| DIMENSION | METRIC | METRIC/ENGLISH |
|---|---|---|
| Power, heat transfer rate | 1 W = 1 J/s<br>1 kW = 1000 W = 1.341 hp<br>1 hp‡ = 745.7 W | 1 kW = 3412.14 Btu/h<br>= 737.56 lbf · ft/s<br>1 hp = 550 lbf · ft/s = 0.7068 Btu/s<br>= 42.41 Btu/min = 2544.5 Btu/h<br>= 0.74570 kW<br>1 Btu/h = 1.055056 kJ/h<br>1 ton of refrigeration = 200 Btu/min |
| Pressure | 1 Pa = 1 $N/m^2$<br>1 kPa = $10^3$ Pa = $10^{-3}$ MPa<br>1 atm = 101.325 kPa<br>= 1.01325 bars<br>= 760 mmHg at 0°C<br>= 1.03323 $kgf/cm^2$<br>1 mmHg = 0.1333 kPa | 1 Pa = $1.4504 \times 10^{-4}$ psia<br>= 0.020886 $lbf/ft^2$<br>1 psia = 144 $lbf/ft^2$ = 6.894757 kPa<br>1 atm = 14.696 psia<br>= 29.92 inHg at 30°F<br>1 inHg = 3.387 kPa |
| Specific heat | 1 kJ/(kg · °C) = 1 kJ/(kg · K)<br>= 1 J/(g · °C) | 1 Btu/(lbm · °F) = 4.1868 kJ/(kg · °C)<br>1 Btu/(lbmol · R) = 4.1868 kJ/(kmol · K)<br>1 kJ/(kg · °C) = 0.23885 Btu/(lbm · °F)<br>= 0.23885 Btu/(lbm · R) |
| Specific volume | 1 $m^3$/kg = 1000 L/kg<br>= 1000 $cm^3$/g | 1 $m^3$/kg = 16.02 $ft^3$/lbm<br>1 $ft^3$/lbm = 0.062428 $m^3$/kg |
| Temperature | $T$ (K) = $T$ (°C) + 273.15<br>$\Delta T$ (K) = $\Delta T$ (°C) | $T$ (R) = $T$ (°F) + 459.67 = 1.8 T (K)<br>$T$ (°F) = 1.8 $T$ (°C) + 32<br>$\Delta T$ (°F) = $\Delta T$ (R)<br>= 1.8* $\Delta T$ (K) |
| Velocity | 1 m/s = 3.60 km/h | 1 m/s = 3.2808 ft/s<br>= 2.237 mi/h<br>1 mi/h = 1.609 km/h |
| Volume | 1 $m^3$ = 1000 L = $10^6$ $cm^3$ (cc) | 1 $m^3$ = $6.1024 \times 10^4$ $in^3$ = 35.315 $ft^3$<br>= 264.17 gal (U.S.)<br>1 U.S. gallon = 231 $in^3$ = 3.7854 L |

‡ Mechanical horsepower. The electrical horsepower is taken to be exactly 746 W.

# Some Physical Constants

| | |
|---|---|
| Universal gas constant | $R_u = 8.31434\ \text{kJ/(kmol} \cdot \text{K)}$<br>$= 8.31434\ \text{kPa} \cdot \text{m}^3\text{/(kmol} \cdot \text{K)}$<br>$= 0.0831434\ \text{bar} \cdot \text{m}^3\text{/(kmol} \cdot \text{K)}$<br>$= 82.05\ \text{L} \cdot \text{atm/(kmol} \cdot \text{K)}$<br>$= 1.9858\ \text{Btu/(lbmol} \cdot \text{R)}$<br>$= 1545.35\ \text{ft} \cdot \text{lbf/(lbmol} \cdot \text{R)}$<br>$= 10.73\ \text{psia} \cdot \text{ft}^3\text{/(lbmol} \cdot \text{R)}$ |
| Standard acceleration of gravity | $g = 9.80665\ \text{m/s}^2$<br>$= 32.174\ \text{ft/s}^2$ |
| Standard atmospheric pressure | $1\ \text{atm} = 101.325\ \text{kPa}$<br>$= 1.01325\ \text{bar}$<br>$= 14.696\ \text{psia}$<br>$= 760\ \text{mmHg}\ (0°\text{C})$<br>$= 29.9213\ \text{inHg}\ (32°\text{F})$<br>$= 10.3323\ \text{mH}_2\text{O}\ (4°\text{C})$ |
| Stefan–Boltzmann constant | $\sigma = 5.66961 \times 10^{-8}\ \text{W/(m}^2 \cdot \text{K}^4\text{)}$<br>$= 0.1714 \times 10^{-8}\ \text{Btu/(h} \cdot \text{ft}^2 \cdot \text{R}^4\text{)}$ |
| Boltzmann's constant | $k = 1.380622 \times 10^{-23}\ \text{kJ/(kmol} \cdot \text{K)}$ |
| Speed of light in vacuum | $c = 2.9979 \times 10^8\ \text{m/s}$<br>$= 9.836 \times 10^8\ \text{ft/s}$ |
| Speed of sound in dry air at 0°C and 1 atm | $\text{C} = 331.36\ \text{m/s}$<br>$= 1089\ \text{ft/s}$ |
| Heat of fusion of water at 1 atm | $h_{\text{if}} = 333.7\ \text{kJ/kg}$<br>$= 143.5\ \text{Btu/lbm}$ |
| Heat of vaporization of water at 1 atm | $h_{\text{fg}} = 2257.1\ \text{kJ/kg}$<br>$= 970.4\ \text{Btu/lbm}$ |

# INTRODUCTION TO THERMODYNAMICS AND HEAT TRANSFER

**McGRAW-HILL SERIES IN MECHANICAL ENGINEERING**

CONSULTING EDITORS
Jack P. Holman, *Southern Methodist University*
John R. Lloyd *Michigan State University*

| | |
|---|---|
| Anderson: | *Computational Fluid Dynamics: The Basics with Applications* |
| Anderson: | *Modern Compressible Flow: With Historical Perspective* |
| Arora: | *Introduction to Optimum Design* |
| Bray and Stanley: | *Nondestructive Evaluation: A Tool for Design, Manufacturing, and Service* |
| Burton: | *Introduction to Dynamic Systems Analysis* |
| Çengel: | *Introduction to Thermodynamics and Heat Transfer* |
| Culp: | *Principles of Energy Conversion* |
| Dally: | *Packaging of Electronic Systems: A Mechanical Engineering Approach* |
| Dieter: | *Engineering Design: A Materials & Processing Approach* |
| Doebelin: | *Engineering Experimentation: Planning, Execution, Reporting* |
| Driels: | *Linear Control Systems Engineering* |
| Eckert and Drake: | *Analysis of Heat and Mass Transfer* |
| Edwards and McKee: | *Fundamentals of Mechanical Component Design* |
| Gebhart: | *Heat Conduction and Mass Diffusion* |
| Gibson: | *Principles of Composite Material Mechanics* |
| Hamrock: | *Fundamentals of Fluid Film Lubrication* |
| Heywood: | *Internal Combustion Engine Fundamentals* |
| Hinze: | *Turbulence* |
| Holman: | *Experimental Methods for Engineers* |
| Howell and Buckius: | *Fundamentals of Engineering Thermodynamics* |
| Hutton: | *Applied Mechanical Vibrations* |
| Juvinall: | *Engineering Considerations of Stress, Strain, and Strength* |
| Kane and Levinson: | *Dynamics: Theory and Applications* |
| Kays and Crawford: | *Convective Heat and Mass Transfer* |
| Kelly: | *Fundamentals of Mechanical Vibrations* |
| Kimbrell: | *Kinematics Analysis and Synthesis* |
| Kreider and Rabl: | *Heating and Cooling of Buildings* |
| Martin: | *Kinematics and Dynamics of Machines* |
| Mattingly: | *Elements of Gas Turbine Propulsion* |
| Modest: | *Radiative Heat Transfer* |
| Norton: | *Design of Machinery* |
| Oosthuizen and Carscallen: | *Compressible Fluid Flow* |
| Phelan: | *Fundamentals of Mechanical Design* |
| Raven: | *Automatic Control Engineering* |
| Reddy: | *An Introduction to the Finite Element Method* |
| Rosenberg and Karnopp: | *Introduction to Physical Systems Dynamics* |
| Schlichting: | *Boundary-Layer Theory* |
| Shames: | *Mechanics of Fluids* |
| Sherman: | *Viscous Flow* |
| Shigley: | *Kinematic Analysis of Mechanisms* |
| Shigley and Mischke: | *Mechanical Engineering Design* |
| Shigley and Uicker: | *Theory of Machines and Mechanisms* |
| Stiffler: | *Design with Microprocessors for Mechanical Engineers* |
| Stoecker and Jones: | *Refrigeration and Air Conditioning* |

| | |
|---|---|
| Turns: | *An Introduction to Combustion: Concepts and Applications* |
| Ullman: | *The Mechanical Design Process* |
| Vanderplaats: | *Numerical Optimization: Techniques for Engineering Design, with Applications* |
| Wark: | *Advanced Thermodynamics for Engineers* |
| White: | *Viscous Fluid Flow* |
| Zeid: | *CAD/CAM Theory and Practice* |

# INTRODUCTION TO THERMODYNAMICS AND HEAT TRANSFER

Dr. Yunus A. Çengel
University of Nevada, Reno

The McGraw-Hill Companies, Inc.

New York St. Louis San Francisco Auckland Bogotá Caracas
Lisbon London Madrid Mexico City Milan Montreal New Delhi
San Juan Singapore Sydney Tokyo Toronto

*McGraw-Hill*

*A Division of The* ***McGraw-Hill*** *Companies*

**INTRODUCTION TO THERMODYNAMICS AND HEAT TRANSFER**

Copyright © 1997 by The McGraw-Hill Companies, Inc. All rights reserved. Printed in the United States of America. Except as permitted under the United States Copyright Act of 1976, no part of this publication may be reproduced or distributed in any form or by any means, or stored in a data base or retrieval system, without the prior written permission of the publisher.

This book is printed on acid-free paper.

1234567890 DOC DOC 909876

ISBN 0-07-011498-6

This book was set in Times Roman by The Universities Press, Ltd.
The editors were Debra Riegert and John M. Morriss;
the production supervisor was Elizabeth J. Strange.
The cover was designed by Wanda Lubelska.
Project supervision was done by The Universities Press, Ltd.
R. R. Donnelley & Sons Company was printer and binder.

Çengel, Yunus A.
Introduction to thermodynamics and heat transfer / Yunus A. Çengel.
p. cm. — (McGraw-Hill series in mechanical engineering)
Includes bibliographical references and index.
ISBN 0-07-011498-6 (text). — ISBN 0-07-011499-4 (SM)
1. Thermodynamics. 2. Heat—Transmission. I. Title. II. Series.
QC311.C42 1997
621.402′1—dc20 96-36053

INTERNATIONAL EDITION

Copyright © 1997. Exclusive rights by The McGraw-Hill Companies, Inc. for manufacture and export. This book cannot be re-exported from the country to which it is consigned by McGraw-Hill. The International Edition is not available in North America.

When ordering this title, use ISBN 0-07-114109-X.

http://www.mhcollege.com

# About the Author

**Yunus A. Çengel** received his Ph.D. in mechanical engineering from North Carolina State University, and joined the faculty of mechanical engineering at the University of Nevada, Reno, where he has been teaching undergraduate and graduate courses in thermodynamics and heat transfer while conducting research. He has published primarily in the areas of thermodynamics, radiation heat transfer, natural convection, solar energy, geothermal energy, energy conservation, and engineering education. He has led teams of engineering students to numerous manufacturing facilities in Northern Nevada to conduct energy audits, and prepared energy conservation reports for them. Dr. Çengel has been voted the outstanding teacher by the ASME student sections in both North Carolina State University and the University of Nevada, Reno. He is a member of the American Society of Mechanical Engineers (ASME) and the American Society for Engineering Education (ASEE). Dr. Çengel is also the recepient of ASEE Meriam/Wiley Distinguished Author Award.

# Contents

## 5 ■ THE SECOND LAW OF THERMODYNAMICS 183



## 6 ■ ENTROPY 237



## 7 ■ POWER AND REFRIGERATION CYCLES 283

# Preface

Thermodynamics and Heat Transfer are two closely related basis sciences that have long been an essential part of engineering core curricula all over the world. Most engineering students take an introductory course in thermodynamics, while others such as mechanical engineering students take a second course in thermodynamics as well as a course in heat transfer. Students who take the traditional thermodynamics course only, such as civil, electrical, and mining engineering students, are exposed to little or no heat transfer. To fix this deficiency, several schools have opened separate sections of thermodynamics for these students, and supplemented some material on heat transfer. This introductory text is intended for use in an undergraduate thermodynamics or thermal science course for students who can take only *one course* in thermal sciences. The text covers the basic principles of thermodynamics and heat transfer with engineering applications. It contains sufficient material to give instructors fiexibility, and to accommodate their preferences on the right blend of thermodynamics and heat transfer for their students. By careful selection of topics, an instructor can spend a third, a half, or two-thirds of the course on thermodynamics, and the rest on selected topics of heat transfer.

The text is intended for traditional thermodynamics students, who are typically Sophomores or Juniors in engineering with adequate background in calculus and physics. No background is assumed in differential equations, and no mathematics is involved beyond the level of simple integration. The emphasis is kept on the physics and the physical arguments in order to develop an *intuitive understanding* of the subject matter.

There is a growing need for the engineering students to understand the mechanisms of heat transfer, since heat transfer plays a crucial role in the design of circuit boards, electronic devices, buildings, and even bridges, but most programs do not have room for another thermal science course. To see how other universities are coping with the situation, we surveyed the Mechanical Engineering Departments in the U.S., and received responses from about 150 of them. The responses to the survey confirmed what I already suspected: About 75% of the respondents agreed that a combined thermodynamics and heat transfer course will better serve the needs of students who do not take a separate course in heat transfer than a course on thermodynamics alone. Several schools indicated that they are already doing that, using a standard text for thermodynamics and supplying notes for heat transfer, or using a second text for heat transfer. Some schools are allowing their students to take regular heat transfer instead of thermodynamics. The 25% of the respondents who did not favor the combined course idea stated staffing problems, the lack of a suitable textbook for such a course, and the difficulty in condensing two already condensed courses into a single course as reasons. The survey indicates that there is a need for a text more responsive to the needs of most engineering students, and it was the primary motivation for this undertaking.

It is recognized that all topics of thermodynamics and heat transfer cannot be covered adequately in a typical three semester hour course, and therefore sacrifices must be made from depth if not from breadth. Selecting the right topics and finding the proper levels of depth and breadth is no small challenge for instructors, and this text is intended to serve as the ground for such selection. Students in a combined thermodynamics and heat transfer course can gain a basic understanding of energy and energy interactions, and various mechanisms of heat transfer. Such a course can also instill in students the confidence and the background to do further reading of their own, and to be able to communicate effectively with specialists in thermal sciences.

The text is an abbreviated version of the standard thermodynamics and heat transfer texts, covering topics that engineering students are most likely to need in their professional life. The thermodynamics portion of this text (Chapters 1–7) is based on the text *Thermodynamics: An Engineering Approach*, which I co-authored with M. A. Boles. Chapter 1 on the basic concepts of thermodynamics has remained essentially unchanged. Chapter 2 on the properties of pure substances has largely remained unchanged, but the use of property tables at this stage is limited to the saturation tables only to facilitate discussion of phase-change processes. The compressibility factors and other equations of state are presented for completeness, but can be skipped if desired.

Chapters 3 and 4 on the first law of thermodynamics emphasize energy interactions for liquids, solids, and gases (but no two-phase mixtures) in closed and steady-flow systems. The ideal gas approximation with constant specific heats is used in the analysis of gases, and solids and liquids are approximated as incompressible substances. The last section of

Chapter 3 is devoted to the popular topic of dieting and exercise, together with tables of metabolizable energy content of common foods and energy consumption during common activities.

Chapter 5 on the second law of thermodynamics has remained essentially unchanged, but Chapter 6 on entropy is condensed considerably. The emphasis in this chapter is placed on the physical significance of entropy and the isentropic relations of ideal gases using constant specific heats. Chapter 7 on thermodynamic cycles describes basic gas and vapor power cycles and thermoelectric power generation and refrigeration. Chapters 6 and 7 can be skipped if desired to cover more material on heat transfer.

Chapters 8 and 9 deal with steady and transient heat conduction, respectively. A practical approach is used throughout, without any differential equations and associated boundary conditions, and the thermal resistance concept is emphasized. Chapters 10 and 11 deal with forced and natural convection, respectively. Again a practical engineering approch is used, with a wealth of physical explanations and empirical correlations. Radiation heat transfer is presented in Chapter 12, followed by heat exchangers in Chapter 13. An overview of the cooling techniques for electronic equipment is presented in Chapter 14. The chapters on heat transfer are practically independent of each other, and can be covered in any order. Also, the later sections of each chapter can be skipped if desired.

The philosophy that has contributed to the popularity of our thermodynamics book remains unchanged in this text: talk directly to the minds of tomorrow's engineers in a simple yet precise manner, and encourage creative thinking and developing a deeper understanding of the subject matter. The goal throughout this project has been to offer an engineering textbook that is read by students with interest and enthusiasm instead of one that is used as a reference book to solve problems. Special effort is made to touch curious minds and take them on a pleasant journey in the wonderful world of thermodynamics and heat transfer and explore the wonders of these exciting subjects.

Thermodynamics and heat transfer are often perceived by students as difficult subjects, but an observant mind should have no difficulty understanding them. After all, the principles of thermodynamics and heat transfer are based on our everyday experiences and experimental observations. Both thermodynamics and heat transfer are mature basic sciences, and the topics covered in introductory textbooks are well established. Textbooks differ primarily in the approach taken. A more physical, intuitive approach is used throughout this text. Frequently, parallels are drawn between the subject matter and students' everyday experiences, so that they can relate the subject matter to what they already know.

Yesterday's engineer spent a major portion of his or her time substituting values into formulas and obtaining numerical results. But all the formula manipulations and number crunching are now being left to the computers. Tomorrow's engineer will have to have a clear understanding and a firm grasp of the basic principles so that he or she can understand even the most complex problems, formulate them, and interpret the results. A conscious effort is made to lead students in this direction.

The material in the text is introduced at a level that an average student can follow comfortably. It speaks to students, not over them. In fact, it is self-instructive. This frees the instructor to use class time more productively.

Figures are important learning tools that help students to "get the picture." The text makes effective use of graphics. It probably contains more figures and illustrations than any other thermodynamics or heat transfer book. Figures attract attention, and stimulate curiosity and interest. A popular cartoon character, "Blondie", is also used to make some important points in a humorous way and also to break the ice and ease nerves.

Each chapter contains numerous worked-out examples that clarify the material and illustrate the use of the basic principles. An intuitive and systematic approach is used in the solution of the example problems, with particular attention to the proper use of units. A summary is included at the end of each chapter for a quick overview of basic concepts and important relations.

The end-of-chapter problems are grouped under specific topics in the order in which they are covered to make problem selection easier for both instructors and students. The problems within each group start with concept questions, indicated by "C", to check students' level of understanding of basic concepts. The problems under *Review Problems* are more comprehensive in nature, and are not directly tied to any specific section of a chapter. The problems under the *Computer, Design, and Essay Problems* title are intended to encourage students to use computers in problem solving, to make engineering judgments, to conduct independent searches on topics of interest, and to communicate their findings in a professional manner. Some safety-related problems are incorporated throughout to enhance safety awareness among engineering students. Answers to selected problems are listed immediately following the problem for convenience to the students. The *Solutions Manual,* available to instructors only, provides complete and detailed solutions of end-of-chapter problems.

In recognition of the fact that English units are still widely used in some industries, both SI and English units are used in this text, with an emphasis on SI. The material in this text can be covered using combined SI/English units or SI units alone, depending on the preference of the instructor. All property tables and charts in the Appendix are presented in both units, except those that involve dimensionless quantities. Problems, tables, and charts in English units are designated by "E" after the number for easy recognition. Frequently used conversion factors and physical constants are listed on the inner cover pages of the text for each reference.

I would like to acknowledge with appreciation the numerous and valuable comments, suggestions, criticism, and praise of the following academic reviewers: Dr. J. L. Goddis, Clemson University, Dr. A. Beyene, San Diego State University, Dr. J. E. Drummond, University of Akron, Dr. W. G. Rieder, North Dakota State University, Dr. J. L. Johnsen, University of Portland, Dr. A. Aziz, Gonzaga University, L. Witte, University of Houston. Their suggestions have greatly helped to improve the quality of this text. I also would like to thank the administration of the University of Nevada, Reno for granting me a faculty development leave to prepare this text. Finally, I would like to express my appreciation to my wife Zehra and my children for their continued patience, understanding, encouragement, and support throughout the preparation of this text.

Yunus A. Çengel

# Nomenclature

| | |
|---|---|
| $a$ | Acceleration, $m/s^2$ |
| $A$ | Area, $m^2$ |
| $A_c$ | Cross-sectional area |
| Bi | Biot number |
| $C$ | Specific heat, kJ/(kg·K) |
| $C_D$ | Drag coefficient |
| $C_f$ | Friction coefficient |
| $C_p$ | Constant pressure specific heat, kJ/(kg · K) |
| $C_v$ | Constant volume specific heat, kJ/(kg · K) |
| COP | Coefficient of performance |
| $COP_R$ | Coefficient of performance of a refrigerator |
| $COP_{HP}$ | Coefficient of performance of a heat pump |
| $d$ | Exact differential |
| $d, D$ | Diameter, m |
| $D_h$ | Hydraulic diameter, m |
| $e$ | Specific total energy, kJ/kg |
| $E$ | Total energy, kJ |
| EER | Energy efficiency rating |
| $f$ | Friction factor |
| $F$ | Force, N |
| $g$ | Gravitational acceleration, $m/s^2$ |

| | |
|---|---|
| $G$ | Incident radiation, $W/m^2$ |
| Gr | Grashof number |
| $h$ | Height, m |
| $h$ | Specific enthalpy, $u + Pv$, kJ/kg |
| $h$ | Convection heat transfer coefficient, $W/(m^2 \cdot °C)$ |
| $h_{fg}$ | Latent heat of vaporization, kJ/kg |
| $h_{if}$ | Latent heat of fusion, kJ/kg |
| $H$ | Total enthalpy, $U + PV$, kJ |
| $I$ | Electric current, A |
| $J$ | Radiosity, $W/m^2$ |
| $k$ | Specific heat ratio, $C_p/C_v$ |
| $k_s$ | Spring constant |
| $k_t$ | Thermal conductivity |
| $ke$ | Specific kinetic energy, $V^2/2$, kJ/kg |
| KE | Total kinetic energy, $mV^2/2$, kJ |
| $L$ | Length |
| $L_h$ | Hydrodynamic entry length |
| $L_t$ | Thermal entry length |
| $m$ | Mass, kg |
| $\dot{m}$ | Mass flow rate, kg/s |
| $M$ | Molar mass, kg/kmol |
| MEP | Mean effective pressure, kPa |
| $n$ | Polytropic exponent |
| $N$ | Number of moles, kmol |
| Nu | Nusselt number |
| NTU | Number of transfer units |
| $p$ | Perimeter, m |
| $P$ | Pressure, kPa |
| $P_{cr}$ | Critical pressure, kPa |
| $P_R$ | Reduced pressure |
| pe | Specific potential energy, $gz$, kJ/kg |
| Pr | Prandtl number |
| PE | Total potential energy, $mgz$, kJ |
| $q$ | Heat transfer per unit mass, kJ/kg |
| $Q$ | Total heat transfer, kJ |
| $Q_H$ | Heat transfer with high-temperature body, kJ |
| $Q_L$ | Heat transfer with low-temperature body, kJ |
| $\dot{q}$ | Heat flux, $W/m^2$ |
| $\dot{Q}$ | Heat transfer rate, kW |
| $r$ | Compression ratio |

| | |
|---|---|
| $r_p$ | Pressure ratio |
| $r_c$ | Cutoff ratio |
| $R$ | Gas constant, kJ/(kg · K) |
| $R_f$ | Fouling factor |
| $R_u$ | Universal gas constant, kJ/(kmol · K) |
| Re | Reynolds number |
| $s$ | Specific entropy, kJ/(kg · K) |
| $s_{gen}$ | Specific entropy generation, kJ/(kg · K) |
| $S$ | Total entropy, kJ/K |
| $S_{gen}$ | Total entropy generation, kJ/K |
| $\dot{S}_{gen}$ | Entropy generation rate |
| $t$ | Time, s |
| $T$ | Temperature, °C or K |
| $T_{cr}$ | Critical temperature, K |
| $T_f$ | Film temperature, °C |
| $T_R$ | Reduced temperature |
| $T_H$ | Temperature of high-temperature body, K |
| $T_L$ | Temperature of low-temperature body, K |
| $u$ | Specific internal energy, kJ/kg |
| $U$ | Overall heat transfer coefficient, W/(m$^2$ · °C) |
| $v$ | Specific volume, m$^3$/kg |
| $v_{cr}$ | Critical specific volume, m$^3$/kg |
| $v_R$ | Pseudo-reduced specific volume |
| $V$ | Total volume, m$^3$ |
| $\mathcal{V}$ | Velocity, m/s |
| $w$ | Work per unit mass, kJ/kg |
| $W$ | Total work, kJ |
| $\dot{W}$ | Power, kW |
| $W_{in}$ | Work input, kJ |
| $W_{out}$ | Work output, kJ |
| $W_{rev}$ | Reversible work, kJ |
| $x$ | Quality |
| $z$ | Elevation, m |
| $Z$ | Compressibility factor |

## Greek Letters

| | |
|---|---|
| $\beta$ | Volume expansivity, 1/K |
| $\Delta$ | Finite change in quantity |
| $\delta$ | Characteristic length |

| | |
|---|---|
| $\varepsilon$ | Emissivity; heat exchanger effectiveness |
| $\eta_{th}$ | Thermal efficiency |
| $\theta$ | Total energy of a flowing fluid, kJ/kg |
| $\alpha$ | Absorptivity |
| $\mu_{JT}$ | Joule-Thomson coefficient, K/kPa |
| $\mu$ | Dynamic viscosity, kg/(m · s) |
| $v$ | Kinematic viscosity, $m^2/s$ |
| $\rho$ | Density, $kg/m^3$ |
| $\rho_s$ | Specific weight or relative density |
| $\sigma$ | Stefan-Boltzmann constant |
| $\sigma_s$ | Surface tension, N/m |
| $\sigma_n$ | Normal stress, $N/m^2$ |
| $\tau$ | Transmissivity; Fourier number |
| $\phi$ | Relative humidity |

## Subscripts

| | |
|---|---|
| *a* | Air |
| abs | Absolute |
| act | Actual |
| atm | Atmospheric |
| av | Average |
| cr | Critical point property |
| cv | Control volume |
| e | Exit conditions |
| *f* | Saturated liquid |
| *fg* | Difference in property between saturated liquid and saturated vapor |
| *g* | Saturated vapor |
| gen | Generation |
| *H* | High temperture (as in $T_H$ and $Q_H$) |
| i | Inlet conditions |
| *i* | *i*th component |
| *L* | Low temperature (as in $T_L$ and $Q_L$) |
| *m* | Mixture |
| *R* | Reduced |
| rev | Reversible |
| *s* | Isentropic |
| *s* | Surface |
| sat | Saturated |
| surr | Surrounding surfaces |

| | |
|---|---|
| sys | System |
| $v$ | Water vapor |
| 1 | Initial or inlet state |
| 2 | Final or exit state |

## Superscripts

| | |
|---|---|
| $\dot{}$ (over dot) | Quantity per unit time |
| $\bar{}$ (over bar) | Quantity per unit mole |

# Basic Concepts of Thermodynamics

CHAPTER

1

Every science has a unique vocabulary associated with it, and thermodynamics is no exception. Precise definition of basic concepts forms a sound foundation for the development of a science and prevents possible misunderstandings. In this chapter, the unit systems that will be used are reviewed, and the basic concepts of thermodynamics such as system, energy, property, state, process, cycle, pressure, and temperature are explained. Careful study of these concepts is essential for a good understanding of the topics in the following chapters.

## 1-1 ■ THERMODYNAMICS AND ENERGY

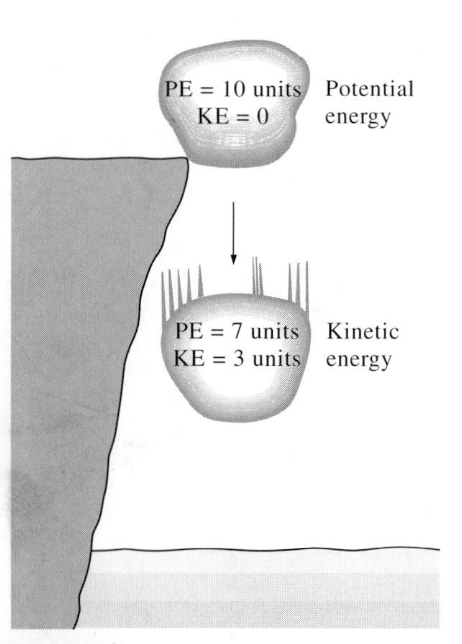


**FIGURE 1-1**

Energy cannot be created or destroyed; it can only change forms (the first law).

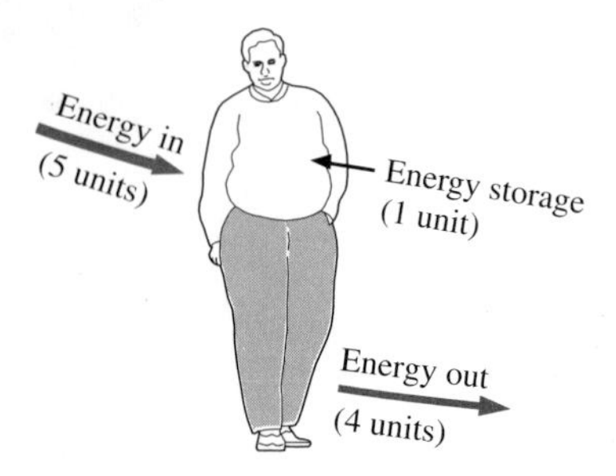


**FIGURE 1-2**

Conservation of energy principle for the human body.

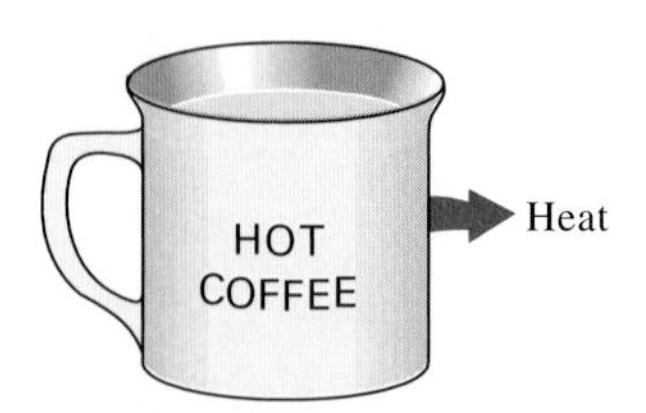


**FIGURE 1-3**

Heat can flow only from hot to cold bodies (the second law).

Thermodynamics can be defined as the science of energy. Although everybody has a feeling of what energy is, it is difficult to give a precise definition for it. Energy can be viewed as the ability to cause changes.

The name *thermodynamics* stems from the Greek words *therme* (heat) and *dynamis* (power), which is most descriptive of the early efforts to convert heat into power. Today the same name is broadly interpreted to include all aspects of energy and energy transformations, including power production, refrigeration, and relationships among the properties of matter.

One of the most fundamental laws of nature is the **conservation of energy** principle. It simply states that during an interaction, energy can change from one form to another but the total amount of energy remains constant. That is, energy cannot be created or destroyed. A rock falling off a cliff, for example, picks up speed as a result of its potential energy being converted to kinetic energy (Fig. 1-1). The conservation of energy principle also forms the backbone of the diet industry: a person who has a greater energy input (food) than energy output (exercise) will gain weight (store energy in the form of fat), and a person who has a smaller energy input than output will lose weight (Fig. 1-2).

The **first law of thermodynamics** for example, is simply an expression of the conservation of energy principle, and it asserts that *energy* is a thermodynamic property. The **second law of thermodynamics** asserts that energy has *quality* as well as *quantity,* and actual processes occur in the direction of decreasing quality of energy. For example, a cup of hot coffee left on a table eventually cools, but a cup of cool coffee on the same table never gets hot by itself (Fig. 1-3). The high-temperature energy of the coffee is degraded (transformed into a less useful form at a lower temperature) once it is transferred to the surrounding air.

Although the principles of thermodynamics have been in existence since the creation of universe, thermodynamics did not emerge as a science until the construction of the first successful atmospheric steam engines in England by Thomas Savery in 1697 and Thomas Newcomen in 1712. These engines were very slow and inefficient, but they opened the way for the development of a new science.

The first and second laws of thermodynamics emerged simultaneously in the 1850s primarily out of the works of William Rankine, Rudolph Clausius, and Lord Kelvin (formerly William Thomson). The term *thermodynamics* was first used in a publication by Lord Kelvin in 1849. The first thermodynamic textbook was written in 1859 by William Rankine, a professor at the University of Glasgow.

It is well known that a substance consists of a large number of particles called *molecules.* The properties of the substance naturally depend on the behavior of these particles. For example, the pressure of a gas in a container is the result of momentum transfer between the molecules and the walls of the container. But one does not need to know the behavior of the gas particles to determine the pressure in the container. It would be sufficient to attach a pressure gage to the container. This macroscopic approach to the study of thermodynamics which does not require a knowledge of the behavior of individual particles is called **classical thermodynamics**. It provides a direct and easy way to the solution of engineering problems. A more elaborate approach, based on the average behavior of large groups of individual particles, is called **statistical thermodynamics**. This microscopic approach is rather involved and is used in this text only in the supporting role.

## Application Areas of Thermodynamics

Every engineering activity involves an interaction between energy and matter; thus it is hard to imagine an area which does not relate to thermodynamics in some respect. Therefore, developing a good understanding of thermodynamic principles has long been an essential part of engineering education.

One does not need to go very far to see some application areas of thermodynamics. In fact, one does not need to go anywhere. These areas are right where one lives. An ordinary house is, in some respects, an exhibition hall filled with thermodynamic wonders. Many ordinary household utensils and appliances are designed, in whole or in part, by using the principles of thermodynamics. Some examples include the electric or gas range, the heating and air-conditioning systems, the refrigerator, the humidifier, the pressure cooker, the water heater, the shower, the iron, and even the computer, the TV, and the VCR set. On a larger scale, thermodynamics plays a major part in the design and analysis of automotive engines, rockets, jet engines, and conventional or nuclear power plants (Fig. 1-4). We should also mention the human body as an interesting application area of thermodynamics.

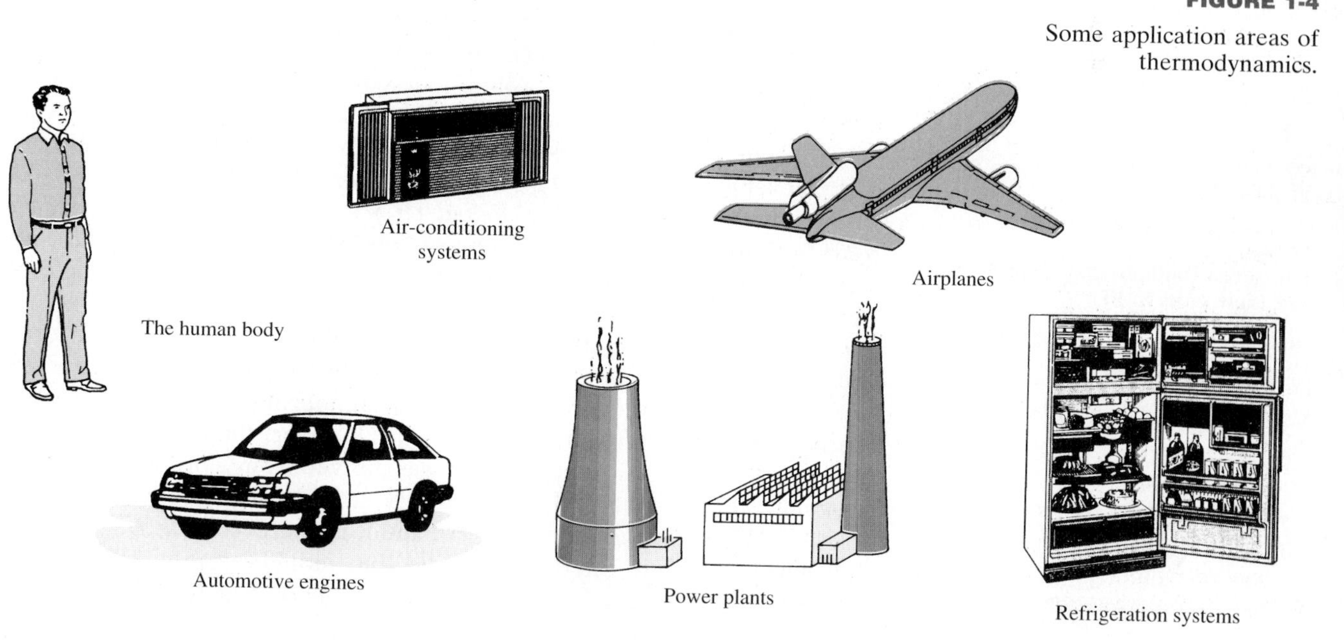


**FIGURE 1-4**
Some application areas of thermodynamics.

## 1-2 A NOTE ON DIMENSIONS AND UNITS

Any physical quantity can be characterized by **dimensions**. The arbitrary magnitudes assigned to the dimensions are called **units**. Some basic dimensions such as mass *m*, length *L*, time *t*, and temperature *T* are selected as **primary** or **fundamental dimensions**, while others such as velocity $\mathscr{V}$, energy *E*, and volume *V* are expressed in terms of the primary dimensions and are called **secondary dimensions**, or **derived dimensions**.

A number of unit systems have been developed over the years. Despite strong efforts in the scientific and engineering community to unify the world with a single unit system, two sets of units are still in common use today: the **English system**, which is also known as the *United States Customary System* (USCS), and the metric **SI** (from *Le Système International d'Unités*), which is also known as the *International System*. The SI is a simple and logical system based on a decimal relationship between the various units, and it is being used for scientific and engineering work in most of the industrialized nations, including England. The English system, however, has no numerical base, and various units in this system are related to each other rather arbitrarily (12 in in 1 ft, 16 oz in 1 lb, 4 qt in 1 gal, etc.) which makes it confusing and difficult to learn. The United States is the only industrialized country that has not yet fully converted to the metric system.

The systematic efforts to develop a universally acceptable system of units dates back to 1790 when the French National Assembly charged the French Academy of Sciences to come up with such a unit system. An early version of the metric system was soon developed in France, but it did not find much universal acceptance until 1875 when *The Metric Convection Treaty* was prepared and signed by 17 nations, including the United States. In this international treaty, meter and gram were established as the metric units for length and mass, respectively, and a *General Conference of Weights and Measures* (CGPM) was established which was to meet every six years. In 1960, the CGPM produced the SI, which was based on six fundamental quantities and their units adopted in 1954 at the Tenth General Conference of Weights and Measures: *meter* (m) for length, *kilogram* (kg) for mass, *second* (s) for time, *ampere* (A) for electrical current, degree *Kelvin* (°K) for temperature, and *candela* (cd) for luminous intensity (amount of light). In 1971, the CGPM added a seventh fundamental quantity and unit: *mole* (mol) for the amount of matter.

**TABLE 1-1**
**The seven fundamental dimensions and their units in SI**

| Dimension | Unit |
|---|---|
| Length | meter (m) |
| Mass | kilogram (kg) |
| Time | second (s) |
| Temperature | kelvin (K) |
| Electric current | ampere (A) |
| Amount of light | candela (c) |
| Amount of matter | mole (mol) |

Based on the notational scheme introduced in 1967, the degree symbol was officially dropped from the absolute temperature unit, and all unit names were to be written without capitalization even if they were derived from proper names (Table 1-1). However, the abbreviation of a unit was to be capitalized if the unit was derived from a proper name. For example, the SI unit of force which is named after Sir Isaac Newton (1647–1723) is *newton* (not Newton), and it is abbreviated as *N*. Also, the full name of a unit may be pluralized, but its abbreviation cannot. For example, the length of an object can be 5 m or 5 meters, *not* 5 ms or

5 meter. Finally, no period is to be used in unit abbreviations unless they appear at the end of a sentence. For example, the proper abbreviation of meter is m (not m.).

The recent trend toward the metric system in the United States seems to have started in 1968 when Congress, in response to what was happening in the rest of the world, passed a Metric Study Act. Congress continued to promote a voluntary switch to the metric system by passing the Metric Conversion Act in 1975. A trade bill passed by Congress in 1988 set a September 1992 deadline for all federal agencies to convert to the metric system.

The industries that are heavily involved in international trade (such as the automotive, soft drink, and liquor industries) have been quick in converting to the metric system for economic reasons (having a single worldwide design, fewer sizes, smaller inventories, etc.). Today, nearly all the cars manufactured in the United States are metric. Most car owners probably do not realize this until they try an inch socket wrench on a metric bolt. Most industries, however, resisted the change, thus slowing down the conversion process.

Presently the United States is a dual-system society, and it will stay that way until the transition to the metric system is completed. This puts an extra burden on today's engineering students, since they are expected to retain their understanding of the English system while learning, thinking, and working in terms of the SI. Given the position of the engineers in the transition period, both unit systems are used in this text with particular emphasis on SI units.

As pointed out earlier, the SI is based on a decimal relationship between units. The prefixes used to express the multiples of the various units are listed in Table 1-2. They are standard for all units, and the student is encouraged to memorize them because of their widespread use (Fig. 1-5).

**TABLE 1-2**
**Standard prefixes in SI units**

| Multiple | Prefix |
|---|---|
| $10^{12}$ | tera, T |
| $10^{9}$ | giga, G |
| $10^{6}$ | mega, M |
| $10^{3}$ | kilo, k |
| $10^{-2}$ | centi, c |
| $10^{-3}$ | milli, m |
| $10^{-6}$ | micro, $\mu$ |
| $10^{-9}$ | nano, n |
| $10^{-12}$ | pico, p |

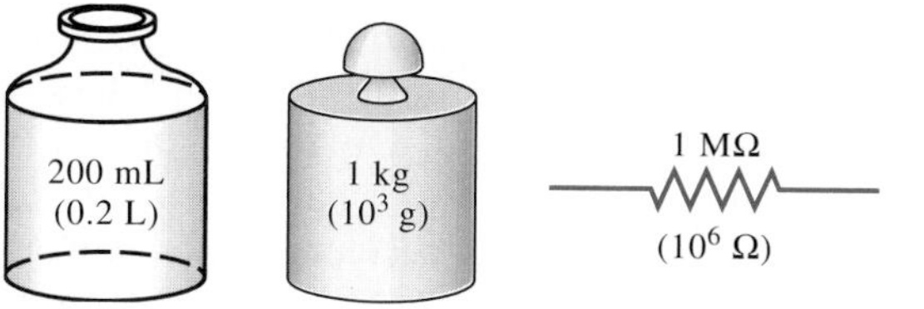


**FIGURE 1-5**
The SI unit prefixes are used in all branches of engineering.

## Some SI and English Units

In SI, the units of mass, length, and time are the *kilogram* (kg), *meter* (m), and *second* (s), respectively. The respective units in the English system are the *pound-mass* (lbm), *foot* (ft), and *second* (s). The pound symbol *lb* is actually the abbreviation of *libra* which was the ancient Roman unit of weight. The English retained this symbol even after the end of the Roman occupation of Britain in A.D. 410. The mass and length units in the two systems are related to each other by

$$1 \text{ lbm} = 0.45359 \text{ kg}$$
$$1 \text{ ft} = 0.3048 \text{ m}$$

In the English system, force is usually considered to be one of the primary dimensions and is assigned a nonderived unit. This is a source of confusion and error that necessitates the use of a conversion factor ($g_c$) in many formulas. To avoid this nuisance, we consider force to be a secondary dimension whose unit is derived from Newton's second law, i.e.,

$$\text{force} = (\text{mass})(\text{acceleration})$$

or

$$F = ma \tag{1-1}$$

In SI, the force unit is the **newton** (N), and it is defined as the *force required to accelerate a mass of* 1 *kg at a rate of* 1 *m/s*$^2$. In the English system, the force unit is the **pound-force** (lbf) and is defined as *the force required to accelerate a mass of 32.174 lbm (1 slug) at a rate of* 1 *ft/s*$^2$. That is,

$$1 \text{ N} = 1 \text{ kg} \cdot \text{m/s}^2$$
$$1 \text{ lbf} = 32.174 \text{ lbm} \cdot \text{ft/s}^2$$

This is illustrated in Fig. 1-6.

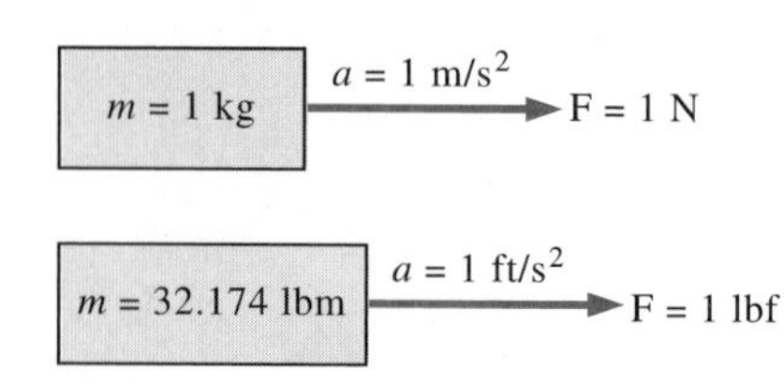


**FIGURE 1-6**
The definition of the force units.

The term **weight** is often incorrectly used to express mass, particularly by the "weight watchers." Unlike mass, weight $W$ is a *force.* It is the gravitational force applied to a body (Fig. 1-7), and its magnitude is determined from Newton's second law,

**FIGURE 1-7**
The ordinary bathroom scale measures the gravitational force applied on a body.

$$W = mg \qquad (\text{N}) \tag{1-2}$$

where $m$ is the mass of the body and $g$ is the local gravitational acceleration ($g$ is 9.807 m/s$^2$ or 32.174 ft/s$^2$ at sea level and 45° latitude). The weight of a unit volume of a substance is called the **specific weight** $w$ and is determined from $w = \rho g$, where $\rho$ is density.

The mass of a body will remain the same regardless of its location in the universe. Its weight, however, will change with a change in gravitational acceleration. A body will weigh less on top of a mountain since $g$ decreases with altitude. On the surface of the moon, an astronaut will weigh about one-sixth of what she or he normally weighs on earth (Fig. 1-8). The gravitational acceleration $g$ changes from 9.807 m/s$^2$ at sea level to 9.804, 9.800, 9.791, 9.776, and 9.745 m/s$^2$ at altitudes of 1000, 2000, 5000, 10,000, and 20,000 meters, respectively. Therefore, for most practical purposes, the gravitational acceleration can be assumed to be a constant at 9.8 m/s$^2$.

**FIGURE 1-8**
A body weighing 150 pounds on earth will weigh only 25 pounds on the moon.

At sea level a mass of 1 kg will weigh 9.807 N, as illustrated in Fig. 1-9. A mass of 1 lbm, however, will weigh 1 lbf, which misleads people to believe that pound-mass and pound-force can be used interchangeably as pound (lb), which is a major source of error in the English system.

*Work,* which is a form of energy, can simply be defined as force times distance; therefore, it has the unit "newton-meter (N · m)," which is called a **joule** (J). That is,

$$1\,\text{J} = 1\,\text{N} \cdot \text{m}$$

A more common unit for energy in SI is the kilojoule ($1\,\text{kJ} = 10^3\,\text{J}$). In the English system, the energy unit is the Btu (British thermal unit), which is defined as the energy required to raise the temperature of 1 lbm of water at 68°F by 1°F. The magnitudes of the kilojoule and Btu are almost identical (1 Btu = 1·055 kJ).

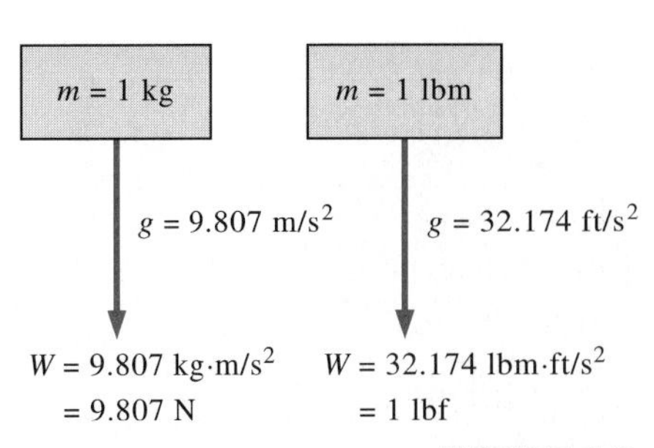


**FIGURE 1-9**
The weight of a unit mass at sea level.

## Dimensional Homogeneity

We all know from grade school that apples and oranges do not add. But we somehow manage to do it (by mistake, of course). In engineering, all equations must be *dimensionally homogeneous.* That is, every term in an equation must have the same units (Fig. 1-10). If, at some stage of an analysis, we find ourselves in a position to add two quantities that have different units, it is a clear indication that we have made an error at an earlier stage. So checking units can serve as a valuable tool to spot errors.

**EXAMPLE 1-1**

While solving a problem, a person ended up with the following equation at some stage:

$$E = 25\,\text{kJ} + 7\,\text{kJ/kg}$$

where $E$ is the total energy and has the unit of kilojoules. Determine the error that may have caused it.

**Solution** The two terms on the right-hand side do not have the same units, and therefore they cannot be added to obtain the total energy. Multiplying the last term by mass will eliminate the kilograms in the denominator, and the whole equation will become dimensionally homogeneous, i.e., every term in the equation will have the same unit. Obviously this error was caused by forgetting to multiply the last term by mass at an earlier stage.

**FIGURE 1-10**
To be dimensionally homogeneous, all the terms in an equation must have the same units.

We all know from experience that units can give terrible headaches if they are not used carefully in solving a problem. But with some attention and skill, units can be used to our advantage. They can be used to check formulas; they can even be used to derive formulas, as explained in the following example.

**EXAMPLE 1-2**

A tank is filled with oil whose density is $\rho = 850\,\text{kg/m}^3$. If the volume of the tank is $V = 2\,\text{m}^3$, determine the amount of mass $m$ in the tank.

**Solution** A sketch of the system described above is given in Fig. 1-11. Suppose we forgot the formula that relates mass to density and volume. But we know that mass has the unit of kilograms. That is, whatever calculations we do,

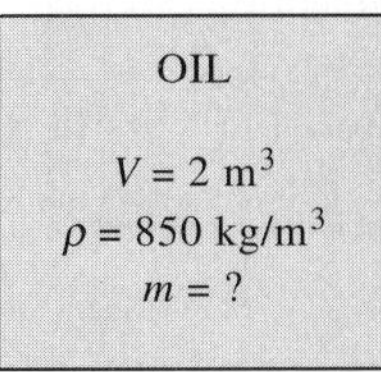


**FIGURE 1-11**
Sketch for Example 1-2.

*BLONDIE cartoons are reprinted with special permission of King Features syndicate, Inc.

we should end up with the unit of kilograms. Putting the given information into perspective, we have

$$\rho = 850\ \text{kg/m}^3 \qquad V = 2\ \text{m}^3$$

It is obvious that we can eliminate $m^3$ and end up with kg by multiplying these two quantities. Therefore, the formula we are looking for is

$$m = \rho V$$

Thus,

$$m = (850\ \text{kg/m}^3)(2\ \text{m}^3) = 1700\ \text{kg}$$

The student should keep in mind that a formula that is not dimensionally homogeneous is definitely wrong, but a dimensionally homogeneous formula is not necessarily right.

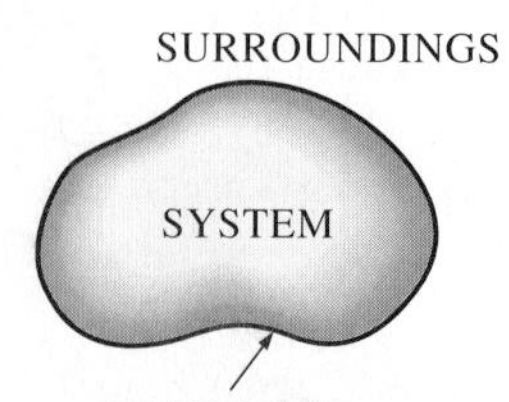


**FIGURE 1-12**
System, surroundings, and boundary.

## 1-3 ■ CLOSED AND OPEN SYSTEMS

A **thermodynamic system**, or simply a **system**, is defined as a *quantity of matter or a region in space chosen for study.* The mass or region outside the system is called the **surroundings**. The real or imaginary surface that separates the system from its surrounding is called the **boundary**. These terms are illustrated in Fig. 1-12. The boundary of a system can be *fixed* or *movable.* Note that the boundary is the contact surface shared by both the system and the surroundings. Mathematically speaking, the boundary has zero thickness, and thus it can neither contain any mass nor occupy any volume in space.

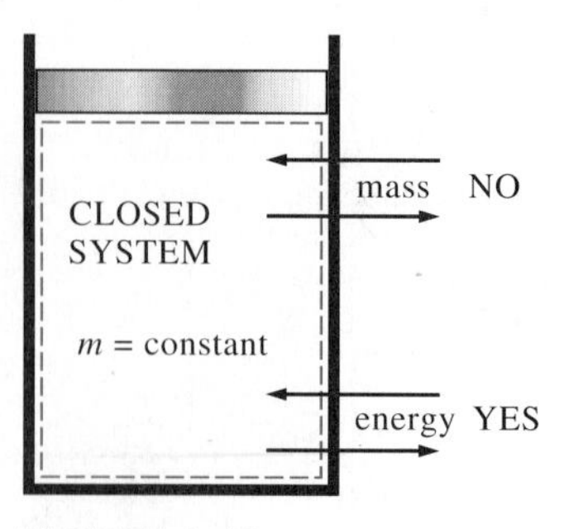


**FIGURE 1-13**
Mass cannot cross the boundaries of a closed system, but energy can.

Systems may be considered to be *closed* or *open,* depending on whether a fixed mass or a fixed volume in space is chosen for study. A **closed system** (also known as a **control mass**) consists of a fixed amount of mass, and no mass can cross its boundary. That is, no mass can enter or leave a closed system, as shown in Fig. 1-13. But energy, in the form of heat or work, can cross the boundary, and the volume of a closed system does not have to be fixed. If, as a special case, even energy is not allowed to cross the boundary, that system is called an **isolated system**.

Consider the piston–cylinder device shown in Fig. 1-14. Let us say that we would like to find out what happens to the enclosed gas when it is heated. Since we are focusing our attention on the gas, it is our system. The inner surfaces of the piston and the cylinder form the boundary, and since no mass is crossing this boundary, it is a closed system. Notice that energy may cross the boundary, and part of the boundary (the inner surface of the piston, in this case) may move. Everything outside the gas, including the piston and the cylinder, is the surroundings.

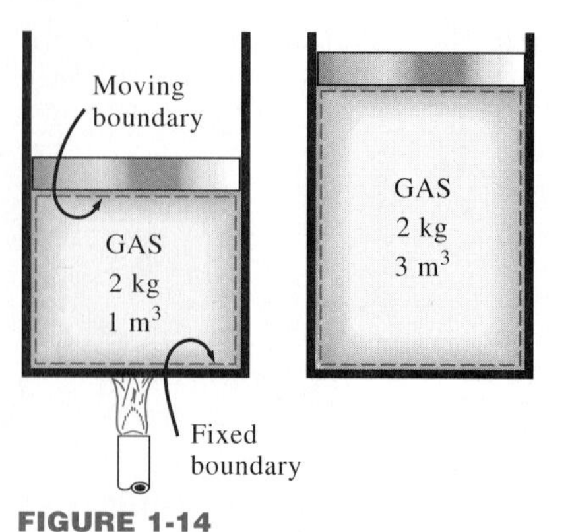


**FIGURE 1-14**
A closed system with a moving boundary.

An **open system**, or a **control volume**, as it is often called, is a properly selected region in space. It usually encloses a device that involves mass flow such as a compressor, turbine, or nozzle. Flow through these devices is best studied by selecting the region within the device as the control volume. Both mass and energy can cross the boundary of a control volume, which is called a **control surface**. This is illustrated in Fig. 1-15.

As an example of an open system, consider the water heater shown in Fig. 1-16. Let us say that we would like to determine how much heat we must transfer to the water in the tank in order to supply a steady stream of hot water. Since hot water will leave the tank and be replaced by cold water, it is not convenient to choose a fixed mass as our system for the analysis. Instead, we can concentrate our attention on the volume formed by the interior surfaces of the tank and consider the hot and cold water streams as mass leaving and entering the control volume. The interior surfaces of the tank form the control surface for this case, and mass is crossing the control surface at two locations.

*The thermodynamic relations that are applicable to closed and open systems are different. Therefore, it is extremely important that we recognize the type of system we have before we start analyzing it.*

In all thermodynamic analyses, the system under study *must* be defined carefully. In most cases, the system investigated is quite simple and obvious, and defining the system may seem like a tedious and unnecessary task. In other cases, however, the system under study may be rather involved, and a proper choice of the system may greatly simplify the analysis.

## 1-4 ■ FORMS OF ENERGY

Energy can exist in numerous forms such as thermal, mechanical, kinetic, potential, electric, magnetic, chemical, and nuclear, and their sum constitutes the **total energy** $E$ of a system. The total energy of a system on a *unit mass* basis is denoted by $e$ and is defined as

$$e = \frac{E}{m} \quad \text{(kJ/kg)} \tag{1-3}$$

Thermodynamics provides no information about the absolute value of the total energy of a system. It only deals with the *change* of the total energy, which is what matters in engineering problems. Thus the total energy of a system can be assigned a value of zero ($E = 0$) at some convenient reference point. The change in total energy of a system is independent of the reference point selected. The decrease in the potential energy of a falling rock, for example, depends on only the elevation difference and not the reference level chosen.

In thermodynamic analysis, it is often helpful to consider the various forms of energy that make up the total energy of a system in two groups: *macroscopic* and *microscopic*. The **macroscopic** forms of energy, on one hand, are those a system possesses as a whole with respect to some outside reference frame, such as kinetic and potential energies (Fig. 1-17). The **microscopic** forms of energy, on the other hand, are those related to the molecular structure of a system and the degree of the



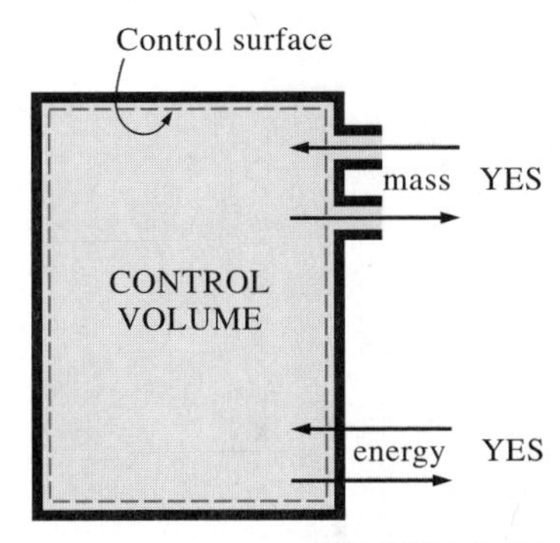


**FIGURE 1-15**

Both mass and energy can cross the boundaries of a control volume.

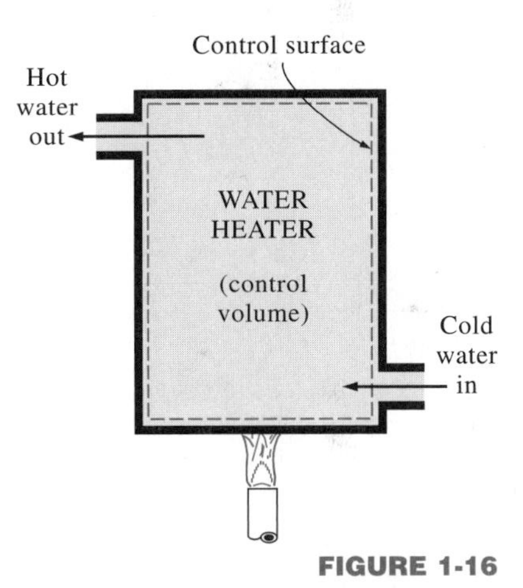


**FIGURE 1-16**

An open system (a control volume) with one inlet and one exit.

**FIGURE 1-17**

The macroscopic energy of an object changes with velocity and elevation.

molecular activity, and they are independent of outside reference frames. The sum of all the microscopic forms of energy is called the **internal energy** of a system and is denoted by $U$. The term *energy* was coined in 1807 by Thomas Young, and its use in thermodynamics was proposed in 1852 by Lord Kelvin. The term *internal energy* and its symbol $U$ first appeared in the works of Rudolph Clausius and William Rankine in the second half of the nineteenth century, and it eventually replaced the alternative terms *inner work, internal work,* and *intrinsic energy* commonly used at the time.

The macroscopic energy of a system is related to motion and the influence of some external effects such as gravity, magnetism, electricity, and surface tension. The energy that a system possesses as a result of its motion relative to some reference frame is called **kinetic energy** KE. When all parts of a system move with the same velocity, the kinetic energy is expressed as

$$\text{KE} = \frac{m\mathscr{V}^2}{2} \qquad \text{(kJ)} \tag{1-4}$$

or, on a unit mass basis,

$$\text{ke} = \frac{\mathscr{V}^2}{2} \qquad \text{(kJ/kg)} \tag{1-5}$$

where the script $\mathscr{V}$ denotes the velocity of the system relative to some fixed reference frame.

The energy that a system possesses as a result of its elevation in a gravitational field is called **potential energy** PE and is expressed as

$$\text{PE} = mgz \qquad \text{(kJ)} \tag{1-6}$$

or, on a unit mass basis,

$$\text{pe} = gz \qquad \text{(kJ/kg)} \tag{1-7}$$

where $g$ is the gravitational acceleration and $z$ is the elevation of the center of gravity of a system relative to some arbitrarily selected reference plane.

The magnetic, electric, and surface tension effects are significant in some specialized cases only and are not considered in this text. In the absence of these effects, the total energy of a system consists of the kinetic, potential, and internal energies and is expressed as

$$E = U + \text{KE} + \text{PE} = U + \frac{m\mathscr{V}^2}{2} + mgz \qquad \text{(kJ)} \tag{1-8}$$

or, on a unit mass basis,

$$e = u + \text{ke} + \text{pe} = u + \frac{\mathscr{V}^2}{2} + gz \qquad \text{(kJ/kg)} \tag{1-9}$$

Most closed systems remain stationary during a process and thus experience no change in their kinetic and potential energies. Closed systems whose velocity and elevation of center of gravity remain constant during a process are frequently referred to as **stationary systems**. The change in the total energy $\Delta E$ of a stationary system is identical to the change in its internal energy $\Delta U$. In this text, a closed system is assumed to be stationary unless it is specifically stated otherwise.



## Some Physical Insight into Internal Energy

Internal energy is defined above as the sum of all the microscopic forms of energy of a system. It is related to the molecular structure and the degree of molecular activity, and it may be viewed as the sum of the kinetic and potential energies of the molecules.

To have a better understanding of internal energy, let us examine a system at the molecular level. The individual molecules of a system, in general, will move around with some velocity, vibrate about each other, and rotate about an axis during their random motion. Associated with these motions are translational, vibrational, and rotational kinetic energies, the sum of which constitutes the kinetic energy of a molecule. The portion of the internal energy of a system associated with the kinetic energies of the molecules is called the **sensible energy** (Fig. 1-18). The average velocity and the degree of activity of the molecules are proportional to the temperature of the gas. Thus, at higher temperatures, the molecules will possess higher kinetic energies, and as a result the system will have a higher internal energy.

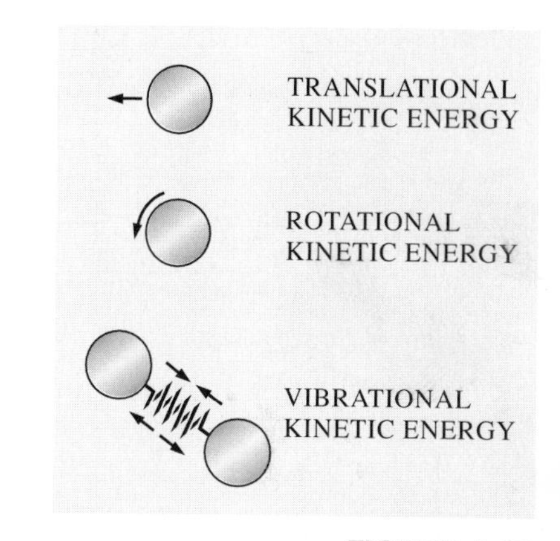


**FIGURE 1-18**
The various forms of molecular energy that make up sensible internal energy.

The internal energy is also associated with the intermolecular forces between the molecules of a system. These are the forces that bind the molecules to each other, and, as one would expect, they are strongest in solids and weakest in gases. If sufficient energy is added to the molecules of a solid or liquid, they will overcome these molecular forces and break away, turning the system to a gas. This is a phase-change process. Because of this added energy, a system in the gas phase is at a higher internal energy level than it is in the solid or the liquid phase. The internal energy associated with the phase of a system is called **latent energy**.

The changes mentioned above can occur without a change in the chemical composition of a system. Most thermodynamic problems fall into this category, and one does not need to pay any attention to the forces binding the atoms in a molecule. The internal energy associated with the atomic bonds in a molecule is called **chemical** (or **bond**) **energy**. During a chemical reaction, such as a combustion process, some chemical bonds are destroyed while others are formed. As a result, the internal energy changes.

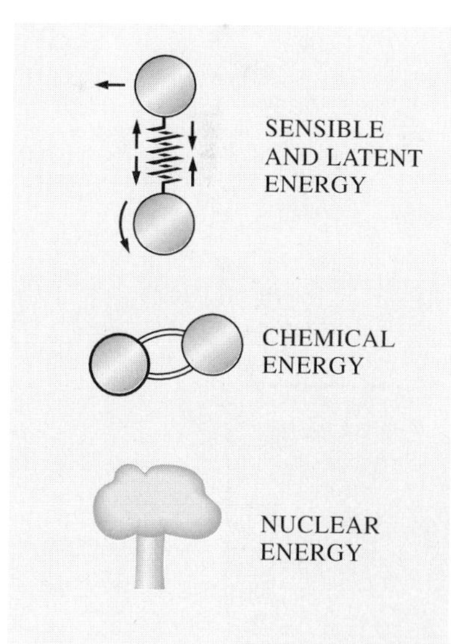


**FIGURE 1-19**
The internal energy of a system is the sum of all forms of the microscopic energies.

We should also mention the tremendous amount of internal energy associated with the bonds within the nucleus of the atom itself (Fig. 1-19). This energy is called **nuclear energy** and is released during nuclear reactions. Obviously, we need not be concerned with nuclear energy in thermodynamics unless, of course, we have a fusion or fission reaction on our hands.

The forms of energy discussed above that constitute the total energy of a system can be *contained* or *stored* in a system, and thus can viewed as the **static** forms of energy. The forms of energy that are not stored in a system can be viewed as the **dynamic** forms of energy, or as **energy interactions**. The dynamic forms of energy are recognized at the system boundary as they cross it, and they represent the energy gained or lost by a system during a process. The only two forms of energy interactions associated with a closed system are **heat transfer** and **work**. An energy interaction is heat transfer if its driving force is a temperature difference. Otherwise it is work, as explained in detail in Chap. 3. (The open systems can also exchange energy via mass transfer, since any time mass is transferred into or out of a system, the energy contained in the mass is also transferred with it).

In daily life, we frequently refer to the sensible and latent forms of internal energy as **heat**, and we talk about heat content of bodies. In thermodynamics, however, we usually refer to those forms of energy as **thermal energy** to prevent any confusion with *heat transfer.*

## 1-5 PROPERTIES OF A SYSTEM

Any characteristic of a system is called a **property**. Some familiar examples are pressure $P$, temperature $T$, volume $V$, and mass $m$. The list can be extended to include less familiar ones such as viscosity, thermal conductivity, modulus of elasticity, thermal expansion coefficient, electric resistivity, and even velocity and elevation.

Not all properties are independent, however. Some are defined in terms of other ones. For example, **density** is defined as *mass per unit volume*:

$$\rho = \frac{m}{V} \qquad (\text{kg/m}^3) \tag{1-10}$$

Sometimes the density of a substance is given relative to the density of a better known substance. Then it is called **specific gravity**, or **relative density**, and is defined as *the ratio of the density of a substance to the density of some standard substance at a specified temperature* (usually water at 4°C, for which $\rho_{H_2O} = 1000\ \text{kg/m}^3$). That is,

$$\rho_s = \frac{\rho}{\rho_{H_2O}} \tag{1-11}$$

Note that specific gravity is a dimensionless quantity.

A more frequently used property in thermodynamics is the **specific volume**. It is the reciprocal of density (Fig. 1-20) and is defined as *the volume per unit mass*:

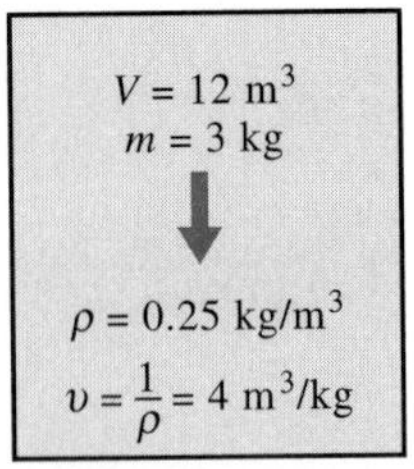

$$v = \frac{V}{m} = \frac{1}{\rho} \qquad (\text{m}^3/\text{kg}) \tag{1-12}$$

$V = 12\ \text{m}^3$
$m = 3\ \text{kg}$

$\rho = 0.25\ \text{kg/m}^3$
$v = \frac{1}{\rho} = 4\ \text{m}^3/\text{kg}$

**FIGURE 1-20**
Density is mass per unit volume; specific volume is volume per unit mass.

Note that in classical thermodynamics, the atomic structure of a substance (thus, the spaces between and within the molecules) is disregarded, and the substance is viewed to be a continuous, homogeneous matter with no microscopic holes, i.e., a **continuum**. This idealization is valid as long as we work with volumes, areas, and lengths that are large relative to the intermolecular spacings.

Properties are considered to be either *intensive* or *extensive*. **Intensive properties** are those that are independent of the size of a system, such as temperature, pressure, and density (Fig. 1-21). **Extensive properties** are those whose values depend on the size—or extent—of the system. Mass $m$, volume $V$, and total energy $E$ are some examples of extensive properties. As easy way to determine whether a property is intensive or extensive is to divide the system into two equal parts with a partition, as shown in Fig. 1-22. Each part will have the same value of intensive properties as the original system, but half the value of the extensive properties.

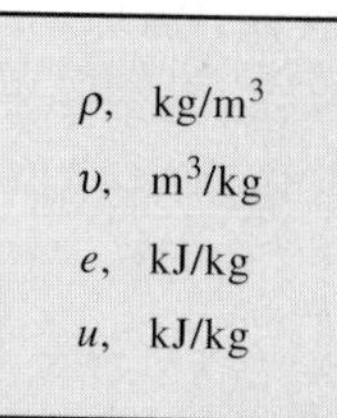


**FIGURE 1-21**
Intensive properties are independent of the size of the system.

Generally, uppercase letters are used to denote extensive properties (with mass $m$ being a major exception), and lowercase letters are used for intensive properties (with pressure $P$ and temperature $T$ being the obvious exceptions).

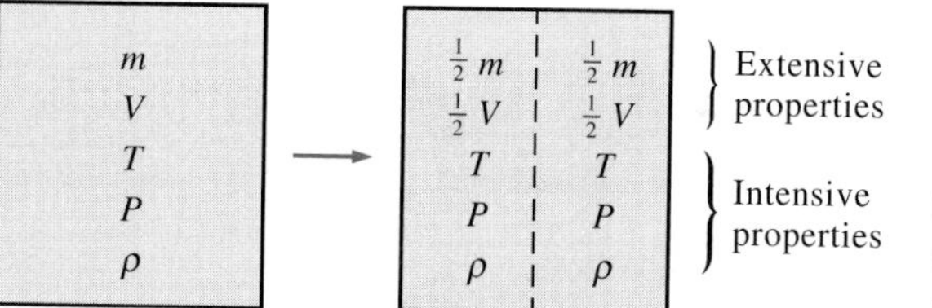

**FIGURE 1-22**
Criteria to differentiate intensive and extensive properties.

Extensive properties per unit mass are called **specific properties**. Some examples of specific properties are specific volume ($v = V/m$), specific total energy ($e = E/m$), and specific internal energy ($u = U/m$).

## 1-6 ■ STATE AND EQUILIBRIUM

Consider a system that is not undergoing any change. At this point, all the properties can be measured or calculated throughout the entire system, which gives us a set of properties that completely describe the condition, or the **state**, of the system. At a given state, all the properties of a system have fixed values. If the value of even one property changes, the state will change to a different one. In Fig. 1-23, a system is shown at two different states.

Thermodynamics deals with **equilibrium** states. The word *equilibrium* implies a state of balance. In an equilibrium state, there are no unbalanced potentials (or driving forces) within the system. A system that is in equilibrium experiences no changes when it is isolated from its surroundings.

There are many types of equilibrium, and a system is not in thermodynamic equilibrium unless the conditions of all the relevant types of equilibrium are satisfied (Fig. 1-24). For example, a system is in **thermal equilibrium** if the temperature is the same throughout the entire

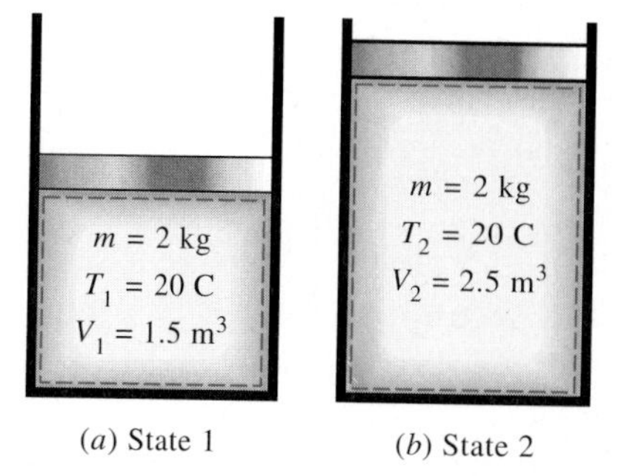


(*a*) State 1 (*b*) State 2

**FIGURE 1-23**
A system at two different states.

**FIGURE 1-24**

A system that involves changes with time is not in equilibrium.

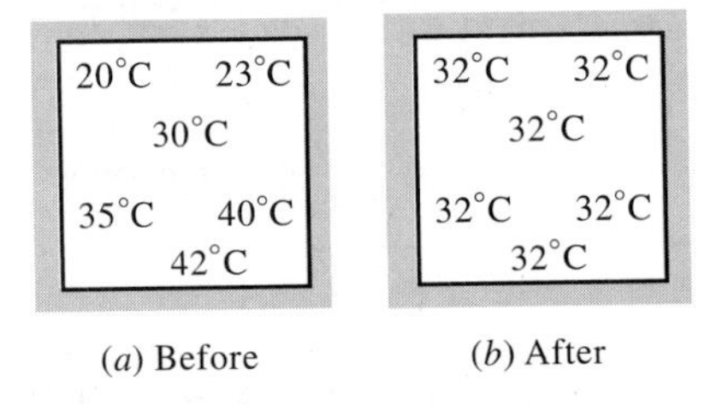


**FIGURE 1-25**

A closed system reaching thermal equilibrium.

system, as shown in Fig. 1-25*b*. That is, the system involves no temperature differentials, which is the driving force for heat flow. **Mechanical equilibrium** is related to pressure, and a system is in mechanical equilibrium if there is no change in pressure at any point of the system with time. However, the pressure may vary within the system with elevation as a result of gravitational effects. But the higher pressure at a bottom layer is balanced by the extra weight it must carry, and, therefore, there is no imbalance of forces. The variation of pressure as a result of gravity in most thermodynamic systems is relatively small and usually disregarded. If a system involves two phases, it is **phase equilibrium** when the mass of each phase reaches an equilibrium level and stays there. Finally, a system is in **chemical equilibrium** if its chemical composition does not change with time, i.e., no chemical reactions occur. A system will not be in equilibrium unless all the relevant equilibrium criteria are satisfied.

## 1-7 ■ PROCESSES AND CYCLES

Any change that a system undergoes from one equilibrium state to another is called a **process**, and the series of states through which a system passes during a process is called the **path** of the process (Fig. 1-26). To describe a process completely, one should specify the initial and final states of the process, as well as the path it follows, and the interactions with the surroundings.

When a process proceeds in such a manner that the system remains infinitesimally close to an equilibrium state at all times, it is called a **quasi-static**, or **quasi-equilibrium, process**. A quasi-equilibrium process can be viewed as a sufficiently slow process that allows the system to adjust itself internally so that properties in one part of the system do not change any faster than those at other parts.

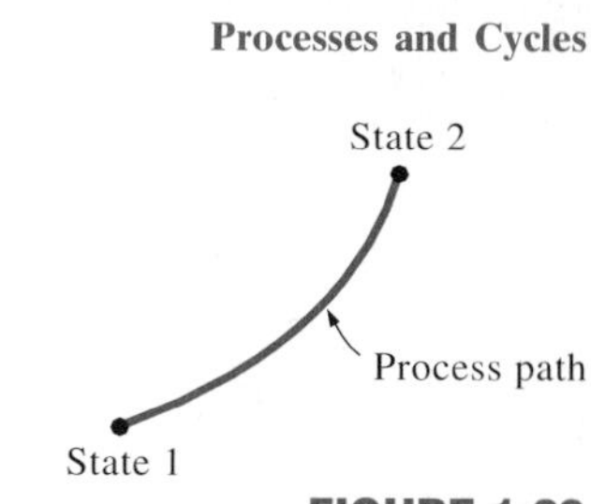


**FIGURE 1-26**
A process between states 1 and 2 and the process path.

This is illustrated in Fig. 1-27. When a gas in a piston–cylinder device is compressed suddenly, the molecules near the face of the piston will not have enough time to escape and they will have to pile up in a small region in front of the piston, thus creating a high-pressure region there. Because of this pressure difference, the system can no longer be said to be in equilibrium, and this makes the entire process non-quasi-equilibrium. However, if the piston is moved slowly, the molecules will have sufficient time to redistribute and there will not be a molecule pileup in front of the piston. As a result, the pressure inside the cylinder will always be uniform and will rise at the same rate at all locations. Since equilibrium is maintained at all times, this is a quasi-equilibrium process.

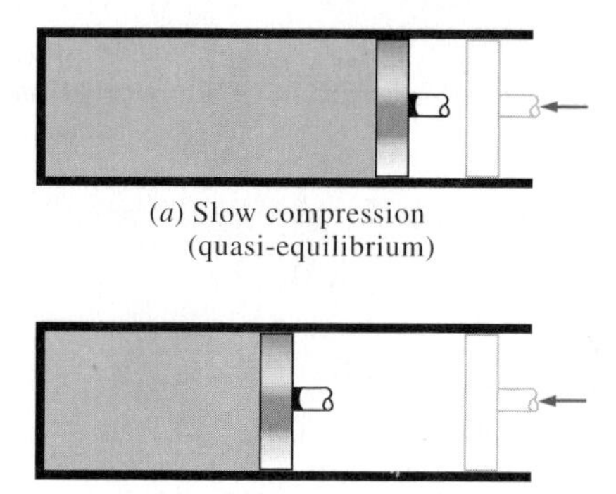

(*a*) Slow compression (quasi-equilibrium)

(*b*) Very fast compression (non-quasi-equilibrium)

**FIGURE 1-27**
Quasi-equilibrium and non-quasi-equilibrium compression processes.

It should be pointed out that a quasi-equilibrium process is an idealized process and is not a true representation of an actual process. But many actual processes closely approximate it, and they can be modeled as quasi-equilibrium with negligible error. Engineers are interested in quasi-equilibrium processes for two reasons. First, they are easy to analyze; second, work-producing devices deliver the most work when they operate on quasi-equilibrium processes (Fig. 1-28). Therefore, quasi-equilibrium processes serve as standards to which actual processes can be compared.

Process diagrams that are plotted by employing thermodynamic properties as coordinates are very useful in visualizing the processes. Some common properties that are used as coordinates are temperature $T$, pressure $P$, and volume $V$ (or specific volume $v$). Figure 1-29 shows the $P$-$V$ diagram of a compression process of a gas.

Note that the process path indicates a series of equilibrium states through which the system passes through during a process and has significance for quasi-equilibrium processes only. For non-quasi-equilibrium processes, we are not able to specify the states through which the system passes during the process and so we cannot speak of a process path. A non-quasi-equilibrium process is denoted by a dashed line between the initial and final states instead of a solid line.

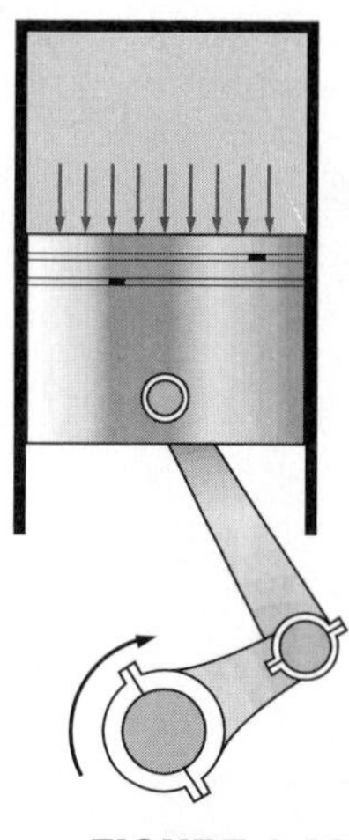

**FIGURE 1-28**
Work-producing devices operating in a quasi-equilibrium manner deliver the most work

The prefix *iso-* is often used to designate a process for which a particular property remains constant. An **isothermal process**, for example, is a process during which the temperature $T$ remains constant, an **isobaric process** is a process during which the pressure $P$ remains constant, and an **isochoric** (or **isometric**) **process** is a process during which the specific volume $v$ remains constant.

A system is said to have undergone a **cycle** if it returns to its initial state at the end of the process. That is, for a cycle the initial and final

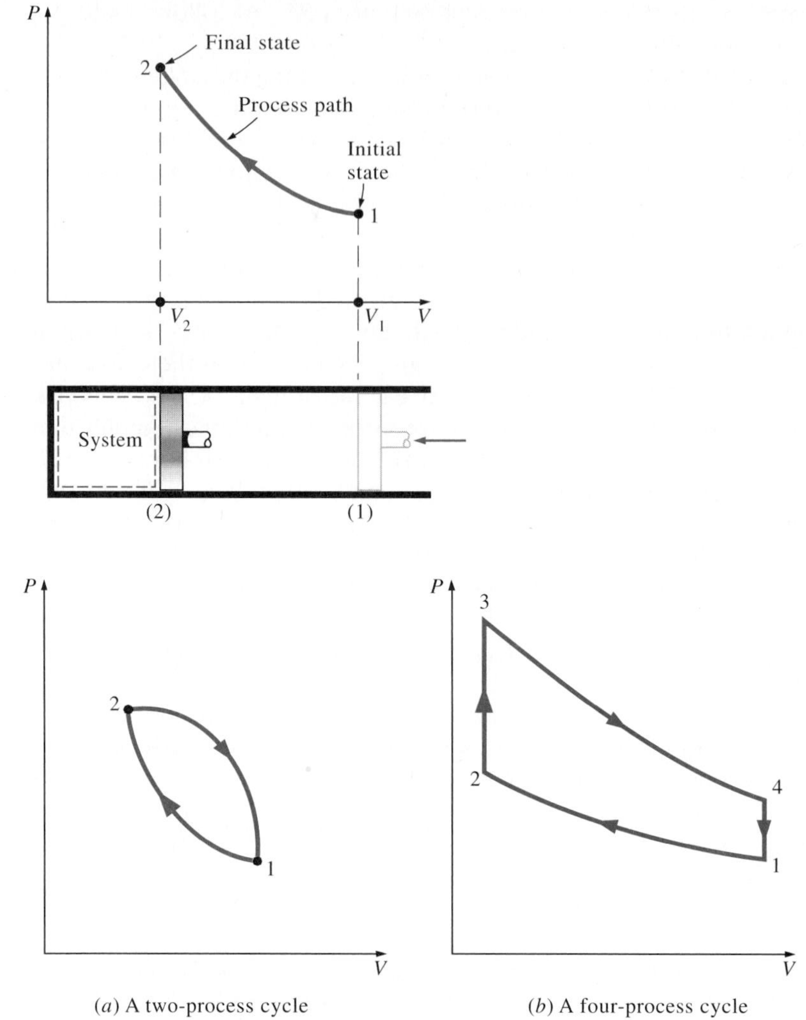


**FIGURE 1-29**
The $P$-$V$ diagram of a compression process.

**FIGURE 1-30**
Two thermodynamic cycles.

($a$) A two-process cycle

($b$) A four-process cycle

states are identical. The cycle in Fig. 1-30$a$ consists of two processes, and the one in Fig. 1-30$b$ consists of four processes (this is the ideal cycle for gasoline engines and is analyzed in Chap. 7).

## 1-8 ■ THE STATE POSTULATE

As noted earlier, the state of a system is described by its properties. But we know from experience that we do not need to specify all the properties in order to fix a state. Once a sufficient number of properties are specified, the rest of the properties assume certain values automatically. That is, specifying a certain number of properties is sufficient to fix a state. The number of properties required to fix the state of a system is given by the **state postulate**:

> *The state of a simple compressible system is completely specified by two independent, intensive properties.*

A system is called a **simple compressible system** in the absence of electrical, magnetic, gravitational, motion, and surface tension effects. These effects are due to external force fields and are negligible for most engineering problems. Otherwise, an additional property needs to be specified for each effect which is significant. If the gravitational effects are to be considered, for example, the elevation $z$ needs to be specified in addition to the two properties necessary to fix the state.

The state postulate requires that the two properties specified be **independent** to fix the state. Two properties are independent if one property can be varied while the other one is held constant. Temperature and specific volume, for example, are always independent properties, and together they can fix the state of a simple compressible system (Fig. 1-31). Temperature and pressure, however, are independent properties for single-phase systems, but are dependent properties for multiphase systems. At sea level ($P = 1$ atm), water boils at 100°C, but on a mountaintop where the pressure is lower, water boils at a lower temperature. That is, $T = f(P)$ during a phase-change process; thus, temperature and pressure are not sufficient to fix the state of a two-phase system. Phase-change processes are discussed in detail in the next chapter.

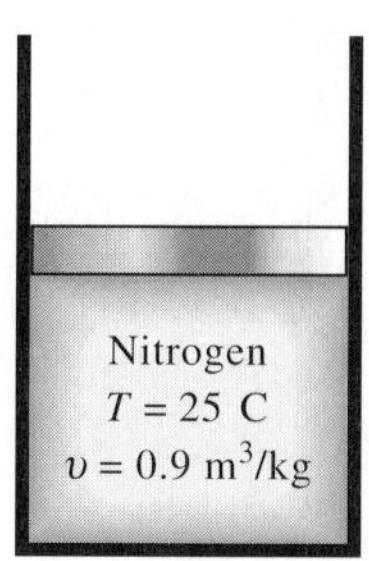

**FIGURE 1-31**
The state of nitrogen is fixed by two independent, intensive properties.

## 1-9 PRESSURE

Pressure is *the force exerted by a fluid per unit area.* We speak of pressure only when we deal with a gas or a liquid. The counterpart of pressure in solids is *stress.* For a fluid at rest, the pressure at a given point is the same in all directions. The pressure in a fluid increases with depth as a result of the weight of the fluid, as shown in Fig. 1-32. This is due to the fluid at lower levels carrying more weight than the fluid at upper levels. The pressure varies in the vertical direction as a result of gravitational effects, but there is no variation in the horizontal direction. The pressure in a tank containing a gas may be considered to be uniform since the weight of the gas is too small to make a significant difference (Fig. 1-33).

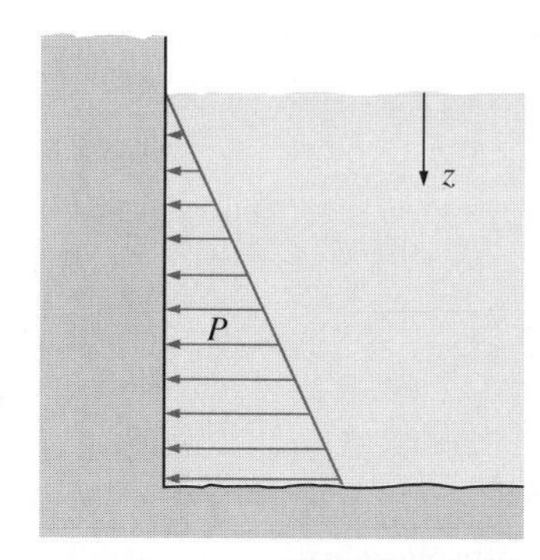


**FIGURE 1-32**
The pressure of a fluid at rest increases with depth (as a result of added weight).

Since pressure is defined as force per unit area, it has the unit of newtons per square meter (N/m$^2$), which is called a *pascal* (Pa). That is,

$$1 \text{ Pa} = 1 \text{ N/m}^2$$

The pressure unit pascal is too small for pressures encountered in practice: therefore, its multiples *kilopascal* (1 kPa $= 10^3$ Pa) and *megapascal* (1 MPa $= 10^6$ Pa) are commonly used. Two other common pressure units are the *bar* and *standard atmosphere*:

$$1 \text{ bar} = 10^5 \text{ Pa} = 0.1 \text{ MPa} = 100 \text{ kPa}$$

$$1 \text{ atm} = 101{,}325 \text{ Pa} = 101.325 \text{ kPa} = 1.01325 \text{ bars}$$

In the English system, the pressure unit is *pound-force per square inch* (lbf/in$^2$, or psi), and 1 atm $=$ 14.696 psi.

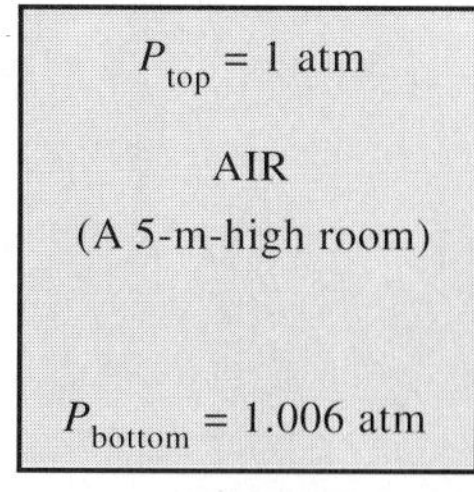


**FIGURE 1-33**
In a container filled with a gas the variation of pressure with height is negligible.

The actual pressure at a given position is called the **absolute pressure**, and it is measured relative to absolute vacuum, i.e., absolute zero pressure. Most pressure-measuring devices, however, are calibrated to read zero in the atmosphere (Fig. 1-34), and so they indicate the difference between the absolute pressure and the local atmospheric pressure. This difference is called the **gage pressure**. Pressures below atmospheric pressure are called **vacuum pressures** and are measured by vacuum gages which indicate the difference between the atmospheric pressure and the absolute pressure. Absolute, gage, and vacuum pressures are all positive quantities and are related to each other by

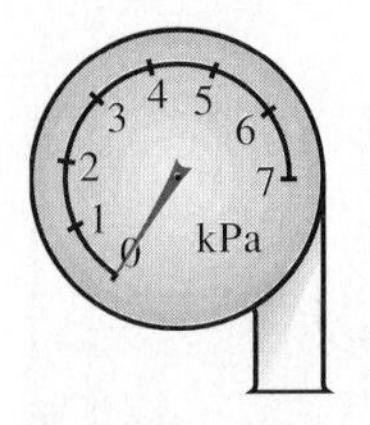


**FIGURE 1-34**
A pressure gage that is open to the atmosphere reads zero.

$$P_{\text{gage}} = P_{\text{abs}} - P_{\text{atm}} \quad \text{(for pressures above } P_{\text{atm}}) \tag{1-13}$$

$$P_{\text{vac}} = P_{\text{atm}} - P_{\text{abs}} \quad \text{(for pressures below } P_{\text{atm}}) \tag{1-14}$$

This is illustrated in Fig. 1-35.

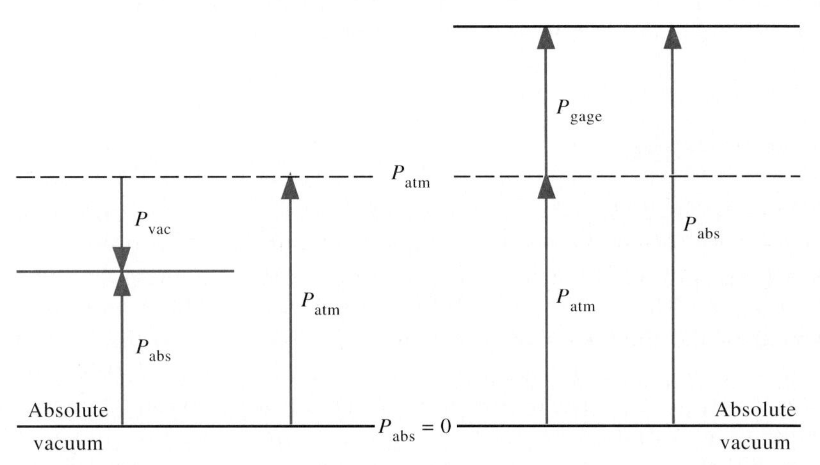


**FIGURE 1-35**
Absolute, gage, and vacuum pressures.

**EXAMPLE 1-3**

A vacuum gage connected to a chamber reads 5.8 psi at a location where the atmospheric pressure is 14.5 psi. Determine the absolute pressure in the chamber.

**Solution** The absolute pressure is easily determined from Eq. 1-14:

$$P_{\text{abs}} = P_{\text{atm}} - P_{\text{vac}} = (14.5 - 5.8)\text{ psi} = \mathbf{8.7\text{ psi}}$$

In thermodynamic relations and tables, absolute pressure is almost always used. Throughout this text, the pressure $P$ will denote *absolute pressure* unless it is otherwise specified. Often the letters "a" (for absolute pressure) and "g" (for gage pressure) are added to pressure units (such as psia and psig) in order to clarify what is meant.

### Manometer

Small and moderate pressure differences are often measured by using a device known as a **manometer**, which mainly consists of a glass or plastic U-tube containing a fluid such as mercury, water, alcohol, or oil. To keep the size of the manometer at a manageable level, heavy fluids such as mercury are used if large pressure differences are anticipated.

Consider the manometer shown in Fig. 1-36 which is used to measure the pressure in the tank. Since the gravitational effects of gases are negligible, the pressure anywhere in the tank and at position 1 has the same value. Furthermore, since pressure in a fluid does not vary in the horizontal direction within a fluid, the pressure at 2 is the same as the pressure at 1, or $P_2 = P_1$.

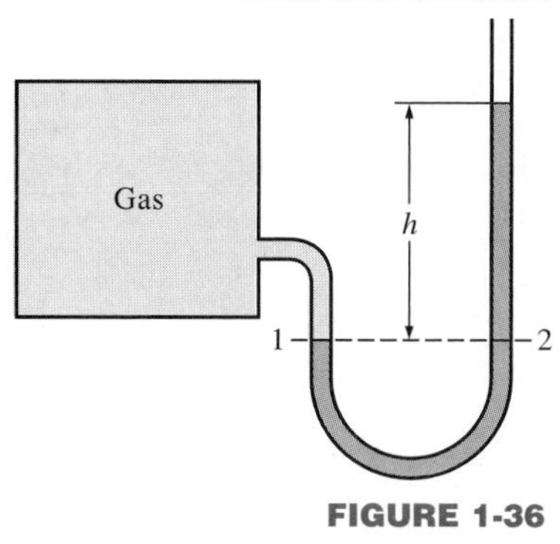


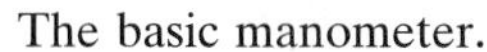
**FIGURE 1-36**
The basic manometer.

The differential fluid column of height $h$ is in static equilibrium, and its free-body diagram is shown in Fig. 1-37. A force balance in the vertical direction gives

$$AP_1 = AP_{\text{atm}} + W$$

where $$W = mg = \rho Vg = \rho Ahg$$

Thus,

$$P_1 = P_{\text{atm}} + \rho gh \qquad \text{(kPa)} \qquad (1\text{-}15)$$

In the above relations, $W$ is the weight of the fluid column, $\rho$ is the density of the fluid and is assumed to be constant, $g$ is the local gravitational acceleration, $A$ is the cross-sectional area of the tube, and $P_{\text{atm}}$ is the atmospheric pressure. The pressure difference can be expressed as

$$\Delta P = P_1 - P_{\text{atm}} = \rho gh \qquad \text{(kPa)} \qquad (1\text{-}16)$$

Note that the cross-sectional area of the tube has no effect on the height differential $h$, and thus the pressure exerted by the fluid.

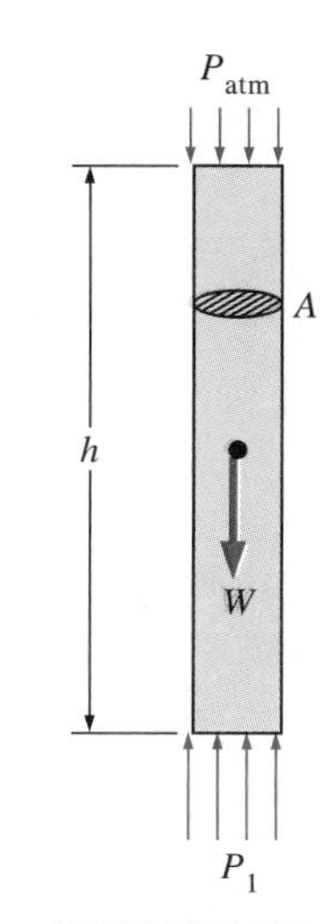


**FIGURE 1-37**
The free-body diagram of a fluid column of height $h$.

**EXAMPLE 1-4**

A manometer is used to measure the pressure in a tank. The fluid used has a specific gravity of 0.85, and the manometer column height is 55 cm, as shown in Fig. 1-38. If the local atmospheric pressure is 96 kPa, determine the absolute pressure within the tank.

**Solution** The gravitational acceleration is not specified, so we assume the standard value of 9.807 m/s$^2$. The density of the fluid is obtained by multiplying its specific gravity by the density of water, which is taken to be 1000 kg/m$^3$:

$$\rho = (\rho_s)(\rho_{H_2O}) = (0.85)(1000\ \text{kg/m}^3) = 850\ \text{kg/m}^3$$

From Eq. 1-15,

$$\begin{aligned} P &= P_{\text{atm}} + \rho gh \\ &= 96\ \text{kPa} + (850\ \text{kg/m}^3)(9.807\ \text{m/s}^2)(0.55\ \text{m})\left(\frac{1\ \text{kPa}}{1000\ \text{N/m}^2}\right) \\ &= 100.6\ \text{kPa} \end{aligned}$$

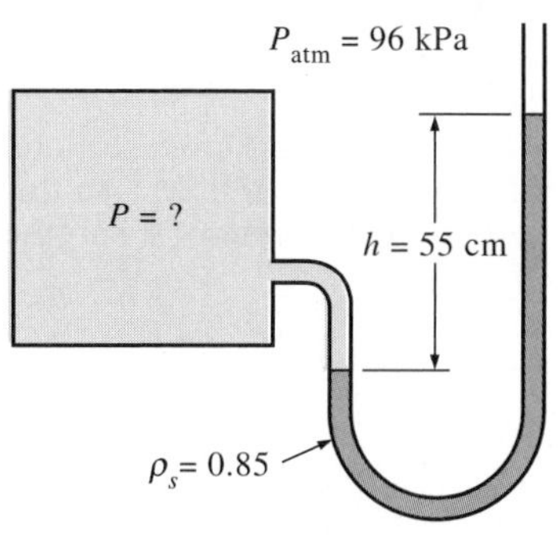


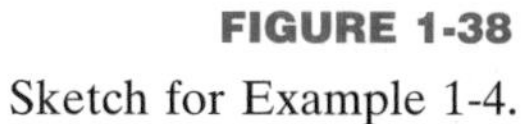
**FIGURE 1-38**
Sketch for Example 1-4.

## Barometer

The atmospheric pressure is measured by a device called a **barometer**; thus, the atmospheric pressure is often called the *barometric pressure.*

As Torricelli (1608–1647) discovered some centuries ago, the atmospheric pressure can be measured by inverting a mercury-filled tube into a mercury container that is open to the atmosphere, as shown in Fig. 1-39. The pressure at point $B$ is equal to the atmospheric pressure, and the pressure at $C$ can be taken to be zero since there is only mercury vapor above point $C$ and the pressure it exerts is negligible. Writing a force balance in the vertical direction gives

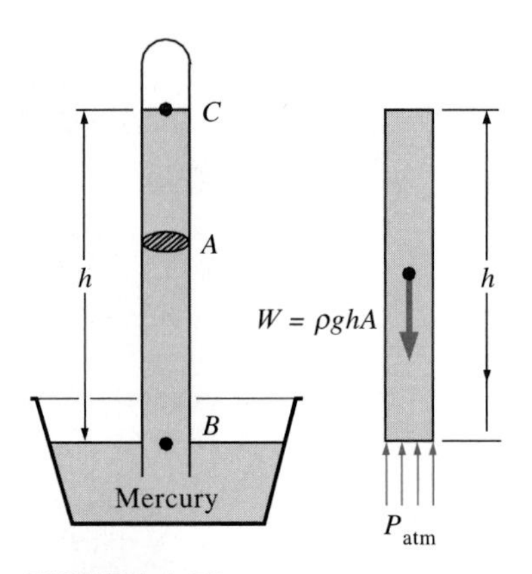


**FIGURE 1-39**
The basic barometer.

$$P_{\text{atm}} = \rho g h \qquad \text{(kPa)} \tag{1-17}$$

where $\rho$ is the density of mercury, $g$ is the local gravitational acceleration, and $h$ is the height of the mercury column above the free surface. Note that the length and the cross-sectional area of the tube have no effect on the height of the fluid column of a barometer (Fig. 1-40).

A frequently used pressure unit is the *standard atmosphere,* which is defined as the pressure produced by a column of mercury 760 mm in height at 0°C ($\rho_{\text{Hg}} = 13{,}595\ \text{kg/m}^3$) under standard gravitational acceleration ($g = 9.807\ \text{m/s}^2$). If water instead of mercury were used to measure the standard atmospheric pressure, a water column of about 10.3 m would be needed. Pressure is sometimes expressed (especially by weather forecasters) in terms of the height of the mercury column. The standard atmospheric pressure, for example, is 760 mmHg (29.92 inHg) at 0°C.

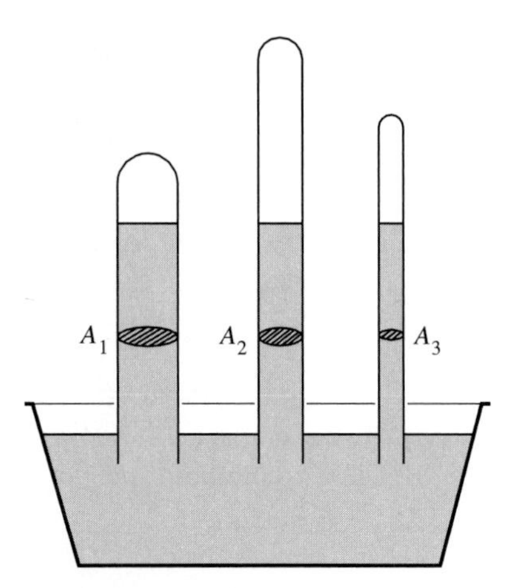


**FIGURE 1-40**
The length or the cross-sectional area of the tube has no effect on the height of the fluid column of a barometer.

The average atmospheric pressure $P_{\text{atm}}$ changes from 101.325 kPa at sea level to 89.88, 79.50, 54.05, 26.5, and 5.53 kPa at altitudes of 1000, 2000, 5000, 10,000, and 20,000 m, respectively. The average atmospheric pressure in Denver (elevation = 1610 m), for example, is 83.4 kPa.

Remember that the atmospheric pressure at a location is simply the weight of the air above that location per unit surface area. Therefore, it changes not only with elevation but also with weather conditions.

**EXAMPLE 1-5**

Determine the atmospheric pressure at a location where the barometric reading is 740 mmHg and the gravitational acceleration is $g = 9.7\ \text{m/s}^2$. Assume the temperature of mercury to be 10°C, at which its density is 13,570 kg/m³.

**Solution** From Eq. 1-17, the atmospheric pressure is determined to be

$$\begin{aligned} P_{\text{atm}} &= \rho g h \\ &= (13{,}570\ \text{kg/m}^3)(9.7\ \text{m/s}^2)(0.74\ \text{m})\left(\frac{1\ \text{N}}{1\ \text{kg}\cdot\text{m/s}^2}\right)\left(\frac{2\ \text{kPa}}{1000\ \text{N/m}^2}\right) \\ &= 97.41\ \text{kPa} \end{aligned}$$

**EXAMPLE 1-6**

The piston of a piston–cylinder device containing a gas has a mass of 60 kg and a cross-sectional area of $0.04\ \mathrm{m}^2$, as shown in Fig. 1-41. The local atmospheric pressure is 0.97 bar, and the gravitational acceleration is $9.8\ \mathrm{m/s^2}$.

(*a*) Determine the pressure inside the cylinder.

(*b*) If some heat is transferred to the gas and its volume doubles, do you expect the pressure inside the cylinder to change?

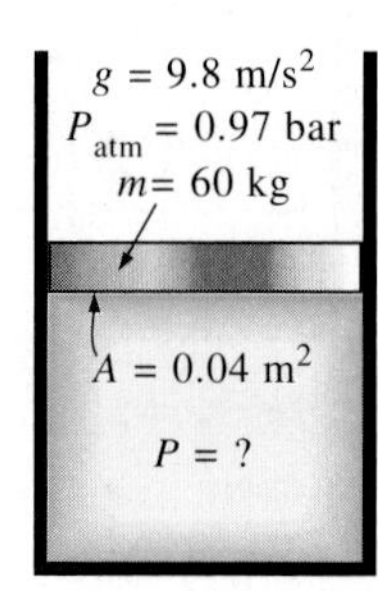


**FIGURE 1-41**
Sketch for Example 1-6.

**Solution** (*a*) The gas pressure in the piston–cylinder device depends on the atmospheric pressure and the weight of the piston. Drawing the free-body diagram of the piston (Fig. 1-42) and balancing the vertical forces yield

$$PA = P_{\mathrm{atm}}A + W$$

$$P = P_{\mathrm{atm}} + \frac{mg}{A}$$

$$= 0.97\ \mathrm{bar} + \frac{(60\ \mathrm{kg})(9.8\ \mathrm{m/s^2})}{0.04\ \mathrm{m^2}}\left(\frac{1\ \mathrm{N}}{1\ \mathrm{kg\cdot m/s^2}}\right)\left(\frac{1\ \mathrm{bar}}{10^5\ \mathrm{N/m^2}}\right)$$

$$= 1.117\ \mathrm{bars}$$

(*b*) The volume change will have no effect on the free-body diagram drawn in part (*a*), and therefore the pressure inside the cylinder will remain the same.

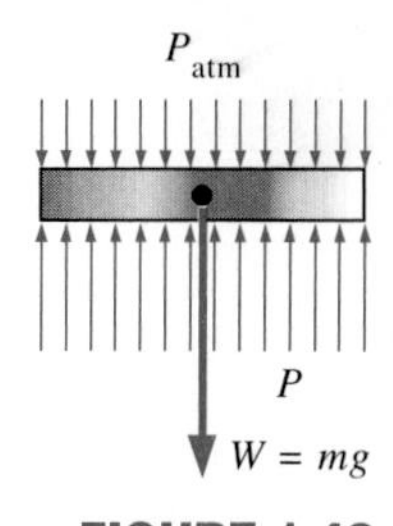


**FIGURE 1-42**
Free-body diagram of the piston.

## 1-10 ■ TEMPERATURE AND THE ZEROTH LAW OF THERMODYNAMICS

Although we are familiar with **temperature** as a measure of "hotness" or "coldness," it is not easy to give an exact definition for it. Based on our physiological sensations, we express the level of temperature qualitatively with words like *freezing cold, cold, warm, hot,* and *red-hot.* However, we cannot assign numerical values to temperatures based on our sensations alone. Furthermore, our senses may be misleading. A metal chair, for example, will feel much colder than a wooden one even when both are at the same temperature.

Fortunately, several properties of materials change with temperature in a repeatable and predictable way, and this forms the basis for accurate temperature measurement. The commonly used mercury-in-glass thermometer, for example, is based on the expansion of mercury with temperature. Temperature is also measured by using several other temperature-dependent properties.

It is common experience that a cup of hot coffee left on the table eventually cools off and a cold drink eventually warms up. That is, when a body is brought into contact with another body which is at a different temperature, heat is transferred from the body at higher temperature to the one at lower temperature until both bodies attain the same temperature (Fig. 1-43). At the point, the heat transfer stops, and the two bodies are said to have reached **thermal equilibrium**. The equality of temperature is the only requirement for thermal equilibrium.

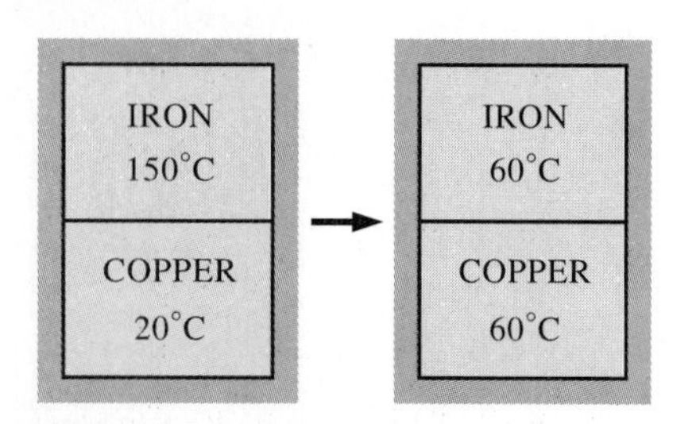


**FIGURE 1-43**
Two bodies reaching thermal equilibrium after being brought into contact in an isolated enclosure.

The **zeroth law of thermodynamics** states that if two bodies are in thermal equilibrium with a third body, they are also in thermal equilibrium with each other. It may seem silly that such an obvious fact is called one of the basic laws of thermodynamics. However, it cannot be concluded from the other laws of thermodynamics, and it serves as a basis for the validity of temperature measurement. By replacing the third body with a thermometer, the zeroth law can be restated as *two bodies are in thermal equilibrium if both have the same temperature reading even if they are not in contact.*

The zeroth law was first formulated and labeled by R. H. Fowler in 1931. As the name suggests, its value as a fundamental physical principle was recognized more than half a century after the formulation of the first and the second laws of thermodynamics. It was named the zeroth law since it should have preceded the first and the second laws of thermodynamics.

## Temperature Scales

Temperature scales enable scientists to use a common basis for temperature measurements, and several have been introduced throughout history. All temperature scales are based on some easily reproducible states such as the freezing and boiling points of water, which are also called the *ice point* and the *steam point,* respectively. A mixture of ice and water that is in equilibrium with air saturated with vapor at 1-atm pressure is said to be at the ice point, and a mixture of liquid water and water vapor (with no air) in equilibrium at 1-atm pressure is said to be at the steam point.

The temperature scales used in the SI and in the English system today are the **Celsius scale** (formerly called the *centigrade scale*; in 1948 it was renamed after the Swedish astronomer A. Celsius, 1701–1744, who devised it) and the **Fahrenheit scale** (named after the German instrument maker G. Fahrenheit, 1686–1736), respectively. On the Celsius scale, the ice and steam points are assigned the values of 0 and 100°C, respectively. The corresponding values on the Fahrenheit scale are 32 and 212°F. These are often referred to as *two-point scales* since temperature values are assigned at two different points.

In thermodynamics, it is very desirable to have a temperature scale that is independent of the properties of any substance or substances. Such a temperature scale is called a **thermodynamic temperature scale**, which is developed in Chap. 5 in conjunction with the second law of thermodynamics. The thermodynamic temperature scale in the SI is the **Kelvin scale**, named after Lord Kelvin (1824–1907). The temperature unit on this scale is the **kelvin**, which is designated by K (not °K; the degree symbol was officially dropped from kelvin in 1967). The lowest temperature on the Kelvin scale is 0 K. Using nonconventional refrigeration techniques, scientists have approached absolute zero kelvin (they achieved 0.000000002 K in 1989).

The thermodynamic temperature scale in the English system is the **Rankine scale**, named after William Rankine (1820–1872). The temperature unit on this scale is the **rankine**, which is designated by R.

A temperature scale that turns out to be identical to the Kelvin scale is the **ideal gas temperature scale**. The temperatures on this scale are measured using a **constant-volume gas thermometer**, which is basically a rigid vessel filled with a gas, usually hydrogen or helium, at low pressure. This thermometer is based on the principle that *at low pressures, the temperature of a gas is proportional to its pressure at constant volume.* That is, the temperature of a gas of fixed volume varies *linearly* with pressure at sufficiently low pressures. Then the relationship between the temperature and the pressure of the gas in the vessel can be expressed as

$$T = a + bP \tag{1-18}$$

where the values of the constants $a$ and $b$ for a gas thermometer are determined experimentally. Once $a$ and $b$ are known, the temperature of a medium can be calculated from the relation above by immersing the rigid vessel of the gas thermometer into the medium, and measuring the gas pressure when thermal equilibrium between the medium and the gas in the vessel whose volume is held constant is established.

An ideal gas temperature scale can be developed by measuring the pressures of the gas in the vessel at two reproducible points (such as the ice and the steam points), and assigning suitable values to temperatures at those two points. Considering that only one straight line passes through two fixed points on a plane, these two measurements are sufficient to determine the constants $a$ and $b$ in Eq. 1-18. Then the unknown temperature $T$ of a medium corresponding to a pressure reading $P$ can be determined from that equation by a simple calculation. The values of the constants will be different for each thermometer, depending on the type and the amount of the gas in the vessel, and the temperature values assigned at the two reference points. If the ice and steam points are assigned the values 0 and 100, respectively, then the gas temperature scale will be identical to the Celsius scale. In this case the value of the constant $a$ (which corresponds to an absolute pressure of zero) is determined to be −273.15°C regardless of the type and the amount of the gas in the vessel of the gas thermometer. That is, on a $P$-$T$ diagram, all the straight lines passing through the data points in this case will intersect the temperature axis at −273.15°C when extrapolated, as shown in Fig. 1-44. This is the lowest temperature that can be obtained by a gas thermometer, and thus we can obtain an *absolute gas temperature scale* by assigning a value of zero to the constant $a$ in Eq. 1-18. In that case Eq. 1-18 reduces to $T = bP$, and thus we need to specify the temperature at only *one* point to define an absolute gas temperature scale.

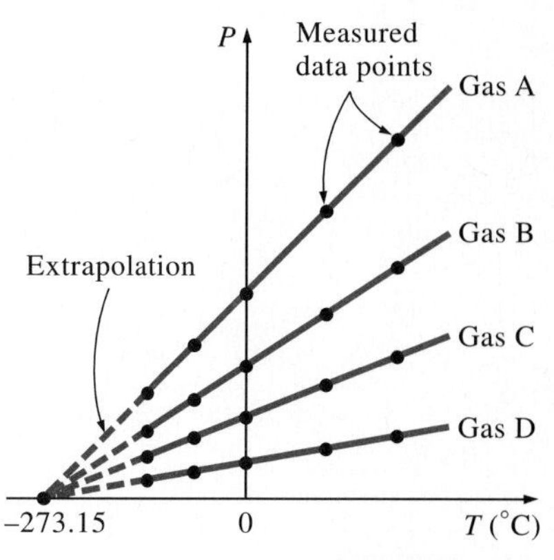


**FIGURE 1-44**

$P$ versus $T$ plots of the expermental data obtained from a constant-volume gas thermometer using four different gases at different (but low) pressures.

It should be noted that the absolute gas temperature scale is not a thermodynamic temperature scale, since it cannot be used at very low temperatures (due to condensation) and at very high temperatures (due to dissociation and ionization). However, absolute gas temperature is identical to the thermodynamic temperature in the temperature range in which the gas thermometer can be used, and thus we can view the thermodynamic temperature scale at this point as an absolute gas temperature scale that utilizes an "ideal" or "imaginary" gas that always acts as a low-pressure gas regardless of the temperature. If such a gas thermometer existed, it would read zero kelvin at absolute zero pressure, which corresponds to −273.15°C on the Celsius scale (Fig. 1-45).

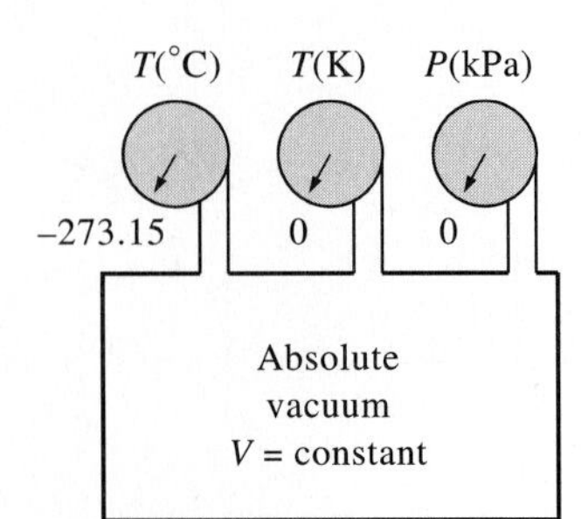


**FIGURE 1-45**
A constant-volume gas thermometer would read −273.15°C at absolute zero pressure.

The Kelvin scale is related to the Celsius scale by

$$T(\mathrm{K}) = T(^\circ\mathrm{C}) + 273.15 \tag{1-19}$$

The Rankine scale is related to the Fahrenheit scale by

$$T(\mathrm{R}) = T(^\circ\mathrm{F}) + 459.67 \tag{1-20}$$

It is common practice to round the constant in Eq. 1-19 to 273 and that in Eq. 1-20 to 460.

The temperature scales in two unit systems are related by

$$T(\mathrm{R}) = 1.8\,T(\mathrm{K}) \tag{1-21}$$

$$T(^\circ\mathrm{F}) = 1.8\,T(^\circ\mathrm{C}) + 32 \tag{1-22}$$

A comparison of various temperature scales is given in Fig. 1-46.

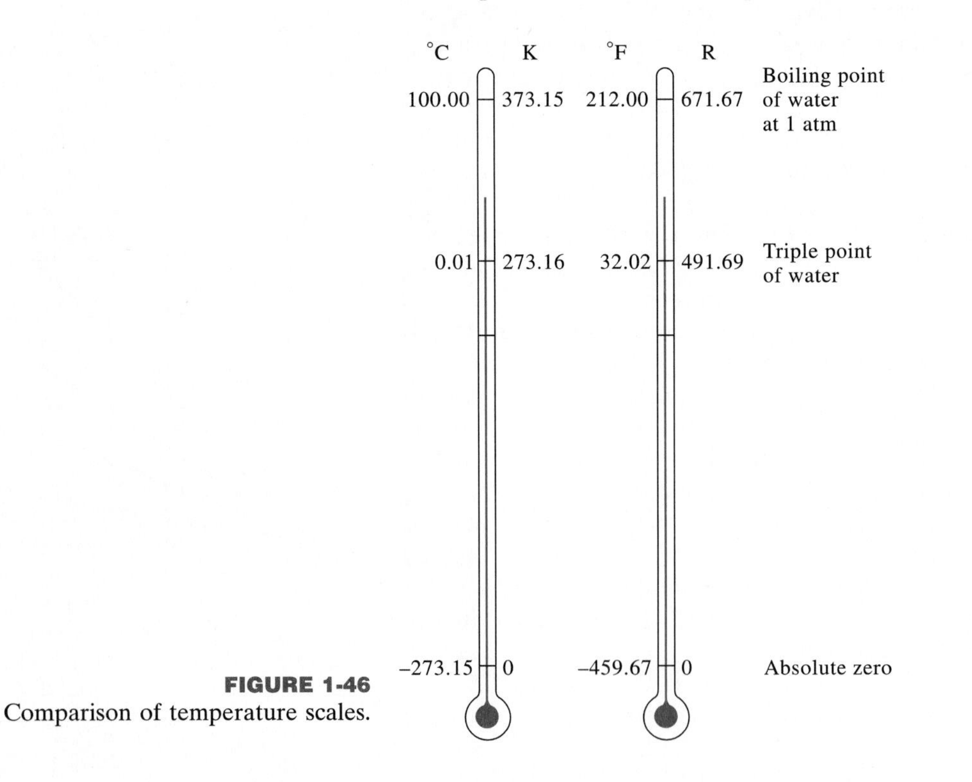


**FIGURE 1-46**
Comparison of temperature scales.

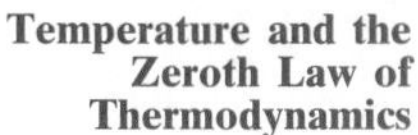

At the Tenth Conference on Weights and Measures in 1954, the Celsius scale was redefined in terms of a single fixed point and the absolute temperature scale. The selected single point is the *triple point* of water (the state at which all three phases of water coexist in equilibrium), which is assigned the value 0.01°C. The magnitude of the degree is defined from the absolute temperature scale. As before, the boiling point of water at 1-atm pressure is 100.00°C. Thus the new Celsius scale is essentially the same as the old one.

On the Kelvin scale, the size of the temperature unit *kelvin* is defined as "the fraction 1/273.16 of the thermodynamic temperature of the triple point of water, which is assigned the value of 273.16 K." The ice points on the Celsius and Kelvin scales are 0°C and 273.15 K, respectively.

Note that the magnitudes of each division of 1 K and 1°C are identical (Fig. 1-47). Therefore, when we are dealing with temperature differences $\Delta T$, the temperature interval on both scales is the same. Raising the temperature of a substance by 10°C is the same as raising it by 10 K. That is,

$$\Delta T(\text{K}) = \Delta T(^\circ\text{C}) \tag{1-23}$$

Similarly,

$$\Delta T(\text{R}) = \Delta T(^\circ\text{F}) \tag{1-24}$$

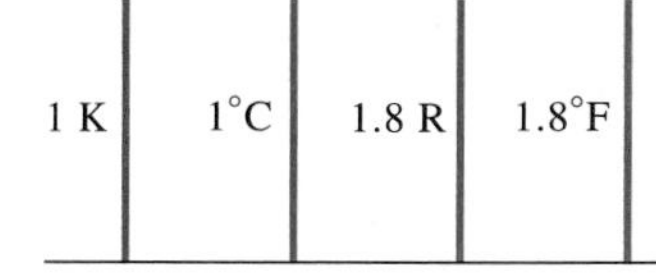


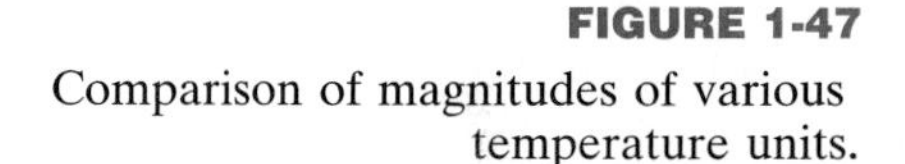

**FIGURE 1-47**
Comparison of magnitudes of various temperature units.

Some thermodynamic relations involve the temperature $T$ and often the question arises of whether it is in K or °C. If the relation involves temperature differences (such as $a = b\,\Delta T$), it makes no difference and either can be used. But if the relation involves temperatures only instead of temperature differences (such as $a = bT$) then K must be used. When in doubt, it is always safe to use K because there are virtually no situations in which the use of K is incorrect, but there are many thermodynamic relations that will yield an erroneous result if °C is used.

**EXAMPLE 1-7**

During a heating process, the temperature of a system rises by 10°C. Express this rise in temperature in K, °F, and R.

**Solution** This problem deals with temperature changes, which are identical in Kelvin and Celsius scales. Then from Eq. 1-23,

$$\Delta T(\text{K}) = \Delta T(^\circ\text{C}) = 10\text{ K}$$

The temperature changes in Fahrenheit and Rankine scales are also identical and are related to the changes in Celsius and Kelvin scales through Eqs. 1-21 and 1-24:

$$\Delta T(\text{R}) = 1.8\,\Delta T(\text{K}) = (1.8)(10) = 18\text{ R}$$

and

$$\Delta T(^\circ\text{F}) = \Delta T(\text{R}) = 18^\circ\text{F}$$

## 1-11 ■ SUMMARY

In this chapter, the basic concepts of thermodynamics are introduced and discussed. *Thermodynamics* is the science that primarily deals with energy. The *first law of thermodynamics* is simply an expression of the conservation of energy principle, and it asserts that *energy* is a thermodynamic property. The *second law of thermodynamics* asserts that energy has *quality* as well as *quantity,* and actual processes occur in the direction of decreasing quality of energy.

A system of fixed mass is called a *closed system,* or *control mass,* and a system that involves mass transfer across its boundaries is called an *open system,* or *control volume.* The mass-dependent properties of a system are called *extensive properties* and the others, *intensive properties. Density* is mass per unit volume, and *specific volume* is volume per unit mass.

The sum of all forms of energy of a system is called *total energy,* which is considered to consist of internal, kinetic, and potential energies. *Internal energy* represents the molecular energy of a system and may exist in sensible, latent, chemical, and nuclear forms.

A system is said to be in *thermodynamic equilibrium* if it maintains thermal, mechanical, phase, and chemical equilibrium. Any change from one state to another is called a *process.* A process with identical end states is called a *cycle.* During a *quasi-static* or *quasi-equilibrium process,* the system remains practically in equilibrium at all times. The state of a simple, compressible system is completely specified by two independent, intensive properties.

Force per unit area is called *pressure,* and its unit is the pascal. The absolute, gage, and vacuum pressures are related by

$$P_{\text{gage}} = P_{\text{abs}} - P_{\text{atm}} \qquad \text{(kPa)}$$
$$P_{\text{vac}} = P_{\text{atm}} - P_{\text{abs}} \qquad \text{(kPa)}$$

Small to moderate pressure differences are measured by a *manometer,* and a differential fluid column of height $h$ corresponds to a pressure difference of

$$\Delta P = \rho g h \qquad \text{(kPa)}$$

where $\rho$ is the fluid density and $g$ is the local gravitational acceleration. The atmospheric pressure is measured by a *barometer* and is determined from

$$P_{\text{atm}} = \rho g h \qquad \text{(kPa)}$$

where $h$ is the height of the liquid column above the free surface.

The *zeroth law of thermodynamics* states that two bodies are in thermal equilibrium if both have the same temperature reading even if they are not in contact.

The temperature scales used in the SI and the English system today are the *Celsius scale* and the *Fahrenheit scale,* respectively. The absolute temperature scale in the SI is the *Kelvin scale,* which is related to the Celsius scale by

$$T(\text{K}) = T(^\circ\text{C}) + 273.15$$

In the English system, the absolute temperature scale is the *Rankine scale,* which is related to the Fahrenheit scale by

$$T(\text{R}) = T(^\circ\text{F}) + 459.67$$

The magnitudes of each division of 1 K and 1°C are identical, and so are the magnitudes of each division of 1 R and 1°F. That is,

$$\Delta T(\text{K}) = \Delta T^\circ(\text{C})$$

and

$$\Delta T(\text{R}) = \Delta T(^\circ\text{F})$$

## REFERENCES AND SUGGESTED READING

**1** American Society for Testing and Materials, *Standards for Metric Practice,* ASTM E 380-79, January 1980.

**2** R. T. Balmer, *Thermodynamics,* West, St. Paul, MN, 1990.

**3** A. Bejan, *Advanced Engineering Thermodynamics,* Wiley, New York, 1988.

**4** W. Z. Black and J. G. Hartley, *Thermodynamics,* Harper and Row, New York, 1985.

**5** J. B. Jones and G. A. Hawkins, *Engineering Thermodynamics,* 2d ed., Wiley, New York, 1986.

**6** G. J. Van Wylen and R. E. Sonntag, *Fundamentals of Classical Thermodynamics,* 3d ed., Wiley, New York, 1985.

**7** K. Wark, *Thermodynamics,* 5th ed., McGraw-Hill, New York, 1988.

## PROBLEMS*

### Thermodynamics and Energy

**1-1C** What is the difference between the classical and the statistical approaches to thermodynamics?

**1-2C** Why does a bicyclist pick up speed on a downhill road even when he is not pedaling? Does this violate the conservation of energy principle?

**1-3C** An office worker claims that a cup of cold coffee on his table warmed up to 80°C by picking up energy from the surrounding air, which

*Students are encouraged to answer *all* the concept "C" questions.

is at 25°C. Is there any truth to his claim? Does this process violate any thermodynamic laws?

### Mass, Force, and Acceleration

**1-4C** What is the difference between pound-mass and pound-force?

**1-5C** What is the net force acting on a car cruising at a constant velocity of 70 km/h (*a*) on a level road and (*b*) on an uphill road?

**1-6** What is the force required to accelerate a mass of 30 kg at a rate of 15 $m/s^2$? *Answer:* 450 N

**1-6E** What is the force required to accelerate a mass of 60 lbm at a rate of 45 $ft/s^2$? *Answer:* 83.92 lbf

**1-7** A 5-kg plastic tank that has a volume of 0.2 $m^3$ is filled with liquid water. Assuming the density of water is 1000 $kg/m^3$, determine the weight of the combined system.

**1-8** Determine the mass and the weight of the air contained in a room whose dimensions are 6 m × 6 m × 8 m. Assume the density of the air is 1.16 $kg/m^3$. *Answer:* 334.1 kg, 3277 N

**1-8E** Determine the mass and the weight of the air contained in a room whose dimensions are 15 ft × 20 ft × 20 ft. Assume the density of the air is 0.0724 $lbm/ft^3$.

**1-9** At 45° latitude, the gravitational acceleration as a function of elevation $z$ above sea level is given by $g = a - bz$, where $a = 9.807\ m/s^2$ and $b = 3.32 \times 10^{-6}\ s^{-2}$. Determine the height above sea level where the weight of a subject will decrease by 1 percent. *Answer:* 29,539 m

**1-10** A 75- kg astronaut took his bathroom scale (a spring scale) and a beam scale (compares masses) to the moon where the local gravity is $g = 1.67\ m/s^2$. Determine how much he will weigh (*a*) on the spring scale and (*b*) on the beam scale.
*Answer:* (*a*) 125 N (or 12.8 kg); (*b*) 735 N (or 75 kg)

**1-10E** A 150-lbm astronaut took his bathroom scale (a spring scale) and a beam scale (compares masses) to the moon where the local gravity is $g = 5.48\ ft/s^2$. Determine how much he will weigh (*a*) on the spring scale and (*b*) on the beam scale.
*Answer:* (*a*) 25.5 lbf; (*b*) 150 lbf

**1-11** The acceleration of high-speed aircraft is sometimes expressed in *g*s (in multiples of the standard acceleration of gravity). Determine the net upward force, in N, that a 70-kg man would experience in an aircraft whose acceleration is 6*g*.

**1-12** A 5-kg rock is thrown upward with a force of 150 N at a location where the local gravitational acceleration is 9.79 $m/s^2$. Determine the acceleration of the rock, in $m/s^2$.

## Forms of Energy, Systems, State, Properties

**1-13C** Most of the energy generated in the engine of a car is rejected to the air by the radiator through the circulating water. Should the radiator be analyzed as a closed system or as an open system? Explain.

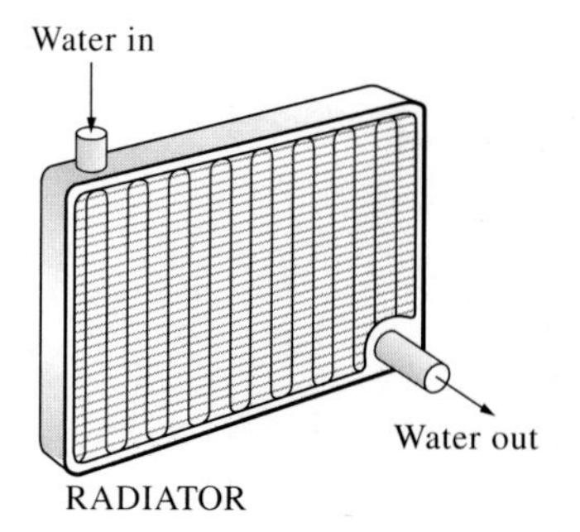


**FIGURE P1-13C**

**1-14C** A can of soft drink at room temperature is put into the refrigerator so that it will cool. Would you model the can of soft drink as a closed system or as an open system? Explain.

**1-15C** Can mass cross the boundaries of a closed system? How about energy?

**1-16C** Portable electric heaters are commonly used to heat small rooms. Explain the energy transformation involved during this heating process.

**1-17C** Consider the process of heating water on top of an electric range. What are the forms of energy involved during this process? What are the energy transformations that take place?

**1-18C** What is the difference between the macroscopic and microscopic forms of energy?

**1-19C** What is total energy? Identify the different forms of energy that constitute the total energy.

**1-20C** List the forms of energy that contribute to the internal energy of a system.

**1-21C** How are heat, internal energy, and thermal energy related to each other?

**1-22C** What is the difference between intensive and extensive properties?

**1-23C** For a system to be in thermodynamic equilibrium, do the temperature and the pressure have to be the same everywhere?

**1-24C** What is a quasi-equilibrium process? What is its importance in engineering?

**1-25C** Define the isothermal, isobaric, and isochoric processes.

**1-26C** What is the state postulate?

**1-27C** Is the state of the air in an isolated room completely specified by the temperature and the pressure? Explain.

## Pressure

**1-28C** What is the difference between gage pressure and absolute pressure?

**1-29C** Explain why some people experience nose bleeding and some others experience shortness of breath at high elevations.

**1-30** A vacuum gage connected to a tank reads 30 kPa at a location where the barometric reading is 755 mmHg. Determine the absolute pressure in the tank. Take $\rho_{Hg}$ = 13,590 kg/m$^3$. *Answer:* 70.6 kPa

**1-30E** A vacuum gage connected to a tank reads 5.4 psi at a location where the barometric reading is 28.5 inHg. Determine the absolute pressure in the tank. Take $\rho_{Hg}$ = 848.4 lbm ft$^3$. *Answer:* 8.6 psia

**1-31** A pressure gage connected to a tank reads 3.5 bars at a location where the barometric reading is 75 cmHg. Determine the absolute pressure in the tank. Take $\rho_{Hg}$ = 13,590 kg/m$^3$. *Answer:* 4.5 bars

**1-31E** A pressure gage connected to a tank read 50 psi at a location where the barometric reading is 29.1 inHg. Determine the absolute pressure in the tank. Take $\rho_{Hg}$ = 848.4 lbm/ft$^3$. *Answer:* 64.29 psia

**1-32** A pressure gage connected to a tank reads 500 kPa at a location where the atmospheric pressure is 94 kPa. Determine the absolute pressure in the tank.

**1-33** The barometer of a mountain hiker reads 930 mbars at the beginning of a hiking trip and 780 mbars at the end. Neglecting the effect of altitude on local gravitational acceleration, determine the vertical distance climbed. Assume an average air density of 1.20 kg/m$^3$ and take $g$ = 9.7 m/s$^2$. *Answer:* 1289 m

**1-33E** The barometer of a mountain hiker reads 13.8 psia at the beginning of a hiking trip and 12.6 psia at the end. Neglecting the effect of altitude on local gravitational acceleration, determine the vertical distance she climbed. Assume an average air density of 0.074 lbm/ft$^3$ and take $g$ = 31.8 ft/s$^2$. *Answer:* 2363 ft

**1-34** The basic barometer can be used to measure the height of a building. If the barometric readings at the top and at the bottom of a building are 730 and 755 mmHg, respectively, determine the height of the building. Assume an average air density of 1.18 kg/m$^3$.

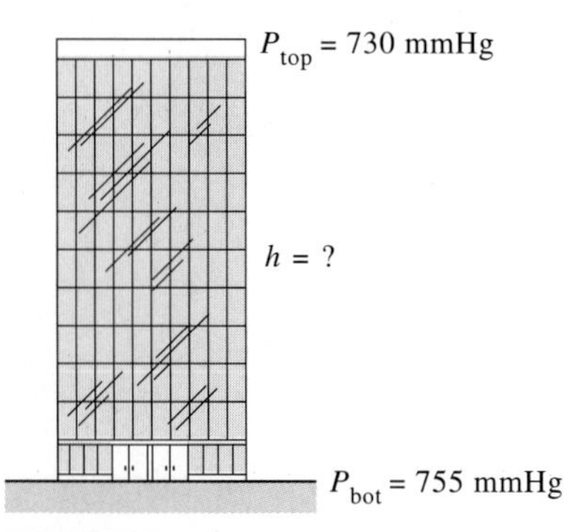


**FIGURE P1-34**

**1-35** Determine the pressure exerted on a diver at 30 m below the free surface of the sea. Assume a barometric pressure of 101 kPa and a specific gravity of 1.03 for seawater. *Answer:* 404.0 KPa

**1-36** Determine the pressure exerted on the surface of a submarine cruising 100 m below the free surface of the sea. Assume that the barometric pressure is 101 kPa and the specific gravity of seawater is 1.03.

**1-36E** Determine the pressure exerted on the surface of a submarine cruising 300 ft below the free surface of the sea. Assume that the barometric pressure is 14.7 psia and the specific gravity of seawater is 1.03.

**1-37** A gas is contained in a vertical, frictionless piston–cylinder device. The piston has a mass of 4 kg and cross-sectional area of 35 cm$^2$.

A compressed spring above the piston exerts a force of 60 N on the piston. If the atmospheric pressure is 95 kPa, determine the pressure inside the 95 cylinder. *Answer:* 123.4 kPa

**1-37E** A gas is contained in a vertical, frictionless piston-cylinder device. The piston has a mass of 8 lbm and a cross-sectional area of 5 in$^2$. A compressed spring above the piston exerts a force of 18 lbf on the piston. If the atmospheric pressure is 14.6 psia, determine the pressure inside the cylinder. *Answer:* 19.8 psia

**FIGURE P1-37**

**1-38** Both a gage and a manometer are attached to a gas tank to measure its pressure. If the reading on the pressure gage is 80 kPa, determine the distance between the two fluid levels of the manometer if the fluid is (*a*) mercury ($\rho$ = 13,600 kg/m$^3$) or (*b*) water ($\rho$ = 1000 kg/m$^3$).

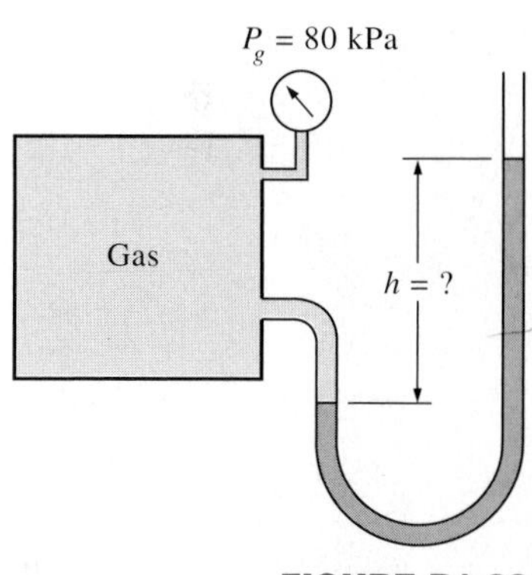

**FIGURE P1-38**

**1-39** A manometer containing oil ($\rho$ = 850 kg/m$^3$) is attached to a tank filled with air. If the oil-level difference between the two columns is 45 cm and the atmospheric pressure is 98 kPa, determine the absolute pressure of the air in the tank. *Answer:* 101.75 kPa

**1-39E** A manometer containing oil ($\rho$ = 53 lbm/ft$^3$) is attached to a tank filled with air. If the oil-level difference between the two columns is 20 in and the atmospheric pressure is 14.6 psia, determine the absolute pressure of the air in the tank. *Answer:* 15.21 psia

**1-40** A mercury manometer ($\rho$ = 13,600 kg/m$^3$) is connected to an air duct to measure the pressure inside. The difference in the manometer levels is 15 mm, and the atmospheric pressure is 100 kPa.

(*a*) Judging from the Fig. P1-40, determine if the pressure in the duct is above or below the atmospheric pressure.

(*b*) Determine the absolute pressure in the duct.

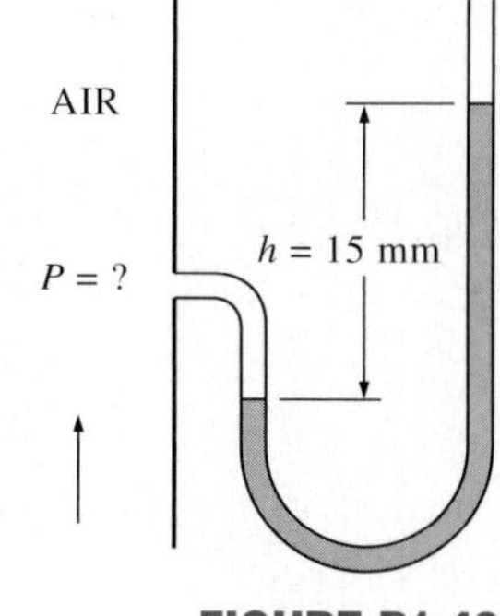

**FIGURE P1-40**

## Temperature

**1-41C** What is the zeroth law of thermodynamics?

**1-42C** What are the ordinary and absolute temperature scales in the SI and the English systems?

**1-43C** Consider an alcohol and a mercury thermometer that read exactly 0°C at the ice point and 100°C at the steam point. The distance between the two points is divided into 100 equal parts in both thermometers. Do you think these thermometers will give exactly the same reading at a temperature of, say, 60°C? Explain.

**1-44** The deep body temperature of a healthy person is 37°C. What is it in kelvins? *Answer:* 310 K

**1-44E** The deep body temperature of a healthy person is 98.6°F. What is it in Rankine? *Answer:* 558.3 R

**1-45** Consider a system whose temperature is 18°C. Express this temperature in kelvins.

**1-45E** Consider a system whose temperature is 18°C. Express this temperature in R, K, and °F.

**1-46** The temperature of a system rises by 30°C during a heating process. Express this rise in temperature in kelvins. *Answer:* 30 K

**1-46E** The temperature of a system rises by 60°F during a heating process. Express this rise in temperature in R, K, and °C.
*Answers:* 60 R, 33.3 K, 33.3°C

**1-47** The temperature of a system drops by 15°C during a cooling process. Express this drop in temperature in kelvins.

**1-47E** The temperature of a system drops by 27°F during a cooling process. Express this drop in temperature in K, R, and °C.

**1-48** Consider two closed systems A and B. System A contains 2000 kJ of thermal energy at 20°C whereas system B contains 200 kJ of thermal energy at 50°C. Now the systems are brought into contact with each other. Determine the direction of any heat transfer between the two systems.

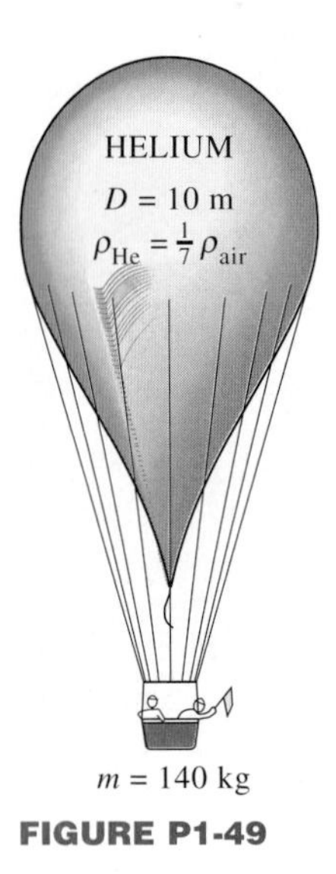


**FIGURE P1-49**

## Review Problems

**1-49** Balloons are often filled with helium gas because it weighs only about one-seventh of what air weighs under identical conditions. The buoyancy force which can be expressed as $F_B = \rho_{air} g V_{balloon}$ will push the balloon upward. If the balloon has a diameter of 10 m and carries two people, 70 kg each, determine the acceleration of the balloon when it is first released. Assume the density of air is $\rho = 1.16$ kg/m$^3$, and neglect the weight of the ropes and the cage. *Answer:* 16.5 m/s$^2$

**1-49E** Ballons are often filled with helium gas because it weighs only about one-seventh of what air weighs under identical conditions. The buoyancy force which can be expressed as $F_B = \rho_{air} g V_{balloon}$ will push the balloon upward. If the balloon has a diameter of 30 ft and carries two people, 140 lbm each, determine the acceleration of the balloon when it is first released. Assume the density of air is $\rho = 0.0724$ lbm/ft$^3$, and neglect the weight of the ropes and the cage. *Answer:* 45.1 ft/s$^2$

**1-50** Determine the maximum amount of load, in kg, the balloon described in Prob. 1-49 can carry. *Answer:* 520.6 kg

**1-51** The basic barometer can be used as an altitude-measuring device in airplanes. The ground control reports a barometric reading of

753 mmHg while the pilot's reading is 690 mmHg. Estimate the altitude of the plane from ground level if the average air density is 1.20 kg/m$^3$ and $g$ = 9.8 m/s$^2$. *Answer:* 714 m

**1-51E** The basic barometer can be used as an altitude-measuring device in airplanes. The ground control reports a barometric reading of 28.9 inHg while the pilot's reading is 26.0 inHg. Estimate the altitude of the plane from ground level if the average air density is 0.075 lbm/ft$^3$ and $g$ = 31.6 ft/s$^2$. *Answer:* 2737 ft

**1-52** The lower half of a 10-m-high cylindrical container is filled with water ($\rho$ = 1000 kg/m$^3$) and the upper half with oil that has a specific gravity of 0.85. Determine the pressure difference between the top and bottom of the cylinder. *Answer:* 90.7 kPa

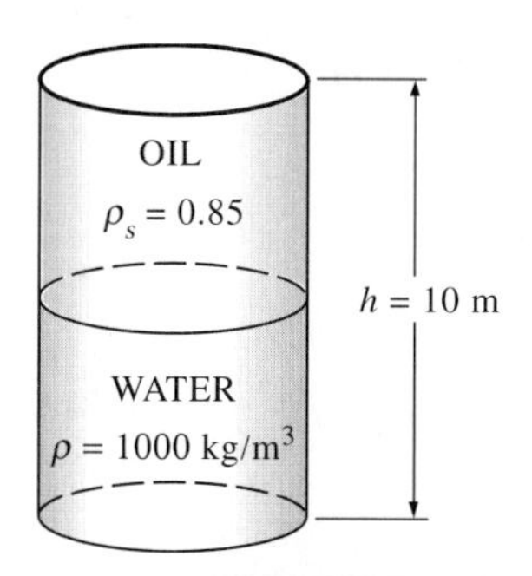

**FIGURE P1-52**

**1-53** A vertical, frictionless piston–cylinder device contains a gas at 500 kPa. The atmospheric pressure outside is 100 kPa, and the piston area is 30 cm$^2$. Determine the mass of the piston. Assume standard gravitational acceleration.

**1-54** A pressure cooker cooks a lot faster than an ordinary pan by maintaining a higher pressure and temperature inside. The lid of a pressure cooker is well sealed, and steam can escape only through an opening in the middle of the lid. A separate piece of certain mass, the petcock, sits on top of this opening and prevents steam from escaping until the pressure force overcomes the weight of the petcock. The periodic escape of the steam in this manner prevents any potentially dangerous pressure buildup and keeps the pressure inside at a constant value.

Determine the mass of the petcock of a pressure cooker whose operation pressure is 100 kPa gage and has an opening cross-sectional area of 4 mm$^2$. Assume an atmospheric pressure of 101 kPa, and draw the freebody diagram of the petcock. *Answer:* 40.8 g

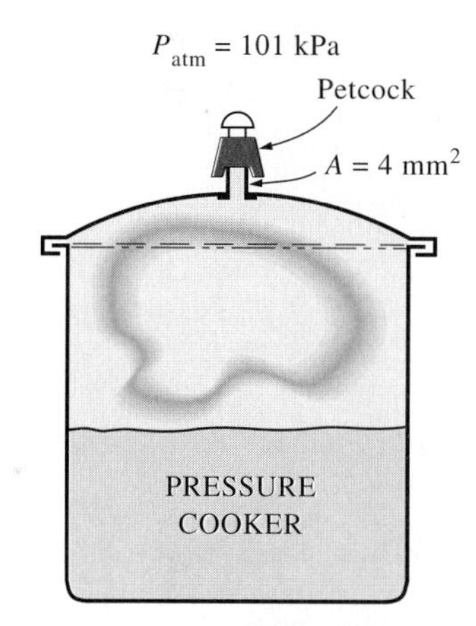

**FIGURE P1-54**

**1-55** A glass tube is attached to a water pipe as shown in Fig. P1-55. If the water pressure at the bottom of the tube is 115 kPa and the local atmospheric pressure is 92 kPa, determine how high the water will rise in the tube, in m. Assume $g$ = 9.8 m/s$^2$ at that location and take the density of water to be 1000 kg/m$^3$.

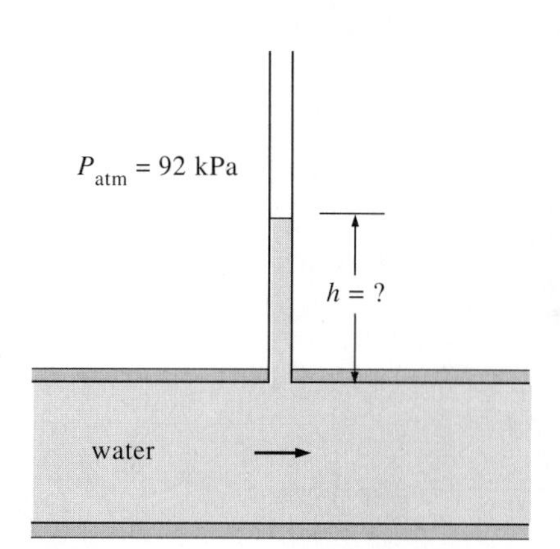


**FIGURE P1-55**

**1-56** The average atmospheric pressure on earth is approximated as a function of altitude by the relation

$$P_{atm} = 101.325(1 - 0.02256z)^{5.256}$$

where $P_{atm}$ is the atmospheric pressure in kPa and $z$ is the altitude in km (1 km = 1000 m) with $z$ = 0 at sea level. Determine the approximate atmospheric pressures at Atlanta ($z$ = 306 m), Denver ($z$ = 1610 m), Mexico City ($z$ = 2309 m), and the top of Mount Everest ($z$ = 8848 m).

**1-57** The weight of bodies may change somewhat from one location to another as a result of the variation of the gravitational acceleration $g$ with elevation. Accounting for this variation using the relation in Prob. 1-9, determine the weight of an 80-kg person at sea level ($z = 0$), in Denver ($z = 1610$ m), and on the top of Mount Everest ($z = 8848$ m).

**1-58** The efficiency of a refrigerator increases by 3 percent for each °C rise in the minimum temperature in the device. What is the increase in the efficiency for each (*a*) K, (*b*) °F, and (*c*) R rise in temperature?

**1-59** The boiling temperature of water decreases by about 3°C for each 1000 m rise in altitude. What is the decrease in the boiling temperature in (*a*) K, (*b*) °F, and (*c*) R for each 1000 m rise in altitude?

**1-60** The average body temperature of a person rises by about 2°C during strenuous exercise. What is the rise in the body temperature in (*a*) K, (*b*) °F, and (*c*) R during strenuous exercise?

**1-61** Hyperthermia of 5°C (i.e., 5°C rise above the normal body temperature) is considered fatal. Express this fatal level of hyperthermia in (*a*) K, (*b*) °F, and (*c*) R.

**1-62** A house is losing heat at a rate of 3000 kJ/h per °C temperature difference between the indoor and the outdoor temperatures. Express the rate of heat loss from this house per (*a*) K, (*b*) °F, and (*c*) R difference between the indoor and the outdoor temperature.

**1-63** The average temperature of the atmosphere in the world is approximated as a function of altitude by the relation

$$T_{\text{atm}} = 288.15 - 6.5z$$

where $T_{\text{atm}}$ is the temperature of the atmosphere in K and $z$ is the altitude in km with $z = 0$ at sea level. Determine the average temperature of the atmosphere outside an airplane that is cruising at an altitude of 12,000 m.

**1-64** Joe Smith, an old-fashioned engineering student, believes that the boiling point of water is best suited for use as the reference point on temperature scales. Unhappy that the boiling point corresponds to some odd number in the current absolute temperature scales, he has proposed a new absolute temperature scale which he called the Smith scale. The temperature unit on this scale is *smith,* denoted by S, and the boiling point of water on this scale is assigned to be 1000 S. From a thermodynamic point of view, discuss if it is an acceptable temperature scale. Also determine the ice point of water on the Smith scale and obtain a relation between the Smith and Celsius scales.

**1-65** A man goes to a traditional market to buy a steak for dinner. He finds a 12-ounce steak (1 lbm = 16 ounces) for $3.15. He then goes to the adjacent international market, and finds a 320-gram steak of identical quality for $2.80. Which steak is a better buy?

**1-66** Milk is to be transported from Texas to California for a distance of 2100 km in a 7-m-long 2-m external-diameter cylindrical tank. The walls of the tank are constructed of 5-cm-thick urethane insulation sandwiched between two metal sheets of negligible thickness. Determine the amount of milk in the tank in kg and in gallons (1 gal = 3.78 L).

**1-67** An engineer who is working on the heat transfer analysis of a brick building in English units needs the thermal conductivity of brick. But the only value he can find from his handbooks is 0.72 W/(m · °C), which is in SI units. To make matters worse, the engineer does not have a direct conversion factor between the two unit systems for thermal conductivity (he should have kept his heat transfer textbook instead of selling it back to the bookstore). Can you help him out?

**1-68** It is well-known that cold air feels much colder in windy weather than what the thermometer reading indicates because of the "chilling effect" of the wind. This effect is due to the increase in the convection heat transfer coefficient with increasing air velocities. The *equivalent wind chill temperature* in °F is given by (1993 ASHRAE Handbook of Fundamentals, Atlanta, GA, p. 8.15)

$$T_{\text{equiv}} = 91.4 - (91.4 - T_{\text{ambient}})(0.475 - 0.0203V + 0.304\sqrt{V})$$

where $V$ is the wind velocity in mph and $T_{\text{ambient}}$ is the ambient air temperature in °F in calm air, which is taken to be air with light winds at speeds up to 4 mph. The constant 91.4°F in the above equation is the mean skin temperature of a resting person in a comfortable environment. Windy air at temperature $T_{\text{ambient}}$ and velocity $V$ will feel as cold as the calm air at temperature $T_{\text{equiv}}$. Using proper conversion factors, obtain an equivalent relation in SI units where $V$ is the wind velocity in km/h and $T_{\text{ambient}}$ is the ambient air temperature in °C.
*Answer*: $T_{\text{equiv}} = 33.0 - (33.0 - T_{\text{ambient}})(0.475 - 0.0126V + 0.240\sqrt{V})$

## Computer, Design, and Essay Problems

**1-69** Write an interactive computer program to express a given temperature in °C, °F, K, and R in terms of the other three units.

**1-70** Write an interactive computer program to express a pressure given in SI units in terms of the height of water and mercury columns.

**1-71** Write an essay on different temperature measurement devices. Explain the operational principle of each device, its advantages and disadvantages, cost, and its range of applicability. Which device would you recommend for use in the following cases: taking the temperatures of patients in a doctor's office, monitoring the variations of temperature of a car engine block at several locations, and monitoring the temperatures in the furnace of a power plant.

**1-72** Write an essay on different pressure-measurement devices. Explain the operational principle of each device, its advantages and disadvantages, cost, and its range of applicability. Give examples of application areas each device is best suited for.

**1-73** Write an essay on the various mass- and volume-measurement devices used through history. Also explain the development of the modern units for mass and volume.

# Properties of Pure Substances

CHAPTER

In this chapter, the concept of pure substances is introduced, and the various phases as well as the physics of phase-change processes are discussed. Various property diagrams and $P$-$v$-$T$ surfaces of pure substances are illustrated and the hypothetical substance "ideal gas" and the ideal-gas equation of state are discussed. The compressibility factor, which accounts for the deviation of real gases from ideal-gas behavior, is introduced, and its use is illustrated. Finally, some of the best-known equations of state are presented.

## 2-1 PURE SUBSTANCE

A substance that has a fixed chemical composition throughout is called a **pure substance**. Water, nitrogen, helium, and carbon dioxide, for example, are all pure substances.

A pure substance does not have to be of a single chemical element or compound, however. A mixture of various chemical elements or compounds also qualifies as a pure substance as long as the mixture is homogeneous. Air, for example, is a mixture of several gases, but it is often considered to be a pure substance because it has a uniform chemical composition (Fig. 2-1). However, a mixture of oil and water is not a pure substance. Since oil is not soluble in water, it will collect on top of the water, forming two chemically dissimilar regions.

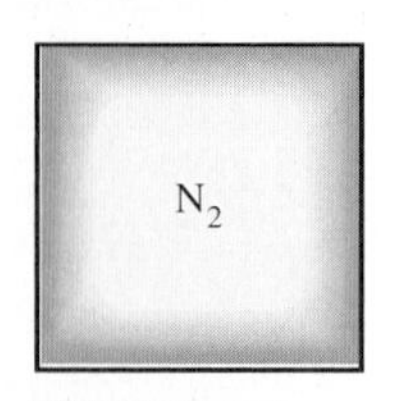

**FIGURE 2-1**

Nitrogen and gaseous air are pure substances.

A mixture of two or more phases of a pure substance is still a pure substance as long as the chemical composition of all phases is the same (Fig. 2-2). A mixture of ice and liquid water, for example, is a pure substance because both phases have the same chemical composition. A mixture of liquid air and gaseous air, however, is not a pure substance since the composition of liquid air is different from the composition of gaseous air, and thus the mixture is no longer chemically homogeneous. This is due to different components in air having different condensation temperatures at a specified pressure.

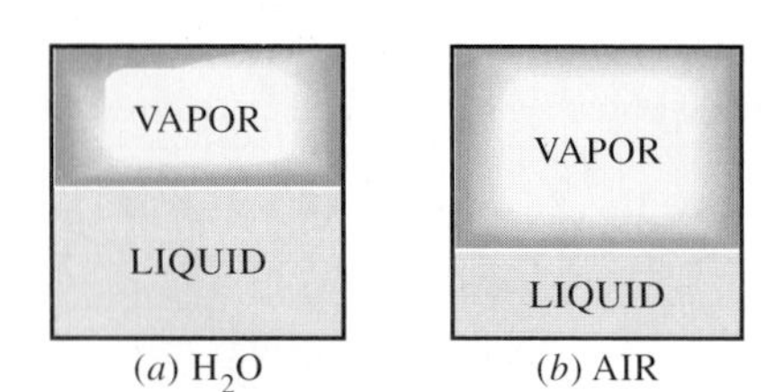


**FIGURE 2-2**

A mixture of liquid and gaseous water is a pure substance, but a mixture of liquid and gaseous air is not.

## 2-2 PHASES OF A PURE SUBSTANCE

We all know from experience that substances exist in different phases. At room temperature and pressure, copper is a solid, mercury is a liquid, and nitrogen is a gas. Under different conditions, each may appear in a different phase. Even though there are three principal phases—solid, liquid, and gas—a substance may have several phases within a principal phase, each with a different molecular structure. Carbon, for example, may exist as graphite or diamond in the solid phase. Helium has two liquid phases; iron has three solid phases. Ice may exist at seven different phases at high pressures. A phase is identified as having a distinct molecular arrangement that is homogeneous throughout and separated from the others by easily identifiable boundary surfaces. The two phases of $H_2O$ in iced water represent a good example of this.

When studying phases or phase changes in thermodynamics, one does not need to be concerned with the molecular structure and behavior of different phases. However, it is very helpful to have some understanding of the molecular phenomena involved in each phase, and a brief discussion of phase transformations is given below.

It is often stated that molecular bonds are the strongest in solids and the weakest in gases. One reason is that molecules in solids are closely packed together, whereas in gases they are separated by great distances.

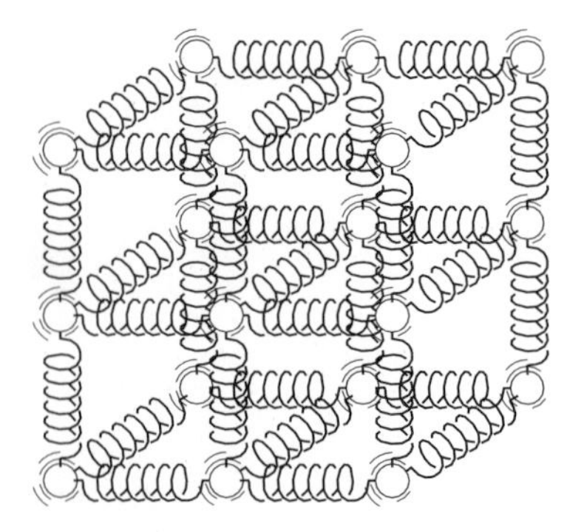

**FIGURE 2-3**

The molecules in a solid are kept at their positions by the large springlike intermolecular forces.

The molecules in a **solid** are arranged in a three-dimensional pattern (lattice) that is repeated throughout the solid (Fig. 2-3). Because of the small distances between molecules in a solid, the attractive forces of molecules on each other are large and keep the molecules at fixed positions within the solid (Fig. 2-4). Note that the attractive forces between molecules turn to repulsive forces as the distance between the molecules approaches zero, thus preventing the molecules from piling up on top of each other. Even though the molecules in a solid cannot move relative to each other, they continually oscillate about their equilibrium position. The velocity of the molecules during these oscillations depends on the temperature. At sufficiently high temperatures, the velocity (and thus the momentum) of the molecules may reach a point where the intermolecular forces are partially overcome and groups of molecules break away (Fig. 2-5). This is the beginning of the melting process.

**FIGURE 2-4**
In a solid, the attractive and the repulsive forces between the molecules tend to maintain them at relatively constant distances from each other.

The molecular spacing in the **liquid** phase is not much different from that of the solid phase, except the molecules are no longer at fixed positions relative to each other. In a liquid, chunks of molecules float about each other; however, the molecules maintain an orderly structure within each chunk and retain their original positions with respect to one another. The distances between molecules generally experience a slight increase as a solid turns liquid, with water being a rare exception.

In the **gas** phase, the molecules are far apart from each other, and a molecular order is nonexistent. Gas molecules move about at random, continually colliding with each other and the walls of the container they are in. Particularly at low densities, the intermolecular forces are very small, and collisions are the only mode of interaction between the molecules. Molecules in the gas phase are at a considerably higher energy level than they are in the liquid or solid phases. Therefore, the gas must release a large amount of its energy before it can condense or freeze.

## 2-3 ■ PHASE-CHANGE PROCESSES OF PURE SUBSTANCES

There are many practical situations where two phases of a pure substance coexist in equilibrium. Water exists as a mixture of liquid and vapor in the boiler and the condenser of a steam power plant. The refrigerant turns from liquid to vapor in the freezer of a refrigerator. Even though many homeowners consider the freezing of water in underground pipes as the most important phase-change process, attention in this section is focused on the liquid and vapor phases and the mixture of these two. As a familiar substance, water will be used to demonstrate the basic principles involved. Remember, however, that all pure substances exhibit the same general behavior.

### Compressed Liquid and Saturated Liquid

Consider a piston–cylinder device containing liquid water at 20°C and 1-atm pressure (state 1, Fig. 2-6). Under these conditions, water exists in

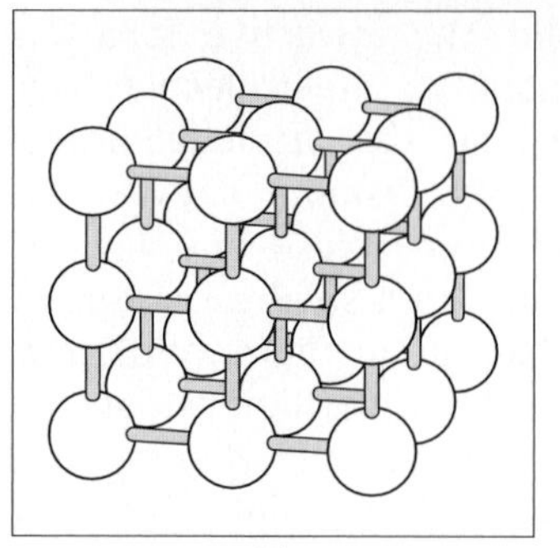

(*a*)

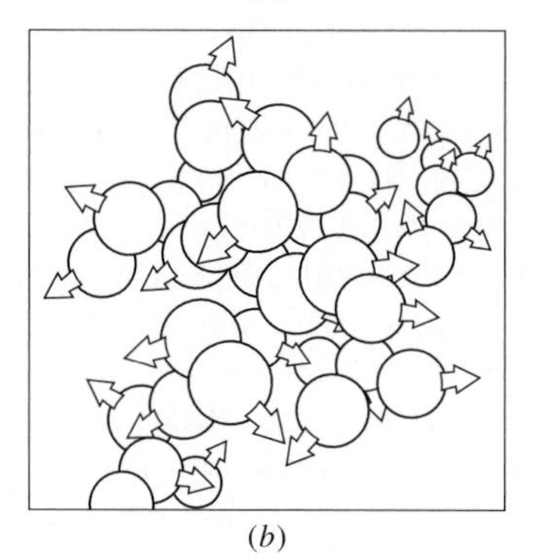

(*b*)

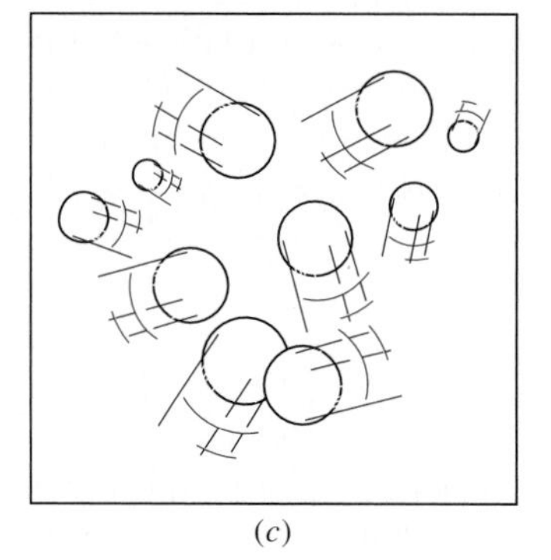

(*c*)

**FIGURE 2-5**
The arrangement of atoms in different phases: (*a*) molecules are at relatively fixed positions in a solid, (*b*) chunks of molecules float about each other in the liquid phase, and (*c*) molecules move about at random in the gas phase.

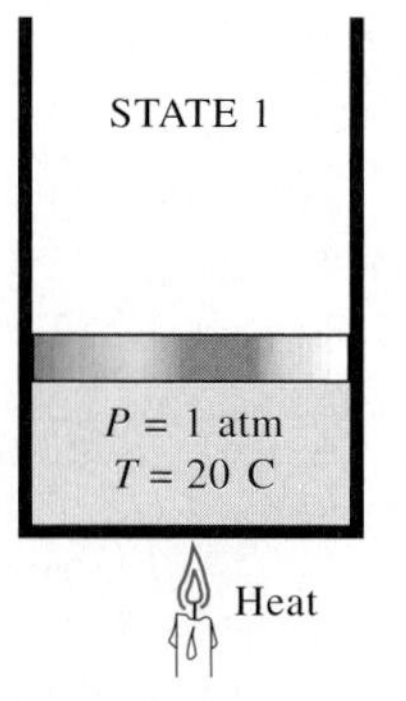


**FIGURE 2-6**
At 1 atm and 20°C water exists in the liquid phase (*compressed liquid*).

the liquid phase, and it is called a **compressed liquid**, or a **subcooled liquid**, meaning that it is *not about to vaporize.* Heat is now added to the water until its temperature rises to, say 40°C. As the temperature rises, the liquid water will expand slightly, and so its specific volume will increase. To accommodate this expansion, the piston will move up slightly. The pressure in the cylinder remains constant at 1 atm during this process since it depends on the outside barometric pressure and the weight of the piston, both of which are constant. Water is still a compressed liquid at this state since it has not started to vaporize.

As more heat is transferred, the temperature will keep rising until it reaches 100°C (state 2, Fig. 2-7). At this point, water is still a liquid, but any heat addition, no matter how small, will cause some of the liquid to

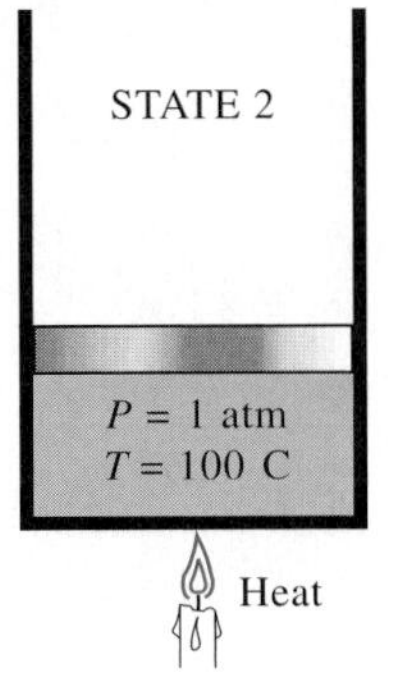


**FIGURE 2-7**
At 1-atm pressure and 100°C, water exists as a liquid which is ready to vaporize (*saturated liquid*).

vaporize. That is, a phase-change process from liquid to vapor is about to take place. A liquid that is *about to vaporize* is called a **saturated liquid**. Therefore, state 2 is a saturated liquid state.

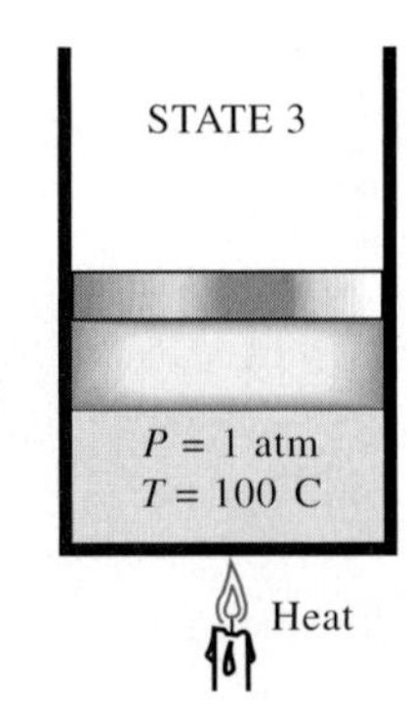

**FIGURE 2-8**
As more heat is transferred, part of saturated liquid vaporizes (*saturated liquid–vapor mixture*).

## Saturated Vapor and Superheated Vapor

Once boiling starts, the temperature will stop rising until the liquid is completely vaporized. That is, the temperature will remain constant during the entire phase-change process if the pressure is held constant. This can easily be verified by placing a thermometer into boiling water on top of a stove. At sea level ($P$ = 1 atm), the thermometer will always read 100°C if the pan is uncovered or covered with a light lid. During a vaporization (boiling) process, the only change we will observe is a large increase in the volume and a steady decline in the liquid level as a result of more liquid turning to vapor.

Midway about the vaporization line (state 3, Fig. 2-8), the cylinder contains equal amounts of liquid and vapor. As we continue adding heat, the vaporization process will continue until the last drop of liquid is vaporized (state 4, Fig. 2-9). At this point, the entire cylinder is filled with vapor that is on the borderline of the liquid phase. Any heat loss from this vapor, no matter how small, will cause some of the vapor to condense (phase change from vapor to liquid). A vapor that is *about to condense* is called a **saturated vapor**. Therefore, state 4 is a saturated vapor state. A substance at states between 2 and 4 is often referred to as a **saturated liquid–vapor mixture** since the *liquid and vapor phases coexist in equilibrium* at these states.

Once the phase-change process is completed, we are back to a single-phase region again (this time vapor), and further transfer of heat will result in an increase in both the temperature and the specific volume (Fig. 2-10). At state 5, the temperature of the vapor is, let us say, 300°C; and if we transfer some heat from the vapor, the temperature may drop somewhat but no condensation will take place as long as the temperature remains above 100°C (for $P$ = 1 atm). A vapor that is *not about to condense* (i.e., not a saturated vapor) is called a **superheated vapor**. Therefore, water at state 5 is a superheated vapor. The constant-pressure

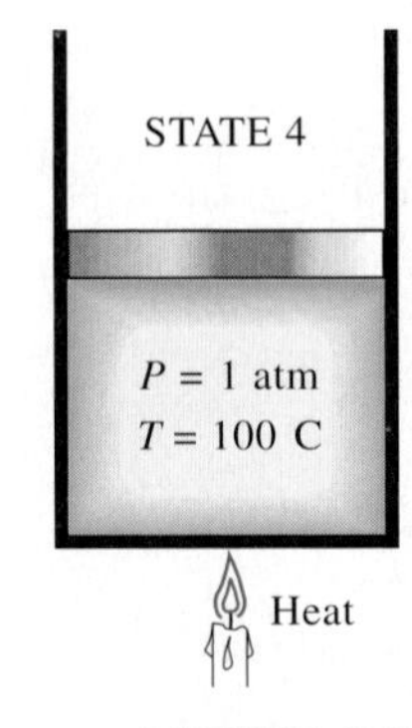


**FIGURE 2-9**
At 1-atm pressure, the temperature remains constant at 100°C until the last drop of liquid is vaporized (*saturated vapor*).

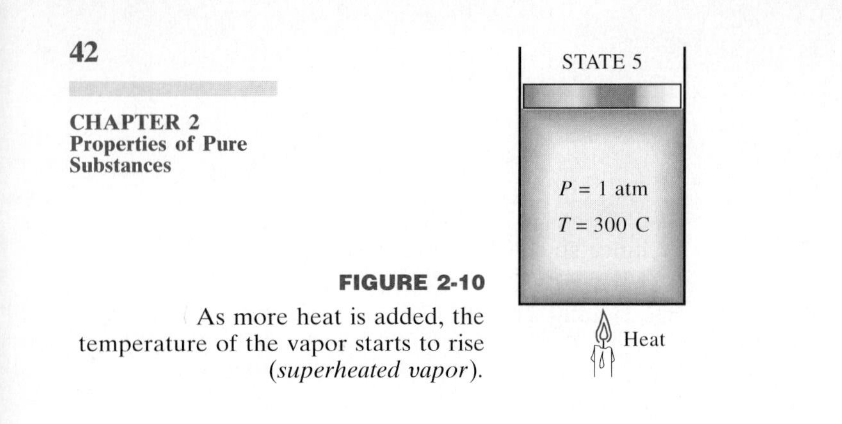


**FIGURE 2-10**
As more heat is added, the temperature of the vapor starts to rise (*superheated vapor*).

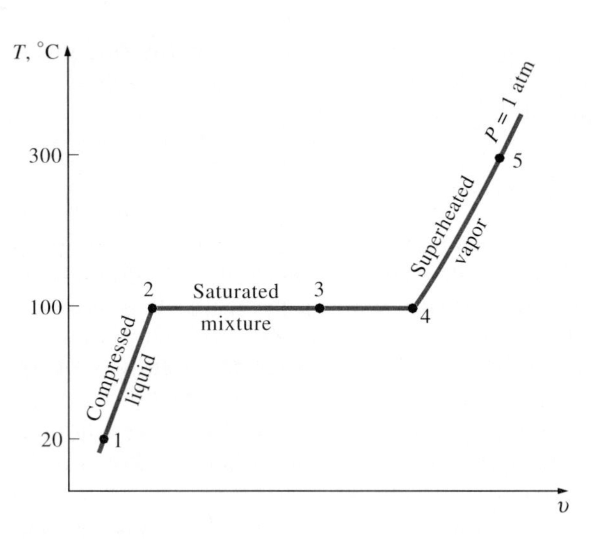

**FIGURE 2-11**
$T$-$v$ diagram for the heating process of water at constant pressure.

phase-change process described above is illustrated on a $T$-$v$ diagram in Fig. 2-11.

If the entire process described above is reversed by cooling the water while maintaining the pressure at the same value, the water will go back to state 1, retracing the same path, and in so doing, the amount of heat released will exactly match the amount of heat added during the heating process.

In our daily life, water implies liquid water and steam implies water vapor. In thermodynamics, however, both water and steam usually mean only one thing: $H_2O$.

## Saturation Temperature and Saturation Pressure

It probably came as no surprise to you that the water started "boiling" at 100°C. Strictly speaking, the statement "water boils at 100°C" is incorrect. The correct statement is "water boils at 100°C at 1-atm pressure." The only reason the water started boiling at 100°C was because we held the pressure constant at 1 atm (101.325 kPa). If the pressure inside the cylinder were raised to 500 kPa by adding weights on top of the piston, the water would start boiling at 151.9°C. That is, *the temperature at which water starts boiling depends on the pressure; therefore, if the pressure is fixed, so is the boiling temperature.*

At a given pressure, the temperature at which a pure substance starts boiling is called the **saturation temperature** $T_{sat}$. Likewise, at a given temperature, the pressure at which a pure substance starts boiling is called the **saturation pressure** $P_{sat}$. At a pressure of 101.325 kPa, $T_{sat}$ is 100°C. Conversely, at a temperature of 100°C, $P_{sat}$ is 101.325 kPa.

During a phase-change process, pressure and temperature are obviously dependent properties, and there is a definite relation between them, that is, $T_{sat} = f(P_{sat})$. A plot of $T_{sat}$ vs. $P_{sat}$, such as the one given for water in Fig. 2-12, is called a **liquid–vapor saturation curve**. A curve of this kind is characteristic of all pure substances.

It is clear from Fig. 2-12 that $T_{sat}$ increases with $P_{sat}$. Thus, a substance at higher pressures will boil at higher temperatures. In the kitchen, higher boiling temperatures mean shorter cooking times and energy savings. A beef stew, for example, may take 1–2 h to cook in a regular pan that operates at 1-atm pressure, but only 20–30 min in a pressure cooker operating at 2-atm absolute pressure (corresponding boiling temperature: 120°C).

The atmospheric pressure, and thus the boiling temperature of water, decrease with elevation. Therefore, it takes longer to cook at higher altitudes than it does at sea level (unless a pressure cooker is used). For example, the standard atmospheric pressure at an elevation of 2000 m is 79.50 kPa, which corresponds to a boiling temperature of 93.2°C as opposed to 100°C at sea level (zero elevation). The variation of the boiling temperature of water with altitude at standard atmospheric conditions is given in Table 2-1. For each 1000 m increase in elevation, the boiling temperature drops by a little over 3°C. Note that the

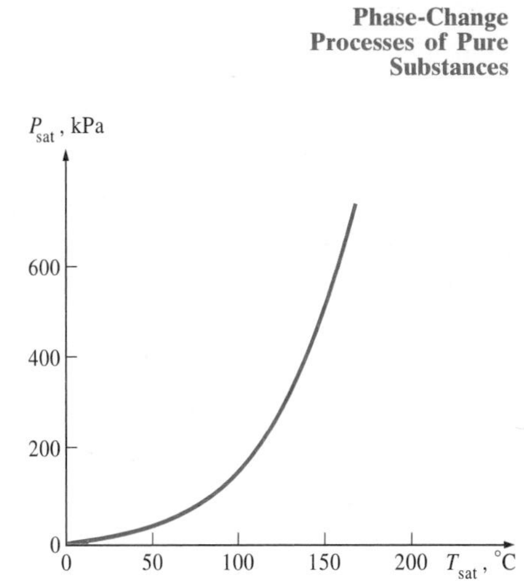

**FIGURE 2-12**

The liquid-vapor saturation curve of a pure substance (numerical values are for water).

**TABLE 2-1**
**Variation of the standard atmospheric pressure and the boiling (saturation) temperature of water with altitude**

| Elevation, m | Atmospheric pressure, kPa | Boiling temperature, °C |
|---|---|---|
| 0 | 101.33 | 100.0 |
| 1,000 | 89.55 | 96.3 |
| 2,000 | 79.50 | 93.2 |
| 5,000 | 54.05 | 83.0 |
| 10,000 | 26.50 | 66.2 |
| 20,000 | 5.53 | 34.5 |

atmospheric pressure at a location, and thus the boiling temperature, changes slightly depending on the weather conditions as a result of the changes in weather. But the change in the boiling temperature is no more than about 1°C.

## 2-4 ■ PROPERTY DIAGRAMS FOR PHASE-CHANGE PROCESSES

The variations of properties during phase-change processes are best studied and understood with the help of property diagrams. Below we develop and discuss the *T-v*, *P-v*, and *P-T* diagrams for pure substances.

### 1 The *T-v* Diagram

The phase-change process of water at 1-atm pressure was described in detail in the last section and plotted on a *T-v* diagram in Fig. 2-11. Now we repeat this process at different pressures to develop the *T-v* diagram for water.

Let us add weights on top of the piston until the pressure inside the cylinder reaches 1 MPa. At this pressure, water will have a somewhat smaller specific volume than it did at 1-atm pressure. As heat is transferred to the water at this new pressure, the process will follow a path that looks very much like the process path at 1-atm pressure, as shown in

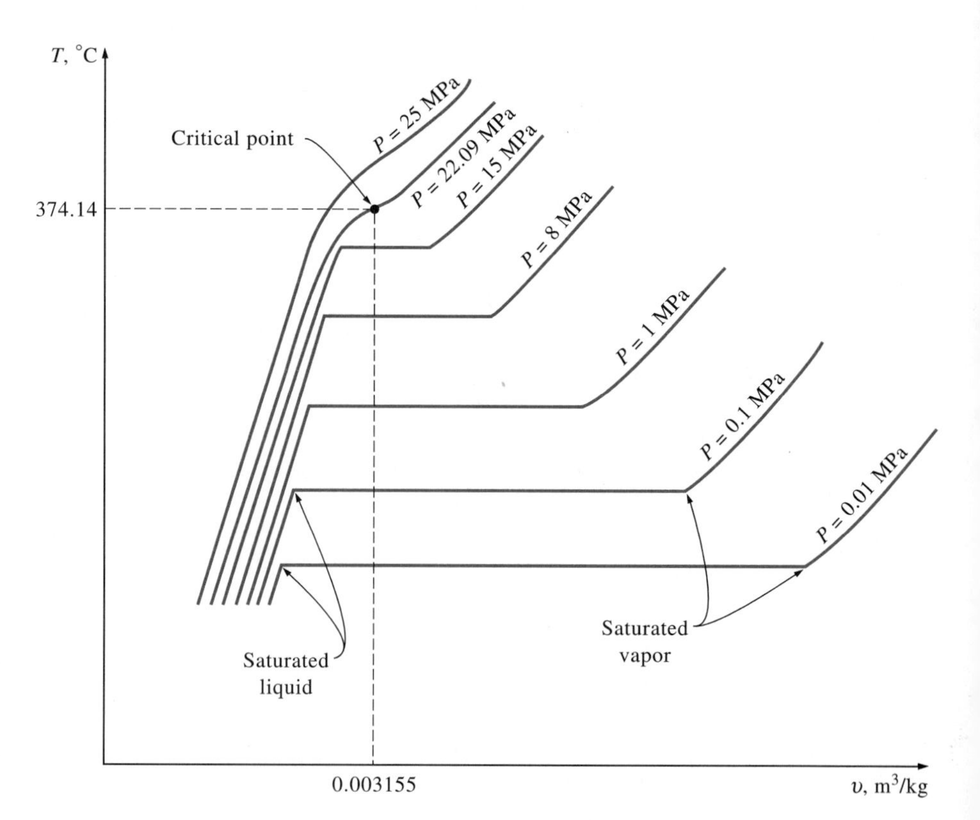


**FIGURE 2-13**

*T-v* diagram of constant-pressure phase-change processes of a pure substance at various pressures (numerical values are for water).

Fig. 2-13, but there are some noticeable differences. First, water will start boiling at a much higher temperature (179.9°C) at this pressure. Second, the specific volume of the saturated liquid is larger, and the specific volume of the saturated vapor is smaller than the corresponding values at 1-atm pressure. That is, the horizontal line that connects the saturated liquid and saturated vapor states is much shorter.

As the pressure is increased further, this saturation line will continue to get shorter, as shown in Fig. 2-13, and it will become a point when the pressure reaches 22.09 MPa for the case of water. This point is called the **critical point**, and it may be defined as *the point at which the saturated liquid and saturated vapor states are identical.*

The temperature, pressure, and specific volume of a substance at the critical point are called, respectively, the *critical temperature* $T_{cr}$, *critical pressure* $P_{cr}$, and *critical specific volume* $v_{cr}$. The critical-point properties of water are $P_{cr} = 22.09$ MPa, $T_{cr} = 374.14$°C, and $v_{cr} = 0.003155$ m$^3$/kg. For helium, they are 0.23 MPa, −267.85°C, and 0.01444 m$^3$/kg. The critical properties for various substances are given in Table A-1 in the Appendix.

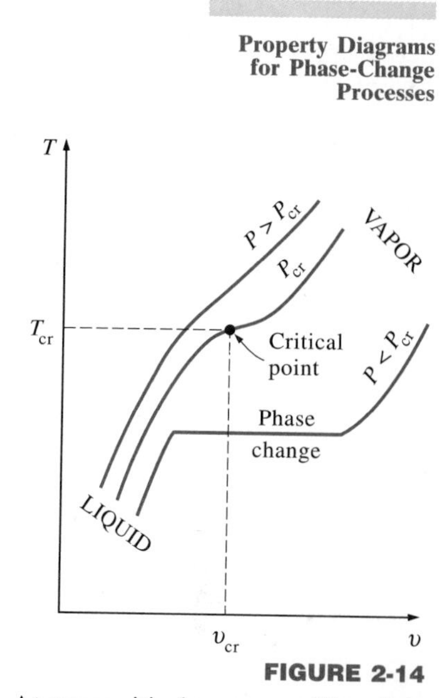


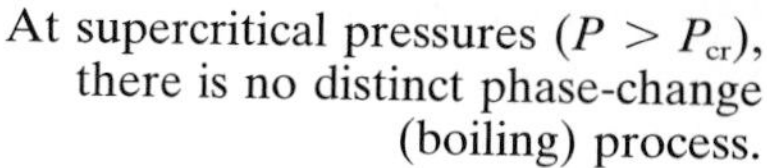

**FIGURE 2-14**
At supercritical pressures ($P > P_{cr}$), there is no distinct phase-change (boiling) process.

At pressures above the critical pressure, there will not be a distinct phase-change process (Fig. 2-14). Instead, the specific volume of the substance will continually increase, and at all times there will be only one phase present. Eventually, it will resemble a vapor, but we can never tell when the change has occurred. Above the critical state, there is no line that separates the compressed liquid region and the superheated vapor region. However, it is customary to refer to the substance as superheated vapor at temperatures above the critical temperature and as compressed liquid at temperatures below the critical temperature.

The saturated liquid states in Fig. 2-13 can be connected by a line, which is called the **saturated liquid line**, and saturated vapor states in the same figure can be connected by another line, which is called the **saturated vapor line**. These two lines meet each other at the critical point, forming a dome as shown in Fig. 2-15. All the compressed liquid states are located in the region to the left of the saturated liquid line, and this is called the **compressed liquid region**. All the superheated vapor states are located to the right of the saturated vapor line, which is called the **superheated vapor region**. In these two regions, the substance exists in a single phase, a liquid or a vapor. All the states that involve both phases in equilibrium are located under the dome, which is called the **saturated liquid–vapor mixture region**, or the **wet region**.

## 2 The $P$-$v$ Diagram

The general shape of the $P$-$v$ diagram of a pure substance is very much like the $T$-$v$ diagram, but the $T$ = constant lines on this diagram have a downward trend, as shown in Fig. 2-16.

Consider again a piston–cylinder device that contains liquid water at 1 MPa and 150°C. Water at this state exists as a compressed liquid. Now the weights on top of the piston are removed one by one so that the

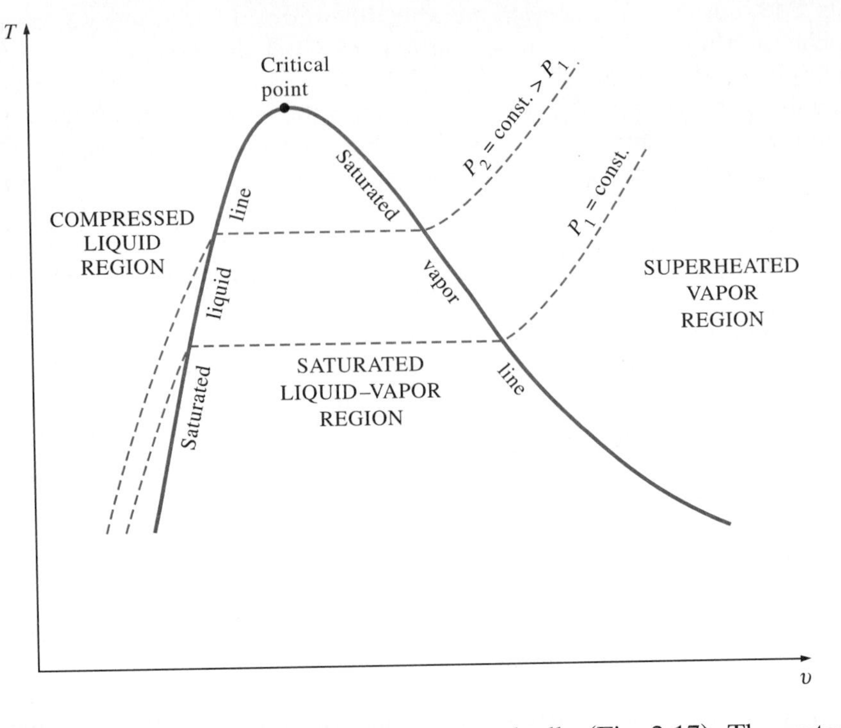


**FIGURE 2-15**

$T$-$\upsilon$ diagram of a pure substance.

pressure inside the cylinder decreases gradually (Fig. 2-17). The water is allowed to exchange heat with the surroundings, so that its temperature remains constant. As the pressure decreases, the volume of the water will increase slightly. When the pressure reaches the saturation-pressure value

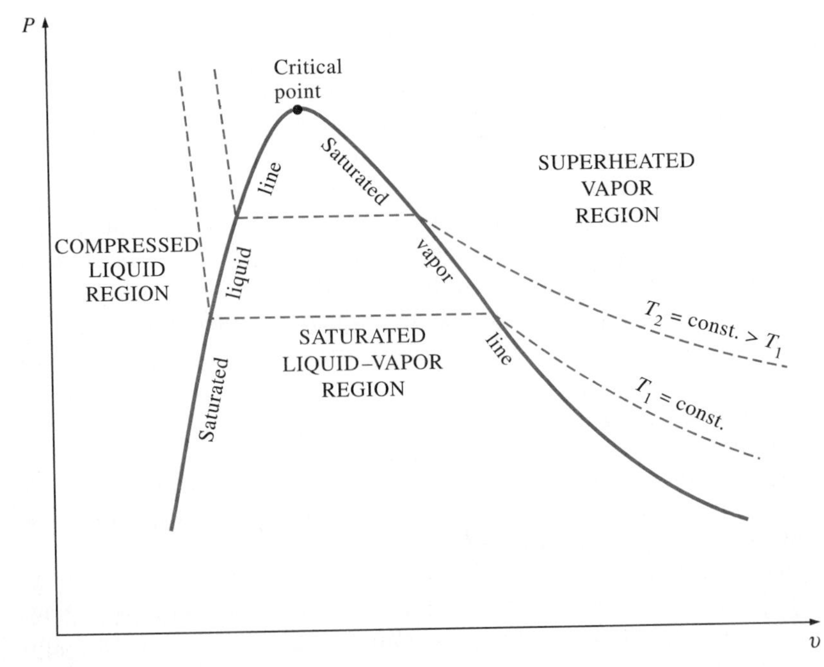


**FIGURE 2-16**

$P$-$\upsilon$ diagram of a pure substance.

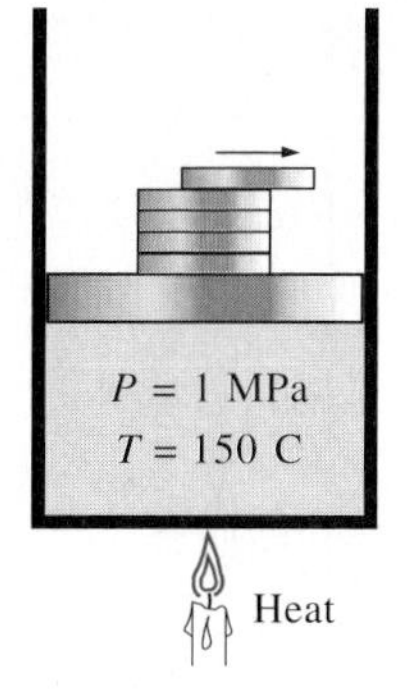


**FIGURE 2-17**

The pressure in a piston–cylinder device can be reduced by reducing the weight of the piston.

at the specified temperature (0.4758 MPa), the water will start to boil. During this vaporization process, both the temperature and the pressure remain constant, but the specific volume increases. Once the last drop of liquid is vaporized, further reduction in pressure results in a further increase in specific volume. Notice that during the phase-change process, we did not remove any weights. Doing so would cause the pressure and therefore the temperature to drop [since $T_{sat} = f(P_{sat})$], and the process would no longer be isothermal.

If the process is repeated for other temperatures, similar paths will be obtained for the phase-change processes. Connecting the saturated liquid and the saturated vapor states by a curve, we obtain the $P$-$v$ diagram of a pure substance, as shown in Fig. 2-16.

## Extending the Diagrams to Include the Solid Phase

The two equilibrium diagrams developed so far represent the equilibrium states involving the liquid and the vapor phases only. But these diagrams can easily be extended to include the solid phase as well as the solid–liquid and the solid–vapor saturation regions. The basic principles discussed in conjunction with the liquid–vapor phase-change process apply equally to the solid–liquid and solid–vapor phase-change processes. Most substances contract during a solidification (i.e., freezing) process. Others, like water, expand as they freeze. The $P$-$v$ diagrams for both groups of substances are given in Figs. 2-18 and 2-19. These two diagrams differ only in the solid–liquid saturation region. The $T$-$v$ diagrams look very much like the $P$-$v$ diagrams, especially for substances that contract on freezing.

The fact that water expands upon freezing has vital consequences in nature. If water contracted on freezing as most other substances do, the ice formed would be heavier than the liquid water, and it would settle to the bottom of rivers, lakes, or oceans instead of floating at the top. The sun's rays would never reach these ice layers, and the bottoms of many

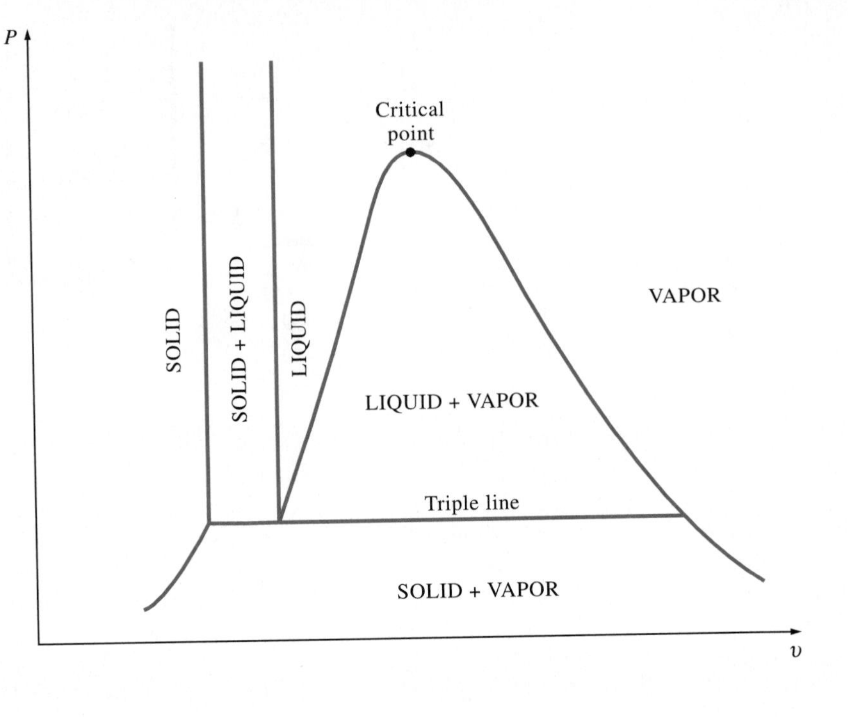


**FIGURE 2-18**

*P*-*υ* diagram of a substance that contracts on freezing.

rivers, lakes, or oceans would be covered with ice year round, seriously disrupting marine life.

We are all familiar with two phases being in equilibrium, but under some conditions all three phases of a pure substance coexist in equilibrium (Fig. 2-20). On *P*-*υ* or *T*-*υ* diagrams, these triple-phase states form a

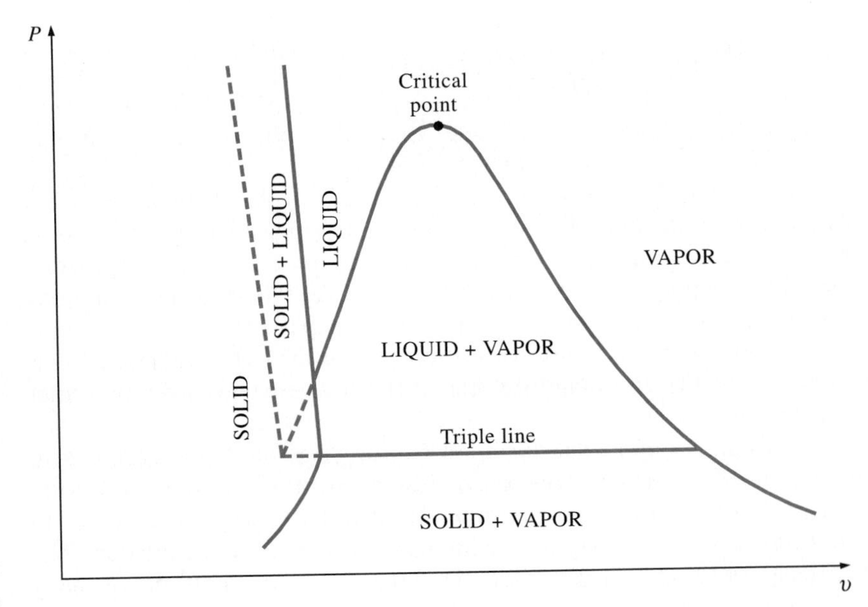


**FIGURE 2-19**

*P*-*υ* diagram of a substance that expands on freezing (such as water).

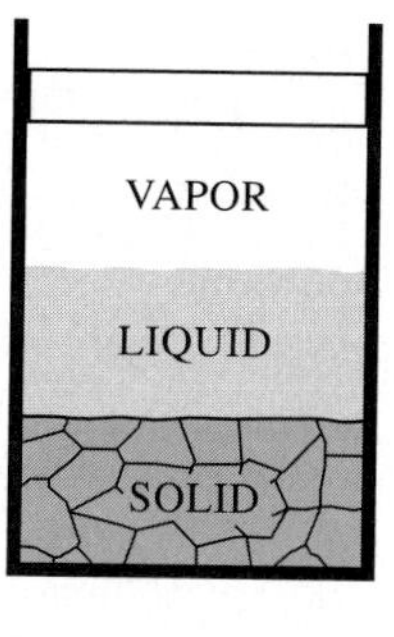


**FIGURE 2-20**
At triple pressure and temperature, a substance exists in three phases in equilibrium.

line called the **triple line**. The states on the triple line of a substance have the same pressure and temperature but different specific volumes. The triple line appears as a point on the $P$-$T$ diagrams and, therefore, is often called the **triple point**. The triple point temperatures and pressures of various substances are given in Table 2-2. For water, the triple-point temperature and pressure values are 0.01°C and 0.6113 kPa, respectively. That is, all three phases of water will exist in equilibrium only if the temperature and pressure have precisely these values. No substance can exist in the liquid phase in stable equilibrium at pressures below the triple-point pressure. The same can be said for temperature for substances that contract on freezing. However, substances at high pressures can exist in the liquid phase at temperatures below the triple-point temperature. For example, water cannot exist in liquid form in equilibrium at atmospheric conditions at temperatures below 0°C, but it can exist as a liquid at −20°C at 200 MPa pressure. Also, ice exists at seven different solid phases at pressures above 100 MPa.

There are two ways a substance can pass from the solid to vapor phase: either it melts first into a liquid and subsequently evaporates, or it evaporates directly without melting first. The latter occurs at pressures below the triple-point pressure, since a pure substance cannot exist in the liquid phase at those pressures (Fig. 2-21). Passing from the solid phase directly into the vapor phase is called **sublimation**. For substances that have a triple-point pressure above the atmospheric pressure such as solid $CO_2$ (dry ice), sublimation is the only way to change from the solid to vapor phase at atmospheric conditions.

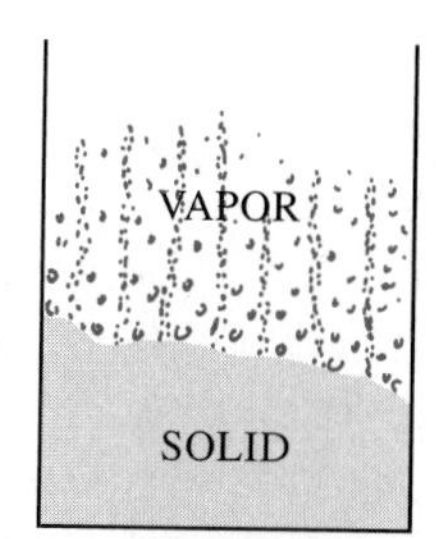


**FIGURE 2-21**
At low pressures (below triple-point value), solids evaporate without melting first (*sublimation*).

## 3 The $P$-$T$ Diagram

Figure 2-22 shows the $P$-$T$ diagram of a pure substance. This diagram is often called the **phase diagram** since all three phases are separated from each other by three lines. The sublimation line separates the solid and the vapor regions, the vaporization line separates the liquid and vapor regions, and the melting (or fusion) line separates the solid and liquid regions. These three lines meet at the triple point, where all three phases coexist in equilibrium. The vaporization line ends at the critical point because no distinction can be made between liquid and vapor phases

**TABLE 2-2**
**Triple-point temperatures and pressures of various substances**

| Substance | Formula | $T_{tp}$, K | $P_{tp}$, kPa |
|---|---|---|---|
| Acetylene | $C_2H_2$ | 192.4 | 120 |
| Ammonia | $NH_3$ | 195.40 | 6.076 |
| Argon | A | 83.81 | 68.9 |
| Carbon (graphite) | C | 3900 | 10,100 |
| Carbon dioxide | $CO_2$ | 216.55 | 517 |
| Cardon monoxide | CO | 68.10 | 15.37 |
| Deuterium | $D_2$ | 18.63 | 17.1 |
| Ethane | $C_2H_6$ | 89.89 | $8 \times 10^{-4}$ |
| Ethylene | $C_2H_4$ | 104.0 | 0.12 |
| Helium 4 ($\lambda$ point) | He | 2.19 | 5.1 |
| Hydrogen | $H_2$ | 13.84 | 7.04 |
| Hydrogen chloride | HCl | 158.96 | 13.9 |
| Mercury | Hg | 234.2 | $1.65 \times 10^{-7}$ |
| Methane | $CH_4$ | 90.68 | 11.7 |
| Neon | Ne | 24.57 | 43.2 |
| Nitric oxide | NO | 109.50 | 21.92 |
| Nitrogen | $N_2$ | 63.18 | 12.6 |
| Nitrous oxide | $N_2O$ | 182.34 | 87.85 |
| Oxygen | $O_2$ | 54.36 | 0.152 |
| Palladium | Pd | 1825 | $3.5 \times 10^{-3}$ |
| Platinum | Pt | 2045 | $2.0 \times 10^{-4}$ |
| Sulfur dioxide | $SO_2$ | 197.69 | 1.67 |
| Titatium | Ti | 1941 | $5.3 \times 10^{-3}$ |
| Uranium hexafluoride | $UF_6$ | 337.17 | 151.7 |
| Water | $H_2O$ | 273.16 | 0.61 |
| Xenon | Xe | 161.3 | 81.5 |
| Zinc | Zn | 692.65 | 0.065 |

*Source:* Data from Natl. Bur. Stand. (U.S.) Circ., 500 (1952).

above the critical point. Substances that expand and contract on freezing differ only in the melting line on the $P$-$T$ diagram.

## 2-5 ■ THE $P$-$v$-$T$ SURFACE

In Chap. 1, we indicated that the state of a simple compressible substance is fixed by any two independent, intensive properties. Once the two appropriate properties are fixed, all the other properties become dependent properties. Remembering that any equation with two independent variables in the form $z = z(x, y)$ represents a surface in space, we can represent the $P$-$v$-$T$ behavior of a substance as a surface in space, as shown in Figs. 2-23 and 2-24. Here $T$ and $v$ may be viewed as the

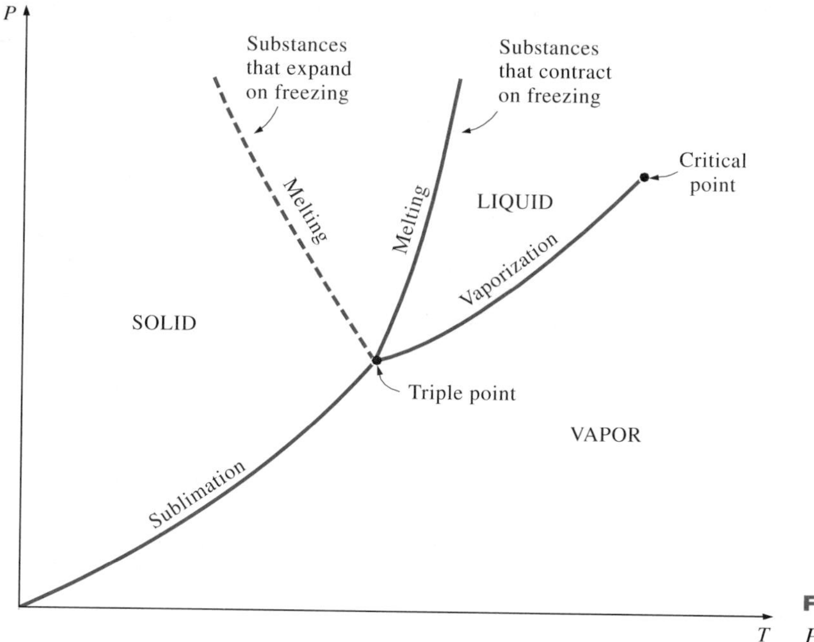


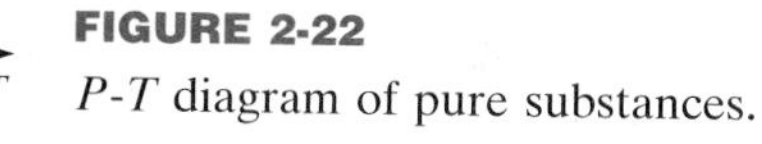
**FIGURE 2-22**
*P-T* diagram of pure substances.

independent variables (the base) and $P$ as the dependent variable (the height).

All the points on the surface represent equilibrium states. All states along the path of a quasi-equilibrium process lie on the $P$-$v$-$T$ surface since such a process must pass through equilibrium states. The single-phase regions appear as curved surfaces on the $P$-$v$-$T$ surface, and the two-phase regions as surfaces perpendicular to the $P$-$T$ plane. This is expected since the projections of two-phase regions on the $P$-$T$ plane are lines.

All the two-dimensional diagrams we have discussed so far are merely projections of this three-dimensional surface onto the appropriate planes. A $P$-$v$ diagram is just a projection of the $P$-$v$-$T$ surface on the $P$-$v$ plane, and a $T$-$v$ diagram is nothing more than the bird's-eye view of this surface. The $P$-$v$-$T$ surfaces present a great deal of information at once, but in a thermodynamic analysis it is more convenient to work with two-dimensional diagrams, such as the $P$-$v$ and $T$-$v$ diagrams.

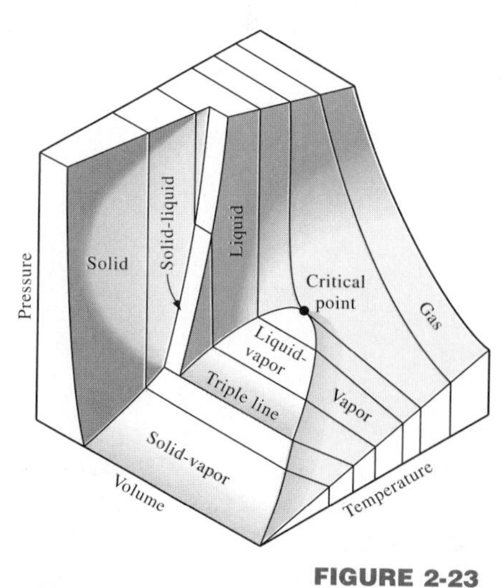


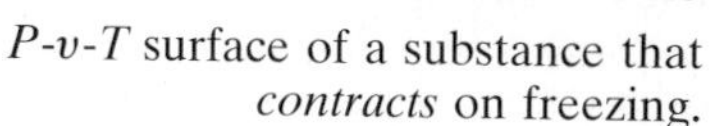
**FIGURE 2-23**
*P-v-T* surface of a substance that *contracts* on freezing.

## 2-6 ■ PROPERTY TABLES

For most substances, the relationships among thermodynamic properties are too complex to be expressed by simple equations. Therefore, properties are frequently presented in the form of tables. Some thermodynamic properties can be measured easily, but others cannot be

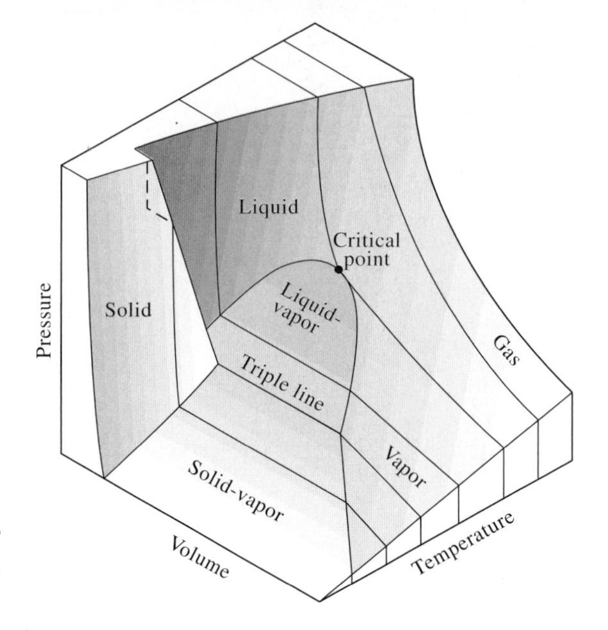


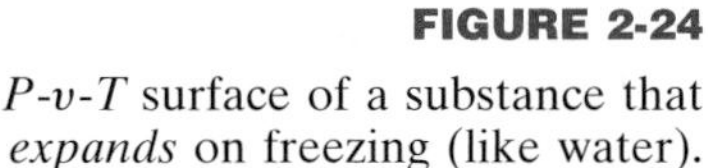

**FIGURE 2-24**

*P-v-T* surface of a substance that *expands* on freezing (like water).

measured directly and are calculated by using the relations that relate them to measurable properties. The results of these measurements and calculations are presented in tables in a convenient format. In the following discussion, the steam tables will be used to demonstrate the use of thermodynamic property tables. Property tables of other substances are used in the same manner.

For each substance, the thermodynamic properties are listed in more than one table. In fact, a separate table is prepared for each region of interest such as the superheated vapor, compressed liquid, and saturated (mixture) regions. Before we get into the discussion of property tables, we will define a new property called *enthalpy*.

## Enthalpy—A Combination Property

A person looking at the tables carefully will notice two new properties: enthalpy $h$ and entropy $s$. Entropy is a property associated with the second law of thermodynamics, and we will not use it until it is properly defined in Chap. 6. However, it is appropriate to introduce enthalpy at this point.

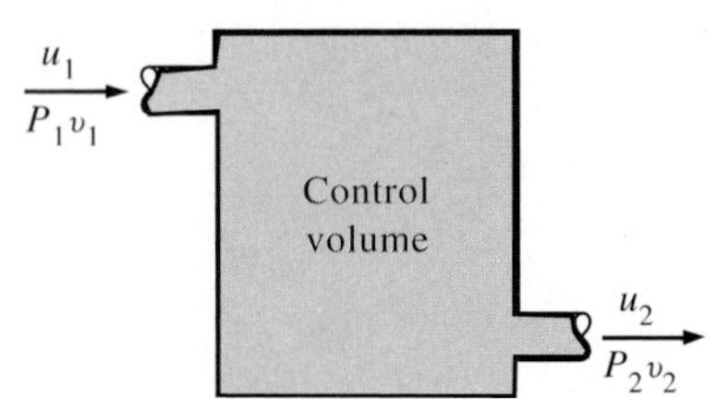


**FIGURE 2-25**

The combination $u + Pv$ is frequently encountered in the analysis of control volumes.

In the analysis of certain types of processes, particularly in power generation and refrigeration (Fig. 2-25), we frequently encounter the combination of properties $U + PV$. For the sake of simplicity and convenience, this combination is defined as a new property, **enthalpy**, and given the symbol $H$:

$$H = U + PV \qquad \text{(kJ)} \qquad (2\text{-}1)$$

or, per unit mass,

$$h = u + Pv \qquad \text{(kJ/kg)} \tag{2-2}$$

Both the total enthalpy $H$ and specific enthalpy $h$ are simply referred to as enthalpy since the context will clarify which one is meant. Notice that the equations given above are dimensionally homogeneous. That is, the unit of the pressure–volume produce may differ from the unit of the internal energy by only a factor (Fig. 2-26). For example, it can be easily shown that $1\ \text{kPa} \cdot \text{m}^3 = 1\ \text{kJ}$. In some tables encountered in practice, the internal energy $u$ is frequently not listed, but it can always be determined from $u = h - Pv$.

The widespread use of the property enthalpy is due to Professor Richard Mollier, who recognized the importance of the group $u + Pv$ in the analysis of steam turbines and in the representation of the properties of steam in tabular and graphical form (as in the famous Mollier chart). Mollier referred to the group $u + Pv$ as *heat contents* and *total heat.* These terms were not quite consistent with the modern thermodynamic terminology and were replaced in 1930s by the term *enthalpy* (from the Greek word *enthalpien* which means *to heat*).

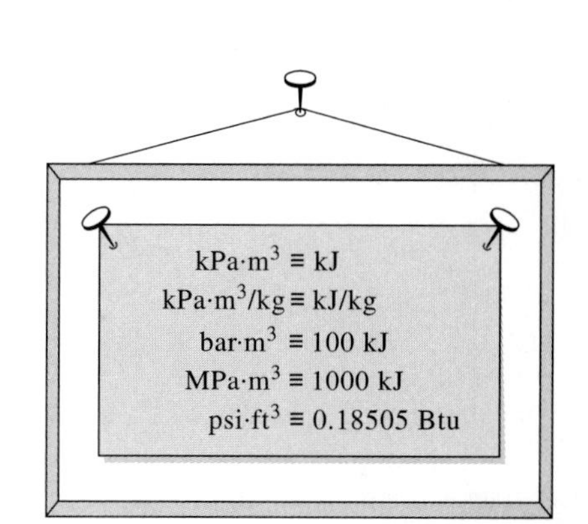


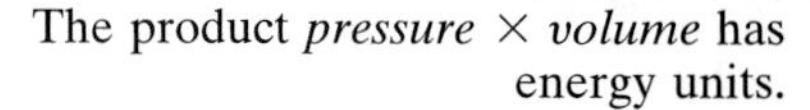

**FIGURE 2-26**
The product *pressure* × *volume* has energy units.

## 1 Saturated Liquid and Saturated Vapor States

The properties of saturated liquid and saturated vapor for water are listed in Table A-4. The use of Table A-4 is illustrated in Fig. 2-27.

The subscript $f$ is used to denote properties of a saturated liquid, and the subscript $g$ to denote the properties of saturated vapor. These symbols are commonly used in thermodynamics and originated from German. Another subscript commonly used is $fg$, which denotes the difference between the saturated vapor and saturated liquid values of the same property. For example,

$$v_f = \text{specific volume of saturated liquid}$$
$$v_g = \text{specific volume of saturated vapor}$$
$$v_{fg} = \text{difference between } v_g \text{ and } v_f \text{ (that is, } v_{fg} = v_g - v_f)$$

The quantity $h_{fg}$ is called the **enthalpy of vaporization** (or latent heat of vaporization). It represents the amount of energy needed to vaporize a unit mass of saturated liquid at a given temperature or pressure. It decreases as the temperature or pressure increases, and it becomes zero at the critical point.

| Temp. °C $T$ | Sat. press kPa $P_{sat}$ | Specific volume m³/kg: Sat. liquid $v_f$ | Specific volume m³/kg: Sat. vapor $v_g$ |
|---|---|---|---|
| 85 | 57.83 | 0.001 033 | 2.828 |
| 90 | 70.14 | 0.001 036 | 2.361 |
| 95 | 84.55 | 0.001 040 | 1.982 |

Specified temperature — Corresponding saturation pressure — Specific volume of saturated liquid — Specific volume of saturated vapor

**FIGURE 2-27**
A partial list of Table A-4.

**EXAMPLE 2-1**

A rigid tank contains 50 kg of saturated liquid water at 90°C. Determine the pressure in the tank and the volume of the tank.

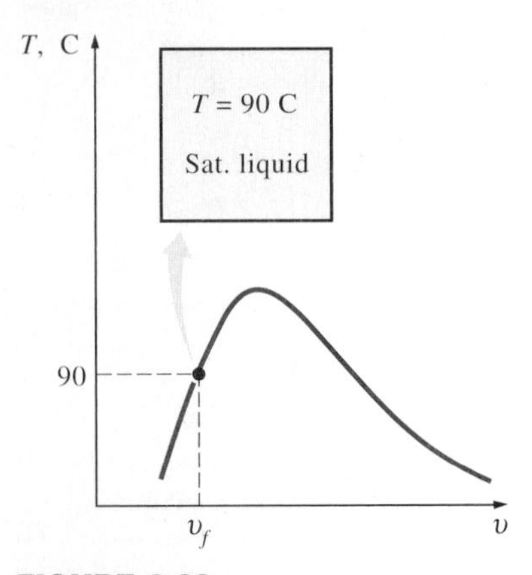


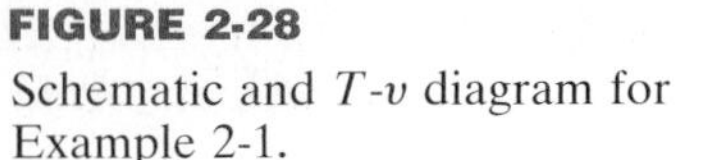

**FIGURE 2-28**
Schematic and $T$-$\upsilon$ diagram for Example 2-1.

**Solution** The state of the saturated liquid water is shown on a $T$-$\upsilon$ diagram in Fig. 2-28. Since saturation conditions exist in the tank, the pressure must be the saturation pressure at 90°C:

$$P = P_{\text{sat @ } 90^\circ\text{C}} = 70.14 \text{ kPa} \qquad \text{(Table A-4)}$$

The specific volume of the saturated liquid at 90°C is

$$\upsilon = \upsilon_{f \text{ @ } 90^\circ\text{C}} = 0.001036 \text{ m}^3/\text{kg} \qquad \text{(Table A-4)}$$

Then the total volume of the tank is determined from

$$V = m\upsilon = (50 \text{ kg})(0.001036 \text{ m}^3/\text{kg}) = \mathbf{0.0518 \text{ m}^3}$$

**EXAMPLE 2-2**

A mass of 200 g of saturated liquid water is completely vaporized at a constant pressure of 100 kPa. Determine (*a*) the volume change and (*b*) the amount of energy added to the water.

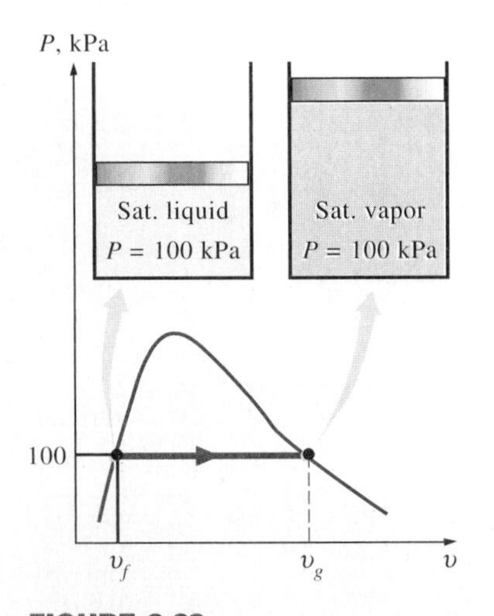


**FIGURE 2-29**
Schematic and $P$-$\upsilon$ diagram for Example 2-2.

**Solution** (*a*) The process described is illustrated on a $P$-$\upsilon$ diagram in Fig. 2-29. The volume change per unit mass during a vaporization process is $\upsilon_{fg}$, which is the difference between $\upsilon_g$ and $\upsilon_f$. Reading these values from Table A-4 at 100 kPa and substituting yield

$$\upsilon_{fg} = \upsilon_g - \upsilon_f = (1.6940 - 0.001043) \text{ m}^3/\text{kg} = 1.6930 \text{ m}^3/\text{kg}$$

Thus, $\quad \Delta V = m\upsilon_{fg} = (0.2 \text{ kg})(1.6930 \text{ m}^3/\text{kg}) = \mathbf{0.3386 \text{ m}^3}$

Note that we have considered the first four decimal digits of $\upsilon_f$ and disregarded the rest. This is because $\upsilon_g$ has significant numbers to the first four decimal places only, and we do not know the numbers in the other decimal places. Taking $\upsilon_f$ as it is would mean that we are assuming $\upsilon_g = 1.694000$, which is not necessarily the case. It could very well be that $\upsilon_g = 1.694038$ since this number, too, would truncate to 1.6940. All the digits in our result (1.6930) are significant. But if we used $\upsilon_f$ as it is, we would obtain $\upsilon_{fg} = 1.692957$, which falsely implies that our result is accurate to the sixth decimal place.

(*b*) The amount of energy needed to vaporize the unit mass of a substance at a given pressure is the enthalpy of vaporization at that pressure which, at 100 kPa, is $h_{fg} = 2258.0$ kJ/kg. Thus the amount of energy added is

$$mh_{fg} = (0.2 \text{ kg})(2258 \text{ kJ/kg}) = \mathbf{451.6 \text{ kJ}}$$

## Saturated Liquid–Vapor Mixture

During a vaporization process, a substance exists as part liquid and part vapor. That is, it is a mixture of saturated liquid and saturated vapor (Fig. 2-30). To analyze this mixture properly, we need to know the proportions of the liquid and vapor phases in the mixture. This is done by defining a new property called the **quality** $x$ as the ratio of the mass of vapor to the total mass of the mixture:



$$x = \frac{m_{\text{vapor}}}{m_{\text{total}}} \quad \text{(2-3)}$$

where $\quad m_{\text{total}} = m_{\text{liquid}} + m_{\text{vapor}} = m_f + m_g$

Quality has significance for saturated mixtures only. It has no meaning in the compressed liquid or superheated vapor regions. Its value is always between 0 and 1. The quality of a system that consists of *saturated liquid* is 0 (or 0 percent), and the quality of a system consisting of *saturated vapor* is 1 (or 100 percent). In saturated mixtures, quality can serve as one of the two independent intensive properties needed to describe a state. Note that *the properties of the saturated liquid are the same whether it exists alone or in a mixture with saturated vapor.* During the vaporization process, only the amount of saturated liquid changes, not its properties. The same can be said about a saturated vapor.

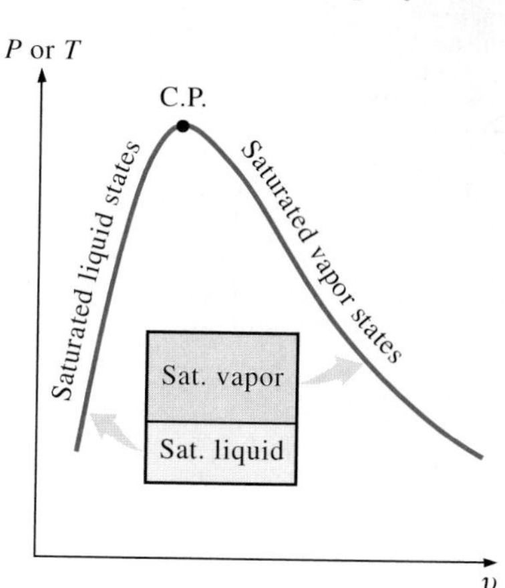


**FIGURE 2-30**

The relative amounts of liquid and vapor phases in a saturated mixture are specified by *quality x*.

A saturated mixture can be treated as a combination of two subsystems: the saturated liquid and the saturated vapor. However, the amount of mass for each phase is usually not known. Therefore, it is often more convenient to imagine that the two phases are mixed very well, forming a homogeneous appearance (Fig. 2-31). Then the properties of this "mixture" will simply be the average properties of the saturated liquid–vapor mixture under consideration. Here is how it is done.

Consider a tank that contains a saturated liquid–vapor mixture. The volume occupied by saturated liquid is $V_f$, and the volume occupied by saturated vapor is $V_g$. The total volume $V$ is the sum of these two:

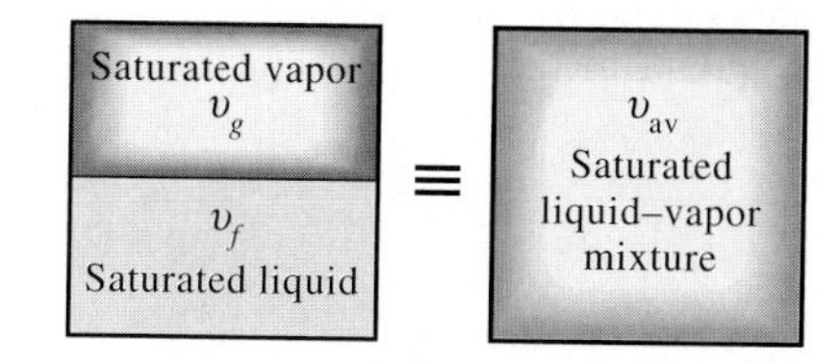


**FIGURE 2-31**

A two-phase system can be treated as a homogeneous mixture for computational purposes.

$$V = V_f + V_g$$

$$V = mv \longrightarrow m_t v_{\text{av}} = m_f v_f + m_g v_g$$

$$m_f = m_t - m_g \longrightarrow m_t v_{\text{av}} = (m_t - m_g)v_f + m_g v_g$$

Dividing by $m_t$ yields

$$v_{\text{av}} = (1 - x)v_f + x v_g$$

since $x = m_g/m_t$. This relation can also be expressed as

$$v_{\text{av}} = v_f + x v_{fg} \qquad (\text{m}^3/\text{kg})$$

where $v_{fg} = v_g - v_f$. Solving for quality, we obtain

$$x = \frac{v_{\text{av}} - v_f}{v_{fg}}$$

Based on this equation, quality can be related to the horizontal distances on a $P$-$v$ or $T$-$v$ diagram. At a given temperature or pressure, the numerator is the distance between the actual state and the saturated liquid state, and the denominator is the length of the entire horizontal line that connects the saturated liquid and saturated vapor states. A state of 50 percent quality will lie in the middle of this horizontal line.

The analysis given above can be repeated for internal energy and enthalpy with the following results:

$$u_{av} = u_f + xu_{fg} \qquad \text{(kJ/kg)}$$

$$h_{av} = h_f + xh_{fg} \qquad \text{(kJ/kg)}$$

All the results are of the same format, and they can be summarized in a single equation as

$$y_{av} = y_f + xy_{fg}$$

where $y$ is $v$, $u$, or $h$. The subscript "av" (for "average") is usually dropped for simplicity. The values of the average properties of the mixtures are always *between* the values of the saturated liquid and the saturated vapor properties (Fig. 2-32). That is,

$$y_f \leqslant y_{av} \leqslant y_g$$

The above relations are used when evaluating the properties of saturated mixture from tables instead of charts for better accuracy.

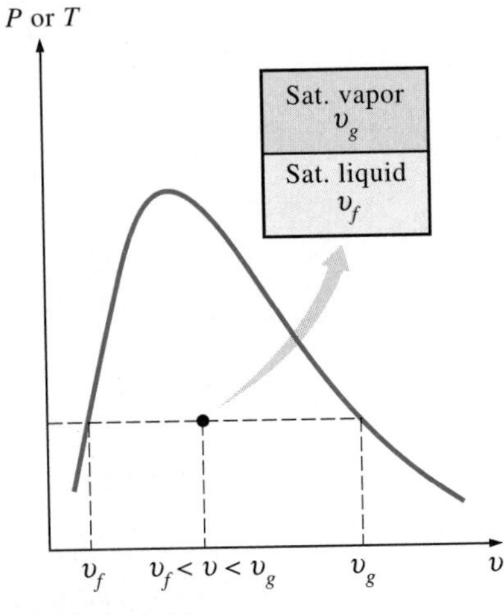


**FIGURE 2-32**

The $v$ values of a saturated liquid–vapor mixture lies between the $v_f$ and $v_g$ values at the spacified $T$ or $P$.

## 2 Superheated Vapor

In the region to the right of the saturated vapor line, a substance exists as superheated vapor. Since the superheat region is a single-phase region (vapor phase only), temperature and pressure are no longer dependent properties and they can conveniently be used as the two independent properties. At pressures sufficiently below the critical pressure or temperatures sufficiently above the critical temperature, a superheated vapor can be approximated as an *ideal gas* which is discussed in the next section. Otherwise property tables or charts should be used.

## 3 Compressed Liquid

The properties of compressed liquid are relatively independent of pressure. Increasing the pressure of a compressed liquid 100 times often causes properties to change less than 1 percent. The property most affected by pressure is enthalpy.

In the absence of compressed liquid data, a general approximation is *to treat compressed liquid as saturated liquid at the given temperature* (Fig. 2-33). This is because the compressed liquid properties depend on temperature more strongly than they do on pressure. Thus,

$$y \approx y_{f@T}$$

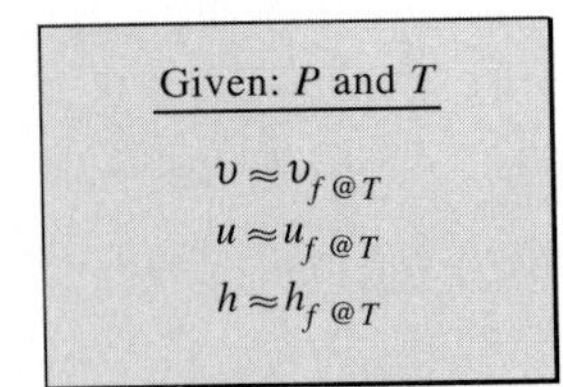


**FIGURE 2-33**

A compressed liquid may be approximated as a saturated liquid at the same temperature.

for compressed liquids, where $y$ is $v$, $u$, or $h$. Of these three properties, the property whose value is most sensitive to variations in the pressure is the enthalpy $h$. Although the above approximation results in negligible error in $v$ and $u$, the error in $h$ may reach undesirable levels. However, the error in $h$ at very high pressures can be reduced significantly by evaluating it from

$$h \approx h_{f @ T} + v_f(P - P_{sat})$$

instead of taking it to be just $h_f$. Here $P_{sat}$ is the saturation pressure at the given temperature.

### Reference State and Reference Values

The values of $u$, $h$, and $s$ cannot be measured directly, and they are calculated from measurable properties using the relations between thermodynamic properties. However, those relations give the *changes* in properties, not the values of properties at specified states. Therefore, we need to choose a convenient *reference state* and assign a value of *zero* for a convenient property or properties at that state. For water, the state of saturated liquid at 0.01°C is taken as the reference state, and the internal energy and entropy are assigned zero values at that state. Note that some properties may have negative values as a result of the reference state chosen.

It should be mentioned that sometimes different tables list different values for some properties at the same state as a result of using a different reference state. However, in thermodynamics we are concerned with the *changes* in properties, and the reference state chosen is of no consequence in calculations as long as we use values from a single consistent set of tables or charts.

## 2-7 ■ THE IDEAL-GAS EQUATION OF STATE

One way of reporting property data for pure substances is to list values of properties at various states. The property tables provide very accurate information about the properties, but they are very bulky and vulnerable to typographical errors. A more practical and desirable approach would be to have some simple relations among the properties that are sufficiently general and accurate.

Any equation that relates the pressure, temperature, and specific volume of a substance is called an **equation of state**. Property relations that involve other properties of a substance at equilibrium states are also referred to as equations of state. There are several equations of state, some simple and other very complex. The simplest and best known equation of state for substances in the gas phase is the ideal-gas equation of state. This equation predicts the $P$-$v$-$T$ behavior of a gas quite accurately within some properly selected region.

*Gas* and *vapor* are often used as synonymous words. The vapor phase of a substance is customarily called a *gas* when it is above the critical temperature. *Vapor* usually implies a gas that is not far from a state of condensation.

In 1662, Robert Boyle, an Englishman, observed during his experiments with a vacuum chamber that the pressure of gases is inversely proportional to their volume. In 1802, J. Charles and J. Gay-Lussac, Frenchmen, experimentally determined that at low pressures the volume of a gas is proportional to its temperature. That is,

$$P = R\left(\frac{T}{v}\right)$$

or

$$Pv = RT \qquad (2\text{-}4)$$

**TABLE 2-3**
**Different substances have different gas constants**

| Substance | $R$, kJ/(kg·K) |
|---|---|
| Air | 0.2870 |
| Helium | 2.0769 |
| Argon | 0.2081 |
| Nitrogen | 0.2968 |

where the constant of proportionality $R$ is called the **gas constant**. Equation 2-4 is called the **ideal-gas equation of state**, or simply the **ideal-gas relation**, and a gas that obeys this relation is called an **ideal gas**. In this equation, $P$ is the absolute pressure, $T$ is the absolute temperature, and $v$ is the specific volume.

The gas constant $R$ is different for each gas (Table 2-3) and is determined from

$$R = \frac{R_u}{M} \qquad [\text{kJ/(kg}\cdot\text{K) or kPa}\cdot\text{m}^3\text{/(kg}\cdot\text{K)}] \qquad (2\text{-}5)$$

where $R_u$ is the **universal gas constant** and $M$ is the molar mass (also called *molecular weight*) of the gas. The constant $R_u$ is the same for all substances, and its value is

$$R_u = \begin{cases} 8.314\ \text{kJ/(kmol}\cdot\text{K)} \\ 8.314\ \text{kPa}\cdot\text{m}^3\text{/(kmol}\cdot\text{K)} \\ 0.08314\ \text{bar}\cdot\text{m}^3\text{/(kmol}\cdot\text{K)} \\ 1.986\ \text{Btu/(lbmol}\cdot\text{R)} \\ 10.73\ \text{psia}\cdot\text{ft}^3\text{/(lbmol}\cdot\text{R)} \\ 1545\ \text{ft lbf/(lbmol}\cdot\text{R)} \end{cases} \qquad (2\text{-}6)$$

The **molar mass** $M$ can simply be defined as *the mass of one mole* (also called a *gram-mole,* abbreviated gmol) *of a substance in grams,* or, *the mass of one kmol* (also called a *kilogram-mole,* abbreviated kgmol) *in kilograms.* In English units, it is the mass of 1 lbmol (1 pound-mole = 0.4536 kmol) in lbm (1 pound-mass = 0.4536 kg). Notice that the molar mass of a substance has the same numerical value in both unit systems because of the way it is defined. When we say the molar mass of nitrogen is 28, it simply means the mass of 1 kmol of nitrogen is 28 kg, or the mass of 1 lbmol of nitrogen is 28 lbm. That is, $M = 28$ kg/kmol = 28 lbm/lbmol. The mass of a system is equal to the product of its molar mass $M$ and the mole number $N$:

$$m = MN \qquad (\text{kg}) \qquad (2\text{-}7)$$

The values of $R$ and $M$ for several substances are given in Table A-1.

The ideal-gas equation of state can be written in several different forms:

$$V = mv \longrightarrow PV = mRT \tag{2-8}$$

$$mR = (MN)R = NR_u \longrightarrow PV = NR_uT \tag{2-9}$$

$$V = N\bar{v} \longrightarrow P\bar{v} = r_uT \tag{2-10}$$

where $\bar{v}$ is the molar specific volume, i.e., the volume per unit mole (in $m^3$/kmol or $ft^3$/lbmol). A bar above a property will denote values on a *unit-mole basis* throughout this text (Fig. 2-34).

By writing Eq. 2-8 twice for a fixed mass and simplifying, the properties of an ideal gas at two different states are related to each other by

$$\frac{P_1V_1}{T_1} = \frac{P_2V_2}{T_2} \tag{2-11}$$

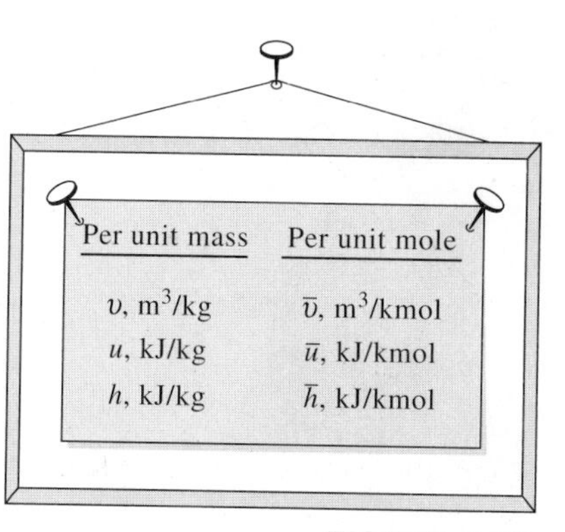


**FIGURE 2-34**
Properties per unit mole are denoted with a bar on the top.

An ideal gas is an *imaginary* substance that obeys the relation $Pv = RT$ (Fig. 2-35). It has been experimentally observed that the ideal-gas relation given above closely approximates the $P$-$v$-$T$ behavior of real gases at low densities. At low pressures and high temperatures, the density of a gas decreases, and the gas behaves as an ideal gas under these conditions. What constitutes low pressure and high temperature is explained in the next section.

In the range of practical interest, many familiar gases such as air, nitrogen, oxygen, hydrogen, helium, argon, neon, krypton, and even heavier gases such as carbon dioxide can be treated as ideal gases with negligible error (often less than 1 percent). Dense gases such as water vapor in steam power plants and refrigerant vapor in refrigerators, however, should not be treated as ideal gases. Instead, the property tables should be used for these substances.

**FIGURE 2-35**
The ideal-gas relation often is not applicable to real gases; thus, care should be exercised when using it.

**EXAMPLE 2-3**

Determine the mass of the air in a room whose dimensions are 4 m × 5 m × 6 m at 100 kPa and 25°C.

**Solution** A sketch of the room is given in Fig. 2-36. Air at specified conditions can be treated as an ideal gas. From Table A-1, the gas constant of air is $R = 0.287\ \text{kPa} \cdot \text{m}^3(\text{kg} \cdot \text{K})$, and the absolute temperature is $T = 25°\text{C} + 273 = 298\ \text{K}$. The volume of the room is

$$V = (4\ \text{m})(5\ \text{m})(6\ \text{m}) = 120\ \text{m}^3$$

By substituting these values into Eq. 2-8, the mass of air in the room is determined to be

$$m = \frac{PV}{RT} = \frac{(100\ \text{kPa})(120\ \text{m}^3)}{[0.287\ \text{kPa} \cdot \text{m}^3/(\text{kg} \cdot \text{K})](198\ \text{K})} = 140.3\ \text{kg}$$

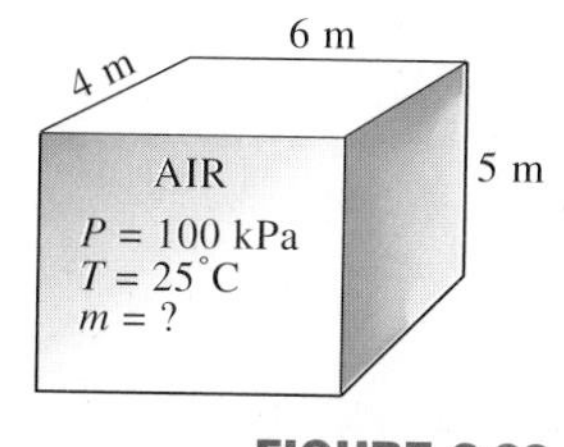


**FIGURE 2-36**
Schematic for Example 2-3.

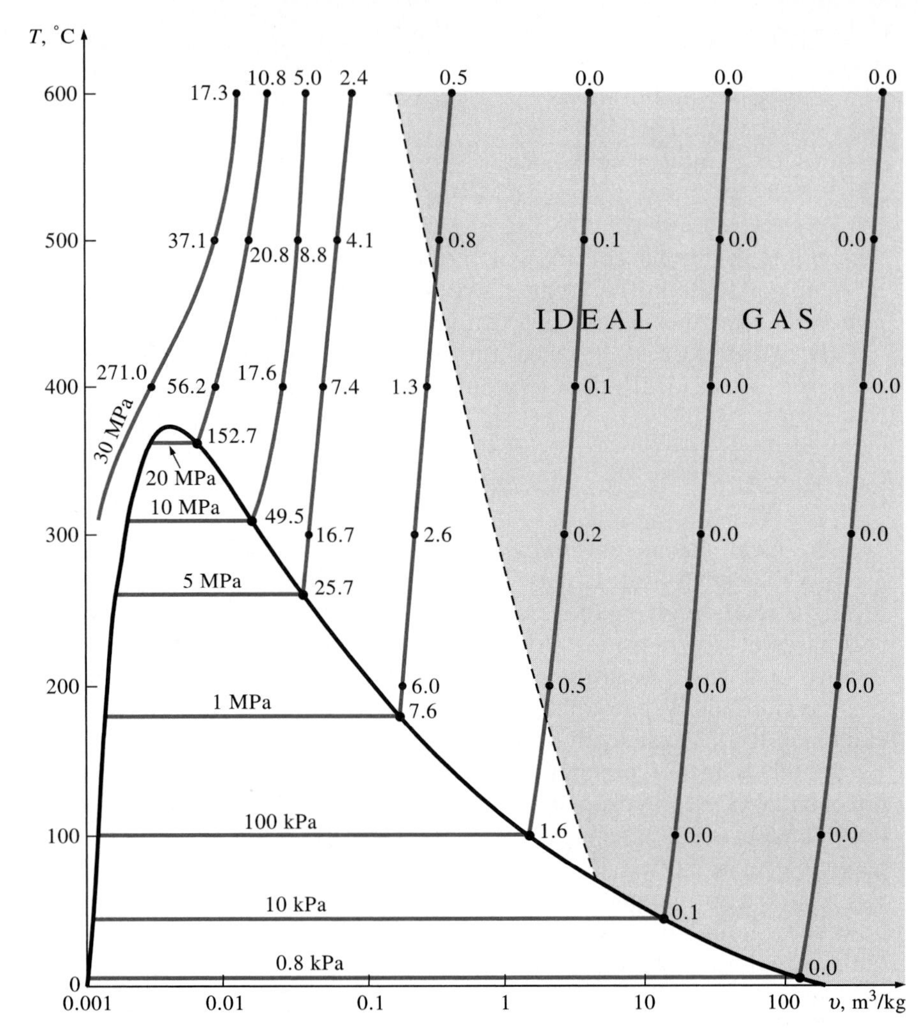


**FIGURE 2-37**
Percentage of error $[(|v_{\text{table}} - v_{\text{ideal}}|/v_{\text{table}}) \times 100]$ involved in assuming steam to be an ideal gas, and the region where steam can be treated as an ideal gas with less than 1-percent error.

### Is Water Vapor an Ideal Gas?

This question cannot be answered with a simple yes or no. The error involved in treating water vapor as an ideal gas is calculated and plotted in Fig. 2-37. It is clear from this figure that at pressures below 10 kPa, water vapor can be treated as an ideal gas, regardless of its temperature, with negligible error (less than 0.1 percent). But at higher pressures, the ideal-gas assumption yields unacceptable errors, particularly in the vicinity of the critical point and the saturated vapor line (over 100 percent). Therefore, in air-conditioning applications, the water vapor in the air can be treated as an ideal gas with essentially no error since the pressure of the water vapor is very low. In steam power plant applications, however, the pressures involved are usually very high; therefore, ideal-gas relations should not be used.

## 2-8 ■ COMPRESSIBILITY FACTOR—A MEASURE OF DEVIATION FROM IDEAL-GAS BEHAVIOR*

The ideal-gas equation is very simple and thus very convenient to use. But, as illustrated in Fig. 2-37, gases deviate from ideal-gas behavior significantly at states near the saturation region and the critical point. This deviation from ideal-gas behavior at a given temperature and pressure can accurately be accounted for by the introduction of a correction factor called the **compressibility factor** $Z$. It is defined as

$$Z = \frac{Pv}{RT} \tag{2-12}$$

or

$$Pv = ZRT \tag{2-13}$$

It can also be expressed as

$$Z = \frac{v_{\text{actual}}}{v_{\text{ideal}}} \tag{2-14}$$

where $v_{\text{ideal}} = RT/P$. Obviously, $Z = 1$ for ideal gases. For real gases $Z$ can be greater than or less than unity (Fig. 2-38). The farther away $Z$ is from unity, the more the gas deviates from ideal-gas behavior.

IDEAL GAS

$Z = 1$

REAL GASES

$Z \begin{cases} > 1 \\ = 1 \\ < 1 \end{cases}$

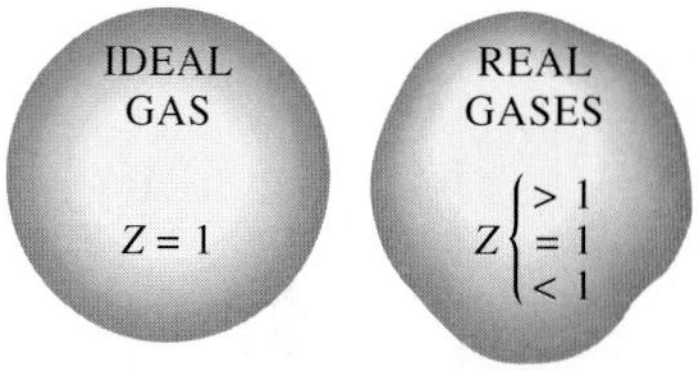

**FIGURE 2-38**
The compressibility factor is unity for ideal gases.

We have repeatedly said that gases follow the ideal-gas equation closely at low pressures and high temperatures. But what exactly constitutes low pressure or high temperature? Is −100°C a low temperature? It definitely is for most substances, but not for air. Air (or nitrogen) can be treated as an ideal gas at this temperature and atmospheric pressure with an error under 1 percent. This is because nitrogen is well over its critical temperature (−147°C) and away from the saturation region. But at this temperature and pressure, most substances would exist in the solid phase. Therefore, the pressure or temperature of a substance is high or low relative to its critical temperature or pressure.

Gases behave differently at a given temperature and pressure, but they behave very much the same at temperatures and pressures normalized with respect to their critical temperatures and pressures. The normalization is done as

$$P_R = \frac{P}{P_{\text{cr}}} \quad \text{and} \quad T_R = \frac{T}{T_{\text{cr}}} \tag{2-15}$$

Here $P_R$ is called the **reduced pressure** and $T_R$ the **reduced temperature**. The $Z$ factor for all gases is approximately the same at the same reduced pressure and temperature (Fig. 2-39). This is called the **principle of corresponding states**. In Fig. 2-40, the experimentally determined $Z$ values are plotted against $P_R$ and $T_R$ for several gases. The gases seem to obey the principle of corresponding states reasonably well. By curve-fitting all the data, we obtain the **generalized compressibility chart** which can be used for all gases. This chart is given in the Appendix in Fig. A-13.

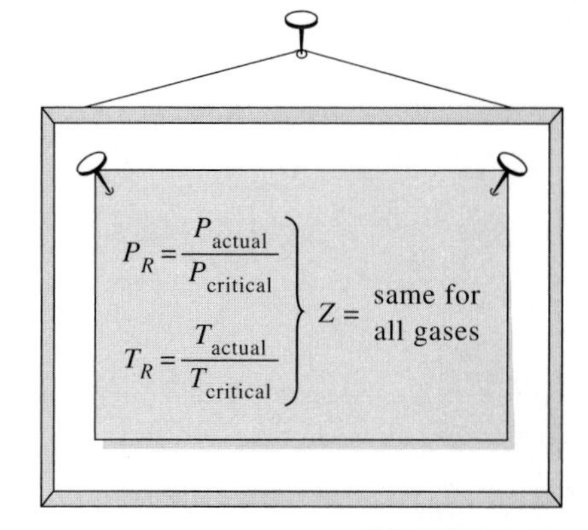


**FIGURE 2-39**
The compressibility factor is the same for all gases at the same reduced pressure and temperature (*principle of corresponding states*).

*This section can be skipped without a loss in continuity.

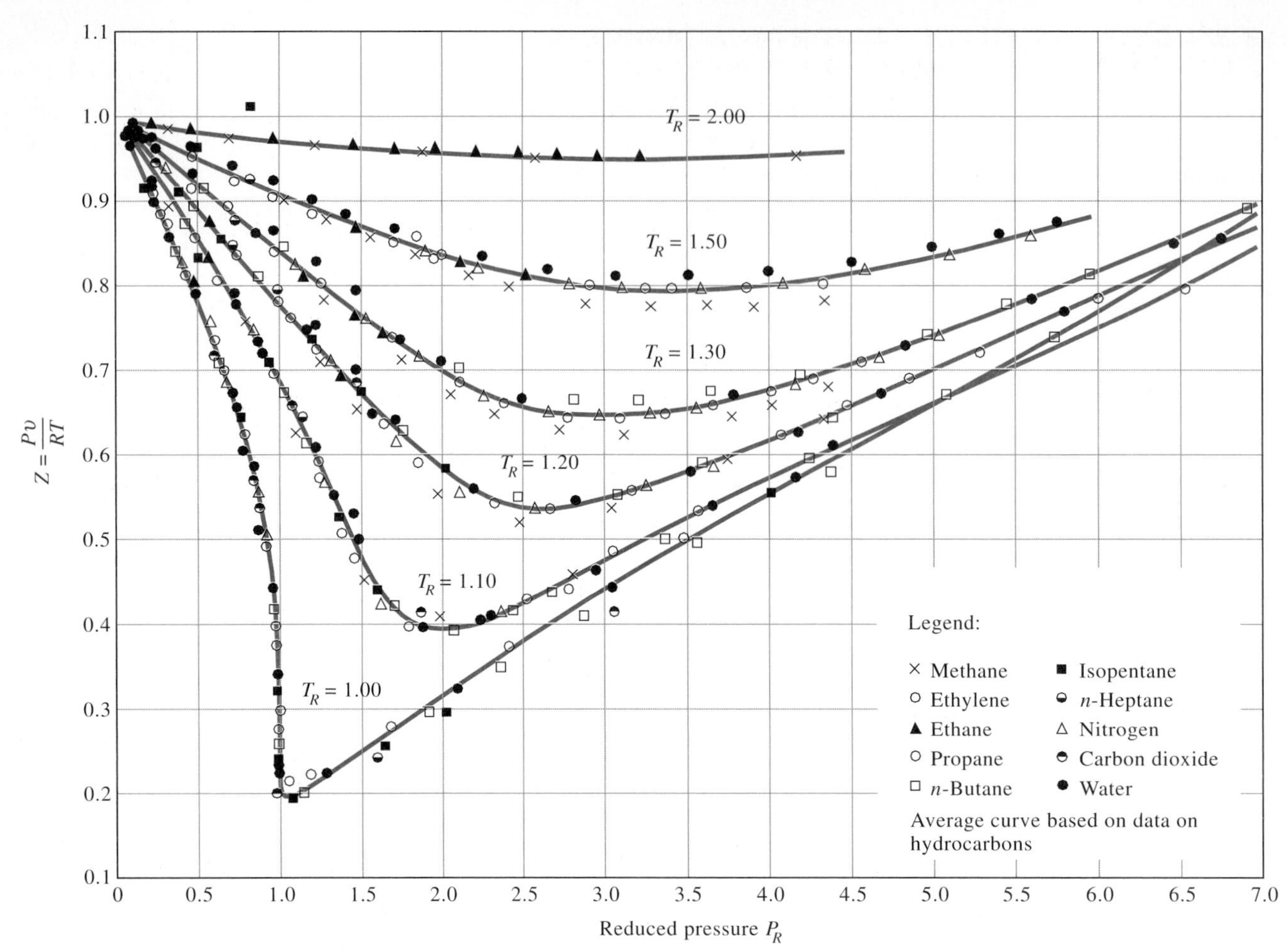


**FIGURE 2-40**

Comparison of $Z$ factors for various gases. [*Source:* Gour-Jen Su, "Modified Law of Corresponding States," *Ind. Eng. Chem.* (Intern. ed.), 38:803 (1946).]

The use of a compressibility chart requires a knowledge of critical-point data, and the results obtained are accurate to within a few percent.

The following observations can be made from the generalized compressibility chart:

1. At very low pressures ($P_R \ll 1$), the gases behave as an ideal gas regardless of temperature (Fig. 2-41),
2. At high temperatures ($T_R > 2$), ideal-gas behavior can be assumed with good accuracy regardless of pressure (except when $P_R \gg 1$).
3. The deviation of a gas from ideal-gas behavior is greatest in the vicinity of the critical point (Fig. 2-42).

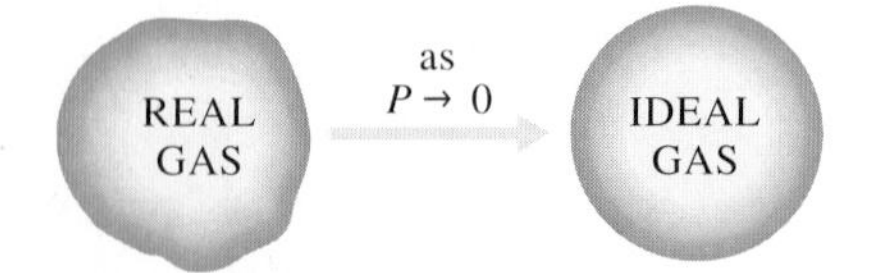


**FIGURE 2-41**

At very low pressures, all gases approach ideal-gas behavior (regardless of their temperature).

**EXAMPLE 2-4**

Determine the specific volume of refrigerant-134a at 1 MPa and 50°C, using (*a*) the ideal gas equation of state and (*b*) the generalized compressibility chart. Compare the values obtained to the actual value of 0.02171 $m^3/kg$, and determine the error involved in each case.

**Solution** The gas constant, the critical pressure, and the critical temperature of refrigerant-134a are determined from Table A-1 to be

$$R = 0.0815\ \text{kPa}\cdot\text{m}^3/(\text{kg}\cdot\text{K})$$
$$P_{cr} = 4.067\ \text{MPa}$$
$$T_{cr} = 374.3\ \text{K}$$

(*a*) The specific volume of the refrigerant-134a under the ideal-gas assumption is determined from the ideal-gas relation to be

$$v = \frac{RT}{P} = \frac{[0.0815\ \text{kPa}\cdot\text{m}^3/(\text{kg}\cdot\text{K})](323\ \text{K})}{1000\ \text{kPa}} = 0.02632\ \text{m}^3/\text{kg}$$

Therefore, treating the refrigerant-134a vapor as an ideal gas would result in an error of (0.02632 − 0.02171)/0.02171 = 0.212, or 21.2 percent in this case.

(*b*) To determine the correction factor $Z$ from the compressibility chart, we first need to calculate the reduced pressure and temperature:

$$\left.\begin{aligned} P_R &= \frac{P}{P_{cr}} = \frac{1\ \text{MPa}}{4.067\ \text{MPa}} = 0.246 \\ T_R &= \frac{T}{T_{cr}} = \frac{323\ \text{K}}{374.3\ \text{K}} = 0.863 \end{aligned}\right\}\quad Z = 0.84 \qquad \text{(Fig. A-13)}$$

Thus, $\quad v = Zv_{\text{ideal}} = (0.84)(0.02632\ \text{m}^3/\text{kg}) = 0.02211\ \text{m}^3/\text{kg}$

The error in this result is less than 2 percent. Therefore, in the absence of exact tabulated data, the generalized compressibility chart can be used with confidence.

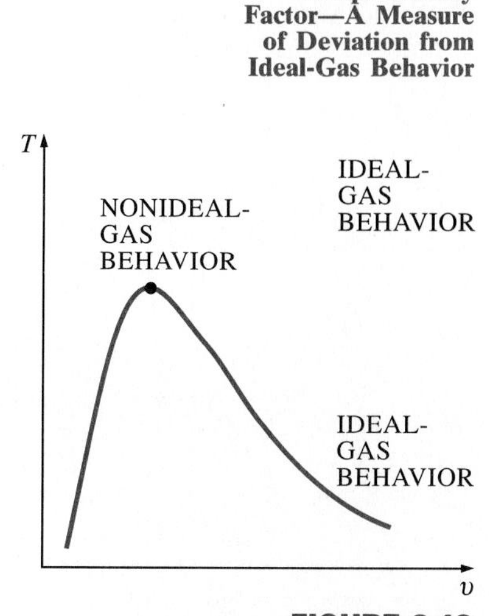


**FIGURE 2-42**
Gases deviate from the ideal-gas behavior most in the neighborhood of the critical point.

When $P$ and $v$, or $T$ and $v$, are given instead of $P$ and $T$, the generalized compressibility chart can still be used to determine the third property, but it would involve tedious trial and error. Therefore, it is very convenient to define one more reduced property called the **pseudo-reduced specific volume** $v_R$ as

$$v_R = \frac{v_{\text{actual}}}{RT_{cr}/P_{cr}} \tag{2-16}$$

Note that $v_R$ is defined differently from $P_R$ and $T_R$. It is related to $T_{cr}$ and $P_{cr}$ instead of $v_{cr}$. Lines of constant $v_R$ are also added to the compressibility charts, and this enables one to determine $T$ of $P$ without having to resort to time-consuming iterations (Fig. 2-43).

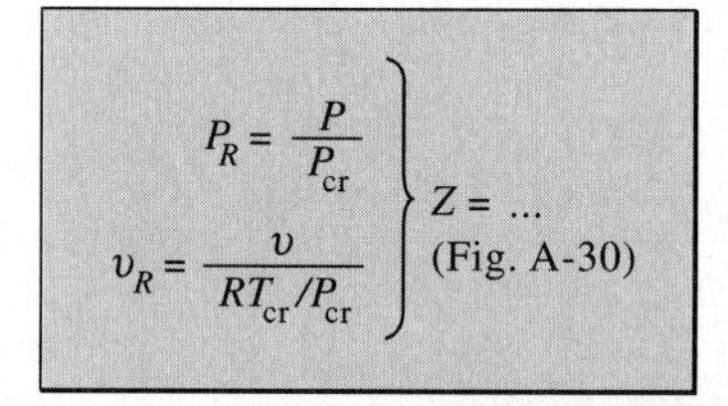


**FIGURE 2-43**
The compressibility factor can also be determined from a knowledge of $P_R$ and $v_R$.

**EXAMPLE 2-5**

Determine the pressure of water vapor at 600°F and 0.514 ft³/lbm, using (*a*) the ideal gas equation of state and (*b*) the generalized compressibility chart. Compare the values obtained to the actual value of 1000 psia.

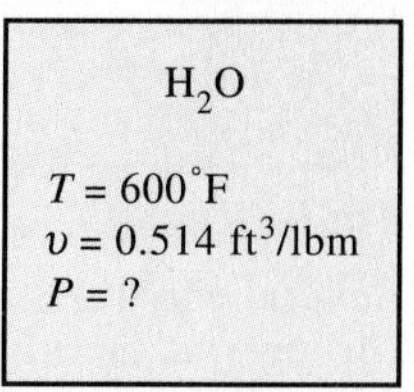


**FIGURE 2-44**
Schematic for Example 2-5.

**Solution** A sketch of the system is given in Fig. 2-44. The gas constant, the critical pressure, and the critical temperature of steam are determined from Table A-1E to be

$$R = 0.5956 \text{ psia} \cdot \text{ft}^3/(\text{lbm} \cdot \text{R})$$
$$P_{cr} = 3204 \text{ psia}$$
$$T_{cr} = 1165.3 \text{ R}$$

(*a*) The pressure of steam under the ideal-gas assumption is determined from the ideal-gas relation (Eq. 2-9) to be

$$P = \frac{RT}{v} = \frac{[0.5956 \text{ psia} \cdot \text{ft}^3/(\text{lbm} \cdot \text{R})](1060 \text{ R})}{0.514 \text{ ft}^3/\text{lbm}} = 1228.3 \text{ psia}$$

Therefore, treating the steam as an ideal gas would result in an error of $(1228.3 - 1000)/1000 = 0.228$, or 22.8 percent in this case.

(*b*) To determine the correction factor $Z$ from the compressibility chart (Fig. A-13), we first need to calculate the pseudo-reduced specific volume and the reduced temperature:

$$\left.\begin{aligned} v_R &= \frac{v_{\text{actual}}}{RT_{cr}/P_{cr}} = \frac{(0.514 \text{ ft}^3/\text{lbm})(3204 \text{ psia})}{[0.5956 \text{ psia} \cdot \text{ft}^3/(\text{lbm} \cdot \text{R})](1165.3 \text{ R})} \\ &= 2.373 \\ T_R &= \frac{T}{T_{cr}} = \frac{1060 \text{ R}}{1165.3 \text{ R}} = 0.91 \end{aligned}\right\} \quad P_R = 0.33$$

Thus, $$P = P_R P_{cr} = (0.33)(3204 \text{ psia}) = 1057.3 \text{ psia}$$

Using the compressibility chart reduced the error from 22.8 to 5.7 percent, which is acceptable for most engineering purposes (Fig. 2-45). A bigger chart, of course, would give better resolution and reduce the reading errors. Notice that we did not have to determine $Z$ in this problem since we could read $P_R$ directly from the chart.

| | *P*, psia |
|---|---|
| Exact | 1000.0 |
| *Z* chart | 1057.3 |
| Ideal gas | 1228.3 |

(from Example 2-5)

**FIGURE 2-45**
Results obtained by using the compressibility chart are usually within a few percent of the experimentally determined values.

## 2-9 OTHER EQUATIONS OF STATE*

The ideal-gas equation of state is very simple, but its range of applicability is limited. It is desirable to have equations of state that represent the *P-v-T* behavior of substances accurately over a larger region with no limitations. Such equations are naturally more complicated. Several equations have been proposed for this purpose (Fig. 2-46), but we shall discuss only three: the *van der Waals* equation because it is one of the earliest, the *Beattie–Bridgeman* equation of state because it is one of the best known and is reasonably accurate, and the *Benedict–Webb–Rubin* equation because it is one of the more recent and is very accurate.

*van der Waals*
*Berthelet*
*Redlich–Kwang*
*Beattie–Bridgeman*
*Benedict–Webb–Rubin*
*Strobridge*
*Virial*

**FIGURE 2-46**
Several equations of state are proposed throughout the history.

*This section can be skipped without a loss in continuity.

## Van der Waals Equation of State

The van der Waals equation of state was proposed in 1873, and it has two constants, which are determined from the behavior of a substance at the critical point. The van der Waals equation of state is given by

$$\left(P + \frac{a}{v^2}\right)(v - b) = RT \tag{2-17}$$

Van der Waals intended to improve the ideal-gas equation of state by including two of the effects not considered in the ideal-gas model: the intermolecular attraction forces and the volume occupied by the molecules themselves. The term $a/v^2$ accounts for the intermolecular attraction forces, and $b$ accounts for the volume occupied by the gas molecules. In a room at atmospheric pressure and temperature, the volume actually occupied by molecules is only about one-thousandth of the volume of the room. As the pressure increases, the volume occupied by the molecules becomes an increasingly significant part of the total volume. Van der Waals proposed to correct this by replacing $v$ in the ideal-gas relation with the quantity $v - b$, where $b$ represents the volume occupied by the gas molecules per unit mass.

The determination of the two constants appearing in this equation is based on the observation that the critical isotherm on a $P$-$v$ diagram has a horizontal inflection point at the critical point (Fig. 2-47). Thus the first and the second derivatives of $P$ with respect to $v$ at the critical point must be zero. That is,

$$\left(\frac{\partial P}{\partial v}\right)_{T=T_{cr}=\text{const}} = 0 \quad \text{and} \quad \left(\frac{\partial^2 P}{\partial v^2}\right)_{T=T_{cr}=\text{const}} = 0 \tag{2-18}$$

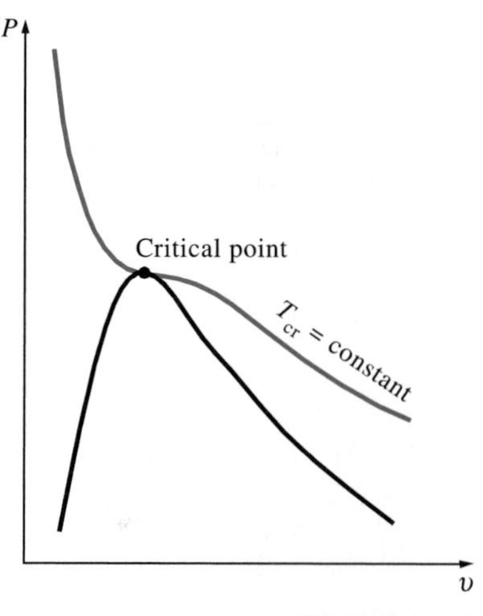


**FIGURE 2-47**
Critical isotherm of a pure substance has an inflection point at the critical state.

By performing the differentiations and eliminating $v_{cr}$, the constants $a$ and $b$ are determined to be

$$a = \frac{27R^2T_{cr}^2}{64P_{cr}} \quad \text{and} \quad b = \frac{RT_{cr}}{8P_{cr}} \tag{2-19}$$

The constants $a$ and $b$ can be determined for any substance from the critical-point data alone (Table A-1).

The accuracy of the van der Waals equation of state is often inadequate, but it can be improved by using the values of $a$ and $b$ that are based on the actual behavior of the gas over a wider range instead of a single point. Despite its limitations, the van der Waals equation of state has a historical value in that it was one of the first attempts to model the behavior of real gases. The van der Waals equation of state can also be expressed on a unit-mole basis by replacing the $v$ in Eq. 2-17 by $\bar{v}$, and the $R$ in Eqs. 2-17 and 2-19 by $R_u$.

### Beattie–Bridgeman Equation of State

The Beattie–Bridgeman equation, proposed in 1928, is an equation of state based on five experimentally determined constants. It was proposed in the form

$$P = \frac{R_u T}{\bar{v}^2}\left(1 - \frac{c}{\bar{v}T^3}\right)(\bar{v} + B) - \frac{A}{\bar{v}^2} \tag{2-20}$$

where $A = A_0\left(1 - \frac{a}{\bar{v}}\right)$ and $B = B_0\left(1 - \frac{b}{\bar{v}}\right)$ (2-21)

The constants appearing in the above equation are given in Table A-12*a* for various substances. The Beattie–Bridgeman equation is known to be reasonably accurate for densities up to about $0.8\rho_{cr}$, where $\rho_{cr}$ is the density of the substance at the critical point.

### Benedict–Webb–Rubin Equation of State

Benedict, Webb, and Rubin extended the Beattie–Bridgeman equation in 1940 by raising the number of constants to eight. It is expressed as

$$P = \frac{R_u T}{\bar{v}} + \left(B_0 R_u T - A_0 - \frac{C_0}{T^2}\right)\frac{1}{\bar{v}^2} + \frac{bR_u T - a}{\bar{v}^3} + \frac{a\alpha}{\bar{v}^6} + \frac{c}{\bar{v}^3 T^2}\left(1 + \frac{\gamma}{\bar{v}^2}\right)e^{-\gamma/\bar{v}^2} \tag{2-22}$$

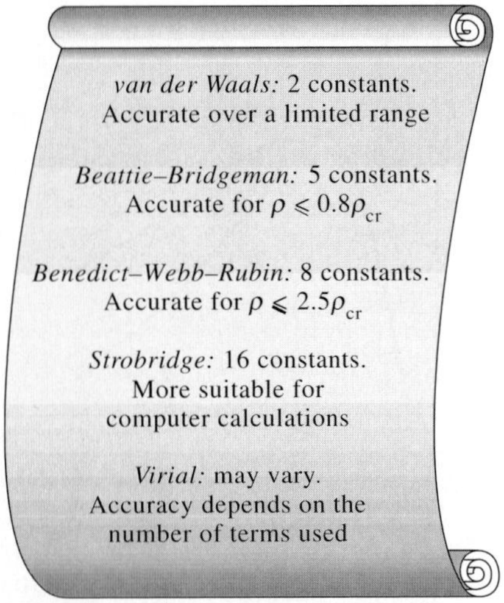

The values of the constants appearing in this equation are given in Table A-12*b*. This equation can handle substances at densities up to about $2.5\rho_{cr}$. In 1962, Strobridge further extended this equation by raising the number of constants to 16 (Fig. 2-48).

*van der Waals:* 2 constants.
Accurate over a limited range

*Beattie–Bridgeman:* 5 constants.
Accurate for $\rho \leqslant 0.8\rho_{cr}$

*Benedict–Webb–Rubin:* 8 constants.
Accurate for $\rho \leqslant 2.5\rho_{cr}$

*Strobridge:* 16 constants.
More suitable for computer calculations

*Virial:* may vary.
Accuracy depends on the number of terms used

**FIGURE 2-48**
Complex equations of state represent the $P$-$v$-$T$ behavior of gases more accurately over a wide range.

### Virial Equation of State

The equation of state of a substance can also be expressed in a series form as

$$P = \frac{RT}{v} + \frac{a(T)}{v^2} + \frac{b(T)}{v^3} + \frac{c(T)}{v^4} + \frac{d(T)}{v^5} + \cdots \tag{2-23}$$

This and similar equations are called the *virial equations of state,* and the coefficients $a(T)$, $b(T)$, $c(T)$, etc., which are functions of temperature alone are called *virial coefficients.* These coefficients can be determined experimentally or theoretically from statistical mechanics. Obviously as the pressure approaches zero, all the virial coefficients will vanish and the equation will reduce to the ideal-gas equation of state. The $P$-$v$-$T$ behavior of a substance can be represented accurately with the virial equation of state over a wider range by including a sufficient number of

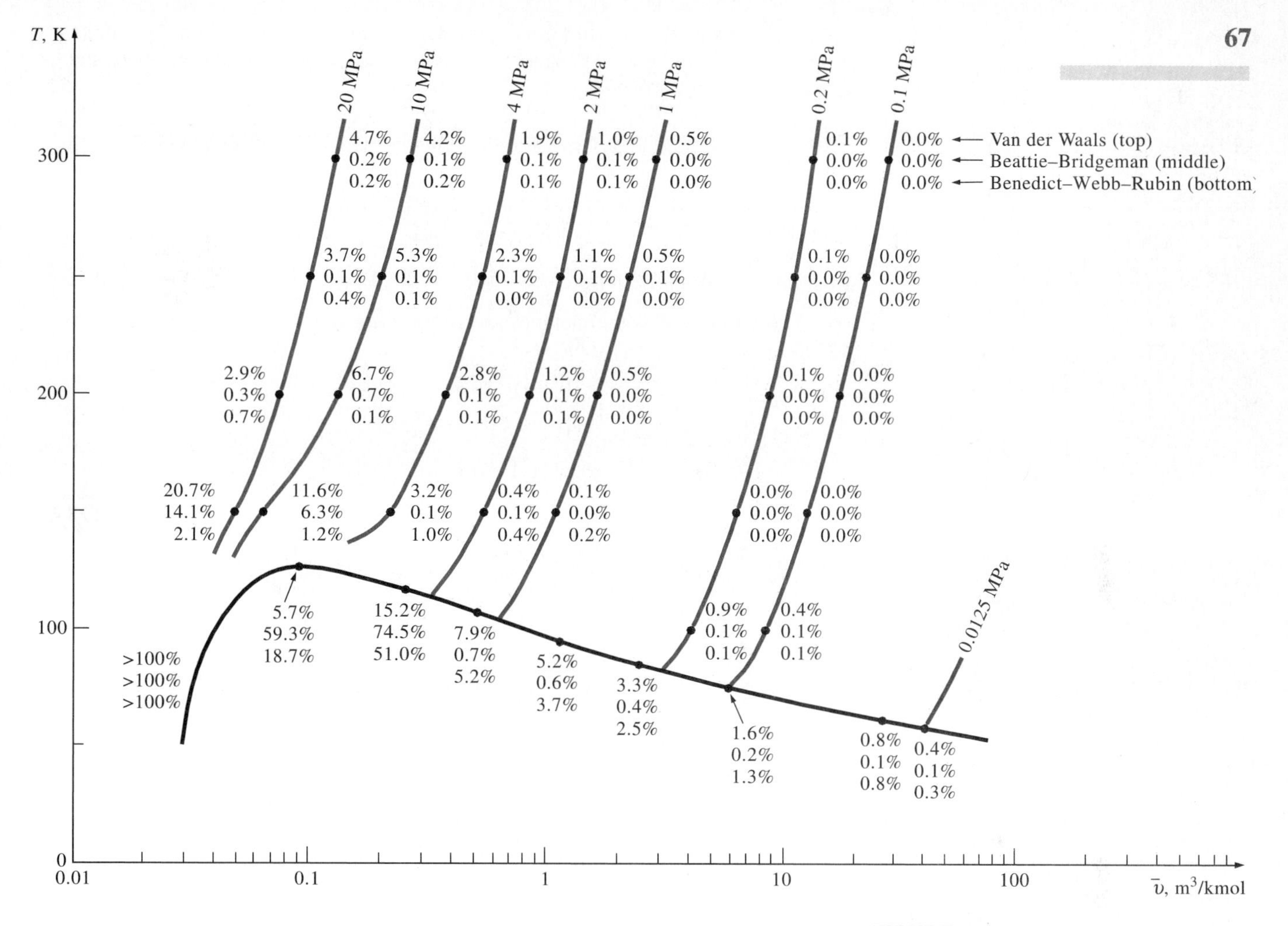


**FIGURE 2-49**

Percentage of error involved in various equations of state for nitrogen (% error $= [(|v_{\text{table}} - v_{\text{equation}}|)/v_{\text{table}}] \times 100$).

terms. All equations of state discussed above are applicable to the gas phase of the substances only, and thus should not be used for liquids or liquid–vapor mixtures.

Complex equations represent the *P-v-T* behavior of substances reasonably well and are very suitable for digital computer applications. For hand calculations, however, it is suggested that the reader use the property tables or the simpler equations of state for convenience. This is particularly true for specific-volume calculations since all the equations above are implicit in $v$ and will require a trial-and-error approach. The accuracy of the van der Waals, Beattie–Bridgeman, and Benedict–Webb–Rubin equations of state is illustrated in Fig. 2-49. It is obvious from this figure that the Benedict–Webb–Rubin equation of state is the most accurate.

**EXAMPLE 2-6**

Predict the pressure of nitrogen gas at $T = 175$ K and $v = 0.00375$ m$^3$/kg on the basis of (*a*) the ideal-gas equation of state, (*b*) the van der Waals equation of state, (*c*) the Beattie-Bridgeman equation of state, and (*d*) the Benedict–Webb–Rubin equation of state. Compare the values obtained with the experimentally determined value of 10,000 kPa.

**Solution** (*a*) By using the ideal-gas equation of state, the pressure is found to be

$$P = \frac{RT}{v} = \frac{[0.279\ \text{kPa}\cdot\text{m}^3/(\text{kg}\cdot\text{K})](175\ \text{K})}{0.00375\ \text{m}^3/\text{kg}} = 13{,}860\ \text{kPa}$$

which is in error by 38.6 percent.

(*b*) The van der Waals constants for nitrogen are determined from Eq. 2-19 to be

$$a = 0.175\ \text{m}^6\cdot\text{kPa/kg}^2$$
$$b = 0.00138\ \text{m}^3/\text{kg}$$

From Eq. 2-17,

$$P = \frac{RT}{v - b} - \frac{a}{v^2} = 9465\ \text{kPa}$$

which is in error by 5.4 percent.

(*c*) The constants in the Beattie–Bridgeman equation are determined from Table A-12*a* to be

$$A = 102.29$$
$$B = 0.05378$$
$$c = 4.2 \times 10^4$$

Also, $\bar{v} = Mv = (28.013\ \text{kg/mol})(0.00375\ \text{m}^3/\text{kg}) = 0.10505\ \text{m}^3/\text{kmol}$. Substituting these values into Eq. 2-20, we obtain

$$P = \frac{R_u T}{\bar{v}^2}\left(1 - \frac{c}{\bar{v}T^3}\right)(\bar{v} + B) - \frac{A}{\bar{v}^2} = 10{,}110\ \text{kPa}$$

which is in error by 1.1 percent.

(*d*) The constants in the Benedict–Webb–Rubin equation are determined from Table A-12*b* to be

$$a = 2.54, \qquad A_0 = 106.73$$
$$b = 0.002328, \qquad B_0 = 0.04074$$
$$c = 7.379 \times 10^4, \qquad C_0 = 8.164 \times 10^5$$
$$\alpha = 1.272 \times 10^{-4}, \qquad \gamma = 0.0053$$

Substituting these values into Eq. 2-22, we obtain

$$P = \frac{R_u T}{\bar{v}} + \left(B_0 R_u T - A_0 - \frac{C_0}{T^2}\right)\frac{1}{\bar{v}^2} + \frac{bR_u T - a}{\bar{v}^3} + \frac{a\alpha}{\bar{v}^6} + \frac{c}{\bar{v}^3 T^2}\left(1 + \frac{\gamma}{\bar{v}^2}\right)e^{-\gamma/\bar{v}^2}$$
$$= 10{,}009\ \text{kPa}$$

which is in error by only 0.09 percent. Thus the accuracy of the Benedict–Webb–Rubin equation of state is rather impressive in this case.

## 2-10 ■ SUMMARY

A substance that has a fixed chemical composition throughout is called a *pure substance.* A pure substance exists in different phases depending on its energy level. In the liquid phase, a substance that is not about to vaporize is called a *compressed* or *subcooled liquid.* In the gas phase, a substance that is not about to condense is called a *superheated vapor.* During a phase-change process, the temperature and pressure of a pure substance are dependent properties. At a given pressure, a substance boils at a fixed temperature, which is called the *saturation temperature.* Likewise, at a given temperature, the pressure at which a substance starts boiling is called the *saturation pressure.* During a phase-change process, both the liquid and the vapor phases coexist in equilibrium, and under this condition the liquid is called *saturated liquid* and the vapor *saturated vapor.*

In a saturated liquid–vapor mixture, the mass fraction of the vapor phase is called the *quality* and is defined as

$$x = \frac{m_{\text{vapor}}}{m_{\text{total}}}$$

The quality may have values between 0 (saturated liquid) and 1 (saturated vapor). It has no meaning in the compressed liquid or superheated vapor regions.

In the absence of compressed liquid data, a general approximation is to treat a compressed liquid as a saturated liquid at the given *temperature,* i.e.,

$$y \approx y_{f\,@\,T}$$

where $y$ stands for $v$, $u$, or $h$.

The state beyond which there is no distinct vaporization process is called the *critical point.* At supercritical pressures, a substance gradually and uniformly expands from the liquid to vapor phase. All three phases of a substance coexist in equilibrium at states along the *triple line* characterized by triple-line temperature and pressure. A compressed liquid has lower $v$, $u$, and $h$ values than the saturated liquid at the same $T$ or $P$. Likewise, superheated vapor has higher $v$, $u$, and $h$ values than the saturated vapor at the same $T$ or $P$.

Any relation among the pressure, temperature, and specific volume of a substance is called an *equation of state.* The simplest and best known equation of state is the *ideal-gas equation of state,* given as

$$Pv = RT$$

where $R$ is the gas constant. Caution should be exercised in using this relation since an ideal gas is a fictitious substance. Real gases exhibit ideal-gas behavior at relatively low pressures and high temperatures.

The deviation from ideal-gas behavior can be properly accounted for by using the *compressibility factor Z,* defined as

$$Z = \frac{Pv}{RT} \quad \text{or} \quad Z = \frac{v_{\text{actual}}}{v_{\text{ideal}}}$$

The $Z$ factor is approximately the same for all gases at the same *reduced temperature* and *reduced pressure,* which are defined as

$$T_R = \frac{T}{T_{\text{cr}}} \quad \text{and} \quad P_R = \frac{P}{P_{\text{cr}}}$$

where $P_{\text{cr}}$ and $T_{\text{cr}}$ are the critical pressure and temperature, respectively. This is known as the *principle of corresponding states.* When either $P$ or $T$ is unknown, it can be determined from the compressibility chart with the help of the *pseudo-reduced specific volume,* defined as

$$v_R = \frac{v_{\text{actual}}}{RT_{\text{cr}}/P_{\text{cr}}}$$

The $P$-$v$-$T$ behavior of substances can be represented more accurately by the more complex equations of state. Three of the best known are

*van der Waals*:

$$\left(P + \frac{a}{v^2}\right)(v - b) = RT$$

where

$$a = \frac{27R^2T_{\text{cr}}^2}{64P_{\text{cr}}} \quad \text{and} \quad b = \frac{RT_{\text{cr}}}{8P_{\text{cr}}}$$

*Beattie–Bridgeman*:

$$P = \frac{R_uT}{\bar{v}^2}\left(1 - \frac{c}{\bar{v}T^3}\right)(\bar{v} + B) - \frac{A}{\bar{v}^2}$$

where

$$A = A_0\left(1 - \frac{a}{\bar{v}}\right) \quad \text{and} \quad B = B_0\left(1 - \frac{b}{\bar{v}}\right)$$

*Benedict–Webb–Rubin*:

$$P = \frac{R_uT}{\bar{v}} + \left(B_0R_uT - A_0 - \frac{C_0}{T^2}\right)\frac{1}{\bar{v}^2} + \frac{bR_uT - a}{\bar{v}^3} + \frac{a\alpha}{\bar{v}^6} + \frac{c}{\bar{v}^3T^2}\left(1 + \frac{\gamma}{\bar{v}^2}\right)e^{-\gamma/\bar{v}^2}$$

The constants appearing in the Beattie–Bridgeman and Benedict–Webb–Rubin equations are given in Table A-12 for various substances.

## REFERENCES AND SUGGESTED READING

**1** A. Bejan, *Advanced Engineering Thermodynamics,* Wiley, New York, 1988.

**2** M. D. Burghardt, *Engineering Thermodynamics with Applications,* Harper and Row, New York, 1986.

**3** J. B. Jones and G. A. Hawkins, *Engineering Thermodynamics,* 2d ed., Wiley, New York, 1986.

**4** J. R. Howell and R. O. Buckius, *Fundamentals of Engineering Thermodynamics,* McGraw-Hill, New York, 1987.

5 G. J. Van Wylen and R. E. Sonntag, *Fundamentals of Classical Thermodynamics,* 3d ed., Wiley, New York, 1985.

6 K. Wark, *Thermodynamics,* 5th ed., McGraw-Hill, New York, 1988.

## PROBLEMS*

### Pure Substances, Phase-Change Processes, Phase Diagrams

**2-1C** Is sweetened tea a pure substance? Explain.

**2-2C** Is iced water a pure substance? Why?

**2-3C** What is the difference between saturated liquid and compressed liquid?

**2-4C** What is the difference between saturated vapor and superheated vapor?

**2-5C** Is there any difference between the properties of saturated vapor at a given temperature and the vapor of a saturated mixture at the same temperature?

**2-6C** Is there any difference between the properties of saturated liquid at a given temperature and the liquid of a saturated mixture at the same temperature?

**2-7C** Is it true that water boils at higher temperatures at higher pressures? Explain.

**2-8C** If the pressure of a substance is increased during a boiling process, will the temperature also increase or will it remain constant? Why?

**2-9C** Why are the temperature and pressure dependent properties in the saturated mixture region?

**2-10C** What is the difference between the critical point and the triple point?

**2-11C** Is it possible to have water vapor at $-10°C$?

**2-12C** A househusband is cooking beef stew for his family in a pan that is (*a*) uncovered, (*b*) covered with a light lid, and (*c*) covered with a heavy lid. For which case will the cooking time be the shortest? Why?

**2-13C** A rigid tank contains some liquid water at 20°C. The rest of the tank is filled with atmospheric air. Is it possible to boil the water in the tank without raising its temperature? How?

**2-14C** A well-sealed rigid tank contains some water and air at atmospheric pressure. The tank is now heated, and the water starts boiling. Will the temperature in the tank remain constant during this boiling process? Why?

**2-15C** How does the boiling process at supercritical pressures differ from the boiling process at subcritical pressures?

---

*Students are encouraged to answer *all* the concept "C" questions.

## Property Tables

**2-16C** Can the enthalpy of a pure substance at a given state be determined from a knowledge of $u$, $P$, and $v$ data? How?

**2-17C** Does the reference point selected for the properties of a substance have any effect in thermodynamic analysis? Why?

**2-18C** What is the physical significance of $h_{fg}$? Can it be obtained from a knowledge of $h_f$ and $h_g$? How?

**2-19C** Is it true that it takes more energy to vaporize 1 kg of saturated liquid water at 100°C than it would to vaporize 1 kg of saturated liquid water at 120°C?

**2-20C** What is quality? Does it have any meaning in the superheated vapor region?

**2-21C** Which process requires more energy: completely vaporizing 1 kg of saturated liquid water at 1-atm pressure or completely vaporizing 1 kg of saturated liquid water at 8-atm pressure?

**2-22C** Does $h_{fg}$ change with pressure? How?

**2-23C** Can quality be expressed as the ratio of the volume occupied by the vapor phase to the total volume?

**2-24C** In the absence of compressed liquid tables, how is the specific volume of a compressed liquid at a given $P$ and $T$ determined?

**2-25** The average atmospheric pressure in Denver (elevation = 1610 m) is 83.4 kPa. Determine the temperature at which water in an uncovered pan will boil in Denver. *Answer:* 94.4°C.

**2-26** Water in a 5-cm deep pan is observed to boil at 98°C. At what temperature will the water in a 40-cm-deep pan boil? Assume both pans are full with water.

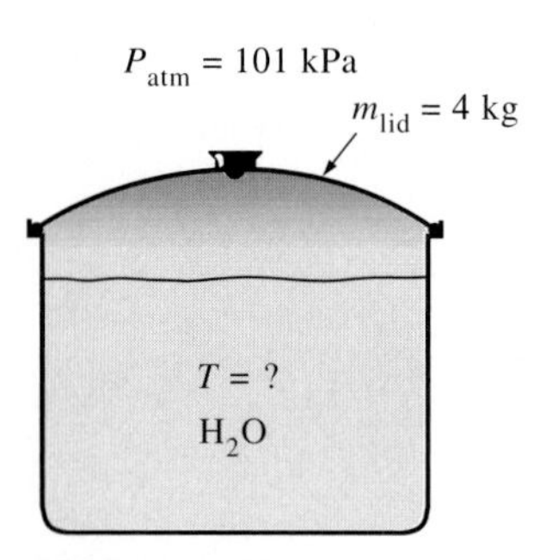


**FIGURE P2-27**

**2-27** A cooking pan whose inner diameter is 20 cm is filled with water and covered with a 4-kg lid. If the local atmospheric pressure is 101 kPa, determine the temperature at which the water will start boiling when it is heated. *Answer:* 100.2°C

**2-28** Water is being heated in a vertical piston–cylinder device. The piston has a mass of 20 kg and a cross-sectional area of 100 cm$^2$. If the local atmospheric pressure is 100 kPa, determine the temperature at which the water will start boiling.

**2-29** A rigid tank with a volume of 2.5 m$^3$ contains 5 kg of saturated liquid–vapor mixture of water at 75°C. Now the water is slowly heated. Determine the temperature at which the liquid in the tank is completely vaporized. Also show the process on a $T$-$v$ diagram with respect to saturation lines. *Answer:* 140.7°C

**2-29E** A rigid tank with a volume of 30 $ft^3$ contains 1.5 lbm of saturated liquid–vapor mixture of water at 190°F. Now the water is slowly heated. Determine the temperature at which the liquid in the tank is completely vaporized. Also show the process on a *T-v* diagram with respect to saturation lines. *Answer:* 228.4°F

**2-30** A 0.5-$m^3$ vessel contains 10 kg of refrigerant-134a at −20°C. Determine (*a*) the pressure, (*b*) the total internal energy, and (*c*) the volume occupied by the liquid phase.
*Answers:* (*a*) 132.99 kPa, (*b*) 889.5 kJ, (*c*) 0.00487 $m^3$

**2-31** A piston–cylinder device initially contains 50 L of liquid water at 25°C and 300 kPa. Heat is added to the water at constant pressure until the entire liquid is vaporized.

(*a*) What is the mass of the water?
(*b*) What is the final temperature?
(*c*) Determine the total enthalpy change.
(*d*) Show the process on a *T-v* diagram with respect to saturation lines.

*Answers:* (*a*) 49.85 kg, (*b*) 133.55°C, (*c*) 130,627 kJ

**2-31E** A piston–cylinder device initially contains 2 $ft^3$ of liquid water at 70°F and 50 psia. Heat is transferred to the water at constant pressure until the entire liquid is vaporized.

(*a*) What is the mass of the water?
(*b*) What is the final temperature?
(*c*) Determine the total enthalpy change.
(*d*) Show the process on a *T-v* diagram with respect to saturation lines.

**2-32** A 0.5-$m^3$ rigid vessel initially contains saturated liquid–vapor mixture of water at 100°C. The water is now heated until it reaches the critical state. Determine the mass of the liquid water and the volume occupied by the liquid at the initial state.
*Answers:* 158.28 kg, 0.165 $m^3$

**2-32E** A 5-$ft^3$ rigid vessel initially contains saturated liquid–vapor mixture of water at 212°F. The water is now heated until it reaches the critical state. Determine the mass of the liquid water and the volume occupied by the liquid at the initial state. *Answers:* 98.8 lbm, 1.65 $ft^3$

**2-33** A 0.5-$m^3$ rigid tank contains saturated mixture of refrigerant-134a at 200 kPa. If the saturated liquid occupies 10 percent of the volume, determine the quality and the total mass of the refrigerant in the tank.

**2-33E** A 15-$ft^3$ rigid tank contains saturated mixture of refrigerant-134a at 30 psia. If the saturated liquid occupies 10 percent of the volume, determine the quality and the total mass of the refrigerant in the tank.

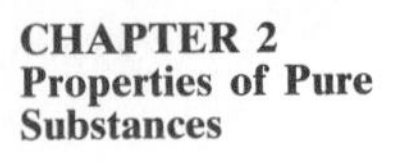



## Ideal Gas

**2-34C** Under what conditions is the ideal-gas assumption suitable for real gases?

**2-35C** What is the difference between $R$ and $R_u$? How are these two related?

**2-36C** What is the difference between mass and molar mass? How are these two related?

**2-37** A spherical balloon with a diameter of 6 m is filled with helium at 20°C and 200 kPa. Determine the mole number and the mass of the helium in the balloon. *Answers:* 9.28 kmol, 37.15 kg

**2-37E** A spherical balloon with a diameter of 25 ft is filled with helium at 70°F and 30 psia. Determine the mole number and the mass of the helium in the balloon. *Answers:* 43.16 lbmol, 172.6 lbm

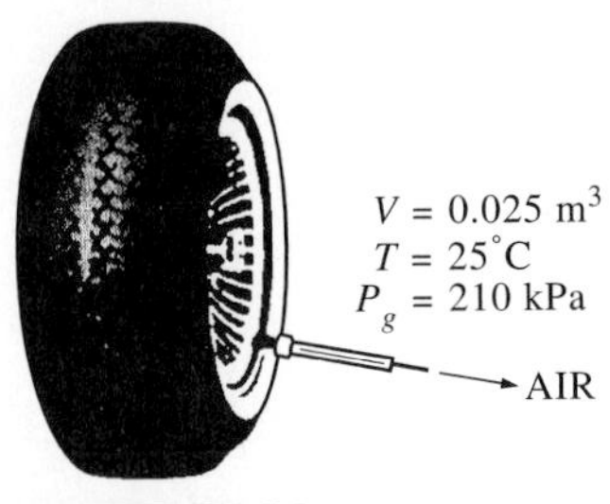


**FIGURE P2-38**

**2-38** The pressure in an automobile tire depends on the temperature of the air in the tire. When the air temperature is 25°C, the pressure gage reads 210 kPa. If the volume of the tire is 0.025 m$^3$, determine the pressure rise in the tire when the air temperature in the tire rises to 50°C. Also determine the amount of air that must be bled off to restore pressure to its original value at this temperature. Assume the atmospheric pressure to be 100 kPa.

**2-39** The air in an automobile tire with a volume of 0.015 m$^3$ is at 30°C and 150 kPa (gage). Determine the amount of air that must be added to raise the pressure to the recommended value of 200 kPa (gage). Assume the atmospheric pressure to be 98 kPa and the temperature and the volume to remain constant. *Answer:* 0.0086 kg

**2-39E** The air in an automobile tire with a volume of 0.53 ft$^3$ is at 90°F and 20 psig. Determine the amount of air that must be added to raise the pressure to the recommended value of 30 psig. Assume the atmospheric pressure to be 14.6 psia and the temperature and the volume to remain constant. *Answer:* 0.0260 lbm

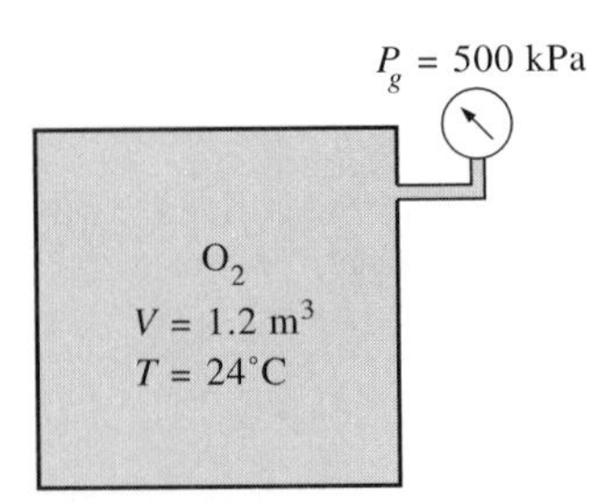


**FIGURE P2-40**

**2-40** The pressure gage on a 1.2-m$^3$ oxygen tank reads 500 kPa. Determine the amount of oxygen in the tank if the temperature is 24°C and the atmospheric pressure is 97 kPa.

**2-40E** The pressure gage on a 30-ft$^3$ oxygen tank reads 50 psig. Determine the amount of oxygen in the tank if the temperature is 73°F and the atmospheric pressure is 14.6 psia.

**2-41** A rigid tank contains 10 kg of air at 150 kPa and 20°C. More air is added to the tank until the pressure and temperature rise to 250 kPa and 30°C, respectively. Determine the amount of air added to the tank. *Answer:* 6.12 kg

**2-41E** A rigid tank contains 20 lbm of air at 20 psia and 70°F. More air is added to the tank until the pressure and temperature rise to 35 psia and 90°F, respectively. Determine the amount of air added to the tank. *Answer:* 13.73 lbm

**2-42** A 800-L rigid tank contains 10 kg of air at 25°C. Determine the reading on the pressure gage if the atmospheric pressure is 97 kPa.

**2-43** A 1-$m^3$ tank containing air at 25°C and 500 kPa is connected through a valve to another tank containing 5 kg of air at 35°C and 200 kPa. Now the valve is opened, and the entire system is allowed to reach thermal equilibrium with the surroundings which are at 20°C. Determine the volume of the second tank and the final equilibrium pressure of air. *Answer:* 284.1 kPa

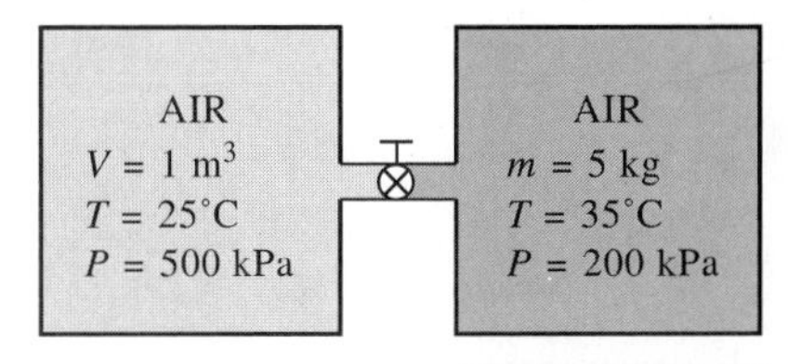


**FIGURE P2-43**

## Compressibility Factor

**2-44C** What is the physical significance of the compressibility factor $Z$?

**2-45C** What is the principle of corresponding states?

**2-46C** What are the reduced pressure and reduced temperature?

**2-47** Determine the specific volume of superheated water vapor at 10 MPa and 400°C, using (*a*) the ideal-gas equation of state and (*b*) the generalized compressibility chart. Compare your results with the actual value of 0.02641 $m^3$/kg, and determine the error involved in each case.
*Answers:* (*a*) 0.03106 $m^3$/kg, 17.6 percent; (*b*) 0.02609 $m^3$/kg, 1.2 percent

**2-48** Determine the specific volume of refrigerant-134a vapor at 1.4 MPa and 140°C, using (*a*) the ideal gas equation of state and (*b*) the generalized compressibility chart. Compare your results with the actual value of 0.02189 $m^3$/kg, and determine the error involved in each case.

**2-49** Determine the specific volume of nitrogen gas at 10 MPa and 150 K based on (*a*) the ideal-gas equation and (*b*) the generalized compressibility chart. Compare these results with the experimental value of 0.002388 $m^3$/kg, and determine the error involved in each case.
*Answers:* (*a*) 0.004452 $m^3$/kg, 86.4 percent; (*b*) 0.002404 $m^3$/kg, 0.7 percent

**2-50** Determine the specific volume of superheated water vapor at 1.6 MPa and 225°C, using (*a*) the ideal-gas equation of state and (*b*) the generalized compressibility chart. Compare your results with the actual value of 0.13287 $m^3$/kg, and determine the error involved in each case.

**2-51** Refrigerant-134a at 1-MPa pressure has a specific volume of 0.02171 $m^3$/kg. Determine the temperature of the refrigerant using (*a*) the ideal gas equation of state and (*b*) the generalized compressibility chart. Compare your results to the actual value of 50°C.

**2-52** A 0.01677-$m^3$ tank contains 1 kg of refrigerant-134a at 110°C. Determine the pressure of the refrigerant using (*a*) the ideal gas equation of state and (*b*) the generalized compressibility chart. Compare your results with the actual value of 1.6 MPa
*Answers:* (*a*) 1.861 MPa, (*b*) 1.586 MPa

**2-53** Somebody claims that oxygen gas at 160 K and 3 MPa can be treated as an ideal gas with an error of less than 10 percent. Is this claim valid?

**2-53E** Somebody claims that oxygen gas at 280 R and 300 psia can be treated as an ideal gas with an error of less than 10 percent. Is this claim valid?

**2-54** What is the percentage of error involved in treating carbon dioxide at 3 MPa and 10°C as an ideal gas? *Answer:* 25 percent

**2-55** What is the percentage of error involved in treating carbon dioxide at 5 MPa and 350 K as an ideal gas?

### Other Equations of state

**2-56C** What is the physical significance of the two constants that appear in the van der Waals equation of state? On what basis are they determined?

**2-57** A 3.27-$m^3$ tank contains 100 kg of nitrogen at 225 K. Determine the pressure in the tank, using (*a*) the ideal-gas equation, (*b*) the van der Waals equation, and (*c*) the Beattie–Bridgeman equation. Compare your results with the actual value of 2000 kPa.

**2-58** A 1-$m^3$ tank contains 2.841 kg of steam at 0.6 MPa. Determine the temperature of the steam using (*a*) the ideal gas equation of state and (*b*) the van der Waals equation of state. Compare your results with the actual value of 473 K. *Answers:* (*a*) 457.6 K, (*b*) 465.9 K

**2-59** Refrigerant-134a at 0.7 MPa has a specific volume of 0.03324 $m^3/kg$. Determine the temperature of the refrigerant using (*a*) the ideal gas equation of state and (*b*) the van der Waals equation of state. Compare your results with the actual value of 50°C.

**2-59E** Refrigerant-134a at 100 psia has a specific volume of 0.4761 $ft^3/lbm$. Determine the temperature of the refrigerant based on (*a*) the ideal-gas equation and (*b*) the van der Waals equation. Compare your results with the actual value of 80°F.

**2-60** Nitrogen at 150 K has a specific volume of 0.041884 $m^3/kg$. Determine the pressure of the nitrogen, using (*a*) the ideal-gas equation and (*b*) the Beattie–Bridgeman equation. Compare your results with the experimental value of 1000 kPa.
*Answers:* (*a*) 1063 kPa, (*b*) 1000.4 kPa

**2-60E** Nitrogen at 350 R has a specific volume of 0.115 $ft^3/lbm$. Determine the pressure of nitrogen, using (*a*) the ideal-gas equation and (*b*) the Beattie–Bridgeman equation. Compare your results with the experimental value of 1000 psia.

**2-61** Although balloons have been around since 1783 when the first balloon took to the skies in France, a real breakthrough in ballooning occurred in 1960 with the design of the modern hot-air balloon fueled by inexpensive propane and constructed of lightweight nylon fabric. Over the years, ballooning has become a sport and a hobby for many people around the world. Unlike balloons filled with the light helium gas, hot-air balloons are open to the atmosphere. Therefore, the pressure in the balloon is always the same as the local atmospheric pressure, and the balloon is never in danger of exploding.

Hot-air balloons range from about 15 to 25 m in diameter. The air in the balloon cavity is heated by a propane burner located at the top of the passenger cage. The flames from the burner that shoot into the balloon heat the air in the balloon cavity, raising the air temperature at top of the balloon from 65°C to over 120°C. The air temperature is maintained at the desired levels by periodically firing the propane burner. The buoyancy force that pushes the balloon upward is proportional to the density of the cooler air outside the balloon and the volume of the balloon, and can be expressed as

**FIGURE P2-61**
A hot-air balloon

$$F_B = \rho_{\text{cool air}} g V_{\text{balloon}}$$

where $g$ is the gravitational acceleration. When air resistance is negligible, the buoyancy force is opposed by (1) the weight of the hot air in the balloon, (2) the weight of the cage, the ropes, and the balloon material, and (3) the weight of the people and other load in the cage. The operator of the balloon can control the height and the vertical motion of the balloon by firing the burner or by letting some hot air in the balloon escape and be replaced by cooler air. The forward motion of the balloon is provided by the winds.

Consider a 20-m-diameter hot-air balloon that, together with its cage, has a mass of 80 kg when empty. This balloon is hanging still in the air at a location where the atmospheric pressure and temperature are 90 kPa and 15°C, respectively, while carrying three 65-kg people. Determine the average temperature of the air in the balloon. What would your response be if the atmospheric air temperature were 30°C?

**2-62** Consider an 18-m-diameter hot-air balloon that, together with its cage, has a mass of 120 kg when empty. The air in the balloon, which is now carrying two 70-kg people, is heated by propane burners at a location where the atmospheric pressure and temperature are 93 kPa and 12°C, respectively. Determine the average temperature of the air in the balloon when the balloon first starts rising. What would your response be if the atmospheric air temperature were 25°C?

**2-63** Water in a pressure cooker is observed to boil at 120°C. What is the absolute pressure in the pressure cooker, in kPa?

**2-63E** Water in a pressure cooker is observed to boil at 250°F. What is the absolute pressure in the pressure cooker, in psia?

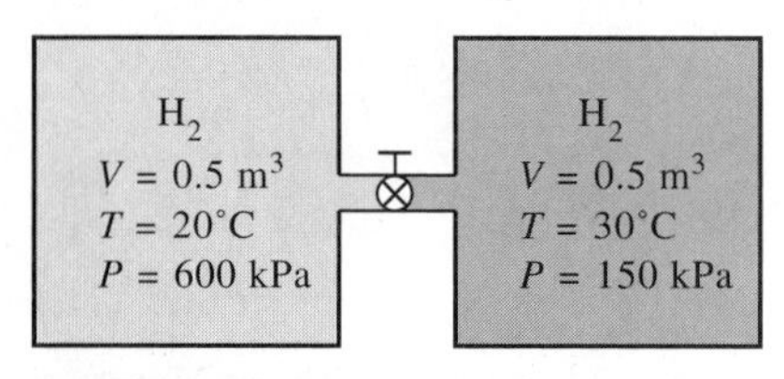


**FIGURE P2-64**

**2-64** A 0.5-m$^3$ rigid tank containing hydrogen at 20°C and 600 kPa is connected by a valve to another 0.5-m$^3$ rigid tank that holds hydrogen at 30°C and 150 kPa. Now the valve is opened, and the system is allowed to reach thermal equilibrium with the surroundings which are at 15°C. Determine the final pressure in the tank.

**2-65** A 20-m$^3$ tank contains nitrogen at 25°C and 800 kPa. Some nitrogen is allowed to escape until the pressure in the tank drops to 600 kPa. If the temperature at this point is 20°C, determine the amount of nitrogen that has escaped. *Answer:* 42.9 kg

**2-66** The boiling temperature of nitrogen at atmospheric pressure at sea level (1 atm pressure) is −196°C. Therefore, nitrogen is commonly used in low-temperature scientific studies, since the temperature of liquid nitrogen in a tank open to the atmosphere will remain constant at −196°C until the liquid nitrogen in the tank is depleted. Any heat transfer to the tank will result in the evaporation of some liquid nitrogen, which has a heat of vaporization of 198 kJ/kg and a density of 810 kg/m$^3$ at 1 atm.

Consider a 3-m internal-diameter spherical tank that is initially filled with liquid nitrogen at 1 atm and −196°C. Determine the mass of nitrogen in the tank, and the amount of heat it will absorb before evaporating completely.

**2-67** Repeat Prob. 2-66 for liquid oxygen, which has a boiling temperature of −183°C, a heat of vaporization of 213 kJ/kg, and a density of 1140 kg/m$^3$ at 1-atm pressure.

**2-68** A fan is to be installed to ventilate a 10 m × 15 m smoking lounge that is 2.4 m high. The ventilation standards indicate that fresh air must be supplied to smoking lounges at a rate of at least 30 L/s per person. If the lounge is to accommodate 25 smokers, determine (*a*) the flow rate of the fan that needs to be installed, and (*b*) how many times the air in the room will be changed by the fan per hour.

**2-69** A fan is to be installed to ventilate a 2 m × 2.5 m bathroom that is 3 m high. It is desired that the fan changes the entire air in the bathroom at least once every 15 min. There is a fan available in storage whose flow rate is rated to be 25 L/s. Determine if this fan is large enough to do the job. Also determine the mass flow rate of air through the fan, in kg/s. Make any reasonable assumptions.

**2-70** The average atmospheric pressure in Denver, Colorado (elevation = 1610 m) is 83.4 kPa, and the average winter temperature is 3°C. The pressurization test of a 2.5-m-high 230-m$^2$ older home revealed that the seasonal average infiltration rate of the house is 1.8 air changes per hour (ACH) at outdoor conditions. That is, on average, the entire air volume of the house is replaced by the outdoor air 1.8 times every hour, which represents a considerable energy loss. Determine the mass of outdoor air which infiltrates into the house per hour.

**2-70E** The average atmospheric pressure in Denver, Colorado (elevation = 5300 ft) is 12.1 psia, and the average winter temperature is 38°F. The pressurization test of a 9-ft-high 2500-$ft^2$ older home revealed that the seasonal average infiltration rate of the house is 1.8 air changes per hour (ACH) at outdoor conditions. That is, on average, the entire air volume of the house is replaced by the outdoor air 1.8 times every hour, which represents a considerable energy loss. Determine the mass of outdoor air which infiltrates into the house per hour.

**2-71** The volume flow rate of a fan or compressor is a measure of volume displaced, and is independent of location. However, the amount of air mass which flows depends on the density, which depends on the temperature and pressure of air. Consider the fan of a car radiator which supplies air at a rate of 0.20 kg/s at sea level to cool the engine. How much air will this fan supply at an elevation of 2000 m at the same environment temperature at the same fan speed?

## Computer, Design, and Essay Problems

**2-72** Write a computer program to express $T_{sat} = f(P_{sat})$ for steam as a fifth-degree polynomial where the pressure is in kPa and the temperature is in °C. Use tabulated data from Table A-4. What is the accuracy of this equation?

**2-73** A solid normally absorbs heat as it melts, but there is a known exception at temperatures close to absolute zero. Find out which solid it is, and give a physical explanation for it.

**2-74** Numerous equations of state have been proposed throughout history. Write an essay on two equations of state not discussed in the chapter, describe them in sufficient detail, and discuss the accuracy and range of applicability of each equation.

**2-75** It is well known that water freezes at 0°C at atmospheric pressure. The mixture of liquid water and ice at 0°C is said to be at stable equilibrium since it cannot undergo any changes when it is isolated from its surrounding. However, when water is free of impurities and the inner surfaces of the container are smooth, the temperature of water can be lowered to −2°C or even lower without any formation of ice at atmospheric pressure. But at that state even a small disturbance can initiate the formation of ice abruptly, and the water temperature stabilizes at 0°C following this sudden process. The water at −2°C is said to be in a *metastable state.* Write an essay on metastable states and discuss how they differ from stable equilibrium states.

**2-76** Using a thermometer, measure the boiling temperature of water and calculate the corresponding saturation pressure. From this information estimate the altitude of your town, and compare it with the actual altitude value.

# The First Law of Thermodynamics: Closed Systems

CHAPTER 3

The first law of thermodynamics is simply a statement of the *conservation of energy principle,* and it asserts that *total energy* is a thermodynamic property. In this chapter, energy transfer with heat and work is introduced, and the mechanisms of heat transfer as well as various work modes are discussed. The first-law relation for closed systems is developed in a step-by-step manner. Specific heats are defined, and relations are obtained for the internal energy and enthalpy of *ideal gases* in terms of specific heats and temperature. This approach is also applied to solids and liquids, which are approximated as *incompressible substances.* Finally, thermodynamic aspects of biological systems are considered to shed some light on dieting and exercise.

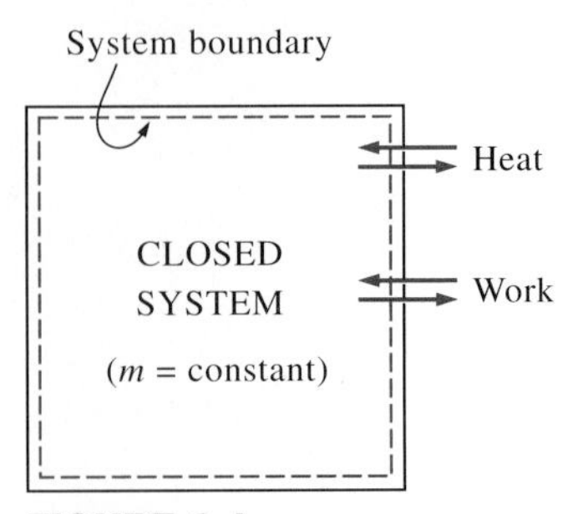


**FIGURE 3-1**

Energy can cross the boundaries of a closed system in the form of heat and work.

## 3-1 ■ INTRODUCTION TO THE FIRST LAW OF THERMODYNAMICS

In Chap. 1, it was pointed out that energy can be neither created nor destroyed; it can only change forms. This principle is based on experimental observations and is known as the *first law of thermodynamics,* or the *conservation of energy principle.* The first law can simply be stated as follows: During an interaction between a system and its surroundings, the amount of energy gained by the system must be exactly equal to the amount of energy lost by the surroundings.

Energy can cross the boundary of a closed system in two distinct forms: *heat* and *work* (Fig. 3-1). It is important to distinguish between these two forms of energy. Therefore, they will be discussed first, to form a sound basis for the development of the first law of thermodynamics.

## 3-2 ■ HEAT TRANSFER

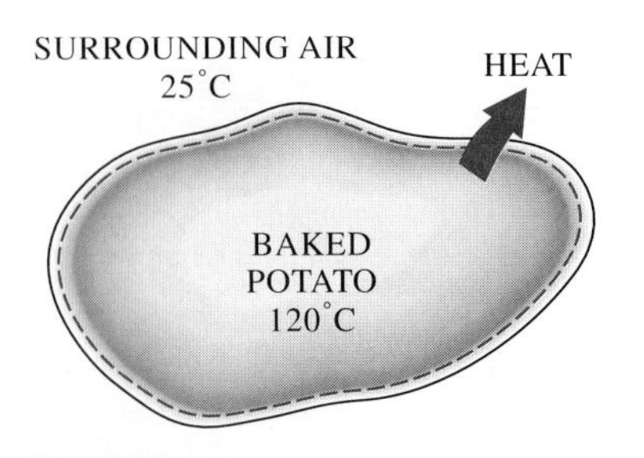


**FIGURE 3-2**

Heat is transferred from hot bodies to colder ones by virtue of a temperature difference.

We know from experience that a can of cold soda left on a table eventually warms up and that a hot baked potato on the same table cools down (Fig. 3-2). That is, when a body is left in a medium that is at a different temperature, energy transfer takes place between the body and the surrounding medium until thermal equilibrium is established, i.e., the body and the medium reach the same temperature. The direction of energy transfer is always from the higher-temperature body to the lower-temperature one. For the case of the baked potato, energy will be leaving the potato until it cools to room temperature. Once the temperature equality is established, the energy transfer stops. In the processes described above, energy is said to be transferred in the form of heat.

**Heat** is defined as *the form of energy that is transferred between two systems (or a system and its surroundings) by virtue of a temperature difference.* That is, an energy interaction is heat only if it takes place because of a temperature difference. Then it follows that there cannot be any heat transfer between two systems that are at the same temperature.

In daily life, we frequently refer to the sensible and latent forms of internal energy as *heat,* and we talk about the heat content of bodies. In thermodynamics, however, we usually refer to those forms of energy as *thermal energy* to prevent any confusion with *heat transfer.*

Several phrases that are in common use today, such as *heat flow, heat addition, heat rejection, heat absorption, heat removal, heat gain, heat loss, heat storage, heat generation, electrical heating, resistance heating, frictional heating, gas heating, heat of reaction, liberation of heat, specific heat, sensible heat, latent heat, waste heat, body heat, process heat, heat sink,* and *heat source,* are not consistent with the strict thermodynamic meaning of the term *heat,* which limits its use to the *transfer* of thermal energy during a process. However, these phrases are deeply rooted in our vocabulary, and they are used by both ordinary people and scientists without causing any misunderstanding since they are usually interpreted properly instead of being taken literally. (Besides, no acceptable alternatives exist for some of these phrases.) For example, the phrase *body heat* is understood to mean the *thermal energy content* of a body. Likewise, *heat flow* is understood to mean *the transfer of thermal energy,* not the flow of a fluidlike substance called heat, although the latter incorrect interpretation, which is based on the caloric theory, is the origin of this phrase. Also, the transfer of heat into a system is frequently referred to as *heat addition* and the transfer of heat out of a system as *heat rejection.* Perhaps there are thermodynamic reasons for being so reluctant to replace *heat* by *thermal energy*: it takes less time and energy to say, write, and comprehend *heat* than it does *thermal energy.*

Heat is energy in transition. It is recognized only as it crosses the boundary of a system. Consider the hot baked potato one more time. The potato contains energy, but this energy is heat transfer only as it passes through the skin of the potato (the system boundary) to reach the air, as shown in Fig. 3-3. Once in the surroundings, the transferred heat becomes part of the internal energy of the surroundings. Thus, in thermodynamics, the term *heat* simply means *heat transfer.*

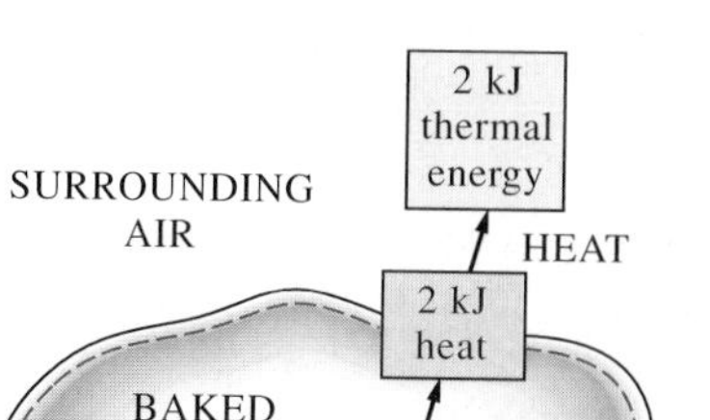


**FIGURE 3-3**
Energy is recognized as heat transfer only as it crosses the system boundary.

A process during which there is no heat transfer is called an **adiabatic process** (Fig. 3-4). The word *adiabatic* comes from the Greek word *adiabatos,* which means *not to be passed.* There are two ways a process can be adiabatic: Either the system is well insulated so that only a negligible amount of heat can pass through the boundary, or both the system and the surroundings are at the same temperature and therefore there is no driving force (temperature difference) for heat transfer. An adiabatic process should not be confused with an *isothermal process.* Even though there is no heat transfer during an adiabatic process, the energy content and thus the temperature of a system can still be changed by other means such as work.

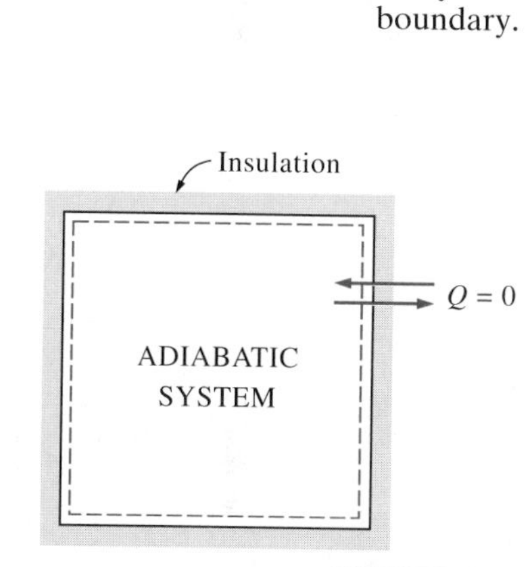

**FIGURE 3-4**
During an adiabatic process, a system exchanges no heat with its surroundings.

As a form of energy, heat has energy units, kJ (or Btu) being the most common one. The amount of heat transferred during the process between two states (states 1 and 2) is denoted by $Q_{12}$, or just $Q$. Heat transfer *per unit mass* of a system is denoted $q$ and is determined from

$$q = \frac{Q}{m} \quad \text{(kJ/kg)} \tag{3-1}$$

Sometimes it is desirable to know the *rate of heat transfer* (the amount of heat transferred per unit time) instead of the total heat transferred over some time interval (Fig. 3-5). The heat transfer rate is denoted $\dot{Q}$, where the overdot stands for the time derivative, or "per unit time." The heat transfer rate $\dot{Q}$ has the unit kJ/s, which is equivalent to kW. When $\dot{Q}$ varies with time, the amount of heat transfer during a process is determined by integrating $\dot{Q}$ over the time interval of the process:

$$Q = \int_{t_1}^{t_2} \dot{Q}\, dt \quad \text{(kJ)} \tag{3-2}$$

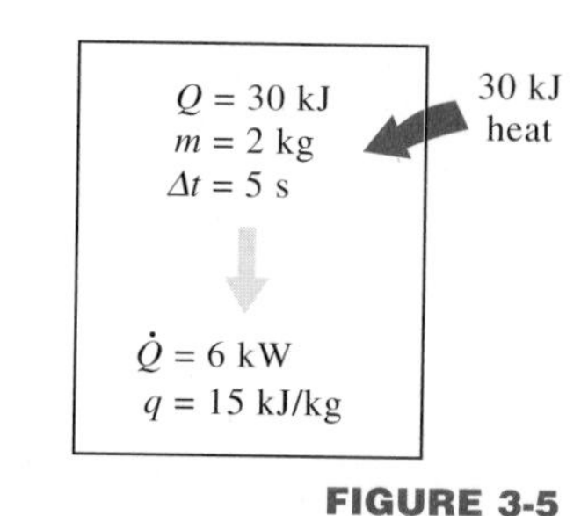


**FIGURE 3-5**
The relationships among $q$, $Q$, and $\dot{Q}$.

When $\dot{Q}$ remains constant during a process, the relation above reduces to

$$Q = \dot{Q}\,\Delta t \qquad \text{(kJ)} \tag{3-3}$$

where $\Delta t = t_2 - t_1$ is the time interval during which the process occurs.

Heat is a directional (or vector) quantity; thus, the equation $Q = 5$ kJ tells us nothing about the direction of heat flow unless we adopt a **sign convention.** The universally accepted sign convention for heat is as follows: *Heat transfer to a system is positive, and heat transfer from a system is negative* (Fig. 3-6). That is, any heat transfer that increases the energy of a system is positive, and any heat transfer that decreases the energy of a system is negative.

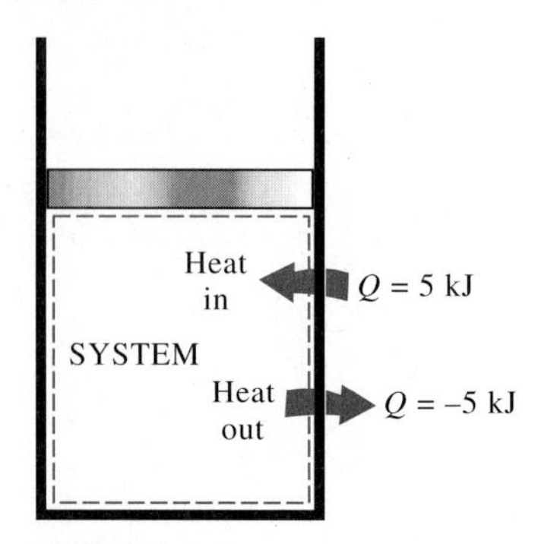


**FIGURE 3-6**

Sign convention for heat: positive if to the system, negative if from the system.

## Historical Background

Heat was always perceived to be something that produces in us a sensation of warmth, and one would think that the nature of heat is one of the first things understood by mankind. But it was only in the middle of the nineteenth century that we had a true physical understanding of the nature of heat, thanks to the development of the **kinetic theory** at that time, which treats the molecules as tiny balls that are in motion and thus possess kinetic energy. Heat is then defined as the energy associated with the random motion of atoms and molecules.

Although it was suggested in the eighteenth and early nineteenth centuries that heat is the manifestation of motion at the molecular level (called the *live force*), the prevailing view of heat until the middle of the nineteenth century was based on the **caloric theory**, which was proposed by the French chemist Antoine Lavoisier (1743–1794) in 1789. The caloric theory asserts that heat is a fluidlike substance called the **caloric** that is a *massless, colorless, odorless,* and *tasteless* substance that can be poured from one body into another (Fig. 3-7). When caloric was added to a body, its temperature increased; and when caloric was removed from a body, its temperature decreased. When a body could not contain any more caloric, in much the same way as when a glass of water could not dissolve any more salt or sugar, the body was said to be *saturated* with caloric. This interpretation gave rise to the terms *saturated liquid* and *saturated vapor,* which are still in use today.

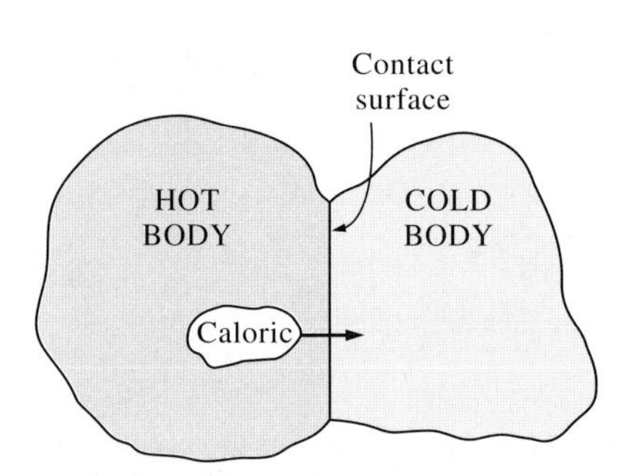


**FIGURE 3-7**

In early nineteenth century, heat was thought to be an invisible fluid called the *caloric* that flowed from warmer bodies to the cooler ones.

Caloric theory came under attack soon after its introduction. The caloric theory maintained that heat is a substance, and it cannot be created or destroyed. Yet it was known that heat can be generated indefinitely by rubbing one's hands together or rubbing two pieces of wood together. In 1798, Benjamin Thompson (Count Rumford) (1753–1814) from the U.S.A. showed that heat can be generated continuously through friction. The validity of the caloric theory was also challenged by several others. But it was the careful experiments of the Englishman James P. Joule (1818–1889) published in 1843 that finally convinced the skeptics that heat was not a substance after all, and put the caloric theory to rest. Although the caloric theory was totally abandoned in the middle of the nineteenth century, it contributed greatly to the development of thermodynamics.

### Modes of Heat Transfer

Heat can be transferred in three different ways: *conduction, convection,* and *radiation.* A detailed study of these heat transfer modes is given later in this text. Below we will give a brief description of each mode to familiarize the reader with the basic mechanisms of heat transfer. All modes of heat transfer require the existence of a temperature difference, and all modes of heat transfer are from the high-temperature medium to a lower-temperature one.

**Conduction** is the transfer of energy from the more energetic particles of a substance to the adjacent less energetic ones as a result of interactions between the particles. Conduction can take place in solids, liquids, or gases. In gases and liquids, conduction is due to the collisions of the molecules during their random motion. In solids, it is due to the combination of vibrations of the molecules in a lattice and the energy transport by free electrons. A cold canned drink in a warm room, for example, eventually warms up to the room temperature as a result of heat transfer from the room to the drink through the aluminum can by conduction (Fig. 3-8).

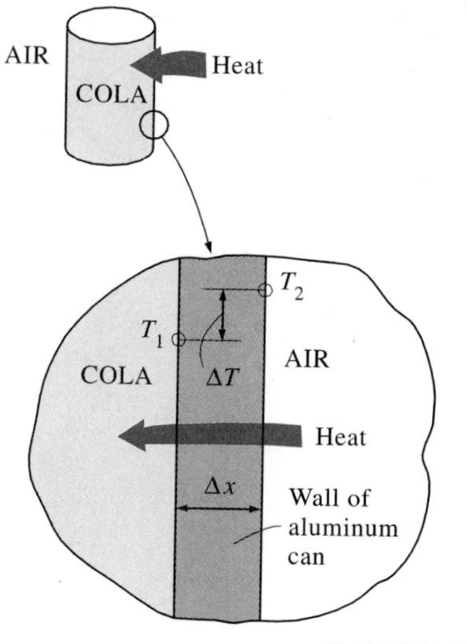


**FIGURE 3-8**
Heat conduction from warm air to a cold canned drink through the wall of the aluminum can.

It is observed that the rate of heat conduction $\dot{Q}_{cond}$ through a layer of constant thickness $\Delta x$ is proportional to the temperature difference $\Delta T$ across the layer and the area $A$ normal to the direction of heat transfer, and is inversely proportional to the thickness of the layer. Therefore,

$$\dot{Q}_{cond} = kA\frac{\Delta T}{\Delta x} \quad \text{(W)} \qquad (3\text{-}4)$$

where the constant of proportionality $k$ is the *thermal conductivity* of the material which is a measure of the ability of a material to conduct heat (Table 3-1). Materials such as copper and silver that are good electric conductors are also good heat conductors, and therefore have high $k$ values. Materials such as rubber, wood, and styrofoam are poor conductors of heat, and therefore have low $k$ values.

**TABLE 3-1**
**Thermal conductivities of some materials at room conditions**

| Material | Thermal conductivity, W/(m · K) |
|---|---|
| Diamond | 2300 |
| Silver | 429 |
| Copper | 401 |
| Gold | 317 |
| Aluminum | 237 |
| Iron | 80.2 |
| Mercury ($\ell$) | 8.54 |
| Glass | 1.4 |
| Brick | 0.72 |
| Water ($\ell$) | 0.613 |
| Human skin | 0.37 |
| Wood (oak) | 0.17 |
| Helium (g) | 0.152 |
| Soft rubber | 0.13 |
| Refrigerant-12 ($\ell$) | 0.072 |
| Glass fiber | 0.043 |
| Air (g) | 0.026 |
| Urethane, rigid foam | 0.026 |

In the limiting case of $\Delta x \to 0$, the equation above reduces to the differential form

$$\dot{Q}_{cond} = -kA\frac{dT}{dx} \quad \text{(W)} \qquad (3\text{-}5)$$

which is known as **Fourier's law** of heat conduction. It indicates that the rate of heat conduction in a direction is proportional to the *temperature gradient* in that direction. Heat is conducted in the direction of decreasing temperature, and the temperature gradient becomes negative when temperature decreases with increasing $x$. Therefore, a negative sign is added in Eq. 3-5 to make heat transfer in the positive $x$ direction a positive quantity.

Temperature is a measure of the kinetic energies of the molecules. In a liquid or gas, the kinetic energy of the molecules is due to the random motion of the molecules as well as the vibrational and rotational motions. When two molecules possessing different kinetic energies collide, part of the kinetic energy of the more energetic (higher-temperature) molecule is transferred to the less energetic (lower-temperature) particle, in much the same way as when two elastic balls of the same mass at different velocities collide, part of the kinetic energy of the faster ball is transferred to the slower one.

In solids, heat conduction is due to two effects: the lattice vibrational waves induced by the vibrational motions of the molecules positioned at relatively fixed positions in a periodic manner called a lattice, and the energy transported via the free flow of electrons in the solid. The thermal conductivity of a solid is obtained by adding the lattice and the electronic components. The thermal conductivity of pure metals is primarily due to the electronic component whereas the thermal conductivity of nonmetals is primarily due to the lattice component. The lattice component of thermal conductivity strongly depends on the way the molecules are arranged. For example, the thermal conductivity of diamond, which is a highly ordered crystalline solid, is much higher than the thermal conductivities of pure metals.

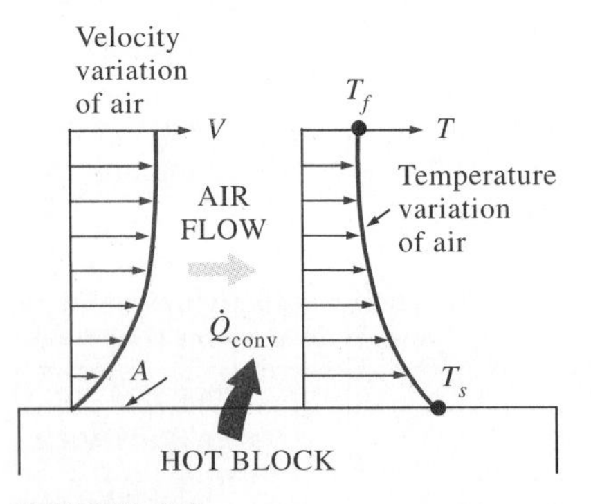


**FIGURE 3-9**
Heat transfer from a hot surface to air by convection.

**Convection** is the mode of energy transfer between a solid surface and the adjacent liquid or gas which is in motion, and it involves the combined effects of *conduction* and *fluid motion.* The faster the fluid motion, the greater the convection heat transfer. In the absence of any bulk fluid motion, heat transfer between a solid surface and the adjacent fluid is by pure conduction The presence of bulk motion of the fluid enhances the heat transfer between the solid surface and the fluid, but it also complicates the determination of heat transfer rates.

Consider the cooling of a hot block by blowing of cool air over its top surface (Fig. 3-9). Energy is first transferred to the air layer adjacent to the surface of the block by conduction. This energy is then carried away from the surface by convection; that is, by the combined effects of conduction within the air, which is due to random motion of air molecules, and the bulk or macroscopic motion of the air, which removes the heated air near the surface and replaces it by the cooler air.

Convection is called *forced convection* if the fluid is *forced* to flow in a tube or over a surface by external means such as a fan, pump, or the wind. In contrast, convection is called *free* (or *natural*) *convection* if the fluid motion is caused by buoyancy forces that are induced by density differences due to the variation of temperature in the fluid (Fig. 3-10). For example, in the absence of a fan, heat transfer from the surface of the hot block in Fig. 3-9 will be by natural convection since any motion in the air in this case will be due to the rise of the warmer (and thus lighter) air near the surface and the fall of the cooler (and thus heavier) air to fill its place. Heat transfer between the block and the surrounding air will be by conduction if the temperature difference between the air and the block is not large enough to overcome the resistance of air to move and thus to initiate natural convection currents.

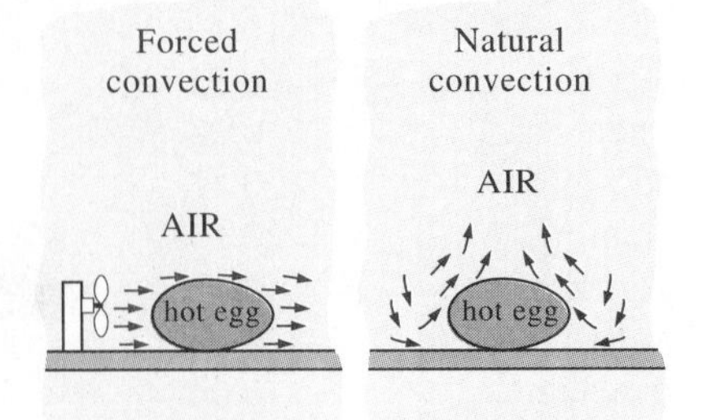


**FIGURE 3-10**
The cooling of a boiled egg by forced and natural convection.

Heat transfer processes that involve *change of phase* of a fluid are also considered to be convection because of the fluid motion induced during the process such as the rise of the vapor bubbles during *boiling* or the fall of the liquid droplets during *condensation.*

The rate of heat transfer by convection $\dot{Q}_{conv}$ is determined from **Newton's law of cooling**, which is expressed as

$$\dot{Q}_{conv} = hA(T_s - T_f) \qquad (W) \qquad (3\text{-}6)$$

where $h$ is the *convection heat transfer coefficient*, $A$ is the surface area through which heat transfer takes place, $T_s$ is the surface temperature, and $T_f$ is bulk fluid temperature away from the surface. (At the surface, the fluid temperature equals the surface temperature of the solid.)

The convection heat transfer coefficient $h$ is not a property of the fluid. It is an experimentally determined parameter whose value depends on all the variables that influence convection such as the surface geometry, the nature of fluid motion, the properties of the fluid, and the bulk fluid velocity. Typical values of $h$, in $W/(m^2 \cdot K)$, are 2–25 for the free convection of gases, 50–1000 for the free convection of liquids, 25–250 for the forced convection of gases, 50–20,000 for the forced convection of liquids, and 2500–100,000 for convection in boiling and condensation processes.

**Radiation** is the energy emitted by matter in the form of electromagnetic waves (or photons) as a result of the changes in the electronic configurations of the atoms or molecules. Unlike conduction and convection, the transfer of energy by radiation does not require the presence of an intervening medium. In fact, energy transfer by radiation is fastest (at the speed of light) and it suffers no attenuation in a vacuum. This is exactly how the energy of the sun reaches the earth.

In heat transfer studies, we are interested in *thermal radiation*, which is the form of radiation emitted by bodies because of their temperature. It differs from other forms of electromagnetic radiation such as X-rays, gamma rays, microwaves, radio waves, and television waves which are not related to temperature. All bodies at a temperature above absolute zero emit thermal radiation.

Radiation is a *volumetric phenomena*, and all solids, liquids, and gases emit, absorb, or transmit radiation to varying degrees. However, radiation is usually considered to be a *surface phenomenon* for solids that are opaque to thermal radiation such as metals, wood, and rocks since the radiation emitted by the interior regions of such material can never reach the surface, and the radiation incident on such bodies is usually absorbed within a few microns from the surface.

The maximum rate of radiation that can be emitted from a surface at an *absolute* temperature $T_s$ is given by the *Stefan–Boltzmann, law* as

$$\dot{Q}_{emit,max} = \sigma A T_s^4 \qquad (W) \qquad (3\text{-}7)$$

where $A$ is the surface area and $\sigma = 5.67 \times 10^{-8}\ W/(m^2 \cdot K^4)$ is the *Stefan–Boltzmann constant.* The idealized surface which emits radiation at this maximum rate is called a **blackbody**, and the radiation emitted by a blackbody is called *blackbody radiation.* The radiation emitted by all *real* surfaces is less than the radiation emitted by a blackbody at the same temperatures and is expressed as

$$\dot{Q}_{emit} = \varepsilon \sigma A T_s^4 \qquad (W) \qquad (3\text{-}8)$$

where $\varepsilon$ is the *emissivity* of the surface. The property emissivity, whose value is in the range $0 \leq \varepsilon \leq 1$, is a measure of how closely a surface approximates a blackbody for which $\varepsilon = 1$. The emissivities of some surfaces are given in Table 3-2.

Another important radiation property of a surface is its *absorptivity*, $\alpha$, which is the fraction of the radiation energy incident on a surface that is absorbed by the surface. Like emissivity, its value is in the range $0 \leq \alpha \leq 1$. A blackbody absorbs the entire radiation incident on it. That is, a blackbody is a perfect absorber ($\alpha = 1$) as well as a perfect emitter.

In general, both $\varepsilon$ and $\alpha$ of a surface depend on the temperature and the wavelength of the radiation. **Kirchhoff's law** of radiation states that the emissivity and the absorptivity of a surface are equal at the same temperature and wavelength. In most practical applications, the dependence of $\varepsilon$ and $\alpha$ on the temperature and wavelength is ignored, and the average absorptivity of a surface is taken to be equal to its average emissivity. The rate at which a surface absorbs radiation is determined from (Fig. 3-11)

$$\dot{Q}_{abs} = \alpha \dot{Q}_{inc} \qquad \text{(W)} \tag{3-9}$$

where $\dot{Q}_{inc}$ is the rate at which radiation is incident on the surface and $\alpha$ is the absorptivity of the surface. For opaque (nontransparent) surfaces, the portion of incident radiation that is not absorbed by the surface is reflected back.

The difference between the rates of radiation emitted by the surface and the radiation absorbed is the *net* radiation heat transfer. If the rate of radiation absorption is greater than the rate of radiation emission, the surface is said to be *gaining* energy by radiation. Otherwise, the surface is said to be *losing* energy by radiation. In general, the determination of the net rate of heat transfer by radiation between two surfaces is a complicated matter since it depends on the properties of the surfaces, their orientation relative to each other, and the interaction of the medium between the surfaces with radiation. However, in the special case of a relatively small surface of emissivity $\varepsilon$ and surface area $A$ at *absolute* temperature $T_s$ that is completely enclosed by a much larger surface at *absolute* temperature $T_{surr}$ separated by a gas (such as air) that does not interact with radiation (i.e., the amount of radiation emitted, absorbed, or scattered by the medium is negligible), the net rate of radiation heat transfer between these two surfaces is determined from (Fig. 3-12)

$$\dot{Q}_{rad} = \varepsilon \sigma A (T_s^4 - T_{surr}^4) \qquad \text{(W)} \tag{3-10}$$

**TABLE 3-2**
**Emissivities of some materials at 300 K**

| Material | Emissivity |
|---|---|
| Aluminum foil | 0.07 |
| Anodized aluminum | 0.82 |
| Polished copper | 0.03 |
| Polished gold | 0.03 |
| Polished silver | 0.02 |
| Polished stainless steel | 0.17 |
| Black paint | 0.98 |
| White paint | 0.90 |
| White paper | 0.92–0.97 |
| Asphalt pavement | 0.85–0.93 |
| Red brick | 0.93–0.96 |
| Human skin | 0.95 |
| Wood | 0.82–0.92 |
| Soil | 0.93–0.96 |
| Water | 0.96 |
| Vegetation | 0.92–0.96 |

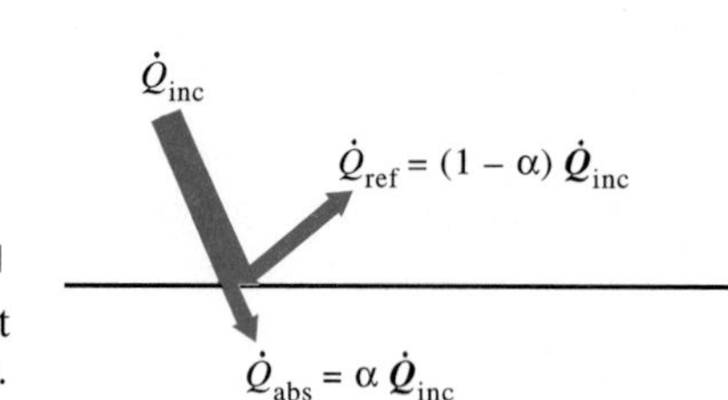


**FIGURE 3-11**
The absorption of radiation incident on an opaque surface of absorptivity $\alpha$.

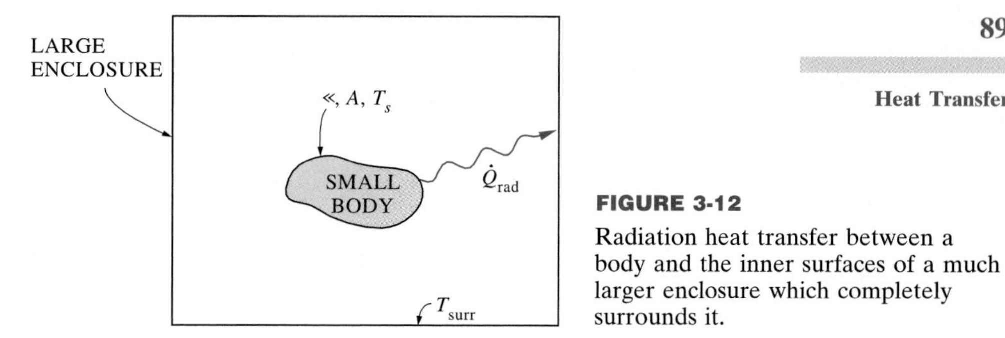


**FIGURE 3-12**
Radiation heat transfer between a body and the inner surfaces of a much larger enclosure which completely surrounds it.

In this special case, the emissivity and the surface area of the surrounding surface do not have any effect on the net radiation heat transfer.

**EXAMPLE 3-1**

Consider a person standing in a breezy room at 20°C. Determine the total rate of heat transfer from this person if the exposed surface area and the average outer surface temperature of the person are 1.6 m$^2$ and 29°C, respectively, and the convection heat transfer coefficient is 6 W/(m$^2$ · °C) (Fig. 3-13).

**Solution** The heat transfer between the person and the air in the room will be by convection (instead of conduction) since it is conceivable that the air in the vicinity of the skin or clothing will warm up and rise as a result of heat transfer from the body, initiating natural convection currents. It appears that the experimentally determined value for the rate of convection heat transfer in this case is 6 W per unit surface area (m$^2$) per unit temperature difference (in K or °C) between the person and the air away from the person. Thus, the rate of convection heat transfer from the person to the air in the room is, from Eq. 3-6,

$$\begin{aligned}\dot{Q}_{conv} &= hA(T_s - T_f)\\ &= [6\ \text{W}/(\text{m}^2 \cdot {}^\circ\text{C})](1.6\ \text{m}^2)(29\text{–}20)^\circ\text{C}\\ &= 86.4\ \text{W}\end{aligned}$$

The person will also lose heat by radiation to the surrounding wall surfaces. We take the temperature of the surfaces of the walls, ceiling, and the floor to be equal to the air temperature in this case for simplicity, but we recognize that this does not need to be the case. These surfaces may be at a higher or lower temperature than the average temperature of the room air, depending on the outdoor conditions and the structure of the walls. Considering that air does not intervene with radiation and the person is completely enclosed by the surrounding surfaces, the net rate of radiation heat transfer from the person to the surrounding walls, ceiling, and the floor is, from Eq. 3-10,

$$\begin{aligned}\dot{Q}_{rad} &= \varepsilon\sigma A(T_s^4 - T_{surr}^4)\\ &= (0.95)[5.67 \times 10^{-8}\ \text{W}/(\text{m}^2 \cdot \text{K}^4)](1.6\ \text{m}^2)[(29 + 273)^4 - (20 + 273)^4]\text{K}^4\\ &= 81.7\ \text{W}\end{aligned}$$

Note that we must use *absolute* temperatures in radiation calculations. Also note that we used the emissivity value for the skin and clothing at room temperature since the emissivity is not expected to change significantly at a slightly higher temperature.

ROOM
AIR
20°C
29°C
$\dot{Q}_{conv}$
$\dot{Q}_{rad}$
$\dot{Q}_{cond}$

**FIGURE 3-13**
Heat and transfer from the person described in Example 3-1.

Then the rate of total heat transfer from the body is determined by adding these two quantities to be

$$\dot{Q}_{\text{total}} = \dot{Q}_{\text{conv}} + \dot{Q}_{\text{rad}} = (86.4 + 81.7)\,\text{W} = \mathbf{168.1\ W}$$

The heat transfer would be much higher if the person were not dressed since the exposed surface temperature would be higher. Thus, an important function of the clothes is to serve as a barrier against heat transfer.

In the above calculations, heat transfer through the feet to the floor by conduction, which is usually very small, is neglected. Heat transfer from the skin by perspiration, which is the dominant mode of heat transfer in hot environments, is not considered here.

## 3-3 ■ WORK

Work, like heat, is an energy interaction between a system and its surroundings. As mentioned earlier, energy can cross the boundary of a closed system in the form of heat or work. Therefore, *if the energy crossing the boundary of a closed system is not heat, it must be work.* Heat is easy to recognize: Its driving force is a temperature difference between the system and its surroundings. Then we can simply say that an energy interaction which is not caused by a temperature difference between a system and its surroundings is work. More specifically, *work is the energy transfer associated with a force acting through a distance.* A rising piston, a rotating shaft, and an electric wire crossing the system boundaries are all associated with work interactions.

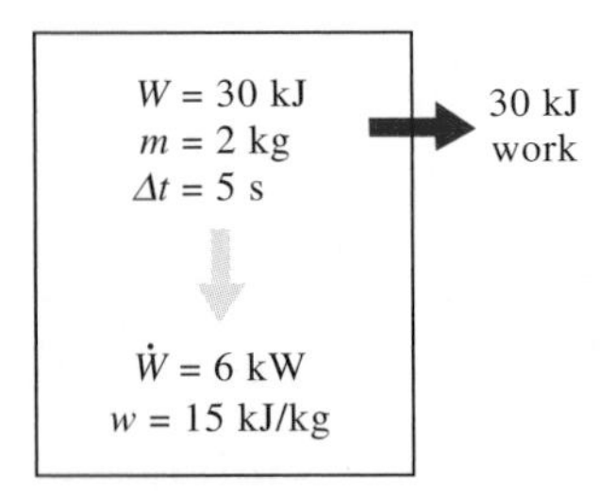


**FIGURE 3-14**
The relationships among $w$, $W$, and $\dot{W}$.

Work is also a form of energy transferred like heat and, therefore, has energy units such as kJ. The work done during a process between states 1 and 2 is denoted $W_{12}$, or simply $W$. The work done *per unit mass* of a system is denoted $w$ and is defined as

$$w = \frac{W}{m} \qquad \text{(kJ/kg)} \tag{3-11}$$

The work done *per unit time* is called **power** and is denoted $\dot{W}$ (Fig. 3-14). The unit of power is kJ/s, or kW.

The production of work by a system is viewed as a desirable, positive effect and the consumption of work as an undesirable, negative effect. The sign convention for work adapted in this text reflects this philosophy: *Work done by a system is positive, and work done on a system is negative* (Fig. 3-15). By this convention, the work produced by car engines, hydraulic, steam, or gas turbines is positive, and the work consumed by compressors, pumps, and mixers is negative. In other words, work *produced* by a system during a process is positive, and work *consumed* is negative. Notice that the energy of a system decreases as it does work and increases as work is done on the system. The reader is reminded that some texts use the opposite sign convention for work.

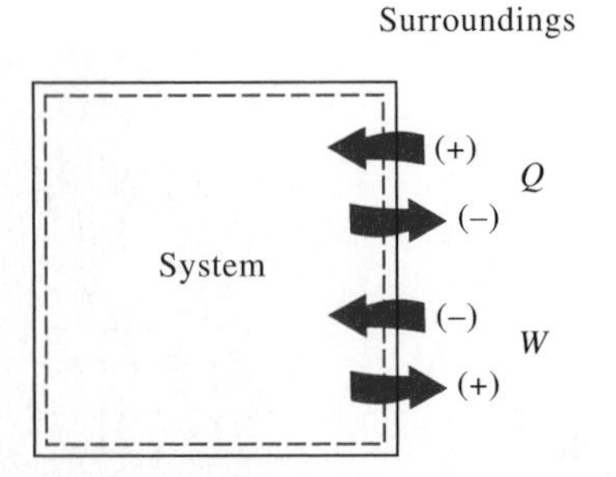


**FIGURE 3-15**
Sign convention for heat and work.

Sometimes the identifiers *in* and *out* are used to indicate the direction of any heat or work interaction in place of the negative and positive signs.

Heat transfer to a system is denoted by $Q_{in}$, and heat transfer from a system, by $Q_{out}$. A heat loss of 5 kJ can be expressed as either $Q = -5$ kJ or $Q_{out} = 5$ kJ. Likewise, a work output of 5 kJ can be expressed as $W = 5$ kJ or $W_{out} = 5$ kJ. When we are dealing with work-consuming devices such as compressors and pumps, the negative sign associated with the work term can conveniently be avoided by speaking of work (or power) input instead of work done (for example, $\dot{W}_{in} = 2$ kW instead of $\dot{W} = -2$ kW).

Heat transfer and work are *interactions* between a system and its surroundings, and there are many similarities between the two:

**1** Both are recognized at the boundaries of the system as they cross them. That is, both heat and work are *boundary* phenomena.

**2** Systems possess energy, but not heat or work. That is, heat and work are *transfer* phenomena.

**3** Both are associated with a *process,* not a state. Unlike properties, heat or work has no meaning at a state.

**4** Both are *path functions* (i.e., their magnitudes depend on the path followed during a process as well as the end states).

**Path functions** have **inexact differentials** designated by the symbol $\delta$. Therefore, a differential amount of heat or work is represented by $\delta Q$ or $\delta W$, respectively, instead of $dQ$ or $dW$. Properties, however, are **point functions** (i.e., they depend on the state only, and not on how a system reaches that state), and they have **exact differentials** designated by the symbol *d.* A small change in volume, for example, is represented by $dV$ and the total volume change during a process between states 1 and 2 is

$$\int_1^2 dV = V_2 - V_1 = \Delta V$$

That is, the volume change during process 1-2 is always the volume at state 2 minus the volume at state 1, regardless of the path followed (Fig. 3-16). The total work done during process 1-2, however, is

$$\int_1^2 \delta W = W_{12} \qquad (\textit{not } \Delta W)$$

That is, the total work is obtained by following the process path and adding the differential amounts of work ($\delta W$) done along the way. The integral of $\delta W$ *is not* $W_2 - W_1$ (i.e., the work at state 2 minus work at state 1), which is meaningless since work is not a property and systems do not possess work at a state.

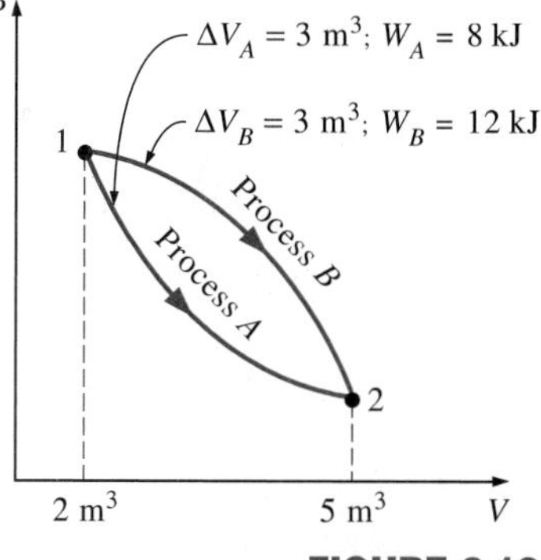


**FIGURE 3-16**

Properties are point functions; but heat and work are path functions (their magnitudes depend on the path followed).

**EXAMPLE 3-2**

A candle is burning in a well-insulated room. Taking the room (the air plus the candle) as the system, determine (*a*) if there is any heat transfer during this burning process and (*b*) if there is any change in the internal energy of the system.

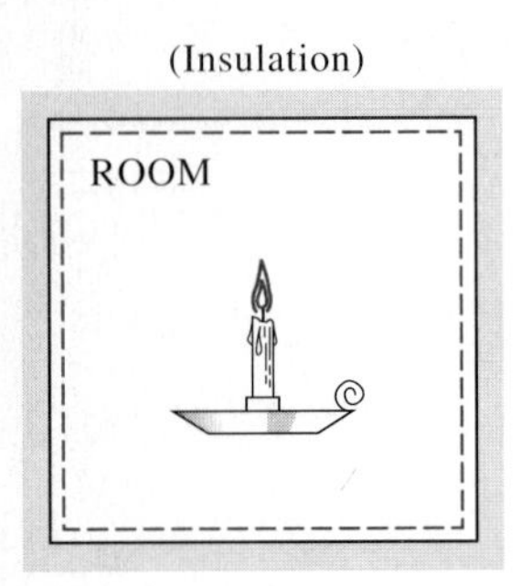


**FIGURE 3-17**
Schematic for Example 3-2.

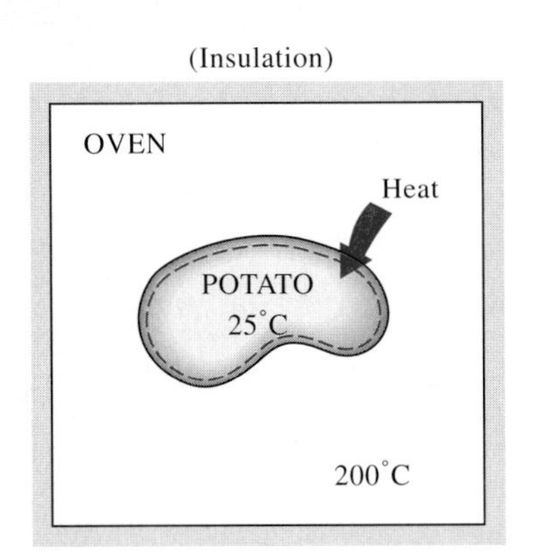


**FIGURE 3-18**
Schematic for Example 3-3.

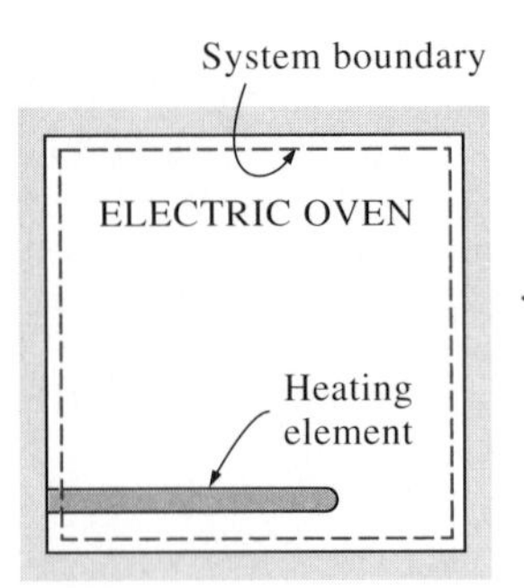


**FIGURE 3-19**
Schematic for Example 3-4.

**Solution** (*a*) The interior surfaces of the room form the system boundary, as indicated by the dashed lines in Fig. 3-17. As pointed out earlier, heat is recognized as it crosses the boundaries. Since the room is well insulated, we have an adiabatic system and no heat will pass through the boundaries. Therefore, $Q = 0$ for this process.

(*b*) As discussed in Chap. 1, the internal energy involves energies that exist in various forms (sensible, latent, chemical, nuclear). During the process described above, part of the chemical energy is converted to sensible energy. That is, part of the internal energy of the system is changed from one form to another. Since there is no increase or decrease in the total internal energy of the system, $\Delta U = 0$ for this process.

**EXAMPLE 3-3**

A potato that is initially at room temperature (25°C) is being baked in an oven which is maintained at 200°C, as shown in Fig. 3-18. Is there any heat transfer during this baking process?

**Solution** This is not a well-defined problem since the system is not specified. Let us assume that we are observing the potato, which will be our system. Then the skin of the potato may be viewed as the system boundary. Part of the energy in the oven will pass through the skin to the potato. Since the driving force for this energy transfer is a temperature difference, this is a heat transfer process.

**EXAMPLE 3-4**

A well-insulated electric oven is being heated through its heating element. If the entire oven, including the heating element, is taken to be the system, determine whether this is a heat or work interaction.

**Solution** For this problem, the interior surfaces of the oven form the system boundary, as shown in Fig. 3-19. The energy content of the oven obviously increases during this process, as evidenced by a rise in temperature. This energy transfer to the oven is not caused by a temperature difference between the oven and the surrounding air. Instead, it is caused by negatively charged particles called *electrons* crossing the system boundary and thus doing work. Therefore, this is a work interaction.

**EXAMPLE 3-5**

Answer the question in Example 3-4 if the system is taken as only the air in the oven without the heating element.

**Solution** This time, the system boundary will include the outer surface of the heating element and will not cut through it, as shown in Fig. 3-20. Therefore, no electrons will be crossing the system boundary at any point. Instead, the energy generated in the interior of the heating element will be transferred to the air around it as a result of the temperature difference between the heating element and the air in the oven. Therefore, this is a heat transfer process.

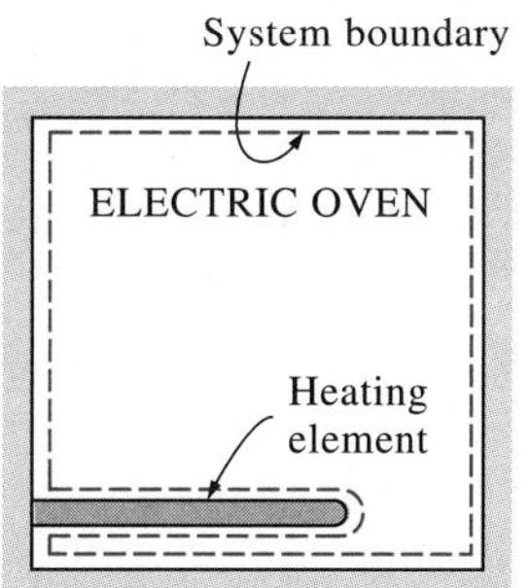


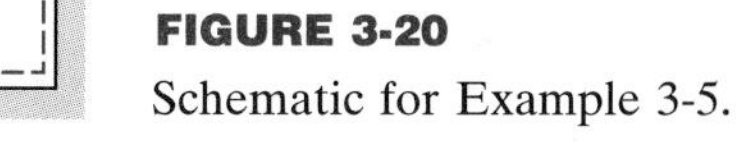

**FIGURE 3-20**
Schematic for Example 3-5.

For both cases, the amount of energy transfer to the air is the same. These two examples show that the same interaction can be heat or work depending on how the system is selected.

## Electrical Work

It was pointed out in Example 3-4 that electrons crossing the system boundary do electrical work on the system. In an electric field, electrons in a wire move under the effect of electromotive forces, doing work. When $N$ coulombs of electrons move through a potential difference $V$, the electrical work done is

$$W_e = VN \qquad \text{(kJ)}$$

which can also be expressed in the rate form as

$$\dot{W}_e = VI \qquad \text{(kW)} \qquad (3\text{-}12)$$

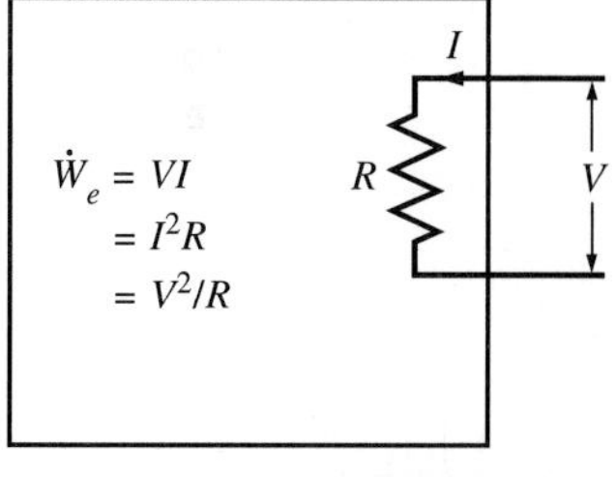


**FIGURE 3-21**
Electrical power in terms of resistance $R$, current $I$, and potential difference $V$.

where $\dot{W}_e$ is the electrical power and $I$ is the number of electrons flowing per unit time, i.e., the current (Fig. 3-21). In general, both $V$ and $I$ vary with time, and the electrical work done during a time interval $\Delta t$ is expressed as

$$W_e = \int_1^2 VI\,dt \qquad \text{(kJ)} \qquad (3\text{-}13)$$

If both $V$ and $I$ remain constant during the time interval $\Delta t$, this equation will reduce to

$$W_e = VI\,\Delta t \qquad \text{(kJ)} \qquad (3\text{-}14)$$

**EXAMPLE 3-6**

A small tank containing iced water at 0°C is placed in the middle of a large, well-insulated tank filled with oil, as shown in Fig. 3-22. The entire system is initially in thermal equilibrium at 0°C. The electric heater in the oil is now turned on, and 10 kJ of electrical work is done on the oil. After a while, it is noticed that the entire system is again at 0°C, but some ice in the small tank has melted. Considering the oil to be system $A$ and the iced water to be system $B$, discuss the heat and work interactions for system $A$, system $B$, and the combined system (oil and iced water).

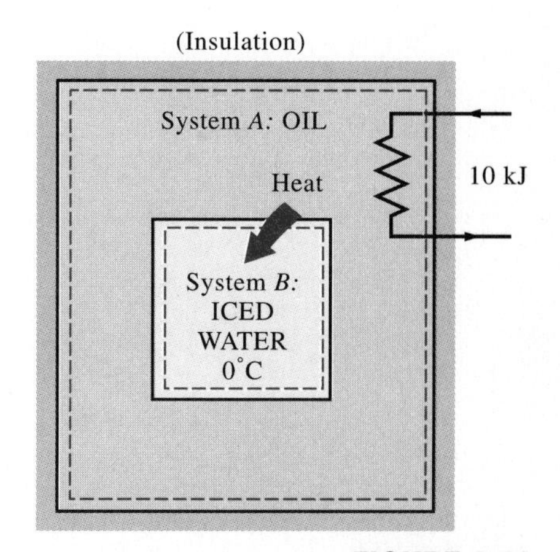


**FIGURE 3-22**
Schematic for Example 3-6.

**Solution** The boundaries of each system are indicated by dashed lines in the figure. Notice that the boundary of system $B$ also forms the inner part of the boundary of system $A$.

*System A*: When the heater is turned on, electrons cross the outer boundary of system $A$, doing electrical work. This work is done on the system, and therefore $W_A = -10$ kJ. Because of this added energy, the temperature of the oil will rise, creating a temperature gradient, which results in a heat flow process from the oil to the iced water through their common boundary. Since the oil is restored to its initial temperature of 0°C, the energy lost as heat must equal the energy gained as work. Therefore, $Q_A = -10$ kJ (or $Q_{A,\text{out}} = 10$ kJ).

*System B*: The only energy interaction at the boundaries of system $B$ is the heat flow from system $A$. All the heat lost by the oil is gained by the iced water. Thus, $W_B = 0$ and $Q_B = +10$ kJ.

*Combined system*: The outer boundary of system $A$ forms the entire boundary of the combined system. The only energy interaction at this boundary is the electrical work. Since the tank is well insulated, no heat will cross this boundary. Therefore, $W_{\text{comb}} = -10$ kJ and $Q_{\text{comb}} = 0$. Notice that the heat flow from the oil to the iced water is an internal process for the combined system and, therefore, is not recognized as heat. It is simply the redistribution of the internal energy.

## 3-4 MECHANICAL FORMS OF WORK

There are several different ways of doing work, each in some way related to a force acting through a distance (Fig. 3-23). In elementary mechanics, the work done by a constant force $F$ on a body that is displaced a distance $s$ in the direction of the force is given by

$$W = Fs \qquad \text{(kJ)} \tag{3-15}$$

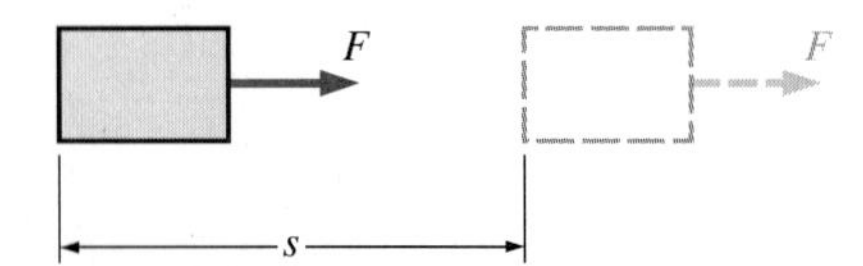


**FIGURE 3-23**
The work done is proportional to the force applied ($F$) and the distance traveled ($s$).

If the force $F$ is not constant, the work done is obtained by adding (i.e., integrating) the differential amounts of work (force times the differential displacement $ds$):

$$W = \int_1^2 F\,ds \qquad \text{(kJ)} \tag{3-16}$$

Obviously one needs to know how the force varies with displacement to perform this integration. Equations 3-15 and 3-16 give only the magnitude of the work. The sign is easily determined from physical considerations: The work done on a system by an external force acting in the direction of motion is negative, and work done by a system against an external force acting in the opposite direction to motion is positive.

There are two requirements for a work interaction between a system and its surroundings to exist: (1) there must be a *force* acting on the boundary, and (2) the boundary must *move*. Therefore, the presence of forces on the boundary without any displacement of the boundary does not constitute a work interaction. Likewise, the displacement of the boundary without any force to oppose or drive this motion (such as the expansion of a gas into an evacuated space) is not a work interaction.

In many thermodynamic problems, mechanical work is the only form of work involved. It is associated with the movement of the boundary of a system or with the movement of the entire system as a whole (Fig. 3-24). Some common forms of mechanical work are discussed below.

FIGURE 3-24
If there is no movement, no work is done.

## Moving Boundary Work

One form of mechanical work frequently encountered in practice is associated with the expansion or compression of a gas in a piston–cylinder device. During this process, part of the boundary (the inner face of the piston) moves back and forth. Therefore, the expansion and compression work is often called **moving boundary work**, or simply **boundary work** (Fig. 3-25). Some prefer to call it the $P\,dV$ work for reasons explained below. Moving boundary work is the primary form of work involved in automobile engines. During their expansion, the combustion gases force the piston to move, which in turn forces the crank shaft to rotate.

The moving boundary work associated with real engines or compressors cannot be determined exactly from a thermodynamic analysis alone because the piston usually moves at very high speeds, making it difficult for the gas inside to maintain equilibrium. Then the states that the system passes through during the process cannot be specified, and no process path can be drawn. Work, being a path function, cannot be determined analytically without a knowledge of the path. Therefore, the boundary work in real engines or compressors is determined by direct measurements.

In this section, we analyze the moving boundary work for a *quasi-equilibrium process,* a process during which the system remains in equilibrium at all times. A quasi-equilibrium process, also called a *quasi-static process*, is closely approximated by real engines, especially when the piston moves at low velocities. Under identical conditions, the work output of the engines is found to be a maximum, and the work input to the compressors to be a minimum, when quasi-equilibrium processes are used in place of non-quasi-equilibrium processes. Below, the work associated with a moving boundary is evaluated for a quasi-equilibrium process.

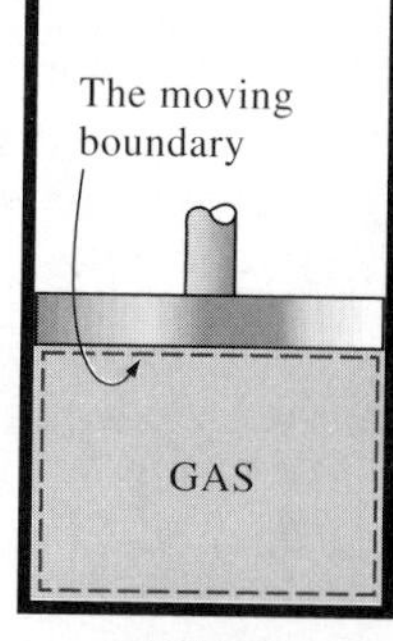


FIGURE 3-25
The work associated with a moving boundary is called *boundary work.*

Consider the gas enclosed in the piston–cylinder device shown in Fig. 3-26. The initial pressure of the gas is $P$, the total volume is $V$, and the cross-sectional area of the piston is $A$. If the piston is allowed to move a distance $ds$ in a quasi-equilibrium manner, the differential work done during this process is

$$\delta W_b = F\,ds = PA\,ds = P\,dV \tag{3-17}$$

That is, the boundary work in the differential form is equal to the product of the absolute pressure $P$ and the differential change in the volume $dV$ of the system. This expression also explains why the moving boundary work is sometimes called the $P\,dV$ work.

Note in Eq. 3-17 that $P$ is the absolute pressure, which is always positive. However, the volume change $dV$ is positive during an expansion

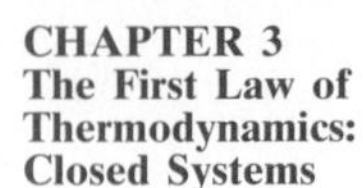

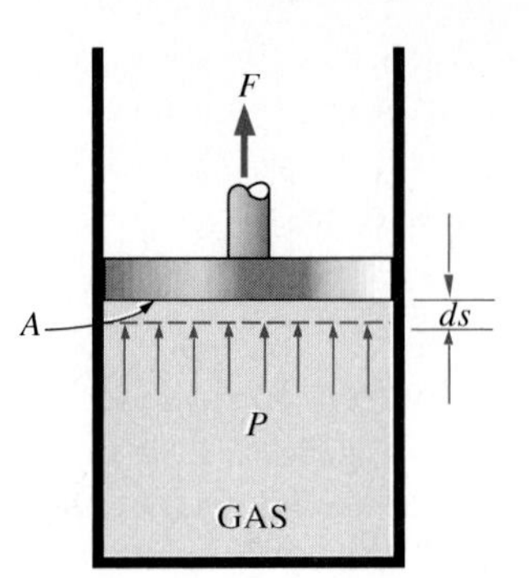


**FIGURE 3-26**

A gas does a differential amount of work $\delta W_b$ as it forces the piston to move by a differential amount $ds$.

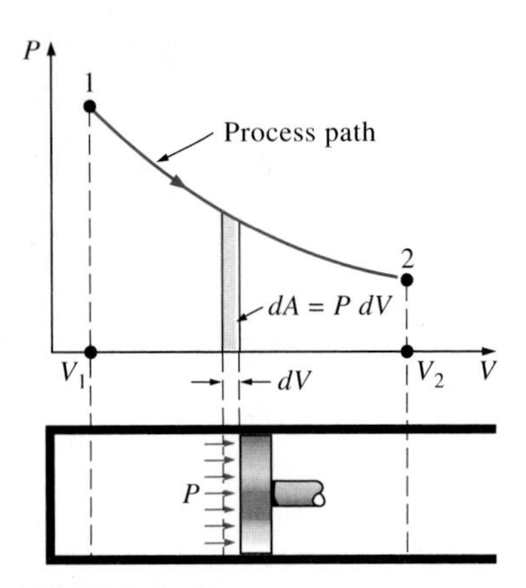


**FIGURE 3-27**

The area under the process curve on a $P$-$V$ diagram represents the boundary work.

process (volume increasing) and negative during a compression process (volume decreasing). Thus, the boundary work is positive during an expansion process and negative during a compression process, which is consistent with the sign convention adopted for work.

The total boundary work done during the entire process as the piston moves is obtained by adding all the differential works from the initial state to the final state:

$$W_b = \int_1^2 P\,dV \qquad \text{(kJ)} \tag{3-18}$$

This integral can be evaluated only if we know the functional relationship between $P$ and $V$ during the process. That is, $P = f(V)$ should be available. Note that $P = f(V)$ is simply the equation of the process path on a $P$-$V$ diagram.

The quasi-equilibrium expansion process described above is shown on a $P$-$V$ diagram in Fig. 3-27. On this diagram, the differential area $dA$ is equal to $P\,dV$, which is the differential work. The total area $A$ under the process curve 1–2 is obtained by adding these differential areas:

$$\text{area} = A = \int_1^2 dA = \int_1^2 P\,dV$$

A comparison of this equation with Eq. 3-18 reveals that *the area under the process curve on a P-V diagram is equal, in magnitude, to the work done during a quasi-equilibrium expansion or compression process of a closed system.* (On the $P$-$v$ diagram, it represents the boundary work done per unit mass.)

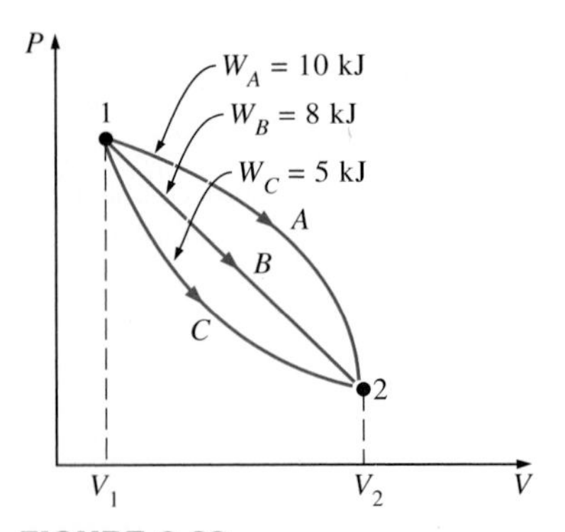


**FIGURE 3-28**

The boundary work done during a process depends on the path followed as well as the end states.

A gas can follow several different paths as it expands from state 1 to state 2. In general, each path will have a different area underneath it, and since this area represents the magnitude of the work, the work done will be different for each process (Fig. 3-28). This is expected, since work is a path function (i.e., it depends on the path followed as well as the end states). If work were not a path function, no cyclic devices (car engines, power plants) could operate as work-producing devices. The work produced by these devices during one part of the cycle would have to be consumed during another part, and there would be no net work output.

The cycle shown in Fig. 3-29 produces a net work output because the work done by the system during the expansion process (area under path $A$) is greater than the work done on the system during the compression part of the cycle (area under path $B$), and the difference between these two is the net work done during the cycle (the colored area).

If the relationship between $P$ and $V$ during an expansion or a compression process is given in terms of experimental data instead of in a functional form, obviously we cannot perform the integration analytically. But we can always plot the $P$-$V$ diagram of the process, using these data points, and calculate the area underneath graphically to determine the work done.

The use of the boundary work relation (Eq. 3-18) is not limited to the quasi-equilibrium processes of gases only. It can also be used for solids and liquids.

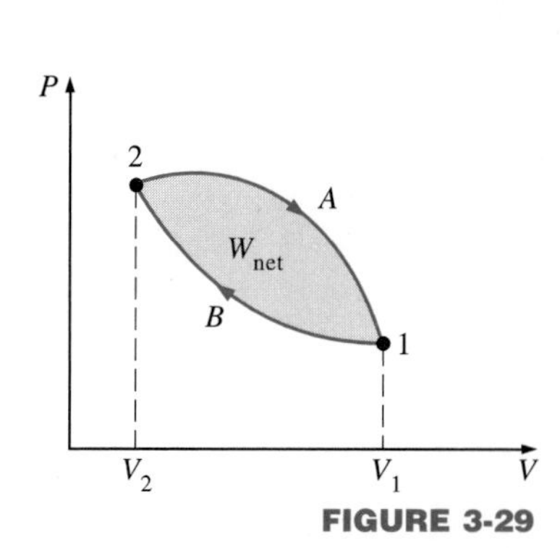


**FIGURE 3-29**

The net work done during a cycle is the difference between the work done by the system and the work done on the system.

**EXAMPLE 3-7**

A rigid tank contains air at 500 kPa and 150°C. As a result of heat transfer to the surroundings, the temperature and pressure inside the tank drop to 65°C and 400 kPa, respectively. Determine the boundary work done during this process.

**Solution** A sketch of the system and the $P$-$V$ diagram of the process are shown in Fig. 3-30. Assuming the process to be quasi-equilibrium, the boundary work can be determined from Eq. 3-18:

$$W_b = \int_1^2 P \overset{0}{\cancel{dV}} = 0$$

This is expected since a rigid tank has a constant volume and $dV = 0$ in the above equation. Therefore, there is no boundary work done during this process. That is, the boundary work done during a constant-volume process is always zero. This is also evident from the $P$-$V$ diagram of the process (the area under the process curve is zero).

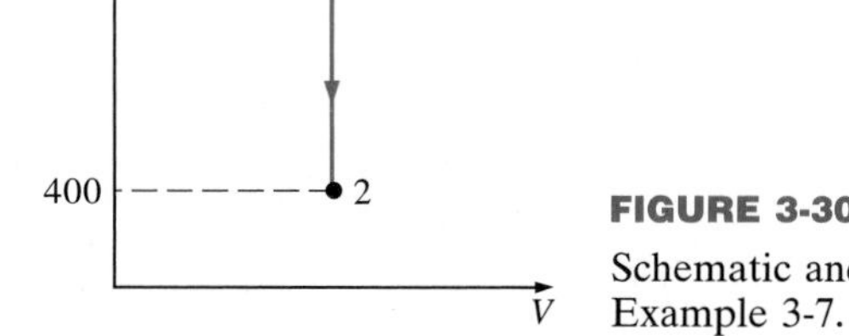


**FIGURE 3-30**

Schematic and $P$-$V$ diagram for Example 3-7.

**EXAMPLE 3-8**

A frictionless piston–cylinder device contains 0.1 lbm of water vapor at 20 psia and 320 °F. Heat is now added to the steam until the temperature reaches 400°F. If the piston is not attached to a shaft and its mass is constant, determine the work done by the steam during this process.

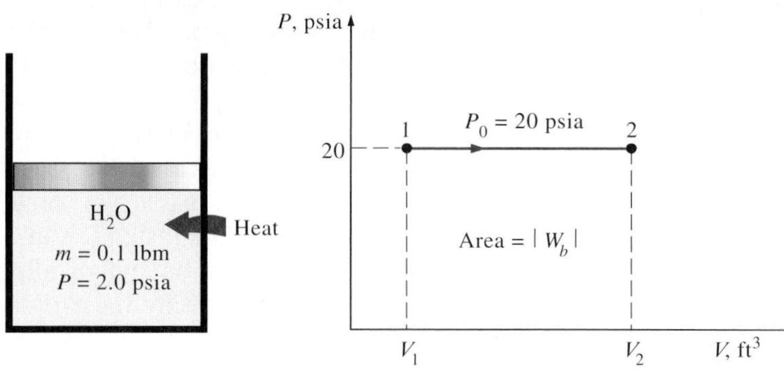


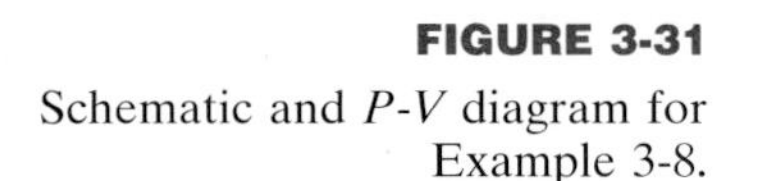

**FIGURE 3-31**
Schematic and *P-V* diagram for Example 3-8.

**Solution** A sketch of the system and the *P-V* diagram of the process are shown in Fig. 3-31. Even though it is not explicitly stated, the pressure of the steam within the cylinder remains constant during this process since both the atmospheric pressure and the weight of the piston remain constant. Therefore, this is a constant-pressure process, and, from Eq. 3-18,

$$W_b = \int_1^2 P\,dV = P_0 \int_1^2 dV = P_0(V_2 - V_1) \tag{3-19}$$

Approximating steam as an ideal gas, its initial and final volumes are determined to be

$$V_1 = \frac{mRT_1}{P_1} = \frac{(0.1\text{ lbm})[0.5956\text{ psia}\cdot\text{ft}^3/(\text{lbm}\cdot\text{R})](320 + 460)\text{ R}}{20\text{ psia}} = 2.32\text{ ft}^3$$

$$V_2 = \frac{mRT_2}{3P_2} = \frac{(0.1\text{ lbm})[0.5956\text{ psia}\cdot\text{ft}^3/(\text{lbm}\cdot\text{R})](400 + 460)\text{ R}}{20\text{ psia}} = 2.56\text{ ft}^3$$

Substituting,

$$W_b = (20\text{ psia})[(2.56 - 2.32)\text{ ft}^3]\left(\frac{1\text{ Btu}}{5.404\text{ psia}\cdot\text{ft}^3}\right) = \mathbf{0.89\text{ Btu}}$$

This answer is only *approximate* since water vapor deviates considerably from ideal gas behavior except at low pressures or high temperatures.

The positive sign indicates that the work is done by the system. That is, the steam used 0.89 Btu of its energy to do this work. The magnitude of this work could also be determined by calculating the area under the process curve on the *P-V* diagram, which is $P_0\,\Delta V$ for this case.

**EXAMPLE 3-9**

A piston-cylinder device initially contains 0.4 m³ of air at 100 kPa and 80°C. The air is now compressed to 0.1 m³ in such a way that the temperature inside the cylinder remains constant. Determine the work done during this process.

**Solution** A sketch of the system and the *P-V* diagram of the process are shown in Fig. 3-32. At the specified conditions, air can be considered to be an ideal gas since it is at a high temperature and low pressure relative to its critical-point values ($T_{cr} = -147$°C, $P_{cr} = 3390$ kPa for nitrogen, the main constituent of air). For an ideal gas at constant temperature $T_0$,

$$PV = mRT_0 = C \qquad \text{or} \qquad P = \frac{C}{V}$$

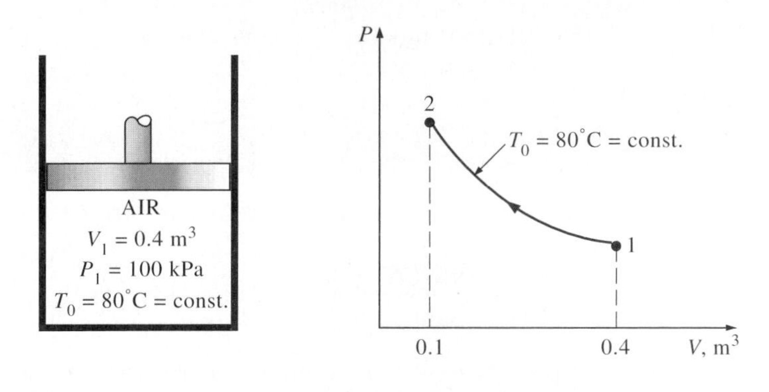


**FIGURE 3-32**

Schematic and *P-V* diagram for Example 3-9.

where $C$ is a constant. Substituting this into Eq. 3-18, we have

$$W_b = \int_1^2 P\,dV = \int_1^2 \frac{C}{V}\,dV = C\int_1^2 \frac{dV}{V} = C\ln\frac{V_2}{V_1} = P_1V_1\ln\frac{V_2}{V_1} \qquad (3\text{-}20)$$

In the above equation, $P_1V_1$ can be replaced by $P_2V_2$ or $mRT_0$. Also, $V_2/V_1$ can be replaced by $P_1/P_2$ for this case since $P_1V_1 = P_2V_2$.

Substituting the numerical values into the above equation yields

$$W_b = (100\text{ kPa})(0.4\text{ m}^3)\left(\ln\frac{0.1}{0.4}\right)\left(\frac{1\text{ kJ}}{1\text{ kPa}\cdot\text{m}^3}\right)$$
$$= -55.45\text{ kJ}$$

The negative sign indicates that this work is done on the system, which is always the case for compression processes.

## Polytropic Process

During expansion and compression processes of real gases, pressure and volume are often related by $PV^n = C$, where $n$ and $C$ are constants. A process of this kind is called a **polytropic process**. Below we develop a general expression for the work done during a polytropic process.

A sketch of the system and the *P-V* diagram of the process are shown in Fig. 3-33. The pressure for a polytropic process can be expressed as

$$P = CV^{-n} \qquad (3\text{-}21)$$

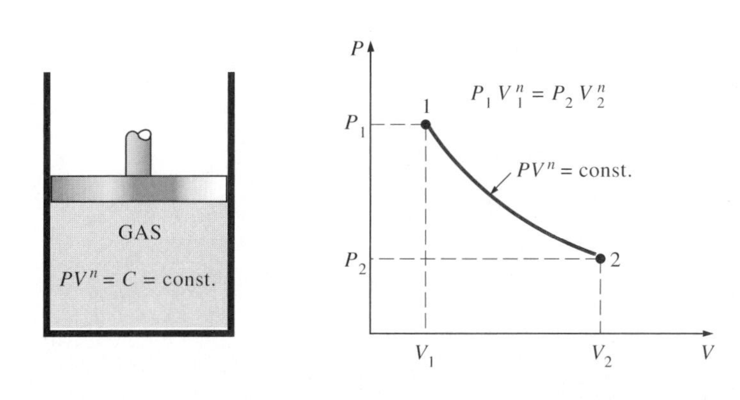

**FIGURE 3-33**

Schematic and *P-V* diagram for a polytropic process.

Substituting this relation into Eq. 3-18, we obtain

$$W_b = \int_1^2 P\,dV = \int_1^2 CV^{-n}\,dV = C\frac{V_2^{-n+1} - V_1^{-n+1}}{-n+1} = \frac{P_2V_2 - P_1V_1}{1-n} \tag{3-22}$$

since $C = P_1V_1^n = P_2V_2^n$. For an ideal gas ($PV = mRT$) it becomes

$$W_b = \frac{mR(T_2 - T_1)}{1-n}, \qquad n \neq 1 \qquad \text{(kJ)} \tag{3-23}$$

The special case of $n = 1$ is equivalent to the isothermal process discussed in the previous example.

### Spring Work

It is common knowledge that when a force is applied on a spring, the length of the spring changes (Fig. 3-34). When the length of the spring changes by a differential amount $dx$ under the influence of a force $F$, the work done is

$$\delta W_{\text{spring}} = F\,dx \tag{3-24}$$

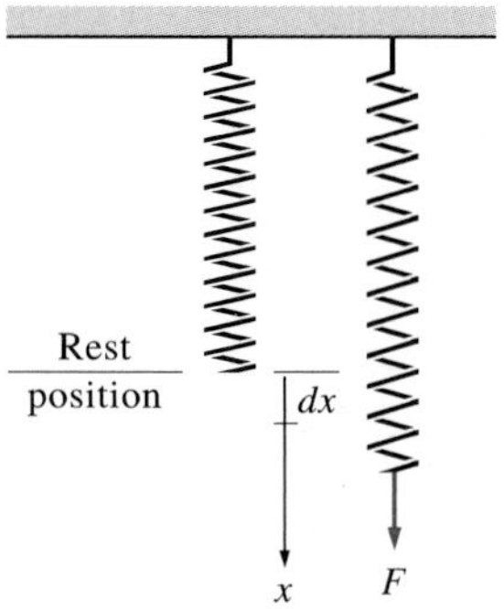

To determine the total spring work, we need to know a functional relationship between $F$ and $x$. For linear elastic springs, the displacement $x$ is proportional to the force applied (Fig. 3-35). That is,

$$F = k_s x \qquad \text{(kN)} \tag{3-25}$$

where $k_s$ is the spring constant and has the unit kN/m. The displacement $x$ is measured from the undisturbed position of the spring (that is, $x = 0$ when $F = 0$). Substituting Eq. 3-25 into Eq. 3-24 and integrating yield

$$W_{\text{spring}} = \tfrac{1}{2}k_s(x_2^2 - x_1^2) \qquad \text{(kJ)} \tag{3-26}$$

where $x_1$ and $x_2$ are the initial and the final displacements of the spring, respectively. Both $x_1$ and $x_2$ are measured from the undisturbed position of the spring. Note that the work done on a spring equals the energy stored in the spring.

**FIGURE 3-34**
Elongation of a spring under the influence of a force.

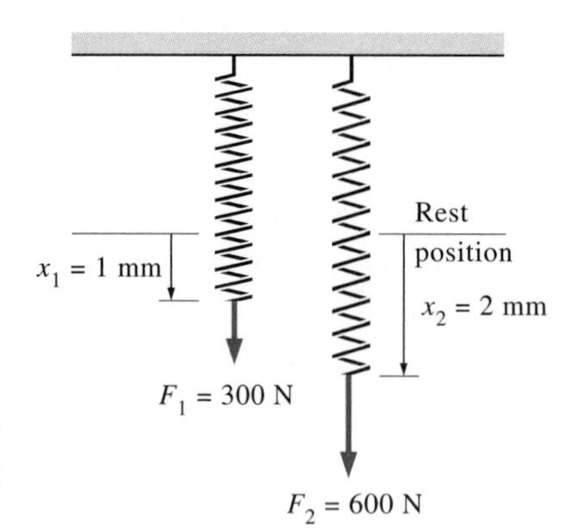


**FIGURE 3-35**
The displacement of a linear spring doubles when the force is doubled.

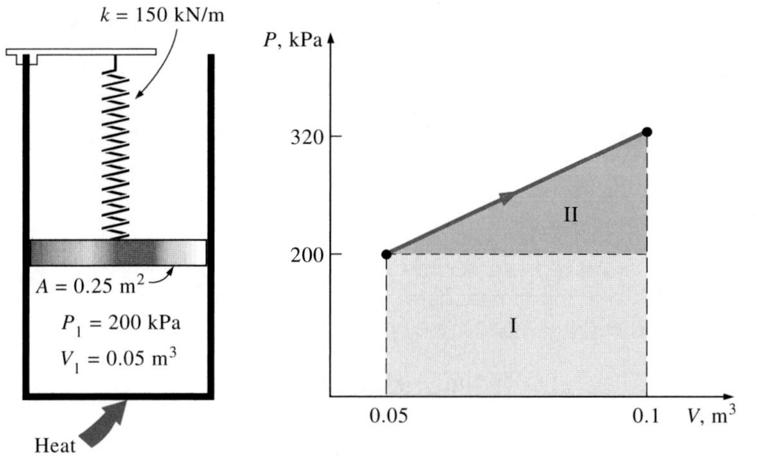


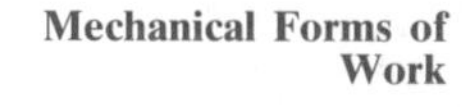

**FIGURE 3-36**
Schematic and *P-V* diagram for Example 3-10.

**EXAMPLE 3-10**

A piston–cylinder device contains $0.05\ \text{m}^3$ of a gas initially at 200 kPa. At this state, a linear spring that has a spring constant of 150 kN/m is touching the piston but exerting no force on it. Now heat is transferred to the gas, causing the piston to rise and to compress the spring until the volume inside the cylinder doubles. If the cross-sectional area of the piston is $0.25\ \text{m}^2$, determine (*a*) the final pressure inside the cylinder, (*b*) the total work done by the gas, and (*c*) the fraction of this work done against the spring to compress it.

**Solution** (*a*) A sketch of the system and the *P-V* diagram of the process are shown in Fig. 3-36. The enclosed volume at the final state is

$$V_2 = 2V_1 = (2)(0.05\ \text{m}^3) = 0.1\ \text{m}^3$$

Then the displacement of the piston (and the spring) becomes

$$x = \frac{\Delta V}{A} = \frac{(0.1 - 0.05)\ \text{m}^3}{0.25\ \text{m}^2} = 0.2\ \text{m}$$

The force applied by the linear spring at the final state is determined from Eq. 3-25 to be

$$F = k_s x = (150\ \text{kN/m})(0.2\ \text{m}) = 30\ \text{kN}$$

The additional pressure applied by the spring on the gas at this state is

$$P = \frac{F}{A} = \frac{30\ \text{kN}}{0.25\ \text{m}^2} = 120\ \text{kPa}$$

Without the spring, the pressure of the gas would remain constant at 200 kPa while the piston is rising. But under the effect of the spring, the pressure rises linearly from 200 kPa to

$$200 + 120 = 320\ \text{kPa}$$

at the final state.

(*b*) An easy way of finding the work done is to plot the process on a *P-V* diagram and find the area under the process curve. From Fig. 3-36 the area under the process curve (a trapezoid) is determined to be

$$|W| = \text{area} = \frac{(200 + 320)\text{ kPa}}{2}[(0.1 - 0.05)\text{ m}^3]\left(\frac{1\text{ kJ}}{1\text{ kPa}\cdot\text{m}^3}\right) = 13\text{ kJ}$$

The sign of the work is determined, by inspection, to be positive since it is done by the system.

(*c*) The work represented by the rectangular area (region I) is done against the piston and the atmosphere, and the work represented by the triangular area (region II) is done against the spring. Thus

$$W_{\text{spring}} = \tfrac{1}{2}[(320 - 200)\text{ kPa}](0.05\text{ m}^3)\left(\frac{1\text{ kJ}}{1\text{ kPa}\cdot\text{m}^3}\right) = 3\text{ kJ}$$

This result could also be obtained from Eq. 3-26:

$$W_{\text{spring}} = \tfrac{1}{2}k_s(x_2^2 - x_1^2) = \tfrac{1}{2}(150\text{ kN/m})[(0.2\text{ m})^2 - 0^2]\left(\frac{1\text{ kJ}}{1\text{ kN}\cdot\text{m}}\right) = 3\text{ kJ}$$

### Nonmechanical Forms of Work

Some work modes encountered in practice are not mechanical in nature. However, these nonmechanical work modes can be treated in a similar manner by identifying a *generalized force F* acting in the direction of a *generalized displacement x*. Then the work associated with the differential displacement under the influence of this force is determined from $\delta W = F \cdot dx$.

Some examples of nonmechanical work modes are **electrical work**, where the generalized force is the *voltage* (the electrical potential) and the generalized displacement is the *electrical charge* as discussed in the last section; **magnetic work**, where the generalized force is the *magnetic field strength* and the generalized displacement is the total *magnetic dipole moment*; and **electrical polarization work**, where the generalized force is the *electric field strength* and the generalized displacement is the *polarization of the medium* (the sum of the electric dipole rotation moments of the molecules). Detailed consideration of these and other nonmechanical work modes can be found in specialized books on these topics.

## 3-5 ■ THE FIRST LAW OF THERMODYNAMICS

So far, we have considered various forms of energy such as heat *Q*, work *W*, and total energy *E* individually, and no attempt has been made to relate them to each other during a process. The *first law of thermodynamics*, also known as *the conservation of energy principle*, provides a sound basis for studying the relationships among the various forms of energy and energy interactions. Based on experimental observations, the first law of thermodynamics states that *energy can be neither created nor destroyed*; *it can only change forms.* Therefore, every bit of energy should be accounted for during a process. The first law cannot be proved mathematically, but no process in nature is known to have violated the first law, and this should be taken as sufficient proof.

We all know that a rock at some elevation possesses some potential energy, and part of this potential energy is converted to kinetic energy as the rock falls (Fig. 3-37). Experimental data show that the decrease in potential energy ($mg\,\Delta z$) exactly equals the increase in kinetic energy $[m(\mathcal{V}_2^2 - \mathcal{V}_1^2)/2]$ when the air resistance is negligible, thus confirming the conservation of energy principle.

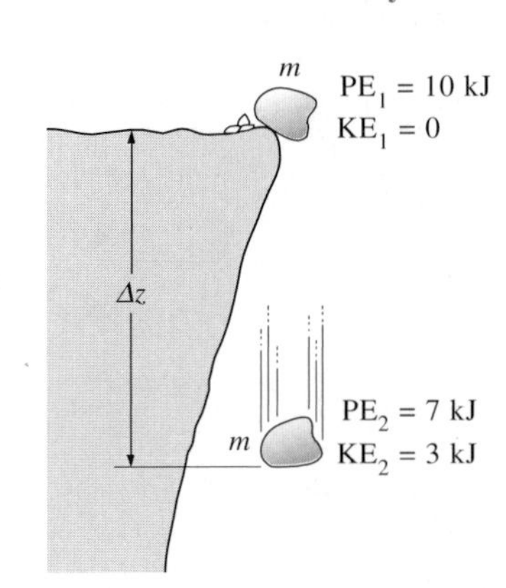


**FIGURE 3-37**
Energy cannot be created or destroyed; it can only change forms.

Consider a system undergoing a series of *adiabatic* processes from a specified state 1 to another specified state 2. Being adiabatic, these processes obviously cannot involve any heat transfer but they may involve several kinds of work interactions. Careful measurements during these experiments indicate the following: *For all adiabatic processes between two specified states of a closed system, the net work done is the same regardless of the nature of the closed system and the details of the process.* Considering that there are an infinite number of ways to perform work interactions under adiabatic conditions, the statement above appears to be very powerful, with a potential for far-reaching implications. This statement, which is largely based on the experiments of Joule in the first half of the nineteenth century, cannot be drawn from any other known physical principle, and is recognized as a fundamental principle. This principle is called the **first law of thermodynamics** or just the **first law**.

A major consequence of the first law is the existence and the definition of the property *total energy E.* Considering that the net work is the same for all adiabatic processes of a closed system between two specified states, the value of the net work must depend on the end states of the system only, and thus it must correspond to a change in a property of the system. This property is the *total energy.* Note that the first law makes no reference to the value of the total energy of a closed system at a state. It simply states that the *change* in the total energy during an adiabatic process must be equal to the net work done. Therefore, any convenient arbitrary value can be assigned to total energy at a specified state to serve as a reference point.

Implicit in the first law statement is the conservation of energy. Although the essence of the first law is the existence of the property *total energy,* the first law is often viewed as a statement of the *conservation of energy* principle. Below we develop the first law or the conservation of energy relation for closed systems with the help of some familiar examples using intuitive arguments.

Let us consider first some processes that involve heat transfer but no work interactions. The potato in the oven that we have discussed previously is a good example for this case (Fig. 3-38). As a result of heat transfer to the potato, the energy of the potato will increase. If we disregard any mass transfer (moisture loss from the potato), the increase in the total energy of the potato becomes equal to the amount of heat

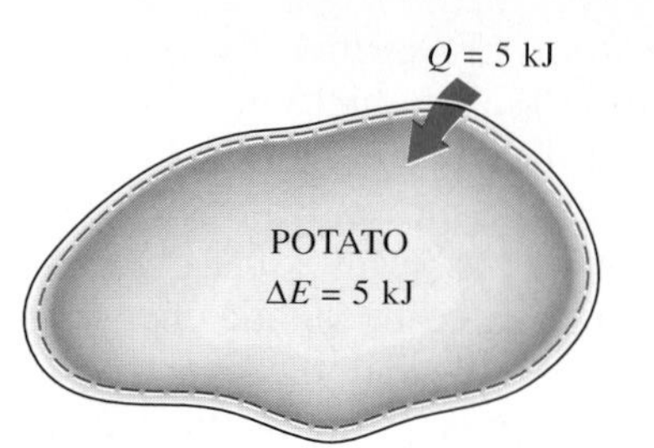


**FIGURE 3-38**
The increase in the energy of a potato in an oven is equal to the amount of heat transferred to it.

transfer. That is, if 5 kJ of heat is transferred to the potato, the energy increase of the potato will also be 5 kJ. Therefore, the conservation of energy principle for this case can be expressed as $Q = \Delta E$.

As another example, consider the heating of water in a pan on top of a range (Fig. 3-39). If 15 kJ of heat is transferred to the water from the heating element and 3 kJ of it is lost from the water to the surrounding air, the increase in energy of the water will be equal to the net heat transfer to the water, which is 12 kJ. That is, $Q = Q_{\text{net}} = \Delta E$.

The above conclusions can be summarized as follows: *In the absence of any work interactions between a system and its surroundings, the amount of net heat transfer is equal to the change in energy of a closed system.* That is,

$$Q = \Delta E \qquad \text{when } W = 0$$

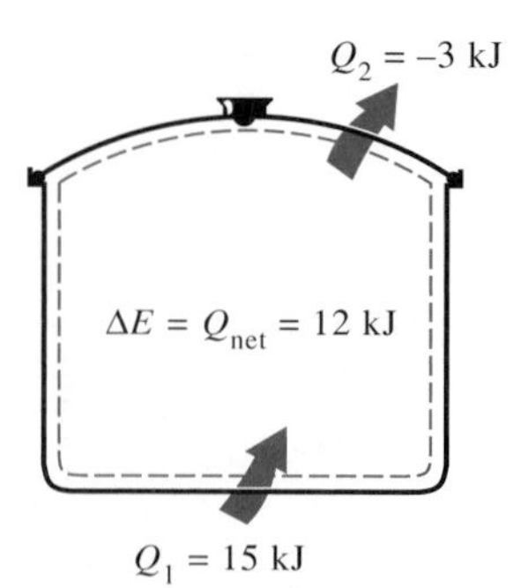


**FIGURE 3-39**
In the absence of any work interactions, energy change of a system is equal to the net heat transfer.

Now consider a well-insulated (i.e., adiabatic) room heated by an electric heater as our system (Fig. 3-40). As a result of electrical work done, the energy of the system will increase. Since the system is adiabatic and cannot have any heat interactions with the surroundings ($Q = 0$), the conservation of energy principle dictates that the electrical work done on the system must equal the increase in energy of the system. That is, $-W_e = \Delta E$.

The negative sign is due to the sign convention that work done on a system is negative. This ensures that work done on a system increases the energy of the system and work done by a system decreases it.

Now let us replace the electric heater with a paddle wheel (Fig. 3-41). As a result of the stirring process, the energy of the system will increase. Again since there is no heat interaction between the system and its surroundings ($Q = 0$), the paddle-wheel work done on the system must show up as an increase of the system. That is, $-W_{\text{pw}} = \Delta E$.

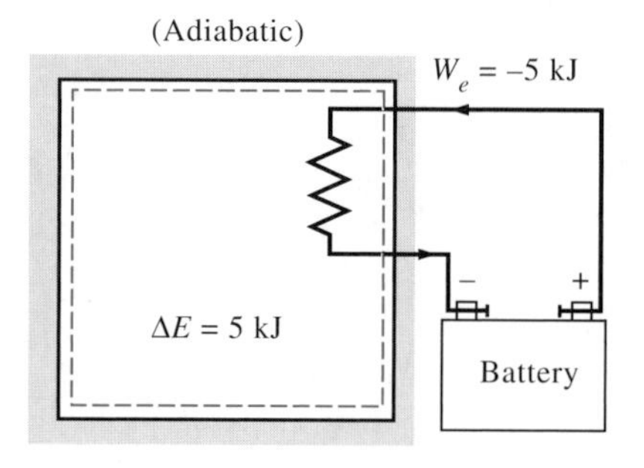


**FIGURE 3-40**
The work (electrical) done on an adiabatic system is equal to the increase in the energy of the system.

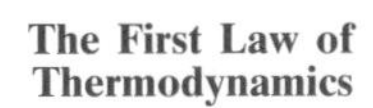

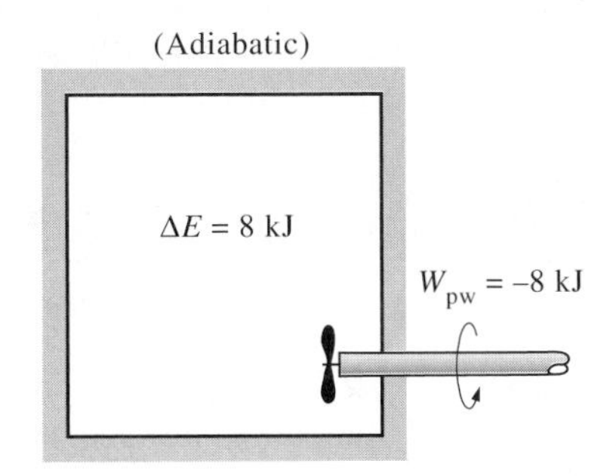


**FIGURE 3-41**

The work (shaft) done on an adiabatic system is equal to the increase in the energy of the system.

Many of you have probably noticed that the temperature of air rises when it is compressed (Fig. 3-42). This is because energy is added to the air in the form of boundary work. In the absence of any heat transfer ($Q = 0$), the entire boundary work will be stored in the air as part of its total energy. The conservation of energy principle again requires that $-W_b = \Delta E$.

It is clear from the foregoing examples that *for adiabatic processes, the amount of work done is equal to the change in the energy of a closed system.* That is,

$$-W = \Delta E \qquad \text{when } Q = 0$$

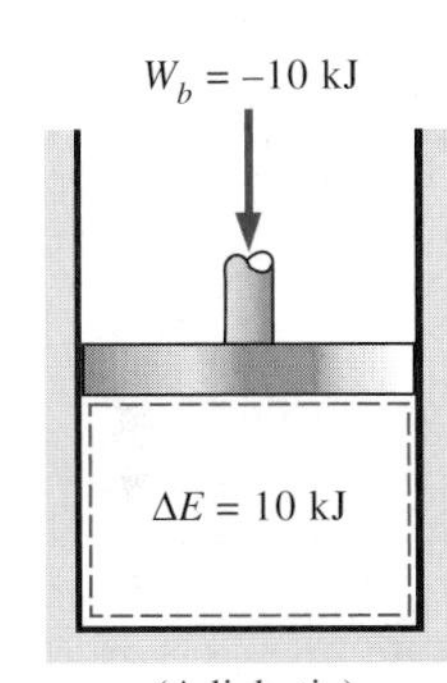


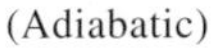


**FIGURE 3-42**

The work (boundary) done on an adiabatic system is equal to the increase in the energy of the system.

Now we are in a position to consider simultaneous heat and work interactions. As you may have already guessed, when a system involves both heat and work interactions during a process, their contributions are simply added. That is, if a system receives 12 kJ of heat while a paddle wheel does 6 kJ of work on the system, the net increase in energy of the system for this process will be 18 kJ (Fig. 3-43).

To generalize our conclusions, the **first law of thermodynamics**, or the **conservation of energy principle** for a closed system or a fixed mass, may be expressed as follows:

$$\begin{bmatrix} \text{net energy transfer} \\ \text{to (or from) the system} \\ \text{as heat and work} \end{bmatrix} = \begin{bmatrix} \text{net increase (or decrease)} \\ \text{in the total energy} \\ \text{of the system} \end{bmatrix}$$

or

$$Q - W = \Delta E \qquad \text{(kJ)} \tag{3-27}$$

where

$Q$ = net heat transfer across system boundaries ($= \sum Q_{\text{in}} - \sum Q_{\text{out}}$)
$W$ = net work done in all forms ($= \sum W_{\text{out}} - \sum W_{\text{in}}$)
$\Delta E$ = net change in total energy of system, $E_2 - E_1$

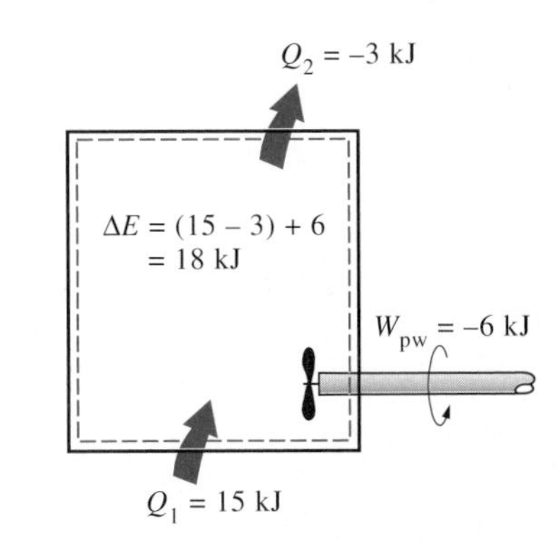


**FIGURE 3-43**

The energy change of a system during a process is equal to the *net* work and heat transfer between the system and its surroundings.

As discussed in Chap. 1, the total energy $E$ of a system is considered to consist of three parts: internal energy $U$, kinetic energy KE, and potential energy PE. Then the change in total energy of a system during a process can be expressed as the sum of the changes in its internal, kinetic, and potential energies:

$$\Delta E = \Delta U + \Delta \text{KE} + \Delta \text{PE} \qquad \text{(kJ)} \tag{3-28}$$

Substituting this relation into Eq. 3-27, we obtain

$$Q - W = \Delta U + \Delta KE + \Delta PE \qquad \text{(kJ)} \tag{3-29}$$

where

$$\Delta U = m(u_2 - u_1)$$
$$\Delta KE = \tfrac{1}{2}m(\mathcal{V}_2^2 - \mathcal{V}_1^2)$$
$$\Delta PE = mg(z_2 - z_1)$$

Stationary Systems

$z_1 = z_2 \rightarrow \Delta PE = 0$

$\mathcal{V}_1 = \mathcal{V}_2 \rightarrow \Delta KE = 0$

$\Delta E = \Delta U$

**FIGURE 3-44**
For stationary systems, $\Delta KE = \Delta PE = 0$; thus, $\Delta E = \Delta U$.

Most closed systems encountered in practice are stationary, i.e., they do not involve any changes in their velocity or the elevation of their center of gravity during a process (Fig. 3-44). Thus, for **stationary closed systems**, the changes in kinetic and potential energies are negligible (that is, $\Delta KE = \Delta PE = 0$), and the first-law relation reduces to

$$Q - W = \Delta U \qquad \text{(kJ)} \tag{3-30}$$

If the initial and final states are specified, the internal energies $u_1$ and $u_2$ can easily be determined from property tables or some thermodynamic relations.

Sometimes it is convenient to consider the work term in two parts: $W_{\text{other}}$ and $W_b$, where $W_{\text{other}}$ represents all forms of work except the boundary work. (This distinction has important bearings with regard to the second law of thermodynamics, as is discussed in later chapters.) Then the first law takes the following form:

$$Q - W_{\text{other}} - W_b = \Delta E \qquad \text{(kJ)} \tag{3-31}$$

It is extremely important that the *sign convention* be observed for heat and work interactions. Heat flow to a system and work done by a system are positive, and heat flow from a system and work done on a system are negative. A system may involve more than one form of work during a process. The only form of work whose sign we do not need to be concerned with is the boundary work $W_b$ as defined by Eq. 3-18. Boundary work calculated by using Eq. 3-18 will always have the correct sign. The signs of other forms of work are determined by inspection.

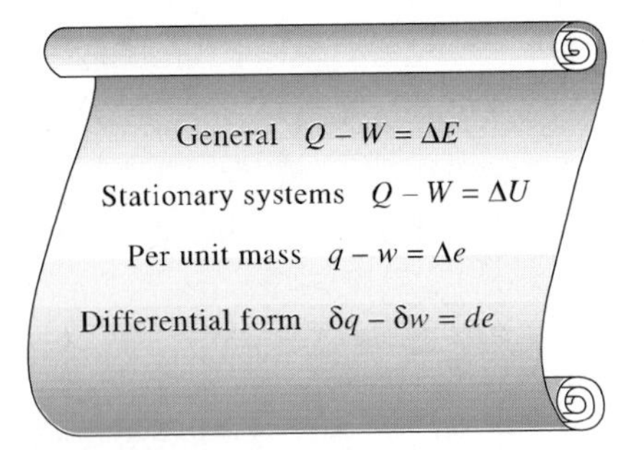

General $Q - W = \Delta E$
Stationary systems $Q - W = \Delta U$
Per unit mass $q - w = \Delta e$
Differential form $\delta q - \delta w = de$


**FIGURE 3-45**
Various forms of the first-law relation for closed systems.

### Other Forms of the First-Law Relation

The first-law relation for closed systems can be written in various forms (Fig. 3-45). Dividing Eq. 3-27 by the mass of the system, for example, gives the first-law relation on a **unit-mass** basis as

$$q - w = \Delta e \qquad \text{(kJ/kg)} \tag{3-32}$$

The **rate form** of the first law is obtained by dividing Eq. 3-27 by the time interval $\Delta t$ and taking the limit as $\Delta t \rightarrow 0$. This yields

$$\dot{Q} - \dot{W} = \frac{dE}{dt} \qquad \text{(kW)} \tag{3-33}$$

where $\dot{Q}$ is the rate of net heat transfer, $\dot{W}$ is the power, and $dE/dt$ is the rate of change of total energy.

Equation 3-27 can be expressed in **differential form** as

$$\delta Q - \delta W = dE \qquad \text{(kJ)} \qquad (3\text{-}34)$$

$$\delta q - \delta w = de \qquad \text{(kJ/kg)} \qquad (3\text{-}35)$$

For a **cyclic process**, the initial and final states are identical, and therefore $\Delta E = E_2 - E_1 = 0$. Then the first-law relation for a cycle simplifies to

$$Q - W = 0 \qquad \text{(kJ)}$$

That is, the net heat transfer and the net work done during a cycle must be equal (Fig. 3-46).

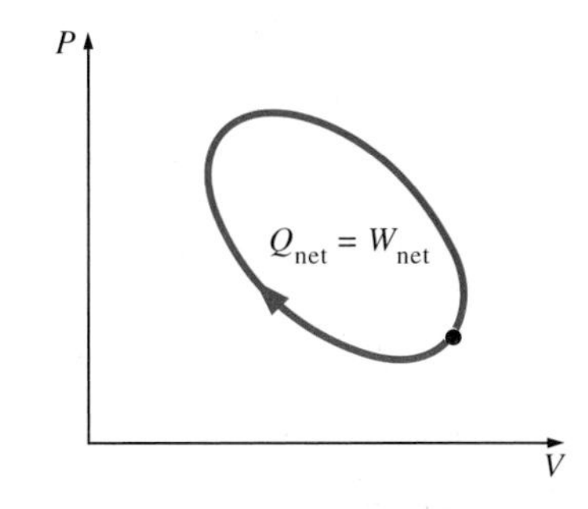


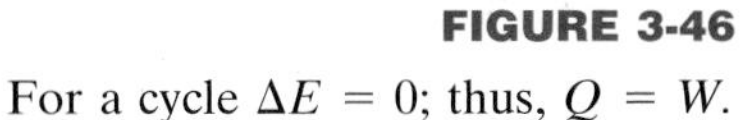

**FIGURE 3-46**

For a cycle $\Delta E = 0$; thus, $Q = W$.

As energy quantities, heat and work are not that different, and you probably wonder why we keep distinguishing them. After all, the change in the energy content of a system is equal to the amount of energy that crosses the system boundaries, and it makes no difference whether the energy crosses the boundary as heat or work. It seems as if the first-law relations would be much simpler if we had just one quantity which we could call *energy interaction* to represent both heat and work. Well, from the first-law point of view, heat and work are not different at all, and the first law can simply be expressed as

$$E_{in} - E_{out} = \Delta E \qquad (3\text{-}36)$$

where $E_{in}$ and $E_{out}$ are the total energy that *enters* and *leaves* the system, respectively, during a process. But from the second-law point of view, heat and work are very different, as you will see in later chapters.

**EXAMPLE 3-11**

A rigid tank contains a hot fluid that is cooled while being stirred by a paddle wheel. Initially, the internal energy of the fluid is 800 kJ. During the cooling process, the fluid loses 500 J of heat, and the paddle wheel does 100 kJ of work on the fluid. Determine the final internal energy of the fluid. Neglect the energy stored in the paddle wheel.

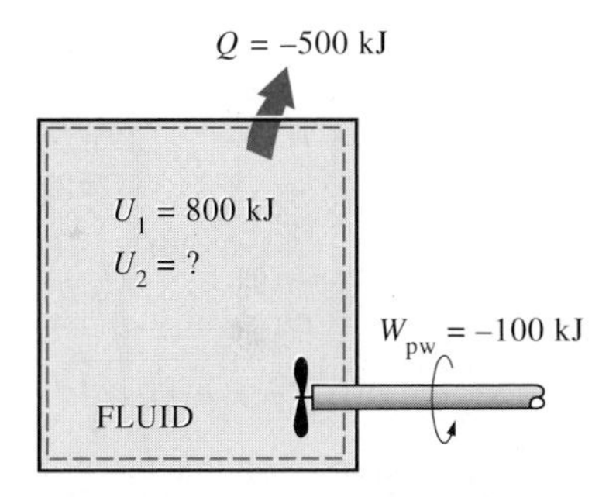


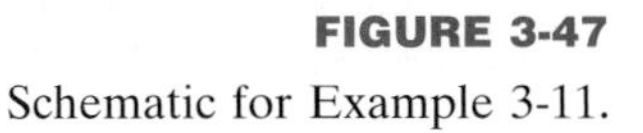

**FIGURE 3-47**

Schematic for Example 3-11.

**Solution** We choose the fluid in the tank as our system. The system boundaries are indicated in Fig. 3-47.

Since no mass crosses the boundary, this is a closed system (a fixed mass). There is no mention of motion. Therefore, we assume that the closed system is stationary and that the potential and kinetic energy changes are zero. By applying the conservation of energy principle as given by Eq. 3-29 to this process, $U_2$ is determined to be

$$Q - W = \Delta U + \Delta KE^{\nearrow 0} + \Delta PE^{\nearrow 0}$$

$$= U_2 - U_1$$

$$-500 \text{ kJ} - (-100 \text{ kJ}) = U_2 - 800 \text{ KJ}$$

$$U_2 = 400 \text{ kJ}$$

Note that the heat transfer is negative since it is from the system; so is the work, since it is done on the system.

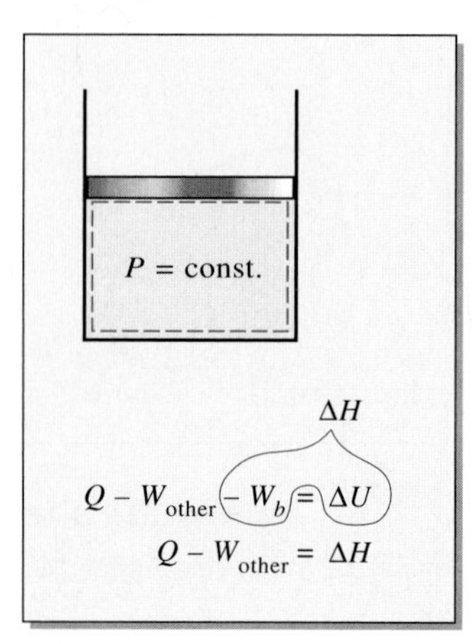


**FIGURE 3-48**

For a closed system undergoing a quasi-equilibrium $P$ = constant process, $\Delta U + W_b = \Delta H$.

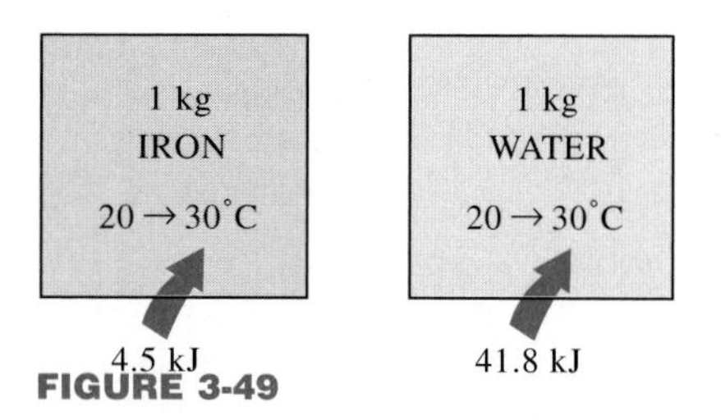


**FIGURE 3-49**

It takes different amounts of energy to raise the temperature of different substances by the same amount.

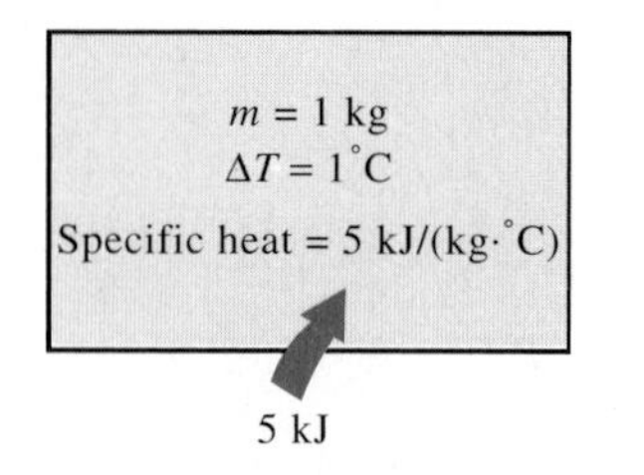


**FIGURE 3-50**

Specific heat is the energy required to raise the temperature of a unit mass of a substance by one degree in a specified way.

**EXAMPLE 3-12**

Consider the quasi-equilibrium expansion or compression of a gas in piston cylinder device. Show that the boundary work $W_b$ and the change internal energy $\Delta U$ in the first law relation can be combined into one term, $\Delta H$, for such a system undergoing a constant-pressure process.

**Solution** Neglecting the changes in kinetic and potential energies and expressing the work as the sum of boundary and other forms of work, Eq. 3-29 simplifies to

$$Q - W = \Delta U + \cancel{\Delta \mathrm{KE}}^{\,0} + \cancel{\Delta \mathrm{PE}}^{\,0}$$

$$Q - W_{\text{other}} - W_b = U_2 - U_1$$

For a constant-pressure process, the boundary work is given by Eq. 3-19 as $W_b = P_0(V_2 - V_1)$. Substituting this into the above relation gives

$$Q - W_{\text{other}} - P_0(V_2 - V_1) = U_2 - U_1$$

But $\quad P_0 = P_2 = P_1 \longrightarrow Q - W_{\text{other}} = (U_2 + P_2V_2) - (U_1 + P_1V_1)$

Also $H = U + PV$, and thus

$$Q - W_{\text{other}} = H_2 - H_1 \qquad \text{(kJ)} \tag{3-37}$$

which is the desired relation (Fig. 3-48). *This equation is very convenient to use in the analysis of closed systems undergoing a constant-pressure quasi-equilibrium process since the boundary work is automatically taken care of by the enthalpy terms, and one no longer needs to determine it separately.*

## 3-6 ■ SPECIFIC HEATS

We know from experience that it takes different amounts of energy to raise the temperature of identical masses of different substances by one degree. For example, we need about 4.5 kJ of energy to raise the temperature of 1 kg iron from 20 to 30°C, whereas it takes about 9 times this energy (41.8 kJ to be exact) to raise the temperature of 1 kg of liquid water by the same amount (Fig. 3-49). Therefore, it is desirable to have a property that will enable us to compare the energy storage capabilities of various substances. This property is the specific heat.

The **specific heat** is defined as *the energy required to raise the temperature of a unit mass of a substance by one degree* (Fig. 3-50). In general, this energy will depend on how the process is executed. In thermodynamics, we are interested in two kinds of specific heats: **specific heat at constant volume** $C_v$ and **specific heat at constant pressure** $C_p$.

Physically, the specific heat at constant volume $C_v$ can be viewed as *the energy required to raise the temperature of the unit mass of a substance by one degree as the volume is maintained constant.* The energy required to do the same as the pressure is maintained constant is the specific heat

at constant pressure $C_p$. This is illustrated in Fig. 3-51. The specific heat at constant pressure $C_p$ is always greater than $C_v$ because at constant pressure the system is allowed to expand and the energy for this expansion work must also be supplied to the system.

Now we will attempt to express the specific heats in terms of other thermodynamic properties. First, consider a stationary closed system undergoing a constant-volume process ($w_b = 0$). The first-law relation for this process can be expressed in the differential form as

$$\delta q - \delta w_{\text{other}} = du$$

The left-hand side of this equation ($\delta q - \delta w_{\text{other}}$) represents the amount of energy transferred to the system in the form of heat and/or work. From the definition of $C_v$, this energy must be equal to $C_v\,dT$, where $dT$ is the differential change in temperature. Thus,

$$C_v\,dT = du \qquad \text{at constant volume}$$

or

$$C_v = \left(\frac{\partial u}{\partial T}\right)_v \tag{3-38}$$

Similarly, an expression for the specific heat at constant pressure $C_p$ can be obtained by considering a constant-pressure process ($w_b + \Delta u = \Delta h$). It yields

$$C_p = \left(\frac{\partial h}{\partial T}\right)_p \tag{3-39}$$

Equations 3-38 and 3-39 are the defining equations for $C_v$ and $C_p$, and their interpretation is given in Fig. 3-52.

Note that $C_v$ and $C_p$ are expressed in terms of other properties; thus, they must be properties themselves. Like any other property, the specific heats of a substance depend on the state, which, in general, is specified by two independent, intensive properties. That is, the energy required to raise the temperature of a substance by one degree will be different at different temperatures and pressures (Fig. 3-53). But this difference is usually not very large.

A few observations can be made from Eqs. 3-38 and 3-39. First, these equations are *property relations* and as such *are independent of the type of processes.* They are valid for *any* substance undergoing *any* process. The only relevance $C_v$ has to a constant-volume process is that $C_v$ happens to be the energy transferred to a system during a constant-volume process

V = const.
$m$ = 1 kg
$\Delta T$ = 1°C
$C_v = 3.13\ \frac{\text{kJ}}{\text{kg}\cdot{}^\circ\text{C}}$
3.13 kJ

$P$ = const.
$m$ = 1 kg
$\Delta T$ = 1°C
$C_p = 5.2\ \frac{\text{kJ}}{\text{kg}\cdot{}^\circ\text{C}}$
5.2 kJ
(1)
(2)

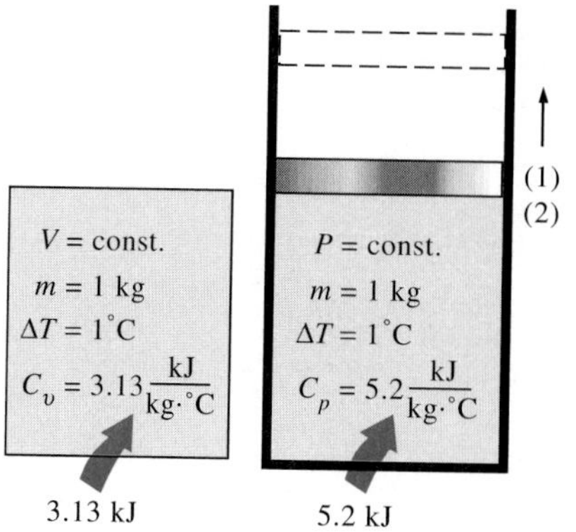

**FIGURE 3-51**
Constant-volume and constant-pressure specific heats $C_v$ and $C_p$ (values given are for helium gas).

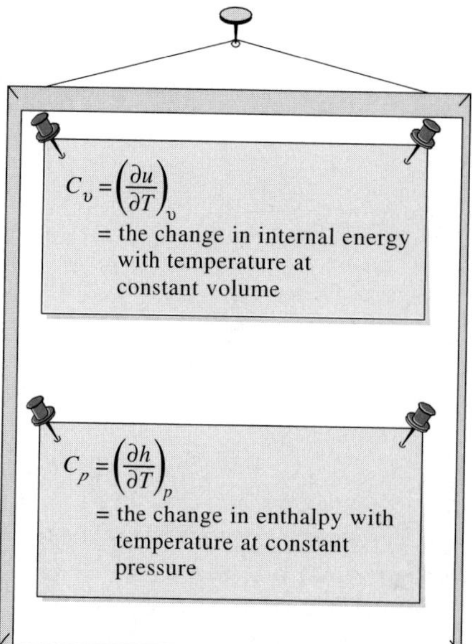


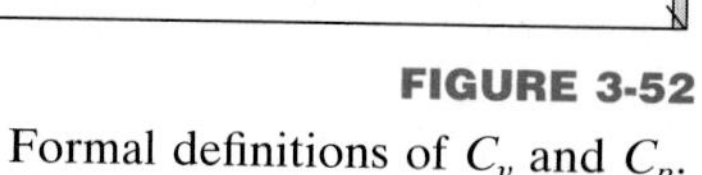

**FIGURE 3-52**
Formal definitions of $C_v$ and $C_p$.

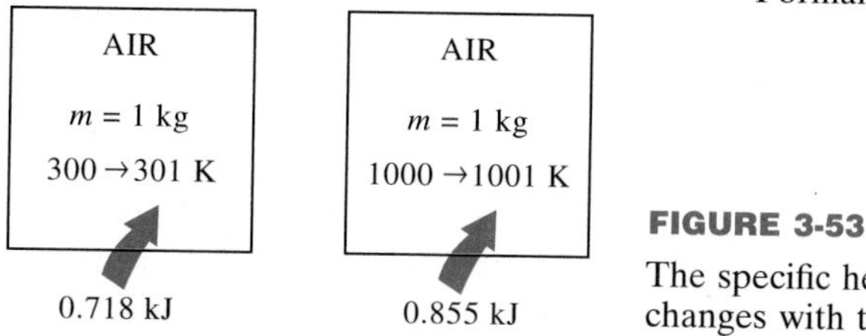


**FIGURE 3-53**
The specific heat of a substance changes with temperature.

per unit mass per unit degree rise in temperature. This is how the values of $C_v$ are determined. This is also how the name *specific heat at constant volume* originated. Likewise, the energy transferred to a system per unit mass per unit temperature rise during a constant-pressure process happens to be equal to $C_p$. This is how the values of $C_p$ can be determined and also explains the origin of the name *specific heat at constant pressure.*

Another observation that can be made from Eqs. 3-38 and 3-39 is that $C_v$ is related to the changes in *internal energy* and $C_p$ to the changes in *enthalpy*. In fact, it would be more proper to define $C_v$ as *the change in the specific internal energy of a substance per unit change in temperature at constant volume.* Likewise, $C_p$ can be defined as *the change in the specific enthalpy of a substance per unit change in temperature at constant pressure.* In other words, $C_v$ is a measure of the variation of internal energy of a substance with temperature, and $C_p$ is a measure of the variation of enthalpy of a substance with temperature.

Both the internal energy and enthalpy of a substance can be changed by the transfer of *energy* in any form, with heat being only one of them. Therefore, the term *specific energy* is probably more appropriate than the term *specific heat,* which implies that energy is transferred (and even stored) in the form of heat.

A common unit for specific heats is kJ/(kg · °C) or kJ/(kg · K). Notice that these two units are *identical* since $\Delta T(°C) = \Delta T(K)$, and 1°C change in temperature is equivalent to a change of 1 K (see Sec. 1-10). The specific heats are sometimes given on a *molar basis.* They are then denoted by $\bar{C}_v$ and $\bar{C}_p$ and have the unit kJ/(kmol · °C) or kJ/(kmol · K).

## 3-7 ■ INTERNAL ENERGY, ENTHALPY, AND SPECIFIC HEATS OF IDEAL GASES

In Chap. 2, we defined an ideal gas as a gas whose temperature, pressure, and specific volume are related by

$$Pv = RT$$

It has been demonstrated mathematically and experimentally (Joule, 1843) that for an ideal gas the internal energy is a function of the temperature only. That is,

$$u = u(T) \tag{3-40}$$

In his classical experiment, Joule submerged two tanks connected with a pipe and a valve in a water bath, as shown in Fig. 3-54. Initially, one tank contained air at a high pressure, and the other tank was evacuated. When thermal equilibrium was attained, he opened the valve to let air pass from one tank to the other until the pressures equalized. Joule observed no change in the temperature of the water bath and assumed that no heat was transferred to or from the air. Since there was also no work done, he concluded that the internal energy of the air did

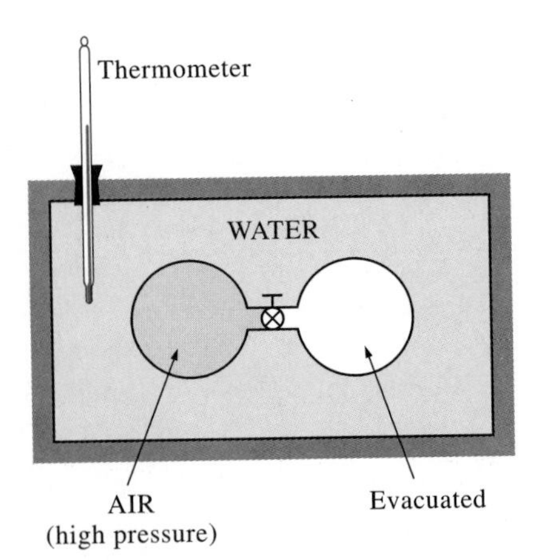


**FIGURE 3-54**
Schematic of the experimental apparatus used by Joule.

not change even though the volume and the pressure changed. Therefore, he reasoned, the internal energy is a function of temperature only and not a function of pressure or specific volume. (Joule later showed that for gases that deviate significantly from ideal-gas behavior, the internal energy is not a function of temperature alone.)

Using the definition of enthalpy and the equation of state of an ideal gas, we have

$$\left.\begin{aligned} h &= u + Pv \\ Pv &= RT \end{aligned}\right\} \qquad h = u + RT$$

Since $R$ is constant and $u = u(T)$, it follows that the enthalpy of an ideal gas is also a function of temperature only:

$$h = h(T) \tag{3-41}$$

Since $u$ and $h$ depend only on temperature for an ideal gas, the specific heats $C_v$ and $C_p$ also depend, at most, on temperature only. Therefore, *at a given temperature, $u$, $h$, $C_v$, and $C_p$ of an ideal gas will have fixed values regardless of the specific volume or pressure* (Fig. 3-55). Thus for ideal gases, the partial derivatives in Eqs. 3-38 and 3-39 can be replaced by ordinary derivatives. Then the differential changes in the internal energy and enthalpy of an ideal gas can be expressed as

$$du = C_v(T)\,dT \tag{3-42}$$

$$dh = C_p(T)\,dT \tag{3-43}$$

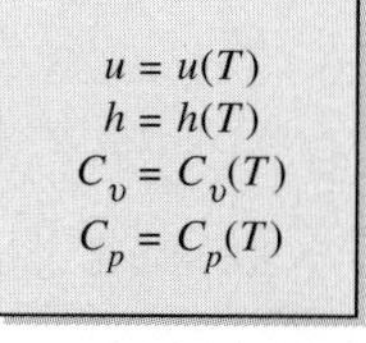


**FIGURE 3-55**
For ideal gases, $u$, $h$, $C_v$, and $C_p$ vary with temperature only.

The change in internal energy or enthalpy for an ideal gas during a process from state 1 to state 2 is determined by integrating these equations:

$$\Delta u = u_2 - u_1 = \int_1^2 C_v(T)\,dT \qquad \text{(kJ/kg)} \tag{3-44}$$

and

$$\Delta h = h_2 - h_1 = \int_1^2 C_p(T)\,dT \qquad \text{(kJ/kg)} \tag{3-45}$$

To carry out these integrations, we need to have relations for $C_v$ and $C_p$ as functions of temperature.

At low pressures, all real gases approach ideal-gas behavior, and therefore their specific heats depend on temperature only. The specific heats of real gases at low pressures are called *ideal-gas specific heats,* or *zero-pressure specific heats,* and are often denoted $C_{p0}$ and $C_{v0}$. Accurate analytical expressions for ideal-gas specific heats, based on direct measurements or calculations from statistical behavior or molecules, are available and are given as a third-degree polynomial in the Appendix (Table A-2*c*) for several gases. A plot of $\bar{C}_{p0}(T)$ data for some common gases is given in Fig. 3-56.

The use of ideal-gas specific heat data is limited to low pressures, but these data can also be used at moderately high pressures with reasonable accuracy as long as the gas does not deviate from ideal-gas behavior significantly.

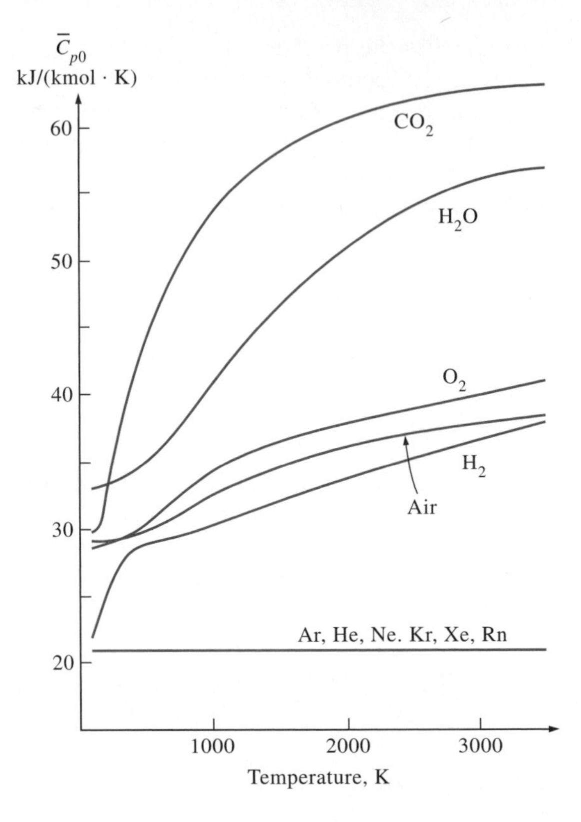


**FIGURE 3-56**

Ideal-gas constant-pressure specific heats for some gases (see Table A-2$c$ for $\bar{C}_{p0}$ equations).

Some observations can be made from Fig. 3-56. First, the specific heats of gases with complex molecules (molecules with two or more atoms) are higher and increase with temperature. Also the variation of specific heats with temperature is smooth and may be approximated as linear over small temperature intervals (a few hundred degrees or less). Then the specific heat functions in Eqs. 3-44 and 3-45 can be replaced by the constant average specific heat values. Now the integrations in these equations can be performed, yielding

$$u_2 - u_1 = C_{v,\text{av}}(T_2 - T_1) \qquad \text{(kJ/kg)} \tag{3-46}$$

$$h_2 - h_1 = C_{p,\text{av}}(T_2 - T_1) \qquad \text{(kJ/kg)} \tag{3-47}$$

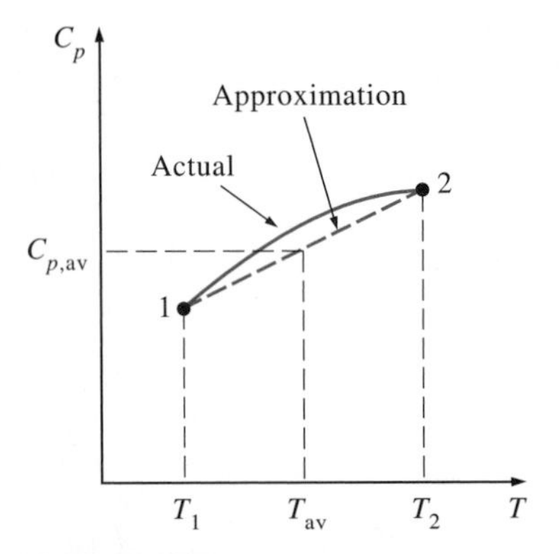


**FIGURE 3-57**

For small temperature intervals, the specific heats may be assumed to vary linearly with temperature.

The specific heat values for some common gases are listed as a function of temperature in Table A-2$b$. The average specific heats $C_{p,\text{av}}$ and $C_{v,\text{av}}$ are evaluated from this table at the average temperature $(T_1 + T_2)/2$, as shown in Fig. 3-57. If the final temperature $T_2$ is not known, the specific heats may be evaluated at $T_1$ or at anticipated average temperature. Then $T_2$ can be determined by using these specific heat values. The value of $T_2$ can be refined, if necessary, by evaluating the specific heats at the new average temperature.

Another way of determining the average specific heats is to evaluate them at $T_1$ and $T_2$ and then take their average. Usually both methods give reasonably good results, and one is not necessarily better than the other.

Another observation that can be made from Fig. 3-56 is that the ideal-gas specific heats of *monatomic gases* such as argon, neon, and helium remain constant over the entire temperature range. Thus, $\Delta u$ and $\Delta h$ of monatomic gases can easily be evaluated from Eqs. 3-46 and 3-47.

Note that the $\Delta u$ and $\Delta h$ relations given above are not restricted to any kind of process. They are valid for all processes. The presence of the constant-volume specific heat $C_v$ in an equation should not lead one to believe that this equation is valid for a constant-volume process only. On the contrary, the relation $\Delta u = C_{u,\text{av}}\,\Delta T$ is valid for *any* ideal gas undergoing *any* process (Fig. 3-58). A similar argument can be given for $C_p$ and $\Delta h$.

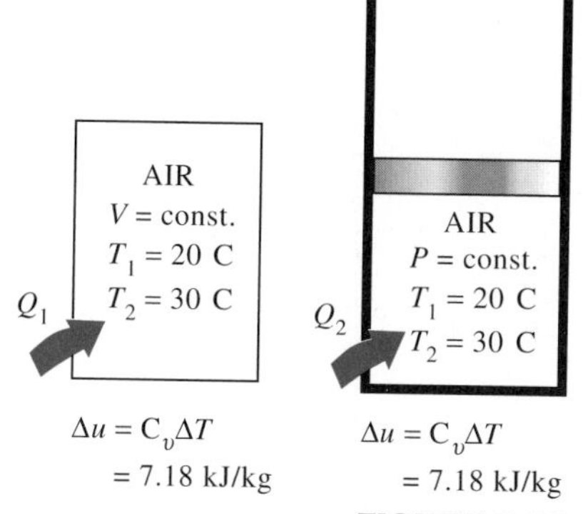


**FIGURE 3-58**

The relation $\Delta u = C_v\,\Delta T$ is valid for *any* kind of process, constant-volume or not.

## Specific-Heat Relations of Ideal Gases

A special relationship between $C_p$ and $C_v$ for ideal gases can be obtained by differentiating the relation $h = u + RT$, which yields

$$dh = du + R\,dT$$

Replacing $dh$ by $C_p\,dT$ and $du$ by $C_v\,dT$ and dividing the resulting expression by $dT$. We obtain

$$C_p = C_v + R \qquad [\text{kJ/(kg}\cdot\text{K)}] \tag{3-48}$$

This is an important relationship for ideal gases since it enables us to determine $C_v$ from a knowledge of $C_p$ and the gas constant $R$.

When the specific heats are given on a molar basis, $R$ in the above equation should be replaced by the universal gas constant $R_u$ (Fig. 3-59). That is,

$$\bar{C}_p = \bar{C}_v + R_u \qquad [\text{kJ/(kmol}\cdot\text{K)}] \tag{3-49}$$

At this point, we introduce another ideal-gas property called the **specific heat ratio** $k$, defined as

$$k = \frac{C_p}{C_v} \tag{3-50}$$

The specific heat ratio also varies with temperature, but this variation is very mild. For monatomic gases, its value is essentially constant at 1.667.

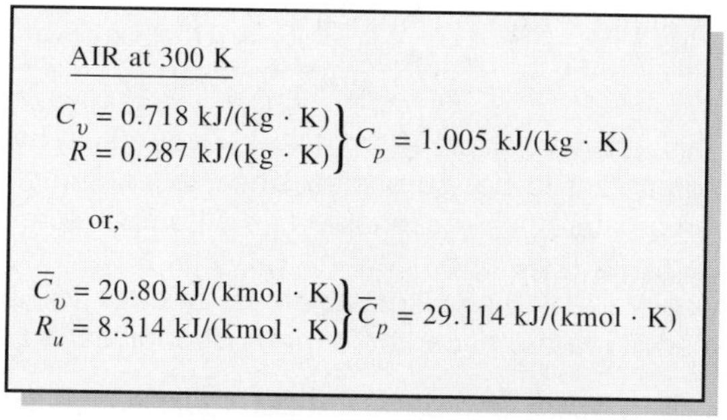

AIR at 300 K

$$\left.\begin{aligned} C_v &= 0.718 \text{ kJ/(kg}\cdot\text{K)} \\ R &= 0.287 \text{ kJ/(kg}\cdot\text{K)} \end{aligned}\right\} C_p = 1.005 \text{ kJ/(kg}\cdot\text{K)}$$

or,

$$\left.\begin{aligned} \bar{C}_v &= 20.80 \text{ kJ/(kmol}\cdot\text{K)} \\ R_u &= 8.314 \text{ kJ/(kmol}\cdot\text{K)} \end{aligned}\right\} \bar{C}_p = 29.114 \text{ kJ/(kmol}\cdot\text{K)}$$

**FIGURE 3-59**

The $C_p$ of an ideal gas can be determined from a knowledge of $C_v$ and $R$.

Many diatomic gases, including air, have a specific heat ratio of about 1.4 at room temperature.

**EXAMPLE 3-13**

Air at 300 K and 200 kPa is heated at constant pressure to 600 K. Determine the change in internal energy of air per unit mass, using (*a*) the functional form of the specific heat (Table A-2*c*), and (*b*) the average specific heat value (Table A-2*b*).

**Solution** At specified conditions, air can be considered to be an ideal gas since it is at a high temperature and low pressure relative to its critical-point values ($T_{cr} = -147°C$, $P_{cr} = 3390$ kPa for nitrogen, the main constituent of air). The internal energy change $\Delta u$ of ideal gases depends on the initial and final temperatures only, and not on the type of process. Thus, the solution given below is valid for any kind of process.

(*a*) The change in internal energy of air, using the functional form of the specific heat, is determined as follows. The $\bar{C}_p(T)$ of air is given in Table A-2*c* in the form of a third-degree polynomial expressed as

$$\bar{C}_p(T) = a + bT + cT^2 + dT^3$$

where $a = 28.11$, $b = 0.1967 \times 10^{-2}$, $c = 0.4802 \times 10^{-5}$, and $d = -1.966 \times 10^{-9}$. From Eq. 3-49,

$$\bar{C}_v(T) = \bar{C}_p - R_u = (a - R_u) + bT + cT^2 + dT^3$$

From Eq. 3-44,

$$\Delta\bar{u} = \int_1^2 \bar{C}_v(T)\,dT$$

$$= \int_{T_1}^{T_2} [(a - R_u) + bT + cT^2 + dT^3]\,dT$$

Performing the integration and substituting the values, we obtain

$$\Delta\bar{u} = 6447.15 \text{ kJ/kmol}$$

The change in the internal energy on a unit-mass basis is determined by dividing this value by the molar mass of air (Table A-1):

$$\Delta u = \frac{\Delta\bar{u}}{M} = \frac{6447.15 \text{ kJ/kmol}}{28.97 \text{ kg/kmol}} = 222.55 \text{ kJ/kg}$$

This result is sufficiently accurate, and can be viewed as the exact result.

(*b*) The average value of the constant-volume specific heat $C_{v,av}$ is determined from Table A-2*b* at the average temperature $(T_1 + T_2)/2 = 450$ K to be

$$C_{v,av} = C_{v\,@\,450\,K} = 0.733 \text{ kJ/(kg}\cdot\text{K)}$$

Thus,

$$\Delta u = C_{v,av}(T_2 - T_1) = [0.733 \text{ kJ/(kg}\cdot\text{K)}][(600 - 300)\text{ K}]$$
$$= 219.9 \text{ kJ/kg}$$

This answer differs from the above result by only 1.2 percent. This close agreement is not surprising since the assumption that $C_v$ varies linearly with temperature is a reasonable one at temperature intervals of only a few hundred degrees. If we had used the $C_v$ value at $T_1 = 300$ K instead of at $T_{av}$, the result would be 215.4 kJ/kg, which is in error by about 3 percent. Errors of this magnitude are acceptable for most engineering purposes.

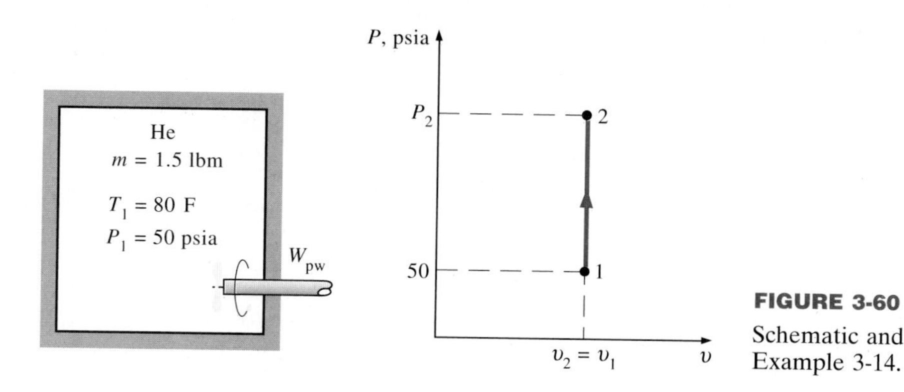


**FIGURE 3-60**
Schematic and $P$-$v$ diagram for Example 3-14.

**EXAMPLE 3-14**

An insulated rigid tank initially contains 1.5 lbm of helium at 80°F and 50 psia. A paddle wheel with a power rating of 0.02 hp is operated within the tank for 30 min. Determine (*a*) the final temperature and (*b*) the final pressure of the helium gas.

**Solution** We take the helium gas within the tank as our system, which is a stationary closed system. A sketch of the system and the $P$-$v$ diagram of the process are shown in Fig. 3-60. The helium gas at the specified conditions can be considered to be an ideal gas since it is at a very high temperature relative to its critical-point temperature ($T_{cr} = -451°F$ for helium).

(*a*) The amount of paddle-wheel work done on the system is

$$W_{pw} = \dot{W}_{pw}\,\Delta t = (-0.02\text{ hp})(0.5\text{ h})\left(\frac{2545\text{ Btu/h}}{1\text{ hp}}\right) = -25.45\text{ Btu}$$

Since we have no moving boundaries, the boundary work is zero ($W_b = 0$), and the heat losses can be neglected since the system is well insulated ($Q = 0$). The system is assumed to be stationary; thus, the changes in kinetic and potential energies are also zero ($\Delta KE = \Delta PE = 0$). Then the conservation of energy equation for this closed system reduces to

$$\cancelto{0}{Q} - W_{pw} - \cancelto{0}{W_b} = \Delta U + \cancelto{0}{\Delta KE} + \cancelto{0}{\Delta PE}$$
$$-W_{pw} = m(u_2 - u_1) \approx mC_{v,av}(T_2 - T_1)$$

As we pointed out earlier, the ideal-gas specific heats of monatomic gases (helium being one of them) are constant. The $C_v$ value of helium is determined from Table A-2E*a* to be $C_v = 0.753\text{ Btu/(lbm}\cdot°\text{F)}$. Substituting this and other known quantities into the above energy equation, we obtain

$$-(-25.45\text{ Btu}) = (1.5\text{ lbm})[0.753\text{ Btu/(lbm}\cdot°\text{F)}](T_2 - 80°\text{F})$$
$$T_2 = 102.5°\text{F}$$

(*b*) The final pressure is determined from the ideal-gas relation

$$\frac{P_1 V_1}{T_1} = \frac{P_2 V_2}{T_2}$$

where $V_1$ and $V_2$ are identical and cancel. Then the final pressure becomes

$$\frac{50 \text{ psia}}{(80 + 460)\text{ R}} = \frac{P_2}{(102.5 + 460)\text{ R}}$$

$$P_2 = 52.1 \text{ psia}$$

**EXAMPLE 3-15**

A piston–cylinder device initially contains 0.5 $m^3$ of nitrogen gas at 400 kPa and 27°C. An electric heater within the device is turned on and is allowed to pass a current of 2 A for 5 min from a 120-V source. Nitrogen expands at constant pressure, and a heat loss of 2800 J occurs during the process. Determine the final temperature of the nitrogen.

**Solution** This time, we take the nitrogen in the piston–cylinder device as our system. A sketch of the system and the *P-V* diagram of the process are given in Fig. 3-61. At the specified conditions, the nitrogen gas can be considered to be an ideal gas since it is at a high temperature and low pressure relative to its critical-point values ($T_{cr} = -147$°C, $P_{cr} = 3390$ kPa).

First, let us determine the electrical work done on the nitrogen:

$$W_e = VI\,\Delta t = (120\text{ V})(2\text{ A})(5 \times 60\text{ s})\left(\frac{1\text{ kJ}}{1000\text{ VA}\cdot\text{s}}\right) = -72\text{ kJ}$$

The negative sign is added because the work is done on the system.

The mass of nitrogen is determined from the ideal-gas relation:

$$m = \frac{P_1 V_1}{RT_1} = \frac{(400\text{ kPa})(0.5\text{ m}^3)}{[0.2968\text{ kPa}\cdot\text{m}^3/(\text{kg}\cdot\text{K})](300\text{ K})} = 2.25\text{ kg}$$

Assuming no changes in kinetic and potential energies ($\Delta$KE $=$ $\Delta$PE $= 0$), the conservation of energy equation for this closed system can be written as

$$Q - W_e - W_b = \Delta U$$

For a constant-pressure process of a closed system, $W_b + \Delta U$ is equivalent to $\Delta H$. Thus,

$$Q - W_e = \Delta H = mC_p(T_2 - T_1)$$

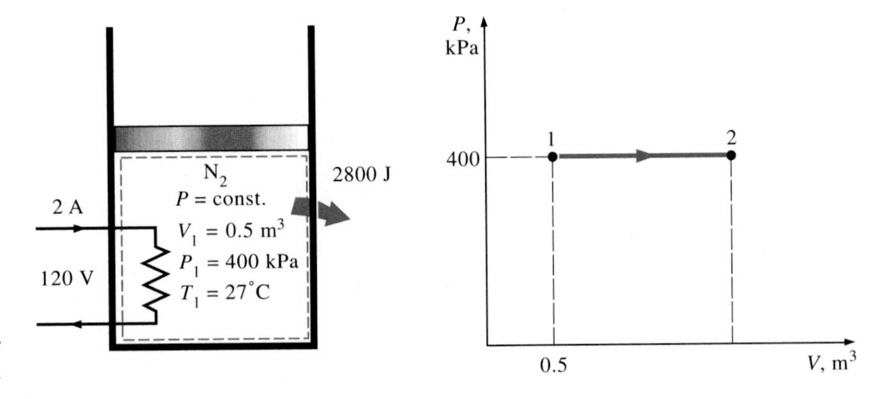


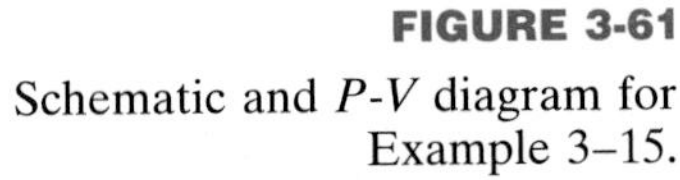

**FIGURE 3-61**

Schematic and *P-V* diagram for Example 3–15.

Using specific heat values at room temperature from Table A-2*a* and substituting the known quantities, the final temperature of nitrogen is determined to be

$$(-2.8\text{ kJ}) - (-72\text{ kJ}) = (2.25\text{ kg})[1.039\text{ kJ/(kg}\cdot{}^\circ\text{C)}](T_2 - 27^\circ\text{C})$$
$$T_2 = 56.6^\circ C$$

**EXAMPLE 3-16**

A piston–cylinder device initially contains air at 150 kPa and 27°C. At this state, the piston is resting on a pair of stops, as shown in Fig. 3-62, and the enclosed volume is 400 L. The mass of the piston is such that a 350-kPa pressure is required to move it. The air is now heated until its volume has doubled. Determine (*a*) the final temperature, (*b*) the work done by the air, and (*c*) the total heat added.

**Solution** The air contained within the piston–cylinder device is the obvious choice for the system, and since no mass is crossing the system boundaries, it is a closed system. Under the given conditions, the air may be assumed to behave as an ideal gas since it is at a high temperature and low pressure relative to its critical-point values ($T_{cr} = -147^\circ$C, $P_{cr} = 3390$ kPa for nitrogen, the main constituent of air). This process can be considered in two parts: a constant-volume process during which the pressure rises to 350 kPa and a constant-pressure process during which the volume doubles.

(*a*) The final temperature can be determined easily by using the ideal-gas relation between states 1 and 3 in the following form:

$$\frac{P_1 V_1}{T_1} = \frac{P_3 V_3}{T_3} \longrightarrow \frac{(150\text{ kPa})(V_1)}{300\text{ K}} = \frac{(350\text{ kPa})(2V_1)}{T_3}$$
$$T_3 = 1400\text{ K}$$

(*b*) The work done could be determined from Eq. 3-18 by integration, but for this case it is much easier to find it from the area under the process curve on a *P-V* diagram, shown in Fig. 3-62:

$$A = (V_2 - V_1)(P_2) = (0.4\text{ m}^3)(350\text{ kPa}) = 140\text{ m}^3\cdot\text{kPa}$$

Therefore,

$$W_{13} = 140\text{ kJ}$$

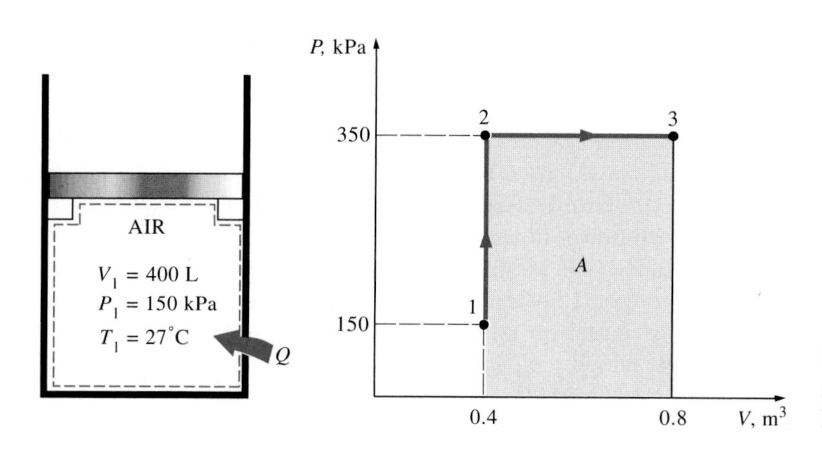


**FIGURE 3-62**

Schematic and *P-V* diagram for Example 3-16.

The work is done by the system (to raise the piston and to push the atmospheric air out of the way), thus it is positive.

(*c*) The total heat transfer can be determined from the first-law relation written between the initial and the final states. Assuming $\Delta KE = \Delta PE = 0$,

$$Q_{13} - W_{13} = U_3 - U_1 = mC_v(T_3 - T_1)$$

The mass of the system can be determined from the ideal-gas equation of state:

$$m = \frac{P_1 V_1}{RT_1} = \frac{(150\text{ kPa})(0.4\text{ m}^3)}{[0.287\text{ kPa}\cdot\text{m}^3/(\text{kg}\cdot\text{K})](300\text{ K})} = 0.697\text{ kg}$$

Using the specific heat of air from Table A-2*a* at the average temperature of $(1400 + 300)/2 = 850$ K and substituting the known quantities, the total heat transfer is determined to be

$$Q_{13} - 140\text{ kJ} = (0.697\text{ kg})[0.823\text{ kJ/(kg4·4K)}](1400 - 300)\text{ K}$$

$$Q_{13} = 771\text{ kJ}$$

## 3-8 INTERNAL ENERGY, ENTHALPY, AND SPECIFIC HEATS OF SOLIDS AND LIQUIDS

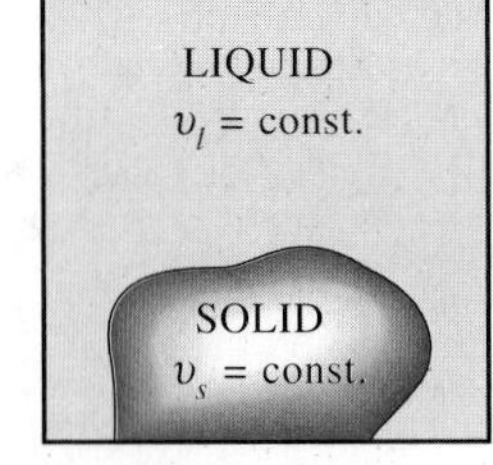


**FIGURE 3-63**
The specific volumes of incompressible substances remain constant during a process.

A substance whose specific volume (or density) is constant is called an **incompressible substance**. The specific volumes of solids and liquids essentially remain constant during a process (Fig. 3-63). Therefore, liquids and solids can be approximated as incompressible substances without sacrificing much in accuracy. The constant-volume assumption should be taken to imply that the energy associated with the volume change, such as the boundary work, is negligible compared with other forms of energy. Otherwise, this assumption would be ridiculous for studying the thermal stresses in solids (caused by volume change with temperature) or analyzing liquid-in-glass thermometers.

It can be mathematically shown that the constant-volume and constant-pressure specific heats are identical for incompressible substances (Fig. 3-64). Therefore, for solids and liquids, the subscripts on $C_p$ and $C_v$ can be dropped, and both specific heats can be represented by a single symbol $C$. That is,

$$C_p = C_v = C \tag{3-51}$$

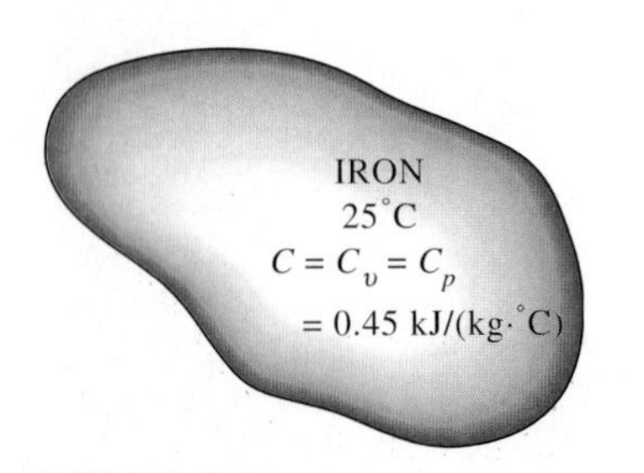


**FIGURE 3-64**
The $C_v$ and $C_p$ values of incompressible substances are identical and are denoted by $C$.

This result could also be deduced from the physical definitions of constant-volume and constant-pressure specific heats. Specific heat values for several common liquids and solids are given in Table A-3.

Like those of ideal gases, the specific heats of incompressible substances depend on temperature only. Thus, the partial differentials in the defining equation of $C_v$ (Eq. 3-38) can be replaced by ordinary differentials, which yields

$$du = C_v\,dT = C(T)\,dT \tag{3-52}$$

The change in internal energy between states 1 and 2 is then obtained by integration:

$$\Delta u = u_2 - u_1 = \int_1^2 C(T)\,dT \qquad \text{(kJ/kg)} \tag{3-53}$$

The variation of specific heat $C$ with temperature should be known before this integration can be carried out. For small temperature intervals, a $C$ value at the average temperature can be used and treated as a constant, yielding

$$\Delta u \approx C_{av}(T_2 - T_1) \qquad \text{(kJ/kg)} \tag{3-54}$$

The enthalpy change of incompressible substances (solids or liquids) during process 1–2 can be determined from the definition of enthalpy ($h = u + Pv$) to be

$$h_2 - h_1 = (u_2 - u_1) + v(P_2 - P_1) \tag{3-55}$$

since $v_1 = v_2 = v$. It can also be expressed in a compact form as

$$\Delta h = \Delta u + v\,\Delta P \qquad \text{(kJ/kg)} \tag{3-56}$$

The second term ($v\,\Delta P$) in Eq. 3-56 is often small compared with the first term ($\Delta u$) and can be neglected without significant loss in accuracy.

**EXAMPLE 3-17**

A 50-kg iron block at 80°C is dropped into an insulated tank that contains 0.5 m$^3$ of liquid water at 25°C. Determine the temperature when thermal equilibrium is reached.

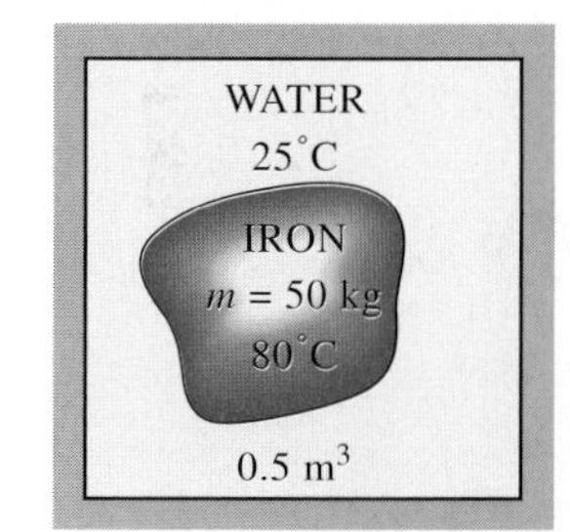


**FIGURE 3-65**
Schematic for Example 3-17.

**Solution** We take the iron block and the water as our system. The inner surfaces of the tank walls form the system boundary, as shown in Fig. 3-65. Since the tank is insulated, no heat will cross these boundaries ($Q = 0$). Also, since there is no movement of the boundary ($W_b = 0$) and no indication of other forms of work ($W_{other} = 0$), the work term for this process is zero ($W = 0$). Then the conservation of energy equation for this process will reduce to

$$\cancelto{0}{Q} - \cancelto{0}{W} = \Delta U \qquad \text{or} \qquad \Delta U = 0$$

The total internal energy $U$ is an extensive property, and therefore it can be expressed as the sum of the internal energies of the parts of the system. Then the total internal energy change of the system is

$$\Delta U_{sys} = \Delta U_{iron} + \Delta U_{water} = 0$$

$$[mC(T_2 - T_1)]_{iron} + [mC(T_2 - T_1)]_{water} = 0$$

The specific volume of liquid water at or about room temperature can be taken to be 0.001 m$^3$/kg. Then the mass of the water is

$$m_{water} = \frac{V}{v} = \frac{0.05\text{ m}^3}{0.001\text{ m}^3/\text{kg}} = 500\text{ kg}$$

The specific heats of iron and liquid water are determined from Table A-3 to be $C_{iron} = 0.45\ kJ/(kg \cdot °C)$ and $C_{water} = 4.184\ kJ/(kg \cdot °C)$. Substituting these values into the energy equation, we obtain

$$(50\ kg)[0.45\ kJ/(kg \cdot °C)](T_2 - 80°C) + (500\ kg)[4.184\ kJ/(kg \cdot °C)](T_2 - 25°C) = 0$$

$$T_2 = 25.6°C$$

Therefore, when thermal equilibrium is established, both the water and iron will be at 25.6°C. The small rise in water temperature is due to its large mass and large specific heat.

## 3-9 ■ THERMODYNAMIC ASPECTS OF BIOLOGICAL SYSTEMS

An important and exciting application area of thermodynamics is to biological systems, which are the sites of rather complex and intriguing energy transfer and transformation processes. Biological systems are not in thermodynamic equilibrium, and thus they are not easy to analyze. Despite their complexity, biological systems are primarily made up of four simple elements: hydrogen, oxygen, carbon, and nitrogen. In the human body, hydrogen accounts for 63 percent, oxygen 25.5 percent, carbon 9.5 percent, and nitrogen 1.4 percent of all the atoms. The remaining 0.6 percent of the atoms come from 20 other elements essential for life. By mass, about 72 percent of the human body is water.

The building blocks of living organisms are *cells,* which resemble miniature factories performing functions that are vital for the survival of organisms. A biological system can be as simple as a single cell. The human body contains about 100 trillion cells with an average diameter of 0.01 mm. The membrane of the cell is a semipermiable wall that allows some substances to pass through it while excluding others.

In a typical cell, thousands of chemical reactions occur every second, during which some molecules are broken down and energy is released and some new molecules are formed. This high level of chemical activity in the cells, which maintains the human body at a temperature of 37°C while performing the necessary bodily tasks, is called **metabolism**. In simple terms, metabolism refers to the burning of foods such as carbohydrates, fat, and protein. The rate of metabolism in the resting state is called the *basal metabolic rate,* which is the rate of metabolism required to keep a body performing the necessary functions (such as breathing and blood circulation) at zero external activity level. The metabolic rate can also be interpreted as the energy consumption rate for a body. For an average male (30 years old, 70 kg, 1.8-$m^2$ body surface area), the basal metabolic rate is 84 W. That is, the body dissipates energy to the environment at a rate of 84 W (joules per second), which means that the body is converting chemical energy of the food (or of the body fat if the person has not eaten) into thermal energy at a rate of

84 W (Fig. 3-66). The metabolic rate increases with the level of activity, and it may exceed 10 times the basal metabolic rate when a body is doing strenuous exercise. That is, two people doing heavy exercising in a room may be supplying more energy to the room than a 1-kW electrical resistance heater (Fig. 3-67). The fraction of sensible heat varies from about 40 percent in the case of heavy work to about 70 percent in the case of light work. The rest of the energy is rejected from the body by perspiration in the form of latent heat.

The basal metabolic rate varies with sex, body size, general health conditions, etc., and decreases considerably with age. This is one of the reasons people tend to put on weight in their late twenties and thirties even though they do not increase their food intake. The brain and the liver are the major sites of metabolic activity. These two organs are responsible for almost 50 percent of the basal metabolic rate of an adult human body although they constitute only about 4 percent of the body mass. In small children, it is remarkable that about half of the basal metabolic activity occurs in the brain alone.

**FIGURE 3-66**
An average person dissipates energy to its surroundings at rate of 84 W when resting.

The metabolic rate of an animal can be measured directly (*direct calorimetry*) or indirectly (*indirect calorimetry*). In direct calorimetry, the animal is placed in a well-insulated closed box equipped with a water circulating system through all sides of the box. The metabolic energy released by the animal is eventually transferred to the water, and the metabolic rate is determined by measuring the temperature rise in water during the period of observation. Although simple in concept, direct calorimetry is difficult to carry out in practice. Therefore, practically all metabolic measurements today are done with indirect calorimetry, which is much simpler and just as accurate as the direct calorimetry.

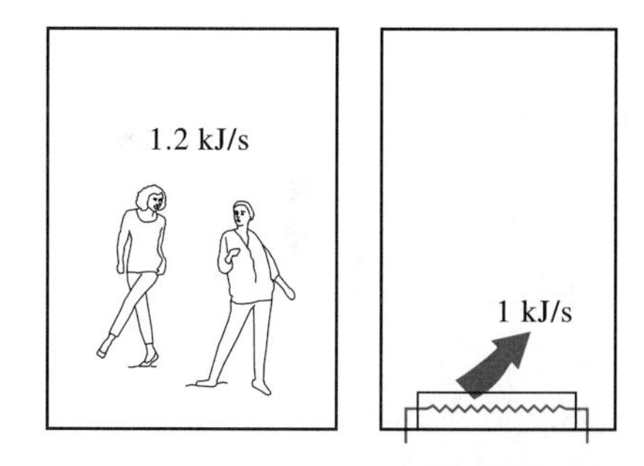

**FIGURE 3-67**
Two fast dancing people supply more energy to a room than a 1-kW electric resistance heater.

In indirect calorimetry, the metabolism rate is determined from the measurements of the rates of $O_2$ consumption and $CO_2$ production of the body. The ratio of the number of moles of $CO_2$ produced to the number of moles of $O_2$ consumed is called the *respiratory quotient* (RQ), whose value depends on the type of food consumed. For example, RQ = 1.0 for glucose ($C_6H_{12}O_6$) since equal number of moles of $O_2$ and $CO_2$ are produced when glucose is oxidized (burned). The RQ is 0.84 for protein and 0.707 for fat. In practice, the protein in the diet is ignored in the calculation of the metabolic rate. The error in ignoring the protein is negligible since the protein forms only a small fraction of the diet, and it has an RQ between those of carbohydrate and fat. Under basal conditions, the RQ of an average adult male is 0.80, which corresponds to a metabolic rate of 20.1 kJ/L of $O_2$ consumed. Thus, a good estimate of the average basal metabolic rate of a person is obtained by measuring the number of liters of $O_2$ the person consumes per unit time, and multiplying this value by 20.1 kJ/L $O_2$. For example, an average resting adult male consumes $O_2$ at a rate of 0.250 L/min, which corresponds to a basal metabolic rate of 84 W. In the absence of any food intake, the starving person consumes his or her own body fat and protein. The average basal metabolic rate in this case is 21.3 kJ/L $O_2$.

The biological reactions in cells occur essentially at constant temperature, pressure, and volume. The temperature of the cell tends to rise when some chemical energy is converted to heat, but this energy is quickly transferred to the circulatory system, which transports it to outer parts of the body, and eventually to the environment through the skin.

The muscle cells function very much like an engine, converting the chemical energy into mechanical energy (work) with a conversion efficiency of close to 20 percent. When the body does no net work on the environment (such as moving some furniture upstairs), the entire work is also converted to heat. In that case, the entire chemical energy in the food released during metabolism in the body is eventually transferred to the environment. This is like a TV set that consumes electricity at a rate of 300 W must reject heat to its environment at a rate of 300 W in steady operation regardless of what goes on in the set. That is, turning on a 300-W TV set or three 100-W light bulbs will produce the same heating effect in a room as a 300-W resistance heater (Fig. 3-68). This is a consequence of the conservation of energy principle, which requires that the energy input into a system must equal the energy output when the total energy of a system remains constant during a process.

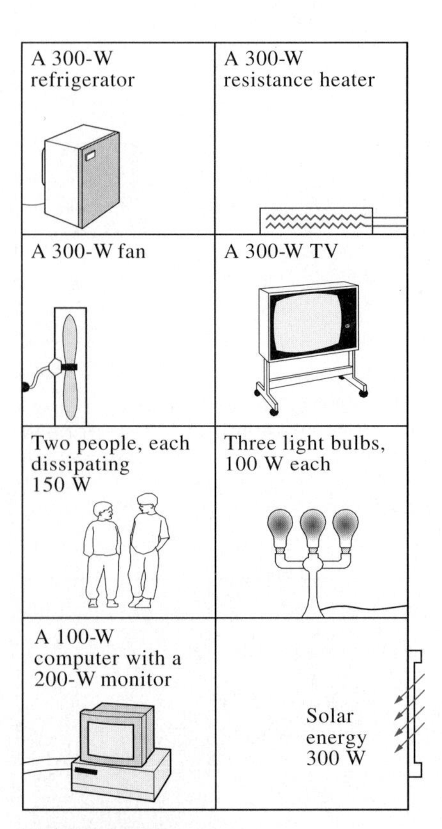


**FIGURE 3-68**
Some arrangements that supply a room the same amount of energy as a 300-W electric resistance heater.

## Food and Exercise

The energy requirements of a body is met by the food we eat. The nutrients in the food are considered in three major groups: carbohydrates, proteins, and fats. *Carbohydrates* are characterized by having hydrogen and oxygen atoms in 2:1 ratio in their molecules. The molecules of carbohydrates range from very simple (as in plain sugar) to very complex or large (as in starch). Bread and plain sugar are the major sources of carbohydrates. *Proteins* are very large molecules that contain carbon, hydrogen, oxygen, and nitrogen, and they are essential for the building and repairing of the body tissues. Proteins are made up of smaller building blocks called *amino acids.* Complete proteins such as meat, milk, and eggs have all the amino acids needed to build body tissues. Plant source proteins such as those in fruits, vegetables, and grains lack one or more amino acids, and are called incomplete proteins. *Fats* are relatively small molecules which consist of carbon, hydrogen, and oxygen. Vegetable oils and animal fats are major sources of fats. Most foods we eat contain all three nutrition groups at varying amounts. The typical average American diet consists of 45 percent carbohydrate, 40 percent fat, and 15 percent protein, although it is recommended that in a healthy diet less than 30 percent of the calories should come from fat.

The energy content of a given food is determined by burning a small sample of the food in a device called *bomb calorimeter* which is basically a well-insulated rigid tank (Fig. 3-69). The tank contains a small combustion chamber surrounded by water. The food is ignited and burned in the combustion chamber in the presence of excess oxygen, and the energy released is transferred to the surrounding water. The energy content of the food is calculated on the basis of the conservation of energy principle from observation of the temperature rise of the water.

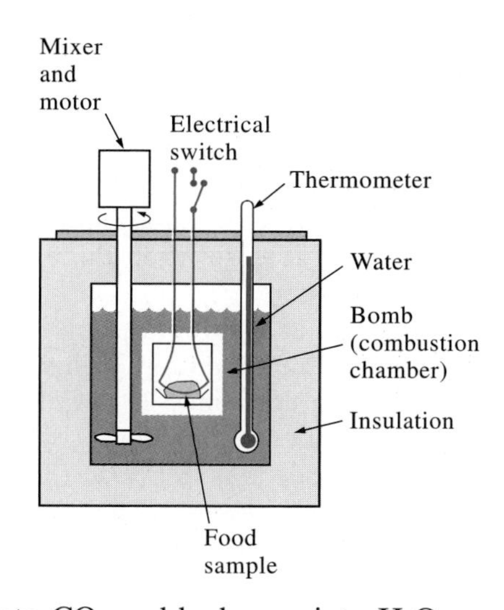


**FIGURE 3-69**

Schematic of a bomb calorimeter used to determine the energy content of food samples.

The carbon in the food is converted into $CO_2$ and hydrogen into $H_2O$ as the food burns. The same chemical reactions occur in the body, and thus the same amount of energy is released.

Using dry (free of water) samples, the average energy contents of the three basic food groups are determined by bomb calorimeter measurements to be 18.0 MJ/kg for carbohydrates, 22.2 MJ/kg for proteins, and 39.8 MJ/kg for fats. These food groups are not entirely metabolized in the human body, however. The fraction of metabolizable energy contents are 95.5 percent for carbohydrates, 77.5 percent for proteins, and 97.7 percent for fats. That is, the fats we eat are almost entirely metabolized in the body, but close to one-quarter of the protein we eat is discarded from the body unburned. This corresponds to 4.1 Calories/g for proteins and carbohydrates and 9.3 Calories/g for fats (Fig. 3-70) commonly seen in nutrition books and on food labels. The energy contents of the foods we normally eat are much lower than the values above because of the large water content (water adds bulk to the food but it cannot be metabolized or burned, and thus it has no energy value). Most vegetables, fruits, and meats, for example, are mostly water. The average metabolizable energy contents of the three basic food groups are 4.2 MJ/kg for carbohydrates, 8.4 MJ/kg for proteins, and 33.1 MJ/kg for fats. Note that 1 kg of natural fat contains almost 8 times the metabolizable energy of 1 kg of natural carbohydrate. Thus, a person who fills his stomach with fatty foods is consuming much more energy than a person who fills his stomach with carbohydrates such as bread or rice.

Fat: (8g)(9.3 Cal/g) = 74.4 Cal
Protein: (2g)(4.1 Cal/g) = 8.2 Cal
Carbohydrates: (21g)(4.1 Cal/g) = 86.1 Cal
Other: (1g) (0 Cal/g) = 0

TOTAL (for 32 g): 169 Cal

**FIGURE 3-70**

Evaluating the calorie content of one serving of chocolate chip cookies (values are for Chips Ahoy cookies made by Nabisco).

The metabolizable energy content of foods is usually expressed by nutritionists in terms of the capitalized *Calories.* One Calorie is equivalent to one *kilocalorie* (1000 calories), which is equivalent to 4.1868 kJ. That is,

$$1 \text{ Calorie} = 1000 \text{ calories} = 1 \text{ kcal (kilocalorie)} = 4.1868 \text{ kJ}$$

The calorie notation often causes confusion since it is not always followed in the tables or articles on nutrition. When the topic is food or fitness, a calorie normally means a kilocalorie whether it is capitalized or not.

Energy or calorie needs of people vary greatly with age, sex, the state of health, the level of activity, and the body size. A small person needs fewer calories than a larger person of the same sex and age. An average man needs about 2400 to 2700 Calories a day. The daily need of an average woman varies from 1800 to 2200 Calories. The extra calories a body consumes are usually stored as fat, which serves as the spare energy of the body for use when the energy intake of the body is less than the needed amount.

Like other natural fat, 1 kg of human body fat contains about 33.1 MJ of metabolizable energy. Therefore, a starving person (zero energy intake) who consumes 2200 Calories (9211 kJ) a day can meet his daily energy intake requirements by burning only 9211/33100 = 0.28 kg of body fat. So it is no surprise that people are known to survive over 100 days without eating. (They still need to drink water, however, to replenish the water loss through the lungs and the skin, to avoid dehydration, which may occur in just a few days.) Although the desire to get rid of the excess fat in a thin world may be overwhelming at times, starvation diets are not recommended because the body soon starts to consume its own muscle tissue in addition to fat. A healthy diet should involve regular exercise while allowing a reasonable amount of calorie intake.

**TABLE 3-3**
**Approximate metabolizable energy content of some common foods. (1 Calorie = 4.1868 kJ = 3.968 Btu)**

| Food | Calories |
|---|---|
| Apple (one, medium) | 70 |
| Baked potato (plain) | 250 |
| Baked potato with cheese | 550 |
| Bread (white, one slice) | 70 |
| Butter (one teaspoon) | 35 |
| Cheeseburger | 325 |
| Chocolate candy bar (20 g) | 105 |
| Cola (200 ml) | 87 |
| Egg (one) | 80 |
| Fish sandwitch | 450 |
| French fries (regular) | 250 |
| Hamburger | 275 |
| Hot dog | 300 |
| Ice cream (100 ml, 10% fat) | 110 |
| Lettuce salad with French dressing | 150 |
| Milk (whole, 200 ml) | 136 |
| Milk (skim, 200 ml) | 76 |
| Peach (one, medium) | 65 |
| Pie (one 1/8 slice, 23 cm diameter) | 300 |
| Pizza (large, cheese, one 1/8 slice) | 350 |

The average metabolizable energy contents of various foods and the energy consumption during various activities are given in Tables 3-3 and 3-4. Considering that no two hamburgers are alike, and that no two people walk exactly the same way, there is some uncertainty in these values, as you would expect. Therefore, you may encounter somewhat different values in other books or magazines for the same items.

The rates of energy consumption listed in Table 3-4 during some activities are for a 68-kg adult. The energy consumed for smaller or larger adults can be determined using the proportionality of the metabolism rate and the body size. For example, the rate of energy consumption by a 68-kg bicyclist is listed in Table 3-4 to be 639 Calories/h. Then the rate of energy consumption by a 50-kg bicyclist is

$$(50\text{ kg})\frac{639\text{ Calories/h}}{68\text{ kg}} = 470\text{ Calories/h}$$

For a 100-kg person, it would be 960 Calories/h.

The thermodynamic analysis of the human body is rather complicated since it involves mass transfer (during breathing, perspiring, etc.) as well as energy transfer. As such, it should be treated as an open system. However, the energy transfer with mass is difficult to quantify. Therefore, the human body is often modeled as a closed system for simplicity by treating energy transported with mass as just energy transfer. For example, eating is modeled as the transfer of energy into the human body in the amount of the metabolizable energy content of the food.

**TABLE 3-4**
**Approximate energy consumption of a 68-kg adult during some activities. (1 Calorie = 4.1868 kJ = 3.968 Btu)**

| Activity | Calories/h |
|---|---|
| Basal metabolism | 72 |
| Basketball | 550 |
| Bicycling (21 km/h) | 639 |
| Cross-country skiing (13 km/h) | 936 |
| Driving a car | 180 |
| Eating | 99 |
| Fast dancing | 600 |
| Fast running (13 km/h) | 936 |
| Jogging (8 km/h) | 540 |
| Swimming (slow) | 288 |
| Swimming (fast) | 860 |
| Tennis (beginner) | 288 |
| Tennis (advanced) | 480 |
| Walking (7.2 km/h) | 432 |
| Watching TV | 72 |

**EXAMPLE 3-18**

A 90-kg man had two hamburgers, a regular serving of french fries, and a 200-ml Coke for lunch (Fig. 3-71). Determine how long it will take for him to burn the lunch calories off (a) by watching TV and (b) by fast swimming. What would your answer be for a 45-kg man?

**FIGURE 3-71**
A typical lunch discussed in Example 3-18.

**Solution** (a) We take the human body as our system and treat it as a closed system whose energy content remains unchanged during the process. Then the conservation of energy principle requires that the energy input into the body must be equal to the energy output. The net energy input in this case is the metabolizable energy content of the food eaten. It is determined from Table 3-3 to be

$$
\begin{aligned}
E_{\text{in}} &= 2 \times E_{\text{hamburger}} + E_{\text{fries}} + E_{\text{cola}} \\
&= (2 \times 275 + 250 + 7)\text{ Calories} \\
&= 887\text{ Calories}
\end{aligned}
$$

Then $E_{\text{out}} = E_{\text{in}} = 887$ Calories. The rate of energy output for a 68-kg man watching TV is given in Table 3-4 to be 72 Calories/h. For a 90-kg man, it becomes

$$\dot{E}_{\text{out}} = (90\text{ kg})\frac{72\text{ Calories/h}}{68\text{ kg}} = 95.3\text{ Calories/h}$$

Therefore, it will take

$$\Delta t = \frac{887\text{ Calories}}{95.3\text{ Calories/h}} = 9.3\text{ h}$$

to burn the lunch calories off by watching TV.

(*b*) It can be shown in a similar manner that it takes only 47 min to burn the lunch calories off by fast swimming.

The 45-kg man is half as large as the 90-kg man. Therefore, expending the same amount of energy will take twice as long in this case: 18.6 h by watching TV, and 94 min by fast swimming.

---

Most diets are based on *calorie counting*; that is, the conservation of energy principle: a person who consumes more calories than his body burns will gain weight whereas a person who consumes less calories than his body burns will loose weight. Yet, people who eat whatever they want whenever they want without gaining any weight are living proofs that the calorie counting technique alone does not work in dieting. Obviously there is more to dieting than keeping track of calories. It should be noted that the phrases *weight gain* and *weight loss* are misnomers. The correct phrases should be *mass gain* and *mass loss.* A man who goes to space loses practically all of his weight but none of his mass. When the topic is food and fitness, *weight* is understood to mean *mass,* and weight is expressed in mass units.

Researchers on nutrition proposed several theories on dieting. One theory suggests that some people have very "food-efficient" bodies. These people need fewer calories than other people do for the same activity, just like a fuel-efficient car needing less fuel for traveling a given distance. It is interesting that we want our cars to be fuel-efficient but we do not want the same high efficiency for our bodies. One thing which frustrates dieters is that the body interprets dieting as *starvation* and starts using the energy reserves of the body more stringently. Shifting from a normal 2000-Calorie daily diet to an 800-Calorie diet without exercise is observed to lower the basal metabolic rate by 10–20 percent. Although the metabolic rate returns to normal once the dieting stops, extended periods of low-Calorie dieting without adequate exercise may result in the loss of considerable muscle tissue together with fat. With less muscle tissue to burn Calories, the metabolic rate of the body declines and stays below normal even after a person starts eating normally. As a result, the person regains the weight he or she has lost in the form of fat, plus more. The basal metabolic rate remains about the same in people who exercise while dieting.

Another theory suggests that people with too many fat cells developed during childhood or adolescence are much more likely to gain weight. Some people believe that the fat content of their bodies are controlled by the setting of a "fat control" mechanism, much like the temperature of a house is controlled by the thermostat setting. Hormone problems are also believed to cause excessive weight gain or loss. Some put the blame for weight problems simply on the genes. Considering that 80 percent of the children of overweight parents are also overweight, heredity may indeed be partly responsible for the way a body stores fat. Whatever the reason, it is clear that the first law of thermodynamics does not give us the complete picture in energy transformation processes, and there is a need for another fundamental principle to complement it. That principle is the second law, which we will study in later chapters.

## 3-10 ■ SUMMARY

The first law of thermodynamics is essentially an expression of the conservation of energy principle. Energy can cross the boundaries of a closed system in the form of heat or work. If the energy transfer is due to a temperature difference between a system and its surroundings, it is *heat*; otherwise, it is *work*. Heat transfer to a system and work done by a system are positive; heat transfer from a system and work done on a system are negative.

Heat is transferred in three different ways: conduction, convection, and radiation. *Conduction* is the transfer of energy from the more energetic particles of a substance to the adjacent less energetic ones as a result of interactions between the particles. *Convection* is the mode of energy transfer between a solid surface and the adjacent liquid or gas which is in motion, and it involves the combined effects of conduction and fluid motion. *Radiation* is the energy emitted by matter in the form of electromagnetic waves (or photons) as a result of the changes in the electronic configurations of the atoms or molecules. The three modes of heat transfer are expressed as

$$\dot{Q}_{\text{cond}} = -kA\frac{dT}{dx} \qquad \text{(W)}$$

$$\dot{Q}_{\text{conv}} = hA(T_s - T_f) \qquad \text{(W)}$$

$$\dot{Q}_{\text{rad}} = \varepsilon\sigma A(T_s^4 - T_{\text{surr}}^4) \qquad \text{(W)}$$

Various forms of work are expressed as follows:

Electrical work $$W_e = VI\,\Delta t \qquad \text{(kJ)}$$

Boundary work $$W_b = \int_1^2 P\,dV \qquad \text{(kJ)}$$

Spring work $$W_{\text{spring}} = \tfrac{1}{2}k_s(x_2^2 - x_1^2) \qquad \text{(kJ)}$$

For the *polytropic process* ($Pv^n$ = constant) of real gases, the boundary work can be expressed as

$$W_b = \frac{P_2V_2 - P_1V_1}{1 - n} \qquad (n \neq 1) \qquad \text{(kJ)}$$

The *first law of thermodynamics* for a closed system is given by

$$Q - W = \Delta U + \Delta\text{KE} + \Delta\text{PE} \qquad \text{(kJ)}$$

where

$$W = W_{\text{other}} + W_b$$

$$\Delta U = m(u_2 - u_1)$$

$$\Delta\text{KE} = \tfrac{1}{2}m(\mathscr{V}_2^2 - \mathscr{V}_1^2)$$

$$\Delta\text{PE} = mg(z_2 - z_1)$$

*For a constant-pressure process,* $W_b + \Delta U = \Delta H$. Thus,

$$Q - W_{\text{other}} = \Delta H + \Delta\text{KE} + \Delta\text{PE} \qquad \text{(kJ)}$$

The amount of energy needed to raise the temperature of a unit mass of a substance by one degree is called the *specific heat at constant volume* $C_v$ for a constant-volume process and the *specific heat at constant pressure* $C_p$ for a constant-pressure process. They are defined as

$$C_v = \left(\frac{\partial u}{\partial T}\right)_v \qquad \text{and} \qquad C_p = \left(\frac{\partial h}{\partial T}\right)_p$$

For ideal gases $u$, $h$, $C_v$, and $C_p$ are functions of temperature alone. The $\Delta u$ and $\Delta h$ of ideal gases can be expressed as

$$\Delta u = u_2 - u_1 = \int_1^2 C_v(T)\,dT \approx C_{v,\text{av}}(T_2 - T_1)$$

$$\Delta h = h_2 - h_1 = \int_1^2 C_p(T)\,dT \approx C_{p,\text{av}}(T_2 - T_1)$$

For ideal gases, $C_v$ and $C_p$ are related by

$$C_p = C_v + R \qquad [\text{kJ/(kg}\cdot\text{K)}]$$

where $R$ is the gas constant. The *specific heat ratio* $k$ is defined as

$$k = \frac{C_p}{C_v}$$

For *incompressible substances* (liquids and solids), both the constant-pressure and constant-volume specific heats are identical and denoted by $C$:

$$C_p = C_v = C \qquad [\text{kJ/(kg}\cdot\text{K)}]$$

The $\Delta u$ and $\Delta h$ of incompressible substances are given by

$$\Delta u = \int_1^2 C(T)\,dT \cong C_{\text{av}}(T_2 - T_1) \qquad \text{(kJ/kg)}$$

$$\Delta h = \Delta u + v\,\Delta P \qquad \text{(kJ/kg)}$$

## REFERENCES AND SUGGESTED READING

**1** R. Balmer, *Thermodynamics,* West, St. Paul, MN, 1990.

**2** A. Bejan, *Advanced Engineering Thermodynamics,* Wiley, New York, 1988.

**3** W. Z. Black and J. G. Hartley, *Thermodynamics,* Harper and Row, New York, 1985.

**4** J. B. Jones and G. A. Hawkins, *Engineering Thermodynamics,* 2d ed., Wiley, New York, 1986.

5 J. R. Howell and R. O. Buckius, *Fundamentals of Engineering Thermodynamics,* McGraw-Hill, New York, 1987.

6 M. Snowman, *Food and Fitness,* New Readers Press, Syracuse, NY, 1986.

7 G. J. Van Wylen and R. E. Sonntag, *Fundamentals of Classical Thermodynamics,* 3d ed., Wiley, New York, 1985.

8 K. Wark, *Thermodynamics,* 5th ed., McGraw-Hill, New York, 1988.

9 W. J. Yang, *Biothermal–Fluid Sciences,* Hemisphere, New York, 1989.

## PROBLEMS*

### Heat Transfer and Work

**3-1C** In what forms can energy cross the boundaries of a system?

**3-2C** When is the energy crossing the boundaries of a system heat and when is it work?

**3-3C** What is an adiabatic process? What is an adiabatic system?

**3-4C** What are the sign conventions for heat and work?

**3-5C** A gas in a piston–cylinder device is compressed, and as a result its temperature rises. Is this a heat or work interaction for the gas?

**3-6C** A room is heated by an iron that is left plugged in. Is this a heat or work interaction? Take the entire room, including the iron, as the system.

**3-7C** A room is heated as a result of solar radiation coming in through the windows. Is this a heat or work interaction for the room?

**3-8C** An insulated room is heated by burning candles. Is this a heat or work interaction? Take the entire room, including the candles, as the system.

**3-9C** What are point and path functions? Give some examples.

**3-10C** What are the mechanisms of heat transfer?

**3-11C** What is the caloric theory? When and why is it abandoned?

**3-12C** Does any of the energy of the sun reach the earth by conduction or convection?

**3-13C** Which is a better heat conductor: diamond or silver?

**3-14C** How does forced convection differ from natural convection?

**3-15C** Define emissivity and absorptivity. What is Kirchhoff's law of radiation?

---

*Students are encouraged to answer *all* the concept "C" questions.

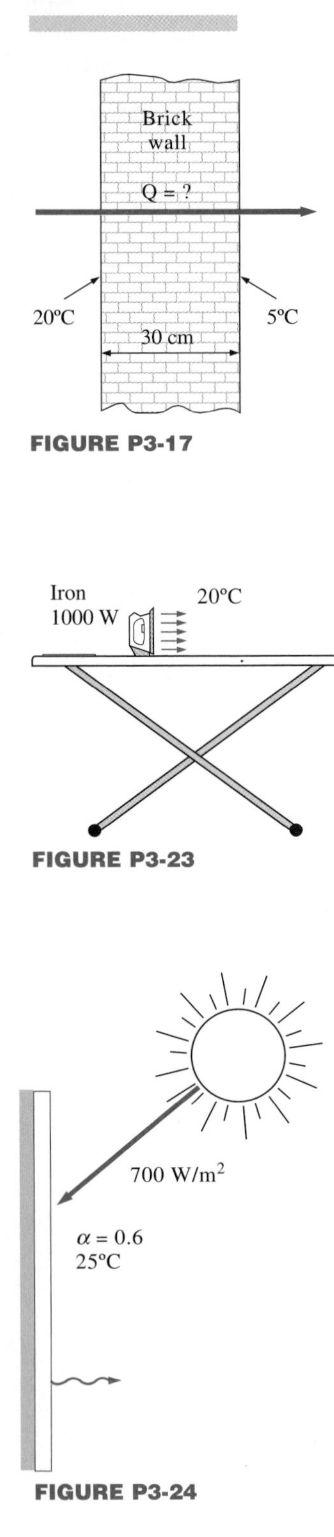


FIGURE P3-17

FIGURE P3-23

FIGURE P3-24

**3-16C** What is a blackbody? How do real bodies differ from a blackbody?

**3-17** The inner and outer surfaces of a 5 m × 6 m brick wall of thickness 30 cm and thermal conductivity 0.69 W/(m · °C) are maintained at temperatures of 20°C and 5°C, respectively. Determine the rate of heat transfer through the wall, in W. *Answer:* 1035 W

**3-18** The inner and outer surfaces of a 0.5-cm-thick 2 m × 2 m window glass in winter are 10°C and 3°C, respectively. If the thermal conductivity of the glass is 0.78 W/(m · °C), determine the amount of heat loss, in kJ, through the glass over a period of 5 h. What would your answer be if the glass were 1 cm thick?

**3-19** An aluminum pan whose thermal conductivity is 237 W/(m · °C) has a flat bottom whose diameter is 20 cm and thickness 0.4 cm. Heat is transferred steadily to boiling water in the pan through its bottom at a rate of 500 W. If the inner surface of the bottom of the pan is 105°C, determine the temperature of the outer surface of the bottom of the pan.

**3-20** For heat transfer purposes, a standing naked man can be modeled as a 30-cm diameter, 170-cm long vertical cylinder with both the top and bottom surfaces insulated and with the side surface at an average temperature of 34°C. For a convection heat transfer coefficient of 15 W/(m$^2$ · °C), determine the rate of his loss from this man by convection in an environment at 20°C. *Answer:* 336 W

**3-21** A 5-cm-diameter spherical ball whose surface is maintained at a temperature of 70°C is suspended in the middle of a room at 20°C. If the convection heat transfer coefficient is 15 W/(m$^2$ · °C) and the emissivity of the surface is 0.8, determine the total rate of heat transfer from the ball.

**3-22** Hot air at 80°C is blown over a 2 m × 4 m flat surface at 30°C. If the convection heat transfer coefficient is 55 W/(m$^2$ · °C), determine the rate of heat transfer from the air to the plate, in kW.

**3-23** A 1000-W iron is left on the iron board with its base exposed to the air at 20°C. The convection heat transfer coefficient between the base surface and the surrounding air is 35 W/(m$^2$ · °C). If the base has an emissivity of 0.6 and a surface area of 0.02 m$^2$, determine the temperature of the base of the iron. *Answer:* 674°C

**3-24** A thin metal plate is insulated at the back and exposed to solar radiation at the front surface. The exposed surface of the plate has an absorptivity of 0.6 for solar radiation. If solar radiation is incident on the plate at a rate of 700 W/m$^2$ and the surrounding air temperature is 25°C, determine the surface temperature of the plate when the heat loss by convection equals the solar energy absorbed by the plate. Assume the convection heat transfer coefficient to be 50 W/(m$^2$ · °C), and disregard heat loss by radiation. *Answer:* 33.4°C

**3-25** A 5-cm-external-diameter, hot water pipe at 80°C is losing heat to the surrounding air at 5°C by natural convection with a heat transfer coefficient of $25\ \mathrm{W/(m^2 \cdot {}^\circ C)}$. Determine the rate of heat loss from the pipe by natural convection, in kW.

**3-26** The outer surface of a spacecraft in space has an emissivity of 0.8 and an absorptivity of 0.3 for solar radiation. If solar radiation is incident on the spacecraft at a rate of $1000\ \mathrm{W/m^2}$, determine the surface temperature of the spacecraft when the radiation emitted equals the solar energy absorbed. *Answer:* 285.2 K

**3-27** A hollow spherical iron container whose outer diameter is 20 cm and thickness 0.4 cm is filled with iced water at 0°C. If the outer surface temperature is 5°C, determine the approximate rate of heat loss from the sphere, and the rate at which ice melts in the container.

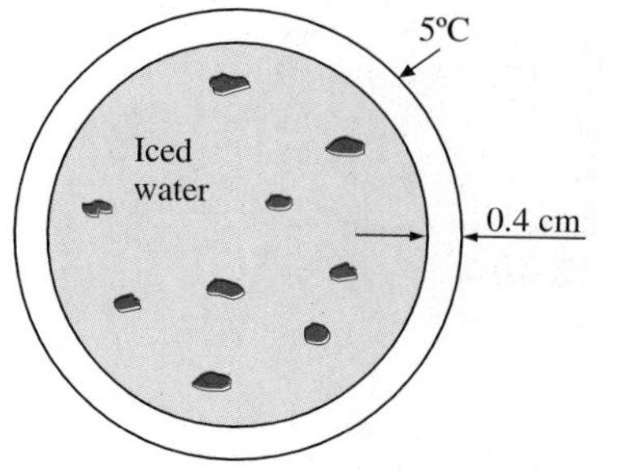


**FIGURE P3-27**

**3-28** The inner and outer glasses of a 2 m × 2 m double pane window are at 18°C and 6°C, respectively. If the 1-cm space between the two glasses is filled with still air, determine the rate of heat transfer through the window, in kW. *Answer:* 124.6 W

**3-29** The two surfaces of a 2-cm-thick plate are maintained at 0°C and 100°C, respectively. If it is determined that heat is transferred through the plate at a rate of $500\ \mathrm{W/m^2}$, determine its thermal conductivity.

## Boundary Work

**3-30C** On a *P-v* diagram, what does the area under the process curve represent?

**3-31C** Explain why the work output is greater when a process takes place in a quasi-equilibrium manner.

**3-32C** Is the boundary work associated with constant-volume systems always zero?

**3-33C** An ideal gas at a given state expands to a fixed final volume first at constant pressure and then at constant temperature. For which case is the work done greater?

**3-34C** Show that $1\ \mathrm{kPa \cdot m^3} = 1\ \mathrm{kJ}$.

**3-35** A mass of 1.2 kg of air of 150 kPa and 12°C is contained in a gas-tight, frictionless piston-cylinder device. The air is now compressed to a final pressure of 600 kPa. During the process heat is transferred from the air such that the temperature inside the cylinder remains constant. Calculate the work done during this process. *Answer:* −136.1 kJ

**3-36** Nitrogen at an initial state of 300 K, 150 kPa, and $0.2\ \mathrm{m^3}$ is compressed slowly in an isothermal process to a final pressure of 800 kPa. Determine the work done during this process.

**3-36E** Nitrogen at an initial state of 70°F, 20 psia, and 5 $ft^3$ is compressed slowly in an isothermal process to a final pressure of 100 psia. Determine the work done during this process.

**3-37** A gas is compressed from an initial volume of 0.42 $m^3$ to a final volume of 0.12 $m^3$. During the quasi-equilibrium process, the pressure changes with volume according to the relation $P = aV + b$, where $a = -1200\ kPa/m^3$ and $b = 600$ kPa. Calculate the work done during this process (*a*) by plotting the process on a *P-V* diagram and finding the area under the process curve and (*b*) by performing the necessary integrations.

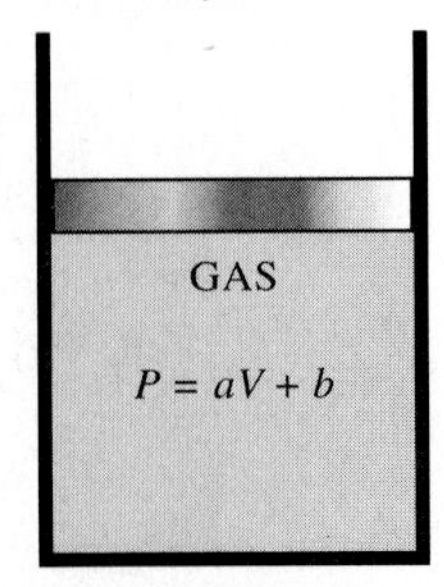


**FIGURE P3-37**

**3-38** During an expansion process, the pressure of a gas changes from 100 to 900 kPa according to the relation $P = aV + b$, where $a = 1\ MPa/m^3$ and $b$ is a constant. If the initial volume of the gas is 0.2 $m^3$, calculate the work done during the process. *Answer:* 400 kJ

**3-38E** During an expansion process, the pressure of a gas changes from 15 to 100 psia according to the relation $P = aV + b$, where $a = 5\ psia/ft^3$ and $b$ is a constant. If the initial volume of the gas is 7 $ft^3$, calculate the work done during the process. *Answer:* 180.9 Btu

**3-39** During some actual expansion and compression processes in piston-cylinder devices, the gases have been observed to satisfy the relationship $PV^n = C$, where $n$ and $C$ are constants. Calculate the work done when a gas expands from a state of 150 kPa and 0.03 $m^3$ to a final volume of 0.2 $m^3$ for the case of $n = 1.3$.

**3-40** A frictionless piston–cylinder device contains 2 kg of nitrogen at 100 kPa and 300 K. Nitrogen is now compressed slowly according to the relation $PV^{1.4}$ = constant until it reaches a final temperature of 360 K. Calculate the work done during this process. *Answer:* −89.0 kJ

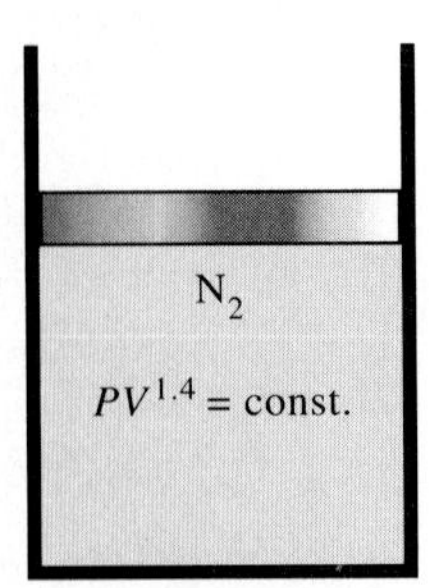


**FIGURE P3-40**

**3-40E** A frictionless piston–cylinder device contains 5 lbm of nitrogen at 14.7 psia and 550 R. Nitrogen is now compressed slowly according to the relation $PV^{1.4}$ = constant until it reaches a final temperature of 700 R. Calculate the work done during this process. *Answer:* −132.9 Btu

**3-41** The equation of state of a gas is given as $\bar{v}(P + 10/\bar{v}^2) = R_u T$, where the units of $\bar{v}$ and $P$ are $m^3/kmol$ and kPa, respectively. Now 0.5 kmol of this gas is expanded in a quasi-equilibrium manner from 2 to 4 $m^3$ at a constant temperature of 300 K. Determine (*a*) the unit of the quantity 10 in the equation and (*b*) the work done during this isothermal expansion process.

**3-42** Carbon dioxide contained in a piston–cylinder device is compressed from 0.3 to 0.1 $m^3$. During the process, the pressure and volume are related by $P = aV^{-2}$, where $a = 8\ kPa \cdot m^6$. Calculate the work done on the carbon dioxide during this process. *Answer:* −53.3 kJ

**3-42E** Carbon dioxide contained in a piston–cylinder device is compressed from 10 to 3 $ft^3$. During the process, the pressure and volume are related by $P = aV^{-2}$, where $a = 175 \text{ psia} \cdot \text{ft}^6$. Calculate the work done on the carbon dioxide during this process.

**3-43** Hydrogen is contained in a piston–cylinder device at 100 kPa and $1 \text{ m}^3$. At this state, a linear spring ($F \propto x$) with a spring constant of 200 kN/m is touching the piston but exerts no force on it. The cross-sectional area of the piston is $0.8 \text{ m}^2$. Heat is transferred to the hydrogen, causing it to expand until its volume doubles. Determine (*a*) the final pressure, (*b*) the total work done by the hydrogen, and (*c*) the fraction of this work done against the spring. Also show the process on a *P-V* diagram.

**3-43E** Hydrogen is contained in a piston–cylinder device at 14.7 psia and $15 \text{ ft}^3$. At this state, a linear spring ($F \propto x$) with a spring constant of 15,000 lbf/ft is touching the piston but exerts no force on it. The cross-sectional area of the piston is $3 \text{ ft}^2$. Heat is transferred to the hydrogen, causing it to expand until its volume doubles. Determine (*a*) the final pressure, (*b*) the total work done by the hydrogen, and (*c*) the fraction of this work done against the spring. Also show the process on a *P-V* diagram.

## Closed-System Energy Analysis: General Systems

**3-44C** For a cycle, is the net work necessarily zero? For what kind of systems will this be the case?

**3-45C** Under what conditions is the relation $Q - W_{\text{other}} = H_2 - H_1$ valid for a closed system?

**3-46C** On a hot summer day, a student turns his fan on when he leaves his room in the morning. When he returns in the evening, will the room be warmer or cooler than the neighboring rooms? Why? Assume all the doors and windows are kept closed.

**3-47C** Consider two identical rooms, one with a refrigerator in it and the other without one. If all the doors and windows are closed, will the room that contains the refrigerator be cooler or warmer than the other room? Why?

**3-48C** Consider a can of soft drink that is dropped from the top of a tall building. Will the temperature of the soft drink increase as it falls, as a result of decreasing potential energy?

**3-49** Water is being heated in a closed pan on top of a range while being stirred by a paddle wheel. During the process, 30 kJ of heat is added to the water, and 5 kJ of heat is lost to the surrounding air. The paddle-wheel work amounts to $500 \text{ N} \cdot \text{m}$. Determine the final energy of the system if its initial energy is 10 kJ. *Answer:* 35.5 kJ

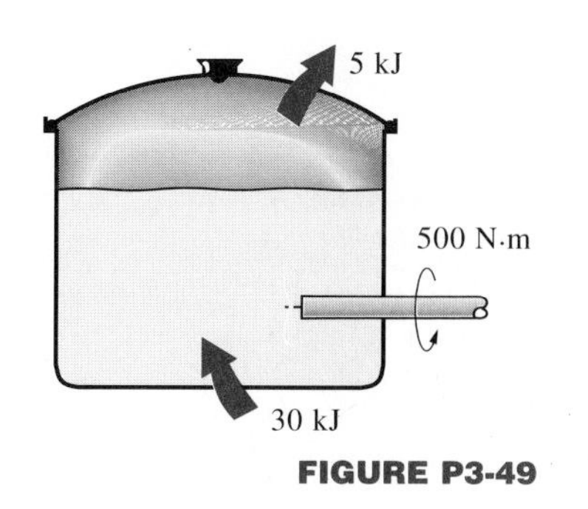

**FIGURE P3-49**

**3-50** A vertical piston–cylinder device contains water and is being heated on top of a range. During the process, 50 kJ of heat is transferred to the water, and heat losses from the side walls amount to 8 kJ. The piston rises as a result of evaporation, and 5 kJ of boundary work is done. Determine the change in the energy of the water for this process.

**3-50E** A vertical piston–cylinder device contains water and is being heated on top of a range. During the process, 50 Btu of heat is transferred to the water, and heat losses from the side walls amount to 8 Btu. The piston rises as a result of evaporation, and 5 Btu of boundary work is done. Determine the change in the energy of the water for this process. *Answer:* 37 Btu

**3-51** Fill in the missing data for each of the following processes of a closed system between states 1 and 2. (Everything is in kJ.)

| | $Q$ | $W$ | $E_1$ | $E_2$ | $\Delta E$ |
|---|---|---|---|---|---|
| (*a*) | 18 | −6 | | 35 | |
| (*b*) | −10 | | | 4 | −15 |
| (*c*) | | 12 | 3 | | 32 |
| (*d*) | 25 | | 14 | | 10 |

**3-52** Fill in the missing data for each of the following processes of a closed system between states 1 and 2. (Everything is in kJ.)

| | $Q$ | $W$ | $E_1$ | $E_2$ | $\Delta E$ |
|---|---|---|---|---|---|
| (*a*) | | 18 | 6 | | 20 |
| (*b*) | 5 | | 20 | | 35 |
| (*c*) | 25 | −10 | | 40 | |
| (*d*) | −9 | | | 12 | −15 |

**3-53** A closed system undergoes a cycle consisting of two processes. During the first process, 40 kJ of heat is transferred to the system while the system does 60 kJ of work. During the second process, 45 kJ of work is done on the system.

(*a*) Determine the heat transfer during the second process.
(*b*) Calculate the net work and net heat transfer for the cycle.
*Answers:* (*a*) −25 kJ; (*b*) 15 kJ, 15 kJ

**3-53E** A closed system undergoes a cycle consisting of two processes. During the first process, 40 Btu of heat is transferred to the system while the system does 60 Btu of work. During the second process, 45 Btu of work is done on the system.

(*a*) Determine the heat transfer during the second process.
(*b*) Calculate the net work and net heat transfer for the cycle.

**3-54** A closed system undergoes a cycle consisting of three processes. During the first process, which is adiabatic, 50 kJ of work is done on the system. During the second process, 200 kJ of heat is transferred to the system while no work interaction takes place. And during the third process, the system does 90 kJ of work as it returns to its initial state.

(*a*) Determine the heat transfer during the last process.

(*b*) Determine the net work done during this cycle.

**3-55** A classroom that normally contains 40 people is to be air-conditoned with window air-conditioning units of 5-kW rating. A person at rest may be assumed to dissipate heat at a rate of about 360 kJ/h. There are 10 light bulbs in the room, each with a rating of 100 W. The rate of heat transfer to the classroom through the walls and the windows is estimated to be 15,000 kJ/h. If the room air is to be maintained at a constant temperature of 21°C, determine the number of window air-conditioning units required. *Answer:* 2 units

### Specific Heats, $\Delta u$, and $\Delta h$ of ideal Gases

**3-56C** Is the relation $\Delta U = mC_{v,\text{av}}\,\Delta T$ restricted to constant-volume processes only, or can it be used for any kind of process of an ideal gas?

**3-57C** Is the relation $\Delta H = mC_{p,\text{av}}\,\Delta T$ restricted to constant-pressure processes only, or can it be used for any kind of process of an ideal gas?

**3-58C** Show that for an ideal gas $\bar{C}_p = \bar{C}_v + R_u$.

**3-59C** Is the energy required to heat air from 295 to 305 K the same as the energy required to heat it from 345 to 355 K? Assume the pressure remains constant in both cases.

**3-60C** In the relation $\Delta U = mC_v\,\Delta T$, what is the correct unit of $C_v$: kJ/(kg · °C) or kJ/(kg · K)?

**3-61C** A fixed mass of an ideal gas is heated from 50 to 80°C at a constant pressure of (*a*) 1 atm and (*b*) 3 atm. For which case do you think the energy required will be greater? Why?

**3-62C** A fixed mass of an ideal gas is heated from 50 to 80°C at a constant volume of (*a*) 1 $m^3$ and (*b*) 3 $m^3$. For which case do you think the energy required will be greater? Why?

**3-63C** A fixed mass of an ideal gas is heated from 50 to 80°C (*a*) at constant volume and (*b*) at constant pressure. For which case do you think the energy required will be greater? Why?

**3-64** Determine the enthalpy change $\Delta h$ of nitrogen, in kJ/kg, as it is heated from 600 to 1000 K, using (*a*) the empirical specific heat equation as a function of temperature (Table A-2*c*), (*b*) the $C_p$ value at the average temperature (Table A-2*b*), and (*c*) the $C_p$ value at room temperature (Table A-2*a*).
*Answers:* (*a*) 447.8 kJ/kg, (*b*) 448.4 kJ/kg, (*c*) 415.6 kJ/kg

**3-65** Determine the enthalpy change $\Delta h$ of oxygen, in kJ/kg, as it is heated from 500 to 800 K, using (*a*) the emprical specific heat equation as a function of temperature (Table A-2*c*), (*b*) the $C_p$ value at the average

temperature (Table A-2*b*), and (*c*) the $C_p$ value at room temperature (Table A-2*a*).
*Answers:* (*a*) 170.1 Btu/lbm, (*b*) 178.5 Btu/lbm, (*c*) 153.3 Btu/lbm

**3-65E** Determine the enthalpy change $\Delta h$ of oxygen, in Btu/lbm, as it is heated from 800 to 1500 R, using (*a*) the empirical specific heat equation as a function of temperature (Table A-2E*c*), (*b*) the $C_p$ value at the average temperature (Table A-2E*b*), and (*c*) the $C_p$ value at room temperature (Table A-2E*a*).
*Answers:* (*a*) 170.1 Btu/lbm, (*b*) 178.5 Btu/lbm, (*c*) 153.3 Btu/lbm

**3-66** Determine the internal energy change $\Delta u$ of hydrogen, in kJ/kg, as it is heated from 400 to 1000 K, using (*a*) the empirical specific heat equation as a function of temperature (Table A-2*c*), (*b*) the $C_v$ value at average temperature (Table A-2*b*), and (*c*) the $C_v$ value at room temperature (Table A-2*a*).

**3-66E** Determine the internal energy change $\Delta u$ of hydrogen, in Btu/lbm, as it is heated from 700 to 1500 R, using (*a*) the empirical specific heat equation as a function of temperature (Table A-2E*c*), (*b*) the $C_v$ value at average temperature (Table A-2E*b*), and (*c*) the $C_v$ value at room temperature (Table A-2E*a*).

### Closed-System Energy Analysis: Ideal Gases

**3-67C** Is is possible to compress an ideal gas isothermally in an adiabatic piston–cylinder device? Explain.

**3-68** A rigid tank contains 10 kg of air at 200 kPa and 27°C. The air is now heated until its pressure doubles. Determine (*a*) the volume of the tank and (*b*) the amount of heat transfer.
*Answers:* (*a*) 4.305 $m^3$, (*b*) 2199 kJ

**3-68E** A rigid tank contains 20 lbm of air at 50 psia and 80°F. The air is now heated until its pressure doubles. Determine (*a*) the volume of the tank and (*b*) the amount of heat transfer.
*Answers:* (*a*) 80 $ft^3$, (*b*) 2035 Btu

**3-69** A 1-$m^3$ rigid tank contains hydrogen at 250 kPa and 500 K. The gas is now cooled until its temperature drops to 300 K. Determine (*a*) the final pressure in the tank and (*b*) the amount of heat transfer.

**3-70** A 4 m × 5 m × 6 m room is to be heated by a baseboard resistance heater. It is desired that the resistance heater be able to raise the air temperature in the room from 7 to 23°C within 15 min. Assuming no heat losses from the room and an atmospheric pressure of 100 kPa, determine the required power of the resistance heater. Assume constant specific heats at room temperature. *Answer:* 1.91 kW

**3-71** A 4 m × 5 m × 7 m room is heated by the radiator of a steam-heating system. The steam radiator transfers heat at a rate of 10,000 kJ/h, and a 100-W fan is used to distribute the warm air in the room. The rate of heat loss from the room is estimated to be about 5000 kJ/h. If the initial temperature of the room air is 10°C, determine how long it will take for the air temperature to rise to 20°C. Assume constant specific heats at room temperature.

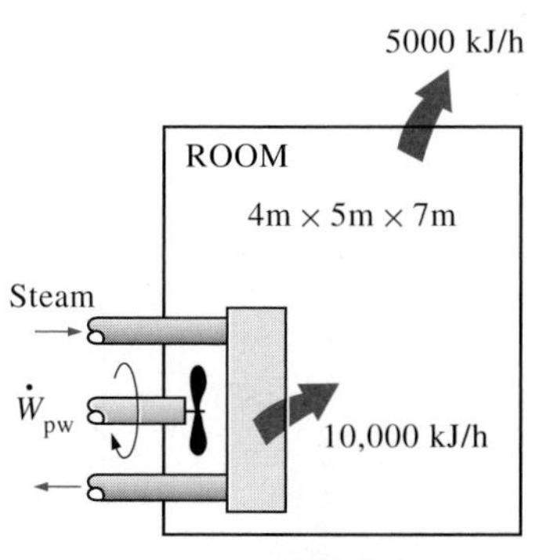

FIGURE P3-71

**3-71E** A 12 ft × 15 ft × 20 ft room is heated by the radiator of a steam-heating system. The steam radiator transfers heat at a rate of 10,000 Btu/h, and a 100-W fan is used to distribute the warm air in the room. The heat losses from the room are estimated to be at a rate of about 5000 Btu/h. If the initial temperature of the room air is 45°F, determine how long it will take for the air temperature to rise to 75°F. Assume constant specific heats at room temperature. *Answer:* 978 s

**3-72** A student living in a 4 m × 6 m × 6 m dormitory room turns on her 150-W fan before she leaves the room on a summer day, hoping that the room will be cooler when she comes back in the evening. Assuming all the doors and windows are tightly closed and disregarding any heat transfer through the walls and the windows, determine the temperature in the room when she comes back 10 h later. Use specific heat values at room temperature, and assume the room to be at 100 kPa and 15°C in the morning when she leaves. *Answer:* 58.2°C

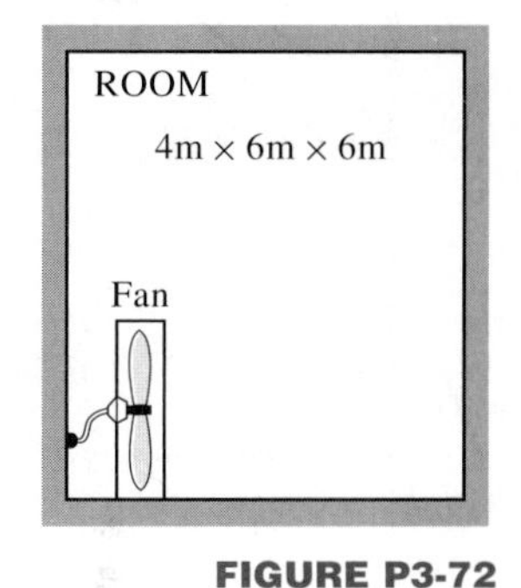


FIGURE P3-72

**3-73** A 0.3-m$^3$ tank contains oxygen initially at 100 kPa and 27°C. A paddle wheel within the tank is rotated until the pressure inside rises to 150 kPa. During the process 2 kJ of heat is lost to the surroundings. Determine the paddle-wheel work done. Neglect the energy stored in the paddle wheel. *Answer:* −40.94 kJ

**3-73E** A 10-ft$^3$ tank contains oxygen initially at 14.7 psia and 80°F. A paddle wheel within the tank is rotated until the pressure inside rises to 20 psia. During the process 20 Btu of heat is lost to the surroundings. Determine the paddle-wheel work done. Neglect the energy stored in the paddle wheel.

**3-74** An insulated rigid tank is divided into two equal parts by a partition. Initially, one part contains 3 kg of an ideal gas at 800 kPa and 50°C, and the other part is evacuated. The partition is now removed, and the gas expands into the entire tank. Determine the final temperature and pressure in the tank.

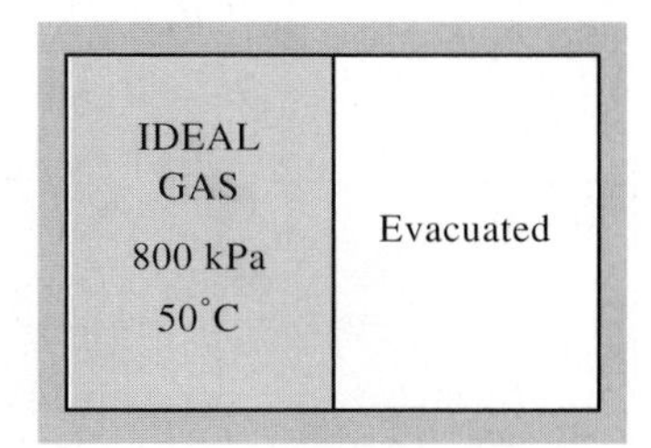


FIGURE P3-74

**3-74E** An insulated rigid tank is divided into two equal parts by a partition. Initially, one part contains 8 lbm of an ideal gas at 80 psia and 200°F, and the other part is evacuated. The partition is now removed, and the gas expands into the entire tank. Determine the final temperature and pressure in the tank.

**3-75** A piston–cylinder device, whose piston is resting on top of a set of stops, initially contains 0.5 kg of helium gas at 100 kPa and 25°C. The mass of the piston is such that 500 kPa of pressure is required to raise it. How much heat must be transferred to the helium before the piston starts rising? *Answer:* 1857 kJ

**3-76** An insulated piston–cylinder device contains 100 L of air at 400 kPa and 25°C. A paddle wheel within the cylinder is rotated until 15 kJ of work is done on the air while the pressure is held constant. Determine the final temperature of the air. Neglect the energy stored in the paddle wheel.

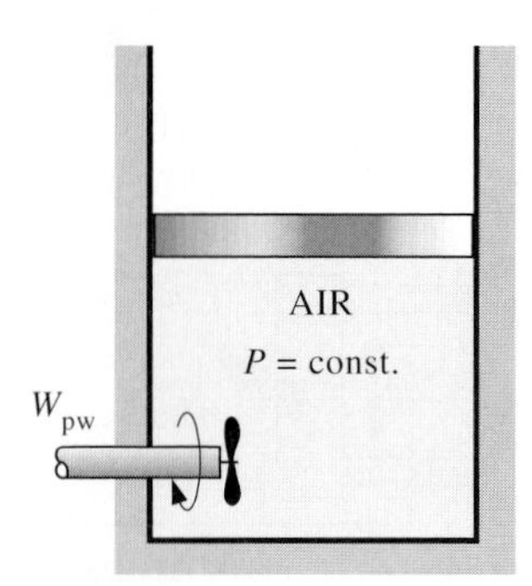


**FIGURE P3-76**

**3-77** A piston–cylinder device contains 0.8 $m^3$ of nitrogen at 300 kPa and 327°C. The nitrogen is now allowed to cool at constant pressure until the temperature drops to 77°C. Determine the heat transfer.
*Answer:* −355.2 kJ

**3-77E** A piston–cylinder device contains 25 $ft^3$ of nitrogen at 50 psia and 700°F. Nitrogen is now allowed to cool at constant pressure until the temperature drops to 140°F. Determine the heat transfer.

**3-78** A mass of 15 kg of air in a piston–cylinder device is heated from 25 to 77°C by passing current through a resistance heater inside the cylinder. The pressure inside the cylinder is held constant at 300 kPa during the process, and a heat loss of 60 kJ occurs. Determine the electric energy supplied, in kWh. *Answer:* 0.235 kWh

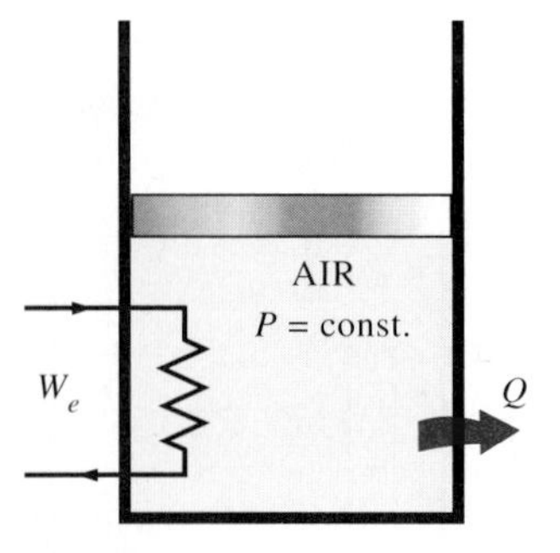


**FIGURE P3-78**

**3-79** An insulated piston–cylinder device initially contains 0.3 $m^3$ of carbon dioxide at 200 kPa and 27°C. An electric switch is turned on, and a 110-V source supplies current to a resistance heater inside the cylinder for a period of 10 min. The pressure is held constant during the process, while the volume is doubled. Determine the current that passes through the resistance heater.

**3-79E** An insulated piston–cylinder device initially contains 10 $ft^3$ of carbon dioxide at 30 psia and 80°F. An electric switch is turned on, and a 110-V source supplies current to a resistance heater inside the cylinder for 10 min. The pressure is held constant during the process, while the volume is doubled. Determine the current that passes through the resistance heater. *Answer:* 4.6 A

**3-80** A piston–cylinder device contains 0.8 kg of nitrogen initially at 100 kPa and 27°C. The nitrogen is now compressed slowly in a polytropic process during which $PV^{1.3}$ = constant until the volume is reduced by one-half. Determine the work done and the heat transfer for this process.

**3-81** A room is heated by a baseboard resistance heater. When the heat losses from the room on a winter day amount to 8000 kJ/h, the air temperature in the room remains constant even though the heater operates continuously. Determine the power rating of the heater, in kW.

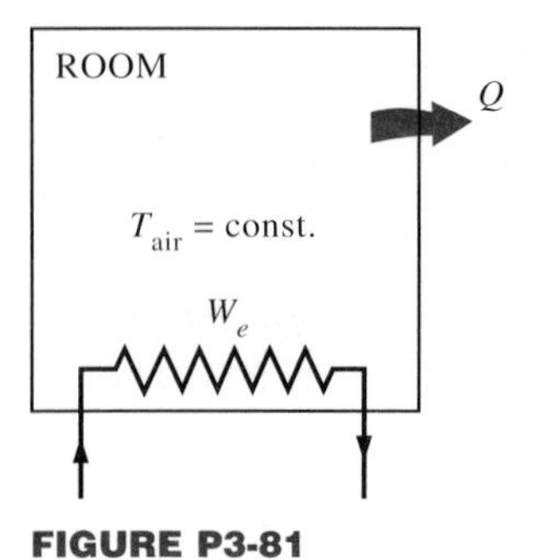


**FIGURE P3-81**

**3-82** A piston–cylinder device contains 0.1 $m^3$ of air at 400 kPa and 50°C. Heat is transferred to the air in the amount of 40 kJ as the air expands isothermally. Determine the amount of boundary work done during this process. *Answer:* 40 kJ

**3-82E** A piston–cylinder device contains 3 $ft^3$ of air at 60 psia and 150°F. Heat is transferred to the air in the amount of 40 Btu as the air expands isothermally. Determine the amount of boundary work done during this process.

**3-83** A piston–cylinder device contains 5 kg of argon at 400 kPa and 30°C. During a quasi-equilibrium, isothermal expansion process, 15 kJ of boundary work is done by the system, and 3 kJ of paddle-wheel work is done on the system. Determine the heat transfer for this process. *Answer:* 12 kJ

**3-83E** A piston–cylinder device contains 8 lbm of argon at 75 psia and 70°F. During a quasi-equilibrium, isothermal expansion process, 15 Btu of boundary work is done by the system, and 3 Btu of paddle-wheel work is done on the system. Determine the heat transfer for this process.

**3-84** A piston–cylinder device, whose piston is resting on a set of stops, initially contains 3 kg of air at 200 kPa and 27°C. The mass of the piston is such that a pressure of 400 kPa is required to move it. Heat is now transferred to the air until its volume doubles. Determine the work done by the air and the total heat transferred to the air during this process. Also show the process on a *P-v* diagram. *Answers:* 516 kJ, 2674 kJ

**3-85** A piston–cylinder device, with a set of stops on the top, initially contains 3 kg of air at 200 kPa and 27°C. Heat is now transferred to the air, and the piston rises until it hits the stops, at which point the volume is twice the initial volume. More heat is transferred until the pressure inside the cylinder also doubles. Determine the work done and the amount of heat transfer for this process. Also show the process on a *P-v* diagram.

**3-85E** A piston–cylinder device, with a set of stops on the top, initially contains 5 lbm of air at 30 psia and 80°F. Heat is now transferred to the air, and the piston rises until it hits the stops, at which point the volume is twice the initial volume. More heat is transferred until the pressure inside the cylinder also doubles. Determine the work done and the amount of heat transfer for this process. Also show the process on a *P-v* diagram. *Answers*: 184.9 Btu, 1731 Btu

**3-86** A rigid tank containing 0.4 $m^3$ of air at 400 kPa and 30°C is connected by a valve to a piston–cylinder device with zero clearance. The mass of the piston is such that a pressure of 200 kPa is required to raise the piston. The valve is now opened slightly, and air is allowed to flow into the cylinder until the pressure in the tank drops to 200 kPa. During this process, heat is exchanged with the surroundings such that the entire air remains at 30°C at all times. Determine the heat transfer for this process.

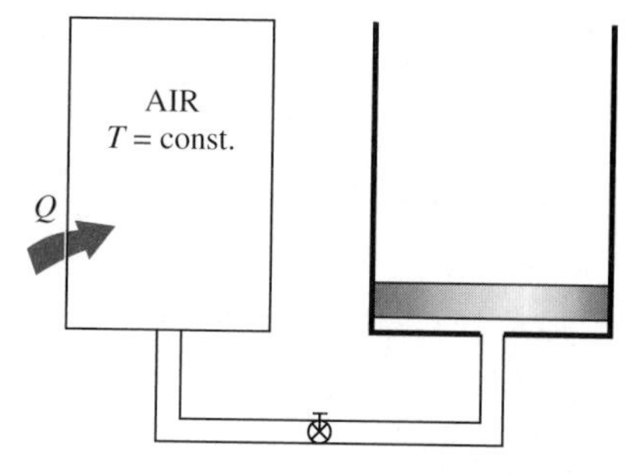


**FIGURE P3-86**

FIGURE P3-88

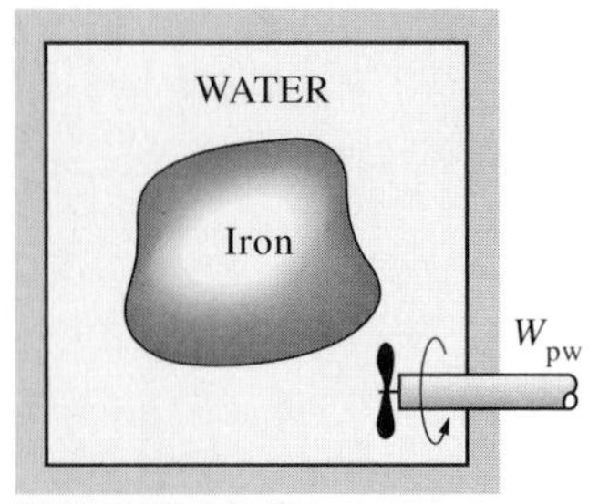


FIGURE P3-90

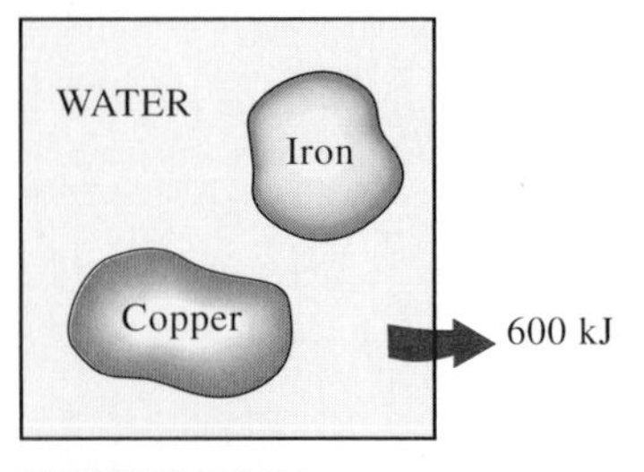


FIGURE P3-91

## Closed-System Energy Analysis: Solids and Liquids

**3-87** An unknown mass of aluminum at 60°C is dropped into an insulated tank that contains 40 L of water at 25°C and atmospheric pressure. If the final equilibrium temperature is 30°C, determine the mass of the aluminum. Assume the density of liquid water to be 1000 kg/m$^3$. *Answer:* 31.0 kg

**3-88** A 50-kg mass of copper at 70°C is dropped into an insulated tank that contains 80 kg of water at 25°C. Determine the final equilibrium temperature in the tank.

**3-88E** A 90-lbm mass of copper at 180°F is dropped into an insulated tank that contains 120 lbm of water at 70°F. Determine the final equilibrium temperature in the tank. *Answer:* 77.1°F

**3-89** A 20-kg mass of iron at 100°C is brought into contact with 20 kg of aluminum at 200°C in an insulated enclosure. Determine the final equilibrium temperature of the combined system.

**3-90** An unknown mass of iron at 90°C is dropped into an insulated tank that contains 80 L of water at 20°C. At the same time, a paddle wheel driven by a 200-W motor is activated to stir the water. Thermal equilibrium is established after 25 min with a final temperature of 27°C. Determine the mass of the iron. Neglect the energy stored in the paddle wheel, and take the water density to be 1000 kg/m$^3$. *Answer:* 72.1 kg

**3-91** A 50-kg mass of copper at 70°C and a 20-kg mass of iron at 80°C are dropped into a tank containing 150 kg of water at 20°C. If 600 kJ of heat is lost to the surroundings during the process, determine the final equilibrium temperature. *Answer:* 21.4°C

**3-91E** A 90-lbm mass of copper at 160°F and a 50-lbm mass of iron at 200°F are dropped into a tank containing 180 lbm of water at 70°F. If 600 Btu of heat is lost to the surroundings during the process, determine the final equilibrum temperature.

## Biological Systems

**3-92C** What is metabolism? What is basal metabolic rate? What is the value of basal metabolic rate for an average man?

**3-93C** What is the energy released during metabolism in humans used for?

**3-94C** Is the metabolizable energy content of a food the same as the energy released when it is burned in a bomb calorimeter? If not, how does it differ?

**3-95C** Is the number of prospective occupants an important consideration in the design of heating and cooling systems of classrooms? Explain.

**3-96C** What do you think of a diet program which allows for generous amounts of bread and rice provided that no butter or margarine is added?

**3-97** Consider two identical rooms, one with a 2-kW electric resistance heater and the other with three couples fast dancing. In which room will the air temperature rise faster?

**3-98** The average specific heat of the human body is 3.6 kJ/(kg · °C). If the body temperature of a 65-kg woman rises by 3°C as a result of fever, determine the amount of body fat that must be consumed to provide for the energy required for the rise in body temperature.

**3-99** Consider two identical 80-kg men who are eating identical meals and doing identical things except that one of them jogs for 30 min every day while the other watches TV. Determine the weight difference between the two in a month. *Answer:* 1.025 kg

**3-100** Consider a classroom that is losing heat to the outdoors at a rate of 20,000 kJ/h. If there are 30 students in class, each dissipating sensible heat at a rate of 100 W, determine if it is necessary to turn the heater in the classroom on to prevent the room temperature from dropping.

**3-101** A 68-kg woman is planning to bicycle for an hour. If she is to meet her entire energy needs while bicycling by eating 30-g chocolate candy bars, determine how many candy bars she needs to take with her.

**3-102** A 55-kg man gives in to temptation, and eats an entire 1-L box of ice cream. How long does this man need to jog to burn off the calories he consumed from the ice cream?

**3-103** Consider a man who has 20 kg of body fat when he goes on a hunger strike. Determine how long he can survive on his body fat alone.

**3-104** Consider two identical 50-kg women, Candy and Wendy, who are doing identical things and eating identical food except that Candy eats her baked potato with four spoons of butter while Wendy eats hers plain every evening. Determine the difference in the weights of Candy and Wendy after one year.

**3-105** A women who used to have about one liter of regular cola every day switches to diet cola (zero calorie) and starts eating two slices of apple pie every day. Is she now consuming fewer or more calories?

**3-106** A 60-kg man used to have an apple every day after dinner without losing or gaining any weight. He now eats a 200-ml serving of ice cream instead of apple, and walks 20 min every day. On this new diet, how much weight will he lose or gain?

**3-107** The average specific heat of the human body is 3.6 kJ/(kg · °C). If the body temperature of a 80-kg man rises from 37°C to 39°C during strenuous exercise, determine the increase in the thermal energy of the body as a result of this rise in body temperature.

## Review Problems

**3-108** Consider a vertical elevator whose cabin has a total mass of 800 kg when fully loaded, and 150 kg when empty. The weight of the elevator cabin is partially balanced by a 400-kg counterweight, which is connected to the top of the cabin by cables which pass through a pulley located on top of the elevator well. Neglecting the weight of the cables and assuming the guide rails and the pulleys to be frictionless, determine (*a*) the power required while the fully loaded cabin is rising at a constant speed of 2 m/s and (*b*) the power required while the empty cabin is descending at a constant speed of 2 m/s.

What would your answer be to (*a*) if no counterweight were used? What would your answer be to (*b*) if a friction force of 1500 N has developed between the cabin and the guide rails?

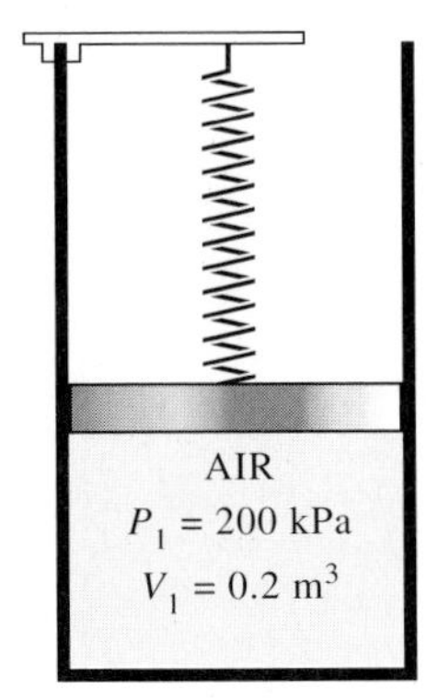


**FIGURE P3-109**

**3-109** A frictionless piston–cylinder device initially contains air at 200 kPa and 0.2 m$^3$. At this state, a linear spring ($F \propto x$) is touching the piston but exerts no force on it. The air is now heated to a final state of 0.5 m$^3$ and 800 kPa. Determine (*a*) the total work done by the air and (*b*) the work done against the spring. Also show the process on a *P-v* diagram. *Answers:* (*a*) 150 kJ, (*b*) 90 kJ

**3-110** A spherical balloon contains 5 kg of air at 200 kPa and 500 K. The balloon material is such that the pressure inside is always proportional to the square of the diameter. Determine the work done when the volume of the balloon doubles as a result of heat transfer. *Answer:* 936 kJ

**3-110E** A spherical balloon contains 10 lbm of air at 30 psia and 800 R. The balloon material is such that the pressure inside is always proportional to the square of the diameter. Determine the work done when the volume of the balloon doubles as a result of heat transfer.
*Answer:* 715.3 Btu

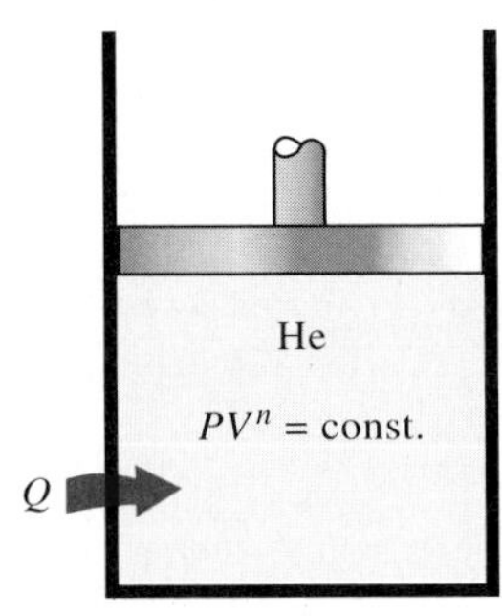


**FIGURE P3-111**

**3-111** A piston–cylinder device contains helium gas initially at 150 kPa, 20°C, and 0.5 m$^3$. The helium is now compressed in a polytropic process ($PV^n$ = constant) to 400 kPa and 140°C. Determine the heat transfer for this process. *Answer:* −11.2 kJ

**3-111E** A piston–cylinder device contains helium gas initially at 25 psia, 70°F, and 15 ft$^3$. The helium is now compressed in a polytropic process ($PV^n$ = constant) to 60 psia and 300°F. Determine the heat transfer for this process.

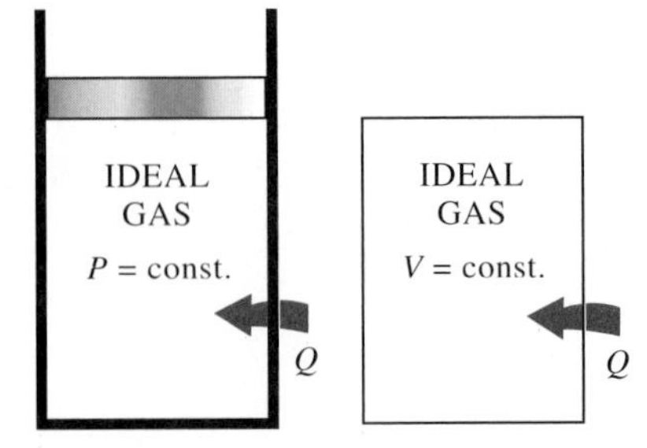


**FIGURE P3-112**

**3-112** A frictionless piston–cylinder device and a rigid tank initially contain 12 kg of an ideal gas each at the same temperature, pressure, and volume. It is desired to raise the temperatures of both systems by 15°C. Determine the amount of extra heat that must be supplied to the gas in the cylinder which is maintained at constant pressure to achieve this result. Assume the molar mass of the gas is 25.

**3-113** A piston–cylinder device contains 0.5 m$^3$ of helium gas initially at 100 kPa and 25°C. At this position, a linear spring is touching the piston

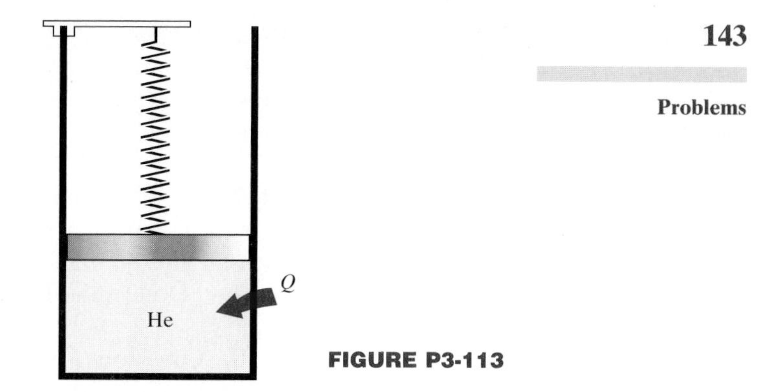


FIGURE P3-113

but exerts no force on it. Heat is now transferred to helium until both the pressure and the volume triple. Determine (*a*) the work done and (*b*) the amount of heat transfer for this process. Also show the process on a *P-v* diagram. *Answers:* (*a*) 200 kJ, (*b*) 801. 4kJ

**3-114** A passive solar house that is losing heat to the outdoors at an average rate of 50,000 kJ/h is maintained at 22°C at all times during a winter night for 10 h. The house is to be heated by 50 glass containers each containing 20 L of water, which is heated to 80°C during the day by absorbing solar energy. A thermostat controlled 15-kW back-up electric resistance heater turns on whenever necessary to keep the house at 22°C. (*a*) How long did the electric heating system run that night? (*b*) How long would the electric heater run that night if the house incorporated no solar heating? *Answers:* (*a*) 4.77 h, (*b*) 9.26 h

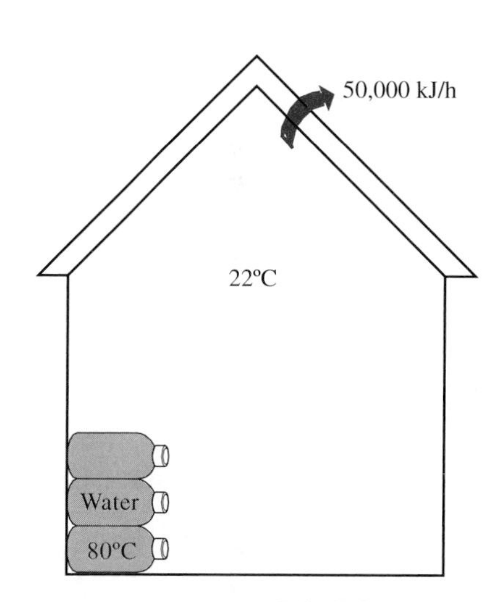


FIGURE P3-114

**3-115** It is well known that wind makes cold air feel much colder as a result of the *wind-chill* effect, which is due to the increase in the convection heat transfer coefficient as a result of the increase in air velocity. The wind-chill effect is usually expressed in terms of the *wind-chill factor,* which is the difference between the actual air temperature and the equivalent calm-air temperature. For example, a wind-chill factor of 20°C for an actual air temperature of 5°C means that the windy air at 5°C feels as cold as the still air at −15°C. In other words, a person will lose as much heat to air at 5°C with a wind-chill factor of 20°C as he or she would in calm air at −15°C.

For heat transfer purposes, a standing naked man can be modeled as a 30-cm-diameter, 170-cm-long vertical cylinder with both the top and bottom surfaces insulated and with the side surface at an average temperature of 34°C. For a convection heat transfer coefficient of 15 W/(m$^2$ · °C), determine the rate of heat loss from this man by convection in still air at 20°C. What would your answer be if the convection heat transfer coefficient was increased to 50 W/(m$^2$ · °C) as a result of winds? What is the wind-chill factor in this case?

**3-116** A 50-cm-long, 800-W electric resistance heating element whose diameter is 0.5 cm and surface temperature 120°C is immersed in 40 kg of water initially at 20°C. Determine how long it will take for this heater to raise the water temperature to 80°C. Also determine the convection heat transfer coefficients at the beginning and at the end of the heating process.

**3-117** A 100-kg man decides to lose 5 kg without cutting down his intake of 3000 Calories a day. Instead, he starts fast swimming, fast dancing, jogging, and biking for an hour every day. He sleeps or relaxes the rest of the day. Determine how long it will take for him to lose 5 kg.

**3-118** One ton (1000 kg) of liquid water at 80°C is brought into a well-insulated and well-sealed 4 m × 5 m × 6 m room initially at 22°C and 100 kPa. Assuming constant specific heats for both air and water at room temperature, determine the final equilibrium temperature in the room. *Answer:* 78.6°C

**3-119** A 4 m × 5 m × 6 m room is to be heated by one ton (1000 kg) of liquid water contained in a tank that is placed in the room. The room is losing heat to the outside at an average rate of 10,000 kJ/h. The room is initially at 20°C and 100 kPa, and is maintained at an average temperature of 20°C at all times. If the hot water is to meet the heating requirements of this room for a 24-h period, determine the minimum temperature of the water when it is first brought into the room. Assume constant specific heats for both air and water at room temperature.

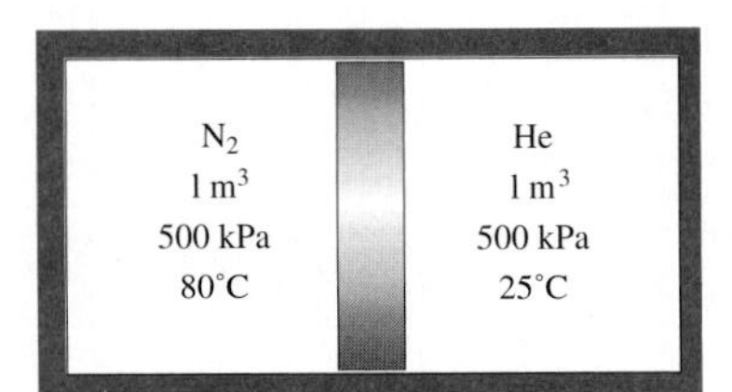


**FIGURE P3-120**

**3-120** Consider a well-insulated horizontal rigid cylinder that is divided into two compartments by a piston that is free to move but does not allow either gas to leak into the other side. Initially, one side of the piston contains 1 $m^3$ of $N_2$ gas at 500 kPa and 80°C while the other side contains 1 $m^3$ of He gas at 500 kPa and 25°C. Now thermal equilibrium is established in the cylinder as a result of heat transfer through the piston. Using constant specific heats at room temperature, determine the final equilibrium temperature in the cylinder. What would your answer be if the piston were not free to move?

**3-121** Repeat Prob. 3-120 assuming the piston is made of 5 kg of copper initially at the average temperature of the two gases on both sides.

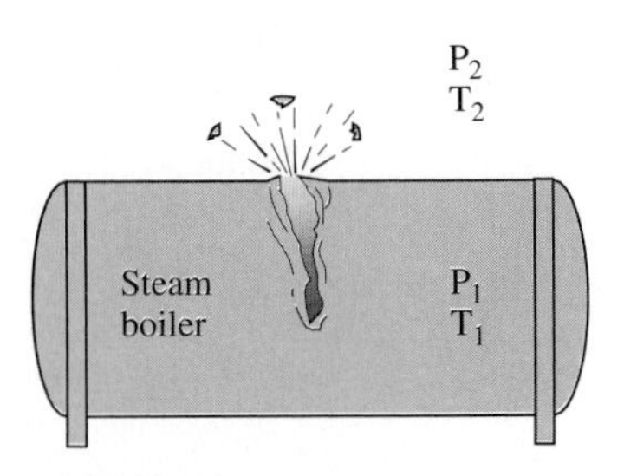


**FIGURE P3-122**

**3-122** Catastrophic explosions of steam boilers in the 1800s and early 1900s resulted in hundreds of deaths, which prompted the development of the ASME Boiler and Pressure Vessel Code in 1915. Considering that the pressurized fluid in a vessel eventually reaches equilibrium with its surroundings shortly after the explosion, the work that a pressurized fluid would do if allowed to expand adiabatically to the state of the surroundings can be viewed as the *explosive energy* of the pressurized fluid. Because of very short time period of the explosion and the apparent stability afterward, the explosion process can be considered to be adiabatic with no changes in kinetic and potential energies. The closed-system conservation of energy relation in this case reduces to $-W = m(u_2 - u_1)$. Then the explosive energy $E_{exp}$ becomes

$$E_{exp} = m(u_1 - u_2)$$

where the subscripts 1 and 2 refer to the state of the fluid before and after the explosion, respectively. The specific explosion energy $e_{exp}$ is usually expressed *per unit volume,* and it is obtained by dividing the quantity above by the total volume $V$ of the vessel:

$$e_{exp} = \frac{u_1 - u_2}{v_1}$$

where $v_1$ is the specific volume of the fluid before the explosion.

Show that the specific explosion energy of an ideal gas with constant specific heats is

$$e_{exp} = \frac{P_1}{k - 1}\left(1 - \frac{T_2}{T_1}\right)$$

Also determine the total explosion energy of 20 m$^3$ of air at 5 MPa and 100°C when the surroundings are at 20°C.

**3-123** The energy content of a certain food is to be determined in a bomb calorimeter that contains 3 kg of water by burning a 2-g sample of it in the presence of 100 g of air in the reaction chamber. If the water temperature rises by 3.2°C when equilibrium is established, determine the energy content of the food, in kJ/kg, by neglecting the thermal energy stored in the reaction chamber and the energy supplied by the mixer. What is a rough estimate of the error involved in neglecting the thermal energy stored in the reaction chamber?

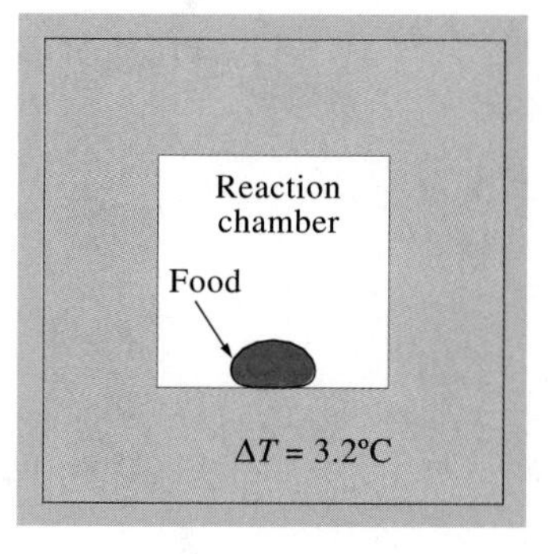


**FIGURE P3-123**

**3-124** A 68-kg man whose average body temperature is 38°C drinks 1 L of cold water at 3°C in an effort to cool down. Taking the average specific heat of human body to be 3.6 kJ/(kg · °C), determine the drop in the average body temperature of this person under the influence of this cold water.

**3-125** A 0.2-L glass of water at 20°C is to be cooled with ice to 5°C. Determine how much ice needs to be added to the water, in grams, if the ice is at (*a*) 0°C and (*b*) −8°C. Also determine how much water would be needed if the cooling is to be done with cold water at 0°C. The melting temperature and the heat of fusion of ice at atmospheric pressure are 0°C and 333.7 kJ/kg, respectively, and the density of water is 1 kg/L.

**3-126** In order to cool 1 ton (1000 kg) of water at 20°C in a insulated tank, a person pours 80 kg of ice at −5°C into the water. Determine the final equilibrium temperature in the tank. The melting temperature and the heat of fusion of ice at atmospheric pressure are 0°C and 333.7 kJ/kg, respectively. *Answer:* 12.4°C

**3-127** An insulated piston–cylinder device initially contains 0.01 m$^3$ of saturated liquid–vapor mixture with a quality of 0.2 at 100°C. Now some ice at 0°C is added into the cylinder. If the cylinder contains saturated liquid at 100°C when thermal equilibrium is established, determine the amount of ice added. The melting temperature and the heat of fusion of ice at atmospheric pressure are 0°C and 333.7 kJ/(kg · °C), respectively.

## Computer, Design, and Essay Problems

**3-128** Write a computer program to express the variation of specific heat $\bar{C}_p$ of air with temperature as a third-degree polynomial, using the data in Table A-2*b*. Compare your result with that given in Table A-2*c*.

**3-129** Write a computer program to determine the boundary work numerically using the trapezoidal rule when $n$ measured values of $P$ are given at the $n$ equally spaced values of $V$.

**3-130** Prepare a nutritional analysis of everything you have eaten and drunk during the last 24 h. Determine your total calorie intake and compare it to your calorie needs for that day. Also determine the percentage of calories that came from fats. Do you consider your diet during the last 24 h to be a healthy one? What changes would make it a healthier diet?

**3-131** Find out how the specific heats of gases, liquids, and solids are determined in national laboratories. Describe the experimental apparatus and the procedures used.

**3-132** Using information from the utility bills for the coldest month last year, estimate the average rate of heat loss from your house for that month. In your analysis, consider the contribution of the internal heat sources such as people, lights, and appliances. Identify the primary sources of heat losses from your house, and propose ways of improving the energy efficiency of your house.

**3-133** Design an experiment complete with instrumentation to determine the specific heats of a gas using a resistance heater. Discuss how the experiment will be conducted, what measurements need to taken, and how the specific heats will be determined. What are the sources of error in your system? How can you minimize the experimental error?

**3-134** Design an experiment complete with instrumentation to determine the specific heats of a liquid using a resistance heater. Discuss how the experiment will be conducted, what measurements need to taken, and how the specific heats will be determined. What are the sources of error in your system? How can you minimize the experimental error? How would you modify this system to determine the specific heat of a solid?

**3-135** Design a reciprocating compressor capable of supplying compressed air at 800 kPa at a rate of 15 kg/min. Also specify the size of the electric motor capable of driving this compressor. The compressor is to operate at no more than 2000 rpm (revolutions per minute).

# The First Law of Thermodynamics: Control Volumes

CHAPTER 4

In Chap. 3, we discussed the energy interactions between a system and its surroundings, and the conservation of energy principle for closed (nonflow) systems. In this chapter, we extend the analysis to systems that involve mass flow across their boundaries, i.e., *control volumes.* The conservation of energy equation for a general control volume can be rather involved. In this chapter, we first give a general description of the conservation of mass and energy equations for a general control volume. We then concentrate on the *steady-flow* process, which is the model process for many engineering devices such as turbines, compressors, heat exchangers and flow through ducts and pipes.

## 4-1 THERMODYNAMIC ANALYSIS OF CONTROL VOLUMES

A large number of engineering problems involve mass flow in and out of a system and, therefore, are modeled as *control volumes* (Fig. 4-1). A water heater, a car radiator, a turbine, and a compressor all involve mass flow and should be analyzed as control volumes (open systems) instead of as control masses (closed systems). In general, *any arbitrary region in space* can be selected as a control volume. There are no concrete rules for the selection of control volumes, but the proper choice certainly makes the analysis much easier. If we were to analyze the flow of air through a nozzle, for example, a good choice for the control volume would be the region within the nozzle.

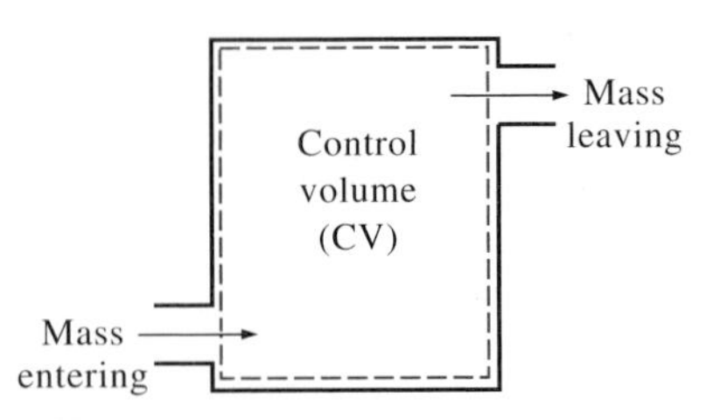


**FIGURE 4-1**

Mass may flow into and out of a control volume.

The boundaries of a control volume are called a *control surface,* and they can be real or imaginary. In the case of a nozzle, the inner surface of the nozzle forms the real part of the boundary, and the entrance and exit areas form the imaginary part, since there are no physical surfaces there (Fig. 4-2).

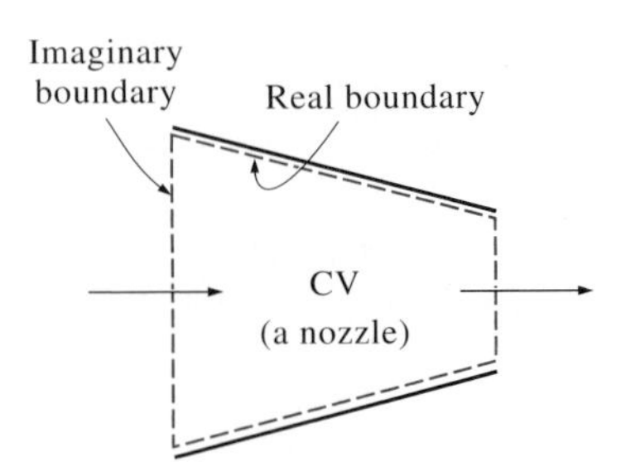


**FIGURE 4-2**

Real and imaginary boundaries of a control volume.

A control volume can be fixed in size and shape, as in the case of a nozzle, or it may involve a moving boundary, as shown in Fig. 4-3. Most control volumes, however, have fixed boundaries and thus do not involve any moving boundary work. A control volume may also involve heat and work interactions just as a closed system, in addition to mass interaction.

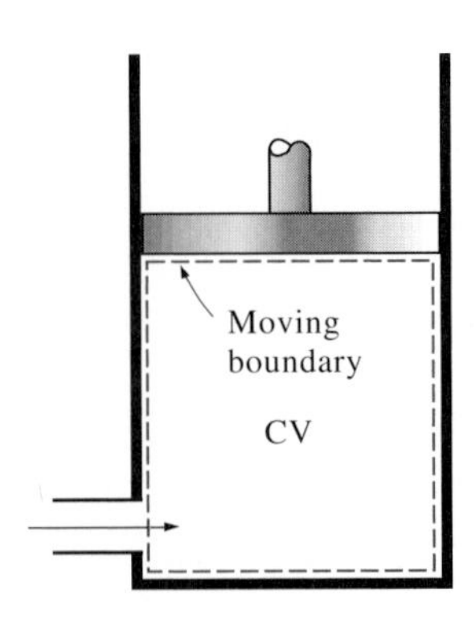


**FIGURE 4-3**

Some control volumes involve moving boundaries.

A large variety of thermodynamic problems may be solved by the control volume analysis. Even though it is possible to derive the relevant equations for the most general case and simplify them for special cases, many students are intimidated early in the analysis by the complexities involved and find them difficult to digest. The method that we use is a step-by-step approach, starting with the simplest case and adding complexities one at a time.

The terms *steady* and *uniform* are used extensively in this chapter, and thus it is important to have a clear understanding of their meanings. The term *steady* implies *no change with time.* The opposite of steady is *unsteady,* or *transient.* The term *uniform,* however, implies *no change with location* over a specified region. These meanings are consistent with their everyday use (steady girlfriend, uniform distribution, etc.).

An overview of the conservation of mass and the conservation of energy principles for control volumes is given below.

### Conservation of Mass Principle

The conservation of mass is one of the most fundamental principles in nature. We are all familiar with this principle, and it is not difficult to understand. As the saying goes, you cannot have your cake and eat it, too! A person does not have to be an engineer to figure out how much vinegar-and-oil dressing she is going to have if she mixes 100 g of oil with 25 g of vinegar. Even chemical equations are balanced on the basis of the conservation of mass principle (Fig. 4-4). When 16 kg of oxygen reacts with 2 kg of hydrogen, 18 kg of water is formed. In an electrolysis process, the water will separate back to 2 kg of hydrogen and 16 kg of oxygen.

Mass, like energy, is a conserved property, and it cannot be created or destroyed. However, mass $m$ and energy $E$ can be converted to each other according to the famous formula proposed:

$$E = mc^2 \tag{4-1}$$

where $c$ is the speed of light. This equation suggests that the mass of a system will change when its energy changes. However, for all energy interactions encountered in practice, with the exception of nuclear reactions, the change in mass is extremely small and cannot be detected by even the most sensitive devices. For example, when 1 kg of water is formed from oxygen and hydrogen, the amount of energy released is 15,879 kJ, which corresponds to a mass of $1.76 \times 10^{-10}$ kg. A mass of this magnitude is beyond the accuracy required by practically all engineering calculations and thus can be disregarded.

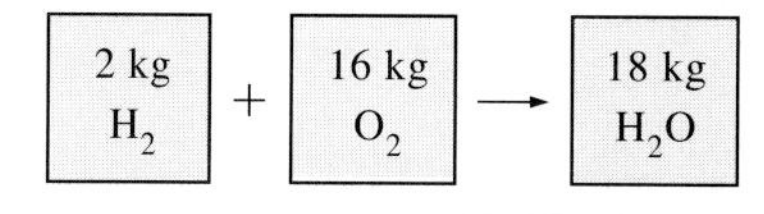


**FIGURE 4-4**

Mass is conserved even during chemical reactions.

For closed systems, the conservation of mass principle is implicitly used by requiring that the mass of the system remain constant during a process. For control volumes, however, mass can cross the boundaries, and so we must keep track of the amount of the mass entering and leaving the control volume (Fig. 4-5). The **conservation of mass principle** for a control volume (CV) undergoing a process can be expressed as

$$\begin{pmatrix}\text{total}\\ \text{mass entering}\\ \text{CV}\end{pmatrix} - \begin{pmatrix}\text{total}\\ \text{mass leaving}\\ \text{CV}\end{pmatrix} = \begin{pmatrix}\text{net change}\\ \text{in mass within}\\ \text{CV}\end{pmatrix}$$

or

$$\sum m_i - \sum m_e = \Delta m_{\text{CV}} \tag{4-2}$$

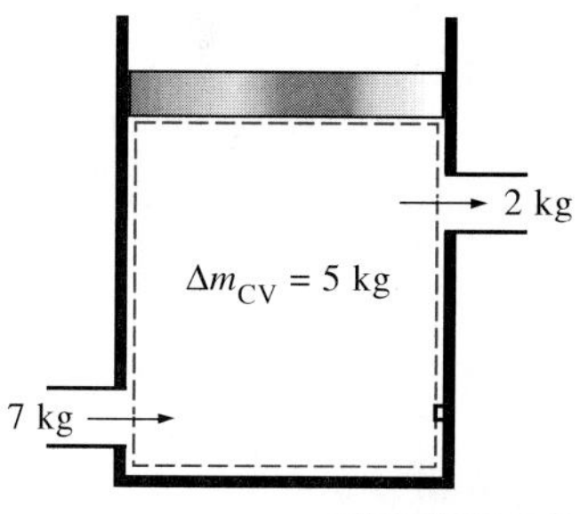


**FIGURE 4-5**

Conservation of mass principle for a control volume.

where the subscripts $i$, $e$, and CV stand for *inlet, exit,* and *control volume,* respectively. The conservation of mass equation could also be expressed in the rate form by expressing the quantities per unit time. Equation 4-2 is a verbal statement of the conservation of mass principle for a general control volume undergoing *any* process.

A person who can balance a checkbook (by keeping track of deposits and withdrawals, or simply by observing the "conservation of money" principle) should have no difficulty in applying the conservation of mass principle to thermodynamic systems. The conservation of mass equation is often referred to as the *continuity equation* in fluid mechanics.

## Mass and Volume Flow Rates

The amount of mass flowing through a cross section per unit time is called the **mass flow rate** and is denoted $\dot{m}$. As before, the dot over a symbol is used to indicate a quantity per unit time.

A liquid or a gas flows in and out of a control volume through pipes or ducts. The mass flow rate of a fluid flowing in a pipe or duct is proportional to the cross-sectional area $A$ of the pipe or duct, the density $\rho$, and the velocity $\mathcal{V}$ of the fluid. The mass flow rate through a differential area $dA$ can be expressed as

$$d\dot{m} = \rho \mathcal{V}_n \, dA \tag{4-3}$$

where $\mathcal{V}_n$ is the velocity component normal to $dA$. The mass flow rate through the entire cross-sectional area of the pipe or duct is obtained by integration:

$$\dot{m} = \int_A \rho \mathcal{V}_n \, dA \qquad \text{(kg/s)} \tag{4-4}$$

In most practical applications, the flow of a fluid through a pipe or duct can be approximated to be **one-dimensional flow**. That is, the properties can be assumed to vary in *one* direction only (the direction of flow). As a result, all properties are *uniform* at any cross-section normal to the flow direction, and the properties are assumed to have *bulk average values* over the cross-section. But the values of the properties at a cross section *may* change with time.

The one-dimensional-flow approximation has little impact on most properties of a fluid flowing in a pipe or duct such as temperature, pressure, and density since these properties usually remain constant over the cross section. But this is not the case for *velocity,* whose value varies from zero at the wall to a maximum at the center because of the viscous effects (friction between fluid layers). Under the one-dimensional-flow assumption, the velocity is assumed to be constant across the entire cross-section at some equivalent average value (Fig. 4-6). Then the integration in Eq. 4-4 can be performed for one-dimensional flow to yield

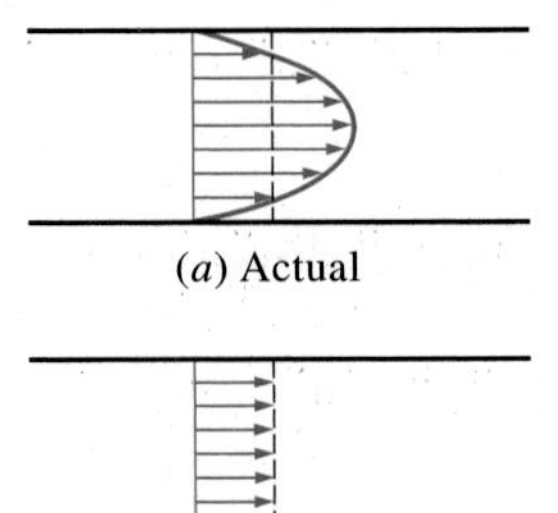

(*a*) Actual

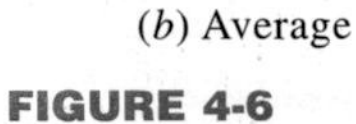

(*b*) Average

**FIGURE 4-6**
Actual and average velocity profiles for flow in a pipe (the mass flow rate is the same for both cases).

$$\dot{m} = \rho \mathcal{V}_{av} A \qquad \text{(kg/s)} \tag{4-5}$$

where $\rho$ = density, kg/m$^3$ ($= 1/v$)
$\mathcal{V}_{av}$ = average fluid velocity normal to $A$, m/s
$A$ = cross-sectional area normal to flow direction, m$^2$

The volume of the fluid flowing through a cross-section per unit time is called the **volume flow rate** $\dot{V}$ (Fig. 4-7) and is given by

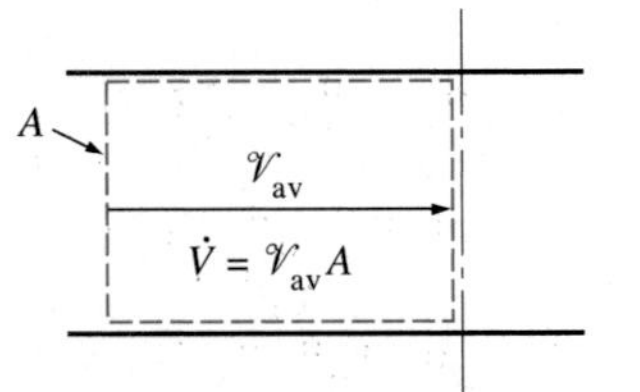


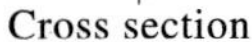


**FIGURE 4-7**
The volume flow rate is the volume of fluid flowing through a cross-section per unit time.

$$\dot{V} = \int_A \mathcal{V}_n \, dA = \mathcal{V}_{av} A \qquad \text{(m}^3\text{/s)} \tag{4-6}$$

The mass and volume flow rates are related by

$$\dot{m} = \rho \dot{V} = \frac{\dot{V}}{v} \tag{4-7}$$

This relation is analogous to $m = V/v$, which is the relation between the mass and the volume of a fluid.

For simplicity, we drop the subscript on the average velocity. Unless otherwise stated, $\mathcal{V}$ denotes the average velocity in the flow direction. Also $A$ denotes the cross-sectional area normal to the flow direction.

We have already discussed the conservation of energy principle, or the first law of thermodynamics, in relation to a closed system. As we pointed out in Chap. 3, the energy of a closed system can be changed by heat or work interactions only, and the change in the energy of a closed system during a process is equal to the net heat and work transfer across the system boundary. This was expressed as

$$Q - W = \Delta E$$

For control volumes, however, an additional mechanism can change the energy of a system: *mass flow in and out of the control volume* (Fig. 4-8). When mass enters a control volume, the energy of the control volume increases because the entering mass carries some energy with it. Likewise, when some mass leaves the control volume, the energy contained within the control volume decreases because the leaving mass takes *out* some energy with it. For example, when some hot water is taken out of a water heater and is replaced by the same amount of cold water, the energy content of the hot-water tank (the control volume) decreases as a result of this mass interaction.

Then the conservation of energy equation for a control volume undergoing a process can be expressed as

$$\begin{pmatrix}\text{total energy}\\ \text{crossing boundary}\\ \text{as heat and work}\end{pmatrix} + \begin{pmatrix}\text{total energy}\\ \text{of mass}\\ \text{entering CV}\end{pmatrix} - \begin{pmatrix}\text{total energy}\\ \text{of mass}\\ \text{leaving CV}\end{pmatrix} = \begin{pmatrix}\text{net change}\\ \text{in energy}\\ \text{of CV}\end{pmatrix}$$

or
$$Q - W + \sum E_{\text{in,mass}} - \sum E_{\text{out,mass}} = \Delta E_{\text{CV}} \qquad (4\text{-}8)$$

Obviously if no mass is entering or leaving the control volume, the second and third terms drop out and the above equation reduces to the one given for a closed system. Despite its simple appearance, Eq. 4-8 is applicable to *any* control volume undergoing *any* process. This equation can also be expressed in the rate form by expressing the quantities above per unit time.

Heat transfer to or from a control volume should not be confused with the energy transported with mass into and out of a control volume.

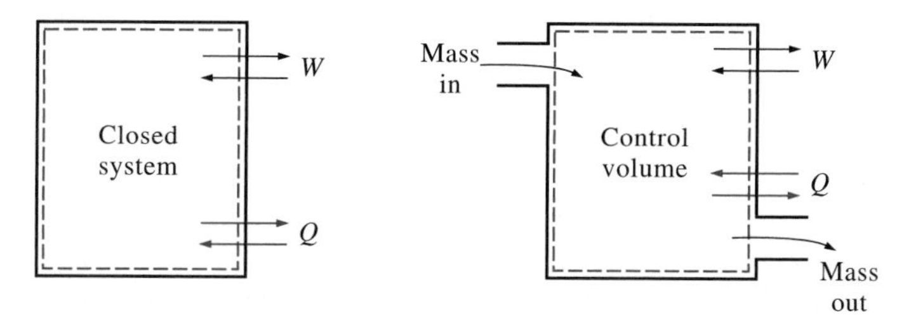


**FIGURE 4-8**
The energy content of a control volume can be changed by mass flow as well as heat and work interactions.

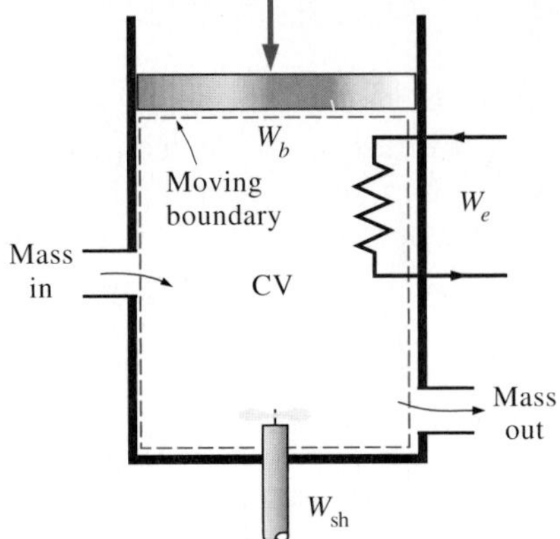


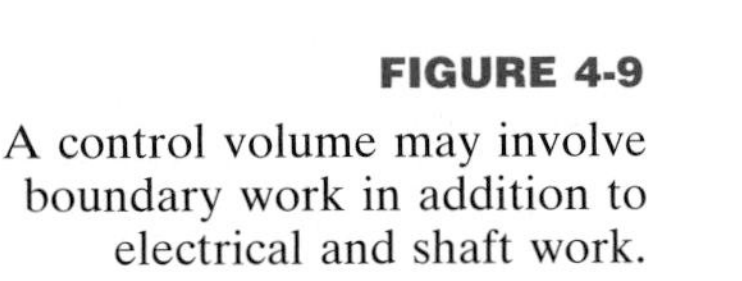

**FIGURE 4-9**
A control volume may involve boundary work in addition to electrical and shaft work.

Remember that heat is the form of energy transferred as a result of a temperature difference between the control volume and the surroundings.

A control volume, like a closed system, may involve one or more forms of work at the same time (Fig. 4-9). If the boundary of the control volume is stationary, as is often the case, the moving boundary work is zero. Then the work term will involve, at most, shaft work and electrical work for simple compressible systems. As before, when the control volume is insulated, the heat transfer term becomes zero.

The energy required to push fluid into or out of a control volume is called the *flow work,* or *flow energy.* It is considered to be part of the energy transported with the fluid and is discussed below.

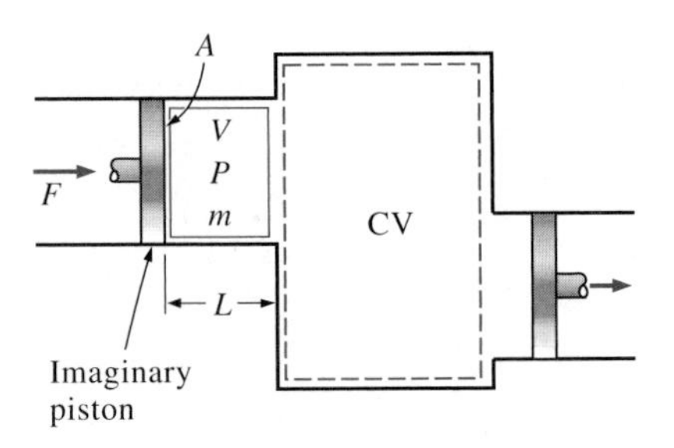


**FIGURE 4-10**
Schematic for flow work.

### Flow Work

Unlike closed systems, control volumes involve mass flow across their boundaries, and some work is required to push the mass into or out of the control volume. This work is known as the **flow work**, or **flow energy**, and is necessary for maintaining a continuous flow through a control volume.

To obtain a relation for flow work, consider a fluid element of volume $V$ as shown in Fig. 4-10. The fluid immediately upstream will force this fluid element to enter the control volume; thus, it can be regarded as an imaginary piston. The fluid element can be chosen to be sufficiently small so that it has uniform properties throughout.

If the fluid pressure is $P$ and the cross-sectional area of the fluid element is $A$ (Fig. 4-11), the force applied on the fluid element by the imaginary piston is

$$F = PA \tag{4-9}$$

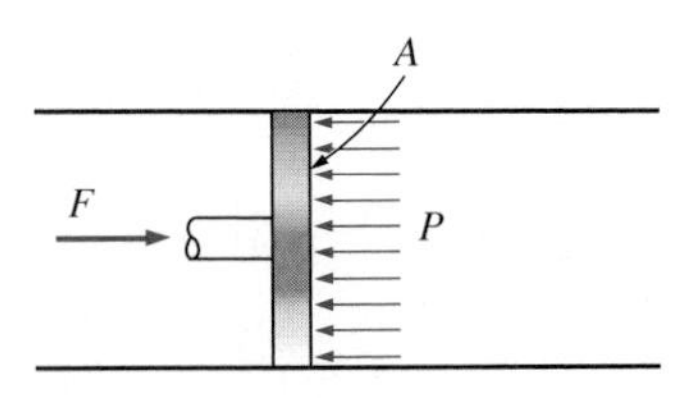


**FIGURE 4-11**
In the absence of acceleation, the force applied on a fluid by a piston is equal to the force applied on the piston by the fluid.

To push the entire fluid element into the control volume, this force must act through a distance $L$. Thus, the work done in pushing the fluid element across the boundary (i.e., the flow work) is

$$W_{\text{flow}} = FL = PAL = PV \qquad \text{(kJ)} \tag{4-10}$$

The flow work per unit mass is obtained by dividing both sides of this equation by the mass of the fluid element:



$$w_{\text{flow}} = Pv \qquad (\text{kJ/kg}) \tag{4-11}$$

The flow work relation is the same whether the fluid is pushed into or out of the control volume (Fig. 4-12).

It is interesting that unlike other work quantities, flow work is expressed in terms of the properties. In fact, it is the product of two properties of the fluid. For that reason, some people view it as a *combination property* (like enthalpy) and refer to it as *flow energy, convected energy,* or *transport energy* instead of flow work. Others, however, argue rightfully that the product $Pv$ represents energy for flowing fluids only and does not represent any form of energy for nonflow (closed) systems. Therefore, it should be treated as work. This controversy is not likely to end, but it is comforting to know that both arguments yield the same result for the energy equation. In the discussions that follow, we consider the flow energy to be part of the energy of a flowing fluid, since this greatly simplifies the derivation of the energy equation for control volumes.

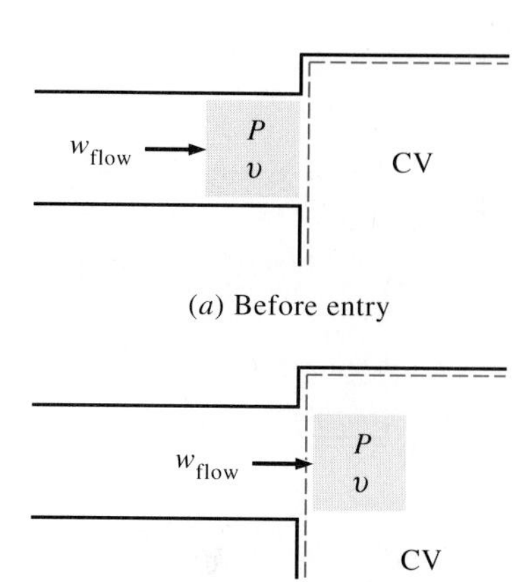


**FIGURE 4-12**

Flow work is the energy needed to push a fluid into or out of a control volume, and it is equal to $Pv$.

### Total Energy of a Flowing Fluid

As we discussed in Chap. 1, the total energy of a simple compressible system consists of three parts: internal, kinetic, and potential energies (Fig. 4-13). On a unit-mass basis, it is expressed as

$$e = u + \text{ke} + \text{pe} = u + \frac{\mathscr{V}^2}{2} + gz \qquad (\text{kJ/kg}) \tag{4-12}$$

where $\mathscr{V}$ is the velocity and $z$ is the elevation of the system relative to some external reference point.

The fluid entering or leaving a control volume possesses an additional form of energy—the *flow energy* $Pv$, as discussed above. Then the total energy of a **flowing fluid** on a unit-mass basis (denoted $\theta$) becomes

$$\theta = Pv + e = Pv + (u + \text{ke} + \text{pe})$$

But the combination $Pv + u$ has been previously defined as the enthalpy

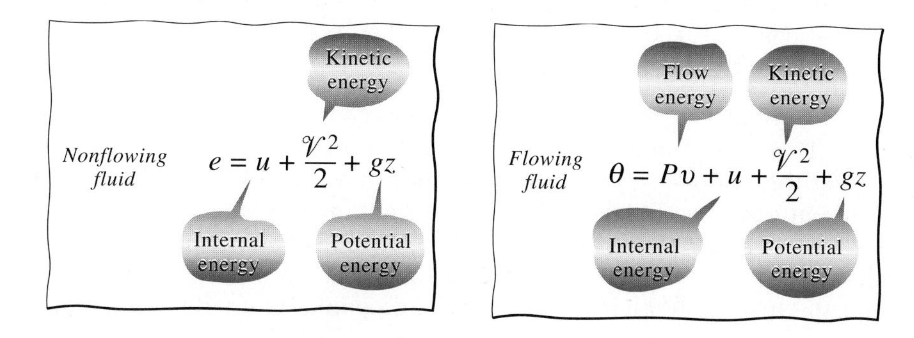


**FIGURE 4-13**

The total energy consists of three parts for a nonflowing fluid and four parts for a flowing fluid.

*h*. So the above relation reduces to

$$\theta = h + \text{ke} + \text{pe} = h + \frac{\mathscr{V}^2}{2} + gz \qquad \text{(kJ/kg)} \qquad (4\text{-}13)$$

Professor J. Kestin proposed in 1966 that the term $\theta$ be called **methalpy** (from *metaenthalpy,* which means *beyond enthalpy*).

By using the enthalpy instead of the internal energy to represent the energy of a flowing fluid, one does not need to be concerned about the flow work. The energy associated with pushing the fluid into or out of the control volume is automatically taken care of by enthalpy. In fact, this is the main reason for defining the property enthalpy. From now on, the energy of a fluid stream flowing into or out of a control volume is represented by Eq. 4-13, and no reference will be made to flow work or flow energy. Thus, the work term $W$ in the control-volume energy equations will represent all forms of work (boundary, shaft, electrical, etc.) except flow work.

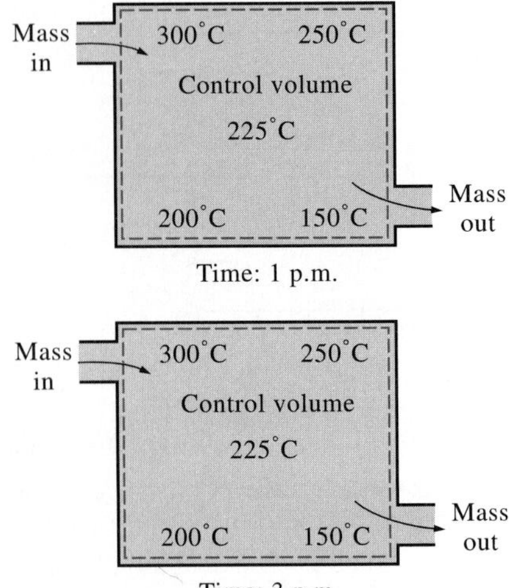


**FIGURE 4-14**
During a steady-flow process fluid properties within the control volume may change with position, but not with time.

## 4-2 THE STEADY-FLOW PROCESS

A large number of engineering devices such as turbines, compressors, and nozzles operate for long periods of time under the same conditions, and they are classified as *steady-flow devices.*

Processes involving steady-flow devices can be represented reasonably well by a somewhat idealized process, called the **steady-flow process.** A steady-flow process can be defined as *a process during which a fluid flows through a control volume steadily* (Fig. 4-14). That is, the fluid properties can change from point to point within the control volume, but at any fixed point they remain the same during the entire process. (Remember, *steady* means *no change with time.*) A steady-flow process is characterized by the following:

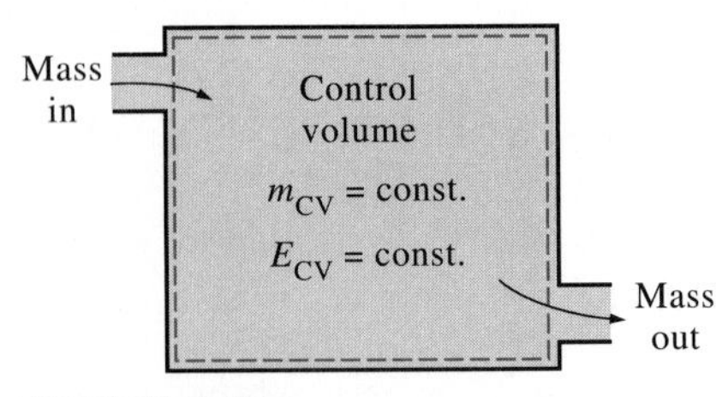


**FIGURE 4-15**
Under steady-flow conditions, the mass and energy contents of a control volume remain constant.

1 No properties (intensive or extensive) *within the control volume* change with time. Thus, the volume $V$, the mass $m$, and the total energy content $E$ of the control volume remain constant during a steady flow process (Fig. 4-15). As a result, the boundary work is zero for steady flow systems (since $V_{CV}$ = constant), and the total mass or energy entering the control volume must be equal to the total mass or energy leaving it (since $m_{CV}$ = constant and $E_{CV}$ = constant). These observations greatly simplify the analysis.

2 No properties change at the *boundaries* of the control volume with time. Thus, the fluid properties at an inlet or an exit will remain the same during the entire process. The properties may, however, be different at different openings (inlets and exits). They may even vary over the cross-section of an inlet or an exit. But all properties, including the velocity and elevation, must remain constant with time at a fixed position. It follows that the mass flow rate of the fluid at an opening must remain constant

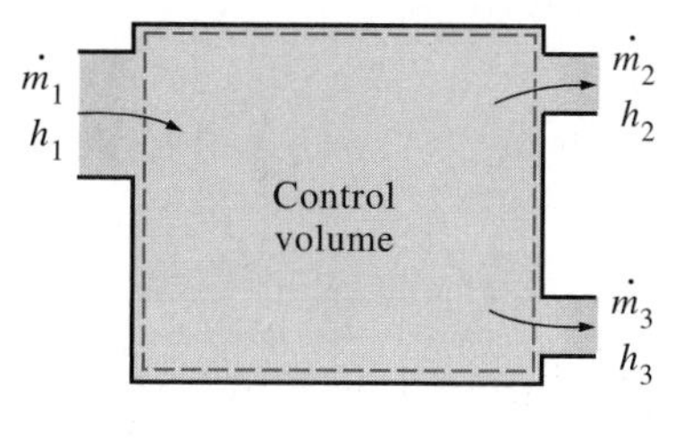


**FIGURE 4-16**

Under steady-flow conditions, the fluid properties at an inlet or exit remain constant (do not change with time).

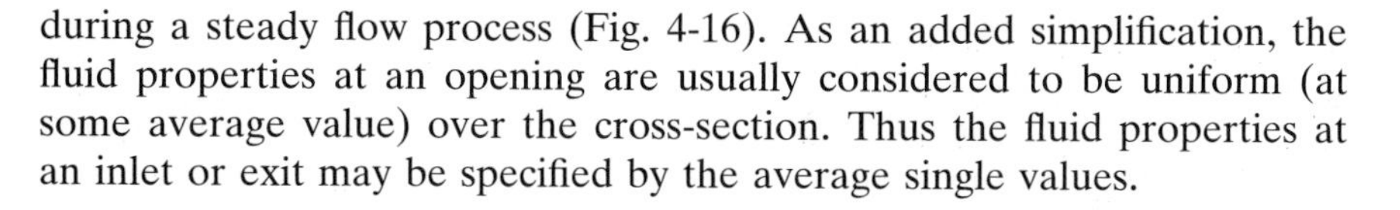

during a steady flow process (Fig. 4-16). As an added simplification, the fluid properties at an opening are usually considered to be uniform (at some average value) over the cross-section. Thus the fluid properties at an inlet or exit may be specified by the average single values.

**3** The *heat* and *work* interactions between a steady-flow system and its surroundings do not change with time. Thus the power delivered by a system and the rate of heat transfer to or from a system remain constant during a steady flow process.

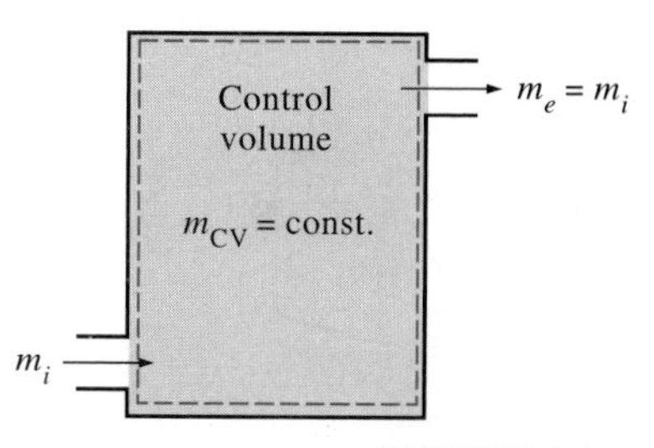


**FIGURE 4-17**

During a steady-flow process, the amount of mass entering the control volume equals the amount of mass leaving.

Some cyclic devices, such as reciprocating engines or compressors, do not satisfy any of the conditions stated above since the flow at the inlets and the exits will be pulsating and not steady. However, the fluid properties vary with time in a periodic manner, and the flow through these devices can still be analyzed as a steady flow process by using time-averaged values for the properties and the heat flow rates through the boundaries.

Steady-flow conditions can be closely approximated by devices that are intended for continuous operation such as turbines, pumps, boilers, condensers, and heat exchangers of steam power plants. The equations that are developed later in this section can be used for these and similar devices once the transient start-up period is completed and a steady operation is established.

## Conservation of Mass

During a steady-flow process, the total amount of mass contained within a control volume does not change with time ($m_{CV}$ = constant). Then the conservation of mass principle requires that the total amount of mass entering a control volume equal the total amount of mass leaving it (Fig. 4-17). For a garden hose nozzle, for example, the amount of water entering the nozzle is equal to the amount of water leaving it under steady operation.

When dealing with steady-flow processes, we are not interested in the amount of mass that flows in and out of a device over time; instead, we are interested in the amount of mass flowing per unit time, i.e., *the mass flow rate* $\dot{m}$. The **conservation of mass principle** for a general steady-flow system with multiple inlets and exits (Fig. 4-18) can be expressed in the

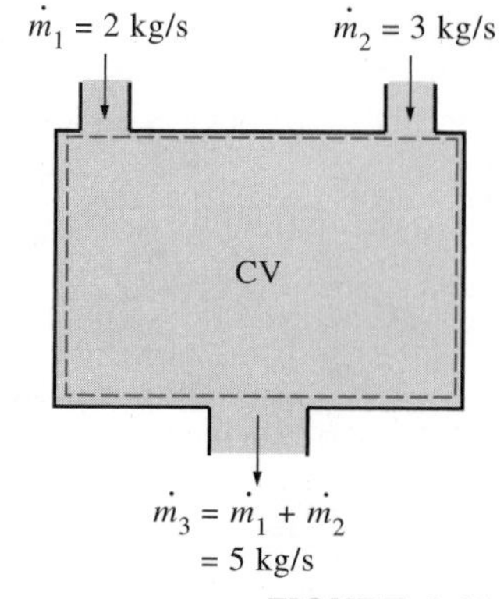


**FIGURE 4-18**

Conservation of mass principle for a two-inlet one-exit steady-flow system.

rate form as

$$\begin{pmatrix}\text{total mass}\\ \text{entering CV}\\ \text{per unit time}\end{pmatrix} = \begin{pmatrix}\text{total mass}\\ \text{leaving CV}\\ \text{per unit time}\end{pmatrix}$$

or

$$\sum \dot{m}_i = \sum \dot{m}_e \qquad \text{(kg/s)} \tag{4-14}$$

where the subscript $i$ stands for *inlet* and $e$ for *exit.* Most engineering devices such as nozzles, diffusers, turbines, compressors, and pumps involve a single stream (only one inlet and one exit). For these cases, we denote the inlet state by the subscript 1 and the exit state by the subscript 2. We also drop the summation signs. Then Eq. 4-14 reduces, for single-stream steady-flow systems, to

$$\dot{m}_1 = \dot{m}_2 \qquad \text{(kg/s)} \tag{4-15}$$

or

$$\rho_1 \mathcal{V}_1 A_1 = \rho_2 \mathcal{V}_2 A_2 \tag{4-16}$$

or

$$\frac{1}{v_1} \mathcal{V}_1 A_1 = \frac{1}{v_2} \mathcal{V}_2 A_2 \tag{4-17}$$

where $\rho$ = density, kg/m$^3$
$v$ = specific volume, m$^3$/kg ($= 1/\rho$)
$\mathcal{V}$ = average flow velocity in flow direction, m/s
$A$ = cross-sectional area normal to flow direction, m$^2$

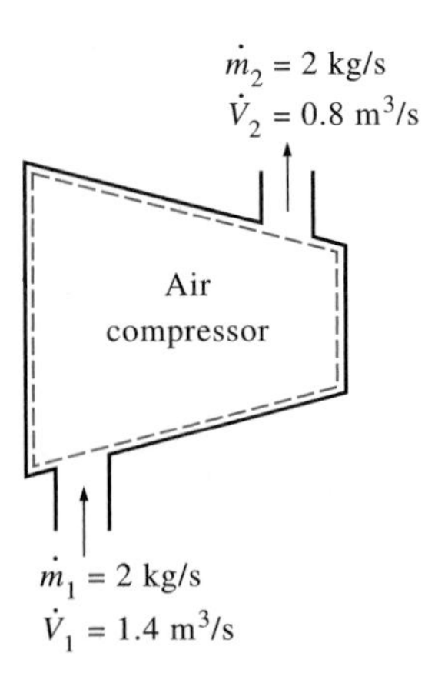


**FIGURE 4-19**
During a steady-flow process, volume flow rates are not necessarily conserved.

The reader is reminded that there is no such thing as a "conservation of volume" principle. Therefore, the volume flow rates ($\dot{V} = \mathcal{V}A$, m$^3$/s) into and out of a steady-flow device may be different. The volume flow rate at the exit of an air compressor will be much less than that at the inlet even though the mass flow rate of air through the compressor is constant (Fig. 4-19). This is due to the higher density of air at the compressor exit. For liquid flow, however, the volume flow rates, as well as the mass flow rates, remain constant since liquids are essentially incompressible (constant-density) substances. Water flow through the nozzle of a garden hose is a good example for the latter case.

## Conservation of Energy

It was pointed out earlier that during a steady-flow process the total energy content of a control volume remains constant ($E_{\text{CV}}$ = constant). That is, the change in the total energy of the control volume during such a process is zero ($\Delta E_{\text{CV}} = 0$). Thus, the amount of energy entering a control volume in all forms (heat, work, mass transfer) must be equal to the amount of energy leaving it for a steady-flow process.

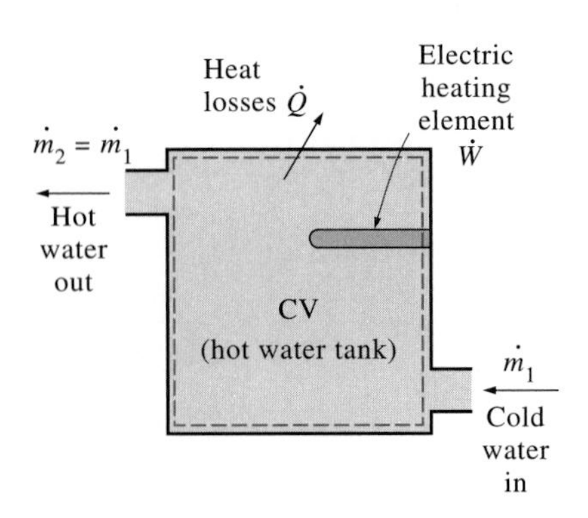


**FIGURE 4-20**
A water heater under steady operation.

Consider, for example, an ordinary electric hot-water heater under steady operation, as shown in Fig. 4-20. A cold-water stream with a mass flow rate $\dot{m}$ is continuously flowing into the water heater, and a hot-water stream of the same mass flow rate is continuously flowing out of it. The water heater (the control volume) is losing heat to the surrounding air at a rate of $\dot{Q}$, and the electric heating element is doing electrical work (heating) on the water at a rate of $\dot{W}$. On the basis of the conservation of energy principle, we can say that *the water stream will experience an increase in its total energy as it flows through the water heater, which is equal to the electric energy supplied to the water minus the heat losses.*

By this line of reasoning, the first law of thermodynamics or **conservation of energy principle** for a general steady-flow system with multiple inlets and exits can be expressed verbally as

$$\begin{pmatrix}\text{total energy}\\ \text{crossing boundary}\\ \text{as heat and work}\\ \text{per unit time}\end{pmatrix} = \begin{pmatrix}\text{total energy}\\ \text{transported out of}\\ \text{CV with mass}\\ \text{per unit time}\end{pmatrix} - \begin{pmatrix}\text{total energy}\\ \text{transported into}\\ \text{CV with mass}\\ \text{per unit time}\end{pmatrix}$$

or

$$\dot{Q} - \dot{W} = \sum \dot{m}_e\theta_e - \sum \dot{m}_i\theta_i \tag{4-18}$$

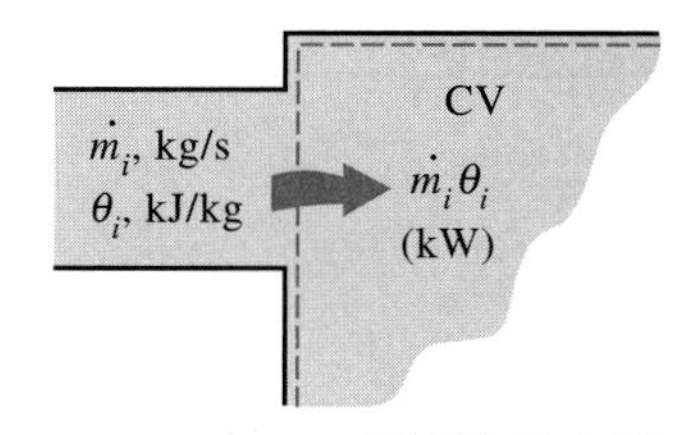


**FIGURE 4-21**
The product $\dot{m}_i\theta_i$ is the energy transported into the control volume by mass per unit time.

where $\theta$ is the total energy of the flowing fluid, including the flow work, per unit mass (Fig. 4-21). It can also be expressed as

$$\dot{Q} - \dot{W} = \underbrace{\sum \dot{m}_e\left(h_e + \frac{\mathcal{V}_e^2}{2} + gz_e\right)}_{\text{for each exit}} - \underbrace{\sum \dot{m}_i\left(h_i + \frac{\mathcal{V}_i^2}{2} + gz_i\right)}_{\text{for each inlet}} \tag{4-19}$$

since $\theta = h + \text{ke} + \text{pe}$ (Eq. 4-13). Equation 4-19 is the general form of the first-law relation for steady-flow processes. The first law for steady-flow systems first appeared in 1859 in a German thermodynamics book written by Gustav Zeuner.

For single-stream (one-inlet, one-exit) systems, the summations over the inlets and the exits drop out, and the inlet and exit states in this case are denoted by subscripts 1 and 2, respectively, for simplicity. The mass flow rate through the entire control volume remains constant ($\dot{m}_1 = \dot{m}_2$) and is denoted $\dot{m}$. Then the conservation of energy equation for *single-stream steady-flow systems* becomes

$$\dot{Q} - \dot{W} = \dot{m}\left[h_2 - h_1 + \frac{\mathcal{V}_2^2 - \mathcal{V}_1^2}{2} + g(z_2 - z_1)\right] \qquad \text{(kW)} \tag{4-20}$$

or

$$\dot{Q} - \dot{W} = \dot{m}(\Delta h + \Delta\text{ke} + \Delta\text{pe}) \qquad \text{(kW)} \tag{4-21}$$

Dividing these equations by $\dot{m}$, we obtain the first-law relation on a unit-mass basis as

$$q - w = h_2 - h_1 + \frac{\mathcal{V}_2^2 - \mathcal{V}_1^2}{2} + g(z_2 - z_1) \qquad \text{(kJ/kg)} \tag{4-22}$$

or

$$q - w = \Delta h + \Delta\text{ke} + \Delta\text{pe} \qquad \text{(kJ/kg)} \tag{4-23}$$

where

$$q = \frac{\dot{Q}}{\dot{m}} \qquad \text{(heat transfer per unit mass, kJ/kg)} \tag{4-24}$$

and

$$w = \frac{\dot{W}}{\dot{m}} \qquad \text{(work done per unit mass, kJ/kg)} \tag{4-25}$$

If the fluid experiences a negligible change in its kinetic and potential energies as it flows through the control volume (that is, $\Delta\text{ke} \approx 0$, $\Delta\text{pe} \approx 0$) then the energy equation for a single-stream steady-flow system reduces further to

$$q - w = \Delta h \qquad \text{(kJ/kg)} \tag{4-26}$$

This is the simplest form of the first-law relation for control volumes. Its form resembles the first-law relation for closed systems except that $\Delta u$ is replaced by $\Delta h$ in this case.

The various terms appearing in the above equations are as follows:

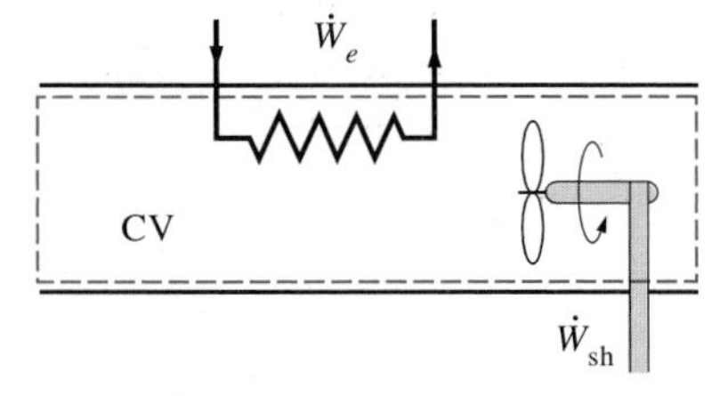


**FIGURE 4-22**
Under steady operation, shaft work and electrical work are the only forms of work a simple compressible system may involve.

$\dot{Q}$ = **rate of heat transfer between the control volume and its surroundings.** When the control volume is losing heat (as in the case of the water heater), $\dot{Q}$ is negative. If the control volume is well insulated (i.e., adiabatic), then $\dot{Q} = 0$.

$\dot{W}$ = **power.** For steady-flow sevices, the volume of the control volume is constant; thus, there is no boundary work involved. The work required to push mass into and out of the control volume is also taken care of by using enthalpies for the energy of fluid streams instead of internal energies. Then $\dot{W}$ represents the remaining forms of work done per unit time (Fig. 4-22). Many steady-flow devices, such as turbines, compressors, and pumps, transmit power through a shaft, and $\dot{W}$ simply becomes the shaft power for those devices. If the control surface is crossed by electric wires (as in the case of an electric water heater), $\dot{W}$ will represent the electrical work done per unit time. If neither is present then $\dot{W} = 0$.

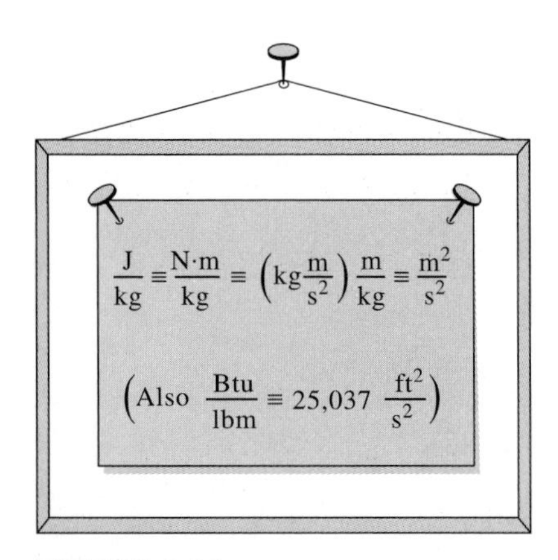


**FIGURE 4-23**
The units $m^2/s^2$ and J/kg are equivalent.

$\Delta h = h_{\text{exit}} - h_{\text{inlet}}$. The enthalpy change of a fluid can easily be determined by reading the enthalpy values at the exit and inlet states from the tables. For ideal gases, it may be approximated by $\Delta h = C_{p,\text{av}}(T_2 - T_1)$. Note that (kg/s)(kJ/kg) $\equiv$ kW.

$\Delta\text{ke} = (\mathscr{V}_2^2 - \mathscr{V}_1^2)/2$. The unit of kinetic energy is $m^2/s^2$, which is equivalent ot J/kg (Fig. 4-23). The enthalpy is usually given in kJ/kg. To add these two quantities, the kinetic energy should be expressed in kJ/kg. This is easily accomplished by dividing it by 1000.

A velocity of 45 m/s corresponds to a kinetic energy of only 1 kJ/kg, which is a very small value compared with the enthalpy values encountered in practice. Thus, the kinetic energy term at low velocities can be neglected. When a fluid stream enters and leaves a steady-flow device at about the same velocity ($\mathscr{V}_1 \approx \mathscr{V}_2$), the change in the kinetic energy is close to zero regardless of the velocity. Caution should be exercised at high velocities, however, since small changes in velocities may cause significant changes in kinetic energy (Fig. 4-24).

| $\mathscr{V}_1$ m/s | $\mathscr{V}_2$ m/s | $\Delta$ke kJ/kg |
|---|---|---|
| 0 | 40 | 1 |
| 50 | 67 | 1 |
| 100 | 110 | 1 |
| 200 | 205 | 1 |
| 500 | 502 | 1 |

**FIGURE 4-24**
At very high velocities, even small changes in velocities may cause significant changes in the kinetic energy of the fluid.

$\Delta\text{pe} = g(z_2 - z_1)$. A similar argument can be given for the potential energy term. A potential energy change of 1 kJ/kg corresponds to an elevation difference of 102 m. The elevation difference between

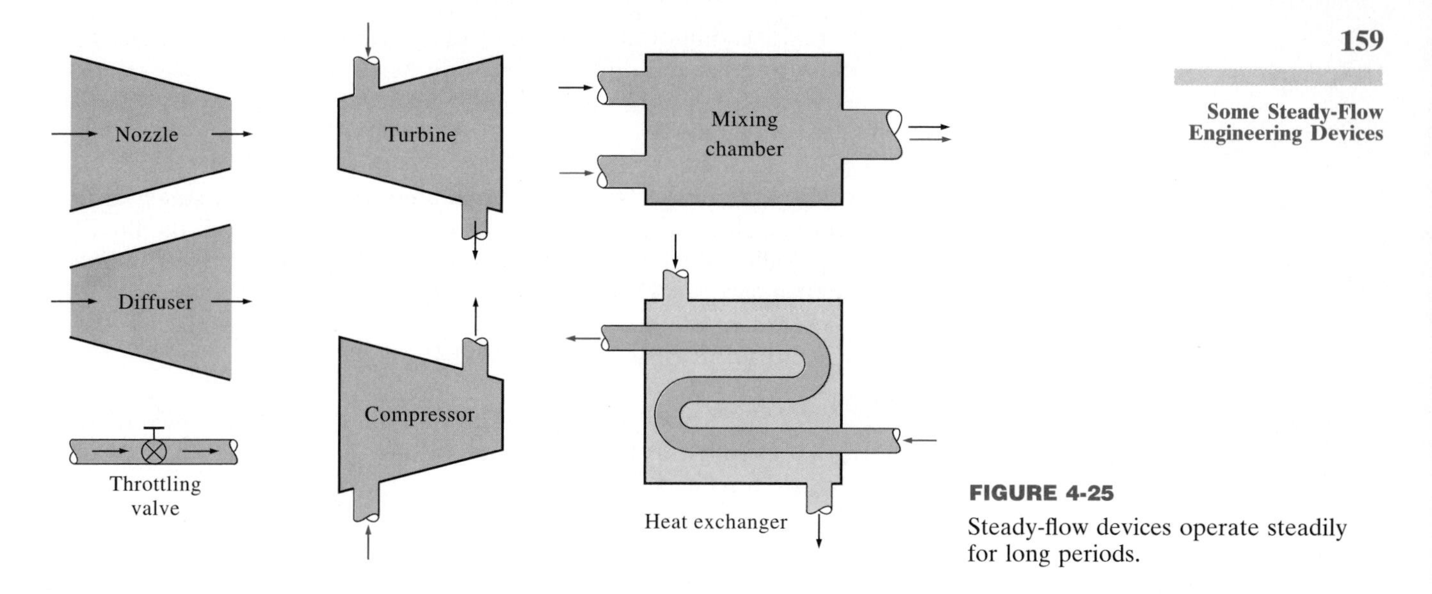


**FIGURE 4-25**
Steady-flow devices operate steadily for long periods.

the inlet and exit of most industrial devices such as turbines and compressors is well below this value, and the potential energy term is always neglected for these devices. The only time the potential energy term is significant is when a process involves pumping a fluid to high elevations. This is particularly true for systems involving negligible heat transfer.

## 4-3 ■ SOME STEADY-FLOW ENGINEERING DEVICES

Many engineering devices operate essentially under the same conditions for long periods of time. The components of a steam power plant (turbines, compressors, heat exchangers, and pumps), for example, operate nonstop for months before the system is shut down for maintenance (Fig. 4-25). Therefore, these devices can be conveniently analyzed as steady-flow devices.

In this section, some common steady-flow devices are described, and the thermodynamic aspects of the flow through them are analyzed. The conservation of mass and the conservation of energy principles for these devices are illustrated with examples.

### 1 Nozzles and Diffusers

Nozzles and diffusers are commonly utilized in jet engines, rockets, spacecraft, and even garden hoses. A **nozzle** is a device that *increases the velocity of a fluid* at the expense of pressure. A **diffuser** is a device that *increases the pressure of a fluid* by slowing it down. That is, nozzles and diffusers perform opposite tasks. The cross-sectional area of a nozzle decrease in the flow direction for subsonic flows and increases for supersonic flows. The reverse is true for diffusers.

The relative importance of the terms appearing in the energy equation for nozzles and diffusers is as follows:

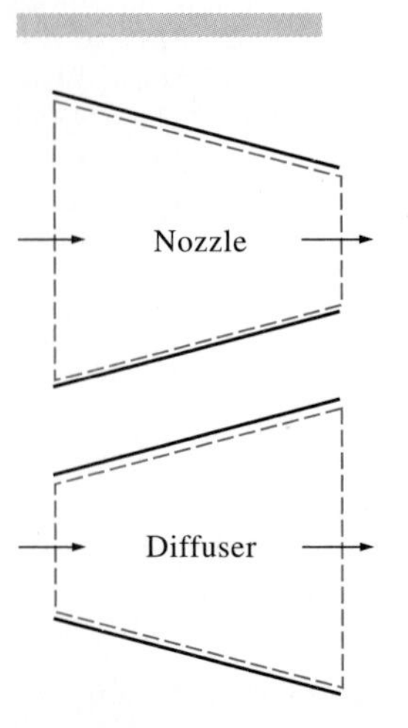


**FIGURE 4-26**
Schematic of a nozzle and diffuser for subsonic flows (velocities under the speed of sound).

$\dot{Q} \approx 0.$ The rate of heat transfer between the fluid flowing through a nozzle or a diffuser and the surroundings is usually very small, even when these devices are not insulated. This is mainly due to the fluid's having high velocities and thus not spending enough time in the device for any significant heat transfer to take place. Therefore, in the absence of any heat transfer data, the flow through nozzles and diffusers may be assumed to be adiabatic.

$\dot{W} = 0.$ The work term for nozzles and diffusers is zero since these devices basically are properly shaped ducts an they involve no shaft or electric resistance wires.

$\Delta \text{ke} \neq 0.$ Nozzles and diffusers usually involve very high velocities, and as a fluid passes through a nozzle or diffuser, it experiences large changes in its velocity (Fig. 4-27). Therefore, the kinetic energy changes must be accounted for in analyzing the flow through these devices.

$\Delta \text{pe} \approx 0.$ The fluid usually experiences little or no change in its elevation as it flows through a nozzle or a diffuser, and therefore the potential energy term can be neglected.

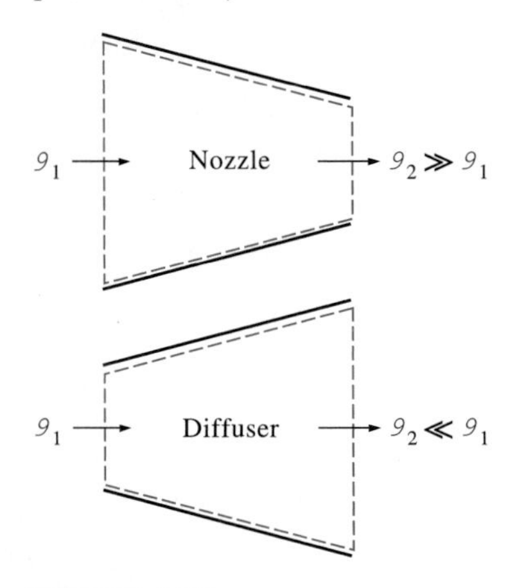


**FIGURE 4-27**
Nozzles and diffusers are shaped so that they cause large changes in fluid velocities and thus kinetic energies.

**EXAMPLE 4-1**

Air at 10°C and 80 kPa enters the diffuser of a jet engine steadily with a velocity of 200 m/s. The inlet area of the diffuser is 0.4 $m^2$. The air leaves the diffuser with a velocity that is very small compared with the inlet velocity. Determine (*a*) the mass flow rate of the air and (*b*) the temperature of the air leaving the diffuser.

**Solution** The region within the diffuser is selected as the system, and its boundaries are shown in Fig. 4-28. Mass is crossing the boundaries, thus it is a *control volume.* And since there is no observable change within the control volume with time, it is a *steady-flow system.* At the specified conditions, the

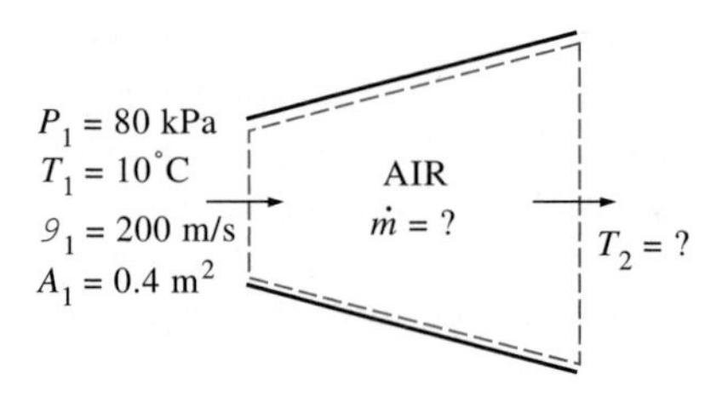


**FIGURE 4-28**
Schematic for Example 4-1.

air can be treated as an *ideal gas* since it is at a high temperature and low pressure relative to its critical values ($T_{cr} = -147°C$ and $P_{cr} = 3390$ kPa for nitrogen, the main constituent or air).

(*a*) To determine the mass flow rate, we need to find the specific volume of the air first. This is determined from the ideal-gas relation at the inlet conditions:

$$v_1 = \frac{RT_1}{P_1} = \frac{[0.287\ \text{kPa}\cdot\text{m}^3/(\text{kg}\cdot\text{K})](283\ \text{K})}{80\ \text{kPa}} = 1.015\ \text{m}^3/\text{kg}$$

Then from Eq. 4-5,

$$\dot{m} = \frac{1}{v_1}\mathscr{V}_1 A_1 = \frac{1}{1.015\ \text{m}^3/\text{kg}}(200\ \text{m/s})(0.4\ \text{m}^2) = 78.8\ \text{kg/s}$$

Since the flow is steady, the mass flow rate through the entire diffuser will remain constant at this value.

(*b*) A diffuser normally involves no shaft or electrical work ($w = 0$), negligible heat transfer ($q \approx 0$), and a small (if any) elevation change between the inlet and the exit ($\Delta\text{pe} \approx 0$). Then the conservation of energy relation on a unit-mass basis for this single-stream steady-flow system (Eq. 4-23) reduces to

$$\cancelto{0}{q} - \cancelto{0}{w} = \Delta h + \Delta\text{ke} + \cancelto{0}{\Delta\text{pe}}$$

$$0 = h_2 - h_1 + \frac{\mathscr{V}_2^2 - \mathscr{V}_1^2}{2}$$

$$0 = C_p(T_2 - T_1) + \frac{\mathscr{V}_2^2 - \mathscr{V}_1^2}{2}$$

The exit velocity of a diffuser is usually small compared with the inlet velocity ($\mathscr{V}_2 \ll \mathscr{V}_1$); thus the kinetic energy at the exit can be neglected. Using specific heats at room temperature, the exit temperature of air is determined to be

$$T_2 = T_1 - \frac{\mathscr{V}_2^2 - \mathscr{V}_1^2}{2C_p} = 10°\text{C} - \frac{0 - (200\ \text{m/s})^2}{2 \times [1.005\ \text{kJ/(kg}\cdot°\text{C)}]}\left(\frac{1\ \text{kJ/kg}}{1000\ \text{m}^2/\text{s}^2}\right) = 29.9°\text{C}$$

which shows that the temperature of the air increased by about 20°C as it was slowed down in the diffuser. The temperature rise of the air is mainly due to the conversion of kinetic energy to internal energy.

## 2 Turbines and Compressors

In steam, gas, or hydroelectric power plants, the device that drives the electric generator is the turbine. As the fluid passes through the turbine, work is done against the blades which are attached to the shaft. As a result, the shaft rotates, and the turbine produces work. The work done in a turbine is positive since it is done by the fluid.

Compressors, as well as pumps and fans, are devices used to increase the pressure of a fluid. Work is supplied to these devices from an external source through a rotating shaft. Therefore, the work term for compressors is negative since work is done on the fluid. Even though these three devices function similarly, they do differ in the tasks they perform. A *fan* increases the pressure of a gas slightly and is mainly used to move a gas around. A *compressor* is capable of compressing the gas to very high pressures. *Pumps* work very much like compressors except that they handle liquids instead of gases.

For turbines and compressors, the relative magnitudes of the various terms appearing in the energy equation are as follows:

$\dot{Q} \approx 0$. The heat transfer for these devices is generally small relative to the shaft work unless there is intentional cooling (as for the case of a compressor). An estimated value based on the experimental studies can be used in the analysis, or the heat transfer may be neglected if there is no intentional cooling.

$\dot{W} \neq 0$. All these devices involve rotating shafts crossing their boundaries, therefore the work term is important. For turbines, $\dot{W}$ represents the power output; for pumps and compressors, it represents the power input.

$\Delta \text{pe} \approx 0$. The potential energy change that a fluid experiences as it flows through turbines, compressors, fans, and pumps is usually very small and is normally neglected.

$\Delta \text{ke} \approx 0$. The velocities involved with these devices, with the exception of turbines, are usually too low to cause any significant change in the kinetic energy. The fluid velocities encountered in most turbines are very high, and the fluid experiences a significant change in its kinetic energy. However, this change is usually very small relative to the change in enthalpy, and thus it is often disregarded.

**EXAMPLE 4-2**

Air at 100 kPa and 280 K is compressed steadily to 600 kPa and 400 K. The mass-flow rate of the air is 0.02 kg/s, and a heat loss of 16 kJ/kg occurs during the process. Assuming the changes in kinetic and potential energies are negligible, determine the necessary power input to the compressor.

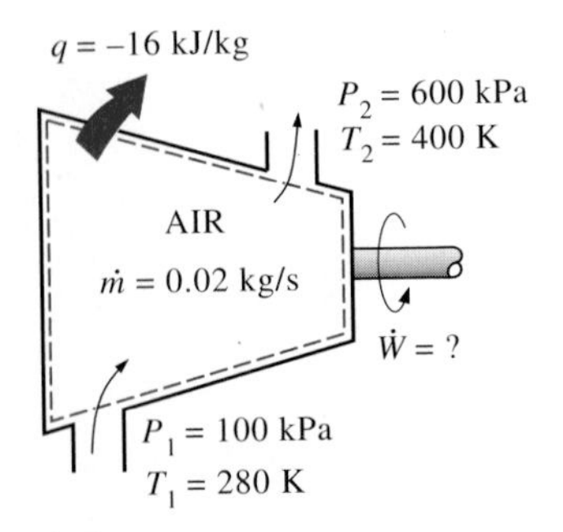


**FIGURE 4-29**
Schematic for Example 4-2.

**Solution** We choose the region within the compressor as our system, and its boundaries are indicated by dashed lines in Fig. 4-29. Mass is crossing the boundaries, thus it is a *control volume*; and there is no observable change within the control volume with time, so it is a *steady-flow system.* At the specified conditions, the air can be treated as an *ideal gas* since it is at a high temperature and low pressure relative to its critical values ($T_{cr}$ = −147°C and $P_{cr}$ = 3390 kPa for nitrogen, the main constituent of air).

It is stated that $\Delta ke \approx 0$ and $\Delta pe \approx 0$. Then the conservation of energy equation (Eq. 4-23) for this single-stream steady-flow system reduces to

$$q - w = \Delta h + \cancelto{0}{\Delta ke} + \cancelto{0}{\Delta pe} = C_p(T_2 - T_1)$$

Using the specific heat of air from Table A-2*a* at the average temperature of $(400 + 280)/2 = 340$ K and substituting the known quantities, the compressor work is determined to be

$$\begin{aligned} w &= q - C_p(T_2 - T_1) \\ &= (-16 \text{ kJ/kg}) - [1.007 \text{ kJ/(kg} \cdot \text{K)}](400 - 280) \text{ K} \\ &= 136.8 \text{ kJ/kg} \end{aligned}$$

This is the work done on the air per unit mass. The power input to the compressor is determined by multiplying this value by the mass flow rate:

$$\dot{W} = \dot{m}w = (0.02 \text{ kg/s})(-136.8 \text{ kJ/kg}) = -2.74 \text{ kW}$$

**EXAMPLE 4-3**

The power output of an adiabatic gas turbine is 5 MW, and the inlet and the exit conditions of the hot gases are as indicated in Fig. 4-30. The gases can be treated as air.

(*a*) Compare the magnitudes of $\Delta h$, $\Delta ke$, and $\Delta pe$.

(*b*) Determine the work done per unit mass of hot gases.

(*c*) Calculate the mass flow rate of the steam.

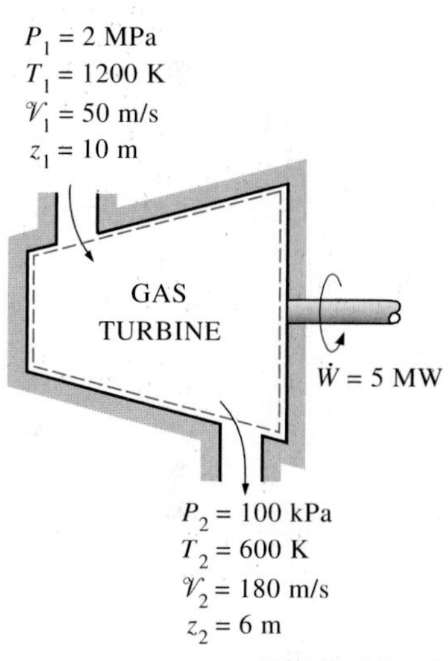


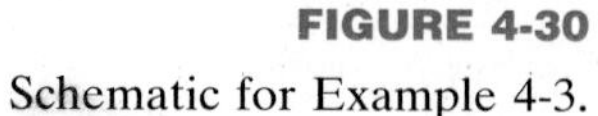

**FIGURE 4-30**
Schematic for Example 4-3.

**Solution** This time we select the region within the turbine as our system. Its boundaries are indicated by dashed lines in Fig. 4-30. Mass is crossing the boundaries, thus it is a *control volume.* There is no indication of any change within the control volume with time; thus, it is a *steady-flow system.*

(*a*) The changes in enthalpy, kinetic energy, and potential energy are determined from their definitions to be

$$\begin{aligned} \Delta h &= h_2 - h_1 = C_p(T_2 - T_1) = [1.121 \text{ kJ/(kg} \cdot \text{K)}](600 - 1200) \text{ K} \\ &= 672.6 \text{ kJ/kg} \end{aligned}$$

$$\Delta ke = \frac{\mathscr{V}_2^2 - \mathscr{V}_1^2}{2} = \frac{(180 \text{ m/s})^2 - (50 \text{ m/s})^2}{2}\left(\frac{1 \text{ kJ/kg}}{1000 \text{ m}^2/\text{s}^2}\right) = 14.95 \text{ kJ/kg}$$

$$\Delta pe = g(z_2 - z_1) = (9.807 \text{ m/s}^2)[(6 - 10)\text{m}]\left(\frac{1 \text{ kJ/kg}}{1000 \text{ m}^2/\text{s}^2}\right) = -0.04 \text{ kJ/kg}$$

Two observations can be made from the above results. First, the change in potential energy is insignificant in comparison to the changes in enthalpy and kinetic energy. This is typical for most engineering devices. Second, as a result of low pressure and thus high specific volume, the gas velocity at the turbine exit can be very high. Yet the change in kinetic energy is a small fraction of the change in enthalpy and is therefore often neglected.

(*b*) The work done per unit mass for this one-inlet, one-exit device can be determined from the first-law relation on a unit-mass basis (Eq. 4-23):

$$\cancelto{0}{q} - w = \Delta h + \Delta \text{ke} + \Delta \text{pe}$$

$$w = -(-672.6 + 14.95 - 0.04)\ \text{kJ/kg} = \mathbf{657.61\ kJ/kg}$$

(*c*) The required mass flow rate for a 5-MW power output is determined from

$$\dot{m} = \frac{\dot{W}}{w} = \frac{5000\ \text{kJ/s}}{657.61\ \text{kJ/kg}} = \mathbf{7.60\ kg/s}$$

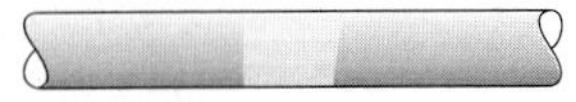

(*a*) An adjustable valve

(*b*) A porous plug

(*c*) A capillary tube

**FIGURE 4-31**
Throttling valves are devices that cause large pressure drops in the fluid.

## 3 Throttling Valves

Throttling valves are *any kind of flow-restricting devices* that cause a significant pressure drop in the fluid. Some familiar examples are ordinary adjustable valves, capillary tubes, and porous plugs (Fig. 4-31). Unlike turbines, they produce a pressure drop without involving any work. The pressure drop in the fluid is often accompanied by a *large drop in temperature,* and for that reason throttling devices are commonly used in refrigeration and air-conditioning applications. The magnitude of the temperature drop (or, sometimes, the temperature rise) during a throttling process is governed by a property called the *Joule–Thomson coefficient.*

Throttling valves are usually small devices, and the flow through them may be assumed to be adiabatic ($q \approx 0$) since there is neither sufficient time nor large enough area for any effective heat transfer to take place. Also, there is no work done ($w = 0$), and the change in potential energy, if any, is very small ($\Delta\text{pe} \approx 0$). Even though the exit velocity is often considerably higher than the inlet velocity, in many cases, the increase in kinetic energy is insignificant ($\Delta\text{ke} \approx 0$). Then the conservation of energy equation for this single-stream steady-flow device reduces to

$$h_2 \approx h_1 \qquad \text{(kJ/kg)} \tag{4-27}$$

That is, enthalpy values at the inlet and exit of a throttling valve are the same. For this reason, a throttling valve is sometimes called an *isenthalpic device.*

To gain some insight into how throttling affects fluid properties, let us express Eq. 4-27 as follows:

$$u_1 + P_1 v_1 = u_2 + P_2 v_2$$

or

$$\text{internal energy} + \text{flow energy} = \text{constant}$$

Thus the final outcome of a throttling process depends on which of the two quantities increases during the process. If the flow energy increases during the process ($P_2 v_2 > P_1 v_1$), it can do so at the expense of the internal energy. As a result, internal energy decreases, which is usually accompanied by a drop in temperature. If the product $Pv$ decreases, the internal energy and the temperature of a fluid will increase during a throttling process. In the case of an ideal gas $h = h(T)$, and thus the temperature has to remain constant during a throttling process (Fig. 4-32).

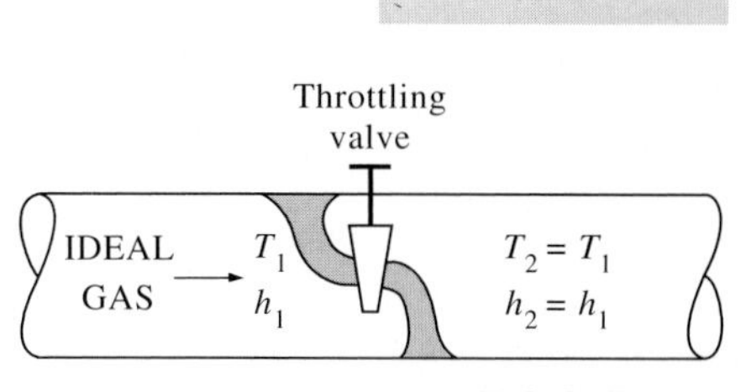


**FIGURE 4-32**
The temperature of an ideal gas does not change during a throttling ($h$ = constant) process since $h = h(T)$.

### EXAMPLE 4-4

Refrigerant-134a enters the capillary tube of a refrigerator as saturated liquid at 0.8 MPa and is throttled to a pressure of 0.12 MPa. Part of the refrigerant evaporates during this process and the refrigerant exists as a saturated liquid–vapor mixture at the final state. Determine the temperature drop of the refrigerant during this process.

**Solution** A capillary tube is a simple flow-restricting device that is commonly used in refrigeration applications to cause a large pressure drop in the refrigerant. Flow through a capillary tube is a throttling tube is a throttling process, and thus the enthalpy of the refrigerant remains constant during this process

The refrigerant is a saturated liquid at the initial state, and thus its initial temperature must be the saturation temperature at the initial pressure, which is determined from Table 4-1 to be

$$T_1 = T_{\text{sat@0.8 MPa}} = 31.33°\text{C}$$

The exit temperature of the refrigerant must also be the saturation temperature at the exit pressure, since it exists as a saturated mixture at the final state. That is,

$$T_2 = T_{\text{sat@0.12 MPa}} = -22.36°\text{C}$$

Therefore, the temperature change the refrigerant experiences during this throttling process is

$$\Delta T = T_2 - T_1 = -22.36 - 31.33 = -53.69°\text{C}$$

That is, the temperature of the refrigerant drops by 53.69°C during this throttling process. About one third of the refrigerant vaporizes during this throttling process, and the energy needed to vaporize this refrigerant is absorbed from the refrigerant itself.

**TABLE 4-1**
**Saturation temperature of refrigerant-134a at various pressures**

| $P$, MPa | $T_{sat}$, °C |
|---|---|
| 0.10 | −26.43 |
| 0.12 | −22.36 |
| 0.14 | −18.80 |
| 0.16 | −15.62 |
| 0.18 | −12.73 |
| 0.20 | −10.09 |
| 0.24 | −5.37 |
| 0.28 | −1.23 |
| 0.32 | 2.48 |
| 0.4 | 8.93 |
| 0.5 | 15.74 |
| 0.6 | 21.58 |
| 0.7 | 26.72 |
| 0.8 | 31.33 |
| 0.9 | 35.53 |
| 1.0 | 39.39 |
| 1.2 | 46.32 |

## 4a Mixture Chambers

In engineering applications, mixing two streams of fluids is not a rare occurrence. The section where the mixing process takes place is commonly referred to as a **mixing chamber**. The mixing chamber does not have to be a distinct "chamber." An ordinary T-elbow or a Y-elbow in a shower, for example, serves as the mixing chamber for the cold- and hot-water streams (Fig. 4-33).

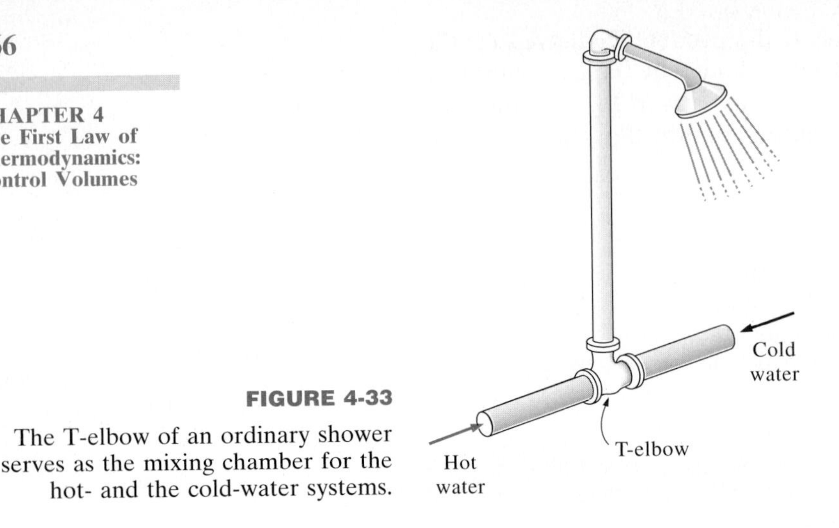


**FIGURE 4-33**
The T-elbow of an ordinary shower serves as the mixing chamber for the hot- and the cold-water systems.

The conservation of mass principle for a mixing chamber requires that the sum of the incoming mass flow rates equal the mass flow rate of the outgoing mixture.

Mixing chambers are usually well insulated ($q \approx 0$) and do not involve any kind of work ($w = 0$). Also, the kinetic and potential energies of the fluid streams are usually negligible (ke $\approx 0$, pe $\approx 0$). Then all there is left in the energy equation (Eq. 4-19) is the total energies of the incoming streams and the outgoing mixture. The conservation of energy principle requires that these two equal each other. Therefore, the conservation of energy equation becomes analogous to the conservation of mass equation for this case.

**EXAMPLE 4-5**

Consider an ordinary shower where hot water at 140°F is mixed with cold water at 50°F. If it is desired that a steady stream of warm water at 110°F be supplied, determine the ratio of the mass flow rates of the hot to cold water. Assume the heat losses from the mixing chamber to be negligible and the mixing to take place at a pressure of 20 psia.

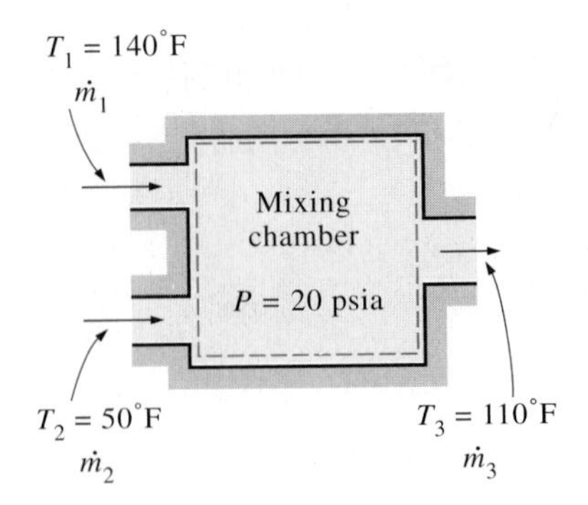


**FIGURE 4-34**
Schematic for Example 4-5.

**Solution** We take the mixing chamber as our system whose boundaries are indicated by dashed lines in Fig. 4-34. Mass is crossing the boundaries; thus, it is a *control volume.* And there is no indication of any change within the control volume with time; thus, it is a *steady-flow system.*

The conservation of mass equation (Eq. 4-14) for this multiple-stream steady-flow system is

$$\sum \dot{m}_i = \sum \dot{m}_e$$

or

$$\dot{m}_1 + \dot{m}_2 = \dot{m}_3$$

No heat or work is crossing the boundaries ($\dot{Q} \approx 0$, $\dot{W} = 0$), and the kinetic and potential energies are considered to be negligible (ke $\approx 0$, pe $\approx 0$). Then the conservation of energy equation for this steady-flow system reduces to

$$\dot{Q}^{\nearrow 0} - \dot{W}^{\nearrow 0} = \sum \dot{m}_e\left(h_e + \frac{\mathscr{V}_e^{2\nearrow 0}}{2} + gz_e^{\nearrow 0}\right) - \sum \dot{m}_i\left(h_i + \frac{\mathscr{V}_i^{2\nearrow 0}}{2} + gz_i^{\nearrow 0}\right)$$

$$\sum \dot{m}_i h_i = \sum \dot{m}_e h_e$$

$$\dot{m}_1 h_1 + \dot{m}_2 h_2 = \dot{m}_3 h_3$$

or

$$\dot{m}_1 h_1 + \dot{m}_2 h_2 = (\dot{m}_1 + \dot{m}_2)h_3$$

Dividing this equation by $\dot{m}_2$ yields

$$yh_1 + h_2 = (y + 1)h_3$$

where $y = \dot{m}_1/\dot{m}_2$ is the desired mass flow rate ratio. Solving for $y$ gives

$$y = \frac{h_3 - h_2}{h_1 - h_3}$$

Assuming constant specific heats for water, the relation above simplifies to

$$y = \frac{C_p(T_3 - T_2)}{C_p(T_1 - T_3)} = \frac{T_3 - T_2}{T_1 - T_3} = \frac{(110 - 50)°F}{(140 - 110)°F} = 2.0$$

Thus, the mass flow rate of the hot water must be twice the mass flow rate of the cold water for the mixture to leave at 110°F.

## 4b Heat Exchangers

As the name implies, **heat exchangers** are devices where two moving fluid streams exchange heat without mixing. Heat exchangers are widely used in various industries, and they come in various designs.

The simplest form of a heat exchanger is a *double-tube* (also called *tube-and-shell*) *heat exchanger,* shown in Fig. 4-35. It is composed of two concentric pipes of different diameters. One fluid flows in the inner pipe, and the other in the annular space between the two pipes. Heat is transferred from the hot fluid to the cold one through the wall separating them. Sometimes the inner tube makes a couple of turns inside the shell to increase the heat transfer area, and thus the rate of heat transfer. The mixing chambers discussed earlier are sometimes classified as *direct-contact* heat exchangers.

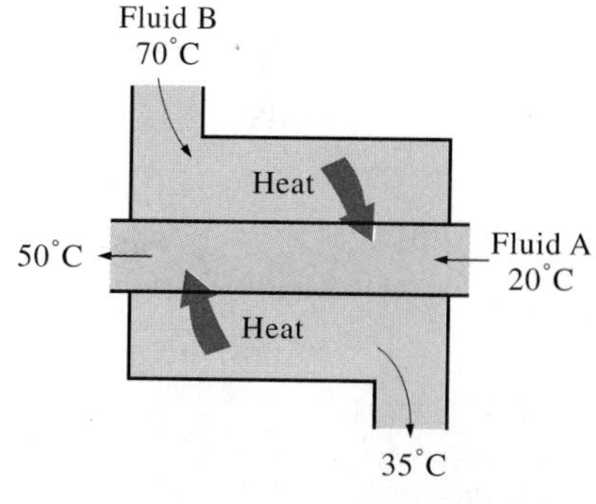

**FIGURE 4-35**

A heat exchanger can be as simple as two concentric pipes of different diameters.

The conservation of mass principle for a heat exchanger in steady operation requires that the sum of the inbound mass flow rates equal the sum of the outbound mass flow rates. This principle can also be expressed as follows: *Under steady operation, the mass flow rate of each fluid stream flowing through a heat exchanger remains constant.*

Heat exchangers typically involve no work interactions ($w = 0$) and negligible kinetic and potential energy changes ($\Delta ke \approx 0$, $\Delta pe \approx 0$) for each fluid stream. The heat transfer rate associated with heat exchangers depends on how the control volume is selected. Heat exchangers are intended for heat transfer between two fluids *within* the device, and the outer shell is usually well insulated to prevent any heat loss to the surrounding medium.

When the entire heat exchanger is selected as the control volume, $\dot{Q}$ becomes zero, since the boundary for this case lies just beneath the insulation and little or no heat crosses the boundary (Fig. 4-36). If, however, only one of the fluids is selected as the control volume then heat will cross this boundary as it flows from one fluid to the other and $\dot{Q}$ will not be zero. In fact, $\dot{Q}$ in this case will be the amount of heat transfer between the two fluids.

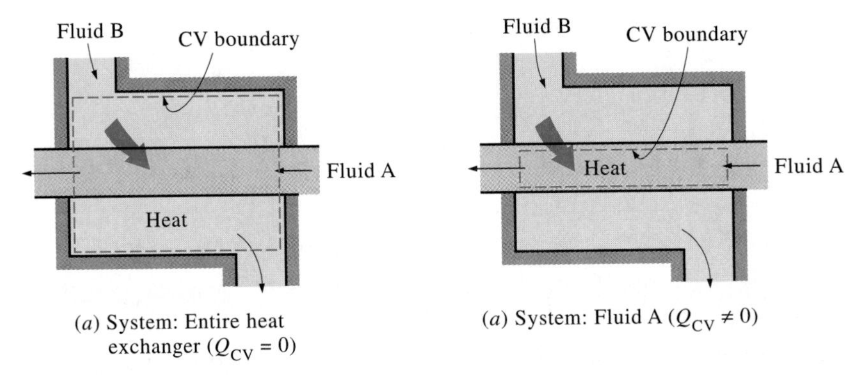

**FIGURE 4-36**
The heat transfer associated with a heat exchanger may be zero or nonzero depending on how the system is selected.

**EXAMPLE 4-6**

Engine oil is to be cooled by water in a condenser. The engine oil enters the consenser with a mass flow rate of 6 kg/min at 1 MPa and 70°C and leaves at 35°C. The cooling water enters at 300 kPa and 15°C and leaves at 25°C. Neglecting any pressure drops, determine (*a*) the mass flow rate of the cooling water required and (*b*) the heat transfer rate from the engine oil to water.

**Solution** There are several possibilities for selecting the control volume for multiple-stream steady-flow devices. The proper choice for the control volume will depend on the situation at hand. Again no work crosses the boundary, and the kinetic and potential energies are considered negligible.

(*a*) To determine the mass flow rate of water, we choose the entire heat exchanger as our control volume, as shown in Fig. 4-37. This is a good choice since the equations will involve only one unknown—the mass flow rate. The conservation of mass principle requires that the mass flow rate of each fluid stream remain constant:

$$\dot{m}_1 = \dot{m}_2 = \dot{m}_w$$
$$\dot{m}_3 = \dot{m}_4 = \dot{m}_{\text{oil}}$$

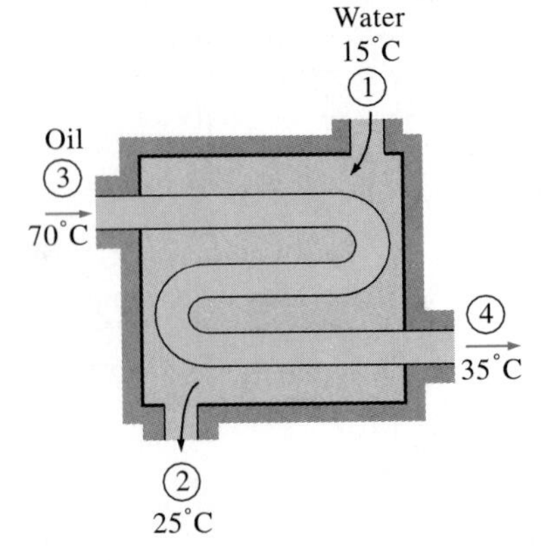


**FIGURE 4-37**
Schematic for Example 4-6.

Heat exchangers are ordinarily well insulated. The boundaries of the control volume selected lie just beneath the insulation, so no heat will cross the boundaries. Then the conservation of energy equation reduces to

$$\dot{Q}^{\nearrow 0} - \dot{W}^{\nearrow 0} = \sum \dot{m}_e\left(h_e + \frac{\mathscr{V}_e^2}{2}^{\nearrow 0} + gz_e^{\nearrow 0}\right) - \sum \dot{m}_i\left(h_i + \frac{\mathscr{V}_i^2}{2}^{\nearrow 0} + gz_i^{\nearrow 0}\right)$$

$$\sum \dot{m}_i h_i = \sum \dot{m}_e h_e$$

or

$$\dot{m}_w h_1 + \dot{m}_{oil} h_3 = \dot{m}_w h_2 + \dot{m}_{oil} h_4$$

Solving for $\dot{m}_w$ gives

$$\dot{m}_w = \frac{h_3 - h_4}{h_2 - h_1}\dot{m}_{oil}$$

Assuming constant specific heats for both the oil and the water at their average temperature, the mass flow rate of the oil is determined to be

$$\dot{m}_w = \frac{C_{p,oil}(T_3 - T_4)}{C_{p,w}(T_2 - T_1)}\dot{m}_{oil} = \frac{[2.016 \text{ kJ/(kg}\cdot{}^\circ\text{C)}](70 - 35)^\circ\text{C}}{[4.18 \text{ kJ/(kg}\cdot{}^\circ\text{C)}](25 - 15)^\circ\text{C}}(6 \text{ kg/min})$$
$$= 10.1 \text{ kg/min}$$

(*b*) To determine the heat transfer from the refrigerant to the water, we have to choose a control volume whose boundary lies on the path of the heat flow, since heat is recognized as it crosses the boundaries. We can choose the volume occupied by either fluid as our control volume. For no particular reason, we choose the volume occupied by the water. All the assumptions stated earlier apply, except that the heat flow is no longer zero. Then the conservation of energy equation (Eq. 4-21) for this single-stream steady-flow system reduces to

$$\dot{Q} - \dot{W}^{\nearrow 0} = \dot{m}_w(\Delta h + \Delta ke^{\nearrow 0} + \Delta pe^{\nearrow 0})$$

$$\dot{Q} = \dot{m}_w(h_2 - h_1) = \dot{m}_w C_{p,w}(T_2 - T_1)$$
$$= (10.1 \text{ kg/min})[4.18 \text{ kJ/(kg}\cdot{}^\circ\text{C)}](25 - 15)^\circ\text{C}$$
$$= 422.2 \text{ kJ/kg}$$

Had we chosen the volume occupied by the oil as the control volume (Fig. 4-38), we would have obtained the same result with the opposite sign since the heat gained by the water is equal to the heat lost by the oil.

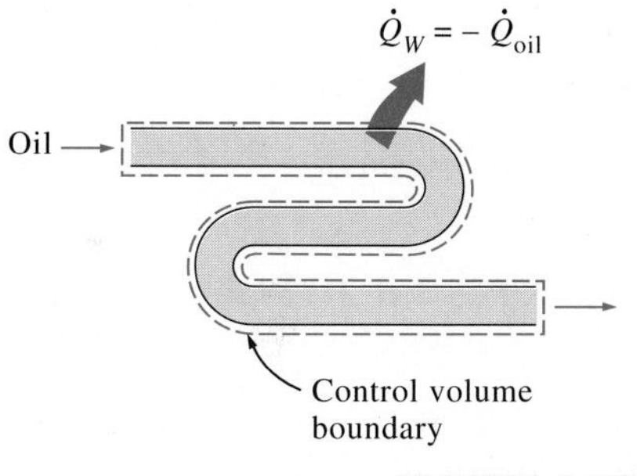


**FIGURE 4-38**
In a heat exchanger, the heat transfer depends on the choice of the control volume.

## 5 Pipe and Duct Flow

The transport of liquids or gases in pipes and ducts is of great importance in many engineering applications. Flow through a pipe or a duct usually satisfies the steady-flow conditions and thus can be analyzed as a steady-flow process. This, of course, excludes the transient start-up and shut-down periods. The control volume can be selected to coincide with the interior surfaces of the portion of the pipe or the duct that we are interested in analyzing.

When flow through pipes or ducts is analyzed, the following points should be considered:

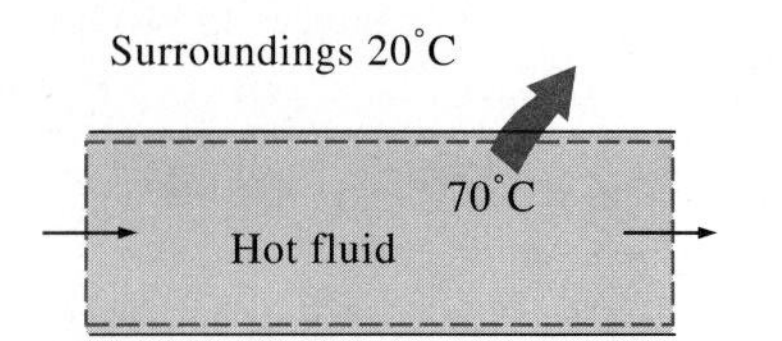


**FIGURE 4-39**

Heat losses from a hot fluid flowing through an uninsulated pipe or duct to he cooler environment may be very significant.

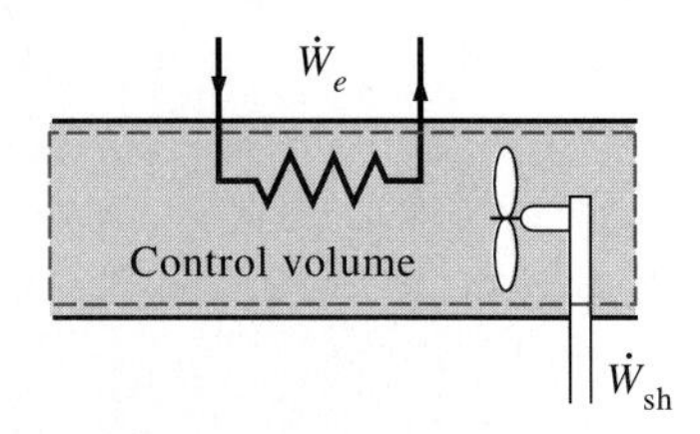


**FIGURE 4-40**

Pipe or duct flow may involve more than one form of work at the same time.

$\dot{Q} \neq 0$. Under normal operating conditions, the amount of heat gained or lost by the fluid may be very significant, particularly if the pipe or duct is long (Fig. 4-39). Sometimes heat transfer is desirable and is the sole purpose of the flow. Water flow through the pipes in the furnace of a power plant, the flow of refrigerant in a freezer, and the flow in heat exchangers are some examples of this case. At other times, heat transfer is undesirable, and the pipes or ducts are insulated to prevent any heat loss or gain, particularly when the temperature difference between the flowing fluid and the surroundings is large. Heat transfer in this case is negligible.

$\dot{W} \neq 0$. If the control volume involves a heating section (electric wires), a fan, or a pump (shaft), the work term interactions should be considered (Fig. 4-40). Of these, fan work is usually small and often neglected. If the control volume involves none of these work devices, the work term is zero.

$\Delta ke \approx 0$. The velocities involved in pipe and duct flow are relatively low, and the kinetic energy changes are usually insignificant. This is particularly true when the pipe or duct diameter is constant and the heating effects are negligible. Kinetic energy changes may be significant, however, for gas flow in ducts with variable cross-sectional areas.

$\Delta pe \neq 0$. In pipes and ducts, the fluid may undergo a considerable elevation change. Thus, the potential energy term may be significant. This is particularly true for flow through insulated pipes and ducts where the heat transfer does not overshadow other effects.

**EXAMPLE 4-7**

The electric heating systems used in many houses consist of a simple duct with resistance wires. Air is heated as it flows over resistance wires. Consider a 15-kW electric heating system. Air enters the heating section at 100 kPa and 17°C with a volume flow rate of 150 m$^3$/min. If heat is lost from the air in the duct to the surroundings at a rate of 200 W, determine the exit temperature of air.

**Solution** We select the heating section portion of the duct, shown in Fig. 4-41, as our control volume. By neglecting the kinetic and potential energy changes, the conservation of energy equation for this single-stream steady-flow system simplifies to

$$\dot{Q} - \dot{W} = \dot{m}(\Delta h + \cancelto{0}{\Delta ke} + \cancelto{0}{\Delta pe}) = \dot{m}(h_2 - h_1)$$

At the specified conditions, the air can be treated as an *ideal gas* since it is at a high temperature and low pressure relative to its critical values

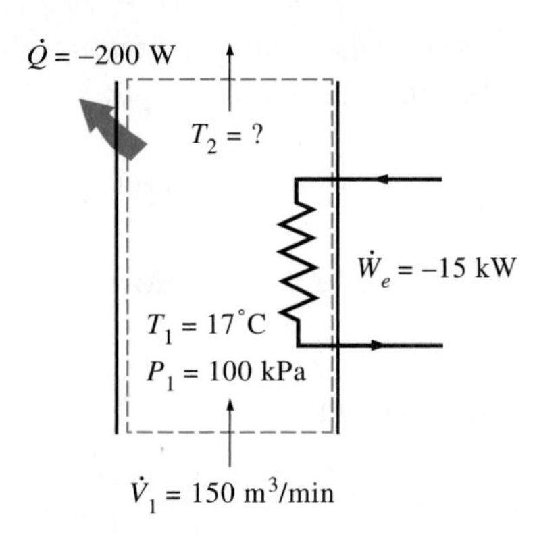


**FIGURE 4-41**
Schematic for Example 4-7.

($T_{cr}$ = −147°C and $P_{cr}$ = 3390 kPa for nitrogen, the main constituent of air). Then the specific volume of the air at the inlet becomes

$$v_1 = \frac{RT_1}{P_1} = \frac{[0.287\ \text{kPa}\cdot\text{m}^3/(\text{kg}\cdot\text{K})](290\ \text{K})}{100\ \text{kPa}} = 0.832\ \text{m}^3/\text{kg}$$

The mass flow rate of the air through the duct is determined from

$$\dot{m} = \frac{\dot{V}_1}{v_1} = \frac{150\ \text{m}^3/\text{min}}{0.832\ \text{m}^3/\text{kg}}\left(\frac{1\ \text{min}}{60\ \text{s}}\right) = 3.0\ \text{kg/s}$$

At the temperatures encountered in heating and air-conditioning systems, $\Delta h$ can be replaced by $C_p\,\Delta T$, where $C_p$ = 1.005 kJ/(kg · °C) (the value at room temperature) with negligible error (Fig. 4-42). Then the energy equation takes the following form:

$$\dot{Q} - \dot{W} = \dot{m}C_p(T_2 - T_1)$$

Substituting the known quantities, we see that the exit temperature of the air is

$$-0.2\ \text{kJ/s} - (-15\ \text{kJ/s}) = (3\ \text{kg/s})[1.005\ \text{kJ/(kg}\cdot{}^\circ\text{C)}](T_2 - 17^\circ\text{C})$$
$$T_2 = 21.9^\circ\text{C}$$

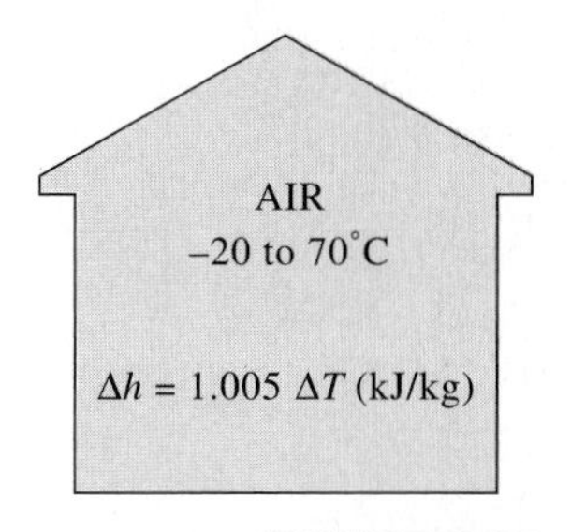


**FIGURE 4-42**
The error involved in $\Delta h = C_p\,\Delta T$, where $C_p$ = 1.005 kJ/(kg · °C), is less than 0.5 percent for air in the temperature range −20 to 70°C.

## 4-4 ■ SUMMARY

In this chaper, we have discussed the conservation of mass and the conservation of energy principles for *control volumes*. Mass carries energy with it, thus the energy content changes when mass enters or leaves the control volume.

Mass flow through a cross section per unit time is called the *mass flow rate* and is denoted $\dot{m}$. It is expressed as

$$\dot{m} = \rho \mathscr{V}_{av} A \qquad \text{(kg/s)}$$

where $\rho$ = density, kg/m³ (= $1/v$)
$\mathscr{V}_{av}$ = average fluid velocity normal to $A$, m/s
$A$ = cross-sectional area, m²

The fluid volume flowing through a cross section per unit time is called the *volume flow rate* $\dot{V}$. It is given by

$$\dot{V} = \int_A \mathscr{V}_n \, dA = \mathscr{V}_{av} A \qquad (\mathrm{m^3/s})$$

The mass and volume flow rates are related by

$$\dot{m} = \rho \dot{V} = \frac{\dot{V}}{v}$$

Thermodynamic processes involving control volumes can be considered in two groups: steady-flow processes and unsteady-flow processes. During a *steady-flow process*, the fluid flows through the control volume steadily, experiencing no change with time at a fixed position. The mass and energy content of the control volume remains constant during a steady-flow process. The conservation of mass and energy equations for steady-flow processes are expressed as

$$\sum \dot{m}_i = \sum \dot{m}_e \qquad (\mathrm{kg/s})$$

$$\dot{Q} - \dot{W} = \underbrace{\sum \dot{m}_e \left( h_e + \frac{\mathscr{V}_e^2}{2} + g z_e \right)}_{\text{for each exit}} - \underbrace{\sum \dot{m}_i \left( h_i + \frac{\mathscr{V}_i^2}{2} + g z_i \right)}_{\text{for each inlet}} \qquad (\mathrm{kW})$$

where the subscript $i$ stands for *inlet* and $e$ for *exit.* These are the most general forms of the equations for steady-flow processes. For single-stream (one-inlet, one-exit) systems such as nozzles, diffusers, turbines, compressors, and pumps, they simplify to

$$\dot{m}_1 = \dot{m}_2 \qquad (\mathrm{kg/s})$$

or

$$\frac{1}{v_1} \mathscr{V}_1 A_1 = \frac{1}{v_2} \mathscr{V}_2 A_2$$

and

$$\dot{Q} - \dot{W} = \dot{m} \left[ h_2 - h_1 + \frac{\mathscr{V}_2^2 - \mathscr{V}_1^2}{2} + g(z_2 - z_1) \right] \qquad (\mathrm{kW})$$

$$q - w = h_2 - h_1 + \frac{\mathscr{V}_2^2 - \mathscr{V}_1^2}{2} + g(z_2 - z_1) \qquad (\mathrm{kJ/kg})$$

or

$$q - w = \Delta h + \Delta \mathrm{ke} + \Delta \mathrm{pe} \qquad (\mathrm{kJ/kg})$$

where

$$q = \frac{\dot{Q}}{\dot{m}} \qquad \text{(heat transfer per unit mass, kJ/kg)}$$

and

$$w = \frac{\dot{W}}{\dot{m}} \qquad \text{(work done per unit mass, kJ/kg)}$$

In the above relations, subscripts 1 and 2 denote the inlet and exit states, respectively.

The steady-flow process is the model process for flow through nozzles, diffusers, turbines, compressors, fans, pumps, pipes, throttling valves, mixing chambers, and heat exchangers.

## REFERENCES AND SUGGESTED READING

**1** A. Bejan, *Advanced Engineering Thermodynamics,* Wiley, New York, 1988.

**2** W. Z. Black and J. G. Hartley, *Thermodynamics,* Harper & Row, New York, 1985.

**3** J. R. Howell and R. O. Buckius, *Fundamentals of Engineering Thermodynamics,* McGraw-Hill, New York, 1987.

**4** J. B. Jones and G. A. Hawkins, *Engineering Thermodynamics,* 2d ed., Wiley, New York, 1986.

**5** W. C. Reynolds and H. C. Perkins, *Engineering Thermodynamics,* 2d ed., McGraw-Hill, New York, 1977.

**6** G. J. Van Wylen and R. E. Sonntag, *Fundamentals of Classical Thermodynamics,* 3d ed., Wiley, New York, 1985.

**7** K. Wark, *Thermodynamics,* 5th ed., McGraw-Hill, New York, 1988.

## PROBLEMS*

### General Control Volume Analysis

**4-1C** Express in words the conservation of mass principle for a control volume.

**4-2C** Define mass and volume flow rates. How do they differ?

**4-3C** Express the conservation of energy principle for a control volume verbally.

**4-4C** What are the different mechanisms for transferring energy to or from a control volume?

**4-5C** What is flow energy? Do fluids at rest possess any flow energy?

**4-6C** How do the energies of flowing fluids and a fluid at rest compare? Name the specific forms of energy associated with each case.

**4-7C** Consider a room filled with warm air. Some cold air leaks into the room now, and the air temperature drops somewhat. No air leaks out during the process. Does the room contain more or less energy now? Explain.

### Steady-Flow Processes

**4-8C** When is the flow through a control volume steady?

**4-9C** How is a steady-flow system characterized?

**4-10C** Can a steady-flow system involve boundary work?

---

*Students are encouraged to answer *all* the concept "C" questions.

## Nozzles and Diffusers

**4-11C** A diffuser is an adiabatic device that decreases the kinetic energy of the fluid by slowing it down. What happens to this *lost* kinetic energy?

**4-12C** The kinetic energy of a fluid increases as it is accelerated in an adiabatic nozzle. Where does this energy come from?

**4-13C** Is heat transfer to or from the fluid desirable as it flows through a nozzle? How will heat transfer affect the fluid velocity at the nozzle exit?

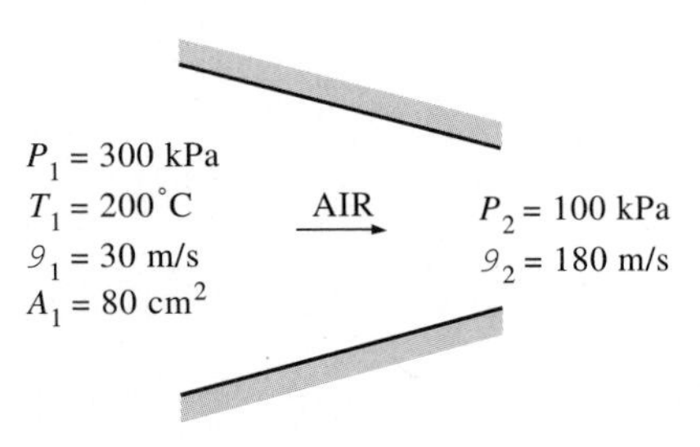


FIGURE P4-14

**4-14** Air enters an adiabatic nozzle steadily at 300 kPa, 200°C, and 30 m/s and leaves at 100 kPa and 180 m/s. The inlet area of the nozzle is 80 cm$^2$. Determine (*a*) the mass flow rate through the nozzle, (*b*) the exit temperature of the air, and (*c*) the exit area of the nozzle.
*Answers:* (*a*) 0.5304 kg/s, (*b*) 184.60°C, (*c*) 38.7 cm$^2$

**4-14E** Air enters an adiabatic nozzle steadily at 75 psia, 400°F, and 100 ft/s and leaves at 15 psia and 500 ft/s. The inlet area of the nozzle is 20 in$^2$. Determine (*a*) the mass flow rate through the nozzle, (*b*) the exit temperature of the air and (*c*) the exit area of the nozzle.

**4-15** Carbon dioxide enters an adiabatic nozzle steadily at 1 MPa and 500°C with a mass flow rate of 6000 kg/h and leaves at 100 kPa and 450 m/s. The inlet area of the nozzle is 40 cm$^2$. Determine (*a*) the inlet velocity and (*b*) the exit temperature.
*Answers:* (*a*) 60.8 m/s, (*b*) 685.8 K

**4-16** Air enters a nozzle steadily at 300 kPa, 77°C, and 50 m/s and leaves at 100 kPa and 320 m/s. The heat loss from the nozzle is estimated to be 3.2 kJ/kg of air flowing. The inlet area of the nozzle is 100 cm$^2$. Determine (*a*) the exit temperature of air and (*b*) the exit area of the nozzle. *Answers:* (*a*) 24.2°C, (*b*) 39.7 cm$^2$

**4-16E** Air enters a nozzle steadily at 50 psia, 140°F, and 150 ft/s and leaves at 14.7 psia and 900 ft/s. The heat loss from the nozzle is estimated to be 6.5 Btu/lbm of air flowing. The inlet area of the nozzle is 0.1 ft$^2$. Determine (*a*) the exit temperature of air and (*b*) the exit area of the nozzle. *Answers:* (*a*) 507.4 R, (*b*) 0.0479 ft$^2$

**4-17** Air at 600 kPa and 500 K enters an adiabatic nozzle that has an inlet-to-exit area ratio of 2:1 with a velocity of 120 m/s and leaves with a velocity of 380 m/s. Determine (*a*) the exit temperature and (*b*) the exit pressure of the air. *Answers:* (*a*) 436.5 K, (*b*) 330.8 kPa

**4-18** Air at 80 kPa and 127°C enters an adiabatic diffuser steadily at a rate of 6000 kg/h and leaves at 100 kPa. The velocity of the airstream is decreased from 230 to 30 m/s as it passes through the diffuser. Find (*a*) the exit temperature of the air and (*b*) the exit area of the diffuser.

**4-19** Air at 80 kPa and −8°C enters an adiabatic diffuser steadily with a velocity of 200 m/s and leaves with a low velocity at a pressure of 95 kPa. The exit area of the diffuser is 5 times the inlet area. Determine (*a*) the exit temperature and (*b*) the exit velocity of the air.

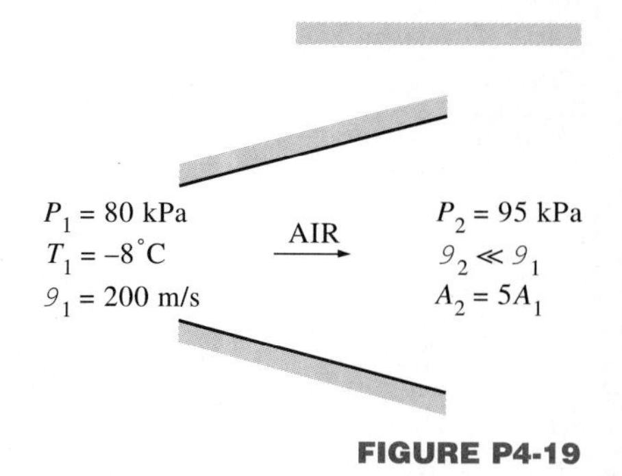


**FIGURE P4-19**

**4-19E** Air at 13 psia and 20°F enters an adiabatic diffuser steadily with a velocity of 600 ft/s and leaves with a low velocity at a pressure of 14.5 psia. The exit area of the diffuser is 5 times the inlet area. Determine (*a*) the exit temperature and (*b*) the exit velocity of the air.

**4-20** Air at 80 kPa, 27°C, and 220 m/s enters a diffuser at a rate of 2.5 kg/s and leaves at 42°C. The exit area of the diffuser is 400 cm$^2$. The air is estimated to lose heat at a rate of 18 kJ/s during this process. Determine (*a*) the exit velocity and (*b*) the exit pressure of the air.
*Answers:* (*a*) 62.0 m/s, (*b*) 91.1 kPa

**4-21** Nitrogen gas at 60 kPa and 7°C enters an adiabatic diffuser steadily with a velocity of 200 m/s and leaves at 85 kPa and 22°C. Determine (*a*) the exit velocity of the nitrogen and (*b*) the ratio of the inlet to exit area $A_1/A_2$.

### Turbines and Compressors

**4-22C** Consider an adiabatic turbine operating steadily. Does the work output of the turbine have to be equal to the decrease in the energy of the steam flowing through it?

**4-23C** Consider a steam turbine operating on a steady-flow process. Would you expect the temperatures at the turbine inlet and exit to be the same?

**4-24C** Consider an air compressor operating on a steady-flow process. Would you expect the air density to be the same at the compressor inlet and exit?

**4-25C** Consider an air compressor operating on a steady-flow process. How would you compare the volume flow rates of the air at the compressor inlet and exit?

**4-26C** Will the temperature of air rise as it is compressed by an adiabatic compressor? Why?

**4-27C** Somebody proposes the following system to cool a house in the summer: Compress the regular outdoor air, let it cool back to the outdoor temperature, pass it through a turbine, and discharge the cold air leaving the turbine into the house. From a thermodynamic point of view, is the proposed system sound?

**4-28** Argon gas enters steadily an adiabatic turbine at 900 kPa and 450°C with a velocity of 80 m/s and leaves at 150 kPa with a velocity of 150 m/s. The inlet area of the turbine is 60 cm$^2$. If the power output of the turbine is 250 kW, determine the exit temperature of the argon.
*Answer:* 267°C

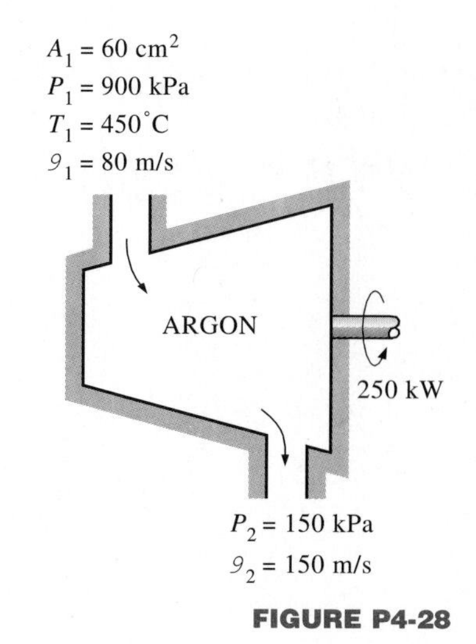


**FIGURE P4-28**

**4-29** Air flows steadily through an adiabatic turbine, entering at 1 MPa, 500°C, and 120 m/s and leaving at 150 kPa, 150°C, and 250 m/s. The inlet area of the turbine is 80 cm$^2$. Determine (*a*) the mass flow rate of the air and (*b*) the power output of the turbine.

**4-29E** Air flows steadily through an adiabatic turbine, entering at 150 psia, 900°F, and 350 ft/s and leaving at 20 psia, 300°F, and 700 ft/s. The inlet area of the turbine is 0.1 ft$^2$. Determine (*a*) the mass flow rate of the air and (*b*) the power output of the turbine.

**4-30** Air enters the compressor of a gas-turbine plant at ambient conditions of 100 kPa and 25°C with a low velocity and exits at 1 MPa and 347°C with a velocity of 90 m/s. The compressor is cooled at a rate of 1500 kJ/min, and the power input to the compressor is 250 kW. Determine the mass flow rate of air through the compressor.
*Answer:* 0.675 kg/s

**4-31** Air is compressed from 100 kPa and 22°C to a pressure of 1 MPa while being cooled at a rate of 16 kJ/kg by circulating water through the compressor casing. The volume flow rate of the air at the inlet conditions is 150 m$^3$/min, and the power input to the compressor is 500 kW. Determine (*a*) the mass flow rate of the air and (*b*) the temperature at the compressor exit. *Answers:* (*a*) 2.95 kg/s, (*b*) 174°C

**4-31E** Air is compressed from 14.7 psia and 60°F to a pressure of 150 psia while being cooled at a rate of 10 Btu/lbm by circulating water through the compressor casing. The volume flow rate of the air at the inlet conditions is 5000 ft$^3$/min, and the power input to the compressor is 700 hp. Determine (*a*) the mass flow rate of the air and (*b*) the temperature at the compressor exit.
*Answers:* (*a*) 6.36 lbm/s, (*b*) 801 R

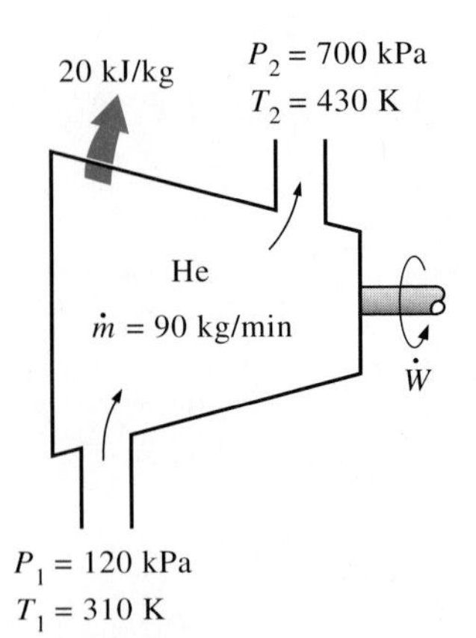


**FIGURE P4-32**

**4-32** Helium is to be compressed from 120 kPa and 310 K to 700 kPa and 430 K. A heat loss of 20 kJ/kg occurs during the compression process. Neglecting kinetic energy changes, determine the power input required for a mass flow rate of 90 kg/min.

**4-33** Carbon dioxide enters an adiabatic compressor at 100 kPa and 300 K at a rate of 0.5 kg/s and leaves at 600 kPa and 450 K. Neglecting kinetic energy changes, determine (*a*) the volume flow rate of the carbon dioxide at the compressor inlet and (*b*) the power input to the compressor. *Answers:* (*a*) 0.28 m$^3$/s, (*b*) 68.8 kW

## Throttling Valves

**4-34C** Why are throttling devices commonly used in refrigeration and air-conditioning applications?

**4-35C** During a throttling process, the temperature of a fluid drops from 30 to −20°C. Can this process occur adiabatically?

**4-36C** Would you expect the temperature of air to drop as it undergoes a steady-flow throttling process?

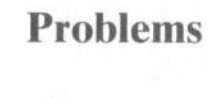

4-37C Would you expect the temperature of a liquid to change as it is throttled? How?

**4-38** Refrigerant-134a is throttled from the saturated liquid state at 800 kPa to a pressure of 140 kPa. Determine the temperature drop during this process. *Answer:* 50.13°C

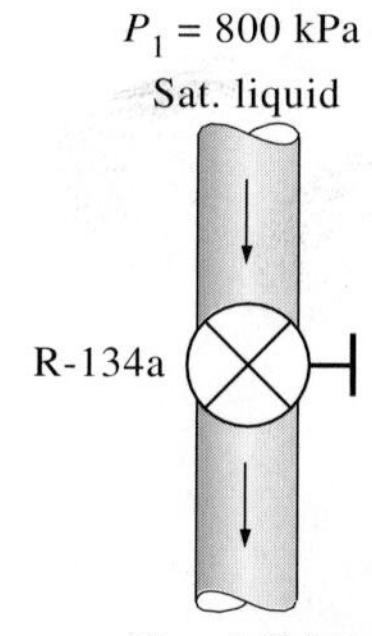


**FIGURE P4-38**

**4-39** Refrigerant-134a is throttled from the saturated liquid state at 800 kPa to a temperature of −20°C. Determine the pressure of the refrigerant at the final state. *Answer:* 133 kPa

**4-40** Air at 2 MPa and 30°C is throttled to the atmospheric pressure of 100 kPa. Determine the final temperature of the air.

**4-40E** Air at 200 psia and 90°F is throttled to the atmospheric pressure of 14.7 psia. Determine the final temperature of the air.

## Mixing Chambers and Heat Exchangers

4-41C When two fluid streams are mixed in a mixing chamber, can the mixture temperature be lower than the temperature of both streams? How?

4-42C Consider a steady-flow mixing process. Under what conditions will the energy transported into the control volume by the incoming streams be equal to the energy transported out of it by the outgoing stream?

4-43C Consider a steady-flow heat exchanger involving two different fluid streams. Under what conditions will the amount of heat lost by one fluid be equal to the amount of heat gained by the other?

**4-44** A hot-water stream at 80°C enters a mixing chamber with a mass flow rate of 0.5 kg/s where it is mixed with a stream of cold water at 20°C. If it is desired that the mixture leave the chamber at 42°C, determine the mass flow rate of the cold-water stream. Assume all the streams are at a pressure of 250 kPa. *Answer:* 0.864 kg/s

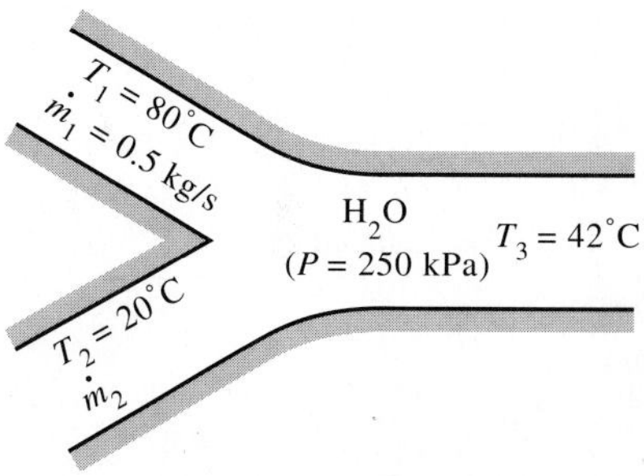

**FIGURE P4-44**

**4-44E** A hot-water stream at 180°F enters a mixing chamber with a mass flow rate of 2 lbm/s, where it is mixed with a stream of cold water at 60°F. If it is desired that the mixture leave the chamber at 110°F, determine the mass flow rate of the cold-water stream. Assume all the streams are at a pressure of 50 psia. *Answer:* 2.80 lbm/s

**4-45** Liquid water at 300 kPa and 20°C is heated in a chamber by mixing it with hot water at 300 kPa and 90°C. Cold water enters the chamber at a rate of 1.8 kg/s. If the mixture leaves the mixing chamber at 60°C, determine the mass flow rate of the hot water required.

**4-46** Water at 25°C and 300 kPa is heated in a chamber by mixing it with hot water at 80°C and 300 kPa. If both streams enter the mixing chamber at the same mass flow rate, determine the temperature of the exiting stream.

**4-46E** Water at 50°F and 50 psia is heated in a chamber by mixing it with hot water at 200°F and 50 psia. If both streams enter the mixing chamber at the same mass flow rate, determine the temperature of the exiting stream.

**4-47** A stream of cold air at 1 MPa and 12°C is mixed with another stream at 1 MPa and 60°C. If the mass flow rate of the cold stream is twice that of the hot one, determine the temperature of the exit stream. Use specific heats at room temperature.

**4-48** Hot water at 1 MPa and 80°C is to be cooled to 1 MPa and 30°C in a condenser by air. The air enters at 100 kPa and 27°C with a volume flow rate of 800 $m^3/min$ and leaves at 95 kPa and 60°C. Determine the mass flow rate of the water. *Answer:* 147 kg/min

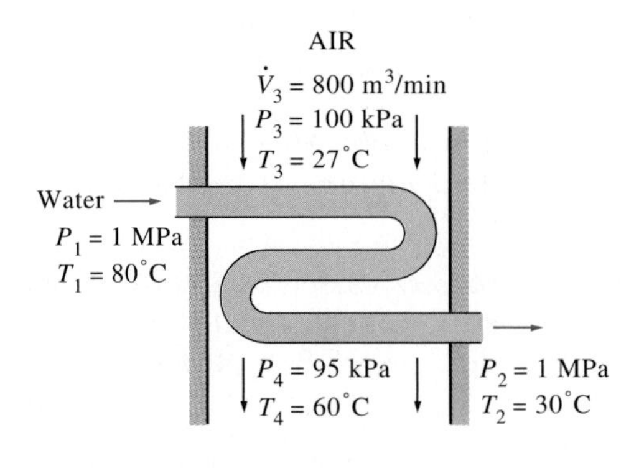


**FIGURE P4-48**

**4-49** Extruded aluminum sheets enter a cooling section at 200°C at a steady velocity of 30 m/min and are cooled to 60°C by water. The cooling water enters the cooling section at 15°C and leaves at 30°C. If the mass of the aluminum sheets is 0.4 kg per meter length, determine the mass flow rate of the cooling water required to cool the aluminum sheets.

**4-50** In a water based heating system, air is heated by passing it through the fins of a radiator. Hot water enters the radiator at 90°C at a rate of 8 kg/min and leaves at 70°C. Air enters at 100 kPa and 25°C and leaves at 47°C. Determine the volume flow rate of air at the inlet.

**4-50E** In a water based heating system, air is heated by passing it through the fins of a radiator. Hot water enters the radiator at 200°F at a rate of 15 lbm/min and leaves at 170°F. Air enters at 14.5 psia and 65°F and leaves at 120°F. Determine the volume flow rate of air at the inlet.

**4-51** Steam enters the condenser of a steam power plant at 20 kPa as saturated vapor with a mass flow rate of 20,000 kg/h. It is to be cooled by water from a nearby river by circulating the water through the tubes within the condenser. To prevent thermal pollution, the river water is not allowed to experience a temperature rise above 10°C. If the steam is to leave the condenser as saturated liquid at 20 kPa, determine the mass flow rate of the cooling water required.

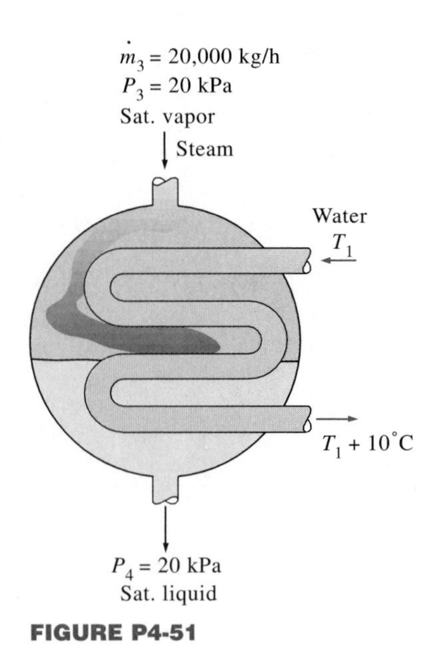


**FIGURE P4-51**

**4-51E** Steam enters the condenser of a steam power plant at 3 psia as saturated vapor with a mass flow rate of 40,000 lbm/h. It is to be cooled by water from a nearby river by circulating the water through the tubes within the condenser. To prevent thermal pollution, the river water is not allowed to experience a temperature rise above 18°F. If the steam is to leave the condenser as saturated liquid at 3 psia, determine the mass flow rate of the cooling water required.

## Pipe and Duct Flow

**4-52** A 5 m × 6 m × 8 m room is to be heated by an electric resistance heater placed in a short duct in the room. Initially, the room is at 15°C, and the local atmospheric pressure is 98 kPa. The room is losing heat steadily to the outside at a rate of 200 kJ/min. A 200-W fan circulates the air steadily through the duct and the electric heater at an average mass flow rate of 50 kg/min. The duct can be assumed to be adiabatic, and there is no air leaking in or out of the room. If it takes 15 min for the room air to reach an average temperature of 25°C, find (*a*) the power rating of the electric heater and (*b*) the temperature rise that the air experiences each time it passes through the heater.

**4-53** A house has an electric heating system that consists of a 300-W fan and an electric resistance heating element placed in a duct. Air flows steadily through the duct at a rate of 0.6 kg/s and experiences a temperature rise of 5°C. The rate of heat loss from the air in the duct is estimated to be 400 W. Determine the power rating of the electric resistance heating element. *Answer:* 3.12 kW

**4-53E** A house has an electric heating system that consists of a 300-W fan and an electric resistance heating element placed in a duct. Air flows through the duct at a rate of 1 lbm/s and experiences a temperature rise of 10°F. The rate of heat loss from the air in the duct is estimated to be 0.2 Btu/s. Determine the power rating of the electric heating element.

**4-54** A hair dryer is basically a duct in which a few layers of electric resistors are placed. A small fan pulls the air in and forces it through the resistors where it is heated. Air enters a 1200-W hair dryer at 100 kPa and 22°C and leaves at 47°C. The cross-sectional area of the hair dryer at the exit is 60 cm². Neglecting the power consumed by the fan and the heat losses through the walls of the hair dryer, determine (*a*) the volume flow rate of air at the inlet and (*b*) the velocity of the air at the exit.
*Answers:* (*a*) 0.0404 m³/s, (*b*) 7.31 m/s

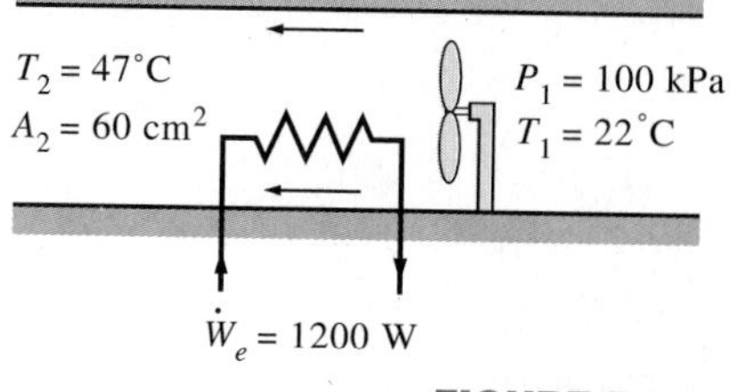


**FIGURE P4-54**

**4-55** The ducts of an air heating system pass through an unheated area. As a result of heat losses, the temperature of the air in the duct drops by 4°C. If the mass flow rate of air is 120 kg/min, determine the rate of heat loss from the air to the cold environment.

**4-56** Air enters the duct of an air-conditioning system at 105 kPa and 12°C at a volume flow rate of 12 m³/min. The diameter of the duct is 20 cm, and heat is transferred to the air in the duct from the surroundings

at a rate of 2 kJ/s. Determine (*a*) the velocity of the air at the duct inlet and (*b*) the temperature of the air at the exit.
*Answers:* (*a*) 6.37 m/s, (*b*) 19.74°C

**4-56E** Air enters the duct of an air-conditioning system at 15 psia and 50°F at a volume flow rate of 450 $ft^3$/min. The diameter of the duct is 10 in, and heat is transferred to the air in the duct from the surroundings at a rate of 2 Btu/s. Determine (*a*) the velocity of the air at the duct inlet and (*b*) the temperature of the air at the exit.

**4-57** Water is heated in an insulated, constant-diameter tube by a 7-kW electric resistance heater. If the water enters the heater steadily at 15°C and leaves at 70°C, determine the mass flow rate of water.

**4-57E** Water is heated in an insulated, constant-diameter tube by a 10-kW electric resistance heater. If the water enters the heater steadily at 50°F and leaves at 170°F, determine the mass flow rate of water.

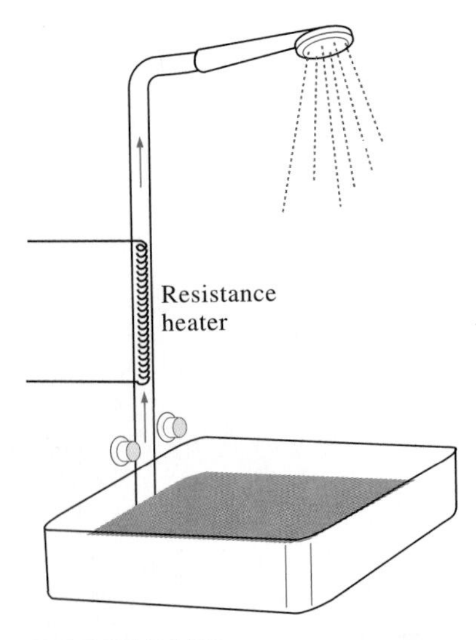


FIGURE P4-58

## Review Problems

**4-58** Water flows through a shower head steadily at a rate of 10 L/min. An electric resistance heater placed in the water pipe heats the water from 16°C to 43°C. Taking the density of water to be 1 kg/L, determine the electric power input to the heater, in kW.

In an effort to conserve energy, it is proposed to pass the drained warm water at a temperature of 39°C through a heat exchanger to preheat the incoming cold water. If the heat exchanger has an effectiveness of 0.50 (that is, it recovers only half of the energy which can possibly be transferred from the drained water to incoming cold water), determine the electric power input required in this case. If the price of the electric energy is 8.5 ¢/kWh, determine how much money is saved during a 10 min shower as a result of installing this heat exchanger.

**4-59** In large gas-turbine power plants, air is preheated by the exhaust gases in a heat exchanger called the *regenerator* before it enters the combustion chamber. Air enters the regenerator at 1 MPa and 550 K at a mass flow rate of 800 kg/min. Heat is transferred to the air at a rate of 3200 kJ/s. Exhaust gases enter the regenerator at 140 kPa and 800 K and leave at 130 kPa and 600 K. Treating the exhaust gases as air, determine (*a*) the exit temperature of the air and (*b*) the mass flow rate of exhaust gases. *Answers:* (*a*) 775 K, (*b*) 14.9 kg/s

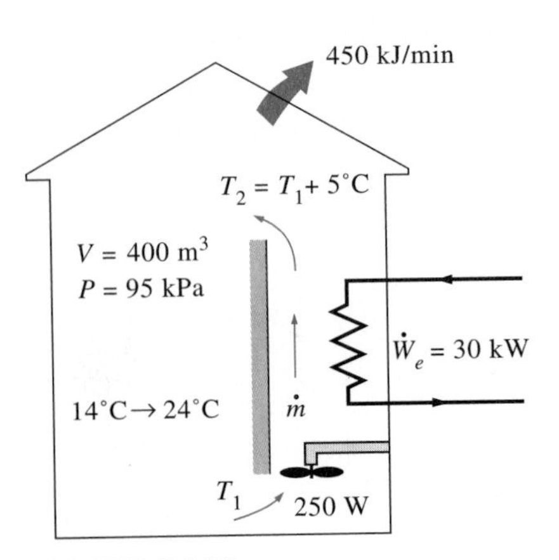


FIGURE P4-60

**4-60** A building with an internal volume of 400 $m^3$ is to be heated by a 30-kW electric resistance heater placed in the duct inside the building. Initially, the air in the building is at 14°C, and the local atmospheric pressure is 95 kPa. The building is losing heat to the surroundings at a steady rate of 450 kJ/min. Air is forced to flow through the duct and the heater steadily by a 250-W fan, and it experiences a temperature rise of 5°C each time it passes through the duct, which may be assumed to be adiabatic.

(*a*) How long will it take for the air inside the building to reach an average temperature of 24°C?
(*b*) Determine the average mass flow rate of air through the duct.
*Answers:* (*a*) 146 s, (*b*) 6.02 kg/s

**4-61** It is proposed to have a water heater that consists of an insulated pipe of 5-cm diameter and an electric resistor inside. Cold water at 15°C enters the heating section steadily at a rate of 30 L/min. If water is to be heated to 50°C, determine (*a*) the power rating of the resistance heater and (*b*) the average velocity of the water in the pipe.

**4-62** The average atmospheric pressure in Spokane, Washington (elevation = 718 m) is 93.0 kPa, and the average winter temperature is 2.5°C. The pressurization test of a 3-m-high 300-$m^2$ older home revealed that the seasonal average infiltration rate of the house is 2.2 air changes per hour (ACH). That is, on average, the entire air volume of the house is replaced by the outdoor air 2.2 times every hour, which represents considerable energy loss. It is suggested that the infiltration rate of the house can be reduced by half to 1.1 ACH by winterizing the doors and the windows. If the house is heated by natural gas whose unit cost is $0.62/therm and the heating season can be taken to be 6 months, determine how much the home owner will save from the heating costs per year by this winterization project. Assume the house is maintained at 22°C at all times, and the efficiency of the furnace is 0.65.

**4-63** Determine the rate of heat loss from a building due to infiltration if outdoor air at −10°C and 90 kPa enters the building at a rate of 35 L/s when the indoors is maintained at 22°C.

**4-64** The ventilating fan of the bathroom of a building has a volume flow rate of 30 L/s, and runs continuously. The building is located in San Francisco, California where the average winter temperature is 12.2°C, and is maintained at 22°C at all times. The building is heated by electricity whose unit cost is $0.09/kWh. Determine the amount and cost of the heat "vented out" per month in winter.

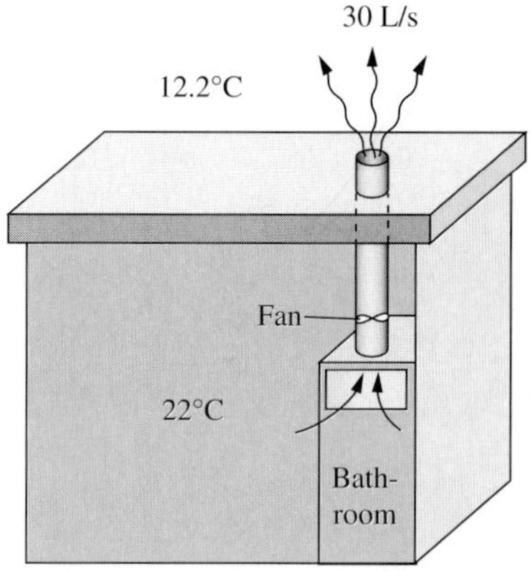

**FIGURE P4-64**

**4-64E** The ventilating fan of the bathroom of a building has a volume flow rate of 150 cfm, and runs continuously. The building is located in San Francisco, California where the average winter temperature is 53.4°F, and is maintained at 72°F. The building is heated by electricity whose unit cost is $0.09/kWh. Determine the amount and cost of the heat "vented out" per month in winter.

**4-65** In a dairy plant, milk at 4°C is pasteurized continuously at 72°C at a rate of 12 L/s for 24 h a day 365 days a year. The milk is heated to the pasteurizing temperature by hot water heated in a natural gas fired boiler which has an efficiency of 82 percent. The pasteurized milk is then cooled by cold water at 18°C before it is finally refrigerated back to 4°C. To save energy and money, the plant installs a regenerator which has an effectiveness of 82 percent. If the cost of natural gas is $0.52/therm (1 therm = 105,500 kJ), determine how much energy and money the

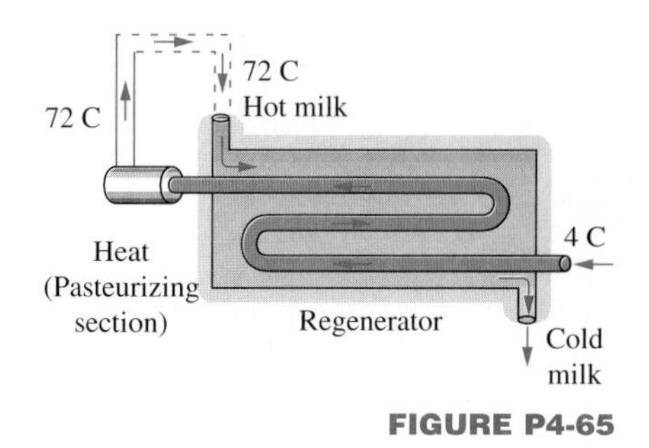


**FIGURE P4-65**

regenerator will save this company per year. The average density and specific heat of milk can be taken to be $\rho_{\text{milk}} \approx \rho_{\text{water}} = 1000\,\text{kg/m}^3$ and $C_{p,\text{milk}} = 3.98\,\text{kJ/(kg}\cdot{}^\circ\text{C)}$.

**4-66** Chickens with an average mass of 2.2 kg and average specific heat of 3.54 kJ/(kg · °C) are to be cooled by chilled water which enters a continuous flow type immersion chiller at 0.5°C. Chickens are dropped into the chiller at a uniform temperature of 15°C at a rate of 500 chickens per hour, and are cooled to an average temperature of 3°C before they are taken out. The chiller gains heat from the surroundings at a rate of 200 kJ/h. Determine (*a*) the rate of heat removal from the chicken, in kW, and (*b*) the mass flow rate of water, in kg/s, if the temperature rise of water is not to exceed 2°C.

## Computer, Design, and Essay Problems

**4-67** Design a 1200 W electric hair dryer such that the air temperature and velocity in the dryer will not exceed 50°C and 3 m/s, respectively.

**4-68** Design an electric hot water heater for a family of four in your area. The maximum water temperature in the tank and the power consumption are not to exceed 60°C and 4 kW, respectively. There are two showers in the house, and the flow rate of water through each of the shower heads is about 10 L/min. Each family member takes a 5-min shower every morning. Explain why a hot water tank is necessary, and determine the proper size of the tank for this family.

**4-69** Write an essay on the classification of heat exchangers. Describe the operation of each type of heat exchanger, and discuss how different types differ from each other.

**4-70** Explain how you can determine the average velocity of air at the exit of your hair dryer at its highest power setting, using a thermometer and a tape measure only.

**4-71** A 1982 U.S. Department of Energy article (FS 204) states that a leak of one drip of hot water per second can cost $1.00 per month. Making reasonable assumptions about the drop size and the unit cost of energy, determine if this claim is reasonable.

# The Second Law of Thermodynamics

CHAPTER 5

Up to this point, we have focused our attention on the first law of thermodynamics, which requires that energy be conserved during a process. In this chapter, we introduce the second law of thermodynamics, which asserts that processes occur in a certain direction and that energy has quality as well as quantity. A process cannot take place unless it satisfies both the first and second laws of thermodynamics. In this chapter, the thermal energy reservoirs, reversible and irreversible processes, heat engines, refrigerators, and heat pumps are introduced first. Various statements of the second law are followed by a discussion of perpetual-motion machines and the absolute thermodynamic temperature scale. The Carnot cycle is introduced next, and the Carnot principles are examined. Finally, idealized Carnot heat engines, refrigerators, and heat pumps are discussed.

## 5-1 ■ INTRODUCTION TO THE SECOND LAW OF THERMODYNAMICS

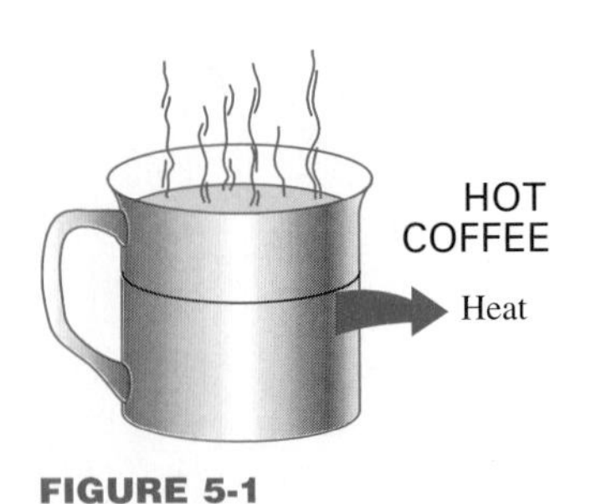


**FIGURE 5-1**

A cup of hot coffee does not get hotter in a cooler room.

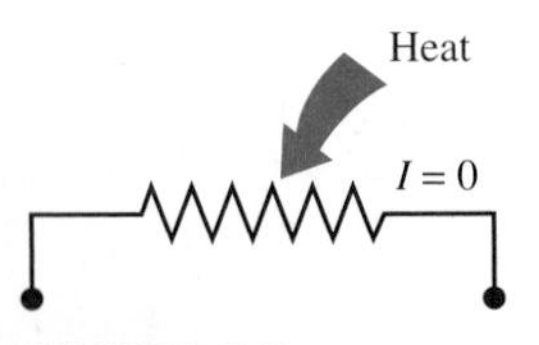


**FIGURE 5-2**

Transferring heat to a wire will not generate electricity.

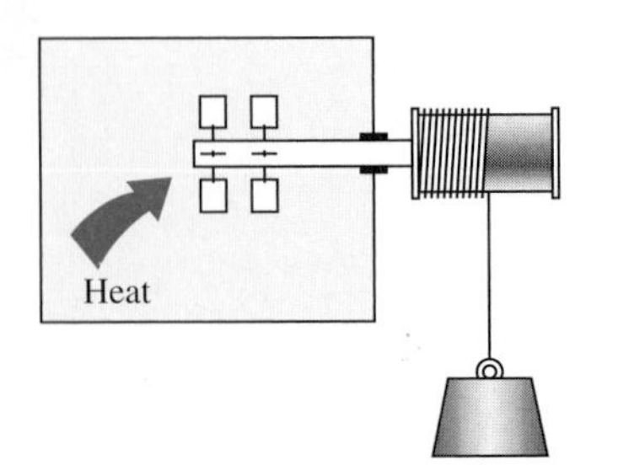


**FIGURE 5-3**

Transferring heat to a paddle wheel will not cause it to rotate.

**FIGURE 5-4**

Processes occur in a certain direction, and not in the reverse direction.

In the preceding two chapters, we applied the *first law of thermodynamics,* or the *conservation of energy principle,* to processes involving closed and open systems. As pointed out repeatedly in those chapters, energy is a conserved property, and no process is known to have taken place in violation of the first law of thermodynamics. Therefore, it is reasonable to conclude that a process must satisfy the first law to occur. However, as explained below, satisfying the first law alone does not ensure that the process will actually take place.

It is common experience that a cup of hot coffee left in a cooler room eventually cools off (Fig. 5-1). This process satisfies the first law of thermodynamics since the amount of energy lost by the coffee is equal to the amount gained by the surrounding air. Now let us consider the reverse process—the hot coffee getting even hotter in a cooler room as a result of heat transfer from the room air. We all know that this process never takes place. Yet, doing so would not violate the first law as long as the amount of energy lost by the air is equal to the amount gained by the coffee.

As another familiar example, consider the heating of a room by the passage of current through an electric resistor (Fig. 5-2). Again, the first law dictates that the amount of electric energy supplied to the resistance wires be equal to the amount of energy transferred to the room air as heat. Now let us attempt to reverse this process. It will come as no surprise that transferring some heat to the wires will not cause an equivalent amount of electric energy to be generated in the wires, even though doing so would not violate the first law.

Finally, consider a paddle-wheel mechanism that is operated by the fall of a mass (Fig. 5-3). The paddle wheel rotates as the mass falls and stirs a fluid within an insulated container. As a result, the potential energy of the mass decreases, and the internal energy of the fluid increases in accordance with the conservation of energy principle. However, the reverse process, raising the mass by transferring heat from the fluid to the paddle wheel, does not occur in nature, although doing so would not violate the first law of thermodynamics.

It is clear from the above that processes proceed in a *certain direction* and not in the reverse direction (Fig. 5-4). The first law places no restriction on the direction of a process, but satisfying the first law does not ensure that that process will actually occur. This inadequacy of the first law to identify whether a process can take place is remedied by introducing another general principle, the *second law of thermodynamics.* We show later in this chapter that the reverse processes discussed above violate the second law of thermodynamics. This violation is easily detected with the help of a property, called *entropy,* defined in the next chapter. *A process will not occur unless it satisfies both the first and the second laws of thermodynamics* (Fig. 5-5).

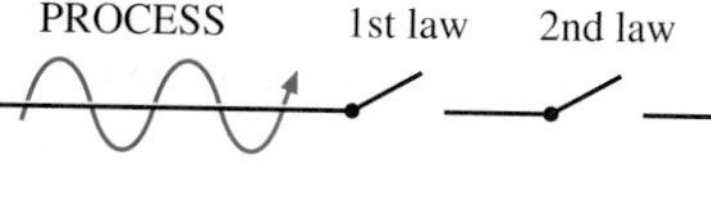


**FIGURE 5-5**
A process must satisfy both the first and second laws of thermodynamics to proceed.

There are numerous valid statements of the second law of thermodynamics. Two such statements are presented and discussed later in this chapter in relation to some engineering devices that operate on cycles.

The use of the second law of thermodynamics is not limited to identifying the direction of processes, however. The second law also asserts that energy has *quality* as well as quantity. The first law is concerned with the quality of energy and the transformations of energy from one form to another with no regard to its quality. Preserving the quality of energy is a major concern to engineers, and the second law provides the necessary means to determine the quality as well as the degree of degradation of energy during a process. As discussed later in this chapter, more of high-temperature energy can be converted to work, and thus it has a higher quality than the same amount of energy at a lower temperature.

The second law of thermodynamics is also used in determining the *theoretical limits* for the performance of commonly used engineering systems, such as heat engines and refrigerators, as well as predicting the *degree of completion* of chemical reactions.

## 5-2 ■ THERMAL ENERGY RESERVOIRS

In the development of the second law of thermodynamics, it is very convenient to have a hypothetical body with a relatively large *thermal energy capacity* (mass × specific heat) that can supply or absorb finite amounts of heat without undergoing any change in temperature. Such a body is called a **thermal energy reservoir**, or just a **reservoir**. In practice, large bodies of water such as oceans, lakes, and rivers as well as the atmospheric air can be modeled accurately as thermal energy reservoirs because of their large thermal energy storage capabilities or thermal masses (Fig. 5-6). The *atmosphere,* for example, does not warm up as a result of heat losses from residential buildings in winter. Likewise, megajoules of waste energy dumped in large rivers by power plants do not cause any significant change in water temperature.

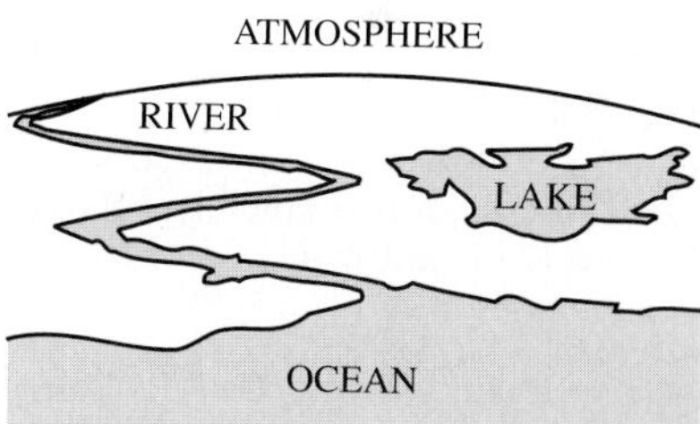


**FIGURE 5-6**
Bodies with relatively large thermal masses can be modeled as thermal energy reservoirs.

A *two-phase system* can be modeled as a reservoir also since it can absorb and release large quantities of heat while remaining at constant temperature. Another familiar example of a thermal energy reservoir is the *industrial furnace.* The temperatures of most furnaces are carefully controlled, and they are capable of supplying large quantities of thermal energy as heat in an essentially isothermal manner. Therefore, they can be modeled as reservoirs.

A body does not actually have to be very large to be considered a reservoir. Any physical body whose thermal energy capacity is large relative to the amount of energy it supplies or absorbs can be modeled as one. The air in a room, for example, can be treated as a reservoir in the analysis of the heat dissipation from a TV set in the room, since the amount of heat transfer from the TV set to the room air is not large enough to have a noticeable effect on the room air temperature.

A reservoir that supplies energy in the form of heat is called a **source**, and one that absorbs energy in the form of heat is called a **sink** (Fig. 5-7). Thermal energy reservoirs are often referred to as **heat reservoirs** since they supply or absorb energy in the form of heat.

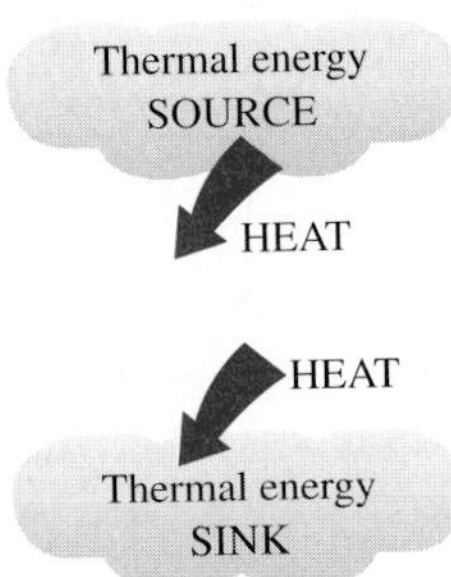


**FIGURE 5-7**
A source supplies energy in the form of heat, and a sink absorbs it.

Heat transfer from industrial sources to the environment is of major concern to environmentalists as well as to engineers. Irresponsible management of waste energy can significantly increase the temperature of portions of the environment, causing what is called *thermal pollution.* If it is not carefully controlled, thermal pollution can seriously disrupt marine life in lakes and rivers. However, by careful design and management, the waste energy dumped into large bodies of water can be used to significantly improve the quality of marine life by keeping the local temperature increases within safe and desirable levels.

## 5-3 HEAT ENGINES

As pointed out in Sec. 5-1, work can easily be converted to other forms of energy, but converting other forms of energy to work is not that easy. The mechanical work done by the shaft shown in Fig. 5-8, for example, is first converted to the internal energy of the water. This energy may then leave the water as heat. We know from experience that any attempt to reverse this process will fail. That is, transferring heat to the water will not cause the shaft to rotate. From this and other observations, we conclude that work can be converted to heat directly and completely, but converting heat to work requires the use of some special devices. These devices are called **heat engines**.

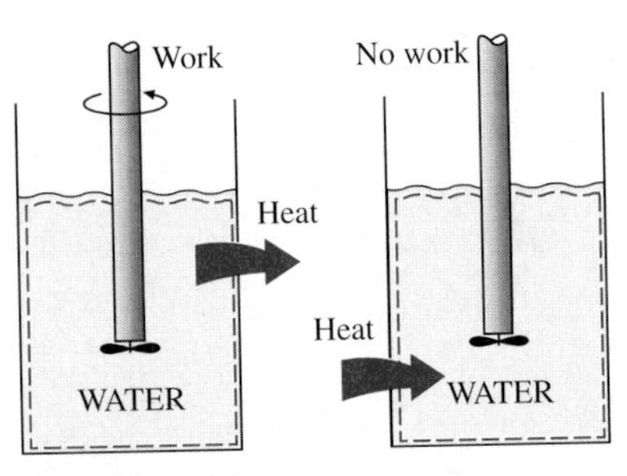


**FIGURE 5-8**
Work can always be converted to heat directly and completely, but the reverse is not true.

Heat engines differ considerably from one another, but all can be characterized by the following (Fig. 5-9):

**1** They receive heat from a high-temperature source (solar energy, oil furnace, nuclear reactor, etc.).

**2** They convert part of this heat to work (usually in the form of a rotating shaft).

**3** They reject the remaining waste heat to a low-temperature sink (the atmosphere, rivers, etc.).

**4** They operate on a cycle.

Heat engines and other cyclic devices usually involve a fluid to and from which heat is transferred while undergoing a cycle. This fluid is called the **working fluid**.

The term *heat engine* is often used in a broader sense to include

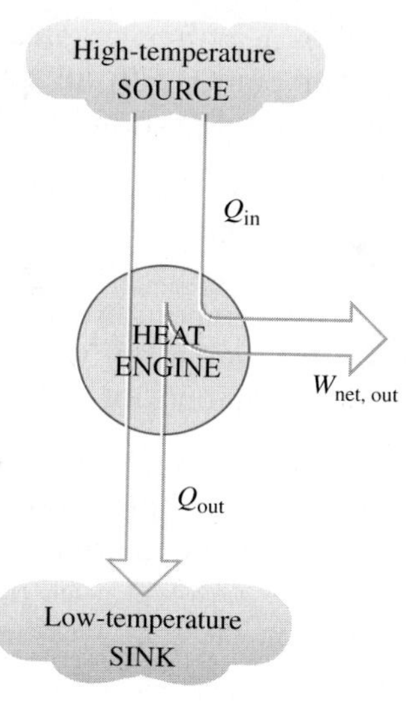


**FIGURE 5-9**

Part of the heat received by a heat engine is converted to work, while the rest is rejected to a sink.

work-producing devices that do not operate in a thermodynamic cycle. Engines that involve internal combustion such as gas turbines and car engines fall into this category. These devices operate in a mechanical cycle but not in a thermodynamic cycle since the working fluid (the combustion gases) does not undergo a complete cycle. Instead of being cooled to the initial temperature, the exhaust gases are purged and replaced by fresh air-and-fuel mixture at the end of the cycle.

The work-producing device that best fits into the definition of a heat engine is the *steam power plant,* which is an external-combustion engine. That is, the combustion process takes place outside the engine, and the thermal energy released during this process is transferred to the steam as heat. The schematic of a basic steam power plant is shown in Fig. 5-10.

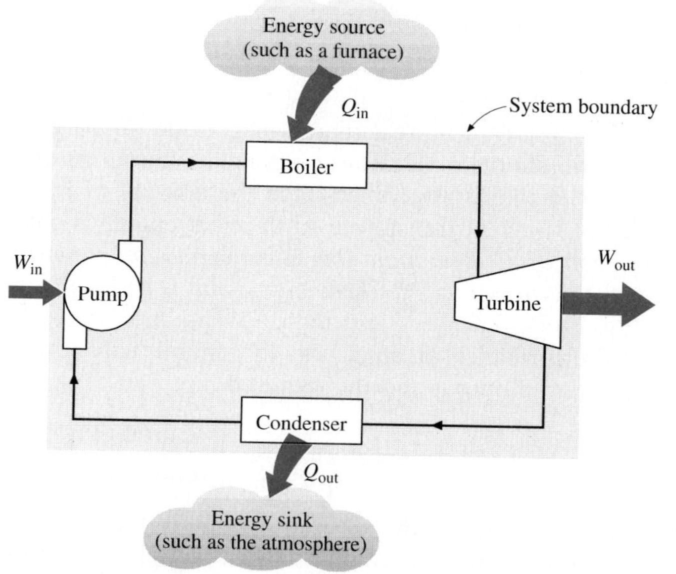


**FIGURE 5-10**

Schematic of a steam power plant.

This is a rather simplified diagram, and the discussion of actual steam power plants with all their complexities is left to Chap. 9. The various quantities shown on this figure are as follows:

$Q_{in}$ = amount of heat supplied to steam in boiler from a high-temperature source (furnace)
$Q_{out}$ = amount of heat rejected from steam in condenser to a low-temperature sink (the atmosphere, a river, etc.)
$W_{out}$ = amount of work delivered by steam as it expands in turbine
$W_{in}$ = amount of work required to compress water to boiler pressure

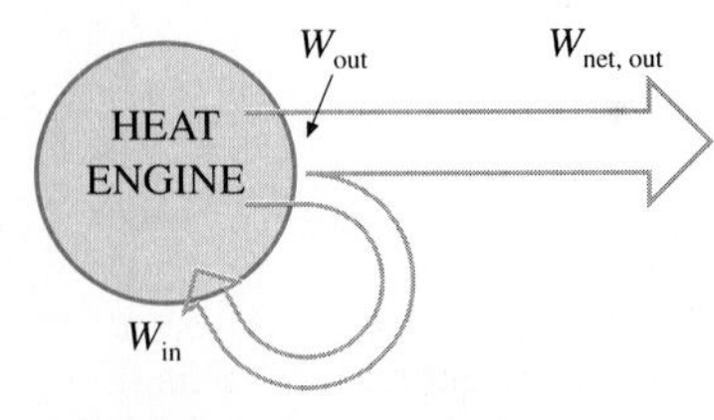


**FIGURE 5-11**
A portion of the work output of a heat engine is consumed internally to maintain continuous operation.

Notice that the directions of the heat and work interactions are indicated by the subscripts *in* and *out.* Therefore, all four quantities described above are always *positive.*

The net work output of this power plant is simply the difference between the total work output of the plant and the total work input (Fig. 5-11):

$$W_{net,out} = W_{out} - W_{in} \quad \text{(kJ)} \tag{5-1}$$

The net work can also be determined from the heat transfer data alone. The four components of the steam power plant involve mass flow in and out, and therefore they should be treated as open systems. These components, together with the connecting pipes, however, always contain the same fluid (not counting the steam that may leak out, or course). No mass enters or leaves this combination system, which is indicated by the shaded area on Fig. 5-11; thus it can be analyzed as a closed system. Recall that for a closed system undergoing a cycle, the change in internal energy $\Delta U$ is zero, and therefore the net work output of the system is also equal to the net heat transfer to the system:

$$W_{net,out} = Q_{in} - Q_{out} \quad \text{(kJ)} \tag{5-2}$$

### Thermal Efficiency

In Eq. 5-2, $Q_{out}$ represents the magnitude of the energy wasted in order to complete the cycle. But $Q_{out}$ is never zero; thus, the net work output of a heat engine is always less than the amount of heat input. That is, only part of the heat transferred to the heat engine is converted to work. *The fraction of the heat input that is converted to net work output is a measure of the performance of a heat engine and is called the* **thermal efficiency** $\eta_{th}$ (Fig. 5-12).

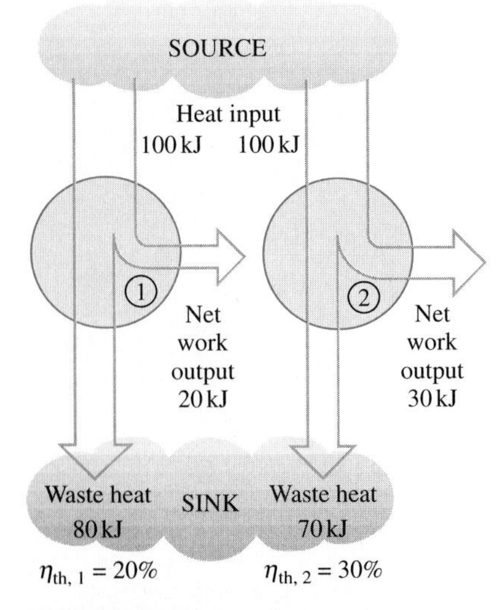


**FIGURE 5-12**
Some heat engines perform better than others (convert more of the heat they receive to work).

Performance or efficiency, in general, can be expressed in terms of the desired output and the required input as (Fig. 5-13)

$$\text{performance} = \frac{\text{desired output}}{\text{required input}} \tag{5-3}$$

For heat engines, the desired output is the net work output, and the required input is the amount of heat supplied to the working fluid. Then

**FIGURE 5-13**
The definition of performance is not limited to thermodynamics only.

the thermal efficiency of a heat engine can be expressed as

$$\text{thermal efficiency} = \frac{\text{net work output}}{\text{total heat input}}$$

or

$$\eta_{\text{th}} = \frac{W_{\text{net,out}}}{Q_{\text{in}}} \tag{5-4}$$

It can also be expressed as

$$\eta_{\text{th}} = 1 - \frac{Q_{\text{out}}}{Q_{\text{in}}} \tag{5-5}$$

since $W_{\text{net,out}} = Q_{\text{in}} - Q_{\text{out}}$.

Cyclic devices of practical interest such as heat engines, refrigerators, and heat pumps operate between a high-temperature medium (or reservoir) at temperature $T_H$ and a low-temperature medium (or reservoir) at temperature $T_L$. To bring uniformity to the treatment of heat engines, refrigerators, and heat pumps, we define the following two quantities:

$Q_H$ = *magnitude* of heat transfer between cyclic device and high-temperature medium at temperature $T_H$

$Q_L$ = *magnitude* of heat transfer between cyclic device and low-temperature medium at temperature $T_L$

Notice that both $Q_L$ and $Q_H$ are defined as *magnitudes* and therefore are *positive quantities.* The direction of $Q_H$ and $Q_L$ is easily determined by inspection, and we do not need to be concerned about their signs. Then the net work output and thermal efficiency relations for any heat engine (shown in Fig. 5-14) can also be expressed as

$$W_{\text{net,out}} = Q_H - Q_L \tag{5-6}$$

and

$$\eta_{\text{th}} = \frac{W_{\text{net,out}}}{Q_H} \tag{5-7}$$

or

$$\eta_{\text{th}} = 1 - \frac{Q_L}{Q_H} \tag{5-8}$$

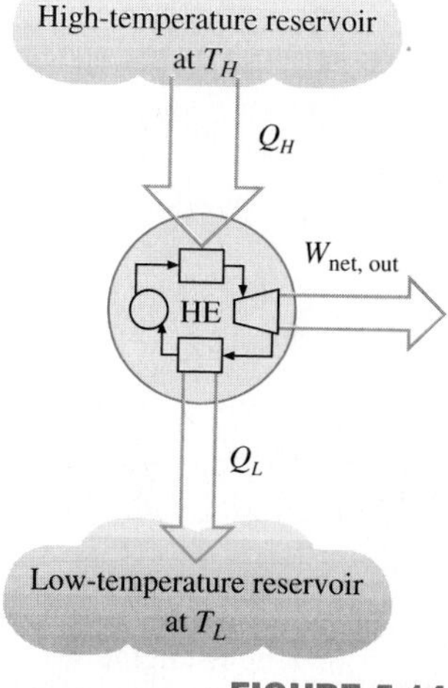


**FIGURE 5-14**
Schematic of a heat engine.

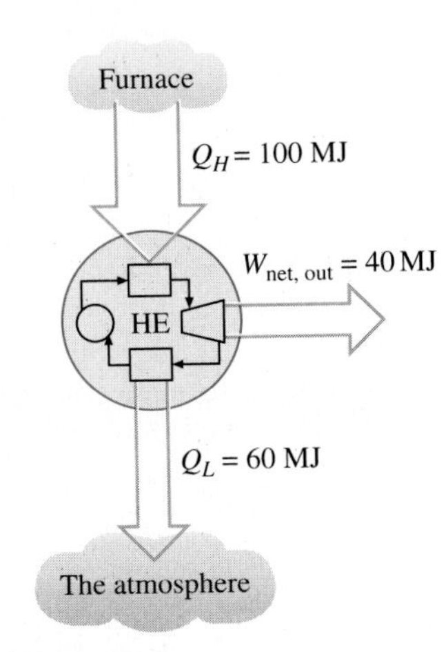


**FIGURE 5-15**
Even the most efficient heat engines reject most of the heat they receive as waste heat.

The thermal efficiency of a heat engine is always less than unity since both $Q_L$ and $Q_H$ are defined as positive quantities.

Thermal efficiency is a measure of how efficiently a heat engine converts the heat that it receives to work. Heat engines are built for the purpose of converting heat to work, and engineers are constantly trying to improve the efficiencies of these devices since increased efficiency means less fuel consumption and thus lower fuel bills.

The thermal efficiencies of work-producing devices are amazingly low. Ordinary spark-ignition automobile engines have a thermal efficiency of about 20 percent. That is, an automobile engine converts, at an average, about 20 percent of the chemical energy of the gasoline to mechanical work. This number is about 30 percent for diesel engines and large gas-turbine plants and 40 percent for large steam power plants. Thus, even with the most efficient heat engines available today, more than one-half of the energy supplied ends up in the rivers, lakes, or the atmosphere as waste or unusable energy (Fig. 5-15).

### Can We Save $Q_{out}$?

In a steam power plant, the condenser is the device where large quantities of waste heat is rejected to rivers, lakes, or the atmosphere. Then one may ask, can we not just take the condenser out of the plant and save all that waste energy? The answer to this question is, unfortunately, a firm *no* for the simple reason that without the cooling process in a condenser the cycle cannot be completed. (Cyclic devices such as steam power plants cannot run continuously unless the cycle is completed.) This is demonstrated below with the help of a simple heat engine.

Consider the simple heat engine shown in Fig. 5-16 that is used to lift weights. It consists of a piston–cylinder device with two sets of stops. The working fluid is the gas contained within the cylinder. Initially, the gas

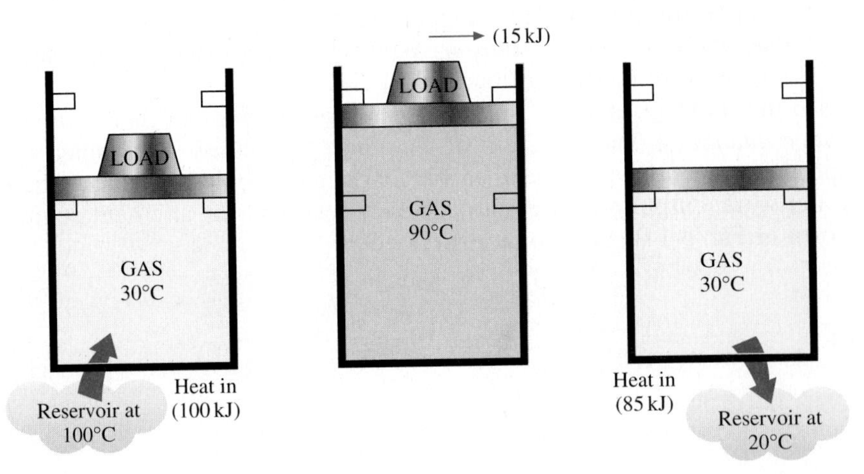


**FIGURE 5-16**
A heat-engine cycle cannot be completed without rejecting some heat to low-temperature sink.

temperature is 30°C. The piston, which is loaded with the weights, is resting on top of the lower stops. Now 100 kJ of heat is transferred to the gas in the cylinder from a source at 100°C, causing it to expand and to raise the loaded piston until the piston reaches the upper stops, as shown in the figure. At this point, the load is removed, and the gas temperature is observed to be 90°C.

The work done on the load during this expansion process is equal to the increase in its potential energy, say 15 kJ. Even under ideal conditions (weightless piston, no friction, no heat losses, and quasi-equilibrium expansion), the amount of heat supplied to the gas is greater than the work done since part of the heat supplied is used to raise the temperature of the gas.

Now let us try to answer the following question: *Is it possible to transfer the 85 kJ of excess heat at 90°C back to the reservoir at 100°C for later use?* If it is, then we will have a heat engine that can have a thermal efficiency of 100 percent under ideal conditions. The answer to this question is again *no,* for the very simple reason that heat always flows from a high-temperature medium to a low-temperature one, and never the other way around. Therefore, we cannot cool this gas from 90 to 30°C by transferring heat to a reservoir at 100°C. Instead, we have to bring the system into contact with a low-temperature reservoir, say at 20°C, so that the gas can return to its initial state by rejecting its 85 kJ of excess energy as heat to this reservoir. This energy cannot be recycled, and it is properly called *waste energy.*

We conclude from the above discussion that every heat engine must *waste* some energy by transferring it to a low-temperature reservoir in order to complete the cycle, even under idealized conditions. The requirement that a heat engine exchange heat with at least two reservoirs for continuous operation forms the basis for the Kelvin–Planck expression of the second law of thermodynamics discussed later in this section.

**EXAMPLE 5-1**

Heat is transferred to a heat engine from a furnace at a rate of 80 MW. If the rate of waste heat rejection to a nearby river is 50 MW, determine the net power output and the thermal efficiency for this heat engine.

**Solution** A schematic of the heat engine is given in Fig. 5-17. The furnace serves as the high-temperature reservoir for this heat engine and the river as the low-temperature reservoir. Then the given quantities can be expressed in rate form as

$$\dot{Q}_H = 80\text{ MW} \qquad \text{and} \qquad \dot{Q}_L = 50\text{ MW}$$

Neglecting the heat losses that may occur from the working fluid as it passes through the pipes and other components, the net power output of this heat engine is determined from Eq. 5-5 to be

$$\dot{W}_{\text{net,out}} = \dot{Q}_H - \dot{Q}_L = (80 - 50)\text{ WM} = 30\text{ MW}$$

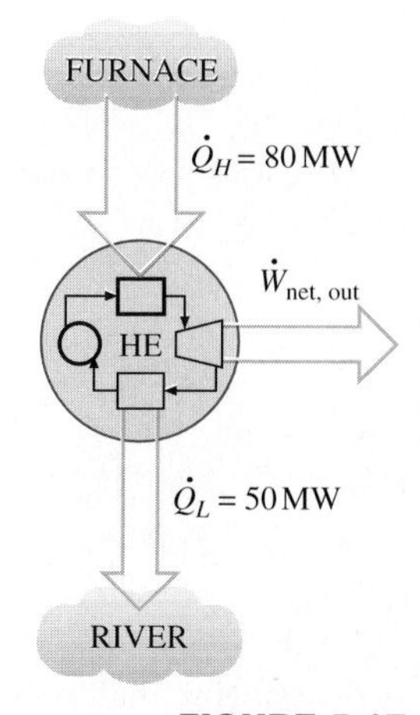


**FIGURE 5-17**
Schematic for Example 5-1.

Then the thermal efficiency is easily determined from Eq. 5-7:

$$\eta_{th} = \frac{\dot{W}_{net,out}}{\dot{Q}_H} = \frac{30 \text{ MW}}{80 \text{ MW}} = 0.375 \text{ (or 37.5\%)}$$

That is, the heat engine converts 37.5 percent of the heat it receives to work.

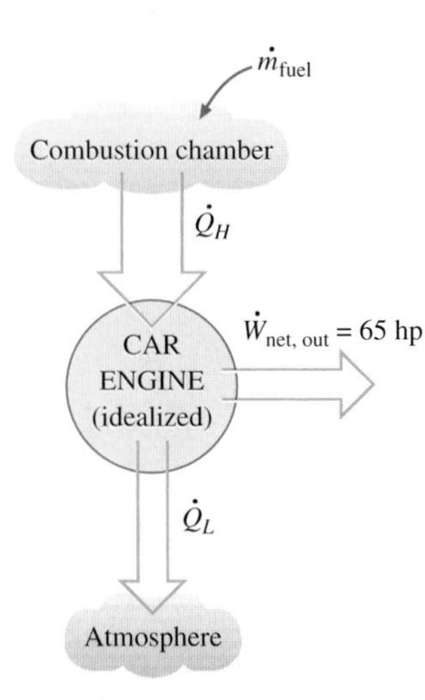


**FIGURE 5-18**
Schematic for Example 5-2.

**EXAMPLE 5-2**

A car engine with a power output of 65 hp has a thermal efficiency of 24 percent. Determine the fuel consumption rate of this car if the fuel has a heating value of 19,000 Btu/lbm (that is, 19,000 Btu of energy is released for each lbm of fuel burned).

**Solution** A schematic of the car engine is given in Fig. 5-18. The car engine is powered by converting 24 percent of the chemical energy released during the combustion process to work. The amount of energy input required to produce a power output of 65 hp is determined from the definition of thermal efficiency (Eq. 5-7):

$$\dot{Q}_H = \frac{\dot{W}_{net,out}}{\eta_{th}} = \frac{65 \text{ hp}}{0.24}\left(\frac{2545 \text{ Btu/h}}{1 \text{ hp}}\right) = 689{,}262 \text{ Btu/h}$$

To supply energy at this rate, the engine must burn fuel at a rate of

$$\dot{m} = \frac{689{,}262 \text{ Btu/h}}{19{,}000 \text{ Btu/lbm}} = 36.3 \text{ lbm/h}$$

since 19,000 Btu of thermal energy is released for each lbm of fuel burned.

## The Second Law of Thermodynamics: Kelvin–Planck Statement

We have demonstrated earlier with reference to the heat engine shown in Fig. 5-16 that, even under ideal conditions, a heat engine must reject some heat to a low-temperature reservoir in order to complete the cycle. That is, no heat engine can convert all the heat it receives to useful work. This limitation on the thermal efficiency of heat engines forms the basis for the Kelvin–Planck statement of the second law of thermodynamics, which is expressed as follows:

> *It is impossible for any device that operates on a cycle to receive heat from a single reservoir and produce a net amount of work.*

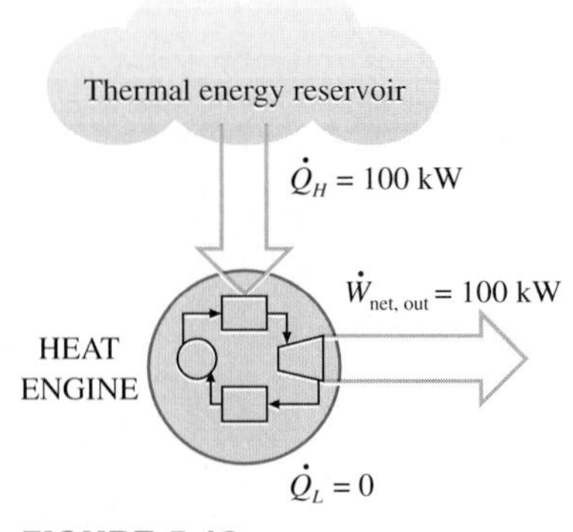


**FIGURE 5-19**
A heat engine that violates the Kelvin–Planck statement of the second law.

That is, a heat engine must exchange heat with a low-temperature sink as well as a high-temperature source to keep operating. The Kelvin–Planck statement can also be expressed as follows: *No heat engine can have a thermal efficiency of 100 percent* (Fig. 5-19), or *for a power plant to operate, the working fluid must exchange heat with the environment as well as the furnace.*

Note that the impossibility of having a 100 percent efficient heat engine is not due to friction or other dissipative effects. It is a limitation that applies to both the idealized and the actual heat engines. Later in this chapter, we develop a relation for the maximum thermal efficiency of a heat engine. We also demonstrate that this maximum value depends on the reservoir temperatures only.

## 5-4 ■ REFRIGERATORS AND HEAT PUMPS

We all know from experience that heat flows in the direction of decreasing temperature, i.e., from high-temperature mediums to low-temperature ones. This heat transfer process occurs in nature without requiring any devices. The reverse process, however, cannot occur by itself. The transfer of heat from a low-temperature medium to a high-temperature one requires special devices called **refrigerators**

Refrigerators, like heat engines, are cyclic devices. The working fluid used in the refrigeration cycle is called a **refrigerant**. The most frequently used refrigeration cycle is the *vapor-compression refrigeration cycle* which involves four main components: a compressor, a condenser, an expansion valve, and an evaporator, as shown in Fig. 5-20.

The refrigerant enters the compressor as a vapor and is compressed to the condenser pressure. It leaves the compressor at a relatively high temperature and cools down and condenses as it flows through the coils of the condenser by rejecting heat to the surrounding medium. It then enters a capillary tube where its pressure and temperature drop drastically due to the throttling effect. The low-temperature refrigerant then enters the evaporator, where it evaporates by absorbing heat from the refrigerated space. The cycle is completed as the refrigerant leaves the evaporator and reenters the compressor.

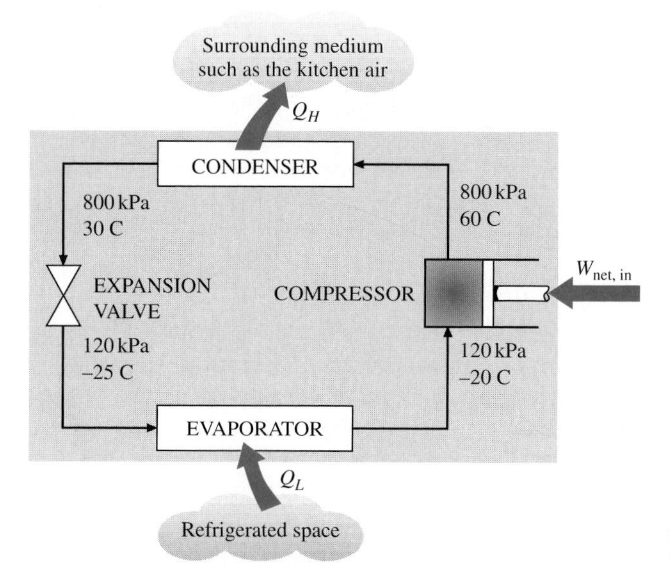


**FIGURE 5-20**

Basic components of a refrigeration system and typical operating conditions.

In a household refrigerator, the freezer compartment where heat is picked up by the refrigerant serves as the evaporator, and the coils behind the refrigerator where heat is dissipated to the kitchen air as the condenser.

A refrigerator is shown schematically in Fig. 5-21. Here $Q_L$ is the magnitude of the heat removed from the refrigerated space at temperature $T_L$, $Q_H$ is the magnitude of the heat rejected to the warm environment at temperature $T_H$, and $W_{net,in}$ is the net work input to the refrigerator. As discussed before, $Q_L$ and $Q_H$ represent magnitudes and so are positive quantities.

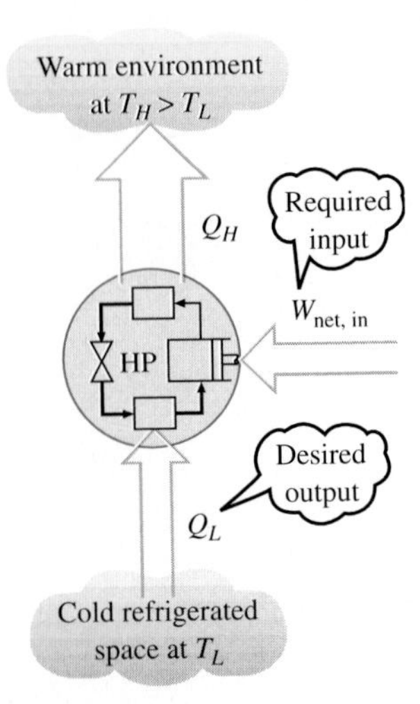


**FIGURE 5-21**
The objective of a refrigerator is to remove $Q_L$ from the cooled space.

### Coefficient of Performance

The *efficiency* of a refrigerator is expressed in terms of the **coefficient of performance** (COP), denoted by $COP_R$. The objective of a refrigerator is to remove heat ($Q_L$) from the refrigerated space. To accomplish this objective, it requires a work input of $W_{net,in}$. Then the COP of a refrigerator can be expressed as

$$COP_R = \frac{\text{desired output}}{\text{required input}} = \frac{Q_L}{W_{net,in}} \tag{5-9}$$

This relation can also be expressed in rate form by replacing $Q_L$ by $\dot{Q}_L$ and $W_{net,in}$ by $\dot{W}_{net,in}$.

The conservation of energy principle for a cyclic device requires that

$$W_{net,in} = Q_H - Q_L \quad \text{(kJ)} \tag{5-10}$$

Then the COP relation can also be expressed as

$$COP_R = \frac{Q_L}{Q_H - Q_L} = \frac{1}{Q_H/Q_L - 1} \tag{5-11}$$

Notice that the value of $COP_R$ can be *greater than unity.* That is, the amount of heat removed from the refrigerated space can be greater than the amount of work input. This is in contrast to the thermal efficiency, which can never be greater than 1. In fact, one reason for expressing the efficiency of a refrigerator by another term—the coefficient of performance—is the desire to avoid the oddity of having efficiencies greater than unity.

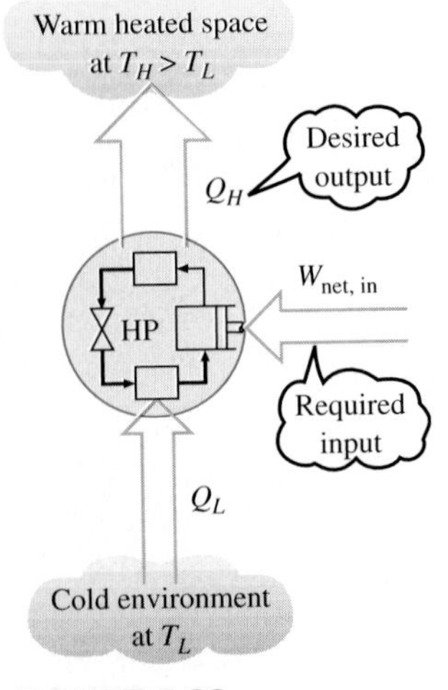


**FIGURE 5-22**
The objective of a heat pump is to supply heat $Q_H$ into the warmer space.

### Heat Pumps

Another device that transfers heat from a low-temperature medium to a high-temperature one is the **heat pump**, shown schematically in Fig. 5-22. Refrigerators and heat pumps operate on the same cycle but differ in their objectives. The objective of a refrigerator is to maintain the refrigerated space at a low temperature by removing heat from it. Discharging this heat to a higher-temperature medium is merely a necessary part of the operation, not the purpose. The objective of a heat pump, however, is to maintain a heated space at a high temperature. This is accomplished by absorbing heat from a low-temperature source, such as well water or cold outside air in winter, and supplying this heat to the high-temperature medium such as a house (Fig. 5-23).

An ordinary refrigerator that is placed in the window of a house with its door open to the cold outside air in winter will function as a heat pump since it will try to cool the outside by absorbing heat from it and rejecting this heat into the house through the coils behind it (Fig. 5-24).

The measure of performance of a heat pump is also expressed in terms of the **coefficient of performance** $COP_{HP}$, defined as

$$COP_{HP} = \frac{\text{desired output}}{\text{required input}} = \frac{Q_H}{W_{net,in}} \tag{5-12}$$

which can also be expressed as

$$COP_{HP} = \frac{Q_H}{Q_H - Q_L} = \frac{1}{1 - Q_L/Q_H} \tag{5-13}$$

A comparison of Eqs. 5-9 and 5-12 reveals that

$$COP_{HP} = COP_R + 1 \tag{5-14}$$

for fixed values of $Q_L$ and $Q_H$. This relation implies that the coefficient of performance of a heat pump is always greater than unity since $COP_R$ is a positive quantity. That is, a heat pump will function, at worst, as a resistance heater, supplying as much energy to the house as it consumes. In reality, however, part of $Q_H$ is lost to the outside air through piping and other devices, and $COP_{HP}$ may drop below unity when the outside air temperature is too low. When this happens, the system usually switches to a resistance heating mode. Most heat pumps in operation today have seasonally averaged COP of 2 to 3.

**Air conditioners** are basically refrigerators whose refrigerated space is a room or a building instead of the food compartment. A window air conditioning unit cools a room by absorbing heat from the room air and discharging it to the outside. The same air conditioning unit can be used as a heat pump in winter by installing it backward. In this mode, the unit will pick up heat from the cold outside and deliver it to the room. Air conditioning systems that are equipped with proper controls and a reversing valve operate as air conditioners in summer and as heat pumps in winter.

The performance of refrigerators and air conditioners in the U.S. is often expressed in terms of the **Energy Efficiency Rating** (EER), which is the amount of heat removed from the cooled space in Btu's for 1 Wh (watt-hour) of electricity consumed. Considering that 1 kWh = 3412 Btu and thus 1 Wh = 3.412 Btu, a unit that removes 1 kWh of heat from the cooled space for each kWh of electricity it consumes (COP = 1) will have an EER of 3.412. Therefore, the relation between EER and COP is

$$EER = 3.412\ COP_R$$

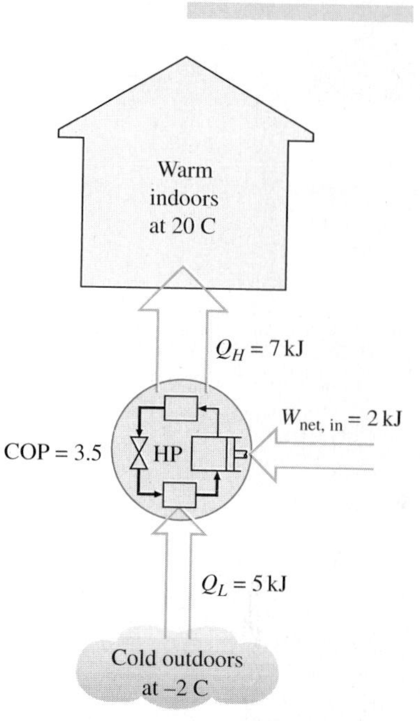


**FIGURE 5-23**

The work supplied to a heat pump is used to extract energy from the cold outdoors and carry it into the warm indoors.

**FIGURE 5-24**

When installed backward, an air conditioner will function as a heat pump.

Most air conditioners have an EER between 8 and 12 (a COP of 2.3 to 3.5). A high-efficiency heat pump recently manufactured by the Trane Company using a reciprocating variable-speed compressor is reported to have a COP of 3.3 in the heating mode, and an EER of 16.9 (COP of 5.0) in the air conditioning mode. Variable-speed compressors and fans allow the unit to operate at maximum efficiency for varying heating/cooling needs and weather conditions as determined by a microprocessor. In the air conditioning mode, for example, they operate at higher speeds on hot days and at lower speeds on cooler days, enhancing both efficiency and comfort.

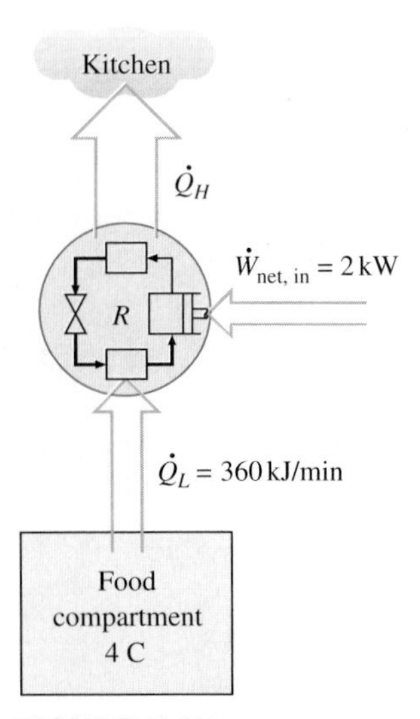


**FIGURE 5-25**
Schematic for Example 5-3.

**EXAMPLE 5-3**

The food compartment of a refrigerator, shown in Fig. 5-25, is maintained at 4°C by removing heat from it at a rate of 360 kJ/min. If the required power input to the refrigerator is 2 kW, determine (*a*) the coefficient of performance of the refrigerator and (*b*) the rate of heat discharge to the room that houses the refrigerator.

**Solution** (*a*) The coefficient of performance of a refrigerator is defined by Eq. 5-9, which can be expressed in rate form as

$$\text{COP}_R = \frac{\dot{Q}_L}{\dot{W}_{net,in}} = \frac{360 \text{ kJ/min}}{2 \text{ kW}}\left(\frac{1 \text{ kW}}{60 \text{ kJ/min}}\right) = 3$$

That is, 3 kJ of heat is removed from the refrigerated space for each kJ of work supplied.

(*b*) The rate at which heat is discharged to the room that houses the refrigerator is determined from the conservation of energy relation for cyclic devices (Eq. 5-10), expressed in rate form as

$$\dot{Q}_H = \dot{Q}_L + \dot{W}_{net,in} = 360 \text{ kJ/min} + (2 \text{ kW})\left(\frac{60 \text{ kJ/min}}{1 \text{ kW}}\right) = 480 \text{ kJ/min}$$

Notice that both the energy removed from the refrigerated space as heat and the energy supplied to the refrigerator as electrical work eventually show up in the room air and become part of the internal energy of the air. This demonstrates that energy can change from one form to another, can move from one place to another, but is never destroyed during a process.

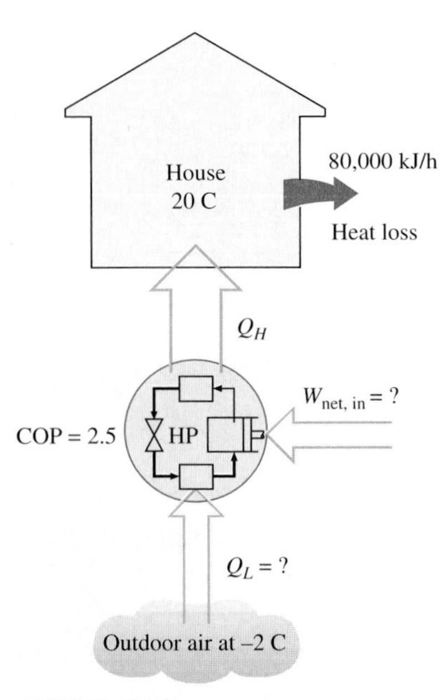


**FIGURE 5-26**
Schematic for Example 5-4.

**EXAMPLE 5-4**

A heat pump is used to meet the heating requirements of a house and maintain it at 20°C. On a day when the outdoor air temperature drops to −2°C, the house is estimated to lose heat at a rate of 80,000 kJ/h. If the heat pump under these conditions has a COP of 2.5, determine (*a*) the power consumed by the heat pump and (*b*) the rate at which heat is extracted from the cold outdoor air.

**Solution** (*a*) The power consumed by this heat pump, shown in Fig. 5-26, can be determined from the definition of the coefficient of performance of a heat pump (Eq. 5-12), expressed in rate form as

$$\dot{W}_{net,in} = \frac{\dot{Q}_H}{\text{COP}_{HP}} = \frac{80{,}000 \text{ kJ/h}}{2.5} = 32{,}000 \text{ kJ/h (or 8.9 kW)}$$

(*b*) The house is losing heat at a rate of 80,000 kJ/h. If the house is to be maintained at a constant temperature of 20°C, the heat pump must deliver heat to the house at the same rate, i.e., at a rate of 80,000 kJ/h. Then the rate of heat transfer from the outdoor air is determined from the conservation of energy principle for a cyclic device (Eq. 5-10):

$$\dot{Q}_L = \dot{Q}_H - \dot{W}_{\text{net,in}} = (80{,}000 - 32{,}000)\ \text{kJ/h} = \mathbf{48{,}000\ kJ/h}$$

That is, 48,000 of the 80,000 kJ/h heat delivered to the house is actually extracted from the cold outdoor air. Therefore, we are paying only for the 32,000-kJ/h energy, which is supplied as electrical work to the heat pump. If we were to use an electric resistance heater instead, we would have to supply the entire 80,000 kJ/h to the resistance heater as electric energy. This would mean a heating bill which is 2.5 times higher. This explains the popularity of heat pumps as heating systems and why they are preferred to simple electric resistance heaters despite their considerably higher initial cost.

## The Second Law of Thermodynamics: Clausius Statement

There are two classical statements of the second law: the Kelvin–Planck statement, which is related to heat engines and discussed in the preceding section, and the Clausius statement, which is related to refrigerators or heat pumps. The Clausius statement is expressed as follows:

> *It is impossible to construct a device that operates in a cycle and produces no effect other than the transfer of heat from a lower-temperature body to a higher-temperature body.*

It is common knowledge that heat does not, of its own volition, flow from a cold medium to a warmer one. The Clausius statement does not imply that a cyclic device that transfers heat from a cold medium to a warmer one is impossible to construct. In fact, this is precisely what a common household refrigerator does. It simply states that a refrigerator will not operate unless its compressor is driven by an external power source, such as an electric motor (Fig. 5-27). This way, the net effect on the surroundings involves the consumption of some energy in the form of work, in addition to the transfer of heat from a colder body to a warmer one. That is, it leaves a trace in the surroundings. Therefore, a household refrigerator is in complete compliance with the Clausius statement of the second law.

Both the Kelvin–Planck and the Clausius statements of the second law are negative statements, and a negative statement cannot be proved. Like any other physical law, the second law of thermodynamics is based on experimental observations. To date, no experiment has been conducted that contradicts the second law, and this should be taken as sufficient evidence of its validity.

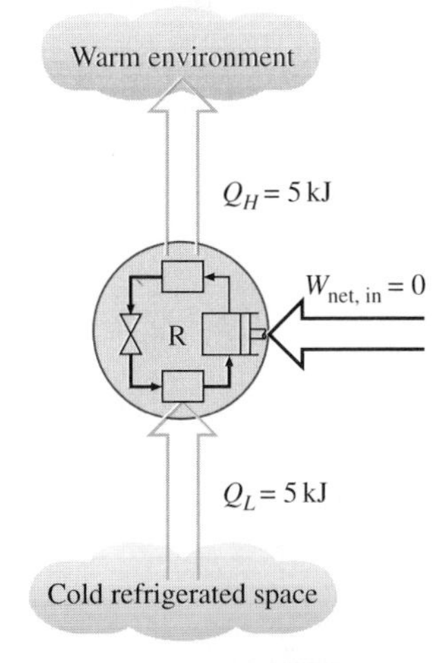


**FIGURE 5-27**
A refrigerator that violates the Clausius statement of the second law.

## Equivalence of the Two Statements

The Kelvin–Planck and the Clausius statements are equivalent in their consequences, and either statement can be used as the expression of the second law of thermodynamics. Any device that violates the Kelvin–Planck statement also violates the Clausius statement, and vice versa. This can be demonstrated as follows:

Consider the heat-engine–refrigerator combination shown in Fig. 5-28*a*, operating between the same two reservoirs. The heat engine is assumed to have, in violation of the Kelvin–Planck statement, a thermal efficiency of 100 percent, and therefore it converts all the heat $Q_H$ it receives to work $W$. This work is now supplied to a refrigerator that removes heat in the amount of $Q_L$ from the low-temperature reservoir and rejects heat in the amount of $Q_L + Q_H$ to the high-temperature reservoir. During this process, the high-temperature reservoir recieves a net amount of heat $Q_L$ (the difference between $Q_L + Q_H$ and $Q_H$). Thus the combination of these two devices can be viewed as a refrigerator, as shown in Fig. 5-28*b*, that transfers heat in an amount of $Q_L$ from a cooler body to a warmer one without requiring any input from outside. This is clearly a violation of the Clausius statement. Therefore, a violation of the Kelvin–Planck statement results in the violation of the Clausius statement.

It can also be shown in a similar manner that a violation of the Clausius statement leads to the violation of the Kelvin–Planck statement. Therefore, the Clausius and the Kelvin–Planck statements are two equivalent expressions of the second law of thermodynamics.

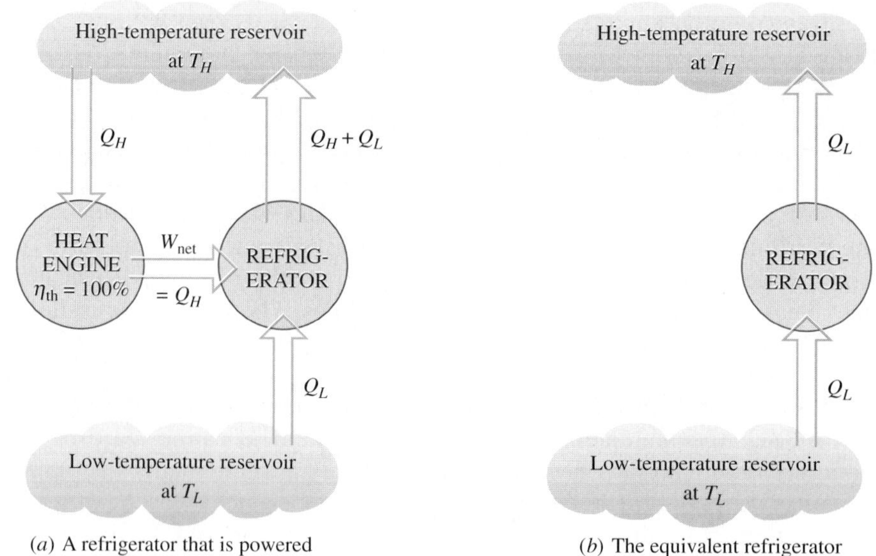


**FIGURE 5-28**

Proof that the violation of the Kelvin–Planck statement leads to the violation of the Clausius statement.

**Perpetual-Motion Machines**

We have repeatedly stated that a process cannot take place unless it satisfies both the first and second laws of thermodynamics. Any device that violates either law is called a **perpetual-motion machine**, and despite numerous attempts, no perpetual-motion machine is known to have worked. But this has not stopped inventors from trying to create new ones.

A device that violates the first law of thermodynamics (by *creating* energy) is called a **perpetual-motion machine of the first kind** (PMM1), and a device that violates the second law of thermodynamics is called a **perpetual-motion machine of the second kind** (PMM2).

Consider the steam power plant shown in Fig. 5-29. It is proposed to heat the steam by resistance heaters placed inside the boiler, instead of by the energy supplied from fossil or nuclear fuels. Part of the electricity generated by the plant is to be used to power the resistors as well as the pump. The rest of the electric energy is to be supplied to the electric network as the net work output. The inventor claims that once the system is started, this power plant will produce electricity indefinitely without requiring any energy input from the outside.

Well, here is an invention that could solve the world's energy problem—if it works, of course. A careful examination of this invention reveals that the system enclosed by the shaded area is continuously supplying energy to the outside at a rate of $\dot{Q}_{out} + \dot{W}_{net,out}$ without receiving any energy. That is, this system is creating energy at a rate of $\dot{Q}_{out} + \dot{W}_{net,out}$, which is clearly a violation of the first law. Therefore, this wonderful device is nothing more than a PMM1 and does not warrant any further consideration.

Now let us consider another novel idea by the same inventor. Convinced that energy cannot be created, the inventor suggests the following modification which will greatly improve the thermal efficiency of that power plant without violating the first law. Aware that more than

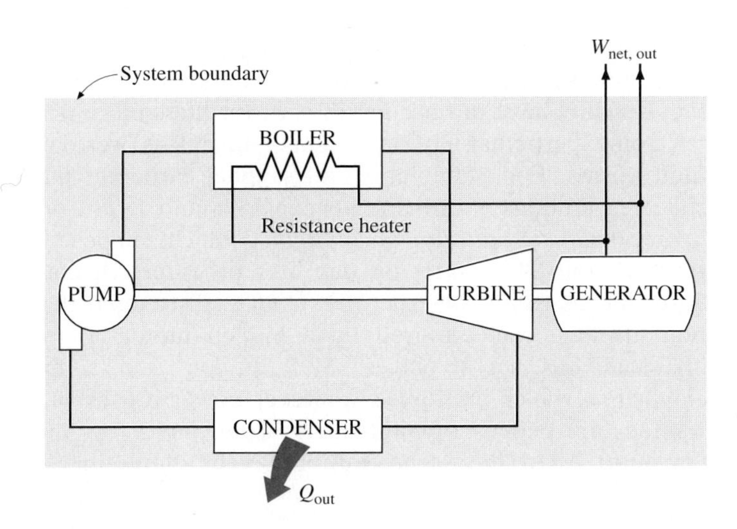


**FIGURE 5-29**
A perpetual-motion machine that violates the first law of thermodynamics (PMM1).

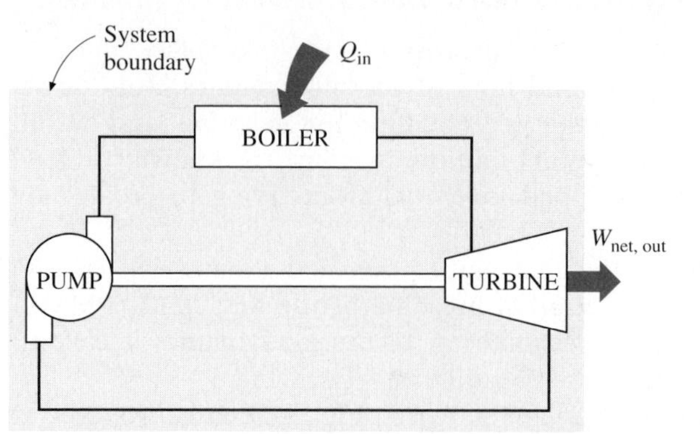


**FIGURE 5-30**
A perpetual-motion machine that violates the second law of thermodynamics (PMM2).

one-half of the heat transferred to the steam in the furnace is discarded in the condenser to the environment, the inventor suggests getting rid of this wasteful component and sending the steam to the pump as soon as it leaves the turbine, as shown in Fig. 5-30. This way, all the heat transferred to the steam in the boiler will be converted to work, and thus the power plant will have a theoretical efficiency of 100 percent. The inventor realizes that some heat losses and friction between the moving components are unavoidable and that these effects will hurt the efficiency somewhat, but still expects the efficiency to be no less than 80 percent (as opposed to 40 percent in actual power plants) for a carefully designed system.

Well, the possibility of doubling the efficiency would certainly be very tempting to plant managers and, if not properly trained, they would probably give this idea a chance, since intuitively they see nothing wrong with it. A student of thermodynamics, however, will immediately label this device as a PMM2, since it works on a cycle and does a net amount of work while exchanging heat with a single reservoir (the furnace) only. It satisfies the first law but violates the second law, and therefore it will not work.

Countless perpetual-motion machines have been proposed throughout history, and many more are being proposed. Some proposers have even gone so far as patenting their inventions, only to find out that what they actually have in their hands is a worthless piece of paper.

Some perpetual-motion machine inventors were very successful in fund raising. For example, a Philadelphia carpenter named J. W. Kelly collected millions of dollars between 1874 and 1898 from investors in his *hydropneumatic-pulsating-vacu-engine,* which supposedly could push a railroad train 3000 miles on one liter of water. Of course it never did. After his death in 1898, the investigators discovered that the demonstration machine was powered by a hidden motor. Recently a group of investors was set to invest $2.5 million into a mysterious *energy augmentor,* which multiplied whatever power it took in, but their lawyer wanted an expert opinion first. Confronted by the scientists, the "inventor" fled the scene without even attempting to run his demo machine.

Tired of applications for perpetual-motion machines, the U.S. Patent Office decreed in 1918 that it would no longer even consider any perpetual-motion applications. But several such patent applications were still filed, and some made it through the patent office undetected. Some applicants whose patent applications were denied sought legal action. For example, in 1982 the U.S. Patent Office dismissed a huge device that involves several hundred kilograms of rotating magnets and kilometers of copper wire and is supposed to be generating more electricity than it is consuming from a battery pack as just another perpetual-motion machine. But the inventor challenged the decision, and in 1985 the National Bureau of Standards finally tested the machine just to certify that it is battery-operated. But it did not convince the inventor that his machine will not work.

The proposers of perpetual-motion machines generally have innovative minds, but they usually lack formal engineering training, which is very unfortunate. No one is immune from being deceived by an innovative perpetual-motion mechanism. But, as the saying goes, if something sounds too good to be true, it probably is.

## 5-6 ■ REVERSIBLE AND IRREVERSIBLE PROCESSES

The sound law of thermodynamics states that no heat engine can have an efficiency of 100 percent. Then one may ask, What is the highest efficiency that a heat engine *can* possibly have? Before we can answer this question, we need to define an idealized process first, which is called the *reversible process.*

The processes that were discussed in Sec. 5-1 occurred in a certain direction. Once having taken place, these processes cannot reverse themselves spontaneously and restore the system to its initial state. For this reason, they are classified as *irreversible processes.* Once a cup of hot coffee cools, it will not heat up retrieving the heat it lost from the surroundings. If it could, the surroundings, as well as the system (coffee), would be restored to their original condition, and this would be a reversible process.

A **reversible process** is defined as a *process that can be reversed without leaving any trace on the surroundings* (Fig. 5-31). That is, both the system *and* the surroundings are returned to their initial states at the end of the reverse process. This is possible only if the net heat *and* net work exchange between the system and the surroundings is zero for the combined (original and reverse) process. Processes that are not reversible are called **irreversible processes**.

It should be pointed out that a system can be restored to its initial state following a process, regardless of whether the process is reversible or irreversible. But for reversible processes, this restoration is made without leaving any net change on the surroundings, whereas for irreversible processes, the surroundings usually do some work on the system and therefore will not return to their original state.

(*a*) Frictionless pendulum

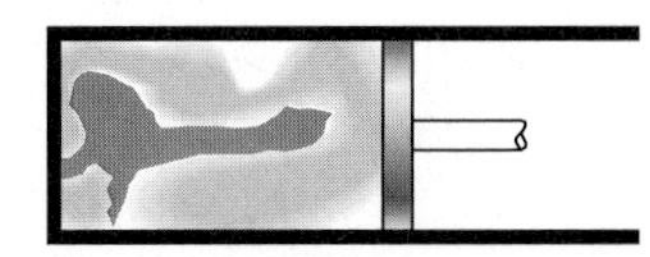

(*b*) Quasi-equilibrium expansion and compression of a gas

**FIGURE 5-31**
Two familiar reversible processes.

Reversible processes actually do not occur in nature. They are merely *idealizations* of actual processes. Reversible processes can be approximated by actual devices, but they can never be achieved. That is, all the processes occurring in nature are irreversible. You may be wondering, then, *why* we are bothering with such fictitious processes. There are two reasons. First, they are easy to analyze, since a system passes through a series of equilibrium states during a reversible process; second, they serve as idealized models to which actual processes can be compared.

In daily life, the concepts of Mr. Right and Ms. Right are also idealizations, just like the concept of reversible (perfect) process. People who insist on finding Mr. or Ms. Right to settle down are bound to remain Mr. or Ms. Single for the rest of their lives. The possibility of finding the perfect prospective mate is no higher than the possibility of finding a perfect (reversible) process. Likewise, a person who insists on perfection in friends is bound to have no friends.

Engineers are interested in reversible processes because work-producing devices such as car engines and gas or steam turbines *deliver the most work,* and work-consuming devices such as compressors, fans, and pumps *require least work* when reversible processes are used instead of irreversible ones (Fig. 5-32).

Reversible processes can be viewed as *theoretical limits* for the corresponding irreversible ones. Some processes are more irreversible than others. We may never be able to have a reversible process, but we may certainly approach it. The more closely we approximate a reversible process, the more work delivered by a work-producing device or the less work required by a work-consuming device.

The concept of reversible processes leads to the definition of *second-law efficiency* for actual processes, which is the degree of approximation to the corresponding reversible processes. This enables us to compare the performance of different devices that are designed to do the same task on the basis of their efficiencies. The better the design, the lower the irreversibilities and the higher the second-law efficiency.

## Irreversibilities

The factors that cause a process to be irreversible are called **irreversibilities**. They include friction, unrestrained expansion, mixing of two gases, heat transfer across a finite temperature difference, electric resistance,

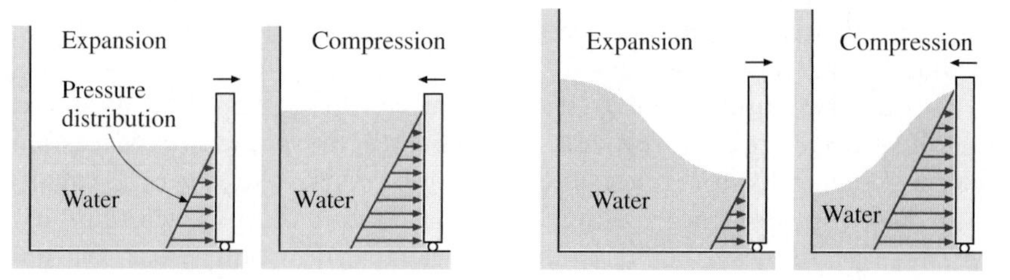


**FIGURE 5-32**
Reversible processes deliver the most work and consume the least.

inelastic deformation of solids, and chemical reactions. The presence of any of these effects renders a process irreversible. A reversible process involves none of these. Some of the frequently encountered irreversibilities are discussed briefly below.



### Friction

Friction is a familiar form of irreversibility associated with bodies in motion. When two bodies in contact are forced to move relative to each other (a piston in a cylinder for example, as shown in Fig. 5-33), a friction force that opposes the motion develops at the interface of these two bodies, and some work is needed to overcome this friction force. The energy supplied as work is eventually converted to heat during the process and is transferred to the bodies in contact, as evidenced by a temperature rise at the interface. When the direction of the motion is reversed, the bodies will be restored to their original position, but the interface will not cool, and heat will not be converted back to work. Instead, more of the work will be converted to heat while overcoming the friction forces which also oppose the reverse motion. Since the system (the moving bodies) and the surroundings cannot be returned to their original states, this process is irreversible. Therefore, any process that involves friction is irreversible. The larger the friction forces involved, the more irreversible the process is.

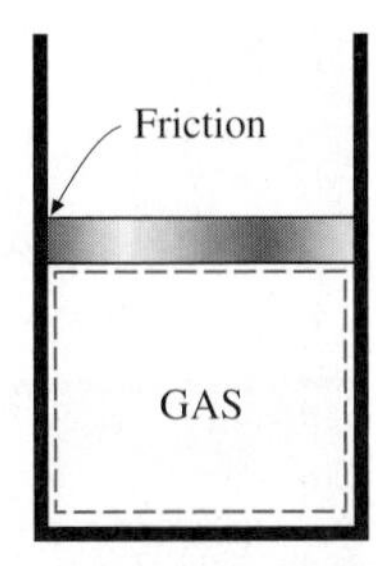


**FIGURE 5-33**
Friction renders a process irreversible.

Friction does not always involve two solid bodies in contact. It is also encountered between a fluid and solid and even between the layers of a fluid moving at different velocities. A considerable fraction of the power produced by a car engine is used to overcome the friction (the drag force) between the air and the external surfaces of the car, and it eventually becomes part of the internal energy of the air. It is not possible to reverse this process and recover that lost power, even though doing so would not violate the conservation of energy principle.

### Non-Quasi-Equilibrium Expansion and Compression

In Chap. 1, we defined a quasi-equilibrium process as one during which the system remains infinitesimally close to a state of equilibrium at all times. Consider a frictionless adiabatic piston–cylinder device that contains a gas. Now the piston is pushed into the cylinder, compressing the gas. If the piston velocity is not very high, the pressure and the temperature will increase uniformly throughout the gas. Since the system is always maintained at a state close to equilibrium, this is a quasi-equilibrium process.

Now the external force on the piston is slightly decreased, allowing the gas to expand. The expansion process will also be *quasi-equilibrium* if the gas is allowed to expand slowly. When the piston returns to its original position, all the boundary ($P\,dV$) work done on the gas during compression is returned to the surroundings during expansion. That is,

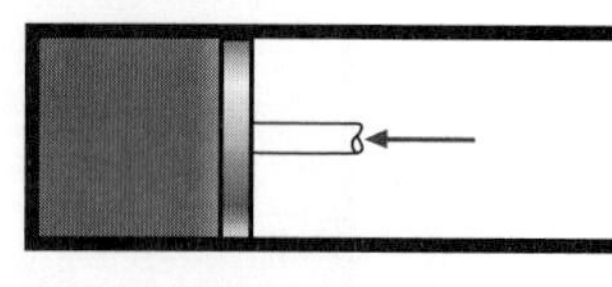
(*a*) Fast compression

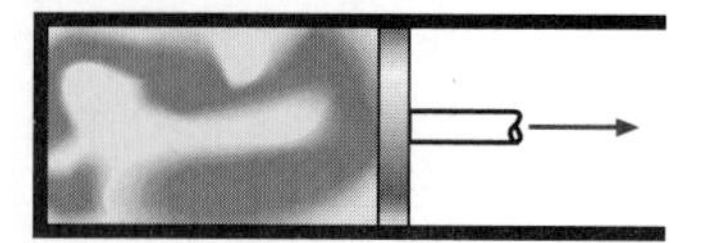
(*b*) Fast expansion

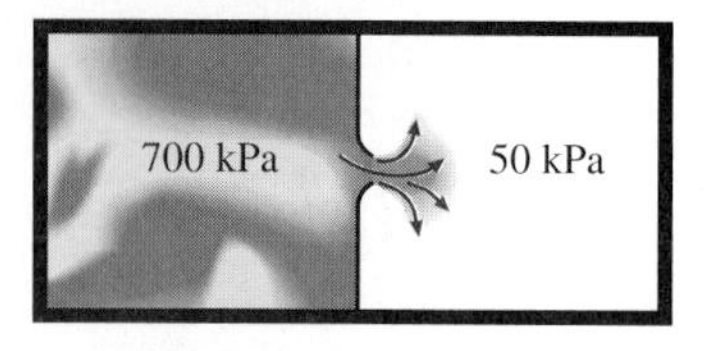

(*c*) Unrestrained expansion

**FIGURE 5-34**
Irreversible compression and expansion processes.

the net work for the combined process is zero. Also, there has been no heat transfer involved during this process, and thus both the system and the surroundings will return to their initial states at the end of the reverse process. Therefore, the slow frictionless adiabatic expansion or compression of a gas is a reversible process.

Now let us repeat this adiabatic process in a *non-quasi-equilibrium* manner, as shown in Fig. 5-34. If the piston is pushed in very rapidly, the gas molecules near the piston face will not have sufficient time to escape, and they will pile up in front of the piston. This will raise the pressure near the piston face, and as a result, the pressure there will be higher than the pressure in other parts of the cylinder. The nonuniformity of pressure will render this process non-quasi-equilibrium. The actual boundary work is a function of pressure, as measured at the piston face. Because of this higher pressure value at the piston face, a non-quasi-equilibrium compression process will require a larger work input than the corresponding quasi-equilibrium one. When the process is reversed by letting the gas expand rapidly, the gas molecules in the cylinder will not be able to follow the piston as fast, thus creating a low-pressure region before the piston face. Because of this low-pressure value at the piston face, a non-quasi-equilibrium process will deliver less work than a corresponding reversible one. Consequently, the work done by the gas during expansion is less than the work done by the surroundings on the gas during compression, and thus the surroundings have a net work deficit. When the piston returns to its initial position, the gas will have excess internal energy, equal in magnitude to the work deficit of the surroundings.

The system can easily be returned to its initial state by transferring this excess internal energy to the surroundings as heat. But the only way the surroundings can be returned to their initial condition is by completely converting this heat to work, which can only be done by a heat engine that has an efficiency of 100 percent. This, however, is impossible to do, even theoretically, since it would violate the second law of thermodynamics. Since only the system, not both the system and the surroundings, can be returned to its initial state, we conclude that the adiabatic non-quasi-equilibrium expansion or compression of a gas is irreversible.

Another example of non-quasi-equilibrium expansion processes is the unrestrained expansion of a gas separated from a vacuum by a membrane, as shown in Fig. 5-34*c*. When the membrane is ruptured, the gas fills the entire tank. The only way to restore the system to its original state is to compress it to its initial volume, while transferring heat from the gas until it reaches its initial temperature. From the conservation of energy considerations, it can easily be shown that the amount of heat transferred from the gas equals the amount of work done on the gas by the surroundings. The restoration of the surroundings involves conversion of this heat completely to work, which would violate the second law. Therefore, unrestrained expansion of a gas is an irreversible process.

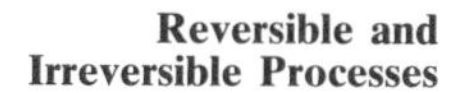



### Heat Transfer

Another form of irreversibility familiar to us all is heat transfer through a finite temperature difference. Consider a can of cold soda left in a warm room, for example, as shown in Fig. 5-35. Heat will flow from the warmer room air to the cooler soda. The only way this process can be reversed and the soda restored to its original temperature is to provide refrigeration, which requires some work input. At the end of the reverse process, the soda will be restored to its initial state, but the surroundings will not be. The internal energy of the surroundings will increase by an amount equal in magnitude to the work supplied to the refrigerator. The restoration of the surroundings to its initial state can be done only by converting this excess internal energy completely to work, which is impossible to do without violating the second law. Since only the system, not both the system and the surroundings, can be restored to its initial condition, heat transfer through a finite temperature difference is an irreversible process.

Heat transfer can occur only when there is a temperature difference between a system and its surroundings. Therefore, it is physically impossible to have a reversible heat transfer process. But a heat transfer process becomes less and less irreversible as the temperature difference between the two bodies approaches zero. Then heat transfer through a differential temperature difference $dT$ can be considered to be reversible. As $dT$ approaches zero, the process can be reversed in direction (at least theoretically) without requiring any refrigeration. Notice that reversible heat transfer is a conceptual process and cannot be duplicated in the laboratory.

The smaller the temperature difference between two bodies, the smaller the heat transfer rate will be. When the temperature difference is small, any significant heat transfer will require a very large surface area and a very long time. Therefore, even though approaching reversible heat transfer is desirable from a thermodynamic point of view, it is impractical and not economically feasible.

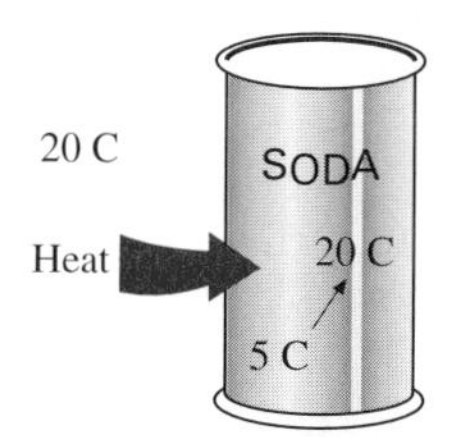


(*a*) An irreversible heat transfer process

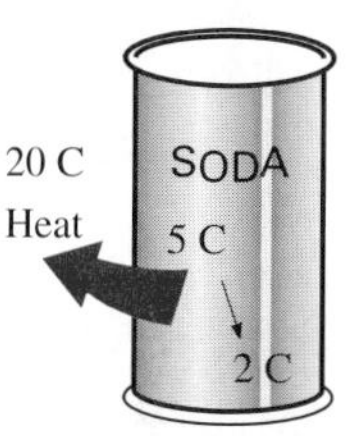


(*b*) An impossible heat transfer process

**FIGURE 5-35**

(*a*) Heat transfer through a temperature difference is irreversible, and (*b*) the reverse process is impossible.

## Internally and Externally Reversible Processes

A process is an interaction between a system and its surroundings, and a reversible process involves no irreversibilities associated with either of them.

A process is called **internally reversible** if no irreversibilities occur within the boundaries of the system during the process. During an internally reversible process, a system proceeds through a series of equilibrium states, and when the process is reversed, the system passes through exactly the same equilibrium states while returning to its initial state. That is, the paths of the forward and reverse processes coincide for an internally reversible process. The quasi-equilibrium process discussed earlier is an example of an internally reversible process.

A process is called **externally reversible** if no irreversibilities occur outside the system boundaries during the process. Heat transfer between a reservoir and a system is an externally reversible process if the surface of contact between the system and the reservoir is at the temperature of the reservoir.

A process is called **totally reversible**, or simply **reversible**, if it involves no irreversibilities within the system or its surroundings (Fig. 5-36). A totally reversible process involves no heat transfer through a finite temperature difference, no non-quasi-equilibrium changes, and no friction or other dissipative effects.

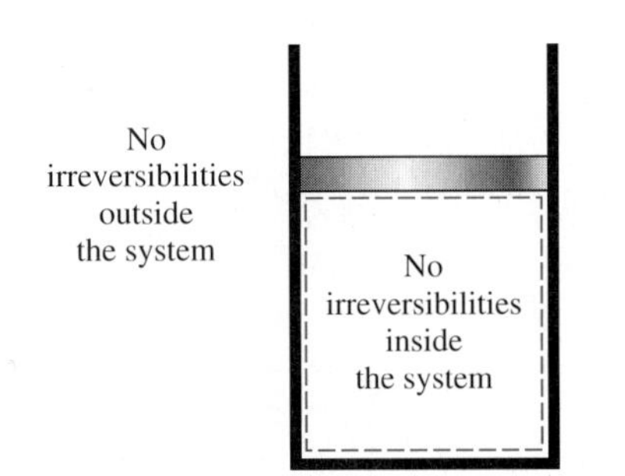


**FIGURE 5-36**
A reversible process involves no internal and external irreversibilities.

As an example, consider the transfer of heat to two identical systems that are undergoing a constant-pressure (thus constant-temperature) phase-change process, as shown in Fig. 5-37. Both processes are internally reversible, since both take place isothermally and both pass through exactly the same equilibrium states. The first process shown is externally reversible also, since heat transfer for this process takes place through an infinitesimal temperature difference $dT$. The second process, however, is externally irreversible, since it involves heat transfer through a finite temperature difference $\Delta T$.

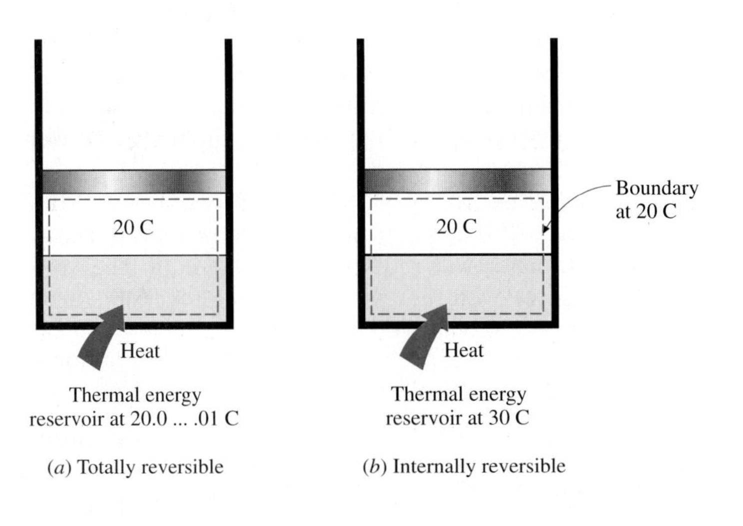


**FIGURE 5-37**
Totally and internally reversible heat transfer processes.

## 5-7 ■ THE CARNOT CYCLE

We mentioned earlier that heat engines are cyclic devices and that the working fluid of a heat engine returns to its initial state at the end of each cycle. Work is done by the working fluid during one part of the cycle and on the working fluid during another part. The difference between these two is the net work delivered by the heat engine. The efficiency of a heat-engine cycle greatly depends on how the individual processes that make up the cycle are executed. The net work, thus the cycle efficiency, can be maximized by using processes that require the least amount of work and deliver the most, that is, by using *reversible processes*.

Therefore, it is no surprise that the most efficient cycles are reversible cycles, i.e., cycles that consist entirely of reversible processes.

Reversible cycles cannot be achieved in practice because the irreversibilities associated with each process cannot be eliminated. However, reversible cycles provide upper limits on the performance of real cycles. Heat engines and refrigerators that work on reversible cycles serve as models to which actual heat engines and refrigerators can be compared. Reversible cycles also serve as starting points in the development of actual cycles and are modified as needed to meet certain requirements.

Probably the best known reversible cycle is the **Carnot cycle**, first proposed in 1824 by a French engineer Sadi Carnot. The theoretical heat engine that operates on the Carnot cycle is called the **Carnot heat engine**. The Carnot cycle is composed of four reversible processes—two isothermal and two adiabatic—and it can be executed either in a closed or a steady-flow system.

Consider a closed system that consists of a gas contained in an adiabatic piston–cylinder device, as shown in Fig. 5-38. The insulation of the cylinder head is such that it may be removed to bring the cylinder into contact with reservoirs to provide heat transfer. The four reversible processes that make up the Carnot cycle are as follows:

**Reversible isothermal expansion** (process 1-2, $T_H$ = constant). Initially (state 1) the temperature of the gas is $T_H$, and the cylinder head is in close contact with a source at temperature $T_H$. The gas is allowed to expand slowly, doing work on the surroundings. As the gas expands, the temperature of the gas tends to decrease. But as soon as the temperature drops by an infinitesimal amount $dT$, some heat flows from the reservoir into the gas, raising the gas temperature to $T_H$. Thus, the gas temperature is kept constant at $T_H$. Since the temperature difference between the gas and the reservoir never exceeds a differential amount $dT$, this is a reversible heat transfer process. It continues until the piston reaches position 2. The amount of total heat transferred to the gas during this process is $Q_H$.

**Reversible adiabatic expansion** (process 2-3, temperature drops from $T_H$ to $T_L$). At state 2, the reservoir that was in contact with the cylinder head is removed and replaced by insulation so that the system becomes adiabatic. The gas continues to expand slowly, doing work on the surroundings until its temperature drops from $T_H$ to $T_L$ (state 3). The piston is assumed to be frictionless and the process to be quasi-equilibrium, so the process is reversible as well as adiabatic.

**Reversible isothermal compression** (process 3-4, $T_L$ = constant). At state 3, the insulation at the cylinder head is removed, and the cylinder is brought into contact with a sink at temperature $T_L$. Now the piston is pushed inward by an external force, doing work on the gas. As the gas is compressed, its temperature tends to rise. But as soon as it rises by an infinitesimal amount $dT$, heat flows from the gas to the sink, causing the gas temperature to drop to $T_L$. Thus, the

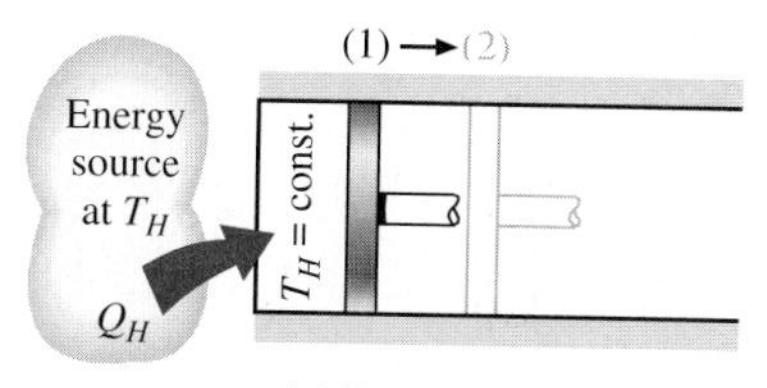


(*a*) Process 1-2

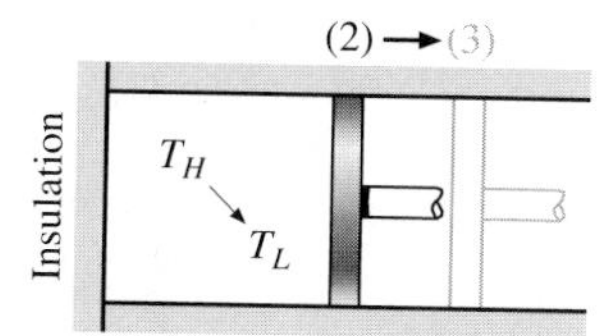


(*b*) Process 2-3

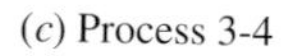


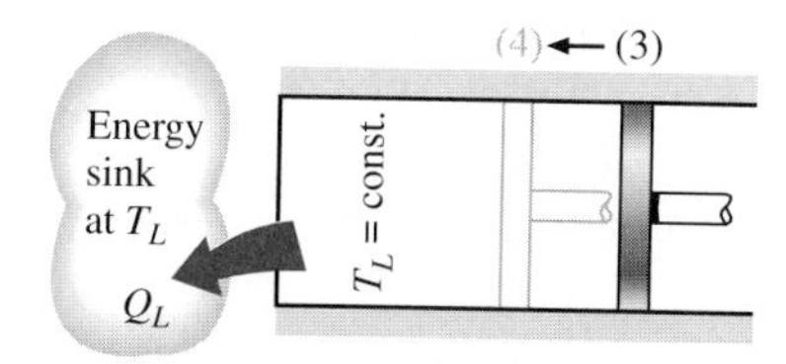


(*c*) Process 3-4

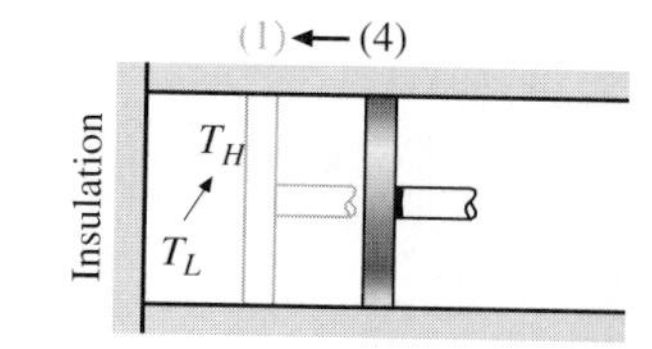


(*d*) Process 4-1

**FIGURE 5-38**
Execution of the Carnot cycle in a closed system.

gas temperature is maintained constant at $T_L$. Since the temperature difference between the gas and the sink never exceeds a differential amount $dT$, this is a reversible heat transfer process. It continues until the piston reaches position 4. The amount of heat rejected from the gas during this process is $Q_L$.

**Reversible adiabatic compression** (process 4-1, temperature rises from $T_L$ to $T_H$). State 4 is such that when the low-temperature reservoir is removed and the insulation is put back on the cylinder head and the gas is compressed in a reversible manner, the gas returns to its initial state (state 1). The temperature rises from $T_L$ to $T_H$ during this reversible adiabatic compression process, which completes the cycle.

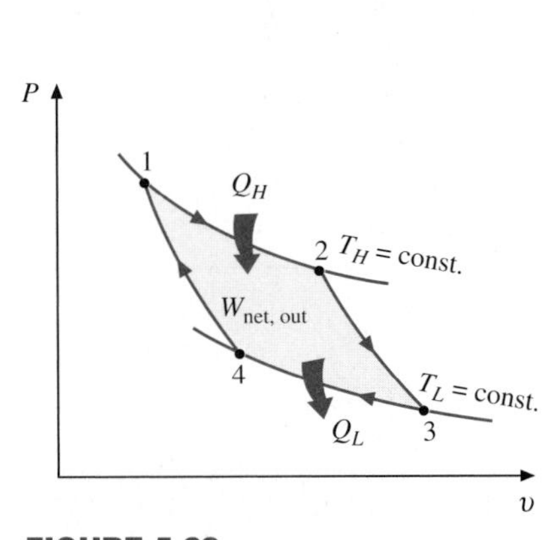


**FIGURE 5-39**
*P-v* diagram of the Carnot cycle.

The *P-v* diagram of this cycle is shown in Fig. 5-39. Remembering that on a *P-v* diagram the area under the process curve represents the boundary work for quasi-equilibrium (internally reversible) processes, we see that the area under curve 1-2-3 is the work done by the gas during the expansion part of the cycle, and the area under curve 3-4-1 is the work done on the gas during the compression part of the cycle. The area enclosed by the path of the cycle (area 1-2-3-4-1) is the difference between these two and represents the net work done during the cycle.

Notice that if we acted stingily and compressed the gas at state 3 adiabatically instead of isothermally in an effort *to save* $Q_L$, we would end up back at state 2, retracing the process path 3-2. By doing so we would save $Q_L$, but we would not be able to obtain any net work output from this engine. This illustrates once more the necessity of a heat engine exchanging heat with at least two reservoirs at different temperatures to operate in a cycle and produce a net amount of work.

The Carnot cycle can also be executed in a steady-flow system. It is discussed in Chap. 7 in conjunction with other power cycles.

Being a reversible cycle, the Carnot cycle is the most efficient cycle operating between two specified temperature limits. Even though the Carnot cycle cannot be achieved in reality, the efficiency of actual cycles can be improved by attempting to approximate the Carnot cycle more closely.

## The Reversed Carnot Cycle

The Carnot heat-engine cycle described above is a totally reversible cycle. Therefore, all the processes that comprise it can be *reversed*, in which case it becomes the **Carnot refrigeration cycle**. This time, the cycle remains exactly the same, except that the directions of any heat and work interactions are reversed: Heat in the amount of $Q_L$ is absorbed from the low-temperature reservoir, heat in the amount of $Q_H$ is rejected to a high-temperature reservoir, and a work input of $W_{net,in}$ is required to accomplish all this.

The $P$-$v$ diagram of the reversed Carnot cycle is the same as the one given for the Carnot cycle, except that the directions of the processes are reversed, as shown in Fig. 5-40.

## 5-8 ■ THE CARNOT PRINCIPLES

The second law of thermodynamics places limitations on the operation of cyclic devices as expressed by the Kelvin–Planck and Clausius statements. A heat engine cannot operate by exchanging heat with a single reservoir, and a refrigerator cannot operate without a net work input from an external source.

We can draw valuable conclusions from these statements. Two conclusions pertain to the thermal efficiency of reversible and irreversible (i.e., actual) heat engines, and they are known as the **Carnot principles** (Fig. 5-41). They are expressed as follows:

1 *The efficiency of an irreversible heat engine is always less than the efficiency of a reversible one operating between the same two reservoirs.*

2 *The efficiencies of all reversible heat engines operating between the same two reservoirs are the same.*

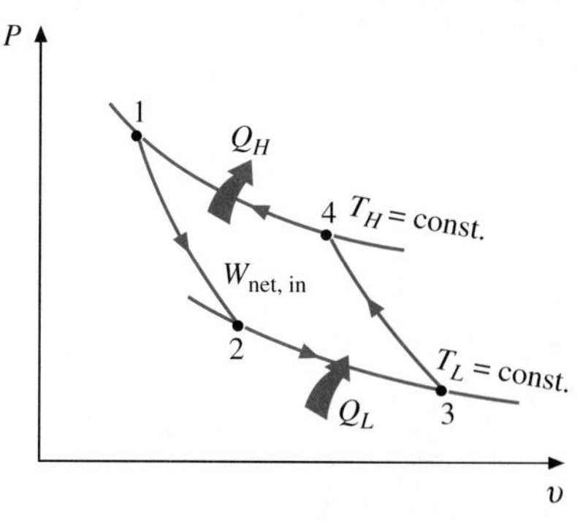


**FIGURE 5-40**

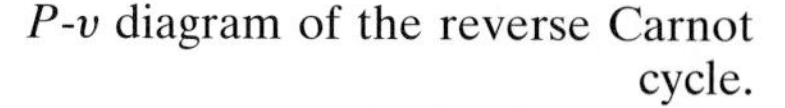

$P$-$v$ diagram of the reverse Carnot cycle.

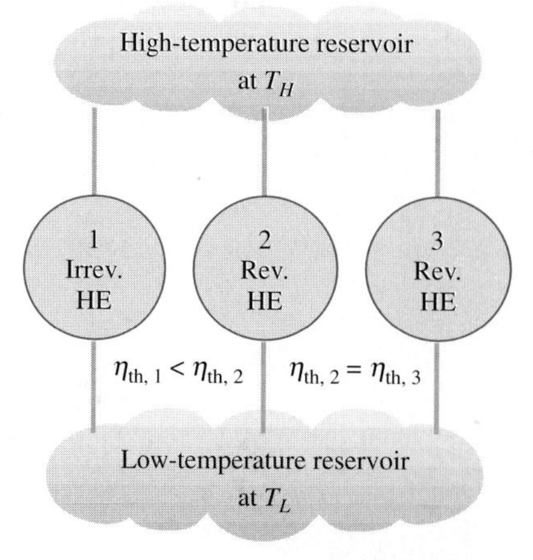


**FIGURE 5-41**

The Carnot principles.

These two statements can be proved by demonstrating that the violation of either statement results in the violation of the second law of thermodynamics.

To prove the first statement, consider two heat engines operating between the same reservoirs, as shown in Fig. 5-42. One engine is reversible, and the other is irreversible. Now each engine is supplied with the same amount of heat $Q_H$. The amount of work produced by the reversible heat engine is $W_{rev}$, and the amount produced by the irreversible one is $W_{irrev}$.

In violation of the first Carnot principle, we assume that the irreversible heat engine is more efficient than the reversible one (that is, $\eta_{th,irrev} > \eta_{th,rev}$) and thus delivers more work than the reversible one. Now let the reversible heat engine be reversed and operate as a refrigerator. This refrigerator will receive a work input of $W_{rev}$ and reject heat to the high-temperature reservoir. Since the refrigerator is rejecting heat in the amount of $Q_H$ to the high-temperature reservoir and the irreversible heat engine is receiving the same amount of heat from this reservoir, the net heat exchange for this reservoir is zero. Thus it could be eliminated by having the refrigerator discharge $Q_H$ directly into the irreversible heat engine.

Now considering the refrigerator and the irreversible engine together, we have an engine that produces a net work in the amount of $W_{irrev} - W_{rev}$ while exchanging heat with a single reservoir—a violation of the Kelvin–Planck statement of the second law. Therefore, our initial assumption that $\eta_{th,irrev} > \eta_{th,rev}$ is incorrect. Then we conclude that no

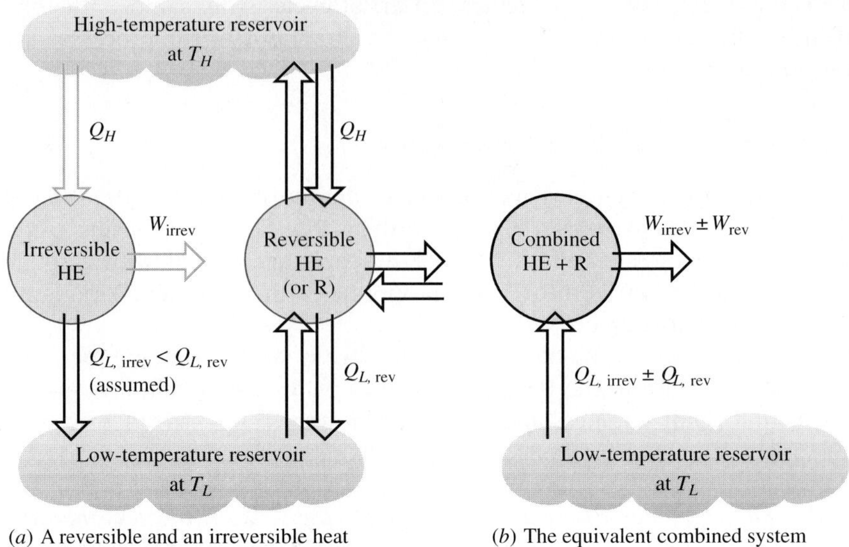


(*a*) A reversible and an irreversible heat engine operating between the same two reservoirs (the reversible heat engine is then reversed to run as a refrigerator)

(*b*) The equivalent combined system

**FIGURE 5-42**
Proof of the first Carnot principle.

heat engine can be more efficient than a reversible heat engine operating between the same reservoirs.

The second Carnot principle can also be proved in a similar manner. This time, let us replace the irreversible engine by another reversible engine that is more efficient and thus delivers more work than the first reversible engine. By following through the same reasoning as above, we will end up having an engine that produces a new amount of work while exchanging heat with a single reservoir, which is a violation of the second law. Therefore we conclude that no reversible heat engine can be more efficient than another reversible heat engine operating between the same two reservoirs, regardless of how the cycle is completed or the kind of working fluid used.

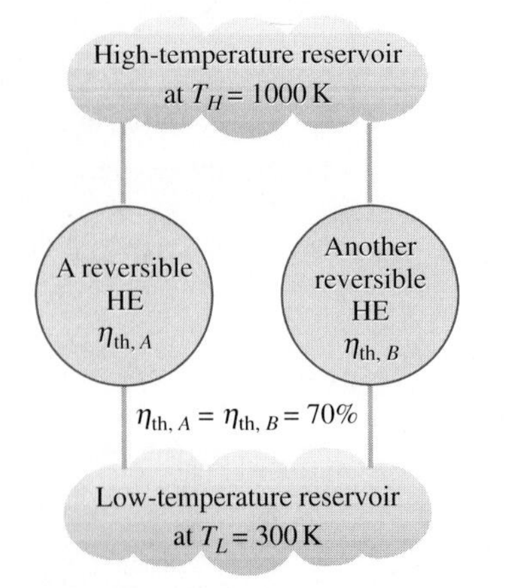


**FIGURE 5-43**
All reversible heat engines operating between the same two reservoirs have the same efficiency (the second Carnot principle.

## 5.9 THE THERMODYNAMIC TEMPERATURE SCALE

A temperature scale that is independent of the properties of the substances that are used to measure temperature is called an **thermodynamic temperature scale**. Such a temperature scale offers great conveniences in thermodynamic calculations, and its derivation is given below using some reversible heat engine.

The second Carnot principle discussed in Sec. 5-8 states that all reversible heat engines have the same thermal efficiency when operating between the same two reservoirs (Fig. 5-43). That is, the efficiency of a reversible engine is independent of the working fluid employed and its properties, the way the cycle is executed, or the type of reversible engine used. Since energy reservoirs are characterized by their temperatures, the thermal efficiency of reversible heat engines is a function of the reservoir temperatures only. That is,

$$\eta_{\text{th,rev}} = g(T_h, T_L)$$

or

$$\frac{Q_H}{Q_L} = f(T_H, R_L) \tag{5-15}$$

since $\eta_{\text{th}} = 1 - Q_L/Q_H$. In these relations $T_H$ and $T_L$ are the temperatures of the high- and low-temperature reservoirs, respectively.

The functional form of $f(T_H, T_L)$ can be developed with the help of the three reversible heat engines shown in Fig. 5-44. Engines $A$ and $C$ are supplied with the same amount of heat $Q_1$ from the high-temperature reservoir at $T_1$. Engine $C$ rejects $Q_3$ to the low-temperature reservoir at $T_3$. Engine $B$ receives the heat $Q_2$ rejected by engine $A$ at temperature $T_2$ and rejects heat in the amount of $Q_3$ to a reservoir at $T_3$.

The amounts of heat rejected by engines $B$ and $C$ must be the same since engines $A$ and $B$ can be combined into one reversible engine operating between the same reservoirs as engine $C$ and thus the combined engine will have the same efficiency as engine $C$. Since the heat input to engine $C$ is the same as the heat input to the combined engines $A$ and $B$, both systems must reject the same amount of heat.

Applying Eq. 5-15 to all three engines separately, we obtain

$$\frac{Q_1}{Q_2} = f(T_1, T_2), \qquad \frac{Q_2}{Q_3} = f(T_2, T_3), \qquad \text{and} \qquad \frac{Q_1}{Q_3} = f(T_1, T_3)$$

Now consider the identity

$$\frac{Q_1}{Q_3} = \frac{Q_1}{Q_2}\frac{Q_2}{Q_3}$$

which corresponds to

$$f(T_1, T_3) = f(T_1, T_2) f(T_2, T_3)$$

A careful examination of this equation reveals that the left-hand side is a function of $T_1$ and $T_3$, and therefore the right-hand side must also be a function of $T_1$ and $T_3$ only, and not $T_2$. That is, the value of the product on the right-hand side of this equation is independent of the value of $T_2$. This condition will be satisfied only if the function $f$ has the following form:

$$f(T_1, T_2) = \frac{\phi(T_1)}{\phi(T_2)} \qquad \text{and} \qquad f(T_2, T_3) = \frac{\phi(T_2)}{\phi(T_3)}$$

so that $\phi(T_2)$ will cancel from the products of $f(T_1, T_2)$ and $f(T_2, T_3)$, yielding

$$\frac{Q_1}{Q_3} = f(T_1, T_3) = \frac{\phi(T_1)}{\phi(T_3)} \tag{5-16}$$

This relation is much more specific than Eq. 5-15 for the functional form of $Q_1/Q_3$ in terms of $T_1$ and $T_3$.

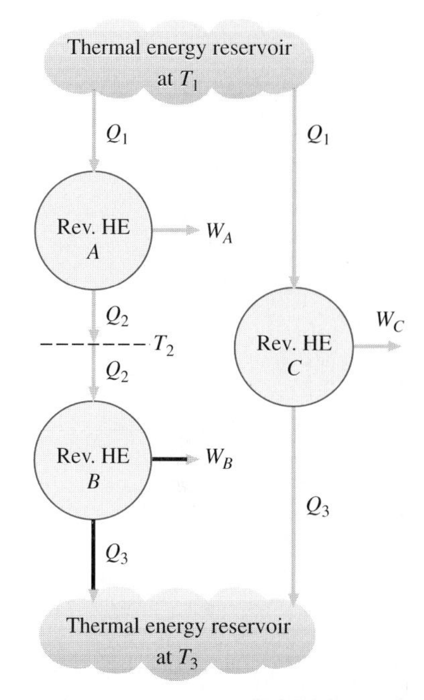


**FIGURE 5-44**
The arrangement of heat engines used to develop the absolute temperature scale.

For a reversible heat engine operating between two reservoirs at temperatures $T_H$ and $T_L$, Eq. 5-16 can be written as

$$\frac{Q_H}{Q_L} = \frac{\phi(T_H))}{\phi(T_L)} \tag{5-17}$$

This is the only requirement that the second law places on the ratio of heat flows to and from the reversible heat engines. Several functions $\phi(T)$ will satisfy this equation, and the choice is completely arbitrary. Lord Kelvin first proposed taking $\phi(T) = T$ to define a thermodynamic temperature scale as (Fig. 5-45)

$$\left(\frac{Q_H}{Q_L}\right)_{\text{rev}} = \frac{T_H}{T_L} \tag{5-18}$$

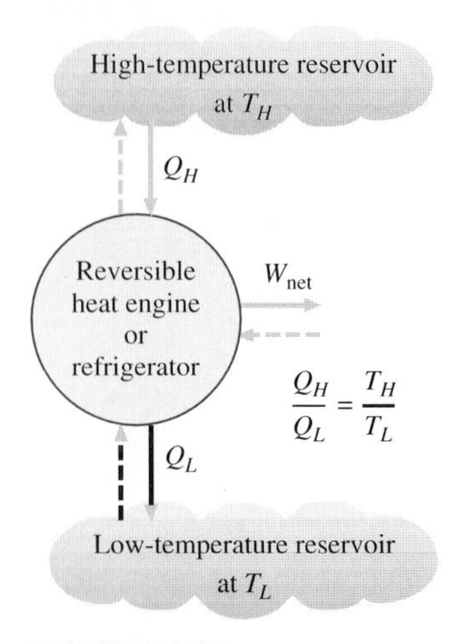


**FIGURE 5-45**
For reversible cycles, the heat transfer ratio $Q_H/Q_L$ can be replaced by the absolute temperature ratio $T_H/T_L$.

This temperature scale is called the **Kelvin scale**, and the temperatures on this scale are called **absolute temperatures**. On the Kelvin scale, the temperature ratios depend on the ratios of heat transfer between a reversible heat engine and the reservoirs and are independent of the physical properties of any substance. On this scale, temperatures vary between zero and infinity.

The thermodynamic temperature scale is not completely defined by Eq. 5-18 since it gives us only a ratio of absolute temperatures. We also need to know the magnitude of a kelvin degree. At the International Conference on Weights and Measures held in 1954, the triple point of water (the state at which all three phases of water exist in equilibrium) was assigned the value 273.16 K. The *magnitude of a kelvin* is defined as 1/273.16 of the temperature interval between absolute zero and the triple-point temperature of water. The magnitudes of temperature units on the Kelvin and Celsius scales are identical (1 K ≡ 1°C). The temperatures on these two scales differ by a constant 273.15:

$$T(^\circ\text{C}) = T(\text{K}) - 273.15 \tag{5-19}$$

Even though the thermodynamic temperature scale is defined with the help of the reversible heat engines, it is not possible, nor is it practical, to actually operate such an engine to determine numerical values on the absolute temperature scale. Absolute temperatures can be measured accurately by other means, such as the constant-volume ideal-gas thermometer discussed in Chap. 1 together with extrapolation techniques. The validity of Eq. 5-18 can be demonstrated from physical considerations for a reversible cycle using an ideal gas as the working fluid.

## 5-10 ■ THE CARNOT HEAT ENGINE

The hypothetical heat engine that operates on the reversible Carnot cycle is called the **Carnot heat engine**. The thermal efficiency of any heat engine, reversible or irreversible, is given by Eq. 5-8 as

$$\eta_{\text{th}} = 1 - \frac{Q_L}{Q_H}$$

where $Q_H$ is heat transferred to the heat engine from a high-temperature reservoir at $T_H$, and $Q_L$ is heat rejected to a low-temperature reservoir at $T_L$. For reversible heat engines, the heat transfer ratio in the above relation can be replaced by the ratio of the absolute temperatures of the two reservoirs, as given by Eq. 5-18. Then the efficiency of a Carnot engine, or any reversible heat engine, becomes

$$\eta_{th,rev} = 1 - \frac{T_L}{T_H} \qquad (5\text{-}20)$$

This relation is often referred to as the **Carnot efficiency** since the Carnot heat engine is the best known reversible engine. *This is the highest efficiency a heat engine operating between the two thermal energy reservoirs at temperatures $T_L$ and $T_H$ can have* (Fig. 5-46). All irreversible (i.e., actual) heat engines operating between these temperature limits ($T_L$ and $T_H$) will have lower efficiencies. An actual heat engine cannot reach this maximum theoretical efficiency value because it is impossible to completely eliminate all the irreversibilities associated with the actual cycle.

Note that $T_L$ and $T_H$ in Eq. 5-20 are *absolute temperatures.* Using °C or °F for temperatures in this relation will give results grossly in error.

The thermal efficiencies of actual and reversible heat engines operating between the same temperature limits compare as follows (Fig. 5-47):

$$\eta_{th} \begin{cases} < \eta_{th,rev} & \text{irreversible heat engine} \\ = \eta_{th,rev} & \text{reversible heat engine} \\ > \eta_{th,rev} & \text{impossible heat engine} \end{cases} \qquad (5\text{-}21)$$

Most work-producing devices (heat engines) in operation have efficiencies under 40 percent, which appear low relative to 100 percent. However, when the performance of actual heat engines is assessed, the efficiencies should not be compared to 100 percent; instead, they should be compared to the efficiency of a reversible heat engine operating

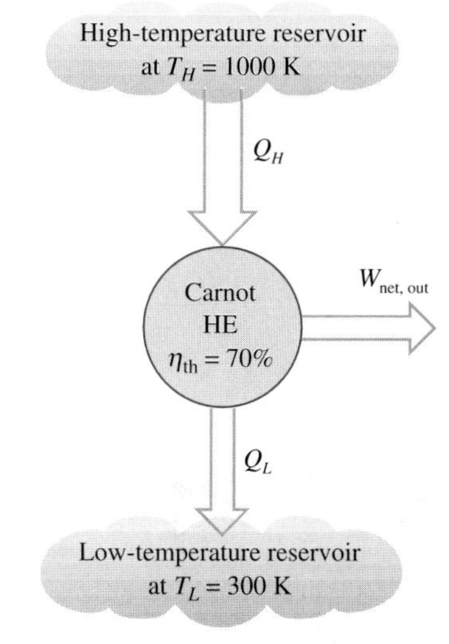


**FIGURE 5-46**

The Carnot heat engine is the most efficient of all heat engines operating between the same high- and low-temperature reservoirs.

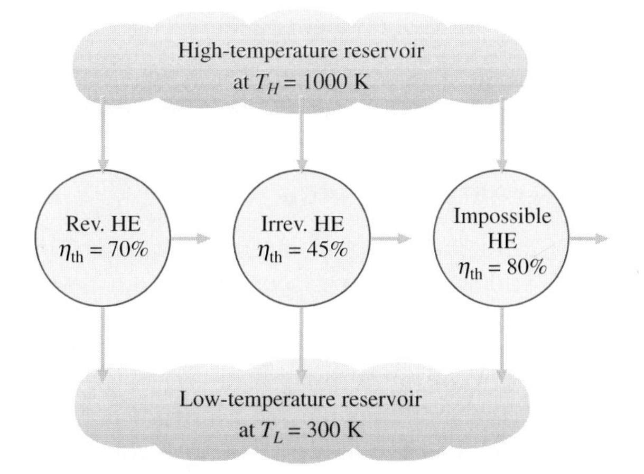


**FIGURE 5-47**

No heat engine can have a higher efficiency than a reversible heat engine operating between the same high- and low-temperature reservoirs.

between the same temperature limits—because this is the true theoretical upper limit for the efficiency, not the 100 percent.

The maximum efficiency of a steam power plant operating between $T_H = 750$ K and $T_L = 300$ K is 60 percent, as determined from Eq. 5-20. Compared with this value, an actual efficiency of 40 percent does not seem so bad, even though there is still plenty of room for improvement.

It is obvious from Eq. 5-20 that the efficiency of a Carnot heat engine increases as $T_H$ is increased, or as $T_L$ is decreased. This is to be expected since as $T_L$ decreases, so does the amount of heat rejected, and as $T_L$ approaches zero, the Carnot efficiency approaches unity. This is also true for actual heat engines. *The thermal efficiency of actual heat engines can be maximized by supplying heat to the engine at the highest possible temperature* (limited by material strength) *and rejecting heat from the engine at the lowest possible temperature* (limited by the temperature of the cooling medium such as rivers, lakes, or the atmosphere).

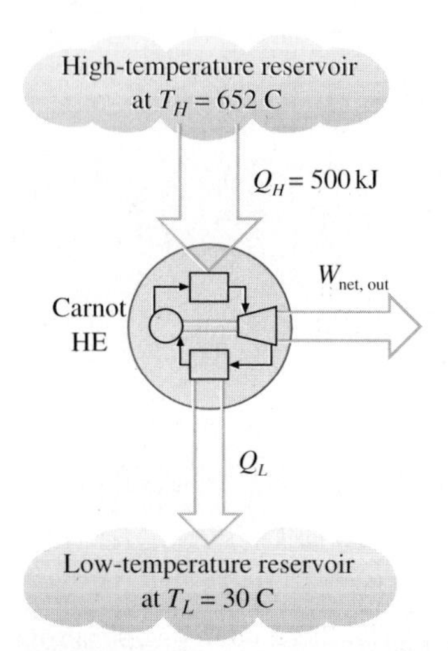

**FIGURE 5-48**
Schematic for Example 5-5.

**EXAMPLE 5-5**

A Carnot heat engine, shown in Fig. 5-48, receives 500 kJ of heat per cycle from a high-temperature source at 652°C and rejects heat to a low-temperature sink at 30°C. Determine (*a*) the thermal efficiency of this Carnot engine and (*b*) the amount of heat rejected to the sink per cycle.

**Solution** (*a*) The Carnot heat engine is a reversible heat engine, and so its efficiency can be determined from Eq. 5-20:

$$\eta_{th,C} = \eta_{th,rev} = 1 - \frac{T_L}{T_H} = 1 - \frac{(30 + 273)\text{ K}}{(652 + 273)\text{ k}} = 0.672$$

That is, this Carnot heat engine converts 67.2 percent of the heat it receives to work.

(*b*) The amount of heat rejected $Q_L$ by this reversible heat engine is easily determined from Eq. 5-18:

$$Q_{L,rev} = \frac{T_L}{T_H} Q_{H,rev} = \frac{(30 + 273)\text{ K}}{(652 + 273)\text{ K}}(500\text{ kJ}) = 163.8\text{ kJ}$$

Therefore, this Carnot heat engine discharges 163.8 kJ of the 500 kJ of heat it receives during each cycle to a sink.

## The Quality of Energy

The Carnot heat engine in Example 5-5 receives heat from a source at 925 K and converts 67.2 percent of it to work while rejecting the rest (32.8 percent) to a sink at 303 K. Now let us examine how the thermal efficiency varies with the source temperature when the sink temperature is held constant.

The thermal efficiency of a Carnot heat engine that rejects heat to a sink at 303 K is evaluated at various source temperatures using Eq. 5-20

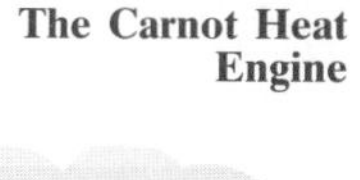

and is listed in Fig. 5-49. Clearly the thermal efficiency decreases as the source temperature is lowered. When heat is supplied to the heat engine at 500 instead of 925 K, for example, the thermal efficiency drops from 67.2 to 39.4 percent. That is, the fraction of heat that can be converted to work drops to 39.4 percent when the temperature of the source drops to 500 K. When the source temperature is 350 K, this fraction becomes a mere 13.4 percent.

These efficiency values show that energy has **quality** as well as quantity. It is clear from the thermal efficiency values in Fig. 5-49 that *more of the high-temperature thermal energy can be converted to work. Therefore, the higher the temperature, the higher the quality of the energy* (Fig. 5-50).

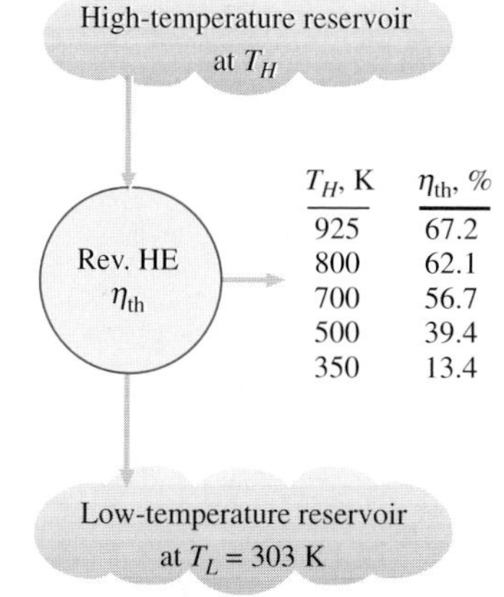


**FIGURE 5-49**
The fraction of heat that can be converted to work as a function of source temperature (for $T_L = 303$ K).

Large quantities of solar energy, for example, can be stored in large bodies of water called *solar ponds* at about 350 K. This stored energy can then be supplied to a heat engine to produce work (electricity). However, the efficiency of solar pond power plants is very low (under 5 percent) because of the low quality of the energy stored in the source, and the construction and maintenance costs are relatively high. Therefore, they are not competitive even though the energy supply of such plants is free. The temperature (and thus the quality) of the solar energy stored could be raised by utilizing concentrating collectors, but the equipment cost in the case becomes very high.

Work is a more valuable form of energy than heat since 100 percent of work can be converted to heat but only a fraction of heat can be converted to work. When heat is transferred from a high-temperature body to a lower-temperature one, it is degraded since less of it now can be converted to work. For example, if 100 kJ of heat is transferred from a body at 1000 K to a body at 300 K, at the end we will have 100 kJ of thermal energy stored at 300 K, which has no practical value. But if this conversion is made through a heat engine, up to $1 - 300/1000 = 70$ percent of it could be converted to work, which is a more valuable form of energy. Thus 70 kJ of work potential is wasted as a result of this heat transfer, and energy is degraded. The degradation of energy during a process is discussed more fully in Chap. 7.

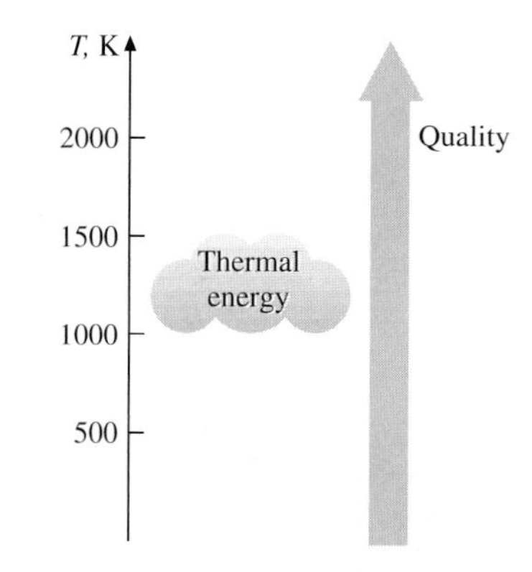


**FIGURE 5-50**
The higher the temperature of the thermal energy, the higher its quality.

## Quantity Versus Quality in Daily Life

At times of energy crisis we are bombarded with speeches and articles on how to "conserve" energy. Yet we all know that the *quantity* of energy is already conserved. What is not conserved is the *quality* of energy, or the work potential of energy. Wasting energy is synonymous to converting it to a less useful form. One unit of high-quality energy can be more valuable than three units of lower-quality energy. For example, a finite amount of heat energy at high temperature is more attractive to power plant engineers than a vast amount of heat energy at low temperature, such as the energy stored in the upper layers of the oceans at tropical climates.

As part of our culture, we seem to be fascinated by quantity, and little attention is given to quality. But quantity alone cannot give the whole picture, and we need to consider the quality also. That is, we need to look at something from both the first- and second-law points of view when evaluating something, even in nontechnical areas. Below we present some ordinary events and show their relevance to the second law of thermodynamics.

Those who shop for a diamond engagement ring quickly realize that not all diamonds are created equal. For example, a half carat fine-quality diamond ring is more valuable (and thus more expensive) than a full carat standard-quality diamond ring. Thus bigger is not necessarily better.

As another example, consider two students Andy and Wendy. Andy has 10 friends who never miss his parties and are always around during fun times. But they seem to be busy when Andy needs their help. Wendy, on the other hand, has five friends. But they are never too busy for her, and she can count on them at times of need. Let us now try to answer the question *Who has more friends*? From the first law point of view, which considers quantity only, it is obvious that Andy has more friends. But from the second-law point of view, which considers quality as well, there is no doubt that Wendy is the one with more friends.

Another example most people will identify with is the multibillion-dollar diet industry, which is primarily based on the first law of thermodynamics. But considering that 90 percent of the people who loose weight gain it back shortly, with interest, suggests that the first law alone does not give the whole picture. This is also confirmed by recent work, which shows that calories that come from fat are more likely to be stored as fat than the calories that come from carbohydrates and protein. A Stanford study found that the body weight was related to fat calories consumed and not calories per se. A Harvard study found no correlation between calories eaten and degree of obesity. A major Cornell University survey involving 6500 people in nearly all provinces of China found that the Chinese eat more—gram for gram, calorie for calorie—than Americans do, but they weigh less, with less body fat. Studies indicate that the metabolism rates and hormone levels change noticeably in the mid 30s. Some researchers concluded that prolonged dieting teaches a body to survive on fewer calories, making it more *fuel efficient.* This probably explains why the dieters gain more weight than they lost once they go back to their normal eating levels.

People who seem to be eating whatever they want, whenever they want, are living proofs that calorie-counting technique (the first law) leaves many questions on dieting unanswered. Obviously more research focused on the second-law effects of dieting is needed before we can fully understand the weight gain and weight loss process.

Those holding academic positions know it well that "quantity" is still used as the basis in the annual merit reviews. That is, faculty members are judged on the basis of the *number* of their papers, and the *dollar amount* of their research grants. Skillful faculty members play this game well by diluting their work over two or more papers, instead of publishing their findings in one sterling paper. The result is a huge increase in the number of publications, but little increase in new knowledge. The emphasis on numbers (the *quantity*) encourages faculty to publish many papers of questionable quality instead of few high-quality ones, and discourages them from undertaking long-term high-risk studies. The issue of declining quality of publications is frequently raised, but no serious effort is made to find a workable solution. Maybe it is time that publications and other scholarly works are examined from the second-law point of view to assess their value realistically.

It is tempting to judge things on the basis of their *quantity* instead of their *quality* since assessing quality is much more difficult than assessing quantity. However, assessments made on the basis of quantity only (the first law) may be grossly inadequate and misleading. The above discussions on the quality of energy and the examples that follow indicate that there is a need for a further in-depth study of the second law of thermodynamics. This is what we will do in the following two chapters.

## 5-11 ■ THE CARNOT REFRIGERATOR AND HEAT PUMP

A refrigerator or a heat pump that operates on the reversed Carnot cycle is called a **Carnot refrigerator**, or a **Carnot heat pump**. The coefficient of performance of any refrigerator or heat pump, reversible or irreversible, is given by Eqs. 5-11 and 5-13 as

$$\text{COP}_\text{R} = \frac{1}{Q_H/Q_L - 1} \quad \text{and} \quad \text{COP}_\text{HP} = \frac{1}{1 - Q_L/Q_H}$$

where $Q_L$ is the amount of heat absorbed from the low-temperature medium and $Q_H$ is the amount of heat rejected to the high-temperature medium. The COPs of all reversible (such as Carnot) refrigerators or heat pumps can be determined by replacing the heat transfer ratios in the above relations by the ratios of the absolute temperatures of the high- and low-temperature media, as expressed by Eq. 5-18. Then the COP relations for reversible refrigerators and heat pumps become

$$\text{COP}_\text{R,rev} = \frac{1}{T_H/T_L - 1} \tag{5-22}$$

and

$$\text{COP}_\text{HP,rev} = \frac{1}{1 - T_L/T_H} \tag{5-23}$$

*These are the highest coefficients of performance that a refrigerator or a heat pump operating between the temperature limits of $T_L$ and $T_H$ can have.* All actual refrigerators or heat pumps operating between these temperature limits ($T_L$ and $T_H$) will have lower coefficients of performance (Fig. 5-51).

The coefficients of performance of actual and reversible (such as Carnot) refrigerators operating between the same temperature limits can

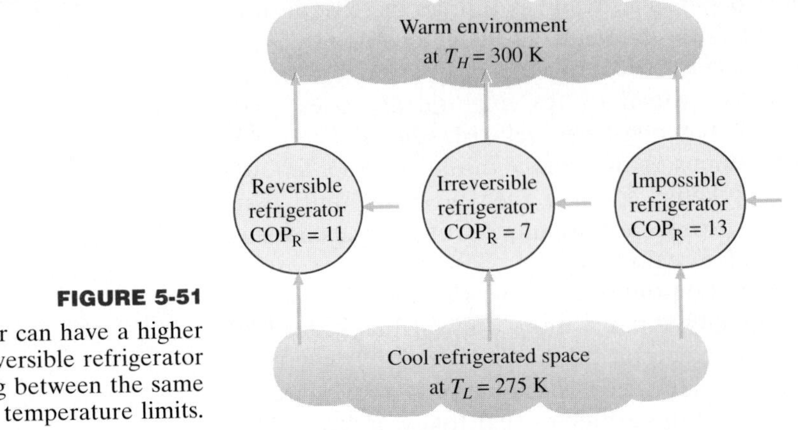


**FIGURE 5-51**
No refrigerator can have a higher COP than a reversible refrigerator operating between the same temperature limits.

be compared as follows:

$$COP_R \begin{cases} < COP_{R,rev} & \text{irreversible refrigerator} \\ = COP_{R,rev} & \text{reversible refrigerator} \\ > COP_{R,rev} & \text{impossible refrigerator} \end{cases} \tag{5-24}$$

A similar relation can be obtained for heat pumps by replacing all values of $COP_R$ in Eq. 5-24 by $COP_{HP}$.

The COP of a reversible refrigerator or heat pump is the maximum theoretical value for the specified temperature limits. Actual refrigerators or heat pumps may approach these values as their designs are improved, but they can never reach them.

As a final note, the COPs of both the refrigerators and the heat pumps decrease as $T_L$ decreases. That is, it requires more work to absorb heat from lower-temperature media. As the temperature of the refrigerated space approaches zero, the amount of work required to produce a finite amount of refrigeration approaches infinity and $COP_R$ approaches zero.

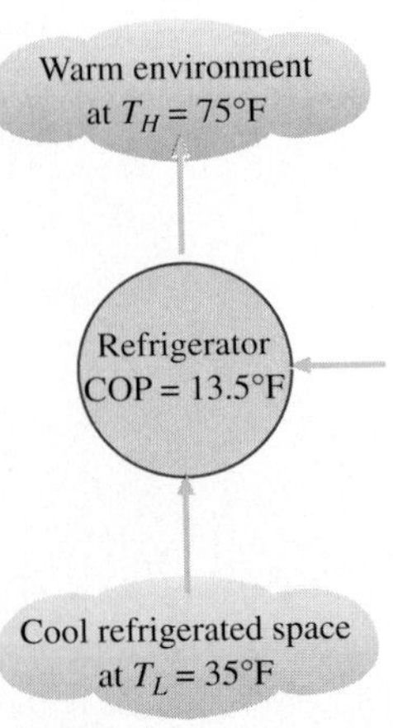


**FIGURE 5-52**
Schematic for Example 5-6.

**EXAMPLE 5-6**

An inventor claims to have developed a refrigerator that maintains the refrigerated space at 35°F while operating in a room where the temperature is 75°F and that has a COP of 13.5. Is there any truth to this claim?

**Solution** The performance of this refrigerator (shown in Fig. 5-52) can be evaluated by comparing it with a Carnot or any other reversible refrigerator operating between the same temperature limits:

$$COP_{R,max} = COP_{R,rev} = \frac{1}{T_H/T_L - 1}$$

$$= \frac{1}{(75 + 460\text{ R})/(35 + 460\text{ R}) - 1} = 12.4$$

This is the highest COP a refrigerator can have when removing heat from a cool medium at 35°F to a warmer medium at 75°F. Since the COP claimed by the inventor is above this maximum value, his claim is *false.*

**EXAMPLE 5-7**

A heat pump is to be used to heat a house during the winter, as shown in Fig. 5-53. The house is to be maintained at 21°C at all times. The house is estimated to be losing heat at a rate of 135,000 kJ/h when the outside temperature drops to −5°C. Determine the minimum power required to drive this heat pump unit.

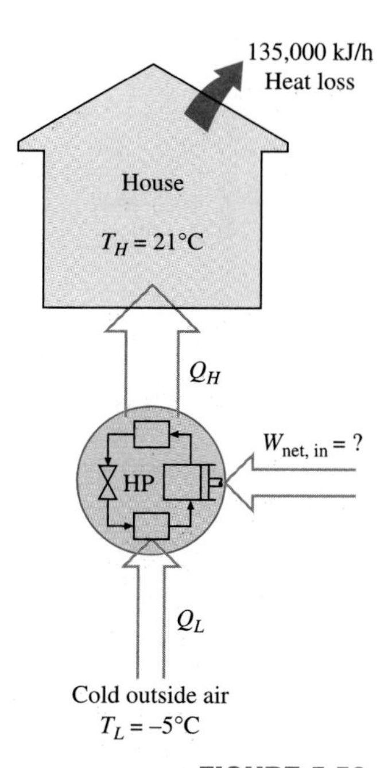


**FIGURE 5-53**
Schematic for Example 5-7.

**Solution** To maintain the house at a fixed temperature, the heat pump must supply the house with as much heat as it is losing. That is, the heat pump must reject heat to the house (the high-temperature medium) at a rate of $Q_H$ = 135,000 kJ/h = 37.5 kW.

The power requirements will be minimum if a reversible heat pump is used to do the job. The COP of a reversible heat pump operating between the house ($T_H$ = 21 + 273 = 294 K) and the outside air ($T_L$ = −5 + 273 = 268 K) is, from Eq. 5-23.

$$\text{COP}_{\text{HP,rev}} = \frac{1}{1 - T_L/T_H} = \frac{1}{1 - (268\text{ K}/294\text{ K})} = 11.3$$

Then the required power input to this reversible heat pump is determined from the definition of the COP, Eq. 5-12:

$$\dot{W}_{\text{net,in}} = \frac{\dot{Q}_H}{\text{COP}_{\text{HP}}} = \frac{37.5\text{ kw}}{11.3} = 3.32\text{ kW}$$

That is, this heat pump can meet the heating requirements of this house by consuming electric power at a rate of 3.32 kW only. If this house were to be heated by electric resistance heaters instead, the power consumption rate would jump up to 11.3 times to 37.5 kW. This is because in resistance heaters the electric energy is converted to heat at a one-to-one ratio. With a heat pump, however, energy is absorbed from the outside and carried to the inside using a refrigeration cycle that consumes only 3.32 kW. Notice that the heat pump does not create energy. It merely transports it from one medium (the cold outdoors) to another (the warm indoors).

## 5-12 ■ SUMMARY

The *second law of thermodynamics* states that processes occur in a certain direction, not in any direction. A process will not occur unless it satisfies both the first and the second laws of thermodynamics. Bodies that can absorb or reject finite amounts of heat isothermally are called *thermal energy reservoirs* or just *thermal reservoirs.*

Work can be converted to heat directly, but heat can be converted to work only by some devices called heat engines. The *thermal efficiency* of a heat engine is defined as

$$\eta_{th} = \frac{W_{net,out}}{Q_H} = 1 - \frac{Q_L}{Q_H}$$

where $W_{net,out}$ is the net work output of the heat engine, $Q_H$ is the amount of heat supplied to the engine, and $Q_L$ is the amount of heat rejected by the engine.

Refrigerators and heat pumps are devices that absorb heat from low-temperature media and reject it to higher temperature ones. The performance of a refrigerator or a heat pump is expressed in terms of the *coefficient of performance,* which is defined as

$$\text{COP}_R = \frac{Q_L}{W_{net,in}} = \frac{1}{Q_H/Q_L - 1}$$

$$\text{COP}_{HP} = \frac{Q_H}{W_{net,in}} = \frac{1}{1 - Q_L/Q_H}$$

The *Kelvin–Planck statement* of the second law of thermodynamics states that no heat engine can produce a net amount of work while exchanging heat with a single reservoir only. The *Clausius statement* of the second law states that no device can transfer heat from a cooler body to a warmer one without leaving an effect on the surroundings.

Any device that violates the first or the second law of thermodynamics is called a *perpetual-motion machine.*

A process is said to be *reversible* if both the system and the surroundings can be restored to their original conditions. Any other process is *irreversible.* The effects, such as friction, not-quasi-equilibrium expansion or compression, and heat transfer through a finite temperature difference render a process irreversible and are called *irreversibilities.*

The *Carnot cycle* is a reversible cycle that is composed of four reversible processes, two isothermal and two adiabatic. The *Carnot principles* state that the thermal efficiencies of all reversible heat engines operating between the same two reservoirs are the same, and that no heat engine is more efficient than a reversible one operating between the same two reservoirs. These statements form the basis for establishing an *absolute temperature scale,* also called the *Kelvin scale,* related to the heat transfers between a reversible device and the high- and low-temperature reservoirs by

$$\left(\frac{Q_H}{Q_L}\right)_{rev} = \frac{T_H}{T_L}$$

Therefore, the $Q_H/Q_L$ ratio can be replaced by $T_H/T_L$ for reversible devices, where $T_H$ and $T_L$ are the absolute temperatures of the high- and low-temperature reservoirs, respectively.

A heat engine that operates on the reversible Carnot cycle is called the *Carnot heat engine.* The thermal efficiency of a Carnot heat engine, as well as all other reversible heat engines, is given by

$$\eta_{\text{th,rev}} = 1 - \frac{T_L}{T_H}$$

This is the maximum efficiency a heat engine operating between two reservoirs at temperatures $T_H$ and $T_L$ can have.

The COPs of reversible refrigerators and heat pumps are given in a similar manner as

$$\text{COP}_{\text{R,rev}} = \frac{1}{T_H/T_L - 1}$$

and

$$\text{COP}_{\text{HP,rev}} = \frac{1}{1 - T_L/T_H}$$

Again, these are the highest COPs a refrigerator or a heat pump operating between the temperature limits of $T_H$ and $T_L$ can have.

## REFERENCES AND SUGGESTED READING

**1** W. Z. Black and J. G. Hartley, *Thermodynamics,* Harper & Row, New York, 1985.

**2** J. R. Howell and R. O. Buckius, *Fundamentals of Engineering Thermodynamics,* McGraw-Hill, New York, 1987.

**3** D. Stewart, "Wheels go round and round, but always run down," *Smithsonian,* pp. 193–208, November 1986.

**4** G. J. Van Wylen and R. E. Sonntag, *Fundamentals of Classical Thermodynamics,* 3d ed., Wiley, New York, 1985.

**5** K. Wark, *Thermodynamics,* 5th ed., McGraw-Hill, New York, 1988.

## PROBLEMS*

### Introduction to the Second Law of Thermodynamics

**5-1C** A mechanic claims to have developed a car engine that runs on water instead of gasoline. What is your response to this claim?

**5-2C** Does satisfying the first law of thermodynamics ensure that the process can actually take place? Explain.

**5-3C** Describe an imaginary process that satisfies the first law but violates the second law of thermodynamics.

**5-4C** Describe an imaginary process that satisfies the second law but violates the first law of thermodynamics.

---

*Students are encouraged to answer *all* the concept "C" questions.

**5-5C** Describe an imaginary process that violates both the first and the second laws of thermodynamics.

**5-6C** An experimentalist claims to have raised the temperature of a small amount of water to 150°C by transferring heat from high-pressure steam at 120°C. Is this a reasonable claim? Why? Assume no refrigerator or heat pump is used in the process.

### Thermal Energy Reservoirs

**5-7C** Consider the energy dissipated by a computer in a room. What is a suitable choice for a thermal energy reservoir?

**5-8C** What is a thermal energy reservoir? Give some examples.

**5-9C** Study the process of boiling eggs. Can the boiling water be treated as a thermal energy reservoir? Explain.

**5-10C** Study the process of baking potatoes in a conventional oven. Can the hot air in the oven be treated as a thermal energy reservoir? Explain.

**5-11C** Study the energy generated by a TV set. What is a suitable choice for a thermal energy reservoir?

### Heat Engines and Thermal Efficiency

**5-12C** Is it possible for a heat engine to operate without rejecting any waste heat to a low-temperature reservoir? Explain.

**5-13C** What are the characteristics of all heat engines?

**5-14C** What is the physical significance of the thermal efficiency of a heat engine, and how is it determined?

**5-15C** Describe two ways to determine the net work output of a heat engine.

**5-16C** Consider the process of baking potatoes in a conventional oven. How would you define the efficiency of the oven for this baking process?

**5-17C** Consider a pan of water being heated (*a*) by placing it on an electric range and (*b*) by placing a heating element in the water. Which method is a more efficient way of heating water? Explain.

**5-18C** Which one of these is a more efficient way of heating a house—burning wood in a fireplace or burning wood in a stove in the middle of the house?

**5-19C** Which is a more efficient way of converting electricity to light—using a light bulb or using a fluorescent tube?

**5-20C** Baseboard heaters are basically electric resistance heaters and are frequently used in space heating. A homeowner claims that her

5-year-old baseboard heaters have a conversion efficiency of 100 percent. Is this claim in violation of any thermodynamic laws? Explain.

**5-21C** What is the Kelvin–Planck expression of the second law of thermodynamics?

**5-22C** Does a heat engine that has a thermal efficiency of 100 percent necessarily violate (*a*) the first law and (*b*) the second law of thermodynamics? Explain.

**5-23C** In the absence of any friction and other irreversibilities, can a heat engine have an efficiency of 100 percent? Explain

**5-24C** Are the efficiencies of all the work-producing devices, including the hydroelectric power plants, limited by the Kelvin–Planck statement of the second law? Explain.

**5-25** An 800-MW steam power plant, which is cooled by a nearby river, has a thermal efficiency of 40 percent. Determine the rate of heat transfer to the river water. Will the actual heat transfer rate be higher or lower than this value? Why?

**5-26** A steam power plant receives heat from a furnace at a rate of 280 GJ/h. Heat losses to the surrounding air from the steam as it passes through the pipes and other components are estimated to be about 8 GJ/h. If the waste heat is transferred to the cooling water at a rate of 145 GJ/h, determine (*a*) the net power output and (*b*) the thermal efficiency of this power plant.
*Answers:* (*a*) 35.3 MW, (*b*) 45.4 percent

**5-27** A car engine with a power output of 90 kW has a thermal efficiency of 28 percent. Determine the rate of fuel consumption if the value of the fuel is 44,000 kJ/kg.

**5-27E** A car engine with a power output of 95 hp has a thermal efficiency of 28 percent. Determine the rate of fuel consumption if the energy content of the fuel is 19,000 Btu/lbm.

**5-28** A steam power plant with a power output of 150 MW consumes coal at a rate of 60 tons/h. If the energy content of the coal is 30,000 kJ/kg, determine the thermal efficiency of this plant (1 ton ≡ 1000 kg).
*Answer:* 30.0 percent

**5-29** An automobile engine consumes fuel at a rate of 20 L/h and delivers 60 kW of power to the wheels. If the fuel has a heating value of 44,000 kJ/kg and a density of 0.8 g/cm$^3$, determine the efficiency of this engine. *Answer:* 30.7 percent

**5-29E** An automobile engine consumes fuel at a rate of 5 gal/h and delivers 70 hp of power to the wheels. If the fuel has an energy content of 19,000 Btu/lbm and a density of 50 lbm/ft$^3$, determine the efficiency of this engine. *Answer:* 28.1 percent

**5-30** Solar energy stored in large bodies of water, called solar ponds, is being used for generating electricity. If such a solar power plant has an efficiency of 3 percent and a net power output of 100 kW, determine the average value of the required solar energy collection rate, in kJ/h. *Answer:* $1.2 \times 10^7$ kJ/h

**5-30E** Solar energy stored in large bodies of water, called solar ponds, is being used for generating electricity. If such a solar power plant has an efficiency of 4 percent and a net power output of 300 kW, determine the average value of the required solar energy collection rate, in Btu/h.

### Refrigerators and Heat Pumps

**5-31C** What is the difference between a refrigerator and a heat pump?

**5-32C** What is the difference between a refrigerator and an air conditioner?

**5-33C** In a refrigerator, heat is transferred from a lower-temperature medium (the refrigerated space) to a higher-temperature one (the kitchen air). Is this a violation of the second law of thermodynamics? Explain.

**5-34C** A heat pump is a device that absorbs energy from the cold outdoor air and transfers it to the warmer indoors. Is this a violation of the second law of thermodynamics? Explain.

**5-35C** Define the coefficient of performance of a refrigerator in words. Can it be greater than unity?

**5-36C** Define the coefficient of performance of a heat pump in words. Can it be greater than unity?

**5-37C** A heat pump that is used to heat a house has a COP of 2.5. That is, the heat pump delivers 2.5 kWh of energy to the house for each 1 kWh of electricity it consumes. Is this a violation of the first law of thermodynamics? Explain.

**5-38C** A refrigerator has a COP of 1.5. That is, the refrigerator removes 1.5 kWh of energy from the refrigerated space for each 1 kWh of electricity it consumes. Is this a violation of the first law of thermodynamics? Explain.

**5-39C** What is the Clausius expression of the second law of thermodynamics?

**5-40C** Show that the Kelvin–Planck and the Clausius expressions of the second law are equivalent.

**5-41** A household refrigerator with a COP of 1.8 removes heat from the refrigerated space at a rate of 90 kJ/min. Determine (*a*) the electric power consumed by the refrigerator and (*b*) the rate of heat transfer to the kitchen air. *Answers:* (*a*) 0.83 kW, (*b*) 140 kJ/min

**5-41E** A household refrigerator with a COP of 1.8 removes heat from the refrigerated space at a rate of 55 Btu/min. Determine (*a*) the electric power consumed by the refrigerator and (*b*) the rate of heat transfer to the kitchen air. *Answers:* (*a*) 0.72 hp, (*b*) 85.56 Btu/min

**5-42** An air conditioner removes heat steadily from a house at a rate of 750 kJ/min while drawing electric power at a rate of 6 kW. Determine (*a*) the COP of this air conditioner and (*b*) the rate of heat discharge to the outside air. *Answers:* (*a*) 2.08, (*b*) 1110 kJ/min

**5-43** A household refrigerator runs one-fourth of the time and removes heat from the food compartment at an average rate of 1200 kJ/h. If the COP of the refrigerator is 2.5 determine the power the refrigerator draws when running.

FIGURE P5-43

**5-43E** A household refrigerator runs one-fourth of the time and removes heat from the food compartment at an average rate of 800 Btu/h. If the COP of the refrigerator is 2.2, determine the power the refrigerator draws when running.

**5-44** Water enters an ice machine at 15°C and leaves as ice at −5°C. If the COP of the ice machine is 2.4 during this operation, determine the required power input for an ice production rate of 12 kg/h. (Note that 384 kJ of energy needs to be removed from each 1 kg of water at 15°C to turn it into ice at −5°C.)

**5-44E** Water enters an ice machine at 55°F and leaves as ice at 25°F. If the COP of the ice machine is 2.4 during this operation, determine the required power input for an ice production rate of 20 lbm/h. (Note that 169 Btu of energy needs to be removed from each lbm of water at 55°F to turn it into ice at 25°F.)

**5-45** A household refrigerator that has a power input of 450 W and a COP of 2.5 is to cool five large watermelons, 10 kg each, to 8°C. If the watermelons are initially at 20°C, determine how long it will take for the refrigerator to cool them. The watermelons can be treated as water whose specific heat is 4.2 kJ/(kg · °C). Is your answer realistic or optimistic? Explain. *Answer:* 2240 s

**5-46** When a man returns to his well-sealed house on a summer day, he finds that the house is at 32°C. He turns on the air conditioner, which cools the entire house to 20°C in 15 min. If the COP of the air conditioning system is 2.5, determine the power drawn by the air conditioner. Assume the entire mass within the house is equivalent to 800 kg of air for which $C_v = 0.72$ kJ/(kg · °C) and $C_p = 1.0$ kJ/(kg · °C).

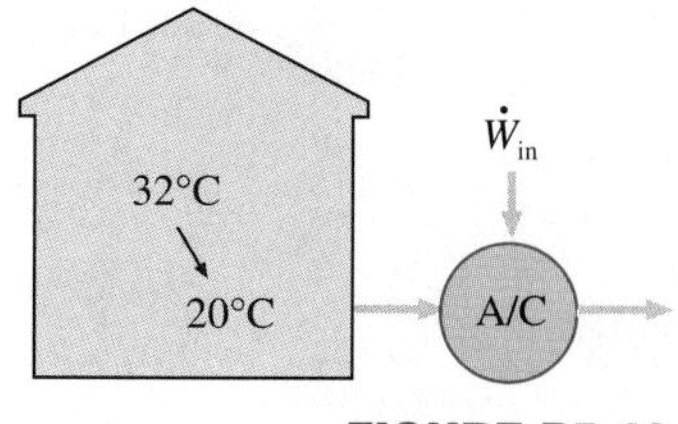

FIGURE P5-46

**5-46E** When a man returns to his well-sealed house on a summer day, he finds that the entire house is at 90°F. He turns on the air conditioner, which cools the entire house to 70°F in 15 min. If the COP of the air conditioning system is 2.5, determine the power drawn by the air conditioner. Assume the entire mass within the house is equivalent to 1800 lbm of air for which $C_v = 0.17$ Btu/(lbm · °F) and $C_p = 0.24$ Btu/(lbm · °F).

**5-47** Determine the COP of a refrigerator that removes heat from the food compartment at a rate of 8000 kJ/h for each 1 kW of power it consumes. Also determine the rate of heat rejection to the surrounding air.

**5-48** Determine the COP of a heat pump that supplies energy to a house at a rate of 8000 kJ/h for each kW of electric power it draws. Also determine the rate of energy absorption from the outdoor air.
*Answers:* 2.22, 4400 kJ/h

**5-49** A house that was heated by electric resistance heaters consumed 1200 kWh of electric energy in a winter month. If this house were heated by a heat pump, instead, that has an average COP of 2.4, determine how much money the homeowner would have saved that month. Assume a price of 8.5 ¢/kWh for electricity.

**5-50** A heat pump with a COP of 1.8 supplies energy to a house at a rate of 75,000 kJ/h. Determine (*a*) the electric power drawn by the heat pump and (*b*) the rate of heat removal from the outside air.
*Answers:* (*a*) 11.6 kW, (*b*) 33,333 kJ/h

**5-50E** A heat pump with a COP of 2.5 supplies energy to a house at a rate of 60,000 Btu/h. Determine (*a*) the electric power drawn by the heat pump and (*b*) the rate of heat removal from the outside air.
*Answers:* (*a*) 9.43 hp, (*b*) 36,000 Btu/h

**5-51** A heat pump that is used to heat a house runs about one-third of the time. The house is losing heat at an average rate of 15,000 kJ/h. If the COP of the heat pump is 3.5, determine the power the heat pump draws when running.

**5-51E** A heat pump that is used to heat a house runs about one-third of the time. The house is losing heat at an average rate of 15,000 Btu/h. If the COP of the heat pump is 3.5, determine the power the heat pump draws when running.

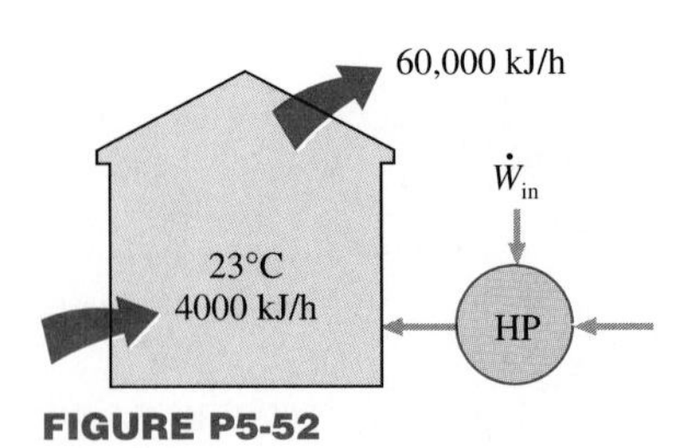

FIGURE P5-52

**5-52** A heat pump is used to maintain a house at a constant temperature of 23°C. The house is losing heat to the outside air through the walls and the windows at a rate of 60,000 kJ/h while the energy generated within the house from people, lights, and appliances amounts to 4000 kJ/h. For a COP of 2.5, determine the required power input to the heat pump.
*Answer:* 6.22 kW

**5-52E** A heat pump is used to maintain a house at a constant temperature of 70°F. The house is losing heat to the outside air through the walls and the windows at a rate of 50,000 Btu/h while the energy generated within the house from the people, lights, and appliances amounts to 8000 Btu/h. For a COP of 2.5, determine the required power input to the heat pump.

### Perpetual-Motion Machines

**5-53C** An inventor claims to have developed a resistance heater that supplies 1.2 kWh of energy to a room for each 1 kWh of electricity it consumes. Is this a reasonable claim, or has she developed a perpetual-motion machine? Explain.

**5-54C** It is common knowledge that the temperature of air rises as it is compressed. An inventor thought about using this high-temperature air to heat buildings. He used a compressor driven by an electric motor. The inventor claims that the compressed hot-air system is 12 percent more efficient than a resistance heating system that provides an equivalent amount of heating. Is this claim valid, or is this just another perpetual-motion machine? Explain.

### Reversible and Irreversible Processes

**5-55C** A cold canned drink is left in a warmer room where its temperature rises as a result of heat transfer. Is this a reversible process? Explain.

**5-56C** A hot baked potato is left on a table where it cools to the room temperature. Is this a reversible or an irreversible process? Explain.

**5-57C** Why are engineers interested in reversible processes even though they can never be achieved?

**5-58C** Air is compressed from 20°C and 100 kPa to 300°C and 800 kPa first in a reversible manner and then in an irreversible manner. Which case do you think will require more work input?

**5-59C** Why does a non-quasi-equilibrium compression process require a larger work input than the corresponding quasi-equilibrium one?

**5-60C** Why does a non-quasi-equilibrium expansion process deliver less work than the corresponding quasi-equilibrium one?

**5-61C** How do you distinguish between internal and external irreversibilities?

**5-62C** Is a reversible expansion or compression process necessarily quasi-equilibrium? Is a quasi-equilibrium expansion or compression process necessarily reversible? Explain.

## The Carnot Cycle and Carnot Principles

**5-63C** What are the four processes that make up the Carnot cycle?

**5-64C** What are the four processes that make up the reversed Carnot cycle?

**5-65C** What are the two statements known as the Carnot principles?

**5-66C** Somebody claims to have developed a new reversible heat-engine cycle that has a higher theoretical efficiency than the Carnot cycle operating between the same temperature limits. How do you evaluate this claim?

**5-67C** Somebody claims to have developed a new reversible heat-engine cycle that has the same theoretical efficiency as the Carnot cycle operating between the same temperature limits. Is this a reasonable claim?

**5-68C** Is it possible to develop (*a*) an actual and (*b*) a reversible heat-engine cycle that is more efficient than a Carnot cycle operating between the same temperature limits? Explain.

## Carnot Heat Engines

**5-69C** Is there any way to increase the efficiency of a Carnot heat engine other than by increasing $T_H$ or decreasing $T_L$?

**5-70C** Consider two actual power plants operating with solar energy. Energy is supplied to one plant from a solar pond at 80°C and to the other from concentrating collectors that raise the water temperature to 600°C. Which of these power plants will have a higher efficiency and why?

**5-71** A Carnot heat engine operates between a source at 1000 K and a sink at 300 K. If the heat engine is supplied with heat at a rate of 800 kJ/min, determine (*a*) the thermal efficiency and (*b*) the power output of this heat engine. *Answers:* (*a*) 70 percent, (*b*) 9.33 kW

**5-71E** A Carnot heat engine operates between a source at 1800 R and a sink at 440 R. If the heat engine is supplied with heat at a rate of 1200 Btu/min, determine (*a*) the thermal efficiency and (*b*) the power output of this heat engine. *Answers:* (*a*) 75.6 percent, (*b*) 21.4 hp

**5-72** A Carnot heat engine receives 500 kJ of heat from a source of unknown temperature and rejects 200 kJ of it to a sink at 17°C. Determine (*a*) the temperature of the source and (*b*) the thermal efficiency of the heat engine.

**5-73** A heat engine operates between a source at 550°C and a sink at 25°C. If heat is supplied to the heat engine at a steady rate of 1200 kJ/min, determine the maximum power output of this heat engine.

**5-74** A heat engine is operating on Carnot cycle and has a thermal efficiency of 55 percent. The waste heat from this engine is rejected to a nearby lake at 15°C at a rate of 800 kJ/min. Determine (*a*) the power output of the engine and (*b*) the temperature of the source.
*Answers:* (*a*) 16.3 kW, (*b*) 640 K

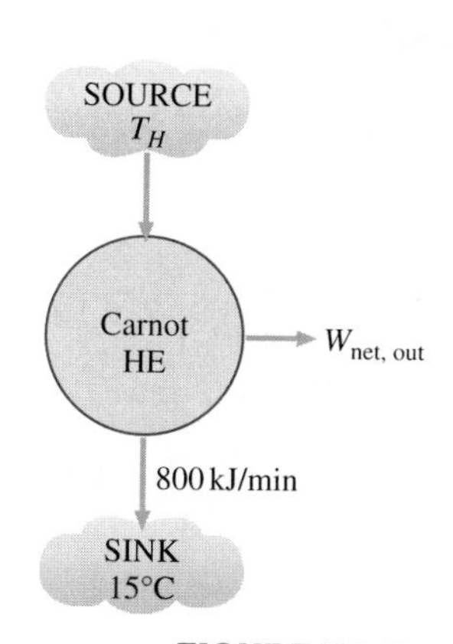


**FIGURE P5-74**

**5-74E** A heat engine is operating on Carnot cycle and has a thermal efficiency of 55 percent. The waste heat from this engine is rejected to a nearby lake at 60°F at a rate of 800 Btu/min. Determine (*a*) the power output of the engine and (*b*) the temperature of the source.
*Answers:* (*a*) 23.1 hp, (*b*) 1155.6 R

**5-75** In tropical climates, the water near the surface of the ocean remains warm throughout the year as a result of solar energy absorption. In the deeper parts of the ocean, however, the water remains at a relatively low temperature since the sun's rays cannot penetrate very far. It is proposed to take advantage of this temperature difference and construct a power plant that will absorb heat from the warm water near the surface and reject the waste heat to the cold water a few hundred meters below. Determine the maximum thermal efficiency of such a plant if the water temperatures at the two respective locations are 24 and 4°C.

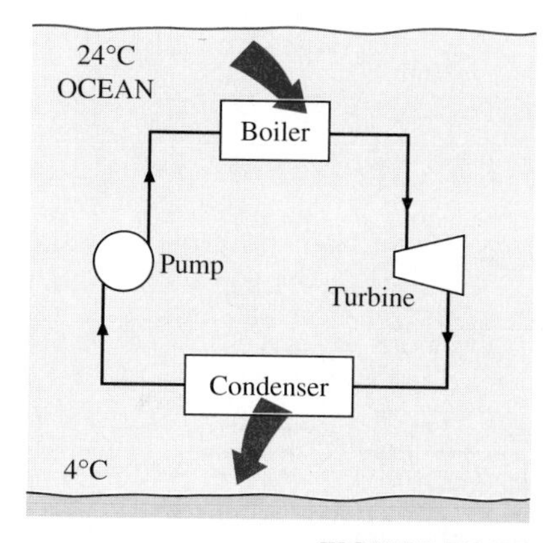

**FIGURE P5-75**

**5-76** An innovative way of power generation involves the utilization of geothermal energy—the energy of hot water that exists naturally underground—as the heat source. If a supply of hot water at 140°C is discovered at a location where the environmental temperature is 20°C, determine the maximum thermal efficiency a geothermal power plant built at that location can have. *Answer:* 29.1 percent

**5-77** An inventor claims to have developed a heat engine that receives 800 kJ of heat from a source at 400 K and produces 250 kJ of net work while rejecting the waste heat to a sink at 300 K. Is this a reasonable claim? Why?

**5-77E** An inventor claims to have developed a heat engine that receives 600 Btu of heat from a source at 750 R and produces 200 Btu of net work while rejecting the waste heat to a sink at 550 R. Is this a reasonable claim? Why?

**5-78** An experimentalist claims that, based on his measurements, a heat engine receives 320 kJ of heat from a source at 500 K, converts 180 kJ of it to work, and rejects the rest as waste heat to a sink at 300 K. Are these measurements reasonable? Why?

**5-78E** An experimentalist claims that, based on his measurements, a heat engine receives 300 Btu of heat from a source of 900 R, converts 160 Btu of it to work, and rejects the rest as waste heat to a sink at 540 R. Are these measurements reasonable? Why?

## Carnot Refrigerators and Heat Pumps

**5-79C** How can we increase the COP of a Carnot refrigerator?

**5-80C** What is the highest COP that a refrigerator operating between temperature levels $T_L$ and $T_H$ can have?

**5-81C** What is the highest COP that a heat pump operating between temperature levels $T_L$ and $T_H$ can have?

**5-82C** In an effort to conserve energy in a heat-engine cycle, somebody suggests incorporating a refrigerator that will absorb some of the waste energy $Q_L$ and transfer it to the energy source of the heat engine. Is this a smart idea? Explain.

**5-83C** It is well established that the thermal efficiency of a heat engine increases as the temperature at which heat is rejected from the heat engine $T_L$ decreases. In an effort to increase the efficiency of a power plant, somebody suggests refrigerating the cooling water before it enters the condenser, where heat rejection takes place. Would you be in favor of this idea? Why?

**5-84C** It is well known that the thermal efficiency of heat engines increases as the temperature of the energy source increases. In an attempt to improve the efficiency of a power plant, somebody suggests transferring heat from the available energy source to a higher temperature medium by a heat pump before energy is supplied to the power plant. What do you think of this suggestion? Explain.

**5-85** A Carnot refrigerator operates in a room in which the temperature is 25°C and consumes 2 kW of power when operating. If the food compartment of the refrigerator is to be maintained at 3°C, determine the rate of heat removal from the food compartment.

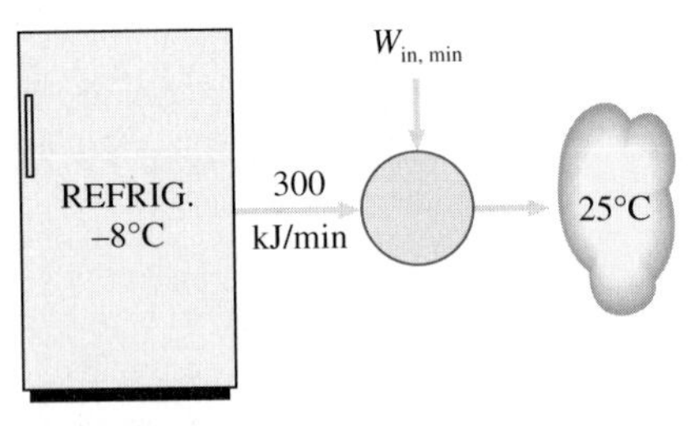


**FIGURE P5-86**

**5-86** A refrigerator is to remove heat from the cooled space at a rate of 300 kJ/min to maintain its temperature at −8°C. If the air surrounding the refrigerator is at 25°C, determine the minimum power input required for this refrigerator. *Answer:* 0.623 kW

**5-86E** A refrigerator is to remove heat from the cooled space at a rate of 150 Btu/min to keep its temperature at 25°F. If the air surrounding the refrigerator is at 80°F, determine the minimum power input required for this refrigerator. *Answer:* 0.4 hp

**5-87** An air conditioning system operating on the reversed Carnot cycle is required to transfer heat from a house at a rate of 750 kJ/min, to maintain its temperature at 20°C. If the outdoor air temperature is 35°C, determine the power required to operate this air conditioning system. *Answer:* 0.64 kW

**5-88** An air conditioning system is used to maintain a house at 20°C when the temperature outside is 35°C. If this air conditioning system draws 5 kW of power when operating, determine the maximum rate of heat removal from the house that it can provide.

**5-88E** An air conditioning system is used to maintain a house at 70°F when the temperature outside is 90°F. If this air conditioning system draws 5 hp of power when operating, determine the maximum rate of heat removal from the house that it can provide.

**5-89** A Carnot refrigerator operates in a room in which the temperature is 25°C. The refrigerator consumes 500 W of power when operating and had a COP of 4.5. Determine (*a*) the rate of heat removal from the refrigerated space and (*b*) the temperature of the refrigerated space. *Answers:* (*a*) 135 kJ/min, (*b*) −29.2°C

**5-90** An inventor claims to have developed a refrigeration system that removes heat from the closed region at −5°C and transfers it to the surrounding air at 22°C while maintaining a COP of 8.2. Is this claim reasonable? Why?

**5-90E** An inventor claims to have developed a refrigeration system that removes heat from the cooled region at 20°F and transfers it to the surrounding air at 75°F while maintaining a COP of 6.7. Is this claim reasonable? Why?

**5-91** During an experiment conducted in a room at 25°C, a laboratory assistant measures that a refrigerator that draws 2 kW of power has removed 30,000 kJ of heat from the refrigerated space, which is maintained at −30°C. The running time of the refrigerator during the experiment was 20 min. Determine if these measurements are reasonable.

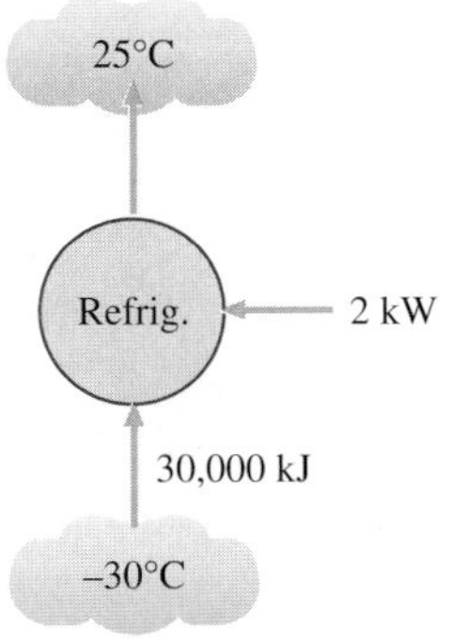


**FIGURE P5-91**

**5-92** An air conditioning system is used to maintain a house at 22°C when the temperature outside is 33°C. The house is gaining heat through the walls and the windows at a rate of 600 kJ/min, and the heat generation rate within the house from people, lights, and appliances amounts to 120 kJ/min. Determine the minimum power input required for this air conditioning system. *Answer:* 0.45 kW

**5-92E** An air conditioning system is used to maintain a house at 75°F when the temperature outside is 95°F. The house is gaining heat through the walls and the windows at a rate of 750 Btu/min, and the heat generation rate within the house from people, lights, and appliances amounts to 150 Btu/min. Determine the minimum power input required for this air conditioning system. *Answer:* 0.79 hp

**5-93** A heat pump is used to heat a house and maintain it at 20°C. On a winter day when the outdoor air temperature is −5°C, the house is estimated to lose heat at a rate of 75,000 kJ/h. Determine the minimum power required to operate this heat pump.

**5-93E** A heat pump is used to heat a house and maintain it at 70°F. On a winter day when the outdoor air temperature is 20°F, the house is estimated to lose heat at a rate of 55,000 Btu/h. Determine the minimum power required to operate this heat pump.

**5-94** A heat pump is used to maintain a house at 22°C by extracting heat from the outside air on a day when the outside air temperature is 2°C. The house is estimated to lose heat at a rate of 110,000 kJ/h, and the heat pump consumes 8 kW of electric power when operating. Is this heat pump powerful enough to do the job?

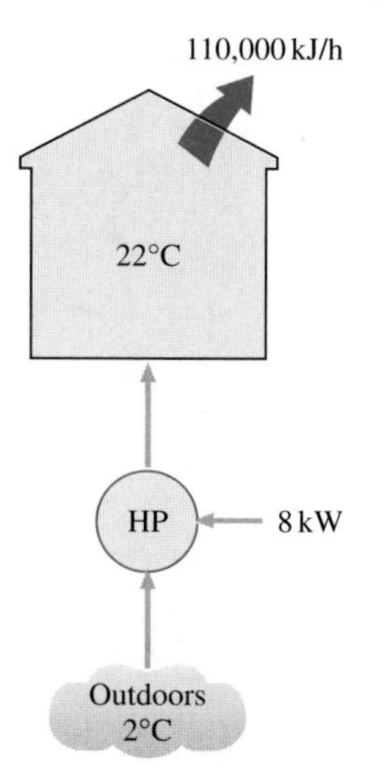

**FIGURE P5-94**

**5-95** The structure of a house is such that it loses heat at a rate of 5400 kJ/h per °C difference between the indoors and outdoors. A heat pump that requires a power input of 6 kW is used to maintain this house at 21°C. Determine the lowest outdoors temperature for which the heat pump can meet the heating requirements of this house.
*Answer:* −13.3°C

**5-96** The performance of a heat pump degrades (i.e., its COP decreases) as the temperature of the heat source decreases. This makes using heat pumps at locations with severe weather conditions unattractive. Consider a house that is heated and maintained at 20°C by a heat pump during the winter. What is the maximum COP for this heat pump if heat is extracted from the outdoor air at (*a*) 10°C, (*b*) −5°C, and (*c*) −30°C?

**5-97** A heat pump is to be used for heating a house in winter. The house is to be maintained at 21°C at all times. When the temperature outdoors drops to −2°C, the heat losses from the house are estimated to be 80,000 kJ/h. Determine the minimum power required to run this heat pump if heat is extracted from (*a*) the outdoor air at −2°C and (*b*) the well water at 12°C.

**5-97E** A heat pump is to be used for heating a house in the winter. The house is to be maintained at 78°F at all times. When the temperature outdoors drops to 25°F, the heat losses from the house are estimated to be 80,000 Btu/h. Determine the minimum power required to run this heat pump if heat is extracted from (*a*) the outdoor air at 25°F and (*b*) the well water at 50°F.

**5-98** A Carnot heat pump is to be used for heating a house and maintaining it at 20°C during the winter. On a day when the average outdoor temperature remains at about 2°C, the house is estimated to lose heat at a steady rate of 82,000 kJ/h. If the heat pump consumes 8 kW of power while operating, determine (*a*) how long the heat pump ran on that day; (*b*) the total heating costs, assuming an average price of 8.5¢/kWh for electricity; and (*c*) the heating cost for the same day if resistance heating is used instead of a heat pump.
*Answers:* (*a*) 4.19 h, (*b*) \$2.85, (*c*) \$46.47

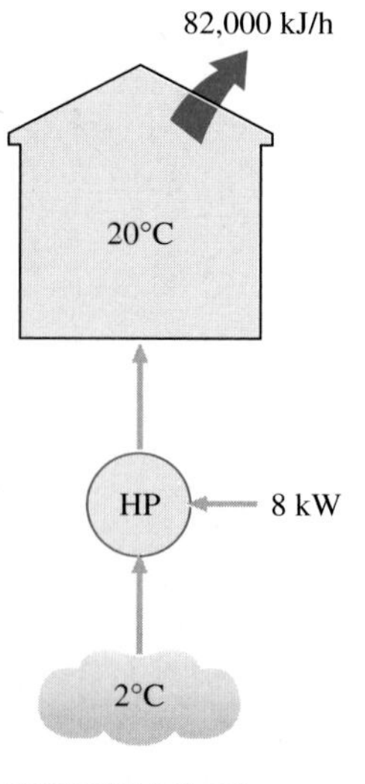


**FIGURE P5-98**

**5-99** A Carnot heat engine receives heat from a reservoir at 900°C at a rate of 800 kJ/min and rejects the waste heat to the ambient air at 27°C. The entire work output of the heat engine is used to drive a refrigerator that removes heat from the refrigerated space at −5°C and transfers it to the same ambient air at 27°C. Determine (*a*) the maximum rate of heat removal from the refrigerated space and (*b*) the total rate of heat rejection to the ambient air. *Answers:* (*a*) 4982 kJ/min, (*b*) 5782 kJ

**5-99E** A Carnot heat engine receives heat from a reservoir at 1700°F at a rate of 700 Btu/min and rejects the waste heat to the ambient air at 80°F. The entire work output of the heat engine is used to drive a refrigerator that removes heat from the refrigerated space at 20°F and transfers it to the same ambient air at 80°F. Determine (*a*) the maximum rate of heat removal from the refrigerated space and (*b*) the total rate of heat rejection to the ambient air.
*Answers:* (*a*) 4262 Btu/min, (*b*) 4962 Btu/min

## Review Problems

**5-100** Consider a Carnot heat engine cycle executed in a steady-flow system using steam as the working fluid. The cycle has a thermal efficiency of 30 percent, and steam changes from saturated liquid to saturated vapor at 300°C during the heat addition process. If the mass flow rate of the steam is 5 kg/s, determine the net power output of this engine, in kW.

**5-101** A heat pump with a COP of 2.4 is used to heat a house. When running, the heat pump consumes 8 kW of electric power. If the house is loosing heat to the outside at an average rate of 4000 kJ/h and the temperature of the house is 3°C when the heat pump is turned on, determine how long it will take for the temperature in the house to rise to 22°C. Assume the house is well sealed (i.e., no air leaks) and take the entire mass within the house (air, furniture, etc.) to be equivalent to 2000 kg of air.

**5-102** A gas turbine has an efficiency of 17 percent and develops a power output of 6000 kW. Determine the fuel consumption rate of this gas turbine, in L/min, if the fuel has a heating value of 46,000 kJ/kg and a density of 0.8 g/cm$^3$.

**5-103** Show that $COP_{HP} = COP_R + 1$ when both the heat pump and the refrigerator have the same $Q_L$ and $Q_H$ values.

**5-104** An air conditioning system is used to maintain a house at a constant temperature of 20°C. The house is gaining heat from outdoors at a rate of 20,000 kJ/h, and the heat generated in the house from the people, lights, and appliances amounts to 8000 kJ/h. For a COP of 2.5, determine the required power input to this air conditioning system.
*Answer:* 3.11 kW

**5-105** Consider a Carnot heat engine cycle executed in a closed system using 0.01 kg of steam as the working fluid. The cycle has a thermal efficiency of 15 percent, and the steam changes from saturated liquid to saturated vapor at 70°C during the heat addition process. Determine the net work output of this engine, in kJ.

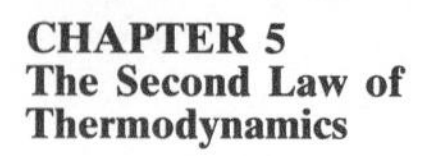


**5-106** A heat pump with a COP of 3.2 is used to heat a house. When running, the heat pump consumes power at a rate of 5 kW. If the temperature in the house is 7°C when the heat pump is turned on, how long will it take for the heat pump to raise the temperature of the house to 22°C? Is this answer realistic or optimistic? Explain. Assume the entire mass within the house (air, furniture, etc.) is equivalent to 1500 kg of air for which $C_v = 0.72\ \text{kJ/(kg}\cdot{}^\circ\text{C)}$ and $C_p = 1.0\ \text{kJ/(kg}\cdot{}^\circ\text{C)}$.
*Answer:* 1012 s

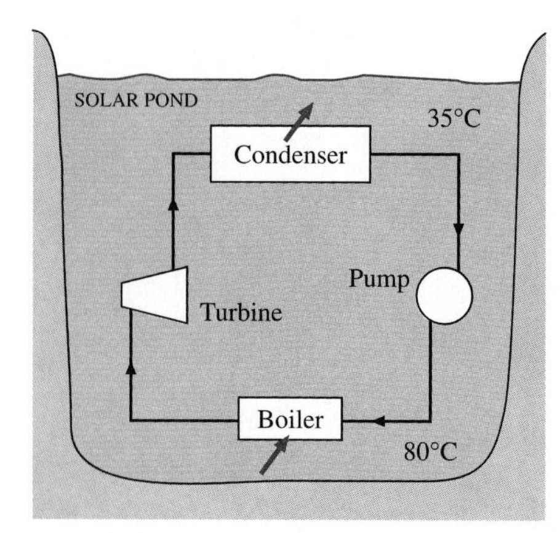

**FIGURE P5-107**

**5-107** A promising method of power generation involves collecting and storing solar energy in large artificial lakes a few meters deep, called solar ponds. Solar energy is absorbed by all parts of the pond, and the water temperature rises everywhere. The top part of the pond, however, loses much of the heat it absorbs to the atmosphere, and as a result, its temperature drops. This cool water serves as insulation for the bottom part of the pond and helps trap the energy there. Usually, salt is planted at the bottom of the pond to prevent the rise of this hot water to the top. A power plant that uses an organic fluid, such as alcohol, as the working fluid can be operated between the top and the bottom portions of the pond. If the water temperature is 35°C near the surface and 80°C near the bottom of the pond, determine the maximum thermal efficiency that this power plant can have. Is it realistic to use 35 and 85°C for temperatures in the calculations? Explain.
*Answer:* 12.7 percent

**5-108** Consider a Carnot heat engine cycle executed in a closed system using 0.0103 kg of steam as the working fluid. It is known that the maximum absolute temperature in the cycle is twice the minimum absolute temperature, and the net work output of the cycle is 25 kJ. If the steam changes from saturated liquid to saturated vapor during the heat rejection process, determine the temperature of the steam during the heat rejection process, in °C.

**5-109** Consider two Carnot heat engines operating in series. The first engine receives heat from the reservoir at 1200 K and rejects the waste heat to another reservoir at temperature $T$. The second engine receives this energy rejected by the first one, converts some of it to work, and rejects the rest to a reservoir at 300 K. If the thermal efficiencies of both engines are the same, determine the temperature $T$. *Answer:* 600 K

**5-110** The COP of a refrigerator decreases as the temperature of the refrigerated space is decreased. That is, removing heat from a medium at a very low temperature will require a large work input. Determine the minimum work input required to remove 1 kJ of heat from liquid helium at 3 K when the outside temperature is 300 K. *Answer:* 99 kJ

**5-111** A Carnot heat pump is used to heat and maintain a residential building at 22°C. An energy analysis of the house reveals that it loses heat at a rate of 2500 kJ/h per °C temperature difference between the indoors and the outdoors. For an outdoor temperature of 4°C, determine (*a*) the coefficient of performance and (*b*) the required power input to the heat pump. *Answers:* (*a*) 16.4, (*b*) 0.762 kW

**5-111E** A Carnot heat pump is used to heat and maintain a residential building at 75°F. An energy analysis of the house reveals that it loses heat at a rate of 2500 Btu/h per °F temperature difference between the indoors and the outdoors. For an outdoor temperature of 35°F, determine (*a*) the coefficient of performance and (*b*) the required power input to the heat pump. *Answers:* (*a*) 13.4, (*b*) 2.93 hp

**5-112** A Carnot heat engine receives heat at 750 K and rejects the waste heat to the environment at 300 K. The entire work output of the heat engine is used to drive a Carnot refrigerator that removes heat from the cooled space at −15°C at a rate of 400 kJ/min and rejects it to the same environment at 300 K. Determine (*a*) the rate of heat supplied to the heat engine and (*b*) the total rate of heat rejection to the environment.

**5-113** A heat engine operates between two reservoirs at 800 and 20°C. One-half of the work output of the heat engine is used to drive a Carnot heat pump that removes heat from the cold surroundings at 2°C and transfers it to a house maintained at 22°C. If the house is losing heat at a rate of 95,000 kJ/h, determine the minimum rate of heat supply to the heat engine required to keep the house at 22°C.

**5-114** A Carnot heat engine is operating between a source at $T_H$ and a sink at $T_L$. If it is desired to double the thermal efficiency of this engine, what should the new source temperature be? Assume the sink temperature is held constant.

**5-115** When discussing Carnot engines, it is assumed that the engine is in thermal equilibrium with the source and the sink during the heat addition and heat rejection processes, respectively. That is, it is assumed that $T_H^* = T_H$ and $T_L^* = T_L$ so that there is no external irreversibility. In that case the thermal efficiency of the Carnot engine is $\eta_C = 1 - T_L/T_H$.

In reality, however, we must maintain a reasonable temperature difference between the two heat transfer mediums in order to have an acceptable heat transfer rate through a finite heat exchanger surface area. The heat transfer rates in that case can be expressed as

$$\dot{Q}_H = (hA)_H(T_H - T_H^*)$$
$$\dot{Q}_L = (hA)_L(T_L^* - T_L)$$

where $h$ and $A$ are heat transfer coefficient and heat transfer surface area, respectively. When the values of $h$, $A$, $T_H$, and $T_L$ are fixed, show that the power output will be a maximum when

$$\frac{T_L^*}{T_H^*} = \left(\frac{T_L}{T_H}\right)^{1/2}$$

Also show that the maximum net power output in this case is

$$\dot{W}_{C,\max} = \frac{(hA)_H T_H}{1 + (hA)_H/(hA)_L}\left[1 - \left(\frac{T_L}{T_H}\right)^{1/2}\right]^2$$

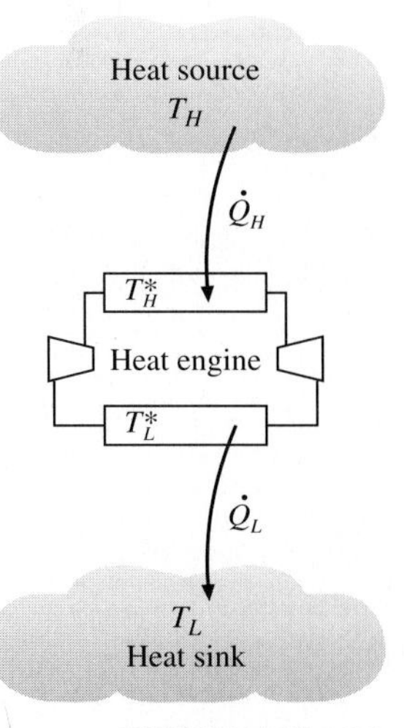


**FIGURE P5-115**

## Computer, Design, and Essay Problems

**5-116** Write a computer program to determine the maximum work that can be extracted from a pond containing $10^5$ kg of water at 350 K when the temperature of the surroundings is 300 K. Notice that the temperature of water in the pond will be gradually decreasing as energy is extracted from it; therefore, the efficiency of the engine will be decreasing. Use temperature intervals of (*a*) 5 K, (*b*) 2 K, and (*c*) 1 K until the pond temperature drops to 300 K. Also solve this problem exactly by integration, and compare the results.

**5-117** Find out the prices of heating oil, natural gas, and electricity in your area, and determine the cost of each per kWh energy supplied to the house as heat. Go through your utility bills and determine how much money you spent for heating last January. Also determine how much your January heating bill would be for each of the heating systems if you had the latest and most efficient system installed.

**5-118** Prepare a report on the residual heating systems available in your area for residential buildings. Discuss the advantages and disadvantages of each system, and compare their initial and operating costs. What are the important factors in the selection of a heating system? Give some guidelines. Identify the conditions under which each heating system would be the best choice in your area.

**5-119** The performance of a cyclic device is defined as the ratio of the desired output to the required input, and this definition can be extended to nontechnical fields. For example, your performance in this course can be viewed as the grade you earn relative to the effort you put in. If you have been investing a lot of time in this course and your grades do not reflect it, you are performing poorly. In that case, perhaps you should try to find out the underlying cause, and how to correct the problem. Give three other definitions of performance from nontechnical fields, and discuss them

**5-120** Devise a Carnot heat engine using steady-flow components, and describe how the Carnot cycle is executed in that engine. What happens when the directions of heat and work interactions are reversed?

**5-121** Consider a house whose annual air-conditioning load (heat gain plus the internal heat generation from people, lights, appliances, etc.) is estimated to be 40,000 kWh in an area where the unit cost of electricity is \$0.09/kWh. Two air-conditioners are considered for the house. Air-conditioner A has a seasonal average COP of 2.5, and costs \$2500 to purchase and install. Air-conditioner B has a seasonal average COP of 5.0, and costs \$4000 to purchase and install. If all else is equal, which air-conditioner would you buy? Justify your selection. Under what conditions would you select the other air-conditioner?

# Entropy

CHAPTER

# 6

In Chap. 5, we introduced the second law of thermodynamics and applied it to cycles and cyclic devices. In this chapter, we apply the second law to processes. The first law of thermodynamics deals with the property *energy* and the conservation of it. The second law leads to the definition of a new property called *entropy.* Entropy is a somewhat abstract property, and it is difficult to give a physical description of it. Entropy is best understood and appreciated by studying its uses in commonly encountered engineering processes, and this is precisely what we intend to do.

This chapter starts with a discussion of the Clausius inequality, which forms the basis for the definition of entropy, and continues with the increase of entropy principle. Unlike energy, entropy is a nonconserved property, and there is no such thing as *conservation of entropy principle.* Next, the entropy changes that take place during processes for pure substances, incompressible substances, and ideal gases are discussed, and a special class of idealized processes, called *isentropic processes,* are examined. Finally, reversible steady-flow work is studied.

## 6-1 ■ THE CLAUSIUS INEQUALITY

The second law of thermodynamics often leads to expressions that involve inequalities. An irreversible (i.e., actual) heat engine, for example, is less efficient than a reversible one operating between the same two thermal energy reservoirs. Likewise, an irreversible refrigerator or a heat pump has a lower coefficient of performance (COP) than a reversible one operating between the same temperature limits. Another important inequality that has major consequences in thermodynamics is the **Clausius inequality**. It is first stated by the German physicist R. J. E. Clausius (1822–1888), one of the founders of thermodynamics, and is expressed as

$$\oint \frac{\delta Q}{T} \leqslant 0$$

That is, *the cyclic integral of $\delta Q/T$ is always less than or equal to zero.* This inequality is valid for all cycles, reversible or irreversible. The symbol $\oint$ (integral symbol with a circle in the middle) is used to indicate that the integration is to be performed over the entire cycle. Any heat transfer to or from a system can be considered to consist of differential amounts of heat transfer. Then the cyclic integral of $\delta Q/T$ can be viewed as the sum of all these differential amounts of heat transfer divided by the absolute temperature at the boundary.

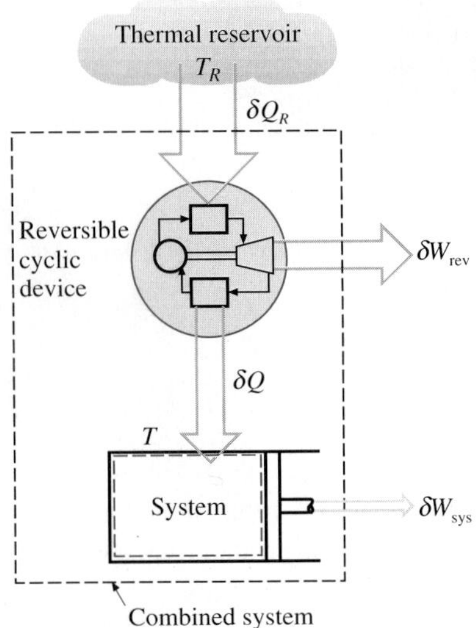


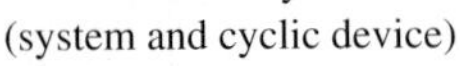


**FIGURE 6-1**
The system considered in the development of Clausius inequality.

To demonstrate the validity of Clausius inequality, consider a system connected to a thermal energy reservoir at a constant absolute temperature of $T_R$ through a *reversible* cyclic device (Fig. 6-1). The cyclic device receives heat $\delta Q_R$ from the reservoir and supplies heat $\delta Q$ to the system whose absolute temperature at that part of the boundary is $T$ (a variable) while producing work $\delta W_{rev}$. The system produces work $\delta W_{sys}$ as a result of this heat transfer. Applying the conservation of energy principle to the combined system identified by dashed lines yields

$$\delta W_C = \delta Q_R - dE_C$$

where $\delta W_C$ is the total work of the combined system ($\delta W_{rev} + \delta W_{sys}$) and $dE_C$ is the change in the total energy of the combined system. Considering that the cyclic device is a *reversible* one, we have (Eq. 5-18)

$$\frac{\delta Q_R}{T_R} = \frac{\delta Q}{T}$$

where the sign of $\delta Q$ is determined with respect to the system (positive if *to* the system and negative if *from* the system) and the sign of $\delta Q_R$ is determined with respect to the reversible cyclic device. Eliminating $\delta Q_R$ from the two relations above yields

$$\delta W_C = T_R \frac{\delta Q}{T} - dE_C$$

We now let the system undergo a cycle while the cyclic device undergoes an integral number of cycles. Then the relation above becomes

$$W_C = T_R \oint \frac{\delta Q}{T}$$

since the cyclic integral of energy (the net change in the energy, which is a property, during a cycle) is zero. Here $W_C$ is the cyclic integral of $\delta W_C$, and it represents the net work for the combined cycle.

It appears that the combined system is exchanging heat with a single thermal energy reservoir while involving (producing or consuming) work $W_C$ during a cycle. On the basis of the Kelvin–Planck statement of the second law, which states that *no system can produce a net amount of work while operating in a cycle and exchanging heat with a single thermal energy reservoir,* we reason that $W_C$ cannot be a work output, and thus it cannot be a positive quantity. Considering that $T_R$ is an absolute temperature and thus a positive quantity, we must have

$$\oint \frac{\delta Q}{T} \leqslant 0 \tag{6-1}$$

which is the *Clausius inequality.* This inequality is valid for all thermodynamic cycles, reversible or irreversible, including the refrigeration cycles.

If no irreversibilities occur within the system as well as the reversible cyclic device, then the cycle undergone by the combined system will be internally reversible. As such, it can be reversed. In the reversed cycle case, all the quantities will have the same magnitude but the opposite sign. Therefore, the work $W_C$, which could not be a positive quantity in the regular case, cannot be a negative quantity in the reversed case. Then it follows that $W_{C,\,\text{int rev}} = 0$ since it cannot be a positive or negative quantity, and therefore

$$\oint \left(\frac{\delta Q}{T}\right)_{\text{int rev}} = 0 \tag{6-2}$$

for internally reversible cycles. Thus we conclude that *the equality in the Clausius inequality (Eq. 6-1) holds for totally or just internally reversible cycles and the inequality for the irreversible ones.*

## 6-2 ■ ENTROPY

The Clausius inequality discussed in Sec. 6-1 forms the basis for the definition of a new property called *entropy.*

To develop a relation for the definition of entropy, let us examine Eq. 6-2 more closely. Here we have a quantity whose cyclic integral is zero. Let us think for a moment what kind of quantities can have this characteristic. We know that the cyclic integral of *work* is not zero. (It is a good thing that it is not. Otherwise, heat engines that work on a cycle, such as steam power plants, would produce zero net work.) Neither is the cyclic integral of heat. As you may recall, these two quantities were defined in Chap. 3 as *path functions* since their magnitudes depend on the process path followed.

Now consider the volume occupied by a gas in a piston-cylinder device undergoing a cycle, as shown in Fig. 6-2. When the piston returns to its initial position at the end of a cycle, the volume of the gas also returns to its initial value. Thus the net change in volume during a cycle is zero. This is also expressed as

$$\oint dV = 0 \tag{6-3}$$

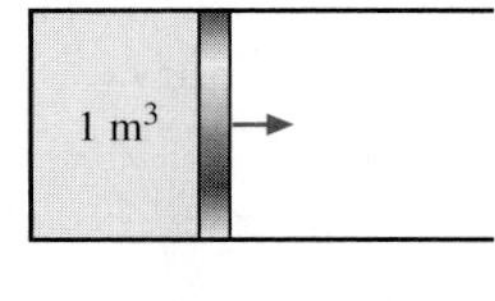


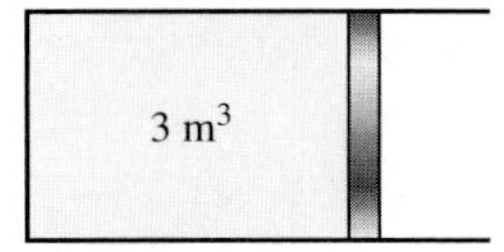


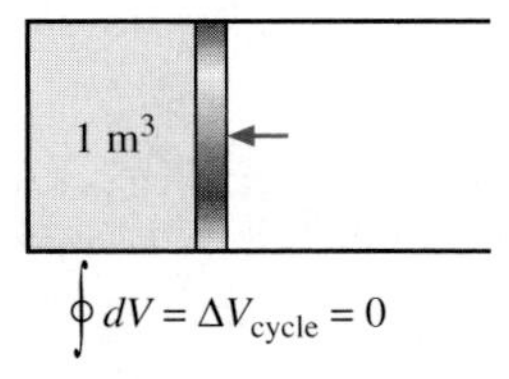


**FIGURE 6-2**
The net change in volume (a property) during a cycle is always zero.

That is, the cyclic integral of volume (or any other property) is zero. Conversely, a quantity whose cyclic integral is zero depends on the *state* only and not the process path, and thus it is a property. Therefore the quantity $(\delta Q/T)_{\text{int rev}}$ must represent a property in the differential form.

Clausius realized in 1865 that he had discovered a new thermodynamic property, and he chose to name this property **entropy**. It is designated $S$ and is defined as

$$dS = \left(\frac{\delta Q}{T}\right)_{\text{int rev}} \qquad (\text{kJ/K}) \tag{6-4}$$

Entropy is an extensive property of a system and sometimes is referred to as *total entropy*. Entropy per unit mass, designated $s$, is an intensive property and has the unit kJ/(kg · K). The term *entropy* is generally used to refer to both total entropy and entropy per unit mass since the context usually clarifies which one is meant.

The entropy change of a system during a process can be determined by integrating Eq. 6-4 between the initial and the final states:

$$\Delta S = S_2 - S_1 = \int_1^2 \left(\frac{\delta Q}{T}\right)_{\text{int rev}} \qquad (\text{kJ/K}) \tag{6-5}$$

Notice that we have actually defined the *change* in entropy instead of entropy itself, just as we defined the change in energy instead of energy when we developed the first-law relation for closed systems in Chap. 3. Absolute values of entropy are determined on the basis of the third law of thermodynamics, which is discussed later in this chapter. Engineers are usually concerned with the *changes* in entropy. Therefore, the entropy of a substance can be assigned a zero value at some arbitrarily selected reference state, and the entropy values at other states can be determined from Eq. 6-5 by choosing state 1 to be the reference state ($S = 0$) and state 2 to be the state at which entropy is to be determined.

To perform the integration in Eq. 6-5, one needs to know the relation between $Q$ and $T$ during a process. This relation is often not available, and the integral in Eq. 6-5 can be performed for a few cases only. For the majority of cases we have to rely on tabulated data for entropy.

Note that entropy is a property, and like all other properties, it has fixed values at fixed states. Therefore, the entropy change $\Delta S$ between two specified states is the same no matter what path, reversible or irreversible, is followed during a process (Fig. 6-3).

Also note that the integral of $\delta Q/T$ will give us the value of entropy change *only if* the integration is carried out along an *internally reversible* path between the two states. The integral of $\delta Q/T$ along an irreversible path is not a property, and in general, different values will be obtained when the integration is carried out along different irreversible paths. Therefore, even for irreversible processes, the entropy change should be determined by carrying out this integration along some convenient *imaginary* internally reversible path between the specified states.

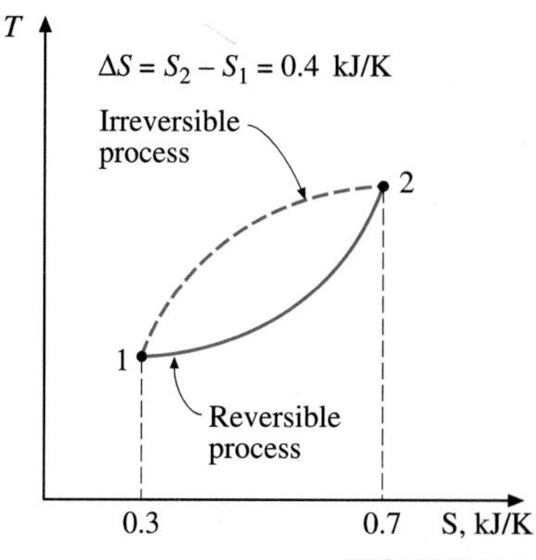

**FIGURE 6-3**

The entropy change between two specified states is the same whether the process is reversible or irreversible.

## 6-3 ■ THE INCREASE OF ENTROPY PRINCIPLE

Consider a cycle that is made up of two processes: process 1-2, which is arbitrary (reversible or irreversible), and process 2-1, which is internally reversible, as shown in Fig. 6-4. From Clausius inequality,

$$\oint \frac{\delta Q}{T} \leqslant 0$$

or

$$\int_1^2 \frac{\delta Q}{T} + \int_1^2 \left(\frac{\delta Q}{T}\right)_{\text{int rev}} \leqslant 0$$

The second integral in the above relation is readily recognized as the entropy change $S_1 - S_2$. Therefore,

$$\int_1^2 \frac{\delta Q}{T} + S_1 - S_2 \leqslant 0 \qquad (6\text{-}6)$$

which can be rearranged as

$$\Delta S = S_2 - S_1 \geqslant \int_1^2 \frac{\delta Q}{T} \qquad (6\text{-}7)$$

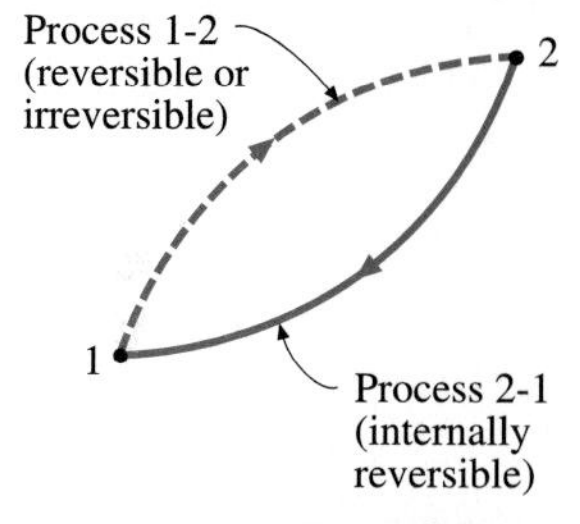

**FIGURE 6-4**

A cycle composed of a reversible and an irreversible process.

Equation 6-7 can be viewed as a mathematical statement of the second law of thermodynamics for a closed mass. It can also be expressed in differential form as

$$dS \geqslant \frac{\delta Q}{T} \qquad (6\text{-}8)$$

where the equality holds for an internally reversible process and the inequality for an irreversible process. We may conclude from these equations that *the entropy change of a closed system during an irreversible process is greater than the integral of $\delta Q/T$ evaluated for that process. In the limiting case of a reversible process, these two quantities become equal.* We again emphasize that $T$ in the above relations is the *absolute temperature* at the *boundary* where the differential heat $\delta Q$ is transferred between the system and the surroundings.

The quantity $\Delta S = S_2 - S_1$ represents the *entropy change* of the system which, for a reversible process, becomes equal to $\int_1^2 \delta Q/T$, which represents the *entropy transfer* with heat.

The inequality sign in the relations above is a constant reminder that the entropy change of a closed system during an irreversible process is always greater than the entropy transfer. That is, some entropy is *generated* or *created* during an irreversible process, and this generation is due entirely to the presence of irreversibilities. The entropy generated during a process is called **entropy generation**, and is denoted by $S_{gen}$. Noting that the difference between the entropy change of a closed system and the entropy transfer is equal to entropy generation, Eq. 6-7 can be rewritten as an *equality* as

$$S_2 - S_1 = \int_1^2 \frac{\delta Q}{T} + S_{gen} \tag{6-9}$$

Note that entropy generation $S_{gen}$ is always a *positive* quantity or zero. Its value depends on the process, and thus it is *not* a property of the system.

Equation 6-7 has far-reaching implications in thermodynamics. For an isolated system (or just an adiabatic closed system), the heat transfer is zero, and Eq. 6-7 reduces to

$$\Delta S_{isolated} \geqslant 0 \tag{6-10}$$

This equation can be expressed as *the entropy of an isolated system during a process always increases or, in the limiting case of a reversible process, remains constant.* In other words, it *never* decreases. This is known as the **increase of entropy principle**. Note that in the absence of any heat transfer, entropy change is due to irreversibilities only, and their effect is always to increase the entropy.

Since no actual process is truly reversible, we can conclude that some entrophy is generated during a process, and therefore the entropy of the universe, which can be considered to be an isolated system, is continuously increasing. The more irreversible a process is, the larger the entropy generated during that process. No entropy is generated during reversible processes ($S_{gen} = 0$).

Entropy increase of the universe is a major concern not only to engineers but also to philosophers and theologians since entropy is viewed as a measure of the disorder (or "mixed-up-ness") in the universe.

The increase of entropy principle does not imply that the entropy of a system or the surroundings cannot decrease. The entropy change of a system or its surroundings *can* be negative during a process (Fig. 6-5); but entropy generation cannot. The increase of entropy principle can be summarized as follows:

$$S_{gen} \begin{cases} > 0 & \text{irreversible process} \\ = 0 & \text{reversible process} \\ < 0 & \text{impossible process} \end{cases}$$

This relation serves as a criterion in determining whether a process is reversible, irreversible, or impossible.

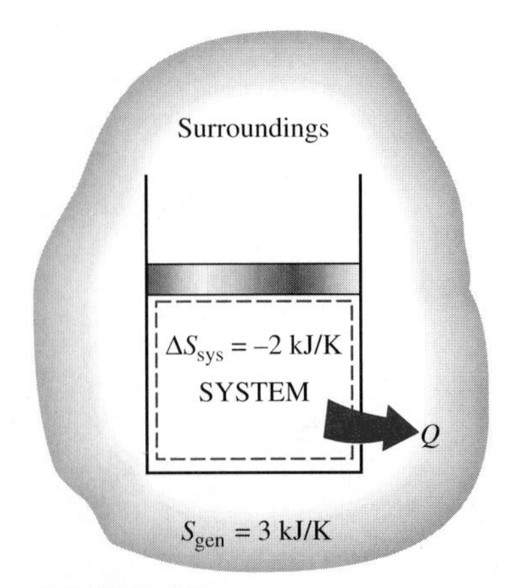

**FIGURE 6-5**
The entropy change of a system can be negative; but the sum entropy generation cannot.

The things in nature have a tendency to change until they attain a state of equilibrium. The increase of entropy principle dictates that the entropy of an isolated system will increase until the entropy of the system reaches a maximum value. At that point, the system is said to have reached an equilibrium state since the increase of entropy principle prohibits the system from undergoing any change of state that will result in a decrease in entropy.

### Some Remarks about Entropy

In the light of the preceding discussions, we can draw the following conclusions:

**1** Processes can occur in a *certain* direction only, not in *any* direction. A process must proceed in the direction that complies with the increase of entropy principle, that is, $S_{gen} \geq 0$. A process that violates this principle is impossible. This principle often forces chemical reactions to come to a halt before reaching completion.

**2** Entropy is a *nonconserved property,* and there is *no* such thing as the *conservation of entropy principle.* Entropy is conserved during the idealized reversible processes only and increases during *all* actual processes. Therefore, the entropy of the universe is continuously increasing.

**3** The performance of engineering systems is degraded by the presence of irreversibilities, and the *entropy generation* is a measure of the magnitudes of the irreversibilities present during that process. The greater the extent of irreversibilities, the greater the entropy generation. Therefore, entropy can be used as a quantitative measure of irreversibilities associated with a process. It is also used to establish criteria for the performance of engineering devices.

## 6-4 ■ ENTROPY BALANCE

The property *entropy* is a measure of molecular disorder or randomness of a system, and the second law of thermodynamics states that entropy can be created but it cannot be destroyed. Therefore, the entropy change of a system during a process is greater than the entropy transfer by an amount equal to the entropy generated during the process within the system, and the *increase of entropy principle* is expressed as

$$\text{Entropy change} = \text{Entropy transfer} + \text{Entropy generation}$$

or

$$\Delta S_{\text{system}} = S_{\text{transfer}} + S_{\text{gen}} \qquad (6\text{-}11)$$

which is a verbal statement of Eq. 6-9. This relation is often referred to as the **entropy balance**, and is applicable to any kind of system undergoing any kind of process. The entropy balance relation above can be stated as *the entropy change of a system during a process is equal to the sum of the entropy transfer through the system boundary and the entropy generated* *within the system as a result of irreversibilities.* Next we discuss the various terms in that relation.

## 1 Entropy Change

Despite the reputation of entropy as being vague and abstract and the intimidation associated with it, entropy balance is actually easier to deal with than energy balance since, unlike energy, entropy does not exist in various forms. Therefore, the determination of entropy change of a system during a process involves the evaluation of the entropy of the system at the beginning and at the end of the process, and taking their difference. That is,

Entropy change = Entropy at final state − Entropy at initial state

or
$$\Delta S_{\text{system}} = S_{\text{final}} - S_{\text{initial}} \tag{6-12}$$

Note that entropy is a property, and the value of a property does not change unless the state of the system changes. Therefore, the entropy change of a system is zero if the state of the system does not change during the process. For example, the entropy change of steady flow devices such as nozzles, compressors, turbines, pumps, and heat exchangers is zero during steady operation.

## 2 Mechanisms of Entropy Transfer

Entropy can be transferred to or from a system in two forms: *heat transfer* and *mass flow* (in contrast, energy is transferred by work also). Entropy transfer is recognized at the system boundary as entropy crosses the boundary, and it represents the entropy gained or lost by a system during a process. The only form of entropy interaction associated with a fixed mass or closed system is *heat transfer,* and thus the entropy transfer for an adiabatic closed system is zero.

**Heat Transfer** Heat is, in essence, a form of disorganized energy, and some disorganization (entropy) will flow with heat. Heat transfer to a system increases the entropy of that system and thus the level of molecular disorder or randomness, and heat transfer from a system decreases it. In fact, heat rejection is the only way the entropy of a fixed mass can be decreased. The ratio of the heat transfer $Q$ at a location to the absolute temperature $T$ at that location is called the *entropy flow* or *entropy transfer,* and is expressed as

Entropy transfer with heat
$$S_{\text{heat}} = \frac{Q}{T} \tag{6-13}$$

The quantity $Q/T$ represents the entropy transfer accompanied by heat transfer, and the direction of entropy transfer is the same as the direction of heat transfer since absolute temperature $T$ is always a positive quantity.

Therefore, the sign of entropy transfer is the same as the sign of heat transfer *positive* if *into* the system, and *negative* if *out of* the system.

When two systems are in contact, the entropy transfer from the warmer system is equal to the entropy transfer into the cooler one at the point of contact. That is, no entropy can be created or destroyed at the boundary since the boundary has no thickness and occupies no volume.

Note that **work** is entropy-free, and no entropy is transferred with work. Energy is transferred with both heat and work whereas entropy is transferred only with heat. The first law of thermodynamics makes no distinction between heat transfer and work; it considers them as *equals*. The distinction between heat transfer and work is brought out by the second law: *an energy interaction which is accompanied by entropy transfer is heat transfer, and an energy interaction which is not accompanied by entropy transfer is work.* That is, no entropy is exchanged during a work interaction between a system and its surroundings. Thus only *energy* is exchanged during work interaction whereas both *energy* and *entropy* are exchanged during heat transfer (Fig. 6-6).

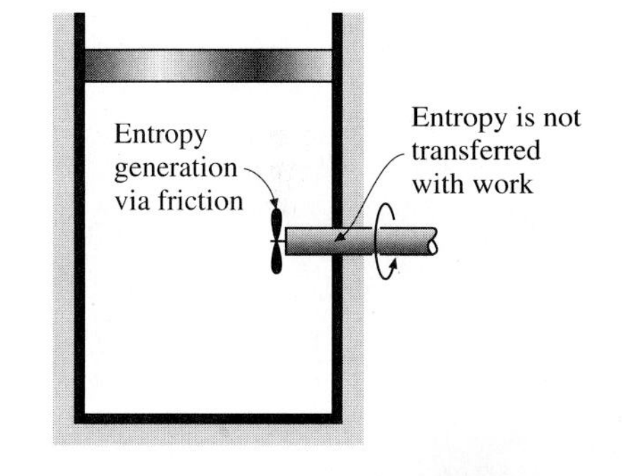


**FIGURE 6-6**
No entropy accompanies work as it crosses the system boundary. But entropy may ge generated within the system as work energy is dissipated into a less useful form of energy.

**Mass Flow** Mass contains entropy as well as energy, and the entropy and energy contents of a system are proportional to the mass. (When the mass of a system is doubled, so are the entropy and energy contents of the system). Both entropy and energy are carried into or out of a system by streams of matter, and the rates of entropy and energy transport into or out of a system are proportional to the mass flow rate. Closed systems do not involve any mass flow and thus any entropy transport. When a mass in the amount of $m$ enters or leaves a system, entropy in the amount of $ms$, where $s$ is the specific entropy (entropy per unit mass), accompanies it. Therefore, the entropy of a system increases by $ms$ when mass in the amount of $m$ enters, and decreases by the same amount when the same amount of mass at the same state leaves the system.

## 3 Entropy Generation

Irreversibilities such as friction, mixing, chemical reactions, heat transfer through a finite temperature difference, unrestrained expansion, non-quasiequilibrium compression or expansion always cause the entropy of a system to increase, and entropy generation $S_{\text{gen}}$ is a measure of the entropy created by such affects during a process.

For a *reversible process* (a process that involves no irreversibilities), the entropy generation is zero and thus the *entropy change* of a system is equal to the *entropy transfer*. Therefore, the entropy balance relation in the reversible case becomes analogous to the energy balance relation, which states that *energy change* of a system during a process is equal to the *energy transfer* during that process. However, note that the energy change of a system equals the energy transfer for *any* process, but the entropy change of a system equals the entropy transfer only for a *reversible* process.

### Entropy Balance for Closed Systems

A closed system involves no mass flow across its boundaries, and its entropy change is simply the difference between the initial and final entropies of the system. The *entropy change* of a closed system is due to the *entropy transfer* accompanying heat transfer and the *entropy generation* within the system boundaries, and Eq. 6-9 is an expression for the entropy balance of a closed system. When heat in the amounts of $Q_k$ is transferred through the boundary at constant temperatures $T_k$ at several locations, the entropy transfer term can be expressed more conveniently as a *sum* instead of an integral to give

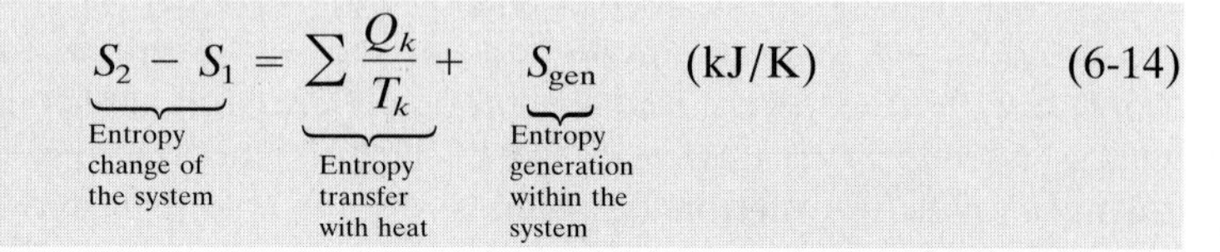

$$\underbrace{S_2 - S_1}_{\substack{\text{Entropy}\\ \text{change of}\\ \text{the system}}} = \underbrace{\sum \frac{Q_k}{T_k}}_{\substack{\text{Entropy}\\ \text{transfer}\\ \text{with heat}}} + \underbrace{S_{\text{gen}}}_{\substack{\text{Entropy}\\ \text{generation}\\ \text{within the}\\ \text{system}}} \qquad \text{(kJ/K)} \tag{6-14}$$

The entropy balance relation above can be stated as *the entropy change of a closed system during a process is equal to the sum of the entropy transferred through the system boundary by heat transfer and the entropy generated within the system boundaries.* It can also be expressed in *rate* form as

$$\frac{dS}{dt} = \sum \frac{\dot{Q}_k}{T_k} + \dot{S}_{\text{gen}} \qquad \text{(kW/K)} \tag{6-15}$$

where $dS/dt$ is the *rate of change of entropy* of the system, and $\dot{Q}_k$ is the rate of heat transfer through the boundary at temperature $T_k$. For an *adiabatic process* ($Q = 0$), the entropy transfer terms in the above relations drop out and entropy change of the closed system becomes equal to the entropy generation within the system. That is,

$$\Delta S_{\text{adiabatic}} = S_{\text{gen}} \tag{6-16}$$

Note that $S_{\text{gen}}$ represents the entropy generation *within the system boundary* only, and not the entropy generation that may occur outside the system boundary during the process as a result of external irreversibilities. Therefore, a process for which $S_{\text{gen}} = 0$ is *internally reversible,* but it is not necessarily *totally* reversible. The *total* entropy generated during a process can be determined by applying the entropy balance to an *extended system* that includes the system itself and its immediate surroundings where external irreversibilities might be occurring (Fig. 6-7). Also, the entropy change in this case is equal to the sum of the entropy change of the system and the entropy change of the immediate surroundings. Note that under steady conditions, the state and thus the entropy of the immediate surroundings (let us call it the "buffer zone") at any point will not change during the process, and the entropy change of the buffer zone will be zero. The entropy change of the buffer zone is usually small relative to the entropy change of the system, and thus it is usually disregarded.

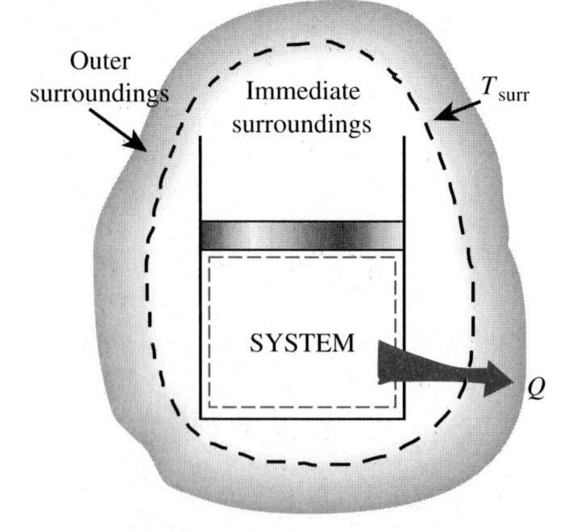


**FIGURE 6-7**
Entropy generation outside system boundaries can be accounted for by writing an entropy balance on an extended system that includes the system and its immediate surroundings.

When evaluating the entropy transfer between an extended system and the surroundings, the boundary temperature of the extended system is simply taken to be the temperature of the surroundings.

## Entropy Balance for Control Volumes

The entropy balance relations for control volumes are similar to the ones given earlier for closed systems, but this time we have to consider one more mechanism of entropy exchange: *mass flow across the control volume boundaries.* As mentioned earlier, mass possesses entropy as well as energy, and the amounts of these two extensive properties are proportional to the amount of mass (Fig. 6-8). Because a control volume is essentially a closed system that allows for mass transfer, the rate form of the **entropy balance** for a control volume can be obtained from the closed-system entropy balance relation (Eq. 6-1) by modifying it to allow for entropy transport with mass. Assuming one-dimensional flow, it can be expressed as

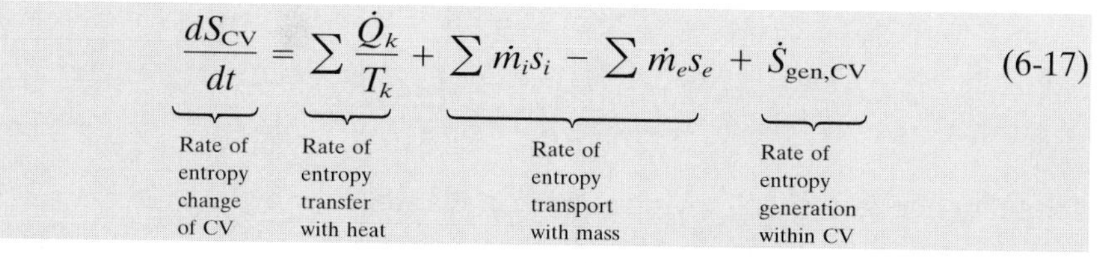

$$\frac{dS_{CV}}{dt} = \sum \frac{\dot{Q}_k}{T_k} + \sum \dot{m}_i s_i - \sum \dot{m}_e s_e + \dot{S}_{gen,CV} \qquad (6\text{-}17)$$



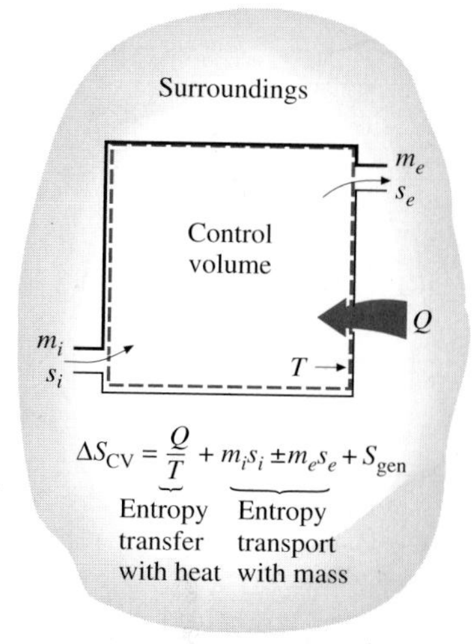



**FIGURE 6-8**

The entropy of a control volume changes as a result of mass flow as well as heat flow.

The entropy balance relation above can be stated as *the rate of entropy change within the control volume $dS_{CV}/dt$ during a process is equal to the sum of the rate of entropy transfer through the control volume boundary by heat transfer, the net rate of entropy transport into the control volume by mass flow, and the rate of entropy generation within the boundaries of the control volume as a result of irreversibilities.*

Most control volumes encountered in practice, such as turbines, compressors, nozzles, diffusers, heat exchangers, pipes, and ducts, operate steadily, and thus they experience no change in their entropy. Therefore, the entropy balance relation for a general **steady-flow process** can be obtained from Eq. 6-17, by setting $dS/dt = 0$,

$$\dot{S}_{gen} = \sum \dot{m}_e s_e - \sum \dot{m}_i s_i - \sum \frac{\dot{Q}_k}{T_k} \qquad (6\text{-}18)$$

For single-stream (one-inlet and one-exit) steady-flow devices, the entropy balance relation simplifies to

$$\dot{S}_{gen} = \dot{m}(s_e - s_i) - \sum \frac{\dot{Q}_k}{T_k} \qquad (6\text{-}19)$$

For the case of an adiabatic single stream device, the entropy balance relation further simplifies to $\dot{S}_{gen} = \dot{m}(s_e - s_i)$ which indicates that the specific entropy of the fluid must increase as it flows through an adiabatic device since $\dot{S}_{gen} \geqslant 0$ (Fig. 6-9).

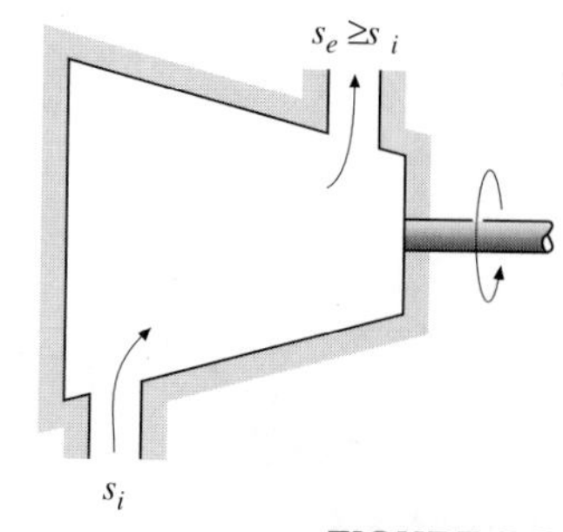



**FIGURE 6-9**

The entropy of a substance always increases (or remains constant in the case of reversible process) as it flows through a single-stream adiabatic steady-flow device.

**EXAMPLE 6-1**

Consider steady heat flow through a 5 m × 6 m brick wall of thickness 30 cm and thermal conductivity 0.69 W/(m.°C). On a day when the temperature of the outdoors is 0°C, the house is maintained at 27°C. The temperatures of the inner and outer surfaces of the brick wall are measured to be 20°C and 5°C, respectively. Determine the rate of heat transfer through the wall, the rate of entropy generation in the wall, and the rate of total entropy generation associated with this heat transfer process (Fig. 6-10).

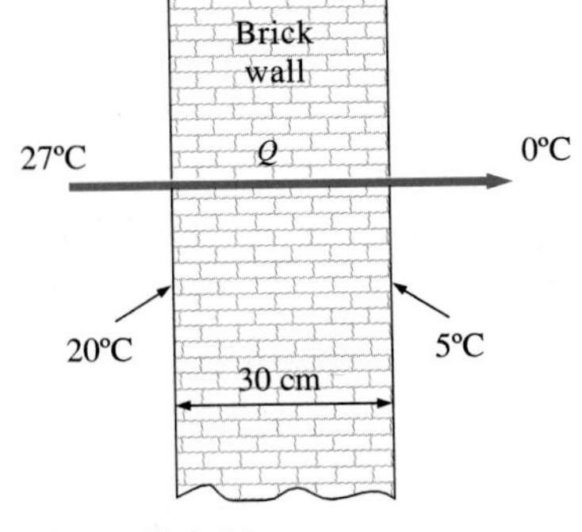


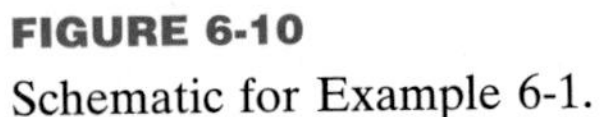
**FIGURE 6-10**
Schematic for Example 6-1.

**Solution** Knowing the wall surface temperatures, the rate of heat transfer through the wall is determined from Fourier's law of heat conduction to be

$$\dot{Q} = kA\left(\frac{\Delta T}{L}\right)_{\text{wall}} = [0.69\ \text{W}/(\text{m.°C})](5 \times 6\ \text{m}^2)\frac{(20-5)\text{°C}}{0.3\ \text{m}} = 1035\ \text{W}$$

Taking the wall as the system, the entropy balance can be expressed in the rate form as

$$dS_{\text{wall}}/dt = \dot{S}_{\text{transfer}} + \dot{S}_{\text{gen,wall}}$$

$$0 = \sum \frac{\dot{Q}}{T} + \dot{S}_{\text{gen,wall}}$$

$$0 = \frac{1035\ \text{W}}{29315\ \text{K}} - \frac{1035\ \text{W}}{27815\ \text{K}} + \dot{S}_{\text{gen,wall}}$$

Therefore, the rate of entropy generation in the wall is

$$\dot{S}_{\text{gen,wall}} = 0.191\ \text{W/K}$$

Note that the entropy change of the wall is zero during this process, since the state and thus the entropy of the wall does not change anywhere in the wall. Also, entropy transfer with heat at any location is $Q/T$ at that location, and the direction of entropy transfer is the same as the direction of heat transfer.

To determine the rate of total entropy generation during this heat transfer process, we extend the system to include the regions on both sides of the wall that experience a temperature change. Then one side of the system boundary becomes at room temperature while the other side becomes at the temperature of the outdoors. The entropy balance for this extended system can be written as

$$dS_{\text{total}}/dt = \dot{S}_{\text{transfer}} + \dot{S}_{\text{gen,total}}$$

$$0 = \sum \frac{\dot{Q}}{T} + \dot{S}_{\text{gen,total}}$$

$$0 = \frac{1035\ \text{W}}{300.15\ \text{K}} - \frac{1035\ \text{W}}{273.15\ \text{K}} + \dot{S}_{\text{gen,total}}$$

Therefore, the rate of total entropy generation becomes

$$\dot{S}_{\text{gen,total}} = 0.341\ \text{W/K}$$

Note that the entropy change of this extended system is zero also, since the state of air at any point does not change during the process. The differences between the two entropy generations is 0.150 W/K, and it represents the entropy generated in the air layers on both sides of the wall. The entropy generation in this case is entirely due to irreversible heat transfer through a finite temperature difference.

It is clear from the previous discussion that entropy is a useful property and serves as a valuable tool in the second-law analysis of engineering devices. But this does not mean that we know and understand entropy well. Because we do not. In fact, we cannot even give an adequate answer to the question, What is entropy? Not being able to describe entropy fully, however, does not take anything away from its usefulness. In Chap. 1, we could not define *energy* either, but it did not interfere with our understanding of energy transformations and the conservation of energy principle. Granted, entropy is not a household word like energy. But with continued use, our understanding of entropy will deepen, and our appreciation of it will grow. The discussion below will shed some light on the physical meaning of entropy by considering the microscopic nature of matter.

Entropy can be viewed as a measure of *molecular disorder,* or *molecular randomness.* As a system becomes more disordered, the positions of the molecules become less predictable and the entropy increases. Thus, it is not surprising that the entropy of a substance is lowest in the solid phase and highest in the gas phase (Fig. 6-11). In the solid phase, the molecules of a substance continually oscillate about their equilibrium positions, but they cannot move relative to each other, and their position at any instant can be predicted with good certainty. In the gas phase, however, the molecules move about at random, collide with each other, and change direction, making it extremely difficult to predict accurately the microscopic state of a system at any instant. Associated with this molecular chaos is a high value of entropy.

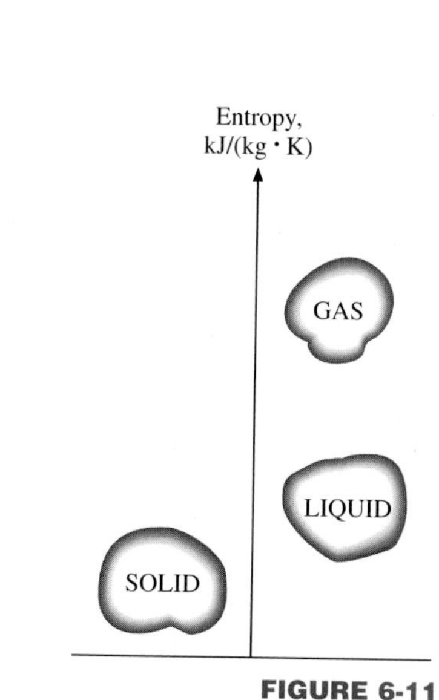


**FIGURE 6-11**
The level of molecular disorder (entropy) of a substance increases as it melts or evaporates.

When viewed microscopically (from a statistical thermodynamics point of view), an isolated system that appears to be at a state of equilibrium may exhibit a high level of activity because of the continual motion of the molecules. To each state of macroscopic equilibrium there corresponds a large number of possible microscopic states or molecular configurations. The entropy of a system is related to the total number of possible microscopic states of that system, called *thermodynamic probability p,* by the **Boltzmann relation** expressed as

$$S = k \ln p$$

where $k = 1.3806 \times 10^{-23}\ \text{kJ/(kmol} \cdot \text{K)}$ is the Boltzmann constant. Therefore, from a microscopic point of view, the entropy of a system increases whenever the molecular randomness or uncertainty (i.e., molecular probability) of a system increases. Thus, entropy is a measure of molecular disorder, and the molecular disorder of an isolated system increases anytime it undergoes a process.

Molecules in the gas phase possess a considerable amount of kinetic energy. But we know that no matter how large their kinetic energies are, the gas molecules will not rotate a paddle wheel inserted into the container and produce work. This is because the gas molecules, and the energy they carry with them, are disorganized. Probably the number of

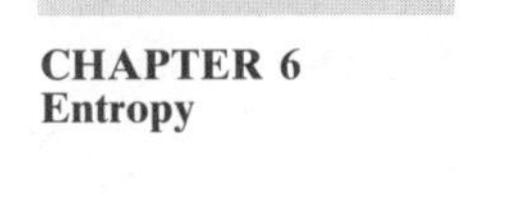

molecules trying to rotate the wheel in one direction at any instant is equal to the number of molecules that are trying to rotate it in the opposite direction, causing the wheel to remain motionless. Therefore, we cannot extract any useful work directly from disorganized energy (Fig. 6-12).

**FIGURE 6-12**
Disorganized energy does not create much useful effect, no matter how large it is.

Now consider a rotating shaft shown in Fig. 6-13. This time, the energy of the molecules is completely organized since the molecules of the shaft are rotating in the same direction together. This organized energy can readily be used to perform useful tasks such as raising a weight or generating electricity. Being an organized form of energy, work is free of disorder or randomness and thus free of entropy. *There is no entropy transfer associated with energy transfer as work.* Therefore, in the absence of any friction, the process of raising a weight by a rotating shaft (or a flywheel) will not produce any entropy. Any process that does not produce a net entropy is reversible, and thus the process described above can be reversed by lowering the weight. Therefore, energy is not degraded during this process, and no potential to do work is lost.

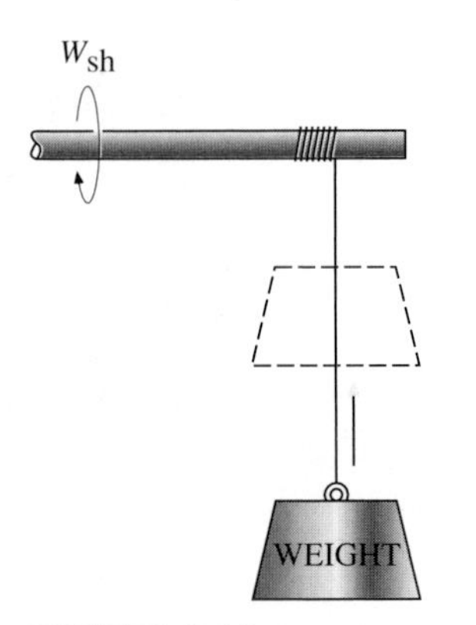


**FIGURE 6-13**
In the absence of friction, raising a weight by a rotating shaft does not create any disorder (entropy), and thus energy is not degraded during this process.

Instead of raising a weight, let us operate the paddle wheel in a container filled with a gas, as shown in Fig. 6-14. The paddle-wheel work in this case will be converted to the internal energy of the gas, as evidenced by a rise in gas temperature, creating a higher level of molecular chaos and disorder in the container. This process is quite different from raising a weight since the organized paddle-wheel energy is now converted to a highly disorganized form of energy, which cannot be converted back to the paddle wheel as the rotational kinetic energy. Only a portion of this energy can be converted to work by partially reorganizing it through the use of a heat engine. Therefore, energy is degraded during this process, the ability to do work is reduced, molecular disorder is produced, and associated with all this is an increase in entropy.

The *quantity* of energy is always preserved during an actual process (the first law), but the *quality* is bound to decrease (the second law). This decrease in quality is always accompanied by an increase in entropy. As an example, consider the transfer of 10 kJ of energy as heat from a hot medium to a cold one. At the end of the process, we will still have the 10 kJ of energy, but at a lower temperature and thus at a lower quality.

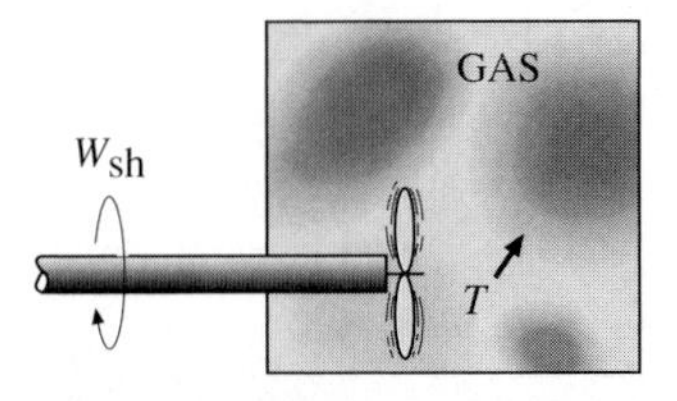


**FIGURE 6-14**
The paddle-wheel work done on a gas increases the level of disorder (entropy) of the gas, and thus energy is degraded during this process.

Heat is, in essence, a form of disorganized energy, and some disorganization (entropy) will flow with heat (Fig. 6-15). As a result, the entropy and the level of molecular disorder or randomness of the hot body will decrease with the entropy and the level of molecular disorder of the cold body increase. The second law requires that the increase in entropy of the cold body be greater than the decrease in entropy of the hot body, and thus the net entropy of the combined system (the cold body and the hot body) increases. That is, the combined system is at a state of greater disorder at the final state. Thus we can conclude that processes can occur only in the direction of increased overall entropy or molecular disorder. That is, the entire universe is getting more and more chaotic every day. This is a major concern not only to engineers but also to philosophers.

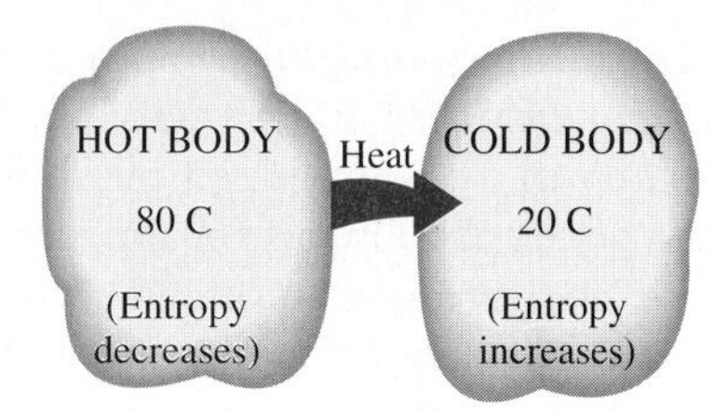


**FIGURE 6-15**

During a heat transfer process, the net disorder (entropy) increases. (The increase in the disorder of the cold body more than offsets the decrease in the disorder of the hot body.)

From a statistical point of view, entropy is a measure of molecular randomness, i.e., the uncertainty about the positions of molecules at any instant. Even in the solid phase, the molecules of a substance continually oscillate, creating an uncertainty about their position. These oscillations, however, fade as the temperature is decreased, and the molecules become completely motionless at absolute zero. This represents a state of ultimate molecular order (and minimum energy). Therefore, *the entropy of a pure crystalline substance at absolute zero temperature is zero* since there is no uncertainty about the state of the molecules at that instant (Fig. 6-16). This statement is known as the **third law of thermodynamics**. The third law of thermodynamics provides an absolute reference point for the determination of entropy. The entropy determined relative to this point is called **absolute entropy**, and it is extremely useful in the thermodynamic analysis of chemical reactions. Notice that the entropy of a substance that is not pure crystalline (such as a solid solution) is not zero at absolute zero temperature. This is because more than one molecular configuration exists for such substances, which introduces some uncertainty about the microscopic state of the substance.

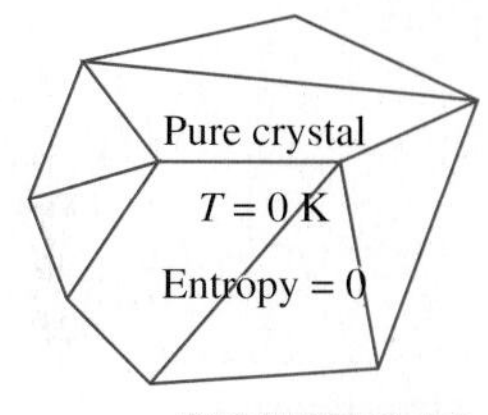


**FIGURE 6-16**

A pure substance at absolute zero temperature is in perfect order, and its entropy is zero (the third law of thermodynamics).

The concept of entropy as a measure of disorganized energy can also be applied to other areas. Iron molecules, for example, create a magnetic field around themselves. In ordinary iron, molecules are randomly aligned, and they cancel each other's magnetic effect. When iron is treated and the molecules are realigned, however, that piece of iron turns into a piece of magnet, creating a powerful magnetic field around it.

## Entropy and Entropy Generation in Daily Life

Entropy can be viewed as a measure of disorder or disorganization in a system. Likewise, entropy generation can be viewed as a measure of disorder or disorganization generated during a process. The concept of entropy is not used in daily life nearly as extensively as the concept of energy, even though entropy is readily applicable to various aspects of daily life. The extension of the entropy concept to nontechnical fields is not a novel idea. It has been the topic of several articles, and even some books. Below we present several ordinary events and show their relevance to the concept of entropy and entropy generation.

Efficient people lead low-entropy (highly organized) lives. They have a place for everything (minimum uncertainty), and it takes minimum energy for them to locate something. Inefficient people, on the other hand, are disorganized and lead high-entropy lives. It takes them minutes (if not hours) to find something they need, and they are likely to create a bigger disorder as they are searching since they will probably conduct the search in a disorganized manner (Fig. 6-17). People leading high-entropy life styles are always on the run, and never seem to catch up.

DO YOU REALIZE YOU ARE INCREASING THE ENTROPY EVERY TIME YOU DO SOMETHING IN THIS HOUSE??


**FIGURE 6-17**
The use of entropy (disorganization, uncertainty) is not limited to thermodynamics.

You probably noticed (with frustration) that some people seem to learn fast and remember what they learn well. We can call this type of learning organized or low-entropy learning. These people make a conscientious effort to file the new information properly by relating it to their existing knowledge base and creating a solid information network in their minds. On the other hand, people who throw the information into their minds as they study, with no effort to secure it, may *think* they are learning. They are bound to discover otherwise when they need to locate the information, for example during a test. It is not easy to retrieve information from a data base which is, in a sense, in the gas phase. Students who have blackouts during the tests should reexamine their study habits.

A library with a good shelving and indexing system can be viewed as a low-entropy library because of the high level of organization. Likewise, a library with a poor shelving and indexing system can be viewed as a high-entropy library because of the high level of disorganization. A library with no indexing system is like no library, since a book is of no value if it cannot be found.

Consider two identical buildings, each containing identical copies of one million books. In the first building, the books are *piled* on top of each other whereas in the second building they are *highly organized, shelved, and indexed* for easy reference. Probably there is no doubt about which building a student will go to check out a certain book. Yet, some may argue from the first law point of view that these two buildings are equivalent since the mass and energy content of the two buildings are identical, despite the high level of disorganization (entropy) in the first building. This example illustrates that any realistic comparisons should involve the second law point of view.

Two textbooks that seem to be identical because both cover basically the same topics and present the same information may actually be *very* different on *how* they cover the topics. After all, two seemingly identical cars are not so identical if one goes only half as many miles as the other one on the same amount of fuel. Likewise, two seemingly identical books are not so identical if it takes twice as long to learn a topic from one of them than it does from the other. Thus, comparisons made on the basis of the first law only may be highly misleading.

Having a disorganized (high-entropy) army is like having no army at all. It is no coincidence that command centers of any armed forces are among the primary targets during a war. One army that consists of ten divisions is ten times more powerful than ten armies each consisting of a single division. Likewise, one country that consists of ten states is more powerful than ten countries, each consisting of a single state. The United States would not be such a powerful country if there were fifty independent countries in its place instead of a single country with fifty states. The new European common market has the potential to be a new economic super power. The old cliché "divide and conquer" can be rephrased as "increase the entropy and conquer."

We know that mechanical friction is always accompanied by entropy generation, and thus reduced performance. We can generalize this to daily life: friction in the work place with fellow workers is bound to generate entropy, and thus adversely affect performance. It will result in reduced productivity. Hopefully, someday we will be able to come up with some procedures to quantify entropy generated during nonmechanical activities, and maybe even pinpoint its primary sources and magnitude.

We also know that unrestrained expansion (or explosion) and uncontrolled electron exchange (chemical reactions) generate entropy and are highly irreversible. Likewise, unrestrained opening of the mouth to scatter angry words is highly irreversible since this generates entropy, and it can cause considerable damage. A person who gets up in anger is bound to sit down at a loss.

## 6-6 THE $T$-$s$ DIAGRAM

Property diagrams serve as great visual aids in the thermodynamic analysis of processes. We have used $P$-$v$ and $T$-$v$ diagrams extensively in previous chapters in conjunction with the first law of thermodynamics. In the second-law analysis, it is very helpful to plot the processes on diagrams for which one of the coordinates is entropy. The two diagrams used most extensively in the second-law analysis are the *temperature–entropy* and the *enthalpy–entropy* diagrams.

Consider the defining equation of entropy (Eq. 6-4). It can be rearranged as

$$\delta Q_{\text{int rev}} = T\,dS \qquad \text{(kJ)} \qquad (6\text{-}20)$$

As shown in Fig. 6-18, $\delta Q_{\text{rev}}$ corresponds to a differential area on a $T$-$S$ diagram. The total heat transfer during an internally reversible process is determined by integration to be

$$Q_{\text{int rev}} = \int_1^2 T\,dS \qquad \text{(kJ)} \qquad (6\text{-}21)$$

which corresponds to the area under the process curve on a $T$-$S$ diagram. Therefore, we conclude that *the area under the process curve on a $T$-$S$ diagram represents the internally reversible heat transfer.* This is somewhat analogous to reversible boundary work being represented by the area under the process curve on a $P$-$V$ diagram. Note that the area under the process curve represents heat transfer for processes that are internally (or totally) reversible. It has no meaning for irreversible processes.

Equations 6-20 and 6-21 can also be expressed on a unit-mass basis as

$$\delta q_{\text{int rev}} = T\,ds \qquad \text{(kJ/kg)} \qquad (6\text{-}22a)$$

and

$$q_{\text{int rev}} = \int_1^2 T\,ds \qquad \text{(kJ/kg)} \qquad (6\text{-}22b)$$

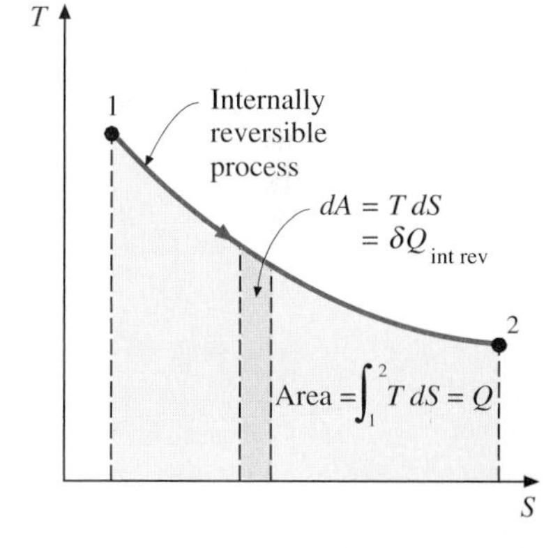

**FIGURE 6-18**
On a $T$-$S$ diagram, the area under the process curve represents the heat transfer for internally reversible processes.

To perform the integrations in Eqs. 6-21 and 6-22*b*, one needs to know the relationship between $T$ and $s$ during a process. One special case for which these integrations can be performed easily is the *internally reversible isothermal process*. It yields

$$Q_{\text{int rev}} = T_0\,\Delta S \qquad \text{(kJ)}$$

or

$$q_{\text{int rev}} = T_0\,\Delta s \qquad \text{(kJ/kg)}$$

where $T_0$ is the constant temperature and $\Delta S$ is the entropy change of the system during the process.

In the relations above, $T$ is the absolute temperature, which is always positive. Therefore, heat transfer during internally reversible processes is positive when entropy increases and negative when entropy decreases. An isentropic process on a $T$-$s$ diagram is easily recognized as a vertical-line segment. This is expected since an isentropic process involves no heat transfer, and therefore the area under the process path must be zero (Fig. 6-19). The $T$-$s$ diagrams serve as valuable tools for visualizing the second-law aspects of processes and cycles, and thus they are frequently used in thermodynamics. The $T$-$s$ diagram of water is given in the Appendix in Fig. A-7.

The general characteristics of a $T$-$s$ diagram for the liquid and vapor regions of a pure substance are shown in Fig. 6-20. The following

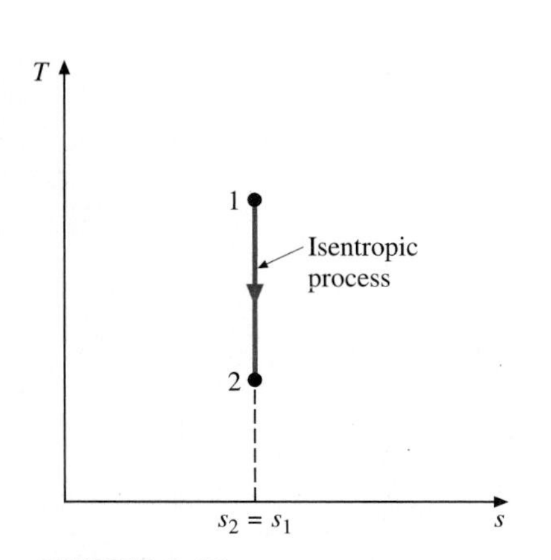


**FIGURE 6-19**
During an internally reversible, adiabatic (isentropic) process, the entropy of a system remains constant.

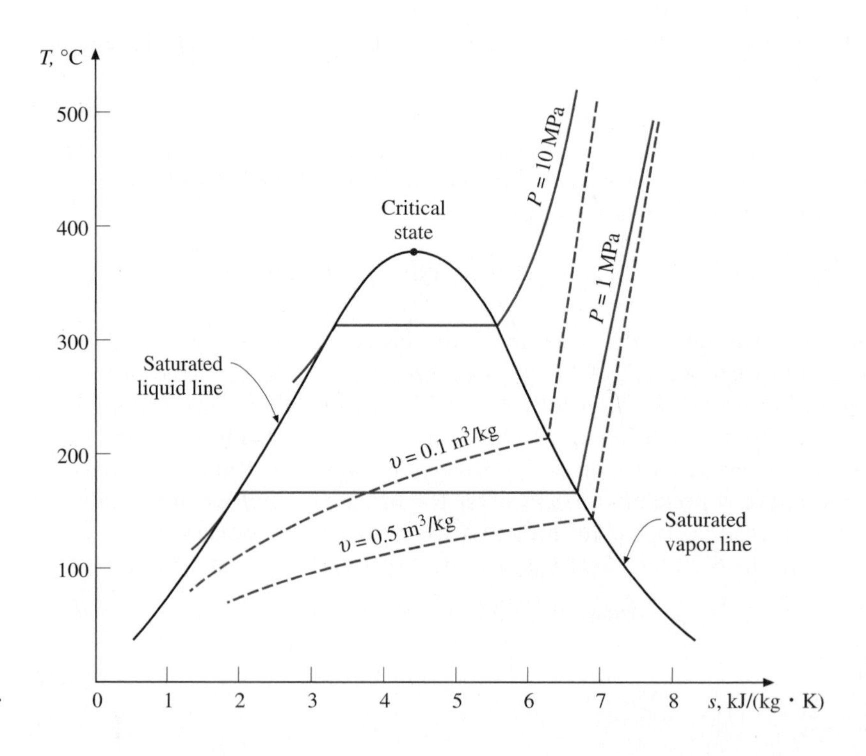


**FIGURE 6-20**
Schematic of a $T$-$s$ diagram for water.

observations can be made:



1 At any point in a single-phase region, the constant-volume lines are steeper than the constant-pressure lines.

2 In the saturated liquid-mixture region, the constant-pressure lines are parallel to the constant-temperature lines.

3 In the compressed liquid region, the constant-pressure lines almost coincide with the saturated liquid line.

**EXAMPLE 6-2**

Show the Carnot cycle on a *T-S* diagram and indicate the areas that represent the heat added $Q_H$, heat rejected $Q_L$, and the net work output $W_{net.\,out}$ on this diagram.

**Solution** You will recall from Chap. 5 that the Carnot cycle is made up of two reversible isothermal ($T$ = constant) processes and two isentropic ($s$ = constant) processes. These four processes form a rectangle on a *T-S* diagram, as shown in Fig. 6-21.

On a *T-S* diagram, the area under the process curve rrepresents the heat transfer for that process. Thus the area *A*12*B* represents $Q_H$, the area *A*43*B* represents $Q_L$, and the difference between these two (the area in color) represents the net work since

$$W_{net,\,out} = Q_H - Q_L$$

Therefore, the area enclosed by the path of a cycle (area 1234) on a *T-S* diagram represents the net work. Recall from Chap. 3 that the area enclosed by the cycle also represents the net work on a *P-V* diagram.

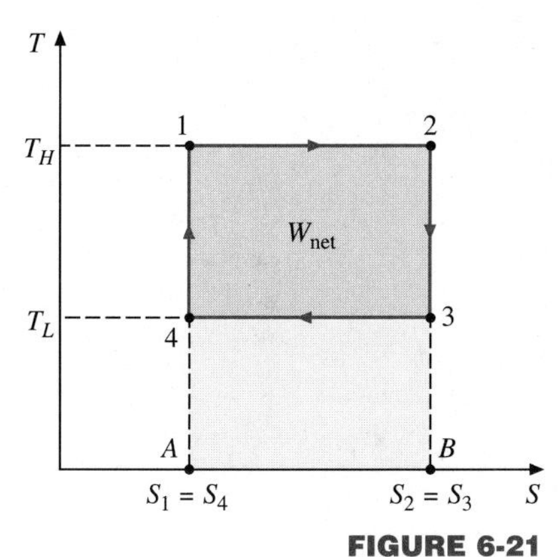


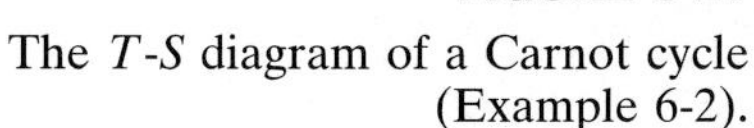

**FIGURE 6-21**
The *T-S* diagram of a Carnot cycle (Example 6-2).

## 6-7 EVALUATION OF THE ENTROPY CHANGE

Earlier in this chapter, it was shown that the quantity $(\delta Q/T)_{int\,rev}$ corresponds to a differential change in a property, called entropy. The entropy change for a process, then, was evaluated by integrating $\delta Q/T$ along some imaginary internally reversible path between the actual end states. For isothermal internally reversible processes, this integration is straightforward. But when the temperature varies during the process, we have to have a relation between $\delta Q$ and $T$ to perform this integration. Finding such relations is what we intend to do in this section.

The differential form of the conservation of energy equation for a closed stationary system (a fixed mass) containing a simple compressible substance can be expressed for an internally reversible process as

$$\delta Q_{int\,rev} - \delta W_{int\,rev} = dU$$

But

$$\delta Q_{int\,rev} = T\,dS$$

$$\delta W_{int\,rev} = P\,dV$$

Thus,

$$T\,dS = dU + P\,dV$$

or

$$T\,ds = du + P\,dv \tag{6-23}$$

per unit mass. This equation is known as the first *T ds,* or *Gibbs, equation.* Notice that the only type of work interaction a simple compressible system may involve as it undergoes an internally reversible process is the quasi-equilibrium boundary work.

The second *T ds* equation is obtained by eliminating *du* from Eq. 6-23 by using the definition of enthalpy ($h = u + Pv$):

$$\left.\begin{aligned} h = u + Pv &\longrightarrow \quad dh = du + P\,dv + v\,dP \\ \text{Eq. 6-23} &\longrightarrow \quad T\,ds = du + P\,dv \end{aligned}\right\} \qquad T\,ds = dh - v\,dP \tag{6-24}$$

Equations 6-23 and 6-24 are extremely valuable since they relate entropy changes of a system to the changes in other properties. Unlike Eq. 6-4, they are property relations and therefore are independent of the type of the processes.

The *T ds* relations above are developed with an internally reversible process in mind since the entropy change between two states must be evaluated along a reversible path. But the results obtained are valid for both reversible and irreversible processes since entropy is a property and the change in a property between two states is independent of the type of process the system undergoes. Equations 6-23 and 6-24 are relations between the properties of a unit mass of a simple compressible system as it undergoes a change of state, and they are applicable whether the change occurs in a closed or an open system (Fig. 6-22).

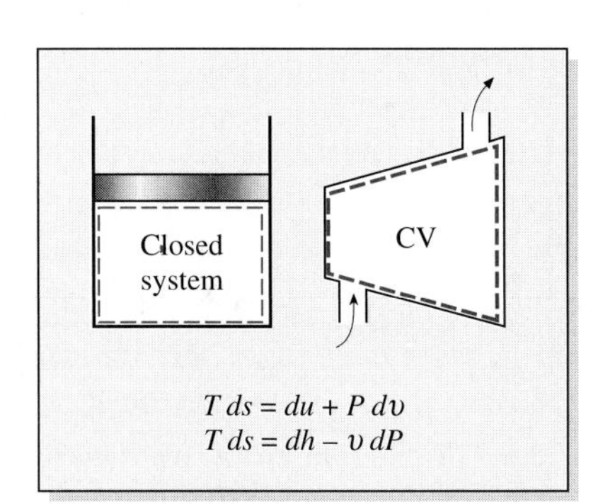


**FIGURE 6-22**
The *T ds* relations are valid for both reversible and irreversible processes and for both closed and open systems.

Explicit relations for differential changes in entropy are obtained by solving for *ds* in Eqs. 6-23 and 6-24:

$$ds = \frac{du}{T} + \frac{P\,dv}{T} \tag{6-25}$$

and 

$$ds = \frac{dh}{T} - \frac{v\,dP}{T} \tag{6-26}$$

The entropy change during a process can be determined by integrating either of these equations between the initial and the final states. To perform these integrations, however, we must know the relationship between *du* or *dh* and the temperature (such as $du = C_p\,dT$ and $dh = C_p\,dT$ for ideal gases) as well as the equation of state for the substance (such as the ideal-gas equation of state $Pv = RT$). For substances for which such relations exist, such as ideal gases and incompressible substances, the integration of Eq. 6-25 or 6-26 is straightforward. This is done later in this chapter. For other substances we have to rely on tabulated data.

The *T ds* relations for nonsimple systems, i.e., systems that involve more than one mode of quasi-equilibrium work interactions, can be obtained in a similar fashion by including all the relevant quasi-equilibrium work modes.

The *T ds* relations developed above are not limited to a particular substance in a particular phase. They are valid for all pure substances at any phase or combination of phases. The successful use of these relations, however, depends on the availability of the property relations between $T$ and $du$ or $dh$ and $P$-$v$-$T$ behavior of the substance. For a pure substance, in general, these relations are too complicated, and this makes it impossible to obtain simple relations for entropy changes. The values of $s$, therefore, are determined from measurable property data following rather involved computations and are tabulated or plotted in exactly the same manner as the properties $v$, $u$, and $h$.

The entropy values in the property tables or charts are given relative to an arbitrary reference state. In steam tables, the entropy of saturated liquid $s_f$ at 0.01°C is assigned the value of zero. For refrigerant-134a, the zero value is assigned to saturated liquid at −40°C. The entropy values become negative at temperatures below the reference value.

The entropy change of a pure substance during a process is simply the difference between the entropy values at the final and initial states:

$$\Delta S = m(s_2 - s_1) \qquad \text{(kJ/K)} \tag{6-27}$$

or

$$\Delta s = s_2 - s_1 \qquad [\text{kJ/(kg}\cdot\text{K)}] \tag{6-28}$$

Equation 6-28 is applicable to a closed system as well as to a unit mass passing through a control volume.

## Isentropic Processes

We pointed out earlier that two factors can change the entropy of a fixed mass: heat transfer and irreversibilities. Then it follows that the entropy of a fixed mass will not change during an internally reversible, adiabatic process, which is called an *isentropic* (constant-entropy) *process.* An isentropic process appears as a vertical line on a $T$-$s$ diagram.

Many engineering systems or devices such as pumps, turbines, nozzles, and diffusers are essentially adiabatic in their operation, and they perform best when the irreversibilities, such as the friction associated with the process, are minimized. Therefore, an isentropic process can serve as an appropriate model for actual processes. Also isentropic processes enables us to define efficiencies for processes to compare the actual performance of these devices to the performance under idealized conditions. This should be sufficient motivation for studying the isentropic processes.

No special relations exist for the isentropic processes of pure substances other than

$$s_2 = s_1 \qquad [\text{kJ/(kg}\cdot\text{K)}] \tag{6-29}$$

except for some idealized cases that are discussed in the next two sections. A substance will have the same entropy value at the final state as it does at the initial state if the process is carried out in an isentropic manner.

It should be recognized that a reversible adiabatic process is necessarily isentropic ($s_2 = s_1$) but an isentropic process is not necessarily reversible adiabatic. (The entropy increase of a substance during a process as a result of irreversibilities may be offset by a decrease in entropy as a result of heat losses, for example.) However, the term *isentropic process* is customarily used in thermodynamics to imply an *internally reversible, adiabatic process.*

## 6-8 THE ENTROPY CHANGE OF SOLIDS AND LIQUIDS

We mentioned in Sec. 3-8 that solids and liquids can be idealized as *incompressible substances* since their volumes remain essentially constant during a process. Thus, $dv = 0$ for solids and liquids, and Eq. 6-25 for this case reduces to

$$ds = \frac{du}{T} = \frac{C\,dT}{T} \tag{6-30}$$

since $C_p = C_v = C$ for incompressible substances and $du = C\,dt$. The entropy change for a process is determined by integration:

$$s_2 - s_1 = \int_1^2 C(T)\frac{dT}{T} \qquad [\text{kJ/(kg}\cdot\text{K)}] \tag{6-31}$$

The specific heat $C$ of liquids and solids, in general, depends on temperature, and we need a relation for $C$ as a function of temperature to perform this integration. In many cases, however, $C$ may be treated as a constant at some average value over the given temperature range. Then the integration in Eq. 6-31 can be performed, to yield

$$s_2 - s_1 = C_{av} \ln\frac{T_2}{T_1} \qquad [\text{kJ/(kg}\cdot\text{K)}] \tag{6-32}$$

Note that the entropy change of a truly incompressible substance depends on temperature only. Equation 6-32 can be used to determine the entropy changes of solids and liquids with reasonable accuracy.

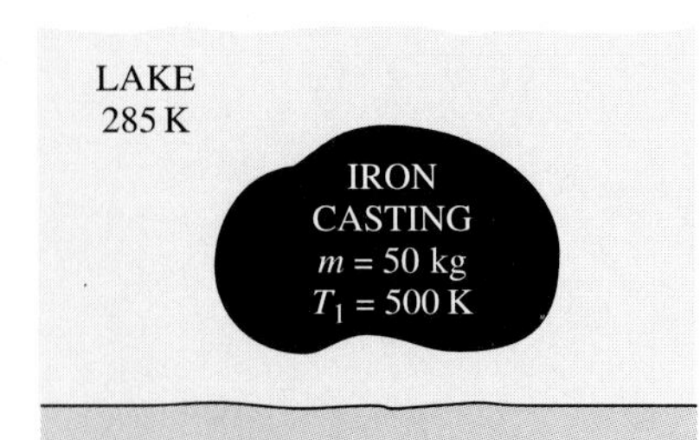


**FIGURE 6-23**
Schematic for Example 6-3.

**EXAMPLE 6-3**

A 50-kg block of iron casting at 500 K is thrown into a large lake that is at a temperature of 285 K, as shown in Fig. 6-23. The iron block eventually reaches thermal equilibrium with the lake water. Assuming an average specific heat of 0.45 kJ/(kg · K) for the iron, determine (*a*) the entropy change of the iron block, (*b*) the entropy change of the lake water, and (*c*) the total entropy change for this process.

**Solution** To determine the entropy change for the iron block and for the lake, first we need to know the final equilibrium temperature. Given that the thermal energy capacity of the lake is very large relative to that of the iron block, the lake will absorb all the heat rejected by the iron block without experiencing any change in its temperature. Therefore, the iron block will cool to 285 K during this process while the lake temperature remains constant at 285 K. Then the entropy changes are determined as follows:

(*a*) Like all solids, the iron block can be approximated as an incompressible substance, and thus its entropy change can be determined from Eq. 6-32:

$$\begin{aligned}\Delta S_{\text{iron}} &= m(s_2 - s_1) = mC_{\text{av}} \ln \frac{T_2}{T_1} \\ &= (50\text{ kg})[0.45\text{ kJ/(kg}\cdot\text{K)}] \ln \frac{285\text{ K}}{500\text{ K}} \\ &= -12.65\text{ kJ/K}\end{aligned}$$

(*b*) The lake water in this problem acts as a thermal energy reservoir, and its entropy change can be determined from Eq. 6-6. But first we need to determine the heat transfer to the lake. Taking the iron block as our system and disregarding the changes in the kinetic and potential energies, we see that the conservation of energy equation for this closed system reduces to

$$Q - \cancelto{0}{W} = \Delta U + \cancelto{0}{\Delta KE} + \cancelto{0}{\Delta PE}$$

$$\begin{aligned}Q_{\text{iron}} &= mC_{\text{av}}(T_2 - T_1) = (50\text{ kg})[0.45\text{ kJ/(kg}\cdot\text{K)}][(285 - 500)\text{ K}] \\ &= -4837.5\text{ kJ}\end{aligned}$$

Then $$Q_{\text{lake}} = -Q_{\text{iron}} = +4837.5\text{ kJ}$$

and $$\Delta S_{\text{lake}} = \frac{Q_{\text{lake}}}{T_{\text{lake}}} = \frac{+4837.5\text{ kJ}}{285\text{ K}} = 16.97\text{ kJ/K}$$

(*c*) The total entropy change for this process is the sum of these two since the iron block and the lake together form an adiabatic system:

$$\begin{aligned}\Delta S_{\text{total}} &= \Delta S_{\text{iron}} + \Delta S_{\text{lake}} = (-12.65 + 16.97)\text{ kJ/K} \\ &= 4.32\text{ kJ/K}\end{aligned}$$

The positive sign for the total entropy change indicates that this is an irreversible process.

---

### Isentropic Processes of Solids and Liquids

A relation for isentropic (constant-entropy) processes of solids and liquids is obtained by setting the entropy-change relation (Eq. 6-32) equal to zero:

$$C_{\text{av}} \ln \frac{T_2}{T_1} = 0 \qquad (6\text{-}33)$$

which gives $$T_2 = T_1 \qquad (6\text{-}34)$$

That is, the temperature of a truly incompressible substance remains constant during an isentropic process. Therefore, the isentropic process of an incompressible substance is also isothermal. This behavior is closely approximated by solids and liquids.

**FIGURE 6-24**

A broadcast from channel IG.

## 6-9 ■ THE ENTROPY CHANGE OF IDEAL GASES

An expression for the entropy change of an ideal gas can be obtained from Eq. 6-25 or 6-26 by employing the property relations for ideal gases (Fig. 6-24). By substituting $du = C_v\, dT$ and $P = RT/v$ into Eq. 6-25, the differential entropy change of an ideal gas becomes

$$ds = C_v \frac{dT}{T} + R \frac{dv}{v}$$

The entropy change for a process is obtained by integrating this relation between the end states:

$$s_2 - s_1 = \int_1^2 C_v(T) \frac{dT}{T} + R \ln \frac{v_2}{v_1} \tag{6-35}$$

A second relation for the entropy change of an ideal gas is obtained in a similar manner by substituting $dh = C_p\, dT$ and $v = RT/P$ into Eq. 6-26 and integrating. The result is

$$s_2 - s_1 = \int_1^2 C_p(T) \frac{dT}{T} - R \ln \frac{P_2}{P_1} \tag{6-36}$$

The specific heats of ideal gases, with the exception of monatomic gases, depend on temperature, and the integrals in Eqs. 6-35 and 6-36 cannot be performed unless the dependence of $C_v$ and $C_p$ on temperature is known. Even when the $C_v(T)$ and $C_p(T)$ functions are available, performing long integrations every time entropy change is calculated is not practical. Then two reasonable choices are left: either perform these integrations by simply assuming constant specific heats, or evaluate those integrals once and tabulate the results.

Assuming constant specific heats for ideal gases is a common approximation, and we used this assumption before on several occasions. It usually simplifies the analysis greatly, and the price we pay for this convenience is some loss in accuracy. The magnitude of the error introduced by this assumption depends on the situation on hand. For example, for monatomic ideal gases such as helium, the specific heats are independent of temperature, and therefore the constant-specific-heat assumption introduces no error. For ideal gases whose specific heats vary almost linearly in the temperature range of interest, the possible error is minimized by using specific-heat values evaluated at the average temperature (Fig. 6-25). The results obtained in this way usually are sufficiently accurate for most ideal gases if the temperature range is not greater than a few hundred degrees.

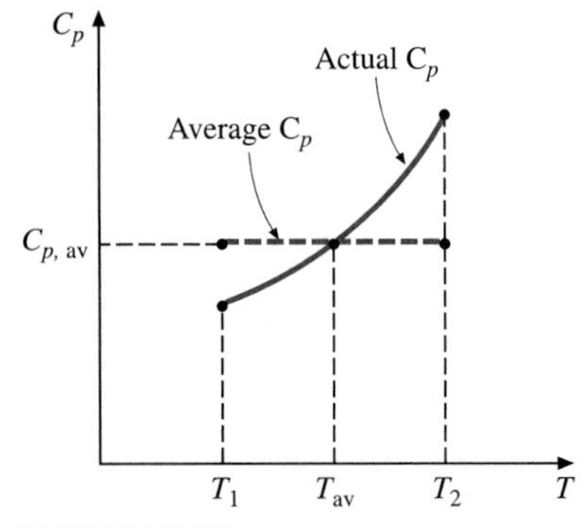


**FIGURE 6-25**

Under the constant-specific-heat assumption, the specific heat is assumed to be constant at some average value.

The entropy-change relations for ideal gases under the constant-specific-heat assumption are easily obtained by replacing $C_v(T)$ and $C_p(T)$ in Eqs. 6-35 and 6-36 by $C_{v,\,av}$ and $C_{p,\,av}$, respectively, and performing the integrations. We obtain

$$s_2 - s_1 = C_{v,\,av} \ln \frac{T_2}{T_1} + R \ln \frac{v_2}{v_1} \qquad [\text{kJ/(kg} \cdot \text{K)}] \tag{6-37}$$

and

$$s_2 - s_1 = C_{p,\,av} \ln \frac{T_2}{T_1} - R \ln \frac{P_2}{P_1} \qquad [\text{kJ/(kg} \cdot \text{K)}] \tag{6-38}$$

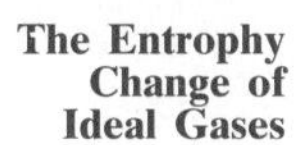

**EXAMPLE 6-4**
Nitrogen gas is compressed from an initial state of 100 kPa and 17°C to a final state of 600 kPa and 57°C. Determine the entropy change of the nitrogen during this compression process by using average specific heats.

**Solution** A sketch of the system and the *T-s* diagram for the process are given in Fig. 6-26. At specific conditions, nitrogen can be treated as an ideal gas since it is at a high temperature and low pressure relative to its critical values ($T_{cr} = -147°C$ and $P_{cr} = 3390$ kPa for nitrogen). Therefore, the entropy-change relations developed under the ideal-gas assumption are applicable.

The entropy change of the nitrogen during this process can also be determined approximately from Eq. 6-38 by using a $C_p$ value at the average temperature of 37°C (Table A-2*b*) and treating it as a constant:

$$s_2 - s_1 = C_{p,\text{av}} \ln \frac{T_2}{T_1} - R \ln \frac{P_2}{P_1}$$

$$= [1.0394 \text{ kJ/(kg} \cdot \text{K)}] \ln \frac{330 \text{ K}}{290 \text{ K}} - [0.297 \text{ kJ/(kg} \cdot \text{K)}] \ln \frac{600 \text{ kPa}}{100 \text{ kPa}}$$

$$= -0.3978 \text{ kJ/(kg} \cdot \text{K)}$$

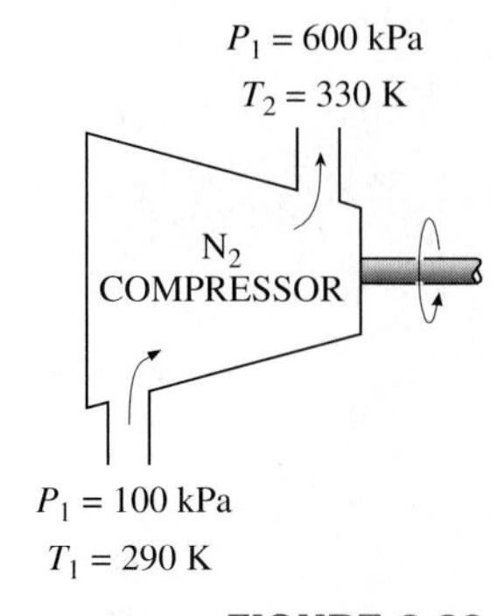


**FIGURE 6-26**
Schematic for Example 6.4.

## Isentropic Processes of Ideal Gases

Several relations for the isentropic processes of ideal gases can be obtained by setting the entropy-change relations developed above equal to zero.

When the constant-specific-heat assumption is valid, the isentropic relations for ideal gases are obtained by setting Eqs. 6-37 and 6-38 equal to zero. From Eq. 6-37,

$$\ln \frac{T_2}{T_1} = -\frac{R}{C_v} \ln \frac{v_2}{v_1} \tag{6-39}$$

which can be rearranged as

$$\ln \frac{T_2}{T_1} = \ln \left(\frac{v_1}{v_2}\right)^{R/C_v} \tag{6-40}$$

or

$$\left(\frac{T_2}{T_1}\right)_{s=\text{const.}} = \left(\frac{v_1}{v_2}\right)^{k-1} \tag{6-41}$$

Since $R = C_p - C_v$, $k = C_p/C_v$, and thus $R/C_v = k - 1$.

Equation 6-41 is the *first isentropic relation* for ideal gases under the constant-specific-heat assumption. The *second isentropic relation* is obtained in a similar manner from Eq. 6-38 with the following result:

$$\left(\frac{T_2}{T_1}\right)_{s=\text{const.}} = \left(\frac{P_2}{P_1}\right)^{(k-1)/k} \tag{6-42}$$

The *third isentropic relation* is obtained by substituting Eq. 6-42 into Eq. 6-41 and simplifying:

$$\left(\frac{P_2}{P_1}\right)_{s=\text{const.}} = \left(\frac{v_1}{v_2}\right)^{k} \tag{6-43}$$

Equations 6-41 through 6-43 can also be expressed in a compact form as

$$Tv^{k-1} = \text{constant} \tag{6-44}$$

$$TP^{(1-k)/k} = \text{constant} \tag{6-45}$$

$$Pv^{k} = \text{constant} \tag{6-46}$$

The specific heat ratio $k$, in general, varies with temperature, and so in the isentropic relations above an average $k$ value for the given temperature range should be used.

Note that the isentropic relations above, as the name implies, are strictly valid for isentropic processes only when the constant-specific-heat assumption is appropriate (Fig. 6-27).

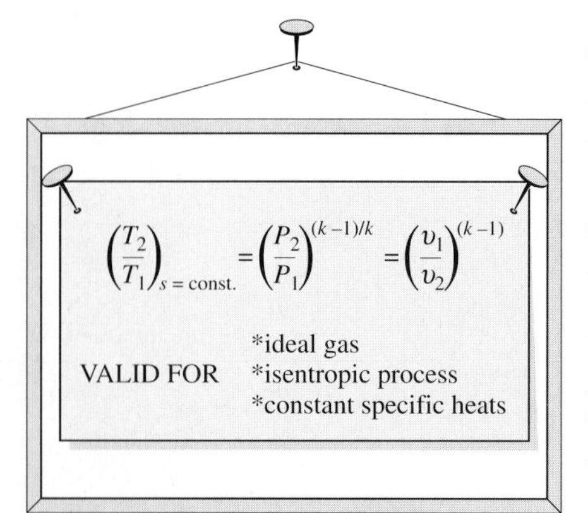


**FIGURE 6-27**
The isentropic relations of ideal gases are valid for the isentropic processes of ideal gases only.

**EXAMPLE 6-5**

Air is compressed in an adiabatic piston–cylinder device from 22°C and 95 kPa in a reversible manner. If the compression ratio $V_1/V_2$ of this piston–cylinder device is 8, determine the final temperature of the air.

**Solution** A sketch of the system and the *T-s* diagram for the process are given in Fig. 6-28. At specified conditions, air can be treated as an *ideal gas* since it is at a high temperature and low pressure relative to its critical values ($T_{cr} = -147°C$ and $P_{cr} = 3390$ kPa for nitrogen, the main constituent of air). Therefore, the isentropic relations developed above for ideal gases are applicable.

This process is easily recognized as isentropic since it is both reversible and adiabatic. The final temperature for this isentropic process can be determined from Eq. 6-41 by assuming constant specific heats for the air:

$$\left(\frac{T_2}{T_1}\right)_{s=\text{const.}} = \left(\frac{v_1}{v_2}\right)^{k-1}$$

The specific heat ratio $k$ also varies with temperature, and we need to use the value of $k$ corresponding to the average temperature. However, the final temperature is not given, and so we cannot determine the average temperature in advance. For such cases, calculations can be started with a $k$ value at the

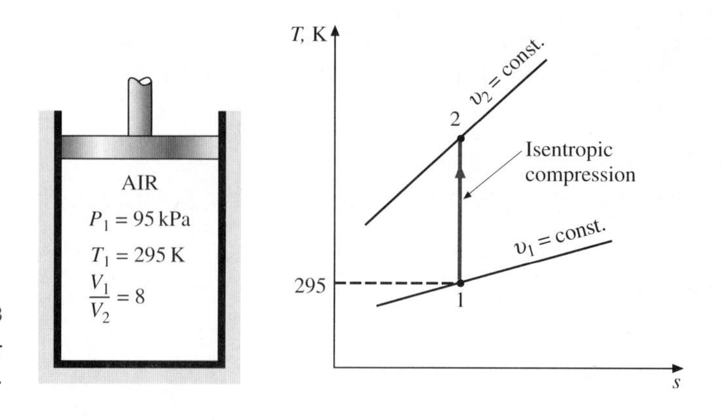


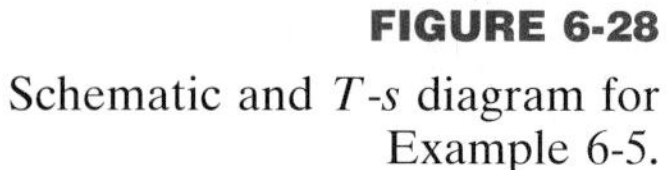

**FIGURE 6-28**
Schematic and *T-s* diagram for Example 6-5.

initial or the anticipated average temperature. This value could be refined later, if necessary, and the calculations can be repeated. We know that the temperature of the air will rise considerably during this adiabatic compression process, so we guess that the average temperature will be about 450 K. The $k$ value at this anticipated average temperature is determined from Table A-2*b* to be 1.391. Then the final temperature of air becomes

$$T_2 = (295\text{ K})(8)^{1.391-1} = 665.2\text{ K}$$

This will give an average temperature value of 480.1 K, which is sufficiently close to the assumed value of 450 K. Therefore, it is not necessary to repeat the calculations by using the $k$ value at this average temperature.

**EXAMPLE 6-6**

Helium gas is compressed in an adiabatic compressor from an initial state of 14 psia and 50°F to a final temperature of 320°F in a reversible manner. Determine the exit pressure of helium.

**Solution** A sketch of the system and the $T$-$s$ diagram for the process are given in Fig. 6-29. At the specified conditions, helium can be treated as an

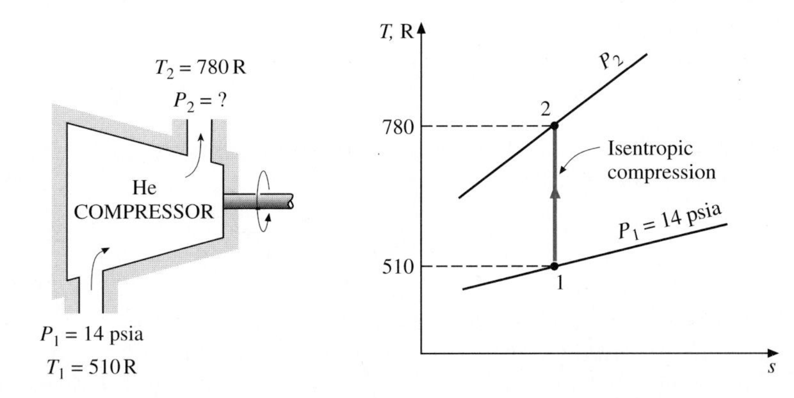


**FIGURE 6-29**
Schematic and $T$-$s$ diagram for Example 6-6.

*ideal gas* since it is at a very high temperature relative to its critical temperature ($T_{cr} = -450$°F for helium). Therefore, the specific relations developed above for ideal gases are applicable. The specific heat ratio $k$ of helium is 1.667 and is independent of temperature in the region where it behaves as an ideal gas. Thus the final pressure of helium can be determined from Eq. 6-42:

$$P_2 = P_1\left(\frac{T_2}{T_1}\right)^{k/(k-1)} = (14\text{ psia})\left(\frac{780\text{ R}}{510\text{ R}}\right)^{1.667/0.667} = 40.5\text{ psia}$$

## 6-10 ■ REVERSIBLE STEADY-FLOW WORK

Work and heat, in general, are path functions, and the heat transfer or the work done during a process depends on the path followed as well as on the properties at the end states. In Chap. 3, we discussed reversible (quasi-equilibrium) moving boundary work associated with closed systems and expressed it in terms of the fluid properties as:

$$W_b = \int_1^2 P\,dV$$

We mentioned that the quasi-equilibrium work interactions lead to the maximum work output for work-producing devices and the minimum work input for work-consuming devices.

It would also be very desirable and insightful to express the work associated with steady-flow devices in terms of fluid properties.

The conservation of energy equation for a steady-flow device undergoing an internally reversible process can be expressed in differential form as

$$\delta q_{\text{rev}} - \delta w_{\text{rev}} = dh + d\text{ke} + d\text{pe}$$

But

$$\left.\begin{aligned} \delta q_{\text{rev}} &= T\,ds \\ T\,ds &= dh - v\,dP \end{aligned}\right\} \quad \delta q_{\text{rev}} = dh - v\,dP$$

Substituting this into the relation above and canceling $dh$ yields

$$-\delta w_{\text{rev}} = v\,dP + d\text{ke} + d\text{pe} \tag{6-47}$$

Integrating, we find

$$w_{\text{rev}} = -\int_1^2 v\,dP - \Delta\text{ke} - \Delta\text{pe} \qquad \text{(kJ/kg)} \tag{6-48}$$

When the changes in kinetic and potential energies are negligible, this equation reduces to

$$w_{\text{rev}} = -\int_1^2 v\,dP \qquad \text{(kJ/kg)} \tag{6-49}$$

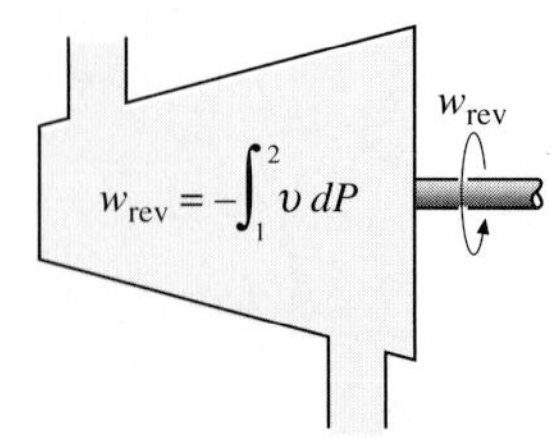

(*a*) Steady-flow system

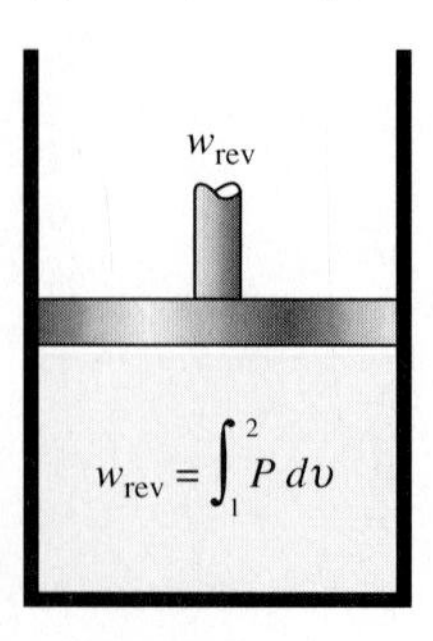

(*b*) Closed system

**FIGURE 6-30**
Reversible work relations for closed and steady-flow systems.

Equations 6-48 and 6-49 are relations for the reversible work associated with an internally reversible process in a steady-flow device. The resemblance between the $v\,dP$ in these relations and $P\,dv$ is striking. They should not be confused with each other, however, since $P\,dv$ is associated with reversible boundary work in closed systems (Fig. 6-30).

Obviously, one needs to know $v$ as a function of $P$ for the given process to perform the integration in Eq. 6-48. When the working fluid is an *incompressible fluid,* the specific volume $v$ remains constant during the process and can be taken out of the integration. Then Eq. 6-48 simplifies to

$$w_{\text{rev}} = v(P_1 - P_2) - \Delta\text{ke} - \Delta\text{pe} \qquad \text{(kJ/kg)} \tag{6-50}$$

For the steady flow of a liquid through a device that involves no work interactions (such as nozzle or a pipe section), the work term is zero, and the equation above can be expressed as

$$v(P_2 - P_1) + \frac{\mathcal{V}_2^2 - \mathcal{V}_1^2}{2} + g(z_2 - z_1) = 0 \tag{6-51}$$

which is known as the **Bernoulli equation** in fluid mechanics. Equation 6-51 is developed for an internally reversible process and thus is applicable to incompressible fluids that involve no irreversibilities such as friction or shock waves. This equation can be modified, however, to incorporate these effects.

Equation 6-49 has far-reaching implications in engineering regarding devices that produce or consume work steadily such as turbines, compressors, and pumps. It is obvious from this equation that the reversible steady-flow work is closely associated with the specific volume of the fluid flowing through the device. *The larger the specific volume, the larger the reversible work produced or consumed by the steady-flow device* (Fig. 6-31). This conclusion is equally valid for actual steady-flow devices. Therefore, every effort should be made to keep the specific volume of a fluid as small as possible during a compression process to minimize the work input and as large as possible during an expansion process to maximize the work output.

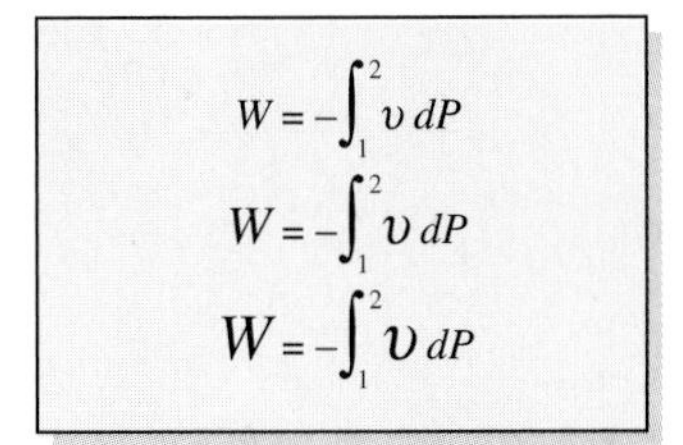


**FIGURE 6-31**
The larger the specific volume, the greater the work produced (or consumed) by a steady-flow device.

In steam or gas power plants, the pressure rise in the pump or compressor is equal to the pressure drop in the turbine if we disregard the pressure losses in various other components. In steam power plants, the pump handles liquid, which has a very small specific volume, and the turbine handles vapor, whose specific volume is many times larger. Therefore, the work output of the turbine is much larger than the work input to the pump. This is one of the reasons for the overwhelming popularity of steam power plants in electric power generation.

If we were to compress the steam exiting the turbine back to the turbine inlet pressure before cooling it first in the condenser in order to "save" the heat rejected, we would have to supply all the work produced by the turbine back to the compressor. In reality, the required work input would be even greater than the work output of the turbine because of the irreversibilities present in both processes.

In gas power plants, the working fluid (typically air) is compressed in the gas phase, and a considerable portion of the work output of the turbine is consumed by the compressor. As a result, a gas power plant delivers less net work per unit mass of the working fluid.

**EXAMPLE 6-7**

Determine the work required to compress steam isentropically from 100 kPa to 1 MPa, assuming that the steam exists as (*a*) saturated liquid and (*b*) saturated vapor at the initial state. Neglect the changes in kinetic and potential energies.

**Solution** Sketches of the pump and the compressor are given in Fig. 6-32. We expect the work input to the pump to be considerably smaller since it handles a liquid. The compression process is stated to be reversible, and the kinetic and potential energy changes are negligible. Thus Eq. 6-49 is applicable.

(*a*) In this case, the steam is a saturated liquid initially, and its specific volume is

$$v_1 = v_{f\,@\,100\text{ kPa}} = 0.001043\text{ m}^3/\text{kg}$$

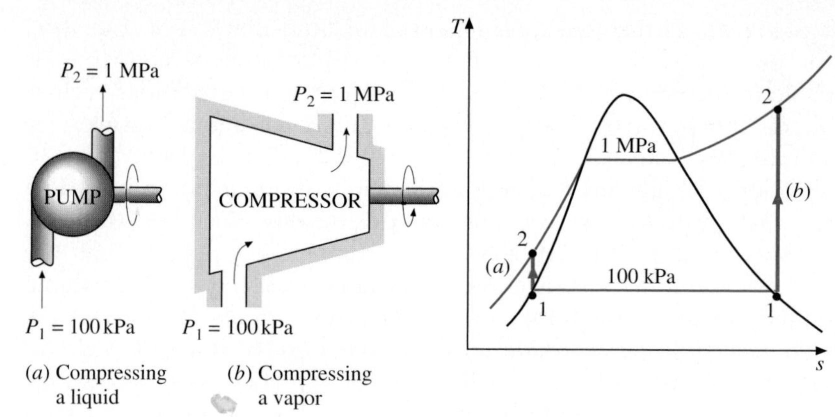

**FIGURE 6-32**
Schematic and $T$-$s$ diagram for Example 6-7.

which remains essentially constant during the process. Thus,

$$w_{rev} = -\int_1^2 v\,dP \approx v_1(P_1 - P_2)$$

$$= (0.001043\ \text{m}^3/\text{kg})[(100 - 1000)\ \text{kPa}]\left(\frac{1\ \text{kJ}}{1\ \text{kPa}\cdot\text{m}^3}\right)$$

$$= -0.94\ \text{kJ/kg}$$

(*b*) This time, the steam is a saturated vapor initially and remains a vapor during the entire compression process. Since the specfic volume of a gas changes considerably during a compression process, we need to know how $v$ varies with $P$ to perform the integration in Eq. 6-49. This relation, in general, is not readily available. But for an isentropic process, it is easily obtained from the second $T\,ds$ relation by setting $ds = 0$:

$$\left.\begin{array}{l} T\,ds = dh - v\,dP \\ ds = 0 \quad \text{(isentropic process)} \end{array}\right\} \quad v\,dP = dh$$

Thus,

$$w_{rev} = -\int_1^2 v\,dP = -\int_1^2 dh = h_1 - h_2$$

This result could also be obtained from the first-law analysis of an isentropic steady-flow process. Next we determine the enthalpies:

*State 1:* $\left.\begin{array}{l} P_1 = 100\ \text{kPa} \\ \text{(sat. vapor)} \end{array}\right\}\quad h_1 = 2675.5\ \text{kJ/kg}$ (Table A-5)

*State 2:* $\left.\begin{array}{l} P_2 = 1\ \text{MPa} \\ s_2 = s_1 \end{array}\right\}\quad h_2 = 3195.5\ \text{kJ/kg}$ (Table A-6)

Thus,

$$w_{rev} = (2675.5 - 3195.5)\ \text{kJ/kg} = -520\ \text{kJ/kg}$$

That is, compressing steam in the vapor form would require over 500 times more work than compressing it in the liquid form between the same pressure limits.

### Minimizing the Compressor Work

We have shown that the work input to a compressor is minimized when the compression process is executed in an internally reversible manner. When the changes in kinetic and potential energies are negligible, the compressor work is given by (Eq. 6-49)

$$w_{\text{rev}} = -\int_1^2 v\,dP$$

Obviously one way of minimizing the compressor work is to approach an internally reversible process as much as possible by minimizing the irreversibilities such as friction, turbulence, and non-quasi-equilibrium compression. The extent to which this can be accomplished is limited by economic considerations. A second (and more practical) way of reducing the compressor work is to keep the specific volume of the gas as small as possible during the compression process. This is done by maintaining the temperature of the gas as low as possible during compression since the specific volume of a gas is proportional to temperature. Therefore, reducing the work input to a compressor requires that the gas be cooled as it is compressed.

## 6-11 ■ SUMMARY

The second law of thermodynamics leads to the definition of a new property called *entropy,* which is a quantitative measure of microscopic disorder for a system. The definition of entropy is based on the *Clausius inequality,* given by

$$\oint \frac{\delta Q}{T} \leqslant 0 \qquad \text{(kJ/K)}$$

where the equality holds for internally or totally reversible processes and the inequality for irreversible processes. Any quantity whose cyclic integral is zero is a property, and entropy is defined as

$$dS = \left(\frac{\delta Q}{T}\right)_{\text{int rev}} \qquad \text{(kJ/K)}$$

The *entropy change* during a process is obtained by integrating this relation:

$$\Delta S = S_2 - S_1 = \int_1^2 \left(\frac{\delta Q}{T}\right)_{\text{int rev}} \qquad \text{(kJ/K)}$$

The inquality part of the Clausius inequality combined with the definition of entropy yields an inequality that is known as the *increase of entropy principle*:

$$dS \geqslant \frac{\delta Q}{T}$$

or

$$\Delta S_{\text{isolated}} \geqslant 0$$

Thus the entropy generation during a process is positive (for actual processes) or zero (for reversible processes). The total entropy change for a process is the amount of entropy generated during that process ($S_{gen}$), and it is equal to the sum of the entropy change of the system and of the surroundings. The entropy of a system (closed system or control volume) or its surroundings may decrease during a process, but the sum of these two can never decrease.

Entropy change is caused by heat transfer, mass flow, and irreversibilities. Heat transfer to a system increases the entropy, and heat transfer from a system decreases it. The effect of irreversibilities is always to increase the entropy.

The *increase of entropy principle for a closed system* is expressed as

$$S_2 - S_1 = \sum \frac{Q_k}{T_k} + S_{gen} \qquad \text{(kJ/K)}$$

or, in the rate form as

$$\frac{dS}{dt} = \sum \frac{\dot{Q}_k}{T_k} + \dot{S}_{gen} \qquad \text{(kW/K)}$$

where $dS/dt$ is the rate of change of entropy of the system, and $\dot{Q}_k$ is the rate of heat transfer through the boundary at temperature $T_k$.

The increase of entropy principle for control volumes is expressed as

$$\frac{dS_{CV}}{dt} = \sum \frac{\dot{Q}_k}{T_k} + \sum \dot{m}_i s_i - \sum \dot{m}_e s_e + \dot{S}_{gen} \qquad \text{(kW/K)}$$

For single-stream (one-inlet and one-exit) steady-flow devices it simplifies to

$$\dot{S}_{gen} = \dot{m}(s_e - s_i) - \sum \frac{\dot{Q}_k}{T_k}$$

The value of $S_{gen}$ can be used to determine whether a process is reversible, irreversible, or impossible:

$$S_{gen} \begin{cases} > 0 & \text{irreversible process} \\ = 0 & \text{reversible process} \\ < 0 & \text{impossible process} \end{cases}$$

The third law of thermodynamics states that the entropy of a pure crystalline substance at absolute zero temperature is zero. This law provides an absolute reference point for the determination of entropy. The entropy determined relative to this point is called *absolute entropy*.

Entropy is a property, and it can be expressed in terms of more familiar properties through the $T\,ds$ relations, expressed as

$$T\,ds = du + P\,dv$$

and

$$T\,ds = dh - v\,dP$$

These two relations have many uses in thermodynamics and serve as the starting point in developing entropy-change relations for processes. The successful use of $T\,ds$ relations depends on the availability of property relations. Such relations do not exist for a general pure substance but are available for incompressible substances (solids, liquids) and ideal gases.

The *entropy-change* and *isentropic relations* for a process can be summarized as follows:

**1** *General*:

Any process: $\Delta s = s_2 - s_1 \quad [\text{kJ/(kg} \cdot \text{K)}]$

Isentropic process: $s_2 = s_1$

**2** *Incompressible substances*:

Any process: $s_2 - s_1 = C_{\text{av}} \ln \dfrac{T_2}{T_1} \quad [\text{kJ/(kg} \cdot \text{K)}]$

Isentropic process: $T_2 = T_1$

**3** *Ideal gases (constant specific heats)*:

Any process:

$$s_2 - s_1 = C_{v,\text{ av}} \ln \frac{T_2}{T_1} + R \ln \frac{v_2}{v_1} \quad [\text{kJ/(kg} \cdot \text{K)}]$$

and

$$s_2 - s_1 = C_{p,\text{ av}} \ln \frac{T_2}{T_1} - R \ln \frac{P_2}{P_1} \quad [\text{kJ/(kg} \cdot \text{K)}]$$

Isentropic process:

$$\left(\frac{T_2}{T_1}\right)_{s=\text{const.}} = \left(\frac{v_1}{v_2}\right)^{k-1}$$

$$\left(\frac{T_2}{T_1}\right)_{s=\text{const.}} = \left(\frac{P_2}{P_1}\right)^{(k-1)/k}$$

$$\left(\frac{P_2}{P_1}\right)_{s=\text{const.}} = \left(\frac{v_1}{v_2}\right)^{k}$$

The *steady-flow work* for a reversible process can be expressed in terms of the fluid properties as

$$w_{\text{rev}} = -\int_1^2 v\, dP - \Delta\text{ke} - \Delta\text{pe} \quad (\text{kJ/kg})$$

For incompressible substances ($v$ = constant), it simplifies to

$$w_{\text{rev}} = v(P_1 - P_2) - \Delta\text{ke} - \Delta\text{pe} \quad (\text{kJ/kg})$$

The work done during a steady-flow process is proportional to the specific volume. Therefore, $v$ should be kept as small as possible during a compression process to minimize the work input and as large as possible during an expansion process to maximize the work output.

## REFERENCES AND SUGGESTED READING

**1** A. Bejan, *Advanced Engineering Thermodynamics,* Wiley, New York, 1988.

**2** A. Bejan, *Entropy Generation through Heat and Fluid Flow,* Wiley-Interscience, New York, 1982.

**3** W. Z. Black and J. G. Hartley, *Thermodynamics,* Harper & Row, New York, 1985.

**4** J. B. Jones and G. A. Hawkins, *Engineering Thermodynamics,* 2d ed., Wiley, New York, 1986.

**5** M. J. Moran and H. N. Shapiro, *Fundamentals of Engineering Thermodynamics,* Wiley, New York, 1988.

**6** W. C. Reynolds and H. C. Perkins, *Engineering Thermodynamics,* 2d ed., McGraw-Hill, New York, 1977.

**7** J. Rifkin, *Entropy,* The Viking Press, New York, 1980.

**8** G. J. Van Wylen and R. E. Sonntag, *Fundamentals of Classical Thermodynamics,* 3d ed., Wiley, New York, 1985.

**9** K. Wark, *Thermodynamics,* 5th ed., McGraw-Hill, New York, 1988.

## PROBLEMS*

### Clausius Inequality, Entropy, and the Increase of Entropy Principle

**6-1C** Does the temperature in the Clausius inequality relation have to be absolute temperature? Why?

**6-2C** Does a cycle for which $\oint \delta Q > 0$ violate the Clausius inequality? Why?

**6-3C** Is a quantity whose cyclic integral is zero necessarily a property?

**6-4C** Does the cyclic integral of heat have to be zero (i.e., does a system have to reject as much heat as it receives to complete a cycle)? Explain.

**6-5C** Does the cyclic integral of work have to be zero (i.e., does a system have to produce as much work as it consumes to complete a cycle)? Explain.

**6-6C** A system undergoes a process between two fixed states first in a reversible manner and then in an irreversible manner. For which case is the entropy change of the system greater? Why?

**6-7C** Is the value of the integral $\int_1^2 \delta Q/T$ the same for all processes between states 1 and 2? Explain.

*Students are encouraged to answer *all* the concept "C" questions.

**6-8C** Is the value of the integral $\int_1^2 \delta Q/T$ the same for all reversible processes between states 1 and 2?

**6-9C** To determine the entropy change for an irreversible process between states 1 and 2, should the integral $\int_1^2 \delta Q/T$ be performed along the actual process path or an imaginary reversible path? Explain.

**6-10C** Is an isothermal process necessarily internally reversible? Explain your answer with an example.

**6-11C** How do the values of the integral $\int_1^2 \delta Q/T$ compare for a reversible and irreversible process between the same end states?

**6-12C** The entropy of a hot baked potato decreases as it cools. Is this a violation of the increase-in-entropy principle? Explain.

**6-13C** Is it possible to create entropy? Is it possible to destroy it?

**6-14C** A piston–cylinder device contains helium gas. During a reversible, isothermal process, the entropy of the helium will (*never, sometimes, always*) increase.

**6-15C** A piston–cylinder device contains nitrogen gas. During a reversible, adiabatic process the entropy of the nitrogen will (*never, sometimes, always*) increase.

**6-16C** A piston-cylinder device contains superheated steam. During an actual adiabatic process, the entropy of the steam will (*never, sometimes, always*) increase.

**6-17C** The entropy of steam will (*increase, decrease, remain the same*) as it flows through an actual adiabatic turbine.

**6-18C** The entropy of the working fluid of the ideal Carnot cycle (*increases, decreases, remains the same*) during the isothermal heat addition process.

**6-19C** The entropy of the working fluid of the ideal Carnot cycle (*increases, decreases, remains the same*) during the isothermal heat rejection process.

**6-20C** During a heat transfer process, the entropy of a system (*always, sometimes, never*) increases.

**6-21C** Is it possible for the entropy change of a closed system to be zero during an irreversible process? Explain.

**6-22C** What three different mechanisms can cause the entropy of a control volume to change?

**6-23C** Steam is accelerated as it flows through an actual adiabatic nozzle. The entropy of the steam at the nozzle exit will be (*greater than, equal to, less than*) the entropy at the nozzle inlet.

**6-24C** Consider a person who organizes his room, and thus decreases the entropy of the room. Does this process violate the second law of thermodynamics?

**6-25C** Consider a fruit tree that makes highly organized fruits out of the water and highly disorganized soil, and thus decreases the entropy of its locality. Does this process violate the second law of thermodynamics?

**6-26C** Consider an army unit whose soldiers are walking around at random in a field. Suddenly an order is issued and the soldiers align in a highly organized manner, decreasing the entropy. Does this process violate the second law of thermodynamics?

**6-27** The inner and outer surfaces of a 5 m × 6 m brick wall of thickness 30 cm and thermal conductivity 0.69 W/m°C are maintained at temperatures of 15°C and 5°C, respectively. Determine (*a*) the rate of heat transfer through the wall, in W, and (*b*) the rate of entropy generation within the wall.

**6-28** For heat transfer purposes, a standing man can be modeled as a 30-cm diameter, 170-cm long vertical cyclinder with both the top and bottom surfaces insulated and with the side surface at an average temperature of 34°C. For a convection heat transfer coefficient of 15 W/(m$^2$ · °C), determine the rate of his loss from this man by convection in an environment at 20°C. Also determine the rate of entropy transfer from the body of this person accompanying heat transfer, in W/K.

**6-29** A 1000 W iron is left on the iron board with its base exposed to the air at 20°C. The convection heat transfer coefficient between the base surface and the surrounding air is 80 W/(m$^2$ · °C). If the base has a surface area of 0.02 m$^2$, determine (*a*) the temperature of the base of the iron and (*b*) the rate of entropy generation during this process in steady operation. How much of this entropy generation occurs within the iron? Disregard heat transfer by radiation.

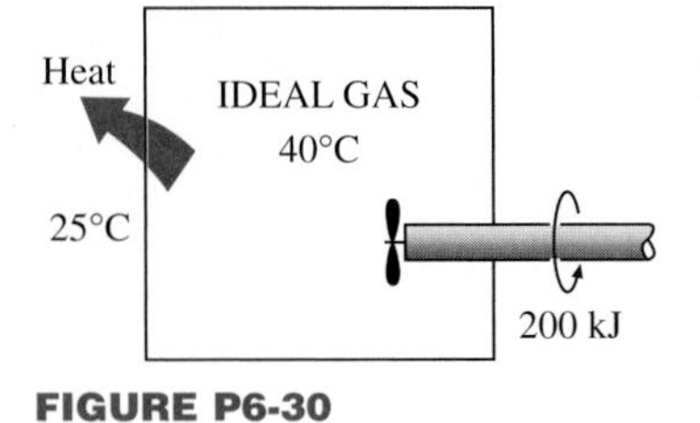


**FIGURE P6-30**

**6-30** A rigid tank contains an ideal gas at 40°C that is being stirred by a paddle wheel. The paddle wheel does 200 kJ of work on the ideal gas. It is observed that the temperature of the ideal gas remains constant during this process as a result of heat transfer between the system and the surroundings at 25°C. Determine (*a*) the entropy change of the ideal gas and (*b*) the total entropy generation. Is the increase of entropy principle satisfied during this process? *Answers:* (*a*) 0, (*b*) 0.671 kJ/K

**6-30E** A rigid tank contains an ideal gas at 85°F that is being stirred by a paddle wheel. The paddle wheel does 200 Btu of work on the ideal gas. It is observed that the temperature of the ideal gas remains constant during this process as a result of heat transfer between the system and the surroundings at 70°F. Determine (*a*) the entropy change of the ideal gas and (*b*) the total entropy generation. Is the increase of entropy principle satisfied during this process?

**6-31** Air is compressed by a 8-kW compressor from $P_1$ to $P_2$. The air temperature is maintained constant at 25°C during this process as a result of heat transfer to the surrounding medium at 10°C. Determine the rate

of entropy change of the air and the rate of total entropy generation. State the assumptions made in solving this problem. Does this process satisfy the second law of thermodynamics?
*Answers:* −0.0268 kW/K, 0.0015 kW/K

**6-32** A frictionless piston–cylinder device contains saturated liquid water at 200-kPa pressure. Now 450 kJ of heat is transferred to water from a source at 500°C, and part of the liquid vaporizes at constant pressure. Determine the total entropy generation for this process, in kJ/K. Is this process reversible, irreversible, or impossible?
*Answers:* 0.562 kJ/K, irreversible

**6-32E** A frictionless piston–cylinder device contains saturated liquid water at 20-psia pressure. Now 600 Btu of heat is transferred to water from a source at 900°F, and part of the liquid vaporizes at constant pressure. Determine the total entropy generation for this process, in Btu/R. Is this process reversible, irreversible, or impossible?

**6-33** During the isothermal heat addition process of a Carnot cycle, 900 kJ of heat is added to the working fluid from a source at 400°C. Determine (*a*) the entropy change of the working fluid, (*b*) the entropy change of the source, and (*c*) the total entropy generation for the process.

**6-34** During the isothermal heat rejection process of a Carnot cycle, the working fluid experiences an entropy change of −0.6 kJ/K. If the temperature of the energy sink is 30°C, determine (*a*) the amount of heat transfer to the sink, (*b*) the entropy change of the sink, and (*c*) the total entropy generation for this process.
*Answers:* (*a*) 181.8 kJ, (*b*) 0.6 kJ/K, (*c*) 0

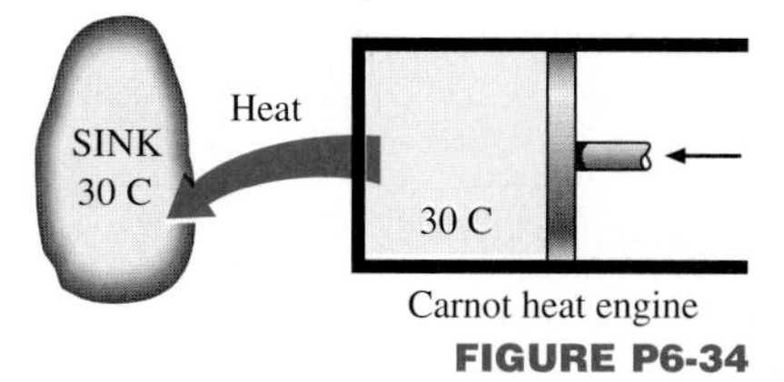

Carnot heat engine
**FIGURE P6-34**

**6-34E** During the isothermal heat rejection process of a Carnot cycle, the working fluid experiences an entropy change of −0.7 Btu/R. If the temperature of the energy sink is 95°F, determine (*a*) the amount of heat transfer, (*b*) the entropy change of the sink, and (*c*) the total entropy generation for this process. *Answers:* (*a*) −388.5 Btu, (*b*) 0.7 Btu/R, (*c*) 0

**6-35** Refrigerant-134a enters the coils of the evaporator of a refrigeration system as a saturated liquid–vapor mixture at pressure of 200 kPa. The refrigerant absorbs 120 kJ of heat from the cooled space which is maintained at −5°C and leaves as saturated vapor at the same pressure. Determine (*a*) the entropy change of the refrigerant, (*b*) the entropy change of the cooled space, and (*c*) the total entropy generation for this process.

### Entropy Changes of Incompressible Substances

**6-36C** Equation 6-32 [that is, $\Delta S = mC_{av} \ln(T_2/T_1)$] is developed for incompressible substances. Can this relation be used to determine the entropy changes of ideal gases? If so, under what conditions?

**6-37C** Are the $T\,ds$ relations developed in Sec. 6-7 limited to reversible processes only, or are they valid for all processes, reversible or irreversible? Explain.

**6-38C** Is a process which is internally reversible and adiabatic necessarily isentropic? Explain.

**6-39C** Consider two solid blocks, one hot and the other cold, brought into contact in an adiabatic container. After a while, thermal equilibrium is established in the container as a result of heat transfer. The first law requires that the amount of energy lost by the hot solid be equal to the energy gained by the cold one. Does the second law require that the decrease in entropy of the hot solid be equal to the increase in entropy of the cold one?

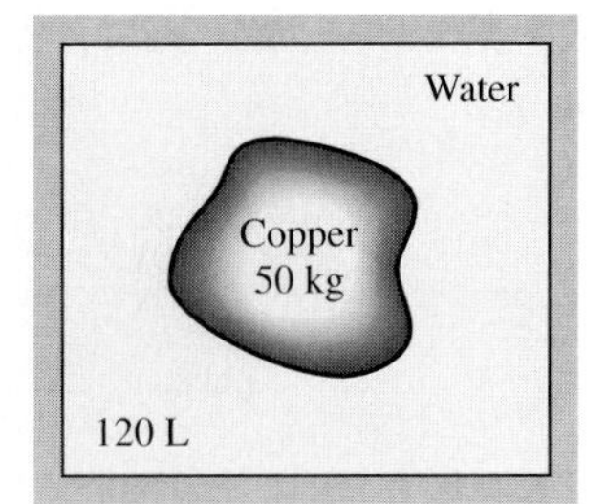


**FIGURE P6-40**

**6-40** A 50-kg copper block initially at 80°C is dropped into an insulated tank that contains 120 L of water at 25°C. Determine the final equilibrium temperature and the total entropy generation for this process.

**6-40E** A 70-lbm copper block initially at 200°F is dropped into an insulated tank that contains 1 $ft^3$ of water at 75°F. Determine the final equilibrium temperature and the total entropy generation for this process.

**6-41** A 5-kg iron block initially at 350°C is quenched in an insulated tank that contains 100 kg of water at 22°C. Assuming the water that vaporizes during the process condenses back in the tank, determine the amount of entropy generated during this process.

**6-42** A 20-kg aluminum block initially at 200°C is brought into contact with a 20-kg block of iron at 100°C in an insulated enclosure. Determine the final equilibrium temperature and the total entropy generation for this process. *Answers:* 168.4°C, 0.169 kJ/K

**6-43** An iron block of unknown mass at 85°C is dropped into an insulated tank that contains 100 L of water at 20°C. At the same time, a paddle wheel driven by a 200-W motor is activated to stir the water. It is observed that the thermal equilibrium is established after 20 min with a final temperature of 24°C. Determine the mass of the iron block and the entropy generated during this process. *Answers:* 52.2 kg, 1.285 kJ/K

**6-43E** An iron block of unknown mass at 185°F is dropped into an insulated tank that contains 0.8 $ft^3$ of water at 70°F. At the same time, a paddle wheel driven by a 200-W motor is activated to stir the water. Thermal equilibrium is established after 10 min with a final temperature of 75°F. Determine the mass of the iron block and the entropy generated during this process.

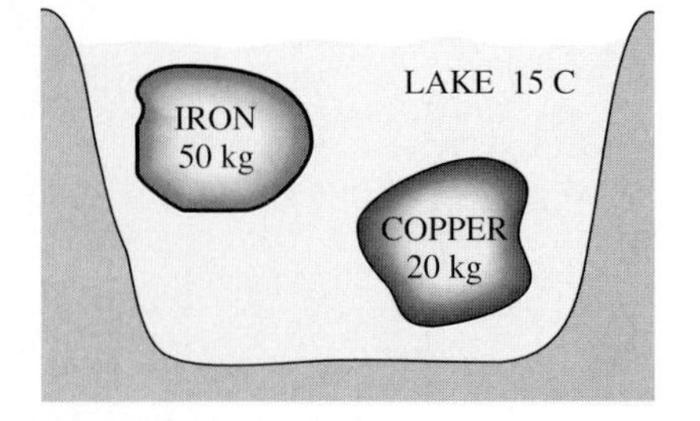


**FIGURE P6-44**

**6-44** A 50-kg iron block and a 20-kg copper block, both initially at 80°C, are dropped into a large lake at 15°C. Thermal equilibrium is established after a while as a result of heat transfer between the blocks and the lake water. Determine the total entropy generation for this process.

### Entropy Changes of Ideal Gases

**6-45C** Prove that the two relations for entropy changes of ideal gases under the constant-specific-heat assumption (Eqs. 6-37 and 6-38) are equivalent.

**6-46C** Starting with the second $T\,ds$ relation (Eq. 6-26), obtain Eq. 6-38 for the entropy change of ideal gases under the constant-specific-heat assumption.

**6-47C** Does the K of an ideal gas vary with temperature?

**6-48C** Some properties of ideal gases such as internal energy and enthalpy vary with temperature only [that is, $u = u(T)$ and $h = h(T)$]. Is this also the case for entropy?

**6-49C** Starting with Eq. 6-38, obtain Eq. 6-42.

**6-50C** Can the entropy of an ideal gas change during an isentropic process?

**6-51C** An ideal gas undergoes a process between two specified temperatures, first at constant pressure and then at constant volume. For which case will the ideal gas experience a larger entropy change? Explain.

**6-52** Oxygen gas is compressed in a piston–cylinder device from an initial state of 0.8 $m^3$/kg and 25°C to a final state of 0.1 $m^3$/kg and 287°C. Determine the entropy change of the oxygen during this process.

**6-53** A 0.5-$m^3$ insulated rigid tank contains 0.9 kg of carbon dioxide at 100 kPa. Now paddle-wheel work is done on the system until the pressure in the tank rises to 120 kPa. Determine the entropy change of carbon dioxide during this process in kJ/K. Assume constant specific heats. *Answer:* 0.108 kJ/K

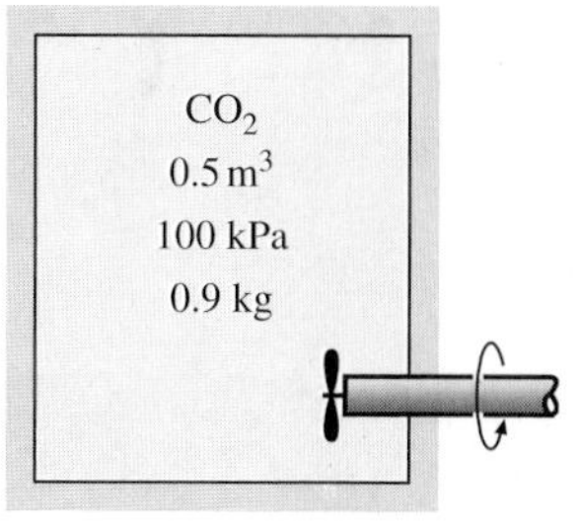


**FIGURE P6-53**

**6-54** An insulated piston–cylinder device initially contains 300 L of air at 120 kPa and 17°C. Air is now heated for 15 min by a 200-W resistance heater placed inside the cylinder. The pressure of air is maintained constant during this process. Determine the entropy change of air.

**6-54E** An insulated piston–cylinder device initially contains 10 $ft^3$ of air at 20 psia and 60°F. Air is now heated for 15 min by a 200-W resistance heater placed inside the cylinder. The pressure of air is maintained constant during this process. Determine the entropy change of air.

**6-55** A piston–cylinder device contains 1.2 kg of nitrogen gas at 120 kPa and 27°C. The gas is now compressed slowly in a polytropic process during which $PV^{1.3}$ = constant. The process ends when the volume is reduced by one-half. Determine the entropy change of nitrogen during this process. *Answer:* −0.0615 kJ/K

**6-56** A mass of 3 kg of helium undergoes a process from an initial state of 3 $m^3$/kg and 20°C to a final state of 0.5 $m^3$/kg and 120°C. Determine the entropy change of helium during this process, assuming (*a*) the process is reversible and (*b*) the process is irreversible.

**6-56E** A mass of 8 lbm of helium undergoes a process from an initial state of 50 $ft^3$/lbm and 80°F to a final state of 10 $ft^3$/lbm and 200°F. Determine the entropy change of helium during this process, assuming (*a*) the process is reversible and (*b*) the process is irreversible.

**6-57** Air is compressed in a piston–cylinder device from 90 kPa and 20°C to 400 kPa in a reversible isothermal process. Determine (*a*) the entropy change of air and (*b*) the work done.

**6-58** Air is compressed steadily by a 5-kW compressor from 100 kPa and 17°C to 600 kPa and 167°C at a rate of 1.6 kg/min. During this process, some heat transfer takes place between the compressor and the surrounding medium at 17°C. Determine (*a*) the rate of entropy change of air and (*b*) the rate of entropy generation during this process. Does this process satisfy the second law?
*Answers:* (*a*) −0.0025 kW/K, (*b*) 0.00083 kW/K

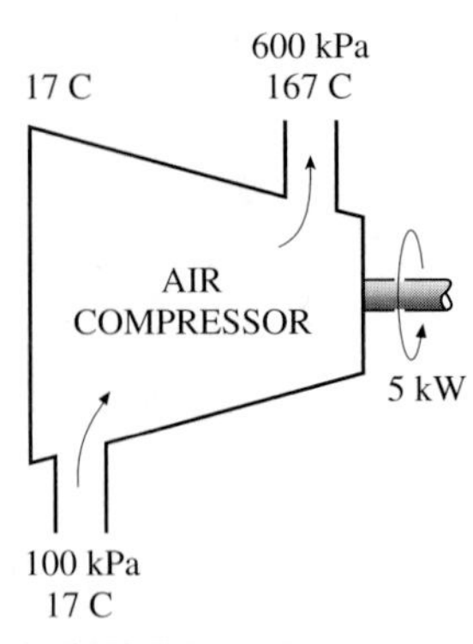


**FIGURE P6-58**

**6-58E** Air is compressed steadily by a 25-hp compressor from 15 psia and 60°F to 90 psia and 340°F at a rate of 3.5 lbm/min. During the process, some heat transfer takes place between the compressor and the surrounding medium at 60°F. Determine (*a*) the rate of entropy change of air and (*b*) the rate of entropy generation during this process. Does this process satisfy the second law?

**6-59** An insulated rigid tank is divided into two equal parts by a partition. Initially, one part contains 5 kmol of an ideal gas at 400 kPa and 50°C, and the other side is evacuated. The partition is now removed, and the gas fills the entire tank. Determine the entropy generation during this process. *Answer:* 28.81 kJ/K

**6-60** Air is compressed in a piston–cylinder device from 100 kPa and 17°C to 800 kPa in a reversible, adiabatic process. Determine the final temperature and the work done during this process.
*Answers:* 525.3 K, 171.1 kJ/kg

**6-61** Helium gas is compressed from 100 kPa and 30°C to 500 kPa in a reversible, adiabatic process. Determine the final temperature and the work done, assuming the process takes place (*a*) in a piston–cylinder device and (*b*) in a steady-flow compressor.

**6-61E** Helium gas is compressed from 15 psia and 90°F to 75 psia in a reversible, adiabatic process. Determine the final temperature and the work done, assuming the process takes place (*a*) in a piston–cylinder device and (*b*) in a steady-flow compressor.

**6-62** An insulated, rigid tank contains 4 kg of argon gas at 450 kPa and 30°C. A valve is now opened, and argon is allowed to escape until the pressure inside drops to 150 kPa. Assuming the argon remaining inside the tank has undergone a reversible, adiabatic process, determine the final mass in the tank. *Answer:* 2.07 kg

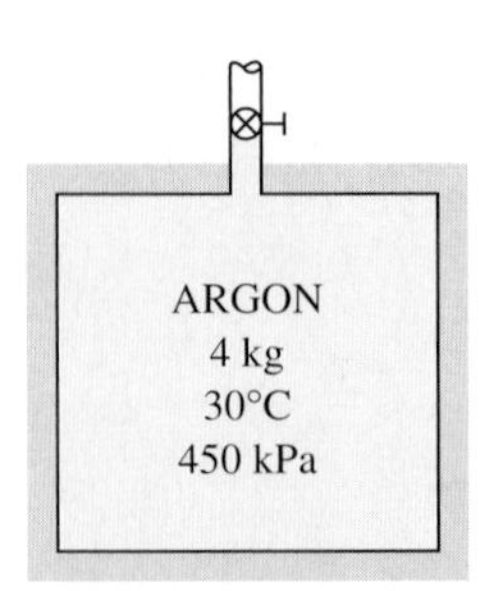


**FIGURE P6-62**

**6-62E** An insulated, rigid tank contains 8 lbm of argon gas at 120 psia and 80°F. A valve is now opened, and argon is allowed to escape until the pressure inside drops to 40 psia. Assuming the argon remaining inside the tank has undergone a reversible, adiabatic process, determine the final mass in the tank.

**6-63** Air enters an adiabatic nozzle at 400 kPa, 247°C, and 60 m/s and exits at 80 kPa. Disregarding any irreversibilities, determine the exit velocity of air.

**6-63E** Air enters an adiabatic nozzle at 60 psia, 540°F, and 200 ft/s and exits at 12 psia. Disregarding any irreversibilities, determine the exit velocity of air.

**6-64** Air enters a nozzle steadily at 280 kPa and 77°C with a velocity of 50 m/s and exits at 85 kPa and 320 m/s. The heat losses from the nozzle to the surrounding medium at 20°C are estimated to be 3.2 kJ/kg. Determine (*a*) the exit temperature and (*b*) the total entropy generated during this process.

**6-65** Air enters a compressor at ambient conditions of 96 kPa and 17°C with a low velocity and exits at 1 MPa, 327°C, and 120 m/s. The compressor is cooled by the ambient air at 17°C at a rate of 1500 kJ/min. The power input to the compressor is 300 kW. Determine (*a*) the mass flow rate of air and (*b*) the rate of entropy generation.
*Answers:* (*a*) 0.851 kg/s, (*b*) 0.144 kW/K

**6-65E** Air enters a compressor at ambient conditions of 15 psia and 60°F with a low velocity and exits at 150 psia, 620°F, and 350 ft/s. The compressor is cooled by the ambient air at 60°F at a rate of 1500 Btu/min. The power input to the compressor is 400 hp. Determine (*a*) the mass flow rate of air and (*b*) the rate of entropy generation.

### Reversible Steady-Flow Work

**6-66C** In large compressors, the gas is frequently cooled while being compressed to reduce the power consumed by the compressor. Explain how cooling the gas during a compression process reduces the power consumption.

**6-67C** The turbines in steam power plants operate essentially under adiabatic conditions. A plant engineer suggests to end this practice. She proposes to run cooling water through the outer surface of the casing to cool the steam as it flows through the turbine. This way, she reasons, the entropy of the steam will decrease, the performance of the turbine will improve, and as a result the work output of the turbine will increase. How would you evaluate this proposal?

**6-68C** It is well known that the power consumed by a compressor can be reduced by cooling the gas during compression. Inspired by this, somebody proposes to cool the liquid as it flows through a pump, in order to reduce the power consumption of the pump. Would you support this proposal? Explain.

**6-99** Water enters the pump of a steam power plant as saturated liquid at 20 kPa at a rate of 20 kg/s and exits at 6 MPa. Neglecting the changes in kinetic and potential energies and assuming the process to be reversible, determine the power input to the pump.

**6-69E** Water enters the pump of a steam power plant as saturated liquid at 5 psia at a rate of 50 lbm/s and exits at 800 psia. Neglecting the changes in kinetic and potential energies and assuming the process to be reversible, determine the power input to the pump. *Answer:* 170.8 hp

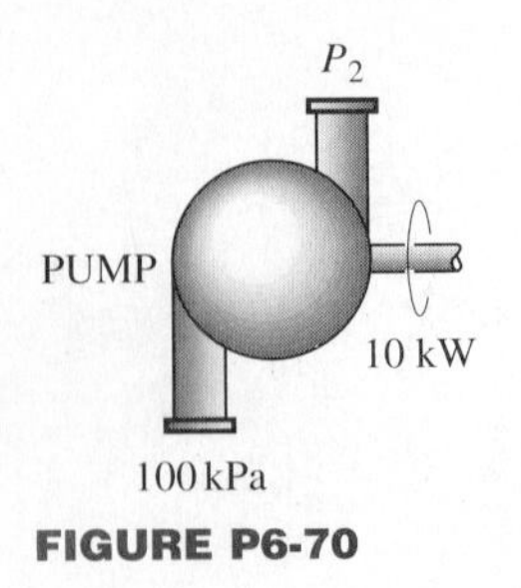


**FIGURE P6-70**

**6-70** Liquid water enters a 10-kW pump at 100-kPa pressure at a rate of 5 kg/s. Determine the highest pressure the liquid water can have at the exit of the pump. Neglect the kinetic and potential energy changes of water, and assume the specific volume of water to be 0.001 $m^3$/kg.
*Answer:* 2100 kPa

**6-71** Liquid water at 120 kPa enters a 15-kW pump where its pressure is raised to 3 MPa. If the elevation difference between the exit and the inlet levels is 10 m, determine the highest mass flow rate of liquid water this pump can handle. Neglect the kinetic change of water, and assume the specific volume of water to be 0.001 $m^3$/kg.

**6-72** Helium gas is compressed from 80 kPa and 20°C to 600 kPa at a rate of 0.2 $m^3$/s. Determine the power input to the compressor, assuming the compression process to be (*a*) isentropic, (*b*) polytropic with $n = 1.2$, and (*c*) isothermal.

**6-72E** Helium gas is compressed from 14 psia and 70°F to 120 psia at a rate of 5 $ft^3$/s. Determine the power input to the compressor, assuming the compression process to be (*a*) isentropic, (*b*) polytropic with $n = 1.2$, and (*c*) isothermal.

**6-73** Nitrogen gas is compressed from 80 kPa and 27°C to 480 kPa by a 10-kW compressor. Determine the mass flow rate of nitrogen through the compressor, assuming the compression process to be (*a*) isentropic, (*b*) polytropic with $n = 1.3$, and (*c*) isothermal.
*Answers:* (*a*) 0.048 kg/s, (*b*) 0.05 kg/s, (*c*) 0.063 kg/s.

## Review Problems

**6-74** Show that the difference between the reversible steady-flow work and reversible moving boundary work is equal to the flow energy.

**6-75** The inner and outer surfaces of a 0.5-cm thick 2 m × 2 m window glass in winter are 10°C and 3°C, respectively. If the thermal conductivity of the glass is 0.78 W/(m · °C), determine the amount of heat loss, in kJ, through the glass over a period of 5 h. Also determine the amount of entropy generated during this process within the glass.

**6-76** An aluminum pan [$k_t = 237\ \mathrm{W/(m \cdot K)}$] has a flat bottom whose diameter is 20 cm and thickness 0.4 cm. Heat is transferred steadily to boiling water in the pan through its bottom at a rate of 500 W. If the inner surface of the bottom of the pan is 105°C, determine (*a*) the temperature of the outer surface of the bottom of the pan, and (*b*) the rate of entropy generation within bottom of the pan, in W/K.

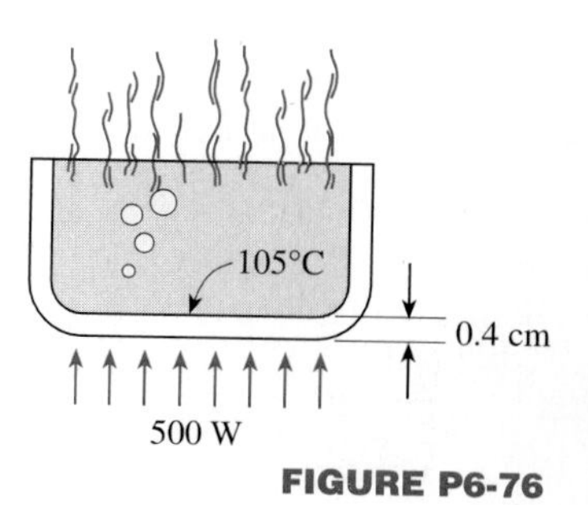


FIGURE P6-76

**6-77** A 50-cm-long 800-W electric resistance heating element whose diameter is 0.5 cm is immersed in 40 kg of water initially at 20°C. Assuming the water container is well-insulated, determine how long it will take for this heater to raise the water temperature to 80°C. Also determine the entropy generated during this process, in kJ/K.

**6-78** A 5-cm-external-diameter 10-m-long hot water pipe at 80°C is losing heat to the surrounding air at 5°C by natural convection with a heat transfer coefficient of 25 $\mathrm{W/(m^2 \cdot °C)}$. Determine the rate of heat loss from the pipe by natural convection, in W, and the rate of entropy generation in the surrounding air, in W/K.

**6-79** During a heat transfer process, the entropy change of incompressible substances, such as liquid water, can be determined from Eq. 6-33, that is, $\Delta S = mC_{\mathrm{av}} \ln(T_2/T_1)$. Show that for thermal energy reservoirs, such as large lakes, this relation reduces to Eq. 6-6, that is, $\Delta S = Q/T$.

**6-80** The inner and outer glasses of a 2 m × 2 m double-paned window area at 18°C and 6°C, respectively. If the 1-cm space between the two glasses is filled with still air [$k_t = 0.026\ \mathrm{W/(m \cdot K)}$], and the glasses are very nearly isothermal, determine the rate of heat transfer through the window, in W. Also determine the rates of entropy transfer through both sides of the window, and the rate of entropy generation within the window, in W/K.

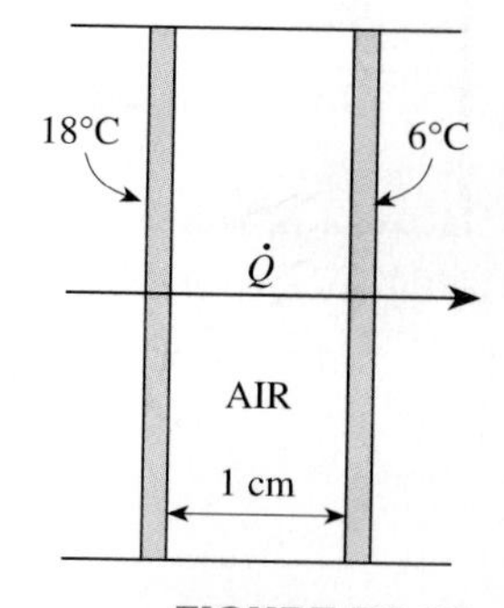

FIGURE P6-80

**6-81** A passive solar house that is losing heat to the outdoors at an average rate of 50,000 kJ/h is maintained at 22°C at all times during a winter night for 10 h. The house is to be heated by 50 glass containers each containing 20 L of water, which is heated to 80°C during the day by absorbing solar energy. A thermostat controlled 15 kW back-up electric resistance heater turns on whenever necessary to keep the house at 22°C. Determine how long the electric heating system was on that night, and the amount of entropy generated during the night.

**6-82** A 0.2-m³ steel container that has a mass of 30 kg when empty is filled with liquid water. Initially, both the steel tank and the water are at 50°C. Now heat is transferred, and the entire system cools to the surrounding air temperature of 25°C. Determine the total entropy generated during this process. *Answers:* 2.83 kJ/K

**6-82E** A 15-ft³ steel container that has a mass of 40 lbm when empty is filled with liquid water. Initially, both the steel tank and the water are

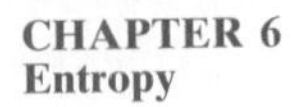



at 120°F. Now heat is transferred, and the entire system cools to the surrounding air temperature of 70°F. Determine the total entropy generated during this process.

**6-83** One ton (1000 kg) of liquid water at 80°C is brought into a well-insulated and well-sealed 4 m × 5 m × 6 m room initially at 22°C and 100 kPa. Assuming constant specific heats for both air and water at room temperature, determine (*a*) the final equilibrium temperature in the room and (*b*) the entropy generated during this process, in kJ/K.

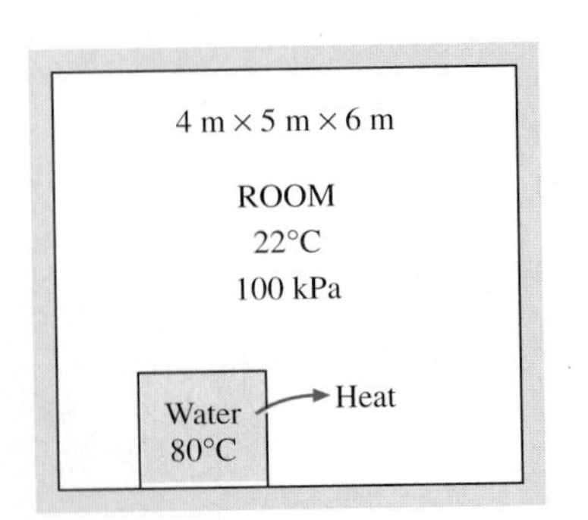


**FIGURE P6-83**

**6-84** A piston–cylinder device initially contains 0.5 m³ of helium gas at 150 kPa and 20°C. Helium is now compressed in a polytropic process ($PV^n$ = constant) to 400 kPa and 140°C. Determine the entropy change of helium and whether this process is reversible, irreversible, or impossible. Assume the surroundings are at 20°C.

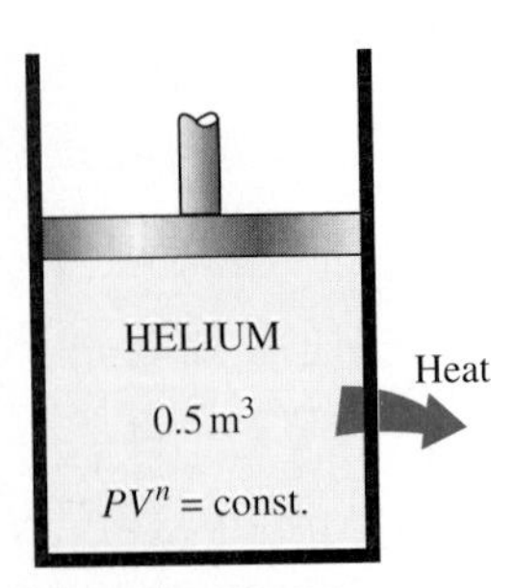


**FIGURE P6-84**

**6-84E** A piston–cylinder device initially contains 15 ft³ of helium gas at 25 psia and 70°F. Helium is now compressed in a polytropic process ($PV^n$ = constant) to 70 psia and 300°F. Determine the entropy change of helium and whether this process is reversible, irreversible, or impossible. Assume the surroundings are at 70°F.
*Answers:* −0.016 Btu/R, irreversible

**6-85** Air is compressed steadily by a compressor from 100 kPa and 17°C to 700 kPa at a rate of 2 kg/min. Determine the minimum power input required if the process is (*a*) adiabatic and (*b*) isothermal. Assume air to be an ideal gas with constant specific heats, and neglect the changes in kinetic and potential energies. *Answers:* (*a*) 7.29 kW, (*b*) 5.4 kW.

**6-86** A 4 m × 5 m × 6 m well-sealed room is to be heated by one ton (1000 kg) of liquid water contained in a tank that is placed in the room. The room is losing heat to the outside air at 5°C at an average rate of 10,000 kJ/h. The room is initially at 20°C and 100 kPa, and is maintained at an average temperature of 20°C at all times. If the hot water is to meet the heating requirements of this room for a 24-h period, determine (*a*) the minimum temperature of the water when it is first brought into the room and (*b*) the entropy generated during a 24-h period. Assume constant specific heats for both air and water at room temperature.

**6-87** Consider a well-insulated horizontal rigid cylinder that is divided into two compartments by a piston that is free to move but does not allow either gas to leak into the other side. Initially, one side of the piston contains 1 m³ of $N_2$ gas at 500 kPa and 80°C while the other side contains 1 m³ of He gas at 500 kPa and 25°C. Now thermal equilibrium is established in the cylinder as a result of heat transfer through the piston. Using constant specific heats at room temperature, determine (*a*) the final equilibrium temperature in the cylinder and (*b*) the entropy generation during this process. What would your answer be if the piston were not free to move?

**6-88** Repeat the problem above by assuming the piston is made of 5 kg of copper initially at the average temperature of the two gases on both sides.

**6-89** In order to cool 1-ton (1000 kg) of water at 20°C in an insulated tank, a person pours 80 kg of ice at −5°C into the water. Determine (*a*) the final equilibrium temperature in the tank and (*b*) the entropy generation during this process. The melting temperature and the heat of fusion of ice at atmospheric pressure are 0°C and 333.7 kJ, respectively.

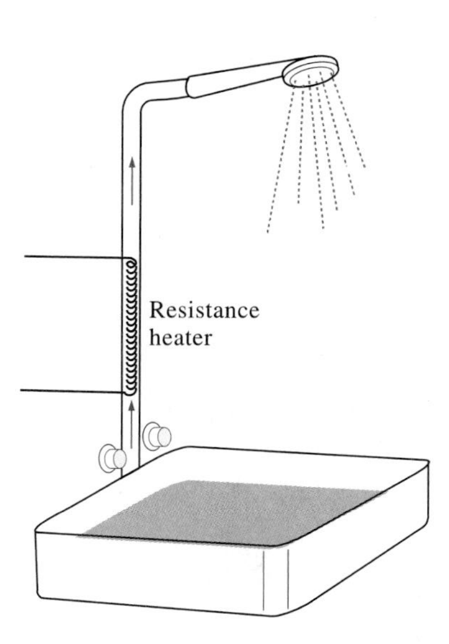

FIGURE P6-90

**6-90** (*a*) Water flows through a shower head steadily at a rate of 10 L/min. An electric resistance heater placed in the water pipe heats the water from 16°C to 43°C. Taking the density of water to be 1 kg/L, determine the electric power input to the heater, in kW, and the rate of entropy generation during this process, in kW/K.

(*b*) In an effort to conserve energy, it is proposed to pass the drained warm water at a temperature of 39°C through a heat exchanger to preheat the incoming cold water. If the heat exchanger has an effectiveness of 0.50 (that is, it recovers only half of the energy that can possibly be transferred from the drained water to incoming cold water), determine the electric power input required in this case as the rate of entropy generation.

**6-91** Consider two bodies of identical mass $m$ and specific heat $C$ used as thermal reservoirs (source and sink) for a heat engine. The first body is initially at an absolute temperature $T_1$ while the second one is at a lower absolute temperature $T_2$. Heat is transferred from the first body to the heat engine which rejects the waste heat to the second body. The process continues until the final temperatures of the two bodies $T_f$ become equal. Show that $T_f = \sqrt{T_1 T_2}$ when the heat engine produces the maximum possible work.

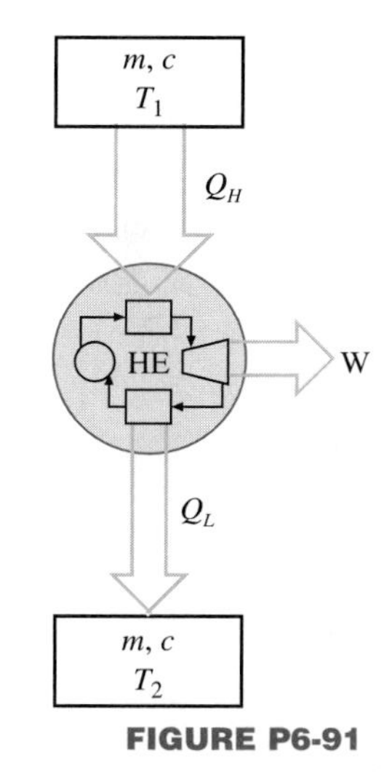


FIGURE P6-91

**6-92** The explosion of a hot water tank in a school in Spencer, Oklahoma, in 1982 killed 7 people while injuring 33 others. Although the number of such explosions has decreased drastically since the development of the ASME Pressure Vessel Code, which requires the tanks to be designed to withstand four times the normal operating pressures, they still occur as a result of the failure of the pressure relief valves and thermostats. When a tank filled with a high-pressure and high-temperature liquid ruptures, the sudden drop of the pressure of the liquid to the atmospheric level causes part of the liquid to flash into vapor, and thus to experience a huge rise in its volume. The resulting pressure wave that propagates rapidly can cause considerable damage.

Considering that the pressurized liquid in the tank eventually reaches equilibrium with its surroundings shortly after the explosion, the work that a pressurized liquid would do if allowed to expand reversibly and adiabatically to the pressure of the surroundings can be viewed as the *explosive energy* of the pressurized liquid. Because of the very short time period of the explosion and the apparent calm afterward, the explosion process can be considered to be adiabatic with no changes in kinetic and potential energies, and no mixing with the air.

Consider a 100-L hot water tank that has a working pressure of 0.5 MPa. As a result of some malfunction, the pressure in the tank rises to 2 MPa, at which point the tank explodes. Taking the atmospheric pressure to be 100 kPa and assuming the liquid in the tank to be saturated at the time of explosion, determine the total explosion energy of the tank in terms of the TNT equivalence. (The explosion energy of TNT is about 3250 kJ/kg, and 5 kg of TNT can cause total destruction of unreinforced structures within about a 7-m radius.) *Answer:* 2.467 kg TNT

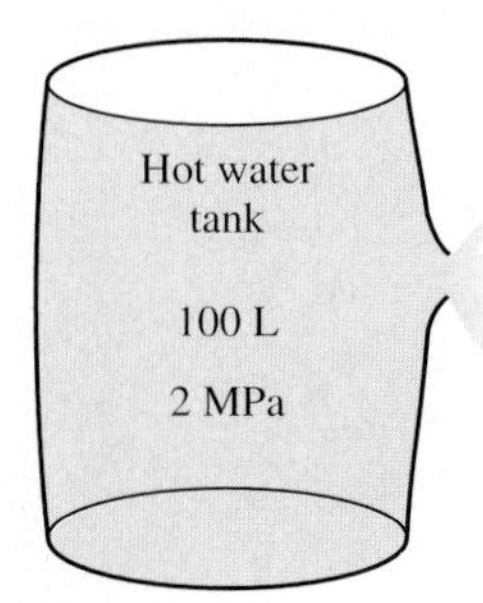


**FIGURE P6-92**

**6-93** Using the arguments in the problem above, determine the total explosion energy of a 0.2-L cannot drink that explodes at a pressure of 1 MPa. How many kg of TNT is this explosion energy equivalent to?

## Computer, Design, and Essay Problems

**6-94** It is well known that the temperature of a gas rises while it is compressed as a result of the energy input in the form of compression work. At high compression ratios, the air temperature may rise above the autoignition temperature of some hydrocarbons, including some lubricating oil. Therefore, the presence of some lubricating oil vapor in high-pressure air raises the possibility of an explosion, creating a fire hazard. The concentration of the oil within the compressor is usually too low to create a real danger. However, the oil that collects on the inner walls of exhaust piping of the compressor may cause an explosion. Such explosions have largely been eliminated by using the proper lubricating oils, carefully designing the equipment, intercooling between compressor stages, and keeping the system clean.

A compressor is to be designed for an industrial application in Los Angeles. If the compressor exit temperature is not to exceed 250°C for safety consideration, determine the maximum allowable compression ratio which is safe for all possible weather conditions for that area.

**6-95** Identify the major sources of entropy generation in your house, and propose ways of reducing them.

**6-96** Obtain the following information about the power plant that is closest to your town: the net power output, the type and amount of fuel, the power consumed by the pumps, fans, and other auxiliary equipment, stack gas losses, temperatures at several locations, and the rate of heat rejection at the condenser. Using this and other relevant data, determine the rate of entropy generation in that power plant.

**6-97** Think about all of your activities all day yesterday. List three of the activities during which you contributed considerably to the entropy increase of the universe. Explain why those activities are irreversible and how they generate entropy.

# Power and Refrigeration Cycles

CHAPTER 7

Two important areas of application for thermodynamics are power generation and refrigeration. Both power generation and refrigeration are usually accomplished by systems that operate on a thermodynamic cycle. Thermodynamic cycles can be divided into general categories: *power cycles,* and *refrigeration cycles*

The devices or systems used to produce a net power output are often called *engines,* and the thermodynamic cycles they operate on are called *power cycles.* The devices or systems used to produce refrigeration are called *refrigerators, air conditioners,* or *heat pumps,* and the cycles they operate on are called *refrigeration cycles.*

Thermodynamic cycles can also be categorized as *gas cycles* or *vapor cycles,* depending on the *phase* of the working fluid—the substance that circulates through the cyclic device. In gas cycles, the working fluid remains in the gaseous phase throughout the entire cycle, whereas in vapor cycles the working fluid exists in the vapor phase during one part of the cycle and in the liquid phase during another part.

Thermodynamic cycles can be categorized yet another way: *closed* and *open cycles.* In closed cycles, the working fluid (such as the steam in steam power plants) is returned to the initial state at the end of the cycle and is recirculated. In open cycles, the working fluid is renewed at the end of each cycle instead of being recirculated. In automobile engines, for example, the combustion gases are exhausted and replaced by fresh air-fuel mixture at the end of each cycle. The engine operates on a mechanical cycle, but the working fluid in this type of device does not go through a complete thermodynamic cycle.

Heat engines are categorized as *internal combustion* or *external combustion engines,* depending on how the heat is supplied to the working fluid. In external combustion engines (such as steam power plants), energy is supplied to the working fluid from an external source such as a furnace, a geothermal well, a nuclear reactor, or even the sun. In internal combustion engines (such as automobile engines), this is done by burning the fuel within the system boundary. In this chapter, various gas power cycles are analyzed under some simplifying assumptions.

Steam is the most common working fluid used in vapor power cycles because of its many desirable characteristics, such as low cost, availability, and high enthalpy of vaporization. Other working fluids used include sodium, potassium, and mercury for high-temperature applications and some organic fluids such as benzene and the freons for low-temperature applications.

Steam power plants are commonly referred to as *coal plants, nuclear plants,* or *natural gas plants,* depending on the type of fuel used to supply heat to the steam. But the steam goes through the same basic cycle in all of them. Therefore, all can be analyzed in the same manner.

The most frequently used refrigeration cycle is the *vapor-compression refrigeration cycle* in which the refrigerant is vaporized and condensed alternately and is compressed in the vapor phase. We also discuss *thermoelectric refrigeration,* where refrigeration is produced by the passage of electric current through two dissimilar materials.

## 7-1 BASIC CONSIDERATIONS IN THE ANALYSIS OF POWER CYCLES

Most power-producing devices operate on cycles, and the study of power cycles is an exciting and important part of thermodynamics. The cycles encountered in actual devices are difficult to analyze because of the presence of complicating effects, such as friction, and the absence of sufficient time for establishment of the equilibrium conditions during the cycle. The make an analytical study of a cycle feasible, we have to keep the complexities at a manageable level and utilize some idealizations (Fig. 7-1). When the actual cycle is stripped of all the internal irreversibilities and complexities, we end up with a cycle that resembles the actual cycle closely but is made up entirely of internally reversible processes. Such a cycle is called an **ideal cycle** (Fig. 7-2).

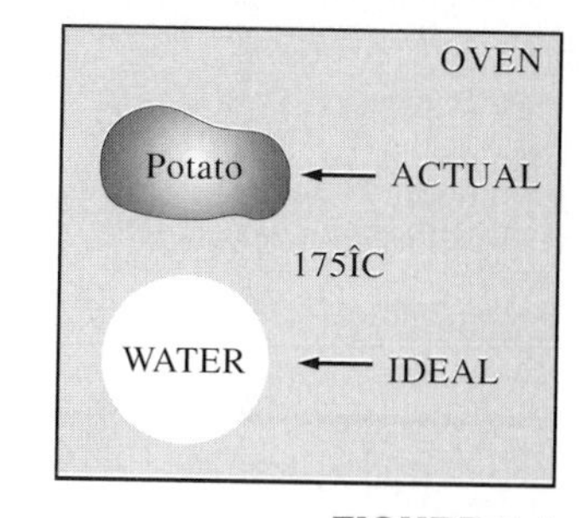


**FIGURE 7-1**

Modeling is a powerful engineering tool that provides great insight and simplicity at the expense of some loss in accuracy.

A simple idealized model enables engineers to study the effects of the major parameters that dominate the cycle without getting bogged down in the details. The cycles discussed in this chapter are somewhat idealized, but they still retain the general characteristics of the actual cycles they represent. The conclusions reached from the analysis of ideal cycles are often applicable to actual cycles. The thermal efficiency of the Otto cycle, the ideal cycle for spark-ignition automobile engines, for example, increases with the compression ratio. This is also the case for actual automobile engines. The numerical values obtained from the analysis of an ideal cycle, however, are not necessarily representative of the actual cycles, and care should be exercised in their interpretation (Fig. 7-3). The simplified analysis presented in this chapter for various power cycles of practical interest may also serve as the starting point for a more in-depth study.

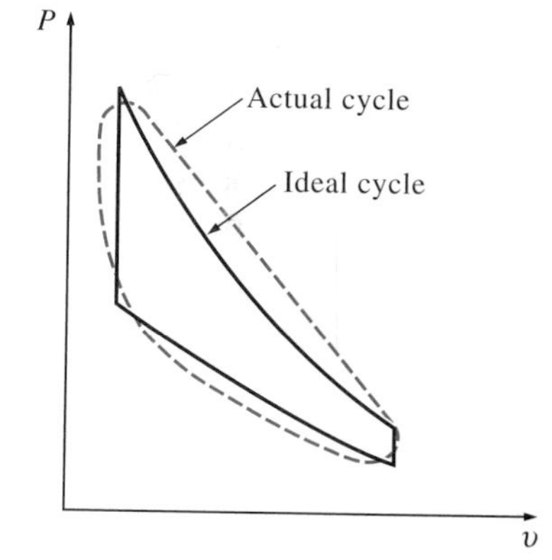


**FIGURE 7-2**

The analysis of many complex processes can be reduced to a manageable level by utilizing some idealizations.

Heat engines are designed for the purpose of converting other forms of energy (usually in the form of heat) to work, and their performance is expressed in terms of the **thermal efficiency** $\eta_{th}$, which is the ratio of the net work produced by the engine to the total heat input:

$$\eta_{th} = \frac{W_{net}}{Q_{in}} \quad \text{or} \quad \eta_{th} = \frac{w_{net}}{q_{in}} \tag{7-1a, b}$$

It was pointed out in Chap. 5 that heat engines that operate on a totally reversible cycle, such as the Carnot cycle, have the highest thermal efficiency of all heat engines operating between the same temperature levels. That is, nobody can develop a cycle more efficient than the Carnot cycle. Then the following question arises naturally: If the Carnot cycle is the best possible cycle, why do we not use it as the model cycle for all the heat engines instead of bothering with several so-called *ideal cycles*? The answer to this equestion is hardware-related. Most cycles encountered in practice differ significantly from the Carnot cycle, which makes it unsuitable as a realistic model. Each ideal cycle discussed in this chapter is related to a specific work-producing device and is an *idealized* version of the actual cycle.

**FIGURE 7-3**

Care should be exercised in the interpretation of the results from ideal cycles.

The ideal cycles are *internally reversible,* but, unlike the Carnot cycle, they are not necessarily externally reversible. That is, they may involve irreversibilities external to the system such as heat transfer through a finite temperature difference. Therefore, the thermal efficiency of an ideal cycle, in general, is less than that of a totally reversible cycle operating between the same temperature limits. However, it is still considerably higher than the thermal efficiency of an actual cycle because of the idealizations utilized.

The idealizations and simplifications commonly employed in the analysis of power cycles can be summarized as follows:

**1** The cycle does not involve any friction. Therefore, the working fluid does not experience any pressure drop as it flows in pipes or devices such as heat exchangers.

**2** All expansion and compression processes take place in a quasi-equilibrium manner (Fig. 7-4).

**3** The pipes connecting the various components of a system are well insulated, and heat transfer through them is negligible.

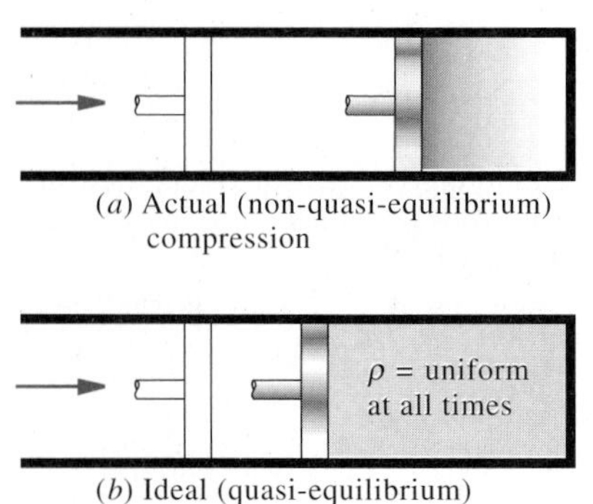


(*a*) Actual (non-quasi-equilibrium) compression

(*b*) Ideal (quasi-equilibrium) compression

**FIGURE 7-4**

All compression and expansion processes in ideal cycles are assumed to be quasi-equilibrium (internally reversible).

Neglecting the changes in *kinetic* and *potential energies* of the working fluid is another commonly utilized simplification in the analysis of power cycles. This is a reasonable assumption since in devices that involve shaft work, such as turbines, compressors, and pumps, the kinetic and potential energy terms are usually very small relative to the other terms in the energy equation. Fluid velocities encountered in devices such as condensers, boilers, and mixing chambers are typically low, and the fluid streams experience little change in their velocities, again making kinetic energy changes negligible. The only devices where the changes in kinetic energy are significant are the nozzles and diffusers, which are specifically designed to create large changes in velocity.

In the preceding chapters, property diagrams such as the $P$-$v$ and $T$-$s$ diagrams have served as valuable aids in the analysis of thermodynamic processes. On both the $P$-$v$ and $T$-$s$ diagrams, the area enclosed by the process curves of a cycle represents the net work produced during the cycle (Fig. 7-5). It is also equivalent to the net heat transfer for that cycle. The $T$-$s$ diagram is particularly useful as a visual aid in the analysis of ideal power cycles. An ideal power cycle does not involve any internal irreversibilities, and so the only effect that can change the entropy of the working fluid during a process is heat transfer.

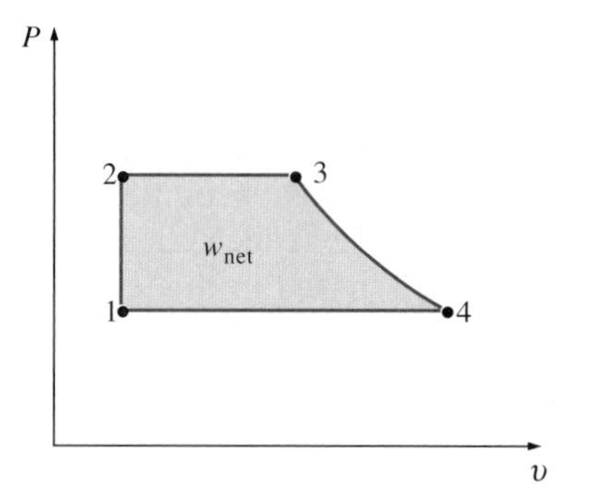


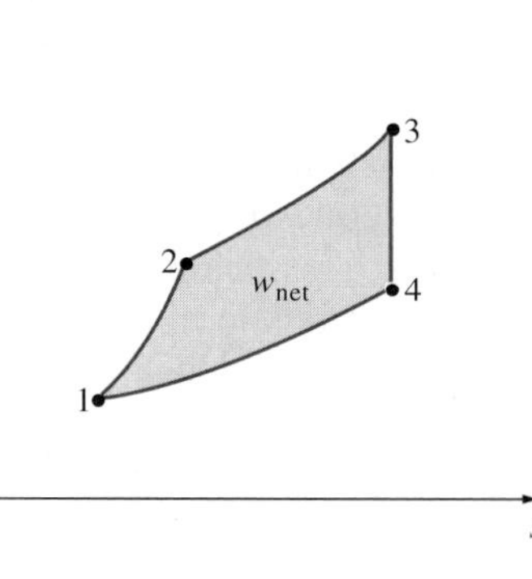


**FIGURE 7-5**

On both $P$-$v$ and $T$-$s$ diagrams, the area enclosed by the process curve represents the net work of the cycle.

On a $T$-$s$ diagram, a *heat addition* process proceeds in the direction of increasing entropy, a *heat rejection* process proceeds in the direction of decreasing entropy, and an *isentropic* (internally reversible, adiabatic) process proceeds at constant entropy. The area under the process curve on a $T$-$s$ diagram represents the heat transfer for that process. The area under the heat addition process on a $T$-$s$ diagram is a geometric measure of the total heat added during the cycle $q_{in}$, and the area under the heat rejection process is a measure of the total heat rejected $q_{out}$. The difference between these two (the area enclosed by the cyclic curve) is the net heat transfer, which is also the net work produced during the cycle. Therefore, on a $T$-$s$ diagram, the ratio of the area enclosed by the cyclic curve to the area under the heat-addition process curve represents the thermal efficiency of the cycle. *Any modification that will increase the ratio of these two areas will also improve the thermal efficiency of the cycle.*

Although the working fluid in an ideal power cycle operates on a closed loop, the type of individual processes that composes the cycle depends on the individual devices used to execute the cycle. In the Rankine cycle, which is the ideal cycle for steam power plants, the working fluid flows through a series of steady-flow devices such as the turbine and condenser, whereas in the Otto cycle, which is the ideal cycle for the spark-ignition automobile engine, the working fluid is alternately expanded and compressed in a piston-cylinder device. Therefore, equations pertaining to steady-flow systems should be used in the analysis of the Rankine cycle, and equations pertaining to closed systems should be used in the analysis of the Otto cycle.

## 7-2 ■ THE CARNOT CYCLE AND ITS VALUE IN ENGINEERING

The Carnot cycle, which was introduced and discussed in Chap. 5, is composed of four totally reversible processes: isothermal heat addition, isentropic expansion, isothermal heat rejection, and isentropic compression. The $P$-$v$ and $T$-$s$ diagrams of a Carnot cycle are replotted in Fig. 7-6. The Carnot cycle can be executed in a closed system (a piston–cylinder device) or a steady-flow system (utilizing two turbines and two compressors, as shown in Fig. 7-7), and either a gas or a vapor can be utilized as the working fluid. The Carnot cycle is the most efficient cycle that can be executed between a thermal energy source at temperature $T_H$ and a sink at temperature $T_L$, and its thermal efficiency is expressed as

$$\eta_{\text{th,Carnot}} = 1 - \frac{T_L}{T_H} \tag{7-2}$$

Reversible isothermal heat transfer is very difficult to achieve in reality because it would require very large heat exchangers and it would take a very long time (a power cycle in a typical engine is completed in a fraction of a second). Therefore, it is not practical to build an engine that would operate on a cycle that closely approximates the Carnot cycle.

The real value of the Carnot cycle comes from its being a standard

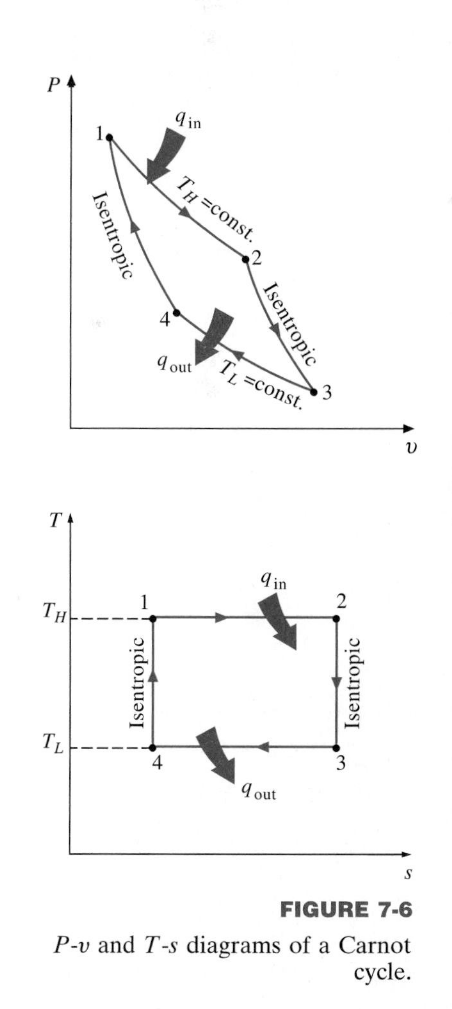


**FIGURE 7-6**

$P$-$v$ and $T$-$s$ diagrams of a Carnot cycle.

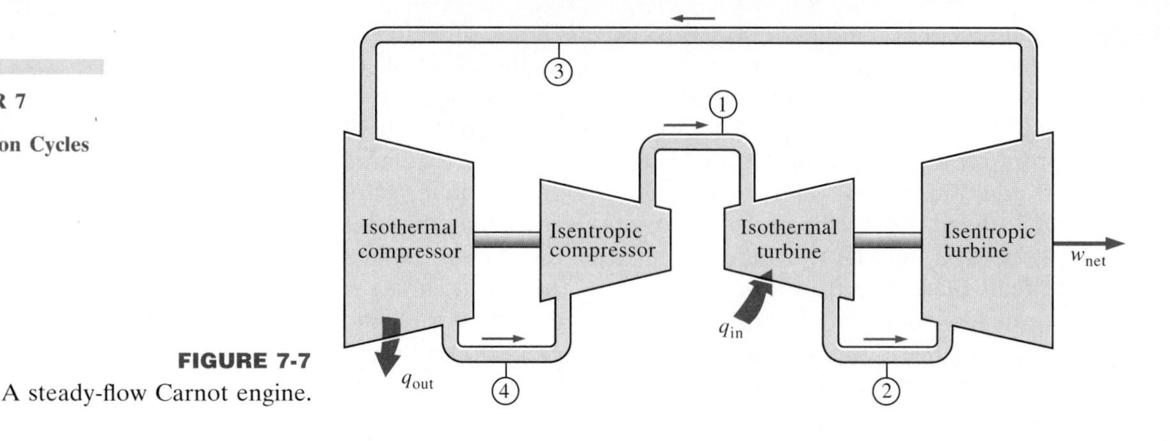


**FIGURE 7-7**
A steady-flow Carnot engine.

against which the actual or other ideal cycles can be compared. The thermal efficiency of the Carnot cycle is a function of the sink and source temperatures only, and the thermal efficiency relation for the Carnot cycle (Eq. 7-2) conveys an important message that is equally applicable to both ideal and actual cycles: *Thermal efficiency increases with an increase in the average temperature at which heat is added to the system or with a decrease in the average temperature at which heat is rejected from the system.*

The source and sink temperatures that can be used in practice are not without limits, however. The highest temperature in the cycle is limited by the maximum temperature that the components of the heat engine, such as the piston or the turbine blades, can withstand. The lowest temperature is limited by the temperature of the cooling medium utilized in the cycle such as a lake, a river, or the atmospheric air.

**EXAMPLE 7-1**

Show that the thermal efficiency of a Carnot cycle operating between the temperature limits of $T_H$ and $T_L$ is solely a function of these two temperatures and is given by Eq. 7-2.

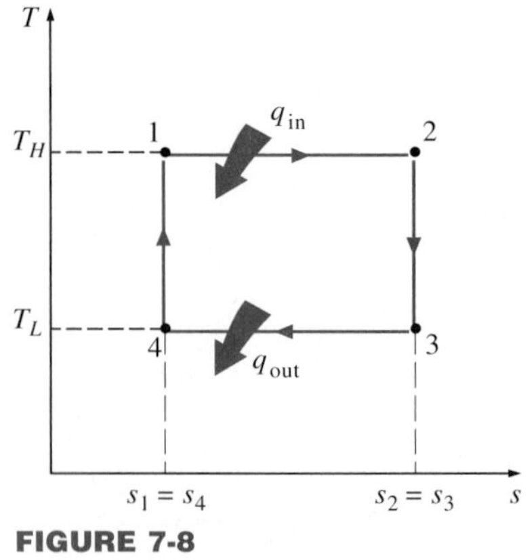


**FIGURE 7-8**
*T-s* diagram for Example 7-1.

**Solution** The *T-s* diagram of a Carnot cycle is redrawn in Fig. 7-8. All four processes that compose the Carnot cycle are reversible, and thus the area under each process curve represents the heat transfer for that process. Heat is transferred to the system during process 1-2 and rejected during process 3-4. Therefore, the amount of heat input and heat output for the cycle can be expressed as

$$q_{in} = T_H(s_2 - s_1) \qquad \text{and} \qquad q_{out} = T_L(s_3 - s_4) = T_L(s_2 - s_1)$$

since processes 2-3 and 4-1 are isentropic, and thus $s_2 = s_3$ and $s_4 = s_1$. Substituting these into Eq. 7-1*b*, we see that the thermal efficiency of a Carnot cycle is

$$\eta_{th} = \frac{w_{net}}{q_{in}} = 1 - \frac{q_{out}}{q_{in}} = 1 - \frac{T_L(s_2 - s_1)}{T_H(s_2 - s_1)} = 1 - \frac{T_L}{T_H}$$

which is the desired result. Notice that the thermal efficiency of a Carnot cycle is independent of the type of the working fluid used (an ideal gas, steam, etc.) or whether the cycle is executed in a closed or steady-flow system.



## 7-3 AIR-STANDARD ASSUMPTIONS

In gas power cycles, the working fluid remains a gas throughout the entire cycle. Spark-ignition automobile engines, diesel engines, and conventional gas turbines are familiar examples of devices that operate on gas cycles. In all these engines, energy is provided by burning a fuel within the system boundaries. That is, they are *internal combustion engines.* Because of this combustion process, the composition of the working fluid changes from air and fuel to combustion products during the course of the cycle. However, considering that air is predominantly nitrogen, which undergoes hardly any chemical reactions in the combustion chamber, the working fluid closely resembles air at all times.

Even though internal combustion engines operate on a mechanical cycle (the piston returns to its starting position at the end of each revolution), the working fluid does not undergo a complete thermodynamic cycle. It is thrown out of the engine at some point in the cycle (as exhaust gases) instead of being returned to the initial state. Working on an open cycle is the characteristic of all internal combustion engines.

The actual gas power cycles are rather complex. To reduce the analysis to a manageable level, we utilize the following approximations, commonly known as the **air-standard assumptions**.

1 The working fluid is air that continuously circulates in a closed loop and always behaves as an ideal gas.

2 All the processes that make up the cycle are internally reversible.

3 The combustion process is replaced by a heat addition process from an external source (Fig. 7-9).

4 The exhaust process is replaced by a heat rejection process that restores the working fluid to its initial state.

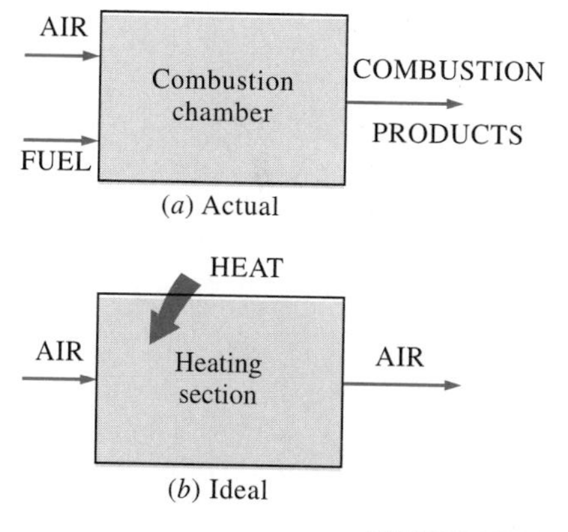


**FIGURE 7-9**
The combustion process is replaced by a heat addition process in ideal cycles.

Another assumption that is often utilized to simplify the analysis even more is that the air has constant specific heats whose values are determined at *room temperature* (25°C, or 77°F). When this assumption is utilized, the air-standard assumptions are called the **cold-air-standard assumptions**. A cycle for which the air-standard assumptions are applicable is frequently referred to as an **air-standard cycle**.

The air-standard assumptions stated above provide considerable simplification in the analysis without significantly deviating from the actual cycles. This simplified model enables us to study qualitatively the influence of major parameters on the performance of the actual engines.

## 7-4 ■ BRIEF OVERVIEW OF RECIPROCATING ENGINES

Despite its simplicity, the reciprocating engine (basically a piston–cylinder device) is one of the rare inventions that has proved to be very versatile and to have a wide range of applications. It is the powerhouse of the vast majority of automobiles, trucks, light aircraft, ships, and electric power generators, as well as many other devices.

The basic components of a reciprocating engine are shown in Fig. 7-10. The piston reciprocates in the cylinder between two fixed positions called the **top dead center** (TDC)—the position of the piston when it forms the smallest volume in the cylinder—and the **bottom dead center** (BDC)—the position of the piston when it forms the largest volume in the cylinder. The distance between the TDC and the BDC is the largest distance that the piston can travel in one direction, and it is called the **stroke** of the engine. The diameter of the piston is called the **bore**. The air or air-fuel mixture is drawn into the cylinder through the **intake valve**, and the combustion products are expelled from the cylinder through the **exhaust valve**.

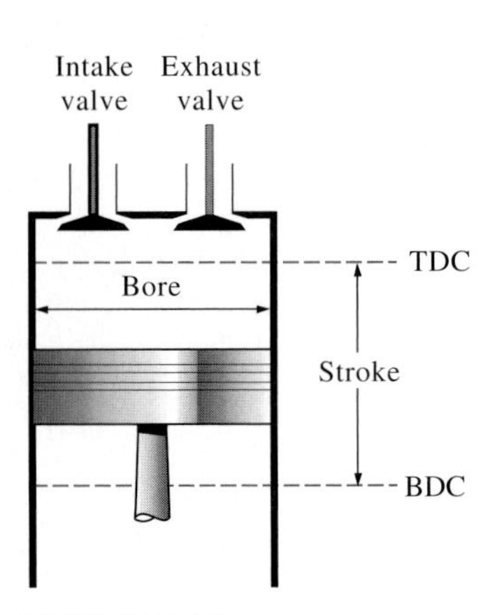


**FIGURE 7-10**

Nomenclature for reciprocating engines.

The minimum volume formed in the cylinder when the piston is at TDC is called the **clearance volume** (Fig. 7-11). The volume displaced by the piston as it moves between TDC and BDC is called the **displacement volume**. The ratio of the maximum volume formed in the cylinder to the minimum (clearance) volume is called the **compression ratio** $r$ of the engine:

$$r = \frac{V_{\max}}{V_{\min}} = \frac{V_{\mathrm{BDC}}}{V_{\mathrm{TDC}}} \tag{7-3}$$

Notice that the compression ratio is a *volume ratio* and should not be confused with the pressure ratio.

Another term frequently used in conjunction with reciprocating engines is the **mean effective pressure** (MEP). It is a fictitious pressure which, if it acted on the piston during the entire power stroke, would

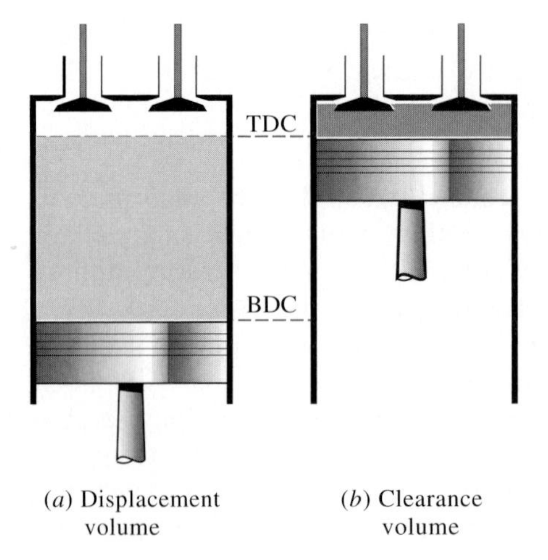


**FIGURE 7-11**

Displacement and clearance volumes of a reciprocating engine.

produce the same amount of net work as that produced during the actual cycle (Fig. 7-12). That is,

$$W_{net} = \text{MEP} \times \text{piston area} \times \text{stroke} = \text{MEP} \times \text{displacement volume}$$

or

$$\text{MEP} = \frac{W_{net}}{V_{max} - V_{min}} = \frac{w_{net}}{v_{max} - v_{min}} \quad \text{(kPa)} \qquad (7\text{-}4)$$

The mean effective pressure can be used as a parameter to compare the performances of reciprocating engines of equal size. The engine that has a larger value of MEP will deliver more net work per cycle and thus will perform better.

Reciprocating engines are classified as **spark-ignition** (SI) **engines** or **compression-ignition** (CI) **engines**, depending on how the combustion process in the cylinder is initiated. In SI engines, the combustion of the air-fuel mixture is initiated by a spark plug. In CI engines, the air-fuel mixture is self-ignited as a result of compressing the mixture above its self-ignition temperature. In the next two sections, we discuss the *Otto* and *Diesel cycles,* which are the ideal cycles for the SI and CI reciprocating engines, respectively.

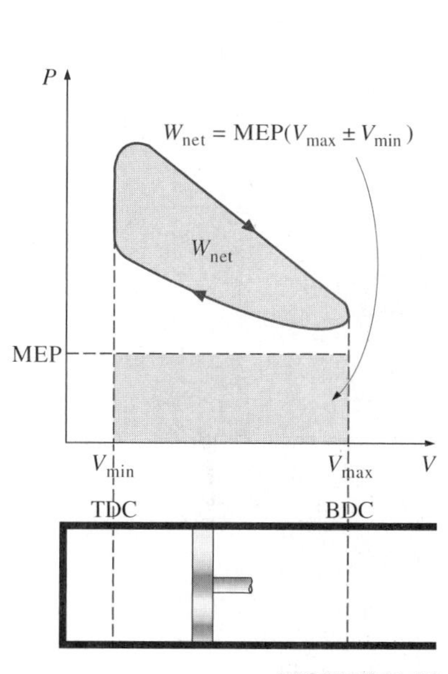

**FIGURE 7-12**
The net work output of a cycle is equivalent to the product of the mean effective pressure and the displacement volume.

## 7-5 ■ OTTO CYCLE—THE IDEAL CYCLE FOR SPARK-IGNITION ENGINES

The Otto cycle is the ideal cycle for spark-ignition reciprocating engines. It is named after Nikolaus A. Otto, who built a successful four-stroke engine in 1876 in Germany using the cycle proposed by Frenchman Beau de Rochas in 1862. In most spark-ignition engines, the piston executes four complete strokes (two mechanical cycles) within the cylinder, and the crankshaft completes 2 revolutions for each thermodynamic cycle. These engines are called **four-stroke** internal combustion engines. A schematic of each stroke as well as a *P-v* diagram for an actual four-stroke spark-ignition engine is given in Fig. 7-13*a.*

Initially, both the intake and the exhaust valves are closed, and the piston is at its lowest position (BDC). During the *compression stroke,* the piston moves upward, compressing the air–fuel mixture. Shortly before the piston reaches its highest position (TCD), the spark plug fires and the mixture ignites, increasing the pressure and temperature of the system. The high-pressure gases force the piston down, which in turn forces the crankshaft to rotate, producing a useful work output during the *expansion* or *power stroke.* At the end of this stroke, the piston is at its lowest position (the completion of the first mechanical cycle), and the cylinder is filled with combustion products. Now the piston moves upward one more time, purging the exhaust gases through the exhaust valve (the *exhaust stroke*), and down a second time, drawing in fresh air-fuel mixture through the intake valve (the *intake stroke*). Notice that the pressure in the cylinder is slightly above the atmospheric value during the exhaust stroke and slightly below during the intake stroke.

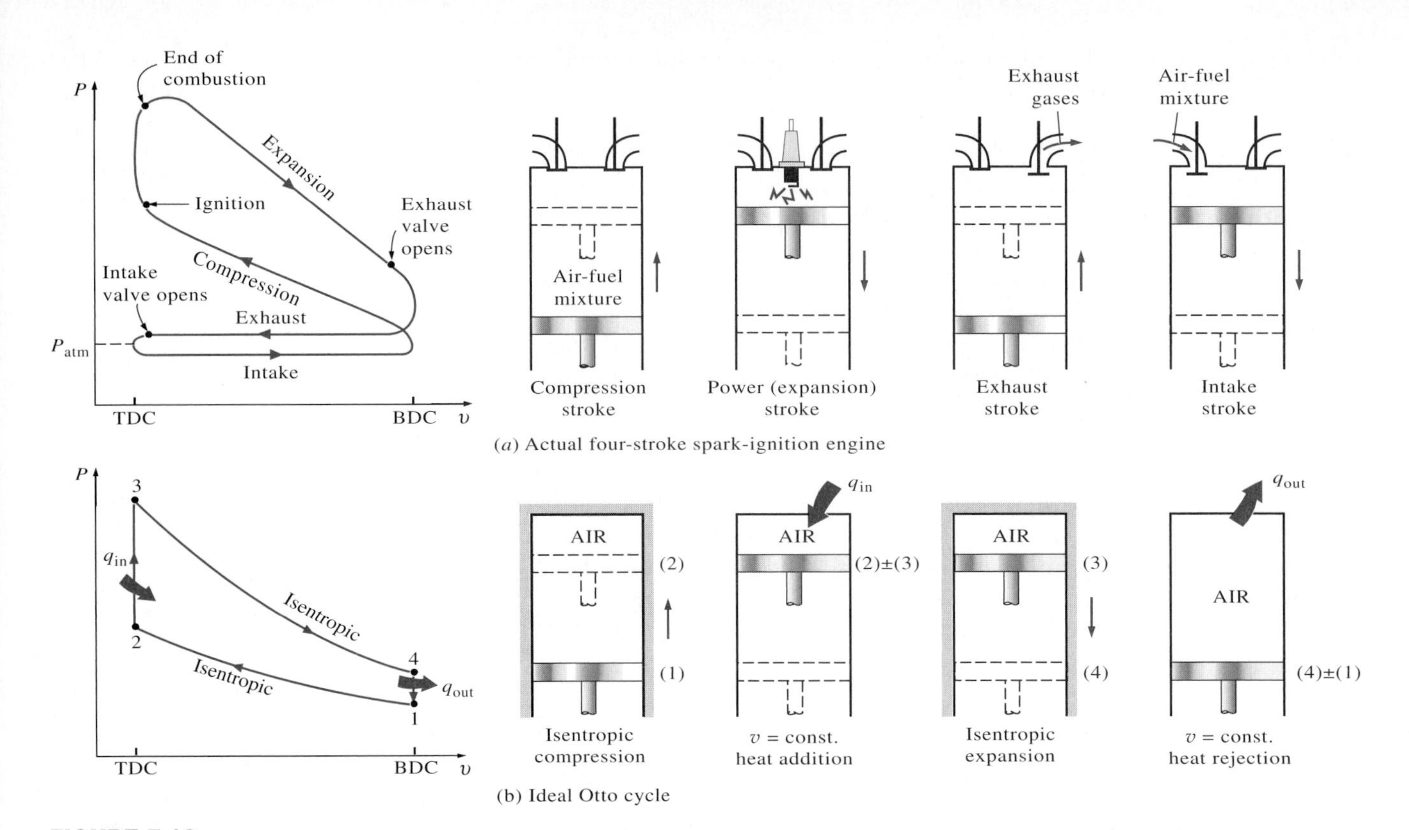


**FIGURE 7-13**

Actual and ideal cycles in spark-ignition engines and their *P-v* diagrams.

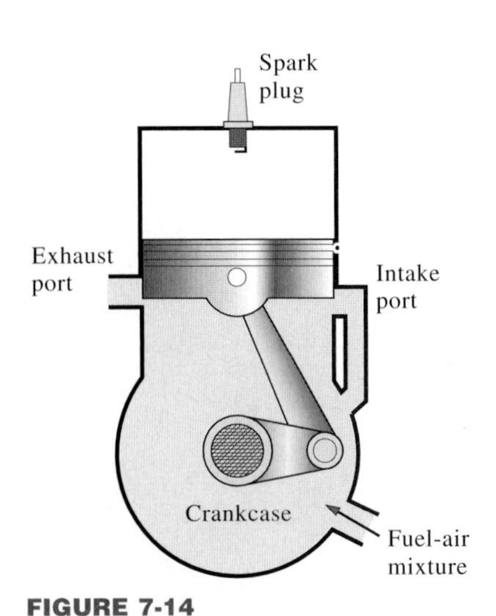


**FIGURE 7-14**

Schematic of a two-stroke reciprocating engine.

In **two-stroke engines**, all four functions described above are executed in just two strokes: the power stroke and the compression stroke. In these engines, the crankcase is sealed, and the outward motion of the piston is used to slightly pressurize the air–fuel mixture in the crankcase, as shown in Fig. 7-14. Also, the intake and exhaust valves are replaced by openings in the lower portion of the cylinder wall. During the latter part of the power stroke, the piston uncovers first the exhaust port, allowing the exhaust gases to be partially expelled, and then the intake port, allowing the fresh air–fuel mixture to rush in and drive most of the remaining exhaust gases out of the cylinder. This mixture is then compressed as the piston moves upward during the compression stroke and is subsequently ignited by a spark plug.

The two-stroke engines are generally less efficient than their four-stroke counterparts because of the incomplete expulsion of the exhaust gases and the partial expulsion of the fresh air–fuel mixture with the exhaust gases. However, they are relatively simple and inexpensive, and they have high power-to-weight and power-to-volume ratios, which make them suitable for applications requiring small size and weight such as for motorcycles, chain saws, and lawn mowers.

Advances in several technologies—such as direct fuel injection, stratified charge combustion, and electronic controls—brought about a renewed interest in two-stroke engines, which can offer high performance and fuel economy while satisfying the future stringent emission requirements. For a given weight and displacement, a well-designed two-stroke engine can provide significantly more power than its four-stroke counterpart because two-stroke engines produce power on every engine revolution instead of every other one. In the new two-stroke engines under development, the highly atomized fuel spray that is injected with compressed air into the combustion chamber towards the end of the compression stroke burns much more completely. The fuel is sprayed after the exhaust valve is closed, which prevents unburned fuel from being ejected into the atmosphere. With stratified combustion, the flame, which is initiated by igniting a small amount of rich fuel/air mixture near the spark plug, propagates through the combustion chamber filled with much leaner mixture, and this results in much cleaner combustion. Also, the advances in electronics made it possible to ensure the optimum operation under varying engine load and speed conditions. Major car companies have research programs underway on two-stroke engines which are expected to make a comeback in the near future.

The thermodynamic analysis of the actual four-stroke or two-stroke cycles described above is not a simple task. However, the analysis can be simplified significantly if the air-standard assumptions are utilized. The resulting cycle which closely resembles the actual operating conditions is the ideal Otto cycle. It consists of four internally reversible processes:

- 1-2 Isentropic compression
- 2-3 $v$ = constant heat addition
- 3-4 Isentropic expansion
- 4-1 $v$ = constant heat rejection

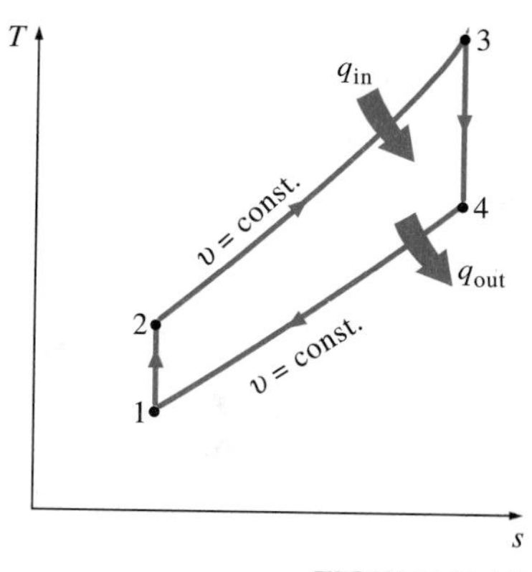

**FIGURE 7-15**

$T$-$s$ diagram for the ideal Otto cycle.

The execution of the Otto cycle in a piston–cylinder device together with a $P$-$v$ diagram is illustrated in Fig. 7-13$b$. The $T$-$s$ diagram of the Otto cycle is given in Fig. 7-15.

The Otto cycle is executed in a closed system, and thus the first-law relation for any of the processes is expressed, on a unit-mass basis, as

$$q - w = \Delta u \qquad \text{(kJ/kg)} \tag{7-5}$$

No work is involved during the two heat transfer processes since both take place at constant volume. Therefore, heat transfer to and from the working fluid can be expressed, under the cold-air-standard assumptions, as

$$q_{in} = q_{23} = u_3 - u_2 = C_v(T_3 - T_2) \tag{7-6a}$$

and

$$q_{out} = -q_{41} = -(u_1 - u_4) = C_v(T_4 - t_1) \tag{7-6b}$$

Then the thermal efficiency of the ideal-air-standard Otto cycle becomes

$$\eta_{th,Otto} = \frac{w_{net}}{q_{in}} = 1 - \frac{q_{out}}{q_{in}} = 1 - \frac{T_4 - T_1}{T_3 - T_2} = 1 - \frac{T_1(T_4/T_1 - 1)}{T_2(T_3/T_2 - 1)}$$

Processes 1-2 and 3-4 are isentropic, and $v_2 = v_3$ and $v_4 = v_1$. Thus,

$$\frac{T_1}{T_2} = \left(\frac{v_2}{v_1}\right)^{k-1} = \left(\frac{v_3}{v_4}\right)^{k-1} = \frac{T_4}{T_3} \tag{7-7}$$

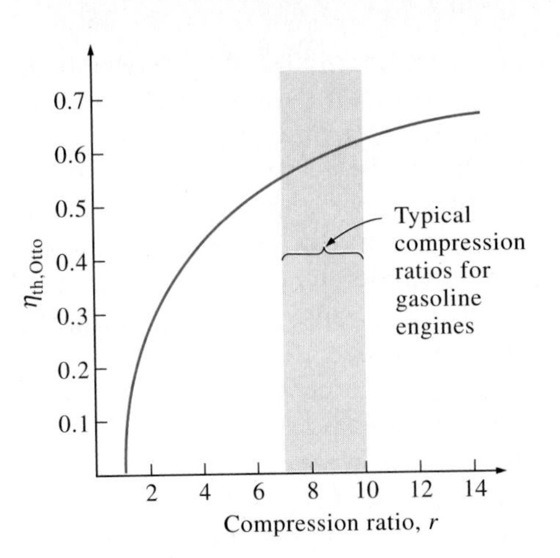


**FIGURE 7-16**

Thermal efficiency of the ideal Otto cycle as a function of compression ratio ($k = 1.4$).

Substituting these equations into the thermal efficiency relation and simplifying gives

$$\eta_{th,Otto} = 1 - \frac{1}{r^{k-1}} \tag{7-8}$$

where

$$r = \frac{V_{max}}{V_{min}} = \frac{V_1}{V_2} = \frac{v_1}{v_2} \tag{7-9}$$

is the compression ratio and $k$ is the specific heat ratio $C_p/C_v$.

Equation 7-8 shows that under the cold-air-standard assumptions, the thermal efficiency of an ideal Otto cycle depends on the compression ratio of the engine and the specific heat ratio of the working fluid (if different from air). The thermal efficiency of the ideal Otto cycle increases with both the compression ratio and the specific heat ratio. This is also true for actual spark-ignition internal combustion engines. A plot of thermal efficiency versus the compression ratio is given in Fig. 7-16 for $k = 1.4$, which is the specific-heat-ratio value of air at room temperature. For a given compression ratio, the thermal efficiency of an actual spark-ignition engine will be less than that of an ideal Otto cycle because of the irreversibilities, such as friction, and other factors such as incomplete combustion.

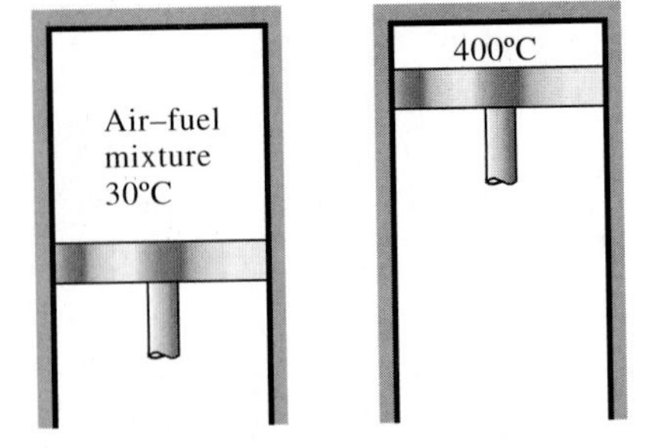


**FIGURE 7-17**

At high compression ratios, the air-fuel mixture temperature rises above the self-ignition temperature of the fuel during the compression process.

We can observe from Fig. 7-16 that the thermal efficiency curve is rather steep at low compression ratios but flattens out starting with a compression ratio value of about 8. Therefore, the increase in thermal efficiency with the compression ratio is not that pronounced at high compression ratios. Also, when high compression ratios are used, the temperature of the air-fuel mixture rises above the autoignition temperature of the fuel (the temperature at which the fuel ignites without the help of a spark) during the combustion process, causing an early and rapid burn of the fuel at some point or points ahead of the flame front, followed by almost instantaneous inflammation of the end gas (Fig. 7-17). This premature ignition of the fuel, called **autoignition**, produces an audible noise, which is called **engine knock**. Autoignition in spark-ignition engines cannot be tolerated because it hurts performance and can cause engine damage. The requirement that autoignition not be allowed places an upper limit on the compression ratios that can be used in spark-ignition internal combustion engines.

Improvement of the thermal efficiency of gasoline engines by utilizing higher compression ratios (up to about 12) without facing the autoignition problem has been made possible by using gasoline blends that have good antiknock characteristics, such as gasoline mixed with tetraethyl lead. Tetraethyl lead has been added to gasoline since the 1920s because it is the cheapest method of raising the *octane rating,* which is a measure of the engine knock resistance of a fuel. Leaded gasoline, however, has a very undesirable side effect: it forms compounds during the combustion process that are hazardous to health and pollute the environment. In an effort to combat air pollution, the government adopted a policy in the mid-1970s that resulted in the eventual phase-out of the leaded gasoline. Unable to use lead, the refiners developed other, more elaborate techniques to improve the antiknock characteristics of the gasoline. Most cars made since 1975 have been designed to use unleaded gasoline, and the compression ratios had to be lowered to avoid engine knock. The thermal efficiency of car engines has decreased somewhat as a result of decreased compression ratios. But, owing to the improvements in other areas (reduction in overall automobile weight, improved aerodynamic design, etc.), today's cars have better fuel economy and consequently get more miles per gallon of fuel. This is an example of how engineering decisions involve compromises, and efficiency is only one of the considerations in reaching a final decision.

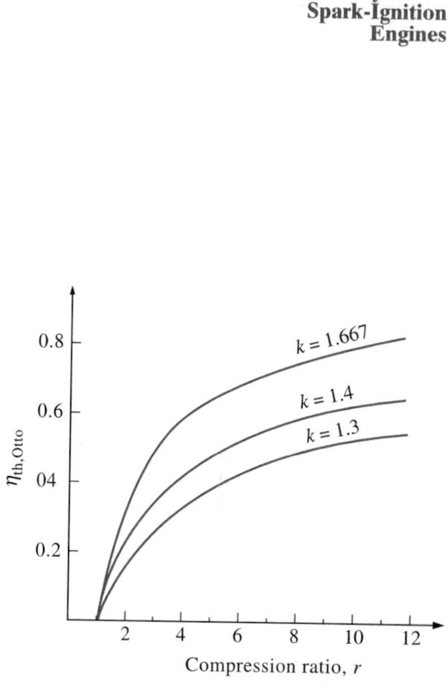


**FIGURE 7-18**

The thermal efficiency of the Otto cycle increases with the specific heat ratio $k$ of the working fluid.

The second parameter affecting the thermal efficiency of an ideal Otto cycle is the specific heat ratio $k$. For a given compression ratio, an ideal Otto cycle using a monatomic gas (such as argon or helium, $k = 1.667$) as the working fluid will have the highest thermal efficiency. The specific heat ratio $k$, and thus the thermal efficiency of the ideal Otto cycle, decreases as the molecules of the working fluid get larger (Fig. 7-18). At room temperature it is 1.4 for air, 1.3 for carbon dioxide, and 1.2 for ethane. The working fluid in actual engines contains larger molecules such as carbon dioxide, and the specific heat ratio decreases with temperature, which is one of the reasons that the actual cycles have lower thermal efficiencies than the ideal Otto cycle. The thermal efficiencies of actual spark-ignition engines range from about 25 to 30 percent.

**EXAMPLE 7-2**

An ideal Otto cycle has a compression ratio of 8. At the beginning of the compression process, the air is at 100 kPa and 17°C, and 800 kJ/kg of heat is transferred to air during the constant-volume heat addition process. Determine (*a*) the maximum temperature and pressure that occur during the cycle, (*b*) the net work output, (*c*) the thermal efficiency, and (*d*) the mean effective pressure for the cycle.

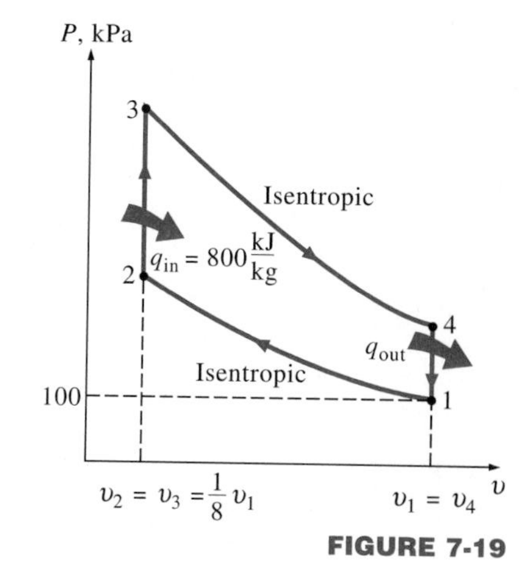


**FIGURE 7-19**

*P-v* diagram for the Otto cycle discussed in Example 7-2.

**Solution** The Otto cycle described is shown on a *P-v* diagram in Fig. 7-19.

The air contained in the cylinder forms a closed system, which should be remembered in applying the thermodynamic laws to the individual processes. We assume constant specific heat for air at room temperature and thus take $C_p = 1.005\ \text{kJ/(kg}\cdot\text{K)}$, $C_v = 0.718\ \text{kg}\cdot\text{K)}$, and $k = 1.4$.

(*a*) The maximum temperature and pressure in an Otto cycle occur at the end of the constant-volume heat addition process (state 3). But first we need to determine the temperature and pressure of air at the end of the isentropic compression process (state 2):

$$T_2 = T_1\left(\frac{v_1}{v_2}\right)^{k-1} = T_1(r)^{k-1} = (290\ \text{K})(8)^{1.4-1} = 666.2\ \text{K}$$

$$\frac{P_2 v_2}{T_2} = \frac{P_1 v_1}{T_1} \rightarrow P_2 = P_1\left(\frac{T_2}{T_1}\right)\left(\frac{v_1}{v_2}\right)$$

$$= (100\ \text{kPa})\left(\frac{666.2\ \text{K}}{290\ \text{K}}\right)(8) = 1837.8\ \text{kPa}$$

Process 2-3 ($v$ = constant heat addition):

$$q_{23} - \cancelto{0}{w_{23}} = u_3 - u_2 = C_v(T_3 - T_2)$$

$$T_3 = T_2 + \frac{q_{23}}{C_v} = 666.2\ \text{K} + \frac{800\ \text{kJ/kg}}{0.718\ \text{kJ/(kg}\cdot\text{K)}} = 1780.4\ \text{K}$$

$$\frac{P_3 v_3}{T_3} = \frac{P_2 v_2}{T_2} \rightarrow P_3 = P_2\left(\frac{T_3}{T_2}\right)\left(\frac{v_2}{v_3}\right)$$

$$= (1873.8\ \text{MPa})\left(\frac{1780.4\ \text{K}}{666.2\ \text{K}}\right)(1) = 4911.5\ \text{kPa}$$

(*b*) The net work output for the cycle is determined either by finding the boundary ($P\,dV$) work involved in each process by integration and adding them or by finding the net heat transfer which is equivalent to the net work done during the cycle. We take the latter approach. But first we need to find the internal energy of the air at state 4:

Process 3-4 (isentropic expansion of an ideal gas):

$$T_4 = T_3\left(\frac{v_3}{v_4}\right)^{k-1} = T_3\left(\frac{1}{r}\right)^{k-1} = (1780.4\ \text{K})\left(\frac{1}{8}\right)^{1.4-1} = 775.0\ \text{K}$$

Process 4-1 ($v$ = constant heat rejection):

$$q_{41} - \cancelto{0}{w_{41}} = u_1 - u_4 = C_v(T_1 - T_4)$$

or

$$q_{out} = -q_{41} = C_v(T_4 - T_1) = [0.718\ \text{kJ/(kg}\cdot\text{K)}](775 - 290)\ \text{K} = 348.2\ \text{kJ/kg}$$

Thus, $\quad w_{net} = q_{net} = q_{in} - q_{out} = (800 - 348.2)\ \text{kJ/kg} = 451.8\ \text{kJ/kg}$

(*c*) The thermal efficiency of the cycle is determined from its definition, Eq. 7-1:

$$\eta_{th} = \frac{w_{net}}{q_{in}} = \frac{451.8\ \text{kJ/kg}}{800\ \text{kJ/kg}} = 0.565 \text{ or } 56.5\%$$

Under the cold-air-standard assumptions (constant specific heat values at room temperature, the thermal efficiency could also be determined from

$$\eta_{\text{th,Otto}} = 1 - \frac{1}{r^{k-1}} = 1 - r^{1-k} = 1 - (8)^{1-1.4} = 0.565 \text{ or } 56.5\%$$

which is identical to the value obtained above.

(*d*) The mean effective pressure is determined from its definition, Eq. 7-4:

$$\text{MEP} = \frac{w_{\text{net}}}{v_1 - v_2} = \frac{w_{\text{net}}}{v_1 - v_1/r} = \frac{w_{\text{net}}}{v_1(1 - 1/r)}$$

where $$v_1 = \frac{RT_1}{P_1} = \frac{[0.287 \text{ kPa}\cdot\text{m}^3/(\text{kg}\cdot\text{K})](290 \text{ K})}{100 \text{ kPa}} = 0.832 \text{ m}^3/\text{kg}$$

Thus, $$\text{MEP} = \frac{451.8 \text{ kJ/kg}}{(0.832 \text{ m}^3/\text{kg})(1 - \frac{1}{8})}\left(\frac{1 \text{ kPa}\cdot\text{m}^3}{1 \text{ kJ}}\right) = 620.6 \text{ kPa}$$

Therefore, a constant pressure of 620.6 kPa during the power stroke would produce the same net work output as the entire cycle.

## 7-6 ■ DIESEL CYCLE—THE IDEAL CYCLE FOR COMPRESSION-IGNITION ENGINES

The Diesel cycle is the ideal cycle for CI reciprocating engines. The CI engine, first proposed by Rudolph Diesel in the 1890s, is very similar to the SI engine discussed in the last section, differing mainly in the method of initiating combustion. In spark–ignition engines (also known as *gasoline engines*), the air–fuel mixture is compressed to a temperature that is below the autoignition temperature of the fuel, and the combustion process is initiated by firing a spark plug. In CI engines (also known as *diesel engines*), the air is compressed to a temperature which is above the autoignition temperature of the fuel, and combustion starts on contact as the fuel is injected into this hot air. Therefore, the spark plug and carburetor are replaced by a fuel injector in diesel engines (Fig. 7-20).

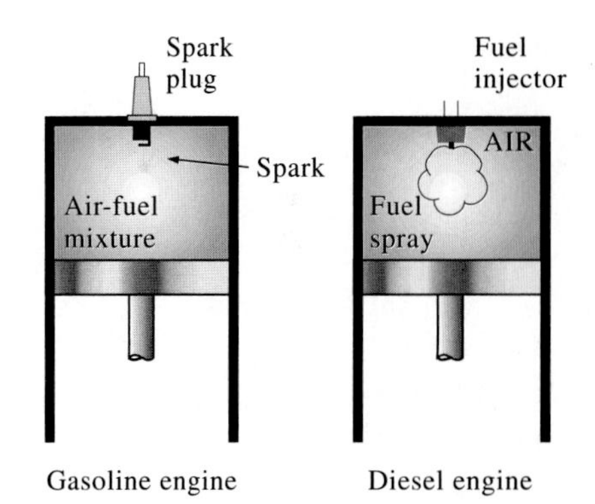


**FIGURE 7-20**
In diesel engines, the spark plug is replaced by a fuel injector, and only air is compressed during the compression process.

In gasoline engines, a mixture of air and fuel is compressed during the compression stroke, and the compression ratios are limited by the onset of autoignition or engine knock. In diesel engines, only air is compressed during the compression stroke, eliminating the possibility of autoignition. Therefore, diesel engines can be designed to operate at much higher compression ratios, typically between 12 and 24. Not having to deal with the problem of autoignition has another benefit: many of the stringent requirements placed on the gasoline can now be removed, and fuels that are less refined (thus less expensive) can be used in diesel engines.

The fuel injection process in diesel engines starts when the piston approaches TDC and continues during the first part of the power stroke. Therefore, the combustion process in these engines takes place over a longer interval. Because of this longer duration, the combustion process in the ideal Diesel cycle is approximated as a constant-pressure heat addition process. In fact, this is the only process where the Otto and the Diesel cycles differ. The remaining three processes are the same for both ideal cycles. That is, process 1-2 is isentropic compression, 3-4 is isentropic expansion, and 4-1 is constant-volume heat rejection. The similarity between the two cycles is also apparent from the $P$-$v$ and $T$-$s$ diagrams of the Diesel cycle, shown in Fig. 7-21.

A measure of performance for any power cycle is its thermal efficiency. Below we develop a relation for the thermal efficiency of a Diesel cycle, utilizing the cold-air-standard assumptions. Such a relation will enable us to examine the effects of major parameters on the performance of diesel engines.

The Diesel style cycle, like the Otto cycle, is executed in a piston–cylinder device, which forms a closed system. Therefore, equations developed for closed systems should be used in the analysis of individual processes. Under the cold-air-standard assumptions, the amount of heat added to the working fluid at constant pressure and rejected from it at constant volume can be expressed as

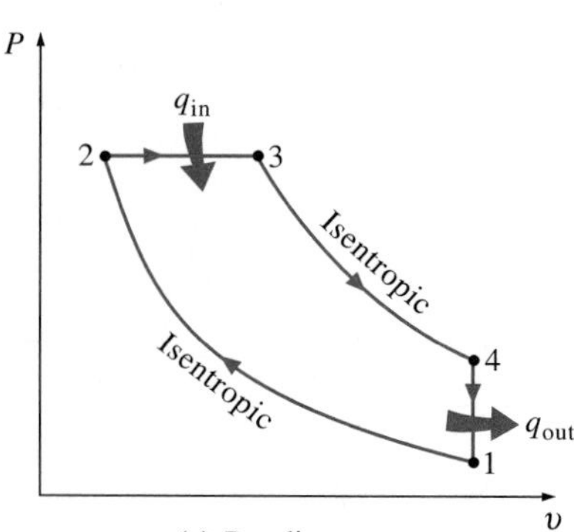

(a) $P$-$v$ diagram

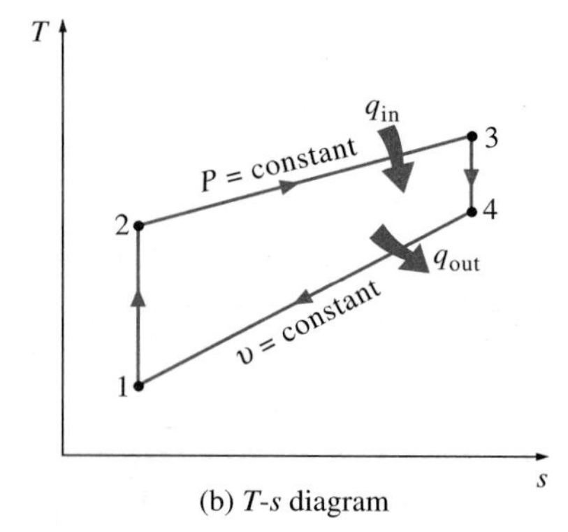

(b) $T$-$s$ diagram

**FIGURE 7-21**
$T$-$s$ and $P$-$v$ diagrams for the ideal Diesel cycle.

$$\begin{aligned} q_{in} = q_{23} &= w_{23} + (\Delta u)_{23} = P_2(v_3 - v_2) + (u_3 - u_2) \\ &= h_3 - h_2 = C_p(T_3 - T_2) \end{aligned} \tag{7-10a}$$

and

$$\begin{aligned} q_{out} = -q_{41} &= -\overset{\nearrow 0}{\cancel{w_{41}}} - (\Delta u)_{41} = u_4 - u_1 \\ &= C_v(T_4 - T_1) \end{aligned} \tag{7-10b}$$

Then the thermal efficiency of the ideal Diesel cycle under the cold-air-standard assumptions becomes

$$\eta_{th,Diesel} = \frac{w_{net}}{q_{in}} = 1 - \frac{q_{out}}{q_{in}} = 1 - \frac{T_4 - T_1}{k(T_3 - T_2)} = 1 - \frac{T_1(T_4/T_1 - 1)}{kT_2(T_3/T_2 - 1)}$$

We now define a new quantity, the **cutoff ratio** $r_c$, as the ratio of the cylinder volumes after and before the combustion process:

$$r_c = \frac{V_3}{V_2} = \frac{v_3}{v_2} \tag{7-11}$$

Utilizing this definition and the isentropic ideal-gas relations for processes 1-2 and 3-4, we see that the thermal efficiency relation reduces to

$$\eta_{th,Diesel} = 1 - \frac{1}{r^{k-1}}\left[\frac{r_c^k - 1}{k(r_c - 1)}\right] \tag{7-12}$$

where $r$ is the compression ratio defined by Eq. 7-9. Looking at Eq. 7-12 carefully, one would notice that under the cold-air-standard assumptions, the efficiency of a Diesel cycle differs from the efficiency of an Otto cycle by the quantity in the brackets. This quantity is always greater than 1. Therefore,

$$\eta_{th,Otto} > \eta_{th,Diesel} \tag{7-13}$$

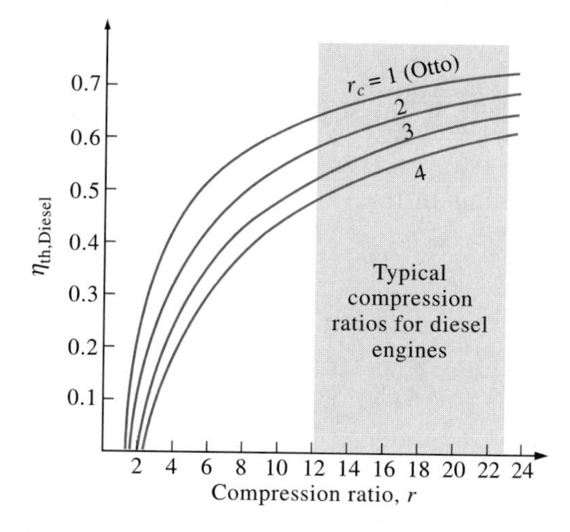


**FIGURE 7-22**

Thermal efficiency of the ideal Diesel cycle as a function of compression and cutoff ratios ($k$ = 1.4).

when both cycles operate on the same compression ratio. Also as the cutoff ratio decreases, the efficiency of the Diesel cycle increases (Fig. 7-22). For the limiting case of $r_c = 1$, the quantity in the brackets becomes unity (can you prove it?), and the efficiencies of the Otto and Diesel cycles become identical. Remember, though, that diesel engines operate at much higher compression ratios and thus are usually more efficient than the spark-ignition (gasoline) engines. The diesel engines also burn the fuel more completely since they usually operate at lower revolutions per minute than spark-ignition engines. Thermal efficiencies of diesel engined range from about 35 to 40 percent.

The higher efficiency and lower fuel costs of diesel engines make them the clear choice in applications, requiring relatively large amounts of power, such as in locomotive engines, emergency power generation units, large ships, and heavy trucks. As an example of how large a diesel engine can be, a 12-cylinder diesel engine built in 1964 by the Fiat Corporation of Italy had a normal power output of 25,200 hp (18.8 MW) at 122 rpm, a cylinder bore of 90 cm, and a stroke of 91 cm.

Approximating the combustion process in internal combustion engines as a constant-volume or a constant-pressure heat addition process is overly simplistic and not quite realistic. Probably a better (but slightly more complex) approach would be to model the combustion process in both gasoline and diesel engines as a combination of two heat transfer processes, one occurring at constant volume and the other at constant pressure. The ideal cycle based on this concept is called the **dual cycle**, and a $P$-$v$ diagram for it is given in Fig. 7-23. The relative amounts of heat added during each process can be adjusted to approximate the actual cycle more closely. Note that both the Otto and the Diesel cycles can be obtained as special cases of the dual cycle.

P
X
3
$q_{in}$
2
Isentropic
Isentropic
4
$q_{out}$
1
$v$

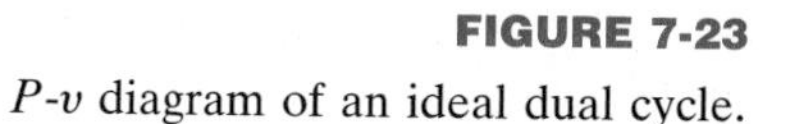

**FIGURE 7-23**

$P$-$v$ diagram of an ideal dual cycle.

**EXAMPLE 7-3**

An ideal Diesel cycle with air as the working fluid has a compression ratio of

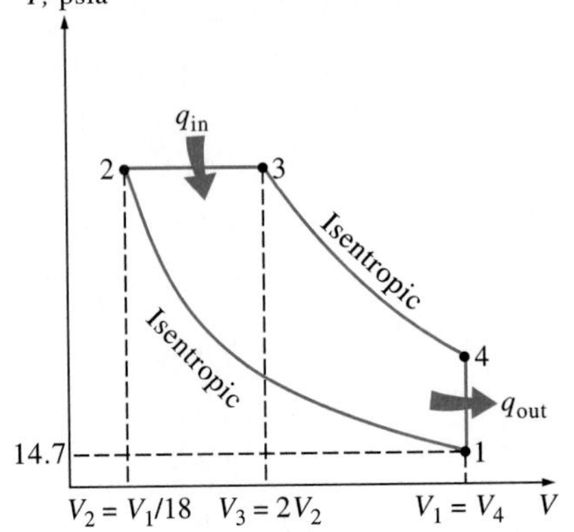


**FIGURE 7-24**

*P-v* diagram for the ideal Diesel cycle discussed in Example 7-3.

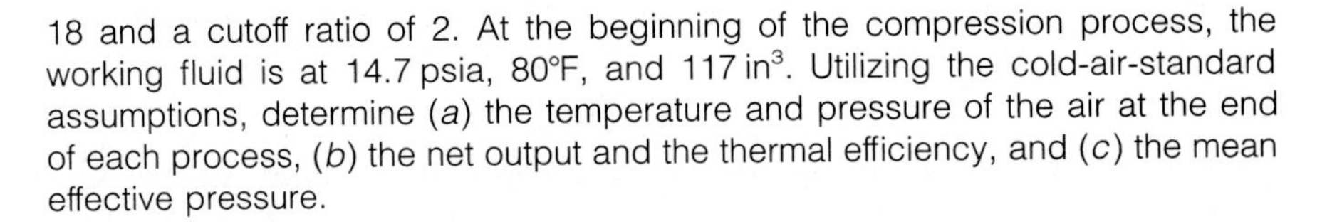

18 and a cutoff ratio of 2. At the beginning of the compression process, the working fluid is at 14.7 psia, 80°F, and 117 in$^3$. Utilizing the cold-air-standard assumptions, determine (*a*) the temperature and pressure of the air at the end of each process, (*b*) the net output and the thermal efficiency, and (*c*) the mean effective pressure.

**Solution** The *P-v* diagram of the cycle is shown in Fig. 7-24. The ideal Diesel cycle is executed in a piston–cylinder device with a fixed amount of air, and thus all four processes should be analyzed as closed-system processes. Under the cold-air-standard assumptions, the working fluid (air) is assumed to be an ideal gas and to have constant specific heat evaluated at room temperature. The gas constant of air is $R = 0.06855$ Btu/(lbm · R), and its specific heats at room temperature are $C_p = 0.240$ Btu/(lbm · R) and $C_v = 0.171$ Btu/(lbm · R) (Table A-2E*a*).

(*a*) The temperature and pressure values at the end of each process can be determined by utilizing the ideal-gas isentropic relations for processes 1-2 and 3-4. But first we determine the volumes at the end of each process from the definitions of the compression ratio and the cutoff ratio:

$$V_2 = \frac{V_1}{r} = \frac{117 \text{ in}^3}{18} = 6.5 \text{ in}^3$$

$$V_3 = r_c V_2 = (2)(6.5 \text{ in}^3) = 13 \text{ in}^3$$

$$V_4 = V_1 = 117 \text{ in}^3$$

Process 1-2 (isentropic compression of an ideal gas, constant specific heats):

$$T_2 = T_1\left(\frac{V_1}{V_2}\right)^{k-1} = (540 \text{ R})(18)^{1.4-1} = 1716 \text{ R}$$

$$P_2 = P_1\left(\frac{V_1}{V_2}\right)^{k} = (14.7 \text{ psia})(18)^{1.4} = 841 \text{ psia}$$

Process 2-3 ($P$ = constant heat addition to an ideal gas):

$$P_3 = P_2 = 841 \text{ psia}$$

$$\frac{P_2 V_2}{T_2} = \frac{P_3 V_3}{T_3} \longrightarrow T_3 = T_2\left(\frac{V_3}{V_2}\right) = (1716 \text{ R})(2) = 3432 \text{ R}$$

Process 3-4 (isentropic expansion of an ideal gas, constant specific heats):

$$T_4 = T_3\left(\frac{V_3}{V_4}\right)^{k-1} = (3432\text{ R})\left(\frac{13\text{ in}^3}{117\text{ in}^3}\right)^{1.4-1} = 1425\text{ R}$$

$$P_2 = P_3\left(\frac{V_3}{V_4}\right)^{k} = (841\text{ psia})\left(\frac{13\text{ in}^3}{117\text{ in}^3}\right)^{1.4} = 38.8\text{ psia}$$

(*b*) The net work for a cycle is equivalent to the net heat transfer, i.e., the difference between the total heat supplied and the total heat rejected. But first we find the mass of air:

$$m = \frac{P_1V_1}{RT_1} = \frac{(14.7\text{ psia})(117\text{ in}^3)}{[0.3704\text{ psia}\cdot\text{ft}^3/(\text{lbm}\cdot\text{R})](540\text{ R})}\left(\frac{1\text{ ft}^3}{1728\text{ in}^3}\right)$$
$$= 0.00498\text{ lbm}$$

Process 2-3 is a constant-pressure heat addition process, for which the boundary work and $\Delta u$ terms can be combined into $\Delta h$. Thus,

$$Q_{in} = Q_{23} = m(h_3 - h_2) = mC_p(T_3 - T_2)$$
$$= (0.00498\text{ lbm})[0.240\text{ Btu}/(\text{lbm}\cdot\text{R})][(3432 - 1716)\text{ R}]$$
$$= 2.051\text{ Btu}$$

Process 4-1 is a constant-volume heat rejection process (it involves no work interactions), and the amount of heat rejected is

$$Q_{out} = -Q_{41} = m(u_4 - u_1) = mC_v(T_4 - T_1)$$
$$= (0.00498\text{ lbm})[0.171\text{ Btu}/(\text{lbm}\cdot\text{R})][(1425 - 540)\text{ R}]$$
$$= 0.758\text{ Btu}$$

Thus, $\quad W_{net} = Q_{in} - Q_{out} = (2.051 - 0.758)\text{ Btu} = 1.293\text{ Btu}$

Then the thermal efficiency becomes

$$\eta_{th} = \frac{W_{net}}{Q_{in}} = \frac{1.293\text{ Btu}}{2.051\text{ Btu}} = 0.630\text{ (or 63.0\%)}$$

The thermal efficiency of this Diesel cycle under the cold-air-standard assumptions could also be determined from Eq. 7-12.

(*c*) The mean effective pressure is determined from its definition, Eq. 7-4:

$$\text{MEP} = \frac{W_{net}}{V_{max} - V_{min}} = \frac{W_{net}}{V_1 - V_2} = \frac{1.293\text{ Btu}}{(117 - 6.5)\text{ in}^3}\left(\frac{778.17\text{ lbf}\cdot\text{ft}}{1\text{ Btu}}\right)\left(\frac{12\text{ in}}{1\text{ ft}}\right)$$
$$= 109.3\text{ psia}$$

Therefore, a constant pressure of 109.3 psia during the power stroke would produce the same net work output as the entire Diesel cycle.

## 7-7 ■ BRAYTON CYCLE—THE IDEAL CYCLE FOR GAS-TURBINE ENGINES

The Brayton cycle was first proposed by George Brayton for use in the reciprocating oil-burning engine that he developed around 1870. Today, it

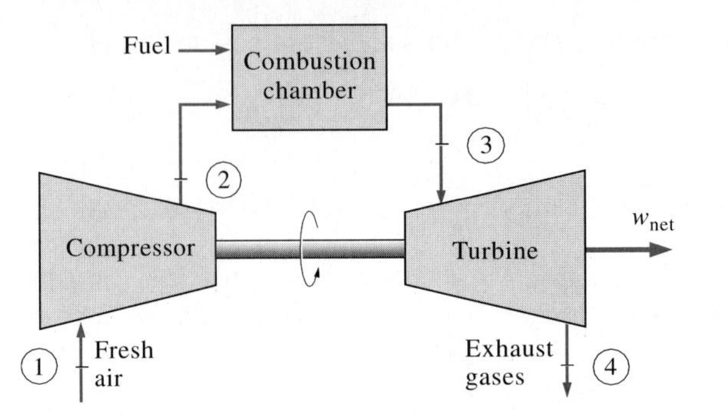


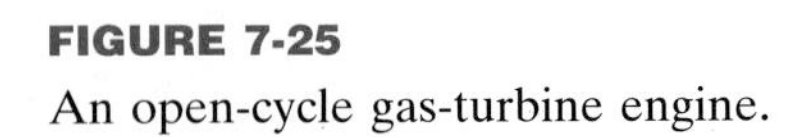

**FIGURE 7-25**
An open-cycle gas-turbine engine.

is used for gas turbines only where both the compression and expansion processes take place in rotating machinery. Gas turbines usually operate on an *open cycle,* as shown in Fig. 7-25. Fresh air at ambient conditions is drawn into the compressor, where its temperature and pressure are raised. The high-pressure air proceeds into the combustion chamber, where the fuel is burned at constant pressure. The resulting high-temperature gases then enter the turbine, where they expand to the atmospheric pressure, thus producing power. The exhaust gases leaving the turbine are thrown out (not recirculated), causing the cycle to be classified as an open cycle.

The open gas-turbine cycle described above can be modeled as a *closed cycle,* as shown in Fig. 7-26, by utilizing the air-standard assumptions. Here the compression and expansion processes remain the same, but the combustion process is replaced by a constant-pressure heat addition process from an external source, and the exhaust process is replaced by a constant-pressure heat rejection process to the ambient air. The ideal cycle that the working fluid undergoes in this closed loop is the Brayton cycle, which is made up of four internally reversible processes:

1-2 Isentropic compression (in a compressor)

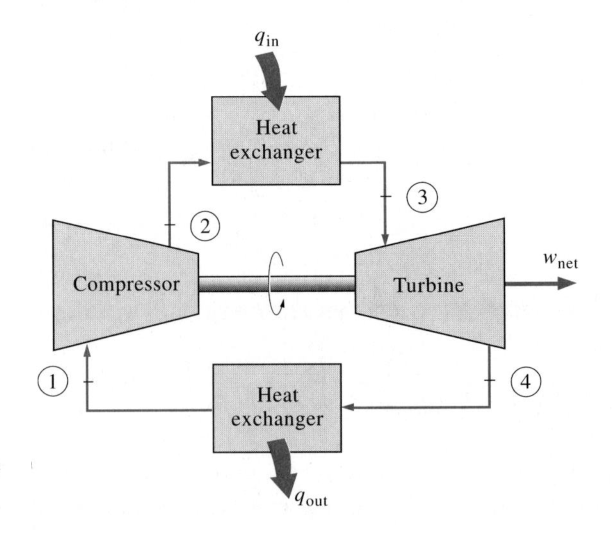


**FIGURE 7-26**
A closed-cycle gas-turbine engine.

2-3 $\quad P$ = constant heat addition

3-4 $\quad$ Isentropic expansion (in a turbine)

4-1 $\quad P$ = constant heat rejection

The $T$-$s$ and $P$-$v$ diagrams of an Ideal Brayton cycle are shown in Fig. 7-27. Notice that all four processes of the Brayton cycle are executed in steady-flow devices; thus, they should be analyzed as steady-flow processes. When the changes in kinetic and potential energies are neglected, the conservation of energy equation for a steady-flow process can be expressed, on a unit-mass basis, as

$$q - w = h_{\text{exit}} - h_{\text{inlet}} \tag{7-14}$$

Assuming constant specific heats at room temperature (cold-air-standard assumption), heat transfer to and from the working fluid becomes

$$q_{\text{in}} = q_{23} = h_3 - h_2 = C_p(T_3 - T_2) \tag{7-15a}$$

and

$$q_{\text{out}} = -q_{41} = h_4 - h_1 = C_p(T_r - T_1) \tag{7-15b}$$

Then the thermal efficiency of the ideal Brayton cycle becomes

$$\eta_{\text{th,Brayton}} = \frac{w_{\text{net}}}{q_{\text{in}}} = 1 - \frac{q_{\text{out}}}{q_{\text{in}}} = 1 - \frac{C_p(T_4 - T_1)}{C_p(T_3 - T_2)} = 1 - \frac{T_1(T_4/T_1 - 1)}{T_2(T_3/T_2 - 1)}$$

Processes 1-2 and 3-4 are isentropic, and $P_2 = P_3$ and $P_4 = P_1$. Thus,

$$\frac{T_2}{T_1} = \left(\frac{P_2}{P_1}\right)^{(k-1)/k} = \left(\frac{P_3}{P_4}\right)^{(k-1)/k} = \frac{T_3}{T_4}$$

Substituting these equations into the thermal efficiency relation and simplifying give

$$\eta_{\text{th,Brayton}} = 1 - \frac{1}{r_p^{(k-1)/k}} \tag{7-16}$$

where

$$r_p = \frac{P_2}{P_1} \tag{7-17}$$

is the **pressure ratio** and $k$ is the specific heat ratio. Equation 7-16 shows that under the cold-air-standard assumptions, the thermal efficiency of an ideal Brayton cycle depends on the pressure ratio of the gas turbine and the specific heat ratio of the working fluid (if different from air). The thermal efficiency increases with both these parameters, which is also the case for actual gas turbines. A plot of thermal efficiency versus the pressure ratio is given in Fig. 7-28 for $k = 1.4$, which is the specific-heat-ratio value of air at room temperature.

The highest temperature in the cycle occurs at the end of the combustion process (state 3), and it is limited by the maximum temperature that the turbine blades can withstand. This also limits the pressure ratios that can be used in the cycle. For a fixed turbine inlet temperature $T_3$, the net work output per cycle increases with the pressure ratio, reaches a maximum, and then starts to decrease, as shown in

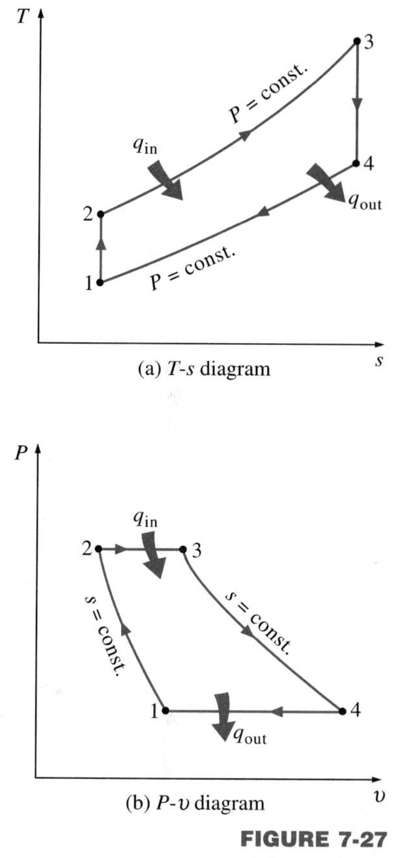


(a) $T$-$s$ diagram

(b) $P$-$v$ diagram

**FIGURE 7-27**

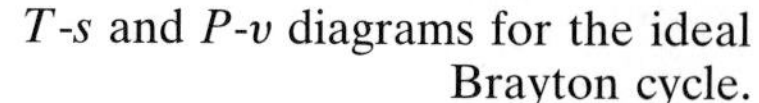

$T$-$s$ and $P$-$v$ diagrams for the ideal Brayton cycle.

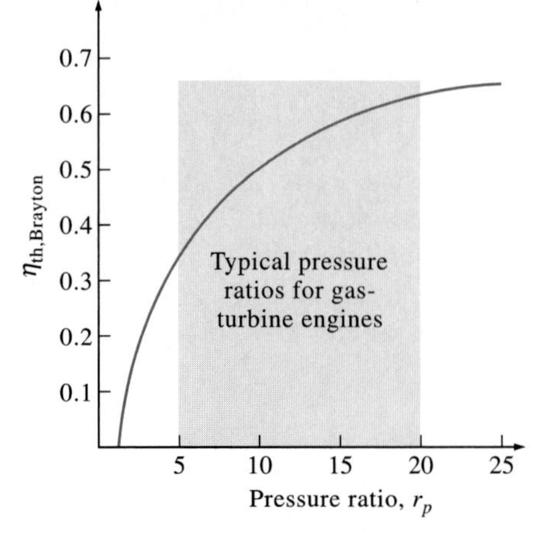


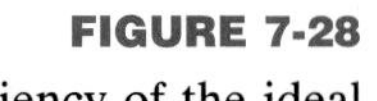

**FIGURE 7-28**

Thermal efficiency of the ideal Brayton cycle as a function of the pressure ratio.

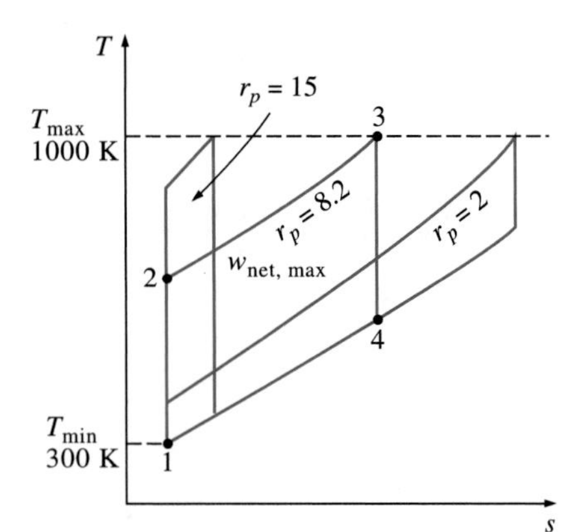


**FIGURE 7-29**

For fixed values of $T_{min}$ and $T_{max}$, the net work of the Brayton cycle first increases with the pressure ratio, then reaches a maximum at $r_p = (T_{max}/T_{min})^{k/[2(k-1)]}$ and finally decreases.

Fig. 7-29. Therefore, there should be a compromise between the pressure ratio (thus the thermal efficiency) and the net work output. With less work output per cycle, a larger mass flow rate (thus a larger system) is needed to maintain the same power output which may not be economical. In most common designs, the pressure ratio ranges from 11 to 16.

The air in gas turbines performs two important functions: It supplies the necessary oxidant for the combustion of the fuel, and it serves as a coolant to keep the temperature of various components within safe limits. The second function is accomplished by drawing in more air than is needed for the complete combustion of the fuel. In gas turbines, an air–fuel mass ratio of 50 or above is not uncommon. Therefore, in a cycle analysis, treating the combustion gases as air will not cause any appreciable error. Also, the mass flow rate through the turbine will be greater than that through the compressor, the difference being equal to the mass flow rate of the fuel. Thus, assuming a constant mass flow rate throughout the cycle will yield conservative results for open-loop gas-turbine engines.

The thermal efficiency of a gas-turbine engine depends on the allowable maximum gas temperature at the turbine inlet. Raising the turbine inlet temperature from 900°C to 1200°C increases the power output by 71 percent and the thermal efficiency by 26 percent. Significant advances, such as coating the turbine blades with ceramic layers and cooling the blades with the discharge air from the compressor, have been made during the last two decades. As a result, today's gas turbines can withstand temperatures as high as 1425°C (2600°F) at the turbine inlet, and gas turbine power plants have efficiencies well over 30%.

A gas-turbine engine manufactured recently by General Electric has a pressure ratio of 13.5 and generates 135.7 MW of net power at a thermal efficiency of 33 percent in simple cycle operation. Air in this engine is compressed by an 18-stage axial-flow compressor. The combustion gases at 1260°C expand in a three-stage turbine and exhaust at 593°C.

The two major application areas of gas-turbine engines are aircraft propulsion and electric power generation. When it is used for aircraft propulsion, the gas turbine produces just enough power to drive the compressor and a small generator to power the auxiliary equipment. The high-velocity exhaust gases are responsible for producing the necessary thrust to propel the aircraft. Gas turbines are also used as stationary power plants to generate electricity. Electricity is predominantly generated by large steam power plants which are discussed in Secs. 7-11 and 7-12. Gas-turbine power plants are mostly utilized in the power generation industry to cover emergencies and peak periods because of their relatively low cost and quick response time. Gas turbines are also used in conjunction with steam power plants on the high-temperature side, forming a dual cycle. In these plants, the exhaust gases of the gas turbine serve as the heat source for the steam. The gas-turbine cycle can also be executed as a closed cycle for use in nuclear power plants. This time the working fluid is not limited to air, and a gas with more desirable characteristics (such as helium) can be used.

The majority of the Western world's naval fleets already use gas-turbine engines for propulsion and electric power generation. Compared with steam turbine and diesel propulsion systems, the gas turbine offers greater power for a given size and weight, high reliability, long life, and more convenient operation. The engine start-up time has been reduced from 4 h required for a typical steam propulsion system to less than 2 min for a gas turbine. Many modern marine propulsion systems use gas turbines together with diesel engines because of the high fuel consumption of simple cycle gas-turbine engines. In combined diesel and gas-turbine systems, diesel is used to provide for efficient low-power and cruise operation, and gas turbine is used when high speeds are needed.

In gas-turbine power plants, the ratio of the compressor work to the turbine work, called the **back work ratio**, is very high (Fig. 7-30). Usually more than one-half of the turbine work output is used to drive the compressor. The situation is even worse when the adiabatic efficiencies of the compressor and the turbine are low. This is quite in contrast to steam power plants, where the back work ratio is only a few percent. This is not surprising, however, since a liquid is compressed in steam power plants instead of a gas, and the reversible steady-flow work is proportional to the specific volume of the working fluid ($w = -\int v\,dP$) when the kinetic and potential energy changes are negligible.

A power plant with a high back work ratio requires a larger turbine to provide the additional power requirements of the compressor. Therefore, the turbines used in gas-turbine power plants are larger than those used in steam power plants of the same power rating.

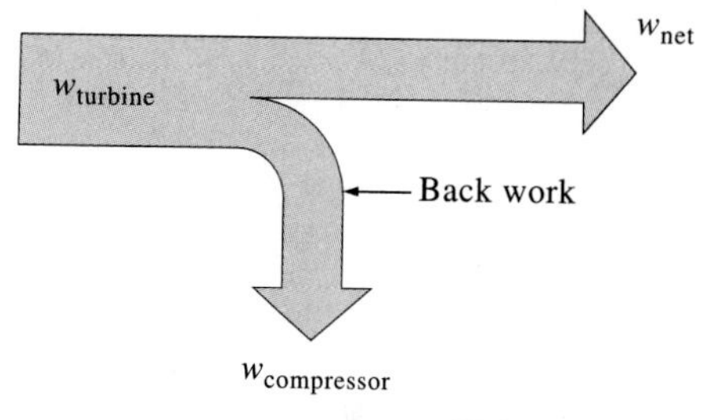


**FIGURE 7-30**
The fraction of the turbine work used to drive the compressor is called the back work ratio.

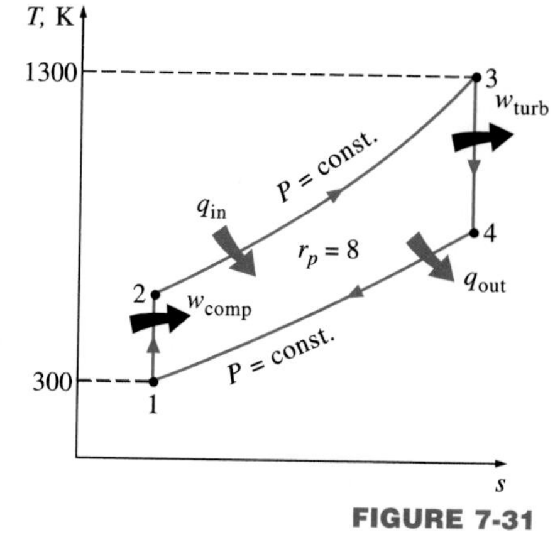


**FIGURE 7-31**
*T-s* diagram for the Brayton cycle discussed in Example 7-4.

**EXAMPLE 7-4**

A stationary power plant operating on an ideal Brayton cycle has a pressure ratio of 8. The gas temperature is 300 K at the compressor inlet and 1300 K at the turbine inlet. Utilizing the cold-air-standard assumptions, determine (*a*) the gas temperature at the exits of the compressor and the turbine, (*b*) the back work ratio, and (*c*) the thermal efficiency.

**Solution** The cycle is shown on a *T-s* diagram in Fig. 7-31. Under the air-standard assumptions, the working fluid is assumed to be air, which behaves as an ideal gas, and all four processes that make up the cycle are internally reversible. Furthermore, the combustion and exhaust processes are replaced by heat addition and heat rejection processes, respectively. Also, when changes in kinetic and potential energies are neglected, the energy equation for a steady-flow device simplifies to Eq. 7-14.

We assume constant specific heat for air at room temperature and thus take $C_p = 1.005 \text{ kJ/(kg}\cdot\text{K)}$, $C_v = 0.718 \text{ kJ/(kg}\cdot\text{K)}$, and $k = 1.4$.

(*a*) The air temperatures at the compressor and turbine exits are determined by applying the energy equation to processes 1-2 and 3-4:

Process 1-2 (isentropic compression of an ideal gas);

$$T_2 = T_1\left(\frac{P_2}{P_1}\right)^{(k-1)/k} = (300\text{ K})(8)^{0.4/0.8} = \mathbf{543.4\ K} \text{ (at compressor exit)}$$

Process 3-4 (isentropic expansion of an ideal gas):

$$T_4 = T_3\left(\frac{P_4}{P_3}\right)^{(k-1)/k} = (1300\text{ K})\left(\frac{1}{8}\right)^{0.4/1.4} = \mathbf{717.7\ K} \text{ (at turbine exit)}$$

(*b*) To find the back work ratio, we need to find the work input to the compressor and the work output of the turbine:

$$w_{\text{comp,in}} = h_2 - h_1 = C_p(T_2 - T_1) = [1.005\text{ kJ/(kg}\cdot\text{K)}](543.4 - 300)\text{ K} = 244.6\text{ kJ/kg}$$
$$w_{\text{turb,out}} = h_3 - h_4 = C_p(T_3 - T_4) = [1.005\text{ kJ/(kg}\cdot\text{K)}](1300 - 717.7)\text{ K} = 585.2\text{ kJ/kg}$$

Thus, back work ratio $r_{\text{bw}} = \dfrac{w_{\text{comp,in}}}{w_{\text{turb,out}}} = \dfrac{244.16\text{ kJ/kg}}{585.2\text{ kJ/kg}} = \mathbf{0.418}$

That is, 41.8 percent of the turbine work output is used just to drive the compressor.

(*c*) The thermal efficiency of the cycle is the ratio of the net power output to the total heat input:

$$q_{\text{in}} = h_3 - h_2 = C_p(T_3 - T_2) = [1.005\text{ kJ/(kg}\cdot\text{K)}](1300 - 543.4) = 760.4\text{ kJ/kg}$$
$$w_{\text{net}} = w_{\text{out}} - w_{\text{in}} = (585.2 - 244.6)\text{ kJ/kg} = 362.7\text{ kJ/kg}$$

Thus,

$$\eta_{\text{th}} = \frac{w_{\text{net}}}{q_{\text{in}}} = \frac{362.7\text{ kJ/kg}}{851.62\text{ kJ/kg}} = 0.448 \text{ (or 44.8\%)}$$

The thermal efficiency could also be determined from

$$\eta_{\text{th,Brayton}} = 1 - \frac{1}{r_p^{(k-1)/k}} = 1 - \frac{1}{8^{(1.4-1)/1.4}} = 0.448$$

which is identical to the value obtained above.

## 7-8 THE BRAYTON CYCLE WITH REGENERATION

In gas-turbine engines, the temperature of the exhaust gas leaving the turbine is often considerably higher than the temperature of the air

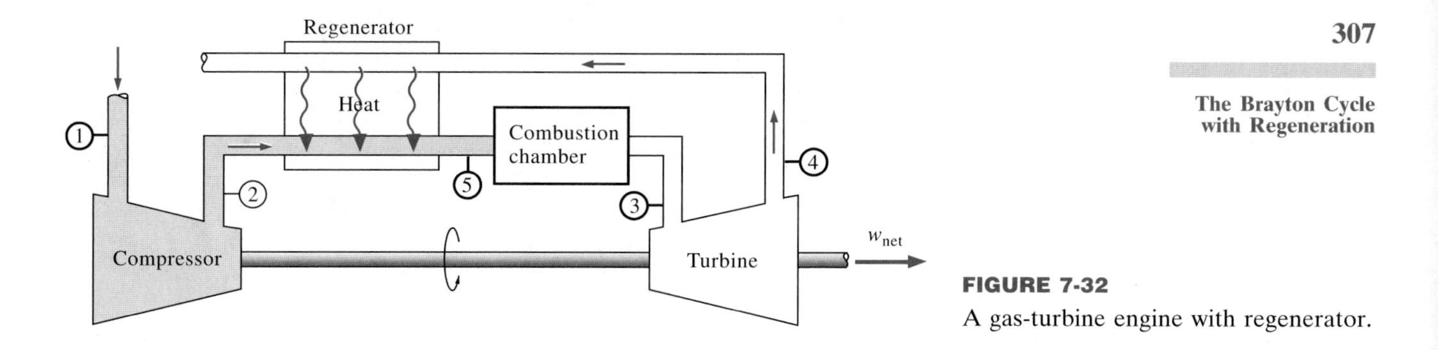


**FIGURE 7-32**
A gas-turbine engine with regenerator.

leaving the compressor. Therefore, the high-pressure air leaving the compressor can be heated by transferring heat to it from the hot exhaust gases in a counter-flow heat exchanger, which is also known as a *regenerator,* or a *recuperator.* A sketch of the gas-turbine engine utilizing a regenerator and the $T$-$s$ diagram of the new cycle are shown in Figs. 7-32 and 7-33, respectively.

The thermal efficiency of the Brayton cycle increases as a result of regeneration since the portion of energy of the exhaust gases that is normally rejected to the surroundings is now used to preheat the air entering the combustion chamber. This, in turn, decreases the heat input (thus fuel) requirements for the same net work output. Note, however, that the use of a regenerator is recommended only when the turbine exhaust temperature is higher than the compressor exit temperature. Otherwise, heat will flow in the reverse direction (*to* the exhaust gases), decreasing the efficiency. This situation is encountered in gas-turbine engines operating at very high pressure ratios.

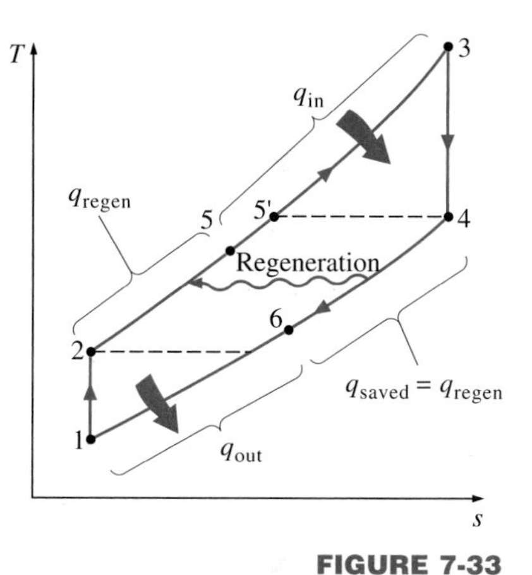


**FIGURE 7-33**
$T$-$s$ diagram of a Brayton cycle with regeneration.

The highest temperature occurring within the regenerator is $T_4$, the temperature of the exhaust gases leaving the turbine and entering the regenerator. Under no conditions can the air be preheated in the regenerator to a temperature above this value. Air normally leaves the regenerator at a lower temperature, $T_5$. In the limiting (ideal) case, the air will exit the regenerator at the inlet temperature of the exhaust gases $T_4$. Assuming the regenerator to be well insulated and any changes in kinetic and potential energies to be negligible, the actual and maximum heat transfers from the exhaust gases to the air can be expressed as

$$q_{regen,act} = h_5 - h_2 \tag{7-18}$$

and

$$q_{regen,max} = h_{5'} - h_2 = h_4 - h_2 \tag{7-19}$$

The extent to which a regenerator approaches an ideal regenerator is called the **effectiveness** $\varepsilon$ and is defined as

$$\varepsilon = \frac{q_{regen,act}}{q_{regen,max}} = \frac{h_5 - h_2}{h_4 - h_2} \tag{7-20}$$

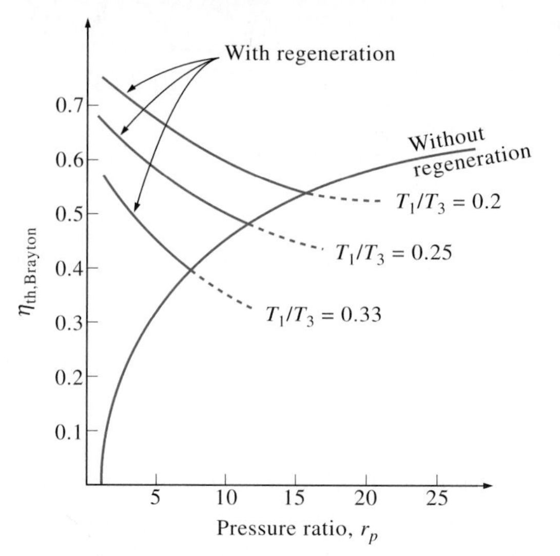


**FIGURE 7-34**
Thermal efficiency of the ideal Brayton cycle with and without regeneration.

When the cold-air-standard assumptions are utilized, it reduces to

$$\varepsilon \approx \frac{T_5 - T_2}{T_4 - T_2} \tag{7-21}$$

It is obvious that a regenerator with a higher effectiveness will save a greater amount of fuel since it will preheat the air to a higher temperature prior to combustion. However, achieving a higher effectiveness requires the use of a larger regenerator, which carries a higher price tag and causes a larger pressure drop. Therefore, the use of a regenerator with a very high effectiveness cannot be justified economically unless the savings from the fuel costs exceed the additional expenses involved. Most regenerators used in practice have an effectiveness below 0.85.

Under the cold-air-standard assumptions, the thermal efficiency of an ideal Brayton cycle with regeneration is

$$\eta_{\text{th,regen}} = 1 - \left(\frac{T_1}{T_3}\right)(r_p)^{(k-1)/k} \tag{7-22}$$

Therefore, the thermal efficiency of an ideal Brayton cycle with regeneration depends on the ratio of the minimum to maximum temperatures as well as the pressure ratio. The thermal efficiency is plotted in Fig. 7-34 for various pressure ratios and minimum-to-maximum temperature ratios. This figure shows that regeneration is most effective at lower pressure ratios and low minimum-to-maximum temperature ratios.

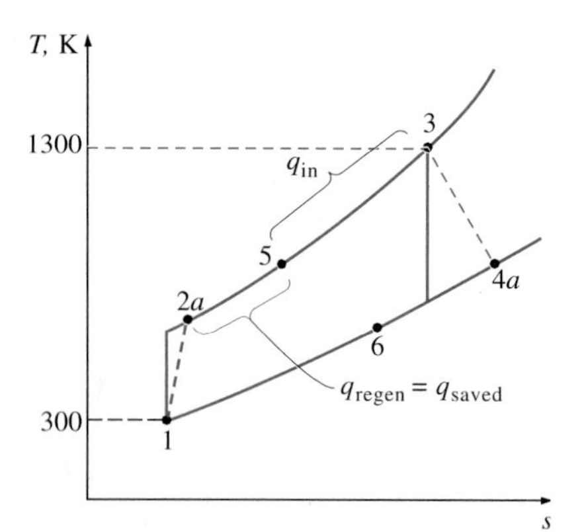


**FIGURE 7-35**
$T$-$s$ diagram of the regenerative Brayton cycle described in Example 7-5.

**EXAMPLE 7-5**

Determine the thermal efficiency of the gas-turbine power plant described in Example 7-4 if a regenerator having an effectiveness of 80 percent is installed.

**Solution** The $T$-$s$ diagram of the cycle is shown in Fig. 7-35. We first determine the enthalpy of the air at the exit of the regenerator, using the definition of effectiveness:

$$\varepsilon = \frac{h_5 - h_2}{h_4 - h_2} = \frac{C_p(T_5 - T_2)}{C_p(T_4 - T_2)} = \frac{T_5 - T_2}{T_4 - T_2}$$

$$0.80 = \frac{(T_5 - 543.4)\text{ K}}{(717.7 - 543.4)\text{ K}} \longrightarrow T_5 = 682.8\text{ K}$$

Thus,

$$q_{in} = h_3 - h_5 = C_p(T_3 - T_5) = [1005\text{ kJ/(kg}\cdot\text{K)}](1300 - 682.8) = 620.3\text{ kJ/kg}$$

This represents a savings of 140.1 kJ/kg from the heat input requirements. The addition of a regenerator (assumed to be frictionless) does not affect the net work output of the plant. Thus,

$$\eta_{th} = \frac{w_{net}}{q_{in}} = \frac{340.6\text{ kJ/kg}}{620.3\text{ kJ/kg}} = 0.549\text{ (or 54.9\%)}$$

That is, the thermal efficiency of the power plant has gone up from 44.8 to 54.9 percent as a result of installing a regenerator, which helps to recuperate some of the excess energy of the exhaust gases.

The net work of a gas-turbine cycle is the difference between the turbine work output and the compressor work input, and it can be increased by either decreasing the compressor work or increasing the turbine work, or both. The work required to compress a gas between two specified pressures can be decreased by carrying out the compression process in stages and cooling the gas in between—that is, using *multistage compression with intercooling.* As the number of stages is increased, the compression process becomes isothermal at the compressor inlet temperature, and the compression work decreases.

Likewise, the work output of a turbine operating between two pressure levels can be increased by expanding the gas in stages and reheating it in between—that is, utilizing *multistage expansion with reheating.* This is accomplished without raising the maximum temperature in the cycle. As the number of stages is increased, the expansion process becomes isothermal. The foregoing argument is based on a simple principle: The steady-flow compression or expansion work is proportional to the specific volume of the fluid. Therefore, the specific volume of the working fluid should be as low as possible during a compression process and as high as possible during an expansion process. This is precisely what intercooling and reheating accomplish.

The working fluid leaves the compressor at a lower temperature, and the turbine at a higher temperature, when intercooling and reheating are utilized. This makes regeneration more attractive since a greater potential for regeneration exists. Also, the gases leaving the compressor can be heated to a higher temperature before they enter the combustion chamber because of the higher temperature of the turbine exhaust.

A schematic of the physical arrangement of an two-stage gas-turbine cycle with intercooling, reheating, and regeneration is shown in Fig. 7-36.

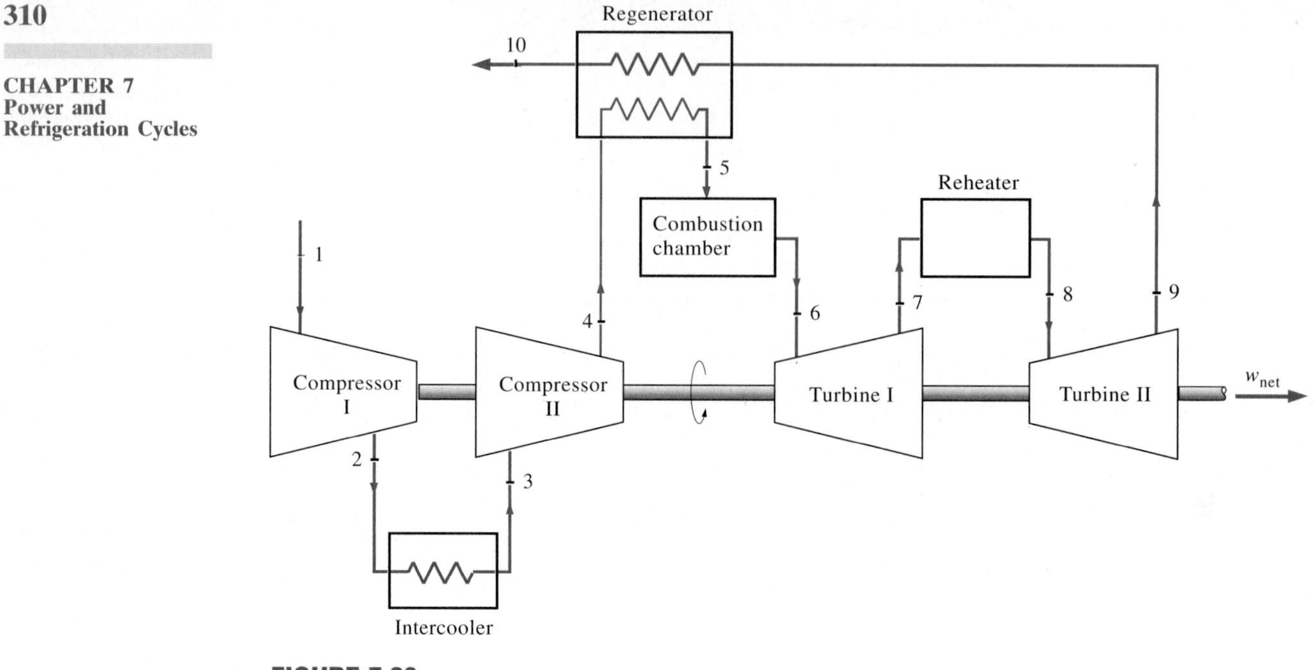


**FIGURE 7-36**
A gas-turbine engine with two-stage compression with intercooling, two-stage expansion with reheating, and regeneration.

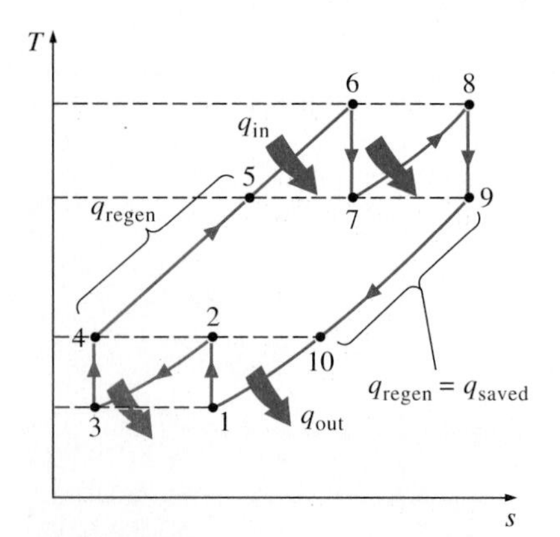


**FIGURE 7-37**
As the number of compression and expansion stages increases, the gas-turbine cycle with intercooling reheating, and regeneration approaches the Ericsson cycle.

The gas enters the first stage of the compressor at state 1, is compressed isentropically to an intermediate pressure $P_2$, is cooled at constant pressure to state 3 and is compressed in the second stage isentropically to the final pressure $P_4$. At state 4, the gas enters the regenerator, where it is heated to $T_5$ at constant pressure. In an ideal regenerator, the gas will leave the regenerator at the temperature of the turbine exhaust, that is, $T_5 = T_9$. The primary heat addition (or combustion) process takes place between states 5 and 6. The gas enters the first stage of the turbine at state 6 and expands isentropically to state 7, where it enters the reheater. It is reheated at constant pressure to state 8, where it enters the second stage of the turbine. The gas exits the turbine at state 9 and enters the regenerator, where it is cooled to state 10 at constant pressure. The cycle is completed by cooling the gas to the initial state (or purging the exhaust gases).

Therefore, in gas-turbine power plants, intercooling and reheating are always used in conjunction with regeneration.

If the number of compression and expansion stages is increased, the ideal gas-turbine cycle with intercooling, reheating, and regeneration will approach the Ericsson cycle, as illustrated in Fig. 7-37, and the thermal efficiency will approach the theoretical limit (the Carnot efficiency). However, the contribution of each additional stage to the thermal efficiency is less and less, and the use of more than two or three stages cannot be justified economically.

Gas-turbine engines are widely used to power aircraft because they are light and compact and have a high power-to-weight ratio. Aircraft gas turbines operate on an open cycle called a **jet-propulsion cycle**. The ideal jet-propulsion cycle differs from the simple ideal Brayton cycle in that the gases are not expanded to the ambient pressure in the turbine. Instead, the gases are expanded to a pressure such that the power produced by the turbine is just sufficient to drive the compressor and the auxiliary equipment, such as a small generator and hydraulic pumps. That is, the net work output of a jet-propulsion cycle is zero. The gases that exit the turbine at a relatively high pressure are subsequently accelerated in a nozzle to provide the thrust to propel the aircraft (Fig. 7-38). Also, aircraft gas turbines operate at higher pressure ratios (typically between 10 and 25), and the fluid passes through a diffuser first, where it is decelerated and its pressure is increased before it enters the compressor.

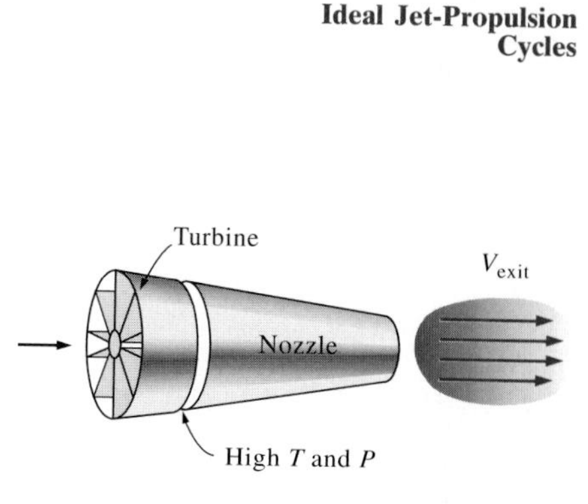


**FIGURE 7-38**

In jet engines, the high-temperature and high-pressure gases leaving the turbine are accelerated in a nozzle to provide thrust.

Aircraft are propelled by accelerating a fluid in the opposite direction to motion. This is accomplished by either slightly accelerating a large mass of fluid (*propeller-driven engine*) or greatly accelerating a small mass of fluid (*jet* or *turbojet engine*) or both (*turboprop engine*).

A schematic of a turbojet engine and the $T$-$s$ diagram of the ideal turbojet cycle are shown in Fig. 7-39. The pressure of air rises slightly as it is decelerated in the diffuser. Air is compressed in the compressor. It is mixed with fuel in the combustion chamber, where the mixture is burned at constant pressure. The high-pressure and high-temperature combustion gases partially expand in the turbine, producing enough power to drive the compressor and other equipment. Finally, the gases expand in a nozzle to the ambient pressure and leave the aircraft at a high velocity.

In the ideal case, the turbine work is assumed to equal the compressor work. Also, the processes in the diffuser, the compressor, the turbine, and the nozzle are assumed to be isentropic. In the analysis of

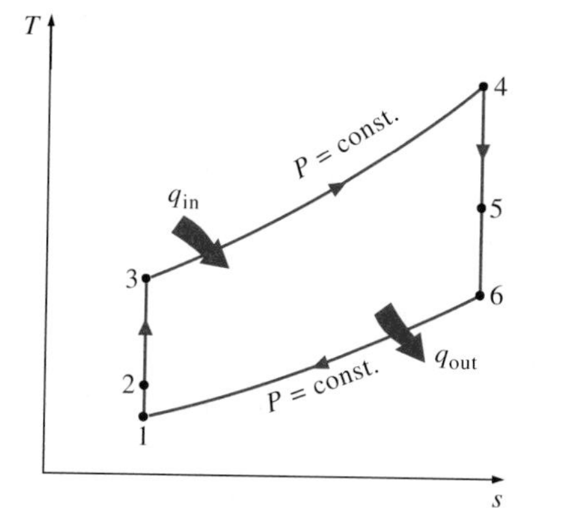


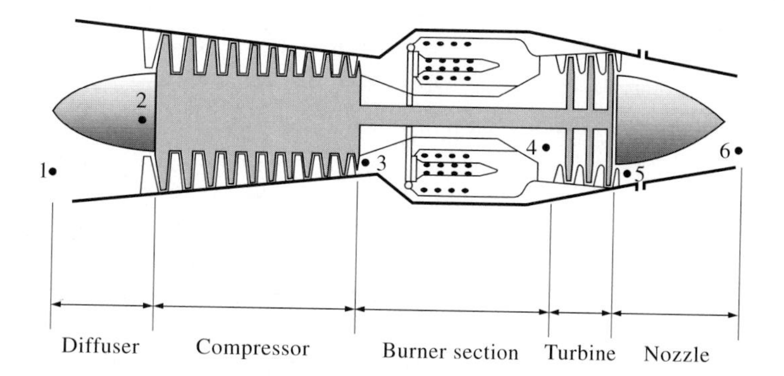


**FIGURE 7-39**

Basic components of a turbojet engine and the $T$-$s$ diagram for the ideal turbojet cycle. [*Source:* The Aircraft Gas Turbine Engine and Its Operation. © United Aircraft Corporation (Now United Technologies Corp.), 1951, 1974.]

actual cycles, however, the irreversibilities associated with these devices should be considered. The effect of the irreversibilities is to reduce the thrust that can be obtained from a turbojet engine.

The **thrust** developed in a turbojet engine is the unbalanced force which is caused by the difference in the momentum of the low-velocity air entering the engine and the high-velocity exhaust gases leaving the engine, and it is determined from Newton's second law. The pressures at the inlet and the exit of a turbojet engine are identical (the ambient pressure), thus the net thrust developed by the engine is

$$F = (\dot{m}\mathcal{V})_{\text{exit}} - (\dot{m}\mathcal{V})_{\text{inlet}} = \dot{m}(\mathcal{V}_{\text{exit}} - \mathcal{V}_{\text{inlet}}) \quad \text{(N)} \qquad (7\text{-}23)$$

where $\mathcal{V}_{\text{exit}}$ is the exit velocity of the exhaust gases and $\mathcal{V}_{\text{inlet}}$ is the inlet velocity of the air, both relative to the aircraft. Thus, for an aircraft cruising in still air, $\mathcal{V}_{\text{inlet}}$ is the aircraft velocity. In reality, the mass flow rates of the gases at the engine exit and the inlet are different, the difference being equal to the combustion rate of the fuel. But the air–fuel mass ratio used in jet-propulsion engines is usually very high, making this difference very small. Thus, $\dot{m}$ in Eq. 7-23 is taken as the mass flow rate of air through the engine. For an aircraft cruising at a steady speed, the thrust is used to overcome the fluid drag, and the net force acting on the body of the aircraft is zero. Commercial airplanes save fuel by flying at higher altitudes during long trips since the air at higher altitudes is thinner and exerts a smaller drag force on aircraft.

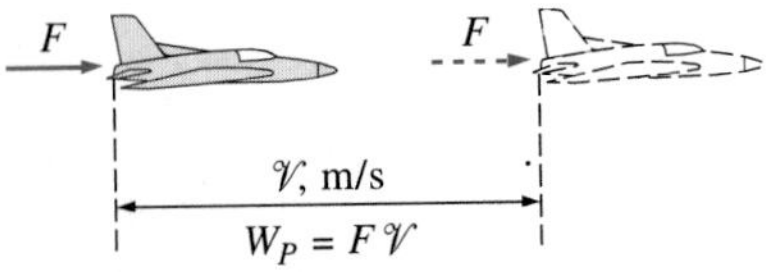

The power developed from the thrust of the engine is called the **propulsive power** $\dot{W}_P$, which is the *propulsive force* (*thrust*) times the *distance* this force acts on the aircraft per unit time, i.e., the thrust times the aircraft velocity (Fig. 7-40):

**FIGURE 7-40**
Propulsive power is the thrust acting on the aircraft through a distance per unit time.

$$\dot{W}_P = (F)\mathcal{V}_{\text{aircraft}} = \dot{m}(\mathcal{V}_{\text{exit}} - \mathcal{V}_{\text{inlet}})\mathcal{V}_{\text{aircraft}} \quad \text{(kW)} \qquad (7\text{-}24)$$

The net work developed by a turbojet engine is zero. Thus, we cannot define the efficiency of a turbojet engine in the same way as stationary gas-turbine engines. Instead, we should use the general definition of efficiency, which is the ratio of the desired output to the required input. The desired output in a turbojet engine is the *power produced* to propel the aircraft $\dot{W}_P$, and the required input is the *thermal energy of the fuel* released during the combustion process $\dot{Q}_{\text{in}}$. The ratio of these two quantities is called the **propulsive efficiency** and is given by

$$\eta_P = \frac{\text{propulsive power}}{\text{energy input rate}} = \frac{\dot{W}_P}{\dot{Q}_{\text{in}}} \qquad (7\text{-}25)$$

Propulsive efficiency is a measure of how efficiently the energy released during the combustion process is converted to propulsive energy. The remaining part of the energy released will show up as the kinetic energy of the exhaust gases relative to a fixed point on the ground and as an increase in the enthalpy of the air leaving the engine.



**EXAMPLE 7-6**

A turbojet aircraft files with a velocity of 850 ft/s at an altitude where the air is at 5 psia and −40°F. The compressor has a pressure ratio of 10, and the temperature of the gases at the turbine inlet is 2000°F. Air enters the compressor at a rate of 100 lbm/s. Utilizing the cold-air-standard assumptions, determine (*a*) the temperature and pressure of the gases at the turbine exit, (*b*) the velocity of the gases at the nozzle exit, and (*c*) the propulsive efficiency of the cycle.

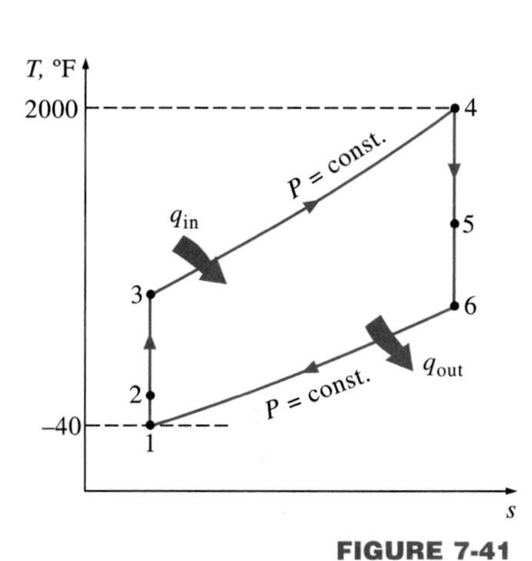


**FIGURE 7-41**
*T-s* diagram for the turbojet cycle described in Example 7-6.

**Solution** The *T-s* diagram of the cycle is shown in Fig. 7-41. Under the cold-air-standard assumptions, all the processes are assumed to be internally reversible, and the working fluid is air, which behaves as an ideal gas and has constant specific heats evaluated at room temperature [$C_p$ = 0.240 Btu/(lbm · R) and $k$ = 1.4, from Table A-2E*a*]. The combustion process is also replaced by a heat addition process.

(*a*) Before we can determine the temperature and pressure at the turbine exit. We need to find the temperatures and pressures at other states:

*Process 1-2* (isentropic compression of an ideal gas in a diffuser): For convenience, we can assume that the aircraft is stationary and the air is moving toward the aircraft at a velocity of $\mathscr{V}_1$ = 850 ft/s. Ideally, the air will leave the diffuser with a negligible velocity ($\mathscr{V}_2 \approx 0$):

$$q_{12}^{\nearrow 0} - w_{12}^{\nearrow 0} = h_2 - h_1 + \frac{\mathscr{V}_2^{2\nearrow 0} - \mathscr{V}_1^2}{2}$$

$$0 = C_p(T_2 - T_1) - \frac{\mathscr{V}_1^2}{2}$$

$$T_2 = T_1 + \frac{\mathscr{V}_1^2}{2C_p}$$

$$= 420\text{ R} + \frac{(850\text{ ft/s})^2}{2[0.240\text{ Btu/(lbm}\cdot\text{R)}]}\left(\frac{1\text{ Btu/lbm}}{25{,}037\text{ ft}^2/\text{s}}\right)$$

$$= 480.1\text{ R}$$

$$P_2 = P_1\left(\frac{T_2}{T_1}\right)^{k/(k-1)} = (5\text{ psia})\left(\frac{480.1\text{ R}}{420\text{ R}}\right)^{1.4/(1.4-1)} = 8.0\text{ psia}$$

*Process 2-3* (isentropic compression of an ideal gas in a compressor):

$$P_3 = (r_p)(P_2) = (10)(8.0\text{ psia}) = 80\text{ psia } (= P_4)$$

$$T_3 = T_2\left(\frac{P_3}{P_2}\right)^{(k-1)/k} = (480.1\text{ R})(10)^{(1.4-1)/1.4} = 926.9\text{ R}$$

*Process 4-5* (isentropic expansion of an ideal gas in a turbine): Neglecting the kinetic energy changes across the compressor and the turbine and assuming the turbine work to be equal to the compressor work, we find the temperature and pressure at the turbine exit to be

$$w_{comp,in} = w_{turb,out}$$
$$h_3 - h_2 = h_4 - h_5$$
$$C_p(T_3 - T_2) = C_p(T_4 - T_5)$$
$$T_5 = T_4 - T_3 + T_2 = (2460 - 926.9 + 480.1)\text{ R} = 2013.2\text{ R}$$
$$P_5 = P_4\left(\frac{T_5}{T_4}\right)^{k/(k-1)} = (80\text{ psia})\left(\frac{2013.2\text{ R}}{2460\text{ R}}\right)^{1.4/(1.4-1)} = 39.7\text{ psia}$$

(*b*) To find the air velocity at the nozzle exit, we need to first determine the nozzle exit temperature and then apply the steady-flow energy equation.

*Process 5-6* (isentropic expansion of an ideal gas in a nozzle):

$$T_6 = T_5\left(\frac{P_6}{P_5}\right)^{(k-1)/k} = (2013.2\text{ R})\left(\frac{5\text{ psia}}{39.7\text{ psia}}\right)^{(1.4-1)/1.4} = 1113.8\text{ R}$$

$$q_{56}^{\nearrow 0} - w_{56}^{\nearrow 0} = h_6 - h_5 + \frac{\mathcal{V}_6^2 - \mathcal{V}_5^{2\nearrow 0}}{2}$$
$$0 = C_p(T_6 - T_5) + \frac{\mathcal{V}_6^2}{2}$$
$$\mathcal{V}_6 = \sqrt{2C_p(T_5 - T_6)}$$
$$= \sqrt{2[0.240\text{ Btu/(lbm}\cdot\text{R)}][(2013.2 - 1113.8)\text{ R}]\left(\frac{25.037\text{ ft}^2/\text{s}^2}{1\text{ Btu/lbm}}\right)}$$
$$= 3287.7\text{ ft/s}$$

(*c*) The propulsive, efficiency of a turbojet engine is the ratio of the propulsive power developed $\dot{W}_p$ to the total heat transfer rate to the working fluid:

$$\dot{W}_P = \dot{m}(\mathcal{V}_{exit} - \mathcal{V}_{inlet})\mathcal{V}_{aircraft}$$
$$= (100\text{ lbm/s})[(3287.7 - 850)\text{ ft/s}](850\text{ ft/s})\left(\frac{1\text{ Btu/lbm}}{25{,}037\text{ ft}^2/\text{s}^2}\right)$$
$$= 8276\text{ Btu/s} \quad (11{,}707\text{ hp})$$
$$\dot{Q}_{in} = \dot{m}(h_4 - h_3) = \dot{m}C_p(T_4 - T_3)$$
$$= (100\text{ lbm/s})[0.240\text{ Btu/(lbm}\cdot\text{R)}][(2460 - 926.9)\text{ R}]$$
$$= 36{,}794\text{ Btu/s}$$
$$\eta_P = \frac{\dot{W}_P}{\dot{Q}_{in}} = \frac{8276\text{ Btu/s}}{36{,}794\text{ Btu/s}} = 22.5\%$$

That is, 22.5 percent of the energy input is used to propel the aircraft and to overcome the drag force exerted by the air.

For those who are wondering what happened to the rest of the energy, here is a brief account:

$$\dot{KE}_{out} = \dot{m}\frac{\mathcal{V}_g^2}{2} = (100\text{ lbm/s})\left\{\frac{[(3287.7 - 850)\text{ ft/s}]^2}{2}\right\}\left(\frac{1\text{ Btu/lbm}}{25{,}037\text{ ft}^2/\text{s}^2}\right)$$
$$= 11{,}867\text{ Btu/s } (32.2\%)$$
$$\dot{Q}_{out} = \dot{m}(h_6 - h_1) = \dot{m}C_p(T_6 - T_1)$$
$$= (100\text{ lbm/s})[0.24\text{ Btu/(lbm}\cdot\text{R)}][(1113.8 - 420)\text{ R}]$$
$$= 16{,}651\text{ Btu/s } (54.3\%)$$

Thus, 32.2 percent of the energy shows up as excess kinetic energy (kinetic energy of the gases relative to a fixed point on the ground). Notice that for the highest propulsion efficiency, the velocity of the exhaust gases relative to the ground $\mathcal{V}_g$ should be zero. That is, the exhaust gases should leave the nozzle at the velocity of the aircraft. The remaining 45.3 percent of the energy shows up as an increase in enthalpy of the gases leaving the engine. These last two forms of energy eventually become part of the internal energy of the atmospheric air (Fig. 7-42).

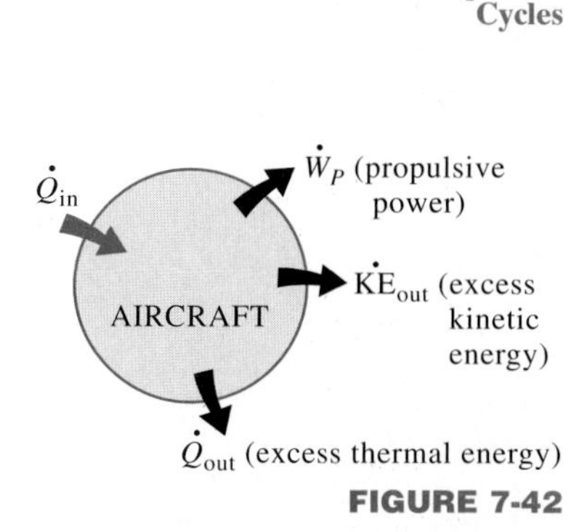


**FIGURE 7-42**

Energy supplied to an aircraft (from the burning of a fuel) manifests itself in various forms.

## Modifications to Turbojet Engines

The first airplanes built were all propeller-driven, with propellers powered by engines essentially identical to automobile engines. The major breakthrough in commercial aviation occurred with the introduction of the turbojet engine in 1952. Both propeller-driven engines and jet-propulsion-driven engines have their own strengths and limitations, and several attempts have been made to combine the desirable characteristics of both in one engine. Two such modifications are the *propjet engine* and the *turbofan engine.*

The most widely used engine in aircraft propulsion is the **turbofan** (or *fanjet*) engine wherein a large fan driven by the turbine forces a considerable amount of air through a duct (cowl) surrounding the engine, as shown in Figs. 7-43 and 7-44. The fan exhaust leaves the duct at a higher velocity, enhancing the total thrust of the engine significantly. A higher velocity, enhancing the total thrust of the engine significantly. A turbofan engine is based on the principle that for the same power, a large volume of slower moving air will produce more thrust than a small volume of fast moving air. The first commercial turbofan engine was successfully tested in 1955.

The turbofan engine on an airplane can be distinguished from the less efficient turbojet engine by its fat cowling covering the large fan. All the thrust of a turbojet engine is due to the exhaust gases leaving the engine at about twice the speed of sound. In a turbofan engine, the high-speed exhaust gases are mixed with the lower-speed air, which results in a considerable reduction in noise.

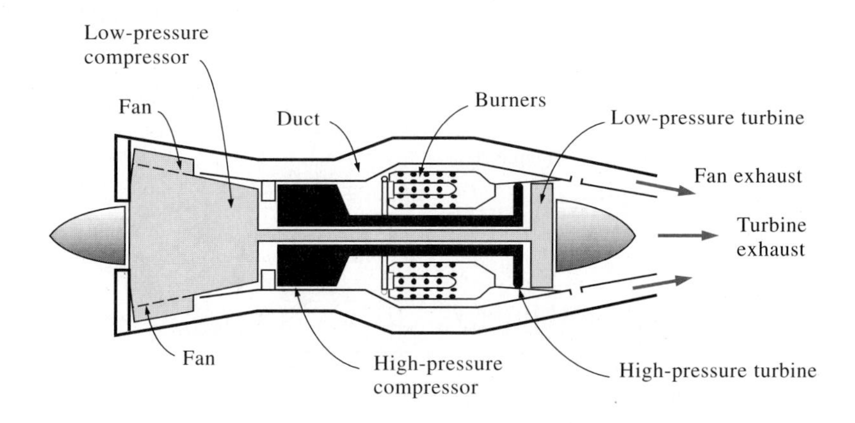


**FIGURE 7-43**

A turbofan engine. [*Source:* The Aircraft Gas Turbine and Its Operation. © United Aircraft Corporation (Now United Technologies Corporation), 1951, 1974.]

**FIGURE 7-44**
A turbofan engine. (Courtesy of Allied-Signal Aerospace Company, Garrett Engine Division.)

New cooling techniques have resulted in considerable increases in efficiencies by allowing gas temperatures at the burner exit to reach over 1500°C, which is more than 100°C above the melting point of the turbine blade materials. Turbofan engines deserve most of the credit for the success of jumbo jets, which weigh almost 400,000 kg and are capable of carrying over 400 passengers for up to 10,000 km at speeds over 950 km/h with less fuel per passenger mile.

The ratio of the mass flow rate of air bypassing the combustion chamber to that of air flowing through it is called the *bypass ratio.* The first commercial high-bypass ratio engines had a bypass ratio of five. Increasing the bypass ratio of a turbofan engine increases thrust. Thus, it makes sense to remove the cowl from the fan. The result is a **propjet** engine, as shown in Fig. 7-45. Turbofan and propjet engines differ primarily in their bypass ratios: 5 or 6 for turbofans and as high as 100 for propjets. As a general rule, propellers are more efficient than jet engines; but they are limited to low-speed and low-altitude operation since their efficiency decreases at high speeds and altitudes. The old propjet engines (*turboprops*) were limited to speeds about Mach 0.62 and to altitudes of around 9100 m. The new propjet engines (*propfans*) under development are expected to achieve speeds of about Mach 0.82 and altitudes of about

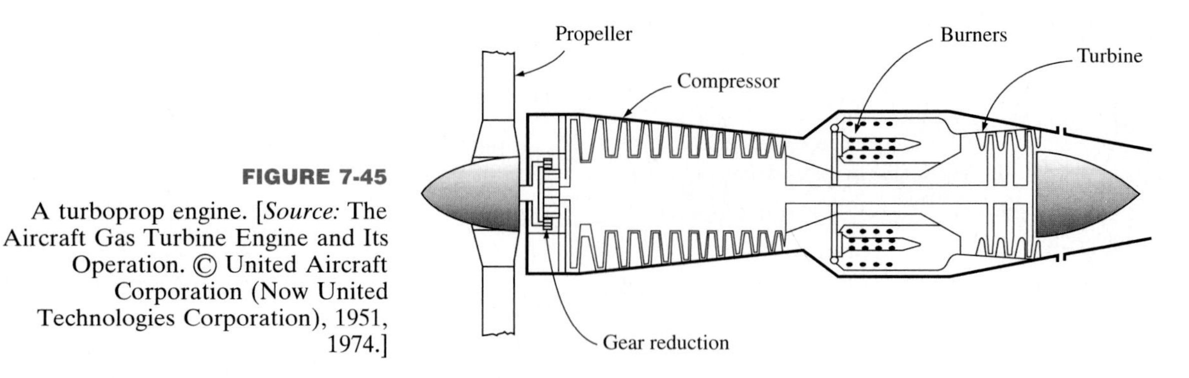


**FIGURE 7-45**
A turboprop engine. [*Source:* The Aircraft Gas Turbine Engine and Its Operation. © United Aircraft Corporation (Now United Technologies Corporation), 1951, 1974.]

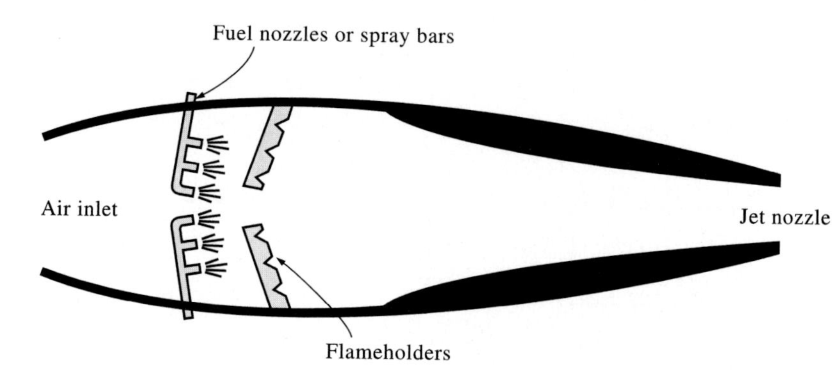


**FIGURE 7-46**
A ramjet engine. [*Source:* The Aircraft Gas Turbine Engine and Its Operation. © United Aircraft Corporation (Now United Technologies Corporation), 1951, 1974.]

12,200 m. Commercial airplanes of medium size and range propelled by propfans are expected to fly as high and as fast as the planes propelled by turbofans, and to do so on 30 percent less fuel. Research is underway to realize these goals.

Another modification that is popular in military aircraft is the addition of an **afterburner** section between the turbine and the nozzle. Whenever a need for extra thrust arises, such as for short takeoffs or combat conditions, additional fuel is injected into the oxygen-rich combustion gases leaving the turbine. As a result of this added energy, the exhaust gases leave at a higher velocity, providing a greater thrust.

A **ramjet** engine is a properly shaped duct with no compressor or turbine, as shown in Fig. 7-46, and is sometimes used for high-speed propulsion of missiles and aircraft. The pressure rise in the engine is provided by the ram effect of the incoming high-speed air being rammed against a barrier. Therefore, a ramjet engine needs to be brought to a sufficiently high speed by an external source before it can be fired.

The ramjet performs best in aircraft flying above Mach 2 or 3 (2 or 3 times the speed of sound). In a ramjet, the air is slowed down to about Mach 0.2, fuel is added to the air and burned at this low velocity, and the combustion gases are expended and accelerated in a nozzle.

A **scramjet** engine is essentially a ramjet in which air flows through at supersonic speeds (above the speed of sound). Ramjets that convert to scramjet configurations at speeds above Mach 6 are successfully tested at speeds of about Mach 8.

Finally, a **rocket** is a device where a solid or liquid fuel and an oxidizer react in the combustion chamber. The high-pressure combustion gases are then expanded in a nozzle. The gases leave the rocket at very high velocities, producing the thrust to propel the rocket.

## 7-10 ■ THE CARNOT VAPOR CYCLE

We have mentioned many time that the Carnot cycle is the most efficient cycle operating between two specified temperature levels. Thus, it is natural to look at the Carnot cycle first as a prospective ideal cycle for vapor power plants. If we could, we would certainly adopt it as the ideal cycle. But as explained below, the Carnot cycle is an unsuitable model for vapor power cycles. Throughout the discussions, we assume steam to be

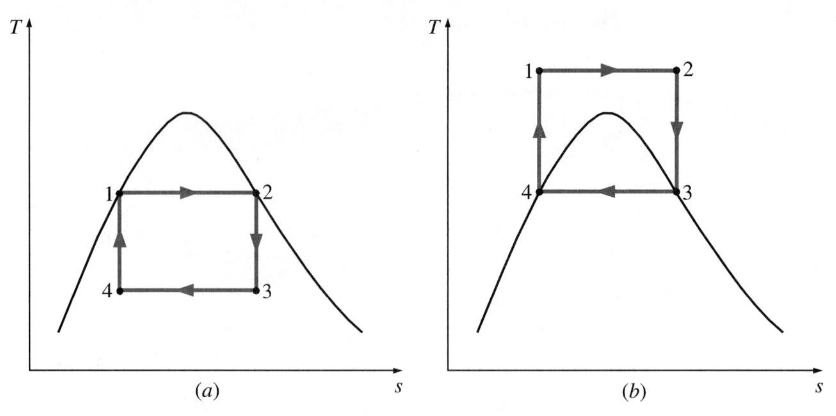


**FIGURE 7-47**
*T*-*s* diagram of two Carnot vapor cycles.

the working fluid since it is the working fluid predominantly used in vapor power cycles.

Consider a steady-flow *Carnot cycle* executed within the saturation dome of a pure substance such as water, as shown in Fig. 7-47*a*. Water is heated reversibly and isothermally in a boiler (process 1-2), expanded isentropically in a turbine (process 2-3), condensed reversibly and isothermally in a condenser (process 3-4), and compressed isentropically by a compressor to the initial state (process 4-1).

Several impracticalities are associated with this cycle:

**1** Isothermal heat transfer to or from a two-phase system is not difficult to achieve in practice since maintaining a constant pressure in the device will automatically fix the temperature at the saturation value. Therefore, processes 1-2 and 3-4 can be approached closely in actual boilers and condensers. Limiting the heat transfer processes to two-phase systems, however, severely limits the maximum temperature that can be used in the cycle (it has to remain under the critical-point value, which is 374°C for water). Limiting the maximum temperature in the cycle also limits the thermal efficiency. Any attempt to raise the maximum temperature in the cycle will involve heat transfer to the working fluid in a single phase, which is not easy to accomplish isothermally.

**2** The isentropic expansion process (process 2-3) can be approximated closely by a well-designed turbine. However, the quality of the steam decreases during this process, as shown on the *T*-*s* diagram in Fig. 7-47*a*. Thus, the turbine will have to handle steam with low quality, i.e., steam with a high moisture content. The impingement of liquid droplets on the turbine blades causes erosion and is a major source of wear. Thus, steam with qualities less than about 90 percent cannot be tolerated in the operation of power plants. This problem could be eliminated by using a working fluid with a very steep saturated vapor line.

**3** The isentropic compression process (process 4-1) involves the compression of a liquid–vapor mixture to a saturated liquid. There are two difficulties associated with this process. First, it is not easy to control the condensation process so precisely as to end up with the desired quality at state 4. Second, it is not practical to design a compressor that will handle two phases.



Some of these problems could be eliminated by executing the Carnot cycle in a different way, as shown in Fig. 7-47*b*. This cycle, however, presents other problems such as isentropic compression to extremely high pressures and isothermal heat transfer at variable pressures. Thus, we conclude that the Carnot cycle cannot be approximated in actual devices and is not a realistic model for vapor power cycles.

## 7-11 ■ RANKINE CYCLE—THE IDEAL CYCLE FOR VAPOR POWER CYCLES

Many of the impracticalities associated with the Carnot cycle can be eliminated by superheating the steam in the boiler and condensing it completely in the condenser, as shown schematically on a $T$-$s$ diagram in Fig. 7-48. The cycle that results is the **Rankine cycle**, which is the ideal cycle for vapor power plants. The ideal Rankine cycle does not involve any internal irreversibilities and consists of the following four processes:

1-2 Isentropic compression in a pump
2-3 $P$ 5 constant heat addition in a boiler
3-4 Isentropic expansion in a turbine
4-1 $P$ 5 constant heat rejection in a condenser

Water enters the *pump* at state 1 as saturated liquid and is compressed isentropically to the operating pressure of the boiler. The water temperature increases somewhat during this isentropic compression process due to a slight decrease in the specific volume of the water. The vertical distance between states 1 and 2 on the $T$-$s$ diagram is greatly exaggerated for clarity. (If water were truly incompressible, would the temperature change at all during this process?)

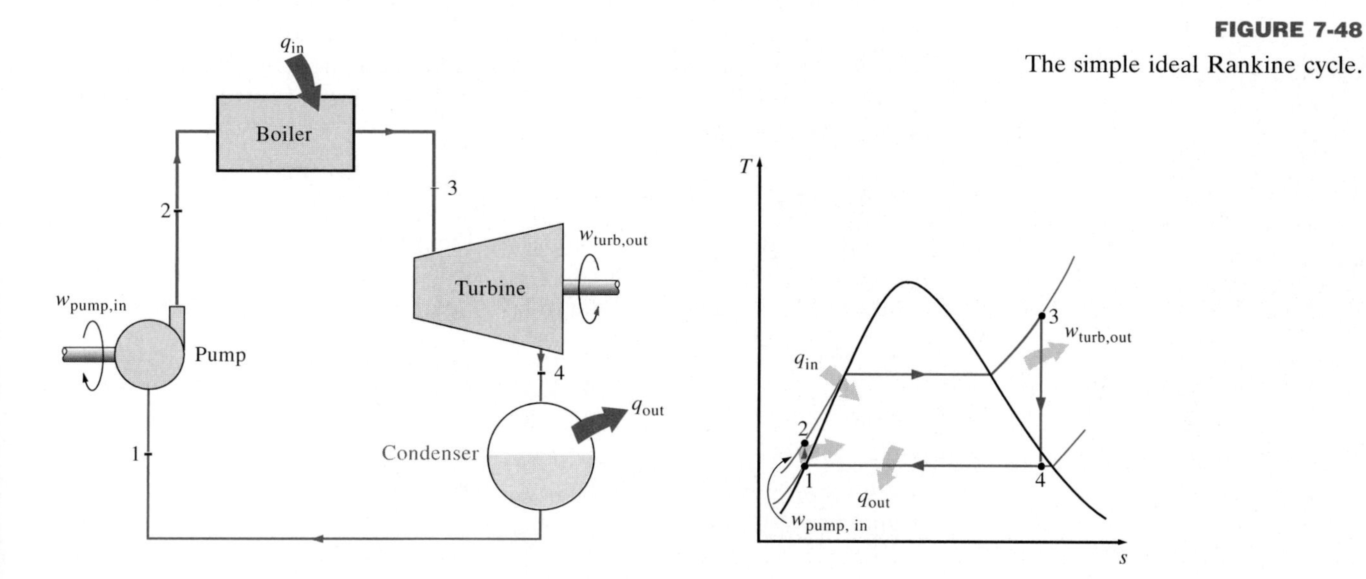


**FIGURE 7-48**
The simple ideal Rankine cycle.

Water enters the *boiler* as a compressed liquid at state 2 and leaves as a superheated vapor at state 3. The boiler is basically a large heat exchanger where the heat originating from combustion gases, nuclear reactors, or other sources is transferred to the water essentially at constant pressure. The boiler, together with the section where the steam is superheated (the superheater), is often called the *steam generator.*

The superheated vapor at state 3 enters the *turbine,* where it expands isentropically and produces work by rotating the shaft connected to an electric generator. The pressure and the temperature of the steam drop during this process to the values at state 4, where steam enters the *condenser.* At this state, steam is usually a saturated liquid–vapor mixture with a high quality. Steam is condensed at constant pressure in the condenser, which is basically a large heat exchanger, by rejecting heat to a cooling medium such as a lake, a river, or the atmosphere. Steam leaves the condenser as saturated liquid and enters the pump, completing the cycle. In areas where water is precious, the power plants are cooled by air instead of water. This method of cooling is called *dry cooling.* Several power plants in the world and a few in the United States are already using dry cooling to conserve water.

Remembering that the area under the process curve on a $T$-$s$ diagram represents the heat transfer for internally reversible processes, we see that the area under process curve 2-3 represents the heat transferred to the water in the boiler and the area under the process curve 4-1 represents the heat rejected in the condenser. The difference between these two (the area enclosed by the cycle) is the net work produced during the cycle.

## Energy Analysis of the Ideal Rankine Cycle

All four components associated with the Rankine cycle (the pump, boiler, turbine, and condenser) are steady-flow devices, and thus all four processes that make up the Rankine cycle can be analyzed as steady-flow processes. The kinetic and potential energy changes of the steam are usually small relative to the work and heat transfer terms and are therefore usually neglected. Then the *steady-flow energy equation* per unit mass of steam reduces to

$$q - w = h_e - h_i \qquad \text{(kJ/kg)} \tag{7-26}$$

The boiler and the condenser do not involve any work, and the pump and the turbine are assumed to be isentropic. Then the conservation of energy relation for each device can be expressed as follows:

*Pump* ($q = 0$): $$w_{\text{pump,in}} = h_2 - h_1 \tag{7-27}$$

or, $$w_{\text{pump,in}} = v(P_2 - P_1) \tag{7-28}$$

where $h_1 = h_{f\,@\,P_1}$ and $v \approx v_1 = v_{f\,@\,P_1}$ (7-29)

*Boiler* ($w = 0$): $$q_{\text{in}} = h_3 - h_2 \tag{7-30}$$

*Turbine* ($q = 0$): $$w_{\text{turb,out}} = h_3 - h_4 \tag{7-31}$$

*Condenser* ($w = 0$): $$q_{\text{out}} = h_4 - h_1 \tag{7-32}$$

The *thermal efficiency* of the Rankine cycle is determined from

$$\eta_{th} = \frac{w_{net}}{q_{in}} = 1 - \frac{q_{out}}{q_{in}} \tag{7-33}$$

where $$w_{net} = q_{in} - q_{out} = w_{turb,out} - w_{pump,in}$$

The converstion efficiency of power plants in the United States is often expressed in terms of **heat rate**, which is the amount of heat supplied, in Btu's, to generate 1 kWh of electricity. The smaller the heat rate, the greater the efficiency. Considering that 1 kWh = 3412 Btu, the relation between the heat rate and the thermal efficiency can be expressed as

$$\eta_{th} = \frac{3412 \text{ (Btu/kWh)}}{\text{heat rate (Btu/kWh)}} \tag{7-34}$$

For example, a heat rate of 11,363 Btu/kWh is equivalent to 30-percent thermal efficiency.

The thermal efficiency can also be interpreted as the ratio of the area enclosed by the cycle on a $T$-$s$ diagram to the area under the heat addition process.

The analysis of vapor power or refrigeration cycles requires the use properties of the working fluid at various states. Such properties can be obtained exactly from property tables, or approximately from property charts presented in the Appendix. The use of charts is straightforward, but is subject to reading errors. The use of property tables is demonstrated below with an example.

**EXAMPLE 7-7**

Determine the missing properties and the phase descriptions in the following table for water:

| | $T$, °C | $P$, kPa | $h$, kJ/kg | $x$ | Phase description |
|---|---|---|---|---|---|
| (*a*) | | 200 | | 0.6 | |
| (*b*) | 125 | | 1600 | | |
| (*c*) | | 1000 | 2950 | | |
| (*d*) | 75 | 500 | | | |
| (*e*) | | 850 | | 0.0 | |

**Solution** (*a*) The quality is given to be $x = 0.6$, which implies that 60 percent of the mass is in the vapor phase and the remaining 40 percent is in liquid phase. Therefore, we have saturated liquid–vapor mixture at a pressure of 200 kPa. Then the temperature must be the saturation temperature at the given pressure:

$$T = T_{sat\,@\,200\,kPa} = 120.23°C \qquad \text{(Table A-5)}$$

At 200 kPa, we also read from Table A-5 that $h_f = 504.70$ kJ/kg and $h_{fg} = 2706.7$ kJ/kg. Then the average internal energy of the mixture is determined from

$$h = h_f + xh_{fg}$$
$$= 504.70\ \text{kJ/kg} + (0.6)(2706.7\ \text{kJ/kg})$$
$$= \mathbf{2128.72\ kJ/kg}$$

(*b*) This time the temperature and the internal energy are given, but we do not know which table to use to determine the missing properties because we have no clue on whether we have saturated mixture, compressed liquid, or superheated vapor. To determine the region we are in, we first go to the saturation table (Table A-4) and determine the $h_f$ and $h_g$ values at the given temperature. At 125°C, we read $h_f = 524.99$ kJ/kg and $h_g = 2713.5$ kJ/kg. Next we compare the given $h$ value to these $h_f$ and $h_g$ values, keeping in mind that

if $h < h_f$, we have *compressed liquid*
if $h_f \leqslant h \leqslant h_g$, we have *saturated mixture*
if $h > h_g$, we have *superheated vapor*

The same can be said about the specific volume, internal energy, and entropy. In our case the given $h$ value is 1600, which falls between the $h_f$ and $h_g$ values at 125°C. Therefore, we have saturated liquid–vapour mixture. Then the pressure must be the saturation pressure at the given temperature:

$$P = P_{\text{sat @ 125°C}} = \mathbf{232.1\ kPa} \qquad \text{(Table A-4)}$$

The quality is determined from

$$x = \frac{h - h_f}{h_{fg}} = \frac{1600 - 524.99}{2188.5} = 0.491$$

(*c*) This case is similar to case (*b*), except pressure is given instead of temperature. Following the argument given above, we read the $h_f$ and $h_g$ values at the specified pressure. At 1 MPa, we have $h_f = 762.81$ kJ/kg and $h_g = 2778.1$ kJ/kg. The specified $h$ value is 2950 kJ/kg, which is greater than the $u_g$ value at 1 MPa. Therefore, we have **superheated vapor**, and the temperature at this state is determined from the superheated vapor table by interpolation to be

$$T = 254.2°\text{C} \qquad \text{(Table A-6)}$$

(*d*) In this case the temperature and pressure are given, but again we cannot tell which table to use to determine the missing properties because we do not know whether we have saturated mixture, compressed liquid, or superheated vapor. To determine the region we are in, we go to the saturation table (Table A-5) and determine the saturation temperature value at the given pressure. At 500 kPa, we have $T_{sat} = 151.86$°C. We then compare the given $T$ value to this $T_{sat}$ value, keeping in mind that

if $T < T_{\text{sat @ given } P}$, we have *compressed liquid*
if $T = T_{\text{sat @ given } P}$, we have *saturated mixture*
if $T > T_{\text{sat @ given } P}$, we have *superheated vapor*

In our case the given $T$ value is 75°C, which is less than the $T_{sat}$ value at the specified pressure. Therefore, we have **compressed liquid**, and normally we would determine the enthalpy value from the compressed liquid table. But in normally available in the compressed liquid tables, and therefore we are justified to treat the compressed liquid as saturated liquid at the given temperature (*not* pressure):

$$h \approx h_{f\,@\,75°C} = 313.93\ \text{kJ/kg} \qquad \text{(Table A-4)}$$

We would leave the quality column blank in this case since quality has no meaning in the compressed liquid region.

(*e*) The quality is given to be $x = 0.0$, and thus we have saturated liquid at the specified pressure of 850 kPa. Then the temperature must be the saturation temperature at the given pressure, and the enthalpy must have the saturated liquid value:

$$T = T_{\text{sat}\,@\,850\ \text{kPa}} = 172.96°C$$
$$h = h_{f\,@\,850\ \text{kPa}} = 732.22\ \text{kJ/kg} \qquad \text{(Table A-5)}$$

**EXAMPLE 7-8**

Consider a steam power plant operating on the simple ideal Rankine cycle. The steam enters the turbine at 3 MPa and 350°C and is condensed in the condenser at a pressure of 75 kPa. Determine the thermal efficiency of this cycle.

**Solution** The schematic of the power plant and the *T-s* diagram of the cycle are shown in Fig. 7-49. Since the power plant operates on the ideal Rankine cycle, it is assumed that the turbine and the pump are isentropic, there are no pressure drops in the boiler and the condenser, and steam leaves the condenser and enters the pump as saturated liquid at the condenser pressure.

First, we determine the enthalpies at various points in the cycle, using data from steam tables (Tables A-4, A-5, and Fig. A-6):

*State 1:* $\left.\begin{array}{l} P_1 = 75\ \text{kPa} \\ \text{sat. liquid} \end{array}\right\}$ $\begin{array}{l} h_1 = h_{f\,@\,75\ \text{kPa}} = 384.39\ \text{kJ/kg} \\ v_1 = v_{f\,@\,75\ \text{kPa}} = 0.001037\ \text{m}^3\text{/kg} \end{array}$

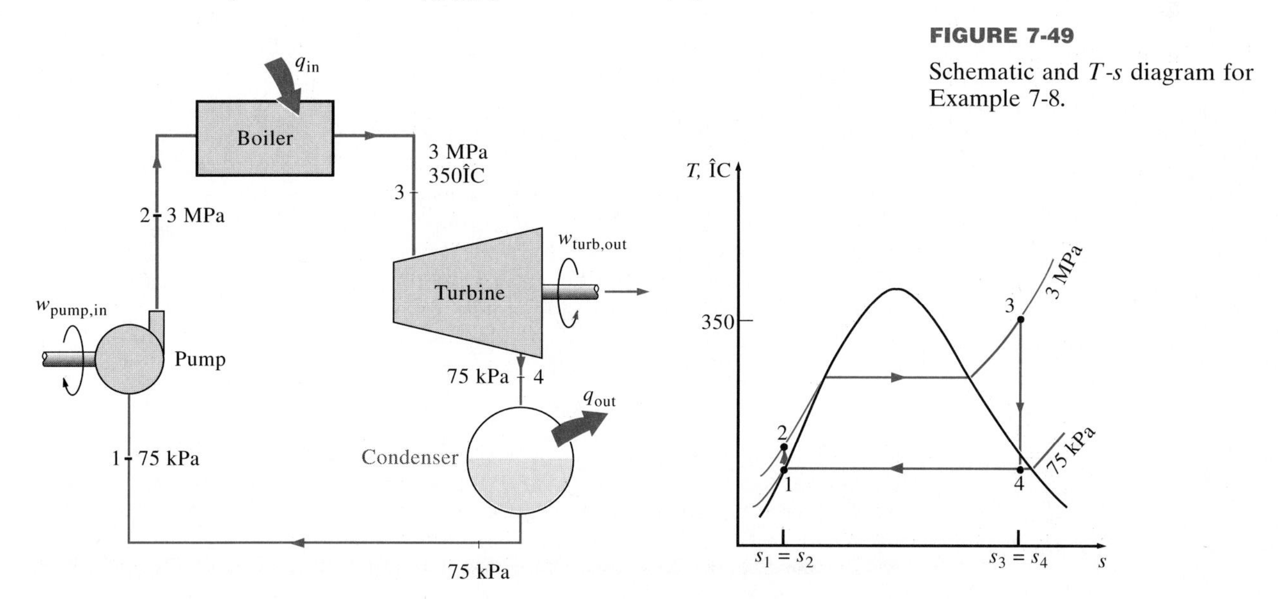


**FIGURE 7-49**

Schematic and *T-s* diagram for Example 7-8.

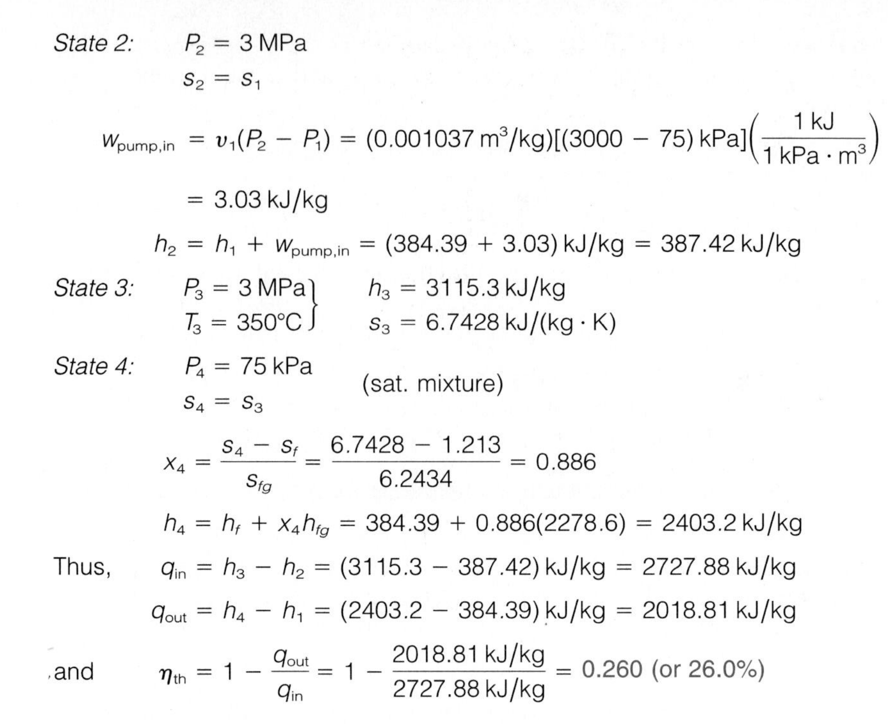

*State 2:* $P_2 = 3\text{ MPa}$
$s_2 = s_1$

$$w_{\text{pump,in}} = v_1(P_2 - P_1) = (0.001037\text{ m}^3/\text{kg})[(3000 - 75)\text{ kPa}]\left(\frac{1\text{ kJ}}{1\text{ kPa}\cdot\text{m}^3}\right)$$

$$= 3.03\text{ kJ/kg}$$

$$h_2 = h_1 + w_{\text{pump,in}} = (384.39 + 3.03)\text{ kJ/kg} = 387.42\text{ kJ/kg}$$

*State 3:* $\left.\begin{array}{l} P_3 = 3\text{ MPa} \\ T_3 = 350°\text{C} \end{array}\right\}$ $\begin{array}{l} h_3 = 3115.3\text{ kJ/kg} \\ s_3 = 6.7428\text{ kJ/(kg}\cdot\text{K)} \end{array}$

*State 4:* $\begin{array}{l} P_4 = 75\text{ kPa} \\ s_4 = s_3 \end{array}$ (sat. mixture)

$$x_4 = \frac{s_4 - s_f}{s_{fg}} = \frac{6.7428 - 1.213}{6.2434} = 0.886$$

$$h_4 = h_f + x_4 h_{fg} = 384.39 + 0.886(2278.6) = 2403.2\text{ kJ/kg}$$

Thus, $$q_{\text{in}} = h_3 - h_2 = (3115.3 - 387.42)\text{ kJ/kg} = 2727.88\text{ kJ/kg}$$

$$q_{\text{out}} = h_4 - h_1 = (2403.2 - 384.39)\text{ kJ/kg} = 2018.81\text{ kJ/kg}$$

and $$\eta_{\text{th}} = 1 - \frac{q_{\text{out}}}{q_{\text{in}}} = 1 - \frac{2018.81\text{ kJ/kg}}{2727.88\text{ kJ/kg}} = 0.260\text{ (or 26.0\%)}$$

The thermal efficiency could also be determined from

$$w_{\text{turb,out}} = h_3 - h_4 = (3115.3 - 2403.2)\text{ kJ/kg} = 712.1\text{ kJ/kg}$$

$$w_{\text{net}} = w_{\text{turb,out}} - w_{\text{pump,in}} = (712.1 - 3.03)\text{ kJ/kg} = 709.07\text{ kJ/kg}$$

or $$w_{\text{net}} = q_{\text{in}} - q_{\text{out}} = (2727.88 - 2018.81)\text{ kJ/kg} = 709.07\text{ kJ/kg}$$

and $$\eta_{\text{th}} = \frac{w_{\text{net}}}{q_{\text{in}}} = \frac{709.07\text{ kJ/kg}}{2727.88\text{ kJ/kg}} = 0.260\text{ (or 26.0\%)}$$

That is, this power plant converts 26 percent of the heat it receives in the boiler to net work. An actual power plant operating between the same temperature and pressure limits will have a lower efficiency because of the irreversibilities such as friction.

Notice that the back work ratio ($r_{\text{bw}} = w_{\text{in}}/w_{\text{out}}$) of this power plant is 0.004. That, is, only 0.4 percent of the turbine work output is required to operate the pump. Having low back work ratios (usually under 1 percent) is characteristic of vapor power cycles. This is in contrast to the gas power cycles, which typically have very high back work ratios (40–80 percent).

It is also interesting to note the thermal efficiency of a Carnot cycle operating between the same temperature limits

$$\eta_{\text{th,Carnot}} = 1 - \frac{T_{\text{min}}}{T_{\text{max}}} = 1 - \frac{(91.78 + 273)\text{ K}}{(350 + 273)\text{ K}} = 0.414$$

The difference between the two efficiencies is due to the large temperature difference between the steam and the combustion gases during the heat addition process of the Rankine cycle.

## How Can We Increase the Efficiency of the Rankine Cycle?

Steam power plants are responsible for the production of most of the electric power in the world, and even small increases in thermal efficiency can mean large savings from the fuel requirements. Therefore, every effort is made to improve the efficiency of the cycle on which steam power plants operate.

The basic idea behind all the modifications to increase the thermal efficiency of a power cycle is the same: *Increase the average temperature at which heat is transferred to the working fluid in the boiler, or decrease the average temperature at which heat is rejected from the working fluid in the condenser.* That is, the average fluid temperature should be as high as possible during heat addition and as low as possible during rejection. Next we discuss three ways of accomplishing this for the simple ideal Rankine cycle.

### 1 Lowering the Condenser Pressure

*(Lowers $T_{low,av}$)*

Steam exists as a saturated mixture in the condenser at the saturation temperature corresponding to the pressure inside. Therefore, lowering the operating pressure of the condenser automatically lowers the temperature of the steam and thus the temperature at which heat is rejected.

The effect of lowering the condenser pressure on the Rankine cycle efficiency is illustrated on a *T-s* diagram in Fig. 7-50. For comparison purposes, the turbine inlet state is maintained the same. The colored area on this diagram represents the increase in net work output as a result of lowering the condenser pressure from $P_4$ to $P_4'$. The heat input requirements also increase (represented by the area under curve 2′-2), but this increase is very small. Thus, the overall effect of lowering the condenser pressure is an increase in the thermal efficiency of the cycle.

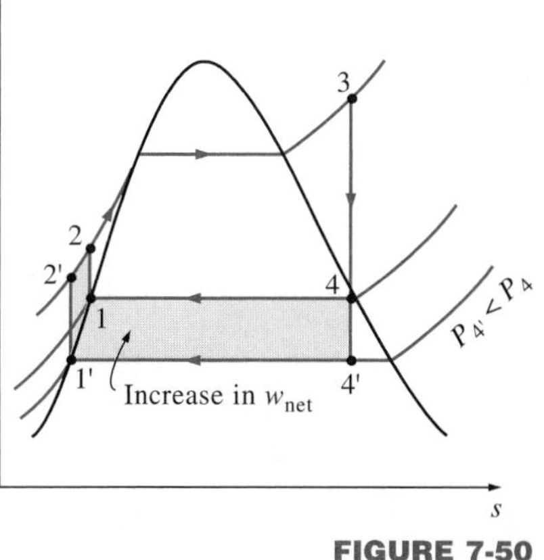


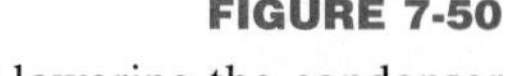

**FIGURE 7-50**

The effect of lowering the condenser pressure on the ideal Rankine cycle.

To take advantage of the increased efficiencies at low pressures, the condensers of steam power plants usually operate well below the atmospheric pressure. This does not present a major problem, since the vapor power cycles operate in a closed loop. However, there is a lower limit on the condenser pressure that can be used. It cannot be lower than the saturation pressure corresponding to the temperature of the cooling medium. Consider, for example, a condenser which is to be cooled by a nearby river at 15°C. Allowing a temperature difference of 10°C for effective heat transfer, the steam temperature in the condenser must be above 25°C; thus, the condenser pressure must be above 3.2 kPa, which is the saturation pressure at 25°C.

Lowering the condenser pressure is not without any side effects, however. For one thing, it creates the problem of air leakage into the condenser. More importantly, it increases the moisture content of the steam at the last stages of the turbine, as can be seen from Fig. 7-50. The presence of large quantities of moisture is highly undesirable in turbines because it decreases the turbine efficiency and erodes the turbine blades.

Fortunately, this problem can be corrected, as discussed later in this chapter.

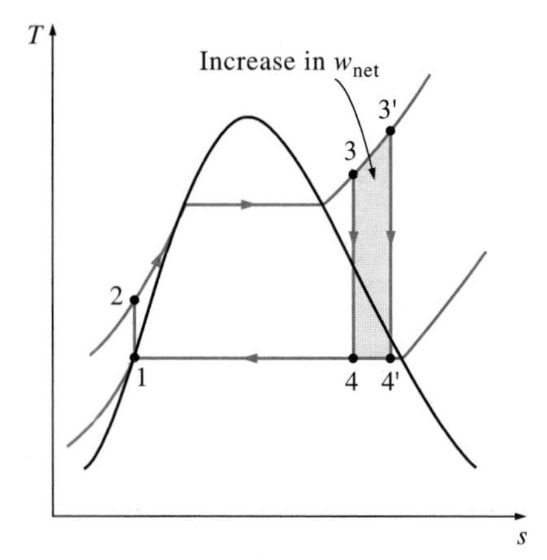


**FIGURE 7-51**

The effect of superheating the steam to higher temperatures on the ideal Rankine cycle.

## 2 Superheating the Steam to High Temperatures
*(Increases $T_{high,av}$)*

The average temperature at which heat is added to the steam can be increased without increasing the boiler pressure by superheating the steam to high temperatures. The effect of superheating on the performance of vapor power cycles is illustrated on a *T-s* diagram in Fig. 7-51. The colored area on this diagram represents the increase in the net work. The total area under the process curve 3-3′ represents the increase in the heat input. Thus both the net work and heat input increases as a result of superheating the steam to a higher temperature. The overall effect is an increase in thermal efficiency, however, since the average temperature at which heat is added increases.

Superheating the steam to higher temperatures has another very desirable effect: it decreases the moisture content of the steam at the turbine exit, as can be seen from the *T-s* diagram (the quality at state 4′ is higher than that at state 4).

The temperature to which steam can be superheated is limited, however, by metallurgical considerations. Presently the highest steam temperature allowed at the turbine inlet is about 620°C (1150°F). Any increase in this value depends on improving the present materials or finding new ones that can withstand higher temperatures. Ceramics are very promising in this regard.

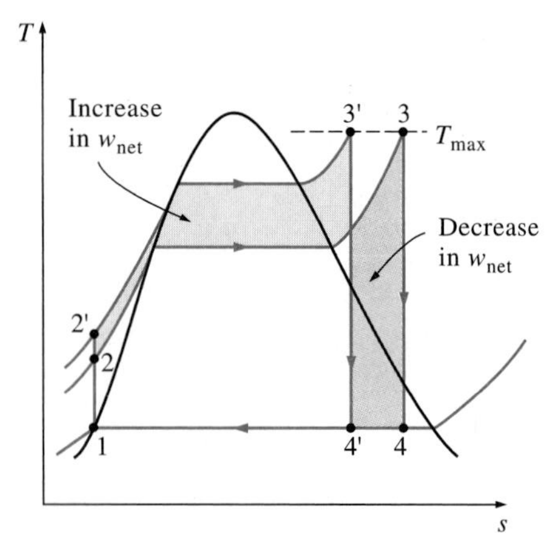


**FIGURE 7-52**

The effect of increasing the boiler pressure on the ideal Rankine cycle.

## 3 Increasing the Boiler Pressure
*(Increases $T_{high,av}$)*

Another way of increasing the average temperature during the heat addition process is to increase the operating pressure of the boiler, which automatically raises the temperature at which boiling takes place. This, in turn, raises the average temperature at which heat is added to the steam and thus raises the thermal efficiency of the cycle.

The effect of increasing the boiler pressure on the performance of vapor power cycles is illustrated on a *T-s* diagram in Fig. 7-52. Notice that for a fixed turbine inlet temperature, the cycle shifts to the left and the moisture content of steam at the turbine exit increases. This undesirable side effect can be corrected, however, by reheating the steam, as discussed in the next section.

Operating pressures of boilers have gradually increased over the years from about 2.7 MPa (400 psia) in 1922 to over 30 MPa (4500 psia) today, generating enough steam to produce a net power output of 1000 MW or more. Today many modern steam power plants operate at supercritical pressures ($P > 22.09$ MPa) and have thermal efficiencies of about 40 percent for fossil-fuel plants and 34 percent for nuclear plants. There are about 170 supercritical pressure steam power plants in operation in the United States. The lower efficiencies of nuclear power plants is due to the lower maximum temperatures used in those plants for safety reasons. The $T$-$s$ diagram of a supercritical Rankine cycle is shown in Fig. 9-9. The United States has 112 nuclear power plants which generate about 21 percent of the nation's electricity. (In contrast, 75 percent of the electricity of France comes from nuclear plants.)

## 7-12 ■ THE IDEAL REHEAT RANKINE CYCLE

We noted in the last section that increasing the boiler pressure increases the thermal efficiency of the Rankine cycle, but it also increases the moisture content of the steam to unacceptable levels. Then it is natural to ask the following question:

*How can we take advantage of the increased efficiencies at higher boiler pressures without facing the problem of excessive moisture at the final stages of the turbine?*

Two possibilities come to mind:

**1** Superheat the steam to very high temperatures before it enters the turbine. This would be the desirable solution since the average temperature at which heat is added would also increase, thus increasing the cycle efficiency. This is not a viable solution, however, since it will require raising the steam temperature to metallurgically unsafe levels.

**2** Expand the steam in the turbine in two stages, and reheat it in between. In other words, modify the simple ideal Rankine cycle with a **reheat** process. Reheating is a practical solution to the excessive moisture problem in turbines, and it is used frequently in modern steam power plants.

The $T$-$s$ diagram of the ideal reheat Rankine cycle and the schematic of the power plant operating on this cycle are shown in Fig. 7-53. The ideal reheat Rankine cycle differs from the simple ideal Rankine cycle in that the expansion process takes place in two stages. In the first stage (the high-pressure turbine), steam is expanded isentropically to an intermediate pressure and sent back to the boiler where it is reheated at constant pressure, usually to the inlet temperature of the first turbine stage. Steam then expands isentropically in the second stage (low-pressure turbine) to the condenser pressure. Thus, the total heat input and the total turbine work output for a reheat cycle become

$$q_{in} = q_{primary} + q_{reheat} = (h_3 - h_2) + (h_5 - h_4) \qquad \text{(kJ/kg)} \qquad (7\text{-}35)$$

and $$w_{turb,out} = w_{turb,I} + w_{turb,II} = (h_3 - h_4) + (h_5 - h_6) \qquad \text{(kJ/kg)} \qquad (7\text{-}36)$$

The incorporation of the single reheat in a modern power plant improves the cycle efficiency by 4–5 percent by increasing the average temperature at which heat is added to the steam.

**FIGURE 7-53**
The ideal reheat Rankine cycle.

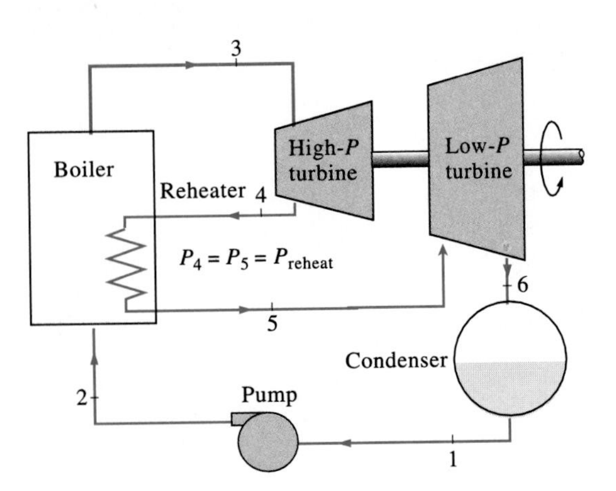


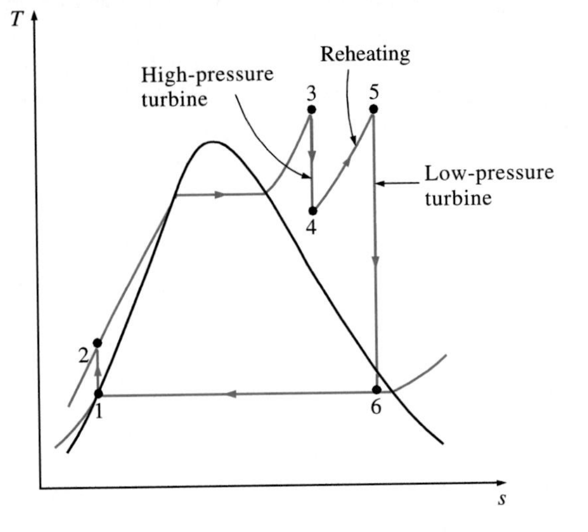


The average temperature during the reheat process can be increased by increasing the number of expansion and reheat stages. As the number of stages is increased, the expansion and reheat processes approach an isothermal process at the maximum temperature, as shown in Fig. 7-54. The use of more than two reheat stages, however, is not practical. The theoretical improvement in efficiency from the second reheat is about half of that which results from a single reheat. If the turbine inlet pressure is not high enough, double reheat would result in superheated exhaust. This is undesirable as it would cause the average temperature for heat rejection to increase and thus the cycle efficiency to decrease. Therefore, double reheat is used only on supercritical pressure ($P > 22.09$ MPa) power plants. A third reheat stage would increase the cycle efficiency by about half of the improvement attained by the second reheat. This gain is so small that it does not justify the added cost and complexity.

The reheat cycle was introduced in the mid 1920s, but it was

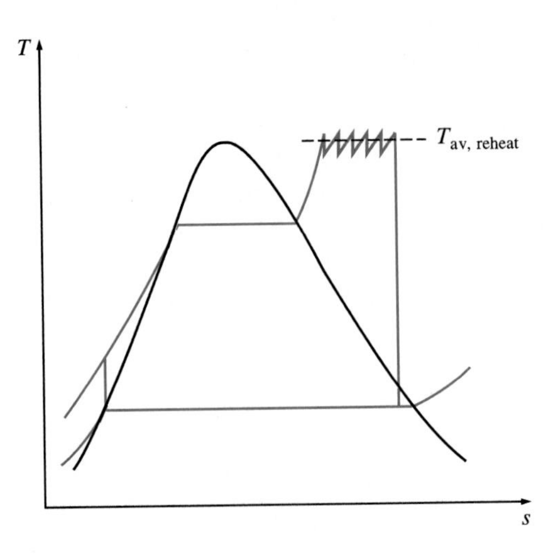


**FIGURE 7-54**
The average temperature at which heat is added during reheating increases as the number of reheat stages is increased.

abandoned in the 1930s because of the operational difficulties. The steady increase in boiler pressures over the years made it necessary to reintroduce single reheat in late 1940s, and double reheat in early 1950s.

The reheat temperatures are very close or equal to the turbine inlet temperature. The optimum reheat pressure is about one-fourth of the maximum cycle pressure. For example, the optimum reheat pressure for a cycle with a boiler pressure of 12 MPa is about 3 MPa.

Remember that the sole purpose of the reheat cycle is to reduce the moisture content of the steam at the final stages of the expansion process. If we had materials that could withstand sufficiently high temperatures, there would be no need for the reheat cycle.

**EXAMPLE 7-9**

Consider a steam power plant operating on the ideal reheat Rankine cycle. The steam enters the high-pressure turbine at 15 MPa and 600°C and is condensed in the condenser at a pressure of 10 kPa. If the moisture content of the steam at the exit of the low-pressure turbine is not to exceed 10.4 percent, determine (*a*) the pressure at which the steam should be reheated and (*b*) the thermal efficiency of the cycle. Assume the steam is reheated to the inlet temperature of the high-pressure turbine.

**Solution** The schematic of the power plant and the *T*-*s* diagram of the cycle are shown in Fig. 7-55. Since the power plant operates on the ideal reheat Rankine cycle, we assume that both stages of the turbine and the pump are isentropic, there are no pressure drops in the boiler and the condenser, and steam leaves the condenser and enters the pump as a saturated liquid at the condenser pressure.

(*a*) The reheat pressure is determined from the requirement that the entropies at states 5 and 6 be the same:

*State 6:*

$$\left.\begin{array}{l} P_6 = 10\text{ kPa} \\ x_6 = 0.896\text{ (sat. mixture)} \end{array}\right\} \begin{array}{l} h_6 = 2335.8\text{ kJ/kg} \\ s_6 = 7.370\text{ kJ/(kg}\cdot\text{K)} \end{array}$$

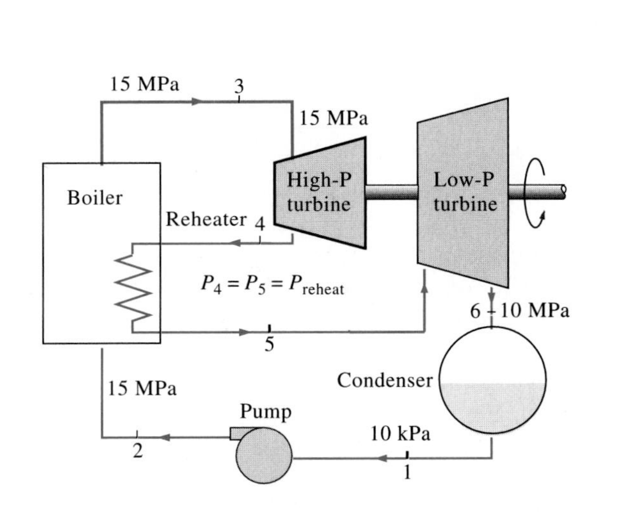


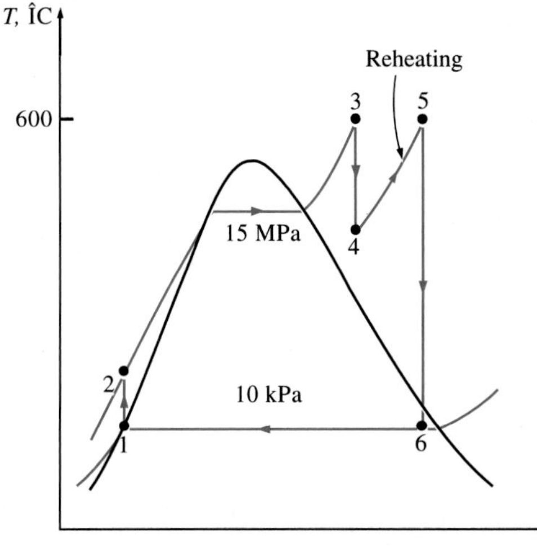


**FIGURE 7-55**
Schematic and *T*-*s* diagram for Example 7-9.

Thus,

*State 5:* $T_5 = 600°C$, $s_5 = s_6$ → $P_5 = 4.0$ MPa, $h_5 = 3674.4$ kJ/kg

Therefore, steam should be reheated at a pressure of 4 MPa or lower to prevent a moisture content above 10.4 percent.

(*b*) To determine the thermal efficiency, we need to know the enthalpies at all other states:

*State 1:* $P_1 = 10$ kPa, sat. liquid → $h_1 = h_{f @ 10\,\text{kPa}} = 191.83$ kJ/kg, $v_1 = v_{f @ 10\,\text{kPa}} = 0.001010\ \text{m}^3/\text{kg}$

*State 2:* $P_2 = 15$ MPa, $s_2 = s_1$

$$w_{\text{pump,in}} = v_1(P_2 - P_1) = (0.001010\ \text{m}^3/\text{kg})[(15{,}000 - 10)\ \text{kPa}]\left(\frac{1\ \text{kJ}}{1\ \text{kPa}\cdot\text{m}^3}\right)$$
$$= 15.11\ \text{kJ/kg}$$
$$h_2 = h_1 + w_{\text{pump,in}} = (191.83 + 15.11)\ \text{kJ/kg} = 206.94\ \text{kJ/kg}$$

*State 3:* $P_3 = 15$ MPa, $T_3 = 600°C$ → $h_3 = 3582.3$ kJ/kg, $s_3 = 6.6776$ kJ/(kg·K)

*State 4:* $P_4 = 4$ mPa, $s_4 = s_3$ → $h_4 = 3154.3$ kJ/kg ($T_4 = 375.5°C$)

Thus,
$$q_{\text{in}} = (h_3 - h_2) + (h_5 - h_4)$$
$$= (3582.3 - 206.94)\ \text{kJ/kg} + (3674.4 - 3154.3)\ \text{kJ/kg}$$
$$= 3895.46\ \text{kJ/kg}$$
$$q_{\text{out}} = h_6 - h_1 = (2335.8 - 191.83)\ \text{kJ/kg}$$
$$= 2143.97\ \text{kJ/kg}$$

and
$$\eta_{\text{th}} = 1 - \frac{q_{\text{out}}}{q_{\text{in}}} = 1 - \frac{2143.97\ \text{kJ/kg}}{3895.46\ \text{kJ/kg}} = \mathbf{0.450\ (or\ 45.0\%)}$$

## 7-13 ■ REFRIGERATORS AND HEAT PUMPS

We all know from experience that heat flows in the direction of decreasing temperature, i.e., from high-temperature regions to low-temperature ones. This heat transfer process occurs in nature without requiring any devices. The reverse process, however, cannot occur by itself. The transfer of heat from a low-temperature region to a high-temperature one requires special devices called **refrigerators.**

Refrigerators are cyclic devices, and the working fluids used in the refrigeration cycles are called **refrigerants**. A refrigerator is shown schematically in Fig. 7-56*a*. Here $Q_L$ is the magnitude of the heat removed from the refrigerated space at temperature $T_L$, $Q_H$ is the magnitude of the heat rejected to the warm space at temperature $T_H$, and $W_{\text{net, in}}$ is the net work input to the refrigerator. As discussed in Chap. 5, $Q_L$ and $Q_H$ represent magnitudes and thus are positive quantities.

Another device that transfers heat from a low-temperature medium to a high-temperature one is the **heat pump**. Refrigerators and heat pumps are essentially the same devices; they differ in their objectives only. The objective of a refrigerator is to maintain the refrigerated space at a low temperature by removing heat from it. Discharging this heat to a higher-temperature medium is merely a necessary part of the operation, not the purpose. The objective of a heat pump, however, is to maintain a heated space at a high temperature. This is accomplished by absorbing heat from a low-temperature source, such as well water or cold outside air in winter, and supplying this heat to a warmer medium such as a house (Fig. 7-56*b*).

The performance of refrigerators, and heat pumps is expressed in terms of the **coefficient of performance** (COP), which is defined as

$$\text{COP}_\text{R} = \frac{\text{desired output}}{\text{required input}} = \frac{\text{cooling effect}}{\text{work input}} = \frac{Q_L}{W_{\text{net, in}}} \tag{7-37}$$

$$\text{COP}_\text{HP} = \frac{\text{desired output}}{\text{required input}} = \frac{\text{heating effect}}{\text{work input}} = \frac{Q_H}{W_{\text{net, in}}} \tag{7-38}$$

These relations can also be expressed in rate form by replacing the quantities $Q_L$, $Q_H$, and $W_{\text{net, in}}$ by $\dot{Q}_L$, $\dot{Q}_H$, and $\dot{W}_{\text{net, in}}$, respectively. Notice that both $\text{COP}_\text{R}$ and $\text{COP}_\text{HP}$ can be greater than 1. A comparison of Eqs. 7-37 and 7-38 reveals that

$$\text{COP}_\text{HP} = \text{COP}_\text{R} + 1 \tag{7-39}$$

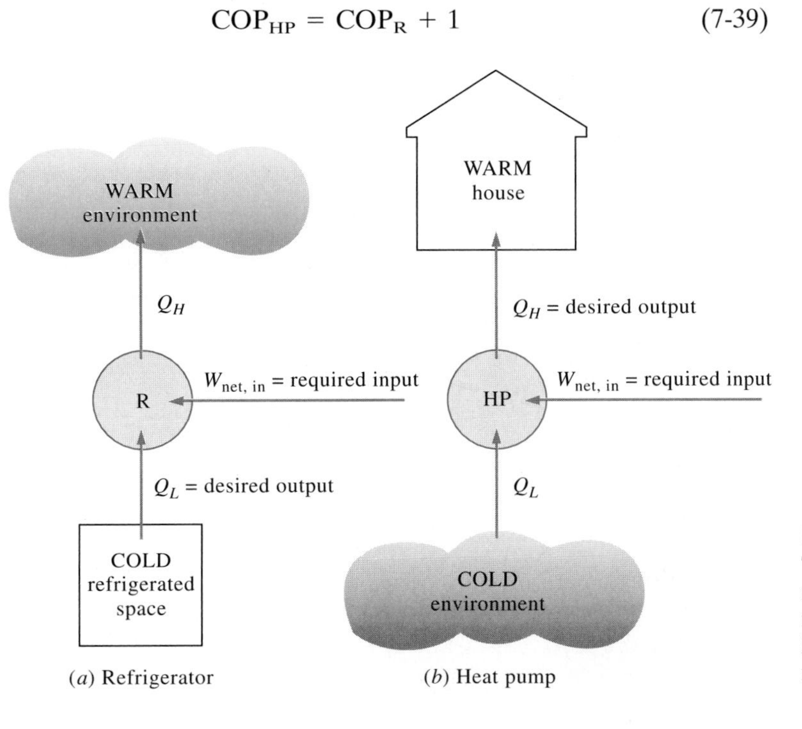


**FIGURE 7-56**
The objective of a refrigerator is to remove heat ($Q_L$) from the cold medium; the objective of a heat pump is to supply heat ($Q_H$) to a warm medium.

for fixed values of $Q_L$ and $Q_H$. This relation implies that $COP_{HP} > 1$ since $COP_R$ is a positive quantity. That is, a heat pump will function, at worst, as a resistance heater, supplying as much energy to the house as it consumes. In reality, however, part of $Q_H$ is lost to the outside air through piping and other devices, and $COP_{HP}$ may drop below unity when the outside air temperature is too low. When this happens, the system usually switches to a resistance heating mode.

The *cooling capacity* of a refrigeration system—that is, the rate of removal from the refrigerated space—is often expressed in terms of **tons of refrigeration**. The capacity of a refrigeration system that can freeze 1 ton (2000 lbm) of liquid water at 0°C (32°F) into ice at 0°C in 24 h is said to be 1 ton. One ton of refrigeration is equivalent to 211 kJ/min or 200 Btu/min. The cooling load of a typical 200 $m^2$ residence is in the 3-ton (10-kW) range.

## 7-14 ■ THE REVERSED CARNOT CYCLE

You will recall from the preceding chapters that the Carnot cycle is a totally reversible cycle that consists of two reversible isothermal and two isentropic processes. It has the maximum thermal efficiency for given temperature limits, and it serves as a standard against which actual power cycles can be compared. The Carnot cycle proved to be a valuable tool in the study of gas and vapor power cycles discussed earlier.

Since it is a reversible cycle, all four processes that form the Carnot cycle can be reversed. Reversing the cycle will also reverse the directions of any heat and work interactions. The result is a cycle that operates in

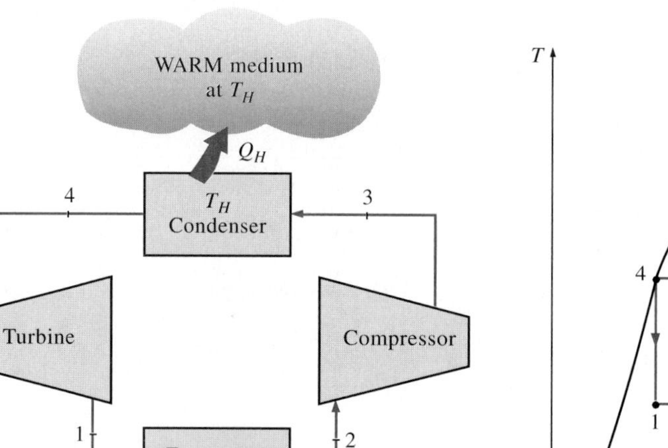


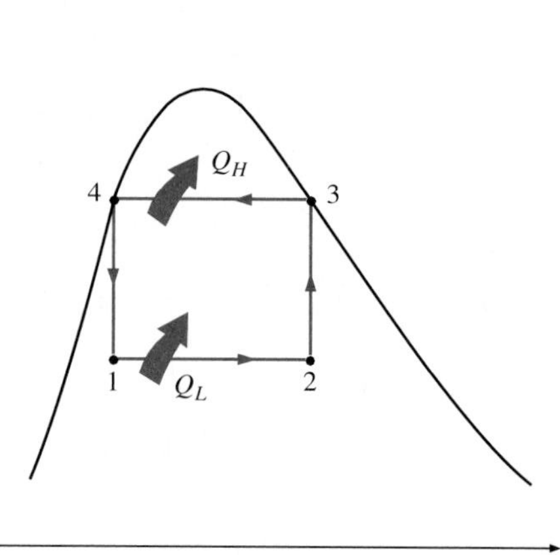


**FIGURE 7-57**

Schematic of a Carnot refrigerator and *T-s* diagram of the reversed Carnot cycle.

the counterclockwise direction, which is called the **reversed Carnot cycle.** A refrigerator or heat pump that operates on the reversed Carnot cycle is called a **Carnot refrigerator** or a **Carnot heat pump.**

Consider a reversed Carnot cycle executed within the saturation dome of a refrigerant, as shown in Fig. 7-57. The refrigerant absorbs heat isothermally from a low-temperature source at $T_L$ in the amount of $Q_L$ (process 1-2), is compressed isentropically to state 3 (temperature rises to $T_H$), rejects heat isothermally to a high-temperature sink at $T_H$ in the amount of $Q_H$ (process 3-4), and expands isentropically to state 1 (temperature drops to $T_L$). The refrigerant changes from a saturated vapor state to a saturated liquid state in the condenser during process 3-4.

The coefficients of performance of Carnot refrigerators and heat pumps were determined in Sec. 5-11 to be

$$\mathrm{COP}_{\mathrm{R,\,Carnot}} = \frac{1}{T_H/T_L - 1} \tag{7-40}$$

and

$$\mathrm{COP}_{\mathrm{HP,\,Carnot}} = \frac{1}{1 - T_L/T_H} \tag{7-41}$$

Notice that both COPs increase as the difference between the two temperatures decreases, i.e., as $T_L$ rises or $T_H$ falls.

The reversed Carnot cycle is the *most efficient* refrigeration cycle operating between two specific temperature levels. Therefore, it is natural to look at it first as a prospective ideal cycle for refrigerators and heat pumps. If we could, we certainly would adapt it as the ideal cycle. But as explained below, the reversed Carnot cycle is an unsuitable model for refrigeration cycles.

The two isothermal heat transfer processes are not difficult to achieve in practice since maintaining a constant pressure automatically fixes the temperature of a two-phase mixture at the saturation value. Therefore, processes 1-2 and 3-4 can be approached closely in actual evaporators and condensers. However, processes 2-3 and 4-1 cannot be approximated closely in practice. This is because process 2-3 involves the compression of a liquid–vapor mixture, which requires a compressor that will handle two phases, and process 4-1 involves the expansion of high-moisture-content refrigerant.

It seems as if these problems could be eliminated by executing the reversed Carnot cycle outside the saturation region. But in this case, we will have difficulty in maintaining isothermal conditions during the heat absorption and heat rejection processes. Therefore, we conclude that the reversed Carnot cycle cannot be approximated in actual devices and is not a realistic model for refrigeration cycles. However, the reversed Carnot cycle can serve as a standard against which actual refrigeration cycles are compared.

## 7-15 ■ THE IDEAL VAPOR-COMPRESSION REFRIGERATION CYCLE

Many of the impracticalities associated with the reversed Carnot cycle can be eliminated by vaporizing the refrigerant completely before it is compressed and by replacing the turbine with a throttling device, such as an expansion valve or capillary tube. The cycle that results is called the **ideal vapor-compression refrigeration cycle**, and it is shown schematically and on a *T-s* diagram in Fig. 7-58. The vapor-compression refrigeration cycle is the most widely used cycle for refrigerators, air conditioning systems, and heat pumps. It consists of four processes:

1-2 Isentropic compression in a compressor

2-3 $P$ = constant heat rejection in a condenser

3-4 Throttling in an expansion device

4-1 $P$ = constant heat absorption in an evaporator

In an ideal vapor-compression refrigeration cycle, the refrigerant enters the compressor at state 1 as saturated vapor and is compressed isentropically to the condenser pressure. The temperature of the refrigerant increases during this isentropic compression process to well above the temperature of the surrounding medium, such as atmospheric air. The refrigerant then enters the condenser as superheated vapor at state 2 and leaves as saturated liquid at state 3 as a result of heat rejection to the surroundings. The temperature of the refrigerant at this state is still above the temperature of the surroundings.

The saturated liquid refrigerant at state 3 is throttled to the evaporator pressure by passing it through an expansion valve or capillary tube. The temperature of the refrigerant drops below the temperature of the refrigerated space during this process. The refrigerant enters the evaporator at state 4 as a low-quality saturated mixture, and it completely

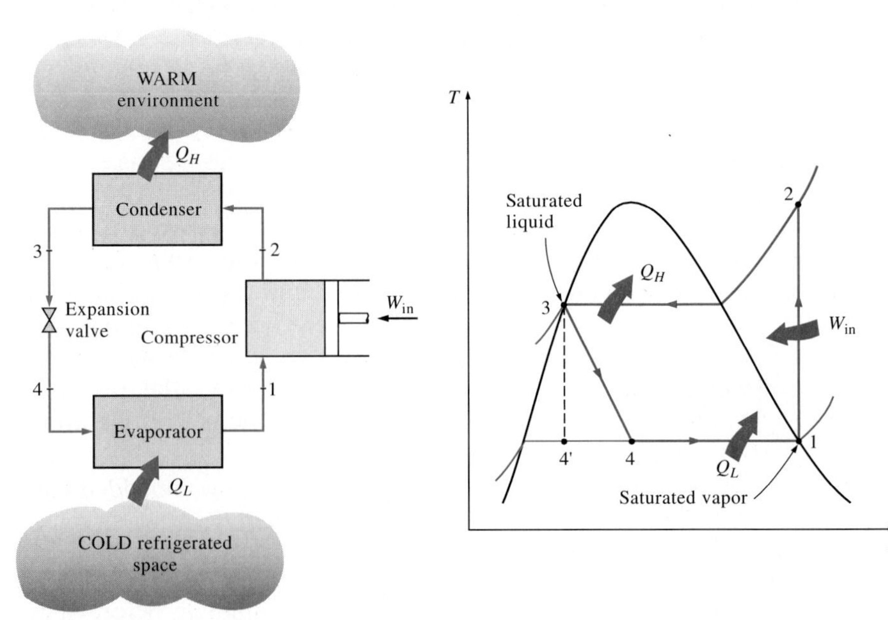


**FIGURE 7-58**

Schematic and *T-s* diagram for the ideal vapor-compression refrigeration cycle.

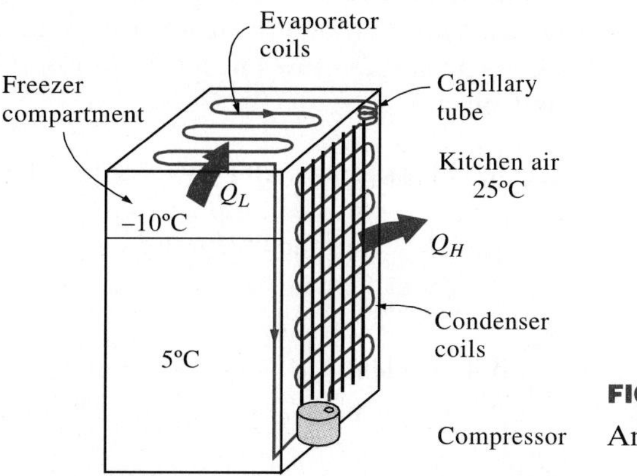

**FIGURE 7-59**
An ordinary household refrigerator.

evaporates by absorbing heat from the refrigerated space. The refrigerant leaves the evaporator as saturated vapor and reenters the compressor, completing the cycle.

In a household refrigerator, the freezer compartment where heat is absorbed by the refrigerant, serves as the evaporator. The coils behind the refrigerator where heat is dissipated to the kitchen air serve as the condenser (Fig. 7-59).

Remember that the area under the process curve on a $T$-$s$ diagram represents the heat transfer for internally reversible processes. The area under process curve 4-1 represents the heat absorbed by the refrigerant in the evaporator, and the area under process curve 2-3 represents the heat rejected in the condenser. A rule of thumb is that the *COP improves by* 2 *to* 4 *percent for each* °*C the evaporating temperature is raised or the condensing temperature is lowered.*

Another diagram frequently used in the analysis of vapor-compression refrigeration cycles is the $P$-$h$ diagram, as shown in Fig. 7-60. On this diagram, three of the four processes appear as straight lines, and the heat transfer in the condenser and the evaporator is proportional to the lengths of the corresponding process curves.

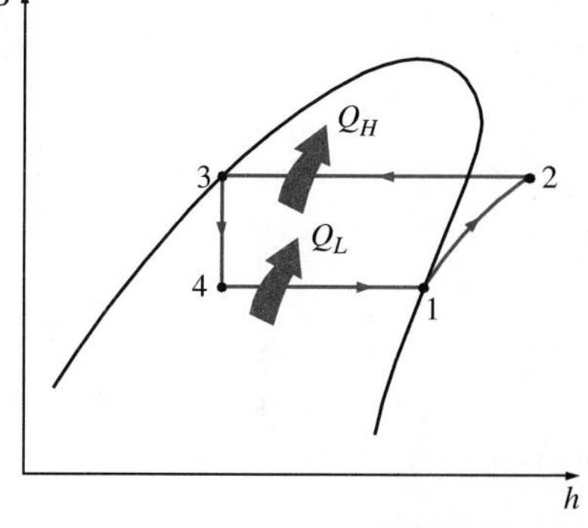

**FIGURE 7-60**
The $P$-$h$ diagram of an ideal vapor-compression refrigeration cycle.

Notice that, unlike the ideal cycles discussed before, the ideal vapor-compression refrigeration cycle is not an internally reversible cycle since it involves an irreversible (throttling) process. This process is maintained in the cycle to make it a more realistic model for the actual vapor-compression refrigeration cycle. If the throttling device were replaced by an isentropic turbine, the refrigerant would enter the evaporator at state 4′ instead of state 4. As a result, the refrigeration capacity would increase (by the area under process curve 4′-4 in Fig. 7-58) and the net work input would decrease (by the amount of work output of the turbine). Replacing the expansion valve by a turbine is not practical, however, since the added benefits cannot justify the added cost and complexity.

All four components associated with the vapor-compression refrigeration cycle are steady-flow devices, and thus all four processes that make up the cycle can be analyzed as steady-flow processes. The kinetic and potential energy changes of the refrigerant are usually small relative to the work and heat transfer terms, and therefore they can be neglected. Then the steady-flow energy equation on a unit-mass basis reduces to

$$q - w = h_e - h_i \tag{7-42}$$

The condenser and the evaporator do not involve any work, and the compressor can be approximated as adiabatic. Then the COPs of refrigerators and heat pumps operating on the vapor-compression refrigeration cycle can be expressed as

$$\text{COP}_\text{R} = \frac{q_L}{w_{\text{net, in}}} = \frac{h_1 - h_4}{h_2 - h_1} \tag{7-43}$$

and

$$\text{COP}_\text{HP} = \frac{q_H}{w_{\text{net, in}}} = \frac{h_2 - h_3}{h_2 - h_1} \tag{7-44}$$

where $h_1 = h_{g @ P_1}$ and $h_3 = h_{f @ P_3}$ for the ideal case.

Vapor compression refrigeration dates back to 1834, when Englishman Jacob Perkins received a patent for a closed cycle ice machine using ether or other volatile fluids as refrigerants. A working model of this machine was built, but it was never produced commercially. In 1850, Alexander Twinning began to design and build vapor compression ice machines using ethyl ether, which is the commercially used refrigerant in vapor compression systems. Initially, vapor compression refrigeration systems were large and were mainly used for ice making, brewing, and cold storage. They lacked automatic control, and were steam engine-driven. In 1890s, electric motor-driven smaller machines equipped with automatic control started to replace the older units, and refrigeration systems began to appear in butcher shops and households. By 1930, the continued improvements made it possible to have vapor compression refrigeration systems that were relatively efficient, reliable, small, and inexpensive.

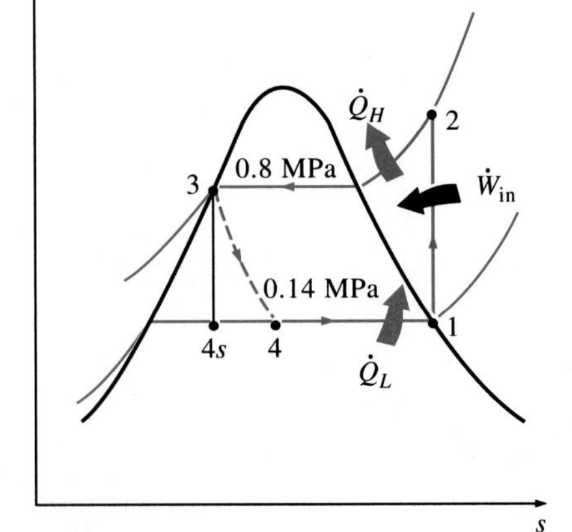


**FIGURE 7-61**

*T-s* diagram of the ideal vapor-compression refrigeration cycle described in Example 7-10.

**EXAMPLE 7-10**

A refrigerator uses refrigerant-134a as the working fluid and operates on an ideal vapor-compression refrigeration cycle between 0.14 and 0.8 MPa. If the mass flow rate of the refrigerant is 0.05 kg/s, determine (*a*) the rate of heat removal from the refrigerated space and the power input to the compressor, (*b*) the heat rejection rate to the environment, and (*c*) the COP of the refrigerator.

**Solution** The refrigeration cycle is shown on a *T-s* diagram in Fig. 7-61. In an ideal vapor-compression refrigeration cycle, the compression process is isentropic, and the refrigerant enters the compressor as a saturated vapor at the evaporator pressure. Also, the refrigerant leaves the condenser as saturated liquid at the condenser pressure.

From the refrigerant-134a tables, the enthalpies of the refrigerant at all four states are determined as follows:

$$P_1 = 0.14\text{ MPa} \longrightarrow h_1 = h_{g @ 0.14\text{ MPa}} = 236.04\text{ kJ/kg}$$
$$s_1 = s_{g @ 0.14\text{ MPa}} = 0.9322\text{ kJ/(kg}\cdot\text{K)}$$

$$\left.\begin{array}{l} P_2 = 0.8\text{ MPa} \\ s_2 = s_1 \end{array}\right\} \quad h_2 = 272.05\text{ kJ/kg}$$

$$P_3 = 0.8\text{ MPa} \longrightarrow h_3 = h_{f@0.8\text{ MPa}} = 93.42\text{ kJ/kg}$$
$$h_4 \cong h_3\text{ (throttling)} \longrightarrow h_4 = 93.42\text{ kJ/kg}$$

(*a*) The rate of heat removal from the refrigerated space and the power input to the compressor are determined from their definitions:

$$\dot{Q}_L = \dot{m}(h_1 - h_4) = (0.05\text{ kg/s})[(236.04 - 93.42)\text{ kJ/kg}] = 7.13\text{ kW}$$

and $$\dot{W}_{in} = \dot{m}(h_2 - h_1) = (0.05\text{ kg/s})[(272.05 - 236.04)\text{ kJ/kg}] = 1.80\text{ kW}$$

(*b*) The rate of heat rejection from the refrigerant to the environment is determined from

$$\dot{Q}_H = \dot{m}(h_2 - h_3) = (0.05\text{ kg/s})[(272.05 - 93.42)\text{ kJ/kg}] = 8.93\text{ kW}$$

It could also be determined from

$$\dot{Q}_H = \dot{Q}_L + \dot{W}_{in} = 7.13 + 1.80 = 8.93\text{ kW}$$

(*c*) The coefficient of performance of the refrigerator is determined from its definition:

$$\text{COP}_R = \frac{\dot{Q}_L}{\dot{W}_{in}} = \frac{7.13\text{ kW}}{1.80\text{ kW}} = 4.0$$

That is, this refrigerator removes 4 units of energy from the refrigerated space for each unit of electric energy it consumes.

## 7-16 ■ SELECTING THE RIGHT REFRIGERANT

When designing a refrigeration system, there are several refrigerants to choose from, such as chlorofluorocarbons (CFCs), ammonia, hydrocarbons (propane, ethane, ehtylene, etc.), carbon dioxide, air (in the air conditioning of aircraft), and even water (in applications above the freezing point). The right choice of refrigerant depends on the situation at hand. Of these, CFCs such as R-11, R-12, R-22, and R-502 account for over 90 percent of the market in the United States.

*Ethyl ether* was the first commercially used refrigerant in vapor compression systems in 1850, followed by ammonia, carbon dioxide, methyl chloride, sulphur dioxide, butane, ethane, propane, isobutane, gasoline, and chlorofluorocarbons, among others.

The *industrial* and the heavy commercial sectors were very satisfied with *ammonia*, and it still are, although ammonia is toxic. The advantages of ammonia over other refrigerants are its low cost, higher COPs (and thus lower energy cost), more favorable thermodynamic and transport properties and thus higher heat transfer coefficients (requires smaller and lower cost heat exchangers), greater detectability in the event of a leak, and no effect on the ozone layer. The major drawback of ammonia is its being toxicity which makes it unsuitable for domestic use. Ammonia is predominantly used in food refrigeration facilities such as the cooling of fresh fruits, vegetables, meat, and fish; refrigeration of beverages and dairy products such as beer, wine, milk, and cheese; freezing of ice cream and other foods; ice production; and low-temperature refrigeration in the pharmaceutical and other process industries.

It is remarkable that the refrigerants used in the light commercial and household sectors such as sulfur dioxide, ethyl chloride, and methyl chloride were highly toxic. The widespread publicity of a few instances of serious leaks that resulted in illnesses and death in the 1920s caused a public cry to ban or limit the use of these refrigerants, creating a need for the development of a safe refrigerant for household use. At the request of Frigidaire Corporation, General Motors research laboratory developed R-21, the first member of the CFC family of refrigerants, within three days in 1928. Of several CFCs developed, the research team settled on R-12 as the refrigerant most suitable for commercial use, and gave the CFC family the trade name "Freon." Commercial production of R-11 and R-12 was started in 1931 by a company jointly formed by General Motors and E. I. duPont deNemours and Co. Inc. The versatility and low cost of CFCs made them the refrigerants of choice. CFCs were also widely used in aerosols, foam insulations, and in the electronic industry as solvents to clean computer chips.

R-11 is used primarily in large-capacity water chillers serving air conditioning systems in buildings. R-12 and R-134a are used in domestic refrigerators and freezers, as well as automotive air conditioners. R-23 is used in window air conditioners, heat pumps, air conditioners of commercial buildings, and large industrial refrigeration systems, and offers strong competition to ammonia. R-502 (a blend of R-115 and R-22) is the dominant refrigerant used in commercial refrigeration systems such as those used in supermarkets because it allows low temperatures at evaporators while operating at single-stage compression.

The ozone crisis has caused a major stir in the refrigeration and air conditioning industry and has triggered a critical look at the refrigerants in use. It was realized in the mid-1970s that CFCs allow more ultraviolet radiation into the earth's atmosphere while preventing the infrared radiation from escaping the earth and thus contributing to the greenhouse effect which causes global warming. As a result, the use of some CFCs is banned and being phased out in many countries. Fully halogenated CFCs (such as R-11, R-12, and R-115) do the most damage to the ozone layer. The nonfully halogenated refrigerants such as R-22 have about 5 percent of the ozone-depleting capability of R-12. CFCs that are friendly to the ozone layer that protects the earth from harmful ultraviolet rays and at the same time do not contribute to the greenhouse effect are under development. The recently developed chlorine-free R-134a is presently replacing R-12.

Two important parameters that need to be considered in the selection of a refrigerant are the temperatures of the two media (the refrigeration space and the environment) with which the refrigerant exchanges heat.

To have heat transfer at a reasonable rate, a temperature difference of 5 to 10°C should be maintained between the refrigerant and the medium it is exchanging heat with. If a refrigerated space is to be maintained at $-10$°C, for example, the temperature of the refrigerant should remain at about −20°C while it absorbs heat in the evaporator. The lowest pressure in a refrigeration cycle occurs in the evaporator, and this pressure should be above atmospheric pressure to prevent any air leakage into the refrigeration system. Therefore, a refrigerant should have a saturation pressure of 1 atm or higher at −20°C in this particular case. Ammonia, refrigerant-134a, and refrigerant-22 (usually marketed under the trade name Freon-22) are three such substances.

The temperature (and thus the pressure) of the refrigerant on the condenser side depends on the medium to which heat is rejected. Lower temperatures in the condenser (thus higher COPs) can be maintained if the refrigerant is cooled by liquid water instead of air. The use of water cooling cannot be justified economically, however, except in large industrial refrigeration systems. The temperature of the refrigerant in the condenser cannot fall below the temperature of the cooling medium (about 20°C for a household refrigerator), and the saturation pressure of the refrigerant at this temperature should be well below its critical pressure if the heat rejection process is to be approximately isothermal. If no single refrigerant can meet the temperature requirements then two or more refrigeration cycles with different refrigerants can be used in series. Such a refrigeration system is called a *cascade system,* and it is discussed later in this chapter.

Other desirable characteristics of a refrigerant include being nontoxic, noncorrosive, nonflammable, and chemically stable; having a high enthalpy of vaporization (minimizes the mass flow rate); and, of course, being available at low cost.

In the case of heat pumps, the minimum temperature (and pressure) for the refrigerant may be considerably higher since heat is usually extracted from media that are well above the temperatures encountered in refrigeration systems.

## 7-17 ■ HEAT PUMP SYSTEMS

Heat pumps are generally more expensive to purchase and install than other heating systems, but they save money in the long run in some areas because they lower the heating bill. Despite their relatively higher initial costs, the popularity of heat pumps is increasing. About one-third of all single-family homes built in the United States in the last decade are heated by heat pumps.

The most common energy source for heat pumps is atmospheric air (air-to-air systems), although water and soil are also used. The major problem with air-source systems is *frosting,* which occurs in humid climates when the temperature falls below 2 to 5°C. The frost accumulation on the evaporator coils is highly undesirable since it seriously disrupts the heat transfer. The coils can be defrosted, however, by reversing the heat pump cycle (running it as an air conditioner). This results in a reduction in the efficiency of the system. Water-source systems usually use well water from depths of up to 80 m in the temperature range of 5–18°C, and they do not have a frosting problem.

They typically have higher COPs but are more complex and require easy access to a large body of water such as underground water. Soil-source systems are also rather involved since they require long tubing placed deep in the ground where the soil temperature is relatively constant. The *COP of heat pumps usually ranges between* 1.5 *and* 4, depending on the particular system used and the temperature of the source. A new class of recently developed heat pumps that use variable-speed electric motor drive are at least twice as energy-efficient as their predecessors.

Both the capacity and the efficiency of a heat pump fall significantly at low temperatures. Therefore, most air-source heat pumps require a supplementary heating system such as electric resistance heaters or an oil or gas furnace. Since water and soil temperatures do not fluctuate much, supplementary heating may not be required for water-source or soil-source systems. But the heat pump system must be large enough to meet the maximum heating load.

Heat pumps and air conditioners have the same mechanical components. Therefore, it is not economical to have two separate systems to meet the heating and cooling requirements of a building or a house. One system can be used as a heat pump in winter and air conditioner in summer. This is accomplished by adding a reversing valve to the cycle, as shown in Fig. 7-62. As a result of this modification, the condenser of the heat pump (located indoors) functions as the evaporator of the air conditioner in summer. Also, the evaporator of the heat pump (located

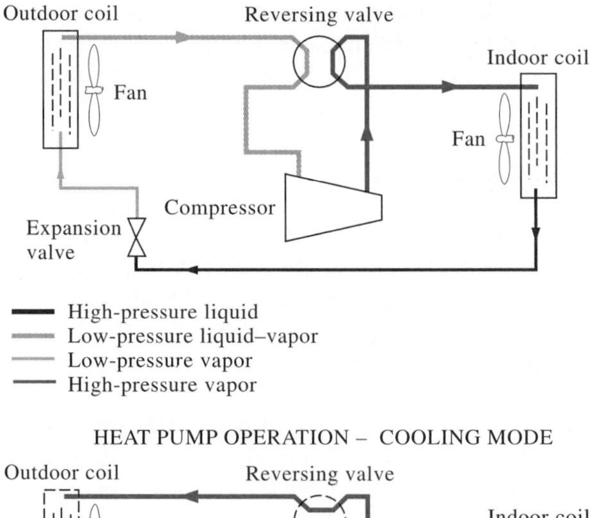


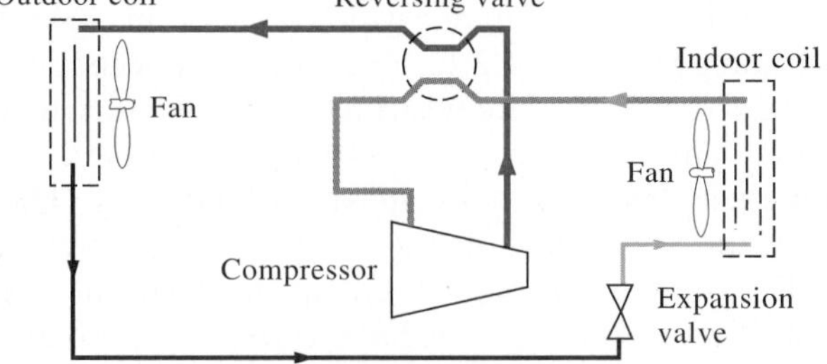


**FIGURE 7-62**

A heat pump can be used to heat a house in winter and to cool it in summer.

outdoors) serves as the condenser of the air conditioner. This feature increases the competitiveness of the heat pump. Such dual-purpose window units are commonly used in motels.

Heat pumps are most competitive in areas that have a large cooling load during the cooling season and a relatively small heating load during the heating season, such as in the southern parts of the United States. In these areas, the heat pump can meet the entire cooling and heating needs of residential or commercial buildings. The heat pump is least competitive in areas where the heating load is significant and the cooling load is small or nonexistent, such as in the northern parts of the United States.

## 7-18 THERMOELECTRIC POWER GENERATION AND REFRIGERATION SYSTEMS

All the refrigeration systems discussed above involve many moving parts and bulky, complex components. Then this question comes to mind: Is it really necessary for a refrigeration system to be so complex? Can we not achieve the same effect in a more direct way? The answer to this question is *yes.* It is possible to use electric energy more directly to produce cooling without involving any refrigerants and moving parts. Below we discuss one such system, called a *thermoelectric refrigerator.*

Consider two wires made from different metals joined at both ends (junctions), forming a closed circuit. Ordinarily, nothing will happen. But when one of the ends is heated, something interesting happens: A current flows continuously in the circuit, as shown in Fig. 7-63. This is called the **Seebeck effect**, in honor of Thomas Seebeck, who made this discovery in 1821. A circuit that incorporates both thermal and electrical effects is called a **thermoelectric circuit**, and a device that operates using such a circuit is called a **thermoelectric device**.

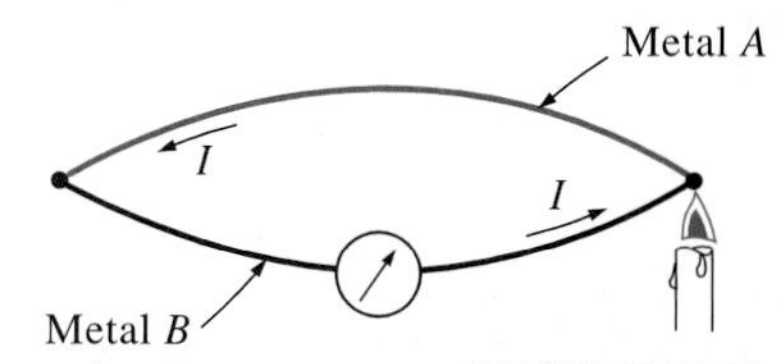


**FIGURE 7-63**

When one of the junctions of two dissimilar metals is heated, a current $I$ flows through the closed circuit.

The Seebeck effect has two major applications: temperature measurement and power generation. When the thermoelectric circuit is broken, as shown in Fig. 7-64, the current ceases to flow, and we can measure the driving force (the electromotive force) or the voltage generated in the circuit by a voltmeter. The voltage generated is a function of the temperature difference and the materials of the two wires used. Therefore, temperature can be measured by simply measuring voltages. The two wires used to measure the temperature in this manner form a *thermocouple,* which is the most versatile and most widely used temperature measurement device. A common T-type thermocouple, for example, consists of copper and constantan wires, and it produces about 40 $\mu$V per degree Celsius temperature difference.

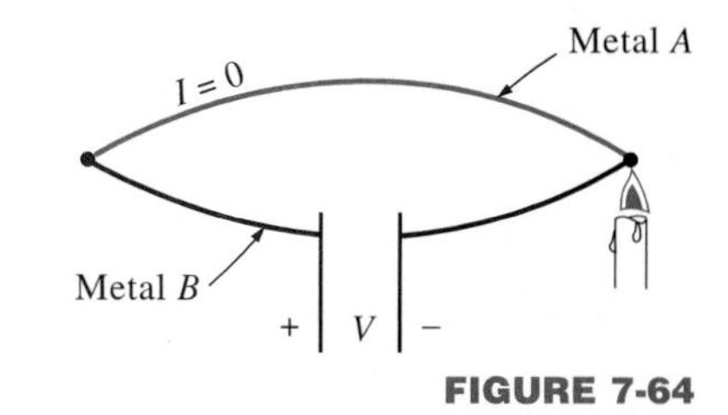


**FIGURE 7-64**

When a thermoelectric circuit is broken, a potential difference is generated.

The Seebeck effect also forms the basis for thermoelectric power generation. The schematic diagram of a **thermoelectric generator** is shown in Fig. 7-65. Heat is transferred from a high-temperature source to the hot junction in the amount of $Q_H$, and it is rejected to a low-temperature sink from the cold junction in the amount of $Q_L$. The difference between these two quantities is the net electrical work

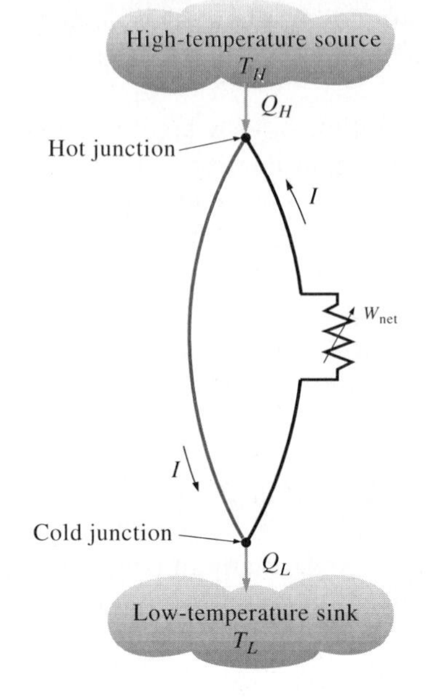


**FIGURE 7-65**
Schematic of a simple thermoelectric power generator.

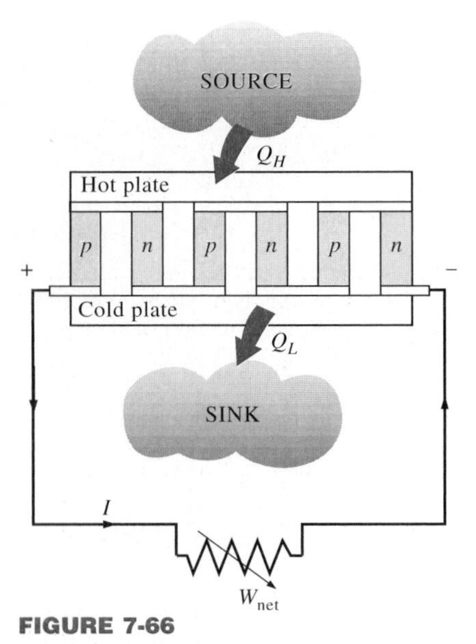


**FIGURE 7-66**
A thermoelectric power generator.

produced, that is, $W_e = Q_H - Q_L$. It is evident from Fig. 7-65 that the thermoelectric power cycle closely resembles an ordinary heat engine cycle, with electrons serving as the working fluid. Therefore, the thermal efficiency of a thermoelectric generator operating between the temperature limits of $T_H$ and $T_L$ is limited by the efficiency of a Carnot cycle operating between the same temperature limits. Thus, in the absence of any irreversibilities (such as $I^2R$ heating, where $R$ is the total electrical resistance of the wires), the thermoelectric generator will have the Carnot efficiency.

The major drawback of thermoelectric generators is their low efficiency. The future success of these devices depends on the finding materials with more desirable characteristics. For example, the voltage output of thermoelectric devices has been increased several times by switching from metal pairs to semiconductors. A practical thermoelectric generator using $n$-type (heavily doped to create excess electrons) and $p$-type (heavily doped to create a deficiency of electrons) materials connected in series is shown in Fig. 7-66. Despite their low efficiencies, thermoelectric generators have a definite weight and reliability advantage and are presently used in rural areas and in space applications. For example, the silicon–germanium-based thermoelectric generators of *Voyagers* have been powering these spacecraft since 1980, and are expected to power them for many more years.

If Seebeck had been fluent in thermodynamics, he would probably have tried reversing the direction of flow of electrons in the thermoelectric circuit (by externally applying a potential difference in the reverse direction), to create a refrigeration effect. But this honor belongs to Jean Charles Athanase Peltier, who discovered this phenomenon in 1834. He noticed during his experiments that when a small current was passed through the junction of two dissimilar wires, the junction was cooled, as shown in Fig. 7-67. This is called the **Peltier effect**, and it forms the basis for **thermoelectric refrigeration**. A practical thermoelectric refrigeration circuit using semiconductor materials is shown in Fig. 7-68. Heat is absorbed from the refrigerated space in the amount of $Q_L$ and rejected to the warmer environment in the amount of $Q_H$. The difference between these two quantities is the net electrical work that needs to be supplied, that is, $W_e = Q_H - Q_L$. Thermoelectric refrigerators presently cannot complete with vapor-compression refrigeration systems because of their low coefficient of performance. They are available in the market, however, and are preferred in some applications, because of their small size, simplicity, quietness, and reliability.

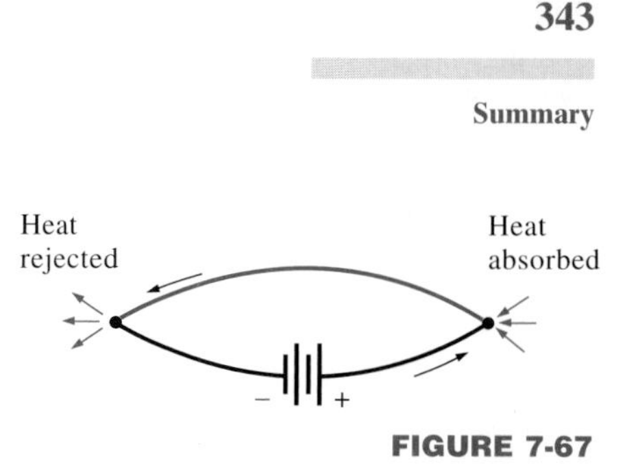


**FIGURE 7-67**
When a current is passed through the junction of two dissimilar materials, the junction is cooled.

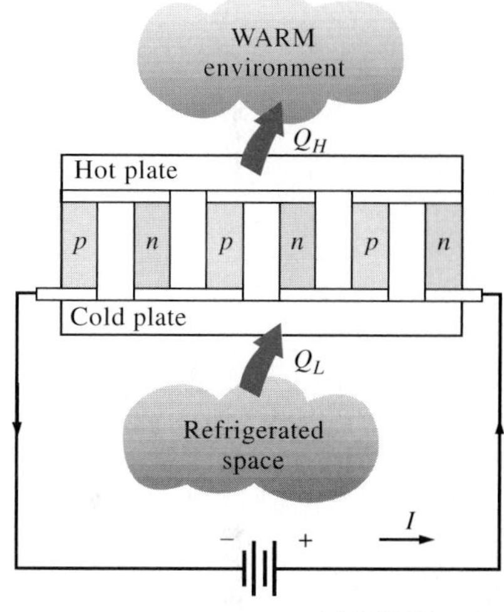


**FIGURE 7-68**
A thermoelectric refrigerator.

## 7-19 SUMMARY

A cycle during which a net amount of work is produced is called a *power cycle,* and a power cycle during which the working fluid remains a gas throughout is called a *gas power cycle.* The most efficient cycle operating between a source at temperature $T_H$ and a sink at temperature $T_L$ is the *Carnot cycle,* and its thermal efficiency is given by

$$\eta_{\text{th,Carnot}} = 1 - \frac{T_L}{T_H}$$

The actual gas cycles are rather complex. The approximations used to simplify the analysis are known as the *air-standard assumptions.* Under these assumptions, all the processes are assumed to be internally reversible; the working fluid is assumed to be air that behaves as an ideal gas; and the combustion and exhaust processes are replaced by heat addition and heat rejection processes, respectively. The air-standard assumptions are called *cold-air-standard assumptions* if, in addition, air is assumed to have constant specific heats at room temperature.

In reciprocating engines, the *compression ratio* $r$ and the *mean effective pressure* MEP are defined as

$$r = \frac{V_{\max}}{V_{\min}} = \frac{V_{\text{BDC}}}{V_{\text{TDC}}}$$

$$\text{MEP} = \frac{w_{\text{net,cycle}}}{v_{\max} - v_{\min}}$$

The *Otto cycle* is the ideal cycle for the spark-ignition reciprocating engines, and it consists of four internally reversible processes: isentropic compression, $v$ = constant heat addition, isentropic expansion, and $v$ = constant heat rejection. Under cold-air-standard assumptions, the thermal efficiency of the ideal Otto cycle is

$$\eta_{\text{th,Otto}} = 1 - \frac{1}{r^{k-1}}$$

where $r$ is the compression ratio and $k$ is the specific heat ratio $C_p/C_v$.

The *Diesel cycle* is the ideal cycle for the compression-ignition reciprocating engines. It is very similar to the Otto cycle, except that the $v$ = constant heat addition process is replaced by a $P$ = constant heat addition process. Its thermal efficiency under cold-air-standard assumptions is

$$\eta_{\text{th,diesel}} = 1 - \frac{1}{r^{k-1}}\left[\frac{r_c^k - 1}{k(r_c - 1)}\right]$$

where $r_c$ is the *cutoff* ratio, defined as the ratio of the cylinder volumes after and before the combustion process.

*Stirling* and *Ericsson cycles* are two totally reversible cycles that involve an isothermal heat addition process at $T_H$ and an isothermal heat rejection process at $T_L$. They differ from the Carnot cycle in that the two isentropic processes are replaced by two $v$ = constant regeneration processes in the Stirling cycle and by two $P$ = constant regeneration processes in the Ericsson cycle. Both cycles utilize *regeneration,* a process during which heat is transferred to a thermal energy storage device (called a *regenerator*) during one part of the cycle which is transferred back to the working fluid during another part of the cycle.

The ideal cycle for modern gas-turbine engines is the *Brayton cycle,* which is made up of four internally reversible processes: isentropic compression (in a compressor), $P$ = constant heat addition, isentropic expansion (in a turbine), and $P$ = constant heat rejection. Under cold-air-standard assumptions, its thermal efficiency is

$$\eta_{\text{th,Brayton}} = 1 - \frac{1}{r_p^{(k-1)/k}}$$

where $r_p = P_{\max}/P_{\min}$ is the pressure ratio and $k$ is the specific heat ratio. The thermal efficiency of the simple Brayton cycle increases with the pressure ratio.

In gas-turbine engines, the temperature of the exhaust gas leaving the turbine is often considerably higher than the temperature of the air leaving the compressor. Therefore, the high-pressure air leaving the compressor can be heated by transferring heat to it from the hot exhaust gases in a counter-flow heat exchanger, which is also known as a *regenerator.* The extent to which a regenerator approaches an ideal regenerator is called the *effectiveness* $\varepsilon$ and is defined as

$$\varepsilon = \frac{q_{\text{regen,act}}}{q_{\text{regen,max}}}$$

Under cold-air-standard assumptions, the thermal efficiency of an ideal Brayton cycle with regeneration becomes

$$\eta_{\text{th,regen}} = 1 - \left(\frac{T_1}{T_3}\right)(r_p)^{(k-1)/k}$$

where $T_1$ and $T_3$ are the minimum and maximum temperatures, respectively, in the cycle.

The thermal efficiency of the Brayton cycle can also be increased by utilizing *multistage compression with intercooling, regeneration, and multistage expansion with reheating.* The work input to the compressor is minimized when equal pressure ratios are maintained across each stage. This procedure also maximizes the turbine work output.

Gas-turbine engines are widely used to power aircraft because they are light and compact and have a high power-to-weight ratio. The ideal *jet-propulsion cycle* differs from the simple ideal Brayton cycle in that the gases are partially expanded in the turbine. The gases that exit the turbine at a relatively high pressure are subsequently accelerated in a nozzle to provide the thrust needed to propel the aircraft.

The *net thrust* developed by the engine is

$$F = \dot{m}(\mathcal{V}_{\text{exit}} - \mathcal{V}_{\text{inlet}}) \qquad \text{(N)}$$

where $\dot{m}$ is the mass flow rate of gases, $\mathcal{V}_{\text{exit}}$ is the exit velocity of the exhaust gases, and $\mathcal{V}_{\text{inlet}}$ is the inlet velocity of the air, both relative to the aircraft.

The power developed from the thrust of the engine is called the *propulsive power* $\dot{W}_P$, and it is given by

$$\dot{W}_P = \dot{m}(\mathcal{V}_{\text{exit}} - \mathcal{V}_{\text{inlet}})\mathcal{V}_{\text{aircraft}} \qquad \text{(kW)}$$

*Propulsive efficiency* is a measure of how efficiently the energy released during the combustion process is converted to propulsive energy, and it is defined as

$$\eta_P = \frac{\text{propulsive power}}{\text{energy input rate}} = \frac{\dot{W}_P}{\dot{Q}_{\text{in}}}$$

The energy not utilized for propulsion shows up as excess enthalpy and kinetic energy relative to a fixed point on the ground.

The *Carnot cycle* is not a suitable model for vapor power cycles, because it cannot be approximated in practice. The model cycle for vapor power cycles is the *Rankine cycle,* which is composed of four internally reversible processes: constant-pressure heat addition in a boiler, isentropic expansion in a turbine, constant-pressure heat rejection in a condenser, and isentropic compression in a pump. Steam leaves the condenser as a saturated liquid at the condenser pressure.

The thermal efficiency of the Rankine cycle can be increased by increasing the average temperature at which heat is added to the working fluid and/or by decreasing the average temperature at which heat is rejected to the cooling medium, such as a lake or a river. The average temperature during heat rejection can be decreased by lowering the turbine exit pressure. Consequently, the condenser pressure of most vapor power plants is well below the atmospheric pressure. The average temperature during heat addition can be increased by raising the boiler pressure or by superheating the fluid to high temperatures. There is a limit to the degree of superheating, however, since the fluid temperature is not allowed to exceed a metallurgically safe value.

Superheating has the added advantage of decreasing the moisture content of the steam at the turbine exit. Lowering the exhaust pressure or raising the boiler pressure, however, increases the moisture content. To take advantage of the improved efficiencies at higher boiler pressures and lower condenser pressures, steam is usuallly *reheated* after expanding partially in the high-pressure turbine. This is done by extracting the steam after partial extraction in the high-pressure turbine, sending it back to the boiler where it is reheated at constant pressure, and returning it to the low-pressure turbine for complete expansion to the condenser pressure. The transfer of heat from lower-temperature regions to higher temperature ones is called *refrigeration.* Devices that produce refrigeration are called *refrigerators,* and the cycles on which they operate are called *refrigeration cycles.* The working fluids used in the refrigeration cycles are called *refrigerants.* Refrigerators used for the purpose of heating a space by transferring heat from a cooler medium are called *heat pumps.*

The performance of refrigerators and heat pumps is expressed in terms of the *coefficient of performance* (COP), defined as

$$\mathrm{COP_R} = \frac{\text{cooling effect}}{\text{work input}} = \frac{Q_L}{W_{\text{net, in}}}$$

$$\mathrm{COP_{HP}} = \frac{\text{heating effect}}{\text{work input}} = \frac{Q_H}{W_{\text{net, in}}}$$

The standard of comparison for refrigeration cycles is the *reversed Carnot cycle.* A refrigerator or heat pump that operates on the reversed Carnot cycle is called a *Carnot refrigerator* or a *Carnot heat pump,* and their COPs are

$$\mathrm{COP_{R,\,Carnot}} = \frac{1}{T_H/T_L - 1}$$

and

$$\mathrm{COP_{HP,\,Carnot}} = \frac{1}{1 - T_L/T_H}$$

The most widely used refrigeration cycle is the *vapor-compression refrigeration cycle.* In an ideal vapor-compression refrigeration cycle, the refrigerant enters the compressor as a saturated vapor and is cooled to the saturated liquid state in the condenser. It is then throttled to the evaporator pressure and vaporizes as it absorbs heat from the refrigerated space.

A refrigeration effect can also be achieved without using any moving parts by simply passing a small current through a closed circuit made up of two dissimilar materials. This effect is called the *Peltier effect,* and a refrigerator that works on this principle is called a *thermoelectric refrigerator.*

## REFERENCES AND SUGGESTED READING

**1** *ASHRAE Handbook of Fundamentals,* American Society of Heating, Refrigerating, and Air-Conditioning Engineers, Atlanta, 1985.

**2** R. L. Bannister and G. J. Silvestri, "The Evolution of Central Station Steam Turbines," *Mechanical Engineering,* pp. 70–78, Feb. 1989.

**3** R. L. Bannister, G. J. Silvestri, A. Hizume, and T. Fujikawa, "High Temperature Supercritical Steam Turbines," *Mechanical Engineering,* pp. 60–65, Feb. 1987.

**4** W. Z. Black and J. G. Hartley, *Thermodynamics,* Harper & Row, New York, 1985.

**5** M. D. Burghard, *Engineering Thermodynamics with Applications,* 3d ed., Harper & Row, New York, 1986.

**6** V. D. Chase, "Propfans: A New Twist for the Propeller," *Mechanical Engineering,* pp. 47–50, Nov. 1986.

**7** M. M. El-Wakil, *Powerplant Technology,* McGraw-Hill, New York, 1984.

**8** R. C. Fellinger and W. J. Cook, *Introduction to Engineering Thermodynamics,* WCB, Dubuque, IA, 1985.

**9** J. B. Jones and G. A. Hawkins, *Engineering Thermodynamics,* 2d ed., New York, 1986.

**10** R. A. Harmon, "The Keys to Cogeneration and Combined Cycles," *Mechanical Engineering,* pp. 64–73, Feb. 1988.

**11** B. V. Karlekar, *Thermodynamics for Engineers,* Prentice-Hall, Englewood Cliffs, NJ, 1983.

**12** K. W. Li and A. P. Priddy, *Power Plant System Design,* Wiley, New York, 1985.

**13** L. C. Lichty, *Combustion Engine Processes,* McGraw-Hill, New York, 1967.

**14** D. C. Look, Jr., and H. J. Sauer, Jr., *Engineering Thermodynamics,* PWS Engineering, Boston, 1986.

**15** H. McIntosh, "Jumbo Jet," *10 Outstanding Achievements 1964–1989,* National Academy of Engineering, pp. 30–33, Washington, D.C., 1989.

**16** B. Nagengast, "A Historical Look at CFC Refrigerants," *ASHRAE Journal,* Vol. 30, Part 11, pp. 37–39, Nov. 1988.

**17** W. Siuru, "Two-stroke engines: Cleaner and Meaner," *Mechanical Engineering,* pp. 66–69, June 1990.

**18** H. Sorensen, *Energy Conversion Systems,* Wiley, New York, 1983.

**19** *Steam, Its Generation and Use,* 39th ed., Babcock and Wilcox Co., New York, 1978.

**20** W. F. Stoecker, "Growing Opportunities for Ammonia Refrigeration," *Precedings of the Meeting of the International Institute of Ammonia Refrigeration,* Austin, Texas, 1989.

**21** W. F. Stoecker and J. W. Jones, *Refrigeration and Air Conditioning,* 2d ed., McGraw-Hill, New York, 1982.

**22** C. F. Taylor, *The Internal Combustion Engine in Theory and Practice,* M.I.T. Press, Cambridge, MA, 1968.

**23** *Turbomachinery,* vol. 28, no. 2, Business Journals, Inc., Norwalk, CT, March/April 1987.

**24** G. J. Van Wylen and R. E. Sonntag, *Fundamentals of Classical Thermodynamics,* 3d ed., Wiley, New York, 1985.

**25** K. Wark, *Thermodynamics,* 5th ed., McGraw-Hill, New York, 1988.

## PROBLEMS*

### Actual and Ideal Cycles, Carnot Cycle, Air-Standard Assumptions, Reciprocating Engines

**7-1C** The efficiency of a Carnot cycle is given by $\eta_{\text{th},C} = 1 - T_L/T_H$. Does this relation suggest any ways of improving the thermal efficiency of actual cycles?

**7-2C** How do gas power cycles differ from vapor power cycles?

**7-3C** Why is the Carnot cycle not suitable as an ideal cycle for all power-producing cyclic devices?

**7-4C** How does the thermal efficiency of an ideal cycle, in general, compare to that of a Carnot cycle operating between the same temperature limits?

**7-5C** What does the area enclosed by the cycle represent on a $P$-$v$ diagram? How about on a $T$-$s$ diagram?

**7-6C** What is the difference between air-standard assumptions and the cold-air-standard assumptions?

**7-7C** Do internal combustion engines operate on a closed or an open cycle? Why?

**7-8C** How are the combustion and exhaust processes modeled under the air-standard assumptions?

**7-9C** What are the air-standard assumptions?

**7-10C** What is the difference between the clearance volume and the displacement volume of reciprocating engines?

**7-11C** Define the compression ratio for reciprocating engines.

---

*Students are encouraged to answer *all* the concept "C" questions.

**7-12C** How is the mean effective pressure for reciprocating engines defined?

**7-13C** Can the mean effective pressure of an automobile engine in operation be less than the atmospheric pressure?

**7-14C** As a car gets older, will its compression ratio change? How about the mean effective pressure?

**7-15C** What is the difference between spark-ignition and compression-ignition engines?

**7-16C** Define the following terms related to reciprocating engines: stroke, bore, top dead center, and clearance volume.

**7-17** An air-standard cycle is executed in a closed system and is composed of the following four processes:

1-2 Isentropic compression from 100 kPa and 27°C to 800 kPa
2-3 $v$ = constant heat addition to 1800 K
3-4 Isentropic expansion to 100 kPa
4-1 $P$ = constant heat rejection to initial state

(*a*) Show the cycle on $P$-$v$ and $T$-$s$ diagrams.
(*b*) Calculate the net work output per unit mass.
(*c*) Determine the thermal efficiency.

**7-17E** An air-standard cycle is executed in a closed system and is composed of the following four processes:

1-2 Isentropic compression from 14.7 psia and 80°F to 120 psia
2-3 $v$ = constant heat addition to 3000 R
3-4 Isentropic expansion to 14.7 psia
4-1 $P$ = constant heat rejection to initial state

(*a*) Show the cycle on $P$-$v$ and $T$-diagrams.
(*b*) Calculate the net work output per unit mass.
(*c*) Determine the thermal efficiency.

**7-18** An air-standard cycle is executed in a closed system and is composed of the following four processes:

1-2 Isentropic compression from 100 kPa and 27°C to 1 MPa
2-3 $P$ = constant heat addition in amount of 2840 kJ/kg
3-4 $v$ = constant heat rejection to 100 kPa
4-1 $P$ = constant heat rejection to initial state

(*a*) Show the cycle on $P$-$v$ and $T$-$s$ diagrams.
(*b*) Calculate the maximum temperature in the cycle.
(*c*) Determine the thermal efficiency.

Assume constant specific heats at room temperature.

*Answers:* (*b*) 3405.1 K, (*c*) 21.1 percent

**7-19** An air-standard cycle is executed in a closed system and is composed of the following four processes:

1-2 $v$ = constant heat addition from 100 kPa and 27°C in the amount of 701.5 kJ/kg
2-2 $P$ = constant heat addition to 2000 K
3-4 Isentropic expansion to 100 kPa
4-1 $P$ = constant heat rejection to initial state

(*a*) Show the cycle on $P$-$v$ and $T$-$s$ diagrams.
(*b*) Calculated the total heat input per unit mass.
(*c*) Determine the thermal efficiency.

Account for the variation of specific heats with temperature.

**7-19E** An air-standard cycle is executed in a closed system and is composed of the following four processes:

1-2 $v$ = constant heat addition from 14.7 psia and 80°F In amount of 300 Btu/lbm
2-3 $P$ = constant heat addition to 3200 R
3-4 Isentropic expansion to 14.7 psia
4-1 $P$ = constant heat rejection to initial state

(*a*) Show the cycle on $P$-$v$ and $T$-$s$ diagrams.
(*b*) Calculate the total heat input per unit mass.
(*c*) Determine the thermal efficiency.

Account for the variation of the specific heats with temperature.

**7-20** An air-standard cycle is executed in a closed system with 0.001 kg of air and consists of the following three processes:

1-2 Isentropic compression from 100 kPa and 27°C to 1 MPa
2-3 $P$ = constant heat addition in amount of 1.84 kJ
3-1 $P = c_1 v + c_2$ heat rejection to initial state ($c_1$ and $c_2$ are constants)

(*a*) Show the cycle on $P$-$v$ and $T$-$s$ diagrams.
(*b*) Calculate the heat rejected.
(*c*) Determine the thermal efficiency.

Assume constant specific heats at room temperature.

**7-21** An air-standard cycle is executed in a closed system with 0.003 kg of air and consists of the following three processes:

1-2 $v$ = constant heat addition from 95 kPa and 17°C to 380 kPa
2-3 Isentropic expansion to 95 kPa
3-1 $P$ = constant heat rejection to initial state

(*a*) Show the cycle on $P$-$v$ and $T$-$s$ diagrams.
(*b*) Calculate the net work per cycle, in kJ.
(*c*) Determine the thermal efficiency.

**7-22** Consider a Carnot cycle executed in a closed system with 0.004 kg of air. The temperature limits of the cycle are 300 and 1000 K, and the minimum and maximum pressures that occur during the cycle are 20 and 1800 kPa. Assuming constant specific heats, determine the net work output per cycle.

**7-23** An air-standard Carnot cycle is executed in a closed system

between the temperature limits of 350 and 1200 K. The pressures before and after the isothermal compression are 150 and 300 kPa, respectively. If the net work output per cycle is 0.5 kJ, determine (*a*) the maximum pressure in the cycle, (*b*) the heat transfer to air, and (*c*) the mass of air. Assume variable specific heats for air.
*Answers:* (*a*) 32.4 MPa, (*b*) 0.706 kJ, (*c*) 0.00296 kg

**7-23E** An air-standard Carnot cycle is executed in a closed system between the temperature limits of 540 and 2000 R. The pressures before and after the isothermal compression are 20 and 40 psia, respectively. If the net work output per cycle is 0.4 Btu, determine (*a*) the maximum pressure in the cycle, (*b*) the heat transfer to air, and (*c*) the mass of air. Assume variable specific heats for air.

**7-24** Repeat Prob. 7-23 using helium as the working fluid.

## Otto Cycle

**7-25C** What four processes make up the ideal Otto cycle?

**7-26C** How do the efficiencies of the ideal Otto cycle and the Carnot cycle compare for the same temperature limits? Explain.

**7-27C** How is the rpm (revolutions per minute) of an actual four-stroke gasoline engine related to the number of thermodynamic cycles? What would your answer be for a two-stroke engine?

**7-28C** Are the processes that make up the Otto cycle analyzed as closed-system or steady-flow processes? Why?

**7-29C** How does the thermal efficiency of an ideal Otto cycle change with the compression ratio of the engine and the specific heat ratio of the working fluid?

**7-30C** Why are high compression ratios not used in spark-ignition engines?

**7-31C** An ideal Otto cycle with a specified compression ratio is executed using (*a*) air, (*b*) argon, and (*c*) ethane as the working fluid. For which case will the thermal efficiency be the highest? Why?

**7-32C** What is the difference between fuel injected gasoline engines and diesel engines?

**7-33** An ideal Otto cycle has a compression ratio of 8. At the beginning of the compression process, air is at 95 kPa and 27°C, and 750 kJ/kg of heat is transferred to air during the constant-volume heat addition process. Using constant specific heats at room temperature, determine (*a*) the pressure and temperature at the end of the heat addition process, (*b*) the net work output, (*c*) the thermal efficiency, and (*d*) the mean effective pressure for the cycle.
*Answers:* (*a*) 4392 kPa, 1734 K; (*b*) 423.5 kJ/kg; (*c*) 56.5 percent; (*d*) 534 kPa

**7-33E** An ideal Otto cycle has a compression ratio of 8. At the beginning of the compression process, air is at 14.5 psia and 80°F, and 450 Btu/lbm of heat is transferred to the air during the constant-volume heat addition process. Using constant specific heats at room temperature, determine (*a*) the pressure and temperature at the end of the heat addition process, (*b*) the net work output, (*c*) the thermal efficiency, and (*d*) the mean effective pressure for the cycle.

**7-34** The compression ratio of an air-standard Otto cycle is 9.5. Prior to the isentropic compression process, the air is at 100 kPa, 17°C, and 600 $cm^3$. The temperature at the end of the isentropic expansion process is 800 K. Using specific heat values at room temperature, determine (*a*) the highest temperature and pressure in the cycle, (*b*) the amount of heat transferred during heat addition, in kJ, (*c*) the thermal efficiency, and (*d*) the mean effective pressure.
*Answers:* (*a*) 1969 K, 6449 kPa; (*b*) 0.65 kJ; (*c*) 59.4 percent; (*d*) 719 kPa

**7-34E** The compression ratio of an air-standard Otto cycle is 9.5. Prior to the isentropic compression process, the air is at 14.7 psia, 60°F, and 35 $in^3$. The temperature at the end of the isentropic expansion process is 1400 R. Using specific heat values at room temperature, determine (*a*) the highest temperature and pressure in the cycle, (*b*) the amount of heat transferred during heat addition, in Btu, (*c*) the thermal efficiency, and (*d*) the mean effective pressure.

**7-35** Repeat Prob. 7-34, but replace the isentropic expansion process by a polytropic expansion process with the polytropic exponent $n = 1.35$.

**7-36** An ideal Otto cycle with air as the working fluid has a compression ratio of 8. The minimum and maximum temperatures in the cycle are 310 and 1600 K. Using constant specific heats at room temperature, determine (*a*) the amount of heat transferred to air during the heat addition process, (*b*) the thermal efficiency, and (*c*) the thermal efficiency of a Carnot cycle operating between the same temperature limits.
*Answers:* (*a*) 637 kJ/kg, (*b*) 60 percent, (*c*) 80.6 percent

**7-36E** An ideal Otto cycle with air as the working fluid has a compression ratio of 8. The minimum and maximum temperatures in the cycle are 540 and 2200 R. Using constant specific heats at room temperature, determine (*a*) the amount of heat transferred to the air during the heat addition process, (*b*) the thermal efficiency, and (*c*) the thermal efficiency of a Carnot cycle operating between the same temperature limits.
*Answers:* (*a*) 164 Btu/lbm, (*b*) 56.5 percent, (*c*) 75.5 percent

**7-37** Repeat Prob. 7-36 using argon as the working fluid.

### Diesel Cycle

**7-38C** What is the dual cycle? How does it differ from the Otto and Diesel cycles?

**7-39C** How does a diesel engine differ from a gasoline engine?

**7-40C** How does the ideal Diesel cycle differ from the ideal Otto cycle?

**7-41C** For a specified compression ratio, is a diesel or gasoline engine more efficient?

**7-42C** Do diesel or gasoline engines operate at higher compression ratios? Why?

**7-43C** What is the cutoff ratio? How does it affect the thermal efficiency of a Diesel cycle?

**7-44** An air-standard Diesel cycle has a compression ratio of 16 and a cutoff ratio of 2. At the beginning of the compression process, air is at 95 kPa and 27°C. Using constant specific heats at room temperature, determine (*a*) the temperature after the heat addition process, (*b*) the thermal efficiency, and (*c*) the mean effective pressure.
*Answers:* (*a*) 1819 K, (*b*) 61.4 percent, (*c*) 660.5 kPa

**7-44E** An air-standard Diesel cycle has a compression ratio of 16 and a cutoff ratio of 2. At the beginning of the compression process, air is at 14.5 psia and 80°F. Using constant specific heats at room temperature, determine (*a*) the temperature after the heat addition process, (*b*) the thermal efficiency, and (*c*) the mean effective pressure.

**7-45** An air-standard Diesel cycle has a compression ratio of 18.2. Air is at 27°C and 0.1 MPa at the beginning of the compression process and at 2000 K at the end of the heat addition process. Using constant specific heats at room temperature, determine (*a*) the cutoff ratio, (*b*) the heat rejection per unit mass, and (*c*) the thermal efficiency.

**7-45E** An air-standard Diesel cycle has a compression ratio of 18.2. Air is at 80°F and 14.7 psia at the beginning of the compression process and at 3400 R at the end of the heat addition process. Using constant specific heats at room temperature, determine (*a*) the cutoff ratio, (*b*) the heat rejection per unit mass, and (*c*) the thermal efficiency.
*Answers:* (*a*) 1.97, (*b*) 147 Btu/lbm, (*c*) 63.4 percent

**7-46** An ideal diesel engine has a compression ratio of 20 and uses air as the working fluid. The state of air at the beginning of the compression process is 95 kPa and 20°C. If the maximum temperature in the cycle is not to exceed 2200 K, determine (*a*) the thermal efficiency and (*b*) the mean effective pressure. Assume constant specific heats for air at room temperature. *Answers:* (*a*) 63.5 percent, (*b*) 933 kPa

**7-47** Repeat Prob. 7-46, but replace the isentropic expansion process by polytropic expansion process with the polytropic exponent $n = 1.35$.

**7-48** A four-cylinder 4.5-L diesel engine that operates on an ideal Diesel cycle has a compression ratio of 17 and a cutoff ratio of 2.2. Air is at 27°C and 97 kPa at the beginning of the compression process. Using the cold-air-standard assumptions, determine how much power the engine will deliver at 1500 rpm.

**7-49** Repeat Prob. 7-48 using nitrogen as the working fluid.

**7-50** The compression ratio of an ideal dual cycle is 14. Air is at 100 kPa and 300 K at the beginning of the compression process and at 2200 K at the end of the heat addition process. Heat transfer to air takes place partly at constant volume and partly at constant pressure, and it amounts to 1520.4 kJ/kg. Assuming variable specific heats for air, determine (*a*) the fraction of heat transferred at constant volume and (*b*) the thermal efficiency of the cycle.

### Gas-Turbine (Brayton) Cycle

**7-51C** Why are the back work ratios relatively high in gas-turbine engines?

**7-52C** What four processes make up the simple ideal Brayton cycle?

**7-53C** For fixed maximum and minimum temperatures, what is the effect of the pressure ratio on (*a*) the thermal efficiency and (*b*) the net work output of a simple ideal Brayton cycle?

**7-54C** Why are gas turbines operated at very high air–fuel mass ratios?

**7-55C** Should the processes that make up the Brayton cycle be analyzed as closed-system or steady-flow processes? Why?

**7-56C** What is the back work ratio? What are typical back work ratio values for gas-turbine engines?

**7-57C** How can the irreversibilities in the turbine and compressor of gas-turbine engines be properly accounted for?

**7-58C** How do the inefficiencies of the turbine and the compressor affect (*a*) the back work ratio and (*b*) the thermal efficiency of a gas-turbine engine?

**7-59** A simple ideal Brayton cycle with air as the working fluid has a pressure ratio of 10. The air enters the compressor at 300 K and the turbine at 1200 K. Using constant specific heats at room temperature, determine (*a*) the air temperature at the compressor exit, (*b*) the back work ratio, and (*c*) the thermal efficiency.

**7-59E** A simple ideal Brayton cycle with air as the working fluid has a pressure ratio of 10. The air enters the compressor at 520 R and the turbine at 2000 R. Using constant specific heats at room temperature, determine (*a*) the air temperature at the compressor exit, (*b*) the back work ratio, and (*c*) the thermal efficiency.

**7-60** Air is used as the working fluid in a simple ideal Brayton cycle that has a pressure ratio of 12, a compressor inlet temperature of 300 K, and a turbine inlet temperature of 1000 K. Determine the required mass flow rate of air for a net power output of 30 MW, assuming both the compressor and the turbine to be isentropic. Assume constant specific heats at room temperature. *Answer:* 150.7 kg/s

**7-61** A stationary gas-turbine power plant operates on a simple ideal Brayton cycle with air as the working fluid. The air enters the compressor at 95 kPa and 290 K and the turbine at 760 kPa and 1100 K. Heat is transferred to air at a rate of 50,000 kJ/s. Determine the power delivered by this plant, assuming constant specific heats at room temperature.

**7-62** Air enters the compressor of a gas-turbine engine at 300 K and 100 kPa, where it is compressed to 700 kPa and 580 K. Heat is transferred to air in the amount of 950 kJ/kg before it enters the turbine. Assuming the turbine to be isentropic, determine (*a*) the fraction of the turbine work output used to drive the compressor and (*b*) the thermal efficiency.

**7-62E** Air enters the compressor of a gas-turbine engine at 540 R and 14.5 psia, where it is compressed to 116 psia and 1000 R. Heat is transferred to air in the amount of 420 Btu/lbm before it enters the turbine. Assuming the turbine to be isentropic, determine (*a*) the fraction of the turbine work output used to drive the compressor and (*b*) the thermal efficiency.

**7-63** A gas-turbine power plant operates on a simple Brayton cycle with air as the working fluid. The air enters the turbine at 1 MPa and 1000 K and leaves at 125 kPa and 600 K. Heat is rejected to the surroundings at a rate of 7922 kJ/s, and air flows through the cycle at a rate of 2.5 kg/s. Assuming the compressor to be isentropic and using specific heats at room temperature, determine the net power output of the plant.

**7-63E** A gas-turbine power plant operates on a simple Brayton cycle with air as the working fluid. The air enters the turbine at 120 psia and 2000 R and leaves at 15 psia and 1200 R. Heat is rejected to the surroundings at a rate of 6400 Btu/s, and air flows through the cycle at a rate of 40 lbm/s. Assuming the compressor to be isentropic, determine the net power output of the plant. *Answer:* 3726 kW

## Brayton Cycle with Regeneration

**7-64C** How does regeneration affect the efficiency of a Brayton cycle, and how does it accomplish it?

**7-65C** Somebody claims that at very high pressure ratios, the use of regeneration actually decreases the thermal efficiency of a gas-turbine engine. Is there any truth in this claim? Explain.

**7-66C** Define the effectiveness of a regenerator used in gas-turbine cycles.

**7-67C** In an ideal regenerator, is the air leaving the compressor heated to the temperature at (*a*) turbine inlet, (*b*) turbine exit, (*c*) slightly above turbine exit?

**7-68** An ideal Brayton cycle with regeneration has a pressure ratio of 10. The air enters the compressor at 300 K and the turbine at 1200 K. If the effectiveness of the regenerator is 100 percent, determine the net work output and the thermal efficiency of the cycle.

**7-68E** An ideal Brayton cycle with regeneration has a pressure ratio of 10. The air enters the compressor at 520 R and the turbine at 2000 R. If the effectiveness of the regenerator is 100 percent, determine the net work output and the thermal efficiency.

**7-69** A Brayton cycle with regeneration using air as the working fluid has a pressure ratio of 8. The minimum and maximum temperatures in the cycle are 310 and 1150 K. Assuming that both the compressor and the turbine are isentropic and that the regenerator has an effectiveness of 65 percent, determine (*a*) the air temperature at the turbine exit, (*b*) the net work output, and (*c*) the thermal efficiency.
*Answers:* (*a*) 635 K, (*b*) 263.5 kJ/kg, (*c*) 48.5 percent

**7-70** A stationary gas-turbine power plant operates on an ideal regenerative Brayton cycle ($\varepsilon$ = 100 percent) with air as the working fluid. Air enters the compressor at 95 kPa and 290 K and the turbine at 760 kPa and 1100 K. Heat is transferred to air from an external source at a rate of 60,000 kJ/s. Determine the power delivered by this plant, assuming constant specific heats for air at room temperature.

**7-71** Air enters the compressor of a regenerative gas-turbine engine at 300 K and 100 kPa, where it is compressed to 800 kPa and 580 K. The regenerator has an effectiveness of 72 percent, and the air enters the turbine at 1200 K. Assuming the turbine to be isentropic, determine (*a*) the amount of heat transfer in the regenerator and (*b*) the thermal efficiency. Account for the variation of specific heats with temperature.

**7-71E** Air enters the compressor of a regenerative gas-turbine engine at 540 R and 14.5 psia, where it is compressed to 116 psia and 1080 R. The regenerator has an effectiveness of 65 percent, and the air enters the turbine at 2000 R. Assuming the turbine to be isentropic, determine (*a*) the amount of heat transfer in the regenerator and (*b*) the thermal efficiency. Account for the variation of specific heats with temperature.

**7-72** Repeat Prob. 7-59, assuming that a regenerator of 75 percent effectiveness is added to the gas-turbine power plant.

### Jet-Propulsion Cycles

**7-73C** How does the ideal jet-propulsion cycle differ from the ideal Brayton cycle?

**7-74C** What is the function of the nozzle in turbojet engines?

**7-75C** What is propulsive power? How is it related to thrust?

**7-76C** What is propulsive efficiency? How is it determined?

**7-77C** Is the effect of turbine and compressor irreversibilities of a turbojet engine to reduce (*a*) the net work, (*b*) the thrust, or (*c*) the fuel consumption rate?

**7-78** A turbojet aircraft is flying with a velocity 280 m/s at an altitude of 6100 m, where the ambient conditions are 48 kPa and −13°C. The pressure ratio across the compressor is 13, and the temperature at the turbine inlet is 1300 K. Assuming ideal operation for all components and constant specific heats for air at room temperature, determine (*a*) the pressure at the turbine exit, (*b*) the velocity of the exhaust gases, and (*c*) the propulsive efficiency.
*Answers:* (*a*) 374.3 kPa, (*b*) 933.6 m/s, (*c*) 26.9 percent

**7-78E** A turbojet is flying with a velocity 900 ft/s at an altitude of 20,000 ft, where the ambient conditions are 7 psia and 10°F. The pressure ratio across the compressor is 13, and the temperature at the turbine inlet is 2400 R. Assuming ideal operation for all components and constant specific heats for air at room temperature, determine (*a*) the pressure at the turbine exit, (*b*) the velocity of the exhaust gases, and (*c*) the propulsive efficiency.

**7-79** A turbojet aircraft is flying with a velocity 320 m/s at an altitude of 9150 m, where the ambient conditions are 32 kPa and −32°C. The pressure ratio across the compressor is 12, and the temperature at the turbine inlet is 1400 K. Air enters the compressor at a rate of 40 kg/s, and the jet fuel has a heating value of 42,700 kJ/kg. Assuming ideal operation for all components and constant specific heats for air at room temperature, determine (*a*) the velocity of the exhaust gases, (*b*) the propulsive power developed, and (*c*) the rate of fuel consumption.

**7-79E** A turbojet aircraft is flying with a velocity 950 ft/s at an altitude of 30,000 ft, where the ambient conditions are 5 psia and −26°F. The pressure ratio across the compressor is 12, and the temperature of the gases at the turbine inlet is 2300 R. Air enters the compressor at a rate of 75 lbm/s, and the jet fuel has a heating value of 18,400 Btu/lbm. Assuming ideal operation for all components and constant specific heats for air at room temperature, determine (*a*) the velocity of the exhaust gases, (*b*) the propulsive power developed, and (*c*) the rate of fuel consumption.

**7-80** Consider an aircraft powered by a turbojet engine that has a pressure ratio of 12. The aircraft is stationary on the ground, held in position by its brakes. The ambient air is at 27°C and 95 kPa and enters the engine at a rate of 10 kg/s. The jet fuel has a heating value of 42,700 kJ/kg, and it is burned completely at a rate of 0.2 kg/s. Neglecting the effect of the diffuser and disregarding the slight increase in mass at the engine exit as well as the inefficiencies of engine components, determine the force that must be applied on the brakes to hold the plane stationary. Account for the variation of specific heats with temperature.

**7-81** Air at 7°C enters a turbojet engine at a rate of 20 kg/s and at a velocity of 300 m/s (relative to the engine). Air is heated in the combustion chamber at a rate 20,000 kJ/s and it leaves the engine at 427°C. Determine the thrust produced by this turbojet engine. (*Hint:* Choose the entire engine as your control volume.)

### Carnot Vapor Cycle

**7-82C** Why is excessive moisture in steam undesirable in steam turbines? What is the highest moisture content allowed?

**7-83C** How is moisture content related to quality?

**7-84C** Why is the Carnot cycle not a realistic model for steam power plants?

**7-85** A steady-flow Carnot cycle uses water as the working fluid. Water changes from saturated liquid to saturated vapor as heat is transferred to it from a source at 250°C. Heat rejection takes place at a pressure of 20 kPa. Show the cycle on a $T$-$s$ diagram relative to the saturation lines, and determine ($a$) the thermal efficiency, ($b$) the amount of heat rejected, in kJ/kg, and ($c$) the net work output.

**7-86** Consider a steady-flow Carnot cycle with water as the working fluid. The maximum and minimum temperatures in the cycle are 350 and 60°C. The quality of water is 0.891 at the beginning of the heat rejection process and 0.1 at the end. Show the cycle on a $T$-$s$ diagram relative to the saturation lines, and determine ($a$) the thermal efficiency, ($b$) the pressure at the turbine inlet, and ($c$) the net work output.
*Answers:* ($a$) 0.465, ($b$) 1.40 MPa, ($c$) 1624 kJ/kg

### The Simple Rankine Cycle

**7-87C** What four processes make up the simple ideal Rankine cycle?

**7-88C** Consider a simple ideal Rankine cycle with fixed turbine inlet conditions. What is the effect of lowering the condenser pressure on

| | |
|---|---|
| Pump work input: | ($a$) increases, ($b$) decreases, ($c$) remains the same |
| Turbine work output: | ($a$) increases, ($b$) decreases, ($c$) remains the same |
| Heat added: | ($a$) increases, ($b$) decreases, ($c$) remains the same |
| Heat rejected: | ($a$) increases, ($b$) decreases, ($c$) remains the same |
| Cycle efficiency: | ($a$) increases, ($b$) decreases, ($c$) remains the same |
| Moisture content at turbine exit: | ($a$) increases, ($b$) decreases, ($c$) remains the same |

**7-89C** Consider a simple ideal Rankine cycle with fixed turbine inlet temperature and condenser pressure. What is the effect of increasing the boiler pressure on

Pump work input: (*a*) increases, (*b*) decreases, (*c*) remains the same
Turbine work
output: (*a*) increases, (*b*) decreases, (*c*) remains the same
Heat added: (*a*) increases, (*b*) decreases, (*c*) remains the same
Heat rejected: (*a*) increases, (*b*) decreases, (*c*) remains the same
Cycle efficiency: (*a*) increases, (*b*) decreases, (*c*) remains the same
Moisture content
at turbine exit: (*a*) increases, (*b*) decreases, (*c*) remains the same

**7-90C** Consider a simple ideal Rankine cycle with fixed boiler and condenser pressures. What is the effect of superheating the steam to a higher temperature on

Pump work input: (*a*) increases, (*b*) decreases, (*c*) remains the same
Turbine work
output: (*a*) increases, (*b*) decreases, (*c*) remains the same
Heat added: (*a*) increases, (*b*) decreases, (*c*) remains the same
Heat rejected: (*a*) increases, (*b*) decreases, (*c*) remains the same
Cycle efficiency: (*a*) increases, (*b*) decreases, (*c*) remains the same
Moisture content
at turbine exit: (*a*) increases, (*b*) decreases, (*c*) remains the same

**7-91C** How do actual vapor power cycles differ from the idealized ones?

**7-92C** Compare the pressures at the inlet and the exit of the boiler for (*a*) actual and (*b*) ideal cycles.

**7-93C** The entropy of steam increases in actual steam turbines as a result of irreversibilities. In an effort to control entropy increase, it is proposed to cool the steam in the turbine by running cooling water around the turbine casing. It is argued that this will reduce the entropy and the enthalpy of the steam at the turbine exit and thus increase the work output. How would you evaluate this proposal?

**7-94C** Is it possible to maintain a pressure of 10 kPa in a condenser which is being cooled by river water entering at 20°C?

**7-95** A steam power plant operates on a simple ideal Rankine cycle between the pressure limits of 3 MPa and 50 kPa. The temperature of the steam at the turbine inlet is 400°C, and the mass flow rate of steam through the cycle is 25 kg/s. Show the cycle on a *T-s* diagram with respect to saturation lines, and determine (*a*) the thermal efficiency of the cycle and (*b*) the net power output of the power plant.

**7-96** Consider a 300-MW steam power plant that operates on a simple ideal Rankine cycle. Steam enters the turbine at 10 MPa and 500°C and is cooled in the condenser at a pressure of 10 kPa. Show the cycle on a *T-s* diagram with respect to saturation lines, and determine (*a*) the quality of the steam at the turbine exit, (*b*) the thermal efficiency of the cycle, and (*c*) the mass flow rate of the steam.
*Answers:* (*a*) 0.793, (*b*) 40.2 percent, (*c*) 235.4 kg/s

**7-96E** Consider a 150-MW steam power plant that operates on a simple ideal Rankine cycle. Steam enters the turbine at 1000 psia and 1000°F and is cooled in the condenser at a pressure of 2 psia. Show the cycle on a *T-s* diagram with respect to saturation lines, and determine (*a*) the quality of the steam at the turbine exit, (*b*) the thermal efficiency of the cycle, and (*c*) the mass flow rate of the steam.

**7-97** A steam power plant operates on a simple ideal Rankine cycle between the pressure limits of 9 MPa and 10 kPa. The mass flow rate of steam through the cycle is 60 kg/s. The moisture content of the steam at the turbine exit is not to exceed 10 percent. Show the cycle on a *T-s* diagram with respect to saturation lines, and determine (*a*) the minimum turbine inlet temperature, (*b*) the rate of heat input in the boiler, and (*c*) the thermal efficiency of the cycle.

**7-97E** A steam power plant operates on a simple ideal Rankine cycle between the pressure limits of 1250 and 2 psia. The mass flow rate of steam through the cycle is 75 lbm/s. The moisture content of the steam at the turbine exit is not to exceed 10 percent. Show the cycle on a *T-s* diagram with respect to saturation lines, and determine (*a*) the minimum turbine inlet temperature, (*b*) the rate of heat input in the boiler, and (*c*) the thermal efficiency of the cycle.

**7-98** Consider a coal-fired steam power plant that produces 300 MW of electric power. The power plant operates on a simple ideal Rankine cycle with turbine inlet conditions of 5 MPa and 450°C and a condenser pressure of 25 kPa. The coal used has a heating value (energy released when the fuel is burned) of 29,300 kJ/kg. Assuming that 75 percent of this energy is transferred to the steam in the boiler and that electric generator has an efficiency of 96 percent, determine (*a*) the overall plant efficiency (the ratio of net electric power output to the energy input as fuel) and (*b*) the required rate of coal supply, in t/h [1 metric ton (t) = 1000 kg]. *Answers:* (*a*) 24.6 percent, (*b*) 150.1 t/h

**7-98E** Consider a coal-fired steam power plant that produces 300 MW of electric power. The power plant operates on a simple ideal Rankine cycle with turbine inlet conditions of 700 psia and 800°F and a condenser pressure of 3 psia. The coal used has a heating value (energy released when the fuel is burned) of 12,600 Btu/lbm. Assuming that 75 percent of this energy is transferred to the steam in the boiler and that the electric generator has an efficiency of 96 percent, determine (*a*) the overall plant efficiency (the ratio of net electric power output to the energy input as fuel) and (*b*) the required rate of coal supply, in tons/h (1 ton = 2000 lbm). *Answers:* (*a*) 24.6 percent, (*b*) 183.8 tons/h

**7-99** Consider a solar-pond power plant that operates on a simple ideal Rankine cycle with refrigerant-134a as the working fluid. The refrigerant enters the turbine as a saturated vapor at 1.6 MPa and leaves at 0.7 MPa. The mass flow rate of the refrigerant is 6 kg/s. Show the cycle on a *T-s*

diagram with respect to saturation lines, and determine (*a*) the thermal efficiency of the cycle and (*b*) the power output of this plant.

**7-100** Consider a steam power plant that operates on a simple ideal Rankine cycle and has a net power output of 30 MW. Steam enters the turbine at 7 MPa and 500°C and is cooled in the condenser at a pressure of 10 kPa by running cooling water from a lake through the tubes of the condenser at a rate of 2000 kg/s. Show the cycle on a *T*-*s* diagram with respect to saturation lines, and determine (*a*) the thermal efficiency of the cycle, (*b*) the mass flow rate of the steam, and (*c*) the temperature rise of the cooling water. *Answers:* (*a*) 38.9 percent, (*b*) 24.0 kg/s, (*c*) 5.63°C

**7-100E** Consider a steam power plant that operates on a simple Rankine cycle and has a net power output of 50 MW. Steam enters the turbine at 1000 psia and 1000°F and is cooled in the condenser at a pressure of 2 psia by running cooling water from a lake through the tubes of the condenser at a rate of 7000 lbm/s. The turbine and the pump have adiabatic efficiencies of 84 percent. Show the cycle on a *T*-s diagram with respect to saturation lines, and determine (*a*) the thermal efficiency of the cycle, (*b*) the mass flow rate of the steam, and (*c*) the temperature rise of the cooling water.

### The Reheat Rankine Cycle

**7-101C** How do the following quantities change when a simple ideal Rankine cycle is modified with reheating? Assume the mass flow rate is maintained the same.

| | |
|---|---|
| Pump work input: | (*a*) increases, (*b*) decreases, (*c*) remains the same |
| Turbine work output: | (*a*) increases, (*b*) decreases, (*c*) remains the same |
| Heat added: | (*a*) increases, (*b*) decreases, (*c*) remains the same |
| Heat rejected: | (*a*) increases, (*b*) decreases, (*c*) remains the same |
| Moisture content at turbine exit: | (*a*) increases, (*b*) decreases, (*c*) remains the same |

**7-102C** Show the ideal Rankine cycle with three stages of reheating on a *T*-*s* diagram. Assume the turbine inlet temperature is the same for all stages. How does the cycle efficiency vary with the number of reheat stages?

**7-103C** Consider a simple Rankine cycle and an ideal Rankine cycle with three reheat stages. Both cycles operate between the same pressure limits. The maximum temperature is 700°C in the simple cycle and 500°C in the reheat cycle. Which cycle do you think will have a higher thermal efficiency?

**7-104** A steam power plant operates on the ideal reheat Rankine cycle. Steam enters the high-pressure turbine at 8 MPa and 500°C and leaves at 3 MPa. Steam is then reheated at constant pressure to 500°C before it expands to 20 kPa in the low-pressure turbine. Determine the turbine work output, in kJ/kg, and the thermal efficiency of the cycle. Also show the cycle on a $T$-$s$ diagram with respect to saturation lines.

**7-105** Consider a steam power plant that operates on the ideal reheat Rankine cycle and has a net power output of 150 MW. Steam enters the high-pressure turbine at 10 MPa and 500°C and the low-pressure turbine at 1 MPa and 500°C. Steam leaves the condenser as a saturated liquid at a pressure of 10 kPa. Show the cycle on a $T$-$s$ diagram with respect to saturation lines, and determine (*a*) the quality (or temperature, if superheated) of the steam at the turbine exit, (*b*) the thermal efficiency of the cycle, and (*c*) the mass flow rate of the steam.
*Answers:* (*a*) 0.948, (*b*) 41.4 percent, (*c*) 93.8 kg/s

**7-105E** Consider a steam power plant that operates on the ideal reheat Rankine cycle and has a net power output of 175 MW. Steam enters the high-pressure turbine at 1500 psia and 1100°F and the low-pressure turbine at 500 psia and 1000°F. Steam leaves the condenser as a saturated liquid at a pressure of 2 psia. Show the cycle on a $T$-$s$ diagram with respect to saturation lines, and determine (*a*) the quality (or temperature, if superheated) of the steam at the turbine exit, (*b*) the thermal efficiency of the cycle, and (*c*) the mass flow rate of the steam.

**7-106** Steam enters the high-pressure turbine of a steam power plant that operates on the ideal reheat Rankine cycle at 6 MPa and 450°C and leaves as saturated vapor. Steam is then reheated to 400°C before it expands to a pressure of 7.5 kPa. Heat is transferred to the steam in the boiler at a rate of $4 \times 10^4$ kJ/s. Steam is cooled in the condenser by the cooling water from a nearby river, which enters the condenser at 15°C. Show the cycle on a $T$-$s$ diagram with respect to saturation lines, and determine (*a*) the pressure at which reheating takes place, (*b*) the net power output and thermal efficiency, and (*c*) the minimum mass flow rate of the cooling water required.

**7-106E** Steam enters the high-pressure turbine of a steam power plant that operates on the ideal reheat Rankine cycle at 800 psia and 900°F and leaves as saturated vapor. Steam is then reheated to 800°F before it expands to a pressure of 1 psia. Heat is transferred to the steam in the boiler at a rate of $6 \times 10^4$ Btu/s. Steam is cooled in the condenser by the cooling water from a nearby river, which enters the condenser at 45°F. Show the cycle on a $T$-$s$ diagram with respect to saturation lines, and determine (*a*) the pressure at which reheating takes place, (*b*) the net power output and thermal efficiency, and (*c*) the minimum mass flow rate of the cooling water required.

**7-107** A steam power plant operates on an ideal reheat Rankine cycle between the pressure limits of 9 MPa and 10 kPa. The mass flow rate of steam through the cycle is 25 kg/s. Steam enters both stages of the turbine at 500°C. If the moisture content of the steam at the exit of the low-pressure turbine is not to exceed 10 percent, determine (*a*) the pressure at which reheating takes place, (*b*) the total rate of heat input in the boiler, and (*c*) the thermal efficiency of the cycle. Also show the cycle on a *T*-*s* diagram with respect to saturation lines.

### Vapor-Compression Refrigeration Cycles

**7-108C** Draw the *T*-*s* diagrams of a reversed Carnot cycle and a Carnot cycle operating between the same temperature limits. How do they differ?

**7-109C** Why is the reversed Carnot cycle executed within the saturation dome not a realistic model for refrigeration cycles?

**7-110C** What is the difference between a refrigerator and a heat pump?

**7-111C** Does the ideal vapor-compression refrigeration cycle involve any internal irreversibilities?

**7-112C** Why is the throttling valve not replaced by an isentropic turbine in the ideal vapor-compression refrigeration cycle?

**7-113C** It is proposed to use water instead of refrigerant-134a as the working fluid in air conditioning applications, where the minimum temperature never falls below the freezing point. Would you support this proposal? Explain.

**7-114C** In a refrigeration system, would you recommend condensing the refrigerant-12 at a pressure of 0.7 or 1.0 MPa if heat is to be rejected to a cooling medium at 15°C? Why?

**7-115C** Does the area enclosed by the cycle on a *T*-*s* diagram represent the net work input for the reversed Carnot cycle? How about for the ideal vapor-compression refrigeration cycle?

**7-116C** Consider two vapor-compression refrigeration cycles. The refrigerant enters the throttling valve as a saturated liquid at 30°C in one cycle and as subcooled liquid at 30°C in the other one. The evaporator pressure for both cycles is the same. Which cycle do you think will have a higher COP?

**7-117** A refrigerator uses refrigerant-134a as the working fluid and operates on an ideal vapor-compression refrigeration cycle between 0.12 and 0.7 MPa. The mass flow rate of the refrigerant is 0.05 kg/s. Show the cycle on a *T*-*s* diagram with respect to saturation lines. Determine (*a*) the rate of heat removal from the refrigerated space and the power input to the compressor, (*b*) the rate of heat rejection to the environment, and (*c*) the coefficient of performance.
*Answers:* (*a*) 7.35 kW, 1.85 kW; (*b*) 9.20 kW; (*c*) 3.97

**7-118** Consider a 300 kJ/min refrigeration system that operates on an ideal vapor-compression refrigeration cycle with refrigerant-134a as the working fluid. The refrigerant enters the compressor as saturated vapor at 140 kPa and is compressed to 800 kPa. Show the cycle on a *T*-*s* diagram with respect to saturation lines, and determine (*a*) the quality of the refrigerant at the end of the throttling process, (*b*) the coefficient of performance, and (*c*) the power input to the compressor.

**7-118E** Consider a 3-ton refrigeration system that operates on an ideal vapor-compression refrigeration cycle with refrigerant-134a as the working fluid. The refrigerant enters the compressor as saturated vapor at 20 psia and is compressed to a pressure of 140 psia. Show the cycle on a $T$-$s$ diagram with respect to saturation lines, and determine (*a*) the quality of the refrigerant at the end of the throttling process, (*b*) coefficient of performance, and (*c*) the power input to the compressor. *Answers:* (*a*) 0.371, (*b*) 3.29, (*c*) 4.30 hp

**7-119** An ice-making machine operates on the ideal vapor-compression cycle, using refrigerant-134a. The refrigerant enters the compressor as saturated vapor at 160 kPa and leaves the condenser as saturated liquid at 700 kPa. Water enters the ice machine at 15°C and leaves as ice at −5°C. For an ice production rate of 12 kg/h, determine the power input to the ice maker (384 kJ of heat needs to be removed from each kilogram of water at 15°C to turn it into ice at −5°C). *Answer:* 0.258 kW

**7-119E** An ice-making machine operates on the ideal vapor-compression cycle, using refrigerant-134a. The refrigerant enters the compressor as saturated vapor at 20 psia and leaves the condenser as saturated liquid at 100 psia. Water enters the ice machine at 55°F and leaves as ice at 25°F. For an ice production rate of 20 lbm/h, determine the power input to the ice machine (169 Btu of heat needs to be removed from each lbm of water at 55°F to turn it into ice at 25°F).

### Selecting the Right Refrigerant

**7-120C** When selecting a refrigerant for a certain application, what qualities would you look for in the refrigerant?

**7-121C** Consider a refrigeration system using refrigerant-12 as the working fluid. If this refrigerator is to operate in an environment at 30°C, what is the minimum pressure to which the refrigerant should be compressed? Why?

**7-122C** A refrigerant-12 is to maintain the refrigerated space at −10°C. Would you recommend an evaporator pressure of 0.12 or 0.14 MPa for this system? Why?

**7-123** A refrigerator that operates on the ideal vapor-compression cycle with refrigerant-12 is to maintain the refrigerated space at −10°C while rejecting heat to the environment at 25°C. Select reasonable pressures for the evaporator and the condenser, and explain why you chose those values.

**7-124** A heat pump that operates on the ideal vapor-compression cycle with refrigerant-12 is used to heat a house and maintain it at 20°C by using underground water at 10°C as the heat source. Select reasonable pressures for the evaporator and the condenser, and explain why you chose those values.



## Heat Pump Systems

**7-125C** What are the advantages and disadvantages of heat pumps? How do they compare with other heating systems?

**7-126C** Do you think a heat pump system will be more cost-effective in New York or in Miami? Why?

**7-127C** What is a water-source heat pump? How does the COP of a water-source heat pump system compare with that of an air-source system?

**7-128** A heat pump that operates on the ideal vapor-compression cycle with refrigerant-134a is used to heat a house and maintain it at 20°C, using underground water at 10°C as the heat source. The house is losing heat at a rate of 75,000 kJ/h. The evaporator and condenser pressures are 320 and 800 kPa, respectively. Determine the power input to the heat pump and the electric power saved by using a heat pump instead of a resistance heater. *Answers:* 2.27 kW, 18.56 kW

**7-128E** A heat pump that operates on the ideal vapor-compression cycle with refrigerant-134a is used to heat a house and maintain it at 75°F by using underground water at 50°F as the heat source. The house is losing heat at a rate of 90,000 Btu/h. The evaporator and condenser pressures are 50 and 120 psia, respectively. Determine the power input to the heat pump and the electric power saved by using a heat pump instead of a resistance heater. *Answers:* 3.68 hp, 31.7 hp

**7-129** A heat pump that operates on the ideal vapor-compression cycle with refrigerant-134a is used to heat water from 15 to 54°C at a rate of 0.18 kg/s. The condenser and evaporator pressures are 1.4 and 0.32 MPa. respectively. Determine the power input to the heat pump.

**7-130** A heat pump using refrigerant-12 heats a house by using underground water at 8°C as the heat source. The house is losing heat at a rate of 60,000 kJ/h. The refrigerant enters the compressor at 280 kPa and 0°C, and it leaves at 1 MPa and 60°C. The refrigerant leaves the condenser at 30°C. Determine (*a*) the power input to the heat pump, (*b*) the rate of heat absorption from the water, and (*c*) the increase in electric power input if an electric resistance heater is used instead of a heat pump. *Answers:* (*a*) 3.65 kW, (*b*) 13.0 kW, (*c*) 13.0 kW

**7-130E** A heat pump using refrigerant-134a heats a house by using underground water at 45°F as the heat source. The house is losing heat at a rate of 70,000 Btu/h. The refrigerant enters the compressor at 30 psia and 20°F and leaves at 120 psia and 140°F. The refrigerant leaves the condenser at 90°F. Determine (*a*) the power input to the heat pump, (*b*) the rate of heat absorption from the water, and (*c*) the increase in electric power input if an electric resistance heater is used instead of a heat pump.

### Thermoelectric Power Generation and Refrigeration Systems

**7-131C** What is a thermoelectric circuit?

**7-132C** Describe the Seebeck and the Peltier effects.

**7-133C** Consider a circular copper wire formed by connecting the two ends of a copper wire. The connection point is now heated by a burning candle. Do you expect any current to flow through the wire?

**7-134C** An iron and a constantan wire are formed into a closed circuit by connecting the ends. Now both junctions are heated and are maintained at the same temperature. Do you expect any electric current to flow through this circuit?

**7-135C** A copper and a constantan wire are formed into a closed circuit by connecting the ends. Now one junction is heated by a burning candle while the other is maintained at room temperature. Do you expect any current to flow through this circuit?

**7-136C** How does a thermocouple work as a temperature measurement device?

**7-137C** Why are semiconductor materials preferable to metals in thermoelectric refrigerators?

**7-138C** Is the efficiency of a thermoelectric generator limited by the Carnot efficiency? Why?

**7-139** A thermoelectric generator receives heat from a source at 150°C and rejects the waste heat to the environment at 25°C. What is the maximum thermal efficiency this thermoelectric generator can have?
*Answer:* 29.6 percent

**7-139E** A thermoelectric generator receives heat from a source at 240°F and rejects the waste heat to the environment at 80°F. What is the maximum thermal efficiency this thermoelectric generator can have?
*Answer:* 22.9 percent

**7-140** A thermoelectric refrigerator removes heat from a refrigerated space at −5°C at a rate of 130 W and rejects it to an environment at 20°C. Determine the maximum coefficient of performance this thermoelectric refrigerator can have and the minimum required power input.
*Answers:* 10.72, 12.1 W

**7-141** A thermoelectric cooler has a COP of 0.6 and removes heat from a refrigerated space at a rate of 180 W. Determine the required power input to the thermoelectric cooler, in W.

**7-141E** A thermoelectric cooler has a COP of 0.8 and removes heat from a refrigerated space at a rate of 35 Btu/min. Determine the required power input to the thermoelectric cooler, in hp.



## Review Problems

**7-142** Consider a simple ideal Brayton cycle operating between the temperature limits of 290 K and 1500 K. Using constant specific heats at room temperature, determine the pressure ratio for which the compressor and the turbine exit temperatures of air are equal.

**7-143** An air-standard cycle is executed in a closed system and is composed of the following four processes:

1-2 $v$ = constant heat addition from 100 kPa and 27°C to 300 kPa
2-3 $P$ = constant heat addition to 1027°C
3-4 Isentropic expansion to 100 kPa
4-1 $P$ = constant heat rejection to initial state

(*a*) Show the cycle on *P*-*v* and *T*-*s* diagrams.
(*b*) Calculate the net work output per unit mass.
(*c*) Determine the thermal efficiency.

**7-144** An air-standard cycle is executed in a closed system with 0.002 kg of air and consists of the following three processes:

1-2 Isentropic compression from 100 kPa and 27°C to 700 kPa
2-3 $P$ = constant heat addition to initial specific volume
3-1 $v$ = constant heat rejection to initial state

(*a*) Show the cycle on *P*-*v* and *T*-*s* diagrams.
(*b*) Calculate the maximum temperature in the cycle.
(*c*) Determine the thermal efficiency.

Assume constant specific heats at room temperature.

**7-144E** An air-standard cycle is executed in a closed system with 0.005 lbm of air and consists of the following three processes:

1-2 Isentropic compression from 14.7 psia and 80°F to 120 psia
2-3 $P$ = constant heat addition to initial specific volume
3-1 $v$ = constant heat rejection to initial state

(*a*) Show the cycle on *P*-*v* and *T*-*s* diagrams.
(*b*) Calculate the maximum temperature in the cycle.
(*c*) Determine the thermal efficiency.

**7-145** A Carnot cycle is executed in a closed system and uses 0.002 kg of air as the working fluid. The cycle efficiency is 70 percent, and the lowest temperature in the cycle is 300 K. The pressure at the beginning of the isentropic expansion is 700 kPa, and at the end of the isentropic compression it is 1 MPa. Determine the net work output per cycle.

**7-146** A four-cylinder spark-ignition engine has a compression ratio of 8, and each cylinder has a maximum volume of 0.6 L. At the beginning of the compression process, the air is at 98 kPa and 17°C, and the maximum temperature in the cycle is 1800 K. Assuming the engine to operate on the ideal Otto cycle, determine (*a*) the amount of heat supplied per cylinder, (*b*) the thermal efficiency, and (*c*) the number of revolutions per minute required for a net power output of 60 kW. Assume constant specific heats at room temperature.

**7-147** An ideal Otto cycle has a compression ratio of 9.2 and uses air as the working fluid. At the beginning of the compression process, air is at 98 kPa and 27°C. The pressure is doubled during the constant-volume heat addition process. Using constant specific heats at room temperature, determine (*a*) the amount of heat transferred to the air, (*b*) the net work output, (*c*) the thermal efficiency, and (*d*) the mean effective pressure.

**7-148** Consider an engine operating on the ideal Diesel cycle with air as the working fluid. The volume of the cylinder is 1200 $cm^3$ at the beginning of the compression process, 75 $cm^3$ at the end, and 150 $cm^3$ after the heat addition process. Air is at 17°C and 100 kPa at the beginning of the compression process. Determine (*a*) the pressure at the beginning of the heat rejection process, (*b*) the net work per cycle, in kJ, and (*c*) the mean effective pressure.

**7-149** Repeat Prob. 7-148 using argon as the working fluid.

**7-150** An ideal dual cycle has a compression ratio of 12 and uses air as the working fluid. At the beginning of the compression process, air is at 100 kPa and 30°C, and occupies a volume of 1.2 L. During the heat addition process, 0.3 kJ of heat is transferred to air at constant volume and 1.1 kJ at constant pressure. Using constant specific heats evaluated at room temperature, determine the thermal efficiency of the cycle.

**7-150E** An ideal dual cycle has a compression ratio of 12 and uses air as the working fluid. At the beginning of the compression process, air is at 14.7 psia and 90°F, and occupies a volume of 75 $in^3$. During the heat addition process, 0.3 Btu of heat is transferred to air at constant volume and 1.1 Btu at constant pressure. Using constant specific heats evaluated at room temperature, determine the thermal efficiency of the cycle.

**7-151** Consider a simple ideal Brayton cycle with air as the working fluid. The pressure ratio of the cycle is 6, and the minimum and maximum temperatures are 300 and 1300 K, respectively. Now the pressure ratio is doubled without changing the minimum and maximum temperatures in the cycle. Determine the change in (*a*) the net work output per unit mass and (*b*) the thermal efficiency of the cycle as a result of this modification. Assume constant specific heats at room temperature.

**7-152** Helium is used as the working fluid in a Brayton cycle with regeneration. The pressure ratio of the cycle is 8, the compressor inlet temperature is 300 K, and the turbine inlet temperature is 1800 K. The effectiveness of the regenerator is 75 percent. Determine the thermal efficiency and the required mass flow rate of helium for a net power output of 30 MW, assuming both the compressor and the turbine to be isentropic.

**7-153** Consider the ideal regenerative Brayton cycle. Determine the pressure ratio that maximizes the thermal efficiency of the cycle and compare this value with the pressure ratio that maximizes the cycle net work. For the same maximum to minimum temperature ratios, explain why the pressure ratio for maximum efficiency is less than the pressure ratio for maximum work.

**7-154** Consider a steam power plant operating on the ideal Rankine cycle with reheat between the pressure limits of 25 MPa and 10 kPa with a maximum cycle temperature of 600°C and a moisture content of 12 percent at the turbine exit. For a reheat temperature of 600°C, determine the reheat pressures of the cycle for the cases of (*a*) single and (*b*) double reheat.

**7-155** The Stillwater geothermal power plant in Nevada, which started full commercial power operation in 1986, is designed to operate with seven identical units. Each of these seven units consists of a pair of power cycles, labeled Level I and Level II, operating on the simple Rankine cycle using an organic fluid as the working fluid.

The heat source for the plant is geothermal water (brine) entering

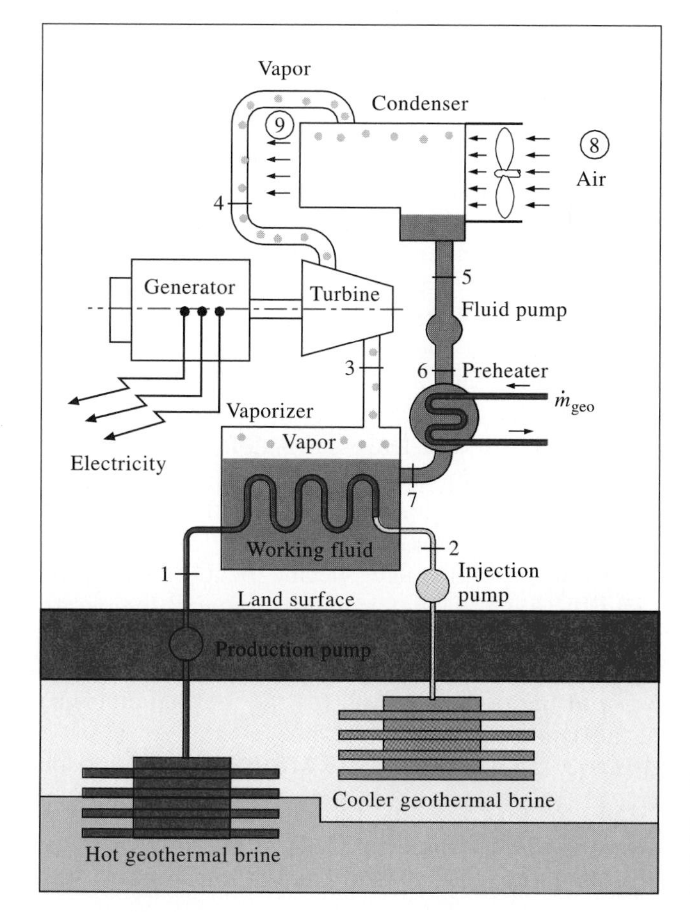


**FIGURE P7-155**

Schematic of a binary geothermal power plant. (*Courtesy of ORMAT Energy Systems, Inc.*)

the vaporizer (boiler) of Level I of each unit at 162.8°C at a rate of 174,309 kg/h and delivering 24,044 MJ/h. The organic fluid that enters the vaporizer at 94.6°C at a rate of 71,620 kg/h leaves it at 139.1°C and 1.557 MPa as saturated vapor. This saturated vapor expands in the turbine to 35.4°C and 131 kPa and produces 1271 kW of electric power. About 200 kW of this power is used by the pumps, the auxiliaries, and the 6 fans of the condenser. Subsequently, the organic working fluid is condensed in an air-cooled condenser by air which enters the condenser at 12.8°C at a rate of 1,903,000 kg/h and leaves at 29.2°C. The working fluid is pumped and then preheated in a preheater to 94.6°C by absorbing 11,750 MJ of heat from the geothermal water (coming from the vaporizer of Level II) entering the preheater at 99.9°C and leaving at 67.8°C.

Taking the average specific heat of the geothermal water to be 4.31 kJ/(kg · °C), determine (*a*) the exit temperature of the geothermal water from the vaporizer, (*b*) the rate of heat rejection from the working fluid to the air in the condenser, (*c*) the mass flow rate of the geothermal water at the preheater, and (*d*) the thermal efficiency of the Level I cycle of this geothermal power plant.

**7-155E** The Stillwater geothermal power plant in Nevada, which started full commercial power operation in 1986, is designed to operate with seven identical units. Each of these seven units consists of a pair of power cycles, labeled Level I and Level II, operating on the simple Rankine cycle using an organic fluid as the working fluid.

The heat source for the plant is geothermal water (brine) entering the vaporizer (boiler) of Level I of each unit at 325°F at a rate of 384,286 lbm/h and delivering 22.79 MBtu/h ("M" stands for "million"). The organic fluid that enters the vaporizer at 202.2°F at a rate of 157,895 lbm/h leaves it at 282.4°F and 225.8 psia as saturated vapor. This saturated vapor expands in the turbine to 95.8°F and 19.0 psia and produces 1271 kW of electric power. About 200 kW of this power is used by the pumps, the auxiliaries, and the 6 fans of the condenser. Subsequently, the organic working fluid is condensed in an air-cooled condenser by air which enters the condenser at 55°F at a rate of 4,195,100 lbm/h and leaves at 84.5°F. The working fluid is pumped and then preheated in a preheater to 202.4°F by absorbing 11.14 MBtu/h of heat from the geothermal water (coming from the vaporizer of Level II) entering the preheater at 211.8°F and leaving at 154.0°F.

Taking the average specific heat of the geothermal water to be 1.03 Btu/(lbm · °F), determine (*a*) the exit temperature of the geothermal water from the vaporizer, (*b*) the rate of heat rejection from the working fluid to the air in the condenser, (*c*) the mass flow rate of the geothermal water at the preheater, and (*d*) the thermal efficiency of the Level I cycle of this geothermal power plant.
*Answers:* (*a*) 267.4°F, (*b*) 29.7 MBtu/h, (*c*) 187,120 lbm/h, (*d*) 10.8 percent.

**7-156** Steam enters the turbine of a steam power plant that operates on a simple ideal Rankine cycle at a pressure of 6 MPa, and it leaves as a saturated vapor at 7.5 kPa. Heat is transferred to the steam in the boiler at a rate of $4 \times 10^4$ kJ/s. Steam is cooled in the condenser by the cooling water from a nearby river, which enters the condenser at 18°C. Show the cycle on a $T$-$s$ diagram with respect to saturation lines, and determine ($a$) the turbine inlet temperature, ($b$) the net power output and thermal efficiency, and ($c$) the minimum mass flow rate of the cooling water required.

**7-157** A large refrigeration plant is to be maintained at −15°C, and it requires refrigeration at a rate of 100 kW. The condenser of the plant is to be cooled by liquid water, which experiences a temperature rise of 8°C as it flows over the coils of the condenser. Assuming the plant operates on the ideal vapor-compression cycle using refrigerant-134a between the pressure limits of 120 and 700 kPa, determine ($a$) the mass flow rate of the refrigerant, ($b$) the power input to the compressor, and ($c$) the mass flow rate of the cooling water.

**7-158** A heat pump that operates on the ideal vapor-compression cycle with refrigerant-134a is used to heat a house. The mass flow rate of the refrigerant is 0.15 kg/s. The condenser and evaporator pressures are 900 and 240 kPa, respectively. Show the cycle on a $T$-$s$ diagram with respect to saturation lines, and determine ($a$) the rate of heat supply to the house, ($b$) the volume flow rate of the refrigerant at the compressor inlet, and ($c$) the COP for this heat pump.

**7-159** It is proposed to run a thermoelectric generator in conjunction with a solar pond which can supply heat at a rate of $10^6$ kJ/h at 80°C. The waste heat is to be rejected to the environment at 30°C. What is the maximum power this thermoelectric generator can produce?

## Computer, Design, and Essay Problems

**7-160** Write a computer program to determine the effects of pressure ratio, maximum cycle temperature, and compressor and turbine inefficiencies on the net work output per unit mass and the thermal efficiency of a simple Brayton cycle. Assume the working fluid is air that is at 100 kPa and 300 K at the compressor inlet. Also assume constant specific heats for air at room temperature. Determine the net work output and the thermal efficiency for all combinations of the following parameters:

Pressure ratio: 5, 8, 14
Maximum cycle temperature: 800, 1200, and 1600 K

Draw conclusions from the results.

**7-161** Repeat Prob. 7-160 using helium as the working fluid.

**7-162** Write a computer program to determine the effects of pressure ratio, maximum cycle temperature, regenerator effectiveness, and compressor and turbine efficiencies on the net work output per unit mass and on the thermal efficiency of a regenerative Brayton cycle. Assume the working fluid is air which is at 100 kPa and 300 K at the compressor inlet.

Also assume constant specific heats for air at room temperature. Determine the net work output and the thermal efficiency for all combinations of the following parameters:

Pressure ratio: 5, 8, 14
Maximum cycle temperature: 1000, 1400, 1600 K
Regenerator effectiveness: 70, 80 percent

**7-163** Repeat Prob. 7-162 using helium as the working fluid.

**7-164** Write an essay on the most recent developments on the two-stroke engines, and find out when we might be seeing cars powered by two-stroke engines in the market. Why do the major car manufacturers have a renewed interest in two-stroke engines?

**7-165** In response to concerns about the environment, some major car manufacturers are currently marketing electric cars. Write an essay on the advantages and disadvantages of electric cars, and discuss when it is advisable to purchase an electric car instead of a traditional internal combustion car.

**7-166** Intense research is underway to develop adiabatic engines that require no cooling of the engine block. Such engines are based on ceramic materials because of the ability of such materials to withstand high temperatures. Write an essay on the current status of adiabatic engine development. Also determine the highest possible efficiencies with these engines, and compare them to the highest possible efficiencies of current engines.

**7-167** Several geothermal power plants are in operation in the United States and more are being built since the heat source of a geothermal plant is the hot geothermal water which is "free energy." An 8-MW geothermal power plant is being considered at a location where geothermal water at 160°C is available. Geothermal water is to serve as the heat source for a closed Rankine power cycle with refrigerant-134a as the working fluid. Specify suitable temperatures and pressures for the cycle, and determine the thermal efficiency of the cycle. Justify your selections.

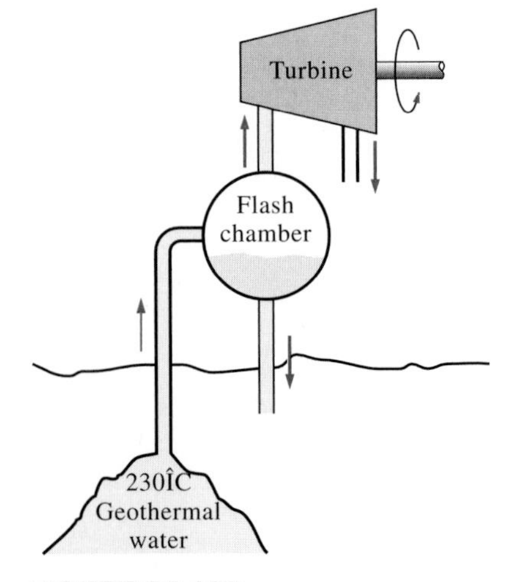


**FIGURE P7-168**

**7-168** A 10-MW geothermal power plant is considered at a site where geothermal water at 230°C is available. Geothermal water is to be flashed into a chamber to a lower pressure where part of the water evaporates. The liquid is returned to the ground while the vapor is used to drive the steam turbine. The pressures at the turbine inlet and the turbine exit are to remain above 200 kPa and 8 kPa, respectively. High-pressure flash chambers yield a small amount of steam with high availability whereas lower pressure flash chambers yield considerably more steam but at a lower availability. By trying several pressures, determine the optimum pressure of the flash chamber to maximize the power production per unit mass of geothermal water withdrawn. Also determine the thermal efficiency for each case assuming 10 percent of the power produced is used to drive the pumps and other auxiliary equipment.

**7-169** Write an essay on the latest developments of OTEC (Ocean Thermal Energy Conversion) plants. Discuss the major problems associated with OTEC systems and possible solutions.

**7-170** Design a vapor-compression refrigeration system that will maintain the refrigerated space at −15°C while operating in an environment at 20°C using refrigerant-134a as the working fluid.

**7-171** Write an essay on air-, water-, and soil-based heat pumps. Discuss the advantages and the disadvantages of each system. For each system, identify the conditions under which that system is preferable over the other two. In what situations would you not recommend a heat pump heating system?

**7-172** Consider a solar pond power plant operating on a closed Rankine cycle. Using refrigerant-134a as the working fluid, specify the operating temperatures and pressures in the cycle, and estimate the required mass flow rate of the refrigerant-12 for a net power output of 50 kW. Also estimate the surface area of the pond for this level of continuous power production. Assume that the solar energy is incident on the pond at a rate of 500 W per m$^2$ of pond area at noon time, and that the pond is capable of storing 15 percent of the incident solar energy in the storage zone.

**7-173** Design a thermoelectric refrigerator that is capable of cooling a canned drink in a car. The refrigerator is to be powered by the cigarette lighter of the car. Draw a sketch of your design. Semiconductor components for building thermoelectric power generators or refrigerators are available from several manufacturers. Using data from one of these manufacturers, determine how many of these components you need in your design, and estimate the coefficient of performance of you system. A critical problem in the design of thermoelectric refrigerators is the effective rejection of waste heat. Discuss how you can enhance the rate of heat rejection without using any devices with moving parts such as a fan.

**7-174** It is proposed to use a solar powered thermoelectric system installed on the roof to cool residential buildings. The system consists of a thermoelectric refrigerator that is powered by a thermoelectric power generator whose top surface is a solar collector. Discuss the feasibility and the cost of such a system, and determine if the proposed system installed on one side of the roof can meet a significant portion of the cooling requirements of a typical house in your area.

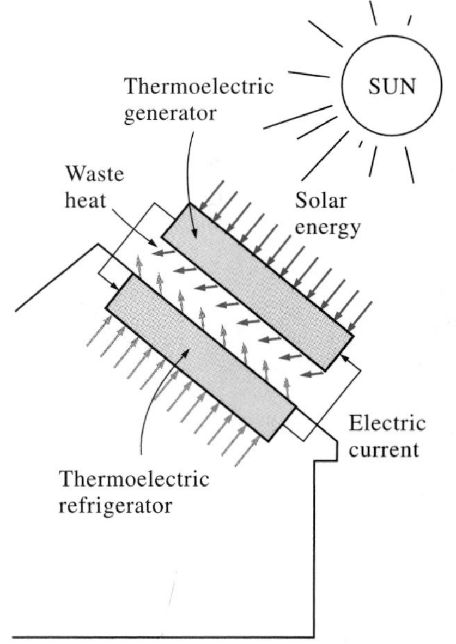


**FIGURE P7-174**

# Steady Heat Conduction

CHAPTER

So far, we have considered *thermodynamics,* which deals with the *amount* of heat transfer as a system undergoes a process from one equilibrium state to another, and makes no reference to *how long* the process takes. But often we are interested in the *rate* of heat transfer as well as the temperature distribution within the system at a specified time. The science that deals with these topics is *heat transfer,* which we will discuss in this and the following chapters.

Heat transfer has *direction* as well as *magnitude.* The rate of heat conduction in a specified direction is proportional to the *temperature gradient,* which is the change in temperature per unit length in that direction. Heat conduction in a medium, in general, is three-dimensional and time-dependent. That is, $T = T(x, y, z; t)$ and the temperature in a medium varies with position as well as time. Heat conduction in a medium is said to be *steady* when the temperature does not vary with time, and *unsteady* or *transient* when it does. Heat conduction in a medium is said to be *one-dimensional* when conduction is significant in one dimension only and negligible in the other two dimensions. It is said to be *two-dimensional* when conduction in the third dimension is negligible, and *three-dimensional* when conduction in all dimensions is significant.

We start this chapter with a physical description of the *conduction* mechanism and the thermal conductivity of materials. We then present *one-dimensional steady* heat conduction in a plane wall, cylinder, and a sphere, and introduce the concept of *thermal resistance.* We apply this concept to heat conduction problems in *multilayer* plane walls, cylinders, and spheres, as well as multidimensional geometries. We continue with the analysis of heat conduction problems that involve *heat generation.* Finally, we discuss steady heat transfer from *finned surfaces* and some complex geometries commonly encountered in practice through the use of *conduction shape factors.*

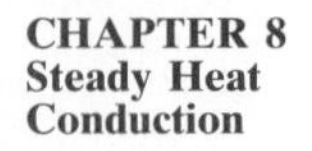



## 8-1 ■ THERMODYNAMICS AND HEAT TRANSFER

Perhaps you are wondering why we are about to undertake a major study on heat transfer. After all, we can determine the amount of heat transfer for any kind of system for any kind of process using a thermodynamic analysis alone. The reason is that thermodynamics is concerned with the *amount* of heat transfer as a system undergoes a process from one equilibrium state to another, and it gives no indication about *how long* the process takes. A thermodynamic analysis simply tells us how much heat must be transferred to realize a specified change of state to satisfy the conservation of energy principle.

In practice, we are more concerned about the *rate* of heat transfer (heat transfer per unit time) than we are with the *amount* of it. For example, we can determine the amount of heat transferred from a thermos bottle as the hot coffee inside cools from 90°C to 80°C by a thermodynamic analysis alone. But a typical user or designer of a thermos is primarily interested in *how long* it will be before the hot coffee inside cools to 80°C, and a thermodynamic analysis cannot answer this question. Determining the rates of heat transfer to or from a system and thus the times of cooling or heating, as well as the variation of the temperature is the subject of *heat transfer* (Fig. 8-1).

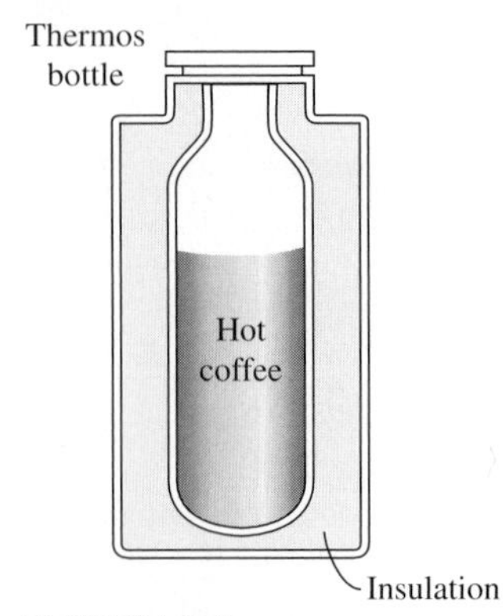


**FIGURE 8-1**

We are usually interested in how long it takes for the hot coffee in a thermos to cool to a certain temperature, which cannot be determined from a thermodynmic analysis alone.

The laws of thermodynamics lay the framework for the science of heat transfer. The *first law* requires that the rate of heat transfer into a system be *equal* to the rate of increase of the energy of that system. The *second law* requires that heat be transferred in the direction of *decreasing temperature*; that is, from a high-temperature medium to a lower-temperature one (Fig. 8-2). This is like the fact that a car parked on an inclined road must go down hill in the direction of *decreasing elevation* when its brakes are released. It is also analogous to the electric current flowing in the direction of *decreasing voltage* or the fluid flowing in the direction of *decreasing pressure*.

The basic requirement for heat transfer is the presence of a *temperature difference.* There can be no net heat transfer between two mediums that are at the same temperature. The temperature difference is the *driving force* for heat transfer, just as the *voltage difference* is the driving force for electric current flow and *pressure difference* is the driving force for fluid flow. The *rate* of heat transfer in a certain direction depends on the *magnitude* of the temperature difference per unit length in that direction. The larger the temperature difference, the higher the rate of heat transfer.

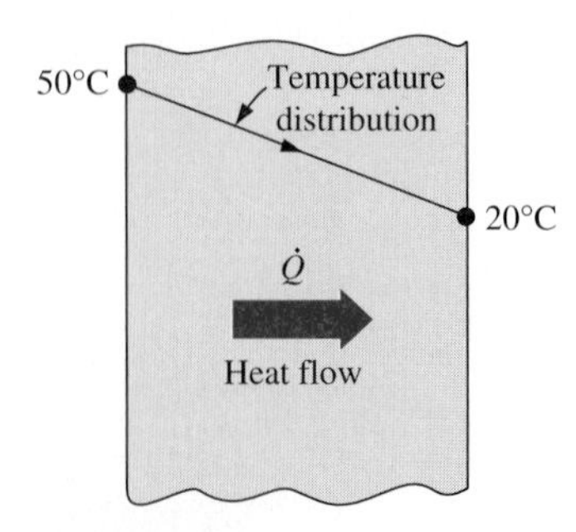


**FIGURE 8-2**

Heat flows in the direction of decreasing temperature.

In daily life, we frequently refer to the sensible and latent forms of internal energy as heat, and we talk about the heat content of bodies. In thermodynamics, however, those forms of energy are usually referred to as thermal energy to prevent any confusion with heat transfer. The term *heat* and the associated phrases such as *heat flow, head addition, heat rejection, heat absorption, heat gain, heat loss, heat storage, heat generation, electrical heating, latent heat, body heat,* and *heat source* are in common use today, and the attempt to replace *heat* in these phrases by *thermal energy* has had only limited success. These phrases are deeply rooted in our vocabulary, and they are used by both ordinary people and scientists without causing any misunderstanding. For example, the phrase *body heat* is understood to mean the thermal energy content of a body. Likewise, *heat flow* is understood to mean the transfer of thermal energy, not the flow of a fluidlike substance. Also, the transfer of heat into a system is frequently referred to as *heat addition* and the transfer of heat out of a system as *heat rejection.* Thus we will refer to the thermal energy as *heat* and the transfer of thermal energy as *heat flow* or *heat transfer.*

When the *rate* of heat transfer $\dot{Q}$ is available, the *total amount* of heat transfer $Q$ during a time interval $\Delta t$ can be determined from

$$Q = \int_0^{\Delta t} \dot{Q}\, dt \quad \text{(kJ)} \tag{8-1}$$

provided that the variation of $\dot{Q}$ with time is known. For the special case of $\dot{Q}$ = *constant,* Eq. 8-1 reduces to

$$Q = \dot{Q}\, \Delta t \quad \text{(kJ)} \tag{8-2}$$

The rate of heat transfer per unit *surface area* is called *heat flux,* and the average heat flux on a surface is expressed as (Fig. 8-3)

$$\dot{q} = \frac{\dot{Q}}{A} \quad (\text{W/m}^2) \tag{8-3}$$

Note that the heat flux may vary with time as well as position on the heat transfer surface.

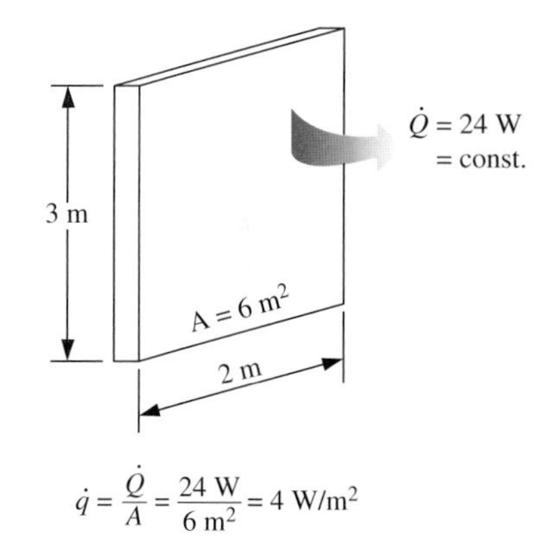


**FIGURE 8-3**
Heat flux is heat transfer *per unit time* and *per unit area,* and it is equal to $\dot{q} = \dot{Q}/A$ when $\dot{Q}$ is constant over the surface area $A$.

## 8-2 ■ AN OVERVIEW OF HEAT TRANSFER MECHANISMS

In Chap. 3 we defined *heat transfer* as *the form of energy transferred from one system to another as a result of temperature difference between the two systems.* We also mentioned that heat can be transferred in three different ways: *conduction, convection,* and *radiation.* All modes of heat transfer require the existence of a temperature difference, and all modes of heat transfer are from the high-temperature medium to a lower-temperature one. Below we give a brief description of each mode to refresh the memory of the reader.

**Conduction** is the transfer of energy from the more energetic particles of a substance to the adjacent less energetic ones as a result of interactions between the particles. Conduction can take place in solids, liquids, or gases. In *gases* and *liquids,* conduction is due to the *collisions* and *diffusion* of the molecules during their random motion. In *solids,* it is due to the combination of *vibrations* of the molecules in a lattice and the energy transport by *free electrons.* A cold canned drink in a warm room, for example, eventually warms up to the room temperature as a result of heat transfer from the room through the aluminum can by conduction.

The *rate* of heat conduction through a medium depends on the *geometry* of the medium, its *thickness,* and the *material* of the medium, as well as the *temperature difference* across the medium. We know that wrapping a hot water tank with glass wool (an insulating material) reduces the rate of heat loss from the tank. The thicker the insulation, the smaller the heat loss. We also know that a hot water tank will lose heat at a higher rate when the temperature of the room which houses the tank is lowered. Further, the larger the tank, the larger the surface area and thus the rate of heat loss.

Consider steady heat conduction through a large plane wall of thickness $\Delta x = L$ and surface area $A$, as shown in Fig. 8-4. The temperature difference across the wall is $\Delta T = T_2 - T_1$. Experiments have shown that the rate of heat transfer $\dot{Q}$ through the wall is *doubled* when the temperature difference $\Delta T$ across the wall or the area $A$ normal to the direction of heat transfer is doubled, but is *halved* when the wall thickness $L$ is doubled. Thus we conclude that *the rate of heat conduction through a layer is proportional to the temperature difference across the layer and the heat transfer area, but is inversely proportional to the thickness of the layer.* That is,

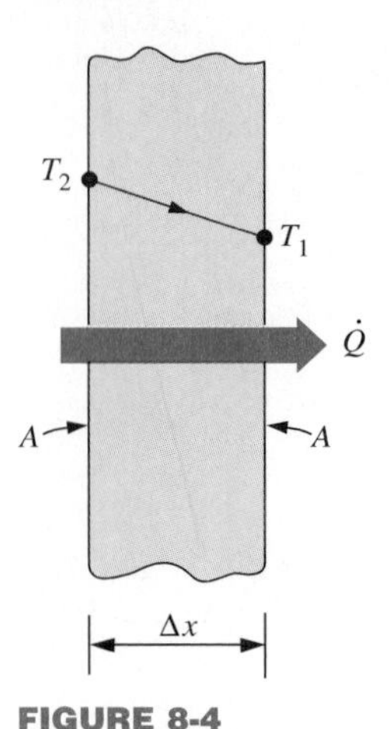


**FIGURE 8-4**
Heat conduction through a large plane wall of thickness $\Delta x$ and area $A$.

$$\text{rate of heat conduction} \propto \frac{\text{(surface area)(temperature difference)}}{\text{thickness}}$$

or

$$\dot{Q}_{\text{cond}} = kA\frac{\Delta T}{\Delta x} \quad \text{(W)} \tag{8-4}$$

where the constant of proportionality $k$ is the **thermal conductivity** of the material, which is a *measure of the ability of a material to conduct heat* (Fig. 8-5). In the limiting case of $\Delta x \to 0$, the equation above reduces to the differential form

$$\dot{Q}_{\text{cond}} = -kA\frac{dT}{dx} \quad \text{(W)} \tag{8-5}$$

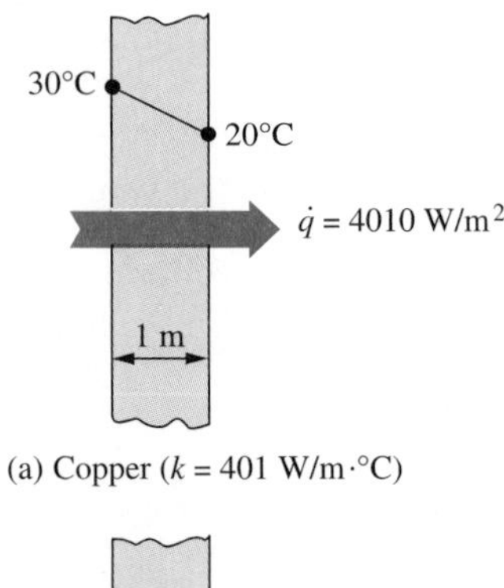


(a) Copper ($k$ = 401 W/m·°C)

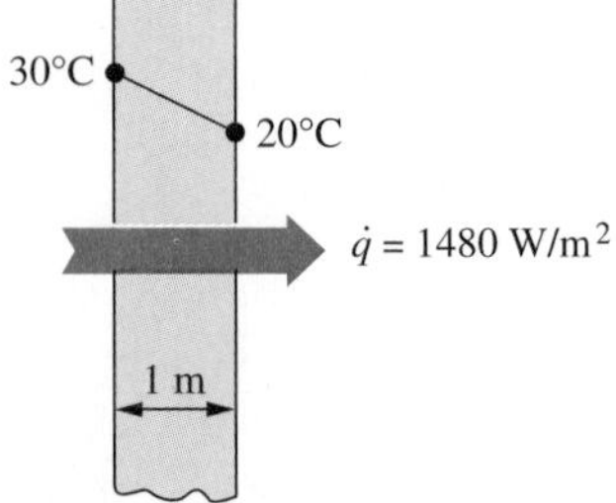


(b) Silicon ($k$ = 148 W/m·°C)

**FIGURE 8-5**
The rate of heat conduction through a solid is directly proportional to its thermal conductivity.

which is called **Fourier's law of heat conduction** after J. Fourier, who expressed it first in his heat transfer text in 1822. Here $dT/dx$ is the *temperature gradient,* which is the slope of the temperature curve on a $T$-$x$ diagram (the rate of change of $T$ with $x$), at location $x$. The relation above indicates that the rate of heat conduction in a direction is proportional to the temperature gradient in that direction. Heat is conducted in the direction of decreasing temperature, and the temperature gradient becomes negative when temperature decreases with increasing $x$. Therefore, a *negative sign* is added to Eq. 8-5 to make heat transfer in the positive $x$ direction a positive quantity.

The heat transfer surface area $A$ is always *normal* to the direction of heat transfer. For heat loss through a 5-m-long, 3-m-high, and 25-cm-thick wall, for example, the heat transfer area is $A = 15\text{ m}^2$. Note that the thickness of the wall has no effect on $A$ (Fig. 8-6).

**Convection** is the mode of energy transfer between a solid surface and the adjacent liquid or gas that is in motion, and it involves the combined effects of *conduction* and *fluid motion.* The faster the fluid motion, the greater the convection heat transfer. In the absence of any bulk fluid motion, heat transfer between a solid surface and the adjacent fluid is by pure conduction. The presence of bulk motion of the fluid enhances the heat transfer between the solid surface and the fluid, but it also complicates the determination of heat transfer rates. Convection is called *forced convection* if the fluid is forced to flow over the surface by external means such as a fan, pump, or the wind. In contrast, convection is called *natural* (or *free*) *convection* if the fluid motion is caused by buoyancy forces induced by density differences due to the variation of temperature in the fluid.

Despite the complexity of convection, the rate of *convection heat transfer* is observed to be proportional to the temperature difference, and is conveniently expressed by **Newton's law of cooling** as

$$\dot{Q}_{\text{convection}} = hA(T_s - T_\infty) \quad \text{(W)} \tag{8-6}$$

where $h$ is the *convection heat transfer coefficient* in W/(m$^2 \cdot$ °C), $A$ is the *surface area* through which convection heat transfer takes place, $T_s$ is the *surface temperature,* and $T_\infty$ is the *temperature of the fluid* sufficiently far from the surface. Not that at the surface, the fluid temperature equals the surface temperature of the solid.

The convection heat transfer coefficient $h$ is not a property of the fluid. It is an experimentally determined parameter whose value depends on all the variables that influence convection, such as the surface geometry, the nature of fluid motion, the properties of the fluid, and the bulk fluid velocity. Typical values of $h$ are given in Table 8-1.

**Radiation** is the energy emitted by matter in the form of *electromagnetic waves* (or *photons*) as a result of the changes in the electronic configurations of the atoms or molecules. Unlike conduction and convection, the transfer of energy by radiation does not require the presence of an intervening medium. In fact, energy transfer by radiation is fastest (at the speed of light), and it suffers no attenuation in a vacuum. This is exactly how the energy of the sun reaches the earth.

In heat transfer studies we are intesested in *thermal radiation,* which is the form of radiation emitted by bodies because of their temperature. All bodies at a temperature above absolute zero emit thermal radiation. The maximum rate of radiation which can be emitted from a surface at an absolute temperature $T_s$ is given by the *Stefan–Boltzman law* as $\dot{q}_{\text{max}} = \sigma T_s^4$, where $\sigma = 5.67 \times 10^{-8}$ W/(m$^2 \cdot$ K$^4$) is the Stefan–Boltzmann constant. The idealized surface that emits radiation at this maximum rate is called a *blackbody* (Fig. 8-7). The radiation emitted by all real surfaces is less than the radiation emitted by a blackbody at the same temperature, and is expressed as $\dot{q} = \varepsilon\sigma T_s^4$ where $\varepsilon$ is the *emissivity* of the surface. The property emissivity, whose value is in the range $0 \leqslant \varepsilon \leqslant 1$ is a measure of how closely a surface approximates a blackbody.

The difference between the rates of radiation emitted by the surface and the radiation absorbed is the net radiation heat transfer. If the rate of radiation absorption is greater than the rate of radiation emission, the surface is said to be *gaining* energy by radiation. Otherwise, the surface is said to be *losing* energy by radiation. In general, the determination of

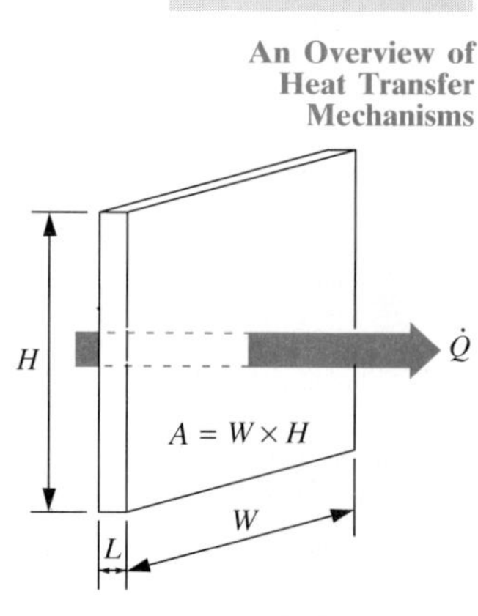


**FIGURE 8-6**
In heat conduction analysis, $A$ represents the area *normal* to the direction of heat transfer.

**TABLE 8-1**

**Typical values of convection heat transfer coefficient**

| Type of convection | $h$, W/(m$^2 \cdot$ °C) |
|---|---|
| Free convection of gases | 2–25 |
| Free convection of liquids | 10–1000 |
| Forced convection of gases | 25–250 |
| Forced convection of liquids | 50–20,000 |
| Boiling and condensation | 2500–100,000 |

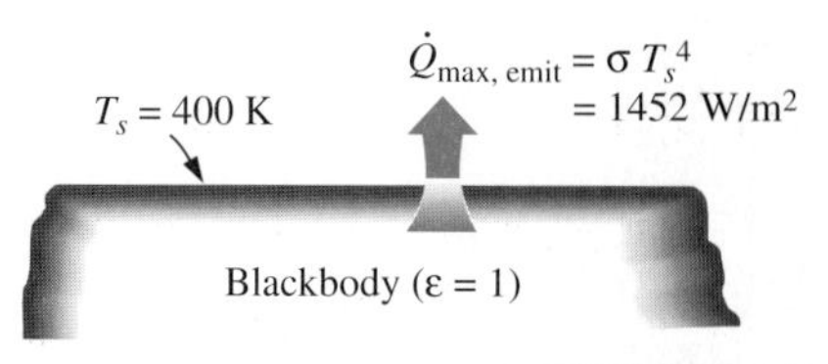


**FIGURE 8-7**
Blackbody radiation represents the *maximum amount of radiation that can be emitted from a surface at a specified temperature.*

the net rate of heat transfer by radiation between two surfaces is a complicated matter, since it depends on the properties of the surfaces, their orientation relative to each other, and the interaction of the medium between the surfaces with radiation.

When a surface of emissivity $\varepsilon$ and surface area $A_s$ at an *absolute temperature* $T_s$ is *completely enclosed* by a much larger (or black) surface at absolute temperature $T_{surr}$ separated by a gas (such as air) that does not intervene with radiation, the net rate of radiation heat transfer between these two surfaces is given by (Fig. 8-8)

$$\dot{Q}_{rad} = \varepsilon\sigma A(T_s^4 - T_{surr}^4) \quad \text{(W)} \tag{8-7}$$

where $\sigma = 5.67 \times 10^{-8}\ \text{W/(m}^2 \cdot \text{K}^4) = 0.1714 \times 10^{-8}\ \text{Btu/(h} \cdot \text{ft}^2 \cdot \text{R}^4)$.

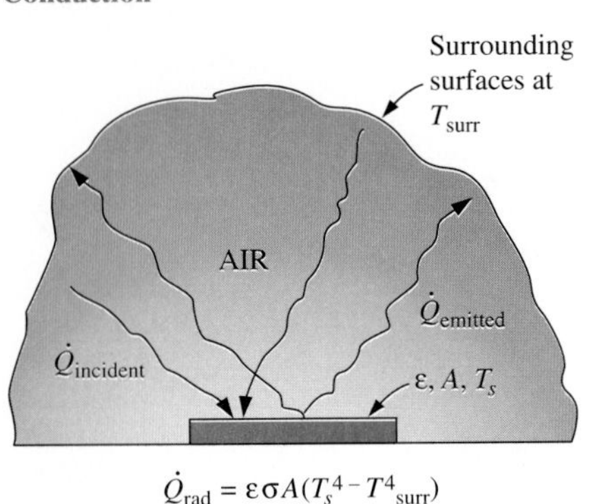


**FIGURE 8-8**
Radiation heat transfer between a surface and the surfaces surrounding it.

Radiation heat transfer to or from a surface surrounded by a gas such as air occurs *parallel* to conduction (or convention if there is bulk gas motion) between the surface and the gas. Thus, the total heat transfer is determined by *adding* the contributions of both heat transfer mechanisms. Radiation is usually significant relative to conduction or natural convection, but negligible relative to forced convection. Thus radiation in forced convection applications is normally disregarded, especially when the surfaces involved have high emissivities and low or moderate temperatures.

There are three mechanisms of heat transfer, but not all three can exist simultaneously in a medium. For example, heat transfer is only by conduction in *opaque solids,* but by conduction and radiation in *semitransparent solids.* Thus a solid may involve conduction and radiation, but not convection. Heat transfer is by conduction and possibly by radiation in a *still fluid* (no bulk fluid motion), and by convection and radiation in a *flowing fluid.* In the absence of radiation, heat transfer through a fluid is either by conduction or convection, depending on the presence of any bulk fluid motion. Convection can be viewed as combined conduction and fluid motion, and conduction in a fluid can be viewed as a special case of convection in the absence of any fluid motion (Fig. 8-9).

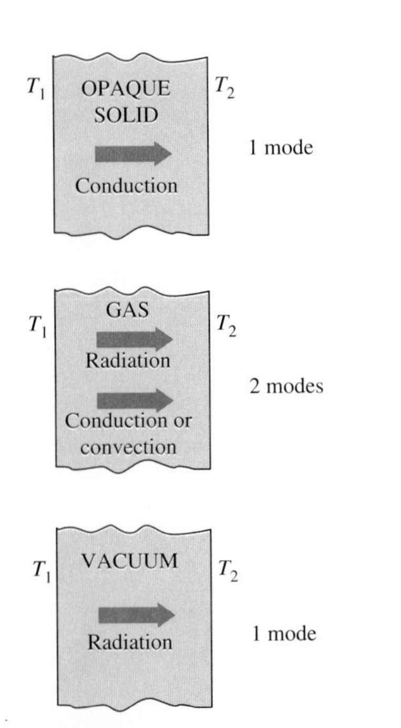


**FIGURE 8-9**
Although there are three mechanisms of heat transfer, a medium may involve only two of them simultaneously.

Thus when we deal with heat transfer through a *fluid,* we have either *conduction* or *convection,* but not both. Also, gases are practically transparent to radiation, except that some gases are known to absorb radiation strongly at certain wavelengths. Ozone, for example, strongly absorbs ultraviolet radiation. But, in most cases, a gas between two solid surfaces does not interfere with radiation, and acts effectively as a vacuum. Liquids, on the other hand, are usually strong absorbers of radiation.

Finally, heat transfer through a *vacuum* is by radiation only, since conduction or convection requires the presence of a material medium.

**EXAMPLE 8-1**

A 10-cm-diameter copper ball shown in Fig. 8-10 is observed to cool from 150°C to an average temperature of 100°C in 30 min in atmospheric air at 25°C. Determine (*a*) the total amount of heat transferred from the copper ball, (*b*) the

average rate of heat transfer from the ball, (*c*) the average heat flux, and (*d*) the convection heat transfer coefficient at the beginning of the cooling process.

**Solution** The thermophysical properties of most materials vary with temperature. Therefore, properties at the average temperature are usually used to account for this variation. The average temperature of the copper ball during this cooling process is

$$T_{ave} = \frac{T_1 + T_2}{2} = \frac{150 + 100}{2} = 125°C = 398 \text{ K}$$

The specific heat of copper at 400 K is listed in Table A-14 to be $C_p = 0.393$ kJ/(kg · K). The density of copper is given at 20°C only, and is listed to be $\rho = 8950 \text{ kg/m}^3$. Thus it is reasonable to use these property values for copper and to treat them as constants.

(*a*) Noting that heat transfer is the only energy interaction involved here, the conservation of energy principle requires that the amount of heat transfer from the copper ball must be equal to the decrease in the energy content of the ball,

$$Q = mC_p(T_2 - T_1)$$

where $$m = \rho V = \frac{\pi}{6}\rho D^3 = \frac{\pi}{6}(8950 \text{ kg/m}^3)(0.1 \text{ m})^3 = 4.69 \text{ kg}$$

Substituting,

$$Q = (4.69 \text{ kg})[0.393 \text{ kJ/(kg} \cdot \text{K)}](100 - 150)°C = -92.2 \text{ kJ}$$

The negative sign indicates that the copper ball is losing heat. In heat transfer analysis, we often take the heat transfer to be a positive quantity for convenience, but indicate the direction of heat transfer by clearly stating it. In this case, we can say $Q = 92.2$ kJ from the copper ball.

(*b*) The *rate* of heat transfer normally changes during a process with time in the same way that the velocity of a car changes with time in normal driving. However, we can determine the *average* rate of heat transfer by dividing the amount of heat transfer by the time interval, in the same way that we can determine the average velocity of a car by dividing the distance traveled by the driving time. Therefore,

$$\dot{Q}_{ave} = \frac{Q}{\Delta t} = \frac{92.2 \text{ kJ}}{1800 \text{ s}} = 0.0512 \text{ kJ/s} = 51.2 \text{ W}$$

(*c*) *Heat flux* is defined as the heat transfer per unit time per unit surface area, or the rate of heat transfer per unit surface area. Therefore, the average heat flux in this case is

$$\dot{q}_{ave} = \frac{\dot{Q}_{ave}}{A} = \frac{51.2 \text{ W}}{0.0314 \text{ m}^2} = 1631 \text{ W/m}^2$$

since $$A = \pi D^2 = \pi(0.1)^2 = 0.0314 \text{ m}^2$$

We recognize that heat flux may vary with location on the surface. The value calculated above is the *average* heat flux over the entire surface of the ball.

(*d*) Newton's law of cooling for convection heat transfer is expressed as

$$\dot{Q} = hA(T_s - T_\infty)$$

Disregarding any heat transfer by radiation, and thus assuming all the heat

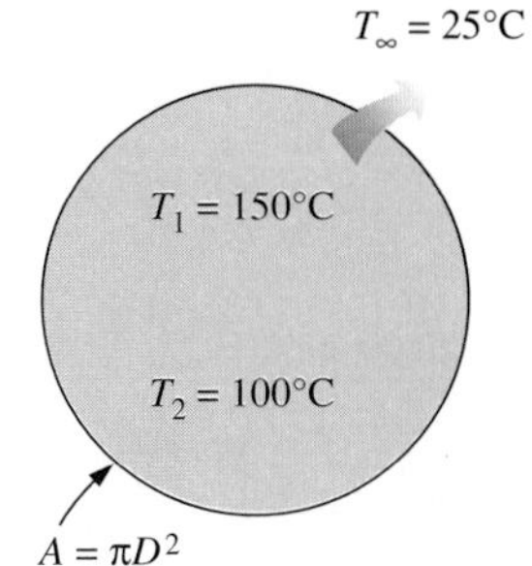


**FIGURE 8-10**
Schematic for Example 8-1.

loss from the ball to occur by convection, the convection heat transfer coefficient at the beginning of the cooling process is determined to be

$$h = \frac{\dot{Q}}{A(T_1 - T_\infty)} = \frac{51.2\,\text{W}}{(0.0314\,\text{m}^2)(150\text{–}25)°\text{C}} = 13.0\,\text{W/(m}^2\cdot°\text{C)}$$

The convection heat transfer coefficient at the *end* of the cooling process can be determined by using 100°C for the surface temperature instead of 150°C to be 21.7 W/(m$^2$ · °C). The accuracy of these values depends on the accuracy of the constant heat flux assumption.

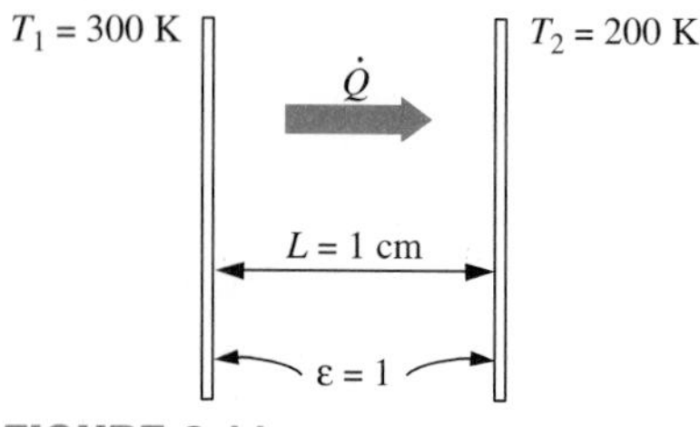


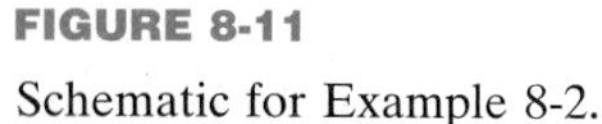
**FIGURE 8-11**
Schematic for Example 8-2.

### EXAMPLE 8-2

Consider steady heat transfer between two large parallel plates at constant temperatures of $T_1$ = 300 K and $T_2$ = 200 K that are $L$ = 1 cm apart, as shown in Fig. 8-11. Assuming the surfaces to be black (emissivity $\varepsilon$ = 1), determine the rate of heat transfer between the plates per unit surface area assuming the gap between the plates is (*a*) filled with atmospheric air, (*b*) evacuated, (*c*) filled with urethane insulation, and (*d*) filled with superinsulation that has an apparent thermal conductivity of 0.00002 W/(m · °C).

**Solution** (*a*) The thermal conductivity of air at the average temperature of 250 K is 0.0223 W/(m · °C). Disregarding any natural convection currents that may occur in the air, the rates of conduction and radiation heat transfer between the plates through the air layer are

$$\dot{Q}_{\text{cond}} = kA\frac{T_1 - T_2}{L} = [0.0223\,\text{W/(m}\cdot°\text{C)}](1\,\text{m}^2)\frac{(300 - 200)°\text{C}}{0.01\,\text{m}} = 223\,\text{W}$$

and
$$\dot{Q}_{\text{rad}} = \varepsilon\sigma A(T_1^4 - T_2^4)$$
$$= (1)[5.67 \times 10^{-8}\,\text{W/(m}^2\cdot\text{K}^4)](1\,\text{m}^2)[(300\,\text{K})^4 - (200\,\text{K})^4] = 386\,\text{W}$$

Therefore,
$$\dot{Q}_{\text{total}} = \dot{Q}_{\text{cond}} + \dot{Q}_{\text{rad}} = 223 + 368 = 591\,\text{W}$$

The heat transfer rate in reality will be higher because of the natural convection currents that are likely to occur in the air space between the plates.

(*b*) When the air space between the plates is evacuated, there will be no conduction or convection, and the only heat transfer between the plates will be by radiation. Therefore,

$$\dot{Q}_{\text{total}} = \dot{Q}_{\text{rad}} = 368\,\text{W}$$

(*c*) An opaque solid material placed between two plates blocks direct radiation heat transfer between the plates. Also, the thermal conductivity of an insulating material accounts for the radiation heat transfer that may be occurring through the voids in the insulating material. The thermal conductivity of urethane insulation is listed in the Table A-16 to be 0.026 W/(m · °C). Then the rate of heat transfer through the urethane insulation becomes

$$\dot{Q}_{\text{total}} = \dot{Q}_{\text{cond}} = kA\frac{T_1 - T_2}{L}[0.026\,\text{W/(m}\cdot°\text{C)}](1\,\text{m}^2)\frac{(300 - 200)°\text{C}}{0.01\,\text{m}} = 260\,\text{W}$$

Note that heat transfer through the urethane material is less than the heat transfer through the air determined in (*a*), although the thermal conductivity of

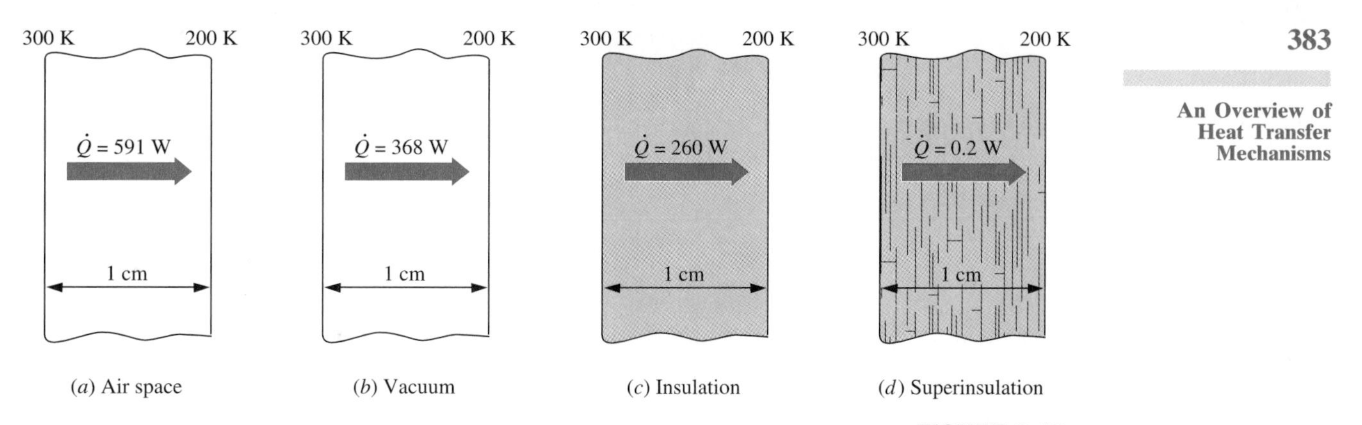


**FIGURE 8-12**

Different ways of reducing heat transfer between two isothermal plates, and their effectiveness.

the insulation is much higher than that of air. This is because the insulation blocks the radiation whereas air transmits it.

(*d*) The layers of the superinsulation prevents any direct radiation heat transfer between the plates. However, radiation heat transfer between the sheets of superinsulation does occur, and the apparent thermal conductivity of the superinsulation accounts for this effect. Therefore,

$$\dot{Q}_{\text{total}} = kA\frac{T_1 - T_2}{L} = [0.00002\ \text{W/(m}\cdot\text{°C)}](1\ \text{m}^2)\frac{(300 - 200)\text{°C}}{0.01\ \text{m}} = 0.2\ \text{W}$$

which is 1/1840 of the heat transfer through the vacuum. The results of this example are summarized in Fig. 8-12 to put them into perspective.

This example demonstrates the effectiveness of superinsulations, which are discussed in the next section, and explains why they are the insulation of choice in critical applications despite their high cost.

**EXAMPLE 8-3**

The fast and efficient cooking of microwave ovens made them one of the essential appliances in modern kitchens (Fig. 8-13). Discuss the heat transfer mechanisms associated with the cooking of a chicken in microwave and conventional ovens, and explain why cooking in a microwave oven is more efficient.

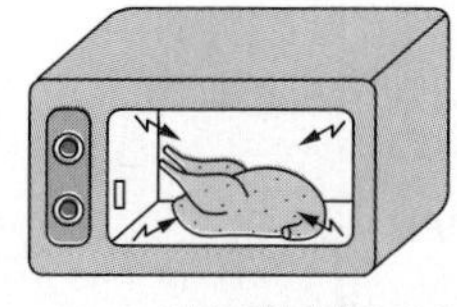

**FIGURE 8-13**

A chicken being cooked in a microwave oven (Example 8-3).

**Solution** Food is cooked in a microwave oven by absorbing the electromagnetic radiation energy generated by the microwave tube called the magnetron. The radiation emitted by the magnetron is not thermal radiation, since its emission is not due to the temperature of the magnetron; rather, it is due to the conversion of electrical energy into electromagnetic radiation at a specified wavelength. The wavelength of the microwave radiation is such that it is *reflected* by metal surfaces, *transmitted* by the cookware made of glass, ceramic, or plastic, and *absorbed* and converted to internal energy by food (especially by the water, sugar, and fat) molecules.

In a microwave oven, the *radiation* that strikes the chicken is absorbed by the skin of the chicken and the outer layers. As a result, the temperature of the chicken at and near the skin rises. Heat is then *conducted* towards the inner parts of the chicken from its outer parts. Of course, some of the heat absorbed by the outer surface of the chicken is lost to the air in the oven by *convection.*

In a conventional oven, the air in the oven is first heated to the desired temperature by the electric or gas heating element. This preheating may take several minutes. The heat is then transferred from the air to the skin of the chicken by *natural convection* in most ovens or by *forced convection* in the newer convection ovens that utilize an air mover. The air motion in convection ovens increases the convection heat transfer coefficient and thus decreases the cooking time. Heat is then *conducted* towards the inner parts of the chicken from its outer parts as in microwave ovens.

Microwave ovens replace the slow convection heat transfer process in conventional ovens by the instantaneous radiation heat transfer. As a result, microwave ovens transfer energy to the food at full capacity the moment they are turned on, and thus they cook faster while consuming less energy.

## 8-3 THERMAL CONDUCTIVITY

We have seen in Chap. 3 that different materials store heat differently, and we have defined the property specific heat $C_p$ as a measure of a material's ability to store heat. For example, $C_p = 4.18\ \text{kJ/(kg} \cdot {}^\circ\text{C)}$ for water and $C_p = 0.45\ \text{kJ/(kg} \cdot {}^\circ\text{C)}$ for iron at room temperature, which indicates that water can store almost 10 times the energy that iron can per unit mass. Likewise, the thermal conductivity $k$ is a measure of a material's ability to conduct heat. For example, $k = 0.608\ \text{W/(m} \cdot {}^\circ\text{C)}$ for water and $k = 60\ \text{W/(m} \cdot {}^\circ\text{C)}$ for iron at room temperature, which indicates that iron conducts heat almost 100 times faster than water can. Thus we say that water is a poor heat conductor relative to iron, although water is an excellent medium to store heat.

**TABLE 8-2**

**The thermal conductivities of some materials at room temperature**

| Material | $k$, W/(m · °C) |
|---|---|
| Diamond | 2300 |
| Silver | 429 |
| Copper | 401 |
| Gold | 317 |
| Aluminum | 237 |
| Iron | 80.2 |
| Mercury (l) | 8.54 |
| Glass | 0.78 |
| Brick | 0.72 |
| Water (l) | 0.613 |
| Human skin | 0.37 |
| Wood (oak) | 0.17 |
| Helium (g) | 0.152 |
| Soft rubber | 0.13 |
| Refrigerant-12 | 0.072 |
| Glass fiber | 0.043 |
| Air (g) | 0.026 |
| Urethane, rigid foam | 0.026 |

Equation 8-4 for the rate of conduction heat transfer under steady conditions can also be viewed as the defining equation for thermal conductivity. Thus the **thermal conductivity** of a material can be defined as *the rate of heat transfer through a unit thickness of the material per unit area per unit temperature difference.* The thermal conductivity of a material is a measure of how fast heat will flow in that material. A large value of a thermal conductivity indicates that the material is a good heat conductor, and a low value indicates that the material is a poor heat conductor or *insulator.* The thermal conductivities of some common materials at room temperature are given in Table 8-2. The thermal conductivity of pure copper at room temperature is $k = 401\ \text{W/(m} \cdot {}^\circ\text{C)}$, which indicates that a 1-m-thick copper wall will conduct heat at a rate of 401 W per $m^2$ area per °C temperature difference across the wall. Note that materials such as copper and silver that are good electric conductors are also good heat conductors, and have high values of thermal conductivity. Materials such as rubber, wood, and styrofoam are poor conductors of heat, and have low conductivity values.

Equation 8-4 can also be used to determine the thermal conductivity of a material. A layer of a material of known thickness and area can be heated from one side by an electric resistance heater of known output. If the outer surface of the heater is well insulated, all the heat generated by the resistance heater will be transferred through the material whose conductivity is to be determined. Then measuring the two surface temperatures of the material when steady heat transfer is reached and substituting them into Eq. 8-4 together with other known quantities give the thermal conductivity (Fig. 8-14).

The thermal conductivities of materials vary over a wide range, as shown in Fig. 8-15. The thermal conductivities of gases such as air vary by a factor of $10^4$ from those of pure metals such as copper. Note that the metals have the highest thermal conductivities, and gases and insulating materials the lowest.

Temperature is a measure of the kinetic energies of the particles such as the molecules or atoms of a substance. In a liquid or gas, the kinetic energy of the molecules is due to the random motion of the molecules as well as the vibrational and rotational motions. When two molecules possessing different kinetic energies collide, part of the kinetic energy of

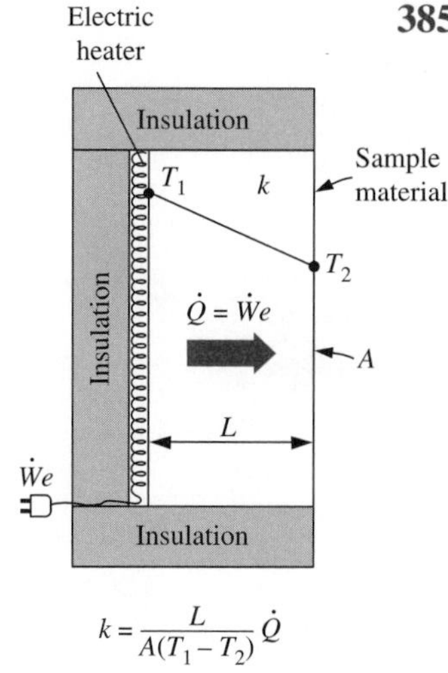


**FIGURE 8-14**
A simple experimental setup to determine the thermal conductivity of a material.

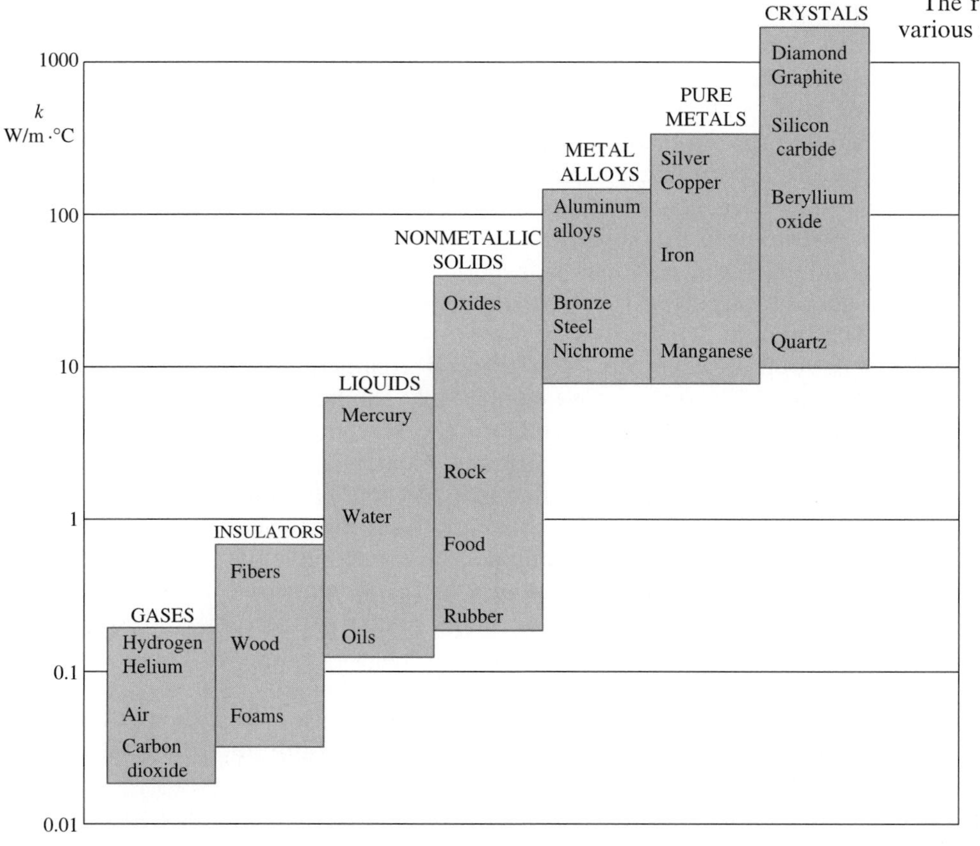


**FIGURE 8-15**
The range of thermal conductivity of various materials at room temperature.

the more energetic (higher-temperature) molecule is transferred to the less energetic (lower-temperature) molecule, much the same way when two elastic balls of the same mass at different velocities collide, part of the kinetic energy of the faster ball is transferred to the slower one. The higher the temperature, the faster the molecules move and the higher the number of such collisions, and the better the heat transfer.

The *kinetic theory* of gases predicts and the experiments confirm that the thermal conductivity of gases is proportional to the *square root of the absolute temperature T,* and inversely proportional to the *square root of the molar mass M.* Therefore, the thermal conductivity of a gas increases with increasing temperature and decreasing molar mass. So it is not surprising that the thermal conductivity of helium ($M = 4$) is much higher than those of air ($M = 29$) and argon ($M = 40$).

The thermal conductivities of *gases* at 1 atm pressure are listed in Table A-19. However, they can be used at pressures other than 1 atm, since the thermal conductivity of gases is *independent of pressure* at most pressures encountered in practice.

The mechanism of heat conduction in a *liquid* is complicated by the fact that the molecules are more closely spaced, and they exert a stronger force field. The thermal conductivities of liquids usually lie between those of solids and gases. The thermal conductivity of a substance is normally highest in the solid phase and lowest in the gas phase. Unlike gases, the thermal conductivities of most liquids decrease with increasing temperature, with water being a notable exception. Like gases, the conductivity of liquids decreases with increasing molar mass. Liquid metals such as mercury and sodium have high thermal conductivites, and are very suitable for use in applications where a high heat transfer rate to a liquid is desired, as in nuclear power plants.

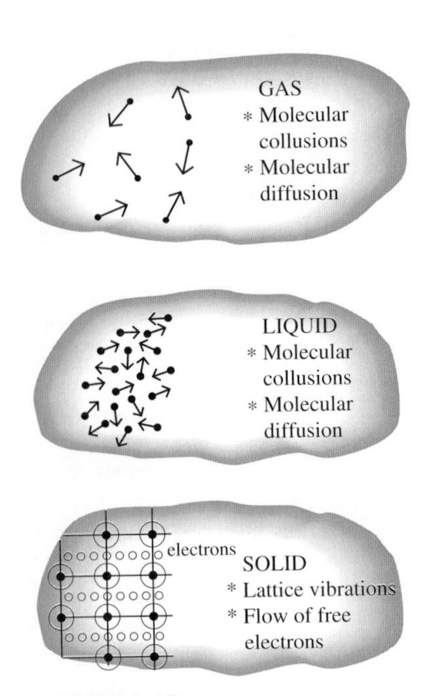


**FIGURE 8-16**

The mechanisms of heat conduction in different phases of a substance.

In *solids,* heat conduction is due to two effects: the *lattice vibrational waves* induced by the vibrational motions of the molecules positioned at relatively fixed positions in a periodic manner called a lattice, and the energy transported via the *free flow of electrons* in the solid (Fig. 8-16). The thermal conductivity of a solid is obtained by adding the lattice and electronic components. The thermal conductivities of pure metals are primarily due to the lattice component. The lattice component of thermal conductivity strongly depends on the way the molecules are arranged. Diamond, which is a highly ordered crystalline solid, for example, has the highest known thermal conductivity at room temperature.

**TABLE 8-3**

**The thermal conductivity of an alloy is usually much lower than the thermal conductivity of either metal of which it is composed**

| Pure metal or alloy | $k$, W/(m · °C), at 300 K |
|---|---|
| Copper | 401 |
| Nickel | 91 |
| *Constantan* (55% Cu, 45% Ni) | 23 |
| Copper | 401 |
| Aluminum | 237 |
| *commercial bronze* (90% Cu, 10% Al) | 52 |

Unlike metals, which are good electrical and heat conductors, *crystalline solids* such as diamond and semiconductors such as silicon are good heat conductors but poor electrical conductors. As a result, such materials find widespread use in the electronics industry. Despite their higher price, diamond heat sinks are used in the cooling of sensitive electronic components because of the excellent thermal conductivity of diamond. Silicon oils and gaskets are commonly used in the packaging of electronic components because they provide both good thermal contact and good electrical insulation.

Pure metals have high thermal conductivities, and one would think that *metal alloys* should also have high conductivities. One would expect an alloy made of two metals of thermal conductivities $k_1$ and $k_2$ to have a conductivity $k$ between $k_1$ and $k_2$. But this turns out not to be the case. The thermal conductivity of an alloy of two metals is usually much lower than that of either metal, as shown in Table 8-3. Even small amounts in a pure metal of "foreign" molecules that are good conductors themselves seriously disrupt the flow of heat in that metal. For example, the thermal conductivity of steel containing just 1% of chrome is 62 W/(m · °C), while the thermal conductivities of iron and chromium are 83 and 95 W/(m · °C), respectively.

The thermal conductivities of materials vary with temperature (Table 8-4). The variation of thermal conductivity over certain temperature ranges is negligible for some materials, but significant for others, as shown in Fig. 8-17. The thermal conductivities of certain solids exhibit dramatic increases at temperatures near absolute zero, when these solids become *superconductors.* For example, the conductivity of copper reaches a maximum value of about 20 000 W/(m · °C) at 20 K, which is about 50 times the conductivity at toom temperature. The thermal conductivities of various materials are given in Tables A-14 to A-20.

The temperature dependence of thermal conductivity causes con

**TABLE 8-4**

**Thermal conductivities of materials vary with temperature**

| | $k$, W/(m · °C) | |
|---|---|---|
| $T$, K | **Copper** | **Aluminum** |
| 100 | 482 | 302 |
| 200 | 413 | 237 |
| 300 | 401 | 237 |
| 400 | 393 | 240 |
| 600 | 379 | 231 |
| 800 | 366 | 218 |

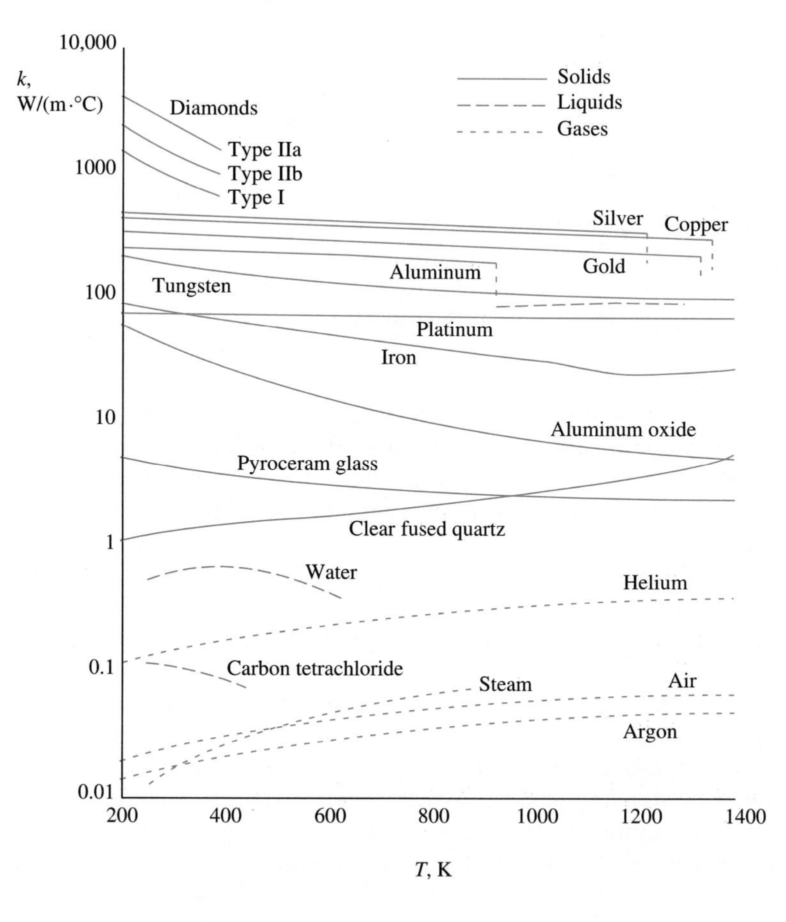


**FIGURE 8-17**
The variation of the thermal conductivity of various solids, liquids, and gases with temperature.

siderable complexity in conduction analysis. Therefore, it is common practice to evaluate the thermal conductivity $k$ at the average temperature and treat it as a constant in calculations.

In heat transfer analysis, a material is normally assumed to be *isotropic*; that is, to have uniform properties in all directions. This assumption is realistic for most materials, except those that exhibit different structural characteristics in different directions, such as laminated composite materials and wood. The thermal conductivity of wood across the grain, for example, is different than that parallel to the grain.

## Insulation Materials

Thermal insulations are materials used primarily to provide resistance to heat flow. You are probably familiar with the several kinds of insulating materials available in the market. Most insulations are heterogeneous materials made of materials of low thermal conductivity, and they involve air pockets. This is not surprising, since air has one of the lowest thermal conductivities, and it is freely available. The styrofoam that is commonly used as a packaging material for TVs, VCRs, computers, and just about anything, because of its light weight is also an excellent insulator.

Heat flow between two mediums at different temperatures can be slowed down by putting "barriers" on the path of heat flow. Thermal insulations serve as such barriers, and they play a major role in the design and manufacture of all energy-efficient devices or systems, and they are usually the cornerstone of all energy conservation projects. A 1991 Drexel University study of the energy-intensive U.S. industries revealed that insulation saves the U.S. industry nearly 2 billion barrels of oil per year valued at $60 billion a year in energy costs, and more can be saved by practicing better insulation techniques and retrofitting the older industrial facilities. Insulation *pays for itself* from the energy it saves. Insulating properly requires a one-time capital investment, but its effects are dramatic and long term. The payback period of insulation is usually under two years. That is, the money insulation saves during the first two years is usually greater than its initial material and installation cost. On a broader perspective, insulation also helps the environment and fights air pollution and the greenhouse effect by reducing the amount of fuel burned and thus the amount of $CO_2$ and other gases released to the atmosphere.

Saving energy by insulation is not limited to hot surfaces. We can also save energy and money by insulating *cold surfaces* (surfaces whose temperature is below the ambient temperature) such as chilled water lines, cryogenic storage tanks, refrigerated trucks, and air conditioning ducts. The source of "coldness" is *refrigeration,* which requires energy input, usually electricity. In this case, heat is transferred from the surroundings to the cold surfaces, and the refrigeration unit must now work harder and longer to make up for this heat gain and thus it must consume more electrical energy. A cold canned drink can be kept cold much longer by wrapping it in a blanket. A refrigerator with well-insulated walls will consume much less electricity than a similar refrigerator with little or no insulation. Insulating a house well will result in reduced cooling load, and thus reduced electricity consumption for air-conditioning.

Thermal insulation in the form *mud, clay, straw, rags,* and *wood strips* was first used in the 18th Century on steam engines to keep workmen from being burned by hot surfaces. As a result, boiler room temperatures dropped and it was noticed that fuel consumption was also reduced. The realization of improved engine efficiency and energy savings prompted the search for materials with improved thermal efficiency. One of the first such material was *mineral wool* insulation, which, like many materials, was discovered by accident. At about 1840, an iron producer in Wales aimed a stream of high-pressure steam at the slag flowing from a blast furnace, and the manufactured mineral wool was born. In the early 1860s, this slag wool was a by-product of manufacturing cannons for the Civil War, and quickly found its way into many industrial uses. By 1880, builders began installing mineral wool in houses.

The energy crisis of 1970s had a tremendous impact on the public awareness of energy and limited energy reserves, and brought an emphasis on *energy conservation.* We have also seen the development of new and more effective insulation materials since then, and a considerable increase in the use of insulation. Thermal insulation is used in more places than you may be aware of. The walls of your house is probably filled with some kind of insulation, and the roof is likely to have a thick layer of insulation. The "thickness" of the walls of your refrigerator is due to the insulation layer sandwiched between two layers of sheet metal (Fig. 3-36). The walls of your range is also insulated to conserve energy, and your hot water tank contains less water than you think because of the 2- to 4-cm-thick insulation in the walls of the tank. Also, your hot water pipe may look much thicker than the cold water pipe because of insulation.

Conserving energy by reducing the rate of heat flow is the primary reason for insulating surfaces. Insulation materials which will perform satisfactorily in the temperature range of −268°C to 1000°C (−450°F to 1800°F) are widely available. Other reasons for insulating surfaces include personnel protection and comfort, maintaining process temperature, reducing temperature variation and fluctuations, condensation and corrosion prevention, fire protection, freezing protection, and reducing noise and vibration.

Most ordinary insulations are obtained by mixing fibers, powders or flakes of insulating material with air. The properties of such loose insulations depend on the volumetric fraction of air space. Heat transfer through such insulations is by conduction through the solid material, and conduction or convection through the air space as well as radiation. Such systems are characterized by an *apparent thermal conductivity* that accounts for all mechanisms of heat transfer (Fig. 8-18). Insulations with extremely low apparent thermal conductivities (about one-thousandth of that of air), called *superinsulations,* are obtained by using layers of highly

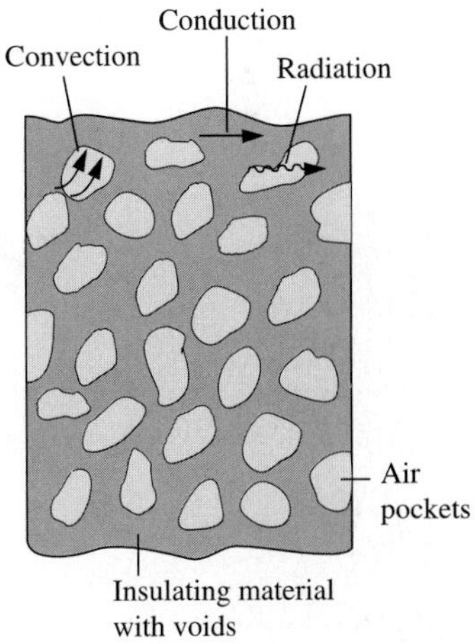

Conduction
Convection
Radiation
Air pockets
Insulating material with voids

**FIGURE 8-18**

The *apparent thermal conductivity* of an insulating material accounts for the conduction through the solid material, and conduction or convection through the air space as well as radiation.

reflective sheets separated by glass fibers in an evacuated space. Such insulations are used in space and cryogenic applications.

## Superinsulators

You may be tempted to think that the most effective way to reduce heat transfer is to use insulating materials that are known to have very low thermal conductivities, such as urethane or rigid foam ($k = 0{\cdot}026\ \mathrm{W/(m \cdot {}^\circ C)}$) or glass fiber ($k = 0{\cdot}035\ \mathrm{W/(m \cdot {}^\circ C)}$). After all, they are widely available, inexpensive, and easy to install. Looking at the thermal conductivities of materials, you may also notice that the thermal conductivity of air at room temperature is $0.026\ \mathrm{W/(m \cdot {}^\circ C)}$, which is lower than the conductivities of practically all of the ordinary insulating materials. Thus you may think that a layer of enclosed air space is as much effective as any of the common insulating materials of the same thickness. Of course, heat transfer through the air will probably be higher than what a pure conduction analysis alone would indicate, because of the natural convection currents that are likely to occur in the air layer. Besides, air is transparent to radiation, and thus heat will also be lost from the surface by radiation.

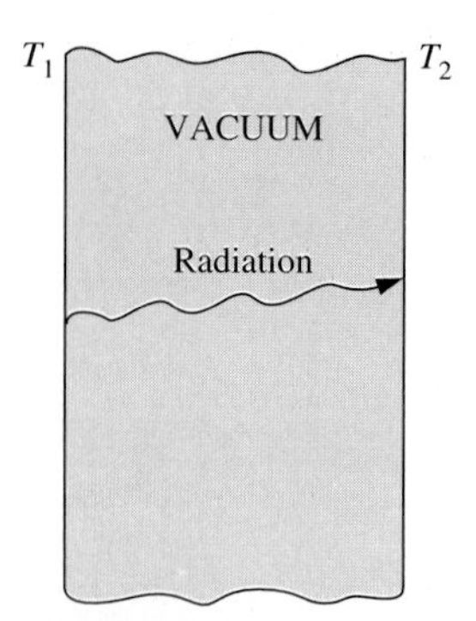


**FIGURE 8-19**

Evacuating the space between two surfaces completely eliminates heat transfer by conduction or convection, but leaves the door wide open for radiation.

The thermal conductivity of air is practically independent of pressure unless the pressure is extremely high or extremely low. Therefore, we can reduce the thermal conductivity of air and thus the conduction heat transfer through the air by evacuating the air space. In the limiting case of absolute vacuum, the thermal conductivity will be zero since there will be no particles in this case to "conduct" heat from one surface to the other, and thus the conduction heat transfer will be zero. Noting that the thermal conductivity cannot be negative, an absolute vacuum must be the ultimate insulator, right? Well, not quite.

The purpose of insulation is to reduce "total" heat transfer from a surface, not just conduction. A vacuum totally eliminates conduction, but offers zero resistance to radiation whose magnitude can be comparable to conduction or natural convection in gases (Fig. 8-19). Thus a vacuum is no more effective in reducing heat transfer than sealing off one of the lanes of a two-lane road is in reducing the flow of traffic in a one-way road.

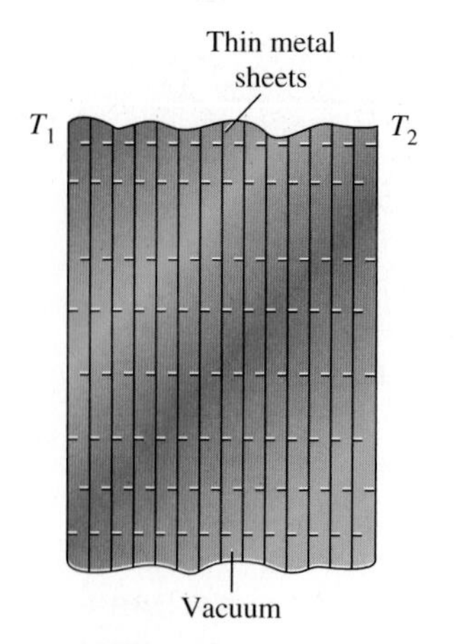


**FIGURE 8-20**

Superinsulators are built by closely packing layers of highly reflective thin metal sheets and evacuating the space between them.

Insulation against radiation heat transfer between two surfaces is achieved by placing "barriers" between the two surfaces that are highly reflective thin metal sheets. Radiation heat transfer between two surfaces is inversely proportional to the number of such sheets placed between the surfaces. Very effective insulations are obtained by using closely packed layers of highly reflective thin metal sheets such as aluminum foil (usually 25 sheets per cm) separated by fibers made of insulating material such as glass fiber (Fig. 8-20). Further, the space between the layers is evacuated to form a vacuum under 0.000001 atm pressure to minimize conduction or convection heat transfer through the air space between the layers. The result is an insulating material whose apparent thermal conductivity is below $2 \times 10^{-5}\ \mathrm{W/(m \cdot {}^\circ C)}$, which is one thousand times less than the conductivity of air or any common insulating material. These specially built insulators are called **superinsulators**, and they are commonly used in space applications and *cryogenics*, which is the branch of heat transfer dealing with temperatures below 200 K (−173°C) such as those encountered in the liquifaction, storage, and transportation of gases, with helium, hydrogen, nitrogen, and oxygen being the most common ones.

### The *R*-value of Insulations

The effectiveness of insulation materials in the construction industry is expressed in terms of their *R-value,* which is the *thermal resistance* of the material for a unit area. That is,

$$R_{\text{value}} = \frac{L}{k} \tag{8-8}$$

where $L$ is the thickness and $k$ is the thermal conductivity of the material. Note that doubling the thickness $L$ doubles the $R$-value of insulation.

In the United States, the $R$-value of insulation is expressed without any dimension, such as $R$-19 and $R$-30. These $R$-values are obtained by dividing the thickness of the material in *feet* by its thermal conductivity in the unit Btu/(h · ft · °F) so that the $R$-values actually have the unit h · ft² · °F/Btu. For example, the $R$-value of 6-inch-thick glass fiber insulation whose thermal conductivity is 0.025 Btu/(h · ft · °F) is (Fig. 8-21)

$$R_{\text{value}} = \frac{L}{k} = \frac{0.5\ \text{ft}}{0.025\ \text{Btu/(h}\cdot\text{ft}\cdot{}^\circ\text{F)}} = 20\ \text{h}\cdot\text{ft}^2\cdot{}^\circ\text{F/Btu}$$

Thus the 6-inch-thick glass fiber insulation would be referred to as $R$-20 insulation by builders.

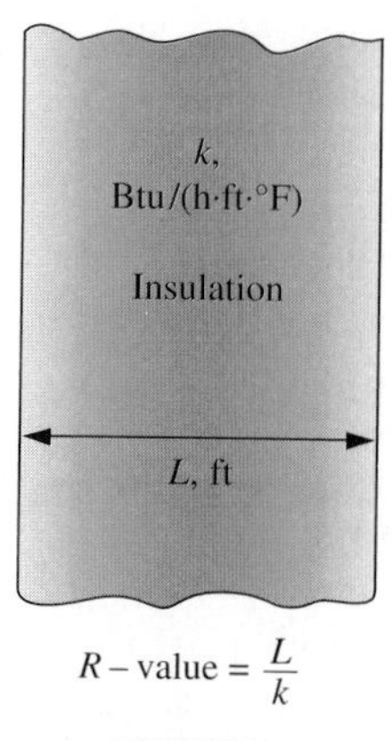


$R - \text{value} = \frac{L}{k}$

**FIGURE 8-21**
The $R$-value of an insulating material is simply the ratio of the thickness of the material to its thermal conductivity in proper units.

### Thermal Diffusivity

The product $\rho C_p$, which is frequently encountered in heat transfer analysis, is called the **heat capacity** of a material. Both the specific heat $C_p$ and the heat capacity $\rho C_p$ represent the heat storage capability of a material. But $C_p$ expresses it *per unit mass* whereas $\rho C_p$ expresses it *per unit volume,* as can be noticed from their units J/(kg · °C) and J/(m³ · °C), respectively.

Another material property that appears in the transient heat conduction analysis is the **thermal diffusivity**, which is defined as

$$\alpha = \frac{\text{heat conducted}}{\text{heat stored}} = \frac{k}{\rho C_p} \quad (\text{m}^2/\text{s}) \tag{8-9}$$

where $\rho$ is the density, $k$ is the thermal conductivity, and $C_p$ is the specific heat of the material.

Note that the thermal conductivity $k$ represents how well a material conducts heat, and the heat capacity $\rho C_p$ represents how much energy a material stores per unit volume. Therefore, the thermal diffusivity of a material can be viewed as the ratio of the *heat conducted* through the material to the *heat stored* per unit volume. A material that has a high thermal conductivity or a low heat capacity will obviously have a large thermal diffusivity. The larger the thermal diffusivity, the faster the propagation of heat into the medium. A small value of thermal diffusivity means that heat is mostly absorbed by the material and a small amount of heat will be conducted further.

The thermal diffusivities of some common materials at 20°C are given in Table 8-5. Note that the thermal diffusivity ranges from $\alpha = 0.14 \times 10^{-6}$ m²/s for water to $174 \times 10^{-6}$ m²/s for silver, which is a difference of more than a thousand times. Also note that the thermal diffusivities of beef and water are the same. This is not surprising, since meat as well as fresh vegetables and fruits are mostly water, and thus they possess the thermal properties of water.

**TABLE 8-5**

**The thermal diffusivities of some materials at room temperature**

| Material | $\alpha$, m²/s |
|---|---|
| Silver | $149 \times 10^{-6}$ |
| Gold | $127 \times 10^{-6}$ |
| Copper | $113 \times 10^{-6}$ |
| Aluminum | $97.5 \times 10^{-6}$ |
| Iron | $22.8 \times 10^{-6}$ |
| Mercury (l) | $4.7 \times 10^{-6}$ |
| Marble | $1.2 \times 10^{-6}$ |
| Ice | $1.2 \times 10^{-6}$ |
| Concrete | $0.75 \times 10^{-6}$ |
| Brick | $0.52 \times 10^{-6}$ |
| Heavy soil (dry) | $0.52 \times 10^{-6}$ |
| Glass | $0.34 \times 10^{-6}$ |
| Glass wool | $0.23 \times 10^{-6}$ |
| Water (l) | $0.14 \times 10^{-6}$ |
| Beef | $0.14 \times 10^{-6}$ |
| Wood (oak) | $0.13 \times 10^{-6}$ |

**EXAMPLE 8-4**

A common way of measuring the thermal conductivity of a material is to sandwich an electric thermofoil heater between two identical samples of the material, as shown in Fig. 8-22. The thickness of the resistance heater, including its cover, which is made of thin silicon rubber, it usually less than 0.5 mm. A circulating fluid such as tap water keeps the exposed ends of the samples at constant temperature. The lateral surfaces of the samples are well insulated to ensure that heat transfer through the samples is one-dimensional. Two thermocouples are embedded into each sample some distance $L$ apart, and a differential thermometer reads the temperature drop $\Delta T$ across this distance along each sample. When steady operating conditions are reached, the total rate of heat transfer through both samples becomes equal to the electric power drawn by the heater, which is determined by multiplying the electric current by the voltage.

In a certain experiment, cylindrical samples of diameter 5 cm and length 10 cm are used. The two thermocouples in each sample are placed 3 cm apart. After initial transients, the electric heater is observed to draw 0.4 A at 110 V, and both differential thermometers read a temperature difference of 15°C. Determine the thermal conductivity of the sample.

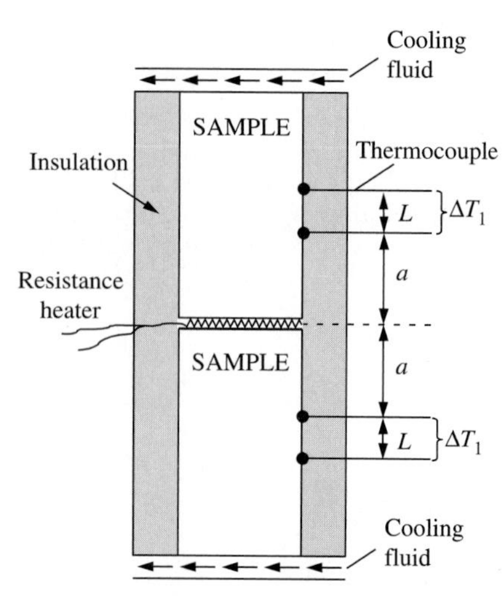


**FIGURE 8-22**

Apparatus to measure the thermal conductivity of a material using two identical samples and a thin resistance heater (Example 8-4).

**Solution** It appears that measurements are taken when the experimental apparatus has reached steady operation, since it is indicated that the readings do not change with time. The electrical power consumed by the resistance heater and converted to heat is

$$\dot{W}_e = VI = (110\text{ V})(0.4\text{ A}) = 44\text{ W}$$

We assume the entire heat generated by the heater is conducted through the samples, since the lateral surfaces of the whole apparatus are heavily insulated. Also, the rate of heat flow through each sample is

$$\dot{Q} = \tfrac{1}{2}\dot{W}_e = \tfrac{1}{2} \times 44\text{ W} = 22\text{ W}$$

since only half of the heat generated will flow through each sample because of symmetry. Reading the same temperature difference across the same distance in each samples also confirms that the apparatus possesses thermal symmetry. The heat transfer area is the area normal to the direction of heat flow, which is the cross-sectional area of the cylinder in this case:

$$A = \tfrac{1}{4}\pi D^2 = \tfrac{1}{4}\pi(0.05\text{ m})^2 = 0.00196\text{ m}^2$$

Noting that the temperature drops by 15°C within 3 cm in the direction of heat flow, the thermal conductivity of the sample is determined to be

$$\dot{Q} = kA\frac{\Delta T}{L} \rightarrow k = \frac{\dot{Q}L}{A\,\Delta T} = \frac{(22\text{ W})(0.03\text{ m})}{(0.00196\text{ m}^2)(15°\text{C})} = 22.4\text{ W/(m}\cdot°\text{C)}$$

Perhaps you are wondering if we really need to use two samples in the apparatus, since the measurements on the second sample do not give any additional information. It seems like we can replace the second sample by insulation. Indeed, we do not need the second sample; however, it enables us to verify the temperature measurements on the first sample, and provides thermal symmetry, which reduces experimental error.



## 8-4 ■ STEADY HEAT CONDUCTION IN PLANE WALLS

Consider steady heat conduction through the walls of a house in a winter day. We know that heat is continuously lost to the outdoors through the wall. We intuitively feel that heat transfer through the wall is in the *normal direction* to the wall surface, and no significant heat transfer takes place in the wall in other directions (Fig. 8-23).

Recall that heat transfer in a certain direction is driven by the *temperature gradient* in that direction. There will be no heat transfer in a direction in which there is no change in temperature. Temperature measurements at several locations on the inner or outer wall surface will confirm that a wall surface is nearly *isothermal.* That is, the temperatures at the top and bottom of a wall surface as well as at the right or left ends are almost the same. Therefore, there will be no heat transfer through the wall from the top to the bottom, or from left to right. but there will be considerable temperature difference between the inner and the outer surfaces of the wall, and thus significant heat transfer in the direction from the inner surface to the outer one.

The small thickness of the wall causes the temperature gradient in that direction to be large. Further, if the air temperatures in and outside the house remain constant then heat transfer through the wall of a house can be modeled as *steady* and *one-dimensional.* The temperature of the wall in this case will depend on one direction only (say the $x$-direction), and can be expressed as $T(x)$.

Noting that heat transfer is the only energy interaction involved in this case and there is no heat generation, the *energy balance* for the wall can be expressed as

$$\begin{pmatrix}\text{rate of}\\ \text{heat transfer}\\ \text{into the wall}\end{pmatrix} - \begin{pmatrix}\text{rate of}\\ \text{heat transfer}\\ \text{out of the wall}\end{pmatrix} = \begin{pmatrix}\text{rate of}\\ \text{change of the}\\ \text{energy content}\\ \text{of the wall}\end{pmatrix}$$

or

$$\dot{Q}_{\text{in}} - \dot{Q}_{\text{out}} = \frac{dE_{\text{wall}}}{dt} \qquad (8\text{-}10)$$

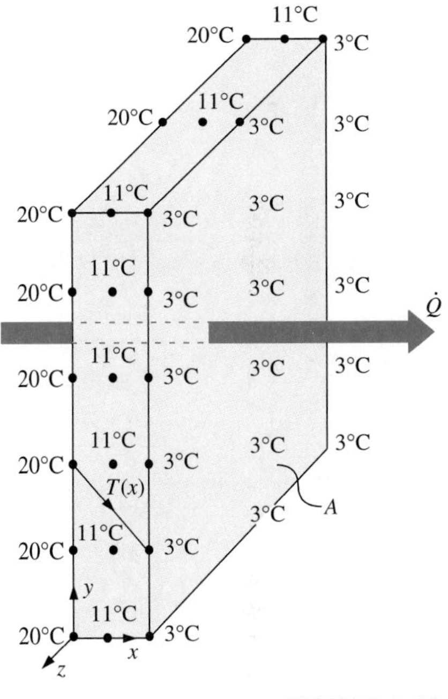


**FIGURE 8-23**

Heat flow through a wall is one-dimensional when the temperature of the wall varies in one direction only.

But $d\dot{E}_{wall}/dt = 0$ for *steady* operation, since there is no change in the temperature of the wall with time at any point. Therefore, the rate of heat transfer into the wall must be equal to the rate of heat transfer out of it. In other words, *the rate of heat transfer through the wall must be constant,* $\dot{Q}_{cond,wall}$ = constant.

Consider a plane wall of thickness $L$ and average thermal conductivity $k$. The two surfaces of the wall are maintained at constant temperatures of $T_1$ and $T_2$. For one-dimensional steady heat conduction through the wall, we have $T(x)$. Then the Fourier's law of heat conduction for the wall can be expressed as

$$\dot{Q}_{cond,wall} = -kA\frac{dT}{dx} \quad \text{(W)} \tag{8-11}$$

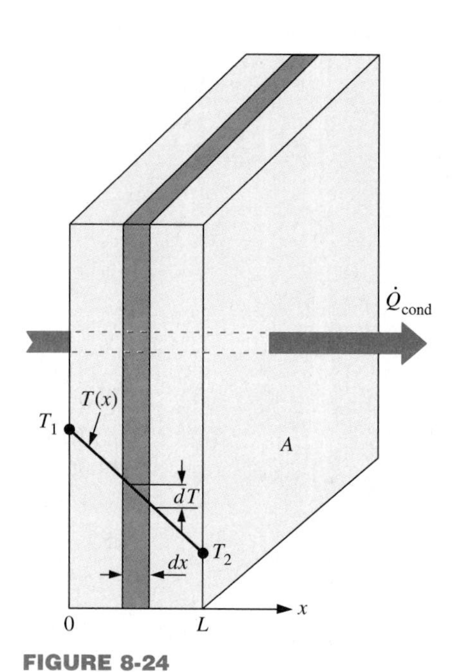


**FIGURE 8-24**
Under steady conditions, the temperature distribution in a plane wall is a straight line.

where the rate of conduction heat transfer $\dot{Q}_{cond,wall}$ and the surface area $A$ are constant. Thus we have $dT/dx$ = constant, which means that *the temperature through the wall varies linearly with x*. That is, the temperature distribution in the wall under steady conditions in a *straight line* (Fig. 8-24).

Separating the variables in the above equation and integrating from $x = 0$ where $T(0) = T_1$ to $x = L$, where $T(L) = T_2$, we get

$$\int_{x=0}^{L} \dot{Q}_{cond,wall}\, dx = -\int_{T=T_1}^{T_2} kA\, dT$$

Performing the integration and rearranging gives

$$\dot{Q}_{cond,wall} = kA\frac{T_1 - T_2}{L} \quad \text{(W)} \tag{8-12}$$

which is identical to Eq. 8-4. Again, *the rate of heat conduction through a plane wall is proportional to the average thermal conductivity, the wall area, and the temperature difference, but is inversely proportional to the wall thickness.* Also, once the rate of heat conduction is available, the temperature $T(x)$ at any location $x$ can be determined by replacing $T_2$ in Eq. 8-12 by $T$, and $L$ by $x$.

## The Thermal Resistance Concept

Equation 8-12 for heat conduction through a plane wall can be rearranged as

$$\dot{Q}_{cond,wall} = \frac{T_1 - T_2}{R_{wall}} \quad \text{(W)} \tag{8-13}$$

where

$$R_{wall} = \frac{L}{kA} \quad (°\text{C/W}) \tag{8-14}$$

is the *thermal resistance* of the wall against heat conduction or simply the **conduction resistance** of the wall. Note that the thermal resistance of a medium depends on the *geometry* and the *thermal properties* of the medium.

The equation above for heat flow is analogous to the relation for *electric current flow I* expressed as

$$I = \frac{\mathscr{V}_1 - \mathscr{V}_2}{R_e} \tag{8-15}$$

where $R_e = L/\sigma_e A$ is the *electric resistance* and $\mathscr{V}_1 - \mathscr{V}_2$ is the *voltage difference* across the resistance ($\rho_e$ is the electrical conductivity). Thus, the *rate of heat transfer* through a layer corresponds to the *electric current,* the *thermal resistance* corresponds to *electrical resistance,* and the *temperature difference* corresponds to *voltage difference* across the layer (Fig. 8-25).

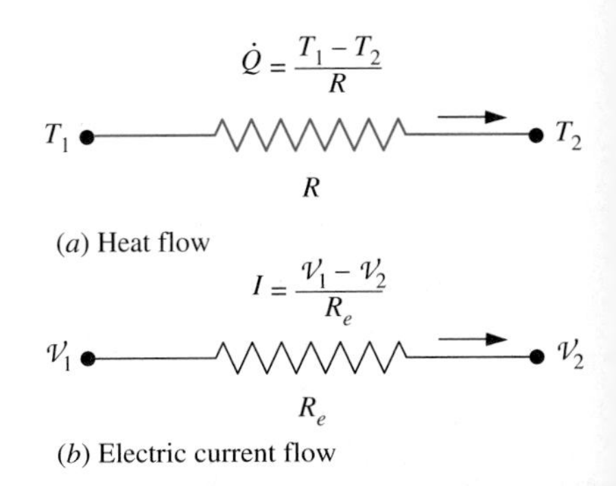


**FIGURE 8-25**
Analogy between thermal and electrical resistance concepts.

Consider convection heat transfer from a solid surface of area $A$ and temperature $T_s$ to a fluid whose temperature sufficiently far from the surface is $T_\infty$, with a convection heat transfer coefficient $h$. Newton's law of cooling for convection heat transfer rate $\dot{Q}_{\text{conv}} = hA(T_s - T_\infty)$ can be rearranged as

$$\dot{Q}_{\text{conv}} = \frac{T_s - T_\infty}{R_{\text{conv}}} \quad \text{(W)} \tag{8-16}$$

where

$$R_{\text{conv}} = \frac{1}{hA} \quad \text{(°C/W)} \tag{8-17}$$

is the *thermal resistance* of the surface against heat convection or simply the **convection resistance** of the surface (Fig. 8-26). Note that when the convection heat transfer coefficient is very large ($h \to \infty$), the convection resistance becomes *zero* and $T_s \approx T_\infty$. That is, the surface offers *no resistance to convection,* and thus it does not slow down the heat transfer process. This situation is approached in practice at surfaces where boiling and condensation occurs. Also note that the surface does not have to be a plane surface. Equation 8-17 for convection resistance is valid for surfaces of any shape, provided that the assumption of $h$ = constant and uniform is reasonable.

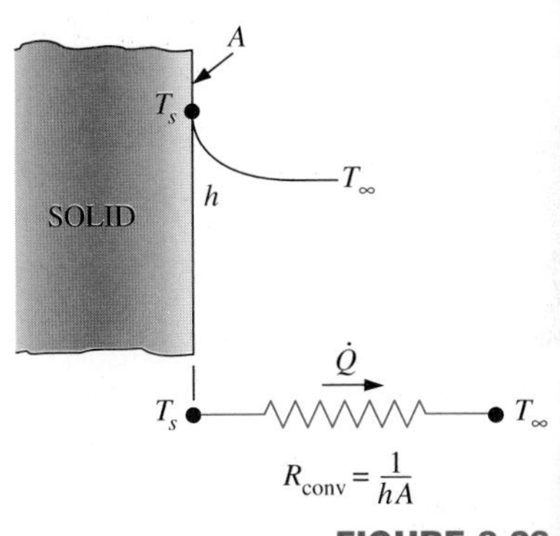


**FIGURE 8-26**
Schematic for convection resistance at a surface.

When the wall is surrounded by a gas, the *radiation effects,* which we have ignored so far, can be significant, and may need to be considered. The rate of radiation heat transfer between a surface of emissivity $\varepsilon$ and area $A$ at temperature $T_s$ and the surrounding surfaces at some average temperature $T_{\text{surr}}$ can be expressed as

$$\dot{Q}_{\text{rad}} = \varepsilon\sigma A(T_s^4 - T_{\text{surr}}^4) = h_{\text{rad}}A(T_s - T_{\text{surr}}) = \frac{T_s - T_{\text{surr}}}{R_{\text{rad}}} \quad \text{(W)} \tag{8-18}$$

where

$$R_{\text{rad}} = \frac{1}{h_{\text{rad}}A} \quad \text{(K/W)} \tag{8-19}$$

is the *thermal resistance* of a surface against radiation or the **radiation resistance** and

$$h_{\text{rad}} = \varepsilon\sigma(T_s^2 + T_{\text{surr}}^2)(T_s + T_{\text{surr}}) \quad [\text{W/(m}^2 \cdot \text{K)}] \tag{8-20}$$

is the **radiation heat transfer coefficient.** Note that both $T_s$ and $T_{\text{surr}}$ *must* be in K in the evaluation of $h_{\text{rad}}$. The definition of the radiation heat transfer coefficient enables us to express radiation conveniently in an analogous manner to convection in terms of a temperature difference. But $h_{\text{rad}}$ depends strongly on temperature while $h_{\text{conv}}$ usually does not.

A surface exposed to the surrounding air involves convection and radiation simultaneously, and the total heat transfer at the surface is determined by adding (or subtracting if in the opposite directions) the radiation and convection components. The convection and radiation resistances are parallel to each other, as shown in Fig. 8-27, and may cause some complication in the thermal resistance network. When $T_{\text{surr}} \approx T_\infty$, the radiation effect can properly be accounted for by replacing $h$ in the convection resistance relation by $h = h_{\text{conv}} + h_{\text{rad}}$. This way, all the complications associated with radiation are avoided.

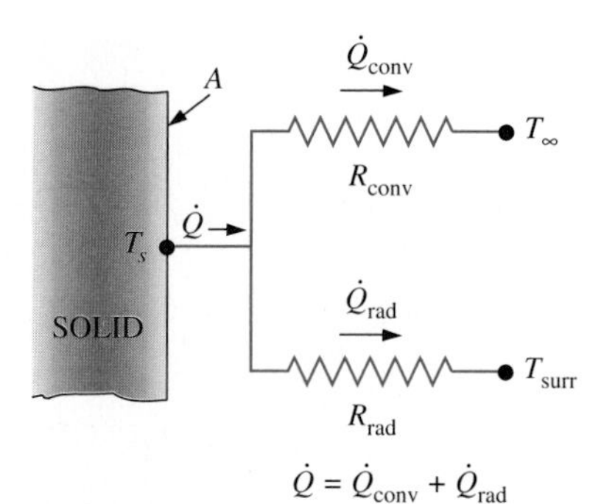


**FIGURE 8-27**

Schematic for convection and radiation resistances at a surface.

## Thermal Resistance Network

Now consider steady one-dimensional heat flow through a plane wall of thickness $L$ and thermal conductivity $k$ that is exposed to convection on both sides to fluids at temperatures $T_{\infty 1}$ and $T_{\infty 2}$ with heat transfer coefficients $h_1$ and $h_2$, respectively, as shown in Fig. 8-28. Assuming $T_{\infty 2} < T_{\infty 1}$, the variation of temperature will be as shown in the figure. Note that the temperature varies linearly in the wall, and asymptotically approaches $T_{\infty 1}$ and $T_{\infty 2}$ in the fluids as we move away from the wall.

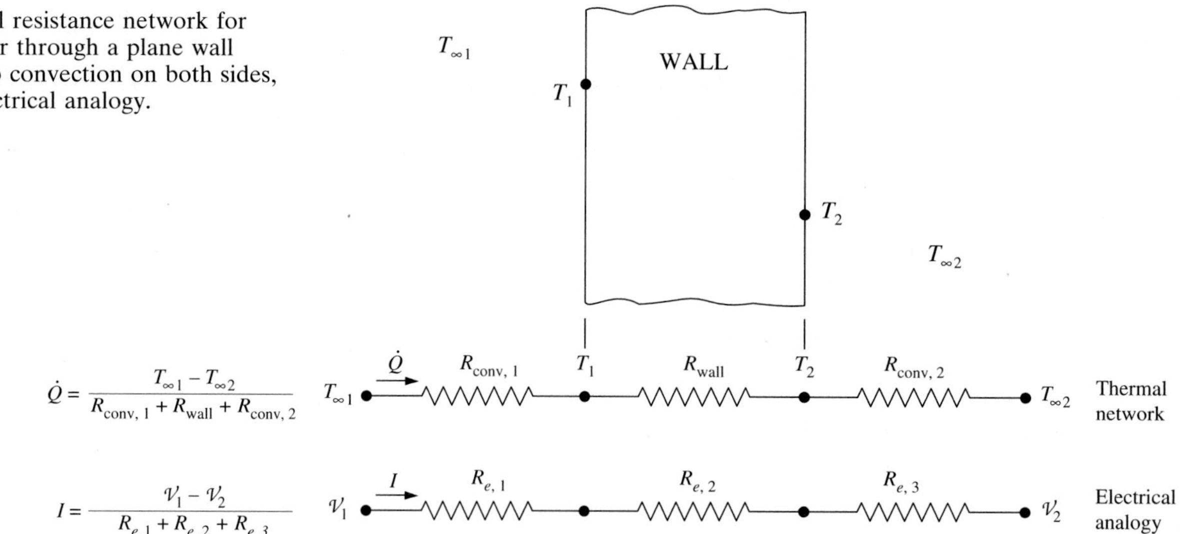


**FIGURE 8-28**

The thermal resistance network for heat transfer through a plane wall subjected to convection on both sides, and the electrical analogy.

Under steady conditions, we have

$$\begin{pmatrix}\text{rate of}\\ \textit{heat convection}\\ \text{into the wall}\end{pmatrix} = \begin{pmatrix}\text{rate of}\\ \textit{heat conduction}\\ \text{through the wall}\end{pmatrix} = \begin{pmatrix}\text{rate of}\\ \textit{heat convection}\\ \text{from the wall}\end{pmatrix}$$

or,

$$\dot{Q} = h_1 A(T_{\infty 1} - T_1) = kA\frac{T_1 - T_2}{L} = h_2 A(T_2 - T_{\infty 2}) \qquad (8\text{-}21)$$

which can be rearranged as

$$\dot{Q} = \frac{T_{\infty 1} - T_1}{1/h_1 A} = \frac{T_1 - T_2}{L/kA} = \frac{T_2 - T_{\infty 2}}{1/h_1 A}$$
$$= \frac{T_{\infty 1} - T_1}{R_{\text{conv},1}} = \frac{T_1 - T_2}{R_{\text{wall}}} = \frac{T_2 - T_{\infty 2}}{R_{\text{conv},2}} \qquad (8\text{-}22)$$

Adding the numerators and denominators yields (Fig. 8-29)

$$\dot{Q} = \frac{T_{\infty 1} - T_{\infty 2}}{R_{\text{total}}} \quad \text{(W)} \qquad (8\text{-}23)$$

where

$$R_{\text{total}} = R_{\text{conv},1} + R_{\text{wall}} + R_{\text{conv},2} = \frac{1}{h_1 A} + \frac{L}{kA} + \frac{1}{h_2 A} \quad (°\text{C/W}) \qquad (8\text{-}24)$$

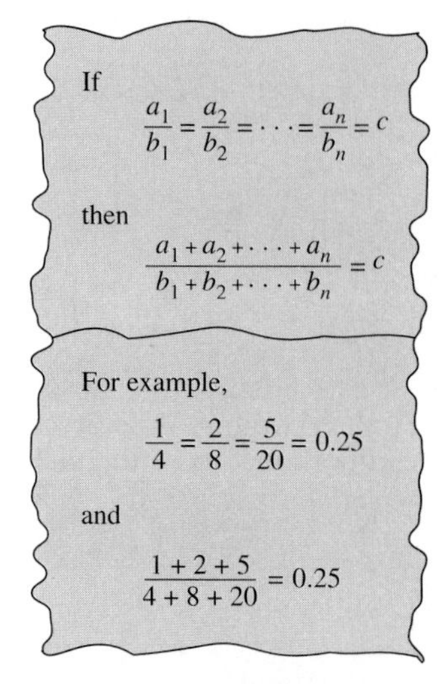

If
$\frac{a_1}{b_1} = \frac{a_2}{b_2} = \cdots = \frac{a_n}{b_n} = c$
then
$\frac{a_1 + a_2 + \cdots + a_n}{b_1 + b_2 + \cdots + b_n} = c$
For example,
$\frac{1}{4} = \frac{2}{8} = \frac{5}{20} = 0.25$
and
$\frac{1+2+5}{4+8+20} = 0.25$


**FIGURE 8-29**
A useful mathematical identity.

Note that the heat transfer area $A$ is constant for a plane wall, and the rate of heat transfer through a wall separating two mediums is equal to the temperature difference divided by the total thermal resistance between the mediums. Also note that the thermal resistances are in *series,* and the equivalent thermal resistance is determined by simply *adding* the individual resistances, just like the electrical resistances connected in series. Thus the electrical analogy still applies. We summarize this as *the rate of steady heat transfer between two surfaces is equal to the temperature difference divided by the total thermal resistance between those two surfaces.*

Another observation which can be made from Eq. 8-23 is that the ratio of the temperature drop to the thermal resistance across any layer is constant, and thus the temperature drop across any layer is proportional to the thermal resistance of the layer. The larger the resistance, the larger the temperature drop. In fact, the equation $\dot{Q} = \Delta T/R$ can be rearranged as

$$\Delta T = \dot{Q}R \quad (°\text{C}) \qquad (8\text{-}25)$$

which indicates that the *temperature drop* across any layer is equal to the *rate of heat transfer* times the *thermal resistance* across that layer (Fig. 8-30). You may recall that this is also true for voltage drop across an electrical resistance when the electric current is constant.

Note that we do not need to know the surface temperatures of the wall in order to evaluate the rate of heat transfer through it. All we need to know is the convection heat transfer coefficients and the fluid

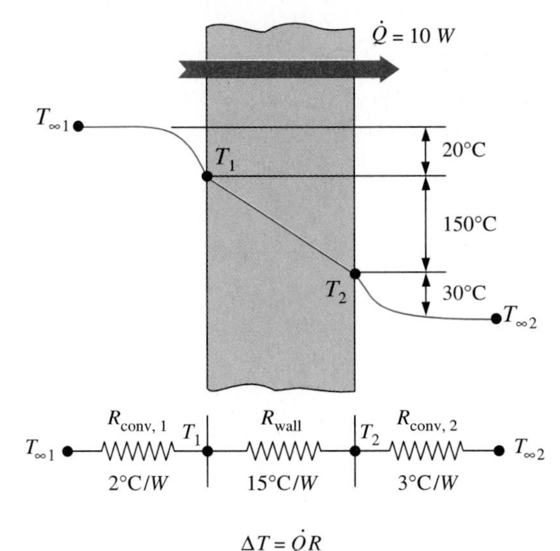


**FIGURE 8-30**
The temperature drop across a layer is proportional to its thermal resistance.

temperatures on both sides of the wall. The *surface temperature* of the wall can be determined as described above using the thermal resistance concept, but by taking the surface at which the temperature is to be determined as one of the terminal surfaces. For example, once $\dot{Q}$ is evaluated, the surface temperature $T_1$ can be determined from

$$\dot{Q} = \frac{T_{\infty 1} - T_1}{R_{\text{conv},1}} = \frac{T_{\infty 1} - T_1}{1/h_1 A} \tag{8-26}$$

## Multilayer Plane Walls

In practice we often encounter plane walls which consist of several layers of different materials. The thermal resistance concept can still be used to determine the rate of steady heat transfer through such *composite* walls. As you may have already guessed, this is done by simply noting that the conduction resistance of each wall is $L/kA$ connected in series, and using the electrical analogy. That is, by dividing the *temperature difference* between two surfaces at known temperatures by the *total thermal resistance* between them.

Consider a plane wall which consists of two layers (such as a brick wall with a layer of insulation). The rate of steady heat transfer through this two-layer composite wall can be expressed as (Fig. 8-31)

$$\dot{Q} = \frac{T_{\infty 1} - T_{\infty 2}}{R_{\text{total}}} \tag{8-27}$$

where $R_{\text{total}}$ is the *total thermal resistance,* expressed as

$$\begin{aligned} R_{\text{total}} &= R_{\text{conv},1} + R_{\text{wall},1} + R_{\text{wall},2} + R_{\text{conv},2} \\ &= \frac{1}{h_1 A} + \frac{L_1}{k_1 A} + \frac{L_2}{k_2 A} + \frac{1}{h_2 A} \end{aligned} \tag{8-28}$$

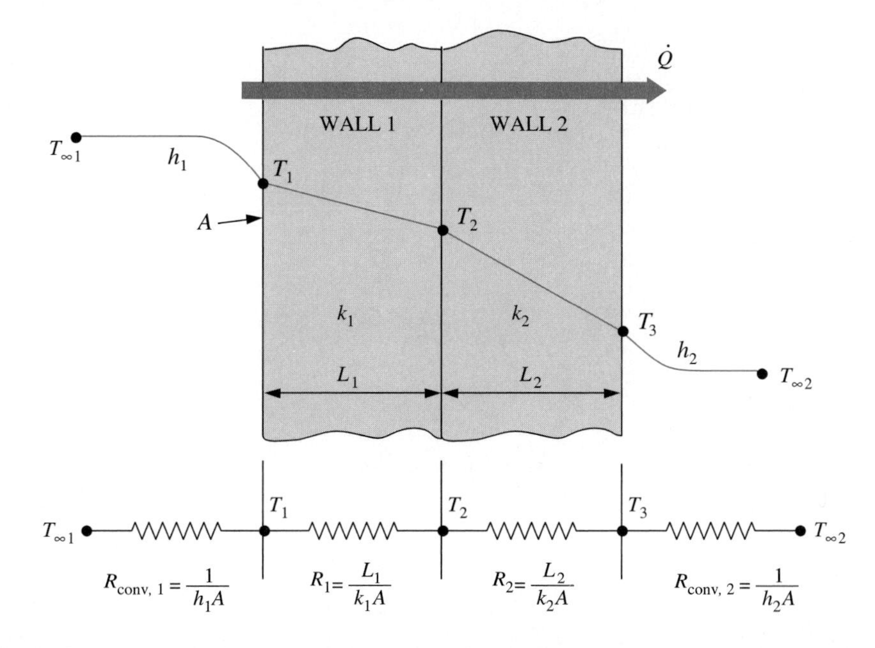


**FIGURE 8-31**
The thermal resistance network for heat transfer through a two-layer plane wall subjected to convection on both sides.

The subscripts 1 and 2 in the $R_{\text{wall}}$ relations above indicate the first and the second layers, respectively. We could also obtain this result by following the approach used above for the single-layer case by noting that the rate of steady heat transfer $\dot{Q}$ through a multilayer medium is constant, and thus it must be the same through each layer. Note from the thermal resistance network that the resistances are *in series,* and thus the *total thermal resistance* is simply the *arithmetic sum* of the individual thermal resistances in the path of heat flow.

The result above for the *two-layer* case is analogous to the *single-layer* case, except that an *additional resistance* is added for the *additional layer.* This result can be extended to plane walls that consist of *three* or *more layers* by adding an *additional resistance* for each *additional layer.*

Once $\dot{Q}$ is *known,* an unknown surface temperature $T_j$ at any surface or interface $j$ can be determined from

$$\dot{Q} = \frac{T_i - T_j}{R_{\text{total},i-j}} \tag{8-29}$$

where $T_i$ is a *known* temperature at location $i$ and $R_{\text{total},i-j}$ is the total thermal resistance between locations $i$ and $j$. For example, when the fluid temperatures $T_{\infty 1}$ and $T_{\infty 2}$ for the two-layer case shown in Fig. 8-30 are available and $\dot{Q}$ is calculated from Eq. 8-29, the interface temperature $T_2$ between the two walls can be determined from (Fig. 8-32)

$$\dot{Q} = \frac{T_{\infty 1} - T_2}{R_{\text{covn},1} + R_{\text{wall},1}} = \frac{T_{\infty 1} - T_2}{\dfrac{1}{h_1A} + \dfrac{L_1}{k_1A}} \tag{8-30}$$

*The temperature drop across a layer is easily determined from Eq.* 8-30 *by multiplying* $\dot{Q}$ by the thermal resistance of that layer.

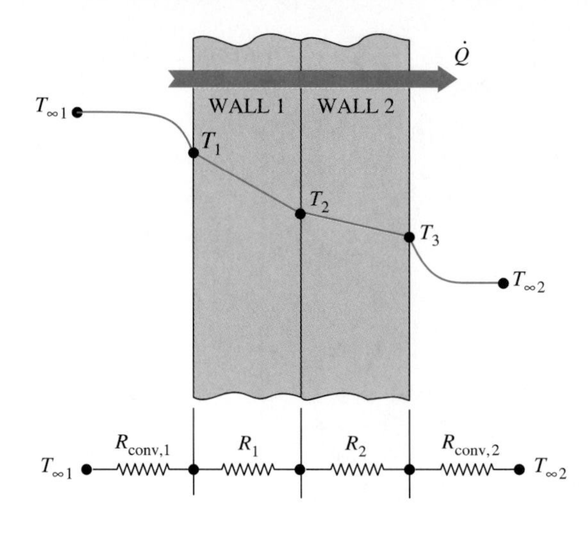


$$\text{To find } T_1: \quad \dot{Q} = \frac{T_{\infty 1} - T_1}{R_{\text{conv},1}}$$

$$\text{To find } T_2: \quad \dot{Q} = \frac{T_{\infty 1} - T_1}{R_{\text{conv},1} + R_1}$$

$$\text{To find } T_3: \quad \dot{Q} = \frac{T_3 - T_{\infty 2}}{R_{\text{conv},2}}$$

**FIGURE 8-32**

The evaluation of the surface and interface temperatures when $T_{\infty 1}$ and $T_{\infty 2}$ are given and $\dot{Q}$ is calculated.

The thermal resistance concept is widely used in practice because it is intuitively easy to understand and it has proven to be a powerful tool in the solution of a wide range of heat transfer problems. But its use is limited to systems through which the rate of heat transfer $\dot{Q}$ remains *constant*; that is, to systems involving *steady* heat transfer with no *heat generation* (such as resistance heating or chemical reactions) within the medium.

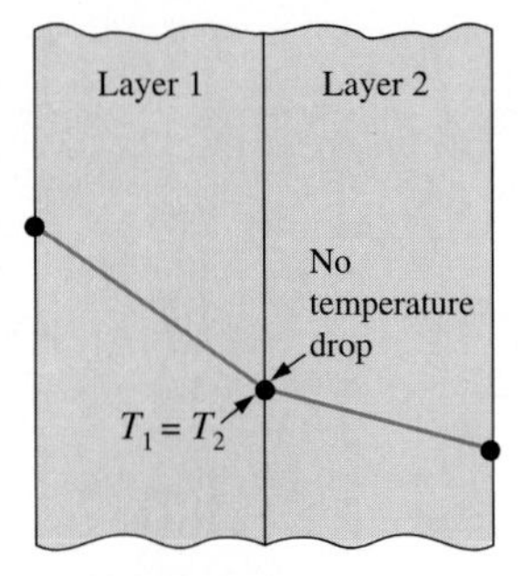


(*a*) Perfect thermal contact

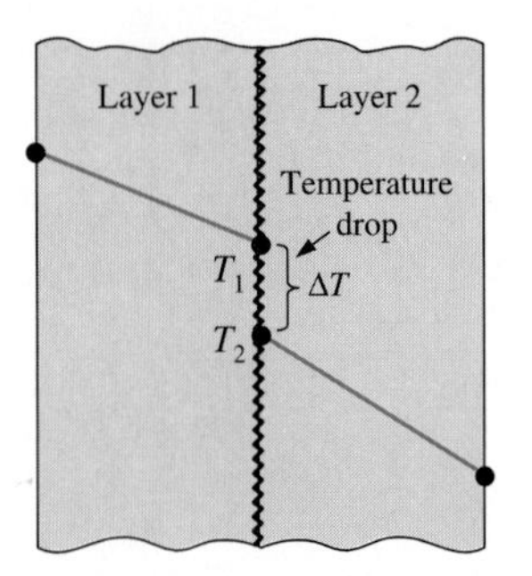


(*b*) Actual (imperfect) thermal contact

**FIGURE 8-33**

Imperfect contact at an interface produces a thermal contact resistance and an associated temperature drop at the interface.

## Thermal Contact Resistance

In the analysis of heat conduction through multilayer solids, we assumed "perfect contact" at the interface of two layers, and thus no temperature drop at the interface. This would be the case when the surfaces are perfectly smooth and are in perfect contact at each point. In reality, however, even surfaces that appear smooth to the eye appear rather rough when examined under a microscope, with numerous peaks and valleys.

When two such surfaces are pressed against each other, the peaks will form good material contact, but the valleys will form voids filled with air. Thus, an interface will contain numerous air gaps of varying sizes, which act as insulation because of the low thermal conductivity of air. Thus an interface offers some resistance to heat transfer, and this resistance is called the **thermal contact resistance** $R_c$ (Fig. 8-33). The value of $R_c$ is determined experimentally, and there is considerable scatter of data because of the difficulty in characterizing the surfaces. As expected, the contact resistance is observed to *decrease* with *decreasing surface rough-*

*ness* and *increasing interface pressure.* Noting that $\dot{Q} = \Delta T/R$ for any layer, the thermal contact resistance can be determined by measuring the temperature drop $\Delta T$ at the interface and multiplying it by the steady heat transfer rate $\dot{Q}$. Most experimentally determined values of the thermal contact resistance fall between 0.00001 and 0.001 $m^2 \cdot K/W$.

The thermal contact resistance can be minimized by applying a thermally conducting liquid called a *thermal grease* such as silicon oil on the surfaces before they are pressed against each other. This is commonly done when attaching electronic components such as power transistors to heat sinks.

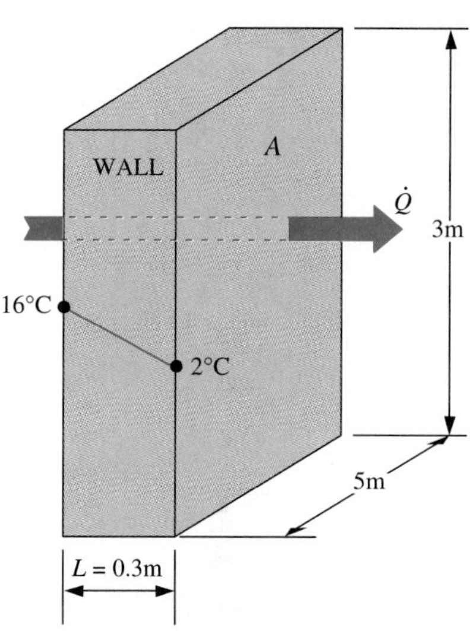


**FIGURE 8-34**
Schematic for Example 8-5.

**EXAMPLE 8-5**

Consider a 3-m-high, 5-m-wide, and 0.3-m-thick wall whose thermal conductivity is $k = 0.9$ W/(m · °C) (Fig. 8-34). On a certain day, the temperatures of the inner and the outer surfaces of the wall are measured to be 16°C and 2°C, respectively. Determine the rate of heat loss through the wall on that day.

**Solution** We assume the temperatures of the wall surfaces remained at the given values long enough so that heat transfer through the wall can be considered to be steady. We also assume heat transfer through the wall to be one-dimensional, since any significant temperature gradients in this case will exist in the direction from the indoors to the outdoors.

Noting that the heat transfer through the wall is by conduction and the surface of the wall is $A = 3\text{ m} \times 5\text{ m} = 15\text{ m}^2$, the steady rate of heat transfer through the wall can be determined from Eq. 8-4 to be

$$\dot{Q} = kA\frac{T_1 - T_2}{L} = [0.9\text{ W/(m} \cdot \text{°C)}](15\text{ m}^2)\frac{(16-2)\text{°C}}{0.3\text{ m}} = 630\text{ W}$$

We could also determine the steady rate of heat transfer through the wall by making use of the thermal resistance concept from

$$\dot{Q} = \frac{\Delta T_{\text{wall}}}{R_{\text{wall}}}$$

where $$R_{\text{wall}} = \frac{L}{kA} = \frac{0.3\text{ m}}{[0.9\text{ W/(m} \cdot \text{°C)}](15\text{ m}^2)} = 0.02222\text{°C/W}$$

Substituting, we get

$$\dot{Q} = \frac{(16-2)\text{°C}}{0.02222\text{°C/W}} = 630\text{ W}$$

which is the result obtained earlier. Note that heat conduction through a plane wall with specified surface temperatures can be determined directly and easily without utilizing the thermal resistance concept. However, the thermal resistance concept serves as a valuable tool in more complex heat transfer problems, as you will see in the following examples.

**EXAMPLE 8-6**

Consider a 0.8-m-high and 1.5 m-wide glass window whose thickness is 8 mm and which has a thermal conductivity of $k = 0.78$ W/(m · °C). Determine the steady rate of heat transfer through this glass window and the temperature of its inner surface for a day during which the room is maintained at 20°C while the temperature of the outdoors is −10°C. Take the heat transfer coefficients on the inner and outer surfaces of the window to be $h_1 = 10\ \text{W/(m}^2\cdot{}^\circ\text{C)}$ and $h_2 = 40\ \text{W/(m}^2\cdot{}^\circ\text{C)}$, which includes the effects of radiation.

**Solution** We assume the indoor and outdoor air temperatures to have remained at the given values long enough that heat transfer through the glass window can be considered to be steady. We also assume heat transfer through the window to be one-dimensional, since any significant temperature gradients in this case will exist in the direction from the indoors to the outdoors.

This problem involves conduction through the glass window and convection at its surfaces, and can best be handled by making use of the thermal resistance concept and drawing the thermal resistance network, as shown in Fig. 8-35. Noting that the surface area of the window is $A_s = 0.8\ \text{m} \times 1.5\ \text{m} = 1.2\ \text{m}^2$, the individual resistances are evaluated from their definitions to be

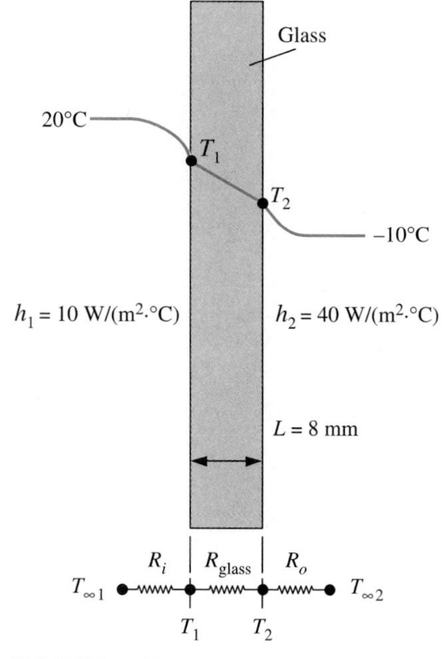


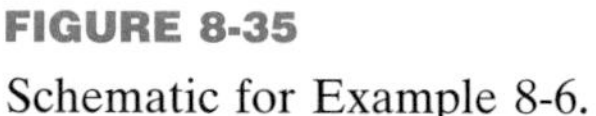


**FIGURE 8-35**
Schematic for Example 8-6.

$$R_i = R_{\text{conv},1} = \frac{1}{h_1 A} = \frac{1}{[10\ \text{W/(m}^2\cdot{}^\circ\text{C)}](1.2\ \text{m}^2)} = 0.08333^\circ\text{C/W}$$

$$R_{\text{glass}} = \frac{L}{kA} = \frac{0.008\ \text{m}}{[0.78\ \text{W/(m}\cdot{}^\circ\text{C)}](1.2\ \text{m}^2)} = 0.00855^\circ\text{C/W}$$

$$R_o = R_{\text{conv},2} = \frac{1}{h_2 A} = \frac{1}{[40\ \text{W/(m}^2\cdot{}^\circ\text{C)}](1.2\ \text{m}^2)} = 0.02083^\circ\text{C/W}$$

Noting that all three resistances are in series, the total resistance is determined to be

$$\begin{aligned} R_{\text{total}} &= R_{\text{conv},1} + R_{\text{glass}} + R_{\text{conv},2} = 0.08333 + 0.00855 + 0.02083 \\ &= 0.1127^\circ\text{C/W} \end{aligned}$$

Then the steady rate of heat transfer through the window becomes

$$\dot{Q} = \frac{T_{\infty 1} - T_{\infty 2}}{R_{\text{total}}} = \frac{[20 - (-10)]^\circ\text{C}}{0.1127^\circ\text{C/W}} = 266\ \text{W}$$

Knowing the rate of heat transfer, the inner surface temperature of the window glass can be determined from

$$\dot{Q} = \frac{T_{\infty 1} - T_1}{R_{\text{conv},1}} \longrightarrow T_1 = T_{\infty 1} - \dot{Q}R_{\text{conv},1}$$

$$= 20^\circ\text{C} - (266\ \text{W})(0.08333^\circ\text{C/W}) = -2.2^\circ\text{C}$$

Therefore, the inner surface temperature of the window glass will be −2.2°C, even though the temperature of the air in the room is maintained at 20°C. Such low surface temperatures are highly undesirable, since they cause the formation of fog or even frost on the inner surfaces of the glass when the humidity in the room is high.

**EXAMPLE 8-7**

Consider a 0.8-m-high and 1.5-m-wide double-pane window consisting of two 4-mm-thick layers of glass $[k = 0.78\ \text{W/(m}\cdot{}^\circ\text{C)}]$ separated by a 10-mm-wide stagnant air space $[k = 0.026\ \text{W/(m}\cdot{}^\circ\text{C)}]$. Determine the steady rate of heat transfer through this double-pane window and the temperature of its inner surface for a day during which the room is maintained at 20°C while the temperature of the outdoors is −10°C. Take the convection heat transfer coefficients on the inner and outer surfaces of the window to be $h_1 = 10\ \mathrm{W/(m^2 \cdot {}^\circ C)}$ and $h_2 = 40\ \mathrm{W/(m^2 \cdot {}^\circ C)}$, which includes the effects of radiation.

**Solution** This example problem is identical to the previous one except that the single 8-mm-thick window glass is replaced by two 4-mm-thick glasses enclosing a 10-mm-wide stagnant air space. Therefore, the thermal resistance of this problem will involve two additional conduction resistances corresponding to the two additional layers, as shown in Fig. 3-36. Noting that the surface area of the window is again $A = 0.8\ \mathrm{m} \times 1.5\ \mathrm{m} = 1.2\ \mathrm{m^2}$, the individual resistances are evaluated from their definitions to be

$$R_i = R_{\text{conv},1} = \frac{1}{h_1 A} = \frac{1}{[10\ \mathrm{W/(m^2 \cdot {}^\circ C)}](1.2\ \mathrm{m^2})} = 0.08333^\circ\mathrm{C/W}$$

$$R_1 = R_3 = R_{\text{glass}} = \frac{L_1}{k_1 A} = \frac{0.004\ \mathrm{m}}{[0.78\ \mathrm{W/(m \cdot {}^\circ C)}](1.2\ \mathrm{m^2})} = 0.00427^\circ\mathrm{C/W}$$

$$R_2 = R_{\text{air}} = \frac{L_2}{k_2 A} = \frac{0.01\ \mathrm{m}}{[0.026\ \mathrm{W/(m \cdot {}^\circ C)}](1.2\ \mathrm{m^2})} = 0.3205^\circ\mathrm{C/W}$$

$$R_o = R_{\text{conv},2} = \frac{1}{h_2 A} = \frac{1}{[40\ \mathrm{W/(m^2 \cdot {}^\circ C)}](1.2\ \mathrm{m^2})} = 0.02083^\circ\mathrm{C/W}$$

Noting that all three resistances are in series, the total resistance is determined to be

$$\begin{aligned} R_{\text{total}} &= R_{\text{conv},1} + R_{\text{glass},1} + R_{\text{air}} + R_{\text{glass},2} + R_{\text{conv},2} \\ &= 0.08333 + 0.00427 + 0.3205 + 0.00427 + 0.02083 \\ &= 0.4332^\circ\mathrm{C/W} \end{aligned}$$

Then the steady rate of heat transfer through the window becomes

$$\dot{Q} = \frac{T_{\infty 1} - T_{\infty 2}}{R_{\text{total}}} = \frac{[20 - (-10)]^\circ\mathrm{C}}{0.4332^\circ\mathrm{C/W}} = 69.2\ \mathrm{W}$$

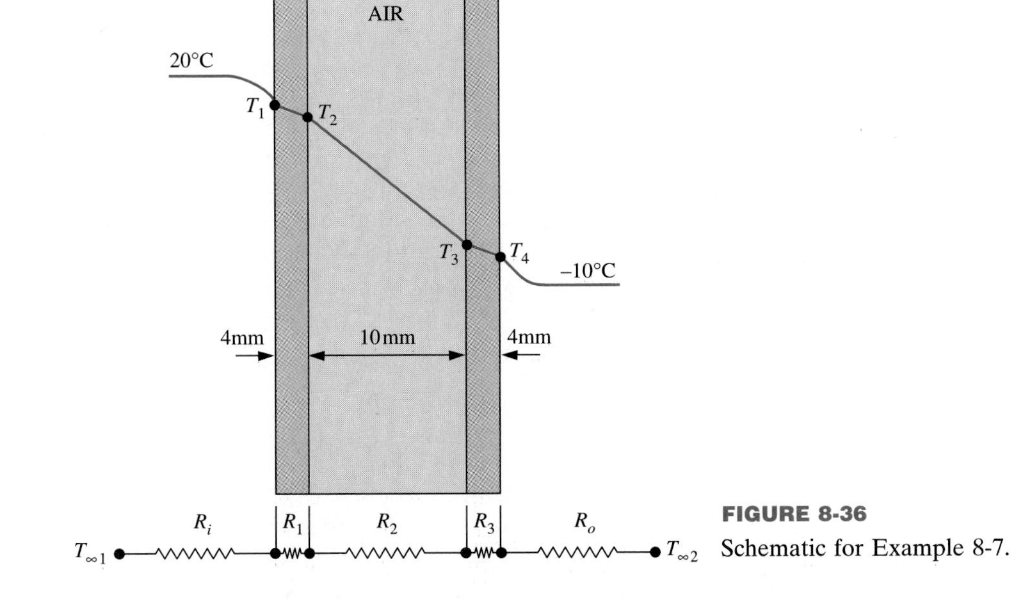


**FIGURE 8-36**
Schematic for Example 8-7.

which is about one-fourth of the result obtained in the previous example. This explains the popularity of double- and even triple-pane windows in cold climates. The drastic reduction in the heat transfer rate in this case is due to the large thermal resistance of the air layer between the glasses. In reality, the thermal resistance of the air layer will be somewhat lower because of the natural convection currents that are likely to occur in the air space.

The inner surface temperature of the window in this case will be

$$T_1 = T_{\infty 1} - \dot{Q}R_{\text{conv},1} = 20°\text{C} - (69.2\text{ W})(0.08333°\text{C/W}) = 14.2°\text{C}$$

which is considerably higher than the −2.2°C obtained in the previous example. Therefore, a double-pane window will rarely get fogged. A double-pane window will also reduce the heat gain in summer, and thus reduce the air-conditioning costs.

## 8-5 ■ GENERALIZED THERMAL RESISTANCE NETWORKS

The *thermal resistance* concept or the *electrical analogy* can also be used to solve steady heat transfer problems which involve parallel layers or combined series-parallel arrangements. Although such problems are often two or even three dimensional, approximate solutions can be obtained by assuming one dimensional heat transfer and using the thermal resistance network.

Consider the composite wall shown in Fig. 8-37, which consists of two parallel layers. The thermal resistance network, which consists of two parallel resistances, can be represented as shown in the figure. Noting that the total heat transfer is the sum of the heat transfers through each layer, we have

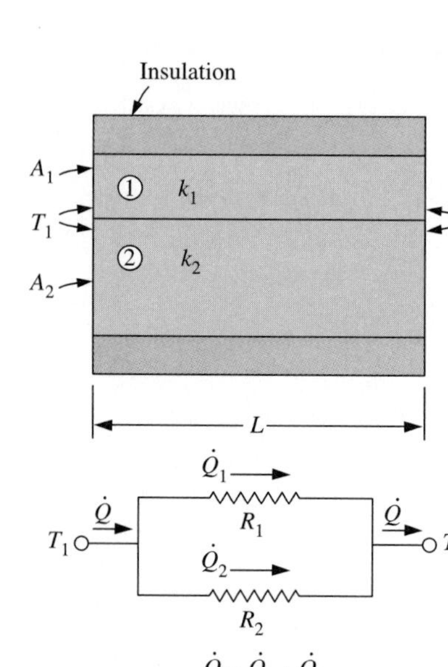


**FIGURE 8-37**
Thermal resistance network for two parallel layers.

$$\dot{Q} = \dot{Q}_1 + \dot{Q}_2 = \frac{T_1 - T_2}{R_1} + \frac{T_1 - T_2}{R_2} = (T_1 - T_2)\left(\frac{1}{R_1} + \frac{1}{R_2}\right) \tag{8-31}$$

Utilizing electrical analogy, we get

$$\dot{Q} = \frac{T_1 - T_2}{R_{\text{total}}} \tag{8-32}$$

where

$$\frac{1}{R_{\text{total}}} = \frac{1}{R_1} + \frac{1}{R_2} \longrightarrow R_{\text{total}} = \frac{R_1 R_2}{R_1 + R_2} \tag{8-33}$$

since the resistances are in parallel.

Now consider the combined series–parallel arrangement shown in Fig. 8-38. The total rate of heat transfer through this composite system can again be expressed as

$$\dot{Q} = \frac{T_1 - T_\infty}{R_{\text{total}}} \tag{8-34}$$

where

$$R_{\text{total}} = R_{12} + R_3 + R_{\text{conv}} = \frac{R_1 R_2}{R_1 + R_2} + R_3 + R_{\text{conv}} \tag{8-35}$$

and

$$R_1 = \frac{L_1}{k_1 A_1} \qquad R_2 = \frac{L_2}{k_2 A_2}, \qquad R_3 = \frac{L_3}{k_3 A_3}, \qquad R_{\text{conv}} = \frac{1}{hA_3} \tag{8-36}$$

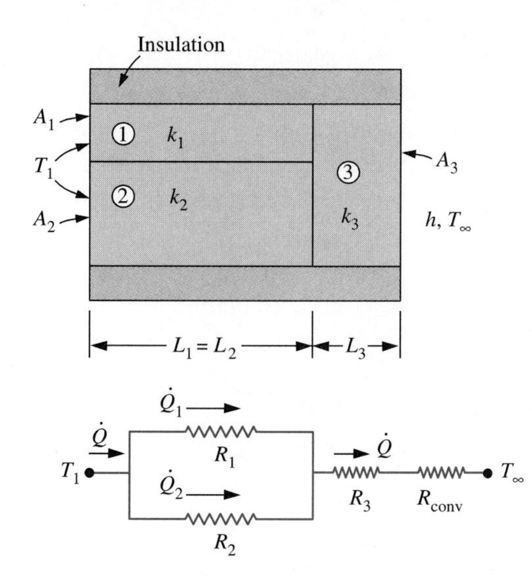


**FIGURE 8-38**
Thermal resistance network for combined series–parallel arrangement.

Once the individual thermal resistances are evaluated, the total resistance and the total rate of heat transfer can easily be determined from the relations above.

The result obtained will be somewhat approximate, since the surfaces of the third layer will probably not be isothermal, and heat transfer between the first two layers is likely to occur.

Two approximations commonly used in solving complex multi-dimensional heat transfer problems by treating them as one-dimensional (say, in the $x$-direction) using the thermal resistance network are (1) to assume any plane wall normal to the $x$-axis to be *isothermal* (i.e., to assume the temperature to vary in the $x$-direction only), and (2) to assume any plane parallel to the $x$-axis to be *adiabatic* (i.e., to assume heat transfer to occur in the $x$-direction only). These two assumptions result in different resistance networks, and thus different (but usually close) values for the total thermal resistance and thus heat transfer. The actual result lies between these two values. In geometries in which heat transfer occurs predominantly in one direction, either approach gives satisfactory results.

**EXAMPLE 8-8**

A 3-m-high and 5-m-wide wall consists of long 16 cm × 22 cm cross-section horizontal bricks [$k = 0.72$ W/(m · °C)] separated by 3-cm thick plaster layers [$k = 0.22$ W/(m · °C)]. There are also 2-cm-thick plaster layers on each side of the brick, and a 3-cm-thick rigid foam [$k = 0.026$ W/(m · °C)] on the inner side of the wall, as shown in Fig. 8-39. The indoor and the outdoor temperatures are 20°C and −10°C, and the convection heat transfer coefficients on the inner and the outer sides are $h_1 = 10$ W/(m² · °C) and $h_2 = 25$ W/(m² · °C), respectively. Assuming one-dimensional heat transfer and disregarding radiation, determine the rate of heat transfer through the wall.

**Solution** Heat transfer in this case is multi-dimensional, but it can be approximated as being one-dimensional, since heat transfer is predominantly in the $x$-direction. There is a pattern in the construction of this wall that repeats

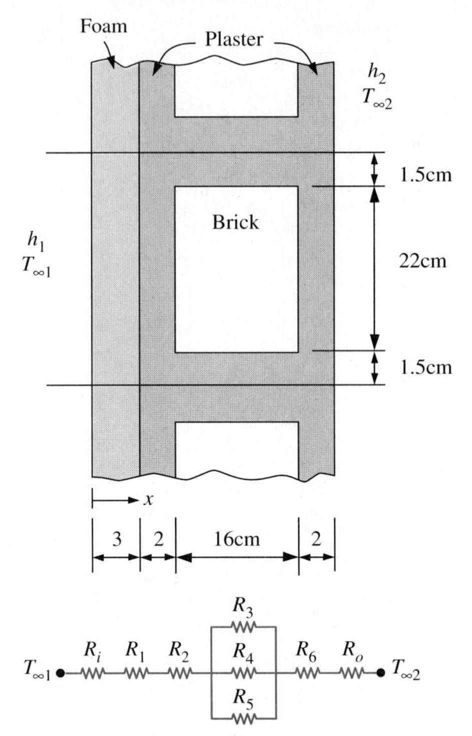


**FIGURE 8-39**
Schematic for Example 8-8.

itself every 25-cm distance in the vertical direction. There is no variation in the horizontal direction. Therefore, we consider a 1-m-deep and 0.25-m-high portion of the wall, since it is representative of the entire wall.

Assuming any cross-section of the wall normal to the $x$-direction to be *isothermal,* the thermal resistance network for the representative section of the wall becomes as shown in Fig. 8-39. The individual resistances are evaluated as follows:

$$R_i = R_{\text{conv},1} = \frac{1}{h_1 A} = \frac{1}{[10\ \text{W/(m}^2\cdot{}^\circ\text{C)}](0.25 \times 1\ \text{m}^2)} = 0.4^\circ\text{C/W}$$

$$R_1 = R_{\text{foam}} = \frac{L}{kA} = \frac{0.03\ \text{m}}{[0.026\ \text{W/(m}\cdot{}^\circ\text{C)}](0.25 \times 1\ \text{m}^2)} = 4.6^\circ\text{C/W}$$

$$R_2 = R_6 = R_{\text{plaster,side}} = \frac{L}{kA} = \frac{0.02\ m}{[0.22\ \text{W/(m}\cdot{}^\circ\text{C)}](0.25 \times 1\ \text{m}^2)}$$
$$= 0.36^\circ\text{C/W}$$

$$R_3 = R_5 = R_{\text{plaster,center}} = \frac{L}{kA} = \frac{0.16\ \text{m}}{[0.22\ \text{W/(m}\cdot{}^\circ\text{C)}](0.015 \times 1\ \text{m}^2)}$$
$$= 48.48^\circ\text{C/W}$$

$$R_4 = R_{\text{brick}} = \frac{L}{kA} = \frac{0.16\ \text{m}}{([0.72\ \text{W/(m}\cdot{}^\circ\text{C)}](0.22 \times 1\ \text{m}^2)} = 1.01^\circ\text{C/W}$$

$$R_o = R_{\text{conv},2} = \frac{1}{h_2 A} = \frac{1}{[25\ \text{W/(m}^2\,.\,{}^\circ\text{C)}](0.25 \times 1\ \text{m}^2)} = 0.16^\circ\text{C/W}$$

The three resistances $R_3$, $R_4$, and $R_5$ in the middle are parallel, and their equivalent resistance is determined from

$$\frac{1}{R_{mid}} = \frac{1}{R_3} + \frac{1}{R_4} + \frac{1}{R_5} = \frac{1}{48.48} + \frac{1}{1.01} + \frac{1}{48.48} = 1.03 \text{ W/°C}$$

which gives $$R_{mid} = 0.97\text{°C/W}$$

Now all the resistances are in series, and the total resistance is determined to be

$$\begin{aligned} R_{total} &= R_i + R_1 + R_2 + R_{mid} + R_6 + R_o \\ &= 0.4 + 4.6 + 0.36 + 0.97 + 0.36 + 0.16 \\ &= 6.85\text{°C/W} \end{aligned}$$

Then the steady rate of heat transfer through the wall becomes

$$\dot{Q} = \frac{T_{\infty 1} - T_{\infty 2}}{R_{total}} = \frac{[20 - (10)]\text{°C}}{6.85\text{°C/W}} = 4.38 \text{ W} \quad (\text{per } 0.25 \text{ m}^2 \text{ surface area})$$

or $4.38/0.25 = 17.5$ W per m$^2$ surface area. The total surface area of the wall is $A = 3 \text{ m} \times 5 \text{ m} = 15 \text{ m}^2$. Then the rate of heat transfer through the entire wall becomes

$$\dot{Q}_{total} = (17.5 \text{ W/m}^2)(15 \text{ m}^2) = 262.5 \text{ W}$$

Of course, this result is approximate, since we assumed the temperature within the wall to vary in one direction only and ignored any temperature change (and thus heat transfer) in the other two directions.

**Discussion** In the above solution, we assumed the temperature at any cross-section of the wall normal to the $x$-direction to be *isothermal.* We could also solve this problem by going to the other extreme and assuming the surfaces parallel to the $x$-direction to be *adiabatic.* The thermal resistance network in this case will be as shown in Fig. 8-40. By following the approach outlined above, the total thermal resistance in this case is determined to be $R_{total} = 6.97$°C/W, which is almost identical to the value 6.85°C/W obtained before. Thus, either approach would give roughly the same result in this case. This example demonstrates that either approach can be used in practice to obtain satisfactory results.

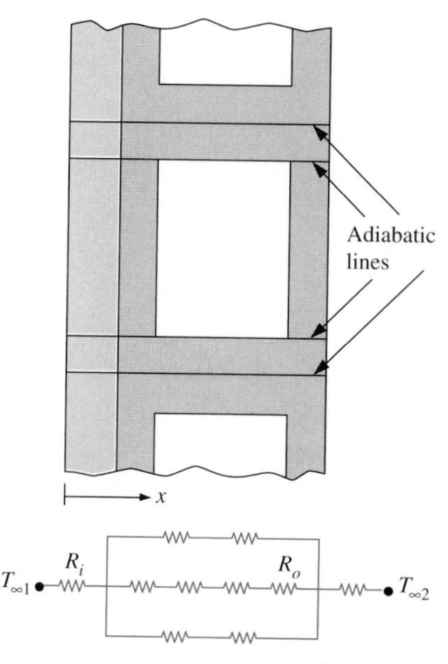


**FIGURE 8-40**

Alternative thermal resistance network for Example 8-8 for the case of surfaces parallel to the primary direction of heat transfer being adiabatic.

## 8-6 HEAT CONDUCTION IN CYLINDERS AND SPHERES

Consider steady heat conduction through a hot water pipe. Heat is continuously lost to the outdoors through the wall of the pipe, and we intuitively feel that heat transfer through the pipe is in the normal direction to the wall surface, and no significant heat transfer takes place in the pipe in other directions (Fig. 8-41). The wall of the pipe, whose thickness is rather small, separates two fluids at different temperatures and thus the temperature gradient in the radial direction will be relatively large. Further, if the fluid temperatures in an outside the pipe remain constant, then heat transfer through the pipe can be modeled as *steady.* Thus heat transfer through the wall of a pipe can be modeled as *steady* and *one-dimensional.* The temperature of the pipe in this case will depend on one direction only (the radial $r$-direction), and can be

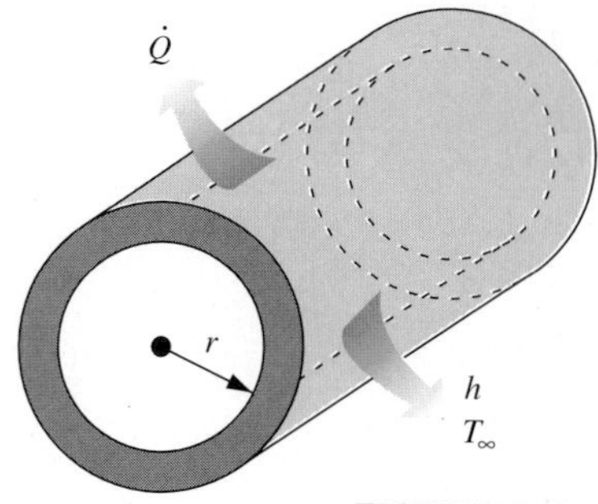


**FIGURE 8-41**

Heat is lost from a hot water pipe to the air outside in the radial direction, and thus heat transfer from a long pipe is very nearly one-dimensional.

expressed as $T = T(r)$. The temperature is independent of the azimuthal angle or the axial distance. This situation is approximated in practice in long cylindrical pipes and spherical containers.

In *steady* operation, there is no change in the temperature of the pipe with time at any point. Therefore, the rate of heat transfer into the pipe must be equal to the rate of heat transfer out of it. In other words, the rate of heat transfer through the pipe must be constant, $\dot{Q}_{\text{cond,cyl}} =$ constant.

Consider a long cylindrical layer (such as a circular pipe) of inner radius $r_1$, outer radius $r_2$, length $L$, and average thermal conductivity $k$ (Fig. 8-42). The two surfaces of the cylindrical layer are maintained at constant temperatures $T_1$ and $T_2$. There is no heat generation in the layer and the thermal conductivity is constant. For one-dimensional heat conduction through the cylindrical layer, we have $T(r)$. Then Fourier's law of heat conduction for heat transfer through the cylindrical layer can be expressed as

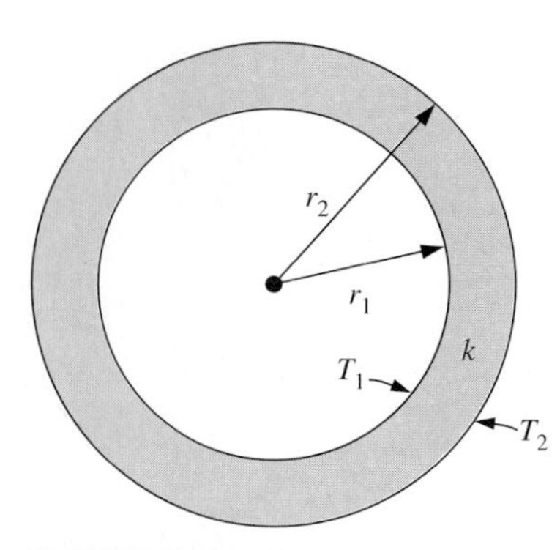


**FIGURE 8-42**

A long cylindrical pipe (or spherical shell) with specified inner and outer surface temperatures $T_1$ and $T_2$.

$$\dot{Q}_{\text{cond,cyl}} = -kA\frac{dT}{dr} \quad \text{(W)} \tag{8-37}$$

where $A = 2\pi rL$ is the heat transfer surface area at location $r$. Note that $A$ depends on $r$, and thus it *varies* in the direction of heat transfer. Separating the variables in the above equation and integrating from $r = r_1$ where $T(r_1) = T_1$ to $r = r_2$ where $T(r_2) = T_2$ gives

$$\int_{r=r_1}^{r_2} \frac{\dot{Q}_{\text{cond,cyl}}}{A}\,dr = -\int_{T=T_1}^{T_2} k\,dT \tag{8-28}$$

Substituting $A = 2\pi rL$ and performing the integrations give

$$\dot{Q}_{\text{cond,cyl}} = 2\pi Lk\frac{T_1 - T_2}{\ln(r_2/r_1)} \quad \text{(W)} \tag{8-39}$$

since $\dot{Q}_{\text{cond,cyl}} =$ constant. This equation can be rearranged as

$$\dot{Q}_{\text{cond,cyl}} = \frac{T_1 - T_2}{R_{\text{cyl}}} \quad \text{(W)} \tag{8-40}$$

where

$$R_{\text{cyl}} = \frac{\ln(r_2/r_1)}{2\pi Lk} = \frac{\ln(\text{outer radius/inner radius})}{2\pi \times (\text{length}) \times (\text{thermal conductivity})} \tag{8-41}$$

is the *thermal resistance* of the cylindrical layer against heat conduction or simply the **conduction resistance** of the cylinder layer.

We can repeat the analysis above for a *spherical layer* by taking $A = 4\pi r^2$ and performing the integrations in Eq. 8-38. The result can be expressed as

$$\dot{Q}_{\text{cond,sph}} = \frac{T_1 - T_2}{R_{\text{sph}}} \tag{8-42}$$

where

$$R_{\text{sph}} = \frac{r_2 - r_1}{4\pi r_1 r_2 k} = \frac{\text{outer radius} - \text{inner radius}}{4\pi(\text{outer radius})(\text{inner radius})(\text{thermal conductivity})} \tag{8-43}$$

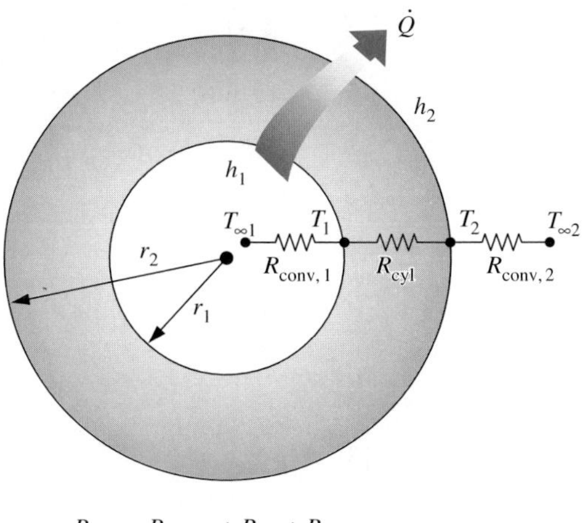


**FIGURE 8-43**
The thermal resistance network for cylindrical (or spherical) shell subjected to convection from both the inner and the outer sides.

is the *thermal resistance* of the spherical layer against heat conduction or simply the **conduction resistance** of the spherical layer.

Now consider steady one-dimensional heat flow through a cylindrical or spherical layer that is exposed to convection on both sides to fluids at temperatures $T_{\infty 1}$ and $T_{\infty 2}$ with heat transfer coefficients $h_1$ and $h_2$, respectively, as shown in Fig. 8-43. The thermal resistance network in this case consists of one conduction and two convection resistances in series, just like the one for the plane wall, and the rate of heat transfer under steady conditions can be expressed as

$$\dot{Q} = \frac{T_{\infty 1} - T_{\infty 2}}{R_{total}} \tag{8-44}$$

where

$$R_{total} = R_{conv,1} + R_{cyl} + R_{conv,2} = \frac{1}{(2\pi r_1 L)h_1} + \frac{\ln(r_2/r_1)}{2\pi Lk} + \frac{1}{(2\pi r_2 L)h_2} \tag{8-45}$$

for a *cylindrical* layer, and

$$R_{total} = R_{conv,1} + R_{sph} + R_{conv,2} = \frac{1}{(4\pi r_1^2)h_1} + \frac{r_2 - r_1}{4\pi r_1 r_2 k} + \frac{1}{(4\pi r_2^2)h_2} \tag{8-46}$$

for a *spherical* layer. Note that $A$ in the convection resistance relation $R_{conv} = 1/hA$ is the *surface area at which convection occurs.* It is equal to $A = 2\pi rL$ for a cylindrical surface and $A = 4\pi r^2$ for a spherical surface of radius $r$. Also note that the thermal resistances are in series, and thus the total thermal resistance is determined by simply adding the individual resistances, just like the electrical resistances connected in series.

## Multilayered Cylinders and Spheres

Steady heat transfer through multilayered cylindrical or spherical shells can be handled just like multilayered plane walls discussed earlier by simply adding an *additional resistance* in series for each *additional layer.*

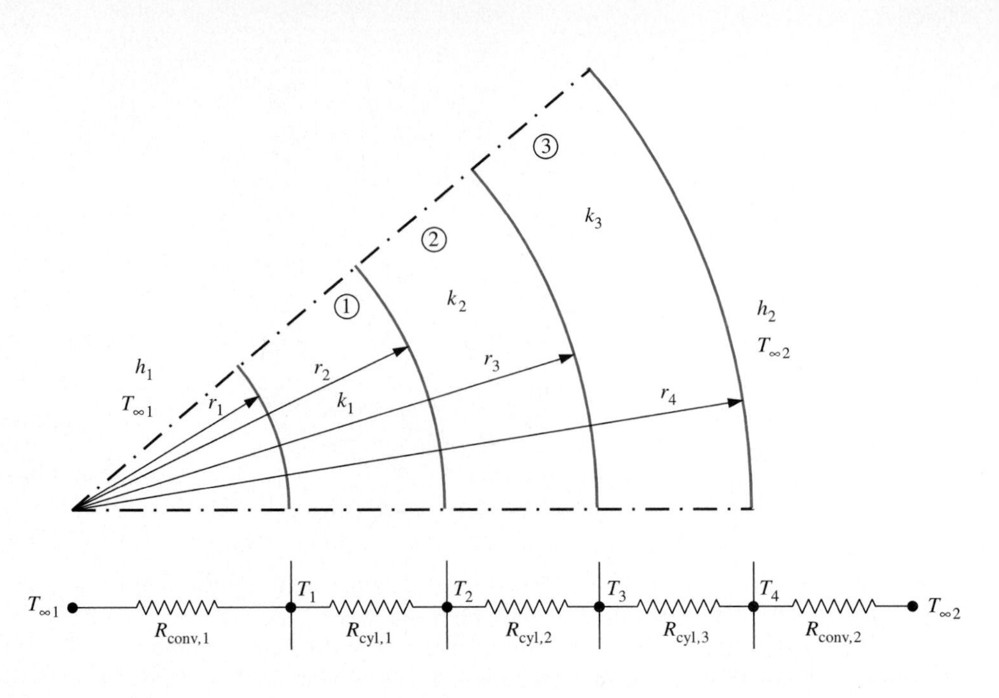


**FIGURE 8-44**

The thermal resistance network for heat transfer through a three-layered composite cylinder subjected to convection on both sides.

For example, the steady heat transfer rate through the three-layered composite cylinder of length $L$ shown in Fig. 8-44 with convection on both sides can be expressed as

$$\dot{Q} = \frac{T_{\infty 1} - T_{\infty 2}}{R_{\text{total}}} \tag{8-47}$$

where $R_{\text{total}}$ is the *total thermal resistance* expressed as

$$\begin{aligned} R_{\text{total}} &= R_{\text{conv},1} + R_{\text{cyl},1} + R_{\text{cyl},2} + R_{\text{cyl},3} + R_{\text{conv},2} \\ &= \frac{1}{h_1 A_1} + \frac{\ln(r_2/r_1)}{2\pi L k_1} + \frac{\ln(r_3/r_2)}{2\pi L k_2} + \frac{\ln(r_4/r_3)}{2\pi L k_3} + \frac{1}{h_2 A_4} \end{aligned} \tag{8-48}$$

where $A_1 = 2\pi r_1 L$ and $A_4 = 2\pi r_4 L$. Equation 8-48 can also be used for a three-layered spherical shell by replacing the thermal resistances of cylindrical layers by the corresponding spherical ones. Again note from the thermal resistance network that the resistances are in series, and thus the total thermal resistance is simply the *arithmetic sum* of the individual thermal resistances in the path of heat flow.

Once $\dot{Q}$ is known, we can determine any intermediate temperature $T_j$ by applying the relation $\dot{Q} = (T_i - T_j)/R_{\text{total},i-j}$ across any layer or layers such that $T_i$ is a *known* temperature at location $i$ and $R_{\text{total},i-j}$ is the total thermal resistance between locations $i$ and $j$ (Fig. 8-45). For example, once $\dot{Q}$ has been calculated, the interface temperature $T_2$ between the first and second cylindrical layers can be determined from

$$\dot{Q} = \frac{T_{\infty 1} - T_2}{R_{\text{conv},1} + R_{\text{cyl},1}} = \frac{T_{\infty 1} - T_2}{\dfrac{1}{h_1(2\pi r_1 L)} + \dfrac{\ln(r_2/r_1)}{2\pi L k_1}} \tag{8-49}$$

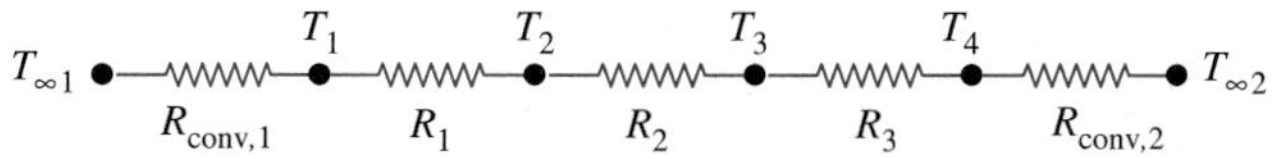

$T_{\infty 1}$ — $R_{conv,1}$ — $T_1$ — $R_1$ — $T_2$ — $R_2$ — $T_3$ — $R_3$ — $T_4$ — $R_{conv,2}$ — $T_{\infty 2}$

$$\dot{Q} = \frac{T_{\infty 1} - T_1}{R_{conv,1}} = \frac{T_{\infty 1} - T_2}{R_{conv,1} + R_1} = \frac{T_1 - T_3}{R_1 + R_2} = \frac{T_3 - T_4}{R_3} = \frac{T_3 - T_{\infty 2}}{R_3 + R_{conv,2}} = \cdots$$

**FIGURE 8-45**
The ratio $\Delta T/R$ across any layer is equal to $\dot{Q}$, which remains constant in one-dimensional steady conduction.

We could also calculate $T_2$ from

$$\dot{Q} = \frac{T_2 - T_{\infty 2}}{R_2 + R_3 + R_{conv,2}} = \frac{T_2 - T_{\infty 2}}{\dfrac{\ln(r_3/r_2)}{2\pi L k_2} + \dfrac{\ln(r_4/r_3)}{2\pi L k_3} + \dfrac{1}{h_o(2\pi r_4 L)}} \tag{8-50}$$

Although both relations will give the same result, we prefer the first one, since it involves fewer terms and thus less work.

The thermal resistance concept can also be used for *other geometries*, provided that the proper conduction resistances and the proper surface areas in convection resistances are used.

**EXAMPLE 8-9**

A 3-m internal diameter spherical tank made of 2-cm-thick stainless steel [$k = 15$ W/(m · °C)] is used to store iced water at $T_{\infty 1} = 0$°C. The tank is located in a room whose temperature is $T_{\infty 2} = 22$°C. The walls of the room are also at 22°C. The outer surface of the tank is black (emissivity $\varepsilon = 1$), and heat transfer between the outer surface of the tank and the surroundings is by natural convection and radiation. The convection heat transfer coefficients at the inner and the outer surfaces of the tank are $h_1 = 80$ W/(m² · °C) and $h_2 = 10$ W/(m² · °C), respectively. Determine (*a*) the rate of heat transfer to the iced water in the tank, and (*b*) the amount of ice at 0°C that melts during a 24-h period. The heat of fusion of water at atmospheric pressure is $h_{if} = 333.7$ kJ/kg.

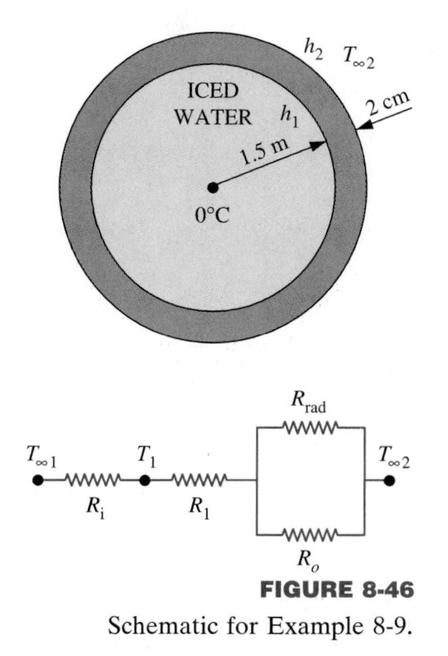


**FIGURE 8-46**
Schematic for Example 8-9.

**Solution** (*a*) We assume the specified thermal conditions do not change with time, and are uniform for the entire tank, so that heat transfer is steady and one-dimensional. The thermal resistance network for this problem is given in Fig. 8-46.

Noting that the inner diameter of the tank is $D_1 = 3$ m and the outer diameter is $D_2 = 3.04$ m, the inner and the outer surface areas of the tank are

$$A_1 = \pi D_1^2 = \pi(3\text{ m})^2 = 28.3\text{ m}^2$$

$$A_2 = \pi D_2^2 = \pi(3.04\text{ m})^2 = 29.0\text{ m}^2$$

Also, the radiation heat transfer coefficient is given by

$$h_{rad} = \varepsilon\sigma(T_2^2 + T_{\infty 2}^2)(T_2 + T_{\infty 2})$$

But we do not know the outer surface temperature $T_2$ of the tank, and thus we cannot calculate $h_{rad}$. Therefore, we need to assume a $T_2$ value now and check the accuracy of this assumption later. We will repeat the calculations if necessary using a revised value for $T_2$.

We note that $T_2$ must be between 0°C and 22°C, but it must be closer to 0°C, since the heat transfer coefficient inside the tank is much larger. Taking $T_2 = 5°C = 278$ K, the radiation heat transfer coefficient is determined to be

$$\begin{aligned} h_{rad} &= (1)[5.67 \times 10^{-8} \text{ W/(m}^2 \cdot \text{K}^4)][(295 \text{ K})^2 + (278 \text{ K})^2][(295 + 278) \text{ K}] \\ &= 5.34 \text{ W/(m}^2 \cdot \text{K}) = 5.34 \text{ W/(m}^2 \cdot °\text{C}) \end{aligned}$$

Then the individual thermal resistances become

$$R_i = R_{conv,1} = \frac{1}{h_1 A_1} = \frac{1}{[80 \text{ W/(m}^2 \cdot °\text{C})](28.3 \text{ m}^2)} = 0.000442°\text{C/W}$$

$$\begin{aligned} R_1 = R_{sphere} &= \frac{r_2 - r_1}{4\pi k r_1 r_2} = \frac{(1.52 - 1.50) \text{ m}}{4\pi[15 \text{ W/(m} \cdot °\text{C})](1.52 \text{ m})(1.50 \text{ m})} \\ &= 0.000047°\text{C/W} \end{aligned}$$

$$R_o = R_{conv,2} = \frac{1}{h_2 A_2} = \frac{1}{[10 \text{ W/(m}^2 \cdot °\text{C})](29.0 \text{ m}^2)} = 0.00345°\text{C/W}$$

$$R_{rad} = \frac{1}{h_{rad} A_2} = \frac{1}{[5.34 \text{ W/(m}^2 \cdot °\text{C})](29.0 \text{ m}^2)} = 0.00646°\text{C/W}$$

The two parallel resistances $R_o$ and $R_{rad}$ can be replaced by an equivalent resistance $R_{equiv}$ determined from

$$\frac{1}{R_{equiv}} = \frac{1}{R_o} = \frac{1}{R_{rad}} = \frac{1}{0.00345} + \frac{1}{0.00646} = 444.7 \text{ W/°C}$$

which gives $\qquad R_{equiv} = 0.00225°\text{C/W}$

Now all the resistances are in series, and the total resistance is determined to be

$$R_{total} = R_i + R_1 + R_{equiv} = 0.000442 + 0.000047 + 0.00225 = 0.00274°\text{C/W}$$

Then the steady rate of heat transfer to the iced water becomes

$$\dot{Q} = \frac{T_{\infty 1} - T_{\infty 2}}{R_{total}} = \frac{(22 - 0)°\text{C}}{0.00274°\text{C/W}} = 8029 \text{ W} \qquad (\text{or } \dot{Q} = 8.029 \text{ kJ/s}).$$

To check the validity of our original assumption, we now determine the outer surface temperature from

$$\dot{Q} = \frac{T_\infty - T_2}{R_{equiv}} \longrightarrow T_2 = T_\infty - \dot{Q}R_{equiv} = 22°\text{C} - (8029 \text{ W})(0.00225°\text{C/W}) = 4°\text{C}$$

which is sufficiently close to the 5°C assumed in the determination of the radiation heat transfer coefficient. Therefore, there is no need to repeat the calculations using 4°C for $T_2$.

(*b*) The total amount of heat transfer during a 24-h period is

$$Q = \dot{Q}\,\Delta t = (8.029 \text{ kJ/s})(24 \times 3600 \text{ s}) = 673{,}700 \text{ kJ}$$

Noting that it takes 333.7 kJ of energy to melt 1 kg of ice at 0°C, the amount of ice that will melt during a 24-h period is

$$m_{\text{ice}} = \frac{Q}{h_{if}} = \frac{673{,}700\ \text{kJ}}{333.7\ \text{kJ/kg}} = 2079\ \text{kg}$$

Therefore, about 2 metric tons of ice will melt in the tank every day.

**Discussion** An easier way to deal with combined convection and radiation at a surface when the surrounding medium and surfaces are at the same temperature is to add the radiation and convection heat transfer coefficients and to treat the result as the convection heat transfer coefficient; that is, to take $h = 10 + 5.34 = 15.34\ \text{W/(m}^2\cdot{}^\circ\text{C)}$ in this case. This way, we can ignore radiation, since its contribution is accounted for in the convection heat transfer coefficient. The convection resistance of the outer surface in this case would be

$$R_{\text{combined}} = \frac{1}{h_{\text{combined}}A_2} = \frac{1}{(15.34\ \text{W/(m}^2\cdot{}^\circ\text{C)})(29.0\ \text{m}^2)} = 0.00225^\circ\text{C/W}$$

which is identical to the value obtained for the equivalent resistance for the parallel convection and the radiation resistances.

**EXAMPLE 8-10**

Steam at $T_{\infty 1} = 320^\circ\text{C}$ flows in a cast iron pipe $[k = 80\ \text{W/(m}\cdot{}^\circ\text{C)}]$ whose inner and outer diameters are $D_1 = 5$ cm and $D_2 = 5.5$ cm, respectively. The pipe is covered with 3-cm-thick glass wool insulation $[k = 0.05\ \text{W/(m}\cdot{}^\circ\text{C)}]$. Heat is lost to the surroundings at $T_\infty = 5^\circ\text{C}$ by natural convection and radiation, with a combined heat transfer coefficient of $h_2 = 18\ \text{W/(m}^2\cdot{}^\circ\text{C)}$. Taking the heat transfer coefficient inside the pipe to be $h_1 = 60\ \text{W/(m}^2\cdot{}^\circ\text{C)}$, determine the rate of heat loss from the steam per unit length of the pipe. Also determine the temperature drops across the pipe shell and the insulation.

**Solution** (*a*) We assume steady and one-dimensional heat transfer through the pipe. The thermal resistance network for this problem involves four resistances in series, and is given in Fig. 8-47. Taking $L = 1$ m, the areas of the surfaces exposed to convection are determined to be

$$A_1 = 2\pi r_1 L = 2\pi(0.025\ \text{m})(1\ \text{m}) = 0.157\ \text{m}^2$$

$$A_3 = 2\pi r_3 L = 2\pi(0.0575\ \text{m})(1\ \text{m}) = 0.361\ \text{m}^2$$

Then the individual thermal resistances become

$$R_i = R_{\text{conv},1} = \frac{1}{h_1 A} = \frac{1}{[60\ \text{W/(m}^2\cdot{}^\circ\text{C)}](0.157\ \text{m}^2)} = 0.106^\circ\text{C/W}$$

$$R_1 = R_{\text{pipe}} = \frac{\ln(r_2/r_1)}{2\pi k_1 L} = \frac{\ln(2.75/2.5)}{2\pi[80\ \text{W/(m}\cdot{}^\circ\text{C)}](1\ \text{m})} = 0.0002^\circ\text{C/W}$$

$$R_2 = R_{\text{insulation}} = \frac{\ln(r_3/r_2)}{2\pi k_2 L} = \frac{\ln(5.75/2.75)}{2\pi[0.05\ \text{W/(m}\cdot{}^\circ\text{C)}](1\ \text{m})} = 2.35^\circ\text{C/W}$$

$$R_o = R_{\text{conv},2} = \frac{1}{h_2 A_3} = \frac{1}{[18\ \text{W/(m}^2\cdot{}^\circ\text{C)}](0.361\ \text{m}^2)} = 0.154^\circ\text{C/W}$$

Noting that all resistances are in series, the total resistance is determined to be

$$R_{\text{total}} = R_i + R_1 + R_2 + R_o = 0.106 + 0.0002 + 2.35 + 0.154 = 2.61^\circ\text{C/W}$$

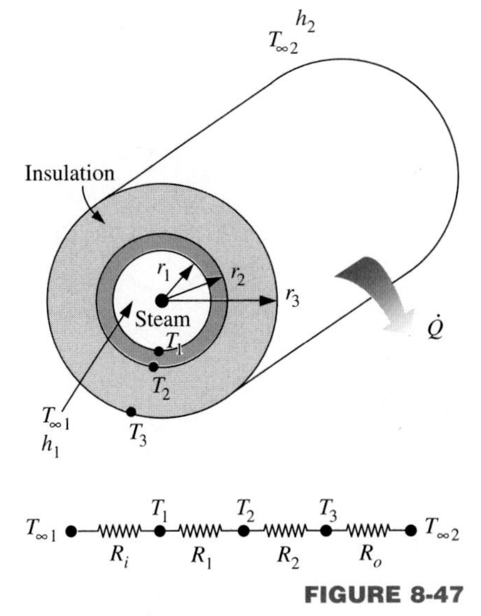


**FIGURE 8-47**

Schematic for Example 8-10.

Then the steady rate of heat loss from the steam becomes

$$\dot{Q} = \frac{T_{\infty 1} - T_{\infty 2}}{R_{\text{total}}} = \frac{(320 - 5)°\text{C}}{2.61°\text{C/W}} = 120.7\ \text{W} \quad \text{(per m pipe length)}$$

The heat loss for a given pipe length can be determined by multiplying the above quantity by the pipe length $L$.

The temperature drops across the pipe and the insulation are determined from Eq. 8-25 to be

$$\Delta T_{\text{pipe}} = \dot{Q}R_{\text{pipe}} = (120.7\ \text{W})(0.0002°\text{C/W}) = 0.02°\text{C}$$
$$\Delta T_{\text{insulation}} = \dot{Q}R_{\text{insulation}} = (120.7\ \text{W})(2.35°\text{C/W}) = 284°\text{C}$$

That is, the temperatures between the inner and the outer surfaces of he pipe differ by 0.02°C, whereas the temperatures between the inner and the outer surfaces of the insulation differ by 284°C.

Note that the thermal resistance of the pipe is too small relative to the other resistances, and can be neglected without causing any significant error. Also note that the temperature drop across the pipe is practically zero, and thus the pipe can be assumed to be isothermal. The resistance to heat flow in insulated pipes is primarily due to the insulation.

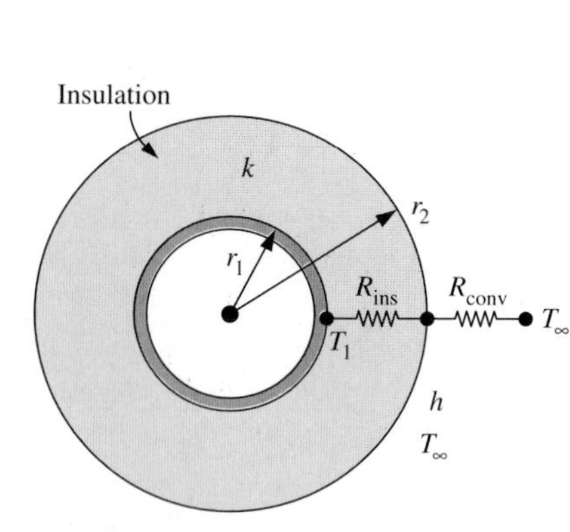


**FIGURE 8-48**
An insulated cylindrical pipe exposed to convection from the outer surface and the thermal resistance network associated with it.

## 8-7 CRITICAL RADIUS OF INSULATION

We know that adding more insulation to a wall or to the attic always decreases heat transfer. The thicker the insulation, the lower the heat transfer rate. This is expected, since the heat transfer area $A$ is constant, and adding insulation always increases the thermal resistance of the wall without effecting the convection resistance.

Adding insulation to a cylindrical pipe or a spherical shell, however, is a different matter. The additional insulation increases the conduction resistance of the insulation layer, but decreases the convection resistance of the surface because of the increase in the outer surface area for convection. The heat transfer from the pipe may increase or decrease, depending on which effect dominates.

Consider a cylindrical pipe of outer radius $r_1$ whose outer surface temperature $T_1$ is maintained constant (Fig. 8-48). The pipe is now insulated with an insulating material whose thermal conductivity is $k$ and outer radius is $r_2$. Heat is lost from the pipe to the surrounding medium at temperature $T_\infty$, with a convection heat transfer coefficient $h$. The rate of heat transfer from the insulated pipe to the surrounding air can be expressed as (see Fig. 8-49)

$$\dot{Q} = \frac{T_1 - T_\infty}{R_{\text{ins}} + R_{\text{conv}}} = \frac{T_1 - T_\infty}{\dfrac{\ln(r_2/r_1)}{2\pi Lk} + \dfrac{1}{h(2\pi r_2 L)}} \tag{8-51}$$

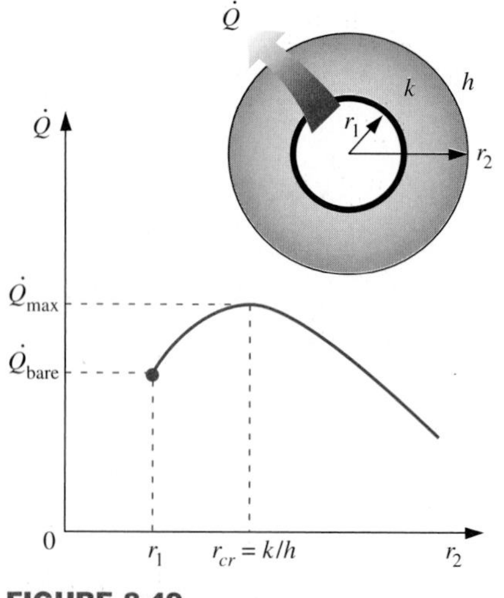


**FIGURE 8-49**

The variation of $\dot{Q}$ with the outer radius of the insulation $r_2$ is also plotted in Fig. 8-49. The value of $r_2$ at which $\dot{Q}$ reaches a maximum is determined from the requirement that $d\dot{Q}/dr_2 = 0$ (zero slope). Performing the differentiation and solving for $r_2$ yields the **critical radius of insulation** for a cylindrical body to be

$$r_{cr,cylinder} = \frac{k}{h} \quad \text{(m)} \tag{8-52}$$

Note that the critical radius of insulation depends on the thermal conductivity of the insulation $k$ and the external convection heat transfer coefficient $h$. The rate of heat transfer from the cylinder increases with the addition of insulation for $r_2 < r_{cr}$, reaches a maximum when $r_2 = r_{cr}$, and starts to decrease for $r_2 > r_{cr}$. Thus, insulating the pipe may actually increase the rate of heat transfer from the pipe instead of decreasing it when $r_2 < r_{cr}$.

The important question to answer at this point is whether we need to be concerned about the critical radius of insulation when insulating hot water pipes or even hot water tanks. Should we always check and make sure that the outer radius of insulation exceeds the critical radius before we install any insulation? Probably not, as explained below.

The value of the critical radius $r_{cr}$ will be the largest when $k$ is large and $h$ is small. Noting that the lowest value of $h$ encountered in practice is about 5 W/(m$^2$ · °C) for the case of natural convection of gases, and that the thermal conductivity of common insulating materials is about 0.05 W/(m · °C), the largest value of the critical radius we are likely to encounter is

$$r_{cr,max} = \frac{k_{max,insulation}}{h_{min}} \approx \frac{0.05 \text{ W/(m} \cdot {}^\circ\text{C)}}{5 \text{ W/(m}^2 \cdot {}^\circ\text{C)}} = 0.01 \text{ m} = 1 \text{ cm}$$

This value would be even smaller when the radiation effects are considered. The critical radius would be much less in forced convection, often less than 1 mm, because of the much larger $h$ values associated with forced convection. Therefore, we can insulate hot water or steam pipes freely without worrying about the possibility of increasing the heat transfer by insulating the pipes.

The radius of electric wires may be smaller than the critical radius. Therefore, plastic electrical insulation may actually *enhance* the heat transfer from electric wires and thus keep their steady operating temperatures at lower and thus safer levels.

The discussions above can be repeated for a sphere, and it can be shown in a similar manner that the critical radius of insulation for a spherical shell is

$$r_{cr,sphere} = \frac{2k}{h} \tag{8-53}$$

where $k$ is the thermal conductivity of the insulation and $h$ is the convection heat transfer coefficient on the outer surface.

**EXAMPLE 8-11**

A 3-mm-diameter and 5-m-long electric wire is tightly wrapped with a 2-mm-thick plastic cover whose thermal conductivity is $k = 0.15$ W/(m · °C). Electrical measurements indicate that a current of 10 A passes through the wire and there is a voltage drop of 8 V along the wire. If the insulated wire is

exposed to a medium at $T_\infty = 30°C$ with a heat transfer coefficient of $h = 12\ W/(m^2 \cdot °C)$, determine the temperature at the interface of the wire and the plastic cover in steady operation. Also determine whether doubling the thickness of the plastic cover will increase or decrease this interface temperature.

**Solution** We assume the thermal contact resistance at the interface is negligible, and the heat transfer coefficient incorporates the radiation effects, if any. Heat is generated in the wire, and its temperature rises as a result of resistance heating. We assume heat is generated uniformly throughout the wire, and is transferred to the surrounding medium in the radial direction. In steady operation, the rate of heat transfer becomes equal to the heat generated within the wire, which is determined from

$$\dot{Q} = \dot{W}_e = VI = (8\ V)(10\ A) = 80\ W$$

The thermal resistance network for this problem involves a conduction resistance for the plastic cover and a convection resistance for the outer surface in series, as shown in Fig. 8-50. The values of these two resistances are determined to be

$$A_2 = (2\pi r_2)L = 2\pi(0.0035\ m)(5\ m) = 0.110\ m^2$$

$$R_{conv} = \frac{1}{hA_2} = \frac{1}{(12\ W/m^2 \cdot °C)(0.110\ m^2)} = 0.76°C/W$$

$$R_{plastic} = \frac{\ln(r_2/r_1)}{2\pi kL} = \frac{\ln(3.5/1.5)}{2\pi[0.15\ W/(m \cdot °C)](5\ m)} = 0.18°C/W$$

and $\quad R_{total} = R_{plastic} + R_{conv} = 0.76 + 0.18 = 0.94°C/W$

Then the interface temperature can be determined from

$$\dot{Q} = \frac{T_1 - T_\infty}{R_{total}} \longrightarrow T_1 = T_\infty + \dot{Q}R_{total} = 30°C + (80\ W)(0.94°C/W) = 105°C$$

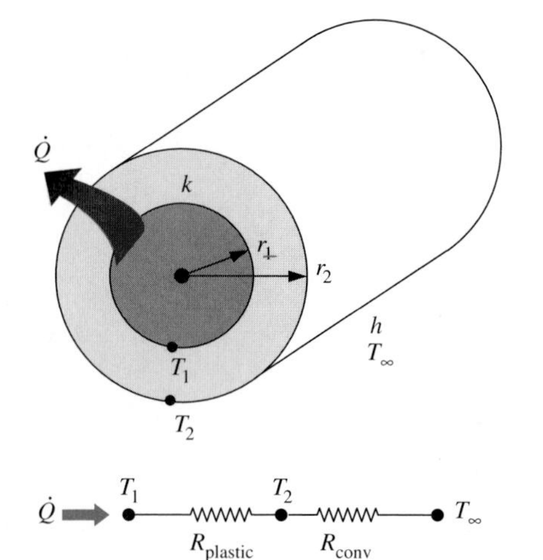


**FIGURE 8-50**
Schematic for Example 8-11.

Note that we did not involve the electrical wire directly in the thermal resistance network, since the wire involves heat generation.

To answer the second part of the question, we need to know the critical radius of insulation of the plastic cover. It is determined from Eq. 8-52 to be

$$r_{cr} = \frac{k}{h} = \frac{0.15\ W/(m \cdot °C)}{12\ W/(m^2 \cdot °C)} = 0.0125\ m = 12.5\ mm$$

which is larger than the radius of the plastic cover. Therefore, increasing the thickness of the plastic cover will *enhance* heat transfer until the outer radius of the cover reaches 12.5 mm. As a result, the rate of heat transfer $\dot{Q}$ will *increase* when the interface temperature $T_1$ is held constant, or $T_1$ will *decrease* when $\dot{Q}$ is held constant which is the case here.

It can be shown by repeating the calculations above for a 4-mm-thick plastic cover that the interface temperature drops to 90.6°C when the thickness of the plastic cover is doubled. It can also be shown in a similar manner that the interface reaches a minimum temperature of 83°C when the outer radius of the plastic cover equals the critical radius.

## 8-8 ■ HEAT GENERATION IN A SOLID

Many practical heat transfer applications involve the conversion of some form of energy into *heat* energy in the medium. Such mediums are said to involve internal *heat generation,* which manifests itself as a rise in temperature throughout the medium. Some examples of heat generation are *resistance heating* in wires, exothermic *chemical reactions* in a solid, and *nuclear reactions* in nuclear fuel rods where electrical, chemical, and nuclear energies are converted to heat, respectively (Fig. 8-51). The absorption of radiation throughout the volume of a semitransparent medium such as water can also be considered as heat generation within the medium.

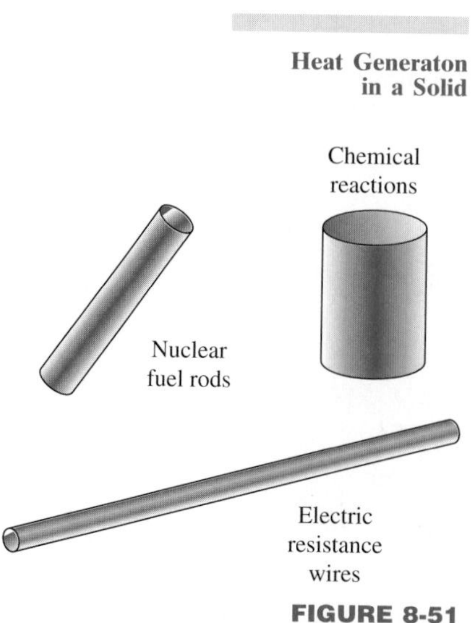


**FIGURE 8-51**
Heat generation in solids is commonly encountered in practice.

Heat generation is usually expressed *per unit volume* of the medium, and is denoted by $\dot{g}$, whose unit is W/m$^3$. For example, heat generation in an electrical wire of outer radius $r_o$ and length $L$ can be expressed as

$$\dot{g} = \frac{\dot{E}_{g,\text{electric}}}{V_{\text{wire}}} = \frac{I^2 R_e}{\pi r_o^2 L} \quad (\text{W/m}^3) \tag{8-54}$$

where $I$ is the electric current and $R_e$ is the electrical resistance of the wire.

The temperature of a medium *rises* during heat generation as a result of the absorption of the generated heat by the medium during transient start-up period. As the temperature of the medium increases, so does the heat transfer from the medium to its surroundings. This continues until steady operating conditions are reached, and the rate of heat generation equals the rate of heat transfer to the surroundings. Once steady operation has been established, the temperature of the medium at any point no longer changes.

The *maximum temperature* $T_{\text{max}}$ in a solid that involves uniform heat generation will occur at a location *furthest away* from the outer surface when the outer surface of the solid is maintained at a constant temperature $T_s$. For example, the maximum temperature occurs at the *midplane* in a plain wall, at the *centerline* in a long cylinder, and at the *midpoint* in a sphere. The temperature distribution within the solid in these cases will be *symmetrical* about the center of symmetry.

The quantities of major interest in a medium with heat generation are the surface temperture $T_s$ and the maximum temperature $T_{\text{max}}$ that occurs in the medium in *steady* operation. Below we develop expressions for these two quantities for common geometries for the case of *uniform* heat generation ($\dot{g}$ = constant) within the medium.

Consider a solid medium of surface area $A$, volume $V$, and constant thermal conductivity $k$, where heat is generated at a constant rate of $\dot{g}$ per unit volume. Heat is transferred from the solid to the surrounding medium at $T_\infty$, with a constant heat transfer coefficient of $h$. All the surfaces of the solid are maintained at a common temperature $T_s$. Under *steady* conditions, the energy balance for this solid can be expressed as (Fig. 8-52)

$$\begin{pmatrix}\text{rate of} \\ \textit{heat transfer} \\ \text{from the solid}\end{pmatrix} = \begin{pmatrix}\text{rate of} \\ \textit{energy generation} \\ \text{within the solid}\end{pmatrix} \tag{8-55}$$

or

$$\dot{Q} = \dot{g}V \quad (\text{W}) \tag{8-56}$$

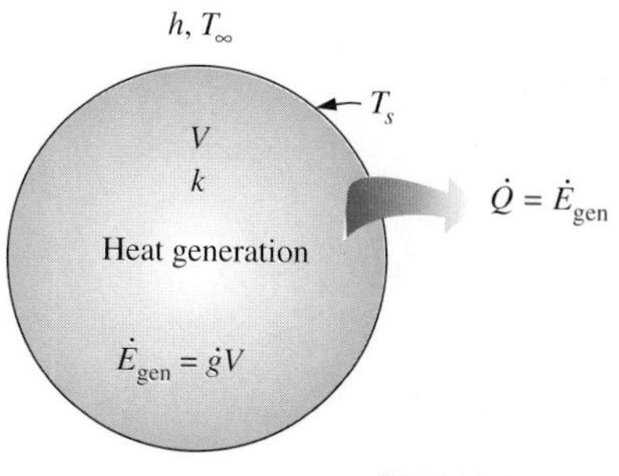


**FIGURE 8-52**
At steady conditions, the entire heat generated in a solid must leave the solid through its outer surface.

Disregarding radiation (or incorporating it in the heat transfer coefficient $h$), the heat transfer rate can also be expressed from the Newton's law of cooling as

$$\dot{Q} = hA(T_s - T_\infty) \quad \text{(W)} \tag{8-57}$$

Combining Eqs. 8-55 and 8-57 and solving for the surface temperature $T_s$ gives

$$T_s = T_\infty + \frac{\dot{g}V}{hA} \tag{8-58}$$

For a large *plane wall* of thickness $2L$ ($A = 2A_{\text{wall}}$ and $V = 2LA_{\text{wall}}$), a long solid *cylinder* of radius $r_o$ ($A = 2\pi r_o L$ and $V = \pi r_o^2 L$), and a solid *sphere* of radius $r_o$ ($A = 4\pi r_o^2$ and $V = \frac{4}{3}\pi r_o^3$), Eq. 8-58 reduces to

$$T_{s,\,\text{plane wall}} = T_\infty + \frac{\dot{g}L}{h}$$

$$T_{s,\,\text{cylinder}} = T_\infty + \frac{\dot{g}r_o}{2h} \tag{8-59}$$

$$T_{s,\,\text{sphere}} = T_\infty + \frac{\dot{g}r_o}{3h}$$

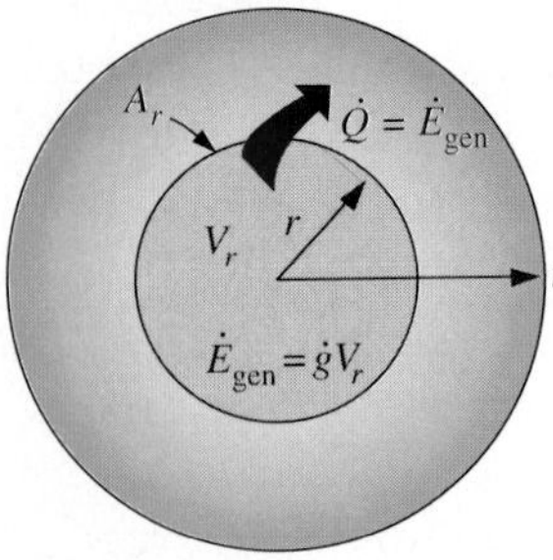

Note that the rise in surface temperature $T_s$ is due to heat generation in the solid.

Consider heat transfer from the long solid cylinder with heat generation one more time. We mentioned above that, under *steady* conditions, the entire heat generated within the medium is conducted through the outer surface of the cylinder. Now consider an imaginary inner cylinder of radius $r$ within the cylinder (Fig. 8-53). Again the *heat generated* within this inner cylinder must be equal to the *heat conducted* through the outer surface of this inner cylinder. That is, from Fourier's law of heat conduction,

**FIGURE 8-53**
Heat conducted through a cylindrical shell of radius $r$ is equal to the heat generated within a shell.

$$-kA_r\frac{dT}{dr} = \dot{g}V_r \tag{8-60}$$

where $A_r = 2\pi rL$ and $V_r = \pi r^2 L$ at any location $r$. Substituting these expressions into Eq. 8-60, separating the variables, and integrating from $r = 0$ where $T(0) = T_o$ to $r = r_o$ where $T(r_o) = T_s$ yield

$$\Delta T_{\text{max, cylinder}} = T_o - T_s = \frac{\dot{g}r_o^2}{4k} \tag{8-61}$$

where $T_o$ is the centerline temperature of the cylinder, which is the *maximum temperature,* and $\Delta T_{\max}$ is the difference between the centerline and the surface temperatures of the cylinder, which is the *maximum temperature rise* in the cylinder above the surface temperature. Once $\Delta T_{\max}$ is available, the centerline temperature can easily be determined from (Fig. 8-54)

$$T_{\text{center}} = T_o = T_s + \Delta T_{\max} \tag{8-62}$$

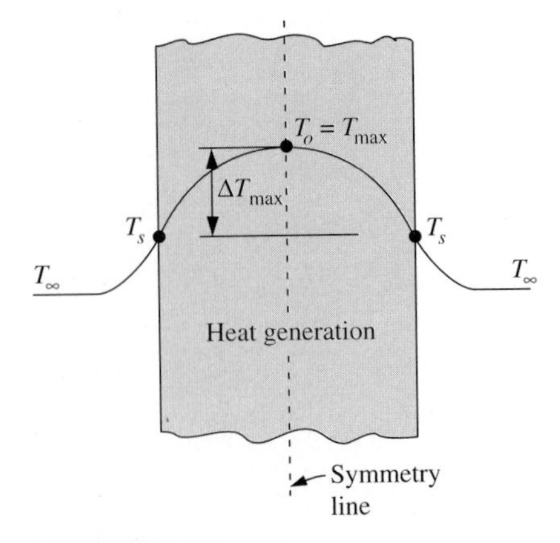


**FIGURE 8-54**
The maximum temperature in a symmetrical solid with uniform heat generation occurs at its center.

The approach outlined above can also be used to determine the *maximum temperature rise* in a plane wall of thickness $2L$ and a solid sphere of radius $r_o$, with the following results:

$$\Delta T_{\text{max, plane wall}} = \frac{\dot{g}L^2}{2k}$$
$$\Delta T_{\text{max, sphere}} = \frac{\dot{g}r_o^2}{6k} \tag{8-63}$$

Again the maximum temperature at the center can be determined from Eq. 8-62 by adding the maximum temperature rise to the surface temperature of the solid.

Note that the thermal resistance concept discussed earlier cannot be used when there is heat generation in the medium since the heat transfer rate through the medium in this case is no longer constant.

**EXAMPLE 8-12**

A 2-kW resistance heater wire whose thermal conductivity is $k = 15\ \text{W/(m}\cdot\text{°C)}$ has a diameter of $D = 4$ mm and a length of $L = 0.5$ m, and is used to boil water (Fig. 8-55). If the outer surface temperature of the resistance wire is $T_s = 105$°C, determine the temperature at the center of the wire.

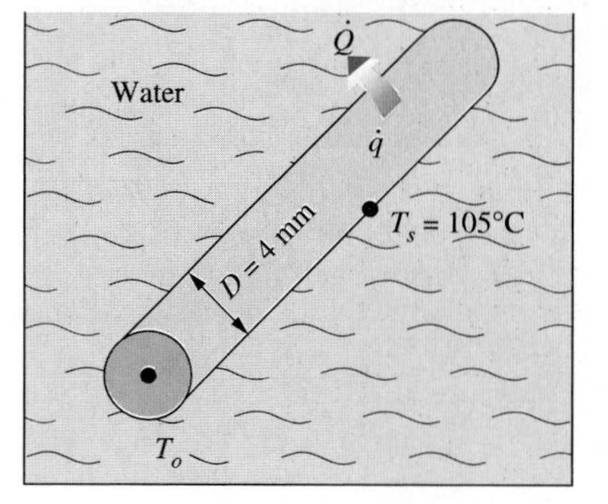


**FIGURE 8-55**
Schematic for Example 8-12.

**Solution** The 2-kW resistance heater converts electric energy into heat at a rate of 2 kW. Assuming heat is generated uniformly throughout the wire, the heat generation per unit volume of the wire becomes

$$\dot{g} = \frac{\dot{Q}_{\text{gen}}}{V_{\text{wire}}} = \frac{\dot{Q}_{\text{gen}}}{\pi r_o^2 L} = \frac{2000\ \text{W}}{\pi(0.002\ \text{m})^2(0.5\ \text{m})} = 0.318 \times 10^9\ \text{W/m}^3$$

Then the center temperature of the wire is determined from Eq. 8-61 to be

$$T_o = T_s + \frac{\dot{g}r_o^2}{4k} = 105\text{°C} + \frac{(0.318 \times 10^9\ \text{W/m}^3)(0.002\ \text{m})^2}{4 \times [15\ \text{W/(m}\cdot\text{°C)}]} = 126.2\text{°C}$$

Note that the temperature difference between the center and the surface of the wire is 21.2°C.

## 8-9 ■ HEAT TRANSFER FROM FINNED SURFACES

The rate of heat transfer from a surface at a temperature $T_s$ to the surrounding medium at $T_\infty$ is given by Newton's law of cooling as

$$\dot{Q}_{\text{conv}} = hA(T_s - T_\infty) \tag{8-64}$$

where $A$ is the heat transfer surface area and $h$ is the convection heat transfer coefficient. When the temperatures $T_s$ and $T_\infty$ are fixed by design considerations, as is often the case, there are *two ways* to increase the rate of heat transfer: to increase the *convection heat transfer coefficient h* or to increase the *surface area A*. Increasing $h$ may require the installation of a pump or fan, or the replacement of the existing one with a larger one, and this approach may or may not be practical. Besides, it may not be adequate. The alternative is to increase the surface area by

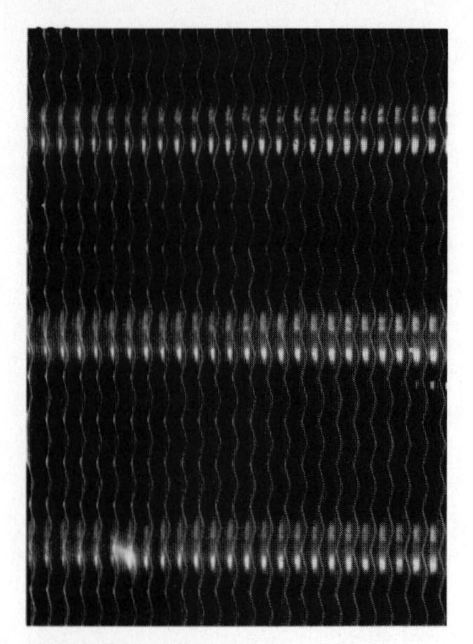

**FIGURE 8-56**

The thin plate fins of a car radiatior greatly increase the rate of heat transfer to the air. (Courtesy of James Kleiser.)

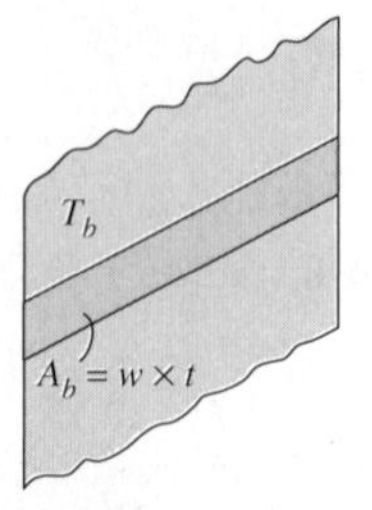


(*a*) Surface without fins

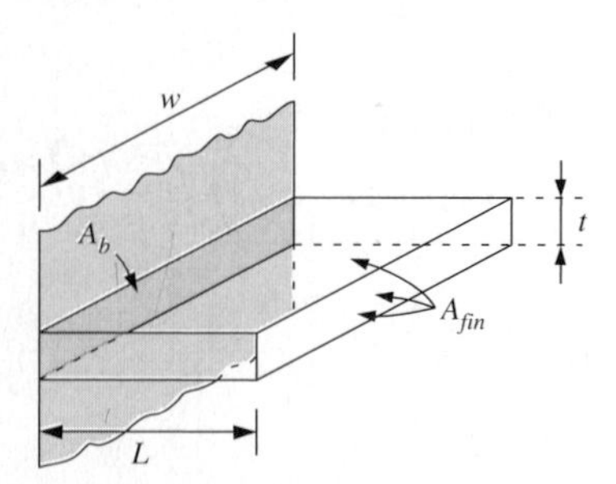


(*b*) Surface with a fin

$$A_{\text{fin}} = 2 \times w \times L + w \times t$$
$$\cong 2 \times w \times L$$

**FIGURE 8-57**

Fins enhance heat transfer from a surface by enhancing surface area.

attaching to the surface *extended surfaces* called *fins* made of highly conductive materials such as aluminum. Finned surfaces are manufactured by extruding, welding, or wrapping a thin metal sheet on a surface. Fins enhance heat transfer from a surface by exposing a larger surface area to convection.

Finned surfaces are commonly used in practice to enhance heat transfer, and they often increase the rate of heat transfer from a surface several folds. The *car radiator* shown in Fig. 8-56 is an example of a finned surface. The closely packed thin metal sheets attached to the hot water tubes increase the surface area for convection and thus the rate of convection heat transfer from the tubes to the air many times.

In the analysis of the fins we consider *steady* operation with *no heat generation* in the fin, and we assume the thermal conductivity $k$ of the material to be constant. We also assume the convection heat transfer coefficient $h$ to be *constant* and *uniform* over the entire surface of the fin for convenience in calculations. We recognize that the convection heat transfer coefficient $h$, in general, varies along the fin as well as its circumference, and its value at a point is a strong function of the *fluid motion* at that point. The value of $h$ is usually much lower at the *fin base* than it is at the *fin tip,* because the fluid is surrounded by solid surfaces near the base, which seriously disrupt its motion to the point of "suffocating" it, while the fluid near the fin tip has little contact with a solid surface, and thus encounters little resistance to flow. Therefore, adding too many fins on a surface may actually decrease the overall heat transfer when the decrease in $h$ offsets any gain resulting from the increase in the surface area.

Consider the surface of a *plane wall* at temperature $T_b$ exposed to a medium at temperature $T_\infty$. Heat is lost from the surface to the surrounding medium by convection with a heat transfer coefficient of $h$. Disregrading radiation or accounting for its contribution in the convection coefficient $h$, heat transfer from a surface area $A_b$ is as given by Eq. 8-64. Now let us consider a fin of constant cross-sectional area $A_c = A_b$ and length $L$ that is attached to the surface with a perfect contact (Fig. 8-57). This time, heat will flow from the surface to the fin *by conduction,* and from the fin to the surrounding medium *by convection* with the same heat transfer coefficient $h$. The temperature of the fin will be $T_b$ at the fin base and gradually decrease towards the fin tip. Convection from the fin surface causes the temperature at any cross-section to drop somewhat from the midsection towards the outer surfaces. However, the cross-sectional area of the fins is usually very small, and thus the temperature at any cross-section can be considered to be uniform. For the same reason, the area of the *fin tip* is usually negligible relative to the lateral surface area when determining the total heat transfer surface area $A_{\text{fin}}$ of the fin. For example, for a pin fin of diameter $D$ and length $L$, the fin surface area is

$$A_{\text{fin}} = A_{\text{lateral}} + A_{\text{fin tip}} = pL + A_c = \pi DL + \tfrac{1}{4}\pi D^2 \approx \pi DL \quad (8\text{-}65)$$

when $L \gg D$, which is usually the case.

In the limiting case of *zero thermal resistance* ($k \to \infty$), the temperature of the fin will be uniform at the base value of $T_b$. The heat transfer from the fin will be *maximum* in this case, and can be expressed as

$$\dot{Q}_{\text{fin, max}} = hA_{\text{fin}}(T_b - T_\infty) \tag{8-66}$$

In reality, however, the temperature of the fin will drop along the fin, and thus the heat transfer from the fin will be less because of the decreasing temperature difference $T(x) - T_\infty$ towards the fin tip (Fig. 8-58). To account for the effect of this decrease in temperature on heat transfer, we define a **fin efficiency** as

$$\eta_{\text{fin}} = \frac{\dot{Q}_{\text{fin}}}{\dot{Q}_{\text{fin, max}}} = \frac{\text{actual heat transfer rate from the fin}}{\substack{\text{ideal heat transfer rate from the fin} \\ \text{if the entire fin were at base temperature}}} \tag{8-67}$$

or

$$\dot{Q}_{\text{fin}} = \eta_{\text{fin}}\dot{Q}_{\text{fin, max}} = \eta_{\text{fin}} hA_{\text{fin}}(T_b - T_\infty) \tag{8-68}$$

where $A_{\text{fin}}$ is the total surface area of the fin. This relation enables us to determine the heat transfer from a fin when its efficiency is known. Fin efficiency relations are developed for fins of various profiles, and are plotted in Fig. 8-59 for fins on a *plain surface,* and in Fig. 8-60 for *circular fins* of constant thickness. The fin surface area associated with each profile is also given on each figure. For most fins of constant thickness encountered in practice, the fin thickness $t$ is too small relative to the fin length $L$, and thus the fin tip area is negligible.

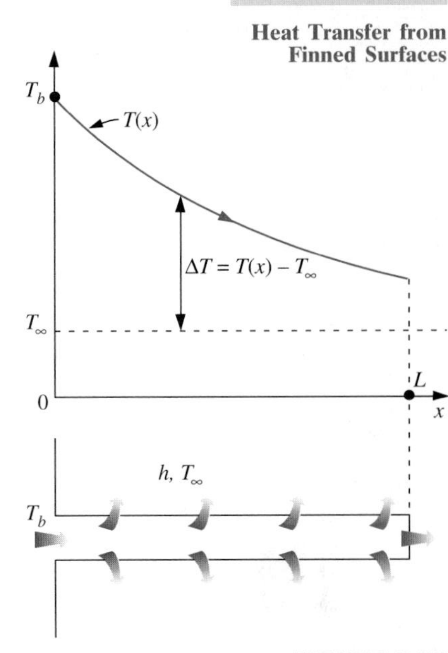

**FIGURE 8-58**
The temperature of a fin drops gradually along the fin from its base towards its tip.

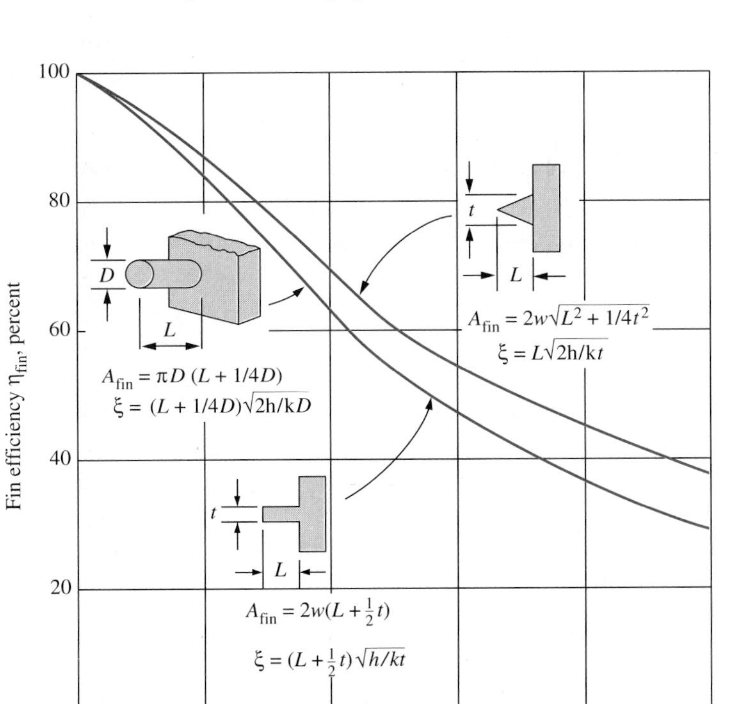

**FIGURE 8-59**
Efficiency of circular, rectangular and triangular fins on a plain surface of width $w$ (from Gardner).

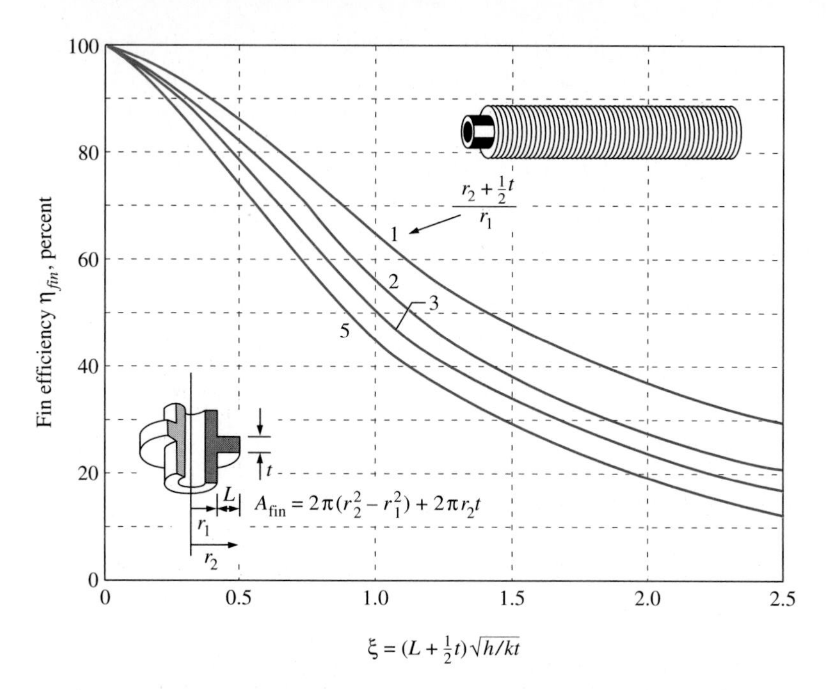

**FIGURE 8-60**
Efficiency of circular fins of length $L$ and constant thickness $t$ (from Gardner).

Note that fins with triangular and parabolic profiles contain less material and are more efficient than the ones with rectangular profiles, and thus are more suitable for applications that require minimum weight such as space applications.

An important consideration in the design of finned surfaces is the selection of the proper *fin length L*. Normally the *longer* the fin, the *larger* the heat transfer area and thus the *higher* the rate of heat transfer from the fin. But also, the larger the fin, the bigger the mass, the higher the price, and the larger the fluid friction. Therefore, increasing the length of the fin beyond a certain value cannot be justified unless the added benefits outweigh the added cost. Also, the fin efficiency decreases with increasing fin length because of the decrease in fin temperature with length. Fin lengths that cause the fin efficiency to drop below 60 percent usually cannot be justified economically, and should be avoided.

## Fin Effectiveness

Fins are used to *enhance* heat transfer, and the use of fins on a surface cannot be recommended unless the enhancement in heat transfer justifies the added cost and complexity associated with the fins. In fact, there is no assurance that adding fins on a surface will *enhance* heat transfer. The performance of the fins is judged on the basis of the enhancement in heat transfer relative to the no-fin case, and expressed in terms of the *fin effectiveness* $\varepsilon_{fin}$ defined as (Fig. 8-61)

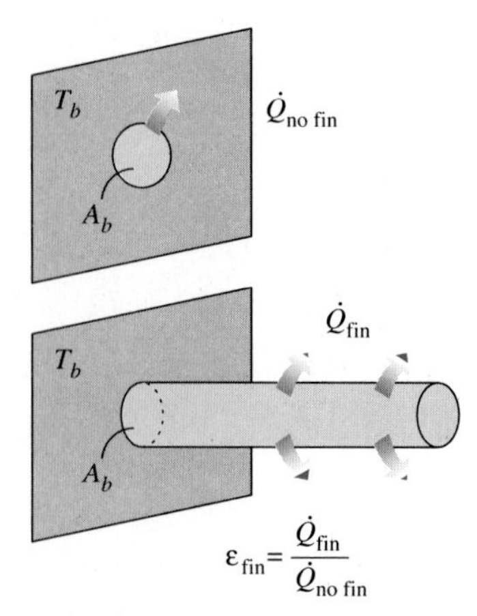


**FIGURE 8-61**
The effectiveness of a fin.

$$\varepsilon_{fin} = \frac{\dot{Q}_{fin}}{\dot{Q}_{no\,fin}} = \frac{\dot{Q}_{fin}}{hA_b(T_b - T_\infty)} = \frac{\text{heat transfer rate from the fin of } \textit{base area } A_b}{\text{heat transfer rate from the surface of } \textit{area } A_b} \tag{8-69}$$

Here $A_b$ is the cross-sectional area of the fin at the base, and $\dot{Q}_{\text{no fin}}$ represents the rate of heat transfer from this area if no fins are attached to the surface. An effectiveness of $\varepsilon_{\text{fin}} = 1$ indicates that the addition of fins to the surface does not affect heat transfer at all. That is, heat conducted to the fin through the base area $A_b$ is equal to the heat transferred from the same area $A_b$ to the surrounding medium. An effectiveness $\varepsilon_{\text{fin}} < 1$ indicates that the fin actually acts as *insulation,* slowing down the heat transfer from the surface. This situation can occur when fins made of low thermal conductivity materials are used. An effectiveness $\varepsilon_{\text{fin}} > 1$ indicates that fins are *enhancing* heat transfer from the surface, as they should. However, the use of fins cannot be justified unless $\varepsilon_{\text{fin}}$ is sufficiently larger than 1. Finned surfaces are designed on the basis of *maximizing* effectiveness for a specified cost, or *minimizing cost* for a desired effectiveness.

For a sufficiently *long* fin of *uniform* cross-section ($A_c$ = constant), the temperature of the fin at the fin tip will approach the environment temperature $T_\infty$. By writing an energy balance and solving the resulting differential equation, it can be shown that the temperature along the fin decreases *exponentially* from $T_b$ to $T_\infty$. The variation of *temperature* along the fin and the steady rate of *heat transfer* from the entire fin in this case are given by (Fig. 8-62)

$$\frac{T(x) - T_\infty}{T_b - T_\infty} = e^{-x\sqrt{hp/kA_c}} \tag{8-70}$$

and

$$\dot{Q}_{\text{long fin}, A_c=\text{const.}} = \sqrt{hpkA_c}\,(T_b - T_\infty) \tag{8-71}$$

where $p$ is the perimeter, $A_c$ is the cross-sectional area of the fin, and $x$ is the distance from the fin base. Then the effectiveness of this long fin becomes

$$\varepsilon_{\text{long fin}, A_c=\text{const.}} = \frac{\dot{Q}_{\text{fin}}}{\dot{Q}_{\text{no fin}}} = \frac{\sqrt{hpkA_c}\,(T_b - T_\infty)}{hA_b(T_b - T_\infty)} = \sqrt{\frac{kp}{hA_c}} \tag{8-72}$$

We can draw several important conclusions from the fin effectiveness relation above for consideration in the design and selection of the fins:

- The *thermal conductivity* $k$ of the fin material should be as high as possible. Thus it is no coincidence that fins are made from metals, with copper, aluminum and iron being the most common ones. Perhaps the most widely used fins are made of aluminum because of its low cost and weight and its resistance to corrosion.
- The ratio of the *perimeter* to the *cross-sectional area* of the fin, $p/A_c$, should be as high as possible. This criteria is satisfied by *thin* plate fins or *slender* pin fins.
- The use of fins is *most effective* in applications that involve *low convection heat transfer coefficient.* Thus, the use of fins is more easily justified when the medium is a *gas* instead of a liquid, and the heat transfer is by *natural convection* instead of by forced convection. Thus, it is no coincidence that in liquid-to-gas heat exchangers such as the car radiator, fins are placed on the *gas* side.

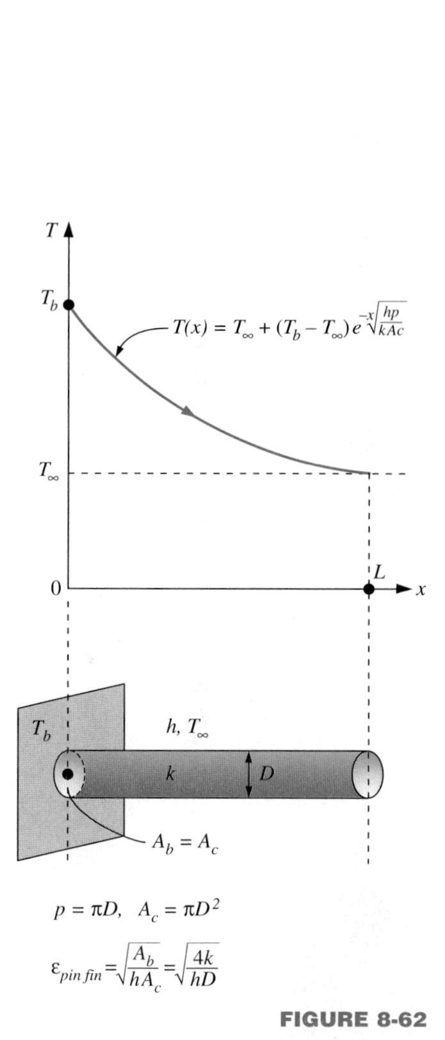


**FIGURE 8-62**
A long circular fin of uniform cross-section and the variation of temperature along it.

When determining the rate of heat transfer from a finned surface, we must consider the *unfinned portion* of the surface as well as the *fins.* Therefore, the rate of heat transfer for a surface that contains $n$ fins can be expressed as

$$\begin{aligned}\dot{Q}_{\text{total,fin}} &= \dot{Q}_{\text{unfin}} + \dot{Q}_{\text{fin}} \\ &= hA_{\text{unfin}}(T_b - T_\infty) + \eta_{\text{fin}} hA_{\text{fin}}(T_b - T_\infty) \\ &= h(A_{\text{unfin}} + n\eta_{\text{fin}}A_{\text{fin}})(T_b - T_\infty)\end{aligned} \quad (8\text{-}73)$$

We can also define an **overall effectiveness** for a finned surface as the ratio of the total heat transfer from the finned surface to the heat transfer from the same surface if there were no fins,

$$\varepsilon_{\text{fin, overall}} = \frac{\dot{Q}_{\text{total, fin}}}{\dot{Q}_{\text{total, no fin}}} = \frac{h(A_{\text{unfin}} + \eta_{\text{fin}}A_{\text{fin}})(T_b - T_\infty)}{hA_{\text{no fin}}(T_b - T_\infty)} \quad (8\text{-}74)$$

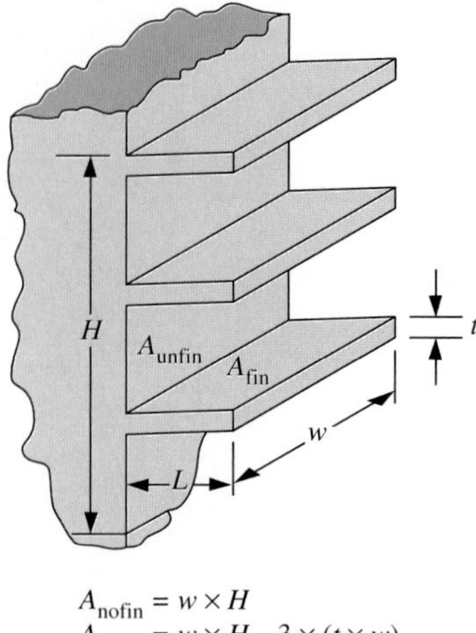


**FIGURE 8-63**
Various surface areas associated with a rectangular surface with three fins.

where $A_{\text{no fin}}$ is the area of the surface when there are no fins, $A_{\text{fin}}$ is the total surface area of all the fins on the surface, and $A_{\text{unfin}}$ is the area of the unfinned portion of the surface (Fig. 8-63). Note that the overall fin effectiveness depends on the fin density (number of fins per unit length), as well as the effectiveness of the individual fins. The overall effectiveness is a better measure of the performance of a finned surface than the effectiveness of the individual fins.

Specially designed finned surfaces called *heat sinks,* which are commonly used in the cooling of electronic equipment involve one-of-a-kind complex geometries, as shown in Table 8-6. The heat transfer performance of heat sinks is usually expressed in terms of their *thermal resistances,* in °C/W. A small value of thermal resistance indicates a small temperature drop across the heat sink, and thus a high fin efficiency.

**EXAMPLE 8-13**

Power transistors that are commonly used in electronic devices consume large amounts of electric power. The failure rate of electronic components increases almost exponentially with operating temperature. As a rule of thumb, the failure rate of electronic components is halved for each 10°C reduction in the junction operating temperature. Therefore, the operating temperature of electronic components is kept below a safe level to minimize the risk of failure.

The sensitive electronic circuitry of a power transistor at the junction is protected by its case, which is a rigid metal enclosure. Heat transfer characteristics of a power transistor are usually specified by the manufacturer in terms of the case-to-ambient thermal resistance, which accounts for both the natural convection and radiation heat transfers.

The case-to-ambient thermal resistance of a power transistor that has a maximum power rating of 10 W is given as 20°C/W. If the case temperature of the transistor is not to exceed 85°C, determine the power at which this transistor can be operated safely in an environment at 25°C.

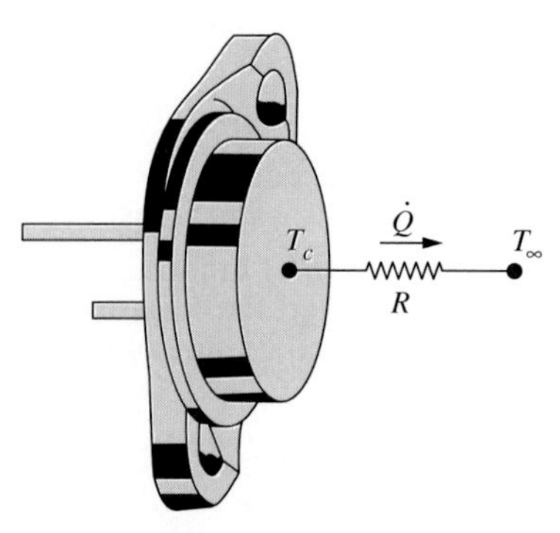


**FIGURE 8-64**
Schematic for Example 8-13.

**Solution** The power transistor and the thermal resistance network associated with it are shown in Fig. 8-64. The transistor case is assumed to be isothermal at 85°C. We notice from the thermal resistance network than there is

**TABLE 8-6**

**Combined natural convection and radiation thermal resistance of various heat sinks used in the cooling of electronic devices between the heat sink and the surroundings. All fins are made of aluminum 6063T-5, are black anodized, and are 76 mm (3 in.) long. (Courtesy of Vemaline Products, Inc.)**

| | |
|---|---|
| HS 5030 | $R = 0.9°C/W$ (vertical)<br>$R = 1.2°C/W$ (horizontal)<br>Dimensions: 76 mm × 105 mm × 44 mm<br>Surface area: 677 $cm^2$ |
| HS 6065 | $R = 5°C/W$<br>Dimensions: 76 mm × 38 mm × 24 mm<br>Surface area: 387 $cm^2$ |
| HS 6071 | $R = 1.4°C/W$ (vertical)<br>$R = 1.8°C/W$ (horizontal)<br>Dimensions: 76 mm × 92 mm × 26 mm<br>Surface area: 968 $cm^2$ |
| HS 6105 | $R = 1.8°C/W$ (vertical)<br>$R = 2.1°C/W$ (horizontal)<br>Dimensions: 76 mm × 127 mm × 19 mm<br>Surface area: 677 $cm^2$ |
| HS 6115 | $R = 1.1°C/W$ (vertical)<br>$R = 1.3°C/W$ (horizontal)<br>Dimensions: 76 mm × 102 mm × 25 mm<br>Surface area: 929 $cm^2$ |
| HS 7030 | $R = 2.9°C/W$ (vertical)<br>$R = 3.1°C/W$ (horizontal)<br>Dimensions: 76 mm × 97 mm × 19 mm<br>Surface area: 290 $cm^2$ |

a single resistance of 20°C/W between the case at $T_c$ = 85°C and the ambient at $T_\infty$ = 25°C, and thus the rate of heat transfer is

$$\dot{Q} = \left(\frac{\Delta T}{R}\right)_{\text{case-ambient}} = \frac{T_c - T_\infty}{R_{\text{case-ambient}}} = \frac{(85 - 25)°\text{C}}{20°\text{C/W}} = 3\text{ W}$$

Therefore, this power transistor should not be operated at power levels above 3 W if its case temperature is not to exceed 85°C.

This transistor can be used at higher power levels by attaching it to a heat sink (which lowers the thermal resistance by increasing the heat transfer surface area, as discussed in the next example) or by using a fan (which lowers the thermal resistance by increasing the convection heat transfer coefficient).

**EXAMPLE 8-14**

A 60-W power transistor is to be cooled by attaching it to one of the commercially available heat sinks shown in Table 8-6. Select a heat sink that will allow the case temperature of the transistor not to exceed 90°C in the ambient air at 30°C.

**Solution** The rate of heat transfer from a 60-W transistor at full power is $\dot{Q}$ = 60 W. Disregarding the contact resistance between the transistor and the heat sink, the thermal resistance between the transistor attached to the heat sink and the ambient air for the specified temperature difference is determined to be

$$\dot{Q} = \frac{\Delta T}{R} \longrightarrow R = \frac{\Delta T}{\dot{Q}} = \frac{(90 - 30)°\text{C}}{60\text{ W}} = 1.0°\text{C/W}$$

Therefore, the thermal resistance of the heat sink should be below 1.0°C/W. An examination of Table 8-6 reveals that the HS 5030, whose thermal resistance is 0.9°C/W in the vertical position is the only heat sink that will meet this requirement.

**EXAMPLE 8-15**

Steam in a heating system flows through tubes whose outer diameter is $D_1$ = 3 cm and whose walls are maintained at a temperature of 120°C. Circular aluminum fins [$k$ = 180 W/(m · °C)] of outer diameter $D_2$ = 6 cm and constant thickness $t$ = 2 mm are attached to the tube, as shown in Fig. 8-65. The space between the fins is 3 mm, and thus there are 200 fins per meter length of the tube. Heat is transferred to the surrounding air at $T_\infty$ = 25°C, with a combined heat transfer coefficient of $h$ = 60 W/(m² · °C). Determine the increase in heat transfer from the tube per meter of its length as a result of adding fins.

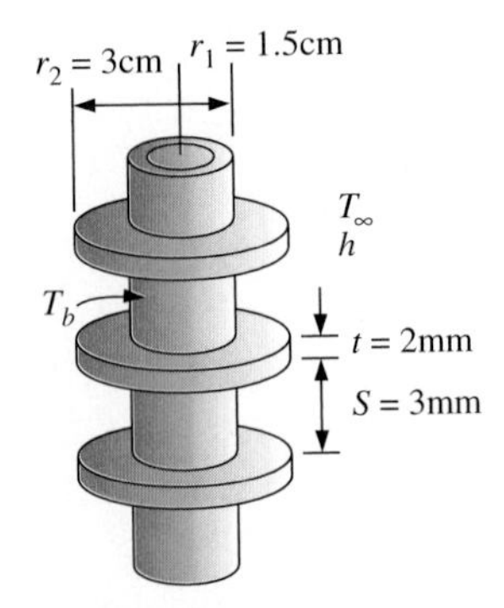


**FIGURE 8-65**
Schematic for Example 8-15.

**Solution** In the case of no fins, heat transfer from the tube per meter of its length is determined from Newton's law of cooling to be

$$A_{\text{no fin}} = \pi D_1 L = \pi(0.03\text{ m})(1\text{ m}) = 0.0942\text{ m}^2$$

$$\begin{aligned}\dot{Q}_{\text{no fin}} &= hA_{\text{no fin}}(T_b - T_\infty)\\ &= [60\text{ W/(m}^2\cdot°\text{C)}](0.0942\text{ m}^2)(120 - 25)°\text{C}\\ &= 537\text{ W}\end{aligned}$$

The efficiency of the circular fins attached to a circular tube is plotted in Fig. 8-60. Noting that $L = \frac{1}{2}(D_2 - D_1) = \frac{1}{2}(0.06 - 0.03) = 0.015$ m in this case, we have

$$\left.\begin{aligned} \frac{r_2 + \frac{1}{2}t}{r_1} &= \frac{(0.03 + \frac{1}{2} \times 0.002)\text{ m}}{0.015\text{ m}} = 2.07 \\ (L + \tfrac{1}{2}t)\sqrt{\frac{h}{kt}} &= (0.015 + \tfrac{1}{2} \times 0.002)\text{ m} \\ &\quad \times \sqrt{\frac{60\text{ W/(m}^2\cdot{}^\circ\text{C)}}{[180\text{ W/(m}\cdot{}^\circ\text{C)}](0.002\text{ m})}} = 0.207 \end{aligned}\right\} \quad \eta_{\text{fin}} = 0.95$$

Also

$$\begin{aligned} A_{\text{fin}} &= 2\pi(r_2^2 - r_1^2) + 2\pi r_2 t \\ &= 2\pi[(0.03\text{ m})^2 - (0.015\text{ m})^2] + 2\pi(0.03\text{ m})(0.002\text{ m}) \\ &= 0.00462\text{ m}^2 \end{aligned}$$

$$\begin{aligned} \dot{Q}_{\text{fin}} &= \eta_{\text{fin}}\dot{Q}_{\text{fin,max}} = \eta_{\text{fin}} h A_{\text{fin}}(T_b - T_\infty) \\ &= 0.95[60\text{ W/(m}^2\cdot{}^\circ\text{C)}](0.00462\text{ m}^2)(120 - 25)^\circ\text{C} \\ &= 25.0\text{ W} \end{aligned}$$

Noting that the space between the two fins is 3 mm, heat transfer from the unfinned portion of the tube is

$$\begin{aligned} A_{\text{unfin}} &= \pi D_1 S = \pi(0.03\text{ m})(0.003\text{ m}) = 0.000283\text{ m}^2 \\ \dot{Q}_{\text{unfin}} &= h A_{unfin}(T_b - T_\infty) \\ &= [60\text{ W/(m}^2\cdot{}^\circ\text{C)}](0.000283\text{ m}^2)(120 - 25)^\circ\text{C} \\ &= 1.60\text{ W} \end{aligned}$$

Noting that there are 200 fins and thus 200 interfin spacings per meter length of the tube, the total heat transfer from the finned tube becomes

$$\dot{Q}_{\text{total,fin}} = n(\dot{Q}_{\text{fin}} + \dot{Q}_{\text{unfin}}) = 200(25.0 + 1.6)\text{ W} = 5320\text{ W}$$

Therefore, the increase in heat transfer from the tube per meter of its length as a result of the addition of fins is

$$\dot{Q}_{\text{increase}} = \dot{Q}_{\text{total,fin}} - \dot{Q}_{\text{no fin}} = 5320 - 537 = 4783\text{ W (per m tube length)}$$

**Discussion** The overall effectiveness of the finned tube is

$$\varepsilon_{\text{fin, overall}} = \frac{\dot{Q}_{\text{total, fin}}}{\dot{Q}_{\text{total, no fin}}} = \frac{5320\text{ W}}{537\text{ W}} = 9.9$$

That is, the rate of heat transfer from the steam tube increases by a factor of almost 10 as a result of adding fins. This explains the widespread use of finned surfaces.

## 8-10 ■ HEAT TRANSFER IN COMMON CONFIGURATIONS

So far, we have considered heat transfer in *simple* geometries such as large plane walls, long cylinders, and spheres. This is because heat transfer in such geometries can be approximated as *one-dimensional,* and simple analytical solutions can be obtained easily. But many problems encountered in practice are two- or three-dimensional, and involve rather complicated geometries for which no simple solutions are available.

An important class of heat transfer problems for which simple solutions are obtained involve two surfaces maintained at *constant* temperatures $T_1$ and $T_2$. The steady rate of heat transfer between these two surfaces is expressed as

$$\dot{Q} = Sk(T_1 - T_2) \tag{8-75}$$

where $S$ is the **conduction shape factor**, which has the dimensions of *length*, and $k$ is the thermal conductivity of the medium between the surfaces. The conduction shape factor depends on the *geometry* of the system only.

Conduction shape factors have been determined for a number of configurations encountered in practice, and are given in Table 8-7 for some common cases. More comprehensive tables are available in the literature. Once the value of the shape factor is known for a specific geometry, the total steady heat transfer rate can be determined from the equation above using the specified two constant temperatures of the two surfaces and the thermal conductivity of the medium between them. Note that conduction shape factors are applicable only when heat transfer between the two surfaces is by *conduction*. Therefore, they cannot be used when the medium between the surfaces is a liquid or gas which involve natural or forced convection currents.

A comparison of Eqs. 8-13 and 8-75 reveals that the conduction shape factor $S$ is related to the thermal resistance $R$ by $R = 1/kS$ or $S = 1/kR$. Thus, these two quantities are the inverse of each other when the thermal conductivity of the medium is unity. The use of the conduction shape factors is illustrated in the following examples.

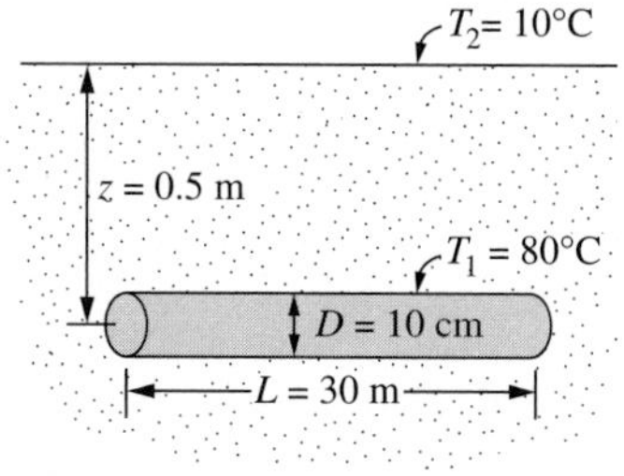

$T_2 = 10°C$

$z = 0.5$ m

$T_1 = 80°C$

$D = 10$ cm

$L = 30$ m

**FIGURE 8-66**
Schematic for Example 8-16.

**EXAMPLE 8-16**

A 30-m-long, 10-cm-diameter hot water pipe of a district heating system is buried in the soil 50 cm below the ground surface, as shown in Fig. 8-66. The outer surface temperature of the pipe is 80°C. Taking the surface temperature of the earth to be 10°C and the thermal conductivity of the soil at that location to be 0.9 W/(m · °C), determine the rate of heat loss from the pipe.

**Solution** This is a two-dimensional heat transfer problem (no change in the axial direction), and the shape factor for this configuration is given in Table 8-7 to be

$$S = \frac{2\pi L}{\ln(4z/D)}$$

since $z > 1.5D$, where $z$ is the distance of the pipe from the ground surface and $D$ is the diameter of the pipe. Substituting,

$$S = \frac{2\pi \times (30\text{ m})}{\ln(4 \times 0.5/0.1)} = 62.9\text{ m}$$

Then the steady rate of heat transfer from the pipe becomes

$$Q = Sk(T_1 - T_2) = (62.9\text{ m})[0.9\text{ W/m} \cdot °\text{C})](80 - 10)°\text{C} = 3963\text{ W}$$

Note that this heat is conducted from the pipe surface to the surface of the earth through the soil, and then transferred to the atmosphere by convection and radiation.

**TABLE 8-7**

**Conduction shape factors $S$ for several configurations for use in $\dot{Q} = kS(T_1 - T_2)$ to determine the steady rate of heat transfer through a medium of thermal conductivity $k$ between the surfaces at temperatures $T_1$ and $T_2$**

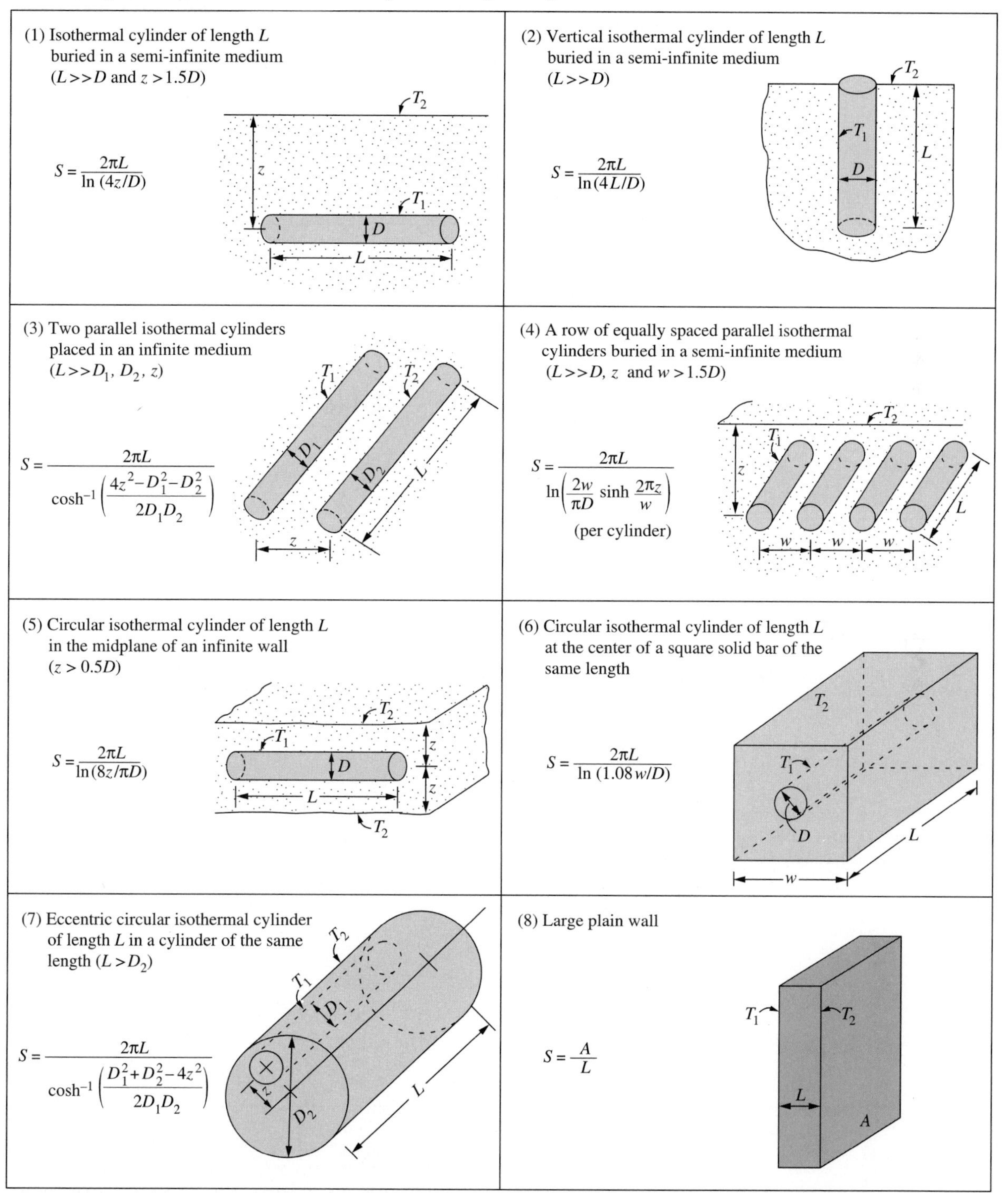

(1) Isothermal cylinder of length L buried in a semi-infinite medium (L>>D and z>1.5D)
$S = \frac{2\pi L}{\ln(4z/D)}$
T2
T1
z
D
L
(2) Vertical isothermal cylinder of length L buried in a semi-infinite medium (L>>D)
$S = \frac{2\pi L}{\ln(4L/D)}$
T2
T1
D
L
(3) Two parallel isothermal cylinders placed in an infinite medium (L>>D1, D2, z)
$S = \frac{2\pi L}{\cosh^{-1}\left(\frac{4z^2 - D_1^2 - D_2^2}{2D_1D_2}\right)}$
T1
T2
D1
D2
L
z
(4) A row of equally spaced parallel isothermal cylinders buried in a semi-infinite medium (L>>D, z and w>1.5D)
$S = \frac{2\pi L}{\ln\left(\frac{2w}{\pi D}\sinh\frac{2\pi z}{w}\right)}$
(per cylinder)
T2
T1
z
w
w
w
L
(5) Circular isothermal cylinder of length L in the midplane of an infinite wall (z>0.5D)
$S = \frac{2\pi L}{\ln(8z/\pi D)}$
T2
T1
D
L
z
z
T2
(6) Circular isothermal cylinder of length L at the center of a square solid bar of the same length
$S = \frac{2\pi L}{\ln(1.08w/D)}$
T2
T1
D
L
w
(7) Eccentric circular isothermal cylinder of length L in a cylinder of the same length (L>D2)
$S = \frac{2\pi L}{\cosh^{-1}\left(\frac{D_1^2 + D_2^2 - 4z^2}{2D_1D_2}\right)}$
T2
T1
D1
z
D2
L
(8) Large plain wall
$S = \frac{A}{L}$
T1
T2
L
A

**TABLE 8-7 (Continued)**

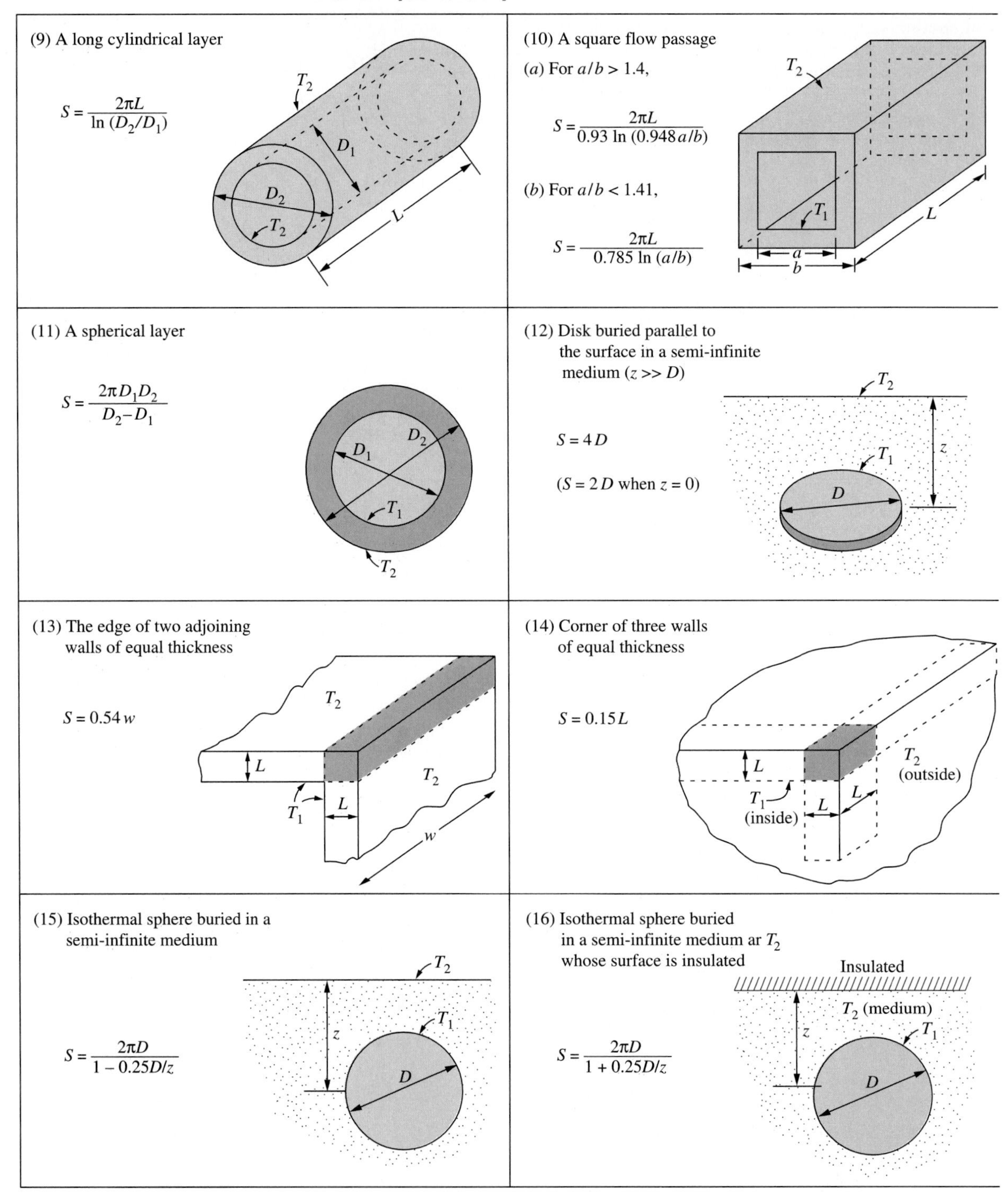

(9) A long cylindrical layer

$$S = \frac{2\pi L}{\ln (D_2/D_1)}$$

(10) A square flow passage

(*a*) For $a/b > 1.4$,

$$S = \frac{2\pi L}{0.93 \ln (0.948\, a/b)}$$

(*b*) For $a/b < 1.41$,

$$S = \frac{2\pi L}{0.785 \ln (a/b)}$$

(11) A spherical layer

$$S = \frac{2\pi D_1 D_2}{D_2 - D_1}$$

(12) Disk buried parallel to the surface in a semi-infinite medium ($z >> D$)

$$S = 4D$$

($S = 2D$ when $z = 0$)

(13) The edge of two adjoining walls of equal thickness

$$S = 0.54\, w$$

(14) Corner of three walls of equal thickness

$$S = 0.15\, L$$

(15) Isothermal sphere buried in a semi-infinite medium

$$S = \frac{2\pi D}{1 - 0.25D/z}$$

(16) Isothermal sphere buried in a semi-infinite medium ar $T_2$ whose surface is insulated

$$S = \frac{2\pi D}{1 + 0.25D/z}$$

**EXAMPLE 8-17**

A 5-m-long section of hot and cold water pipes run parallel to each other in a thick concrete layer, as shown in Fig. 8-67. The diameters of both pipes is 5 cm, and the distance between the centerline of the pipes is 30 cm. The surface temperature of the hot and cold pipes are 70°c and 15°C, respectively. Taking the thermal conductivity of the concrete to be $k = 0.75$ W/(m · °C), determine the rate of heat transfer between the pipes.

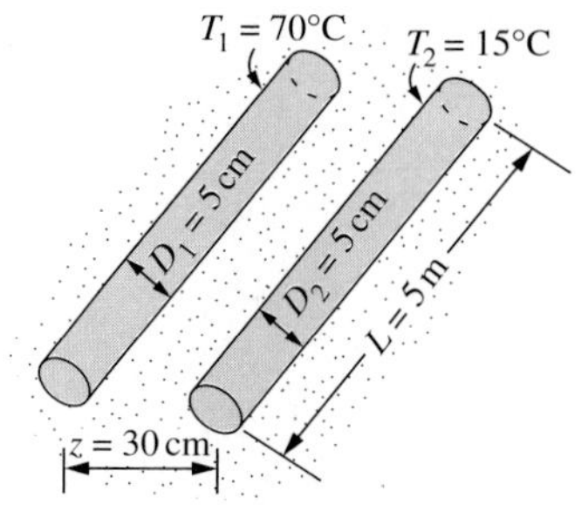

**FIGURE 8-67**
Schematic for Example 8-17.

**Solution** This is also a two-dimensional heat transfer problem (no change in the axial direction along the pipes), and the shape factor for this configuration is given in Table 8-7 to be

$$S = \frac{2\pi L}{\cosh^{-1}\left(\dfrac{4z^2 - D_1^2 - D_2^2}{2D_1D_2}\right)}$$

where $z$ is the distance between the centerlines of the pipes, and $L$ is their length. Substituting,

$$S = \frac{2\pi \times (5\text{ m})}{\cosh^{-1}\left(\dfrac{4 \times 0.3^2 - 0.05^2 - 0.05^2}{2 \times 0.05 \times 0.05}\right)} = 6.34\text{ m}$$

Then the steady rate of heat transfer between the pipes becomes

$$Q = Sk(T_1 - T_2) = (6.34\text{ m})[0.75\text{ W/(m}\cdot{}^\circ\text{C)}](70 - 15)^\circ\text{C} = 262\text{ W}$$

We can reduce this heat loss by placing the hot and cold water pipes further away from each other.

It is well known that insulation reduces heat transfer and saves energy and money. Decisions on the right amount of insulation are based on a heat transfer analysis, followed by an economic analysis to determine the "monetary value" of energy loss. This is illustrated below with an example.

**EXAMPLE 8-18**

Consider an electrically heated house whose walls are 9 ft high and have an $R$-value of insulation of 13 (i.e., a thickness-to-thermal conductivity ratio of $L/k = 13$ h · ft² · °F/Btu). Two of the walls of the house are 40 ft long and the others are 30 ft long. The house is maintained at 75°F at all times, while the temperature of the outdoors varies. Determine the amount of heat lost through the walls of the house on a certain day during which the average temperature of the outdoors is 45°F. Also determine the cost of this heat loss to the homeowner if the unit cost of electricity is \$0.075/kWh. For combined convection and radiation heat transfer coefficients, use the ASHRAE (American Society of Heating, Refrigeration, and Air Conditioning Engineers) recommended values of $h_i = 1.46$ Btu/(h · ft² · °F) for the inner surfaces of the walls and $h_o = 4.0$ Btu/(h · ft² · °F) for the inner and outer surfaces of the walls, respectively, under 15 mph wind conditions in winter.

**Solution** We assume the indoor and outdoor air temperatures to have remained at the given values for the entire day, so that heat transfer through the walls can be considered to be steady. We also assume heat transfer through

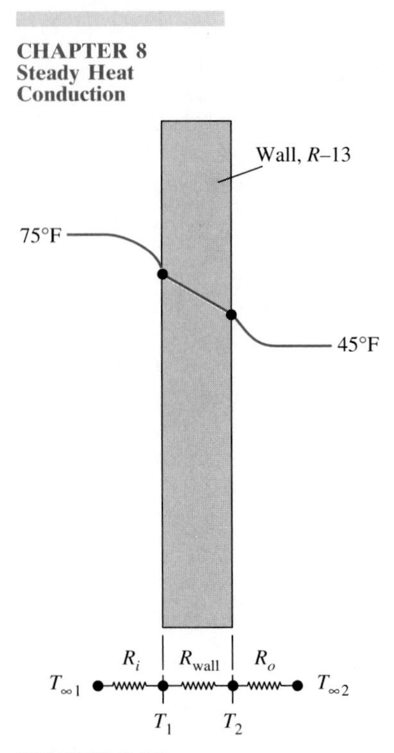

**FIGURE 8-68**
Schematic for Example 8-18.

the walls to be one-dimensional, since any significant temperature gradients in this case will exist in the direction from the indoors to the outdoors. The radiation effects are accounted for in the heat transfer coefficients.

This problem involves conduction through the wall and convection at its surfaces, and can best be handled by making use of the thermal resistance concept and drawing the thermal resistance network, as shown in Fig. 8-68. The heat transfer area of the walls is

$$A = \text{circumference} \times \text{height} = (2 \times 30\text{ ft} + 2 \times 40\text{ ft})(9\text{ ft}) = 1260\text{ ft}^2$$

Then the individual resistances are evaluated from their definitions to be

$$R_i = R_{\text{conv},i} = \frac{1}{h_i A} = \frac{1}{[1.46\text{ Btu/(h}\cdot\text{ft}^2\cdot{}^\circ\text{F)}](1260\text{ ft}^2)} = 0.00054\text{ h}\cdot{}^\circ\text{F/Btu}$$

$$R_{\text{wall}} = \frac{L}{kA} = \frac{13\text{ h}\cdot\text{ft}^2\cdot{}^\circ\text{F/Btu}}{1260\text{ ft}^2} = 0.01032\text{ h}\cdot{}^\circ\text{F/Btu}$$

$$R_o = R_{\text{conv},o} = \frac{1}{h_o A} = \frac{1}{[4.0\text{ Btu/(h}\cdot\text{ft}^2\cdot{}^\circ\text{F)}](1260\text{ ft}^2)} = 0{\cdot}00020\text{ h}\cdot{}^\circ\text{F/Btu}$$

Noting that all three resistances are in series, the total resistance is determined to be

$$R_{\text{total}} = R_i + R_{\text{wall}} + R_o = 0.00054 + 0.01032 + 0.00020 = 0.01106\text{ h}\cdot{}^\circ\text{F/Btu}$$

Then the steady rate of heat transfer through the walls of the house becomes

$$\dot{Q} = \frac{T_{\infty 1} - T_{\infty 2}}{R_{\text{total}}} = \frac{(75 - 45)^\circ\text{F}}{0.01106\text{ h}\cdot{}^\circ\text{F/Btu}} = 2712\text{ Btu/h}$$

Finally, the total amount of heat lost through the walls during a 24-h period and its cost to the home owner are

$$Q = \dot{Q}\,\Delta t = (2712\text{ Btu/h})(24\text{-h/day}) = 65\,099\text{ Btu/day} = 19.1\text{ kWh/day}$$

since 1 kWh = 3412 Btu, and

$$\begin{aligned}\text{heating cost} &= (\text{energy lost})(\text{cost of energy}) = (19.1\text{ kWh/day})(\$0.075/\text{kWh})\\ &= \$1.43/\text{day}\end{aligned}$$

Therefore, the heat losses through the walls of the house will cost the home owner than day $1.43 worth of electricity.

## 8-11 ■ SUMMARY

*Heat* is the form of energy transferred from one system to another as a result of temperature difference, and the science of heat transfer deals with the rate of this energy transfer. Heat conduction in medium is said to be *steady* when the temperature does not vary with time, and *unsteady* when it does. Heat conduction in a medium is said to be *one-dimensional* when conduction is significant in one dimension only and negligible in the other two dimensions. *Conduction* is the transfer of energy from the more energetic particles of a substance to the adjacent less energetic ones as a result of interactions between the particles. The rate of heat conduction through a layer is proportional to the temperature difference across the layer and the heat transfer area, but is inversely proportional to the thickness of the layer, and is expressed by *Fourier's law of heat conduction* as

$$\dot{Q}_{\text{cond}} = -kA\frac{dT}{dx} \quad \text{(W)}$$

where $k$ is the *thermal conductivity* of the material. *Convection* is the mode of energy transfer between a solid surface and adjacent liquid or gas that is in motion, and it involves the combined effects of conduction and fluid motion. Convection is observed to be proportional to the temperature difference, and is expressed by *Newton's law of cooling* as

$$\dot{Q}_{\text{conv}} = hA(T_s - T_\infty) \quad \text{(W)}$$

where $h$ is the *convection heat transfer coefficient. Radiation* is the energy emitted by matter in the form of electromagnetic waves (or photons) as a result of the changes in the electronic configurations of the atoms or molecules. The net rate of radiation heat transfer between a body of emissivity $\varepsilon$ and surface area $A$ and its surroundings is given by

$$\dot{Q}_{\text{rad}} = \varepsilon\sigma A(T_s^4 - T_{\text{surr}}^4) \quad \text{(W)}$$

where $\sigma = 5.67 \times 10^{-8}\ \text{W/(m}^2 \cdot \text{K}^4)$ is the Stefan-Boltzman constant.

The *thermal conductivity* of a material is defined as the rate of heat transfer through a unit thickness of the material per unit surface area per unit temperature difference. The product $\rho C_p$ is called the *heat capacity* of a material, and the property defined as $\alpha = k/\rho C_p$ is the *thermal diffusivity.*

The one-dimensional heat transfer through a simple or composite body exposed to convection from both sides to mediums at temperatures $T_{\infty 1}$ and $T_{\infty 2}$ can be expressed as

$$\dot{Q} = \frac{T_{\infty 1} - T_{\infty 2}}{R_{\text{total}}} \quad \text{(W)}$$

where $R_{\text{total}}$ is the total thermal resistance between the two mediums. For a plane wall exposed to convection on both sides, the total resistance is expressed as

$$R_{\text{total}} = R_{\text{conv},1} + R_{\text{wall}} + R_{\text{conv},2} = \frac{1}{h_1 A} + \frac{L}{kA} + \frac{1}{h_2 A} \quad (^\circ\text{C/W})$$

This relation can be extended to plane walls consisting of two or more layers by adding an additional resistance for each additional layer. The elementary thermal resistance relations can be expressed as follows:

*conduction resistance* (*plane wall*): $R_{\text{wall}} = \dfrac{L}{kA}$

*conduction resistance* (*cylinder*): $R_{\text{cyl}} = \dfrac{\ln(r_2/r_1)}{2\pi Lk}$

*conduction resistance* (*sphere*): $R_{\text{sphere}} = \dfrac{r_2 - r_1}{4\pi r_1 r_2 k}$

*convection resistance*: $$R_{\text{conv}} = \frac{1}{hA}$$

*radiation resistance*: $$R_{\text{rad}} = \frac{1}{h_{\text{rad}}A}$$

where the radiation heat transfer coefficient is defined as

$$h_{\text{rad}} = \varepsilon\sigma(T_s^2 + T_{\text{surr}}^2)(T_s + T_{\text{surr}}) \quad [\text{W/(m}^2 \cdot \text{K)}]$$

Once the rate of heat transfer is available, the *temperature drop* across any layer can be determined from

$$\Delta T = \dot{Q}R \quad (°\text{C})$$

The thermal resistance concept can also be used to solve steady heat transfer problems involving parallel layers or combined series-parallel arrangements.

Adding insulation to a cylindrical pipe or a spherical shell will increase the rate of heat transfer if the outer radius of the insulation is less than the *critical radius of inuslation* defined as

$$r_{\text{cr, cylinder}} = \frac{k}{h}$$

$$r_{\text{cr, sphere}} = \frac{2k}{h}$$

For a large *plane wall* of thickness $2L$, a long solid *cylinder* of radius $r_o$, and a solid *sphere* of radius $r_o$ that involve uniform heat generation at a rate of $\dot{g}$ per unit volume, the surface temperatures $T_s$ is expressed as

$$T_{s,\text{ plane wall}} = T_\infty + \frac{\dot{g}L}{h}$$

$$T_{s,\text{ cylinder}} = \frac{\dot{g}r_o}{2h}$$

$$T_{s,\text{ sphere}} = T_\infty + \frac{\dot{g}r_o}{3h}$$

The maximum temperature rise in these geometries, which is the difference between the center temperature $T_0$ and the surface temperature $T_s$, is given by

$$\Delta T_{\text{max, plane wall}} = \frac{\dot{g}L^2}{2k}$$

$$\Delta T_{\text{max, cylinder}} = \frac{\dot{g}r_o^2}{4k}$$

$$\Delta T_{\text{max, sphere}} = \frac{\dot{g}r_o^2}{6k}$$

Finned surfaces are commonly used in practice to enhance heat transfer. Fins enhance heat transfer from a surface by exposing a larger surface area to convection. The fin temperature drops along the fin. To account for the effect of this decrease in temperature on heat transfer, we define a *fin efficiency* as

$$\eta_{\text{fin}} = \frac{\dot{Q}_{\text{fin}}}{\dot{Q}_{\text{fin, max}}} = \frac{\text{actual heat transfer rate from the fin}}{\substack{\text{ideal heat transfer rate from the fin}\\ \text{if the entire fin were at base temperature}}}$$

When the fin efficiency is available, the rate of heat transfer from a fin can be determined from

$$\dot{Q}_{\text{fin}} = \eta_{\text{fin}}\dot{Q}_{\text{fin, max}} = \eta_{\text{fin}} h A_{\text{fin}}(T_b - T_\infty)$$

The performance of the fins is judged on the basis of the enhancement in heat transfer relative to the no-fin case, and is expressed in terms of the *fin effectiveness* $\varepsilon_{\text{fin}}$, defined as

$$\varepsilon_{\text{fin}} = \frac{\dot{Q}_{\text{fin}}}{\dot{Q}_{\text{no fin}}} = \frac{\dot{Q}_{\text{fin}}}{hA_b(T_b - T_\infty)} = \frac{\substack{\text{heat transfer rate from}\\ \text{the fin of } \textit{base area } A_b}}{\substack{\text{heat transfer rate from}\\ \text{the surface of } \textit{area } A_b}}$$

Here $A_b$ is the cross-sectional area of the fin at the base and $\dot{Q}_{\text{no fin}}$ represents the rate of heat transfer from this area if no fins are attached to the surface. The *overall effectiveness* for a finned surface is defined as the ratio of the total heat transfer from the finned surface to the heat transfer from the same surface if there were no fins,

$$\varepsilon_{\text{fin, overall}} = \frac{\dot{Q}_{\text{total, fin}}}{\dot{Q}_{\text{total, no fin}}} = \frac{h(A_{\text{unfin}} + \eta_{\text{fin}}A_{\text{fin}})(T_b - T_\infty)}{hA_{\text{no fin}}(T_b - T_\infty)}$$

Certain multidimensional heat transfer problems involve two surfaces that are maintained at *constant* temperatures $T_1$ and $T_2$. The steady rate of heat transfer between these two surfaces is expressed as

$$\dot{Q} = Sk(T_1 - T_2)$$

where $S$ is the *conduction shape factor,* which has the dimensions of *length,* and $k$ is the thermal conductivity of the medium between the surfaces.

## REFERENCES AND SUGGESTED READING

**1** *ASHRAE Handbook of Fundamentals,* American Society of Heating, Refrigeration, and Air Conditioning Engineers, Atlanta, 1993.

**2** R. V. Andrews, "Solving Conductive Heat Transfer Problems with Electrical-Analogue Shape Factors," *Chemical Engineering Progress,* Vol. 5, p. 67, 1955.

**3** R. Barron, *Cryogenic Systems,* McGraw-Hill, New York, 1967.

**4** Y. Bayazitoglu and M. N. Özışık, *Elements of Heat Transfer,* McGraw-Hill, New York, 1988.

**5** H. S. Carslaw and J. C. Jaeger, *Conduction of Heat in Solids,* Oxford University Press, London, 1959.

**6** L. S. Fletcher, "Recent Developments in Contact Conductance Heat Transfer," *Journal of Heat Transfer,* Vol. 110, No. 4B, p. 1059, 1988.

**7** K. A. Gardner, "Efficiency of Extended Surfaces," *Transactions of the ASME,* Vol. 67, pp. 621–631, 1945.

**8** E. Hahne and U. Grigull, "Formfactor und Formwiderstand der stationaren mehrdimensionalen Warmeleteitung," *International Journal of Heat and Mass Transfer,* Vol. 18, p. 75, 1975.

**9** J. P. Holman, *Heat Transfer,* 7th ed., McGraw-Hill, New York, 1990.

**10** F. P. Incropera and D. P. DeWitt, *Introduction to Heat Transfer,* 2nd ed., Wiley, 1990.

**11** M. Jakob, *Heat Transfer,* Vol. 1, Wiley, New York, 1949.

**12** D. Q. Kern and A. D. Kraus, *Extended Surface Heat Transfer,* McGraw-Hill, New York, 1972.

**13** P. G. Klemens, "Theory of Thermal Conductivity of Solids," in R. P. Tye, Ed., *Thermal Conductivity,* Vol. 1, Academic Press, London, 1969.

**14** F. Kreith and M. S. Bohn, *Principles of Heat Transfer,* 5th ed., West, St. Paul, MN, 1993.

**15** S. S. Kutateladze, *Fundamentals of Heat Transfer,* Academic Press, New York, 1963.

**16** E. McLaughlin, "Theory of Thermal Conductivity of Fluids," in R. P. Tye, Ed., *Thermal Conductivity,* Vol. 2, Academic Press, London, 1969.

**17** M. N. Özışık, *Heat Transfer—A Basic Approach,* McGraw-Hill, New York, 1985.

**18** J. E. Sunderland and K. R. Johnson, "Shape Factors for Heat Conduction Through Bodies with Isothermal or Convective Boundary Conditions," *Transactions of the ASHRAE,* Vol. 10, pp. 237–241, 1964.

**19** L. C. Thomas, *Heat Transfer,* Prentice-Hall, Englewood Cliffs, NJ, 1992.

**20** Y. S. Touloukian and C. Y. Ho, Eds., *Thermophysical Properties of Matter: The TPRC Data Series,* 13 Volumes, Plenum Press, New York, 1970–1977.

**21** F. M. White, *Heat and Mass Transfer,* Addison-Wesley, Reading, MA, 1988.

## PROBLEMS*

### Thermodynamics and Heat Transfer Mechanisms

**8-1C** How does the science of heat transfer differ from the science of thermodynamics? What is the driving force for heat transfer?

---

*Students are encouraged to answer *all* the concept "C" questions.

**8-2C** How does transient heat conduction differ from steady conduction?

**8-3C** Under what conditions can heat transfer be considered to be one-dimensional? Give examples of one-, two-, and three-dimensional heat transfer.

**8-4C** What is heat flux? How is it related to the heat transfer rate?

**8-5C** What are the mechanisms of energy transfer to a system? How is heat transfer distinguished from the other forms of energy transfer?

**8-6C** What are the mechanisms of heat transfer? How are they distinguished from each other?

**8-7C** What is the physical mechanism of heat conduction in a solid, liquid, and a gas?

**8-8C** Consider heat transfer through a windowless wall of a house in a winter day. Discuss the parameters that affect the rate of heat conduction through the wall.

**8-9C** Write down the expressions for the physical laws that govern each mode of heat transfer, and identify the variables involved in each relation.

**8-10C** How does heat conduction differ from convection? How does natural convection differ from forced convection?

**8-11C** Can all three modes of heat transfer occur simultaneously (in parallel) in a medium?

**8-12C** Can a medium involve (*a*) conduction and convection, (*b*) conduction and radiation, or (*c*) convection and radiation simultaneously? Give examples for the "yes" answers.

**8-13C** The deep body temperature of a healthy person remains constant at 37°C while the temperature and humidity of the environment changes with time. Discuss the heat transfer mechanisms between the human body and the environment in both summer and winter, and explain how a person can keep cooler in the summer and warmer in winter.

**8-14C** We often turn the fan on in summer to help us cool. Explain how a fan makes us feel cooler in the summer. Also explain why some people use ceiling fans also in winter.

**FIGURE P8-14C**

**8-15C** Consider two identical people wearing jeans whose ends are tucked into their shoes. One of the people wears his jeans tight while the other wears them loose. Explain which person will feel colder in winter.

**8-16** A 15-cm-diameter brass ball [$k = 111\ \text{W/(m} \cdot \text{°C)}$, $\rho = 8520\ \text{kg/m}^3$, $C_p = 0.38\ \text{kJ/(kg} \cdot \text{°C)}$] is observed to cool from 130°C to an average temperature of 70°C in 20 min in atmospheric air at 30°C. Determine (*a*) the total amount of heat transferred from the copper ball, (*b*) the average rate of heat transfer from the ball, (*c*) the average heat flux, and (*d*) the convection heat transfer coefficient at the beginning of the cooling process.

**8-17** Consider steady heat transfer between two large parallel plates at constant temperatures of $T_1 = 290$ K and $T_2 = 150$ K that are $L = 2$ cm apart. Assuming the surfaces to be black (emissivity $\varepsilon = 1$), determine the rate of heat transfer between the plates per unit surface area, if the gap between the plates is (*a*) filled with atmospheric air, (*b*) evacuated, (*c*) filled with fiber glass insulation, and (*d*) filled with superinsulation that has an apparent thermal conductivity of $0.00015\ \text{W/(m} \cdot \text{°C)}$.

**8-18** A logic chip used in a computer dissipates 3 W of power in an environment at 60°C, and has a heat transfer surface area of $0.34\ \text{cm}^2$. Assuming the heat transfer from the surface to be uniform, determine (*a*) the amount of heat this chip dissipates during an eight-hour work day, in kWh, and (*b*) the heat flux on the surface of the chip, in $\text{W/m}^2$.

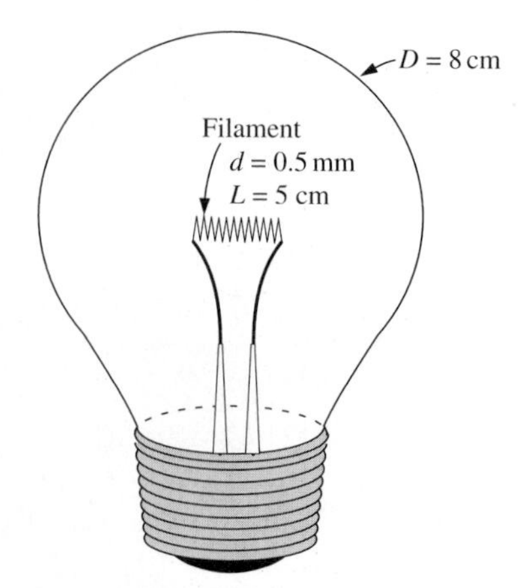


**FIGURE P8-19**

**8-19** Consider a 150-W incandescent lamp. The filament of the lamp is 5 cm long and has a diameter of 0.5 mm. The diameter of the glass bulb of the lamp is 8 cm. Determine the heat flux, in $\text{W/m}^2$, (*a*) on the surface of the filament and (*b*) on the surface of the glass bulb, and (*c*) calculate how much it will cost per year to keep that light on for eight hours a day every day if the price of electricity is \$0.08/kWh.

**8-20** A 35-cm-diameter watermelon is to be cooled from 25°C to 10°C in a refrigerator. Previous observations indicate that heat is removed from the watermelon at an average rate 200 kJ/h. Using the properties of water for the watermelon, determine (*a*) the average heat flux on the surface of the watermelon, in $\text{W/m}^2$, and (*b*) how long it will take to cool the watermelon. *Answers:* (*a*) $144\ \text{W/m}^2$, (*b*) 7 h

**8-20E** A 15-in-diameter watermelon is to be cooled from 75°F to 50°F in a refrigerator. Previous observations indicate that heat is removed from the watermelon at an average rate 200 Btu/h. Using the properties of water for the watermelon, determine (*a*) the average heat flux on the surface of the watermelon, in $\text{Btu/(h} \cdot \text{ft}^2)$, and (*b*) how long it will take to cool the watermelon. *Answers:* (*a*) $40.8\ \text{Btu/(h} \cdot \text{ft}^2)$, (*b*) 7.9 h

**8-21** A 1200-W iron is left on the ironing board with its base exposed to the air. About 90 percent of the heat generated in the iron is dissipated through its base whose surface area is $150\ \text{cm}^2$, and the remaining 10 percent through other surfaces. Assuming the heat transfer from the surface to be uniform, determine (*a*) the amount of heat the iron dissipates during a two-hour period, in kWh, (*b*) the heat flux on the surface of the iron base, in $\text{W/m}^2$, and (*c*) the cost of electricity consumed during this two-hour period for an electricity price of \$0.07/kWh.

**8-22** A 15 cm × 20 cm circuit board houses 120 closely spaced logic chips, each dissipating 0.1 W, on its surface. If the heat transfer from the back surface of the board is negligible, determine (*a*) the amount of heat this circuit board dissipates during a 10-hour period, in kWh, and (*b*) the heat flux on the surface of the circuit board, in W/m$^2$.

**8-22E** A 6 in. × 9 in. circuit board houses 120 closely spaced logic chips, each dissipating 0.1 W, on its surface. If the heat transfer from the back surface of the board is negligible, determine (*a*) the amount of heat this circuit board dissipates during a ten-hour period, in kWh, and (*b*) the heat flux on the surface of the circuit board, in Btu/(h · ft$^2$).

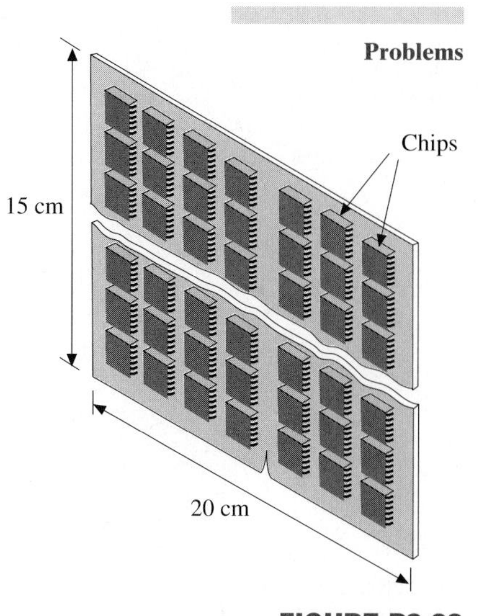


**FIGURE P8-22**

**Thermal conductivity**

**8-23C** Define thermal conductivity and explain its significance in heat transfer.

**8-24C** Judging from its unit W/(m · °C), can we define thermal conductivity of a material as the rate of heat transfer through the material per unit thickness per unit temperature difference? Explain.

**8-25C** Consider heat loss through the two walls of a house in a winter night. The walls are identical, except that one of them has a tightly fit glass window. Through which wall will the house lose more heat? Explain.

**8-26C** Consider two walls of a house, which are identical except that one is made of 10-cm-thick wood while the other is made of 25-cm-thick brick. Through which wall will the house lose more heat in winter?

**8-27C** How do the thermal conductivities of gases and liquids vary with temperature?

**8-28C** Why are the thermal conductivities of superinsulations order of magnitude lower than those of ordinary insulations?

**8-29C** Why do we characterize the heat conduction ability of insulators in terms of their apparent thermal conductivity instead of the ordinary thermal conductivity?

**8-30C** Consider an alloy of two metals whose thermal conductivites are $k_1$ and $k_2$. Will the thermal conductivity of the alloy be less than $k_1$, greater than $k_2$, or between $k_1$ and $k_2$?

**8-31C** What is the *R*-value of insulation? How is it determined? Will doubling the thickness of the insulation double its *R*-value?

**8-32C** How is the heat capacity of a material defined? How does it differ from specific heat?

**8-33C** What is the physical significance of thermal diffusivity? How does it differ from thermal conductivity?

**8-34C** Two identical spherical balls made of materials having different thermal diffusivities are placed in a hot oven. Compare the heat conduction processes through each ball.

**8-35** In a certain experiment cylindrical samples of diameter 4 cm and length 7 cm are used. The two thermocouples in each sample are placed 3 cm apart. After initial transients the electric heater is observed to draw 0.6 A at 110 V, and both differential thermometers read a temperature difference of 10°C. Determine the thermal conductivity of the sample.

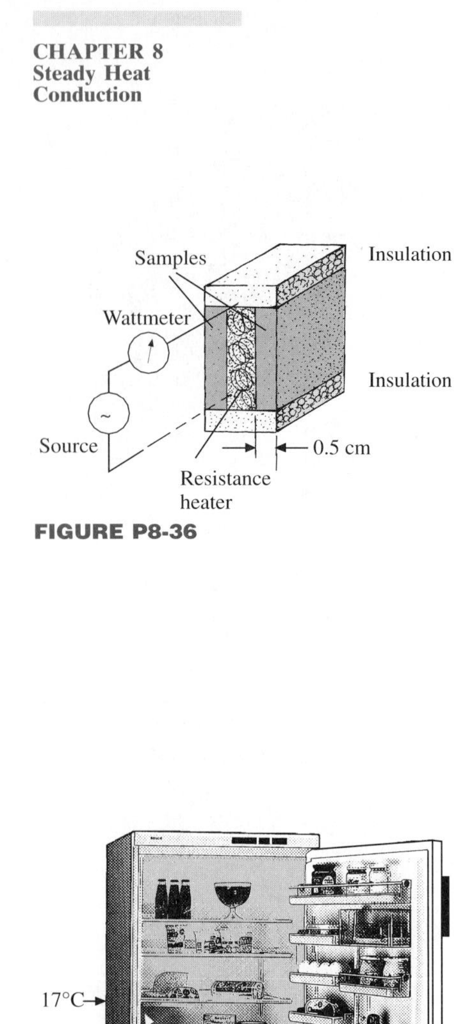


**FIGURE P8-36**

**FIGURE P8-39**

**8-36** One way of measuring the thermal conductivity of a material is to sandwich an electric thermofoil heater between two identical rectangular samples of the material and to heavily insulate the four outer edges, as shown in Fig. P8-36. Thermocouples attached to the inner and outer surfaces of the samples record the temperature.

During an experiment, two 0.5-cm-thick samples 10 cm × 10 cm in size are used. When steady operation is reached, the heater is observed to draw 35 W of electric power, and the temperature of each sample is observed to drop from 82°C at the inner surface to 74°C at the outer surface. Determine the thermal conductivity of the material at the average temperature.

**8-37** Repeat Prob. 8-36 for an electric power consumption of 20 W.

**8-38** A heat flux meter attached to the inner surface of a 3-cm-thick refrigerator door indicates a heat flux of 25 $W/m^2$ through the door. Also, the temperatures of the inner and the outer surfaces of the door are measured to be 7°C and 15°C, respectively. Determine the average thermal conductivity of the refrigerator door.
*Answer:* 0.0938 W/(m · °C)

**8-39** Consider a refrigerator whose dimensions are 1.8 m × 1.2 m × 0.8 m and whose walls are 3 cm thick. The refrigerator consumes 600 W of power when operating, and has a COP of 2.5. It is observed that the motor of the refrigerator remains on for 5 min and then off for 15 min periodically. If the average temperatures at the inner and outer surfaces of the refrigerator are 6°C and 17°C, respectively, determine the average thermal conductivity of the refrigerator walls. Also determine the annual cost of operating this refrigerator if the price of electricity is $0.06/kWh.

**8-39E** Consider a refrigerator whose dimensions are 6 ft × 4 ft × 3 ft and whose walls are 1 in. thick. The refrigerator consumes 600 W of power when operating, and has a COP of 2.5. It is observed that the motor of the refrigerator remains on for 5 min and then off for 15 min periodically. If the average temperatures at the inner and outer surfaces of the refrigerator are 50°F and 65°F, respectively, determine the average thermal conductivity of the refrigerator walls. Also determine the annual cost of operating this refrigerator if the price of electricity is $0.06/kWh.

### Steady Heat Conduction in Plane Walls

**8-40C** Consider one-dimensional heat conduction through a cylindrical rod of diameter $D$ and length $L$. What is the heat transfer area of the rod if (*a*) the lateral surfaces of the rod are insulated, and (*b*) the top and bottom surfaces of the rod are insulated?



**8-41C** Consider a 1.5 m × 2 m glass window whose thickness is 0.01 m. What is the heat transfer area of the window?

**8-42C** Consider heat conduction through a plane wall. Does the energy content of the wall change during steady heat conduction? How about during transient conduction? Explain.

**8-43C** Consider heat conduction through a wall of thickness $L$ and surface area $A$. Under what conditions will the temperature distributions in the wall be a straight line?

**8-44C** What does the thermal resistance of a medium represent?

**8-45C** How does the $R$-value of an insulation differ from its thermal resistance?

**8-46C** Can we define the convection resistance per unit surface area as the inverse of the convection heat transfer coefficient?

**8-47C** Why are the convection and the radiation resistances at a surface are in parallel instead of being in series?

**8-48C** Consider a surface of area $A$ at which the convection and radiation heat transfer coefficients are $h_{conv}$ and $h_{rad}$, respectively. Explain how you would determine (*a*) the single equivalent heat transfer coefficient and (*b*) the equivalent thermal resistance. Assume the medium and the surrounding surfaces are at the same temperature.

**8-49C** How does the thermal resistance network associated with a single-layer plane wall differ from the one associated with a five-layer composite wall?

**8-50C** Consider steady one-dimensional heat transfer through a multilayer medium. If the rate of heat transfer $\dot{Q}$ is known, explain how you would determine the temperature drop across each layer.

**8-51C** Consider steady one-dimensional heat transfer through a plane wall exposed to convection from both sides to environments at known temperatures $T_{\infty 1}$ and $T_{\infty 2}$ with known heat transfer coefficients $h_1$ and $h_2$. Once the rate of heat transfer $\dot{Q}$ has been evaluated, explain how you would determine the temperature of each surface.

**8-52C** Someone comments that a microwave oven can be viewed as a conventional oven with zero convection resistance at the surface of the food. Is this an accurate statement?

**8-53C** Consider a window glass consisting of two 4-mm-thick glass sheets pressed tightly against each other. Compare the heat transfer rate through this window with that of one consisting of a single 8-mm-thick glass sheet under identical conditions.

**8-54C** Consider steady heat transfer through an outer wall of a room in winter. The convection heat transfer coefficient at the outer surface of the wall is three times that of the inner surface as a result of the winds. On which surface of the wall do you think the temperature will be closer to the surrounding air temperature? Explain.

**8-55C** The bottom of a pan is made of a 4-mm-thick aluminum layer. In order to increase the rate of heat transfer through the bottom of the pan, someone proposes a design for the bottom which consists of a 3-mm-thick copper layer sandwiched between two 2-mm-thick aluminum layers. Will the new design conduct heat better? Explain. Assume perfect contact between the layers.

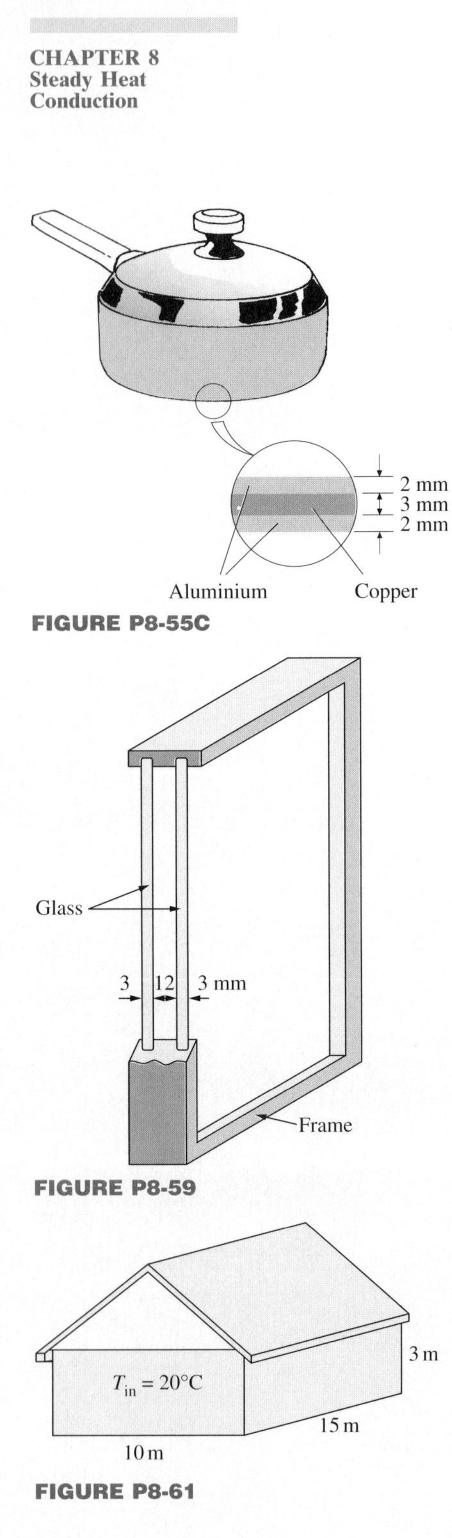


FIGURE P8-55C

FIGURE P8-59

FIGURE P8-61

**8-56C** Will the thermal contact resistance be greater for smooth or rough surfaces?

**8-57** Consider a 4-m-high, 6-m-wide, and 0.3-m-thick brick wall whose thermal conductivity is $k = 0.8\ \mathrm{W/(m \cdot {}^\circ C)}$. On a certain day, the temperatures of the inner and the outer surfaces of the wall are measured to be 14°C and 6°C, respectively. Determine the rate of heat loss through the wall on that day.

**8-58** Consider a 1.2-m-high and 2-m-wide glass window whose thickness is 6 mm and which has a thermal conductivity of $k = 0.78\ \mathrm{W/(m \cdot {}^\circ C)}$. Determine the steady rate of heat transfer through this glass window and the temperature of its inner surface for a day during which the room is maintained at 24°C while the temperature of the outdoors is −5°C. Take the convection heat transfer coefficients on the inner and outer surfaces of the window to be $h_1 = 10\ \mathrm{W/(m^2 \cdot {}^\circ C)}$ and $h_2 = 25\ \mathrm{W/(m^2 \cdot {}^\circ C)}$, and disregard any heat transfer by radiation.

**8-59** Consider a 1.2-m-high and 2-m-wide double-pane window consisting of two 3-mm-thick layers of glass [$k = 0.78\ \mathrm{W/(m \cdot {}^\circ C)}$] separated by a 12-mm-wide stagnant air space [$k = 0.026\ \mathrm{W/(m \cdot {}^\circ C)}$]. Determine the steady rate of heat transfer through this double-paned window and the temperature of its inner surface for a day during which the room is maintained at 24°C while the temperature of the outdoors is −5°C. Take the convection heat transfer coefficients on the inner and outer surfaces of the window to be $h_1 = 10\ \mathrm{W/(m^2 \cdot {}^\circ C)}$ and $h_2 = 25\ \mathrm{W/(m^2 \cdot {}^\circ C)}$, and disregard any heat transfer by radiation. *Answers:* 113 W, 19.2°C

**8-60** Repeat Prob. 8-59, assuming the space between the two glass layers is evacuated.

**8-61** Consider an electrically heated brick house [$k = 0.69\ \mathrm{W/(m \cdot {}^\circ C)}$] whose walls are 3 m high and 0.3 m thick. Two of the walls of the house are 15 m long and the other two are 10 m long. The house is maintained at 20°C at all times, while the temperature of the outdoors varies. On a certain day, the temperature of the inner surface of the walls is measured to be at 14°C, while the average temperature of the outer surface is observed to remain at 10°C during the day for 10 h and at 6°C at night for 14 h. Determine the amount of heat lost from the house that day. Also determine the cost of that heat loss to the homeowner for an electricity price of \$0.075/kWh. *Answers:* 52.4 kWh, \$3.93

**8-61E** Consider an electrically heated brick house ($k =$ 0.40 Btu/(h · ft · °F)] whose walls are 9 ft high and 1 ft thick. Two of the walls of the house are 40 ft long and the other two are 30 ft long. The house is maintained at 70°F at all times, while the temperature of the outdoors varies. On a certain day, the temperature of the inner surface of the walls is measured to be at 55°F, while the average temperature of the outer surface is observed to remain at 45°F during the say for 10 h and at 35°F at night for 14 h. Determine the amount of heat lost from the house that day. Also determine the cost of that heat loss to the homeowner for an electricity price of \$0.075/kWh.

**8-62** A cyclindrical resistor element on a circuit board dissipates 0.15 W of power in an environment at 40°C. The resistor is 1.2 cm long and has a diameter of 0.3 cm. Assuming heat to be transferred uniformly from all surfaces, determine (*a*) the amount of heat this resistor dissipates during a 24-h period, (*b*) the heat flux on the surface of the resistor, in W/m$^2$, and (*c*) the surface temperature of the resistor for a combined convection and radiation heat transfer coefficient of 9 W/(m$^2$ · °C).

**8-63** Consider a power transistor which dissipates 0.2 W of power in an in environment at 30°C. The transistor is 0.4 cm long, and has a diameter of 0.5 cm. Assuming heat to be transferred uniformly from all surfaces, determine (*a*) the amount of heat this resistor dissipates during a 24-h period, in kWh, (*b*) the heat flux on the surface of the transistor, in W/m$^2$, and (*c*) the surface temperature of the resistor for a combined convection and radiation heat transfer coefficient of 12 W/(m$^2$ · °C).

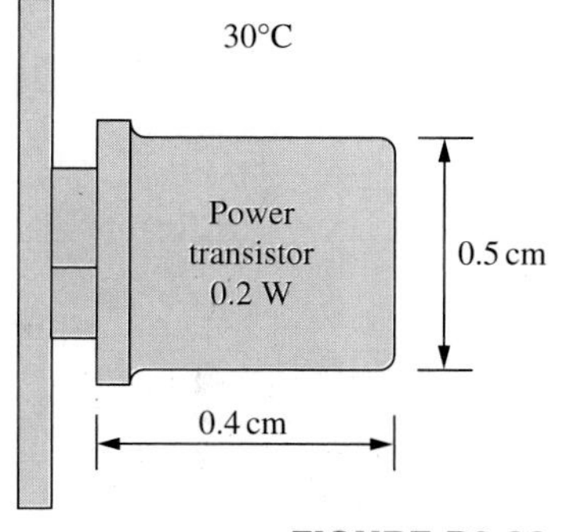

FIGURE P8-63

**8-64** A 12 cm × 18 cm circuit board houses 100 closely spaced logic chips, each dissipating 0.07 W, on its surface. The heat transfer from the back surface of the board is negligible. If the heat transfer coefficient on the surface of the board is 10 W/(m$^2$ · °C), determine (*a*) the heat flux on the surface of the circuit board, in W/m$^2$, (*b*) the surface temperature of the chips, and (*c*) the thermal resistance between the surface of the circuit board and the cooling medium, in °C/W.

**8-65** Consider a naked person standing in a room at 20°C with an exposed surface area of 1.7 m$^2$. The deep body temperature of the human body is 37°C, and the thermal conductivity of the human tissue near the skin is about 0.3 W/(m · °C). The body is losing heat at a rate of 150 W by natural convection and radiation to the surroundings. Taking the body temperature 0.5 cm beneath the skin to be 37°C, determine the skin temperature of the person. *Answer:* 35.5°C

**8-66** Water is boiling in a 25-cm-diameter aluminum pan [$k =$ 237 W/(m · °C)] at 95°C. Heat is transferred steadily to the boiling water in the pan through its 0.5 cm thick flat bottom at a rate of 600 W. If the

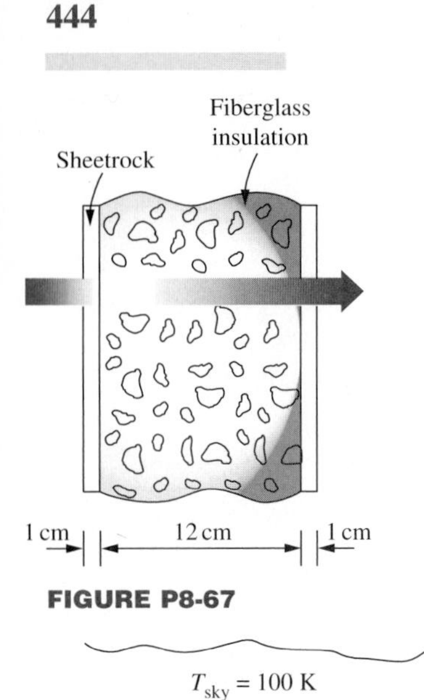


FIGURE P8-67

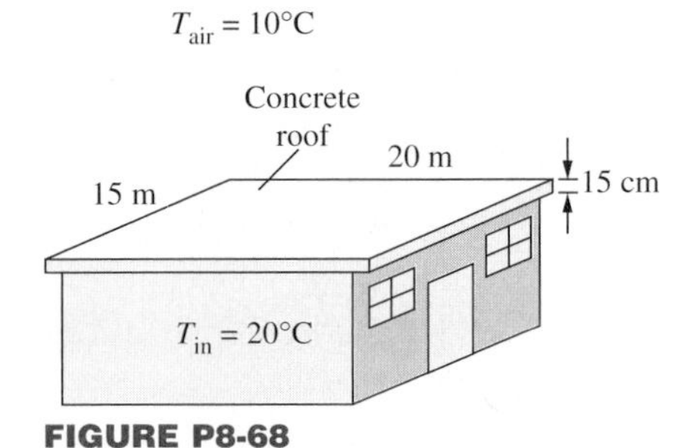


FIGURE P8-68

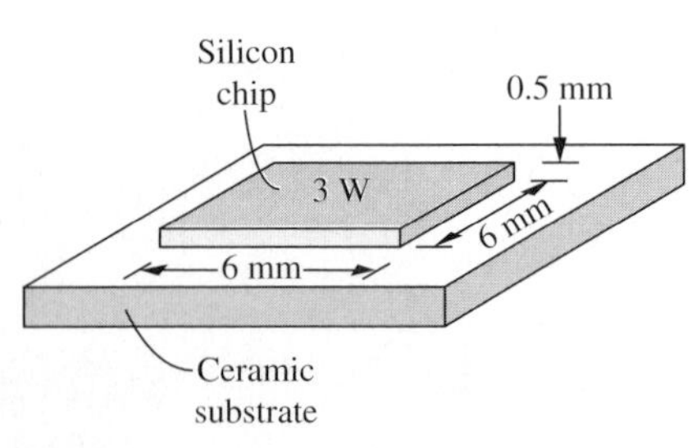

FIGURE P8-69

inner surface temperature of the bottom of the pan is 108°C, determine (*a*) the boiling heat transfer coefficient on the inner surface of the pan and (*b*) the outer surface temperature of the bottom of the pan.

**6-67** A wall is constructed of two layers of 1-cm-thick sheetrock [$k$ = 0.17 W/(m · °C)], which is a plasterboard made of two layers of heavy paper separated by a layer of gypsum, placed 12 cm apart. The space between the sheetrocks is filled with fiberglass insulation [$k$ = 0.035 W/(m · °C)]. Determine (*a*) the thermal resistance of the wall and (*b*) its $R$-value of insulation in SI units.

**8-67E** A wall is constructed of two layers of 0.5-in.-thick sheetrock [$k$ = 0.10 Btu/(h · ft · °F)], which is a plasterboard made of two layers of heavy paper separated by a layer of gypsum, placed 5 in. apart. The space between the sheetrocks is filled with fiberglass insulation [$k$ = 0.020 Btu/(h · ft · °F)]. Determine (*a*) the thermal resistance of the wall and (*b*) its $R$-value of insulation in English units.

**8-68** The roof of a house consists of 15-cm-thick concrete slab ($k$ = 2 W/(m · °C)] that is 15 m wide and 20 m long. The convection heat transfer coefficients on the inner and outer surfaces of the roof are 5 and 12 W/($m^2$ · °C), respectively. On a clear winter night, the ambient air is reported to be at 10°C, while the night sky temperature is 100 K. The house and the interior surfaces of the wall are maintained at a constant temperature of 20°C. The emissivity of both surfaces of the concrete roof is 0.9. Considering both radiation and convection heat transfers, determine the rate of heat transfer through the roof, and the inner surface temperature of the roof.

If the house is heated by a furnace burning natural gas with an efficiency of 80%, and the price of natural gas is $0.60/therm (1 therm = 105,500 kJ of energy content), determine the money lost through the roof that night during a 14-h period. *Answers:* 37,440 W, 7.3°C, $12.6

**8-69** The heat generated in the circuitry on the surface of a silicon chip [$k$ = 130 W/(m · °C)] is conducted to the ceramic substrate to which it is attached. The chip is 6 mm × 6 mm in size and 0.5 mm thick, and dissipates 3 W of power. Determine the temperature difference between the front and back surfaces of the chip in steady operation.

**8-69E** The heat generated in the circuitry on the surface of a silicon chip [$k$ = 75 Btu/(h · ft · °F)] is conducted to the ceramic substrate to which it is attached. The chip is 0.25 in. × 0.25 in. in size and 0.02 in. thick, and dissipates 3 W of power. Determine the temperature difference between the front and back surfaces of the chip in steady operation.

**8-70** A 2 m × 1.5 m section wall of an industrial furnace burning natural gas is not insulated, and the temperature at the outer surface of this section is measured to be 80°C. The temperature of the furnace room is 30°C, and the combined convection and radiation heat transfer coefficient at the surface of the outer furnace is 10 W/($m^2$ · C). It is proposed to insulate this section of the furnace wall with glass wool insulation

[$k$ = 0.038 W/(m · °C)] in order to reduce the heat loss by 90 percent. Assuming the outer surface temperature of the metal section still remains at about 80°C, determine the thickness of the insulation which needs to be used.

The furnace operates continuously, and has an efficiency of 78%. The price of the natural gas is \$0.55/therm (1 therm = 105,500 kJ of energy content). If the installation of the insulation will cost \$250 for materials and labor, determine how long it will take for the insulation to pay for itself from the energy it saves.

**8-71** Consider a house whose walls are 4 m high and 12 m long. Two of the walls of the house have no windows, while each of the other two walls have four windows made of a 0.6-cm-thick glass [$k$ = 0.78 W/(m · °C)], 1.2 m × 2 m in size. The walls are certified to have an $R$-value of 3.38 $m^2$ · °C/W. Disregarding any direct radiation gain or loss through the windows and taking the heat transfer coefficients at the inner and outer surfaces of the house to be 10 and 20 W/($m^2$ · °C), respectively, determine the ratio of the heat transfer through the walls with and without windows. *Answer:* 6.24

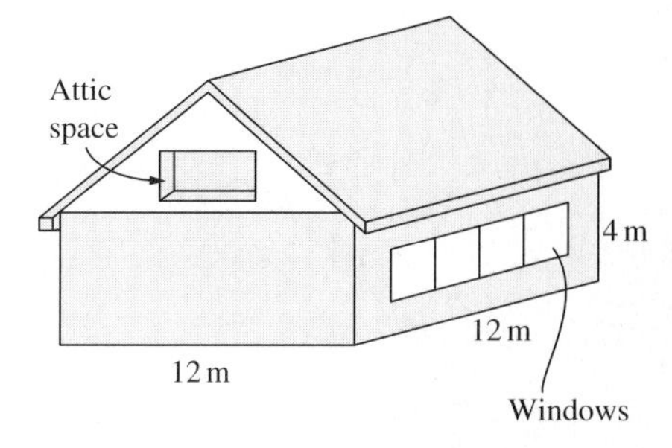

**FIGURE P8-71**

**8-71E** Consider a house whose walls are 12 ft high and 40 ft long. Two of the walls of the house have no windows, while each of the other two walls have four windows made of 0.25-in-thick glass [$k$ = 0.45 Btu/(h · ft · °F)], 3 ft × 5 ft in size. The walls are certified to have an $R$-value of 19 (i.e., an $L/k$ value of 19 h · $ft^2$ · °F/Btu). Disregarding any direct radiation gain or loss through the windows and taking the heat transfer coefficients at the inner and outer surfaces of the house to be 2 and 4 Btu/(h · $ft^2$ · °F), respectively, determine the ratio of the heat transfer through the walls with and without windows.

**8-72** Consider a house that has a 10 m × 20 m base and a 4-m-high walls. All four walls of the house have an $R$-value of 2.31 $m^2$ · °C/W. The two 10 m × 4 m walls have no windows. The third wall has five windows made of 0.5-cm-thick glass [$k$ = 0.78 W/(m · °C)], 1.2 m × 1.8 m in size. The fourth wall has the same size and number of windows, but they are double-paned with a 1.5-cm-thick stagnant air space [$k$ = 0.026 W/(m · °C)] enclosed between two 0.5-cm-thick glass layers. The thermostat in the house is set at 22°C and the average temperature outside at that location is 5°C during the seven-month-long heating season. Disregarding any direct radiation gain or loss through the windows and taking the heat transfer coefficients at the inner and outer surfaces of the house to be 7 and 15 W/($m^2$ · °C), respectively, determine the average rate of heat transfer through each wall.

If the house is electrically heated and the price of electricity is given to be \$0.09/kWh, determine the amount of money this household will save per heating season by converting the single pane windows to double pane windows.

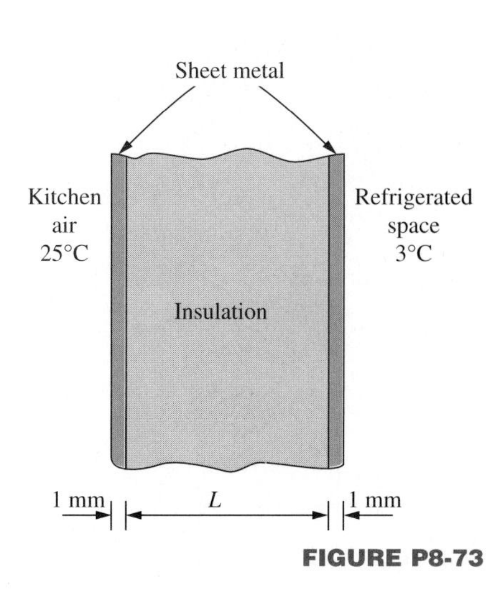


**FIGURE P8-73**

**8-73** The wall of a refrigerator is constructed of fiberglass insulation [$k$ = 0.035 W/(m · °C)] sandwiched between two layers of 1-mm-thick sheet metal [$k = 15.1$ W/(m · °C)]. The refrigerated space is maintained at 3°C, and the average heat transfer coefficients at the inner and outer surfaces of the wall are 4 W/(m$^2$ · °C) and 9 W/(m$^2$ · °C), respectively. The kitchen temperature averages 25°C. It is observed that condensation occurs on the outer surfaces of the refrigerator when the temperature of the outer surface drops to 20°C. Determine the minimum thickness of fiberglass insulation that needs to be used in the wall in order to avoid condensation on the outer surfaces.

**8-74** Repeat Prob. 8-73 for a condensation temperature of 15°C on the outer surfaces of the refrigerator when the kitchen temperature is 19°C.

**8-75** Heat is to be conducted along a circuit board that has a copper layer on one side. The circuit board is 15 cm long and 15 cm wide, and the thicknesses of the copper and epoxy layers are 0.1 mm and 1.2 mm, respectively. Disregarding heat transfer from side surfaces, determine the percentages of heat conduction along the copper [$k = 386$ W/(m · °C)] and epoxy [$k = 0.26$ W/(m · °C)] layers. Also determine the effective thermal conductivity of the board.
*Answers:* 0.8%, 99.2%, and 29.9 W/(m · °C)

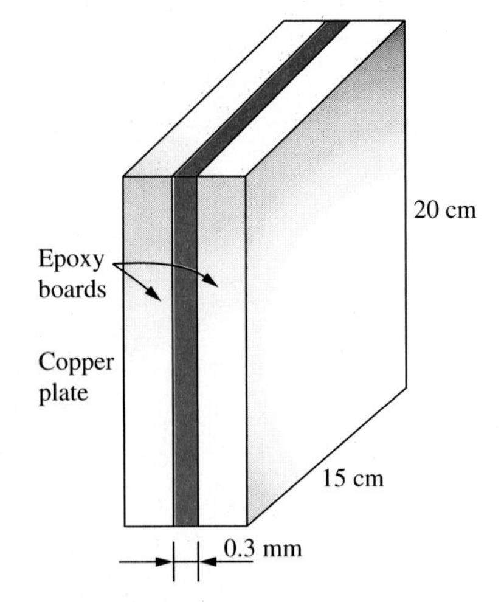

**FIGURE P8-76**

**8-76** A 0.3-mm-thick copper plate [$k = 386$ W/(m · °C)] is sandwiched between two 4-mm-thick epoxy boards [$k = 0.26$ W/(m · °C)] that are 15 cm × 20 cm in size. Determine the effective thermal conductivity of the board along its 20-cm-long side. What fraction of the heat transferred along that side is conducted through copper?

**8-76E** A 0.03-in.-thick copper plate [$k = 223$ Btu/(h · ft · °F)] is sandwiched between two 0.1-in.-thick epoxy boards [$k = 0.15$ Btu/(h · ft · °F)] that are 7 in. × 9 in. in size. Determine the effective thermal conductivity of the board along its 9-in.-long side. What fraction of the heat transferred along that side is conducted through copper?

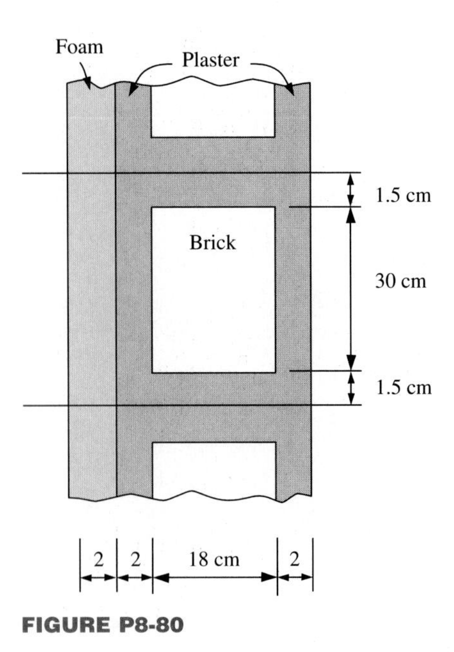


**FIGURE P8-80**

### Generalized Thermal Resistance Networks

**8-77C** When plotting the thermal resistance network associated with a heat transfer problem, explain when two resistances are taken to be in series and when they are in parallel.

**8-78C** The thermal resistance networks can also be used approximately for multidimensional problems. For what kind of multidimensional problems will the thermal resistance approach give adequate results?

**8-79C** What are the two approaches used in the development of the thermal resistance network for two-dimensional problems?

**8-80** A 4-m-high and 6-m-wide wall consists of long 18 cm × 30 cm cross-section horizontal bricks [$k = 0.72$ W/(m · °C)] separated by 3-cm-thick plaster layers [$k = 0.22$ W/(m · °C)]. There are also 2-cm-thick plaster layers on each side of the wall, and a 2-cm-thick rigid foam ($k = 0.026$ W/(m · °C)] on the inner side of the wall. The indoor and the outdoor temperatures are 22°C and −4°C, and the convection heat transfer coefficients on the inner and the outer sides are $h_1 = 10\ \text{W/(m}^2 \cdot {}^\circ\text{C)}$ and $h_2 = 20\ \text{W/(m}^2 \cdot {}^\circ\text{C)}$, respectively. Assuming one-dimensional heat transfer and disregarding radiation, determine the rate of heat transfer through the wall.

**8-81** A 10-cm-thick wall is to be constructed with 2.5-m-long wood studs [$k = 0.11\ \text{W/(m} \cdot {}^\circ\text{C)}$] that have a cross-section of 10 cm × 10 cm. At some point, the builder has run out of those studs, and started using pairs of 2.5-m-long wood studs that have a cross-section of 5 cm × 10 cm nailed to each other instead. The manganese steel nails [$k = 50\ \text{W/(m} \cdot {}^\circ\text{C)}$] are 10 cm long, and have a diameter of 0.4 cm. A total of 50 nails are used to connect the two studs, which are mounted to the wall such that the nails cross the wall. The temperature difference between the inner and outer surfaces of the wall is 15°C. Assuming the thermal contact resistance between the two layers to be negligible, determine the rate of heat transfer (*a*) through a solid stud and (*b*) through a stud pair of equal length and width nailed to each other. (*c*) Also determine the effective conductivity of the nailed stud pair.

**8-82** A 12-m-long and 5-m-high wall is constructed of two layers of 1-cm-thick sheetrock [$k = 0.17\ \text{W/(m} \cdot {}^\circ\text{C)}$] spaced 12 cm by wood studs [$k = 0.11\ \text{W/(m} \cdot {}^\circ\text{C)}$] whose cross-section is 12 cm × 5 cm. The studs are placed vertically 60 cm apart, and the space between them is filled with fiberglass insulation [$k = 0.034\ \text{W/(m} \cdot {}^\circ\text{C)}$]. The house is maintained at 20°C and the ambient temperature outside is −5°C. Taking the heat transfer coefficients at the inner and outer surfaces of the house to be 8.3 and 34 $\text{W/(m}^2 \cdot \text{C)}$, respectively, determine (*a*) the thermal resistance of the wall considering a representative section of it and (*b*) the rate of heat transfer through the wall.

**8-83** A 22-cm-thick, 10-m-long, and 3.5-m-high wall is to be constructed using 20-cm-long solid bricks [$k = 0.7\ \text{W/(m} \cdot {}^\circ\text{C)}$] of cross-section 16 cm × 16 cm, or identical size bricks with nine square air holes [$k = 0.026\ \text{W/(m} \cdot {}^\circ\text{C)}$] that are 20 cm long and have a cross-section of 4 cm × 4 cm, as shown in Fig. P8-83. There is a 1-cm-thick plaster layer [$k = 0.22\ \text{W/(m} \cdot {}^\circ\text{C)}$] between two adjacent bricks on all four sides, and on both sides of the wall. The house is maintained at 24°C and the ambient temperature outside is 2°C. Taking the heat transfer coefficients at the inner and outer surfaces of the wall to be 8 and 24 $\text{W/(m}^2 \cdot {}^\circ\text{C)}$, respectively, determine the rate of heat transfer through the wall constructed of (*a*) solid bricks and (*b*) bricks with air holes.
*Answers:* (*a*) 1326 W, (*b*) 840 W

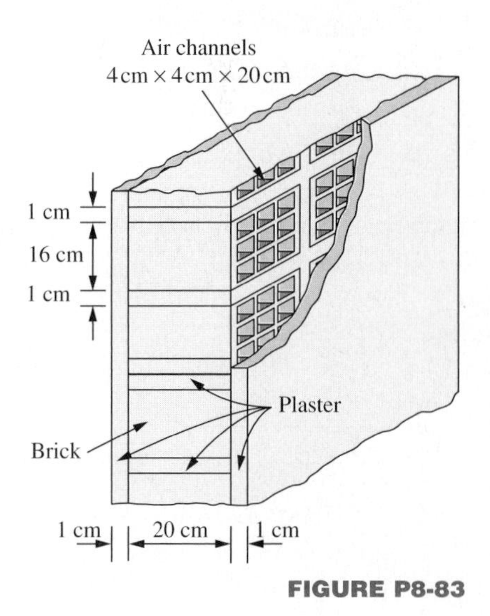


**FIGURE P8-83**

**8-83E** A 10-in.-thick, 30-ft-long, and a 10-ft-high wall is to be constructed using 9-in.-long solid bricks [$k = 0.40\ \text{Btu/(h} \cdot \text{ft} \cdot {}^\circ\text{F)}$] of cross-section 7 in. × 7 in., or identical size bricks with nine square air holes [$k = 0.015\ \text{Btu/(h} \cdot \text{ft} \cdot {}^\circ\text{F)}$] that are 9 in. long and have a cross-section of 1.5 in. × 1.5 in. There is a 0.5-in.-thick plaster layer [$k = 0.10\ \text{Btu/(h} \cdot \text{ft} \cdot {}^\circ\text{F)}$] between two adjacent bricks on all four sides, and on both sides of the wall. The house is maintained at 75°F and the ambient temperature outside is 35°F. Taking the heat transfer coefficients at the inner and outer surfaces of the wall to be 1.5 and 4 Btu/(h · ft$^2$ · °F), respectively, determine the rate of heat transfer through the wall constructed of (*a*) solid bricks and (*b*) bricks with air holes.

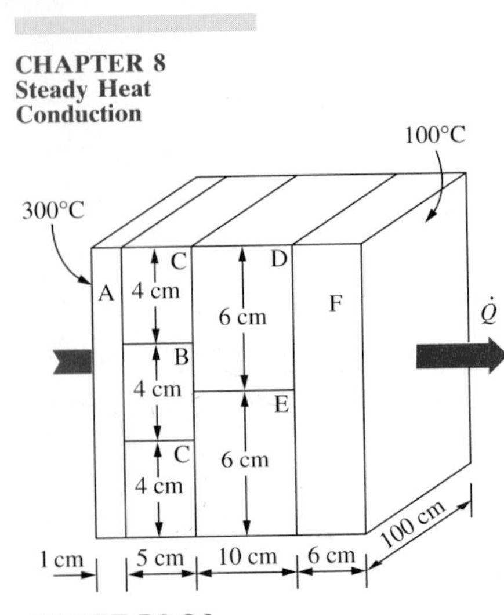

**FIGURE P8-84**

**8-84** Consider a 5-m-high, 8-m-long, and 0.22-m-thick wall whose representative cross-section is as given in Fig. P8-84. The thermal conductivities of various materials used, in W/(m · °C), are $k_A = k_F = 2$, $k_B = 8$, $k_C = 20$, $k_D = 15$, and $k_E = 35$. The left and right surfaces of the wall are maintained at uniform temperatures of 300°C and 100°C, respectively. Assuming heat transfer through the wall to be one-dimensional, determine (*a*) the rate of heat transfer through the wall, (*b*) the temperature at the point where the sections B, D, and E meet, and (*c*) the temperature drop across the section F. Disregard any contact resistances at the interfaces.

**8-85** Repeat Prob. 8-84 assuming that the thermal contact resistance at the interfaces D–F and E–F is 0.00012 m$^2$ · °C/W.

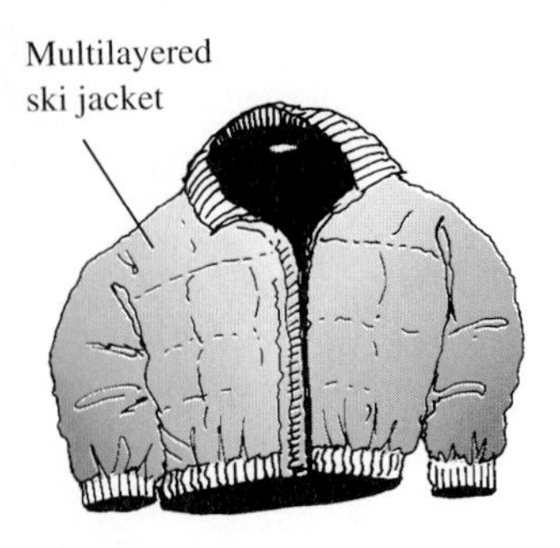

**FIGURE P8-86**

**8-86** Clothing made of several thin layers of fabric with trapped air in between, often called ski clothing, is commonly used in cold climates because it is light, fashionable, and a very effective thermal insulator. So it is no surprise that such clothing has largely replaced thick and heavy old-fashioned coats.

Consider coat made of five layers of 0.1-mm-thick synthetic fabric [$k$ = 0.13 W/(m · °C)] with 1.5-mm-thick air space [$k$ = 0.026 W/(m · °C)] between the layers. Assuming the inner surface temperature of the jacket to be 28°C and the surface area to be 1.1 m$^2$, determine the rate of heat loss through the jacket when the temperature of the outdoors is −5°C and the heat transfer coefficient at the outer surface is 25 W/(m$^2$ · °C).

What would your response be if the jacket were made of a single layer of 0.5-mm-thick synthetic fabric? What should the thickness of a wool fabric [$k$ = 0.035 W/(m · °C)] be if the person is to achieve the same level of thermal comfort wearing a thick wool coat instead of a ski jacket?

**8-87** Repeat Prob. 8-86 assuming that the layers of the jacket are made of cotton fabric [$k$ = 0.06 W/(m · °C)].

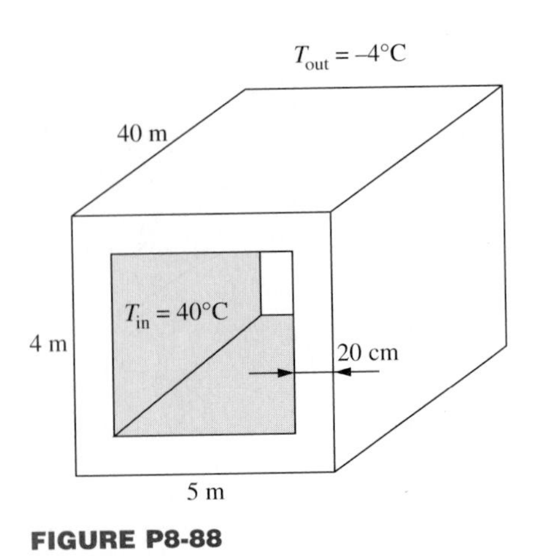


**FIGURE P8-88**

**8-88** A 5-m-wide, 4-m-high, and 40-m-long kiln used to cure concrete pipes is made of 20-cm-thick concrete walls and ceiling [$k$ = 0.9 W/(m · °C)]. The kiln is maintained at 40°C by injecting hot steam into it. The two ends of the kiln, 4 × 5 m in size, are made of a 3-mm-thick sheet metal covered with 2-cm-thick styrofoam [$k$ = 0.033 W/(m · °C)]. The convection heat transfer coefficients on the inner and the outer surfaces of the kiln are 3000 W/(m$^2$ · °C) and 25 W/(m$^2$ · °C), respectively. Disregarding any heat loss through the floor, determine the rate of heat loss from the kiln when the ambient air is at −4°C.

**8-89** Consider a 15 cm × 18 cm epoxy glass laminate [$k$ = 0.26 W/(m · °C)] whose thickness is 1.4 mm. In order to reduce the thermal resistance across its thickness, cylindrical copper fillings [$k$ = 386 W/(m · °C)] of 1 mm diameter are to be planted throughout the board, with a center-to-center distance of 3 mm. Determine the new value of the thermal resistance of the epoxy board for heat conduction across its thickness as a result of this modification.
*Answer:* 0.00153 °C/W

**8-89E** Consider a 6 in. × 8 in. epoxy glass laminate [$k$ = 0.10 Btu/(h · ft · °F)] whose thickness is 0.05 in. In order to reduce the thermal resistance across its thickness, cylindrical copper fillings [$k$ = 223 Btu/(h · ft · °F)] of 0.02 in diameter are to be planted throughout the board, with a center-to-center distance of 0.06 in. Determine the new value of the thermal resistance of the epoxy board for heat conduction across its thickness as a result of this modification.
*Answer:* 0.00065 h · °F/Btu

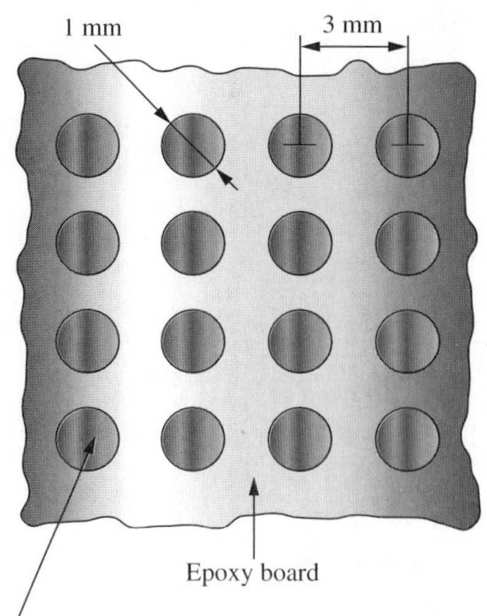

**FIGURE P8-89**

### Heat Conduction in Cylinders and Spheres

**8-90C** Consider one-dimensional heat conduction through a plane wall, a long cylinder, and a sphere. For which of these geometries is the heat transfer area constant, and for which of them is variable? Explain.

**8-91C** What is an infinitely long cylinder? When is it proper to treat an actual cylinder as being infinitely long, and when is it not?

**8-92C** Consider a short cylinder whose top and bottom surfaces are insulated. The cylinder is initially at a uniform temperature $T_i$, and is subjected to convection from its side surface to a medium at temperature $T_\infty$, with a heat transfer coefficient of $h$. Is the heat transfer in this short cylinder one- or two-dimensional? Explain.

**8-93C** Can the thermal resistance concept be used for a solid cylinder or sphere in steady operation? Explain.

**8-94** A 5-m internal diameter spherical tank made of 1.5-cm-thick stainless steel [$k$ = 15 W/(m · °C)] is used to store iced water at 0°C. The tank is located in a room whose temperature is 20°C. The walls of the room are also at 20°C. The outer surface of the tank is black (emissivity $\varepsilon$ = 1), and heat transfer between the outer surface of the tank and the surroundings is by natural convection and radiation. The convection heat transfer coefficients at the inner and the outer surfaces of the tank are 80 W/(m$^2$ · °C) and 10 W/(m$^2$ · °C), respectively. Determine (*a*) the rate of heat transfer to the iced water in the tank and (*b*) the amount of ice at 0°C that melts during a 24-h period. The heat of fusion of water at atmospheric pressure is $h_{if}$ = 333.7 kJ/kg.

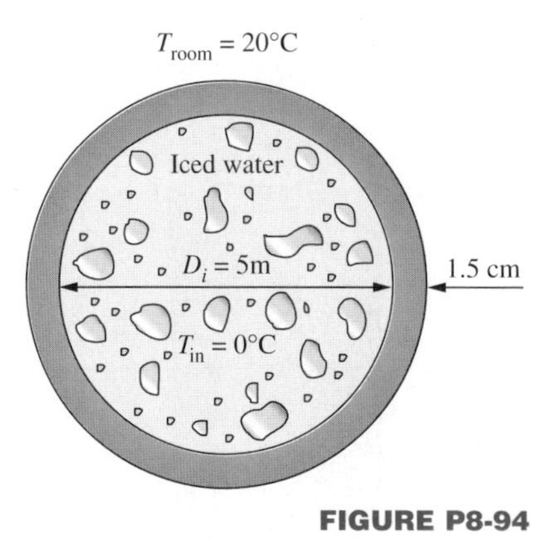

**FIGURE P8-94**

**8-95** Steam at 320°C flows in a stainless steel pipe [$k$ = 15 W/(m · °C)] whose inner and outer diameters are 5 cm and 5.5 cm, respectively. The pipe is covered with 3-cm-thick glass wool insulation [$k$ = 0.038 W/(m · °C)]. Heat is lost to the surroundings at 5°C by natural

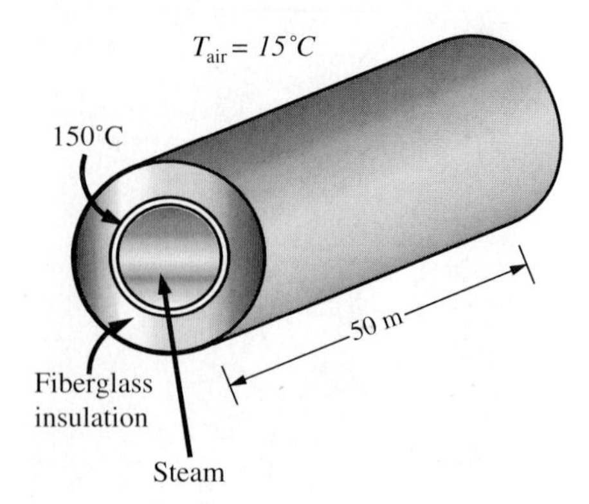


FIGURE P8-96

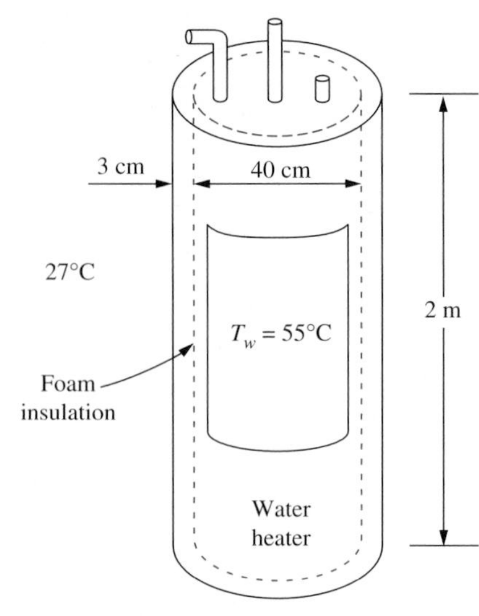

FIGURE P8-97

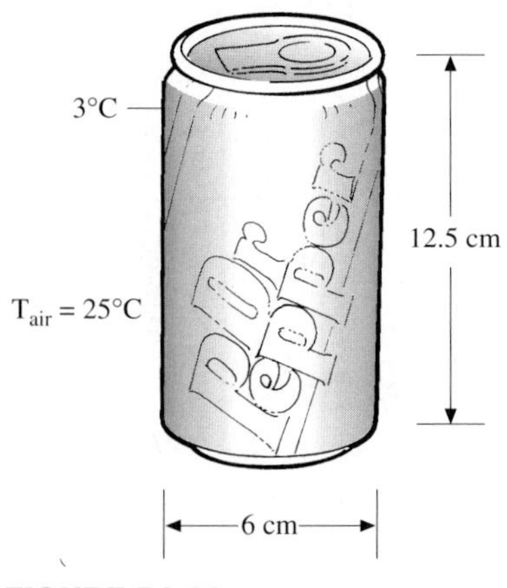


FIGURE P8-98

convection and radiation, with a combined natural convection and radiation heat transfer coefficient of 15 W/(m$^2 \cdot$ C). Taking the heat transfer coefficient inside the pipe to be 80 W/(m$^2 \cdot$ °C), determine the rate of heat loss from the steam per unit length of the pipe. Also determine the temperature drops across the pipe shell and the insulation.

**8-96** A 50-m-long section of a steam pipe whose outer diameter is 10 cm passes through an open space at 15°C. The average temperature of the outer surface of the pipe is measured to be 150°C. If the combined heat transfer coefficient on the outer surface of the pipe is 20 W/(m$^2 \cdot$ °C), determine (*a*) the rate of heat loss from the steam pipe, (*b*) the annual cost of this energy lost if steam is generated in a natural gas furnace that has an efficiency of 75 percent, and the price of natural gas is $0.52/therm (1 therm = 105,500 kJ), and (*c*) the thickness of fiberglass insulation [$k$ = 0.035 W/(m · °C)] needed in order to save 90 percent of the heat lost. Assume the pipe temperature to remain constant at 150°C.

**8-96E** A 150-ft-long section of a steam pipe whose outer diameter is 4 in. passes through an open space at 60°F. The average temperature of the outer surface of the pipe is measured to be 300°F. If the combined heat transfer coefficient on the outer surface of the pipe is 3.5 Btu/(h · ft$^2$ · °F), determine (*a*) the rate of heat loss from the steam pipe, (*b*) the annual cost of this energy lost if steam is generated in a natural gas furnace that has an efficiency of 75 percent, and the price of natural gas is $0.52/therm (1 therm = 100,000 Btu), and (*c*) the thickness of fiberglass insulation [$k$ = 0.020 Btu/(h · ft · °F)] needed in order to save 90 percent of the heat lost. Assume the pipe temperature to remain constant at 300°F.

**8-97** Consider a 2-m-high electric hot water heater that has a diameter of 40 cm and maintains the hot water at 55°C. The tank is located in a small room whose average temperature is 27°C, and the heat transfer coefficients on the inner and outer surfaces of the heater are 50 and 12 W/(m$^2 \cdot$ °C), respectively. The tank is placed in another 46-cm-diameter sheet metal tank of negligible thickness, and the space between the two tanks is filled with foam insulation [$k$ = 0.03 W/(m · °C)]. The thermal resistances of the water tank and the outer thin sheet metal shell are very small, and can be neglected. The price of electricity is $0.08/kWh, and the homeowner pays $280 a year for water heating. Determine the fraction of the hot water energy cost of this household that is due to the heat loss from the tank.

Hot water tank insulation kits consisting of 3-cm-thick fiber glass insulation [$k$ = 0.035 W/(m · °C)] large enough to wrap the entire tank are available in the market for about $30. If such an insulation is installed on this water tank by the homeowner himself, how long will it take for this additional insulation to pay for itself?
*Answers:* 17.5 percent, 1.5 year

**8-98** Consider an aluminum cold drink can that is initially at a uniform temperature of 3°C. The can is 12.5 cm high and has a diameter of 6 cm.

If the combined convection/radiation heat transfer coefficient between the can and the surrounding air at 25°C is 10 W/(m$^2$ · C), determine how long it will take for the average temperature of the drink to rise to 10°C.

In an effort to slow down the warming of the cold drink a person puts the can in a perfectly fitting 1-cm-thick cylindrical rubber insulation [$k$ = 0.13 W/(m · °C)]. Now how long will it take for the average temperature of the drink to rise to 10°C? Assume the top of the can is not covered.

**8-99** Repeat Prob. 8-98, assuming a thermal contact resistance of 0.00008 m$^2$ · °C/W between the can and the insulation.

**8-100** Steam at 300°C is flowing through a steel pipe [$k$ = 15.1 W/(m · °C)] whose inner and outer diameters are 8 and 8.8 cm, respectively, in an environment at 15°C. The pipe is insulated with 3-cm-thick fiberglass insulation [$k$ = 0.035 W/(m · °C)]. If the heat transfer coefficients on the inside and the outside of the pipe are 150 and 25 W/(m$^2$ · °C), respectively, determine the rate of heat loss from the steam per meter length of the pipe. What is the error involved in neglecting the thermal resistance of the steel pipe in calculations?

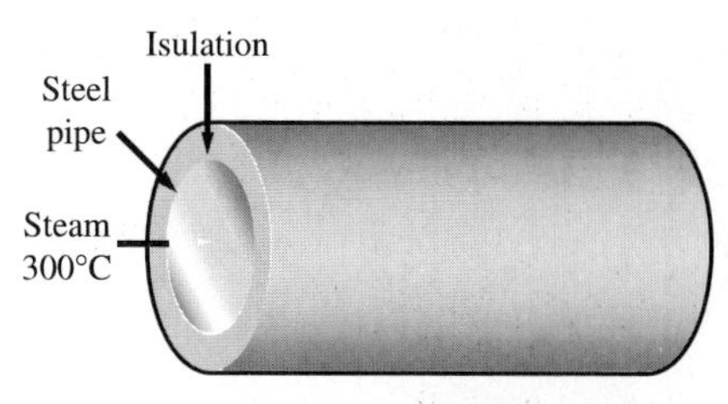

FIGURE P8-100

**8-100E** Steam at 600°F is flowing through a steel pipe [$k$ = 8.7 Btu/(h · ft · °F)] whose inner and outer diameters are 3.5 and 4.0 in., respectively, in an environment at 60°F. The pipe is insulated with 2-in.-thick fiberglass insulation [$k$ = 0.020 Btu/(h · ft · °F)]. If the heat transfer coefficients on the inside and the outside of the pipe are 30 and 5 Btu/(h · ft$^2$ · °F), respectively, determine the rate of heat loss from the steam per foot length of the pipe. What is the error involved in neglecting the thermal resistance of the steel pipe in calculations?

**8-101** Hot water at an average temperature of 90°C is flowing through a 15-m section of a cast iron pipe [$k$ = 52 W/(m · °C)] whose inner and outer diameters are 4 and 4.6 cm, respectively. The outer surface of the pipe, whose emissivity is 0.7, is exposed to the cold air at 10°C in the basement, with a heat transfer coefficient of 15 W/(m$^2$ · °C). The heat transfer coefficient at the inner surface of the pipe is 120 W/(m$^2$ · °C). Taking the walls of the basement to be 10°C also, determine the rate of heat loss from the hot water. Also determine the average velocity of the water in the pipe if the temperature of the water drops by 3°C as it passes through the basement.

**8-102** Repeat Prob. 8-101 for a pipe made of copper [$k$ = 386 W/(m · °C)] instead of cast iron.

**8-103** Steam exiting the turbine of a steam power plant at 35°C is to be condensed in a large condenser by cooling water flowing through copper pipes [$k$ = 386 W/(m · °C)] of inner diameter 1 cm and outer diameter 1.4 cm at an average temperature of 20°C. The heat of vaporization of water at 35°C is 2419 kJ/kg. The heat transfer coefficients are 8000 W/(m$^2$ · °C) on the steam side and 160 W/(m$^2$ · °C) on the water side. Determine the length of the tube required to condense steam at a rate of 200 kg/h. *Answer:* 1773 m

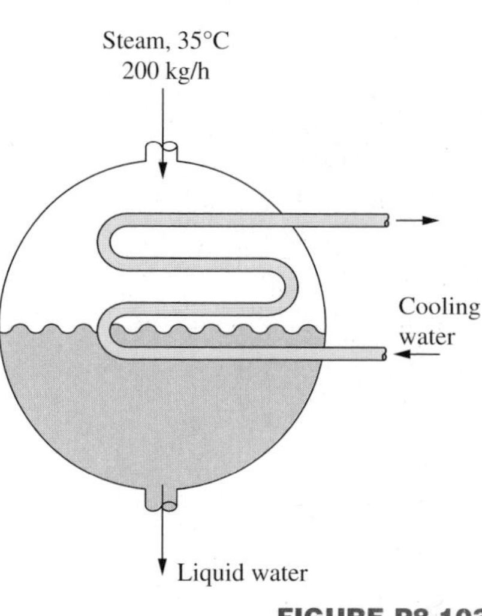

FIGURE P8-103

**8-103E** Steam exiting the turbine of a steam power plant at 100°F is to be condensed in a large condenser by cooling water flowing through copper pipes [$k = 223$ Btu/(h · ft · °F)] of inner diameter 0.4 in. and outer diameter 0.6 in. at an average temperature of 70°F. The heat of vaporization of water at 100°F is 1037 Btu/lbm. The heat transfer coefficients are 1500 Btu/(h · ft² · °F) on the steam side and 35 Btu/(h · ft² · °F) on the water side. Determine the length of the tube required to condense steam at a rate of 400 lbm/h. *Answer:* 3830 ft

**8-104** Repeat Prob. 8-103, assuming that a 0.15-mm-thick layer of mineral deposit [$k = 3$ W/(m · °C)] is formed on the inner surface of the pipe.

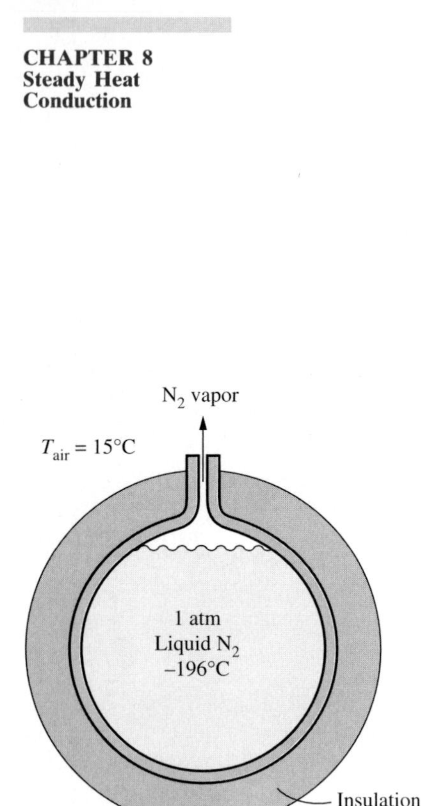


**FIGURE P8-105**

**8-105** The boiling temperature of nitrogen at atmospheric pressure at sea level (1-atm pressure) is −196°C. Therefore, nitrogen is commonly used in low-temperature scientific studies, since the temperature of liquid nitrogen in a tank open to the atmosphere will remain constant at −196°C until it is depleted. Any heat transfer to the tank will result in the evaporation of some liquid nitrogen which has a heat of vaporization of 198 kJ/kg and a density of 810 kg/m³ at 1 atm.

Consider a 3-m-diameter spherical tank that is initially filled with liquid nitrogen at 1 atm and −196°C. The tank is exposed to ambient air at 15°C, with a combined convection and radiation heat transfer coefficient of 35 W/(m² · °C). The temperature of the thin-shelled spherical tank is observed to be almost the same as the temperature of the nitrogen inside. Determine the rate of evaporation of the liquid nitrogen in the tank as a result of the heat transfer from the ambient air if the tank is (*a*) not insulated, (*b*) insulated with 5-cm-thick fiberglass insulation [$k = 0.035$ W/(m · °C)], and (*c*) insulated with 2-cm-thick superinsulation that has an effective thermal conductivity of 0.00005 W/(m · °C).

**8-106** Repeat Prob. 8-105 for liquid oxygen, which has a boiling temperature of −183°C, a heat of vaporization of 213 kJ/kg, and a density of 1140 kg/m³ at 1-atm pressure.

## Critical Radius of Insulation

**8-107C** What is the critical radius of insulation? How is defined for a cylindrical layer?

**8-108C** A pipe is insulated such that the outer radius of the insulation is less than the critical radius. Now the insulation is taken off. Will the rate of heat transfer from the pipe increase or decrease for the same pipe surface temperature?

**8-109C** A pipe is insulated to reduce the heat loss from it. However, measurements indicate that the rate of heat loss has increased instead of decreasing. Can the measurements be right?

**8-110C** Consider a pipe at a constant temperature whose radius is greater than the critical radius of insulation. Someone claims that the

rate of heat loss from the pipe has increased when some insulation is added to the pipe. Is this claim valid?

**8-111C** Consider an insulated pipe exposed to the atmosphere. Will the critical radius of insulation be greater on calm days or on windy days? Why?

**8-112** A 2-mm-diameter and 10-m-long electric wire is tightly wrapped with a 1-mm-thick plastic cover whose thermal conductivity is $k = 0.15\ \text{W/(m} \cdot °\text{C)}$. Electrical measurements indicate that a current of 10 A passes through the wire and there is a voltage drop of 8 V along the wire. If the insulated wire is exposed to a medium at $T_\infty = 30°\text{C}$, with a heat transfer coefficient of $h = 18\ \text{W/(m}^2 \cdot °\text{C)}$, determine the temperature at the interface of the wire and the plastic cover in steady operation. Also determine if doubling the thickness of the plastic cover will increase or decrease this interface temperature.

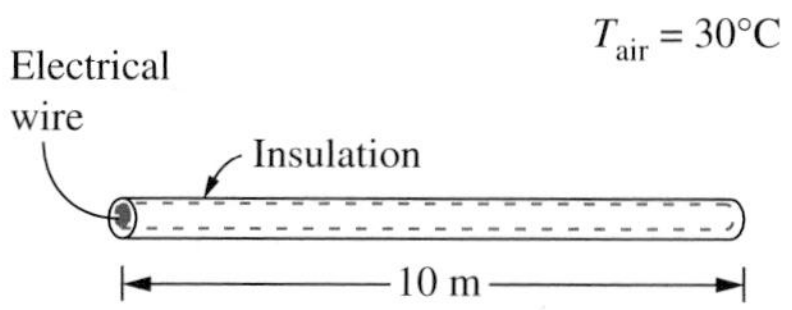

$T_{air} = 30°\text{C}$
Electrical wire
Insulation
10 m

**FIGURE P8-112**

**8-113** A 2-mm-diameter electrical wire at 45°C is covered by 0.5-mm-thick plastic insulation $[k = 0.13\ \text{W/(m} \cdot °\text{C)}]$. The wire is exposed to a medium at 10°C, with a combined convection and radiation heat transfer coefficient of $12\ \text{W/(m}^2 \cdot °\text{C)}$. Determine if the plastic insulation on the wire will help or hurt heat transfer from the wire. *Answer:* It helps

**8-113E** A 0.08-in.-diameter electrical wire at 115°F is covered by 0.02-in.-thick plastic insulation $[k = 0.075\ \text{Btu/(h} \cdot \text{ft} \cdot °\text{F)}]$. The wire is exposed to a medium at 50°F, with a combined convection and radiation heat transfer coefficient of $2.5\ \text{Btu/(h} \cdot \text{ft}^2 \cdot °\text{F)}$. Determine if the plastic insulation on the wire will help or hurt heat transfer from the wire.
*Answer:* It helps

**8-114** Repeat Prob. 8-113, assuming a thermal contact resistance of $0.0002\ \text{m}^2 \cdot °\text{C/W}$ at the interface of the wire and the insulation.

**8-115** A 5-mm-diameter spherical ball at 50°C is covered by 1-mm-thick plastic insulation $[k = 0.13\ \text{W/(m} \cdot °\text{C)}]$. The ball is exposed to a medium at 15°C, with a combined convection and radiation heat transfer coefficient of $20\ \text{W/(m}^2 \cdot °\text{C)}$. Determine if the plastic insulation on the ball will help or hurt heat transfer from the ball.

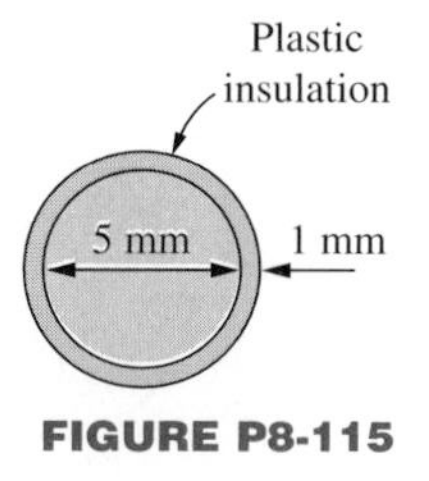


**FIGURE P8-115**

## Heat Generation in a Solid

**8-116C** Does heat generation in a solid violate the first law of thermodynamics which states that energy cannot be created or destroyed? Explain.

**8-117C** What is heat generation? Give some examples.

**8-118C** An iron is left unattended, and its base temperature rises as a result of resistance heating inside. When will the rate of heat generation inside the iron be equal to the rate of heat loss from the iron?

**8-119C** Consider the uniform heating of a plate in an environment at a constant temperature. Is it possible for part of the heat generated in the left half of the plate to leave the plate through the right surface? Explain.

**8-120C** Consider uniform heat generation in a cylinder and a sphere of equal radius made of the same material in the same environment. Which geometry will have a higher temperature at its center? Why?

**8-121** A 2-kW resistance heater wire whose thermal conductivity is $k = 12\ \text{W/(m} \cdot \text{°C)}$ has a diameter of 3 mm and a length of 0.8 m, and is used to boil water. If the outer surface temperature of the resistance wire is 110°C, determine the temperature at the center of the wire.

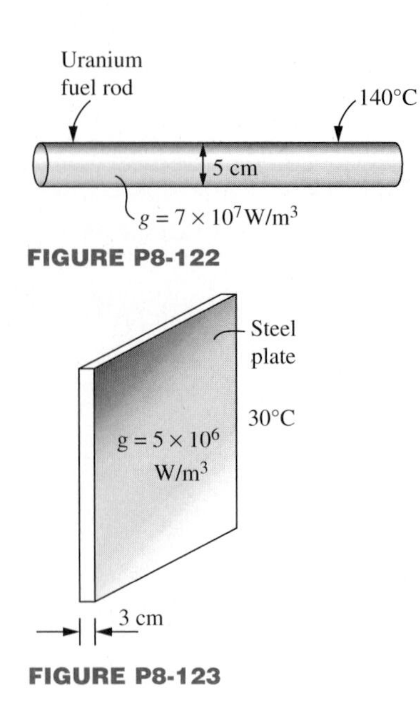


FIGURE P8-122

FIGURE P8-123

**8-122** In a nuclear reactor, 5-cm-diameter cylindrical uranium rods cooled by water from outside serve as the fuel. Heat is generated uniformly in the rods $[k = 29.5\ \text{W/(m} \cdot \text{°C)}]$ at a rate of $7 \times 10^7\ \text{W/m}^3$. If the outer surface temperature of the rods is 140°C, determine the temperature at their centers.

**8-123** Consider a large 3-cm-thick stainless steel plate $[k = 15.1\ \text{W/(m} \cdot \text{°C)}]$ in which heat is generated uniformly at a rate of $5 \times 10^6\ \text{W/m}^3$. Both sides of the plate are exposed to an environment at 30°C, with a heat transfer coefficient of $600\ \text{W/(m}^2 \cdot \text{°C)}$. Explain where in the plate the highest and the lowest temperatures will occur, and determine their values. *Answers:* 155°C, 192°C

**8-123E** Consider a large 1-in.-thick stainless steel plate $[k = 8.7\ \text{Btu/(h} \cdot \text{ft} \cdot \text{°F)}]$ in which heat is generated uniformly at a rate of $5 \times 10^5\ \text{Btu/(h} \cdot \text{ft}^3)$. Both sides of the plate are exposed to an environment at 90°F, with a heat transfer coefficient of $120\ \text{Btu/(h} \cdot \text{ft}^2 \cdot \text{°F)}$. Explain where in the plate the highest and the lowest temperatures will occur, and determine their values. *Answers:* 437°F, 637°F

**8-124** Consider a large 5-cm-thick brass plate $[k = 111\ \text{W/(m} \cdot \text{°C)}]$ in which heat is generated uniformly at a rate of $2 \times 10^5\ \text{W/m}^3$. One side of the plate is insulated, while the other side is exposed to an environment at 25°C, with a heat transfer coefficient of $44\ \text{W/(m}^2 \cdot \text{°C)}$. Explain where in the plate the highest and the lowest temperatures will occur, and determine their values.

**8-125** A 6-m-long 2-kW electrical resistance wire is made of 0.2-mm-diameter stainless steel $[k = 15.1\ \text{W/(m} \cdot \text{°C)}]$. The resistance wire operates in an environment at 30°C, with a heat transfer coefficient of $140\ \text{W/(m}^2 \cdot \text{°C)}$ at the outer surface. Determine the surface temperature of the wire.

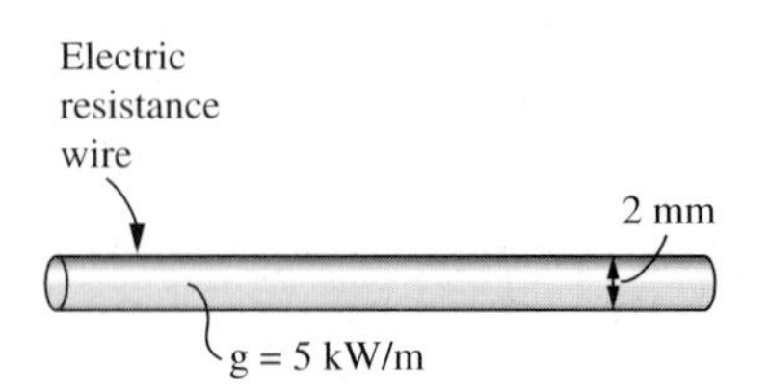

FIGURE P8-126

**8-126** Heat is generated in a 2-mm-diameter electric resistance wire made of nickel steel $[k = 10\ \text{W/(m} \cdot \text{°C)}]$ uniformly at a rate of 5 kW per m length. Determine the temperature difference between the centerline and the surface of the wire.

**8-126E** Heat is generated in a 0.08-in.-diameter electric resistance wire made of nickel steel $[k = 5.8\ \text{Btu/(h} \cdot \text{ft} \cdot \text{°F)}]$ uniformly at a rate of 5 kW per m length. Determine the temperature difference between the centerline and the surface of the wire.

**8-127** Repeat Prob. 8-126 for a manganese wire [$k = 7.8$ W/(m · °C)].

### Heat Transfer From Finned Surfaces

**8-128C** What is the reason for the widespread use of fins on surfaces?

**8-129C** What is the difference between the fin effectiveness and the fin efficiency.

**8-130C** The fins attached to a surface are determined to have an effectiveness of 0.9. Do you think the rate of heat transfer from the surface has increased or decreased as a result of the addition of these fins?

**8-131C** Explain how the fins enhance heat transfer from a surface. Also explain how the additon of fins may actually decrease heat transfer from a surface.

**8-132C** How does the overall effectiveness of a finned surface differ from the effectiveness of a single fin?

**8-133C** Hot water is to be cooled as it flows through the tubes exposed to atmospheric air. Fins are to be attached in order to enhance heat transfer. Would you recommend attaching the fins inside or outside the tubes? Why?

**8-134C** Hot air is to be cooled as it is forced to flow through the tubes exposed to atmospheric air. Fins are to be added in order to enhance heat transfer. Would you recommend attaching the fins inside or outside the tubes? Why? When would you recommend attaching fins both inside and outside the tubes?

**8-135C** Consider two finned surfaces that are identical except that the fins on the first surface are formed by casting or extrusion, whereas they are attached to the second surface afterwards by welding or tight fitting. For which case do you think the fins will provide greater enhancement in heat transfer? Explain.

**8-136C** The heat transfer surface area of a fin is equal to the sum of all surfaces of the fin exposed to the surrounding medium, including the surface are of the fin tip. Under what conditions can we neglect heat transfer from the fin tip?

**8-137C** Do (*a*) the efficiency and (*b*) the effectiveness of a fin increase or decrease as the fin length is increased?

**8-138C** Two pin fins are identical, except that the diameter of one of them is twice the diameter of the other. For which fin will (*a*) the fin effectiveness and (*b*) the fin efficiency be higher? Explain.

**8-139C** Two plate fins of constant rectangular cross-section are identical, except that the thickness of one of them is twice the thickness of the other. For which fin will (*a*) the fin effectiveness and (*b*) the fin efficiency be higher? Explain.

**8-140C** Two finned surfaces are identical, except that the convection heat transfer coefficient of one of them is twice that of the other. For which finned surface will (*a*) the fin effectiveness and (*b*) the fin efficiency be higher? Explain.

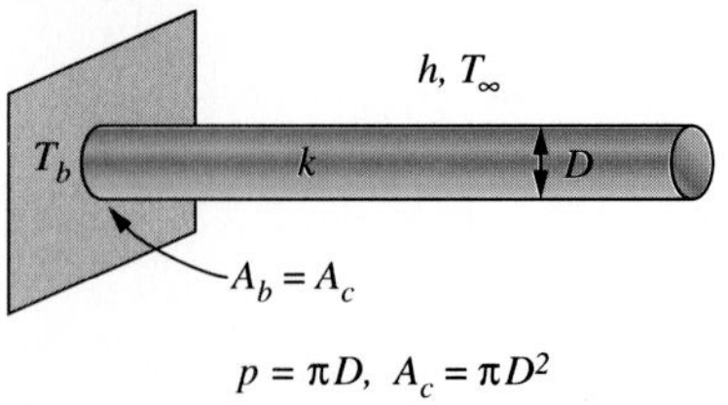

**8-141** Obtain a relation for the fin efficiency for a fin of constant cross-sectional area $A_c$, perimeter $p$, length $L$, and thermal conductivity $k$ exposed to convection to medium at $T_\infty$ with a heat transfer coefficient $h$. Assume the fins are sufficiently long that the temperature of the fin at the tip is nearly $T_\infty$. Take the temperature of the fin at the base to be $T_b$ and neglect heat transfer from the fin tips. Simplify the relation for (*a*) a circular fin of diameter $D$ and (*b*) a rectangular fin of thickness $t$.

**8-142** The case-to-ambient thermal resistance of a power transistor that has a maximum power rating of 15 W is given as 25°C/W. If the case temperature of the transistor is not to exceed 80°C, determine the power at which this transistor can be operated safely in an environment at 30°C.

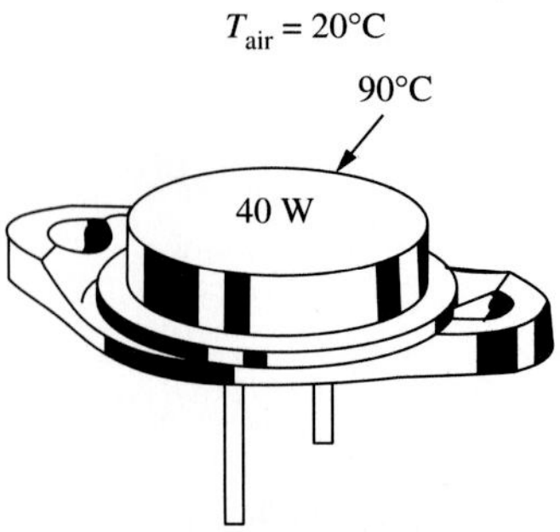

**8-143** A 40-W power transistor is to be cooled by attaching it to one of the commercially available heat sinks shown in Table 8-6. Select a heat sink that will allow the case temperature of the transistor not to exceed 90°C in the ambient air at 20°C.

**8-144** A 30-W power transistor is to be cooled by attaching it to one of the commercially available heat sinks shown in Table 8-6. Select a heat sink that will allow the case temperature of the transistor not to exceed 80°C in the ambient air at 35°C.

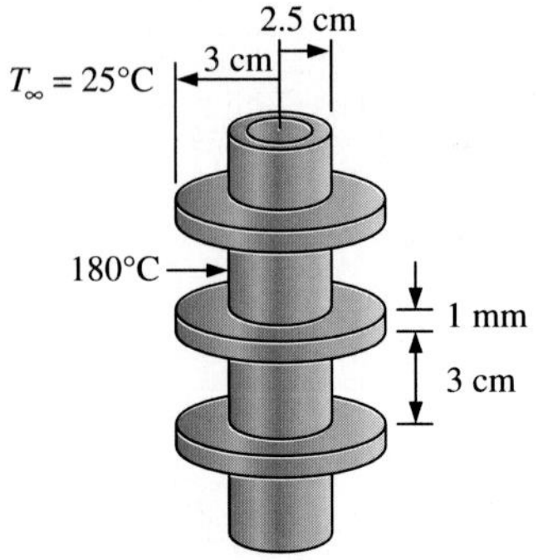

**8-145** Steam in a heating system flows through tubes whose outer diameter is 5 cm and whose walls are maintained at a temperature of 180°C. Circular aluminum alloy 2024-T6 fins [$k$ = 186 W/(m · °C)] of outer diameter 6 cm and constant thickness 1 mm are attached to the tube. The space between the fins is 3 mm, and thus there are 250 fins per m length of the tube. Heat is transferred to the surrounding air at $T_\infty$ = 25°C, with a heat transfer coefficient of 40 W/(m$^2$ · °C). Determine the increase in heat transfer from the tube per meter of its length as a result of adding fins. *Answer:* 2274 W

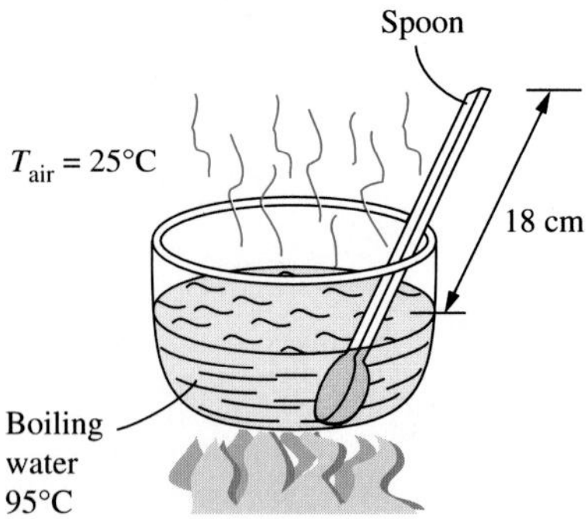

**8-146** Consider a stainless steel spoon [$k$ = 15.1 W/(m · °C)] partially immersed in boiling water at 95°C in a kitchen at 25°C. The handle of the spoon has a cross-section of 0.2 cm × 1 cm, and extends 18 cm in the air from the free surface of the water. If the heat transfer coefficient at the exposed surfaces of the spoon handle is 15 W/(m$^2$ · C), determine the temperature difference across the exposed surface of the spoon handle. State your assumptions. *Answer:* 69.9°C

**8-146E** Consider a stainless steel spoon [$k$ = 8.7 Btu/(h · ft · °F)] partially immersed in boiling water at 200°F in a kitchen at 75°F. The handle

**FIGURE P8-141**

**FIGURE P8-143**

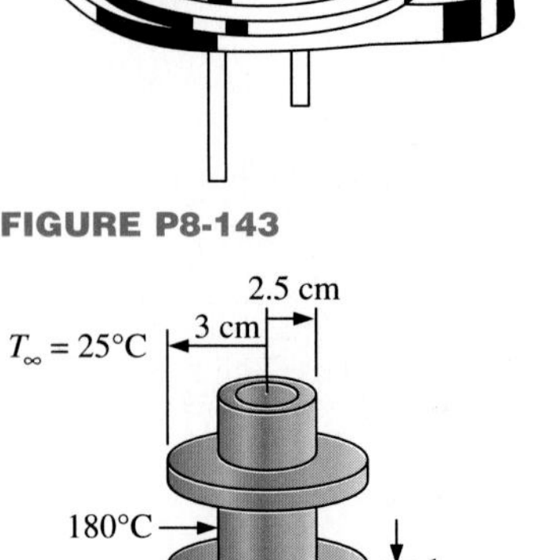

**FIGURE P8-145**

**FIGURE P8-146**

of the spoon has a cross-section of 0·08 in. × 0.5 in., and extends 7 in. in the air from the free surface of the water. If the heat transfer coefficient at the exposed surfaces of the spoon handle is 3 Btu/(h · ft$^2$ · °F), determine the temperature difference across the exposed surface of the spoon handle. State your assumptions. *Answer:* 124.8°F

**8-147** Repeat Prob. 8-146 for a silver spoon [$k$ = 427 W/(m · °C)].

**8-148** A 0.3-cm-thick, 12-cm-high, and 18-cm-long circuit board houses 80 closely spaced logic chips on one side, each dissipating 0.04 W. The board is impregnated with copper fillings, and has an effective thermal conductivity of 20 W/(m · °C). All the heat generated in the chips is conducted across the circuit board, and is dissipated from the back side of the board to a medium at 40°C, with a heat transfer coefficient of 50 W/(m$^2$ · °C). (*a*) Determine the temperatures on the two sides of the circuit board. (*b*) Now a 0.2-cm-thick, 12-cm-high, and 18-cm-long aluminum plate [$k$ = 237 W/(m · °C)] with 864 2-cm-long aluminum pin fins of diameter 0.25 cm is attached to the back side of the circuit board with a 0.02-cm-thick epoxy adhesive [$k$ = 1.8 W/(m · °C)]. Determine the new temperatures on the two sides of the circuit board.

**8-149** Repeat Prob. 8-148 using a copper plate with copper fins [$k$ = 386 W/(m · °C)] instead of aluminum ones.

**8-150** A hot surface at 100°C is to be cooled by attaching 3-cm-long, 0.25-cm-diameter aluminum pin fins [$k$ = 237 W/(m · °C)] to it, with a center-to-center distance of 0·6 cm. The temperature of the surrounding medium is 30°C, and the heat transfer coefficient on the surfaces is 35 W/(m$^2$ · °C). Determine the rate of heat transfer from the surface for a 1 m × 1 m section of the plate. Also determine the overall effectiveness of the fins.

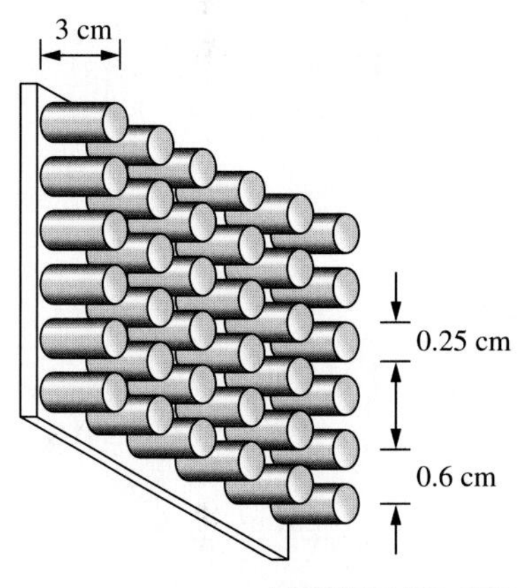

FIGURE P8-150

**8-150E** A hot surface at 250°F is to be cooled by attaching 1-in.-long, 0.1-in.-diameter aluminum pin fins [$k$ = 137 Btu/(h · ft · °F)] to it with a center-to-center distance of 0.25 in. The temperature of the surrounding medium is 85°F, and the heat transfer coefficient on the surfaces is 7 Btu/(h · ft$^2$ · °F). Determine the rate of heat transfer from the surface for a 1 ft section of the plate. Also determine the overall effectiveness of the fins.

**8-151** Repeat Prob. 8-150 using copper fins [$k$ = 386 W/(m · °C)] instead of aluminum ones.

**8-152** Two 3-m-long and 0.4-cm-thick cast iron [$k$ = 52 W/(m · °C)] steam pipes of outer diameter 10 cm are connected to each other through two 1-cm-thick flanges of outer diameter 20 cm, as shown in Fig. P8-152. The steam flows inside the pipe at an average temperature of 200°C with a heat transfer coefficient of 180 W/(m$^2$ · C). The outer surface of the pipe is exposed to an ambient at 8°C, with a heat transfer coefficient of 25 W/(m$^2$ · °C). (*a*) Disregarding the flanges, determine the average

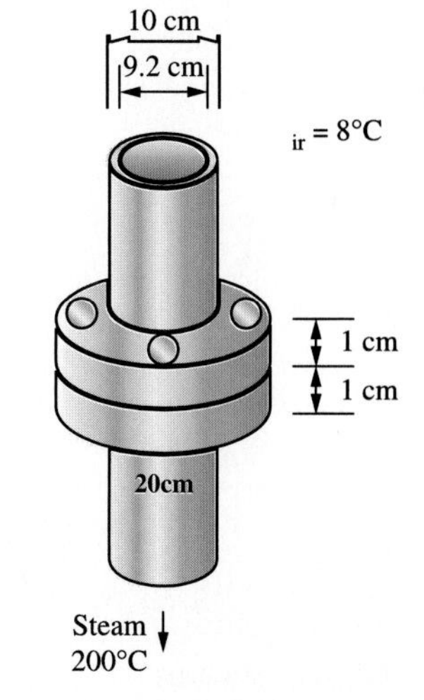


FIGURE P8-152

outer surface temperature of the pipe. (*b*) Using this temperature for the base of the flange and treating the flanges as the fins, determine the fin efficiency and the rate of heat transfer from the flanges. (*c*) What length of pipe is the flange section equivalent to for heat transfer purposes?

## Heat Transfer in Common Configurations

**8-153C** What is a conduction shape factor? How is it related to the thermal resistance?

**8-154C** What is the value of conduction shape factors in engineering?

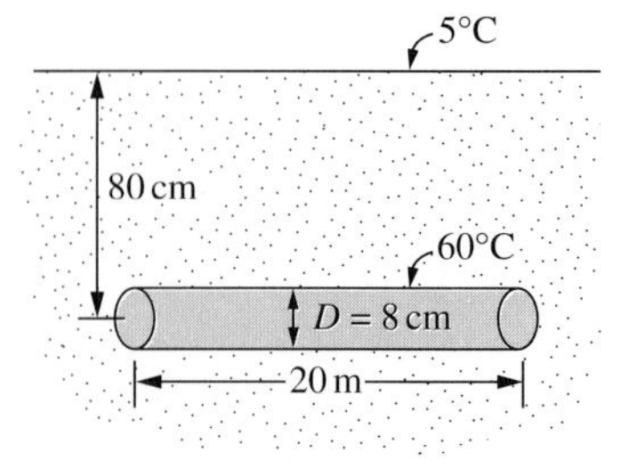

FIGURE P8-155

**8-155** A 20-m-long and 8-cm diameter hot water pipe of a district heating system is buried in the soil 80 cm below the ground surface. The outer surface temperature of the pipe is 60°C. Taking the surface temperature of the earth to be 5°C and the thermal conductivity of the soil at that location to be 0.9 W/(m · °C), determine the rate of heat loss from the pipe.

**8-156** An 8-m-long section of hot and cold water pipes run parallel to each other in a thick concrete layer. The diameters of both pipes are 5 cm, and the distance between the centerlines of the pipes is 40 cm. The surface temperatures of the hot and cold pipes are 60°C and 15°C, respectively. Taking the thermal conductivity of the concrete to be $k = 0.75$ W/(m · °C), determine the rate of heat transfer between the pipes. *Answer:* 306 W

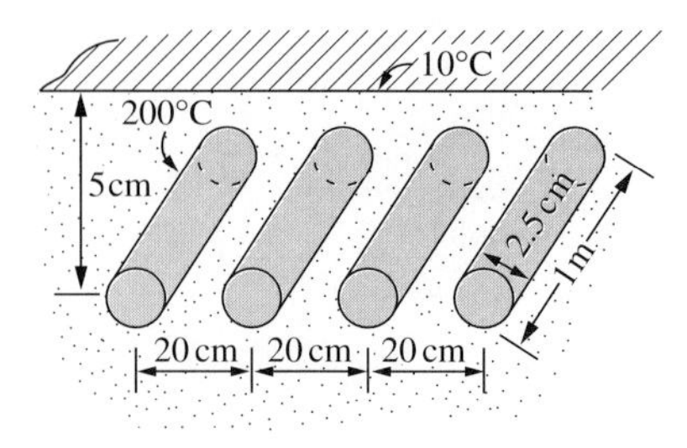

FIGURE P8-157

**8-157** A row of 1-m-long and 2.5-cm-diameter used uranium fuel rods that are still radioactive are buried in the ground parallel to each other, with a center-to-center distance of 20 cm at a depth 5 m from the ground surface at a location where the thermal conductivity of the soil is 1.1 W/(m · °C). If the surface temperatures of the rods and the ground are 200°C and 10°C, respectively, determine the rate of heat transfer from the fuel rods to the atmosphere through the soil.

**8-157E** A row of 3-ft-long and 1-in.-diameter used uranium fuel rods that are still radioactive are buried in the ground parallel to each other, with a center-to-center distance of 8 in. at a depth 15 ft from the ground surface at a location where the thermal conductivity of the soil is 0.6 Btu/(h · ft · °F). If the surface temperatures of the rods and the ground are 400°F and 50°F, respectively, determine the rate of heat transfer from the fuel rods to the atmosphere through the soil.

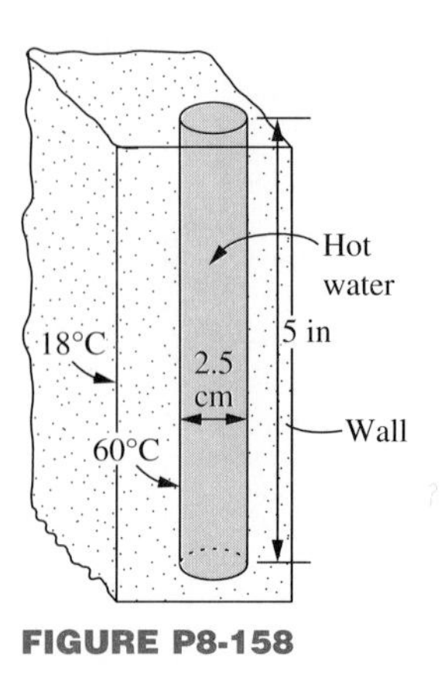

FIGURE P8-158

**8-158** Hot water at an average temperature of 60°C and an average velocity of 0.6 m/s is flowing through a 5-m section of a thin walled hot water pipe that has an outer diameter of 2.5 cm. The pipe passes through the center of a 14-cm-thick wall filled with fiberglass insulation [$k$ = 0.035 W/(m · °C)]. If the surfaces of the wall are at 18°C, determine (*a*) the rate of heat transfer from the pipe to the air in the rooms and (*b*) the temperature drop of the hot water as it flows through this 5-m-long section of the wall. *Answers:* 23.5 W, 0.02°C

**8-159** Hot water at an average temperature of 80°C and an average velocity of 1.5 m/s is flowing through a 25-m section of a pipe that has an outer diameter of 5 cm. The pipe extends 2 m in the ambient air above the ground, dips into the ground [$k$ = 1.5 W/(m · °C)] vertically for 3 m, and continues horizontally at this depth for 20 m more before it enters the next building. The first section of the pipe is exposed to the ambient air at 8°C, with a heat transfer coefficient of 22 W/(m$^2$ · °C). If the surface of the ground is covered with snow at 0°C, determine (*a*) the total rate of heat loss from the hot water, and (*b*) the temperature drop of the hot water as it flows through this 25-m-long section of the pipe.

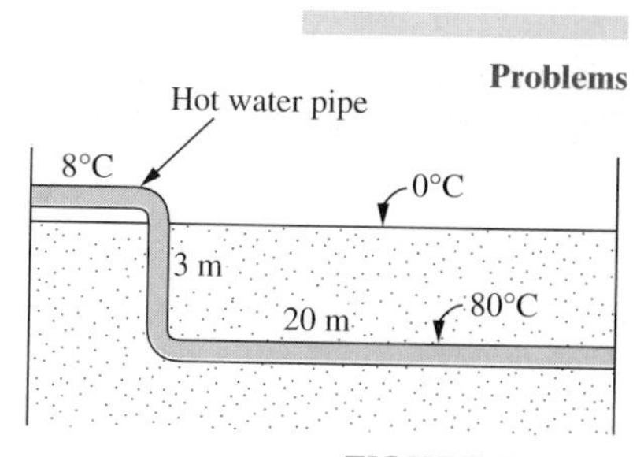


**FIGURE P8-159**

**8-160** Consider a house with a flat roof whose outer dimensions are 12 m × 12 m. The outer walls of the house are 6 m high. The walls and the roof of the house are made of 20-cm-thick concrete [$k$ = 0.75 W/(m · °C)]. The temperatures of the inner and outer surfaces of the house are 15°C and 3°C, respectively. Accounting for the effects of the edges of adjoining surfaces, determine the rate of heat loss from the house through its walls and the roof. What is the error involved in ignoring the effects of the edges and corners, and treating the roof as a 12 m × 12 m surface and the walls as 6 m × 12 m surfaces for simplicity?

**8-161** Consider a 10-m-long thick-walled concrete duct [$k$ = 0.75 W/(m · °C)] of square cross-section. The outer dimensions of the duct are 20 cm × 20 cm, and the thickness of the duct wall is 2 cm. If the inner and outer surfaces of the duct are at 100°C and 15°C, respectively, determine the rate of heat transfer through the walls of the duct.
*Answer:* 22.9 kW

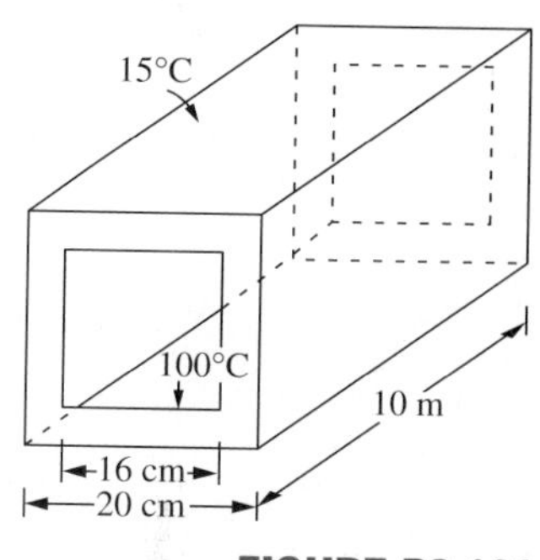


**FIGURE P8-161**

**8-161E** Consider a 30-ft-long thick-walled concrete duct [$k$ = 0.45 Btu/(h · ft · °F)] of square cross-section. The outer dimensions of the duct are 10 in. × 10 in., and the thickness of the duct wall is 1 in. If the inner and outer surfaces of the duct are at 200°F and 40°F, respectively, determine the rate of heat transfer through the walls of the duct.
*Answer:* 77,500 Btu/h

**8-162** A 3-m-diameter spherical tank containing some radioactive material is buried in the ground [$k$ = 1.4 W/(m · °C)]. The distance between the top surface of the tank and the ground surface is 4 m. If the surface temperatures of the tank and the ground are 170°C and 15°C, respectively, determine the rate of heat transfer from the tank.

**8-163** Hot water at an average temperature of 85°C passes through a row of eight parallel pipes that are 4 m long and have an outer diameter of 3 cm, located vertically in the middle of a concrete wall [$k$ = 0.75 W/(m · °C)] that is 4-m-high, 8-m-long, and 15-cm-thick. If the surfaces of the concrete walls are exposed to a medium at 20°C, with a heat transfer coefficient of 8 W/(m$^2$ · °C), determine the rate of heat loss from the hot water and the surface temperature of the wall.

## Review Problems

**8-164** Steam is produced in the copper tubes [$k$ = 386 W/(m · °C)] of a heat exchanger at a temperature of 120°C by another fluid condensing on the outside surfaces of the tubes at 170°C. The inner and outer diameters of the tube are 2 cm and 2.6 cm, respectively. When the heat exchanger was new, the rate of heat transfer per meter length of the tube was 7500 W. Determine the rate of heat transfer per meter length of the tube when a 0.2-mm-thick layer of limestone [$k$ = 2.9 W/(m · °C)] is formed on the inner surface of the tube after extended use.

**8-164E** Steam is produced in the copper tubes [$k$ = 223 Btu/(h · ft · °F)] of a heat exchanger at a temperature of 250°F by another fluid condensing on the outside surfaces of the tubes at 350°F. The inner and outer diameters of the tube are 1 in. and 1.3 in., respectively. When the heat exchanger was new, the rate of heat transfer per foot length of the tube was $2 \times 10^4$ Btu/h. Determine the rate of heat transfer per foot length of the tube when a 0.01-in.-thick layer of limestone [$k$ = 1.7 Btu/(h · ft · °F)] is formed on the inner surface of the tube after extended use.

**8-165** Repeat Prob. 8-164, assuming that a 0.25-mm-thick limestone layer is formed on both the inner and outer surfaces of the tube.

**8-166** A 1.2-m-diameter and 6-m-long cylindrical propane tank is initially filled with liquid propane whose density is 581 kg/m$^3$. The tank is exposed to the ambient air at 15°C, with a heat transfer coefficient of 20 W/(m$^2$ · °C). Now a crack develops at the top of the tank, and the pressure inside drops to 1 atm while the temperature drops to −42°C, which is the boiling temperature of propane at 1 atm. The heat of vaporization of propane at 1 atm is 425 kJ/kg. The propane is slowly vaporized as a result of the heat transfer from the ambient air into the tank, and the propane vapor escapes the tank at −42°C through the crack. Assuming the propane tank to be at about the same temperature as the propane inside at all times, determine how long it will take for the propane tank to empty if the tank is (*a*) not insulated, and (*b*) insulated with 7.5-cm-thick glass wool insulation [$k$ = 0.038 W/(m · °C)].

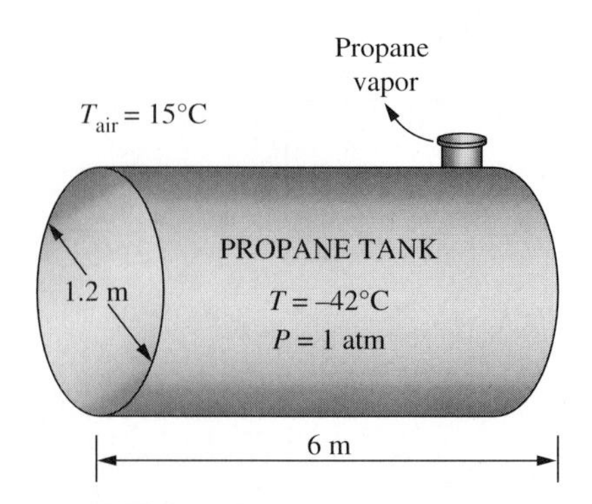


**FIGURE P8-166**

**8-167** Hot water is flowing at an average velocity of 1.5 m/s through a cast iron pipe [$k$ = 52 W/(m · °C)] whose inner and outer diameters are 3 and 3.5 cm, respectively. The pipe passes through a 15-m-long section of a basement whose temperature is 15°C. If the temperature of the water drops from 70°C to 67°C as it passes through the basement and the heat transfer coefficient on the inner surface of the pipe is 400 W/(m$^2$ · °C), determine the combined convection and radiation heat transfer coefficient at the outer surface of the pipe. *Answer:* 272.5 W/(m$^2$ · °C)

**8-167E** Hot water is flowing at an average velocity of 5 ft/s through a cast iron pipe [$k$ = 30 Btu/(h · ft · °F)] whose inner and outer diameters are 1.2 and 1.4 in., respectively. The pipe passes through a 50-ft-long section of a basement whose temperature is 60°F. If the temperature of the water drops from 160°F to 155°F as it passes through the basement and the heat transfer coefficient on the inner surface of the pipe is 30 Btu/(h · ft$^2$ · °F), determine the combined convection and radiation heat transfer coefficient at the outer surface of the pipe.

**8-168** Newly formed concrete pipes are usually cured first overnight by steam in a curing kiln maintained at a temperature of 45°C before the pipes are cured for several days outside. The heat and moisture to the kiln is provided by steam flowing in a pipe whose outer diameter is 12 cm. During a plant inspection, it was noticed that the pipe passes through a 10-m section that is completely exposed to the ambient air before it reaches the kiln. The temperature measurements indicate that the average temperature of the outer surface of the steam pipe is 82°C when the ambient temperature is 5°C. The combined convection and radiation heat transfer coefficient at the outer surface of the pipe is estimated to be 25 W/(m$^2$ · °C). Determine the amount of heat lost from the steam during a 10-h curing process that night.

Steam is supplied by a gas-fired steam generator that has an efficiency of 80 percent, and the plant pays \$0.60/therm of natural gas (1 therm = 105,500 kJ). If the pipe is insulated and 90 percent of the heat loss is saved as a result, determine the amount of money this facility will save a year as a result of insulating the steam pipes. Assume that the concrete pipes are cured 110 nights a year. State your assumptions.

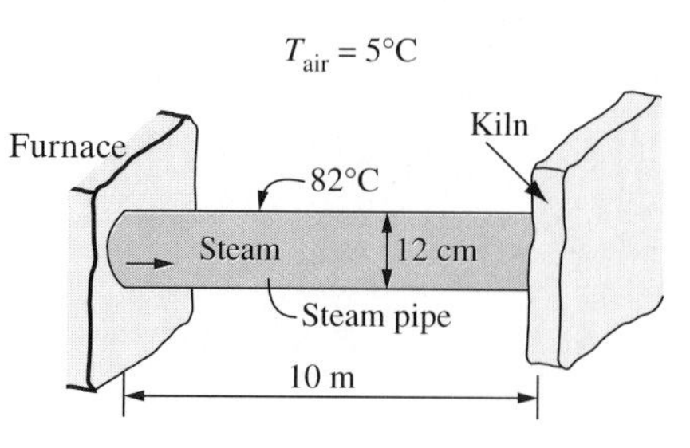


FIGURE P8-168

**8-169** Consider an 18 cm × 18 cm multilayer circuit board dissipating 27 W of heat. The board consists of four layers of 0.2-mm-thick copper [$k$ = 386 W/(m · °C)] and three layers of 1.5-mm-thick epoxy glass [$k$ = 0.26 W/(m · °C)] sandwiched together, as shown in the figure. The circuit board is attached to a heat sink from both ends, and the temperature of the board at those ends is 35°C. Heat is considered to be uniformly generated in the epoxy layers of the board at a rate of 0.5 W per 1 cm × 18 cm epoxy laminate strip (or 1.5 W per 1 cm × 18 cm strip of the board). Considering only a portion of the board because of symmetry, determine the magnitude and location of the maximum temperature that occurs in the board. Assume heat transfer from the top and bottom faces of the board to be negligible.

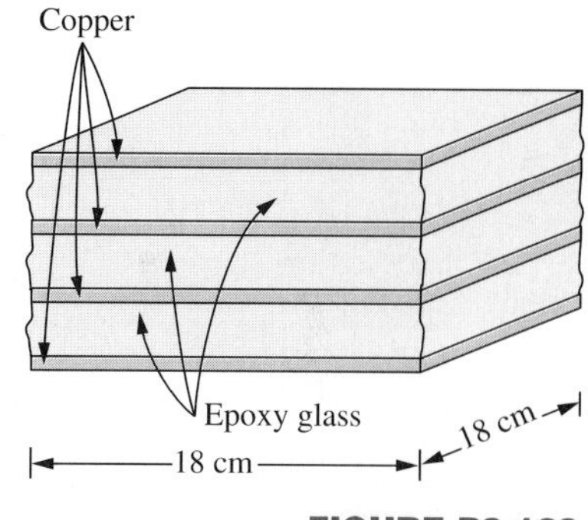

FIGURE P8-169

**8-170** The plumbing system of a house involves a 0.5-m section of a plastic pipe [$k$ = 0.16 W/(m · °C)] of inner diameter 2 cm and outer diameter 2.4 cm, exposed to the ambient air. During a cold and windy night, the ambient air temperature remains at about −5°C for a period of 14 h. The combined convection and radiation heat tansfer coefficient on the outer surface of the pipe is estimated to be 40 W/(m$^2$ · °C), and the heat of fusion of water is 333.7 kJ/kg. Assuming the pipe to contain stationary water initially at 0°C, determine if the water in that section of the pipe will completely freeze that night.

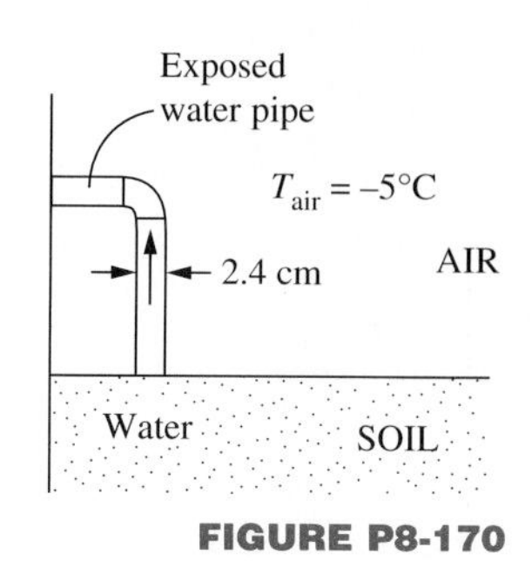


FIGURE P8-170

**8-171** Repeat Prob. 8-170 for the case of a heat transfer coefficient of 10 W/(m$^2$ · °C) on the outer surface as a result of putting a fence around the pipe that blocks the wind.

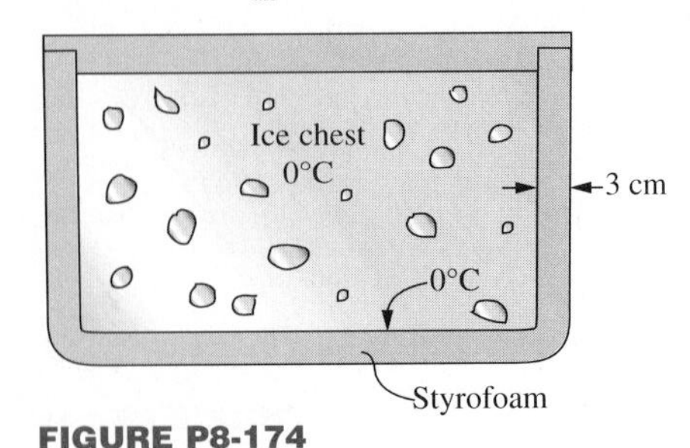


FIGURE P8-174

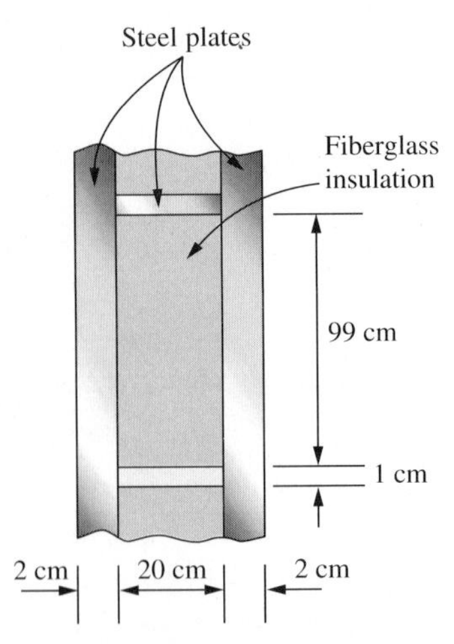

FIGURE P8-175

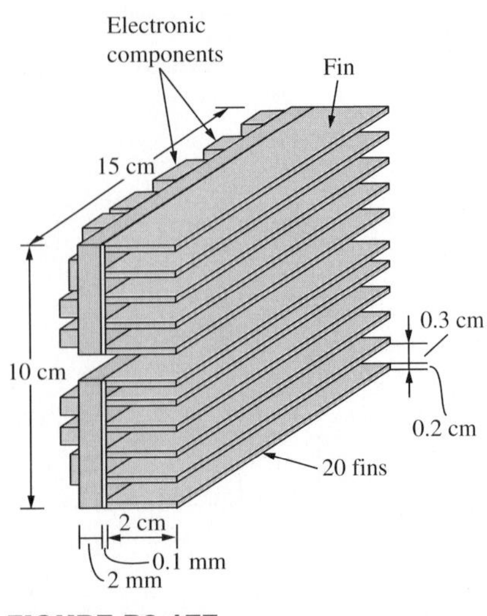


FIGURE P8-177

**8-172** The surface temperature of an 8-cm-diameter baked potato is observed to drop from 150°C to 100°C in 5 min in an environment at 25°C. Determine the average heat transfer coefficient between the potato and its surroundings. Using this heat transfer coefficient and the same surface temperature, determine how long it will take for the potato to experience the same temperature drop if it is wrapped completely in a 3-mm-thick towel [$k$ = 0.06 W/(m · °C)]. You may use the properties of water for potato. *Answers:* 92.9 W/(m$^2$ · °C), 25.9 min

**8-172E** The surface temperature of an 3-in.-diameter baked potato is observed to drop from 300°F to 200°F in 5 min in an environment at 70°F. Determine the average heat transfer coefficient between the potato and its surroundings. Using this heat transfer coefficient and the same surface temperature, determine how long it will take for the potato to experience the same temperature drop if it is wrapped completely in a 0.12-in.-thick towel [$k$ = 0.035 Btu/(h · ft · °F)]. You may use the properties of water for potato.

**8-173** Repeat Prob. 8-172 assuming there is a 0.4-mm-thick air space [$k$ = 0.026 W/(m · °C)] between the potato and the towel.

**8-174** An ice chest whose outer dimensions are 30 cm × 40 cm × 50 cm is made of 3-cm-thick Styrofoam [$k$ = 0.033 W/(m · °C)]. Initially, the chest is filled with 45 kg of ice at 0°C, and the inner surface temperature of the ice chest can be taken to be 0°C at all times. The heat of fusion of ice at 0°C is 333.7 kJ/kg, and the heat transfer coefficient between the outer surface of the ice chest and surrounding air at 30°C is 20 W/(m$^2$ · °C). Disregarding any heat transfer from the 40 cm × 50 cm base of the ice chest, determine how long it will take for the ice in the chest to melt completely.

**8-175** A 4-m-high and 6-m-long wall is constructed of two large 2-cm-thick steel plates [$k$ = 15 W/(m · °C)] separated by 1-cm-thick and 20-cm-wide steel bars placed 99 cm apart. The remaining space between the steel plates is filled with fiberglass insulation [$k$ = 0.035 W/(m · °C)]. If the temperature difference between the inner and the outer surfaces of the wall is 15°C, determine the rate of heat transfer through the wall. Can we ignore the steel bars between the plates in heat transfer analysis, since they occupy only 1 percent of the heat transfer surface area?

**8-176** In a nuclear reactor, heat is generated in 2.5-cm-diameter cylindrical uranium fuel rods at a rate of 4 × 10$^7$ W/m$^3$. Determine the temperature difference between the center and the surface of the fuel rods. *Answer:* 56.6°C

**8-176E** In a nuclear reactor, heat is generated in 1-in.-diameter cylindrical uranium fuel rods at a rate of 5 × 10$^6$ Btu/(h · ft$^3$). Determine the temperature difference between the center and the surface of the fuel rods. *Answer:* 135.6°F

**8-177** A 0.2-cm-thick, 10-cm-high, and 15-cm-long circuit board houses electronic components on one side that dissipate a total of 15 W of heat uniformly. The board is impregnated with conducting metal fillings, and has an effective thermal conductivity of 12 W/(m · °C). All the heat generated in the components is conducted across the circuit board, and is dissipated from the back side of the board to a medium at 37°C, with a heat transfer coefficient of 45 W/(m$^2$ · °C). (*a*) Determine the surface temperatures on the two sides of the circuit board. (*b*) Now a 0.1-cm-thick, 10-cm-high, and 15-cm-long aluminum plate [$k$ = 237 W/(m · °C)] with 20 0.2-cm-thick, 2-cm-long, and 15-cm-wide aluminum fins of rectangular profile are attached to the back side of the circuit board with a 0.015-cm-thick epoxy adhesive [$k$ = 1.8 W/(m · °C)]. Determine the new temperatures on the two sides of the circuit board.

**8-178** Repeat Prob. 8-177 using a copper plate with copper fins [$k$ = 386 W/(m · °C)] instead of aluminum ones.

**8-179** A row of 10 parallel pipes that are 5 m long and have an outer diameter of 6 cm are used to transport steam at 150°C through the concrete floor [$k$ = 0.75 W/(m · °C)] of a 10 m × 5 m room that is maintained at 25°C. The combined convection and radiation heat transfer coefficient at the floor is 12 W/(m$^2$ · °C). If the surface temperature of the concrete floor is not to exceed 40°C, determine how deep the steam pipes should be buried below the surface of the concrete floor.

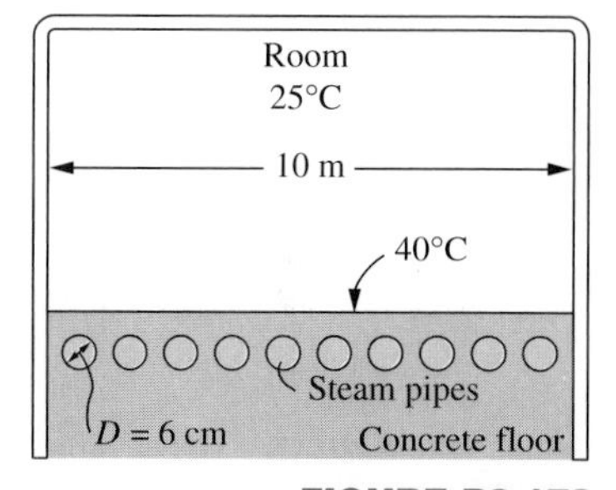

**FIGURE P8-179**

**8-179E** A row of 10 parallel pipes that are 15 ft long and have an outer diameter of 2.5 in. are used to transport steam at 300°F through the concrete floor [$k$ = 0.45 btu/(h · ft · °F)] of a 30 ft × 15 ft room that is maintained at 70°F. The combined convection and radiation heat transfer coefficient at the floor is 2.4 Btu/(h · ft$^2$ · °F). If the surface temperature of the concrete floor is not to exceed 105°F, determine how deep the steam pipes should be buried below the surface of the concrete floor.

## Computer, Design, and Essay Problems

**8-180** The temperature in deep space is close to absolute zero, which presents thermal challenges for astronauts who do space walks. Propose a design for the clothing of the astronauts that will be most suitable for the thermal environment in space. Defend the selections in your design.

**8-181** In the design of electronic components, it is very desirable to attach the electronic circuitry to a substrate material that is very good thermal conductor but also a very effective electrical insulator. If the high cost is not a major concern, what material would you propose for the substrate?

**8-182** Using cylindrical samples of the same material, device an experiment to determine the thermal contact resistance. Cylindrical samples are available at any length, and the thermal conductivity of the material is known.

**8-183** What are the considerations in determining the right number of fins attached to a surface?

**8-184** What are the considerations in determining the proper length of the fins attached to a surface?

**8-185** Find out about the wall construction of the cabins of large commercial airplanes, the range of ambient conditions under which they operate, typical heat transfer coefficients on the inner and outer surfaces of the wall, and the heat generation rates inside. Determine the size of the heating and air-conditioning system that will be able to maintain the cabin at 20°C at all times for an airplane capable of carrying 400 people.

**8-186** Repeat Prob. 8-195 for a submarine with a crew of 60 people.

**8-187** A house with 200-$m^2$ floor space is to be heated with geothermal water flowing through pipes laid in the ground under the floor. The walls of the house are 4 m high, and there are 10 single-paned windows in the house, which are 1.2 m wide and 1.8 m high. The house has R-19 (in $h \cdot ft^2 \cdot °F/Btu$) insulation in the walls, and R-30 on the ceiling. The floor temperature is not to exceed 40°C. Hot geothermal water is available at 90°C, and the inner and outer diameter of the pipes to be used are 2.4 cm and 3.0 cm. Design such a heating system for this house in your area.

# Transient Heat Conduction

CHAPTER 9

The temperature of a body, in general, varies with time as well as position. In rectangular coordinates, this variation is expressed as $T(x, y, z; t)$, where $(x, y, z)$ indicates variation in the $x$, $y$, and $z$ directions, respectively, and $t$ indicates variation with time. In the preceding chapter, we considered heat conduction under *steady* conditions, for which the temperature of a body at any point does not change with time. This certainly simplified the analysis, especially when the temperature varied in one direction only, and we were able to obtain analytical solutions. In this chapter, we consider the variation of temperature with *time* as well as *position* in one- or multidimensional systems.

We start this chapter with the analysis of *lumped systems* in which the temperature of a solid varies with time but remains uniform throughout the solid at any time. Then we consider the variation of temperature with time as well as position for one-dimensional heat conduction problems such as those associated with a large plane wall, a long cylinder, a sphere, and a semi-infinite medium using *transient temperature charts* and analytical solutions. Finally, we consider transient heat conduction in multidimensional systems by utilizing the *product solution.*

## 9-1 ■ LUMPED SYSTEM ANALYSIS

In heat transfer analysis, some bodies are observed to behave like a "lump" whose interior temperature remains essentially uniform at all times during a heat transfer process. The temperature of such bodies can be taken to be a function of time only, $T(t)$. Heat transfer analysis that utilizes this idealization is known as **lumped system analysis**, which provides great simplification in certain classes of heat transfer problems without much sacrifice from accuracy.

Consider a hot copper ball coming out of an oven (Fig. 9-1). Measurements indicate that the temperature of the copper ball changes with time, but it does not change with position at any given time. Thus the temperature of the ball remains uniform at all times, and we can talk about the temperature of the ball with no reference to a specific location.

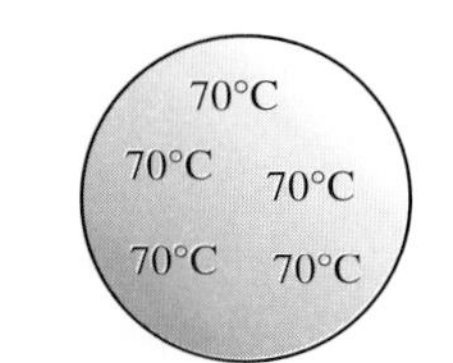


(*a*) Copper ball

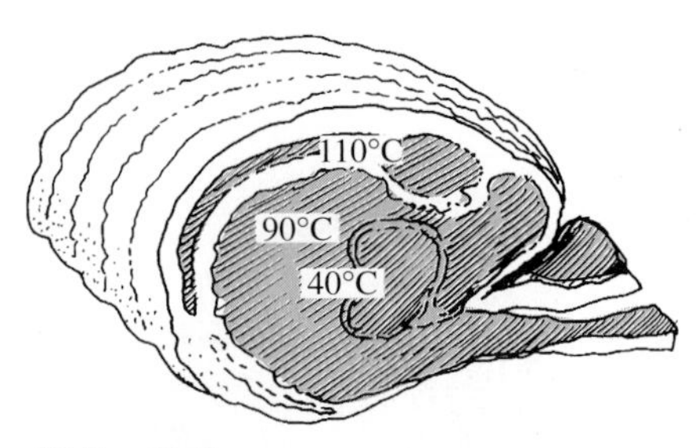


(*b*) Roast ball

**FIGURE 9-1**

A small copper ball can be modeled as a lumped system, but a roast beef cannot.

Now let us go to the other extreme, and consider a large roast in an oven. If you have done any roasting, you must have noticed that the temperature distribution within the roast is not even close to being uniform. You can easily verify this by taking the roast out before it is completely done, and cutting it in half. You will see that the outer parts of the roast are well done while the center part if barely warm. Thus, lumped system analysis is not applicable in this case. Before presenting a criteria about applicability of lumped system analysis, we develop the formulation associated with it.

Consider a body of arbitrary shape of mass $m$, volume $V$, surface area $A$, density $\rho$, and specific heat $C_p$ initially at a uniform temperature $T_i$ (Fig. 9-2). At time $t = 0$, the body is placed into a medium at temperature $T_\infty$, and heat transfer takes place between the body and its environment, with a heat transfer coefficient $h$. For the sake of discussion, we will assume that $T_\infty > T_i$, but the analysis is equally valid for the opposite case. We assume lumped system analysis to be applicable, so that the temperature remains uniform within the body at all times, and changes with time only, $T = T(t)$.

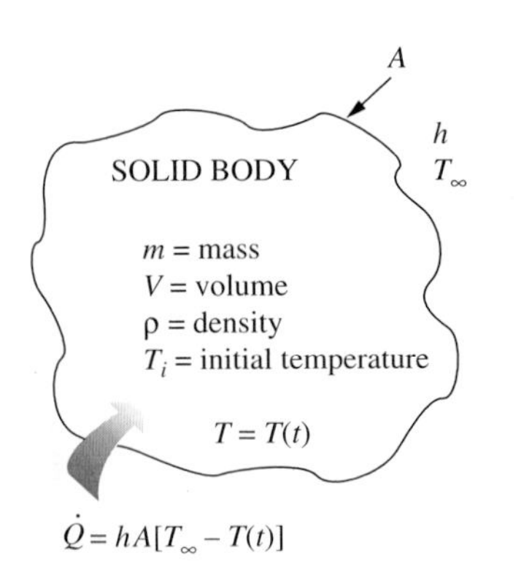


**FIGURE 9-2**

The geometry and parameters involved in the lumped system analysis.

During a differential time interval $dt$, the temperature of the body rises by a differential amount $dT$. An energy balance of the solid for the time interval $dt$ can be expressed as

$$\begin{pmatrix}\text{heat transfer into the body}\\ \text{during } dt\end{pmatrix} = \begin{pmatrix}\text{the increase in the}\\ \text{energy of the body}\\ \text{during } dt\end{pmatrix}$$

or

$$hA(T_\infty - T)\,dt = mC_p\,dT \tag{9-1}$$

Noting that $m = \rho V$ and $dT = d(T - T_\infty)$ since $T_\infty =$ constant, Eq. 9-1 can be rearranged as

$$\frac{d(T - T_\infty)}{T - T_\infty} = -\frac{hA}{\rho V C_p}\,dt \tag{9-2}$$

Integrating from $t = 0$ at which $T = T_i$ to any time $t$ at which $T = T(t)$ gives

$$\ln \frac{T(t) - T_\infty}{T_i - T_\infty} = -\frac{hA}{\rho V C_p} t \tag{9-3}$$

Taking the exponential of both sides and rearranging, we obtain

$$\frac{T(t) - T_\infty}{T_i - T_\infty} = e^{-bt} \tag{9-4}$$

where

$$b = \frac{hA}{\rho V C_p} \quad (1/\text{s}) \tag{9-5}$$

is a positive quantity whose dimension is $(\text{time})^{-1}$. Equation 9-4 is plotted in Fig. 9-3 for different values of $b$. There are two observations that can be made from this figure and the relation above:

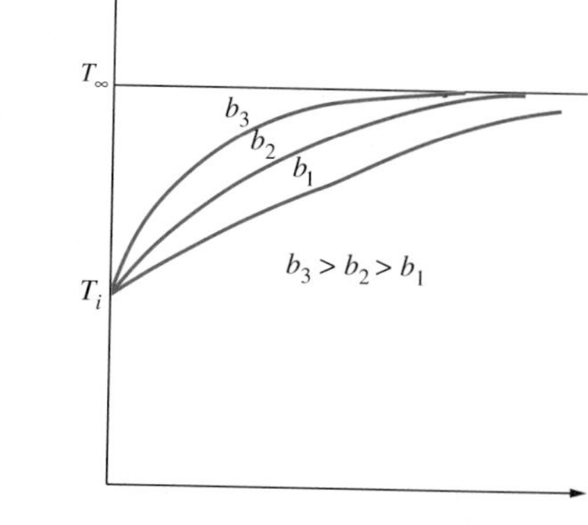


**FIGURE 9-3**
The temperature of a lumped system approaches the environment temperature as time gets larger.

**1** Equation 9-4 enables us to determine the temperature $T(t)$ of a body at time $t$, or alternately, the time $t$ required for the temperature to reach a specified value $T(t)$.

**2** The temperature of a body approaches the ambient temperature $T_\infty$ exponentially. The temperature of the body changes rapidly at the beginning, but rather slowly later on. A large value of $b$ indicates that the body will approach the environment temperature in a short time. The larger the value of the exponent $b$, the higher the rate of decay in temperature. Note that $b$ is proportional to the surface area, but inversely proportional to the mass and the specific heat of the body. This is not surprising, since it takes longer to heat or cool a larger mass, especially when it has a large specific heat.

Once the temperature $T(t)$ at time $t$ is available from Eq. 9-4, the *rate* of convection heat transfer between the body and its environment at that time can be determined from Newton's law of cooling as

$$\dot{Q}(t) = hA[T(t) - T_\infty] \quad (\text{W}) \tag{9-6}$$

The *total amount* of heat transfer between the body and the surrounding medium over time time interval $t = 0$ to $t$ is simply the change in the energy content of the body:

$$Q = mC_p[T(t) - T_i] \quad (\text{kJ}) \tag{9-7}$$

The amount of heat transfer reaches its *upper limit* when the body reaches the surroundings temperature $T_\infty$. Therefore, the *maximum* heat transfer between the body and its surroundings is (Fig. 9-4)

$$Q_{\max} = mC_p(T_\infty - T_i) \quad (\text{kJ}) \tag{9-8}$$

We could also obtain this equation by substituting the $T(t)$ relation from Eq. 9-4 into the $\dot{Q}(t)$ relation in Eq. 9-6, and integrating it from $t = 0$ to $t \to \infty$.

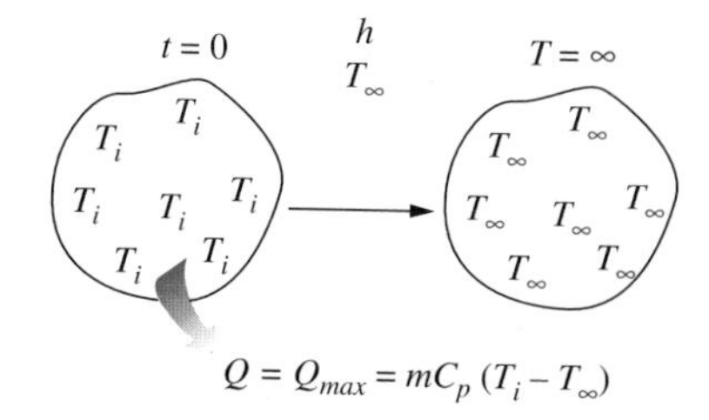


**FIGURE 9-4**
Heat transfer to or from a body reaches its maximum value when the body reaches the environment temperature.

## Criteria for Lumped System Analysis

The lumped system analysis certainly provides great convenience in heat transfer analysis, and naturally we would like to know when it is appropriate to use it. The first step in establishing a criteria for the applicability of the lumped system analysis is to define a **characteristic length** as

$$L_c = \frac{V}{A}$$

and a **Biot number** Bi as

$$\text{Bi} = \frac{hL_c}{k} \qquad \text{(9-9)}$$

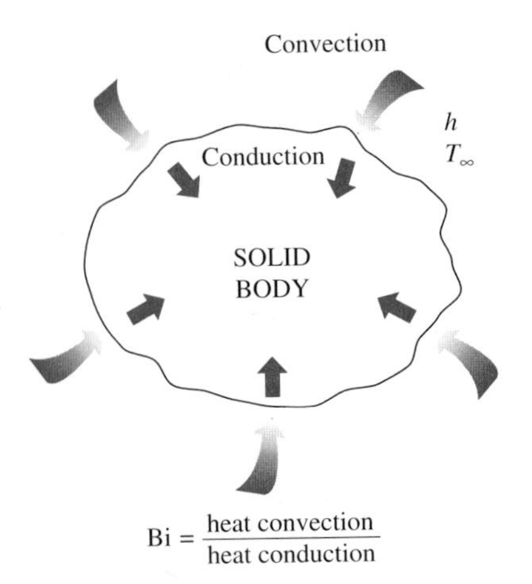


**FIGURE 9-5**

The Biot number can be viewed as the ratio of the convection at the surface to conduction within the body.

It can also be expressed as (Fig. 9-5)

$$\text{Bi} = \frac{h}{k/L_c}\frac{\Delta T}{\Delta T} = \frac{\text{convection at the surface of the body}}{\text{conduction within the body}}$$

or

$$\text{Bi} = \frac{L_c/k}{1/h} = \frac{\text{conduction resistance within the body}}{\text{convection resistance at the surface of the body}}$$

When a solid body is being heated by the hotter fluid surrounding it (such as a potato being baked in an oven), heat is first *convected* to the body, and subsequently *conducted* within the body. The Biot number is the *ratio* of the internal resistance of a body to *heat conduction* to its external resistance to *heat convection.* Thus, a small Biot number represents small resistance to heat conduction, and thus small temperature gradients within the body.

Lumped system analysis assumes a *uniform* temperature distribution throughout the body, which will be the case only when the thermal resistance of the body to heat conduction (the *conduction resistance*) is zero. Thus, lumped system analysis is *exact* when Bi $= 0$ and *approximate* when Bi $> 0$. Of course, the smaller the Bi number, the more accurate is the lumped system analysis. Then the question we must answer is how much accuracy we are willing to sacrifice for the convenience of the lumped system analysis.

Before answering this question, we should mention that a 20 percent uncertainty in the convection heat transfer coefficient $h$ in most cases is considered "normal" and "expected." Assuming $h$ to be *constant* and *uniform* is also an approximation of questionable validity, especially for irregular geometries. Therefore, in the absence of sufficient experimental data for the specific geometry under consideration, we cannot claim our results to be better than $\pm 20$ percent, even when Bi $= 0$. This being the case, introducing another source of uncertainty in the problem will hardly have any effect on the overall uncertainty, provided that it is minor. It is generally accepted that lumped system analysis is *applicable* if

$$\text{Bi} \leq 0.1$$

When this criteria is satisfied, the temperatures within the body relative to the surroundings (i.e., $T - T_\infty$) remain within 5 percent of each other even for well-rounded geometries such as a spherical ball. Thus, when Bi < 0.1, the variation of temperature with location within the body will be slight, and can reasonably be approximated as being uniform.

The first step in the application of lumped system analysis is the calculation of the *Biot number,* and the assessment of the applicability of this approach. One may still wish to use lumped system analysis even when the criteria Bi < 0.1 is not satisfied, if high accuracy is not a major concern.

Note that the Biot number is the ratio of the *convection* at the surface to *conduction* within the body, and this number should be as small as possible for lumped system analysis to be applicable. Therefore, *small bodies* with *high thermal conductivity* are good candidates for lumped system analysis, especially when they are in a medium that is a poor conductor of heat (such as air or another gas) and motionless. Thus, the hot small copper ball placed in quiescent air discussed earlier is most likely to satisfy the criteria for lumped system analysis (Fig. 9-6).

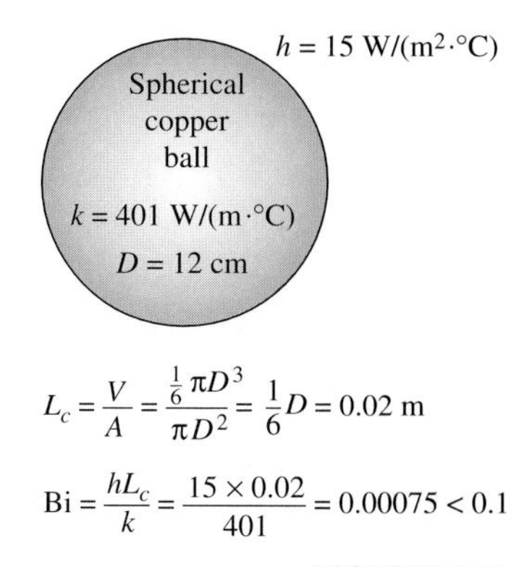

$h = 15$ W/(m²·°C)
Spherical copper ball
$k = 401$ W/(m·°C)
$D = 12$ cm
$L_c = \frac{V}{A} = \frac{\frac{1}{6}\pi D^3}{\pi D^2} = \frac{1}{6}D = 0.02$ m
$\text{Bi} = \frac{hL_c}{k} = \frac{15 \times 0.02}{401} = 0.00075 < 0.1$


**FIGURE 9-6**
Small bodies with high thermal conductivities and low convection coefficients are most likely to satisfy the criteria for lumped system analysis.

## Some Remarks on Heat Transfer in Lumped Systems

To understand the heat transfer mechanism during the heating or cooling of a solid by the fluid surrounding it, and the criteria for lumped system analysis, consider the following analogy (Fig. 9-7). People from the mainland are to go *by boat* to an island whose entire shore is a harbor, and from the harbor to their destinations in the island *by bus*. The overcrowding of people at the harbor depends on the boat traffic to the island and the ground transportation system in the island. If there is an excellent ground transporation system with plenty of busses, there will be no overcrowding at the harbor, especially when the boat traffic is light. But when the opposite is true, there will be a huge overcrowding at the harbor, creating a large difference between the populations at the harbor and the inland. The chance of overcrowding is much lower in a small island with plenty of fast busses.

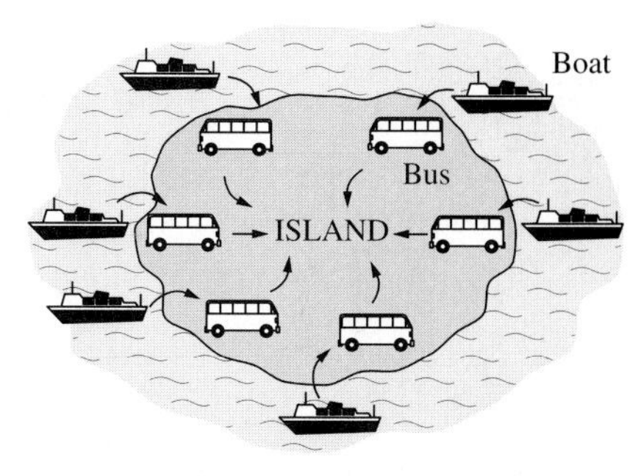

**FIGURE 9-7**
Analogy between heat transfer to a solid and passenger traffic to an island.

In heat transfer, a poor ground transportation system corresponds to poor heat conduction in a body, and overcrowding at the harbor to the accumulation of heat and the subsequent rise in temperature near the surface of the body relative to its inner parts. Lumped system analysis is obviously not applicable when there is overcrowding at the surface. Of course, we have disregarded radiation in this analogy and thus the air traffic to the island. Like passengers at the harbor, heat changes *vehicles* at the surface from *convection* to *conduction.* Noting that a surface has zero thickness and thus cannot store any energy, heat reaching the surface of a body by convection must continue its journey within the body by conduction.

Consider heat transfer from a hot body to its cooler surroundings. Heat will be transferred from the body to the surrounding fluid as a result of a temperature difference. But this energy will come from the region near the surface, and thus the temperature of the body near the surface will drop. This creates a *temperature gradient* between the inner and outer regions of the body, and initiates heat flow by conduction from the interior of the body towards the outer surface.

When the convection heat transfer coefficient $h$ and thus convection heat transfer from the body are high, the temperature of the body near the surface will drop quickly (Fig. 9-8). This will create a larger temperature difference between the inner and outer regions unless the body is able to transfer heat from the inner to the outer regions just as fast. Thus, the magnitude of the maximum temperature difference within the body depends strongly on the ability of a body to conduct heat towards its surface relative to the ability of the surrounding medium to convect this heat away from the surface. The Biot number is a measure of the relative magnitudes of these two competing effects.

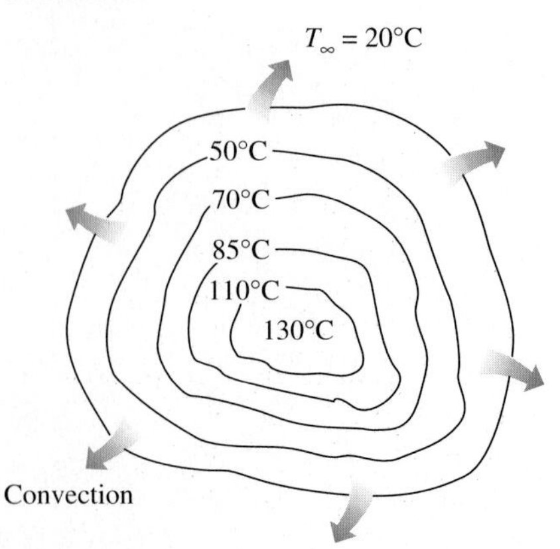


**FIGURE 9-8**

When the convection coefficient $h$ is high and $k$ is low, large temperature differences occur between the inner and outer regions of a large solid.

Recall that heat conduction in a specified direction $n$ per unit surface area is expressed as $\dot{q} = -k\partial T/\partial n$, where $\partial T/\partial n$ is the temperature gradient and $k$ is the thermal conductivity of the solid. Thus, the temperature distribution in the body will be *uniform* only when its thermal conductivity is *infinity,* and no such material is known to exist. Therefore, temperature gradients and thus temperature differences must exist within the body, no matter how small, in order for heat conduction to take place. Of course, the temperature gradient and the thermal conductivity are inversely proportional for a given heat flux. Therefore, the larger the thermal conductivity, the smaller the temperature gradient.

**EXAMPLE 9-1**

The temperature of a gas stream is to be measured by a thermocouple whose junction can be approximated as a 1-mm-diameter sphere, as shown in Fig. 9-9. The properties of the junction are $k = 35$ W/(m · °C), $\rho = 8500$ kg/m$^3$, and $C_p = 320$ J/(kg · °C), and the convection heat transfer coefficient between the junction and the gas is $h = 210$ W/(m$^2$ · °C). Determine how long it will take for the thermocouple to read 99 percent of the initial temperature difference.

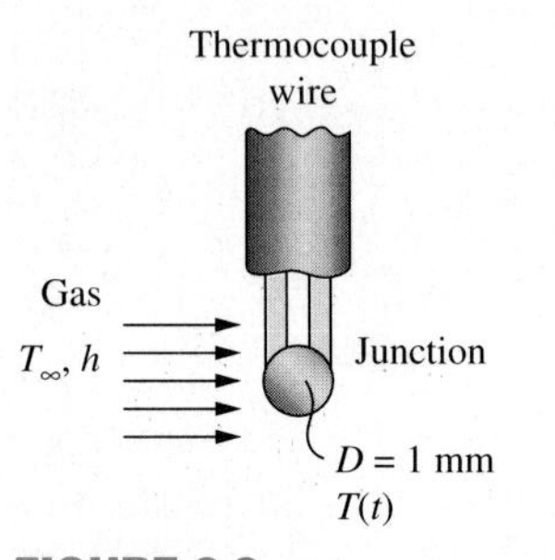


**FIGURE 9-9**

Schematic for Example 9-1.

**Solution** We assume the heat transfer coefficient $h$ to remain uniform and constant for the entire surface, and to include any radiation effects. The characteristic length of the junction is

$$L_c = \frac{V}{A} = \frac{\frac{1}{6}\pi D^3}{\pi D^2} = \frac{1}{6}D = \frac{1}{6} \times 0.001 \text{ m} = 1.67 \times 10^{-4} \text{ m}$$

Then the Biot number becomes

$$\text{Bi} = \frac{hL_c}{k} = \frac{[210 \text{ W/(m}^2 \cdot {}^\circ\text{C)}](1.67 \times 10^{-4} \text{ m})}{35 \text{ W/(m} \cdot {}^\circ\text{C)}} = 0.001 < 0.1$$

Therefore, lumped system analysis is applicable, and the error involved in this approximation is negligible.

In order to read 99 percent of the initial temperature difference $T_i - T_\infty$ between the junction and the gas, we must have

$$\frac{T(t) - T_\infty}{T_i - T_\infty} = 0.01$$

For example, when $T_i = 0$°C and $T_\infty = 100$°C, a thermocouple is considered to have read 99 percent of this applied temperature difference when its reading indicates $T(t) = 99°C$.

The value of the exponent $b$ is

$$b = \frac{hA_s}{\rho C_p V} = \frac{h}{\rho C_p L_c} = \frac{210\ \text{W/(m}^2 \cdot °\text{C)}}{(8500\ \text{kg/m}^3)[320\ \text{J/(kg} \cdot °\text{C)}](1.67 \times 10^{-4}\ \text{m})} = 0.462\ \text{s}^{-1}$$

We now substitute these values into Eq. 9-4 and obtain

$$\frac{T(t) - T_\infty}{T_i - T_\infty} = e^{-bt} \longrightarrow 0.01 = e^{-(0.462\ \text{s}^{-1})t}$$

which yields

$$t = 10\ \text{s}$$

Therefore, we must wait at least 10 s for the temperature of the thermocouple junction to approach within 1 percent of the initial junction–gas temperature difference.

**EXAMPLE 9-2**

A person is found dead at 5 p.m. in a room whose temperature is 20°C. The temperature of the body is measured to be 25°C when found, and the heat transfer coefficient is estimated to be $h = 8\ \text{W/(m}^2 \cdot °\text{C)}$. Modeling the body as a 30-cm-diameter, 1.70-m-long cylinder, estimate the time of death of that person (Fig. 9-10).

**FIGURE 9-10**
Schematic for Example 9-2.

**Solution** The average human body is 72 percent water by mass, and thus we can assume the body to have the properties of water at room temperature. That is, $k = 0.608\ \text{W/(m} \cdot °\text{C)}$, $\rho = 1000\ \text{kg/m}^3$, and $C_p = 4180\ \text{J/(kg} \cdot °\text{C)}$. We assume the convection coefficient $h$ to remain constant and uniform for the entire surface, and disregard any radiation effects. We also assume the person to have been healthy(!) when he or she died with a body temperature of 37°C.

Modeling the body as a 0.30-m-diameter, 1.70-m-long cylinder, its characteristic length is determined to be

$$L_c = \frac{V}{A} = \frac{\pi r_o^2 L}{2\pi r_o L + 2\pi r_o^2} = \frac{\pi(0.15\ \text{m})^2(1.7\ \text{m})}{2\pi(0.15\ \text{m})(1.7\ \text{m}) + 2\pi(0.15\ \text{m})^2} = 0.0689\ \text{m}$$

Then the Biot number becomes

$$\text{Bi} = \frac{hL_c}{k} = \frac{[8\ \text{W/(m}^2 \cdot °\text{C)}](0.0689\ \text{m})}{0.608\ \text{W/(m} \cdot °\text{C)}} = 0.92 > 0.1$$

Therefore, lumped system analysis is *not* applicable. However, we can still use it to get a "rough" estimate of the time of death. The exponent $b$ in this case is

$$b = \frac{hA}{\rho C_p V} = \frac{h}{\rho C_p L_c} = \frac{8\ \text{W/(m}^2 \cdot °\text{C)}}{(1000\ \text{kg/m}^3)[4180\ \text{J/(kg} \cdot °\text{C)}](0.0689\ \text{m})}$$
$$= 2.78 \times 10^{-5}\ \text{s}^{-1}$$

We now substitute these values into Eq. 9-4,

$$\frac{T(t) - T_\infty}{T_i - T_\infty} = e^{-bt} \longrightarrow \frac{25 - 20}{37 - 20} = e^{-(2.78 \times 10^{-5}\ \text{s}^{-1})t}$$

which yields

$$t = 31{,}825\ \text{s} = 8.8\ \text{h}$$

Therefore, as a rough estimate, the person died about 9 h before the body was found, and thus the time of death is 8 a.m. This example demonstrates how to obtain "ball park" values using a simple analysis.

## 9-2 ■ TRANSIENT HEAT CONDUCTION IN LARGE PLANE WALLS, LONG CYLINDERS, AND SPHERES

In the preceding section, we considered bodies in which the variation of temperature within the body was negligible; that is, bodies that remain nearly *isothermal* during a process. Relatively *small* bodies of *highly conductive* materials approximate this behavior. In general, however, the temperature within a body will change from point to point as well as with time. In this section, we consider the variation of temperature with *time* and *position* in one-dimensional problems such as those associated with a large plane wall, a long cylinder, and a sphere.

Consider a plane wall of thickness $2L$, a long cylinder of radius $r_o$, and a sphere of radius $r_o$ initially at a *uniform temperature* $T_i$, as shown in Fig. 9-11. At time $t = 0$, each geometry is placed in a large medium that is at a constant temperature $T_\infty$ and kept in that medium for $t > 0$. Heat transfer takes place between these bodies and their environments by convection with a *uniform* and *constant* heat transfer coefficient $h$. Note that all three cases posses geometric and thermal symmetry: the plane wall is symmetric about its *center plane* ($x = 0$), the cylinder is symmetric about its *center line* ($r = 0$), and the sphere is symmetric about its *center point* ($r = 0$). We neglect *radiation* heat transfer between these bodies and their surrounding surfaces, or incorporate the radiation effect into the convection heat transfer coefficient $h$.

The variation of the temperature profile with *time* in the plane wall is

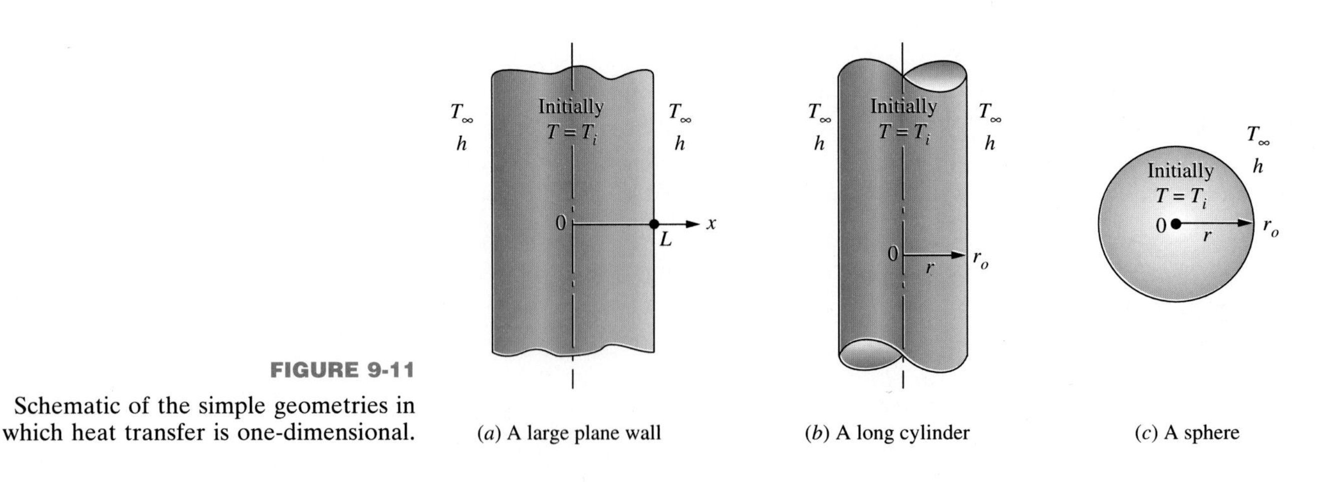


**FIGURE 9-11**
Schematic of the simple geometries in which heat transfer is one-dimensional.

illustrated in Fig. 9-12. When the wall is first exposed to the surrounding medium at $T_\infty < T_i$ at $t = 0$, the entire wall is at its initial temperature $T_i$. But the wall temperature at and near the surfaces starts to drop as a result of convection heat transfer from the wall to the surrounding medium. This creates a *temperature gradient* in the wall, and initiates heat conduction from the inner parts of the wall towards its outer surfaces. Note that the temperature at the center of the wall remains at $T_i$ until $t = t_2$, and that the temperature profile within the wall remains symmetric at all times about the center plane. The temperature profile gets flatter and flatter as time passes as a result of heat transfer, and eventually becomes uniform at $T = T_\infty$. That is, the wall reaches *thermal equilibrium* with its surroundings. At that point, the heat transfer stops since there is no longer a temperature difference. Similar discussions can be given for the long cylinder or sphere.

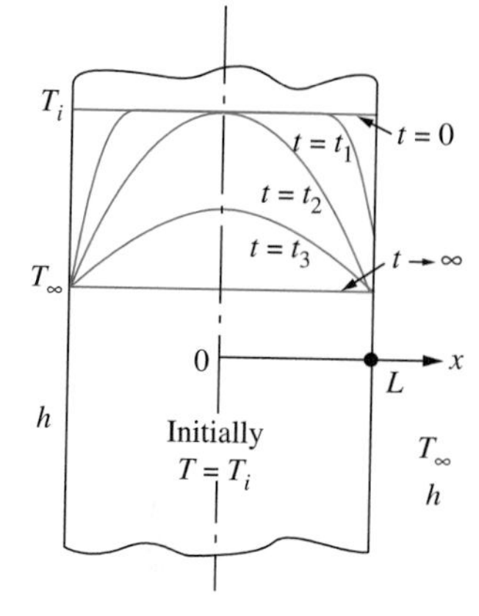


**FIGURE 9-12**
Transient temperature profiles in a plane wall exposed to convection from its surfaces for $T_i > T_\infty$.

The formulation of the problems for the determination of the one-dimensional transient temperature distribution $T(x, t)$ in a wall results in a partial differential equation, which can be solved using advanced mathematical techniques. The solution, however, normally involves infinite series, which are inconvenient and time-consuming to evaluate. Therefore, there is clear motivation to present the solution in *tabular* or *graphical* form. However, the solution involves the parameters $x$, $L$, $t$, $k$, $\alpha$, $h$, $T_i$, and $T_\infty$, which are too many to make any graphical presentation of the results practical. In order to reduce the number of the parameters, we nondimensionalize the problem by defining the following dimensionless quantities:

*dimensionless temperature* $$\theta(x, t) = \frac{T(x, t) - T_\infty}{T_i - T_\infty}$$

*dimensionless distance from the center* $$X = \frac{x}{L}$$

*dimensionless heat transfer coefficient* $$\text{Bi} = \frac{hL}{k}$$ **(Biot number)**

*dimensionless time* $$\tau = \frac{\alpha t}{L^2}$$ **(Fourier number)**

The nondimensionalization enables us to present the temperature in terms of three parameters only: $x$, Bi, and $\tau$. This makes it practical to present the solution in graphical form. The dimensionless quantities defined above for a plane wall can also be used for a *cylinder* or *sphere* by replacing the space variable $x$ by $r$, and the half-thickness $L$ by the outer radius $r_o$. Note that the characteristic length in the definition of the Biot number is taken to be the *half-thickness* $L$ for the plane wall, and the *radius* $r_o$ for the long cylinder and sphere.

The one-dimensional transient heat conduction problem described above can be solved exactly for any of the three geometries, but the solution involves infinite series, which are difficult to deal with. However, the terms in the solutions converge rapidly with increasing time, and for $\tau > 0.2$, keeping the first term and neglecting all the remaining terms in the series results in an error under 2 percent. We are usually interested in the solution for times greater than $\tau > 0.2$, and thus it is very convenient to express the solution using this **one-term approximation**, which is given as

$$\text{Plane wall:} \quad \theta(x, t)_{\text{wall}} = \frac{T(x, t) - T_\infty}{T_i - T_\infty} = A_1 e^{-\lambda_1^2 \tau} \cos(\lambda_1 x/L), \quad \tau > 0.2 \tag{9-10}$$

**TABLE 9-1**
**Coefficients used in the one-term approximate solution of transient one-dimensional heat conduction in plane walls, cylinders, and spheres (Bi = $hL/k$ for a plane wall of thickness $2L$, and Bi = $hr_0/k$ for a cylinder or sphere of radius $r_0$)**

| | Plane slab | | Cylinder | | Sphere | |
|---|---|---|---|---|---|---|
| Bi | $\lambda_1$ | $A_1$ | $\lambda_1$ | $A_1$ | $\lambda_1$ | $A_1$ |
| 0.01 | 0.0998 | 1.0017 | 0.1412 | 1.0025 | 0.1730 | 1.0030 |
| 0.02 | 0.1410 | 1.0033 | 0.1995 | 1.0050 | 0.2445 | 1.0060 |
| 0.04 | 0.1987 | 1.0066 | 0.2814 | 1.0099 | 0.3450 | 1.0120 |
| 0.06 | 0.2425 | 1.0098 | 0.3438 | 1.0148 | 0.4217 | 1.0179 |
| 0.08 | 0.2791 | 1.0130 | 0.3960 | 1.0197 | 0.4860 | 1.0239 |
| 0.1 | 0.3111 | 1.0161 | 0.4417 | 1.0246 | 0.5423 | 1.0298 |
| 0.2 | 0.4328 | 1.0311 | 0.6170 | 1.0483 | 0.7593 | 1.0592 |
| 0.3 | 0.5218 | 1.0450 | 0.7465 | 1.0712 | 0.9208 | 1.0880 |
| 0.4 | 0.5932 | 1.0580 | 0.8516 | 1.0931 | 1.0528 | 1.1164 |
| 0.5 | 0.6533 | 1.0701 | 0.9408 | 1.1143 | 1.1656 | 1.1441 |
| 0.6 | 0.7051 | 1.0814 | 1.0184 | 1.1345 | 1.2644 | 1.1713 |
| 0.7 | 0.7506 | 1.0918 | 1.0873 | 1.1539 | 1.3525 | 1.1978 |
| 0.8 | 0.7910 | 1.1016 | 1.1490 | 1.1724 | 1.4320 | 1.2236 |
| 0.9 | 0.8274 | 1.1107 | 1.2048 | 1.1902 | 1.5044 | 1.2488 |
| 1.0 | 0.8603 | 1.1191 | 1.2558 | 1.2071 | 1.5708 | 1.2732 |
| 2.0 | 1.0769 | 1.1785 | 1.5995 | 1.3384 | 2.0288 | 1.4793 |
| 3.0 | 1.1925 | 1.2102 | 1.7887 | 1.4191 | 2.2889 | 1.6227 |
| 4.0 | 1.2646 | 1.2287 | 1.9081 | 1.4698 | 2.4556 | 1.7202 |
| 5.0 | 1.3138 | 1.2403 | 1.9898 | 1.5029 | 2.5704 | 1.7870 |
| 6.0 | 1.3496 | 1.2479 | 2.0490 | 1.5253 | 2.6537 | 1.8338 |
| 7.0 | 1.3766 | 1.2532 | 2.0937 | 1.5411 | 2.7165 | 1.8673 |
| 8.0 | 1.3978 | 1.2570 | 2.1286 | 1.5526 | 2.7654 | 1.8920 |
| 9.0 | 1.4149 | 1.2598 | 2.1566 | 1.5611 | 2.8044 | 1.9106 |
| 10.0 | 1.4289 | 1.2620 | 2.1795 | 1.5677 | 2.8363 | 1.9249 |
| 20.0 | 1.4961 | 1.2699 | 2.2880 | 1.5919 | 2.9857 | 1.9781 |
| 30.0 | 1.5202 | 1.2717 | 2.3261 | 1.5973 | 3.0372 | 1.9898 |
| 40.0 | 1.5325 | 1.2723 | 2.3455 | 1.5993 | 3.0632 | 1.9942 |
| 50.0 | 1.5400 | 1.2727 | 2.3572 | 1.6002 | 3.0788 | 1.9962 |
| 100.0 | 1.5552 | 1.2731 | 2.3809 | 1.6015 | 3.1102 | 1.9990 |
| ∞ | 1.5708 | 1.2732 | 2.4048 | 1.6021 | 3.1416 | 2.0000 |

Cylinder: $$\theta(x, t)_{cyl} = \frac{T(r, t) - T_\infty}{T_i - T_\infty} = A_1 e^{-\lambda_1^2 \tau} J_0(\lambda_1 r/\lambda_0), \quad \tau > 0.2 \tag{9-11}$$

Sphere: $$\theta(x, t)_{sph} = \frac{T(r, t) - T_\infty}{T_i - T_\infty} = A_1 e^{-\lambda_1^2 \tau} \frac{\sin(\lambda_1 r/r_0)}{\lambda_1 r/r_0}, \quad \tau > 0.2 \tag{9-12}$$

where the constant $A_1$ and $\lambda_1$ are functions of the Bi number only, and their values are listed in Table 9-1 against the Bi number for all three goemetries. The function $J_0$ is the zeroth-order Bessel function of the first kind, whose value can be determined from Table 9-2. Noting that $\cos(0) = J_0(0) = 1$ and the limit of $(\sin x)/x$ is also 1, the above relations simplify to the following at the center of a plane wall, cylinder, or sphere:

Center of plane wall ($x = 0$): $$\theta_{0,wall} = \frac{T_0 - T_\infty}{T_1 - T_\infty} = A_1 e^{-\lambda_1^2 \tau} \tag{9-13}$$

Center of cylinder ($r = 0$): $$\theta_{0,cyl} = \frac{T_0 - T_\infty}{T_i - T_\infty} = A_1 e^{-\lambda_1^2 \tau} \tag{9-14}$$

Center of sphere: ($r = 0$): $$\theta_{0,sph} = \frac{T_0 - T_\infty}{T_i - T_\infty} = A_1 e^{-\lambda_1^2 \tau} \tag{9-15}$$

Once the Bi number is known, the above relations can be used to determine the temperature anywhere in the medium. The determination of the constants $A_1$ and $\lambda_1$ usually requires interpolation. For those who prefer reading charts to interpolations, the relations above are plotted and the one-term approximation solutions are presented in graphical form, which are known as the *transient temperature charts.* Note that the charts are sometimes difficult to read, and they are subject to reading errors. Therefore, the relations above should be preferred to the charts.

The transient temperature charts in Figs. 9-13, 9-14, and 9-15 for a large plane wall, long cylinder, and sphere were presented by M. P. Heisler in 1947, and are called **Heisler charts**. They were supplemented in 1961 with transient heat transfer charts by H. Gröber. There are *three* charts associated with each geometry: the first chart is to determine the temperature $T_o$ at the *center* of the geometry at a given time *t*. The second chart is to determine the temperature at *other locations* at the same time in terms of $T_o$. The third chart is to determine the total amount of *heat transfer* up to the time *t*. These plots are valid for $\tau > 0.2$.

Note that the case $1/\text{Bi} = k/hL = 0$ corresponds to $h \to \infty$, which corresponds to the case of *specified surface temperature* $T_\infty$. That is, the case in which the surfaces of the body are suddenly brought to the temperature $T_\infty$ at $t = 0$ and kept at $T_\infty$ at all times can be handled by setting $h$ to infinity (Fig. 9-16).

The temperature of the body changes from the initial temperature $T_i$ to the temperature of the surroundings $T_\infty$ at the end of the transient heat conduction process. Thus, the *maximum* amount of heat that a body can

**TABLE 9-2**
**The zeroth- and first-order Bessel functions of the first kind**

| $\xi$ | $J_0(\xi)$ | $J_1(\xi)$ |
|---|---|---|
| 0.0 | 1.0000 | 0.0000 |
| 0.1 | 0.9975 | 0.0499 |
| 0.2 | 0.9900 | 0.0995 |
| 0.3 | 0.9776 | 0.1483 |
| 0.4 | 0.9604 | 0.1960 |
| 0.5 | 0.9385 | 0.2423 |
| 0.6 | 0.9120 | 0.2867 |
| 0.7 | 0.8812 | 0.3290 |
| 0.8 | 0.8463 | 0.3688 |
| 0.9 | 0.8075 | 0.4059 |
| 1.0 | 0.7652 | 0.4400 |
| 1.1 | 0.7196 | 0.4709 |
| 1.2 | 0.6711 | 0.4983 |
| 1.3 | 0.6201 | 0.5220 |
| 1.4 | 0.5669 | 0.5419 |
| 1.5 | 0.5118 | 0.5579 |
| 1.6 | 0.4554 | 0.5699 |
| 1.7 | 0.3980 | 0.5778 |
| 1.8 | 0.3400 | 0.5815 |
| 1.9 | 0.2818 | 0.5812 |
| 2.0 | 0.2239 | 0.5767 |
| 2.1 | 0.1666 | 0.5683 |
| 2.2 | 0.1104 | 0.5560 |
| 2.3 | 0.0555 | 0.5399 |
| 2.4 | 0.0025 | 0.5202 |

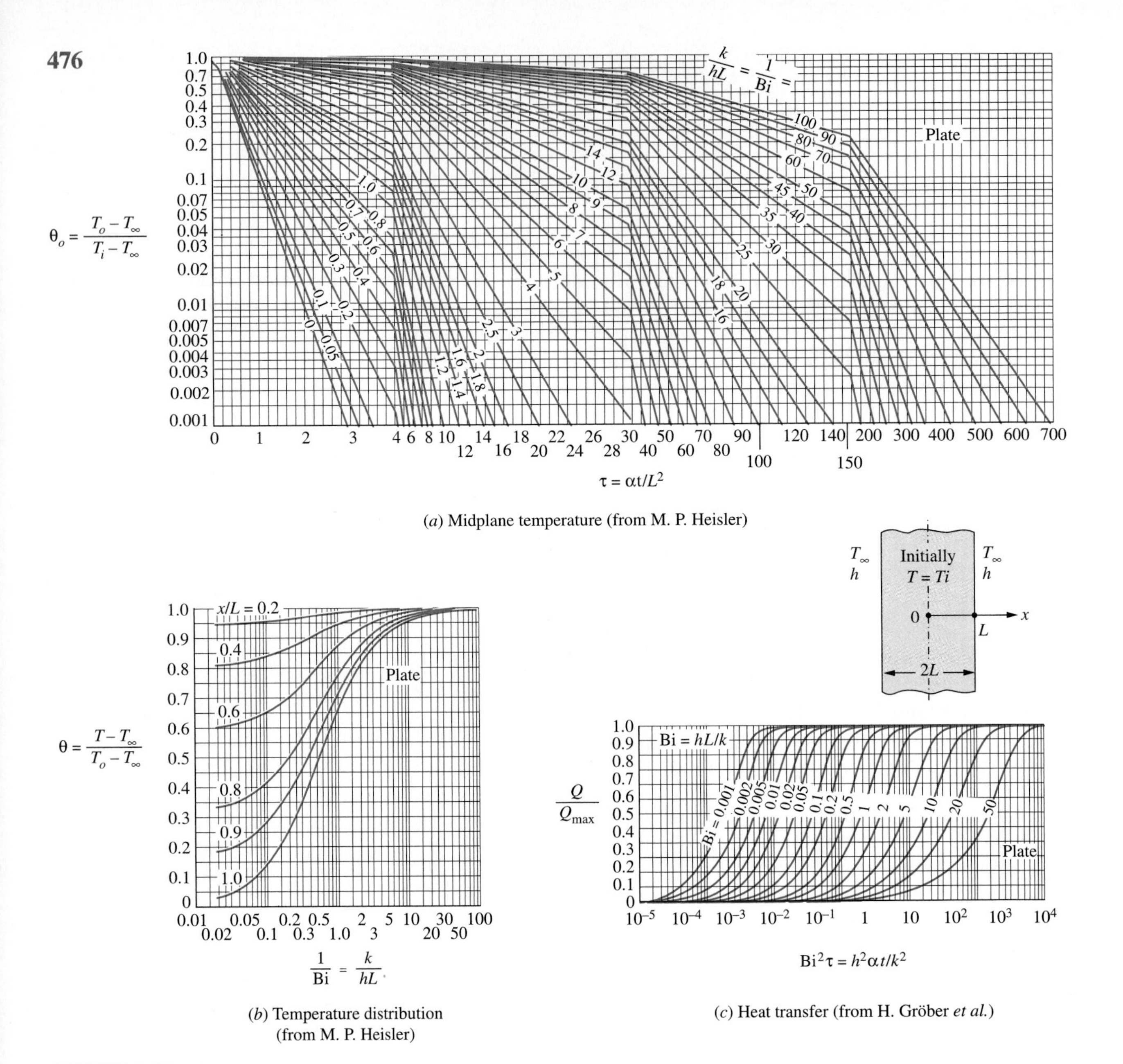


**FIGURE 9-13**

Transient temperature and heat transfer charts for a plane wall of thickness $2L$ initially at a uniform temperature $T_i$ subjected to convection from both sides to an environment at temperature $T_\infty$ with a convection coefficient of $h$.

gain (or lose if $T_i > T_\infty$) is simply the *change* in the *energy content* of the body. That is,

$$Q_{max} = mC_p(T_\infty - T_i) = \rho VC_p(T_\infty - T_i) \quad \text{(kJ)} \tag{9-16}$$

where $m$ is the mass, $V$ is the volume, $\rho$ is the density, and $C_p$ is the specific heat of the body. Thus, $Q_{max}$ represents the amount of heat transfer for $t \to \infty$. The amount of heat transfer $Q$ at a finite time $t$ will obviously be less than this maximum. The ratio $Q/Q_{max}$ is plotted in

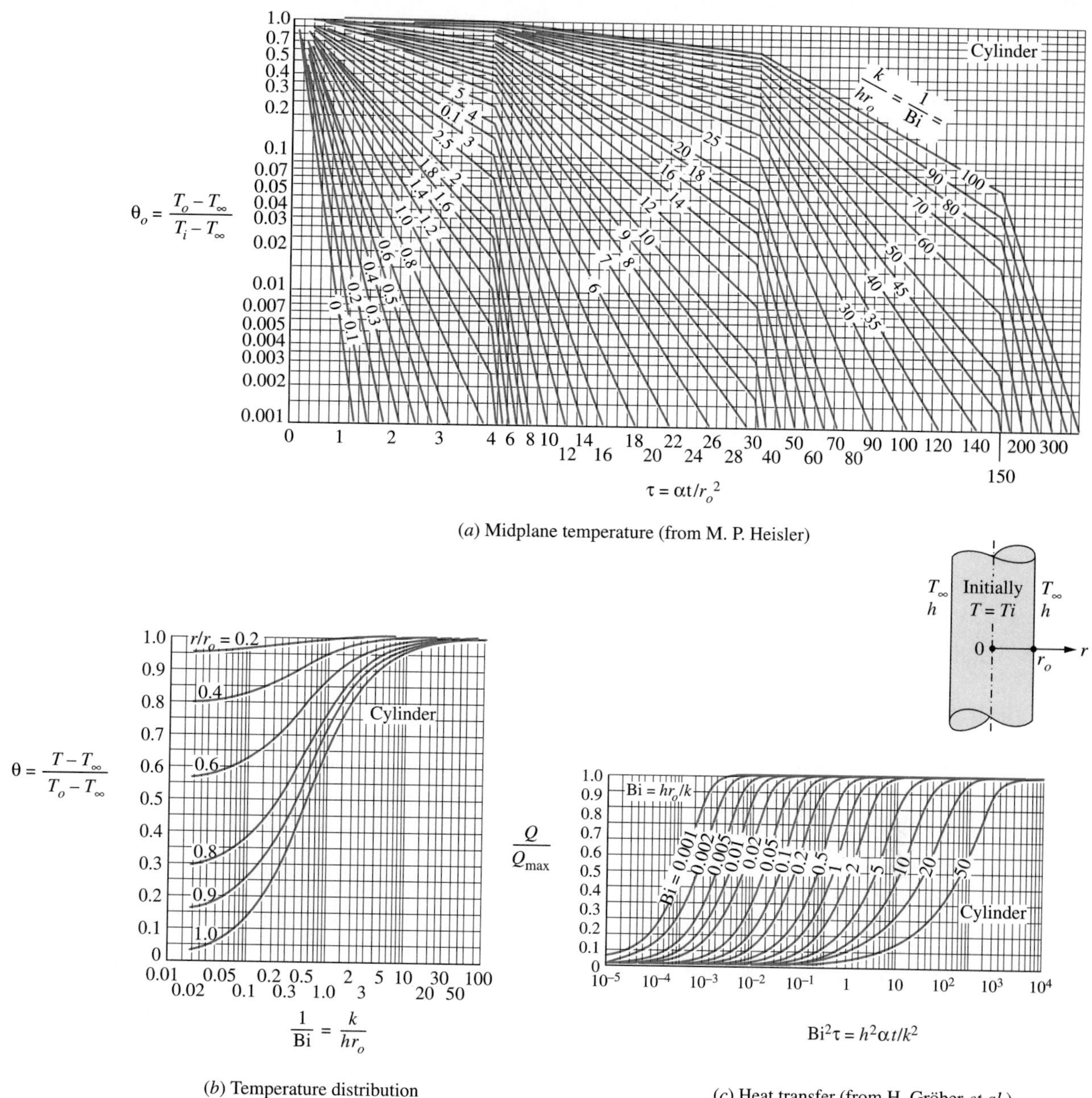


(*a*) Midplane temperature (from M. P. Heisler)

(*b*) Temperature distribution (from M. P. Heisler)

(*c*) Heat transfer (from H. Gröber *et al.*)

**FIGURE 9-14**

Transient temperature and heat transfer charts for a long cylinder of radius $r_o$ initially at a uniform temperature $T_i$ subjected to convection from all sides to an environment at temperature $T_\infty$ with a convection coefficient of $h$.

Figs. 9-13*c*, 9-14*c*, and 9-15*c* against the variables Bi and $h^2\alpha t/k^2$ for the large plane wall, long cylinder, and sphere, respectively. Note that once the *fraction* of heat transfer $Q/Q_{max}$ has been determined from these charts for the given *t*, the actual amount of heat transfer by that time can be evaluated by multiplying this fraction by $Q_{max}$. A *negative* sign for $Q_{max}$ indicates that heat is *leaving* the body (Fig. 9-17).

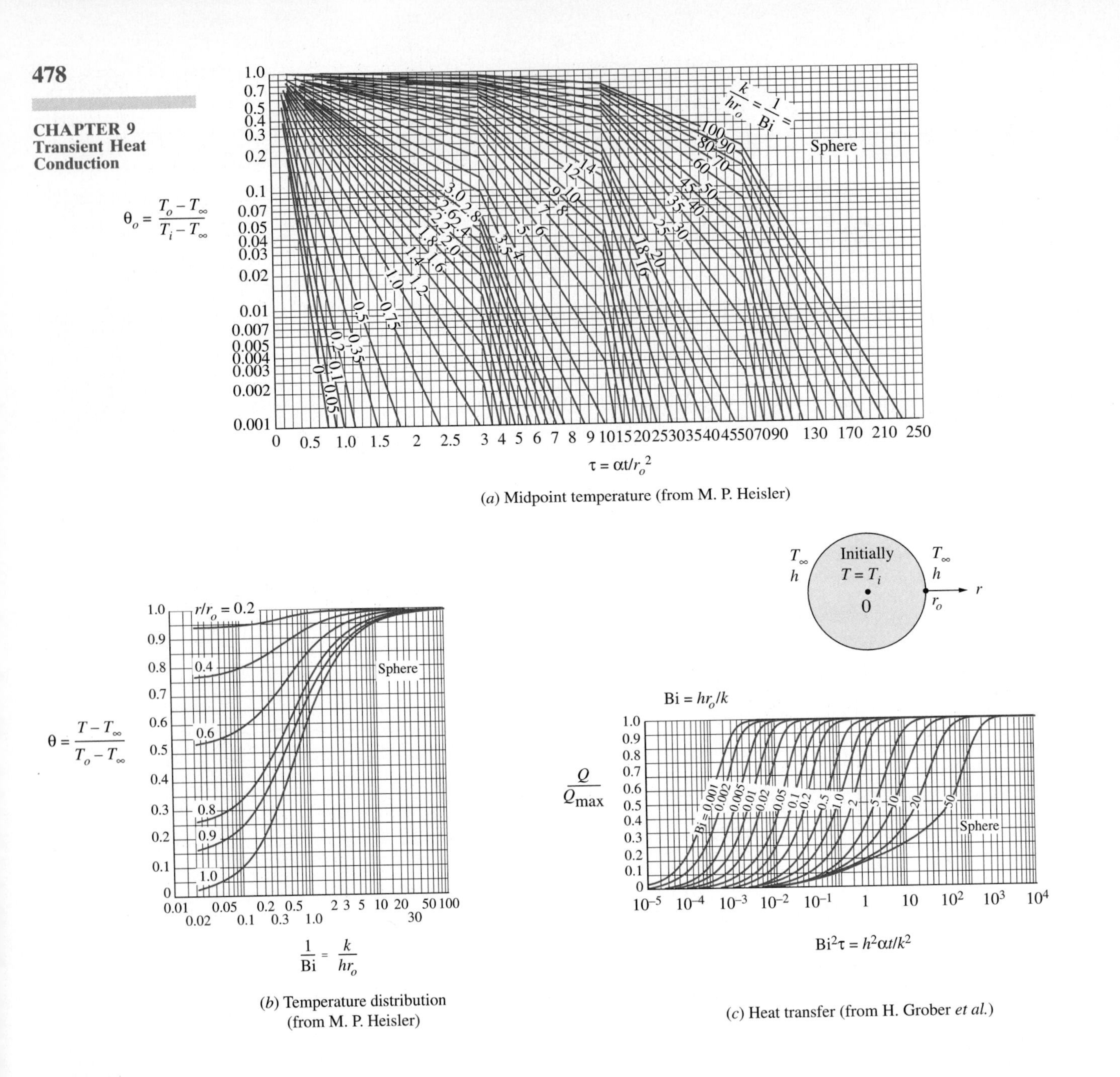


**FIGURE 9-15**

Transient temperature and heat transfer charts for a sphere of radius $r_o$ initially at a uniform temperature $T_i$ subjected to convection from all sides to an environment at temperature $T_\infty$ with a convection coefficient of $h$.

The fraction of heat transfer can also be determined from the following relations, which are based on the one-term approximations discussed above:

$$\text{Plane wall:} \quad \left(\frac{Q}{Q_{max}}\right)_{wall} = 1 - \theta_{0,\text{wall}} \frac{\sin \lambda_1}{\lambda_1} \tag{9-17}$$

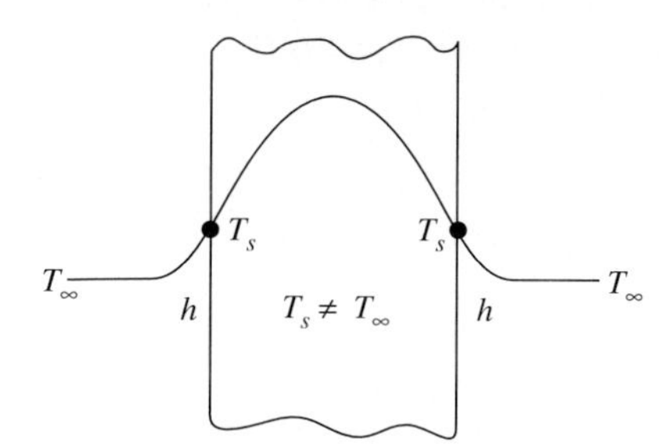


(*a*) Finite convection coefficient

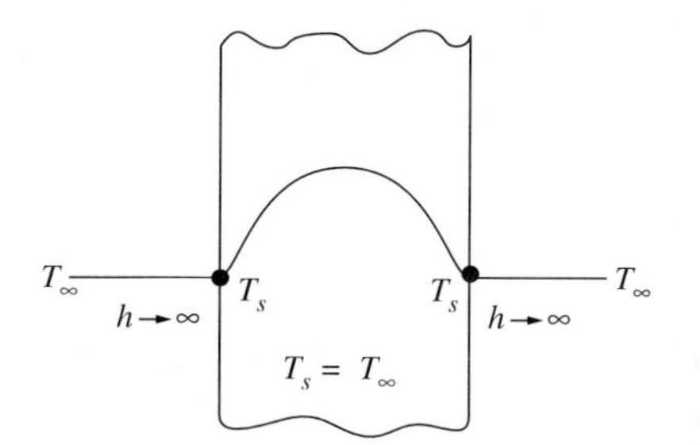


(*b*) Infinite convection coefficient

**FIGURE 9-16**

The specified surface temperature corresponds to the case of convection to an environment at $T_\infty$ with a convection coefficient $h$ that is *infinite*.

Cylinder: $$\left(\frac{Q}{Q_{\max}}\right)_{\text{cyl}} = 1 - 2\theta_{0,\text{cyl}}\frac{J_1(\lambda_1)}{\lambda_1} \tag{9-18}$$

Sphere: $$\left(\frac{Q}{Q_{\max}}\right)_{\text{sph}} = 1 - 3\theta_{0,\text{cyl}}\frac{\sin\lambda_1 - \lambda_1\cos\lambda_1}{\lambda_1^3} \tag{9-19}$$

The use of the Heisler/Gröber charts and the one-term solutions discussed above are limited to the conditions specified at the beginning of this section: the body is initially at a *uniform* temperature, the temperature of the medium surrounding the body and the convection heat transfer coefficient are *constant* and *uniform,* and there is no *energy generation* in the body.

We discussed the physical significance of the *Biot number* Bi $= h/kL$ earlier and indicated that it is a measure of the relative magnitudes of the two heat transfer mechanisms: *convection* at the surface, and *conduction* through the solid. A *small* value of Bi indicates that the inner resistance of the body to heat conduction is *small* relative to the resistance to convection between the surface and the fluid. As a result, the temperature distribution within the solid becomes fairly uniform, and lumped system analysis becomes applicable. Recall that when Bi $< 0.1$, the error in assuming the temperature within the body to be *uniform* is less than 5 percent.

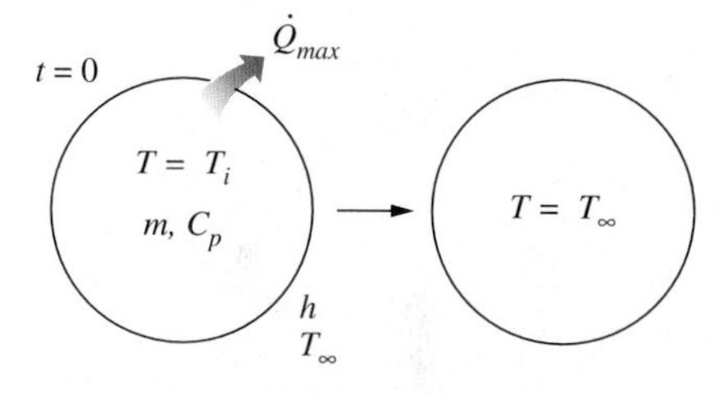


(*a*) Maximum heat transfer ($t \to \infty$)

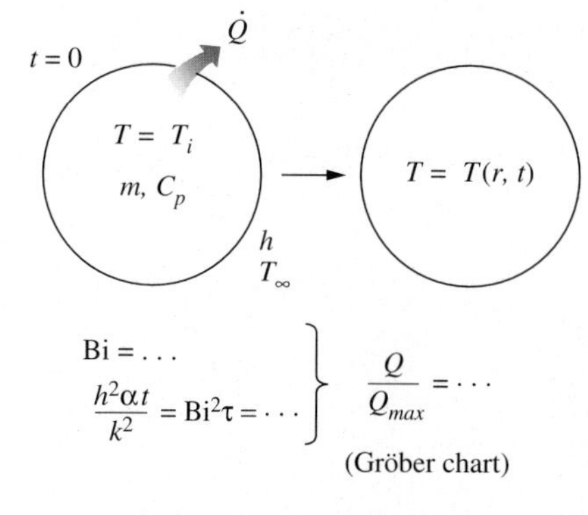


(*b*) Actual heat transfer for time *t*

**FIGURE 9-17**

The fraction of total heat transfer $Q/Q_{\max}$ up to a specified time *t* is determined using the Gröber charts.

To understand the physical significance of the *Fourier number* $\tau$, we express it as (Fig. 9-18)

$$\tau = \frac{\alpha t}{L^2} = \frac{kL^2(1/L)\,\Delta T}{\rho C_p L^3/t\ \Delta T} = \frac{\text{the rate at which heat is } conducted \text{ across } L \text{ of a body of volume } L^3}{\text{the rate at which heat is } stored \text{ in a body of volume } L^3} \tag{9-20}$$

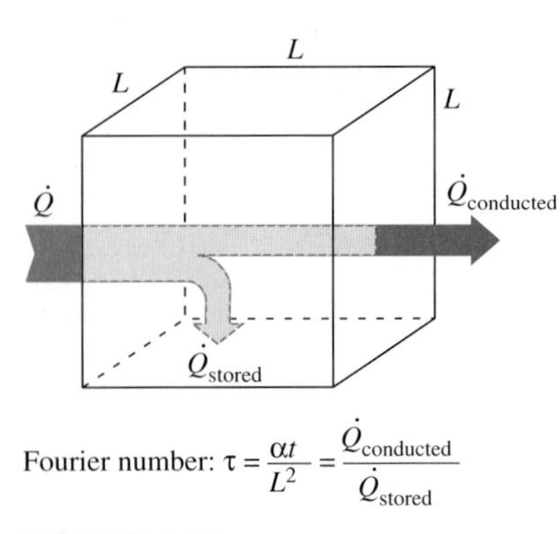


Fourier number: $\tau = \dfrac{\alpha t}{L^2} = \dfrac{\dot{Q}_{\text{conducted}}}{\dot{Q}_{\text{stored}}}$

**FIGURE 9-18**
Fourier number at time $t$ can be viewed as the ratio of the rate of heat conducted to the rate of heat stored at that time.

Therefore, the Fourier number is a measure of *heat conducted* through a body relative to *heat stored.* Thus a large value of Fourier number indicates faster propagation of heat through a body.

Perhaps you are wondering about what constitutes an infinitely large plate or an infinitely long cylinder. After all, nothing in this world is infinite. A plate whose thickness is small relative to the other dimensions can be modeled as an infinitely large plate, except very near the outer edges. But the edge effects on large bodies are usually negligible, and thus a large plane wall such as the wall of a house can be modeled as an infinitely large wall for heat transfer purposes. Similarly, a long cylinder whose diameter is small relative to its length can be analyzed as an infinitely long cylinder. The use of the transient temperature charts and the one-term solutions are illustrated in the following examples.

**EXAMPLE 9-3**

An ordinary egg can be approximated as a 5-cm-diameter sphere whose properties are roughly those of water at room temperature, $k = 0.6\ \text{W/(m}\cdot{}^\circ\text{C)}$ and $\alpha = 0.14 \times 10^{-6}\ \text{m}^2/\text{s}$ (Fig. 9-19). The egg is initially at a uniform temperature of 5°C, and is dropped into boiling water at 95°C. Taking the convection heat transfer coefficient to be $h = 1200\ \text{W/(m}^2\cdot{}^\circ\text{C)}$, determine how long it will take for the center of the egg to reach 70°C.

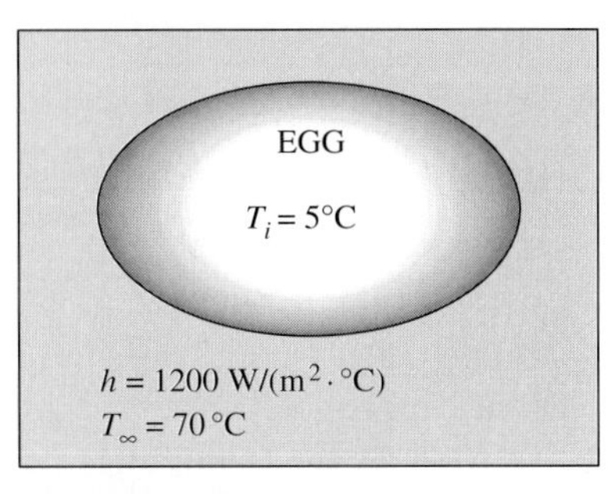


**FIGURE 9-19**
Schematic for Example 9-3.

**Solution** The temperature within the egg varies with radial distance as well as time, and the temperature at a specified location at a given time can be determined from the Heisler charts or the one-term solutions. Here we will use the latter to demonstrate their use. We assume the convection heat transfer coefficient to be uniform, and the properties of the egg to be constant. The Biot number for this problem is

$$\text{Bi} = \frac{hr_0}{k} = \frac{[1200\ \text{W/(m}^2\cdot\text{C)}](0.025\ \text{m})}{0.6\ \text{W/(m}\cdot{}^\circ\text{C)}} = 50$$

which is much greater than 0.1, and thus the lumped system analysis is not applicable. The coefficients $\lambda_1$ and $A_1$ for a sphere corresponding to this Bi are, from Table 9-1,

$$\lambda_1 = 3.0788, \qquad A_1 = 1.9962$$

Substituting these and other values into Eq. 9-15 and solving for $\tau$ gives

$$\frac{T_o - T_\infty}{T_i - T_\infty} = A_1 e^{-\lambda_1^2 \tau} \longrightarrow \frac{70 - 95}{5 - 95} = 1.9962 e^{-(3.0788)^2 \tau} \longrightarrow \tau = 0.208$$

which is greater than 0.2, and thus the one-term solution is applicable with an error of less than 2 percent. Then the cooking time is determined from the definition of the Fourier number to be

$$t = \frac{\tau r_o^2}{\alpha} = \frac{(0.208)(0.025\ \text{m})^2}{0.14 \times 10^{-6}\ \text{m}^2/\text{s}} = 929\ \text{s} \approx 15.5\ \text{min}$$

Therefore, it will take about 15 min for the center of the egg to be heated from 5°C to 70°C.

**EXAMPLE 9-4**

In a production facility, large brass plates of 4-cm thickness [$k$ = 110 W/(m · °C), $\rho$ = 8530 kg/m³, $C_p$ = 380 J/(kg · °C), and $\alpha = 33.9 \times 10^{-6}$ m²/s] that are initially at a uniform temperature of 20°C are heated by passing them through an oven that is maintained at 500°C (Fig. 9-20). the plates remain in the oven for a period of 7 min. Taking the combined convection and radiation heat transfer coefficient to be $h$ = 120 W/(m² · °C), determine the surface temperature of the plates when they come out of the oven.

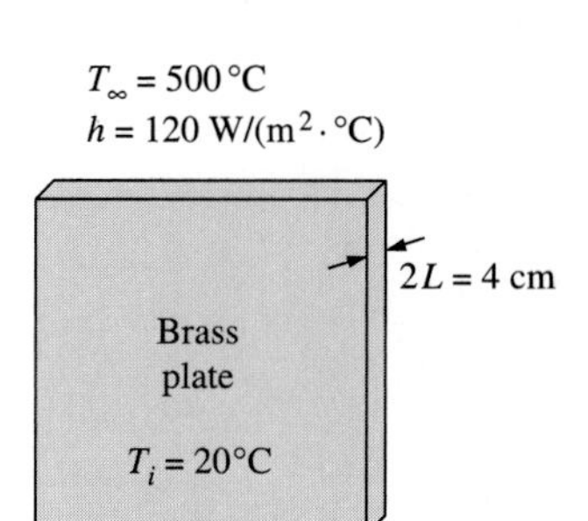


**FIGURE 9-20**
Schematic for Example 9-4.

**Solution** The temperature within the large plate may vary with the distance $x$ as well as time, and the temperature at a specified location at a given time can be determined from the Heisler charts or one-term solutions. Here we will use the charts to demonstrate their use. We assume the heat transfer coefficient to be uniform, and the properties of the plate to be constant. Noting that the half-thickness of the plate is $L$ = 0.02 m, from Fig. 9-13 we have

$$\left.\begin{aligned} \frac{1}{\text{Bi}} &= \frac{k}{hL} = \frac{110\ \text{W/(m}\cdot{}^\circ\text{C)}}{[120\ \text{W/(m}^2\cdot{}^\circ\text{C)}](0.02\ \text{m})} = 45.8 \\ \tau &= \frac{\alpha t}{L^2} = \frac{(33.9 \times 10^{-6}\ \text{m}^2/\text{s})(7 \times 60\ \text{s})}{(0.02\ \text{m})^2} = 35.6 \end{aligned}\right\} \frac{T_o - T_\infty}{T_i - T_\infty} = 0.46$$

Also,

$$\left.\begin{aligned} \frac{1}{\text{Bi}} &= \frac{k}{hL} = 45.8 \\ \frac{x}{L} &= \frac{L}{L} = 1 \end{aligned}\right\} \frac{T - T_\infty}{T_o - T_\infty} = 0.99$$

Therefore,

$$\frac{T - T_\infty}{T_i - T_\infty} = \frac{T - T_\infty}{T_o - T_\infty}\frac{T_o - T_\infty}{T_i - T_\infty} = 0.46 \times 0.99 = 0.455$$

and $$T = T_\infty + 0.455(T_i - T_\infty) = 500 + 0.455(20 - 500) = 282^\circ\text{C}$$

Therefore, the surface temperature of the plates will be 282°C when they leave the oven.

**Discussion** We notice that the Biot number in this case is Bi = 1/45.8 = 0.022, which is much less than 0.1. Therefore, we expect the lumped system analysis to be applicable. This is also evident from $(T - T_\infty)/(T_o - T_\infty) = 0.99$, which indicates that the temperatures at the center and the surface of the plate relative to the surroundings temperature are within 1 percent of each other. Noting that the error involved in reading the Heisler charte is typically at least a few percent, the lumped system analysis in this case may yield just as accurate results with less effort.

Noting that the heat transfer surface area of the plate is $2A$, where $A$ the face area of the plate (the plate transfers heat through both of its surfaces), and the volume of the plate is $V = (2L)A$, where $L$ is the half-thickness of the plate, the exponent $b$ used in the lumped system analysis is determined to be

$$b = \frac{hA}{\rho C_p V} = \frac{h(2A)}{\rho C_p(2LA)} = \frac{h}{\rho C_p L}$$

$$= \frac{120\ \text{W/(m}^2\cdot{}^\circ\text{C)}}{(8530\ \text{kg/m}^3)[380\ \text{J/(kg}\cdot{}^\circ\text{C)}](0.02\ \text{m})} = 0.00185\ \text{s}^{-1}$$

Then the temperature of the plate at $t = 7$ min $= 420$ s is determined from

$$\frac{T(t) - T_\infty}{T_i - T_\infty} = e^{-bt} \longrightarrow \frac{T(t) - 500}{20 - 500} = e^{-(0.00185\ \text{s}^{-1})(420\ \text{s})}$$

It yields

$$T(t) = 279^\circ\text{C}$$

which is practically identical to the result obtained above using the Heisler charts. Therefore, we can use lumped system analysis with confidence when the Biot number is sufficiently small.

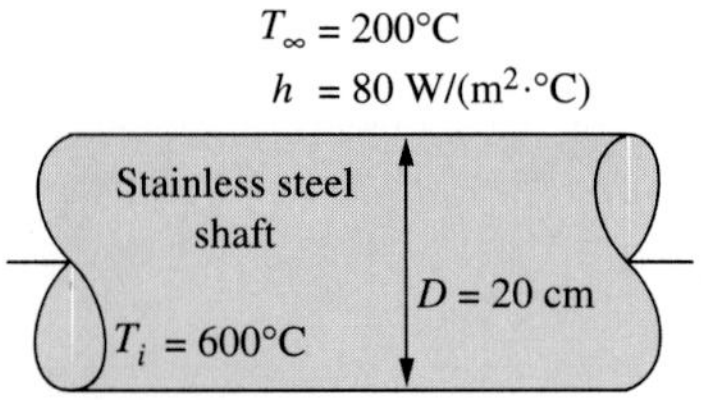

$T_\infty = 200°C$
$h = 80$ W/(m²·°C)
Stainless steel shaft
$T_i = 600°C$
$D = 20$ cm

**FIGURE 9-21**
Schematic for Example 9-5.

**EXAMPLE 9-5**

A long 20-cm-diameter cylindrical shaft made of stainless steel 304 [$k$ = 14.9 W/(m · °C), $\rho$ = 7900 kg/m³, $C_p$ = 477 J/(kg · °C), and $\alpha = 3.95 \times 10^{-6}$ m²/s] comes out of an oven at a uniform temperature of 600°C (Fig. 9-21). The shaft is then allowed to cool slowly in an environment chamber at 200°C with an average heat transfer coefficient of $h$ = 80 W/(m² · °C). Determine the temperature at the center of the shaft 45 min after the start of the cooling process. Also determine the heat transfer per unit length of the shaft during this time period.

**Solution** The shaft is said to be very long, and is subjected to uniform thermal conditions. Therefore, it can be modeled as an infinite cylinder in which heat transfer is one-dimensional. The temperature within the shaft may vary with the radial distance $r$ as well as time, and the temperature at a specified location at a given time can be determined from the Heisler charts. We assume the heat transfer coefficient to be uniform, and the properties of the shaft to be constant. Noting that the radius of the shaft is $r_o = 0.1$ m, from Fig. 9-14 we have

$$\left.\begin{aligned} \frac{1}{\text{Bi}} &= \frac{k}{hr_o} = \frac{14.9\ \text{W/(m}\cdot{}^\circ\text{C)}}{[80\ \text{W/(m}^2\cdot{}^\circ\text{C)}](0.1\ \text{m})} = 1.86 \\ \tau &= \frac{\alpha t}{r_o^2} = \frac{(3.95\times 10^{-6}\ \text{m}^2/\text{s})(45\times 60\ \text{s})}{(0.1\ \text{m})^2} = 1.07 \end{aligned}\right\}\ \frac{T_o - T_\infty}{T_i - T_\infty} = 0.40$$

and $\quad T_o = T_\infty + 0.4(T_i - T_\infty) = 200 + 0.4(600 - 200) = 360^\circ\text{C}$

Therefore, the center temperature of the shaft will drop from 600°C to 360°C in 45 min.

To determine the actual heat transfer, we first need to calculate the maximum heat that can be transferred from the cylinder, which is the sensible energy of the cylinder relative to its environment. Taking $L = 1$ m,

$$m = \rho V = \rho \pi r_o^2 L = (7900\ \text{kg/m}^3)\pi(0.1\ \text{m})^2(1\ \text{m}) = 248.2\ \text{kg}$$

$$Q_{max} = mC_p(T_\infty - T_i) = (248.2\ \text{kg})[0.477\ \text{kJ/(kg}\cdot{}^\circ\text{C)}](600 - 200)^\circ\text{C}$$
$$= 47{,}354\ \text{kJ}$$

The dimensionless heat transfer ratio is determined from Fig. 9-14*c* for a long cylinder to be

$$\left.\begin{aligned} \text{Bi} &= \frac{1}{1/\text{Bi}} = \frac{1}{1.86} = 0.537 \\ \frac{h^2\alpha t}{k^2} &= \text{Bi}^2\tau = (0.537)^2(1.07) = 0.309 \end{aligned}\right\} \quad \frac{Q}{Q_{max}} = 0.62$$

Therefore,

$$Q = 0.62Q_{max} = 0.62 \times (47{,}354\ \text{kJ}) = 29{,}360\ \text{kJ}$$

which is the total heat transfer from the shaft during the first 45 min of the cooling.

**Alternative solution** We could also solve this problem using the one-term solution relation instead of the transient charts. First we find the Biot number

$$\text{Bi} = \frac{hr_o}{k} = \frac{[80\ \text{W/(m}^2\cdot{}^\circ\text{C)}](0.1\ \text{m})}{14.9\ \text{W/(m}\cdot{}^\circ\text{C)}} = 0.537$$

The coefficients $\lambda_1$ and $A_1$ for a cylinder corresponding to this Bi are determined from Table 9-1 to be

$$\lambda_1 = 0.970, \qquad A_1 = 1.122$$

Substituting these values into Eq. 9-14 gives

$$\theta_0 = \frac{T_o - T_\infty}{T_i - T_\infty} = A_1 e^{-\lambda_1^2\tau} = 1.122e^{-(0.970)^2(1.07)} = 0.41$$

and thus $T_o = T_\infty + 0.41(T_i - T_\infty) = 200 + 041(600 - 200) = 364^\circ\text{C}$

The value of $J_1(\lambda_1)$ for $\lambda_1 = 0.970$ is determined from Table 9-2 to be 0.430. Then the fractional heat transfer is determined from Eq. 9-18 to be

$$\frac{Q}{Q_{max}} = 1 - 2\theta_0 \frac{J_1(\lambda_1)}{\lambda_1} = 1 - 2 \times 0.41 \frac{0.430}{0.970} = 0.636$$

and thus $Q = 0.636Q_{max} = 0.636 \times (47{,}354\ \text{kJ}) = 30{,}120\ \text{kJ}$

The slight difference between the two results is due to the reading error of the charts.

## 9-3 ■ TRANSIENT HEAT CONDUCTION IN SEMI-INFINITE SOLIDS

A semi-infinite solid is an idealized body that has a *single plane surface* and extends to infinity in all directions, as shown in Fig. 9-22. This idealized body is used to indicate that the temperature change in the part of the body we are interested in (the region close to the surface) is due to the thermal conditions on a single surface. The earth, for example, can be considered to be a semi-infinite medium in determining the variation of temperature near its surface. Also, a thick wall can be modeled as a semi-infinite medium if all we are interested in is the variation of temperature in the region near one of the surfaces, and the other surface is too far to have any impact on the region of interest during the time of observation.

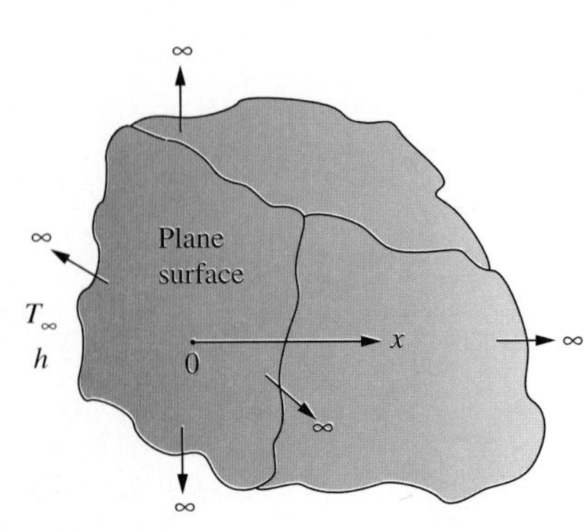


**FIGURE 9-22**
Schematic of a semi-infinite body.

Consider a semi-infinite solid that is at a uniform temperature $T_i$. At time $t = 0$, the surface of the solid at $x = 0$ is exposed to convection by a fluid at a constant temperature $T_\infty$, with a heat transfer coefficient $h$. This problem can be formulated as a partial differential equation, which can be solved analytically for the transient temperature distribution $T(x, t)$. The solution obtained is presented in Fig. 9-23 graphically for the *nondimensionalized temperature* defined as

$$1 - \theta(x, t) = \frac{T(x, t) - T_\infty}{T_i - T_\infty} = \frac{T(x, t) - T_i}{T_\infty - T_i} \tag{9-21}$$

against the dimensionless variable $x/(2\sqrt{\alpha t})$ for various values of the parameter $h\sqrt{\alpha t}/k$.

Note that the values on the vertical axis correspond to $x = 0$, and thus represent the surface temperature. The curve $h\sqrt{\alpha t}/k = \infty$ corresponds to $h \to \infty$, which corresponds to the case of *specified temperature* $T_\infty$ at the surface at $x = 0$. That is, the case in which the surface of the semi-infinite body is suddenly brought to temperature $T_\infty$ at $t = 0$ and kept at $T_\infty$ at all times can be handled by setting $h$ to infinity. The specified surface temperature case is closely approximated in practice when condensation or boiling takes place on the surface. For a *finite* heat transfer coefficient $h$, the surface temperature approaches the fluid temperature $T_\infty$ as the time $t$ approaches infinity.

The exact solution of the transient one-dimensional heat conduction problem in a semi-infinite medium that is initially at a uniform temperature of $T_i$ and is suddenly subjected to convection at time $t = 0$ has been obtained, and is expressed as

$$\frac{T(x, t) - T_i}{T_\infty - T_i} = \operatorname{erfc}\left(\frac{x}{2\sqrt{\alpha t}}\right) - \exp\left(\frac{hx}{k} + \frac{h^2\alpha t}{k^2}\right)\left[\operatorname{erfc}\left(\frac{x}{2\sqrt{\alpha t}} + \frac{h\sqrt{\alpha t}}{k}\right)\right] \tag{9-22}$$

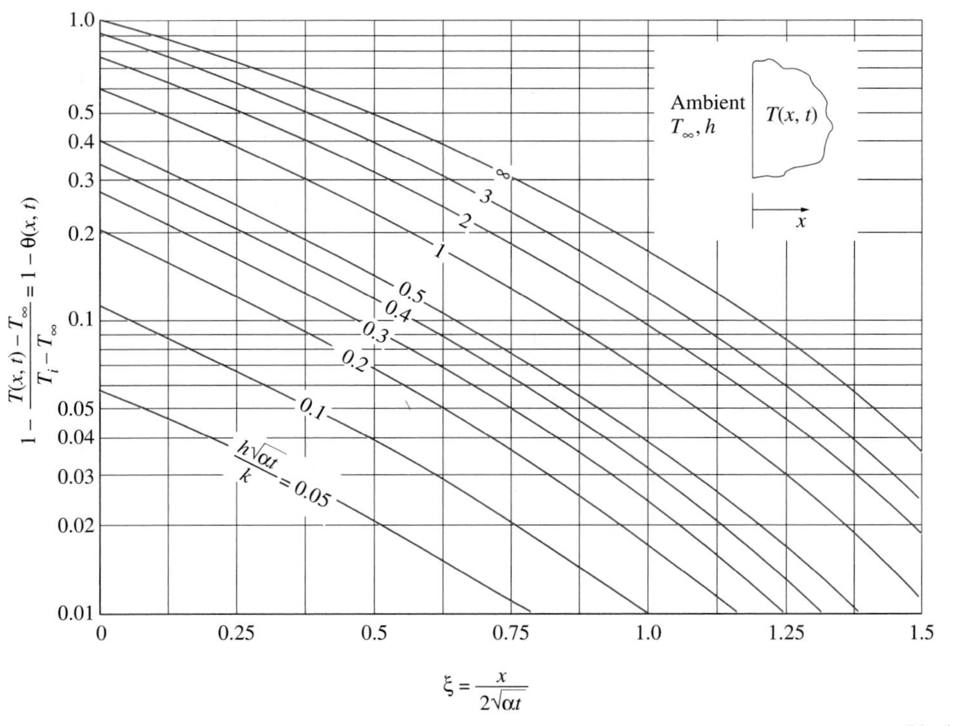


**FIGURE 9-23**

Variation of temperature with position and time in a semi-infinite solid initially at $T_i$ subjected to convection to an environment at $T_\infty$ with a convection heat transfer coefficient of $h$ (from P. J. Schneider).

where the quantity erfc ($\xi$) is the **complementary error function** defined as

$$\text{erfc}(\xi) = 1 - \frac{2}{\sqrt{\pi}} \int_0^\xi e^{-u^2}\, du \tag{9-23}$$

Despite its simple appearance, the integral that appears in the above relation cannot be performed analytically. Therefore, it is evaluated numerically for different values of $\xi$, and the results are listed in Table 9-3. For the special case of $h \to \infty$, the surface temperature $T_s$ becomes equal to the fluid temperature $T_\infty$, and Eq. 9-22 reduces to

$$\frac{T(x, t) - T_i}{T_s - T_i} = \text{erfc}\left(\frac{x}{2\sqrt{\alpha t}}\right) \tag{9-24}$$

This solution corresponds to the case when the temperature of the exposed surface of the medium is suddenly raised (or lowered) to $T_s$ at $t = 0$ and is maintained at that value at all times. Although the graphical solution given in Fig 9-23 is simply a plot of the exact analytical solution given by Eq. 9-23, it is subject to reading errors, and thus is of limited accuracy.

**TABLE 9-3**
**The complementary error function**

| $\xi$ | erfc ($\xi$) | $\xi$ | erfc ($\xi$) | $\xi$ | erfc ($\xi$) | $\xi$ | erfc ($\xi$) | $\xi$ | erfc ($\xi$) | $\xi$ | erfc ($\xi$) |
|---|---|---|---|---|---|---|---|---|---|---|---|
| 0.00 | 1.00000 | 0.38 | 0.5910 | 0.76 | 0.2825 | 1.14 | 0.1069 | 1.52 | 0.03159 | 1.90 | 0.00721 |
| 0.02 | 0.9774 | 0.40 | 0.5716 | 0.78 | 0.2700 | 1.16 | 0.10090 | 1.54 | 0.02941 | 1.92 | 0.00662 |
| 0.04 | 0.9549 | 0.42 | 0.5525 | 0.80 | 0.2579 | 1.18 | 0.09516 | 1.56 | 0.02737 | 1.94 | 0.00608 |
| 0.06 | 0.9324 | 0.44 | 0.5338 | 0.82 | 0.2462 | 1.20 | 0.08969 | 1.58 | 0.02545 | 1.96 | 0.00557 |
| 0.08 | 0.9099 | 0.46 | 0.5153 | 0.84 | 0.2349 | 1.22 | 0.08447 | 1.60 | 0.02365 | 1.98 | 0.00511 |
| 0.10 | 0.8875 | 0.48 | 0.4973 | 0.86 | 0.2239 | 1.24 | 0.07950 | 1.62 | 0.02196 | 2.00 | 0.00468 |
| 0.12 | 0.8652 | 0.50 | 0.4795 | 0.88 | 0.2133 | 1.26 | 0.07476 | 1.64 | 0.02038 | 2.10 | 0.00298 |
| 0.14 | 0.8431 | 0.52 | 0.4621 | 0.90 | 0.2031 | 1.28 | 0.07027 | 1.66 | 0.01890 | 2.20 | 0.00186 |
| 0.16 | 0.8210 | 0.54 | 0.4451 | 0.92 | 0.1932 | 1.30 | 0.06599 | 1.68 | 0.01751 | 2.30 | 0.00114 |
| 0.18 | 0.7991 | 0.56 | 0.4284 | 0.94 | 0.1837 | 1.32 | 0.06194 | 1.70 | 0.01612 | 2.40 | 0.00069 |
| 0.20 | 0.7773 | 0.58 | 0.4121 | 0.96 | 0.1746 | 1.34 | 0.05809 | 1.72 | 0.01500 | 2.50 | 0.00041 |
| 0.22 | 0.7557 | 0.60 | 0.3961 | 0.98 | 0.1658 | 1.36 | 0.05444 | 1.74 | 0.01387 | 2.60 | 0.00024 |
| 0.24 | 0.7343 | 0.62 | 0.3806 | 1.00 | 0.1573 | 1.38 | 0.05098 | 1.76 | 0.01281 | 2.70 | 0.00013 |
| 0.26 | 0.7131 | 0.64 | 0.3654 | 1.02 | 0.1492 | 1.40 | 0.04772 | 1.78 | 0.01183 | 2.80 | 0.00008 |
| 0.28 | 0.6921 | 0.66 | 0.3506 | 1.04 | 0.1413 | 1.42 | 0.04462 | 1.80 | 0.01091 | 2.90 | 0.00004 |
| 0.30 | 0.6714 | 0.68 | 0.3362 | 1.06 | 0.1339 | 1.44 | 0.04170 | 1.82 | 0.01006 | 3.00 | 0.00002 |
| 0.32 | 0.6509 | 0.70 | 0.3222 | 1.08 | 0.1267 | 1.46 | 0.03895 | 1.84 | 0.00926 | 3.20 | 0.00001 |
| 0.34 | 0.6306 | 0.72 | 0.3086 | 1.10 | 0.1198 | 1.48 | 0.03635 | 1.86 | 0.00853 | 3.40 | 0.00000 |
| 0.36 | 0.6107 | 0.74 | 0.2953 | 1.12 | 0.1132 | 1.50 | 0.03390 | 1.88 | 0.00784 | 3.60 | 0.00000 |

**EXAMPLE 9-6**

In areas where the air temperature remains below 0°C for prolonged periods of time, the freezing of water in underground pipes is a major concern. Fortunately, the soil remains relatively warm during those periods, and it takes weeks for the subfreezing temperatures to reach the water mains in the ground. Thus, the soil effectively serves as an insulation to protect the water from the freezing atmospheric temperatures in winter.

The ground at a particular location is covered with snow pack at −10°C for a continuous period of three months, and the average soil properties at that location are $k = 0.4$ W/(m · °C) and $\alpha = 0.15 \times 10^{-6}$ m²/s (Fig. 9-24). Assuming an initial uniform temperature of 15°C for the ground, determine the minimum burial depth to prevent the water pipes from freezing.

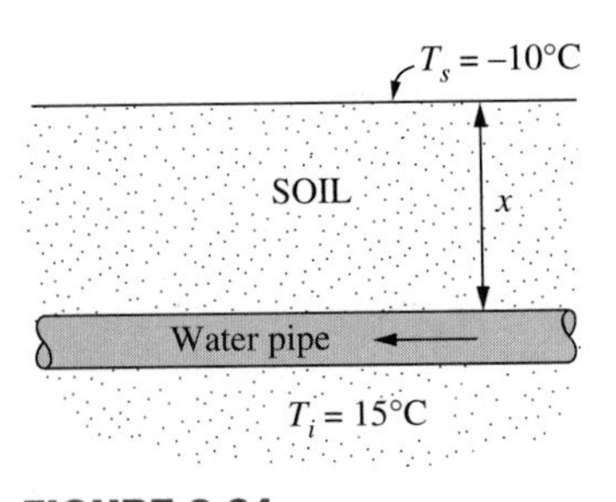


**FIGURE 9-24**
Schematic for Example 9-6.

**Solution** The temperature in the soil is affected by the thermal conditions at one surface only, and thus the soil can be considered to be a semi-infinite medium with a specified surface temperature of −10°C, which is equivalent to an environment temperature of $T_\infty = -10$°C, with a convection heat transfer coefficient of $h \to \infty$. The temperature of the soil surrounding the pipes will be 0°C after three months in the case of minimum burial depth. Therefore, from Fig. 9-23, we have

$$\left.\begin{aligned} &\frac{h\sqrt{\alpha t}}{k} = \infty \quad (\text{since } h \to \infty) \\ &1 - \frac{T(x,t) - T_\infty}{T_i - T_\infty} = 1 - \frac{0 - (-10)}{15 - (-10)} = 0.6 \end{aligned}\right\} \quad \xi = \frac{x}{2\sqrt{\alpha t}} = 0.36$$

We note that

$$t = (90 \text{ days})(24 \text{ h/day})(3600 \text{ s/h}) = 7.78 \times 10^6 \text{ s}$$

and thus

$$x = 2\xi\sqrt{\alpha t} = 2 \times 0.36\sqrt{(0.15 \times 10^{-6} \text{ m}^2\text{/s})(7.78 \times 10^6 \text{ s})} = 0.77 \text{ m}$$

Therefore, the water pipes must be buried to a depth of at least 77 cm to avoid freezing under the specified harsh winter conditions.

**Alternative solution** The solution of this problem could also be determined from Eq. 9-24:

$$\frac{T(x, t) - T_i}{T_s - T_i} = \text{erfc}\left(\frac{x}{2\sqrt{\alpha t}}\right) \longrightarrow \frac{0 - 15}{-10 - 15} = \text{erfc}\left(\frac{x}{2\sqrt{\alpha t}}\right) = 0.60$$

The argument that corresponds to this value of the complimentary error function is determined from Table 9-3 to be $\xi = 0.37$. Therefore,

$$x = 2\xi\sqrt{\alpha t} = 2 \times 0.37\sqrt{(0.15 \times 10^{-6} \text{ m}^2\text{/s})(7.78 \times 10^{-6} \text{ s})} = 0.80 \text{ m}$$

Again the slight difference is due to the reading error of the chart.

## 9-4 ■ TRANSIENT HEAT CONDUCTION IN MULTIDIMENSIONAL SYSTEMS

The transient temperature charts presented earlier can be used to determine the temperature distribution and heat transfer in *one-dimensional* heat conduction problems associated with a large plane wall, a long cylinder, a sphere, and a semi-infinite medium. Using a clever superposition principle called the **product solution**, these charts can also be used to construct solutions for the *two-dimensional* transient heat conduction problems encountered in geometries such as a short cylinder, a long rectangular bar, a semi-infinite cylinder or plate, and even *three-dimensional* problems associated with geometries such as a rectangular prism or a semi-infinite rectangular bar, provided that *all* surfaces of the solid are subjected to convection to the *same* fluid at temperature $T_\infty$, with the *same* heat transfer coefficient $h$, and the body involves no heat generation (Fig. 9-25). The solution in such multidimensional geometries can be expressed as the *product* of the solutions for the one-dimensional geometries whose intersection is the multidimensional geometry.

Consider a *short cylinder* of height $a$ and radius $r_o$ initially at a uniform temperature $T_i$. There is no heat generation in the cylinder. At time $t = 0$, the cylinder is subjected to convection from all surfaces to a medium at temperature $T_\infty$ with a heat transfer coefficient $h$. The temperature within the cylinder will change with the axial variable $x$ as well as the radial variable $r$ and time $t$ since heat transfer will occur from the top and bottom of the cylinder as well as its side surfaces. That is,



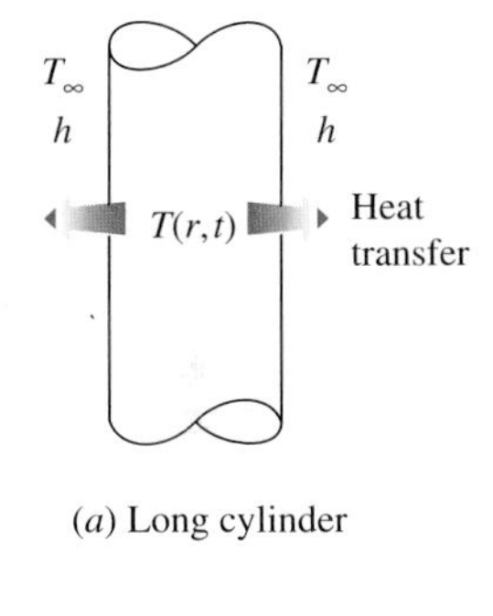


(*a*) Long cylinder

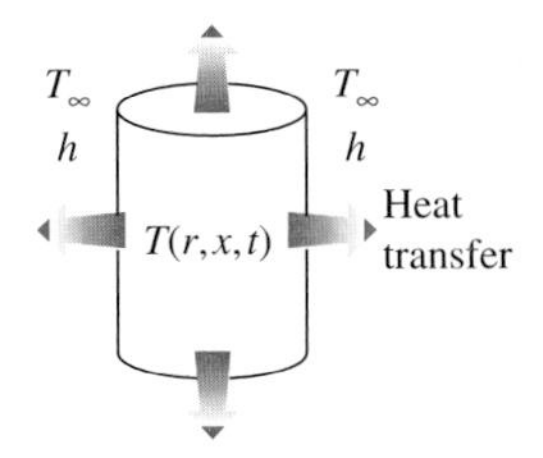


(*b*) Short cylinder (two-dimensional)

**FIGURE 9-25**

The temperature in a short cylinder exposed to convection from all surfaces varies in both the radial and axial directions, and thus heat is transferred in both directions.

$T = T(r, x, t)$ and thus this is a two-dimensional transient heat conduction problem. When the properties are assumed to be constant, it can be shown that the solution of this two-dimensional problem can be expressed as

$$\left(\frac{T(r, x, t) - T_\infty}{T_i - T_\infty}\right)_{\substack{\text{short}\\\text{cylinder}}} = \left(\frac{T(x, t) - T_\infty}{T_i - T_\infty}\right)_{\substack{\text{plane}\\\text{wall}}} \left(\frac{T(r, t) - T_\infty}{T_i - T_\infty}\right)_{\substack{\text{infinite}\\\text{cylinder}}} \quad (9\text{-}25)$$

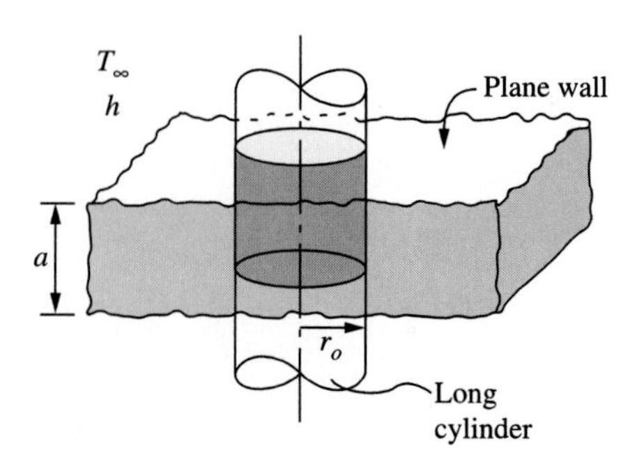


**FIGURE 9-26**
A short cylinder of radius $r_o$ and height $a$ is the *intersection* of a long cylinder of radius $r_o$ and a plane wall of thickness $a$.

That is, the solution for the two-dimensional short cylinder of height $a$ and radius $r_0$ is equal to the *product* of the non-dimensionalized solutions for the one-dimensional plane wall of thickness $a$ and the long cylinder of radius $r_o$, which are the two geometries whose intersection is the short cylinder, as shown in Fig. 9-26. We generalize this as follows: *the solution for a multidimensional geometry is the product of the solutions of the one-dimensional geometries whose intersection is the multidimensional body.*

For convenience, the one-dimensional solutions are denoted by

$$\theta_{\text{wall}}(x, t) = \left(\frac{T(x, t) - T_\infty}{T_i - T_\infty}\right)_{\substack{\text{plane}\\\text{wall}}}$$

$$\theta_{\text{cyl}}(r, t) = \left(\frac{T(r, t) - T_\infty}{T_i - T_\infty}\right)_{\substack{\text{infinite}\\\text{cylinder}}} \quad (9\text{-}26)$$

$$\theta_{\text{seminf}}(x, t) = \left(\frac{T(x, t) - T_\infty}{T_i - T_\infty}\right)_{\substack{\text{semi-infinite}\\\text{solid}}}$$

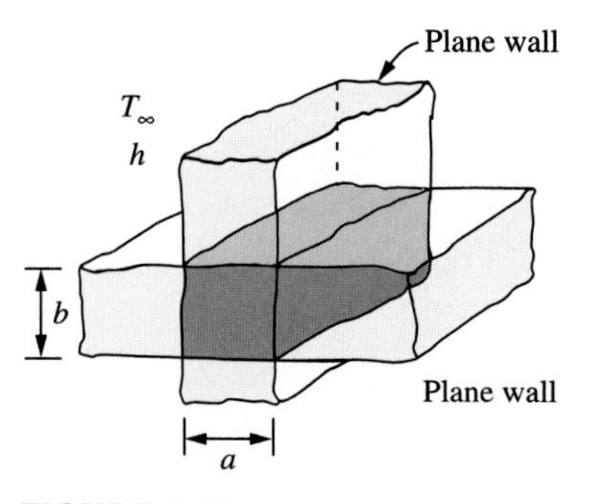


**FIGURE 9-27**
A long solid bar of rectangular profile $a \times b$ is the *intersection* of two plane walls of thicknesses $a$ and $b$.

For example, the solution for a long solid bar whose cross-section is an $a \times b$ rectangle is the intersection of the two infinite plane walls of thicknesses $a$ and $b$, as shown in Fig. 9-27, and thus the transient temperature distribution for this rectangular bar can be expressed as

$$\left(\frac{T(x_1, x_2, t) - T_\infty}{T_i - T_\infty}\right)_{\substack{\text{rectangular}\\\text{bar}}} = \theta_{\text{wall}}(x_1, t)\theta_{\text{wall}}(x_2, t) \quad (9\text{-}27)$$

The proper forms of the product solutions for some other geometries are given in Table 9-4. It is important to note that the $x$-coordinate is measured from the *surface* in a semi-infinite solid, but from the *midplane* in a plane wall. The radial distance $r$ is always measured from the centerline.

Note that the solution of a *two-dimensional* problem involves the product of *two* one-dimensional solutions whereas the solution of a *three-dimensional* problem involves the product of *three* one-dimensional solutions.

A modified form of the product solution can also be used to determine the total transient heat transfer to or from a multidimensional geometry by using the one-dimensional values, as shown by L. S. Langston in 1982. The transient heat transfer for a two-dimensional

**TABLE 9-4**

**Multidimensional solutions expressed as products of one-dimensional solutions for bodies that are initially at a uniform temperature $T_i$ and exposed to convection from all surfaces to a medium at $T_\infty$**

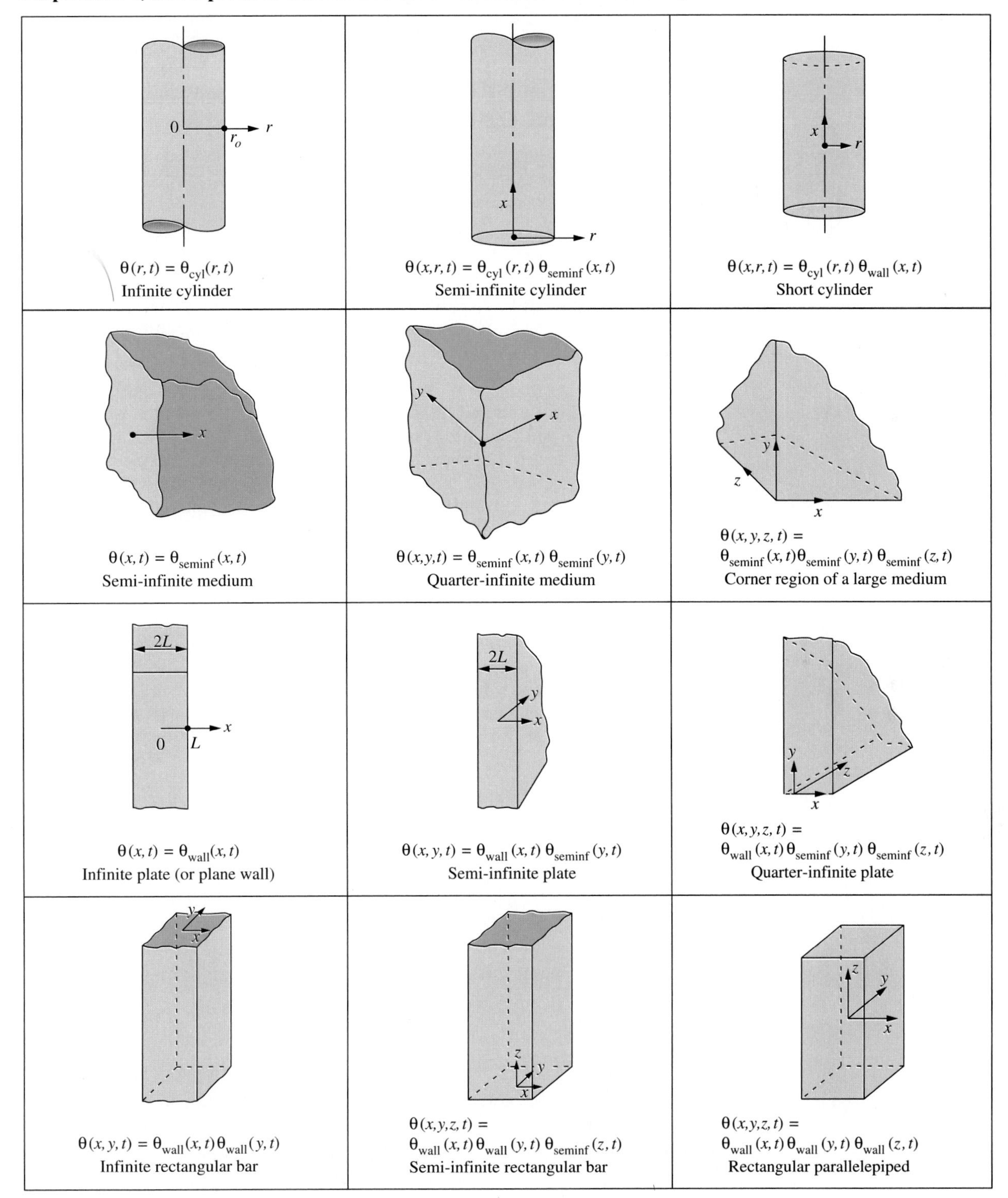

geometry formed by the intersection of two one-dimensional geometries 1 and 2 is

$$\left(\frac{Q}{Q_{\max}}\right)_{\text{total,2D}} = \left(\frac{Q}{Q_{\max}}\right)_1 + \left(\frac{Q}{Q_{\max}}\right)_2\left[1 - \left(\frac{Q}{Q_{\max}}\right)_1\right] \tag{9-28}$$

Transient heat transfer for a three-dimensional body formed by the intersection of three one-dimensional bodies 1, 2, and 3 is given by

$$\left(\frac{Q}{Q_{\max}}\right)_{\text{total,3D}} = \left(\frac{Q}{Q_{\max}}\right)_1 + \left(\frac{Q}{Q_{\max}}\right)_2\left[1 - \left(\frac{Q}{Q_{\max}}\right)_1\right] + \left(\frac{Q}{Q_{\max}}\right)_3\left[1 - \left(\frac{Q}{Q_{\max}}\right)_1\right]\left[1 - \left(\frac{Q}{Q_{\max}}\right)_2\right] \tag{9-29}$$

The use of the product solution in transient two- and three-dimensional heat conduction problems is illustrated in the following examples.

**EXAMPLE 9-7**

A short brass cylinder [$k = 110\ \text{W/(m}\cdot{}^\circ\text{C)}$ and $\alpha = 3.39 \times 10^{-5}\ \text{m}^2/\text{s}$] of diameter $D = 10$ cm and height $H = 12$ cm is initially at a uniform temperature $T_i = 120^\circ$C. The cylinder is now placed in atmospheric air at 25°C, where heat transfer takes place by convection, with a heat transfer coefficient of $h = 60\ \text{W/(m}^2\cdot{}^\circ\text{C)}$. Calculate the temperature at (*a*) the center of the cylinder and (*b*) the center of the top surface of the cylinder 15 min after the start of the cooling.

$T_\infty = 25^\circ$C
$h = 60\ \text{W/(m}^2\cdot{}^\circ\text{C)}$

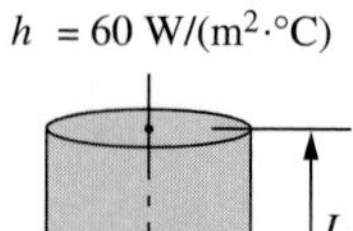
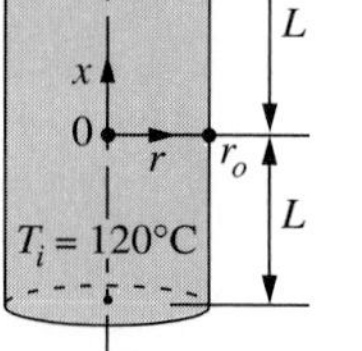


**FIGURE 9-28**
Schematic for Example 9-7.

**Solution** (*a*) This is a two-dimensional transient heat conduction problem, and thus the temperature will vary in both the axial *x*- and the radial *r*-directions within the cylinder, as well as with time *t*. This short cylinder can physically be formed by the intersection of a long cylinder of radius $r_o = \frac{1}{2}D = 5$ cm and a plane wall of thickness $2L = 12$ cm, as shown in Fig. 9-28. Also, we measure *x* from the midplane.

The dimensionless temperature at the center of the plane wall is determined from Fig. 9-13*a* to be

$$\left.\begin{aligned} \tau &= \frac{\alpha t}{L^2} = \frac{(3.39\times10^{-5}\ \text{m}^2/\text{s})(900\ \text{s})}{(0.06\ \text{m})^2} = 8.48 \\ \frac{1}{\text{Bi}} &= \frac{k}{hL} = \frac{110\ \text{W/(m}\cdot{}^\circ\text{C)}}{[60\ \text{W/(m}^2\cdot{}^\circ\text{C)}](0.06\ \text{m})} = 30.6 \end{aligned}\right\} \quad \theta_{\text{wall}}(0,t) = \frac{T(0,t) - T_\infty}{T_i - T_\infty} = 0.8$$

Similarly, at the center of the cylinder, we have

$$\left.\begin{aligned} \tau &= \frac{\alpha t}{r_o^2} = \frac{(3.39\times10^{-5}\ \text{m}^2/\text{s})(900\ \text{s})}{(0.05\ \text{m})^2} = 12.2 \\ \frac{1}{\text{Bi}} &= \frac{k}{hr_o} = \frac{110\ \text{W/(m}\cdot{}^\circ\text{C)}}{[60\ \text{W/(m}^2\cdot{}^\circ\text{C)}](0.05\ \text{m})} = 36.7 \end{aligned}\right\} \quad \theta_{\text{cyl}}(0,t) = \frac{T(0,t) - T_\infty}{T_i - T_\infty} = 0.5$$

Therefore,

$$\left(\frac{T(0,0,t) - T_\infty}{T_i - T_\infty}\right)_{\substack{\text{short}\\\text{cylinder}}} = \theta_{\text{wall}}(0,t) \times \theta_{\text{cyl}}(0,t) = 0.8 \times 0.5 = 0.4$$

and
$$T(0,0,t) = T_\infty + 0.4(T_i - T_\infty) = 25 + 0.4(120 - 25) = 63^\circ\text{C}$$

This is the temperature at the center of the short cylinder, which is also the center of both the long cylinder and the plate.

(*b*) The center of the top surface of the cylinder is still at the center of the long cylinder ($r = 0$), but at the outer surface of the plane wall ($x = L$). Therefore, we first need to find the surface temperature of the wall. Noting that $x = L = 0.06$ m,

$$\left.\begin{aligned} \frac{x}{L} &= \frac{0.06\text{ m}}{0.06\text{ m}} = 1 \\ \frac{1}{\text{Bi}} &= \frac{k}{hL} = \frac{110\text{ W/(m}\cdot{}^\circ\text{C)}}{[60\text{ W/(m}^2\cdot{}^\circ\text{C)}](0.06\text{ m})} = 30.6 \end{aligned}\right\} \quad \frac{T(L,t) - T_\infty}{T_o - T_\infty} = 0.98$$

Then

$$\theta_{\text{wall}}(L,t) = \frac{T(L,t) - T_\infty}{T_i - T_\infty} = \left(\frac{T(L,t) - T_\infty}{T_o - T_\infty}\right)\left(\frac{T_o - T_\infty}{T_i - T_\infty}\right) = 0.98 \times 0.8 = 0.784$$

Therefore,

$$\left(\frac{T(L,0,t) - T_\infty}{T_i - T_\infty}\right)_{\substack{\text{short}\\\text{cylinder}}} = \theta_{\text{wall}}(L,t)\theta_{\text{cyl}}(0,t) = 0.784 \times 0.5 = 0.392$$

and $$T(L,0,t) = T_\infty + 0.392(T_i - T_\infty) = 25 + 0.392(120 - 25) = 62.2^\circ\text{C}$$

which is the temperature at the center of the top surface of the cylinder.

---

**EXAMPLE 9-8**

Determine the total heat transfer from the short brass cylinder [$\rho = 8530$ kg/m$^3$, $C_p = 0.380$ kJ/(kg · °C)] discussed in Example 9-7.

**Solution** We first determine the maximum heat that can be transferred from the cylinder, which is the sensible energy content of the cylinder relative to its environment:

$$m = \rho V = \rho\pi r_o^2 L = (8530\text{ kg/m}^3)\pi(0.05\text{ m})^2(0.06\text{ m}) = 4.02\text{ kg}$$

$$Q_{\max} = mC_p(T_i - T_\infty) = (4.02\text{ kg})[0.380\text{ kJ/(kg}\cdot{}^\circ\text{C)}](120 - 25)^\circ\text{C} = 145.1\text{ kJ}$$

Then we determine the dimensionless heat transfer ratios for both geometries. For the plane wall, it is determined from Fig. 9-13*c* to be

$$\left.\begin{aligned} \text{Bi} &= \frac{1}{1/\text{Bi}} = \frac{1}{30.6} = 0.0327 \\ \frac{h^2\alpha t}{k^2} &= \text{Bi}^2\tau = (0.0327)^2(8.48) = 0.0091 \end{aligned}\right\} \quad \left(\frac{Q}{Q_{\max}}\right)_{\substack{\text{plane}\\\text{wall}}} = 0.23$$

Similarly, for the cylinder, we have

$$\left.\begin{aligned} \text{Bi} &= \frac{1}{1/\text{Bi}} = \frac{1}{36.7} = 0.0272 \\ \frac{h^2\alpha t}{k^2} &= \text{Bi}^2\tau = (0.0272)^2(12.2) = 0.0090 \end{aligned}\right\} \quad \left(\frac{Q}{Q_{\max}}\right)_{\substack{\text{infinite}\\\text{cylinder}}} = 0.47$$

Then the heat transfer ratio for the short cylinder is, from Eq. 9-28,

$$\begin{aligned} \left(\frac{Q}{Q_{\max}}\right)_{\text{short cyl}} &= \left(\frac{Q}{Q_{\max}}\right)_1 + \left(\frac{Q}{Q_{\max}}\right)_2\left[1 - \left(\frac{Q}{Q_{\max}}\right)_1\right] \\ &= 0.23 + 0.47(1 - 0.23) = 0.592 \end{aligned}$$

Therefore, the total heat transfer from the cylinder during the first 15 min of cooling is

$$Q = 0.592Q_{max} = 0.592 \times (145.1 \text{ kJ}) = 85.9 \text{ kJ}$$

**EXAMPLE 9-9**

A semi-infinite aluminum cylinder [$k = 237$ W/(m · °C), $\alpha = 9.71 \times 10^{-5}$ m²/s] of diameter $D = 20$ cm is initially at a uniform temperature $T_i = 200$°C. The cylinder is now placed in water at 15°C where heat transfer takes place by convection, with a heat transfer coefficient of $h = 120$ W/(m² · °C). Determine the temperature at the center of the cylinder 15 cm from the end surface 5 min after the start of the cooling.

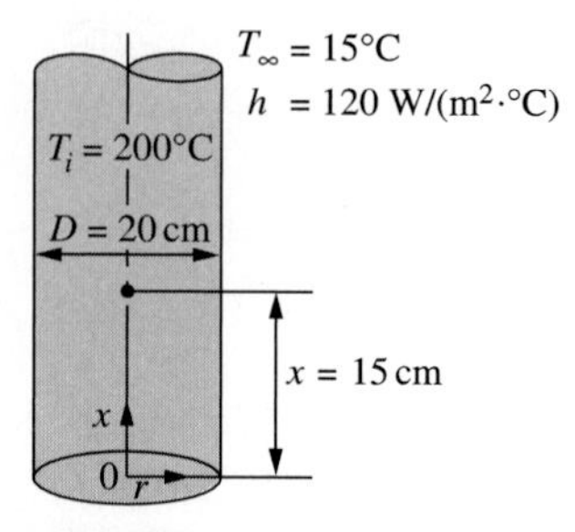


**FIGURE 9-29**
Schematic for Example 9-9.

**Solution** This is a two-dimensional transient heat conduction problem, and thus the temperature will vary in both the axial $x$- and the radial $r$-directions within the cylinder, as well as with time $t$. This semi-infinite cylinder can physically be formed by the intersection of an infinite cylinder of radius $r_o = \frac{1}{2}D = 10$ cm and a semi-infinite medium, as shown in Fig. 9-29. Note that we measure $x$ from the bottom surface of the cylinder.

We will solve this problem using the one-term solution relation for the cylinder, and the analytic solution for the semi-infinite medium. First we consider the infinitely long cylinder and evaluate the Biot number

$$\text{Bi} = \frac{hr_o}{k} = \frac{[120 \text{ W/(m}^2 \cdot \text{°C)}](0.1 \text{ m})}{237 \text{ W/(m} \cdot \text{°C)}} = 0.05$$

The coefficients $\lambda_1$ and $A_1$ for a cylinder corresponding to this Bi are determined from Table 9-1 to be $\lambda_1 = 0.3126$ and $A_1 = 1.0124$. Substituting these values into Eq. 9-14 gives

$$\theta_0 = \theta_{cyl}(0, t) = A_1 e^{-\lambda_1^2 \tau} = 1.0124 e^{-(0.3126)^2(2.91)} = 0.762$$

Note that the Fourier number in this case is

$$\tau = \frac{\alpha t}{r_o^2} = \frac{(9.71 \times 10^{-5} \text{ m}^2\text{/s})(1 \times 60 \text{ s})}{(0.1 \text{ m})^2} = 2.91 > 0.2$$

and thus the one-term approximation is applicable. The solution for the semi-infinite solid can be determined from

$$1 - \theta_{\text{semi-inf}}(x, t) = \text{erfc}\left(\frac{x}{2\sqrt{\alpha t}}\right) - \exp\left(\frac{hx}{k} + \frac{h^2 \alpha t}{k^2}\right)\left[\text{erfc}\left(\frac{x}{2\sqrt{\alpha t}} + \frac{h\sqrt{\alpha t}}{k}\right)\right]$$

First we determine the various quantities in parentheses:

$$\xi = \frac{x}{2\sqrt{\alpha t}} = \frac{0.15 \text{ m}}{2\sqrt{(9.71 \times 10^{-5} \text{ m}^2\text{/s})(5 \times 60 \text{ s})}} = 0.88$$

$$\frac{h\sqrt{\alpha t}}{k} = \frac{[120 \text{ W/(m}^2 \cdot \text{°C)}]\sqrt{(9.71 \times 10^{-5} \text{ m}^2\text{/s})(900 \text{ s})}}{237 \text{ W/(m} \cdot \text{°C)}} = 0.086$$

$$\frac{hx}{k} = \frac{[120 \text{ W/(m}^2 \cdot \text{C)}](0.51 \text{ m})}{237 \text{ W(m} \cdot \text{°C)}} = 0.0759$$

$$\frac{h^2 \alpha t}{k^2} = \left(\frac{h\sqrt{\alpha t}}{k}\right)^2 = (0.086)^2 = 0.0074$$

Substituting and evaluating the complimentary error functions from Table 9-3,

$$\begin{aligned}\theta_{\text{semi-inf}}(x, t) &= 1 - \text{erfc}\,(0.88) + \exp\,(0.0759 + 0.0074)\,\text{erfc}\,(0.88 + 0.086)\\ &= 1 - 0.2133 + \exp\,(0.0833) \times 0.1720\\ &= 0.974\end{aligned}$$

Now we apply the product solution to get

$$\left(\frac{T(x, 0, t) - T_\infty}{T_i - T_\infty}\right)_{\substack{\text{semi-infinite}\\ \text{cylinder}}} = \theta_{\text{semi-inf}}(x, t)\theta_{\text{cyl}}(0, t) = 0.974 \times 0.762 = 0.742$$

and $T(x, 0, t) = T_\infty + 0.742(T_i - T_\infty) = 15 + 0.742(200 - 15) = 152°C$

which is the temperature at the center of the cylinder 15 cm from the exposed bottom surface.

**EXAMPLE 9-10**

In a meat processing plant, 1-in-thick steaks initially at 75°F are to be cooled in the racks of a large refrigerator that is maintained at 5°F. The steaks are placed close to each other, so that heat transfer from the 1-in-thick edges is negligible. The entire steak is to be cooled below 45°F, but its temperature is not to drop below 35°F at any point during refrigeration, to avoid "frostbite". The convection heat transfer coefficient and thus the rate of heat transfer from the steak can be controlled by varying the speed of a circulating fan inside. Determine the heat transfer coefficient $h$ that will enable us to meet both temperature constraints while keeping the refrigeraion time to a minimum. The steak can be treated as a homogeneous layer having the properties $\rho = 74.9\ \text{lbm/ft}^3$, $C_p = 0.98\ \text{Btu/(lbm} \cdot \text{°F)}$, $k = 0.26\ \text{Btu/(h} \cdot \text{ft} \cdot \text{°F)}$, and $\alpha = 0.0035\ \text{ft}^2\text{/h}$.

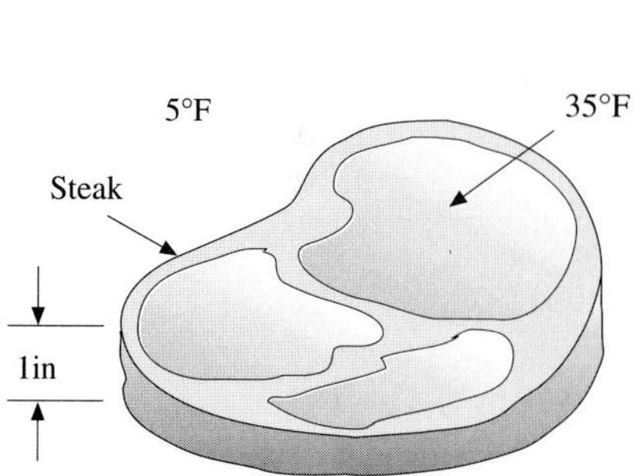


**FIGURE 9-30**
Schematic for Example 9-10.

**Solution** The steaks form a plane layer of uniform thickness, and the heat transfer from the edges is said to be negligible. Therefore, we can treat this problem as a one-dimensional transient heat conduction problem in a plane wall of thickness $2L = 1$ in.

The lowest temperature in the steak will occur at the surfaces and the highest temperature at the center at a given time, since the inner part will be last place to be cooled. In the limiting case, the surface temperature at $x = L = 0.5$ in from the center will be 35°F, while the midplane temperature is 45°F in an environment at 5°F. Then, from Fig. 9-13$b$, we obtain

$$\left.\begin{aligned}\frac{x}{L} &= \frac{0.5\ \text{in.}}{0.5\ \text{in.}} = 1\\ \frac{T(L, t) - T_\infty}{T_o - T_\infty} &= \frac{35 - 5}{45 - 5} = 0.75\end{aligned}\right\} \quad \frac{1}{\text{Bi}} = \frac{k}{hL} = 1.5$$

which gives

$$h = \frac{1}{1.5}\frac{k}{L} = \frac{0.26\ \text{Btu/(h} \cdot \text{ft} \cdot \text{°F)}}{1.5(0.5/12\ \text{ft})} = 4.16\ \text{Btu/(h} \cdot \text{ft}^2 \cdot \text{°F)}$$

Therefore, the convection heat transfer coefficient should be kept below this value to satisfy the constraints on the temperature of the steak during refrigeration. We can also meet the constraints by using a lower heat transfer coefficient, but doing so would extend the refrigeration time unnecessarily.

## 9-5 SUMMARY

In this chapter we considered the variation of temperature with time as well as position in one or multidimensional systems. We first considered the *lumped systems* in which the temperature varies with time but remains uniform throughout the system at any time. The temperature of a lumped body of arbitrary shape of mass $m$, volume $V$, surface area $A_s$, density $\rho$, and specific heat $C_p$ initially at a uniform temperature $T_i$ that is exposed to convection at time $t = 0$ in a medium at temperature $T_\infty$ with a heat transfer coefficient $h$ is expressed as

$$\frac{T(t) - T_\infty}{T_i - T_\infty} = e^{-bt}$$

where
$$b = \frac{hA}{\rho C_p V} = \frac{h}{\rho C_p L_c} \quad (1/\text{s})$$

is a positive quantity whose dimension is $(\text{time})^{-1}$. This relation can be used to determine the temperature $T(t)$ of a body at time $t$, or alternately, the time $t$ required for the temperature to reach a specified value $T(t)$. Once the temperature $T(t)$ at time $t$ is available, the *rate* of convection heat transfer between the body and its environment at that time can be determined from Newton's law of cooling as

$$\dot{Q}(t) = hA[T(t) - T_\infty] \quad (\text{W})$$

The *total amount* of heat transfer between the body and the surrounding medium over the time interval $t = 0$ to $t$ is simply the change in the energy content of the body,

$$Q = mC_p[T(t) - T_i] \quad (\text{kJ})$$

The amount of heat transfer reaches its upper limit when the body reaches the surroundings temperature $T_\infty$. Therefore, the *maximum* heat transfer between the body and its surroundings is

$$Q_{\text{max}} = mC_p(T_\infty - T_i) \quad (\text{kJ})$$

The error involved in lumped system analysis is negligible when

$$\text{Bi} = \frac{hL_c}{k} < 0.1$$

where Bi is the *Biot number* and $L_c = V/A$ is the *characteristic length*.

When the lumped system analysis is not applicable, the variation of temperature with position as well as time can be determined using the *transient temperature charts* given in Figs. 9-13, 9-14, 9-15, and 9-23 for a large plane wall, long cylinder, a sphere, and a semi-infinite medium, respectively. These charts are applicable for one-dimensional heat transfer in those geometries. Therefore, their use is limited to situations in which the body is initially at a uniform temperature, all surfaces are subjected to the same thermal conditions, and the body does not involve any heat generation. These charts can also be used to determine the total heat transfer from the body up to a specified time $t$.

Using a *one-term approximation,* the solution of one-dimensional transient heat conduction problems are expressed analytically as

Plane wall: $$\theta(x, t)_{\text{wall}} = \frac{T(x, t) - T_\infty}{T_i - T_\infty} = A_1 e^{-\lambda_1^2 \tau} \cos(\lambda_1 x/L), \quad \tau > 0.2$$

Cylinder: $$\theta(x, t)_{\text{cyl}} = \frac{T(r, t) - T_\infty}{T_i - T_\infty} = A_1 e^{-\lambda_1^2 \tau} J_0(\lambda_1 r/r_0), \quad \tau > 0.2$$

Sphere: $$\theta(x, t)_{\text{sph}} = \frac{T(r, t) - T_\infty}{T_i - T_\infty} = A_1 e^{-\lambda_1^2 \tau} \frac{\sin(\lambda_1 r/r_0)}{\lambda_1 r/r_0}, \quad \tau > 0.2$$

where the constants $A_1$ and $\lambda_1$ are functions of the Bi number only, and their values are listed in Table 9-1 against the Bi number for all three geometries. The error involved in one-term solutions is less than 2 percent when $\tau > 0.2$.

Using the one-therm solutions, the fractional heat transfers in different geometries are expressed as

Plane wall: $$\left(\frac{Q}{Q_{\max}}\right)_{\text{wall}} = 1 - \theta_{0,\text{wall}} \frac{\sin \lambda_1}{\lambda_1}$$

Cylinder: $$\left(\frac{Q}{Q_{\max}}\right)_{\text{cyl}} = 1 - 2\theta_{0,\text{cyl}} \frac{J_1(\lambda_1)}{\lambda_1}$$

Sphere: $$\left(\frac{Q}{Q_{\max}}\right)_{\text{sph}} = 1 - 3\theta_{0,\text{cyl}} \frac{\sin \lambda_1 - \lambda_1 \cos \lambda_1}{\lambda_1^3}$$

The analytic solution for one-dimensional transient heat conduction in a semi-infinite solid subjected to convection is given by

$$\frac{T(x, t) - T_i}{T_\infty - T_i} = \text{erfc}\left(\frac{x}{2\sqrt{\alpha t}}\right) - \exp\left(\frac{hx}{k} + \frac{h^2 \alpha t}{k^2}\right)\left[\text{erfc}\left(\frac{x}{2\sqrt{\alpha t}} + \frac{h\sqrt{\alpha t}}{k}\right)\right]$$

where the quantity erfc $(\xi)$ is the *complementary error function.* For the special case of $h \to \infty$, the surface temperature $T_s$ becomes equal to the fluid temperature $T_\infty$, and the above equation reduces to

$$\frac{T(x, t) - T_i}{T_s - T_i} = \text{erfc}\left(\frac{x}{2\sqrt{\alpha t}}\right) \quad (T_s = \text{constant})$$

Using a clever superposition principle called the *product solution,* these charts can also be used to construct solutions for the *two-dimensional* transient heat conduction problems encountered in geometries such as a short cylinder, a long rectangular bar, a semi-infinite cylinder or plate, and even *three-dimensional* problems associated with geometries such as a rectangular prism or a semi-infinite rectangular bar, provided that all surfaces of the solid are subjected to convection to the same fluid at temperature $T_\infty$, with the same convection heat transfer coefficient $h$, and the body involves no heat generation. The solution in such multidimensional geometries can be expressed as the product of the solutions for the one-dimensional geometries whose intersection is the multidimensional geometry.

The total heat transfer to or from a multidimensional geometry can also be determined by using the one-dimensional values. The transient heat transfer for a two-dimensional geometry formed by the intersection of two one-dimensional geometries 1 and 2 is

$$\left(\frac{Q}{Q_{\max}}\right)_{\text{total,2D}} = \left(\frac{Q}{Q_{\max}}\right)_1 + \left(\frac{Q}{Q_{\max}}\right)_2\left[1 - \left(\frac{Q}{Q_{\max}}\right)_1\right]$$

Transient heat transfer for a three-dimensional body formed by the intersection of three one-dimensional bodies 1, 2, and 3 is given by

$$\left(\frac{Q}{Q_{\max}}\right)_{\text{total,3D}} = \left(\frac{Q}{Q_{\max}}\right)_1 + \left(\frac{Q}{Q_{\max}}\right)_2\left[1 - \left(\frac{Q}{Q_{\max}}\right)_1\right] + \left(\frac{Q}{Q_{\max}}\right)_3\left[1 - \left(\frac{Q}{Q_{\max}}\right)_1\right]\left[1 - \left(\frac{Q}{Q_{\max}}\right)_2\right]$$

## REFERENCES AND SUGGESTED READING

**1** Y. Bayazitoglu and M. N. Özışık, *Elements of Heat Transfer,* McGraw-Hill, New York, 1988.

**2** H. S. Carslaw and J. C. Jaeger, *Conduction of Heat in Solids,* 2nd ed., Oxford University Press, London, 1959.

**3** H. Gröber, S. Erk, and U. Grigull, *Fundamentals of Heat Transfer,* McGraw-Hill, New York, 1961.

**4** M. P. Heisler, "Temperature Charts for Induction and Constant Temperature Heating," *ASME Transactions,* Vol. 69, pp. 227–236, 1947.

**5** J. P. Holman, *Heat Transfer,* 7th ed., McGraw-Hill, New York, 1990.

**6** F. P. Incropera and D. P. DeWitt, *Introduction to Heat Transfer,* 2nd ed., Wiley, 1990.

**7** M. Jakob, *Heat Transfer,* Vol. 1, Wiley, New York, 1949.

**8** F. Kreith and M. S. Bohn, *Principles of Heat Transfer,* 5th ed., West, St. Paul, MN, 1993.

**9** L. S. Langston, "Heat Transfer from Multidimensional Objects Using One-Dimensional Solutions for Heat Loss," *International Journal of Heat and Mass Transfer,* Vol. 25, pp. 149–150, 1982.

**10** M. N. Özışık, *Heat Transfer—A Basic Approach,* McGraw-Hill, New York, 1985.

**11** P. J. Schneider, *Conduction Heat Transfer,* Addison-Wesley, Reading, Mass., 1955.

**12** L. C. Thomas, *Heat Transfer,* Prentice-Hall, Englewood Cliffs, NJ, 1992.

**13** F. M. White, *Heat and Mass Transfer,* Addison-Wesley, Reading, MA, 1988.

### Lumped System Analysis

**9-1C** How does transient heat conduction differ from steady conduction? How does two-dimensional heat transfer problems differ from one-dimensional ones?

**9-2C** What is lumped system analysis? When is it applicable?

**9-3C** Consider heat transfer between two identical hot solid bodies and the air surrounding them. The first solid is being cooled by a fan while the second one is allowed to cool naturally. For which solid is the lumped system analysis more likely to be applicable? Why?

**9-4C** Consider heat transfer between two identical hot solid bodies and their environments. The first solid is dropped in a large container filled with water, while the second one is allowed to cool naturally in the air. For which solid is the lumped system analysis more likely to be applicable? Why?

**FIGURE P9-5C**

**9-5C** Consider a hot baked potato on a plate. The temperature of the potato is observed to drop by 4°C during the first minute. Will the temperature drop during the second minute be less than, equal to, or more than 4°C? Why?

**9-6C** Consider a potato being baked in an oven that is maintained at a constant temperature. The temperature of the potato is observed to rise by 5°C during the first minute. Will the temperature rise during the second minute be less than, equal to, or more than 5°C? Why?

**9-7C** What is the physical significance of the Biot number? Is the Biot number more likely to be larger for highly conducting solids or poorly conducting ones?

**9-8C** Consider two identical 4-kg pieces of roast beef. The first piece is baked as a whole, while the second is baked after being cut into two equal pieces in the same oven. Will there be any difference between the cooking times of the whole and cut roasts? Why?

**9-9C** Consider a sphere and a cylinder of equal volume made of copper. Both the sphere and the cylinder are initially at the same temperature, and are exposed to convection in the same environment. Which do you think will cool faster, the cylinder or the sphere? Why?

**9-10C** In what medium is the lumped system analysis more likely to be applicable: in water or in air? Why?

**9-11C** For which solid is the lumped system analysis more likely to be applicable: an actual apple or a golden apple of the same size? Why?

**9-12C** For which kind of bodies made of the same material is the lumped system analysis more likely to be applicable: slender ones or well-rounded ones of the same volume? Why?

* Students are encouraged to answer *all* the concept "C" questions.

**9-13** Obtain relations for the characteristic lengths of a large plane wall of thickness $2L$, a very long cylinder of radius $r_o$, and a sphere of radius $r_o$.

**9-14** Obtain a relation for the time required for a lumped system to reach the average temperature $\frac{1}{2}(T_i + T_\infty)$, where $T_i$ is the initial temperature and $T_\infty$ is the temperature of the environment.

**9-15** The temperature of a gas stream is to be measured by a thermocouple whose junction can be approximated as a 1.2-mm-diameter sphere. The properties of the junction are $k = 35\ \text{W/(m} \cdot {}^\circ\text{C)}$, $\rho = 8500\ \text{kg/m}^3$, and $C_p = 320\ \text{J/(kg} \cdot {}^\circ\text{C)}$, and the heat transfer coefficient between the junction and the gas is $h = 65\ \text{W/(m}^2 \cdot {}^\circ\text{C)}$. Determine how long it will take for the thermocouple to read 99 percent of the initial temperature difference. *Answer:* 38.5 s

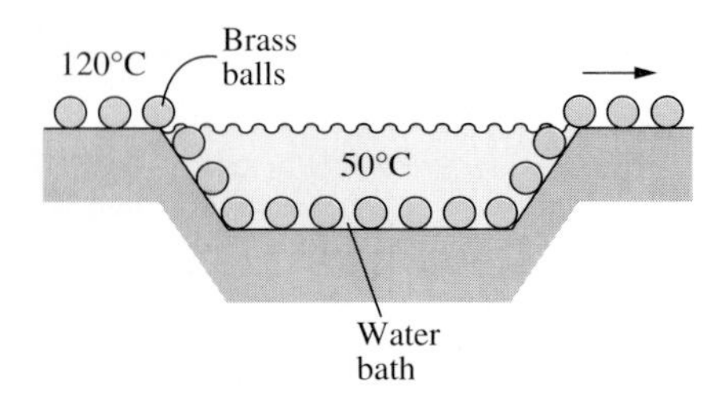

FIGURE P9-16

**9-16** In a manufacturing facility, 5-cm-diameter brass balls [$k = 111\ \text{W/(m} \cdot {}^\circ\text{C)}$, $\rho = 8522\ \text{kg/m}^3$, and $C_p = 0.385\ \text{kJ/(kg} \cdot {}^\circ\text{C)}$] initially at 120°C are quenched in a water bath at 50°C for a period of 2 min at a rate of 100 balls per minute. If the convection heat transfer coefficient is $240\ \text{W/(m}^2 \cdot {}^\circ\text{C)}$, determine (*a*) the temperature of the balls after quenching and (*b*) the rate at which heat needs to be removed from the water in order to keep its temperature constant at 50°C.

**9-16E** In a manufacturing facility, 2-in-diameter brass balls [$k = 64.1\ \text{Btu/(h} \cdot \text{ft} \cdot {}^\circ\text{F)}$, $\rho = 532\ \text{lbm/ft}^3$, and $C_p = 0.092\ \text{But/(lbm} \cdot {}^\circ\text{F)}$] initially at 250°F are quenched in a water bath at 120°F for a period of 2 min at a rate of 100 balls per minute. If the convection heat transfer coefficient is $42\ \text{Btu/(h} \cdot \text{ft}^2 \cdot {}^\circ\text{F)}$, determine (*a*) the temperature of the balls after quenching and (*b*) the rate at which heat needs to be removed from the water in order to keep its temperature constant at 120°F.

**9-17** Repeat Prob. 9-16 for aluminum balls.

**9-18** To warm up some milk for a baby, a mother pours in milk into a thin walled glass whose diameter is 6 cm. The height of the milk in the glass is 7 cm. She then places the glass into a large pan filled with hot water at 60°C. The milk is stirred constantly, so that its temperature is uniform at all times. If the heat transfer coefficient between the water and the glass is $120\ \text{W/(m}^2 \cdot {}^\circ\text{C)}$, determine how long it will take for the milk to warm up from 3°C to 38°C. Take the properties of the milk to be the same as those of water. Can the milk in this case be treated as a lumped system? Why? *Answer:* 5.9 min

**9-19** Repeat Prob. 9-18 for the case of water also being stirred, so that the heat transfer coefficient is doubled to $240\ \text{W/(m}^2 \cdot {}^\circ\text{C)}$.

FIGURE P9-20

**9-20** During a picnic on a hot summer day all the cold drinks disappeared quickly, and the the only available drinks were those at the ambient temperature of 25°C. In an effort tó cool a 335-ml drink in a can, which is 12.5 cm high and has a diameter of 6.5 cm, a person grabs the can and starts shaking it in the iced water of the chest at 0°C. The temperature of the drink can be assumed to be uniform at all times, and the heat transfer coefficient between the iced water and the aluminum can is 170 W/(m$^2 \cdot$ °C). Using the properties of water for the drink, estimate how long it will take for the canned drink to cool to 8°C.

**9-20E** During a picnic on a hot summer day, all the cold drinks disappeared quickly, and the only available drinks were those at the ambient temperature of 75°F. In an effort to cool a 12-fluid-oz drink in a can, which is 5 in. high and has a diameter of 2.5 in., a person grabs the can and starts shaking it in the iced water of the chest at 32°F. The temperature of the drink can be assumed to be uniform at all times, and the heat transfer coefficient between the iced water and the aluminum can is 30 Btu/(h $\cdot$ ft$^2 \cdot$ °F). Using the properties of water for the drink, estimate how long it will take for the canned drink to cool to 45°F.

**9-21** Consider a 1000-W iron whose base plate is made of 0.5-cm-thick aluminum alloy 2024-T6 [$\rho$ = 2770 kg/m$^3$, $C_p$ = 875 J/(kg $\cdot$ °C), $\alpha$ = 7.3 × 10$^{-5}$ m$^2$/s]. The base plate has a surface area of 0.03 m$^2$. Initially, the iron is in thermal equilibrium with the ambient air at 22°C. Taking the heat transfer coefficient at the surface of the base plate to be 12 W/(m$^2 \cdot$ °C) and assuming 85 percent of the heat generated in the resistance wires is transferred to the plate, determine how long it will take for the plate temperature to reach 140°C. Is it realistic to assume the plate temperature to be uniform at all times?

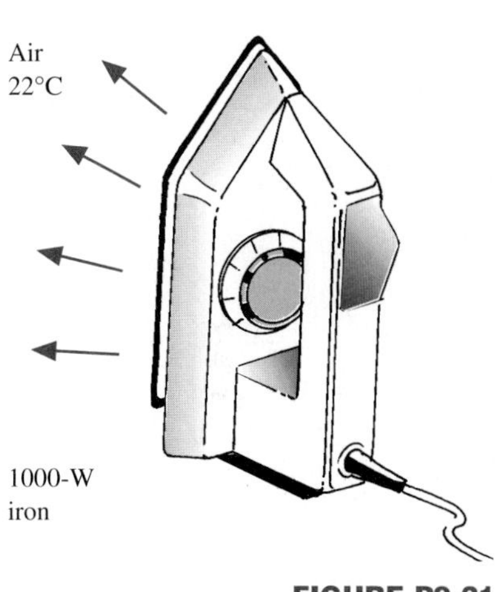


FIGURE P9-21

**9-22** Stainless steel ball bearings [$\rho$ = 8085 kg/m$^3$, $k$ = 15.1 W/(m $\cdot$ °C), $C_p$ = 0.480 kJ/(kg $\cdot$ °C), and $\alpha$ = 3.91 × 10$^{-6}$ m$^2$/s] having a diameter of 1.2 cm are to be quenched in water. The balls leave the oven at a uniform temperature of 900°C, and are exposed to air at 30°C for a while before they are dropped into the water. If the temperature of the balls are not to fall below 850°C prior to quenching and the heat transfer coefficient in the air is 125 W/(m$^2 \cdot$ °C), determine how long they can stand in the air before being dropped into the water. *Answer:* 3.7 s

**9-23** Carbon steel balls [$\rho$ = 7833 kg/m$^3$, $k$ = 54 W/(m $\cdot$ °C), $C_p$ = 0.465 kJ/(kg $\cdot$ °C), and $\alpha$ = 1.474 × 10$^{-6}$ m$^2$/s] 8 mm in diameter are annealed by heating them first to 900°C in a furnace, and then allowing them to cool slowly to 100°C in ambient air at 35°C. If the average heat transfer coefficient is 75 W/(m$^2 \cdot$ °C), determine how long the annealing process will take. If 2500 balls are to be annealed per hour, determine the total rate of heat transfer from the balls to the ambient air.

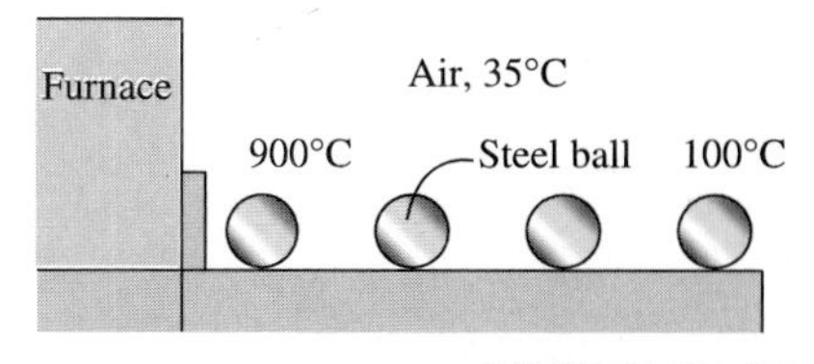


FIGURE P9-23

**9-23E** Carbon steel balls [$\rho$ = 489 lbm/ft$^3$, $k$ = 31.2 Btu/(h $\cdot$ ft $\cdot$ °F), $C_p$ = 0.111 Btu/(lbm $\cdot$ °F), and $\alpha$ = 0.0571 ft$^2$/h] 0.4 in. in diameter are annealed by heating them first to 1700°F in a furnace, and then allowing them to cool slowly to 200°F in ambient air at 95°F. If the average heat transfer coefficient is 15 Btu/(h $\cdot$ ft$^2 \cdot$ °F), determine how long the annealing process will take. If 2500 balls are to be annealed per hour, determine the total rate of heat transfer from the balls to the ambient air.

**9-24** An electronic device dissipating 30 W has a mass of 20 g, a specific heat of 850 J/(kg $\cdot$ °C), and a surface area of 5 cm$^2$. The device is lightly used, and it is on for 5 min and then off for several hours, during which it cools to the ambient temperature of 25°C. Taking the heat transfer coefficient to be 12 W/(m$^2\cdot$°C), determine the temperature of the device at the end of the 5-min operating period. What would your answer be if the device were attached to an aluminum heat sink which has a mass of 200 g and a surface area of 50 cm$^2$? Assume the device and the heat sink to be nearly isothermal.

### Transient Heat Conduction in Large Plane Walls, Long Cylinders, and Spheres

**9-25C** What is an infinitely long cylinder? When is it proper to treat an actual cylinder as being infinitely long, and when is it not? For example, is it proper to use this model when finding the temperatures near the bottom or top surfaces of a cylinder? Explain.

**9-26C** Can the transient temperature charts in Fig. 9-13 for a plane wall exposed to convection on both sides be used for a plane wall whose one side is exposed to convection while the other side is insulated? Explain.

**9-27C** Why are the transient temperature charts prepared using non-dimensionalized quantities such as the Biot and Fourier numbers instead of the actual variables such as thermal conductivity and time?

**9-28C** What is the physical significance of the Fourier number? Will the Fourier number for a specified heat transfer problem double when the time is doubled?

**9-29C** How can we use the transient temperature charts when the surface temperature of the geometry is specified instead of the temperature of the surrounding medium and the convection heat transfer coefficient?

**9-30C** A body at an initial temperature of $T_i$ is brought into a medium at a constant temperature of $T_\infty$. How can you determine the maximum possible amount of heat transfer between the body and the surrounding medium?

**9-31C** The Biot number during a heat transfer process between a sphere and its surroundings is determined to be 0.02. Would you use lumped system analysis or the transient temperature charts when determining the midpoint temperature of the sphere? Why?

**9-32** A student calculates that the total heat transfer from a spherical copper ball of diameter 15 cm initially at 200°C and its environment at a constant temperature of 25°C during the first 20 min of cooling is 4200 kJ. Is this result reasonable? Why?

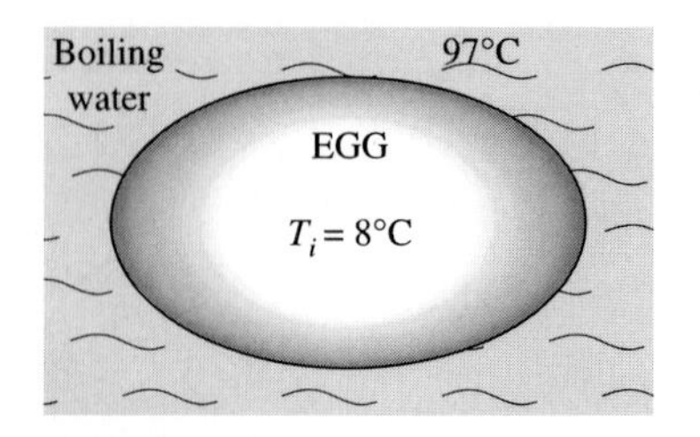


**FIGURE P9-33**

**9-33** An ordinary egg can be approximated as a 5.5-cm-diameter sphere whose properties are roughly those of water at room temperature [$k = 0.6$ W/(m$\cdot$°C) and $\alpha = 0.14\times10^{-6}$ m$^2$/s]. The egg is initially at a uniform temperature of 8°C, and is dropped into boiling water at 97°C.

Taking the convection heat transfer coefficient to be $h = 1400\ \text{W/(m}^2 \cdot {}^\circ\text{C)}$, determine how long it will take for the center of the egg to reach 70°C.

**9-34** In a production facility, 3-cm-thick large brass plates [$k = 110\ \text{W/(m} \cdot {}^\circ\text{C)}$, $\rho = 8530\ \text{kg/m}^3$, $C_p = 380\ \text{J/(kg} \cdot {}^\circ\text{C)}$, and $\alpha = 33.9 \times 10^{-6}\ \text{m}^2\text{/s}$] that are initially at a uniform temperature of 25°C are heated by passing them through an oven maintained at 700°C. The plates remain in the oven for a period of 10 min. Taking the convection heat transfer coefficient to be $h = 80\ \text{W/(m}^2 \cdot {}^\circ\text{C)}$, determine the surface temperature of the plates when they come out of the oven.

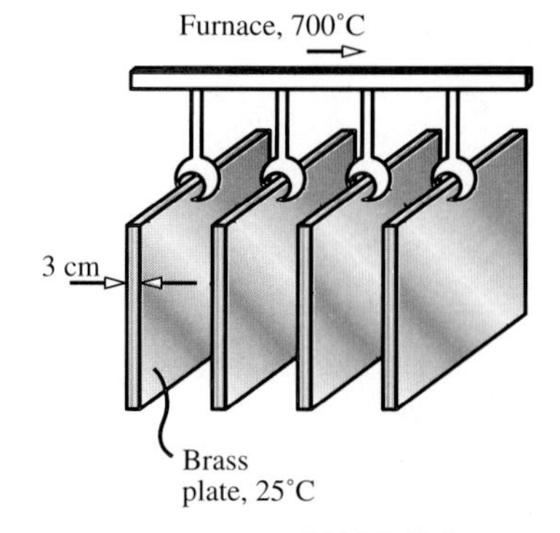

FIGURE P9-34

**9-34E** In a production facility, 1.2-in-thick large brass plates [$k = 63.5\ \text{Btu/(h} \cdot \text{ft} \cdot {}^\circ\text{F)}$, $\rho = 532.5\ \text{lbm/ft}^3$, $C_p = 0.091\ \text{Btu/(lbm} \cdot {}^\circ\text{F)}$, and $\alpha = 1.31\ \text{ft}^2\text{/h}$] that are initially at a uniform temperature of 75°F are heated by passing them through an oven that is maintained at 1300°F. The plates remain in the oven for a period of 10 min. Taking the convection heat transfer coefficient to be $h = 20\ \text{Btu/(h} \cdot \text{ft}^2 \cdot {}^\circ\text{F)}$, determine the surface temperature of the plates when they come out of the oven.

**9-35** A long 35-cm-diameter cylindrical shaft made of stainless steel 304 [$k = 14.9\ \text{W/(m} \cdot {}^\circ\text{C)}$, $\rho = 7900\ \text{kg/m}^3$, $C_p = 477\ \text{J/(kg} \cdot {}^\circ\text{C)}$, and $\alpha = 3.95 \times 10^{-6}\ \text{m}^2\text{/s}$] comes out of an oven at a uniform temperature of 400°C. The shaft is then allowed to cool slowly in a chamber at 150°C with an average convection heat transfer coefficient of $h = 60\ \text{W/(m}^2 \cdot {}^\circ\text{C)}$. Determine the temperature at the center of the shaft 20 min after the start of the cooling process. Also determine the heat transfer per unit length of the shaft during this time period.
*Answers:* 390°C, 15,680 kJ

**9-36** Long cylindrical AISI stainless steel rods [$k = 13.4\ \text{W/(m} \cdot {}^\circ\text{C)}$ and $\alpha = 3.48 \times 10^{-6}\ \text{m}^2\text{/s}$] of 10-cm diameter are heat treated by drawing them at a velocity of 3 m/min through a 9-m-long oven maintained at 900°C. The heat transfer coefficient in the oven is $90\ \text{W/(m}^2 \cdot {}^\circ\text{C)}$. If the rods enter the oven at 30°C, determine their centerline temperature when they leave.

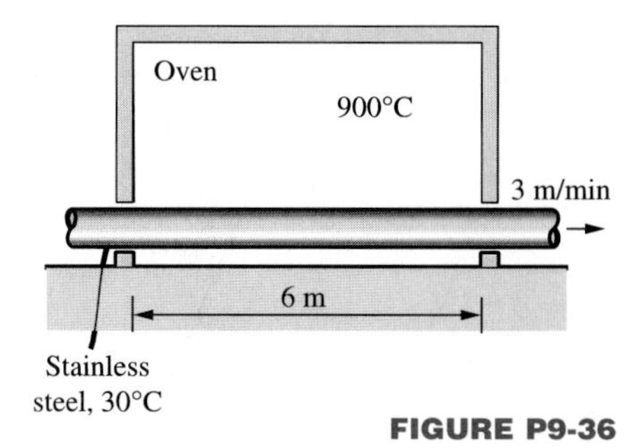


FIGURE P9-36

**9-36E** Long cylindrical AISI stainless steel rods [$k = 7.74\ \text{Btu/(h} \cdot \text{ft} \cdot {}^\circ\text{F)}$ and $\alpha = 0.135\ \text{ft}^2\text{/h}$] of 4-in diameter are heat-treated by drawing them at a velocity of 10 ft/min through a 30-ft-long oven maintained at 1700°F. The heat transfer coefficient in the oven is $20\ \text{Btu/(h} \cdot \text{ft}^2 \cdot {}^\circ\text{F)}$. If the rods enter the oven at 85°F, determine their centerline temperature when they leave.

**9-37** In a meat processing plant, 2-cm-thick steaks [$k = 0.45\ \text{W/(m} \cdot {}^\circ\text{C)}$ and $\alpha = 0.91 \times 10^{-7}\ \text{m}^2\text{/s}$] that are initially at 25°C are to be cooled by passing them through a refrigeration room at −10°C. The heat transfer coefficient on both sides of the steaks is $9\ \text{W/(m}^2 \cdot {}^\circ\text{C)}$. If both surfaces of the steaks are to be cooled to 3°C, determine how long the stakes should be kept in the refrigeration room.

**9-38** A long cylindrical wood log [$k = 0.17\ \text{W/(m} \cdot {}^\circ\text{C)}$ and $\alpha = 1.28 \times$

$10^{-7}$ m$^2$/s] is 10 cm in diameter and is initially at a uniform temperature of 10°C. It is exposed to hot gases at 700°C in a fireplace with a heat transfer coefficient of 40 W/(m$^2$ · °C) on the surface. If the ignition temperature of the wood is 420°C, determine how long it will be before the log ignites.

**9-39** In Betty Crocker's Cookbook, it is stated that it takes 2 h 45 min to roast a 3.2-kg rib initially at 4.5°C "rare" in an oven maintained at 163°C. It is recommended that a meat thermometer be used to monitor the cooking, and the rib is considered rare done when the thermometer inserted into the center of the thickest part of the meat registers 60°C. The rib can be treated as a homogeneous spherical object with the properties $\rho$ = 1200 kg/m$^3$, $C_p$ = 4.1 kJ/(kg · °C), $k$ = 0.45 W/(m · °C), and $\alpha = 0.91 \times 10^{-7}$ m$^2$/s. Determine (*a*) the heat transfer coefficient at the surface of the rib, (*b*) the temperature of the outer surface of the rib when it is done, and (*c*) the amount of heat transferred to the rib. (*d*) Using the values obtained, predict how long it will take to roast this rib to "medium" level, which occurs when the innermost temperature of the rib reaches 71°C. Compare your result to the listed value of 3 h 20 min.

If the roast rib is to be set on the counter for about 15 min before it is sliced, it is recommended that the rib be taken out of the oven when the thermometer registers about 4°C below the indicated value because the rib will continue cooking even after it is taken out of the oven. Do you agree with this recommendation?
*Answers:* (*a*) 156.9 W/(m$^2$ · °C), (*b*) 159.5°C, (*c*) 1829 kJ, (*d*) 3 h 1 min

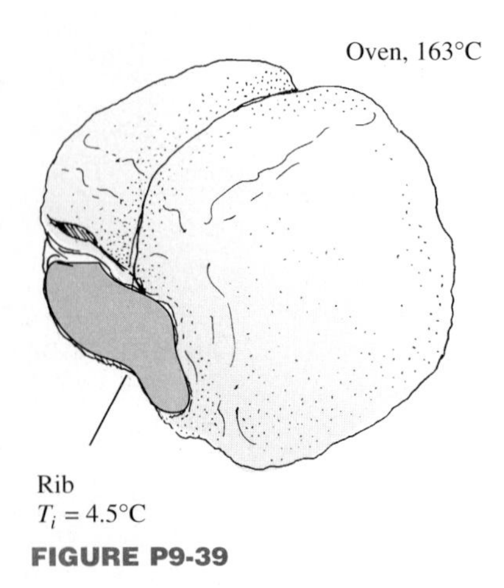

**FIGURE P9-39**

**9-40** Repeat Prob. 9-39 for a roast rib which is to be "well-done" instead of "rare". A rib is considered to be well-done when its center temperature reaches 77°C, and the roasting in this case takes about 4 h 15 min.

**9-41** For heat transfer purposes, an egg can be considered to be a 5.5-cm-diameter sphere having the properties of water. An egg that is initially at 8°C is dropped into the boiling water at 100°C. The heat transfer coefficient at the surface of the egg is estimated to be 500 W/(m$^2$ · °C). If the egg is considered cooked when its center temperature reaches 60°C, determine how long the egg should be kept in the boiling water.

**9-42** Repeat Prob. 9-41 for a location at 1610-m elevation such as Denver, Colorado where the boiling temperature of water is 94.4°C.

**9-43** The author and his 6-year-old son have conducted the following experiment to determine the thermal conductivity of a hot dog. They first boiled water in a large pan, and measured the temperature of the boiling water to be 94°C, which is not surprising, since they live at an elevation of about 1650 m in Reno, Nevada. They then took a hot dog that is 12.5 cm long and 2.2 cm in diameter, and inserted a thermocouple into the midpoint of the hot dog and another thermocouple just under the skin. They waited until both thermocouples read 20°C, which is the ambient

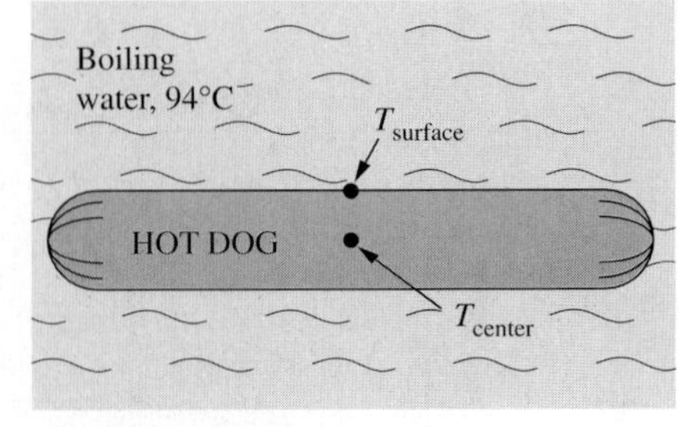

**FIGURE P9-43**

temperature. They then dropped the hot dog into boiling water, and observed the changes in both temperatures. Exactly 2 min after the hot dog was dropped into the boiling water, they recorded the center and the surface temperatures to be 59°C and 88°C, respectively. The density of the hot dog can be taken to be 980 kg/m$^3$, which is slightly less than the density of water, since the hot dog was observed to be floating in water while being almost completely immersed. The specific heat of hot dog can be taken to be 3900 J/(kg · °C), which is slightly less than that of water, since a hot dog is mostly water. Using transient temperature charts, determine (*a*) the thermal diffusivity of water, (*b*) the thermal conductivity of water, and (*c*) the convection heat transfer coefficient.
*Answers:* (*a*) $2 \times 10^{-7}$ m$^2$/s, (*b*) 0.76 W/(m · °C), (*c*) 658 W/(m$^2$ · °C).

**9-44** Using the data and the answers given in Prob. 9-43, determine the center and the surface temperatures of the hot dog 4 min after the start of the cooking. Also determine the amount of heat transferred to the hot dog.

**9-45** In a chicken processing plant, whole chickens averaging 2 kg each and initially at 22°C are to be cooled in the racks of a large refrigerator that is maintained at −15°C. The entire chicken is to be cooled below 6°C, but the temperature of the chicken is not to drop below 2°C at any point during refrigeration. The convection heat transfer coefficient and thus the rate of heat transfer from the chicken can be controlled by varying the speed of a circulating fan inside. Determine the heat transfer coefficient that will enable us to meet both temperature constraints while keeping the refrigeration time to a minimum. The chicken can be treated as a homogeneous spherical object having the properties $\rho = 1200$ kg/m$^3$, $C_p = 4.1$ kJ/(kg · °C), $k = 0.45$ W/(m · °C), and $\alpha = 0.91 \times 10^{-7}$ m$^2$/s.

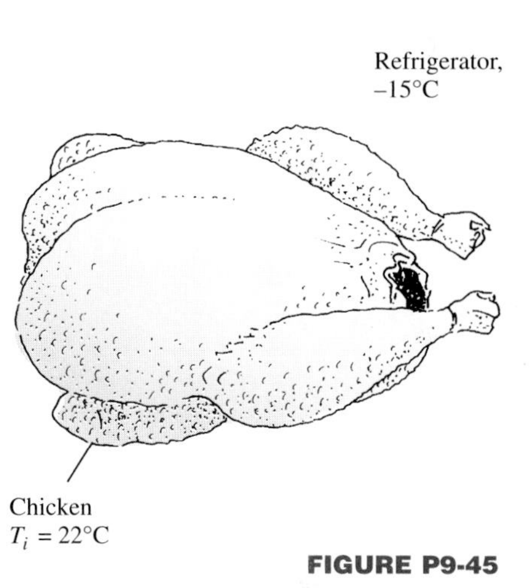

**FIGURE P9-45**

**9-45E** In a chicken processing plant, whole chickens averaging 5 lb each and initially at 72°F are to be cooled in the racks of a large refrigerator that is maintained at 5°F. The entire chicken is to be cooled below 45°F, but the temperature of the chicken is not to drop below 35°F at any point during refrigeration. The convection heat transfer coefficient and thus the rate of heat transfer from the chicken can be controlled by varying the speed of a circulating fan inside. Determine the heat transfer coefficient that will enable us to meet both temperature constraints while keeping the refrigeration time to a minimum. The chicken can be treated as a homogeneous spherical object having the properties $\rho = 74.9$ lbm/ft$^3$, $C_p = 0.98$ Btu/(lbm · °F), $k = 0.26$ Btu/(h · ft · °F), and $\alpha = 0.0035$ ft$^2$/h.

**9-46** A person puts a few apples into the freezer at −15°C to cool them quickly for guests who are about to arrive. Initially, the apples are at a uniform temperature of 20°C, and the heat transfer coefficient on the surfaces is 8 W/(m$^2$ · °C). Treating the apples as 9-cm-diameter spheres and taking their properties to be $\rho = 840$ kg/m$^3$, $C_p = 3.6$ kJ/(kg · °C), $k = 0.513$ W/(m · °C), and $\alpha = 1.3 \times 10^{-7}$ m$^2$/s, determine the center and surface temperatures of the apples in 1 h. Also determine the amount of heat transfer from each apple.

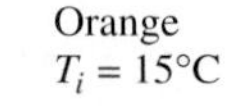


**FIGURE P9-47**

**9-47** Citrus fruits are very susceptible to cold weather, and extended exposure to subfreezing temperatures can destroy them. Consider an 8-cm-diameter orange that is initially at 15°C. A cold front moves in one night, and the ambient temperature suddenly drops to −6°C, with a heat transfer coefficient of 15 W/(m$^2$ · °C). Using the properties of water for the orange and assuming the ambient conditions to remain constant for 4 h before the cold front moves out, determine if any part of the orange will freeze that night.

**9-47E** Citrus fruits are very susceptible to cold weather, and extended exposure to subfreezing temperatures can destroy them. Consider a 3.5-in.-diameter orange that is initially at 60°F. A cold front moves in one night, and the ambient temperature suddenly drops to 20°F with a heat transfer coefficient of 3 Btu/(h · ft$^2$ · °F). Using the properties of water for the orange and assuming the ambient conditions to remain constant for 4 h before the cold front moves out, determine if any part of the orange will freeze that night.

**9-48** An 8-cm-diameter potato [$\rho$ = 1100 kg/m$^3$, $C_p$ = 3900 J/(kg · °C), $k$ = 0.6 W/(m · °C), and $\alpha = 1.4 \times 10^{-7}$ m$^2$/s] that is initially at a uniform temperature of 25°C is baked in an oven at 170°C until a temperature sensor inserted to the center of the potato indicates a reading of 70°C. The potato is then taken out of the oven and is wrapped into thick towels so that almost no heat is lost from the baked potato. Assuming the heat transfer coefficient in the oven to be 25 W/(m$^2$ · °C), determine (*a*) how long the potato is baked in the oven and (*b*) the final equilibrium temperature of the potato after it is wrapped.

### Transient Heat Conduction in Semi-Infinite Solids

**9-49C** What is a semi-infinite medium? Give examples of solid bodies that can be treated as semi-infinite mediums for heat transfer purposes.

**9-50C** Under what conditions can a plane wall be treated as a semi-infinite medium?

**9-51C** Consider a hot semi-infinite solid at an initial temperature of $T_i$ that is exposed to convection to a cooler medium at a constant temperature of $T_\infty$, with a heat transfer coefficient of $h$. Explain how you can determine the total amount of heat transfer from the solid up to a specified time $t_o$.

**9-52** In areas where the air temperature remains below 0°C for prolonged periods of time, the freezing of water in undergound pipes is a major concern. Fortunately, the soil remains relatively warm during those periods, and it takes weeks for the subfreezing temperatures to reach the water mains in the ground. Thus, the soil effectively serves as an insulation to protect the water from the freezing atmospheric temperatures in winter.

The ground at a particular location is covered with snow pack at $-8$°C for a continuous period of 60 days, and the average soil properties at that location are $k = 0.4$ W/(m · °C) and $\alpha = 0.15 \times 10^{-6}$ m$^2$/s. Assuming an initial uniform temperature of 10°C for the ground, determine the minimum burial depth to prevent the water pipes from freezing.

**9-53** The soil temperature in the upper layers of the earth varies with the variations in the atmospheric conditions. Before a cold front moves in, the earth at a location is initially at a uniform temperature of 10°C. Then the area is subjected to a temperature of $-10$°C and high winds which resulted in a convection heat transfer coefficient of 40 W/(m$^2$ · °C) on the earth's surface for a period of 10 h. Taking the properties of the soil at that location to be $k = 0.9$ W/(m · °C) and $\alpha = 1.6 \times 10^{-5}$ m$^2$/s, determine the soil temperature at distances 0, 10, 20 and 50 cm from the earth's surface at the end of this 10-h period.

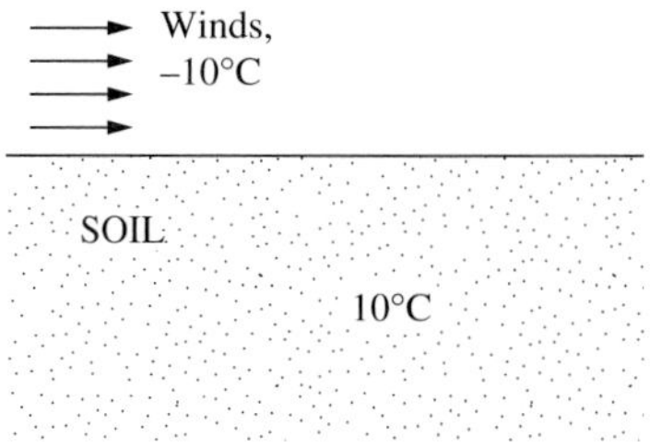

Winds, –10°C
SOIL
10°C

**FIGURE P9-53**

**9-54** The walls of a furnace are made of 40-cm-thick concrete [$k = 1.1$ W/(m · °C) and $\alpha = 0.60 \times 10^{-6}$ m$^2$/s]. Initially, the furnace and the surrounding air are in thermal equilibrium at 20°C. The furnace is then fired, and the inner surfaces of the furnace are subjected to hot gases at 1000°C with a very large heat transfer coefficient. Determine how long it will take for the temperature of the outer surface of the furnace walls to rise to 20.1°C. *Answer:* 146 min

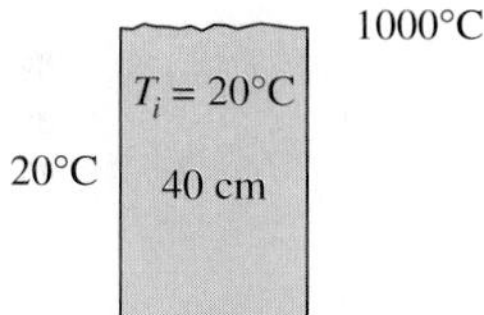


**FIGURE P9-54**

**9-54E** The walls of a furnace are made of 1.5-ft-thick concrete [$k = 0.64$ Btu/(h · ft · °F) and $\alpha = 0.023$ ft$^2$/h]. Initially, the furnace and the surrounding air are in thermal equilibrium at 70°F. The furnace is then fired, and the inner surfaces of the furnace are subjected to hot gases at 1800°F with a very large heat transfer coefficient. Determine how long it will take for the temperature of the outer surface of the furnace walls to rise to 70.1°F. *Answer:* 181 min

**9-55** A thick wood slab [$k = 0.17$ W/(m · °C) and $\alpha = 1.28 \times 10^{-7}$ m$^2$/s] that is initially at a uniform temperature of 25°C is exposed to hot gases at 550°C for a period of 5 minutes. The heat transfer coefficient between the gases and the wood slab is 35 W/(m$^2$ · °C). If the ignition temperature of the wood is 420°C, determine if the wood will ignite.

**9-56** A large cast iron container [$k = 52$ W/(m · °C) and $\alpha = 1.70 \times 10^{-5}$ m$^2$/s] with 5-cm-thick walls is initially at a uniform temperature of 0°C and is filled with ice at 0°C. Now the outer surfaces of the container are exposed to hot water at 60°C with a very large heat transfer coefficient. Determine how long it will be before the ice inside the container starts melting. Also, taking the heat transfer coefficient on the inner surface of the container to be 250 W/(m$^2$ · °C), determine the rate of heat transfer to the ice through a 1.2-m-wide and 2-m-high section of the wall when steady operating conditions are reached. Assume the ice starts melting when its inner surface temperature rises to 0.1°C.

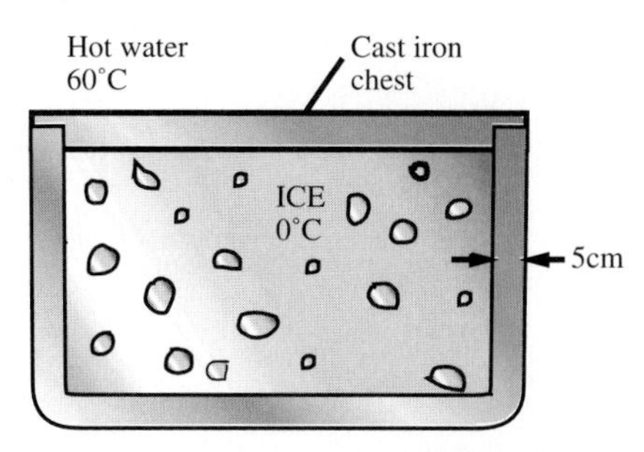


**FIGURE P9-56**

**9-57C** What is the product solution method? How is it used to determine the transient temperature distribution in a two-dimensional system?

**9-58C** How is the product solution used to determine the variation of temperature with time and position in three-dimensional systems?

**9-59C** A short cylinder initially at a uniform temperature $T_i$ is subjected to convection from all of its surfaces to a medium at temperature $T_\infty$. Explain how you can determine the temperature of the midpoint of the cylinder at a specified time $t$.

**9-60C** Consider a short cylinder whose top and bottom surfaces are insulated. The cylinder is initially at a uniform temperature $T_i$, and is subjected to convection from its side surface to a medium at temperature $T_\infty$ with a heat transfer coefficient of $h$. Is the heat transfer in this short cylinder one- or two-dimensional? Explain.

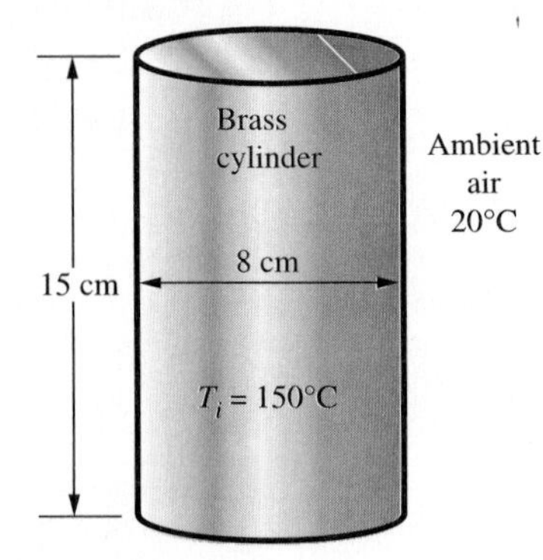


**FIGURE P9-61**

**9-61** A short brass cylinder [$\rho$ = 8530 kg/m$^3$, $C_p$ = 0.389 kJ/(kg · °C), $k$ = 110 W/(m · °C), and $\alpha = 3.39 \times 10^{-5}$ m$^2$/s] of diameter $D$ = 8 cm and height $H$ = 15 cm is initially at a uniform temperature of $T_i$ = 150°C. The cylinder is now placed in atmospheric air at 20°C, where heat transfer takes place by convection with a heat transfer coefficient of $h$ = 40 W/(m$^2$ · °C). Calculate (*a*) the center temperature of the cylinder, (*b*) the center temperature of the top surface of the cylinder, and (*c*) the total heat transfer from the cylinder 15 min after the start of the cooling.

**9-62** A semi-infinite aluminum cylinder [$k$ = 237 W/(m · °C), $\alpha = 9.71 \times 10^{-5}$ m$^2$/s] of diameter $D$ = 15 cm is initially at a uniform temperature of $T_i$ = 150°C. The cylinder is now placed in water at 10°C, where heat transfer takes place by convection with a heat transfer coefficient of $h$ = 140 W/(m$^2$ · °C). Determine the temperature at the center of the cylinder 10 cm from the end surface 8 min after the start of the cooling.

**9-63** A hot dog can be considered to be a cylinder 12 cm long and 2 cm in diameter whose properties are $\rho$ = 980 kg/m$^3$, $C_p$ = 3.9 kJ/(kg · °C), $k$ = 0.76 W/(m · °C), and $\alpha = 2 \times 10^{-7}$ m$^2$/s. A hot dog initially at 5°C is dropped into boiling water at 100°C. If the heat transfer coefficient at the surface of the hot dog is estimated to be 600 W/(m$^2$ · °C), determine the center temperature of the hot dog after 5, 10, and 15 min by treating the hot dog as (*a*) a finite cylinder and (*b*) an infinitely long cylinder.

**9-63E** A hot dog can be considered to be a cylinder 5 in long and 0.8 in in diameter whose properties are $\rho$ = 61.2 lbm/ft$^3$, $C_p$ = 0.93 Btu/(lbm · °F), $k$ = 0.44 Btu/(ft · °F), and $\alpha$ = 0.0077 ft$^2$/h. A hot dog initially at 40°F is dropped into boiling water at 212°F. If the heat transfer coefficient at the surface of the hot dog is estimated to be 120 Btu/(h · ft$^2$ · °F), determine the center temperature of the hot dog after 5, 10, and 15 min by treating the hot dog as (*a*) a finite cylinder and (*b*) an infinitely long cylinder.

**9-64** Repeat Prob. 9-63 for a location at 1610 m elevation such as Denver, Colorado where the boiling temperature of water is 94.4°C.

**9-65** A 5-cm-high rectangular ice block [$k = 2.22$ W/(m · °C) and $\alpha = 0.124 \times 10^{-7}$ m$^2$/s] initially at −20°C is placed on a table on its square base 4 cm × 4 cm in size in a room at 18°C. The heat transfer coefficient on the exposed surfaces of the ice block is 12 W/(m$^2$ · °C). Disregarding any heat transfer from the base to the table, determine how long it will be before the ice block starts melting. Where on the ice block will the first liquid droplets appear?

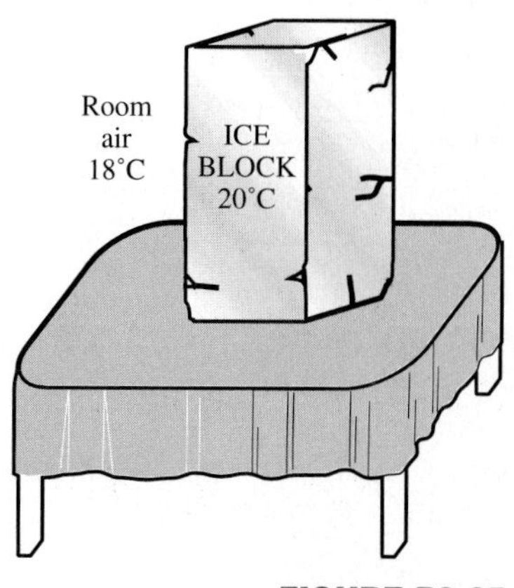


**FIGURE P9-65**

**9-66** A 2-cm-high cylindrical ice block [$k = 2.22$ W/(m · °C) and $\alpha = 0.124 \times 10^{-7}$ m$^2$/s] is placed on a table on its base of diameter 2 cm in a room at 20°C. The heat transfer coefficient on the exposed surfaces of the ice block is 13 W/(m$^2$ · °C), and heat transfer from the base of the ice block to the table is negligible. If the ice block is not to start melting at any point for at least 2 h, determine what the initial temperature of the ice block should be.

**9-67** Consider a cubic block whose sides are 5 cm long and a cylindrical block whose height and diameter are also 5 cm. Both blocks are initially at 20°C and are made of granite [$k = 2.5$ W/(m · °C) and $\alpha = 1.15 \times 10^{-6}$ m$^2$/s]. Now both blocks are exposed to hot gases at 500°C in a furnace on all of their surfaces with a heat transfer coefficient of 40 W/(m$^2$ · °C). Determine the center temperature of each geometry after 10, 20, and 60 min.

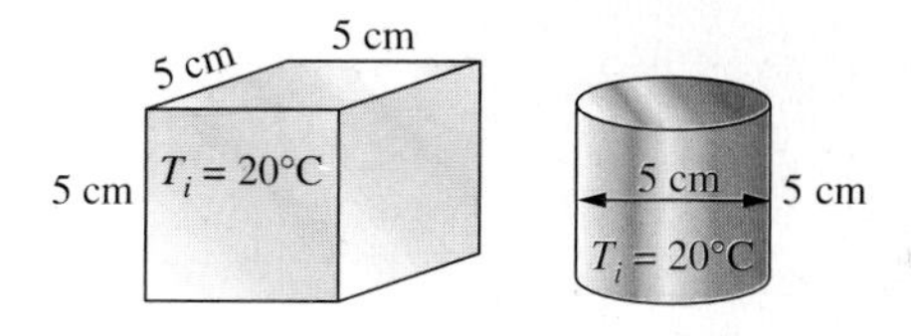


**FIGURE P9-67**

**9-67E** Consider a cubic block whose sides are 2 in long and a cylindrical block whose height and diameter are also 2 in. Both blocks are initially at 70°F and are made of granite [$k = 1.44$ Btu/(h · ft · °F) and $\alpha = 0.0426$ ft$^2$/h]. Now both blocks are exposed to hot gases at 850°F in a furnace on all of their surfaces with a heat transfer coefficient of 8 Btu/(h · ft$^2$ · °F). Determine the center temperature of each geometry after 10, 20, and 60 min.

**9-68** Repeat Prob. 9-67 with the heat transfer coefficient at the top and the bottom surfaces of each block being doubled to 80 W/(m$^2$ · °C).

**9-69** A 20-cm-long cylindrical aluminum block [$\rho = 2702$ kg/m$^3$, $C_p = 0.896$ kJ/(kg · °C), $k = 236$ W/(m · °C), and $\alpha = 9.75 \times 10^{-5}$ m$^2$/s], 15 cm in diameter, is initially at a uniform temperature of 20°C. The block is to be heated in a furnace at 1200°C until its center temperature rises to 300°C. If the heat transfer coefficient on all surfaces of the block is 50 W/(m$^2$ · °C), determine how long the block should be kept in the furnace. Also determine the amount of heat transfer from the aluminum block if it is allowed to cool in the room until its temperature drops to 20°C throughout.

**9-70** Repeat Prob. 9–69 for the case where the aluminum block is inserted into the furnace on a low-conductivity material so that the heat transfer to or from the bottom surface of the block is negligible.

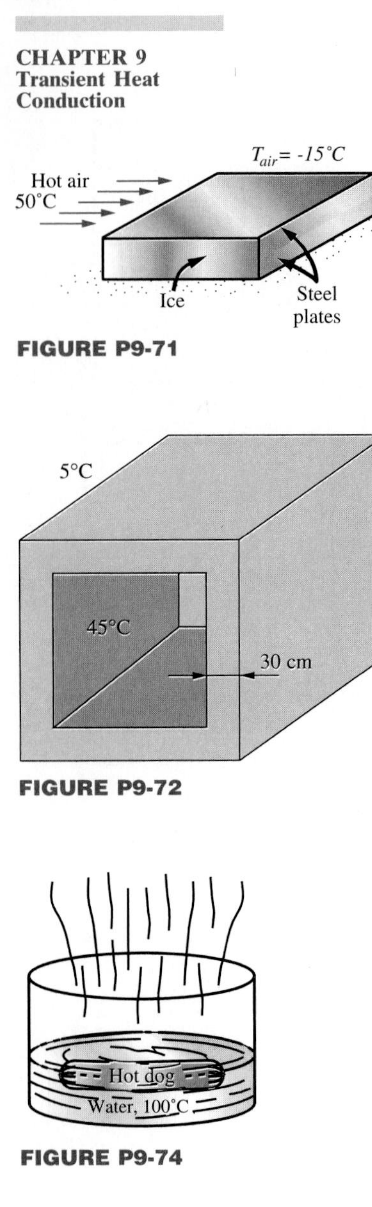


FIGURE P9-71

FIGURE P9-72

FIGURE P9-74

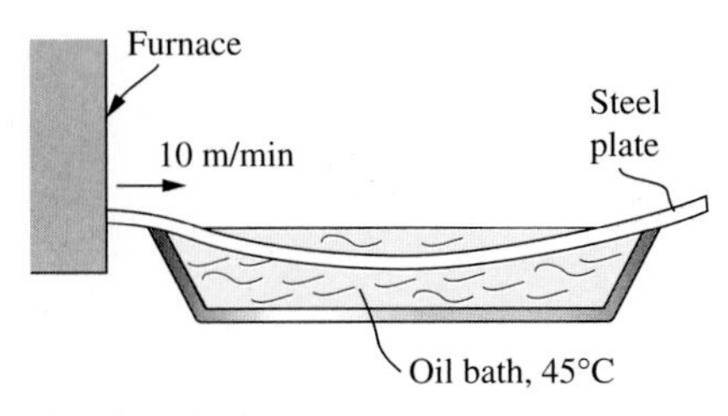


FIGURE P9-75

## Review Problems

**9-71** Consider two 2-cm-thick large steel plates [$k = 43$ W/(m · °C) and $\alpha = 1.17 \times 10^{-5}$ m$^2$/s] that were put on top of each other while wet and left outside during a cold winter night at −15°C. The next day, a worker needs one of the plates, but the plates are stuck together because of the freezing of the water between the two plates which bonded them together. In an effort to melt the ice between the plates and separate them, the worker takes a large hairdryer, and blows hot air at 50°C all over the exposed surface of the plate on the top. The convection heat transfer coefficient at the top surface is estimated to be 40 W/(m$^2$ · °C). Determine how long the worker must keep blowing hot air before the two plates separate. *Answer:* 507 s

**9-72** Consider a curing kiln whose walls are made of 30-cm-thick concrete whose properties are $k = 0.9$ W/(m · °C) and $\alpha = 0.23 \times 10^{-5}$ m$^2$/s. Initially, the kiln and its walls are in equilibrium with the surroundings at 5°C. Then all the doors are closed and the kiln is heated by steam so that the temperature of the inner surface of the walls is raised to 45°C and is maintained at that level for 3 h. The curing kiln is then opened, and is exposed to the atmospheric air after the steam flow is turned off. If the outer surfaces of the walls of the kiln were insulated, would it save any energy that day during the period the kiln was used for curing for 3 h only, or would it make no difference? Base your answer on calculations.

**9-73** The water main in the cities must be placed at sufficient depth below the earth's surface to avoid freezing during extended periods of subfreezing temperatures. Determine the minimum depth at which the water main must be placed at a location where the soil is initially at 15°C and the earth's surface temperature under the worst conditions is expected to remain at −10°C for a period of 75 days. Take the properties of soil at that location to be $k = 0.7$ W/(m · °C) and $\alpha = 1.4 \times 10^{-5}$ m$^2$/s. *Answer:* 7.05 m

**9-74** A hot dog can be considered to be a 12-cm-long cylinder whose diameter is 2 cm and whose properties are $\rho = 980$ kg/m$^3$, $C_p = 3.9$ kJ(kg · °C), $k = 0.76$ W/(m · °C), and $\alpha = 2 \times 10^{-7}$ m$^2$/s. A hot dog initially at 5°C is dropped into boiling water at 100°C. The heat transfer coefficient at the surface of the hot dog is estimated to be 600 W/(m$^2$ · °C). If the hot dog is considered cooked when its center temperature reaches 80°C, determine how long it will take to cook it in the boiling water.

**9-75** A long roll of 2-m-wide and 0.5-cm-thick 1-Mn manganese steel plate coming off a furnace at 820°C is to be quenched in an oil bath [$C_p = 2.0$ kJ/(kg · °C)] at 45°C. The metal sheet is moving at a steady velocity of 10 m/min, and the oil bath is 8 m long. Taking the convection heat transfer coefficient on both sides of the plate to be 860 W/(m$^2$ · °C), determine the temperature of the sheet metal when it leaves the oil bath.

Also determine the required rate of heat removal from the oil to keep its temperature constant at 45°C.

**9-75E** A long roll of 6-ft-wide and 0.25-in-thick 1-Mn manganese steel plate coming off a furnace at 1500°F is to be quenched in an oil bath [$C_p = 0.48$ Btu/(lbm · °F)] at 115°F. The metal sheet is moving at a steady velocity of 25 ft/min, and the oil bath is 20 ft long. Taking the convection heat transfer coefficient on both sides of the plate to be 150 Btu/(h · ft$^2$ · °F), determine the temperature of the sheet metal when it leaves the oil bath. Also determine the required rate of heat removal from the oil to keep its temperature constant at 115°F.

**9-76** In Betty Crocker's Cookbook, it is stated that it takes 5 h to roast a 6.4-kg stuffed turkey that is initially at 4.5°C in an oven maintained at 163°C. It is recommended that a meat thermometer be used to monitor the cooking, and the turkey is considered done when the thermometer inserted deep into the thickest part of the breast or thigh without touching the bone registers 85°C. The turkey can be treated as a homogeneous spherical object with the properties $\rho = 1200$ kg/m$^3$, $C_p =$ 4.1 kJ/(kg · °C), $k = 0.45$ W/(m · °C), and $\alpha = 0.91 \times 10^{-7}$ m$^2$/s. Assuming the tip of the thermometer is at one-third radial distance from the center of the turkey, determine (*a*) the average heat transfer coefficient at the surface of the turkey, (*b*) the temperature of the skin of the turkey when it is done, and (*c*) the total amount of heat transferred to the turkey in the oven. Will the reading of the thermometer be more or less than 85°C 5 min after the turkey is taken out of the oven?

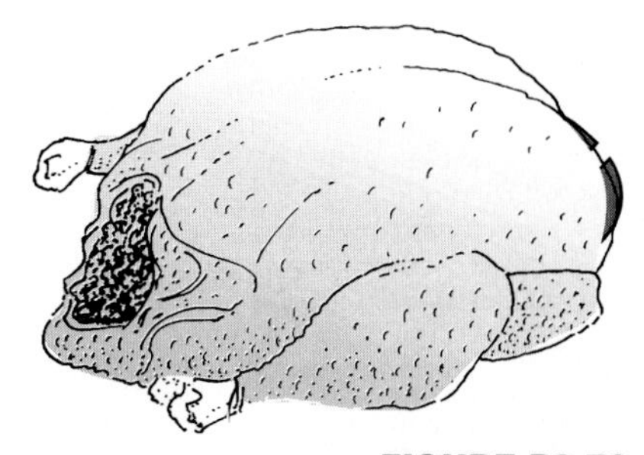

**FIGURE P9-76**

**9-76E** In Betty Crocker's Cookbook, it is stated that it takes 5 h to roast a 14-lb stuffed turkey initially at 40°F in an oven maintained at 325°F. It is recommended that a meat thermometer be used to monitor the cooking, and the turkey is considered done when the thermometer inserted deep into the thickest part of the breast or thigh without touching the bone registers 185°F. The turkey can be treated as a homogeneous spherical object with the properties $\rho = 75$ lbm/ft$^3$, $C_p = 0.98$ Btu/(lbm · °F), $k =$ 0.26 Btu/(h · ft · °F), and $\alpha = 0.0035$ ft$^2$/h. Assuming the tip of the thermometer is at one-third radial distance from the center of the turkey, determine (*a*) the average heat transfer coefficient at the surface of the turkey, (*b*) the temperature of the skin of the turkey when it is done, and (*c*) the total amount of heat transferred to the turkey in the oven. Will the reading of the thermometer be more or less than 185°F 5 min after the turkey is taken out of the oven?

**9-77** During a fire, the trunks of some dry oak trees [$k =$ 0.17 W/(m · °C) and $\alpha = 1.28 \times 10^{-7}$ m$^2$/s] that are initially at a uniform temperature of 30°C are exposed to hot gases at 450°C for a period of 4 h, with a heat transfer coefficient of 65 W/(m$^2$ · °C) on the surface. The ignition temperature of the trees is 410°C. Treating the trunks of the trees as long cylindrical rods of diameter 20 cm, determine if these dry trees will ignite as the fire sweeps through them.

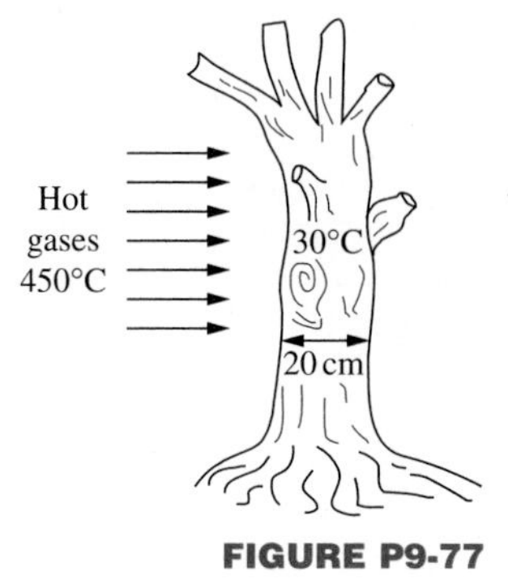


**FIGURE P9-77**

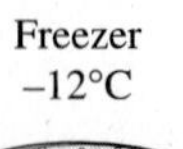

**FIGURE P9-78**

**9-78** We often cut a watermelon in half, and put it into the freezer to cool it quickly. But usually we forget to check on it, and end up having a watermelon with a frozen layer on the top. To avoid this potential problem, a person wants to set the timer such that it will go off when the temperature of the exposed surface of the watermelon drops to 3°C.

Consider a 30-cm-diameter spherical watermelon that is cut into two equal parts and put into a freezer at −12°C. Initially, the entire watermelon is at a uniform temperature of 25°C, and the heat transfer coefficient on the surfaces is 30 W/(m$^2$ · °C). Assuming the watermelon to have the properties of water, determine how long it will take for the center of the exposed cut surfaces of the watermelon to drop to 3°C.

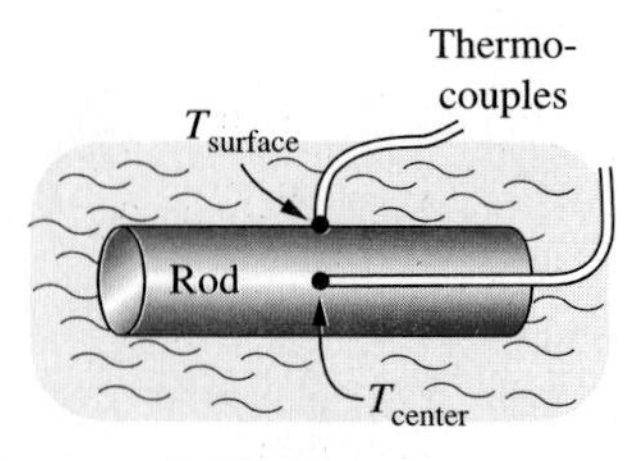


**FIGURE P9-79**

**9-79** The thermal conductivity of a solid whose density and specific heat are known can be determined from the relation $k = \alpha/\rho C_p$ after evaluating the thermal diffusivity $\alpha$.

Consider a 2-cm-diameter cylindrical rod made of a sample material whose density and specific heat are 3700 kg/m$^3$ and 920 J/(kg · °C), respectively. The sample is initially at a uniform temperature of 25°C. In order to measure the temperatures of the sample at its surface and its center, a thermocouple is inserted to the center of the sample along the center line, and another thermocouple is welded into a small hole drilled on the surface. The sample is dropped into boiling water at 100°C. After 3 min, the surface and the center temperatures are recorded to be 93°C and 75°C, respectively. Determine the thermal diffusivity and the thermal conductivity of the material.

**9-80** In desert climates, rainfall is not a common occurrence, since the rain droplets formed in the upper layer of the atmosphere often evaporate before they reach the ground. Consider a raindrop that is initially at a temperature of 5°C and has a diameter of 5 mm. Determine how long it will take for the diameter of the raindrop to reduce to 3 mm as it falls through ambient air at 25°C with a heat transfer coefficient of 400 W/(m$^2$ · °C). The water temperature can be assumed to remain constant and uniform at 5°C at all times, and the heat of vaporization of water at 5°C is 2490 kJ/kg.

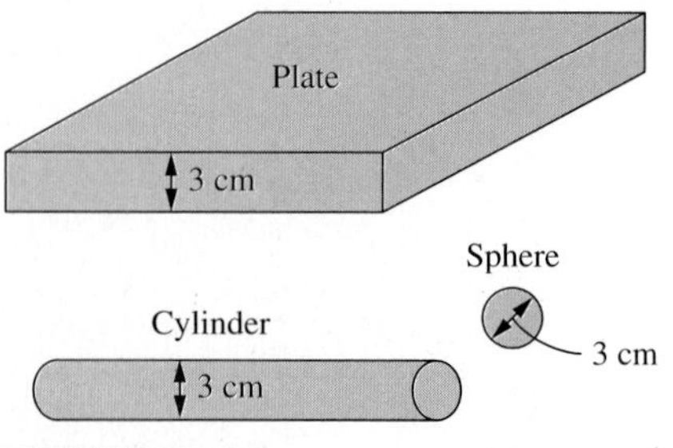


**FIGURE P9-81**

**9-81** Consider a plate of thickness 3 cm, a long cylinder of diameter 3 cm, and a sphere of diameter 3 cm, all initially at 200°C and all made of bronze [$k$ = 26 W/(m · °C) and $\alpha$ = 0.859 × 10$^{-5}$ m$^2$/s]. Now all three of these geometries are exposed to cool air at 25°C on all of their surfaces, with a heat transfer coefficient of 35 W/(m$^2$ · °C). Determine the center temperature of each geometry after 5, 10, and 30 min. Explain why the center temperature of the sphere is always the lowest.

**9-81E** Consider a plate of thickness 1 in, a long cylinder of diameter 1 in, and a sphere of diameter 1 in., all initially at 400°F and all made of bronze [$k$ = 15.0 Btu/(h · ft · °F) and $\alpha$ = 0.333 ft$^2$/h]. Now all three of these geometries are exposed to cool air at 75°F on all of their surfaces, with a heat transfer coefficient of 7 Btu/(h · ft$^2$ · °F). Determine the center temperature of each geometry after 5, 10, and 30 min. Explain why the center temperature of the sphere is always the lowest.

**9-82** Repeat Prob. 9-81 for cast iron geometries [$k$ = 52 W/(m · °C) and $\alpha$ = 1.70 × $10^{-5}$ $m^2$/s].

**9-83** Long aluminum wires of diameter 3 mm [$\rho$ = 2702 kg/$m^3$, $C_p$ = 0.896 kJ/(kg · °C), $k$ = 236 W/(m · °C), and $\alpha$ = 9.75 × $10^{-5}$ $m^2$/s] are extruded at a temperature of 350°C, and are exposed to atmospheric air at 30°C with a heat transfer coefficient of 35 W/($m^2$ · °C). (*a*) Determine how long it will take for the wire temperature to drop to 50°C. (*b*) If the wire is extruded at a velocity of 10 m/min, determine how far the wire travels after extrusion by the time its temperature drops to 50°C. What change in the cooling process would you propose to shorten this distance? (*c*) Assuming the aluminum wire leaves the extrusion room at 50°C determine the rate of heat transfer from the wire to the extrusion room. *Answers:* (*a*) 144 s, (*b*) 24 m, (*c*) 855 W

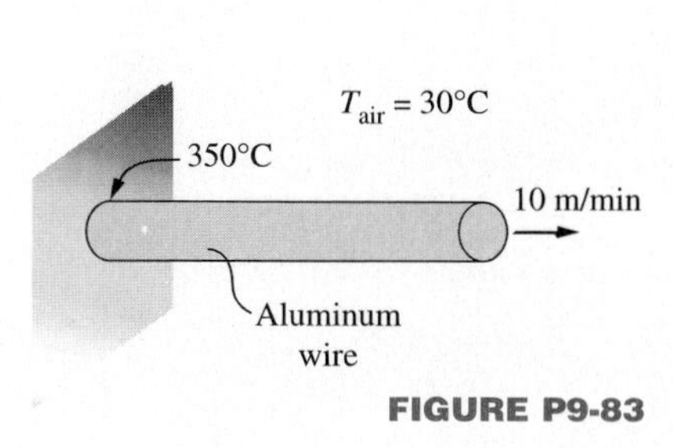


**FIGURE P9-83**

**9-84** Repeat Prob. 9-83 for a copper wire [$\rho$ = 8950 kg/$m^3$, $C_p$ = 0.383 kJ/(kg · °C), $k$ = 386 W/(m · °C), and $\alpha$ = 1.13 × $10^{-4}$ $m^2$/s].

**9-85** Consider a brick house [$k$ = 0.72 W/(m · °C) and $\alpha$ = 0.45 × $10^{-6}$ $m^2$/s] whose walls are 10 m long, 3 m high and 0.3 m thick. The heater of the house broke down one night, and the entire house, including its walls, was observed to be 5°C throughout in the morning. The outdoors warmed up as the day progressed, but no change was felt in the house, which was tightly sealed. Assuming the the outer surface temperature of the house to remain constant at 18°C, determine how long it would take for the temperature of the inner surfaces of the walls to rise to 5.1°C.

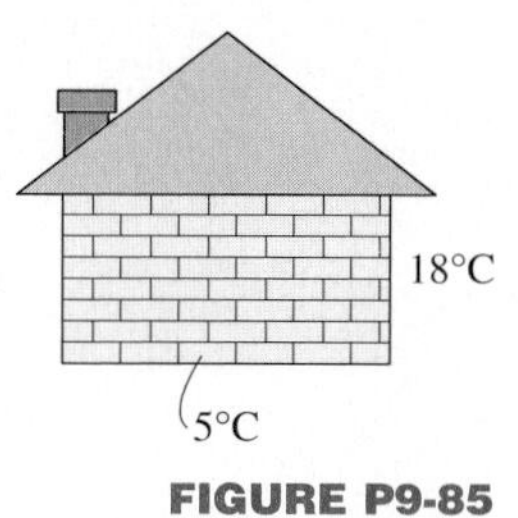


**FIGURE P9-85**

## Computer, Design, and Essay Problems

**9-86** Conduct the following experiment at home to determine the combined convection and radiation heat transfer coefficient at the surface of an apple [$\rho$ = 840 kg/$m^3$, $C_p$ = 3.6 kJ/(kg · °C), $k$ = 0.513 W/(m · °C), and $\alpha$ = 0.17 × $10^{-7}$ $m^2$/s] exposed to the room air. You will need two thermometers and a clock.

First weigh the apple and measure its diameter. You may measure its volume by placing it in a large measuring cup halfway filled with water, and measuring the change in volume when it is completely immersed in the water. Refrigerate the apple overnight so that it is at a uniform temperature in the morning, and measure the air temperature in the kitchen. Then take the apple out, and stick one of the thermometers to its middle and the other just under the skin. Record both temperatures every 5 min for an hour. Using these two temperatures, calculate the heat transfer coefficient for each interval, and take their average. The result is the combined convection and radiation heat transfer coefficient for this heat transfer process. Using your experimental data, also calculate the thermal conductivity and thermal diffusivity of the apple, and compare them to the values given above.

**9-87** Repeat Prob. 9-86 using a banana instead of an apple. The thermal properties of banana are practically the same as those of apple.

**9-88** Conduct the following experiment to determine the lumped exponent $b$ in Eq. 9-5 for a can of drink, and then predict the temperature of the drink at different times. Leave the drink in the refrigerator overnight. Measure the air temperature in the kitchen and the temperature of the drink while it is still in the refrigerator by taping the sensor of the thermometer to the outer surface of the can. Then take the drink out and measure its temperature again in 5 min. Using these values calculate the exponent $b$. Using this $b$-value, predict the temperatures of the drink in 10, 15, 20, 30, and 60 min, and compare the results with the actual temperature measurements. Do you think the lumped system analysis is valid in this case?

**9-89** Whole ready to cook turkeys range from about 2 to 11 kg. Roasted turkeys are considered done when a thermometer inserted deep into the turkey registers 85°C. From your favourite cookbook, obtain the instructions to bake a large stuffed turkey, and evaluate the average heat transfer coefficient during baking. Using this heat transfer coefficient, estimate the baking time for a turkey which is only half as large.

**9-90** Citrus trees are very susceptible to cold weather, and extended exposure to subfreezing temperatures can destroy the crop. In order to protect the trees from occasional cold fronts with subfreezing temperatures, tree growers in Florida usually install water sprinklers on the trees. When the temperature drops below a certain level, the sprinklers spray water on the tree and its fruits to protect them against the damage the subfreezing temperatures can cause. Explain the basic mechanism behind this protection measure, and write an essay on how the system works in practice.

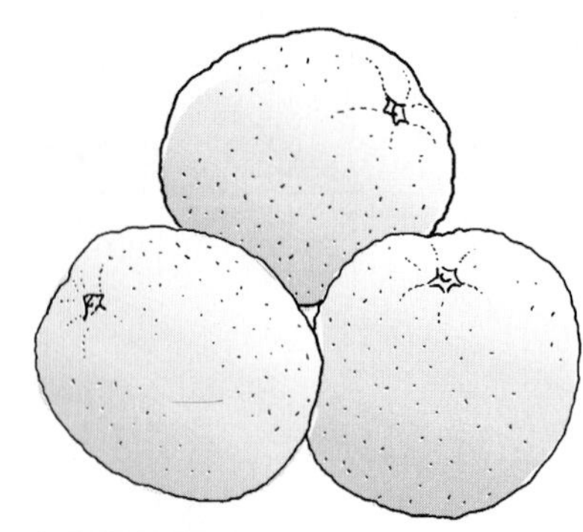

**FIGURE P9-90**

# Forced Convection

CHAPTER

# 10

So far, we have considered *conduction,* which is the mechanism of heat transfer through a solid or fluid in the absence of any fluid motion. We now consider *convection,* which is the mechanism of heat transfer through a fluid in the presence of bulk fluid motion.

Convection is classified as *natural* (or *free*) or *forced convection,* depending on how the fluid motion is initiated. In forced convection, the fluid is forced to flow over a surface or in a tube by external means such as a pump or a fan. In natural convection, any fluid motion is caused by natural means such as the buoyancy effect, which manifests itself as the rise of warmer fluid and the fall of the cooler fluid. Convection is also classified as *external* and *internal,* depending on whether the fluid is forced to flow over a surface or in a channel. Convection in external and internal flows exhibits very different characteristics.

In this chapter, we consider both *external* and *internal forced convection.* In Chap. 11 we will consider *natural convection,* which is associated with naturally caused fluid motions.

We start this chapter with a general physical description of the *convection* mechanism and the *velocity* and *thermal boundary layers.* We continue with the discussion of the dimensionless *Reynolds, Prandtl,* and *Nusselt numbers,* and their physical significance. We then present empirical relations for *friction* and *heat transfer coefficients* for flow over various geometries such as a flat plate, cylinder, and sphere, for both laminar and turbulent flow conditions. Finally, we discuss the characteristics of flow inside tubes, and present the pressure drop and heat transfer correlations associated with it.

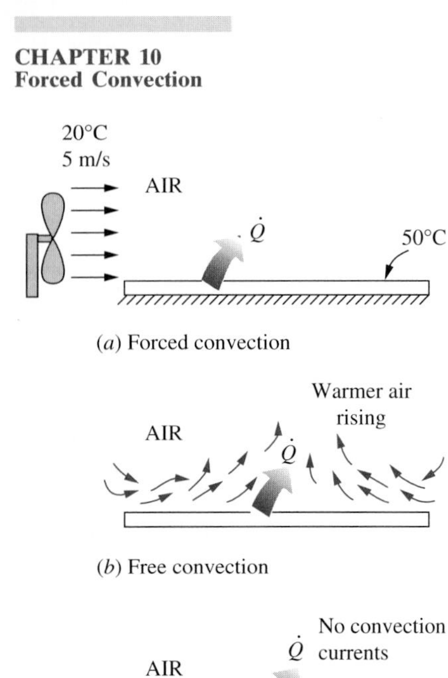


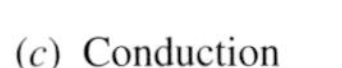


**FIGURE 10-1**
Heat transfer from a hot surface to the surrounding fluid by convection and conduction.

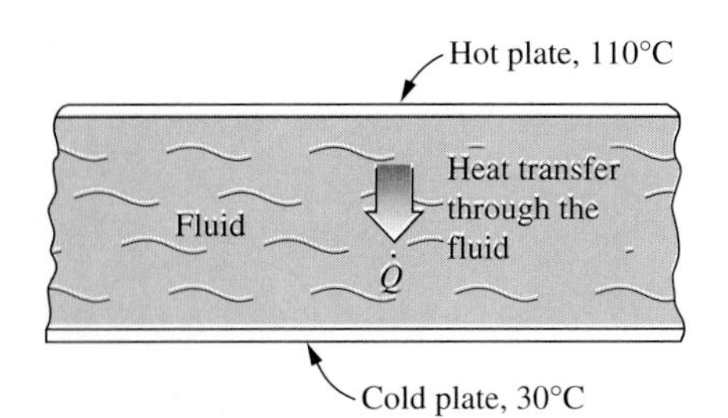


**FIGURE 10-2**
Heat transfer through a fluid sandwiched between two parallel plates.

## 10-1 ■ PHYSICAL MECHANISM OF FORCED CONVECTION

We mentioned earlier that there are three basic mechanisms of heat transfer: conduction, convection, and radiation. Conduction and convection are similar in that both mechanisms require the presence of a material medium. But they are different in that convection requires the presence of *fluid motion.*

Heat transfer through a *solid* is always by *conduction,* since the molecules of a solid remain at relatively fixed positions. Heat transfer through a *liquid* or *gas,* however, can be by *conduction* or *convection,* depending on the presence of any bulk fluid motion. Heat transfer through a fluid is by *convection* in the presence of bulk fluid motion, and by *conduction* in the absence of it. Therefore, conduction in a fluid can be viewed as the *limiting case* of convection, corresponding to the case of quiescent fluid (Fig. 10-1).

Convection heat transfer is complicated by the fact that it involves *fluid motion* as well as *heat conduction.* The fluid motion *enhances* heat transfer, since it brings hotter and cooler chunks of fluid into contact, initiating higher rates of conduction at a greater number of sites in a fluid. Therefore, the rate of heat transfer through a fluid is much higher by convection than it is by conduction. In fact, the higher the *fluid velocity,* the higher the rate of *heat transfer.*

To clarify this point further, consider steady heat transfer through a fluid contained between two *parallel plates* maintained at different temperatures, as shown in Fig. 10-2. The temperatures of the fluid and the plate will be the same at the points of contact because of the *continuity of temperature.* Assuming no fluid motion, the energy of the hotter fluid molecules near the hot plate will be transferred to the adjacent cooler fluid molecules. This energy will then be transferred to the next layer of the cooler fluid molecules. This energy will then be transferred to the next layer of the cooler fluid, and so on, until it is finally transferred to the other plate. This is what happens during *conduction* through a fluid. Now let us use a syringe to draw some fluid near the hot plate and inject it near the cold plate repeatedly. You can imagine that this will speed up the heat transfer process considerably, since some energy is *carried* to the other side as a result of fluid motion.

Consider the cooling of a *hot iron block* with a fan blowing air over its top surface, as shown in Fig. 10-3. We know that heat will be transferred from the hot block to the surrounding cooler air, and the block will eventually cool. We also know that the block will cool faster if the fan is switched to a higher speed. Replacing air by water will enhance the convection heat transfer even more.

Experience shows that convection heat transfer strongly depends on the fluid properties *dynamic viscosity* $\mu$, *thermal conductivity* $k$, *density* $\rho$, and *specific heat* $C_p$, as well as the *fluid velocity* $\mathcal{V}$. It also depends on the *geometry* and *roughness* of the solid surface, in addition to the *type of fluid flow* (such as being streamlined or turbulent). Thus, we expect the

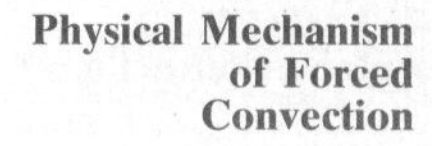

convection heat transfer relations to be rather complex, because of the dependence of convection on so many variables. This is not surprising, since convection is the most complex mechanism of heat transfer.

Despite the complexity of convection, the rate of *convection heat transfer* is observed to be proportional to the temperature difference, and is conveniently expressed by **Newton's law of cooling** as

$$\dot{q}_{\text{conv}} = h(T_s - T_\infty) \quad (\text{W/m}^2) \tag{10-1}$$

or

$$\dot{Q}_{\text{conv}} = hA(T_s - T_\infty) \quad (\text{W}) \tag{10-2}$$

where $h$ = convection heat transfer coefficient, W/(m$^2 \cdot$ °C)
$A$ = heat transfer surface area, m$^2$
$T_s$ = temperature of the surface, °C
$T_\infty$ = temperature of the fluid sufficiently far from the surface, °C

Judging from its units, the **convection heat transfer coefficient** can be defined as *the rate of heat transfer between a solid surface and a fluid per unit surface area per unit temperature difference.*

You should not be deceived by the simple appearance of this relation, because the convection heat transfer coefficient $h$ depends on the several variables mentioned above, and thus is difficult to determine.

When a fluid is forced to flow over a solid surface, it is observed that the fluid layer in contact with the solid surface "sticks" to the surface. That is, a very thin layer of fluid assumes *zero velocity* at the wall. In fluid flow, this phenomena is known as the *no-slip* condition. An implication of the no-slip condition is that heat transfer from the solid surface to the fluid layer adjacent to the surface is by *pure conduction,* since the fluid layer is motionless, and can be expressed as

$$\dot{q}_{\text{conv}} = \dot{q}_{\text{cond}} = -k_{\text{fluid}} \left.\frac{\partial T}{\partial y}\right|_{y=0} \quad (\text{W/m}^2) \tag{10-3}$$

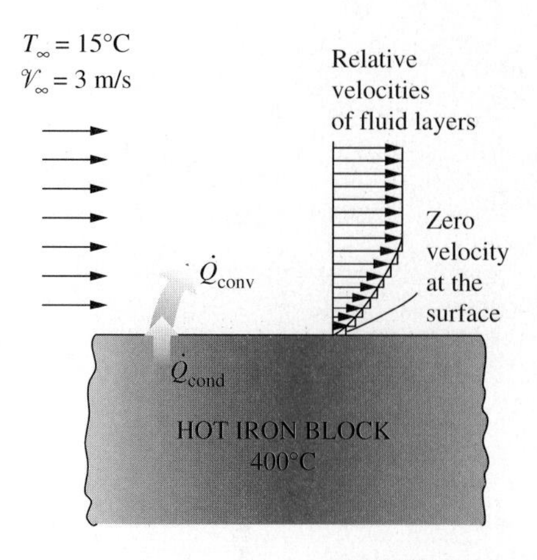


**FIGURE 10-3**
The cooling of a hot block by forced convection.

where $T$ represents the temperature distribution in the fluid and $(\partial T/\partial y)_{y=0}$ is the *temperature gradient* at the surface. This heat is then *convected away* from the surface as a result of fluid motion. Note that convection heat transfer from a solid surface to a fluid is merely the conduction heat transfer from the solid surface to the fluid layer adjacent to the surface. Therefore, we can equate the expressions 10-1 and 10-3 for the heat flux to obtain

$$h = \frac{-k_{\text{fluid}}(\partial T/\partial y)_{y=0}}{T_s - T_\infty} \quad [\text{W/(m}^2 \cdot \text{°C)}] \tag{10-4}$$

for the determination of the *convection heat transfer coefficient* when the temperature distribution within the fluid is known.

The convection heat transfer coefficient, in general, varies along the flow (or $x$) direction. The *average* or *mean* convection heat transfer coefficient for a surface in such cases is determined by properly averaging the *local* convection heat transfer coefficients over the entire surface.

In convection studies, it is common practice to nondimensionalize the governing equations and combine the variables, which group together into *dimensionless numbers* in order to reduce the number of total variables. It is also common practice to *nondimensionalize* the heat transfer coefficient $h$, with **Nusselt number** defined as

$$\mathrm{Nu} = \frac{h\delta}{k} \qquad (10\text{-}5)$$

where $k$ is the thermal conductivity of the fluid and $\delta$ is the *characteristic length.* The Nusselt number is named after Wilhelm Nusselt, who made significant contributions to convective heat transfer in the first half of the 20th century, and it is viewed as the *dimensionless convection heat transfer coefficient.*

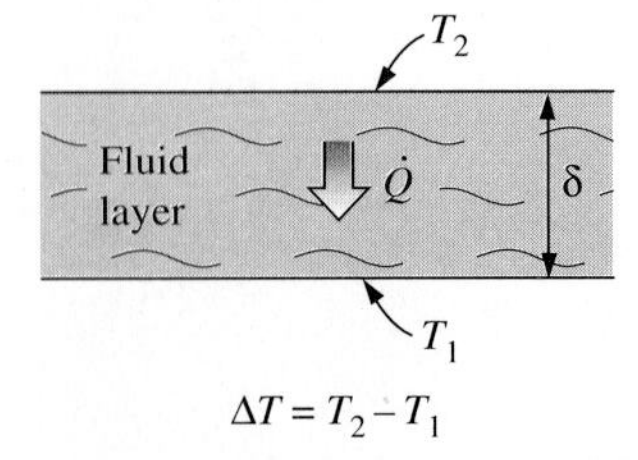

**FIGURE 10-4**
Heat transfer through a fluid layer of thickness $\delta$ and temperature difference $\Delta T$.

To understand the physical significance of the Nusselt number, consider a fluid layer of thickness $\delta$ and temperature difference $\Delta T = T_2 - T_1$, as shown in Fig. 10-4. Heat transfer through the fluid layer will be by *convection* when the fluid involves some motion, and by *conduction* when the fluid layer is motionless. Heat flux (the rate of heat transfer per unit time per unit surface area) in either case will be

$$\dot{q}_{\mathrm{conv}} = h\,\Delta T$$

and

$$\dot{q}_{\mathrm{cond}} = k\frac{\Delta T}{\delta}$$

Taking their ratio gives

$$\frac{\dot{q}_{\mathrm{conv}}}{\dot{q}_{\mathrm{cond}}} = \frac{h\,\Delta T}{k\,\Delta T/\delta} = \frac{h\delta}{k} = \mathrm{Nu}$$

which is the *Nusselt number.* Therefore, the Nusselt number represents the enhancement of heat transfer through a fluid layer as a result of convection relative to conduction across the same fluid layer. The larger the Nusselt number, the more effective is the convection. A Nusselt number $\mathrm{Nu} = 1$ for a fluid layer represents heat transfer by pure conduction.

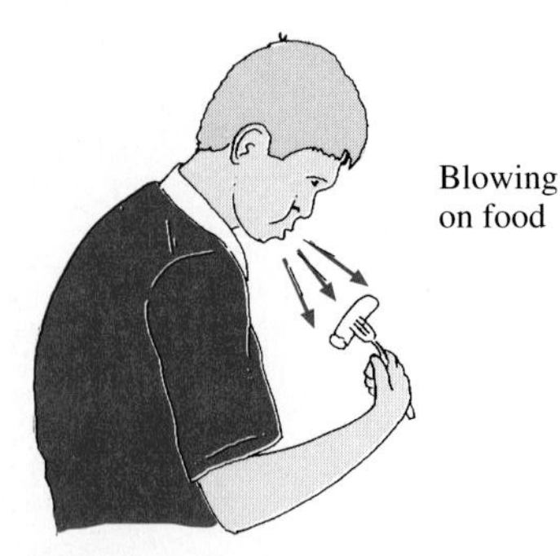


**FIGURE 10-5**
We resort to forced convection whenever we need to increase the rate of heat transfer.

We use forced convection in daily life more often than you might think (Fig. 10-5). We resort to forced convection whenever we want to increase the rate of heat transfer from a hot object. For example, we turn the *fan* on in hot summer days to help our body cool more effectively. The higher the fan speed, the better we feel. We *stir* our soup and *blow* on a hot slice of pizza to make them cool faster. The air on *windy* winter days feels much colder than it actually is. The simplest solution to heating problems in electronics packaging is to use a large enough fan.

## 10-2 ■ VELOCITY BOUNDARY LAYER

Consider the flow of a fluid over a *flat plate,* as shown in Fig. 10-6. The $x$-coordinate is measured along the plate surface from the *leading edge* of

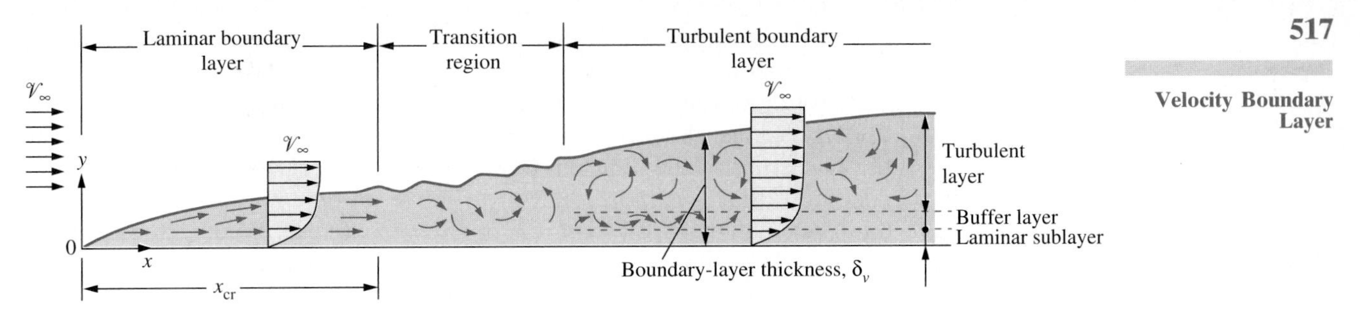


**FIGURE 10-6**

The development of the boundary layer for flow over a flat plate, and the different flow regimes.

the plate in the direction of the flow, and $y$ is measured from the surface in the normal direction. The fluid approaches the plate in the $x$-direction with a uniform velocity $\mathscr{V}_\infty$. For the sake of discussion, we can consider the fluid to consist of *adjacent layers* piled on top of each other. The velocity of the particles in the first fluid layer adjacent to the plate becomes *zero* because of the no-slip condition. This motionless layer *slows down* the particles of the neighboring fluid layer as a result of friction between the particles of these two adjoining fluid layers at different velocities. This fluid layer then slows down the molecules of the next layer, and so on. Thus, the presence of the plate is felt up to some distance $\delta_v$ from the plate beyond which the fluid velocity $\mathscr{V}_\infty$ remains essentially unchanged. As a result, the fluid velocity at any $x$-location will vary from 0 at $y = 0$ to nearly $\mathscr{V}_\infty$ at $y = \delta_v$.

The region of the flow above the plate bounded by $\delta_v$ in which the effects of the viscous shearing forces caused by fluid viscosity are felt is called the **velocity boundary layer** or just the **boundary layer**. The *thickness* of the boundary layer, $\delta_v$, is arbitrarily defined as the distance from the surface at which $\mathscr{V} = 0.99\mathscr{V}_\infty$.

The hypothetical line of $\mathscr{V} = 0.99\mathscr{V}_\infty$ divides the flow over a plate into two regions: the **boundary layer region** in which the viscous effects and the velocity changes are significant, and the **inviscid flow region** in which the frictional effects are negligible and the velocity remains essentially constant.

Consider two adjacent fluid layers in the boundary layer. The faster layer will try to drag along the slower one because of the friction between the two layers, exerting a *drag force* (or *friction force*) on it. The drag force per unit area is called the **shear stress**, and is denoted by $\tau$. Experimental studies indicate that the shear stress for most fluids is proportional to the *velocity gradient,* and the shear stress at the wall surface is expressed as

$$\tau_s = \mu \frac{\partial \mathscr{V}}{\partial y}\bigg|_{y=0} \qquad (\text{N/m}^2) \qquad (10\text{-}6)$$

where the constant of proportionality $\mu$ is the *dynamic viscosity* of the fluid, whose unit is kg/(m · s), or equivalently N · s/m$^2$.

The viscosity of a fluid is a measure of its *resistance to flow,* and it is a strong function of temperature. The viscosities of liquids *decrease* with temperature, whereas the viscosities of gases *increase* with temperature. The viscosities of some fluids at 20°C are listed in Table 10-1. Note that the viscosities of the different fluids differ by several orders of magnitude. Also note that it is more difficult to move an object in a higher-viscosity fluid such as engine oil than it is in a lower-viscosity fluid such as water.

**TABLE 10-1**
**The dynamic viscosities of some liquids and air at 20°C**

| Fluid | $\mu$, kg/(m · s) |
|---|---|
| Glycerin | 1.49 |
| Engine oil | 0.800 |
| Ethyl alcohol | 0.00120 |
| Water | 0.00106 |
| Freon-12 | 0.000262 |
| Air | 0.0000182 |

The determination of the surface shear stress $\tau_s$ from Eq. 10-6 is not practical, since it requires a knowledge of the flow velocity profile. A more practical approach in external flow is to relate $\tau_s$ to the free-stream velocity $\mathcal{V}_\infty$ as

$$\tau_s = C_f \frac{\rho \mathcal{V}_\infty^2}{2} \quad (\text{N/m}^2) \tag{10-7}$$

where $C_f$ is the **friction coefficient** or the **drag coefficient**, whose value in most cases is determined experimentally, and $\rho$ is the density of the fluid. Note that the friction coefficient, in general, will vary with location along the surface. Once the average friction coefficient over a given surface is available, the drag or friction force over the entire surface is determined from

$$F_D = C_f A \frac{\rho \mathcal{V}_\infty^2}{2} \quad (\text{N}) \tag{10-8}$$

where $A$ is the surface area.

The friction coefficient is an important parameter in heat transfer studies since it is directly related to the heat transfer coefficient and the power requirements of the pump or fan.

## Laminar and Turbulent flows

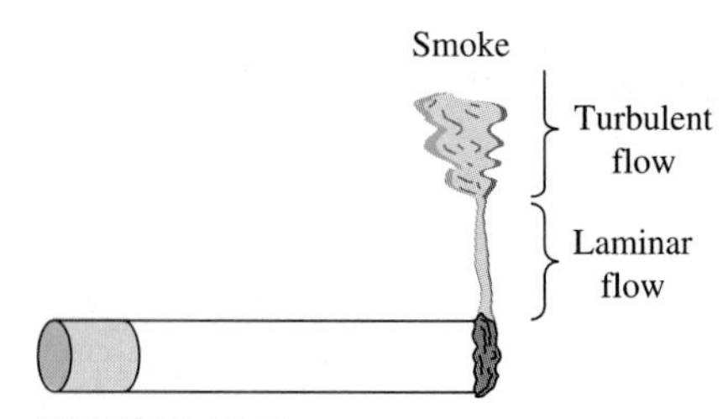


**FIGURE 10-7**
Laminar and turbulent flow regimes of cigarette smoke.

If you have been around smokers, you probably noticed that the cigarette smoke rises in a smooth plume for the first few centimeters, and then starts fluctuating randomly in all directions as it continues its journey towards the lungs of nonsmokers (Fig. 10-7). Likewise, a careful inspection of flow over a flat plate reveals that the fluid flow in the boundary layer starts out as flat and streamlined, but turns chaotic after some distance from the leading edge, as shown in Fig. 10-6. The flow regime in the first case is said to be **laminar**, and is characterized by *smooth streamlines* and *highly ordered motion,* and **turbulent** in the second case where it is characterized by *velocity fluctuations* and *highly disordered motion.* The **transition** from laminar to turbulent flow does not occur suddenly; rather, it occurs over some region in which the flow hesitates between laminar and turbulent flows before it becomes fully turbulent.

We can verify the existence of these laminar, transition, and turbulent flow regimes by injecting some dye into the flow stream. We will observe that the dye streak will form a *smooth line* when the flow is laminar, will have *bursts of fluctuations* in the transition regime, and will *zigzag rapidly and randomly* when the flow becomes fully turbulent.

Typical velocity profiles in laminar and turbulent flow are also given in Fig. 10-6. Note that the velocity profile is approximately parabolic in laminar flow, and becomes flatter in turbulent flow, with a sharp drop near the surface. The turbulent boundary layer can be considered to consist of three layers. The very thin layer next to the wall where the viscous effects are dominant is the **laminar sublayer**. The velocity profile in this layer is nearly linear, and the flow is streamlined. Next to the laminar sublayer is the **buffer layer**, in which the turbulent effects are significant but not dominant of the diffusion effects, and next to it is the **turbulent layer**, in which the turbulent effects dominate.

The *intense mixing* of the fluid in turbulent flow as a result of rapid fluctuations enhances heat and momentum transfer between fluid particles, which increases the friction force on the surface and the convection heat transfer rate (Fig. 10-8). It also causes the boundary layer to enlarge. Both the friction and heat transfer coefficients reach maximum values when the flow becomes *fully turbulent.* So it will come as no surprise that a special effort is made in the design of heat transfer equipment to achieve turbulence in order to take advantage of the high heat transfer coefficients associated with turbulent flow. The enhancement in heat transfer in turbulent flow does not come for free, however. It may be necessary to use a larger pump or fan in turbulent flow to overcome the larger friction forces accompanying the higher heat transfer rate.

(*a*) Before turbulence (°C)

(*b*) After turbulence (°C)

**FIGURE 10-8**

The intense mixing in turbulent flow brings fluid particles at different temperatures into close contact, and thus enhances heat transfer.

## The Reynolds Number

The transition from laminar to turbulent flow depends on the *surface geometry, surface roughness,* the *free-stream velocity,* the *surface temperature,* and the *type of fluid,* among other things. After exhaustive experiments in 1880s, Osborn Reynolds discovered that the flow regime depends mainly on the ratio of the *inertia forces* to *viscous forces* in the fluid. This ratio is called the **Reynolds number**, and is expressed for external flow as (Fig. 10-9)

$$\text{Re} = \frac{\text{inertia forces}}{\text{viscous forces}} = \frac{\mathcal{V}_\infty \delta}{\nu} \tag{10-9}$$

where $\mathcal{V}_\infty$ = free-stream velocity, m/s
$\delta$ = characteristic length of the geometry, m
$\nu = \mu/\rho$ = kinematic viscosity of the fluid, $m^2/s$

$$\text{Re} = \frac{\text{Inertia forces}}{\text{Viscous forces}} = \frac{\rho \mathcal{V}^2/\delta}{\mu \mathcal{V}/\delta^2} = \frac{\rho \mathcal{V} \delta}{\mu} = \frac{\mathcal{V} \delta}{\nu}$$

**FIGURE 10-9**

Reynolds number can be viewed as the ratio of the inertia forces to viscous forced acting on a fluid volume element.

Note that the Reynolds number is a *dimensionless* quantity. Also note that *kinematic viscosity* $\nu$ differs from dynamic viscosity $\mu$ by the factor $\rho$. Kinematic viscosity has the unit $m^2/s$, which is identical to the unit of thermal diffusivity, and can be viewed as *viscous diffusivity*. The characteristic length is the distance from the leading edge $x$ in the flow direction for a flat plate, and the diameter $D$ for a circular cylinder or sphere.

At *large* Reynolds numbers, the inertia forces, which are proportional to the density and the velocity of the fluid, are large relative to the viscous forces, and thus the viscous forces cannot prevent the random and rapid fluctuations of the fluid. At *small* Reynolds numbers, however, the viscous forces are large enough to overcome the inertia forces and to keep the fluid "in line." Thus the flow is *turbulent* in the first case, and *laminar* in the second.

The Reynolds number at which the flow becomes turbulent is called the **critical Reynolds number**. The value of the critical Reynolds number is different for different geometries. For flow over a *flat plate,* transition from laminar to turbulent occurs at the critical Reynolds number of

$$\text{Re}_{\text{critical, flat plate}} \approx 5 \times 10^5$$

This generally accepted value of the critical Reynolds number for a flat plate may vary somewhat depending on the surface roughness, the turbulence level, and the variation of pressure along the surface.

## 10-3 ■ THERMAL BOUNDARY LAYER

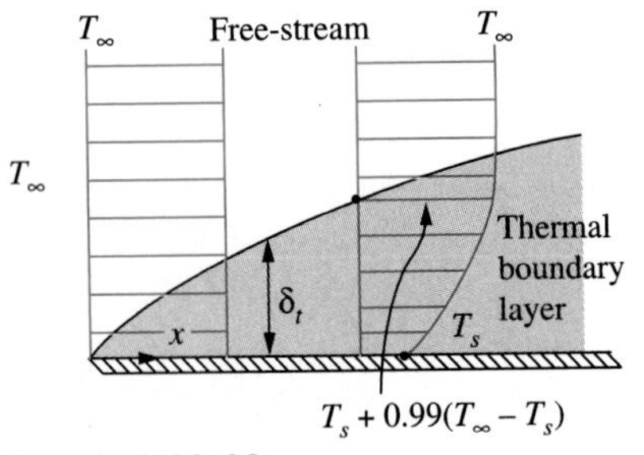


**FIGURE 10-10**
Thermal boundary layer on a flat plate (the fluid is hotter than the plate surface).

We have seen that a velocity boundary layer develops when a fluid flows over a surface as a result of the fluid layer adjacent to the surface assuming the surface velocity (i.e., zero velocity relative to the surface). Also, we defined the velocity boundary layer as the region in which the fluid velocity varies from zero to $0.99\mathcal{V}_\infty$. Likewise, a *thermal boundary layer* develops when a fluid at a specified temperature flows over a surface which is at a different temperature, as shown in Fig. 10-10.

Consider the flow of a fluid at a uniform temperature of $T_\infty$ over an isothermal flat plate at a temperature $T_s$. The fluid particles in the layer adjacent to the surface will reach thermal equilibrium with the plate and assume the surface temperature $T_s$. These fluid particles will then exchange energy with the particles in the adjoining fluid layer, and so on. As a result, a temperature profile will develop in the flow field which ranges from $T_s$ at the surface to $T_\infty$ sufficiently far from the surface. The flow region over the surface in which the temperature variation in the direction normal to the surface is significant is the **thermal boundary layer**. The *thickness* of the thermal boundary layer $\delta_t$ at any location along the surface is defined as *the distance from the surface at which the temperature difference* $T - T_s$ *equals* $0.99(T_\infty - T_s)$. Note that for the special case of $T_s = 0$, we have $T = 0.99T_\infty$ at the outer edge of the thermal boundary layer, which is analogous to $\mathcal{V} = 0.99\mathcal{V}_\infty$ for the velocity boundary layer.

The thickness of the thermal boundary layer increases in the flow direction, since the effects of heat transfer are felt at greater distances from the surface further down stream.

The convection heat transfer rate anywhere along the surface is directly related to the *temperature gradient* at that location. Therefore, the shape of the temperature profile in the thermal boundary layer dictates the convection heat transfer between a solid surface and the fluid flowing over it. In flow over a heated (or cooled) surface, both velocity and thermal boundary layers will develop simultaneously. Noting that the fluid velocity will have a strong influence on the temperature profile, the development of the velocity boundary layer relative to the thermal boundary layer will have a strong effect on the convection heat transfer.

The relative thickness of the velocity and the thermal boundary layers is best described by the *dimensionless* parameter **Prandtl number** defined as

$$\mathrm{Pr} = \frac{\text{molecular diffusivity of momentum}}{\text{molecular diffusivity of heat}} = \frac{\nu}{\alpha} = \frac{\mu C_p}{k} \qquad (10\text{-}10)$$

It is named after Ludwig Prandtl, who introduced the concept of boundary layer in 1904 and made significant contributions to boundary layer theory. The Prandtl numbers of fluids range from less than 0.01 for liquid metals to more than 100,000 for heavy oils (Table 10-2). Note that the Prandtl number is in the order of 10 for water.

The Prandtl numbers of gases are about 1, which indicates that both momentum and heat dissipate through the fluid at about the same rate. Heat diffuses very quickly in liquid metals ($\mathrm{Pr} \ll 1$), and very slowly in oils ($\mathrm{Pr} \gg 1$) relative to momentum. Consequently, the thermal boundary layer is much thicker for liquid metals, and much thinner for oils relative to the velocity boundary layer (Fig. 10-11).

## 10-4 ■ FLOW OVER FLAT RATES

So far we have discussed the physical aspects of forced convection over surfaces. In this section, we will discuss the determination of the *heat transfer rate* to or from a flat plate, as well as the *drag force* exerted on the plate by the fluid for both laminar and turbulent flow cases. Surfaces that are slightly contoured such as turbine blades can also be approximated as flat plates with reasonable accuracy.

The friction and the heat transfer coefficients for a flat plate can be determined theoretically by solving the conservation of mass, momentum, and energy equations approximately or numerically. They can also be determined experimentally, and expressed by empirical correlations. In either approach, it is found that the *average* Nusselt number can be expressed in terms of the Reynolds and Prandtl numbers in the form

$$\mathrm{Nu} = \frac{hL}{k} = C\,\mathrm{Re}_L^m\,\mathrm{Pr}^n \qquad (10\text{-}11)$$

**TABLE 10-2**
**Typical ranges of Prandtl numbers for common fluids**

| Fluid | Pr |
|---|---|
| Liquid metals | 0.004–0.030 |
| Gases | 0.7–1.0 |
| Water | 1.7–13.7 |
| Light organic fluids | 5–50 |
| Oils | 50-100,000 |
| Glycerin | 2000–100,000 |

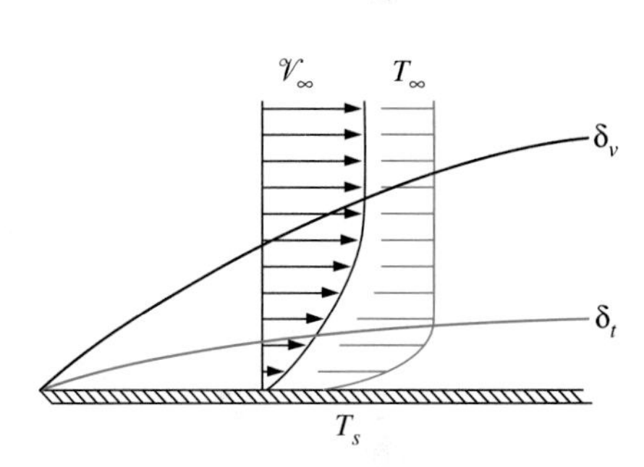

(*a*) Oils

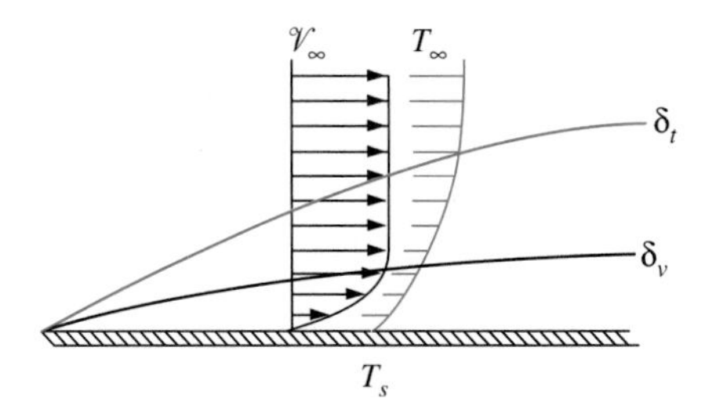


(*b*) Liquid metals (like mercury)

**FIGURE 10-11**
The relative thicknesses of the velocity and thermal boundary layers for liquid metals and oils.

where $C$, $m$, and $n$ are constants and $L$ is the *length* of the plate in the flow direction. The *local* Nusselt number at any point on the plate will depend on the distance of that point from the leading edge.

The fluid temperature in the thermal boundary layer varies from $T_s$ at the surface to about $T_\infty$ at the outer edge of the boundary. The fluid properties also vary with temperature, and thus with position across the boundary layer. In order to account for the variation of the properties with temperature properly, the fluid properties are usually evaluated at the so called **film temperature** defined as

$$T_f = \frac{T_s + T_\infty}{2} \qquad (10\text{-}12)$$

which is the *arithmetic average* of the surface and the free-stream temperatures. The fluid properties are then assumed to remain constant at those values during entire flow.

The local friction and heat transfer coefficients *vary* along the surface of the flat plate as a result of the changes in the velocity and thermal boundary layers in the flow direction. We are usually interested in the heat transfer and drag force on the *entire* surface, which can be determined using the *average* heat transfer and friction coefficients. But sometimes we are also interested in the heat flux and the drag force at a certain location. In such cases, we need to know the *local* values of the heat transfer and friction coefficients. With this in mind, below we present correlations for both local and average friction and heat transfer coefficients. The *local* quantities are identified with the subscript $x$.

Recall that the *average* friction and heat transfer coefficients for the entire plate can be determined from the corresponding *local* values by integration from

$$C_f = \frac{1}{L}\int_0^L C_{f,x}\,dx \qquad (10\text{-}13)$$

and

$$h = \frac{1}{L}\int_0^L h_x\,dx \qquad (10\text{-}14)$$

Once $C_f$ and $h$ are available, the drag force and the heat transfer rate can be determined from Eqs. 10-2 and 10-8. Next we discuss the local and average friction and heat transfer coefficients over a flat plate for *laminar, turbulent,* and *combined laminar and turbulent* flow conditions.

## 1 Laminar Flow

The *local* friction coefficient and the Nusselt number at location $x$ for laminar flow over a flat plate are given by

$$C_{f,x} = \frac{0.664}{\mathrm{Re}_x^{1/2}} \qquad (10\text{-}15)$$

and

$$\mathrm{Nu}_x = \frac{h_x x}{k} = 0.332\,\mathrm{Re}_x^{1/2}\mathrm{Pr}^{1/3} \qquad (\mathrm{Pr} \geqslant 0.6) \qquad (10\text{-}16)$$

where $x$ is the distance from the leading edge of the plate, and $\mathrm{Re}_x = u_\infty x/\nu$ is the Reynolds number at location $x$. Note that $C_{f,x}$ is proportional to $1/\mathrm{Re}_x^{1/2}$ and thus to $x^{-1/2}$. Likewise, $\mathrm{Nu}_x = h_x x/k$ is proportional to $x^{1/2}$ and thus $h_x$ is proportional to $x^{-1/2}$. Therefore, both $C_{f,x}$ and $h_x$ are supposedly *infinite* at the leading edge ($x = 0$), and decrease by a factor of $x^{-1/2}$ in the flow direction. The variation of the boundary layer thickness $\delta$, the friction coefficient $C_f$ and the convection heat transfer coefficient $h$ along an isothermal flat plate is shown in Fig. 10-12.

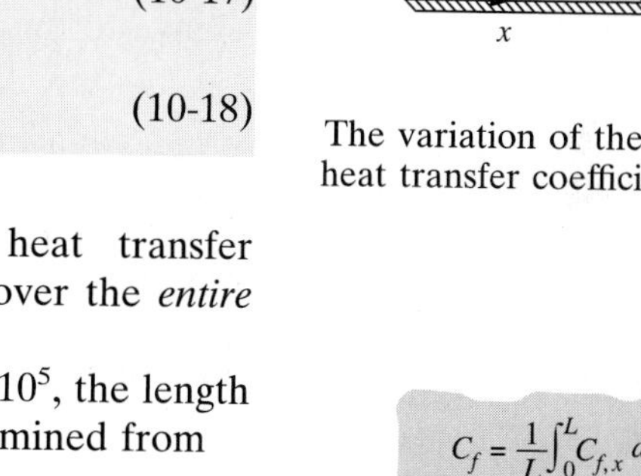



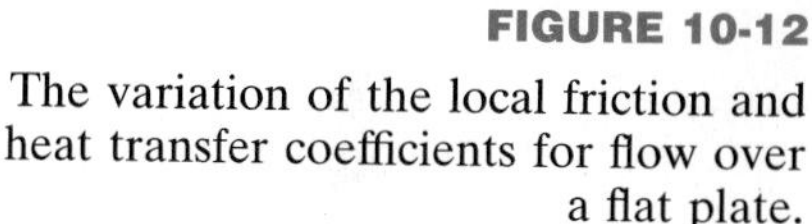

**FIGURE 10-12**
The variation of the local friction and heat transfer coefficients for flow over a flat plate.

The *average* friction coefficient and the Nusselt number over the entire plate are determined by substituting the relations above into Eqs. 10-13 and 10-14 and performing the simple integrations (Fig. 10-13). We get

$$C_f = \frac{1.328}{\mathrm{Re}_L^{1/2}} \tag{10-17}$$

and

$$\mathrm{Nu} = \frac{hL}{k} = 0.664\,\mathrm{Re}_L^{1/2}\,\mathrm{Pr}^{1/3} \qquad (\mathrm{Pr} \geq 0.6) \tag{10-18}$$

The relations above give the average friction and heat transfer coefficients for the entire plate when the flow is *laminar* over the *entire* plate.

Taking the critical Reynolds number to be $\mathrm{Re}_{cr} = 5 \times 10^5$, the length of the plate $x_{cr}$ over which the flow is laminar can be determined from

$$\mathrm{Re}_{cr} = 5 \times 10^5 = \frac{\mathscr{V}_\infty x_{cr}}{\nu} \tag{10-19}$$

Thus, the relations above can be used for $x \leq x_{cr}$.

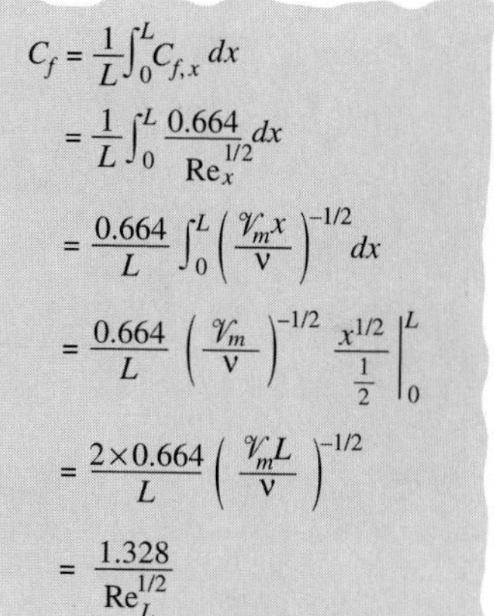



**FIGURE 10-13**
The average friction coefficient over a surface is determined by integrating the local friction coefficient over the entire surface.

## 2 Turbulent Flow

The *local* friction coefficient and the Nusselt number at location $x$ for turbulent flow over a flat plate are given by

$$C_{f,x} = \frac{0.0592}{\mathrm{Re}_x^{1/5}} \qquad (5 \times 10^5 \leq \mathrm{Re}_x \leq 10^7) \tag{10-20}$$

and

$$\mathrm{Nu}_x = \frac{h_x x}{k} = 0.0296\,\mathrm{Re}_x^{4/5}\,\mathrm{Pr}^{1/3} \qquad \begin{pmatrix} 0.6 \leq \mathrm{Pr} \leq 60 \\ 5 \times 10^5 \leq \mathrm{Re}_x \leq 10^7 \end{pmatrix} \tag{10-21}$$

where again $x$ is the distance from the leading edge of the plate, and $\mathrm{Re}_x = \mathscr{V}_\infty x/\nu$ is the Reynolds number at location $x$. The local friction and heat transfer coefficients are higher in turbulent flow than they are in laminar flow because of the intense mixing that occurs in the turbulent boundary layer. Note that both $C_{f,x}$ and $h_x$ reach their highest values when the flow becomes fully turbulent, and then decrease by a factor of $x^{-1/5}$ in the flow direction, as shown in Fig. 10-12.

The *average* friction coefficient and the Nusselt number over the entire plate in turbulent flow are determined by substituting the relations above into Eqs. 10-13 and 10-14 and performing the simple integrations. We get

$$C_f = \frac{0.074}{\mathrm{Re}_L^{1/5}} \qquad (5 \times 10^5 \leq \mathrm{Re}_L \leq 10^7) \tag{10-22}$$

and

$$\mathrm{Nu} = \frac{hL}{k} = 0.037\,\mathrm{Re}_L^{4/5}\,\mathrm{Pr}^{1/3} \qquad \begin{pmatrix} 0.6 \leq \mathrm{Pr} \leq 60 \\ 5 \times 10^5 \leq \mathrm{Re}_L \leq 10^7 \end{pmatrix} \tag{10-23}$$

The two relations above give the average friction and heat transfer coefficients for the entire plate only when the flow is *turbulent* over the *entire* plate, or when the laminar flow region of the plate is too small relative to the turbulent flow region (that is, $x_{cr} \ll L$).

## 3 Combined Laminar and Turbulent Flow

In some cases, a flat plate is sufficiently long for the flow to become turbulent, but not long enough to disregard the laminar flow region. In such cases, the *average* friction coefficient and the Nusselt number over the entire plate are determined by performing the integrations in Eqs. 10-13 and 10-14 over two parts: the laminar region $0 \leq x \leq x_{cr}$ and the turbulent region $x_{cr} < x \leq L$ as

$$C_f = \frac{1}{L}\left(\int_0^{x_{cr}} C_{f,x,\mathrm{laminar}}\,dx + \int_{x_{cr}}^{L} C_{f,x,\mathrm{turbulent}}\,dx\right) \tag{10-24}$$

and

$$h = \frac{1}{L}\left(\int_0^{x_{cr}} h_{x,\mathrm{laminar}}\,dx + \int_{x_{cr}}^{L} h_{x,\mathrm{turbulent}}\,dx\right) \tag{10-25}$$

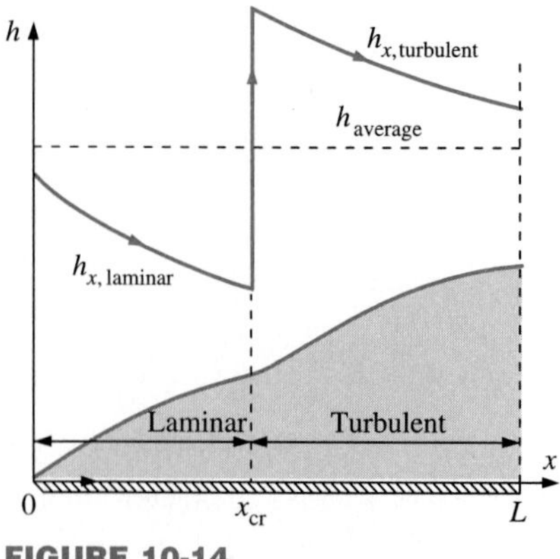


**FIGURE 10-14**

Graphical representation of the average heat transfer coefficient for a flat plate with combined laminar and turbulent flow.

Note that we included the transition region with the turbulent region. Again taking the critical Reynolds number to be $\mathrm{Re}_{cr} = 5 \times 10^5$ and performing the integrations above after substituting the indicated expressions, the *average* friction coefficient and the Nusselt number over the *entire* plate are determined to be (Fig. 10-14)

$$C_f = \frac{0.074}{\mathrm{Re}_L^{1/5}} - \frac{1742}{\mathrm{Re}_L} \qquad (5 \times 10^5 \leq \mathrm{Re}_L \leq 10^7) \tag{10-26}$$

and

$$\mathrm{Nu} = \frac{hL}{k} = (0.037\,\mathrm{Re}_L^{4/5} - 871)\,\mathrm{Pr}^{1/3} \qquad \begin{pmatrix} 0.6 \leq \mathrm{Pr} \leq 60 \\ 5 \times 10^5 \leq \mathrm{Re}_L \leq 10^7 \end{pmatrix} \tag{10-27}$$

The constants in the two relations above will be different for different critical Reynolds numbers.

When solving a forced convection problem over a flat plate, the first thing we do is determine $\mathrm{Re}_L$, which is the Reynolds number at the rear end of the plate, using the fluid properties at the film temperature. If $\mathrm{Re}_L < 5 \times 10^5$, the flow over the entire plate is laminar, and we use the laminar flow relations. If $\mathrm{Re}_L > 5 \times 10^5$, then we use the turbulent flow or the combined laminar and turbulent flow relations, as appropriate. Once the average friction and heat transfer coefficients are calculated, we can determine the heat transfer rate and the drag force from

$$\dot{Q} = hA(T_s - T_\infty) \quad \text{(W)} \tag{10-28}$$

and

$$F_D = C_f A \frac{\rho \mathcal{V}_\infty^2}{2} \quad \text{(N)} \tag{10-29}$$

where $A = wL$ is the surface area of a flat plate of length $L$ and width $w$.

The relations above have been obtained for the case of *isothermal* surfaces, but could also be used approximately for the case of nonisothermal surfaces by assuming the surface temperature to be constant at some average value. Also, the surfaces are assumed to be *smooth*, and the free stream to be *turbulent free*.

When a flat plate is subjected to *uniform heat flux* instead of uniform temperature, the local Nusselt number is given by

$$\mathrm{Nu}_x = 0.453\, \mathrm{Re}_x^{0.5}\, \mathrm{Pr}^{1/3} \tag{10-30}$$

for *laminar* flow, and

$$\mathrm{Nu}_x = 0.0308\, \mathrm{Re}_x^{0.8}\, \mathrm{Pr}^{1/3} \tag{10-31}$$

for *turbulent* flow. The relations above give values that are 36 percent higher for laminar flow, and 4 percent higher for turbulent flow relative to the isothermal plate case.

**EXAMPLE 10-1**

Engine oil at 60°C flows over a 5-m-long flat plate whose temperature is 20°C with a velocity of 2 m/s (Fig. 10-15). Determine the total drag force and the rate of heat transfer per unit width of the entire plate.

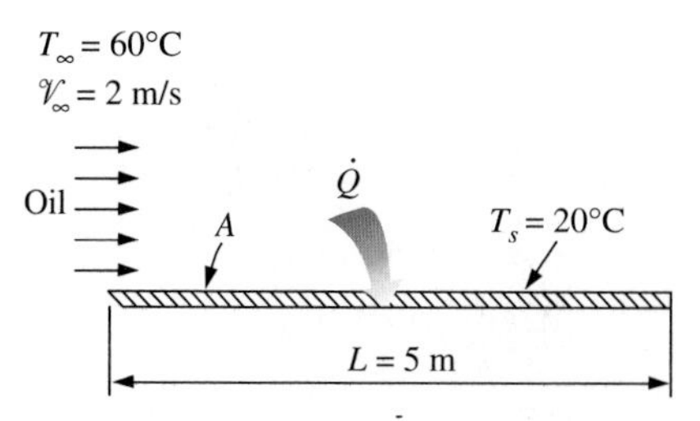


**FIGURE 10-15**
Schematic for Example 10-1.

**Solution** We assume the critical Reynolds number is $\mathrm{Re}_{cr} = 5 \times 10^5$. The properties of the engine oil at the film temperature of

$$T_f = \frac{T_s + T_\infty}{2} = \frac{20 + 60}{2} = 40°\mathrm{C}$$

are

$$\rho = 876 \text{ kg/m}^3$$
$$k = 0.144 \text{ W/(m} \cdot °\text{C)}$$
$$\mathrm{Pr} = 2870$$
$$\nu = 242 \times 10^{-6} \text{ m}^2/\text{s}$$

Noting that $L = 5$ m, and the Reynolds number at the end of the plate becomes

$$\text{Re}_L = \frac{\mathscr{V}_\infty L}{\nu} = \frac{(2\text{ m/s})(5\text{ m})}{242 \times 10^{-6}\text{ m}^2/\text{s}} = 4.13 \times 10^4$$

which is less than the critical Reynolds number. Thus we have *laminar flow* over the entire plate, and the average friction coefficient is determined from

$$C_f = 1.328\,\text{Re}_L^{-0.5} = 1.328 \times (4.13 \times 10^4)^{-0.5} = 0.00653$$

Then the drag force acting on the plate per unit width becomes

$$F_D = C_f A \frac{\rho \mathscr{V}_\infty^2}{2} = 0.00653 \times (5 \times 1\text{ m}^2)\frac{(876\text{ kg/m}^3)(2\text{ m/s})^2}{2} = 57.2\text{ N}$$

This force corresponds to the weight of a mass of about 6 kg. Therefore, a person who applies an equal and opposite force to the plate to keep it from moving will feel like he or she is spending as much power as is necessary to hold a 6 kg mass from dropping.

Similarly, the Nusselt number is determined using the laminar flow relations for a flat plate,

$$\text{Nu} = \frac{hL}{k} = 0.664\,\text{Re}_L^{0.5}\,\text{Pr}^{1/3} = 0.664 \times (4.13 \times 10^4)^{0.5} \times 2870^{1/3} = 1918$$

Then,

$$h = \frac{k}{L}\text{Nu} = \frac{0.144\text{ W/(m}\cdot\text{°C)}}{5\text{ m}}(1918) = 55.2\text{ W/(m}^2\cdot\text{°C)}$$

and

$$\dot{Q} = hA(T_\infty - T_s) = [55.2\text{ W/(m}^2\cdot\text{°C)}](5 \times 1\text{ m}^2)(60 - 20)\text{°C} = 11{,}040\text{ W}$$

Note that heat transfer is always from the higher-temperature medium to the lower-temperature one. In this case, it is from the oil to the plate. Both the drag force and the heat transfer rate are per m width of the plate. The total quantities for the entire plate can be obtained by multiplying these quantities by the actual width of the plate.

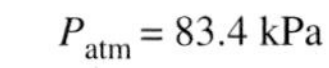


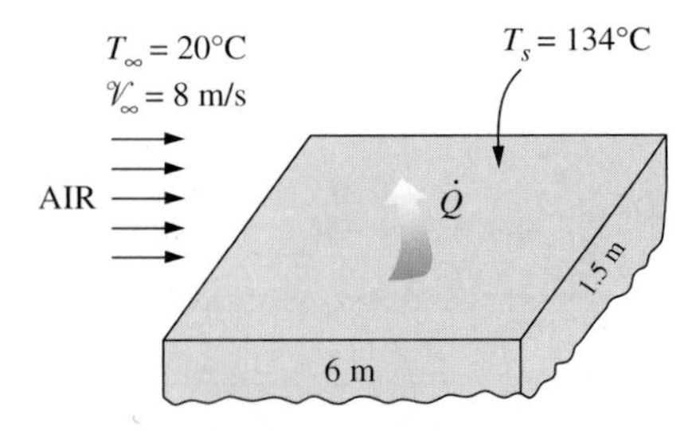


**FIGURE 10-16**
Schematic for Example 10-2.

**EXAMPLE 10-2**

The local atmospheric pressure in Denver, Colorado (elevation 1610 m) is 83.4 kPa. Air at this pressure and 20°C flows with a velocity of 8 m/s over a 1.5 m × 6 m flat plate whose temperature is 134°C (Fig. 10-16). Determine the rate of heat transfer from the plate if the air flows parallel to the (*a*) 6-m-long side and (*b*) the 1.5-m side.

**Solution** We assume steady-state conditions, and disregard any radiation effects. We also assume the critical Reynolds number is $\text{Re}_{cr} = 5 \times 10^5$, and treat air as an ideal gas. The properties of air are to be evaluated at 83.4 kPa and the film temperature of

$$T_f = \frac{T_s + T_\infty}{2} = \frac{134 + 20}{2} = 77\text{°C} = 350\text{ K}$$

The atmospheric pressure in Denver in atm is

$$P = (83.4\text{ kPa})\frac{1\text{ atm}}{101.325\text{ kPa}} = 0.823\text{ atm}$$

For an ideal gas, the thermal conductivity $k$ and the Prandtl number Pr are independent of pressure, but the kinematic viscosity $\nu$ is inversely proportional to pressure. Then the properties of air at 0.823 atm and 350 K become

$$k = 0.0297 \text{ W/(m} \cdot \text{°C)}$$
$$\text{Pr} = 0.706$$
$$\nu = \nu_{1\text{ atm}}/P\text{ (atm)} = (2.06 \times 10^{-5} \text{ m}^2/\text{s})/0.823 = 2.50 \times 10^{-5} \text{ m}^2/\text{s}$$

(*a*) When air flow is parallel to the long side, we have $L = 6$ m, and the Reynolds number at the end of the plate becomes

$$\text{Re}_L = \frac{\mathcal{V}_\infty L}{\nu} = \frac{(8 \text{ m/s})(6 \text{ m})}{2.50 \times 10^{-5} \text{ m}^2/\text{s}} = 1.92 \times 10^6$$

which is greater than the critical Reynolds number. Thus, we have combined laminar and turbulent flow, and the average Nusselt number for the entire plate is determined from

$$\text{Nu} = \frac{hL}{k} = (0.037 \text{ Re}_L^{0.8} - 871) \text{ Pr}^{1/3} = [0.037(1.92 \times 10^6)^{0.8} - 871]0.706^{1/3} = 2727$$

Then

$$h = \frac{k}{L}\text{Nu} = \frac{0.0297 \text{ W/(m} \cdot \text{°C)}}{6 \text{ m}}(2727) = 13.5 \text{ W/(m}^2 \cdot \text{°C)}$$
$$A = wL = (1.5 \text{ m})(6 \text{ m}) = 9 \text{ m}^2$$

and

$$\dot{Q} = hA(T_s - T_\infty) = [13.5 \text{ W/(m}^2 \cdot \text{°C)}](9 \text{ m}^2)(134 - 20)\text{°C} = 13{,}850 \text{ W}$$

Note that if we disregarded the laminar region and assumed turbulent flow over the entire plate, we would get Nu = 3520 from Eq. 10-23, which is 28 percent higher than the value we calculated above. Therefore, that assumption would cause a 28-percent error in heat transfer calculation in this case.

(*b*) When air flow is parallel to the short side, we have $L = 1.5$ m, and the Reynolds number at the end of the plate becomes

$$\text{Re}_L = \frac{\mathcal{V}_\infty L}{\nu} = \frac{(8 \text{ m/s})(1.5 \text{ m})}{2.50 \times 10^{-5} \text{ m}^2/\text{s}} = 4.80 \times 10^5$$

which is less than the critical Reynolds number. Thus we have laminar flow over the entire plate, and the average Nusselt number is determined from

$$\text{Nu} = \frac{hL}{k} = 0.664 \text{ Re}_L^{0.5} \text{ Pr}^{1/3} = 0.664 \times (4.8 \times 10^5)^{0.5} \times 0.706^{1/3} = 410$$

Then

$$h = \frac{k}{L}\text{Nu} = \frac{0.0297 \text{ W/(m} \cdot \text{°C)}}{1.5 \text{ m}}(410) = 8.12 \text{ W/(m}^2 \cdot \text{°C)}$$

and

$$\dot{Q} = hA(T_s - T_\infty) = [8.12 \text{ W/(m}^2 \cdot {}^\circ\text{C)}](9 \text{ m}^2)(134 - 20)^\circ\text{C} = 8330 \text{ W}$$

which is considerably less than the heat transfer rate determined in case (*a*). Therefore, the *direction* of fluid flow can have significant effect on convection heat transfer to or from a surface (Fig. 10-17). In this case, we can increase the heat transfer rate by 67 percent by simply blowing the air parallel to the long side of the rectangular plate instead of the short side.

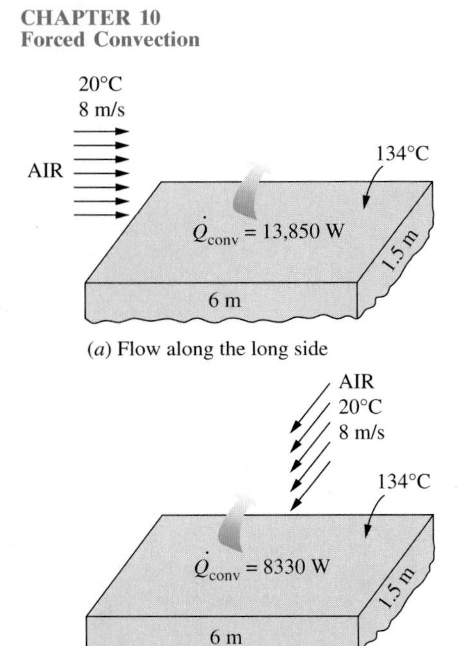


**FIGURE 10-17**
The direction of fluid flow can have significant effect on convection heat transfer.

## 10-5 ■ FLOW ACROSS CYLINDERS AND SPHERES

In the preceding section, we considered fluid flow over *flat* surfaces. In this section, we consider flow over cylinders and spheres, which is frequently encountered in practice. For example, the tubes in a tube-and-shell heat exchanger involve both *internal flow* through the tubes, and *external flow* over the tubes, and both flows must be considered in the analysis of heat transfer between the two fluids. Below we consider external flow only.

The characteristic length for a circular cylinder or sphere is taken to be the *external diameter D*. Thus, the Reynolds number is defined as

$$\text{Re} = \frac{\mathcal{V}_\infty D}{\nu}$$

where $\mathcal{V}_\infty$ is the uniform velocity of the fluid as it approaches the cylinder or sphere. The critical Reynolds number for flow across a circular cylinder or sphere is $\text{Re}_{cr} \approx 2 \times 10^5$. That is, the boundary layer remains laminar for $\text{Re} < 2 \times 10^5$, and becomes turbulent for $\text{Re} > 2 \times 10^5$.

Cross-flow over a cylinder exhibits complex flow patterns, as shown in Fig. 10-18. The fluid approaching the cylinder will branch out and encircle the cylinder, forming a boundary layer that wraps around the cylinder. The fluid particles on the midplane will strike the cylinder at the stagnation point, bringing the fluid to a complete stop and thus raising the pressure at that point. The pressure decreases in the flow direction while the fluid velocity increases.

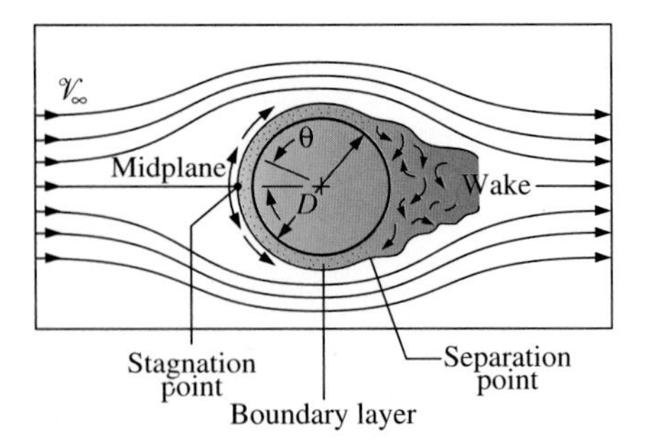


**FIGURE 10-18**
Typical flow patterns in cross-flow over a cylinder.

At very low free-stream velocities ($\text{Re} < 4$), the fluid completely wraps around the cylinder and the two arms of the fluid meet on the rear side of the cylinder in an orderly manner. Thus the fluid follows the curvature of the cylinder. At higher velocities, the fluid still hugs the cylinder on the frontal side, but it is too fast to remain attached to the surface as it approaches the top of the cylinder. As a result, the boundary layer detaches from the surface, forming a wake behind the cylinder. This point is called the **separation point**. Flow in the wake region is characterized by random vortex formation and pressures much lower than the stagnation point pressure.

The flow separation phenomena is analogous to the fast vehicles jumping off on the hills. At low velocities, the wheels of the vehicle always remain in contact with the road surface. But at high velocities, the

vehicle is too fast to follow the curvature of the road, and takes off at the hill losing contact with the road.

Flow separation occurs at about $\theta \approx 80°$ when the boundary layer is *laminar,* and at about $\theta \approx 140°$ when it is *turbulent* (Fig. 10–19). The delay of separation in turbulent flow is caused by the rapid fluctuations of the fluid in transverse direction.

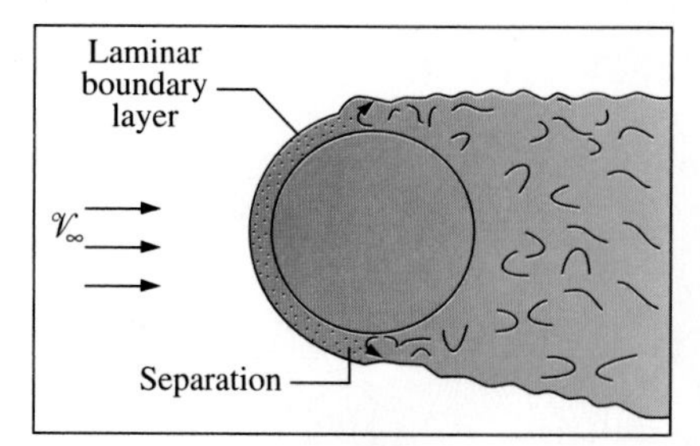


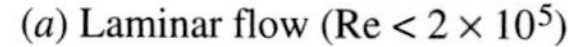
($a$) Laminar flow (Re < 2 × $10^5$)

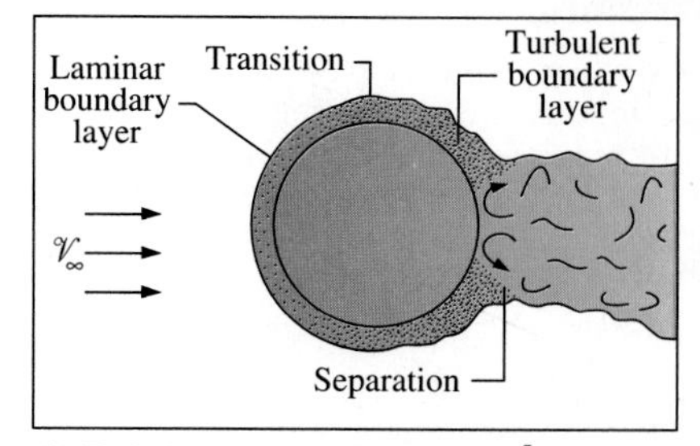


($b$) Turbulence occurs (Re > 2 × $10^5$)

**FIGURE 10-19**

Turbulence delays flow separation.

## The Drag Coefficient

The nature of the flow across a cylinder or sphere strongly effect the drag coefficient $C_D$ and the heat transfer coefficient $h$. The *drag force* acting on a body in cross flow is caused by two effects: the *friction drag,* which is due to the shear stress at the surface, and the *pressure drag,* which is due to pressure differential between the front and the rear sides of the body when a wake is formed in the rear. The high pressure in the vicinity of the stagnation point and the low pressure at the opposite side in the wake produce a net force on the body in the direction of flow. The drag force is primarily due to friction drag at low Reynolds numbers (Re $< 4$), and to pressure drag at high Reynolds numbers (Re $> 5000$). Both effects are significant at intermediate Reynolds numbers.

The average drag coefficients $C_D$ for cross-flow over a single circular cylinder and a sphere are given in Fig. 10-20. The large reduction in $C_D$ for Re $> 2 \times 10^5$ is caused by the transition to *turbulent* flow, which moves the separation point further on the rear of the body, reducing the size of the wake and thus the magnitude of the pressure drag. Once the drag coefficient is available, the drag force acting on a body in cross flow can be determined from

$$F_D = C_D A_N \frac{\rho \mathscr{V}_\infty^2}{2} \quad \text{(N)}$$

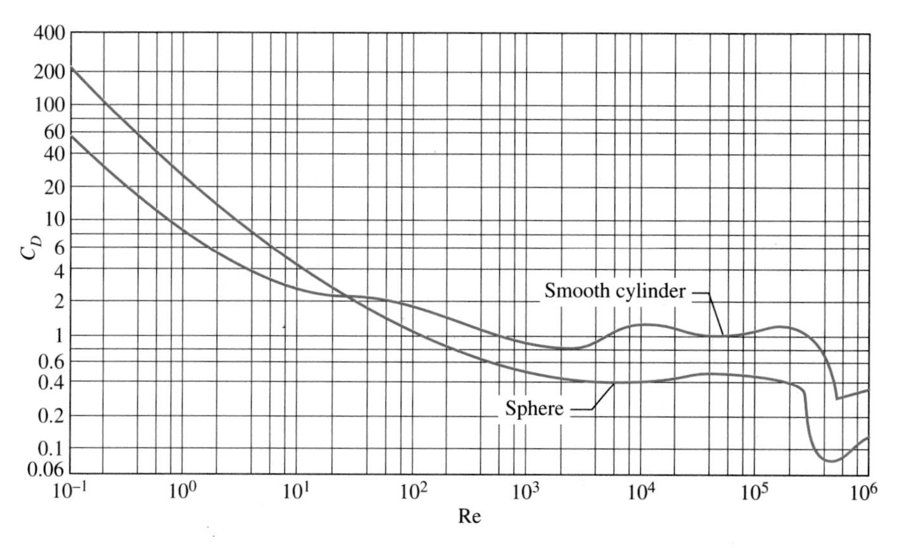

**FIGURE 10-20**

Average drag coefficient for cross flow over a smooth circular cylinder and a smooth sphere (from Schlichting).

where $A_N$ is the *frontal area* (area normal to the direction of flow). It is equal to $LD$ for a cylinder of length $L$, and $\frac{1}{4}\pi D^2$ for a sphere.

The term *drag coefficient* is also used commonly in daily life. Car manufacturers try to attract consumers by pointing out the *low drag coefficients* of their cars (Fig. 10-21). The surfaces of golf balls are intentionally roughened to induce *turbulence* at a lower Reynolds number to take advantage of the sharp *drop* in the drag coefficient at the onset of turbulence in the boundary layer. *Airplanes* are built to resemble birds, and *submarines* to resemble fish in order to minimize the drag coefficients, and thus the fuel consumption. The airplanes retract their wheels, just like the birds retracting their feet, in order to reduce the drag coefficients.

**FIGURE 10-21**

Modern vehicles are shaped as to minimize the drag coefficient and thus to maximize the fuel efficiency.

## The Heat Transfer Coefficient

Flows across cylinders and spheres, in general, involve *flow separation,* which is difficult to handle analytically. Therefore, such flows must be studied experimentally. Indeed, flow across cylinders and spheres has been studied experimentally by numerous investigators, and several empirical correlations are developed for the heat transfer coefficient.

The complicated flow pattern across a cylinder discussed earlier greatly influences heat transfer. The variation of the local Nusselt number $Nu_\theta$ around the periphery of a cylinder subjected to cross-flow of air is given in Fig. 10-22. Note that, for all cases, the value of $Nu_\theta$ starts out relatively high at the stagnation point ($\theta = 0°$), but decreases with increasing $\theta$ as a result of the thickeneing of the laminar boundary layer. On the two curves at the bottom corresponding to Re = 70,000 and 101,300, $Nu_\theta$ reaches a minimum at $\theta \approx 80°$, which is the separation point in laminar flow. Then $Nu_\theta$ increases with increasing $\theta$ as a result of the intense mixing in the separated flow region (the wake). The curves at the top corresponding to Re = 140,000–219,000 differ from the first two curves in that they have *two* minima for $Nu_\theta$. The sharp increase in $Nu_\theta$ at about $\theta \approx 90°$ is due to transition from laminar to turbulent flow. The later decrease in $Nu_\theta$ is again due to the thickening of the boundary layer. $Nu_\theta$ reaches its second minimum at about $\theta \approx 140°$, which is the flow separation point in turbulent flow, and increases with $\theta$ as a result of the intense mixing in the turbulent wake region.

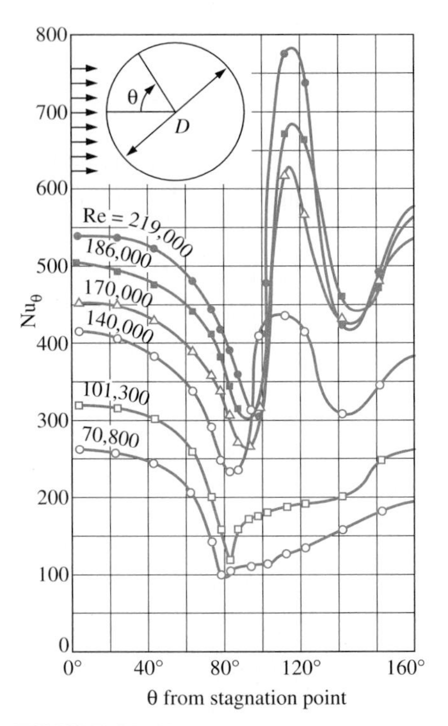


**FIGURE 10-22**

Variation of the local heat transfer coefficient along the circumference of a circular cylinder in cross flow of air (from Giedt).

The discussions above on the local heat transfer coefficients are insightful; however, they are of little value in heat transfer calculations, since the calculation of heat transfer requires the *average* heat transfer coefficient over the entire surface. Of the several such relations available in the literature for the average Nusselt number for cross-flow over a *cylinder,* we present the one proposed by Churchill and Bernstein:

$$Nu_{cyl} = \frac{hD}{k} = 0.3 + \frac{0.62\, Re^{1/2} Pr^{1/3}}{[1 + (0.4/Pr)^{2/3}]^{1/4}} \left[1 + \left(\frac{Re}{28{,}200}\right)^{5/8}\right]^{4/5} \tag{10-32}$$

This relation is quite comprehensive in that it correlates all available data well for Re Pr > 0.2. The fluid properties are evaluated at the *film temperature* $T_f = \frac{1}{2}(T_\infty + T_s)$, which is the average of the free-stream and surface temperatures.

For flow over a *sphere,* Whitaker recommends the following comprehensive correlation:

$$\text{Nu}_{\text{sph}} = \frac{hD}{k} = 2 + [0.4\,\text{Re}^{1/2} + 0.06\,\text{Re}^{2/3}]\,\text{Pr}^{0.4}\left(\frac{\mu_\infty}{\mu_s}\right)^{1/4} \tag{10-33}$$

which is valid for $3.5 \leq \text{Re} \leq 80{,}000$ and $0.7 \leq \text{Pr} \leq 380$. The fluid properties in this case are evaluated at the free-stream temperature $T_\infty$, except for $\mu_s$, which is evaluated at the surface temperature $T_s$. Although the two relations above are considered to be quite accurate, the results obtained from them can be off by as much as 30 percent.

The average Nusselt number for flow across cylinders can be expressed compactly as

$$\text{Nu}_{\text{cyl}} = \frac{hD}{k} = C\,\text{Re}^m\,\text{Pr}^n \tag{10-34}$$

where $n = \frac{1}{3}$ and the experimentally determined constants $C$ and $m$ are given in Table 10-3 for circular as well as various noncircular cylinders. The characteristic length $D$ for use in the calculation of the Reynolds and the Nusselt numbers for different geometries are as indicated on the figure. All fluid properties are evaluated at the *film temperature* $T_f = \frac{1}{2}(T_\infty + T_s)$.

The relations for cylinders above are for *single* cylinders, or cylinders oriented such that the flow over them is not affected by the presence of others. Also, they are applicable to *smooth* surfaces. *Surface roughness* and the *free-stream turbulence* may affect the drag and heat transfer coefficients significantly. Drag and heat transfer coefficients for flow over *tube bundles* can be obtained from some of the standard heat transfer texts listed at the end of this chapter. Equation 10-34 provides a simpler alternative to Eq. 10-32 for flow over cylinders. However, Eq. 10-32 is more accurate, and thus should be preferred in calculations whenever possible.

**EXAMPLE 10-3**

A long 10-cm-diameter steam pipe whose external surface temperature is 110°C passes through some open area that is not protected against the winds (Fig. 10-23). Determine the rate of heat loss from the pipe per unit of its length when the air is at 1-atm pressure and 4°C and the wind is blowing across the pipe at a velocity of 8 m/s.

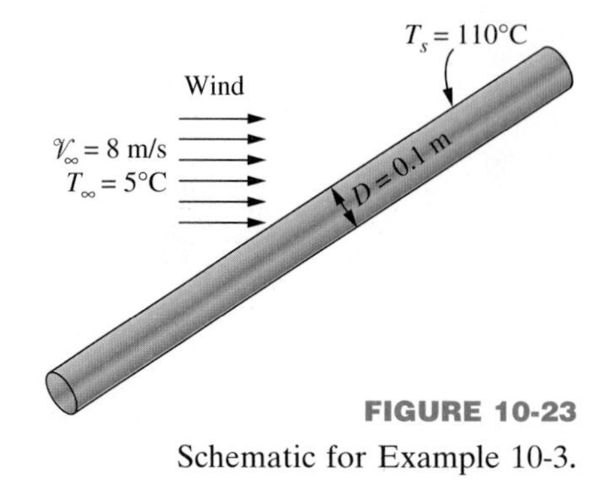


**FIGURE 10-23**
Schematic for Example 10-3.

**Solution** This is an *external flow* problem, since we are interested in the heat transfer from the pipe to the air that is flowing outside the pipe. The

**TABLE 10-3**
**Empirical correlations for the average Nusselt number for forced convection over circular and noncircular cylinders in cross-flow (from Zhukauskas and Jakob)**

| Cross-section of the cylinder | Fluid | Range of Re | Nusselt number |
|---|---|---|---|
| Circle | Gas or liquid | 0.4–4<br>4–40<br>40–4000<br>4000–40,000<br>40,000–400,000 | $Nu = 0.989Re^{0.330} Pr^{1/3}$ (10–35)<br>$Nu = 0.911Re^{0.385} Pr^{1/3}$ (10–36)<br>$Nu = 0.683Re^{0.466} Pr^{1/3}$ (10–37)<br>$Nu = 0.193Re^{0.618} Pr^{1/3}$ (10–38)<br>$Nu = 0.027Re^{0.805} Pr^{1/3}$ (10–39) |
| Square | Gas | 5000–100,000 | $Nu = 0.102Re^{0.675} Pr^{1/3}$ (10–40) |
| Square (tilted 45°) | Gas | 5000–100,000 | $Nu = 0.246Re^{0.588} Pr^{1/3}$ (10–41) |
| Hexagon | Gas | 5000–100,000 | $Nu = 0.153Re^{0.638} Pr^{1/3}$ (10–42) |
| Hexagon (tilted 45°) | Gas | 5000–19,500<br>19,500–100,000 | $Nu = 0.160Re^{0.638} Pr^{1/3}$ (10–43)<br>$Nu = 0.0385Re^{0.782} Pr^{1/3}$ (10–44) |
| Vertical plate | Gas | 4000–15,000 | $Nu = 0.228Re^{0.731} Pr^{1/3}$ (10–45) |
| Ellipse | Gas | 2500–15,000 | $Nu = 0.248Re^{0.612} Pr^{1/3}$ (10–46) |

properties of air at 1-atm pressure and the film temperature of $T_f = \frac{1}{2}(T_\infty + T_s) = \frac{1}{2}(4 + 110) = 57°C = 330$ K are

$$k = 0.0283 \text{ W/(m} \cdot \text{°C)}$$

$$\nu = 1.86 \times 10^{-5} \text{ m}^2\text{/s}$$

$$\text{Pr} = 0.708$$

The Reynolds number of the flow is

$$\text{Re} = \frac{\mathcal{V}_\infty D}{\nu} = \frac{(8 \text{ m/s})(0.1 \text{ m})}{1.86 \times 10^{-5} \text{ m}^2\text{/s}} = 43{,}011$$

Then the Nusselt number in this case can be determined from

$$\mathrm{Nu} = \frac{hD}{k} = 0.3 + \frac{0.62\,\mathrm{Re}^{1/2}\,\mathrm{Pr}^{1/3}}{[1 + (0.4/\mathrm{Pr})^{2/3}]^{1/4}}\left[1 + \left(\frac{\mathrm{Re}}{28{,}200}\right)^{5/8}\right]^{4/5}$$

$$= 0.3 + \frac{0.62(43{,}011)^{1/2}(0.708)^{1/3}}{[1 + (0.4/0.708)^{2/3}]^{1/4}}\left[1 + \left(\frac{43{,}011}{28{,}200}\right)^{5/8}\right]^{4/5}$$

$$= 196.3$$

and

$$h = \frac{k}{D}\mathrm{Nu} = \frac{0.0283\ \mathrm{W/(m \cdot {}^\circ C)}}{0.1\ \mathrm{m}}(196.3) = 55.6\ \mathrm{W/(m^2 \cdot {}^\circ C)}$$

Then the rate of heat transfer from the pipe per unit of its length becomes

$$A = pL = \pi DL = \pi(0.1\ \mathrm{m})(1\ \mathrm{m}) = 0.314\ \mathrm{m^2}$$

$$\dot{Q} = hA(T_s - T_\infty) = [55.6\ \mathrm{W/(m^2 \cdot C)}](0.314\ \mathrm{m^2})(110 - 4){}^\circ\mathrm{C} = 1851\ \mathrm{W}$$

The rate of heat loss from the entire pipe can be obtained by multiplyihng the value above by the length of the pipe in m.

The simpler Nusselt number relation (Eq. 10-39) in this case would give Nu = 129, which is 34 percent lower han the value obtained above using Eq. 10-32.

---

**EXAMPLE 10-4**

A 25-cm-diameter stainless steel ball [$\rho = 8055\ \mathrm{kg/m^3}$, $C_p = 480\ J/(kg \cdot {}^\circ\mathrm{C})$] is removed from the oven at a uniform temperature of 300°C (Fig. 10-24). The ball is then subjected to the flow of air at 1-atm pressure and 27°C with a velocity of 3 m/s. The surface temperature of the ball eventually drops to 200°C. Determine the average convection heat transfer coefficient during this cooling process, and estimate how long this cooling process will take.

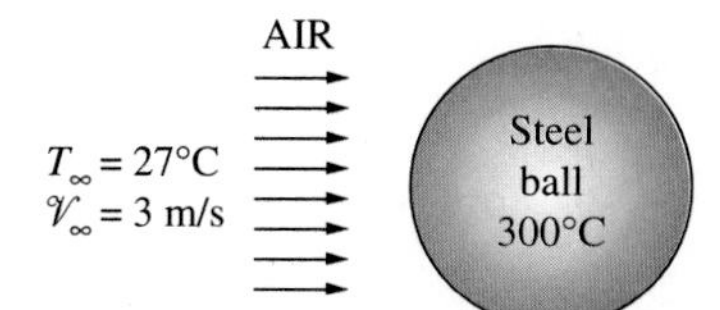


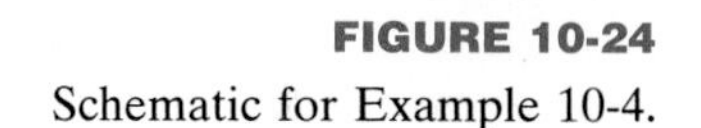
**FIGURE 10-24**
Schematic for Example 10-4.

**Solution** This is an *external flow* problem, since the air flows outside the ball. We assume the outer surface temperature of the ball to be uniform at all times, and disregard any heat loss by radiation. The surface temperature of the ball during cooling is changing. Therefore, the convection heat transfer coefficient between the ball and the air will also change. To avoid this complexity, we assume the surface temperature of the ball to be constant at the average temperature of $\frac{1}{2}(300 + 200) = 250$°C in the evaluation of the heat transfer coefficient.

The properties of air at 1-atm pressure and the free-stream temperature of 27°C are

$$k = 0.0261\ \mathrm{W/(m \cdot {}^\circ C)}$$
$$\mu = 1.85 \times 10^{-5}\ \mathrm{kg/(m \cdot s)}$$
$$\nu = 1.57 \times 10^{-5}\ \mathrm{m^2/s}$$
$$\mathrm{Pr} = 0.712$$

The Reynolds number of the flow is

$$\mathrm{Re} = \frac{\mathscr{V}_\infty D}{\nu} = \frac{(3\ \mathrm{m/s})(0.25\ \mathrm{m})}{1.57 \times 10^{-5}\ \mathrm{m^2/s}} = 47{,}800$$

Then the Nusselt number can be determined from

$$\text{Nu} = \frac{hD}{k} = 2 + [0.4\,\text{Re}^{1/2} + 0.06\,\text{Re}^{2/3}]\,\text{Pr}^{0.4}\left(\frac{\mu_\infty}{\mu_s}\right)^{1/4}$$

$$= 2 + [0.4(47{,}800)^{1/2} + 0.06(47{,}800)^{2/3}](0.712)^{0.4}\left(\frac{1.85 \times 10^{-5}}{2.96 \times 10^{-5}}\right)^{1/4}$$

$$= 131$$

where we substituted $\mu_s = \mu_{250^\circ\text{C}} = 2.96 \times 10^{-5}\,\text{kg/(m}\cdot\text{s)}$ for the dynamic viscosity of air at the surface temperature . Then the average convection heat transfer coefficient becomes

$$h = \frac{k}{D}\text{Nu} = \frac{0.0261\,\text{W/(m}\cdot{}^\circ\text{C)}}{0.25\,\text{m}}(131) = 13.6\,\text{W/(m}^2\cdot{}^\circ\text{C)}$$

In order to estimate the time of cooling of the ball from 300°C to 200°C, we determine the *average* rate of heat transfer from Newton's law of cooling by using the *average* surface temperature. That is,

$$A = \pi D^2 = \pi(0.25\,\text{m})^2 = 0.196\,\text{m}^2$$

$$\dot{Q}_{\text{ave}} = hA(T_{s,\text{ave}} - T_\infty) = [13.6\,\text{W/(m}^2\cdot{}^\circ\text{C)}](0.196\,\text{m}^2)(250 - 27)^\circ\text{C} = 594\,\text{W}$$

Next we determine the *total* heat transferred from the ball, which is simply the change in the energy of the ball as it cools from 300°C to 200°C:

$$m = \rho V = \rho \tfrac{1}{6}\pi D^3 = (8085\,\text{kg/m}^3)\tfrac{1}{6}\pi(0.25\,\text{m})^3 = 66.1\,\text{kg}$$

$$Q_{\text{total}} = mC_p(T_2 - T_1) = (66.1\,\text{kg})[480\,\text{J/(kg}\cdot{}^\circ\text{C)}](300 - 200)^\circ\text{C} = 3{,}172{,}800\,\text{J}$$

In the above calculation, we assumed that the *entire ball* is at 200°C, which is not necessarily true. The inner region of the ball will probably be at a higher temperature than its surface. With this assumption, the time of cooling is determined to be

$$\Delta t \approx \frac{Q}{\dot{Q}_{\text{ave}}} = \frac{3{,}172{,}800\,\text{J}}{594\,\text{J/s}} = 5341\,\text{s} = 1\,\text{h}\;29\,\text{min}$$

The time of cooling could also be determined more accurately using the transient temperature charts or relations introduced in Chap. 9. But the simplifying assumptions we made above can be justified if all we need is a ball-park value. It will be naive to expect the time of cooling to be exactly 1 h 29 min, but, using our engineering judgement, it is realistic to expect the time of cooling to be somewhere between one and two hours.

## 10-6 ■ FLOW IN TUBES

Liquid or gas flow through *pipes* or *ducts* is commonly used in practice in heating and cooling applications. The fluid in such applications is forced to flow by a fan or pump through a tube that is sufficiently long to accomplish the desired heat transfer. In this section, we will discuss the *friction* and *heat transfer coefficients,* which are directly related to the *pressure drop* and *heat flux* for flow through tubes. These quantities are then used to determine the pumping power requirement and the length of the tube.

There is a *fundamental* difference between external and internal flows. In *external flow,* which we have considered so far, the fluid had a free surface, and thus the boundary layer over the surface was free to grow indefinitely. In *internal flow,* however, the fluid is completely confined by the inner surfaces of the tube, and thus there is a limit on how much the boundary layer can grow.

## General Considerations

The fluid velocity in a tube changes from *zero* at the surface to a *maximum* at the tube center. In fluid flow, it is convenient to work with an *average* or *mean* velocity $\mathcal{V}_m$, which remains constant in incompressible flow when the cross-sectional area of the tube is constant. The mean velocity in actual heating and cooling applications may change somewhat because of the changes in density with temperature. But, in practice, we evaluate the fluid properties at some average temperature, and treat them as constants. The convenience in working with constant properties usually more than justifies the slight loss in accuracy.

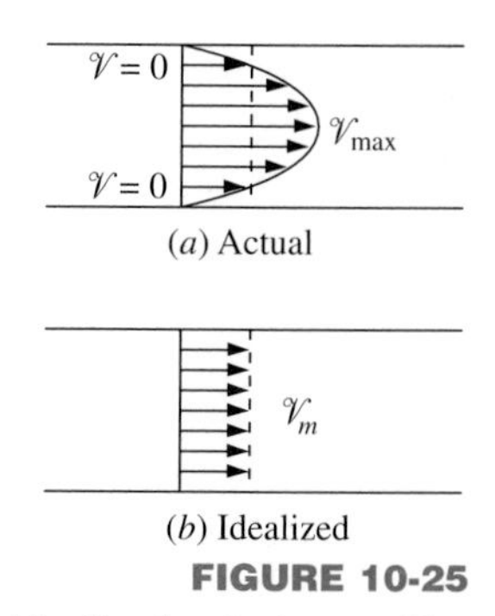


**FIGURE 10-25**

Actual and idealized velocity profiles for flow in a tube (the mass flow rate of the fluid is the same for both cases).

The value of the mean velocity $\mathcal{V}_m$ is determined from the requirement that the *conservation of mass* principle be satisfied (Fig. 10-25). That is, the mass flow rate through the tube evaluated using the mean velocity $\mathcal{V}_m$ from

$$\dot{m} = \rho \mathcal{V}_m A_c \qquad \text{(kg/s)}$$

be equal to the actual mass flow rate. Here $\rho$ is the density of the fluid and $A_c$ is the cross-sectional area, which is equal to $A_c = \frac{1}{4}\pi D^2$ for a circular tube.

When a fluid is heated or cooled as it flows through a tube, the temperature of a fluid at any cross-section changes from $T_s$ at the surface of the wall at that cross-section to some maximum (or minimum in the case of heating) at the tube center. In fluid flow it is convenient to work with an *average* or *mean* tempeature $T_m$ that remains constant at a cross section. The mean temperature $T_m$ *will change* in the flow direction, however, whenever the fluid is heated or cooled.

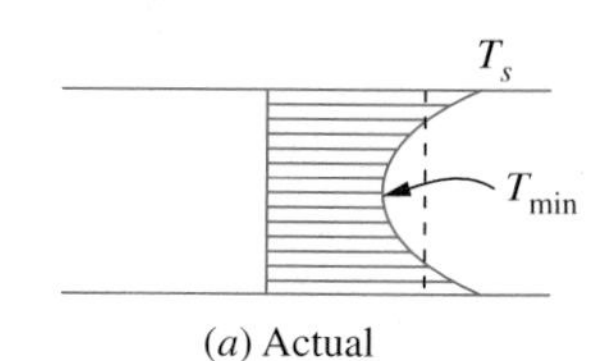


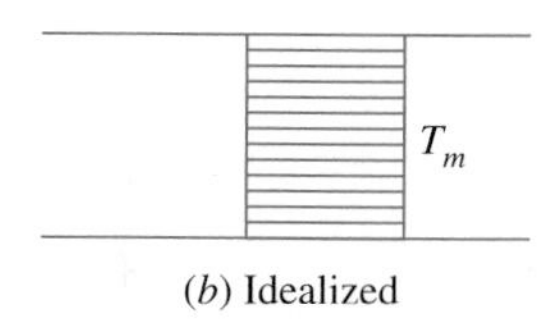


**FIGURE 10-26**

Actual and idealized temperature profiles for flow in a tube (the rate at which energy is transported with the fluid is the same for both cases).

The value of the mean temperature $T_m$ is determined from the requirement that the *conservation of energy* principle be satisfied. That is, the energy transported by the fluid through a cross-section in actual flow be equal to the energy that would be transported through the same cross-section if the fluid were at a constant temperature $T_m$. This can be expressed mathematically as (Fig. 10-26)

$$\dot{E}_{\text{fluid}} = \dot{m}C_pT_m = \int_{\dot{m}} C_pT\,\delta\dot{m} = \int_{A_c} C_pT(\rho\mathcal{V}\,dA_c) \qquad \text{(kJ/s)} \quad (10\text{-}47)$$

where $C_p$ is the specific heat of the fluid and $\dot{m}$ is the mass flow rate. Note that the product $\dot{m}C_pT_m$ at any cross-section along the tube represents the *energy flow* with the fluid at that cross-section. You will recall from Chap. 4 that in the absence of any work interactions (such as electric resistance heating), the conservation of energy equation for the steady flow of a fluid in a tube can be expressed as (Fig. 10-27)

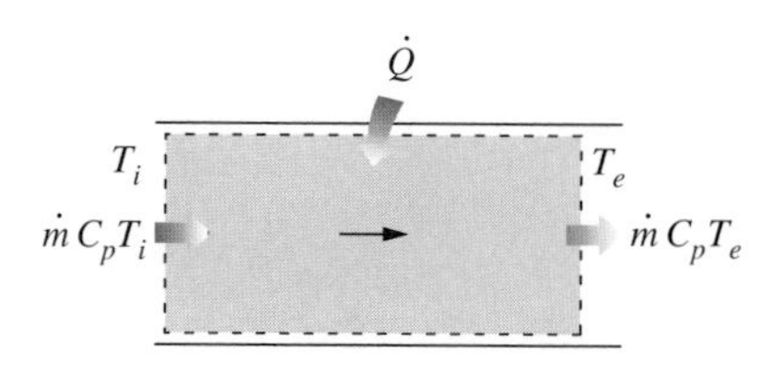


Energy balance:

$\dot{Q} = \dot{m}\,C_p(T_e - T_i)$

**FIGURE 10-27**

The heat transfer to a fluid flowing in a tube is equal to the increase in the energy of the fluid.

$$\dot{Q} = \dot{m}C_p(T_e - T_i) \qquad \text{(kJ/s)} \qquad (10\text{-}48)$$

where $T_i$ and $T_e$ are the mean fluid temperatures at the inlet and exit of the tube, respectively, and $\dot{Q}$ is the rate of heat transfer to or from the fluid. Note that the temperature of a fluid flowing in a tube remains constant in the absence of any energy interactions through the wall of the tube.

Perhaps we should mention that the friction between the fluid layers in a tube do cause a slight rise in fluid temperature as a result of the mechanical energy being converted to sensible heat energy. But this *frictional heating* is too small to warrant any consideration in calculations, and thus is disregarded. For example, in the absence of any heat transfer, no noticeable difference will be detected between the inlet and exit temperatures of a fluid flowing in a tube. Thus, it is reasonable to assume that any temperature change in the fluid is due to heat transfer.

The thermal conditions at the surface of a tube can usually be approximated with reasonable accuracy to be *constant surface temperature* ($T_s$ = constant) or *constant surface heat flux* ($\dot{q}_s$ = constant). For example, the constant surface temperature condition is realized when a phase change process such as boiling or condensation occurs at the outer surface of a tube. The constant surface heat flux condition is realized when the tube is subjected to radiation or electric resistance heating uniformly from all directions.

The convection heat flux at any location on the tube can be expressed as

$$\dot{q} = h(T_s - T_m) \quad \text{(W/m}^2\text{)} \tag{10-49}$$

where $h$ is the *local* heat transfer coefficient and $T_s$ and $T_m$ are the surface and the mean fluid temperatures at that location. Note that the mean fluid temperature $T_m$ of a fluid flowing in a tube must change during heating or cooling. Therefore, when $h$ = constant, the surface temperature $T_s$ must change when $\dot{q}_s$ = constant, and the surface heat flux $\dot{q}_s$ must change when $T_s$ = constant. Thus we may have either $T_s$ = constant or $\dot{q}_s$ = constant at the surface of a tube, but not both. Below we consider convection heat transfer for these two common cases.

### Constant Surface Heat Flux ($\dot{q}_s$ = constant)

In the case of $\dot{q}_s$ = constant, the rate of heat transfer can also be expressed as

$$\dot{Q} = \dot{q}_s A = \dot{m}C_p(T_e - T_i) \quad \text{(W)} \tag{10-50}$$

Then the mean fluid temperature at the tube exit becomes

$$T_e = T_i + \frac{\dot{q}_s A}{\dot{m}C_p} \tag{10-51}$$

Note that the mean fluid temperature increases *linearly* in the flow direction in the case of constant surface heat flux, since the surface area increases linearly in the flow direction ($A$ is equal to the perimeter, which is constant, times the tube length).

The surface temperature in this case can be determined from $\dot{q} = h(T_s - T_m)$. Note that when $h$ is constant, $T_s - T_m$ = constant, and thus the surface temperature will also increase *linearly* in the flow direction (Fig. 10-28). Of course, this is true when the variation of the specific heat $C_p$ with $T$ is disregarded and $C_p$ is assumed to remain constant.

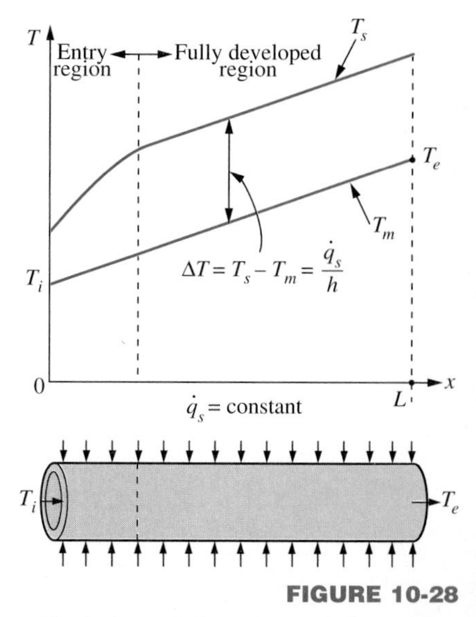


**FIGURE 10-28**

Variation of the *tube surface* and the *mean fluid* temperatures along the tube for the case of constant surface heat flux.

## Constant Surface Temperature ($T_s$ = constant)

From Newton's law of cooling, the rate of heat transfer to or from a fluid flowing in a tube can be expressed as

$$\dot{Q} = hA\,\Delta T_{\text{ave}} = hA(T_s - T_m)_{\text{ave}} \tag{10-52}$$

where $h$ is the average convection heat transfer coefficient, $A$ is the heat transfer surface area (it is equal to $\pi DL$ for a circular pipe of length $L$), and $\Delta T_{\text{ave}}$ is some appropriate *average* temperature difference between the fluid and the surface. Below we discuss two suitable ways of expressing $\Delta T_{\text{ave}}$.

In the constant surface temperature $T_s$ = constant case, $\Delta T_{\text{ave}}$ can be expressed *approximately* by the **arithmetic mean temperature difference** $\Delta T_{\text{am}}$ as

$$\Delta T_{\text{ave}} \approx \Delta T_{\text{am}} = \frac{\Delta T_i + \Delta T_e}{2} = \frac{(T_s - T_i) + (T_s - T_e)}{2}$$

$$= T_s - \frac{T_i + T_e}{2} = T_s - T_b \tag{10-53}$$

where $$T_b = \frac{T_i + T_e}{2}$$

is the *bulk mean fluid temperature,* which is the *arithmetic average* of the mean fluid temperatures at the inlet and the exit of the tube.

Note that the *arithmetic mean temperature difference* $\Delta T_{\text{am}}$ is simply the *average* of the *temperature differences* between the surface and the fluid at the inlet and the exit of the tube. Inherent in this definition is the assumption that the mean fluid temperature varies linearly along the tube, which is hardly ever the case when $T_s$ = constant. This simple approximation often gives acceptable results, but not always. Therefore, we need a better way to evaluate $\Delta T_{\text{ave}}$.

Consider the heating of a fluid in a tube of constant cross-section whose inner surface is maintained at a constant temperature of $T_s$. We know that the mean temperature of the fluid $T_m$ will increase in the flow direction as a result of heat transfer. The energy balance on a differential control volume shown in Fig. 10-29 gives

$$\dot{m}C_p\,dT_m = h(T_s - T_m)\,dA \tag{10-54}$$

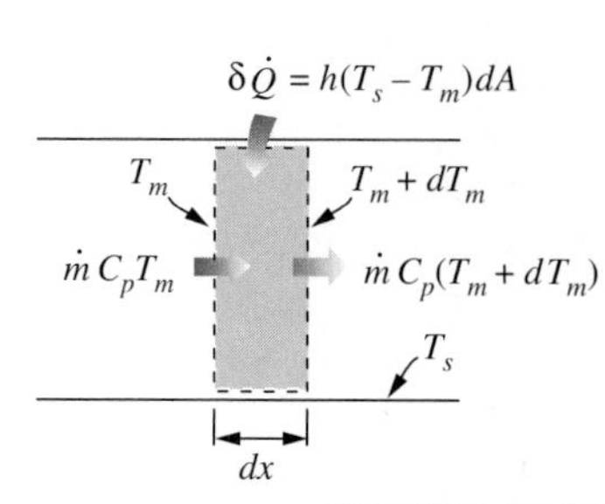


**FIGURE 10-29**

Energy interactions for a differential control volume in a tube.

That is, the increase in the energy of the fluid (represented by an increase in its mean temperature by $dT_m$) is equal to the heat transferred to the fluid from the tube surface by convection. Noting that the differential surface area is $dA = p\,dx$ where $p$ is the perimeter of the tube, and that $dT_m = -d(T_s - T_m)$, since $T_s$ is constant, the relation above can be rearranged as

$$\frac{d(T_s - T_m)}{T_s - T_m} = -\frac{hp}{\dot{m}C_p}\,dx \tag{10-55}$$

Integrating from $x = 0$ (tube inlet where $T_m = T_i$) to $x = L$ (tube exit where $T_m = T_e$) gives

$$\ln\frac{T_s - T_e}{T_s - T_i} = -\frac{hA}{\dot{m}C_p} \tag{10-56}$$

where $A = pL$ is the surface area of the tube and $h$ is the constant *average* convection heat transfer coefficient. Taking the exponential of both sides and solving for $T_e$ gives the following very useful relation for the determination of the *mean fluid temperature at the tube exit*:

$$T_e = T_s - (T_s - T_i)e^{-hA/(\dot{m}C_p)} \tag{10-57}$$

This relation can also be used to determine the mean fluid temperature $T_m(x)$ at any $x$ by replacing $A = pL$ by $px$.

Note that the temperature difference between the fluid and the surface *decays exponentially* in the flow direction, and the rate of decay depends on the mangitude of the exponent $hA/\dot{m}C_p$, as shown in Fig. 10-30. This dimensionless parameter is called the *number of transfer units,* denoted by NTU, and is a measure of the effectiveness of the heat transfer systems. For NTU > 5, the exit temperature of the fluid becomes almost equal to the surface temperature, $T_e \approx T_s$ (Fig. 10-31). Noting that the fluid temperature can approach the surface temperature but cannot cross it, an NTU of about 5 indicates that the limit is reached for heat transfer, and the heat transfer will not increase no matter how much we extend the length of the tube. A small value of NTU, on the other hand, indicates more opportunities for heat transfer, and the heat transfer will continue increasing as the tube length is increased. A large NTU and thus a large heat transfer surface area (which means a large tube) may be desirable from heat transfer point of view, but it may be unacceptable from an economic point of view. The selection of heat transfer equipment usually reflects a compromise between heat transfer performance and cost.

Solving Eq. 10-56 for $\dot{m}C_p$ gives

$$\dot{m}C_p = \frac{hA}{\ln\dfrac{T_s - T_e}{T_s - T_i}} \tag{10-58}$$

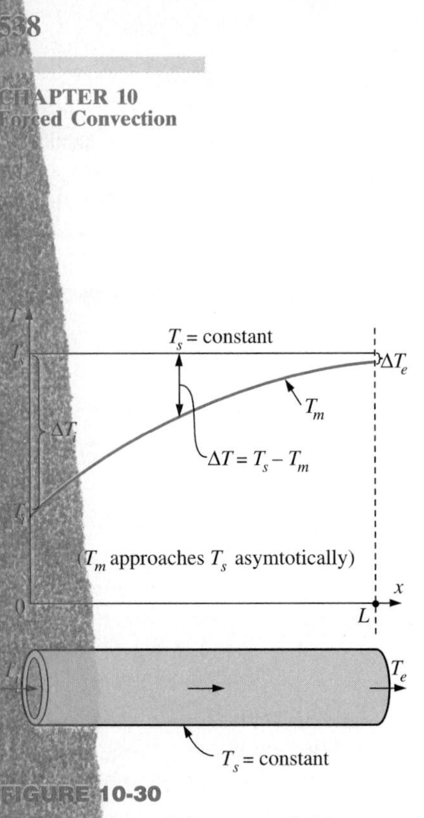


FIGURE 10-30
The variation of the *mean fluid temperature* along the tube for the case of constant surface temperature.

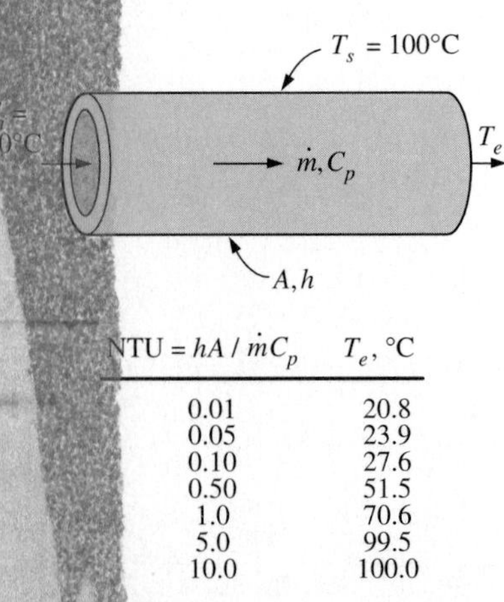


| NTU = $hA/\dot{m}C_p$ | $T_e$, °C |
|---|---|
| 0.01 | 20.8 |
| 0.05 | 23.9 |
| 0.10 | 27.6 |
| 0.50 | 51.5 |
| 1.0 | 70.6 |
| 5.0 | 99.5 |
| 10.0 | 100.0 |

FIGURE 10-31
An NTU greater than 5 indicates that the fluid flowing in a tube will reach the surface temperature at the exit regardless of the inlet temperature.

Substituting this into Eq. 10-50, we obtain

$$\dot{Q} = hA\,\Delta T_{\ln} \tag{10-59}$$

where

$$\Delta T_{\ln} = \frac{T_e - T_i}{\ln \dfrac{T_s - T_e}{T_s - T_i}} = \frac{\Delta T_e - \Delta T_i}{\ln (\Delta T_e/\Delta T_i)} \tag{10-60}$$

is the **logarithmic mean temperature difference.** Note that $\Delta T_i = T_s - T_i$ and $\Delta T_e = T_s - T_e$ are the temperature differences between the surface and the fluid at the inlet and the exit of the tube, respectively. The $\Delta T_{\ln}$ relation above appears to be prone to misuse, but it is practically failsafe, since using $T_i$ in place of $T_e$ and vice versa in the numerator and/or the denominator will, at most, affect the sign, not the magnitude. Also, it can be used for both heating ($T_s > T_i$ and $T_e$) and cooling ($T_s < T_i$ and $T_e$) of a fluid in a tube.

The logarithmic mean temperature difference $\Delta T_{\ln}$ is obtained by tracing the actual temperature profile of the fluid along the tube, and is an *exact* representation of the *average temperature difference* between the fluid and the surface. It truly reflects the exponential decay of the local temperature difference. When $\Delta T_e$ differs from $\Delta T_i$ by no more that 40 percent, the error in using the arithmetic mean temperature difference is less that 1 percent. But the error increases to undesirable levels when $\Delta T_e$ differs from $\Delta T_i$ by greater amounts. Therefore, we should always use the logarithmic mean temperature difference when determining the convection heat transfer in a tube whose surface is maintained at a constant temperature $T_s$.

## Pressure Drop

A quantity of interest in the analysis of tube flow is the *pressure drop* $\Delta P$ along the flow, since this quantity is directly related to the *power requirements* of the fan or pump to maintain the flow. The pressure drop during flow in a tube of length $L$ is expressed as (Fig. 10-32)

$$\Delta P = f\frac{L}{D}\frac{\rho \mathcal{V}_m^2}{2} \quad (\text{N/m}^2) \tag{10-61}$$

where $f$ is the **friction factor.** The required **pumping power** to overcome a specified pressure drop $\Delta P$ is determined from

$$\dot{W}_{\text{pump}} = \dot{V}\,\Delta P = \frac{\dot{m}\,\Delta P}{\rho} \quad (\text{W}) \tag{10-62}$$

where $\dot{V} = \mathcal{V}_m A_c = \dot{m}/\rho$ is the *volume flow rate* of the fluid through the tube.

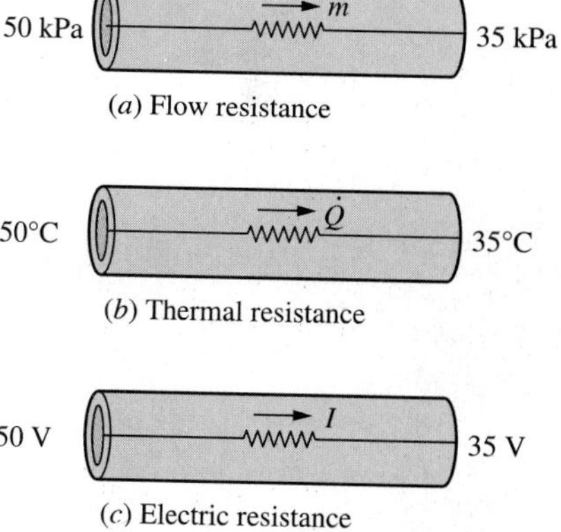


**FIGURE 10-32**
Resistance to *fluid flow, heat flow,* and *current flow* causes, respectively, the *pressure,* the *temperature,* and the *voltage* to drop in the flow direction.

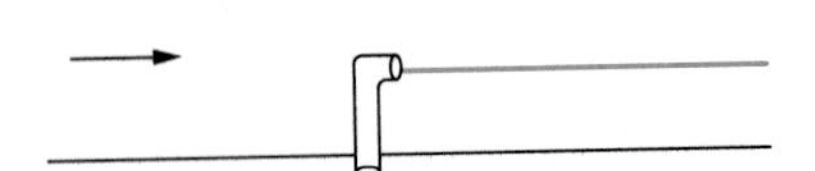

(*a*) Laminar flow

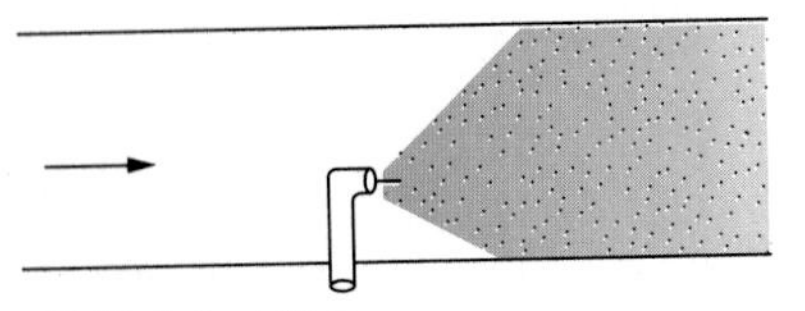

(*b*) Turbulent flow

**FIGURE 10-33**
The behavior of colored fluid injected into the flow in laminar and turbulent flows in a tube.

## Flow Regimes in a Tube

The flow in a tube can be laminar or turbulent, depending on the flow conditions. The type of flow in a tube can be verified experimentally by injecting a small amount of colored fluid into the main flow, as Reynolds did in 1880s. The streak of colored fluid is *straight and smooth* in laminar flow, as shown in Fig. 10-33, but *fluctuates rapidly and randomly* in turbulent flow.

The *Reynolds number* for flow in a circular tube of diameter $D$ is defined as

$$\text{Re} = \frac{\mathscr{V}_m D}{\nu} \tag{10-63}$$

where $\mathscr{V}_m$ is the mean fluid velocity and $\nu$ is the kinematic viscosity of the fluid. The Reynolds number again provides a convenient criteria for determining the flow regime in a tube, although the *roughness* of the tube surface and the *fluctuations* in the flow have considerable influence. The critial Reynolds number for flow in a tube is generally accepted to be 2300. Therefore,

$$\begin{aligned} \text{Re} &< 2300 && \text{laminar flow} \\ 2300 < \text{Re} &< 4000 && \text{transition to turbulence} \\ \text{Re} &> 4000 && \text{turbulent flow} \end{aligned}$$

The transition from laminar to turbulent flow in a tube is quite different than it is over a flat plate, where the Reynolds number is zero at the leading edge, and increases linearly in the flow direction. Consequently, the flow starts out laminar, and becomes turbulent when a critical Reynolds number is reached. Therefore, the flow over a flat plate is partly laminar and partly turbulent. In flow in a tube, however, the Reynolds number is constant. Therefore, the flow is either laminar or turbulent over practically the entire length of the tube.

## Hydrodynamic and Thermal Entry Lengths

Consider a fluid entering a circular tube at a uniform velocity. As in external flow, the fluid particles in the layer in contact with the surface of the tube will come to a complete stop. This layer will also cause the fluid particles in the adjacent layers to slow down gradually as a result of friction. To make up for this velocity reduction, the velocity of the fluid at the mid-section of the tube will have to increase to keep the mass flow rate through the tube constant. As a result, a *velocity boundary layer* develops along the tube. The thickness of this boundary layer increases in the flow direction until the boundary layer reaches the tube center and thus fills the entire tube, as shown in Fig. 10-34. The region from the tube inlet to the point at which the boundary layer merges at the centerline is

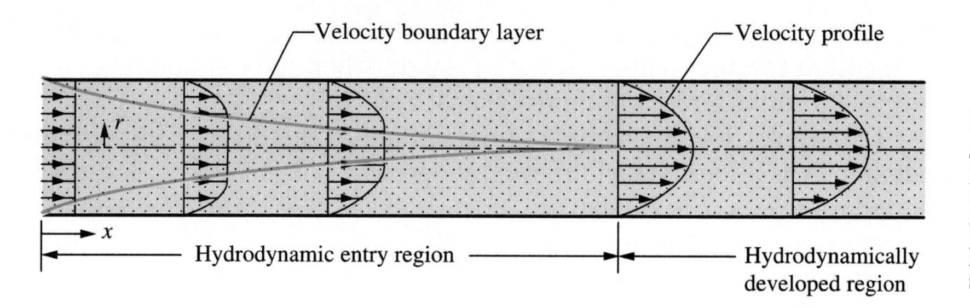


**FIGURE 10-34**

The development of the velocity boundary layer in a tube. (The developed velocity profile will be parabolic in laminar flow as shown, but somewhat blunt in turbulent flow.)

called the **hydrodynamic entry region**, and the length of this region is called the **hydrodynamic entry length** $L_h$. The region beyond the hydrodynamic entry region in which the velocity profile is fully developed and remains unchanged is called the **hydrodynamically developed region**. The velocity profile in the hydrodynamically developed region is *parabolic* in laminar flow, and somewhat *flatter* in turbulent flow.

Now consider a fluid at a uniform temperature entering a circular tube which is at a different temperature. This time, the fluid particles in the layer in contact with the surface of the tube will assume the surface temperature. This will initiate convection heat transfer in the tube, and the development of a *thermal boundary layer* along the tube. The thickness of this boundary layer also increases in the flow direction until the boundary layer reaches the tube center and thus fills the entire tube, as shown in Fig. 10-35. The region of flow over which the thermal boundary layer develops and reaches the tube center is called the **thermal entry region**, and the length of this region is called the **thermal entry length** $L_t$. The region beyond the thermal entry region in which the dimensionless temperature profile expressed as $(T - T_s)/(T_m - T_s)$ remains unchanged is called the **thermally developed region**. The region in which the flow is both hydrodynamically and thermally developed is called the **fully developed flow**.

Note that the *temperature profile* in the thermally developed region may *vary* with $x$ in the flow direction. That is, unlike the velocity profile, the temperature profile can be different at different cross-sections of the tube in the developed region, and it usually is. However, it can be shown

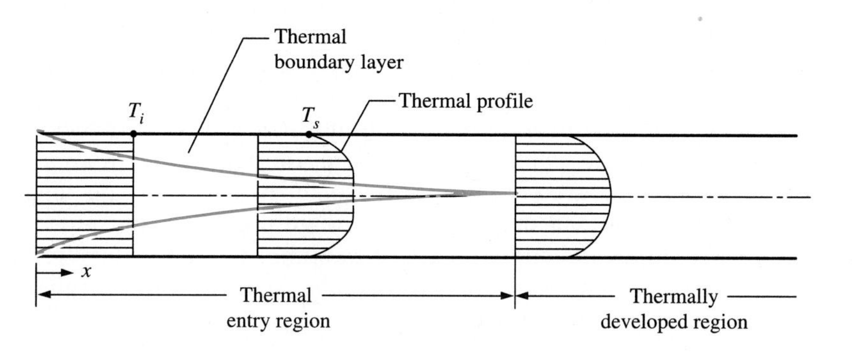


**FIGURE 10-35**

The development of the thermal boundary layer in a tube. (The fluid in the tube is being cooled.)

that the dimensionless temperature profile defined above remains unchanged in the thermally developed region when the temperature or heat flux at the tube surface remains constant.

In laminar flow in a tube, the magnitude of the dimensionless Prandtl number Pr is a measure of the relative growth of the velocity and thermal boundary layers. For fluids with Pr $\approx$ 1, such as gases, the two boundary layers essentially coincide with each other. For fluids with Pr $\gg$ 1, such as oils, the velocity boundary layer outgrows the thermal boundary layer. As a result, the hydrodynamic entry length is smaller than the thermal entry length. The opposite is true for fluids with Pr $\ll$ 1 such as liquid metals.

The hydrodynamic and thermal entry lengths in *laminar flow* are given approximately as

$$L_{h,\text{laminar}} \approx 0.05 \text{ Re } D$$
$$L_{t,\text{laminar}} \approx 0.05 \text{ Re Pr } D \qquad (10\text{-}64)$$

In *turbulent flow*, the hydrodynamic and thermal entry lengths are known to be independent of Re or Pr, and are generally taken to be

$$L_{h,\text{turbulent}} \approx L_{t,\text{turbulent}} \approx 10D \qquad (10\text{-}65)$$

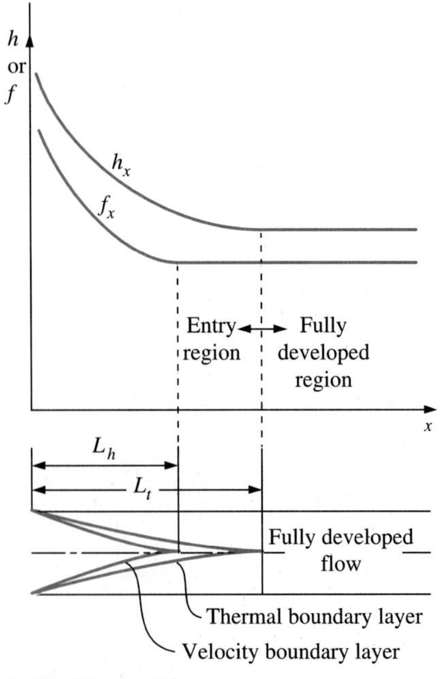


**FIGURE 10-36**
Variation of the friction factor and the convection heat transfer coefficient in the flow direction for flow in a tube (Pr > 1).

The friction coefficient is related to the shear stress at the surface, which is related to the slope of the velocity profile at the surface. Noting that the velocity profile remains unchanged in the hydrodynamically developed region, the friction coefficient also remains constant in that region. A similar argument can be given for the heat transfer coefficient in the thermally developed region. Thus, we conclude that *the friction and the heat transfer coefficients in the fully developed flow region remain constant.*

Consider a fluid that is being heated (or cooled) in a tube as it flows through it. The friction factor and the heat transfer coefficient are *highest* at the tube inlet where the thickness of the boundary layers is zero, and decrease gradually to the fully developed values, as shown in Fig. 10-36. Therefore, the pressure drop and heat flux are *higher* in the entry regions of a tube, and the effect of the entry region is always to *enhance* the average friction and heat transfer coefficients for the entire tube. This enhancement can be significant for short tubes, but negligible for long ones.

Precise correlations for the friction and heat transfer coefficients for the entry regions are available in the literature. However, the tubes used in practice in forced convection are usually many times the length of either entry region, and thus the flow through the tubes is assumed to be fully developed for the entire length of the tube. This approach, which we will also use for simplicity, gives *reasonable* results for long tubes, and *conservative* results for short ones.

### Laminar Flow in Tubes

We mentioned earlier that flow in smooth tubes is laminar for Re < 2300. The theory for laminar flow is well developed, and both the friction and heat transfer coefficients for fully developed laminar flow in smooth circular tubes can be determined analytically by solving the governing differential equations. Combining the conservation of *mass* and *momentum* equations in the axial direction for a tube and solving them subject to the no-slip condition at the boundary and the condition that the velocity profile is symmetric about the tube center gives the following *parabolic* velocity profile for the hydrodynamically developed laminar flow:

$$\mathcal{V}(r) = 2\mathcal{V}_m\left(1 - \frac{r^2}{R^2}\right) \tag{10-66}$$

where $\mathcal{V}_m$ is the *mean fluid velocity* and $R$ is the *radius* of the tube. Note that the maximum velocity occurs at the tube center ($r = 0$), and it is $\mathcal{V}_{max} = 2\mathcal{V}_m$. Knowing the velocity profile, the shear stress at the wall becomes

$$\tau_s = -\mu \left.\frac{d\mathcal{V}}{dr}\right|_{r=R} = -2\mu\mathcal{V}_m\left(-\frac{2r}{R^2}\right)_{r=R} = \frac{8\mu\mathcal{V}_m}{D} \tag{10-67}$$

But we also have the following practical definition of shear stress:

$$\tau_s = C_f \frac{\rho\mathcal{V}_m^2}{2} \tag{10-68}$$

where $C_f$ is the friction coefficient. Combining Eqs. 10-67 and 10-68 and solving for $C_f$ gives

$$C_f = \frac{8\mu\mathcal{V}_m}{D}\frac{2}{\rho\mathcal{V}_m^2} = \frac{16\mu}{\rho\mathcal{V}_m D} = \frac{16}{\text{Re}} \tag{10-69}$$

The *friction factor f*, which is the parameter of interest in the pressure drop calculations, is related to the friction coefficient $C_f$ by $f = 4C_f$. Therefore,

$$f = \frac{64}{\text{Re}} \quad \text{(laminar flow)} \tag{10-70}$$

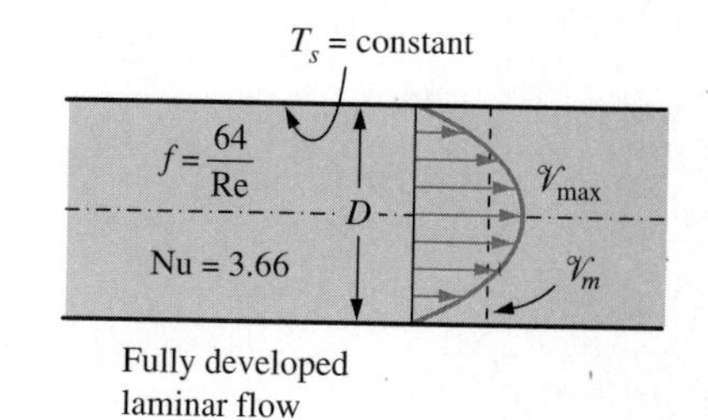


**FIGURE 10-37**
In laminar flow in a tube with constant surface temperature, both the *friction factor* and the *heat transfer coefficient* remain constant in the fully developed region.

Note that the friction factor $f$ is related to the *pressure drop* in the fluid, whereas the friction coefficient $C_f$ is related to the *drag force* on the surface directly. Of course, these two coefficients are simply a constant multiple of each other.

The Nusselt number in the fully developed laminar flow region in a circular tube is determined in a similar manner from the conservation of energy equation to be (Fig. 10-37)

$$\text{Nu} = 3.66 \quad \text{for } T_s = \text{constant} \quad \text{(laminar flow)}$$

and

$$\text{Nu} = 4.36 \quad \text{for } \dot{q}_s = \text{constant} \quad \text{(laminar flow)}$$

A general relation for the *average* Nusselt number for the hydrodynamically and/or thermally *developing laminar flow* in a *circular* tube is given by Sieder and Tate as

$$\mathrm{Nu} = 1.86\left(\frac{\mathrm{Re\,Pr}\,D}{L}\right)^{1/3}\left(\frac{\mu_b}{\mu_s}\right)^{0.14} \qquad (\mathrm{Pr} > 0.5) \tag{10-71}$$

All properties are evaluated at the bulk mean fluid temperature, except for $\mu_s$, which is evaluated at the surface temperature.

The Nusselt number Nu and the friction factor $f$ are given in Table 10-4 for *fully developed laminar flow* in tubes of various cross-sections. The Reynolds and Nusselt numbers for flow in these tubes are based on the **hydraulic diameter** $D_h$ defined as

$$D_h = \frac{4A_c}{p} \tag{10-72}$$

where $A_c$ is the cross-sectional area of the tube and $p$ is its perimeter. The hydraulic diameter is defined such that it reduces to ordinary diameter $D$ for circular tubes since $A_c = \pi D^2/4$ and $p = \pi D$. Once the Nusselt number is available, the convection heat transfer coefficient is determined from $h = k\,\mathrm{Nu}/D_h$. It turns out that for a fixed surface area, the *circular tube* gives the most heat transfer for the least pressure drop, which explains the overwhelming popularity of circular tubes in heat transfer equipment.

The effect of *surface roughness* on the friction factor and the heat transfer coefficient in laminar flow is negligible.

## Turbulent Flow in Tubes

We mentioned earlier that flow in smooth tubes is turbulent at Re > 4000. Turbulent flow is commonly utilized in practice because of the higher heat transfer coefficients associated with it. Most correlation for the friction and heat transfer coefficients in turbulent flow are based on experimental studies because of the difficulty in dealing with turbulent flow theoretically.

For *smooth* tubes, the friction factor in fully developed turbulent flow can be determined from

$$f = 0.184\,\mathrm{Re}^{-0.2} \qquad \text{(smooth tubes)} \tag{10-73}$$

The friction factor for flow in tubes with *smooth* as well as *rough surfaces* over a wide range of Reynolds numbers is given in Fig. 10-38, which is known as the **Moody diagram**. Note that the friction factor and thus the

**TABLE 10-4**
**Nusselt number and friction factor for fully developed laminar flow in tubes of various cross-sections ($D_h = 4A_c/p$, $\mathrm{Re} = \mathcal{V}_m D_h/\nu$ and $\mathrm{Nu} = hD_h/k$)**

| Cross-section of tube | $a/b$ or $\theta°$ | Nusselt number | | Friction factor $f$ |
|---|---|---|---|---|
| | | $T_s$ = const. | $\dot{q}_s$ = const. | |
| Circle | — | 3.66 | 4.36 | 64.00/Re |
| Hexagon | — | 3.35 | 4.00 | 60.20/Re |
| Square | — | 2.98 | 3.61 | 56.92/Re |
| Rectangle | $a/b$ | | | |
| | 1 | 2.98 | 3.61 | 56.92/Re |
| | 2 | 3.39 | 4.12 | 62.20/Re |
| | 3 | 3.96 | 4.79 | 68.36/Re |
| | 4 | 4.44 | 5.33 | 72.92/Re |
| | 6 | 5.14 | 6.05 | 78.80/Re |
| | 8 | 5.60 | 6.49 | 82.32/Re |
| | ∞ | 7.54 | 8.24 | 96.00/Re |
| Ellipse | $a/b$ | | | |
| | 1 | 3.66 | 4.36 | 64.00/Re |
| | 2 | 3.74 | 4.56 | 67.28/Re |
| | 4 | 3.79 | 4.88 | 72.96/Re |
| | 8 | 3.72 | 5.09 | 76.60/Re |
| | 16 | 3.65 | 5.18 | 78.16/Re |
| Triangle | $\theta$ | | | |
| | 10° | 1.61 | 2.45 | 50.80/Re |
| | 30° | 2.26 | 2.91 | 52.28/Re |
| | 60° | 2.47 | 3.11 | 53.32/Re |
| | 90° | 2.34 | 2.98 | 52.60/Re |
| | 120° | 2.00 | 2.68 | 50.96/Re |

pressure drop for flow in a tube can vary several times as a result of surface roughness.

The Nusselt number in turbulent is related to the friction factor through the famous **Chilton–Colburn analogy** expressed as

$$\mathrm{Nu} = 0.125 f\,\mathrm{Re}\,\mathrm{Pr}^{1/3} \quad \text{(turbulent flow)} \tag{10-74}$$

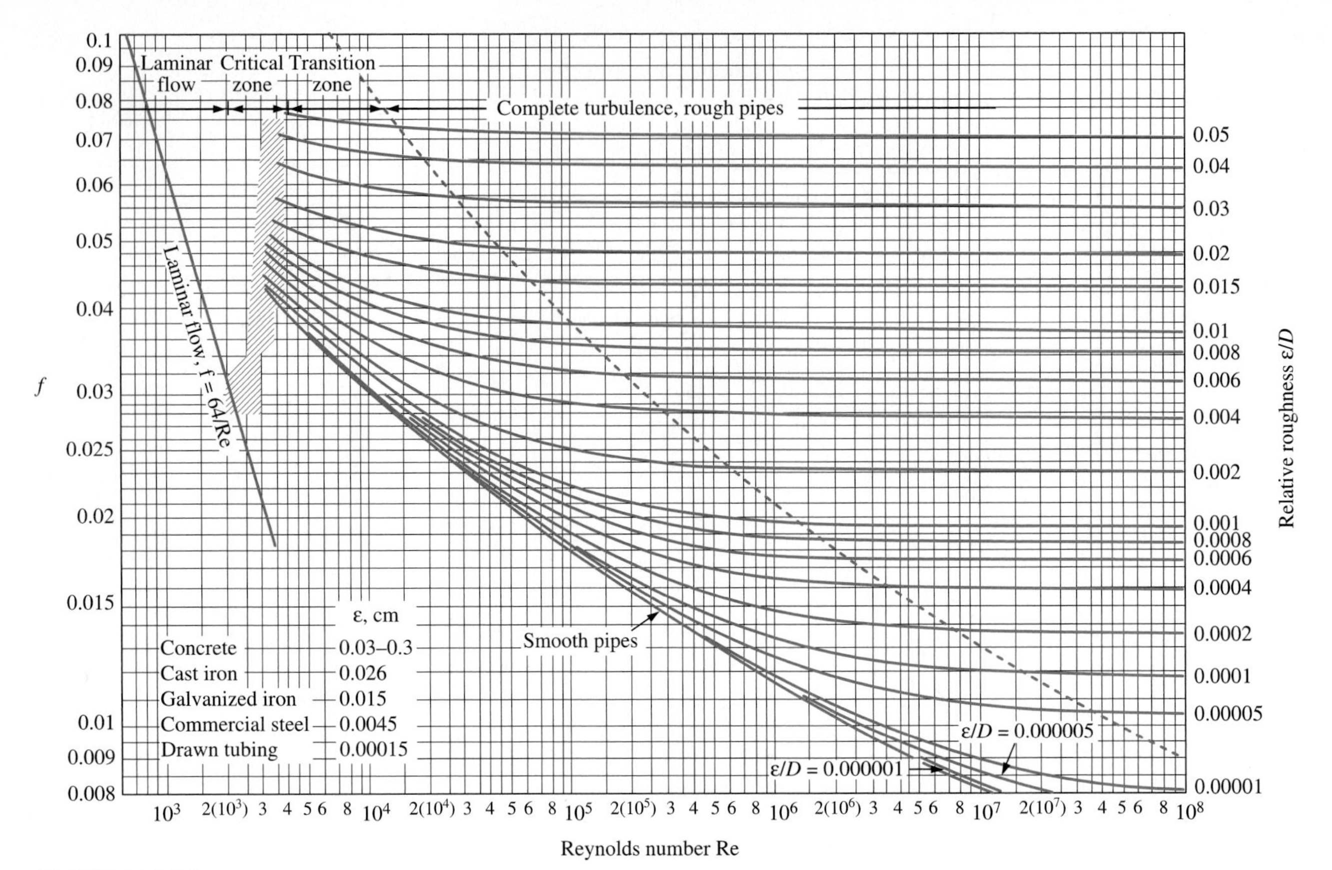


**FIGURE 10-38**

Friction factor for fully developed flow in circular tubes (The Moody chart).

Substituting the $f$ relation from Eq. 10-73 into Eq. 10-74 gives the following relation for the Nusselt number for *fully developed turbulent flow in smooth tubes*:

$$\text{Nu} = 0.023\ \text{Re}^{0.8}\ \text{Pr}^{1/3} \qquad \begin{pmatrix} 0.7 \leq \text{Pr} \leq 160 \\ \text{Re} > 10{,}000 \end{pmatrix} \qquad (10\text{-}75)$$

which is known as the **Colburn equation**. The accuracy of this equation can be improved by modifying it as

$$\text{Nu} = 0.023\ \text{Re}^{0.8}\ \text{Pr}^{n} \qquad \begin{pmatrix} 0.7 \leq \text{Pr} \leq 160 \\ \text{Re} > 10{,}000 \end{pmatrix} \qquad (10\text{-}76)$$

where $n = 0.4$ for *heating* and 0.3 for *cooling* of the fluid flowing through the tube. This equation is known as the **Dittus–Boulter equation**, and it is preferred to the Calburn equation. The fluid properties are evaluated at the *bulk mean fluid temperature* $T_b = \frac{1}{2}(T_i + T_e)$, which is the arithmetic average of the mean fluid temperatures at the inlet and the exit of the tube.

The relations above are not very sensitive to the *thermal conditions* at the tube surfaces, and can be used for both $T_s$ = constant and $\dot{q}_s$ = constant cases. Despite their simplicity, the correlations above give sufficiently accurate results for most engineering purposes. They can also be used to obtain rough estimates of the friction factor and the heat transfer coefficients in the transition region $2300 < \text{Re} < 4000$, especially when the Reynolds number is closer to 4000 than it is to 2300.

The Nusselt number for *rough surfaces* can also be determined from Eq. 10-74 by substituting the friction factor $f$ value from the Moody chart. Note that tubes with rough surfaces have much higher heat transfer coefficients than tubes with smooth surfaces. Therefore, tube surfaces are often intentionally *roughened, corrugated,* or *finned* in order to *enhance* the convection heat transfer coefficient and thus the convection heat transfer rate (Fig. 10-39). Heat transfer in turbulent flow in a tube has been increased by as much as 400 percent by roughening the surface. Roughening the surface, of course, also increases the friction factor and thus the power requirement for the pump or the fan.

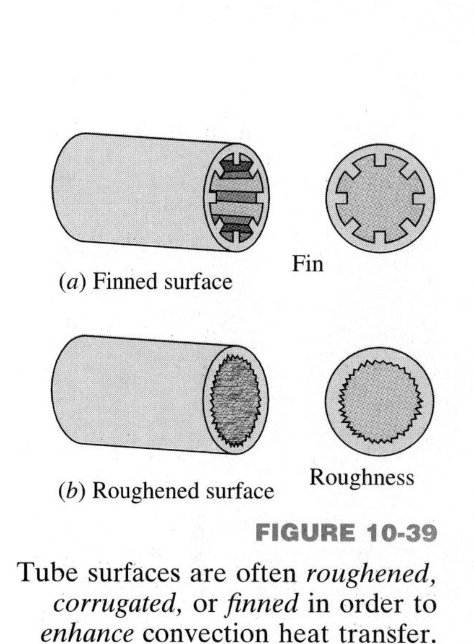


**FIGURE 10-39**

Tube surfaces are often *roughened, corrugated,* or *finned* in order to *enhance* convection heat transfer.

The turbulent flow relations above can also be used for *noncircular tubes* with reasonable accuracy by replacing the diameter $D$ in the evaluation of the Reynolds number by the hydraulic diameter $D_h = 4A_c/p$.

**EXAMPLE 10-5**

Water is to be heated from 15°C to 65°C as it flows through a 3-cm internal-diameter 5-m-long tube (Fig. 10-40). The tube is equipped with an electric resistance heater that provides uniform heating throughout the surface of the tube. The outer surface of the heater is well insulated, so that in steady operation all the heat generated in the heater is transferred to the water in the tube. If the system is to provide hot water at a rate of 10 L/min, determine the power rating of the resistance heater. Also estimate the inner surface temperature of the pipe at the exit.

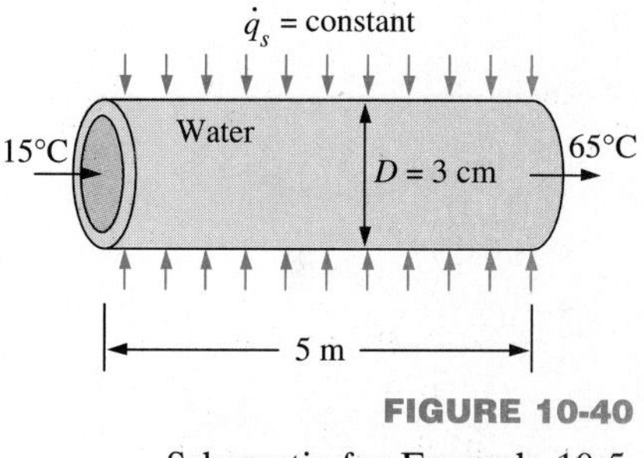


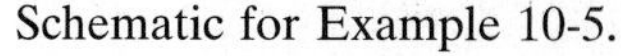


**FIGURE 10-40**

Schematic for Example 10-5.

**Solution** This is an *internal flow* problem, since the water is flowing in a pipe. We assume the tube to be smooth, and the operation to be steady. The properties of water at the bulk mean temperature of $T_b = \frac{1}{2}(T_i + T_e) = \frac{1}{2}(15 + 65) = 40°\text{C}$ are

$$\rho = 994 \text{ kg/m}^3$$
$$k = 0.628 \text{ W/(m} \cdot °\text{C)}$$
$$\nu = 0.658 \times 10^{-6} \text{ m}^2\text{/s}$$
$$C_p = 4178 \text{ J/(kg} \cdot °\text{C)}$$
$$\text{Pr} = 4.34$$

The cross-sectional and heat transfer surface areas are

$$A_c = \tfrac{1}{4}\pi D^2 = \tfrac{1}{4}\pi(0.03 \text{ m})^2 = 7.069 \times 10^{-4} \text{ m}^2$$
$$A = pL = \pi DL = \pi(0.03 \text{ m})(5 \text{ m}) = 0.471 \text{ m}^2$$

The volume flow rate of water is given as $\dot{V} = 10\ \text{L/min} = 0.01\ \text{m}^3\text{/min}$. Then the mass flow rate of water becomes

$$\dot{m} = \rho\dot{V} = (994.6\ \text{kg/m}^3)(0.01\ \text{m}^3\text{/min}) = 9.946\ \text{kg/min} = 0.1658\ \text{kg/s}$$

To heat the water at this mass flow rate from 15°C to 65°C, heat must be supplied to the water at a rate of

$$\begin{aligned}\dot{Q} &= \dot{m}C_p(T_e - T_i)\\ &= (0.1658\ \text{kg/s})[4.178\ \text{kJ/(kg}\cdot\text{°C)}](65 - 15)\text{°C}\\ &= 34.6\ \text{kJ/s} = 34.6\ \text{kW}\end{aligned}$$

All of this energy must come from the resistance heater. Therefore, the power rating of the heater must be 34.6 kW.

The surface temperature $T_s$ of the tube at any location can be determined from

$$\dot{q}_s = h(T_s - T_m) \rightarrow T_s = T_m + \frac{\dot{q}_s}{h}$$

where $h$ is the heat transfer coefficient and $T_m$ is the mean temperature of the fluid at that location. The surface heat flux is constant in this case, and its value can be determined from

$$\dot{q}_s = \frac{\dot{Q}}{A} = \frac{34.6\ \text{kW}}{0.471\ \text{m}^2} = 73.46\ \text{kW/m}^2$$

To determine the heat transfer coefficient, we first need to find the mean velocity of water and the Reynolds number:

$$\mathscr{V}_m = \frac{\dot{V}}{A_c} = \frac{0.010\ \text{m}^3\text{/min}}{7.069 \times 10^{-4}\ \text{m}^2} = 14.15\ \text{m/min} = 0.236\ \text{m/s}$$

$$\text{Re} = \frac{\mathscr{V}_m D}{\nu} = \frac{(0.236\ \text{m/s})(0.03\ \text{m})}{0.658 \times 10^{-6}\ \text{m}^2\text{/s}} = 10{,}760$$

which is greater than 4000. Therefore the flow is turbulent in this case, and the entry lengths are roughly

$$L_h \approx L_t \approx 10D = 10 \times (0.03\ \text{m}) = 0.3\ \text{m}$$

which is much shorter than the total length of the pipe. Therefore, we can assume fully developed turbulent flow in the entire pipe, and determine the Nusselt number from

$$\text{Nu} = \frac{hD}{k} = 0.023\ \text{Re}^{0.8}\,\text{Pr}^{0.4} = 0.023(10{,}760)^{0.8}(4.34)^{0.4} = 69.5$$

Then,

$$h = \frac{k}{D}\text{Nu} = \frac{0.628\ \text{W/(m}\cdot\text{°C)}}{0.03\ \text{m}}(69.5) = 1455\ \text{W/(m}^2\cdot\text{°C)}$$

and the surface temperature of the pipe at the exit becomes

$$T_s = T_m + \frac{\dot{q}_s}{h} = 65\text{°C} + \frac{73{,}460\ \text{kW/m}^2}{1455\ \text{W/(m}^2\cdot\text{°C)}} = 115\text{°C}$$

Therefore, the inner surface temperature of the pipe will be 50°C higher than the mean water temperature at the pipe exit. This temperature difference of

50°C between the water and the surface will remain constant throughout the fully developed flow region.

**EXAMPLE 10-6**

Hot air at atmospheric pressure and 80°C enters an 8-m-long uninsulated square duct of cross-section 0.2 m × 0.2 m that passes through the attic of a house at a rate of 0.15 m³/s (Fig. 10-41). The duct is observed to be nearly isothermal at 60°C. Determine the exit temperature of the air and the rate of heat loss from the duct to the attic space.

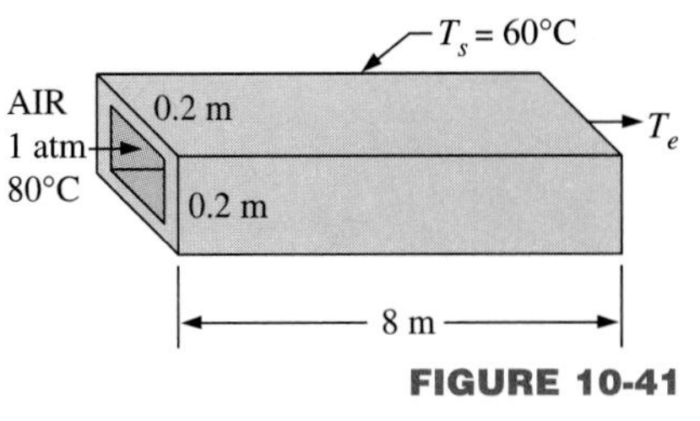


**FIGURE 10-41**
Schematic for Example 10-6.

**Solution** This is an *internal flow* problem, since the air is flowing in a duct. We do not know the exit temperature of the air in the duct, and thus we cannot determine the bulk mean temperature of air, which is the temperature at which the properties are to be determined. The mean temperature of air at the inlet is 80°C, or 353 K, and we expect this temperature to drop somewhat as a result of heat loss through the duct, whose surface is at a lower temperature. Thus it is reasonable to assume a bulk mean temperature of 350 K for air (we will check this assumption later) for the purpose of evaluating the properties of air. At this temperature we read

$$\rho = 1.009 \text{ kg/m}^3$$
$$k = 0.0297 \text{ W/(m} \cdot \text{°C)}$$
$$\nu = 2.06 \times 10^{-5} \text{ m}^2\text{/s}$$
$$C_p = 1008 \text{ J/(kg} \cdot \text{°C)}$$
$$\text{Pr} = 0.706$$

The characteristic length (which is the hydraulic diameter), the mean velocity, and the Reynolds number in this case are

$$D_h = \frac{4A_c}{p} = \frac{4a^2}{4a} = a = 0.2 \text{ m}$$
$$\mathscr{V}_m = \frac{\dot{V}}{A_c} = \frac{0.15 \text{ m}^3\text{/s}}{(0.2 \text{ m})^2} = 3.75 \text{ m/s}$$

and

$$\text{Re} = \frac{\mathscr{V}_m D_h}{\nu} = \frac{(3.75 \text{ m/s})(0.2 \text{ m})}{2.06 \times 10^{-5} \text{ m}^2\text{/s}} = 36{,}408$$

which is greater than 4000. Therefore, the flow is turbulent, and the entry lengths in this case are roughly

$$L_h \approx L_t \approx 10D_h = 10 \times (0.2 \text{ m}) = 2 \text{ m}$$

which is much shorter than the total length of the duct. Therefore, we can assume fully developed turbulent flow in the entire duct, and determine the Nusselt number from

$$\text{Nu} = \frac{hD_h}{k} = 0.023 \text{ Re}^{0.8} \text{ Pr}^{0.3} = 0.023(36{,}408)^{0.8}(0.706)^{0.3} = 92.3$$

Then,

$$h = \frac{k}{D_h}\text{Nu} = \frac{0.0297 \text{ W/(m} \cdot \text{°C)}}{0.2 \text{ m}}(92.3) = 13.7 \text{ W/(m}^2 \cdot \text{°C)}$$
$$A = pL = 4aL = 4 \times (0.2 \text{ m})(8 \text{ m}) = 6.4 \text{ m}^2$$
$$\dot{m} = \rho\dot{V} = (1.009 \text{ kg/m}^3)(0.15 \text{ m}^3\text{/s}) = 0.151 \text{ kg/s}$$

Next, we determine the exit temperature of air from

$$\begin{aligned} T_e &= T_s - (T_s - T_i)e^{-hA/(\dot{m}C_p)} \\ &= 60°C - [(60 - 80)°C] \exp\left\{-\frac{(13.7\ W/(m^2 \cdot °C))(6.4\ m^2)}{(0.151\ kg/s)[1008\ J(kg \cdot °C)]}\right\} \\ &= 71.2°C \end{aligned}$$

Then the logarithmic mean temperature difference and the rate of heat loss from the air become

$$\Delta T_{ln} = \frac{T_e - T_i}{\ln \dfrac{T_s - T_e}{T_s - T_i}} = \frac{71.2 - 80}{\ln \dfrac{60 - 71.2}{60 - 80}} = 15.2°C$$

$$\dot{Q} = hA\,\Delta T_{ln} = [13.7\ W/(m^2 \cdot °C)](6.4\ m^2)(15.2°C) = \mathbf{1368\ W}$$

Therefore, the air will lose heat at a rate of 1368 W as it flows through the duct in the attic.

Having calculated the exit temperature of the air, we can now determine the actual bulk mean fluid temperature from

$$T_b = \frac{T_i + T_e}{2} = \frac{80 + 71.2}{2} = 75.6°C = 348.6\ K$$

which is sufficiently close to the assumed value of 350 K at which we evaluated the properties of air. Therefore, it is not necessary to re-evaluate the properties at this $T_b$ and to repeat the calculations.

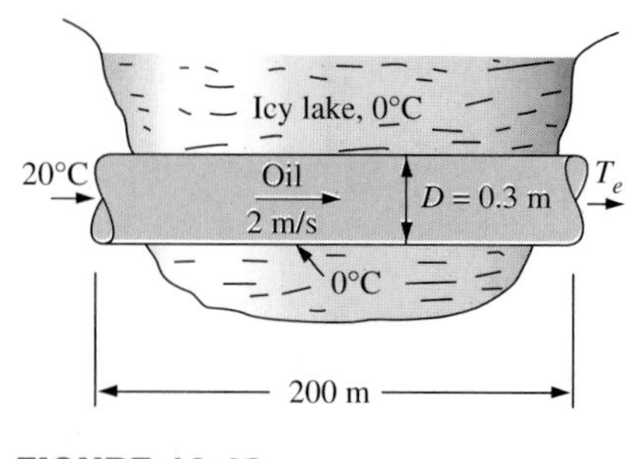


**FIGURE 10-42**
Schematic for Example 10-7.

**EXAMPLE 10-7**

Consider the flow of oil at 20°C in a 30-cm-diameter pipeline at an average velocity of 2 m/s (Fig. 10-42). A 200-m-long section of the pipeline passes through icy waters of a lake at 0°C. Some measurements indicate that the surface temperature of the pipe is very nearly 0°C. Disregarding the thermal resistance of the pipe material, determine (*a*) the temperature of the oil when the pipe leaves the lake, (*b*) the rate of heat transfer from the oil, and (*c*) the pumping power required to overcome the pressure losses and to maintain the flow of the oil in the pipe.

**Solution** (*a*) This is an *internal flow* problem, since the oil is flowing in a pipe. the thermal resistance of the pipe material is given to be negligible. Therefore, there will be no temperature drop through the thickness of the pipe, and the inner surface temperature of the pipe will also be nearly 0°C. Also, the oil has been flowing in the pipe long before the pipe has reached the lake, so it is reasonable to assume the flow to be hydrodynamically developed.

We do not know the exit temperature of the oil, and thus we cannot determine the bulk mean temperature, which is the temperature at which the properties of oil are to be determined. The mean temperature of oil at the inlet is 20°C, and we expect this temperature to drop somewhat as a result of heat loss to the icy waters of the lake. We evaluate the properties of the oil at the inlet temperature, but we will repeat the calculations, if necessary, using properties at the evaluated bulk mean temperature. At 20°C, we read

$$\begin{aligned} \rho &= 888 \text{ kg/m}^3 \\ k &= 0.145 \text{ W/(m} \cdot \text{°C)} \\ \mu &= 0.800 \text{ kg/(m} \cdot \text{s)} \\ \nu &= 901 \times 10^{-6} \text{ m}^2\text{/s} \\ C_p &= 1880 \text{ J/(kg} \cdot \text{°C)} \\ \text{Pr} &= 10{,}400 \end{aligned}$$

The Reynolds number in this case is

$$\text{Re} = \frac{\mathcal{V}_m D_h}{\nu} = \frac{(2 \text{ m/s})(0.3 \text{ m})}{901 \times 10^{-6} \text{ m}^2\text{/s}} = 666$$

which is less than the critical Reynolds number of 2300. Therefore, the flow is laminar, and the thermal entry length in this case is roughly

$$L_t \approx 0.05 \text{ Re Pr } D = 0.05 \times 666 \times 10{,}400 \times (0.3 \text{ m}) \approx 104{,}000 \text{ m}$$

which is much greater than the total length of the pipe. This is typical of fluids with high Prandtl numbers. Therefore, we assume thermally developing flow, and determine the Nusselt number from

$$\begin{aligned} \text{Nu} = \frac{hD}{k} &= 1.86\left(\frac{\text{Re Pr } D}{L}\right)^{1/3}\left(\frac{\mu_b}{\mu_s}\right)^{0.14} \\ &= 1.86\left(\frac{666 \times 10{,}400 \times 0.3 \text{ m}}{200 \text{ m}}\right)^{1/3}\left(\frac{0.8}{3.85}\right)^{0.14} = 32.6 \end{aligned}$$

where the dynamic viscosity $\mu_s$ is determined at the surface temperature of 0°C. Note that this Nusselt number is considerably higher than the fully developed value of 3.66. Then,

$$h = \frac{k}{D}\text{Nu} = \frac{0.145 \text{ W/(m} \cdot \text{°C)}}{0.3 \text{ m}}(32.6) = 15.8 \text{ W/(m}^2 \text{ °C)}$$

Also,

$$\begin{aligned} A &= pL = \pi DL = \pi(0.3 \text{ m})(200 \text{ m}) = 188.5 \text{ m}^2 \\ \dot{m} &= \rho A_c \mathcal{V}_m = (888 \text{ kg/m}^3)[\tfrac{1}{4}\pi(0.3 \text{ m})^2](2 \text{ m/s}) = 125.5 \text{ kg/s} \end{aligned}$$

Next we determine the exit temperature of oil from

$$\begin{aligned} T_e &= T_s - (T_s - T_i)e^{-hA/(\dot{m}C_p)} \\ &= 0\text{°C} - [(0 - 20)\text{°C}] \exp\left\{-\frac{[15.8 \text{ W/(m}^2 \cdot \text{°C)}](188.5 \text{ m}^2)}{(125.5 \text{ kg/s})[1880 \text{ J/(kg} \cdot \text{°C)}]}\right\} \\ &= 19.75\text{°C} \end{aligned}$$

Thus, the mean temperature of oil drops by a mere 0.25°C as it crosses the lake. This makes the bulk mean oil temperature 19.875°C, which is practically identical to the inlet mean temperature of 20°C. Therefore, we do not need to re-evaluate the properties at this bulk temperature and repeat the calculations.

(*b*) The logarithmic mean temperature difference and the rate of heat loss from the oil are

$$\Delta T_{ln} = \frac{T_e - T_i}{\ln \dfrac{T_s - T_e}{T_s - T_i}} = \frac{19.75 - 20}{\ln \dfrac{0 - 19.75}{0 - 20}} = 19.875°\text{C}$$

$$\dot{Q} = hA\,\Delta T_{ln} = [15.8\ \text{W/(m}^2\cdot°\text{C)}](188.5\ \text{m}^2)(19.875°\text{C}) = 59{,}190\ \text{W}$$

Therefore, the oil will lose heat at a rate of 59,190 W as it flows through the pipe in the icy waters of the lake. Note that $\Delta T_{ln}$ is identical to the arithmetic mean temperature in this case, since $\Delta T_i \approx \Delta T_e$.

(*c*) The laminar flow of oil is hydrodynamically developed. Therefore, the friction factor can be determined from

$$f = \frac{64}{\text{Re}} = \frac{64}{666} = 0.0961$$

Then the pressure drop in the pipe and the required pumping power become

$$\Delta P = f\frac{L}{D}\frac{\rho \mathcal{V}_m^2}{2} = 0.0961\frac{200\ \text{m}}{0.3\ \text{m}}\frac{(888\ \text{kg/m}^3)(2\ \text{m/s})^2}{2} = 113{,}780\ \text{N/m}^2$$

$$\dot{W}_{pump} = \frac{\dot{m}\,\Delta P}{\rho} = \frac{(125.5\ \text{kg/s})(113{,}780\ \text{N/m}^2)}{888\ \text{kg/m}^3} = 16.1\ \text{kW}$$

Thus, we will need a 16.1-kW pump just to overcome the friction in the pipe as the oil flows in the 200-m-long pipe through the lake.

---

We will close this chapter with the following external flow problem based on actual practical experience.

**EXAMPLE 10-8**

The forming section of a plastics plant puts out a continuous sheet of plastic that is 4 ft wide and 0.04 in thick at a rate of 30 ft/min. The temperature of the plastic sheet is 200°F when it is exposed to the surrounding air, and a 2-ft-long section of the plastic sheet is subjected to air flow at 80°F at a velocity of 10 ft/s on both sides along its surfaces normal to the direction of motion of the sheet, as shown in Fig. 10-43. Determine (*a*) the rate of heat transfer from the plastic sheet to the air by forced convection and radiation, and (*b*) the temperature of the plastic sheet at the end of the cooling section. Take the density, specific heat, and emissivity of the plastic sheet to be $\rho = 75\ \text{lbm/ft}^3$, $C_p = 0.4\ \text{Btu/(lbm}\cdot°\text{F)}$, and $\varepsilon = 0.9$.

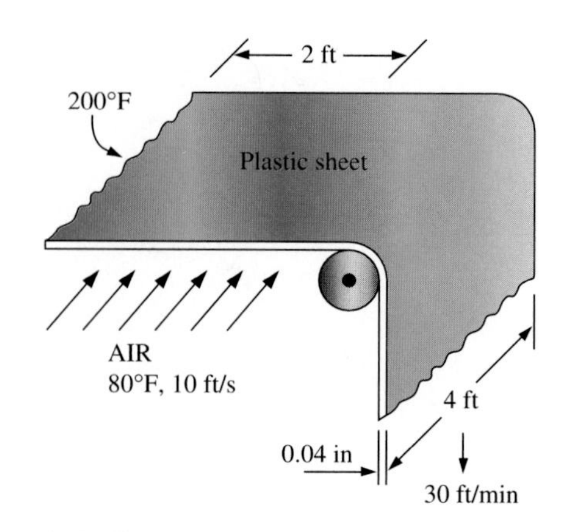


**FIGURE 10-43**
Schematic for Example 10-8.

**Solution** (*a*) We expect the temperature of the plastic sheet to drop somewhat as it flows through the 2-ft-long cooling section, but at this point we do not know the magnitude of that drop. Therefore, we assume the plastic sheet to be isothermal at 200°F to get started. We will repeat the calculations if necessary to account for the temperature drop of the plastic sheet.

This is a forced flow problem over a flat plate. We take the critical Reynolds number to be $\text{Re}_{cr} = 5 \times 10^5$, and evaluate the properties of air at the film temperature of

$$T_f = \frac{T_s + T_\infty}{2} = \frac{200 + 80}{2} = 140°\text{F}$$

which are

$$k = 0.0162 \text{ Btu/(h} \cdot \text{ft} \cdot °\text{F)}$$
$$\text{Pr} = 0.72$$
$$\nu = 0.204 \times 10^{-3} \text{ ft}^2/\text{s}$$

Noting that $L = 4$ ft, the Reynolds number at the end of the air flow across the plastic sheet becomes

$$\text{Re}_L = \frac{\mathscr{V}_\infty L}{\nu} = \frac{(10 \text{ ft/s})(4 \text{ ft})}{0.204 \times 10^{-3} \text{ ft}^2/\text{s}} = 1.96 \times 10^5$$

which is less than the critical Reynolds number. Thus, we have *laminar flow* over the entire sheet, and the Nusselt number is determined, using the laminar flow relations for a flat plate, to be

$$\text{Nu} = \frac{hL}{k} = 0.664 \text{Re}_L^{0.5} \text{Pr}^{1/3} = 0.664 \times (1.96 \times 10^5)^{0.5} \times 0.72^{1/3} = 263.5$$

Then,

$$h = \frac{k}{L}\text{Nu} = \frac{0.0162 \text{ Btu/(h} \cdot \text{ft} \cdot °\text{F)}}{4 \text{ ft}}(263.5) = 1.07 \text{ Btu/(h} \cdot \text{ft}^2 \cdot °\text{F)}$$

$$A = (2 \text{ ft})(4 \text{ ft})(2 \text{ sides}) = 16 \text{ ft}^2$$

and

$$\dot{Q}_{conv} = hA(T_s - T_\infty)$$
$$= [1.07 \text{ Btu/(h . ft}^2 \text{ . °F)}](16 \text{ ft}^2)(200 - 80)°\text{F}$$
$$= 2054 \text{ Btu/h}$$

$$\dot{Q}_{rad} = \varepsilon\sigma A(T_s^4 - T_{surr}^4)$$
$$= (0.9)[0.1714 \text{ Btu/(h} \cdot \text{ft}^2 \cdot \text{R}^4)](16 \text{ ft}^2)[(660 \text{ R})^4 - (540 \text{ R})^4]$$
$$= 2584 \text{ Btu/h}$$

Therefore, the rate of cooling of the plastic sheet by combined convection and radiation is

$$\dot{Q}_{total} = \dot{Q}_{conv} + \dot{Q}_{radv} = 2584 + 2054 = 4638 \text{ Btu/h}$$

(*b*) To find the temperature of the plastic sheet at the end of the cooling section, we need to know the mass of the plastic rolling out per unit time (or the mass flow rate), which is determined from

$$\dot{m} = \rho A_c \mathscr{V} = (75 \text{ lbm/ft}^3)\left(\frac{4 \times 0.04}{12} \text{ ft}^2\right)\left(\frac{30}{60} \text{ ft/s}\right) = 0.5 \text{ lbm/s}$$

Then, an energy balance on the cooled section of the plastic sheet yields

$$\dot{Q} = \dot{m}C_p(T_2 - T_1) \rightarrow T_2 = T_1 + \frac{\dot{Q}}{\dot{m}C_p}$$

Noting that $\dot{Q}$ is a negative quantity (heat loss) for the plastic sheet and substituting, the temperature of the plastic sheet as it leaves the cooling section is determined to be

$$T_2 = 200°\text{F} + \frac{-4638 \text{ Btu/h}}{(0.5 \text{ lbm/s})[0.4 \text{ Btu/(lbm} \cdot °\text{F)}]}\left(\frac{1 \text{ h}}{3600 \text{ s}}\right) = 193.6°\text{F}$$

That is, the average temperature of the plastic sheet drops by about 6.4°F as it passes through the cooling section. The calculations now can be repeated by taking the average temperature of the plastic sheet to be 196.8°F instead of 200°F for better accuracy, but the change in the results will be insignificant because of the small change in temperature.

## 10-7 ■ SUMMARY

Convection is the mode of heat transfer that involves conduction as well as bulk fluid motion. The rate of convection heat transfer in external flow is expressed by *Newton's law of cooling* as

$$\dot{Q}_{\text{conv}} = hA(T_s - T_\infty) \quad \text{(W)}$$

where $T_s$ is the surface temperature and $T_\infty$ is the free-stream temperature. The heat transfer coefficient $h$ is usually expressed in the dimensionless form as the *Nusselt number*

$$\text{Nu} = \frac{h\delta}{k}$$

where $\delta$ is the *characteristic length.* The region of the flow in which the effects of the viscous shearing forces caused by fluid viscosity are felt is called the *velocity boundary layer* or just the *boundary layer.* The *drag force* acting over the entire surface in external flow is determined from

$$F_D = C_f A \left.\frac{\rho \mathcal{V}_\infty^2}{2}\right| \quad \text{(N)}$$

where $C_f$ is the average friction coefficient and $A$ is the surface area for flow over a flat plate, but the frontal area for flow over a cylinder or sphere.

Fluid flow over a flat plate starts out as smooth and streamlined, but turns chaotic after some distance from the leading edge. The flow regime in the first case is said to be *laminar,* and is characterized by smooth streamlines and highly ordered motion, and *turbulent* in the second case, where it is characterized by velocity fluctuations and highly disordered motion. The intense mixing in turbulent flow enhances both the drag force and the heat transfer. The flow regime depends mainly on the ratio of the inertia forces to viscous forces in the fluid. This ratio is called the *Reynolds number,* and is expressed as

$$\text{Re} = \frac{\mathcal{V}\delta}{\nu}$$

where $\mathcal{V}$ is the free stream velocity $\mathcal{V}_\infty$ for external flow and the mean fluid velocity $\mathcal{V}_m$ for internal flow. Also, $\delta$ is the characteristic length of the geometry, which is the distance from the leading edge for a flat plate, the outer diameter for flow over cylinders or spheres, and the inner diameter for flow inside circular tubes. The characteristic length for noncircular tubes is the *hydraulic diameter* $D_h$ defined as

$$D_h = \frac{4A_c}{p}$$

where $A_c$ is the cross-sectional area of the tube and $p$ is its perimeter. The Reynolds number at which the flow becomes turbulent is called the *critical Reynolds number.* The value of the critical Reynolds number is about $5 \times 10^5$ for flow over a flat plate, $2 \times 10^5$ for flow over cylinders and spheres, and 2300 for flow inside tubes.

When the flow over the entire flat plate is laminar, the average friction coefficient and the Nusselt number can be determined from

$$C_f = \frac{1.328}{\mathrm{Re}_L^{1/2}}$$

and

$$\mathrm{Nu} = \frac{hL}{k} = 0.664\,\mathrm{Re}_L^{1/2}\,\mathrm{Pr}^{1/3} \qquad (\mathrm{Pr} \geq 0.6)$$

After transition to turbulent flow at a critical Reynolds number $\mathrm{Re}_{cr} = 5 \times 10^5$, the average friction coefficient and the Nusselt number over the entire plate become

$$C_f = \frac{0.074}{\mathrm{Re}_L^{1/5}} - \frac{1742}{\mathrm{Re}_L} \qquad (5 \times 10^5 \leq \mathrm{Re}_L \leq 10^7)$$

and

$$\mathrm{Nu} = \frac{hL}{k} = (0.037\,\mathrm{Re}_L^{4/5} - 871)\,\mathrm{Pr}^{1/3} \qquad \begin{pmatrix} 0.6 \leq \mathrm{Pr} \leq 60 \\ 5 \times 10^5 \leq \mathrm{Re}_L \leq 10^7 \end{pmatrix}$$

In order to account for the variation of the properties with temperature properly, the fluid properties are usually evaluated at the *film temperature* defined as $T_f = \frac{1}{2}(T_s + T_\infty)$, which is the *arithmetic average* of the surface and the free-stream temperatures.

The average Nusselt numbers for cross-flow over a *cylinder* and *sphere* can be determined from

$$\mathrm{Nu}_{cyl} = \frac{hD}{k} = 0.3 + \frac{0.62\,\mathrm{Re}^{1/2}\,\mathrm{Pr}^{1/3}}{[1 + (0.4/\mathrm{Pr})^{2/3}]^{1/4}}\left[1 + \left(\frac{\mathrm{Re}}{28{,}200}\right)^{5/8}\right]^{4/5}$$

which is valid for $\mathrm{Re\,Pr} > 0.2$, and

$$\mathrm{Nu}_{sph} = \frac{hD}{k} = 2 + [0.4\,\mathrm{Re}^{1/2} + 0.06\,\mathrm{Re}^{2/3}]\,\mathrm{Pr}^{0.4}\left(\frac{\mu_\infty}{\mu_s}\right)^{1/4}$$

which is valid for $3.5 \leq \mathrm{Re} \leq 80{,}000$ and $0.7 \leq \mathrm{Pr} \leq 380$. The fluid properties are evaluated at the film temperature $T_f = \frac{1}{2}(T_\infty + T_s)$ in the case of cylinder, and at the free-stream temperature $T_\infty$ (except for $\mu_s$, which is evaluated at the surface temperature $T_s$) in the case of a sphere.

For flow in a tube, the mean velocity $\mathcal{V}_m$ is the average velocity of the fluid. The mean temperature $T_m$ at a cross-section can be viewed as the average temperature at that cross section. The mean velocity $\mathcal{V}_m$ remains constant, but the mean temperature $T_m$ changes along the tube unless the fluid is not heated or cooled. The heat transfer to a fluid during steady flow in a tube can be expressed as

$$\dot{Q} = \dot{m}C_p(T_e - T_i) \quad \text{(kJ/s)}$$

where $T_i$ and $T_e$ are the mean fluid temperatures at the inlet and exit of the tube.

The conditions at the surface of a tube can usually be approximated with reasonable accuracy to be *constant surface temperature* ($T_s$ = constant) or *constant surface heat flux* ($\dot{q}_s$ = constant). In the case of $\dot{q}_s$ = constant, the rate of heat transfer can be expressed as

$$\dot{Q} = \dot{q}_s A = \dot{m}C_p(T_e - T_i) \quad \text{(W)}$$

Then mean fluid temperature at the tube exit becomes

$$T_e = T_i + \frac{\dot{q}_s A}{\dot{m}C_p}$$

In the case of $T_s$ = constant, the rate of heat transfer is expressed as

$$\dot{Q} = hA\,\Delta T_{\ln}$$

where

$$\Delta T_{\ln} = \frac{T_e - T_i}{\ln \dfrac{T_s - T_e}{T_s - T_i}} = \frac{\Delta T_e - \Delta T_i}{\ln(\Delta T_e/\Delta T_i)}$$

is the *logarithmic mean temperature difference.* Note that $\Delta T_i = T_s - T_i$ and $\Delta T_e = T_s - T_e$ are the temperature differences between the surface and the fluid at the inlet and the exit of the tube, respectively. Then the mean fluid temperature at the tube exit in this case can be determined from

$$T_e = T_s - (T_s - T_i)e^{-hA/(\dot{m}C_p)}$$

The pressure drop during flow in a tube of length $L$ is expressed as

$$\Delta P = f\frac{L}{D}\frac{\rho \mathscr{V}_m^2}{2} \quad \text{(N/m}^2\text{)}$$

where $f$ is the *friction factor.* The required *pumping power* to overcome a specified pressure drop $\Delta P$ is determined from

$$\dot{W}_{\text{pump}} = \dot{V}\,\Delta P = \frac{\dot{m}\,\Delta P}{\rho} \quad \text{(W)}$$

where $\dot{V} = \mathscr{V}_m A_c = \dot{m}/\rho$ is the volume flow rate of the fluid through the tube.

The region from the tube inlet to the point at which the boundary layer merges at the centerline is called the *hydrodynamic entry region,* and the length of this region is called the *hydrodynamic entry length* $L_h$. The region beyond the hydrodynamic entry region in which the velocity profile is fully developed and remains unchanged is called the *hydrodynamically developed region.* The region of flow over which the thermal boundary layer develops and reaches the tube center is called the *thermal entry region,* and the length of this region is called the *thermal entry length* $L_t$. The region beyond the thermal entry region in which the dimensionless temperature profile expressed as $(T - T_s)/(T_m - T_s)$ remains unchanged is called the *thermally developed region.* The region in which the flow is both hydrodynamically and thermally developed is called the *fully developed flow.* The hydrodynamic and thermal entry lengths are given approximately as

$$L_{h,\text{laminar}} \approx 0.05 \text{ Re } D$$

$$L_{t,\text{laminar}} = 0.05 \text{ Re Pr } D$$

and

$$L_{h,\text{turbulent}} \approx L_{t,\text{turbulent}} \approx 10D$$

The friction and the heat transfer coefficients in the fully developed flow region remain constant.

In fully developed laminar flow, the friction factor is determined to be $f = 64/\text{Re}$. The Nusselt number is determined to be $\text{Nu} = 3.66$ for the case of $T_s$ = constant, and $\text{Nu} = 4.36$ for the case of $\dot{q}_s$ = constant. The average Nusselt number for the hydrodynamically and/or thermally developing laminar flow in a circular tube is given as

$$\text{Nu} = 1.86\left(\frac{\text{Re Pr } D}{L}\right)^{1/3}\left(\frac{\mu_b}{\mu_s}\right)^{0.14} \qquad (\text{Pr} > 0.5)$$

The recommended relations for the friction factor $f$ and the Nusselt number for fully developed turbulent flow in smooth circular tubes are

$$f = 0.184 \text{ Re}^{-0.2}$$

and

$$\text{Nu} = 0.023 \text{ Re}^{0.8} \text{ Pr}^n \qquad \left(\begin{matrix} 0.7 \leq \text{Pr} \leq 160 \\ \text{Re} > 10{,}000 \end{matrix}\right)$$

where $n = 0.4$ for *heating* and 0.3 for *cooling* of the fluid flowing through the tube. The fluid properties are evaluated at the *bulk mean fluid temperature* $T_b = \frac{1}{2}(T_i + T_e)$, which is the arithmetic average of the mean fluid temperatures at the inlet and the exit of the tube.

## REFERENCES AND SUGGESTED READING

**1** Y. Bayazitoglu and M. N. Özışık, *Elements of Heat Transfer,* McGraw-Hill, New York, 1988.

**2** S. W. Churchill and M. Bernstein, "A Correlating Equation for Forced Convection from Gases and Liquids to a Circular Cylinder in Cross Flow," *Journal of Heat Transfer,* Vol. 99, pp. 300–306, 1977.

**3** A. P. Colburn, *Transactions of the AIChE,* Vol. 26, p. 174, 1933.

**4** F. W. Dittus and L. M. K. Boelter, University of California Publications on Engineering, Vol. 2, p. 433, 1930.

**5** W. H. Giedt, "Investigation of Variation of Point Unit-Heat Transfer Coefficient Around a Cylinder Normal to an Air Stream," *Transactions of the ASME,* Vol. 71, pp. 375–381, 1949.

**6** J. P. Holman, *Heat Transfer,* 7th ed., McGraw-Hill, New York, 1990.

**7** F. P. Incropera and D. P. DeWitt, *Introduction to Heat Transfer,* 2nd ed., Wiley, 1990.

**8** M. Jakob, *Heat Transfer,* Vol. 1, Wiley, New York, 1949.

**9** F. Kreith and M. S. Bohn, *Principles of Heat Transfer,* 5th ed., West, St. Paul, MN, 1993.

**10** L. F. Moody, "Friction Factor for Pipe Flow," *Transactions of the ASME,* Vol. 66, pp. 671–684, 1944.

**11** M. N. Özışık, *Heat Transfer—A Basic Approach,* McGraw-Hill, New York, 1985.

**12** O. Reynolds, "On the Experimental Investigation of the Circumstances which Determine Whether the Motion of Water Shall be Direct or Sinuous, and the Law of Resistance in Parallel Channels," *Philosophical Transactions of the Royal Society of London,* Vol. 174, pp. 935–982, 1883.

**13** H. Schlichting, *Boundary Layer Theory,* 7th ed., McGraw-Hill, New York, 1979.

**14** E. N. Sieder and G. E. Tate, "Heat Transfer and Pressure Drop of Liquids in Tubes," *Industrial Engineering Chemistry* Vol. 28, pp. 1429–1435, 1936.

**15** L. C. Thomas, *Heat Transfer,* Prentice-Hall, Englewood Cliffs, NJ, 1992.

**16** S. Whitaker, "Forced Convection Heat Transfer Correlations for Flow in Pipe, Past Flat Plates, Single Cylinders, and for Flow in Packed Beds and Tube Bundles," *AIChE Journal,* Vol. 18, pp. 361–371, 1972.

**17** F. M. White, *Heat and Mass Transfer,* Addison-Wesley, Reading, MA, 1988.

**18** A. Zhukauskas, "Heat Transfer from Tubes in Cross Flow," in J. P. Hartnett and T. F. Irvine, Jr., Eds., *Advances in Heat Transfer,* Vol. 8, Academic Press, New York, 1972.

## PROBLEMS*

### Physical Mechanism of Forced Convection

**10-1C** What is forced convection? How does it differ from natural convection? Is convection caused by winds forced or natural convection?

*Students are encouraged to answer *all* the concept "C" questions.

**10-2C** What is external forced convection? How does it differ from internal forced convection? Can a heat transfer system involve both internal and external convection at the same time? Give an example.

**10-3C** In which mode of heat transfer the convection heat transfer coefficient is usually higher, natural convection or forced convection? Why?

**10-4C** Consider a hot baked potato. Will the potato cool faster or slower when we blow the warm air coming from our lungs on it instead of letting it cool naturally in the cooler air in the room? Explain.

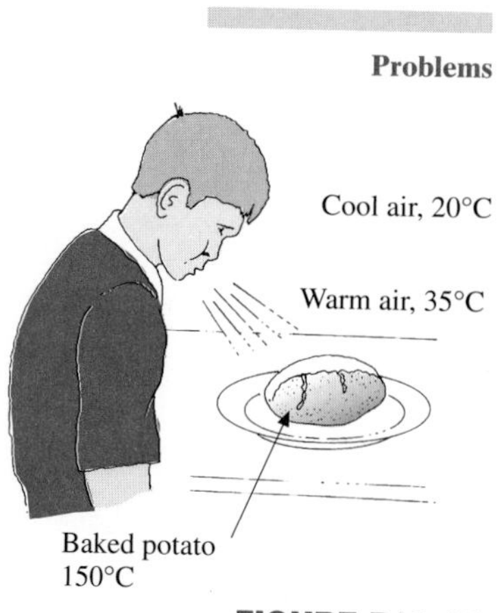

FIGURE P10-4C

**10-5C** What is the physical significance of the Prandtl number? Does the value of the Prandtl number depend on the type of flow or the flow geometry? Does the Prandtl number of air change with pressure? Does it change with temperature?

**10-6C** What is the physical significance of the Reynolds number? How is it defined for (*a*) flow over a flat plate of length $L$, (*b*) flow over a cylinder of outer diameter $D_o$, (*c*) flow in a circular tube of inner diameter $D_i$, and (*d*) flow in a rectangular tube of cross-section $a \times b$?

**10-7C** What is the physical significance of the Nusselt number? How is it defined for (*a*) flow over a flat plate of length $L$, (*b*) flow over a cylinder of outer diameter $D_o$, (*c*) flow in a circular tube of inner diameter $D_i$, and (*d*) flow in a rectangular tube of cross section $a \times b$?

**10-8C** When is heat transfer through a fluid conduction and when is it convection? For what case is the rate of heat transfer higher? How does the convection heat transfer coefficient differ from the thermal conductivity of a fluid?

**10-9C** How does turbulent flow differ from laminar flow? For which flow is the (*a*) friction coefficient and (*b*) the heat transfer coefficient higher?

**10-10C** Will a thermal boundary layer develop in flow over a surface even if both the fluid and the surface are at the same temperature?

## Flow over Flat Plates

**10-11C** What fluid property is responsible for the development of the velocity boundary layer? For what kind of fluids will there be no velocity boundary layer on a flat plate?

**10-12C** Under what conditions can a curved surface be treated as a flat plate in convection calculations?

**10-13C** What is the no-slip condition on a surface?

**10-14C** Consider laminar forced convection from a horizontal flat plate. Will the heat flux be higher at the leading edge or at the tail of the plate? Why?

**10-15C** What does the friction coefficient represent in flow over a flat plate? How is it related to the drag force acting on the plate?

**10-16C** For flow over a flat plate, how does the flow in the thermal boundary layer differ than the flow outside the thermal boundary layer?

**10-17C** Consider laminar flow over a flat plate. Will the friction coefficient change with position? How about the heat transfer coefficient?

**10-18C** How are the average friction and heat transfer coefficients determined in flow over a flat plate?

**10-19** Engine oil at 80°C flows over a 6-m-long flat plate whose temperature is 30°C with a velocity of 3 m/s. Determine the total drag force and the rate of heat transfer over the entire plate per unit width.

**10-20** The local atmospheric pressure in Denver, Colorado (elevation 1610 m) is 83.4 kPa. Air at this pressure and 30°C flows with a velocity of 6 m/s over a 2.5 m × 8 m flat plate whose temperature is 120°C. Determine the rate of heat transfer from the plate if the air flows parallel to the (*a*) 8-m-long side and (*b*) the 2.5-m side.

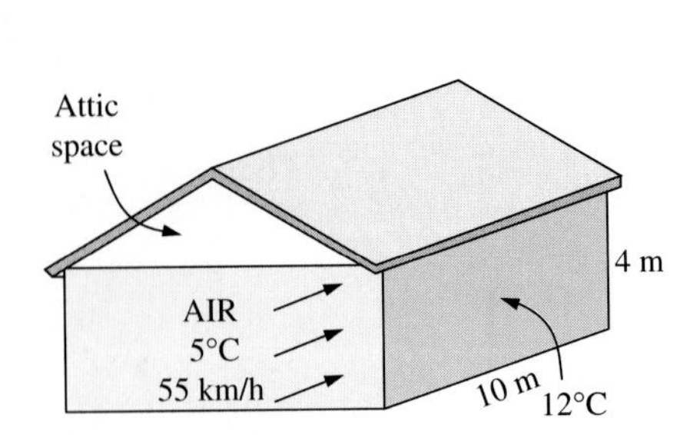


**FIGURE P10-21**

**10-21** During a cold winter day, wind at 55 km/h is blowing parallel to a 4-m-high and 10-m-long wall of a house. If the air outside is at 5°C and the surface temperature of the wall is 12°C, determine the rate of heat loss from that wall by convection. What would your answer be if the wind velocity has doubled? *Answers:* 9212 *W*, 16,408 *W*

**10-21E** During a cold winter day, wind at 40 mph is blowing parallel to a 12-ft-high and 30-ft-long wall of a house. If the air outside is at 40°F and the surface temperature of the wall is 55°F, determine the rate of heat loss from that wall by convection. What would your answer be if the wind velocity has doubled? *Answers:* 36,018 Btu/h, 64,152 Btu/h

**10-22** Air at 17°C flows over a 3-m-long flat plate at 2 m/s. Determine the local friction and heat transfer coefficients at intervals of 25 cm, and plot the results against the distance from the leading edge.

**10-22E** Air at 65°F flows over a 10-ft-long flat plate at 7 ft/s. Determine the local friction and heat transfer coefficients at intervals of 1 ft, and plot the results against the distance from the leading edge.

**10-23** Consider a hot automotive engine, which can be approximated as a 0.5-m-high, 0.40-m-wide, and 0.8-m-long rectangular block. The bottom surface of the block is at a temperature of 80°C and has an emissivity of 0.95. The ambient air is at 30°C, and the road surface is at 25°C. Determine the rate of heat transfer from the bottom surface of the engine block by convection and radiation as the car travels at a velocity of 80 km/h. Assume the flow to be turbulent over the entire surface because of the constant agitation of the engine block.

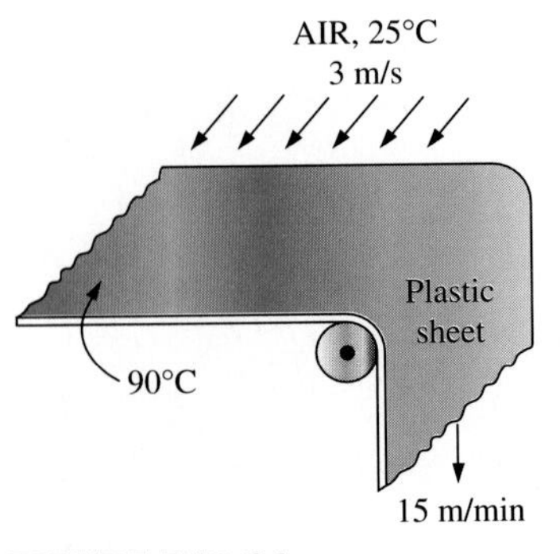


**FIGURE P10-24**

**10-24** The forming section of a plastics plant puts out a continuous sheet of plastic that is 1.2 m wide and 2 mm thick at a rate of 15 m/min. The temperature of the plastic sheet is 90°C when it is exposed to the surrounding air, and the sheet is subjected to air flow at 25°C at a velocity of 3 m/s on both sides along its surfaces normal to the direction of motion of the sheet. The width of the air cooling section is such that a fixed point on the plastic sheet passes through that section in 2 s. Determine the rate of heat transfer from the plastic sheet to the air and the drag force the air exerts on the plastic sheet in the direction of air flow.

**10-25** The top surface of the passenger car of a train moving at a velocity of 70 km/h is 2.8 m wide and 8 m long. The top surface is absorbing solar radiation at a rate of 200 W/m$^2$, and the temperature of the ambient air is 30°C. Assuming the roof of the car to be perfectly insulated and the radiation heat exchange with the surroundings to be small relative to convection, determine the equilibrium temperature of the top surface of the car. *Answer:* 35°C

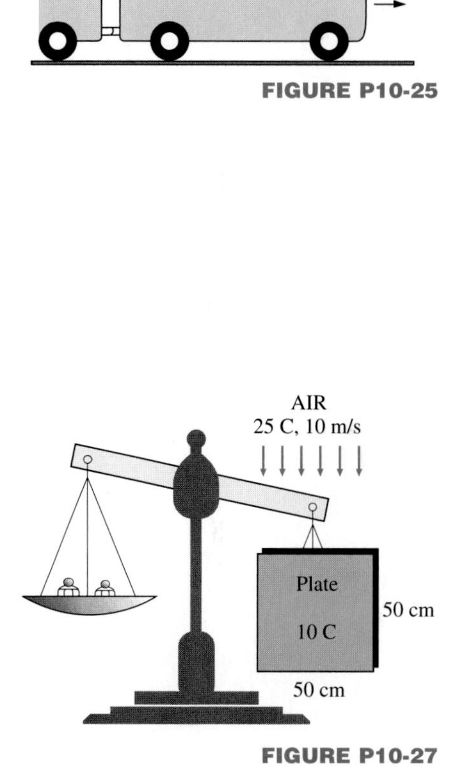


**FIGURE P10-25**

**FIGURE P10-27**

**10-25E** The top surface of the passenger car of a train moving at a velocity of 60 mph is 9 ft wide and 25 ft long. The top surface is absorbing solar radiation at a rate of 70 Btu/(h · ft$^2$), and the temperature of the ambient air is 80°F. Assuming the roof of the car to be perfectly insulated and the radiation heat exchange with the surroundings to be small relative to convection, determine the equilibrium temperature of the top surface of the car. *Answer:* 87.6°F

**10-26** A 15 cm × 15 cm circuit board dissipating 15 W of power uniformly is cooled by air, which approaches the circuit board at 50°C with a velocity of 5 m/s. Disregarding any heat transfer from the back surface of the board, determine the surface temperature of the electronic components (*a*) at the leading edge and (*b*) at the end of the board. Assume the flow to be turbulent, since the electronic components are expected to act as turbulators.

**10-27** The weight of a thin flat plate 50 cm × 50 cm in size is balanced by a counterweight that has a mass of 2 kg, as shown in Fig. P10-27. Now a fan is turned on, and air at 25°C flows downwards over both surfaces of the plate with a free-stream velocity of 10 m/s. Determine the mass of the counterweight that needs to be added in order to balance the plate in this case. Also determine the initial rate of heat transfer to the plate if the plate was initially at a uniform temperature of 10°C.

**10-28** Consider laminar flow of a fluid over a flat plate maintained at a constant temperature. Now the free-stream velocity of the fluid is doubled. Determine the change in the drag force on the plate and the rate of heat transfer between the fluid and the plate. Assume the flow to remain laminar.

**10-29** Consider a refrigeration truck traveling at 85 km/h at a location where the air temperature is 30°C. The refrigerated compartment of the truck can be considered to be a 2.6-m-wide, 2.4-m-high, and 7-m-long rectangular box. The refrigeration system of the truck can provide 3 tons of refrigeration (i.e., it can remove heat at a rate of 633 kJ/min). The

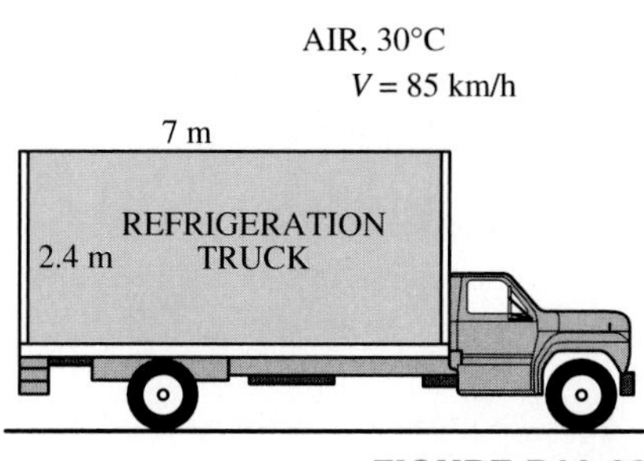


**FIGURE P10-29**

outer surface of the truck is coated with a low-emissivity material, and thus radiation heat transfer is very small. Determine the average temperature of the outer surface of the refrigeration compartment of the truck if the refrigeration system is observed to be operating at half the capacity. Assume the air flow over the entire outer surface to be turbulent, and the heat transfer coefficient at the front and rear surfaces to be equal to that on side surfaces.

**10-29E** Consider a refrigeration truck traveling at 55 mph at a location where the air temperature is 80°F. The refrigerated compartment of the truck can be considered to be a 9-ft-wide, 8-ft-high, and 20-ft-long rectangular box. The refrigeration system of the truck can provide 3 tons of refrigeration (i.e., it can remove heat at a rate of 600 Btu/min). The outer surface of the truck is coated with a low-emissivity material, and thus radiation heat transfer is very small. Determine the average temperature of the outer surface of the refrigeration compartment of the truck if the refrigeration system is observed to be operating at half the capacity. Assume the air flow over the entire outer surface to be turbulent, and the heat transfer coefficient at the front and rear surfaces to be equal to that on side surfaces.

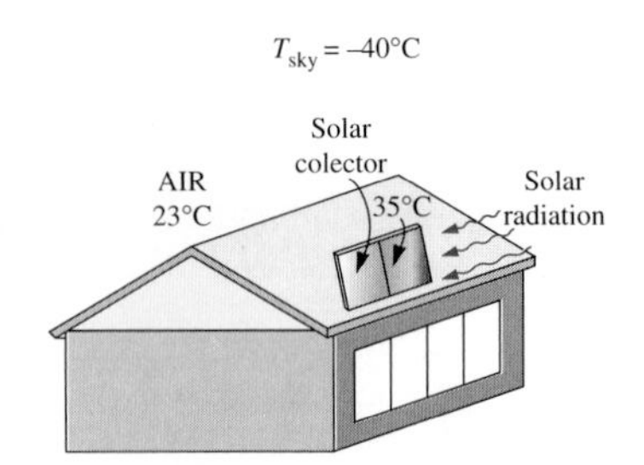

**FIGURE P10-30**

**10-30** Solar radiation is incident on the glass cover of a solar collector at a rate of 700 W/m$^2$. The glass transmits 88 percent of the incident radiation, and has an emissivity of 0.90. The entire hot water needs of a family in summer can be met by two collectors 1.2 m high and 1 m wide. The two collectors are attached to each other on one side so that they appear like a single collector 1.2 m × 2 m in size. The temperature of the glass cover is measured to be 35°C on a day when the surrounding air temperature is 23°C and the wind is blowing at 30 km/h. The effective sky temperature for radiation exchange between the glass cover and the open sky is −40°C. Water enters the tubes attached to the absorber plate at a rate of 1 kg/min. Assuming the backsurface of the absorber plate to be heavily insulated and the only heat loss occurs through the glass cover, determine (*a*) the total rate of heat loss from the collector, (*b*) the collector efficiency, which is the ratio of the amount of heat transferred to the water to the solar energy incident on the collector, and (*c*) the temperature rise of water as it flows through the collector.
*Answers:* (*a*) 1262 W, (*b*) 0.15, (*c*) 3.1°C

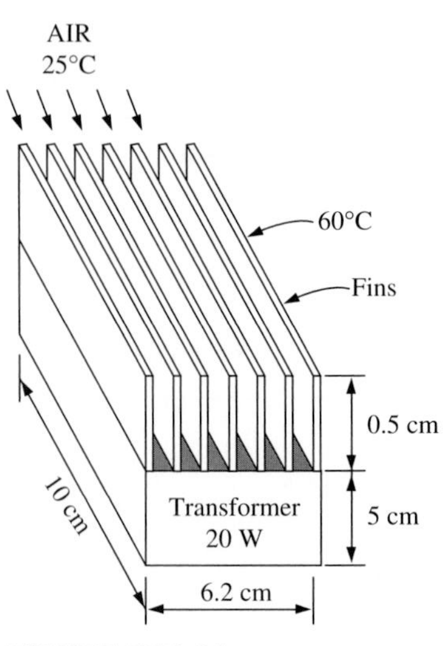


**FIGURE P10-31**

**10-31** A transformer that is 10 cm long, 6.2 cm wide, and 5 cm high is to be cooled by attaching a 10 cm × 6.2 cm wide polished aluminum heat sink (emissivity = 0.03) to its top surface. The heat sink has seven fins, which are 5 mm high, 2 mm thick, and 10 cm long. A fan blows air at 25°C parallel to the passages between the fins. The heat sink is to dissipate 20 W of heat and the base temperature of the heat sink is not to exceed 60°C. Assuming the fins and the base plate to be nearly isothermal and the radiation heat transfer to be negligible, determine the minimum free-stream velocity the fan needs to supply to avoid overheating.

**10-31E** A transformer that is 4 in long, 2.5 in wide, and 2 in high is to be cooled by attaching a 4 in × 2.5 in wide polished aluminum heat sink (emissivity = 0.03) to its top surface. The heat sink has seven fins, which are 0.2 in high, 0.1 in thick, and 4 in long. A fan blows air at 75°F parallel to the passages between the fins. The heat sink is to dissipate 20 W of heat and the base temperature of the heat sink is not to exceed 130°F. Assuming the fins and the base plate to be nearly isothermal and the radiation heat transfer to be negligible, determine the minimum free-stream velocity the fan needs to supply to avoid overheating.

**10-32** Repeat Prob. 10-31 assuming the heat sink to be black-anodized and thus to have an effective emissivity of 0.90. Note that in radiation calculations the base area (10 cm × 6.2 cm) is to be used, not the total surface area.

**10-33** An array of power transistors, dissipating 3 W of power each, are to be cooled by mounting them on a 25 cm × 25 cm square aluminum plate and blowing air at 35°C over the plate with a fan at a velocity of 4 m/s. The average temperature of the plate is not to exceed 65°C. Assuming the heat transfer from the back side of the plate to be negligible and disregarding radiation, determine the number of transistors that can be placed on this plate.

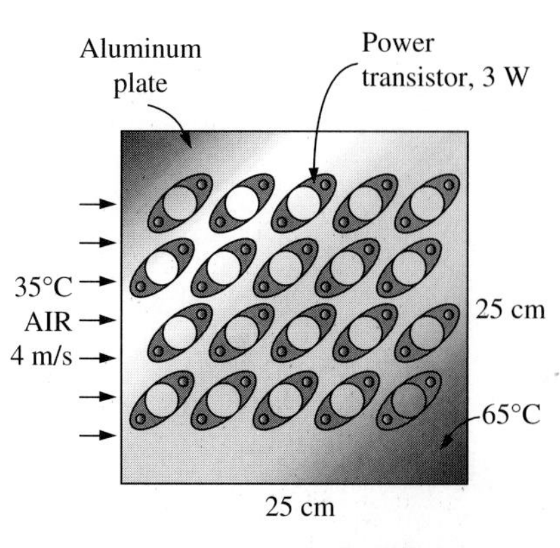

**FIGURE P10-33**

**10-33E** An array of power transistors, dissipating 3 W of power each, are to be cooled by mounting them on a 10 in × 10 in square aluminum plate and blowing air at 90°F over the plate with a fan at a velocity of 15 ft/s. The average temperature of the plate is not to exceed 130°F. Assuming the heat transfer from the back side of the plate to be negligible and disregarding radiation, determine the number of transistors that can be placed on this plate.

**10-34** Repeat Prob. 10-33 for a location at an elevation of 1610 m where the atmospheric pressure is 83.4 kPa. *Answer:* 8

### Flow Across Cylinders and Spheres

**10-35C** How is the film temperature for flow over an isothermal cylinder defined?

**10-36C** In flow over cylinders, why does the drag coefficient suddenly drop when the flow becomes turbulent? Isn't turbulence supposed to increase the drag coefficient instead of decreasing it?

**10-37C** In flow over blunt bodies such as a cylinder, how does the pressure drag differ from the friction drag?

**10-38C** Why is flow separation in flow over cylinders delayed in turbulent flow?

**10-39C** Which bicyclist is more likely to go faster: the one who keeps his head and his body in the most upright position or the one who leans down and brings his body closer to his knees? Why?

**10-40C** Which car is more likely to be more fuel-efficient: the one with sharp corners, or the one that is contoured to resemble an ellipse? Why?

**10-41C** Consider laminar flow of air across a hot circular cylinder. At what point on the cylinder will the heat transfer be highest? What would your answer be if the flow were turbulent?

**10-42** A long 8-cm-diameter steam pipe whose external surface temperature is 90°C passes through some open area that is not protected against the winds. Determine the rate of heat loss from the pipe per unit of its length when the air is at 1 atm pressure and 7°C and the wind is blowing across the pipe at a velocity of 50 km/h.

**10-43** A stainless steel ball [$\rho$ = 8055 kg/m$^3$, $C_p$ = 480 J/(kg · °C)] of diameter $D$ = 15 cm is removed from the oven at a uniform temperature of 350°C. The ball is then subjected to the flow of air at 1-atm pressure and 30°C with a velocity of 6 m/s. The surface temperature of the ball eventually drops to 250°C. Determine the average convection heat transfer coefficient during this cooling process, and estimate how long this process has taken.

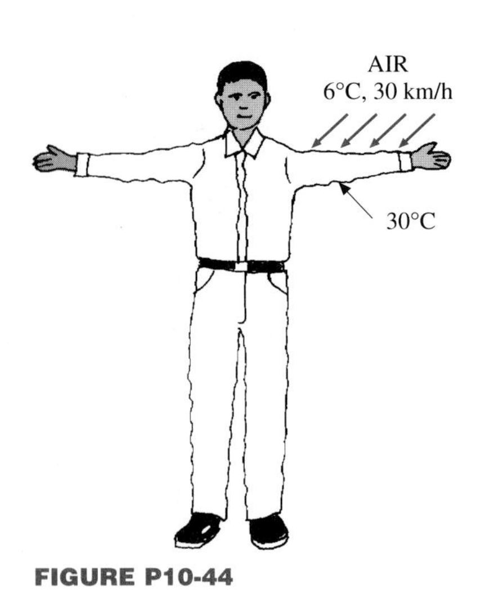


FIGURE P10-44

**10-44** A person extends his uncovered arms into the windy air outside at 6°C and 30 km/h in order to feel nature closely. Initially, the skin temperature of the arm is 30°C. Treating the arm as a 60-cm-long and 7.5-cm-diameter cylinder, determine the rate of heat loss from the arm.

**10-44E** A person extends his uncovered arms into the windy air outside at 40°F and 20 mph in order to feel nature closely. Initially, the skin temperature of the arm is 86°F. Treating the arm as a 2-ft-long and 3-in-diameter cylinder, determine the rate of heat loss from the arm.

**10-45** An average person generates heat at a rate of 84 W while resting. Assuming one-quarter of this heat is lost from the head and disregarding radiation, determine the average surface temperature of the head when it is not covered and is subjected to winds at 10°C and 35 km/h. The head can be approximated as a 30-cm-diameter sphere. *Answer:* 12.7°C

**10-46** Consider the flow of a fluid across a cylinder maintained at a constant temperature. Now the free-stream velocity of the fluid is doubled. Determine the change in the drag force on the cylinder and the rate of heat transfer between the fluid and the cylinder.

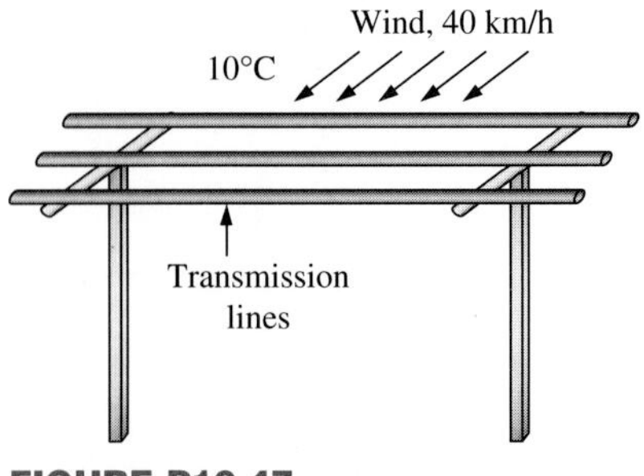


FIGURE P10-47

**10-47** A 6-mm-diameter electrical transmission line carries an electric current of 50 A and has a resistance of 0.002 ohm per meter length. Determine the surface temperature of the wire during a windy day when the air temperature is 10°C and the wind is blowing across the transmission line at 40 km/h. Also determine the drag force exerted on the wire by the wind.

**10-47E** A 0.25-in-diameter electrical transmission line carries an electric current of 50 A and has a resistance of 0.001 ohm per ft length. Determine the surface temperature of the wire during a windy day when the air temperature is 50°F and the wind is blowing across the transmission line at 30 mph. Also determine the drag force exerted on the wire by the wind.

**10-48** A heating system is to be designed to keep the wings of an aircraft cruising at a velocity of 900 km/h above freezing temperatures during flight at 12,200-m altitude where the standard atmospheric conditions are −55.4°C and 18.8 kPa. Approximating the wing as a cylinder of elliptical cross-section whose minor axis is 30 cm and disregarding radiation, determine the average convection heat transfer coefficient on the wing surface and the average rate of heat transfer per unit surface area.

**10-49** A long aluminum wire of diameter 3 mm is extruded at a temperature of 350°C. The wire is subjected to cross air flow at 35°C at a velocity of 6 m/s. Determine the rate of heat transfer from the wire to the air per meter length when it is first exposed to the air.

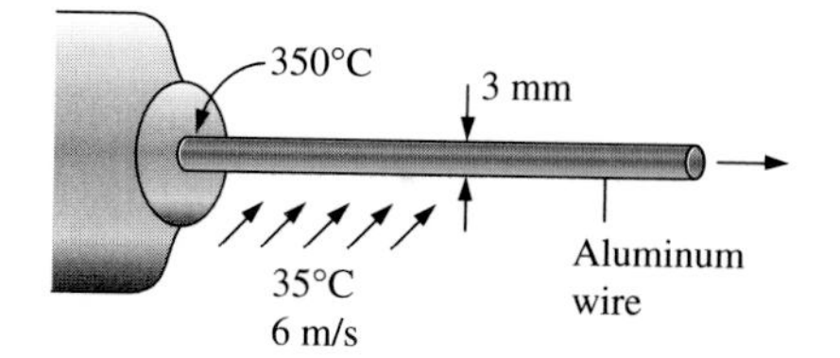

FIGURE P10-49

**10-50** Consider a person who is trying to keep cool in a hot summer day by turning a fan on and exposing his entire body to air flow. The air temperature is 30°C and the fan is blowing air at a velocity of 2 m/s. If the person is doing light work and generating sensible heat at a rate of 100 W, determine the average temperature of the outer surface (skin or clothing) of the person. The average human body can be treated as a 30-cm-diameter cylinder with an exposed surface area of 1.7 m$^2$. Disregard any heat transfer by radiation. What would your answer be if the air velocity were doubled? *Answers:* 33.8°C, 32.2°C

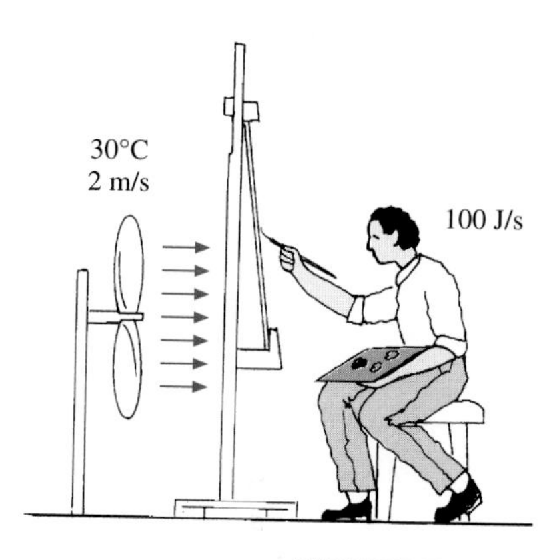


FIGURE P10-50

**10-50E** Consider a person who is trying to keep cool in a hot summer day by turning a fan on and exposing his entire body to air flow. The air temperature is 85°F and the fan is blowing air at a velocity of 6 ft/s. If the person is doing light work and generating sensible heat at a rate of 300 Btu/h, determine the average temperature of the outer surface (skin or clothing) of the person. The average human body can be treated as a 1-ft-diameter cylinder with an exposed surface area of 18 ft$^2$. Disregard any heat transfer by radiation. What would your answer be if the air velocity were doubled? *Answers:* 91.7°F, 88.9°F

**10-51** An incandescent light bulb is an inexpensive but a highly inefficient device which converts electrical energy into light. It converts about 10 percent of the electrical energy it consumes into light while converting the remaining 90 percent into heat. (A fluorescent light bulb will give the same amount of light while consuming only one-fourth of the electrical energy, and it will last 10 times longer than an incandescent light bulb). The glass bulb of the lamp heats up very quickly as a result of absorbing all that heat and dissipating it to the surroundings by convection and radiation.

Consider a 10-cm diameter 100-W light bulb cooled by a fan that blows air at 25°C to the bulb at a velocity of 2 m/s. The surrounding surfaces are also at 25°C, and the emissivity of the glass is 0.9. Assuming

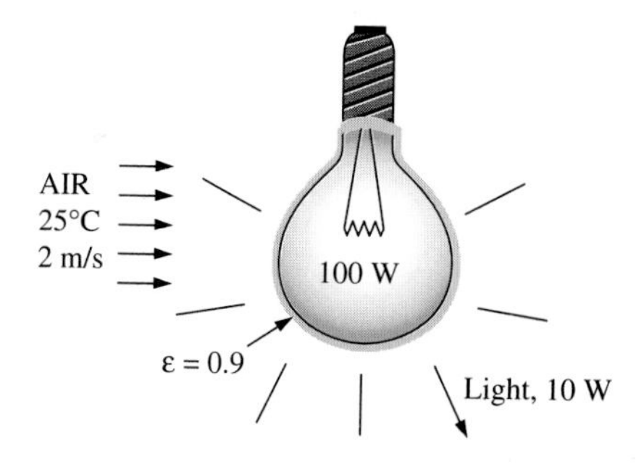

FIGURE P10-51

10 percent of the energy passes through the glass bulb as light with negligible absorption and the rest of the energy is absorbed and dissipated by the bulb itself, determine the equilibrium temperature of the glass bulb.

**10-52** Consider a 3-mm-diameter raindrop at 8°C that is falling freely in atmospheric air at 20°C. Determine the terminal velocity of the raindrop, which is the velocity at which the drag force equals the weight of the drop, and the rate of heat transfer by convection as the raindrop descends at the teminal velocity. Assume there is no evaporation or condensation.

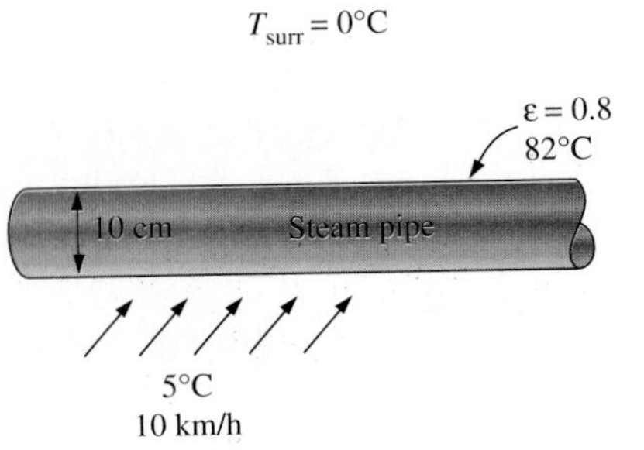

**10-53** During a plant visit, it was noticed that 12-m-long section of a 10-cm-diameter steam pipe is completely exposed to the ambient air. The temperature measurements indicate that the average temperature of the outer surface of the steam pipe is 82°C when the ambient temperature is 5°C. There are also light winds in the area at 10 km/h. The emissivity of the outer surface of the pipe is 0.8, and the average temperature of the surfaces surrounding the pipe, including the sky, is estimated to be 0°C. Determine the amount of heat lost from the steam during a 10-h-long work day.

Steam is supplied by a gas-fired steam generator that has an efficiency of 80 percent, and the plant pays $0.54/therm of natural gas (1 therm = 105,500 kJ). If the pipe is insulated and 90 percent of the heat loss is saved, determine the amount of money this facility will save a year as a result of insulating the steam pipes. Assume the plant operates every day of the year for 10 h. State your assumptions.

$T_{surr} = 0°C$

ε = 0.8
82°C
10 cm
Steam pipe
5°C
10 km/h

**FIGURE P10-53**

**10-54** Reconsider Prob. 10-53. There seems to be some uncertainty about the average temperature of the surfaces surrounding the pipe used in radiation calculations, and you are asked to determine if it makes any significant difference in overall heat transfer. Repeat the calculations in Prob. 10-53 for average surrounding surface temperatures of −20°C and 25°C, and determine the change in the values obtained.

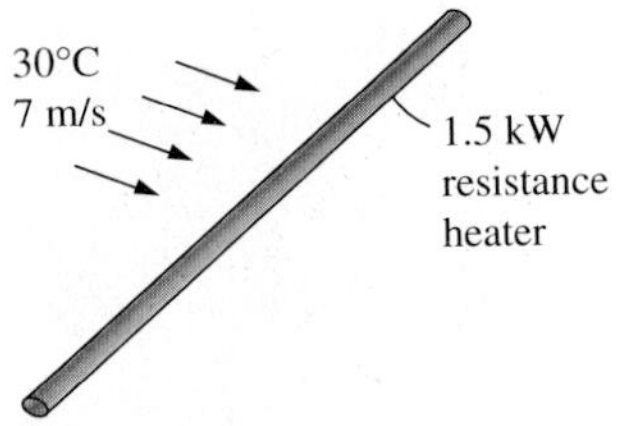

**10-55** A 4-m-long 1.5-kW electrical resistance wire is made of 0.25-cm-diameter stainless steel [$k$ = 15.1 W/(m · °C)]. The resistance wire operates in an environment at 30°C. Determine the surface temperature of the wire if it is cooled by a fan blowing air at a velocity of 7 m/s.

30°C
7 m/s
1.5 kW
resistance
heater

**FIGURE P10-55**

**10-55E** A 12-ft-long 1.5-kW electrical resistance wire is made of 0.1-in-diameter stainless steel [$k$ = 8.7 Btu/(h · ft · °F)]. The resistance wire operates in an environment at 85°F. Determine the surface temperature of the wire if it is cooled by a fan blowing air at a velocity of 20 ft/s.

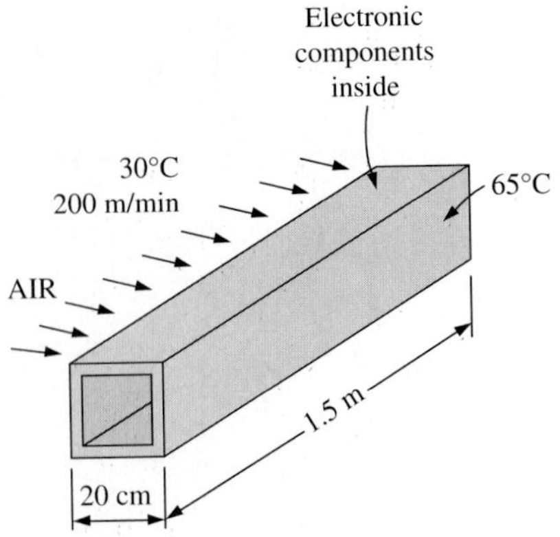

**10-56** The components of an electronic system are located in a 1.5-m-long horizontal duct whose cross-section is 20 cm × 20 cm. The components in the duct are not allowed to come into direct contact with cooling air, and thus are cooled by air at 30°C flowing over the duct with a velocity of 200 m/min. If the surface temperature of the duct is not to exceed 65°C, determine the total power rating of the electronic devices that can be mounted into the duct. *Answer:* 643 W

Electronic
components
inside
30°C
200 m/min
65°C
AIR
1.5 m
20 cm

**FIGURE P10-56**

**10-56E** The components of an electronic system are located in a 5-ft-long horizontal duct whose cross-section is 6 in × 6 in. The components in the duct are not allowed to come into direct contact with cooling air, and thus are cooled by air flowing over the duct at 80°F with a velocity of 600 ft/min. If the surface temperature of the duct is not to exceed 130°F, determine the total power rating of the electronic devices that can be mounted into the duct. *Answer:* 1400 Btu/h

**10-57** Repeat Prob. 10-56 for a location at 4000-m altitude where the atmospheric pressure is 61.66 kPa.

**10-58** A 0.4-W cylindrical electronic component whose diameter is 0.3 cm and length is 1.8 cm mounted on a circuit board is cooled by air flowing across it at a velocity of 150 m/min. If the air temperature is 45°C, determine the surface temperature of the component.

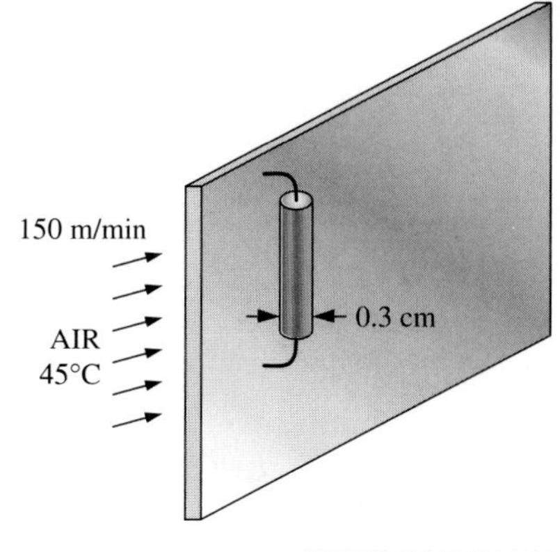


FIGURE P10-58

**10-58E** A 0.4-W cylindrical electronic component whose diameter is 0.12 in and length is 1.2 in mounted on a circuit board is cooled by air flowing across it at a velocity of 500 ft/min. If the air temperature is 120°F, determine the surface temperature of the component.

## Flow in tubes

**10-59C** Show that the Reynolds number for flow in a circular tube of diameter $D$ can be expressed as $\text{Re} = 4\dot{m}/\pi D\mu$.

**10-60C** Which fluid requires a larger pump to move at a specified velocity in a specified tube: water or engine oil? Why?

**10-61C** What are the generally accepted values of the critical Reynolds numbers for (*a*) flow over a flat plate, (*b*) flow over a circular cylinder, and (*c*) flow in a tube?

**10-62C** In the fully developed region of flow in a circular tube, will the velocity profile change in the flow direction? How about the temperature profile?

**10-63C** Consider the flow of oil in a tube. How will the hydrodynamic and thermal entry lengths compare if the flow is laminar? How would they compare if the flow were turbulent?

**10-64C** Consider the flow of mercury (a liquid metal) in a tube. How will the hydrodynamic and thermal entry lengths compare if the flow is laminar? How would they compare if the flow were turbulent?

**10-65C** What is the difference between the friction factor and the friction coefficient?

**10-66C** What do the mean velocity $\mathcal{V}_m$ and the mean temperature $T_m$ represent in flow through circular tubes of constant diameter?

**10-67C** Consider fluid flow in a tube whose surface temperature remains constant. What is the appropriate temperature difference for use in Newton's law of cooling with an average heat transfer coefficient?

**10-68C** What is the physical significance of the number of transfer units NTU $= hA/\dot{m}C_p$? What does a small and a large NTU tell the heat transfer engineer about a heat transfer system?

**10-69C** How is the friction factor for flow in a tube related to the pressure drop? How is the pressure drop related to the pumping power requirement for a given mass flow rate?

**10-70C** What does the logarithmic mean temperature difference represent for flow in a tube whose surface temperature is constant? Why do we use the logarithmic mean temperature instead of the arithmetic mean temperature?

**10-71C** How is the hydrodynamic entry length defined for flow in a tube? How about the thermal entry length? In what region is the flow in a tube fully developed?

**10-72C** Consider laminar forced convection in a circular tube. Will the friction factor be higher near the inlet of the tube or near the exit? Why? What would your response be if the flow were turbulent?

**10-73C** Consider laminar forced convection in a circular tube. Will the heat flux be higher near the inlet of the tube or near the exit? Why?

**10-74C** Consider turbulent forced convection in a circular tube. Will the heat flux be higher near the inlet of the tube or near the exit? Why?

**10-75C** How does surface roughness effect the pressure drop and the heat transfer in a tube if the fluid flow is turbulent? What would your response be if the flow in the tube were laminar?

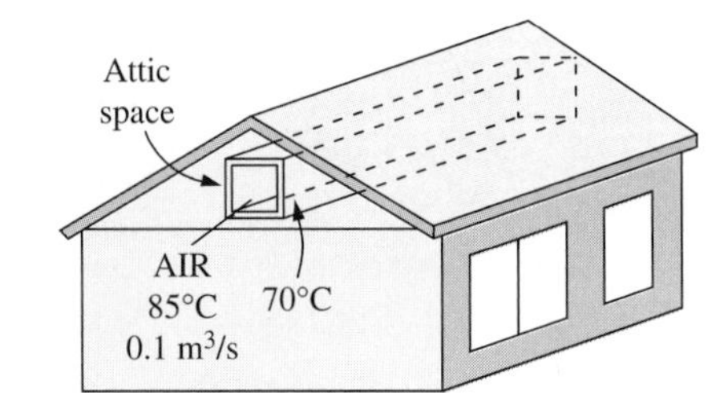


**FIGURE P10-77**

**10-76** Water is to be heated from 12°C to 70°C as it flows through a 2-cm internal-diameter 7-m-long tube. The tube is equipped with an electric resistance heater, which provides uniform heating throughout the surface of the tube. The outer surface of the heater is well insulated, so that in steady operation all the heat generated in the heater is transferred to the water in the tube. If the system is to provide hot water at a rate of 8 L/min, determine the power rating of the resistance heater. Also estimate the inner surface temperature of the pipe at the exit.

**10-77** Hot air at atmospheric pressure and 85°C enters a 10-m-long uninsulated square duct of cross-section 0.15 m × 0.15 m that passes through the attic of a house at a rate of 0.10 m$^3$/s. The duct is observed to be nearly isothermal at 70°C. Determine the exit temperature of the air and the rate of heat loss from the duct to the air space in the attic. *Answers:* 75.6°C, 946 W

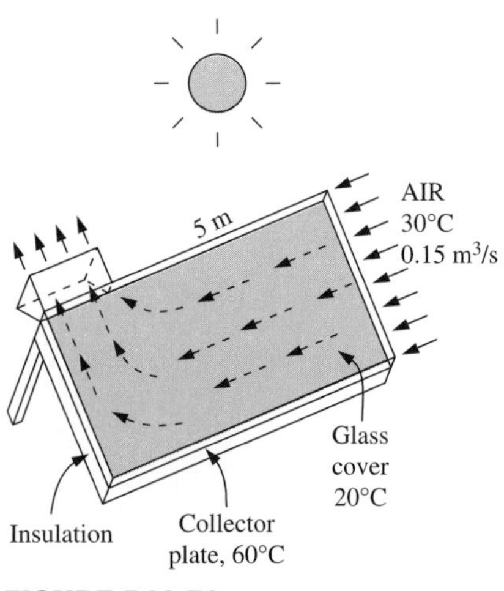


**FIGURE P10-78**

**10-78** Consider an air solar collector that is 1 m wide and 5 m long and has a constant spacing of 3 cm between the glass cover and the collector plate. Air enters the collector at 30°C at a rate of 0.15 m$^3$/s through the

1-m-wide edge, and flows along the 5-m-long passage way. If the average temperatures of the glass cover and the collector plate are 20°C and 60°C, respectively, determine (*a*) the net rate of heat transfer to the air in the collector and (*b*) the temperature rise of air as it flows through the collector.

**10-78E** Consider an air solar collector that is 3 ft wide and 15 ft long and has a constant spacing of 1 in. between the glass cover and the collector plate. Air enters the collector at 85°F at a rate of 5 $ft^3/s$ through the 3-ft-wide edge, and flows along the 15-ft-long passage way. If the average temperatures of the glass cover and the collector plate are 70°F and 130°F, respectively, determine (*a*) the net rate of heat transfer to the air in the collector and (*b*) the temperature rise of air as it flows through the collector.

**10-79** Consider the flow of oil at 10°C in a 40-cm-diameter pipeline at an average velocity of 0.5 m/s. A 300-m-long section of the pipeline passes through icy waters of a lake at 0°C. Measurements indicate that the surface temperature of the pipe is very nearly 0°C. Disregarding the thermal resistance of the pipe material, determine (*a*) the temperature of the oil when the pipe leaves the lake, (*b*) the rate of heat transfer from the oil, and (*c*) the pumping power required to overcome the pressure losses and to maintain the flow of the oil in the pipe.

**10-80** Consider laminar flow of a fluid through a square channel maintained at a constant temperature. Now the free-stream velocity of the fluid is doubled. Determine the change in the pressure drop of the fluid and the rate of heat transfer between the fluid and the walls of the channel. Assume the flow regime remains unchanged.
*Answers:* 2 and 1.26

**10-81** Repeat Prob. 10-80 for turbulent flow.

**10-82** The hot water needs of a household are to be met by heating water at 15°C to 90°C by a parabolic solar collector at a rate of 2 kg/s. Water flows through a 3-cm-diameter thin aluminum tube whose outer surface is black-anodized in order to maximize its solar absorption ability. The centerline of the tube coincides with the focal line of the collector, and a glass sleeve is placed outside the tube to minimize the heat losses. If solar energy is transferred to water at a net rate of 300 W per m length of the tube, determine the required length of the parabolic collector to meet the hot water requirements of this house. Also determine the surface temperature of the tube at the exit.

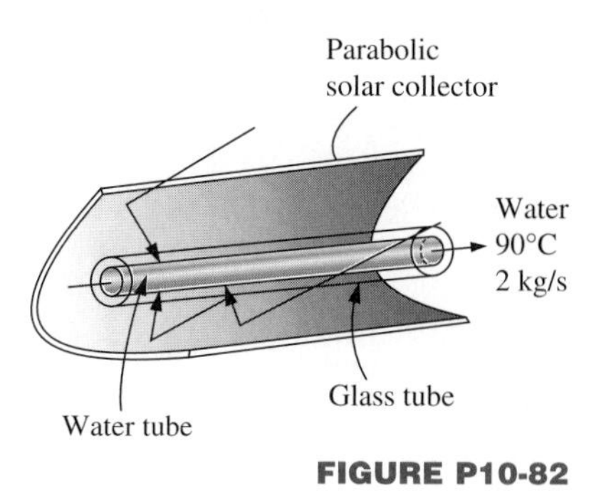


**FIGURE P10-82**

**10-82E** The hot water needs of a household are to be met by heating water at 55°F to 200°F by a parabolic solar collector at a rate of 4 lbm/s. Water flows through a 1.25-in-diameter thin aluminum tube whose outer surface is black-anodized in order to maximize its solar absorption ability. The centerline of the tube coincides with the focal line of the collector, and a glass sleeve is placed outside the tube to minimize the heat losses. If solar energy is transferred to water at a net rate of 350 Btu/h per ft length of the tube, determine the required length of the parabolic collector to meet the hot water requirements of this house. Also determine the surface temperature of the tube at the exit.

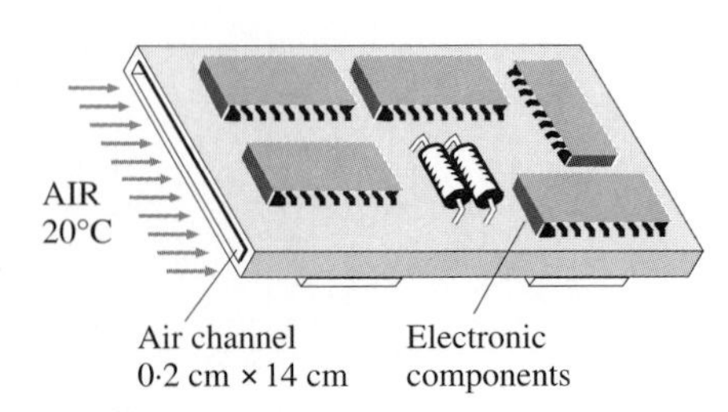


**FIGURE P10-83**

**10-83** A 15 cm × 20 cm printed circuit board whose components are not allowed to come into direct contact with air for reliability reasons is to be cooled by passing cool air through a 20-cm-long channel of rectangular cross-section 0.2 cm × 14 cm drilled into the board. The heat generated by the electronic components is conducted across the thin layer of the board to the channel, where it is removed by air that enters the channel at 20°C. The heat flux at the top surface of the channel can be considered to be uniform, and heat transfer through other surfaces is negligible. If the velocity of the air at the inlet of the channel is not to exceed 4 m/s and the surface temperature of the channel is to remain under 50°C, determine the maximum total power of the electronic components which can safely be mounted on this circuit board.

**10-84** Repeat Prob. 10-83 by replacing air with helium, which has six times the thermal conductivity of air.

**10-85** Air enters a 7-m-long section of a rectangular duct of cross-section 15 cm × 20 cm at 50°C at an average velocity of 7 m/s. If the walls of the duct are maintained at 10°C, determine (*a*) the outlet temperature of the air, (*b*) the rate of heat transfer from the air, and (*c*) the fan power needed to overcome the pressure losses in this section of the duct.
*Answers:* (*a*) 32.8°C, (*b*) 3674 W, (*c*) 4.22 W

**10-85E** Air enters a 20-ft-long section of a rectangular duct of cross-section 6 in. × 6 in. at 120°F at an average velocity of 18 ft/s. If the walls of the duct are maintained at 50°F, determine (*a*) the outlet temperature of the air, (*b*) the rate of heat transfer from the air, and (*c*) the fan power needed to overcome the pressure losses in this section of the duct.

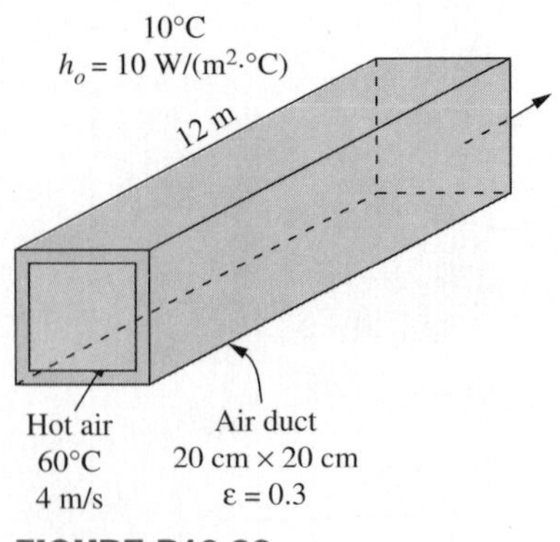

**FIGURE P10-86**

**10-86** Hot air at 60°C leaving the furnace of a house enters a 12-m long section of a sheet metal duct of rectangular cross-section 20 cm × 20 cm at an average velocity of 4 m/s. The thermal resistance of the duct is negligible, and the outer surface of the duct, whose emissivity is 0.3, is exposed to the cold air at 10°C in the basement, with a convection heat transfer coefficient of 10 W/(m$^2$ · °C). Taking the walls of the basement to be at 10°C also, determine (*a*) the temperature at which the hot air will leave the basement and (*b*) the rate of heat loss from the hot air in the duct to the basement.

**10-87** The components of an electronic system dissipating 90 W are located in a 1-m-long horizontal duct whose cross-section is 16 cm × 16 cm. The components in the duct are cooled by forced air, which enters at 32°C at a rate of 0.65 m$^3$/min. Assuming 85 percent of the heat generated inside is transferred to air flowing through the duct and the remaining 15 percent is lost through the outer surfaces of the duct, determine (*a*) the exit temperature of air and (*b*) the highest component surface temperature in the duct.

**10-88** Repeat Prob. 10-87 for a circular horizontal duct of 15-cm diameter.

**10-89** Consider a hollow-core printed circuit board 12 cm high and 18 cm long, dissipating a total of 20 W. The width of the air gap in the middle of the PCB is 0.25 cm. The cooling air enters the 12-cm-wide core at 32°C at a rate of 0.8 L/s. Assuming the heat generated to be uniformly distributed over the two side surfaces of the PCB, determine (*a*) the temperature at which the air leaves the hollow core and (*b*) the highest temperature on the inner surface of the core.
*Answers:* (*a*) 53.7°C, (*b*) 72.0°C

**10-90** Repeat Prob. 10-89 for a hollow-core PCB dissipating 35 W.

**10-91** Water at 12°C is heated by passing it through 1.5-cm internal-diameter thin-walled copper tubes. Heat is supplied to the water by steam, which condenses outside the copper tubes at 120°C. If water is to be heated to 60°C at a rate of 0.3 kg/s, determine (*a*) the length of the copper tube that needs to be used and (*b*) the pumping power required to maintain this flow at the specified rate. Assume the entire copper tube to be at the steam temperature of 120°C.

**10-91E** Water at 54°F is heated by passing it through 0.75-in internal-diameter thin-walled copper tubes. Heat is supplied to the water by steam which condenses outside the copper tubes at 250°F. If water is to be heated to 140°F at a rate of 0.7 lbm/s, determine (*a*) the length of the copper tube that needs to be used and (*b*) the pumping power required to maintain this flow at the specified rate. Assume the entire copper tube to be at the steam temperature of 250°F.

**10-92** A computer cooled by a fan contains eight PCBs, each dissipating 10 W of power. The height of the PCBs is 12 cm and the length is 18 cm. The clearance between the tips of the components on the PCB and the back surface of the adjacent PCB is 0.3 cm. The cooling air is supplied by a 25-W fan mounted at the inlet. If the temperature rise of air as it flows through the case of the computer is not to exceed 10°C, determine (*a*) the flow rate of the air that the fan needs to deliver, (*b*) the fraction of the temperature rise of air that is due to the heat generated by the fan and its motor, and (*c*) the highest allowable inlet air temperature if the surface temperature of the components is not to exceed 70°C anywhere in the system. *Answers:* (*a*) 0.008 kg/s, (*b*) 31 percent, (*c*) 8.3°C

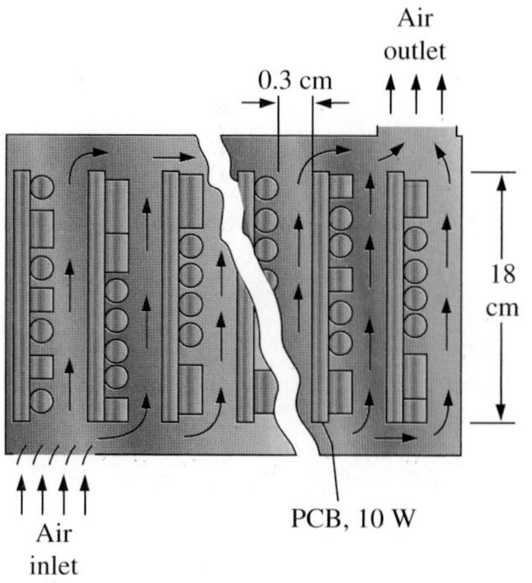

**FIGURE P10-92**

## Review Problems

**10-93** Consider a house that is maintained at 22°C at all times. The walls of the house have R-3.38 insulation in SI units (i.e., they have an $L/k$ value or a thermal resistance of 3.38 $m^2 \cdot °C/W$). During a cold winter night, the outside air temperature is 4°C and wind at 50 km/h is blowing parallel to a 3-m-high and 8-m-long wall of the house. If the heat transfer

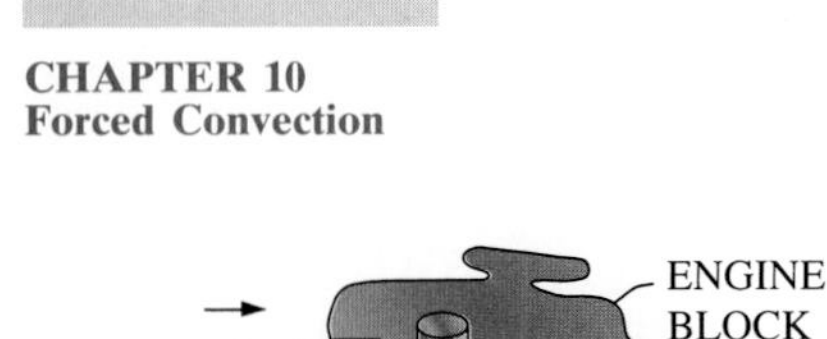


coefficient on the interior surface of the wall is 8 W/(m$^2$ · C), determine the rate of heat loss from that wall of the house. Draw the thermal resistance network and disregard radiation heat transfer.
*Answer:* 122 W

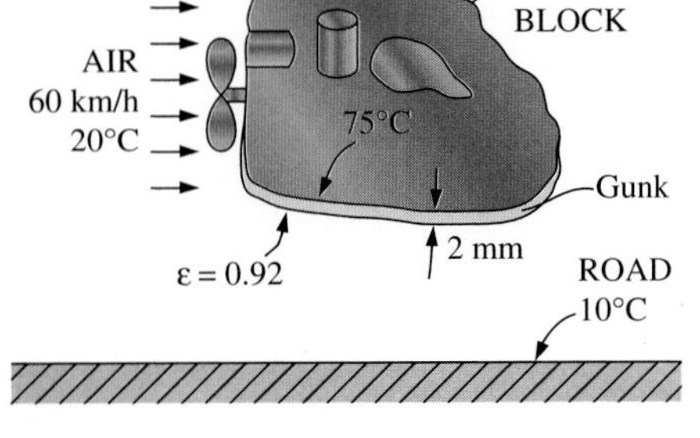

**FIGURE P10-94**

**10-94** An automotive engine can be approximated as a 0.4-m-high, 0.60-m-wide, and 0.7-m-long rectangular block. The bottom surface of the block is a temperature of 75°C and has an emissivity of 0.92. The ambient air is at 20°C, and the road surface is at 10°C. Determine the rate of heat transfer from the bottom surface of the engine block by convection and radiation as the car travels at a velocity of 60 km/h. Assume the flow to be turbulent over the entire surface because of the constant agitation of the engine block. How will the heat transfer be affected when a 2-mm-thick gunk [$k$ = 3 W/(m · °C)] is formed at the bottom surface as a result of the dirt and oil collected at that surface over time? Assume the metal temperature under the gunk to be still 75°C.

**10-95** The thickness of the velocity boundary layer over a flat plate increases in the flow direction, and is given by

$$\delta = \frac{5x}{\sqrt{\text{Re}_x}}$$

where $\text{Re}_x = V_\infty x/\nu$ is the Reynolds number at a distance $x$ from the leading edge of the plate. Calculate the thickness of the boundary layer during flow over a 3-m-long flat plate at intervals of 25 cm, and plot the boundary layer over the plate for the flow of (*a*) air, (*b*) water, and (*c*) engine oil at 20°C at a free-stream velocity of 3 m/s.

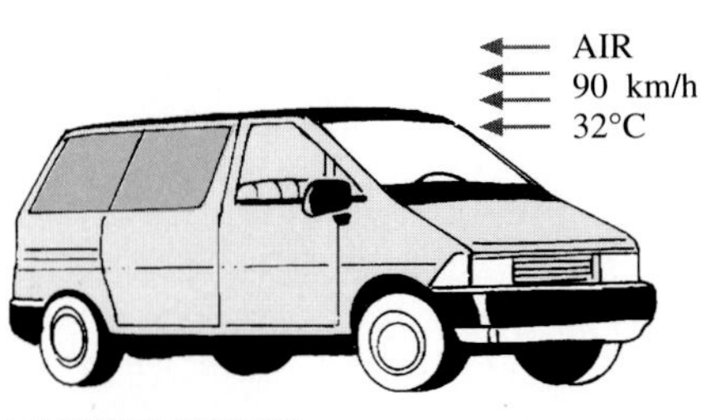

**FIGURE P10-96**

**10-96** The passenger compartment of a minivan traveling at 90 km/h can be modeled as 1.1-m-high, 1.8-m-wide, and 3.5-m-long rectangular box whose walls have an insulating value of R-0.5 in SI units (i.e., a wall thickness-to-thermal conductivity ratio of 0.5 m$^2$ · °C/W). The interior of a minivan is maintained at an average temperature of 20°C during a trip at night while the outside air temperature is 32°. The average heat transfer coefficient on the interior surfaces of the van is 6 W/(m$^2$ · °C). The air flow over the exterior surfaces can be assumed to be turbulent because of the intense vibrations involved, and the heat transfer coefficient on the front and back surfaces can be taken to be equal to that on the top surface. Disregarding any heat gain or loss by radiation, determine the rate of heat transfer from the ambient air to the van.

**10-96E** The passenger compartment of a minivan traveling at 60 mph can be modeled as 3.2-ft-high, 6-ft-wide, and 11-ft-long rectangular box whose walls have an insulating value of R-3 (i.e., a wall thickness-to-thermal conductivity ratio of 3 h · ft$^2$ · °F/Btu). The interior of a minivan is maintained at an average temperature of 70°F during a trip at night while the outside air temperature is 90°F. The average heat transfer coefficient on the interior surfaces of the van is 1.2 Btu/(h · ft$^2$ · °F). The air flow over the exterior surfaces can be assumed to be turbulent because of the intense vibrations involved, and the heat transfer coefficient on the front and back surfaces can be taken to be equal to that on the top surface. Disregarding any heat gain or loss by radiation, determine the rate of heat transfer from the ambient air to the van.

**10-97** Consider a house which is maintained at a constant temperature of 22°C. One of the walls of the house has three single-pane glass windows that are 1.5 m high and 1.2 m long. The glass [$k$ = 0.78 W/(m · °C)] is 0.5 cm thick, and the heat transfer coefficient on the inner surface of the glass is 8 W/(m$^2$ · C). Now winds at 60 km/h start to blow parallel to the surface of this wall. If the air temperature outside is −2°C, determine the rate of heat loss through the windows of this wall. Assume radiation heat transfer to be negligible.

**10–98** The compressed air requirements of a manufacturing facility are met by a 150-hp compressor located in a room that is maintained at 25°C. In order to minimize the compressor work, the intake port of the compressor is connected to the outside through an 8-m-long, 20-cm-diameter duct made of thin aluminum sheet. The compressor takes in air at a rate of 0.27 m$^3$/s at the outdoor conditions of 10°C and 95 kPa. Disregarding the thermal resistance of the duct and taking the heat transfer coefficient on the outer surface of the duct to be 10 W/(m$^2$ · °C), determine ($a$) the power used by the compressor to overcome the pressure drop in this duct, ($b$) the rate of heat transfer to the incoming cooler air, and ($c$) the temperature rise of air as it flows through the duct.

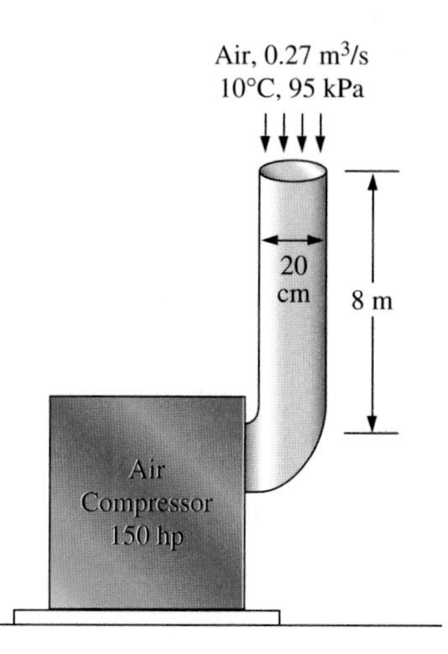

**FIGURE P10-98**

**10-98E** The compressed air requirements of a manufacturing facility are met by a 150-hp compressor located in a room that is maintained at 75°F. In order to minimize the compressor work, the intake port of the compressor is connected to the outside through a 20-ft-long, 8-in-diameter duct made of thin aluminum sheet. The compressor takes in air at a rate of 600 ft$^3$/min at the outdoor conditions of 50°F and 14.0 psia. Disregarding the thermal resistance of the duct and taking the heat transfer coefficient on the outer surface of the duct to be 2 Btu/(h · ft$^2$ · °F), determine ($a$) the power used by the compressor to overcome the pressure drop in this duct, ($b$) the rate of heat transfer to the incoming cooler air, and ($c$) the temperature rise of air as it flows through the duct.

**10-99** Consider a person who is trying to keep cool in a hot summer day by turning a fan on and exposing his body to air flow. The air temperature is 32°C, and the fan is blowing air at a velocity of 5 m/s. The surrounding surfaces are at 40°C, and the emissivity of the person can be taken to be 0.9. If the person is doing light work and generating sensible heat at a rate of 90 W, determine the average temperature of the outer surface (skin or clothing) of the person. The average human body can be treated as a 30-cm-diameter cylinder with an exposed surface area of 1.7 m$^2$. *Answer:* 35°C

**10-100** A house built on a riverside is to be cooled in summer by

utilizing the cool water of the river, which flows at an average temperature of 15°C. A 15-m-long section of a circular duct of 20-cm diameter passes through the water. Air enters underwater section of the duct at 25°C at a velocity of 3 m/s. Assuming the surface of the duct to be at the temperature of the water, determine the outlet temperature of air as it leaves the underwater portion of the duct. Also determine the fan power needed to overcome the flow resistance in this section of the duct.

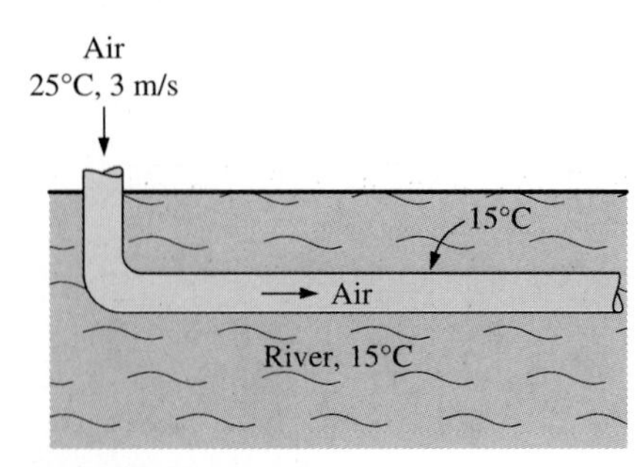


**FIGURE P10-100**

**10-101** Repeat Prob. 10-100 assuming that a 0.15-mm-thick layer of mineral deposit [$k$ = 3 W/(m · °C)] is formed on the inner surface of the pipe.

**10-102** The exhaust gases of an automotive engine leave the combustion chamber and enter a 2.5-m-long and 8-cm-diameter thin-walled steel exhaust pipe at 800 K and 110 kPa at a rate of 0.1 kg/s. The surrounding ambient air is at a temperature of 25°C, and the heat transfer coefficient on the outer surface of the exhaust pipe is 15 W/(m$^2$ · °C). Assuming the exhaust gases to have the properties of air, determine (*a*) the velocity of the exhaust gases at the inlet of the exhaust pipe and (*b*) the temperature at which the exhaust gases will leave the pipe and enter the air.

**10-102E** The exhaust gases of an automotive engine leave the combustion chamber and enter a 8-ft-long and 3.5-in-diameter thin-walled steel exhaust pipe at 940°F and 16.1 psia at a rate of 0.2 lbm/s. The surrounding ambient air is at a temperature of 75°F, and the heat transfer coefficient on the outer surface of the exhaust pipe is 3 Btu/(h · ft$^2$ · °F). Assuming the exhaust gases to have the properties of air, determine (*a*) the velocity of the exhaust gases at the inlet of the exhaust pipe and (*b*) the temperature at which the exhaust gases will leave the pipe and enter the air.

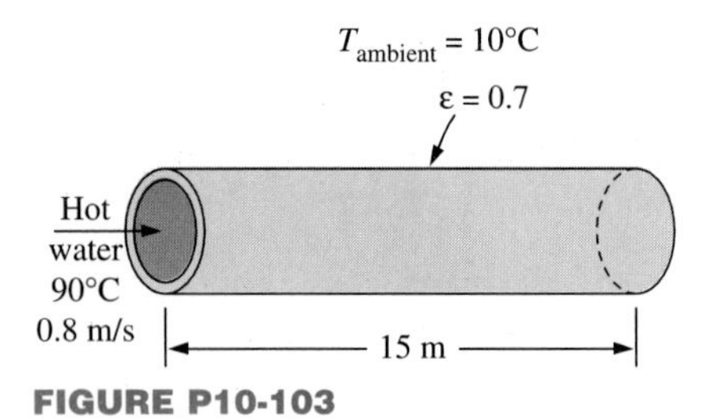


**FIGURE P10-103**

**10-103** Hot water at 90°C enters a 15-m section of a cast iron pipe [$k$ = 52 W/(m · °C)] whose inner and outer diameters are 4 and 4.6 cm, respectively, at an average velocity of 0.8 m/s. The outer surface of the pipe, whose emissivity is 0.7, is exposed to the cold air at 10°C in a basement, with a convection heat transfer coefficient of 15 W/(m$^2$ · °C). Taking the walls of the basement to be at 10°C also, determine (*a*) the rate of heat loss from the water and (*b*) the temperature at which the water leaves the basement.

**10-104** Repeat Prob. 10-103 for a pipe made of copper [$k$ = 386 W/(m · °C)] instead of cast iron.

**10-105** Four power transistors, each dissipating 15 W, are mounted on a thin vertical aluminum plate [$k$ = 237 W/(m · °C)] 22 cm × 22 cm in size. The heat generated by the transistors is to be dissipated by both surfaces of the plate to the surrounding air at 25°C, which is blown over the plate by a fan at a velocity of 250 m/min. The entire plate can be assumed to be nearly isothermal, and the exposed surface area of the transistor can be taken to be equal to its base area. Determine the temperature of the aluminum plate.

**10-106** A 3-m internal-diameter spherical tank made of 1-cm-thick stainless steel [$k = 15\ \text{W/(m}\cdot{}^\circ\text{C)}$] is used to store iced water at 0°C. The tank is located outdoors at 30°C and is subjected to winds at 25 km/h. Assuming the entire steel tank to be at 0°C and thus its thermal resistance to be negligible, determine (*a*) the rate of heat transfer to the iced water in the tank and (*b*) the amount of ice at 0°C that melts during a 24-h period. The heat of fusion of water at atmospheric pressure is $h_{if} =$ 333.7 kJ/kg. Disregard any heat transfer by radiation.

**10-106E** A 10-ft internal-diameter spherical tank made of 0.5-in-thick stainless steel [$k = 8.7\ \text{Btu/(h}\cdot\text{ft}\cdot{}^\circ\text{F)}$] is used to store iced water at 32°F. The tank is located outdoors at 85°F and is subjected to winds at 20 mph. Assuming the entire steel tank to be at 32°F and thus its thermal resistance to be negligible, determine (*a*) the rate of heat transfer to the iced water in the tank and (*b*) the amount of ice at 32°F that melts during a 24-h period. The heat of fusion of water at atmospheric pressure is $h_{if} =$ 143.5 Btu/lbm. Disregard any heat transfer by radiation.

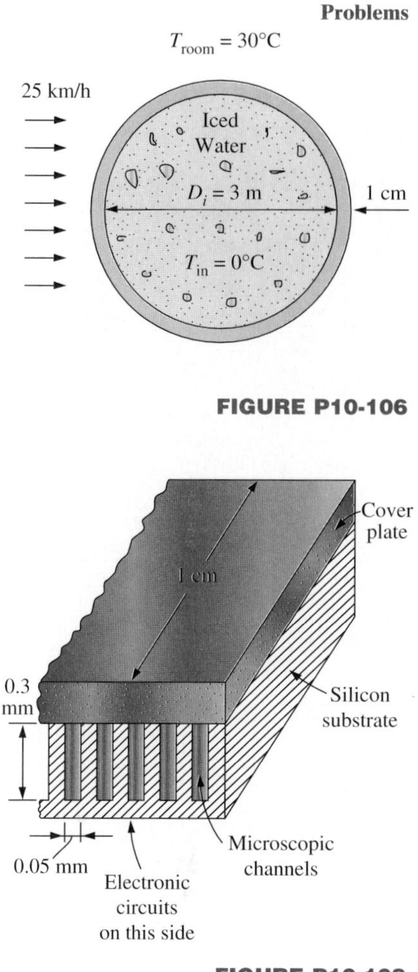


**FIGURE P10-106**

**10-107** Repeat Prob. 10-106, assuming the inner surface of the tank to be at 0°C but by taking the thermal resistance of the tank and heat transfer by radiation into consideration. Assume the average surrounding surface temperature for radiation exchange to be 15°C, and the outer surface of the tank to have an emissivity of 0.9.
*Answers:* (*a*) 13,630 W, (*b*) 3529 kg

**10-108** D. B. Tuckerman and R. F. Pease of Stanford University demonstrated in the early 1980s that integrated circuits can be cooled very effectively by fabricating a series of microscopic channels 0.3 mm high and 0.05 mm wide in the back of the substrate, and covering them with a plate to confine the fluid flow within the channels. They were able to dissipate 790 W of power generated in a 1-cm$^2$ silicon chip at a junction-to-ambient temperature difference of 71°C using water as the coolant flowing at a rate of 0.01 L/s through 100 such channels under a 1 cm × 1 cm silicon chip. Heat is transferred primarily through the base area of the channel, and it was found that the increased surface area and thus the fin effect is of lesser importance. Disregarding the entrance effects and ignoring any heat transfer from the side and cover surfaces, determine (*a*) the temperature rise of water as it flows through the microchannels and (*b*) the average surface temperature of the base of the microchannels for a power dissipation of 50 W. Assume the water enters the channels at 20°C.

**FIGURE P10-108**

**10-109** A transistor with a height of 0.55 cm and a diameter of 0.6 cm is mounted on a circuit board. The transistor is cooled by air flowing over it at a velocity of 200 m/min. If the air temperature is 55°C and the transistor case temperature is not to exceed 70°C, determine the amount of power this transistor can dissipate safely.

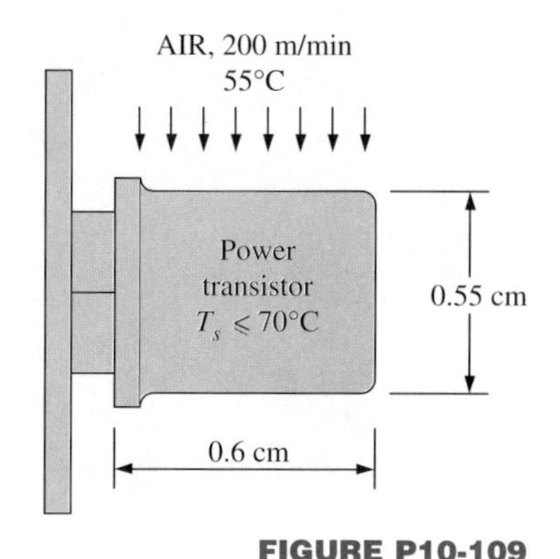


**FIGURE P10-109**

**10-109E** A transistor with a height of 0.25 in. and a diameter of 0.22 in. is mounted on a circuit board. The transistor is cooled by air flowing over it at a velocity of 500 ft/min. If the air temperature is 130°F and the transistor case temperature is not to exceed 165°F, determine the amount of power this transistor can dissipate safely.

**10-110** Liquid-cooled systems have high heat transfer coefficients associated with them, but they have the inherent disadvantage that they present potential leakage problems. Therefore, air is proposed to be used as the microchannel coolant. Repeat Prob. 10-108 using air as the cooling fluid instead of water, entering at a rate of 0.5 L/s.

**10-111** A desktop computer is to be cooled by a fan. The electronic components of the computer consume 45 W of power under full-load conditions. The computer is to operate in environments at temperatures up to 50°C and at elevations up to 3000 m where the atmospheric pressure is 70.12 kPa. The exit temperature of air is not to exceed 60°C to meet the reliability requirements. Also, the average velocity of air is not to exceed 120 m/min at the exit of the computer case, where the fan is installed to keep the noise level down. Determine the flow rate of the fan which needs to be installed, and the diameter of the casing of the fan.

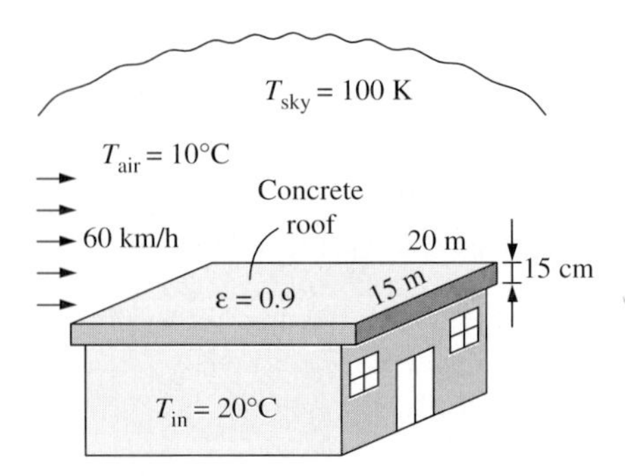


FIGURE P10-112

**10-112** The roof of a house consists of 15-cm-thick concrete slab [$k$ = 2 W/(m · °C)] 15 m wide and 20 m long. The convection heat transfer coefficient on the inner surface of the roof is 5 W/(m$^2$ · °C). On a clear winter night, the ambient air is reported to be at 10°C, while the night sky temperature is 100 K. The house and the interior surfaces of the wall are maintained at a constant temperature of 20°C. The emissivity of both surfaces of the concrete roof is 0.9. Considering both radiation and convection heat transfer, determine the rate of heat transfer through the roof when wind at 60 km/h is blowing over the roof.

If the house is heated by a furnace burning natural gas with an efficiency of 85%, and the price of natural gas is \$0.60/therm (1 therm = 105,500 kJ of energy content), determine the money lost through the roof that night during a 14-h period. *Answers:* 30.4 kW, \$10.30

**10-113** Steam at 250°C flows in a stainless steel pipe [$k$ = 15 W/(m · °C)] whose inner and outer diameters are 4 cm and 4.6 cm, respectively. The pipe is covered with 3.5-cm-thick glass wool insulation [$k$ = 0.038 W/(m · °C)] whose outer surface has an emissivity of 0.3. Heat is lost to the surrounding air and surfaces at 3°C by convection and radiation. Taking the heat transfer coefficient inside the pipe to be 80 W/(m$^2$ · °C), determine the rate of heat loss from the steam per unit length of the pipe when air is flowing across the pipe at 4 m/s.

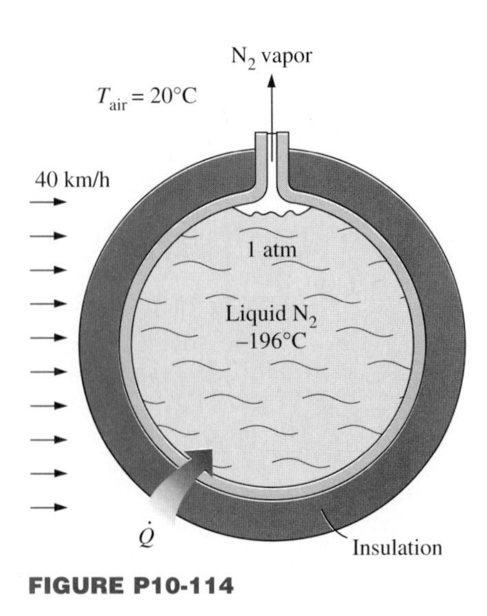


FIGURE P10-114

**10-114** The boiling temperature of nitrogen at atmospheric pressure at sea level (1-atm pressure) is −196°C. Therefore, nitrogen is commonly used in low-temperature scientific studies, since the temperature of liquid nitrogen in a tank open to the atmosphere will remain constant at −196°C until it is depleted. Any heat transfer to the tank will result in the evaporation of some liquid nitrogen, which has a heat of vaporization of 198 kJ/kg and a density of 810 kg/m$^3$ at 1 atm.

Consider a 4-m-diameter spherical tank that is initially filled with liquid nitrogen at 1 atm and −196°C. The tank is exposed to 20°C ambient air and 40 km/h winds. The temperature of the thin-shelled spherical tank is observed to be almost the same as the temperature of the nitrogen inside. Disregarding any radiation heat exchange, determine the rate of evaporation of the liquid nitrogen in the tank as a result of heat transfer from the ambient air if the tank is (*a*) not insulated, (*b*) insulated with 5-cm-thick fiberglass insulation [$k = 0.035\ \text{W/(m} \cdot °\text{C)}$], and (*c*) insulated with 2-cm-thick superinsulation that has an effective thermal conductivity of 0.00005 W/(m · °C).

**10-115** Repeat Prob. 10-114 for liquid oxygen, which has a boiling temperature of −183°C, a heat of vaporization of 213 kJ/kg, and a density of 1140 kg/m$^3$ at 1-atm pressure.

**10-116** A 0.3-cm-thick, 12-cm-high, and 18-cm-long circuit board houses 80 closely spaced logic chips on one side, each dissipating 0.04 W. The board is impregnated with copper fillings, and has an effective thermal conductivity of 16 W/(m · °C). All the heat generated in the chips is conducted across the circuit board, and is dissipated from the back side of the board to the ambient air at 40°C, which is forced to flow over the surface by a fan at a free-stream velocity of 400 m/min. Determine the temperatures on the two sides of the circuit board.
*Answers:* 46.28°C, 46.31°C

**10-117** It is well known that cold air feels much colder in windy weather than what the thermometer reading indicates, because of the "chilling effect" of the wind. This effect is due to the increase in the convection heat transfer coefficient with increasing air velocities. The *equivalent wind chill temperature* in °C is given by (1993 *ASHRAE Handbook of Fundamentals,* Atlanta, GA, p. 8.15)

$$T_{\text{equiv}} = 33.0 - (33.0 - T_{\text{ambient}})(0.475 - 0.0126V + 0.240\sqrt{V})$$

where $V$ is the wind velocity in km/h and $T_{\text{ambient}}$ is the ambient air temperature in °C in calm air, which is taken to be air with light winds at speeds up to 6.5 km/h. The constant 33°C in the above equation is the mean skin temperature of a resting person in a comfortable environment. Windy air at a temperature $T_{\text{ambient}}$ and velocity $V$ will feel as cold as calm air at a temperature $T_{\text{equiv}}$. The equation above is valid for winds up to 70 km/h. Winds at higher velocities produce little additional chilling effect. Determine the equivalent wind chill temperature of an environment at −15°C at wind speeds of 15, 30, 45, and 60 km/h. Exposed flesh can freeze within one minute at a temperature below −30°C in calm weather. Does a person need to be concerned about this possibility in any of the cases above?

**FIGURE P10-117**

**10-117E** It is well known that cold air feels much colder in windy weather than what the thermometer reading indicates because of the "chilling effect" of the wind. This effect is due to the increase in the convection heat transfer coefficient with increasing air velocities. The *equivalent wind chill temperature* in °F is given by (1993 *ASHRAE Handbook of Fundamentals,* Atlanta, GA, p. 8.15)

$$T_{\text{equiv}} = 91.4 - (91.4 - T_{\text{ambient}})(0.475 - 0.0203V + 0.304\sqrt{V})$$

where $V$ is the wind velocity in mph and $T_{\text{ambient}}$ is the ambient air temperature in °F in calm air, which is taken to be air with light winds at speeds up to 4 mph. The constant 91.4°F in the above equation is the mean skin temperature of a resting person in a comfortable environment. Windy air at a temperature $T_{\text{ambient}}$ and velocity $V$ will feel as cold as calm air at a temperature $T_{\text{equiv}}$. The equation above is valid for winds up to 43 mph. Winds at higher velocities produce little additional chilling effect. Determine the equivalent wind chill temperature of an environment at 10°F at wind speeds of 10, 20, 30, and 40 mph. Exposed flesh can freeze within one minute at a temperature below −25°F in calm weather. Does a person need to be concerned about this possibility in any of the cases above?

### Computer, Design, and Essay Problems

**10-118** Electronic boxes such as computers are commonly cooled by a fan. Write an essay on forced air cooling of electronic boxes, and on the selection of the fan.

**10-119** Obtain information on frostbite and the conditions under which it occurs. Using the relation in Prob. 10-117, prepare a table that shows how long people can stay in cold and windy weather for specified temperatures and wind speeds before the exposed flesh is in danger of experiencing frostbite.

**10-120** Write an article on forced convection cooling with air, helium, water, and a dielectric liquid. Discuss the advantages and disadvantages of each fluid in heat transfer. Explain the circumstances under which a certain fluid will be most suitable for the cooling job.

**10-121** Write a computer program that determines the drag force and the rate of heat transfer for air flow over a flat plate when the free-stream velocity and temperature of the air as well as the dimensions and the temperature of the surface are specified. The program should also check if the flow over the plate becomes turbulent.

**10-122** Write a computer program that determines the drag force and the rate of heat transfer for water flow across a long cylinder or sphere when the free-stream velocity and temperature of the air as well as the dimensions and the temperature of the surface are specified.

# Natural Convection

CHAPTER 11

In the preceding chapter, we considered heat transfer by *forced convection* where a fluid was *forced* to move over a surface or in a tube by external means such as a pump or a fan. In this chapter, we consider *natural convection* where any fluid motion occurs by natural means such as buoyancy. The fluid motion in forced convection is quite *noticeable*, since a fan or a pump can transfer enough momentum to the fluid to move it in a certain direction. The fluid motion in natural convection, however, is often not noticeable because of the low velocities involved.

Convection heat transfer coefficient is a strong function of *velocity*: the higher the velocity, the higher the convection heat transfer coefficient. The fluid velocities associated with natural convection are low, typically under 1 m/s. Therefore, the *heat transfer coefficients* encountered in natural convection are usually *much lower* than those encountered in forced convection. Yet several types of heat transfer equipment are designed to operate under natural convection conditions instead of forced convection, because natural convection does not require the use of a fluid mover.

Many familiar heat transfer applications involve *natural convection* as the *primary* mechanism of heat transfer. Some examples are cooling of electronic equipment such as power transistors, TVs and VCRs, heat transfer from electric baseboard heaters or steam radiators, heat transfer from the refrigeration coils and power transmission lines, and heat transfer from the bodies of animals and human beings. Natural convection in gases is usually accompanied by radiation of comparable magnitude except for low-emissivity surfaces.

We start this chapter with a discussion of the physical mechanism of *natural convection* and the *Grashof number.* We then present the correlations to evaluate heat transfer by natural convection for various geometries, including enclosures and finned surfaces. Finally, we discuss simultaneous forced and natural convection.

## 11-1 ■ PHYSICAL MECHANISM OF NATURAL CONVECTION

**Physical Mechanism of Natural Convection**

We know that a hot *boiled egg* (or a hot *baked potato*) in a plate eventually cools to the surrounding air temperature (Fig. 11-1). The egg is cooled by transferring heat by *convection* to the air and by radiation to the surrounding surfaces. Disregarding heat transfer by radiation, the physical mechanism of cooling of a hot egg (or any hot object) in a cooler environment can be explained as follows:

As soon as the hot egg is exposed to cooler air, the temperature of the outer surface of the egg shell will drop somewhat, and the temperature of the air adjacent to the shell will rise as a result of heat conduction from the shell to the air. Consequently, the egg will soon be surrounded by a thin layer of warmer air, and heat will then be transferred from this warmer layer to the outer layers of air. The cooling process in this case would be rather *slow*, since the egg would always be *blanketed* by warm air, and it would have no direct contact with the cooler air farther away. We may not notice any *air motion* in the vicinity of the egg, but careful measurements indicate otherwise.

**FIGURE 11-1**
The cooling of a boiled egg in a cooler environment by natural convection.

The temperature of the air adjacent to the egg is higher, and thus its density is lower, since at constant pressure the *density* of a gas is inversely proportional to its *temperature*. Thus, we have a situation in which some low-density or "light" gas is surrounded by a high-density or "heavy" gas, and the natural laws dictate that *the light gas rise*. This is no different than the oil in a vinegar-and-oil salad dressing rising to the top (note that $\rho_{oil} < \rho_{vinegar}$). This phenomena is characterized incorrectly by the phrase "heat rises," which is understood to mean *heated air rises*. The space vacated by the warmer air in the vicinity of the egg is replaced by the cooler air nearby, and the presence of cooler air in the vicinity of the egg speeds up the cooling process. The rise of warmer air and the flow of cooler air into its place continues until the egg is cooled to the temperature of the surrounding air. The motion that results by the continual replacement of the heated air in the vicinity of the egg by the cooler air nearby is called a **natural convection current**, and the heat transfer that is enhanced as a result of this natural convection current is called **natural convection heat transfer**. Note that in the absence of natural convection currents, heat transfer from the egg to the air surrounding it would be by *conduction* only, and the rate of heat transfer from the egg would be much *lower*.

Natural convection is just as effective in the heating of cold surfaces in a warmer environment as it is in the cooling of hot surfaces in a cooler environment, as shown in Fig. 11-2. Note that the direction of fluid motion is reversed in this case.

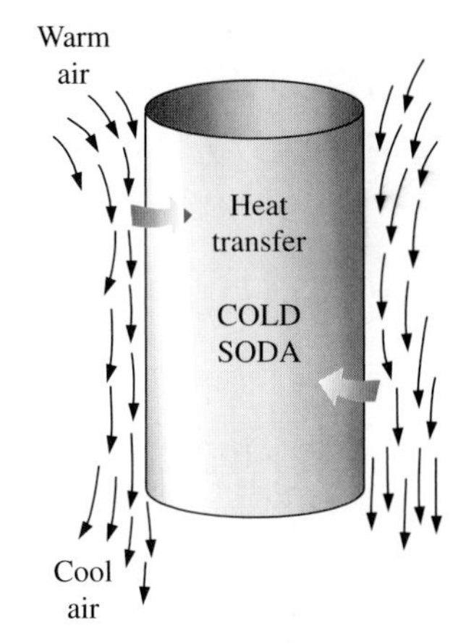


**FIGURE 11-2**
The warming up of a cold drink in a warmer environment by natural convection.

In a gravitational field, there seems to be a net force that pushes a light fluid placed in a heavier fluid upwards. The upward force exerted by a fluid on a body completely or partially immersed in it is called the **buoyancy force**. The magnitude of the buoyancy force is equal to the weight of the *fluid displaced* by the body. That is,

$$F_{\text{buoyancy}} = \rho_{\text{fluid}} g V_{\text{body}} \tag{11-1}$$

where $\rho_{\text{fluid}}$ is the average density of the *fluid* (not the body), $g$ is the gravitational acceleration, and $V_{\text{body}}$ is the volume of the portion of the body immersed in the fluid (for bodies completely immersed in the fluid, it is the total volume of the body). In the absence of other forces, the net vertical force acting on a body is the difference between the weight of the body and the buoyancy force. That is,

$$\begin{aligned} F_{\text{net}} &= W - F_{\text{buoyancy}} \\ &= \rho_{\text{body}} g V_{\text{body}} - \rho_{\text{fluid}} g V_{\text{body}} \\ &= (\rho_{\text{body}} - \rho_{\text{fluid}}) g V_{\text{body}} \end{aligned} \tag{11-2}$$

Note that this force is *proportional* to the difference in the *densities* of the fluid and the body immersed in it. Thus, a body immersed in a fluid will experience a "weight loss" in an amount equal to the weight of the fluid it displaces. This is known as *Archimedes' principle.*

To have a better understanding of the buoyancy effect, consider an egg dropped into water. If the average density of the egg is greater than the density of water (a sign of freshness), the egg will settle at the bottom of the container. Otherwise, it will rise to the top. When the density of the egg equals to the density of water, the egg will settle somewhere in the water while remaining completely immersed, acting like a "weightless object" in space. This occurs when the upward buoyancy force acting on the egg equals the weight of the egg which acts downwards.

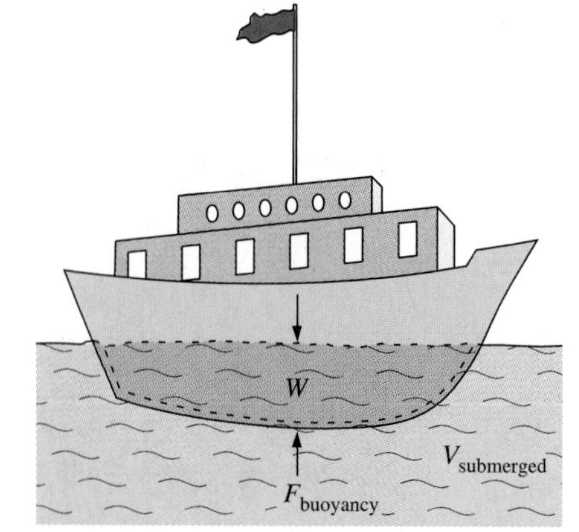


**FIGURE 11-3**
It is the buoyancy force which keeps the ships afloat in water ($W = F_{\text{Buoyancy}}$ for floating objects).

The *buoyancy effect* has far-reaching implications in life. For one thing, without buoyancy, heat transfer between a hot (or cold) surface and the fluid surrounding it would be by *conduction* instead of by *natural convection.* The natural convection currents encountered in the oceans, lakes, and the atmosphere owe their existence to buoyancy. Also, light boats as well as heavy warships made of steel float on water because of buoyancy (Fig. 11-3). Ships are designed on the basis of the principle that the entire weight of a ship and its contents is equal to the weight of the water that the submerged volume of the ship can contain. (Note that a larger portion of the hull of a ship will sink in fresh water than it does in salty water.) The "chimney effect" that induces the upward flow of hot combustion gases through a chimney is also due to buoyancy effect, and the upward force acting on the gases in the chimney is proportional to the difference between the densities of the hot gases in the chimney and the cooler air outside. Note that there is *no gravity* in space, and thus there will be no natural convection heat transfer in a spacecraft, even if the spacecraft is filled with atmospheric air.

In heat transfer studies, the primary variable is *temperature*, and it is desirable to express the net buoyancy force (Eq. 11-2) in terms of temperature differences. But this requires expressing the density difference in terms of a temperature difference, which requires a knowledge of a property which represents the *variation of the density of a fluid with temperature at constant pressure.* The property which provides

that information is the **volume expansion coefficient** $\beta$ defined as (Fig. 11-4)

$$\beta = -\frac{1}{\rho}\left(\frac{\partial \rho}{\partial T}\right)_P \quad (1/\text{K}) \qquad (11\text{-}3)$$

It can also be expressed approximately by replacing derivatives by differences as

$$\beta \approx -\frac{1}{\rho}\frac{\Delta \rho}{\Delta T} \longrightarrow \Delta \rho \approx -\rho\beta\Delta T \qquad (\text{at constant } P)$$

We can show easily that the volume expansion coefficient $\beta$ of an *ideal gas* ($P = \rho RT$) at a temperature $T$ is equivalent to the inverse of the temperature:

$$\beta_{\text{ideal gas}} = \frac{1}{T} \quad (1/\text{K}) \qquad (11\text{-}4)$$

where $T$ is the *absolute* temperature. Note that a large value of $\beta$ for a fluid means a large change in density with temperature, and that the product $\beta\,\Delta T$ represents the fraction of volume change of a fluid that corresponds to a temperature change $\Delta T$ at constant pressure. Also note that the buoyancy force is proportional to the *density difference,* which is proportional to the *temperature difference* at constant pressure. Therefore, the *larger* the temperature difference between the fluid adjacent to a hot (or cold) surface and the fluid away from it, the *larger* the buoyancy force, and the *stronger* the natural convection currents, and thus the *higher* the heat transfer rate.

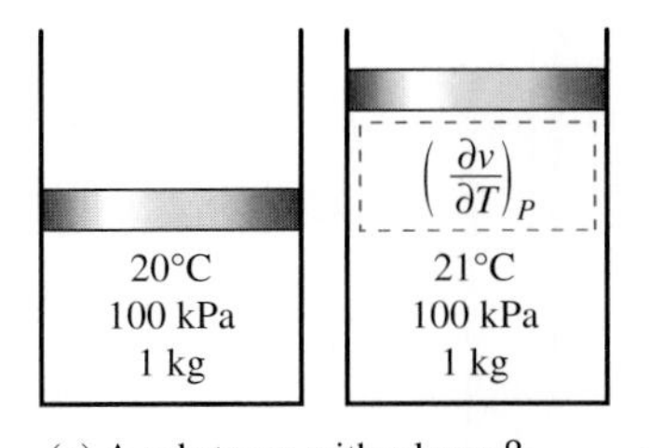



(*a*) A substance with a large β

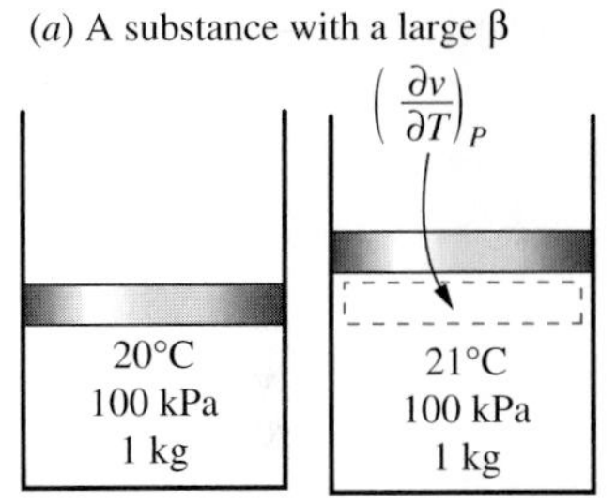



(*b*) A substance with a small β

**FIGURE 11-4**

The coefficient of volume expansion is a measure of the change in volume of a substance with temperature at constant pressure.

The magnitude of the natural convection heat transfer between a surface and a fluid is directly related to the *mass flow rate* of the fluid. The higher the mass flow rate, the higher is the heat transfer rate. In fact, it is the very high flow rates that increase the heat transfer coefficient by orders of magnitude when forced convection is used. In natural convection, no blowers are used, and therefore the flow rate cannot be controlled externally. The flow rate in this case is established by the dynamic balance of *buoyancy* and *friction*.

As we have discussed earlier, the *buoyancy force* is caused by the density difference between the heated (or cooled) fluid adjacent to the surface and the fluid surrounding it, and is proportional to this density difference and the volume occupied by the warmer fluid. It is also well known that whenever two bodies in contact (solid–solid, solid–fluid, or fluid–fluid) move relative to each other, a *friction force* develops at the contact surface in the direction opposite to that of the motion. This opposing force slows down the fluid and thus reduces the flow rate of the fluid. Under steady conditions, the air flow rate driven by buoyancy is established at the point where these two effects *balance* each other. The friction force increases as more and more solid surfaces are introduced, seriously disrupting the fluid flow and heat transfer. For that reason, heat sinks with closely spaced fins are not suitable for natural convection cooling.

Most heat transfer correlations in natural convection are based on experimental measurements. The instrument used in natural convection experiments most often is the *Mach–Zehnder interferometer,* which gives a plot of isotherms in the fluid in the vicinity of a surface. The operation principle of interferometers is based on the fact that at low pressure, the lines of constant temperature for a gas correspond to the lines of constant density, and that the index of refraction of a gas is a function of its density. Therefore, the degree of refraction of light at some point in a gas is a measure of the temperature gradient at that point. An interferometer produces a map of interference fringes, which can be interpreted as lines of *constant temperature*, as shown in Fig. 11-5. The smooth and parallel lines in ($a$) indicate that the flow is *laminar*, whereas the eddies and irregularities in ($b$) indicate that the flow is *turbulent.* Note that the lines are closest near the surface, indicating a *higher temperature gradient.*

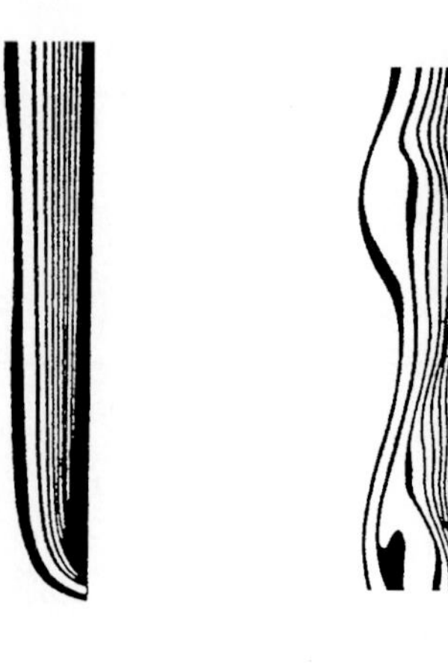

($a$) Laminar flow ($b$) Turbulent flow

**FIGURE 11-5**
Isotherms in natural convection over a hot plate in air.

## The Grashof Number

We mentioned in the preceding chapter that the flow regime in forced convection is governed by the dimensionless *Reynolds number* which represents the ratio of inertial forces to viscous forces acting on the fluid. The flow regime in natural convection is governed by another dimensionless number, called the **Grashof number**, which represents the ratio of the *buoyancy force* to the *viscous force* acting on the fluid. That is,

$$\text{Gr} = \frac{\text{buoyancy forces}}{\text{viscous forces}} = \frac{g\,\Delta\rho\,V}{\rho\nu^2} = \frac{g\beta\,\Delta T\,V}{\rho\nu}$$

It is formally expressed as (Fig. 11-6)

$$\text{Gr} = \frac{g\beta(T_s - T_\infty)\delta^3}{\nu^2} \tag{11-5}$$

where $g$ = gravitational acceleration, $\text{m/s}^2$
$\beta$ = coefficient of volume expansion, 1/K ($\beta = 1/T$ for ideal gases)
$T_s$ = temperature of the surface, °C
$T_\infty$ = temperature of the fluid sufficiently far from the surface, °C
$\delta$ = characteristic length of the geometry, m
$\nu$ = kinematic viscosity of the fluid, $\text{m}^2/\text{s}$

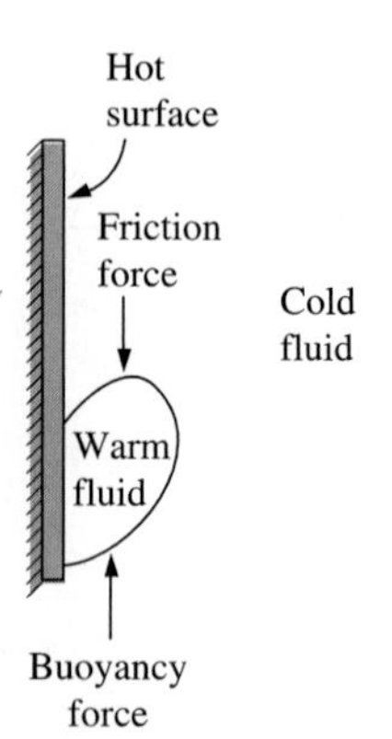


**FIGURE 11-6**
The Grashof number Gr is a measure of the relative magnitudes of the *buoyancy force* and the opposing *friction force* acting on the fluid.

The role played by the *Reynolds number* in forced convection is played by the *Grashof number* in natural convection. As such, the Grashof number provides the main criteria in determining whether the fluid flow is laminar or turbulent in natural convection. For vertical plates, for example, the critical Grashof number is observed to be about $10^9$. Therefore, the flow regime on a vertical plate becomes turbulent at Grashoff numbers greater than $10^9$.

The heat transfer rate in natural convection from a solid surface to the surrounding fluid is expressed by Newton's law of cooling as

$$\dot{Q}_{\text{conv}} = hA(T_s - T_\infty) \quad \text{(W)} \tag{11-6}$$

where $A$ is the heat transfer surface area and $h$ is the average heat transfer coefficient on the surface.

## 11-2 ■ NATURAL CONVECTION OVER SURFACES

Natural convection heat transfer on a surface depends on the geometry of the surface as well as its orientation. It also depends on the variation of temperature on the surface and the thermophysical properties of the fluid involved.

The *velocity* and *temperature profiles* for natural convection over a vertical hot plate immersed in a quiescent fluid body are given in Fig. 11-7. As in forced convection, the thickness of the boundary layer increases in the flow direction. Unlike forced convection, however, the fluid velocity is *zero* at the *outer edge* of the velocity boundary layer as well as at the surface of the plate. This is expected, since the fluid beyond the boundary layer is stationary. Thus, the fluid velocity increases with distance from the surface, reaches a maximum, and gradually decreases to zero at a distance sufficiently far from the surface. The *temperature* of the fluid will equal the plate temperature at the surface, and gradually decrease to the temperature of the surrounding fluid at a distance sufficiently far from the surface, as shown in the figure. In the case of *cold surfaces,* the shape of the velocity and temperature profiles remain the same but their direction is reversed.

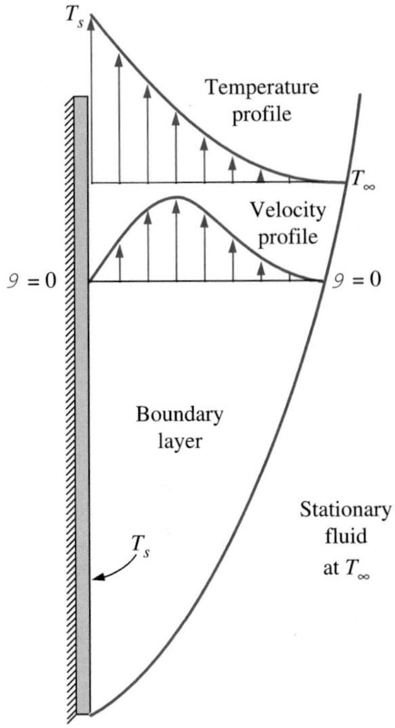

**FIGURE 11-7**
Typical velocity and temperature profiles for natural convection flow over a hot vertical plate at temperature $T_s$ inserted in a fluid at temperature $T_\infty$.

### Natural Convection Correlations

Although we understand the mechanism of natural convection well, the complexities of fluid motion make it very difficult to obtain simple analytical relations for heat transfer by solving the governing equations of motion and energy. Some analytical solutions exist for natural convection, but such solutions lack generality, since they are obtained for simple geometries under some simplifying assumptions. Therefore, with the exception of some simple cases, heat transfer relations in natural convection are based on experimental studies. Of the numerous such correlations of varying complexity and claimed accuracy available in the literature for any given geometry, we present below the *simpler* ones for two reasons: first, the accuracy of simpler relations is usually within the range of uncertainty associated with a problem, and second, we would like to keep the emphasis on the physics of the problems instead of formula manipulation.

The simple empirical correlations for the average *Nusselt number Nu* in natural convection are of the form (Fig. 11-8)

$$\text{Nu} = \frac{h\delta}{k} = C(\text{Gr Pr})^n = C\,\text{Ra}^n \tag{11-7}$$

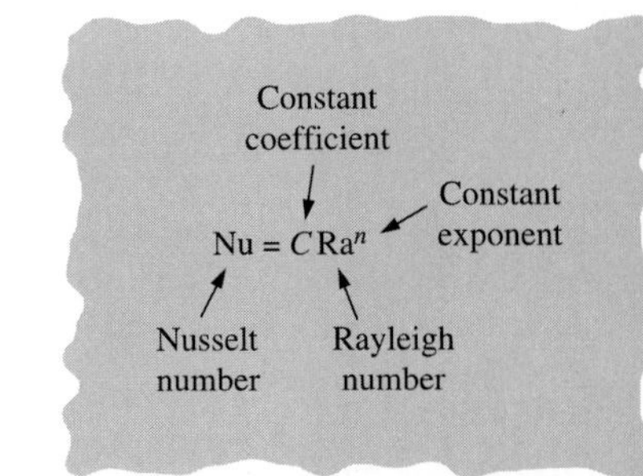


**FIGURE 11-8**
Natural convection heat transfer correlations are usually expressed in terms of the Rayleigh number raised to a constant $n$ multiplied by another constant $C$, both of which determined experimentally.

where Ra is the **Rayleigh number**, which is the product of the Grashof and Prandtl numbers:

$$\mathrm{Ra} = \mathrm{Gr\,Pr} = \frac{g\beta(T_s - T_\infty)\delta^3}{\nu^2}\mathrm{Pr} \tag{11-8}$$

The values of the constants $C$ and $n$ depend on the *geometry* of the surface and the *flow regime*, which is characterized by the range of the Rayleigh number. The value of $n$ is usually $\frac{1}{4}$ for laminar flow and $\frac{1}{3}$ for turbulent flow. The value of the constant $C$ is normally less than 1.

Simple relations for the average Nusselt number for various geometries are given in Table 11-1, together with sketches of the geometries. Also given in this table are the characteristic lengths of the geometries and the ranges of Rayleigh number in which the relation is applicable. All fluid properties are to be evaluated at the film temperature $T_f = \frac{1}{2}(T_s + T_\infty)$.

These relations have been obtained for the case of isothermal surfaces, but could also be used approximately for the case of nonisothermal surfaces by assuming the surface temperature to be constant at some average value. The use of these relations is illustrated below with examples.

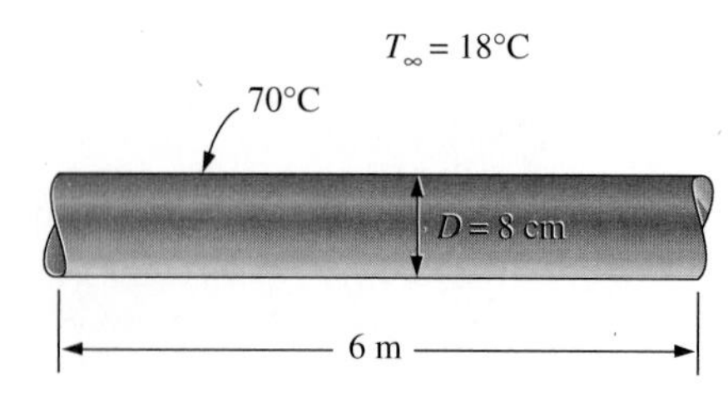

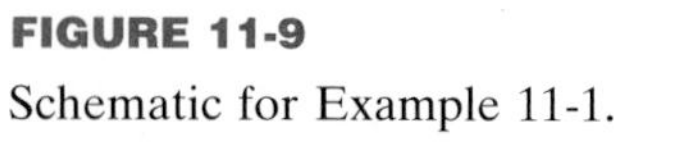


**FIGURE 11-9**
Schematic for Example 11-1.

**EXAMPLE 11-1**

A 6-m-long section of an 8-cm-diameter horizontal hot water pipe shown in Fig. 11-9 passes through a large room whose temperature is 18°C. If the outer surface temperature of the pipe is 70°C, determine the rate of heat loss from the pipe by natural convection.

**Solution** We assume the air pressure in the room to be 1 atm. The properties of air are to be evaluated at the film temperature

$$T_f = \frac{T_s + T_\infty}{2} = \frac{70 + 18}{2} = 44°\mathrm{C} = 317\ \mathrm{K}$$

At this temperature, we read

$$k = 0.0273\ \mathrm{W/(m \cdot °C)}$$
$$\nu = 1.74 \times 10^{-5}\ \mathrm{m^2/s}$$
$$\mathrm{Pr} = 0.710$$
$$\beta = \frac{1}{T_f} = \frac{1}{317\ \mathrm{K}} = 0.00315\ \mathrm{K^{-1}}$$

The characteristic length in this case is the outer diameter of the pipe, $\delta = D = 0.08$ m. Then the Rayleigh number becomes

$$\begin{aligned}\mathrm{Ra} &= \frac{g\beta(T_s - T_\infty)\delta^3}{\nu^2}\mathrm{Pr} \\ &= \frac{(9.8\ \mathrm{m/s^2})(0.00315\ \mathrm{K^{-1}})[(70 - 18)\ \mathrm{K}](0.08\ \mathrm{m})^3}{(1.74 \times 10^{-5}\ \mathrm{m^2/s})^2}(0.710) = 1.930 \times 10^6\end{aligned}$$

**TABLE 11-1**
**Empirical correlations for the average Nusselt number for natural convection over surfaces**

| Geometry | Characteristic length $\delta$ | Range of Ra | Nu |
|---|---|---|---|
| Vertical plate 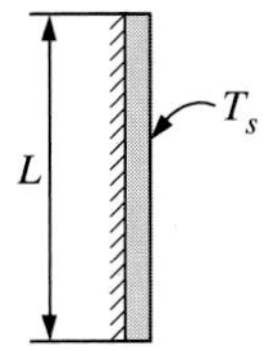  | $L$ | $10^4$–$10^9$<br>$10^9$–$10^{13}$<br>Entire range | $\text{Nu} = 0.59\,\text{Ra}^{1/4}$ (11-8)<br>$\text{Nu} = 0.1\,\text{Ra}^{1/3}$ (11-9)<br>$\text{Nu} = \left\{0.825 + \dfrac{0.387\text{Ra}^{1/6}}{[1 + (0.492/\text{Pr})^{9/16}]^{8/27}}\right\}^2$ (11-10)<br>(complex but more accurate) |
| Inclined plate 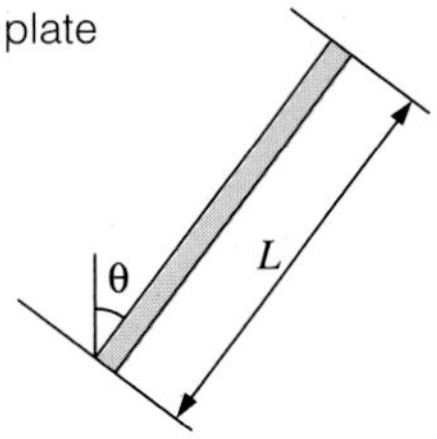  | $L$ | | Use vertical plate equations as a first degree of approximation<br>Replace $g$ by $g\cos\theta$ for $\text{Ra} < 10^9$ |
| Horizontal plate<br>(Surface area $A$ and perimeter $p$)<br>(*a*) Upper surface of a hot plate<br>(or lower surface of a cold plate) 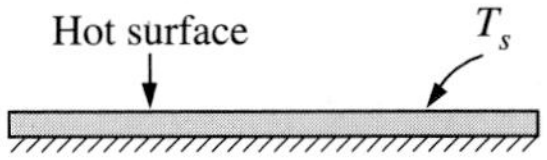 <br>(*b*) Lower surface of a hot plate<br>(or upper surface of a cold plate) 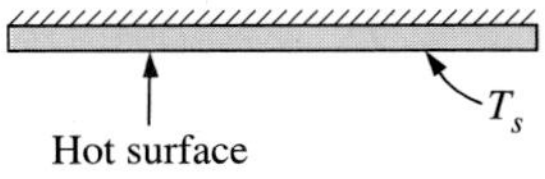  | $A/p$ | $10^4$–$10^7$<br>$10^7$–$10^{11}$<br><br>$10^5$–$10^{11}$ | $\text{Nu} = 0.54\,\text{Ra}^{1/4}$ (11-11)<br>$\text{Nu} = 0.15\,\text{Ra}^{1/3}$ (11-12)<br><br>$\text{Nu} = 0.27\text{Ra}^{1/4}$ (11-13) |
| Vertical cylinder 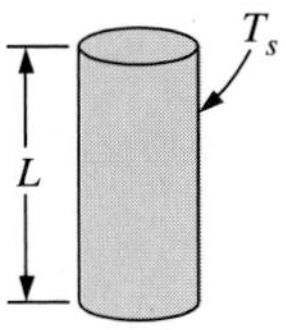  | $L$ | | A vertical clyinder can be treated as a vertical plate when<br>$D \geqslant \dfrac{35L}{\text{Gr}^{1/4}}$ (11-14) |
| Horizontal cylinder 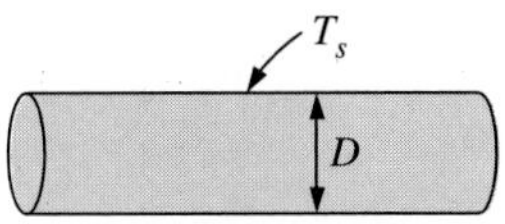  | $D$ | $10^{-5}$–$10^{12}$ | $\text{Nu} = \left\{0.6 + \dfrac{0.387\,\text{Ra}^{1/6}}{[1 + (0.559/\text{Pr})^{9/16}]^{8/27}}\right\}^2$ (11-15) |
| Sphere 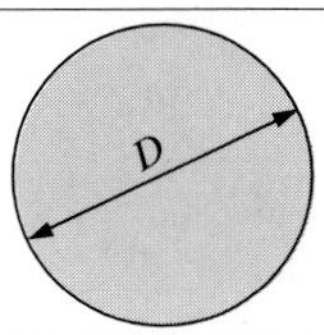  | $\frac{1}{2}\pi D$ | $\text{Ra} \leqslant 10^{11}$<br>$(\text{Pr} \geqslant 0.7)$ | $\text{Nu} = 2 + \dfrac{0.589\,\text{Ra}^{1/4}}{[1 + (0.469/\text{Pr})^{9/16}]^{4/9}}$ (11-16) |

Then the natural convection Nusselt number in this case can be determined from Eq. 11-15 to be

$$\text{Nu} = \left\{0.6 + \frac{0.387\,\text{Ra}^{1/6}}{[1 + (0.559/\text{Pr})^{9/16}]^{8/27}}\right\}^2$$

$$= \left\{0.6 + \frac{0.387(1.930 \times 10^6)^{1/6}}{[1 + (0.559/0.710)^{8/27}}\right\}^2 = 17.2$$

Then

$$h = \frac{k}{D}\text{Nu} = \frac{0.0273\ \text{W/(m}\cdot{}^\circ\text{C)}}{0.08\ \text{m}}(17.2) = 5.9\ \text{W/(m}^2\cdot{}^\circ\text{C)}$$

$$A = \pi DL = \pi(0.08\ \text{m})(6\ \text{m}) = 1.51\ \text{m}^2$$

and

$$\dot{Q} = hA(T_s - T_\infty) = [5.9\ \text{W/(m}^2\cdot{}^\circ\text{C)}](1.51\ \text{m}^2)(70 - 18)^\circ\text{C} = 463\ \text{W}$$

Therefore, the pipe will lose heat to the air in the room at a rate of 463 W by natural convection.

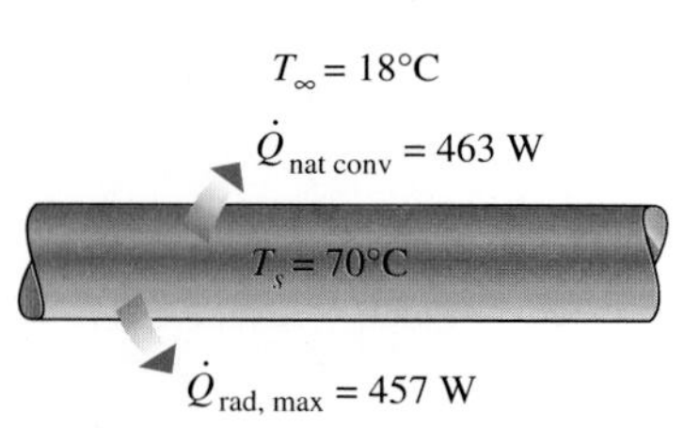


**FIGURE 11-10**
Radiation heat transfer is usually comparable to natural convection in magnitude, and should be considered in heat transfer analysis.

**Discussion** The pipe will lose heat to the surroundings by radiation as well as by natural convection. Assuming the outer surface of the pipe to be black (emissivity $\varepsilon$ = 1) and the inner surfaces of the walls of the room to be at room temperature, the radiation heat transfer in this case is determined to be (Fig. 11-10)

$$\begin{aligned}\dot{Q}_{\text{rad}} &= \varepsilon A\sigma(T_s^4 - T_\infty^4)\\ &= (1)(1.51\ \text{m}^2)[5.67 \times 10^{-8}\ \text{W/(m}^2\cdot\text{K}^4)]\\ &\quad \times [(70 + 273\ \text{K})^4 - (18 + 273\ \text{K})^4]\\ &= 457\ \text{W}\end{aligned}$$

which is as large as that for natural convection. The emissivity of a real surface is less than 1, and thus the radiation heat transfer for a real surface will be less. But radiation will still be significant for most systems cooled by natural convection. Therefore, a radiation analysis should normally accompany a natural convection analysis unless the emissivity of the surface.

**EXAMPLE 11-2**

Consider a 0.6 m × 0.6 m thin square plate in a room at 30°C. One side of the plate is maintained at a temperature of 74°C, while the other side is insulated, as shown in Fig. 11-11. Determine the rate of heat transfer from the plate by natural convection if the plate (*a*) vertical, (*b*) horizontal with hot surface facing up, and (*c*) horizontal with hot surface facing down.

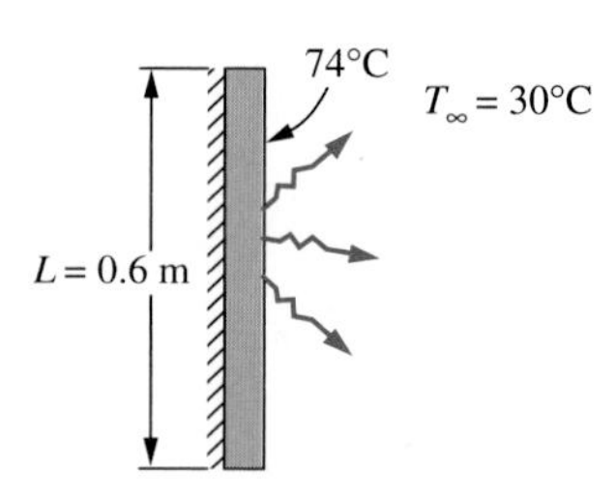


(*a*) Vertical

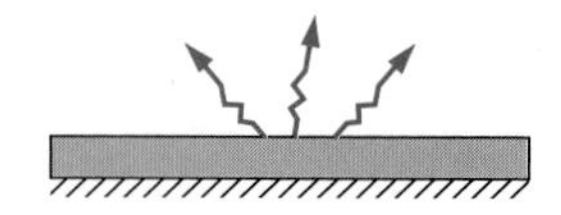

(*b*) Hot surface facing up

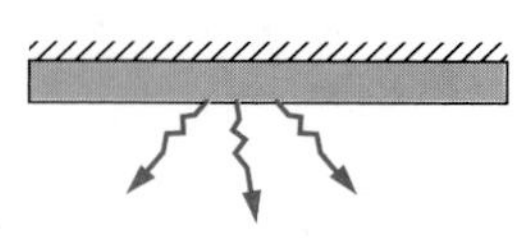

(*c*) Hot surface facing down

**FIGURE 11-11**
Schematic for Example 11-2.

**Solution** We assume the air pressure in the room to be 1 atm. The properties of air are to be evaluated at the film temperature of

$$T_f = \frac{T_s + T_\infty}{2} = \frac{74 + 30}{2} = 52^\circ\text{C} = 325\ \text{K}$$

At this temperature, we read

$$k = 0.0279 \text{ W/(m} \cdot \text{°C)}$$
$$\nu = 1.815 \times 10^{-5} \text{ m}^2\text{/s}$$
$$\text{Pr} = 0.709$$
$$\beta = \frac{1}{T_f} = \frac{1}{325 \text{ K}} = 0.00308 \text{ K}^{-1}$$

(*a*) *Vertical.* The characteristic length in this case is the height of the plate, which is $\delta = 0.6$ m. The Rayleigh number is

$$\text{Ra} = \frac{g\beta(T_s - T_\infty)\delta^3}{\nu^2}\text{Pr}$$
$$= \frac{(9.8 \text{ m/s}^2)(0.00308 \text{ K}^{-1})[(74 - 30) \text{ K}](0.6 \text{ m})^3}{(1.815 \times 10^{-5} \text{ m}^2\text{/s})^2}(0.708) = 6.034 \times 10^8$$

Then the natural convection Nusselt number can be determined from Eq. 11-8 to be

$$\text{Nu} = 0.59 \text{ Ra}^{1/4} = 0.59(6.034 \times 10^8)^{1/4} = 92.5$$

Then

$$h = \frac{k}{\delta}\text{Nu} = \frac{0.0279 \text{ W/(m} \cdot \text{°C)}}{0.6 \text{ m}}(92.5) = 4.3 \text{ W/(m}^2 \cdot \text{°C)}$$
$$A = L^2 = (0.6 \text{ m})^2 = 0.36 \text{ m}^2$$

and

$$\dot{Q} = hA(T_s - T_\infty) = [4.3 \text{ W/(m}^2 \cdot \text{°C)}](0.36 \text{ m}^2)(74 - 30)\text{°C} = 68.1 \text{ W}$$

(*b*) *Horizontal with hot surface facing up.* The characteristic length length and the Rayleigh number in this case are

$$\delta = \frac{A}{p} = \frac{L^2}{4L} = \frac{L}{4} = \frac{0.6}{4} = 0.15 \text{ m}$$

The Rayleigh number in this case is

$$\text{Ra} = \frac{g\beta(T_s - T_\infty)\delta^3}{\nu^2}\text{Pr}$$
$$= \frac{(9.8 \text{ m/s}^2)(0.00308 \text{ K}^{-1})[(74 - 30) \text{ K}](0.15 \text{ m})^3}{(1.815 \times 10^{-5} \text{ m}^2\text{/s})^2}(0.708) = 9.43 \times 10^6$$

Then the natural convection Nusselt number can be determined from Eq. 11–6 to be

$$\text{Nu} = 0.54 \text{ Ra}^{1/4} = 0.54(9.43 \times 10^6)^{1/4} = 29.9$$

Therefore,

$$h = \frac{k}{\delta}\text{Nu} = \frac{0.0279 \text{ W/(m} \cdot \text{°C)}}{0.15 \text{ m}}(29.9) = 3.9 \text{ W/(m}^2 \cdot \text{°C)}$$
$$A = L^2 = (0.6 \text{ m})^2 = 0.36 \text{ m}^2$$

and

$$\dot{Q} = hA(T_s - T_\infty) = [3.9 \text{ W/(m}^2 \cdot \text{°C)}](0.36 \text{ m}^2)(74 - 30)\text{°C} = 88.1 \text{ W}$$

(*c*) *Horizontal with hot surface facing down.* The characteristic length $\delta$, the heat transfer surface area $A$, and the Rayleigh number in this case are the same as those determined in (*b*). But the natural convection Nusselt number is to be determined from Eq. 11–7:

$$\text{Nu} = 0.27 \text{ Ra}^{1/4} = 0.27(9.43 \times 10^6)^{1/4} = 15.0$$

Then,

$$h = \frac{k}{\delta}\text{Nu} = \frac{0.0279\ \text{W/(m}\cdot\text{°C)}}{0.15\ \text{m}}(15.0) = 2.8\ \text{W/(m}^2\cdot\text{°C)}$$

and

$$\dot{Q} = hA(T_s - T_\infty) = [2.8\ \text{W/(m}^2\cdot\text{°C)}](0.36\ \text{m}^2)(74 - 30)\text{°C} = 44.4\ \text{W}$$

Note that the natural convection heat transfer is the lowest in the case of the hot surface facing down. This is not surprising, since the hot air is "trapped" under the plate in this case, and cannot get away from the plate easily. As a result, the cooler air in the vicinity of the plate will have difficulty reaching the plate, which results in a reduced rate of heat transfer.

**Discussion** The plate will lose heat to the surroundings by radiation as well as by natural convection. Assuming the surface of the plate to be black (emissivity $\varepsilon = 1$) and the inner surfaces of the walls of the room to be at room temperature, the radiation heat transfer in this case is determined to be

$$\begin{aligned}\dot{Q}_{\text{rad}} &= \varepsilon A\sigma(T_s^4 - T_\infty^4)\\ &= (1)(0.36\ \text{m}^2)[5.67 \times 10^{-8}\ \text{W/(m}^2\cdot\text{K}^4)]\\ &\quad \times [(74 + 273\ \text{K})^4 - (30 + 273\ \text{K})^4]\\ &= 124\ \text{W}\end{aligned}$$

which is larger than that for natural convection heat transfer for each case. The emissivity of a real surface is less than 1, and thus the radiation heat transfer for a real surface will be less. But radiation can still be significant, and needs to be considered in surfaces cooled by natural convection.

## 11-3 ■ NATURAL CONVECTION INSIDE ENCLOSURES

A considerable portion of heat loss from a typical residence occurs through the windows. We certainly would insulate the windows, if we could, in order to conserve energy. The problem is finding an insulating material which is transparent. An examination of the thermal conductivities of the insulting materials reveals that *air* is a *better insulator* than most common insulating materials. Besides, it is transparent. Therefore, it makes sense to insulate the windows with a layer of air. Of course, we need to use another sheet of glass to trap the air. The result is an *enclosure,* which is known as a *double-paned window* in this case. Other examples of enclosures include wall cavities, solar collectors, and cryogenic chambers involving concentric cylinders or spheres.

Enclosures are frequently encountered in practice, and heat transfer through them is of practical interest. Heat transfer in enclosed spaces is complicated by the fact that the fluid in the enclosure, in general, does not remain stationary. In a vertical enclosure, the fluid adjacent to the hotter surface rises and the fluid adjacent to the cooler one falls, setting off a rotationary motion within the enclosure that enhances heat transfer

through the enclosure. Typical flow patterns in vertical and horizontal rectangular enclosures are shown in Fig. 11-12.

The characteristics of heat transfer through a horizontal enclosure depend on whether the hotter plate is at the top or at the bottom, as shown in Fig. 11-13. When the *hotter plate* is at the *top,* no convection currents will develop in the enclosure, since the lighter fluid will always be on top of the heavier fluid. Heat transfer in this case will be by *pure conduction,* and we will have Nu = 1. When the *hotter plate* is at the *bottom,* the heavier fluid will be on top of the lighter fluid, and there will be a tendency for the lighter fluid to topple the heavier fluid and rise to the top, where it will come in contact with the cooler plate and cool down. Until that happens, however, the heat transfer is still by *pure conduction,* and Nu = 1. When Ra > 1708, the buoyant force overcomes the fluid resistance and initiates natural convection currents, which are observed to be in the form of hexagonal cells called *Bénard cells.* For $\text{Ra} > 3 \times 10^5$, the cells break down and the fluid motion becomes turbulent.

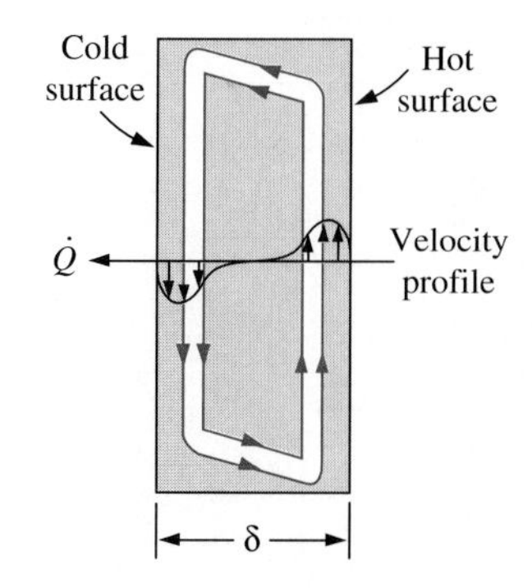


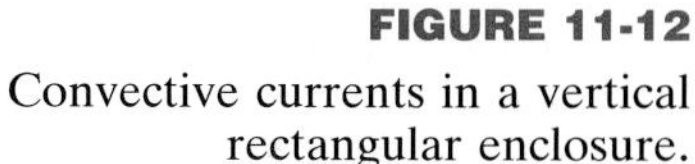

**FIGURE 11-12**
Convective currents in a vertical rectangular enclosure.

The Rayleigh number for an enclosure is determined from

$$\text{Ra} = \frac{g\beta(T_1 - T_2)\delta^3}{\nu^2}\text{Pr} \tag{11-17}$$

where the characteristic length $\delta$ is the distance between the hot and cold surfaces, and $T_1$ and $T_2$ are the temperatures of the hot and cold surfaces, respectively. All fluid properties are to be evaluated at the average fluid temperature $T_{av} = \frac{1}{2}(T_1 + T_2)$.

Simple empirical correlations for the Nusselt number for various enclosures are given in Table 11-2. Once the Nusselt number is available, the heat transfer coefficient and the rate of heat transfer through the enclosure can be determined from

$$h = \frac{k}{\delta}\text{Nu} \tag{11-18}$$

and

$$\dot{Q} = hA_s(T_1 - T_2) = k\,\text{Nu}\,A_s\frac{T_1 - T_2}{\delta} \tag{11-19}$$

where

$$A = \begin{cases} HL & \text{rectangular enclosures} \\ \dfrac{\pi L(D_2 - D_1)}{\ln(D_2/D_1)} & \text{concentric cylinders} \\ \pi D_1 D_2 & \text{concentric spheres} \end{cases} \tag{11–20}$$

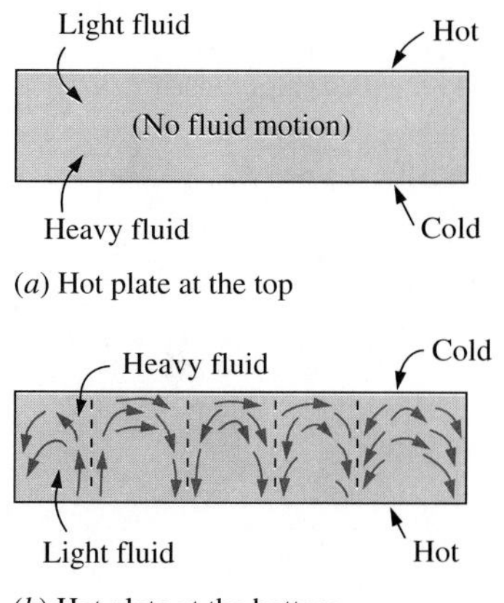


**FIGURE 11-13**
Convective currents in a horizontal enclosure with (*a*) hot plate at the top and (*b*) hot plate at the bottom.

For *inclined rectangular enclosures,* highly accurate but complex correlations are available in the literature. In the absence of such relations, the Nusselt number correlations for vertical enclosures can be used for inclined enclosures heated from below for inclination angles up

**TABLE 11-2**

**Empirical correlations for the average Nusselt number for natural convection in enclosures (the characteristic length $\delta$ is as indicated on the respective diagram)**

| Geometry | Fluid | $H/\delta$ | Range of Pr | Range of Ra | Nusselt number | |
|---|---|---|---|---|---|---|
| Vertical rectangular enclosure (or vertical cylindrical enclosure) | Gas or liquid | — | — | $\text{Ra} < 2000$ | $\text{Nu} = 1$ | (11-23) |
| | Gas | 11-42 | 0.5–2 | $2 \times 10^3–2 \times 10^5$ | $\text{Nu} = 0.197\,\text{Ra}^{1/4}\left(\frac{H}{\delta}\right)^{-1/9}$ | (11-24) |
| | | 11-42 | 0.5–2 | $2 \times 10^5–10^7$ | $\text{Nu} = 0.073\,\text{Ra}^{1/3}\left(\frac{H}{\delta}\right)^{-1/9}$ | (11-25) |
| | Liquid | 10–40 | 1–20,000 | $10^4–10^7$ | $\text{Nu} = 0.42\text{Pr}^{0.012}\,\text{Ra}^{1/4}\left(\frac{H}{\delta}\right)^{-0.3}$ | (11–26) |
| | | 1–40 | 1–20 | $10^6–10^9$ | $\text{Nu} = 0.046\,\text{Ra}^{1/3}$ | (11–27) |
| Inclined rectangular enclosure | | | | | Use the correlations for vertical enclosures as a first-degree approximation for $\theta \leq 20°$ by replacing $g$ in the Ra relation by $g \cos \theta$ | |
| Horizontal rectangular enclosure (hot surface at the top) | Gas or liquid | — | — | — | $\text{Nu} = 1$ | (11-28) |
| Horizontal rectangular enclosure (hot surface at the bottom) | Gas or liquid | — | — | $\text{Ra} < 1700$ | $\text{Nu} = 1$ | (11–29) |
| | Gas | — | 0.5–2 | $1.7 \times 10^3–7 \times 10^3$ | $\text{Nu} = 0.059\,\text{Ra}^{0.4}$ | (11–30) |
| | | — | 0.5–2 | $7 \times 10^3–3.2 \times 10^5$ | $\text{Nu} = 0.212\,\text{Ra}^{1/4}$ | (11–31) |
| | | — | 0.5–2 | $\text{Ra} > 3.2 \times 10^5$ | $\text{Nu} = 0.061\,\text{Ra}^{1/3}$ | (11–32) |
| | Liquid | — | 1–5000 | $1.7 \times 10^3–6 \times 10^3$ | $\text{Nu} = 0.012\,\text{Ra}^{0.6}$ | (11-33) |
| | | — | 1–5000 | $6 \times 10^3–3.7 \times 10^4$ | $\text{Nu} = 0.375\,\text{Ra}^{0.2}$ | (11–34) |
| | | — | 1–20 | $3.7 \times 10^4–10^8$ | $\text{Nu} = 0.13\,\text{Ra}^{0.3}$ | (11-35) |
| | | — | 1–20 | $\text{Ra} > 10^8$ | $\text{Nu} = 0.057\,\text{Ra}^{1/3}$ | (11-36) |
| Concentric horizontal cylinders | Gas or liquid | — | 1–5000 | $6.3 \times 10^3–10^6$ | $\text{Nu} = 0.11\,\text{Ra}^{0.29}$ | (11-37) |
| | | | 1–5000 | $10^6–10^8$ | $\text{Nu} = 0.40\,\text{Ra}^{0.20}$ | (11-38) |
| Concentric spheres | Gas or liquid | — | 0.7–4000 | $10^2–10^9$ | $\text{Nu} = 0.228\,\text{Ra}^{0.226}$ | (11-39) |

to about $\theta = 20°$ from the vertical by replacing $g$ in the Ra relation by $g \cos\theta$ (Fig. 11-14).

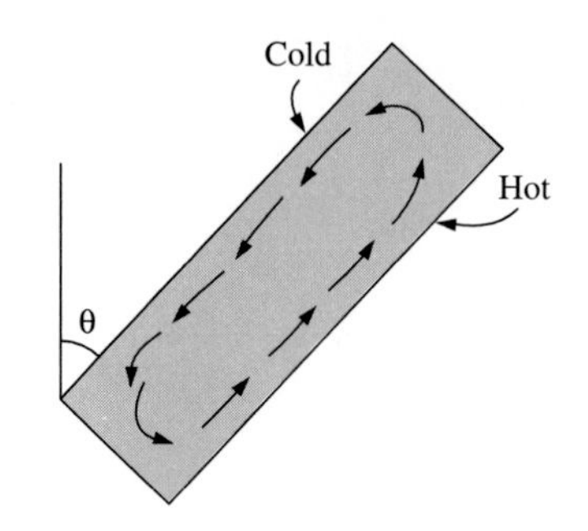


**FIGURE 11-14**

An inclined rectangular enclosure heated from below.

## Effective Thermal Conductivity

You will recall from Chap. 8 that the rate of steady heat conduction across a layer of thickness $\delta$, surface area $A$, and thermal conductivity $k$ is

$$\dot{Q}_{\text{cond}} = kA\frac{T_1 - T_2}{\delta} \tag{11-21}$$

where $T_1$ and $T_2$ are the temperatures on the two sides of the layer. A comparison of this relation with Eq. 11-19 reveals that the convection heat transfer in an enclosure is analogous to heat conduction across the fluid layer in the enclosure provided that the thermal conductivity $k$ is replaced by $k$ Nu. That is, the fluid in an enclosure behaves like a fluid whose thermal conductivity is $k$ Nu as a result of convection currents. Therefore, the quantity $k$ Nu is called the **effective thermal conductivity** of the enclosure. That is,

$$k_{\text{eff}} = k\,\text{Nu} \tag{11-22}$$

Note that for the special case of Nu = 1, the effective thermal conductivity of the enclosure becomes equal to the conductivity of the fluid. This is expected, since this case corresponds to pure conduction (Fig. 11-15).

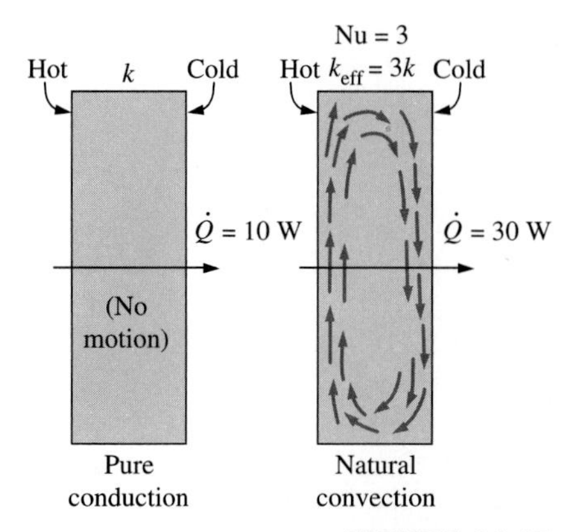


**FIGURE 11-15**

A Nusselt number of 3 for an enclosure indicates that heat transfer through the enclosure by *natural convection* is 3 times that by *pure conduction.*

**EXAMPLE 11-3**

The vertical 0.8-m-high, 2-m-wide double-paned window shown in Fig. 11-16 consists of two sheets of glass separated by a 2-cm air gap at atmospheric pressure. If the glass surface temperatures across the air gap are measured to be 12°C and 2°C, determine the rate of heat transfer through the window.

**Solution** We have a rectangular enclosure filled with air. The properties of air are to be evaluated at the average temperature of

$$T_{\text{ave}} = \frac{T_1 + T_2}{2} = \frac{12 + 2}{2} = 7°\text{C} = 280\text{ K}$$

At this temperature, we read

$$k = 0.0246\text{ W/(m}\cdot°\text{C)}$$
$$\nu = 1.40 \times 10^{-5}\text{ m}^2\text{/s}$$
$$\text{Pr} = 0.717$$
$$\beta = \frac{1}{T_f} = \frac{1}{280\text{ K}} = 0.00357\text{ K}^{-1}$$

The characteristic length in this case is the distance between the two glasses, $\delta = 0.02$ m. Then the Rayleigh number becomes

$$\text{Ra} = \frac{g\beta(T_1 - T_2)\delta^3}{\nu^2}\text{Pr}$$
$$= \frac{(9.8\text{ m/s}^2)(0.00357\text{ K}^{-1})[(12 - 2)\text{ K}](0.02\text{ m})^3}{(1.40 \times 10^{-5}\text{ m}^2\text{/s})^2}(0.717) = 1.024 \times 10^4$$

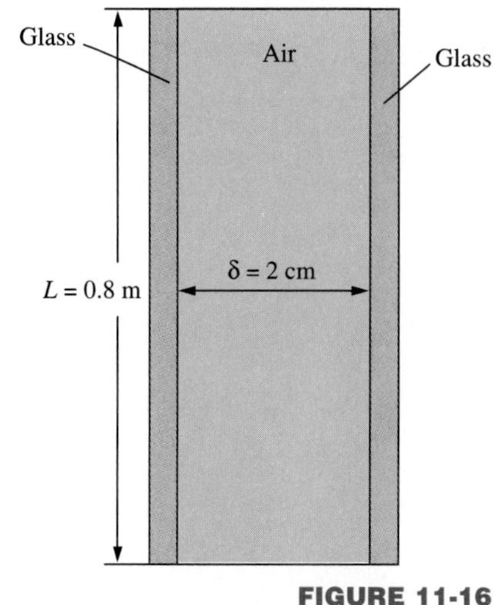


**FIGURE 11-16**

Schematic for Example 11-3.

Then the natural convection Nusselt number in this case can be determined from Eq. 11-24 to be

$$\text{Nu} = 0.197\,\text{Ra}^{1/4}\left(\frac{H}{\delta}\right)^{-1/9} = 0.197(1.024 \times 10^4)^{1/4}\left(\frac{0.8\text{ m}}{0.02\text{ m}}\right)^{-1/9} = 1.32$$

Then $A = H \times L = (0.8\text{ m})(2\text{ m}) = 1.6\text{ m}^2$

and

$$\dot{Q} = k\,\text{Nu}\,A\frac{T_1 - T_2}{\delta}$$

$$= [0.0246\text{ W/(m}\cdot{}^\circ\text{C)}](1.32)(1.6\text{ m}^2)\frac{(12 - 2)^\circ\text{C}}{0.02\text{ m}} = 25.9\text{ W}$$

Therefore, heat will be lost through the window at a rate of 25.9 W.

**Discussion** Recall that a Nusselt number of Nu = 1 for an enclosure corresponds to pure conduction heat transfer through the enclosure. The air in the enclosure in this case remains still, and no natural convection currents occur in the enclosure. The Nusselt number in our case is 1.32, which indicates that heat transfer through the enclosure is 1.32 times that by pure conduction. The increase in heat transfer is due to the natural convection currents that develop in the enclosure.

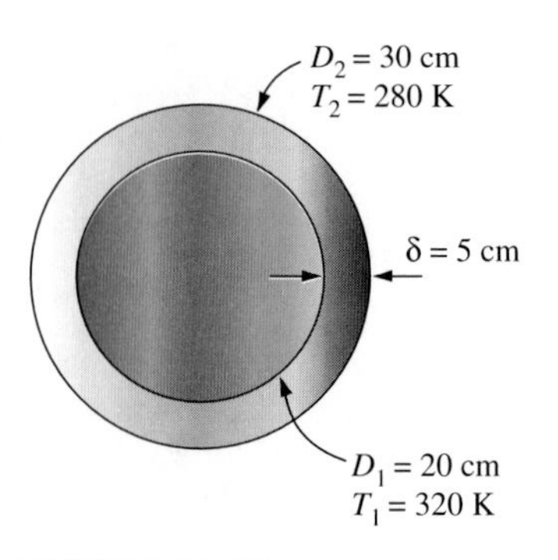


**FIGURE 11-17**
Schematic for Example 11-4.

**EXAMPLE 11-4**

The two concentric spheres of diameters $D_1 = 20$ cm and $D_2 = 30$ cm shown in Fig. 11-17 are separated by air at 1-atm pressure. The surface temperatures of the two spheres enclosing the air are $T_1 = 320$ K and $T_2 = 280$ K, respectively. Determine the rate of heat transfer from the inner sphere to the outer sphere by natural convection.

**Solution** We have a spherical enclosure filled with air. The properties of air are to be evaluated at the average temperature

$$T_{\text{ave}} = \frac{T_1 + T_2}{2} = \frac{320 + 280}{2} = 300\text{ K}$$

At this temperature, we read

$$k = 0.0261\text{ W/(m}\cdot{}^\circ\text{C)}$$

$$\nu = 1.57 \times 10^{-5}\text{ m}^2\text{/s}$$

$$\text{Pr} = 0.712$$

$$\beta = \frac{1}{T_f} = \frac{1}{300\text{ K}} = 0.00333\text{ K}^{-1}$$

The characteristic length in this case is the distance between the two spheres, which is determined to be

$$\delta = \tfrac{1}{2}(D_2 - D_1) = \tfrac{1}{2}(0.3 - 0.2)\text{ m} = 0.05\text{ m}$$

Then the Rayleigh number becomes

$$\begin{aligned}\text{Ra} &= \frac{g\beta(T_1 - T_2)\delta^3}{\nu^2}\text{Pr}\\ &= \frac{(9.8\ \text{m/s}^2)(0.00333\ \text{K}^{-1})[(320 - 280)\ \text{K}](0.05\ \text{m})^3}{(1.57 \times 10^{-5}\ \text{m}^2/\text{s})^2}(0.712)\\ &= 4.713 \times 10^5\end{aligned}$$

Then the natural convection Nusselt number in this case can be determined from Eq. 11–39 to be

$$\text{Nu} = 0.228\ \text{Ra}^{0.226} = 0.228(4.713 \times 10^5)^{0.226} = 4.37$$

That is, the air in the spherical enclosure will act like a stationary fluid whose thermal conductivity is 4.37 times that of air as a result of natural convection currents. Then,

$$A = \pi D_1 D_2 = \pi(0.2\ \text{m})(0.3\ \text{m}) = 0.188\ \text{m}^2$$

and

$$\begin{aligned}\dot{Q} &= k\,\text{Nu}\,A\frac{T_1 - T_2}{\delta}\\ &= [0.0261\ \text{W/(m} \cdot {}^\circ\text{C)}](4.37)(0.188\ \text{m}^2)\frac{(320 - 280)\ \text{K}}{0.05\ \text{m}} = 17.2\ \text{W}\end{aligned}$$

Therefore, heat will be lost from the inner sphere to the outer one at a rate of 17.2 W.

**Discussion** Assuming the surfaces of the spheres to be black (emissivity $\varepsilon = 1$), the rate of heat transfer between the two spheres by radiation is

$$\begin{aligned}\dot{Q}_{\text{rad}} &= \varepsilon A_1 \sigma(T_1^4 - T_2^4)\\ &= (1)\pi(0.2\ \text{m})^2[5.67 \times 10^{-8}\ \text{W/(m}^2 \cdot \text{K}^4)][(320\ \text{K})^4 - (280\ \text{K})^4]\\ &= 30.9\ \text{W}\end{aligned}$$

Thus, the maximum heat transfer by radiation is greater than the heat transfer by natural convection in this case. The emissivity of a real surface is less than 1, and thus the radiation heat transfer for a real enclosure will be less. But radiation can still be significant, and needs to be considered.

**EXAMPLE 11-5**

A solar collector consists of a horizontal aluminum tube having an outer diameter of 2 in enclosed in a concentric thin glass tube of 4-in diameter (Fig. 11-18). Water is heated as it flows through the tube, and the annular space between the aluminum and the glass tubes is filled with air at 1-atm pressure. The pump circulating the water fails during a clear day, and the water temperature in the tube starts rising. The aluminum tube absorbs solar radiation at a rate of 30 Btu/h per foot length, and the temperature of the ambient air outside is 70°F. Disregarding any heat loss by radiation, determine the temperature of the aluminum tube when thermal equilibrium is established (i.e., when the rate of heat loss from the tube equals the amount of solar energy gained by the tube).

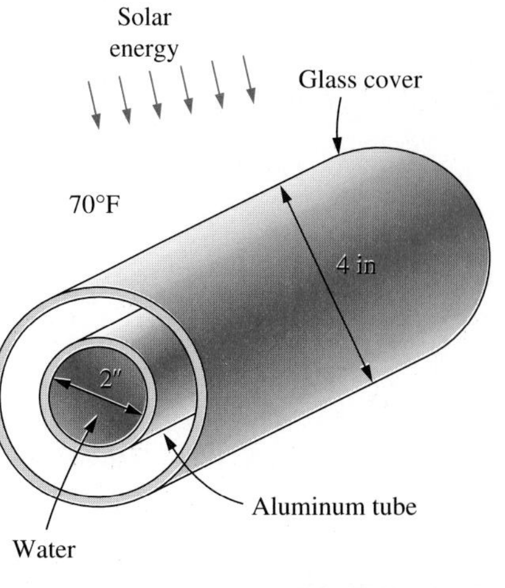


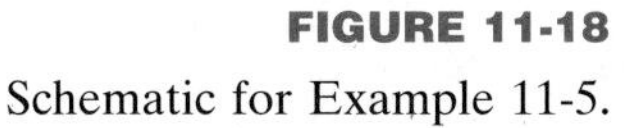
**FIGURE 11-18**
Schematic for Example 11-5.

**Solution** We have a horizontal cylindrical enclosure filled with air at 1-atm pressure. The problem involves heat transfer from the aluminum tube to the glass cover, and from the outer surface of the glass cover to the surrounding ambient air. When thermal equilibrium is established and steady operation is reached, these two heat transfer rates must equal the rate of heat gain. That is,

$$\dot{Q}_{\text{tube-glass}} = \dot{Q}_{\text{glass-ambient}} = \dot{Q}_{\text{solar gain}} = 30 \text{ Btu/h} \quad \text{(per foot of tube)}$$

The properties of air should be evaluated at the average temperature. But we do not know the tube and glass cover temperatures at this point, and thus we cannot evaluate the average temperatures. Therefore, we will use the properties at an anticipated average temperature of 100°F, which are

$$k = 0.0154 \text{ Btu/(h} \cdot \text{ft} \cdot \text{°F)}$$
$$\nu = 0.18 \times 10^{-3} \text{ ft}^2\text{/s}$$
$$\text{Pr} = 0.72$$
$$\beta = \frac{1}{T_f} = \frac{1}{(100 + 460) \text{ R}} = 0.001786 \text{ R}^{-1}$$

The results obtained can then be refined for better accuracy, if necessary, using the evaluated surface temperatures. The heat transfer surface area of the glass cover is

$$A_2 = A_{\text{glass}} = (\pi DL)_{\text{glass}} = \pi\left(\frac{4}{12} \text{ ft}\right)(1 \text{ ft})$$
$$= 1.047 \text{ ft}^2 \quad \text{(per foot length of tube)}$$

To determine the Rayleigh number, we need to know the surface temperature of the glass which is not available. Therefore, it is clear that the solution will require a trial-and-error approach. Assuming the glass cover temperature to be 100°F, the Rayleigh number, the Nusselt number, the convection heat transfer coefficient, and the rate of natural convection heat transfer from the glass cover to the ambient air are determined to be

$$\text{Ra} = \frac{g\beta(T_s - T_\infty)\delta^3}{\nu^2}\text{Pr}$$
$$= \frac{(32.2 \text{ ft/s}^2)(0.001786 \text{ R}^{-1})[(100 - 70) \text{ R}](\frac{4}{12} \text{ ft})^3}{(0.18 \times 10^{-3} \text{ ft}^2\text{/s})^2}(0.72) = 1.420 \times 10^6$$
$$\text{Nu} = \left\{0.6 + \frac{0.387 \text{ Ra}^{1/6}}{[1 + (0.559/\text{Pr})^{9/16}]^{8/27}}\right\}^2$$
$$= \left\{0.6 + \frac{0.387(1.420 \times 10^6)^{1/6}}{[1 + (0.559/0.72)^{9/16}]^{8/27}}\right\}^2 = 16.1$$
$$h = \frac{k}{D}\text{Nu} = \frac{0.0154 \text{ Btu/(ft} \cdot \text{°F)}}{\frac{4}{12} \text{ ft}}(16.1) = 0.743 \text{ Btu/(h} \cdot \text{ft}^2 \cdot \text{°F)}$$
$$\dot{Q}_{\text{glass}} = hA_2(T_2 - T_\infty) = [0.743 \text{ Btu/(h} \cdot \text{ft}^2 \cdot \text{°F)}](1.047 \text{ ft}^2)(100 - 70)\text{°F}$$
$$= 23.3 \text{ Btu/h}$$

which is less than 30 Btu/h. Therefore, the assumed temperature of 100°F for the glass cover is low. Repeating the calculations for a temperature of 110°F gives 33.8 Btu/h, which is high. Then the glass cover temperature corresponding to 30 Btu/h is determined by interpolation to be 107°F.

The temperature of the aluminum tube is determined in a similar manner using the natural convection relations for two horizontal concentric cylinders.

The characteristic length in this case is the distance between the two cylinders, which is determined to be

$$\delta = \tfrac{1}{2}(D_2 - D_1) = \tfrac{1}{2}(4 - 2)\text{ in} = 1\text{ in}$$

Also,

$$A = \frac{\pi L(D_2 - D_1)}{\ln(D_2/D_1)} = \frac{\pi(1\text{ ft})(\frac{4}{12} - \frac{2}{12})\text{ ft}}{\ln(\frac{4}{2})} = 0.755\text{ ft}^2$$

We start the calculations by assuming the tube temperature to be 200°F. This gives

$$\text{Ra} = \frac{g\beta(T_1 - T_2)\delta^3}{\nu^2}\text{Pr}$$

$$= \frac{(32.2\text{ ft/s}^2)(0.001786\text{ R}^{-1})[(200 - 107)\text{ R}](\frac{1}{12}\text{ ft})^3}{(0.18 \times 10^{-3}\text{ ft}^2/\text{s})^2}(0.72) = 6.88 \times 10^4$$

$$\text{Nu} = 0.11\,\text{Ra}^{0.29} = 0.11(6.88 \times 10^4)^{0.29} = 2.78$$

$$\dot{Q}_{\text{tube}} = k\,\text{Nu}\,A\frac{T_1 - T_2}{\delta}$$

$$= [0.0154\text{ Btu/(h}\cdot\text{ft}\cdot\text{°F)}](2.78)(0.755\text{ ft}^2)\frac{(200 - 107)\text{ R}}{\frac{1}{12}\text{ ft}} = 36.1\text{ Btu/h}$$

which is more than 30 Btu/h. Therefore, the assumed temperature of 200°F for the tube is high. Repeating the calculations for a temperature of 180°F gives 26.4 Btu/h, which is low. Then the tube temperature corresponding to 30 Btu/h is determined by interpolation to be 188°F. Therefore, the tube will reach an equilibrium temperature of 188°F when the pump fails.

This result above is obtained by using air properties at 100°F. It appears that this result can be improved by repeating the calculations above using air properties at the average temperature of 88.5°F for heat transfer from the glass cover to the ambient air, and at 147.5°F for heat transfer from the tube to the glass cover. Also, we have not considered heat loss by radiation in the calculations, and thus the tube temperature determined above is probably too high. This problem is considered again in the next chapter by accounting for the effect of radiation heat transfer.

## 11-4 NATURAL CONVECTION FROM FINNED SURFACES

Finned surfaces of various shapes, called *heat sinks,* are frequently used in the cooling of electronic devices. Energy dissipated by these devices is transferred to the heat sinks by *conduction,* and from the heat sinks to the ambient air by *natural* or *forced convection,* depending on the power dissipation requirements. Natural convection is the preferred mode of heat transfer, since it involves *no moving parts,* like the electronic components themselves. However, in the natural convection mode, the components are more likely to run at a *higher temperature* and thus undermine reliability. A properly selected heat sink may considerably *lower* the operation temperature of the components, and thus reduce the risk of failure.

A question that often arises in the selection of a heat sink is whether to select one with *closely packed* fins or *widely spaced* fins for a given

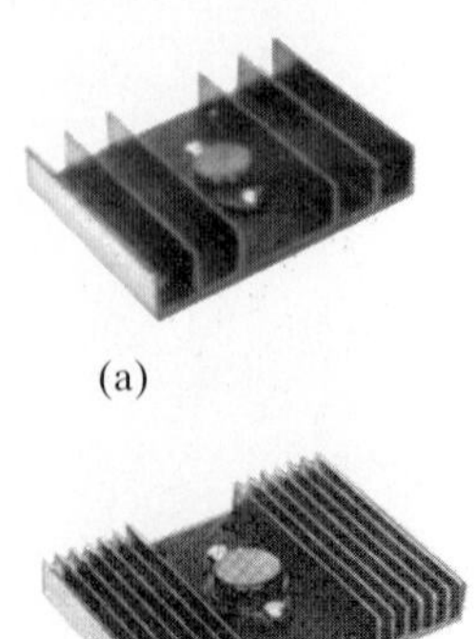
(a)

(b)

**FIGURE 11-19**
Heat sinks with (*a*) widely spaced and (*b*) closely packed fins (courtesy of Vemaline Products).

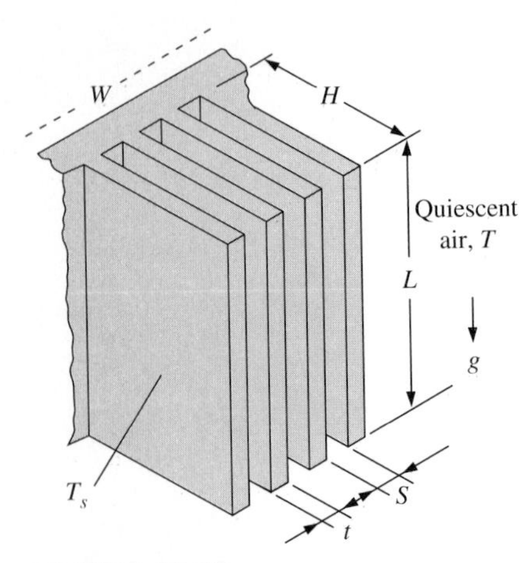


**FIGURE 11-20**
Various dimensions of a finned surface oriented vertically.

base area (Fig. 11-19). A heat sink with closely packed fins will have greater surface area for heat transfer, but a smaller heat transfer coefficient because of the extra resistance the additional fins will introduce to fluid flow through the interfin passages. A heat sink with widely spaced fins, on the other hand, will have a higher heat transfer coefficient but a smaller surface area. Therefore, there must be an *optimum spacing* that maximizes the natural convection heat transfer from the heat sink for a given base area $WL$, where $W$ and $L$ are the width and height of the base of the heat sink, respectively, as shown in Fig. 11-20. When the fins are essentially isothermal and the fin thickness $t$ is small relative to the fin spacing $S$, the optimum fin spacing for a vertical heat sink is determined by Rohsenow and Bar-Cohen to be

$$S_{\text{opt}} = 2.714 \frac{L}{\text{Ra}^{1/4}} \tag{11-40}$$

where the fin length $L$ in the vertical direction is taken to be the characteristic length in the evaluation of the Rayleigh number. The heat transfer coefficient for the optimum spacing case was determined to be

$$h = 1.31 \frac{k}{S_{\text{opt}}} \tag{11-41}$$

Then the rate of heat transfer by natural convection form the fins can be determined from

$$\dot{Q} = h(2nLH)(T_s - T_\infty) \tag{11-42}$$

where $n = W/(S + t) \approx W/S$ is the number of fins on the heat sink and $T_s$ is the surface temperature of the fins.

As we mentioned earlier, the magnitude of the natural convection heat transfer is directly related to the *mass flow rate* of the fluid, which is established by the dynamic balance of two opposing effects: *buoyancy* and *friction.*

The fins of a heat sink introduce both effects: *inducing extra buoyancy* as a result of the elevated temperature of the fin surfaces, and *slowing down the fluid* by acting as an added obstacle on the flow path. As a result, increasing the number of fins on a heat sink can either enhance or reduce natural convection, depending on which effect is dominant. The buoyancy-driven air flow rate is established at the point where these two effects balance each other. The friction force increases as more and more solid surfaces are introduced, seriously disrupting fluid flow and heat transfer. Under some conditions, the increase in friction may more than offset the increase in buoyancy. This in turn will tend to reduce the flow rate and thus the heat transfer. For that reason, heat sinks with closely spaced fins are not suitable for natural convection cooling.

When the heat sink involves closely spaced fins, the narrow channels formed tend to block or "suffocate" the fluid, especially when the heat sink is long. As a result, the blocking action produced overwhelms the extra buoyancy and downgrades the heat transfer characteristics of the heat sink. Then, at a fixed power setting, the heat sink runs at a higher temperature relative to the no-shroud case. When the heat sink involves widely spaced fins, the shroud does not introduce a significant increase in resistance to flow, and the buoyancy effects dominate. As a result, heat transfer by natural convection may improve, and at a fixed power level the heat sink may run at a lower temperature.

When extended surfaces such as fins are used to enhance natural convection heat transfer between a solid and a fluid, the flow rate of the fluid in the vicinity of the solid adjusts itself to incorporate the changes in buoyancy and friction. It is obvious that this enhancement technique will work to advantage only when the increase in buoyancy is greater than the additional friction introduced. One does not need to be concerned with pressure drop or pumping power when studying natural convection, since no pumps or blowers are used in this case. Therefore, an enhancement technique in natural convection is evaluated on heat transfer performance alone.

The failure rate of an electronic component increases almost exponentially with operating temperature. The cooler the electronic device operates, the more reliable it is. A rule of thumb is that semiconductor failure rate is halved for each 10°C reduction in junction operating temperature. The desire to lower the operating temperature without having to resort to forced convection has motivated researchers to investigate enhancement techniques for natural convection. Sparrow and Prakash have demonstrated that, under certian conditions, the use of discrete plates in lieu of continuous plates of the same surface area increases heat transfer considerably. In other experimental work, using transistors as the heat source, Çengel and Zing have demonstrated that ,,,,,,temperature recorded on the transistor case dropped by as much as ,,,,,,,,temperature recorded on the transistor case dropped by as much as 30°C when a shroud was used as opposed to the corresponding no-shroud case.

**EXAMPLE 11-6**

A 12-cm-wide and 18-cm-high vertical hot surface in 25°C air is to be cooled by a heat sink with equally spaced fins of rectangular profile. The fins are 0.1 cm thick, 18 cm long in the vertical direction, and have a height of 2.4 cm from the base. Determine the optimum fin spacing, and the rate of heat transfer by natural convection from the heat sink if the base temperature is 80°C.

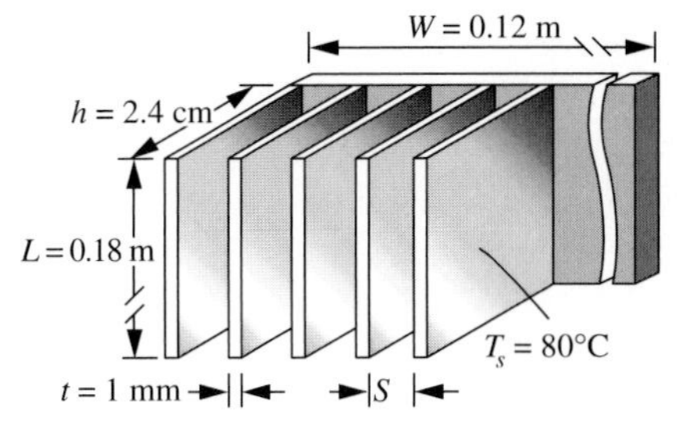


**FIGURE 11-21**
Schematic for Example 11-6.

**Solution** We assume the thickness $t$ of the fins to be very small relative to the fin spacing $S$, and thus Eqs. 11-40 and 11-41 for optimum fin spacing to be applicable. The properties of air are to be evaluated at the film temperature of

$$T_f = \frac{T_s + T_\infty}{2} = \frac{80 + 25}{2} = 52.5°\text{C} = 325.5\ \text{K}$$

At this temperature, we read

$$k = 0.0279\ \text{W/(m} \cdot °\text{C)}$$
$$\nu = 1.82 \times 10^{-5}\ \text{m}^2/\text{s}$$
$$\text{Pr} = 0.709$$
$$\beta = \frac{1}{T_f} = \frac{1}{325.5\ \text{K}} = 0.003\,072\ \text{K}^{-1}$$

The characteristic length in this case is the length of the fins in the vertical direction, which is given to be $L = 0.18$ m. Then the Rayleigh number becomes

$$\text{Ra} = \frac{g\beta(T_s - T_\infty)\delta^3}{\nu^2}\text{Pr}$$

$$= \frac{(9.8\ \text{m/s}^2)(0.003072\ \text{K}^{-1})(80 - 25)\ \text{K}(0.18\ \text{m})^3}{(1.82 \times 10^{-5}\ \text{m}^2/\text{s})^2}(0.709) = 2.067 \times 10^7$$

The optimum fin spacing is determined from Eq. 11-40 to be

$$S_{\text{opt}} = 2.714\frac{L}{\text{Ra}^{1/4}} = 2.714\frac{0.18\ \text{m}}{(2.067 \times 10^7)^{1/4}} = 0.0072\ \text{m} = 7.2\ \text{mm}$$

which is about 7 times the thickness of the fins. Therefore, the assumption of negligible fin thickness in this case is acceptable for practical purposes. The number of fins and the heat transfer coefficient for this optimum fin spacing case are

$$n = \frac{W}{S + t} = \frac{0.12\ \text{m}}{(0.0072 + 0.001)\ \text{m}} \approx 15\ \text{fins}$$

$$h = 1.31\frac{k}{S_{\text{opt}}} = 1.31\frac{0.0279\ \text{W/(m}\cdot{}^\circ\text{C)}}{0.0072\ \text{m}} = 5.08\ \text{W/(m}^2\cdot{}^\circ\text{C)}$$

Then the rate of natural convection heat transfer becomes

$$\dot{Q} = h(2nLH)(T_s - T_\infty)$$

$$= [5.08\ \text{W/(m}^2\cdot{}^\circ\text{C)}][2 \times 15 \times (0.18\ \text{m})(0.12\ \text{m})](80 - 25)^\circ\text{C} = 181\ \text{W}$$

Therefore, the heat sink can dissipate heat by natural convection at a rate of 181 W.

## 11-5 ■ COMBINED NATURAL AND FORCED CONVECTION

The presence of a temperature gradient in a fluid in a gravity field always gives rise to natural convection currents, and thus heat transfer by natural convection. Therefore, forced convection is always accompanied by natural convection.

We mentioned earlier that the convection heat transfer coefficient, natural or forced, is a strong function of the fluid velocity. Heat transfer coefficients encountered in forced convection are typically much higher than those encountered in natural convection because of the higher fluid velocities associated with forced convection. As a result, we tend to ignore natural convection in heat transfer analyses that involve forced convection, although we recognize that natural convection always accompanies forced convection. The error involved in ignoring natural convection is negligible at high velocities, but may be considerable at low velocities associated with forced convection. Therefore, it is desirable to have a criteria to assess the relative magnitude of natural convection in the presence of forced convection.

For a given fluid, it is observed that the parameter $Gr/Re^2$ represents the importance of natural convection relative to forced convection. This is not surprising, since the convection heat transfer coefficient is a strong function of the Reynolds number Re in forced convection and the Grashof number Gr in natural convection.

A plot of the nondimensionalized heat transfer coefficient for combined natural and forced convection on a vertical plate is given in Fig. 11-22 for different fluids. We note from this figure that natural convection is negligible when $Gr/Re^2 < 0.1$, forced convection is negligible when $Gr/Re^2 > 10$, and neither is negligible when $0.1 < Gr/Re^2 < 10$. Therefore, both natural and forced convection must be considered in heat transfer calculations when the Gr and $Re^2$ are of the same order of magnitude (one is within a factor of 10 times the other). Note that forced convection is small relative to natural convection only in the rare case of extremely low forced flow velocities.

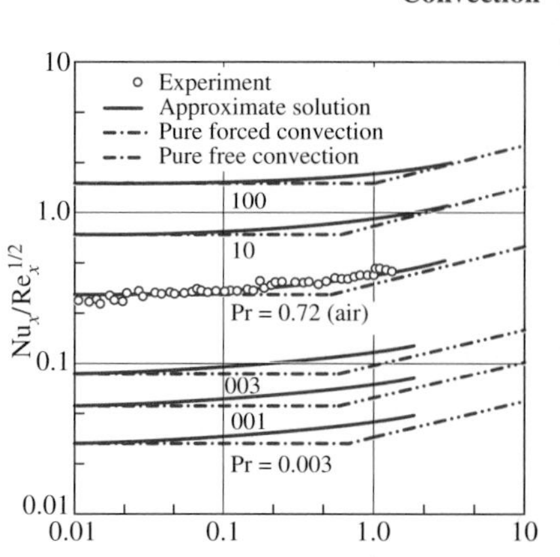

**FIGURE 11-22**

Variation of the local Nusselt number $Nu_x$ for combined natural and forced convection from a hot isothermal vertical plate (from Lloyd and Sparrow).

Natural convection may *help* or *hurt* forced convection heat transfer, depending on the relative directions of *buoyancy induced* and the *forced convection* motions (Fig. 11-23):

1 In *assisting flow,* the buoyant motion is in the *same* direction as the forced motion. Therefore, natural convection assists forced convection, and *enhances* heat transfer. An example is upward forced flow over a hot surface.

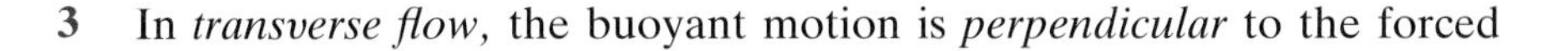

2 In *opposing flow,* the buoyant motion is in the *opposite* direction to the forced motion. Therefore, natural convection resists forced convection, and *decreases* heat transfer. An example is upward forced flow over a cold surface.

3 In *transverse flow,* the buoyant motion is *perpendicular* to the forced

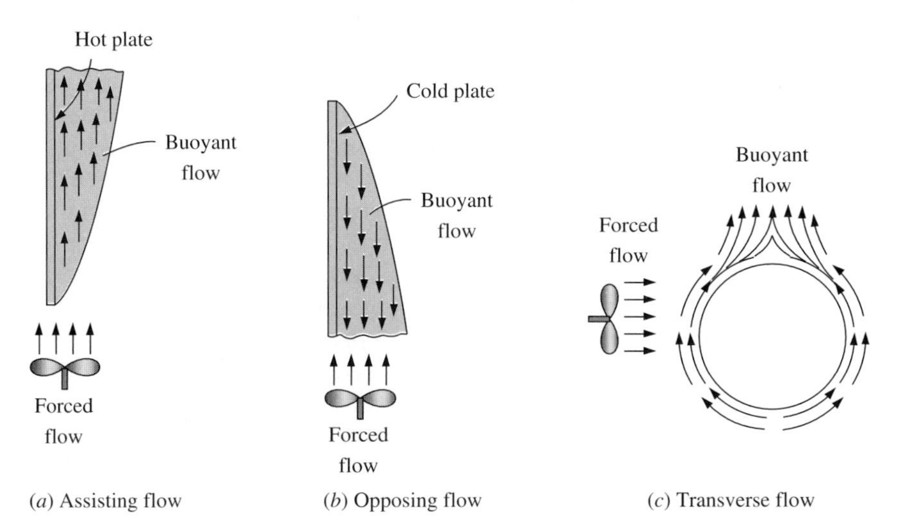


**FIGURE 11-23**

Natural convection can *enhance* or *inhibit* heat transfer, depending on the relative directions of *buoyancy-induced motion* and the *forced convection motion.*

motion. Transverse flow enhances fluid mixing, and thus *enhances* heat transfer. An example is horizontal forced flow over a hot or cold cylinder or sphere.

When determining heat transfer under combined natural and forced convection conditions, it is tempting to add the contributions of natural and forced convection in assisting flows, and to subtract them in opposing flows. However, the evidence indicates differently. A review of experimental data suggests a correlation of the form

$$\text{Nu}_{\text{combined}} = (\text{Nu}^n_{\text{forced}} \pm \text{Nu}^n_{\text{natural}})^{1/n} \tag{11-43}$$

where $\text{Nu}_{\text{forced}}$ and $\text{Nu}_{\text{natural}}$ are determined from the correlations for *pure forced* and *pure natural convection,* respectively. The plus sign is for *assisting* and *transverse* flows, and the minus sign is for *opposing* flows. The value of the exponent $n$ varies between 3 and 4, depending on the geometry involved. It is observed that $n = 3$ correlates experimental data for vertical surfaces well. Larger values of $n$ are better suited for horizontal surfaces.

A question that frequently arises in the cooling of heat-generating equipment such as electronic components is whether to use a fan (or a pump if the cooling medium is a liquid). That is, whether to utilize *natural* or *forced* convection in the cooling of the equipment. The answer depends on the maximum allowable operating temperature. Recall that the convection heat transfer rate from a surface at temperature $T_s$ in a medium at $T_\infty$ is given by

$$\dot{Q}_{\text{conv}} = hA(T_s - T_\infty)$$

where $h$ is the convection heat transfer coefficient and $A$ is the surface area. Note that for a fixed value of power dissipation and surface area, $h$ and $T_s$ are *inversely proportional.* Therefore, the device will operate at a *higher* temperature when $h$ is low (typical of natural convection), and at a *lower* temperature when $h$ is high (typical of forced convection).

Natural convection is the preferred mode of heat transfer, since no blowers or pumps are needed and thus all the problems associated with these, such as noise, vibration, power consumption, and malfunctioning, are avoided. Natural convection is adequate for cooling *low-power-output* devices, especially when they are attached to extended surfaces such as heat sinks. For *high-power-output* devices, however, we have no choice but use a blower or a pump to keep the operating temperature below the maximum allowable level. For *very high-power-output* devices, even forced convection may not be sufficient to keep the surface temperature at the desirable levels. In such cases, we may have to use *boiling* and *condensation,* to take advantage of the very high heat transfer coefficients associated with phase change processes.

In this chapter, we have considered *natural convection* heat transfer where any fluid motion occurs by natural means such as buoyancy. The fluid velocities associated with natural convection are low. Therefore, the heat transfer coefficients encountered in natural convection are usually much lower than those encountered in forced convection.

The upward force exerted by a fluid on a body completely or partially immersed in it is called the *buoyancy force,* whose magnitude is equal to the weight of the fluid displaced by the body. The *volume expansion coefficient* $\beta$ of a substance represents the variation of the density of that substance with temperature at constant pressure, and is defined as

$$\beta = -\frac{1}{\rho}\left(\frac{\partial \rho}{\partial T}\right)_P \qquad (1/\text{K})$$

For an ideal gas, it reduces to

$$\beta_{\text{ideal gas}} = \frac{1}{T} \qquad (1/K)$$

where $T$ is the absolute temperature. The instrument used in natural convection experiments most often is the *Mach–Zehnder interferometer,* which gives a plot of isotherms in the fluid in the vicinity of a surface.

The flow regime in natural convection is governed by a dimensionless number called the *Grashof number,* which represents the ratio of the buoyancy force to the viscous force acting on the fluid, and is expressed as

$$\text{Gr} = \frac{g\beta(T_s - T_\infty)\delta^3}{\nu^2}$$

where $g$ = gravitational acceleration, $\text{m/s}^2$
$\beta$ = coefficient of volume expansion, 1/K ($\beta = 1/T$ for ideal gases)
$T_s$ = temperature of the surface, °C
$T_\infty$ = temperature of the fluid sufficiently far from the surface, °C
$\delta$ = characteristic length of the geometry, m
$\nu$ = kinematic viscosity of the fluid, $\text{m}^2/\text{s}$

The Grashof number provides the main criteria in determining whether the fluid flow is laminar or turbulent in natural convection. The heat transfer rate in natural convection from a solid surface to the surrounding fluid is expressed by Newton's law of cooling as

$$\dot{Q}_{\text{conv}} = hA(T_s - T_\infty) \qquad (\text{W})$$

where $A$ is the heat transfer surface area and $h$ is the average heat transfer coefficient on the surface.

Most heat transfer relations in natural convection are based on experimental studies, and the simple empirical correlations for the average *Nusselt number* Nu in natural convection are of the form

$$\text{Nu} = \frac{h\delta}{k} = C(\text{Gr Pr})^n = C\,\text{Ra}^n$$

where Ra is the *Rayleigh number,* which is the product of the Grashof and Prandtl numbers:

$$\text{Ra} = \text{Gr Pr} = \frac{g\beta(T_s - T_\infty)\delta^3}{\nu^2}\text{Pr}$$

The values of the constants $C$ and $n$ depend on the geometry of the surface and the flow regime, which is characterized by the range of the Rayleigh number. Simple relations for the average Nusselt number for various geometries are given in Table 11-1 together with a sketch of the geometry. All fluid properties are to be evaluated at the film temperature $T_f = \frac{1}{2}(T_s + T_\infty)$.

Simple empirical correlations for the Nusselt number for various enclosures are given in Table 11-2. Once the Nusselt number is available, the rate of heat transfer through the enclosure can be determined from

$$\dot{Q} = hA(T_1 - T_2) = k\,\text{Nu}\,A\frac{T_1 - T_2}{\delta}$$

where
$$A = \begin{cases} HL & \text{rectangular enclosures} \\ \dfrac{\pi L(D_2 - D_1)}{\ln(D_2/D_1)} & \text{concentric cylinders} \\ \pi D_1 D_2 & \text{concentric spheres} \end{cases}$$

The quantity $k\,\text{Nu}$ is called the *effective thermal conductivity* of the enclosure, since a fluid in an enclosure behaves like a quiescent fluid whose thermal conductivity is $k\,\text{Nu}$ as a result of convection currents.

For a given fluid, the parameter $\text{Gr}/\text{Re}^2$ represents the importance of natural convection relative to the forced convection. Natural convection is negligible when $\text{Gr}/\text{Re}^2 < 0.1$, forced convection is negligible when $\text{Gr}/\text{Re}^2 > 10$, and neither is negligible when $0.1 < \text{Gr}/\text{Re}^2 < 10$.

## REFERENCES AND SUGGESTED READING

**1** A. Bar-Cohen and W. M. Rohsenow, "Thermally Optimum Spacing of Vertical Natural Convection Cooled Parallel Plates," *Journal of Heat Transfer,* Vol. 106, pp. 116–123, 1984.

**2** A. Bar-Cohen, "Fin Thickness for an Optimized Natural Convection Array of Rectangular Fins," *Journal of Heat Transfer,* Vol. 101, pp. 564–566, 1979.

**3** Y. Bayazitoglu and M. N. Özışık, *Elements of Heat Transfer,* McGraw-Hill, New York, 1988.

**4** Y. A. Cengel and P. T. L. Zing, "Enhancement of Natural Convection Heat Transfer from Heat Sinks by Shrouding", *Proceedings of ASME/JSME Thermal Engineering Conference,* Honolulu, Hawaii, March 22–27, Vol. 3, pp. 451–457, 1987.

**5** S. W. Churchill, "Combined Free and Forced Convection Around Immersed bodies," *Heat Exchanger Design Handbook,* Section 2.5.9, Hemisphere Publishing, New York, 1986.

**6** S. W. Churchill, "A Comprehensive Correlating Equation for Laminar Assisting Forced and Free Convection, "*AIChE Journal,* Vol. 23, pp. 10–16, 1977.

**7** E. R. G. Eckerd and E. Soehngen, "Interferometric Studies on the Stability and Transition to Turbulence of a Free Convection Boundary Layer," *Proceedings of General Discussion, Heat Transfer ASME–IME,* London, 1951.

**8** E. R. G. Eckerd and E. Soehngen, "Studies on Heat Transfer in Laminar Free Convection with Zehnder–Mach Interferometer," *USAF Technical Report* 5747, December 1948.

**9** J. P. Holman, *Heat Transfer,* 7th ed., McGraw-Hill, New York, 1990.

**10** F. P. Incropera and D. P. DeWitt, *Introduction to Heat Transfer,* 2nd ed., Wiley, New York, 1990.

**11** F. Kreith and M. S. Bohn, *Principles of Heat Transfer,* 5th ed., West, St. Paul, MN, 1993.

**12** J. R. Lloyd and E. M. Sparrow, "Combined Forced and Free Convection Flow on Vertical Surfaces," *International Journal of Heat and Mass Transfer,* Vol. 13, p. 434, 1970.

**13** M. N. Özışık, *Heat Transfer—A Basic Approach,* McGraw-Hill, New York, 1985.

**14** E. M. Sparrow and C. Prakash, "Enhancement of Natural Convection Heat Transfer by a Staggered Array of Vertical Plates," *Journal of Heat Transfer,* Vol. 102, pp. 215–220, 1980.

**15** E. M. Sparrow and S. B. Vemuri., "Natural Convection/Radiation Heat Transfer From Highly Populated Pin Fin Arrays," *Journal of Heat Transfer,* Vol. 107, pp. 190–197, 1985.

**16** L. C. Thomas, *Heat Transfer,* Prentice-Hall, Englewood Cliffs, NJ, 1992.

**17** F. M. White, *Heat and Mass Transfer,* Addison-Wesley, Reading, MA, 1988.

## PROBLEMS*

### Physical Mechanism of Natural Convection

**11-1C** What is natural convection? How does it differ from forced convection? What force causes natural convection currents?

**11-2C** In which mode of heat transfer the convection heat transfer coefficient is usually higher, natural convection or forced convection? Why?

**11-3C** Consider a hot boiled egg in a spacecraft that is filled with air at atmospheric pressure and temperature at all times. Will the egg cool faster or slower when the spacecraft is in space instead of on the ground? Explain.

**11-4C** What is buoyancy force? Compare the relative magnitudes of the buoyancy force acting on a body immersed in the following mediums: (*a*) air, (*b*) water, (*c*) mercury, and (*d*) an evacuated chamber.

**11-5C** When will the hull of a ship sink in water deeper: when the ship is sailing in fresh water or in sea water? Why?

**11-6C** A person weighs himself on a waterproof spring scale placed at the bottom of a 1-m-deep swimming pool. Will the person weigh more or less in water? Why?

**11-7C** Consider two fluids, one with a large coefficient of volume expansion and the other with a small one. In what fluid will a hot surface initiate stronger natural convection currents? Why? Assumc the viscosity of the fluids to be the same.

**11-8C** Consider a fluid whose volume does not change with temperature at constant pressure. What can you say abut natural convection heat transfer in this medium?

**11-9C** What do the lines on an interferometer photograph represent? What do closely packed lines on the same photograph represent?

**11-10C** Physically, what does the Grashof number represent? How does the Grashof number differ from the Reynolds number?

**11-11** Show that the volume expsnaion coefficient of an ideal gas is $\beta = 1/T$ where $T$ is the absolute temperature.

### Natural Convection over Surfaces

**11-12C** How does Rayleigh number differ from Grashof number?

**11-13C** Under what conditions can the outer surface of a vertical cylinder be treated as a vertical plate in natural convection calculations?

**11-14C** Will a hot horizontal plate whose back side is insulated cool faster or slower when its hot surface is facing down instead of up?

* Students are encouraged to answer *all* the concept "C" questions.

**11-15C** Consider laminar natural convection from a vertical hot plate. Will the heat flux be higher at the top or at the bottom of the plate? Why?

**11-16** An 8-m-long section of a 6-cm-diameter horizontal hot water pipe passes through a large room whose temperture is 22°C. If the temperature and the emissivity of the outer surface of the pipe are 65°C and 0.8, respectively, determine the rate of heat loss from the pipe by (*a*) natural convection and (*b*) radiation.

**11-17** Consider a wall-mounted power transistor that dissipates 0.18 W of power in an environment at 35°C. The transistor is 0.45 cm long, and has a diameter of 0.4 cm. The emissivity of the outer surface of the transistor is 0.1, and the average temperature of the surrounding surfaces is 25°C. Disregarding any heat transfer from the base surface, determine the surface temperature of the transistor. *Answer:* 60°C

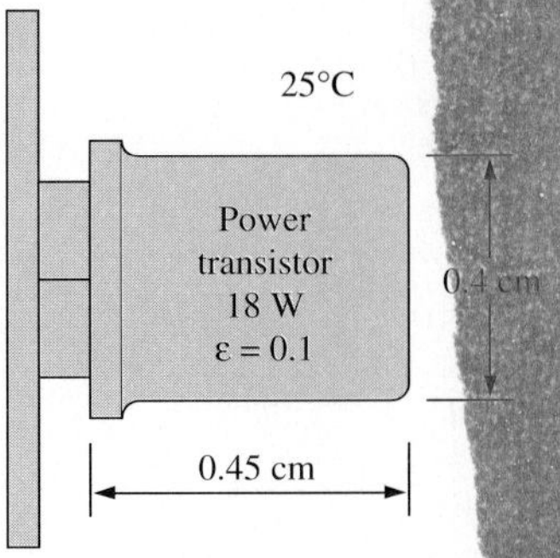

**FIGURE P11-17**

**11-18** Consider a 0.8 m × 0.8 m thin square plate in a room at 25°C. One side of the plate is maintained at a temperature of 60°C, while the other side is insulated. Determine the rate of heat transfer from the plane by natural convection if the plate is (*a*) vertical, (*b*) horizontal with hot surface facing up, an (*c*) horizontal with hot surface facing down.

**11-18E** Consider a 2 ft × 2 ft thin square plate in a room at 75°F. One side of the plate is maintained at a temperature of 130°F, while the other side is insulated. Determine the rate of heat transfer from the plate by natural convection if the plate is (*a*) vertical, (*b*) horizontal with hot surface facing up, and (*c*) horizontal with hot surface facing down.

**11-19** A 500-W cylindrical resistance heater is 1 m long and 0.5 cm in diameter. The resistance wire is placed horizontally in a fluid at 20°C. Determine the outer surface temperature of the resistance wire in steady operation if the fluid is (*a*) air and (*b*) water. Ignore any heat transfer by radiation. For water take $\beta = 0.000365\ \text{K}^{-1}$.

**11-20** Water is boiling in a 12-cm-deep pan that has an outer diameter of 25 cm placed on top of a stove. The ambient air and the surrounding surfaces are at a temperature of 25°C, and the emissivity of the outer surface of the pan is 0.95. Assuming the entire pan to be at an average temperature of 98°C, determine the rate of heat loss from the cylindrical side surface of the pan to the surroundings by (*a*) natural convection and (*b*) radiation. (*c*) If water is boiling at a rate of 2 kg/h at 100°C, determine the ratio of the heat lost from the side surfaces of the pan to that by the evaporation of water. The heat of vaporization of water at 100°C is 2257 kJ/kg. *Answers:* 50 W, 56.1 W, 0.085

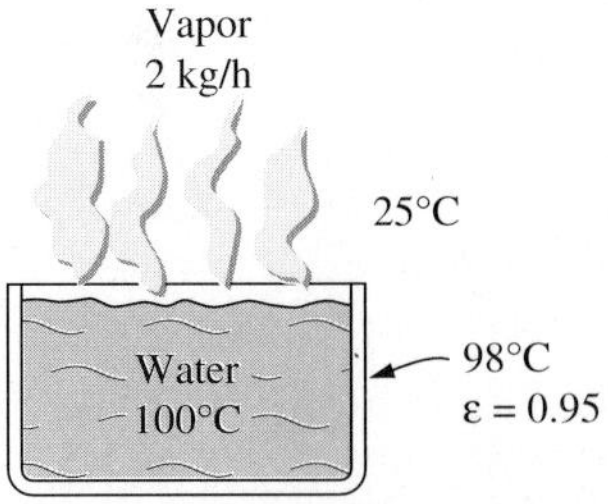

**FIGURE P11-20**

**11-21** Repeat Prob. 11-20 for a pan whose outer surface is polished and has an emissivity of 0.1.

**11-22** In a plant that manufactures canned aerosol paints, the cans are temperature-tested in water baths at 55°C before they are shipped to ensure that they will withstand temperatures up to 55°C during transportation and shelving. The cans, moving on a conveyor, enter the open hot

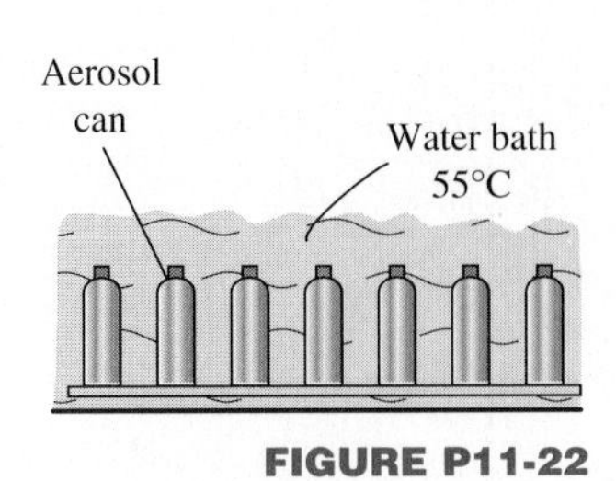


**FIGURE P11-22**

water bath, which is 0.5 m deep, 1 m wide, and 3.5 m long, and move slowly in the hot water towards the other end. Some of the cans fail the test, and explode in the water bath. The water container is made of sheet metal, and the entire container is at about the same temperature as the hot water. The emissivity of the outer surface of the container is 0.7. If the temperature of the surrounding air and surfaces is 20°C, determine the rate of heat loss from the four side surfaces of the container (disregard the top surface which is open).

The water is heated electrically by resistance heaters, and the cost of electricity is \$0.085/kWh. If the plant operates 24 h a day 365 days a year and thus 8760 h a year, determine the annual cost of the heat losses from the container for this facility.

**11-22E** In a plant that manufactures canned aerosol paints, the cans are temperature-tested in water baths at 130°F before they are shipped to ensure that the cans will withstand temperatures up to 130°F during transportation and shelving. The cans moving on a conveyor enter the open hot water bath, which is 1.5 ft deep, 3 ft wide, and 10 ft long, and move slowly in the hot water towards the other end. Some of the cans fail the test and explode in the water bath. The water container is made of sheet metal, and the entire container is at about the same temperature as the hot water. The emissivity of the outer surface of the container is 0.7. If the temperature of the surrounding air and surfaces is 70°F, determine the rate of heat loss from the four side surfaces of the container (disregard the top surface which is open).

The water is heated electrically by resistance heaters, and the cost of electricity is \$0.085/kWh. If the plant operates 24 h a day 365 days a year and thus 8760 h a year, determine the annual cost of the heat losses from the container for this facility.

**11-23** Reconsider Prob. 11-22. In order to reduce the heating cost of the hot water, it is proposed to insulate the side and bottom surfaces of the container with 5-cm-thick fiberglass insulation [$k = 0.035$ W/(m · °C)], and to wrap the insulation with aluminum foil ($\varepsilon = 0.1$) in order to minimize the heat loss by radiation. An estimate is obtained from a local insulation contractor who proposes to do the insulation job for \$350, including materials and labor. Would you support this proposal? How long will it take for the insulation to pay for itself from the energy it saves?

**11-23E** Reconsider Prob. 11-22E. In order to reduce the heating cost of the hot water, it is proposed to insulate the side and bottom surfaces of the container with 2-in-thick fiberglass insulation [$k = 0.020$ Btu/(h · ft · °F)], and to wrap the insulation with aluminum foil ($\varepsilon = 0.1$) in order to minimize the heat losses by radiation. An estimate is obtained from a local insulation contractor who proposes to do the insulation job for \$350, including materials and labor. Would you support this proposal? How long will it take for the insulation to pay for itself from the energy it saves?

**11-24** Consider a 15 cm × 20 cm printed circuit board (PCB) that has electronic components on one side. The board is placed in a room at 20°C. The heat loss from the back surface of the board is negligible. If the circuit board is dissipating 8 W of power in steady operation, determine the average temperature of the hot surface of the board, assuming the board is (*a*) vertical, (*b*) horizontal with hot surface facing up, and (*c*) horizontal with hot surface facing down. Take the emissivity of the surface of the board to be 0.8, and assume the surrounding surfaces to be at the same temperature as the air in the room.
*Answers:* (*a*) 46°C, (*b*) 42°C, (*c*) 50°C

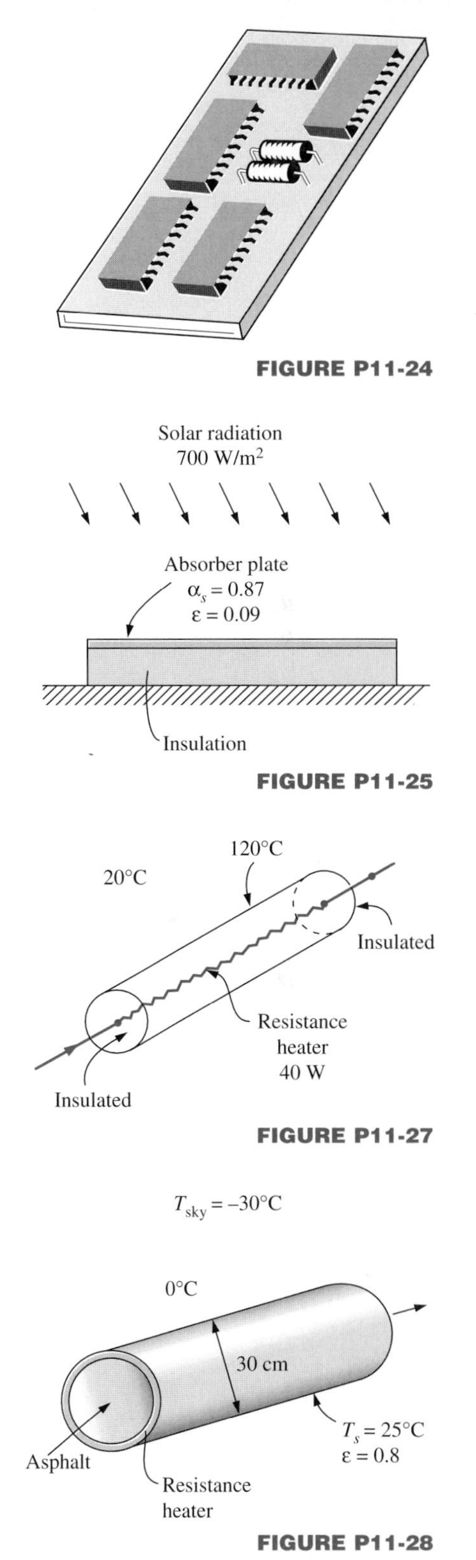


FIGURE P11-24

FIGURE P11-25

FIGURE P11-27

FIGURE P11-28

**11-25** A manufacturer makes absorber plates that are 1.2 m × 0.8 m in size for use in solar collectors. The back side of the plate is heavily insulated, while its front surface is coated with black chrome, which has an absorptivity of 0.87 for solar radiation and an emissivity of 0.09. Consider such a plate placed horizontally outdoors in calm air at 25°C. Solar radiation is incident on the plate at a rate of 700 W/m$^2$. Taking the effective sky temperature to be 10°C, determine the equilibrium temperature of the absorber plate. What would your answer be if the absorber plate is made of ordinary aluminum plate which has a solar absorptivity of 0.28 and an emissivity of 0.07?

**11-26** Repeat Prob. 11-25 for an aluminum plate painted flat black (solar absorptivity 0.98 and emissivity 0.98) and also for a plate painted white (solar absorptivity 0.26 and emissivity 0.90).

**11-27** The following experiment is conducted to determine the natural convection heat transfer coefficient for a horizontal cylinder that is 50 cm long and 2 cm in diameter. A 50-cm-long resistance heater is placed along the center line of the cylinder, and the surfaces of the cylinder are polished to minimize the radiation effect. The two circular side surfaces of the cylinder are well insulated. The resistance heater is turned on, and the power dissipation is maintained constant at 40 W. If the average surface temperature of the cylinder is measured to be 120°C in the 20°C room air when steady operation is reached, determine the natural convection heat transfer coefficient. If the emissivity of the outer surface of the cylinder is 0.1 and a 5 percent error is acceptable, do you think we need to do any correction for the radiation effect? Assume the surrounding surfaces to be at 20°C also.

**11-28** Thick fluids such as asphalt and waxes and the pipes they flow in are often heated in order to reduce the viscosity of the fluids and thus to reduce the pumping costs. Consider the flow of such a fluid through a 100-m-long pipe of outer diameter 30 cm in calm ambient air at 0°C. The pipe is heated electrically, and a thermostat keeps the outer surface temperature of the pipe constant at 25°C. The emissivity of the outer surface of the pipe is 0.8, and the effective sky temperature is −30°C, Determine the power rating of the electric resistance heater, in kW, that needs to be used. Also determine the cost of electricity associated with

heating the pipe during a 10-h period under the above conditions if the price of electricity is \$0.09/kWh. *Answers:* 29.2 kW, \$26.2

**11-29** Reconsider Prob. 11-28. To reduce the heating cost of the pipe, it is proposed to insulate it with sufficiently thick fiberglass insulation [$k$ = 0.035 W/(m · °C)] wrapped with aluminium foil ($\varepsilon$ = 0.1) to cut down the heat losses by 85 percent. Assuming the pipe temperature to remain constant at 25°C, determine the thickness of the insulation that needs to be used. How much money will the insulation save during this 10-h period? *Answers:* 1.3 cm, \$22.3

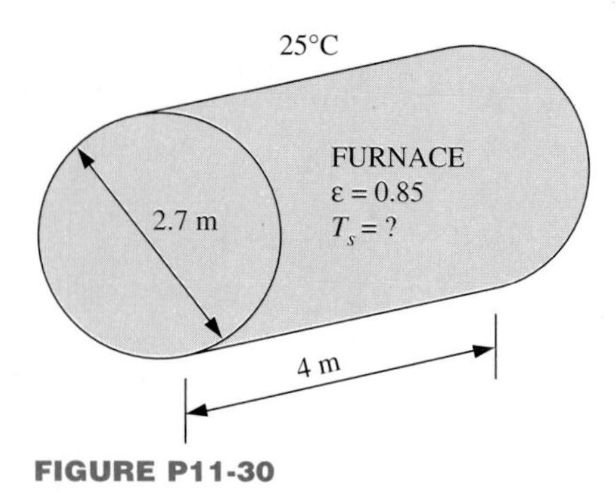


**FIGURE P11-30**

**11-30** Consider an industrial furnace that resembles a 4-m-long horizontal cylindrical enclosure 2.7 m in diameter whose end surfaces are well insulated. The furnace burns natural gas at a rate of 48 therms/h (1 therm = 105,500 kJ). The combustion efficiency of the furnace is 82 percent (i.e., 18 percent of the chemical energy of the fuel is lost through the flue gases as a result of incomplete combustion and the flue gases leaving the furnace at high temperature). If the heat loss from the outer surfaces of the furnace by natural convection and radiation is not to exceed 1 percent of the heat generated inside, determine the highest allowable surface temperature of the furnace. Assume the air and the wall surface temperature of the room to be 25°C, and take the emissivity of the outer surface of the furnace to be 0.85. If the cost of natural gas is \$0.48/therm and the furnace operates 3000 h per year, determine the annual cost of this heat loss to the plant.

**11-30E** Consider an industrial furnace that resembles a 13-ft-long horizontal cylindrical enclosure 8 ft in diameter whose end surfaces are well insulated. The furnace burns natural gas at a rate of 48 therms/h (1 therm = 100,000 Btu). The combustion efficiency of the furnace is 82 percent (i.e., 18 percent of the chemical energy of the fuel is lost through the flue gases as a result of incomplete combustion and the flue gases leaving the furnace at high temperature). If the heat loss from the outer surfaces of the furnace by natural convection and radiation is not to exceed 1 percent of the heat generated inside, determine the highest allowable surface temperature of the furnace. Assume the air and wall surface temperature of the room to be 75°F, and take the emissivity of the outer surface of the furnace to be 0.85. If the cost of natural gas is \$0.48/therm and the furnace operates 3000 h per year, determine the annual cost of this heat loss to the plant.

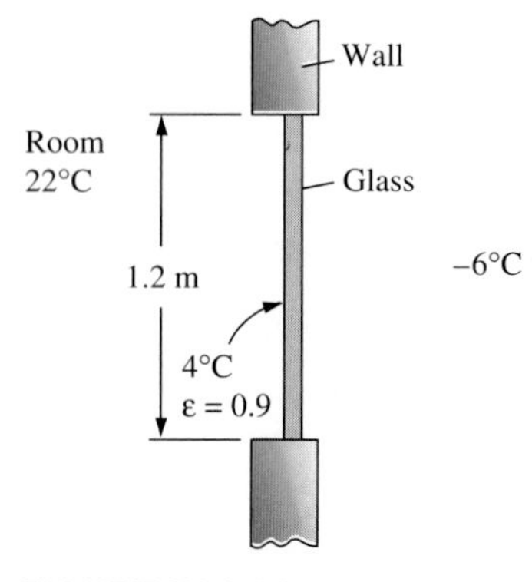


**FIGURE P11-31**

**11-31** Consider a 1.2-m-high and 2-m-wide glass window whose thickness is 6 mm, and has a thermal conductivity $k$ = 0.78 W/(m ·°C) and emissivity $\varepsilon$ = 0.9. The room and the walls which face the window are maintained at 22°C, and the average temperature of the inner surface of the window is measured to be 4°C. If the temperature of the outdoors is −6°C, determine (*a*) the convection heat transfer coefficient on the inner surface of the window, (*b*) the rate of total heat transfer through the window, and (*c*) the combined natural convection and radiation heat transfer coefficient on the outer surface of the window. Is it reasonable to neglect the thermal resistance of the glass in this case?



**11-32** A 3-mm-diameter and 12-m-long electric wire is tightly wrapped with a 1.5-mm-thick plastic cover whose thermal conductivity and emissivity are $k = 0.15\ \text{W/(m} \cdot {}^\circ\text{C)}$ and $\varepsilon = 0.9$. Electrical measurements indicate that a current of 10 A passes through the wire and there is a voltage drop of 8 V along the wire. If the insulated wire is exposed to calm atmospheric air at $T_\infty = 30°\text{C}$, determine the temperature at the interface of the wire and the plastic cover in steady operation. Take the surrounding surfaces to be at about the same temperature as the air. *Answer:* 57.9°C

**11-33** During a visit to a plastic sheeting plant, it was observed that a 60-m-long section of a 2-in nominal (6.03-cm outer-diameter) steam pipe extended from one end of the plant to the other with no insulation on it. The temperature measurements at several locations revealed that the average temperature of the exposed surfaces of the steam pipe was 170°C, while the temperature of the surrounding air was 20°C. The outer surface of the pipe appeared to be oxidized, and its emissivity can be taken to be 0.7. Taking the temperature of the surrounding surfaces to be 20°C also, determine the rate of heat loss from the steam pipe.

Steam is generated in a gas furnace that has an efficiency of 78 percent, and the plant pays $0.538 per therm (1 therm = 105,500 kJ) of natural gas. The plant operates 24 h a day 365 days a year, and thus 8760 h a year. Determine the annual cost of the heat losses from the steam pipe for this facility.

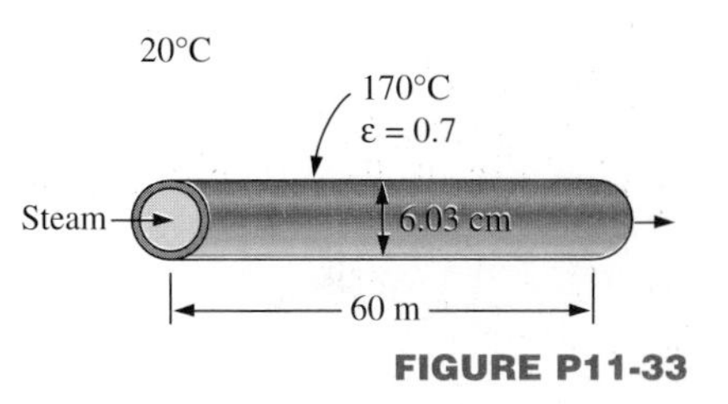


FIGURE P11-33

**11-33E** During a visit to a plastic sheeting plant, it was observed that a 200-ft-long section of a 2-in nominal (2.375-in outer-diameter) steam pipe extended from one end of the plant to the other with no insulation on it. The temperature measurements at several locations revealed that the average temperature of the exposed surfaces of the steam pipe was 340°F, while the temperature of the surrounding air was 70°F. The outer surface of the pipe appeared to be oxidized, and its emissivity can be taken to be 0.7. Taking the temperature of the surrounding surfaces to be 70°F also, determine the rate of heat loss from the steam pipe.

Steam is generated in a natural gas fired furnace that has an efficiency of 78 percent, and the plant pays $0.538 per therm (1 therm = 100,000 Btu) of natural gas. The plant operates 24 h a day 365 days a year, and thus 8760 h a year. Determine the annual cost of the heat losses from the steam pipe for this facility.

**11-34** Reconsider Prob. 11-33. In order to reduce heat losses, it is proposed to insulate the steam pipe with 5-cm-thick fiberglass insulation [$k = 0.038\ \text{W/(m} \cdot {}^\circ\text{C)}$], and to wrap it with aluminum foil ($\varepsilon = 0.1$) in order to minimize the radiation losses. Also, an estimate is obtained from a local insulation contractor, who proposed to do the insulation job for $750, including materials and labor. Would you support this proposal? How long will it take for the insulation to pay for itself from the energy

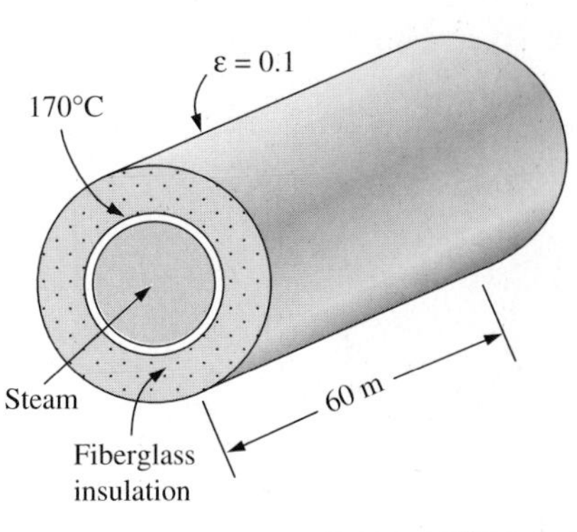


FIGURE P11-34

it saves? Assume the temperature of the steam pipe to remain constant at 170°C.

**11-34E** Reconsider Prob. 11-33E. In order to reduce heat losses, it is proposed to insulate the steam pipe with 2-in-thick fiberglass insulation [$k = 0.022$ Btu/(h · ft · °F)], and to wrap it with aluminum foil ($\varepsilon = 0.1$) in order to minimize the radiation losses. Also, an estimate is obtained from a local insulation contractor, who proposed to do the insulation job for $750, including materials and labor. Would you support this proposal? How long will it take for the insulation to pay for itself from the energy it saves? Assume the temperature of the steam pipe to remain constant at 340°F.

**11-35** A 30 cm × 30 cm circuit board that contains 121 square chips on one side is to be cooled by combined natural convection and radiation by mounting it on a vertical surface in a room at 25°C. Each chip dissipates 0.05 W of power, and the emissivity of the chip surfaces is 0.7. Assuming the heat transfer from the back side of the circuit board to be negligible, and the temperature of the surrounding surfaces to be the same as the air temperature of the room, determine the surface temperature of the chips. *Answer:* 33°C

**11-36** Repeat Prob. 11-35 assuming the circuit board to be positioned horizontally with (*a*) chips facing up, and (*b*) chips facing down.

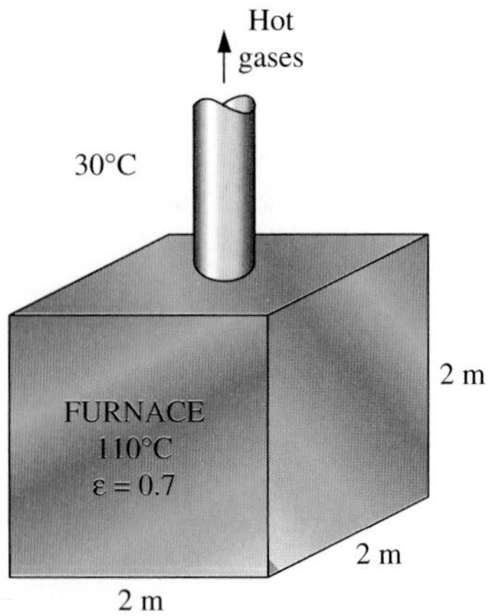

**11-37** The side surfaces of a 2-m-high cubic industrial furnace burning natural gas is not insulated, and the temperature at the outer surface of this section is measured to be 110°C. The temperature of the furnace room, including its surfaces, is 30°C, and the emissivity of the outer surface of the furnace is 0.7. It is proposed to insulate this section of the furnace wall with glass wool insulation [$k = 0.038$ W/(m · °C)] wrapped by a reflective sheet ($\varepsilon = 0.2$) in order to reduce the heat loss by 90 percent. Assuming the outer surface temperature of the metal section still remains at about 110°C, determine the thickness of the insulation that needs to be used.

The furnace operates continuously throughout the year, and has an efficiency of 78%. The price of the natural gas is $0.55/therm (1 therm = 105,500 kJ of energy content). If the installation of the insulation will cost $550 for materials and labor, determine how long it will take for the insulation to pay for itself from the energy it saves.

**FIGURE P11-37**

**11-38** A 1.5-m-diameter, 5-m-long cylindrical propane tank is initially filled with liquid propane, whose density is 581 kg/m$^3$. The tank is exposed to the ambient air at 25°C in calm weather. The outer surface of the tank is polished so that the radiation heat transfer is negligible. Now a crack develops at the top of the tank, and the pressure inside drops to 1 atm while the temperature drops to −42°C which is the boiling temperature of propane at 1 atm. The heat of vaporization of propane at 1 atm is 425 kJ/kg. The propane is slowly vaporized as a result of the heat transfer from the ambient air into the tank, and the propane vapor

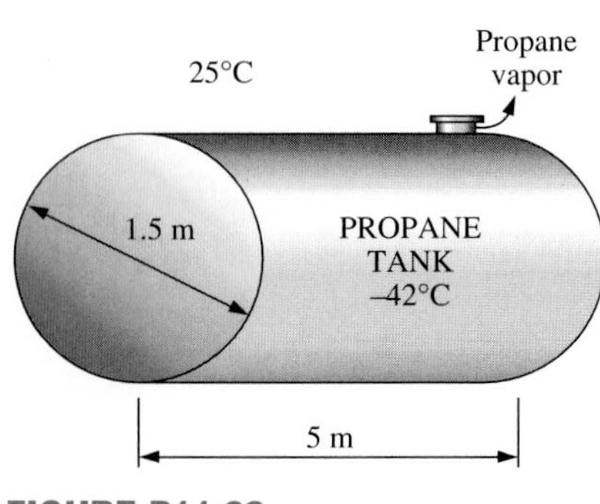

**FIGURE P11-38**

escapes the tank at $-42°C$ through the crack. Assuming the propane tank to be at about the same temperature as the propane inside at all times, determine how long it will take for the tank to empty if it is not insulated.

**11-39** An average person generates heat at a rate of 84 W while resting in a room at 25°C. Assuming one-quarter of this heat is lost from the head and taking the emissivity of the skin to be 0.9, determine the average surface temperature of the head when it is not covered. The head can be approximated as a 30-cm-diameter sphere, and the interior surfaces of the room can be assumed to be at the room temperature.

**11-39E** An average person generates heat at a rate of 287 Btu/h while resting in a room at 77°F. Assuming one-quarter of this heat is lost from the head and taking the emissivity of the skin to be 0.9, determine the average surface temperature of the head when it is not covered. The head can be approximated as a 12-in-diameter sphere, and the interior surfaces of the room can be assumed to be at the room temperature.

**11-40** An incandescent light bulb is an inexpensive but highly inefficient device that converts electrical energy into light. It converts about 10 percent of the electrical energy it consumes into light while converting the remaining 90 percent into heat. The glass bulb of the lamp heats up very quickly as a result of absorbing all that heat and dissipating it to the surroundings by convection and radiation. Consider an 8-cm-diameter 60 W light bulb in a room at 25°C. The emissivity of the glass is 0.9. Assuming that 10 percent of the energy passes through the glass bulb as light with negligible absorption and the rest of the energy is absorbed and dissipated by the bulb itself by natural convection and radiation, determine the equilibrium temperature of the glass bulb. Assume the interior surfaces of the room to be at room temperature.
*Answer:* 175°C

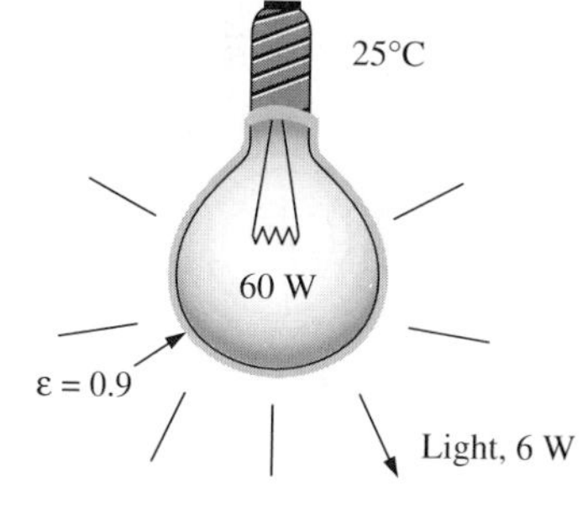


**FIGURE P11-40**

## Natural Convection inside Enclosures

**11-41C** The upper and lower compartments of a well-insulated container are separated by two parallel sheets of glass with an air space between them. One of the compartments is to be filled with a hot fluid and the other with a cold fluid. If it is desired that heat transfer between the two compartments be minimum, would you recommend putting the hot fluid into the upper or the lower compartment of the container? Why?

**11-42C** Someone claims that the air space in a double-pane window enhances the heat transfer from a house because of the natural convection currents that occur in the air space, and recommends that the double-pane window be replaced by a single sheet of glass whose thickness is equal to the sum of the thicknesses of the two glasses of the double pane window to save energy. Do you agree with this claim?

**11-43C** Consider a double-pane window consisting of two glass sheets separated by a 1-cm-wide air space. Someone suggests inserting a thin vinyl sheet in the middle of the two glasses to form two 0.5-cm-wide compartments in the window in order to reduce natural convection heat transfer through the window. From heat transfer point of view, would you be in favor of this idea to reduce heat losses through the window?

**11-44C** What does the effective conductivity of an enclosure represent? How is the ratio of the effective conductivity to thermal conductivity related to the Nusselt number?

**11-45** Show that the thermal resistance of a rectangular enclosure can be expressed as $R = \delta/(Ak \text{ Nu})$, where $k$ is the thermal conductivity of the fluid in the enclosure.

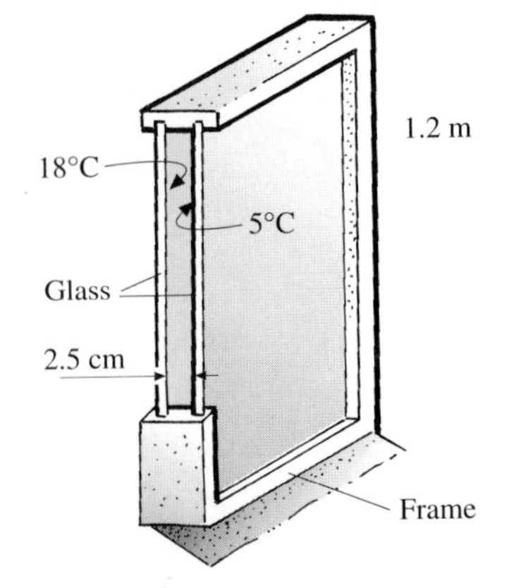


**FIGURE P11-46**

**11-46** A vertical 1.2-m-high and 2-m-wide double-pane window consists of two sheets of glass separated by a 2.5-cm air gap at atmospheric pressure. If the glass surface temperatures across the air gap are measured to be 18°C and 5°C, determine the rate of heat transfer through the window by (*a*) natural convection and (*b*) radiation. Also determine the effective thermal conductivity of the air space of this double-paned window, which also accounts for the radiation effect. The effective emissivity for use in radiation calculations between two large parallel glass plates can be taken to be 0.82.
*Answers:* (*a*) 49.6 W, (*b*) 134 W, 0.147 W/(m · °C)

**11-46E** A vertical 4-ft-high and 6-ft-wide double-pane window consists of two sheets of glass separated by a 1-in air gap at atmospheric pressure. If the glass surface temperatures across the air gap are measured to be 65°F and 40°F, determine the rate of heat transfer through the window by (*a*) natural convection and (*b*) radiation. Also determine the *R*-value of insulation of this window such that multiplying the inverse of the *R*-value by the surface area and the temperature difference gives the total rate of heat transfer through the window. The effective emissivity for use in radiation calculations between two large parallel glass plates can be taken to be 0.82.

**11-47** Two concentric spheres of diameters $D_1 = 15$ cm and $D_2 = 25$ cm are separated by air at 1-atm pressure. The surface temperatures of the two spheres enclosing the air are $T_1 = 350$ K and $T_2 = 275$ K, respectively. Determine the rate of heat transfer from the inner sphere to the outer sphere by natural convection.

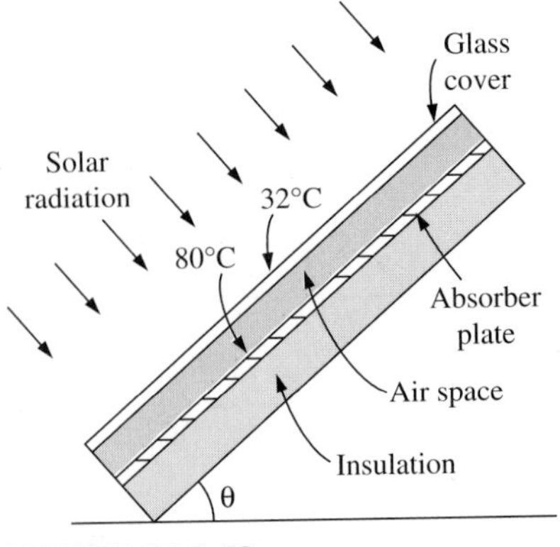


**FIGURE P11-48**

**11-48** Flat-plate solar collectors are often tilted up towards the sun in order to intercept a greater amount of direct solar radiation. The tilt angle from the horizontal also effects the rate of heat loss from the collector. Consider a 2-m-high and 3-m-wide solar collector that is tilted at an angle $\theta$ from the horizontal. The distance between the glass cover and the absorber plate is 3 cm, and the back side of the absorber is heavily insulated. The absorber plate and the glass cover, which are spaced 2.5 cm from each other, are maintained at temperatures of 80°C and 32°C, respectively. Determine the rate of heat loss from the absorber plate by natural convection for $\theta = 0°$, 20°, and 90°.

**11-49** A simple solar collector is built by placing a 5-cm-diameter clear plastic tube around a garden hose whose outer diameter is 1.6 cm. The hose is painted black to maximize solar absorption, and some plastic rings are used to keep the spacing between the hose and the clear plastic cover constant. During a clear day, the temperature of the hose is measured to be 65°C, while the ambient air temperature is 26°C. Determine the rate of heat loss from the water in the hose per meter of its length by natural convection. Also discuss how the performance of this solar collector can be improved. *Answer:* 6.5 W

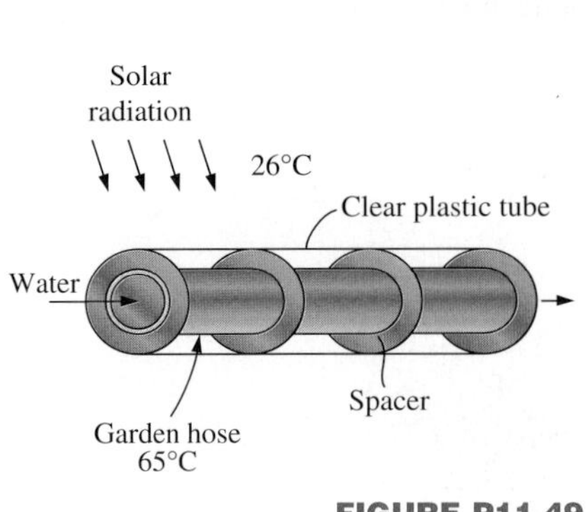

FIGURE P11-49

### Natural Convection from Finned Surfaces

**11-50C** Why are finned surfaces frequently used in practice? Why are the finned surfaces referred to as heat sinks in the electronics industry?

**11-51C** Why are heat sinks with closely packed fins not suitable for natural convection heat transfer, although they increase the heat transfer surface area more?

**11-52C** Consider a heat sink with optimum fin spacing. Explain how heat transfer from this heat sink will be effected by (*a*) removing some of the fins on the heat sink, and (*b*) doubling the number of fins on the heat sink by reducing the fin spacing. The base area of the heat sink remains unchanged at all times.

**11-53** Aluminum heat sinks of rectangular profile are commonly used to cool electronic components. Consider a 7.62-cm-long and 9.68-cm-wide commercially available heat sink whose cross-section and dimensions are as shown in Fig. P11-53. The heat sink is oriented vertically, and is used to cool a power transistor that can dissipate up to 125 W of power. The back surface of the heat sink is insulated. The surfaces of the heat sink are untreated, and thus they have a low emissivity (under 0.1). Therefore, radiation heat transfer from the heat sink can be neglected. During an experiment conducted in room air at 22°C, the base temperature of the heat sink was measured to be 120°C when the power dissipation of the transistor was 15 W. Assuming the entire heat sink to be at the base temperature, determine the average natural convection heat transfer coefficient for this case. *Answer:* 7.1 W/(m$^2$ · °C)

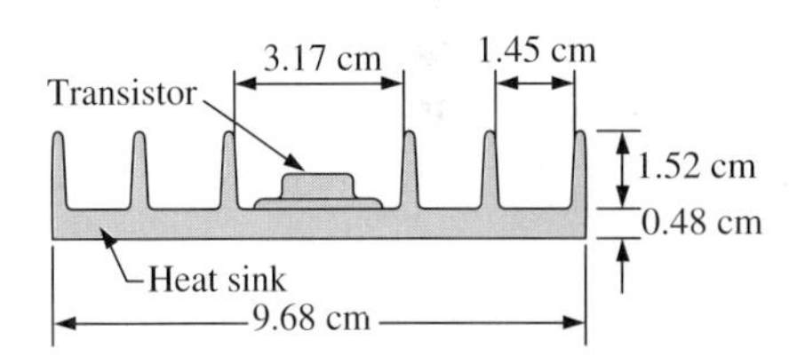

FIGURE P11-53

**11-53E** Aluminum heat sinks of rectangular profile are commonly used to cool electronic components. Consider a 3-in-long and 3.8-in-wide commercially available heat sink whose cross-section and dimensions are as shown in Fig. P11-53E. The heat sink is oriented vertically, and is used to cool a power transistor that can dissipate up to 125 W of power. The back surface of the heat sink is insulated. The surfaces of the heat sink are untreated, and thus they have a low emissivity (under 0.1). Therefore, radiation heat transfer from the heat sink can be neglected. During an

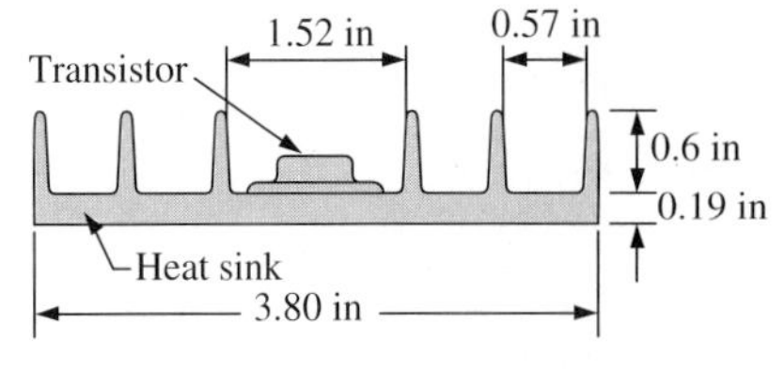

FIGURE P11-53E

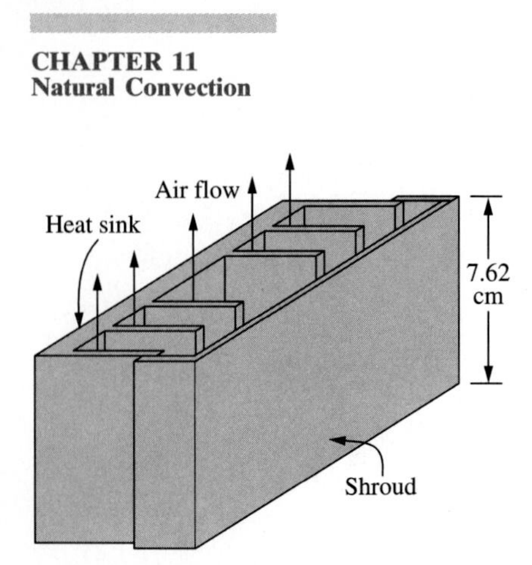


FIGURE P11-54

experiment conducted in room air at 72°F, the base temperature of the heat sink was measured to be 248°F when the power dissipation of the transistor was 15 W. Assuming the entire heat sink to be at the base temperature, determine the average natural convection heat transfer coefficient for this case. *Answer:* 1.32 Btu/(h · ft$^2$ · °F)

**11-54** Reconsider the heat sink in Prob. 11-53. In order to enhance heat transfer, a shroud (a thin rectangular metal plate) whose surface area is equal to the base area of the heat sink is placed very close to the tips of the fins such that the interfin spaces are converted into rectangular channels. The base temperature of the heat sink in this case was measured to be 108°C. Noting that the shroud looses heat to the ambient air from both sides, determine the average natural convection heat transfer coefficient in this shrouded case. (For complete details, see Ref. 4).

**11-55** A 15.2-cm-wide and 20-cm-high vertical hot surface in 25°C air is to be cooled by a heat sink with equally spaced fins of rectangular profile. The fins are 0.2 cm thick, 20 cm long in the vertical direction, and have a height of 3 cm from the base. Determine the optimum fin spacing, and the rate of heat transfer by natural convection from the heat sink if the base temperature is 80°C.

**11-55E** A 6-in-wide and 8-in-high vertical hot surface in 78°F air is to be cooled by a heat sink with equally spaced fins of rectangular profile. The fins are 0.08 in thick, 8 in long in the vertical direction, and have a height of 1.2 in from the base. Determine the optimum fin spacing, and the rate of heat transfer by natural convection from the heat sink if the base temperature is 180°F.

**11-56** A 12.1-cm-wide and 18-cm-high vertical hot surface in 25°C air is to be cooled by a heat sink with 25 equally spaced fins of rectangular profile. The fins are 0.1 cm thick and 18 cm long in the vertical direction. Determine the optimum fin height, and the rate of heat transfer by natural convection from the heat sink if the base temperature is 70°C.

### Combined Natural and Forced Convection

**11-57C** When is natural convection negligible, and when is it not in forced convection heat transfer?

**11-58C** Under what conditions does natural convection enhance forced convection, and under what conditions does it hurt forced convection?

**11-59C** When neither natural nor forced convection is negligible, is it correct to calculate each independently and add them to determine the total convection heat transfer?

**11-60** Consider a 5-m-long vertical plate at 85°C in air at 30°C. Determine the forced motion velocity above which natural convection heat transfer from this plate is negligible. *Answer:* 9.05 m/s

**11-61** Consider a 3-m-long vertical plate at 60°C in water at 25°C. Determine the forced motion velocity above which natural convection heat transfer from this plate is negligible. Take $\beta = 0.0004\ \text{K}^{-1}$ for water.

**11-62** In a production facility, thin square plates 2 m × 2 m in size coming out of the oven at 300°C are cooled by blowing ambient air at 30°C horizontally parallel to their surfaces. Determine the air velocity above which the natural convection effects on heat transfer are less than 10 percent and thus are negligible.

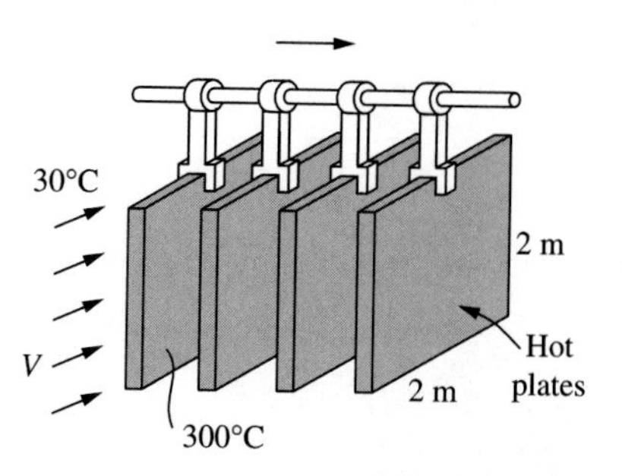


**FIGURE P11-62**

**11-62E** In a production facility, thin square plates 6 ft × 6 ft in size coming out of the oven at 500°F are cooled by blowing ambient air at 90°F horizontally parallel to their surfaces. Determine the air velocity above which the natural convection effects on heat transfer are less than 10 percent and thus are negligible. *Answer:* 32.4 ft/s

**11-63** A 12-cm-high and 20-cm-wide circuit board houses 100 closely spaced logic chips on its surface, each dissipating 0.05 W. The board is cooled by a fan that blows air over the hot surface of the board at 35°C at a velocity of 0.5 m/s. The heat transfer from the back surface of the board is negligible. Determine the average temperature on the surface of the circuit board assuming the air flows vertically upwards along the 12-cm-long side by (*a*) ignoring natural convection and (*b*) considering the contribution of natural convection. Disregard any heat transfer by radiation.

## Review Problems

**11-64** A 0.1-W small cylindrical resistor mounted on the lower part of a vertical circuit board is 0.8 cm long and has a diameter of 0.4 cm. The view of the resistor is largely blocked by other circuit board facing it, and the heat transfer through the connecting wires is negligible. The air is free to flow through the large parallel flow passages between the boards as a result of natural convection currents. If the air temperature at the vicinity of the resistor is 40°C, determine the approximate surface temperature of the resistor. *Answer:* 96°C

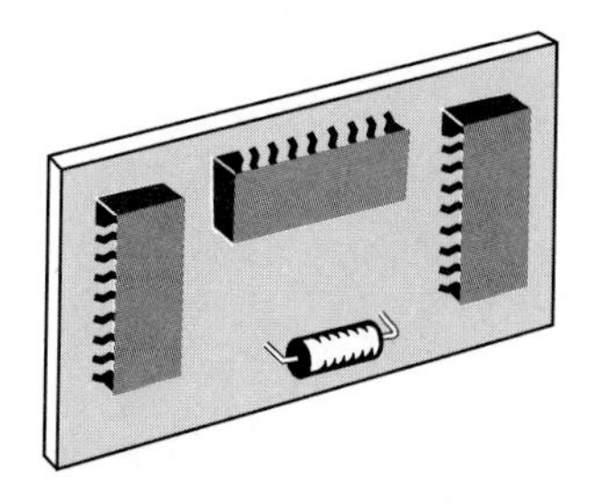

**FIGURE P11-64**

**11-64E** A 0.1-W small cylindrical resistor mounted on a lower part of a vertical circuit board is 0.3 in long and has a diameter of 0.2 in. The view of the resistor is largely blocked by other circuit board facing it, and the heat transfer through the connecting wires is negligible. The air is free to flow through the large parallel flow passages between the boards as a result of natural convection currents. If the air temprature at the vicinity of the resistor is 120°F, determine the approximate surface temperature of the resistor. *Answer:* 212°F

**11-65** An ice chest whose outer dimensions are 30 cm × 40 cm × 40 cm is made of 3-cm-thick styrofoam [$k = 0.033\ \text{W/(m} \cdot {}^\circ\text{C)}$]. Initially, the chest is filled with 40 kg of ice at 0°C, and the inner surface temperature of the ice chest can be taken to be 0°C at all times. The heat of fusion water at 0°C is 333.7 kJ/kg, and the surrounding ambient air is at 20°C. Disregarding any heat transfer from the 40 cm × 40 cm base of the ice chest, determine how long it will take for the ice in the chest to melt completely if the ice chest is subjected to (*a*) calm air and (*b*) winds at 50 km/h. Assume the heat transfer coefficient on the front and back surfaces to be the same as that on the side surfaces.

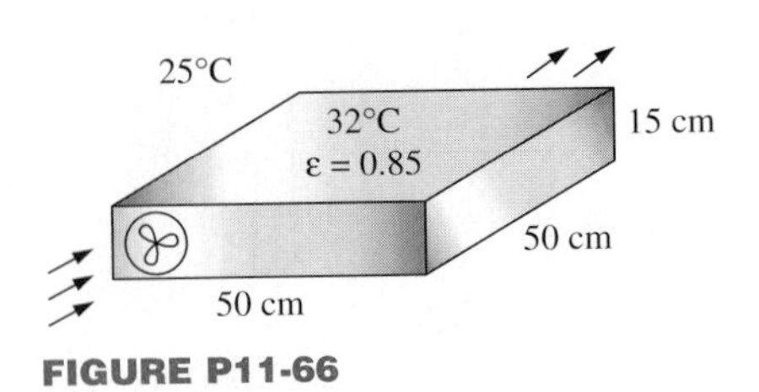


**FIGURE P11-66**

**11-66** An electronic box that consumes 180 W of power is cooled by a fan blowing air into the box enclosure. The dimensions of the electronic box are 15 cm × 50 cn × 50 cm, and all surfaces of the box are exposed to the ambient except the base surface. Temperature measuremnts indicate that the box is at an average temperature of 32°C when the ambient temperature and the temperature of the surrounding walls are 25°C. If the emissivity of the outer surface of the box is 0.85, determine the fraction of the heat lost from the outer surfaces of the electronic box.

**11-66E** An electronic box that consumes 180 W of power is cooled by a fan blowing air into the box enclosure. The dimensions of the electronic box are 6 in × 20 in × 20 in, and all surfaces of the box are exposed to the ambient except the base surface. Temperature measurements indicate that the box is at an average temperature of 90°F when the ambient temperature and the temperature of the surrounding walls are 85°F. If the emissivity of the outer surface of the box is 0.85, determine the fraction of the heat lost from the outer surfaces of the electronic box.

**11-67** A 6-m internal-diameter spherical tank made of 1.5-cm-thick stainless steel [$k$ = 15 W/(m · °C)] is used to store iced water at 0°C. The walls of the room are also at 20°C. The outer surface of the tank is black (emissivity $\varepsilon$ = 1), and heat transfer between the outer surface of the tank and the surroundings is by natural convection and radiation. Assuming the entire steel tank to be at 0°C and thus the thermal resistance of the tank to be negligible, determine (*a*) the rate of heat transfer to the iced water in the tank and (*b*) the amount of ice at 0°C that melts during a 24-h period. The heat of fusion of water at atmospheric pressure is $h_{if}$ = 333.7 kJ/kg.
*Answers:* (*a*) 15,044 W, (*b*) 3895 kg

**11-68** Consider a 1.2-m-high and 2-m-wide double-pane window consisting of two 3-mm-thick layers of glass [$k$ = 0.78 W/(m · °C)] separated by a 11-mm-wide air space. Determine the steady rate of heat transfer through this window and the temperature of its inner surface for a day during which the room is maintained at 24°C while the temperature of the outdoors is −5°C. Take the heat transfer coefficients on the inner and outer surfaces of the window to be $h_1$ = 10 W/(m² · °C) and $h_2$ = 25 W/(m² · °C) and disregard any heat transfer by radiation.

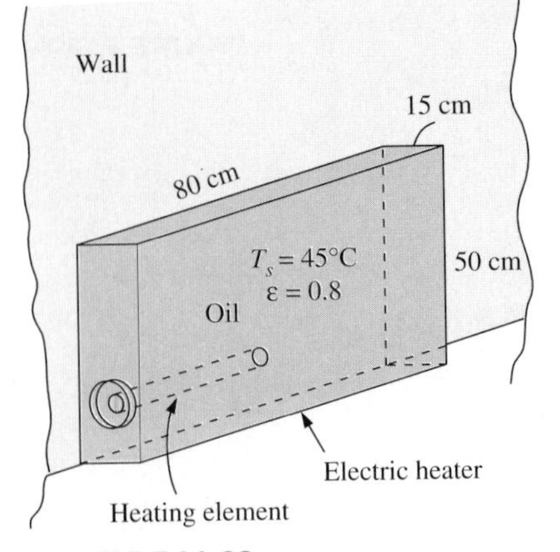


**FIGURE P11-69**

**11-69** An electric resistance space heater is designed such that it resembles a rectangular box 50 cm high, 80 cm long, and 15 cm wide filled with 45 kg of oil. The heater is to be placed against a wall, and thus heat transfer from its back surface is negligible for safety considerations. The surface temperature of the heater is not to exceed 45°C in a room at 25°C. Disregarding heat transfer from the bottom and top surfaces of the heater in anticipation that the top surface will be used as a shelf, determine the power rating of the heater in W. Take the emissivity of the outer surface of the heater to be 0.8 and the average temperature of the ceiling and wall surfaces to be the same as the room air temperature.

Also determine how long it will take for the heater to reach steady operation when it is first turned on (i.e., for the oil temperature to rise from 25°C to 45°C). State your assumptions in calculations.

**11-70** Skylights or "roof windows" are commonly used in homes and manufacturing facilities, since they let natural light in during day time and thus reduce the lighting costs. However, they offer little resistance to heat transfer, and large amounts of energy are lost through them in winter unless they are equipped with a motorized insulating cover that can be used in cold weather and at nights to reduce heat losses. Consider a 1-m-wide and 2.5-m-long horizontal skylight on the roof of a house that is kept at 20°C. The glazing of the skylight is made of a single layer of 0.5-cm-thick glass [$k = 0.78$ W/(m · °C) and $\varepsilon = 0.9$]. Determine the rate of heat loss through the skylight when the air temperature outside is −8°C and the effective sky temperature is −30°C. Compare your result with the rate of heat loss through an equivalent surface area of the roof that has a common R-5.34 construction in SI units (i.e., an effective thermal conductivity to thickness ratio of 5.34 m$^2$ · °C/W).

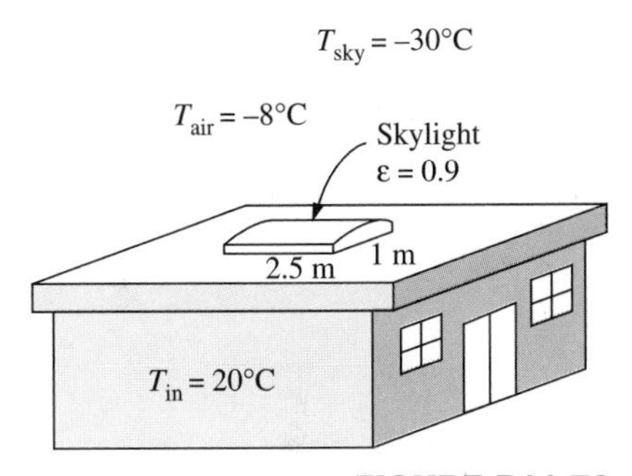


FIGURE P11-70

**11-70E** Skylights or "roof windows" are commonly used in homes and manufacturing facilities since they let natural light in during day time and thus reduce the lighting costs. However, they offer little resistance to heat transfer, and large amounts of energy are lost through them in winter unless they are equipped with a motorized insulating cover that can be used in cold weather and at nights to reduce heat losses. Consider a 3-ft-wide and 8-ft-long horizontal skylight on the roof of a house that is kept at 70°F. The glazing of the skylight is made of a single layer of 0.25-in-thick glass [$k = 0.45$ Btu/(h · ft · °F) and $\varepsilon = 0.9$]. Determine the rate of heat loss through the skylight when the air temperature outside is 20°F and the effective sky temperature is −30°F. Compare your result with the rate of heat loss through an equivalent surface area of the roof that has a common R-30 construction in English units (i.e., an effective thermal conductivity to thickness ratio of 30 ft$^2$ · h · °F/Btu).

**11-71** A solar collector consists of a horizontal copper tube of outer diameter 5 cm enclosed in a concentric thin glass tube of 9 cm diameter. Water is heated as it flows through the tube, and the annular space between the copper and the glass tubes is filled with air at 1-atm pressure. During a clear day, the temperatures of the tube surface and the glass cover are measured to be 60°C and 32°C, respectively. Determine the rate of heat loss from the collector by natural convection per meter length of the tube. *Answer:* 19.6 W

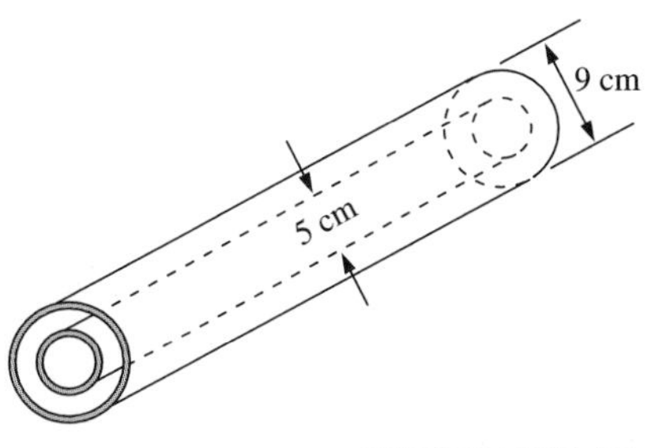


FIGURE P11-71

**11-72** A solar collector consists of a horizontal aluminum tube of outer diameter 4 cm enclosed in a concentric thin glass tube of 7 cm diameter. Water is heated as it flows through the tube, and the annular space between the aluminum and the glass tubes is filled with air at 1-atm pressure. The pump circulating the water fails during a clear day, and the water temprature in the tube starts rising. The aluminum tube absorbs solar radiation at a rate of 20 W per meter length, and the temperature of the ambient air outside is 30°C. Approximating the surfaces of the tube and the glass cover as being black (emissivity $\varepsilon = 1$) in radiation calculations and taking the effective sky temperature to be 10°C, determine the temperature of the aluminum tube when thermal equilibrium is established (i.e., when the net heat loss from the tube by convection and radiation equals the amount of solar energy absorbed by the tube).

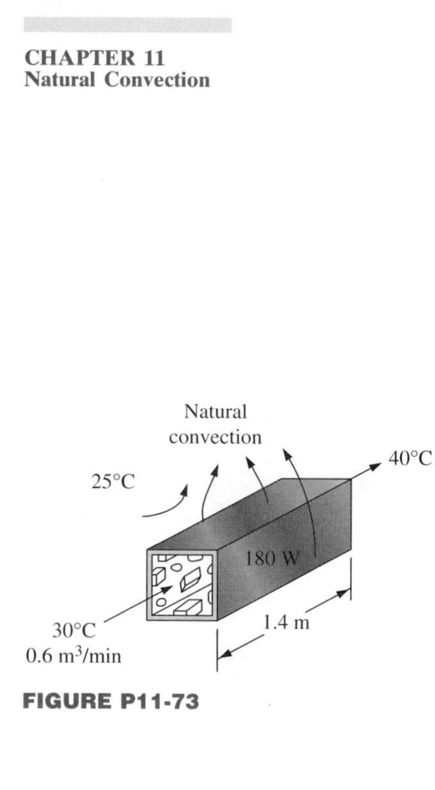

**FIGURE P11-73**

**11-73** The components of an electronic system dissipating 180 W are located in a 1.4-m-long horizontal duct whose cross-section is 20 cm × 20 cm. The components in the duct are cooled by forced air, which enters the duct at 30°C at a rate of 0.6 $m^3$/min, and leaves at 40°C. The surfaces of the sheet metal duct are not painted, and thus radiation heat transfer from the outer surfaces is negligible. If the ambient air temperature is 25°C, determine (*a*) the heat transfer from the outer surfaces of the duct to the ambient air by natural convection and (*b*) the average temperature of the duct. *Answers:* (*a*) 64.3 W, (*b*) 40°C

**11-73E** The components of an electronic system dissipating 180 W are located in a 4-ft-long horizintal duct whose cross-section is 6 in × 6 in. The components in the duct are cooled by forced air, which enters at 85°F at a rate of 22 cfm, and leaves at 100°F. The surfaces of the sheet metal duct are not painted, and thus radiation heat transfer from the outer surfaces is negligible. If the ambient air temperature is 80°F, determine (*a*) the heat transfer from the outer surfaces of the duct to the ambient air by natural convection and (*b*) the average temperature of the duct.

**11-74** Repeat Prob. 11-73 for a circular horizontal duct of diameter 10 cm.

**11-74E** Repeat Prob. 11-73E for a circular horizontal duct of diameter 4 in.

**11-75** Repeat Prob. 11-73 assuming the fan fails and thus the entire heat generated inside the duct must be rejected to the ambient air by natural convection through the outer surfaces of the duct.

**11-76** Consider a cold aluminum cold drink that is initially at a uniform temperature of 5°C. The can is 12.5 cm high, and has a diameter of 6 cm. The emissivity of the outer surface of the can is 0.6. Disregarding any heat transfer from the bottom surface of the can, determine how long it will take for the average temperature of the drink to rise to 7°C if the surrounding air and surfaces are at 25°C. *Answer:* 11.4 min

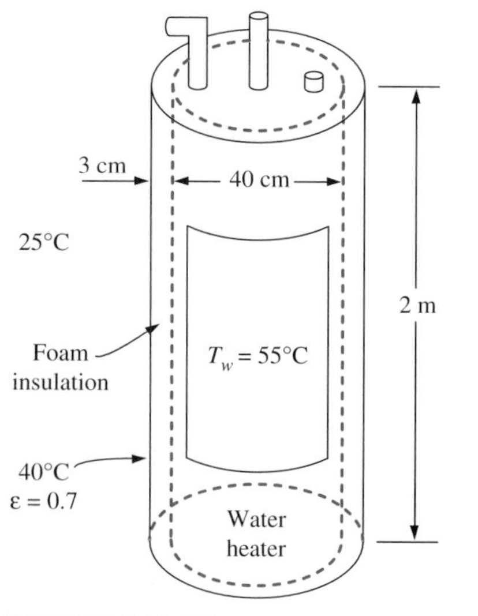

**FIGURE P11-77**

**11-77** Consider a 2-m-high electric hot water heater that has a diameter of 40 cm and maintains the hot water at 55°C. The tank is located in a small room at 25°C whose walls and the ceiling are at about the same temperature. The tank is placed in a 46-cm-diameter sheet metal shell of negligible thickness, and the space between the tank and the shell is filled with foam insulation. The average temperature and emissivity of the outer surface of the shell are 40°C and 0.7, respectively. The price of electricity is \$0.08/kWh. Hot water tank insulation kits large enough to wrap the entire tank are available on the market for about \$30. If such an insulation is installed on this water tank by the homeowner himself, how long will it take for this additional insulation to pay for itself? Disregard any heat loss from the top and bottom surfaces, and assume the insulation to reduce the heat losses by 80 percent.

**11-78** During a plant visit, it was observed that a 1.5-m-high and 1-m-wide section of the vertical front section of a natural gas furnace wall was too hot to touch. The temperature measurements on the surface revealed that the average temperature of the exposed hot surface was 110°C, while the temperature of the surrounding air was 25°C. The surface appeared to be oxidized, and its emissivity can be taken to be 0.7. Taking the temperature of the surrounding surfaces to be 25°C also, determine the rate of heat loss from this furnace.

The furnace has an efficiency of 79 percent, and the plant pays \$0.58 per therm (1 therm = 105,500 kJ) of natural gas. If the plant operates 10 h a day, 260 days a year, and thus 2600 h a year, determine the annual cost of the heat loss from this vertical hot surface on the front section of the furnace wall.

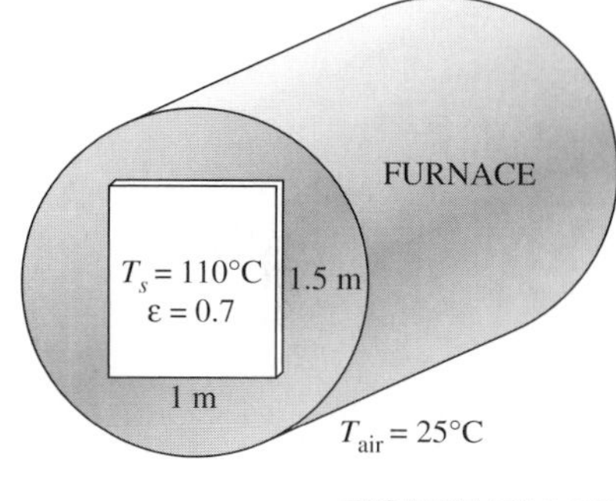


FIGURE P11-78

**11-78E** During a plant visit, it was observed that a 5-ft-high and 3-ft-wide section of the vertical front section of a natural gas furnace wall was too hot to touch. The temperature measurements on the surface revealed that the average temperature of the exposed hot surface was 230°F, while the temperature of the surrounding air was 80°F. The surface appeared to be oxidized, and its emissivity can be taken to be 0.7. Taking the temperature of the surrounding surfaces to be 80°F also, determine the rate of heat loss from this furnace.

The furnace has an efficiency of 79 percent, and the plant pays \$0.58 per therm (1 therm = 100,000 Btu) of natural gas. If the plant operates 10 h a day, 260 days a year, and thus 2600 h a year, determine the annual cost of the heat loss from this vertical hot surface on the front section of the furnace wall.

**11-79** A group of 25 power transistors, dissipating 1.5 W each, are to be cooled by attaching them to a black-anodized square aluminum plate and mounting the plate on the wall of a room at 30°C. The emissivity of the transistor and the plate surfaces is 0.9. Assuming the heat transfer from the black side of the plate to be negligible and the temperature of the surrounding surfaces to be the same as the air temperature of the room, determine the size of the plate if the average surface temperature of the plate is not to exceed 50°C. *Answer:* $L$ = 77.5 cm

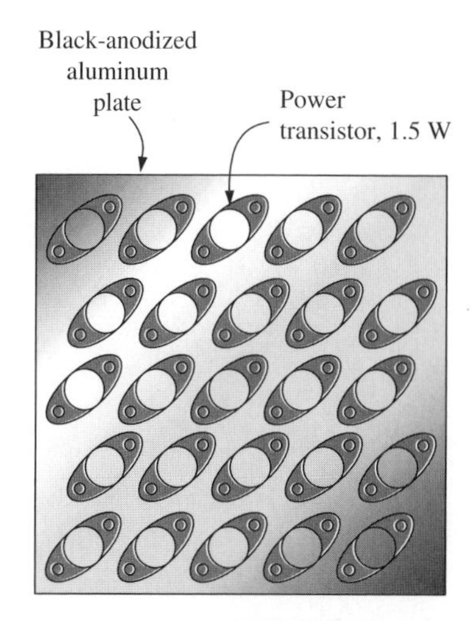


FIGURE P11-79

**11-80** Repeat Prob. 11-79 assuming the plate to be positioned horizontally with (*a*) transistors facing up and (*b*) transistors facing down.

**11-81** Hot water is flowing at an average velocity of 0.8 m/s through a cast iron pipe [$k = 52$ W/(m · °C)] whose inner and outer diameters are 2.5 cm and 3.0 cm, respectively. The pipe passes through a 15-m-long section of a basement whose temperature is 10°C. The emissivity of the outer surface of the pipe is 0.5, and the walls of the basement are also at about 10°C. If the inlet temperature of the water is 70°C and the heat transfer coefficient on the inner surface of the pipe is 150 W/(m$^2$ · °C), determine the temperature drop of water as it passes through the basement.

**11-81E** Hot water is flowing at an average velocity of 4 ft/s through a cast iron pipe [$k = 30$ Btu/(h · ft · °F)] whose inner and outer diameters are 1.0 in and 1.2 in, respectively. The pipe passes through a 50-ft-long section of a basement whose temperature is 60°F. The emissivity of the outer surface of the pipe is 0.5, and the walls of the basement are also at about 60°F. If the inlet temperature of the water is 150°F and the heat transfer coefficient on the inner surface of the pipe is 30 Btu/(h · ft$^2$ · °F), determine the temperature drop of water as it passes through the basement.

**11-82** Consider a flat-plate solar collector placed horizontally on the flat roof of a house. The collector is 1.5 m wide and 6 m long, and the average temperature of the exposed surface of the collector is 42°C. Determine the rate of heat loss from the collector by natural convection during a calm day when the ambient air temperature is 15°C. Also determine the heat loss by radiation by taking the emissivity of the collector surface to be 0.9 and the effective sky temperature to be −30°C.
*Answer:* 12.95 W, 2921 W

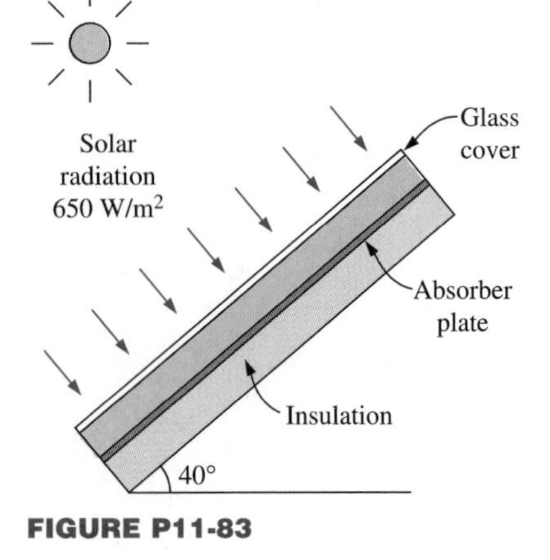


**FIGURE P11-83**

**11-83** Solar radiation is incident on the glass cover of a solar collector at a rate of 650 W/m$^2$. The glass transmits 88 percent of the incident radiation, and has an emissivity of 0.90. The hot water needs of a family in summer can be met completely by a collector 1.5 m high and 2 m wide, and tilted 40° from the horizontal. The temperature of the glass cover is measured to be 35°C on a calm day when the surrounding air temperature is 23°C. The effective sky temperature for radiation exchange between the glass cover and the open sky is −40°C. Water enters the tubes attached to the absorber plate at a rate of 1 kg/min. Assuming the back surface of the absorber plate to be heavily insulated and the only heat loss occurs through the glass cover, determine (*a*) the total rate of heat loss from the collector, (*b*) the collector efficiency, which is the ratio of the amount of heat transferred to the water to the solar energy incident on the collector, and (*c*) the temprature rise of water as it flows through the collector.

**11-84** Write a computer program to evaluate the variation of temperature with time of thin square metal plates that are removed from an oven at a specified temperature, and placed vertically in a large room. The thickness, the size, the initial temperature, the emissivity, and thermophysical properties of the plate as well as the room temperature are to be specified by the user. The program should evaluate the temperature of the plate at specified intervals, and tabulate the results against time. The computer should list the assumptions made during calculations before printing the results.

For each step or time interval, assume the surface temperature to be constant, evaluate the heat loss during that time interval and the temperature drop of the plate as a result of this heat loss. This gives the temperature of the plate at the end of a time interval, which is to serve as the initial temperature of the plate for the beginning of the next time interval.

Try your program for 0.2-cm-thick vertical copper plates of 40 cm × 40 cm in size initially at 300°C cooled in a room at 25°C. Take the surface emissivity to be 0.9. Use a time interval of 1 s in calculations, but print the results at 10-s intervals for a total cooling period of 15 min.

**11-85** Repeat Prob. 11-84 for a vertical slender cylindrical metal object instead of a square plate. The height and the diameter of the cylinder are to be specified by the user.

**11-86** Write a computer program to optimize the spacing between the two glasses of a double-paned window. Assume the spacing is filled with dry air at atmospheric pressure. The program should evaluate the recommended practical value of the spacing to minimize the heat losses and list it when the size of the window (the height and the width) and the temperatures of the two glasses are specified.

**11-87** Contact a manufacturer of aluminum heat sinks, and obtain their product catalog for cooling electronic components by natural convection and radiation. Write an essay on how to select a suitable heat sink for an electronic component when its maximum power dissipation and maximum allowable surface temperature are specified.

**11-88** The top surfaces of practically all flat-plate solar collectors are covered with glass in order to reduce the heat losses from the absorber plate underneath. Although the glass cover reflects or absorbs about 15 percent of the incident solar radiation, it saves much more from the potential heat losses from the absorber plate, and thus it is considered to be an essential part of a well-designed solar collector. Inspired by the energy efficiency of double-paned windows, someone proposes to use double glazing on solar collectors instead of a single glass. Investigate if this is a good idea for the town you live in. Use local weather data, and base your conclusion on heat transfer analysis and economic considerations.

# Radiation Heat Transfer

CHAPTER 12

So far, we have considered the conduction and convection modes of heat transfer, which are related to the nature of the materials involved and the presence of fluid motion, among other things. We now turn our attention to a third mechanism of heat transfer: *radiation,* which is characteristically different from the other two.

We start this chapter with a discussion of *electromagnetic waves* and the *electromagnetic spectrum,* with particular emphasis on *thermal radiation.* Then we introduce the idealized *blackbody, blackbody radiation,* and the *blackbody radiation function,* together with the *Stefan–Boltzmann law, Plank's distribution law* and *Wien's displacement law.* This is followed by a discussion of radiation properties of materials such as *emissivity, absorptivity, reflectivity,* and *transmissivity,* and their dependence on wavelength and temperature. The *greenhouse effect* is presented as an example of the consequences of the wavelength dependence of radiation properties. A separate section is devoted to the discussions of *atmospheric* and *solar radiation* because of their importance.

The second part of this chapter starts with a discussion of *view factors* and the rules associated with them. View factor *expressions* and *charts* for some common configurations are given, and the *crossed-strings method* is presented. We then discuss *radiation heat transfer,* first between black surfaces and then between nonblack surfaces using the *radiation network* approach. Finally, we consider *radiation shields,* and discuss the *radiation effect* on temperature measurements and comfort.

## 12-1 INTRODUCTION

Consider a hot object that is placed in an evacuated chamber whose walls are at room temperature (Fig. 12-1). Our experience tells us that the hot object will eventually cool down and reach thermal equilibrium with its surroundings. That is, it will lose heat until its temperature reaches the temperature of the walls of the chamber. Heat transfer between the object and the chamber could not have taken place by conduction or convection, because these two mechanisms cannot occur in a vacuum. Therefore, heat transfer must have occurred through another mechanism that involves the emission of the sensible internal energy of the object. This mechanism is *radiation.*

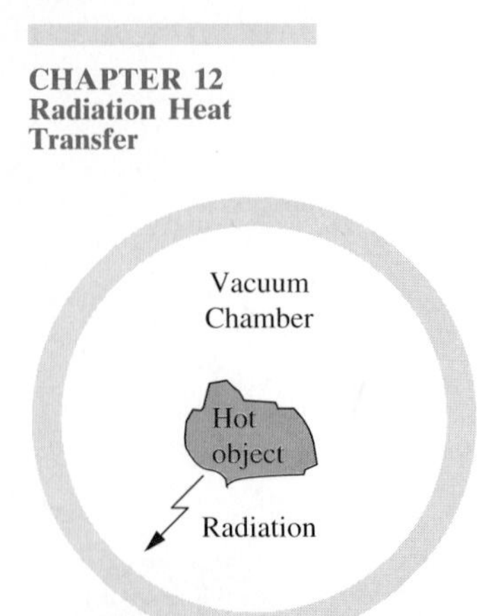


**FIGURE 12-1**
A hot object in a vacuum chamber loses heat by radiation only.

Radiation differs from the other two heat transfer mechanisms in that it does not require the presence of a material medium to take place. In fact, energy transfer by radiation is fastest (at the speed of light) and it suffers no attenuation in a *vacuum.* Also, radiation transfer occurs in solids as well as liquids and gases. In most practical applications, all three modes of heat transfer occur concurrently at varying degrees. But heat transfer through an evacuated space can occur only by radiation. For example, the energy of the sun reaches the earth by radiation.

You will recall that heat transfer by conduction or convection takes place in the direction of decreasing temperature; that is, from a high-temperature medium to a lower-temperature one. It is interesting that radiation heat transfer can occur between two bodies separated by a medium colder than both bodies (Fig. 12-2). For example, solar radiation reaches the surface of the earth after passing through extremely cold air layers at high altitudes. Also, the radiation absorbing surfaces inside a green house reach high temperatures even when its plastic or glass cover remains relatively cool.

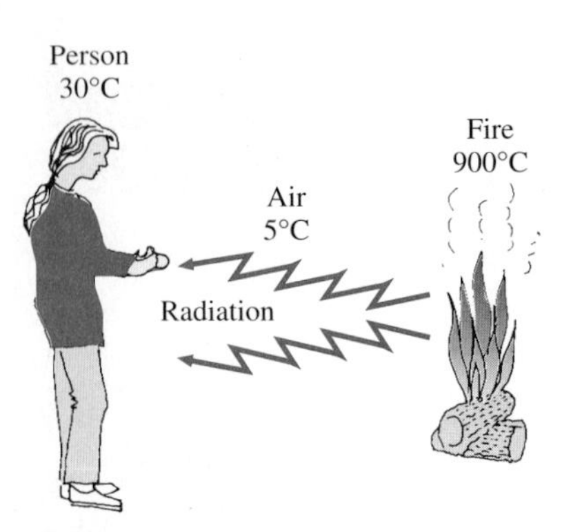


**FIGURE 12-2**
Unlike conduction and convection, heat transfer by radiation can occur between two bodies, even when they are separated by a medium colder than both of them.

The theoretical foundation of radiation was established in 1864 by physicist James Clerk Maxwell, who postulated that accelerated charges or changing electric currents give rise to electric and magnetic fields. These rapidly moving fields are called **electromagnetic waves** or **electromagnetic radiation**, and they represent the energy emitted by matter as a result of the changes in the electronic configurations of the atoms or molecules. In 1887, Heinrich Hertz experimentally demonstrated the existence of such waves. Electromagnetic waves transport energy just like other waves, and all electromagnetic waves travel at the *speed of light*. Electromatnetic waves are characterized by their *frequency* $\nu$ and *wavelength* $\lambda$. These two properties in a medium are related by

$$\lambda = \frac{c}{\nu} \tag{12-1}$$

where $c$ is the speed of light in that medium. In a vacuum, $c = c_0 = 2.998 \times 10^8$ m/s. The speed of light in a medium is related to the speed of light in a vacuum by $c = c_0/n$, where $n$ is the *index of refraction* of that medium. The index of refraction is essentially unity for air and most gases, and about 1.5 for water and glass. The commonly used unit of wavelength is the *micrometer* ($\mu$m), where $1\ \mu\text{m} = 10^{-6}$ m. Unlike the wavelength and the speed of propagation, the frequency of an electromagnetic wave depends only on the source, and is independent of the medium through which the wave travels. The *frequency* (the number of oscillations per second) of an electromagnetic wave can range from a few cycles to millions of cycles and even higher per second, depending on the source. Note from Eq. 12-1 that the wavelength and the frequency of electromagnetic radiation are inversely proportional.

In radiation studies, it has proven useful to view electromagnetic radiation as the propagation of a collection of discrete packets of energy called **photons** or **quanta**, as proposed by Max Planck in 1900 in conjunction with his *quantum theory.* In this view, each photon of frequency $\nu$ is considered to have an energy of

$$e = h\nu = \frac{hc}{\lambda} \tag{12-2}$$

where $h = 6.625 \times 10^{-34}$ J · s is *Planck's constant.* Note from the second part of Eq. 12-2 that $h$ and $c$ are constants, and thus the energy of a photon is inversely proportional to its wavelength. Therefore, shorter-wavelength radiation possesses larger photon energies. It is no wonder that we try to avoid the very short-wavelength radiation such as gamma rays and X-rays, since they are highly destructive.

## 12-2 THERMAL RADIATION

Although all electromagnetic waves have the same general features, waves of different wavelength differ significantly in their behavior. The electromagnetic radiation encountered in practice cover a wide range of wavelengths, varying from less than $10^{-10}$ $\mu$m for cosmic rays to more than $10^{10}$ $\mu$m for electrical power waves. The **electromagnetic spectrum** also includes gamma rays, X-rays, ultraviolet radiation, visible light, infrared radiation, thermal radiation, microwaves, and radio waves, as shown in Fig. 12-3.

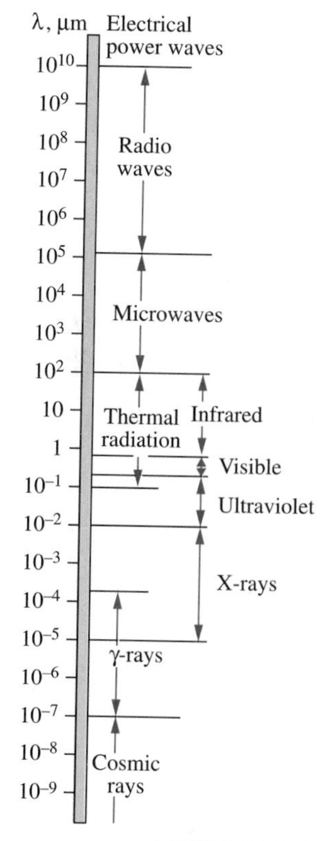

**FIGURE 12-3**
The electromagnetic wave spectrum.

Different types of electromagnetic radiation are produced differently through different mechanisms. For example, *gamma rays* are produced by nuclear reactions, *X-rays* by the bombardment of metals with high-energy electrons, *microwaves* by special types of electron tubes such as klystrons and magnetrons, and *radio waves* by the excitation of some crystals or by the flow of alternating current through electric conductors.

The short-wavelength gamma rays and X-rays are primarily of concern to nuclear engineers, while the long-wavelength microwaves and radio waves are of concern to electrical engineers. The type of electromagnetic radiation that is pertinent to heat transfer is the **thermal radiation** emitted as a result of vibrational and rotational motions of molecules, atoms, and electrons of a substance. Temperature is a measure of the strength of these activities at the microscopic level, as discussed in Chapter 1, and the rate of thermal radiation emission increases with

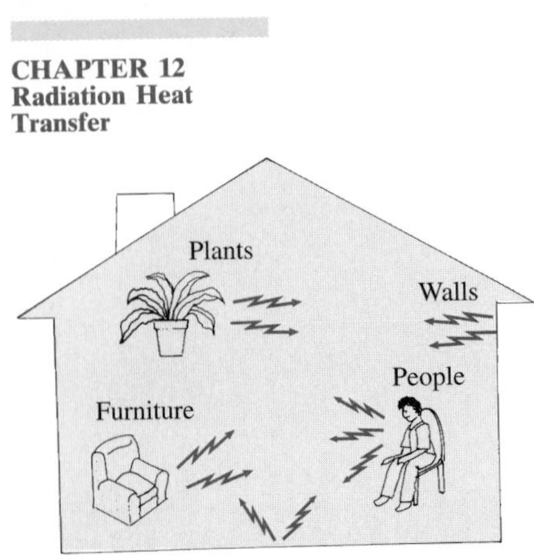


**FIGURE 12-4**

Everything around us constantly emits thermal radiation.

**TABLE 12-1**
**The wavelength ranges of different colors**

| Color | Wavelength band |
|---|---|
| Violet | 0.40–0.44 $\mu$m |
| Blue | 0.44–0.49 $\mu$m |
| Green | 0.49–0.54 $\mu$m |
| Yellow | 0.54–0.60 $\mu$m |
| Orange | 0.60–0.63 $\mu$m |
| Red | 0.63–0.76 $\mu$m |

increasing temperature. Thermal radiation is continuously emitted by all matter whose temperature is above absolute zero. That is, everything around us such as walls, furniture, and our friends constantly emit (and absorb) radiation (Fig. 12-4). Thermal radiation is also defined as the portion of the electromagnetic spectrum that extends from about 0.1 to 100 $\mu$m, since the radiation emitted by bodies because of their temperature falls almost entirely into this wavelength range. Thus, thermal radiation includes the entire visible and infrared (IR) radiation as well as a portion of the ultraviolet (UV) radiation.

What we call **light** is simply the *visible* portion of the electromagnetic spectrum that lies between 0.40 and 0.76 $\mu$m. Light is characteristically no different than other electromagnetic radiation, except that it happens to trigger the sensation of seeing in the human eye. Light or the visible spectrum consists of narrow bands of color from violet (0.40–0.44 $\mu$m) to red (0.63–0.76 $\mu$m), as shown in Table 12-1. The color of a surface depends on its ability to *reflect* certain wavelengths. For example, a surface that reflects radiation in the wavelength range 0.63–0.76 $\mu$m while absorbing the rest of the visible radiation appears red to the eye. A surface that reflects all of the light appears *white,* while a surface that absorbs all of the light incident on it appears *black.*

A body that emits some radiation in the visible range is called a light source. The sun is obviously our primary light source. The electromagnetic radiation emitted by the sun is known as the **solar radiation**, and nearly all of it falls into the wavelength band 0.1–3 $\mu$m. Almost half of solar radiation is light (i.e., it falls into the visible range), with the remaining being ultraviolet and infrared.

The radiation emitted by bodies at room temperature fall into the **infrared** region of the spectrum, which extends from 0.76 to 100 $\mu$m. Bodies start emitting noticeable visible radiation at temperatures above 800 K. The tungsten filament of a light bulb must be heated to temperatures above 2000 K before it can emit any significant amount of radiation in the visible range.

The **ultraviolet** radiation occupies the low-wavelength end of the thermal radiation spectrum, and lies between the wavelengths 0.01 and 0.40 $\mu$m. Ultraviolet rays are to be avoided, since they can kill microorganisms and cause serious damage to humans and other living organisms. About 12 percent of solar radiation is in the ultraviolet range, and it would be devastating if it were to reach the surface of the earth. Fortunately, the ozone ($O_3$) layer in the atmosphere acts as a protective blanket, and absorbs most of this ultraviolet radiation. The ultraviolet rays that remain in sunlight are still sufficient to cause serious sunburns in sun worshippers, and prolonged exposure to direct sunlight is the leading cause of skin cancer, which can be lethal. Recent discoveries of "holes" in the ozone layer has prompted the international community to ban the use of ozone-destroying chemicals such as the widely used refrigerant Freon-12 in order to save the earth. Ultraviolet radiation is also produced artificially in fluorescent lamps for use in medicine as a bacteria killer and in tanning parlors as an artificial tanner. The connection between the skin cancer and ultraviolet rays has caused dermatologists to issue strong warnings against its use for tanning.

Microwave ovens utilize electromagnetic radiation in the **microwave** region of the spectrum generated by microwave tubes called *magnetrons.* Microwaves in the range of $10^2$–$10^5$ $\mu$m are very suitable for use in cooking since they are *reflected* by metals, *transmitted* by glass and plastics, and *absorbed* by food (especially water) molecules. Thus, the electric energy converted to radiation in a microwave oven eventually becomes part of the internal energy of the food. The fast and efficient cooking of microwave ovens has made them some of the essential appliances in modern kitchens (Fig. 12-5).

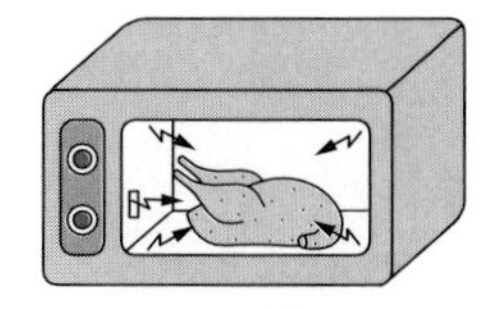

**FIGURE 12-5**

Food is heated or cooked in a microwave oven by absorbing the electromagnetic radiation energy generated by the magnetron of the oven.

Radars and cordless telephones also use electromagnetic radiation in the microwave region. The wavelength of the electromagnetic waves used in radio and TV broadcasting usually range between 1 and 1000 m in the **radio wave** region of the spectrum.

In heat transfer studies, we are interested in the energy emitted by bodies because of their temperature only. Therefore, we will limit our consideration to *thermal radiation*, which we will simply call *radiation.* The relations developed below are restricted to thermal radiation only, and may not be applicable to other forms of electromagnetic radiation.

The electrons, atoms, and molecules of all solids, liquids and gases above absolute zero temperature are constantly in motion, and thus radiation is constantly emitted, as well as being absorbed or transmitted throughout the entire volume of matter. That is, radiation is a **volumetric phenomena**. However, for opaque (nontransparent) solids such as metals, wood, and rocks, radiation is considered to be a **surface phenomena**, since the radiation emitted by the interior regions can never reach the surface, and the radiation incident on such bodies is usually absorbed within a few microns from the surface (Fig. 12-6). Note that the radiation characteristics of surfaces can be changed completely by applying thin layers of coatings on them.

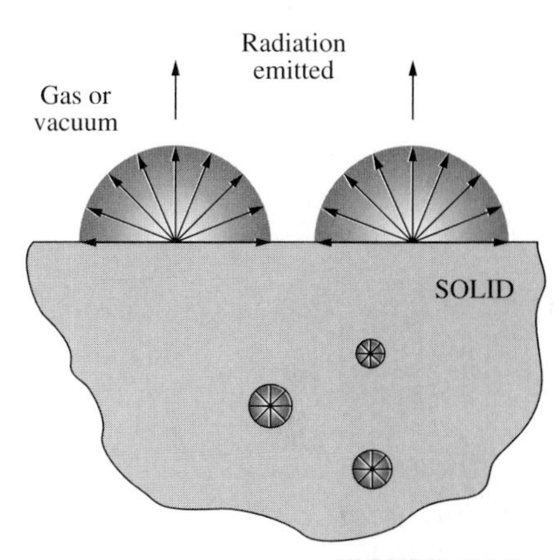


**FIGURE 12-6**

Radiation in opaque solids is considered a surface phenomena since the radiation emitted only by the molecules at the surface can escape the solid.

## 12-3 ■ BLACKBODY RADIATION

A body at a temperature above absolute zero emits radiation in all directions over a wide range of wavelengths. The amount of radiation energy emitted from a surface at a given wavelength depends on the material of the body and the condition of its surface as well as the surface temperature. Therefore, different bodies may emit different amounts of radiation per unit surface area, even when they are at the same temperature. Then it is natural to be curious about the *maximum* amount of radiation that can be emitted by a surface at a given temperature. Satisfying this curiosity requires the definition of an idealized body, called a *blackbody,* to serve as a standard against which the radiative properties of real surfaces may be compared.

A **blackbody** is defined as *a perfect emitter and absorber of radiation.* At a specified temperature and wavelength, no surface can emit more energy than a blackbody. A blackbody absorbs *all* incident radiation, regardless of wavelength and direction. Also, a blackbody emits radiation energy uniformly in all directions (Fig. 12-7). That is, a blackbody is a *diffuse* emitter. The term *diffuse* means "independent of direction".

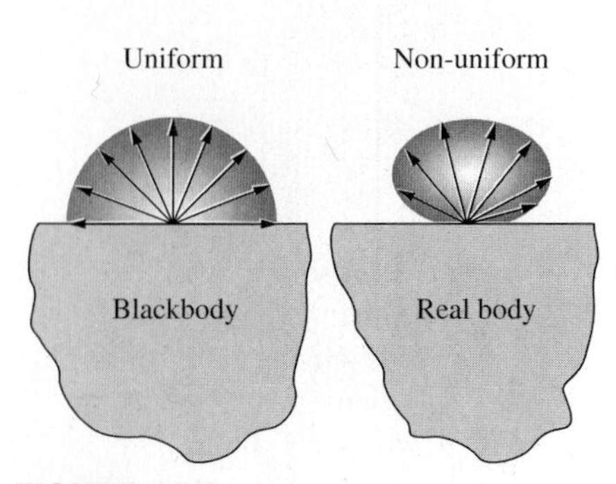


**FIGURE 12-7**

A blackbody is said to be a *diffuse* emitter since it emits radiation energy uniformly in all directions.

The radiation energy emitted by a blackbody per unit time and per unit surface area was determined experimentally by Joseph Stefan in 1879, and is expressed as

$$E_b = \sigma T^4 \quad (\text{W/m}^2) \tag{12-3}$$

where $\sigma = 5.67 \times 10^{-8}\ \text{W/(m}^2 \cdot \text{K}^4)$ is the *Stefan–Boltzmann constant* and $T$ is the absolute temperature of the surface in K. This relation is theroetically verified in 1884 by Ludwig Boltzmann. Equation 12-3 is known as the **Stefan–Boltzmann law** and $E_b$ is called the **blackbody emissive power**. Note that the emission of thermal radiation is proportional to the *fourth power* of the absolute temperature.

Although a blackbody would appear *black* to the eye, a distinction should be made between the idealized blackbody and an ordinary black surface. Any surface that absorbs light (the visible portion of radiation) would appear black to the eye, and a surface that reflects it completely would appear white. Considering that visible radiation occupies a very narrow band of the spectrum from 0.4 to 0.76 $\mu$m, we cannot make any judgments about the blackness of a surface on the basis of visual observations. For example, snow and white paint reflect light and thus appear white. But they are essentially black for infrared radiation, since they strongly absorb long-wavelength radiation. Surfaces coated with lampblack paint approach idealized blackbody behavior.

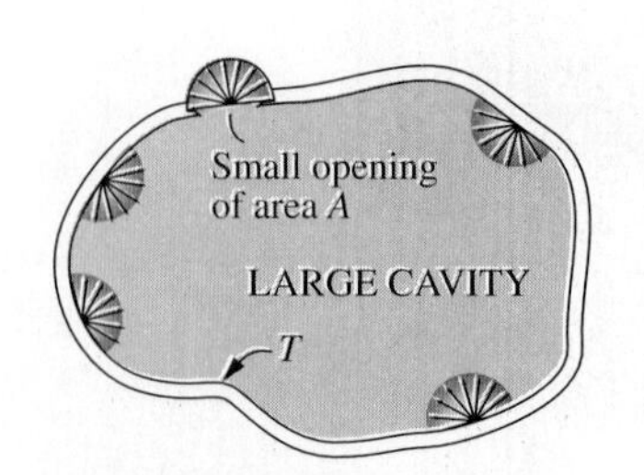


**FIGURE 12-8**

A large isothermal cavity at temperature $T$ with a small opening of area $A$ closely resembles a blackbody of surface area $A$ at the same temperature.

Another type of body that closely resembles a blackbody is a *large cavity with a small opening,* as shown in Fig. 12-8. Radiation coming in through the opening of area $A$ will undergo multiple reflections, and thus it will have several chances to be absorbed by the interior surfaces of the cavity before any part of it can possibly escape. Also, if the surface of the cavity is isothermal at temperature $T$, the radiation emitted by the interior surfaces will stream through the opening after undergoing multiple reflections, and thus it will have a diffuse nature. Therefore, the cavity will act as a perfect absorber and perfect emitter, and the opening will resemble a blackbody of surface area $A$ at temperature $T$, regardless of the actual radiative properties of the cavity.

The Stefan–Boltzmann law in Eq. 12-3 gives the *total* blackbody emissive power $E_b$, which is the sum of the radiation emitted over all wavelengths. Sometimes we need to know the **spectral blackbody emissive power**, which is *the amount of radiation energy emitted by a blackbody at an absolute temperature T per unit time, per unit surface area, and per unit wavelength about the wavelength* $\lambda$. For example, we are more interested in the amount of radiation an incandescent light bulb emits in the visible wavelength spectrum than we are in the total amount of radiation that the light bulb emits.

The relation for the spectral blackbody emissive power $E_{b\lambda}$ was developed by Max Planck in 1901 in conjunction with his famous

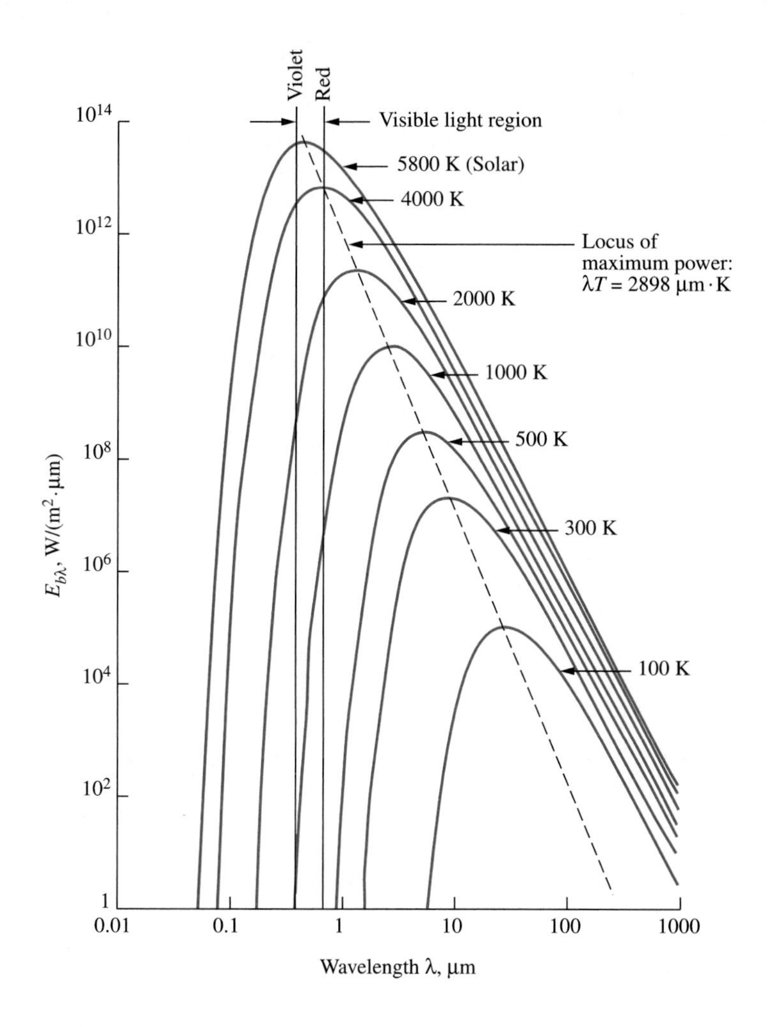

**FIGURE 12-9**

The variation of the blackbody emissive power with wavelength for several temperatures.

quantum theory. This relation is known as **Planck's distribution law**, and is expressed as

$$E_{b\lambda}(T) = \frac{C_1}{\lambda^5[\exp(C_2/\lambda T) - 1]} \qquad [\mathrm{W/(m^2 \cdot \mu m)}] \qquad (12\text{-}4)$$

where $C_1 = 2\pi h c_0^2 = 3.742 \times 10^8 \ \mathrm{W \cdot \mu m^4/m^2}$ and $C_2 = hc_0/k = 1.439 \times 10^4 \ \mu\mathrm{m \cdot K}$. Also, $T$ is the absolute temperature of the surface, $\lambda$ is the wavelength of the radiation emitted, and $k = 1.3805 \times 10^{-23}$ J/K is *Boltzmann's constant.* This relation is valid for a surface in a *vacuum* or a *gas.* For other mediums, it needs to be modified by replacing $C_1$ by $C_1/n^2$, where $n$ is the index of refraction of the medium. Note that the term *spectral* indicates dependence on wavelength.

The variation of the blackbody emissive power with wavelength is plotted in Fig. 12-9 for selected temperatures. Several observations can be made from this figure:

1 The emitted radiation is a continuous function of *wavelength.* At any

specified temperature, it increases with wavelength, reaches a peak, and then decreases with increasing wavelength.

**2** At any wavelength, the amount of emitted radiation *increases* with increasing temperature.

**3** As temperature increases, the curves get steeper and shift to the left to the shorter-wavelength region. Consequently, a larger fraction of the readiation is emitted at *shorter wavelengths* at higher temperatures.

**4** The radiation emitted by the *sun,* which is considered to be a blackbody at 5762 K (or roughly at 5800 K), reaches its peak in the visible region of the spectrum. Therefore, the sun is in tune with our eyes. On the other hand, surfaces at $T \leqslant 800$ K emit almost entirely in the infrared region, and thus are not visible to the eye unless they reflect light coming from other sources.

As the temperature increases, the peak of the curve in Fig. 12-9 shifts towards shorter wavelengths. The wavelength at which the peak occurs for a specified temperature is given by **Wien's displacement law** as

$$(\lambda T)_{\text{max power}} = 2897.8\ \mu\text{m}\cdot\text{K} \qquad (12\text{-}5)$$

This relation was originally developed by Willy Wien in 1894 using classical thermodynamics, but it can also be obtained by differentiating Eq. 12-4 with respect to $\lambda$ while holding $T$ constant, and setting the result equal to zero. A plot of Wien's displacement law, which is the locus of the peaks of the radiation emission curves, is also given in Fig. 12-9.

The peak of the solar radiation, for example, occurs at $\lambda = 2897.8/5762 = 0.50\ \mu$m, which is near the middle of the visible range. The peak of the radiation emitted by a surface at room temperature ($T = 298$ K) occurs at 9.72 $\mu$m, which is well into the infrared region of the spectrum.

An electrical resistance heater starts radiating heat soon after it is plugged in, and we can feel the emitted radiation energy by holding our hands against the heater. But this radiation is entirely in the infrared region, and thus cannot be sensed by our eyes. The heater would appear dull red when its temperature reaches about 1000 K, since it will start emitting a detectable amount [about 1 W/(m$^2\cdot\mu$m)] of visible red radiation at that temperature. As the temperature rises even more, the heater appears bright red, and is said to be *red hot.* When the temperature reaches about 1500 K, the heater emits enough radiation in the entire visible range of the spectrum to appear almost *white* to the eye, and it is called *white hot.*

Although it cannot be sensed directly by the human eye, infrared radiation can be detected by infrared cameras, which transmit the information to microprocessors to display visual images of objects at night. *Rattlesnakes* can sense the infrared radiation or the "body heat" coming off warm-blooded animals, and thus they can see at night without using any instruments. Similarly, honeybees are sensitive to ultraviolet radiation.

It should be clear from the discussion above that the color of an object is not due to emission, which is primarily in the infrared region, unless the surface temperature of the object exceeds about 1000 K. Instead, the color of a surface depends on the absorption and reflection characteristics of the surface, and is due to selective absorption and reflection of the incident visible radiation coming from a light source such as the sun or an incandescent light bulb. A piece of clothing containing a pigment that reflects red while absorbing the remaining parts of the incident light appears "red" to the eye (Fig. 12-10). Leaves appear "green" because their cells contain the pigment chlorophyll, which strongly reflects green while absorbing other colors.

It is left as an exercise to show that integration of the *spectral* blackbody emissive power $E_{b\lambda}$ over the entire wavelength spectrum gives the *total* blackbody emissive power $E_b$:

$$E_b(T) = \int_0^{\infty} E_{b\lambda}(T)\, d\lambda = \sigma T^4 \quad (\text{W/m}^2) \tag{12-6}$$

Thus, we obtained the Stefan–Boltzmann law (Eq. 12-3) by integrating Planck's distribution law (Eq. 12-4) over all wavelengths. Note that on a $E_{b\lambda}$-$\lambda$ chart, $E_{b\lambda}$ corresponds to any value on the curve, whereas $E_b$ corresponds to the area under the entire curve for a specified temperature (Fig. 12-11). Also, the term *total* means "integrated over all wavelengths."

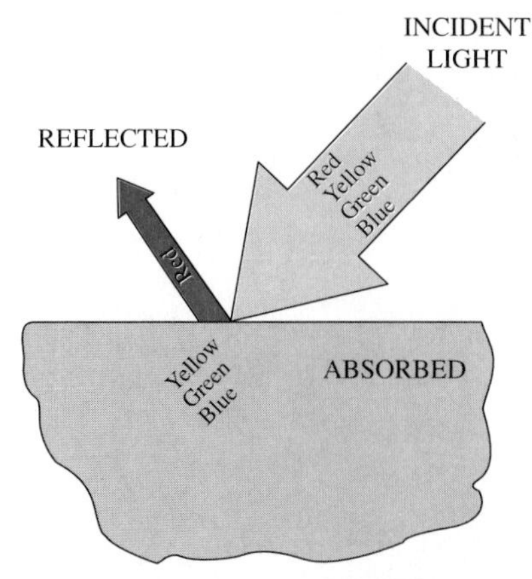


**FIGURE 12-10**

A surface that reflects red while absorbing the remaining parts of the incident light appears red to the eye.

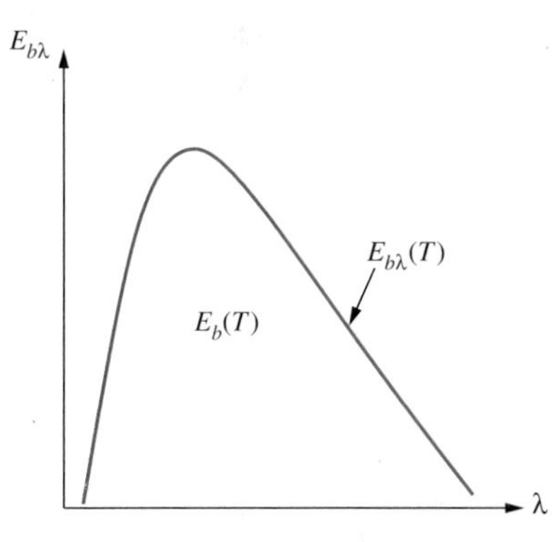


**FIGURE 12-11**

On a $E_{b\lambda}$-$\lambda$ chart, the area under a curve for a given temperature represents the total radiation energy emitted by a blackbody at that temperature.

**EXAMPLE 12-1**

Consider a 20-cm-diameter spherical ball at 800 K suspended in the air as shown in Fig. 12-12. Assuming that the ball closely approximates a blackbody, determine (*a*) the total blackbody emissive power, (*b*) the total amount of radiation emitted by the ball in 5 min, and (*c*) the spectral blackbody emissive power at a wavelength of 3 $\mu$m.

**Solution** (*a*) The total blackbody emissive power is determined from the Stefan–Boltamann law to be

$$E_b = \sigma T^4 = [5.67 \times 10^{-8}\ \text{W/(m}^2 \cdot \text{K}^4)](800\ \text{K}^4) = 23{,}224\ \text{W/m}^2$$

That is, the ball emits 23,224 J of energy in the form of electromagnetic radiation per second per m$^2$ of the surface area of the ball.

(*b*) The total amount of radiation energy emitted from the entire ball in 5 min is determined by multiplying the blackbody emissive power obtained above by the total surface area of the ball and the given time interval:

$$A = \pi D^2 = \pi(0.2\ \text{m}^2) = 0.1257\ \text{m}^2$$

$$\Delta t = (5\ \text{min})\left(\frac{60\ \text{s}}{1\ \text{min}}\right) = 300\ \text{s}$$

$$Q_{\text{rad}} = E_b A\, \Delta t = (23{,}224\ \text{W/m}^2)(0.1257\ \text{m}^2)(300\ \text{s})\left(\frac{1\ \text{kJ}}{1000\ \text{W} \cdot \text{s}}\right)$$

$$= 875.8\ \text{kJ}$$

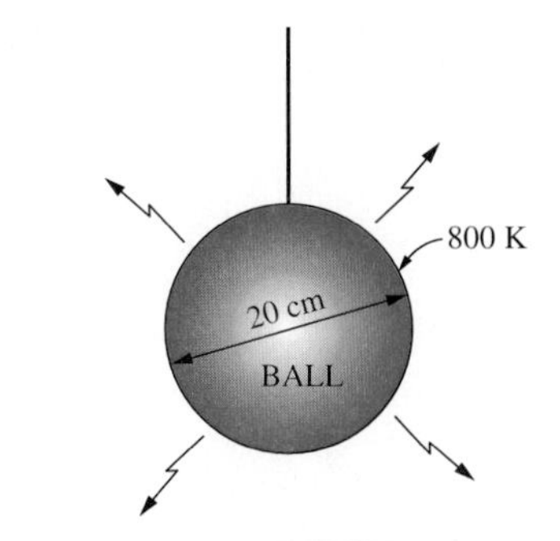


**FIGURE 12-12**

The spherical ball considered in Example 12-1.

That is, the ball loses 875.8 kJ of its internal energy in the form of electromagnetic waves to the surroundings in 5 min, which is enough energy to raise the temperature of 1 kg of water by 50°C. Note that the surface temperature of the ball cannot remain constant at 800 K unless there is an equal amount of energy flow to the surface from the surroundings or from the interior regions of the ball through some mechanisms such as chemical or nuclear reactions.

(*c*) The spectral blackbody emissive power at a wavelength of 3 μm is determined from Planck's distribution law to be

$$E_{b\lambda} = \frac{C}{\lambda^5\left[\exp\left(\frac{C_2}{\lambda T}\right) - 1\right]} = \frac{3.743 \times 10^8 \text{ W} \cdot \mu\text{m}^4/\text{m}^2}{(3\ \mu\text{m})^5\left[\exp\left(\frac{1.4387 \times 10^4\ \mu\text{m} \cdot \text{K}}{(3\ \mu\text{m})(800\text{ K})}\right) - 1\right]}$$

$$= 3848.4 \text{ W/(m}^2 \cdot \mu\text{m)}$$

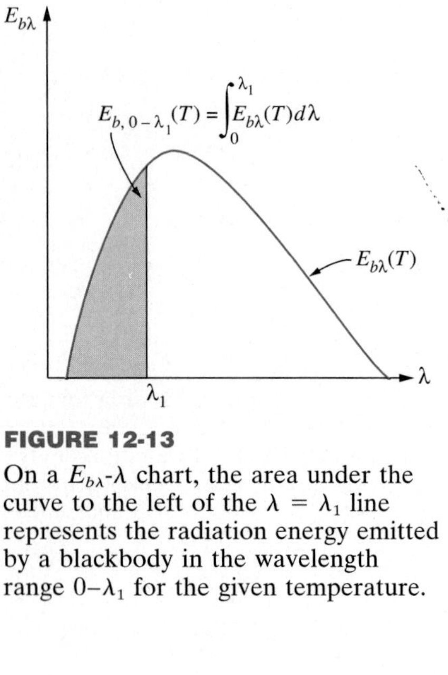

The Stefan–Boltzmann law $E_b(T) = \sigma T^4$ gives the *total* radiation emitted by a blackbody at all wavelengths from $\lambda = 0$ to $\lambda = \infty$. But we are often interested in the amount of radiation emitted over *some wavelength band*. For example, an incandescent light bulb is judged on the basis of the radiation it emits in the visible range rather than the radiation it emits at all wavelengths.

The radiation energy emitted by a blackbody per unit area over a wavelength band from $\lambda = 0$ to $\lambda$ is determined from (Fig. 12-13)

$$E_{b,0-\lambda}(T) = \int_0^{\lambda} E_{b\lambda}(T)\, d\lambda \qquad (\text{W/m}^2) \tag{12-7}$$

**FIGURE 12-13**
On a $E_{b\lambda}$-$\lambda$ chart, the area under the curve to the left of the $\lambda = \lambda_1$ line represents the radiation energy emitted by a blackbody in the wavelength range $0$–$\lambda_1$ for the given temperature.

It looks like we can determine $E_{b,0-\lambda}$ by substituting the $E_{b\lambda}$ relation from Eq. 12-4 and performing this integration. But it turns out that this integration does not have a simple closed-form solution, and performing a numerical integration each time we need a value of $E_{b,0-\lambda}$ is not practical. Therefore, we define a dimensionless quantity $f_\lambda$ called the **blackbody radiation function** as

$$f_\lambda(T) = \frac{\int_0^{\lambda} E_{b\lambda}(T)\, d\lambda}{\sigma T^4} \tag{12-8}$$

*The function* $f_\lambda$ represents *the fraction of radiation emitted from a blackbody at temperature T in the wavelength band from* $\lambda = 0$ *to* $\lambda$. The values of $f_\lambda$ are listed in Table 12-2 as a function of $\lambda T$, where $\lambda$ is in μm and $T$ is in K.

**TABLE 12-2**
**Blackbody radiation functions $f_\lambda$**

| $\lambda T$, μm · K | $f_\lambda$ | $\lambda T$, μm · K | $f_\lambda$ |
|---|---|---|---|
| 200 | 0.000000 | 6,200 | 0.754140 |
| 400 | 0.000000 | 6,400 | 0.769234 |
| 600 | 0.000000 | 6,600 | 0.783199 |
| 800 | 0.000016 | 6,800 | 0.796129 |
| 1000 | 0.000321 | 7,000 | 0.808109 |
| 1200 | 0.002134 | 7,200 | 0.819217 |
| 1400 | 0.007790 | 7,400 | 0.829527 |
| 1600 | 0.019718 | 7,600 | 0.839102 |
| 1800 | 0.039341 | 7,800 | 0.848005 |
| 2000 | 0.066728 | 8,000 | 0.856288 |
| 2200 | 0.100888 | 8,500 | 0.874608 |
| 2400 | 0.140256 | 9,000 | 0.890029 |
| 2600 | 0.183120 | 9,500 | 0.903085 |
| 2800 | 0.227897 | 10,000 | 0.914199 |
| 3000 | 0.273232 | 10,500 | 0.923710 |
| 3200 | 0.318102 | 11,000 | 0.931890 |
| 3400 | 0.361735 | 11,500 | 0.939959 |
| 3600 | 0.403607 | 12,000 | 0.945098 |
| 3800 | 0.443382 | 13,000 | 0.955139 |
| 4000 | 0.480877 | 14,000 | 0.962898 |
| 4200 | 0.516014 | 15,000 | 0.969981 |
| 4400 | 0.548796 | 16,000 | 0.973814 |
| 4600 | 0.579280 | 18,000 | 0.980860 |
| 4800 | 0.607559 | 20,000 | 0.985602 |
| 5000 | 0.633747 | 25,000 | 0.992215 |
| 5200 | 0.658970 | 30,000 | 0.995340 |
| 5400 | 0.680360 | 40,000 | 0.997967 |
| 5600 | 0.701046 | 50,000 | 0.998953 |
| 5800 | 0.720158 | 75,000 | 0.999713 |
| 6000 | 0.737818 | 100,000 | 0.999905 |

The fraction of radiation energy emitted by a blackbody at temperature $T$ over a finite wavelength band from $\lambda = \lambda_1$ to $\lambda = \lambda_2$ is determined from (Fig. 12-14)

$$f_{\lambda_1-\lambda_2}(T) = f_{\lambda_2}(T) - f_{\lambda_1}(T) \tag{12-9}$$

where $f_{\lambda_1}(T)$ and $f_{\lambda_2}(T)$ are blackbody radiation functions corresponding to $\lambda_1 T$ and $\lambda_2 T$, respectively.

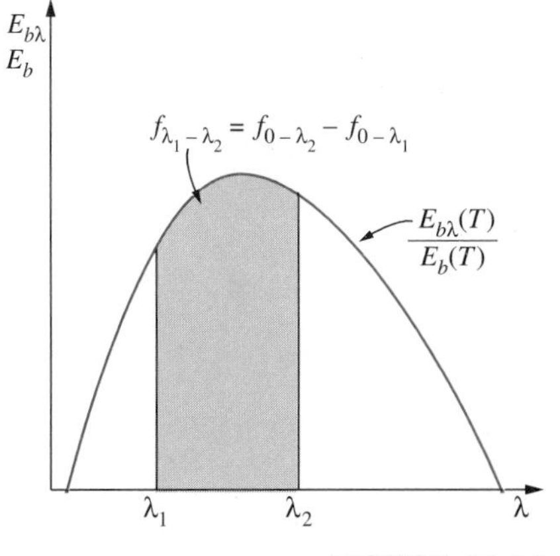


**FIGURE 12-14**
Graphical representation of the fraction of radiation emitted in the wavelength band from $\lambda_1$ to $\lambda_2$.

**EXAMPLE 12-2**

The temperature of the filament of an incandescent light bulb is 2500 K. Assuming the filament to be a blackbody, determine the fraction of the radiant energy emitted by the filament that falls in the visible range. Also determine the wavelength at which the emission of radiation from the filament peaks.

**Solution** The visible range of the electromagnetic spectrum extends from $\lambda_1 = 0.4\ \mu m$ to $\lambda_2 = 0.76\ \mu m$. Noting that $T = 2500$ K, the blackbody radiation functions corresponding to $\lambda_1 T$ and $\lambda_2 T$ are determined from Table 12-2 to be

$$\lambda_1 T = (0.40\ \mu m)(2500\ K) = 1000\ \mu m \cdot K \longrightarrow f_{\lambda_1} = 0.000321$$

$$\lambda_2 T = (0.76\ \mu m)(2500\ K) = 1900\ \mu m \cdot K \longrightarrow f_{\lambda_2} = 0.053035$$

That is, 0.03% of the radiation is emitted at wavelengths less than 0.4 $\mu$m and 5.3% at wavelengths less than 0.76 $\mu$m. Then the fraction of radiation emitted between these two wavelengths is (Fig. 12-15)

$$f_{\lambda_1 - \lambda_2} = f_{\lambda_2} - f_{\lambda_1} = 0.053035 - 0.000321 = 0.0527135$$

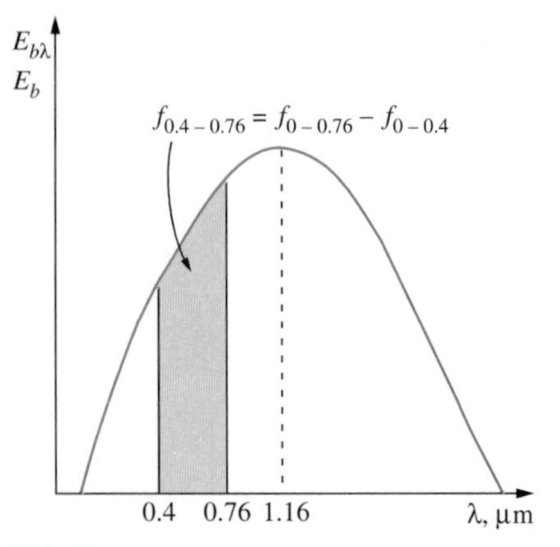


**FIGURE 12-15**

Graphical representation of the fraction of radiation emitted in the visible range in Example 12-2.

That is, only about 5% of the radiation emitted by the filament of the light bulb falls in the visible range. The remaining 95% of the radiation appears in the infrared region in the form of radiant heat or "invisible light," as it used to be called. This is certainly a very inefficient way of converting electrical energy to light, and explains why florescent tubes are a wiser choice for lighting.

The wavelength at which the emission of radiation from the filament peaks is easily determined from Wien's displacement law (Eq. 12-5) to be

$$(\lambda T)_{\text{max power}} = 2897.8\ \mu m \cdot K \rightarrow \lambda_{\text{max power}} = \frac{2897.8\ \mu m \cdot K}{2500\ K} = 1.16\ \mu m$$

Note that the radiation emitted from the filament peaks in the infrared region.

## 2-4 RADIATION PROPERTIES

Most materials encountered in practice, such as metals, wood, and bricks, are *opaque* to thermal radiation, and radiation is considered to be a *surface phenomena* for such materials. That is, thermal radiation is emitted or absorbed within the first few microns of the surface, and we speak of radiation properties of *surfaces* for opaque materials.

Some other materials, such as glass and water, allow visible radiation to penetrate to considerable depths before any significant absorption takes place. Radiation through such *semitransparent* materials obviously cannot be considered to be a surface phenomena, since the entire volume of the material interacts with radiation. On the other hand, both glass and water are practically opaque to infrared radiation. Therefore, materials can exhibit different behavior at different wavelengths, and the dependence on wavelength is an important consideration in the study of radiation properties such as emissivity, absorptivity, reflectivity, and transmissivity of materials.

In the preceding section, we defined a *blackbody* as a perfect emitter and absorber of radiation, and said that no body can emit more radiation than a blackbody at the same temperature. Therefore, a blackbody can serve as a convenient *reference* in describing the emission and absorption characteristics of real surfaces.

## Emissivity

The **emissivity** of a surface is defined as *the ratio of the radiation emitted by the surface to the radiation emitted by a blackbody at the same temperature.* The emissivity of a surface is denoted by $\varepsilon$, and it varies between zero and one, $0 \leqslant \varepsilon \leqslant 1$. Emissivity is a measure of how closely a surface approximates a blackbody for which $\varepsilon = 1$.

The emissivity of a real surface is not a constant. Rather, it varies with the *temperature* of the surface as well as the *wavelength* and the *direction* of the emitted radiation. Therefore, different emissivities can be defined for a surface, depending on the effects considered. For example, the emissivity of a surface at a specified wavelength is called *spectral emissivity,* and is denoted by $\varepsilon_\lambda$. Likewise, the emissivity in a specified direction is called *directional emissivity,* denoted by $\varepsilon_\theta$, where $\theta$ is the angle between the direction of radiation and the normal of the surface. The emissivity of a surface averaged over all directions is called the *hemispherical emissivity,* and the emissivity averaged over all wavelengths is called the *total emissiviity.* Thus the *total hemispherical emissivity* $\varepsilon$ of a surface is simply the average emissivity over all directions and wavelengths, and can be expressed as

$$\varepsilon(T) = \frac{E(T)}{E_b(T)} = \frac{E(T)}{\sigma T^4} \tag{12-10}$$

where $E(T)$ is the total emissive power of the real surface. Equation 12-10 can be rearranged as

$$E(T) = \varepsilon(T)\sigma T^4 \quad (\mathrm{W/m^2}) \tag{12-11}$$

Thus, the radiation emitted by the unit area of a real surface at temperature $T$ is obtained by multiplying the radiation emitted by a blackbody at the same temperature by the emissivity of the surface.

Spectral emissivity is defined in a similar manner as

$$\varepsilon_\lambda(T) = \frac{E_\lambda(T)}{E_{b\lambda}(T)} \tag{12-12}$$

where $E_\lambda(T)$ is the spectral emissive power of the real surface.

Radiation is a complex phenomena as it is, and the consideration of wavelength and direction dependence of properties, assuming sufficient data exists, makes it even more complicated. Therefore, the *gray* and *diffuse* approximations are commonly utilized in radiation calculations. A surface is said to be *diffuse* if its properties are *independent of direction,* and *gray* if its properties are *independent of wavelength.* Therefore, the emissivity of a gray, diffuse surface is simply the total hemispherical emissivity of that surface because of independence of direction and wavelength (Fig. 12-16).

*Real surface*:
$\varepsilon_\theta \neq$ constant
$\varepsilon_\lambda \neq$ constant

*Diffuse surface*:
$\varepsilon_\theta =$ constant

*Gray surface*:
$\varepsilon_\lambda =$ constant

*Diffuse, gray surface*:
$\varepsilon = \varepsilon_\lambda = \varepsilon_\theta =$ constant

**FIGURE 12-16**
The effect of diffuse and gray approximations on the emissivity of a surface.

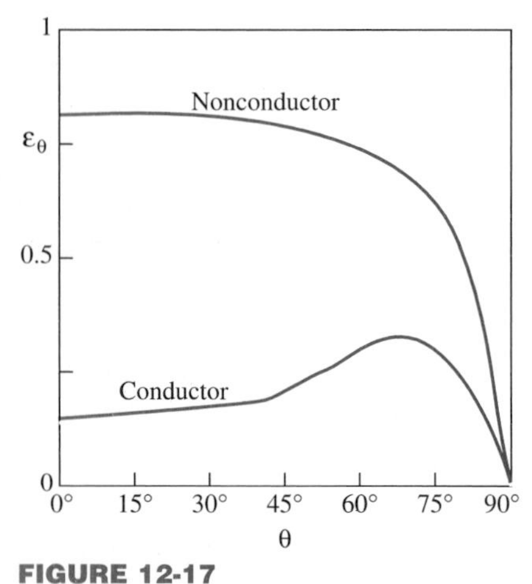


**FIGURE 12-17**

Typical variations of emissivity with direction for electrical conductors and nonconductors.

A few comments about the validity of the diffuse approximation are in order. Although real surfaces do not emit radiation in a perfectly diffuse manner as a blackbody does, they usually come close. The variation of emissivity with direction for both conductors and nonconductors is given in Fig. 12-17. Here $\theta$ is the angle measured from the normal of the surface, and thus $\theta = 0$ for radiation emitted in a direction normal to the surface. Note that $\varepsilon_\theta$ remains nearly constant for about $\theta < 40°$ for conductors such as metals, and for $\theta < 70°$ for nonconductors such as plastics. Therefore, the directional emissivity of a surface in the normal direction is representative of the hemispherical emissivity of the surface. In radiation analysis, it is common practice to assume the surfaces to be diffuse emitters with an emissivity equal to the value in the normal ($\theta = 0$) direction.

The effect of the gray approximation on emissivity and emissive power of a real surface is illustrated in Fig. 12-18. Note that the radiation emission from a real surface, in general, differs from Planck distribution, and the emission curve may have several peaks and valleys.

A gray surface should emit as much radiation as the real surface it represents at the same temperature. Therefore, the areas under the emission curves of the real and gray surfaces must be equal. That is, $\varepsilon(T)\sigma T^4 = \int_0^\infty \varepsilon_\lambda(T)E_{b\lambda}(T)\,d\lambda$. This requirement yields the following expression for the average emissivity:

$$\varepsilon(T) = \frac{\displaystyle\int_0^\infty \varepsilon_\lambda(T)E_{b\lambda}(T)\,d\lambda}{\sigma T^4} \tag{12-13}$$

To perform this integration, we need to know the variation of spectral emissivity with wavelength at the specified temperature. The integrand is usually a complicated function, and the integration has to be performed numerically. However, the integration can be performed quite easily by

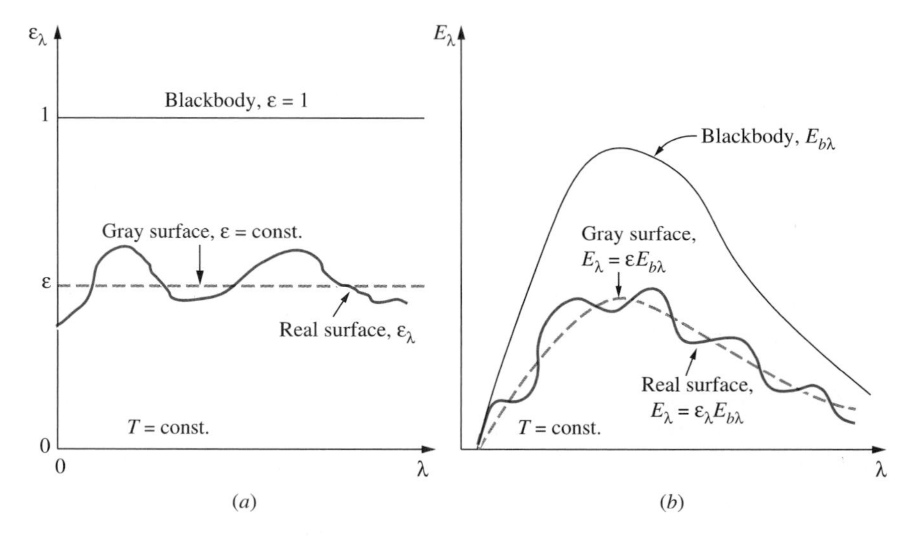


**FIGURE 12-18**

Comparison of the emissivity (*a*) and emissive power (*b*) of a real surface with those of a gray surface and a blackbody at the same temperature.

dividing the spectrum into a sufficient number of *wavelength bands* and assuming the emissivity to remain constant over each band; that is, by expressing the function $\varepsilon(T)$ as a step function. This simplification offers great convenience for little sacrifice of accuracy, since it allows us to transform the integration into a summation in terms of blackbody emission functions.

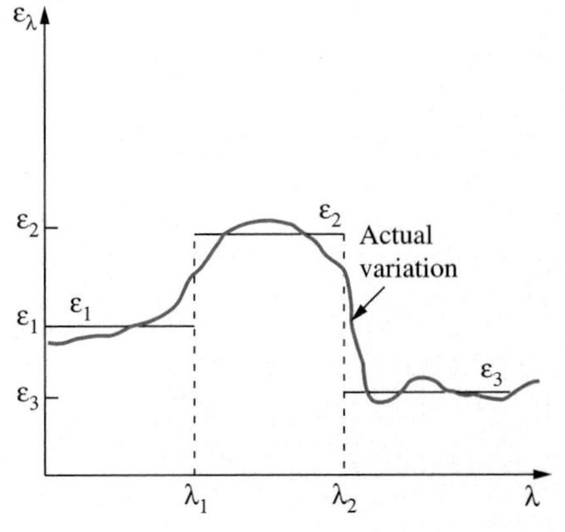

As an example, consider the emissivity function plotted in Fig. 12-19. It seems like this function can be approximated reasonably well by a step function of the form

$$\varepsilon_\lambda = \begin{cases} \varepsilon_1 = \text{constant}, & 0 \leqslant \lambda < \lambda_1 \\ \varepsilon_2 = \text{constant}, & \lambda_1 \leqslant \lambda < \lambda_2 \\ \varepsilon_2 = \text{constant}, & \lambda_2 \leqslant \lambda < \infty \end{cases} \tag{12-14}$$

**FIGURE 12-19**

Approximating the actual variation of emissivity with wavelength by a step function.

Then the average emissivity can be determined from Eq. 12-13 by breaking the integral into three parts and utilizing the definition of the blackbody radiation function as

$$\varepsilon(T) = \frac{\varepsilon_1 \int_0^{\lambda_1} E_{b\lambda}(T)\, d\lambda}{\sigma T^4} + \frac{\varepsilon_2 \int_{\lambda_1}^{\lambda_2} E_{b\lambda}(T)\, d\lambda}{\sigma T^4} + \frac{\varepsilon_3 \int_{\lambda_2}^{\infty} E_{b\lambda}(T)\, d\lambda}{\sigma T_4}$$

$$= \varepsilon_1 f_{0-\lambda_1}(T) + \varepsilon_2 f_{\lambda_1-\lambda_2}(T) + \varepsilon_3 f_{\lambda_2-\infty}(T) \tag{12-15}$$

The emissivities of common materials are listed in Table A-21 in the Appendix, and the variation of emissivity with wavelength and temperature is illustrated in Fig. 12-20. Typical ranges of emissivity of various materials are given in Fig. 12-21. Note that metals generally have low emissivities, as low as 0.02 for polished surfaces, and nonmetals such as ceramics and organic materials have high ones. The emissivity of metals

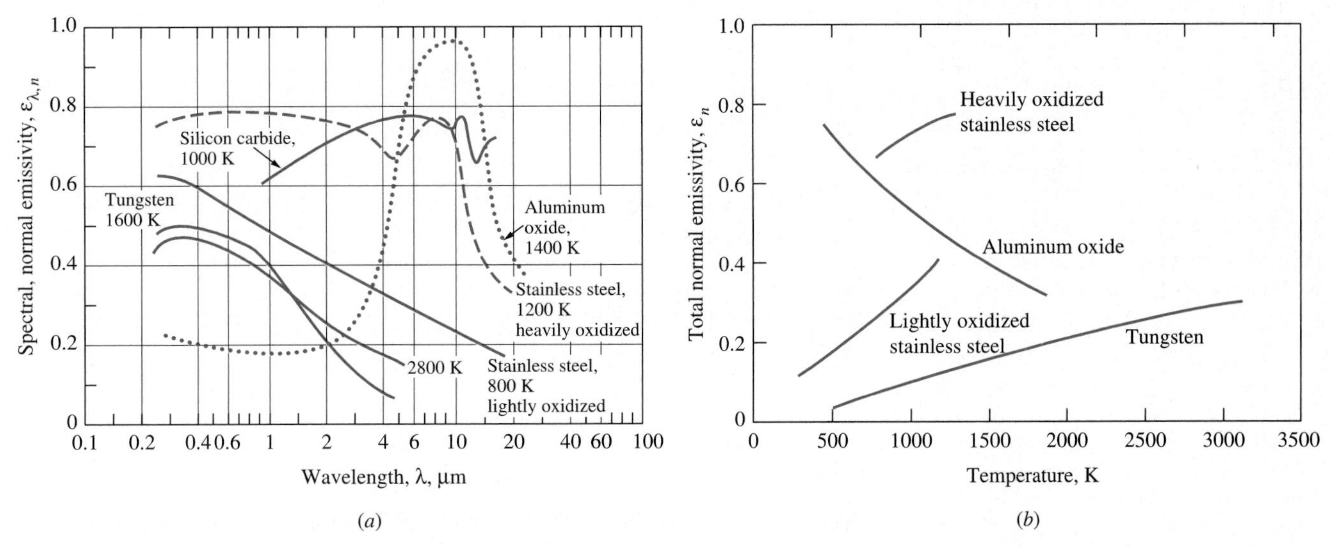


**FIGURE 12-20**

The variation of normal emissivity with (*a*) wavelength and (*b*) temperature for various materials.

increases with temperature. Also, oxidation causes significant increases in the emissivity of metals. Heavily oxidized metals can have emissivities comparable to those of nonmetals.

Care should be exercised in the use and interpretation of radiation property data reported in the literature, since the properties strongly depend on the surface conditions such as oxidation, roughness, type of finish, and cleanliness. Consequently, there is considerable discrepancy and uncertainty in the reported values. This uncertainty is largely due to the difficulty in characterizing and describing the surface conditions precisely.

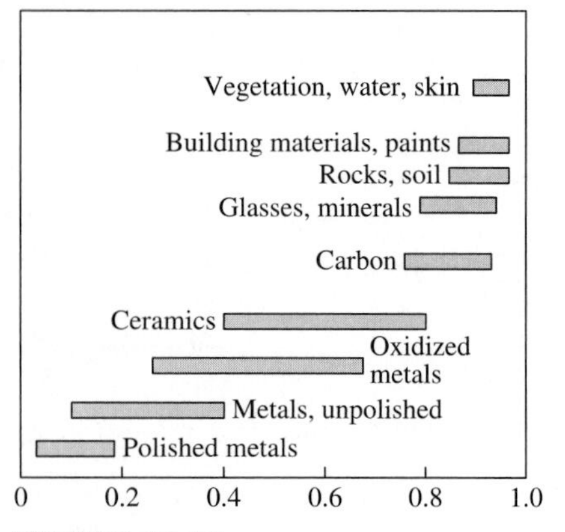


**FIGURE 12-21**

Typical ranges of emissivity for various materials.

**EXAMPLE 12-3**

The spectral emissivity function of an opaque surface at 800 K is approximated as (Fig. 12-22)

$$\varepsilon_\lambda = \begin{cases} \varepsilon_1 = 0.3, & 0 \leq \lambda < 3\ \mu\text{m} \\ \varepsilon_2 = 0.8, & 3\ \mu\text{m} \leq \lambda < 7\ \mu\text{m} \\ \varepsilon_3 = 0.1, & 7\ \mu\text{m} \leq \lambda < \infty \end{cases}$$

Determine the average emissivity of the surface and its emissive power.

**Solution** The variation of the emissivity of the surface with wavelength is given as a step function. Therefore, the average emissivity of the surface can be determined from Eq. 12-13 by breaking the integral into three parts as shown in Eq. 12-15:

$$\begin{aligned} \varepsilon(T) &= \frac{\varepsilon_1 \int_0^{\lambda_1} E_{b\lambda}(T)\,d\lambda}{\sigma T^4} + \frac{\varepsilon_2 \int_{\lambda_1}^{\lambda_2} E_{b\lambda}(T)\,d\lambda}{\sigma T^4} + \frac{\varepsilon_3 \int_{\lambda_2}^{\infty} E_{b\lambda}(T)\,d\lambda}{\sigma T^4} \\ &= \varepsilon_1 f_{0-\lambda_1}(T) + \varepsilon_2 f_{\lambda_1-\lambda_2}(T) + \varepsilon_3 f_{\lambda_2-\infty}(T) \\ &= \varepsilon_1 f_{\lambda_1} + \varepsilon_2(f_{\lambda_2} - f_{\lambda_1}) + \varepsilon_3(1 - f_{\lambda_2}) \end{aligned}$$

where $f_{\lambda_1}$ and $f_{\lambda_2}$ are blackbody radiation functions corresponding to $\lambda_1 T$ and $\lambda_2 T$. These functions are determined from Table 12-2 to be

$$\lambda_1 T = (3\ \mu\text{m})(800\ \text{K}) = 2400\ \mu\text{m}\cdot\text{K} \rightarrow f_{\lambda_1} = 0.140256$$
$$\lambda_2 T = (7\ \mu\text{m})(800\ \text{K}) = 5600\ \mu\text{m}\cdot\text{K} \rightarrow f_{\lambda_2} = 0.701046$$

Note that $f_{0-\lambda_1} = f_{\lambda_1} - f_0 = f_{\lambda_1}$, since $f_0 = 0$, and $f_{\lambda_2-\infty} = f_\infty - f_{\lambda_2} = 1 - f_{\lambda_2}$, since $f_\infty = 1$. Substituting,

$$\begin{aligned} \varepsilon &= 0.3 \times 0.140256 + 0.8(0.701046 - 0.140256) + 0.1(1 - 0.701046) \\ &= 0.521 \end{aligned}$$

That is, the surface will emit as much radiation energy at 800 K as a gray surface having a constant emissivity $\varepsilon = 0.521$. The emissive power of the surface is

$$E = \varepsilon\sigma T^4 = 0.521[5.67 \times 10^{-8}\ \text{W/(m}^2\cdot\text{K}^4)](800\ \text{K}^4) = 12{,}100\ \text{W/m}^2$$

That is, the surface emits 12,100 J of radiation energy per second per m² area of the surface.

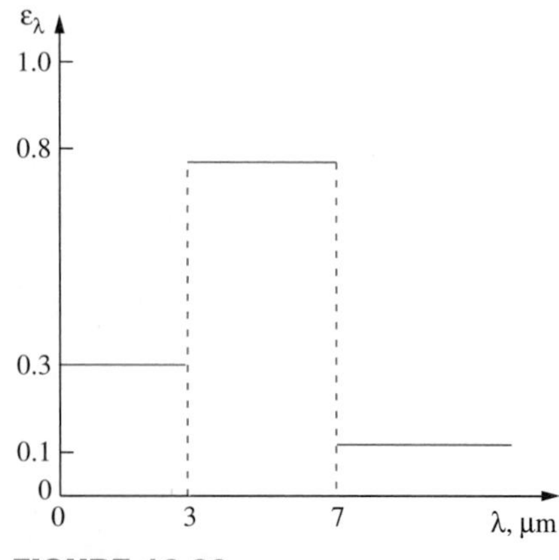


**FIGURE 12-22**

The spectral emissivity of the surface considered in Example 12-3.

Everything around us constantly emits radiation, and the emissivity represent the emission characteristics of those bodies. This means that every body, including our own, is constantly bombarded by radiation coming from all directions over a range of wavelengths. *The radiation energy incident on a surface per unit surface area per unit time* is called **irradiation**, and is denoted by $G$.

When radiation strikes a surface, part of it is absorbed, part of it is reflected, and the remaining part, if any, is transmitted, as illustrated in Fig. 12-23. *The fraction of irradiation absorbed by the surface* is called the **absorptivity** $\alpha$, *the fraction reflected by the surface* is called the **reflectivity** $\rho$, and *the fraction transmitted* is called the **transmissivity** $\tau$. That is,

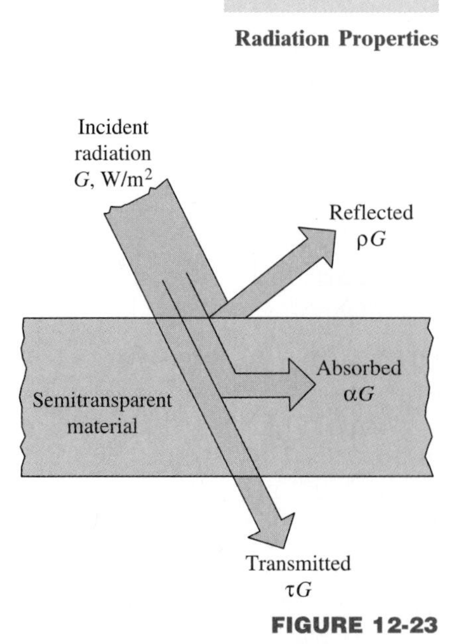


**FIGURE 12-23**
The absorption, reflection, and transmission of incident radiation by a semitransparent material.

absorptivity: $$\alpha = \frac{\text{absorbed radiation}}{\text{incident radiation}} = \frac{G_{\text{abs}}}{G}, \qquad 0 \leqslant \alpha \leqslant 1 \tag{12-16a}$$

reflectivity: $$\rho = \frac{\text{reflected radiation}}{\text{incident radiation}} = \frac{G_{\text{ref}}}{G}, \qquad 0 \leqslant \rho \leqslant 1 \tag{12-16b}$$

transmissivity: $$\tau = \frac{\text{transmitted radiation}}{\text{incident radiation}} = \frac{G_{\text{tr}}}{G}, \qquad 0 \leqslant \tau \leqslant 1 \tag{12-16c}$$

where $G$ is the radiation energy incident on the surface, and $G_{\text{abs}}$, $G_{\text{ref}}$, and $G_{\text{tr}}$ are the absorbed, reflected, and transmitted portions of it, respectively. The first law of thermodynamics requires that the sum of the absorbed, reflected, and transmitted radiation energy be equal to the incident radiation. That is,

$$G_{\text{abs}} + G_{\text{ref}} + G_{\text{tr}} = G$$

Dividing each term of this relation by $G$ yields

$$\alpha + \rho + \tau = 1 \tag{12-17}$$

For opaque surfaces $\tau = 0$, and thus

$$\alpha + \rho = 1 \tag{12-18}$$

This is an important property relation, since it allows us to determine both the absorptivity and reflectivity of an opaque surface by measuring either of these properties.

The definitions above are for *total hemispherical* properties, since $G$ represents the radiation energy incident on the surface from all directions over the hemispherical space and over all wavelengths. Thus, $\alpha$, $\rho$, and $\tau$ are the *average* properties of a medium for all directions and all wavelengths. However, like emissivity, these properties can also be defined for a specific wavelength or direction. For example, the *spectral* absorptivity, reflectivity, and transmissivity of a surface are defined in a similar manner as

$$\alpha_\lambda = \frac{G_{\lambda,\text{abs}}}{G_\lambda}, \qquad \rho_\lambda = \frac{G_{\lambda,\text{ref}}}{G_\lambda}, \qquad \tau_\lambda = \frac{G_{\lambda,\text{tr}}}{G_\lambda} \tag{12-19}$$

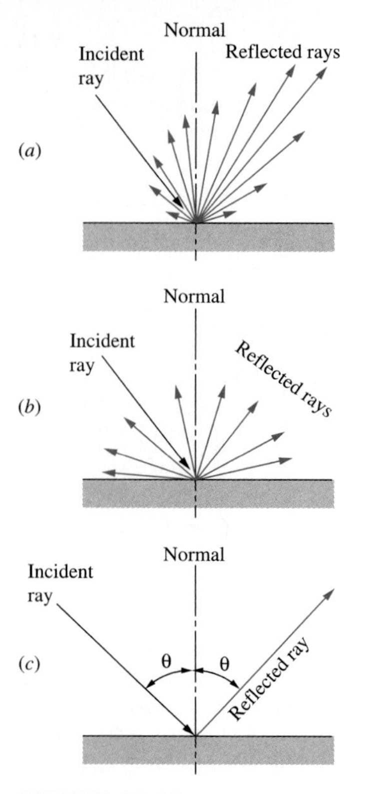


**FIGURE 12-24**

Different types of reflection from a surface: (*a*) actual or irregular, (*b*) diffuse, and (*c*) specular or mirrorlike.

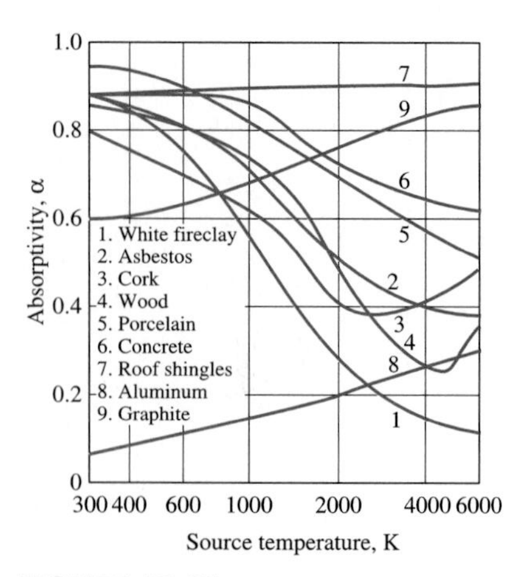


**FIGURE 12-25**

Variation of absorptivity with the temperature of the source of irradiation for various common materials at toom temperature.

where $G_\lambda$ is the radiation energy incident at the wavelength $\lambda$, and $G_{\lambda,\text{abs}}$, $G_{\lambda,\text{ref}}$, and $G_{\lambda,\text{tr}}$ are the absorbed, reflected, and transmitted portions of it, respectively. Similar definitions can be given for *directional* properties in direction $\theta$ by replacing all occurrences of the subscripts $\lambda$ in Eq. 12-19 by $\theta$.

The average absorptivity, reflectivity, and transmissivity of a surface can also be defined in terms of their spectral counterparts as

$$\alpha = \frac{\int_0^\infty \alpha_\lambda G_\lambda \, d\lambda}{\int_0^\infty G_\lambda \, d\lambda}, \qquad \rho = \frac{\int_0^\infty \rho_\lambda G_\lambda \, d\lambda}{\int_0^\infty G_\lambda \, d\lambda}, \qquad \tau = \frac{\int_0^\infty \tau_\lambda G_\lambda \, d\lambda}{\int_0^\infty G_\lambda \, d\lambda} \tag{12-20}$$

The reflectivity differs somewhat from the other properties in that it is *bidirectional* in nature. That is, the value of the reflectivity of a surface depends not only on the direction of the incident radiation but also the direction of reflection. Therefore, the reflected rays of a radiation beam incident on a real surface in a specified direction will form an irregular shape, as shown in Fig. 12-24. Such detailed reflectivity data does not exist for most surfaces, and even if it did, it would be of little value in radiation calculations, since it would usually add more complication to the analysis than it is worth.

In practice, for simplicity, surfaces are assumed to reflect in a perfectly *specular* or *diffuse* manner. In **specular** (or *mirrorlike*) **reflection**, *the angle or reflection equals the angle of incidence of the radiation beam.* In **diffuse reflection**, *radiation is reflected equally in all directions,* as shown in Fig. 12-24. Reflection from smooth and polished surfaces approximates specular reflection, whereas reflection from rough surfaces approximates diffuse reflection. In radiation analysis, smoothness is defined relative to wavelength. A surface is said to be *smooth* if the height of the surface roughness is much smaller than the wavelength of the incident radiation.

Unlike emissivity, the absorptivity of a material is practically independent of surface temperature. However, the absorptivity depends strongly on the temperature of the source at which the incident radiation is originating. This is also evident from Fig. 12-25, which shows the absorptivities of various materials at room temperature as functions of the temperature of the radiation source. For example, the absorptivity of the concrete roof of a house is about 0.6 for solar radiation (source temperature 5762 K) and 0.9 for radiation originating from the surrounding trees and buildings (source temperature 300 K), as illustrated in Fig. 12-26.

Notice that the absorptivity of aluminum increases with temperature, a characteristic for metals, and the absorptivity of electric nonconductors, in general, decreases with temperature. This decrease is most pronounced for surfaces that appear white to the eye. For example, the absorptivity of a white painted surface is low for solar radiation, although it is rather high for infrared radiation.

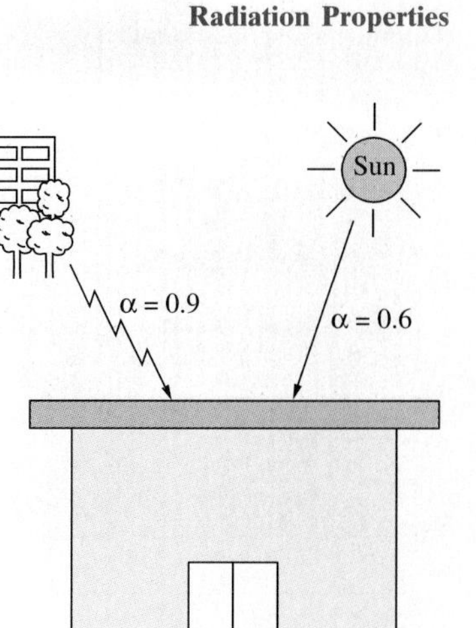


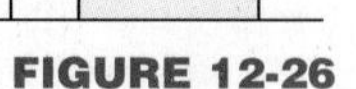
**FIGURE 12-26**
The absorptivity of a material may be quite different for radiation originating from sources at different temperatures.

## Kirchhoff's Law

Consider a small body of surface area $A$, emissivity $\varepsilon$, and absorptivity $\alpha$ at temperature $T$ contained in a large isothermal enclosure at the same temperature, as shown in Fig. 12-27. Recall that a large isothermal enclosure forms a blackbody cavity regardless of the radiative properties of the enclosure surface, and the body in the enclosure is too small to interfere with the blackbody nature of the cavity. Therefore, the radiation incident on any part of the surface of the small body is equal to the radiation emitted by a blackbody at temperature $T$. That is, $G = E_b(T) = \sigma T^4$, and the radiation absorbed by the small body per unit of its surface area is

$$G_{\text{abs}} = \alpha G = \alpha \sigma T^4$$

The radiation emitted by the small body is (Eq. 12-3)

$$E_{\text{emit}} = \varepsilon \sigma T^4$$

Considering that the small body is in thermal equilibrium with the enclosure, the net rate of heat transfer to the body must be zero. Therefore, the radiation emitted by the body must be equal to the radiation absorbed by it:

$$A\varepsilon\sigma T^4 = A\alpha\sigma T^4$$

Thus, we conclude that

$$\varepsilon(T) = \alpha(T) \tag{12-21}$$

That is, *the total hemispherical emissivity of a surface at temperature T is equal to its total hemispherical absorptivity for radiation coming from a blackbody at the same temperature.* This relation, which greatly simplifies the radiation analysis, was first developed by Gustav Kirchhoff in 1860, and is now called **Kirchhoff's law**. Note that this relation is derived under the condition that the surface temperature is equal to the temperature of the source of irradiation, and the reader is cautioned against using it when considerable difference (more than a few hundred degrees) exists between the surface temperature and the temperature of the source of irradiation.

The derivation above can also be repeated for radiation at a specified wavelength to obtain the *spectral* form of Kirchhoff's law:

$$\varepsilon_\lambda(T) = \alpha_\lambda(T) \tag{12-22}$$

This relation is valid when the irradiation or the emitted radiation is independent of direction. The form of Kirchhoff's law that involves no restrictions is the *spectral directional* form expressed as $\varepsilon_{\lambda,\theta}(T) = \alpha_{\lambda,\theta}(T)$. That is, the emissivity of a surface at a specified wavelength, direction, and temperature is always equal to its absorptivity at the same wavelength, direction, and temperature.

**FIGURE 12-27**
The small body contained in a large isothermal enclosure used in the development of Kirchhoff's law.

It is very tempting to use Kirchhoff's law in radiation analysis since the relation $\varepsilon = \alpha$ together with $\rho = 1 - \alpha$ enables us to determine all three properties of an opaque surface from a knowledge of only *one* property. Although using Kirchhoff's law gives acceptable results in many cases, in practice, care should be exercised when there is considerable difference between the surface temperature and the temperature of the source of incident radiation.

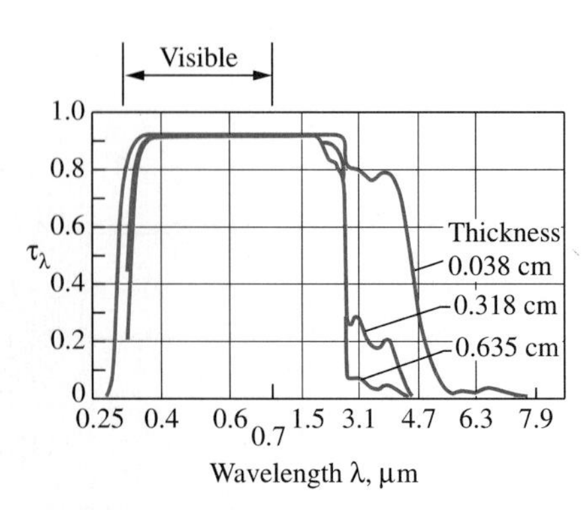


**FIGURE 12-28**

The spectral transmissivity of low-iron glass at room temperature for different thicknesses.

## The Greenhouse Effect

You probably noticed that when you leave yout car under direct sunlight on a sunny day, the interior of the car gets much warmer than the air outside, and you may have wondered why the car acts like a *heat trap.* The answer lies in the spectral transmissivity curve of the *glass,* which resembles an inverted U, as shown in Fig. 12-28. We observe from this figure that glass at thicknesses encountered in practice transmits over 90% of radiation in the visible range, and is practically opaque (nontransparent) to radiation in the longer-wavelength infrared regions of the electromagnetic spectrum (roughly $\lambda > 3\ \mu$m). Therefore, glass has a transparent window in the wavelength range $0.3\ \mu\text{m} < \lambda < 3\ \mu\text{m}$ in which over 90% of solar radiation is emitted. On the other hand, the entire radiation emitted by surfaces at room temperature falls in the infrared region. Consequently, glass allows the solar radiation to enter but does not allow the infrared radiation from the interior surfaces to exit. This causes a rise in the interior temperature as a result of the energy build-up in the car. This heating effect, which is due to the nongray characteristic of glass (or clear plastics), is known as the **greenhouse effect**, since it is utilized primarily in greenhouses (Fig. 12-29).

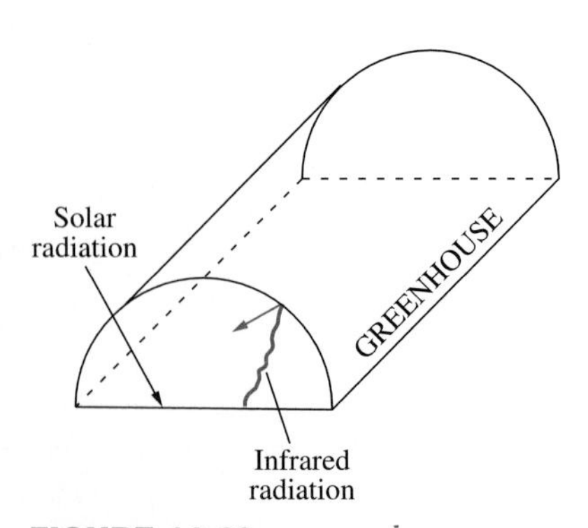


**FIGURE 12-29**

A greenhouse traps energy by allowing the solar radiation to come in but not allowing the infrared radiation to go out.

The greenhouse effect is also experienced on a larger scale on earth. The surface of the earth, which warms up during the day as a result of the absorption of solar energy, cools down at night by radiating its energy into deep space as infrared radiation. The combustion gases such as $CO_2$ and water vapor in the atmosphere transmit the bulk of the solar radiation but absorb the infrared radiation emitted by the surface of the earth. Thus, there is concern that the energy trapped on earth will eventually cause global warming, and thus drastic changes in weather patterns.

In *humid* places such as coastal areas, there is not a drastic change between the daytime and nighttime temperatures, because the humidity acts as a barrier on the path of the infrared radiation coming from the earth, and thus slows down the cooling process at night. In areas with clear skies such as deserts, there is a large swing between the daytime and nighttime temperatures, because of the absence of such barriers for infrared radiation.

## 12-5 ATMOSPHERIC AND SOLAR RADIATION

The sun is our primary source of energy. The energy coming off the sun, called *solar energy,* reaches us in the form of electromagnetic waves after experiencing considerable interactions with the atmosphere. The radiation energy emitted or reflected by the constituents of the atmosphere form the *atmospheric radiation.* Below we give an overview of the solar and atmospheric radiation because of their importance and relevance to daily life. Also, our familiarity with solar energy makes it an effective tool in developing a better understanding for some of the new concepts introduced earlier. Detailed tratment of this exciting subject can be found in numerous books devoted to this topic.

The *sun* is a nearly spherical body which has a diameter of $D \approx 1.39 \times 10^9$ m and a mass of $m \approx 2 \times 10^{30}$ kg, and is located at a mean distance of $L = 1.50 \times 10^{11}$ m from the earth. It emits radiation energy continuously at a rate of $E_{\text{sun}} \approx 3.8 \times 10^{26}$ W. Less than a billionth of this energy (about $1.7 \times 10^{17}$ W) strikes the earth, which is sufficient to keep the earth warm and to maintain life through the photosynthesis process. The energy of the sun is due to the continuous *fusion* reaction during which two hydrogen atoms fuse to form one atom of helium. Therefore, the sun is essentially a *nuclear reactor,* with temperatures as high as 40,000,000 K in its core region. The temperature drops to about 6000 K in the outer region of the sun, called the convective zone, as a result of the dissipation of this energy by radiation.

The solar energy reaching the earth's atmosphere is determined by a series of measurements taken in late 1960s by using high-altitude aircraft, balloons, and spacecraft to be 1353 W/m². This quantity is called the **solar constant** $G_s$:

$$G_s = 1353 \text{ W/m}^2 \tag{12-23}$$

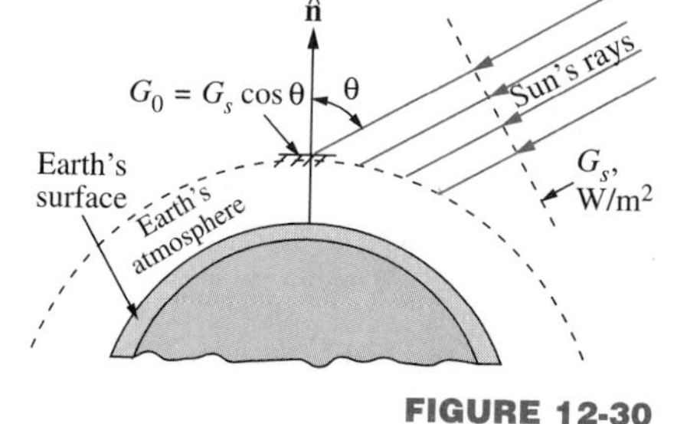


**FIGURE 12-30**
Solar radiation reaching the earth's atmosphere and the solar constant.

The **solar constant** represents *the rate at which solar energy is incident on a surface normal to sun's rays at the outer edge of the atmosphere when the earth is at its mean distance from the sun* (Fig. 12-30). Owing to the ellipticity of the earth's orbit, the distance between the sun and earth, and thus the actual value of the solar constant, changes throughout the year. It varies from a maximum of 1399 W/m² on December 21 to a minimum of 1310 W/m² on June 21. (Note that the earth is farthest away from the sun in summer in the northern hemisphere.) However, this variation, which remains within ±3.4 percent of the mean value is considered negligible for most practical purposes, and $G_s$ is taken to be a *constant* at its mean value of 1353 W/m².

The measured value of the solar constant can be used to estimate the effective surface temperature of the sun from the requirement that

$$(4\pi L^2)G_s = (4\pi r^2)\sigma T_{\text{sun}}^4 \tag{12-24}$$

where $L$ is the mean distance between the sun and the earth and $r$ is the radius of the sun. The left-hand side of this equation represents the total solar energy passing through a spherical surface whose radius is the mean earth–sun distance, and the right-hand side represents the total energy

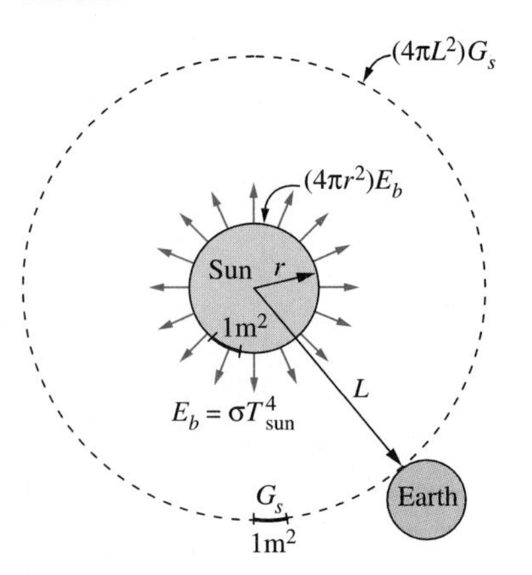


**FIGURE 12-31**

The total solar energy passing through concentric spheres remains constant, but the energy falling per unit area decreases with increasing radius.

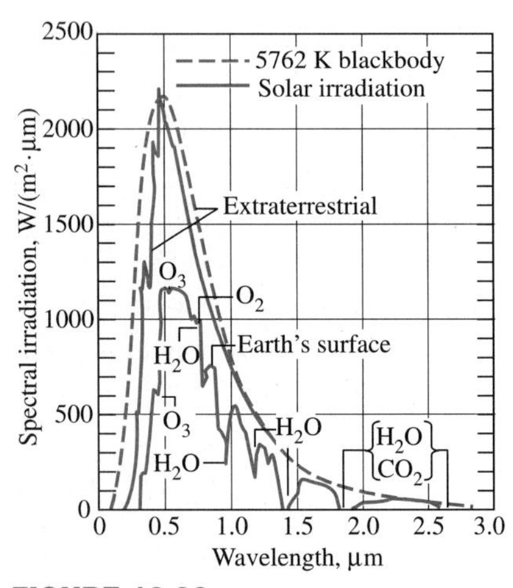


**FIGURE 13-32**

Spectral distribution of solar radiation just outside the atmosphere, at the surface of the earth on a typical day, and comparison with blackbody radiation at 5762 K.

that leaves the sun's outer surface. The conservation of energy principle requires that these two quantities be equal to each other, since the solar energy experiences no attenuation (or enhancement) on its way through the vacuum (Fig. 12-31). The **effective surface temperature** of the sun is determined from Eq. 12-24 to be $T_{sun}$ = 5762 K. That is, the sun can be treated as a blackbody at a temperature of 5762 K. This is also confirmed by the measurements of the spectral distribution of the solar radiation just outside the atmosphere plotted in Fig. 12-32, which shows only small deviations from the idealized blackbody behavior.

The spectral distribution of solar radiation on the ground plotted in Fig. 12-32 shows that the solar radiation undergoes considerable *attenuation* as it passes through the atmosphere as a result of *absorption* and scattering. About 99 percent of the atmosphere is contained within a distance of 30 km from the earth's surface. The several dips on the spectral distribution of radiation on the earth's surface are due to *absorption* by the gases $O_2$, $O_3$ (ozone), $H_2O$, and $CO_2$. Absorption by *oxygen* occurs in a narrow band about $\lambda$ = 0.76 $\mu$m. The *ozone* absorbs *ultraviolet* radiation at wavelengths below 0.3 $\mu$m almost completely, and radiation in the range 0.3–0.4 $\mu$m considerably. Thus, the ozone layer in the upper regions of the atmosphere protects biological systems on earth from harmful ultraviolet radiation. In turn, we must protect the ozone layer from the destructive chemicals commonly used as refrigerants, cleaning agents, and propellants in aerosol cans. The use of these chemicals is now banned in many countries. The ozone gas also absorbs some radiation in the visible range. Absorption in the infrared region is domimated by *water vapor* and *carbon dioxide.* The dust particles and other pollutants in the atmosphere also absorb radiation at various wavelengths.

As a result of these absorptions, the solar energy reaching the *earth's surface* is weakened considerably; to about 950 W/m² on a clear day and much less on cloudy or smoggy days. Also, practically all of the solar radiation reaching the earth's surface falls in the wavelength band from 0.3 to 2.5 $\mu$m.

Another mechanism that attenuates solar radiation as it passes through the atmosphere is *scattering* or *reflection* by air molecules and the many other kinds of particles such as dust, smog, and water droplets suspended in the atmosphere. Scattering is mainly governed by the size of the particle relative to the wavelength of radiation. The oxygen and nitrogen molecules primarily scatter radiation at very short wavelengths, comparable to the size of the molecules themselves. Therefore, radiation at wavelengths corresponding to violet and blue colors is scattered the most. This molecular scattering in all directions is what gives the sky its bluish color. The same phenomena is responsible for red sunrises and sunsets. Early in the morning and late in the afternoon, the sun's rays pass through a greater thickness of the atmosphere than they do at midday, when the sun is at the top. Therefore, the violet and blue colors of the light encounter a greater number of molecules by the time they reach the earth's surface, and thus a greater fraction of them is scattered

(Fig. 12-33). Consequently, the light that reaches the earth's surface consists primarily of colors corresponding to longer wavelengths such as red, orange, and yellow. The clouds appear in reddish-orange color during sunrise and sunset because the light they reflect is reddish-orange at those times. For the same reason, a red traffic light is visible from a longer distance than is a green light under the same circumstances.

The solar energy incident on a surface on earth is considered to consist of *direct* and *diffuse* parts. The part of solar radiation that reaches the earth's surface without being scattered or absorbed by the atmosphere is called **direct solar radiation** $G_D$. The scattered radiation is assumed to reach the earth's surface uniformly from all directions, and is called **diffuse solar radiation** $G_d$. Then the *total solar energy* incident on the unit area of a *horizontal surface* on the ground is (Fig. 12-34)

$$G_{\text{solar}} = G_D \cos\theta + G_d \qquad (\text{W/m}^2) \tag{12-25}$$

where $\theta$ is the angle of incidence of direct solar radiation (the angle that the sun's rays make with the normal of the surface). The diffuse radiation varies from about 10 percent of the total radiation on a clear day to nearly 100 percent on a totally cloudy day.

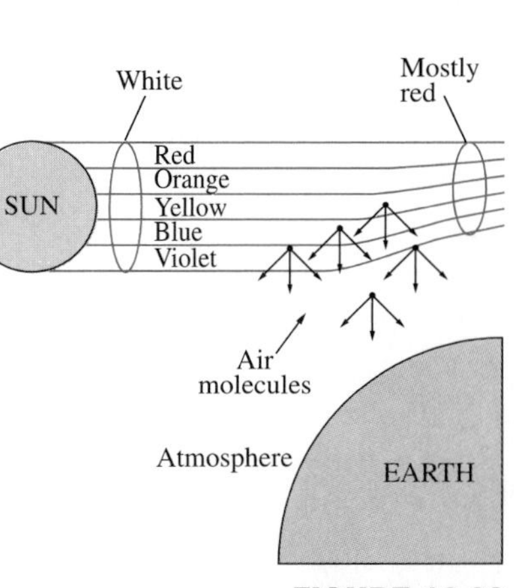


**FIGURE 12-33**

Air molecules scatter blue light much more than they do red light. At sunset, the light travels through a thicker layer of atmosphere, which removes much of the blue from the natural light, and letting the red dominate.

The gas molecules and the suspended particles in the atmosphere *emit radiation* as well as absorbing it. The atmospheric emission is primarily due to the $CO_2$ and $H_2O$ molecules, and is concentrated in the regions from 5 to 8 $\mu$m and above 13 $\mu$m. Although this emission is far from resembling the distribution of radiation from a blackbody, it is found convenient in radiation calculations to treat the atmosphere as a blackbody at some lower fictitious temperature which emits an equivalent amount of radiation energy. This fictitious temperature is called the **effective sky temperature** $T_{\text{sky}}$. Then the radiation emission from the atmosphere to the earth's surface is expressed as

$$G_{\text{sky}} = \sigma T_{\text{sky}}^4 \qquad (\text{W/m}^2) \tag{12-26}$$

The value of $T_{\text{sky}}$ depends on the atmospheric conditions. It ranges from about 230 K for cold, clear-sky conditions to about 285 K for warm, cloudy-sky conditions.

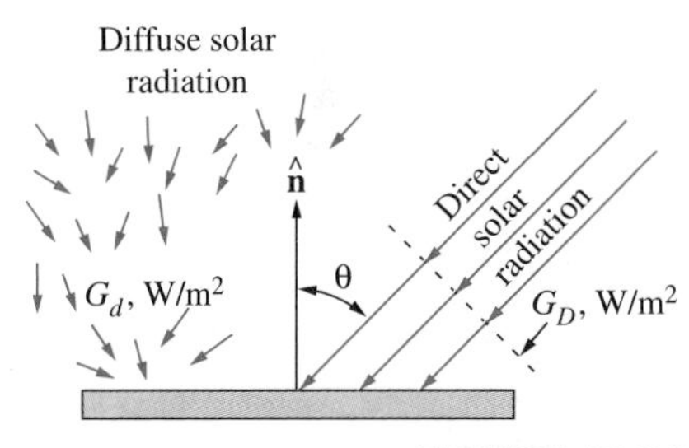


**FIGURE 12-34**

The direct and diffuse radiation incident on a horizontal surface at the earth's surface.

Note that the effective sky temperature does not deviate much from the room temperature. Thus, in the light of Kirchhoff's law, we can take the absorptivity of a surface to be equal to its emissivity at room temperature, $\alpha = \varepsilon$. Then the sky radiation absorbed by a surface can be expressed as

$$E_{\text{sky, absorbed}} = \alpha G_{\text{sky}} = \alpha\sigma T_{\text{sky}}^4 = \varepsilon\sigma T_{\text{sky}}^4 \qquad (\text{W/m}^2) \tag{12-27}$$

The net rate of radiation heat transfer to a surface exposed to solar and atmospheric radiation is determined from an energy balance (Fig. 12-35):

$$\begin{aligned} \dot{q}_{\text{net, rad}} &= \sum E_{\text{absorbed}} - \sum E_{\text{emitted}} \\ &= E_{\text{solar, absorbed}} + E_{\text{sky, absorbed}} - E_{\text{emitted}} \\ &= \alpha_s G_{\text{solar}} + \varepsilon\sigma T_{\text{sky}}^4 - \varepsilon\sigma T_s^4 \\ &= \alpha_s G_{\text{solar}} + \varepsilon\sigma(T_{\text{sky}}^4 - T_s^4) \qquad (\text{W/m}^2) \end{aligned} \tag{12-28}$$

where $T_s$ is the temperature of the surface in K and $\varepsilon$ is its emissivity at room temperature. A positive result for $\dot{q}_{\text{net,rad}}$ indicates a heat gain by the surface and a negative result indicates a heat loss.

The absorption and emission of radiation by the *elementary gases* such as $H_2$, $O_2$, and $N_2$ at moderate temperatures are negligible, and a medium filled with these gases can be treated as a *vacuum* in radiation analysis. The absorption and emission of gases with *larger molecules* such as $H_2O$ and $CO_2$, however, can be *significant* and may need to be considered when considerable amounts of such gases are present in a medium. For example, a 1-m-thick layer of water vapor at 1-atm pressure and 100°C emits more than 50 percent of the energy that a blackbody would emit at the same temperature.

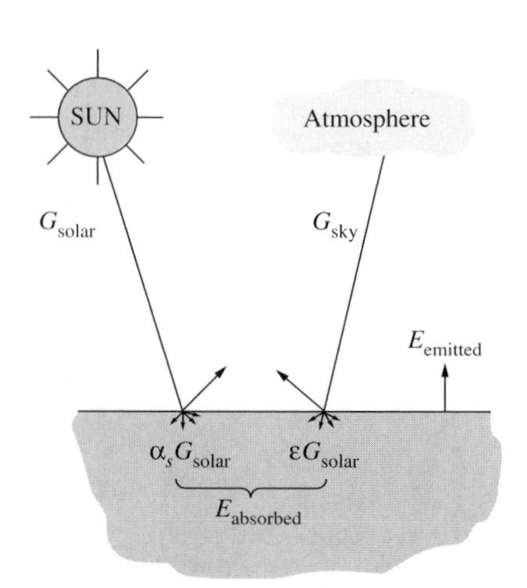


**FIGURE 12-35**
Radiation interactions of a surface exposed to solar and atmospheric radiation.

In solar energy applications, the spectral distribution of incident solar radiation is very different than the spectral distribution of emitted radiation by the surfaces, since the former is concentrated in the short-wavelength region and the latter in the infrared region. Therefore, the radiation properties of surfaces will be quite different for the incident and emitted radiation, and the surfaces cannot be assumed to be gray. Instead, the surfaces are assumed to have two sets of properties: one for solar radiation, and another for infrared radiation at room temperature. Table 12-3 lists the *emissivity* $\varepsilon$ and the *solar absorptivity* $\alpha_s$ of the surfaces of some common materials. Surfaces that are intended to *collect solar energy,* such as the absorber surfaces of solar collectors, are desired to have high $\alpha_s$ but low $\varepsilon$ values to maximize the absorption of solar radiation and to minimize the emission of radiation. Surfaces that are intended to *remain cool* under the sun, such as the outer surfaces of fuel tanks and refrigerator trucks, are desired to have just the opposite properties. Surfaces are often given the desired properties by coating them with thin layers of *selective* materials. A surface can be kept cool, for example, by simply painting it white.

We close this section by pointing out that what we call *renewable energy* is usually nothing more than the manifestation of solar energy in different froms. Such energy sources include wind energy, hydroelectric power, ocean thermal energy, ocean wave energy, and wood. For example, no hydroelectric power plant can generate electricity year after year unless the water evaporates by absorbing solar energy, and comes back as a rainfall to replensih the water source (Fig. 12-36). Although solar energy is sufficient to meet the entire energy needs of the world, currently it is not economical to do so because of the low concentration of solar energy on earth and the high capical cost of harnessing it.

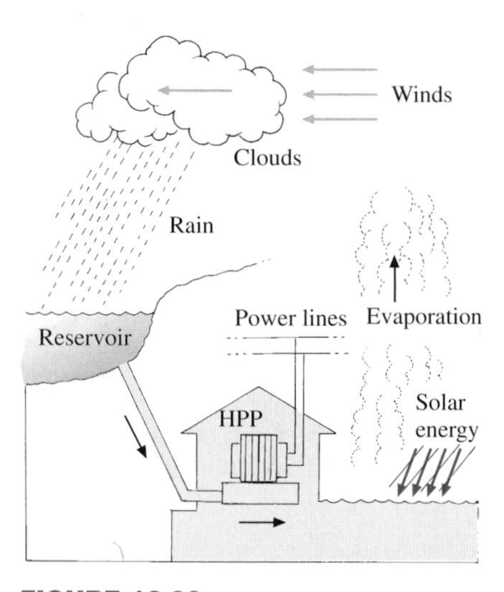


**FIGURE 12-36**
The cycle that water undergoes in a hydroelectric power plant.

**EXAMPLE 12-4**

Consider a surface exposed to solar radiation. At some time, the direct and diffuse components of solar radiation are $G_D$ = 400 and $G_d$ = 300 W/m², and the direct radiation makes a 20° angle with the normal of the surface. The surface temperature is observed to be 320 K at that time. Assuming an effective sky temperature of 260 K, determine the net rate of radiation heat transfer for the following cases (Fig. 12-37):

**TABLE 12-3**
**Comparison of the solar absorptivity $\alpha_s$ of some surfaces with their emissivity $\varepsilon$ at room temperature**

| Surface | $\alpha_s$ | $\varepsilon$ |
|---|---|---|
| Aluminum | | |
| Polished | 0.09 | 0.03 |
| anodized | 0.14 | 0.84 |
| Foil | 0.15 | 0.05 |
| Copper | | |
| Polished | 0.18 | 0.03 |
| Tarnished | 0.65 | 0.75 |
| Stainless steel | | |
| Polished | 0.37 | 0.60 |
| Dull | 0.50 | 0.21 |
| Plated metals | | |
| Black nickel oxide | 0.92 | 0.08 |
| Black chrome | 0.87 | 0.09 |
| Concrete | 0.60 | 0.88 |
| White marble | 0.46 | 0.95 |
| Red brick | 0.63 | 0.93 |
| Asphalt | 0.90 | 0.90 |
| Black paint | 0.97 | 0.97 |
| White paint | 0.14 | 0.93 |
| Snow | 0.28 | 0.97 |
| Human skin (caucasian) | 0.62 | 0.97 |

(*a*) $\alpha_s = 0.9$ and $\varepsilon = 0.9$ (gray absorber surface)
(*b*) $\alpha_s = 0.1$ and $\varepsilon = 0.1$ (gray reflector surface)
(*c*) $\alpha_s = 0.9$ and $\varepsilon = 0.1$ (selective absorber surface)
(*d*) $\alpha_s = 0.1$ and $\varepsilon = 0.9$ (selective reflector surface)

**Solution** The total solar energy incident on the surface is determined from Eq. 12-25 to be

$$\begin{aligned} G_{\text{solar}} &= G_D \cos\theta + G_d \\ &= (400\ \text{W/m}^2)\cos 20^\circ + (300\ \text{W/m}^2) \\ &= 675.9\ \text{W/m}^2 \end{aligned}$$

Then the net rate of radiation heat transfer for each of the four cases is determined from Eq. 12-25:

$$\dot{q}_{\text{net, rad}} = \alpha_s G_{\text{solar}} + \varepsilon\sigma(T_{\text{sky}}^4 - T_s^4)$$

(*a*) $\alpha_s = 0.9$ and $\varepsilon = 0.9$ (gray absorber surface):

$$\begin{aligned} \dot{q}_{\text{net, rad}} &= 0.9(675.9\ \text{W/m}^2) + 0.9[5.67\times10^{-8}\ \text{W/(m}^2\cdot\text{K}^4)][(260\ \text{K})^4 - (320\ \text{K})^4] \\ &= 306.5\ \text{W/m}^2 \end{aligned}$$

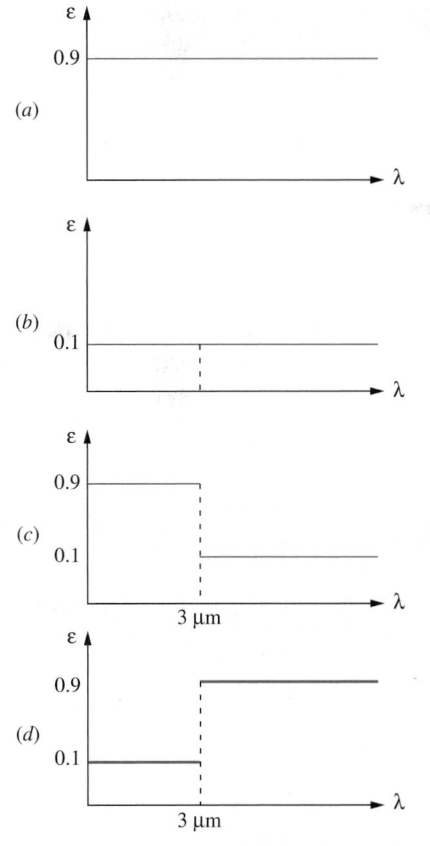


**FIGURE 12-37**
Graphical representation of the spectral emissivities of the four surfaces considered in Example 12-4.

(*b*) $\alpha_s = 0.1$ and $\varepsilon = 0.1$ (gray reflector surface):

$$\dot{q}_{\text{net, rad}} = 0.1(675.9\ \text{W/m}^2) + 0.1[5.67 \times 10^{-8}\ \text{W/(m}^2 \cdot \text{K}^4)][(260\ \text{K})^4 - (320\ \text{K})^4]$$
$$= 34.1\ \text{W/m}^2$$

(*c*) $\alpha_s = 0.9$ and $\varepsilon = 0.1$ (selective absorber surface):

$$\dot{q}_{\text{net, rad}} = 0.9(675.9\ \text{W/m}^2) + 0.1[5.67 \times 10^{-8}\ \text{W/(m}^2 . \text{K}^4)][(260\ \text{K})^4 - (320\ \text{K})^4]$$
$$= 574.8\ \text{W/m}^2$$

(*d*) $\alpha_s = 0.1$ and $\varepsilon = 0.9$ (selective reflector surface):

$$\dot{q}_{\text{net, rad}} = 0.1(675.9\ \text{W/m}^2) + 0.9[5.67 \times 10^{-8}\ \text{W/(m}^2 \cdot \text{K}^4)][(260\ \text{K})^4 - (320\ \text{K})^4]$$
$$= -234.3\ \text{W/m}^2$$

Note that the surface of an ordinary gray material of high absorptivity gains heat at a rate of 306.5 W/m$^2$. The amount of heat gain increases to 574.8 W/m$^2$ when the surface is coated with a selective material that has the same absorptivity for solar radiation bu a low emissivity for infrared radiation. Also note that the surface of an ordinary gray material of high reflectivity still gains heat at a rate of 34.1 W/m$^2$. When the surface is coated with a selective material which has the same reflectivity for solar radiation but a high emissivity for infrared radiation, the surface loses 234.3 W/m$^2$ instead. Therefore, the temperature of the surface will decrease when a selective reflector surface is used.

## 12-6 THE VIEW FACTOR

So far, we have considered the radiation properties of surfaces, and the radiation interactions of a single surface. We are now in a position to consider *radiation heat transfer* between two or more surfaces, which is the primary quantity of interest in practice. Radiation heat transfer between surfaces depends on the *orientation* of the surfaces relative to each other as well as their radiation properties and temperatures, as illustrated in Fig. 12-38. for example, a camper will make the most use of a campfire in a cold night by standing as close to the fire as possible and by blocking as much of the radiation coming from the fire by turning his front to the fire instead his side. Likewise, a person will maximize the amount of solar radiation incident on him by lying down on his back instead of standing up on his feet.

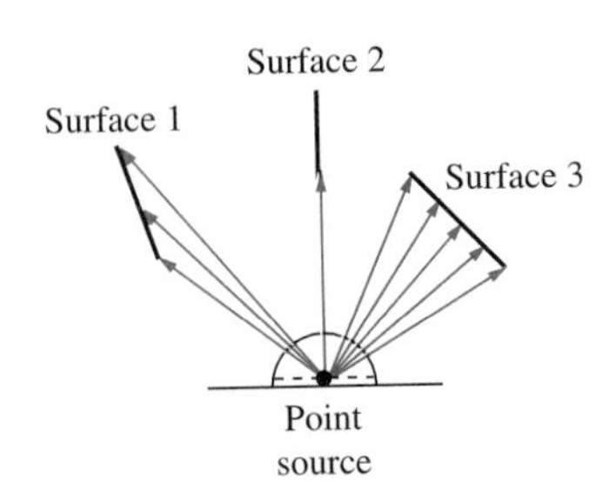


**FIGURE 12-38**
Radiation heat exchange between surfaces depends on the *orientation* of the surfaces relative to each other, and this dependence on orientation is accounted by the *view factor*.

To account for the effects of orientation on radiation heat transfer between two surfaces, we define a new parameter called the *view factor*, which is a purely geometric quantity and is independent of the surface properties and temperature. It is also called the *shape factor, configuration factor,* and *angle factor*. The view factor based on the assumption that the surfaces are diffuse emitters and diffuse reflectors is called the *diffuse view factor,* and the view factor based on the assumption that the surfaces are diffuse emitters but specular reflectors is called the *specular view factor*. In this book, we will consider radiation exchange between diffuse surfaces only, and thus the term *view factor* will simply imply *diffuse view factor*.

The **view factor** from a surface $i$ to a surface $j$ is denoted by $F_{i \to j}$, and is defined as

$F_{i \to j}$ = *the fraction of the radiation leaving surface i that strikes surface j directly*

Therefore, the view factor $F_{1 \to 2}$ represents the fraction of the radiation leaving surface 1 that strikes surface 2 and $F_{2 \to 1}$ represents the fraction of the radiation leaving surface 2 that strikes surface 1 directly. Note that the radiation that strikes a surface does not need to be absorbed by that surface. Also, radiation which strikes a surface after being reflected by other surfaces is not considered in the evaluation of the view factors. For the special case of $j = i$, we have

$F_{i \to i}$ = *the fraction of radiation leaving surface i that strikes itself directly*

Noting that in the absence of strong electromagnetic fields radiation beams travel in straight paths, the view factor from a surface to itself will be zero unless the surface "sees" itself. Therefore, $F_{i \to i} = 0$ for *plane* or *convex* surfaces, and $F_{i \to i} \neq 0$ for *concave* surfaces, as illustrated in Fig. 12-39.

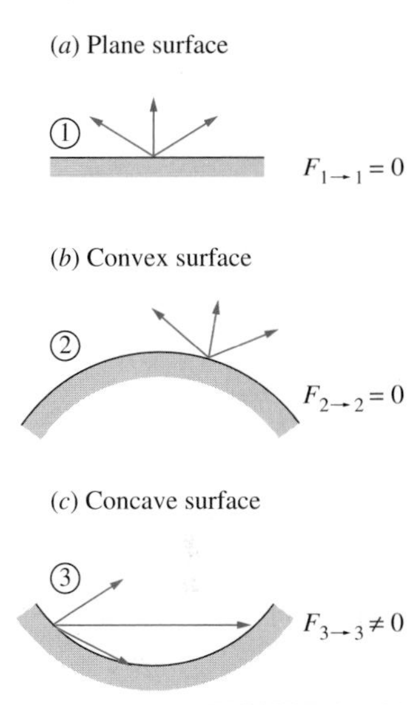


**FIGURE 12-39**
The view factor from a surface to itself is *zero* for *plane* or *convex* surfaces, and *nonzero* for *concave* surfaces.

The value of the view factor ranges between *zero* and *one.* The limiting case $F_{i \to j} = 0$ indicates that the two surfaces do not have a direct view of each other, and thus radiation leaving surface $i$ cannot strike surface $j$ directly. The other limiting case $F_{i \to j} = 1$ indicates that surface $j$ completely surrounds surface $i$, so that the entire radiation leaving surface $i$ is intercepted by surface $j$. For example, in a geometry consisting of two concentric spheres, the entire radiation leaving the surface of the smaller sphere (surface 1) will strike the larger sphere (surface 2), and thus $F_{1 \to 2} = 1$, as illustrated in Fig. 12-40.

The view factor has proven to be very useful in radiation analysis because it allows us to express the *fraction of radiation* leaving a surface that strikes another surface in terms of the orientation of these two surfaces relative to each other. The underlying assumption in this process is that the radiation a surface receives from a source is directly proportional to the angle the surface subtends when viewed from the source. This would be the case only if the radiation coming off the source is *uniform* in all directions throughout its surface, and the medium between the surfaces does not *absorb, emit,* or *scatter* radiation. That is, it will be the case when the surfaces are *isothermal* and *diffuse* emitters and reflectors, and the surfaces are separated by a *nonparticipating* medium such as a vacuum or air.

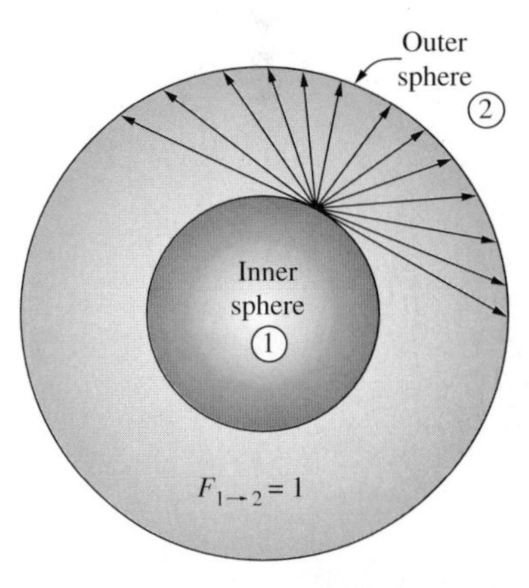


**FIGURE 12-40**
In a geometry that consists of two concentric spheres, the view factor $F_{1 \to 2} = 1$, since the entire radiation leaving the surface of the smaller sphere will be intercepted by the larger sphere.

The view factor $F_{1 \to 2}$ between two surfaces $A_1$ and $A_2$ can be determined in a systematic manner by first expressing the view factor between two differential areas $dA_1$ and $dA_2$ in terms of the spatial variables, and then by performing the necessary integrations. However, this approach is not practical, since, even for simple geometries, the resulting integrations are usually very complex and difficult to perform.

**TABLE 12-4**
**View factor expressions for some common geometries of finite size (3D)**

| Geometry | Relation |
|---|---|
| Aligned parallel rectangles | $\bar{X} = X/L,\ \bar{Y} = Y/L$ <br> $F_{i \to j} = \dfrac{2}{\pi \bar{X}\bar{Y}} \left\{ \ln \left[ \dfrac{(1 + \bar{X}^2)(1 + \bar{Y}^2)}{1 + \bar{X}^2 + \bar{Y}^2} \right]^{1/2} + \bar{X}(1 + \bar{Y}^2)^{1/2} \tan^{-1} \dfrac{\bar{X}}{(1 + \bar{Y}^2)^{1/2}} + \bar{Y}(1 + \bar{X}^2)^{1/2} \tan^{-1} \dfrac{\bar{Y}}{(1 + \bar{X}^2)^{1/2}} - \bar{X} \tan^{-1} \bar{X} - \bar{Y} \tan^{-1} \bar{Y} \right\}$ |
| Coaxial parallel disks | $R_i = r_i/L,\ R_j = r_j/L$ <br> $S = 1 + \dfrac{1 + R_j^2}{R_i^2}$ <br> $F_{i \to j} = \dfrac{1}{2} \left\{ S - \left[ S^2 - 4\left(\dfrac{r_j}{r_i}\right)^2 \right]^{1/2} \right\}$ |
| Perpendicular rectangles with a common edge | $H = Z/X,\ W = Y/X$ <br> $F_{i \to j} = \dfrac{1}{\pi W} \left( W \tan^{-1} \dfrac{1}{W} + H \tan^{-1} \dfrac{1}{H} - (H^2 + W^2)^{1/2} \tan^{-1} \dfrac{1}{(H^2 + W^2)^{1/2}} + \dfrac{1}{4} \ln \left\{ \dfrac{(1 + W^2)(1 + H^2)}{1 + W^2 + H^2} \times \left[ \dfrac{W^2(1 + W^2 + H^2)}{(1 + W^2)(W^2 + H^2)} \right]^{W^2} \times \left[ \dfrac{H^2(1 + H^2 + W^2)}{(1 + H^2)(H^2 + W^2)} \right]^{H^2} \right\} \right)$ |

View factors for hundreds of common geometries are evaluated and the results are given in analytical, graphical, and tabular form in several publications. View factors for selected geometries are given in Tables 12-4 and 12-5 in *analytical* form, and in Figs 12-41 to 12-44 in *graphical* form. The view factors in Table 12-4 are for geometries that are *infinitely long* in the direction perpendicular to the plane of the paper, and are therefore two-dimensional. The view factors in Table 12-5, on the other hand, are for *three-dimensional* geometries.

**TABLE 12-5**
**View factor expressions for some infinitely long (2D) geometries**

| Geometry | Relation |
|---|---|
| Parallel plates with midlines connected by perpendicular 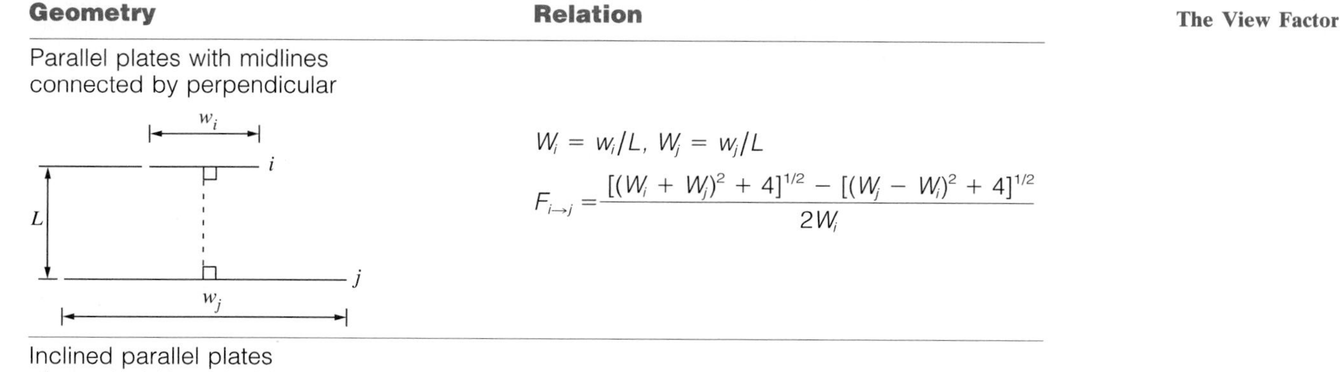 | $W_i = w_i/L,\ W_j = w_j/L$ <br> $F_{i\to j} = \dfrac{[(W_i + W_j)^2 + 4]^{1/2} - [(W_j - W_i)^2 + 4]^{1/2}}{2W_i}$ |
| Inclined parallel plates of equal width and with a common edge 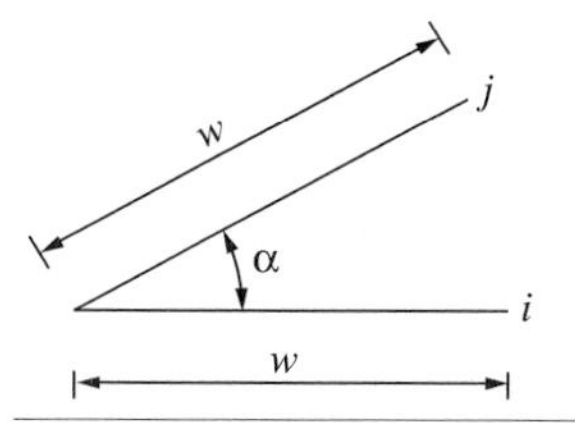 | $F_{i\to j} = 1 - \sin \frac{1}{2}\alpha$ |
| Perpendicular plates with a common edge 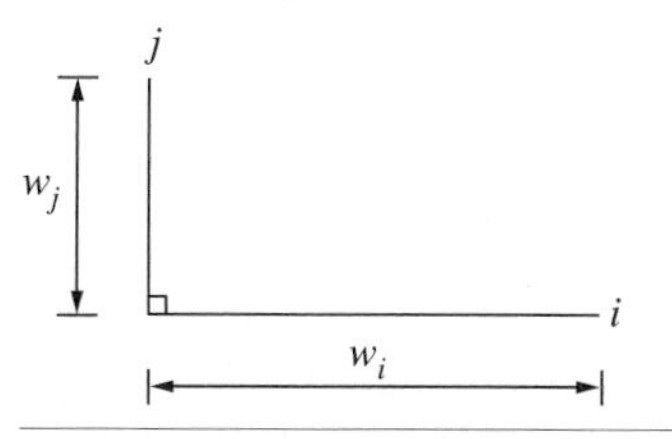 | $F_{i\to j} = \dfrac{1}{2}\left\{1 + \dfrac{w_j}{w_i} - \left[1 + \left(\dfrac{w_j}{w_i}\right)^2\right]^{1/2}\right\}$ |
| Three-sided enclosure 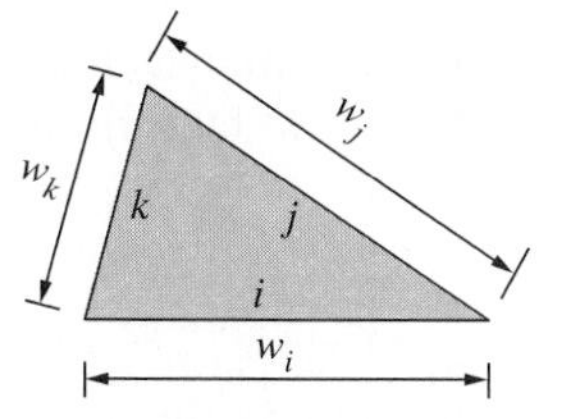 | $F_{i\to j} = \dfrac{w_i + w_j - w_k}{2w_i}$ |
| Infinite plane and row of cylinders | $F_{i\to j} = 1 - \left[1 - \left(\dfrac{D}{s}\right)^2\right]^{1/2} + \dfrac{D}{s}\tan^{-1}\left(\dfrac{s^2 - D^2}{D^2}\right)^{1/2}$ |

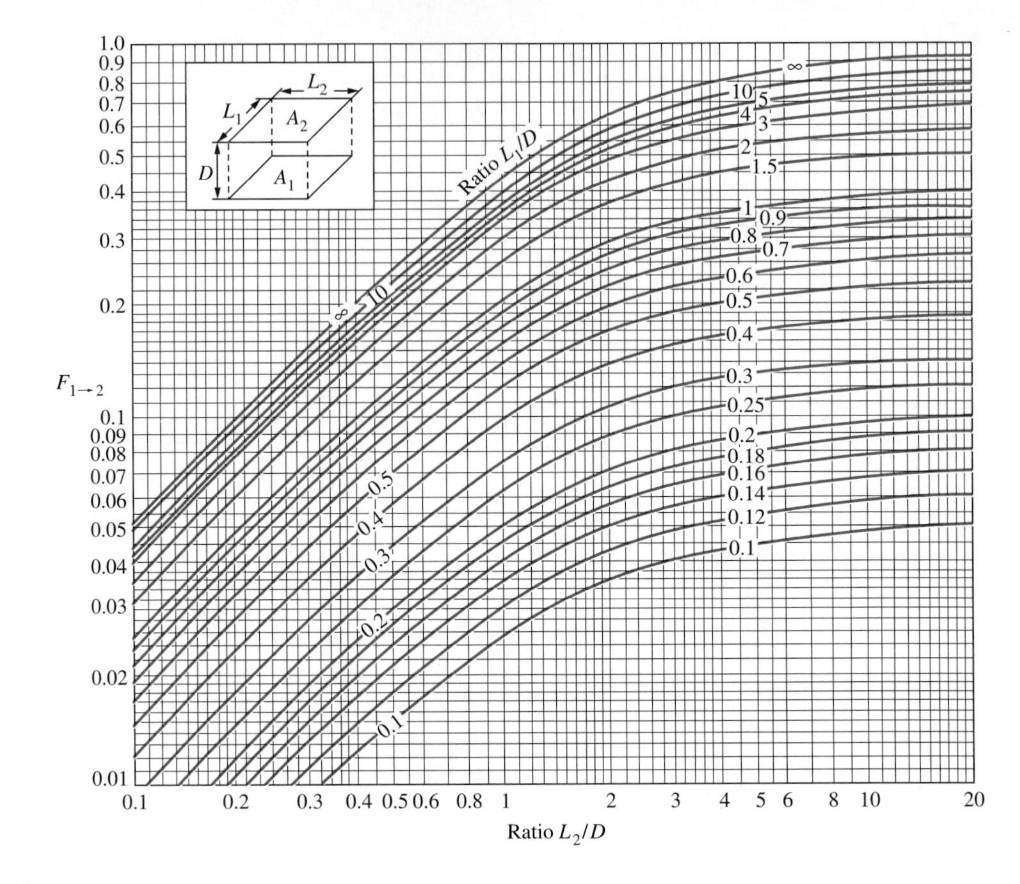

**FIGURE 12-41**

View factor between two aligned parallel rectangles of equal size.

## View Factor Relations

Radiation analysis of an enclosure consisting of $N$ surfaces requires the evaluation of $N^2$ view factors, and this evaluation process is probably the most time-consuming part of a radiation analysis. However, it is neither practical nor necessary to evaluate all of the view factors directly. Once a sufficient number of view factors are available, the rest of them can be determined by utilizing some fundamental relations for view factors, as discussed below.

### 1 The Reciprocity Rule

The view factors $F_{i\to j}$ and $F_{j\to i}$ are *not* equal to each other unless the areas of the two surfaces are equal. That is,

$$F_{j\to i} = F_{i\to j} \quad \text{when} \quad A_i = A_j$$
$$F_{j\to i} \neq F_{i\to j} \quad \text{when} \quad A_i \neq A_j$$

Using the radiation intensity concept and going through some manipulations, it can be shown that the pair of view factors $F_{i\to j}$ and $F_{j\to i}$ are related to each other by

$$A_i F_{i\to j} = A_j F_{j\to i} \tag{12-29}$$

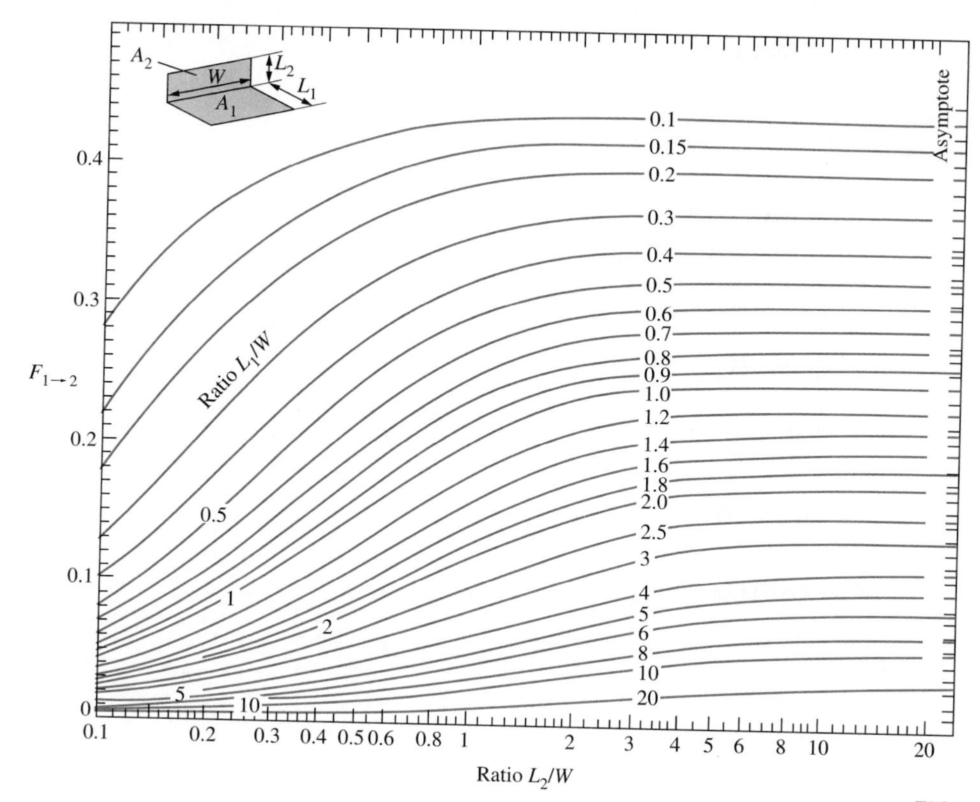


**FIGURE 12-42**

View factor between two perpendicular rectangles with a common edge.

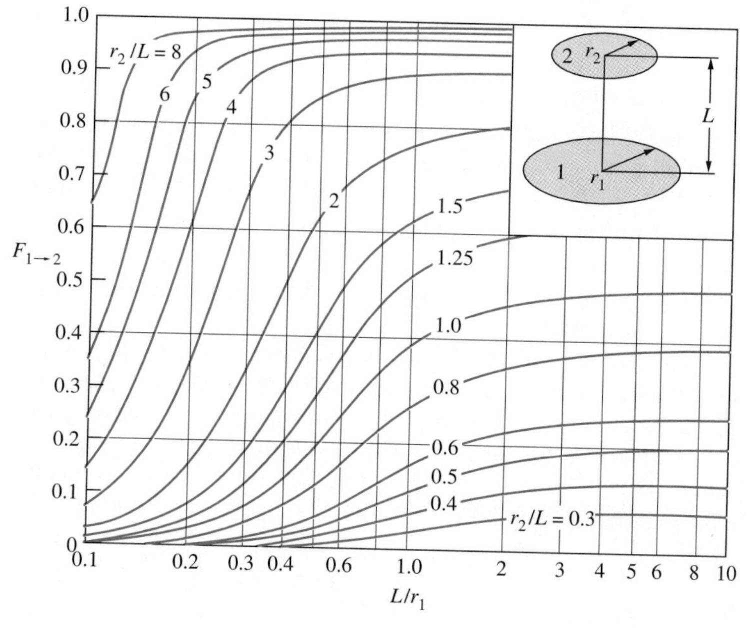


**FIGURE 12-43**

View factor between two coaxial parallel disks.

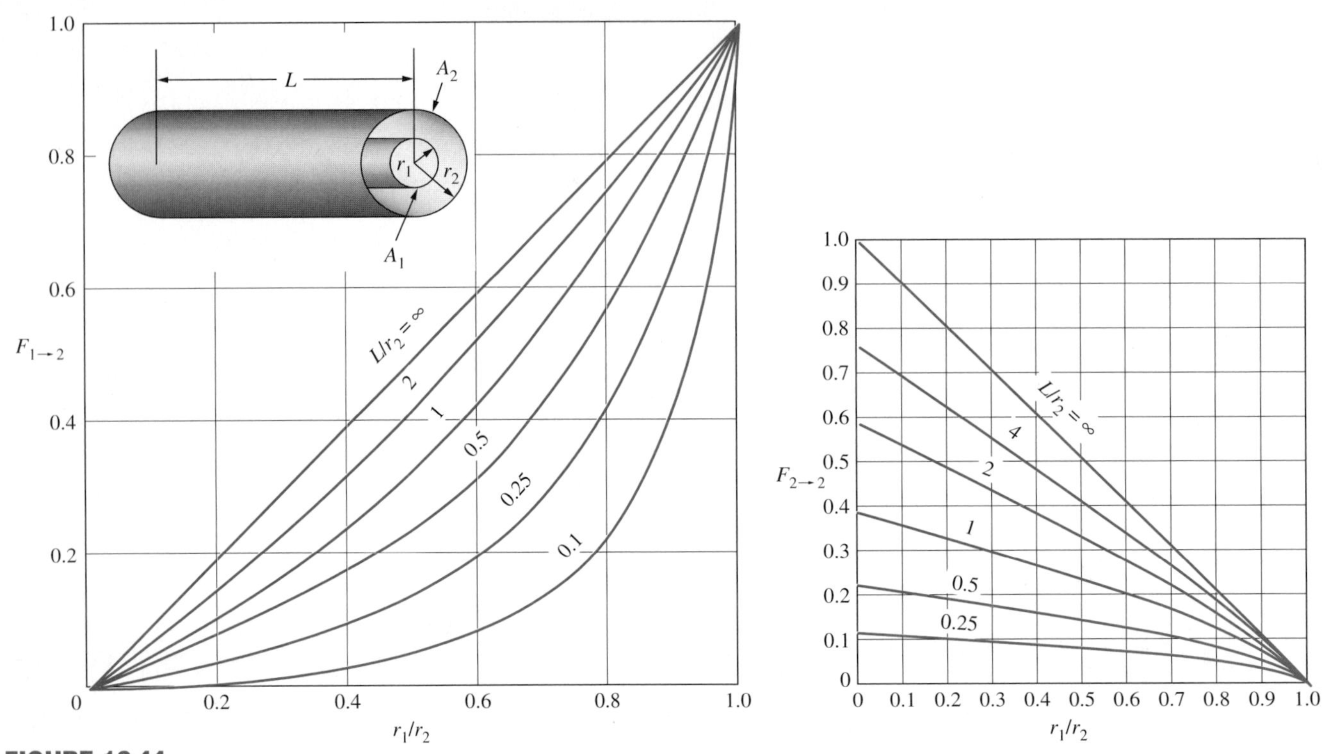


**FIGURE 12-44**

View factors for two concentric cylinders of finite length: (*a*) outer cylinder to inner cylinder; (*b*) outer cylinder to itself.

This relation is known as the **reciprocity rule**, and it enables us to determine the counterpart of a view factor from a knowledge of the view factor itself and the areas of the two surfaces. When determining the pair of view factors $F_{i\to j}$ and $F_{j\to i}$, it makes sense to evaluate first the easier one directly and then the harder one by applying the reciprocity rule.

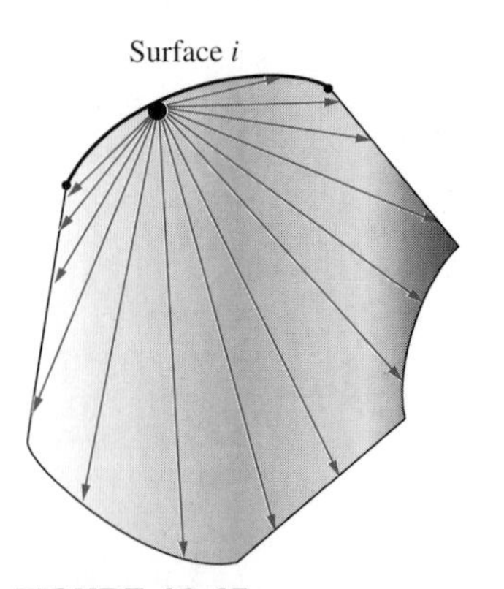


**FIGURE 12-45**

Radiation leaving any surface *i* of an enclosure must be intercepted completely by the surfaces of the enclosure. Therefore, the sum of the view factors from surface *i* to each one of the surfaces of the enclosure must be unity.

## 2 The Summation Rule

The radiation analysis of a surface normally requires the consideration of the radiation coming in or going out in all directions. Therefore, most radiation problems encountered in practice involve enclosed spaces. When formulating a radiation problem, we usually form an *enclosure* consisting of the surfaces interacting radiatively. Even openings are treated as imaginary surfaces with radiation properties equivalent to those of the opening.

The conservation of energy principle requires that the entire radiation leaving any surface *i* of an enclosure be intercepted by the surfaces of the enclosure. Therefore, *the sum of the view factors from surface i of an enclosure to all surfaces of the enclosure, including to itself, must equal unity*. This is known as the **summation rule** for an enclosure, and is expressed as (Fig. 12-45)

$$\sum_{j=1}^{N} F_{i\to j} = 1 \tag{12-30}$$

where $N$ is the number of surfaces of the enclosure. For example, applying the summation rule to surface 1 of a three-surface enclosure yields

$$\sum_{j=1}^{3} F_{1\to j} = F_{1\to 1} + F_{1\to 2} + F_{1\to 3} = 1$$

The notation $F_{i\to j}$ is *instructive* for beginners, since it emphasizes that the view factor is for radiation that travels from surface $i$ to surface $j$. However, this notation becomes rather awkward when it has to be used many times in a problem. In such cases, it is convenient to replace it by its *shorthand* version $F_{ij}$.

The summation rule can be applied to each surface of an enclosure by varying $i$ from 1 to $N$. Therefore, the summation rule applied to each of the $N$ surfaces of an enclosure gives $N$ relations for the determination of the view factors. Also, the reciprocity rule gives $\frac{1}{2}N(N-1)$ additional relations. Then the total number of view factors that need to be evaluated directly for an $N$-surface enclosure becomes

$$N^2 - [N + \tfrac{1}{2}N(N-1)] = \tfrac{1}{2}N(N-1)$$

For example, for a six-surface enclosure, we need to determine only $\frac{1}{2} \times 6(6-1) = 15$ of the $6^2 = 36$ view factors directly. The remaining 21 view factors can be determined from the 21 equations that are obtained by applying the reciprocity and the summation rules.

**EXAMPLE 12-5**

Determine the view factors associated with an enclosure formed by two spheres shown in Fig. 12-46.

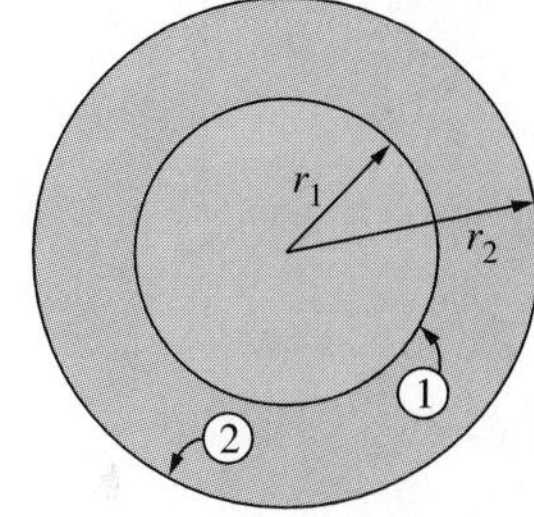


**FIGURE 12-46**
The geometry considered in Example 12-5.

**Solution** The outer surface of the smaller sphere (surface 1) and inner surface of the larger sphere (surface 2) form a two-surface enclosure. Therefore, $N = 2$, and this enclosure involves $N^2 = 2^2 = 4$ view factors, which are $F_{11}$, $F_{12}$, $F_{21}$, and $F_{22}$. In this two-surface enclosure, we need to determine only

$$\tfrac{1}{2}N(N-1) = \tfrac{1}{2} \times 2(2-1) = 1$$

view factor directly. The remaining three view factors can be determined by the application of the summation and reciprocity rules. But it turns out that we can determine not only one but *two* view factors directly in this case by a simple *inspection*:

$F_{11} = 0$, since no radiation leaving surface 1 strikes itself

and $F_{12} = 1$, since all radiation leaving surface 1 strikes surface 2

Actually it would be sufficient to determine only one of these view factors by inspection, since we could always determine the other one from the summation rule applied to surface 1 as $F_{11} + F_{12} = 1$.

The view factor $F_{21}$ is determined by applying the reciprocity rule to surfaces 1 and 2:

$$A_1F_{12} = A_2F_{21}$$

which yields

$$F_{21} = \frac{A_1}{A_2}F_{12} = \frac{4\pi r_1^2}{4\pi r_2^2} \times 1 = \left(\frac{r_1}{r_2}\right)^2$$

Finally, the view factor $F_{22}$ is determined by applying the summation rule to surface 2:

$$F_{21} + F_{22} = 1$$

and thus
$$F_{22} = 1 - F_{21} = 1 - \left(\frac{r_1}{r_2}\right)^2$$

Note that when the outer sphere is much larger than the inner sphere ($r_2 \gg r_1$), $F_{22}$ approaches one. This is expected, since the fraction of radiation leaving the outer sphere that is intercepted by the inner sphere will be negligible in that case. Also note that the two spheres considered above do not need to be concentric. However, the radiation analysis will be most accurate for the case of concentric spheres, since the radiation is most likely to be uniform on the surfaces in that case.

## 3 The Superposition Rule

Sometimes the view factor associated with a given geometry is not available in standard tables and charts. In such cases, it is desirable to express the given geometry as the sum or differences of some geometries with known view factors, and then to apply the **superposition rule**, which can be expressed as follows: *the view factor from a surface i to a surface j is equal to the sum of the view factors from surface i to the parts of surface j*. Note that the reverse of this is not true. That is, the view factor from a surface *j* to a surface *i* is *not* equal to the sum of the view factors from the parts of surface *j* to surface *i*.

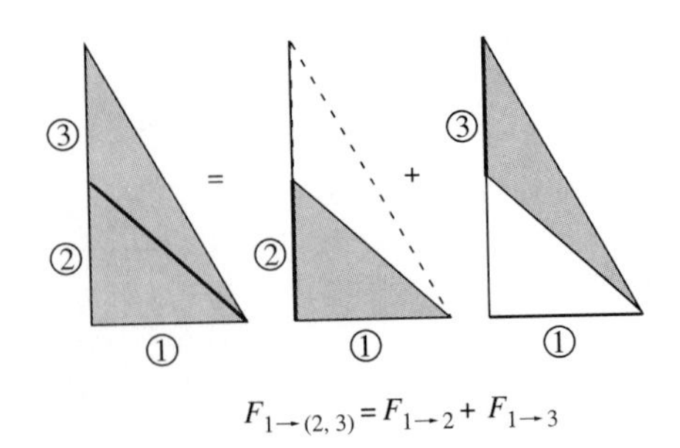


**FIGURE 12-47**

The view factor from a surface to a composite surface is equal to the sum of the view factors from the surface to the parts of the composite surface.

Consider the geometry in Fig. 12-47, which is infinitely long in the direction perpendicular to the plane of the paper. The radiation that leaves surface 1 and strikes the combined surfaces 2 and 3 is equal to the sum of the radiation that strikes surfaces 2 and 3. Therefore, the view factor from surface 1 to the combined surfaces of 2 and 3 is

$$F_{1\to(2,3)} = F_{1\to2} + F_{1\to3} \qquad \text{(12-31)}$$

Suppose we need to find the view factor $F_{1\to3}$. A quick check of the view factor expressions and charts in this section will reveal that such a view factor cannot be evaluated directly. However, the view factor $F_{1\to(2,3)}$ can be determined from Eq. 12-31 after determining both $F_{1\to2}$ and $F_{1\to(2,3)}$ from the chart in Fig. 12-42. Therefore, it may be possible to determine some difficult view factors with relative ease by expressing one or both of the areas as the sum or differences of areas and then applying the superposition rule.

To obtain a relation for the view factor $F_{(2,3)\to1}$, we multiply Eq. 12-31 by $A_1$,

$$A_1F_{1\to(2,3)} = A_1F_{1\to2} + A_1F_{1\to3}$$

and apply the reciprocity rule to each term to get

$$(A_2 + A_3)F_{(2,3)\to1} = A_2F_{2\to1} + A_2F_{3\to1}$$

or
$$F_{(2,3)\to1} = \frac{A_2F_{2\to1} + A_3F_{3\to1}}{A_2 + A_3} \qquad \text{(12-32)}$$

Areas that are expressed as the sum of more than two parts can be handled in a similar manner.



**EXAMPLE 12-6**

Determine the fraction of the radiation leaving the base of the cylindrical enclosure shown in Fig. 12-48 that escapes through a coaxial ring opening at its top surface. The radius and the length of the enclosure are $r_1 = 10$ cm and $L = 10$ cm, while the inner and outer radii of the ring are $r_2 = 5$ cm and $r_3 = 8$ cm, respectively.

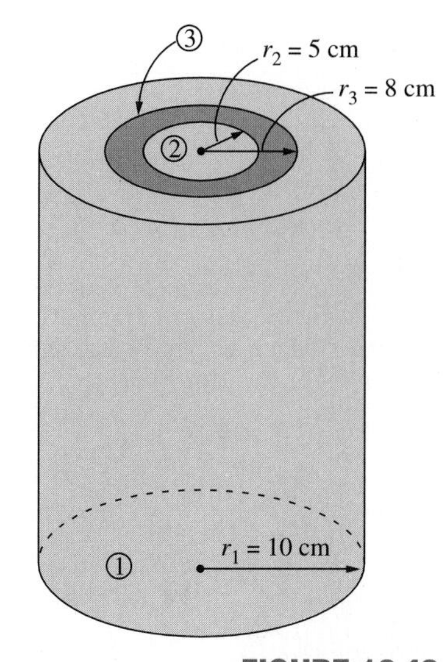


**FIGURE 12-48**
The cylindrical enclosure considered in Example 12-6.

**Solution** We are asked to determine the fraction of the radiation leaving the base of the enclosure that escapes through an opening at the top surface. Assuming the base surface is a diffuse emitter and diffuse reflector, what we are asked to determine is simply the *view factor* $F_{1\to\text{ring}}$ from the base of the enclosure to the ring-shaped surface at the top.

We do not have an analytical expression or chart for view factors between a circular area and a coaxial ring, and so we cannot determine $F_{1\to\text{ring}}$ directly. However, we do have a chart for view factors between two coaxial parallel disks, and we can always express a ring in terms of disks.

Let the base surface of radius $r_1 = 10$ cm be surface 1, the circular area of $r_2 = 5$ cm at the top be surface 2, and the circular area of $r_3 = 8$ cm be surface 3. Using the superposition rule, the view factor from surface 1 to surface 3 can be expressed as

$$F_{1\to3} = F_{1\to2} + F_{1\to\text{ring}}$$

since surface 3 is the sum of surface 2 and the ring area. The view factors $F_{1\to2}$ and $F_{1\to3}$ are determined from the chart in Fig. 12-43 as follows:

$$\frac{L}{r_1} = \frac{10\text{ cm}}{10\text{ cm}} = 1 \quad\text{and}\quad \frac{r_2}{L} = \frac{5\text{ cm}}{10\text{ cm}} = 0.5 \xrightarrow{\text{(Fig. 12-43)}} F_{1\to2} = 0.11$$

$$\frac{L}{r_1} = \frac{10\text{ cm}}{10\text{ cm}} = 1 \quad\text{and}\quad \frac{r_3}{L} = \frac{8\text{ cm}}{10\text{ cm}} = 0.8 \xrightarrow{\text{(Fig. 12-43)}} F_{1\to3} = 0.28$$

Therefore, $$F_{1\to\text{ring}} = F_{1\to3} - F_{1\to2} = 0.28 - 0.11 = 0.17$$

which is the desired result. Note that $F_{1\to2}$ and $F_{1\to3}$ represent the fractions of radiation leaving the base that strike the circular surfaces 2 and 3, respectively, and their difference gives the fraction that strikes the ring area.

## 4 The Symmetry Rule

The determination of the view factors in a problem can be simplified further if the geometry involved possesses some sort of symmetry. Therefore, it is good practice to check for the presence of any *symmetry* in a problem before attempting to determine the view factors directly. The presence of symmetry can be determined *by inspection,* keeping the definition of the view factor in mind. Identical surfaces that are oriented in an identical manner with respect to another surface will intercept identical amounts of radiation leaving that surface. Therefore, the **symmetry rule** can be expressed as follows: *two (or more) surfaces that*

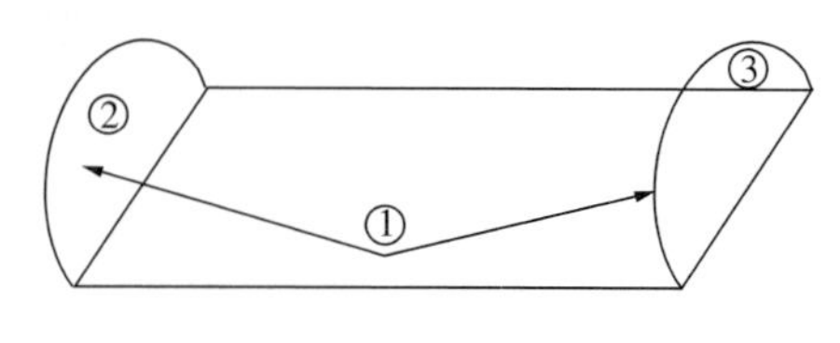


$F_{1\to2} = F_{1\to3}$

(Also, $F_{2\to1} = F_{3\to1}$)

**FIGURE 12-49**

Two surfaces that are symmetric about a third surface will have the same view factor from the third surface.

*possess symmetry about a third surface will have identical view factors from that surface* (Fig. 12-49).

The symmetry rule can also be expressed as follows: *if the surfaces j and k are symmetric about the surface i then* $F_{i\to j} = F_{i\to k}$. Using the reciprocity rule, we can show that the relation $F_{j\to i} = F_{k\to i}$ is also true in this case.

**EXAMPLE 12-7**

Determine the view factors from the base of the pyramid shown in Fig. 12-50 to each of its four side surfaces. The base of the pyramid is a square, and its side surfaces are isosceles triangles.

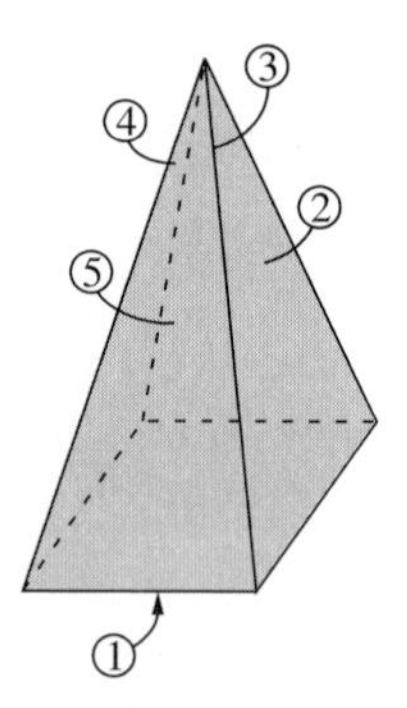


**FIGURE 12-50**

The pyramid considered in Example 12-7.

**Solution** The base of the pyramid (surface 1) and its four side surfaces (surfaces 2, 3, 4, and 5) form a five-surface enclosure. The first thing we notice about this enclosure is its symmetry. The four side surfaces are symmetric about the base surface. Then, from the *symmetry rule,* we have

$$F_{12} = F_{13} = F_{14} = F_{15}$$

Also, the *summation rule* applied to surface 1 yields

$$\sum_{j=1}^{5} F_{1j} = F_{11} + F_{12} + F_{13} + F_{14} + F_{15} = 1$$

However, $F_{11} = 0$, since the base is a *flat* surface. Then the two relations above yield

$$F_{12} = F_{13} = F_{14} = F_{15} = 0.25$$

That is, each of the four side surfaces of the pyramid receive one fourth of the entire radiation leaving the base surface, as expected. Note that the presence of symmetry greatly simplified the determination of the view factors.

**EXAMPLE 12-8**

Determine the view factor from any one side of the infinitely long triangular duct whose cross-section is given in Fig. 12-51 to any other side.

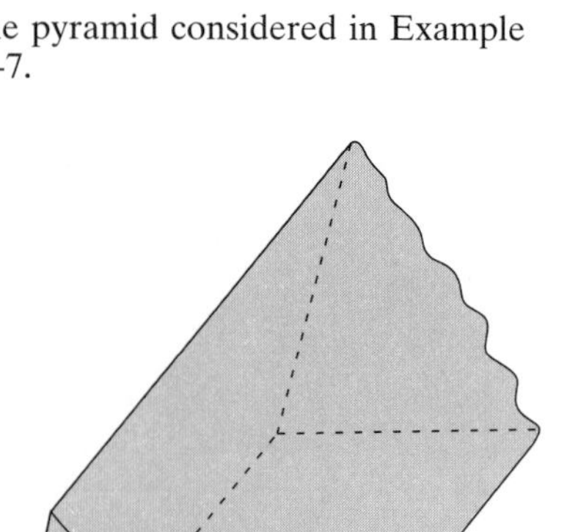

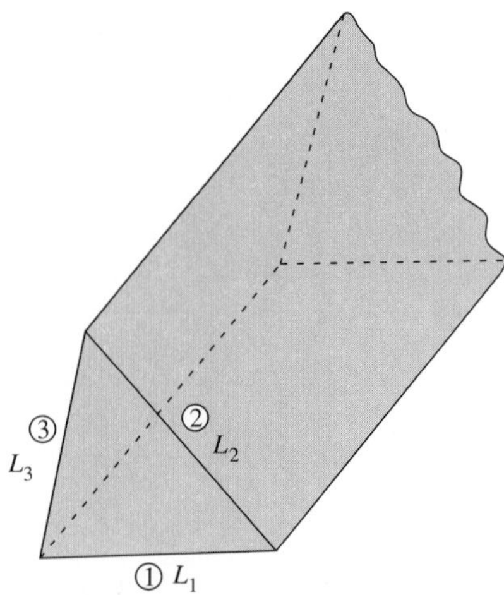


**FIGURE 12-51**

The infinitely long triangular duct considered in Example 12-8.

**Solution** The widths of the sides of the triangular cross-section of the duct are $L_1$, $L_2$, and $L_3$, and the surface areas corresponding to them are $A_1$, $A_2$, and $A_3$, respectively. Since the duct is infinitely long, the fraction of radiation leaving any surface that escapes through the ends of the duct is negligible. Therefore, the infinitely long duct can be considered to be a three-surface enclosure, $N = 3$.

This enclosure involves $N^2 = 3^2 = 9$ view factors, and we need to determine

$$\tfrac{1}{2}N(N-1) = \tfrac{1}{2} \times 3(3-1) = 3$$

of these view factors directly. Fortunately, we can determine all three of them by inspection to be

$$F_{11} = F_{22} = F_{33} = 0$$

since all three surfaces are flat. The remaining six view factors can be determined by the application of the summation and reciprocity rules.

Applying the summation rule to each of the three surfaces gives

$$F_{11} + F_{12} + F_{13} = 1$$
$$F_{21} + F_{22} + F_{23} = 1$$
$$F_{31} + F_{32} + F_{33} = 1$$

Noting that $F_{11} = F_{22} = F_{33} = 0$ and multiplying the first equation by $A_1$, the second by $A_2$, and the third by $A_3$ gives

$$A_1F_{12} + A_1F_{13} = A_1$$
$$A_2F_{21} + A_2F_{23} = A_2$$
$$A_3F_{31} + A_3F_{32} = A_3$$

Finally, applying the three reciprocity rules $A_1F_{12} = A_2F_{21}$, $A_1F_{13} = A_3F_{31}$, and $A_2F_{23} = A_3F_{32}$ gives

$$A_1F_{12} + A_1F_{13} = A_1$$
$$A_1F_{12} + A_2F_{23} = A_2$$
$$A_1F_{13} + A_2F_{23} = A_3$$

This is a set of three algebraic equations with three unknowns, which can be solved to obtain

$$F_{12} = \frac{A_1 + A_2 - A_3}{2A_1} = \frac{L_1 + L_2 - L_3}{2L_1}$$
$$F_{13} = \frac{A_1 + A_3 - A_2}{2A_1} = \frac{L_1 + L_3 - L_2}{2L_1} \qquad (12\text{-}33)$$
$$F_{23} = \frac{A_2 + A_3 - A_1}{2A_2} = \frac{L_2 + L_3 - L_1}{2L_2}$$

Note that we have replaced the areas of the side surfaces by their corresponding widths for simplicity, since $A = Ls$, and the length $s$ can be factored out and cancelled. We can generalize this result as follows: *the view factor from a surface of a very long triangular duct to another surface is equal to the sum of the widths of these two surfaces minus the width of the third surface, divided by twice the width of the first surface.*

## View Factors between Infinitely Long Surfaces: The Crossed-Strings Method

Many problems encountered in practice involve geometries of constant cross-section such as channels and ducts that are *very long* in one direction relative to the other directions. Such geometries can conveniently be considered to be *two-dimensional,* since any radiation interaction through their end surfaces will be negligible. Then they can be modeled as being *infinitely long,* and the view factor between their surfaces can be determined by the amazingly simple *crossed-strings method* developed by H. C. Hottel in the 1950s. The surfaces of the geometry do not need to be flat; they can be convex, concave, or any irregular shape.

To demonstrate the method, consider the geometry shown in Fig. 12-52, and let us try to find the view factor $F_{1\to 2}$ between surfaces 1 and

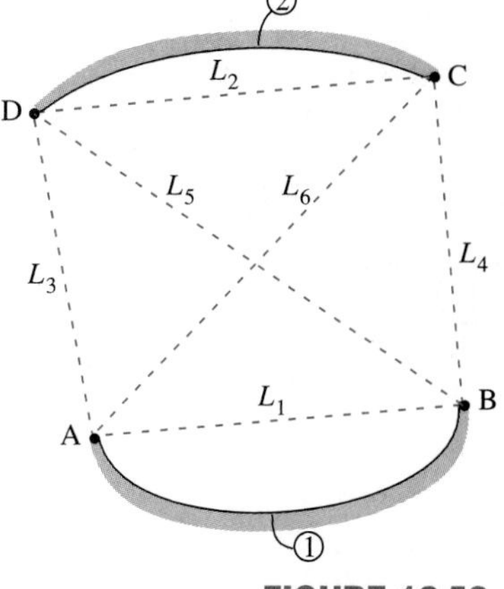

**FIGURE 12-52**
Determination of the view factor $F_{1\to 2}$ by the application of the crossed-strings method.

2. The first thing we do is identify the end points of the surfaces (the points A, B, C, and D) and connect them to each other with tightly stretched strings, which are indicated by dashed lines. Hottel has shown that the view factor $F_{1\to 2}$ can be expressed in terms of the lengths of these stretched strings, which are straight lines, as

$$F_{1\to 2} = \frac{(L_5 + L_6) - (L_3 + L_4)}{2L_1} \tag{12-34}$$

Note that $L_5 + L_6$ is the sum of the lengths of the *crossed strings,* and $L_3 + L_4$ is the sum of the lengths of the *uncrossed strings* attached to the endpoints. Therefore, Hottel's crossed-string method can be expressed verbally as

$$F_{i\to j} = \frac{\Sigma\,(\text{crossed strings}) - \Sigma\,(\text{uncrossed strings})}{2 \times (\text{string on surface } i)} \tag{12-35}$$

The crossed-strings method is applicable even when the two surfaces considered share a common edge, as in a triangle. In such cases, the common edge can be treated as an imaginary string of zero length. The method can also be applied to surfaces that are partially blocked by other surfaces by allowing the strings to bend around the blocking surfaces.

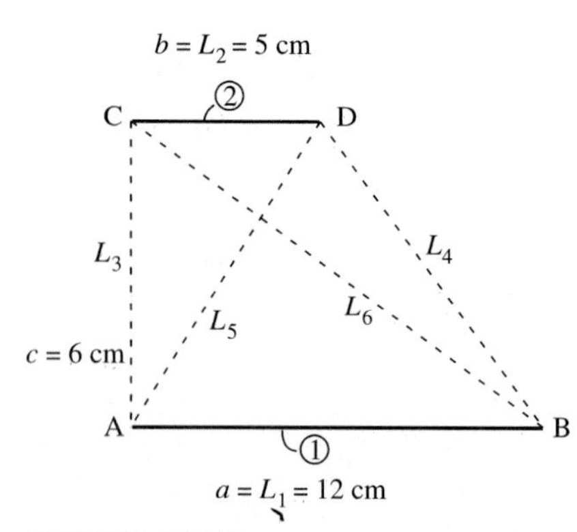

**FIGURE 12-53**

The two infinitely long parallel plates considered in Example 12-9.

**EXAMPLE 12-9**

Two infinitely long parallel plates of widths $a = 12$ cm and $b = 5$ cm are located a distance $c = 6$ cm apart, as shown in Fig. 12-53. (*a*) Determine the view factor $F_{1\to 2}$ from surface 1 to surface 2 by using the crossed-strings method. (*b*) Derive the crossed-strings formula by forming triangles on the given geometry and using Eq. 12-33 for view factors between the sides of triangles.

**Solution** (*a*) First we label the endpoints of both surfaces and draw straight dashed lines between the endpoints, as shown in Fig. 12-53. Then we identify the crossed and uncrossed strings, and apply the crossed-strings method (Eq. 12-35) to determine the view factor $F_{1\to 2}$:

$$F_{1\to 2} = \frac{\Sigma\,(\text{crossed strings}) - \Sigma\,(\text{uncrossed strings})}{2 \times (\text{string on surface 1})} = \frac{(L_5 + L_6) - (L_3 + L_4)}{2L_1}$$

where

$$L_1 = a = 12 \text{ cm}, \quad L_4 = \sqrt{7^2 + 6^2} = 9.22 \text{ cm}$$
$$L_2 = b = 5 \text{ cm}, \quad L_5 = \sqrt{5^2 + 6^2} = 7.81 \text{ cm}$$
$$L_3 = c = 6 \text{ cm} \quad L_6 = \sqrt{12^2 + 6^2} = 13.42 \text{ cm}$$

Substituting,

$$F_{1\to 2} = \frac{[(7.81 + 13.42) - (6 + 9.22)] \text{ cm}}{2 \times 12 \text{ cm}} = 0.250$$

(*b*) The geometry is infinitely long in the direction perpendicular to the plane of the paper, and thus the two plates (surfaces 1 and 2) and the two openings (imaginary surfaces 3 and 4) form a four-surface enclosure. Then applying the summation rule ot surface 1 yields

$$F_{11} + F_{12} + F_{13} + F_{14} = 1$$

But $F_{11} = 0$, since it is a flat surface. Therefore,

$$F_{12} = 1 - F_{13} - F_{14}$$

where the view factors $F_{13}$ and $F_{14}$ can be determined by considering the triangles ABC and ABD, respectively, and applying Eq. 12-33 for view factors between the sides of triangles. We obtain

$$F_{13} = \frac{L_1 + L_3 - L_6}{2L_1}, \qquad F_{14} = \frac{L_1 + L_4 - L_5}{2L_1}$$

Substituting,

$$F_{12} = 1 - \frac{L_1 + L_3 - L_6}{2L_1} - \frac{L_1 + L_4 - L_5}{2L_1}$$

$$= \frac{(L_5 + L_6) - (L_3 + L_4)}{2L_1}$$

which is the desired result. This is also a mini-proof of the crossed-strings method for the case of two infinitely long plain parallel surfaces.

## 12-7 ■ RADIATION HEAT TRANSFER: BLACK SURFACES

So far, we have considered the nature of radiation, the radiation properties of materials, and the view factors, and we are now in a position to consider the rate of heat transfer between surfaces by radiation. The analysis of radiation exchange between surfaces, in general, is complicated because of reflection: a radiation beam leaving a surface may be reflected several times, with partial reflection occurring at each surface, before it is completely absorbed. The analysis is simplified greatly when the surfaces involved can be approximated as black bodies, because of the absence of reflection. In this section, we consider radiation exchange between *black surfaces* only; we will extend the analysis to reflecting surfaces in the next section.

Consider two black surfaces of arbitrary shape maintained at uniform temperatures $T_1$ and $T_2$, as shown in Fig. 12-54. Recognizing that radiation leaves a black surface at a rate of $E_b = \sigma T^4$ per unit surface area and that the view factor $F_{1\rightarrow 2}$ represents the fraction of radiation leaving surface 1 that strikes surface 2, the *net* rate of radiation heat transfer from surface 1 to surface 2 can be expressed as

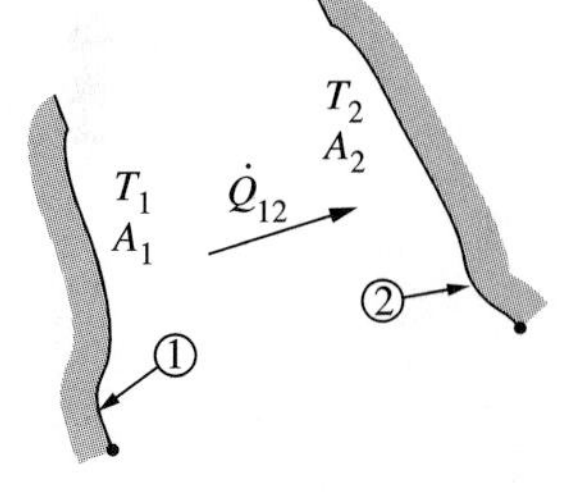


**FIGURE 12-54**
Two general black surfaces maintained at uniform temperatures $T_1$ and $T_2$.

$$\dot{Q}_{1\rightarrow 2} = \begin{pmatrix} \text{radiation leaving} \\ \text{the entire surface 1} \\ \text{that strikes surface 2} \end{pmatrix} - \begin{pmatrix} \text{radiation leaving} \\ \text{the entire surface 2} \\ \text{that strikes surface 1} \end{pmatrix}$$

$$= A_1 E_{b1} F_{1\rightarrow 2} - A_2 E_{b2} F_{2\rightarrow 1} \quad \text{(W)} \qquad (12\text{-}36)$$

Applying the reciprocity rule $A_1 F_{1\rightarrow 2} = A_2 F_{2\rightarrow 1}$ yields

$$\dot{Q}_{1\rightarrow 2} = A_1 F_{1\rightarrow 2} \sigma (T_1^4 - T_2^4) \quad \text{(W)} \qquad (12\text{-}37)$$

which is the desired relation. A negative value for $\dot{Q}_{1\rightarrow 2}$ indicates that net radiation heat transfer is from surface 2 to surface 1.

Now consider an *enclosure* consisting of *N black* surfaces maintained at specified temperatures. The *net* radiation heat transfer *from* any surface *i* of this enclosure is determined by adding up the net radiation heat transfers from surface *i* to each of the surfaces of the enclosure:

$$\dot{Q}_i = \sum_{j=1}^{N} \dot{Q}_{i\to j} = \sum_{j=1}^{N} A_i F_{i\to j}\sigma(T_i^4 - T_j^4) \quad \text{(W)} \tag{12-38}$$

Again a negative value for $\dot{Q}$ indicates that net radiation heat transfer is *to* surface *i* (i.e., surface *i gains* radiation energy instead of losing). Also, the net heat transfer from a surface to itself is zero, regardless of the shape of the surface.

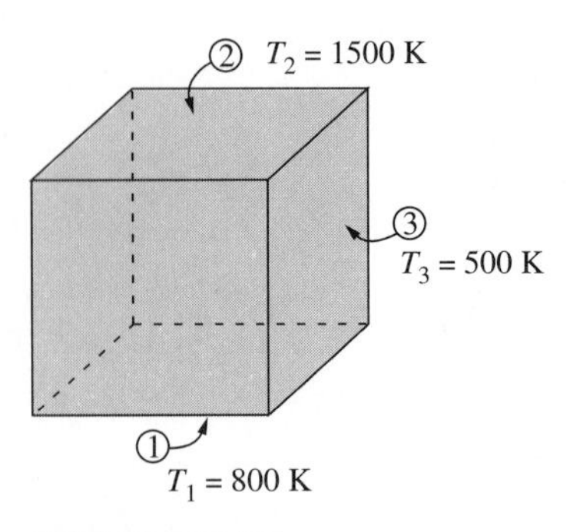


**FIGURE 12-55**
The cubical furnace of black surfaces considered in Example 12-10.

**EXAMPLE 12-10**

Consider the 5 m × 5 m × 5 m cubical furnace shown in Fig. 12-55, whose surfaces closely approximate black surfaces. The base, top, and side surfaces of the furnace are maintained at uniform temperatures of 800 K, 1500 K, and 500 K, respectively. Determine (*a*) the net rate of radiation heat transfer between the base and the side surfaces, (*b*) the net rate of radiation heat transfer between the base and the top surface, and (*c*) the net radiation heat transfer from the base surface.

**Solution** (*a*) Considering that the geometry involves six surfaces, we may be tempted at first to treat the furnace as a six-surface enclosure. However, the four side surfaces possess the same properties, and thus we can treat them as a single side surface in radiation analysis. We consider the base surface to be surface 1, the top surface to be surface 2, and the side surfaces to be surface 3. Then the problem reduces to determining $\dot{Q}_{1\to 3}$, $\dot{Q}_{1\to 2}$, and $\dot{Q}_1$.

The net rate of radiation heat transfer $\dot{Q}_{1\to 3}$ from surface 1 to surface 3 can be determined from Eq. 12-37, since both surfaces involved are black, by replacing the subscript 2 by 3:

$$\dot{Q}_{1\to 3} = A_1 F_{1\to 3}\sigma(T_1^4 - T_3^4)$$

But first we need to evaluate the view factor $F_{1\to 3}$. After checking the view factor charts and tables, we realize that we cannot determine this view factor directly. However, we can determine the view factor $F_{1\to 2}$ directly from Fig. 12-41 to be $F_{1\to 2} = 0.2$, and we know that $F_{1\to 1} = 0$, since surface 1 is plane. Then applying the summation rule to surface 1 yields

$$F_{1\to 1} + F_{1\to 2} + F_{1\to 3} = 1$$

or

$$\begin{aligned} F_{1\to 3} &= 1 - F_{1\to 1} - F_{1\to 2} \\ &= 1 - 0 - 0.2 \\ &= 0.8 \end{aligned}$$

Substituting,

$$\begin{aligned} \dot{Q}_{1\to 3} &= (25\ \text{m}^2)(0.8)[5.67 \times 10^{-8}\ \text{W/(m}^2\cdot\text{K}^4)][(800\ \text{K})^4 - (500\ \text{K})^4] \\ &= 393{,}611\ \text{W} \end{aligned}$$

(*b*) The net rate of radiation heat transfer $\dot{Q}_{1\to2}$ from surface 1 to surface 2 is determined in a similar manner from Eq. 12-37 to be

$$\begin{aligned}\dot{Q}_{1\to2} &= A_1F_{1\to2}\sigma(T_1^4 - T_2^4)\\ &= (25\ \text{m}^2)(0.2)[5.67 \times 10^{-8}\ \text{W/(m}^2\cdot\text{K}^4)][(800\ \text{K})^4 - (1500\ \text{K})^4]\\ &= -1{,}319{,}097\ \text{W}\end{aligned}$$

The negative sign indicates that net radiation heat transfer is from surface 2 to surface 1.

(*c*) The net radiation heat transfer from the base surface $\dot{Q}_1$ is determined from Eq. 12-37 by replacing the subscript *i* by 1 and taking $N = 3$:

$$\begin{aligned}\dot{Q}_1 &= \sum_{j=1}^{3}\dot{Q}_{1\to j} = \dot{Q}_{1\to1} + \dot{Q}_{1\to2} + \dot{Q}_{1\to3}\\ &= 0 + (-1{,}319{,}097\ \text{W}) + (393{,}611\ \text{W})\\ &= -925{,}486\ \text{W}\end{aligned}$$

Again the negative sign indicates that net radiation heat transfer is *to* surface 1. That is, the base of the furnace is gaining net radiation at a rate of about 925 kW.

## 12-8 ■ RADIATION HEAT TRANSFER: DIFFUSE, GRAY SURFACES

The analysis of radiation transfer in enclosures consisting of black surfaces is relatively easy, as we have seen above, but most enclosures encountered in practice involve nonblack surfaces, which allow multiple reflections to occur. Radiation analysis of such enclosures becomes very complicated unless some simplifying assumptions are made.

To make a simple radiation analysis possible, it is common to assume the surfaces of an enclosure to be *opaque, diffuse,* and *gray.* That is, the surfaces are nontransparent, they are diffuse emitters and diffuse reflectors, and their radiation properties are independent of wavelength. Also, each surface of the enclosure is *isothermal,* and both the incoming and outgoing radiation are *uniform* over each surface. But first we introduce the concept of radiosity.

### Radiosity

Surfaces emit radiation as well as reflecting it, and thus the radiation leaving a surface consists of emitted and reflected parts. The calculation of radiation heat transfer between surfaces involves the *total* radiation energy streaming away from a surface, with no regard for its origin. Thus, we need to define a new quantity to represent the *total radiation energy leaving a surface per unit time and per unit area.* This quantity is called the **radiosity**, and is denoted by $J$ (Fig. 12-56).

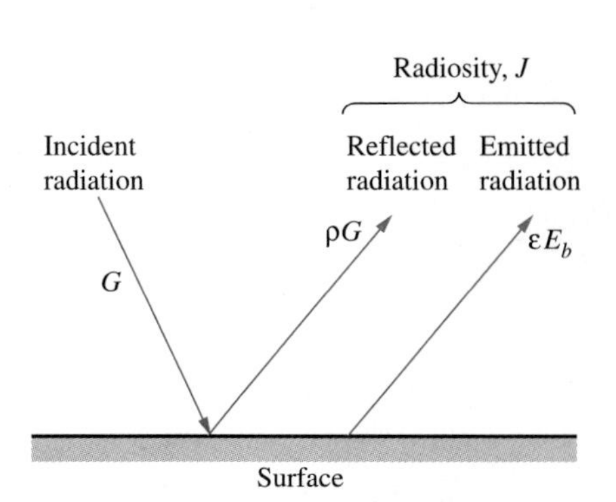


**FIGURE 12-56**
Radiosity represents the sum of the radiation energy emitted and reflected by a surface.

For a surface $i$ that is *gray* and *opaque* ($\varepsilon_i = \alpha_i$ and $\alpha_i + \rho_i = 1$), the radiosity can be expresses as

$$J_i = \begin{pmatrix}\text{radiation emitted}\\ \text{by surface } i\end{pmatrix} + \begin{pmatrix}\text{radiation reflected}\\ \text{by surface } i\end{pmatrix}$$
$$= \varepsilon_i E_{bi} + \rho_i G_i$$
$$= \varepsilon_i E_{bi} + (1 - \varepsilon_i)G_i \quad (\text{W/m}^2) \tag{12-39}$$

where $E_{bi} = \sigma T_i^4$ is the blackbody emissive power of surface $i$, and $G_i$ is irradiation (i.e., the radiation energy incident on surface $i$ per unit time per unit area).

For a surface that can be approximated as a *blackbody* ($\varepsilon_i = 1$), the radiosity relation reduces to

$$J_i = E_{bi} = \sigma T_i^4 \quad (\text{blackbody}) \tag{12-40}$$

That is, *the radiosity of a blackbody is equal to its emissive power.* This is expected, since a blackbody does not reflect any radiation, and thus radiation coming from a blackbody is due to emission only.

## Net Radiation Heat Transfer to (or from) a Surface

During a radiation interaction, a surface *loses* energy by emitting radiation and *gains* energy by absorbing radiation emitted by other surfaces. A surface experiences a net gain or a net loss of energy, depending on which quantity is larger. The *net* rate of radiation heat transfer from a surface $i$ of surface area $A_i$ is denoted by $\dot{Q}_i$, and is expressed as

$$\dot{Q}_i = \begin{pmatrix}\text{radiation leaving}\\ \text{entire surface } i\end{pmatrix} - \begin{pmatrix}\text{radiation incident}\\ \text{on entire surface } i\end{pmatrix}$$
$$= A_i(J_i - G_i) \quad (\text{W}) \tag{12-41}$$

Solving for $G_i$ from Eq. 12-39 and substituting into Eq. 12-41 yields

$$\dot{Q}_i = A_i\left(J_i - \frac{J_i - \varepsilon_i E_{bi}}{1 - \varepsilon_i}\right) = \frac{A_i \varepsilon_i}{1 - \varepsilon_i}(E_{bi} - J_i) \quad (\text{W}) \tag{12-42}$$

In electrical analogy to Ohm's law, this equation can be rearranged as

$$\dot{Q}_i = \frac{E_{bi} - J_i}{R_i} \quad (\text{W}) \tag{12-43}$$

where

$$R_i = \frac{1 - \varepsilon_i}{A_i \varepsilon_i} \tag{12-44}$$

is the **surface resistance** to radiation. The quantity $E_{bi} - J_i$ corresponds to a *potential difference* and the net rate of heat transfer corresponds to *current* in electrical analogy, as illustrated in Fig. 12-57.

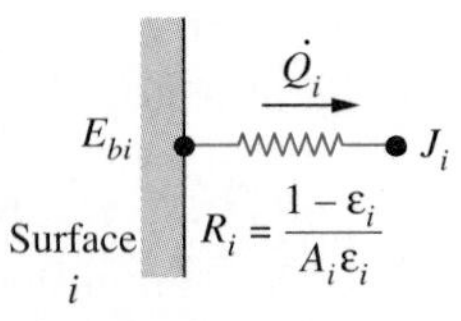


**FIGURE 12-57**
Electrical analogy of surface resistance to radiation.

The direction of the net radiation heat transfer depends on the relative magnitudes of $J_i$ (the radiosity) and $E_{bi}$ (the emissive power of a blackbody at the temperature of the surface). It will be *from* the surface if $E_{bi} > J_i$, and *to* the surface if $J_i > E_{bi}$. A negative value for $\dot{Q}_i$

indicates that heat transfer is *to* the surface. All of this radiation energy gained must be removed from the other side of the surface through some mechanism if the surface temperature is to remain constant.

The surface resistance to radiation for a *blackbody* is *zero,* since $\varepsilon_i = 1$, and $J_i = E_{bi}$. The net rate of radiation heat transfer in this case is determined directly from Eq. 12-41.

Some surfaces encountered in numerous practical heat transfer applications are modeled as being *adiabatic,* since their back sides are well insulated, and the net heat transfer through them is zero. When the convection effects on the front (heat transfer) side of such a surface is negligible and steady-state conditions are reached, the surface must lose as much radiation energy as it gains, and thus $\dot{Q}_i = 0$. In such cases, the surface is said to *reradiate* all the radiation energy it receives, and such a surface is called a **reradiating surface**. Setting $\dot{Q}_i = 0$ in Eq. 12-43 yields

$$J_i = E_{bi} = \sigma T_i^4 \quad \text{(W/m}^2\text{)} \tag{12-45}$$

Therefore, the *temperature* of a reradiating surface under steady conditions can easily be determined from the equation above once its radiosity is known. Note that the temperature of a reradiating surface is *independent of its emissivity.* In radiation analysis, the surface resistance of a reradiating surface is disregarded, since there is no net heat transfer through it. (This is like the fact that there is no need to consider a resistance in an electrical network if no current is flowing through it.)

## Net Radiation Heat Transfer between Any Two Surfaces

Consider two diffuse, gray, and opaque surfaces of arbitrary shape maintained at uniform temperatures, as shown in Fig. 12-58. Recognizing that the radiosity $J$ represents the rate of radiation leaving a surface per unit surface area and that the view factor $F_{i\to j}$ represents the fraction of radiation leaving surface $i$ that strikes surface $j$, the *net* rate of radiation heat transfer from surface $i$ to surface $j$ can be expressed as

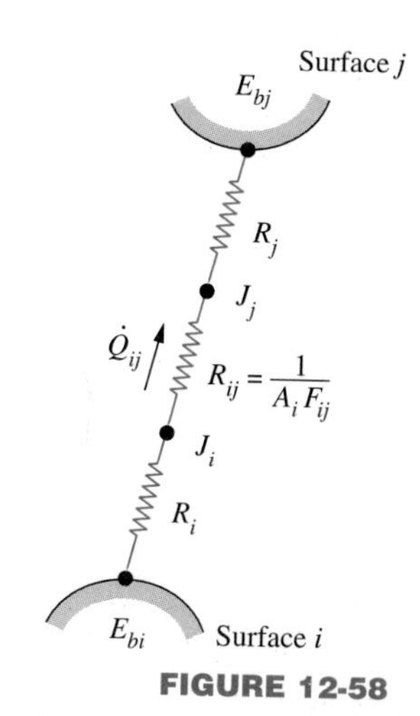


**FIGURE 12-58**
Electrical analogy of space resistance to radiation.

$$\dot{Q}_{i\to j} = \begin{pmatrix}\text{radiation leaving}\\ \text{the entire surface } i\\ \text{that strikes surface } j\end{pmatrix} - \begin{pmatrix}\text{radiation leaving}\\ \text{the entire surface } j\\ \text{that strikes surface } i\end{pmatrix}$$
$$= A_i J_i F_{i\to j} - A_j J_j F_{j\to i} \quad \text{(W)} \tag{12-46}$$

Applying the reciprocity relation $A_i F_{i\to j} = A_j F_{j\to i}$ yields

$$\dot{Q}_{i\to j} = A_i F_{i\to j}(J_i - J_j) \quad \text{(W)} \tag{12-47}$$

Again in analogy to Ohm's law, this equation can be rearranged as

$$\dot{Q}_{i\to j} = \frac{J_i - J_j}{R_{i\to j}} \quad \text{(W)} \tag{12-48}$$

where

$$R_{i\to j} = \frac{1}{A_i F_{i\to j}} \tag{12-49}$$

is the **space resistance** to radiation. Again the quantity $J_i - J_j$ corresponds to a *potential difference,* and the net rate of heat transfer between two surfaces corresponds to *current* in electrical analogy, as illustrated in Fig. 12-58.

The direction of the net radiation heat transfer between two surfaces depends on the relative magnitudes of $J_i$ and $J_j$. A positive value for $\dot{Q}_{i\to j}$ indicates that net heat transfer is *from* surface $i$ *to* surface $j$. A negative value indicates the opposite.

In an $N$-surface enclosure, the conservation of energy principle requires that the net heat transfer from surface $i$ to be equal to the sum of the net heat transfers from surface $i$ to each of the $N$ surfaces of the enclosure. That is,

$$\dot{Q}_i = \sum_{j=1}^{N} \dot{Q}_{i\to j} = \sum_{j=1}^{N} \frac{J_i - J_j}{R_{i\to j}} \quad \text{(W)} \tag{12-50}$$

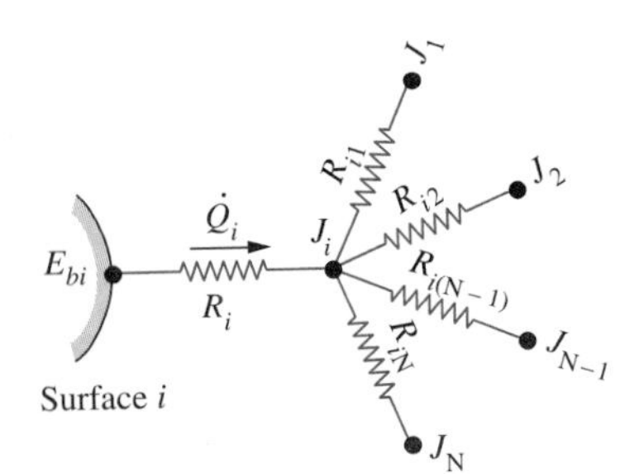


**FIGURE 12-59**
Network representation of net radiation heat transfer from surface $i$ to the remaining surfaces of an $N$-surface enclosure.

The network representation of net radiation heat transfer from surface $i$ to the remaining surfaces of an $N$-surface enclosure is given in Fig. 12-59. Note that $\dot{Q}_{i\to i}$ (the net rate of heat transfer from a surface to itself) is zero, regardless of the shape of the surface. Combining Eqs. 12-42 and 12-50 gives

$$\frac{E_{bi} - J_i}{R_i} = \sum_{j=1}^{N} \frac{J_i - J_j}{R_{i\to j}} \quad \text{(W)} \tag{12-51}$$

which has the electrical analogy interpretation that *the net radiation flow from a surface through its surface resistance is equal to the sum of the radiation flows from that surface to all other surfaces through the corresponding space resistances.*

## Methods of Solving Radiation Problems

In the radiation analysis of an enclosure, either the temperature or the net rate of heat transfer must be given for each of the surfaces to obtain a unique solution for the unknown surface temperatures and heat transfer rates. Equations 12-50 (for surfaces with specified heat transfer rates) and 12-51 (for surfaces with specified temperatures) give $N$ linear algebraic equations for the determination of the $N$ unknown radiosities for an $N$-surface enclosure. Once the radiosities $J_1, J_2, \ldots, J_N$ are available, the unknown surface temperatures and heat transfer rates can be determined from Eqs. 12-51 and 12-50, respectively.

The direct approach described above for solving radiation heat transfer problems normally involves the use of matrices, especially when there are a large number of surfaces, and is known as the **matrix method**. Therefore, the efficient use of this method requires some knowledge of linear algebra. Below we describe an alternative method called the **network method**, which is based on the electrical network analogy.

The network method is first introduced by A. K. Oppenheim in the 1950s, and found widespread acceptance because of its simplicity and its emphasis on the physics of the problem. The application of the method is straightforward: draw a surface resistance associated with each surface of an enclosure, and connect them with space resistances. Then solve the radiation problem by treating it as an electrical network problem where the radiation heat transfer replaces the current and radiosity replaces the potential.

The network method is not practical for enclosures with more than three or four surfaces, however, because of the increased complexity of the network. Below we apply the method to solve radiation problems in two- and three-surface enclosures.

### Radiation Heat Transfer in Two-Surface Enclosures

Consider an enclosure consisting of two opaque surfaces at specified temperatures $T_1$ and $T_2$, as shown in Fig. 12-60, and try to determine the net rate of radiation heat transfer between the two surfaces with the network method. Surfaces 1 and 2 have emissivities $\varepsilon_1$ and $\varepsilon_2$, surface areas $A_1$ and $A_2$, and are maintained at uniform temperatures $T_1$ and $T_2$, respectively. There are only two surfaces in the enclosure, and thus we can write

$$\dot{Q}_{12} = \dot{Q}_1 = -\dot{Q}_2$$

That is, the net rate of radiation transfer from surface 1 to surface 2 must equal to the net rate of radiation transfer *from* surface 1 and the net rate of radiation transfer *to* surface 2.

The radiation network of this two-surface enclosure consists of two surface resistances and one space resistance, as shown in Fig. 12-60. In an electrical network, the electric current flowing through these resistances connected in series would be determined by dividing the potential difference between points A and B by the total resistance between the same two points. The net rate of radiation transfer is determined in the same manner, and is expressed as

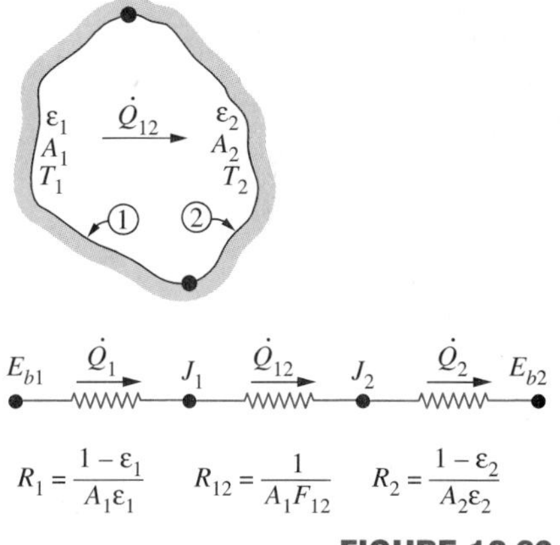


**FIGURE 12-60**

Schematic of a two-surface enclosure and the radiation network associated with it.

$$\dot{Q}_{12} = \frac{E_{b1} - E_{b2}}{R_1 + R_{12} + R_2} = \dot{Q}_1 = -\dot{Q}_2 \quad \text{(W)} \tag{12-52}$$

or

$$\dot{Q}_{12} = \frac{\sigma(T_1^4 - T_2^4)}{\dfrac{1-\varepsilon_1}{A_1\varepsilon_1} + \dfrac{1}{A_1F_{12}} + \dfrac{1-\varepsilon_2}{A_2\varepsilon_2}} \quad \text{(W)} \tag{12-53}$$

This important result is applicable to any two gray, diffuse, opaque surfaces that form an enclosure. The view factor $F_{12}$ depends on the geometry, and must be determined in advance. Simplified forms of Eq. 12-53 for some familiar arrangements that form a two-surface enclosure are given in Table 12-6. Note that $F_{12} = 1$ for all of these special cases.

**EXAMPLE 12-11**

Two very large parallel plates are maintained at uniform temperatures $T_1 = 800$ K and $T_2 = 500$ K, and have emissivities $\varepsilon_1 = 0.2$ and $\varepsilon_2 = 0.7$, respectively, as shown in Fig. 12-61. Determine the net rate of radiation heat transfer between the two surfaces per unit surface area of the plates.

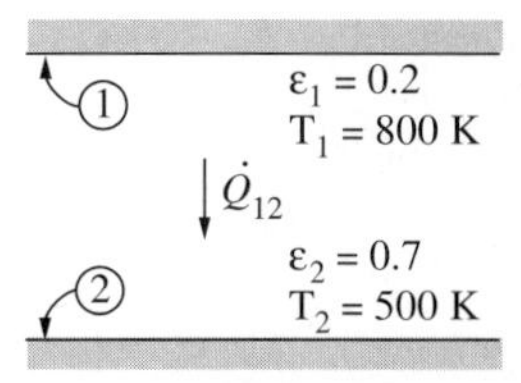


**FIGURE 12-61**

The two parallel plates considered in Example 12-11.

**Solution** Assuming the surfaces to be opaque, diffuse, and gray, the net

**TABLE 12-6**

| Configuration | Relations | Net radiation | Eq. |
|---|---|---|---|
| Small object in a large cavity ($A_1, T_1, \varepsilon_1$; $A_2, T_2, \varepsilon_2$) | $\frac{A_1}{A_2} \approx 0$, $F_{12} = 1$ | $\dot{Q}_{12} = A_1 \sigma \varepsilon_1 (T_1^4 - T_2^4)$ | (12–54) |
| Infinitely large parallel plates ($A_1, T_1, \varepsilon_1$; $A_2, T_2, \varepsilon_2$) | $A_1 = A_2 = A$, $F_{12} = 1$ | $\dot{Q}_{12} = \dfrac{A\sigma(T_1^4 - T_2^4)}{\frac{1}{\varepsilon_1} + \frac{1}{\varepsilon_2} - 1}$ | (12–55) |
| Infinitely long concentric cylinders ($r_1$, $r_2$) | $\frac{A_1}{A_2} = \frac{r_1}{r_2}$, $F_{12} = 1$ | $\dot{Q}_{12} = \dfrac{A_1\sigma(T_1^4 - T_2^4)}{\frac{1}{\varepsilon_1} + \frac{1-\varepsilon_2}{\varepsilon_2}\left(\frac{r_1}{r_2}\right)}$ | (12–56) |
| Concentric spheres ($r_1$, $r_2$) | $\frac{A_1}{A_2} = \left(\frac{r_1}{r_2}\right)^2$, $F_{12} = 1$ | $\dot{Q}_{12} = \dfrac{A_1\sigma(T_1^4 - T_2^4)}{\frac{1}{\varepsilon_1} + \frac{1-\varepsilon_2}{\varepsilon_2}\left(\frac{r_1}{r_2}\right)^2}$ | (12–57) |

rate of radiation heat transfer between the two plates per unit area is readily determined from Eq. 12-55 to be

$$\dot{q}_{12} = \frac{\dot{Q}_{12}}{A} = \frac{\sigma(T_1^4 - T_2^4)}{\frac{1}{\varepsilon_1} + \frac{1}{\varepsilon_2} - 1} = \frac{[5.67 \times 10^{-8}\ \text{W/(m}^2 \cdot \text{K}^4)][(800\ \text{K})^4 - (500\ \text{K})^4]}{\frac{1}{0.2} + \frac{1}{0.7} - 1}$$
$$= 3625\ \text{W/m}^2$$

That is, heat at a net rate of 3625 W is transferred from plate 1 to plate 2 by radiation per unit surface area of either plate.

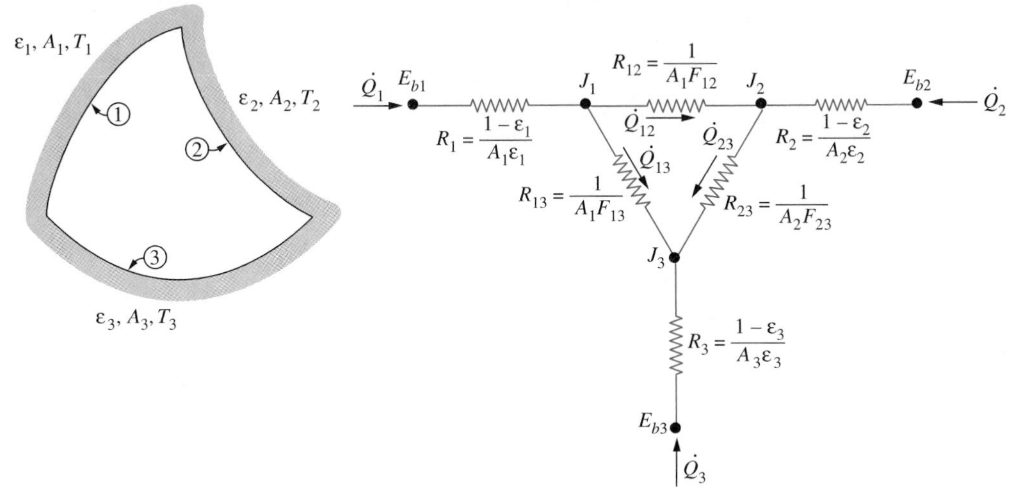


**FIGURE 12-62**
Schematic of a three-surface enclosure and the radiation network associated with it.

## Radiation Heat Transfer in Three-Surface Enclosures

We now consider an enclosure consisting of three opaque, diffuse, gray surfaces, as shown in Fig. 12-62. Surfaces 1, 2, and 3 have surface areas $A_1$, $A_2$, and $A_3$, emissivities $\varepsilon_1$, $\varepsilon_2$, and $\varepsilon_3$, and uniform temperatures $T_1$, $T_2$, and $T_3$, respectively. The radiation network of this geometry is constructed by following the standard procedure: drawing a surface resistance associated with each of the three surfaces, and connecting these surface resistances with space resistances, as shown in the figure. Relations for the surface and space resistances are given by Eqs. 12-44 and 12-49. The three end-point potentials $E_{b1}$, $E_{b2}$, and $E_{b3}$ are considered known, since the surface temperatures are specified. Then all we need to find are the radiosities $J_1$, $J_2$, and $J_3$. The three equations for the determination of these three unknowns are obtained from the requirement that *the algebraic sum of the currents (net radiation heat transfer) at each node must equal zero.* That is,

$$\frac{E_{b1} - j_1}{R_1} + \frac{J_2 - J_1}{R_{12}} + \frac{J_3 - J_1}{R_{13}} = 0$$

$$\frac{J_1 - J_2}{R_{12}} + \frac{E_{b2} - J_2}{R_2} + \frac{J_3 - J_2}{R_{23}} = 0 \qquad (12\text{-}58)$$

$$\frac{J_1 - J_3}{R_{13}} + \frac{J_2 - J_3}{R_{23}} + \frac{E_{b3} - J_3}{R_3} = 0$$

Once the radiosities $J_1$, $J_2$, and $J_3$ are available, the net rate of radiation heat transfer at each surface can be determined from Eq. 12-50.

The set of equations above simplify further if one or more surfaces are "special" in some way. For example, $J_i = E_{bi} = \sigma T_i^4$ for a *black* or

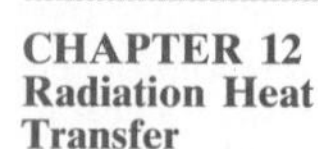

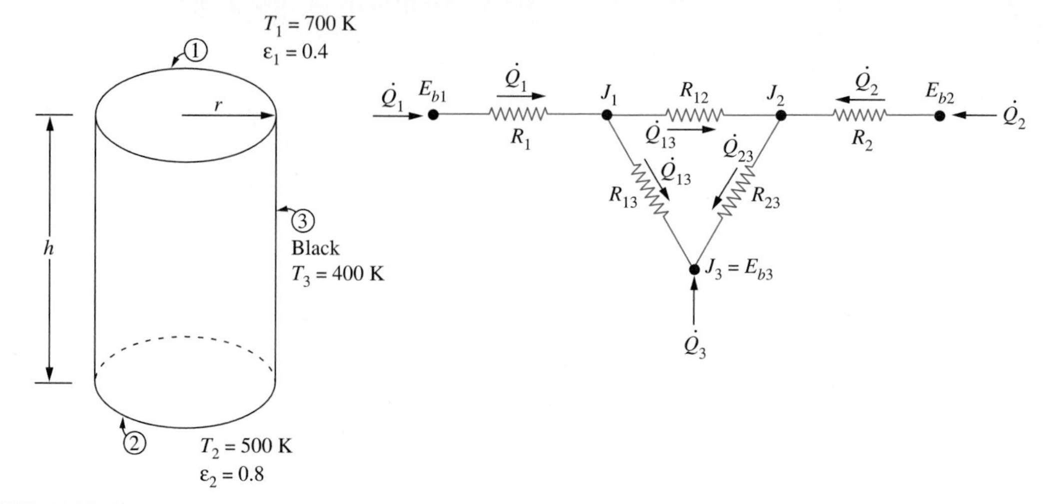


**FIGURE 12-63**

The cylindrical furnace considered in Example 12-12.

*reradiating* surface. Also, $\dot{Q}_i = 0$ for a reradiating surface. Finally, when the net rate of radiation heat transfer $\dot{Q}_i$ is specified at surface $i$ instead of the temperature, the term $(E_{bi} - J_i)/R_i$ should be replaced by the specified $\dot{Q}_i$.

**EXAMPLE 12-12**

Consider a cylindrical furnace with $r = h = 1$ m, as shown in Fig. 12-63. The base (surface 1) and the top (surface 2) of the furnace have emissivities $\varepsilon_1 = 0.4$ and $\varepsilon_2 = 0.8$, respectively, and are maintained at uniform temperatures $T_1 = 700$ K and $T_2 = 500$ K. The side surface closely approximates a blackbody, and is maintained at a temperature of $T_3 = 400$ K. Determine the net rate of radiation heat transfer at each surface during steady-state operation, and explain how these surfaces can be maintained at specified temperatures.

**Solution** The furnace can be considered to be a three-surface enclosure with a radiation network as shown in the figure. We assume that the surfaces are opaque, diffuse, and gray, and that steady-state conditions exist. To determine the net radiation heat transfers, we need to find the radiosites first. The side surface is given as black, and thus its radiosity is simply $J_3 = E_{b3}$. The other two radiosities are determined by setting the net radiation flow (the "current") entering each of the nodes 1 and 2 equal to zero (Eq. 12-58):

$$\frac{E_{b1} - J_1}{R_1} + \frac{J_2 - J_1}{R_{12}} + \frac{J_3 - J_1}{R_{13}} = 0$$

$$\frac{J_1 - J_2}{R_{12}} + \frac{E_{b2} - J_2}{R_2} + \frac{J_3 - J_2}{R_{23}} = 0$$

where

$$E_{b1} = \sigma T_1^4 = [5.67 \times 10^{-8}\ \text{W/(m}^2\cdot\text{K}^4)](700\ \text{K})^4 = 13{,}614\ \text{W/m}^2$$

$$E_{b2} = \sigma T_2^4 = [5.67 \times 10^{-8}\ \text{W/(m}^2\cdot\text{K}^4)](500\ \text{K})^4 = 3544\ \text{W/m}^2$$

$$E_{b3} = \sigma T_3^4 = [5.67 \times 10^{-8}\ \text{W/(m}^2\cdot\text{K}^4)](400\ \text{K})^4 = 1452\ \text{W/m}^2 = J_3$$

and $$A_1 = A_2 = \pi r^2 = 3.14 \times (1\ \text{m})^2 = 3.14\ \text{m}^2$$

The view factor from the base to the top surface is determined from Fig. 12-43 to be $F_{12} = 0.38$. Then the view factor from the base to the side surface is determined by applying the summation rule to be

$$F_{11} + F_{12} + F_{13} = 1 \rightarrow F_{13} = 1 - F_{11} - F_{12} = 1 - 0 - 0.38 = 0.62$$

since the base surface is flat and thus $F_{11} = 0$. Then the radiation resistances that appear in the two equations above become

$$R_1 = \frac{1 - \varepsilon_1}{A_1 \varepsilon_1} = \frac{1 - 0.8}{(3.14\ \text{m}^2) \times 0.8} = 0.0796\ \text{m}^{-2}$$

$$R_2 = \frac{1 - \varepsilon_2}{A_2 \varepsilon_2} = \frac{1 - 0.4}{(3.14\ \text{m}^2) \times 0.4} = 0.4777\ \text{m}^{-2}$$

$$R_{12} = \frac{1}{A_1 F_{12}} = \frac{1}{(3.14\ \text{m}^2) \times 0.38} = 0.8381\ \text{m}^{-2}$$

$$R_{23} = \frac{1}{A_2 F_{23}} = \frac{1}{(3.14\ \text{m}^2) \times 0.62} = 0.5137\ \text{m}^{-2} = R_{13} \quad \text{(symmetry)}$$

Substituting,

$$\frac{13{,}614 - J_1}{0.0796} + \frac{J_2 - J_1}{0.8381} + \frac{1452 - J_1}{0.5137} = 0$$

$$\frac{J_1 - J_2}{0.8381} + \frac{3544 - J_2}{0.4777} + \frac{1542 - J_2}{0.5137} = 0$$

Solving these two equations with two unknowns yields

$$J_1 = 11{,}418\ \text{W/m}^2, \qquad J_2 = 4562\ \text{W/m}^2$$

Then the net rates of radiation heat transfer at the top and the base become

$$\dot{Q}_1 = \frac{E_{b1} - J_1}{R_1} = \frac{(13{,}614 - 11{,}418)\ \text{W/m}^2}{0.0796\ \text{m}^{-2}} = 27{,}588\ \text{W}$$

$$\dot{Q}_2 = \frac{E_{b2} - J_2}{R_2} = \frac{(3544 - 4562)\ \text{W/m}^2}{0.4777\ \text{m}^{-2}} = -2131\ \text{W}$$

The net rate of radiation heat transfer at the side surface is determined by requiring the net radiation flow to node 3 be equal to zero:

$$\dot{Q}_3 + \frac{J_1 - J_3}{R_{13}} + \frac{J_2 - J_3}{R_{23}} = 0$$

or,

$$\dot{Q}_3 = \frac{J_3 - J_1}{R_{13}} + \frac{J_3 - J_2}{R_{23}} = \frac{(1452 - 11{,}418)\ \text{W/m}^2}{0.5137\ \text{m}^{-2}} + \frac{(1452 - 4562)\ \text{W/m}^2}{0.5137\ \text{m}^{-2}}$$
$$= -25{,}455\ \text{W}$$

Note that the direction of net radiation heat transfer is *from* the top surface *to* the base and side surfaces, and the algebraic sum of these three quantities must be equal to zero. That is,

$$\dot{Q}_1 + \dot{Q}_2 + \dot{Q}_3 = 27{,}588 + (-2131) + (-25{,}455) \approx 0$$

The small difference is due to round-off error. To maintain the surfaces at the specified temperatures, we must supply the top surface heat continuously at a

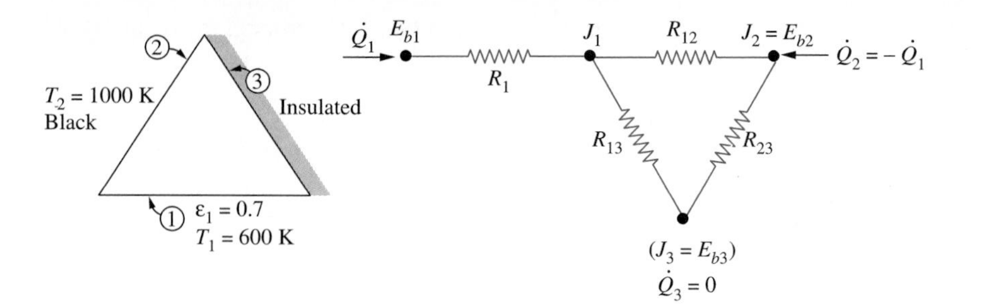


**FIGURE 12-64**
The triangular furnace considered in Example 12-13.

rate of 27,588 W while removing 2131 W from the base and 25,455 W from the side surfaces.

---

**EXAMPLE 12-13**

A furnace is shaped like a long equilateral triangular duct, as shown in Fig. 12-64. The width of each side is 1 m. The base surface has an emissivity of 0.7 and is maintained at a uniform temperature of 600 K. The heated left side surface closely approximates a blackbody at 1000 K. The right side surface is well-insulated. Determine the rate at which energy must be supplied to the heated side externally per unit length of the duct in order to maintain these operating conditions.

**Solution** The furnace can be considered to be a three-surface enclosure with a radiation network as shown in the figure, since the duct is very long and thus the end effects are negligible. We assume that the surfaces are opaque, diffuse, and gray. We observe that the view factor from any surface to any other surface in the enclosure is 0.5 because of symmetry. Surface 3 is a reradiating surface, since the net rate of heat transfer at that surface is zero. Then we must have $\dot{Q}_1 = -\dot{Q}_2$, since the entire heat lost by surface 1 must be gained by surface 2. The radiation network in this case is a simple series–parallel connection, and we can determine $\dot{Q}_1$ directly from

$$\dot{Q}_1 = \frac{E_{b1} - E_{b2}}{R_1 + \left(\dfrac{1}{R_{12}} + \dfrac{1}{R_{13} + R_{23}}\right)^{-1}} = \frac{E_{b1} - E_{b2}}{\dfrac{1 - \varepsilon_1}{A_1\varepsilon_1} + \left(A_1F_{12} + \dfrac{1}{1/A_1F_{13} + 1/A_2F_{23}}\right)^{-1}}$$

where

$$A_1 = A_2 = A_3 = wL = 1\text{ m} \times 1\text{ m} = 1\text{ m}^2 \quad \text{(per unit length of the duct)}$$
$$F_{12} = F_{13} = F_{23} = 0.5 \quad \text{(symmetry)}$$
$$E_{b1} = \sigma T_1^4 = [5.67 \times 10^{-8}\text{ W/(m}^2\cdot\text{K}^4)](600\text{ K})^4 = 7348\text{ W/m}^2$$
$$E_{b2} = \sigma T_2^4 = [5.67 \times 10^{-8}\text{ W/(m}^2\cdot\text{K}^4)](1000\text{ K})^4 = 56{,}700\text{ W/m}^2$$

Substituting,

$$\dot{Q}_1 = \frac{(56{,}700 - 7348)\text{ W/m}^2}{\dfrac{1 - 0.7}{0.7 \times 1\text{ m}^2} + \left[(0.5 \times 1\text{ m}^2) + \dfrac{1}{1/(0.5 \times 1\text{ m}^2) + 1/(0.5 \times 1\text{ m}^2)}\right]^{-1}}$$
$$= 28{,}009\text{ W}$$

Therefore, energy at a rate of 28,009 W must be supplied to the heated surface per unit length of the duct to maintain steady operation in the furnace.



**EXAMPLE 12-14**

A solar collector consists of a horizontal aluminum tube having an outer diameter of 2 in enclosed in a concentric thin glass tube of 4-in diameter, as shown in Fig. 12-65. Water is heated as it flows through the tube, and the annular space between the aluminum and the glass tubes is filled with air at 1-atm pressure. The pump circulating the water fails during a clear day, and the water temperature in the tube starts rising. The aluminum tube absorbs solar radiation at a rate of 30 Btu/h per foot length, and the temperature of the ambient air outside is 70°F. The emissivities of the tube and the glass cover are 0.95 and 0.9, respectively. Taking the effective sky temperature to be 50°F, determine the temperature of the aluminum tube when thermal equilibrium is established (i.e., when the rate of heat loss from the tube equals the amount of solar energy gained by the tube).

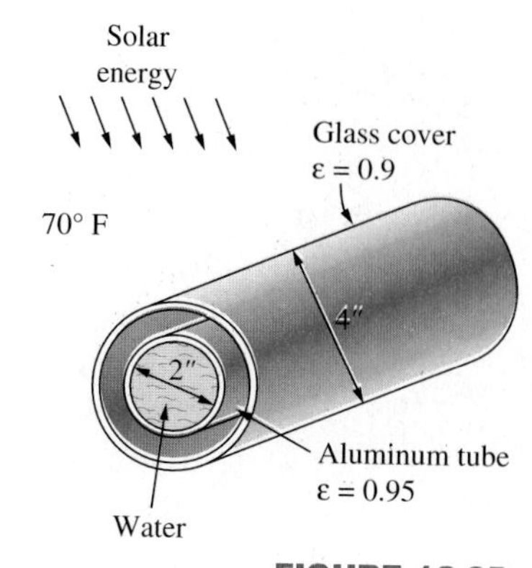


**FIGURE 12-65**
Schematic for Example 12-14.

**Solution** This problem was solved in the previous chapter by disregarding radiation heat transfer. Now we will repeat the solution by considering natural convection and radiation occurring simultaneously. We have a horizontal cylindrical enclosure filled with air at 1-atm pressure. The problem involves heat transfer from the aluminum tube to the glass cover, and from the outer surface of the glass cover to the surrounding ambient air. When thermal equilibrium is established and steady operation is reached, these two heat transfer rates must equal the rate of heat gain. That is,

$$\dot{Q}_{\text{tube-glass}} = \dot{Q}_{\text{glass-ambient}} = \dot{Q}_{\text{solar gain}} = 30 \text{ Btu/h} \quad \text{(per foot of tube)}$$

The properties of air should be evaluated at the average temperature. But we do not know the tube and glass cover temperatures at this point, and thus we cannot evaluate the average temperatures. Therefore, we will use the properties at an anticipated average temperature of 100°F, which are

$$k = 0.0154 \text{ Btu/(h}\cdot\text{ft}\cdot{}^\circ\text{F)}$$

$$\nu = 0.18 \times 10^{-3} \text{ ft}^2\text{/s}$$

$$\text{Pr} = 0.72$$

$$\beta = \frac{1}{T_f} = \frac{1}{(100 + 460)\text{ R}} = 0.001786 \text{ R}^{-1}$$

The results obtained can then be refined for better accuracy, if necessary, using the evaluated surface temperatures. The heat transfer surface area of the glass cover is

$$A_2 = A_{\text{glass}} = (\pi DL)_{\text{glass}} = \pi(\tfrac{4}{12}\text{ ft})(1\text{ ft}) = 1.047 \text{ ft}^2 \quad \text{(per foot length of tube)}$$

To determine the Rayleigh number, we need to know the surface temperature of the glass, which is not available. Therefore, it is clear that the solution will require a trial-and-error approach. Assuming the glass cover temperature to be 80°F, the Raleigh number, the Nusselt number, the convection heat transfer coefficient, and the rate of natural convection heat transfer from the glass cover to the ambient air are determined to be

$$\mathrm{Ra} = \frac{g\beta(T_s - T_\infty)\delta^3}{\nu^2}\mathrm{Pr}$$

$$= \frac{(32.2\ \mathrm{ft/s^2})(0.001786\ \mathrm{R^{-1}})[(80 - 70)\ \mathrm{K}](\frac{4}{12}\ \mathrm{ft})^3}{(0.18 \times 10^{-3}\ \mathrm{ft^2/s})^2}(0.72) = 4.733 = 10^5$$

$$\mathrm{Nu} = \left\{0.6 + \frac{0.387\ \mathrm{Ra}^{1/6}}{[1 + (0.559/\mathrm{Pr})^{9/16}]^{8/27}}\right\}^2$$

$$= \left\{0.6 + \frac{0.387(4.733 \times 10^5)^{1/6}}{[1 + (0.559/0.72)^{9/16}]^{8/27}}\right\}^2 = 11.8$$

$$h = \frac{k}{D}\mathrm{Nu} = \frac{0.0154\ \mathrm{Btu/(ft \cdot {}^\circ F)}}{4/12\ \mathrm{ft}}(11.8) = 0.546\ \mathrm{Btu/(h \cdot ft^2 \cdot {}^\circ F)}$$

$$\dot{Q}_{2,\mathrm{conv}} = hA_2(T_2 - T_\infty) = [0.546\ \mathrm{Btu/(h \cdot ft^2 \cdot {}^\circ F)}](1.047\ \mathrm{ft^2})(80 - 70)^\circ\mathrm{F}$$
$$= 5.7\ \mathrm{Btu/h}$$

Also,

$$\dot{Q}_{2,\mathrm{rad}} = \varepsilon\sigma A_2(T_2^4 - T_{\mathrm{surr}}^4)$$
$$= (0.9)[0.1714\ \mathrm{Btu/(h \cdot ft^2 \cdot R^4)}](1.047\ \mathrm{ft^2})[(540\ \mathrm{R})^4 - (510\ \mathrm{R})^4]$$
$$= 28.1\ \mathrm{Btu/h}$$

and $$\dot{Q}_{2,\mathrm{total}} = \dot{Q}_{2,\mathrm{conv}} + \dot{Q}_{2,\mathrm{rad}} = 5.7 + 28.1 = 33.8\ \mathrm{Btu/h}$$

which is more than 30 Btu/h. Therefore, the assumed temperature of 80°F for the glass cover is high. Repeating the calculations for a temperature of 75°F gives 25.5 Btu/h, which is low. Then the glass cover temperature corresponding to 30 Btu/h is determined by interpolation to be 78°F.

The temperature of the aluminum tube is determined in a similar manner using the natural convection radiation relations for two horizontal concentric cylinders. The characteristic length in this case is the distance between the two cylinders, which is determined to be

$$\delta = \tfrac{1}{2}(D_2 - D_1) = \tfrac{1}{2}(4 - 2)\ \mathrm{in} = 1\ \mathrm{in}$$

Also,

$$A_1 = A_{\mathrm{tube}} = (\pi DL)_{\mathrm{tube}} = \pi(\tfrac{2}{12}\ \mathrm{ft})(1\ \mathrm{ft}) = 0.524\ \mathrm{ft^2} \quad \text{(per foot length of tube)}$$

$$A = \frac{\pi L(D_2 - D_1)}{\ln(D_2/D_1)} = \frac{\pi(1\ \mathrm{ft})(\frac{4}{12} - \frac{2}{12})\ \mathrm{ft}}{\ln(\frac{4}{2})} = 0.755\ \mathrm{ft^2}$$

We start the calculations by assuming the tube temperature to be 120°F. This gives

$$\mathrm{Ra} = \frac{g\beta(T_1 - T_2)\delta^3}{\nu^2}\mathrm{Pr}$$

$$= \frac{(32.2\ \mathrm{ft/s^2})(0.001786\ \mathrm{R^{-1}})[(120 - 78)\ \mathrm{R}](\frac{1}{12}\ \mathrm{ft})^3}{(0.18 \times 10^{-3}\ \mathrm{ft^2/s})^2}(0.72) = 3.11 \times 10^4$$

$$\mathrm{Nu} = 0.11\ \mathrm{Ra}^{0.29} = 0.11(3.11 \times 10^4)^{0.29} = 2.21$$

$$\dot{Q}_{1,\mathrm{conv}} = kNuA\frac{T_1 - T_2}{\delta}$$

$$= [0.0154\ \mathrm{Btu/(h \cdot ft \cdot {}^\circ F)}](2.21)(0.755\ \mathrm{ft^2})\frac{(120 - 78)\ \mathrm{R}}{\frac{1}{12}\ \mathrm{ft}} = 13.0\ \mathrm{Btu/h}$$

Also,

$$\dot{Q}_{1,\text{rad}} = \frac{\sigma A_1(T_1^4 - T_2^4)}{\dfrac{1}{\varepsilon_1} + \dfrac{1-\varepsilon_2}{\varepsilon_2}\left(\dfrac{D_1}{D_2}\right)}$$

$$= \frac{[0.1714 \times 10^{-8}\ \text{Btu/(h}\cdot\text{ft}^2\cdot\text{R}^4)](0.524\ \text{ft}^2)][(580\ \text{R})^4 - (538\ \text{R})^4]}{\dfrac{1}{0.95} + \dfrac{1-0.9}{0.9}\left(\dfrac{2\ \text{in}}{4\ \text{in}}\right)}$$

$$= 23.8\ \text{Btu/h}$$

and $$\dot{Q}_{1,\text{total}} = \dot{Q}_{1,\text{conv}} + \dot{Q}_{1,\text{rad}} = 13.0 + 23.8 = 36.8\ \text{Btu/h}$$

which is more than 30 Btu/h. Therefore, the assumed temperature of 120°F for the tube is high. Repeating the calculations for a temperature of 110°F gives 26.7 Btu/h, which is low. Then the tube temperature corresponding to 30 Btu/h is determined by interpolation to be **113°F**. Therefore, the tube will reach an equilibrium temperature of 113°F when the pump fails. Recall that disregarding radiation in the previous chapter gave a result of 188°F, which seems to be quite unrealistic. Therefore, radiation should always be considered in systems that are heated or cooled by natural convection, unless the surfaces involved are polished and thus have emissivities close to zero.

## 12-9 RADIATION SHIELDS AND THE RADIATION EFFECT

Radiation heat transfer between two surfaces can be reduced greatly by inserting a thin, high-reflectivity (low-emissivity) sheet of material between the two surfaces. Such highly reflective thin plates or shells are called **radiation shields**. Multilayer radiation shields constructed of about 20 sheets per cm thickness separated by evacuated space are commonly used in cryogenic and space applications. Radiation shields are also used in temperature measurements of fluids to reduce the error caused by the radiation effect when the temperature sensor is exposed to surfaces that are much hotter or colder than the fluid itself. The role of the radiation shield is to reduce the rate of radiation heat transfer by placing additional resistances in the path of radiation heat flow. The lower the emissivity of the shield, the higher the resistance.

Radiation heat transfer between two large parallel plates of emissivities $\varepsilon_1$ and $\varepsilon_2$ maintained at uniform temperatures $T_1$ and $T_2$ is given by Eq. 12-55:

$$\dot{Q}_{12,\text{no shield}} = \frac{A\sigma(T_1^4 - T_2^4)}{\dfrac{1}{\varepsilon_1} + \dfrac{1}{\varepsilon_2} - 1}$$

Now consider a radiation shield placed between these two plates, as shown in Fig. 12-66. Let the emissivities of the shield facing plates 1 and 2 be $\varepsilon_{3,1}$ and $\varepsilon_{3,2}$, respectively. Note that the emissivity of different surfaces

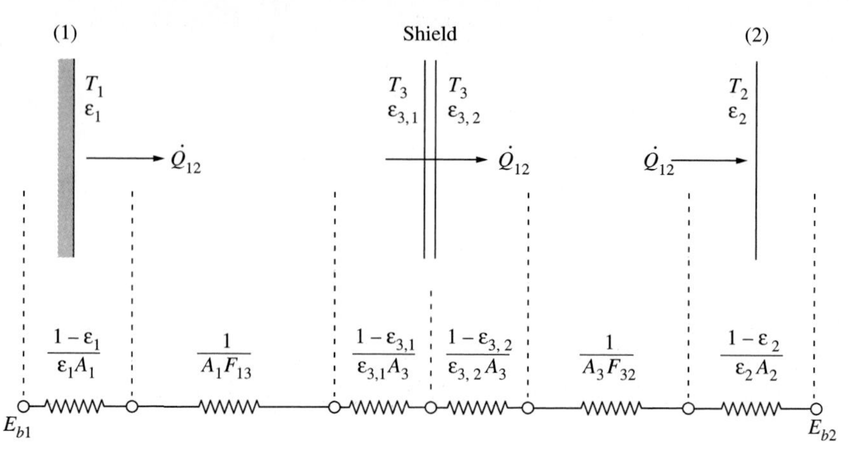

**FIGURE 12-66**

The radiation shield placed between two parallel plates and the radiation network associated with it.

of the shield may be different. The radiation network of this geometry is constructed, as usual, by drawing a surface resistance associated with each surface, and connecting these surface resistances with space resistances, as shown in the figure. The resistances are connected in series, and thus the rate of radiation heat transfer is

$$\dot{Q}_{12,\,\text{one shield}} = \frac{E_{b1} - E_{b2}}{\dfrac{1-\varepsilon_1}{A_1\varepsilon_1} + \dfrac{1}{A_1F_{12}} + \dfrac{1-\varepsilon_{3,1}}{A_3\varepsilon_{3,1}} + \dfrac{1-\varepsilon_{3,2}}{A_3\varepsilon_{3,2}} + \dfrac{1}{A_3F_{32}} + \dfrac{1-\varepsilon_2}{A_2\varepsilon_2}} \qquad \text{(12-59)}$$

Noting that $F_{13} = F_{23} = 1$ and $A_1 = A_2 = A_3 = A$ for parallel plates, Eq. 12-59 simplifies to

$$\dot{Q}_{12,\,\text{one shield}} = \frac{A\sigma(T_1^4 - T_2^4)}{\left(\dfrac{1}{\varepsilon_1} + \dfrac{1}{\varepsilon_2} - 1\right) + \left(\dfrac{1}{\varepsilon_{3,1}} + \dfrac{1}{\varepsilon_{3,2}} - 1\right)} \qquad \text{(12-60)}$$

where the terms in the second set of parentheses in the denominator represent the additional resistance to radiation introduced by the shield. The appearance of the equation above suggests that parallel plates involving multiple radiation shields can be handled by adding a group of terms like those in the second set of parentheses to the denominator for each radiation shield. Then the radiation heat transfer through large parallel plates separated by $N$ radiation shields becomes

$$\dot{Q}_{12,\,N\,\text{shields}} = \frac{A\sigma(T_1^4 - T_2^4)}{\left(\dfrac{1}{\varepsilon_1} + \dfrac{1}{\varepsilon_2} - 1\right) + \left(\dfrac{1}{\varepsilon_{3,1}} + \dfrac{1}{\varepsilon_{3,2}} - 1\right) + \cdots + \left(\dfrac{1}{\varepsilon_{N,1}} + \dfrac{1}{\varepsilon_{N,2}} - 1\right)} \qquad \text{(12-61)}$$

If the emmissivities of all surfaces are equal, Eq. 12-61 reduces to

$$\dot{Q}_{12,\,N\,\text{shields}} = \frac{A\sigma(T_1^4 - T_2^4)}{(N+1)\left(\dfrac{1}{\varepsilon} + \dfrac{1}{\varepsilon} - 1\right)} = \frac{1}{N+1}\dot{Q}_{12,\,\text{no shields}} \qquad (12\text{-}62)$$

Therefore, when all emissivities are equal, 1 shield reduces the rate of radiation heat transfer to one-half, 9 shields reduce it to one-tenth, and 19 shields reduce it to one-twentieth (or 5 percent) of what it was when there were no shields.

The equilibrium temperature of the radiation shield $T_3$ in Fig. 12-66 can be determined by expressing Eq. 12-55 for $\dot{Q}_{13}$ or $\dot{Q}_{23}$ (which involves $T_3$) after evaluating $\dot{Q}_{12}$ from Eq. 12-60 and noting that $\dot{Q}_{12} = \dot{Q}_{13} = \dot{Q}_{23}$ when steady-state conditions are reached.

Radiation shields used to retard the rate of radiation heat transfer between concentric cylinders and spheres can be handled in a similar manner. In case of one shield, Eq. 12-59 can be used by taking $F_{13} = F_{23} = 1$ for both cases and by replacing the $A$'s by the proper area relations.

## Radiation Effect on Temperature Measurements

A temperature measuring device indicates the temperature of its *sensor,* which is supposed to be, but is not necessarily, the temperature of the medium that the sensor is in. When a thermometer (or any other temperature measuring device such as a thermocouple) is placed in a medium, heat transfer takes place between the sensor of the thermometer and the medium by convection until the sensor reaches the temperature of the medium. But when the sensor is surrounded by surfaces that are at a different temperature than the fluid, radiation exchange will take place between the sensor and the surrounding surfaces. When the heat transfers by convection and radiation balance each other, the sensor will indicate a temperature that falls between the fluid and surface temperatures. Below we develop a procedure to account for the radiation effect and to determine the actual fluid temperature.

Consider a thermometer that is used to measure the temperature of a fluid flowing through a large channel whose walls are at a lower temperature than the fluid (Fig. 12-67). Equilibrium will be established and the reading of the thermometer will stabilize when heat gain by the sensor by convection equals heat loss by radiation. That is, on a unit-area basis,

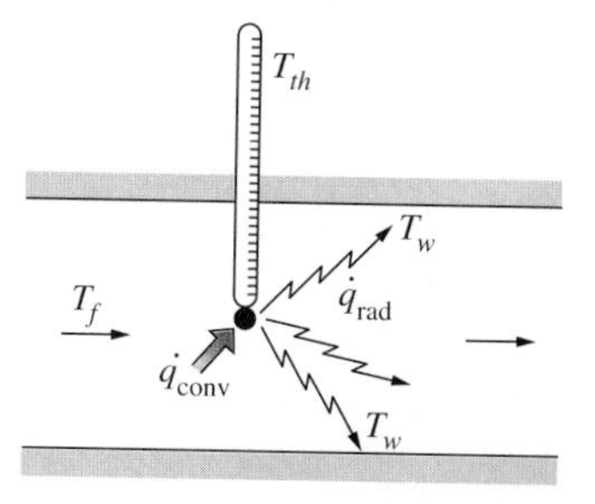


**FIGURE 12-67**
A thermometer used to measure the temperature of a fluid in a channel.

$$\dot{q}_{\text{conv, to sensor}} = \dot{q}_{\text{rad, from sensor}}$$

$$h(T_f - T_{\text{th}}) = \varepsilon_{\text{th}}\sigma(T_{\text{th}}^4 - T_w^4)$$

or

$$T_f = T_{\text{th}} + \frac{\varepsilon_{\text{th}}\sigma(T_{\text{th}}^4 - T_w^4)}{h} \quad \text{(K)} \qquad (12\text{-}63)$$

where $T_f$ = actual temperature of the fluid, K
$T_{th}$ = temperature value measured by the thermometer, K
$T_w$ = temperature of the surrounding surfaces, K
$h$ = convection heat transfer coefficient, $W/(m^2 \cdot K)$
$\varepsilon$ = emissivity of the sensor of the thermometer

The last term in Eq. 12-63 is due to the *radiation effect,* and represents the radiation correction. Note that the radiation correction terms is most significant when the convection heat transfer coefficient is small, and the emissiviity of the surface of the sensor is large. Therefore, the sensor should be coated with a material of high reflectivity (low emissivity) to reduce the radiation effect.

Placing the sensor in a radiation shield without interfering with the fluid flow also reduces the radiation effect. The sensors of temperature measurement devices used outdoors must be protected from direct sunlight, since the radiation effect in that case is sure to reach unaccaptable levels.

The radiation effect is also a significant factor in *human comfort* in heating and air-conditioning applications. A person who feels fine in a room at a specified temperature may feel chilly in another room at the same temperature as a result of the radiation effect if the walls of the second room are at a considerably lower temperature. For example, most people will feel comfortable in a room at 22°C if the walls of the room are also roughly at that temperature. When the wall temperature drops to 5°C for some reason, the interior temperature of the room must be raised to at least 27°C to maintain the same level of comfort. Therefore, well-insulated buildings conserve energy not only by reducing the heat loss or heat gain, but also by allowing the thermostats to be set at a lower temperature in winter and at a higher temperature in summer without compromising the comfort level.

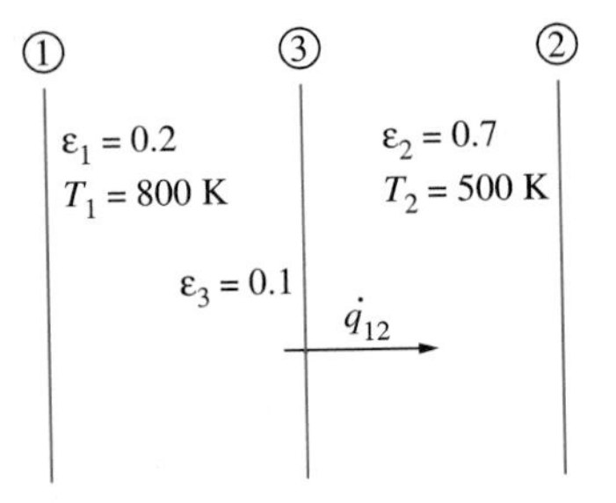


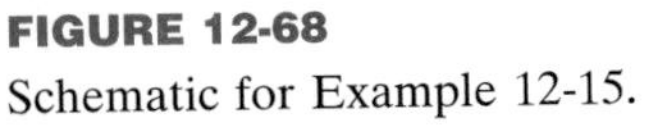
**FIGURE 12-68**
Schematic for Example 12-15.

**EXAMPLE 12-15**

A thin aluminum sheet with an emissivity of 0.1 on both sides is placed between two very large parallel plates that are maintained at uniform temperatures $T_1 = 800$ K and $T_2 = 500$ K, and have emissivities $\varepsilon_1 = 0.2$ and $\varepsilon_2 = 0.7$, respectively, as shown in Fig. 12-68. Determine the net rate of radiation heat transfer between the two plates per unit surface area of the plates, and compare the result to that without the shield.

**Solution** The net rate of radiation heat transfer between these two plates without the shield was determined in Example 12-11 to be 3625 W/m². Heat transfer in the presence of one shield is determined from Eq. 12-60 to be

$$\dot{q}_{12,\text{ one shield}} = \frac{\dot{Q}_{12,\text{ one shield}}}{A} = \frac{\sigma(T_1^4 - T_2^4)}{\left(\dfrac{1}{\varepsilon_1} + \dfrac{1}{\varepsilon_2} - 1\right) + \left(\dfrac{1}{\varepsilon_{3,1}} + \dfrac{1}{\varepsilon_{3,2}} - 1\right)}$$

$$= \frac{[5.67 \times 10^{-8}\ \text{W/(m}^2 \cdot \text{K}^4)][(800\ \text{K})^4 - (500\ \text{K})^4]}{\left(\dfrac{1}{0.2} + \dfrac{1}{0.7} - 1\right) + \left(\dfrac{1}{0.1} + \dfrac{1}{0.1} - 1\right)}$$

$$= 805.6\ \text{W/m}^2$$

Note that the rate of radiation heat transfer reduces to about one-fourth of what it was a result of placing a radiation shield between the two parallel plates.



**EXAMPLE 12-16**

A thermocouple used to measure the temperature of hot air flowing in a duct whose walls are maintained at $T_w$ = 400 K shows a temperature reading of $T_{th}$ = 650 K (Fig. 12-69). Assuming the emissivity of the thermocouple junction to be $\varepsilon$ = 0.6 and the convection heat transfer coefficient to be $h$ = 80 W/(m² · °C), determine the actual temperature of the air.

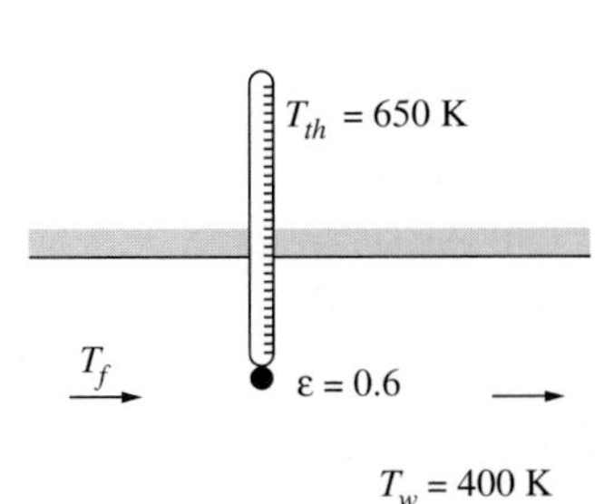


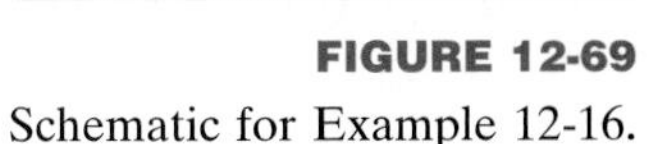

**FIGURE 12-69**
Schematic for Example 12-16.

**Solution** The walls of the duct are at a considerably lower temperature than the air in it, and thus we expect the thermocouple to show a reading lower than the actual air temperature as a result of radiation effect. The actual air temperature is determined from Eq. 12-63 to be

$$\begin{aligned} T_f &= T_{th} + \frac{\varepsilon_{th}\sigma(T_{th}^4 - T_w^4)}{h} \\ &= (650\text{ K}) + \frac{0.6 \times [5.67 \times 10^{-8}\text{ W/(m}^2\cdot\text{K}^4)][(650\text{ K})^4 - (400\text{ K})^4]}{80\text{ W/(m}^2\cdot{}^\circ\text{C)}} \\ &= 715.0\text{ K} \end{aligned}$$

Note that the radiation effect causes a difference of 65°C (or 65 K since °C ≡ K for temperature differences) in temperature reading in this case.

## 12-10 ■ SUMMARY

Radiation propagates in the form of electromagnetic waves. The *frequency* $\nu$ and *wavelength* $\lambda$ of electromagnetic waves in a medium are related by $\lambda = c/\nu$, where $c$ is the speed of light in that medium. All matter whose temperature is above absolute zero continuously emits *thermal radiation* as a result of vibrational and rotational motions of molecules, atoms, and electrons of a substance. Temperature is a measure of the strength of these activities at the microscopic level.

A *blackbody* is defined as a *perfect emitter and absorber of radiation.* At a specified temperature and wavelength, no surface can emit more energy than a blackbody. A blackbody absorbs *all* incident radiation, regardless of wavelength and direction. The radiation energy emitted by a blackbody per unit time and per unit surface area is called the *blackbody emissive power,* and is expressed by the *Stefan–Boltzmann law* as

$$E_b = \sigma T^4 \quad (\text{W/m}^2)$$

where $\sigma = 5.67 \times 10^{-8}$ W/(m² · K⁴) is the *Stefan-Boltzmann constant,* $E_b$ is the blackbody emissive power, and $T$ is the absolute temperature of the surface in K. At any specified temperature, the spectral blackbody emissive power $E_{b\lambda}$ increases with wavelength, reaches a peak, and then decreases with increasing wavelength. The wavelength at which the peak occurs for a specified temperature is given by *Wien's displacement law* as

$$(\lambda T)_{\text{max power}} = 2897.8\ \mu\text{m}\cdot\text{K}$$

The *blackbody radiation function* $f_\lambda$ represents the fraction of radiation emitted by a blackbody at temperature $T$ in the wavelength band from $\lambda = 0$ to $\lambda$. The fraction of radiation energy emitted by a blackbody at temperature $T$ over a finite wavelength band from $\lambda = \lambda_1$ to $\lambda = \lambda_2$ is determined from

$$f_{\lambda_1-\lambda_2}(T) = f_{\lambda_2}(T) - f_{\lambda_1}(T)$$

where $f_{\lambda_1}(T)$ and $f_{\lambda_2}(T)$ are the blackbody radiation functions corresponding to $\lambda_1 T$ and $\lambda_2 T$, respectively.

The *emissivity* of a surface is defined as the ratio of the radiation emitted by the surface to the radiation emitted by a blackbody at the same temperature. The *total hemispherical emissivity* $\varepsilon$ of a surface is simply the average emissivity over all directions and wavelengths, and is expressed as

$$\varepsilon(T) = \frac{E(T)}{E_b(T)} = \frac{E(T)}{\sigma T^4}$$

where $E(T)$ is the total emissive power of the surface.

The consideration of wavelength and direction dependence of properties makes radiation calculations very complicated. Therefore, the *gray* and *diffuse* approximations are commonly utilized in radiation calculations. A surface is said to be *diffuse* if its properties are *independent of direction,* and *gray* if its properties are *independent of wavelength.*

The radiation energy incident on a surface per unit surface area per unit time is called *irradiation,* and is denoted by $G$. When irradiation strikes a surface, part of it is absorbed, part of it is reflected, and the remaining part, if any, is transmitted. The fraction of irradiation absorbed by the surface is called the *absorptivity* $\alpha$. The fraction of irradiation reflected by the surface is called the *reflectivity* $\rho$, and the fraction transmitted is called the *transmissivity* $\tau$. The sum of the aborbed, reflected, and transmitted fractions of radiation energy must be equal to unity,

$$\alpha + \rho + \tau = 1$$

For *opaque* surfaces $\tau = 0$, and thus

$$\alpha + \rho = 1$$

Surfaces are usually assumed to reflect in a perfectly *specular* or *diffuse* manner for simplicity. In *specular* (or *mirrorlike*) *reflection,* the angle of reflection equals the angle of incidence of the radiation beam. In *diffuse reflection,* radiation is reflected equally in all directions. Reflection from smooth and polished surfaces approximates specular reflection, whereas reflection from rough surfaces approximates diffuse reflection.

*Kirchhoff's law* of radiation is expressed as

$$\varepsilon(T) = \alpha(T)$$

That is, the total hemispherical emissivity of a surface at temperature $T$ is equal to its total hemispherical absorptivity for radiation coming from a blackbody at the same temperature.

The *view factor* from a surface $i$ to a surface $j$ is denoted by $F_{i \to j}$, and is defined as the fraction of the radiation leaving surface $i$ that strikes surface $j$ directly. The view factor $F_{i \to i}$ represents the fraction of the radiation leaving surface $i$ that strikes itself directly. $F_{i \to i} = 0$ for *plane* or *convex* surfaces, and $F_{i \to i} \neq 0$ for *concave* surfaces. For view factors, the *reciprocity rule* is expressed as

$$A_i F_{i \to j} = A_j F_{j \to i}$$

The sum of the view factors from surface $i$ of an enclosure to all surfaces of the enclosure, including to itself, must equal unity. This is known as the *summation rule* for an enclosure. The *superposition rule* is expressed as follows: the view factor from a surface $i$ to a surface $j$ is equal to the sum of the view factors from surface $i$ to the parts of surface $j$. The symmetry rule is expressed as follows: if the surfaces $j$ and $k$ are symmetric about the surface $i$ then $F_{i \to j} = F_{i \to k}$.

The rate of net radiation heat transfer between two *black* surfaces is determined from

$$\dot{Q}_{1 \to 2} = A_1 F_{1 \to 2} \sigma (T_1^4 - T_2^4) \quad \text{(W)}$$

The *net* radiation heat transfer from any surface $i$ of a *black* enclosure is determined by adding up the net radiation heat transfers from surface $i$ to each of the surfaces of the enclosure:

$$\dot{Q}_i = \sum_{j=1}^{N} \dot{Q}_{i \to j} = \sum_{j=1}^{N} A_i F_{i \to j} \sigma (T_i^4 - T_j^4) \quad \text{(W)}$$

The total radiation energy leaving a surface per unit time and per unit area is called the *radiosity,* and is denoted by $J$. The *net* rate of radiation heat transfer from a surface $i$ of surface area $A_i$ is expressed as

$$\dot{Q}_i = \frac{E_{bi} - J_i}{R_i} \quad \text{(W)}$$

where

$$R_i = \frac{1 - \varepsilon_i}{A_i \varepsilon_i}$$

is the *surface resistance* to radiation. The *net* rate of radiation heat transfer from surface $i$ to surface $j$ can be expressed as

$$\dot{Q}_{i \to j} = \frac{J_i - J_j}{R_{i \to j}} \quad \text{(W)}$$

where

$$R_{i \to j} = \frac{1}{A_i F_{i \to J}}$$

is the *space resistance* to radiation. The *network method* is applied to radiation enclosure problems by drawing a surface resistance associated with each surface of an enclosure, and connecting them with space resistances. Then the problem is solved by treating it as an electrical network problem where the radiation heat transfer replaces the current and the radiosity replaces the potential. The *matrix method* is based on the following two equations:

$$\dot{Q}_i = \sum_{j=1}^{N} \dot{Q}_{i\to j} = \sum_{j=1}^{N} \frac{J_i - J_j}{R_{i\to j}} \quad \text{(W)}$$

$$\frac{E_{bi} - J_i}{R_i} = \sum_{j=1}^{N} \frac{J_i - J_j}{R_{i\to j}} \quad \text{(W)}$$

The first group (for surfaces with specified heat transfer rates) and the second group (for surfaces with specified temperatures) of equations give $N$ linear algebraic equations for the determination of the $N$ unknown radiosities for an $N$-surface enclosure. Once the radiosities $J_1, J_2, \ldots, J_N$ are available, the unknown surface temperatures and heat transfer rates can be determined from the equations above.

The net rate of radiation transfer between any two gray, diffuse, opaque surfaces that form an enclosure is given by

$$\dot{Q}_{12} = \frac{\sigma(T_1^4 - T_2^4)}{\dfrac{1-\varepsilon_1}{A_1\varepsilon_1} + \dfrac{1}{A_1F_{12}} + \dfrac{1-\varepsilon_2}{A_2\varepsilon_2}} \quad \text{(W)}$$

Radiation heat transfer between two surfaces can be reduced greatly by inserting thin, high-reflectivity (low-emissivity) sheets of material between the two surfaces called *radiation shields.* Radiation heat transfer between two large parallel plates separated by $N$ radiation shields is

$$\dot{Q}_{12,\,N\,\text{shields}} = \frac{A\sigma(T_1^4 - T_2^4)}{\left(\dfrac{1}{\varepsilon_1} + \dfrac{1}{\varepsilon_2} - 1\right) + \left(\dfrac{1}{\varepsilon_{3,1}} + \dfrac{1}{\varepsilon_{3,2}} - 1\right) + \cdots + \left(\dfrac{1}{\varepsilon_{N,1}} + \dfrac{1}{\varepsilon_{N,2}} - 1\right)}$$

The radiation effect in temperature measurements can be properly accounted for by the relation

$$T_f = T_{\text{th}} + \frac{\varepsilon_{\text{th}}\sigma(T_{\text{th}}^4 - T_w^4)}{h} \quad \text{(K)}$$

where $T_f$ is the actual temperature of the fluid, $T_{\text{th}}$ is the temperature value measured by the thermometer, and $T_w$ is the temperature of the surrounding walls, all in K.

## REFERENCES AND SUGGESTED READING

**1** A. G. H. Dietz, "Diathermanous Materials and Properties of Surfaces," in *Space Heating with Solar Energy,* by R. W. Hamilton, MIT Press, Cambridge, MA, 1954.

**2** J. A. Duffy and W. A. Backman, *Solar Energy Thermal Process,* Wiley, New York, 1974.

**3** J. P. Holman, *Heat Transfer,* 7th ed., McGraw-Hill, New York, 1990.

**4** H. C. Hottel, "Radiant Heat Transmission," in W. H. McAdams (Ed.), *Heat Transmission,* 3d ed., McGraw-Hill, New York, 1954.

**5** J. R. Howell, *A Catalog of Radiation Configuration Factors,* McGraw-Hill, New York, 1982.

**6** D. C. Hamilton and W. R. Morgan, "Radiation Interchange Configuration Factors," National Advisory Committee for Aeronautics, Technical Note 2836, 1952.

**7** F. P. Incropera and D. P. DeWitt, *Introduction to Heat Transfer,* 2d ed., Wiley, New York, 1990.

**8** A. K. Oppenheim, "Radiation Analysis by the Network Method", *Transactions, of the ASME,* Vol. 78, pp. 725–735, 1956.

**9** M. N. Özışık, *Heat Transfer—A Basic Approach,* McGraw-Hill, New York, 1985.

**10** W. Sieber, *Zeitschrift für Technische Physics*, Vol. 22, pp. 130–135, 1941.

**11** L. C. Thomas, *Heat Transfer,* Prentice-Hall, Englewood Cliffs, NJ, 1992.

**12** Y. S. Touloukain and D. P. DeWitt, "Nonmetallic Solids," in *Thermal Radiative Properties,* Vol. 8, IFI/Plenum, New York, 1970.

**13** Y. S. Touloukian and D. P. DeWitt, "Metallic Elements and Alloys," in *Thermal Radiative Properties,* Vol. 7, IFI/Plenum, New York, 1970.

## PROBLEMS*

### Electromagnetic and Thermal Radiation

**12-1C** What is an electromagnetic wave? How does it differ from a sound wave?

**12-2C** By what properties is an electromagnetic wave characterized? How are these properties related to each other?

**12-3C** What is visible light? How does it differ from the other forms of electromagnetic radiation?

**12-4C** How does ultraviolet and infrared radiation differ? Do you think your body emits any radiation in the ultraviolet range?

*Students are encouraged to answer *all* the concept "C" questions.

**12-5C** What is thermal radiation? How does it differ from the other forms of electromagnetic radiation?

**12-6C** What is the cause of color? Why do some objects appear blue to the eye while others appear red? Is the color of a surface at room temperature related to the radiation it emits?

**12-7C** Why is radiation usually treated as a surface phenomena?

**12-8C** Why do skiers get sunburn so easily?

**12-9C** How does microwave cooking differ from conventional cooking?

**12-10** Electricity is generated and transmitted in power lines at a frequency of 60 Hz (1 Hz = 1 cycle per second). Determine the wavelength of the electromagnetic waves generated by the passage of electricity in power lines.

**12-11** A microwave oven is designed to operate at a frequency of $2.8 \times 10^9$ Hz. Determine the wavelength of these microwaves, and the energy of each microwave.

**12-12** A radio station is broadcasting radiowaves at a wavelength of 300 m. Determine the frequency of these waves.
*Answer:* $1.0 \times 10^6$ Hz

**12-13** A cordless telephone is designed to operate at a frequency of $8.5 \times 10^8$ Hz. Determine the wavelength of these telephone waves.

## Blackbody Radiation

**12-14C** What is a blackbody? Does a blackbody actually exist?

**12-15C** Define the total and spectral blackbody emissive powers. How are they related to each other? How do they differ?

**12-16C** Why did we define the blackbody radiation function? What does it represent? What is it used for?

**12-17C** Consider two identical bodies, one at 1000 K and the other at 1500 K. Which body emits more radiation in the shorter-wavelength region? Which body emits more radiation at a wavelength of 20 $\mu$m?

**12-18** Consider a 20 cm $\times$ 20 cm $\times$ 20 cm cubical body at 1000 K suspended in the air. Assuming the body closely approximates a blackbody, determine (*a*) the rate at which the cube emits radiation energy, in W, and (*b*) the spectral blackbody emissive power at a wavelength of 4 $\mu$m.

**12-19** The sun can be treated as a blackbody at an effective surface temperature of 5762 K. Determine the fraction of the radiation energy emitted by the sun that falls in (*a*) the ultraviolet range ($\lambda$ = 0.01–0.40 $\mu$m) and (*b*) the visible range ($\lambda$ = 0.40–0.76 $\mu$m). Also determine the wavelength at which the emission of radiation from the sun is a maximum. *Answers:* (*a*) 0.121, (*b*) 0.425, 0.503 $\mu$m

**12-19E** The sun can be treated as a blackbody at an effective surface temperature of 10,372 R. Determine the rate at which infrared radiation energy ($\lambda = 0.76$–$100\ \mu$m) is emitted by the sun, in Btu/(h · ft$^2$).

**12-20** The temperature of the filament of an incandescent light bulb is 3200 K. Treating the filament as a blackbody, determine the fraction of the radiant energy emitted by the filament that falls in the visible range. Also determine the wavelength at which the emission of radiation from the filament peaks.

**21-21** An incandescent light bulb is desired to emit at least 20 percent of its energy at wavelengths shorter than 1 $\mu$m. Determine the minimum temperature to which the filament of the light bulb must be heated.

**12-22** It is desired that the radiation energy emitted by a light source reach a maximum in the blue range ($\lambda = 0.47\ \mu$m). Determine the temperature of this light source, and the fraction of radiation it emits in the visible range ($\lambda = 0.40$–$0.76\ \mu$m).

**12-23** A 3-mm-thick glass window transmits 90 percent of radiation between $\lambda = 0.3$ and $3.0\ \mu$m, and is essentially opaque for radiation at other wavelengths. Determine the rate of radiation transmitted through a 2 m × 2 m glass window from blackbody sources at (*a*) 5800 K and (*b*) 1000 K. *Answers:* (*a*) 98,150 kW, (*b*) 0.00265 kW

## Radiation Properties

**12-24C** Define the properties emissivity and absorptivity. When are these two properties equal to each other?

**12-25C** Define the properties reflectivity and transmissivity, and discuss the different forms of reflection.

**12-26C** What is a graybody? How does it differ from a blackbody? What is a diffuse gray surface?

**12-27C** What is the greenhouse effect? Why is it a matter of great concern among atmospheric scientists?

**12-28C** We can see the inside of a microwave oven during operation through its glass door, which indicates that visible radiation is escaping the oven. Do you think that the harmful microwave radiation might also be escaping?

**12-29** The spectral emissivity function of an opaque surface at 1000 K is approximated as

$$\varepsilon_\lambda = \begin{cases} \varepsilon_1 = 0.4, & 0 \leq \lambda < 2\ \mu\text{m} \\ \varepsilon_2 = 0.7, & 2\ \mu\text{m} \leq \lambda < 6\ \mu\text{m} \\ \varepsilon_3 = 0.3, & 6\ \mu\text{m} \leq \lambda < \infty \end{cases}$$

Determine the average emissivity of the surface and the rate of radiation emission from the surface, in W/m$^2$. *Answers:* 0.575, 32.6 kW/m$^2$

**12-30** The reflectivity of aluminum coated with lead sulfate is 0.35 for radiation at wavelength less than 3 $\mu$m and 0.95 for radiation greater than 3 $\mu$m. Determine the average reflectivity of this surface for solar radiation ($T \approx 5800$ K) and radiation coming from surfaces at room temperature ($T \approx 300$ K). Also determine the emissivity and absorptivity of this surface at both temperatures. Do you think this material is suitable for use in solar collectors?

**12-31** A furnace that has a 20 cm × 20 cm glass window can be considered to be a blackbody at 1200 K. If the transmissivity of the glass is 0.8 for radiation at wavelengths less than 3 $\mu$m and zero for radiation at greater than 3 $\mu$m, determine the fraction and the rate of radiation coming from the furnace and transmitted through the window.

**12-32** The emissivity of a tungsten filament can be approximated to be 0.5 for radiation at wavelengths less than 1 $\mu$m and 0.15 for radiation at greater than 1 $\mu$m. Determine the average emissivity of the filament at (a) 1500 K and (b) 3000 K. Also determine the absorptivity and reflectivity of the filament at both temperatures.

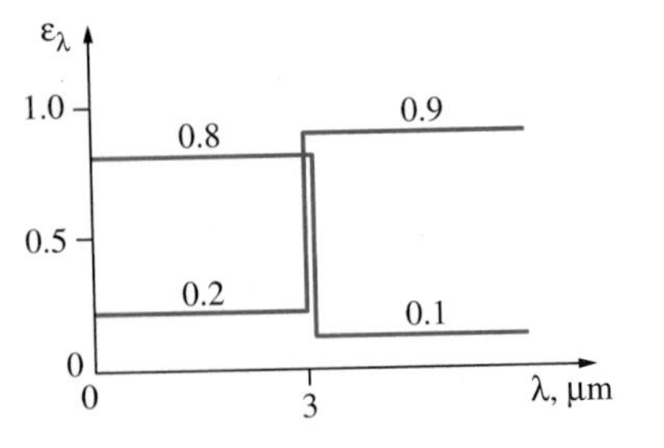

FIGURE P12-33

**12-33** The variations of the spectral emissivity of two surfaces are as given in Fig. P12-33. Determine the average emissivity of each surface at $T = 3000$ K. Also determine the average absorptivity and reflectivity of each surface for radiation coming from a source at 3000 K. Which surface is more suitable to serve as a solar absorber?

**12-34** The emissivity of a surface coated with aluminum oxide can be approximated to be 0.2 for radiation at wavelengths less than 5 $\mu$m and 0.9 for radiation at greater than 5 $\mu$m. Determine the average emissivity of this surface at (*a*) 5800 K and (*b*) 300 K. What can you say about the absorptivity of this surface for radiation coming from sources at 5800 K and 300 K? *Answers:* (*a*) 0.203, (*b*) 0.89

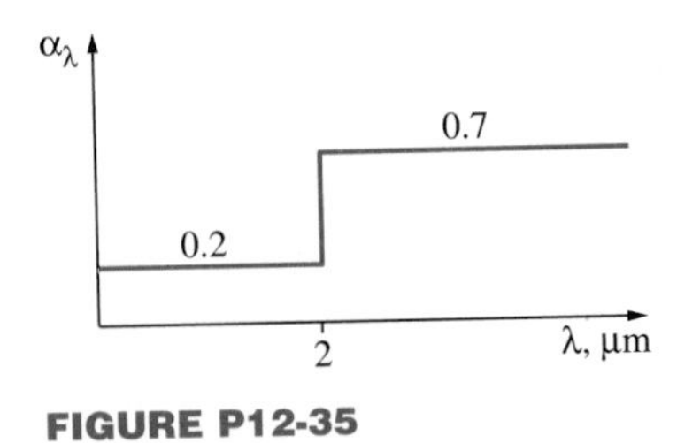


FIGURE P12-35

**12-35** The variation of the spectral absorptivity of a surface is as given in Fig. P12-35. Determine the average absorptivity and reflectivity of the surface for radiation that originates from a source at $T = 2500$ K. Also determine the average emissivity of this surface at 3000 K.

**12-35E** The variation of the spectral absorptivity of a surface is as given in Fig. P12-35. Determine the average absorptivity and reflectivity of the surface for radiation that originates from a source at $T = 4500$ R. Also determine the average emissivity of this surface at 5400 R.

**12-36** A 10-cm-diameter spherical ball is known to emit radiation at a rate of 30 W when its surface temperature is 400 K. Determine the average emissivity of the ball at this temperature.

**12-36E** A 5-in-diameter spherical ball is known to emit radiation at a rate of 120 Btu/h when its surface temperature is 800 R. Determine the average emissivity of the ball at this temperature.

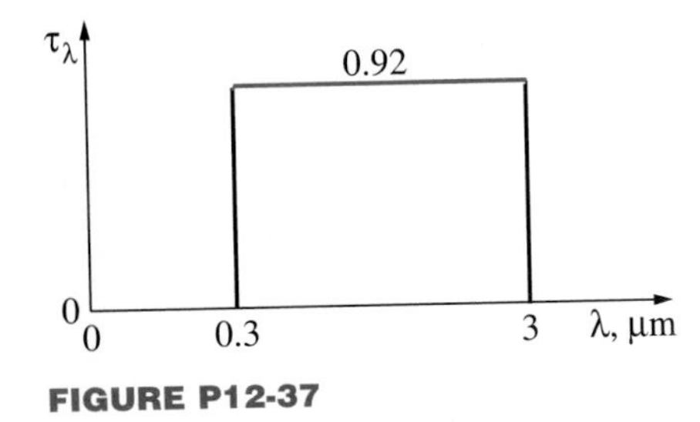


FIGURE P12-37

**12-37** The variation of the spectral transmissivity of a 0.6-cm-thick glass pane is as given in Fig. P12-37. Determine the average transmissivity of this pane for solar radiation ($T \approx 5800$ K) and radiation coming from surfaces at room temperatue ($T \approx 300$ K). Also determine the amount of solar radiation transmitted through the pane for incident solar radiation of 650 W/m$^2$. *Answers:* 0.848, 0.00015, 551.1 W/m$^2$

### Atmospheric and Solar Radiation

**12-38C** What is the solar constant? How is it used to determine the effective surface temperature of the sun? How would the value of the solar constant change if the distance between the earth and the sun doubled?

**12-39C** What changes would you notice if the sun emitted radiation at an effective temperature of 2000 K instead of 5762 K?

**12-40C** Explain why the sky is blue and the sunset is yellow-orange.

**12-41C** When the earth is closest to the sun, we have winter in the northern hemisphere. Explain why. Also explain why we have summer in the northern hemisphere when the earth is farthest away from the sun.

**12-42C** What is the effective sky temperature?

**12-43C** You have probably noticed warning signs on the highways stating that bridges may be icy even when the roads are not. Explain how this can happen.

**12-44C** Unless you live in a warm southern state, you have probably had to scratch ice from the windshield and windows of your car many mornings. You may have noticed, with frustration, that the thickest layer of ice always forms on the windshield instead of the side windows. Explain why this is the case.

**12-45C** Explain why surfaces usually have quite different absorptivities for solar radiation and for radiation originating from the surrounding bodies.

**12-46** A surface has an absorptivity of $\alpha_s = 0.85$ for solar radiation and an emissivity of $\varepsilon = 0.5$ at room temperature. The surface temperature is observed to be 350 K when the direct and the diffuse components of solar radiation are $G_D = 350$ and $G_d = 400$ W/m$^2$, respectively, and the direct radiation makes a 30° angle with the normal of the surface. Taking the effective sky temperature to be 280 K, determine the net rate of radiation to the surface at that time.

**12-47** Solar radiation is incident on the outer surface of a spaceship at a rate of 1320 W/m$^2$. The surface has an absorptivity of $\alpha_s = 0.10$ for solar radiation and an emissivity of $\varepsilon = 0.8$ at room temperature. The outer surface radiates heat into space at 0 K. If there is no net heat transfer into the spaceship, determine the equilibrium temperature of the surface. *Answer:* 232.3 K

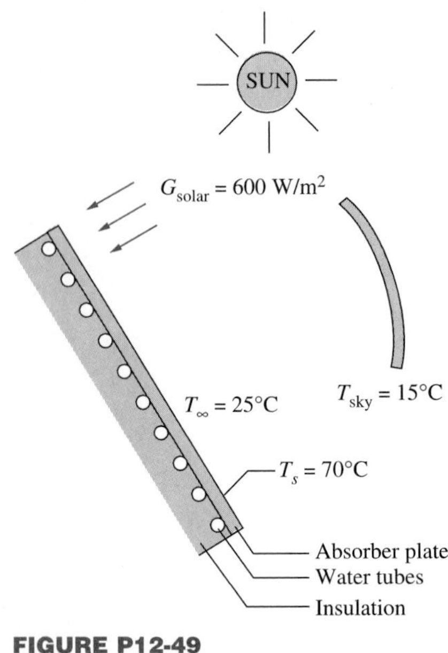


**FIGURE P12-49**

**12-47E** Solar radiation is incident on the outer surface of a spaceship at a rate of 400 Btu/(h · ft²). The surface has an absorptivity of $\alpha_s = 0.10$ for solar radiation and an emissivity of $\varepsilon = 0.8$ at room temperature. The outer surface radiates heat into space at 0 R. If there is no net heat transfer into the spaceship, determine the equilibrium temperature of the surface. *Answer:* 413.3 R

**12-48** The air temperature on a clear night is observed to remain at about 4°C. Yet water is reported to have frozen that night. Taking the convection heat transfer coefficient to be 10 W/(m² · °C), determine the value of the effective sky temperature that night.

**12-49** The absorber surface of a solar collector is made of aluminum coated with black chrome ($\alpha_s = 0.87$ and $\varepsilon = 0.09$). Solar radiation is incident on the surface at a rate of 600 W/m². The air and the effective sky temperatures are 25°C and 15°C, respectively, and the convection heat transfer coefficient is 10 W/(m² · °C). For an absorber surface temperature of 60°C, determine the net rate of solar energy delivered by the absorber plate to the water circulating behind it.

**12-49E** The absorber surface of a solar collector is made of aluminum coated with black chrome ($\alpha_s = 0.87$ and $\varepsilon = 0.09$). Solar radiation is incident on the surface at a rate of 200 Btu/(h · ft²). The air and the effective sky temperatures are 75°F and 60°F, respectively, and the convection heat transfer coefficient is 2 Btu/(h · ft² · °F). For an absorber surface temperature of 140°F, determine the net rate of solar energy delivered by the absorber plate to the water circulating behind it.

**12-50** Determine the equilibrium temperature of the absorber surface in Prob. 12-49 if the back side of the absorber is insulated.

## View Factors

**12-51C** What does the view factor represent? When is the view factor from a surface to itself not zero?

**12-52C** How can you determine the view factor $F_{12}$ when the view factor $F_{21}$ is available?

**12-53C** What are the summation rule and the superposition rule for view factors?

**12-54C** What is the crossed-strings method? For what kind of geometries is the crossed-strings method applicable?

**12-55** Consider an enclosure consisting of seven surfaces. How many view factors does this geometry involve? How many of these view factors can be determined by the application of the reciprocity and the summation rules?

**12-56** Consider an enclosure consisting of five surfaces. How many view factors does this geometry involve? How many of these view factors can be determined by the application of the reciprocity and summation rules?

**12-57** Consider an enclosure consisting of 12 surfaces. How many view factors does this geometry involve? How many of these view factors can be determined by the application of the reciprocity and the summation rules? *Answers:* 144, 78

**12-58** Determine the view factors $F_{13}$ and $F_{23}$ between the rectangular surfaces shown in Fig. P12-58.

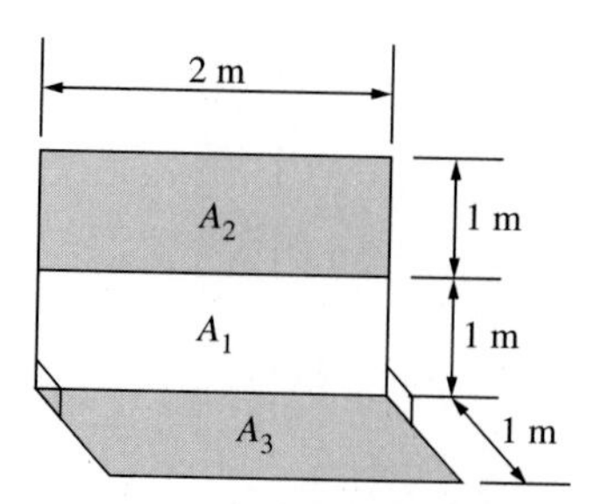


**FIGURE P12-58**

**12-59** Consider a cylindrical enclosure whose height is twice the diameter of its base. Determine the view factor from the side surface of this cylindrical enclosure to its base surface.

**12-60** Consider a hemispherical furnace with a flat circular base of diameter $D$. Determine the view factor from the dome of this furnace to its base. *Answer:* 0.5

**12-61** Determine the view factors $F_{12}$ and $F_{21}$ for the very long ducts shown in Fig. P12-61 without using any view factor tables or charts. Neglect end effects.

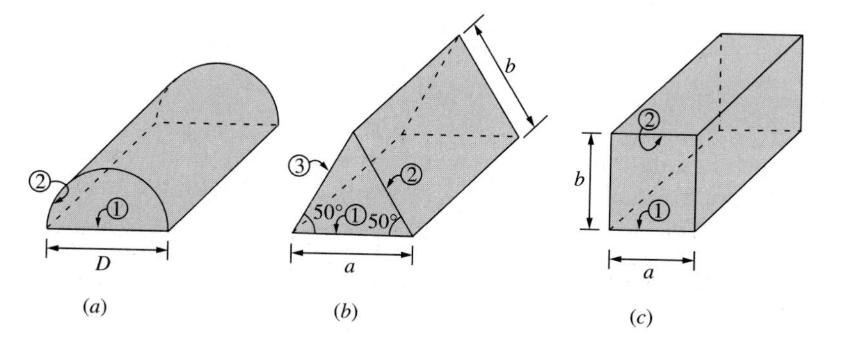


**FIGURE P12-61**

(*a*) Semicylindrical duct.
(*b*) Triangular duct.
(*c*) Rectangular duct.

**12-62** Determine the view factors from the very long grooves shown in Fig. P12-62 to the surroundings without using any view factor tables or charts. Neglect end effects.

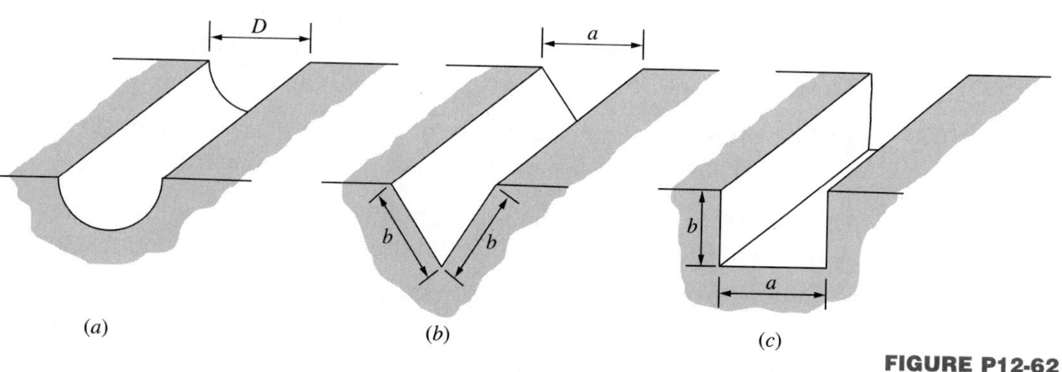

**FIGURE P12-62**

(*a*) Semicylindrical groove.
(*b*) Triangular groove.
(*c*) Rectangular groove.

**12-63** Determine the view factors from the base of a cube to each of the other five surfaces. *Answer:* 0.2

**12-64** Consider a conical enclosure of height $h$ and base diameter $D$. Determine the view factor from the conical side surface to a hole of diameter $d$ located at the center of the base.

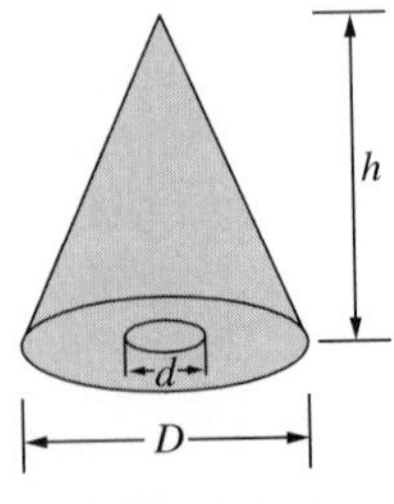


FIGURE P12-64

**12-65** Determine the four view factors associated with an enclosure formed by two very long concentric cylinders of radii $r_1$ and $r_2$. Neglect the end effects.

**12-66** Determine the view factor $F_{12}$ between the rectangular surfaces shown in Fig. P12-66.

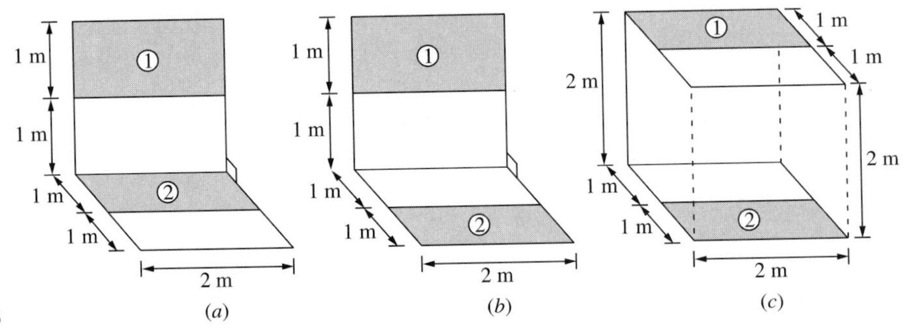


FIGURE P12-66

**12-67** Two infinitely long parallel cylinders of diameter $D$ are located a distance $s$ apart from each other. Determine the view factor $F_{12}$ between these two cylinders.

**12-68** Three infinitely long parallel cylinders of diameter $D$ are located a distance $s$ apart from each other. Determine the view factor between the cylinder in the middle and the surroundings.

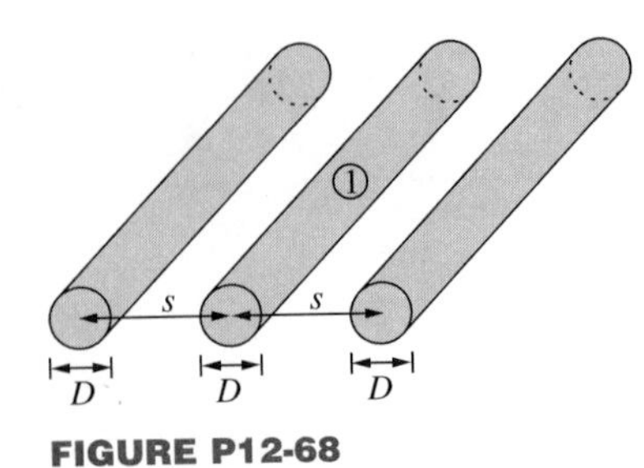


FIGURE P12-68

### Radiation Heat Transfer Between Surfaces

**12-69C** Why is the radiation analysis of enclosures that consist of black surfaces relatively easy? How is the rate of radiation heat transfer between two surfaces expressed in this case?

**12-70C** How does radiosity differ for a surface from the emitted energy? For what kind of surfaces are these two quantities identical?

**12-71C** What are the radiation surface and space resistances? How are they expressed? For what kind of surfaces is the radiation surface resistance zero?

**12-72C** What are the two methods used in radiation analysis? How do these two methods differ?

**12-73C** What is a reradiating surface? What simplifications does a reradiating surface offer in the radiation analysis?

**12-74** Consider a 3 m × 3 m × 3 m cubical furnace whose top and side surfaces closely approximate black surfaces, and whose base surface has an emissivity $\varepsilon = 0.7$. The base, top, and side surfaces of the furnace are maintained at uniform temperatures of 400 K, 800 K, and 1200 K, respectively. Determine the net rate of radiation heat transfer between (*a*) the base and side surfaces and (*b*) the base and top surfaces. Also determine the net rate of radiation heat transfer to the base surface.

**12-74E** Consider a 10 ft × 10 ft × 10 ft cubical furnace whose top and side surfaces closely approximate black surfaces, and whose base surface has an emissivity $\varepsilon = 0.7$. The base, top, and side surfaces of the furnace are maintained at uniform temperatures of 800 R, 1600 R, and 2400 R, respectively. Determine the net rate of radiation heat transfer between (*a*) the base and side surfaces and (*b*) the base and top surfaces. Also determine the net rate of radiation heat transfer to the base surface.

**12-75** Two very large parallel plates are maintained at uniform temperatures of $T_1 = 600$ K and $T_2 = 400$ K, and have emissivities $\varepsilon_1 = 0.5$ and $\varepsilon_2 = 0.9$, respectively. Determine the net rate of radiation heat transfer between the two surfaces per unit area of the plates.

**12-76** A furnace is of cylindrical shape with $r = h = 2$ m. The base, top, and side surfaces of the furnace are all black, and are maintained at uniform temperatures of 500, 700, and 800 K, respectively. Determine the net rate of radiation heat transfer from the top surface during steady operation.

FIGURE P12-76

**12-76E** A furnace is of cylindrical shape with $r = h = 6$ ft. The base, top, and side surfaces of the furnace are all black, and are maintained at uniform temperatures of 900, 1200, and 1400 R, respectively. Determine the net rate of radiation heat transfer from the top surface during steady operation.

**12-77** Consider a hemispherical furnace of diameter $D = 5$ m with a flat base. The dome of the furnace is black, and the base has an emissivity of 0.7. The base and the dome of the furnace are mainteaned at uniform temperatures of 400 and 1000 K, respectively. Determine the net rate of radiation heat transfer from the dome to the base surface during steady operation. *Answer:* 759 kW

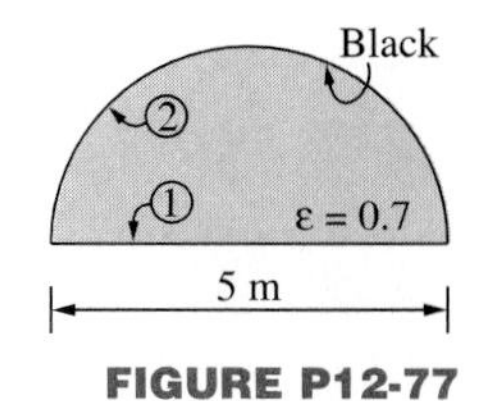


FIGURE P12-77

**12-77E** Consider a hemispherical furnace of diameter $D = 15$ ft with a flat base. The dome of the furnace is black, and the base has an emissivity of 0.7. The base and the dome of the furnace are maintained at uniform temperatures of 800 and 1800 R, respectively. Determine the net rate of radiation heat transfer from the dome to the base surface during steady operation. *Answer:* 2,139,000 Btu/h

**12-78** Two very long concentric cylinders of diameters $D_1 = 0.2$ m and $D_2 = 0.5$ m are maintained at uniform temperatures of $T_1 = 800$ K and $T_2 = 500$ K, and have emissivities $\varepsilon_1 = 1$ and $\varepsilon_2 = 0.7$, respectively.

Determine the net rate of radiation heat transfer between the two cylinders per unit length of the cylinders.

**12-79** The following experiment is conducted to determine the emissivity of a certain material. A long cylindrical rod of diameter $D_1 = 0.01$ m is coated with this new material, and is placed in an evacuated long cylindrical enclosure of diameter $D_2 = 0.1$ m and emissivity $\varepsilon_2 = 0.95$, which is cooled externally and is maintained at a temperature of 200 K at all times. The rod is heated by passing electric current through it. When steady operating conditions are reached, it is observed that the rod is dissipating electric power at a rate of 8 W per unit of its length, and its surface temperature is 500 K. Based on these measurements, determine the emissivity of the coating on the rod.

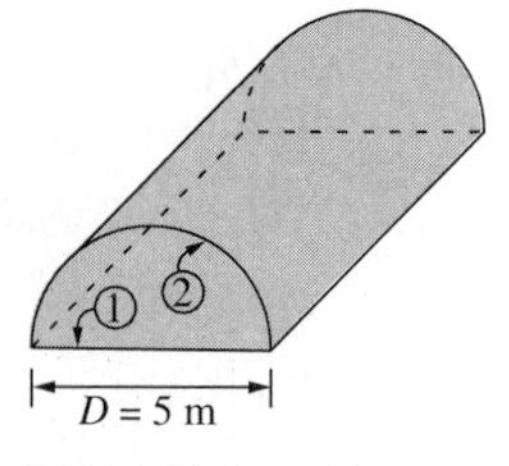


**FIGURE P12-80**

**12-80** A furnace is shaped like a long semicylindrical duct of diameter $D = 5$ m. The base and the dome of the furnace have emissivities of 0.5 and 0.9, and are maintained at uniform temperatures of 400 and 1000 K, respectively. Determine the net rate of radiation heat transfer from the dome to the base surface per unit length during steady operation.

**12-80E** A furnace is shaped like a long semicylindrical duct of diameter $D = 15$ ft. The base and the dome of the furnace have emissivities of 0.5 and 0.9, and are maintained at uniform temperatures of 700 and 1800 R, respectively. Determine the net rate of radiation heat transfer from the dome to the base surface per unit length during steady operation.

**12-81** Two parallel disks of diameter $D = 0.6$ m separated by $L = 0.4$ m are located directly on top of each other. Both disks are black, and are maintained at a temperature of 700 K. The back sides of the disks are insulated, and the environment that the disks are in can be considered to be a blackbody at $T_\infty = 300$ K. Determine the net rate of radiation heat transfer from the disks to the environment. *Answer:* 5505 W

**12-82** A furnace is shaped like a long equilateral-triangular duct where the width of each side is 2 m. Heat is supplied from the base surface, whose emissivity is $\varepsilon_1 = 0.8$, at a rate of 800 W/m$^2$ while the side surfaces, whose emissivities are 0.5, are maintained at 500 K. Neglecting the end effects, determine the temperature of the base surface. Can you treat this geometry as a two-surface enclosure?

**12-83** Consider a 4 m × 4 m × 4 m cubical furnace whose floor and ceiling are black and whose side surfaces are reradiating. The floor and the ceiling of the furnace are maintained at temperatures of 550 K and 1100 K, respectively. Determine the net rate of radiation heat transfer between the floor and the ceiling of the furnace.

**12-84** Two concentric spheres of diameters $D_1 = 0.3$ m and $D_2 = 0.8$ m are maintained at uniform temperatures $T_1 = 700$ K and $T_2 = 400$ K, and have emissivities $\varepsilon_1 = 0.5$ and $\varepsilon_2 = 0.7$, respectively. Determine the net rate of radiation heat transfer between the two spheres. Also determine the convection heat transfer coefficient at the outer surface if both the surrounding medium and the surrounding surfaces are at 25°C. Assume the emissivity of the outer surface is 0.2.

**12-85** A spherical tank of diameter $D = 2$ m that is filled with liquid nitrogen at 100 K is kept in an evacuated cubic enclosure whose sides are 3 m long. The emissivities of the spherical tank and the enclosure are $\varepsilon_1 = 0.1$ and $\varepsilon_2 = 0.8$, respectively. If the temperature of the cubic enclosure is measured to be 240 K, determine the net rate of radiation heat transfer to liquid nitrogen. *Answer:* 228 W

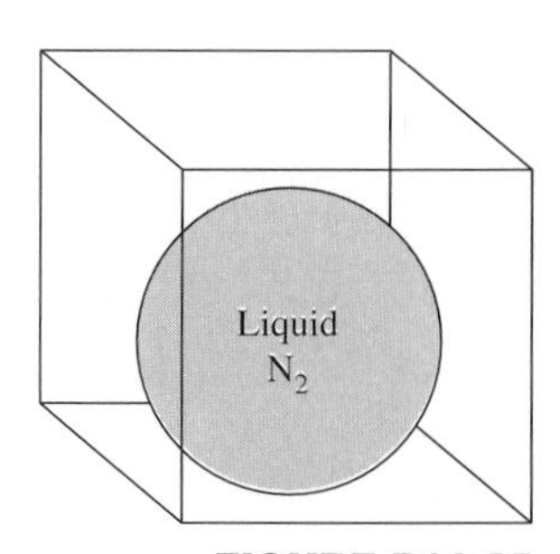


**FIGURE P12-85**

**12-85E** A spherical tank of diameter $D = 6$ ft that is filled with liquid nitrogen at 180 R is kept in an evacuated cubic enclosure whose sides are 9ft long. The emissivities of the spherical tank and the enclosure are $\varepsilon_1 = 0.1$ and $\varepsilon_2 = 0.8$, respectively. If the temperature of the cubic enclosure is measured to be 430 R, determine the net rate of radiation heat transfer to liquid nitrogen. *Answer:* 639 Btu/h

**12-86** Repeat Prob. 12-85 by replacing the cubic enclosure by a spherical enclosure whose diameter is 3 m.

**12-87** Consider a circular grill whose diameter is 0.3 m. The bottom of the grill is covered with hot coal bricks at 1100 K, while the wire mesh on top of the grill is covered with steaks initially at 5°C. The distance between the coal bricks and the steaks is 0.20 m. Treating both the steaks and the coal bricks as blackbodies, determine the initial rate of radiation heat transfer from the coal bricks to the steaks. Also determine the initial rate of radiation heat transfer to the steaks if the side opening of the grill is covered by aluminum foil, which can be approximated as a reradiating surface. *Answer:* 1519 W, 3679 W

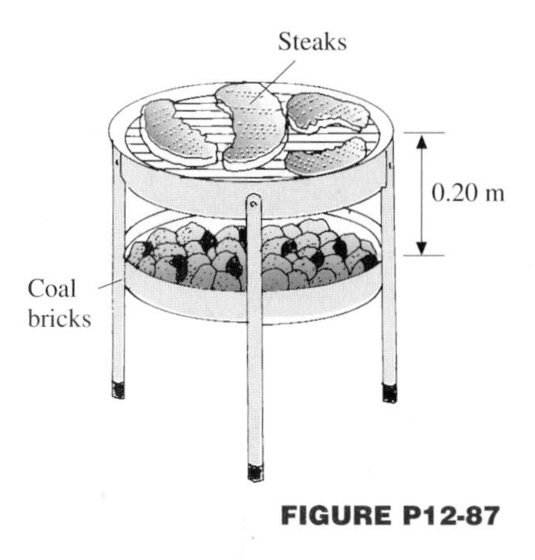


**FIGURE P12-87**

**12-88** A 3-m-high room with a base area of 4 m × 4 m is to be heated by electric resistance heaters placed on the ceiling, which is maintained at a uniform temperature of 32°C at all times. The floor of the room is at 18°C and has an emissivity of 0.8. The side surfaces are well insulated. Treating the ceiling as a blackbody, determine the rate of heat loss from the room through the floor.

**12-88E** A 9-ft-high room with a base area of 12 ft × 12 ft is to be heated by electric resistance heaters placed on the ceiling, which is maintained at a uniform temperature of 90°F at all times. The floor of the room is at 65°F and has an emissivity of 0.8. The side surfaces are well insulated. Treating the ceiling as a blackbody, determine the rate of heat loss from the room through the floor.

### Radiation Shields and the Radiation Effect

**12-89C** What is a radiation shield? Why is it used?

**12-90C** What is the radiation effect? How does it influence the temperature measurements?

**12-91C** Give examples of radiation effects that affect human comfort.

**12-92** Consider a person whose exposed surface area is 1.7 m$^2$, emissivity is 0.7, and surface temperature is 32°C. Determine the rate of heat loss from that person by radiation in a large room whose walls are at a temperature of (*a*) 300 K and (*b*) 280 K.
*Answers:* (*a*) 37.4 W, (*b*) 169.2 W

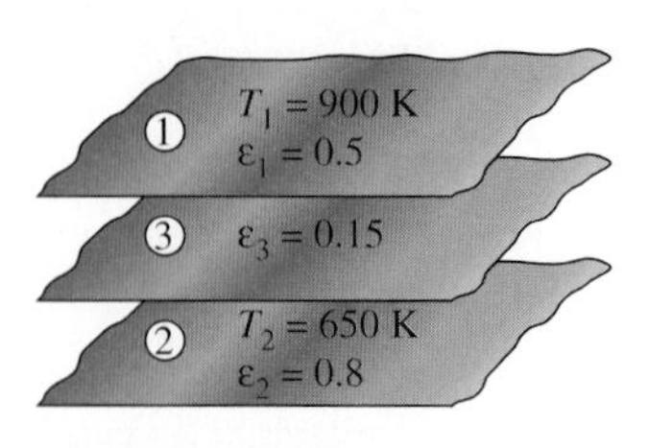


**FIGURE P12-93**

**12-93** A thin aluminum sheet with an emissivity of 0.15 on both sides is placed between two very large parallel plates, which are maintained at uniform temperatures $T_1 = 900$ K and $T_2 = 650$ K, and have emissivities $\varepsilon_1 = 0.5$ and $\varepsilon_2 = 0.8$, respectively. Determine the net rate of radiation heat transfer between the two plates per unit surface are of the plates, and compare the result with that without the shield.

**12-94** Two very large parallel plates are maintained at uniform temperatures of $T_1 = 1000$ K and $T_2 = 500$ K, and have emissivities of $\varepsilon_1 = \varepsilon_2 = 0.2$, respectively. It is desired to reduce the net rate of radiation heat transfer between the two plates to one-fifth by placing thin aluminum sheets with an emissivity of 0.2 on both sides between the plates. Determine the number of sheets that need to be inserted.

**12-95** Five identical thin aluminum sheets with emissivities of 0.1 on both sides are placed between two very large parallel plates, which are maintained at uniform temperatures of $T_1 = 800$ K and $T_2 = 450$ K, and have emissivities of $\varepsilon_1 = \varepsilon_2 = 0.1$, respectively. Determine the net rate of radiation heat transfer between the two plates per unit surface area of the plates, and compare the result to that without the shield.

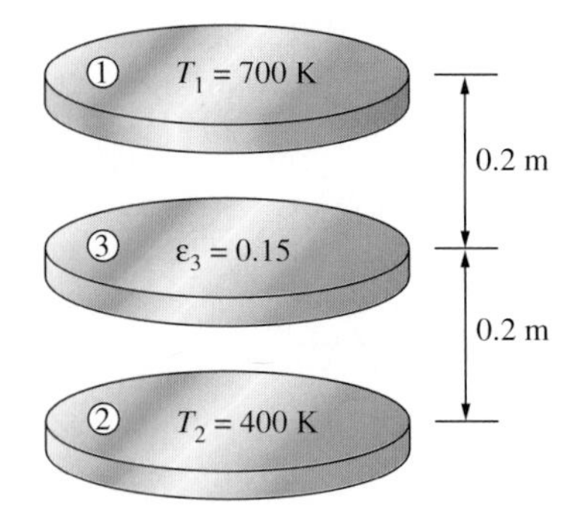


**FIGURE P12-96**

**12-96** Two parallel disks of diameter $D = 0.6$ m separated by $L = 0.4$ m are located directly on top of each other. The disks are separated by a radiation shield whose emissivity is 0.15, as shown in Fig. P12-96. Both disks are black, and are maintained at temperatures of 700 K and 400 K, respectively. The environment that the disks are in can be considered to be a blackbody at $T_\infty = 300$ K. Determine the net rate of radiation heat transfer between the disks under steady conditions. *Answer:* 115 W

**12-96E** Two parallel disks of diameter $D = 3$ ft separated by $L = 2$ ft are located directly on top of each other. The disks are separated by a radiation shield whose emissivity is 0.15. Both disks are black, and are maintained at temperatures of 1200 R and 700 R, respectively. The environment that the disks are in can be considered to be a blackbody at $T_\infty = 540$ R. Determine the net rate of radiation heat transfer between the disks under steady conditions. *Answer:* 799 Btu/h

**12-97** A radiation shield that has the same emissivity $\varepsilon_3$ on both sides is placed between two large parallel plates, which are maintained at uniform temperatures of $T_1 = 650$ K and $T_2 = 400$ K, and have emissivities of $\varepsilon_1 = 0.6$ and $\varepsilon_2 = 0.9$, respectively. Determine the emissivity of the radiation shield if the radiation heat transfer between the plates is to be reduced to 15 percent of that without the radiation shield.

**12-98** Two coaxial cylinders of diameters $D_1 = 0.10$ m and $D_2 = 0.30$ m and emissivities $\varepsilon_1 = 0.7$ and $\varepsilon_2 = 0.4$ are maintained at uniform temperatures of $T_1 = 750$ K and $T_2 = 500$ K, respectively. Now a coaxial radiation shield of diameter $D_3 = 0.20$ m and emissivity $\varepsilon_3 = 0.2$ m is placed between the two cylinders. Determine the net rate of radiation heat transfer between the two cylinders per unit length of the cylinders, and compare the result with that without the shield.

## Review Problems

**12-99** A thermocouple used to measure the temperature of hot air flowing in a duct whose walls are maintained at $T_w = 500$ K shows a temperature reading of $T_{th} = 850$ K. Assuming the emissivity of the thermocouple junction to be $\varepsilon = 0.6$ and the convection heat transfer coefficient to be $h = 60$ W/(m$^2 \cdot$ °C), determine the actual temperature of the air. *Answer:* 1111 K

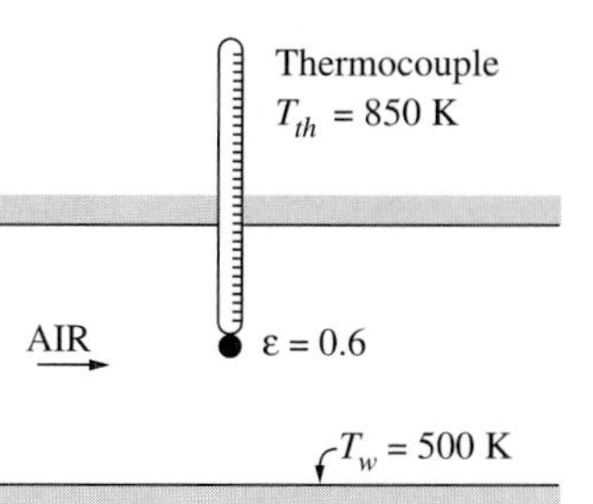

FIGURE P12-99

**12-99E** A thermocouple used to measure the temperature of hot air flowing in a duct whose walls are maintained at $T_w = 900$ R shows a temperature reading of $T_{th} = 1500$ R. Assuming the emissivity of the thermocouple junction to be $\varepsilon = 0.6$ and the convection heat transfer coefficient to be $h = 12$ Btu/(h $\cdot$ ft$^2 \cdot$ °F), determine the actual temperature of the air. *Answer:* 1878 R

**12-100** A thermocouple shielded by aluminum foil of emissivity 0.1 is used to measure the temperature of hot gases flowing in a duct whose walls are maintained at $T_w = 380$ K. The thermometer shows a temperature reading of $T_{th} = 530$ K. Assuming the emissivity of the thermocouple junction to be $\varepsilon = 0.8$ and the convection heat transfer coefficient to be $h = 120$ W/(m$^2 \cdot$ °C), determine the actual temperature of the gas. What would the thermometer reading be if no radiation shield was used?

**12-101** The spectral emissivity function of an opaque surface at 1200 K is approximated as

$$\varepsilon_\lambda = \begin{cases} \varepsilon_1 = 0.0, & 0 \leqslant \lambda < 2\ \mu\text{m} \\ \varepsilon_2 = 0.8, & 2\ \mu\text{m} \leqslant \lambda < 6\ \mu\text{m} \\ \varepsilon_3 = 0.6, & 6\ \mu\text{m} \leqslant \lambda < \infty \end{cases}$$

Determine the average emissivity of the surface, and the rate of radiation emission from the surface, in W/m$^2$.

**12-102** Consider a sealed 20-cm-high electronic box whose base dimensions are 40 cm × 40 cm, placed in a vacuum chamber. The emissivity of the outer surface of the box is 0.95. If the electronic components in the box dissipate a total of 100 W of power and the outer surface temperature of the box is not to exceed 60°C, determine the highest temperature at which the surrounding surfaces must be kept if this box is to be cooled by radiation alone. Assume the heat transfer from the bottom surface of the box to the stand to be negligible. *Answer:* 30°C

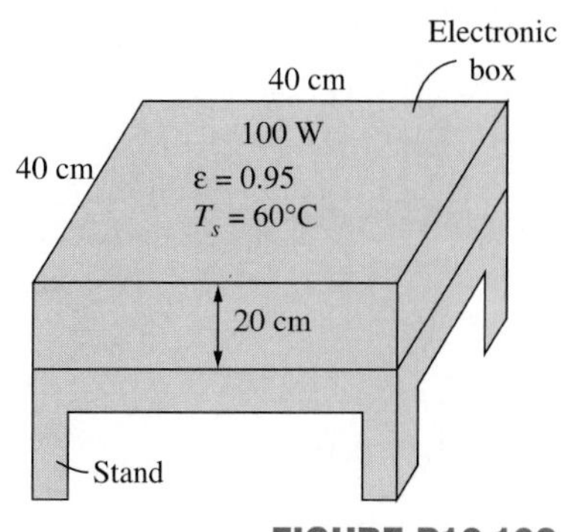


FIGURE P12-102

**12-102E** Consider a sealed 8-in-high electronic box whose base dimensions are 12 in × 12 in, placed in a vacuum chamber. The emissivity of the outer surface of the box is 0.95. If the electronic components in the box dissipate a total of 100 W of power and the outer surface temperature of the box is not to exceed 130°F, determine the highest temperature at which the surrounding surfaces must be kept if this box is to be cooled by radiation alone. Assume the heat transfer from the bottom surface of the box to the stand to be negligible. *Answer:* 43°F

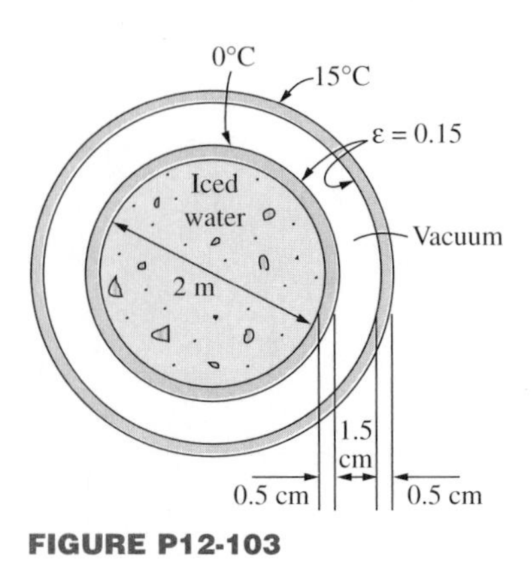


**FIGURE P12-103**

**12-103** A 2-m internal-diameter double-walled spherical tank is used to store iced water at 0°C. Each wall is 0.5 cm thick, and the 1.5-cm-thick air space between the two walls of the tank is evacuated in order to minimize heat transfer. The surfaces surrounding the evacuated space are polished so that each surface has an emissivity of 0.15. The temperature of the outer wall of the tank is measured to be 15°C. Assuming the inner wall of the steel tank to be at 0°C, determine (*a*) the rate of heat transfer to the iced water in the tank and (*b*) the amount of ice at 0°C that melts during a 24-h period. The heat of fusion of water at atmospheric pressure is $h_{if}$ = 333.7 kJ/kg.

**12-103E** A 6-ft internal-diameter double-walled spherical tank is used to store iced water at 32°F. Each wall is 0.2 in thick, and the 0.5-in-thick air space between the two walls of the tank is evacuated in order to minimize the heat transfer. The surfaces surrounding the evacuated space are polished so that each surface has an emissivity of 0.15. The temperature of the outer wall of the tank is measured to be 60°F. Assuming the inner wall of the steel tank to be at 32°F, determine (*a*) the rate of heat transfer to the iced water in the tank and (*b*) the amount of ice at 32°F that melts during a 24-h period. The heat of fusion of water at atmospheric pressure is $h_{if}$ = 143.5 Btu/lbm.

**12-104** Two concentric spheres of diameters $D_1$ = 15 cm and $D_2$ = 25 cm are separated by air at 1-atm pressure. The surface temperatures of the two spheres enclosing the air are $T_1$ = 350 K and $T_2$ = 275 K, respecctivly, and their emissivities are 0.5. Determine the rate of heat transfer from the inner sphere to the outer sphere by (*a*) natural convection and (*b*) radiation.

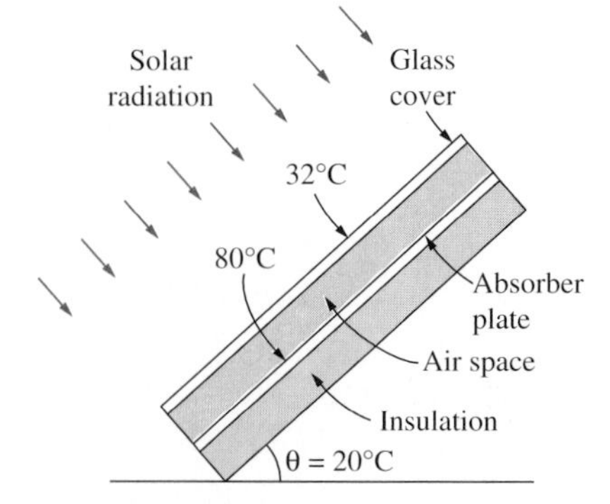


**FIGURE P12-105**

**12-105** Consider a 1.5-m-high and 3-m-wide solar collector that is tilted at an angle 20° from the horizontal. The distance between the glass cover and the absorber plate is 3 cm, and the back side of the absorber is heavily insulated. The absorber plate and the glass cover are maintained at temperatures of 80°C and 32°C, respectively. The emissivity of the glass surface is 0.9 and that of the absorber plate is 0.8. Determine the rate of heat loss from the absorber plate by natural convection and radiation. *Answers:* 713 W, 1289 W

**12-106** A solar collector consists of a horizontal aluminum tube having an outer diameter of 5 cm enclosed in a concentric thin glass tube of diameter 8 cm. Water is heated as it flows through the tube, and the annular space between the aluminum and the glass tube is filled with air at 0.5-atm pressure. The pump circulating the water fails during a clear day, and the water temperature in the tube starts rising. The aluminum tube absorbs solar radiation at a rate of 35 W per meter length, and the temperature of the ambient air outside is 25°C. The emissivities of the tube and the glass cover are 0.9. Taking the effective sky temperature to be 15°C, determine the temperature of the aluminum tube when thermal equilibrium is established (i.e., when the rate of heat loss from the tube equals the amount of solar energy gained by the tube).

**12-106E** A solar collector consists of a horizontal aluminum tube having an outer diameter of 2.5 in. enclosed in a concentric thin glass tube of diameter 5 in. Water is heated as it flows through the tube, and the annular space between the aluminum and the glass tube is filled with air at 0.5-atm pressure. The pump circulating the water fails during a clear day, and the water temperature in the tube starts rising. The aluminum tube absorbs solar radiation at a rate of 30 Btu/h per foot length, and the temperature of the ambient air outside is 75°F. The emissivities of the tube and the glass cover are 0.9. Taking the effective sky temperature to be 60°F, determine the temperature of the aluminum tube when thermal equilibrium is established (i.e., when the rate of heat loss from the tube equals the amount of solar energy gained by the tube).

**12-107** A vertical 1.5-m-high and 3-m-wide double-pane window consists of two sheets of glass separated by a 1.5-cm-thick air gap. In order to reduce heat transfer through the window, the air space between the two glasses is partially evacuated to 0.3-atm pressure. The emissivities of the glass surfaces are 0.9. Taking the glass surface temperatures across the air gap to be 15°C and 5°C, determine the rate of heat transfer through the window by natural convection and radiation.

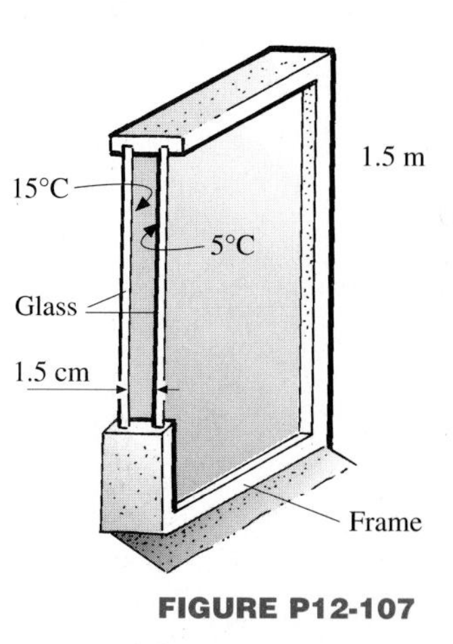


**FIGURE P12-107**

**12-107E** A vertical 6-ft-high and 10-ft-wide double-pane window consists of two sheets of glass separated by a 0.5-in-thick air gap. In order to reduce heat transfer through the window, the air space between the two glasses is partially evacuated to 0.3 atm. The emissivities of the glass surfaces are 0.9. Taking the glass surface temperatures across the air gap to be 60°F and 40°F, determine the rate of heat transfer through the window by natural convection and radiation.

**12-108** A simple solar collector is built by placing a 6-cm-diameter clear plastic tube around a garden hose whose outer diameter is 2 cm. The hose is painted black to maximize solar absorption, and some plastic rings are used to keep the spacing between the hose and the clear plastic cover constant. The emissivities of the hose surface and the glass cover are 0.9, and the effective sky temperature is estimated to be 15°C. The temperature of the plastic tube is measured to be 40°C, while the ambient air temperature is 25°C. Determine the rate of heat loss from the water in the hose by natural convection and radiation per meter of its length under steady conditions. *Answers:* 5.9 W, 26.2 W

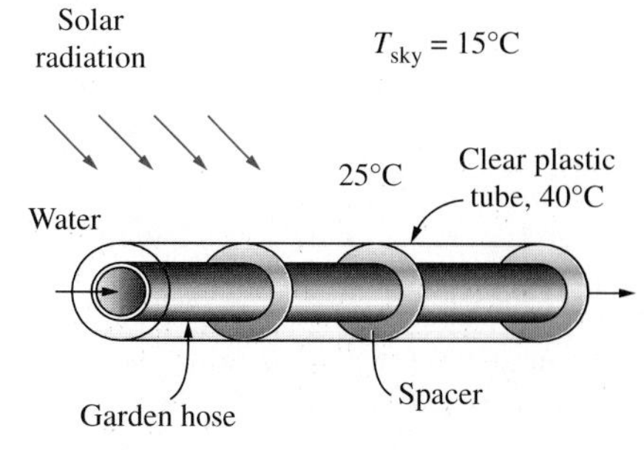


**FIGURE P12-108**

**12-109** A solar collector consists of a horizontal copper tube of outer diameter 5 cm enclosed in a concentric thin glass tube of diameter 9 cm. Water is heated as it flows through the tube, and the annular space between the copper and the glass tubes is filled with air at 1-atm pressure. The emissivities of the tube surface and the glass cover are 0.85 and 0.9, respectively. During a clear day, the temperatures of the tube surface and the glass cover are measured to be 60°C and 32°C, respectively. Determine the rate of heat loss from the collector by natural convection and radiation per meter length of the tube.

## Computer, Design, and Essay Problems

**12-110** Consider an enclosure consisting of *N* diffuse and gray surfaces. The emissivity and temperature of each surface as well as all the view factors between the surfaces are specified. Using the matrix method and obtaining a subroutine for solving a system of *N* linear equations simultaneously, write a program to determine the net rate of radiation heat transfer for each surface.

**12-111** Radiation shields are commonly used in the design of superinsulations for use in space and cryogenic applications. Write an essay on superinsulations and how they are used in different applications.

**12-112** Thermal comfort in a house is strongly effected by the so-called "radiation effect," which is due to radiation heat transfer between the person and surrounding surfaces. A person feels much colder in the morning, for example, because of the lower surface temperature of the walls at that time, although the thermostat setting of the house is fixed. Write an essay on the radiation effect, how it affects human comfort, and how it is accounted for in heating and air conditioning applications.

# Heat Exchangers

CHAPTER 13

Heat exchangers are devices that facilitate the *exchange of heat* between *two fluids* that are at different temperatures while keeping them from mixing with each other. Heat exchangers are commonly used in practice in a wide range of applications; from heating and air conditioning systems in a household, to chemical processing and power production in large plants. Heat exchangers differ from mixing chambers in that they do not allow the two fluids involved to mix. In a car radiator, for example, heat is transferred from the hot water flowing through the radiator tubes to the air flowing through the closely spaced thin plates outside attached to the tubes.

Heat transfer in a heat exchanger usually involves *convection* in each fluid and *conduction* through the wall separating the two fluids. In the analysis of heat exchangers, it is convenient to work with an *overall heat transfer coefficient U* that accounts for the contribution of all these effects on heat transfer. The rate of heat transfer between the two fluids at a location in a heat exchanger depends on the magnitude of the temperature difference at that location, which varies along the heat exchanger. In the analysis of heat exchangers, it is usually convenient to work with the *logarithmic mean temperature difference LMTD,* which is an equivalent mean temperature difference between the two fluids for the entire heat exchanger.

Heat exchangers are manufactured in a variety of types, and thus we start this chapter with the *classification* of heat exchangers. We then discuss the determination of the overall heat transfer coefficient in heat exchangers, and the logarithmic mean temperature difference LMTD for some configurations. We then introduce the *correction factor F* to account for the deviation of the mean temperature difference from the LMTD in complex configurations. Next we discuss the $\varepsilon$-NTU method, which enables us to analyze heat exchangers when the outlet temperatures of the fluids are not known. Finally, we discuss the selection of heat exchangers.

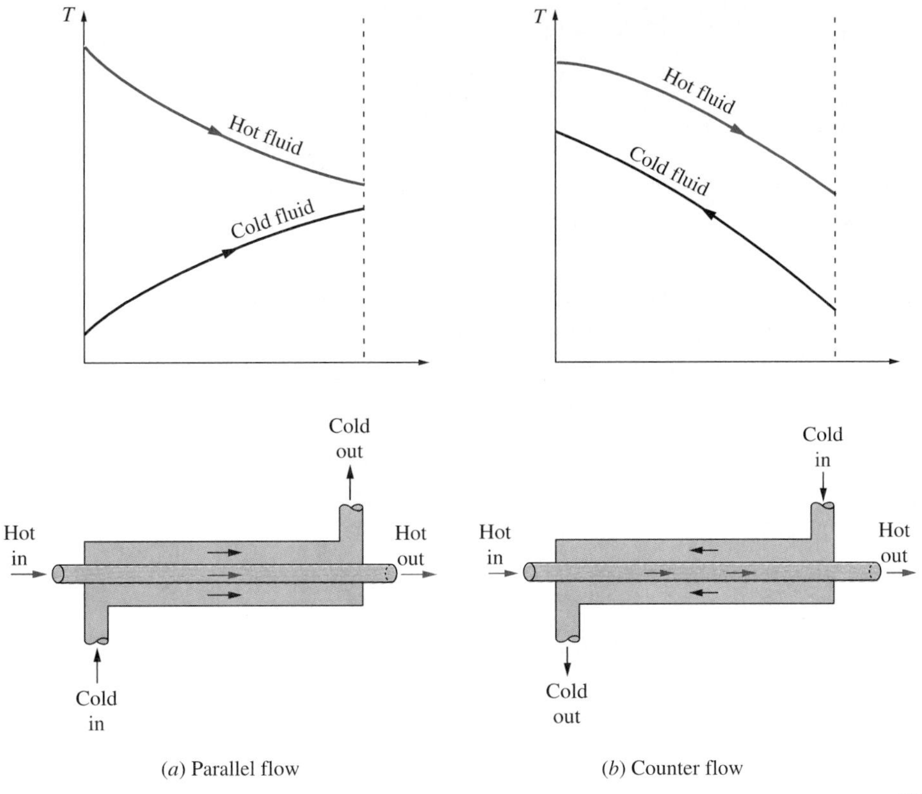


**FIGURE 13-1**

Different flow regimes and associated temperature profiles in a double-pipe heat exchanger.

## 13-1  TYPES OF HEAT EXCHANGERS

Different heat transfer applications require different types of hardware and different configurations of heat transfer equipment. The attempt to match the heat transfer hardware to the heat transfer requirements within the specified constraints has resulted in numerous types of innovative heat exchanger designs.

The simplest type of heat exchanger consist of two concentric pipes of different diameters, as shown in Fig. 13-1, called the **double-pipe** heat exchanger. One fluid in a double-pipe heat exchanger flows through the smaller pipe while the other fluid flows through the annular space between the two pipes. Two types of flow arrangement are possible in a double-pipe heat exchanger: in **parallel flow**, both the hot and cold fluids enter the heat exchanger at the same end, and move in the *same* direction. In **counter flow**, on the other hand, the hot and cold fluids enter the heat exchanger at opposite ends, and flow in *opposite* directions.

Another type of heat exchanger, which is specifically designed to realize a large heat transfer surface area per unit volume, is the **compact** heat exchanger. The ratio of the heat transfer surface area of a heat exchanger to its volume is called the *area density* $\beta$. A heat exchanger

**FIGURE 13-2**
A gas-to-liquid compact heat exchanger for a residential air-conditioning system.

with $\beta > 700\,\mathrm{m^2/m^3}$ (or $200\,\mathrm{ft^2/ft^3}$) is classified as being compact. Examples of compact heat exchangers are car radiators ($\beta \approx 1000\,\mathrm{m^2/m^3}$), glass ceramic gas turbine heat exchangers ($\beta \approx 6000\,\mathrm{m^2/m^3}$), the regenerator of a Stirling engine ($\beta \approx 15{,}000\,\mathrm{m^2/m^3}$), and the human lung ($\beta \approx 20{,}000\,\mathrm{m^2/m^3}$). Compact heat exchangers enable us to achieve high heat transfer rates between two fluids in a small volume, and they are commonly used in applications with strict limitations on the weight and volume of heat exchangers (Fig. 13-2).

The large surface area in compact heat exchangers is obtained by attaching closely spaced *thin plate* or *corrugated fins* to the walls separating the two fluids. Compact heat exchangers are commonly used in gas-to-gas and gas-to-liquid (or liquid-to-gas) heat exchangers to counteract the low heat transfer coefficient associated with gas flow with increased surface area. In a car radiator, which is a water-to-air compact heat exchanger, for example, it is no surprise that fins are attached to the air side of the tube surface.

In compact heat exchangers, the two fluids usually move *perpendicular* to each other, and such flow configuration is called **cross-flow**. The cross-flow is further classified as *unmixed* and *mixed flow,* depending on the flow configuration, as shown in Fig. 13-3. In (*a*) the cross-flow is said to be *unmixed,* since the plate fins force the fluid to flow through a particular interfin spacing, and prevent it from moving in the transverse direction (i.e., parallel to the tubes). The cross-flow in (*b*) is said to be *mixed,* since the fluid now is free to move in the transverse direction. Both fluids are unmixed in a car radiator. The presence of mixing in the fluid can have a significant effect on the heat transfer characteristics of the heat exchanger.

Perhaps the most common type of heat exchanger in industrial

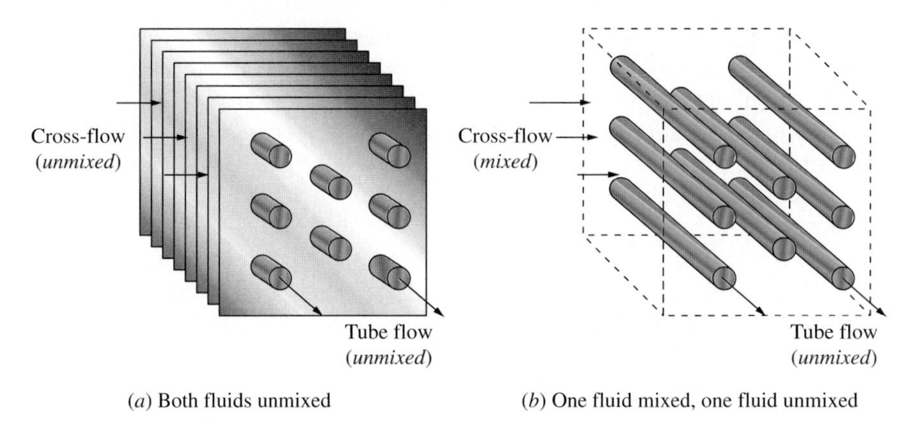


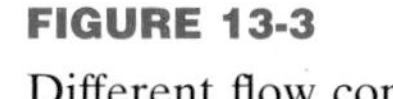
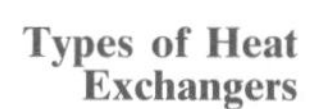

**FIGURE 13-3**

Different flow configurations in cross-flow heat exchangers.

applications is the **shell-and-tube** heat exchanger, shown in Fig. 13-4. Shell-and-tube heat exchangers contain a large number of tubes (sometimes several hundred) packed in a shell with their axes parallel to that of the shell. Heat transfer takes place as one fluid flows inside the tubes while the other fluid flows outside the tubes through the shell. *Baffles* are commonly placed in the shell to force the shell-side fluid to flow across the shell to enhance heat transfer and to maintain uniform spacing between the tubes. Despite their widespread use, shell-and-tube heat exchangers are not suitable for use in automotive, aircraft, and marine applications because of their relatively large size and weight. Note that the tubes in a shell-and-tube heat exchangers open to some large flow areas called *headers* at both ends of the shell, where the tube-side fluid accumulates before entering the tubes and after leaving them.

Shell-and-tube heat exchangers are further classified according to the number of shell and tube passes involved. Heat exchangers in which all the tubes make one U-turn in the shell, for example, are called *one-shell pass and two-tube pass* heat exchangers. Likewise, a heat exchanger that

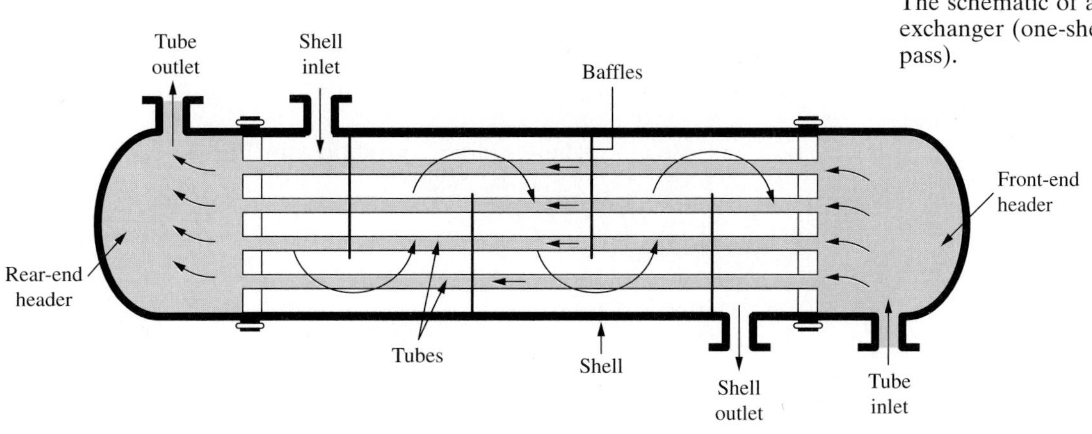


**FIGURE 13-4**

The schematic of a shell-and-tube heat exchanger (one-shell pass and one-tube pass).

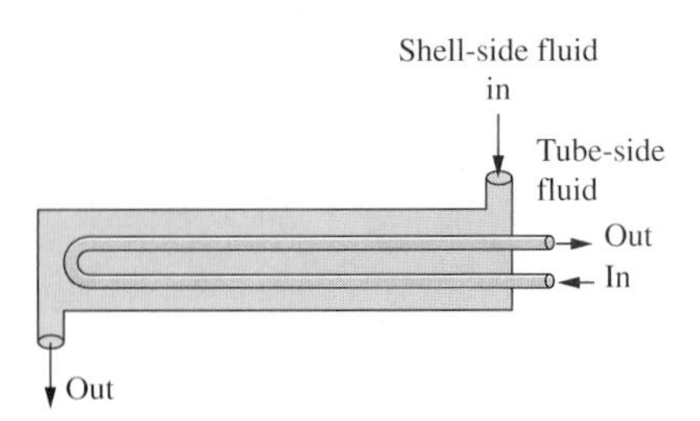
Shell-side fluid
in
Tube-side
fluid
Out
In
Out

(*a*) One-shell pass and two-tube pass

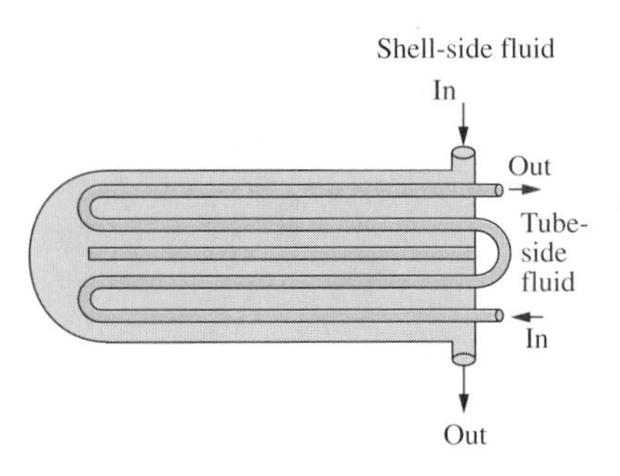
Shell-side fluid
In
Out
Tube-
side
fluid
In
Out

(*b*) Two-shell pass and four-tube pass

**FIGURE 13-5**
Multipass flow arrangements in shell-and-tube heat exchangers.

involves two passes in the shell and four passes in the tubes is called a *two-shell pass and four-tube pass* heat exchanger (Fig. 13-5).

An innovative type of heat exchanger that has found widespread use is the **plate and frame** (or just plate) heat exchanger, which consists of a series of plates with corrogated flat flow passages (Fig. 13-6). The hot and cold fluids flow in alternate passages, and thus each cold fluid stream is surrounded by two hot fluid streams, resulting in very effective heat transfer. Also, plate heat exchangers can grow with growing demand for heat transfer by simply mounting more plates. They are well suited for liquid-to-liquid heat exchange applications, provided that the hot and cold fluid streams are at about the same pressure.

Another type of heat exchanger that involves the alternate passage of the hot and cold fluid streams through the same flow area is the **regenerative** heat exchanger. The *static*-type regenerative heat exchanger is basically a porous mass that has a large heat storage capacity, such as a ceramic wire mesh. Hot and cold fluids flow through this porous mass alternately. Heat is transferred from the hot fluid to the matrix of the regenerator during the flow of the hot fluid, and from the matrix to the cold fluid during the flow of the cold fluid. Thus the matrix serves as a temporary heat storage medium.

The *dynamic*-type regenerator involes a rotating drum and continuous flow of the hot and cold fluid through different portions of the drum so that any portion of the drum passes periodically through the hot stream, storing heat, and then through the cold stream, *rejecting* this

**FIGURE 13-6**
A plate-and-frame liquid-to-liquid heat exchanger (Courtesy of Trante Corp.).

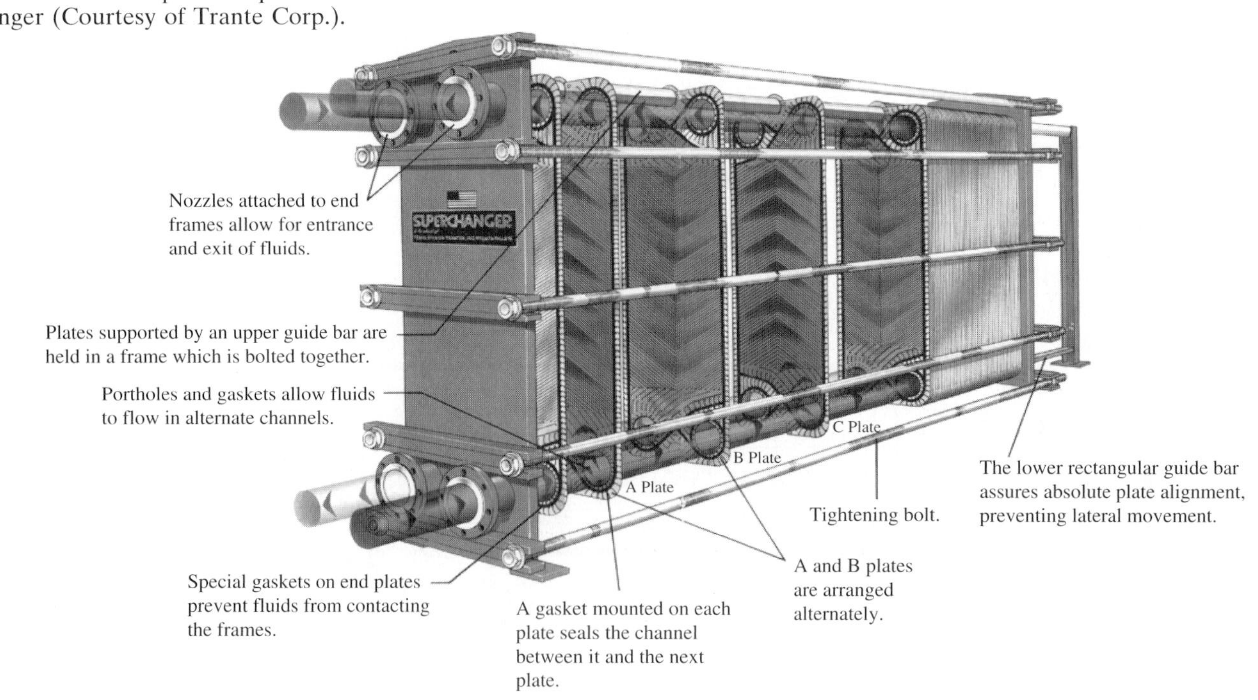
Nozzles attached to end frames allow for entrance and exit of fluids.
SUPERCHANGER
Plates supported by an upper guide bar are held in a frame which is bolted together.
Portholes and gaskets allow fluids to flow in alternate channels.
C Plate
B Plate
A Plate
Tightening bolt.
The lower rectangular guide bar assures absolute plate alignment, preventing lateral movement.
A and B plates are arranged alternately.
Special gaskets on end plates prevent fluids from contacting the frames.
A gasket mounted on each plate seals the channel between it and the next plate.

*stored* heat. Again the drum serves as the medium to transport the heat from the hot to the cold fluid stream.

Heat exchangers are often given specific names to reflect the specific application they are used for. For example, a *condenser* is a heat exchanger in which one of the fluids gives up heat and condenses as it flows through the heat exchanger. A *boiler* is another heat exchanger in which one of the fluids absorbs heat and vaporizes. A *space radiator* is a heat exchanger that transfers heat from the hot fluid to the surrounding space by radiation.

## 13-2 ■ THE OVERALL HEAT TRANSFER COEFFICIENT

A heat exchanger typically involves two flowing fluids separated by a solid wall. Heat is first transferred from the hot fluid to the wall by *convection*, through the wall by *conduction*, and from the wall to the cold fluid again by *convection*. Any radiation effects are usually included in the convection heat transfer coefficients.

The thermal resistance network associated with this heat transfer process involves two convection and one conduction resistances, as shown in Fig. 13-7. Here the subscripts $i$ and $o$ represent the inner and outer surfaces of the inner tube. For a double-pipe heat exchanger, we have $A_i = \pi D_i L$ and $A_o = \pi D_o L$, and the *thermal resistance* of the tube wall in this case is

$$R_{\text{wall}} = \frac{\ln (D_o/D_i)}{2\pi k L} \tag{13-1}$$

where $k$ is the thermal conductivity of the wall material and $L$ is the length of the tube. Then the *total thermal resistance* becomes

$$R = R_{\text{total}} = R_i + R_{\text{wall}} + R_o = \frac{1}{h_i A_i} + \frac{\ln (D_o/D_i)}{2\pi k L} + \frac{1}{h_o A_o} \tag{13-2}$$

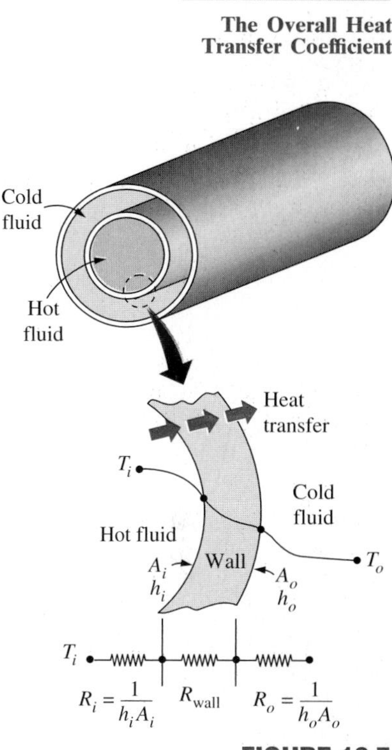

**FIGURE 13-7**

Thermal resistance network associated with heat transfer in a double-pipe heat exchanger.

The $A_i$ is the area of the *inner surface* of the wall that separates the two fluids, and $A_o$ is the area of the outer surface of the wall. In other words, $A_i$ and $A_o$ are surface areas of the separating wall wetted by the inner and the outer fluids, respectively. When one fluid flows inside a circular tube and the other outside of it, we have $A_i = \pi D_i L$ and $A_o = \pi D_o L$ (Fig. 13-8).

In the analysis of heat exchangers, it is convenient to combine all the thermal resistances in the path of heat flow from the hot fluid to the cold one into a single resistance $R$ as we discussed in Chapter 8, and to express the rate of heat transfer between the two fluids as

$$\dot{Q} = \frac{\Delta T}{R} = UA\,\Delta T = U_i A_i\,\Delta T = U_o A_o\,\Delta T \tag{13-3}$$

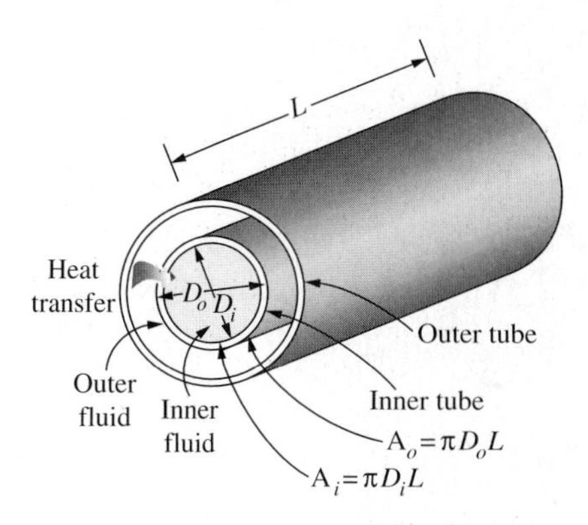


**FIGURE 13-8**

The two heat transfer surface areas associated with a double-pipe heat exchanger (for thin tubes, $D_i \approx D_o$ and thus $A_i \approx A_o$).

where $U$ is the **overall heat transfer coefficient**, whose unit is $\text{W}/(\text{m}^2 \cdot {}^\circ\text{C})$, which is identical to the unit of the ordinary convection coefficient $h$. Canceling $\Delta T$, Eq. 13-3 reduces to

$$\frac{1}{UA} = \frac{1}{U_i A_i} = \frac{1}{U_o A_o} = R = \frac{1}{h_i A_i} + R_{\text{wall}} + \frac{1}{h_o A_o} \tag{13-4}$$

Perhaps you are wondering why we have *two* overall heat transfer coefficients $U_i$ and $U_o$ for a heat exchanger. The reason is that every heat exchanger has two heat transfer surface areas $A_i$ and $A_o$, which, in general, are not equal to each other.

Note that $U_i A_i = U_o A_o$, but $U_i \neq U_o$ unless $A_i = A_0$. Therefore, the overall heat transfer coefficient $U$ of a heat exchanger is meaningless unless the area on which it is based is specified. This is especially the case when one side of the tube wall is finned and the other side is not, since the surface area of the finned side is several times that of the unfinned side.

When the wall thickness of the tube is small and the thermal conductivity of the tube material is high, as is usually the case, the thermal resistance of the tube is negligible ($R_{\text{wall}} \approx 0$) and the inner and outer surfaces of the tube are almost identical ($A_i \approx A_0 \approx A$). Then Eq. 13-4 for the overall heat transfer coefficient simplifies to

$$\frac{1}{U} \approx \frac{1}{h_i} + \frac{1}{h_o} \tag{13-5}$$

where $U \approx U_i \approx U_0$. The individual convection heat transfer coefficients inside and outside the tube, $h_i$ and $h_o$, are determined using the convection relations discussed in earlier chapters.

The overall heat transfer coefficient $U$ in Eq. 13-5 is dominated by the *smaller* convection coefficient, since the inverse of a large number is small. When one of the convection coefficients is *much smaller* than the other (say, $h_i \ll h_o$), we have $1/h_i \gg 1/h_o$, and thus $U \approx h_i$. Therefore, the smaller heat transfer coefficient creates a *bottleneck* on the path of heat flow, and seriously impedes heat transfer. This situation arises frequently when one of the fluids is a gas and the other is a liquid. In such cases, fins are commonly used on the gas side to enhance the product $UA$ and thus the heat transfer on that side.

Representative values of the overall heat transfer coefficient $U$ are given in Table 13-1. Note that the overall heat transfer coefficient ranges from about $10\ \text{W}/(\text{m}^2 \cdot {}^\circ\text{C})$ for gas-to-gas heat exchangers to about $10{,}000\ \text{W}/(\text{m}^2 \cdot {}^\circ\text{C})$ for heat exchangers that involve phase changes. This is not surprising, since gases have very low thermal conductivities, and phase-change processes involve very high heat transfer coefficients.

When the tube is *finned* on one side to enhance heat transfer, the total heat transfer surface area on the finned side becomes

$$A = A_{\text{total}} = A_{\text{fin}} + A_{\text{unfinned}} \tag{13-6}$$

**TABLE 13-1**
**Representative values of the overall heat transfer coefficients in heat exchangers**

| Type of heat exchanger | $U$, W/(m$^2$ · °C)* |
|---|---|
| Water-to-water | 850–1700 |
| Water-to-oil | 100–350 |
| Water-to-gasoline or kerosene | 300–1000 |
| Feedwater heaters | 1000–8500 |
| Steam-to-light fuel oil | 200–400 |
| Steam-to-heavy fuel oil | 50–200 |
| Steam condenser | 1000–6000 |
| Freon condenser (water cooled) | 300–1000 |
| Ammonia condenser (water cooled) | 800–1400 |
| Alcohol condensers (water cooled) | 250–700 |
| Gas-to-gas | 10–40 |
| Water-to-air in finned tubes (water in tubes) | 30–60† |
| | 400–850‡ |
| Steam-to-air in finned tubes (steam in tubes) | 30–300† |
| | 400–4000‡ |

* Multiply the listed values by 0.176 to convert them to Btu/(h · ft$^2$ · °F).
† Based on air-side surface area.
‡ Based on water or steam side surface area.

where $A_{\text{fin}}$ is the surface area of the fins and $A_{\text{unfinned}}$ is the area of unfinned portion of the tube surface. For short fins of high thermal conductivity, we can use this total area in the convection resistance relation $R_{\text{conv}} = 1/hA$, since the fins in this case will be very nearly isothermal. Otherwise, we should determine the effective surface area $A$ from

$$A = A_{\text{unfinned}} + \eta_{\text{fin}} A_{\text{fin}} \tag{13-7}$$

where $\eta_{\text{fin}}$ is the fin efficiency. This way, the temperature drop along the fins is accounted for. Note that $\eta_{\text{fin}} = 1$ for isothermal fins, and thus Eq. 13-7 reduces to Eq. 13-6 in that case.

## Fouling Factor

The performance of heat exchangers usually deteriorates with time as a result of accumulation of *deposits* on heat transfer surfaces. The layer of deposits represents *additional resistance* to heat transfer, and causes the rate of heat transfer in a heat exchanger to decrease. The net effect of these accumulations on heat transfer is represented by a **fouling factor** $R_f$, which is a measure of the *thermal resistance* introduced by fouling.

The most common type of fouling is the *preciptation* of solid deposits in a fluid on the heat transfer surfaces. You can observe this type of fouling even in your house. If you check the inner surfaces of your teapot after prolonged use, you will probably notice a layer of calcium-based

**FIGURE 13-9**
Precipitation fouling of ash particles on superheater tubes (from *Steam, Its Generation, and Use,* Babcock and Wilcox Co., 1978).

deposits on the surfaces at which boiling occurs. This is especially the case in areas where the water is hard. The scales of such deposits come off by scratching, and the surfaces can be cleaned off such deposits by chemical treatment. Now imagine those mineral deposits forming on the inner surfaces of fine tubes in a heat exchanger (Fig. 13-9), and the detrimental effect it may have on the flow passage area and the heat transfer. To avoid this potential problem, water in power and process plants is extensively treated and its solid contents are removed before it is allowed to circulate through the system. The solid ash particles in the flue gases accumulating on the surfaces of air preheaters create similar problems.

Another form of fouling, which is common in the chemical process industry, is *corrosion* and other *chemical fouling.* In this case, the surfaces are fouled by the accumulation of the products of chemical reactions on the surfaces. This form of fouling can be avoided by coating metal pipes with glass or using plastic pipes instead of metal ones. Heat exchangers may also be fouled by the growth of algae in warm fluids. This type of fouling is called *biological fouling,* and can be prevented by chemical treatment.

In applications where it is likely to occur, fouling should be considered in the design and selection of heat exchangers. In such applications, it may be necessary to select a larger and thus more expensive heat exchanger to ensure that it meets the design heat transfer requirements even after fouling occurs. The periodic cleaning of heat exchangers and the resulting down time are additional penalities associated with fouling.

The fouling factor is obviously zero for a new heat exchanger, and increases with time as the solid deposits build up on the heat exchanger surface. The fouling factor depends on the *operating temperature* and the *velocity* of the fluids, as well as the length of service. Fouling increases with *increasing temperature* and *decreasing velocity*.

The overall heat transfer coefficient relation given above is valid for clean surfaces, and needs to be modified to account for the effects of fouling on both the inner and the outer surfaces of the tube. For an unfinned shell-and-tube heat exchanger, it can be expressed as

$$\frac{1}{UA} = \frac{1}{U_iA_i} = \frac{1}{U_oA_o} = R = \frac{1}{h_iA_i} + \frac{R_{f,i}}{A_i} + \frac{\ln(D_o/D_i)}{2\pi kL} + \frac{R_{f,o}}{A_o} + \frac{1}{h_oA_o} \tag{13-8}$$

where $A = \pi D_iL$ and $A_o = \pi D_oL$ are the areas of inner and outer surfaces, and $R_{f,i}$ and $R_{f,o}$ are the fouling factors at those surfaces.

Representative values of fouling factors are given in Table 13-2. More comprehensive tables of fouling factors are available in handbooks. As you would expect, considerable uncertainty exists in these values, and they should be used as a guide in the selection and evaluation of heat exchangers to account for the effects of anticipated fouling on heat transfer. Note that most fouling factors in the table are of the order of $10^{-4}\,\text{m}^2\cdot{}^\circ\text{C/W}$, which is equivalent to the thermal resistance of a 0.2-mm-thick limestone layer [$k = 2.9\,\text{W/(m}\cdot{}^\circ\text{C)}$] per unit surface area. Therefore, in the absence of specific data, we can assume the surfaces to be coated with 0.2 mm of limestone as a starting point to account for the effects of fouling.

**TABLE 13-2**
**Representative fouling factors (thermal resistance due to fouling for a unit surface area).**

| Fluid | $R_f$, $\text{m}^2\cdot{}^\circ\text{C/W}$ |
|---|---|
| Distilled water, sea water, river water, boiler feedwater: | |
| Below 50°C | 0.0001 |
| Above 50°C | 0.0002 |
| Fuel oil | 0.0009 |
| Steam (oil-free) | 0.0001 |
| Refrigerants (liquid) | 0.0002 |
| Refrigerants (vapor) | 0.0004 |
| Alcohol vapors | 0.0001 |
| Air | 0.0004 |

*Source*: Tubular Exchanger Manufacturers Association.

**EXAMPLE 13-1**

Hot oil is to be cooled in a double-tube counterflow heat exchanger. The copper inner tubes have a diameter of 2 cm and negligible thickness. The inner diameter of the outer tube (the shell) is 3 cm. Water flows through the tube at a rate of 0.5 kg/s, and the oil through the shell at a rate of 0.8 kg/s. Taking the average temperatures of the water and the oil to be 47°C and 80°C, respectively, in property evaluations, and assuming fully developed flow, determine the overall heat transfer coefficient of this heat exchanger.

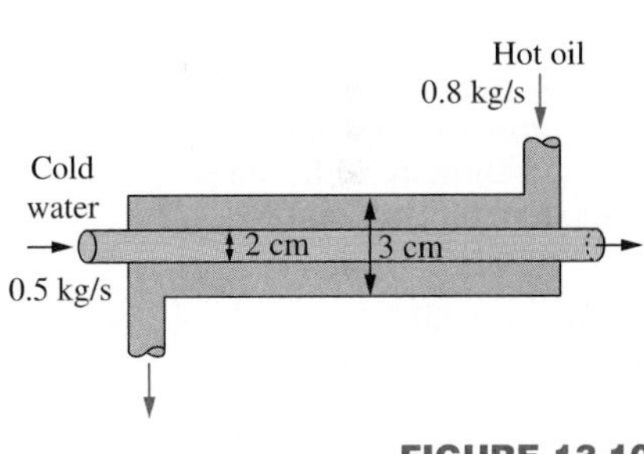


**FIGURE 13-10**
Schematic for Example 13-1.

**Solution** The schematic of the heat exchanger is given in Fig. 13-10. The thermal resistance of the inner tube can be neglected, since the tube material is highly conductive, and its thickness is negligible. Then the overall heat transfer coefficient $U$ can be determined from Eq. 13-5:

$$\frac{1}{U} \approx \frac{1}{h_i} + \frac{1}{h_o}$$

where $h_i$ and $h_o$ are the convection heat tranfer coefficients inside and outside the tube, respectively, which are to be determined using the forced convection relations discussed in Chapter 10.

The properties of water at 47°C (320 K) are

$$\rho = 989\,\text{kg/m}^3$$
$$k = 0.637\,\text{W/(m}\cdot{}^\circ\text{C)}$$
$$\nu = 0.59 \times 10^{-6}\,\text{m}^2\text{/s}$$
$$\text{Pr} = 3.79$$

The hydraulic diameter for a circular tube is the diameter of the tube itself, $D_h = D = 0.02$ m. The average velocity of water in the tube and the Reynolds number are

$$\mathscr{V}_m = \frac{\dot{m}}{\rho A_c} = \frac{\dot{m}}{\rho(\frac{1}{4}\pi D^2)} = \frac{0.5 \text{ kg/s}}{(989 \text{ kg/m}^3)[\frac{1}{4}\pi(0.02 \text{ m})^2]} = 1.61 \text{ m/s}$$

and

$$\text{Re} = \frac{\mathscr{V}_m D_h}{\nu} = \frac{(1.61 \text{ m/s})(0.02 \text{ m})}{0.59 \times 10^{-6} \text{ m}^2/\text{s}} = 54{,}576$$

which is greater than 4000. Therefore, the flow of water is turbulent. Assuming the flow to be fully developed, the Nusselt number can be determined from

$$\text{Nu} = \frac{hD_h}{k} = 0.023 \text{ Re}^{0.8} \text{ Pr}^{0.4} = 0.023(54{,}576)^{0.8}(3.79)^{0.4} = 241.4$$

Then,

$$h = \frac{k}{D_h}\text{Nu} = \frac{0.637 \text{ W/(m} \cdot {}^\circ\text{C)}}{0.02 \text{ m}}(241.4) = 7690 \text{ W/(m}^2 \cdot {}^\circ\text{C)}$$

Now we repeat the analysis above for oil. The properties of oil at 80°C are

$$\rho = 852 \text{ kg/m}^3$$
$$k = 0.138 \text{ W/(m} \cdot {}^\circ\text{C)}$$
$$\nu = 37.5 \times 10^{-6} \text{ m}^2/\text{s}$$
$$\text{Pr} = 490$$

The hydraulic diameter for the annular space is

$$D_h = D_o - D_i = 0.03 - 0.02 = 0.01 \text{ m}$$

The average velocity and the Reyonlds number in this case are

$$\mathscr{V}_m = \frac{\dot{m}}{\rho A_c} = \frac{\dot{m}}{\rho[\frac{1}{4}\pi(D_o^2 - D_i^2)]} = \frac{0.8 \text{ kg/s}}{(852 \text{ kg/m}^3)[\frac{1}{4}(0.03^2 - 0.02^2)] \text{ m}^2} = 2.39 \text{ m/s}$$

and

$$\text{Re} = \frac{\mathscr{V}_m D_h}{\nu} = \frac{(2.39 \text{ m/s})(0.01 \text{ m})}{37.5 \times 10^{-6} \text{ m}^2/\text{s}} = 637$$

which is less than 4000. Therefore, the flow of oil is laminar. Assuming fully developed flow, the Nusselt number on the tube side of the annular space $Nu_i$ corresponding to $D_i/D_o = 0.02/0.03 = 0.667$ can be determined from Table 13-3 by interpolation to be

$$\text{Nu} = 5.45$$

and

$$h_o = \frac{k}{D_h}\text{Nu} = \frac{0.138 \text{ W/(m} \cdot {}^\circ\text{C)}}{0.01 \text{ m}}(5.45) = 75.2 \text{ W/(m}^2 \cdot {}^\circ\text{C)}$$

Then the overall heat transfer coefficient for this heat exchanger becomes

$$U = \frac{1}{\frac{1}{h_i} + \frac{1}{h_o}} = \frac{1}{\frac{1}{7690 \text{ W/(m}^2 \cdot {}^\circ\text{C)}} + \frac{1}{75.2 \text{ W/(m}^2 \cdot {}^\circ\text{C)}}} = 75.1 \text{ W/(m}^2 \cdot {}^\circ\text{C)}$$

Note that $U \approx h_o$ in this case, since $h_i \gg h_o$. This confirms our earlier statement that the overall heat transfer coefficient in a heat exchanger is

**TABLE 13-3**
**Nusselt number for fully developed laminar flow in a circular annulus with one surface insulated and the other isothermal.**

| $D_i/D_o$ | $\text{Nu}_i$ | $\text{Nu}_o$ |
|---|---|---|
| 0.00 | — | 3.66 |
| 0.05 | 17.46 | 4.06 |
| 0.10 | 11.56 | 4.11 |
| 0.25 | 7.37 | 4.23 |
| 0.50 | 5.74 | 4.43 |
| 1.00 | 4.86 | 4.86 |

*Source*: Ref. 8.

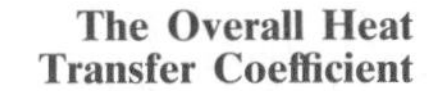

dominated by the smaller heat transfer coefficient when the difference between the two values is large.

To improve the overall heat transfer coefficient and thus the heat transfer in this heat exchanger, we must use some enhancement techniques on the oil side, such as a finned surface.

**EXAMPLE 13-2**

A double-pipe (shell-and-tube) heat exchanger is constructed of a stainless steel [$k = 15.1$ W/(m · °C)] inner tube of inner diameter $D_i = 1.5$ cm and outer diameter$D_o = 1.9$ cm, and an outer tube of diameter 3.2 cm. The convection heat transfer coefficient is given to be $h_i = 800$ W/(m$^2$ · °C) on the inner surface of the tube, and $h_o = 1200$ W/(m$^2$ · °C) on the otuer surface. For a fouling factor of $R_{f,i} = 0.0004$ m$^2$ · °C/W on the tube side and $R_{f,o} = 0.0001$ m$^2$ · °C/W on the shell side, determine (*a*) the thermal resistance of the heat exchanger per unit length and (*b*) the overall heat transfer coefficients $U_i$ and $U_o$ based on the inner and outer surface areas of the tube, respectively.

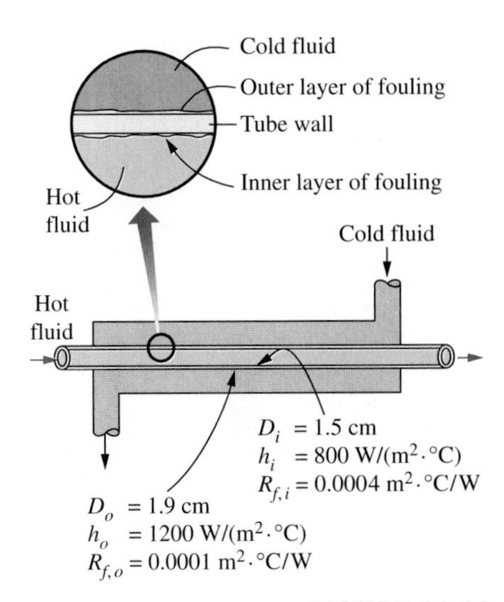

**FIGURE 13-11**

Schematic for Example 13-2.

**Solution** (*a*) The schematic of the heat exchanger is given in Fig. 13-11. The thermal resistance for an unfinned shell-and-tube heat exchanger with fouling on both heat transfer surfaces is given by Eq. 13-8 as

$$R = \frac{1}{UA} = \frac{1}{U_iA_i} = \frac{1}{U_oA_o} = \frac{1}{h_iA_i} + \frac{R_{f,i}}{A_i} + \frac{\ln(D_o/D_i)}{2\pi kL} + \frac{R_{f,o}}{A_o} + \frac{1}{h_oA_o}$$

where
$$A_i = \pi D_i L = \pi(0.015\text{ m})(1\text{ m}) = 0.0471\text{ m}^2$$
$$A_o = \pi D_o L = \pi(0.019\text{ m})(1\text{ m}) = 0.0597\text{ m}^2$$

Substituting, the total thermal resistance is determined to be

$$\begin{aligned} R &= \frac{1}{[800\text{ W/(m}^2\cdot\text{°C)}](0.0471\text{ m}^2)} + \frac{0.0004\text{ W/(m}^2\cdot\text{°C)}}{0.0471\text{ m}^2} \\ &+ \frac{\ln(0.019/0.015)}{2\pi[15.1\text{ W/(m}\cdot\text{°C)}](1\text{ m})} \\ &+ \frac{0.0001\text{ W/(m}^2\cdot\text{°C)}}{0.0597\text{ m}^2} + \frac{1}{[1200\text{ W/(m}^2\cdot\text{°C)}](0.0597\text{ m}^2)} \\ &= (0.02654 + 0.00849 + 0.0025 + 0.00168 + 0.01396)\text{ W/°C} \\ &= 0.0532\text{°C/W} \end{aligned}$$

Note that about 19 percent of the total thermal resistance in this case is due to fouling, and about 5 percent of it is due to the steel tube separating the two fluids. The rest (76 percent) is due to the convection resistances on the two sides of the inner tube.

(*b*) Knowing the total thermal resistance and the heat transfer surface areas, the overall heat transfer coefficient based on the inner and outer surfaces of the tube are determined again from Eq. 13-8 to be

$$U_i = \frac{1}{RA_i} = \frac{1}{(0.0532\text{ W/°C})(0.0471\text{ m}^2)} = 399.1\text{ W/(M}^2\cdot\text{°C)}$$

and

$$U_o = \frac{1}{RA_o} = \frac{1}{(0.0532\text{ W/°C})(0.0597\text{ m}^2)} = 314.9\text{ W/(M}^2\cdot\text{°C)}$$

Note that the two overall heat transfer coefficients differ significantly (by 27 percent) in this case because of the considerable difference between the heat transfer surface areas on the inner and the outer sides of the tube. For tubes of negligible thickness, the difference between the two overall heat transfer coefficients would be negligible.

## 13-3 ■ ANALYSIS OF HEAT EXCHANGERS

Heat exchangers are commonly used in practice, and an engineer often finds himself or herself in a position to *select a heat exchanger* that will achieve a *specified temperature change* in a fluid stream of known mass flow rate, or to *predict the outlet temperatures* of the hot and cold fluid streams in a *specified heat exchanger*.

In the following sections, we will discuss the two methods used in the analysis of heat exchangers. Of these, the *log mean temperature difference* (or LMTD) method is best suited for the first task, and the *effectiveness-NTU* method for the second task stated above. But first we present some general considerations.

Heat exchangers ususally operate for long periods of time with no change in their operating conditions. Therefore, they can be modeled as *steady-flow* devices. As such, the mass flow rate of each fluid remains constant, and the fluid properties such as temperature and velocity at any inlet or outlet remain the same. Also, the fluid streams experience little or no change in their velocities and elevations, and thus the *kinetic* and *potential energy changes* are negligible. The *specific heat* of a fluid, in general, changes with temperature. But, in a specified temperature range, it can be treated as a constant at some average value with little loss in accuracy. *Axial heat conduction* along the tube is usually insignificant, and can be considered negligible. Finally, the outer surface of the heat exchanger is assumed to be *perfectly insulated,* so that there is no heat loss to the surrounding medium, and any heat transfer occurs between the two fluids only.

The idealizations stated above are closely approximated in practice, and they greatly simplify the analysis of a heat exchanger with little sacrifice of accuracy. Therefore, they are commonly used. Under these assumptions, the *first law of thermodynamics* requires that the rate of heat transfer from the hot fluid be equal to the rate of heat transfer to the cold one. That is,

$$\dot{Q} = \dot{m}_c C_{pc}(T_{c,\text{out}} - T_{c,\text{in}}) \tag{13-9}$$

and

$$\dot{Q} = \dot{m}_h C_{ph}(T_{h,\text{in}} - T_{h,\text{out}}) \tag{13-10}$$

where the subscripts $c$ and $h$ stand for *cold* and *hot* fluids, respectively, and

$\dot{m}_c, \dot{m}_h$ = mass flow rates

$C_{pc}, C_{ph}$ = specific heats

$T_{c,\text{out}}, T_{h,\text{out}}$ = outlet temperatures

$T_{c,\text{in}}, \text{T}_{\text{h,in}}$ = inlet temperatures

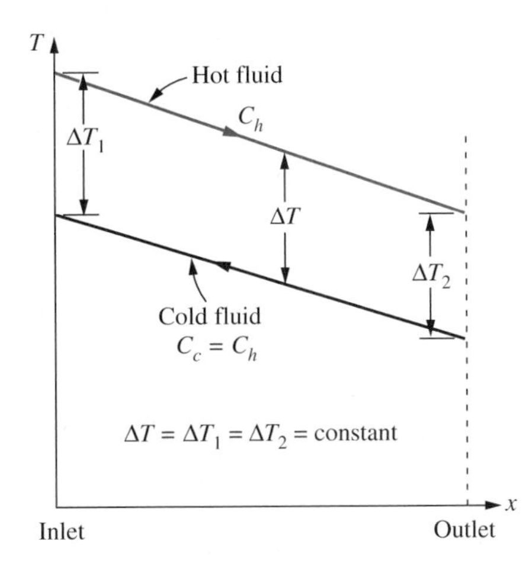


**FIGURE 13-12**

Two fluids that have the same mass flow rate and the same specific heat experience the same temperature change in a well-insulated heat exchanger.

Note that the heat transfer rate $\dot{Q}$ is taken to be a positive quantity, and its direction is understood to be from the hot fluid to the cold one in accordance with the second law of thermodynamics.

In heat exchanger analysis, it is often convenient to combine the product of the *mass flow rate* and the *specific heat* of a fluid into a single quantity. This quantity is called the **heat capacity rate**, and is defined as

$$C = \dot{m}C_p \tag{13-11}$$

The heat capacity rate of a fluid stream represents the rate of heat transfer needed to change the temperature of the fluid stream by 1°C as it flows through a heat exchanger. Note that in a heat exchanger, the fluid with a *large* heat capacity rate will experience a *small* temperature change, and the fluid with a *small* heat capacity rate will experience a *large* temperature change. Therefore, *doubling* the mass flow rate of a fluid while leaving everything else unchanged will *halve* the temperature change of that fluid.

With the definition of the heat capacity rate above, Eqs. 13-9 and 13-10 can also be expressed as

$$\dot{Q} = C_c(T_{c,\text{out}} - T_{c,\text{in}}) \tag{13-12}$$

and

$$\dot{Q} = C_h(T_{h,\text{in}} - T_{h,\text{out}}) \tag{13-13}$$

That is, the heat transfer rate in a heat exchanger is equal to the heat capacity rate of either fluid multiplied by temperature change of that fluid. Note that *the only time the temperature rise of a cold fluid is equal to the temperature drop of the hot fluid is when the heat capacity rates of the two fluids are equal to each other* (Fig. 13-12).

Two special types of heat exchangers commonly used in practice are *condensers* and *boilers.* One of the fluids in a condenser or a boiler undergoes a phase-change process, and the rate of heat transfer is expressed as

$$\dot{Q} = \dot{m}h_{fg} \tag{13-14}$$

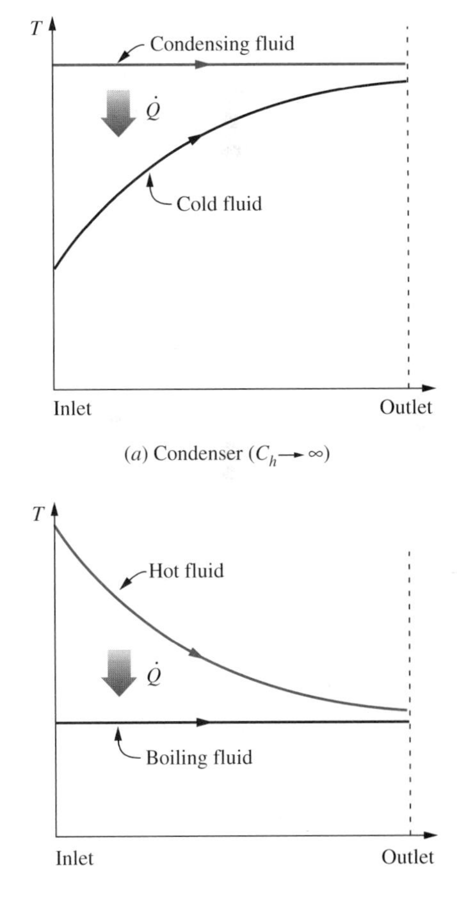


**FIGURE 13-13**

Variation of fluid temperatures in a heat exchanger when one of the fluids condenses or boils.

where $\dot{m}$ is the rate of evaporation or condensation of the fluid and $h_{fg}$ is the enthalpy of vaporization of the fluid at the specified temperature or pressure.

An ordinary fluid absorbs or releases a large amount of heat essentially at constant temperature during a phase-change process, as shown in Fig. 13-13. The heat capacity rate of a fluid during a phase-change process must approach infinity since the temperature change is practically zero. That is, $C = \dot{m}C_p \to \infty$ when $\Delta T \to 0$, so that the heat transfer rate $\dot{Q} = \dot{m}C_p\,\Delta T$ is a finite quantity. Therefore, in heat exchanger analysis, a condensing or boiling fluid is conveniently modeled as a fluid whose heat capacity rate is *infinity*.

The rate of heat transfer in a heat exchanger can also be expressed in an analogous manner to Newton's law of cooling as

$$\dot{Q} = UA\,\Delta T_m \tag{13-15}$$

where $U$ is the overall heat transfer coefficient, $A$ is the heat transfer area, and $\Delta T_m$ is an appropriate average temperature difference between the two fluids. Here the surface area $A$ can be determined precisely using the dimensions of the heat exchanger. However, the overall heat transfer coefficient $U$ and the temperature difference $\Delta T$ between the hot and cold fluids, in general, are not constant and vary along the heat exchanger.

The average value of the overall heat transfer coefficient can be determined as described in the preceding section by using the average convection coefficients for each fluid. It turns out that the appropriate form of the mean temperature difference between the two fluids is *logarithmic* in nature, and its determination is presented in the next section.

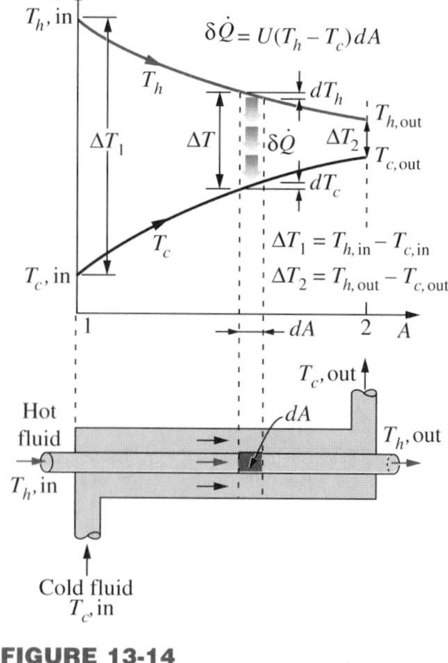


**FIGURE 13-14**

Variation of the fluid temperatures in a parallel-flow double-pipe heat exchanger.

## 13-4 THE LOG MEAN TEMPERATURE DIFFERENCE METHOD

Earlier, we mentioned that the temperature difference between the hot and cold fluids varies along the heat exchanger, and it is convenient to have a *mean temperature difference* $\Delta T_m$ for use in the relation

$$\dot{Q} = UA\,\Delta T_m$$

In order to develop a relation for the equivalent average temperature difference between the two fluids, consider the *parallel-flow double-pipe* heat exchanger shown in Fig. 13-14. Note that the temperature difference $\Delta T$ between the hot and cold fluids is large at the inlet of the heat exchanger, but decreases exponentially towards the outlet. As you would expect, the temperature of the hot fluid decreases and the temperature of the cold fluid increases along the heat exchanger, but the temperature of the cold fluid can never exceed that of the hot fluid, no matter how long the heat exchanger is.

Assuming the outer surface of the heat exchanger to be well insulated so that any heat transfer occurs between the two fluids, and disregarding any changes in kinetic and potential energy, an energy balance on each fluid in a differential section of the heat exchanger can be expressed as

$$\delta\dot{Q} = -\dot{m}_h C_{ph}\, dT_h \qquad (13\text{-}16)$$

and

$$\delta\dot{Q} = \dot{m}_c C_{pc}\, dT_c \qquad (13\text{-}17)$$

That is, the rate of heat loss from the hot fluid at any section of a heat exchanger is equal to the rate of heat gain by the cold fluid in that section. The temperature change of the hot fluid is a *negative* quantity, and so a *negative sign* is added to Eq. 13-16 to make the heat transfer rate $\dot{Q}$ a positive quantity. Solving the equations above for $dT_h$ and $dT_c$ gives

$$dT_h = -\frac{\delta\dot{Q}}{\dot{m}_h C_{ph}} \qquad (13\text{-}18)$$

and

$$dT_c = \frac{\delta\dot{Q}}{\dot{m}_c C_{pc}} \qquad (13\text{-}19)$$

Taking their difference, we get

$$dT_h - dT_c = d(T_h - T_c) = -\delta\dot{Q}\left(\frac{1}{\dot{m}_h C_{ph}} + \frac{1}{\dot{m}_c C_{pc}}\right) \qquad (13\text{-}20)$$

The rate of heat transfer in the differential section of the heat exchanger can also be expressed as

$$\delta\dot{Q} = U(T_h - T_c)\, dA \qquad (13\text{-}21)$$

Substituting this equation into Eq. 13-20 and rearranging gives

$$\frac{d(T_h - T_c)}{T_h - T_c} = -U\, dA\left(\frac{1}{\dot{m}_h C_{ph}} + \frac{1}{\dot{m}_c C_{pc}}\right) \qquad (13\text{-}22)$$

Integrating from the inlet of the heat exchanger to its outlet, we obtain

$$\ln\frac{T_{h,\text{out}} - T_{c,\text{out}}}{T_{h,\text{in}} - T_{c,\text{in}}} = -UA\left(\frac{1}{\dot{m}_h C_{ph}} + \frac{1}{\dot{m}_c C_{pc}}\right) \qquad (13\text{-}23)$$

Finally, solving Eqs. 13-9 and 13-10 for $\dot{m}_c C_{pc}$ and $\dot{m}_h C_{ph}$ and substituting into Eq. 13-23 gives, after some rearrangement,

$$\dot{Q} = UA\, \Delta T_{\text{lm}} \qquad (13\text{-}24)$$

where

$$\Delta T_{\text{lm}} = \frac{\Delta T_1 - \Delta T_2}{\ln(\Delta T_1/\Delta T_2)} \qquad (13\text{-}25)$$

is the **log mean temperature difference**, which is the suitable form of the average temperature difference for use in the analysis of heat exchangers. Here $\Delta T_1$ and $\Delta T_2$ represent the temperature difference between the two fluids at the two ends (inlet and outlet) of the heat exchanger. It makes no difference which end of the heat exchanger is designated as the *inlet* or the *outlet* (Fig. 13-15).

The temperature difference between the two fluids decreases from $\Delta T_1$ at the inlet to $\Delta T_2$ at the outlet. Thus, it is tempting to use the arithmetic mean temperature $\Delta T_{\text{am}} = \frac{1}{2}(\Delta T_1 + \Delta T_2)$ as the average temperature difference. The logarithmic mean temperature difference $\Delta T_{\text{lm}}$ is obtained by tracing the actual temperature profile of the fluids along the

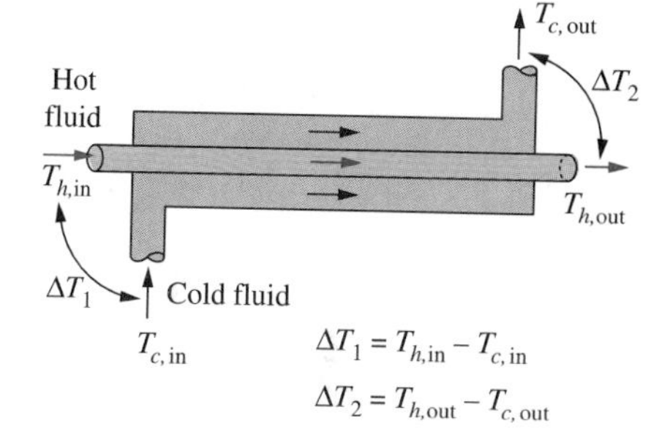


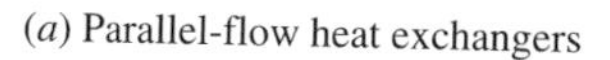

(*a*) Parallel-flow heat exchangers

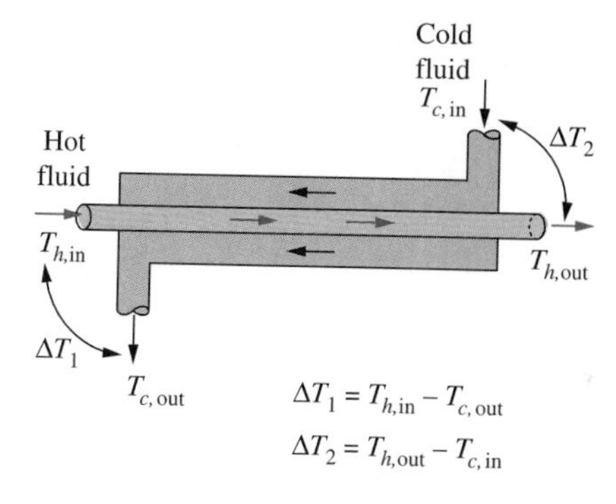


(*b*) Counter-flow heat exchangers

**FIGURE 13-15**

The $\Delta T_1$ and $\Delta T_2$ expressions in parallel-flow and counter-flow heat exchangers.

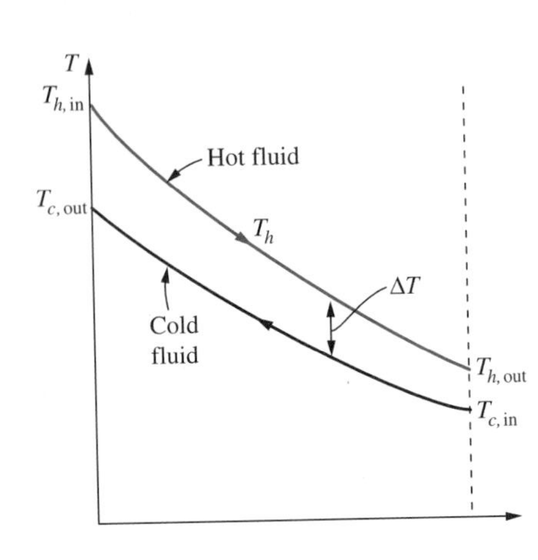


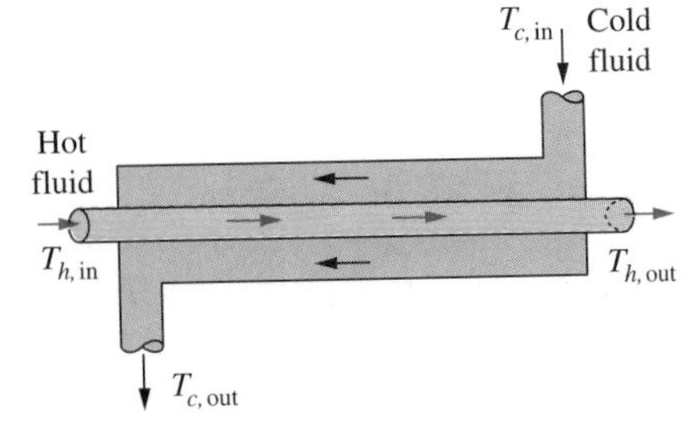


**FIGURE 13-16**

The variation of the fluid temperatures in a counter-flow double-pipe heat exchanger.

heat exchanger, and is an *exact* representation of the *average temperature difference* between the hot and cold fluids. It truly reflects the exponential decay of the local temperature difference.

Note that $\Delta T_{lm}$ is always less than $\Delta T_{am}$. Therefore, using $\Delta T_{am}$ in calculations instead of $\Delta T_{lm}$ will overestimate the rate of heat transfer in a heat exchanger between the two fluids. When $\Delta T_1$ differs from $\Delta T_2$ by no more than 40 percent, the error in using the arithmetic mean temperature difference is less than 1 percent. But the error increases to undesirable levels when $\Delta T_1$ differs from $\Delta T_2$ by greater amounts. Therefore, we should always use the *logarithmic mean temperature difference* when determining the rate of heat transfer in a heat exchanger.

### Counter-flow Heat Exchangers

The variation of temperatures of hot and cold fluids in a counter-flow heat exchanger is given in Fig. 13-16. Note that the hot and cold fluids enter the heat exchanger from opposite ends, and the outlet temperature of the *cold fluid* in this case may exceed the outlet temperature of the *hot fluid.* In the limiting case, the cold fluid will be heated to the inlet temperature of the hot fluid. However, the outlet temperature of the cold fluid can *never* exceed the inlet temperature of the hot fluid, since this would be a violation of the second law of thermodynamics.

The relation above for the log mean temperature difference is developed using a parallel-flow heat exchanger, but we can show by repeating the analysis above for a counter-flow heat exchanger that is also applicable to counter-flow heat exchangers. But this time, $\Delta T_1$ and $\Delta T_2$ are expressed as shown in Fig. 13-15.

For specified inlet and outlet temperatures, the log mean temperature difference for a *counter-flow* heat exchanger is always *greater* than that for a parallel-flow heat exchanger. That is, $\Delta T_{lm,CF} > \Delta T_{lm,PF}$, and thus a smaller parallel-flow heat exchanger. That is, $\Delta T_{lm,CF} > \Delta T_{lm,PF}$, and thus a smaller surface area (and thus a smaller heat exchanger) is needed to achieve a specified heat transfer rate in a counter-flow heat exchanger. Therefore, it is common practice to use counter-flow arrangements in heat exchangers.

In a counter-flow heat exchanger, the temperature difference between the hot and the cold fluids will remain constant along the heat exchanger when the *heat capacity rates* of the two fluids are *equal* (that is, $\Delta T$ = constant when $C_h = C_c$ or $\dot{m}_h C_{ph} = \dot{m}_c C_{pc}$). Then we have $\Delta T_1 = \Delta T_2$, and the log mean temperature difference relation above gives $\Delta T_{lm} = 0/0$, which is indeterminate. It can be shown by the application of l'Hôpital's rule that in this case we have $\Delta T_{lm} = \Delta T_1 = \Delta T_2$, as expected.

A *condenser* or a *boiler* can be considered to be either a parallel or counterflow heat exchanger since both approaches give the same result.

### Multipass and Cross-Flow Heat Exchangers: Use of a Correction Factor

The log mean temperature difference $\Delta T_{lm}$ relation developed earlier is limited to parallel-flow and counter-flow heat exchangers only. Similar relations are also developed for *cross-flow* and *multipass shell-and-tube* heat exchangers, but the resulting expressions are too complicated because of the complex flow conditions.

In such cases, it is convenient to relate the equivalent temperature difference to the log mean temperature diffrence relation for the counter-flow case as

$$\Delta T_{\text{lm}} = F\,\Delta T_{\text{lm},CF} \tag{13-26}$$

where $F$ is the **correction factor**, which depends on the *geometry* of the heat exchanger and the inlet and outlet temperatures of the hot and cold fluid streams. The $\Delta T_{\text{lm},CF}$ is the log mean temperature difference for the case of a *counter-flow* heat exchanger with the same inlet and outlet temperatures, and is determined from Eq. 13-25 by taking $\Delta T_1 = T_{h,\text{in}} - T_{c,\text{out}}$ and $\Delta T_2 = T_{h,\text{out}} - T_{c,\text{in}}$ (Fig. 13-17).

The correction factor is less than unity for a cross-flow and multipass shell-and-tube heat exchanger. That is, $F \leqslant 1$. The limiting value of $F = 1$ corresponds to the counter-flow heat exchanger. Thus the correction factor $F$ for a heat exchanger is *a measure of deviation of the* $\Delta T_{\text{lm}}$ *from the corresponding values for the counter-flow case.*

The correction factor $F$ for common cross-flow and shell-and-tube heat exchanger configurations are given in Fig. 13-18 versus two temperature ratios $P$ and $R$ defined as

$$P = \frac{t_2 - t_1}{T_1 - t_1} \tag{13-27}$$

and

$$R = \frac{T_1 - T_2}{t_2 - t_1} = \frac{(\dot{m}C_p)_{\text{tube side}}}{(\dot{m}C_p)_{\text{shell side}}} \tag{13-28}$$

where the subscripts 1 and 2 represent the *inlet* and *outlet,* respectively. Note that for a shell-and-tube heat exchanger, $T$ and $t$ represent the *shell-* and *tube-side* temperatures, respectively, as shown in the correction factor charts. It makes no difference whether the hot or the cold fluid flows through the shell or the tube. The determination of the correction factor $F$ requires the availability of the *inlet* and the *outlet* temperatures for both the cold and hot fluids.

Note that the value of $P$ ranges from 0 to 1. The value of $R$, on the other hand, ranges from 0 to infinity, with $R = 0$ corresponding to the phase-change (condensation or boiling) on the shell-side, and $R \to \infty$ to phase change on the tube side. The correction factor is $F = 1$ for both of these limiting cases. Therefore, the correction factor for a *condenser* or *boiler* is $F = 1$, regardless of the configuration of the heat exchanger.



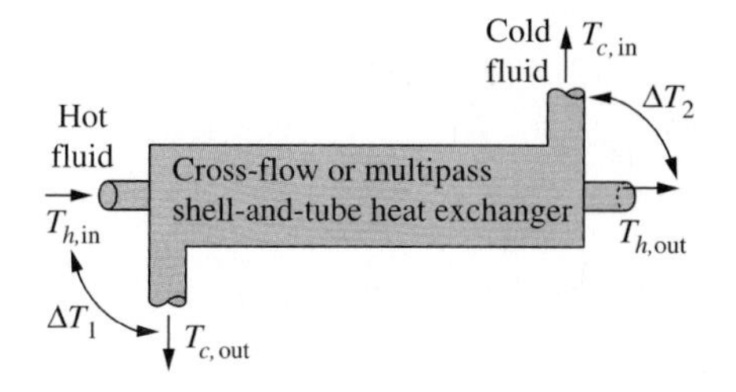


Heat transfer rate:

$$\dot{Q} = UAF\Delta T_{lm,CF}$$

where $\Delta T_{lm,CF} = \dfrac{\Delta T_1 - \Delta T_2}{\ln(\Delta T_1/\Delta T_2)}$

$\Delta T_1 = T_{h,\text{in}} - T_{c,\text{out}}$

$\Delta T_2 = T_{h,\text{out}} - T_{c,\text{in}}$

and $F = \ldots$ (Fig. 13–18)

**FIGURE 13-17**

The determination of the heat transfer rate for crossflow and multipass shell-and-tube heat exchangers using the correction factor.

**EXAMPLE 13-3**

Steam in the condenser of a steam power plant is to be condensed at a temperature of 30°C with cooling water from a nearby lake, which enters the tubes of the condenser at 14°C and leaves at 22°C. The surface area of the tubes is 45 m$^2$, and the overall heat transfer coefficient is 2100 W/(m$^2 \cdot$ °C). Determine the mass flow rate of the cooling water needed and the rate of condensation of the steam in the condenser.

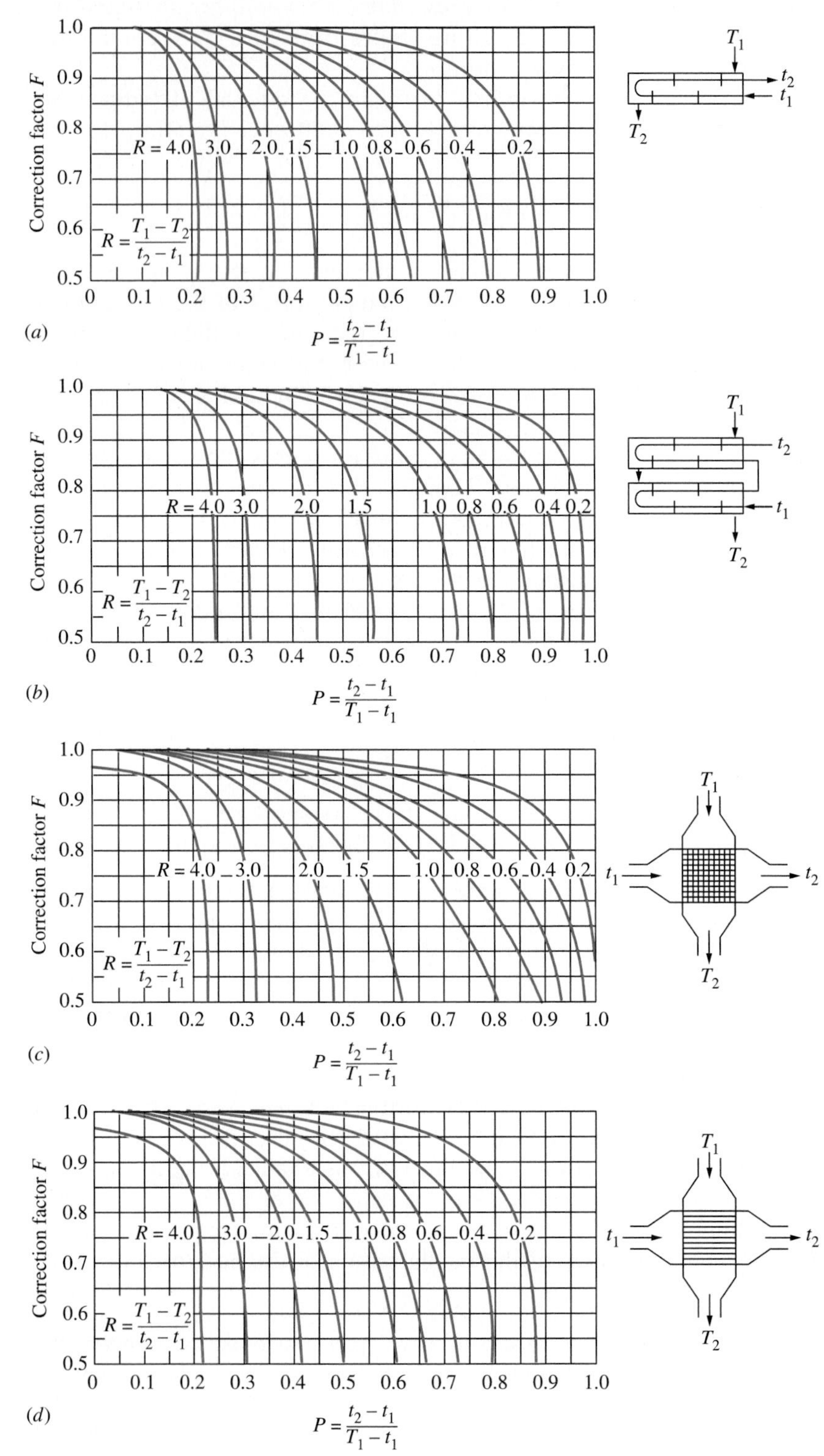


**FIGURE 13-18**

Correction factor $F$ charts for common shell-and-tube and cross-flow heat exchangers (from Bowman, Mueller, and Nagle).
($a$) One-shell pass and two, four, six, etc. (any multiple of two) tube passes.
($b$) Two-shell pass and four, eight, twelve, etc. (any multiple of four) tube passes.
($c$) Single pass crossflow with both fluids *unmixed.*
($d$) Single pass crossflow with one fluid *mixed* and the other *unmixed.*

**Solution** The schematic of the condenser is given in Fig. 13-19. We assume the condenser to be well-insulated, so that heat transfer from the steam is equal to the heat transfer to the cooling water. The condenser can be treated as a counter-flow heat exchanger since the temperature of one of the fluids (the steam) remains constant.

The temperature difference between the steam and the cooling water at the two ends of the condenser is

$$\Delta T_1 = T_{h,\text{in}} - T_{c,\text{out}} = (30 - 22)°\text{C} = 8°\text{C}$$
$$\Delta T_2 = T_{h,\text{out}} - T_{c,\text{in}} = (30 - 14)°\text{C} = 16°\text{C}$$

That is, the temperature difference between the two fluids varies from 8°C at one end to 16°C at the other. The proper average temperature difference between the two fluids is the *logarithmic mean temperature difference* (not the arithmetic), which is determined from

$$\Delta T_{\text{lm}} = \frac{\Delta T_1 - \Delta T_2}{\ln(\Delta T_1/\Delta T_2)} = \frac{8 - 16}{\ln(8/16)} = 11.5°\text{C}$$

This is a little less than the arithmetic mean temperature difference of $\frac{1}{2}(8 + 16) = 12°\text{C}$. Then the heat transfer rate in the condenser is determined from

$$\dot{Q} = UA\,\Delta T_{\text{lm}} = [2100\ \text{W/(m}^2 \cdot °\text{C)}](45\ \text{m}^2)(11.5°\text{C}) = 1{,}086{,}750\ \text{W}$$

Therefore, the steam will lose heat at a rate of 1,086.75 kW as it flows through the condenser, and the cooling water will gain practically all of it, since the condenser is well insulated.

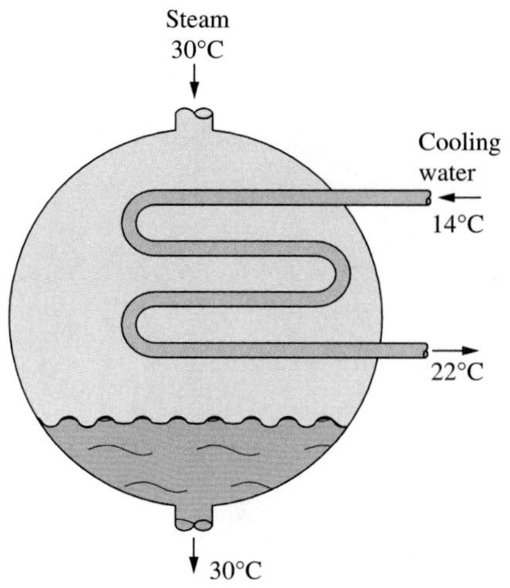


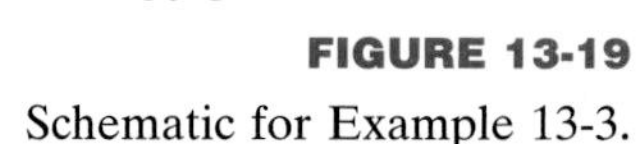
**FIGURE 13-19**
Schematic for Example 13-3.

The specific heat of water at the temperature range of the cooling water is 4.18 kJ/(kg · °C), and the heat of vaporization of water at 30°C is 2430.5 kJ/kg. Then the mass flow rate of the cooling water and the rate of the condensation of the steam are determined from $\dot{Q} = [\dot{m}C_p(T_{\text{out}} - T_{\text{in}})]_{\text{cooling water}} = (\dot{m}h_{fg})_{\text{steam}}$ to be

$$\dot{m}_{\text{cooling water}} = \frac{\dot{Q}}{C_p(T_{\text{out}} - T_{\text{in}})} = \frac{1{,}086.75\ \text{kJ/s}}{[4.18\ \text{kJ/(kg} \cdot °\text{C)}](22 - 14)°\text{C}} = 32.5\ \text{kg/s}$$

and

$$\dot{m}_{\text{steam}} = \frac{\dot{Q}}{h_{fg}} = \frac{1{,}086.75\ \text{kJ/s}}{2430.5\ \text{kJ/kg}} = 0.45\ \text{kg/s}$$

Therefore, we need to circulate about 72 kg of cooling water for each 1 kg of steam condensing to remove the heat released during the condensation process.

---

**EXAMPLE 13-4**

A counter-flow double-pipe heat exchanger is to heat water from 20°C to 80°C at a rate of 1.2 kg/s. The heating is to be accomplished by geoethermal water available at 160°C at a mass flow rate of 2 kg/s. The inner tube is thin-walled, and has a diameter of 1.5 cm. If the overall heat transfer coefficient of the heat exchanger is 640 W/(m² · °C), determine the length of the heat exchanger required to achieve the desired heating. Take the specific heat of the geothermal water to be 4.31 kg/(kg · °C) and that of water to be 4.18 kJ/(kg · °C).

**Solution** The schematic of the heat exchanger is given in Fig. 13-20. We assume the heat exchanger to be well insulated, so that heat transferred from the geothermal water is equal to the heat transferred to the water. Then the rate of heat transfer in the heat exchanger can be determined from

$$\dot{Q} = [\dot{m}C_p(T_{out} - T_{in})]_{water} = (1.2\ \text{kg/s})[4.18\ \text{kJ/(kg}\cdot{}^\circ\text{C)}](80 - 20)^\circ\text{C}$$
$$= 301.0\ \text{kW}$$

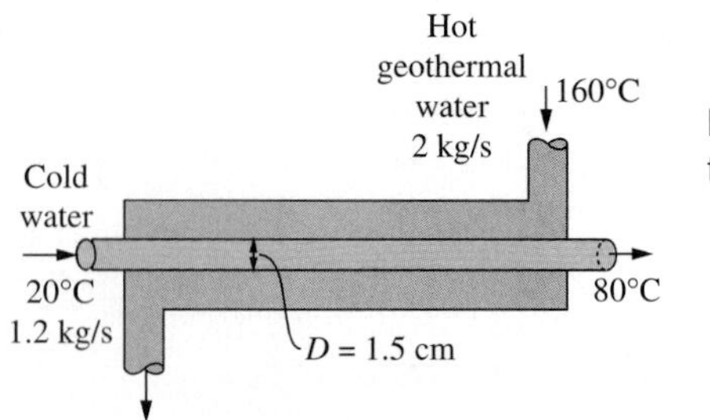


**FIGURE 13-20**
Schematic for Example 13-4.

Noting that all of this heat is supplied by the geothermal water, the outlet temperature of the geothermal water is determined to be

$$\dot{Q} = [\dot{m}C_p(T_{in} - T_{out})]_{geothermal} \longrightarrow T_{out} = T_{in} - \frac{\dot{Q}}{\dot{m}C_p}$$
$$= 160^\circ\text{C} - \frac{301.0\ \text{kW}}{(2\ \text{kg/s})[4.31\ \text{kJ/(kg}\cdot{}^\circ\text{C)}]}$$
$$= 125.1^\circ\text{C}$$

Knowing the inlet and outlet temperatures of both fluids, the logiarithmic mean temperature difference for this counter-flow heat exchanger becomes

$$\Delta T_1 = T_{h,in} - T_{c,out} = (160 - 80)^\circ\text{C} = 80^\circ\text{C}$$
$$\Delta T_2 = T_{h,out} - T_{c,in} = (125.1 - 20)^\circ\text{C} = 105.1^\circ\text{C}$$

and

$$\Delta T_{lm} = \frac{\Delta T_1 - \Delta T_2}{\ln(\Delta T_1/\Delta T_2)} = \frac{80 - 105.1}{\ln(90/105.1)} = 92.0^\circ\text{C}$$

Then the surface area of the heat exchanger is determined to be

$$\dot{Q} = UA\,\Delta T_{lm} \longrightarrow A = \frac{\dot{Q}}{U\,\Delta T_{lm}} = \frac{301{,}000\ \text{W}}{[640\ \text{W/(m}^2\cdot{}^\circ\text{C)}](92.0^\circ\text{C})} = 5.11\ \text{m}^2$$

To provide this much heat transfer surface area, the length of the tube must be

$$A = \pi DL \longrightarrow L = \frac{A}{\pi D} = \frac{5.11\ \text{m}^2}{\pi(0.015\ \text{m})} = 108.4\ \text{m}$$

Thus the inner tube of this counter-flow heat exchanger (and thus the heat exchanger itself) needs to be over 100 m long to achieve the desired heat transfer, which is impractical. In cases like this, we need to use a plate heat exchanger or a multipass shell-and-tube heat exchanger with multiple passes of tube bundles.

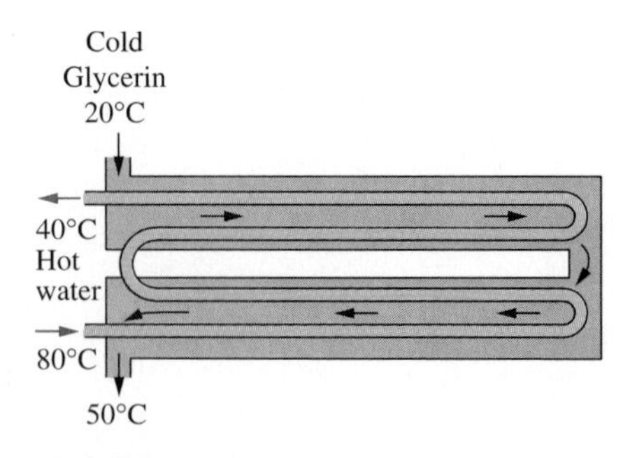


**FIGURE 13-21**
Schematic for Example 13-5.

**EXAMPLE 13-5**

A two-shell and four-tube passes heat exchanger is used to heat glycerin from 20°C to 50°C by hot water, which enters the thin-walled 2-cm-diameter tubes at 80°C and leaves at 40°C (Fig. 13-21). The total length of the tubes in the heat exchanger is 60 m. The convection heat transfer coefficient is 25 W/(m$^2\cdot$°C) on the glycerin (shell) side, and 160 W/(m$^2\cdot$°C) on the water (tube) side. Determine the rate of heat transfer in the heat exchanger (*a*) before any fouling occurs and (*b*) after fouling with a fouling factor of 0.0006 m$^2\cdot$°C/W occurs on the outer surfaces of the tubes.

**Solution** The tubes are said to be thin-walled, and thus it is reasonable to assume the inner surface area of the tubes to be equal to the outer surface area. Then the heat transfer surface area of this heat exchanger becomes

$$A = \pi DL = \pi(0.02\text{ m})(60\text{ m}) = 3.77\text{ m}^2$$

The rate of heat transfer in this heat exchanger can be determined from

$$\dot{Q} = UAF\,\Delta T_{lm,CF}$$

where $F$ is the correction factor and $\Delta T_{lm,CF}$ is the log mean temperature difference for the counterflow arrangement. These two quantities are determined from

$$\Delta T_1 = T_{h,\text{in}} - T_{c,\text{out}} = (80 - 50)°\text{C} = 30°\text{C}$$

$$\Delta T_2 = T_{h,\text{out}} - T_{c,\text{in}} = (40 - 20)°\text{C} = 20°\text{C}$$

$$\Delta T_{lm,CF} = \frac{\Delta T_1 - \Delta T_2}{\ln(\Delta T_1/\Delta T_2)} = \frac{30 - 20}{\ln(30/20)} = 24.7°\text{C}$$

and

$$\left.\begin{aligned} P &= \frac{t_2 - t_1}{T_1 - t_1} = \frac{40 - 80}{20 - 80} = 0.67 \\ R &= \frac{T_1 - T_2}{t_2 - t_1} = \frac{20 - 50}{40 - 80} = 0.75 \end{aligned}\right\} \quad F = 0.87 \quad \text{(Fig. 13-18}b\text{)}$$

($a$) In the case of no fouling, the overall heat transfer coefficient $U$ is determined from

$$U = \frac{1}{\dfrac{1}{h_i} + \dfrac{1}{h_o}} = \frac{1}{\dfrac{1}{160\text{ W/(m}^2\cdot°\text{C)}} + \dfrac{1}{25\text{ W/(m}^2\cdot°\text{C)}}} = 21.6\text{ W/(m}^2\cdot°\text{C)}$$

Then the rate of heat transfer becomes

$$\dot{Q} = UAF\,\Delta T_{lm,CF} = [21.6\text{ W/(m}^2\cdot°\text{C)}](3.77\text{ m}^2)(0.87)(24.6°\text{C}) = 1743\text{ W}$$

($b$) When there is fouling on one of the surfaces, the overall heat transfer coefficient $U$ is determined from

$$U = \frac{1}{\dfrac{1}{h_i} + \dfrac{1}{h_o} + R_f} = \frac{1}{\dfrac{1}{160\text{ W/(m}^2\cdot°\text{C)}} + \dfrac{1}{25\text{ W/(m}^2\cdot°\text{C)}} + 0.0006\text{ m}^2\cdot°\text{C/W}}$$
$$= 21.3\text{ W/(m}^2\cdot°\text{C)}$$

The rate of heat transfer in this case becomes

$$\dot{Q} = UAF\,\Delta T_{lm,CF} = [21.3\text{ W/(m}^2\cdot°\text{C)}](3.77\text{m}^2)(0.87)(24.6°\text{C}) = 1719\text{ W}$$

Note that the rate of heat transfer decreases as a result of fouling, as expected. The decrease is not dramatic, however, because of the relatively low convection heat transfer coefficients involved.

---

**EXAMPLE 13-6**

A test is conducted to determine the overall heat transfer coefficient in an automotive radiator that is a compact cross-flow water-to-air heat exchanger with both fluids (air and water) unmixed (Fig. 13-22). The radiator has 40 tubes of internal diameter 0.5 cm and length 65 cm in a closely spaced plate-finned matrix. Hot water enters the tubes at 90°C at a rate of 0.6 kg/s and leaves at

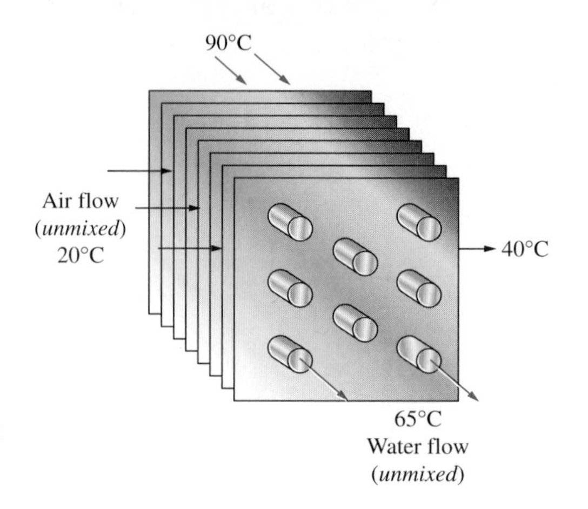


**FIGURE 13-22**
Schematic for Example 13-6.

65°C. Air flows across the radiator through the interfin spaces, and is heated from 20°C to 40°C. Determine the overall heat transfer coefficient $U_i$ of this radiator based on the inner surface area of the tubes.

**Solution** The rate of heat transfer in this radiator from the hot water to the air is determined from an energy balance on water flow. Taking the specific heat of water to be 4.18 kJ/(kg · °C), we have

$$\dot{Q} = [\dot{m}C_p(T_{\text{in}} - T_{\text{out}})]_{\text{water}} = (0.6\text{ kg/s})[4.18\text{ kJ/(kg}\cdot{}^\circ\text{C)}](90 - 65)^\circ\text{C} = 62.5\text{ kW}$$

The tube side heat transfer area is the total surface areas of the tubes, and is determined from

$$A_i = n\pi D_i L = (40)\pi(0.005\text{ m})(0.65\text{ m}) = 0.408\text{ m}^2$$

Knowing the rate of heat transfer and the surface area, the overall heat transfer coefficient in this heat exchanger can be determined from

$$\dot{Q} = U_i A_i F\,\Delta T_{\text{lm},CF} \longrightarrow U_i = \frac{\dot{Q}}{A_i F\,\Delta T_{\text{lm},CF}}$$

where $F$ is the correction factor and $\Delta T_{\text{lm},CF}$ is the log mean temperature difference for the counterflow arrangement. These two quantities are found to be

$$\Delta T_1 = T_{h,\text{in}} - T_{c,\text{out}} = (90 - 40)^\circ\text{C} = 50^\circ\text{C}$$
$$\Delta T_2 = T_{h,\text{out}} - T_{c,\text{in}} = (65 - 20)^\circ\text{C} = 45^\circ\text{C}$$
$$\Delta T_{\text{lm},CF} = \frac{\Delta T_1 - \Delta T_2}{\ln(\Delta T_1/\Delta T_2)} = \frac{50 - 45}{\ln(50/45)} = 47.6^\circ\text{C}$$

and

$$\left.\begin{aligned} P &= \frac{t_2 - t_1}{T_1 - t_1} = \frac{65 - 90}{20 - 90} = 0.36 \\ R &= \frac{T_1 - T_2}{t_2 - t_1} = \frac{20 - 40}{65 - 90} = 0.80 \end{aligned}\right\} \quad F = 0.97 \qquad \text{(Fig 13-18}c\text{)}$$

Substituting, the overall heat transfer coefficient $U_i$ is determined to be

$$U_i = \frac{\dot{Q}}{A_i F\,\Delta T_{\text{lm},CF}} = \frac{62{,}700\text{ W}}{(0.408\text{ m}^2)(0.97)(47.6^\circ\text{C})} = 3328\text{ W/(m}^2\cdot{}^\circ\text{C)}$$

Note that the overall heat transfer coefficient on the air side would be much lower because of the large surface area involved on that side.

The log mean temperature difference (LMTD) method discussed in the previous section is easy to use in heat exchanger analysis when the inlet and the outlet temperatures of the hot and cold fluiuds are known or can be determined from an energy balance. Once $\Delta T_{lm}$, the mass flow rates, and the overall heat transfer coefficient are available, the heat transfer surface area of the heat exchanger can be determined from

$$\dot{Q} = UA\,\Delta T_{lm}$$

Therefore, the LMTD method is very suitable for determining the *size* of a heat exchanger to realize prescribed outlet temperatures when the mass flow rates and the inlet and outlet temperatures of the hot and cold fluids are specified.

With the LMTD method the task is to *select* a heat exchanger that will meet the prescribed heat transfer requirements. The procedure to be followed in the selection process is as follows:

1 select the type of heat exchanger suitable for the application;

2 determine any unknown inlet or outlet temperature and the heat transfer rate using an energy balance;

3 calculate the log mean temperature difference $\Delta T_{lm}$ and the correction factor $F$, if necessary;

4 obtain (select or calculate) the value of the overall heat transfer coefficient $U$; and

5 calculate the heat transfer surface area $A$.

The task is completed by selecting a heat exchanger that has a heat transfer surface area equal to or larger than $A$.

A second kind of problem encountered in heat exchanger analysis is the determination of the *heat transfer rate* and the *outlet temperatures* of the hot and cold fluids for prescribed fluid mass flow rates and inlet temperatures when the *type* and *size* of the heat exchanger are specified. The heat transfer surface area $A$ of the heat exchanger in this case is known, but the *outlet temperatures* are not. Here the task is to determine the heat transfer performance of a specified heat exchanger or to determine if a heat exchanger available in storage will do the job.

The LMTD method could still be used for this alternative problem, but the procedure would require tedious iterations, and thus it is not practical. In an attempt to eliminate the iterations from the solution of such problems, Kays and London came up with a new method in 1955 called the **effectiveness-NTU method**, which greatly simplified heat exchanger analysis.

This new method is based on a dimensionless parameter called the **heat transfer effectiveness** $\varepsilon$ defined as

$$\varepsilon = \frac{\dot{Q}}{\dot{Q}_{max}} = \frac{\text{actual heat transfer rate}}{\text{maximum possible heat transfer rate}} \tag{13-29}$$

The *actual* heat transfer rate in a heat exchanger can be determined from an energy balance on the hot or cold fluids, and can be expressed as

$$\dot{Q} = C_c(T_{c,\text{out}} - T_{c,\text{in}}) = C_h(T_{h,\text{in}} - T_{h,\text{out}}) \tag{13-30}$$

where $C_c = \dot{m}_c C_{pc}$ and $C_h = \dot{m}_c C_{ph}$ are the heat capacity rates of the cold and the hot fluids, respectively.

To determine the maximum possible heat transfer rate in a heat exchanger, we first recognize that the *maximum temperature difference* in a heat exchanger is the difference between the *inlet* temperatures of the hot and cold fluids. That is,

$$\Delta T_{\text{max}} = T_{h,\text{in}} - T_{c,\text{in}} \tag{13-31}$$

The heat transfer in a heat exchanger will reach its maximum value when (1) the cold fluid is heated to the inlet temperature of the hot fluid, or (2) the hot fluid is cooled to the inlet temperature of the cold fluid. These two limiting conditions will not be reached simultaneously unless the heat capacity rates of the hot and cold fluids are identical (i.e., $C_c = C_h$). When $C_c \neq C_h$, which is usually the case, the fluid with the *smaller* heat capacity rate will experience a larger temperature change, and thus it will be the first to experience the maximum temperature at which point the heat transfer will come to a halt. Therefore, the maximum possible heat transfer rate in a heat exchanger is (Fig. 13-23)

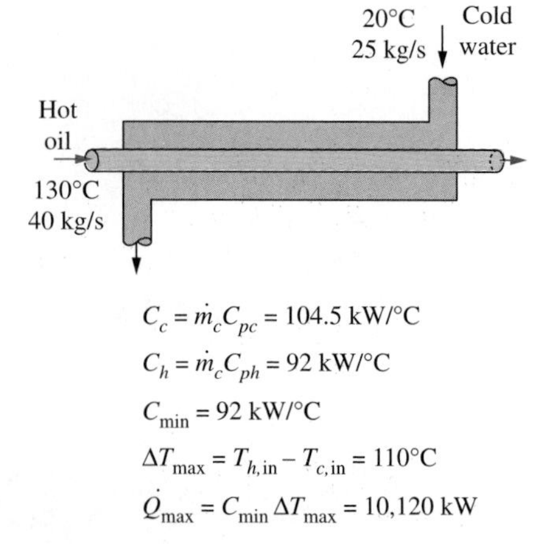


**FIGURE 13-23**
The determination of the maximum rate of heat transfer in a heat exchanger.

$$\dot{Q}_{\text{max}} = C_{\text{min}}(T_{h,\text{in}} - T_{c,\text{in}}) \tag{13-32}$$

where $C_{\text{min}}$ is the smaller of $C_h = \dot{m}_h C_{ph}$ and $C_c = \dot{m}_c C_{pc}$. This is further clarified by the following example.

**EXAMPLE 13-7**

Cold water enters a counterflow heat exchanger at 10°C at a rage of 8 kg/s, where it is heated by a hot water stream that enters the heat exchanger at 70°C at a rate of 2 kg/s. Assuming the specific heat of water to remain constant at $C_p = 4.18$ kJ/(kg · °C), determine the maximum heat transfer rate and the outlet temperatures of the cold and the hot water streams for this limiting case.

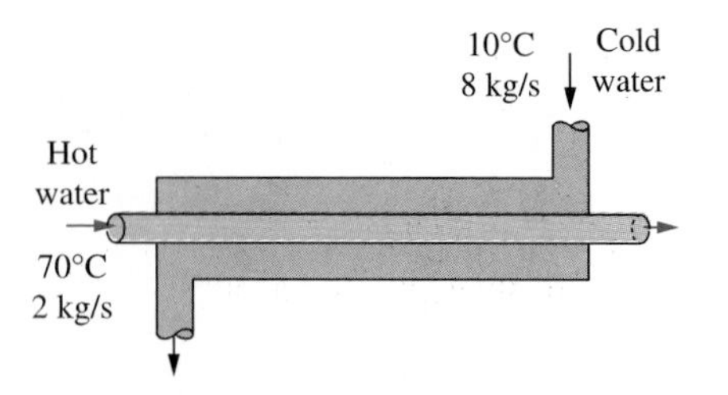


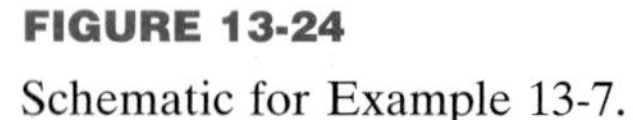

**FIGURE 13-24**
Schematic for Example 13-7.

**Solution** A schematic of the heat exchanger is given in Fig. 13-24. The heat capacity rates of the hot and cold fluids are determined from

$$C_h = \dot{m}_h C_{ph} = (2 \text{ kg/s})[4.18 \text{ kJ/(kg} \cdot {}^\circ\text{C)}] = 8.36 \text{ kW/}^\circ\text{C}$$

and

$$C_c = \dot{m}_c C_{pc} = (8 \text{ kg/s})[4.18 \text{ kJ/(kg} \cdot {}^\circ\text{C)}] = 33.44 \text{ kW/}^\circ\text{C}$$

Therefore

$$C_{\text{min}} = C_h = 8.36 \text{ kW/}^\circ\text{C}$$

which is the smaller of the two heat capacity rates. Then the maximum heat transfer rate is determined from Eq. 13-32 to be

$$\begin{aligned} \dot{Q}_{max} &= C_{\text{min}}(T_{h,\text{in}} - T_{c,\text{in}}) \\ &= (8.36 \text{ kW/}^\circ\text{C})(70 - 10)^\circ\text{C} \\ &= 501.6 \text{ kW} \end{aligned}$$

That is, the maximum possible heat transfer rate in this heat exchanger is 501.6 kW. This value would be approached in a counterflow heat exchanger with a *very large* heat transfer surface area.

The maximum temperature difference in this heat exchanger is $\Delta T_{\max} = T_{h,\text{in}} - T_{c,\text{in}} = (70 - 10)°\text{C} = 60°\text{C}$. Therefore, the hot water cannot be cooled by more than 60°C (to 10°C) in this heat exchanger, and the cold water cannot be heated by more than 60°C (to 70°C), no matter what we do. The outlet temperatures of the cold and the hot streams in this limiting case are determined to be

$$\dot{Q} = C_c(T_{c,\text{out}} - T_{c,\text{in}}) \longrightarrow T_{c,\text{out}} = T_{c,\text{in}} + \frac{\dot{Q}}{C_c} = 10°\text{C} + \frac{501.6\text{ kW}}{33.44\text{ kW/°C}} = 25°\text{C}$$

$$\dot{Q} = C_h(T_{h,\text{in}} - T_{h,\text{out}}) \longrightarrow T_{h,\text{out}} = T_{h,\text{in}} - \frac{\dot{Q}}{C_h} = 70°\text{C} - \frac{501.6\text{ kW}}{8.38\text{ kW/°C}} = 10°\text{C}$$

Note that the hot water is cooled to the limit of 10°C (the inlet temperature of the cold water stream), but the cold water is heated to 25°C only when maximum heat transfer occurs in the heat exchanger. This is not surprising, since the mass flow rate of the hot water is only one-fourth that of the cold water, and, as a result, the temperature of the cold water increases by 0.25°C for each 1°C drop in the temperature of the hot water.

You may be tempted to think that the cold water should be heated to 70°C in the limiting case of maximum heat transfer. But this will require the temperature of the hot water to drop to −170°C (below 10°C), which is impossible. Therefore, heat transfer in a heat exchanger reaches its maximum value when the fluid with the smaller heat capacity rate (or the smaller mass flow rate when both fluids have the same specific heat value) experiences the maximum temperature change. This example explains why we use $C_{\min}$ in the evaluation of $\dot{Q}_{\max}$ instead of $C_{\max}$.

We can show that the hot water will leave at the inlet temperature of the cold water and vice versa in the limiting case of maximum heat transfer when the mass flow rates of the hot and cold water streams are identical (Fig. 13-25). We can also show that the outlet temperature of the cold water will reach the 70°C limit when the mass flow rate of the hot water is greater than that of the cold water.

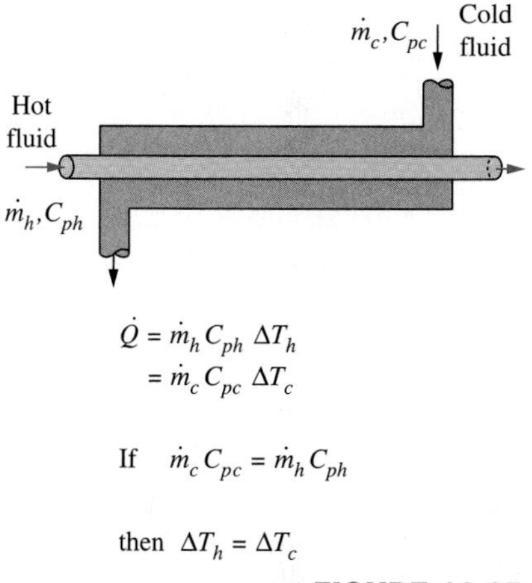


**FIGURE 13-25**
The temperature rise of the cold fluid in a heat exchanger will be equal to the temperature drop of the hot fluid when the mass flow rates and the specific heats of the hot and cold fluids are identical.

The determination of $\dot{Q}_{\max}$ requires the availability of the *inlet temperature* of the hot and cold fluids and their *mass flow rates,* which are usually specified. Then, once the effectiveness of the heat exchanger is known, the actual heat transfer rate $\dot{Q}$ can be determined from

$$\dot{Q} = \varepsilon \dot{Q}_{\max} = \varepsilon C_{\min}(T_{h,\text{in}} - T_{c,\text{in}}) \tag{13-33}$$

Therefore, the effectiveness of a heat exchanger enables us to determine the heat transfer rate without knowing the *outlet temperatures* of the fluids.

The effectiveness of a heat exchanger depends on the *geometry* of the heat exchanger as well as the *flow arrangement.* Therefore, different types of heat exchangers have different effectiveness relations. Below we illustrate the development of the effectiveness $\varepsilon$ relation for the double-pipe *parallel-flow* heat exchanger.

Equation 13-23 developed in the previous section for a parallel-flow heat exchanger can be rearranged as

$$\ln \frac{T_{h,\text{out}} - T_{c,\text{out}}}{T_{h,\text{in}} - T_{c,\text{in}}} = -\frac{UA}{C_c}\left(1 + \frac{C_c}{C_h}\right) \tag{13-34}$$

Also, solving Eq. 13-30 for $T_{h,\text{out}}$ gives

$$T_{h,\text{out}} = \mathrm{T}_{\mathrm{h,in}} - \frac{C_c}{C_h}(T_{c,\text{out}} - T_{c,\text{in}}) \tag{13-35}$$

Substituting this relation into Eq. 13-34 after adding and subtracting $T_{c,in}$ gives

$$\ln \frac{T_{h,\text{in}} - T_{c,\text{in}} + T_{c,\text{in}} - T_{c,\text{out}} - \dfrac{C_c}{C_h}(T_{c,\text{out}} - T_{c,\text{in}})}{T_{h,\text{in}} - T_{c,\text{in}}} = -\frac{UA}{C_c}\left(1 + \frac{C_c}{C_h}\right)$$

which simplifies to

$$\ln\left[1 - \left(1 + \frac{C_c}{C_h}\right)\frac{T_{c,\text{out}} - T_{c,\text{in}}}{T_{h,\text{in}} - T_{c,\text{in}}}\right] = -\frac{UA}{C_c}\left(1 + \frac{C_c}{C_h}\right) \tag{13-36}$$

We now manipulate the definition of effectiveness to obtain

$$\varepsilon = \frac{\dot{Q}}{\dot{Q}_{\max}} = \frac{C_c(T_{c,\text{out}} - T_{c,\text{in}})}{C_{\min}(T_{h,\text{in}} - T_{c,\text{in}})} \longrightarrow \frac{T_{c,\text{out}} - T_{c,\text{in}}}{T_{h,\text{in}} - T_{c,\text{in}}} = \varepsilon\frac{C_{\min}}{C_c}$$

Substituting this result into Eq. 13-36 and solving for $\varepsilon$ gives the following relation for the effectiveness of a *parallel-flow* heat exchanger:

$$\varepsilon_{\text{parallel flow}} = \frac{1 - \exp\left[-\dfrac{UA}{C_c}\left(1 + \dfrac{C_c}{C_h}\right)\right]}{\left(1 + \dfrac{C_c}{C_h}\right)\dfrac{C_{\min}}{C_c}} \tag{13-37}$$

Taking either $C_c$ or $C_h$ to be $C_{\min}$ (both approaches give the same result), the relation above can be expressed more conveniently as

$$\varepsilon_{\text{parallel flow}} = \frac{1 - \exp\left[-\dfrac{UA}{C_{\min}}\left(1 + \dfrac{C_{\min}}{C_{\max}}\right)\right]}{1 + \dfrac{C_{\min}}{C_{\max}}} \tag{13-38}$$

Again $C_{\min}$ is the *smaller* heat capacity ratio and $C_{\max}$ is the larger one, and it makes no difference whether $C_{\min}$ belongs to the hot or cold fluid.

Effectiveness relations of the heat exchangers typically involve the *dimensionless* group $UA/C_{\min}$. This quantity is called the **number of transfer units NTU**, and is expressed as

$$\text{NTU} = \frac{UA}{C_{\min}} = \frac{UA}{(\dot{m}C_p)_{\min}} \tag{13-39}$$

where $U$ is the overall heat transfer coefficient and $A$ is the heat transfer surface area of the heat exchanger. Note that NTU is proportional to $A$. Therefore, for specified values of $U$ and $C_{min}$, the value of NTU *is a measure of the heat transfer surface area A*. Thus, the larger the NTU, the larger is the heat exchanger.

In heat exchanger analysis, it is also convenient to define another dimensionless quantity called **capacity ratio** $C$ as

$$C = \frac{C_{min}}{C_{max}} \tag{13-40}$$

It can be shown that the effectiveness of a heat exchanger is a function of the number of transfer units NTU and the capacity ratio $C$. That is,

$$\varepsilon = \text{function } (UA/C_{min}, C_{min}/C_{max}) = \text{function (NTU, } C)$$

Effectiveness relations have been developed for a large number of heat exchangers, and the results are given in Table 13-4. The effectivenesses of some common types of heat exchangers are also plotted in Fig. 13-26. More extensive effectiveness charts and relations are available in the literature. The dashed lines in Fig. 13-26*f* are for the case of $C_{min}$ unmixed and $C_{max}$ mixed, and the solid lines for the opposite case. The analytic relations for the effectiveness give more accurate results than the charts, since reading errors in charts are unavoidable, and the relations are very suitable for computerized design analysis of heat exchangers

**TABLE 13-4**

**Effectiveness relations for heat exchangers:** NTU $= UA/C_{min}$ **and** $C = C_{min}/C_{max} = (\dot{m}C_p)_{min}/(\dot{m}C_p)_{max}$

| Heat exchanger type | Effectiveness relation |
|---|---|
| 1 *Double pipe*: | |
| Parallel-flow | $\varepsilon = \dfrac{1 - \exp[-\text{NTU}(1 + C)]}{1 + C}$ |
| Counter-flow | $\varepsilon = \dfrac{1 - \exp[-\text{NTU}(1 - C)]}{1 - C\exp[-\text{NTU}(1 - C)]}$ |
| 2 *Shell and tube*: One-shell pass 2, 4, . . . tube passes | $\varepsilon = 2\left\{1 + C + \sqrt{1 + C^2}\dfrac{1 + \exp[-\text{NTU}\sqrt{1 + C^2}]}{1 - C\exp[-\text{NTU}\sqrt{1 + C^2}]}\right\}^{-1}$ |
| 3 *Cross-flow* (*single-pass*) | |
| Both fluids unmixed | $\varepsilon = 1 - \exp\left\{\dfrac{\text{NTU}^{0.22}}{C}[\exp(-C\,\text{NTU}^{0.78}) - 1]\right\}$ |
| $C_{max}$ mixed, $C_{min}$ unmixed | $\varepsilon = \dfrac{1}{C}(1 - \exp\{-C[1 - \exp(-\text{NTU})]\})$ |
| $C_{min}$ mixed, $C_{max}$ unmixed | $\varepsilon = 1 - \exp\left\{-\dfrac{1}{C}[1 - \exp(-C\,\text{NTU})]\right\}$ |
| 4 *All heat exchangers with* $C = 0$ | $\varepsilon = 1 - \exp(-\text{NTU})$ |

*Source*: Ref. 7.

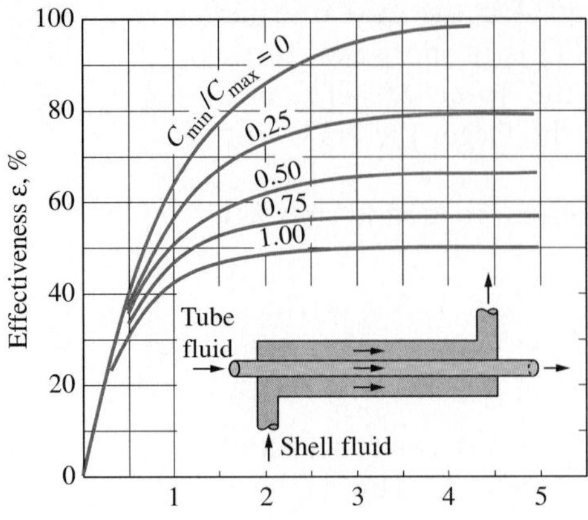


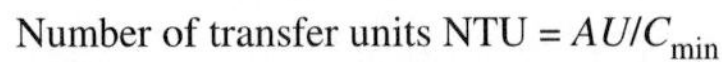
Number of transfer units NTU = $AU/C_{min}$

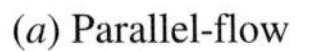
(*a*) Parallel-flow

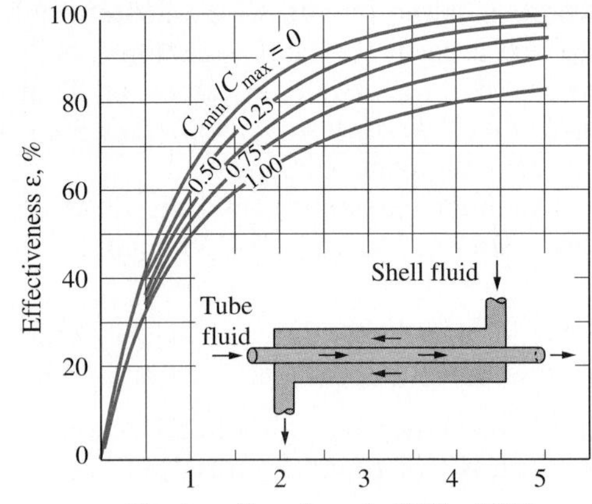


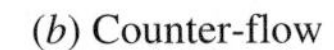
(*b*) Counter-flow

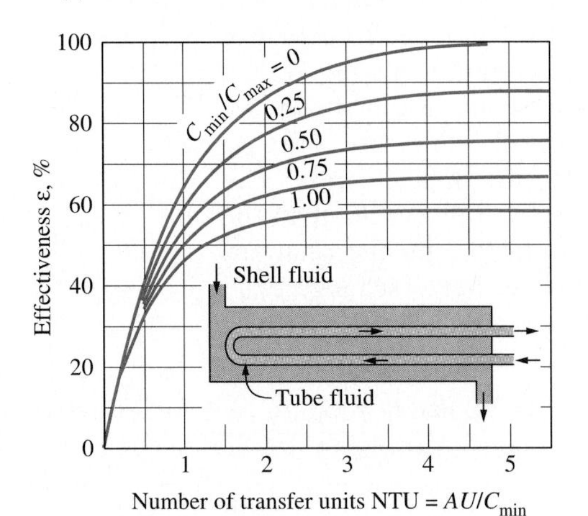


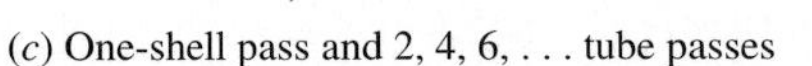
(*c*) One-shell pass and 2, 4, 6, . . . tube passes

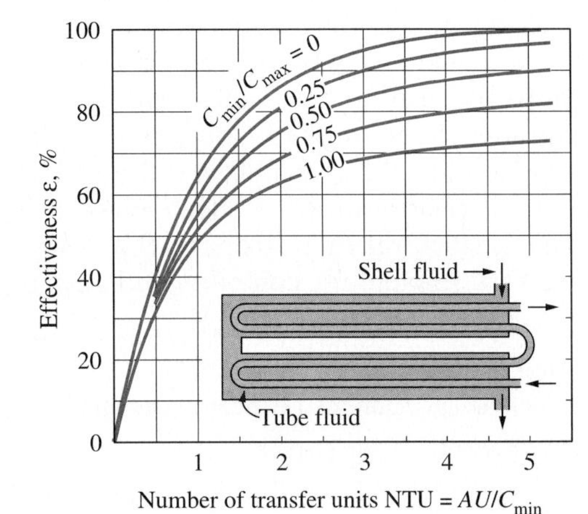


(*d*) Two-shell pass and 4, 8, 12, ... tube passes

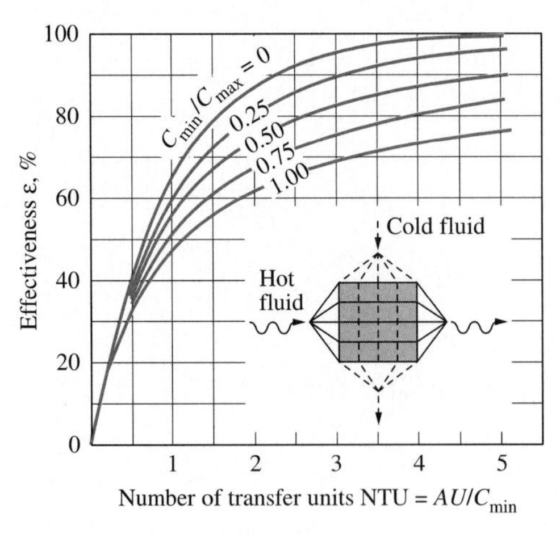


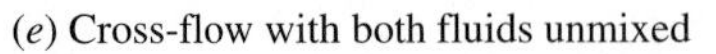
(*e*) Cross-flow with both fluids unmixed

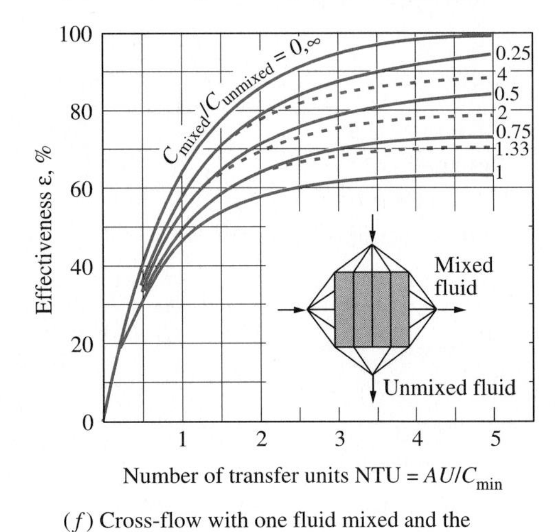


(*f*) Cross-flow with one fluid mixed and the other unmixed

**FIGURE 13-26**

Effectiveness for heat exchangers (from Ref. 7).

We make the following observations from the effectiveness relations and charts given above:

1 The value of the effectiveness ranges from 0 to 1. It increases rapidly with NTU for small values (up to about NTU = 1.5), but rather slowly for larger values. Therefore, the use of a heat exchanger with a large NTU (usually larger than 3) and thus a large size cannot be justified economically, since a large increase in NTU in this case corresponds to a small increase in effectiveness. Thus, a heat exchanger with a very high effectiveness may be highly desirable from a heat transfer point of view, but rather undesirable from an economical point of view

2 For a given NTU and capacity ratio $C = C_{min}/C_{max}$, the *counter-flow* heat exchanger has the *highest* effectiveness, followed closely by the cross-flow heat exchangers with both fluids unmixed. As you might expect, the lowest effectiveness values are encountered in parallelflow heat exchangers (Fig. 13-27).

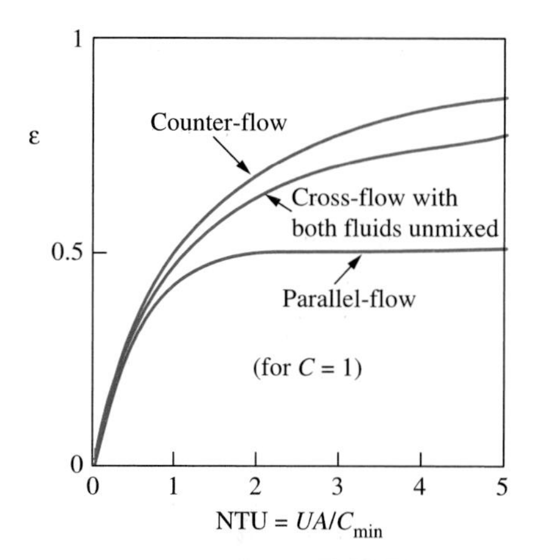


**FIGURE 13-27**

For a specified NTU and capacity ratio $C$, the counter-flow heat exchanger has the highest effectiveness, and the parallel-flow the lowest.

3 The effectiveness of a heat exchanger is independent of the capacity ratio $C$ for NTU values of less than about 0.3.

4 The value of the capacity ratio $C$ ranges between 0 and 1. For a given NTU, the effectiveness becomes a *maximum* for $C = 0$, and a *minimum* for $C = 1$. The case $C = C_{min}/C_{max} \to 0$ corresponds to $C_{max} \to \infty$, which is realized during a phase-change process in a *condenser* or *boiler*. All effectiveness relations in this case reduce to

$$\varepsilon = \varepsilon_{max} = 1 - \exp(-\text{NTU}) \tag{13-41}$$

regardless of the type of heat exchanger (Fig. 13-28). Note that the temperature of the condensing or boiling fluid remains constant in this case. The effectiveness is the *lowest* in the other limiting case of $C = C_{min}/C_{max} = 1$, which is realized when the heat capacity rates of the two fluids are equal.

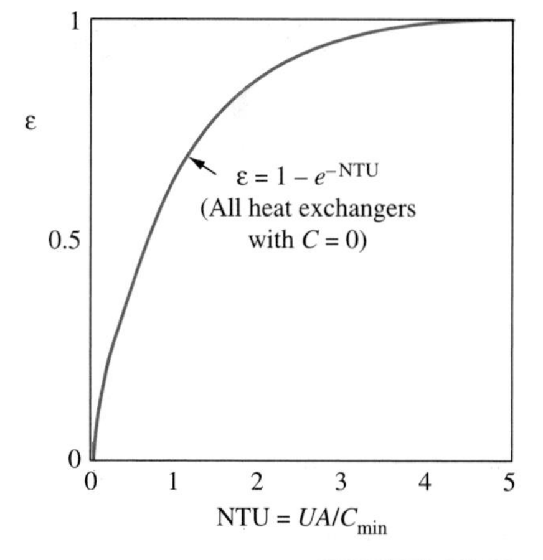


**FIGURE 13-28**

The effectiveness relation reduces to $\varepsilon = \varepsilon_{max} = 1 - \exp(-\text{NTU})$ for all heat exchangers when the capacity ratio $C = 0$.

Once the quantities $C = C_{min}/C_{max}$ and $\text{NTU} = UA/C_{min}$ have been evaluated, the effectiveness $\varepsilon$ can be determined from either the charts or (preferably) from the effectiveness relation for the specified type of heat exchanger. Then the rate of heat transfer $\dot{Q}$ and the outlet temperatures $T_{h,out}$ and $T_{c,out}$ can be determined from Eqs. 13-33 and 13-30, respectively. Note that the analysis of heat exchangers with unknown outlet temperatures is a straightforward matter with the effectiveness-NTU method, but requires rather tedious iterations with the LMTD method.

We mentioned earlier that when all the inlet and outlet temperatures are specified, the *size* of the heat exchanger can easily be determined using the LMTD method. Alternatively, it can also be determined from the effectiveness-NTU method by first evaluating the effectiveness $\varepsilon$

**TABLE 13-5**
**NTU relations for heat exchangers** NTU $= UA/C_{\min}$ **and** $C = C_{\min}/C_{\max} = (\dot{m}C_p)_{\min}/(\dot{m}C_p)_{\max}$

| Heat exchanger type | NTU relation |
|---|---|
| 1 *Double-pipe*: | |
| Parallel-flow | $\text{NTU} = -\dfrac{\ln[1-\varepsilon(1+C)]}{1+C}$ |
| Counter-flow | $\text{NTU} = \dfrac{1}{C-1}\ln\left(\dfrac{\varepsilon-1}{\varepsilon C-1}\right)$ |
| 2 *Shell and tube*: One-shell pass 2,4, . . . tube passes | $\text{NTU} = -\dfrac{1}{\sqrt{1+C^2}}\ln\left(\dfrac{2/\varepsilon-1-C-\sqrt{1+C^2}}{2/\varepsilon-1-C+\sqrt{1+C^2}}\right)$ |
| 3 *Cross-flow (single-pass)* $C_{\max}$ mixed, $C_{\min}$ unmixed | $\text{NTU} = -\ln\left[1+\dfrac{\ln(1-\varepsilon C)}{C}\right]$ |
| $C_{\min}$ mixed, $C_{\max}$ unmixed | $\text{NTU} = -\dfrac{\ln[C\ln(1-\varepsilon)+1]}{C}$ |
| 4 *All heat exchangers with* $C = 0$ | $\text{NTU} = -\ln(1-\varepsilon)$ |

*Source*: Ref. 7.

from its definition (Eq. 13-29), and then the NTU from the appropriate NTU relation in Table 13-5.

Note that the relations in Table 13-5 are equivalent to those in Table 13-4. Both sets of relations are given for convenience. The relations in Table 13-4 give the effectiveness directly when NTU is known, and the relations in Table 13-5 give the NTU directly when the effectiveness $\varepsilon$ is known.

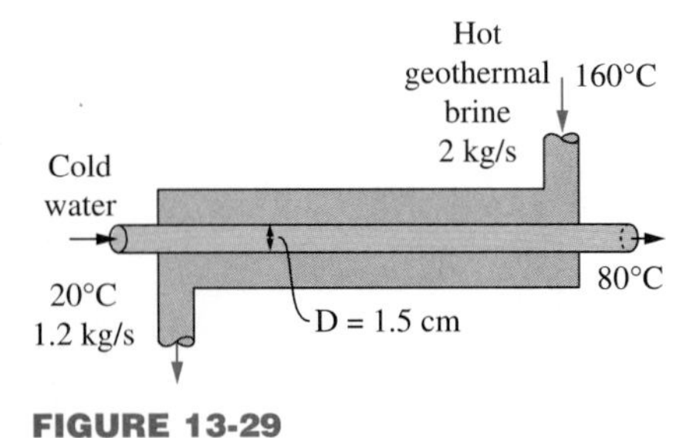


**FIGURE 13-29**
Schematic for Example 13-8.

**EXAMPLE 13-8**

Repeat Example 13-7, which was solved with the LMTD method, using the effectiveness-NTU method.

**Solution** The schematic of the heat exchanger is redrawn in Fig. 13-29. Again we assume the heat exchanger to be well insulated, so that heat transferred from the geothermal water is equal to that heat transferred to the water.

In the effectiveness-NTU method, we first determine the heat capacity rates of the hot and cold fluids, and identify the smaller one:

$$C_h = \dot{m}_h C_{ph} = (2\text{ kg/s})[4.31\text{ kJ/(kg}\cdot{}^\circ\text{C)}] = 8.62\text{ kW/}^\circ\text{C}$$

and

$$C_c = \dot{m}_c C_{pc} = (1.2\text{ kg/s})[4.18\text{ kJ/(kg}\cdot{}^\circ\text{C)}] = 5.02\text{ kW/}^\circ\text{C}$$

Therefore,

$$C_{\min} = C_c = 5.02\text{ kW/}^\circ\text{C}$$

and

$$C = C_{\min}/C_{\max} = 5.02/8.62 = 0.583$$

Then the maximum heat transfer rate is determined from Eq. 13-32 to be

$$\begin{aligned}\dot{Q}_{max} &= C_{min}(T_{h,in} - T_{c,in}) \\ &= (5.02\ \text{kW/°C})(160 - 20)\text{°C} \\ &= 702.8\ \text{kW}\end{aligned}$$

That is, the maximum possible heat transfer rate in this heat exchanger is 702.8 kW. The actual rate of heat transfer in the heat exchanger is

$$\begin{aligned}\dot{Q} = [\dot{m}C_p(T_{out} - T_{in})]_{water} &= (1.2\ \text{kg/s})[4.18\ \text{kJ/(kg}\cdot\text{°C)}](80 - 20)\text{°C} \\ &= 301.0\ \text{kW}\end{aligned}$$

Thus, the effectiveness of the heat exchanger is

$$\varepsilon = \frac{\dot{Q}}{\dot{Q}_{max}} = \frac{301.0\ \text{kW}}{702.8\ \text{kW}} = 0.428$$

Knowing the effectiveness, the NTU of this counter-flow heat exchanger can be determined from Fig. 13-25*b* or the appropriate relation from Table 13-5. We choose the latter approach for greater accuracy:

$$\text{NTU} = \frac{1}{C-1}\ln\left(\frac{\varepsilon - 1}{\varepsilon C - 1}\right) = \frac{1}{0.583 - 1}\ln\left(\frac{0.428 - 1}{0.428 \times 0.583 - 1}\right) = 0.651$$

Then the heat transfer surface area becomes

$$\text{NTU} = \frac{UA}{C_{min}} \longrightarrow A = \frac{\text{NTU}\,C_{min}}{U} = \frac{(0.651)(5020\ \text{W/°C})}{640\ \text{W/(m}^2\cdot\text{°C)}} = 5.11\ \text{m}^2$$

To provide this much heat transfer surface area, the length of the tube must be

$$A = \pi DL \longrightarrow L = \frac{A}{\pi D} = \frac{5.11\ \text{m}^2}{\pi(0.015\ \text{m})} = 108.4\ \text{m}$$

Thus, we obtained the same result with the effectiveness-NTU method in a systematic and straightforward manner.

---

**EXAMPLE 13-9**

Hot oil is to be cooled by water in a one-shell pass and 8-tube-passes heat exchanger. The tubes are thin-walled, and are made of copper with an internal diameter of 1.4 cm. The length of each tube pass in the heat exchanger is 5 m, and the overall heat transfer coefficient is 310 W/(m² · °C). Water flows through the tubes at a rate of 0.2 kg/s, and the oil through the shell at a rate of 0.3 kg/s. The water and the oil enter at temperatures of 20°C and 150°C, respectively. Determine the rate of heat transfer in the heat exchanger, and the outlet temperatures of the water and the oil.

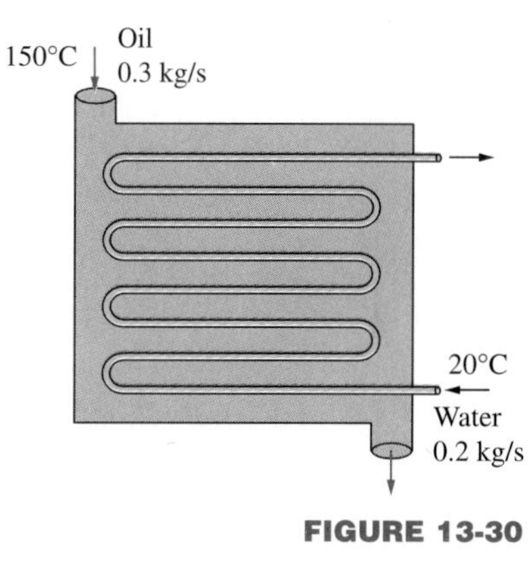


**FIGURE 13-30**
Schematic for Example 13-9.

**Solution** The schematic of the heat exchanger is given in Fig. 13-30. We assume the heat exchanger to be well insulated, so that heat transferred from the hot oil is equal to the heat transferred to the cooling water.

The outlet temperatures are not specified, and they cannot be determined from an energy balance. The use of the LMTD method in this case will involve tedious iterations, and thus the ε-NTU method is indicated. The first step in the

$\varepsilon$-NTU method is to determine the heat capacity rates of the hot and cold fluids, and identify the smaller one:

$$C_h = \dot{m}_h C_{ph} = (0.3\ \text{kg/s})[2.13\ \text{kJ/(kg}\cdot{}^\circ\text{C)}] = 0.639\ \text{kW/}^\circ\text{C}$$

and

$$C_c = \dot{m}_c C_{pc} = (0.2\ \text{kg/s})[4.18\ \text{kJ/(kg}\cdot{}^\circ\text{C)}] = 0.836\ \text{kW/}^\circ\text{C}$$

Therefore,

$$C_{\min} = C_h = 0.639\ \text{kW/}^\circ\text{C}$$

and

$$C = \frac{C_{\min}}{C_{\max}} = \frac{0.639}{0.836} = 0.764$$

Then the maximum heat transfer rate is determined from Eq. 13-32 to be

$$\begin{aligned}\dot{Q}_{\max} &= C_{\min}(T_{h,\text{in}} - T_{c,\text{in}})\\ &= (0.639\ \text{kW/}^\circ\text{C})(150 - 20)^\circ\text{C}\\ &= 83.1\ \text{kW}\end{aligned}$$

That is, the maximum possible heat transfer rate in this heat exchanger is 83.1 kW. The heat transfer surface area is

$$A = n(\pi DL) = 8\pi(0.014\ \text{m})(5\ \text{m}) = 1.76\ \text{m}^2$$

Then the NTU of this heat exchanger becomes

$$\text{NTU} = \frac{UA}{C_{\min}} = \frac{[310\ \text{W/(m}^2\cdot{}^\circ\text{C)}](1.76\ \text{m}^2)}{639\ \text{W/}^\circ\text{C}} = 0.853$$

The effectiveness of this heat exchanger corresponding to $C = 0.764$ and $\text{NTU} = 0.853$ is determined from Fig. 13-26*c* to be

$$\varepsilon = 0.59$$

We could also determine the effectiveness from the third relation in Table 13-4 more accurately but with more labor. Then the actual rate of heat transfer becomes

$$\dot{Q} = \varepsilon\dot{Q}_{\max} = (0.59)(83.1\ \text{kW}) = 49.0\ \text{kW}$$

Finally, the outlet temperatures of the cold and the hot fluid streams are determined to be

$$\dot{Q} = C_c(T_{c,\text{out}} - T_{c,\text{in}}) \longrightarrow T_{c,\text{out}} = T_{c,\text{in}} + \frac{\dot{Q}}{C_c} = 20^\circ\text{C} + \frac{49.0\ \text{kW}}{0.836\ \text{kW/}^\circ\text{C}} = 78.6^\circ\text{C}$$

$$\dot{Q} = C_h(T_{h,\text{in}} - T_{h,\text{out}}) \longrightarrow T_{h,\text{out}} = T_{h,\text{in}} - \frac{\dot{Q}}{C_h} = 150^\circ\text{C} - \frac{49.06\ \text{kW}}{0.639\ \text{kW/}^\circ\text{C}} = 73.3^\circ\text{C}$$

Therefore, the temperature of the cooling water will rise from 20°C to 78.6°C as it cools the hot oil from 150°C to 73.3°C in this heat exchanger.

## 13-6 ■ THE SELECTION OF HEAT EXCHANGERS

Heat exchangers are complicated devices, and the results obtained with simplified approaches presented above should be used with care. For example, we assumed that the overall heat transfer coefficient $U$ is constant throughout the heat exchanger, and that the convection heat transfer coefficients can be predicted using the convection correlations. However, it should be kept in mind that the uncertainty in the predicted value of $U$ can even exceed 30 percent. Thus, it is natural to tend to overdesign the heat exchangers in order to avoid unpleasant surprises.

Heat transfer enhancement in heat exchangers is usually accompanied by *increased pressure drop,* and thus *higher pumping power.* Therefore, any gain from the enhancement in heat transfer should be weighed against the cost of the accompanying pressure drop. Also, some thought should be given to which fluid should pass through the tube side and which through the shell side. Usually, the more viscous fluid is more suitable for the shell side (larger passage area and thus lower pressure drop) and the fluid with the higher pressure for the tube side.

Engineers in industry often find themselves in a position to select heat exchangers to accomplish certain heat transfer tasks. Usually, the goal is to heat or cool a certain fluid at a known mass flow rate and temperature to a desired temperature. Thus, the rate of heat transfer in the prospective heat exchanger is

$$\dot{Q}_{max} = \dot{m}C_p(T_{in} - T_{out})$$

which gives the heat transfer requirement of the heat exchanger before having any idea about the heat exchanger itself.

An engineer going through catalogs of heat exchanger manufacturers will be overwhelmed by the type and number of readily available off-the-shelf heat exchangers. The proper selection depends on several factors.

### Heat Transfer Rate

This is the most important quantity in the selection of a heat exchanger. A heat exchanger should be capable of transferring heat at the specified rate in order to achieve the desired temperature change of the fluid at the specified mass flow rate.

### Cost

Budgetary limitations usually play the most important role in the selection of heat exchangers, except for some specialized cases where "money is no object." An off-the-shelf heat exchanger has a definite cost advantage over those made to order. However, in some cases, none of the existing heat exchangers will do, and it may be necessary to undertake the expensive and time-consuming task of designing and manufacturing a heat exchanger from the scratch to suit the needs. This is often the case when the heat exchanger is an integral part of the overall device to be manufactured.

The operation and maintenance costs of the heat exchanger are also important considerations in assessing the overall cost.

### Pumping Power

In a heat exchanger, both fluids are usually forced to flow by pumps or fans which consume electrical power. The annual cost of electricity associated with the operation of the pumps and fans can be determined from

$$\text{operating cost} = (\text{pumping power, kW}) \times (\text{hours of operation, h}) \times (\text{price of electricity, \$/kWh})$$

where the pumping power is the total electrical power consumed by the motors of the pumps and fans. For example, a heat exchanger that involves a 1-hp pump and a $\frac{1}{3}$-hp fan (1 hp = 0.746 kW) operating 8 h a day and 5 days a week will consume 2017 kWh of electricity per year, which will cost \$161.4 at an electricity cost of 8 cents/kWh.

Minimizing the pressure drop and the mass flow rate of the fluids will *minimize* the operating cost of the heat exchanger, but it will *maximize* the size of the heat exchanger and thus the initial cost. As a rule of thumb, doubling the mass flow rate will reduce the initial cost by *half,* but will increase the pumping power requirements by a factor of roughly *eight.*

Typically, fluid velocities encountered in heat exchangers range between 0.7 and 7 m/s for liquids, and between 3 and 30 m/s for gases. Low velocities are helfpul in avoiding erosion, tube vibrations, and noise as well as pressure drop.

### Size and Weight

Normally, the *smaller* and the *lighter* the heat exchanger, the better it is. This is especially the case in the *automotive* and *aerospace* industries, where size and weight requirements are most stringent. Also, a larger heat exchanger normally carries a higher price tag. The space available for the heat exchanger in some cases limits the length of the tubes that can be used.

### Type

The type of the heat exchanger to be selected depends primarily on the type of *fluids* involved, the *size* and *weight* limitations, and the presence of any *phase-change* processes. For example, a heat exchanger is suitable to cool a liquid by a gas if the surface area on the gas side is many times that on the liquid side. On the other hand, a plate or shell-and-tube heat exchanger is very suitable for cooling a liquid by another liquid.

### Materials

The materials used in the construction of the heat exchanger may be an important consideration in the selection of heat exchangers. For example, the thermal and structural *stress effects* need not be considered at pressures below 15 atm or temperature below 150°C. But these effects are major considerations above 70 atm or 550°C, and seriously limit the acceptable materials of the heat exchanger.

A temperature difference of 50°C or more between the tubes and the shell will probably pose *differential thermal expansion* problems, and needs to be considered. In the case of corrosive fluids, we may have to select expensive *corrosion-resistant* materials such as stainless steel or even titanium if we are not willing to replace low-cost heat exchangers frequently.

### Other Considerations

There are other considerations in the selection of heat exchanger that may or may not be important, depending on the application. For example, being *leak-tight* is an important consideration when *toxic* or *expensive* fluids are involved. Ease of servicing, low maintenance cost, and safety and reliability are some other important considerations in the selection process. Quietness is one of the primary considerations in the selection of liquid-to-air heat exchangers used in heating and air-conditioning applications

**EXAMPLE 13-10**

In a dairy plant, milk is pasteurized by hot water supplied by a natural gas furnace. The hot water is then discharged to an open floor drain at 80°C at a rate of 15 kg/min. The plant operates 24 h a day and 365 days a year. The furnace has an efficiency of 80 percent, and the cost of the natural gas is $0.40 per therm (1 therm = 100,000 Btu = 105,500 kJ). The average temperature of the cold water entering the furnace throughout the year is 15°C. The drained hot water cannot be returned to the furnace and recirculated, because it is contaminated during the process.

In order to save energy, it is proposed to install a water-to-water heat exchanger to preheat the incoming cold water by the drained hot water. Assuming that the heat exchanger will recover 75 percent of the available heat in the hot water, determine the heat transfer rating of the heat exchanger that needs to be purchased, and suggest a suitable type. Also determine the amount of money this heat exchanger will save the company per year from natural gas savings.

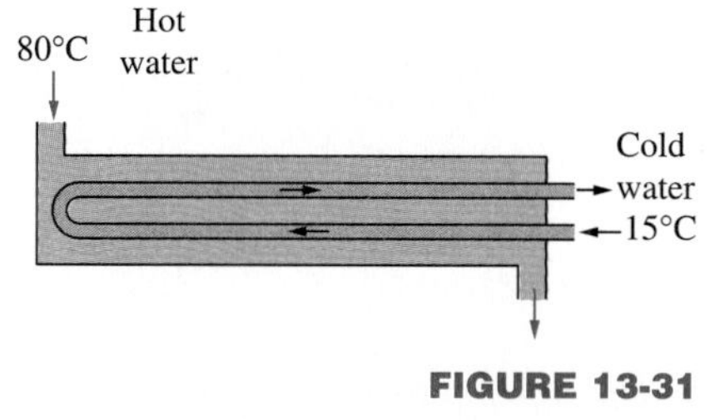


**FIGURE 13-31**
Schematic for Example 13-10.

**Solution** A schematic of the prospective heat exchanger is given in Fig. 13-31. We take the specific heat of water to be $C_p = 4.18$ kJ/(kg · °C) and treat it as constant. The heat recovery from the hot water will be a maximum when it leaves the heat exchanger at the inlet temperature of the cold water. Therefore,

$$\begin{aligned}\dot{Q}_{max} &= \dot{m}_h C_p (T_{h,in} - T_{c,in}) \\ &= \left(\frac{15}{60}\text{ kg/s}\right)[4.18\text{ kJ/(kg}\cdot\text{°C)}](80 - 15)\text{°C} \\ &= 67.9\text{ kJ/s}\end{aligned}$$

That is, the existing hot water stream has the potential to supply heat at a rate of 67.9 kJ/s to the incoming cold water. This value would be approached in a counter-flow heat exchanger with a *very large* heat transfer surface area. A heat exchanger of reasonable size and cost can capture 75 percent of this heat transfer potential. Thus, the heat transfer rating of the prospective heat exchanger must be

$$\dot{Q} = \varepsilon \dot{Q}_{max} = (0.75)(67.9\text{ kJ/s}) = 50.9\text{ kJ/s}$$

That is, the heat exchanger should be able to deliver heat at a rate of 50.9 kJ/s from the hot to the cold water. An ordinary plate or shell-and-tube heat exchanger should be adequate for this purpose, since both sides of the heat exchanger involve the same fluid at comparable flow rates and thus comparable heat transfer coefficients. (Note that if we were heating air with hot water, we would have to specify a heat exchanger that has a large surface area on the air side.)

The heat exchanger will operate 24 h a day and 365 days a year. Therefore, the annual operating hours are

$$\text{operating hours} = (24\ \text{h/day})(365\ \text{days/year}) = 8760\ \text{h/year}$$

Noting that this heat exchanger saves 50.9 kJ of energy per second, the energy saved during an entire year will be

$$\begin{aligned}\text{energy saved} &= (\text{heat transfer rate})(\text{operation time})\\ &= (50.9\ \text{kJ/s})(8760\ \text{h/year})(3600\ \text{s/h})\\ &= 1.605 \times 10^9\ \text{kJ/year}\end{aligned}$$

The furnace is said to be 80-percent efficient. That is, for each 80 units of heat supplied by the furance, natural gas with an energy content of 100 units must be supplied to the furnace. Therefore, the energy savings determined above results in fuel savings in the amount of

$$\begin{aligned}\text{fuel saved} &= \frac{\text{energy saved}}{\text{furnace efficiency}} = \frac{1.605 \times 10^9\ \text{kJ/year}}{0.80}\left(\frac{1\ \text{therm}}{105{,}500\ \text{kJ}}\right)\\ &= 19{,}020\ \text{therms/year}\end{aligned}$$

since 1 therm = 105,500 kJ. Noting that the price of natural gas is \$0.40 per therm, the amount of money saved becomes

$$\begin{aligned}\text{money saved} &= (\text{fuel saved}) \times (\text{price of fuel})\\ &= (19{,}020\ \text{therms/year})(\$0.40/\text{therm})\\ &= \$7607/\text{year}\end{aligned}$$

Therefore, the installation of the proposed heat exchanger will save the company \$7607 a year, and the installation cost of the heat exchanger will probably be paid from the fuel savings in a short time.

## 13-7 ■ SUMMARY

Heat exchangers are devices that allow the exchange of heat between two fluids without allowing them to mix with each other. Heat exchangers are manufactured in a variety of types, the simplest being the *double-pipe* heat exchanger. In a *parallel-flow* type, both the hot and cold fluids enter the heat exchanger at the same end, and move in the same direction, whereas in a *counter-flow* type, the hot and cold fluids enter the heat exchanger at opposite ends, and flow in opposite directions. In *compact* heat exchangers, the two fluids move perpendicular to each other, and such a flow configuration is called *cross-flow*. Other common types of heat exchangers in industrial applications are the *plate* and the *shell-and-tube* heat exchangers.

Heat transfer in a heat exchanger usually involves convection in each fluid, and conduction through the wall separating the two fluids. In the analysis of heat exchangers, it is convenient to work with an *overall heat transfer coefficient* $U$ or a *total thermal resistance* $R$ expressed as

$$\frac{1}{UA} = \frac{1}{U_iA_i} = \frac{1}{U_oA_o} = R = \frac{1}{h_iA_i} + R_{\text{wall}} + \frac{1}{h_oA_o}$$

where the subscripts $i$ and $o$ stand for the inner and outer surfaces of the wall that separates the two fluids, respectively. When the wall thickness of the tube is small and the thermal conductivity of the tube material is high, the above relation simplifies to

$$\frac{1}{U} \approx \frac{1}{h_i} + \frac{1}{h_o}$$

where $U \approx U_i \approx U_o$. The effects of fouling on both the inner and the outer surfaces of the tubes of a heat exchanger can be accounted for by

$$\frac{1}{UA} = \frac{1}{U_iA_i} = \frac{1}{U_oA_o} = R = \frac{1}{h_iA_i} + \frac{R_{f,i}}{A_i} + \frac{\ln(D_o/D_i)}{2\pi kL} + \frac{R_{f,o}}{A_o} + \frac{1}{h_oA_o}$$

where $A_i = \pi D_i L$ and $A_o = \pi D_o L$ are the areas of the inner and outer surfaces, and $R_{f,i}$ and $R_{f,o}$ are the fouling factors at those surfaces.

In a well-insulated heat exchanger, the rate of heat transfer from the hot fluid is equal to the rate of heat transfer to the cold one. That is,

$$\dot{Q} = \dot{m}_c C_{pc}(T_{c,\text{out}} - T_{c,\text{in}}) = C_c(T_{c,\text{out}} - T_{c,\text{in}})$$

and

$$\dot{Q} = \dot{m}_h C_{ph}(T_{h,\text{in}} - T_{h,\text{out}}) = C_h(T_{h,\text{in}} - T_{h,\text{out}})$$

where the subscripts $c$ and $h$ stand for the cold and hot fluids, respectively, and the product of the mass flow rate and the specific heat of a fluid $C = \dot{m}C_p$ is called the *heat capacity rate.*

Of the two methods used in the analysis of heat exchangers, the *log mean temperature difference* (or LMTD) method is best suited for determining the size of a heat exchanger when all the inlet and the outlet temperatures are known. The *effectiveness-NTU* method is best suited to predict the outlet temperatures of the hot and cold fluid streams in a specified heat exchanger. In the LMTD method, the rate of heat transfer is determined from

$$\dot{Q} = UA\,\Delta T_{\text{lm}}$$

where

$$\Delta T_{\text{lm}} = \frac{\Delta T_1 - \Delta T_2}{\ln(\Delta T_1/\Delta T_2)}$$

is the *log mean temperature difference,* which is the suitable form of the average temperature difference for use in the analysis of heat exchangers. Here $\Delta T_1$ and $\Delta T_2$ represent the temperature difference between the two fluids at the two ends (inlet and outlet) of the heat exchanger. For cross-flow and multipass shell-and-tube heat exchangers, the logarithmic mean temperature difference is related to the counterflow one $\Delta T_{\text{lm},CF}$ as

$$\Delta T_{\text{lm}} = F\,\Delta T_{\text{lm},CF}$$

where $F$ is the *correction factor,* which depends on the geometry of the heat exchanger and the inlet and outlet temperatures of the hot and cold fluid streams.

The *effectiveness* of a heat exchanger is defined as

$$\varepsilon = \frac{\dot{Q}}{\dot{Q}_{\max}} = \frac{\text{actual heat transfer rate}}{\text{maximum possible heat transfer rate}}$$

where
$$\dot{Q}_{\max} = C_{\min}(T_{h,\text{in}} - T_{c,\text{in}})$$

and $C_{\min}$ is the smaller of $C_h = \dot{m}_h C_{ph}$ and $C_c = \dot{m}_c C_{pc}$. The effectiveness of heat exchangers can be determined from effectiveness relations or charts.

The selection or design of a heat exchanger depends on several factors such as the heat transfer rate, cost, pressure drop, size, weight, construction type, materials, and operating environment.

## REFERENCES AND SUGGESTED READING

**1** N. Afgan and E. U. Schlunder, *Heat Exchanger: Design and Theory Sourcebook,* McGraw-Hill/Scripta, Washington, DC, 1974.

**2** R. A. Bowman, A. C. Mueller, and W. M. Nagle, "Mean Temperature Difference in Design," *Transactions of the ASME,* Vol. 62, p. 283, 1940.

**3** A. P. Fraas, *Heat Exchanger Design,* 2d ed., Wiley, 1989.

**4** K. A. Gardner, "Variable Heat Transfer Rate Correction in Multipass Exchangers, Shell Side Film Controlling," *Transactions of the ASME,* Vol. 67, pp. 31–38, 1945.

**5** J. P. Holman, *Heat Transfer,* 7th ed., McGraw-Hill, New York, 1990.

**6** F. P. Incropera and D. P. DeWitt, *Introduction to Heat Transfer,* 2nd ed., Wiley, New York, 1990.

**7** W. M. Kays and A. L. London, *Compact Heat Exchangers,* 3rd ed., McGraw-Hill, New York, 1984.

**8** W. M. Kays and H. C. Perkins, Chap. 7, in W. M. Rohsenow and J. P. Hartnett, (Eds.), *Handbook of Heat Transfer,* McGraw-Hill, New York, 1972.

**9** M. N. Özışık, *Heat Transfer—A Basic Approach,* McGraw-Hill, New York, 1985.

**10** A. C. Mueller, "Heat Exchangers," Chap. 18 in W. M. Rohsenow and J. P. Hartnett (Eds.), *Handbook of Heat Transfer,* McGraw-Hill, New York, 1972.

**11** E. U. Schlunder, *Heat Exchanger Design Handbook,* Hemisphere, Washington, 1982.

**12** *Standards of Tubular Exchanger Manufacturers Association,* Latest ed., Tubular Exchanger Manufacturers Association, New York.

**13** R. A. Stevens, J. Fernandes, and J. R. Woolf, "Mean Temperature Difference in One, Two, and Three Pass Crossflow Heat Exchangers," *Transactions of the ASME,* Vol. 79, pp. 287–297, 1957.

**14** J. Taborek, G. F. Hewitt, and N. Afgan, *Heat Exchangers: Theory and Practice,* Hemisphere, New York, 1983.

**15** L. C. Thomas, *Heat Transfer,* Prentice-Hall, Englewood Cliffs, NJ, 1992.

**16** G. Walker, *Industrial Heat Exchangers,* Hemisphere, Washington, 1982.

**17** F. M. White, *Heat and Mass Transfer,* Addison-Wesley, Reading, MA, 1988.

## PROBLEMS*

### Types of Heat Exchangers

**13-1C** Classify heat exchangers according to flow type, and explain the characteristics of each type.

**13-2C** Classify heat exchangers according to construction type, and explain the characteristics of each type.

**13-3C** When is a heat exchanger classified as being compact? Do you think a double-pipe heat exchanger can be classified as a compact heat exchanger?

**13-4C** How does a cross-flow heat exchanger differ from a counter-flow one? What is the difference between mixed and unmixed fluids in cross-flow?

**13-5C** What is the role of the baffles in a shell-and-tube heat exchanger? How does the presence of baffles effect the heat transfer and the pumping power requirements? Explain.

**13-6C** Draw a one-shell pass and six-tube-passes shell-and-tube heat exchanger. What are the advantages and disadvantages of using six tube passes instead of just two of the same diameter?

**13-7C** Draw a two-shell-passes and eight-tube passes shell-and-tube heat exchanger. What is the primary reason for using so many tube passes?

* Students are encouraged to answer *all* the concept "C" questions.

**13-8C** What is a regenerative heat exchanger? How does a static type regenerative heat exchanger differ from a dynamic type?

### The Overall Heat Transfer Coefficient

**13-9C** What are the heat transfer mechanisms involved during heat transfer from the hot to the cold fluid?

**13-10C** Under what conditions is the thermal resistance of the tube in a heat exchanger negligible?

**13-11C** Consider a double-pipe parallel-flow heat exchanger of length $L$. The inner and outer diameters of the inner tube are $D_1$ and $D_2$, respectively, and the inner diameter of the outer tube is $D_3$. Explain how you would determine the two heat transfer surface areas $A_i$ and $A_o$. When is it reasonable to assume $A_i \approx A_o \approx A$?

**13-12C** Is the approximation $h_i \approx h_0 \approx h$ for the convection heat transfer coefficient in a heat exchanger a reasonable one when the thickness of the tube wall is negligible?

**13-13C** Under what conditions can the overall heat transfer coefficient of a heat exchanger be determined from $U = (1/h_i + 1/h_o)^{-1}$?

**13-14C** What are the restrictions on the relation $UA = U_iA_i = U_oA_o$ for a heat exchanger? Here $A$ is the heat transfer surface area and $U$ is the overall heat transfer coefficient.

**13-15C** In a thin-walled double-pipe heat exchanger, when is the approximation $U = h_i$ a reasonable one? Here $U$ is the overall heat transfer coefficient and $h_i$ is the convection heat transfer coefficient inside the tube.

**13-16C** What are the common causes of fouling in a heat exchanger? How does fouling effect heat transfer and pressure drop?

**13-17C** How is the thermal resistance due to fouling in a heat exchanger accounted for? How does the fluid velocity and temperature effect fouling?

**13-18** A double-pipe heat exchanger is constructed of a copper [$k = 380\ \text{W/(m} \cdot {}^\circ\text{C)}$] inner tube of internal diameter $D_i = 1.2$ cm and external diameter $D_o = 1.6$ cm, and an outer tube of diameter 3.0 cm. The convection heat transfer coefficient is reported to be $h_i = 700\ \text{W/(m}^2 \cdot {}^\circ\text{C)}$ on the inner surface of the tube, and $h_o = 1400\ \text{W/(m}^2 \cdot {}^\circ\text{C)}$ on its outer surface. For a fouling factor $R_{f,i} = 0.0005\ \text{m}^2 \cdot {}^\circ\text{C/W}$ on the tube side and $R_{f,o} = 0.0002\ \text{m}^2 \cdot {}^\circ\text{C/W}$ on the shell side, determine ($a$) the thermal resistance of the heat exchanger per unit length and ($b$) the overall heat transfer coefficients $U_i$ and $U_o$ based on the inner and outer surface areas of the tube, respectively.

**13-19** Water at an average temperature of 107°C and an average velocity of 3.5 m/s flows through a 5-m-long stainless steel [$k = 14.2\ \text{W/(m} \cdot {}^\circ\text{C)}$] in a boiler. The inner and outer diameters of the tube are $D_i = 1.0$ cm and $D_o = 1.4$ cm, respectively. If the convection heat transfer coefficient at the outer surface of the tube where boiling is taking place is $h_o = 8400$ W/(m$^2 \cdot$ °C), determine the overall heat transfer coefficient $U_i$ of this boiler based on the inner surface area of the tube.

**13-20** Repeat Problem 13-19, assuming a fouling factor $R_{f,i} = 0.0005$ m$^2 \cdot$ °C/W on the inner surface of the tube.

**13-21** A long thin-walled double-pipe heat exchanger with tube and shell diameters of 1.0 cm and 2.5 cm, respectively, is used to condense refrigerant 134a at 20°C by water. The refrigerant flows through the tube, with a convection heat transfer coefficient of $h_i = 5000$ W/(m$^2 \cdot$ °C). Water flows through the shell at a rate of 0.3 kg/s. Determine the overall heat transfer coefficient of this heat exchanger.
*Answer:* 2100 W/(m$^2 \cdot$ °C)

**13-22** Repeat Problem 13-21 by assuming a 2-mm-thick layer of limestone [$k = 1.3$ W/(m $\cdot$ °C)] forms on the outer surface of the inner tube.

**13-23** Water at an average temperature of 60°C and an average velocity of 2.5 m/s flows through a thin-walled 1.8-cm-diameter tube. The water is cooled by air, which flows across the tube with a velocity of $U_\infty = 6$ m/s at an average temperature of 300 K. Determine the overall heat transfer coefficient.

**13-23E** Water at an average temperature of 140°F and an average velocity of 8 ft/s flows through a thin-walled $\frac{3}{4}$-in-diameter tube. The water is cooled by air which flows across the tube with a velocity of $U_\infty = 20$ ft/s at an average temperature of 80°F. Determine the overall heat transfer coefficient.

## Analysis of Heat Exchangers

**13-24C** What are the common approximations made in the analysis of heat exchangers?

**13-25C** Under what conditions is the heat transfer relation

$$\dot{Q} = \dot{m}_c C_{pc}(T_{c,\text{out}} - T_{c,\text{in}}) = \dot{m}_h C_{ph}(T_{h,\text{in}} - T_{h,\text{out}})$$

valid for a heat exchanger?

**13-26C** What is the heat capacity rate? What can you say about the temperature changes of the hot and cold fluids in a heat exchanger if both fluids have the same capacity rate? What does a heat capacity of infinity for a fluid in a heat exchanger mean?

**13-27C** Consider a condenser in which steam at a specified temperature is condensed by rejecting heat to the cooling water. If the heat transfer rate in the condenser and the temperature rise of the cooling water is known, explain how the rate of condensation of the steam and the mass flow rate of the cooling water can be determined. Also explain how the total thermal resistance $R$ of this condenser can be evaluated in this case.

**13-28C** Under what conditions will the temperature rise of the cold fluid in a heat exchanger be equal to the temperature drop of the hot fluid?

### The Log Mean Temperature Difference Method

**13-29C** In the heat transfer relation $\dot{Q} = UA\,\Delta T_{lm}$ for a heat exchanger, what is $\Delta T_{lm}$ called? How is it determined for a parallelflow and counterflow heat exchanger?

**13-30C** How does the log mean temperature difference for a heat exchanger differ from the arithmetic mean temperature difference (AMTD)? For specified inlet and outlet temperatures, which one of these two quantities is larger?

**13-31C** The temperature difference between the hot and cold fluids in a heat exchanger is given to be $\Delta T_1$ at one end and $\Delta T_2$ at the other end. Can the logarithmic temperature difference $\Delta T_{lm}$ of this heat exchanger be greater than both $\Delta T_1$ and $\Delta T_2$? Explain.

**13-32C** Can the logarithmic mean temperature difference $\Delta T_{lm}$ of a heat exchanger be a negative quantity? Explain.

**13-33C** Can the outlet temperature of the cold fluid in a heat exchanger be higher than the outlet temperature of the hot fluid in a parallel-flow heat exchanger? How about in a counter-flow heat exchanger? Explain.

**13-34C** For specified inlet and outlet temperatures, for what kind of heat exchanger will the $\Delta T_{lm}$ be greatest: double-pipe parallel-flow, double-pipe counter-flow, cross-flow, or multipass shell-and-tube heat exchanger?

**13-35C** In the heat transfer relation $\dot{Q} = UAF\,\Delta T_{lm}$ for a heat exchanger, what is the quantity $F$ called? What does it represent? Can $F$ be greater than one?

**13-36C** When the outlet temperatures of the fluids in a heat exchanger are not known, it is still practical to use the LMTD method? Why?

**13-37C** Explain how the LMTD method can be used to determine the heat transfer surface area of a multipass shell-and-tube heat exchanger when all the necessary information, including the outlet temperatures, is given.

**13-38** Steam in the condenser of a steam power plant is to be condensed at a temperature of 50°C ($h_{fg}$ = 2305 kJ/kg) with cooling water [$C_p$ = 4180 J/(kg · °C)] from a nearby lake, which enters the tubes of the condenser at 18°C and leaves at 27°C. The surface area of the tubes is 58 $m^2$, and the overall heat transfer coefficient is 2400 W/($m^2$ · °C). Determine the mass flow rate of the cooling water needed and the rate of condensation of the steam in the condenser.
*Answer:* 101 kg/s, 1.65 kg/s

**13-39** A double-pipe parallel-flow heat exchanger is to heat water [$C_p$ = 4180 J/(kg · °C)] from 25°C to 60°C at a rate of 0.2 kg/s. The heating is to be accomplished by geothermal water [$C_p$ = 4310 J/(kg · °C)] available at 140°C at a mass flow rate of 0.3 kg/s. The inner tube is thin-walled, and has a diameter of 0.8 cm. If the overall heat transfer coefficient of the heat exchanger is 550 W/(m$^2$ · °C), determine the length of the heat exchanger required to achieve the desired heating.

**13-40** A one-shell-pass and eight-tube-passes heat exchanger is used to heat glycerin [$C_p$ = 2500 J/(kg · °C)] from 25°C to 60°C by hot water [$C_p$ = 4190 J/(kg · °C)], which enters the thin-walled 1.2-cm-diameter tubes at 90°C and leaves at 50°C. The total length of the tubes in the heat exchanger is 150 m. The convection heat transfer coefficient is 20 W/(m$^2$ · °C) on the glycerin (shell) side, and 240 W/(m$^2$ · °C) on the water (tube) side. Determine the rate of heat transfer in the heat exchanger (*a*) before any fouling occurs and (*b*) after fouling with a fouling factor of 0.0004 m$^2$ · °C/W occurs on the outer surfaces of the tubes.

**13-40E** A one-shell-pass and eight-tube-passes heat exchanger is used to heat glycerin [$C_p$ = 0.60 Btu/(lbm · °F)] from 75°F to 140°F by hot water [$C_p$ = 1.0 Btu/(lbm · °F)] which enters the thin-walled 0.5-in-diameter tubes at 190°F and leaves at 120°F. The total length of the tubes in the heat exchanger is 500 ft. The convection heat transfer coefficient is 4 Btu/(h · ft$^2$ · °F) on the glycerin (shell) side, and 50 Btu/(h · ft$^2$ · °F) on the water (tube) side. Determine the rate of heat transfer in the heat exchanger (*a*) before any fouling occurs and (*b*) after fouling with a fouling factor of 0.002 h · ft$^2$ · °F/Btu occurs on the outer surfaces of the tubes.

**13-41** A test is conducted to determine the overall heat transfer coefficient in a shell-and-tube oil-to-water heat exchanger that has 24 tubes of internal diameter 1.2 cm and length 2 m in a single shell. Cold water [$C_p$ = 4180 J/(kg · °C)] enters the tubes at 20°C at a rate of 5 kg/s and leaves at 55°C. Oil [$C_p$ = 2150 J/(kg · °C)] flows through the shell, and is cooled from 120°C to 45°C. Determine the overall heat transfer coefficient $U_i$ of this heat exchanger based on the inner surface area of the tubes. *Answer:* 17.6 kW/(m$^2$ · °C)

**13-42** A double-pipe counter-flow heat exchanger is to cool ethylene glycol [$C_p$ = 2560 J/(kg · °C)] flowing at a rate of 2 kg/s from 80°C to 40°C by water [$C_p$ = 4180 J/(kg · °C)] which enters at 20°C and leaves at 55°C. The overall heat transfer coefficient based on the inner surface area of the tube is 250 W/(m$^2$ · °C). Determine (*a*) the rate of heat transfer, (*b*) the mass flow rate of water, and (*c*) the heat transfer surface area on the inner side of the tube.

**13-43** Water [$C_p$ = 4180 J/(kg · °C)] enters the 2.5-cm internal-diameter tube of a double-pipe counter-flow heat exchanger at 17°C at a rate of 3 kg/s. It is heated by steam condensing at 120°C ($h_{fg}$ = 2203 kJ/kg) in the shell. If the overall heat transfer coefficient of the heat exchanger is 1500 W/(m$^2$ · °C), determine the length of the tube required in order to heat the water to 80°C.

**13-44** A thin-walled double-pipe counter-flow heat exchanger is to be used to cool oil [$C_p$ = 2200 J/(kg · °C)] from 150°C to 40°C at a rate of 2 kg/s by water [$C_p$ = 4180 J/(kg · °C)] which enters at 22°C at a rate of 1.5 kg/s. The diameter of the tube is 2.5 cm, and its length is 6 m. Determine the overall heat transfer coefficient of this heat exchanger.

**13-44E** A thin-walled double-pipe counter-flow heat exchanger is to be used to cool oil [$C_p$ = 0.525 Btu/(lbm · °F)] from 300°F to 100°F at a rate of 5 lbm/s by water [$C_p$ = 1.0 Btu/(lbm · °F)], which enters at 70°F at a rate of 3 lbm/s. The diameter of the tube is 1 in, and its length is 20 ft. Determine the overall heat transfer coefficient of this heat exchanger.

**13-45** Consider a water-to-water double-pipe heat exchanger whose flow arrangement is not known. The temperature measurements indicate that the cold water enters at 20°C, and leaves at 50°C while the hot water enters at 80°C and leaves at 45°C. Do you think this is a parallel-flow or counter-flow heat exchanger? Explain.

**13-46** Cold water [$C_p$ = 4180 J/(kg · °C)] leading to a shower enters a thin-walled double-pipe counter-flow heat exchanger at 15°C at a rate of 0.25 kg/s, and is heated to 45°C by hot water [$C_p$ = 4190 J/(kg · °C)], which enters at 100°C at a rate of 3 kg/s. If the overall heat transfer coefficient is 950 W/(m$^2$ · °C), determine the rate of heat transfer and the heat transfer surface area of the heat exchanger.

**13-47** Engine oil [$C_p$ = 2100 J/(kg · °C)] is to be heated from 20°C to 60°C at a rate of 0.3 kg/s in a 2-cm-diameter thin-walled copper tube by condensing steam outside at a temperature of 130°C ($h_{fg}$ = 2174 kJ/kg). For an overall heat transfer coefficient of 650 W/(m$^2$ · °C), determine the rate of heat transfer and the length of the tube required to achieve it. *Answers:* 25.2 kW, 7.0 m

**13-48** Geothermal Water [$C_p$ = 4310 J/(kg · °C)] is to be used as the heat source to supply heat to the hydronic heating system of a house at a rate of 30 kW in a double-pipe counter-flow heat exchanger. Water [$C_p$ = 4190 J/(kg · °C)] is heated from 60°C to 90°C in the heat exchanger as the geothermal water is cooled from 120°C to 80°C. Determine the mass flow rate of each fluid and the total thermal resistance of this heat exchanger.

**13-48E** Geothermal water [$C_p$ = 1.03 Btu/(lbm · °F)] is to be used as the heat source to supply heat to the hydronic heating system of a house at a rate of 30 Btu/s in a double-pipe counter-flow heat exchanger. Water [$C_p$ = 1.0 Btu/(lbm · °F)] is heated from 140°F to 200°F in the heat exchanger as the geoethermal water is cooled from 250°F to 180°F. Determine the mass flow rate of each fluid and the total thermal resistance of this heat exchanger.

**13-49** Glycerin [$C_p$ = 2400 J/(kg · °C)] at 20°C and 0.3 kg/s is to be heated by ethylene glycol [$C_p$ = 2500 J/(kg · °C)] at 60°C in a thin-walled double-pipe parallelflow heat exchanger. The temperature difference between the two fluids is 15°C at the outlet of the heat exchanger. If the overall heat transfer coefficient is 240 W/(m$^2$ · °C) and the heat transfer surface are is 7.6 m$^2$, determine (*a*) the rate of heat transfer, (*b*) the outlet temperature of the glycerin, and (*c*) the mass flow rate of the ethylene glycol.

**13-50** Air [$C_p$ = 1005 J/(kg · °C)] is to be preheated by hot exhaust gases in a crossflow heat exchanger before it enters the furnace. Air enters the heat exchanger at 95 kPa and 20°C at a rate of 0.8 m$^3$/s. The combustion gases [$C_p$ = 1100 J/(kg · °C)] enter at 180°C at a rate of 1.1 kg/s and leave at 95°C. The product of the overall heat transfer coefficient and the heat transfer surface area is $AU$ = 1200 W/°C. Assuming both fluids to be unmixed, determine the rate of heat transfer and the outlet temperature of the air.

**13-51** A shell-and-tube heat exchanger with two-shell passes and twelve-tube passes is used to heat water [$C_p$ = 4180 J/(kg · °C)] in the tubes from 20°C to 70°C at a rate of 4.5 kg/s. Heat is supplied by hot oil [$C_p$ = 2300 J/(kg · °C)], which enters the shell side at 170°C at a rate of 10 kg/s. For a tube-side overall heat transfer coefficient of 600 W/(m$^2$ · °C), determine the heat transfer surface area on the tube side. *Answer:* 15 m$^2$

**13-52** Repeat Prob. 13-51 for a mass flow rate of 2 kg/s for water.

**13-53** A shell-and-tube heat exchanger with two shell passes and eight tube passes is used to heat ethyl alcohol [$C_p$ = 2670 J/(kg · °C)] in the tubes from 25°C to 70°C at a rate of 2.1 kg/s. The heating is to be done by water [$C_p$ = 4190 J/(kg · °C)], which enters the shell side at 95°C and leaves at 45°C. If the overall heat transfer coefficient is 800 W/(m$^2$ · °C), determine the heat transfer surface area of the heat exchanger.

**13-54** A shell-and-tube heat exchanger with two shell passes and twelve tube passes is used to heat water [$C_p$ = 4180 J/(kg · °C)] with ethylene glycol [$C_p$ = 2680 J/(kg · °C)]. Water enters the tubes at 22°C at a rate of 0.8 kg/s and leaves at 70°C. Ethylene glycol enters the shell at 110°C and leaves at 60°C. If the overall heat transfer coefficient based on the tube side is 280 W/(m$^2$ · °C), determine the rate of heat transfer and the heat transfer surface area on the tube side.

**13-55** Steam is to be condensed on the shell side of a one-shell-pass and eight-tube-pass condenser with 50 tubes in each pass, at 30°C ($h_{fg}$ = 2430 kJ/kg) at a rate of 10 kg/s. Cooling water [$C_p$ = 4180 J/(kg · °C)] enters the tubes at 15°C and leaves at 23°C. The tubes are thin-walled, and have a diameter of 1.5 cm and length of 2 m per mass. If the overall heat transfer coefficient is 3000 W/(m$^2$ · °C), determine (*a*) the rate of heat transfer, (*b*) the rate of condensation of steam, and (*c*) the mass flow rate of cold water. *Answers:* (*a*) 1188 kW, (*b*) 0.489 kg/s, (*c*) 35.5 kg/s

**13-55E** Steam is to be condensed on the shell side of a one-shell-pass and eight-tube-pass condenser, wtih 50 tubes in each pass at 90°F ($h_{fg}$ = 1043 But/lbm) at a rate of 20 lbm/s. Cooling water

[$C_p$ = 1.0 But/(lbm · °F)] enters the tubes at 60°F and leaves at 73°F. The tubes are thin-walled, and have a diameter of $\frac{3}{4}$ in and length of 5 ft per pass. If the overall heat transfer coefficient is 600 But/(h · ft$^2$ · °F), determine (*a*) the rate of heat transfer, (*b*) the rate of condensation of steam, and (*c*) the mass flow rate of cold water.

**13-56** A shell-and-tube heat exchanger with one shell pass and ten tube passes is used to heat glycerin [$C_p$ = 2480 J/(kg · °C)] in the shell, with hot water in the tubes. The tubes are thin-walled, and have a diameter of 1.5 cm and length of 2 m per pass. The water enters the tubes at 100°C at a rate of 5 kg/s and leaves at 55°C. The glycerin enters the shell at 15°C and leaves at 55°C. Determine the mass flow rate of the glycerin and the overall heat transfer coefficient of the heat exchanger.

### The Effectiveness-NTU Method

**13-57C** Under what conditions is the effectiveness-NTU method definitely preferred over the LMTD method in heat exchanger analysis?

**13-58C** What does the effectiveness of a heat exchanger represent? Can effectiveness be greater than one? What factors does the effectiveness of a heat exchanger depend on?

**13-59C** For a specified fluid pair, inlet temperatures, and mass flow rates, what kind of heat exchanger will have the highest effectiveness: double-pipe parallel-flow, double-pipe counter-flow, cross-flow, or multi-pass shell-and-tube heat exchanger?

**13-60C** Explain how you can evaluate the outlet temperatures of the cold and hot fluids in a heat exchanger after its effectiveness is determined.

**13-61C** Can the temperature of the hot fluid drop below the inlet temperature of the cold fluid at any location in a heat exchanger? Explain.

**13-62C** Can the temperature of the cold fluid rise above the inlet temperature of the hot fluid at any location in a heat exchanger? Explain.

**13-63C** Consider a heat exchanger in which both fluids have the same specific heats, but different mass flow rates. Which fluid will experience a larger temperature change: the one with the lower or higher mass flow rate?

**13-64C** Explain how the maximum possible heat transfer rate $\dot{Q}_{max}$ in a heat exchanger can be determined when the mass flow rates, specific heats, and the inlet temperatures of the two fluids are specified. Does the value of $\dot{Q}_{max}$ depend on the type of the heat exchanger?

**13-65C** Consider two double-pipe counter-flow heat exchangers that are identical except that one is twice as long as the other one. Which heat exchanger is more likely to have a higher effectiveness?

**13-66C** Consider a double-pipe counter-flow heat exchanger. In order to enhance heat transfer, the length of the heat exchanger is now doubled. Do you think its effectiveness will also double?

**13-67C** Consider a shell-and-tube water-to-water heat exchanger with identical mass flow rates for both the hot and cold water streams. Now the mass flow rate of the cold water is reduced by half. Will the effectiveness of this heat exchanger increase, decrease, or remain the same as a result of this modification? Explain. Assume the overall heat transfer coefficient and the inlet temperatures remain the same.

**13-68C** Under what conditions can a counter-flow heat exchanger have an effectiveness of one? What would your answer be for a parallel-flow heat exchanger?

**13-69C** How is the NTU of a heat exchanger defined? What does it represent? Is a heat exchanger with a very large NTU (say, 10) necessarily a good one to buy?

**13-70C** Consider a heat exchanger that has an NTU of 4. Someone proposes to double the size of the heat exchanger and thus double the NTU to 8 in order to increase the effectiveness of the heat exchanger and thus save energy. Would you support this proposal?

**13-71C** Consider a heat exchanger that has an NTU of 0.1. Someone proposes to triple the size of the heat exchanger and thus triple the NTU to 0.3 in order to increase the effectiveness of the heat exchanger and thus save energy. Would you support this proposal?

**13-72** Air [$C_p$ = 1005 J/(kg · °C)] enters a cross-flow heat exchanger at 12°C at a rate of 3 kg/s, where it is heated by a hot water stream [$C_p$ = 4190 J/(kg · °C)], which enters the heat exchanger at 90°C at a rate of 1 kg/s. Determine the maximum heat transfer rate and the outlet temperatures of the cold and the hot water streams for that case.

**13-73** Hot oil [$C_p$ = 2200 J/(kg · °C)] is to be cooled by water [$C_p$ = 4180 J/(kg · °C)] in a 2-shell-passes and 12-tube-passes heat exchanger. The tubes are thin-walled, and are made of copper with an internal diameter of 1.8 cm. The length of each tube pass in the heat exchanger is 3 m, and the overall heat transfer coefficient is 340 W/(m$^2$ · °C). Water flows through the tubes at a rate of 0.1 kg/s, and the oil through the shell at a rate of 0.2 kg/s. The water and the oil enter at temperatures 18°C and 160°C, respectively, Determine the rate of heat transfer in the heat exchanger, and the outlet temperatures of the water and the oil.
*Answers:* 36.2 kW, 104.6°C, 77.7°C

**13-74** Consider an oil-to-oil double-pipe heat exchanger whose flow arrangement is not known. The temperature measurements indicate that the cold oil enters at 20°C and leaves at 55°C, while the hot oil enters at 80°C and leaves at 45°C. Do you think this is a parallel-flow or counter-flow heat exchanger? Why? Assuming the mass flow rates of both fluids to be identical, determine the effectiveness of this heat exchanger.

**13-75** Hot water enters a double-pipe counter-flow water-to-oil heat exchanger at 90°C and leaves at 40°C. Oil enters at 20°C and leaves at 55°C. Determine which fluid has the smaller heat capacity rate, and calculate the effectiveness of this heat exchanger.

**13-75E** Hot water enters a double-pipe counter-flow water-to-oil heat exchanger at 200°F and leaves at 100°F. Oil enters at 70°F and leaves at 130°F. Determine which fluid has the smaller heat capacity rate, and calculate the effectiveness of this heat exchanger.

**13-76** A thin-walled double-pipe parallel-flow heat exchanger is used to heat a chemical whose specific heat is 1800 J/(kg · °C) with hot water [$C_p$ = 4180 J/(kg · °C)]. The chemical enters at 20°C at a rate of 3 kg/s, while the water enters at 110°C at a rate of 2 kg/s. The heat transfer surface area of the heat exchanger is 7 $m^2$ and the overall heat transfer coefficient is 1200 W/($m^2$ · °C). Determine the outlet temperatures of the chemical and the water.

**13-77** A cross-flow air-to-water heat exchanger with an effectiveness of 0.65 is used to heat water [$C_p$ = 4180 J/(kg · °C)] with hot air [$C_p$ = 1010 J/(kg · °C)]. Water enters the heat exchanger at 20°C at a rate of 4 kg/s, while air enters at 100°C at a rate of 9 kg/s. If the overall heat transfer coefficient based on the water side is 260 W/($m^2$ · °C), determine the heat transfer surface area of the heat exchanger on the water side. Assume both fluids are unmixed. *Answer:* 52.4 $m^2$

**13-78** Water [$C_p$ = 4180 J/(kg · °C)] enters the 2.5-cm internal-diameter tube of a double-pipe counter-flow heat exchanger at 17°C at a rate of 3 kg/s. Water is heated by steam condensing at 120°C ($h_{fg}$ = 2203 kJ/kg) in the shell. If the overall heat transfer coefficient of the heat exchanger is 900 W/($m^2$ · °C), determine the length of the tube required in order to heat the water to 80°C using (*a*) the LMTD method and (*b*) the $\varepsilon$-NTU method.

**13-79** Ethanol is vaporized at 78°C ($h_{fg}$ = 846 kJ/kg) in a double-pipe parallel-flow heat exchanger at a rate of 0.03 kg/s by hot oil [$C_p$ = 2200 J/(kg · °C)], which enters at 120°C. If the heat transfer surface area and the overall heat transfer coefficients are 7.8 $m^2$ and 210 W/($m^2$ · °C), respectively, determine the outlet temperature and the mass flow rate of the oil using (*a*) the LMTD method and (*b*) the $\varepsilon$-NTU method.

**13-80** Water [$C_p$ = 4180 J/(kg · °C)] is to be heated by solar-heated hot air [$C_p$ = 1010 J/(kg · °C)] in a double-pipe counter-flow heat exchanger. Air enters the heat exchanger at 90°C at a rate of 0.3 kg/s, while water enters at 22°C at a rate of 0.1 kg/s. The overall heat transfer coefficient based on the inner side of the tube is given to be 80 W/($m^2$ · °C). The length of the tube is 12 m and the internal diameter of the tube is 1.2 cm. Determine the outlet temperatures of the water and the air.

**13-81** A thin-walled double-pipe heat exchanger is to be used to cool oil [$C_p$ = 2200 J/(kg · °C)] from 150°C to 40°C at a rate of 2 kg/s by water [$C_p$ = 4180 J/(kg · °C)], which enters at 22°C at a rate of 1.5 kg/s. The diameter of the tube is 2.5 cm and its length is 6 m. Determine the overall heat transfer coefficient of this heat exchanger using (*a*) the LMTD method and (*b*) the ε-NTU method.

**13-81E** A thin-walled double-pipe heat exchanger is to be used to cool oil [$C_p$ = 0.525 Btu/(lbm · °F)] from 300°F to 105°F at a rate of 5 lbm/s by water [$C_p$ = 1.0 Btu/(lbm · °F)], which enters at 70°F at a rate of 3 lbm/s. The diameter of the tube is 1 in, and its length is 20 ft. Determine the overall heat transfer coefficient of this heat exchanger using (*a*) the LMTD method and (*b*) the ε-NTU method.

**13-82** Cold water [$C_p$ = 4180 J/(kg · °C)] leading to a shower enters a thin-walled double-pipe counter-flow heat exchanger at 15°C at a rate of 0.25 kg/s, and is heated to 45°C by hot water [$C_p$ = 4190 J/(kg · °C)], which enters at 100°C at a rate of 3 kg/s. If the overall heat transfer coefficient is 950 W/(m$^2$ · °C), determine the rate of heat transfer and the heat transfer surface area of the heat exchanger using the ε-NTU method. *Answers:* 31.35 kW, 0.482 m$^2$

**13-83** Glycerin [$C_p$ = 2400 J/(kg · °C)] at 20°C and 0.3 kg/s is to be heated by ethylene glycol [$C_p$ = 2500 J/(kg · °C)] at 60°C and the same mass flow rate in a thin-walled double pipe parallel-flow heat exchanger. If the overall heat transfer coefficient is 240 W/(m$^2$ · °C) and the heat transfer surface area is 7.6 m$^2$, determine (*a*) the rate of heat transfer and (*b*) the outlet temperatures of the glycerin and the glycol.

**13-84** A cross-flow heat exchanger consists of 40 thin-walled tubes of 1-cm diameter located in a duct of 1 m × 1 m cross-section. There are no fins attached to the tubes. Cold water [$C_p$ = 4180 J/(kg · °C)] enters the tubes at 18°C with an average velocity of 3 m/s, while hot air [$C_p$ = 1010 J/(kg · °C)] enters the channel at 130°C and 105 kPa at an average velocity of 12 m/s. If the overall heat transfer coefficient is 80 W/(m$^2$ · °C), determine the outlet temperatures of both fluids and the rate of heat transfer.

**13-85** A shell-and-tube heat exchanger with two shell passes and eight tube passes is used to heat ethyl alcohol [$C_p$ = 2670 J/(kg · °C)] in the tubes from 25°C to 70°C at a rate of 2.1 kg/s. The heating is to be done by water [$C_p$ = 4190 J/(kg · °C)], which enters the shell at 95°C and leaves at 45°C. If the overall heat transfer coefficient is 800 W/(m$^2$ · °C), determine the heat transfer surface area of the heat exchanger using (*a*) the LMTD method and (*b*) the ε-NTU method. *Answer:* 17.4 m$^2$

**13-86** Steam is to be condensed on the shell side of a one-shell-pass and eight-tube-passes condenser, with 50 tubes in each pass, at 30°C ($h_{fg}$ = 2430 kJ/kg). Cooling water [$C_p$ = 4810 J/(kg · °C)] enters the tubes at 15°C at a rate of 1800 kg/h. The tubes are thin-walled, and have a diameter of 1.5 cm and length of 2 m per pass. If the overall heat transfer coefficient is 3000 W/(m$^2$ · °C), determine (*a*) the rate of heat transfer and (*b*) the rate of condensation of steam.

**13-87** Cold water [$C_p$ = 4180 J/(kg · °C)] enters the tubes of a heat exchanger with 2 shell passes and 20 tube passes at 20°C at a rate of 3 kg/s, while hot oil [$C_p$ = 2200 J/(kg · °C)] enters the shell at 130°C at the same mass flow rate. The overall heat transfer coefficient based on the outer surface of the tube is 300 W/($m^2$ · °C) and the heat transfer surface area on that side is 20 $m^2$. Determine the rate of heat transfer using (*a*) the LMTD method and (*b*) the $\varepsilon$-NTU method.

### Selection of Heat Exchangers

**13-88C** A heat exchanger is to be selected to cool a hot liquid chemical at a specified rate to a specified temperature. Explain the steps involved in the selection process.

**13-89C** There are two heat exchangers that can meet the heat transfer requirements of a facility. One is smaller and cheaper, but requires a larger pump, while the other is larger and more expensive, but has a smaller pressure drop and thus requires a smaller pump. Both heat exchangers have the same life expectancy, and meet all other requirements. Explain which heat exchanger you would choose under what conditions.

**13-90C** There are two heat exchangers that can meet the heat transfer requirements of a facility. Both have the same pumping power requirements, the same useful life, and the same price tag. But one is heavier and larger in size. Under what conditions would you choose the smaller one?

**13-91** A heat exchanger is to cool oil [$C_p$ = 2200 J/(kg · °C)] at a rate of 20 kg/s from 120°C to 50°C by air. Determine the heat transfer rating of the heat exchanger, and propose a suitable type.

**13-92** A shell-and-tube process heater is to be selected to heat water [$C_p$ = 4190 J/(kg · °C)] from 20°C to 90°C by steam flowing on the shell side. The heat transfer load of the heater is 600 kW. If the inner diameter of the tubes is 1 cm and the velocity of water is not to exceed 3 m/s, determine how many tube passes need to be used in the heat exchanger.

**13-93** The condenser of a large power plant is to remove 500 MW of heat from steam condensing at 30°C ($h_{fg}$ = 2430 J/kg). The cooling is to be accomplished by cooling water [$C_p$ = 4180 J/(kg · °C)] from a nearby river, which enters the tubes at 18°C and leaves at 26°C. The tubes of the heat exchanger have an internal diameter of 2 cm, and the overall heat transfer coefficient is 3500 W/($m^2$ · °C). Determine the total length of the tubes required in the condenser. What type of heat exchanger is suitable for this task? *Answer:* 312.3 km

**13-94** Repeat Prob. 13-93 for a heat transfer load of 300 MW.

### Review Problems

**13-95** Hot oil is to be cooled in a multipass shell-and-tube heat exchanger by water. The oil flows through the shell, with a heat transfer coefficient of $h_o = 35\ W/(m^2 \cdot °C)$, and the water flows through the tube with an average velocity of 3 m/s. The tube is made of brass $[k = 110\ W/(m \cdot °C)]$ with internal and external diameters of 1.3 cm and 1.5 cm, respectively. Using water properties at 300 K, determine the overall heat transfer coefficient of this heat exchanger based on the inner surface.

**13-96** Repeat Problem 13-95 by assuming a fouling factor $R_{f,o} = 0.0004\ m^2 \cdot °C/W$ on the outer surface of the tube.

**13-97** Cold water $[C_p = 4180\ J/(kg \cdot °C)]$ enters the tubes of a heat exchanger with 2 shell passes and 20 tube passes at 20°C at a rate of 3 kg/s, while hot oil $[C_p = 2200\ J/(kg \cdot °C)]$ enters the shell at 130°C at the same mass flow rate and leaves at 60°C. If the overall heat transfer coefficient based on the outer surface of the tube is $300\ W/(m^2 \cdot °C)$, determine (*a*) the rate of heat transfer and (*b*) the heat transfer surface area on the outer side of the tube. *Answers:* (*a*) 462 kW, (*b*) 71.6 $m^2$

**13-98** Water $[C_p = 4180\ J/(kg \cdot °C)]$ is to be heated by solar-heated hot air $[C_p = 1010\ J/(kg \cdot °C)]$ in a double-pipe counter-flow heat exchanger. Air enters the heat exchanger at 90°C at a rate of 0.3 kg/s, and leaves at 50°C. Water enters at 22°C at a rate of 0.1 kg/s. The overall heat transfer coefficient based on the inner side of the tube is given to be $80\ W/(m^2 \cdot °C)$. Determine the length of the tube required for a tube internal diameter of 1.2 cm.

**13-98E** Water $[C_p = 1.0\ Btu/(lbm \cdot °F)]$ is to be heated by solar-heated hot air $[C_p = 0.24\ Btu/(lbm \cdot °F)]$ in a double-pipe counter-flow heat exchanger. Air enters the heat exchanger at 190°F at a rate of 0.7 lbm/s, and leaves at 120°F. Water enters at 70°F at a rate of 0.2 lbm/s. The overall heat transfer coefficient based on the inner side of the tube is given to be $20\ Btu/(h \cdot ft^2 \cdot °F)$. Determine the length of the tube required for a tube internal diameter of 0.5 in.

**13-99** By taking the limit as $\Delta T_2 \rightarrow \Delta T_1$, show that when $\Delta T_1 = \Delta T_2$ for a heat exchanger, the $\Delta T_{lm}$ relation reduces to $\Delta T_{lm} = \Delta T_1 = \Delta T_2$.

**13-100** The condenser of a room air-conditioner is designed to reject heat at a rate of 15,000 kJ/h from Refrigerant-134a as the refrigerant is condensed at a temperature of 40°C. Air $[C_p = 1005\ J/(kg \cdot °C)]$ flows across the finned condenser coils, entering at 25°C and leaving at 35°C. If the overall heat transfer coefficient based on the refrigerant side is $150\ W/(m^2 \cdot °C)$, determine the heat transfer area on the refrigerant side. *Answer:* 3.05 $m^2$.

**13-101** Air $[C_p = 1005\ J/(kg \cdot °C)]$ is to be preheated by hot exhaust gases in a crossflow heat exchanger before it enters the furnace. Air enters the heat exchanger at 95 kPa and 20°C at a rate of 0.8 $m^3/s$. The combustion gases $[C_p = 1100\ J/(kg \cdot °C)]$ enter at 180°C at a rate of 1.1 kg/s and leave at 95°C. The product of the overall heat transfer coefficient and the heat transfer surface area is $AU = 1200\ W/°C$. Assuming both fluids to be unmixed, determine the rate of heat transfer.

**13-102** In a chemical plant, a certain chemical is heated by hot water supplied by a natural gas furnace. The hot water [$C_p$ = 4180 J/(kg · °C)] is then discharged at 60°C at a rate of 8 kg/min. The plant operates 8 h a day, 5 days a week, and 52 weeks a year. The furnace has an efficiency of 78 percent, and the cost of the natural gas is $0.54 per therm (1 therm = 100,000 Btu = 105,500 kJ). The average temperature of the cold water entering the furnace throughout the year is 14°C. In order to save energy, it is proposed to install a water-to-water heat exchanger to preheat the incoming cold water by the drained hot water. Assuming that the heat exchanger will recover 72 percent of the available heat in the hot water, determine the heat transfer rating of the heat exchanger that needs to be purchased, and suggest a suitable type. Also determine the amount of money this heat exchanger will save the company per year from natural gas savings.

## Computer, Design, And Essay Problems

**13-103** Write an interactive computer program that will give the effectiveness of a heat exchanger and the outlet temperatures of both the hot and cold fluids when the type of fluids, the inlet temperatures, the mass flow rates, the heat transfer surface area, the overall heat transfer coefficient, and the type of heat exchanger are specified. The program should allow the user to select from the fluids water, engine oil, glycerin, ethyl alcohol, and ammonia. Assume constant specific heats at about room temperature.

**13-104** Repeat the problem above, accounting for the variation of specific heats with temperature.

**13-105** Water flows through a shower head steadily at a rate of 8 kg/min. The water is heated in an electric water heater from 15°C to 45°C. In an attempt to conserve energy, it is proposed to pass the drained warm water at a temperature of 38°C through a heat exchanger to preheat the incoming cold water. Design a heat exchanger that is suitable for this task, and discuss the potential savings in energy and money for your area.

**13-106** Open the engine compartment of your car, and search for heat exchangers. How many do you have? What type are they? Why do you think those specific types are selected? If you were redesigning the car, would you use different kinds? Explain.

**13-107** Write an essay on the static and dynamic types of regenerative heat exchangers, and compile information about the manufacturers of such heat exchangers. Choose a few models by different manufacturers, and compare their costs and performance.

# Cooling of Electronic Equipment

CHAPTER 14

Electronic equipment has made its way into practically every aspect of modern life, from toys and appliances to high-power computers. The reliability of the electronics of a system is a major factor in the overall reliability of the system. Electronic components depend on the passage of electric current to perform their duties, and they become potential sites for excessive heating, since the current flow through a resistance is always accompanied by heat generation. The continued miniaturization of electronic systems has resulted in a dramatic increase in the amount of heat generated per unit volume, comparable in magnitude to those encountered at nuclear reactors and the surface of the sun. Unless properly designed and controlled, high rates of heat generation results in high operating temperatures for electronic equipment, which jeopardizes its safety and reliability. The failure rate of electronic equipment increases exponentially with temperature. Also, the high thermal stresses in the solder joints of electronic components mounted on circuit boards resulting from temperature variations are major causes of failure. Therefore, thermal control has become increasingly important in the design and operation of electronic equipment.

In an electronic system, the electronic components are the sites of heat generation, and thus they normally form the hottest spots. The essence of thermal design is the removal of this internally generated heat by providing an effective path for heat flow from the components themselves to the surrounding medium. There are several ways of doing this, and the proper way depends on the situation on hand. In this chapter, we discuss several cooling techniques commonly used in electronic equipment such as conduction cooling, natural convection and radiation cooling, forced air cooling, liquid cooling, immersion cooling, and heat pipes. This chapter is intended to familiarize the reader with these techniques, and put them into perspective. The reader interested in an in-depth coverage of any of these topics can consult numerous other sources available, such as those listed in the references.

**Introduction and History**

The field of electronics deals with the construction and utilization of devices that involve current flow through a vacuum, a gas, or a semiconductor. This exciting field of science and engineering dates back to 1883, when Thomas Edison discovered the vacuum diode. The **vacuum tube** served as the foundation of electronics industry until 1950s, and played a central role in the development of radio, TV, radar, and the digital computer. Of the several computers developed in this era, the largest and best known is the ENIAC (Electronic Numerical Integrator and Computer), which was built at the Univeristy of Pennsylvania in 1946. It had over 18,000 vacuum tubes, and occupied a room 7 m × 14 m in size. It consumed a large amount of power, and its reliability was poor because of the high failure rate of the vacuum tubes.

The invention of bipolar **transistor** in 1948 marked the beginning of a new era in electronics industry, and the obsolescence of vacuum tube technology. Transistor circuits performed the functions of the vacuum tubes with greater reliability, while occupying negligible space and consuming negligible power compared with vacuum tubes. The first transistors were made from germanium, which could not function properly at temperatures above 100°C. Soon they were replaced by silicon transistors, which could operate at much higher temperatures.

The next turning point in electronics occurred in 1959 with the introduction of the **integrated circuits** (IC), where several components such as diodes, transistors, resistors, and capacitors are placed in a single chip. The number of components packed in a single chip has been increasing steadily since then at an amazing rate, as shown in Fig. 14-1. The continued miniaturization of electronic components has resulted in *medium-scale integration* (MSI) in the 1960s with 50–1000 components per chip, *large-scale integration* (LSI) in the 1970s with 1000–100,000 components per chip, and *very large-scale integration* (VLSI) in the 1980s with 100,000–10,000,000 components per chip. Today it is not unusual to have a chip 3 cm × 3 cm in size with several million components on it.

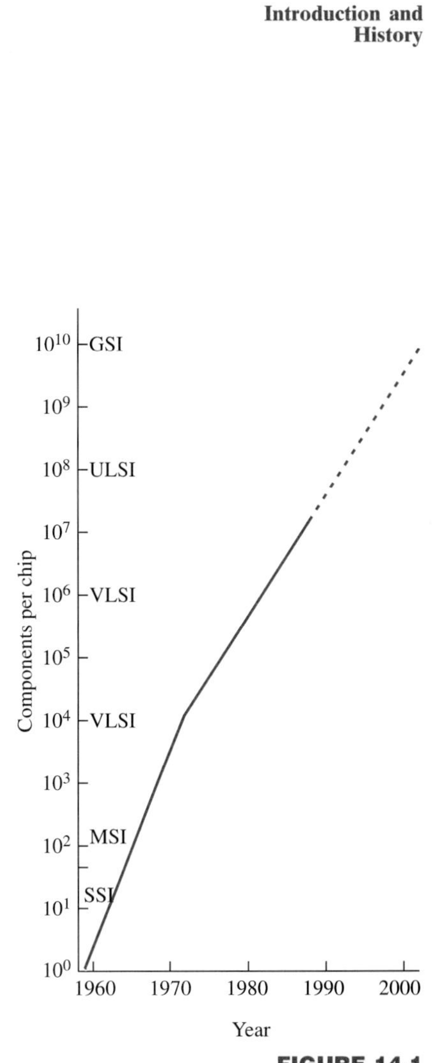


**FIGURE 14-1**

The increase in the number of components packed on a chip over the years.

The development of the **microprocessor** in the early 1970s by the Intel Corporation marked yet another beginning in the electronics industry. The accompanying rapid development of large-capacity memory chips in this decade made it possible to introduce capable personal computers for use at work or at home at an affordable price. Electronics has made its way into practically everything from watches to household appliances to automobiles. Today it is difficult to imagine a new product that does not involve any electronic parts.

The current flow through a resistance is always accompanied by *heat generation* in the amount of $I^2R$, where $I$ is the electric current and $R$ is the resistance. When the transistor was first introduced, it was touted in the newspapers as a device that "produces no heat." This certainly was a fair statement, considering the huge amount of heat generated by vacuum tubes. Obviously, the little heat generated in the transistor was no match to that generated in its prodecessor. But when thousands or even millions of such components are packed in a small volume, the heat generated increases to such high levels that its removal becomes a formidable task and a major concern for the safety and reliability of the electronic devices. The heat fluxes encountered in electronic devices range from less than 1 W/cm$^2$ to more than 100 W/cm$^2$.

Heat is generated in a resistive element for as long as current continues to flow through it. This creates a *heat build-up* and a subsequent *temperature rise* at and around the component. The temperature of the component will continue rising until the component is destroyed unless heat is transferred away from it. The temperature of the component will remain constant when the rate of heat removal from it equals the rate of heat generation.

Individual electronic components have *no moving parts,* and thus nothing to wear out with time. Therefore, they are inherently reliable, and it seems as if they can operate safely for many years. Indeed, this would be the case if components operated at room temperature. But electronic components are observed to fail under prolonged use at high temperatures. Possible causes of failure are *diffusion* in semiconductor materials, *chemical reactions,* and *creep* in the bonding materials, among other things. The failure rate of electronic devices increases almost *exponentially* with the operating temperature, as shown in Fig. 14-2. The cooler the electronic device operates, the more reliable it is. A rule of thumb is that the failure rate of electronic components is halved for each 10°C reduction in their junction temperature.

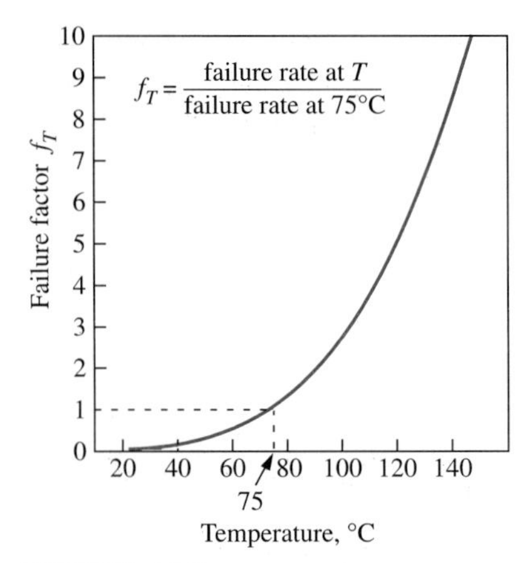


**FIGURE 14-2**

The increase in the failure rate of bipolar digital devices with temperature (from Ref. 17).

## 14-2 ■ MANUFACTURING OF ELECTRONIC EQUIPMENT

The narrow band where two different regions of a semiconductor (such as the p-type and n-type regions) come in contact is called a **junction**. A transistor, for example, involves two such junctions, and a diode, which is the simplest semiconductor device, is based on a single p–n junction. In heat transfer analysis, the circuitry of an electronic component through which electrons flow and thus heat is generated is also referred to as the junction. That is, junctions are the sites of heat generation and thus the hottest spots in a component. In silicon-based semiconductor devices, the junction temperature is limited to 125°C for safe operation. However, lower junction temperatures are desirable for extended life and lower maintenance costs. In a typical application, numerous electronic components, some smaller than 1 $\mu$m in size, are formed from a silicon wafer into a chip.

### The Chip Carrier

The chip is housed in a **chip carrier** or substrate made of ceramic, plastic, or glass in order to protect its delicate circuitry from the detrimental effects of the environment. The chip carrier provides a rugged housing for the safe handling of the chip during the manufacturing process, as well as the connectors between the chip and the circuit board. The various components of the chip carrier are shown in Fig. 14-3. The chip is secured in the carrier by bonding it to the bottom surface. The thermal expansion coefficient of the *plastic* is about 20 times that of silicon. Therefore, bonding the silicon chip directly to the plastic case would result in such large thermal stresses that the reliability would be seriously jeopardized. To avoid this problem, a *lead frame* made of a copper alloy with a thermal expansion coefficient close to that of silicion is used as the bonding surface.

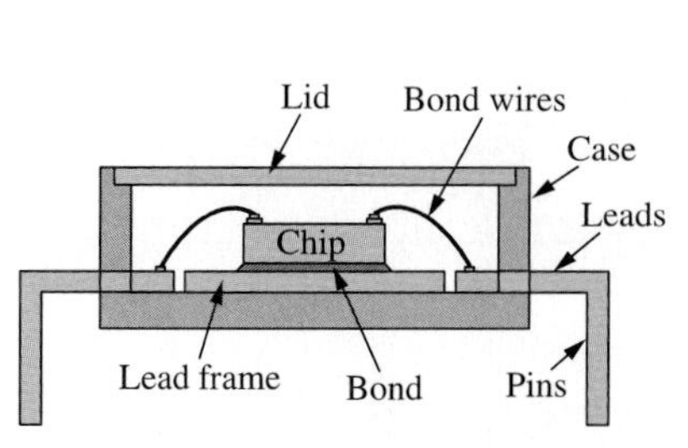


**FIGURE 14-3**
The components of a chip carrier (from Ref. 4).

The design of the chip carrier is the *first level* in the thermal control of electronic devices, since the transfer of heat from the chip to the chip carrier is the first step in the dissipation of the heat generated on the chip. The heat generated on the chip is transferred to the case of the chip carrier by a combination of conduction, convection, and radiation. However, it is obvious from the figure that the common chip carrier is designed with the *electrical aspects* in mind, and little consideration is given to the *thermal aspects.* First of all, the cavity of the chip carrier is filled with a gas, which is a poor heat conductor, and the case is often made of materials that are also poor conductors of heat. This results in a relatively large thermal resistance between the chip and the case, called the **junction-to-case resistance**, and thus a large temperature difference. As a result, the temperature of the chip will be much higher than that of the case for a specified heat dissipation rate. The junction-to-case thermal resistance depends on the *geometry* and the *size* of the chip and the chip carrier as well as the material properties of the *bonding* and the *case.* It varies considerably from one device to another, and ranges from about 10°C/W to more than 100°C/W.

*Moisture* in the cavity of the chip carrier is highly undesirable, since it causes corrosion on the wiring. Therefore, chip carriers are made of materials that prevent the entry of moisture by diffusion, and are *hermetically sealed* in order to prevent the direct entry of moisture through cracks. Materials that outgas are also not permitted in the chip cavity, because such gases can also cause corrosion. In products with strict hermeticity requirements the more expensive ceramic cases are used instead of the plastic ones.

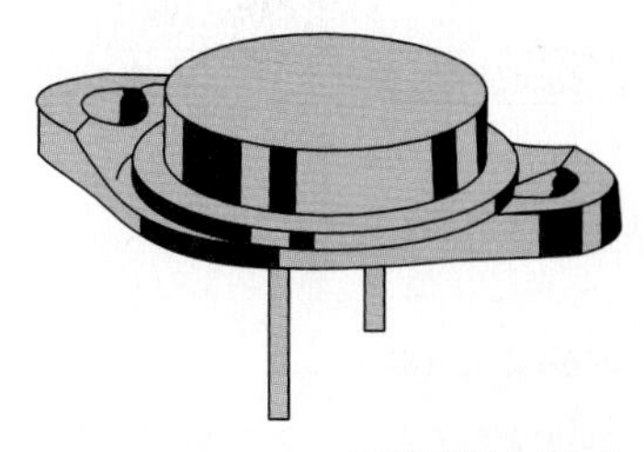

**FIGURE 14-4**
The chip carrier for a high-power transistor attached to a flange for enhanced heat transfer.

A common type of chip carrier for high-power transistors is shown in Fig. 14-4. The transistor is formed on a small silicon chip housed in the disk-shaped cavity, and the I/O pins come out from the bottom. The case of the transistor carrier is usually attached directly to a flange, which provides a large surface area for heat dissipation, and reduces the junction-to-case thermal resistance.

It is often desirable to house more than one chip in a single chip carrier. The result is a *hybrid* or *multichip package.* Hybrid packages house several chips, individual electronic components, and ordinary circuit elements connected to each other in a single chip carrier. The result is improved performance due to the shortening of the wiring lengths, and enhanced reliability. Lower cost would be an added benefit of multichip packages if they are produced in sufficiently large quantity.

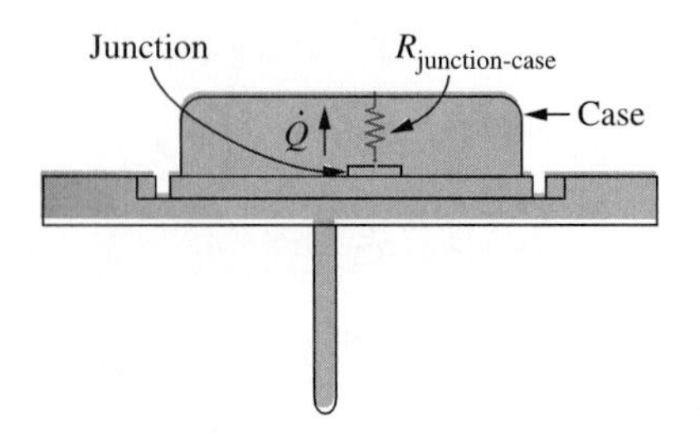


**FIGURE 14-5**

Schematic for Example 14-1.

**EXAMPLE 14-1**

The temperature of the case of a power transistor that is dissipating 3 W is measured to be 50°C. If the junction-to-case resistance of this transistor is specified by the manufacturer to be 15°C/W, determine the temperature at the junction of the transistor.

**Solution** The schematic of the transistor is given in Fig. 14-5. The rate of heat transfer between the junction and the case in steady operation can be expressed as

$$\dot{Q} = \left(\frac{\Delta T}{R}\right)_{\text{junction-case}} = \frac{T_{\text{junction}} - T_{\text{case}}}{R_{\text{junction-case}}}$$

Then the junction temperature becomes

$$\begin{aligned} T_{\text{junction}} &= T_{\text{case}} + \dot{Q}R_{\text{junction-case}} \\ &= 50°\text{C} + (3\text{ W})(15°\text{C/W}) \\ &= 95°\text{C} \end{aligned}$$

Therefore, the temperature of the transistor junction will be 95°C when its case is at 50°C.

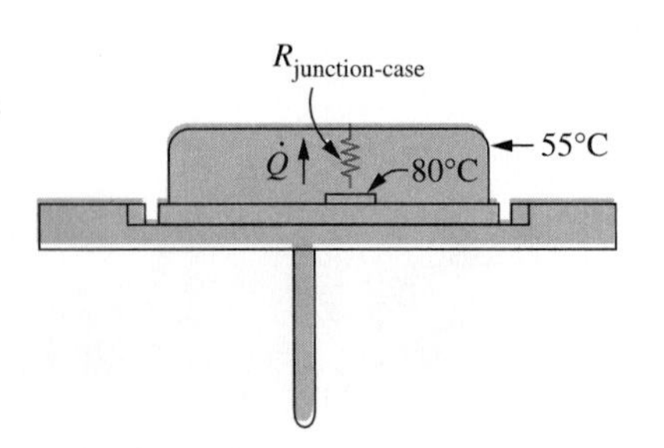


**FIGURE 14-6**

Schematic for Example 14-2.

**EXAMPLE 14-2**

The following experiment is conducted to determine the junction-to-case thermal resistance of an electronic component. Power is supplied to the component from a 15-V source, and the variation in the electric current and in the junction and the case temperatures with time are observed. When things are stabilized, the current is observed to be 0.1 A and the temperatures to be 80°C and 55°C at the junction and the case, respectively. Calculate the junction-to-case resistance of this component.

**Solution** The schematic of the component is given in Fig. 14-6. The electric power consumed by this electronic component is

$$\dot{W}_e = VI = (15\text{ V})(0.1\text{ A}) = 1.5\text{ W}$$

In steady operation, this is equivalent to the heat dissipated by the component. That is,

$$\dot{Q} = \left(\frac{\Delta T}{R}\right)_{\text{junction-case}} = \frac{T_{\text{junction}} - T_{\text{case}}}{R_{\text{junction-case}}} = 1.5\text{ W}$$

Then the junction-to-case resistance is determined to be

$$R_{\text{junction-case}} = \frac{T_{\text{junction}} - T_{\text{case}}}{\dot{Q}} = \frac{(80 - 55)°\text{C}}{1.5\text{ W}} = 16.7°\text{C/W}$$

Therefore, a temperature difference of 16.7°C will occur between the electronic circuitry and the case of the chip carrier for each W of power consumed by the component.

**FIGURE 14-7**
A printed circuit board (PCB) with a variety of components on it (courtesty of Litton Systems, Inc.).

## Printed Circuit Boards

A **printed circuit board** (PCB) is a properly wired plane board made of polymers and glass–epoxy materials on which various electronic components such as the ICs, diodes, transistors, resistors, and capacitors are mounted to perform a certain task, as shown in Fig. 14-7. The PCBs are commonly called *cards,* and they can be replaced easily during a repair. The PCBs are plane boards, usually 10 cm wide and 15 cm long and only a few millimeters thick, and they are not suitable for heavy components such as transformers. Usually a copper cladding is added on one or both sides of the board. The cladding on one side is subjected to an etching process to form wiring strips and attachment pads for the components. The power dissipated by a PCB usually ranges from 5 W to about 30 W.

A typical electronic system involves several layers of PCBs. The PCBs are usually cooled by direct contact with a fluid such as air flowing between the boards. But when the boards are placed in a hermetically sealed enclosure, they must be cooled by a cold plate (a heat exchanger) in contact with the edge of the boards. The *device-to-board edge thermal resistance* of a PCB is usually high (about 20–60°C/W) because of the small thickness of the board and the low thermal conductivity of the board material. In such cases, even a thin layer of copper cladding on one side of the board can decrease the device-to-board edge thermal resistance in the plane of the board and enhance heat transfer in that direction drastically.

In the thermal design of a PCB, it is important to pay particular attention to the components that are not tolerant to high temperatures, such as certain high-performance capacitors, and to ensure their safe operation. Often when one component on a PCB fails, the whole board fails and must be replaced.

Printed circuit boards come in three types: *single-sided, double-sided,* and *multilayer* boards. Each type has its own strengths and weaknesses. Single-sided PCBs have circuitry lines on one side of the board only, and are suitable for low-density electronic devices (10–20 components). Double-sided PCBs have circuits on both sides, and are best suited for intermediate-density devices. Multilayer PCBs contain several layers of circuitry, and are suitable for high-density devices. They are equivalent to several PCBs sandwiched together.

The single-sided PCB has the lowest cost, as expected, and it is easy to maintain. But it occupies a lot of space. The multilayer PCB, on the other hand, allows the placement of a large number of components in a three-dimensional configuration, but it has the highest initial cost and is difficult to repair. Also, temperatures are most likely to be the highest in multilayer PCBs.

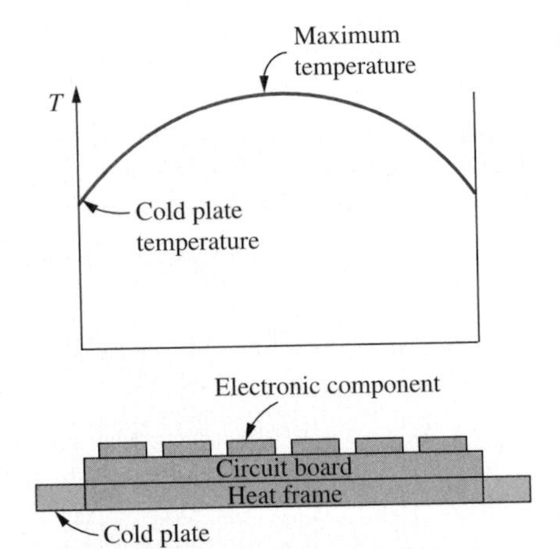


**FIGURE 14-8**
The path of heat flow in a conduction cooled PCB and the temperature distribution.

In cricital applications, the electronic components are placed on boards attached to a conductive metal, called the *heat frame,* which serves as a conduction path to the edge of circuit board and thus to the cold plate for the heat generated in the components. Such boards are said to be *conduction-cooled.* The temperature of the components in this case will depend on the location of the components on the boards: it will be highest for the components in the middle, and lowest for those near the edge, as shown in Fig. 14-8.

Materials used in the fabrication of circuit boards should be (1) effective electrical insulators to prevent electrical breakdown and (2) good heat conductors to conduct the heat generated away. They should also have (3) high material strength to withstand forces and to maintain dimensional stability, (4) thermal expansion coefficients that closely match that of copper, to prevent cracking in the copper cladding during thermal cycling, (5) resistance to moisture absorption, since moisture can effect both mechanical and electrical properties and degrade performance, (6) stability in properties at temperature levels encountered in electronic applications, (7) ready availability and manufacturability, and, of course, (8) low cost. As you might have already guessed, no existing material has all of these desirable characteristics.

Glass–epoxy laminates made of an epoxy or polymide matrix reinforced by several layers of woven glass cloth are commonly used in the production of circuit boards. Polyimide matrices are more expensive than epoxy, but can withstand much higher temperatures. Polymer or polyimide films are also used without reinforcement for flexible circuits.

### The Enclosure



An electronic system is not complete without a rugged enclosure (a case or a cabinet) that will *house* the circuit boards and the necessary peripheral equipment and connectors, *protect* them from the detrimental effects of the environment, and *provide* a cooling mechanism (Fig. 14-9). In a small electronic system such a personal computer, the enclosure can simply be an inexpensive box made of sheet metal with proper connectors and a small fan. But for a large system with several hundred PCBs, the design and construction of the enclosure is a challenge for both electronic and thermal designers. An enclosure must provide *easy access* for service personnel so that they can identify and replace any defective parts easily and quickly in order to minimize down time, which can be very costly. But, at the same time, the enclosure must *prevent* any easy access by unauthorized people in order to protect the sensitive electronics from them as well as the people from possible electrical hazards. Electronic circuits are powered by low voltages (usually under $\pm 15$ V), but the currents involved may be very high (sometimes a few hundred amperes).

**FIGURE 14-9**
A cabinet style enclosure.

Plug-in type circuit boards make it very easy to replace a defective board, and are commonly used in low-power electronic equipment. High-power circuit boards in large systems, however, are tightly attached to the racks of the cabinet with special brackets. A well-designed enclosure also includes switches, indicator lights, a screen to display messages and present information about the operation, and a key pad for user interface.

The printed circuit boards (PCBs) in a large system are plugged into a *back panel* through their edge connectors. The back panel supplies power to the PCBs, and interconnects them to facilitate the passage of current from one board to another. The PCBs are assembled in an orderly manner in card racks or chassis. One or more such assemblies are housed in a *cabinet,* as shown in Fig. 14-10.

Electronic enclosures come in a wide variety of sizes and shapes. Sheet metals such as thin-gauge aluminum or steel sheets are commonly used in the production of enclosures. The thickness of the enclosure walls depends on the shock and vibration requirements. Enclosures made of thick metal sheets or by casting can meet these requirements, but at the expense of increased weight and cost.

Electronic boxes are sometimes sealed to prevent the fluid inside (usually air) from leaking out and the water vapor outside from leaking in. Sealing against moisture migration is very difficult, because of the small size of the water molecule and the large vapor pressure outside the box relative to that within the box. Sealing adds to the size, weight, and cost of an electronic box, especially in space or high-altitude operation, since the box in this case must withstand the larger forces due to the higher pressure differential between the inside and outside the box.

## 14-3 ■ COOLING LOAD OF ELECTRONIC EQUIPMENT

The first step in the selection and design of a cooling system is the determination of the heat dissipation, which constitutes the *cooling load.*

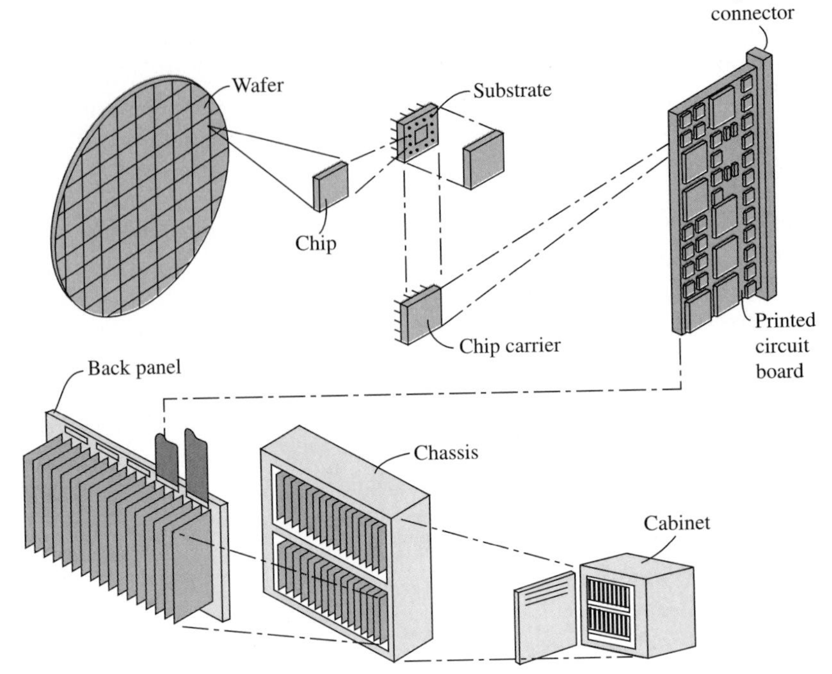


**FIGURE 14-10**

Different stages involved in the production of an electronic system (from Ref. 4).

The easiest way to determine the power dissipation of electronic equipment is to measure the voltage applied $V$ and the electric current $I$ at the entrance of the electronic device under full load conditions, and to substitute them into the relation

$$\dot{W}_e = VI = I^2R \quad \text{(W)} \tag{14-1}$$

where $\dot{W}_e$ is the electric power consumption of the electronic device, which constitutes the *energy input* to the device.

The first law of thermodynamics requires that in *steady* operation the energy input into a system be equal to the energy output from the system. Considering that the only form of energy leaving the electronic device is heat generated as the current flows through resistive elements, we conclude that the heat dissipation or cooling load of an electronic device is equal to its power consumption. That is, $\dot{Q} = \dot{W}_e$, as shown in Fig. 14-11. The exception to this rule is equipment that outputs other forms of energy as well, such as the emitter tubes of a radar, radio, or TV installation emitting radiofrequency (RF) electromagnetic radiation. In such cases, the cooling load will be equal to the difference between the power consumption and the RF power emission. An equivalent but cumbersome way of determining the cooling load of an electronic device is to determine the heat dissipated by each component in the device, and then to add them up.

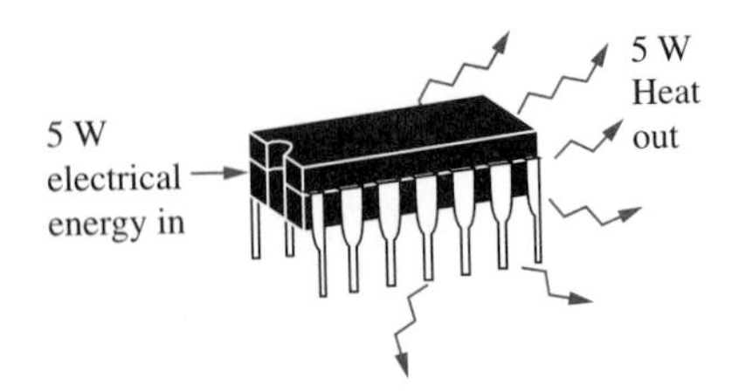


**FIGURE 14-11**

In the absence of other energy interactions, the heat output of an electronic device in steady operation is equal to the power input to the device.

The discovery of *superconductor* materials that can operate at room temperature will cause drastic changes in the design of electronic devices and cooling techniques, since such devices will generate hardly any heat. As a result, more components can be packed into a smaller volume, resulting in enhanced speed and reliability without having to resort to exotic cooling techniques.

Once the cooling load has been determined, it is common practice to inflate this number to leave some *safety margin* or a "cushion," and to make some allowance for future growth. It is not uncommon to add another card to an existing system (such as adding a fax/modem card to a PC) to perform an additional task. But we should not go overboard in being conservative, since an *oversized* cooling system will cost more, occupy more space, be heavier, and consume more power. For example, there is no need to install a large and noisy fan in an electronic system just to be "safe" when a smaller one will do. For the same reason, there is no need to use an expensive and failure-prone liquid cooling system when air cooling is adequate. We should always keep in mind that the most desirable form of cooling is *natural convection cooling,* since it does not require any moving parts, and thus it is inherently reliable, quite, and best of all, free.

The cooling system of an electronic device must be designed considering the actual *field* operating conditions. In critical applications such as those in the military, the electronic device must undergo extensive testing to satisfy stringent requirements for safety and reliability. Several such codes exist to specify the minimum standards to be met in some applications.

The *duty cycle* is another important consideration in the design and selection of a cooling technique. The actual power dissipated by a device can be considerably less than the rated power, depending on its duty cycle (the fraction of time it is on). A 5 W power transistor, for example, will dissipate an average of 2 W of power if it is active only 40 percent of the time. If the chip of this transistor is 1.5 mm wide, 1.5 mm high, and 0.1 mm thick then the heat flux on the chip will be $(2\text{ W})/(0.15\text{ cm})^2 = 89\text{ W/cm}^2$.

An electronic device that is not running is in *thermal equilibrium* with its surroundings, and thus is at the temperature of the surrounding medium. When the device is turned on, the temperature of the components and thus the device starts rising as a result of absorbing the heat generated. The temperature of the device stabilizes at some value when the heat generated equals the heat removed by the cooling mechanism. At this point, the device is said to have reached *steady* operating conditions. The warming-up period during which the component temperature rises is called the *transient* operation stage (Fig. 14-12).

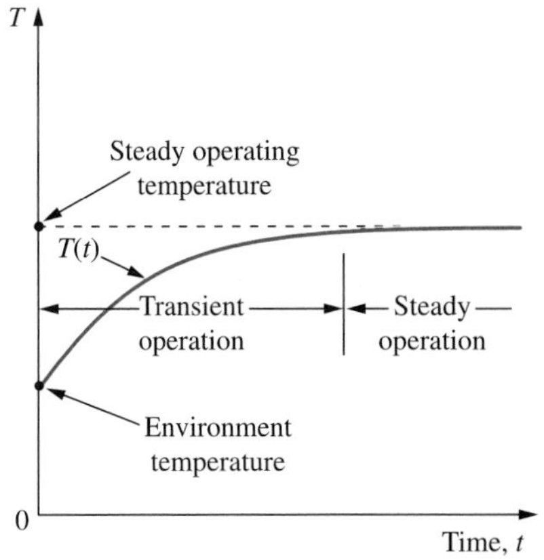

Another thermal factor that undermines the reliability of electronic devices is the thermal stresses caused by *temperature cycling.* In an experimental study (see Ref. 11), the failure rate of electronic devices subjected to deliberate temperature cycling of more than 20°C is observed to increase eightfold. *Shock* and *vibration* are other common

**FIGURE 14-12**
The temperature change of an electronic component with time as it reaches steady operating temperature after it is turned on.

causes of failure for electronic devices, and should be considered in the design and manufacturing process for increased reliability.

Most electronic devices operate for long periods of time, and thus their cooling mechanism is designed for steady operation. But electronic devices in some applications never run long enough to reach steady operation. In such cases, it may be sufficient to use a limited cooling technique, such as thermal storage for a short period, or not to use one at all. Transient operation can also be caused by large swings in the environmental conditions. A common cooling technique for transient operation is to use a double-wall construction for the enclosure of the electronic equipment, with the space between the walls filled with a wax with a suitable melting temperature. As the wax melts, it absorbs a large amount of heat and thus delays overheating of the electronic components considerably. During off periods, the wax solidifies by rejecting heat to the environment.

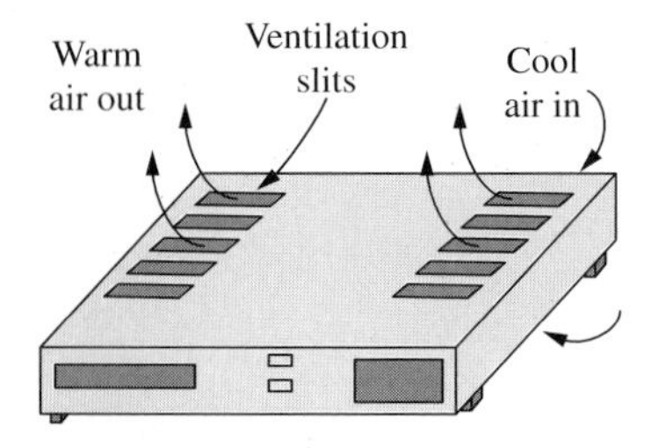


**FIGURE 14-13**
Strategically located ventilation holes are adequate to cool low-power electronics such as a TV or VCR.

## 14-4 ■ THERMAL ENVIRONMENT

An important consideration in the selection of a cooling technique is the *environment* in which the electronic equipment is to operate. Simple *ventilation holes* on the case may be all we need for the cooling of low-power-density electronics such as a TV or a VCR in a room, and a *fan* may be adequate for the safe operation of a home computer (Fig. 14-13). But the thermal control of the electronics of an *aircraft* will challenge thermal designers, since the environmental conditions in this case will swing from one extreme to another in a matter of minutes. The expected *duration of operation* in a hostile environment is also an important consideration in the design process. The thermal design of the electronics for an aircraft that cruises for hours each time it takes off will be quite different than that of a missile that has an operation time of a few minutes.

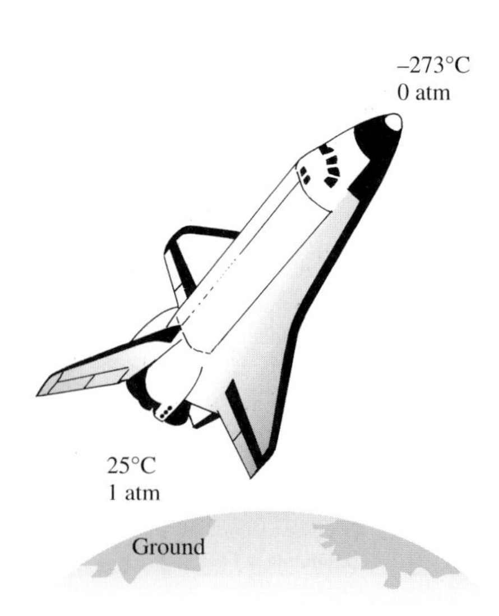


**FIGURE 14-14**
The thermal environment of a spacecraft changes drastically in a short time, and this complicates the thermal control of the electronics.

The thermal environment in *marine applications* is relatively stable, since the ultimate heat sink in this case is water with a temperature range of 0°C to 30°C. For *ground applications,* however, the ultimate heat sink is the atmospheric air, whose temperature varies from −50°C at polar regions to +50°C in desert climates, and whose pressure ranges from about 70 kPa (0.7 atm) at 3000 m elevation to 107 kPa (1.08 atm) at 500 m below sea level. The combined convection and radiation heat transfer coefficient can range from $10\ \text{W}/(\text{m}^2 \cdot {}^\circ\text{C})$ in calm weather to $80\ \text{W}(\text{m}^2 \cdot {}^\circ\text{C})$ in 100 km/h (62 mph) winds. Also, the surfaces of the devices facing the sun directly can be subjected to solar radiation heat flux of $1000\ \text{W/m}^2$ on a clear day.

In *airborne applications,* the thermal environment can change from 1 atm and 35°C on the ground to 19 kPa (0.2 atm) and −60°C at a typical cruising altitude of 12,000 m in minutes (Fig. 14-14). At supersonic velocities, the surface temperature of some part of the aircraft may rise 200°C above the environment temperature.

Electronic devices are rarely exposed to uncontrolled environmental conditions directly, because of the wide variations in the environmental variables. Instead, a conditioned fluid such as air, water, or a dielectric fluid is used to serve as a local heat sink and as an intermediary between the electronic equipment and the environment, just like the air-conditioned air in a building providing thermal comfort to the human body. Conditioned air is the preferred cooling medium, since it is benign, readily available, and not prone to leakage. But its use is limited to equipment with low power densities, because of the low thermal conductivity of air. The thermal design of electronic equipment in military applications must comply with strict military standards in order to satisfy the utmost reliability requirements.



## 14-5 ■ ELECTRONICS COOLING IN DIFFERENT APPLICATIONS

The cooling techniques used in the cooling of electronic equipment vary widely, depending on the particular application. Electronic equipment designed for *airborne applications* such as airplanes, satellites, space vehicles, and missiles offers challenges to designers because it must fit into odd-shaped spaces because of the curved shape of the bodies, yet be able to provide adequate paths for the flow of fluid and heat. Most such electronic equipment is cooled by forced convection using pressurized air bled off a compressor. This compressed air is usually at a high temperature, and thus it is cooled first by expanding it through a turbine. The moisture in the air is also removed before the air is routed to the electronic boxes. But the removal process may not be adequate under rainy conditions. Therefore, electronics in some cases are placed in sealed finned boxes that are externally cooled to eliminate any direct contact with electronic components.

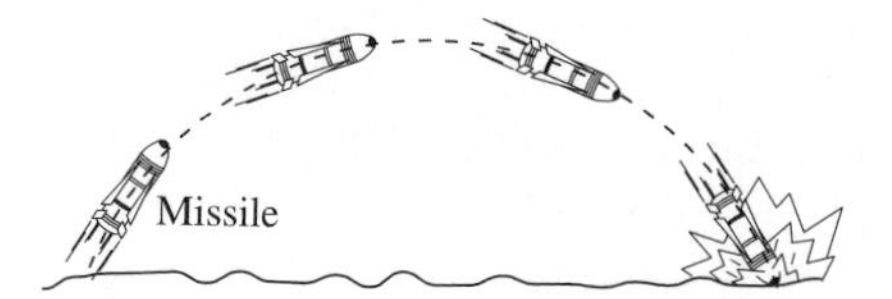

**FIGURE 14-15**
The electronics of short-range missiles may not need any cooling because of the short flight time involved.

The electronics of short-range *missiles* do not need any cooling, because of their short cruising times (Fig. 14-15). The missiles reach their destinations before the electronics reach unsafe temperatures. Long-range missiles such as cruise missiles, however, may have a flight time of several hours. Therefore, they must utilize some form of cooling mechanism. The first thing that comes to mind is to use forced convection with the air that rams the missile by utilizing its large *dynamic pressure.* However, the *dynamic temperature* of air, which is the rise in the temperature of the air as a result of the ramming effect, may be more than 50°C at speeds close to the speed of sound (Fig. 14-16). For example, at a speed of 320 m/s, the dynamic temperature of air is

$$T_{\text{dynamic}} = \frac{\mathcal{V}^2}{2C_p} = \frac{(320\text{ m/s})^2}{2[1005\text{ J/(kg}\cdot°\text{C)}]}\left(\frac{1\text{ J/kg}}{1\text{ m}^2/\text{s}^2}\right) = 51°\text{C} \qquad (14\text{-}2)$$

Therefore, the temperature of air at a velocity of 320 m/s and a temperature of 30°C will rise to 81°C as a result of the conversion of kinetic energy to internal energy. Air at such high temperatures is not suitable for use as a cooling medium. Instead, cruise missiles are often

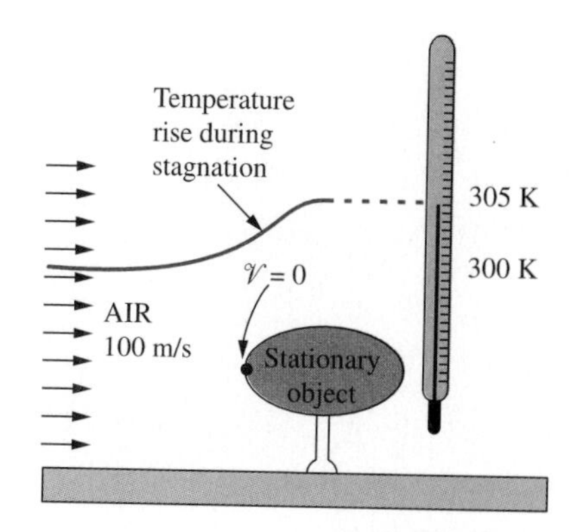


**FIGURE 14-16**
The temperature of a gas having a specific heat $C_p$ flowing at a velocity of $\mathcal{V}$ rises by $\mathcal{V}^2/2C_p$ when it is brought to a complete stop.

cooled by taking advantage of the cooling capacity of the large quantities of *liquid fuel* they carry. The electronics in this case are cooled by passing the fuel through the cold plate of the electronics enclosure as it flows towards the combustion chamber.

Electronic equipment in *space vehicles* is usually cooled by a liquid circulated through the components, where heat is picked up, and then through a space radiator, where the waste heat is radiated into deep space at 0 K. Note that radiation is the only heat transfer mechanism for rejecting heat to the vacuum environment of space, and radiation exchange depends strongly on surface properties. Desirable radiation properties on surfaces can be obtained by special coatings and surface treatments. When electronics in sealed boxes are cooled by a liquid flowing through the outer surface of the electronics box, it is important to run a fan in the box to circulate the air, since there is no natural convection currents in space because of the absence of a gravity field.

Electronic equipment in *ships* and *submarines* is usually housed in rugged cabinets to protect it from vibrations and shock during stormy weather. Because of easy access to water, water-cooled heat exchangers are commonly used to cool seaborn electronics. This is usually done by cooling air in a closed or open loop air-to-water heat exchanger, and forcing the cool air to the electronic cabinet by a fan. When forced air cooling is used, it is important to establish a flow path for air such that no trapped hot air pockets will be formed in the cabinets.

*Communication systems* located at remote locations offer challenges to thermal designers because of the extreme conditions under which they operate. These electronic systems operate for long periods of time under adverse conditions such as rain, snow, high winds, solar radiation, high altitude, high humidity, and extremely high or low temperatures. Large communication systems are housed in specially built shelters. Sometimes it is necessary to air-condition these shelters to safely dissipate the large quantities of heat dissipated by the electronics of communication systems.

Electronic components used in high-power *microwave equipment* such as radars generate enormous amounts of heat because of the low conversion efficiency of electrical energy to microwave energy. Klystron tubes of high-power radar systems where radiofrequency (RF) energy is generated can yield local heat fluxes as high as 2000 W/cm$^2$, which is close to one-third of the heat flux on the sun's surface. The safe and reliable dissipation of these high heat fluxes usually requires the immersion of such equipment in a suitable dielectric fluid that can remove large quantities of heat by boiling.

The manufacturers of electronic devices usually specify the *rate of heat dissipation* and the *maximum allowable component temperature* for reliable operation. These two numbers help us determine the cooling techniques that are suitable for the device under consideration.

The heat fluxes attainable at specified temperature differences are plotted in Fig. 14-17 for some common heat transfer mechanisms. When the power rating of a device or component is given, the heat flux is determined by dividing the power rating by the exposed surface area of

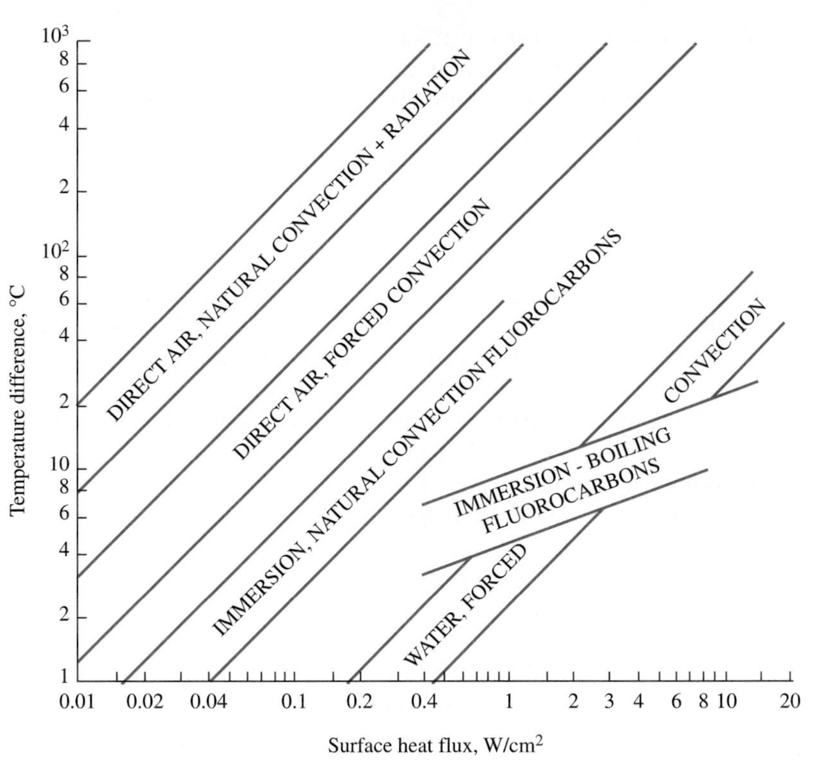


**FIGURE 14-17**
Heat fluxes that can be attained at specified temperature differences with various heat transfer mechanisms and fluids (from Ref. 16, p. 22; reproduced with permission).

the device or component. Then suitable heat transfer mechanisms can be determined from Fig. 14-17 from the requirement that the temperature difference between the surface of the device and the surrounding medium not exceed the allowable maximum value. For example, a heat flux of 0.5 W/cm$^2$ for an electronic component would result in a temperature difference of about 500°C between the component surface and the surrounding air if natural convection in air is used. Considering that the maximum allowable temperature difference is typically under 80°C, the natural convection cooling of this component in air is out of question. But forced convection with air is a viable option if using a fan is acceptable. Note that at heat fluxes greater than 1 W/cm$^2$, even forced convection with air will be inadequate, and we must use a sufficiently large heat sink or switch to a different cooling fluid such as water. Forced convection with water can be used effectively for cooling electronic components with high heat fluxes. Also note that dielectric liquids such as fluorochemicals can remove high heat fluxes by immersing the component directly in them.

## 14-6 ■ CONDUCTION COOLING

Heat is *generated* in electronic components whenever electric current flows through them. The generated heat causes the temperature of the components to rise, and the resulting temperature difference drives the heat away from the components through a path of *least thermal resistance.* The temperature of the components stabilizes when the heat dissipated equals the heat generated. In order to minimize the temperature rise of the components, *effective heat transfer paths* must be established between the components and the ultimate heat sink, which is usually the atmospheric air.

The selection of a cooling mechanism for electronic equipment depends on the *magnitude* of the heat generated, *reliability* requirements, *environmental* conditions, and *cost.* For low-cost electronic equipment, inexpensive cooling mechanisms such as natural or forced convection with air as the cooling medium are commonly used. For high-cost, high-performance electronic equipment, however, it is often necessary to resort to expensive and complicated cooling techniques.

Conduction cooling is based on the *diffusion* of heat through a solid, liquid, or gas as a result of molecular interactions in the absence of any bulk fluid motion. Steady *one-dimensional* heat conduction through a plane medium of thickness $L$, heat transfer surface area $A$, and thermal conductivity $k$ is given by (Fig. 14-18)

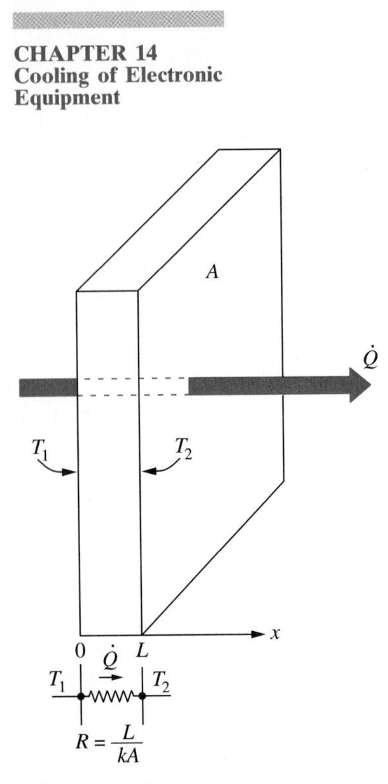


**FIGURE 14-18**
The thermal resistance of a medium is proportional to its length in the direction of heat transfer, and inversely proportional to its heat transfer surface area and thermal conductivity.

$$\dot{Q} = kA\frac{\Delta T}{L} = \frac{\Delta T}{R} \quad \text{(W)} \tag{14-3}$$

where

$$R = \frac{L}{kA} = \frac{\text{length}}{\text{thermal conductivity} \times \text{heat transfer area}} \quad (\text{°C/W}) \tag{14-4}$$

is the *thermal resistance* of the medium and $\Delta T$ is the temperature difference across the medium. Note that this is analogous to *electric current* being equal to the potential difference divided by the electrical resistance.

The thermal resistance concept enables us to solve heat transfer problems in an analogous manner to electric circuit problems using the thermal resistance network, as discussed in Chapter 8. When the rate of heat conduction $\dot{Q}$ is known, the temperature drop along a medium whose thermal resistance is $R$ is simply determined from

$$\Delta T = \dot{Q}R \quad (\text{°C}) \tag{14-5}$$

Therefore, the greatest temperature drops along the path of heat conduction will occur across portions of the heat flow path with the largest thermal resistances.

## Conduction in Chip Carriers

The conduction analysis of an electronic device starts with the *circuitry* or *junction* of a chip, which is the site of *heat generation.* In order to

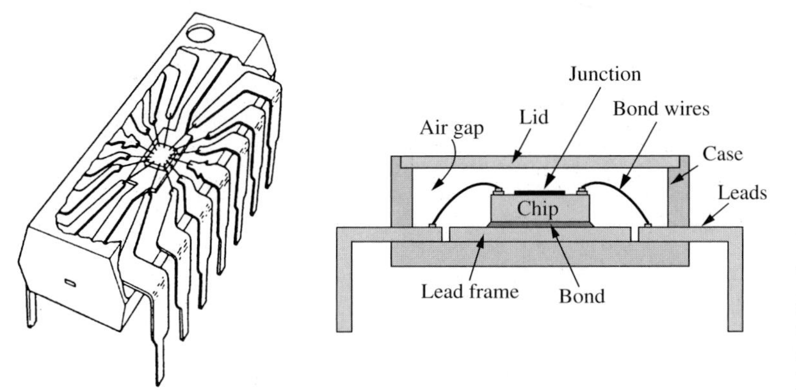


**FIGURE 14-19**

The schematic for the internal geometry and the cross-sectional view of a DIP (dual in-line package) type electronic device with 14 leads (from Ref. 4).

understand the heat transfer mechanisms at the chip level, consider the DIP (dual in-line package) type chip carrier shown in Fig. 14-19.

The heat generated at the junction spreads throughout the chip, and is conducted across the *thickness* of the chip. The spread of heat from the junction to the body of the chip is three-dimensional in nature, but can be approximated as one-dimensional by adding a *constriction thermal resistance* to the thermal resistance network. For a small heat generation area of diameter $d$ on a considerably larger body, the constriction resistance is given by

$$R_{\text{constriction}} = \frac{1}{2\sqrt{\pi}\, dk} \quad (°\text{C/W}) \tag{14-6}$$

where $k$ is the thermal conductivity of the larger body.

The chip is attached to the lead frame with a highly conductive *bonding material* that provides a low-resistance path for heat flow from the chip to the lead frame. There is no metal connection between the *lead frame* and the *leads,* since this would short-circuit the entire chip. Therefore, heat flow from the lead frame to the leads is through the dielectric case material such as *plastic* or *ceramic.* Heat is then transported outside the electronic device through the leads.

When solving a heat transfer problem, it is often necessary to make some simplifying assumptions regarding the *primary* heat flow path and the magnitudes of heat transfer in other directions (Fig. 14-20). In the

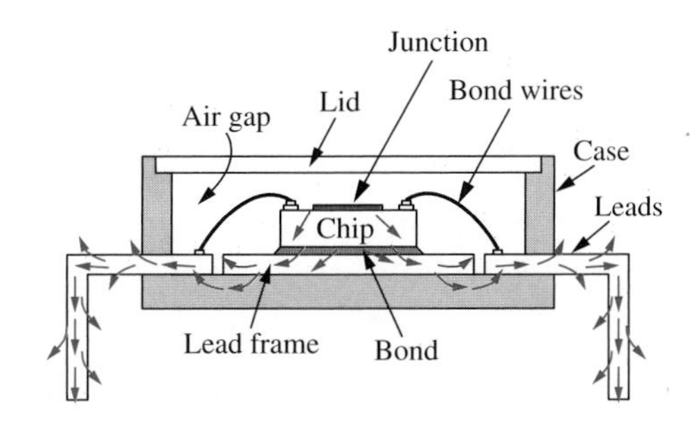


**FIGURE 14-20**

Heat generated at the junction of an electronic device flows through the path of least resistance.

chip carrier discussed above, for example, heat transfer through the top is disregarded, since it is very small because of the large thermal resistance of the stagnant air space between the chip and the lid. Heat transfer from the base of the electronic device is also considered to be negligible, because of the low thermal conductivity of the case material and the lack of effective convection on the base surface.

**EXAMPLE 14-3**

A chip is dissipating 0.6 W of power in a DIP with 12 pin leads. The materials and the dimensions of various sections of this electronic device are as given in the table below. If the temperature of the leads is 40°C, estimate the temperature at the junction of the chip.

| Section and material | Thermal conductivity, W/(m · °C) | Thickness, mm | Heat transfer surface area |
|---|---|---|---|
| Junction constriction | — | — | diameter 0.4 mm |
| Silicon chip | 120† | 0.4 | 3 mm × 3 mm |
| Eutectic bond | 296 | 0.03 | 3 mm × 3 mm |
| Copper lead frame | 386 | 0.25 | 3 mm × 3 mm |
| Plastic separator | 1 | 0.2 | 12 × 1 mm × 0.25 mm |
| Copper leads | 386 | 5 | 12 × 1 mm × 0.25 mm |

† The thermal conductivity of silicon varies greatly with temperature from 153.5 W/(m · °C) at 27°C to 113.7 W/(m · °C) at 100°C, and the value 120 W/(m · °C) reflects the anticipation that the temperature of the silicon chip will be close to 100°C.

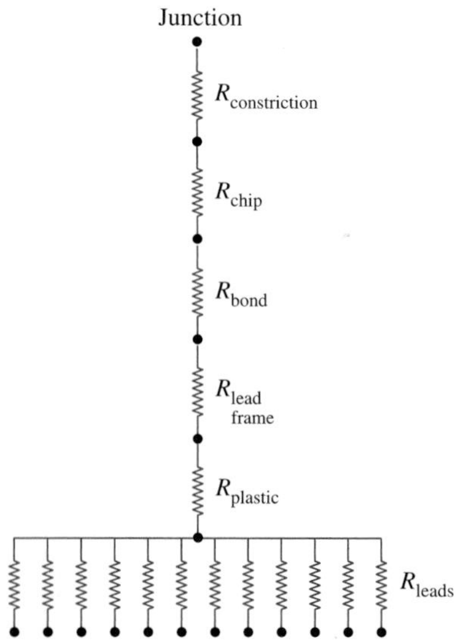

**FIGURE 14-21**
Thermal resistance network for the electronic device considered in Example 14-3.

**Solution** The geometry of the device is as shown in Fig. 14-19. We take the primary heat flow path to be the chip, the eutectic bond, the lead frame, the plastic insulator, and the 12 leads. When the constriction resistance between the junction and the chip is considered, the thermal resistance network for this problem becomes as shown in Fig. 14-21. Note that heat transfer through the air gap and the lid on top of the chip is neglected, because of the very large thermal resistance involved along this path.

The various thermal resistances on the path of primary heat flow are determined as follows.

$$R_{\text{constriction}} = \frac{1}{2\sqrt{\pi}\,dk} = \frac{1}{2\sqrt{\pi}(0.4 \times 10^{-3}\text{ m})[120\text{ W/(m}\cdot{}^\circ\text{C)}]} = 5.88^\circ\text{C/W}$$

$$R_{\text{chip}} = \left(\frac{L}{kA}\right)_{\text{chip}} = \frac{0.4 \times 10^{-3}\text{ m}}{[120\text{ W/(m}\cdot{}^\circ\text{C)}](9 \times 10^{-6}\text{ m}^2)} = 0.37^\circ\text{C/W}$$

$$R_{\text{bond}} = \left(\frac{L}{kA}\right)_{\text{bond}} = \frac{0.03 \times 10^{-3}\text{ m}}{[296\text{ W/(m}\cdot{}^\circ\text{C)}](9 \times 10^{-6}\text{ m}^2)} = 0.01^\circ\text{C/W}$$

$$R_{\text{lead frame}} = \left(\frac{L}{kA}\right)_{\text{lead frame}} = \frac{0.25 \times 10^{-3}\text{ m}}{[386\text{ W/(m}\cdot{}^\circ\text{C)}](9 \times 10^{-6}\text{ m}^2)} = 0.07^\circ\text{C/W}$$

$$R_{\text{plastic}} = \left(\frac{L}{kA}\right)_{\text{plastic}} = \frac{0.2 \times 10^{-3}\text{ m}}{[1\text{ W/(m}\cdot{}^\circ\text{C)}](12 \times 0.25 \times 10^{-6}\text{ m}^2)} = 66.67^\circ\text{C/W}$$

$$R_{\text{leads}} = \left(\frac{L}{kA}\right)_{\text{leads}} = \frac{5 \times 10^{-3}\text{ m}}{[386\text{ W/(m}\cdot{}^\circ\text{C)}](12 \times 0.25 \times 10^{-6}\text{ m}^2)} = 4.32^\circ\text{C/W}$$

Note that for heat transfer purposes, all 12 leads can be considered as a single lead whose cross-sectional area is 12 times as large. The alternative is to find the resistance of a single lead, and to calculate the equivalent resistance for 12 such resistances connected in parallel. Both approaches give the same result.

All the resistances determined above are in series. Thus the total thermal resistance between the junction and the leads is determined by simply adding them up:

$$
\begin{aligned}
R_{\text{total}} &= R_{\text{junction-lead}} = R_{\text{constriction}} + R_{\text{chip}} + R_{\text{bond}} + R_{\text{lead frame}} + R_{\text{plastic}} + R_{\text{leads}} \\
&= (5.88 + 0.37 + 0.01 + 0.07 + 66.67 + 4.32)\text{°C/W} \\
&= 77.32\text{°C/W}
\end{aligned}
$$

Heat transfer through the chip can be expressed as

$$\dot{Q} = \left(\frac{\Delta T}{R}\right)_{\text{junction-leads}} = \frac{T_{\text{junction}} - T_{\text{leads}}}{R_{\text{junction-leads}}}$$

Solving for $T_{\text{junction}}$ and substituting the given values, the junction temperature is determined to be

$$T_{\text{junction}} = T_{\text{leads}} + \dot{Q}R_{\text{junction-leads}} = 40\text{°C} + (0.6\text{ W})(77.32\text{°C/W}) = 86.4\text{°C}$$

Note that the plastic layer between the lead frame and the leads accounts for the 66.67/77.32 = 86 percent of the total thermal resistance and thus the 86 percent of the temperature drop (0.6 × 66.67 = 40°C) between the junction and the leads. In other words, the temperature of the junction would be just 86.5 − 40 = 46.5°C if the thermal resistance of the plastic was eliminated.

The simplified analysis given above points out that any attempt to reduce the thermal resistance in the chip carrier and thus improve the heat flow path should start with the plastic layer. We also notice from the magnitudes of individual resistances that some sections, such as the eutectic bond and the lead frame, have negligible thermal resistances, and any attempt to improve them further will have practically no effect on the junction temperature of the chip.

The analytical determination of the junction-to-case thermal resistance of an electronic device can be rather complicated, and can involve considerable uncertainty, as shown above. Therefore, the manufacturers of electronic devices usually determine this value *experimentally,* and list it as part of their product description. When the thermal resistance is known, the *temperature difference* between the junction and the outer surface of the device can be determined from

$$\Delta T_{\text{junction-case}} = T_{\text{junction}} - T_{\text{case}} = \dot{Q}R_{\text{junction-case}} \quad (\text{°C}) \qquad (14\text{-}7)$$

where $\dot{Q}$ is the power consumed by the device.

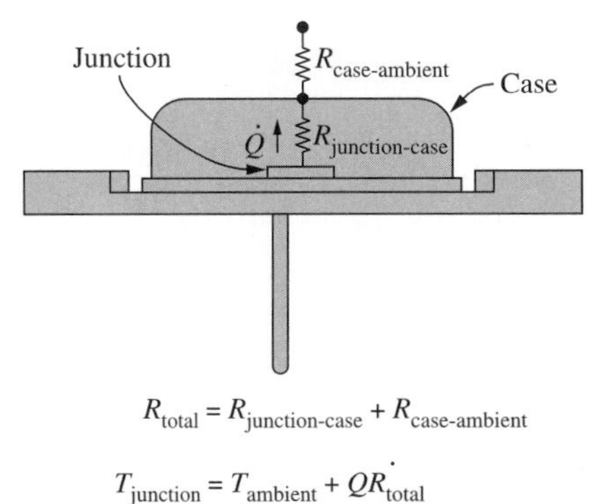


**FIGURE 14-22**

The junction temperature of a chip depends on the external case-to-ambient thermal resistance as well as the internal junction-to-case resistance.

The determination of the *actual junction temperature* depends on the ambient temperature $T_{\text{ambient}}$ as well as the thermal resistance $R_{\text{case-ambient}}$ between the case and the ambient (Fig. 14-22). The magnitude of this resistance depends on the type of ambient (such as air or water) and the fluid velocity. The two thermal resistances discussed above are in series, and the total resistance between the junction and the ambient is determined by simply adding them up:

$$R_{\text{total}} = R_{\text{junction-ambient}} = R_{\text{junction-case}} + R_{\text{case-ambient}} \quad (\text{°C/W}) \qquad (14\text{-}8)$$

Many manufacturers of electronic devices go to the extra step and list the *total resistance* between the junction and the ambient for various chip configurations and ambient conditions likely to be encountered. Once

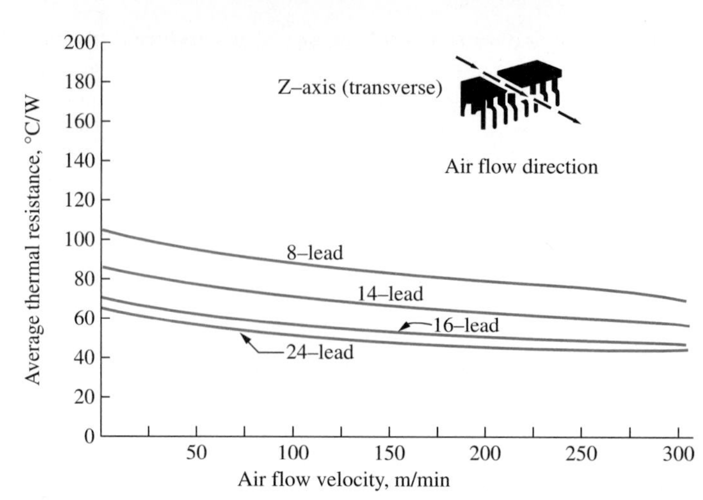


**FIGURE 14-23**
Total thermal resistance between the junction of a plastic DIP device mounted on a circuit board and the ambient air as a function of the air velocity and the number of leads (courtesy of Motorola Semiconductor Products, Inc.).

the total thermal resistance is available, the *junction temperature* corresponding to the specified power consumption (or heat dissipation rate) of $\dot{Q}$ is determined from

$$T_{\text{junction}} = T_{\text{ambient}} + Q\dot{R}_{\text{junction-ambient}} \quad (°\text{C}) \tag{14-9}$$

A typical chart for the total junction-to-ambient thermal resistance for a single DIP type electronic device mounted on a circuit board is given in Fig. 14-23 for various air velocities and lead numbers. The values at the intersections of the curves and the vertical axis represent the thermal resistances corresponding to *natural convection* conditions (zero air velocity). Note that the thermal resistance and thus the junction temperature decrease with increasing air velocity and the number of leads extending from the electronic device, as expected.

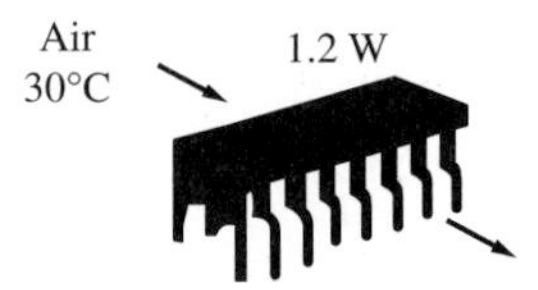


**FIGURE 14-24**
Schematic for Example 14-4.

**EXAMPLE 14-4**

A fan blows air at 30°C and a velocity of 200 m/min over a 1.2-W plastic DIP with 16 leads mounted on a PCB, as shown in Fig. 14-24. Using data from Fig. 14-23, determine the junction temperature of the electronic device. What would the junction temperature be if the fan were to fail?

**Solution** The junction-to-ambient thermal resistance of the device with 16 leads corresponding to an air velocity of 200 m/min is determined from Fig. 14-23 to be

$$R_{\text{junction-ambient}} = 55°\text{C/W}$$

Then the junction temperature can be determined from Eq. 14-9 to be

$$T_{\text{junction}} = T_{\text{ambient}} + \dot{Q}R_{\text{junction-ambient}} = 30°\text{C} + (1.2\ \text{W})(55°\text{C/W}) = 96°\text{C}$$

When the fan fails, the air flow velocity over the device will be zero. The total thermal resistance in this case is determined from the same chart by reading the value at the intersection of the curve and the vertical axis to be

$$R_{\text{junction-ambient}} = 70°\text{C/W}$$

which gives

$$T_{\text{junction}} = T_{\text{ambient}} + \dot{Q}R_{\text{junction-ambient}} = 30°\text{C} + (1.2\ \text{W})(70°\text{C/W}) = 114°\text{C}$$

Therefore, the temperature of the junction will rise by 18°C when the fan fails. Of course this analysis assumes the temperature of the surrounding air to be still 30°C, which may no longer be the case. Any increase in the ambient temperature as a result of inadequate air flow will reflect on the junction temperature, which will serously jeopardize the safety of the electronic device.

## Conduction in Printed Circuit Boards (PCBs)

Heat-generating electronic devices are commonly mounted on thin rectangular *boards,* usually 10 cm × 15 cm in size, made of electrically insulating materials such as *glass–epoxy laminates,* which are also poor conductors of heat. The resulting printed circuit boards (PCBs) are usually cooled by blowing air or passing a dielectric liquid through them. In such cases, the components on the PCBs are cooled directly, and we are not concerned about heat conduction along the PCBs. But in some critical applications such as those encountered in the military, the PCBs are contained in *sealed enclosures,* and the boards provide the only effective heat path between the components and the heat sink attached to the sealed enclosure. In such cases, heat transfer from the side faces of the PCBs is negligible, and the heat generated in the components must be *conducted* along the PCB towards its edges, which are clamped to cold plates for removing the heat externally.

Heat transfer along a PCB is complicated in nature because of the multidimensional effects and non-uniform heat generation on the surfaces. We can still obtain sufficiently accurate results by using the thermal resistance network in one or more dimensions.

Copper or aluminum *cladding, heat frames,* or *cores* are commonly used to enhance heat conduction along the PCBs. The thickness of the copper cladding on the PCB is usually expressed in terms of *ounces of copper,* which is the thickness of 1-ft$^2$ copper sheet made of one ounce of copper. An ounce of copper is equivalent to 0.03556 mm (1.4 mil) thickness of a copper layer.

When analyzing heat conduction along a PCB with *copper* (or aluminum) *cladding* on one or both sides, often the question arises whether heat transfer along the epoxy laminate can be ignored relative to that along the copper layer, since the thermal conductivity of copper is about 1500 times that of epoxy. The answer depends on the relative cross-sectional areas of each layer, since heat conduction is proportional to the cross-sectional area as well as the thermal conductivity.

Consider a copper-cladded PCB of width $w$ and length $L$, across which the temperature difference is $\Delta T$, as shown in Fig. 14–25. Assuming heat conduction is along the length $L$ only and heat conduction in other directions is negligible, the rate of heat conduction along this

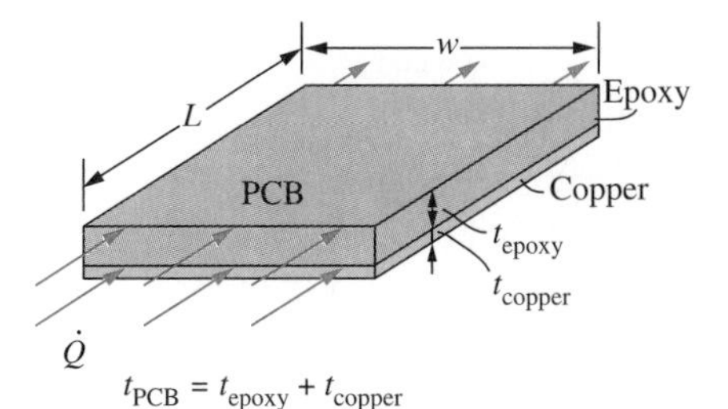


**FIGURE 14-25**
Schematic of a copper cladded epoxy board and heat conduction along it.

PCB is the sum of the heat conduction along the *epoxy board* and the *copper* layer, and is expressed as

$$\begin{aligned}\dot{Q}_{\text{PCB}} &= \dot{Q}_{\text{epoxy}} + \dot{Q}_{\text{copper}} = \left(kA\frac{\Delta T}{L}\right)_{\text{epoxy}} + \left(kA\frac{\Delta T}{L}\right)_{\text{copper}} \\ &= [(kA)_{\text{epoxy}} + (kA)_{\text{copper}}]\frac{\Delta T}{L} \\ &= [(kt)_{\text{epoxy}} + (kt)_{\text{copper}}]\frac{w\,\Delta T}{L} \end{aligned} \qquad (14\text{-}10)$$

where $t$ denotes the thickness. Therefore, the relative magnitudes of heat conduction along the two layers depends on the relative magnitudes of the thermal conductivity–thickness product $kt$ of the layer. Therefore, if the $kt$ product of the copper is 100 times that of epoxy then neglecting heat conduction along the epoxy board will involve an error of just 1 percent, which is negligible.

We can also define an **effectice thermal conductivity** for metal-cladded PCBs as

$$k_{\text{eff}} = \frac{(kt)_{\text{epoxy}} + (kt)_{\text{copper}}}{t_{\text{epoxy}} + t_{\text{copper}}} \quad [\text{W/(m}\cdot{}^\circ\text{C)}] \qquad (14\text{-}11)$$

so that the rate of heat conduction along the PCB can be expressed as

$$\dot{Q}_{\text{PCB}} = k_{\text{eff}}(t_{\text{epoxy}} + t_{\text{copper}})\frac{w\,\Delta T}{L} = k_{\text{eff}}A_{\text{PCB}}\frac{\Delta T}{L} \quad (\text{W}) \qquad (14\text{-}12)$$

where $A_{\text{PCB}} = w(t_{\text{epoxy}} + t_{\text{copper}})$ is the area normal to the direction of heat transfer. When there are holes or discontinuities along the copper cladding, the above analysis needs to be modified to account for thier effect.

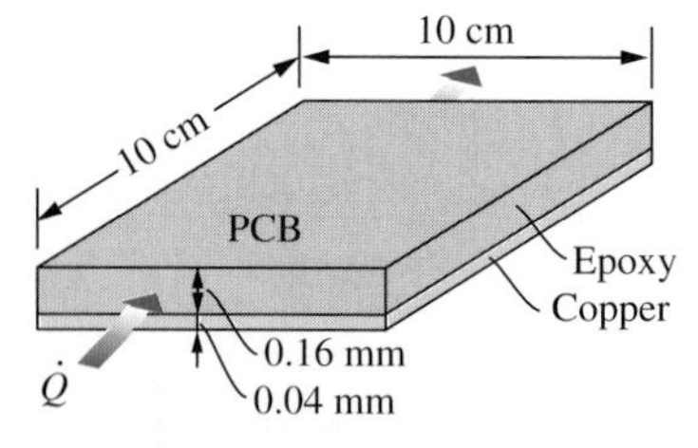


**FIGURE 14-26**
Schematic for Example 14-5.

**EXAMPLE 14-5**

Heat is to be conducted along a PCB with copper cladding on one side. The PCB is 10 cm long and 10 cm wide, and the thickness of the copper and epoxy layers are 0.04 mm and 0.16 mm, respectively, as shown in Fig. 14-26. Disregarding heat transfer from side surfaces, determine the percentages of heat conduction along the copper ($k$ = 386 W/(m · °C)] and epoxy [$k$ = 0.26 W/(m · °C)] layers. Also determine the effective thermal conductivity of the PCB.

**Solution** The length and width of the two layers are the same, and so is the temperature difference across each layer. Heat conduction along a layer is proportional to the thermal conductivity–thickness product $kt$, which is determined for each layer and the entire PCB to be

$$(kt)_{\text{copper}} = [386\ \text{W/(m}\cdot{}^\circ\text{C)}](0.04 \times 10^{-3}\ \text{m}) = 15.44 \times 10^{-3}\ \text{W/}^\circ\text{C}$$
$$(kt)_{\text{epoxy}} = [0.26\ \text{W/(m}\cdot{}^\circ\text{C)}](0.16 \times 10^{-3}\ \text{m}) = 0.04 \times 10^{-3}\ \text{W/}^\circ\text{C}$$

and $$(kt)_{\text{PCB}} = (kt)_{\text{copper}} + (kt)_{\text{epoxy}} = (15.44 + 0.04) = 15.48 \times 10^{-3}\ \text{W/}^\circ\text{C}$$

Therefore, heat conduction along the epoxy board will constitute

$$f = \frac{(kt)_{\text{epoxy}}}{(kt)_{\text{PCB}}} = \frac{0.04 \times 10^{-3}\,\text{W/°C}}{15.48 \times 10^{-3}\,\text{W/°C}} = 0.0026$$

or 0.26 percent of the thermal conduction along the PCB, which is negligible. Therefore, heat conduction along the epoxy layer in this case can be disregarded without any reservations.

The effective thermal conductivity of the board is determined from Eq. 14-11 to be

$$k_{\text{eff}} = \frac{(kt)_{\text{epoxy}} + (kt)_{\text{copper}}}{t_{\text{epoxy}} + t_{\text{copper}}} = \frac{(15.44 + 0.04) \times 10^{-3}\,\text{W/°C}}{(0.16 + 0.04) \times 10^{-3}\,\text{m}} = 77.4\,\text{W/(m} \cdot \text{°C)}$$

That is, the entire PCB can be treated as a 0.20-mm-thick single homogeneous layer whose thermal conductivity is 77.4 W/(m · °C) for heat transfer along its length.

Note that a very thin layer of copper cladding on a PCB improves heat conduction along the PCB drastically, and thus it is commonly used in conduction-cooled electronic devices.

## Heat Frames

In applications where direct cooling of circuit boards by passing air or a dielectric liquid over the electronic components is not allowed, and the junction temperatures are to be maintained relatively low to meet strict safety requirements, a thick *heat frame* is used instead of a thin layer of copper cladding. This is especially the case for multilayer PCBs that are packed with high-power output chips.

The schematic of a PCB that is conduction-cooled via a heat frame is shown in Fig. 14-27. Heat generated in the chips is conducted through the circuit board, through the epoxy adhesive, to the center of the heat frame, along the heat frame, and to a heat sink or cold plate, where heat is externally removed.

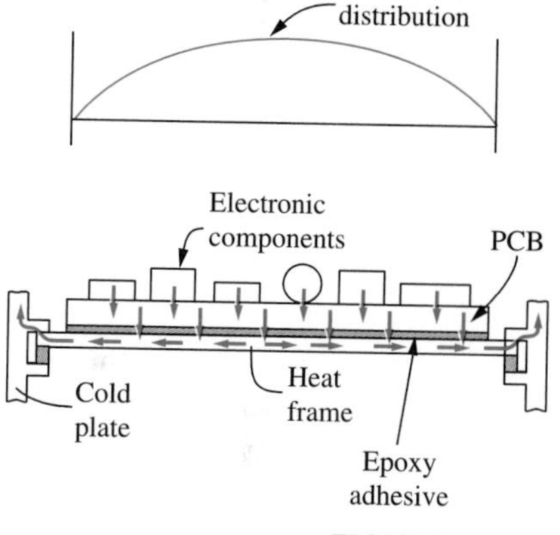


**FIGURE 14-27**
Conduction cooling of a printed circuit board with a heat frame, and the typical temperature distribution along the frame.

The heat frame provides a low-resistance path for the flow of heat from the circuit board to the heat sink. The thicker the heat frame, the lower the thermal resistance, and thus the smaller the temperature difference between the center and the ends of the heat frame. When the heat load is evenly distributed on the PCB, there will be thermal symmetry about the centerline, and the temperature distribution along the heat frame and the PCB will be *parabolic* in nature, with the chips in the middle of the PCB (furthest away from the edges) operating at the highest temperatures, and the chips near the edges operating at the lowest temperatures. Also, when the PCB is cooled from two edges, heat generated in the left half of the PCB will flow towards the left edge, and heat generated in the right half will flow towards the right edge of the heat frame. But when the PCB is cooled from all four edges, the heat transfer along the heat frame as well as the resistance network will be two-dimensional.

When a heat frame is used, heat conduction in the epoxy layer of the PCB is through its *thickness* instead of along its length. The epoxy layer

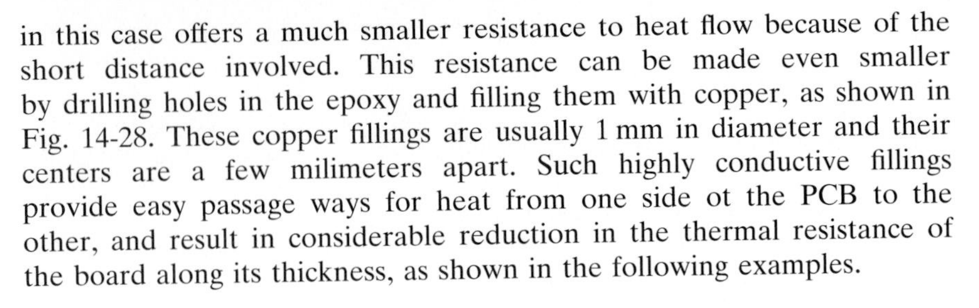

in this case offers a much smaller resistance to heat flow because of the short distance involved. This resistance can be made even smaller by drilling holes in the epoxy and filling them with copper, as shown in Fig. 14-28. These copper fillings are usually 1 mm in diameter and their centers are a few milimeters apart. Such highly conductive fillings provide easy passage ways for heat from one side ot the PCB to the other, and result in considerable reduction in the thermal resistance of the board along its thickness, as shown in the following examples.

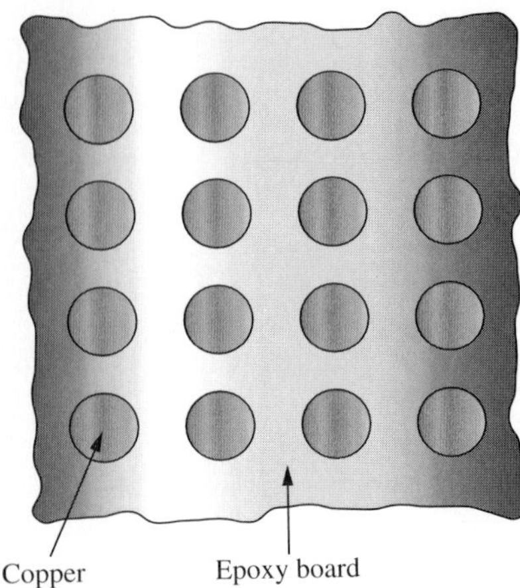


**FIGURE 14-28**

Planting the epoxy board with copper filing decreases the thermal resistance across its thickness considerably.

**EXAMPLE 14-6**

Consider a 10 cm × 15 cm glass–epoxy laminate [$k = 0.26$ W/(m · °C)] whose thickness is 0.8 mm, as shown in Fig. 14-29. Determine the thermal resistance of this epoxy layer for heat flow (*a*) along the 15-cm-long side and (*b*) across its thickness.

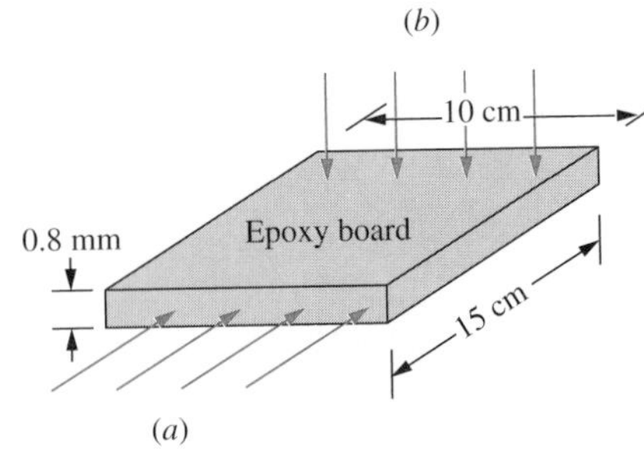


**FIGURE 14-29**

Schematic for Example 14-6.

**Solution** The thermal resistance of a plane parallel medium in the direction of heat conduction is given by

$$R = \frac{L}{kA}$$

where $L$ is the length in the direction of heat flow, $k$ is the thermal conductivity, and $A$ is the area normal to the direction of heat conduction. Substituting the given values, the thermal resistances of the board for both cases are determined to be

(*a*)
$$R_{\text{along length}} = \left(\frac{L}{kA}\right)_{\text{along length}}$$
$$= \frac{0.15\ \text{m}}{[0.26\ \text{W/(m}\cdot\text{°C)}](0.1\ \text{m})(0.8 \times 10^{-3}\ \text{m})} = 7212\text{°C/W}$$

(*b*)
$$R_{\text{across thickness}} = \left(\frac{L}{kA}\right)_{\text{across thickness}}$$
$$= \frac{0.8 \times 10^{-3}\ \text{m}}{[0.26\ \text{W/(m}\cdot\text{°C)}](0.1\ \text{m})(0.15\ \text{m})} = 0.21\text{°C/W}$$

Note that heat conduction at a rate of 1 W along this PCB would cause a temperature difference of 7212°C across a length of 15 cm. But the same rate of heat conduction would cause a temperature difference of only 0.21°C across the thickness of the epoxy board.

**EXAMPLE 14-7**

Reconsider the 10 cm × 15 cm glass–epoxy laminate [$k = 0.26$ W/(m · °C)] of thickness 0.8 mm discussed in Example 14-7. In order to reduce the thermal resistance across its thickness from the current value of 0.21°C/W, cylindrical copper fillings [$k = 386$ W/(m · °C)] of 1-mm diameter are to be planted throughout the board with a center-to-center distance of 2.5 mm, as shown in Fig. 14-30. Determine the new value of the thermal resistance of the epoxy board for heat conduction across its thickness as a result of this modification.

**Solution** Heat flow through the thickness of the board in this case will take place partly through the copper fillings and partly through the epoxy in parallel paths. The thickness of both materials is the same, and is given to be 0.8 mm. But we also need to know the surface area of each material before we can determine the thermal resistances.

It is stated that the distance between the centers of the copper fillings is 2.5 mm. That is, there is only one 1-mm-diameter copper filling in every 2.5 mm × 2.5 mm square section of the board. The number of such squares and thus the number of copper fillings on the board are

$$n = \frac{\text{Area of the board}}{\text{area of one square}} = \frac{(100\text{ mm})(150\text{ mm})}{(2.5\text{ mm})(2.5\text{ mm})} = 2400$$

Then the surface areas of the copper fillings and the remaining epoxy layer become

$$A_{\text{copper}} = n\frac{\pi D^2}{4} = (2400)\frac{\pi(1 \times 10^{-3}\text{ m})^2}{4} = 0.001885\text{ m}^2$$

$$A_{\text{total}} = (\text{length})(\text{width}) = (0.1\text{ m})(0.15\text{ m}) = 0.015\text{ m}^2$$

$$A_{\text{epoxy}} = A_{\text{total}} - A_{\text{copper}} = (0.15 - 0.001885)\text{ m}^2 = 0.013115\text{ m}^2$$

The thermal resistance of each material is

$$R_{\text{copper}} = \left(\frac{L}{kA}\right)_{\text{copper}} = \frac{0.8 \times 10^{-3}\text{ m}}{[386\text{ W/(m}\cdot{}^\circ\text{C)}](0.001885\text{ m}^2)} = 0.0011^\circ\text{C/W}$$

$$R_{\text{epoxy}} = \left(\frac{L}{kA}\right)_{\text{epoxy}} = \frac{0.8 \times 10^{-3}\text{ m}}{[0.26\text{ W/(m}\cdot{}^\circ\text{C)}](0.013115\text{ m}^2)} = 0.2346^\circ\text{C/W}$$

Noting that these two resistances are in parallel, the equivalent thermal resistance of the entire board is determined from

$$\frac{1}{R_{\text{board}}} = \frac{1}{R_{\text{copper}}} + \frac{1}{R_{\text{copper}}} = \frac{1}{0.0011^\circ\text{C/W}} + \frac{1}{0.2346^\circ\text{C/W}}$$

which gives

$$R_{\text{board}} = 0.00109^\circ\text{C/W}$$

Note that the thermal resistance of the epoxy board has dropped from 0.21°C/W by a factor of almost 200 to just 0.00109°C/W as a result of implanting 1-mm-diameter copper fillings into it. Therefore, implanting copper pins into the epoxy laminate has virtually eliminated the thermal resistance of the epoxy across its thickness.

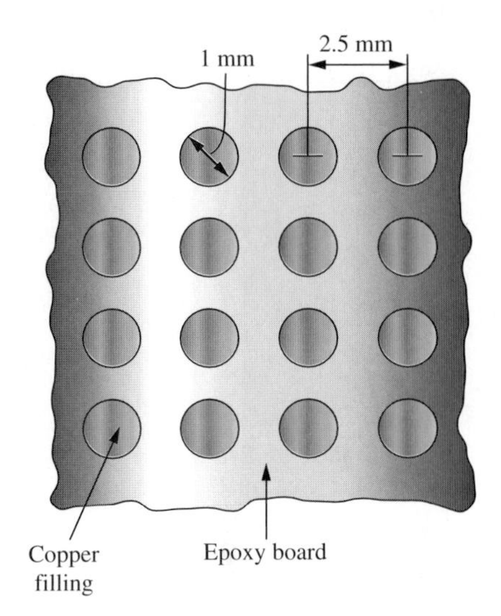


**FIGURE 14-30**
Schematic for Example 14-7.

**EXAMPLE 14-8**

A 10 cm × 12 cm circuit board dissipating 24 W of heat is to be conduction-cooled by a 1.2-mm-thick copper heat frame [$k$ = 386 W/(m · °C)] 10 cm × 14 cm in size. The epoxy laminate [$k$ = 0.26 W/(m · °C)] has a thickness of 0.8 mm, and is attached to the heat frame with conductive epoxy adhesive [$k$ = 1.8 W/(m · °C)] of 0.13-mm thickness, as shown in Fig. 14-31. The PCB is attached to a heat sink by clamping a 5-mm-wide portion of the edge to the heat sink from both ends. The temperature of the heat frame at this point is 20°C. Heat is uniformly generated on the PCB at a rate of 2 W per 1 cm × 10 cm strip. Considering only one half of the PCB board because of symmetry, determine the maximum temperature on the PCB and the temperature distribution along the heat frame.

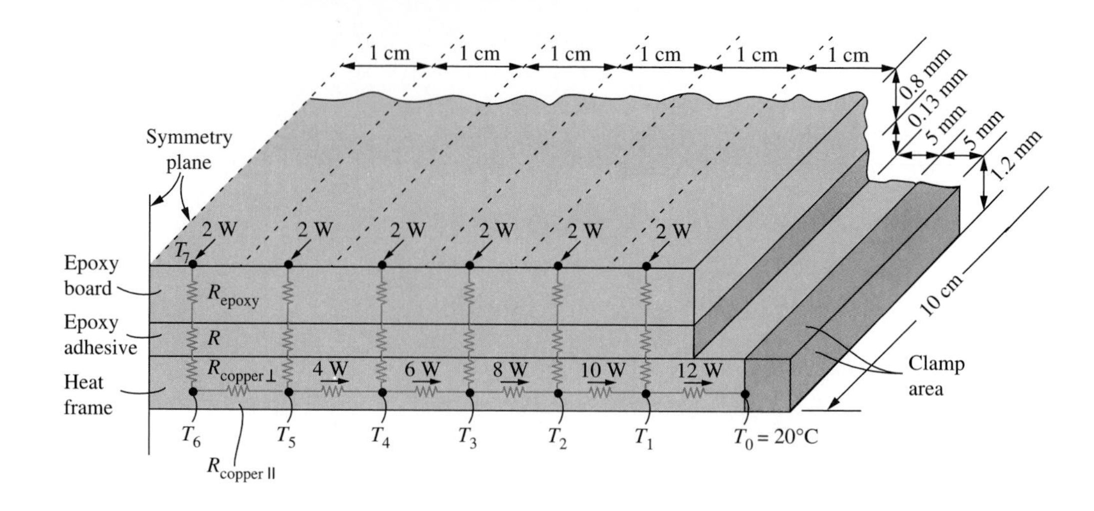


**FIGURE 14-31**

The schematic and thermal resistance network for Example 14-8.

**Solution** The PCB under consideration possesses thermal symmetry about the centerline. Therefore, the heat generated on the left half of the PCB is conducted to the left heat sink, and the heat generated on the right half is conducted to the right heat sink. Thus, we need to consider only half of the board in the analysis.

The maximum temperature will occur at a location furthest away from the heat sinks, which is the symmetry line. Therefore, the temperature of the electronic components located at the center of the PCB will be the highest, and their reliability will be the lowest.

We assume there is no direct heat dissipation from the surfaces of the PCB, and thus all the heat generated is conducted by the heat frame to the heat sink. Heat generated in the components on each strip is conducted through the epoxy layer underneath. Heat is then conducted across the epoxy adhesive and to the middle of the copper heat frame. Finally, heat is conducted along the heat frame to the heat sink.

The thermal resistance network associated with heat flow in the right half of the PCB is also shown in Fig. 14-31. Note that all vertical resistances are identical, and are equal to the sum of the three resistances in series. Also note that heat conduction towards the heat sink is assumed to be predominantly along the heat frame, and conduction along the epoxy adhesive is considered to be negligible. This assumption is quite reasonable, since the conductivity–thickness product of the heat frame is much larger than those of the other two layers.

The properties and dimensions of various sections of the PCB are summarized in the table below.

| Section and material | Thermal conductivity, W/(m · °C) | Thickness, mm | Heat transfer surface area |
|---|---|---|---|
| Epoxy board | 0.26 | 0.8 | 10 mm × 100 mm |
| Epoxy adhesive | 1.8 | 0.13 | 10 mm × 100 mm |
| Copper heat frame, ⊥ (normal to frame) | 386 | 0.6 | 10 mm × 100 mm |
| Copper heat frame, ‖ (along the frame) | 386 | 10 | 1.2 mm × 100 mm |

Using the values in the table, the various thermal resistances are determined to be

$$R_{\text{epoxy}} = \left(\frac{L}{kA}\right)_{\text{epoxy}} = \frac{0.8 \times 10^{-3}\text{ m}}{[0.26\text{ W/(m}\cdot{}^\circ\text{C)}](0.01\text{ m} \times 0.1\text{ m})} = 3.077^\circ\text{C/W}$$

$$R_{\text{adhesive}} = \left(\frac{L}{kA}\right)_{\text{adhesive}} = \frac{0.13 \times 10^{-3}\text{ m}}{[1.8\text{ W/(m}\cdot{}^\circ\text{C)}](0.01\text{ m} \times 0.1\text{ m})} = 0.072^\circ\text{C/W}$$

$$R_{\text{copper},\perp} = \left(\frac{L}{kA}\right)_{\text{copper},\perp} = \frac{0.6 \times 10^{-3}\text{ m}}{[386\text{ W/(m}\cdot{}^\circ\text{C)}](0.01\text{ m} \times 0.1\text{ m})} = 0.002^\circ\text{C/W}$$

$$R_{\text{frame}} = R_{\text{copper},\parallel} = \left(\frac{L}{kA}\right)_{\text{copper},\parallel} = \frac{0.01\text{ m}}{[386\text{ W/(m}\cdot{}^\circ\text{C)}](0.0012\text{ m} \times 0.1\text{ m})}$$
$$= 0.216^\circ\text{C/W}$$

The combined resistance between the electronic components on each strip and the heat frame can be determined, by adding the three resistances in series, to be

$$\begin{aligned} R_{\text{vertical}} &= R_{\text{epoxy}} = R_{\text{adhesive}} + R_{\text{copper},\perp} \\ &= (3.077 + 0.072 + 0.002)^\circ\text{C/W} \\ &= 3.151^\circ\text{C/W} \end{aligned}$$

The various temperatures along the heat frame can be determined from the relation

$$\Delta T = T_{\text{high}} - T_{\text{low}} = \dot{Q}R$$

where $R$ is the thermal resistance between two specified points, $\dot{Q}$ is the heat transfer rate through that resistance, and $\Delta T$ is the temperature difference across that resistance.

The temperature at the location where the heat frame is clamped to the heat sink is given as $T_0 = 20^\circ\text{C}$. Noting that the entire 12 W of heat generated on the right half of the PCB must pass through the last thermal resistance adjacent to the heat sink, the temperature $T_1$ can be determined from

$$T_1 = T_0 + \dot{Q}_{1-0}R_{1-0} = 20^\circ\text{C} + (12\text{ W})(0.216^\circ\text{C/W}) = 22.59^\circ\text{C}$$

Following the same line of reasoning, the temperatures at specified locations along the heat frame are determined to be

$$\begin{aligned} T_2 &= T_1 + \dot{Q}_{2-1}R_{2-1} = 22.59^\circ\text{C} + (10\text{ W})(0.216^\circ\text{C/W}) = 24.75^\circ\text{C} \\ T_3 &= T_2 + \dot{Q}_{3-2}R_{3-2} = 24.75^\circ\text{C} + (8\text{ W})(0.216^\circ\text{C/W}) = 26.48^\circ\text{C} \\ T_4 &= T_3 + \dot{Q}_{4-3}R_{4-3} = 26.68^\circ\text{C} + (6\text{ W})(0.216^\circ\text{C/W}) = 27.78^\circ\text{C} \\ T_5 &= T_4 + \dot{Q}_{5-4}R_{5-4} = 27.78^\circ\text{C} + (4\text{ W})(0.216^\circ\text{C/W}) = 28.64^\circ\text{C} \\ T_6 &= T_5 + \dot{Q}_{6-5}R_{6-5} = 28.64^\circ\text{C} + (2\text{ W})(0.216^\circ\text{C/W}) = 29.07^\circ\text{C} \end{aligned}$$

Finally, $T_7$, which is the maximum temperature on the PCB, is determined from

$$T_7 = T_6 + \dot{Q}_{\text{vertical}}R_{\text{vertical}} = 29.07^\circ\text{C} + (2\text{ W})(3.151^\circ\text{C/W}) = 35.37^\circ\text{C}$$

Therefore, the maximum temperature difference between the PCB and the heat sink is only 15.37°C, which is very impressive considering that the PCB has no direct contact with the cooling medium. The junction temperatures in this case can be determined by calculating the temperature difference between the junction and the leads of the chip carrier at the point of contact to the PCB, and adding 35.37°C to it. The maximum temperature rise of 15.37°C can be reduced, if necessary by using a thicker heat frame.

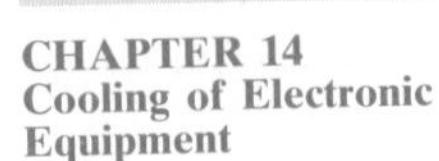



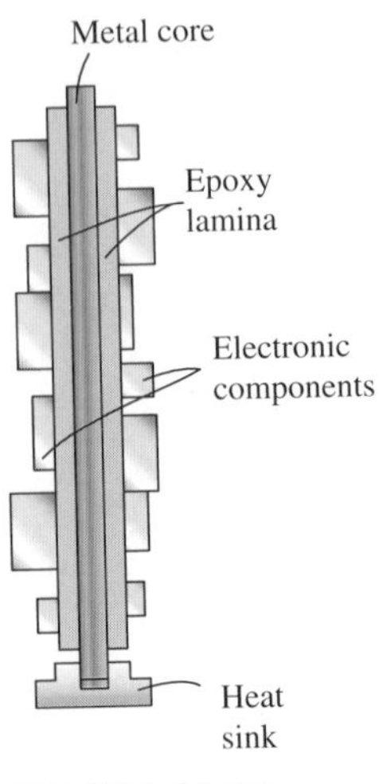


**FIGURE 14-32**

A two-sided printed circuit board with a metal core for conduction cooling.

Conduction cooling can also be used when electronic components are mounted on *both sides* of the PCB by using a copper or aluminum core plate in the middle of the PCB, as shown in Fig. 14-32. The heat load in this case will be twice that of a PCB that has components on one side only. Again heat generated in the components will be conducted through the thickness of the epoxy layer to the *metal core,* which serves as a *channel* for effective heat removal. The thickness of the core is selected such that the maximum component temperatures remain below specified values to meet a prescribed reliability criteria.

The thermal expansion coefficients of aluminum and copper are about twice as large as that of the glass–epoxy. This large difference in *thermal expansion coefficients* can cause *warping* on the PCBs if the epoxy and the metal are not bonded properly. One way of avoiding warping is to use PCBs with components on both sides, as discussed above. Extreme care should be exercised during the bonding and curing process, when components are mounted on only one side of the PCB.

## The Thermal Conduction Module (TCM)

The heat flux for logic chips has been increasing steadily as a result of the increasing circuit density in the chips. For example, the peak flux at the chip level has increased from 2 W/cm$^2$ on IBM System 370 to 20 W/cm$^2$ on IBM System 3081, which was introduced in the early 1980s. The conventional forced-air cooling technique used in earlier machines was inadequate for removing such high heat fluxes, and it was necessary to develop a new and more effective cooling technique. The result was the **thermal conduction module** (TCM) shown in Fig. 14-33. The TCM was different from previous chip packaging designs in that incorporated both electrical and thermal considerations in early stages of chip design.

**FIGURE 14-33**

Cutaway view of the thermal conduction module (TCM), and the thermal resistance network between a single chip and the cooling fluid (courtesy of IBM Corporation).

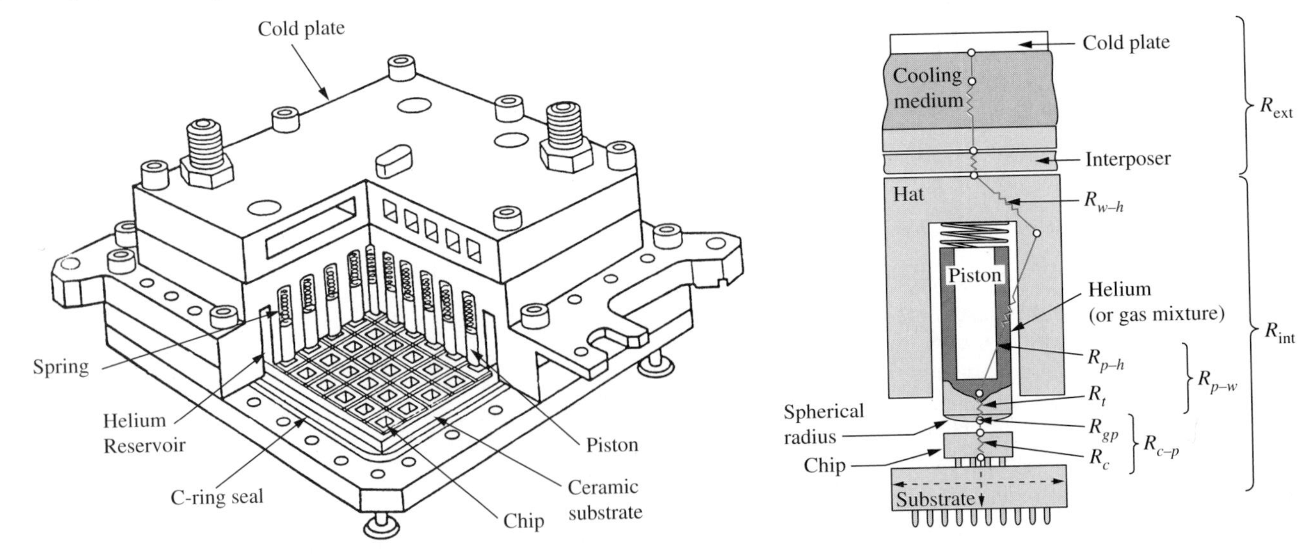

Previously, a chip would be designed primarily by electrical designers, and the thermal designer would be told to come up with a cooling scheme for the chip. That approach resulted in unnecessarily high junction temperatures, and reduced reliability, since the thermal designer had no direct access to the chip. The TCM reflects a new philosophy in electronic packaging in that the thermal and electrical aspects are given equal treatment in the design process, and a successful thermal design starts at the chip level.

In the TCM, one side of the chip is reserved for electrical connections and the other side for heat rejection. The chip is cooled by direct contact to the cooling system to minimize the junction-to-case thermal resistance.

The TCM houses 100–118 logic chips, which are bonded to a multilayer ceramic substrate 90 mm × 90 mm in size with solder balls, which also provide the electrical connections between the chips and the substrate. Each chip dissipates about 4 W of power. The heat flow path from the chip to the metal casing is provided by a *piston,* which is pressed against the back surface of the chip by a spring. The tip of the piston is slightly curved to ensure good thermal contact even when the chip is tilted or misaligned.

Heat conduction between the chip and the piston occurs primarily through the gas space between the chip and the piston because of the limited contact area between them. To maximize heat conduction through the gas, the air in the TCM cavity is evacuated and is replaced by helium gas, whose thermal conductivity is about six times that of air. Heat is then conducted through the piston, across the surrounding helium gas layer, through the module housing, and finally to the cooling water circulating through the cold plate attached to the top surface of the TCM.

The total *internal thermal resistance* $R_{int}$ of the TCM is about 8°C/W, which is rather impressive. This means that the temperature difference between the chip surface and the outer surface of the housing of the module will be only 24°C for a 3-W chip. The *external thermal resistance* $R_{ext}$ between the housing of the module and the cooling fluid is usually comparable in magnitude to $R_{int}$. Also, the thermal resistance between the junction and the surface of the chip can be taken to be 1°C/W.

The compact design of the TCM significantly reduces the *distance* between the chips, and thus the signal transmission time between the chips. This, in turn, increases the *operating speed* of the electronic device.

**EXAMPLE 14-9**

Consider a thermal conduction module with 100 chips, each dissipating 3 W of power. The module is cooled by water at 25°C flowing through the cold plate on top of the module. The thermal resistances in the path of heat flow are $R_{chip}$ = 1°C/W between the juncion and the surface of the chip, $R_{int}$ = 9°C/W between the surface of the chip and the outer surface of the thermal conduction module, and $R_{ext}$ = 6°C/W between the outer surface of the module and the cooling water. Determine the junction temperature of the chip.

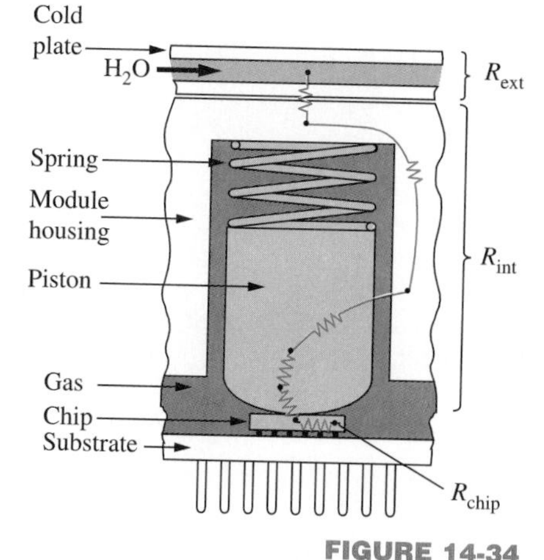


**FIGURE 14-34**
Thermal resistance network for Example 14-9.

**Solution** Because of symmetry, we will consider only one of the chips in our analysis. The thermal resistance network for heat flow is given in Fig. 14-34.

Noting that all resistances are in series, the total thermal resistance between the junction and the cooling water is

$$R_{\text{total}} = R_{\text{junction-water}} = R_{\text{chip}} + R_{\text{int}} + R_{\text{ext}} = (1 + 8 + 6)°\text{C/W} = 15°\text{C/W}$$

Noting that the total power dissipated by the chip is 3 W and the water temperature is 25°C, the junction temperature of the chip in steady operation can be determined from

$$\dot{Q} = \left(\frac{\Delta T}{R}\right)_{\text{junction-water}} = \frac{T_{\text{junction}} - T_{\text{water}}}{R_{\text{junction-water}}}$$

Solving for $T_{\text{junction}}$ and substituting the specified values gives

$$T_{\text{junction}} = T_{\text{water}} + \dot{Q}R_{\text{junction-water}} = 25°\text{C} + (3\text{ W})(15°\text{C/W}) = 70°\text{C}$$

Therefore, the circuits of the chip will operate at about 70°C, which is considered to be a safe operating temperature for silicon chips.

Cold plates are usually made of metal plates with fluid channels running through them, or copper tubes attached to them by brazing. Heat transferred to the cold plate is conducted to the tubes, and from the tubes to the fluid flowing through them. The heat carried away by the fluid is finally dissipated to the ambient in a heat exchanger.

## 14-7 ■ AIR COOLING: NATURAL CONVECTION AND RADIATION

Low-power electronic systems are conveniently cooled by *natural convection* and *radiation.* Natural convection cooling is very desirable, since it does not involve any fans that may break down.

Natural convection is based on the *fluid motion* caused by the density differences in a fluid due to a temperature difference. A fluid expands when heated and becomes less dense. In a gravitational field, this lighter fluid rises and initiates a motion in the fluid which is called *natural convection currents* (Fig. 14-35). Natural convection cooling is most effective when the path of the fluid is relatively free of obstacles, which tend to slow down the fluid, and is least effective when the fluid has to pass through narrow flow passages and over many obstacles.

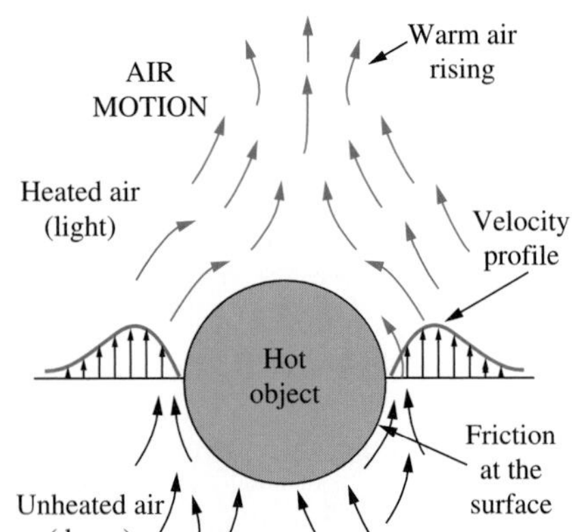


**FIGURE 14-35**

Natural convection currents around a hot object in air.

The magnitude of the natural convection heat transfer between a surface and a fluid is directly related to the *flow rate* of the fluid. The higher the flow rate, the higher is the heat transfer rate. In natural convection, no blowers are used and therefore the flow rate cannot be controlled externally. The flow rate in this case is established by the dynamic balance of *buoyancy* and *friction.* The *larger* the temperature difference between the fluid adjacent to a hot surface and the fluid away from it, the *larger* is the buoyancy force, and the *stronger* is the natural convection currents, and thus the *higher* is the heat transfer rate. Also, whenever two bodies in contact move relative to each other, a *friction force* develops at the contact surface in the direction opposite to that of the motion. This opposing force slows down the fluid, and thus reduces the flow rate of the fluid. Under steady conditions, the air flow rate driven by buoyancy is established at the point where these two effects *balance* each other. The friction force increases as more and more solid surfaces are introduced, seriously disrupting the fluid flow and heat transfer.

Electronic components or PCBs placed in *enclosures* such as a TV or VCR are cooled by natural convection by providing a sufficient number of *vents* on the case to enable the cool air to enter and the heated air to leave the case freely, as shown in Fig. 14-36. From the heat transfer point of view, the vents should be as large as possible to minimize the flow resistance, and should be located at the bottom of the case for air entering, and at the top for air leaving. But equipment and human safety requirements dictate that the vents should be quite narrow to discourage unintended entry into the box. Also, concern about human habits such as putting a cup of coffee on the closest flat surface makes it very risky to place vents on the top surface. The narrow clearance allowed under the case also offers resistance to air flow. Therefore, vents on the enclosures of natural convection cooled electronic equipment are usually placed at the *lower section* of the side or back surfaces for air inlet, and at the *upper section* of those surfaces for air exit.

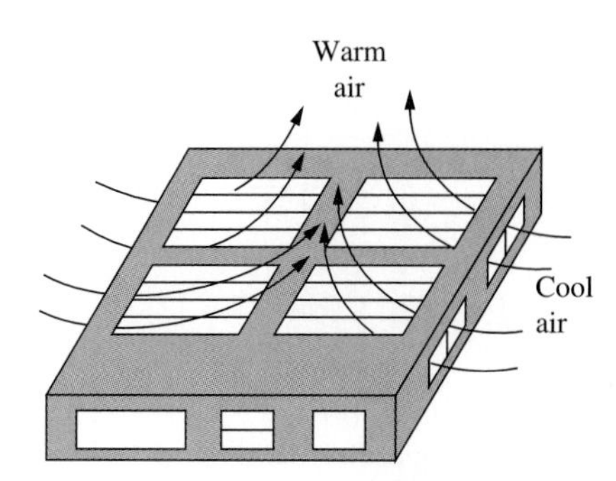


**FIGURE 14-36**
Natural convection cooling of electronic components in an enclosure with air vents.

The heat transfer from a surface at temperature $T_s$ to a fluid at temperature $T_{\text{fluid}}$ by convection is expressed as

$$\dot{Q}_{\text{conv}} = h_{\text{conv}} A \, \Delta T = h_{\text{conv}} A (T_s - T_{\text{fluid}}) \quad \text{(W)} \qquad (14\text{-}13)$$

where $h_{\text{conv}}$ is the convection heat transfer coefficient and $A$ is the heat transfer surface area. The value of $h_{\text{conv}}$ depends on the geometry of the surface and the type of fluid flow, among other things.

Natural convection currents start out as *laminar* (smooth and orderly) and turn *turbulent* when the dimension of the body and the temperature difference between the hot surface and the fluid is large. For air, the flow remains laminar when the temperature differences involved are less than 100°C and the characteristic length of the body is less than 0.5 m, which is almost always the case in electronic equipment. Therefore, the air flow in the analysis of electronic equipment can be assumed to be laminar.

The natural convection heat transfer coefficient for laminar flow of *air* at atmospheric pressure is given by a simplified relation of the form

$$h_{\text{conv}} = K \left( \frac{\Delta T}{L} \right)^{0.25} \quad [\text{W}/(\text{m}^2 \cdot {}^\circ\text{C})] \qquad (14\text{-}14)$$

where $\Delta T = T_s - T_{\text{fluid}}$ is the temperature difference between the surface and the fluid, $L$ is the characteristic length (the length of the body along the heat flow path) and $K$ is a constant whose value depends on the *geometry* and *orientation* of the body.

The heat transfer coefficient relations are given in Table 14-1 for some common geometries encountered in electronic equipment in both SI and English unit systems. Once $h_{\text{conv}}$ has been determined from one of these relations, the rate of heat transfer can be determined from

**TABLE 14-1**

**Simplified relations for natural convection heat transfer coefficients for various geometries in air at atmospheric pressure for laminar flow conditions**

| Geometry | Natural convection heat transfer coefficient, W/(m² · °C) (ΔT in °C, L or D in m) | Natural convection heat transfer coefficient, Btu/(h · ft² · °F) (ΔT in °F, L or D in ft) |
|---|---|---|
| Vertical plate or cylinder 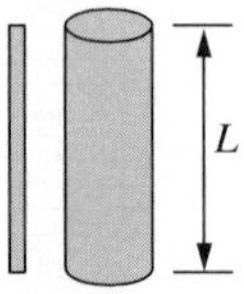 | $h_{conv} = 1.42\left(\frac{\Delta T}{L}\right)^{0.25}$ | $h_{conv} = 0.29\left(\frac{\Delta T}{L}\right)^{0.25}$ |
| Horizontal cylinder  | $h_{conv} = 1.32\left(\frac{\Delta T}{D}\right)^{0.25}$ | $h_{conv} = 0.27\left(\frac{\Delta T}{D}\right)^{0.25}$ |
| Horizontal plate ($L = 4A/p$, where $A$ is surface area and $p$ is perimeter) (*a*) Hot surface facing up 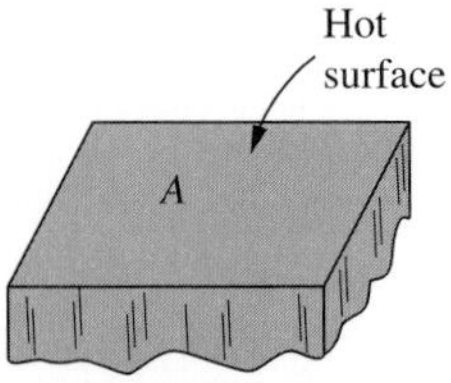 | $h_{conv} = 1.32\left(\frac{\Delta T}{L}\right)^{0.25}$ | $h_{conv} = 0.27\left(\frac{\Delta T}{L}\right)^{0.25}$ |
| (*b*) Hot surface facing down 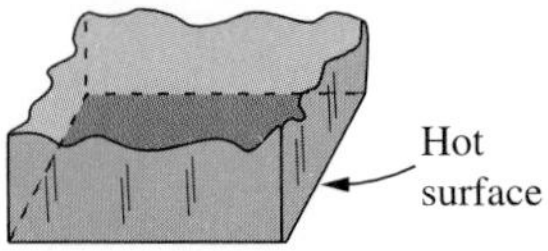 | $h_{conv} = 0.59\left(\frac{\Delta T}{L}\right)^{0.25}$ | $h_{conv} = 0.12\left(\frac{\Delta T}{L}\right)^{0.25}$ |
| Components on a circuit board  | $h_{conv} = 2.44\left(\frac{\Delta T}{L}\right)^{0.25}$ | $h_{conv} = 0.50\left(\frac{\Delta T}{L}\right)^{0.25}$ |
| Small components or wires in free air 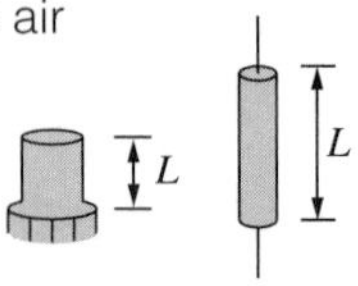 | $h_{conv} = 3.53\left(\frac{\Delta T}{L}\right)^{0.25}$ | $h_{conv} = 0.72\left(\frac{\Delta T}{L}\right)^{0.25}$ |
| Sphere 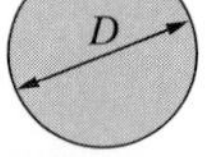 | $h_{conv} = 1.92\left(\frac{\Delta T}{D}\right)^{0.25}$ | $h_{conv} = 0.39\left(\frac{\Delta T}{D}\right)^{0.25}$ |

*Sources*: Refs. 5 and 6.

Eq. 14-13. The relations in Table 14-1 can also be used at pressures other than 1 atm by multiplying them by $\sqrt{P}$, where $P$ is the *air pressure* in *atm* (1 atm = 101.325 kPa = 14.696 psia). That is,

$$h_{\text{conv}, P\ \text{atm}} = h_{\text{conv}, 1\ \text{atm}}\sqrt{P} \quad [\text{W}/(\text{m}^2 \cdot {}^\circ\text{C})] \tag{14-15}$$

When hot surfaces are surrounded by cooler surfaces such as the walls and ceilings of a room or just the sky, the surfaces are also cooled by *radiation,* as shown in Fig. 14-37. The magnitude of radiation heat transfer, in general, is comparable to the magnitude of natural convection heat transfer. This is especially the case for surfaces whose emissivity is close to *unity*, such as plastics and painted surfaces(regardless of color). Radiation heat transfer is negligible for *polished metals* because of their very low emissivity, and for bodies surrounded by surfaces at about the same temperature.

Radiation heat transfer between a surface at temperature $T_s$ completely surrounded by a much larger surface at temperature $T_{\text{surr}}$ can be expressed as

$$\dot{Q}_{\text{rad}} = \varepsilon A \sigma (T_s^4 - T_{\text{surr}}^4) \quad \text{(W)} \tag{14-16}$$

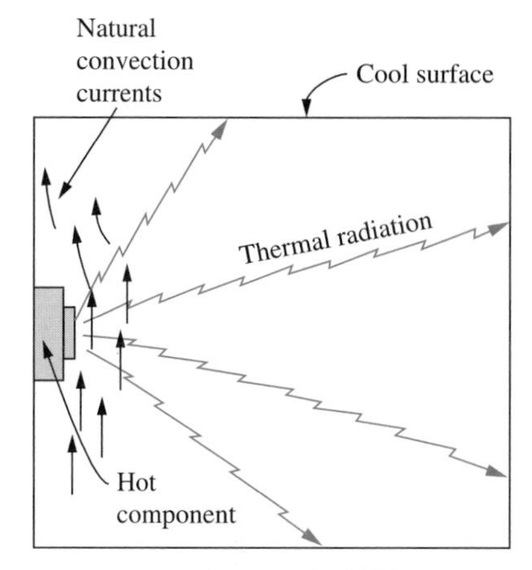


**FIGURE 14-37**

Simultaneous natural convection heat transfer to air and radiation heat transfer to the surrounding surfaces from a hot electronic component mounted on the wall of an enclosure.

where $\varepsilon$ is the emissivity of the surface, $A$ is the heat transfer surface area, and $\sigma$ is the Stefan–Boltzmann constant, whose value is $\sigma = 5.67 \times 10^{-8}\ \text{W}/(\text{m}^2 \cdot \text{K}^4) = 0.1714 \times 10^{-8}\ \text{Btu}/(\text{h} \cdot \text{ft}^2 \cdot \text{R}^4)$. Here both temperatures must be expressed in K or R. Also, if the hot surface analyzed has only a partial view of the surrounding cooler surface at $T_{\text{surr}}$, the result obtained from Eq. 14-16 must be multiplied by a *view factor,* which is the fraction of the view of the hot surface blocked by the cooler surface. The value of view factor ranges from 0 (the hot surface has no direct view of the cooler surface) to 1 (the hot surface is completely surrounded by the cooler surface). In preliminary analysis, the surface is usually assumed to be completely surrounded by a single hypothetical surface whose temperature is the equivalent average temperature of the surrounding surfaces.

Arrays of low-power PCBs are often cooled by natural convection by mounting them within a chassis with adequate openings at the top and at the bottom to facilitate air flow, as shown in Fig. 14-38. The air between the PCBs rises when heated by the electronic components, and is replaced by the cooler air entering from below. This initiates the natural convection flow through the parallel flow passages formed by the PCBs. The PCBs must be placed *vertically* to take advantage of natural convection currents, and to minimize trapped air pockets (Fig. 14-39). Placing the PCBs too far from each other wastes valuable cabinet space, and placing them too close tends to "choke" the flow because of the increased resistance. Therefore, there should be an optimum spacing between the PCBs. It turns out that a distance of about 2 cm between the PCBs provides adequate air flow for effective natural convection cooling.

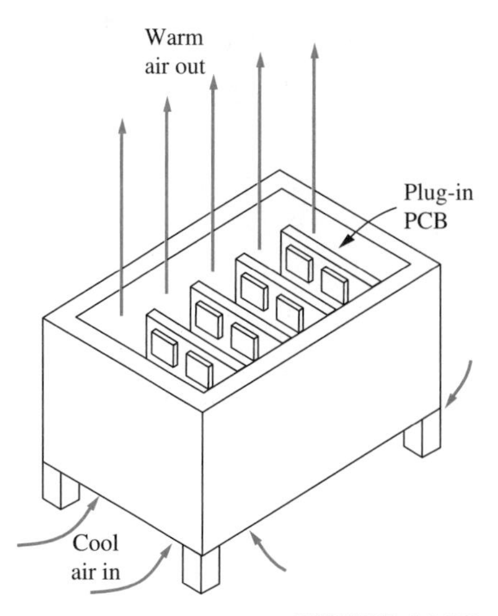


**FIGURE 14-38**

A chassis with an array of vertically oriented PCBs cooled by natural convection.

In the heat transfer analysis of PCBs, *radiation* heat transfer is

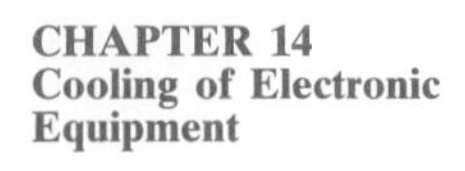



*disregarded,* since the view of the components is largely *blocked* by other heat generating components. As a result, hot components face other hot surfaces instead of a cooler surface. The exceptions are the two PCBs at the ends of the chassis that view the cooler side surfaces. Therefore, it is wise to mount any high-power components on the PCBs facing the walls of the chassis to take advantage of the additional cooling provided by radiation.

Circuit boards that dissipate up to about 5 W of power (or that have a power density of about 0.02 W/cm$^2$) can be cooled effectively by natural convection. Heat transfer from PCBs can be analyzed by treating them as rectangular plates with uniformly distributed heat sources on one side, and insulated on the other side, since heat transfer from the back surfaces of the PCBs is usually small. For PCBs with electronic components mounted on both sides, the rate of heat transfer and the heat transfer surface area will be twice as large.

It should be remembered that natural convection currents occur only in *gravitational fields.* Therefore, there can be no heat transfer in space by natural convection. This will also be the case when the air passageways are blocked and hot air cannot rise. In such cases, there will be no air motion, and heat transfer through the air will be by conduction.

The heat transfer from hot surfaces by natural convection and radiation can be enhanced by attaching fins to the surfaces. The heat transfer in this case can best be determined by using the data supplied by the manufacturers, as discussed in Chapter 8, especially for complex geometries.

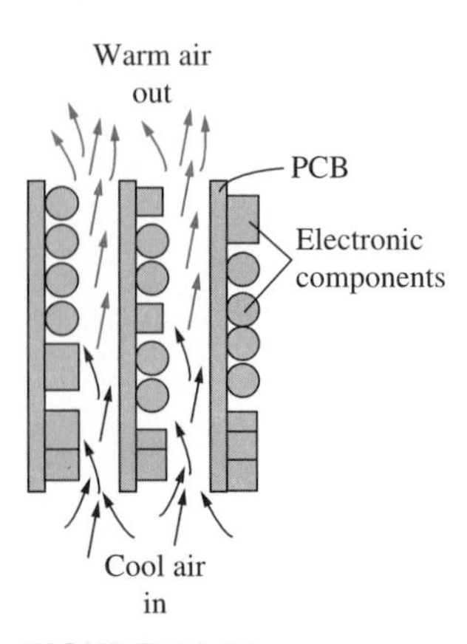


**FIGURE 14-39**
The PCBs in a chassis must be oriented vertically and spaced adequately to maximize heat transfer by natural convection.

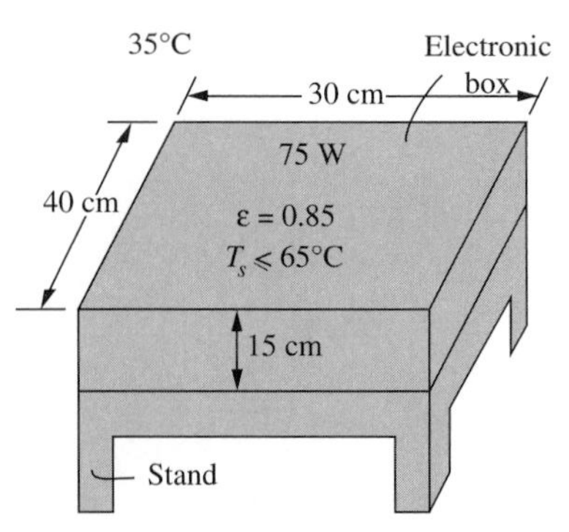


**FIGURE 14-40**
Schematic for Example 14-10.

**EXAMPLE 14-10**

Consider a sealed electronic box whose dimensions are 15 cm × 30 cm × 40 cm placed on top of a stand in a room at 35°C, as shown in Fig. 14-40. The box is painted, and the emissivity of its outer surface is 0.85. If the electronic components in the box dissipate 75 W of power and the outer surface temperature of the box is not to exceed 65°C, determine if this box can be cooled by natural convection and radiation alone. Assume the heat transfer from the bottom surface of the box to the stand to be negligible, and the temperature of the surrounding surfaces to be the same as the air temperature of the room.

**Solution** We assume the box is located at sea level, so that the local atmospheric pressure is 1 atm. The sealed electronic box will lose heat from the top and the side surfaces by natural convection and radiation.

All four side surfaces of the box can be treated as 0.15-m-high vertical surfaces. Then the natural convection heat transfer from these surfaces is determined to be

$$L = 0.15 \text{ m}$$

$$A_{\text{side}} = (2 \times 0.4 \text{ m} + 2 \times 0.3 \text{ m})(0.15 \text{ m}) = 0.21 \text{ m}^2$$

$$h_{\text{conv, side}} = 1.42\left(\frac{\Delta T}{L}\right)^{0.25} = 1.42\left(\frac{65 - 35}{0.15}\right)^{0.25} = 5.34 \text{ W/(m}^2 \cdot \text{°C)}$$

$$\dot{Q}_{\text{conv, side}} = h_{\text{conv, side}} A_{\text{side}}(T_s - T_{\text{fluid}})$$
$$= [535\ \text{W/(m}^2 \cdot {}^\circ\text{C)}](0.21\ \text{m}^2)(65 - 35)^\circ\text{C}$$
$$= 33.6\ \text{W}$$

Similarly, heat transfer from the horizontal top surface by natural convection is determined to be

$$L = \frac{4A}{p} = \frac{4(0.3\ \text{m})(0.4\ \text{m})}{2(0.3 + 0.4)\ \text{m}} = 0.34\ \text{m}$$
$$A_{\text{top}} = (0.3\ \text{m})(0.4\ \text{m}) = 0.12\ \text{m}^2$$
$$h_{\text{conv, top}} = 1.32\left(\frac{\Delta T}{L}\right)^{0.25} = 1.32\left(\frac{65 - 35}{0.34}\right)^{0.25} = 4.05\ \text{W/(m}^2 \cdot {}^\circ\text{C)}$$
$$\dot{Q}_{\text{conv, top}} = h_{\text{conv, top}} A_{\text{top}}(T_s - T_{\text{fluid}})$$
$$= [4.05\ \text{W/(m}^2 \cdot {}^\circ\text{C)}](0.12\ \text{m}^2)(65 - 35)^\circ\text{C}$$
$$= 14.6\ \text{W}$$

Therefore, the natural convection heat transfer from the entire box is

$$\dot{Q}_{\text{conv}} = \dot{Q}_{\text{conv, side}} + \dot{Q}_{\text{conv, top}} = 33.6 + 14.6 = 48.2\ \text{W}$$

The box is completely surrounded by the surfaces of the room, and it is stated that the temperature of the surfaces facing the box is equal to the air temperature in the room. Then the rate of heat transfer from the box by radiation can be determined from

$$\dot{Q}_{\text{rad}} = \varepsilon A \sigma (T_s^4 - T_{\text{surr}}^4)$$
$$= 0.85[(0.21 + 0.12)\ \text{m}^2][5.67 \times 10^{-8}\ \text{W/(m}^2 \cdot \text{K}^4)]$$
$$\times [(65 + 273)^4 - (35 + 273)^4]\ \text{K}^4$$
$$= 64.5\ \text{W}$$

Note that we must use absolute temperatures in radiation calculations. Then the total heat transfer from the box is simply

$$\dot{Q}_{\text{total}} = \dot{Q}_{\text{conv}} + \dot{Q}_{\text{rad}} = 48.2 + 64.5 = 112.7\ \text{W}$$

which is greater than 75 W. Therefore, this box can be cooled by combined natural convection and radiation, and there is no need to install any fans. There is even some safety margin left for occasions when the air temperature rises above 35°C.

**EXAMPLE 14-11**

A 0.2-W small cylindrical resistor mounted on a PCB is 1 cm long and has a diameter of 0.3 cm, as shown in Fig. 14-41. The view of the resistor is largely blocked by the PCB facing it, and the heat transfer from the connecting wires is negligible. The air is free to flow through the parallel flow passages between the PCBs. If the air temperature at the vicinity of the resistor is 50°C, determine the surface temperature of the resistor.

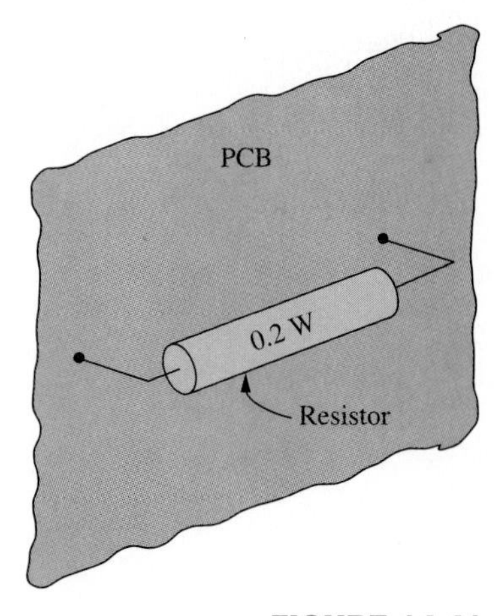


**FIGURE 14-41**
Schematic for Example 14-11.

**Solution** The resistor is to be cooled by natural convection and radiation. However, radiation can be neglected in this case, since the resistor is surrounded by surfaces that are at about the same temperature, and the net radiation heat transfer between two surfaces at the same temperature is zero.

This leaves natural convection as the only mechanism of heat transfer from the resistor.

Using the relation for components on a circuit board from Table 14-1, the natural convection heat transfer coefficient for this cylindrical component can be determined from

$$h_{conv} = 2.44\left(\frac{T_s - T_{fluid}}{D}\right)^{0.25}$$

where the diameter $D = 0.003$ m, which is the length in the heat flow path, is the characteristic length. We cannot determine $h_{conv}$ yet, since we do not know the surface temperature of the component and thus $\Delta T$. But we can substitute this relation into the heat transfer relation to get

$$\dot{Q}_{conv} = h_{conv}A(T_s - T_{fluid}) = 2.44\left(\frac{T_s - T_{fluid}}{D}\right)^{0.25} A(T_s - T_{fluid})$$
$$= 2.44A\frac{(T_s - T_{fluid})^{1.25}}{D^{0.25}}$$

The heat transfer surface area of the component is

$$A = 2 \times \tfrac{1}{4}\pi D^2 + \pi DL = 2 \times \tfrac{1}{4}\pi(0.3\text{ cm})^2 + \pi(0.3\text{ cm})(1\text{ cm}) = 1.084\text{ cm}^2$$

Substituting this and other known quantities in proper units (W for $\dot{Q}$, °C for $T$, m$^2$ for $A$, and m for $D$) into this equation and solving for $T_s$ yields

$$0.2 = 2.44(1.084 \times 10^{-4})\frac{(T_s - 50)^{1.25}}{0.003^{0.25}} \longrightarrow T_s = 113°\text{C}$$

Therefore, the surface temperature of the resistor on the PCB will be 113°C, which is considered to be a safe operating temperature for the resistors. Note that blowing air to the circuit board will lower this temperature considerably as a result of increasing the convection heat transfer coefficient and decreasing the air temperature at the vicinity of the components due to the larger flow rate of air.

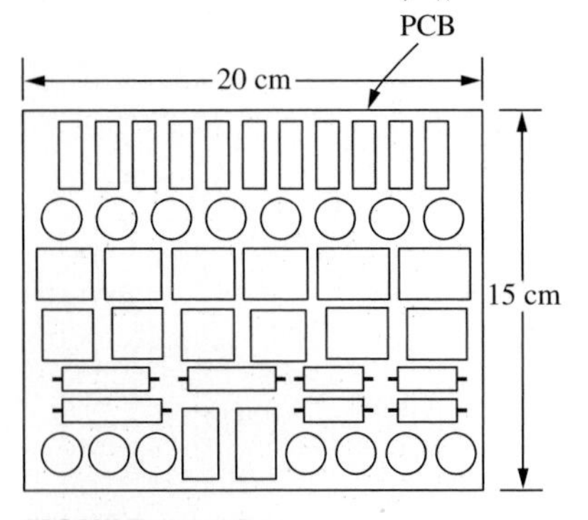


**FIGURE 14-42**
Schematic for Example 14-12.

**EXAMPLE 14-12**

A 15 cm × 20 cm PCB has electronic components on one side, dissipating a total of 7 W, as shown in Fig. 14-42. The PCB is mounted in a rack vertically together with other PCBs. If the surface temperature of the components is not to exceed 100°C, determine the maximum temperature of the environment in which this PCB can operate safely at sea level. What would your answer be if this rack is located at a location at 4000 m altitude where the atmospheric pressure is 61.66 kPa?

**Solution** We disregard any heat transfer by radiation since the PCB is surrounded by other PCBs at about the same temperature. We also neglect any heat transfer from the back surface of the PCB, since it will be small. Then the entire heat load of the PCB must be dissipated to the ambient air by natural convection from its front surface, which can be treated as a vertical flat plate.

Using the simplified relation for a vertical surface from Table 14-1, the natural convection heat transfer coefficient for this PCB can be determined from

$$h_{conv} = 1.42\left(\frac{\Delta T}{L}\right)^{0.25} = 1.42\left(\frac{T_s - T_{fluid}}{L}\right)^{0.25}$$

The characteristic length in this case is the height ($L = 0.15$ m) of the PCB, which is the length in the path of heat flow. We cannot determine $h_{conv}$ yet, since we do not know the ambient temperature and thus $\Delta T$. But we can substitute this relation into the heat transfer relation to get

$$\dot{Q}_{conv} = h_{conv}A(T_s - T_{fluid}) = 1.42\left(\frac{T_s - T_{fluid}}{L}\right)^{0.25} A(T_s - T_{fluid})$$

$$= 1.42A\frac{(T_s - T_{fluid})^{1.25}}{L^{0.25}}$$

The heat transfer surface area of the PCB is

$$A = \text{(width)(height)} = (0.2\text{ m})(0.15\text{ m}) = 0.03\text{ m}^2$$

Substituting this and other known quantities in proper units (W for $\dot{Q}$, °C for $T$, m$^2$ for $A$, and m for $L$) into this equation and solving for $T_{fluid}$ yields

$$7 = 1.42(0.03)\frac{(100 - T_{fluid})^{1.25}}{0.15^{0.25}} \longrightarrow T_{fluid} = 59.5°\text{C}$$

Therefore, the PCB will operate safely in environments with temperatures up to 59.4°C by relying solely on natural convection.

At an altitude of 4000 m, the atmospheric pressure is 61.66 kPa, which is equivalent to

$$P = (61.66\text{ kPa})\frac{1\text{ atm}}{101.325\text{ kPa}} = 0.609\text{ atm}$$

The heat transfer coefficient in this case is obtained by multiplying the value at sea level by $\sqrt{P}$, where $P$ is in atm. Substituting

$$7 = 1.42(0.03)\frac{(100 - T_{fluid})^{1.25}}{0.15^{0.25}}\sqrt{0.609} \longrightarrow T_{fluid} = 50.6°\text{C}$$

which is about 10°C lower than the value obtained at 1-atm pressure. Therefore, the effect of altitude on convection should be considered in high altitude applications.

## 14-8 ■ AIR COOLING: FORCED CONVECTION

We mentioned earlier that convection heat transfer between a solid surface and a fluid is proportional to the velocity of the fluid. The *higher* the velocity, the *larger* the flow rate and the *higher* the heat transfer rate. The fluid velocities associated with natural convection currents are naturally low, and thus natural convection cooling is limited to low-power electronic systems.

When natural convection cooling is not adequate, we simply add a *fan* and blow air through the enclosure that houses the electronic components. In other words, we resort to *forced convection* in order to enhance the velocity and thus the flow rate of the fluid as well as the heat

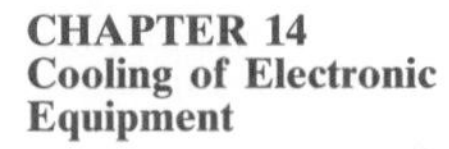



transfer. By doing so, we can increase the heat transfer coefficient by a factor of up to about 10, depending on the size of the fan. This means we can remove heat at much higher rates for a specified temperature difference between the components and the air, or we can reduce the surface temperature of the components considerably for a specified power dissipation.

The *radiation* heat transfer in forced-convection-cooled electronic systems is usually disregarded for two reasons. First, forced convection heat transfer is usually much larger than that due to radiation, and the consideration of radiation causes no significant change in the results. Second, the electronic components and circuit boards in convection-cooled systems are mounted so close to each other that a component is almost entirely surrounded by other components at about the same high temperature. That is, the components have hardly any direct view of a cooler surface. This results in little or no radiation heat transfer from the components. The components near the edges of circuit boards with a large view of a cooler surface may benefit somewhat from the additional cooling by radiation, and it is a good design practice to reserve those spots for high-power components to have a thermally balanced system.

When heat transfer from the outer surface of the enclosure of the electronic equipment is negligible, the amount of heat absorbed by the air becomes equal to the amount of heat given up (or power dissipated) by the electronic components in the enclosure, and can be expressed as (Fig. 14-43)

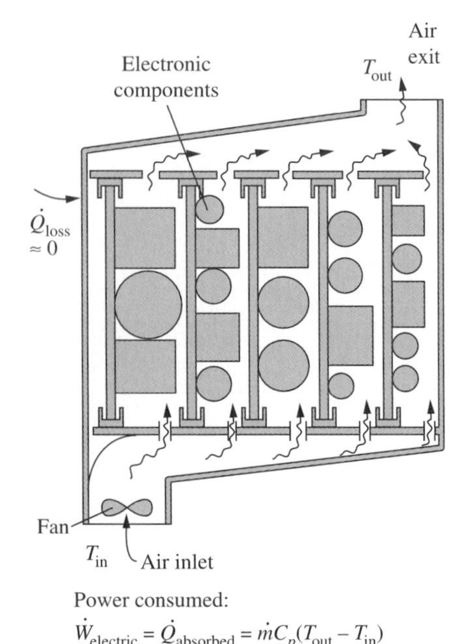


**FIGURE 14-43**

In steady operation, the heat absorbed by air per unit time as it flows through an electronic box is equal to the power consumed by the electronic components in the box.

$$\dot{Q} = \dot{m}C_p(T_{out} - T_{in}) \quad \text{(W)} \tag{14-17}$$

where $\dot{Q}$ is the rate of heat transfer to the air, $C_p$ is the specific heat of air, $T_{in}$ and $T_{out}$ are the average temperatures of air at the inlet and exit of the enclosure, respectively, and $\dot{m}$ is the mass flow rate of air.

Note that for a specified mass flow rate and power dissipation, the *temperature rise* of air, $T_{out} - T_{in}$, remains constant as it flows through the enclosure. Therefore, the higher the inlet temperature of the air, the higher is the exit temperature, and thus the higher is the surface temperature of the components. It is considered a good design practice to limit the temperature rise of air to 10°C, and the maximum exit temperature of air to 70°C. In a properly designed forced air-cooled system, this results in a maximum component surface temperature of under 100°C.

The mass flow rate of air required for cooling an electronic box depends on the temperature of air available for cooling. In cool environments, such as an air-conditioned room, a smaller flow rate will be adequate. However, in hot environments, we may need to use a larger flow rate to avoid overheating of the components and the potential problems associated with it.

Forced convection is covered in detail in Chap. 10. For those who skipped that chapter because of time limitations, below we present a brief review of basic concepts and relations.

The fluid flow over a body such as a transistor is called **external flow**, and flow through a confined space such as inside a tube or through the parallel passage area between two circuit boards in an enclosure is called **internal flow** (Fig. 14-44). Both types of flow are encountered in a typical electronic system.

Fluid flow is also categorized as being **laminar** (smooth and streamlined) or **turbulent** (intense eddy currents and random motion of chunks of fluid). Turbulent flow is desirable in heat transfer applications, since it results in a much larger heat transfer coefficient. But it also comes with a much larger friction coefficient, which requires a much larger fan (or pump for liquids).

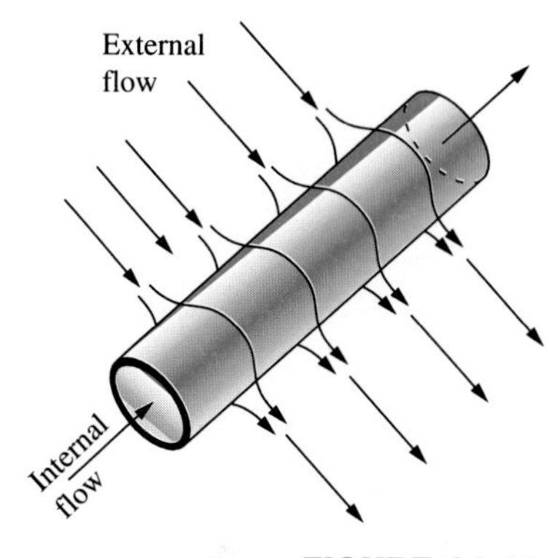


**FIGURE 14-44**
Internal flow through a circular tube and external flow over it.

Numerous experimental studies have shown that turbulence tends to occur at larger velocities, during flow over larger bodies or flow through larger channels, and with fluids having smaller viscosities. These effects are combined into the dimensionless **Reynolds number** defined as

$$\text{Re} = \frac{\mathcal{V} D}{\nu} \tag{14-18}$$

where

$\mathcal{V}$ = velocity of the fluid (*free-stream* velocity for external flow, and *average* velocity for internal flow), m/s
$D$ = characteristic length of the geometry (the length the fluid flows over in external flow, and the equivalent diameter in internal flow), m
$\nu = \mu/\rho$ = kinematic viscosity of the fluid, $m^2/s$.

The Reynolds number at which the flow changes from laminar to turbulent is called the *critical Reynolds number,* whose value is 2300 for internal flow, 500,000 for flow over a flat plate, and 200,000 for flow over a cylinder or sphere.

The *equivalent* (or *hydraulic*) *diameter* for internal flow is defined as

$$D_h = \frac{4A_c}{p} \quad \text{(m)} \tag{14-19}$$

where $A_c$ is the cross-sectional area of the flow passage and $p$ is the perimeter. Note that for a circular pipe the hydraulic diameter is equivalent to the ordinary diameter.

The convection heat transfer is expressed by *Newton's law of cooling* as

$$\dot{Q}_{\text{conv}} = hA(T_s - T_{\text{fluid}}) \quad \text{(W)} \tag{14-20}$$

where

$h$ = average convection heat transfer coefficient, $W/(m^2 \cdot °C)$
$A$ = heat transfer surface area, $m^2$
$T_s$ = temperature of the surface, °C
$T_{\text{fluid}}$ = temperature of the fluid sufficiently far from the surface for *external* flow, and average temperature of the fluid at a specified location in *internal* flow, °C.

When the heat load is distributed uniformly on the surfaces with a constant heat flux $\dot{q}$, the total rate of heat transfer can also be expressed as $\dot{Q} = \dot{q}A$.

In *fully developed flow* through a pipe or duct (i.e., when the entrance effects are negligible) subjected to constant heat flux on the surfaces, the convection heat transfer coefficient $h$ remains *constant.* In this case, both the surface temperature $T_s$ and the fluid temperature $T_{\text{fluid}}$ increase *linearly,* as shown in Fig. 14-45, but the difference between them, $T_s - T_{\text{fluid}}$, remains *constant.* Then the *temperature rise* of the surface above the fluid temperature can be determined from Eq. 14-20 to be

$$\Delta T_{\text{rise, surface}} = T_s - T_{\text{fluid}} = \frac{\dot{Q}_{\text{conv}}}{hA} \quad (°\text{C}) \tag{14-21}$$

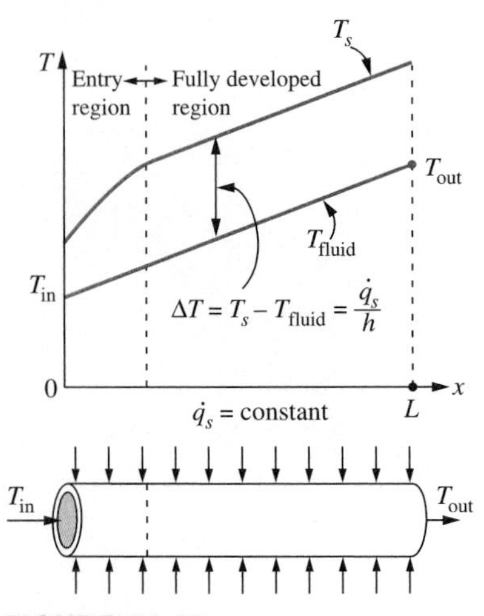


**FIGURE 14-45**
Under constant heat flux conditions, the surface and fluid temperatures increase linearly, but their difference remains constant in the fully developed region.

Note that the temperature rise of the surface is inversely proportional to the convection heat transfer coefficient. Therefore, the greater the convection coefficient, the lower the surface temperature of the electronic components.

When the exit temperature of the fluid, $T_{\text{out}}$, is known, the highest surface temperature that will occur at the end of the flow channel can be determined from Eq. 14-21 to be

$$T_{s,\max} = T_{\text{fluid, max}} + \frac{\dot{Q}}{hA} = T_{\text{out}} + \frac{\dot{Q}}{hA} \quad (°\text{C}) \tag{14-22}$$

If this temperature is within the safe range, then we don't need to worry about temperatures at other locations. But if it is not, it may be necessary to use a larger fan to increase the flow rate of the fluid.

In convection analysis, the convection heat transfer coefficient $h$ is usually expressed in terms of the dimensionless **Nusselt number** Nu as

$$h = \frac{k}{D}\text{Nu} \quad [\text{W/(m}^2 \cdot °\text{C)}] \tag{14.23}$$

where $k$ is the thermal conductivity of the fluid and $D$ is the characteristic length of the geometry. Relations for the average Nusselt number based on experimental data are given in Table 14-2 for external flow, and in Table 14-3 for laminar ($\text{Re} < 2300$) internal flow under a uniform heat flux condition, which is closely approximated by electronic equipment. For *turbulent flow* ($\text{Re} > 2300$) through smooth tubes and channels, the Nusselt number can be determined from

$$\text{Nu} = 0.023\ \text{Re}^{0.8}\ \text{Pr}^{0.4} \tag{14-24}$$

for any geometry. Here Pr is the dimensionless **Prandtl number**, and its value is about 0.7 for air at room temperature.

**TABLE 14-2**

**Empirical correlations for the average Nusselt number for forced convection over a flat plate and circular and noncircular cylinders in cross-flow**

| Cross-section of the cylinder | Fluid | Range of Re | Nusselt number |
|---|---|---|---|
| Circle | Gas or liquid | 0.4–4 | $Nu = 0.989\,Re^{0.330}\,Pr^{1/3}$ |
| | | 4–40 | $Nu = 0.911\,Re^{0.385}\,Pr^{1/3}$ |
| | | 40–4000 | $Nu = 0.683\,Re^{0.466}\,Pr^{1/3}$ |
| | | 4000–40,000 | $Nu = 0.193\,Re^{0.618}\,Pr^{1/3}$ |
| | | 40,000–400,000 | $Nu = 0.027\,Re^{0.805}\,Pr^{1/3}$ |
| Square | Gas | 5000–100,000 | $Nu = 0.102\,Re^{0.675}\,Pr^{1/3}$ |
| Square (tilted 45°) | Gas | 5000–100,000 | $Nu = 0.246\,Re^{0.588}\,Pr^{1/3}$ |
| Flat plate | Gas or liquid | $0–5 \times 10^5$ | $Nu = 0.664\,Re^{1/2}\,Pr^{1/3}$ |
| | | $5 \times 10^5–10^7$ | $Nu = (0.037\,Re^{4/5} - 871)\,Pr^{1/3}$ |
| Vertical plate | Gas | 4000–15,000 | $Nu = 0.228\,Re^{0.731}\,Pr^{1/3}$ |

*Source*: Refs. 14 and 19.

The fluid properties in the above relations are to be evaluated at the *bulk mean fluid temperature* $T_{ave} = \frac{1}{2}(T_{in} + T_{out})$ for internal flow, which is the arithmetic average of the mean fluid temperatures at the inlet and the exit of the tube, and at the *film temperature* $T_{film} = \frac{1}{2}(T_s + T_{fluid})$ for external flow, which is the arithmetic average of the surface temperature and free-stream temperature of the fluid.

The relations in Table 14-3 for internal flow assume fully developed flow over the entire flow section, and disregard the heat transfer enhancement effects of the development region at the entrance. Therefore, the results obtained from these relations are on the conservative

**TABLE 14-3**

**Nusselt number for fully developed laminar flow in circular tubes and rectangular channels**

| Cross-section of the tube | Aspect ratio | Nusselt number |
|---|---|---|
| Circle 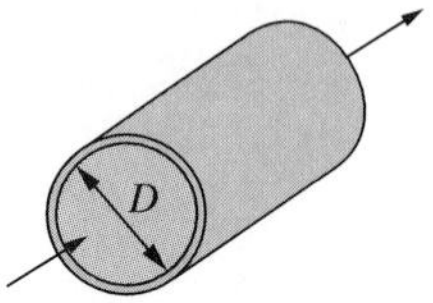 | — | 4.36 |
| Square 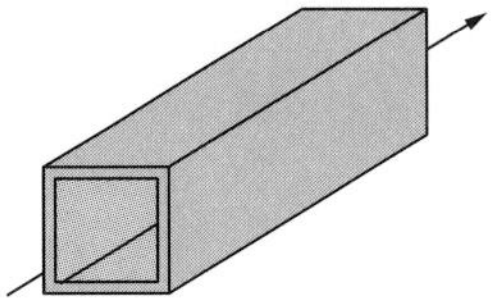 | — | 3.61 |
| Rectangle 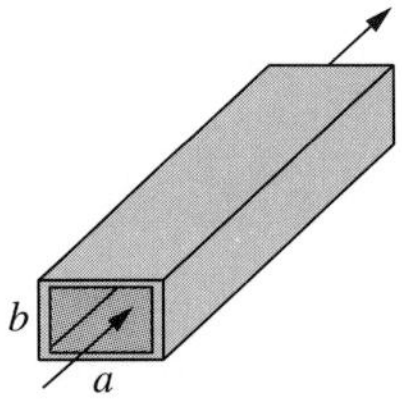 | $a/b$ | |
| | 1 | 3.61 |
| | 2 | 4.12 |
| | 3 | 4.79 |
| | 4 | 5.33 |
| | 6 | 6.05 |
| | 8 | 6.49 |
| | ∞ | 8.24 |

side. We don't mind this much, however, since it is common practice in engineering design to have some safety margin to fall back to "just in case," as long as it does not result in a grossly overdesigned system. Also, it may sometimes be necessary to do some local analysis for critical components with small surface areas to assure reliability and to incorporate solutions to local problems such as attaching heat sinks to high power components.

### Fan Selection

Air is supplied to electronic equipment by one or several fans. Although the air is free and abundant, the fans are not. Therefore, a few words about the fan selection are in order.

A fan at a fixed speed (or fixed rpm) will deliver a fixed volume of air regardless of the altitude and pressure. But the mass flow rate of air will be less at *high altitude* as a result of the lower density of air. For example, the atmospheric pressure of air drops by more than 50 percent at an altitude of 6000 m from its value at sea level. This means that the fan will deliver *half* as much air mass at this altitude at the same rpm and temperature, and thus the temperature rise of air cooling will double. This may create serious reliability problems and catastrophic failures of electronic equipment if proper precautions are not taken. *Variable-speed fans* that automatically increase speed when the air density decreases are available to avoid such problems. Expensive electronic systems are usually equipped with thermal cutoff switches to prevent overheating due to inadequate air flow rate or the failure of the cooling fan.

Fans draw in not only cooling air but also all kinds of *contaminants* that are present in the air, such as lint, dust, moisture, and even oil. If unattended, these contaminants can pile up on the components, and plug up narrow passageways, causing overheating. It should be remembered that the dust that settles on the electronic components acts as a an insulation layer that makes it very difficult for the heat generated in the component to escape. To minimize the contamination problem, air filters are commonly used. It is good practice to use the largest air filter practical to minimize the pressure drop of air and to maximize the dust capacity.

Often the question arises about whether to place the fan at the inlet or the exit of an electronic box. The generally preferred location is the *inlet.* A fan placed at the inlet draws air in and pressurizes the electronic box and prevents air infiltration into the box from cracks or other openings. Having only one location for air inlet makes it practical to install a filter at the inlet to clean the air from all the dust and dirt before they enter the box. This allows the electronic system to operate in a *clean* environment. Also, a fan placed at the inlet handles cooler and thus *denser* air, which results in a higher mass flow rate for the same volume flow rate or rpm. Since the fan is always subjected to cool air this has the added benefit that it increases the reliability and extends the life of the fan. The major disadvantage associated with having a fan mounted at the inlet is that the *heat* generated by the fan and its motor is picked up by air on its way into the box, which adds to the heat load of the system.

When the fan is placed at the *exit,* the heat generated by the fan and its motor is immediately discarded to the atmosphere without getting blown first into the electronic box. However, a fan at the exit creates a *vacuum* inside the box, which draws air into the box through inlet vents as well as any cracks and openings (Fig. 14-46). Therefore, the air is difficult to filter, and the dirt and dust that collects on the components undermines the reliability of the system.

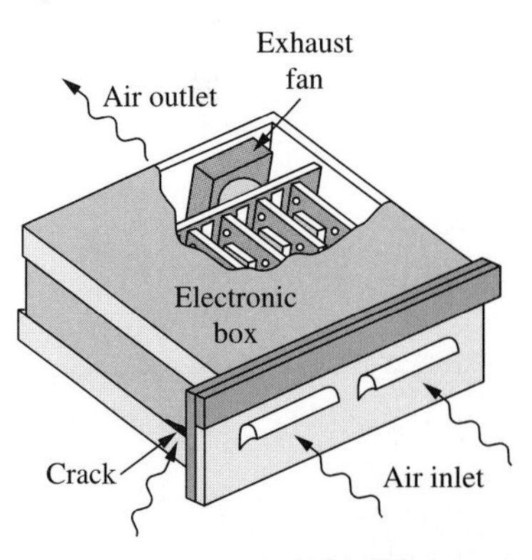


**FIGURE 14-46**
A fan placed at the exit of an electronic box draws in air as well as contaminants in the air through cracks.

There are several types of fans available on the market for cooling electronic equipment, and the right choice depends on the situation on hand. There are two primary considerations in the selection of the fan: the *static pressure head* of the system, which is the total resistance an electronic system offers to air as it passes through, and the *volume flow rate* of the air. *Axial fans* are simple, small, light, and inexpensive, and they can deliver a large flow rate. However, they are suitable for systems with relatively small pressure heads. Also, axial fans usually run at very high speeds, and thus they are noisy. The *radial* or *centrifugal fans,* on the other hand, can deliver moderate flow rates to systems with high static pressure heads at relatively low speeds. But they are larger, heavier, more complex, and more expensive than axial fans.

The performance of a fan is represented by a set of curves called the *characteristic curves,* which are provided by fan manufacturers to help engineers with the selection of fans. A typical static pressure head curve for a fan is given in Fig. 14-47 together with a typical system flow resistance curve plotted against the flow rate of air. Note that a fan creates the *highest* pressure head at *zero* flow rate. This corresponds to the limiting case of blocked exit vents of the enclosure. The flow rate increases with decreasing static head, and reaches its maximum value when the fan meets no flow resistance.

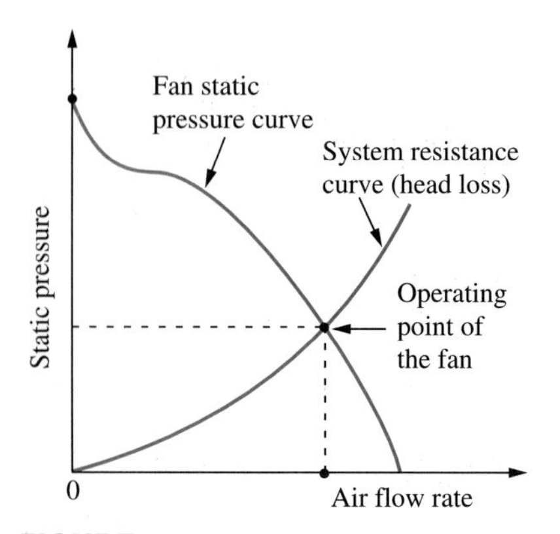


**FIGURE 14-47**

The air flow rate a fan delivers into an electronic enclosure depends on the flow resistance of that system as well as the variation of the static head of the fan with flow rate.

Any electronic enclosure will offer some resistance to flow. The system resistance curve is *parabolic* in shape, and the pressure or head loss due to this resistance is nearly proportional to the *square* of the flow rate. The fan must overcome this resistance to maintain flow through the enclosure. The design of a forced convection cooling system requires the determination of the total system resistance characteristic curve. This curve can be generated accurately by measuring the static pressure drop at different flow rates. It can also be determined approximately by evaluating the pressure drops.

A fan will operate at the *point* where the fan static head curve and the system resistance curve *intersects.* Note that a fan will deliver a higher flow rate to a system with a low flow resistance. The required air flow rate for a system can be determined from heat transfer requirements alone, using the *design heat load* of the system and the *allowable temperature rise* of air. Then the flow resistance of the system at this flow rate can be determined analytically or experimentally. Knowing the flow rate and the needed pressure head, it is easy to select a fan from manufacturers catalogs which will meet both of these requirements.

Below we present some general guidelines associated with the forced air cooling of electronic systems.

**1** Before deciding on forced air cooling, check to see if *natural convection* cooling is adequate. If it is, which may be the case for low-power systems, incorporate it and avoid all the problems associated with fans such as cost, power consumption, noise, complexity, maintenance, and possible failure.

**2** Select a fan that is neither too small nor too large. An *undersized* fan may cause the electronic system to overheat and fail. An *oversized* fan will definitely provide adequate cooling, but it will needlessly be larger, more expensive, and will consume more power.

**3** If the temperature rise of air due to the power consumed by the motor of the fan is acceptable, mount the fan at the *inlet* of the box to pressurize the box and filter the air to keep dirt and dust out (Fig. 14-48).

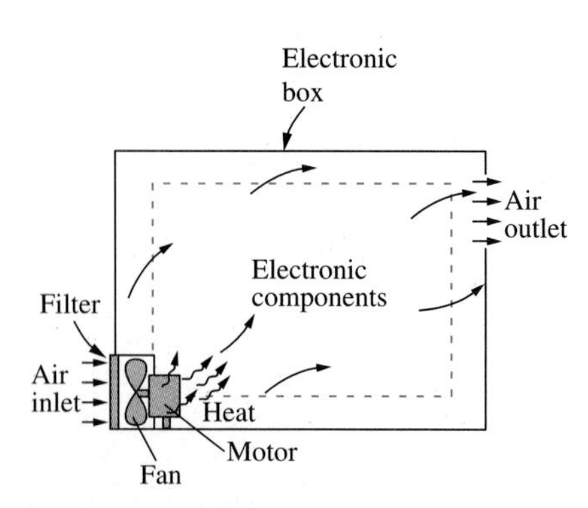


**FIGURE 14-48**

Installing the fan at the inlet keeps the dirt and dust out, but the heat generated by the fan motor in.

**4** Position and size the air exit vents so that there is *adequate air flow* throughout the entire box. More air can be directed to a certain area by

enlarging the size of the vent at that area. The total exit areas should be at least as large as the inlet flow area to avoid the choking of the air flow, which may result in reduced air flow rate.

**5** Place the most critical electronic components near the *entrance,* where the air is coolest. Place the components that consume a lot of power near the exit (Fig. 14-49).

**6** Arrange the circuit boards and the electronic components in the box such that the *resistance* of the box to air flow is *minimized* and thus the flow rate of air through the box is maximized for the same fan speed. Make sure that no hot air pockets are formed during operation.

**7** Consider the effect of *altitude* in high-altitude applications.

**8** Try to avoid any flow sections that increase the flow resistance of the systems, such as unnecessary corners, sharp turns, sudden expansions and contractions, and very high velocities (greater than 7 m/s), since the flow resistance is nearly proportional to the flow rate. Also avoid very low velocities, since these result in a poor heat transfer performance, and allow the dirt and the dust in the air to settle on the components.

**9** Arrange the system such that *natural convection* helps forced convection instead of hurting it. For example, mount the PCBs vertically, and blow the air from the bottom towards the top instead of the other way around.

**10** When the design calls for the use of two or more fans, a decision needs to be made about mounting the fans in parallel or in series. Fans mounted in *series* will boost the pressure head available, and are best suited for systems with a high flow resistance. Fans connected in *parallel* will increase the flow rate of air, and are best suited for systems with small flow resistance.

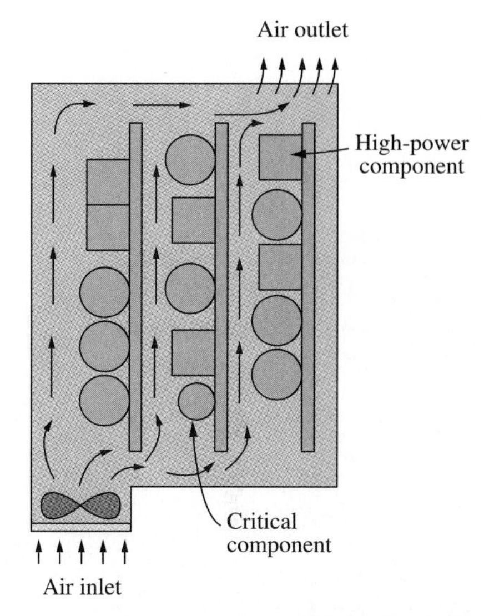


**FIGURE 14-49**

Sensitive components should be located near the inlet, and high-power components near the exit.

## Cooling Personal Computers

The introduction of the 4004 chip, the first general purpose microprocessor, by the Intel Corporation in the early 1970s marked the beginning of the electronics era in consumer goods, from calculators and washing machines to personal computers. The microprocessor, which is the "brain" of the personal computer, is basically a DIP-type LSE package that incorporates a central processing unit (CPU), memory, and some input/output capabilities.

A typical desktop personal computer consists of a few circuit boards plugged into a mother board, which houses the microprocessor and the memory chips as well as the network of interconnections enclosed in a formed sheet metal chassis, which also houses the disk and CD-ROM drives. Connected to this "magic" box are the monitor, a keyboard, a printer, and other auxiliary equipment (Fig. 14-50). The PCBs are normally mounted vertically on a mother board, since this facilitates better cooling.

**FIGURE 14-50**

A desktop personal computer with monitor and keyboard.

A small and quiet fan is usually mounted to the rear or side of the chassis to cool the electronic components. There are also louvers and openings on the side surfaces to facilitate air circulation. Such openings are not placed on the top surface, since many users would block them by putting books or other things there, which will jeopardize safety, and a coffee or soda spill can cause major damage to the system.

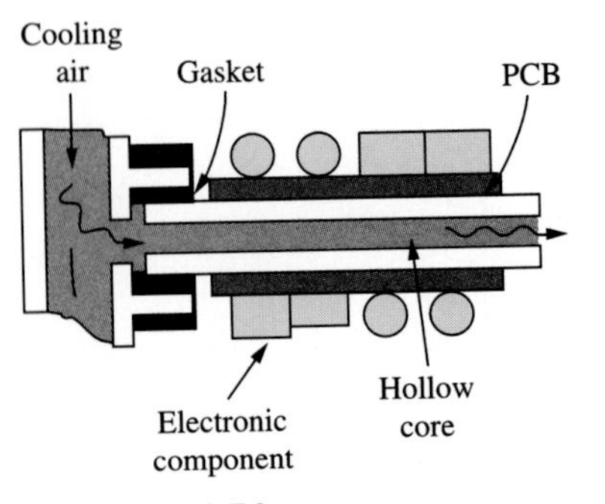


**FIGURE 14-51**
The hollow-core PCB discussed in Example 14-13.

**EXAMPLE 14-13**

Some strict specifications of electronic equipment require that the cooling air not come into direct contact with the electronic components, in order to protect them from exposure to the contaminants in the air. In such cases, heat generated in the components on a PCB must be conducted a long way to the walls of the enclosure through a metal core strip or a heat frame attached to the PCB. An alternative solution is the *hollow-core PCB,* which is basically a narrow duct of rectangular cross-section made of thin glass–epoxy board with electronic components mounted on both sides, as shown in Fig. 14-51. Heat generated in the components is conducted to the hollow core through a thin layer of epoxy board, and is then removed by the cooling air flowing through the core. Effective sealing is provided to prevent air leakage into the component chamber.

Consider a hollow-core PCB 12 cm high and 18 cm long, dissipating a total of 40 W. The width of the air gap between the two sides of the PCB is 0.3 cm. The cooling air enters the core at 20°C at a rate of 0.72 L/s. Assuming the heat generated to be uniformly distributed over the two side surfaces of the PCB, determine (*a*) the temperature at which the air leaves the hollow core and (*b*) the highest temperature on the inner surface of the core.

**Solution** We assume operation at sea level, and take the atmospheric pressure to be 1 atm. We also assume that all the heat generated in electronic components is removed by the air flowing through the hollow core. The temperature of air varies as it flows through the core, and so does its properties. We will perform the calculations using property values at 27°C (or 300 K) given below, since the air enters at 20°C, and its temperature will increase.

$$\rho = 1.177\ \text{kg/m}^3, \qquad k = 0.0261\ \text{W/(m}\cdot{}^\circ\text{C)}$$
$$C_p = 1005\ \text{J/(kg}\cdot{}^\circ\text{C)}, \qquad \nu = 1.57 \times 10^{-5}\ \text{m}^2\text{/s}$$
$$\text{Pr} = 0.712$$

After we calculate the exit temperature of air, we can repeat the calculations, if necessary, using properties at the average temperature.

The cross-sectional area of the channel and its hydraulic diameter are

$$A_c = \text{(height)(width)} = (0.12\ \text{m})(0.003\ \text{m}) = 3.6 \times 10^{-4}\ \text{m}^2$$
$$D_h = \frac{4A_c}{p} = \frac{4 \times (3.6 \times 10^{-4}\ \text{m}^2)}{2 \times (0.12 + 0.003)\ \text{m}} = 0.00585\ \text{m}$$

The average velocity and the mass flow rate of air are

$$\mathcal{V} = \frac{\dot{V}}{A_c} = \frac{0.72 \times 10^{-3}\ \text{m}^3\text{/s}}{3.6 \times 10^{-4}\ \text{m}^2} = 2.0\ \text{m/s}$$
$$\dot{m} = \rho\dot{V} = (1.177\ \text{kg/m}^3)(0.72 \times 10^{-3}\ \text{m}^3\text{/s}) = 0.847 \times 10^{-3}\ \text{kg/s}$$

(*a*) The temperature of air at the exit of the hollow core can be determined from

$$\dot{Q} = \dot{m}C_p(T_{out} - T_{in})$$

Solving for $T_{out}$ and substituting the given values, we obtain

$$T_{out} = T_{in} + \frac{\dot{Q}}{\dot{m}C_p} = 20°C + \frac{40\ J/s}{(0.847 \times 10^{-3}\ kg/s)[1005\ J/(kg\,°C)]} = 67°C$$

(*b*) The surface temperature of the channel at any location can be determined from

$$\dot{Q}_{conv} = hA(T_s - T_{fluid})$$

where the heat transfer surface area is

$$A = 2A_{side} = 2(\text{height})(\text{length}) = 2(0.12\ m)(0.18\ m) = 0.0432\ m^2$$

To determine the convection heat transfer coefficient, we first need to calculate the Reynolds number:

$$Re = \frac{\mathcal{V}D_h}{\nu} = \frac{(2\ m/s)(0.00585\ m)}{1.57 \times 10^{-5}\ m^2/s} = 745 < 2300$$

Therefore, the flow is laminar, and, assuming fully developed flow, the Nusselt number for the air flow in this rectangular cross-section corresponding to the aspect ratio $a/b = (12\ cm)(0.3\ cm) = 40 \approx \infty$ is determined from Table 14-3 to be

$$Nu = 8.24$$

and thus

$$h = \frac{k}{D_h}Nu = \frac{0.0261\ W/(m \cdot °C)}{0.00585\ m}(8.24) = 36.7\ W/(m^2 \cdot °C)$$

Then the surface temperature of the hollow core near the exit is determined to be

$$T_{s,max} = T_{out} + \frac{\dot{Q}}{hA} = 67°C + \frac{40\ W}{[36.7\ W/(m^2 \cdot °C)](0.0432\ m^2)} = 92.2°C$$

Note that the temperature difference between the surface and the air at the exit of the hollow core is 25.2°C. This temperature difference between the air and the surface remains at that value throughout the core, since the heat generated on the side surfaces is uniform, and the convection heat transfer coefficient is constant. Therefore, the surface temperature of the core at the inlet will be 20°C + 25.2°C = 45.2°C. In reality, however, this temperature will be somewhat lower because of the entrance effects, which affect heat transfer favorably. The fully developed flow assumption gives somewhat conservative results, but is commonly used in practice because it provides considerable simplification in calculations.

**EXAMPLE 14-4**

A TO 71 transistor with a height of 0.53 cm and a diameter of 0.44 cm is mounted on a circuit board, as shown in Fig. 14-52. The transistor is cooled by air flowing over it at a velocity of 90 m/min. If the air temperature is 65°C and the transistor case temperature is not to exceed 90°C, determine the amount of power this transistor can dissipate safely.

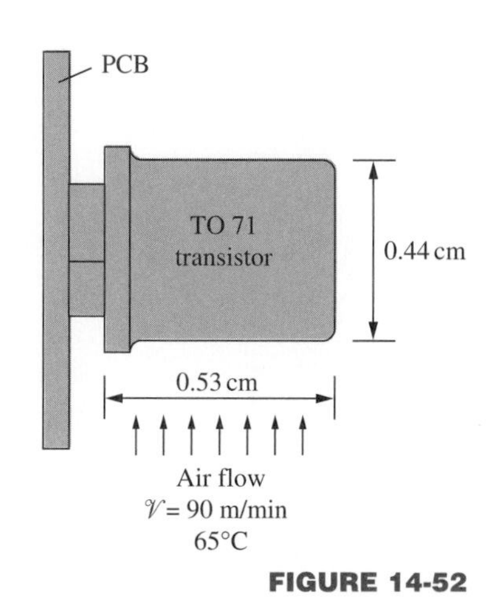


**FIGURE 14-52**
Schematic for Example 14-14.

**Solution** The transistor is cooled by forced convection through its cylindrical surface as well as its flat top and bottom surfaces. We assume operation at sea level, and take the atmospheric pressure to be 1 atm. The properties of air at 1-atm pressure and the film temperaure $T_f = \frac{1}{2}(T_s + T_{\text{fluid}}) = \frac{1}{2}(90 + 65) = 77.5°\text{C} \approx 350\text{ K}$ are

$$\rho = 1.009\text{ kg/m}^3, \qquad k = 0.0297\text{ W/(m}\cdot°\text{C)}$$
$$C_p = 1008\text{ J/(kg}\cdot°\text{C)}, \qquad \nu = 2.06 \times 10^{-5}\text{ m}^2\text{/s}$$
$$\text{Pr} = 0.706$$

The characteristic length for flow over a cylinder is the diameter $D = 0.0044$ m. Then the Reynolds number becomes

$$\text{Re} = \frac{\mathcal{V}D}{\nu} = \frac{(90/60\text{ m/s})(0.0044\text{ m})}{2.06 \times 10^{-5}\text{ m}^2\text{/s}} = 320$$

which falls into the range 40–4000. Using the corresponding relation from Table 14-2 for the Nusselt number, we obtain

$$\text{Nu} = 0.683\text{ Re}^{0.466}\text{ Pr}^{1/3} = 0.683(320)^{0.466}(0.706)^{1/3} = 8.94$$

and $$h = \frac{k}{D}\text{Nu} = \frac{0.0297\text{ W/(m}\cdot°\text{C)}}{0.0044\text{ m}}(8.94) = 60.4\text{ W/(m}^2\cdot°\text{C)}$$

Also, $$A_{\text{cyl}} = \pi DL = \pi(0.0044\text{ m})(0.0053\text{ m}) = 0.733 \times 10^{-4}\text{ m}^2$$

Then the rate of heat transfer from the cylindrical surface becomes

$$\dot{Q}_{\text{cyl}} = hA_{\text{cyl}}(T_s - T_{\text{fluid}}) = [60.4\text{ W/(m}^2\cdot°\text{C)}](0.733 \times 10^{-4}\text{ m}^2)(90\text{-}65)°\text{C} = 0.11\text{ W}$$

We now repeat the calculations for the top and bottom surfaces of the transistor, which can be treated as flat plates of length $L = 0.0044$ m in the flow direction (which is the diameter), and, using the proper relation from Table 14-2,

$$\text{Re} = \frac{\mathcal{V}L}{\nu} = \frac{(\frac{90}{60}\text{ m/s})(0.0044\text{ m})}{2.06 \times 10^{-5}\text{ m}^2\text{/s}} = 320$$
$$\text{Nu} = 0.664\text{ Re}^{1/2}\text{ Pr}^{1/3} = 0.664(320)^{1/2}(0.706)^{1/3} = 10.6$$

and $$h = \frac{k}{D}\text{Nu} = \frac{0.0297\text{ W/(m}\cdot°\text{C)}}{0.0044\text{ m}}(10.6) = 71.6\text{ W/(m}^2\cdot°\text{C)}$$

Also,

$$A_{\text{flat}} = A_{\text{top}} + A_{\text{bottom}} = 2 \times \tfrac{1}{4}\pi D^2 = 2 \times \tfrac{1}{4}\pi(0.0044\text{ m})^2 = 0.30 \times 10^{-4}\text{ m}^2$$
$$\dot{Q}_{\text{flat}} = hA_{\text{flat}}(T_s - T_{\text{fluid}}) = [71.6\text{ W/(m}^2\cdot°\text{C)}](0.30 \times 10^{-4}\text{ m}^2)(90\text{-}65)°\text{C} = 0.05\text{ W}$$

Therefore, the total rate of heat that can be dissipated from all surfaces of the transistor is

$$\dot{Q}_{\text{total}} = \dot{Q}_{\text{cyl}} + \dot{Q}_{\text{flat}} = (0.11 + 0.05)\text{ W} = 0.16\text{ W}$$

which seems to be low. This value can be increased considerably by attaching a heat sink to the transistor to enhance the heat transfer surface area and thus heat transfer, or by increasing the air velocity, which will increase the heat transfer coefficient.

**EXAMPLE 14-15**

The desktop computer shown in Fig. 14-53 is to be cooled by a fan. The electronics of the computer consume 75 W of power under full load conditions. The computer is to operate in environments at temperatures up to 40°C and at elevations up to 2000 m where the atmospheric pressure is 79.50 kPa. The exit temperature of air is not to exceed 70°C, to meet reliability requirements. Also, the average velocity of air is not to exceed 75 m/min at the exit of the computer case, where the fan is installed to keep the noise level down. Determine the flow rate of the fan that needs to be installed, and the diameter of the casing of the fan.

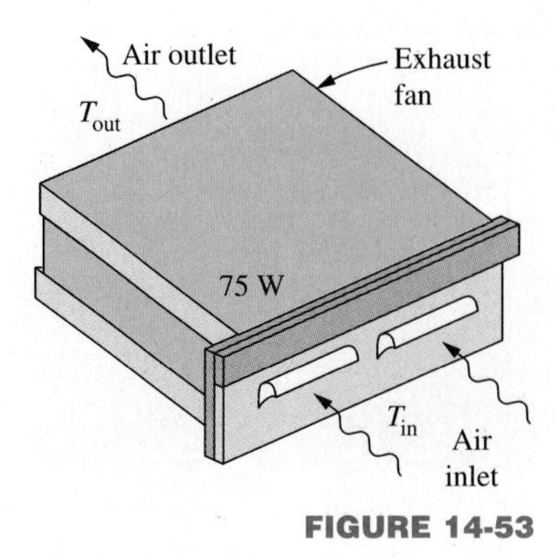


**FIGURE 14-53**
Schematic for Example 14-15.

**Solution** We need to determine the flow rate of air for the worst-case scenerio. Therefore, we assume the inlet temperature of air to be 40°C, the atmospheric pressure to be 79.50 kPa, and disregard any heat transfer from the outer surfaces of the computer case. Note that any direct heat loss from the computer case will provide a safety margin in the design.

The specific heat of air at the average temperature of $\frac{1}{2}(40 + 70) = 55°C = 328$ K is 1006 J/(kg · °C). Noting that all the heat dissipated by the electronic components is absorbed by the air, the required mass flow rate of air to absorb heat at a rate of 75 W can be determined from

$$\dot{Q} = \dot{m}C_p(T_{out} - T_{in})$$

Solving for $\dot{m}$ and substituting the given values, we obtain

$$\dot{m} = \frac{\dot{Q}}{C_p(T_{out} - T_{in})} = \frac{75\ \text{J/s}}{[1006\ \text{J/(kg}\cdot°\text{C)}](70 - 40)°\text{C}}$$
$$= 0.00249\ \text{kg/s} = 0.149\ \text{kg/min}$$

In the worst case, the exhaust fan will handle air at 70°C. Then the density of air entering the fan and the volume flow rate become

$$\rho = \frac{P}{RT} = \frac{79.50\ \text{kPa}}{[0.287\ \text{kPa}\cdot\text{m}^3/(\text{kg}\cdot\text{K})](70 + 273)\ \text{K}} = 0.8076\ \text{kg/m}^3$$

$$\dot{V} = \frac{\dot{m}}{\rho} = \frac{0.149\ \text{kg/min}}{0.8076\ \text{kg/m}^3} = 0.184\ \text{m}^3/\text{min}$$

Therefore, the fan must be able to provide a flow rate of 0.184 m$^3$/min or 6.5 cfm (cubic feet per minute). Note that if the fan were installed at the inlet instead of the exit then we would need to determine the flow rate using the density of air at the inlet temperature of 40°C, and we would need to add the power consumed by the motor of the fan to the heat load of 75 W. The result may be a slightly smaller or larger fan, depending on which effect dominates.

For an average velocity of 75 m/min, the diameter of the duct in which the fan is installed can be determined from

$$\dot{V} = A_c\mathscr{V} = \tfrac{1}{4}\pi D^2\mathscr{V}$$

Solving for $\mathscr{V}$ and substituting the known values, we obtain

$$D = \sqrt{\frac{4\dot{V}}{\pi\mathscr{V}}} = \sqrt{\frac{4 \times (0.184\ \text{m}^3/\text{min})}{\pi(75\ \text{m/min})}} = 0.056\ \text{m} = 5.6\ \text{cm}$$

Therefore, a fan with a casing diameter of 5.6 cm and a flow rate of 0.184 m$^3$/min will meet the design requirements.

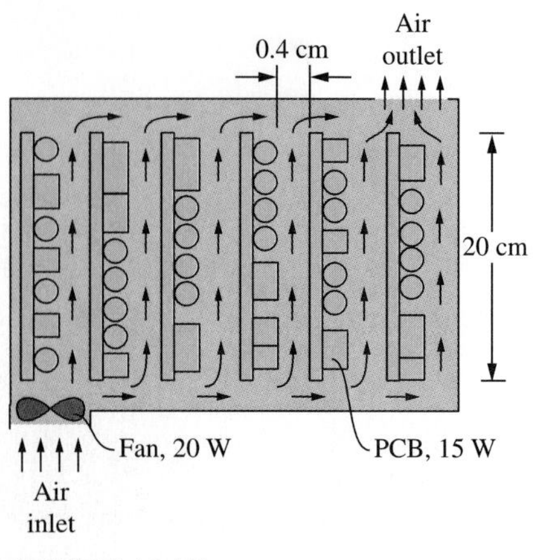


**FIGURE 14-54**

Schematic of the computer discussed in Example 14-16.

**EXAMPLE 14-16**

A computer cooled by a fan contains 6 PCBs, each dissipating 15 W of power, as shown in Fig. 14-54. The height of the PCBs is 15 cm and the length is 20 cm. The clearance between the tips of the components on the PCB and the back surface of the adjacent PCB is 0.4 cm. The cooling air is supplied by a 20-W fan mounted at the inlet. If the temperature rise of air as it flows through the case of the computer is not to exceed 10°C, determine (*a*) the flow rate of the air the fan needs to deliver, (*b*) the fraction of the temperature rise of air due to the heat generated by the fan and its motor, and (*c*) the highest allowable inlet air temperautre if the surface temperature of the components is not to exceed 90°C anywhere in the system.

**Solution** We assume operation at sea level, and take the atmospheric pressure to be 1 atm. We also assume that all the heat generated by the electronic components is removed by the air flowing through the opening between the PCBs. Also, the fan and its motor are usually mounted to the chassis of the electronic system, and some of the heat generated in the motor may be conducted to the chassis through the mounting brackets. However, we will take the conservative approach and assume the entire power consumed by the motor to be transferred as heat to the cooling air as air passes through it.

We use properties of air at 27°C, since the air enters at room temperature (about 20°C), and the temperature rise of air is limited to just 10°C:

$$\rho = 1.177\ \text{kg/m}^3, \qquad k = 0.0261\ \text{W/(m}\cdot{}^\circ\text{C)}$$
$$C_p = 1005\ \text{J/(kg}\cdot{}^\circ\text{C)}, \qquad \nu = 1.57 \times 10^{-5}\ \text{m}^2\text{/s}$$
$$\text{Pr} = 0.712$$

Because of symmetry, we consider the flow area between the two adjacent PCBs only. We assume the flow rate of air through all six channels to be identical, and to be equal to one-sixth of the total flow rate.

(*a*) Noting that the temperature rise of air is limited to 10°C and that the power consumed by the fan is also absorbed by the air, the total mass flow rate of air through the computer can be determined from

$$\dot{Q} = \dot{m}C_p(T_{\text{out}} - T_{\text{in}})$$

Solving for $\dot{m}$ and substituting the given values, we obtain

$$\dot{m} = \frac{\dot{Q}}{C_p(T_{\text{out}} - T_{\text{in}})} = \frac{(6 \times 15 + 20)\ \text{J/s}}{[1005\ \text{J/(kg}\cdot{}^\circ\text{C)}](10^\circ\text{C})} = 0.0109\ \text{kg/s}$$

Then the volume flow rate of air and the air velocity become

$$\dot{V} = \frac{\dot{m}}{\rho} = \frac{0.0109\ \text{kg/s}}{1.177\ \text{kg/m}^3} = 0.00926\ \text{m}^3\text{/s} = 0.556\ \text{m}^3\text{/min}$$

$$\mathcal{V} = \frac{\dot{V}}{A_c} = \frac{\frac{1}{6}(0.00926\ \text{m}^3\text{/s})}{6 \times 10^{-4}\ \text{m}^2} = 2.57\ \text{m/s}$$

Therefore, the fan needs to supply air at a rate of 0.556 m$^3$/min or about 20 cfm.

(*b*) The temperature rise of air due to the power consumed by the fan motor can be determined by assuming the entire 20 W of power consumed by the motor to be transferred to air as heat:

$$\Delta T_{\text{air, rise}} = \frac{\dot{Q}_{\text{fan}}}{\dot{m}C_p} = \frac{20\ \text{J/s}}{(0.0109\ \text{kg/s})[1005\ \text{J/(kg}\cdot{}^\circ\text{C)}]} = 1.8^\circ\text{C}$$

Therefore, 18 percent of the temperature rise of air is due to the heat generated by the fan motor. Note that the fraction of the power consumed by the fan is also 18 percent of the total, as expected.

(*c*) The surface temperature of the channel at any location can be determined from

$$\dot{q}_{\text{conv}} = \dot{Q}_{\text{conv}}/A = h(T_s - T_{\text{fluid}})$$

where the heat transfer surface area is

$$A = A_{\text{side}} = (\text{height})(\text{length}) = (0.15\text{ m})(0.20\text{ m}) = 0.03\text{ m}^2$$

To determine the convection heat transfer coefficient, we first need to calculate the Reynolds number. The cross-sectional area of the channel and its hydraulic diameter are

$$A_c = (\text{height})(\text{width}) = (0.15\text{ m})(0.004\text{ m}) = 6 \times 10^{-4}\text{ m}^2$$

$$D_h = \frac{4A_c}{p} = \frac{4 \times (6 \times 10^{-4}\,m^2)}{2 \times (0.15 + 0.004)\text{ m}} = 0.0078\text{ m}$$

Then the Reynolds number becomes

$$\text{Re} = \frac{\mathcal{V}D_h}{\nu} = \frac{(2.57\text{ m/s})(0.0078\text{ m})}{1.57 \times 10^{-5}\text{ m}^2/\text{s}} = 1277 < 2300$$

Therefore, the flow is laminar, and, assuming fully developed flow, the Nusselt number for the air flow in this rectangular cross-section corresponding to the aspect ratio $a/b = (15\text{ cm})/(0.4\text{ cm}) = 37.5 \approx \infty$ is determined from Table 14-3 to be

$$\text{Nu} = 8.24$$

and thus

$$h = \frac{k}{D_h}\text{Nu} = \frac{0.0261\text{ W/(m}\cdot{}^\circ\text{C)}}{0.0078\text{ m}}(8.24) = 27.6\text{ W/(m}^2\cdot{}^\circ\text{C)}$$

Disregarding the entrance effects, the temperature difference between the surface of the PCB and the air anywhere along the channel is determined to be

$$T_s - T_{\text{fluid}} = \left(\frac{\dot{q}}{h}\right)_{\text{PCB}} = \frac{(15\text{ W})/(0.03\text{ m}^2)}{27.6\text{ W/(m}^2\cdot{}^\circ\text{C)}} = 18.1^\circ\text{C}$$

That is, the surface temperature of the components on the PCB will be 18.1°C higher than the temperature of air passing by.

The highest air and component temperatures will occur at the exit. Therefore, in the limiting case, the component surface temperature at the exit will be 90°C. The air temperature at the exit in this case will be

$$T_{\text{out,max}} = T_{s,\text{max}} - \Delta T_{\text{rise}} = 90^\circ\text{C} - 18.1^\circ\text{C} = 71.9^\circ\text{C}$$

Noting that the air experiences a temperature rise of 10°C between the inlet and the exit, the inlet temperature of air is

$$T_{\text{in,max}} = T_{\text{out,max}} - 10^\circ\text{C} = (71.9 - 10)^\circ\text{C} = \mathbf{61.9^\circ C}$$

This is the highest allowable air inlet temperature if the surface temperature of the components is not to exceed 90°C anywhere in the system.

It should be noted that the analysis presented above is approximate, since we have made some simplifying assumptions. However, the accuracy of the results obtained are adequate for most engineering purposes.

## 14-9 ■ LIQUID COOLING

Liquids normally have much higher thermal conductivities than gases, and thus much higher heat transfer coefficients associated with them. Therefore, liquid cooling is far more effective than gas cooling. However, liquid cooling comes with its own risks and potential problems, such as leakage, corrosion, extra weight, and condensation. Therefore, liquid cooling is reserved for applications involving power densities that are too high for safe dissipation by air cooling.

Liquid cooling systems can be classified as **direct cooling** and **indirect cooling** systems. In *direct cooling* systems, the electronic components are in direct contact with the liquid, and thus the heat generated in the components is transferred directly to the liquid. In *indirect cooling* systems, however, there is no direct contact with the components. The heat generated in this case is first transferred to a medium such as a *cold plate* before it is carried away by the liquid. Liquid cooling systems are also classified as *closed-loop* and *open-loop* systems, depending on whether the liquid is discarded or recirculated after it is heated. In open-loop systems, tap water flows through the cooling system, and is discarded into a drain after it is heated. The heated liquid in closed-loop systems is cooled in a heat exchanger, and is recirculated through the system. Closed-loop systems facilitate better temperature control, while conserving water.

The electronic components in direct cooling systems are usually completely immersed in the liquid. The heat transfer from the components to the liquid can be by natural or forced convection or *boiling,* depending on the temperature levels involved and the properties of the fluids. Immersion cooling of electronic devices usually involves boiling and thus very high heat transfer coefficients, as discussed in the next section. Note that only *dielectric fluids* can be used in immersion or direct liquid cooling. This limitation immediately excludes water from consideration as a prospective fluid in immersion cooling. Fluorocarbon fluids such as FC75 are well suited for direct cooling, and are commonly used in such applications.

Indirect liquid cooling systems of electronic devices operate just like the cooling system of a *car engine,* where the water (actually a mixture of water and ethylene glycol) circulates through the passages around the cylinders of the engine block, picking up heat generated in the cylinders by combustion. The heated water is then routed by the water pump to the car radiator, where it is cooled by air blown through the radiator coils by the cooling fan. The cooled water is then rerouted to the engine to pick up more heat. To appreciate the effectiveness of the cooling system of a car engine, it will suffice to say that the temperatures encountered in the engine cylinders are typically much higher than the melting temperatures of the engine blocks.

In an electronic system, the heat is generated in the *components* instead of the combustion chambers. The components in this case are mounted on a metal plate made of a highly conducting material such as copper or aluminum. The metal plate is cooled by circulating a cooling fluid through tubes attached to it, as shown in Fig. 14-55. The heated liquid is then cooled in a heat exchanger, usually by air (or sea water in marine applications), and is recirculated by a pump through the tubes. The expansion and storage tank accommodates any expansions and contractions of the cooling liquid due to temperature variations while acting as a liquid reservoir.

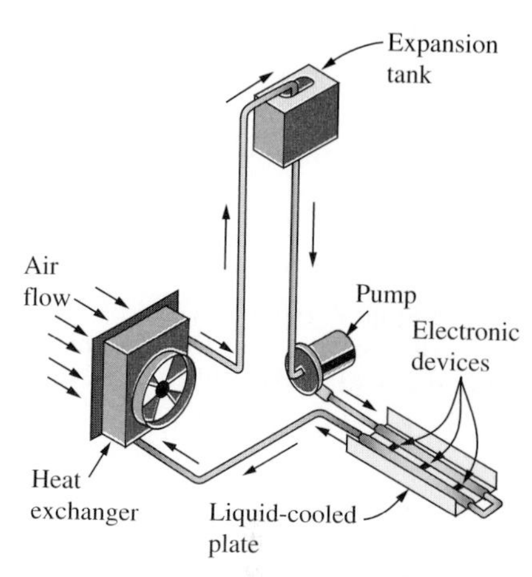

**FIGURE 14-55**

Schematic of an indirect liquid cooling system.

The liquids used in the cooling of electronic equipment must meet several requirements, depending on the specific application. Desirable characteristics of cooling liquids include *high thermal conductivity* (yields high heat transfer coefficients), *high specific heat* (requires smaller mass flow rate), *low viscosity* (causes a smaller pressure drop, and thus requires a smaller pump), *high surface tension* (less likely to cause leakage problems), *high dielectric strength* (a must in direct liquid cooling), *chemical inertness* (does not react with surfaces with which it comes into contact), *chemical stability* (does not decompose under prolonged use), *nontoxic* (safe for personnel to handle), *low freezing and high boiling points* (extends the useful temperature range), and *low cost.* Different fluids may be selected in different applications because of the different priorities set in the selection process.

The heat sinks or cold plates of an electronic enclosure are usually cooled by *water* by passing it through channels made for this purpose or through tubes attached to the cold plate. High heat removal rates can be achieved by circulating water through these channels or tubes. In high-performance systems, a *refrigerant* can be used in place of water to keep the temperature of the heat sink at subzero temperatures and thus reduce the junction temperatures of the electronic components proportionately. The heat transfer and pressure drop calculations in liquid cooling systems can be performed as described in Chap. 10.

Liquid cooling can be used effectively to cool clusters of electronic devices attached to a tubed metal plate (or heat sink), as shown in Fig. 14-56. Here 12 TO-3 cases, each dissipating up to 150 W of power, are mounted on a heat sink equipped with tubes on the back side through which a liquid flows. The thermal resistance between the case of the devices and the liquid is minimized in this case, since the electronic devices are mounted directly over the cooling lines. The case-to-liquid thermal resistance depends on the spacing between the devices, the quality of the thermal contact between the devices and the plate, the thickness of the plate, and the flow rate of the liquid, among other things. The tubed metal plate shown is 15.2 cm $\times$ 18 cm $\times$ 2.5 cm in size, and is capable of dissipating up to 2 kW of power.

The thermal resistance network of a liquid cooling system is shown

**FIGURE 14-56**
Liquid cooling of TO-3 packages placed on top of the coolant line (courtesy of Wakefield Engineering).

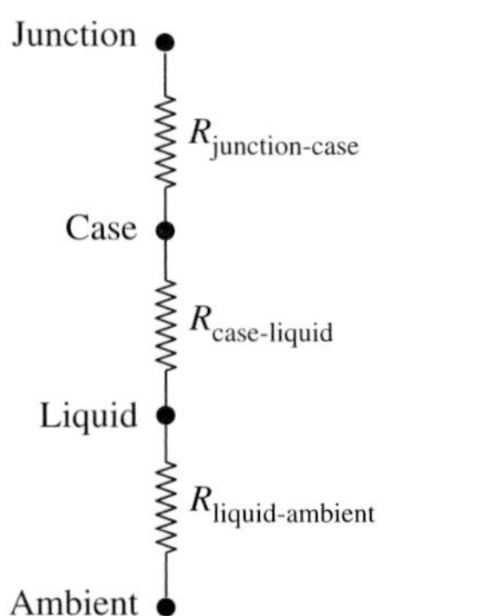


**FIGURE 14-57**
Thermal resistance network for a liquid cooled electronic device.

in Fig. 14-57. The junction temperatures of silicon-based electronic devices are usually limited to 125°C. The *junction-to-case* thermal resistance of a device is provided by the manufacturer. The *case-to-liquid* thermal resistance can be determined experimentally by measuring the temperatures of the case and the liquid, and dividing the difference by the total power dissipated. The *liquid-to-air* thermal resistance at the heat exchanger can be determined in a similar manner. That is,

$$R_{\text{case-liquid}} = \frac{T_{\text{case}} - T_{\text{liquid, device}}}{\dot{Q}}, \qquad R_{\text{liquid-air}} = \frac{T_{\text{liquid, hx}} - T_{\text{air, in}}}{\dot{Q}} \qquad (°\text{C/W})$$

where $T_{\text{liquid, device}}$ and $T_{\text{liquid, hx}}$ are the inlet temperatures of the liquid to the electronic device and the heat exchanger, respectively. The required mass flow rate of the liquid corresponding to a specified temperature rise of the liquid as it flows through the electronic systems can be determined from Eq. 14-17.

Electronic components mounted on liquid-cooled metal plates should be provided with good thermal contact in order to minimize the thermal resistance between the components and the plate. The thermal resistance can be minimized by applying silicone grease or beryllium oxide to the contact surfaces and fastening the components tightly to the metal plate. The liquid cooling of a cold plate with a large number of high-power components attached to it is illustrated in the following example.

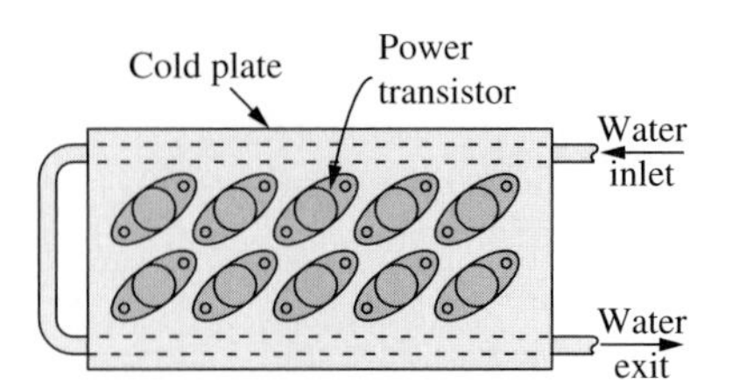


**FIGURE 14-58**
Schematic for Example 14-17.

**EXAMPLE 14-17**

A cold plate that supports 20 power transistors, each dissipating 40 W, is to be cooled with water, as shown in Fig. 14-58. Half of the transistors are attached to the back side of the cold plate. It is specified that the temperature rise of the water is not to exceed 3°C, and the velocity of water is remain under 1 m/s. Assuming that 20 percent of the heat generated is dissipated from the components to the surroundings by convection and radiation, and the remaining 80 percent is removed by the cooling water, determine the mass flow rate of water needed and the diameter of the pipe to satisfy the restriction imposed on the flow velocity. Also determine the case temperature of the devices if the total case-to-liquid thermal resistance is 0.030°C/W and the water enters the cold plate at 35°C.

**Solution** Noting that each of the 20 transistors dissipates 40 W of power and 80 percent of this power must be removed by the water, the amount of heat that must be removed by the water is

$$\dot{Q} = (20 \text{ transistors})(40 \text{ W/transistor})(0.80) = 640 \text{ W}$$

In order to limit the temperature rise of the water to 3°C, the mass flow rate of water must be no less than

$$\dot{m} = \frac{\dot{Q}}{C_p \, \Delta T_{\text{rise}}} = \frac{640 \text{ J/s}}{[4180 \text{ J/(kg} \cdot \text{°C)}](3\text{°C})} = 0.051 \text{ kg/s} = \mathbf{3.06 \text{ kg/min}}$$

The mass flow rate of a fluid through a circular pipe can be expressed as

$$\dot{m} = \rho A \mathcal{V} = \rho \frac{\pi D^2}{4} \mathcal{V}$$

Then taking the density of water to be 1000 kg/m$^3$, the diameter of the pipe to maintain the velocity of water under 1 m/s is determined to be

$$D = \sqrt{\frac{4\dot{m}}{\pi \rho \mathcal{V}}} = \sqrt{\frac{4(0.051 \text{ kg/s})}{\pi(1000 \text{ kg/m}^3)(1 \text{ m/s})}} = 0.0081 \text{ m} = \mathbf{0.81 \text{ cm}}$$

Noting that the total case-to-liquid thermal resistance is 0.030°C/W and the water enters the cold plate at 35°C, the case temperature of the devices is determined from Eq. 14-25 to be

$$T_{\text{case}} = T_{\text{liquid, device}} + \dot{Q}R_{\text{case-liquid}} = 35\text{°C} + (640 \text{ W})(0.03\text{°C/W}) = \mathbf{54.2\text{°C}}$$

The junction temperature of the device can be determined similarly by using the junction-to-case thermal resistance of the device supplied by the manufacturer.

## 14-10 ■ IMMERSION COOLING

High-power electronic components can be cooled effectively by immersing them in a *dielectric liquid* and taking advantage of the very high heat transfer coefficients associated with boiling. Immersion cooling has been used since the 1940s in the cooling of electronic equipment, but for many years its use was largely limited to the electronics of high-power radar systems. The miniaturization of electronic equipment and the resulting high heat fluxes brought about renewed interest in immersion cooling, which had been largely viewed as an exotic cooling technique.

You will recall from thermodynamics that, at a specified pressure, a fluid boils isothermally at the corresponding saturation temperature. A large amount of heat is absorbed during the boiling process, essentially in an isothermal manner. Therefore, immersion cooling also provides a constant-temperature bath for the electronic components, and eliminates hot spots effectively.

The simplest type of immersion cooling system involves an *external reservoir,* which supplies liquid continually to the electronic enclosure. The vapor generated inside is simply allowed to escape to the atmosphere, as shown in Fig. 14-59. A pressure relief valve on the vapor vent line keeps the pressure and thus the temperature inside at the preset value, just like the petcock of a pressure cooker. Note that without a pressure relief valve, the pressure inside the enclosure would be atmospheric pressure, and the temperature would have to be the boiling temperature of the fluid at the atmospheric pressure.

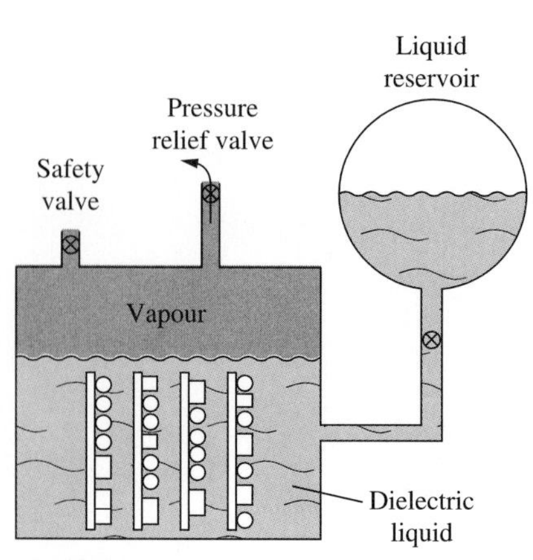

**FIGURE 14-59**

A simple open loop immersion cooling system.

The **open-loop**-type immersion cooling system described above is simple, but there are several impracticalities associated with it. First of all, it is heavy and bulky because of the presence of an external liquid reservoir, and the fluid lost through evaporation needs to be replenished continually, which adds to the cost. Further, the release of the vapor into the atmosphere greatly limits the fluids that can be used in such a system. Therefore, the use of open-loop immersion systems is limited to applications that involve occasional use and thus have a light duty cycle.

More sophisticated immersion cooling systems operate in a **closed loop** in that the vapor is condensed and returned to the electronic enclosure instead of being purged to the atmosphere. Schematics of two such systems are given in Fig. 14-60. The first system involves a condenser *external* to the electronics enclosure, and the vapor leaving the enclosure is cooled by a cooling fluid such as air or water outside the enclosure. The condensate is returned to the enclosure for reuse. The condenser in the second system is actually *submerged* in the electronic enclosure, and is part of the electronic system. The cooling fluid in this case circulates through the condenser tube, removing heat from the vapor. The vapor

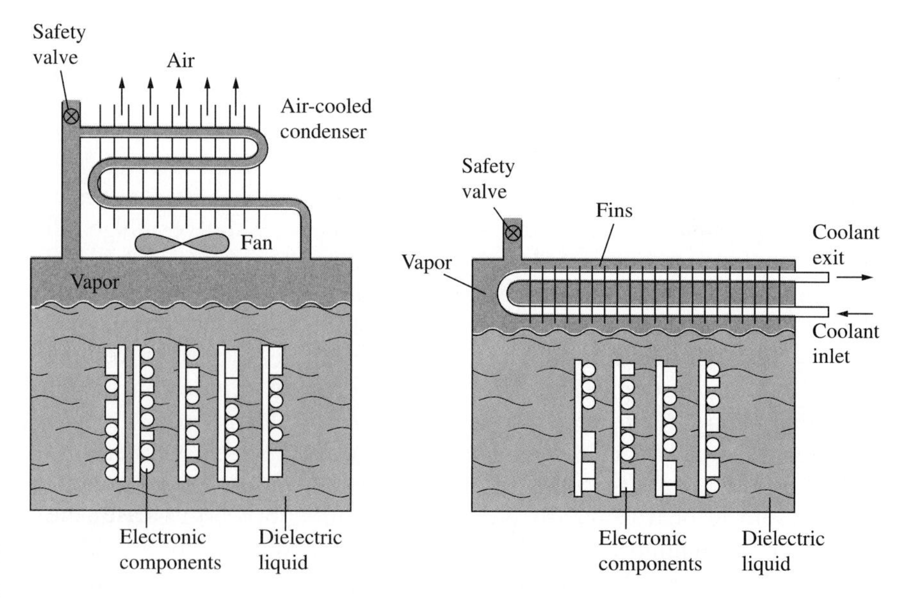


**FIGURE 14-60**

The schematics of two closed loop immersion cooling systems.

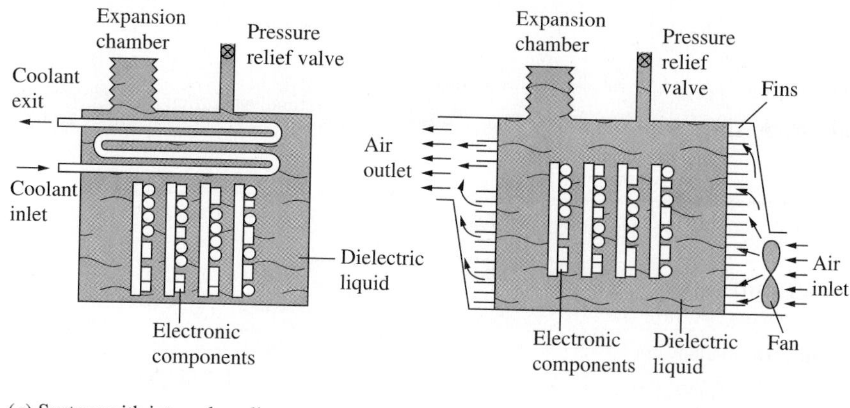


**FIGURE 14-61**
The schematics of two-all-liquid immersion cooling systems.

that condenses drips on top of the liquid in the enclosure, and continues to recirculate.

The performance of closed-loop immersion cooling systems is most susceptible to the presence of *noncondensable gases* such as air in the vapor space. A 0.5 percent of air by mass in the vapor can cause the condensation heat transfer coefficient to drop by a factor of up to 5. Therefore, the fluid used in immersion cooling systems should be degassed as much as practical, and care should be taken during the filling process to avoid introducing any air into the system.

The problems associated with the condensation process and noncondensable gases can be avoided by *submerging* the condenser (actually, heat exchanger tubes in this case) in the *liquid* instead of the vapor in the electronic enclosure, as shown in Fig. 14-61*a*. The cooling fluid, such as water, circulating through the tubes absorbs heat from the dielectric liquid at the top portion of the enclosure, and subcools it. The liquid in contact with the electronic components is heated and may even be vaporized as a result of absorbing heat from the components. But these vapor bubbles collapse as they move up, as a result of transferring heat to the cooler liquid with which they come in contact. This system can still remove heat at high rates from the surfaces of electronic components in an isothermal manner by utilizing the boiling process, but its overall capacity is limited by the rate of heat that can be removed by the external cooling fluid in a *liquid-to-liquid* heat exchanger. Noting that the heat transfer coefficients associated with forced convection are far less than those associated with condensation, this all-liquid immersion cooling system is not suitable for electronic boxes with very high power dissipation rates per unit volume.

A step further in the all-liquid immersion cooling systems is to remove the heat from the dielectric liquid directly from the *outer surface* of the electronics enclosure, as shown in Fig. 14-61*b*. In this case, the dielectric liquid inside the sealed enclosure is heated as a result of absorbing the heat dissipated by the electronic components. The heat is then transferred to the walls of the enclosure, where it is removed by

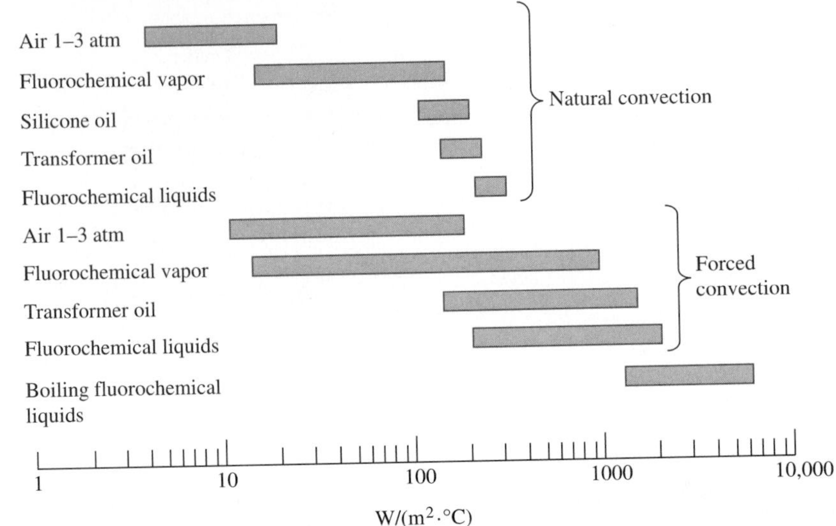


**FIGURE 14-62**

Typical heat transfer coefficients for various dielectric fluids (from Ref. 12).

external means. This immersion cooling technique is the most reliable of all, since it does not involve any penetration into the electronics enclosure, and the components reside in a completely sealed liquid environment. However, the use of this system is limited to applications that involve moderate power dissipation rates. The heat dissipation is limited by the ability of the system to reject the heat from the outer surface of the enclosure. To enhance this ability, the outer surfaces of the enclosures are often *finned,* especially when the enclosure is cooled by air.

Typical ranges of heat transfer coefficients for various dielectric fluids suitable for use in the cooling of electronic equipment are given in Fig. 14-62 for natural convection, forced convection, and boiling. Note that extremely high heat transfer coefficients [from about 1500 to 6000 W/(m$^2$ · °C)] can be attained with the boiling of *fluorocarbon fluids* such as FC78 and FC86 manufactured by the 3M company. Fluorocarbon fluids, not to be confused with the ozone-destroying fluorochloro fluids, are found to be very suitable for immersion cooling of electronic equipment. They have boiling points ranging from 30°C to 174°C, and freezing points below −50°C. They are nonflammable, chemically inert, and highly compatible with materials used in electronic equipment.

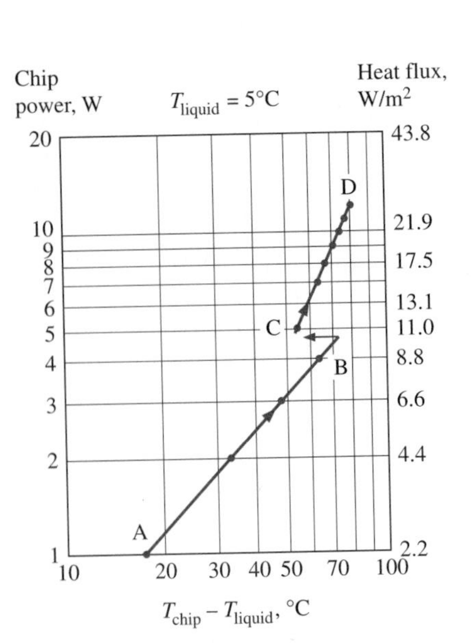


**FIGURE 14-63**

Heat transfer from a chip immersed in the fluorocarbon fluid FC86 (from Ref. 12).

Experimental results for the power dissipation of a chip having a heat transfer area of 0.457 cm$^2$ and its substrate during immersion cooling in an FC86 bath (boiling point −57°C) are given in Fig. 14-63. The FC86 liquid is maintained at a bulk temperature of 5°C during the experiments by the use of a heat exchanger. Heat transfer from the chip is by *natural convection* in regime A–B, and *bubble formation* and thus boiling begins in regime B–C. Note that the chip surface temperature suddenly drops with the onset of *boiling* because of the high heat transfer coefficients associated with boiling. Heat transfer is by nucleate boiling in regime C–D, and very high heat transfer rates can be achieved in this regime with relatively small temperature differences.

**EXAMPLE 14-18**

A logic chip used in an IBM 3081 computer dissipates 4 W of power and has a heat transfer surface area of 0.279 $cm^2$, as shown in Fig. 14-64. If the surface of the chip is to be maintained at 80°C while being cooled by immersion in a dielectric fluid at 20°C, determine the necessary heat transfer coefficient, and the type of cooling mechanism which needs to be used to achieve that heat transfer coefficient.

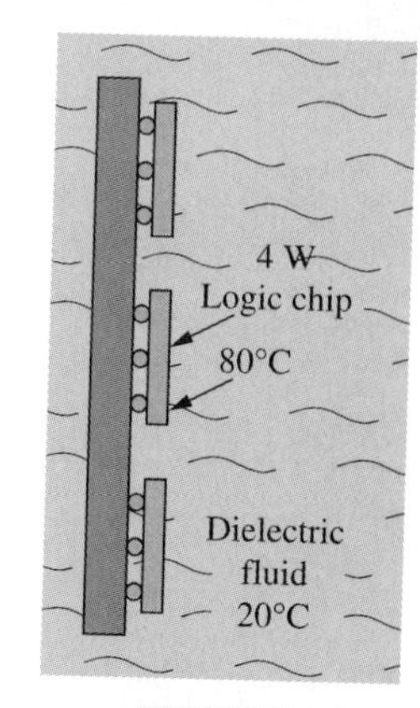


**FIGURE 14-64**
Schematic for Example 14-18.

**Solution** The average heat transfer coefficient over the surface of the chip can be determined from Newton's law of cooling expressed as

$$\dot{Q} = hA(T_{\text{chip}} - T_{\text{fluid}})$$

Solving for $h$ and substituting the given values, the convection heat transfer coefficient is determined to be

$$h = \frac{\dot{Q}}{A(T_{\text{chip}} - T_{\text{fluid}})} = \frac{4\text{ W}}{(0.279 \times 10^{-4}\text{ m}^2)(80 - 20)°\text{C}} = 2390\text{ W/(m}^2 \cdot °\text{C)}$$

which is rather high. Examination of Fig. 14-62 reveals that we can obtain such high heat transfer coefficients with the boiling of fluorocarbon fluids. Therefore, a suitable cooling technique in this case is immersion cooling in such a fluid. A viable alternative to immersion cooling is the thermal conduction module discussed earlier.

**EXAMPLE 14-19**

An 8-W chip having a surface area of 0.6 $cm^2$ is cooled by immersing it in FC86 liquid that is maintained at a temperature of 15°C, as shown in Fig. 14-65. Using the boiling curve in Fig. 14-63, estimate the temperature of the chip surface.

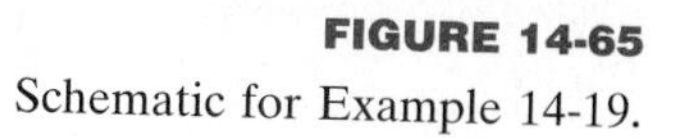

**FIGURE 14-65**
Schematic for Example 14-19.

**Solution** The boiling curve in Fig. 14-63 is prepared for a chip having a surface area of 0.457 $cm^2$ being cooled in FC86 maintained at 5°C. We can still use that chart for similar cases with a good degree of approximation. The heat flux in our case is

$$\dot{q} = \frac{\dot{Q}}{A} = \frac{8\text{ W}}{0.6\text{ cm}^2} = 13.3\text{ W/cm}^2$$

Corresponding to this value on the chart is $T_{\text{chip}} - T_{\text{fluid}} = 60°\text{C}$. Therefore,

$$T_{\text{chip}} = T_{\text{fluid}} + 60 = 15 + 60 = 75°\text{C}$$

That is, the surface of this 8-W chip will be at 75°C as it is cooled by boiling in the dielectric fluid FC86.

A liquid-based cooling system brings with it the possibility of leakage, and associated reliability concerns. Therefore, the consideration of immersion cooling should be limtied to applications that require precise temperature control and those that involve heat dissipation rates that are too high for effective removal by condution or air cooling.

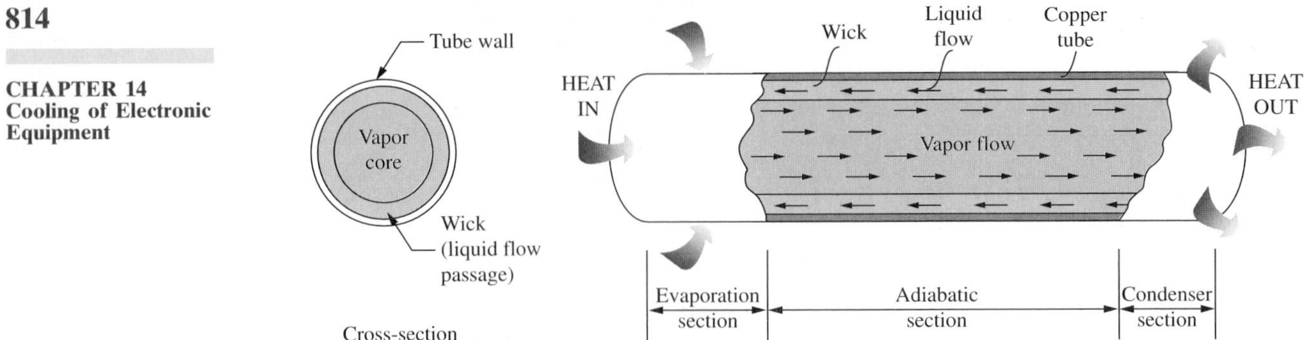


**FIGURE 14-66**
Schematic and operation of a heat pipe.

## 14-11 ■ HEAT PIPES

A **heat pipe** is a simple device with no moving parts that can transfer large quantities of heat over fairly large distances essentially at a constant temperature without requiring any power input. A heat pipe is basically a sealed slender tube containing a wick structure lined on the inner surface and a small amount of fluid such as water at the saturated state, as shown in Fig. 14-66. It is composed of three sections: the *evaporator* section at one end, where heat is absorbed and the fluid is vaporized, a *condenser* section at the other end, where the vapor is condensed and heat is rejected, and the *adiabatic* section in between, where the vapor and the liquid phases of the fluid flow in opposite directions through the core and the wick, respectively, to complete the cycle with no significant heat transfer between the fluid and the surrounding medium.

The *type of fluid* and the *operating pressure* inside the heat pipe depends on the *operating temperature* of the heat pipe. For example, the critical- and triple-point temperatures of water are 0.01°C and 374.1°C, respectively. Therefore, water can undergo a liquid-to-vapor or vapor-to-liquid phase change process in this temperature range only, and thus it will not be a suitable fluid for applications involving temperatures beyond this range. Furthermore, water will undergo a phase-change process at a specified temperature only if its pressure equals the saturation pressure at that temperature. For example, if a heat pipe with water as the working fluid is designed to remove heat at 70°C, the pressure inside the heat pipe must be maintained at 31.2 kPa, which is the boiling pressure of water at this temperature. Note that this value is well below the atmospheric pressure of 101 kPa, and thus the heat pipe operates in a vacuum environment in this case. If the pressure inside is maintained at atmospheric pressure instead, heat transfer would result in an increase in the temperature of the water instead of evaporation.

Although water is a suitable fluid to use in the moderate temperature range encountered in electronic equipment, several other fluids can be used in the construction of heat pipes to enable them to be used in cryogenic as well as high-temperature applications. The suitable temperature ranges for some common heat pipe fluids are given in Table 14-4.

**TABLE 14-4**

**Suitable temperature ranges for some fluids used in heat pipes**

| Fluid | Temperature range, °C |
|---|---|
| Helium | −271 to −268 |
| Hydrogen | −259 to −240 |
| Neon | −248 to −230 |
| Nitrogen | −210 to −150 |
| Methane | −182 to −82 |
| Ammonia | −78 to −130 |
| Water | 5 to 230 |
| Mercury | 200 to 500 |
| Cesium | 400 to 1000 |
| Sodium | 500 to 1200 |
| Lithium | 850 to 1600 |

Note that the overall temperature range extends from almost absolute zero for cryogenic fluids such as helium to over 1600°C for liquid metals such as lithium. The ultimate temperature limits for a fluid are the *triple-* and *critical-point* temperatures. However, a narrower temperature range is used in practice to avoid the extreme pressures and low heats of vaporization that occur near the critical point. Other desirable characteristics of the candidate fluids are having a high surface tension to enhance the capillary effect and being compatible with the wick material, as well as being readily available, chemically stable, nontoxic, and inexpensive.

The concept of heat pipe was originally conceived by R. S. Gaugler of the General Motors Corporation, who filed a patent application for it in 1942. However, it did not receive much attention until 1962, when it was suggested for use in space applications. Since then, heat pipes have found a wide range of applications, including the cooling of electronic equipment.

## The Operation of a Heat Pipe

The operation of a heat pipe is based on the following physical principles:

- At a specified pressure, a liquid will vaporize or a vapor will condense at a certain temperature, called the *saturation temperature.* Thus, fixing the pressure inside a heat pipe fixes the temperature at which phase change will take place.
- At a specified pressure or temperature, the amount of heat *absorbed* as a unit mass of liquid vaporizes is equal to the amount of heat *rejected* as that vapor condenses.
- The capillary pressure developed in a wick will move a liquid in the wick even *against* the gravitational field as a result of the capillary effect.
- A fluid in a channel flows in the direction of *decreasing pressure*.

Initially, the *wick* of the heat pipe is saturated with liquid and the *core section* is filled with vapor. When the evaporator end of the heat pipe is brought into contact with a hot surface or is placed into a hot environment, heat will flow into the heat pipe. Being at a saturated state, the liquid in the evaporator end of the heat pipe will *vaporize* as a result of this heat transfer, causing the vapor pressure there to rise. This resulting pressure difference drives the vapor through the core of the heat pipe from the evaporator towards the condenser section. The condenser end of the heat pipe is in a cooler environment, and thus its surface is slightly cooler. The vapor that comes into contact with this cooler surface *condenses,* releasing the heat of vaporization, which is rejected to the surrounding medium. The liquid then returns to the evaporator end of the heat pipe through the wick as a result of *capillary action* in the wick, completing the cycle. As a result, heat is absorbed at one end of the heat pipe, and is rejected at the other end, with the fluid inside serving as a transport medium for heat.

The boiling and condensation processes are associated with extremely high heat transfer coefficients, and thus it is natural to expect the heat pipe to be an extremely effective heat transfer device, since its operation is based on alternate boiling and condensation of the working fluid. Indeed, heat pipes have effective conductivities *several hundred times* that of copper or silver. That is, replacing a copper bar between two mediums at different temperatures by a heat pipe of equal size can increase the rate of heat transfer between those two mediums by several hundred times. A simple heat pipe with water as the working fluid has an effective thermal conductivity of the order of 100,000 W/(m · °C) compared with about 400 W/(m · °C) for copper. For a heat pipe, it is not unusual to have an effective conductivity of 400,000 W/(m · °C), which is 1000 times that of copper. A 15-cm-long horizontal cylindrical heat pipe with water inside having a diameter of 0.6 cm, for example, can transfer heat at a rate of 300 W. Therefore, heat pipes are preferred in some critical applications, despite their high initial cost.

There is a small pressure difference between the evaporator and condenser ends, and thus a small temperature difference between the two ends of the heat pipe. This temperature difference is usually between 1°C and 5°C.

## The Construction of a Heat Pipe

The wick of a heat pipe provides the means for the return of the liquid to the evaporator. Therefore, the structure of the wick has a strong effect on the performance of a heat pipe, and the design and construction of the wick is the most critical aspect of the manufacturing process.

The wicks are often made of porous ceramic or woven stainless wire mesh. They can also be made together with the tube by extruding axial grooves along its inner surface, but this approach presents manufacturing difficulties.

The performance of a wick depends on its structure. The characteristics of a wick can be changed by changing the *size* and the *number* of the pores per unit volume, and the *continuity* of the passageway. Liquid motion in the wick depends on the dynamic balance between two opposing effects: the *capillary pressure,* which creates the suction effect to draw the liquid, and the *internal resistance to flow* as a result of friction between the mesh surfaces and the liquid. A small pore size increases the capillary action, since the capillary pressure is inversely proportional to the effective capillary radius of the mesh. But decreasing the pore size and thus the capillary radius also increases the friction force opposing the motion. Therefore, the core size of the mesh should be reduced so long as the increase in capillary force is greater than the increase in the friction force.

Note that the *optimum pore size* will be different for different fluids and different orientations of the heat pipe. An improperly designed wick will result in an adequate liquid supply, and eventual failure of the heat pipe.

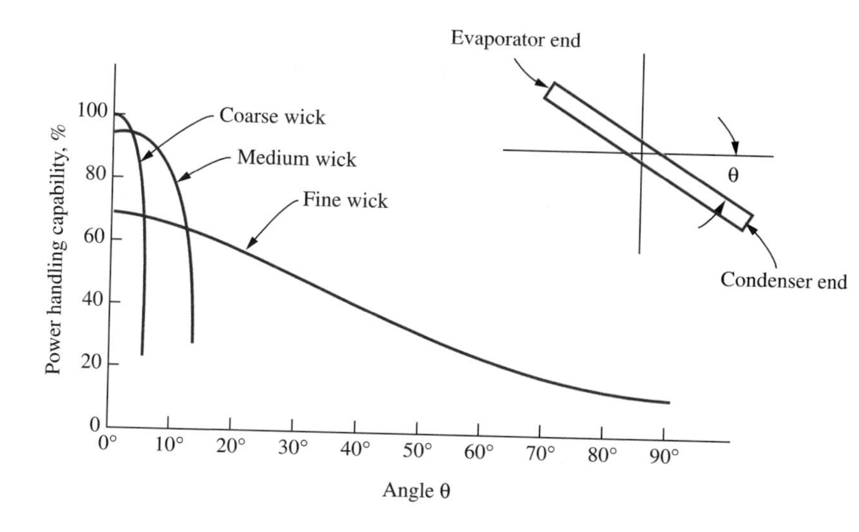


**FIGURE 14-67**
Variation of the heat removal capacity of a heat pipe with tilt angle from the horizontal when the liquid flows in the wick against gravity (from Ref. 18).

Capillary action permits the heat pipe to operate in any orientation in a gravity field. However, the performance of a heat pipe will be best when the capillary and gravity forces act in the same direction (evaporator end down), and will be worst when these two forces act in opposite directions (evaporator end up). Gravity does not effect the capillary force when the heat pipe is in the horizontal position. The heat removal capacity of a horizontal heat pipe can be *doubled* by installing it vertically with evaporator end down so that gravity helps the capillary action. In the opposite case of vertical orientation with evaporator end up, the performance declines considerably relative the horizontal case, since the capillary force in this case must work against the gravity force.

Most heat pipes are cylindrical in shape. However, they can be manufactures in a variety of shapes involving 90° bends, S-turns, or spirals. They can also be made as a flat layer with a thickness of about 0.3 cm. Flat heat pipes are very suitable for cooling high-power-output (say, 50 W or greater) PCBs. In this case, flat heat pipes are attached directly to the back surface of the PCB, and they absorb and transfer the heat to the edges. Cooling fins are usually attached to the condenser end of the heat pipe to improve its effectiveness and to eliminate a bottleneck in the path of heat flow from the components to the environment when the ultimate heat sink is the ambient air.

The decline in the performance of a 122-cm-long water heat pipe with the tilt angle from the horizontal is shown in Fig. 14-67 for heat pipes with coarse, medium, and fine wicks. Note that for the horizontal case, the heat pipe with a coarse wick performs best, but the performance drops off sharply as the evaporator end is raised from the horizontal. The heat pipe with a fine wick does not perform as well in the horizontal position, but maintains its level of performance greatly at tilted positions. It is clear from this figure that heat pipes which work against the gravity must be equipped with *fine* wicks. The heat removal capacities of various heat pipes are given in Table 14-5.

**TABLE 14-5**
**Typical heat removal capacity of various heat pipes**

| Outside diameter, cm (in) | Length, cm (in) | Heat removal rate, W |
|---|---|---|
| 0.635 ($\frac{1}{4}$) | 15.2 (6) | 300 |
| | 30.5 (12) | 175 |
| | 45.7 (18) | 150 |
| 0.95 ($\frac{3}{8}$) | 15.2 (6) | 500 |
| | 30.5 (12) | 375 |
| | 45.7 (18) | 350 |
| 1.27 ($\frac{1}{2}$) | 15.2 (6) | 700 |
| | 30.5 (12) | 575 |
| | 45.7 (18) | 550 |

A major concern about the performance of a heat pipe is degradation with time. Some heat pipes have failed within just a few months after they are put into operation. The major cause of degradation appears to be *contamination* that occurs during the sealing of the ends of the heat pipe tube, and affects the vapor pressure. This form of contamination has been minimized by electron beam welding in clean rooms. Contamination of the wick prior to installation in the tube is another cause of degradation. Cleanliness of the wick is essential for its reliable operation for a long time. Heat pipes usually undergo extensive testing and quality control process before they are put into actual use.

An important consideration in the design of heat pipes is the compatibility of the materials used for the tube, wick, and fluid. Otherwise, reaction between the incompatible materials produces non-condensable gases, which degrades the performance of the heat pipe. For example, the reaction between stainless steel and water in some early heat pipes generated hydrogen gas, which destroyed the heat pipe.

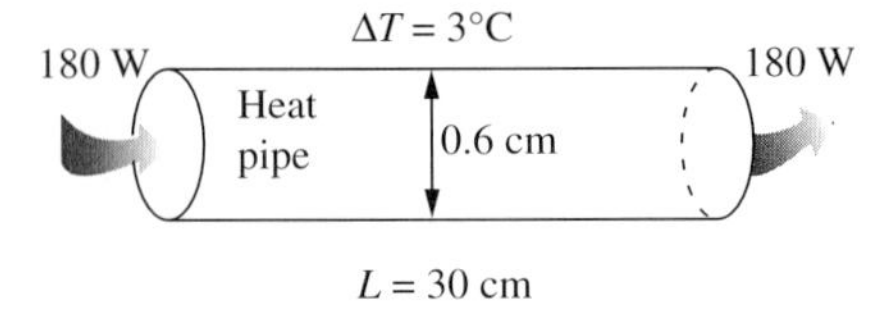


**FIGURE 14-68**
Schematic for Example 14-20.

**EXAMPLE 14-20**

A 30-cm-long cylindrical heat pipe having a diameter of 0.6 cm is dissipating heat at a rate of 180 W, with a temperature difference of 3°C across the heat pipe, as shown in Fig. 14-68. If we were to use a 30-cm-long copper rod [$k$ = 386 W/(m · °C) and $\rho$ = 8950 kg/m$^3$] instead to remove heat at the same rate, determine the diameter and the mass of the copper rod that needs to be installed.

**Solution** The rate of heat transfer $\dot{Q}$ through the copper rod can be expressed as

$$\dot{Q} = kA\frac{\Delta T}{L}$$

where $k$ is the thermal conductivity, $L$ is the length, and $\Delta T$ is the temperature difference across the copper bar. Solving for the cross-sectional area $A$ and substituting the specified values gives

$$A = \frac{L}{k\,\Delta T}\dot{Q} = \frac{0.3\text{ m}}{[386\text{ W/(m}\cdot\text{°C)}](3\text{°C})}(180\text{ W}) = 0.04663\text{ m}^2 = 466.3\text{ cm}^2$$

Then the diameter and the mass of the copper rod becomes

$$A = \tfrac{1}{4}\pi D^2 \longrightarrow D = \sqrt{4A/\pi} = \sqrt{4(466.3\text{ cm}^2)/\pi} = \mathbf{24.4\text{ cm}}$$
$$m = \rho V = \rho AL = (8590\text{ kg/m}^3)(0.04663\text{ m}^2)(0.3\text{ m}) = \mathbf{125.2\text{ kg}}$$

Therefore, the diameter of the copper rod needs to be almost 25 times that of the heat pipe to transfer heat at the same rate. Also, the rod would have a mass of 125.2 kg, which is impossible for an average person to lift.

## 14-12 SUMMARY

Electric current flow through a resistance is always accompanied by heat generation, and the essence of thermal design is the safe removal of this internally generated heat by providing an effective path for heat flow from electronic components to the surrounding medium. In this chapter, we have discussed several *cooling techniques* commonly used in electronic equipment, such as conduction cooling, natural convection and radiation cooling, forced-air convection cooling, liquid cooling, immersion cooling, and heat pipes.

In a chip carrier, heat generated at the junction is conducted through the thickness of the chip, the bonding material, the lead frame, the case material and the leads. The *junction-to-case thermal resistance* $R_{\text{junction-case}}$ represents the total resistance to heat transfer between the junction of a component and its case. This resistance should be as low as possible to minimize the temperature rise of the junction above the case temperature. The epoxy board used in PCBs is a poor heat conductor, and so it is necessary to use copper cladding or to attach the PCB to a heat frame in conduction cooled systems.

Low-power electronic systems can be cooled effectively with natural convection and radiation. The heat transfer from a surface at temperature $T_s$ to a fluid at temperature $T_{\text{fluid}}$ by *convection* is expressed as

$$\dot{Q}_{\text{conv}} = hA(T_s - T_{\text{fluid}}) \quad \text{(W)}$$

where $h$ is the convection heat transfer coefficient and $A$ is the heat transfer surface area. The natural convection heat transfer coefficient for laminar flow of air at atmospheric pressure is given by a simplified relation of the form

$$h = K\left(\frac{\Delta T}{L}\right)^{0.25} \quad [\text{W/(m}^2 \cdot {}^\circ\text{C)}]$$

where $\Delta T = T_s - T_{\text{fluid}}$ is the temperature difference between the surface and the fluid, $L$ is the characteristic length (the length of the body along the heat flow path), and $K$ is a constant, whose value is given in Table 14-1. The relations in Table 14-1 can also bed used at pressures other than 1 atm by multiplying them by $\sqrt{P}$, where $P$ is the air pressure in atm.

Radiation heat transfer between a surface at temperature $T_s$ completely surrounded by a much larger surface at temperature $T_{\text{surr}}$ can be expressed as

$$\dot{Q}_{\text{rad}} = \varepsilon A\sigma(T_s^4 - T_{\text{surr}}^4) \quad \text{(W)}$$

where $\varepsilon$ is the emissivity of the surface, $A$ is the heat transfer surface area, and $\sigma$ is the Stefan–Boltzmann constant, whose value is $\sigma = 5.67 \times 10^{-8}\ \text{W/(m}^2 \cdot \text{K}^4) = 0.1714 \times 10^{-8}\ \text{Btu/(h} \cdot \text{ft}^2 \cdot \text{R}^4)$.

Fluid flow over a body such as a transistor is called *external flow*, and flow through a confined space such as inside a tube or through the parallel passage area between two circuit boards in an enclosure is called *internal flow*. Fluid flow is also categorized as being *laminar* or *turbulent*, depending on the value of the Reynolds number. In convection analysis, the convection heat transfer coefficient is usually expressed in terms of the dimensionless *Nusselt number* Nu as

$$h = \frac{k}{D}\text{Nu} \quad [\text{W/(m}^2 \cdot {}^\circ\text{C)}]$$

where $k$ is thermal conductivity of the fluid and $D$ is the characteristic length of the geometry. Relations for the average Nusselt number based on experimental data are given in Table 14-2 for external flow, and in Table 14-3 for laminar internal flow under the uniform heat flux condition, which is closely approximated by electronic equipment. In forced-air-cooled systems the heat transfer can also be expressed as

$$\dot{Q} = \dot{m}C_p(T_{\text{out}} - T_{\text{in}}) \quad \text{(W)}$$

where $\dot{Q}$ is the rate of heat transfer to the air, $C_p$ is the specific heat of air, $T_{\text{in}}$ and $T_{\text{out}}$ are the average temperatures of air at the inlet and exit of the enclosure, respectively, and $\dot{m}$ is the mass flow rate of air.

The heat transfer coefficients associated with liquids are usually an order of magnitude higher than those associated with gases. Liquid cooling systems can be classified as *direct cooling* and *indirect cooling* systems. In direct cooling systems, the electronic components are in direct contact with the liquid, and thus the heat generated in the components is transferred directly to the liquid. In indirect cooling systems, however, there is no direct contact with the components. Liquid cooling systems are also classified as *closed-loop* and *open-loop* systems, depending on whether the liquid is discharged or recirculated after it is heated. Only *dielectric fluids* can be used in immersion or direct liquid cooling.

High-power electronic components can be cooled effectively by immersing them in a *dielectric liquid* and taking advantage of the very high heat transfer coefficients associated with boiling. The simplest type of immersion cooling system involves an *external reservoir* that supplies liquid continually to the electronic enclosure. This *open-loop*-type immersion cooling system is simple but often impractical. Immersion cooling systems usually operate in a *closed loop,* in that the vapor is condensed and returned to the electronic enclosure instead of being purged to the atmosphere.

A *heat pipe* is basically a sealed slender tube containing a wick structure lined on the inner surface, which can transfer large quantities of heat over fairly large distances essentially at constant temperature without requiring any power input. The type of fluid and the operating pressure inside the heat pipe depends on the operating temperature of the heat pipe.

## REFERENCES AND SUGGESTED READING

**1** E. P. Black and E. M. Daley, "Thermal Design Considerations for Electronic Components," ASME Paper No. 70-DE-17, 1970.

**2** S. W. Chi, *Heat Pipe Theory and Practice,* Hemisphere, Washington, DC, 1976.

**3** R. A. Colclaser, D. A. Neaman, and C. F. Hawkins, *Electronic Circuit Analysis,* Wiley, New York, 1984.

**4** J. W. Dally, *Packaging of Electronic Systems,* McGraw-Hill, New York, 1990.

**5** *Design Manual of Natural Methods of Cooling Electronic Equipment,* NAVSHIPS 900-192, Department of The Navy, Bureau of Ships, November 1956.

**6** *Design Manual of Cooling Methods for Electronic Equipment,* NAVSHIPS 900-190, Department of The Navy, Bureau of Ships, March 1955.

**7** G. N. Ellison, *Thermal Computations for Electronic Equipment,* Van Nostrand Reinhold, New York, 1984.

**8** J. A. Gardner, "Liquid Cooling Safeguards High-Power Semiconductors," *Electronics,* p. 103, Feb. 24, 1974.

**9** R. S. Gaugler, "Heat Transfer Devices," U.S. Patent 2350348, 1944.

**10** G. M. Grover, T. P. Cotter, and G. F. Erickson, "Structures of Very High Thermal Conductivity," *Journal of Applied Physics,* Vol. 35, pp. 1190–1191, 1964.

**11** W. F. Hilbert and F. H. Kube, "Effects on Electronic Equipment Reliability of Temperature Cycling in Equipment," Final Report, Grumman Aircraft Engineering Corporation, Report No. EC-69-400, Bethpage, NY, Feb. 1969.

**12** U. P. Hwang and K. P. Moran, "Boiling Heat Transfer of Silicon Integrated Circuits Chip Mounted on a Substrate," *ASME HTD,* Vol. 20, pp. 53–59, 1981.

**13** R. D. Johnson, "Enclosures—State of the Art," *New Electronics,* p. 29, July 29, 1983.

**14** M. Jakob, *Heat Transfer,* Vol. 1, Wiley, New York, 1949.

**15** R. Kemp, "The Heat Pipe—A New Tune on an Old Pipe," *Electronics and Power,* p. 326, 9 August 1973.

**16** A. D. Kraus and A. Bar-Cohen, *Thermal Analysis and Control of Electronic Equipment,* McGraw-Hill/Hemisphere, New York, 1983.

**17** *Reliability Prediction of Electronic Equipment,* U.S. Department of Defense, MIL-HDBK-2178B, NTIS, Springfield, VA, 1974.

**18** D. S. Steinberg, *Cooling Techniques for Electronic Equipment,* Wiley, New York, 1980.

**19** A. Zhukauskas, "Heat Transfer from Tubes in Cross Flow," in J. P. Hartnett and T. F. Irvine, Jr. (Eds.), *Advances in Heat Transfer,* Vol. 8, Academic Press, New York, 1972.

## PROBLEMS*

### Introduction and History

**14-1C** What invention started the electronic age? Why did the invention of the transistor marked the beginning of a revolution in that age?

**14-2C** What is an integrated circuit? What is its significance in the electronics era? What do the initials MSI, LSI, and VLSI stand for?

**14-3C** When electric current $I$ passes through an electrical element having a resistance $R$, why is heat generated in the element? How is the amount of heat generation determined?

**14-4C** Consider a TV that is wrapped in the blankets from all sides except its screen. Explain what will happen when the TV is turned on and kept on for a long time, and why. What will happen if the TV is kept on for a few minutes only?

**14-5C** Consider an incandescent light bulb that is completely wrapped in a towel. Explain what will happen when the light is turned on and kept on. (P.S. Do not try this at home!)

**14-6C** A businessman ties a large cloth advertisement banner in front of his car such that it completely blocks the air flow to the radiator. What do you think will happen when he starts the car and goes on a trip?

**14-7C** Which is more likely to break: a car or a TV? Why?

**14-8C** Why do electronic components fail under prolonged use at high temperatures?

**14-9** The temperature of the case of a power transistor that is dissipating 12 W is measured to be 60°C. If the junction-to-case thermal resistance of this transistor is specified by the manufacturer to be 5°C/W, determine the junction temperature of the transistor. *Answer:* 120°C

**14-10** Power is supplied to an electronic component from a 12-V source, and the variation in the electric current, the junction temperature, and the case temperatures with time are observed. When everything is stabilized, the current is observed to be 0.15 A and the temperatures to be 90°C and 65°C at the junction and the case, respectively. Calculate the junction-to-case thermal resistance of this component.

**14-11** A logic chip used in a computer dissipates 3.5 W of power in an environment at 55°C, and has a heat transfer surface area of 0.32 $cm^2$. Assuming the heat transfer from the surface to be uniform, determine (*a*) the amount of heat this chip dissipates during an eight-hour work day, in kWh, and (*b*) the heat flux on the surface of the chip, in $W/cm^2$.

**14-12** A 15 cm × 20 cm circuit board houses 90 closely spaced logic chips, each dissipating 0.1 W, on its surface. If the heat transfer from the

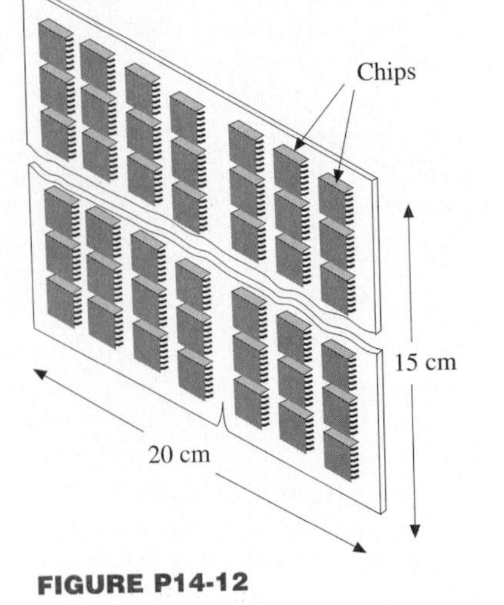


**FIGURE P14-12**

*Students are encouraged to answer *all* the concept "C" questions.

back surface of the board is negligible, determine (*a*) the amount of heat this circuit board dissipates during a 10-h period, in kWh, and (*b*) the heat flux on the surface of the circuit board, in $W/cm^2$.

**14-12E** A 6 in. × 8 in. circuit board houses 90 closely spaced logic chips, each dissipating 0.1 W, on its surface. If the heat transfer from the back surface of the board is negligible, determine (*a*) the amount of heat this circuit board dissipates during a 10-h period, in kWh, and (*b*) the heat flux on the surface of the circuit board, in $W/in.^2$.

**14-13** A resistor on a circuit board has a total thermal resistance of 100°C/W. If the temperature of the resistor is not to exceed 200°C, determine the power at which it can operate safely in an ambient at 50°C.

**14-13E** A resistor on a circuit board has a total thermal resistance of 130°F/W. If the temperature of the resistor is not to exceed 360°F, determine the power at which it can operate safely in an ambient at 120°F.

**14-14** Consider a 0.1-kΩ resistor whose surface-to-ambient thermal resistance is 300°C/W. If the voltage drop across the resistor is 7.5 V and its surface temperature is not to exceed 150°C, determine the power at which it can operate safely in an ambient at 30°C. *Answer:* 0.4 W

### Manufacturing of Electronic Equipment

**14-15C** Why is a chip in a chip carrier bonded to a lead frame instead of the plastic case of the chip carrier?

**14-16C** Draw a schematic of a chip carrier, and explain how heat is transferred from the chip to the medium outside of the chip carrier.

**14-17C** What does the junction-to-case thermal resistance represent? What does it depend on for a chip carrier?

**14-18C** What is a hybrid chip carrier? What are the advantages of hybrid electronic packages?

**14-19C** What is a PCB? What is the board of a PCB made of? What does the "device-to-PCB edge" thermal resistance in conduction-cooled systems represent? Why is this resistance relatively high?

**14-20C** What are the three types of printed circuit boards? What are the advantages and disadvantages of each type?

**14-21C** What are the desirable characteristics of the materials used in the fabrication of the circuit boards?

**14-22C** What is an electronic enclosure? What is the primary function of the enclosure of an electronic system? What materials are the enclosures made of?

## Cooling Load of Electronic Equipment and Thermal Environment

**14-23C** Consider an electronics box that consumes 120 W of power when plugged in. How is the heating load of this box determined?

**14-24C** Why is the development of superconducting materials generating so much excitement among designers of electronic equipment?

**14-25C** How is the duty cycle of an electronic system defined? How does the duty cycle effect the design and selection of a cooling technique for a system?

**14-26C** What is temperature cycling? How does it affect the reliability of electronic equipment?

**14-27C** What is the ultimate heat sink for (*a*) a TV, (*b*) an airplane, and (*c*) a ship? For each case, what is the range of temperature variation of the ambient?

**14-28C** What is the ultimate heat sink for (*a*) a VCR, (*b*) a spacecraft, and (*c*) a communication system on top of a mountain? For each case, what is the range of temperature variation of the ambient?

## Electronics Cooling in Different Applications

**14-29C** How are the electronics of short-range and long-range missiles cooled?

**14-30C** What is dynamic temperature? What causes it? How is it determined? At what velocities is it significant?

**14-31C** How are the electronics of a ship or submarine cooled?

**14-32C** How are the electronics of the communication systems at remote areas cooled?

**14-33C** How are the electronics of high-power microwave equipment such as radars cooled?

**14-34C** How are the electronics of a space vehicle cooled?

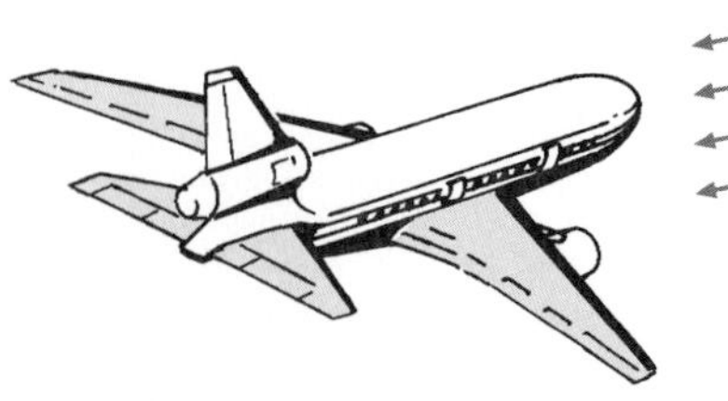


**FIGURE P14-35**

**14-35** Consider an airplane cruising in the air at a temperature of $-10°C$ at a velocity of 850 km/h. Determine the temperature rise of air at the nose of the airplane as a result of the ramming effect of the air.

**14-36** The temperature of air in high winds is measured by a thermometer to be 12°C. Determine the true temperature of air if the wind velocity is 90 km/h. *Answer:* 11.7°C

**14-36E** The temperature of air in high winds is measured by a thermometer to be 60°F. Determine the true temperature of air if the wind velocity is 55 mph. *Answer:* 59.5°F

**14-37** Air at 25°C is flowing in a channel at a velocity of (*a*) 1, (*b*) 10, (*c*) 100, and (*d*) 1000 m/s. Determine the temperature which a stationary probe inserted into the channel will read for each case.

**14-38** An electronic device dissipates 2 W of power and has a surface area of 5 cm$^2$. If the surface temperature of the device is not to exceed the ambient temperature by more than 50°C, determine a suitable cooling technique for this device. Use Fig. 14-17.

**14-39** A stand-alone circuit board, 15 cm × 20 cm in size, dissipates 20 W of power. The surface temperature of the board is not to exceed 80°C in a 30°C environment. Using Fig. 14-17 as a guide, select a suitable cooling mechanism.

**14-39E** A stand alone circuit board, 6 in. × 8 in. in size, dissipates 20 W of power. The surface temperature of the board is not to exceed 170°F in a 85°F environment. Using Fig. 14-17 as a guide, select a suitable cooling mechanism.

## Conduction Cooling

**14-40C** What are the major considerations in the selection of a cooling technique for electronic equipment?

**14-41C** What is thermal resistance? What is it analogous to in electrical circuits? Can thermal resistance networks be analyzed like electrical circuits? Explain.

**14-42C** If the rate of heat conduction through a medium and the thermal resistance of the medium are known, how can the temperature difference between the two sides of the medium be determined?

**14-43C** Consider a wire of electrical resistance $R$, length $L$, and cross-sectional area $A$ through which electric current $I$ is flowing. How is the voltage drop across the wire determined? What happens to the voltage drop when $L$ is doubled while $I$ is held constant?

Now consider heat conduction at a rate of $\dot{Q}$ through the same wire having a thermal resistance of $R$. How is the temperature drop across the wire determined? What happens to the temperature drop when $L$ is doubled while $\dot{Q}$ *is* held constant?

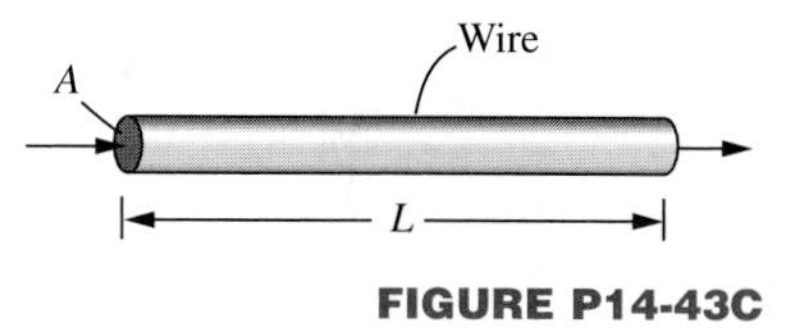


**FIGURE P14-43C**

**14-44C** What is a heat frame? How does it enhance heat transfer along a PCB? Which components on a PCB attached to a heat frame operate at the highest temperatures: those at the middle of the PCB, or those near the edge?

**14-45C** What is constriction resistance in heat flow? What is it analogous to in fluid flow through tubes and electric current flow in wires?

**14-46C** What does the junction-to-case thermal resistance of an electronic component represent? In practice, how is this value determined? How can the junction temperature of a component be determined when the case temperature, the power dissipation of the component, and the junction-to-case thermal resistance are known?

**14-47C** What does the case-to-ambient thermal resistance of an electronic component represent? In practice, how is this value determined?

How can the case temperature of a component be determined when the ambient temperature, the power dissipation of the component, and the case-to-ambient thermal resistance are known?

**14-48C** Consider an electronic component whose junction-to-case thermal resistance $R_{\text{junction-case}}$ is provided by the manufacturer and whose case-to-ambient thermal resistance $R_{\text{case-ambient}}$ is determined by the thermal designer. When the power dissipation of the component and the ambient temperature are known, explain how the junction temperature can be determined. When $R_{\text{junction-case}}$ is greater than $R_{\text{case-ambient}}$, will the case temperature be closer to the junction or ambient temperature?

**14-49C** Why is the rate of heat conduction along a PCB very low? How can heat conduction from the mid-parts of a PCB to its outer edges be improved? How can heat conduction across the thickness of the PCB be improved?

**14-50C** Why is the warping of epoxy boards that are copper-cladded on one side a major concern? What is the cause of this warping? How can the warping of PCBs be avoided?

**14-51C** Why did the thermal conduction module receive so much attention from thermal designers of electronic equipment? How does the design of TCM differ from traditional chip carrier design? Why is the cavity in the TCM filled with helium instead of air?

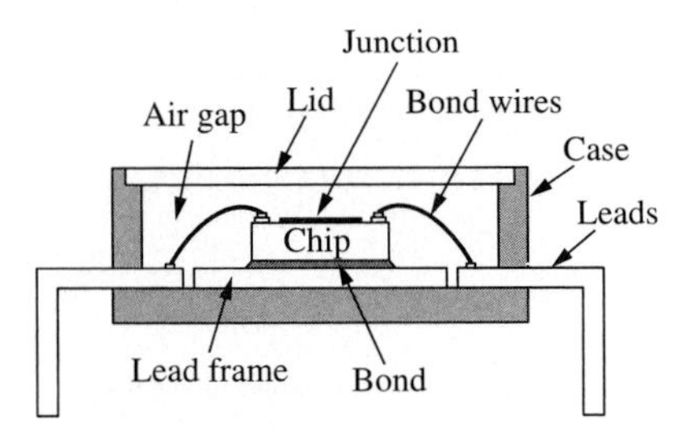


**FIGURE P14-52**

**14-52** Consider a chip dissipating 0.8 W of power in a DIP with 18 pin leads. The materials and the dimensions of various sections of this electronic device are given in the table below. If the temperature of the leads is 50°C, estimate the temperature at the junction of the chip.

| Section and material | Thermal conductivity, W/(m · °C) | Thickness, mm | Heat transfer surface area |
|---|---|---|---|
| Junction constriction | — | — | diameter 0.5 mm |
| Silicon chip | 120 | 0.5 | 4 mm × 4 mm |
| Eutectic bond | 296 | 0.05 | 4 mm × 4 mm |
| Copper lead frame | 386 | 0.25 | 4 mm × 4 mm |
| Plastic separator | 1 | 0.3 | 18 × 1 mm × 0.25 mm |
| Copper leads | 386 | 6 | 18 × 1 mm × 0.25 mm |

**14-53** A fan blows air at 35°C over a 2-W plastic DIP with 16 leads mounted on a PCB at a velocity of 300 m/min. Using data from Fig. 14-23, determine the junction temperature of the electronic device. What would the junction temperature be if the fan were to fail?

**14-53E** A fan blows air at 95°F over a 2-W plastic DIP with 16 leads mounted on a PCB at a velocity of 500 ft/min. Using data from Fig. 14-23, determine the junction temperature of the electronic device. What would the junction temperature be if the fan were to fail?

**14-54** Heat is to be conducted along a PCB with copper cladding on one side. The PCB is 12 cm long and 12 cm wide, and the thicknesses of the copper and epoxy layers are 0.06 mm and 0.5 mm, respectively. Disregarding heat transfer from the side surfaces, determine the percentages of heat conduction along the copper [$k = 386$ W/(m · °C)] and epoxy [$k = 0.26$ W/(m · °C)] layers. Also determine the effective thermal conductivity of the PCB.
*Answers:* 0.6 percent, 99.4 percent, 41.6 W/(m · °C)

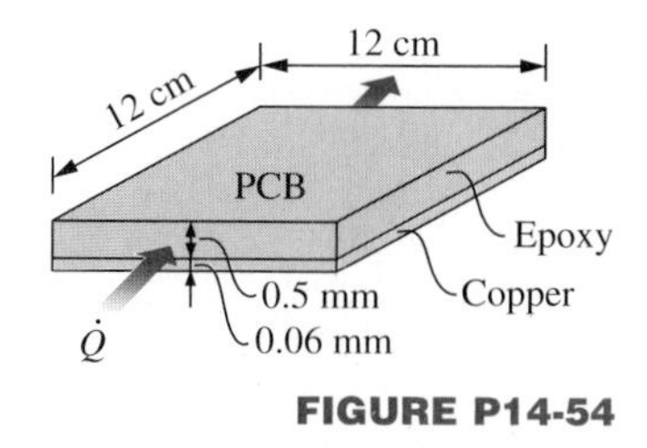


**FIGURE P14-54**

**14-55** The heat generated in the circuitry on the surface of a silicon chip [$k = 130$ W/(m · °C)] is conducted to the ceramic substrate to which it is attached. The chip is 6 mm × 6 mm in size and 0.5 mm thick, and dissipates 2 W of power. Determine the temperature difference between the front and back surfaces of the chip in steady operation.

**14-56** Consider a 15 cm × 18 cm glass–epoxy laminate [$k =$ 0.26 W/(m · °C)] whose thickness is 1.2 mm. Determine the thermal resistance of this epoxy layer for heat flow (*a*) along the 18-cm-long side and (*b*) across its thickness.

**14-56E** Consider a 6 in. × 7 in. glass–epoxy laminate [$k =$ 0.15 Btu/(h · ft · °F)] whose thickness is 0.05 in. Determine the thermal resistance of this epoxy layer for heat flow (*a*) along the 7-in.-long side and (*b*) across its thickness.

**14-57** Consider a 15 cm × 18 cm glass–epoxy laminate [$k =$ 0.26 W(m · °C)] whose thickness is 1.4 mm. In order to reduce the thermal resistance across its thickness, cylindrical copper fillings [$k =$ 386 W/(m · °C)] of diameter 1 mm are to be planted throughout the board with a center-to-center distance of 3 mm. Determine the new value of the thermal resistance of the epoxy board for heat conduction across its thickness as a result of this modification.

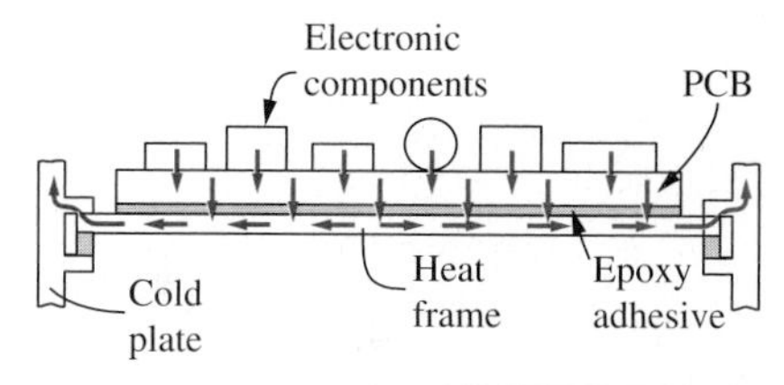

**FIGURE P14-58**

**14-58** A 12 cm × 15 cm circuit board dissipating 45 W of heat is to be conduction-cooled by a 1.5-mm-thick copper heat frame [$k =$ 386 W/(m · °C)] 12 cm × 17 cm in size. The epoxy laminate [$k =$ 0.26 W/(m · °C)] has a thickness of 2 mm, and is attached to the heat frame with conductive epoxy adhesive [$k = 1.8$ W/(m · °C)] of thickness 0.12 mm. The PCB is attached to a heat sink by clamping a 5-mm-wide portion of the edge to the heat sink from both ends. The temperature of the heat frame at this point is 30°C. Heat is uniformly generated on the PCB at a rate of 3 W per 1 cm × 12 cm strip. Considering only one half of the PCB board because of symmetry, determine the maximum surface temperature on the PCB and the temperature distibution along the heat frame.

**14-59** Consider a 15 cm × 20 cm double-sided circuit board dissipating a total of 30 W of heat. The board consists of a 3-mm-thick epoxy layer [$k = 0.26$ W/(m · °C)] with 1-mm-diameter aluminum wires [$k =$ 237 W/(m · °C)] inserted along the 20-cm-long direction, as shown in Fig. P14-59. The distance between the centers of the aluminum wires is 2 mm.

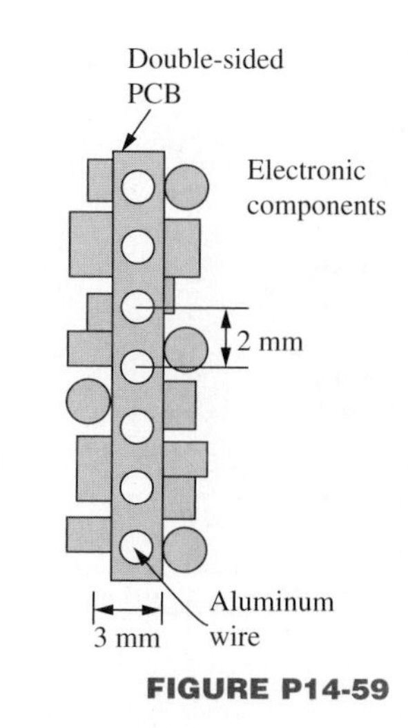


**FIGURE P14-59**

The circuit board is attached to a heat sink from both ends, and the temperature of the board at both ends is 30°C. Heat is considered to be uniformly generated on both sides of the epoxy layer of the PCB. Considering only a portion of the PCB because of symmetry, determine the magnitude and location of the maximum temperature that occurs in the PCB. *Answer:* 136.7°C

**14-59E** Consider a 6 in. × 8 in. double-sided circuit board dissipating a total of 30 W of heat. The board consists of a 0.12-in.-thick epoxy layer [$k$ = 0.15 Btu/(h · ft · °F)] with 0.04-in.-diameter aluminum wires [$k$ = 137 Btu/(h · ft · °F)] inserted along the 8-in.-long direction, as shown in Fig. P14-59. The distance between the centers of the aluminum wires is 0.08 in. The circuit board is attached to a heat sink from both ends, and the temperature of the board at both ends is 90°F. Heat is considered to be uniformly generated on both sides of the epoxy layer of the PCB. Considering only a portion of the PCB because of symmetry, determine the magnitude and location of the maximum temperature that occurs in the PCB. *Answer:* 468°F

**14-60** Repeat Prob. 14-59, replacing the aluminum wires by copper wires [$k$ = 386 W/(m · °C)].

**14-61** Repeat Prob. 14-59 for a center-to-center distance of 4 mm instead of 2 mm between the wires.

**14-62** Consider a thermal conduction module with 80 chips, each dissipating 4 W of power. The module is cooled by water at 22°C flowing through the cold plate on top of the module. The thermal resistances in the path of heat flow are $R_{chip}$ = 12°C/W between the junction and the surface of the chip, $R_{int}$ = 9°C/W between the surface of the chip and the outer surface of the thermal conduction module, and $R_{ext}$ = 7°C/W between the outer surface of the module and the cooling water. Determine the junction temperature of the chip.

**14-63** Consider a 0.3-cm-thick epoxy board [$k$ = 0.26 W/(m · °C)] that is 15 cm × 20 cm in size. Now a 0.1-mm-thick layer of copper [$k$ = 386 W/(m · °C)] is attached to the back surface of the PCB. Determine the effective thermal conductivity of the PCB along its 20-cm-long side. What fraction of the heat conducted along that side is conducted through copper?

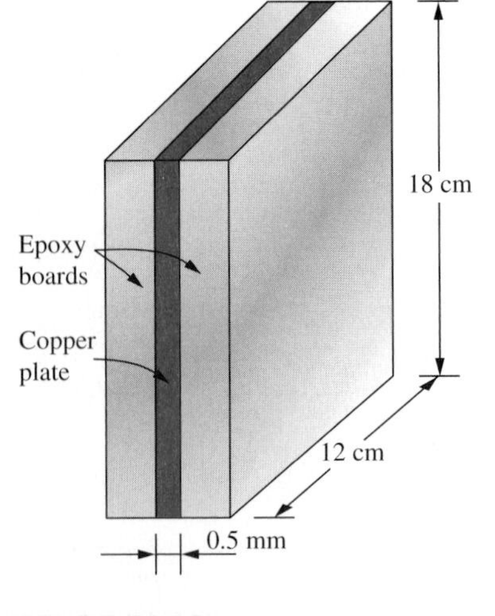


**FIGURE P14-64**

**14-64** A 0.5-mm-thick copper plate [$k$ = 386 W/(m · °C)] is sandwiched between two 3-mm-thick epoxy boards [$k$ = 0.26 W/(m · °C)] that are 12 cm × 18 cm in size. Determine the effective thermal conductivity of the PCB along its 18-cm-long side. What fraction of the heat conducted along that side is conducted through copper?

**14-64E** A 0.02-in.-thick copper plate [$k$ = 223 Btu/(h · ft · °F)] is sandwiched between two 0.12-in.-thick epoxy boards [$k$ = 0.15 Btu/(h · ft · °F)] that are 6 in. × 9 in. in size. Determine the effective thermal conductivity of the PCB along its 9-in.-long side. What fraction of the heat conducted along that side is conducted through the copper plate?

**14-65** A 15 cm × 20 cm × 0.2 cm copper heat frame is used to conduct 20 W of heat generated in a PCB along the 20-cm-long side towards the ends. Determine the temperature difference between the mid-section and either end of the heat frame. *Answer:* 17.3°C

**14-65E** A 6 in. × 8 in. × 0.07 in. copper heat frame is used to conduct 20 W of heat generated in a PCB along the 8-in.-long side towards the ends. Determine the temperature difference between the mid section and either end of the heat frame. *Answer:* 34.8°F

**14-66** A 12-W power transistor is cooled by mounting it on an aluminum bracket [$k$ = 237 W/(m · °C)] that is attached to a liquid-cooled plate by 0.2-mm-thick epoxy adhesive [$k$ = 1.8 W/(m · °C)], as shown in Fig. P14-66. The thermal resistance of the plastic washer is given as 2.5°C/W. Preliminary calculations show that about 20 percent of the heat is dissipated by convection and radiation, and the rest is conducted to the liquid-cooled plate. If the temperature of the cold plate is 50°C, determine the temperature of the transistor case.

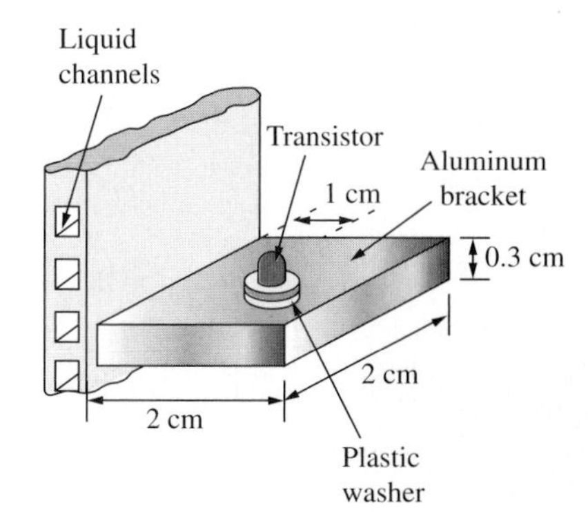


**FIGURE P14-66**

## Air Cooling: Natural Convection and Radiation

**14-67C** A student puts his books on top of a VCR, completely blocking the air vents on the top surface. What do you think will happen as the student watches a rented movie played by that VCR?

**14-68C** Can a low-power electronic system in space be cooled by natural convection? Can it be cooled by radiation? Explain.

**14-69C** Why are there several openings on the various surfaces of a TV, VCR, and other electronic enclosures? What happens if a TV or VCR is enclosed in a cabinet with no free air space around?

**14-70C** Why should radiation heat transfer always be considered in the analysis of natural convection cooled electronic equipment?

**14-71C** How does atmospheric pressure affect natural convection heat transfer? What are the best and worst orientations for heat transfer from a square surface?

**14-72C** What is view factor? How does it affect radiation heat transfer between two surfaces?

**14-73C** What is emissivity? How does it affect radiation heat transfer between two surfaces?

**14-74C** For most effective natural convection cooling of a PCB array, should the PCBs be placed horizontally or vertically? Should they be placed close to each other or far from each other?

**14-75C** Why is radiation heat transfer from the components on the PCBs in an enclosure negligible?

**14-76** Consider a sealed 20-cm-high electronic box whose base dimensions are 35 cm × 50 cm, placed on top of a stand in a room at 30°C. The emissivity of the outer surface of the box is 0.85. If the electronic

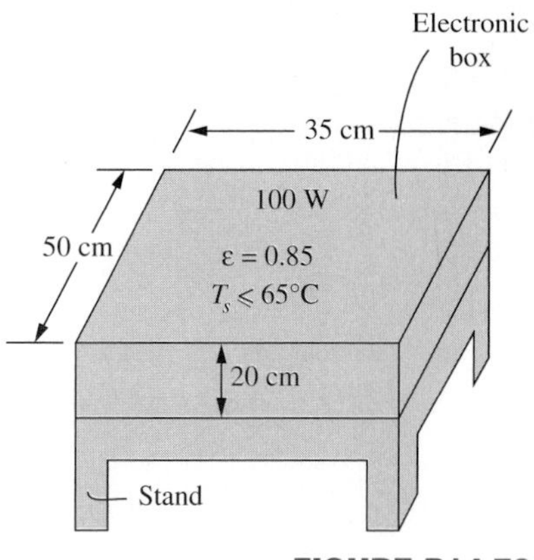


**FIGURE P14-76**

components in the box dissipate a total of 100 W of power and the outer surface temperature of the box is not to exceed 65°C, determine if this box can be cooled by natural convection and radiation alone. Assume the heat transfer from the bottom surface of the box to the stand to be negligible, and the temperature of the surrounding surfaces to be the same as the air temperature of the room.

**14-77** Repeat Prob. 14-76, assuming the box is mounted on a wall instead of a stand such that it is 0.5 m high. Again assume heat transfer from the bottom surface to the wall to be negligible.

**14-78** A 0.15-W small cylindrical resistor mounted on a circuit board is 1.2 cm long and has a diameter of 0.4 cm. The view of the resistor is largely blocked by the circuit board facing it, and the heat transfer from the connecting wires is negligible. The air is free to flow through the parallel flow passages between the PCBs as a result of natural convection currents. If the air temperature at the vicinity of the resistor is 55°C, determine the surface temperature of the resistor. *Answer:* 90.9°C

**14-78E** A 0.15-W small cylindrical resistor mounted on a circuit board is 0.5 in. long and has a diameter of 0.15 in. The view of the resistor is largely blocked by the circuit board facing it, and the heat transfer from the connecting wires is negligible. The air is free to flow through the parallel flow passages between the PCBs as a result of natural convection currents. If the air temperature at the vicinity of the resistor is 130°F, determine the surface temperature of the resistor. *Answer:* 194°F

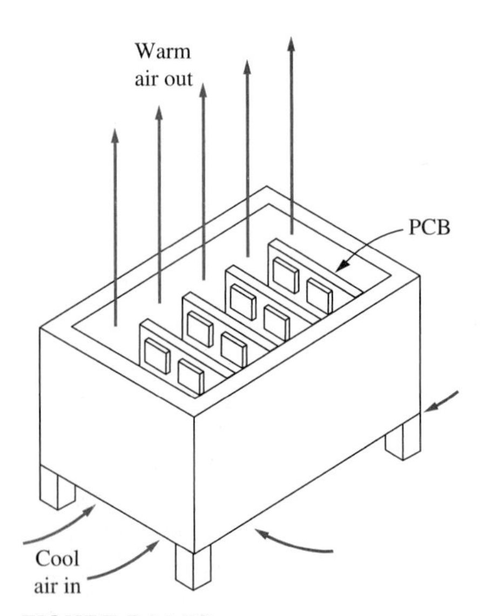


**FIGURE P14-79**

**14-79** A 14 cm × 20 cm PCB has electronic components on one side, dissipating a total of 7 W. The PCB is mounted in a rack vertically (height 14 cm) together with other PCBs. If the surface temperature of the components is not to exceed 90°C, determine the maximum temperature of the environment in which this PCB can operate safely at sea level. What would your answer be if this rack is located at a location at 3000 m altitude where the atmospheric pressure is 70.12 kPa?

**14-80** A cylindrical electronic component whose diameter is 2 cm and length is 4 cm is mounted on a board with its axis in the vertical direction, and is dissipating 3 W of power. The emissivity of the surface of the component is 0.8, and the temperature of the ambient air is 30°C. Assuming the temperature of the surrounding surfaces to be 20°C, determine the average surface temperature of the component under combined natural convection and radiation cooling.

**14-81** Repeat Prob. 14-80, assuming the component is oriented horizontally.

**14-82** Consider a power transistor that dissipates 0.1 W of power in an environment at 30°C. The transistor is 0.4 cm long, and has a diameter of 0.4 cm. Assuming heat to be transferred uniformly from all surfaces, determine ($a$) the heat flux on the surface of the transistor, in W/cm$^2$, and (*b*) the surface temperature of the transistor for a combined convection and radiation heat transfer coefficient of $12\ \mathrm{W/(m^2 \cdot C)}$.

**14-83** The components of an electronic system dissipating 150 W are located in a 1-m-long horizontal duct whose cross-section is 15 cm × 15 cm. The components in the duct are cooled by forced air, which enters at 30°C at a rate of $0.4\ \mathrm{m^3/min}$ and leaves at 45°C. The surfaces of the sheet metal duct are not painted, and thus radiation heat transfer from the outer surfaces is negligible. If the ambient air temperature is 25°C, determine (*a*) the heat transfer from the outer surfaces of the duct to the ambient air by natural convection and (*b*) the average temperature of the duct. *Answers:* (*a*) 31.7 W, (*b*) 40°C

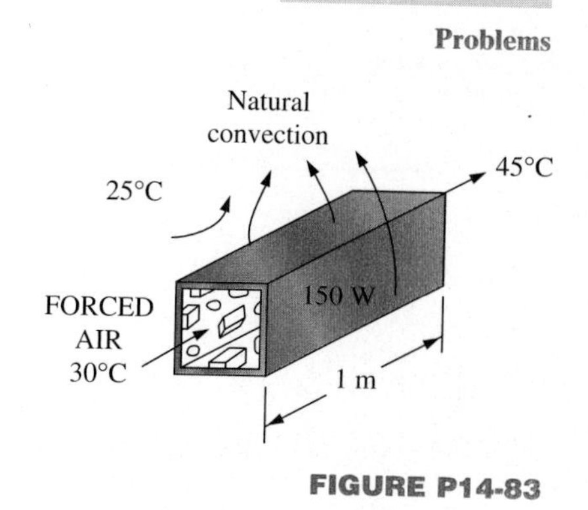


**FIGURE P14-83**

**14-83E** The components of an electronic system dissipating 150 W are located in a 3-ft-long horizontal duct whose cross-section is 6 in. × 6 in. The components in the duct are cooled by forced air, which enters at 85°F at a rate of 15 cfm and leaves at 110°F. The surfaces of the sheet metal duct are not painted, and thus radiation heat transfer from the outer surfaces is negligible. If the ambient air temperature is 80°F, determine (*a*) the heat transfer from the outer surfaces of the duct to the ambient air by natural convection and (*b*) the average temperature of the duct. *Answers:* (*a*) 128.4 Btu/h, (*b*) 113°F

**14-84** Repeat Prob. 14-83 for a circular horizontal duct of diameter 10 cm.

**14-84E** Repeat Prob. 14-83E for a circular horizontal duct of diameter 4 in.

**14-85** Repeat Prob. 14-83, assuming that the fan fails and thus the entire heat generated inside the duct must be rejected to the ambient air by natural convection from the outer surfaces of the duct.

**14-86** A 20 cm × 20 cm circuit board containing 81 square chips on one side is to be cooled by combined natural convection and radiation by mounting it on a vertical surface in a room at 25°C. Each chip dissipates 0.1 W of power, and the emissivity of the chip surfaces is 0.7. Assuming the heat transfer from the back side of the circuit board to be negligible, and the temperature of the surrounding surfaces to be the same as the air temperature of the room, determine the surface temperature of the chips.

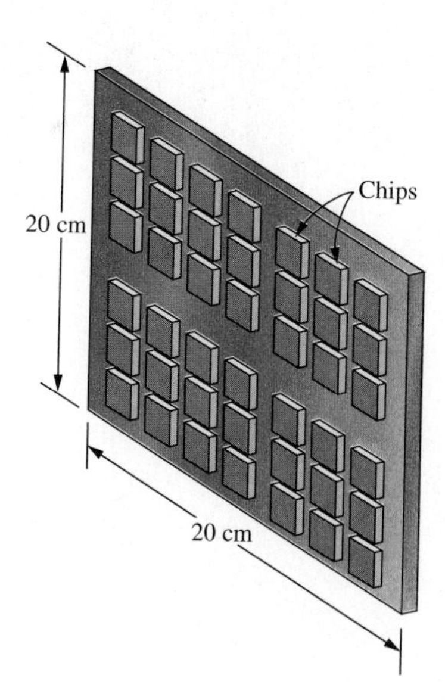


**FIGURE P14-86**

**14-87** Repeat Prob. 14-86, assuming the circuit board to be positioned horizontally with (*a*) chips facing up and (*b*) chips facing down.

## Air Cooling: Forced Convection

**14-88C** Why is radiation heat transfer in forced-air-cooled systems disregarded?

**14-89C** If an electronic system can be cooled adequately by either natural convection or forced air convection, which would you prefer? Why?

**14-90C** Why is forced convection cooling much more effective than natural convection cooling?

**14-91C** Consider a forced-air-cooled electronic system dissipating a fixed amount of power. How will increasing the flow rate of air affect the surface temperature of the components? Explain. How will it affect the exit temperature of the air?

**14-92C** What do internal and external flow refer to in forced convection cooling? Give an example of forced air cooled electronic system which involves both types of flow.

**14-93C** For a specified power dissipation and air inlet temperature, how does the convection heat transfer coefficient affect the surface temperature of the electronic components? Explain.

**14-94C** How does high altitude affect forced convection heat transfer? How would you modify your forced air cooling system to operate at high altitudes safely?

**14-95C** What are the advantages and disadvantages of placing the cooling fan at the inlet or at the exit of an electronic box?

**14-96C** How is the volume flow rate of air in a forced-air-cooled electronic system that has a constant-speed fan established? If a few more PCBs are added to the box while keeping the fan speed constant, will the flow rate of air through the system change? Explain.

**14-97C** What happens if we attempt to cool an electronic system with an undersized fan? What about if we do that with an oversized fan?

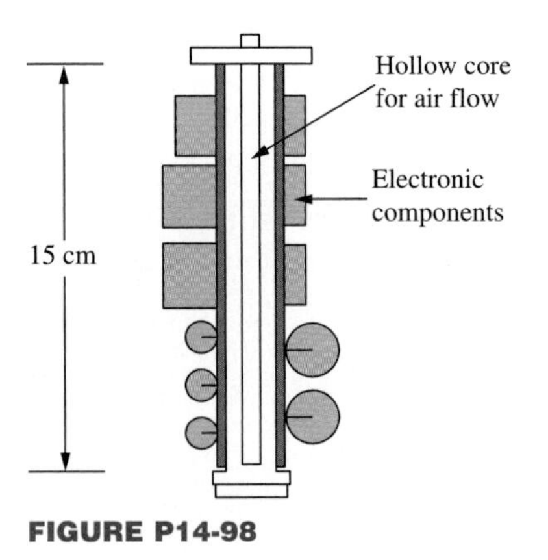


**FIGURE P14-98**

**14-98** Consider a hollow core PCB that is 15 cm high and 20 cm long, dissipating a total of 30 W. The width of the air gap in the middle of the PCB is 0.25 cm. The cooling air enters the core at 30°C at a rate of 1 L/s. Assuming the heat generated to be uniformly distributed over the two side surfaces of the PCB, determine (*a*) the temperature at which the air leaves the hollow core and (*b*) the highest temperature on the inner surface of the core. *Answers:* (*a*) 55°C, (*b*) 66.4°C

**14-99** Repeat Prob. 14-98 for a hollow core PCB dissipating 45 W.

**14-100** A transistor with a height of 0.6 cm and a diameter of 0.5 cm is mounted on a circuit board. The transistor is cooled by air flowing over it at a velocity of 120 m/min. If the air temperature is 60°C and the transistor case temperature is not to exceed 80°C, determine the amount of power this transistor can dissipate safely. *Answer:* 0.15 W

**14-100E** A transistor with a height of 0.25 in. and a diameter of 0.2 in. is mounted on a circuit board. The transistor is cooled by air flowing over it at a velocity of 400 ft/min. If the air temperature is 140°F and the transistor case temperature is not to exceed 175°F, determine the amount of power this transistor can dissipate safely. *Answer:* 0.15 W

**14-101** A desktop computer is to be cooled by a fan. The electronic components of the computer consume 60 W of power under full load conditions. The computer is to operate in environments at temperatures up to 45°C and at elevations up to 3400 m where the atmospheric pressure is 66.63 kPa. The exit temperature of air is not to exceed 60°C to meet reliability requirements. Also, the average velocity of air is not to exceed 110 m/min at the exit of the computer case, where the fan is installed to keep the noise level down. Determine the flow rate of the fan that needs to be installed, and the diameter of the casing of the fan.

**14-102** Repeat the Prob. 14-101 for a computer that consumes 100 W of power.

**14-103** A computer cooled by a fan contains eight PCBs, each dissipating 12 W of power. The height of the PCBs is 12 cm and the length is 18 cm. The clearance between the tips of the components on the PCB and the back surface of the adjacent PCB is 0.3 cm. The cooling air is supplied by a 25-W fan mounted at the inlet. If the temperature rise of air as it flows through the case of the computer is not to exceed 15°C, determine (*a*) the flow rate of the air that the fan needs to deliver, (*b*) the fraction of the temperature rise of air due to the heat generated by the fan and its motor, and (*c*) the highest allowable inlet air temperature if the surface temperature of the components is not to exceed 80°C anywhere in the system.

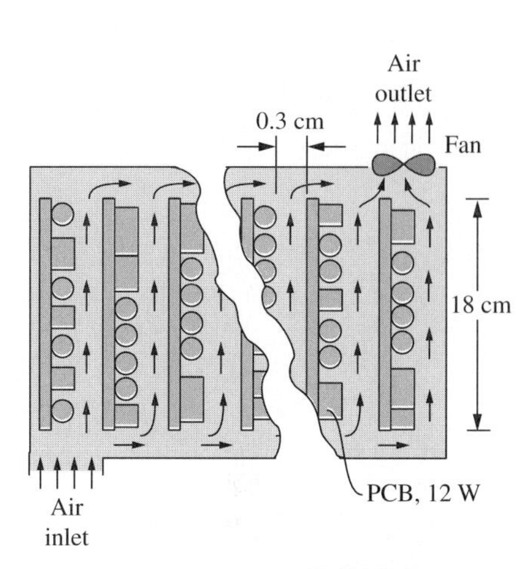


**FIGURE P14-103**

**14-104** An array of power transistors, each dissipating 2 W of power, are to be cooled by mounting them on a 20 cm × 20 cm square aluminum plate and blowing air over the plate with a fan at 30°C with a velocity of 3 m/s. The average temperature of the plate is not to exceed 60°C. Assuming the heat transfer from the back side of the plate to be negligible, determine the number of transistors that can be placed on this plate. *Answer:* 9

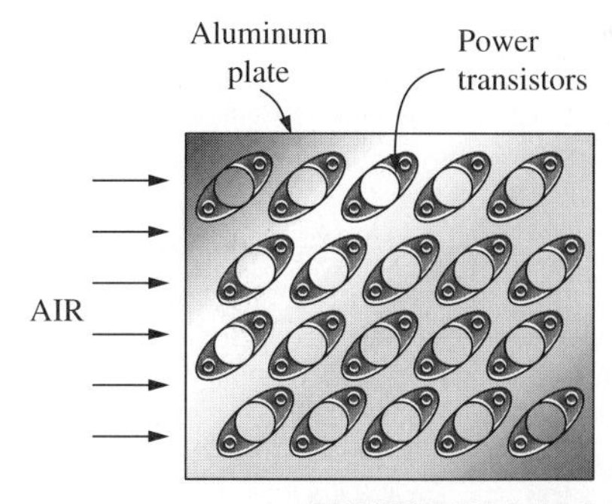


**FIGURE P14-104**

**14-104E** An array of power transistors, each dissipating 2 W of power, are to be cooled by mounting them on a 8 in. × 8 in. square aluminum plate and blowing air over the plate with a fan at 85°F with a velocity of 10 ft/s. The average temperature of the plate is not to exceed 140°F. Assuming the heat transfer from the back side of the plate to be negligible, determine the number of transistors that can be placed on this plate. *Answer:* 9

**14-105** Repeat Prob. 14-104 for a location at an elevation of 1610 m where the atmospheric pressure is 83.4 kPa.

**14-106** An enclosure contains an array of circuit boards, 15 cm high and 20 cm long. The clearance between the tips of the components on the PCB and the back surface of the adjacent PCB is 0.3 cm. Each circuit board contains 75 square chips on one side, each dissipating 0.15 W of power. Air enters the space between the boards through the 0.3 cm × 15 cm cross-section at 40°C with a velocity of 300 m/min. Assuming the heat transfer from the back side of the circuit board to be negligible, determine the exit temperature of the air and the highest surface temperature of the chips.

**14-107** The components of an electronic system dissipating 120 W are located in a 1-m-long horizontal duct whose cross-section is 15 cm × 15 cm. The components in the duct are cooled by forced air, which enters at 30°C at a rate of 0.5 $m^3$/min. Assuming 80 percent of the heat generated inside is transferred to air flowing through the duct and the remaining 20 percent is lost through the outer surfaces of the duct, determine (*a*) the exit temperature of air and (*b*) the highest component surface temperature in the duct.

**14-108** Repeat Prob. 14-107 for a circular horizontal duct of diameter 10 cm.

## Liquid Cooling

**14-109C** If an electronic system can be cooled adequately by either forced air cooling or liquid cooling, which one would you prefer? Why?

**14-110C** Explain how direct and indirect liquid cooling systems differ from each other.

**14-111C** Explain how closed-loop and open-loop liquid cooling systems operate.

**14-112C** What are the properties of a liquid ideally suited for cooling electronic equipment?

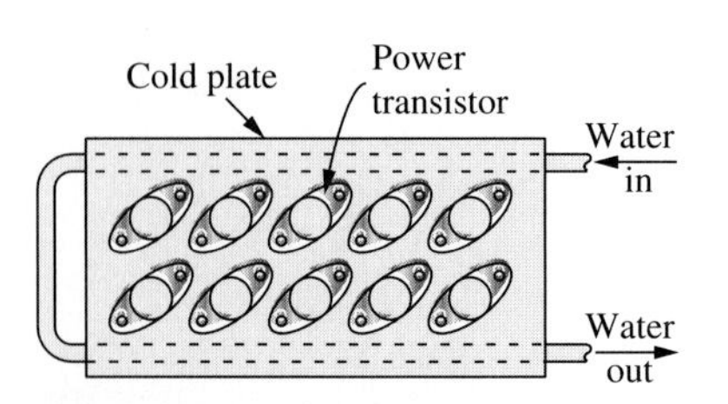


**FIGURE P14-113**

**14-113** A cold plate that supports 10 power transistors, each dissipating 40 W, is to be cooled with water. It is specified that the temperature rise of the water is not to exceed 4°C, and the velocity of water is to remain under 0.5 m/s. Assuming 25 percent of the heat generated is dissipated from the components to the surroundings by convection and radiation, and the remaining 75 percent is removed by the cooling water, determine the mass flow rate of water needed and the diameter of the pipe to satisfy the restriction imposed on the flow velocity. Also determine the case temperature of the devices if the total case-to-liquid thermal resistance is 0.04°C/W and the water enters the cold plate at 25°C.

**14-114** Water enters the tubes of a cold plate at 35°C with an average velocity of 20 m/min, and leaves at 40°C. The diameter of the tubes is 0.5 cm. Assuming 15 percent of the heat generated is dissipated from the components to the surroundings by convection and radiation, and the remaining 85 percent is removed by the cooling water, determine the amount of heat generated by the electronic devices mounted on the cold plate. *Answer:* 160.9 W

**14-114E** Water enters the tubes of a cold plate at 95°F with an average velocity of 60 ft/min, and leaves at 105°F. The diameter of the tubes is 0.25 in. Assuming 15 percent of the heat generated is dissipated from the components to the surroundings by convection and radiation, and the remaining 85 percent is removed by the cooling water, determine the amount of heat generated by the electronic devices mounted on the cold plate. *Answer:* 263 W

**14-115** A sealed electronic box is to be cooled by tap water flowing through channels on two of its sides. It is specified that the temperature rise of the water is not to exceed 4°C. The power dissipation of the box is 2 kW, which is removed entirely by water. If the box operates 24 h a day, 365 days a year, determine the mass flow rate of water flowing through the box, and the amount of cooling water used per year.

**14-116** Repeat Prob. 14-115 for a power dissipation of 3 kW.

## Immersion Cooling

**14-117C** What are the desirable characteristics of a liquid used in immersion cooling of electronic devices?

**14-118C** How does an open-loop immersion cooling system operate? How does it differ from closed-loop cooling systems?

**14-119C** How do immersion cooling systems with internal and external cooling differ? Why are externally cooled systems limited to relatively low-power applications?

**14-120C** Why is boiling heat transfer used in the cooling of very high-power electronic devices instead of forced air or liquid cooling?

**14-121** A logic chip used in a computer dissipates 3 W of power and has a heat transfer surface area of 0.3 cm$^2$. If the surface of the chip is to be maintained at 70°C while being cooled by immersion in a dielectric fluid at 20°C, determine the necessary heat transfer coefficient, and the type of cooling mechanism that needs to be used to achieve that heat transfer coefficient.

**14-122** A 6-W chip having a surface are of 0.5 cm$^2$ is cooled by immersing it into FC86 liquid that is maintained at a temperature of 25°C. Using the boiling curve in Fig. 14-63, estimate the temperature of the chip surface. *Answer:* 82°C

**14-123** A logic chip cooled by immersing it in a dielectric liquid dissipates 3.5 W of power in an environment at 50°C, and has a heat transfer surface area of 0.8 cm$^2$. The surface temperature of the chip is measured to be 95°C. Assuming the heat transfer from the surface to be uniform, determine (*a*) the heat flux on the surface of the chip, in W/cm$^2$, (*b*) the heat transfer coefficient on the surface of the chip, in W/(m$^2$ · °C), and (*c*) the thermal resistance between the surface of the chip and the cooling medium, in °C/W.

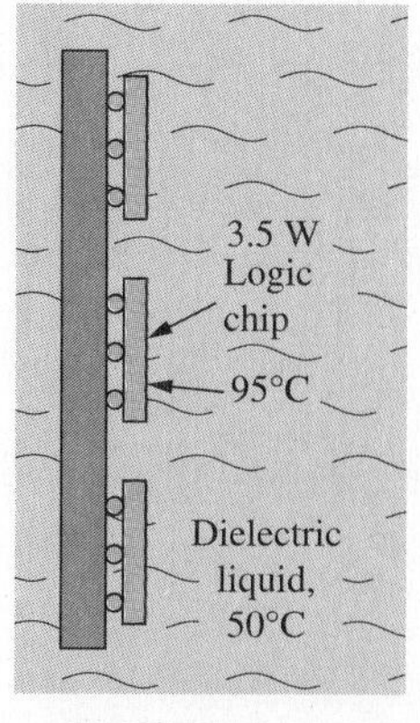


**FIGURE P14-123**

**14-124** A computer chip dissipates 5 W of power and has a heat transfer surface area of 0.4 cm$^2$. If the surface of the chip is to be maintained at 55°C while being cooled by immersion in a dielectric fluid at 10°C, determine the necessary heat transfer coefficient, and the type of cooling mechanism that needs to be used to achieve that heat transfer coefficient.

**14-125** A 3-W chip having a surface area of $0.2\,cm^2$ is cooled by immersing it into FC86 liquid that is maintained at a temperature of 40°C. Using the boiling curve in Fig. 14-63, estimate the temperature of the chip surface. *Answer:* 103°C

**14-126** A logic chip having a surface area of $0.3\,cm^2$ is to be cooled by immersing it into FC86 liquid that is maintained at a temperature of 35°C. The surface temperature of the chip is not to exceed 60°C. Using the boiling curve in Fig. 14-63, estimate the maximum power that the chip can dissipate safely.

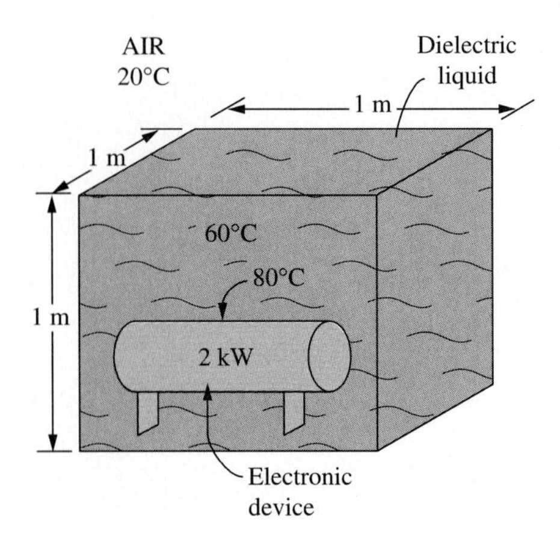


**FIGURE P14-127**

**14-127** A 2-kW electronic device that has a surface area of $120\,cm^2$ is to be cooled by immersing it in a dielectric fiuid with a boiling temperature of 60°C contained in a $1\,m \times 1\,m$ cubic enclosure. Noting that the combined natural convection and the radiation heat transfer coefficients in air are typically about $10\,W/(m^2 \cdot C)$, determine if the heat generated inside can be dissipated to the ambient air at 20°C by natural convection and radiation. If it cannot, explain what modification you could make to allow natural convection cooling.

Also determine the heat transfer coefficients at the surface of the electronic device for a surface temperature of 80°C. Assume the liquid temperature remains constant at 60°C throughout the enclosure.

### Heat Pipes

**14-128C** What is a heat pipe? How does it operate? Does it have any moving parts?

**14-129C** A heat pipe with water as the working fluid is said to have an effective thermal conductivity of $100{,}000\,W/(m \cdot °C)$, which is more than 100,000 times the conductivity of water. How can this happen?

**14-130C** What is the effect of a small amount of noncondensable gas such as air on the performance of a heat pipe?

**14-131C** Why do water-based heat pipes used in the cooling of electronic equipment operate below atmospheric pressure?

**14-132C** What happens when the wick of a heat pipe is too coarse or too fine?

**14-133C** Does the orientation of a heat pipe affect its performance? Does it matter if the evaporator end of the heat pipe is up or down? Explain.

**14-134C** How can the liquid in a heat pipe move up against gravity without a pump? For heat pipes that work against gravity, is it better to have coarse or fine wicks? Why?

**14-135C** What are the important considerations in the design and manufacture of heat pipes?

**14-136C** What is the major cause for the premature degradation of the performance of some heat pipes?

**14-137** A 40-cm-long cylindrical heat pipe having a diameter of 0.5 cm is dissipating heat at a rate of 150 W, with a temperature difference of 4°C across the heat pipe. If we were to use a 40-cm-long copper rod [$k = 386$ W/(m · °C) and $\rho = 8950$ kg/m$^3$] instead to remove heat at the same rate, determine the diameter and the mass of the copper rod that needs to be installed.

**14-138** Repeat Prob. 14-137 for an aluminum rod [$k = 237$ W/(m · °C) and $\rho = 2702$ kg/m$^3$] instead of copper.

**14-139** A plate that supports 10 power transistors, each dissipating 35 W, is to be cooled with 30.5-cm-long heat pipes having a diameter of 0.635 cm. Using Table 14-5, determine how many heat pipes need to be attached to this plate. *Answer:* 2

**14-139E** A plate that supports 10 power transistors, each dissipating 35 W, is to be cooled with 1-ft-long heat pipes having a diameter of $\frac{1}{4}$ in. Using Table 14-5, determine how many pipes need to be attached to this plate. *Answer:* 2

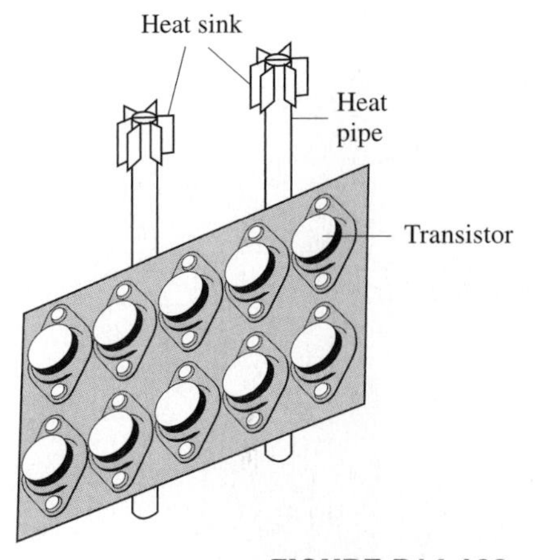

**FIGURE P14-139**

## Review Problems

**14-140C** Several power transistors are to be cooled by mounting them on a water-cooled metal plate. The total power dissipation, the mass flow rate of water through the tube, and the water inlet temperature are fixed. Explain what you would do for the most effective cooling of the transistors.

**14-141C** Consider heat conduction along a vertical copper bar whose sides are insulated. One person claims that the bar should be oriented such that the hot end is at the bottom and the cold end is at the top for better heat transfer, since heat rises. Another person claims that it makes no difference to heat conduction whether heat is conducted downwards or upwards, and thus the orientation of the bar is irrelevant. Which person do you agree with?

**14-142** Consider a 15 cm × 15 cm multilayer circuit board dissipating 22.5 W of heat. The board consists of four layers of 0.1-mm-thick copper [$k = 386$ W/(m · °C)] and three layers of 0.5-mm-thick glass–epoxy [$k = 0.26$ W/(m · °C)] sandwiched together, as shown in Fig. P14-142. The circuit board is attached to a heat sink from both ends, and the temperature of the board at those ends is 35°C. Heat is considered to be uniformly generated in the epoxy layers of the PCB at a rate of 0.5 W per 1 cm × 15 cm epoxy laminate strip (or 1.5 W per 1 cm × 15 cm strip of the board). Considering only a portion of the PCB because of symmetry,

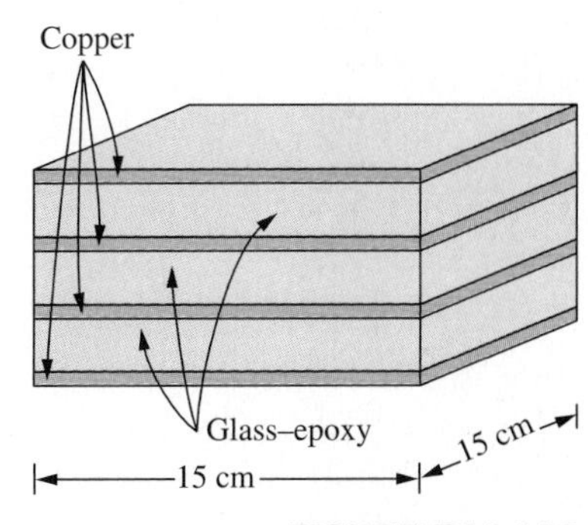

**FIGURE P14-142**

determine the magnitude and location of the maximum temperature that occurs in the PCB. Assume heat transfer from the top and bottom faces of the PCB to be negligible. *Answer:* 108°C

**14-143** Repeat Prob. 14-142, assuming that the board consists of a single 1.5-mm-thick layer of glass–epoxy, with no copper layers.

**14-144** The components of an electronic system that is dissipating 200 W are located in a 1-m-long horizontal duct whose cross-section is 10 cm × 10 cm. The components in the duct are cooled by forced air, which enters at 30°C and 50 m/min and leaves at 45°C. The surfaces of the sheet metal duct are not painted, and so radiation heat transfer from the outer surfaces is negligible. If the ambient air temperature is 30°C, determine (*a*) the heat transfer from the outer surfaces of the duct to the ambient air by natural convection, (*b*) the average temperature of the duct, and (*c*) the highest component surface temperature in the duct.

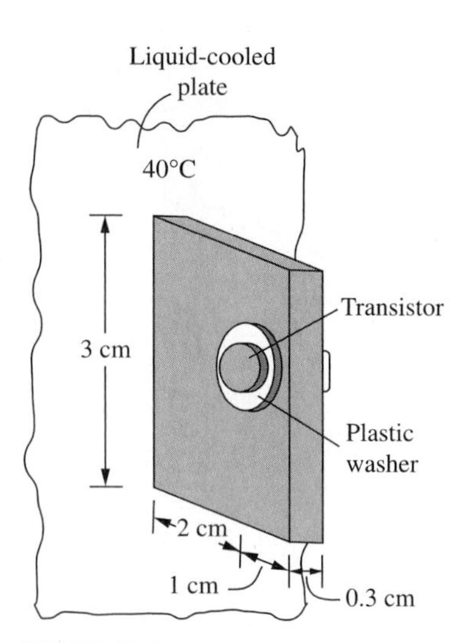

**FIGURE P14-145**

**14-145** Two 10-W power transistors are cooled by mounting them on the two sides of an aluminum bracket [$k = 237$ W/(m · °C)] that is attached to a liquid-cooled plate by 0.2-mm-thick epoxy adhesive [$k = 1.8$ W/(m · °C)], as shown in Fig. P14-145. The thermal resistance of each plastic washer is given as 2°C/W, and the temperature of the cold plate is 40°C. The surface of the aluminum plate is untreated, and thus radiation heat transfer from it is negligible because of the low emissivity of aluminum surfaces. Disregarding heat transfer from the 0.3-cm-wide edges of the aluminum plate, determine the surface temperature of the transistor case. Also determine the fraction of heat dissipation to the ambient air by natural convection, and to the cold plate by conduction.

**14-146** A fan blows air at 30°C and a velocity of 150 m/min over a 1.5-W plastic DIP with 24 leads mounted on a PCB. Using data from Fig. 14-23, determine the junction temperature of the electronic device. What would the junction temperature be if the fan were to fail?

**14-146E** A fan blows air at 85°F and a velocity of 500 ft/min over a 1.5-W plastic DIP with 24 leads mounted on a PCB. Using data from Fig. 14-23, determine the junction temperature of the electronic device. What would the junction temperature be if the fan were to fail?

**14-147** A 15 cm × 18 cm double-sided circuit board dissipating a total of 18 W of heat is to be conduction-cooled by a 1.2-mm-thick aluminum core plate [$k = 237$ W/(m · °C)] sandwiched between two epoxy laminates [$k = 0.26$ W/(m · °C)]. Each epoxy layer has a thickness of 0.5 mm, and is attached to the aluminum core plate with conductive epoxy adhesive [$k = 1.8$ W/(m · °C)] of thickness 0.1 mm. Heat is uniformly generated on each side of the PCB at a rate of 0.5 W per 1 cm × 15 cm epoxy laminate strip. All of the heat is conducted along the 18-cm side, since the PCB is cooled along the two 15-cm-long edges. Considering only part of the PCB board because of symmetry, determine the maximum temperature rise across the 9-cm distance between the center and the sides of the PCB. *Answer:* 19.2°C

**14-148** Ten power transistors, each dissipating 2 W, are attached to a 7 cm × 7 cm × 0.2 cm aluminum plate with a square cutout in the middle in a symmetrical arrangement, as shown in Fig. P14-149. The aluminum plate is cooled from two sides by liquid. If 85 percent of the heat generated by the transistors is estimated to be conducted through the aluminum plate, determine the temperature rise across the 1-cm-wide section of the aluminum plate between the transistors and the heat sink.

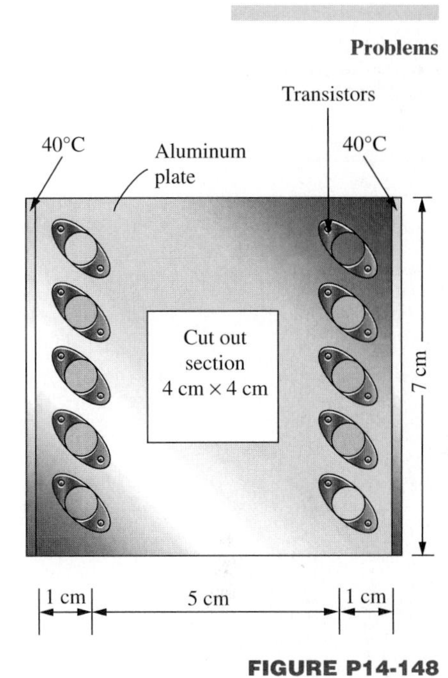


FIGURE P14-148

**14-149** The components of an electronic system are located in a 1.2-m-long horizontal duct whose cross-section is 10 cm × 20 cm. The components in the duct are not allowed to come into direct contact with cooling air, and so are cooled by air flowing over the duct at 30°C with a velocity of 250 m/min. The duct is oriented such that air strikes the 10-cm-high side of the duct normally. If the surface temperature of the duct is not to exceed 60°C, determine the total power rating of the electronic devices that can be mounted in the duct. What would your answer be if the duct is oriented such that air strikes the 20-cm-high side normally? *Answers:* 487 W, 389 W

**14-149E** The components of an electronic system are located in a 4-ft-long horizontal duct whose cross-section is 4 in. × 8 in. The components in the duct are not allowed to come into direct contact with cooling air, and so are cooled by air flowing over the duct at 85°F with a velocity of 750 ft/min. The duct is oriented such that air strikes the 4-in.-high side of the duct normally. If the surface temperature of the duct is not to exceed 140°F, determine the total power rating of the electronic devices that can be mounted in the duct. What would your answer be if the duct is oriented such that air strikes the 8-in.-high side normally? *Answers:* 477 W, 382 W

**14-150** Repeat Prob. 14-149 for a location at altitude 5000 m where the atmospheric pressure is 54.05 kPa.

**14-151** A computer that consumes 80 W of power is cooled by a fan blowing air into the computer enclosure. The dimensions of the computer case are 15 cm × 50 cm × 60 cm, and all surfaces of the case are exposed to the ambient, except for the base surface. Temperature measurements indicate that the case is at an average temperature of 35°C when the ambient temperature and the temperature of the surrounding walls are 25°C. If the emissivity of the outer surface of the case is 0.85, determine the fraction of heat lost from the outer surfaces of the computer case.

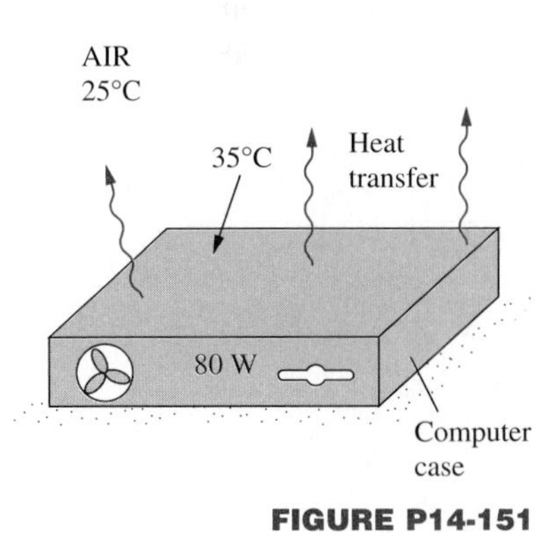


FIGURE P14-151

**14-151E** A computer that consumes 80 W of power is cooled by a fan blowing air into the computer enclosure. The dimensions of the computer case are 6 in. × 20 in. × 24 in., and all surfaces of the case are exposed to the ambient, except for the base surface. Temperature measurements indicate that the case is at an average temperature of 95°F when the ambient temperature and the temperature of the surrounding walls are 80°F. If the emissivity of the outer surface of the case is 0.85, determine the fraction of heat lost from the outer surfaces of the computer case.

## Computer, Design, and Essay Problems

**14-152** Write an interactive computer program that will determine the surface temperature of an electronic component under natural convection and radiation cooling in air when the power dissipation, air temperature, surrounding surface temperature, surface emissivity, and the dimensions and the orientation of the surface are given.

**14-153** Redesign the chip carrier discussed in this chapter such that the thermal performance of the chip is greatly improved while satisfying all the electrical requirements.

**14-154** Bring an electronic device that is cooled by heat sinking to class, and discuss how the heat sink enhances heat transfer.

**14-155** Obtain a catalog from a heat sink manufacturer, and select a heat sink model that can cool a 10-W power transistor safely under (*a*) natural convection and radiation and (*b*) forced convection conditions.

**14-156** Find out about the heat fluxes encountered at nuclear reactors, at the sun's surface, and the solar energy incident on the surface of the earth, and compare the heat fluxes encountered in electronic devices with them.

**14-157** Take the cover off a PC or another electronic box, and identify the individual sections and components that you see. Also identify the cooling mechanisms involved, and discuss how the current design factilitates effective cooling.

# Property Tables and Charts (SI Units)

APPENDIX 1

**TABLE A-1**
**Molar mass, gas constant, and critical-point properties**

| Substance | Formula | $M$ Molar mass kg/kmol | $R$ Gas constant kJ/(kg · K)* | Critical point properties Temperature K | Pressure MPa | Volume $m^3$/kmol |
|---|---|---|---|---|---|---|
| Air | — | 28.97 | 0.2870 | 132.5 | 3.77 | 0.0883 |
| Ammonia | $NH_3$ | 17.03 | 0.4882 | 405.5 | 11.28 | 0.0724 |
| Argon | Ar | 39.948 | 0.2081 | 151 | 4.86 | 0.0749 |
| Benzene | $C_6H_6$ | 78.115 | 0.1064 | 562 | 4.92 | 0.2603 |
| Bromine | $Br_2$ | 159.808 | 0.0520 | 584 | 10.34 | 0.1355 |
| *n*-Butane | $C_4H_{10}$ | 58.124 | 0.1430 | 425.2 | 3.80 | 0.2547 |
| Carbon dioxide | $CO_2$ | 44.01 | 0.1889 | 304.2 | 7.39 | 0.0943 |
| Carbon monoxide | CO | 28.011 | 0.2968 | 133 | 3.50 | 0.0930 |
| Carbon tetrachloride | $CCl_4$ | 153.82 | 0.05405 | 556.4 | 4.56 | 0.2759 |
| Chlorine | $Cl_2$ | 70.906 | 0.1173 | 417 | 7.71 | 0.1242 |
| Chloroform | $CHCl_3$ | 119.38 | 0.06964 | 536.6 | 5.47 | 0.2403 |
| Dichlorodifluoromethane (R-12) | $CCl_2F_2$ | 120.91 | 0.06876 | 384.7 | 4.01 | 0.2179 |
| Dichlorofluoromethane (R-21) | $CHCl_2F$ | 102.92 | 0.08078 | 451.7 | 5.17 | 0.1973 |
| Ethane | $C_2H_6$ | 30.070 | 0.2765 | 305.5 | 4.48 | 0.1480 |
| Ethyl alcohol | $C_2H_5OH$ | 46.07 | 0.1805 | 516 | 6.38 | 0.1673 |
| Ethylene | $C_2H_4$ | 28.054 | 0.2964 | 282.4 | 5.12 | 0.1242 |
| Helium | He | 4.003 | 2.0769 | 5.3 | 0.23 | 0.0578 |
| *n*-Hexane | $C_6H_{14}$ | 86.179 | 0.09647 | 507.9 | 3.03 | 0.3677 |
| Hydrogen (normal) | $H_2$ | 2.016 | 4.1240 | 33.3 | 1.30 | 0.0649 |
| Krypton | Kr | 83.80 | 0.09921 | 209.4 | 5.50 | 0.0924 |
| Methane | $CH_4$ | 16.043 | 0.5182 | 191.1 | 4.64 | 0.0993 |
| Methyl alcohol | $CH_3OH$ | 32.042 | 0.2595 | 513.2 | 7.95 | 0.1180 |
| Methyl chloride | $CH_3Cl$ | 50.488 | 0.1647 | 416.3 | 6.68 | 0.1430 |
| Neon | Ne | 20.183 | 0.4119 | 44.5 | 2.73 | 0.0417 |
| Nitrogen | $N_2$ | 28.013 | 0.2968 | 126.2 | 3.39 | 0.0899 |
| Nitrous oxide | $N_2O$ | 44.013 | 0.1889 | 309.7 | 7.27 | 0.0961 |
| Oxygen | $O_2$ | 31.999 | 0.2598 | 154.8 | 5.08 | 0.0780 |
| Propane | $C_3H_8$ | 44.097 | 0.1885 | 370 | 4.26 | 0.1998 |
| Propylene | $C_3H_6$ | 42.081 | 0.1976 | 365 | 4.62 | 0.1810 |
| Sulfur dioxide | $SO_2$ | 64.063 | 0.1298 | 430.7 | 7.88 | 0.1217 |
| Tetrafluoroethane (R-134a) | $CF_3CH_2F$ | 102.03 | 0.08149 | 374.3 | 4.067 | 0.1847 |
| Trichlorofluoromethane (R-11) | $CCl_3F$ | 137.37 | 0.06052 | 471.2 | 4.38 | 0.2478 |
| Water | $H_2O$ | 18.015 | 0.4615 | 647.3 | 22.09 | 0.0568 |
| Xeon | Xe | 131.30 | 0.06332 | 289.8 | 5.88 | 0.1186 |

*The unit kJ/(kg · K) is equivalent to kPa · $m^3$/(kg · K). The gas constant is calculated from $R = R_u/M$, where $R_u$ = 8.314 kJ/(kmol · K) and $M$ is the molar mass.

*Source:* K. A. Kobe and R. E. Lynn, Jr., *Chemical Review,* Vol. 52, pp. 117–236, 1953 and *ASHRAE Handbook of Fundamentals,* pp. 16.4 and 36.1, American Society of Heating, Refrigerating and Air-Conditioning Engineers, Inc., Atlanta, GA, 1993.

**TABLE A-2**
**Ideal-gas specific heats of various common gases**
($a$) At 300 K

| Gas | Formula | Gas constant $R$ kJ/(kg · K) | $C_{p0}$ kJ/(kg · K) | $C_{v0}$ kJ/(kg · K) | $k$ |
|---|---|---|---|---|---|
| Air | — | 0.2870 | 1.005 | 0.718 | 1.400 |
| Argon | Ar | 0.2081 | 0.5203 | 0.3122 | 1.667 |
| Butane | $C_4H_{10}$ | 0.1433 | 1.7164 | 1.5734 | 1.091 |
| Carbon dioxide | $CO_2$ | 0.1889 | 0.846 | 0.657 | 1.289 |
| Carbon monoxide | $CO$ | 0.2968 | 1.040 | 0.744 | 1.400 |
| Ethane | $C_2H_6$ | 0.2765 | 1.7662 | 1.4897 | 1.186 |
| Ethylene | $C_2H_4$ | 0.2964 | 1.5482 | 1.2518 | 1.237 |
| Helium | He | 2.0769 | 5.1926 | 3.1156 | 1.667 |
| Hydrogen | $H_2$ | 4.1240 | 14.307 | 10.183 | 1.405 |
| Methane | $CH_4$ | 0.5182 | 2.2537 | 1.7354 | 1.299 |
| Neon | Ne | 0.4119 | 1.0299 | 0.6179 | 1.667 |
| Nitrogen | $N_2$ | 0.2968 | 1.039 | 0.743 | 1.400 |
| Octane | $C_8H_{18}$ | 0.0729 | 1.7113 | 1.6385 | 1.044 |
| Oxygen | $O_2$ | 0.2598 | 0.918 | 0.658 | 1.395 |
| Propane | $C_3H_8$ | 0.1885 | 1.6794 | 1.4909 | 1.126 |
| Steam | $H_2O$ | 0.4615 | 1.8723 | 1.4108 | 1.327 |

*Source:* Gordon J. Van Wylen and Richard E. Sonntag, *Fundamentals of Classical Thermodynamics,* English/SI Version, 3d ed., Wiley, New York, 1986, p. 687, Table A.8SI.

**TABLE A-2**
**Ideal-gas specific heats of various common gases (*Continued*)**

(*b*) At various temperatures

| Temperature K | $C_{p0}$ kJ/(kg·K) | $C_{v0}$ kJ/(kg·K) | $k$ | $C_{p0}$ kJ/(kg·K) | $C_{v0}$ kJ/(kg·K) | $k$ | $C_{p0}$ kJ/(kg·K) | $C_{v0}$ kJ/(kg·K) | $k$ |
|---|---|---|---|---|---|---|---|---|---|
| | **Air** | | | **Carbon dioxide, $CO_2$** | | | **Carbon monoxide, CO** | | |
| 250 | 1.003 | 0.716 | 1.401 | 0.791 | 0.602 | 1.314 | 1.039 | 0.743 | 1.400 |
| 300 | 1.005 | 0.718 | 1.400 | 0.846 | 0.657 | 1.288 | 1.040 | 0.744 | 1.399 |
| 350 | 1.008 | 0.721 | 1.398 | 0.895 | 0.706 | 1.268 | 1.043 | 0.746 | 1.398 |
| 400 | 1.013 | 0.726 | 1.395 | 0.939 | 0.750 | 1.252 | 1.047 | 0.751 | 1.395 |
| 450 | 1.020 | 0.733 | 1.391 | 0.978 | 0.790 | 1.239 | 1.054 | 0.757 | 1.392 |
| 500 | 1.029 | 0.742 | 1.387 | 1.014 | 0.825 | 1.229 | 1.063 | 0.767 | 1.387 |
| 550 | 1.040 | 0.753 | 1.381 | 1.046 | 0.857 | 1.220 | 1.075 | 0.778 | 1.382 |
| 600 | 1.051 | 0.764 | 1.376 | 1.075 | 0.886 | 1.213 | 1.087 | 0.790 | 1.376 |
| 650 | 1.063 | 0.776 | 1.370 | 1.102 | 0.913 | 1.207 | 1.100 | 0.803 | 1.370 |
| 700 | 1.075 | 0.788 | 1.364 | 1.126 | 0.937 | 1.202 | 1.113 | 0.816 | 1.364 |
| 750 | 1.087 | 0.800 | 1.359 | 1.148 | 0.959 | 1.197 | 1.126 | 0.829 | 1.358 |
| 800 | 1.099 | 0.812 | 1.354 | 1.169 | 0.980 | 1.193 | 1.139 | 0.842 | 1.353 |
| 900 | 1.121 | 0.834 | 1.344 | 1.204 | 1.015 | 1.186 | 1.163 | 0.866 | 1.343 |
| 1000 | 1.142 | 0.855 | 1.336 | 1.234 | 1.045 | 1.181 | 1.185 | 0.888 | 1.335 |
| | **Hydrogen, $H_2$** | | | **Nitrogen, $N_2$** | | | **Oxygen, $O_2$** | | |
| 250 | 14.051 | 9.927 | 1.416 | 1.039 | 0.742 | 1.400 | 0.913 | 0.653 | 1.398 |
| 300 | 14.307 | 10.183 | 1.405 | 1.039 | 0.743 | 1.400 | 0.918 | 0.658 | 1.395 |
| 350 | 14.427 | 10.302 | 1.400 | 1.041 | 0.744 | 1.399 | 0.928 | 0.668 | 1.389 |
| 400 | 14.476 | 10.352 | 1.398 | 1.044 | 0.747 | 1.397 | 0.941 | 0.681 | 1.382 |
| 450 | 14.501 | 10.377 | 1.398 | 1.049 | 0.752 | 1.395 | 0.956 | 0.696 | 1.373 |
| 500 | 14.513 | 10.389 | 1.397 | 1.056 | 0.759 | 1.391 | 0.972 | 0.712 | 1.365 |
| 550 | 14.530 | 10.405 | 1.396 | 1.065 | 0.768 | 1.387 | 0.988 | 0.728 | 1.358 |
| 600 | 14.546 | 10.422 | 1.396 | 1.075 | 0.778 | 1.382 | 1.003 | 0.743 | 1.350 |
| 650 | 14.571 | 10.447 | 1.395 | 1.086 | 0.789 | 1.376 | 1.017 | 0.758 | 1.343 |
| 700 | 14.604 | 10.480 | 1.394 | 1.098 | 0.801 | 1.371 | 1.031 | 0.771 | 1.337 |
| 750 | 14.645 | 10.521 | 1.392 | 1.110 | 0.813 | 1.365 | 1.043 | 0.783 | 1.332 |
| 800 | 14.695 | 10.570 | 1.390 | 1.121 | 0.825 | 1.360 | 1.054 | 0.794 | 1.327 |
| 900 | 14.822 | 10.698 | 1.385 | 1.145 | 0.849 | 1.349 | 1.074 | 0.814 | 1.319 |
| 1000 | 14.983 | 10.859 | 1.380 | 1.167 | 0.870 | 1.341 | 1.090 | 0.830 | 1.313 |

*Source:* Kenneth Wark, *Thermodynamics,* 4th ed., McGraw-Hill, New York, 1983, p. 783, Table A-4M. Originally published in *Tables of Thermal Properties of Gases,* NBS Circ. 564, 1955.

**TABLE A-2**
**Ideal-gas specific heats of various common gases (*Continued*)**

(*c*) As a function of temperature

$$\bar{C}_{p0} = a + bT + cT^2 + dT^3$$
$$[T \text{ in K, } \bar{C}_{p0} \text{ in kJ/(kmol·K)}]$$

| Substance | Formula | $a$ | $b$ | $c$ | $d$ | Temperature range K | % error Max. | % error Avg. |
|---|---|---|---|---|---|---|---|---|
| Nitrogen | $N_2$ | 28.90 | $-0.1571 \times 10^{-2}$ | $0.8081 \times 10^{-5}$ | $-2.873 \times 10^{-9}$ | 273–1800 | 0.59 | 0.34 |
| Oxygen | $O_2$ | 25.48 | $1.520 \times 10^{-2}$ | $-0.7155 \times 10^{-5}$ | $1.312 \times 10^{-9}$ | 273–1800 | 1.19 | 0.28 |
| Air | | 28.11 | $0.1967 \times 10^{-2}$ | $0.4802 \times 10^{-5}$ | $-1.966 \times 10^{-9}$ | 273–1800 | 0.72 | 0.33 |
| Hydrogen | $H_2$ | 29.11 | $-0.1916 \times 10^{-2}$ | $0.4003 \times 10^{-5}$ | $-0.8704 \times 10^{-9}$ | 273–1800 | 1.01 | 0.26 |
| Carbon monoxide | $CO$ | 28.16 | $0.1675 \times 10^{-2}$ | $0.5372 \times 10^{-5}$ | $-2.222 \times 10^{-9}$ | 273–1800 | 0.89 | 0.37 |
| Carbon dioxide | $CO_2$ | 22.26 | $5.981 \times 10^{-2}$ | $-3.501 \times 10^{-5}$ | $7.469 \times 10^{-9}$ | 273–1800 | 0.67 | 0.22 |
| Water vapor | $H_2O$ | 32.24 | $0.1923 \times 10^{-2}$ | $1.055 \times 10^{-5}$ | $-3.595 \times 10^{-9}$ | 273–1800 | 0.53 | 0.24 |
| Nitric oxide | $NO$ | 29.34 | $-0.09395 \times 10^{-2}$ | $0.9747 \times 10^{-5}$ | $-4.187 \times 10^{-9}$ | 273–1500 | 0.97 | 0.36 |
| Nitrous oxide | $N_2O$ | 24.11 | $5.8632 \times 10^{-2}$ | $-3.562 \times 10^{-5}$ | $10.58 \times 10^{-9}$ | 273–1500 | 0.59 | 0.26 |
| Nitrogen dioxide | $NO_2$ | 22.9 | $5.715 \times 10^{-2}$ | $-3.52 \times 10^{-5}$ | $7.87 \times 10^{-9}$ | 273–1500 | 0.46 | 0.18 |
| Ammonia | $NH_3$ | 27.568 | $2.5630 \times 10^{-2}$ | $0.99072 \times 10^{-5}$ | $-6.6909 \times 10^{-9}$ | 273–1500 | 0.91 | 0.36 |
| Sulfur | $S_2$ | 27.21 | $2.218 \times 10^{-2}$ | $-1.628 \times 10^{-5}$ | $3.986 \times 10^{-9}$ | 273–1800 | 0.99 | 0.38 |
| Sulfur dioxide | $SO_2$ | 25.78 | $5.795 \times 10^{-2}$ | $-3.812 \times 10^{-5}$ | $8.612 \times 10^{-9}$ | 273–1800 | 0.45 | 0.24 |
| Sulfur trioxide | $SO_3$ | 16.40 | $14.58 \times 10^{-2}$ | $-11.20 \times 10^{-5}$ | $32.42 \times 10^{-9}$ | 273–1300 | 0.29 | 0.13 |
| Acetylene | $C_2H_2$ | 21.8 | $9.2143 \times 10^{-2}$ | $-6.527 \times 10^{-5}$ | $18.21 \times 10^{-9}$ | 273–1500 | 1.46 | 0.59 |
| Benzene | $C_6H_6$ | −36.22 | $48.475 \times 10^{-2}$ | $-31.57 \times 10^{-5}$ | $77.62 \times 10^{-9}$ | 273–1500 | 0.34 | 0.20 |
| Methanol | $CH_4O$ | 19.0 | $9.152 \times 10^{-2}$ | $-1.22 \times 10^{-5}$ | $-8.039 \times 10^{-9}$ | 273–1000 | 0.18 | 0.08 |
| Ethanol | $C_2H_6O$ | 19.9 | $20.96 \times 10^{-2}$ | $-10.38 \times 10^{-5}$ | $20.05 \times 10^{-9}$ | 273–1500 | 0.40 | 0.22 |
| Hydrogen chloride | $HCl$ | 30.33 | $-0.7620 \times 10^{-2}$ | $1.327 \times 10^{-5}$ | $-4.338 \times 10^{-9}$ | 273–1500 | 0.22 | 0.08 |
| Methane | $CH_4$ | 19.89 | $5.024 \times 10^{-2}$ | $1.269 \times 10^{-5}$ | $-11.01 \times 10^{-9}$ | 273–1500 | 1.33 | 0.57 |
| Ethane | $C_2H_6$ | 6.900 | $17.27 \times 10^{-2}$ | $-6.406 \times 10^{-5}$ | $7.285 \times 10^{-9}$ | 273–1500 | 0.83 | 0.28 |
| Propane | $C_3H_8$ | −4.04 | $30.48 \times 10^{-2}$ | $-15.72 \times 10^{-5}$ | $31.74 \times 10^{-9}$ | 273–1500 | 0.40 | 0.12 |
| *n*-Butane | $C_4H_{10}$ | 3.96 | $37.15 \times 10^{-2}$ | $-18.34 \times 10^{-5}$ | $35.00 \times 10^{-9}$ | 273–1500 | 0.54 | 0.24 |
| *i*-Butane | $C_4H_{10}$ | −7.913 | $41.60 \times 10^{-2}$ | $-23.01 \times 10^{-5}$ | $49.91 \times 10^{-9}$ | 273–1500 | 0.25 | 0.13 |
| *n*-Pentane | $C_5H_{12}$ | 6.774 | $45.43 \times 10^{-2}$ | $-22.46 \times 10^{-5}$ | $42.29 \times 10^{-9}$ | 273–1500 | 0.56 | 0.21 |
| *n*-Hexane | $C_6H_{14}$ | 6.938 | $55.22 \times 10^{-2}$ | $-28.65 \times 10^{-5}$ | $57.69 \times 10^{-9}$ | 273–1500 | 0.72 | 0.20 |
| Ethylene | $C_2H_4$ | 3.95 | $15.64 \times 10^{-2}$ | $-8.344 \times 10^{-5}$ | $17.67 \times 10^{-9}$ | 273–1500 | 0.54 | 0.13 |
| Propylene | $C_3H_6$ | 3.15 | $23.83 \times 10^{-2}$ | $-12.18 \times 10^{-5}$ | $24.62 \times 10^{-9}$ | 273–1500 | 0.73 | 0.17 |

*Source:* B. G. Kyle, *Chemical and Process Thermodynamics,* Prentice-Hall, Englewood Cliffs, NJ, 1984. Used with permission.

**TABLE A-3**
**Properties of common liquids, solids, and foods**
(*a*) Liquids

| Substance | Boiling data at 1 atm: Normal boiling point °C | Boiling data at 1 atm: Latent heat of vaporization $h_{fg}$ kJ/kg | Freezing data: Freezing point °C | Freezing data: Latent heat of fusion $h_{if}$ kJ/kg | Liquid properties: Temp. °C | Liquid properties: Density $\rho$ kg/m$^3$ | Liquid properties: Specific heat $C_p$ kJ/(kg · °C) |
|---|---|---|---|---|---|---|---|
| Ammonia | −33.3 | 1357 | −77.7 | 322.4 | −33.3 | 682 | 4.43 |
| | | | | | −20 | 665 | 4.52 |
| | | | | | 0 | 639 | 4.60 |
| | | | | | 25 | 602 | 4.80 |
| Argon | −185.9 | 161.6 | −189.3 | 28 | −185.6 | 1394 | 1.14 |
| Benzene | 80.2 | 394 | 5.5 | 126 | 20 | 879 | 1.72 |
| Brine (20% sodium chloride by mass) | 103.9 | — | −17.4 | — | 20 | 1150 | 3.11 |
| *n*-Butane | −0.5 | 385.2 | −138.5 | 80.3 | −0.5 | 601 | 2.31 |
| Carbon dioxide | −78.4* | 230.5 (at 0°C) | −56.6 | | 0 | 298 | 0.59 |
| Ethanol | 78.2 | 838.3 | −114.2 | 109 | 25 | 783 | 2.46 |
| Ethylene glycol | 198.1 | 800.1 | −10.8 | 181.1 | 20 | 1109 | 2.84 |
| Ethyl alcohol | 78.6 | 855 | −156 | 108 | 20 | 789 | 2.84 |
| Glycerine | 179.9 | 974 | 18.9 | 200.6 | 20 | 1261 | 2.32 |
| Helium | −268.9 | 22.8 | — | — | −268.9 | 146.2 | 22.8 |
| Hydrogen | −252.8 | 445.7 | −259.2 | 59.5 | −252.8 | 70.7 | 10.0 |
| Isobutane | −11.7 | 367.1 | −160 | 105.7 | −11.7 | 593.8 | 2.28 |
| Kerosene | 204–293 | 251 | −24.9 | — | 20 | 820 | 2.00 |
| Mercury | 356.7 | 294.7 | −38.9 | 11.4 | 25 | 13560 | 0.139 |
| Methane | −161.5 | 510.4 | −182.2 | 58.4 | −161.5 | 423 | 3.49 |
| | | | | | −100 | 301 | 5.79 |
| Methanol | 64.5 | 1100 | −97.7 | 99.2 | 25 | 787 | 2.55 |
| Nitrogen | −195.8 | 198.6 | −210 | 25.3 | −195.8 | 809 | 2.06 |
| | | | | | −160 | 596 | 2.97 |
| Octane | 124.8 | 306.3 | −57.5 | 180.7 | 20 | 703 | 2.10 |
| Oil (light) | | | | | 25 | 910 | 1.80 |
| Oxygen | −183 | 212.7 | −218.8 | 13.7 | −183 | 1141 | 1.71 |
| Petroleum | — | 230–384 | | | 20 | 640 | 2.0 |
| Propane | −42.1 | 427.8 | −187.7 | 80.0 | −42.1 | 581 | 2.25 |
| | | | | | 0 | 529 | 2.53 |
| | | | | | 50 | 449 | 3.13 |
| Refrigerant-134a | −26.1 | 216.8 | −96.6 | — | −50 | 1443 | 1.23 |
| | | | | | −26.1 | 1374 | 1.27 |
| | | | | | 0 | 1294 | 1.34 |
| | | | | | 25 | 1206 | 1.42 |
| Water | 100 | 2257 | 0.0 | 333.7 | 0 | 1000 | 4.23 |
| | | | | | 25 | 997 | 4.18 |
| | | | | | 50 | 988 | 4.18 |
| | | | | | 75 | 975 | 4.19 |
| | | | | | 100 | 958 | 4.22 |

* Sublimation temperature. (At pressures below the triple point pressure of 518 kPa, carbon dioxide exists as a solid or gas. Also, the freezing point temperature of carbon dioxide is the triple point temperature of −56.5°C.)

**TABLE A-3**

**Properties of common liquids, solids, and foods (*Continued*)**

(*b*) Solids (Values are for room temperature unless indicated otherwise)

| Substance | Density $\rho$ kg/m$^3$ | Specific heat $C_p$ kJ/(kg · °C) | Substance | Density $\rho$ kg/m$^3$ | Specific heat $C_p$ kJ/(kg · °C) |
|---|---|---|---|---|---|
| *Metals* | | | *Nonmetals* | | |
| Aluminum | | | Asphalt | 2110 | 0.920 |
| 200 K | | 0.797 | Brick, common | 1922 | 0.79 |
| 250 K | | 0.859 | Brick, fireclay (500°C) | 2300 | 0.960 |
| 300 K | 2,700 | 0.902 | Concrete | 2300 | 0.653 |
| 350 K | | 0.929 | Clay | 1000 | 0.920 |
| 400 K | | 0.949 | Diamond | 2420 | 0.616 |
| 450 K | | 0.973 | Glass, window | 2700 | 0.800 |
| 500 K | | 0.997 | Glass, pyrex | 2230 | 0.840 |
| Bronze (76% Cu, 2% Zn, 2% Al) | 8,280 | 0.400 | Graphite | 2500 | 0.711 |
| | | | Granite | 2700 | 1.017 |
| Brass, yellow (65% Cu, 35% Zn) | 8,310 | 0.400 | Gypsum or plaster board | 800 | 1.09 |
| Copper | | | Ice | | |
| −173°C | | 0.254 | 200 K | | 1.56 |
| −100°C | | 0.342 | 220 K | | 1.71 |
| −50°C | | 0.367 | 240 K | | 1.86 |
| 0°C | | 0.381 | 260 K | | 2.01 |
| 27°C | 8,900 | 0.386 | 273 K | 921 | 2.11 |
| 100°C | | 0.393 | Limestone | 1650 | 0.909 |
| 200°C | | 0.403 | Marble | 2600 | 0.880 |
| Iron | 7,840 | 0.45 | Plywood (Douglas Fir) | 545 | 1.21 |
| Lead | 11,310 | 0.128 | Rubber (soft) | 1100 | 1.840 |
| Magnesium | 1,730 | 1.000 | Rubber (hard) | 1150 | 2.009 |
| Nickel | 8,890 | 0.440 | Sand | 1520 | 0.800 |
| Silver | 10,470 | 0.235 | Stone | 1500 | 0.800 |
| Steel, mild | 7,830 | 0.500 | Woods, hard (maple, oak, etc) | 721 | 1.26 |
| Tungsten | 19,400 | 0.130 | Woods, soft (fir, pine, etc.) | 513 | 1.38 |

(*c*) Foods

| Food | Water content % (mass) | Freezing point °C | Specific heat kJ/(kg · °C) Above freezing | Specific heat kJ/(kg · °C) Below freezing | Latent heat of fusion kJ/kg | Food | Water content % (mass) | Freezing point °C | Specific heat kJ/(kg · °C) Above freezing | Specific heat kJ/(kg · °C) Below freezing | Latent heat of fusion kJ/kg |
|---|---|---|---|---|---|---|---|---|---|---|---|
| Apples | 84 | −1.1 | 3.65 | 1.90 | 281 | Lettuce | 95 | −0.2 | 4.02 | 2.04 | 317 |
| Bananas | 75 | −0.8 | 3.35 | 1.78 | 251 | Milk, whole | 88 | −0.6 | 3.79 | 1.95 | 294 |
| Beef round | 67 | — | 3.08 | 1.68 | 224 | Oranges | 87 | −0.8 | 3.75 | 1.94 | 291 |
| Broccoli | 90 | −0.6 | 3.86 | 1.97 | 301 | Potatoes | 78 | −0.6 | 3.45 | 1.82 | 261 |
| Butter | 16 | — | — | 1.04 | 53 | Salmon fish | 64 | −2.2 | 2.98 | 1.65 | 214 |
| Cheese, swiss | 39 | −10.0 | 2.15 | 1.33 | 130 | Shrimp | 83 | −2.2 | 3.62 | 1.89 | 277 |
| Cherries | 80 | −1.8 | 3.52 | 1.85 | 267 | Spinach | 93 | −0.3 | 3.96 | 2.01 | 311 |
| Chicken | 74 | −2.8 | 3.32 | 1.77 | 247 | Strawberries | 90 | −0.8 | 3.86 | 1.97 | 301 |
| Corn, sweet | 74 | −0.6 | 3.32 | 1.77 | 247 | Tomatoes, ripe | 94 | −0.5 | 3.99 | 2.02 | 314 |
| Eggs, whole | 74 | −0.6 | 3.32 | 1.77 | 247 | Turkey | 64 | — | 2.98 | 1.65 | 214 |
| Ice cream | 63 | −5.6 | 2.95 | 1.63 | 210 | Watermelon | 93 | −0.4 | 3.96 | 2.01 | 311 |

*Source:* Values are obtained from various handbooks and other sources or are calculated. Water content and freezing point data of foods are from *ASHRAE Handbook of Fundamentals,* SI version, Chap. 30, Table 1, American Society of Heating, Refrigerating and Air-Conditioning Engineers, Inc., Atlanta, GA, 1993. Freezing point is the temperature at which freezing starts for fruits and vegetables, and the average freezing temperature for other foods.

**TABLE A-4**
**Saturated water: temperature table**

| | | Specific volume m³/kg | | Internal energy kJ/kg | | | Enthalpy kJ/kg | | | Entropy kJ/(kg · K) | | |
|---|---|---|---|---|---|---|---|---|---|---|---|---|
| Temp. $T$ °C | Sat. press. $P_{sat}$ kPa | Sat. liquid $v_f$ | Sat. vapor $v_g$ | Sat. liquid $u_f$ | Evap. $u_{fg}$ | Sat. vapor $u_g$ | Sat. liquid $h_f$ | Evap. $h_{fg}$ | Sat. vapor $h_g$ | Sat. liquid $s_f$ | Evap. $s_{fg}$ | Sat. vapor $s_g$ |
| 0.01 | 0.6113 | 0.001000 | 206.14 | 0.0 | 2375.3 | 2375.3 | 0.01 | 2501.3 | 2501.4 | 0.000 | 9.1562 | 9.1562 |
| 5 | 0.8721 | 0.001000 | 147.12 | 20.97 | 2361.3 | 2382.3 | 20.98 | 2489.6 | 2510.6 | 0.0761 | 8.9496 | 9.0257 |
| 10 | 1.2276 | 0.001000 | 106.38 | 42.00 | 2347.2 | 2389.2 | 42.01 | 2477.7 | 2519.8 | 0.1510 | 8.7498 | 8.9008 |
| 15 | 1.7051 | 0.001001 | 77.93 | 62.99 | 2333.1 | 2396.1 | 62.99 | 2465.9 | 2528.9 | 0.2245 | 8.5569 | 8.7814 |
| 20 | 2.339 | 0.001002 | 57.79 | 83.95 | 2319.0 | 2402.9 | 83.96 | 2454.1 | 2538.1 | 0.2966 | 8.3706 | 8.6672 |
| 25 | 3.169 | 0.001003 | 43.36 | 104.88 | 2304.9 | 2409.8 | 104.89 | 2442.3 | 2547.2 | 0.3674 | 8.1905 | 8.5580 |
| 30 | 4.246 | 0.001004 | 32.89 | 125.78 | 2290.8 | 2416.6 | 125.79 | 2430.5 | 2556.3 | 0.4369 | 8.0164 | 8.4533 |
| 35 | 5.628 | 0.001006 | 25.22 | 146.67 | 2276.7 | 2423.4 | 146.68 | 2418.6 | 2565.3 | 0.5053 | 7.8478 | 8.3531 |
| 40 | 7.384 | 0.001008 | 19.52 | 167.56 | 2262.6 | 2430.1 | 167.57 | 2406.7 | 2574.3 | 0.5725 | 7.6845 | 8.2570 |
| 45 | 9.593 | 0.001010 | 15.26 | 188.44 | 2248.4 | 2436.8 | 188.45 | 2394.8 | 2583.2 | 0.6387 | 7.5261 | 8.1648 |
| 50 | 12.349 | 0.001012 | 12.03 | 209.32 | 2234.2 | 2443.5 | 209.33 | 2382.7 | 2592.1 | 0.7038 | 7.3725 | 8.0763 |
| 55 | 15.758 | 0.001015 | 9.568 | 230.21 | 2219.9 | 2450.1 | 230.23 | 2370.7 | 2600.9 | 0.7679 | 7.2234 | 7.9913 |
| 60 | 19.940 | 0.001017 | 7.671 | 251.11 | 2205.5 | 2456.6 | 251.13 | 2358.5 | 2609.6 | 0.8312 | 7.0784 | 7.9096 |
| 65 | 25.03 | 0.001020 | 6.197 | 272.02 | 2191.1 | 2463.1 | 272.06 | 2346.2 | 2618.3 | 0.8935 | 6.9375 | 7.8310 |
| 70 | 31.19 | 0.001023 | 5.042 | 292.95 | 2176.6 | 2469.6 | 292.98 | 2333.8 | 2626.8 | 0.9549 | 6.8004 | 7.7553 |
| 75 | 38.58 | 0.001026 | 4.131 | 313.90 | 2162.0 | 2475.9 | 313.93 | 2321.4 | 2635.3 | 1.0155 | 6.6669 | 7.6824 |
| 80 | 47.39 | 0.001029 | 3.407 | 334.86 | 2147.4 | 2482.2 | 334.91 | 2308.8 | 2643.7 | 1.0753 | 6.5369 | 7.6122 |
| 85 | 57.83 | 0.001033 | 2.828 | 355.84 | 2132.6 | 2488.4 | 355.90 | 2296.0 | 2651.9 | 1.1343 | 6.4102 | 7.5445 |
| 90 | 70.14 | 0.001036 | 2.361 | 376.85 | 2117.7 | 2494.5 | 376.92 | 2283.2 | 2660.1 | 1.1925 | 6.2866 | 7.4791 |
| 95 | 84.55 | 0.001040 | 1.982 | 397.88 | 2102.7 | 2500.6 | 397.96 | 2270.2 | 2668.1 | 1.2500 | 6.1659 | 7.4159 |
| | Sat. press. MPa | | | | | | | | | | | |
| 100 | 0.10133 | 0.001044 | 1.6729 | 418.94 | 2087.6 | 2506.5 | 419.04 | 2257.0 | 2676.1 | 1.3069 | 6.0480 | 7.3549 |
| 105 | 0.12082 | 0.001048 | 1.4194 | 440.02 | 2072.3 | 2512.4 | 440.15 | 2243.7 | 2683.8 | 1.3630 | 5.9328 | 7.2958 |
| 110 | 0.14327 | 0.001052 | 1.2102 | 461.14 | 2057.0 | 2518.1 | 461.30 | 2230.2 | 2691.5 | 1.4185 | 5.8202 | 7.2387 |
| 115 | 0.16906 | 0.001056 | 1.0366 | 482.30 | 2041.4 | 2523.7 | 482.48 | 2216.5 | 2699.0 | 1.4734 | 5.7100 | 7.1833 |
| 120 | 0.19853 | 0.001060 | 0.8919 | 503.50 | 2025.8 | 2529.3 | 503.71 | 2202.6 | 2706.3 | 1.5276 | 5.6020 | 7.1296 |
| 125 | 0.2321 | 0.001065 | 0.7706 | 524.74 | 2009.9 | 2534.6 | 524.99 | 2188.5 | 2713.5 | 1.5813 | 5.4962 | 7.0775 |
| 130 | 0.2701 | 0.001070 | 0.6685 | 546.02 | 1993.9 | 2539.9 | 546.31 | 2174.2 | 2720.5 | 1.6344 | 5.3925 | 7.0269 |
| 135 | 0.3130 | 0.001075 | 0.5822 | 567.35 | 1977.7 | 2545.0 | 567.69 | 2159.6 | 2727.3 | 1.6870 | 5.2907 | 6.9777 |
| 140 | 0.3613 | 0.001080 | 0.5089 | 588.74 | 1961.3 | 2550.0 | 589.13 | 2144.7 | 2733.9 | 1.7391 | 5.1908 | 6.9299 |
| 145 | 0.4154 | 0.001085 | 0.4463 | 610.18 | 1944.7 | 2554.9 | 610.63 | 2129.6 | 2740.3 | 1.7907 | 5.0926 | 6.8833 |
| 150 | 0.4758 | 0.001091 | 0.3928 | 631.68 | 1927.9 | 2559.5 | 632.20 | 2114.3 | 2746.5 | 1.8418 | 4.9960 | 6.8379 |
| 155 | 0.5431 | 0.001096 | 0.3468 | 653.24 | 1910.8 | 2564.1 | 653.84 | 2098.6 | 2752.4 | 1.8925 | 4.9010 | 6.7935 |
| 160 | 0.6178 | 0.001102 | 0.3071 | 674.87 | 1893.5 | 2568.4 | 675.55 | 2082.6 | 2758.1 | 1.9427 | 4.8075 | 6.7502 |
| 165 | 0.7005 | 0.001108 | 0.2727 | 696.56 | 1876.0 | 2572.5 | 697.34 | 2066.2 | 2763.5 | 1.9925 | 4.7153 | 6.7078 |
| 170 | 0.7917 | 0.001114 | 0.2428 | 718.33 | 1858.1 | 2576.5 | 719.21 | 2049.5 | 2768.7 | 2.0419 | 4.6244 | 6.6663 |
| 175 | 0.8920 | 0.001121 | 0.2168 | 740.17 | 1840.0 | 2580.2 | 741.17 | 2032.4 | 2773.6 | 2.0909 | 4.5347 | 6.6256 |
| 180 | 1.0021 | 0.001127 | 0.19405 | 762.09 | 1821.6 | 2583.7 | 763.22 | 2015.0 | 2778.2 | 2.1396 | 4.4461 | 6.5857 |
| 185 | 1.1227 | 0.001134 | 0.17409 | 784.10 | 1802.9 | 2587.0 | 785.37 | 1997.1 | 2782.4 | 2.1879 | 4.3586 | 6.5465 |
| 190 | 1.2544 | 0.001141 | 0.15654 | 806.19 | 1783.8 | 2590.0 | 807.62 | 1978.8 | 2786.4 | 2.2359 | 4.2720 | 6.5079 |
| 195 | 1.3978 | 0.001149 | 0.14105 | 828.37 | 1764.4 | 2592.8 | 829.98 | 1960.0 | 2790.0 | 2.2835 | 4.1863 | 6.4698 |

**TABLE A-4**
**Saturated water: temperature table (*Continued*)**

| Temp. $T$ °C | Sat. press. $P_{sat}$ MPa | Specific volume m³/kg Sat. liquid $v_f$ | Sat. vapor $v_g$ | Internal energy kJ/kg Sat. liquid $u_f$ | Evap. $u_{fg}$ | Sat. vapor $u_g$ | Enthalpy kJ/kg Sat. liquid $h_f$ | Evap. $h_{fg}$ | Sat. vapor $h_g$ | Entropy kJ/(kg · K) Sat. liquid $s_f$ | Evap. $s_{fg}$ | Sat. vapor $s_g$ |
|---|---|---|---|---|---|---|---|---|---|---|---|---|
| 200 | 1.5538 | 0.001157 | 0.12736 | 850.65 | 1744.7 | 2595.3 | 852.45 | 1940.7 | 2793.2 | 2.3309 | 4.1014 | 6.4323 |
| 205 | 1.7230 | 0.001164 | 0.11521 | 873.04 | 1724.5 | 2597.5 | 875.04 | 1921.0 | 2796.0 | 2.3780 | 4.0172 | 6.3952 |
| 210 | 1.9062 | 0.001173 | 0.10441 | 895.53 | 1703.9 | 2599.5 | 897.76 | 1900.7 | 2798.5 | 2.4248 | 3.9337 | 6.3585 |
| 215 | 2.104 | 0.001181 | 0.09479 | 918.14 | 1682.9 | 2601.1 | 920.62 | 1879.9 | 2800.5 | 2.4714 | 3.8507 | 6.3221 |
| 220 | 2.318 | 0.001190 | 0.08619 | 940.87 | 1661.5 | 2602.4 | 943.62 | 1858.5 | 2802.1 | 2.5178 | 3.7683 | 6.2861 |
| 225 | 2.548 | 0.001199 | 0.07849 | 963.73 | 1639.6 | 2603.3 | 966.78 | 1836.5 | 2803.3 | 2.5639 | 3.6863 | 6.2503 |
| 230 | 2.795 | 0.001209 | 0.07158 | 986.74 | 1617.2 | 2603.9 | 990.12 | 1813.8 | 2804.0 | 2.6099 | 3.6047 | 6.2146 |
| 235 | 3.060 | 0.001219 | 0.06537 | 1009.89 | 1594.2 | 2604.1 | 1013.62 | 1790.5 | 2804.2 | 2.6558 | 3.5233 | 6.1791 |
| 240 | 3.344 | 0.001229 | 0.05976 | 1033.21 | 1570.8 | 2604.0 | 1037.32 | 1766.5 | 2803.8 | 2.7015 | 3.4422 | 6.1437 |
| 245 | 3.648 | 0.001240 | 0.05471 | 1056.71 | 1546.7 | 2603.4 | 1061.23 | 1741.7 | 2803.0 | 2.7472 | 3.3612 | 6.1083 |
| 250 | 3.973 | 0.001251 | 0.05013 | 1080.39 | 1522.0 | 2602.4 | 1085.36 | 1716.2 | 2801.5 | 2.7927 | 3.2802 | 6.0730 |
| 255 | 4.319 | 0.001263 | 0.04598 | 1104.28 | 1596.7 | 2600.9 | 1109.73 | 1689.8 | 2799.5 | 2.8383 | 3.1992 | 6.0375 |
| 260 | 4.688 | 0.001276 | 0.04221 | 1128.39 | 1470.6 | 2599.0 | 1134.37 | 1662.5 | 2796.9 | 2.8838 | 3.1181 | 6.0019 |
| 265 | 5.081 | 0.001289 | 0.03877 | 1152.74 | 1443.9 | 2596.6 | 1159.28 | 1634.4 | 2793.6 | 2.9294 | 3.0368 | 5.9662 |
| 270 | 5.499 | 0.001302 | 0.03564 | 1177.36 | 1416.3 | 2593.7 | 1184.51 | 1605.2 | 2789.7 | 2.9751 | 2.9551 | 5.9301 |
| 275 | 5.942 | 0.001317 | 0.03279 | 1202.25 | 1387.9 | 2590.2 | 1210.07 | 1574.9 | 2785.0 | 3.0208 | 2.8730 | 5.8938 |
| 280 | 6.412 | 0.001332 | 0.03017 | 1227.46 | 1358.7 | 2586.1 | 1235.99 | 1543.6 | 2779.6 | 3.0668 | 2.7903 | 5.8571 |
| 285 | 6.909 | 0.001348 | 0.02777 | 1253.00 | 1328.4 | 2581.4 | 1262.31 | 1511.0 | 2773.3 | 3.1130 | 2.7070 | 5.8199 |
| 290 | 7.436 | 0.001366 | 0.02557 | 1278.92 | 1297.1 | 2576.0 | 1289.07 | 1477.1 | 2766.2 | 3.1594 | 2.6227 | 5.7821 |
| 295 | 7.993 | 0.001384 | 0.02354 | 1305.2 | 1264.7 | 2569.9 | 1316.3 | 1441.8 | 2758.1 | 3.2062 | 2.5375 | 5.7437 |
| 300 | 8.581 | 0.001404 | 0.02167 | 1332.0 | 1231.0 | 2563.0 | 1344.0 | 1404.9 | 2749.0 | 3.2534 | 2.4511 | 5.7045 |
| 305 | 9.202 | 0.001425 | 0.019948 | 1359.3 | 1195.9 | 2555.2 | 1372.4 | 1366.4 | 2738.7 | 3.3010 | 2.3633 | 5.6643 |
| 310 | 9.856 | 0.001447 | 0.018350 | 1387.1 | 1159.4 | 2546.4 | 1401.3 | 1326.0 | 2727.3 | 3.3493 | 2.2737 | 5.6230 |
| 315 | 10.547 | 0.001472 | 0.016867 | 1415.5 | 1121.1 | 2536.6 | 1431.0 | 1283.5 | 2714.5 | 3.3982 | 2.1821 | 5.5804 |
| 320 | 11.274 | 0.001499 | 0.015488 | 1444.6 | 1080.9 | 2525.5 | 1461.5 | 1238.6 | 2700.1 | 3.4480 | 2.0882 | 5.5362 |
| 330 | 12.845 | 0.001561 | 0.012996 | 1505.3 | 993.7 | 2498.9 | 1525.3 | 1140.6 | 2665.9 | 3.5507 | 1.8909 | 5.4417 |
| 340 | 14.586 | 0.001638 | 0.010797 | 1570.3 | 894.3 | 2464.6 | 1594.2 | 1027.9 | 2622.0 | 3.6594 | 1.6763 | 5.3357 |
| 350 | 16.513 | 0.001740 | 0.008813 | 1641.9 | 776.6 | 2418.4 | 1670.6 | 893.4 | 2563.9 | 3.7777 | 1.4335 | 5.2112 |
| 360 | 18.651 | 0.001893 | 0.006945 | 1725.2 | 626.3 | 2351.5 | 1760.5 | 720.3 | 2481.0 | 3.9147 | 1.1379 | 5.0526 |
| 370 | 21.03 | 0.002213 | 0.004925 | 1844.0 | 384.5 | 2228.5 | 1890.5 | 441.6 | 2332.1 | 4.1106 | 0.6865 | 4.7971 |
| 374.14 | 22.09 | 0.003155 | 0.003155 | 2029.6 | 0 | 2029.6 | 2099.3 | 0 | 2099.3 | 4.4298 | 0 | 4.4298 |

*Source for Tables A-4 through A-6:* Joseph H. Keenan, Frederick G. Keyes, Philip G. Hill, and Joan G. Moore, *Steam Tables,* SI Units, Wiley, New York, 1978.

**TABLE A-5**
**Saturated water: pressure table**

| Press. $P$ kPa | Sat. temp. $T_{sat}$ °C | Specific volume m³/kg — Sat. liquid $v_f$ | Sat. vapor $v_g$ | Internal energy kJ/kg — Sat. liquid $u_f$ | Evap. $u_{fg}$ | Sat. vapor $u_g$ | Enthalpy kJ/kg — Sat. liquid $h_f$ | Evap. $h_{fg}$ | Sat. vapor $h_g$ | Entropy kJ/(kg · K) — Sat. liquid $s_f$ | Evap. $s_{fg}$ | Sat. vapor $s_g$ |
|---|---|---|---|---|---|---|---|---|---|---|---|---|
| 0.6113 | 0.01 | 0.001000 | 206.14 | 0.00 | 2375.3 | 2375.3 | 0.01 | 2501.3 | 2501.4 | 0.0000 | 9.1562 | 9.1562 |
| 1.0 | 6.98 | 0.001000 | 129.21 | 29.30 | 2355.7 | 2385.0 | 29.30 | 2484.9 | 2514.2 | 0.1059 | 8.8697 | 8.9756 |
| 1.5 | 13.03 | 0.001001 | 87.98 | 54.71 | 2338.6 | 2393.3 | 54.71 | 2470.6 | 2525.3 | 0.1957 | 8.6322 | 8.8279 |
| 2.0 | 17.50 | 0.001001 | 67.00 | 73.48 | 2326.0 | 2399.5 | 73.48 | 2460.0 | 2533.5 | 0.2607 | 8.4629 | 8.7237 |
| 2.5 | 21.08 | 0.001002 | 54.25 | 88.48 | 2315.9 | 2404.4 | 88.49 | 2451.6 | 2540.0 | 0.3120 | 8.3311 | 8.6432 |
| 3.0 | 24.08 | 0.001003 | 45.67 | 101.04 | 2307.5 | 2408.5 | 101.05 | 2444.5 | 2545.5 | 0.3545 | 8.2231 | 8.5776 |
| 4.0 | 28.96 | 0.001004 | 34.80 | 121.45 | 2293.7 | 2415.2 | 121.46 | 2432.9 | 2554.4 | 0.4226 | 8.0520 | 8.4746 |
| 5.0 | 32.88 | 0.001005 | 28.19 | 137.81 | 2282.7 | 2420.5 | 137.82 | 2423.7 | 2561.5 | 0.4764 | 7.9187 | 8.3951 |
| 7.5 | 40.29 | 0.001008 | 19.24 | 168.78 | 2261.7 | 2430.5 | 168.79 | 2406.0 | 2574.8 | 0.5764 | 7.6750 | 8.2515 |
| 10 | 45.81 | 0.001010 | 14.67 | 191.82 | 2246.1 | 2437.9 | 191.83 | 2392.8 | 2584.7 | 0.6493 | 7.5009 | 8.1502 |
| 15 | 53.97 | 0.001014 | 10.02 | 225.92 | 2222.8 | 2448.7 | 225.94 | 2373.1 | 2599.1 | 0.7549 | 7.2536 | 8.0085 |
| 20 | 60.06 | 0.001017 | 7.649 | 251.38 | 2205.4 | 2456.7 | 251.40 | 2358.3 | 2609.7 | 0.8320 | 7.0766 | 7.9085 |
| 25 | 64.97 | 0.001020 | 6.204 | 271.90 | 2191.2 | 2463.1 | 271.93 | 2346.3 | 2618.2 | 0.8931 | 6.9383 | 7.8314 |
| 30 | 69.10 | 0.001022 | 5.229 | 289.20 | 2179.2 | 2468.4 | 289.23 | 2336.1 | 2625.3 | 0.9439 | 6.8247 | 7.7686 |
| 40 | 75.87 | 0.001027 | 3.993 | 317.53 | 2159.5 | 2477.0 | 317.58 | 2319.2 | 2636.8 | 1.0259 | 6.6441 | 7.6700 |
| 50 | 81.33 | 0.001030 | 3.240 | 340.44 | 2143.4 | 2483.9 | 340.49 | 2305.4 | 2645.9 | 1.0910 | 6.5029 | 7.5939 |
| 75 | 91.78 | 0.001037 | 2.217 | 384.31 | 2112.4 | 2496.7 | 384.39 | 2278.6 | 2663.0 | 1.2130 | 6.2434 | 7.4564 |
| **Press. MPa** | | | | | | | | | | | | |
| 0.100 | 99.63 | 0.001043 | 1.6940 | 417.36 | 2088.7 | 2506.1 | 417.46 | 2258.0 | 2675.5 | 1.3026 | 6.0568 | 7.3594 |
| 0.125 | 105.99 | 0.001048 | 1.3749 | 444.19 | 2069.3 | 2513.5 | 444.32 | 2241.0 | 2685.4 | 1.3740 | 5.9104 | 7.2844 |
| 0.150 | 111.37 | 0.001053 | 1.1593 | 466.94 | 2052.7 | 2519.7 | 467.11 | 2226.5 | 2693.6 | 1.4336 | 5.7897 | 7.2233 |
| 0.175 | 116.06 | 0.001057 | 1.0036 | 486.80 | 2038.1 | 2524.9 | 486.99 | 2213.6 | 2700.6 | 1.4849 | 5.6868 | 7.1717 |
| 0.200 | 120.23 | 0.001061 | 0.8857 | 504.49 | 2025.0 | 2529.5 | 504.70 | 2201.9 | 2706.7 | 1.5301 | 5.5970 | 7.1271 |
| 0.225 | 124.00 | 0.001064 | 0.7933 | 520.47 | 2013.1 | 2533.6 | 520.72 | 2191.3 | 2712.1 | 1.5706 | 5.5173 | 7.0878 |
| 0.250 | 127.44 | 0.001067 | 0.7187 | 535.10 | 2002.1 | 2537.2 | 535.37 | 2181.5 | 2716.9 | 1.6072 | 5.4455 | 7.0527 |
| 0.275 | 130.60 | 0.001070 | 0.6573 | 548.59 | 1991.9 | 2540.5 | 548.89 | 2172.4 | 2721.3 | 1.6408 | 5.3801 | 7.0209 |
| 0.300 | 133.55 | 0.001073 | 0.6058 | 561.15 | 1982.4 | 2543.6 | 561.47 | 2163.8 | 2725.3 | 1.6718 | 5.3201 | 6.9919 |
| 0.325 | 136.30 | 0.001076 | 0.5620 | 572.90 | 1973.5 | 2546.4 | 573.25 | 2155.8 | 2729.0 | 1.7006 | 5.2646 | 6.9652 |
| 0.350 | 138.88 | 0.001079 | 0.5243 | 583.95 | 1965.0 | 2548.9 | 584.33 | 2148.1 | 2732.4 | 1.7275 | 5.2130 | 6.9405 |
| 0.375 | 141.32 | 0.001081 | 0.4914 | 594.40 | 1956.9 | 2551.3 | 594.81 | 2140.8 | 2735.6 | 1.7528 | 5.1647 | 6.9175 |
| 0.40 | 143.63 | 0.001084 | 0.4625 | 604.31 | 1949.3 | 2553.6 | 604.74 | 2133.8 | 2738.6 | 1.7766 | 5.1193 | 6.8959 |
| 0.45 | 147.93 | 0.001088 | 0.4140 | 622.77 | 1934.9 | 2557.6 | 623.25 | 2120.7 | 2743.9 | 1.8207 | 5.0359 | 6.8565 |
| 0.50 | 151.86 | 0.001093 | 0.3749 | 639.68 | 1921.6 | 2561.2 | 640.23 | 2108.5 | 2748.7 | 1.8607 | 4.9606 | 6.8213 |
| 0.55 | 155.48 | 0.001097 | 0.3427 | 655.32 | 1909.2 | 2564.5 | 665.93 | 2097.0 | 2753.0 | 1.8973 | 4.8920 | 6.7893 |
| 0.60 | 158.85 | 0.001101 | 0.3157 | 669.90 | 1897.5 | 2567.4 | 670.56 | 2086.3 | 2756.8 | 1.9312 | 4.8288 | 6.7600 |
| 0.65 | 162.01 | 0.001104 | 0.2927 | 683.56 | 1886.5 | 2570.1 | 684.28 | 2076.0 | 2760.3 | 1.9627 | 4.7703 | 6.7331 |
| 0.70 | 164.97 | 0.001108 | 0.2729 | 696.44 | 1876.1 | 2572.5 | 697.22 | 2066.3 | 2763.5 | 1.9922 | 4.7158 | 6.7080 |
| 0.75 | 167.78 | 0.001112 | 0.2556 | 708.64 | 1866.1 | 2574.7 | 709.47 | 2057.0 | 2766.4 | 2.0200 | 4.6647 | 6.6847 |
| 0.80 | 170.43 | 0.001115 | 0.2404 | 720.22 | 1856.6 | 2576.8 | 721.11 | 2048.0 | 2769.1 | 2.0462 | 4.6166 | 6.6628 |
| 0.85 | 172.96 | 0.001118 | 0.2270 | 731.27 | 1847.4 | 2578.7 | 732.22 | 2039.4 | 2771.6 | 2.0710 | 4.5711 | 6.6421 |
| 0.90 | 175.38 | 0.001121 | 0.2150 | 741.83 | 1838.6 | 2580.5 | 742.83 | 2031.1 | 2773.9 | 2.0946 | 4.5280 | 6.6226 |
| 0.95 | 177.69 | 0.001124 | 0.2042 | 751.95 | 1830.2 | 2582.1 | 753.02 | 2023.1 | 2776.1 | 2.1172 | 4.4869 | 6.6041 |
| 1.00 | 179.91 | 0.001127 | 0.19444 | 761.68 | 1822.0 | 2583.6 | 762.81 | 2015.3 | 2778.1 | 2.1387 | 4.4478 | 6.5865 |
| 1.10 | 184.09 | 0.001133 | 0.17753 | 780.09 | 1806.3 | 2586.4 | 781.34 | 2000.4 | 2781.7 | 2.1792 | 4.3744 | 6.5536 |
| 1.20 | 187.99 | 0.001139 | 0.16333 | 797.29 | 1791.5 | 2588.8 | 798.65 | 1986.2 | 2784.8 | 2.2166 | 4.3067 | 6.5233 |
| 1.30 | 191.64 | 0.001144 | 0.15125 | 813.44 | 1777.5 | 2591.0 | 814.93 | 1972.7 | 2787.6 | 2.2515 | 4.2438 | 6.4953 |

**TABLE A-5**

**Saturated water: pressure table (*Continued*)**

| | | Specific volume m³/kg | | Internal energy kJ/kg | | | Enthalpy kJ/kg | | | Entropy kJ/(kg · K) | | |
|---|---|---|---|---|---|---|---|---|---|---|---|---|
| Press. $P$ MPa | Sat. temp. $T_{sat}$ °C | Sat. liquid $v_f$ | Sat. vapor $v_g$ | Sat. liquid $u_f$ | Evap. $u_{fg}$ | Sat. vapor $u_g$ | Sat. liquid $h_f$ | Evap. $h_{fg}$ | Sat. vapor $h_g$ | Sat. liquid $s_f$ | Evap. $s_{fg}$ | Sat. vapor $s_g$ |
| 1.40 | 195.07 | 0.001149 | 0.14084 | 828.70 | 1764.1 | 2592.8 | 830.30 | 1959.7 | 2790.0 | 2.2842 | 4.1850 | 6.4693 |
| 1.50 | 198.32 | 0.001154 | 0.13177 | 843.16 | 1751.3 | 2594.5 | 844.89 | 1947.3 | 2792.2 | 2.3150 | 4.1298 | 6.4448 |
| 1.75 | 205.76 | 0.001166 | 0.11349 | 876.46 | 1721.4 | 2597.8 | 878.50 | 1917.9 | 2796.4 | 2.3851 | 4.0044 | 6.3896 |
| 2.00 | 212.42 | 0.001177 | 0.09963 | 906.44 | 1693.8 | 2600.3 | 908.79 | 1890.7 | 2799.5 | 2.4474 | 3.8935 | 6.3409 |
| 2.25 | 218.45 | 0.001187 | 0.08875 | 933.83 | 1668.2 | 2602.0 | 936.49 | 1865.2 | 2801.7 | 2.5035 | 3.7937 | 6.2972 |
| 2.5 | 223.99 | 0.001197 | 0.07998 | 959.11 | 1644.0 | 2603.1 | 962.11 | 1841.0 | 2803.1 | 2.5547 | 3.7028 | 6.2575 |
| 3.0 | 233.90 | 0.001217 | 0.06668 | 1004.78 | 1599.3 | 2604.1 | 1008.42 | 1795.7 | 2804.2 | 2.6457 | 3.5412 | 6.1869 |
| 3.5 | 242.60 | 0.001235 | 0.05707 | 1045.43 | 1558.3 | 2603.7 | 1049.75 | 1753.7 | 2803.4 | 2.7253 | 3.4000 | 6.1253 |
| 4 | 250.40 | 0.001252 | 0.04978 | 1082.31 | 1520.0 | 2602.3 | 1087.31 | 1714.1 | 2801.4 | 2.7964 | 3.2737 | 6.0701 |
| 5 | 263.99 | 0.001286 | 0.03944 | 1147.81 | 1449.3 | 2597.1 | 1154.23 | 1640.1 | 2794.3 | 2.9202 | 3.0532 | 5.9734 |
| 6 | 275.64 | 0.001319 | 0.03244 | 1205.44 | 1384.3 | 2589.7 | 1213.35 | 1571.0 | 2784.3 | 3.0267 | 2.8625 | 5.8892 |
| 7 | 285.88 | 0.001351 | 0.027437 | 1257.55 | 1323.0 | 2580.5 | 1267.00 | 1505.1 | 2772.1 | 3.1211 | 2.6922 | 5.8133 |
| 8 | 295.06 | 0.001384 | 0.02352 | 1305.57 | 1264.2 | 2569.8 | 1316.64 | 1441.3 | 2758.0 | 3.2068 | 2.5364 | 5.7432 |
| 9 | 303.40 | 0.001418 | 0.02048 | 1350.51 | 1207.3 | 2557.8 | 1363.26 | 1378.9 | 2742.1 | 3.2858 | 2.3915 | 5.6722 |
| 10 | 311.06 | 0.001452 | 0.018026 | 1393.04 | 1151.4 | 2544.4 | 1407.56 | 1317.1 | 2724.7 | 3.3596 | 2.2544 | 5.6141 |
| 11 | 318.15 | 0.001489 | 0.015987 | 1433.7 | 1096.0 | 2529.8 | 1450.1 | 1255.5 | 2705.6 | 3.4295 | 2.1233 | 5.5527 |
| 12 | 324.75 | 0.001527 | 0.014263 | 1473.0 | 1040.7 | 2513.7 | 1491.3 | 1193.3 | 2684.9 | 3.4962 | 1.9962 | 5.4924 |
| 13 | 330.93 | 0.001567 | 0.012780 | 1511.1 | 985.0 | 2496.1 | 1531.5 | 1130.7 | 2662.2 | 3.5606 | 1.8718 | 5.4323 |
| 14 | 336.75 | 0.001611 | 0.011485 | 1548.6 | 928.2 | 2476.8 | 1571.1 | 1066.5 | 2637.6 | 3.6232 | 1.7485 | 5.3717 |
| 15 | 342.24 | 0.001658 | 0.010337 | 1585.6 | 869.8 | 2455.5 | 1610.5 | 1000.0 | 2610.5 | 3.6848 | 1.6249 | 5.3098 |
| 16 | 347.44 | 0.001711 | 0.009306 | 1622.7 | 809.0 | 2431.7 | 1650.1 | 930.6 | 2580.6 | 3.7461 | 1.4994 | 5.2455 |
| 17 | 352.37 | 0.001770 | 0.008364 | 1660.2 | 744.8 | 2405.0 | 1690.3 | 856.9 | 2547.2 | 3.8079 | 1.3698 | 5.1777 |
| 18 | 357.06 | 0.001840 | 0.007489 | 1698.9 | 675.4 | 2374.3 | 1732.0 | 777.1 | 2509.1 | 3.8715 | 1.2329 | 5.1044 |
| 19 | 361.54 | 0.001924 | 0.006657 | 1739.9 | 598.1 | 2338.1 | 1776.5 | 688.0 | 2464.5 | 3.9388 | 1.0839 | 5.0228 |
| 20 | 365.81 | 0.002036 | 0.005834 | 1785.6 | 507.5 | 2293.0 | 1826.3 | 583.4 | 2409.7 | 4.0139 | 0.9130 | 4.9269 |
| 21 | 369.89 | 0.002207 | 0.004952 | 1842.1 | 388.5 | 2230.6 | 1888.4 | 446.2 | 2334.6 | 4.1075 | 0.6938 | 4.8013 |
| 22 | 373.80 | 0.002742 | 0.003568 | 1961.9 | 125.2 | 2087.1 | 2022.2 | 143.4 | 2165.6 | 4.3110 | 0.2216 | 4.5327 |
| 22.09 | 374.14 | 0.003155 | 0.003155 | 2029.6 | 0 | 2029.6 | 2099.3 | 0 | 2099.3 | 4.4298 | 0 | 4.4298 |

**TABLE A-6**
**Superheated water**

| $T$ °C | $v$ m³/kg | $u$ kJ/kg | $h$ kJ/kg | $s$ kJ/(kg · K) | $v$ m³/kg | $u$ kJ/kg | $h$ kJ/kg | $s$ kJ/(kg · K) | $v$ m³/kg | $u$ kJ/kg | $h$ kJ/kg | $s$ kJ/(kg · K) |
|---|---|---|---|---|---|---|---|---|---|---|---|---|
| | $P$ = **0.01 MPa (45.81°C)*** | | | | $P$ = **0.05 MPa (81.33°C)** | | | | $P$ = **0.10 MPa (99.63°C)** | | | |
| Sat.[†] | 14.674 | 2437.9 | 2584.7 | 8.1502 | 3.240 | 2483.9 | 2645.9 | 7.5939 | 1.6940 | 2506.1 | 2675.5 | 7.3594 |
| 50 | 14.869 | 2443.9 | 2592.6 | 8.1749 | | | | | | | | |
| 100 | 17.196 | 2515.5 | 2687.5 | 8.4479 | 3.418 | 2511.6 | 2682.5 | 7.6947 | 1.6958 | 2506.7 | 2676.2 | 7.3614 |
| 150 | 19.512 | 2587.9 | 2783.0 | 8.6882 | 3.889 | 2585.6 | 2780.1 | 7.9401 | 1.9364 | 2582.8 | 2776.4 | 7.6134 |
| 200 | 21.825 | 2661.3 | 2879.5 | 8.9038 | 4.356 | 2659.9 | 2877.7 | 8.1580 | 2.172 | 2658.1 | 2875.3 | 7.8343 |
| 250 | 24.136 | 2736.0 | 2977.3 | 9.1002 | 4.820 | 2735.0 | 2976.0 | 8.3556 | 2.406 | 2733.7 | 2974.3 | 8.0333 |
| 300 | 26.445 | 2812.1 | 3076.5 | 9.2813 | 5.284 | 2811.3 | 3075.5 | 8.5373 | 2.639 | 2810.4 | 3074.3 | 8.2158 |
| 400 | 31.063 | 2968.9 | 3279.6 | 9.6077 | 6.209 | 2968.5 | 3278.9 | 8.8642 | 3.103 | 2967.9 | 3278.2 | 8.5435 |
| 500 | 35.679 | 3132.3 | 3489.1 | 9.8978 | 7.134 | 3132.0 | 3488.7 | 9.1546 | 3.565 | 3131.6 | 3488.1 | 8.8342 |
| 600 | 40.295 | 3302.5 | 3705.4 | 10.1608 | 8.057 | 3302.2 | 3705.1 | 9.4178 | 4.028 | 3301.9 | 3704.4 | 9.0976 |
| 700 | 44.911 | 3479.6 | 3928.7 | 10.4028 | 8.981 | 3479.4 | 3928.5 | 9.6599 | 4.490 | 3479.2 | 3928.2 | 9.3398 |
| 800 | 49.526 | 3663.8 | 4159.0 | 10.6281 | 9.904 | 3663.6 | 4158.9 | 9.8852 | 4.952 | 3663.5 | 4158.6 | 9.5652 |
| 900 | 54.141 | 3855.0 | 4396.4 | 10.8396 | 10.828 | 3854.9 | 4396.3 | 10.0967 | 5.414 | 3854.8 | 4396.1 | 9.7767 |
| 1000 | 58.757 | 4053.0 | 4640.6 | 11.0393 | 11.751 | 4052.9 | 4640.5 | 10.2964 | 5.875 | 4052.8 | 4640.3 | 9.9764 |
| 1100 | 63.372 | 4257.5 | 4891.2 | 11.2287 | 12.674 | 4257.4 | 4891.1 | 10.4859 | 6.337 | 4257.3 | 4891.0 | 10.1659 |
| 1200 | 67.987 | 4467.9 | 5147.8 | 11.4091 | 13.597 | 4467.8 | 5147.7 | 10.6662 | 6.799 | 4467.7 | 5147.6 | 10.3463 |
| 1300 | 72.602 | 4683.7 | 5409.7 | 11.5811 | 14.521 | 4683.6 | 5409.6 | 10.8382 | 7.260 | 4683.5 | 5409.5 | 10.5183 |
| | $P$ = **0.20 MPa (120.23°C)** | | | | $P$ = **0.30 MPa (133.55°C)** | | | | $P$ = **0.40 MPa (143.63°C)** | | | |
| Sat. | 0.8857 | 2529.5 | 2706.7 | 7.1272 | 0.6058 | 2543.6 | 2725.3 | 6.9919 | 0.4625 | 2553.6 | 2738.6 | 6.8959 |
| 150 | 0.9596 | 2576.9 | 2768.8 | 7.2795 | 0.6339 | 2570.8 | 2761.0 | 7.0778 | 0.4708 | 2564.5 | 2752.8 | 6.9299 |
| 200 | 1.0803 | 2654.4 | 2870.5 | 7.5066 | 0.7163 | 2650.7 | 2865.6 | 7.3115 | 0.5342 | 2646.8 | 2860.5 | 7.1706 |
| 250 | 1.1988 | 2731.2 | 2971.0 | 7.7086 | 0.7964 | 2728.7 | 2967.6 | 7.5166 | 0.5951 | 2726.1 | 2964.2 | 7.3789 |
| 300 | 1.3162 | 2808.6 | 3071.8 | 7.8926 | 0.8753 | 2806.7 | 3069.3 | 7.7022 | 0.6548 | 2804.8 | 3066.8 | 7.5662 |
| 400 | 1.5493 | 2966.7 | 3276.6 | 8.2218 | 1.0315 | 2965.6 | 3275.0 | 8.0330 | 0.7726 | 2964.4 | 3273.4 | 7.8985 |
| 500 | 1.7814 | 3130.8 | 3487.1 | 8.5133 | 1.1867 | 3130.0 | 3486.0 | 8.3251 | 0.8893 | 3129.2 | 3484.9 | 8.1913 |
| 600 | 2.013 | 3301.4 | 3704.0 | 8.7770 | 1.3414 | 3300.8 | 3703.2 | 8.5892 | 1.0055 | 3300.2 | 3702.4 | 8.4558 |
| 700 | 2.244 | 3478.8 | 3927.6 | 9.0194 | 1.4957 | 3478.4 | 3927.1 | 8.8319 | 1.1215 | 3477.9 | 3926.5 | 8.6987 |
| 800 | 2.475 | 3663.1 | 4158.2 | 9.2449 | 1.6499 | 3662.9 | 4157.8 | 9.0576 | 1.2372 | 3662.4 | 4157.3 | 8.9244 |
| 900 | 2.705 | 3854.5 | 4395.8 | 9.4566 | 1.8041 | 3854.2 | 4395.4 | 9.2692 | 1.3529 | 3853.9 | 4395.1 | 9.1362 |
| 1000 | 2.937 | 4052.5 | 4640.0 | 9.6563 | 1.9581 | 4052.3 | 4639.7 | 9.4690 | 1.4685 | 4052.0 | 4639.4 | 9.3360 |
| 1100 | 3.168 | 4257.0 | 4890.7 | 9.8458 | 2.1121 | 4256.8 | 4890.4 | 9.6585 | 1.5840 | 4256.5 | 4890.2 | 9.5256 |
| 1200 | 3.399 | 4467.5 | 5147.5 | 10.0262 | 2.2661 | 4467.2 | 5147.1 | 9.8389 | 1.6996 | 4467.0 | 5146.8 | 9.7060 |
| 1300 | 3.630 | 4683.2 | 5409.3 | 10.1982 | 2.4201 | 4683.0 | 5409.0 | 10.0110 | 1.8151 | 4682.8 | 5408.8 | 9.8780 |
| | $P$ = **0.50 MPa (151.86°C)** | | | | $P$ = **0.60 MPa (158.85°C)** | | | | $P$ = **0.80 MPa (170.43°C)** | | | |
| Sat. | 0.3749 | 2561.2 | 2748.7 | 6.8213 | 0.3157 | 2567.4 | 2756.8 | 6.7600 | 0.2404 | 2576.8 | 2769.1 | 6.6628 |
| 200 | 0.4249 | 2642.9 | 2855.4 | 7.0592 | 0.3520 | 2638.9 | 2850.1 | 6.9665 | 0.2608 | 2630.6 | 2839.3 | 6.8158 |
| 250 | 0.4744 | 2723.5 | 2960.7 | 7.2709 | 0.3938 | 2720.9 | 2957.2 | 7.1816 | 0.2931 | 2715.5 | 2950.0 | 7.0384 |
| 300 | 0.5226 | 2802.9 | 3064.2 | 7.4599 | 0.4344 | 2801.0 | 3061.6 | 7.3724 | 0.3241 | 2797.2 | 3056.5 | 7.2328 |
| 350 | 0.5701 | 2882.6 | 3167.7 | 7.6329 | 0.4742 | 2881.2 | 3165.7 | 7.5464 | 0.3544 | 2878.2 | 3161.7 | 7.4089 |
| 400 | 0.6173 | 2963.2 | 3271.9 | 7.7938 | 0.5137 | 2962.1 | 3270.3 | 7.7079 | 0.3843 | 2959.7 | 3267.1 | 7.5716 |
| 500 | 0.7109 | 3128.4 | 3483.9 | 8.0873 | 0.5920 | 3127.6 | 3482.8 | 8.0021 | 0.4433 | 3126.0 | 3480.6 | 7.8673 |
| 600 | 0.8041 | 3299.6 | 3701.7 | 7.3522 | 0.6697 | 3299.1 | 3700.9 | 8.2674 | 0.5018 | 3297.9 | 3699.4 | 8.1333 |
| 700 | 0.8969 | 3477.5 | 3925.9 | 8.5952 | 0.7472 | 3477.0 | 3925.3 | 8.5107 | 0.5601 | 3476.2 | 3924.2 | 8.3770 |
| 800 | 0.9896 | 3662.1 | 4156.9 | 8.8211 | 0.8245 | 3661.8 | 4156.5 | 8.7367 | 0.6181 | 3661.1 | 4155.6 | 8.6033 |
| 900 | 1.0822 | 3853.6 | 4394.7 | 9.0329 | 0.9017 | 3853.4 | 4394.4 | 8.9486 | 0.6761 | 3852.8 | 4393.7 | 8.8153 |
| 1000 | 1.1747 | 4051.8 | 4639.1 | 9.2328 | 0.9788 | 4051.5 | 4638.8 | 9.1485 | 0.7340 | 4051.0 | 4638.2 | 9.0153 |
| 1100 | 1.2672 | 4256.3 | 4889.9 | 9.4224 | 1.0559 | 4256.1 | 4889.6 | 9.3381 | 0.7919 | 4255.6 | 4889.1 | 9.2050 |
| 1200 | 1.3596 | 4466.8 | 5146.6 | 9.6029 | 1.1330 | 4466.5 | 5146.3 | 9.5185 | 0.8497 | 4466.1 | 5145.9 | 9.3855 |
| 1300 | 1.4521 | 4682.5 | 5408.6 | 9.7749 | 1.2101 | 4682.3 | 5408.3 | 9.6906 | 0.9076 | 4681.8 | 5407.9 | 9.5575 |

*The temperature in parentheses is the saturation temperature at the specified pressure.

[†]Properties of saturated vapor at the specified pressure.

**TABLE A-6**

**Superheated water (*Continued*)**

| $T$ °C | $v$ m³/kg | $u$ kJ/kg | $h$ kJ/kg | $s$ kJ/(kg·K) | $v$ m³/kg | $u$ kJ/kg | $h$ kJ/kg | $s$ kJ/(kg·K) | $v$ m³/kg | $u$ kJ/kg | $h$ kJ/kg | $s$ kJ/(kg·K) |
|---|---|---|---|---|---|---|---|---|---|---|---|---|
| | **$P$ = 1.00 MPa (179.91°C)** | | | | **$P$ = 1.20 MPa (187.99°C)** | | | | **$P$ = 1.40 MPa (195.07°C)** | | | |
| Sat. | 0.19444 | 2583.6 | 2778.1 | 6.5865 | 0.16333 | 2588.8 | 2784.8 | 6.5233 | 0.14084 | 2592.8 | 2790.0 | 6.4693 |
| 200 | 0.2060 | 2621.9 | 2827.9 | 6.6940 | 0.16930 | 2612.8 | 2815.9 | 6.5898 | 0.14302 | 2603.1 | 2803.3 | 6.4975 |
| 250 | 0.2327 | 2709.9 | 2942.6 | 6.9247 | 0.19234 | 2704.2 | 2935.0 | 6.8294 | 0.16350 | 2698.3 | 2927.2 | 6.7467 |
| 300 | 0.2579 | 2793.2 | 3051.2 | 7.1229 | 0.2138 | 2789.2 | 3045.8 | 7.0317 | 0.18228 | 2785.2 | 3040.4 | 6.9534 |
| 350 | 0.2825 | 2875.2 | 3157.7 | 7.3011 | 0.2345 | 2872.2 | 3153.6 | 7.2121 | 0.2003 | 2869.2 | 3149.5 | 7.1360 |
| 400 | 0.3066 | 2957.3 | 3263.9 | 7.4651 | 0.2548 | 2954.9 | 3260.7 | 7.3774 | 0.2178 | 2952.5 | 3257.5 | 7.3026 |
| 500 | 0.3541 | 3124.4 | 3478.5 | 7.7622 | 0.2946 | 3122.8 | 3476.3 | 7.6759 | 0.2521 | 3121.1 | 3474.1 | 7.6027 |
| 600 | 0.4011 | 3296.8 | 3697.9 | 8.0290 | 0.3339 | 3295.6 | 3696.3 | 7.9435 | 0.2860 | 3294.4 | 3694.8 | 7.8710 |
| 700 | 0.4478 | 3475.3 | 3923.1 | 8.2731 | 0.3729 | 3474.4 | 3922.0 | 8.1881 | 0.3195 | 3473.6 | 3920.8 | 8.1160 |
| 800 | 0.4943 | 3660.4 | 4154.7 | 8.4996 | 0.4118 | 3659.7 | 4153.8 | 8.4148 | 0.3528 | 3659.0 | 4153.0 | 8.3431 |
| 900 | 0.5407 | 3852.2 | 4392.9 | 8.7118 | 0.4505 | 3851.6 | 4392.2 | 8.6272 | 0.3861 | 3851.1 | 4391.5 | 8.5556 |
| 1000 | 0.5871 | 4050.5 | 4637.6 | 8.9119 | 0.4892 | 4050.0 | 4637.0 | 8.8274 | 0.4192 | 4049.5 | 4636.4 | 8.7559 |
| 1100 | 0.6335 | 4255.1 | 4888.6 | 9.1017 | 0.5278 | 4254.6 | 4888.0 | 9.0172 | 0.4524 | 4254.1 | 4887.5 | 8.9457 |
| 1200 | 0.6798 | 4465.6 | 5145.4 | 9.2822 | 0.5665 | 4465.1 | 5144.9 | 9.1977 | 0.4855 | 4464.7 | 5144.4 | 9.1262 |
| 1300 | 0.7261 | 4681.3 | 5407.4 | 9.4543 | 0.6051 | 4680.9 | 5407.0 | 9.3698 | 0.5186 | 4680.4 | 5406.5 | 9.2984 |
| | **$P$ = 1.60 MPa (201.41°C)** | | | | **$P$ = 1.80 MPa (207.15°C)** | | | | **$P$ = 2.00 MPa (212.42°C)** | | | |
| Sat. | 0.12380 | 2596.0 | 2794.0 | 6.4218 | 0.11042 | 2598.4 | 2797.1 | 6.3794 | 0.09963 | 2600.3 | 2799.5 | 6.3409 |
| 225 | 0.13287 | 2644.7 | 2857.3 | 6.5518 | 0.11673 | 2636.6 | 2846.7 | 6.4808 | 0.10377 | 2628.3 | 2835.8 | 6.4147 |
| 250 | 0.14184 | 2692.3 | 2919.2 | 6.6732 | 0.12497 | 2686.0 | 2911.0 | 6.6066 | 0.11144 | 2679.6 | 2902.5 | 6.5453 |
| 300 | 0.15862 | 2781.1 | 3034.8 | 6.8844 | 0.14021 | 2776.9 | 3029.2 | 6.8226 | 0.12547 | 2772.6 | 3023.5 | 6.7664 |
| 350 | 0.17456 | 2866.1 | 3145.4 | 7.0694 | 0.15457 | 2863.0 | 3141.2 | 7.0100 | 0.13857 | 2859.8 | 3137.0 | 6.9563 |
| 400 | 0.19005 | 2950.1 | 3254.2 | 7.2374 | 0.16847 | 2947.7 | 3250.9 | 7.1794 | 0.15120 | 2945.2 | 3247.6 | 7.1271 |
| 500 | 0.2203 | 3119.5 | 3472.0 | 7.5390 | 0.19550 | 3117.9 | 3469.8 | 7.4825 | 0.17568 | 3116.2 | 3467.6 | 7.4317 |
| 600 | 0.2500 | 3293.3 | 3693.2 | 7.8080 | 0.2220 | 3292.1 | 3691.7 | 7.7523 | 0.19960 | 3290.9 | 3690.1 | 7.7024 |
| 700 | 0.2794 | 3472.7 | 3919.7 | 8.0535 | 0.2482 | 3471.8 | 3918.5 | 7.9983 | 0.2232 | 3470.9 | 3917.4 | 7.9487 |
| 800 | 0.3086 | 3658.3 | 4152.1 | 8.2808 | 0.2742 | 3657.6 | 4151.2 | 8.2258 | 0.2467 | 3657.0 | 4150.3 | 8.1765 |
| 900 | 0.3377 | 3850.5 | 4390.8 | 8.4935 | 0.3001 | 3849.9 | 4390.1 | 8.4386 | 0.2700 | 3849.3 | 4389.4 | 8.3895 |
| 1000 | 0.3668 | 4049.0 | 4635.8 | 8.6938 | 0.3260 | 4048.5 | 4635.2 | 8.6391 | 0.2933 | 4048.0 | 4634.6 | 8.5901 |
| 1100 | 0.3958 | 4253.7 | 4887.0 | 8.8837 | 0.3518 | 4253.2 | 4886.4 | 8.8290 | 0.3166 | 4252.7 | 4885.9 | 8.7800 |
| 1200 | 0.4248 | 4464.2 | 5143.9 | 9.0643 | 0.3776 | 4463.7 | 5143.4 | 9.0096 | 0.3398 | 4463.3 | 5142.9 | 8.9607 |
| 1300 | 0.4538 | 4679.9 | 5406.0 | 9.2364 | 0.4034 | 4679.5 | 5405.6 | 9.1818 | 0.3631 | 4679.0 | 5405.1 | 9.1329 |
| | **$P$ = 2.50 MPa (223.99°C)** | | | | **$P$ = 3.00 MPa (233.90°C)** | | | | **$P$ = 3.50 MPa (242.60°C)** | | | |
| Sat. | 0.07998 | 2603.1 | 2803.1 | 6.2575 | 0.06668 | 2604.1 | 2804.2 | 6.1869 | 0.05707 | 2603.7 | 2803.4 | 6.1253 |
| 225 | 0.08027 | 2605.6 | 2806.3 | 6.2639 | | | | | | | | |
| 250 | 0.08700 | 2662.6 | 2880.1 | 6.4085 | 0.07058 | 2644.0 | 2855.8 | 6.2872 | 0.05872 | 2623.7 | 2829.2 | 6.1749 |
| 300 | 0.09890 | 2761.6 | 3008.8 | 6.6438 | 0.08114 | 2750.1 | 2993.5 | 6.5390 | 0.06842 | 2738.0 | 2977.5 | 6.4461 |
| 350 | 0.10976 | 2851.9 | 3126.3 | 6.8403 | 0.09053 | 2843.7 | 3115.3 | 6.7428 | 0.07678 | 2835.3 | 3104.0 | 6.6579 |
| 400 | 0.12010 | 2939.1 | 3239.3 | 7.0148 | 0.09936 | 2932.8 | 3230.9 | 6.9212 | 0.08453 | 2926.4 | 3222.3 | 6.8405 |
| 450 | 0.13014 | 3025.5 | 3350.8 | 7.1746 | 0.10787 | 3020.4 | 3344.0 | 7.0834 | 0.09196 | 3015.3 | 3337.2 | 7.0052 |
| 500 | 0.13993 | 3112.1 | 3462.1 | 7.3234 | 0.11619 | 3108.0 | 3456.5 | 7.2338 | 0.09918 | 3103.0 | 3450.9 | 7.1572 |
| 600 | 0.15930 | 3288.0 | 3686.3 | 7.5960 | 0.13243 | 3285.0 | 3682.3 | 7.5085 | 0.11324 | 3282.1 | 3678.4 | 7.4339 |
| 700 | 0.17832 | 3468.7 | 3914.5 | 7.8435 | 0.14838 | 3466.5 | 3911.7 | 7.7571 | 0.12699 | 3464.3 | 3908.8 | 7.6837 |
| 800 | 0.19716 | 3655.3 | 4148.2 | 8.0720 | 0.16414 | 3653.5 | 4145.9 | 7.9862 | 0.14056 | 3651.8 | 4143.7 | 7.9134 |
| 900 | 0.21590 | 3847.9 | 4387.6 | 8.2853 | 0.17980 | 3846.5 | 4385.9 | 8.1999 | 0.15402 | 3845.0 | 4384.1 | 8.1276 |
| 1000 | 0.2346 | 4046.7 | 4633.1 | 8.4861 | 0.19541 | 4045.4 | 4631.6 | 8.4009 | 0.16743 | 4044.1 | 4630.1 | 8.3288 |
| 1100 | 0.2532 | 4251.5 | 4884.6 | 8.6762 | 0.21098 | 4250.3 | 4883.3 | 8.5912 | 0.18080 | 4249.2 | 4881.9 | 8.5192 |
| 1200 | 0.2718 | 4462.1 | 5141.7 | 8.8569 | 0.22652 | 4460.9 | 5140.5 | 8.7720 | 0.19415 | 4459.8 | 5139.3 | 8.7000 |
| 1300 | 0.2905 | 4677.8 | 5404.0 | 9.0291 | 0.24206 | 4676.6 | 5402.8 | 8.9442 | 0.20749 | 4675.5 | 5401.7 | 8.8723 |

**TABLE A-6**

**Superheated water (*Continued*)**

| $T$ °C | $v$ $m^3/kg$ | $u$ kJ/kg | $h$ kJ/kg | $s$ kJ/(kg · K) | $v$ $m^3/kg$ | $u$ kJ/kg | $h$ kJ/kg | $s$ kJ/(kg · K) | $v$ $m^3/kg$ | $u$ kJ/kg | $h$ kJ/kg | $s$ kJ/(kg · K) |
|---|---|---|---|---|---|---|---|---|---|---|---|---|
| | $P$ = **4.0 MPa (250.40°C)** | | | | $P$ = **4.5 MPa (257.49°C)** | | | | $P$ = **5.0 MPa (263.99°C)** | | | |
| Sat. | 0.04978 | 2602.3 | 2801.4 | 6.0701 | 0.04406 | 2600.1 | 2798.3 | 6.0198 | 0.03944 | 2597.1 | 2794.3 | 5.9734 |
| 275 | 0.05457 | 2667.9 | 2886.2 | 6.2285 | 0.04730 | 2650.3 | 2863.2 | 6.1401 | 0.04141 | 2631.3 | 2838.3 | 6.0544 |
| 300 | 0.05884 | 2725.3 | 2960.7 | 6.3615 | 0.05135 | 2712.0 | 2943.1 | 6.2828 | 0.04532 | 2698.0 | 2924.5 | 6.2084 |
| 350 | 0.06645 | 2826.7 | 3092.5 | 6.5821 | 0.05840 | 2817.8 | 3080.6 | 6.5131 | 0.05194 | 2808.7 | 3068.4 | 6.4493 |
| 400 | 0.07341 | 2919.9 | 3213.6 | 6.7690 | 0.06475 | 2913.3 | 3204.7 | 6.7047 | 0.05781 | 2906.6 | 3195.7 | 6.6459 |
| 450 | 0.08002 | 3010.2 | 3330.3 | 6.9363 | 0.07074 | 3005.0 | 3323.3 | 6.8746 | 0.06330 | 2999.7 | 3316.2 | 6.8186 |
| 500 | 0.08643 | 3099.5 | 3445.3 | 7.0901 | 0.07651 | 3095.3 | 3439.6 | 7.0301 | 0.06857 | 3091.0 | 3433.8 | 6.9759 |
| 600 | 0.09885 | 3279.1 | 3674.4 | 7.3688 | 0.08765 | 3276.0 | 3670.5 | 7.3110 | 0.07869 | 3273.0 | 3666.5 | 7.2589 |
| 700 | 0.11095 | 3462.1 | 3905.9 | 7.6198 | 0.09847 | 3459.9 | 3903.0 | 7.5631 | 0.08849 | 3457.6 | 3900.1 | 7.5122 |
| 800 | 0.12287 | 3650.0 | 4141.5 | 7.8502 | 0.10911 | 3648.3 | 4139.3 | 7.7942 | 0.09811 | 3646.6 | 4137.1 | 7.7440 |
| 900 | 0.13469 | 3843.6 | 4382.3 | 8.0647 | 0.11965 | 3842.2 | 4380.6 | 8.0091 | 0.10762 | 3840.7 | 4378.8 | 7.9593 |
| 1000 | 0.14645 | 4042.9 | 4628.7 | 8.2662 | 0.13013 | 4041.6 | 4627.2 | 8.2108 | 0.11707 | 4040.4 | 4625.7 | 8.1612 |
| 1100 | 0.15817 | 4248.0 | 4880.6 | 8.4567 | 0.14056 | 4246.8 | 4879.3 | 8.4015 | 0.12648 | 4245.6 | 4878.0 | 8.3520 |
| 1200 | 0.16987 | 4458.6 | 5138.1 | 8.6376 | 0.15098 | 4457.5 | 5136.9 | 8.5825 | 0.13587 | 4456.3 | 5135.7 | 8.5331 |
| 1300 | 0.18156 | 4674.3 | 5400.5 | 8.8100 | 0.16139 | 4673.1 | 5399.4 | 8.7549 | 0.14526 | 4672.0 | 5398.2 | 8.7055 |
| | $P$ = **6.0 MPa (275.64°C)** | | | | $P$ = **7.0 MPa (285.88°C)** | | | | $P$ = **8.0 MPa (295.06°C)** | | | |
| Sat. | 0.03244 | 2589.7 | 2784.3 | 5.8892 | 0.02737 | 2580.5 | 2772.1 | 5.8133 | 0.02352 | 2569.8 | 2758.0 | 5.7432 |
| 300 | 0.03616 | 2667.2 | 2884.2 | 6.0674 | 0.02947 | 2632.2 | 2838.4 | 5.9305 | 0.02426 | 2590.9 | 2785.0 | 5.7906 |
| 350 | 0.04223 | 2789.6 | 3043.0 | 6.3335 | 0.03524 | 2769.4 | 3016.0 | 6.2283 | 0.02995 | 2747.7 | 2987.3 | 6.1301 |
| 400 | 0.04739 | 2892.9 | 3177.2 | 6.5408 | 0.03993 | 2878.6 | 3158.1 | 6.4478 | 0.03432 | 2863.8 | 3138.3 | 6.3634 |
| 450 | 0.05214 | 2988.9 | 3301.8 | 6.7193 | 0.04416 | 2978.0 | 3287.1 | 6.6327 | 0.03817 | 2966.7 | 3272.0 | 6.5551 |
| 500 | 0.05665 | 3082.2 | 3422.2 | 6.8803 | 0.04814 | 3073.4 | 3410.3 | 6.7975 | 0.04175 | 3064.3 | 3398.3 | 6.7240 |
| 550 | 0.06101 | 3174.6 | 3540.6 | 7.0288 | 0.05195 | 3167.2 | 3530.9 | 6.9486 | 0.04516 | 3159.8 | 3521.0 | 6.8778 |
| 600 | 0.06525 | 3266.9 | 3658.4 | 7.1677 | 0.05565 | 3260.7 | 3650.3 | 7.0894 | 0.04845 | 3254.4 | 3642.0 | 7.0206 |
| 700 | 0.07352 | 3453.1 | 3894.2 | 7.4234 | 0.06283 | 3448.5 | 3888.3 | 7.3476 | 0.05481 | 3443.9 | 3882.4 | 7.2812 |
| 800 | 0.08160 | 3643.1 | 4132.7 | 7.6566 | 0.06981 | 3639.5 | 4128.2 | 7.5822 | 0.06097 | 3636.0 | 4123.8 | 7.5173 |
| 900 | 0.08958 | 3837.8 | 4375.3 | 7.8727 | 0.07669 | 3835.0 | 4371.8 | 7.7991 | 0.06702 | 3832.1 | 4368.3 | 7.7351 |
| 1000 | 0.09749 | 4037.8 | 4622.7 | 8.0751 | 0.08350 | 4035.3 | 4619.8 | 8.0020 | 0.07301 | 4032.8 | 4616.9 | 7.9384 |
| 1100 | 0.10536 | 4243.3 | 4875.4 | 8.2661 | 0.09027 | 4240.9 | 4872.8 | 8.1933 | 0.07896 | 4238.6 | 4870.3 | 8.1300 |
| 1200 | 0.11321 | 4454.0 | 5133.3 | 8.4474 | 0.09703 | 4451.7 | 5130.9 | 8.3747 | 0.08489 | 4449.5 | 5128.5 | 8.3115 |
| 1300 | 0.12106 | 4669.6 | 5396.0 | 8.6199 | 0.10377 | 4667.3 | 5393.7 | 8.5475 | 0.09080 | 4665.0 | 5391.5 | 8.4842 |
| | $P$ = **9.0 MPa (303.40°C)** | | | | $P$ = **10.0 MPa (311.06°C)** | | | | $P$ = **12.5 MPa (327.89°C)** | | | |
| Sat. | 0.02048 | 2557.8 | 2742.1 | 5.6772 | 0.018026 | 2544.4 | 2724.7 | 5.6141 | 0.013495 | 2505.1 | 2673.8 | 5.4624 |
| 325 | 0.02327 | 2646.6 | 2856.0 | 5.8712 | 0.019861 | 2610.4 | 2809.1 | 5.7568 | | | | |
| 350 | 0.02580 | 2724.4 | 2956.6 | 6.0361 | 0.02242 | 2699.2 | 2923.4 | 5.9443 | 0.016126 | 2624.6 | 2826.2 | 5.7118 |
| 400 | 0.02993 | 2848.4 | 3117.8 | 6.2854 | 0.02641 | 2832.4 | 3096.5 | 6.2120 | 0.02000 | 2789.3 | 3039.3 | 6.0417 |
| 450 | 0.03350 | 2955.2 | 3256.6 | 6.4844 | 0.02975 | 2943.4 | 3240.9 | 6.4190 | 0.02299 | 2912.5 | 3199.8 | 6.2719 |
| 500 | 0.03677 | 3055.2 | 3386.1 | 6.6576 | 0.03279 | 3045.8 | 3373.7 | 6.5966 | 0.02560 | 3021.7 | 3341.8 | 6.4618 |
| 550 | 0.03987 | 3152.2 | 3511.0 | 6.8142 | 0.03564 | 3144.6 | 3500.9 | 6.7561 | 0.02801 | 3125.0 | 3475.2 | 6.6290 |
| 600 | 0.04285 | 3248.1 | 3633.7 | 6.9589 | 0.03837 | 3241.7 | 3625.3 | 6.9029 | 0.03029 | 3225.4 | 3604.0 | 6.7810 |
| 650 | 0.04574 | 3343.6 | 3755.3 | 7.0943 | 0.04101 | 3338.2 | 3748.2 | 7.0398 | 0.03248 | 3324.4 | 3730.4 | 6.9218 |
| 700 | 0.04857 | 3439.3 | 3876.5 | 7.2221 | 0.04358 | 3434.7 | 3870.5 | 7.1687 | 0.03460 | 3422.9 | 3855.3 | 7.0536 |
| 800 | 0.05409 | 3632.5 | 4119.3 | 7.4596 | 0.04859 | 3628.9 | 4114.8 | 7.4077 | 0.03869 | 3620.0 | 4103.6 | 7.2965 |
| 900 | 0.05950 | 3829.2 | 4364.8 | 7.6783 | 0.05349 | 3826.3 | 4361.2 | 7.6272 | 0.04267 | 3819.1 | 4352.5 | 7.5182 |
| 1000 | 0.06485 | 4030.3 | 4614.0 | 7.8821 | 0.05832 | 4027.8 | 4611.0 | 7.8315 | 0.04658 | 4021.6 | 4603.8 | 7.7237 |
| 1100 | 0.07016 | 4236.3 | 4867.7 | 8.0740 | 0.06312 | 4234.0 | 4865.1 | 8.0237 | 0.05045 | 4228.2 | 4858.8 | 7.9165 |
| 1200 | 0.07544 | 4447.2 | 5126.2 | 8.2556 | 0.06789 | 4444.9 | 5123.8 | 8.2055 | 0.05430 | 4439.3 | 5118.0 | 8.0937 |
| 1300 | 0.08072 | 4662.7 | 5389.2 | 8.4284 | 0.07265 | 4460.5 | 5387.0 | 8.3783 | 0.05813 | 4654.8 | 5381.4 | 8.2717 |

**TABLE A-6**
**Superheated water (*Continued*)**

| $T$ °C | $v$ m³/kg | $u$ kJ/kg | $h$ kJ/kg | $s$ kJ/(kg · K) | $v$ m³/kg | $u$ kJ/kg | $h$ kJ/kg | $s$ kJ/(kg · K) | $v$ m³/kg | $u$ kJ/kg | $h$ kJ/kg | $s$ kJ/(kg · K) |
|---|---|---|---|---|---|---|---|---|---|---|---|---|
| | $P$ = **15.0 MPa (342.24°C)** | | | | $P$ = **17.5 MPa (354.75°C)** | | | | $P$ = **20.0 MPa (365.81°C)** | | | |
| Sat. | 0.010337 | 2455.5 | 2610.5 | 5.3098 | 0.007920 | 2390.2 | 2528.8 | 5.1419 | 0.005834 | 2293.0 | 2409.7 | 4.9269 |
| 350 | 0.011470 | 2520.4 | 2692.4 | 5.4421 | | | | | | | | |
| 400 | 0.015649 | 2740.7 | 2975.5 | 5.8811 | 0.012447 | 2685.0 | 2902.9 | 5.7213 | 0.009942 | 2619.3 | 2818.1 | 5.5540 |
| 450 | 0.018445 | 2879.5 | 3156.2 | 6.1404 | 0.015174 | 2844.2 | 3109.7 | 6.0184 | 0.012695 | 2806.2 | 3060.1 | 5.9017 |
| 500 | 0.02080 | 2996.6 | 3308.6 | 6.3443 | 0.017358 | 2970.3 | 3274.1 | 6.2383 | 0.014768 | 2942.9 | 3238.2 | 6.1401 |
| 550 | 0.02293 | 3104.7 | 3448.6 | 6.5199 | 0.019288 | 3083.9 | 3421.4 | 6.4230 | 0.016555 | 3062.4 | 3393.5 | 6.3348 |
| 600 | 0.02491 | 3208.6 | 3582.3 | 6.6776 | 0.02106 | 3191.5 | 3560.1 | 6.5866 | 0.018178 | 3174.0 | 3537.6 | 6.5048 |
| 650 | 0.02680 | 3310.3 | 3712.3 | 6.8224 | 0.02274 | 3296.0 | 3693.9 | 6.7357 | 0.019693 | 3281.4 | 3675.3 | 6.6582 |
| 700 | 0.02861 | 3410.9 | 3840.1 | 6.9572 | 0.02434 | 3398.7 | 3824.6 | 6.8736 | 0.02113 | 3386.4 | 3809.0 | 6.7993 |
| 800 | 0.03210 | 3610.9 | 4092.4 | 7.2040 | 0.02738 | 3601.8 | 4081.1 | 7.1244 | 0.02385 | 3592.7 | 4069.7 | 7.0544 |
| 900 | 0.03546 | 3811.9 | 4343.8 | 7.4279 | 0.03031 | 3804.7 | 4335.1 | 7.3507 | 0.02645 | 3797.5 | 4326.4 | 7.2830 |
| 1000 | 0.03875 | 4015.4 | 4596.6 | 7.6348 | 0.03316 | 4009.3 | 4589.5 | 7.5589 | 0.02897 | 4003.1 | 4582.5 | 7.4925 |
| 1100 | 0.04200 | 4222.6 | 4852.6 | 7.8283 | 0.03597 | 4216.9 | 4846.4 | 7.7531 | 0.03145 | 4211.3 | 4840.2 | 7.6874 |
| 1200 | 0.04523 | 4433.8 | 5112.3 | 8.0108 | 0.03876 | 4428.3 | 5106.6 | 7.9360 | 0.03391 | 4422.8 | 5101.0 | 7.8707 |
| 1300 | 0.04845 | 4649.1 | 5376.0 | 8.1840 | 0.04154 | 4643.5 | 5370.5 | 8.1093 | 0.03636 | 4638.0 | 5365.1 | 8.0442 |
| | $P$ = **25.0 MPa** | | | | $P$ = **30.0 MPa** | | | | $P$ = **35.0 MPa** | | | |
| 375 | 0.0019731 | 1798.7 | 1848.0 | 4.0320 | 0.0017892 | 1737.8 | 1791.5 | 3.9305 | 0.0017003 | 1702.9 | 1762.4 | 3.8722 |
| 400 | 0.006004 | 2430.1 | 2580.2 | 5.1418 | 0.002790 | 2067.4 | 2151.1 | 4.4728 | 0.002100 | 1914.1 | 1987.6 | 4.2126 |
| 425 | 0.007881 | 2609.2 | 2806.3 | 5.4723 | 0.005303 | 2455.1 | 2614.2 | 5.1504 | 0.003428 | 2253.4 | 2373.4 | 4.7747 |
| 450 | 0.009162 | 2720.7 | 2949.7 | 5.6744 | 0.006735 | 2619.3 | 2821.4 | 5.4424 | 0.004961 | 2498.7 | 2672.4 | 5.1962 |
| 500 | 0.011123 | 2884.3 | 3162.4 | 5.9592 | 0.008678 | 2820.7 | 3081.1 | 5.7905 | 0.006927 | 2751.9 | 2994.4 | 5.6282 |
| 550 | 0.012724 | 3017.5 | 3335.6 | 6.1765 | 0.010168 | 2970.3 | 3275.4 | 6.0342 | 0.008345 | 2921.0 | 3213.0 | 5.9026 |
| 600 | 0.014137 | 3137.9 | 3491.4 | 6.3602 | 0.011446 | 3100.5 | 3443.9 | 6.2331 | 0.009527 | 3062.0 | 3395.5 | 6.1179 |
| 650 | 0.015433 | 3251.6 | 3637.4 | 6.5229 | 0.012596 | 3221.0 | 3598.9 | 6.4058 | 0.010575 | 3189.8 | 3559.9 | 6.3010 |
| 700 | 0.016646 | 3361.3 | 3777.5 | 6.6707 | 0.013661 | 3335.8 | 3745.6 | 6.5606 | 0.011533 | 3309.8 | 3713.5 | 6.4631 |
| 800 | 0.018912 | 3574.3 | 4047.1 | 6.9345 | 0.015623 | 3555.5 | 4024.2 | 6.8332 | 0.013278 | 3536.7 | 4001.5 | 6.7450 |
| 900 | 0.021045 | 3783.0 | 4309.1 | 7.1680 | 0.017448 | 3768.5 | 4291.9 | 7.0718 | 0.014883 | 3754.0 | 4274.9 | 6.9386 |
| 1000 | 0.02310 | 3990.9 | 4568.5 | 7.3802 | 0.019196 | 3978.8 | 4554.7 | 7.2867 | 0.016410 | 3966.7 | 4541.1 | 7.2064 |
| 1100 | 0.02512 | 4200.2 | 4828.2 | 7.5765 | 0.020903 | 4189.2 | 4816.3 | 7.4845 | 0.017895 | 4178.3 | 4804.6 | 7.4037 |
| 1200 | 0.02711 | 4412.0 | 5089.9 | 7.7605 | 0.022589 | 4401.3 | 5079.0 | 7.6692 | 0.019360 | 4390.7 | 5068.3 | 7.5910 |
| 1300 | 0.02910 | 4626.9 | 5354.4 | 7.9342 | 0.024266 | 4616.0 | 5344.0 | 7.8432 | 0.020815 | 4605.1 | 5333.6 | 7.7653 |
| | $P$ = **40.0 MPa** | | | | $P$ = **50.0 MPa** | | | | $P$ = **60.0 MPa** | | | |
| 375 | 0.0016407 | 1677.1 | 1742.8 | 3.8290 | 0.0015594 | 1638.6 | 1716.6 | 3.7639 | 0.0015028 | 1609.4 | 1699.5 | 3.7141 |
| 400 | 0.0019077 | 1854.6 | 1930.9 | 4.1135 | 0.0017309 | 1788.1 | 1874.6 | 4.0031 | 0.0016335 | 1745.4 | 1843.4 | 3.9318 |
| 425 | 0.002532 | 2096.9 | 2198.1 | 4.5029 | 0.002007 | 1959.7 | 2060.0 | 4.2734 | 0.0018165 | 1892.7 | 2001.7 | 4.1626 |
| 450 | 0.003693 | 2365.1 | 2512.8 | 4.9459 | 0.002486 | 2159.6 | 2284.0 | 4.5884 | 0.002085 | 2053.9 | 2179.0 | 4.4121 |
| 500 | 0.005622 | 2678.4 | 2903.3 | 5.4700 | 0.003892 | 2525.5 | 2720.1 | 5.1726 | 0.002956 | 2390.6 | 2567.9 | 4.9321 |
| 550 | 0.006984 | 2869.7 | 3149.1 | 5.7785 | 0.005118 | 2763.6 | 3019.5 | 5.5485 | 0.003956 | 2658.8 | 2896.2 | 5.3441 |
| 600 | 0.008094 | 3022.6 | 3346.4 | 6.0144 | 0.006112 | 2942.0 | 3247.6 | 5.8178 | 0.004834 | 2861.1 | 3151.2 | 5.6452 |
| 650 | 0.009063 | 3158.0 | 3520.6 | 6.2054 | 0.006966 | 3093.5 | 3441.8 | 6.0342 | 0.005595 | 3028.8 | 3364.5 | 5.8829 |
| 700 | 0.009941 | 3283.6 | 3681.2 | 6.3750 | 0.007727 | 3230.5 | 3616.8 | 6.2189 | 0.006272 | 3177.2 | 3553.5 | 6.0824 |
| 800 | 0.011523 | 3517.8 | 3978.7 | 6.6662 | 0.009076 | 3479.8 | 3933.6 | 6.5290 | 0.007459 | 3441.5 | 3889.1 | 6.4109 |
| 900 | 0.012962 | 3739.4 | 4257.9 | 6.9150 | 0.010283 | 3710.3 | 4224.4 | 6.7882 | 0.008508 | 3681.0 | 4191.5 | 6.6805 |
| 1000 | 0.014324 | 3954.6 | 4527.6 | 7.1356 | 0.011411 | 3930.5 | 4501.1 | 7.0146 | 0.009480 | 3906.4 | 4475.2 | 6.9127 |
| 1100 | 0.015642 | 4167.4 | 4793.1 | 7.3364 | 0.012496 | 4145.7 | 4770.5 | 7.2184 | 0.010409 | 4124.1 | 4748.6 | 7.1195 |
| 1200 | 0.016940 | 4380.1 | 5057.7 | 7.5224 | 0.013561 | 4359.1 | 5037.2 | 7.4058 | 0.011317 | 4338.2 | 5017.2 | 7.3083 |
| 1300 | 0.018229 | 4594.3 | 5323.5 | 7.6969 | 0.014616 | 4572.8 | 5303.6 | 7.5808 | 0.012215 | 4551.4 | 5284.3 | 7.4837 |

**FIGURE A-7**

*T-s* diagram for water. (*Source*: Lester Haar, John S. Gallagher, and George S. Kell, *NBS/NRC Steam Tables*, Fig. 9, pp. 256–257, 1984, Hemisphere Publishing Corporation, New York. Reproduced with permission.)

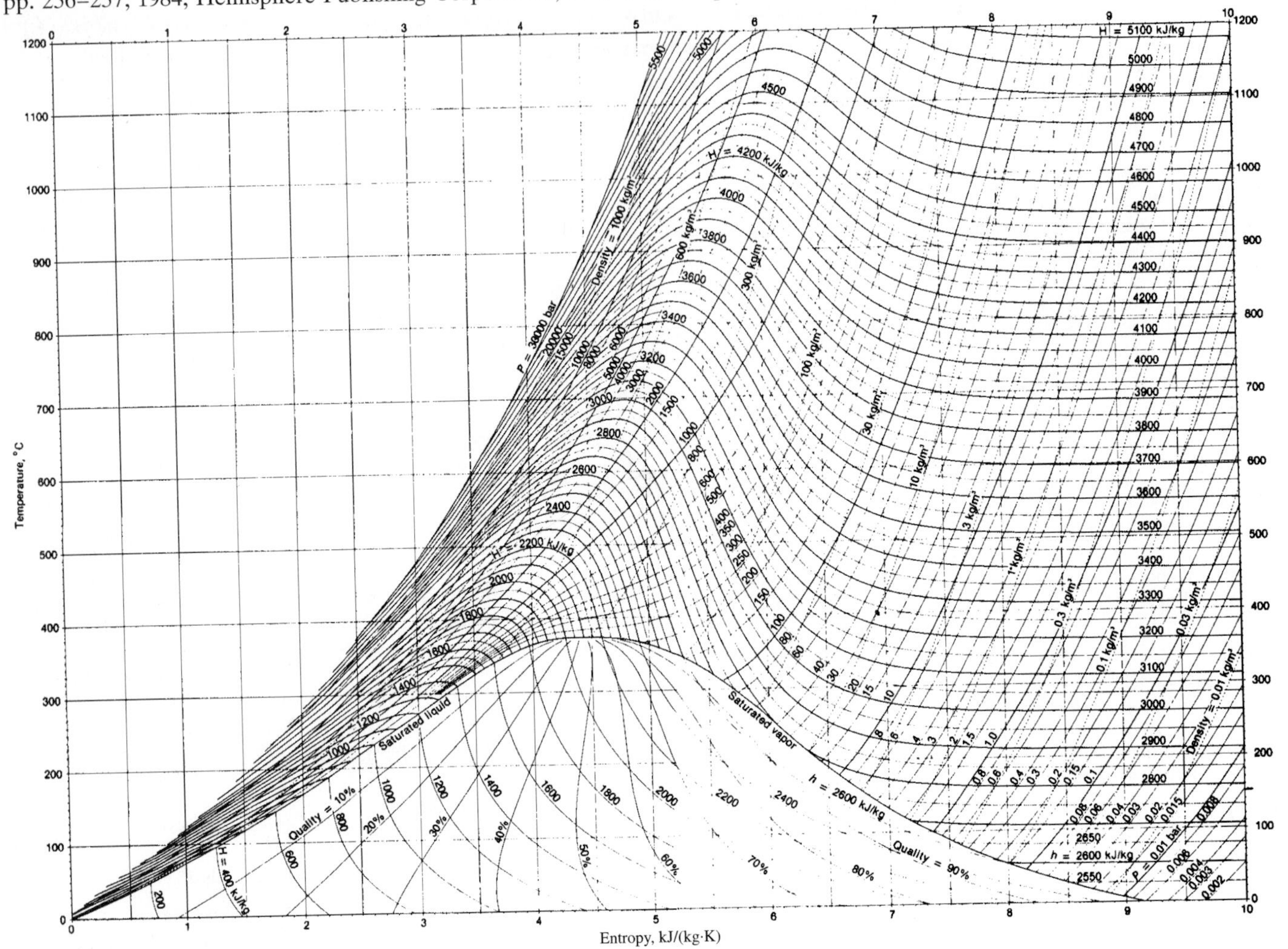

**TABLE A-8**
**Saturated refrigerant-134a: temperature table**

| Temp. $T$ °C | Press. $P_{sat}$ MPa | Specific volume m³/kg Sat. liquid $v_f$ | Sat. vapor $v_g$ | Internal energy kJ/kg Sat. liquid $u_f$ | Sat. vapor $u_g$ | Enthalpy kJ/kg Sat. liquid $h_f$ | Evap. $h_{fg}$ | Sat. vapor $h_g$ | Entropy kJ/(kg · K) Sat. liquid $s_f$ | Sat. vapor $s_g$ |
|---|---|---|---|---|---|---|---|---|---|---|
| −40 | 0.05164 | 0.0007055 | 0.3569 | −0.04 | 204.45 | 0.00 | 222.88 | 222.88 | 0.0000 | 0.9560 |
| −36 | 0.06332 | 0.0007113 | 0.2947 | 4.68 | 206.73 | 4.73 | 220.67 | 225.40 | 0.0201 | 0.9506 |
| −32 | 0.07704 | 0.0007172 | 0.2451 | 9.47 | 209.01 | 9.52 | 218.37 | 227.90 | 0.0401 | 0.9456 |
| −28 | 0.09305 | 0.0007233 | 0.2052 | 14.31 | 211.29 | 14.37 | 216.01 | 230.38 | 0.0600 | 0.9411 |
| −26 | 0.10199 | 0.0007265 | 0.1882 | 16.75 | 212.43 | 16.82 | 214.80 | 231.62 | 0.0699 | 0.9390 |
| −24 | 0.11160 | 0.0007296 | 0.1728 | 19.21 | 213.57 | 19.29 | 213.57 | 232.85 | 0.0798 | 0.9370 |
| −22 | 0.12192 | 0.0007328 | 0.1590 | 21.68 | 214.70 | 21.77 | 212.32 | 234.08 | 0.0897 | 0.9351 |
| −20 | 0.13299 | 0.0007361 | 0.1464 | 24.17 | 215.84 | 24.26 | 211.05 | 235.31 | 0.0996 | 0.9332 |
| −18 | 0.14483 | 0.0007395 | 0.1350 | 26.67 | 216.97 | 26.77 | 209.76 | 236.53 | 0.1094 | 0.9315 |
| −16 | 0.15748 | 0.0007428 | 0.1247 | 29.18 | 218.10 | 29.30 | 208.45 | 237.74 | 0.1192 | 0.9298 |
| −12 | 0.18540 | 0.0007498 | 0.1068 | 34.25 | 220.36 | 34.39 | 205.77 | 240.15 | 0.1388 | 0.9267 |
| −8 | 0.21704 | 0.0007569 | 0.0919 | 39.38 | 222.60 | 39.54 | 203.00 | 242.54 | 0.1583 | 0.9239 |
| −4 | 0.25274 | 0.0007644 | 0.0794 | 44.56 | 224.84 | 44.75 | 200.15 | 244.90 | 0.1777 | 0.9213 |
| 0 | 0.29282 | 0.0007721 | 0.0689 | 49.79 | 227.06 | 50.02 | 197.21 | 247.23 | 0.1970 | 0.9190 |
| 4 | 0.33765 | 0.0007801 | 0.0600 | 55.08 | 229.27 | 55.35 | 194.19 | 249.53 | 0.2162 | 0.9169 |
| 8 | 0.38756 | 0.0007884 | 0.0525 | 60.43 | 231.46 | 60.73 | 191.07 | 251.80 | 0.2354 | 0.9150 |
| 12 | 0.44294 | 0.0007971 | 0.0460 | 65.83 | 233.63 | 66.18 | 187.85 | 254.03 | 0.2545 | 0.9132 |
| 16 | 0.50416 | 0.0008062 | 0.0405 | 71.29 | 235.78 | 71.69 | 184.52 | 256.22 | 0.2735 | 0.9116 |
| 20 | 0.57160 | 0.0008157 | 0.0358 | 76.80 | 237.91 | 77.26 | 181.09 | 258.36 | 0.2924 | 0.9102 |
| 24 | 0.64566 | 0.0008257 | 0.0317 | 82.37 | 240.01 | 82.90 | 177.55 | 260.45 | 0.3113 | 0.9089 |
| 26 | 0.68530 | 0.0008309 | 0.0298 | 85.18 | 241.05 | 85.75 | 175.73 | 261.48 | 0.3208 | 0.9082 |
| 28 | 0.72675 | 0.0008362 | 0.0281 | 88.00 | 242.08 | 88.61 | 173.89 | 262.50 | 0.3302 | 0.9076 |
| 30 | 0.77006 | 0.0008417 | 0.0265 | 90.84 | 243.10 | 91.49 | 172.00 | 263.50 | 0.3396 | 0.9070 |
| 32 | 0.81528 | 0.0008473 | 0.0250 | 93.70 | 244.12 | 94.39 | 170.09 | 264.48 | 0.3490 | 0.9064 |
| 34 | 0.86247 | 0.0008530 | 0.0236 | 96.58 | 245.12 | 97.31 | 168.14 | 265.45 | 0.3584 | 0.9058 |
| 36 | 0.91168 | 0.0008590 | 0.0223 | 99.47 | 246.11 | 100.25 | 166.15 | 266.40 | 0.3678 | 0.9053 |
| 38 | 0.96298 | 0.0008651 | 0.0210 | 102.38 | 247.09 | 103.21 | 164.12 | 267.33 | 0.3772 | 0.9047 |
| 40 | 1.0164 | 0.0008714 | 0.0199 | 105.30 | 248.06 | 106.19 | 162.05 | 268.24 | 0.3866 | 0.9041 |
| 42 | 1.0720 | 0.0008780 | 0.0188 | 108.25 | 249.02 | 109.19 | 159.94 | 269.14 | 0.3960 | 0.9035 |
| 44 | 1.1299 | 0.0008847 | 0.0177 | 111.22 | 249.96 | 112.22 | 157.79 | 270.01 | 0.4054 | 0.9030 |
| 48 | 1.2526 | 0.0008989 | 0.0159 | 117.22 | 251.79 | 118.35 | 153.33 | 271.68 | 0.4243 | 0.9017 |
| 52 | 1.3851 | 0.0009142 | 0.0142 | 123.31 | 253.55 | 124.58 | 148.66 | 273.24 | 0.4432 | 0.9004 |
| 56 | 1.5278 | 0.0009308 | 0.0127 | 129.51 | 255.23 | 130.93 | 143.75 | 274.68 | 0.4622 | 0.8990 |
| 60 | 1.6813 | 0.0009488 | 0.0114 | 135.82 | 256.81 | 137.42 | 138.57 | 275.99 | 0.4814 | 0.8973 |
| 70 | 2.1162 | 0.0010027 | 0.0086 | 152.22 | 260.15 | 154.34 | 124.08 | 278.43 | 0.5302 | 0.8918 |
| 80 | 2.6324 | 0.0010766 | 0.0064 | 169.88 | 262.14 | 172.71 | 106.41 | 279.12 | 0.5814 | 0.8827 |
| 90 | 3.2435 | 0.0011949 | 0.0046 | 189.82 | 261.34 | 193.69 | 82.63 | 276.32 | 0.6380 | 0.8655 |
| 100 | 3.9742 | 0.0015443 | 0.0027 | 218.60 | 248.49 | 224.74 | 34.40 | 259.13 | 0.7196 | 0.8117 |

*Source for Tables A-8 through A-10:* M. J. Moran and H. N. Shapiro, *Fundamentals of Engineering Thermodynamics,* 2d ed., Wiley, New York, 1992, pp. 710–715. Originally based on equations from D. P. Wilson and R. S. Basu, "Thermodynamic Properties of a New Stratospherically Safe Working Fluid—Refrigerant 134a," *ASHRAE Trans.,* Vol. 94, Pt. 2, 1988, pp. 2095–2118. Used with permission.

**TABLE A-9**
**Saturated refrigerant-134: pressure table**

| Press. $P$ MPa | Temp. $T_{sat}$ °C | Specific volume m³/kg Sat. liquid $v_f$ | Sat. vapor $v_g$ | Internal energy kJ/kg Sat. liquid $u_f$ | Sat. vapor $u_g$ | Enthalpy kJ/kg Sat. liquid $h_f$ | Evap. $h_{fg}$ | Sat. vapor $h_g$ | Entropy kJ/(kg · K) Sat. liquid $s_f$ | Sat. vapor $s_g$ |
|---|---|---|---|---|---|---|---|---|---|---|
| 0.06 | −37.07 | 0.0007097 | 0.3100 | 3.41 | 206.12 | 3.46 | 221.27 | 224.72 | 0.0147 | 0.9520 |
| 0.08 | −31.21 | 0.0007184 | 0.2366 | 10.41 | 209.46 | 10.47 | 217.92 | 228.39 | 0.0440 | 0.9447 |
| 0.10 | −26.43 | 0.0007258 | 0.1917 | 16.22 | 212.18 | 16.29 | 215.06 | 231.35 | 0.0678 | 0.9395 |
| 0.12 | −22.36 | 0.0007323 | 0.1614 | 21.23 | 214.50 | 21.32 | 212.54 | 233.86 | 0.0879 | 0.9354 |
| 0.14 | −18.80 | 0.0007381 | 0.1395 | 25.66 | 216.52 | 25.77 | 210.27 | 236.04 | 0.1055 | 0.9322 |
| 0.16 | −15.62 | 0.0007435 | 0.1229 | 29.66 | 218.32 | 29.78 | 208.18 | 237.97 | 0.1211 | 0.9295 |
| 0.18 | −12.73 | 0.0007485 | 0.1098 | 33.31 | 219.94 | 33.45 | 206.26 | 239.71 | 0.1352 | 0.9273 |
| 0.20 | −10.09 | 0.0007532 | 0.0993 | 36.69 | 221.43 | 36.84 | 204.46 | 241.30 | 0.1481 | 0.9253 |
| 0.24 | −5.37 | 0.0007618 | 0.0834 | 42.77 | 224.07 | 42.95 | 201.14 | 244.09 | 0.1710 | 0.9222 |
| 0.28 | −1.23 | 0.0007697 | 0.0719 | 48.18 | 226.38 | 48.39 | 198.13 | 246.52 | 0.1911 | 0.9197 |
| 0.32 | 2.48 | 0.0007770 | 0.0632 | 53.06 | 228.43 | 53.31 | 195.35 | 248.66 | 0.2089 | 0.9177 |
| 0.36 | 5.84 | 0.0007839 | 0.0564 | 57.54 | 230.28 | 57.82 | 192.76 | 250.58 | 0.2251 | 0.9160 |
| 0.4 | 8.93 | 0.0007904 | 0.0509 | 61.69 | 231.97 | 62.00 | 190.32 | 252.32 | 0.2399 | 0.9145 |
| 0.5 | 15.74 | 0.0008056 | 0.0409 | 70.93 | 235.64 | 71.33 | 184.74 | 256.07 | 0.2723 | 0.9117 |
| 0.6 | 21.58 | 0.0008196 | 0.0341 | 78.99 | 238.74 | 79.48 | 179.71 | 259.19 | 0.2999 | 0.9097 |
| 0.7 | 26.72 | 0.0008328 | 0.0292 | 86.19 | 241.42 | 86.78 | 175.07 | 261.85 | 0.3242 | 0.9080 |
| 0.8 | 31.33 | 0.0008454 | 0.0255 | 92.75 | 243.78 | 93.42 | 170.73 | 264.15 | 0.3459 | 0.9066 |
| 0.9 | 35.53 | 0.0008576 | 0.0226 | 98.79 | 245.88 | 99.56 | 166.62 | 266.18 | 0.3656 | 0.9054 |
| 1.0 | 39.39 | 0.0008695 | 0.0202 | 104.42 | 247.77 | 105.29 | 162.68 | 267.97 | 0.3838 | 0.9043 |
| 1.2 | 46.32 | 0.0008928 | 0.0166 | 114.69 | 251.03 | 115.76 | 155.23 | 270.99 | 0.4164 | 0.9023 |
| 1.4 | 52.43 | 0.0009159 | 0.0140 | 123.98 | 253.74 | 125.26 | 148.14 | 273.40 | 0.4453 | 0.9003 |
| 1.6 | 57.92 | 0.0009392 | 0.0121 | 132.52 | 256.00 | 134.02 | 141.31 | 275.33 | 0.4714 | 0.8982 |
| 1.8 | 62.91 | 0.0009631 | 0.0105 | 140.49 | 257.88 | 142.22 | 134.60 | 276.83 | 0.4954 | 0.8959 |
| 2.0 | 67.49 | 0.0009878 | 0.0093 | 148.02 | 259.41 | 149.99 | 127.95 | 277.94 | 0.5178 | 0.8934 |
| 2.5 | 77.59 | 0.0010562 | 0.0069 | 165.48 | 261.84 | 168.12 | 111.06 | 279.17 | 0.5687 | 0.8854 |
| 3.0 | 86.22 | 0.0011416 | 0.0053 | 181.88 | 262.16 | 185.30 | 92.71 | 278.01 | 0.6156 | 0.8735 |

**TABLE A-10**
**Superheated refrigerant 134a**

| $T$ °C | $v$ m³/kg | $u$ kJ/kg | $h$ kJ/kg | $s$ kJ/(kg · K) | $v$ m³/kg | $u$ kJ/kg | $h$ kJ/kg | $s$ kJ/(kg · K) | $v$ m³/kg | $u$ kJ/kg | $h$ kJ/kg | $s$ kJ/(kg · K) |
|---|---|---|---|---|---|---|---|---|---|---|---|---|
| | $P$ = **0.06 MPa** ($T_{sat}$ = **−37.07°C**) | | | | $P$ = **0.10 MPa** ($T_{sat}$ = **−26.43°C**) | | | | $P$ = **0.14 MPa** ($T_{sat}$ = **−18.80°C**) | | | |
| Sat. | 0.31003 | 206.12 | 224.72 | 0.9520 | 0.19170 | 212.18 | 231.35 | 0.9395 | 0.13945 | 216.52 | 236.04 | 0.9322 |
| −20 | 0.33536 | 217.86 | 237.98 | 1.0062 | 0.19770 | 216.77 | 236.54 | 0.9602 | | | | |
| −10 | 0.34992 | 224.97 | 245.96 | 1.0371 | 0.20686 | 224.01 | 244.70 | 0.9918 | 0.14549 | 223.03 | 243.40 | 0.9606 |
| 0 | 0.36433 | 232.24 | 254.10 | 1.0675 | 0.21587 | 231.41 | 252.99 | 1.0227 | 0.15219 | 230.55 | 251.86 | 0.9922 |
| 10 | 0.37861 | 239.69 | 262.41 | 1.0973 | 0.22473 | 238.96 | 261.43 | 1.0531 | 0.15875 | 238.21 | 260.43 | 1.0230 |
| 20 | 0.39279 | 247.32 | 270.89 | 1.1267 | 0.23349 | 246.67 | 270.02 | 1.0829 | 0.16520 | 246.01 | 269.13 | 1.0532 |
| 30 | 0.40688 | 255.12 | 279.53 | 1.1557 | 0.24216 | 254.54 | 278.76 | 1.1122 | 0.17155 | 253.96 | 277.97 | 1.0828 |
| 40 | 0.42091 | 263.10 | 288.35 | 1.1844 | 0.25076 | 262.58 | 287.66 | 1.1411 | 0.17783 | 262.06 | 286.96 | 1.1120 |
| 50 | 0.43487 | 271.25 | 297.34 | 1.2126 | 0.25930 | 270.79 | 296.72 | 1.1696 | 0.18404 | 270.32 | 296.09 | 1.1407 |
| 60 | 0.44879 | 279.58 | 306.51 | 1.2405 | 0.26779 | 279.16 | 305.94 | 1.1977 | 0.19020 | 278.74 | 305.37 | 1.1690 |
| 70 | 0.46266 | 288.08 | 315.84 | 1.2681 | 0.27623 | 287.70 | 315.32 | 1.2254 | 0.19633 | 287.32 | 314.80 | 1.1969 |
| 80 | 0.47650 | 296.75 | 325.34 | 1.2954 | 0.28464 | 296.40 | 324.87 | 1.2528 | 0.20241 | 296.06 | 324.39 | 1.2244 |
| 90 | 0.49031 | 305.58 | 335.00 | 1.3224 | 0.29302 | 305.27 | 334.57 | 1.2799 | 0.20846 | 304.95 | 334.14 | 1.2516 |
| 100 | | | | | | | | | 0.21449 | 314.01 | 344.04 | 1.2785 |
| | $P$ = **0.18 MPa** ($T_{sat}$ = **−12.73°C**) | | | | $P$ = **0.20 MPa** ($T_{sat}$ = **−10.09°C**) | | | | $P$ = **0.24 MPa** ($T_{sat}$ = **−5.37°C**) | | | |
| Sat. | 0.10983 | 219.94 | 239.71 | 0.9273 | 0.09933 | 221.43 | 241.30 | 0.9253 | 0.08343 | 224.07 | 244.09 | 0.9222 |
| −10 | 0.11135 | 222.02 | 242.06 | 0.9362 | 0.09938 | 221.50 | 241.38 | 0.9256 | | | | |
| 0 | 0.11678 | 229.67 | 250.69 | 0.9684 | 0.10438 | 229.23 | 250.10 | 0.9582 | 0.08574 | 228.31 | 248.89 | 0.9399 |
| 10 | 0.12207 | 237.44 | 259.41 | 0.9998 | 0.10922 | 237.05 | 258.89 | 0.9898 | 0.08993 | 236.26 | 257.84 | 0.9721 |
| 20 | 0.12723 | 245.33 | 268.23 | 1.0304 | 0.11394 | 244.99 | 267.78 | 1.0206 | 0.09399 | 244.30 | 266.85 | 1.0034 |
| 30 | 0.13230 | 253.36 | 277.17 | 1.0604 | 0.11856 | 253.06 | 276.77 | 1.0508 | 0.09794 | 252.45 | 275.95 | 1.0339 |
| 40 | 0.13730 | 261.53 | 286.24 | 1.0898 | 0.12311 | 261.26 | 285.88 | 1.0804 | 0.10181 | 260.72 | 285.16 | 1.0637 |
| 50 | 0.14222 | 269.85 | 295.45 | 1.1187 | 0.12758 | 269.61 | 295.12 | 1.1094 | 0.10562 | 269.12 | 294.47 | 1.0930 |
| 60 | 0.14710 | 278.31 | 304.79 | 1.1472 | 0.13201 | 278.10 | 304.50 | 1.1380 | 0.10937 | 277.67 | 303.91 | 1.1218 |
| 70 | 0.15193 | 286.93 | 314.28 | 1.1753 | 0.13639 | 286.74 | 314.02 | 1.1661 | 0.11307 | 286.35 | 313.49 | 1.1501 |
| 80 | 0.15672 | 295.71 | 323.92 | 1.2030 | 0.14073 | 295.53 | 323.68 | 1.1939 | 0.11674 | 295.18 | 323.19 | 1.1780 |
| 90 | 0.16148 | 304.63 | 333.70 | 1.2303 | 0.14504 | 304.47 | 333.48 | 1.2212 | 0.12037 | 304.15 | 333.04 | 1.2055 |
| 100 | 0.16622 | 313.72 | 343.63 | 1.2573 | 0.14932 | 313.57 | 343.43 | 1.2483 | 0.12398 | 313.27 | 343.03 | 1.2326 |
| | $P$ = **0.28 MPa** ($T_{sat}$ = **−1.23°C**) | | | | $P$ = **0.32 MPa** ($T_{sat}$ = **2.48°C**) | | | | $P$ = **0.40 MPa** ($T_{sat}$ = **8.93°C**) | | | |
| Sat. | 0.07193 | 226.38 | 246.52 | 0.9197 | 0.06322 | 228.43 | 248.66 | 0.9177 | 0.05089 | 231.97 | 252.32 | 0.9145 |
| 0 | 0.07240 | 227.37 | 247.64 | 0.9238 | | | | | | | | |
| 10 | 0.07613 | 235.44 | 256.76 | 0.9566 | 0.06576 | 234.61 | 255.65 | 0.9427 | 0.05119 | 232.87 | 253.35 | 0.9182 |
| 20 | 0.07972 | 243.59 | 265.91 | 0.9883 | 0.06901 | 242.87 | 264.95 | 0.9749 | 0.05397 | 241.37 | 262.96 | 0.9515 |
| 30 | 0.08320 | 251.83 | 275.12 | 1.0192 | 0.07214 | 251.19 | 274.28 | 1.0062 | 0.05662 | 249.89 | 272.54 | 0.8937 |
| 40 | 0.08660 | 260.17 | 284.42 | 1.0494 | 0.07518 | 259.61 | 283.67 | 1.0367 | 0.05917 | 258.47 | 282.14 | 1.0148 |
| 50 | 0.08992 | 268.64 | 293.81 | 1.0789 | 0.07815 | 268.14 | 293.15 | 1.0665 | 0.06164 | 267.13 | 291.79 | 1.0452 |
| 60 | 0.09319 | 277.23 | 303.32 | 1.1079 | 0.08106 | 276.79 | 302.72 | 1.0957 | 0.06405 | 275.89 | 301.51 | 1.0748 |
| 70 | 0.09641 | 285.96 | 312.95 | 1.1364 | 0.08392 | 285.56 | 312.41 | 1.1243 | 0.06641 | 284.75 | 311.32 | 1.1038 |
| 80 | 0.09960 | 294.82 | 322.71 | 1.1644 | 0.08674 | 294.46 | 322.22 | 1.1525 | 0.06873 | 293.73 | 321.23 | 1.1322 |
| 90 | 0.10275 | 303.83 | 332.60 | 1.1920 | 0.08953 | 303.50 | 332.15 | 1.1802 | 0.07102 | 302.84 | 331.25 | 1.1602 |
| 100 | 0.10587 | 312.98 | 342.62 | 1.2193 | 0.09229 | 312.68 | 342.21 | 1.2076 | 0.07327 | 312.07 | 341.38 | 1.1878 |
| 110 | 0.10897 | 322.27 | 352.78 | 1.2461 | 0.09503 | 322.00 | 352.40 | 1.2345 | 0.07550 | 321.44 | 351.64 | 1.2149 |
| 120 | 0.11205 | 331.71 | 363.08 | 1.2727 | 0.09774 | 331.45 | 362.73 | 1.2611 | 0.07771 | 330.94 | 362.03 | 1.2417 |
| 130 | | | | | | | | | 0.07991 | 340.58 | 372.54 | 1.2681 |
| 140 | | | | | | | | | 0.08208 | 350.35 | 383.18 | 1.2941 |

**TABLE A-10**
**Superheated refrigerant 134a (*Continued*)**

| $T$ °C | $v$ m$^3$/kg | $u$ kJ/kg | $h$ kJ/kg | $s$ kJ/(kg · K) | $v$ m$^3$/kg | $u$ kJ/kg | $h$ kJ/kg | $s$ kJ/(kg · K) | $v$ m$^3$/kg | $u$ kJ/kg | $h$ kJ/kg | $s$ kJ/(kg · K) |
|---|---|---|---|---|---|---|---|---|---|---|---|---|
| | $P$ = **0.50 MPa** ($T_{sat}$ = **15.74°C**) | | | | $P$ = **0.60 MPa** ($T_{sat}$ = **21.58°C**) | | | | $P$ = **0.70 MPa** ($T_{sat}$ = **26.72°C**) | | | |
| Sat. | 0.04086 | 235.64 | 256.07 | 0.9117 | 0.03408 | 238.74 | 259.19 | 0.9097 | 0.02918 | 241.42 | 261.85 | 0.9080 |
| 20 | 0.04188 | 239.40 | 260.34 | 0.9264 | | | | | | | | |
| 30 | 0.04416 | 248.20 | 270.28 | 0.9597 | 0.03581 | 246.41 | 267.89 | 0.9388 | 0.02979 | 244.51 | 265.37 | 0.9197 |
| 40 | 0.04633 | 256.99 | 280.16 | 0.9918 | 0.03774 | 255.45 | 278.09 | 0.9719 | 0.03157 | 253.83 | 275.93 | 0.9539 |
| 50 | 0.04842 | 265.83 | 290.04 | 1.0229 | 0.03958 | 264.48 | 288.23 | 1.0037 | 0.03324 | 263.08 | 286.35 | 0.9867 |
| 60 | 0.05043 | 274.73 | 299.95 | 1.0531 | 0.04134 | 273.54 | 298.35 | 1.0346 | 0.03482 | 272.31 | 296.69 | 1.0182 |
| 70 | 0.05240 | 283.72 | 309.92 | 1.0825 | 0.04304 | 282.66 | 308.48 | 1.0645 | 0.03634 | 281.57 | 307.01 | 1.0487 |
| 80 | 0.05432 | 292.80 | 319.96 | 1.1114 | 0.04469 | 291.86 | 318.67 | 1.0938 | 0.03781 | 290.88 | 317.35 | 1.0784 |
| 90 | 0.05620 | 302.00 | 330.10 | 1.1397 | 0.04631 | 301.14 | 328.93 | 1.1225 | 0.03924 | 300.27 | 327.74 | 1.1074 |
| 100 | 0.05805 | 311.31 | 340.33 | 1.1675 | 0.04790 | 310.53 | 339.27 | 1.1505 | 0.04064 | 309.74 | 338.19 | 1.1358 |
| 110 | 0.05988 | 320.74 | 350.68 | 1.1949 | 0.04946 | 320.03 | 349.70 | 1.1781 | 0.04201 | 319.31 | 348.71 | 1.1637 |
| 120 | 0.06168 | 330.30 | 361.14 | 1.2218 | 0.05099 | 329.64 | 360.24 | 1.2053 | 0.04335 | 328.98 | 359.33 | 1.1910 |
| 130 | 0.06347 | 339.98 | 371.72 | 1.2484 | 0.05251 | 339.38 | 370.88 | 1.2320 | 0.04468 | 338.76 | 370.04 | 1.2179 |
| 140 | 0.06524 | 349.79 | 382.42 | 1.2746 | 0.05402 | 349.23 | 381.64 | 1.2584 | 0.04599 | 348.66 | 380.86 | 1.2444 |
| 150 | | | | | 0.05550 | 359.21 | 392.52 | 1.2844 | 0.04729 | 358.68 | 391.79 | 1.2706 |
| 160 | | | | | 0.05698 | 369.32 | 403.51 | 1.3100 | 0.04857 | 368.82 | 402.82 | 1.2963 |
| | $P$ = **0.80 MPa** ($T_{sat}$ = **31.33°C**) | | | | $P$ = **0.90 MPa** ($T_{sat}$ = **35.53°C**) | | | | $P$ = **1.00 MPa** ($T_{sat}$ = **39.39°C**) | | | |
| Sat. | 0.02547 | 243.78 | 264.15 | 0.9066 | 0.02255 | 245.88 | 266.18 | 0.9054 | 0.02020 | 247.77 | 267.97 | 0.9043 |
| 40 | 0.02691 | 252.13 | 273.66 | 0.9374 | 0.02325 | 250.32 | 271.25 | 0.9217 | 0.02029 | 248.39 | 268.68 | 0.9066 |
| 50 | 0.02846 | 261.62 | 284.39 | 0.9711 | 0.02472 | 260.09 | 282.34 | 0.9566 | 0.02171 | 258.48 | 280.19 | 0.9428 |
| 60 | 0.02992 | 271.04 | 294.98 | 1.0034 | 0.02609 | 269.72 | 293.21 | 0.9897 | 0.02301 | 268.35 | 291.36 | 0.9768 |
| 70 | 0.03131 | 280.45 | 305.50 | 1.0345 | 0.02738 | 279.30 | 303.94 | 1.0214 | 0.02423 | 278.11 | 302.34 | 1.0093 |
| 80 | 0.03264 | 289.89 | 316.00 | 1.0647 | 0.02861 | 288.87 | 314.62 | 1.0521 | 0.02538 | 287.82 | 313.20 | 1.0405 |
| 90 | 0.03393 | 299.37 | 326.52 | 1.0940 | 0.02980 | 298.46 | 325.28 | 1.0819 | 0.02649 | 297.53 | 324.01 | 1.0707 |
| 100 | 0.03519 | 308.93 | 337.08 | 1.1227 | 0.03095 | 308.11 | 335.96 | 1.1109 | 0.02755 | 307.27 | 334.82 | 1.1000 |
| 110 | 0.03642 | 318.57 | 347.71 | 1.1508 | 0.03207 | 317.82 | 346.68 | 1.1392 | 0.02858 | 317.06 | 345.65 | 1.1286 |
| 120 | 0.03762 | 328.31 | 358.40 | 1.1784 | 0.03316 | 327.62 | 357.47 | 1.1670 | 0.02959 | 326.93 | 356.52 | 1.1567 |
| 130 | 0.03881 | 338.14 | 369.19 | 1.2055 | 0.03423 | 337.52 | 368.33 | 1.1943 | 0.03058 | 336.88 | 367.46 | 1.1841 |
| 140 | 0.03997 | 348.09 | 380.07 | 1.2321 | 0.03529 | 347.51 | 379.27 | 1.2211 | 0.03154 | 346.92 | 378.46 | 1.2111 |
| 150 | 0.04113 | 358.15 | 391.05 | 1.2584 | 0.03633 | 357.61 | 390.31 | 1.2475 | 0.03250 | 357.06 | 389.56 | 1.2376 |
| 160 | 0.04227 | 368.32 | 402.14 | 1.2843 | 0.03736 | 367.82 | 401.44 | 1.2735 | 0.03344 | 367.31 | 400.74 | 1.2638 |
| 170 | 0.04340 | 378.61 | 413.33 | 1.3098 | 0.03838 | 378.14 | 412.68 | 1.2992 | 0.03436 | 377.66 | 412.02 | 1.2895 |
| 180 | 0.04452 | 389.02 | 424.63 | 1.3351 | 0.03939 | 388.57 | 424.02 | 1.3245 | 0.03528 | 388.12 | 423.40 | 1.3149 |
| | $P$ = **1.20 MPa** ($T_{sat}$ = **46.32°C**) | | | | $P$ = **1.40 MPa** ($T_{sat}$ = **52.43°C**) | | | | $P$ = **1.60 MPa** ($T_{sat}$ = **57.92°C**) | | | |
| Sat. | 0.01663 | 251.03 | 270.99 | 0.9023 | 0.01405 | 253.74 | 273.40 | 0.9003 | 0.01208 | 256.00 | 275.33 | 0.8982 |
| 50 | 0.01712 | 254.98 | 275.52 | 0.9164 | | | | | | | | |
| 60 | 0.01835 | 265.42 | 287.44 | 0.9527 | 0.01495 | 262.17 | 283.10 | 0.9297 | 0.01233 | 258.48 | 278.20 | 0.9069 |
| 70 | 0.01947 | 275.59 | 298.96 | 0.9868 | 0.01603 | 272.87 | 295.31 | 0.9658 | 0.01340 | 269.89 | 291.33 | 0.9457 |
| 80 | 0.02051 | 285.62 | 310.24 | 1.0192 | 0.01701 | 283.29 | 307.10 | 0.9997 | 0.01435 | 280.78 | 303.74 | 0.9813 |
| 90 | 0.02150 | 295.59 | 321.39 | 1.0503 | 0.01792 | 293.55 | 318.63 | 1.0319 | 0.01521 | 291.39 | 315.72 | 1.0148 |
| 100 | 0.02244 | 305.54 | 332.47 | 1.0804 | 0.01878 | 303.73 | 330.02 | 1.0628 | 0.01601 | 301.84 | 327.46 | 1.0467 |
| 110 | 0.02335 | 315.50 | 343.52 | 1.1096 | 0.01960 | 313.88 | 341.32 | 1.0927 | 0.01677 | 312.20 | 339.04 | 1.0773 |
| 120 | 0.02423 | 325.51 | 354.58 | 1.1381 | 0.02039 | 324.05 | 352.59 | 1.1218 | 0.01750 | 322.53 | 350.53 | 1.1069 |
| 130 | 0.02508 | 335.58 | 365.68 | 1.1660 | 0.02115 | 334.25 | 363.86 | 1.1501 | 0.01820 | 332.87 | 361.99 | 1.1357 |
| 140 | 0.02592 | 345.73 | 376.83 | 1.1933 | 0.02189 | 344.50 | 375.15 | 1.1777 | 0.01887 | 343.24 | 373.44 | 1.1638 |
| 150 | 0.02674 | 355.95 | 388.04 | 1.2201 | 0.02262 | 354.82 | 386.49 | 1.2048 | 0.01953 | 353.66 | 384.91 | 1.1912 |
| 160 | 0.02754 | 366.27 | 399.33 | 1.2465 | 0.02333 | 365.22 | 397.89 | 1.2315 | 0.02017 | 364.15 | 396.43 | 1.2181 |
| 170 | 0.02834 | 376.69 | 410.70 | 1.2724 | 0.02403 | 375.71 | 409.36 | 1.2576 | 0.02080 | 374.71 | 407.99 | 1.2445 |
| 180 | 0.02912 | 387.21 | 422.16 | 1.2980 | 0.02472 | 386.29 | 420.90 | 1.2834 | 0.02142 | 385.35 | 419.62 | 1.2704 |
| 190 | | | | | 0.02541 | 396.96 | 432.53 | 1.3088 | 0.02203 | 396.08 | 431.33 | 1.2960 |
| 200 | | | | | 0.02608 | 407.73 | 444.24 | 1.3338 | 0.02263 | 406.90 | 443.11 | 1.3212 |

**FIGURE A-11**

*P-h* diagram for refrigerant-134a. (Reprinted by permission of American Society of Heating, Refrigerating, and Air Conditioning Engineers, Inc., Atlanta, GA.)

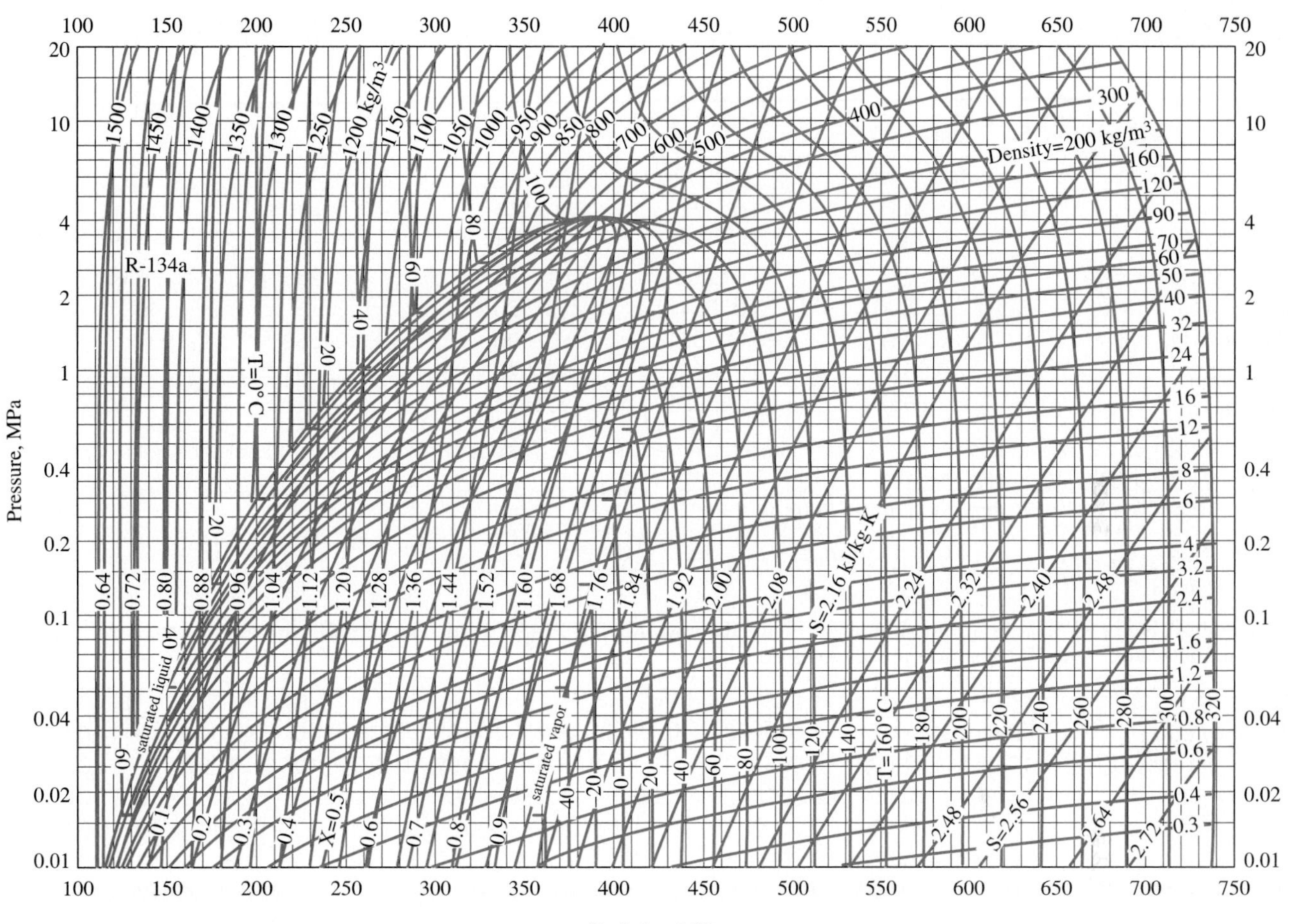

**TABLE A-12**
**Constants that appear in the Beattie–Bridgeman and the Benedict–Webb–Rubin equations of state**

(*a*) The Beattie–Bridgeman equation of state is

$$P = \frac{R_u T}{\bar{v}^2}\left(1 - \frac{c}{\bar{v}T^3}\right)(\bar{v} + B) - \frac{A}{\bar{v}^2}, \quad \text{where } A = A_0\left(1 - \frac{a}{\bar{v}}\right) \quad \text{and} \quad B = B_0\left(1 - \frac{b}{\bar{v}}\right)$$

When $P$ is in kPa, $\bar{v}$ is in $m^3$/kmol, $T$ is in K, and $R_u = 8.314\ \text{kPa} \cdot \text{m}^3/(\text{kmol} \cdot \text{K})$, the five constants in the Beattie–Bridgeman equation are as follows:

| Gas | $A_0$ | $a$ | $B_0$ | $b$ | $c$ |
|---|---|---|---|---|---|
| Air | 131.8441 | 0.01931 | 0.04611 | −0.001101 | $4.34 \times 10^4$ |
| Argon, Ar | 130.7802 | 0.02328 | 0.03931 | 0.0 | $5.99 \times 10^4$ |
| Carbon dioxide, $CO_2$ | 507.2836 | 0.07132 | 0.10476 | 0.07235 | $6.60 \times 10^5$ |
| Helium, He | 2.1886 | 0.05984 | 0.01400 | 0.0 | 40 |
| Hydrogen, $H_2$ | 20.0117 | −0.00506 | 0.02096 | −0.04359 | 504 |
| Nitrogen, $N_2$ | 136.2315 | 0.02617 | 0.05046 | −0.00691 | $4.20 \times 10^4$ |
| Oxygen, $O_2$ | 151.0857 | 0.02562 | 0.04624 | 0.004208 | $4.80 \times 10^4$ |

*Source:* Gordon J. Van Wylen and Richard E. Sonntag, *Fundamentals of of Classical Thermodynamics,* English/SI Version, 3d ed., Wiley, New York, 1986, p. 46, Table 3.3.

(*b*) The Benedict–Webb–Rubin equation of state is

$$P = \frac{R_u T}{\bar{v}} + \left(B_0 R_u T - A_0 - \frac{C_0}{T^2}\right)\frac{1}{\bar{v}^2} + \frac{bR_u T - a}{\bar{v}^3} + \frac{a\alpha}{\bar{v}^6} + \frac{c}{\bar{v}^3 T^2}\left(1 + \frac{\gamma}{\bar{v}^2}\right)e^{-\gamma/\bar{v}^2}$$

When $P$ is in kPa, $\bar{v}$ is in $m^3$/kmol, $T$ is in K, and $R_u = 8.314\ \text{kPa} \cdot \text{m}^3/(\text{kmol} \cdot \text{K})$, the eight constants in the Benedict–Webb–Rubin equation are as follows:

| Gas | $a$ | $A_0$ | $b$ | $B_0$ | $c$ | $C_0$ | $\alpha$ | $\gamma$ |
|---|---|---|---|---|---|---|---|---|
| *n*-Butane, $C_4H_{10}$ | 190.68 | 1021.6 | 0.039998 | 0.12436 | $3.205 \times 10^7$ | $1.006 \times 10^8$ | $1.101 \times 10^{-3}$ | 0.0340 |
| Carbon dioxide, $CO_2$ | 13.86 | 277.30 | 0.007210 | 0.04991 | $1.511 \times 10^6$ | $1.404 \times 10^7$ | $8.470 \times 10^{-5}$ | 0.00539 |
| Carbon monoxide, CO | 3.71 | 135.87 | 0.002632 | 0.05454 | $1.054 \times 10^5$ | $8.673 \times 10^5$ | $1.350 \times 10^{-4}$ | 0.0060 |
| Methane, $CH_4$ | 5.00 | 187.91 | 0.003380 | 0.04260 | $2.578 \times 10^5$ | $2.286 \times 10^6$ | $1.244 \times 10^{-4}$ | 0.0060 |
| Nitrogen, $N_2$ | 2.54 | 106.73 | 0.002328 | 0.04074 | $7.379 \times 10^4$ | $8.164 \times 10^5$ | $1.272 \times 10^{-4}$ | 0.0053 |

*Source:* Kenneth Wark, *Thermodynamics,* 4th ed., McGraw-Hill, New York, 1983, p. 815, Table A-21M. Originally published in H. W. Cooper and J. C. Goldfrank, *Hydrocarbon Processing,* Vol. 46, No. 12, p. 141, 1967.

**FIGURE A-13**

Nelson–Obert generalized compressibility chart—low pressures. (Used with permission of Dr. Edward E. Obert, University of Wisconsin.)

(*a*) $0 < P_r < 1.0$

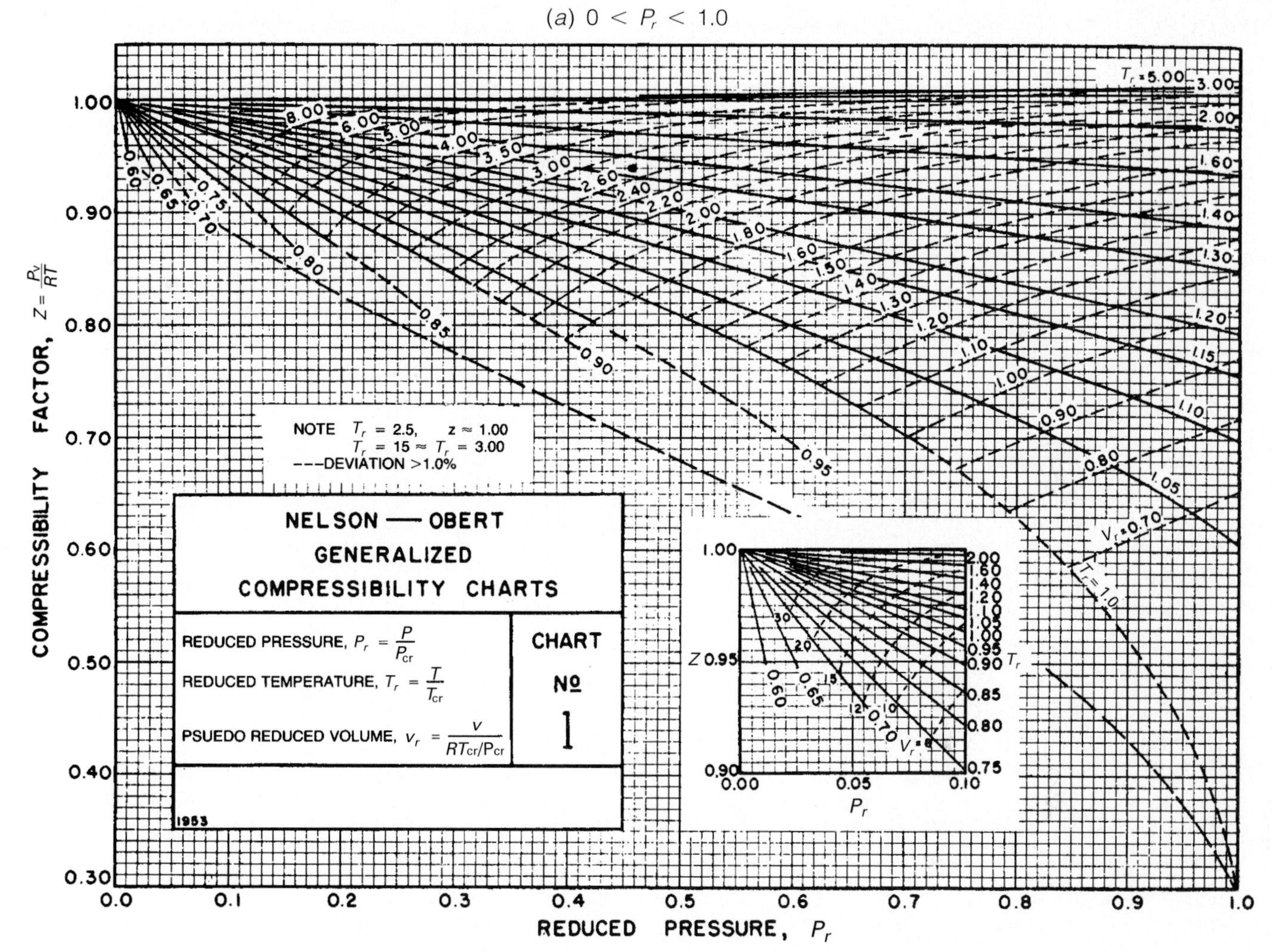

**TABLE A-14**
**Properties of solid metals**

| Composition | Melting point K | Properties at 300 K $\rho$ kg/m³ | $c_p$ J/(kg · K) | $k$ W/(m · K) | $\alpha \times 10^6$ m²/s | Properties at various temperatures (K) $k$[W/(m · K)]/$c_p$[J/(kg · K)] 100 | 200 | 400 | 600 | 800 | 1000 | 1200 |
|---|---|---|---|---|---|---|---|---|---|---|---|---|
| Aluminum | | | | | | | | | | | | |
| Pure | 933 | 2,702 | 903 | 237 | 97.1 | 302 | 237 | 240 | 231 | 218 | | |
| | | | | | | 482 | 798 | 949 | 1033 | 1146 | | |
| Alloy 2024-T6 (4.5% Cu, 1.5% Mg, 0.6% Mn) | 775 | 2,770 | 875 | 177 | 73.0 | 65 | 163 | 186 | 186 | | | |
| | | | | | | 473 | 787 | 925 | 1042 | | | |
| Alloy 195, Cast (4.5% Cu) | | 2,790 | 883 | 168 | 68.2 | | | 174 | 185 | | | |
| | | | | | | | | — | — | | | |
| Beryllium | 1550 | 1,850 | 1825 | 200 | 59.2 | 990 | 301 | 161 | 126 | 106 | 90.8 | 78.7 |
| | | | | | | 203 | 1114 | 2191 | 2604 | 2823 | 3018 | 3227 |
| Bismuth | 545 | 9,780 | 122 | 7.86 | 6.59 | 16.5 | 9.69 | 7.04 | | | | |
| | | | | | | 112 | 120 | 127 | | | | |
| Boron | 2573 | 2,500 | 1107 | 27.0 | 9.76 | 190 | 55.5 | 16.8 | 10.6 | 9.60 | 9.85 | |
| | | | | | | 128 | 600 | 1463 | 1892 | 2160 | 2338 | |
| Cadmium | 594 | 8,650 | 231 | 96.8 | 48.4 | 203 | 99.3 | 94.7 | | | | |
| | | | | | | 198 | 222 | 242 | | | | |
| Chromium | 2118 | 7,160 | 449 | 93.7 | 29.1 | 159 | 111 | 90.9 | 80.7 | 71.3 | 65.4 | 61.9 |
| | | | | | | 192 | 384 | 484 | 542 | 581 | 616 | 682 |
| Cobalt | 1769 | 8,862 | 421 | 99.2 | 26.6 | 167 | 122 | 85.4 | 67.4 | 58.2 | 52.1 | 49.3 |
| | | | | | | 236 | 379 | 450 | 503 | 550 | 628 | 733 |
| Copper | | | | | | | | | | | | |
| Pure | 1358 | 8,933 | 385 | 401 | 117 | 482 | 413 | 393 | 379 | 366 | 352 | 339 |
| | | | | | | 252 | 356 | 397 | 417 | 433 | 451 | 480 |
| Commercial bronze (90% Cu, 10% Al) | 1293 | 8,800 | 420 | 52 | 14 | | 42 | 52 | 59 | | | |
| | | | | | | | 785 | 160 | 545 | | | |
| Phosphor gear bronze (89% Cu, 11% Sn) | 1104 | 8,780 | 355 | 54 | 17 | | 41 | 65 | 74 | | | |
| | | | | | | | — | — | — | | | |
| Cartridge brass (70% Cu, 30% Zn) | 1188 | 8,530 | 380 | 110 | 33.9 | 75 | 95 | 137 | 149 | | | |
| | | | | | | | 360 | 395 | 425 | | | |
| Constantan (55% Cu, 45% Ni) | 1493 | 8.920 | 384 | 23 | 6.71 | 17 | 19 | | | | | |
| | | | | | | 237 | 362 | | | | | |
| Germanium | 1211 | 5,360 | 322 | 59.9 | 34.7 | 232 | 96.8 | 43.2 | 27.3 | 19.8 | 17.4 | 17.4 |
| | | | | | | 190 | 290 | 337 | 348 | 357 | 375 | 395 |
| Gold | 1336 | 19,300 | 129 | 317 | 127 | 327 | 323 | 311 | 298 | 284 | 270 | 255 |
| | | | | | | 109 | 124 | 131 | 135 | 140 | 145 | 155 |
| Iridium | 2720 | 22,500 | 130 | 147 | 50.3 | 172 | 153 | 144 | 138 | 132 | 126 | 120 |
| | | | | | | 90 | 122 | 133 | 138 | 144 | 153 | 161 |
| Iron | | | | | | | | | | | | |
| Pure | 1810 | 7,870 | 447 | 80.2 | 23.1 | 134 | 94.0 | 69.5 | 54.7 | 43.3 | 32.8 | 28.3 |
| | | | | | | 216 | 384 | 490 | 574 | 680 | 975 | 609 |
| Armco (99.75% pure) | | 7.870 | 447 | 72.7 | 20.7 | 95.6 | 80.6 | 65.7 | 53.1 | 42.2 | 32.3 | 28.7 |
| | | | | | | 215 | 384 | 490 | 574 | 680 | 975 | 609 |
| Carbon steels | | | | | | | | | | | | |
| Plain carbon (Mn ≤ 1%, Si ≤ 0.1% | | 7,854 | 434 | 60.5 | 17.7 | | | 56.7 | 48.0 | 39.2 | 30.0 | |
| | | | | | | | | 487 | 559 | 685 | 1169 | |
| AISI 1010 | | 7,832 | 434 | 63.9 | 18.8 | | | 58.7 | 48.8 | 39.2 | 31.3 | |
| | | | | | | | | 487 | 559 | 685 | 1168 | |
| Carbon–silicon (Mn ≤ 1%, 0.1% < Si ≤ 0.6%) | | 7,817 | 446 | 51.9 | 14.9 | | | 49.8 | 44.0 | 37.4 | 29.3 | |
| | | | | | | | | 501 | 582 | 699 | 971 | |
| Carbon–manganese–silicon (1% < Mn ≤ 1.65% 0.1% < Si ≤ 0.6%) | | 8,131 | 434 | 41.0 | 11.6 | | | 42.2 | 39.7 | 35.0 | 27.6 | |
| | | | | | | | | 487 | 559 | 685 | 1090 | |

**TABLE A-14**
**Properties of solid metals (*Continued*)**

| Composition | Melting point K | $\rho$ kg/m$^3$ | $c_p$ J/(kg · K) | $k$ W/(m · K) | $\alpha \times 10^6$ m$^2$/s | 100 | 200 | 400 | 600 | 800 | 1000 | 1200 |
|---|---|---|---|---|---|---|---|---|---|---|---|---|
| | | Properties at 300 K | | | | Properties at various temperatures (K): $k$[W/(m · K)]/$c_p$[J/(kg · K)] | | | | | | |
| *Chromium (low) steels* | | | | | | | | | | | | |
| $\frac{1}{2}$Cr – $\frac{1}{4}$Mo – Si | | 7,822 | 444 | 37.7 | 10.9 | | | 38.2 | 36.7 | 33.3 | 26.9 | |
| (0.18% C, 0.65% Cr, 0.23% Mo, 0.6% Si) | | | | | | | | 492 | 575 | 688 | 969 | |
| 1Cr – $\frac{1}{2}$Mo | | 7,858 | 442 | 42.3 | 12.2 | | | 42.0 | 39.1 | 34.5 | 27.4 | |
| (0.16% C, 1% Cr, 0.54% Mo, 0.39% Si) | | | | | | | | 492 | 575 | 688 | 969 | |
| 1 Cr – V | | 7,836 | 443 | 48.9 | 14.1 | | | 46.8 | 42.1 | 36.3 | 28.2 | |
| (0.2% C, 1.02% Cr, 0.15% V) | | | | | | | | 492 | 575 | 688 | 969 | |
| Stainless steels | | | | | | | | | | | | |
| AISI 302 | | 8,055 | 480 | 15.1 | 3.91 | | | 17.3 | 20.0 | 22.8 | 25.4 | |
| | | | | | | | | 512 | 559 | 585 | 606 | |
| AISI 304 | 1670 | 7,900 | 477 | 14.9 | 3.95 | 9.2 | 12.6 | 16.6 | 19.8 | 22.6 | 25.4 | 28.0 |
| | | | | | | 272 | 402 | 515 | 557 | 582 | 611 | 640 |
| AISI 316 | | 8,238 | 468 | 13.4 | 3.48 | | | 15.2 | 18.3 | 21.3 | 24.2 | |
| | | | | | | | | 504 | 550 | 576 | 602 | |
| AISI 347 | | 7,978 | 480 | 14.2 | 3.71 | | | 15.8 | 18.9 | 21.9 | 24.7 | |
| | | | | | | | | 513 | 559 | 585 | 606 | |
| Lead | 601 | 11,340 | 129 | 35.3 | 24.1 | 39.7 | 36.7 | 34.0 | 31.4 | | | |
| | | | | | | 118 | 125 | 132 | 142 | | | |
| Magnesium | 923 | 1,740 | 1024 | 156 | 87.6 | 169 | 159 | 153 | 149 | 146 | | |
| | | | | | | 649 | 934 | 1074 | 1170 | 1267 | | |
| Molybdenum | 2894 | 10,240 | 251 | 138 | 53.7 | 179 | 143 | 134 | 126 | 118 | 112 | 105 |
| | | | | | | 141 | 224 | 261 | 275 | 285 | 295 | 308 |
| Nickel | | | | | | | | | | | | |
| Pure | 1728 | 8,900 | 444 | 90.7 | 23.0 | 164 | 107 | 80.2 | 65.6 | 67.6 | 71.8 | 76.2 |
| | | | | | | 232 | 383 | 485 | 592 | 530 | 562 | 594 |
| Nichrome | 1672 | 8,400 | 420 | 12 | 3.4 | | | 14 | 16 | 21 | | |
| (80% Ni, 20% Cr) | | | | | | | | 480 | 525 | 545 | | |
| Inconel X-750 | 1665 | 8,510 | 439 | 11.7 | 3.1 | 8.7 | 10.3 | 13.5 | 17.0 | 20.5 | 24.0 | 27.6 |
| (73% Ni, 15% Cr, 6.7% Fe) | | | | | | — | 372 | 473 | 510 | 546 | 626 | — |
| Niobium | 2741 | 8,570 | 265 | 53.7 | 23.6 | 55.2 | 52.6 | 55.2 | 58.2 | 61.3 | 64.4 | 67.5 |
| | | | | | | 188 | 249 | 274 | 283 | 292 | 301 | 310 |
| Palladium | 1827 | 12,020 | 244 | 71.8 | 24.5 | 76.5 | 71.6 | 73.6 | 79.7 | 86.9 | 94.2 | 102 |
| | | | | | | 168 | 227 | 251 | 261 | 271 | 281 | 291 |
| Platinum | | | | | | | | | | | | |
| Pure | 2045 | 21,450 | 133 | 71.6 | 25.1 | 77.5 | 72.6 | 71.8 | 73.2 | 75.6 | 78.7 | 82.6 |
| | | | | | | 100 | 125 | 136 | 141 | 146 | 152 | 157 |
| Alloy 60Pt – 40Rh | 1800 | 16,630 | 162 | 47 | 17.4 | | | 52 | 59 | 65 | 69 | 73 |
| (60% Pt, 40% Rh) | | | | | | | | — | — | — | — | — |
| Rhenium | 3453 | 21,100 | 136 | 47.9 | 16.7 | 58.9 | 51.0 | 46.1 | 44.2 | 44.1 | 44.6 | 45.7 |
| | | | | | | 97 | 127 | 139 | 145 | 151 | 156 | 162 |
| Rhodium | 2236 | 12,450 | 243 | 150 | 49.6 | 186 | 154 | 146 | 136 | 127 | 121 | 116 |
| | | | | | | 147 | 220 | 253 | 274 | 293 | 311 | 327 |
| Silicon | 1685 | 2,330 | 712 | 148 | 89.2 | 884 | 264 | 98.9 | 61.9 | 42.4 | 31.2 | 25.7 |
| | | | | | | 259 | 556 | 790 | 867 | 913 | 946 | 967 |
| Silver | 1235 | 10.500 | 235 | 429 | 174 | 444 | 430 | 425 | 412 | 396 | 379 | 361 |
| | | | | | | 187 | 225 | 239 | 250 | 262 | 277 | 292 |
| Tantalum | 3269 | 16,600 | 140 | 57.5 | 24.7 | 59.2 | 57.5 | 57.8 | 58.6 | 59.4 | 60.2 | 61.0 |
| | | | | | | 110 | 133 | 144 | 146 | 149 | 152 | 155 |
| Thorium | 2023 | 11,700 | 118 | 54.0 | 39.1 | 59.8 | 54.6 | 54.5 | 55.8 | 56.9 | 56.9 | 58.7 |
| | | | | | | 99 | 112 | 124 | 134 | 145 | 156 | 167 |

**TABLE A-14**
**Properties of solid metals (*Continued*)**

| Composition | Melting point K | Properties at 300 K | | | | Properties at various temperatures (K) $k$[W/(m · K)]/$c_p$[J/(kg · K)] | | | | | | |
|---|---|---|---|---|---|---|---|---|---|---|---|---|
| | | $\rho$ kg/m$^3$ | $c_p$ J/(kg · K) | $k$ W/(m · K) | $\alpha \times 10^6$ m$^2$/s | 100 | 200 | 400 | 600 | 800 | 1000 | 1200 |
| Tin | 505 | 7,310 | 227 | 66.6 | 40.1 | 85.2 | 73.3 | 62.2 | | | | |
| | | | | | | 188 | 215 | 243 | | | | |
| Titanium | 1953 | 4,500 | 522 | 21.9 | 9.32 | 30.5 | 24.5 | 20.4 | 19.4 | 19.7 | 20.7 | 22.0 |
| | | | | | | 300 | 465 | 551 | 591 | 633 | 675 | 620 |
| Tungsten | 3660 | 19,300 | 132 | 174 | 68.3 | 208 | 186 | 159 | 137 | 125 | 118 | 113 |
| | | | | | | 87 | 122 | 137 | 142 | 145 | 148 | 152 |
| Uranium | 1406 | 19,070 | 116 | 27.6 | 12.5 | 21.7 | 25.1 | 29.6 | 34.0 | 38.8 | 43.9 | 49.0 |
| | | | | | | 94 | 108 | 125 | 146 | 176 | 180 | 161 |
| Vanadium | 2192 | 6,100 | 489 | 30.7 | 10.3 | 35.8 | 31.3 | 31.3 | 33.3 | 35.7 | 38.2 | 40.8 |
| | | | | | | 258 | 430 | 515 | 540 | 563 | 597 | 645 |
| Zinc | 693 | 7,140 | 389 | 116 | 41.8 | 117 | 118 | 111 | 103 | | | |
| | | | | | | 297 | 367 | 402 | 436 | | | |
| Zirconium | 2125 | 6,570 | 278 | 22.7 | 12.4 | 33.2 | 25.2 | 21.6 | 20.7 | 21.6 | 23.7 | 26.0 |
| | | | | | | 205 | 264 | 300 | 332 | 342 | 362 | 344 |

*Source for Tables A-14 through A-16:* Frank P. Incropera and David P. DeWitt, *Fundamentals of Heat and Mass Transfer,* 3d ed., Wiley, New York, 1990, pp. A3–A14. Originally compiled from various sources. Reprinted by permission of John Wiley & Sons, Inc.

**TABLE A-15**
**Properties of solid nonmetals**

| Composition | Melting point K | Properties at 300 K $\rho$ kg/m$^3$ | $c_p$ J/(kg · K) | $k$ W/(m · K) | $\alpha \times 10^6$ m$^2$/s | Properties at various temperatures (K), $k$[W/(m · K)]/$c_p$[J/(kg · K)] 100 | 200 | 400 | 600 | 800 | 1000 | 1200 |
|---|---|---|---|---|---|---|---|---|---|---|---|---|
| Aluminum oxide, | 2323 | 3970 | 765 | 46 | 15.1 | 450 | 82 | 32.4 | 18.9 | 13.0 | 10.5 | |
| sapphire | | | | | | — | — | 940 | 1110 | 1180 | 1225 | |
| Aluminum oxide, | 2323 | 3970 | 765 | 36.0 | 11.9 | 133 | 55 | 26.4 | 15.8 | 10.4 | 7.85 | 6.55 |
| polycrystalline | | | | | | — | — | 940 | 1110 | 1180 | 1225 | — |
| Beryllium oxide | 2725 | 3000 | 1030 | 272 | 88.0 | | | 196 | 111 | 70 | 47 | 33 |
| | | | | | | | | 1350 | 1690 | 1865 | 1975 | 2055 |
| Boron | 2573 | 2500 | 1105 | 27.6 | 9.99 | 190 | 52.5 | 18.7 | 11.3 | 8.1 | 6.3 | 5.2 |
| | | | | | | — | — | 1490 | 1880 | 2135 | 2350 | 2555 |
| Boron fiber epoxy (30% vol.) composite | 590 | 2080 | | | | | | | | | | |
| $k$, ∥ to fibers | | | | 2.29 | | 2.10 | 2.23 | 2.28 | | | | |
| $k$, ⊥ to fibers | | | | 0.59 | | 0.37 | 0.49 | 0.60 | | | | |
| $c_p$ | | | 1122 | | | 364 | 757 | 1431 | | | | |
| Carbon | | | | | | | | | | | | |
| Amorphous | 1500 | 1950 | — | 1.60 | — | 0.67 | 1.18 | 1.89 | 21.9 | 2.37 | 2.53 | 2.84 |
| | | | | | | — | — | — | — | — | — | — |
| Diamond, type IIa insulator | — | 3500 | 509 | 2300 | | 10,000 | 4000 | 1540 | | | | |
| | | | | | | 21 | 194 | 853 | | | | |
| Graphite, pyrolytic | 2273 | 2210 | | | | | | | | | | |
| $k$, ∥ to layers | | | | 1950 | | 4,970 | 3230 | 1390 | 892 | 667 | 534 | 448 |
| $k$, ⊥ to layers | | | | 5.70 | | 16.8 | 9.23 | 4.09 | 2.68 | 2.01 | 1.60 | 1.34 |
| $c_p$ | | | 709 | | | 136 | 411 | 992 | 1406 | 1650 | 1793 | 1890 |
| Graphite fiber epoxy (25% vol.) composite | 450 | 1400 | | | | | | | | | | |
| $k$, heat flow ∥ to fibers | | | | 11.1 | | 5.7 | 8.7 | 13.0 | | | | |
| $k$, heat flow ⊥ to fibers | | | | 0.87 | | 0.46 | 0.68 | 1.1 | | | | |
| $c_p$ | | | 935 | | | 337 | 642 | 1216 | | | | |
| Pyroceram, | 1623 | 2600 | 808 | 3.98 | 1.89 | 5.25 | 4.78 | 3.64 | 3.28 | 3.08 | 2.96 | 2.87 |
| Corning 9606 | | | | | | — | — | 908 | 1038 | 1122 | 1197 | 1264 |
| Silicon carbide | 3100 | 3160 | 675 | 490 | 230 | | | — | — | — | 87 | 58 |
| | | | | | | | | 880 | 1050 | 1135 | 1195 | 1243 |
| Silicon dioxide, crystalline (quartz) | 1883 | 2650 | | | | | | | | | | |
| $k$, ∥ to $c$ axis | | | | 10.4 | | 39 | 16.4 | 7.6 | 5.0 | 4.2 | | |
| $k$, ⊥ to $c$ axis | | | | 6.21 | | 20.8 | 9.5 | 4.70 | 3.4 | 3.1 | | |
| $c_p$ | | | 745 | | | — | — | 885 | 1075 | 1250 | | |
| Silicon dioxide, | 1883 | 2220 | 745 | 1.38 | 0.834 | 0.69 | 1.14 | 1.51 | 1.75 | 2.17 | 2.87 | 4.00 |
| polycrystalline (fused silica) | | | | | | — | — | 905 | 1040 | 1105 | 1155 | 1195 |
| Silicon nitride | 2173 | 2400 | 691 | 16.0 | 9.65 | — | — | 13.9 | 11.3 | 9.88 | 8.76 | 8.00 |
| | | | | | | — | 578 | 778 | 937 | 1063 | 1155 | 1226 |
| Sulfur | 392 | 2070 | 708 | 0.206 | 0.141 | 0.165 | 0.185 | | | | | |
| | | | | | | 403 | 606 | | | | | |
| Thorium dioxide | 3573 | 9110 | 235 | 13 | 6.1 | | | 10.2 | 6.6 | 4.7 | 3.68 | 3.12 |
| | | | | | | | | 255 | 274 | 285 | 295 | 303 |
| Titanium dioxide, | 2133 | 4157 | 710 | 8.4 | 2.8 | | | 7.01 | 5.02 | 3.94 | 3.46 | 3.28 |
| polycrystalline | | | | | | | | 805 | 880 | 910 | 930 | 945 |

**TABLE A-16**
**Properties of common materials**
**(*a*) Building materials**

| Description/Composition | Typical properties at 300 K | | |
|---|---|---|---|
| | **Density** $\rho$ **kg/m³** | **Thermal conductivity** $k$ **W/(m · K)** | **Specific heat** $c_p$ **J/(kg · K)** |
| Building boards | | | |
| Asbestos–cement board | 1920 | 0.58 | — |
| Gypsum or plaster board | 800 | 0.17 | — |
| Plywood | 545 | 0.12 | 1215 |
| Sheathing, regular-density | 290 | 0.055 | 1300 |
| Acoustic tile | 290 | 0.058 | 1340 |
| Hardboard, siding | 640 | 0.094 | 1170 |
| Hardboard, high-density | 1010 | 0.15 | 1380 |
| Particle board, low-density | 590 | 0.078 | 1300 |
| Particle board, high-density | 1000 | 0.170 | 1300 |
| Woods | | | |
| Hardwoods (oak, maple) | 720 | 0.16 | 1255 |
| Softwoods (fir, pine) | 510 | 0.12 | 1380 |
| Masonry materials | | | |
| Cement mortar | 1860 | 0.72 | 780 |
| Brick, common | 1920 | 0.72 | 835 |
| Brick, face | 2083 | 1.3 | — |
| Clay tile, hollow | | | |
| 1-cell-deep, 10-cm-thick | — | 0.52 | — |
| 3-cells-deep, 30-cm-thick | — | 0.69 | — |
| Concrete block, 3 oval cores | | | |
| sand/gravel, 20-cm-thick | — | 1.0 | — |
| cinder aggregate, 20-cm-thick | — | 0.67 | — |
| Concrete block, rectangular core | | | |
| 2 cores, 20-cm-thick, 16-kg | — | 1.1 | — |
| same with filled cores | — | 0.60 | — |
| Plastering materials | | | |
| Cement plaster, sand aggregate | 1860 | 0.72 | — |
| Gypsum plaster, sand aggregate | 1680 | 0.22 | 1085 |
| Gypsum plaster, vermiculite aggregate | 720 | 0.25 | — |

**TABLE A-16**
**Properties of common materials**
**(*b*) Insulating materials**

| Description/Composition | Density $\rho$ kg/m³ | Thermal conductivity $k$ W/(m · K) | Specific heat $c_p$ J/(kg · K) |
|---|---|---|---|
| | Typical properties at 300 K | | |
| Blanket and batt | | | |
| Glass fiber, paper faced | 16 | 0.046 | — |
| | 28 | 0.038 | — |
| | 40 | 0.035 | — |
| Glass fiber, coated; duct liner | 32 | 0.038 | 835 |
| Board and slab | | | |
| Cellular glass | 145 | 0.058 | 1000 |
| Glass fiber, organic bonded | 105 | 0.036 | 795 |
| Polystyrene, expanded | | | |
| extruded (R-12) | 55 | 0.027 | 1210 |
| molded beads | 16 | 0.040 | 1210 |
| Mineral fiberboard; roofing material | 265 | 0.049 | — |
| Wood, shredded/cemented | 350 | 0.087 | 1590 |
| Cork | 120 | 0.039 | 1800 |
| Loose fill | | | |
| Cork, granulated | 160 | 0.045 | — |
| Diatomaceous silica, coarse | 350 | 0.069 | — |
| powder | 400 | 0.091 | — |
| Diatomaceous silica, fine powder | 200 | 0.052 | — |
| | 275 | 0.061 | — |
| Glass fiber, poured or blown | 16 | 0.043 | 835 |
| Vermiculite, flakes | 80 | 0.068 | 835 |
| | 160 | 0.063 | 1000 |
| Formed/foamed-in-place | | | |
| Mineral wool granules with asbestor/inorganic binders, sprayed | 190 | 0.046 | — |
| Polyvinyl acetate cork mastic; sprayed or troweled | — | 0.100 | — |
| Urethane, two-part mixture; rigid form | 70 | 0.026 | 1045 |
| Reflective | | | |
| Aluminum foil separating fluffy glass mats; 10–12 layers; evacuated; for cryogenic applications (150 K) | 40 | 0.00016 | — |
| Aluminum foil and glass paper laminate; 75–150 layers; evacuated; for cryogenic application (150 K) | 120 | 0.000017 | — |
| Typical silica powder, evacuated | 160 | 0.0017 | — |

**TABLE A-16**
**Properties of common materials**
**(c) Other materials**

| Description/Composition | Temperature K | Density $\rho$ kg/m$^3$ | Thermal conductivity $k$ W/(m · K) | Specific heat $c_p$ J/(kg · K) |
|---|---|---|---|---|
| Asphalt | 300 | 2115 | 0.062 | 920 |
| Bakelite | 300 | 1300 | 1.4 | 1465 |
| Brick, refractory | | | | |
| Carborundum | 872 | — | 18.5 | — |
| | 1672 | — | 11.0 | — |
| Chrome brick | 473 | 3010 | 2.3 | 835 |
| | 823 | | 2.5 | |
| | 1173 | | 2.0 | |
| Diatomaceous | 478 | — | 0.25 | — |
| silica, fired | 1145 | — | 0.30 | |
| Fire clay, burnt 1600 K | 773 | 2050 | 1.0 | 960 |
| | 1073 | — | 1.1 | |
| | 1373 | — | 1.1 | |
| Fire clay, burnt 1725 k | 773 | 2325 | 1.3 | 960 |
| | 1073 | | 1.4 | |
| | 1373 | | 1.4 | |
| Fire clay brick | 478 | 2645 | 1.0 | 960 |
| | 922 | | 1.5 | |
| | 1478 | | 1.8 | |
| Magnesite | 478 | — | 3.8 | 1130 |
| | 922 | — | 2.8 | |
| | 1478 | | 1.9 | |
| Clay | 300 | 1460 | 1.3 | 880 |
| Coal, anthracite | 300 | 1350 | 0.26 | 1260 |
| Concrete (stone mix) | 300 | 2300 | 1.4 | 880 |
| Cotton | 300 | 80 | 0.06 | 1300 |
| Foodstuffs | | | | |
| Banana (75.7% water content) | 300 | 980 | 0.481 | 3350 |
| Apple, red (75% water content | 300 | 840 | 0.513 | 3600 |
| Cake, batter | 300 | 720 | 0.223 | — |
| Cake, fully baked | 300 | 280 | 0.121 | — |
| Chicken meat, white | 198 | — | 1.60 | — |
| (74.4% water content) | 233 | — | 1.49 | |
| | 253 | | 1.35 | |
| | 263 | | 1.20 | |
| | 273 | | 0.476 | |
| | 283 | | 0.480 | |
| | 293 | | 0.489 | |

**TABLE A-16**
**Properties of common materials**
**(*c*) Other materials (*Continued*)**

| Description/Composition | Temperature K | Density $\rho$ kg/m³ | Thermal conductivity $k$ W/(m · K) | Specific heat $c_p$ J/(kg · K) |
|---|---|---|---|---|
| Glass | | | | |
| Plate (soda lime) | 300 | 2500 | 1.4 | 750 |
| Pyrex | 300 | 2225 | 1.4 | 835 |
| Ice | 273 | 920 | 1.88 | 2040 |
| | 253 | — | 2.03 | 1945 |
| Leather (sole) | 300 | 998 | 0.159 | — |
| Paper | 300 | 930 | 0.180 | 1340 |
| Paraffin | 300 | 900 | 0.240 | 2890 |
| Rock | | | | |
| Granite, Barre | 300 | 2630 | 2.79 | 775 |
| Limestone, Salem | 300 | 2320 | 2.15 | 810 |
| Marble, Halston | 300 | 2680 | 2.80 | 830 |
| Quartzite, Sioux | 300 | 2640 | 5.38 | 1105 |
| Sandstone, Berea | 300 | 2150 | 2.90 | 745 |
| Rubber, vulcanized | | | | |
| Soft | 300 | 1100 | 0.13 | 2010 |
| Hard | 300 | 1190 | 0.16 | — |
| Sand | 300 | 1515 | 0.27 | 800 |
| Soil | 300 | 2050 | 0.52 | 1840 |
| Snow | 273 | 110 | 0.049 | — |
| | | 500 | 0.190 | — |
| Teflon | 300 | 2200 | 0.35 | — |
| | 400 | | 0.45 | — |
| Tissue, human | | | | |
| Skin | 300 | — | 0.37 | — |
| Fat layer (adipose) | 300 | — | 0.2 | — |
| Muscle | 300 | — | 0.41 | — |
| Wood, cross-grain | | | | |
| Balsa | 300 | 140 | 0.055 | — |
| Cypress | 300 | 465 | 0.097 | — |
| Fir | 300 | 415 | 0.11 | 2720 |
| Oak | 300 | 545 | 0.17 | 2385 |
| Yellow pine | 300 | 640 | 0.15 | 2805 |
| White pine | 300 | 435 | 0.11 | — |
| Wood, radial | | | | |
| Oak | 300 | 545 | 0.19 | 2385 |
| Fir | 300 | 420 | 0.14 | 2720 |

**TABLE A-17**
**Properties of some liquid metals**

| Temperature $T$ K | Density $\rho$ kg/m$^3$ | Specific heat $C_p$ J/(kg · °C) | Thermal conductivity $k$ W/(m · °C) | Thermal diffusivity $\alpha$ m$^2$/s | Dynamic viscosity $\mu$ kg/m · s | Kinematic viscosity $\nu$ m$^2$/s | Prandtl number Pr |
|---|---|---|---|---|---|---|---|
| | | | *Bismuth* | | | | |
| 545* | 10,069 | 143 | 16.8 | $1.17 \times 10^{-5}$ | $1.75 \times 10^{-3}$ | $1.74 \times 10^{-7}$ | 0.0148 |
| 600 | 9,997 | 145 | 16.4 | $1.13 \times 10^{-5}$ | $1.61 \times 10^{-3}$ | $1.61 \times 10^{-7}$ | 0.0142 |
| 700 | 9,867 | 150 | 15.6 | $1.06 \times 10^{-5}$ | $1.34 \times 10^{-3}$ | $1.36 \times 10^{-7}$ | 0.0128 |
| 800 | 9,752 | 154 | 15.6 | $1.04 \times 10^{-5}$ | $1.12 \times 10^{-3}$ | $1.15 \times 10^{-7}$ | 0.0111 |
| 900 | 9,636 | 159 | 15.6 | $1.02 \times 10^{-5}$ | $0.96 \times 10^{-3}$ | $0.99 \times 10^{-7}$ | 0.0098 |
| 1000 | 9,510 | 163 | 15.6 | $1.01 \times 10^{-5}$ | $0.83 \times 10^{-3}$ | $0.87 \times 10^{-7}$ | 0.0087 |
| | | | *Lead* | | | | |
| 601* | 10,588 | 161 | 15.5 | $0.91 \times 10^{-5}$ | $2.62 \times 10^{-3}$ | $2.47 \times 10^{-7}$ | 0.0272 |
| 700 | 10,476 | 157 | 17.4 | $1.06 \times 10^{-5}$ | $2.15 \times 10^{-3}$ | $2.05 \times 10^{-7}$ | 0.0194 |
| 800 | 10,359 | 153 | 19.0 | $1.20 \times 10^{-5}$ | $2.05 \times 10^{-3}$ | $1.98 \times 10^{-7}$ | 0.0165 |
| 900 | 10,237 | 149 | 20.3 | $1.33 \times 10^{-5}$ | $1.54 \times 10^{-3}$ | $1.50 \times 10^{-7}$ | 0.0113 |
| 1000 | 10,111 | 145 | 21.5 | $1.47 \times 10^{-5}$ | $1.32 \times 10^{-3}$ | $1.30 \times 10^{-7}$ | 0.0089 |
| | | | *Mercury* | | | | |
| 234* | 13,723 | 142 | 7.3 | $3.8 \times 10^{-6}$ | $2.00 \times 10^{-3}$ | $1.46 \times 10^{-7}$ | 0.0389 |
| 273 | 13,628 | 140 | 8·2 | $4.3 \times 10^{-6}$ | $1.69 \times 10^{-3}$ | $1.24 \times 10^{-7}$ | 0.0289 |
| 300 | 13,562 | 139 | 8·9 | $4.7 \times 10^{-6}$ | $1.51 \times 10^{-3}$ | $1.11 \times 10^{-7}$ | 0.0237 |
| 350 | 13,441 | 138 | 10.0 | $5.4 \times 10^{-6}$ | $1.31 \times 10^{-3}$ | $0.98 \times 10^{-7}$ | 0.0181 |
| 400 | 13,320 | 137 | 11.0 | $6.1 \times 10^{-6}$ | $1.18 \times 10^{-3}$ | $0.89 \times 10^{-7}$ | 0.0147 |
| 500 | 13,081 | 136 | 12.7 | $7.1 \times 10^{-6}$ | $1.02 \times 10^{-3}$ | $0.78 \times 10^{-7}$ | 0.0109 |
| 600 | 12,816 | 134 | 14.2 | $8.3 \times 10^{-6}$ | $0.84 \times 10^{-3}$ | $0.66 \times 10^{-7}$ | 0.0080 |
| | | | *Lithium* | | | | |
| 454* | 512 | 4190 | 43 | $2.0 \times 10^{-5}$ | $6.1 \times 10^{-4}$ | $1.18 \times 10^{-6}$ | 0.059 |
| 500 | 508 | 4190 | 44 | $2.1 \times 10^{-5}$ | $5.9 \times 10^{-4}$ | $1.17 \times 10^{-6}$ | 0.056 |
| 600 | 498 | 4190 | 48 | $2.3 \times 10^{-5}$ | $5.7 \times 10^{-4}$ | $1.14 \times 10^{-6}$ | 0.050 |
| 700 | 489 | 4190 | 51 | $2.4 \times 10^{-5}$ | $5.4 \times 10^{-4}$ | $1.11 \times 10^{-6}$ | 0.045 |
| 800 | 480 | 4190 | 54 | $2.7 \times 10^{-5}$ | $5.2 \times 10^{-4}$ | $1.08 \times 10^{-6}$ | 0.040 |
| 900 | 471 | 4190 | 57 | $2.9 \times 10^{-5}$ | $4.9 \times 10^{-4}$ | $1.05 \times 10^{-6}$ | 0.036 |
| 1000 | 462 | 4190 | 60 | $3.1 \times 10^{-5}$ | $4.7 \times 10^{-4}$ | $1.02 \times 10^{-6}$ | 0.033 |
| | | | *Potassium* | | | | |
| 337* | 827 | 802 | 55 | $8.3 \times 10^{-5}$ | $4.7 \times 10^{-4}$ | $5.6 \times 10^{-7}$ | 0.0068 |
| 400 | 812 | 798 | 52 | $8.0 \times 10^{-5}$ | $3.9 \times 10^{-4}$ | $4.9 \times 10^{-7}$ | 0.0061 |
| 500 | 789 | 790 | 48 | $7.7 \times 10^{-5}$ | $3.0 \times 10^{-4}$ | $3.8 \times 10^{-7}$ | 0.0050 |
| 600 | 766 | 783 | 44 | $7.3 \times 10^{-5}$ | $2.3 \times 10^{-4}$ | $3.0 \times 10^{-7}$ | 0.0041 |
| 700 | 742 | 775 | 40 | $7.0 \times 10^{-5}$ | $1.8 \times 10^{-4}$ | $2.4 \times 10^{-7}$ | 0.0034 |
| 800 | 718 | 767 | 37 | $6.7 \times 10^{-5}$ | $1.6 \times 10^{-4}$ | $2.2 \times 10^{-7}$ | 0.0033 |
| 900 | 693 | 760 | 34 | $6.5 \times 10^{-5}$ | $1.4 \times 10^{-4}$ | $2.0 \times 10^{-7}$ | 0.0032 |
| 1000 | 669 | 752 | 31 | $6.2 \times 10^{-5}$ | $1.3 \times 10^{-4}$ | $1.9 \times 10^{-7}$ | 0.0030 |
| | | | *Sodium* | | | | |
| 371* | 929 | 1382 | 88 | $6.9 \times 10^{-5}$ | $7.0 \times 10^{-4}$ | $7.5 \times 10^{-7}$ | 0.0110 |
| 400 | 922 | 1371 | 87 | $6.9 \times 10^{-5}$ | $6.1 \times 10^{-4}$ | $6.7 \times 10^{-7}$ | 0.0097 |
| 500 | 896 | 1334 | 82 | $6.8 \times 10^{-5}$ | $4.1 \times 10^{-4}$ | $4.6 \times 10^{-7}$ | 0.0067 |
| 600 | 871 | 1309 | 76 | $6.7 \times 10^{-5}$ | $3.2 \times 10^{-4}$ | $3.6 \times 10^{-7}$ | 0.0054 |
| 700 | 846 | 1284 | 72 | $6.6 \times 10^{-5}$ | $2.6 \times 10^{-4}$ | $3.0 \times 10^{-7}$ | 0.0046 |
| 800 | 822 | 1259 | 67 | $6.5 \times 10^{-5}$ | $2.1 \times 10^{-4}$ | $2.6 \times 10^{-7}$ | 0.0040 |
| 900 | 797 | 1256 | 63 | $6.2 \times 10^{-5}$ | $1.9 \times 10^{-4}$ | $2.4 \times 10^{-7}$ | 0.0039 |
| 1000 | 773 | 1256 | 58 | $6.0 \times 10^{-5}$ | $1.8 \times 10^{-4}$ | $2.3 \times 10^{-7}$ | 0.0038 |

* Melting point.
*Source:* Tables A-17 through A-19 are adapted from Frank M. White, *Heat and Mass Transfer,* Addison-Wesley, Reading, MA, 1988, pp. 677–688 and 692–694. Originally compiled from various sources. Reprinted by permission of Addison-Wesley Longman Publishing Company, Inc.

## TABLE A-18
## Properties of some liquids

| Temperature $T$ °C | Density $\rho$ kg/m$^3$ | Specific heat $C_p$ J/(kg · ·C) | Thermal conductivity $k$ W/(m · °C) | Thermal diffusivity $\alpha$ m$^2$/s | Dynamic viscosity $\mu$ kg/(m · s) | Kinematic viscosity $\nu$ m$^2$/s | Prandtl number Pr |
|---|---|---|---|---|---|---|---|
| | | | | *Ammonia* | | | |
| −40 | 692 | 4467 | 0.546 | $1.78 \times 10^{-7}$ | $2.81 \times 10^{-4}$ | $4.06 \times 10^{-7}$ | 2.28 |
| −20 | 667 | 4509 | 0.546 | $1.82 \times 10^{-7}$ | $2.54 \times 10^{-4}$ | $3.81 \times 10^{-7}$ | 2.09 |
| 0 | 640 | 4635 | 0.540 | $1.82 \times 10^{-7}$ | $2.39 \times 10^{-4}$ | $3.73 \times 10^{-7}$ | 2.05 |
| 20 | 612 | 4798 | 0.521 | $1.78 \times 10^{-7}$ | $2.20 \times 10^{-4}$ | $3.59 \times 10^{-7}$ | 2.02 |
| 40 | 581 | 4999 | 0.493 | $1.70 \times 10^{-7}$ | $1.98 \times 10^{-4}$ | $3.40 \times 10^{-7}$ | 2.00 |
| | | | | *Ethyl alcohol* ($C_2H_6O$) | | | |
| −40 | 823 | 2037 | 0.186 | $1.11 \times 10^{-7}$ | $4.81 \times 10^{-3}$ | $5.84 \times 10^{-6}$ | 52.7 |
| −20 | 815 | 2124 | 0.179 | $1.03 \times 10^{-7}$ | $2.83 \times 10^{-3}$ | $3.47 \times 10^{-6}$ | 33.6 |
| 0 | 806 | 2249 | 0.174 | $0.960 \times 10^{-7}$ | $1.77 \times 10^{-3}$ | $2.20 \times 10^{-6}$ | 22.9 |
| 20 | 789 | 2395 | 0.168 | $0.889 \times 10^{-7}$ | $1.20 \times 10^{-3}$ | $1.52 \times 10^{-6}$ | 17.0 |
| 40 | 772 | 2572 | 0.162 | $0.816 \times 10^{-7}$ | $0.834 \times 10^{-3}$ | $1.08 \times 10^{-6}$ | 13.2 |
| 60 | 755 | 2781 | 0.156 | $0.743 \times 10^{-7}$ | $0.592 \times 10^{-3}$ | $0.784 \times 10^{-6}$ | 10.6 |
| 80 | 738 | 3026 | 0.150 | $0.672 \times 10^{-7}$ | $0.430 \times 10^{-3}$ | $0.583 \times 10^{-6}$ | 8.7 |
| | | | | *Ethylene glycol* ($C_2H_6O_2$) | | | |
| 0 | 1131 | 2295 | 0.254 | $9.79 \times 10^{-8}$ | $65.1 \times 10^{-3}$ | $57.5 \times 10^{-6}$ | 588 |
| 20 | 1117 | 2386 | 0.257 | $9.64 \times 10^{-8}$ | $21.4 \times 10^{-3}$ | $19.2 \times 10^{-6}$ | 199 |
| 40 | 1101 | 2476 | 0.259 | $9.50 \times 10^{-8}$ | $9.57 \times 10^{-3}$ | $8.69 \times 10^{-6}$ | 91 |
| 60 | 1088 | 2565 | 0.262 | $9.39 \times 10^{-8}$ | $5.17 \times 10^{-3}$ | $4.75 \times 10^{-6}$ | 51 |
| 80 | 1078 | 2656 | 0.265 | $9.26 \times 10^{-8}$ | $3.21 \times 10^{-3}$ | $2.98 \times 10^{-6}$ | 32 |
| 100 | 1059 | 2750 | 0.267 | $9.17 \times 10^{-8}$ | $2.15 \times 10^{-3}$ | $2.03 \times 10^{-6}$ | 22 |
| | | | | *Freon-12 refrigerant* ($CCl_2F_2$) | | | |
| −40 | 1515 | 885 | 0.069 | $5.14 \times 10^{-8}$ | $4.24 \times 10^{-4}$ | $2.80 \times 10^{-7}$ | 5.4 |
| −20 | 1457 | 907 | 0.071 | $5.38 \times 10^{-8}$ | $3.43 \times 10^{-4}$ | $2.35 \times 10^{-7}$ | 4.4 |
| 0 | 1393 | 935 | 0.073 | $5.59 \times 10^{-8}$ | $2.98 \times 10^{-4}$ | $2.14 \times 10^{-7}$ | 3.8 |
| 20 | 1327 | 966 | 0.073 | $5.66 \times 10^{-8}$ | $2.62 \times 10^{-4}$ | $1.97 \times 10^{-7}$ | 3.5 |
| 40 | 1254 | 1002 | 0.069 | $5.46 \times 10^{-8}$ | $2.40 \times 10^{-4}$ | $1.91 \times 10^{-7}$ | 3.5 |
| | | | | *Glycerin* | | | |
| −20 | 1288 | 2143 | 0.282 | $1.02 \times 10^{-7}$ | 134 | $104 \times 10^{-3}$ | 1.020 |
| 0 | 1276 | 2261 | 0.284 | $0.98 \times 10^{-7}$ | 12.1 | $9.5 \times 10^{-3}$ | 0.096 |
| 20 | 1264 | 2386 | 0.287 | $0.95 \times 10^{-7}$ | 1.49 | $1.2 \times 10^{-3}$ | 0.0124 |
| 40 | 1252 | 2513 | 0.290 | $0.92 \times 10^{-7}$ | 0.27 | $0.2 \times 10^{-3}$ | 0.0023 |
| | | | | *Unused engine oil* | | | |
| 0 | 899 | 1796 | 0.147 | $9.11 \times 10^{-8}$ | $3850 \times 10^{-3}$ | $4280 \times 10^{-6}$ | 47100 |
| 20 | 888 | 1880 | 0.145 | $8.72 \times 10^{-8}$ | $800 \times 10^{-3}$ | $901 \times 10^{-6}$ | 10400 |
| 40 | 876 | 1964 | 0.144 | $8.34 \times 10^{-8}$ | $212 \times 10^{-3}$ | $242 \times 10^{-6}$ | 2870 |
| 60 | 864 | 2047 | 0.140 | $8.00 \times 10^{-8}$ | $72.5 \times 10^{-3}$ | $83.9 \times 10^{-6}$ | 1050 |
| 80 | 852 | 2131 | 0.138 | $7.69 \times 10^{-8}$ | $32.0 \times 10^{-3}$ | $37.5 \times 10^{-6}$ | 490 |
| 100 | 840 | 2219 | 0.137 | $7.38 \times 10^{-8}$ | $17.1 \times 10^{-3}$ | $20.3 \times 10^{-6}$ | 276 |
| 120 | 829 | 2307 | 0.135 | $7.10 \times 10^{-8}$ | $10.2 \times 10^{-3}$ | $12.4 \times 10^{-6}$ | 175 |
| 140 | 817 | 2395 | 0.133 | $6.86 \times 10^{-8}$ | $6.53 \times 10^{-3}$ | $8.0 \times 10^{-6}$ | 116 |
| 160 | 806 | 2483 | 0.132 | $6.63 \times 10^{-8}$ | $4.49 \times 10^{-3}$ | $5.6 \times 10^{-6}$ | 84 |
| | | | | *Water* ($T$ in K) | | | |
| 273.2 | 1000 | 4205 | 0.564 | $1.34 \times 10^{-7}$ | $1.79 \times 10^{-3}$ | $1.79 \times 10^{-6}$ | 13.4 |
| 280 | 1000 | 4197 | 0.582 | $1.39 \times 10^{-7}$ | $1.44 \times 10^{-3}$ | $1.44 \times 10^{-6}$ | 10.4 |
| 300 | 997 | 4177 | 0.608 | $1.46 \times 10^{-7}$ | $0.857 \times 10^{-3}$ | $0.86 \times 10^{-6}$ | 5.88 |
| 320 | 989 | 4176 | 0.637 | $1.54 \times 10^{-7}$ | $0.579 \times 10^{-3}$ | $0.59 \times 10^{-6}$ | 3.79 |
| 340 | 980 | 4187 | 0.659 | $1.61 \times 10^{-7}$ | $0.423 \times 10^{-3}$ | $0.43 \times 10^{-6}$ | 2.69 |
| 360 | 967 | 4204 | 0.674 | $1.66 \times 10^{-7}$ | $0.320 \times 10^{-3}$ | $0.33 \times 10^{-6}$ | 2.00 |
| 373.2 | 958 | 4220 | 0.681 | $1.68 \times 10^{-7}$ | $0.282 \times 10^{-3}$ | $0.29 \times 10^{-6}$ | 1.75 |
| 400 | 937 | 4241 | 0.686 | $1.73 \times 10^{-7}$ | $0.219 \times 10^{-3}$ | $0.23 \times 10^{-6}$ | 1.35 |
| 450 | 890 | 4419 | 0.673 | $1.71 \times 10^{-7}$ | $0.153 \times 10^{-3}$ | $0.17 \times 10^{-6}$ | 1.01 |
| 500 | 832 | 4647 | 0.635 | $1.64 \times 10^{-7}$ | $0.118 \times 10^{-3}$ | $0.14 \times 10^{-6}$ | 0.86 |
| 550 | 756 | 5272 | 0.571 | $1.43 \times 10^{-7}$ | $0.095 \times 10^{-3}$ | $0.13 \times 10^{-6}$ | 0.88 |
| 600 | 650 | 6691 | 0.481 | $1.11 \times 10^{-7}$ | $0.076 \times 10^{-3}$ | $0.12 \times 10^{-6}$ | 1.05 |
| 647.3* | 315 | — | — | — | — | — | — |

* Critical point.

**TABLE A-19**
**Properties of some gases at 1-atm pressure**

| Temperature $T$ K | Density $\rho$ kg/m³ | Specific heat $C_p$ J/(kg · °C) | Thermal conductivity $k$ W/(m · °C) | Thermal diffusivity $\alpha$ m²/s | Dynamic viscosity $\mu$ kg/(m · s) | Kinematic viscosity $\nu$ m²/s | Prandtl number Pr |
|---|---|---|---|---|---|---|---|
| | | | | *Air* | | | |
| 200 | 1.766 | 1003 | 0.0181 | $1.02 \times 10^{-5}$ | $1.34 \times 10^{-5}$ | $0.76 \times 10^{-5}$ | 0.740 |
| 250 | 1.413 | 1003 | 0.0223 | $1.57 \times 10^{-5}$ | $1.61 \times 10^{-5}$ | $1.14 \times 10^{-5}$ | 0.724 |
| 280 | 1.271 | 1004 | 0.0246 | $1.95 \times 10^{-5}$ | $1.75 \times 10^{-5}$ | $1.40 \times 10^{-5}$ | 0.717 |
| 290 | 1.224 | 1005 | 0.0253 | $2.08 \times 10^{-5}$ | $1.80 \times 10^{-5}$ | $1.48 \times 10^{-5}$ | 0.714 |
| 300 | 1.177 | 1005 | 0.0261 | $2.21 \times 10^{-5}$ | $1.85 \times 10^{-5}$ | $1.57 \times 10^{-5}$ | 0.712 |
| 310 | 1.143 | 1006 | 0.0268 | $2.35 \times 10^{-5}$ | $1.90 \times 10^{-5}$ | $1.67 \times 10^{-5}$ | 0.711 |
| 320 | 1.110 | 1006 | 0.0275 | $2.49 \times 10^{-5}$ | $1.94 \times 10^{-5}$ | $1.77 \times 10^{-5}$ | 0.710 |
| 330 | 1.076 | 1007 | 0.0283 | $2.64 \times 10^{-5}$ | $1.99 \times 10^{-5}$ | $1.86 \times 10^{-5}$ | 0.708 |
| 340 | 1.043 | 1007 | 0.0290 | $2.78 \times 10^{-5}$ | $2.03 \times 10^{-5}$ | $1.96 \times 10^{-5}$ | 0.707 |
| 350 | 1.009 | 1008 | 0.0297 | $2.92 \times 10^{-5}$ | $2.08 \times 10^{-5}$ | $2.06 \times 10^{-5}$ | 0.706 |
| 400 | 0.883 | 1013 | 0.0331 | $3.70 \times 10^{-5}$ | $2.29 \times 10^{-5}$ | $2.60 \times 10^{-5}$ | 0.703 |
| 450 | 0.785 | 1020 | 0.0363 | $4.54 \times 10^{-5}$ | $2.49 \times 10^{-5}$ | $3.18 \times 10^{-5}$ | 0.700 |
| 500 | 0.706 | 1029 | 0.0395 | $5.44 \times 10^{-5}$ | $2.68 \times 10^{-5}$ | $3.80 \times 10^{-5}$ | 0.699 |
| 550 | 0.642 | 1039 | 0.0426 | $6.39 \times 10^{-5}$ | $2.86 \times 10^{-5}$ | $4.45 \times 10^{-5}$ | 0.698 |
| 600 | 0.589 | 1051 | 0.0456 | $7.37 \times 10^{-5}$ | $3.03 \times 10^{-5}$ | $5.15 \times 10^{-5}$ | 0.698 |
| 700 | 0.504 | 1075 | 0.0513 | $9.46 \times 10^{-5}$ | $3.35 \times 10^{-5}$ | $6.64 \times 10^{-5}$ | 0.702 |
| 800 | 0.441 | 1099 | 0.0569 | $11.7 \times 10^{-5}$ | $3.64 \times 10^{-5}$ | $8.25 \times 10^{-5}$ | 0.704 |
| 900 | 0.392 | 1120 | 0.0625 | $14.2 \times 10^{-5}$ | $3.92 \times 10^{-5}$ | $9.99 \times 10^{-5}$ | 0.705 |
| 1000 | 0.353 | 1141 | 0.0672 | $16.7 \times 10^{-5}$ | $4.18 \times 10^{-5}$ | $11.8 \times 10^{-5}$ | 0.709 |
| 1200 | 0.294 | 1175 | 0.0759 | $22.2 \times 10^{-5}$ | $4.65 \times 10^{-5}$ | $15.8 \times 10^{-5}$ | 0.720 |
| 1400 | 0.252 | 1201 | 0.0835 | $27.6 \times 10^{-5}$ | $5.09 \times 10^{-5}$ | $20.2 \times 10^{-5}$ | 0.732 |
| 1600 | 0.221 | 1240 | 0.0904 | $33.0 \times 10^{-5}$ | $5.49 \times 10^{-5}$ | $24.9 \times 10^{-5}$ | 0.753 |
| 1800 | 0.196 | 1276 | 0.0970 | $38.3 \times 10^{-5}$ | $5.87 \times 10^{-5}$ | $29.9 \times 10^{-5}$ | 0.772 |
| 2000 | 0.177 | 1327 | 0.1032 | $44.1 \times 10^{-5}$ | $6.23 \times 10^{-5}$ | $35.3 \times 10^{-5}$ | 0.801 |
| | | | | *Ammonia* ($NH_3$) | | | |
| 200 | 1.038 | 2199 | 0.0153 | $0.67 \times 10^{-5}$ | $6.89 \times 10^{-6}$ | $0.66 \times 10^{-5}$ | 0.990 |
| 250 | 0.831 | 2248 | 0.0197 | $1.05 \times 10^{-5}$ | $8.53 \times 10^{-6}$ | $1.03 \times 10^{-5}$ | 0.973 |
| 300 | 0.692 | 2298 | 0.0246 | $1.55 \times 10^{-5}$ | $10.27 \times 10^{-6}$ | $1.48 \times 10^{-5}$ | 0.959 |
| 350 | 0.593 | 2349 | 0.0302 | $2.17 \times 10^{-5}$ | $12.06 \times 10^{-6}$ | $2.03 \times 10^{-5}$ | 0.938 |
| 400 | 0.519 | 2402 | 0.0364 | $2.92 \times 10^{-5}$ | $13.90 \times 10^{-6}$ | $2.68 \times 10^{-5}$ | 0.917 |
| 450 | 0.461 | 2455 | 0.0433 | $3.82 \times 10^{-5}$ | $15.76 \times 10^{-6}$ | $3.42 \times 10^{-5}$ | 0.894 |
| 500 | 0.415 | 2507 | 0.0506 | $4.86 \times 10^{-5}$ | $17.63 \times 10^{-6}$ | $4.25 \times 10^{-5}$ | 0.873 |
| 550 | 0.378 | 2559 | 0.0580 | $6.00 \times 10^{-5}$ | $19.5 \times 10^{-6}$ | $5.16 \times 10^{-5}$ | 0.860 |
| 600 | 0.346 | 2611 | 0.0656 | $7.26 \times 10^{-5}$ | $21.4 \times 10^{-6}$ | $6.18 \times 10^{-5}$ | 0.852 |
| 700 | 0.297 | 2710 | 0.0811 | $10.1 \times 10^{-5}$ | $25.1 \times 10^{-6}$ | $8.45 \times 10^{-5}$ | 0.839 |
| 800 | 0.260 | 2810 | 0.0977 | $13.4 \times 10^{-5}$ | $28.8 \times 10^{-6}$ | $11.1 \times 10^{-5}$ | 0.828 |
| | | | | *Argon* | | | |
| 200 | 2.435 | 523.6 | 0.0124 | $0.98 \times 10^{-5}$ | $1.60 \times 10^{-5}$ | $0.66 \times 10^{-5}$ | 0.674 |
| 250 | 1.948 | 522.2 | 0.0152 | $1.49 \times 10^{-5}$ | $1.95 \times 10^{-5}$ | $1.00 \times 10^{-5}$ | 0.672 |
| 300 | 1.623 | 521.6 | 0.0177 | $2.09 \times 10^{-5}$ | $2.27 \times 10^{-5}$ | $1.40 \times 10^{-5}$ | 0.669 |
| 350 | 1.392 | 521.2 | 0.0201 | $2.78 \times 10^{-5}$ | $2.57 \times 10^{-5}$ | $1.85 \times 10^{-5}$ | 0.666 |
| 400 | 1.218 | 521.0 | 0.0223 | $3.52 \times 10^{-5}$ | $2.85 \times 10^{-5}$ | $2.34 \times 10^{-5}$ | 0.665 |
| 450 | 1.082 | 520.9 | 0.0244 | $4.33 \times 10^{-5}$ | $3.12 \times 10^{-5}$ | $2.88 \times 10^{-5}$ | 0.665 |
| 500 | 0.974 | 520.8 | 0.0264 | $5.20 \times 10^{-5}$ | $3.37 \times 10^{-5}$ | $3.45 \times 10^{-5}$ | 0.664 |
| 550 | 0.886 | 520.7 | 0.0283 | $6.14 \times 10^{-5}$ | $3.60 \times 10^{-5}$ | $4.07 \times 10^{-5}$ | 0.662 |
| 600 | 0.812 | 520.6 | 0.0301 | $7.12 \times 10^{-5}$ | $3.83 \times 10^{-5}$ | $4.72 \times 10^{-5}$ | 0.662 |
| 700 | 0.696 | 520.6 | 0.0336 | $9.28 \times 10^{-5}$ | $4.25 \times 10^{-5}$ | $6.11 \times 10^{-5}$ | 0.658 |
| 800 | 0.609 | 520.5 | 0.0369 | $11.6 \times 10^{-5}$ | $4.64 \times 10^{-5}$ | $7.62 \times 10^{-5}$ | 0.655 |
| 900 | 0.541 | 520.5 | 0.0398 | $14.1 \times 10^{-5}$ | $5.01 \times 10^{-5}$ | $9.26 \times 10^{-5}$ | 0.654 |
| 1000 | 0.487 | 520.5 | 0.0427 | $16.8 \times 10^{-5}$ | $5.35 \times 10^{-5}$ | $11.0 \times 10^{-5}$ | 0.652 |
| 1200 | 0.406 | 520.5 | 0.0481 | $22.8 \times 10^{-5}$ | $5.99 \times 10^{-5}$ | $14.8 \times 10^{-5}$ | 0.648 |
| 1400 | 0.348 | 520.4 | 0.0535 | $29.6 \times 10^{-5}$ | $6.56 \times 10^{-5}$ | $18.9 \times 10^{-5}$ | 0.638 |

**Properties of some gases at 1-atm pressure (*Continued*)**

| Temperature $T$ K | Density $\rho$ kg/m$^3$ | Specific heat $C_p$ J/(kg · °C) | Thermal conductivity $k$ W/(m · °C) | Thermal diffusivity $\alpha$ m$^2$/s | Dynamic viscosity $\mu$ kg/(m · s) | Kinematic viscosity $\nu$ m$^2$/s | Prandtl number Pr |
|---|---|---|---|---|---|---|---|
| | | | *Hydrogen* | | | | |
| 200 | 0.1299 | 13,540 | 0.128 | $0.77 \times 10^{-4}$ | $0.68 \times 10^{-5}$ | $0.55 \times 10^{-4}$ | 0.717 |
| 250 | 0.0983 | 14,070 | 0.156 | $1.13 \times 10^{-4}$ | $0.79 \times 10^{-5}$ | $0.80 \times 10^{-4}$ | 0.713 |
| 300 | 0.0819 | 14,320 | 0.182 | $1.55 \times 10^{-4}$ | $0.89 \times 10^{-5}$ | $1.09 \times 10^{-4}$ | 0.705 |
| 350 | 0.0702 | 14,420 | 0.203 | $2.01 \times 10^{-4}$ | $0.99 \times 10^{-5}$ | $1.42 \times 10^{-4}$ | 0.705 |
| 400 | 0.0614 | 14,480 | 0.221 | $2.49 \times 10^{-4}$ | $1.09 \times 10^{-5}$ | $1.78 \times 10^{-4}$ | 0.714 |
| 450 | 0.0546 | 14,500 | 0.239 | $3.02 \times 10^{-4}$ | $1.18 \times 10^{-5}$ | $2.17 \times 10^{-4}$ | 0.719 |
| 500 | 0.0492 | 14,510 | 0.256 | $3.59 \times 10^{-4}$ | $1.27 \times 10^{-5}$ | $2.59 \times 10^{-4}$ | 0.721 |
| 550 | 0.0447 | 14,520 | 0.274 | $4.22 \times 10^{-4}$ | $1.36 \times 10^{-5}$ | $3.04 \times 10^{-4}$ | 0.722 |
| 600 | 0.0410 | 14,540 | 0.291 | $4.89 \times 10^{-4}$ | $1.45 \times 10^{-5}$ | $3.54 \times 10^{-4}$ | 0.724 |
| 700 | 0.0351 | 14,610 | 0.325 | $6.34 \times 10^{-4}$ | $1.61 \times 10^{-5}$ | $4.59 \times 10^{-4}$ | 0.724 |
| 800 | 0.0307 | 14,710 | 0.360 | $7.97 \times 10^{-4}$ | $1.77 \times 10^{-5}$ | $5.76 \times 10^{-4}$ | 0.723 |
| 900 | 0.0273 | 14,840 | 0.394 | $10.8 \times 10^{-4}$ | $1.92 \times 10^{-5}$ | $7.03 \times 10^{-4}$ | 0.723 |
| 1000 | 0.0246 | 14,990 | 0.428 | $11.6 \times 10^{-4}$ | $2.07 \times 10^{-5}$ | $8.42 \times 10^{-4}$ | 0.724 |
| 1200 | 0.0205 | 15,370 | 0.495 | $15.7 \times 10^{-4}$ | $2.36 \times 10^{-5}$ | $11.5 \times 10^{-4}$ | 0.733 |
| | | | *Nitrogen* | | | | |
| 200 | 1.708 | 1043 | 0.0183 | $1.02 \times 10^{-5}$ | $1.29 \times 10^{-5}$ | $0.75 \times 10^{-5}$ | 0.734 |
| 250 | 1.367 | 1042 | 0.0222 | $1.56 \times 10^{-5}$ | $1.55 \times 10^{-5}$ | $1.13 \times 10^{-5}$ | 0.725 |
| 300 | 1.139 | 1040 | 0.0260 | $2.19 \times 10^{-5}$ | $1.79 \times 10^{-5}$ | $1.57 \times 10^{-5}$ | 0.715 |
| 350 | 0.967 | 1041 | 0.0294 | $2.92 \times 10^{-5}$ | $2.01 \times 10^{-5}$ | $2.08 \times 10^{-5}$ | 0.711 |
| 400 | 0.854 | 1045 | 0.0325 | $3.64 \times 10^{-5}$ | $2.21 \times 10^{-5}$ | $2.59 \times 10^{-5}$ | 0.710 |
| 450 | 0.759 | 1050 | 0.0356 | $4.47 \times 10^{-5}$ | $2.41 \times 10^{-5}$ | $3.17 \times 10^{-5}$ | 0.709 |
| 500 | 0.683 | 1057 | 0.0387 | $5.36 \times 10^{-5}$ | $2.59 \times 10^{-5}$ | $3.79 \times 10^{-5}$ | 0.708 |
| 550 | 0.621 | 1065 | 0.0414 | $6.26 \times 10^{-5}$ | $2.76 \times 10^{-5}$ | $4.45 \times 10^{-5}$ | 0.711 |
| 600 | 0.569 | 1075 | 0.0441 | $7.20 \times 10^{-5}$ | $2.93 \times 10^{-5}$ | $5.14 \times 10^{-5}$ | 0.713 |
| 700 | 0.488 | 1098 | 0.0493 | $9.20 \times 10^{-5}$ | $3.24 \times 10^{-5}$ | $6.63 \times 10^{-5}$ | 0.720 |
| 800 | 0.427 | 1122 | 0.0541 | $11.3 \times 10^{-5}$ | $3.52 \times 10^{-5}$ | $8.24 \times 10^{-5}$ | 0.730 |
| 900 | 0.380 | 1146 | 0.0587 | $13.5 \times 10^{-5}$ | $3.79 \times 10^{-5}$ | $9.97 \times 10^{-5}$ | 0.739 |
| 1000 | 0.342 | 1168 | 0.0631 | $15.8 \times 10^{-5}$ | $4.04 \times 10^{-5}$ | $11.8 \times 10^{-5}$ | 0.747 |
| 1200 | 0.285 | 1205 | 0.0713 | $20.8 \times 10^{-5}$ | $4.50 \times 10^{-5}$ | $15.8 \times 10^{-5}$ | 0.761 |
| 1400 | 0.244 | 1233 | 0.0797 | $26.5 \times 10^{-5}$ | $4.92 \times 10^{-5}$ | $20.2 \times 10^{-5}$ | 0.761 |
| | | | *Oxygen* | | | | |
| 200 | 1.951 | 906 | 0.0182 | $1.03 \times 10^{-5}$ | $1.47 \times 10^{-5}$ | $0.75 \times 10^{-5}$ | 0.728 |
| 250 | 1.561 | 914 | 0.0225 | $1.58 \times 10^{-5}$ | $1.78 \times 10^{-5}$ | $1.14 \times 10^{-5}$ | 0.721 |
| 300 | 1.301 | 920 | 0.0267 | $2.23 \times 10^{-5}$ | $2.07 \times 10^{-5}$ | $1.59 \times 10^{-5}$ | 0.711 |
| 350 | 1.115 | 929 | 0.0306 | $2.95 \times 10^{-5}$ | $2.34 \times 10^{-5}$ | $2.10 \times 10^{-5}$ | 0.710 |
| 400 | 0.976 | 942 | 0.0342 | $3.72 \times 10^{-5}$ | $2.59 \times 10^{-5}$ | $2.65 \times 10^{-5}$ | 0.713 |
| 450 | 0.867 | 956 | 0.0377 | $4.55 \times 10^{-5}$ | $2.83 \times 10^{-5}$ | $3.26 \times 10^{-5}$ | 0.717 |
| 500 | 0.780 | 971 | 0.0412 | $5.44 \times 10^{-5}$ | $3.05 \times 10^{-5}$ | $3.91 \times 10^{-5}$ | 0.720 |
| 550 | 0.709 | 987 | 0.0447 | $6.38 \times 10^{-5}$ | $3.27 \times 10^{-5}$ | $4.61 \times 10^{-5}$ | 0.722 |
| 600 | 0.650 | 1003 | 0.0480 | $7.36 \times 10^{-5}$ | $3.47 \times 10^{-5}$ | $5.34 \times 10^{-5}$ | 0.725 |
| 700 | 0.557 | 1032 | 0.0544 | $9.46 \times 10^{-5}$ | $3.85 \times 10^{-5}$ | $6.91 \times 10^{-5}$ | 0.730 |
| 800 | 0.488 | 1054 | 0.0603 | $11.7 \times 10^{-5}$ | $4.21 \times 10^{-5}$ | $8.63 \times 10^{-5}$ | 0.736 |
| 900 | 0.434 | 1074 | 0.0661 | $14.2 \times 10^{-5}$ | $4.54 \times 10^{-5}$ | $10.5 \times 10^{-5}$ | 0.738 |
| 1000 | 0.390 | 1091 | 0.0717 | $16.8 \times 10^{-5}$ | $4.85 \times 10^{-5}$ | $12.4 \times 10^{-5}$ | 0.738 |
| 1200 | 0.325 | 1116 | 0.0821 | $22.6 \times 10^{-5}$ | $5.42 \times 10^{-5}$ | $16.7 \times 10^{-5}$ | 0.737 |
| 1400 | 0.278 | 1136 | 0.0921 | $29.1 \times 10^{-5}$ | $5.95 \times 10^{-5}$ | $21.3 \times 10^{-5}$ | 0.734 |

**TABLE A-19**
**Properties of some gases at 1-atm pressure (*Continued*)**

| Temperature $T$ K | Density $\rho$ kg/m³ | Specific heat $C_p$ J/(kg · °C) | Thermal conductivity $k$ W/(m · °C) | Thermal diffusivity $\alpha$ m²/s | Dynamic viscosity $\mu$ kg/(m · s) | Kinematic viscosity $\nu$ m²/s | Prandtl number Pr |
|---|---|---|---|---|---|---|---|
| | | | *Carbon dioxide* ($CO_2$) | | | | |
| 200 | 2.683 | 759 | 0.0095 | $0.47 \times 10^{-5}$ | $1.02 \times 10^{-5}$ | $0.38 \times 10^{-5}$ | 0.814 |
| 250 | 2.146 | 806 | 0.0129 | $0.75 \times 10^{-5}$ | $1.26 \times 10^{-5}$ | $0.59 \times 10^{-5}$ | 0.790 |
| 300 | 1.789 | 852 | 0.0166 | $1.09 \times 10^{-5}$ | $1.50 \times 10^{-5}$ | $0.84 \times 10^{-5}$ | 0.768 |
| 350 | 1.533 | 897 | 0.0205 | $1.49 \times 10^{-5}$ | $1.73 \times 10^{-5}$ | $1.13 \times 10^{-5}$ | 0.755 |
| 400 | 1.341 | 939 | 0.0244 | $1.94 \times 10^{-5}$ | $1.94 \times 10^{-5}$ | $1.45 \times 10^{-5}$ | 0.747 |
| 450 | 1.192 | 979 | 0.0283 | $2.43 \times 10^{-5}$ | $2.15 \times 10^{-5}$ | $1.80 \times 10^{-5}$ | 0.743 |
| 500 | 1.073 | 1017 | 0.0323 | $2.96 \times 10^{-5}$ | $2.35 \times 10^{-5}$ | $2.19 \times 10^{-5}$ | 0.740 |
| 550 | 0.976 | 1049 | 0.0363 | $3.55 \times 10^{-5}$ | $2.54 \times 10^{-5}$ | $2.60 \times 10^{-5}$ | 0.734 |
| 600 | 0.894 | 1077 | 0.0403 | $4.18 \times 10^{-5}$ | $2.72 \times 10^{-5}$ | $3.04 \times 10^{-5}$ | 0.727 |
| 700 | 0.767 | 1126 | 0.0487 | $5.64 \times 10^{-5}$ | $3.06 \times 10^{-5}$ | $3.99 \times 10^{-5}$ | 0.708 |
| 800 | 0.671 | 1169 | 0.0560 | $7.14 \times 10^{-5}$ | $3.39 \times 10^{-5}$ | $5.05 \times 10^{-5}$ | 0.708 |
| 900 | 0.596 | 1205 | 0.0621 | $8.65 \times 10^{-5}$ | $3.69 \times 10^{-5}$ | $6.19 \times 10^{-5}$ | 0.716 |
| 1000 | 0.537 | 1235 | 0.0680 | $10.25 \times 10^{-5}$ | $3.97 \times 10^{-5}$ | $7.40 \times 10^{-5}$ | 0.721 |
| 1200 | 0.447 | 1283 | 0.0780 | $13.6 \times 10^{-5}$ | $4.49 \times 10^{-5}$ | $10.04 \times 10^{-5}$ | 0.739 |
| 1400 | 0.383 | 1315 | 0.0867 | $17.2 \times 10^{-5}$ | $4.97 \times 10^{-5}$ | $13.0 \times 10^{-5}$ | 0.754 |
| | | | *Carbon monoxide* (CO) | | | | |
| 200 | 1.708 | 1045 | 0.0175 | $0.98 \times 10^{-5}$ | $1.27 \times 10^{-5}$ | $0.75 \times 10^{-5}$ | 0.763 |
| 250 | 1.366 | 1048 | 0.0214 | $1.50 \times 10^{-5}$ | $1.54 \times 10^{-5}$ | $1.13 \times 10^{-5}$ | 0.753 |
| 300 | 1.138 | 1051 | 0.0252 | $2.11 \times 10^{-5}$ | $1.78 \times 10^{-5}$ | $1.56 \times 10^{-5}$ | 0.743 |
| 350 | 0.976 | 1056 | 0.0288 | $2.80 \times 10^{-5}$ | $2.01 \times 10^{-5}$ | $2.05 \times 10^{-5}$ | 0.735 |
| 400 | 0.854 | 1060 | 0.0323 | $3.57 \times 10^{-5}$ | $2.21 \times 10^{-5}$ | $2.59 \times 10^{-5}$ | 0.727 |
| 450 | 0.759 | 1065 | 0.0355 | $4.39 \times 10^{-5}$ | $2.41 \times 10^{-5}$ | $3.18 \times 10^{-5}$ | 0.723 |
| 500 | 0.683 | 1071 | 0.0386 | $5.28 \times 10^{-5}$ | $2.60 \times 10^{-5}$ | $3.80 \times 10^{-5}$ | 0.720 |
| 550 | 0.621 | 1077 | 0.0416 | $6.22 \times 10^{-5}$ | $2.77 \times 10^{-5}$ | $4.46 \times 10^{-5}$ | 0.717 |
| 600 | 0.569 | 1084 | 0.0444 | $7.20 \times 10^{-5}$ | $2.94 \times 10^{-5}$ | $5.17 \times 10^{-5}$ | 0.718 |
| 700 | 0.488 | 1099 | 0.0497 | $9.27 \times 10^{-5}$ | $3.25 \times 10^{-5}$ | $6.66 \times 10^{-5}$ | 0.718 |
| 800 | 0.427 | 1114 | 0.0549 | $11.5 \times 10^{-5}$ | $3.54 \times 10^{-5}$ | $8.29 \times 10^{-5}$ | 0.718 |
| 900 | 0.379 | 1128 | 0.0596 | $13.9 \times 10^{-5}$ | $3.81 \times 10^{-5}$ | $10.04 \times 10^{-5}$ | 0.721 |
| 1000 | 0.342 | 1142 | 0.0644 | $16.5 \times 10^{-5}$ | $4.06 \times 10^{-5}$ | $11.9 \times 10^{-5}$ | 0.720 |
| 1100 | 0.310 | 1155 | 0.0692 | $19.3 \times 10^{-5}$ | $4.30 \times 10^{-5}$ | $13.9 \times 10^{-5}$ | 0.718 |
| 1200 | 0.285 | 1168 | 0.0738 | $22.2 \times 10^{-5}$ | $4.53 \times 10^{-5}$ | $15.9 \times 10^{-5}$ | 0.717 |
| | | | *Helium* | | | | |
| 200 | 0.2440 | 5197 | 0.115 | $0.91 \times 10^{-4}$ | $1.50 \times 10^{-5}$ | $0.61 \times 10^{-4}$ | 0.676 |
| 250 | 0.1952 | 5197 | 0.134 | $1.54 \times 10^{-4}$ | $1.75 \times 10^{-5}$ | $0.90 \times 10^{-4}$ | 0.680 |
| 300 | 0.1627 | 5197 | 0.150 | $1.77 \times 10^{-4}$ | $1.99 \times 10^{-5}$ | $1.22 \times 10^{-4}$ | 0.690 |
| 350 | 0.1394 | 5197 | 0.165 | $2.28 \times 10^{-4}$ | $2.21 \times 10^{-5}$ | $1.59 \times 10^{-4}$ | 0.698 |
| 400 | 0.1220 | 5197 | 0.180 | $2.83 \times 10^{-4}$ | $2.43 \times 10^{-5}$ | $1.99 \times 10^{-4}$ | 0.703 |
| 450 | 0.1085 | 5197 | 0.195 | $3.45 \times 10^{-4}$ | $2.63 \times 10^{-5}$ | $2.43 \times 10^{-4}$ | 0.702 |
| 500 | 0.0976 | 5197 | 0.211 | $4.17 \times 10^{-4}$ | $2.83 \times 10^{-5}$ | $2.90 \times 10^{-4}$ | 0.695 |
| 550 | 0.0887 | 5197 | 0.229 | $4.97 \times 10^{-4}$ | $3.02 \times 10^{-5}$ | $3.40 \times 10^{-4}$ | 0.684 |
| 600 | 0.0813 | 5197 | 0.247 | $5.84 \times 10^{-4}$ | $3.20 \times 10^{-5}$ | $3.93 \times 10^{-4}$ | 0.673 |
| 700 | 0.0697 | 5197 | 0.278 | $7.67 \times 10^{-4}$ | $3.55 \times 10^{-5}$ | $5.09 \times 10^{-4}$ | 0.663 |
| 800 | 0.0610 | 5197 | 0.307 | $9.68 \times 10^{-4}$ | $3.88 \times 10^{-5}$ | $6.37 \times 10^{-4}$ | 0.657 |
| 900 | 0.0542 | 5197 | 0.335 | $11.9 \times 10^{-4}$ | $4.20 \times 10^{-5}$ | $7.75 \times 10^{-4}$ | 0.652 |
| 1000 | 0.0488 | 5197 | 0.363 | $14.3 \times 10^{-4}$ | $4.50 \times 10^{-5}$ | $9.23 \times 10^{-4}$ | 0.645 |
| 1200 | 0.0407 | 5197 | 0.416 | $19.7 \times 10^{-4}$ | $5.08 \times 10^{-5}$ | $12.5 \times 10^{-4}$ | 0.635 |
| 1400 | 0.0349 | 5197 | 0.469 | $25.9 \times 10^{-4}$ | $5.61 \times 10^{-5}$ | $16.1 \times 10^{-4}$ | 0.622 |
| 1600 | 0.0305 | 5197 | 0.521 | $32.9 \times 10^{-4}$ | $6.10 \times 10^{-5}$ | $20.0 \times 10^{-4}$ | 0.608 |
| 1800 | 0.0271 | 5197 | 0.570 | $40.4 \times 10^{-4}$ | $6.57 \times 10^{-5}$ | $24.2 \times 10^{-4}$ | 0.599 |
| 2000 | 0.0244 | 5197 | 0.620 | $48.9 \times 10^{-4}$ | $7.00 \times 10^{-5}$ | $28.7 \times 10^{-4}$ | 0.587 |

**TABLE A-19**

**Properties of some gases at 1-atm pressure (*Continued*)**

| Temperature $T$ K | Density $\rho$ kg/m$^3$ | Specific heat $C_p$ J/(kg · °C) | Thermal conductivity $k$ W/(m · °C) | Thermal diffusivity $\alpha$ m$^2$/s | Dynamic viscosity $\mu$ kg/(m · s) | Kinematic viscosity $\nu$ m$^2$/s | Prandtl number Pr |
|---|---|---|---|---|---|---|---|
| *Water vapor (steam)* | | | | | | | |
| 300 | 0.0253* | 2041 | 0.0181 | $35.1 \times 10^{-5}$* | $0.91 \times 10^{-5}$ | $36.1 \times 10^{-5}$* | 1.03 |
| 350 | 0.258* | 2037 | 0.0222 | $4.22 \times 10^{-5}$* | $1.12 \times 10^{-5}$ | $4.33 \times 10^{-5}$* | 1.02 |
| 400 | 0.555 | 2000 | 0.0264 | $2.38 \times 10^{-5}$ | $1.32 \times 10^{-5}$ | $2.38 \times 10^{-5}$ | 1.00 |
| 450 | 0.491 | 1968 | 0.0307 | $3.17 \times 10^{-5}$ | $1.52 \times 10^{-5}$ | $3.10 \times 10^{-5}$ | 0.98 |
| 500 | 0.441 | 1977 | 0.0357 | $4.09 \times 10^{-5}$ | $1.73 \times 10^{-5}$ | $3.92 \times 10^{-5}$ | 0.96 |
| 550 | 0.401 | 1994 | 0.0411 | $5.15 \times 10^{-5}$ | $1.93 \times 10^{-5}$ | $4.82 \times 10^{-5}$ | 0.94 |
| 600 | 0.367 | 2022 | 0.0464 | $6.25 \times 10^{-5}$ | $2.13 \times 10^{-5}$ | $5.82 \times 10^{-5}$ | 0.93 |
| 700 | 0.314 | 2083 | 0.0572 | $8.74 \times 10^{-5}$ | $2.54 \times 10^{-5}$ | $8.09 \times 10^{-5}$ | 0.93 |
| 800 | 0.275 | 2148 | 0.0686 | $11.6 \times 10^{-5}$ | $2.95 \times 10^{-5}$ | $10.7 \times 10^{-5}$ | 0.92 |
| 900 | 0.244 | 2217 | 0.078 | $14.4 \times 10^{-5}$ | $3.36 \times 10^{-5}$ | $13.7 \times 10^{-5}$ | 0.95 |
| 1000 | 0.220 | 2288 | 0.087 | $17.3 \times 10^{-5}$ | $3.76 \times 10^{-5}$ | $17.1 \times 10^{-5}$ | 0.99 |

* At saturation pressure (less than 1 atm).

For ideal gases, the properties $C_p$, $k$, $\mu$, and Pr are independent of pressure. The properties $\rho$, $\nu$, and $\alpha$ at a pressure $P$ other than 1 atm are determined by multiplying the values of $\rho$ at the given temperature by $P$, and by dividing the values of $\nu$ and $\alpha$ at the given temperature by $P$, where $P$ is in atm (1 atm = 101.325 kPa = 14.696 psi).

**TABLE A-20**
**Properties of the atmosphere at high altitude**

| Altitude m | Temperature °C | Pressure kPa | Gravity, $g$ m/s$^2$ | Speed of sound m/s | Density kg/m$^3$ | Viscosity $\mu$ kg/(m · s) | Thermal conductivity W/(m · °C) |
|---|---|---|---|---|---|---|---|
| 0 | 15.00 | 101.33 | 9.807 | 340.3 | 1.225 | $1.789 \times 10^{-5}$ | 0.0253 |
| 200 | 13.70 | 98.95 | 9.806 | 339.5 | 1.202 | $1.783 \times 10^{-5}$ | 0.0252 |
| 400 | 12.40 | 96.61 | 9.805 | 338.8 | 1.179 | $1.777 \times 10^{-5}$ | 0.0252 |
| 600 | 11.10 | 94.32 | 9.805 | 338.0 | 1.156 | $1.771 \times 10^{-5}$ | 0.0251 |
| 800 | 9.80 | 92.08 | 9.804 | 337.2 | 1.134 | $1.764 \times 10^{-5}$ | 0.0250 |
| 1,000 | 8.50 | 89.88 | 9.804 | 336.4 | 1.112 | $1.758 \times 10^{-5}$ | 0.0249 |
| 1.200 | 7.20 | 87.72 | 9.803 | 335.7 | 1.090 | $1.752 \times 10^{-5}$ | 0.0248 |
| 1,400 | 5.90 | 85.60 | 9.802 | 334.9 | 1.069 | $1.745 \times 10^{-5}$ | 0.0247 |
| 1,600 | 4.60 | 83.53 | 9.802 | 334.1 | 1.048 | $1.739 \times 10^{-5}$ | 0.0245 |
| 1,800 | 3.30 | 81.49 | 9.801 | 333.3 | 1.027 | $1.732 \times 10^{-5}$ | 0.0244 |
| 2,000 | 2.00 | 79.50 | 9.800 | 332.5 | 1.007 | $1.726 \times 10^{-5}$ | 0.0243 |
| 2,200 | 0.70 | 77.55 | 9.800 | 331.7 | 0.987 | $1.720 \times 10^{-5}$ | 0.0242 |
| 2,400 | −0.59 | 75.63 | 9.799 | 331.0 | 0.967 | $1.713 \times 10^{-5}$ | 0.0241 |
| 2,600 | −1.89 | 73.76 | 9.799 | 330.2 | 0.947 | $1.707 \times 10^{-5}$ | 0.0240 |
| 2,800 | −3.19 | 71.92 | 9.798 | 329.4 | 0.928 | $1.700 \times 10^{-5}$ | 0.0239 |
| 3,000 | −4.49 | 70.12 | 9.797 | 328.6 | 0.909 | $1.694 \times 10^{-5}$ | 0.0238 |
| 3,200 | −5.79 | 68.36 | 9.797 | 327.8 | 0.891 | $1.687 \times 10^{-5}$ | 0.0237 |
| 3,400 | −7.09 | 66.63 | 9.796 | 327.0 | 0.872 | $1.681 \times 10^{-5}$ | 0.0236 |
| 3,600 | −8.39 | 64.94 | 9.796 | 326.2 | 0.854 | $1.674 \times 10^{-5}$ | 0.0235 |
| 3,800 | −9.69 | 63.28 | 9.795 | 325.4 | 0.837 | $1.668 \times 10^{-5}$ | 0.0234 |
| 4,000 | −10.98 | 61.66 | 9.794 | 324.6 | 0.819 | $1.661 \times 10^{-5}$ | 0.0233 |
| 4,200 | −12.3 | 60.07 | 9.794 | 323.8 | 0.802 | $1.655 \times 10^{-5}$ | 0.0232 |
| 4,400 | −13.6 | 58.52 | 9.793 | 323.0 | 0.785 | $1.648 \times 10^{-5}$ | 0.0231 |
| 4,600 | −14.9 | 57.00 | 9.793 | 322.2 | 0.769 | $1.642 \times 10^{-5}$ | 0.0230 |
| 4,800 | −16.2 | 55.51 | 9.792 | 321.4 | 0.752 | $1.635 \times 10^{-5}$ | 0.0229 |
| 5,000 | −17.5 | 54.05 | 9.791 | 320.5 | 0.736 | $1.628 \times 10^{-5}$ | 0.0228 |
| 5,200 | −18.8 | 52.62 | 9.791 | 319.7 | 0.721 | $1.622 \times 10^{-5}$ | 0.0227 |
| 5.400 | −20.1 | 51.23 | 9.790 | 318.9 | 0.705 | $1.615 \times 10^{-5}$ | 0.0226 |
| 5,600 | −21.4 | 49.86 | 9.789 | 318.1 | 0.690 | $1.608 \times 10^{-5}$ | 0.0224 |
| 5,800 | −22.7 | 48.52 | 9.785 | 317.3 | 0.675 | $1.602 \times 10^{-5}$ | 0.0223 |
| 6,000 | −24.0 | 47.22 | 9.788 | 316.5 | 0.660 | $1.595 \times 10^{-5}$ | 0.0222 |
| 6,200 | −25.3 | 45.94 | 9.788 | 315.6 | 0.646 | $1.588 \times 10^{-5}$ | 0.0221 |
| 6,400 | −26.6 | 44.69 | 9.787 | 314.8 | 0.631 | $1.582 \times 10^{-5}$ | 0.0220 |
| 6,600 | −27.9 | 43.47 | 9.786 | 314.0 | 0.617 | $1.575 \times 10^{-5}$ | 0.0219 |
| 6,800 | −29.2 | 42.27 | 9.785 | 313.1 | 0.604 | $1.568 \times 10^{-5}$ | 0.0218 |
| 7,000 | −30.5 | 41.11 | 9.785 | 312.3 | 0.590 | $1.561 \times 10^{-5}$ | 0.0217 |
| 8,000 | −36.9 | 35.65 | 9.782 | 308.1 | 0.526 | $1.527 \times 10^{-5}$ | 0.0212 |
| 9,000 | −43.4 | 30.80 | 9.779 | 303.8 | 0.467 | $1.493 \times 10^{-5}$ | 0.0206 |
| 10,000 | −49.9 | 26.50 | 9.776 | 299.5 | 0.414 | $1.458 \times 10^{-5}$ | 0.0201 |
| 12,000 | −56.5 | 19.40 | 9.770 | 295.1 | 0.312 | $1.422 \times 10^{-5}$ | 0.0195 |
| 14,000 | −56.5 | 14.17 | 9.764 | 295.1 | 0.228 | $1.422 \times 10^{-5}$ | 0.0195 |
| 16,000 | −56.5 | 10.53 | 9.758 | 295.1 | 0.166 | $1.422 \times 10^{-5}$ | 0.0195 |
| 18,000 | −56.5 | 7.57 | 9.751 | 295.1 | 0.122 | $1.422 \times 10^{-5}$ | 0.0195 |

*Source:* U.S. Standard Atmosphere Supplements, U.S. Government Printing Office, 1966. Based on year-around mean conditions at 45° latitude, and will vary with time of the year and the weather patterns. The conditions at sea level ($z = 0$) are taken to be $P = 101.325$ kPa, $T = 15$°C, $\rho = 1.2250$ kg/m$^3$, g = 9.80665 m$^2$/s.

**TABLE A-21**
**Emissivities of some surfaces**
**(*a*) Metals**

| Material | Temperature K | Emissivity $\varepsilon$ |
|---|---|---|
| Aluminum | | |
| Polished | 300–900 | 0.04–0.06 |
| Commercial sheet | 400 | 0.09 |
| Heavily oxidized | 400–800 | 0.20–0.33 |
| Anodized | 300 | 0.8 |
| Bismuth, bright | 350 | 0.34 |
| Brass | | |
| Highly polished | 500–650 | 0.03–0.04 |
| Polished | 350 | 0.09 |
| Dull plate | 300–600 | 0.22 |
| Oxidized | 450–800 | 0.6 |
| Chromium, polished | 300–1400 | 0.08–0.40 |
| Copper | | |
| Highly polished | 300 | 0.02 |
| Polished | 300–500 | 0.04–0.05 |
| Commercial sheet | 300 | 0.15 |
| Oxidized | 600–1000 | 0.5–0.8 |
| Black oxidized | 300 | 0.78 |
| Gold | | |
| Highly polished | 300–1000 | 0.03–0.06 |
| Bright foil | 300 | 0.07 |
| Iron | | |
| Highly polished | 300–500 | 0.05–0.07 |
| Case iron | 300 | 0.44 |
| Wrought iron | 300–500 | 0.28 |
| Rusted | 300 | 0.61 |
| Oxidized | 500–900 | 0.64–0.78 |
| Lead | | |
| Polished | 300–500 | 0.06–0.08 |
| Unoxidized, rough | 300 | 0.43 |
| Oxidized | 300 | 0.63 |
| Magnesium, polished | 300–500 | 0.07–0.13 |
| Mercury | 300–400 | 0.09–0.12 |
| Molybdenum | | |
| Polished | 300–2000 | 0.05–0.21 |
| Oxidized | 600–800 | 0.80–0.82 |
| Nickel | | |
| Polished | 500–1200 | 0.07–0.17 |
| Oxidized | 450–1000 | 0.37–0.57 |
| Platinum, polished | 500–1500 | 0.06–0.18 |
| Silver, polished | 300–1000 | 0.02–0.07 |
| Stainless steel | | |
| Polished | 300–1000 | 0.17–0.30 |
| Lightly oxidized | 600–1000 | 0.30–0.40 |
| Highly oxidized | 600–1000 | 0.70–0.80 |
| Steel | | |
| Polished sheet | 300–500 | 0.08–014 |
| Commercial sheet | 500–1200 | 0.20–0.32 |
| Heavily oxidized | 300 | 0.81 |
| Tin, polished | 300 | 0.05 |
| Tungsten | | |
| Polished | 300–2500 | 0.03–0.29 |
| Filament | 3500 | 0.39 |
| Zinc | | |
| Polished | 300–800 | 0.02–0.05 |
| Oxidized | 300 | 0.25 |

**TABLE A-21**
**Emissivities of some surfaces**
**(*b*) Nonmetals**

| Material | Temperature K | Emissivity $\varepsilon$ |
|---|---|---|
| Alumina | 800–1400 | 0.65–0.45 |
| Aluminum oxide | 600–1500 | 0.69–0.41 |
| Asbestos | 300 | 0.96 |
| Asphalt pavement | 300 | 0.85–0.93 |
| Brick | | |
| Common | 300 | 0.93–0.96 |
| Fireclay | 1200 | 0.75 |
| Carbon filament | 2000 | 0.53 |
| Cloth | 300 | 0.75–0.90 |
| Concrete | 300 | 0.88–0.94 |
| Glass | | |
| Window | 300 | 0.90–0.95 |
| Pyrex | 300–1200 | 0.82–0.62 |
| Pyroceram | 300–1500 | 0.85–0.57 |
| Ice | 273 | 0.95–0.99 |
| Magnesium oxide | 400–800 | 0.69–0.55 |
| Masonry | 300 | 0.80 |
| Paints | | |
| Aluminum | 300 | 0.40–0.50 |
| Black, lacquer, shiny | 300 | 0.88 |
| Oils, all colors | 300 | 0.92–0.96 |
| White acrylic | 300 | 0.90 |
| White enamel | 300 | 0.90 |
| Red primer | 300 | 0.93 |
| Paper, white | 300 | 0.90 |
| Plaster, white | 300 | 0.93 |
| Porcelain, glazed | 300 | 0.92 |
| Quartz, rough, fused | 300 | 0.93 |
| Rubber | | |
| Soft | 300 | 0.86 |
| Hard | 300 | 0.93 |
| Sand | 300 | 0.90 |
| Silicon carbide | 600–1500 | 0.87–0.85 |
| Skin, human | 300 | 0.95 |
| Snow | 273 | 0.80–0.90 |
| Soil, earth | 300 | 0.93–0.96 |
| Soot | 300–500 | 0.95 |
| Teflon | 300–500 | 0.85–0.92 |
| Water, deep | 273–373 | 0.95–0.96 |
| Wood | | |
| Beech | 300 | 0.94 |
| Oak | 300 | 0.90 |

**TABLE A-22**
**Solar radiative properties of some materials**

| Description/Composition | Solar absorptivity $\alpha_s$ | Emissivity $\varepsilon$ at 300 K | Ratio $\alpha_s/\varepsilon$ | Solar transmissivity $\tau_s$ |
|---|---|---|---|---|
| Aluminum | | | | |
| Polished | 0.09 | 0.03 | 3.0 | |
| Anodized | 0.14 | 0.84 | 0.17 | |
| Quartz-overcoated | 0.11 | 0.37 | 0.30 | |
| Foil | 0.15 | 0.05 | 3.0 | |
| Brick, red (Purdue) | 0.63 | 0.93 | 0.68 | |
| Concrete | 0.60 | 0.88 | 0.68 | |
| Galvanized sheet metal | | | | |
| Clean, new | 0.65 | 0.13 | 5.0 | |
| Oxidized, weathered | 0.80 | 0.28 | 2.9 | |
| Glass, 3.2-mm thickness | | | | |
| Float or tempered | | | | 0.79 |
| Low iron oxide type | | | | 0.88 |
| Marble, slightly off-white (nonreflective) | 0.40 | 0.88 | 0.45 | |
| Metal, plated | | | | |
| Black sulfide | 0.92 | 0.10 | 9.2 | |
| Black cobalt oxide | 0.93 | 0.30 | 3.1 | |
| Black nickel oxide | 0.92 | 0.08 | 11 | |
| Black chorme | 0.87 | 0.09 | 9.7 | |
| Mylar, 0.13-mm thickness | | | | 0.87 |
| Paints | | | | |
| Black (Parsons) | 0.98 | 0.98 | 1.0 | |
| White, acrylic | 0.26 | 0.90 | 0.29 | |
| White, zinc oxide | 0.16 | 0.93 | 0.17 | |
| Paper, white | 0.27 | 0.83 | 0.32 | |
| Plexiglas, 3.2-mm thickness | | | | 0.90 |
| Porcelain tiles, white (reflective glazed surface) | 0.26 | 0.85 | 0.30 | |
| Roofing tiles, bright red | | | | |
| Dry surface | 0.65 | 0.85 | 0.76 | |
| Wet surface | 0.88 | 0.91 | 0.96 | |
| Sand, dry | | | | |
| Off white | 0.52 | 0.82 | 0.63 | |
| Dull red | 0.73 | 0.86 | 0.82 | |
| Snow | | | | |
| Fine particles, fresh | 0.13 | 0.82 | 0.16 | |
| Ice granules | 0.33 | 0.89 | 0.37 | |
| Steel | | | | |
| mirror-finish | 0.41 | 0.05 | 8.2 | |
| Heavily rusted | 0.89 | 0.92 | 0.96 | |
| Stone (light pink) | 0.65 | 0.87 | 0.74 | |
| Tedlar, 0.10-mm thickness | | | | 0.92 |
| Teflon, 0.13-mm thickness | | | | 0.92 |
| Wood | 0.59 | 0.90 | 0.66 | |

*Source:* V. C. Sharma and A. Sharma, Solar Properties of Some Building Elements, *Energy,* Vol. 14, pp. 805–810, 1989, and other sources.

# Property Tables and Charts (English Units)

APPENDIX

**TABLE A-1E**

**Molar mass, gas constant, and critical-point properties**

| Substance | Formula | $M$ Molar mass lbm/lbmol | Gas constant $R$ Btu/(lbm · R)* | psia · ft³/(lbm · R)* | Critical point properties Temp. R | Pressure psia | Volume ft³/lbmol |
|---|---|---|---|---|---|---|---|
| Air | — | 28.97 | 0.06855 | 0.3704 | 238.5 | 547 | 1.41 |
| Ammonia | $NH_3$ | 17.03 | 0.1166 | 0.6301 | 729.8 | 1636 | 1.16 |
| Argon | Ar | 39.948 | 0.04971 | 0.2686 | 272 | 705 | 1.20 |
| Benzene | $C_6H_6$ | 78.115 | 0.02542 | 0.1374 | 1012 | 714 | 4.17 |
| Bromine | $Br_2$ | 159.808 | 0.01243 | 0.06714 | 1052 | 1500 | 2.17 |
| *n*-Butane | $C_4H_{10}$ | 58.124 | 0.03417 | 0.1846 | 765.2 | 551 | 4.08 |
| Carbon dioxide | $CO_2$ | 44.01 | 0.04513 | 0.2438 | 547.5 | 1071 | 1.51 |
| Carbon monoxide | CO | 28.011 | 0.07090 | 0.3831 | 240 | 507 | 1.49 |
| Carbon tetrachloride | $CCl_4$ | 153.82 | 0.01291 | 0.06976 | 1001.5 | 661 | 4.42 |
| Chlorine | $Cl_2$ | 70.906 | 0.02801 | 0.1517 | 751 | 1120 | 1.99 |
| Chloroform | $CHCl_3$ | 119.38 | 0.01664 | 0.08988 | 965.8 | 794 | 3.85 |
| Dichlorodifluoromethane (R-12) | $CCl_2F_2$ | 120.91 | 0.01643 | 0.08874 | 692.4 | 582 | 3.49 |
| Dichlorofluoromethane (R-21) | $CHCl_2F$ | 102.92 | 0.01930 | 0.1043 | 813.0 | 749 | 3.16 |
| Ethane | $C_2H_6$ | 30.020 | 0.06616 | 0.3574 | 549.8 | 708 | 2.37 |
| Ethyl alcohol | $C_2H_5OH$ | 46.07 | 0.04311 | 0.2329 | 929.0 | 926 | 2.68 |
| Ethylene | $C_2H_4$ | 28.054 | 0.07079 | 0.3825 | 508.3 | 742 | 1.99 |
| Helium | He | 4.003 | 0.4961 | 2.6805 | 9.5 | 33.2 | 0.926 |
| *n*-Hexane | $C_6H_{14}$ | 86.178 | 0.02305 | 0.1245 | 914.2 | 439 | 5.89 |
| Hydrogen (normal) | $H_2$ | 2.016 | 0.9851 | 5.3224 | 59.9 | 188.1 | 1.04 |
| Krypton | Kr | 83.80 | 0.02370 | 0.1280 | 376.9 | 798 | 1.48 |
| Methane | $CH_4$ | 16.043 | 0.1238 | 0.6688 | 343.9 | 673 | 1.59 |
| Methyl alcohol | $CH_3OH$ | 32.042 | 0.06198 | 0.3349 | 923.7 | 1154 | 1.89 |
| Methyl chloride | $CH_3Cl$ | 50.488 | 0.03934 | 0.2125 | 749.3 | 968 | 2.29 |
| Neon | Ne | 20.183 | 0.09840 | 0.5316 | 80.1 | 395 | 0.668 |
| Nitrogen | $N_2$ | 28.013 | 0.07090 | 0.3830 | 227.1 | 492 | 1.44 |
| Nitrous oxide | $N_2O$ | 44.013 | 0.04512 | 0.2438 | 557.4 | 1054 | 1.54 |
| Oxygen | $O_2$ | 31.999 | 0.06206 | 0.3353 | 278.6 | 736 | 1.25 |
| Propane | $C_3H_8$ | 44.097 | 0.04504 | 0.2433 | 665.9 | 617 | 3.20 |
| Propylene | $C_3H_6$ | 42.081 | 0.04719 | 0.2550 | 656.9 | 670 | 2.90 |
| Sulfur dioxide | $SO_2$ | 64.063 | 0.03100 | 1.1675 | 775.2 | 1143 | 1.95 |
| Tetrafluoroethane (R-134a) | $CF_3CH_2F$ | 102.03 | 0.01946 | 0.1052 | 673.7 | 589.9 | 2.96 |
| Trichlorofluoromethane (R-11) | $CCl_3F$ | 137.37 | 0.01446 | 0.07811 | 848.1 | 635 | 3.97 |
| Water | $H_2O$ | 18.015 | 0.1102 | 0.5956 | 1165.3 | 3204 | 0.90 |
| Xeon | Xe | 131.30 | 0.01513 | 0.08172 | 521.55 | 852 | 1.90 |

*Calculated from $R = R_u/M$, where $R_u$ = 1.986 Btu/(lbmol · R) = 10.73 psia · ft³/(lbmol · R) and $M$ is the molar mass.

*Source:* K. A. Kobe and R. E. Lynn, Jr., *Chemical Review,* Vol. 52, pp. 117–236, 1953 and *ASHRAE Handbook of Fundamentals,* pp. 16.4 and 36.1, American Society of Heating, Refrigerating and Air-Conditioning Engineers, Inc., Atlanta, GA, 1993.

**TABLE A-2E**
**Ideal-gas specific heats of various common gases**
($a$) At 80°F

| Gas | Formula | Gas constant $R$ Btu/(lbm · R) | $C_{p0}$ Btu/(lbm · R) | $C_{v0}$ Btu/(lbm · R) | $k$ |
|---|---|---|---|---|---|
| Air | — | 0.06855 | 0.240 | 0.171 | 1.400 |
| Argon | Ar | 0.04971 | 0.1253 | 0.0756 | 1.667 |
| Butane | $C_4H_{10}$ | 0.03424 | 0.415 | 0.381 | 1.09 |
| Carbon dioxide | $CO_2$ | 0.04513 | 0.203 | 0.158 | 1.285 |
| Carbon monoxide | CO | 0.07090 | 0.249 | 0.178 | 1.399 |
| Ethane | $C_2H_6$ | 0.06616 | 0.427 | 0.361 | 1.183 |
| Ethylene | $C_2H_4$ | 0.07079 | 0.411 | 0.340 | 1.208 |
| Helium | He | 0.4961 | 1.25 | 0.753 | 1.667 |
| Hydrogen | $H_2$ | 0.9851 | 3.43 | 2.44 | 1.404 |
| Methane | $CH_4$ | 0.1238 | 0.532 | 0.403 | 1.32 |
| Neon | Ne | 0.09840 | 0.246 | 0.1477 | 1.667 |
| Nitrogen | $N_2$ | 0.07090 | 0.248 | 0.177 | 1.400 |
| Octane | $C_8H_{18}$ | 0.01742 | 0.409 | 0.392 | 1.044 |
| Oxygen | $O_2$ | 0.06206 | 0.219 | 0.157 | 1.395 |
| Propane | $C_3H_8$ | 0.04504 | 0.407 | 0.362 | 1.124 |
| Steam | $H_2O$ | 0.1102 | 0.445 | 0.335 | 1.329 |

*Source:* Gordon J. Van Wylen and Richard E. Sonntag, *Fundamentals of Classical Thermodynamics,* English/SI Version, 3d ed., Wiley, New York, 1986, p. 687, Table A-8E.

**TABLE A-2E**

**Ideal-gas specific heats of various common gases (*Continued*)**

(*b*) At various temperatures

| Temp. °F | $C_{p0}$ Btu/(lbm · R) | $C_{v0}$ Btu/(lbm · R) | $k$ | $C_{p0}$ Btu/(lbm · R) | $C_{v0}$ Btu/(lbm · R) | $k$ | $C_{p0}$ Btu/(lbm · R) | $C_{v0}$ Btu/(lbm · R) | $k$ |
|---|---|---|---|---|---|---|---|---|---|
| | | **Air** | | | **Carbon dioxide, $CO_2$** | | | **Carbon monoxide, CO** | |
| 40 | 0.240 | 0.171 | 1.401 | 0.195 | 0.150 | 1.300 | 0.248 | 0.177 | 1.400 |
| 100 | 0.240 | 0.172 | 1.400 | 0.205 | 0.160 | 1.283 | 0.249 | 0.178 | 1.399 |
| 200 | 0.241 | 0.173 | 1.397 | 0.217 | 0.172 | 1.262 | 0.249 | 0.179 | 1.397 |
| 300 | 0.243 | 0.174 | 1.394 | 0.229 | 0.184 | 1.246 | 0.251 | 0.180 | 1.394 |
| 400 | 0.245 | 0.176 | 1.389 | 0.239 | 0.193 | 1.233 | 0.253 | 0.182 | 1.389 |
| 500 | 0.248 | 0.179 | 1.383 | 0.247 | 0.202 | 1.223 | 0.256 | 0.185 | 1.384 |
| 600 | 0.250 | 0.182 | 1.377 | 0.255 | 0.210 | 1.215 | 0.259 | 0.188 | 1.377 |
| 700 | 0.254 | 0.185 | 1.371 | 0.262 | 0.217 | 1.208 | 0.262 | 0.191 | 1.371 |
| 800 | 0.257 | 0.188 | 1.365 | 0.269 | 0.224 | 1.202 | 0.266 | 0.195 | 1.364 |
| 900 | 0.259 | 0.191 | 1.358 | 0.275 | 0.230 | 1.197 | 0.269 | 0.198 | 1.357 |
| 1000 | 0.263 | 0.195 | 1.353 | 0.280 | 0.235 | 1.192 | 0.273 | 0.202 | 1.351 |
| 1500 | 0.276 | 0.208 | 1.330 | 0.298 | 0.253 | 1.178 | 0.287 | 0.216 | 1.328 |
| 2000 | 0.286 | 0.217 | 1.312 | 0.312 | 0.267 | 1.169 | 0.297 | 0.226 | 1.314 |
| | | **Hydrogen, $H_2$** | | | **Nitrogen, $N_2$** | | | **Oxygen, $O_2$** | |
| 40 | 3.397 | 2.412 | 1.409 | 0.248 | 0.177 | 1.400 | 0.219 | 0.156 | 1.397 |
| 100 | 3.426 | 2.441 | 1.404 | 0.248 | 0.178 | 1.399 | 0.220 | 0.158 | 1.394 |
| 200 | 3.451 | 2.466 | 1.399 | 0.249 | 0.178 | 1.398 | 0.223 | 0.161 | 1.387 |
| 300 | 3.461 | 2.476 | 1.398 | 0.250 | 0.179 | 1.396 | 0.226 | 0.164 | 1.378 |
| 400 | 3.466 | 2.480 | 1.397 | 0.251 | 0.180 | 1.393 | 0.230 | 0.168 | 1.368 |
| 500 | 3.469 | 2.484 | 1.397 | 0.254 | 0.183 | 1.388 | 0.235 | 0.173 | 1.360 |
| 600 | 3.473 | 2.488 | 1.396 | 0.256 | 0.185 | 1.383 | 0.239 | 0.177 | 1.352 |
| 700 | 3.477 | 2.492 | 1.395 | 0.260 | 0.189 | 1.377 | 0.242 | 0.181 | 1.344 |
| 800 | 3.494 | 2.509 | 1.393 | 0.262 | 0.191 | 1.371 | 0.246 | 0.184 | 1.337 |
| 900 | 3.502 | 2.519 | 1.392 | 0.265 | 0.194 | 1.364 | 0.249 | 0.187 | 1.331 |
| 1000 | 3.513 | 2.528 | 1.390 | 0.269 | 0.198 | 1.359 | 0.252 | 0.190 | 1.326 |
| 1500 | 3.618 | 2.633 | 1.374 | 0.283 | 0.212 | 1.334 | 0.263 | 0.201 | 1.309 |
| 2000 | 3.758 | 2.773 | 1.355 | 0.293 | 0.222 | 1.319 | 0.270 | 0.208 | 1.298 |

*Source:* Kenneth Wark, *Thermodynamics*, 4th ed., McGraw-Hill, New York, 1983, p. 830, Table A-4. Originally published in *Tables of Properties of Gases*, NBS *Circular*, 564, 1955.

**TABLE A-2E**
**Ideal-gas specific heats of various common gases (*Continued*)**

(*c*) As a function of temperature

$$\bar{C}_{p0} = a + bT + cT^2 + dT^3$$
**[$T$ in R, $\bar{C}_{p0}$ in Btu/(lbmol · R)]**

| Substance | Formula | $a$ | $b$ | $c$ | $d$ | Temperature range R | % error Max. | % error Avg. |
|---|---|---|---|---|---|---|---|---|
| Nitrogen | $N_2$ | 6.903 | $-0.02085 \times 10^{-2}$ | $0.05957 \times 10^{-5}$ | $-0.1176 \times 10^{-9}$ | 491–3240 | 0.59 | 0.34 |
| Oxygen | $O_2$ | 6.085 | $0.2017 \times 10^{-2}$ | $-0.05275 \times 10^{-5}$ | $0.05372 \times 10^{-9}$ | 491–3240 | 1.19 | 0.28 |
| Air | — | 6.713 | $0.02609 \times 10^{-2}$ | $0.03540 \times 10^{-5}$ | $-0.08052 \times 10^{-9}$ | 491–3240 | 0.72 | 0.33 |
| Hydrogen | $H_2$ | 6.952 | $-0.02542 \times 10^{-2}$ | $0.02952 \times 10^{-5}$ | $-0.03565 \times 10^{-9}$ | 491–3240 | 1.02 | 0.26 |
| Carbon monoxide | $CO$ | 6.726 | $0.02222 \times 10^{-2}$ | $0.03960 \times 10^{-5}$ | $-0.09100 \times 10^{-9}$ | 491–3240 | 0.89 | 0.37 |
| Carbon dioxide | $CO_2$ | 5.316 | $0.79361 \times 10^{-2}$ | $-0.2581 \times 10^{-5}$ | $0.3059 \times 10^{-9}$ | 491–3240 | 0.67 | 0.22 |
| Water vapor | $H_2O$ | 7.700 | $0.02552 \times 10^{-2}$ | $0.07781 \times 10^{-5}$ | $-0.1472 \times 10^{-9}$ | 491–3240 | 0.53 | 0.24 |
| Nitric oxide | $NO$ | 7.008 | $-0.01247 \times 10^{-2}$ | $0.07185 \times 10^{-5}$ | $-0.1715 \times 10^{-9}$ | 491–2700 | 0.97 | 0.36 |
| Nitrous oxide | $N_2O$ | 5.758 | $0.7780 \times 10^{-2}$ | $-0.2596 \times 10^{-5}$ | $0.4331 \times 10^{-9}$ | 491–2700 | 0.59 | 0.26 |
| Nitrogen dioxide | $NO_2$ | 5.48 | $0.7583 \times 10^{-2}$ | $-0.260 \times 10^{-5}$ | $0.322 \times 10^{-9}$ | 491–2700 | 0.46 | 0.18 |
| Ammonia | $NH_3$ | 6.5846 | $0.34028 \times 10^{-2}$ | $0.073034 \times 10^{-5}$ | $-0.27402 \times 10^{-9}$ | 491–2700 | 0.91 | 0.36 |
| Sulfur | $S_2$ | 6.499 | $0.2943 \times 10^{-2}$ | $-0.1200 \times 10^{-5}$ | $0.1632 \times 10^{-9}$ | 491–3240 | 0.99 | 0.38 |
| Sulfur dioxide | $SO_2$ | 6.157 | $0.7689 \times 10^{-2}$ | $-0.2810 \times 10^{-5}$ | $0.3527 \times 10^{-9}$ | 491–3240 | 0.45 | 0.24 |
| Sulfur trioxide | $SO_3$ | 3.918 | $1.935 \times 10^{-2}$ | $-0.8256 \times 10^{-5}$ | $1.328 \times 10^{-9}$ | 491–2340 | 0.29 | 0.13 |
| Acetylene | $C_2H_2$ | 5.21 | $1.2227 \times 10^{-2}$ | $-0.4812 \times 10^{-5}$ | $0.7457 \times 10^{-9}$ | 491–2700 | 1.46 | 0.59 |
| Benzene | $C_6H_6$ | −8.650 | $6.4322 \times 10^{-2}$ | $-2.327 \times 10^{-5}$ | $3.179 \times 10^{-9}$ | 491–2700 | 0.34 | 0.20 |
| Methanol | $CH_4O$ | 4.55 | $1.214 \times 10^{-2}$ | $-0.0898 \times 10^{-5}$ | $-0.329 \times 10^{-9}$ | 491–1800 | 0.18 | 0.08 |
| Ethanol | $C_2H_6O$ | 4.75 | $2.781 \times 10^{-2}$ | $-0.7651 \times 10^{-5}$ | $0.821 \times 10^{-9}$ | 491–2700 | 0.40 | 0.22 |
| Hydrogen chloride | $HCl$ | 7.244 | $-0.1011 \times 10^{-2}$ | $0.09783 \times 10^{-5}$ | $-0.1776 \times 10^{-9}$ | 491–2740 | 0.22 | 0.08 |
| Methane | $CH_4$ | 4.750 | $0.6666 \times 10^{-2}$ | $0.09352 \times 10^{-5}$ | $-0.4510 \times 10^{-9}$ | 491–2740 | 1.33 | 0.57 |
| Ethane | $C_2H_6$ | 1.648 | $2.291 \times 10^{-2}$ | $-0.4722 \times 10^{-5}$ | $0.2984 \times 10^{-9}$ | 491–2740 | 0.83 | 0.28 |
| Propane | $C_3H_8$ | −0.966 | $4.044 \times 10^{-2}$ | $-1.159 \times 10^{-5}$ | $1.300 \times 10^{-9}$ | 491–2740 | 0.40 | 0.12 |
| *n*-Butane | $C_4H_{10}$ | 0.945 | $4.929 \times 10^{-2}$ | $-1.352 \times 10^{-5}$ | $1.433 \times 10^{-9}$ | 491–2740 | 0.54 | 0.24 |
| *i*-Butane | $C_4H_{10}$ | −1.890 | $5.520 \times 10^{-2}$ | $-1.696 \times 10^{-5}$ | $2.044 \times 10^{-9}$ | 491–2740 | 0.25 | 0.13 |
| *n*-Pentane | $C_5H_{12}$ | 1.618 | $6.028 \times 10^{-2}$ | $-1.656 \times 10^{-5}$ | $1.732 \times 10^{-9}$ | 491–2740 | 0.56 | 0.21 |
| *n*-Hexane | $C_6H_{14}$ | 1.657 | $7.328 \times 10^{-2}$ | $-2.112 \times 10^{-5}$ | $2.363 \times 10^{-9}$ | 491–2740 | 0.72 | 0.20 |
| Ethylene | $C_2H_4$ | 0.944 | $2.075 \times 10^{-2}$ | $-0.6151 \times 10^{-5}$ | $0.7326 \times 10^{-9}$ | 491–2740 | 0.54 | 0.13 |
| Propylene | $C_3H_6$ | 0.753 | $3.162 \times 10^{-2}$ | $-0.8981 \times 10^{-5}$ | $1.008 \times 10^{-9}$ | 491–2740 | 0.73 | 0.17 |

*Source:* B. G. Kyle, *Chemical and Process Thermodynamics*, Prentice-Hall, Englewood Cliffs, N.J., 1984. Used with permission.

**TABLE A-3E**
**Properties of common liquids, solids, and foods**
(*a*) Liquids

| Substance | Boiling data at 1 atm | | Freezing data | | Liquid properties | | |
|---|---|---|---|---|---|---|---|
| | Normal boiling point °F | Latent heat of vaporization $h_{fg}$ Btu/lbm | Freezing point °F | Latent heat of fusion $h_{if}$ Btu/lbm | Temp. °F | Density $\rho$ lbm/ft³ | Specific heat $C_p$ Btu/(lbm · °F) |
| Ammonia | −27.9 | 24.54 | −107.9 | 138.6 | −27.9 | 42.6 | 1.06 |
| | | | | | 0 | 41.3 | 1.083 |
| | | | | | 40 | 39.5 | 1.103 |
| | | | | | 80 | 37.5 | 1.135 |
| Argon | −302.6 | 69.5 | −308.7 | 12.0 | −302.6 | 87.0 | 0.272 |
| Benzene | 176.4 | 169.4 | 41.9 | 54.2 | 68 | 54.9 | 0.411 |
| Brine (20% sodium chloride by mass) | 219.0 | — | 0.7 | — | 68 | 71.8 | 0.743 |
| *n*-Butane | 31.1 | 165.6 | −217.3 | 34.5 | 31.1 | 37.5 | 0.552 |
| Carbon dioxide | −109.2* | 99.6 (at 32°F) | −69.8 | — | 32 | 57.8 | 0.583 |
| Ethanol | 172.8 | 360.5 | −173.6 | 46.9 | 77 | 48.9 | 0.588 |
| Ethylene glycol | 388.6 | 344.0 | 12.6 | 77.9 | 68 | 69.2 | 0.678 |
| Ethyl alcohol | 173.5 | 368 | −248.8 | 46.4 | 68 | 49.3 | 0.678 |
| Glycerine | 355.8 | 419 | 66.0 | 86.3 | 68 | 78.7 | 0.554 |
| Helium | −452.1 | 9.80 | — | — | −452.1 | 9.13 | 5.45 |
| Hydrogen | −423.0 | 191.7 | −434.5 | 25.6 | −423.0 | 4.41 | 2.39 |
| Isobutane | 10.9 | 157.8 | −255.5 | 45.5 | 10.9 | 37.1 | 0.545 |
| Kerosene | 399 to 559 | 108 | −12.8 | — | 68 | 51.2 | 0.478 |
| Mercury | 674.1 | 126.7 | −38.0 | 4.90 | 77 | 847 | 0.033 |
| Methane | −258.7 | 219.6 | 296.0 | 25.1 | −258.7 | 26.4 | 0.834 |
| | | | | | −160 | 20.0 | 1.074 |
| Methanol | 148.1 | 473 | −143.9 | 42.7 | 77 | 49.1 | 0.609 |
| Nitrogen | −320.4 | 85.4 | −346.0 | 10.9 | −320.4 | 50.5 | 0.492 |
| | | | | | −260 | 38.2 | 0.643 |
| Octane | 256.6 | 131.7 | −71.5 | 77.9 | 68 | 43.9 | 0.502 |
| Oil (light) | — | — | | | 77 | 56.8 | 0.430 |
| Oxygen | −297.3 | 91.5 | −361.8 | 5.9 | −297.3 | 71.2 | 0.408 |
| Petroleum | — | 99–165 | | | 68 | 40.0 | 0.478 |
| Propane | −43.7 | 184.0 | −305.8 | 34.4 | −43.7 | 36.3 | 0.538 |
| | | | | | 32 | 33.0 | 0.604 |
| | | | | | 100 | 29.4 | 0.673 |
| Refrigerant-134a | −15.0 | 93.2 | −141.9 | — | −40 | 88.5 | 0.283 |
| | | | | | −15 | 86.0 | 0.294 |
| | | | | | 32 | 80.9 | 0.318 |
| | | | | | 90 | 73.6 | 0.348 |
| Water | 212 | 970.5 | 32 | 143.5 | 32 | 62.4 | 1.01 |
| | | | | | 90 | 62.1 | 1.00 |
| | | | | | 150 | 61.2 | 1.00 |
| | | | | | 212 | 59.8 | 1.01 |

* Sublimination temperature. (At pressures below the triple point pressure of 75.1 psia, carbon dioxide exists as a solid or gas. Also, the freezing point temperature of carbon dioxide is the triple point temperature of −69.8°F.)

**TABLE A-3E**

**Properties of common liquids, solids, and foods (*Continued*)**

(*b*) Solids (values are for room temperature unless indicated otherwise)

| Substance | Density $\rho$ lbm/ft³ | Specific heat $C_p$ Btu/(lbm · °F) | Substance | Density $\rho$ lbm/ft³ | Specific heat $C_p$ Btu/(lbm · °F) |
|---|---|---|---|---|---|
| *Metals* | | | *Nonmetals* | | |
| Aluminium | | | Asphalt | 132 | 0.220 |
| −100°F | | 0.192 | Brick, common | 120 | 0.189 |
| 32°F | | 0.212 | Brick, fireclay (500°C) | 144 | 0.229 |
| 100°F | 170 | 0.218 | Concrete | 144 | 0.156 |
| 200°F | | 0.224 | Clay | 62.4 | 0.220 |
| 300°F | | 0.229 | Diamond | 151 | 0.147 |
| 400°F | | 0.235 | Glass, window | 169 | 0.191 |
| 500°F | | 0.240 | Glass, pyrex | 139 | 0.200 |
| Bronze (76% Cu, 2% Zn, 2% Al) | 517 | 0.0955 | Graphite | 156 | 0.170 |
| | | | Granite | 169 | 0.243 |
| Brass, yellow (65% Cu, 35% Zn) | 519 | 0.0955 | Gypsum or plaster board | 50 | 0.260 |
| Copper | | | Ice | | |
| −240°F | | 0.0674 | −100°F | | 0.375 |
| −150°F | | 0.0784 | −50°F | | 0.424 |
| −60°F | | 0.0862 | 0°F | | 0.471 |
| 0°F | | 0.0893 | 20°F | | 0.491 |
| 100°F | 555 | 0.0925 | 32°F | 57.5 | 0.502 |
| 200°F | | 0.0938 | Limestone | 103 | 0.217 |
| 390°F | | 0.0963 | Marble | 162 | 0.210 |
| Iron | 490 | 0.107 | Plywood (Douglas Fir) | 34.0 | |
| Lead | 705 | 0.030 | Rubber (soft) | 68.7 | |
| Magnesium | 108 | 0.239 | Rubber (hard) | 71.8 | |
| Nickel | 555 | 0.105 | Sand | 94.9 | |
| Silver | 655 | 0.056 | Stone | 93.6 | |
| Steel, mild | 489 | 0.119 | Woods, hard (maple, oak, etc.) | 45.0 | |
| Tungsten | 1211 | 0.031 | Woods, soft (fir, pine, etc) | 32.0 | |

(*c*) Foods

| Food | Water content % (mass) | Freezing point °F | Specific heat Btu/(lbm · °F) Above freezing | Specific heat Btu/(lbm · °F) Below freezing | Latent heat of fusion Btu/lbm | Food | Water content % (mass) | Freezing point °F | Specific heat Btu/(lbm · °F) Above freezing | Specific heat Btu/(lbm · °F) Below freezing | Latent heat of fusion Btu/lbm |
|---|---|---|---|---|---|---|---|---|---|---|---|
| Apples | 84 | 30 | 0.873 | 0.453 | 121 | Lettuce | 95 | 32 | 0.961 | 0.487 | 136 |
| Bananas | 75 | 31 | 0.801 | 0.426 | 108 | Milk, whole | 88 | 31 | 0.905 | 0.465 | 126 |
| Beef round | 67 | — | 0.737 | 0.402 | 96 | Oranges | 87 | 31 | 0.897 | 0.462 | 125 |
| Broccoli | 90 | 31 | 0.921 | 0.471 | 129 | Potatoes | 78 | 31 | 0.825 | 0.435 | 112 |
| Butter | 16 | — | — | 0.249 | 23 | Salmon fish | 64 | 28 | 0.713 | 0.393 | 92 |
| Cheese, swiss | 39 | 14 | 0.513 | 0.318 | 56 | Shrimp | 83 | 28 | 0.865 | 0.450 | 119 |
| Cherries | 80 | 29 | 0.841 | 0.441 | 115 | Spinach | 93 | 31 | 0.945 | 0.481 | 134 |
| Chicken | 74 | 27 | 0.793 | 0.423 | 106 | Strawberries | 90 | 31 | 0.921 | 0.471 | 129 |
| Corn, sweet | 74 | 31 | 0.793 | 0.423 | 106 | Tomatoes, ripe | 94 | 31 | 0.953 | 0.484 | 135 |
| Eggs, whole | 74 | 31 | 0.793 | 0.423 | 106 | Turkey | 64 | — | 0.713 | 0.393 | 92 |
| Ice cream | 63 | 22 | 0.705 | 0.390 | 90 | Watermelon | 93 | 31 | 0.945 | 0.481 | 134 |

*Source:* Values are obtained from various handbooks and other sources or are calculated. Water content and freezing point data of foods are from *ASHRAE Handbook of Fundamentals,* I-P version, Chap. 30, Table 1, American Society of Heating, Refrigerating and Air-Conditioning Engineers, Inc., Atlanta, GA, 1993. Freezing point is the temperature at which freezing starts for fruits and vegetables, and the average freezing temperature for other foods.

**TABLE A-4E**
**Saturated water: temperature table**

| | | Specific volume ft³/lbm | | Internal energy Btu/lbm | | | Enthalpy Btu/lbm | | | Entropy Btu/(lbm · R) | | |
|---|---|---|---|---|---|---|---|---|---|---|---|---|
| Temp. $T$ °F | Sat. press. $P_{sat}$ psia | Sat. liquid $v_f$ | Sat. vapor $v_g$ | Sat. liquid $u_f$ | Evap. $u_{fg}$ | Sat. vapor $u_g$ | Sat. liquid $h_f$ | Evap. $h_{fg}$ | Sat. vapor $h_g$ | Sat. liquid $s_f$ | Evap. $s_{fg}$ | Sat. vapor $s_g$ |
| 32.018 | 0.08866 | 0.016022 | 3302 | 0.00 | 1021.2 | 1021.2 | 0.01 | 1075.4 | 1075.4 | 0.00000 | 2.1869 | 2.1869 |
| 35 | 0.09992 | 0.016021 | 2948 | 2.99 | 1019.2 | 1022.2 | 3.00 | 1073.7 | 1076.7 | 0.00607 | 2.1704 | 2.1764 |
| 40 | 0.12166 | 0.016020 | 2445 | 8.02 | 1015.8 | 1023.9 | 8.02 | 1070.9 | 1078.9 | 0.01617 | 2.1430 | 2.1592 |
| 45 | 0.14748 | 0.016021 | 2037 | 13.04 | 1012.5 | 1025.5 | 13.04 | 1068.1 | 1081.1 | 0.02618 | 2.1162 | 2.1423 |
| 50 | 0.17803 | 0.016024 | 1704.2 | 18.06 | 1009.1 | 1027.2 | 18.06 | 1065.2 | 1083.3 | 0.03607 | 2.0899 | 2.1259 |
| 60 | 0.2563 | 0.016035 | 1206.9 | 28.08 | 1002.4 | 1030.4 | 28.08 | 1059.6 | 1087.7 | 0.05555 | 2.0388 | 2.0943 |
| 70 | 0.3632 | 0.016051 | 867.7 | 38.09 | 995.6 | 1033.7 | 38.09 | 1054.0 | 1092.0 | 0.07463 | 1.9896 | 2.0642 |
| 80 | 0.5073 | 0.016073 | 632.8 | 48.08 | 988.9 | 1037.0 | 48.09 | 1048.3 | 1096.4 | 0.09332 | 1.9423 | 2.0356 |
| 90 | 0.6988 | 0.016099 | 467.7 | 58.07 | 982.2 | 1040.2 | 58.07 | 1042.7 | 1100.7 | 0.11165 | 1.8966 | 2.0083 |
| 100 | 0.9503 | 0.016130 | 350.0 | 68.04 | 975.4 | 1043.5 | 68.05 | 1037.0 | 1105.0 | 0.12963 | 1.8526 | 1.9822 |
| 110 | 1.2763 | 0.016166 | 265.1 | 78.02 | 968.7 | 1046.7 | 78.02 | 1031.3 | 1109.3 | 0.14730 | 1.8101 | 1.9574 |
| 120 | 1.6945 | 0.016205 | 203.0 | 87.99 | 961.9 | 1049.9 | 88.00 | 1025.5 | 1113.5 | 0.16465 | 1.7690 | 1.9336 |
| 130 | 2.225 | 0.016247 | 157.17 | 97.97 | 955.1 | 1053.0 | 97.98 | 1019.8 | 1117.8 | 0.18172 | 1.7292 | 1.9109 |
| 140 | 2.892 | 0.016293 | 122.88 | 107.95 | 948.2 | 1056.2 | 107.96 | 1014.0 | 1121.9 | 0.19851 | 1.6907 | 1.8892 |
| 150 | 3.722 | 0.016343 | 96.99 | 117.95 | 941.3 | 1059.3 | 117.96 | 1008.1 | 1126.1 | 0.21503 | 1.6533 | 1.8684 |
| 160 | 4.745 | 0.016395 | 77.23 | 127.94 | 934.4 | 1062.3 | 127.96 | 1002.2 | 1130.1 | 0.23130 | 1.6171 | 1.8484 |
| 170 | 5.996 | 0.016450 | 62.02 | 137.95 | 927.4 | 1065.4 | 137.97 | 996.2 | 1134.2 | 0.24732 | 1.5819 | 1.8293 |
| 180 | 7.515 | 0.016509 | 50.20 | 147.97 | 920.4 | 1068.3 | 147.99 | 990.2 | 1138.2 | 0.26311 | 1.5478 | 1.8109 |
| 190 | 9.343 | 0.016570 | 40.95 | 158.00 | 913.3 | 1071.3 | 158.03 | 984.1 | 1142.1 | 0.27866 | 1.5146 | 1.7932 |
| 200 | 11.529 | 0.016634 | 33.63 | 168.04 | 906.2 | 1074.2 | 168.07 | 977.9 | 1145.9 | 0.29400 | 1.4822 | 1.7762 |
| 210 | 14.125 | 0.016702 | 27.82 | 178.10 | 898.9 | 1077.0 | 178.14 | 971.6 | 1149.7 | 0.30913 | 1.4508 | 1.7599 |
| 212 | 14.698 | 0.016716 | 26.80 | 180.11 | 897.5 | 1077.6 | 180.16 | 970.3 | 1150.5 | 0.31213 | 1.4446 | 1.7567 |
| 220 | 17.188 | 0.016772 | 23.15 | 188.17 | 891.7 | 1079.8 | 188.22 | 965.3 | 1153.5 | 0.32406 | 1.4201 | 1.7441 |
| 230 | 20.78 | 0.016845 | 19.386 | 198.26 | 884.3 | 1082.6 | 198.32 | 958.8 | 1157.1 | 0.33880 | 1.3901 | 1.7289 |
| 240 | 24.97 | 0.016922 | 16.327 | 208.36 | 876.9 | 1085.3 | 208.44 | 952.3 | 1160.7 | 0.35335 | 1.3609 | 1.7143 |
| 250 | 29.82 | 0.017001 | 13.826 | 218.49 | 869.4 | 1087.9 | 218.59 | 945.6 | 1164.2 | 0.36772 | 1.3324 | 1.7001 |
| 260 | 35.42 | 0.017084 | 11.768 | 228.64 | 861.8 | 1090.5 | 228.76 | 938.8 | 1167.6 | 0.38193 | 1.3044 | 1.6864 |
| 270 | 41.85 | 0.017170 | 10.066 | 238.82 | 854.1 | 1093.0 | 238.95 | 932.0 | 1170.9 | 0.39597 | 1.2771 | 1.6731 |
| 280 | 49.18 | 0.017259 | 8.650 | 249.02 | 846.3 | 1095.4 | 249.18 | 924.9 | 1174.1 | 0.40986 | 1.2504 | 1.6602 |
| 290 | 57.53 | 0.017352 | 7.467 | 259.25 | 838.5 | 1097.7 | 259.44 | 917.8 | 1177.2 | 0.42360 | 1.2241 | 1.6477 |
| 300 | 66.98 | 0.017448 | 6.472 | 269.52 | 830.5 | 1100.0 | 269.73 | 910.4 | 1180.2 | 0.43720 | 1.1984 | 1.6356 |
| 310 | 77.64 | 0.017548 | 5.632 | 279.81 | 822.3 | 1102.1 | 280.06 | 903.0 | 1183.0 | 0.45067 | 1.1731 | 1.6238 |
| 320 | 89.60 | 0.017652 | 4.919 | 290.14 | 814.1 | 1104.2 | 290.43 | 895.3 | 1185.8 | 0.464.00 | 1.1483 | 1.6123 |
| 330 | 103.00 | 0.017760 | 4.312 | 300.51 | 805.7 | 1106.2 | 300.43 | 887.5 | 1188.4 | 0.47722 | 1.1238 | 1.6010 |
| 340 | 117.93 | 0.017872 | 3.792 | 310.91 | 797.1 | 1108.0 | 311.30 | 879.5 | 1190.8 | 0.49031 | 1.0997 | 1.5901 |
| 350 | 134.53 | 0.017988 | 3.346 | 321.35 | 788.4 | 1109.8 | 321.80 | 871.3 | 1193.1 | 0.50329 | 1.0760 | 1.5793 |
| 360 | 152.92 | 0.018108 | 2.961 | 331.84 | 779.6 | 1111.4 | 332.35 | 862.9 | 1195.2 | 0.51617 | 1.0526 | 1.5688 |
| 370 | 173.23 | 0.018233 | 2.628 | 342.37 | 770.6 | 1112.9 | 342.96 | 854.2 | 1197.2 | 0.52894 | 1.0295 | 1.5585 |
| 380 | 195.60 | 0.018363 | 2.339 | 352.95 | 761.4 | 1114.3 | 353.62 | 845.4 | 1199.0 | 0.54163 | 1.0067 | 1.5483 |
| 390 | 220.2 | 0.018498 | 2.087 | 363.58 | 752.0 | 1115.6 | 364.34 | 836.2 | 1200.6 | 0.55422 | 0.9841 | 1.5383 |
| 400 | 247.1 | 0.018638 | 1.8661 | 374.27 | 742.4 | 1116.6 | 375.12 | 826.8 | 1202.0 | 0.56672 | 0.9617 | 1.5284 |
| 410 | 276.5 | 0.018784 | 1.6726 | 385.01 | 732.6 | 1117.6 | 385.97 | 817.2 | 1203.1 | 0.57916 | 0.9395 | 1.5187 |
| 420 | 308.5 | 0.018936 | 1.5024 | 395.81 | 722.5 | 1118.3 | 396.89 | 807.2 | 1204.1 | 0.59152 | 0.9175 | 1.5091 |
| 430 | 343.3 | 0.019094 | 1.3521 | 406.68 | 712.2 | 1118.9 | 407.89 | 796.9 | 1204.8 | 0.60381 | 0.8957 | 1.4995 |
| 440 | 381.2 | 0.019260 | 1.2192 | 417.62 | 701.7 | 1119.3 | 418.98 | 786.3 | 1205.3 | 0.61605 | 0.8740 | 1.4900 |
| 450 | 422.1 | 0.019433 | 1.1011 | 428.6 | 690.9 | 1119.5 | 430.2 | 775.4 | 1205.6 | 0.6282 | 0.8523 | 1.4806 |
| 460 | 466.3 | 0.019614 | 0.9961 | 439.7 | 679.8 | 1119.6 | 441.4 | 764.1 | 1205.5 | 0.6404 | 0.8308 | 1.4712 |
| 470 | 514.1 | 0.019803 | 0.9025 | 450.9 | 668.4 | 1119.4 | 452.8 | 752.4 | 1205.2 | 0.6525 | 0.8093 | 1.4618 |
| 480 | 565.5 | 0.020002 | 0.8187 | 462.2 | 656.7 | 1118.9 | 464.3 | 740.3 | 1204.6 | 0.6646 | 0.7878 | 1.4524 |
| 490 | 620.7 | 0.020211 | 0.7436 | 473.6 | 644.7 | 1118.3 | 475.9 | 727.8 | 1203.7 | 0.6767 | 0.7663 | 1.4430 |

**TABLE A-4E**
**Saturated water: temperature table (*Continued*)**

| Temp. $T$ °F | Sat. press. $P_{sat}$ psia | Specific volume ft³/lbm | | Internal energy Btu/lbm | | | Enthalpy Btu/lbm | | | Entropy Btu/(lbm · R) | | |
|---|---|---|---|---|---|---|---|---|---|---|---|---|
| | | Sat. liquid $v_f$ | Sat. vapor $v_g$ | Sat. liquid $u_f$ | Evap. $u_{fg}$ | Sat. vapor $u_g$ | Sat. liquid $h_f$ | Evap. $h_{fg}$ | Sat. vapor $h_g$ | Sat. liquid $s_f$ | Evap. $s_{fg}$ | Sat. vapor $s_g$ |
| 500 | 680.0 | 0.02043 | 0.6761 | 485.1 | 632.3 | 1117.4 | 487.7 | 714.8 | 1202.5 | 0.6888 | 0.7448 | 1.4335 |
| 520 | 811.4 | 0.02091 | 0.5605 | 508.5 | 606.2 | 1114.8 | 511.7 | 687.3 | 1198.9 | 0.7130 | 0.7015 | 1.4145 |
| 540 | 961.5 | 0.02145 | 0.4658 | 532.6 | 578.4 | 1111.0 | 536.4 | 657.5 | 1193.8 | 0.7374 | 0.6576 | 1.3950 |
| 560 | 1131.8 | 0.02207 | 0.3877 | 557.4 | 548.4 | 1105.8 | 562.0 | 625.0 | 1187.0 | 0.7620 | 0.6129 | 1.3749 |
| 580 | 1324.3 | 0.02278 | 0.3225 | 583.1 | 515.9 | 1098.9 | 588.6 | 589.3 | 1178.0 | 0.7872 | 0.5668 | 1.3540 |
| 600 | 1541.0 | 0.02363 | 0.2677 | 609.9 | 480.1 | 1090.0 | 616.7 | 549.7 | 1166.4 | 0.8130 | 0.5187 | 1.3317 |
| 620 | 1784.4 | 0.02465 | 0.2209 | 638.3 | 440.2 | 1078.5 | 646.4 | 505.0 | 1151.4 | 0.8398 | 0.4677 | 1.3075 |
| 640 | 2057.1 | 0.02593 | 0.1805 | 668.7 | 394.5 | 1063.2 | 678.6 | 453.4 | 1131.9 | 0.8681 | 0.4122 | 1.2803 |
| 660 | 2362 | 0.02767 | 0.14459 | 702.3 | 340.0 | 1042.3 | 714.4 | 391.1 | 1105.5 | 0.8990 | 0.3493 | 1.2483 |
| 680 | 2705 | 0.03032 | 0.11127 | 741.7 | 269.3 | 1011.0 | 756.9 | 309.8 | 1066.7 | 0.9350 | 0.2718 | 1.2068 |
| 700 | 3090 | 0.03666 | 0.07438 | 801.7 | 145.9 | 947.7 | 822.7 | 167.5 | 990.2 | 0.9902 | 0.1444 | 1.1346 |
| 705.44 | 3204 | 0.05053 | 0.05053 | 872.6 | 0 | 872.6 | 902.5 | 0 | 902.5 | 1.0580 | 0 | 1.0580 |

*Source for Tables A-4E through A-6E:* Joseph H. Keenan, Frederick G. Keyes, Philip G. Hill, and Joan G. Moore, *Steam Tables,* Wiley, New York, 1969.

**TABLE A-5E**
**Saturated water: pressure table**

| Press. $P$ psia | Sat. temp. $T_{sat}$ °F | Specific volume ft³/lbm | | Internal energy Btu/lbm | | | Enthalpy Btu/lbm | | | Entropy Btu/(lbm · R) | | |
|---|---|---|---|---|---|---|---|---|---|---|---|---|
| | | Sat. liquid $v_f$ | Sat. vapor $v_g$ | Sat. liquid $u_f$ | Evap. $u_{fg}$ | Sat. vapor $u_g$ | Sat. liquid $h_f$ | Evap. $h_{fg}$ | Sat. vapor $h_g$ | Sat. liquid $s_f$ | Evap. $s_{fg}$ | Sat. vapor $s_g$ |
| 1.0 | 101.70 | 0.016136 | 333.6 | 69.74 | 974.3 | 1044.0 | 69.74 | 1036.0 | 1105.8 | 0.13266 | 1.8453 | 1.9779 |
| 2.0 | 126.04 | 0.016230 | 173.75 | 94.02 | 957.8 | 1051.8 | 94.02 | 1022.1 | 1116.1 | 0.17499 | 1.7448 | 1.9198 |
| 3.0 | 141.43 | 0.016300 | 118.72 | 109.38 | 947.2 | 1056.6 | 109.39 | 10131 | 1122.5 | 0.20089 | 1.6852 | 1.8861 |
| 4.0 | 152.93 | 0.016358 | 90.64 | 120.88 | 939.3 | 1060.2 | 120.89 | 1006.4 | 1127.3 | 0.21983 | 1.6426 | 1.8624 |
| 5.0 | 162.21 | 0.016407 | 73.53 | 130.15 | 932.9 | 1063.0 | 130.17 | 1000.9 | 1131.0 | 0.23486 | 1.6093 | 1.8441 |
| 6.0 | 170.03 | 0.016451 | 61.98 | 137.98 | 927.4 | 1065.4 | 138.00 | 996.2 | 1134.2 | 0.24736 | 1.5819 | 1.8292 |
| 8.0 | 182.84 | 0.016526 | 47.35 | 150.81 | 918.4 | 1069.2 | 150.84 | 988.4 | 1139.3 | 0.26754 | 1.5383 | 1.8058 |
| 10 | 193.19 | 0.016590 | 38.42 | 161.20 | 911.0 | 1072.2 | 161.23 | 982.1 | 1143.3 | 0.28358 | 1.5041 | 1.7877 |
| 14.696 | 211.99 | 0.016715 | 26.80 | 180.10 | 897.5 | 1077.6 | 180.15 | 970.4 | 1150.5 | 0.31212 | 1.4446 | 1.7567 |
| 15 | 213.03 | 0.016723 | 26.29 | 181.14 | 896.8 | 1077.9 | 181.19 | 969.7 | 1150.9 | 0.31367 | 1.4414 | 1.7551 |
| 20 | 227.96 | 0.016830 | 20.09 | 196.19 | 885.8 | 1.082.0 | 196.26 | 960.1 | 1156.4 | 0.33580 | 1.3962 | 1.7320 |

**TABLE A-5E**

**Saturated water: pressure table (*Continued*)**

| Press. $P$ psia | Sat. temp. $T_{sat}$ °F | Specific volume ft³/lbm Sat. liquid $v_f$ | Sat. vapor $v_g$ | Internal energy Btu/lbm Sat. liquid $u_f$ | Evap. $u_{fg}$ | Sat. vapor $u_g$ | Enthalpy Btu/lbm Sat. liquid $h_f$ | Evap. $h_{fg}$ | Sat. vapor $h_g$ | Entropy Btu/(lbm · R) Sat. liquid $s_f$ | Evap. $s_{fg}$ | Sat. vapor $s_g$ |
|---|---|---|---|---|---|---|---|---|---|---|---|---|
| 25 | 240.08 | 0.016922 | 16.306 | 208.44 | 876.9 | 1085.3 | 208.52 | 952.2 | 1160.7 | 0.35345 | 1.3607 | 1.7142 |
| 30 | 250.34 | 0.017004 | 13.748 | 218.84 | 869.2 | 1088.0 | 218.93 | 945.4 | 1164.3 | 0.36821 | 1.3314 | 1.6996 |
| 35 | 259.30 | 0.017073 | 11.900 | 227.93 | 862.4 | 1090.3 | 228.04 | 939.3 | 1167.4 | 0.38093 | 1.3064 | 1.6873 |
| 40 | 267.26 | 0.017146 | 10.501 | 236.03 | 856.2 | 1092.3 | 236.16 | 933.8 | 1170.0 | 0.39214 | 1.2845 | 1.6767 |
| 45 | 274.46 | 0.017209 | 9.403 | 243.37 | 850.7 | 1094.0 | 243.51 | 928.8 | 1172.3 | 0.40218 | 1.2651 | 1.6673 |
| 50 | 281.03 | 0.017269 | 8.518 | 250.08 | 845.5 | 1095.6 | 250.24 | 924.2 | 1174.4 | 0.41129 | 1.2476 | 1.6589 |
| 55 | 287.10 | 0.017325 | 7.789 | 256.28 | 840.8 | 1097.0 | 256.46 | 919.9 | 1176.3 | 0.41963 | 1.2317 | 1.6513 |
| 60 | 292.73 | 0.017378 | 7.177 | 262.06 | 836.3 | 1098.3 | 262.25 | 915.8 | 1178.0 | 0.42733 | 1.2170 | 1.6444 |
| 65 | 298.00 | 0.017429 | 6.657 | 267.46 | 832.1 | 1099.5 | 267.67 | 911.9 | 1179.6 | 0.43450 | 1.2035 | 1.6380 |
| 70 | 302.96 | 0.017478 | 6.209 | 272.56 | 828.1 | 1100.6 | 272.79 | 908.3 | 1181.0 | 0.44120 | 1.1909 | 1.6321 |
| 75 | 307.63 | 0.017524 | 5.818 | 277.37 | 824.3 | 1101.6 | 277.61 | 904.8 | 1182.4 | 0.44749 | 1.1790 | 1.6265 |
| 80 | 312.07 | 0.017570 | 5.474 | 281.95 | 820.6 | 1102.6 | 282.21 | 901.4 | 1183.6 | 0.45344 | 1.1679 | 1.6214 |
| 85 | 316.29 | 0.017613 | 5.170 | 286.30 | 817.1 | 1103.5 | 286.58 | 898.2 | 1184.8 | 0.45907 | 1.1574 | 1.6165 |
| 90 | 320.31 | 0.017655 | 4.898 | 290.46 | 813.8 | 1104.3 | 290.76 | 895.1 | 1185.9 | 0.46442 | 1.1475 | 1.6119 |
| 95 | 324.16 | 0.017696 | 4.654 | 294.45 | 810.6 | 1105.0 | 294.76 | 892.1 | 1186.9 | 0.46952 | 1.1380 | 1.6076 |
| 100 | 327.86 | 0.017736 | 4.434 | 298.28 | 807.5 | 1105.8 | 298.61 | 889.2 | 1187.8 | 0.47439 | 1.1290 | 1.6034 |
| 110 | 334.82 | 0.017813 | 4.051 | 305.52 | 801.6 | 1107.1 | 305.88 | 883.7 | 1189.6 | 0.48355 | 1.1122 | 1.5957 |
| 120 | 341.30 | 0.017886 | 3.730 | 312.27 | 796.0 | 1108.3 | 312.67 | 878.5 | 1191.1 | 0.49201 | 1.0966 | 1.5886 |
| 130 | 347.37 | 0.017957 | 3.457 | 318.61 | 790.7 | 1109.4 | 319.04 | 873.5 | 1192.5 | 0.49989 | 1.0822 | 1.5821 |
| 140 | 353.08 | 0.018024 | 3.221 | 324.58 | 785.7 | 1110.3 | 325.05 | 868.7 | 1193.8 | 0.50727 | 1.0688 | 1.5761 |
| 150 | 358.48 | 0.018089 | 3.016 | 330.24 | 781.0 | 1111.2 | 330.75 | 864.2 | 1194.9 | 0.51422 | 1.0562 | 1.5704 |
| 160 | 363.60 | 0.018152 | 2.836 | 335.63 | 776.4 | 1112.0 | 336.16 | 859.8 | 1196.0 | 0.52078 | 1.0443 | 1.5651 |
| 170 | 368.47 | 0.018214 | 2.676 | 340.76 | 772.0 | 1112.7 | 341.33 | 855.6 | 1196.9 | 0.52700 | 1.0330 | 1.5600 |
| 180 | 373.13 | 0.018273 | 2.533 | 345.68 | 767.7 | 1113.4 | 346.29 | 851.5 | 1197.8 | 0.53292 | 1.0223 | 1.5553 |
| 190 | 377.59 | 0.018331 | 2.405 | 350.39 | 763.6 | 1114.0 | 351.04 | 847.5 | 1198.6 | 0.53857 | 1.0122 | 1.5507 |
| 200 | 381.86 | 0.018387 | 2.289 | 354.9 | 759.6 | 1114.6 | 355.6 | 843.7 | 1199.3 | 0.5440 | 1.0025 | 1.5464 |
| 250 | 401.04 | 0.018653 | 1.8448 | 375.4 | 741.4 | 1116.7 | 376.2 | 825.8 | 1202.1 | 0.5680 | 0.9594 | 1.5274 |
| 300 | 417.43 | 0.018896 | 1.5442 | 393.0 | 725.1 | 1118.2 | 394.1 | 809.8 | 1203.9 | 0.5883 | 0.9232 | 1.5115 |
| 350 | 431.82 | 0.019124 | 1.3267 | 408.7 | 710.3 | 1119.0 | 409.9 | 795.0 | 1204.9 | 0.6060 | 0.8917 | 1.4978 |
| 400 | 444.70 | 0.019340 | 1.1620 | 422.8 | 696.7 | 1119.5 | 424.2 | 781.2 | 1205.5 | 0.6218 | 0.8638 | 1.4856 |
| 450 | 456.39 | 0.019547 | 1.0326 | 435.7 | 683.9 | 1119.6 | 437.4 | 768.2 | 1205.6 | 0.6360 | 0.8385 | 1.4746 |
| 500 | 467.13 | 0.019748 | 0.9283 | 447.7 | 671.7 | 1119.4 | 449.5 | 755.8 | 1205.3 | 0.6490 | 0.8154 | 1.4645 |
| 550 | 477.07 | 0.019943 | 0.8423 | 458.9 | 660.2 | 1119.1 | 460.9 | 743.9 | 1204.8 | 0.6611 | 0.7941 | 1.4551 |
| 600 | 486.33 | 0.02013 | 0.7702 | 469.4 | 649.1 | 1118.6 | 471.7 | 732.4 | 1204.1 | 0.6723 | 0.7742 | 1.4464 |
| 700 | 503.23 | 0.02051 | 0.6558 | 488.9 | 628.2 | 1117.0 | 491.5 | 710.5 | 1202.0 | 0.6927 | 0.7378 | 1.4305 |
| 800 | 518.36 | 0.02087 | 0.5691 | 506.6 | 608.4 | 1115.0 | 509.7 | 689.6 | 1199.3 | 0.7110 | 0.7050 | 1.4160 |
| 900 | 532.12 | 0.02123 | 0.5009 | 523.0 | 589.6 | 1112.6 | 526.6 | 669.5 | 1196.0 | 0.7277 | 0.6750 | 1.4027 |
| 1000 | 544.75 | 0.02159 | 0.4459 | 538.4 | 571.5 | 1109.9 | 542.4 | 650.0 | 1192.4 | 0.7432 | 0.6471 | 1.3903 |
| 1200 | 567.37 | 0.02232 | 0.3623 | 566.7 | 536.8 | 1103.5 | 571.7 | 612.3 | 1183.9 | 0.7712 | 0.5961 | 1.3673 |
| 1400 | 587.25 | 0.02307 | 0.3016 | 592.7 | 503.3 | 1096.0 | 598.6 | 575.5 | 1174.1 | 0.7964 | 0.5497 | 1.3461 |
| 1600 | 605.06 | 0.02386 | 0.2552 | 616.9 | 470.5 | 1087.4 | 624.0 | 538.9 | 1162.9 | 0.8196 | 0.5062 | 1.3258 |
| 1800 | 621.21 | 0.02472 | 0.2183 | 640.0 | 437.6 | 1077.7 | 648.3 | 502.1 | 1150.4 | 0.8414 | 0.4645 | 1.3060 |
| 2000 | 636.00 | 0.02565 | 0.18813 | 662.4 | 404.2 | 1066.6 | 671.9 | 464.4 | 1136.3 | 0.8623 | 0.4238 | 1.2861 |
| 2500 | 668.31 | 0.02860 | 0.13059 | 717.7 | 313.4 | 1031.0 | 730.9 | 360.5 | 1091.4 | 0.9131 | 0.3196 | 1.2327 |
| 3000 | 695.52 | 0.03431 | 0.08404 | 783.4 | 185.4 | 968.8 | 802.5 | 213.0 | 1015.5 | 0.9732 | 0.1843 | 1.1575 |
| 3203.6 | 705.44 | 0.05053 | 0.05053 | 872.6 | 0 | 872.6 | 902.5 | 0 | 902.5 | 1.0580 | 0 | 1.0580 |

**TABLE A-6E**
**Superheated water**

| $T$ °F | $v$ ft³/lbm | $u$ Btu/lbm | $h$ Btu/lbm | $s$ Btu/(lbm·R) | $v$ ft³/lbm | $u$ Btu/lbm | $h$ Btu/lbm | $s$ Btu/(lbm·R) | $v$ ft³/lbm | $u$ Btu/lbm | $h$ Btu/lbm | $s$ Btu/(lbm·R) |
|---|---|---|---|---|---|---|---|---|---|---|---|---|
| | $P$ = 1.0 psia (101.70°F)* | | | | $P$ = 5.0 psia (162.21°F) | | | | $P$ = 10.0 psia (193.19°F) | | | |
| Sat.† | 333.6 | 1044.0 | 1105.8 | 1.9779 | 73.53 | 1063.0 | 1131.0 | 1.8441 | 38.42 | 1072.2 | 1143.3 | 1.7877 |
| 200 | 392.5 | 1077.5 | 1150.1 | 2.0508 | 78.15 | 1076.3 | 1148.6 | 1.8715 | 38.85 | 1074.7 | 1146.6 | 1.7927 |
| 240 | 416.4 | 1091.2 | 1168.3 | 2.0775 | 83.00 | 1090.3 | 1167.1 | 1.8987 | 41.32 | 1089.0 | 1165.5 | 1.8205 |
| 280 | 440.3 | 1105.0 | 1186.5 | 2.1028 | 87.83 | 1104.3 | 1185.5 | 1.9244 | 43.77 | 1103.3 | 1184.3 | 1.8467 |
| 320 | 464.2 | 1118.9 | 1204.8 | 2.1269 | 92.64 | 1118.3 | 1204.0 | 1.9487 | 46.20 | 1117.6 | 1203.1 | 1.8714 |
| 360 | 488.1 | 1132.9 | 1223.2 | 2.1500 | 97.45 | 1132.4 | 1222.6 | 1.9719 | 48.62 | 1131.8 | 1221.8 | 1.8948 |
| 400 | 511.9 | 1147.0 | 1241.8 | 2.1720 | 102.24 | 1146.6 | 1241.2 | 1.9941 | 51.03 | 1146.1 | 1240.5 | 1.9171 |
| 440 | 535.8 | 1161.2 | 1260.4 | 2.1932 | 107.03 | 1160.9 | 1259.9 | 2.0154 | 53.44 | 1160.5 | 1259.3 | 1.9385 |
| 500 | 571.5 | 1182.8 | 1288.5 | 2.2235 | 114.20 | 1182.5 | 1288.2 | 2.0458 | 57.04 | 1182.2 | 1287.7 | 1.9690 |
| 600 | 631.1 | 1219.3 | 1336.1 | 2.2706 | 126.15 | 1219.1 | 1335.8 | 2.0930 | 63.03 | 1218.9 | 1335.5 | 2.0164 |
| 700 | 690.7 | 1256.7 | 1384.5 | 2.3142 | 138.08 | 1256.5 | 1384.3 | 2.1367 | 69.01 | 1256.3 | 1384.0 | 2.0601 |
| 800 | 750.3 | 1294.9 | 1433.7 | 2.3550 | 150.01 | 1294.7 | 1433.5 | 2.1775 | 74.98 | 1294.6 | 1433.3 | 2.1009 |
| 1000 | 869.5 | 1373.9 | 1534.8 | 2.4294 | 173.86 | 1373.9 | 1534.7 | 2.2520 | 86.91 | 1373.8 | 1534.6 | 2.1755 |
| 1200 | 988.6 | 1456.7 | 1639.6 | 2.4967 | 197.70 | 1456.6 | 1639.5 | 2.3192 | 98.84 | 1456.5 | 1639.4 | 2.2428 |
| 1400 | 1107.7 | 1543.1 | 1748.1 | 2.5584 | 221.54 | 1543.1 | 1748.1 | 2.3810 | 110.76 | 1543.0 | 1748.0 | 2.3045 |
| | $P$ = 14.696 psia (211.99°F) | | | | $P$ = 20 psia (227.96°F) | | | | $P$ = 40 psia (267.26°F) | | | |
| Sat. | 26.80 | 1077.6 | 1150.5 | 1.7567 | 20.09 | 1082.0 | 1156.4 | 1.7320 | 10.501 | 1092.3 | 1170.0 | 1.6767 |
| 240 | 28.00 | 1087.9 | 1164.0 | 1.7764 | 20.47 | 1086.5 | 1162.3 | 1.7405 | | | | |
| 280 | 29.69 | 1102.4 | 1183.1 | 1.8030 | 21.73 | 1101.4 | 1181.8 | 1.7676 | 10.711 | 1097.3 | 1176.6 | 1.6857 |
| 320 | 31.36 | 1116.8 | 1202.1 | 1.8280 | 22.98 | 1116.0 | 1201.0 | 1.7930 | 11.360 | 1112.8 | 1196.9 | 1.7124 |
| 360 | 33.02 | 1131.2 | 1221.0 | 1.8516 | 24.21 | 1130.6 | 1220.1 | 1.8168 | 11.996 | 1128.0 | 1216.8 | 1.7373 |
| 400 | 34.67 | 1145.6 | 1239.9 | 1.8741 | 25.43 | 1145.1 | 1239.2 | 1.8395 | 12.623 | 1143.0 | 1236.4 | 1.7606 |
| 440 | 36.31 | 1160.1 | 1258.8 | 1.8956 | 26.64 | 1159.6 | 1258.2 | 1.8611 | 13.243 | 1157.8 | 1255.8 | 1.7828 |
| 500 | 38.77 | 1181.8 | 1287.3 | 1.9263 | 28.46 | 1181.5 | 1286.8 | 1.8919 | 14.164 | 1180.1 | 1284.9 | 1.8140 |
| 600 | 42.86 | 1218.6 | 1335.2 | 1.9737 | 31.47 | 1218.4 | 1334.8 | 1.9395 | 15.685 | 1217.3 | 1333.4 | 1.8621 |
| 700 | 46.93 | 1256.1 | 1383.8 | 2.0175 | 34.47 | 1255.9 | 1383.5 | 1.9834 | 17.196 | 1255.1 | 1382.4 | 1.9063 |
| 800 | 51.00 | 1294.4 | 1433.1 | 2.0584 | 37.46 | 1294.3 | 1432.9 | 2.0243 | 18.701 | 1293.7 | 1432.1 | 1.9474 |
| 1000 | 59.13 | 1373.7 | 1534.5 | 2.1330 | 43.44 | 1373.5 | 1534.3 | 2.0989 | 21.70 | 1373.1 | 1533.8 | 2.0223 |
| 1200 | 67.25 | 1456.5 | 1639.3 | 2.2003 | 49.41 | 1456.4 | 1639.2 | 2.1663 | 24.69 | 1456.1 | 1638.9 | 2.0897 |
| 1400 | 75.36 | 1543.0 | 1747.9 | 2.2621 | 55.37 | 1542.9 | 1747.9 | 2.2281 | 27.68 | 1542.7 | 1747.6 | 2.1515 |
| 1600 | 83.47 | 1633.2 | 1860.2 | 2.3194 | 61.33 | 1633.2 | 1860.1 | 2.2854 | 30.66 | 1633.0 | 1859.9 | 2.2089 |
| | $P$ = 60 psia (292.73°F) | | | | $P$ = 80 psia (312.07°F) | | | | $P$ = 100 psia (327.86°F) | | | |
| Sat. | 7.177 | 1098.3 | 1178.0 | 1.6444 | 5.474 | 1102.6 | 1183.6 | 1.6214 | 4.434 | 1105.8 | 1187.8 | 1.6034 |
| 320 | 7.485 | 1109.5 | 1192.6 | 1.6634 | 5.544 | 1106.0 | 1188.0 | 1.6271 | | | | |
| 360 | 7.924 | 1125.3 | 1213.3 | 1.6893 | 5.886 | 1122.5 | 1209.7 | 1.6541 | 4.662 | 1119.7 | 1205.9 | 1.6259 |
| 400 | 8.353 | 1140.8 | 1233.5 | 1.7134 | 6.217 | 1138.5 | 1230.6 | 1.6790 | 4.934 | 1136.2 | 1227.5 | 1.6517 |
| 440 | 8.775 | 1156.0 | 1253.4 | 1.7360 | 6.541 | 1154.2 | 1251.0 | 1.7022 | 5.199 | 1152.3 | 1248.5 | 1.6755 |
| 500 | 9.399 | 1178.6 | 1283.0 | 1.7678 | 7.017 | 1177.2 | 1281.1 | 1.7346 | 5.587 | 1175.7 | 1279.1 | 1.7085 |
| 600 | 10.425 | 1216.3 | 1332.1 | 1.8165 | 7.794 | 1215.3 | 1330.7 | 1.7838 | 6.216 | 1214.2 | 1329.3 | 1.7582 |
| 700 | 11.440 | 1254.4 | 1381.4 | 1.8609 | 8.561 | 1253.6 | 1380.3 | 1.8285 | 6.834 | 1252.8 | 1379.2 | 1.8033 |
| 800 | 12.448 | 1293.0 | 1431.2 | 1.9022 | 9.321 | 1292.4 | 1430.4 | 1.8700 | 7.445 | 1291.8 | 1429.6 | 1.8449 |
| 1000 | 14.454 | 1372.7 | 1533.2 | 1.9773 | 10.831 | 1372.3 | 1532.6 | 1.9453 | 8.657 | 1371.9 | 1532.1 | 1.9204 |
| 1200 | 16.452 | 1455.8 | 1638.5 | 2.0448 | 12.333 | 1455.5 | 1638.1 | 2.0130 | 9.861 | 1455.2 | 1637.7 | 1.9882 |
| 1400 | 18.445 | 1542.5 | 1747.3 | 2.1067 | 13.830 | 1542.3 | 1747.0 | 2.0749 | 11.060 | 1542.0 | 1746.7 | 2.0502 |
| 1600 | 20.44 | 1632.8 | 1859.7 | 2.1641 | 15.324 | 1632.6 | 1859.5 | 2.1323 | 12.257 | 1632.4 | 1859.3 | 2.1076 |
| 1800 | 22.43 | 1726.7 | 1975.7 | 2.2179 | 16.818 | 1726.5 | 1975.5 | 2.1861 | 13.452 | 1726.4 | 1975.3 | 2.1614 |
| 2000 | 24.41 | 1824.0 | 2095.1 | 2.2685 | 18.310 | 1823.9 | 2094.9 | 2.2367 | 14.647 | 1823.7 | 2094.8 | 2.2121 |

*The temperature in parentheses is the saturation temperature at the specified pressure.

†Properties of saturated vapor at the specified pressure.

**TABLE A-6E**
**Superheated water (*Continued*)**

| $T$ °F | $v$ ft³/lbm | $u$ Btu/lbm | $h$ Btu/lbm | $s$ Btu/(lbm·R) | $v$ ft³/lbm | $u$ Btu/lbm | $h$ Btu/lbm | $s$ Btu/(lbm·R) | $v$ ft³/lbm | $u$ Btu/lbm | $h$ Btu/lbm | $s$ Btu/(lbm·R) |
|---|---|---|---|---|---|---|---|---|---|---|---|---|
| | $P$ = **120 psia (341.30°F)** | | | | $P$ = **140 psia (353.08°F)** | | | | $P$ = **160 psia (363.60°F)** | | | |
| Sat. | 3.730 | 1108.3 | 1191.1 | 1.5886 | 3.221 | 1110.3 | 1193.8 | 1.5761 | 2.836 | 1112.0 | 1196.0 | 1.5651 |
| 360 | 3.844 | 1116.7 | 1202.0 | 1.6021 | 3.259 | 1113.5 | 1198.0 | 1.5812 | | | | |
| 400 | 4.079 | 1133.8 | 1224.4 | 1.6288 | 3.466 | 1131.4 | 1221.2 | 1.6088 | 3.007 | 1128.8 | 1217.8 | 1.5911 |
| 450 | 4.360 | 1154.3 | 1251.2 | 1.6590 | 3.713 | 1152.4 | 1248.6 | 1.6399 | 3.228 | 1150.5 | 1246.1 | 1.6230 |
| 500 | 4.633 | 1174.2 | 1277.1 | 1.6868 | 3.952 | 1172.7 | 1275.1 | 1.6682 | 3.440 | 1171.2 | 1273.0 | 1.6518 |
| 550 | 4.900 | 1193.8 | 1302.6 | 1.7127 | 4.184 | 1192.6 | 1300.9 | 1.6944 | 3.646 | 1191.3 | 1299.2 | 1.6784 |
| 600 | 5.164 | 1213.2 | 1327.8 | 1.7371 | 4.412 | 1212.1 | 1326.4 | 1.7191 | 3.848 | 1211.1 | 1325.0 | 1.7034 |
| 700 | 5.682 | 1252.0 | 1378.2 | 1.7825 | 4.860 | 1251.2 | 1377.1 | 1.7648 | 4.243 | 1250.4 | 1376.0 | 1.7494 |
| 800 | 6.195 | 1291.2 | 1428.7 | 1.8243 | 5.301 | 1290.5 | 1427.9 | 1.8068 | 4.631 | 1289.9 | 1427.0 | 1.7916 |
| 1000 | 7.208 | 1371.5 | 1531.5 | 1.9000 | 6.173 | 1371.0 | 1531.0 | 1.8827 | 5.397 | 1370.6 | 1530.4 | 1.8677 |
| 1200 | 8.213 | 1454.9 | 1637.3 | 1.9679 | 7.036 | 1454.6 | 1636.9 | 1.9507 | 6.154 | 1454.3 | 1636.5 | 1.9358 |
| 1400 | 9.214 | 1541.8 | 1746.4 | 2.0300 | 7.895 | 1541.6 | 1746.1 | 2.0129 | 6.906 | 1541.4 | 1745.9 | 1.9980 |
| 1600 | 10.212 | 1632.3 | 1859.0 | 2.0875 | 8.752 | 1632.1 | 1858.8 | 2.0704 | 7.656 | 1631.9 | 1858.6 | 2.0556 |
| 1800 | 11.209 | 1726.2 | 1975.1 | 2.1413 | 9.607 | 1726.1 | 1975.0 | 2.1242 | 8.405 | 1725.9 | 1974.8 | 2.1094 |
| 2000 | 12.205 | 1823.6 | 2094.6 | 2.1919 | 10.461 | 1823.5 | 2094.5 | 2.1749 | 9.153 | 1823.3 | 2094.3 | 2.1601 |
| | $P$ = **180 psia (373.13°F)** | | | | $P$ = **200 psia (381.86°F)** | | | | $P$ = **225 psia (391.87°F)** | | | |
| Sat. | 2.533 | 1113.4 | 1197.8 | 1.5553 | 2.289 | 1114.6 | 1199.3 | 1.5464 | 2.043 | 1115.8 | 1200.8 | 1.5365 |
| 400 | 2.648 | 1126.2 | 1214.4 | 1.5749 | 2.361 | 1123.5 | 1210.8 | 1.5600 | 2.073 | 1119.9 | 1206.2 | 1.5427 |
| 450 | 2.850 | 1148.5 | 1243.4 | 1.6078 | 2.548 | 1146.4 | 1240.7 | 1.5938 | 2.245 | 1143.8 | 1237.3 | 1.5779 |
| 500 | 3.042 | 1169.6 | 1270.9 | 1.6372 | 2.724 | 1168.0 | 1268.8 | 1.6239 | 2.405 | 1165.9 | 1266.1 | 1.6087 |
| 550 | 3.228 | 1190.0 | 1297.5 | 1.6642 | 2.893 | 1188.7 | 1295.7 | 1.6512 | 2.588 | 1187.0 | 1293.5 | 1.6366 |
| 600 | 3.409 | 1210.0 | 1323.5 | 1.6893 | 3.058 | 1208.9 | 1322.1 | 1.6767 | 2.707 | 1207.5 | 1320.2 | 1.6624 |
| 700 | 3.763 | 1249.6 | 1374.9 | 1.7357 | 3.379 | 1248.8 | 1373.8 | 1.7234 | 2.995 | 1247.7 | 1372.4 | 1.7095 |
| 800 | 4.110 | 1289.3 | 1426.2 | 1.7781 | 3.693 | 1288.6 | 1425.3 | 1.7660 | 3.276 | 1287.8 | 1424.2 | 1.7523 |
| 900 | 4.453 | 1329.4 | 1477.7 | 1.8175 | 4.003 | 1328.9 | 1477.1 | 1.8055 | 3.553 | 1328.3 | 1476.2 | 1.7920 |
| 1000 | 4.793 | 1370.2 | 1529.8 | 1.8545 | 4.310 | 1369.8 | 1529.3 | 1.8425 | 3.827 | 1369.3 | 1528.6 | 1.8292 |
| 1200 | 5.467 | 1454.0 | 1636.1 | 1.9227 | 4.918 | 1453.7 | 1635.7 | 1.9109 | 4.369 | 1453.4 | 1635.3 | 1.8977 |
| 1400 | 6.137 | 1541.2 | 1745.6 | 1.9849 | 5.521 | 1540.9 | 1745.3 | 1.9732 | 4.906 | 1540.7 | 1744.9 | 1.9600 |
| 1600 | 6.804 | 1631.7 | 1858.4 | 2.0425 | 6.123 | 1631.6 | 1858.2 | 2.0308 | 5.441 | 1631.3 | 1857.9 | 2.0177 |
| 1800 | 7.470 | 1725.8 | 1974.6 | 2.0964 | 6.722 | 1725.6 | 1974.4 | 2.0847 | 5.975 | 1725.4 | 1974.2 | 2.0716 |
| 2000 | 8.135 | 1823.2 | 2094.2 | 2.1470 | 7.321 | 1823.0 | 2094.0 | 2.1354 | 6.507 | 1822.9 | 2093.8 | 2.1223 |
| | $P$ = **250 psia (401.04°F)** | | | | $P$ = **275 psia (409.52°F)** | | | | $P$ = **300 psia (417.43°F)** | | | |
| Sat. | 1.8448 | 1116.7 | 1202.1 | 1.5274 | 1.6813 | 1117.5 | 1203.1 | 1.5192 | 1.5442 | 1118.2 | 1203.9 | 1.5115 |
| 450 | 2.002 | 1141.1 | 1233.7 | 1.5632 | 1.8026 | 1138.3 | 1230.0 | 1.5495 | 1.6361 | 1135.4 | 1226.2 | 1.5365 |
| 500 | 2.150 | 1163.8 | 1263.3 | 1.5948 | 1.9407 | 1161.7 | 1260.4 | 1.5820 | 1.7662 | 1159.5 | 1257.5 | 1.5701 |
| 550 | 2.290 | 1185.3 | 1291.3 | 1.6233 | 2.071 | 1183.6 | 1289.0 | 1.6110 | 1.8878 | 1181.9 | 1286.7 | 1.5997 |
| 600 | 2.426 | 1206.1 | 1318.3 | 1.6494 | 2.196 | 1204.7 | 1316.4 | 1.6376 | 2.004 | 1203.2 | 1314.5 | 1.6266 |
| 650 | 2.558 | 1226.5 | 1344.9 | 1.6739 | 2.317 | 1225.3 | 1343.2 | 1.6623 | 2.117 | 1224.1 | 1341.6 | 1.6516 |
| 700 | 2.688 | 1246.7 | 1371.1 | 1.6970 | 2.436 | 1245.7 | 1369.7 | 1.6856 | 2.227 | 1244.6 | 1368.3 | 1.6751 |
| 800 | 2.943 | 1287.0 | 1423.2 | 1.7401 | 2.670 | 1286.2 | 1422.1 | 1.7289 | 2.442 | 1285.4 | 1421.0 | 1.7187 |
| 900 | 3.193 | 1327.6 | 1475.3 | 1.7799 | 2.898 | 1327.0 | 1474.5 | 1.7689 | 2.653 | 1326.3 | 1473.6 | 1.7589 |
| 1000 | 3.440 | 1368.7 | 1527.9 | 1.8172 | 3.124 | 1368.2 | 1527.2 | 1.8064 | 2.860 | 1367.7 | 1526.5 | 1.7964 |
| 1200 | 3.929 | 1453.0 | 1634.8 | 1.8858 | 3.570 | 1452.6 | 1634.3 | 1.8751 | 3.270 | 1452.2 | 1633.8 | 1.8653 |
| 1400 | 4.414 | 1540.4 | 1744.6 | 1.9483 | 4.011 | 1540.1 | 1744.2 | 1.9376 | 3.675 | 1539.8 | 1743.8 | 1.9279 |
| 1600 | 4.896 | 1631.1 | 1857.6 | 2.0060 | 4.450 | 1630.9 | 1857.3 | 1.9954 | 4.078 | 1630.7 | 1857.0 | 1.9857 |
| 1800 | 5.376 | 1725.2 | 1974.0 | 2.0599 | 4.887 | 1725.0 | 1973.7 | 2.0493 | 4.479 | 1724.9 | 1973.5 | 2.0396 |
| 2000 | 5.856 | 1822.7 | 2093.6 | 2.1106 | 5.323 | 1822.5 | 2093.4 | 2.1000 | 4.879 | 1822.3 | 2093.2 | 2.0904 |

| $T$ °F | $v$ ft³/lbm | $u$ Btu/lbm | $h$ Btu/lbm | $s$ Btu/(lbm · R) | $v$ ft³/lbm | $u$ Btu/lbm | $h$ Btu/lbm | $s$ Btu/(lbm · R) | $v$ ft³/lbm | $u$ Btu/lbm | $h$ Btu/lbm | $s$ Btu/(lbm · R) |
|---|---|---|---|---|---|---|---|---|---|---|---|---|
| | $P$ = **350 psia (431.82°F)** | | | | $P$ = **400 psia (444.70°F)** | | | | $P$ = **450 psia (456.39°F)** | | | |
| Sat. | 1.3267 | 1119.0 | 1204.9 | 1.4978 | 1.1620 | 1119.5 | 1205.5 | 1.4856 | 1.0326 | 1119.6 | 1205.6 | 1.4746 |
| 450 | 1.3733 | 1129.2 | 1218.2 | 1.5125 | 1.1745 | 1122.6 | 1209.6 | 1.4901 | | | | |
| 500 | 1.4913 | 1154.9 | 1251.5 | 1.5482 | 1.2843 | 1150.1 | 1245.2 | 1.5282 | 1.1226 | 1145.1 | 1238.5 | 1.5097 |
| 550 | 1.5998 | 1178.3 | 1281.9 | 1.5790 | 1.3833 | 1174.6 | 1277.0 | 1.5605 | 1.2146 | 1170.7 | 1271.9 | 1.5436 |
| 600 | 1.7025 | 1200.3 | 1310.6 | 1.6068 | 1.4760 | 1197.3 | 1306.6 | 1.5892 | 1.2996 | 1194.3 | 1302.5 | 1.5732 |
| 650 | 1.8013 | 1221.6 | 1338.3 | 1.6323 | 1.5645 | 1219.1 | 1334.9 | 1.6153 | 1.3803 | 1216.6 | 1331.5 | 1.6000 |
| 700 | 1.8975 | 1242.5 | 1365.4 | 1.6562 | 1.6503 | 1240.4 | 1362.5 | 1.6397 | 1.4580 | 1238.2 | 1359.6 | 1.6248 |
| 800 | 2.085 | 1283.8 | 1418.8 | 1.7004 | 1.8163 | 1282.1 | 1416.6 | 1.6844 | 1.6077 | 1280.5 | 1414.4 | 1.6701 |
| 900 | 2.267 | 1325.0 | 1471.8 | 1.7409 | 1.9776 | 1323.7 | 1470.1 | 1.7252 | 1.7524 | 1322.4 | 1468.3 | 1.7113 |
| 1000 | 2.446 | 1366.6 | 1525.0 | 1.7787 | 2.136 | 1365.5 | 1523.6 | 1.7632 | 1.8941 | 1364.4 | 1522.2 | 1.7495 |
| 1200 | 2.799 | 1451.5 | 1632.8 | 1.8478 | 2.446 | 1450.7 | 1631.8 | 1.8327 | 2.172 | 1450.0 | 1630.8 | 1.8192 |
| 1400 | 3.148 | 1539.3 | 1743.1 | 1.9106 | 2.752 | 1538.7 | 1742.4 | 1.8956 | 2.444 | 1538.1 | 1741.7 | 1.8823 |
| 1600 | 3.494 | 1630.2 | 1856.5 | 1.9685 | 3.055 | 1629.8 | 1855.9 | 1.9535 | 2.715 | 1629.3 | 1855.4 | 1.9403 |
| 1800 | 3.838 | 1724.5 | 1973.1 | 2.0225 | 3.357 | 1724.1 | 1972.6 | 2.0076 | 2.983 | 1723.7 | 1972.1 | 1.9944 |
| 2000 | 4.182 | 1822.0 | 2092.8 | 2.0733 | 3.658 | 1821.6 | 2092.4 | 2.0584 | 3.251 | 1821.3 | 2092.0 | 2.0453 |
| | $P$ = **500 psia (467.13°F)** | | | | $P$ = **600 psia (486.33°F)** | | | | $P$ = **700 psia (503.23°F)** | | | |
| Sat. | 0.9283 | 1119.4 | 1205.3 | 1.4645 | 0.7702 | 1118.6 | 1204.1 | 1.4464 | 0.6558 | 1117.0 | 1202.0 | 1.4305 |
| 500 | 0.9924 | 1139.7 | 1231.5 | 1.4923 | 0.7947 | 1128.0 | 1216.2 | 1.4592 | | | | |
| 550 | 1.0792 | 1166.7 | 1266.6 | 1.5279 | 0.8749 | 1158.2 | 1255.4 | 1.4990 | 0.7275 | 1149.0 | 1243.2 | 1.4723 |
| 600 | 1.1583 | 1191.1 | 1298.3 | 1.5585 | 0.9456 | 1184.5 | 1289.5 | 1.5320 | 0.7929 | 1177.5 | 1280.2 | 1.5081 |
| 650 | 1.2327 | 1214.0 | 1328.0 | 1.5860 | 1.0109 | 1208.6 | 1320.9 | 1.5609 | 0.8520 | 1203.1 | 1313.4 | 1.5387 |
| 700 | 1.3040 | 1236.0 | 1356.7 | 1.6112 | 1.0727 | 1231.5 | 1350.6 | 1.5872 | 0.9073 | 1226.9 | 1344.4 | 1.5661 |
| 800 | 1.4407 | 1278.8 | 1412.1 | 1.6571 | 1.1900 | 1275.4 | 1407.6 | 1.6343 | 1.0109 | 1272.0 | 1402.9 | 1.6145 |
| 900 | 1.5723 | 1321.0 | 1466.5 | 1.6987 | 1.3021 | 1318.4 | 1462.9 | 1.6766 | 1.1089 | 1315.6 | 1459.3 | 1.6576 |
| 1000 | 1.7008 | 1363.3 | 1520.7 | 1.7371 | 1.4108 | 1361.2 | 1517.8 | 1.7155 | 1.2036 | 1358.9 | 1514.9 | 1.6970 |
| 1100 | 1.8271 | 1406.0 | 1575.1 | 1.7731 | 1.5173 | 1404.2 | 1572.7 | 1.7519 | 1.2960 | 1402.4 | 1570.2 | 1.7337 |
| 1200 | 1.9518 | 1449.2 | 1629.8 | 1.8072 | 1.6222 | 1447.7 | 1627.8 | 1.7861 | 1.3868 | 1446.2 | 1625.8 | 1.7682 |
| 1400 | 2.198 | 1537.6 | 1741.0 | 1.8704 | 1.8289 | 1536.5 | 1739.5 | 1.8497 | 1.5652 | 1535.3 | 1738.1 | 1.8321 |
| 1600 | 2.442 | 1628.9 | 1854.8 | 1.9285 | 2.033 | 1628.0 | 1853.7 | 1.9080 | 1.7409 | 1627.1 | 1852.6 | 1.8906 |
| 1800 | 2.684 | 1723.3 | 1971.7 | 1.9827 | 2.236 | 1722.6 | 1970.8 | 1.9622 | 1.9152 | 1721.8 | 1969.9 | 1.9449 |
| 2000 | 2.926 | 1820.9 | 2091.6 | 2.0335 | 2.438 | 1820.2 | 2090.8 | 2.0131 | 2.0887 | 1819.5 | 2090.1 | 1.9958 |
| | $P$ = **800 psia (518.36°F)** | | | | $P$ = **1000 psia (544.75°F)** | | | | $P$ = **1250 psia (572.56°F)** | | | |
| Sat. | 0.5691 | 1115.0 | 1199.3 | 1.4160 | 0.4459 | 1109.9 | 1192.4 | 1.3903 | 0.3454 | 1101.7 | 1181.6 | 1.3619 |
| 550 | 0.6154 | 1138.8 | 1229.9 | 1.4469 | 0.4534 | 1114.8 | 1198.7 | 1.3966 | | | | |
| 600 | 0.6776 | 1170.1 | 1270.4 | 1.4861 | 0.5140 | 1153.7 | 1248.8 | 1.4450 | 0.3786 | 1129.0 | 1216.6 | 1.3954 |
| 650 | 0.7324 | 1197.2 | 1305.6 | 1.5186 | 0.5637 | 1184.7 | 1289.1 | 1.4822 | 0.4267 | 1167.2 | 1266.0 | 1.4410 |
| 700 | 0.7829 | 1222.1 | 1338.0 | 1.5471 | 0.6080 | 1212.0 | 1324.6 | 1.5135 | 0.4670 | 1198.4 | 1306.4 | 1.4767 |
| 750 | 0.8306 | 1245.7 | 1368.6 | 1.5730 | 0.6490 | 1237.2 | 1357.3 | 1.5412 | 0.5030 | 1226.1 | 1342.4 | 1.5070 |
| 800 | 0.8764 | 1268.5 | 1398.2 | 1.5969 | 0.6878 | 1261.2 | 1388.5 | 1.5664 | 0.5364 | 1251.8 | 1375.8 | 1.5341 |
| 900 | 0.9640 | 1312.9 | 1455.6 | 1.6408 | 0.7610 | 1307.3 | 1448.1 | 1.6120 | 0.5984 | 1300.0 | 1438.4 | 1.5820 |
| 1000 | 1.0482 | 1356.7 | 1511.9 | 1.6807 | 0.8305 | 1352.2 | 1505.9 | 1.6530 | 0.6563 | 1346.4 | 1498.2 | 1.6244 |
| 1100 | 1.1300 | 1400.5 | 1567.8 | 1.7178 | 0.8976 | 1396.8 | 1562.9 | 1.6908 | 0.7116 | 1392.0 | 1556.6 | 1.6631 |
| 1200 | 1.2102 | 1444.6 | 1623.8 | 1.7526 | 0.9630 | 1441.5 | 1619.7 | 1.7261 | 0.7652 | 1437.5 | 1614.5 | 1.6991 |
| 1400 | 1.3674 | 1534.2 | 1736.6 | 1.8167 | 1.0905 | 1531.9 | 1733.7 | 1.7909 | 0.8689 | 1529.0 | 1730.0 | 1.7648 |
| 1600 | 1.5218 | 1626.2 | 1851.5 | 1.8754 | 1.2152 | 1624.4 | 1849.3 | 1.8499 | 0.9699 | 1622.2 | 1846.5 | 1.8243 |
| 1800 | 1.6749 | 1721.0 | 1969.0 | 1.9298 | 1.3384 | 1719.5 | 1967.2 | 1.9046 | 1.0693 | 1717.6 | 1965.0 | 1.8791 |
| 2000 | 1.8271 | 1818.8 | 2089.3 | 1.9808 | 1.4608 | 1817.4 | 2087.7 | 1.9557 | 1.1678 | 1815.7 | 2085.8 | 1.9304 |

**TABLE A-6E**
**Superheated water (*Continued*)**

| T °F | v ft³/lbm | u Btu/lbm | h Btu/lbm | s Btu/(lbm · R) | v ft³/lbm | u Btu/lbm | h Btu/lbm | s Btu/(lbm · R) | v ft³/lbm | u Btu/lbm | h Btu/lbm | s Btu/(lbm · R) |
|---|---|---|---|---|---|---|---|---|---|---|---|---|
| | *P* = 1500 psia (596.39°F) | | | | *P* = 1750 psia (617.31°F) | | | | *P* = 2000 psia (636.00°F) | | | |
| Sat. | 0.2769 | 1091.8 | 1168.7 | 1.3359 | 0.2268 | 1080.2 | 1153.7 | 1.3109 | 0.18813 | 1066.6 | 1136.3 | 1.2861 |
| 600 | 0.2816 | 1096.6 | 1174.8 | 1.3416 | | | | | | | | |
| 650 | 0.3329 | 1147.0 | 1239.4 | 1.4012 | 0.2627 | 1122.5 | 1207.6 | 1.3603 | 0.2057 | 1091.1 | 1167.2 | 1.3141 |
| 700 | 0.3716 | 1183.4 | 1286.6 | 1.4429 | 0.3022 | 1166.7 | 1264.6 | 1.4106 | 0.2487 | 1147.7 | 1239.8 | 1.3782 |
| 750 | 0.4049 | 1214.1 | 1326.5 | 1.4767 | 0.3341 | 1201.3 | 1309.5 | 1.4485 | 0.2803 | 1187.3 | 1291.1 | 1.4216 |
| 800 | 0.4350 | 1241.8 | 1362.5 | 1.5058 | 0.3622 | 1231.3 | 1348.6 | 1.4802 | 0.3071 | 1220.1 | 1333.8 | 1.4562 |
| 850 | 0.4631 | 1267.7 | 1396.2 | 1.5320 | 0.3878 | 1258.8 | 1384.4 | 1.5081 | 0.3312 | 1249.5 | 1372.0 | 1.4860 |
| 900 | 0.4897 | 1292.5 | 1428.5 | 1.5562 | 0.4119 | 1284.8 | 1418.2 | 1.5334 | 0.3534 | 1276.8 | 1407.6 | 1.5126 |
| 1000 | 0.5400 | 1340.4 | 1490.3 | 1.6001 | 0.4569 | 1334.3 | 1482.3 | 1.5789 | 0.3945 | 1328.1 | 1474.1 | 1.5598 |
| 1100 | 0.5876 | 1387.2 | 1550.3 | 1.6399 | 0.4990 | 1382.2 | 1543.8 | 1.6197 | 0.4325 | 1377.2 | 1537.2 | 1.6017 |
| 1200 | 0.6334 | 1433.5 | 1609.3 | 1.6765 | 0.5392 | 1429.4 | 1604.0 | 1.6571 | 0.4685 | 1425.2 | 1598.6 | 1.6398 |
| 1400 | 0.7213 | 1526.1 | 1726.3 | 1.7431 | 0.6158 | 1523.1 | 1722.6 | 1.7245 | 0.5368 | 1520.2 | 1718.8 | 1.7082 |
| 1600 | 0.8064 | 1619.9 | 1843.7 | 1.8031 | 0.6896 | 1617.6 | 1841.0 | 1.7850 | 0.6020 | 1615.4 | 1838.2 | 1.7692 |
| 1800 | 0.8899 | 1715.7 | 1962.7 | 1.8582 | 0.7617 | 1713.9 | 1960.5 | 1.8404 | 0.6656 | 1712.0 | 1958.3 | 1.8249 |
| 2000 | 0.9725 | 1814.0 | 2083.9 | 1.9096 | 0.8330 | 1812.3 | 2082.0 | 1.8919 | 0.7284 | 1810.6 | 2080.2 | 1.8765 |
| | *P* = 2500 psia (668.31°F) | | | | *P* = 3000 psia (695.52°F) | | | | *P* = 3500 psia | | | |
| Sat. | 0.13059 | 1031.0 | 1091.4 | 1.2327 | 0.08404 | 968.8 | 1015.5 | 1.1575 | | | | |
| 650 | | | | | | | | | 0.02491 | 663.5 | 679.7 | 0.8630 |
| 700 | 0.16839 | 1098.7 | 1176.6 | 1.3073 | 0.09771 | 1003.9 | 1058.1 | 1.1944 | 0.03058 | 759.5 | 779.3 | 0.9506 |
| 750 | 0.2030 | 1155.2 | 1249.1 | 1.3686 | 0.14831 | 1114.7 | 1197.1 | 1.3122 | 0.10460 | 1058.4 | 1126.1 | 1.2440 |
| 800 | 0.2291 | 1195.7 | 1301.7 | 1.4112 | 0.17572 | 1167.6 | 1265.2 | 1.3675 | 0.13626 | 1134.7 | 1223.0 | 1.3226 |
| 850 | 0.2513 | 1229.5 | 1345.8 | 1.4456 | 0.19731 | 1207.7 | 1317.2 | 1.4080 | 0.15818 | 1183.4 | 1285.9 | 1.3716 |
| 900 | 0.2712 | 1259.5 | 1385.4 | 1.4752 | 0.2160 | 1241.8 | 1361.7 | 1.4414 | 0.17625 | 1222.4 | 1336.5 | 1.4096 |
| 950 | 0.2896 | 1288.2 | 1422.2 | 1.5018 | 0.2328 | 1272.7 | 1402.0 | 1.4705 | 0.19214 | 1256.4 | 1380.8 | 1.4416 |
| 1000 | 0.3069 | 1315.2 | 1457.2 | 1.5262 | 0.2485 | 1301.7 | 1439.6 | 1.4967 | 0.2066 | 1287.6 | 1421.4 | 1.4699 |
| 1100 | 0.3393 | 1366.8 | 1523.8 | 1.5704 | 0.2772 | 1356.2 | 1510.1 | 1.5434 | 0.2328 | 1345.2 | 1496.0 | 1.5193 |
| 1200 | 0.3696 | 1416.7 | 1587.7 | 1.6101 | 0.3036 | 1408.0 | 1576.6 | 1.5848 | 0.2566 | 1399.2 | 1565.3 | 1.5624 |
| 1400 | 0.4261 | 1514.2 | 1711.3 | 1.6804 | 0.3524 | 1508.1 | 1703.7 | 1.6571 | 0.2997 | 1501.9 | 1696.1 | 1.6368 |
| 1600 | 0.4795 | 1610.2 | 1832.6 | 1.7424 | 0.3978 | 1606.3 | 1827.1 | 1.7201 | 0.3395 | 1601.7 | 1821.6 | 1.7010 |
| 1800 | 0.5312 | 1708.2 | 1954.0 | 1.7986 | 0.4416 | 1704.5 | 1949.6 | 1.7769 | 0.3776 | 1700.8 | 1945.4 | 1.7583 |
| 2000 | 0.5820 | 1807.2 | 2076.4 | 1.8506 | 0.4844 | 1803.9 | 2072.8 | 1.8291 | 0.4147 | 1800.6 | 2069.2 | 1.8108 |
| | *P* = 4000 psia | | | | *P* = 5000 psia | | | | *P* = 6000 psia | | | |
| 650 | 0.02447 | 657.7 | 675.8 | 0.8574 | 0.02377 | 648.0 | 670.0 | 0.8482 | 0.01222 | 640.0 | 665.8 | 0.8405 |
| 700 | 0.02867 | 742.1 | 763.4 | 0.9345 | 0.02676 | 721.8 | 746.6 | 0.9156 | 0.02563 | 708.1 | 736.5 | 0.9028 |
| 750 | 0.06331 | 960.7 | 1007.5 | 1.1395 | 0.03364 | 821.4 | 852.6 | 1.0049 | 0.02978 | 788.6 | 821.7 | 0.9746 |
| 800 | 0.10522 | 1095.0 | 1172.9 | 1.2740 | 0.05932 | 987.2 | 1042.1 | 1.1583 | 0.03942 | 896.9 | 940.7 | 1.0708 |
| 850 | 0.12833 | 1156.5 | 1251.5 | 1.3352 | 0.08556 | 1092.7 | 1171.9 | 1.2596 | 0.05818 | 1018.8 | 1083.4 | 1.1820 |
| 900 | 0.14622 | 1201.5 | 1309.7 | 1.3789 | 0.10385 | 1155.1 | 1251.1 | 1.3190 | 0.07588 | 1102.9 | 1187.2 | 1.2599 |
| 950 | 0.16151 | 1239.2 | 1358.8 | 1.4144 | 0.11853 | 1202.2 | 1311.9 | 1.3629 | 0.09008 | 1162.0 | 1262.0 | 1.3140 |
| 1000 | 0.17520 | 1272.9 | 1402.6 | 1.4449 | 0.13120 | 1242.0 | 1363.4 | 1.3988 | 0.10207 | 1209.1 | 1322.4 | 1.3561 |
| 1100 | 0.19954 | 1333.9 | 1481.6 | 1.4973 | 0.15302 | 1310.6 | 1452.2 | 1.4577 | 0.12218 | 1286.4 | 1422.1 | 1.4222 |
| 1200 | 0.2213 | 1390.1 | 1553.9 | 1.5423 | 0.17199 | 1371.6 | 1530.8 | 1.5066 | 0.13927 | 1352.7 | 1507.3 | 1.4752 |
| 1300 | 0.2414 | 1443.7 | 1622.4 | 1.5823 | 0.18918 | 1428.6 | 1603.7 | 1.5493 | 0.15453 | 1413.3 | 1584.9 | 1.5206 |
| 1400 | 0.2603 | 1495.7 | 1688.4 | 1.6188 | 0.20517 | 1483.2 | 1673.0 | 1.5876 | 0.16854 | 1470.5 | 1657.6 | 1.5608 |
| 1600 | 0.2959 | 1597.1 | 1816.1 | 1.6841 | 0.2348 | 1587.9 | 1805.2 | 1.6551 | 0.19420 | 1578.7 | 1794.3 | 1.6307 |
| 1800 | 0.3296 | 1697.1 | 1941.1 | 1.7420 | 0.2626 | 1689.8 | 1932.7 | 1.7142 | 0.21801 | 1682.4 | 1924.5 | 1.6910 |
| 2000 | 0.3625 | 1797.3 | 2065.6 | 1.7948 | 0.2895 | 1790.8 | 2058.6 | 1.7676 | 0.24087 | 1784.3 | 2051.7 | 1.7450 |

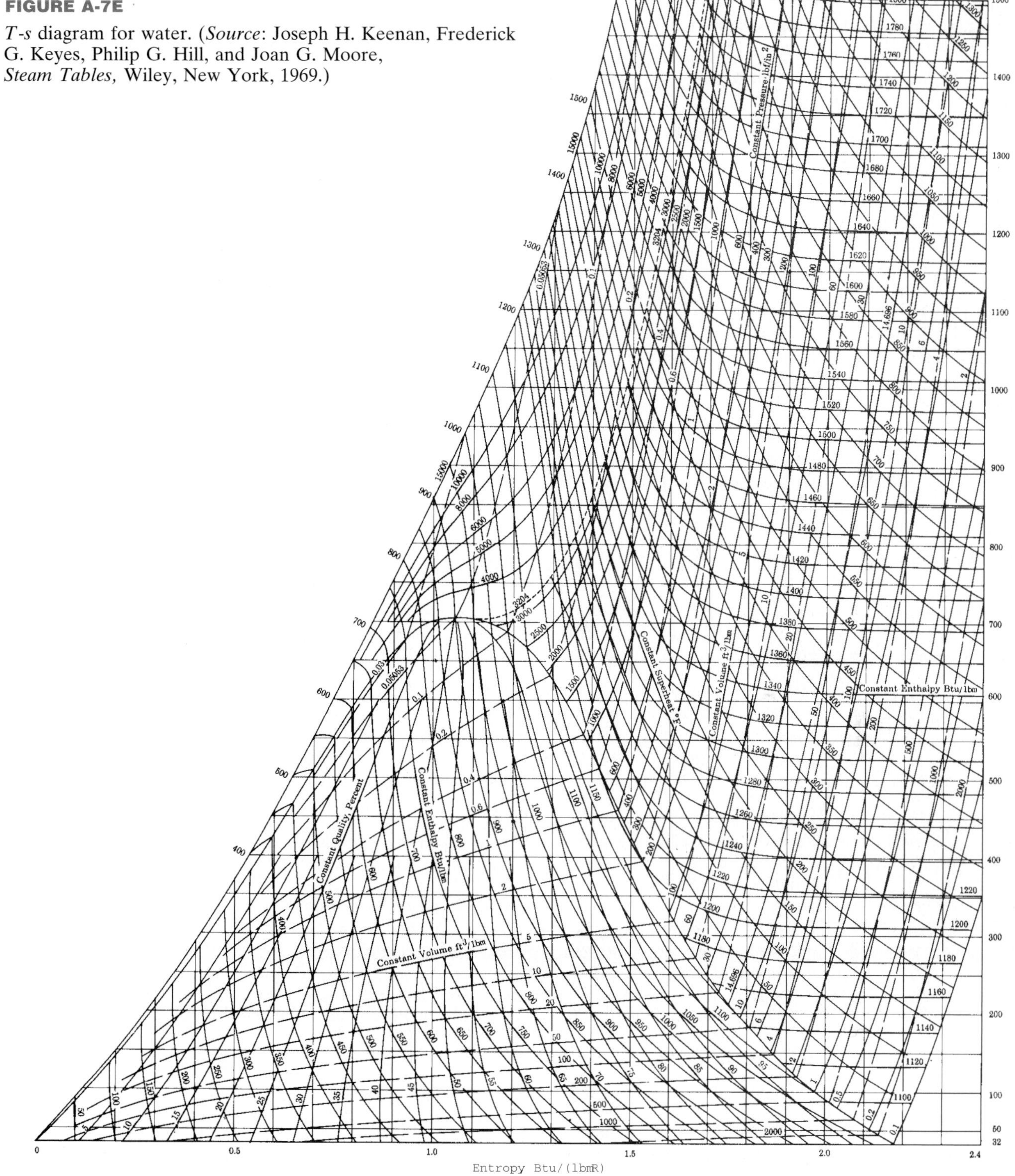


**FIGURE A-7E**

*T-s* diagram for water. (*Source*: Joseph H. Keenan, Frederick G. Keyes, Philip G. Hill, and Joan G. Moore, *Steam Tables,* Wiley, New York, 1969.)

**TABLE A-8E**
**Saturated refrigerant-134a: temperature table**

| Temp. $T$ °F | Press. $P_{sat}$ psia | Specific volume ft³/lbm: Sat. liquid $v_f$ | Sat. vapor $v_g$ | Internal energy Btu/lbm: Sat. liquid $u_f$ | Sat. vapor $u_g$ | Enthalpy Btu/lbm: Sat. liquid $h_f$ | Evap. $h_{fg}$ | Sat. vapor $h_g$ | Entropy Btu/(lbm · R): Sat. liquid $s_f$ | Sat. vapor $s_g$ |
|---|---|---|---|---|---|---|---|---|---|---|
| −40 | 7.490 | 0.01130 | 5.7173 | −0.02 | 87.90 | 0.00 | 95.82 | 95.82 | 0.0000 | 0.2283 |
| −30 | 9.920 | 0.01143 | 4.3911 | 2.81 | 89.26 | 2.8[illegible] | 94.49 | 97.32 | 0.0067 | 0.2266 |
| −20 | 12.949 | 0.01156 | 3.4173 | 5.69 | 90.62 | 5.71 | 93.10 | 98.81 | 0.0133 | 0.2250 |
| −15 | 14.718 | 0.01163 | 3.0286 | 7.14 | 91.30 | 7.17 | 92.38 | 99.55 | 0.0166 | 0.2243 |
| −10 | 16.674 | 0.01170 | 2.6918 | 8.61 | 91.98 | 8.65 | 91.64 | 100.29 | 0.0199 | 0.2236 |
| −5 | 18.831 | 0.01178 | 2.3992 | 10.09 | 92.66 | 10.13 | 90.89 | 101.02 | 0.0231 | 0.2230 |
| 0 | 21.203 | 0.01185 | 2.1440 | 11.58 | 93.33 | 11.63 | 90.12 | 101.75 | 0.0264 | 0.2224 |
| 5 | 23.805 | 0.01193 | 1.9208 | 13.09 | 94.01 | 13.14 | 89.33 | 102.47 | 0.0296 | 0.2219 |
| 10 | 26.651 | 0.01200 | 1.7251 | 14.60 | 94.68 | 14.66 | 88.53 | 103.19 | 0.0329 | 0.2214 |
| 15 | 29.756 | 0.01208 | 1.5529 | 16.13 | 95.35 | 16.20 | 87.71 | 103.90 | 0.0361 | 0.2209 |
| 20 | 33.137 | 0.01216 | 1.4009 | 17.67 | 96.02 | 17.74 | 86.87 | 104.61 | 0.0393 | 0.2205 |
| 25 | 36.809 | 0.01225 | 1.2666 | 19.22 | 96.69 | 19.30 | 86.02 | 105.32 | 0.0426 | 0.2200 |
| 30 | 40.788 | 0.01233 | 1.1474 | 20.78 | 97.35 | 20.87 | 85.14 | 106.01 | 0.0458 | 0.2196 |
| 40 | 49.738 | 0.01251 | 0.9470 | 23.94 | 98.67 | 24.05 | 83.34 | 107.39 | 0.0522 | 0.2189 |
| 50 | 60.125 | 0.01270 | 0.7871 | 27.14 | 99.98 | 27.28 | 81.46 | 108.74 | 0.0585 | 0.2183 |
| 60 | 72.092 | 0.01290 | 0.6584 | 30.39 | 101.27 | 30.56 | 79.49 | 110.05 | 0.0648 | 0.2178 |
| 70 | 85.788 | 0.01311 | 0.5538 | 33.68 | 102.54 | 33.89 | 77.44 | 111.33 | 0.0711 | 0.2173 |
| 80 | 101.37 | 0.01334 | 0.4682 | 37.02 | 103.78 | 37.27 | 75.29 | 112.56 | 0.0774 | 0.2169 |
| 85 | 109.92 | 0.01346 | 0.4312 | 38.72 | 104.39 | 38.99 | 74.17 | 113.16 | 0.0805 | 0.2167 |
| 90 | 118.99 | 0.01358 | 0.3975 | 40.42 | 105.00 | 40.72 | 73.03 | 113.75 | 0.0836 | 0.2165 |
| 95 | 128.62 | 0.01371 | 0.3668 | 42.14 | 105.60 | 42.47 | 71.86 | 114.33 | 0.0867 | 0.2163 |
| 100 | 138.83 | 0.01385 | 0.3388 | 43.87 | 106.18 | 44.23 | 70.66 | 114.89 | 0.0898 | 0.2161 |
| 105 | 149.63 | 0.01399 | 0.3131 | 45.62 | 106.76 | 46.01 | 69.42 | 115.43 | 0.0930 | 0.2159 |
| 110 | 161.04 | 0.01414 | 0.2896 | 47.39 | 107.33 | 47.81 | 68.15 | 115.96 | 0.0961 | 0.2157 |
| 115 | 173.10 | 0.01429 | 0.2680 | 49.17 | 107.88 | 49.63 | 66.84 | 116.47 | 0.0992 | 0.2155 |
| 120 | 185.82 | 0.01445 | 0.2481 | 50.97 | 108.42 | 51.47 | 65.48 | 116.95 | 0.1023 | 0.2153 |
| 140 | 243.86 | 0.01520 | 0.1827 | 58.39 | 110.41 | 59.08 | 59.57 | 118.65 | 0.1150 | 0.2143 |
| 160 | 314.63 | 0.01617 | 0.1341 | 66.26 | 111.97 | 67.20 | 52.58 | 119.78 | 0.1280 | 0.2128 |
| 180 | 400.22 | 0.01758 | 0.0964 | 74.83 | 112.77 | 76.13 | 43.78 | 119.91 | 0.1417 | 0.2101 |
| 200 | 503.52 | 0.02014 | 0.0647 | 84.90 | 111.66 | 86.77 | 30.92 | 117.69 | 0.1575 | 0.2044 |
| 210 | 563.51 | 0.02329 | 0.0476 | 91.84 | 108.48 | 94.27 | 19.18 | 113.45 | 0.1684 | 0.1971 |

*Source for Tables A-8E through A-10E:* M. J. Moran and H. N. Shapiro, *Fundamentals of Engineering Thermodynamics*, 2d ed., Wiley, New York, 1992, pp. 754–758. Originally based on equations from D. P. Wilson and R. S. Basu, "Thermodynamic Properties of a New Stratospherically Safe Working Fluid—Refrigerant 134a," *ASHRAE Trans.*, Vol. 94, Pt. 2, 1988, pp. 2095–2118. Used with permission.

**TABLE A-9E**

**Saturated refrigerant-134a: pressure table**

| | | Specific volume ft³/lbm | | Internal energy Btu/lbm | | Enthalpy Btu/lbm | | | Entropy Btu/(lbm · R) | |
|---|---|---|---|---|---|---|---|---|---|---|
| Press. $P$ psia | Temp. $T_{sat}$ °F | Sat. liquid $v_f$ | Sat. vapor $v_g$ | Sat. liquid $u_f$ | Sat. vapor $u_g$ | Sat. liquid $h_f$ | Evap. $h_{fg}$ | Sat. vapor $h_g$ | Sat. liquid $s_f$ | Sat. vapor $s_g$ |
| 5 | −53.48 | 0.01113 | 8.3508 | −3.74 | 86.07 | −3.73 | 97.53 | 93.79 | −0.0090 | 0.2311 |
| 10 | −29.71 | 0.01143 | 4.3581 | 2.89 | 89.30 | 2.91 | 94.45 | 97.37 | 0.0068 | 0.2265 |
| 15 | −14.25 | 0.01164 | 2.9747 | 7.36 | 91.40 | 7.40 | 92.27 | 99.66 | 0.0171 | 0.2242 |
| 20 | −2.48 | 0.01181 | 2.2661 | 10.84 | 93.00 | 10.89 | 90.50 | 101.39 | 0.0248 | 0.2227 |
| 30 | 15.38 | 0.01209 | 1.5408 | 16.24 | 95.40 | 16.31 | 87.65 | 103.96 | 0.0364 | 0.2209 |
| 40 | 29.04 | 0.01232 | 1.1692 | 20.48 | 97.23 | 20.57 | 85.31 | 105.88 | 0.0452 | 0.2197 |
| 50 | 40.27 | 0.01252 | 0.9422 | 24.02 | 98.71 | 24.14 | 83.29 | 107.43 | 0.0523 | 0.2189 |
| 60 | 49.89 | 0.01270 | 0.7887 | 27.10 | 99.96 | 27.24 | 81.48 | 108.72 | 0.0584 | 0.2183 |
| 70 | 58.35 | 0.01286 | 0.6778 | 29.85 | 101.05 | 30.01 | 79.82 | 109.83 | 0.0638 | 0.2179 |
| 80 | 65.93 | 0.01302 | 0.5938 | 32.33 | 102.02 | 32.53 | 78.28 | 110.81 | 0.0686 | 0.2175 |
| 90 | 72.83 | 0.01317 | 0.5278 | 34.62 | 102.89 | 34.84 | 76.84 | 111.68 | 0.0729 | 0.2172 |
| 100 | 79.17 | 0.01332 | 0.4747 | 36.75 | 103.68 | 36.99 | 75.47 | 112.46 | 0.0768 | 0.2169 |
| 120 | 90.54 | 0.01360 | 0.3941 | 40.61 | 105.06 | 40.91 | 72.91 | 113.82 | 0.0839 | 0.2165 |
| 140 | 100.56 | 0.01386 | 0.3358 | 44.07 | 106.25 | 44.43 | 70.52 | 114.95 | 0.0902 | 0.2161 |
| 160 | 109.56 | 0.01412 | 0.2916 | 47.23 | 107.28 | 47.65 | 68.26 | 115.91 | 0.0958 | 0.2157 |
| 180 | 117.74 | 0.01438 | 0.2569 | 50.16 | 108.18 | 50.64 | 66.10 | 116.74 | 0.1009 | 0.2154 |
| 200 | 125.28 | 0.01463 | 0.2288 | 52.90 | 108.98 | 53.44 | 64.01 | 117.44 | 0.1057 | 0.2151 |
| 220 | 132.27 | 0.01489 | 0.2056 | 55.48 | 109.68 | 56.09 | 61.96 | 118.05 | 0.1101 | 0.2147 |
| 240 | 138.79 | 0.01515 | 0.1861 | 57.93 | 110.30 | 58.61 | 59.96 | 118.56 | 0.1142 | 0.2144 |
| 260 | 144.92 | 0.01541 | 0.1695 | 60.28 | 110.84 | 61.02 | 57.97 | 118.99 | 0.1181 | 0.2140 |
| 280 | 150.70 | 0.01568 | 0.1550 | 62.53 | 111.31 | 63.34 | 56.00 | 119.35 | 0.1219 | 0.2136 |
| 300 | 156.17 | 0.01596 | 0.1424 | 64.71 | 111.72 | 65.59 | 54.03 | 119.62 | 0.1254 | 0.2132 |
| 350 | 168.72 | 0.01671 | 0.1166 | 69.88 | 112.45 | 70.97 | 49.03 | 120.00 | 0.1338 | 0.2118 |
| 400 | 179.95 | 0.01758 | 0.0965 | 74.81 | 112.77 | 76.11 | 43.80 | 119.91 | 0.1417 | 0.2102 |
| 450 | 190.12 | 0.01863 | 0.0800 | 79.63 | 112.60 | 81.18 | 38.08 | 119.26 | 0.1493 | 0.2079 |
| 500 | 199.38 | 0.02002 | 0.0657 | 84.54 | 111.76 | 86.39 | 31.44 | 117.83 | 0.1570 | 0.2047 |

**TABLE A-10E**
**Superheated refrigerant-134a**

| $T$ °F | $v$ ft$^3$/lbm | $u$ Btu/lbm | $h$ Btu/lbm | $s$ Btu/(lbm·R) | $v$ ft$^3$/lbm | $u$ Btu/lbm | $h$ Btu/lbm | $s$ Btu/(lbm·R) | $v$ ft$^3$/lbm | $u$ Btu/lbm | $h$ Btu/lbm | $s$ Btu/(lbm·R) |
|---|---|---|---|---|---|---|---|---|---|---|---|---|
| | **$P$ = 10 psia ($T_{sat}$ = −29.71°F)** | | | | **$P$ = 15 psia ($T_{sat}$ = −14.25°F)** | | | | **$P$ = 20 psia ($T_{sat}$ = −2.48°F)** | | | |
| Sat. | 4.3581 | 89.30 | 97.37 | 0.2265 | 2.9747 | 91.40 | 99.66 | 0.2242 | 2.2661 | 93.00 | 101.39 | 0.2227 |
| −20 | 4.4718 | 90.89 | 99.17 | 0.2307 | | | | | | | | |
| 0 | 4.7026 | 94.24 | 102.94 | 0.2391 | 3.0893 | 93.84 | 102.42 | 0.2303 | 2.2816 | 93.43 | 101.88 | 0.2238 |
| 20 | 4.9297 | 97.67 | 106.79 | 0.2472 | 3.2468 | 97.33 | 106.34 | 0.2386 | 2.4046 | 96.98 | 105.88 | 0.2323 |
| 40 | 5.1539 | 101.19 | 110.72 | 0.2553 | 3.4012 | 100.89 | 110.33 | 0.2468 | 2.5244 | 100.59 | 109.94 | 0.2406 |
| 60 | 5.3758 | 104.80 | 114.74 | 0.2632 | 3.5533 | 104.54 | 114.40 | 0.2548 | 2.6416 | 104.28 | 114.06 | 0.2487 |
| 80 | 5.5959 | 108.50 | 118.85 | 0.2709 | 3.7034 | 108.28 | 118.56 | 0.2626 | 2.7569 | 108.05 | 118.25 | 0.2566 |
| 100 | 5.8145 | 112.29 | 123.05 | 0.2786 | 3.8520 | 112.10 | 122.79 | 0.2703 | 2.8705 | 111.90 | 122.52 | 0.2644 |
| 120 | 6.0318 | 116.18 | 127.34 | 0.2861 | 3.9993 | 116.01 | 127.11 | 0.2779 | 2.9829 | 115.83 | 126.87 | 0.2720 |
| 140 | 6.2482 | 120.16 | 131.72 | 0.2935 | 4.1456 | 120.00 | 131.51 | 0.2854 | 3.0942 | 119.85 | 131.30 | 0.2795 |
| 160 | 6.4638 | 124.23 | 136.19 | 0.3009 | 4.2911 | 124.09 | 136.00 | 0.2927 | 3.2047 | 123.95 | 135.81 | 0.2869 |
| 180 | 6.6786 | 128.38 | 140.74 | 0.3081 | 4.4359 | 128.26 | 140.57 | 0.3000 | 3.3144 | 128.13 | 140.40 | 0.2922 |
| 200 | 6.8929 | 132.63 | 145.39 | 0.3152 | 4.5801 | 132.52 | 145.23 | 0.3072 | 3.4236 | 132.40 | 145.07 | 0.3014 |
| 220 | | | | | | | | | 3.5323 | 136.76 | 149.83 | 0.3085 |
| | **$P$ = 30 psia ($T_{sat}$ = 15.38°F)** | | | | **$P$ = 40 psia ($T_{sat}$ = 29.04°F)** | | | | **$P$ = 50 psia ($T_{sat}$ = 40.27°F)** | | | |
| Sat. | 1.5408 | 95.40 | 103.96 | 0.2209 | 1.1692 | 97.23 | 105.88 | 0.2197 | 0.9422 | 98.71 | 107.43 | 0.2189 |
| 20 | 1.5611 | 96.26 | 104.92 | 0.2229 | | | | | | | | |
| 40 | 1.6465 | 99.98 | 109.12 | 0.2315 | 1.2065 | 99.33 | 108.26 | 0.2245 | | | | |
| 60 | 1.7293 | 103.75 | 113.35 | 0.2398 | 1.2723 | 103.20 | 112.62 | 0.2331 | 0.9974 | 102.62 | 111.85 | 0.2276 |
| 80 | 1.8098 | 107.59 | 117.63 | 0.2478 | 1.3357 | 107.11 | 117.00 | 0.2414 | 1.0508 | 106.62 | 116.34 | 0.2361 |
| 100 | 1.8887 | 111.49 | 121.98 | 0.2558 | 1.3973 | 111.08 | 121.42 | 0.2494 | 1.1022 | 110.65 | 120.85 | 0.2443 |
| 120 | 1.9662 | 115.47 | 126.39 | 0.2635 | 1.4575 | 115.11 | 125.90 | 0.2573 | 1.1520 | 114.74 | 125.39 | 0.2523 |
| 140 | 2.0426 | 119.53 | 130.87 | 0.2711 | 1.5165 | 119.21 | 130.43 | 0.2650 | 1.2007 | 118.88 | 129.99 | 0.2601 |
| 160 | 2.1181 | 123.66 | 135.42 | 0.2786 | 1.5746 | 123.38 | 135.03 | 0.2725 | 1.2484 | 123.08 | 134.64 | 0.2677 |
| 180 | 2.1929 | 127.88 | 140.05 | 0.2859 | 1.6319 | 127.62 | 139.70 | 0.2799 | 1.2953 | 127.36 | 139.34 | 0.2752 |
| 200 | 2.2671 | 132.17 | 144.76 | 0.2932 | 1.6887 | 131.94 | 144.44 | 0.2872 | 1.3415 | 131.71 | 144.12 | 0.2825 |
| 220 | 2.3407 | 136.55 | 149.54 | 0.3003 | 1.7449 | 136.34 | 149.25 | 0.2944 | 1.3873 | 136.12 | 148.96 | 0.2897 |
| 240 | | | | | 1.8006 | 140.81 | 154.14 | 0.3015 | 1.4326 | 140.61 | 153.87 | 0.2969 |
| 260 | | | | | 1.8561 | 145.36 | 159.10 | 0.3085 | 1.4775 | 145.18 | 158.85 | 0.3039 |
| 280 | | | | | 1.9112 | 149.98 | 164.13 | 0.3154 | 1.5221 | 149.82 | 163.90 | 0.3108 |
| | **$P$ = 60 psia ($T_{sat}$ = 49.89°F)** | | | | **$P$ = 70 psia ($T_{sat}$ = 58.35°F)** | | | | **$P$ = 80 psia ($T_{sat}$ = 65.93°F)** | | | |
| Sat. | 0.7887 | 99.96 | 108.72 | 0.2183 | 0.6778 | 101.05 | 109.83 | 0.2179 | 0.5938 | 102.02 | 110.81 | 0.2175 |
| 60 | 0.8135 | 102.03 | 111.06 | 0.2229 | 0.6814 | 101.40 | 110.23 | 0.2186 | | | | |
| 80 | 0.8604 | 106.11 | 115.66 | 0.2316 | 0.7239 | 105.58 | 114.96 | 0.2276 | 0.6211 | 105.03 | 114.23 | 0.2239 |
| 100 | 0.9051 | 110.21 | 120.26 | 0.2399 | 0.7640 | 109.76 | 119.66 | 0.2361 | 0.6579 | 109.30 | 119.04 | 0.2327 |
| 120 | 0.9482 | 114.35 | 124.88 | 0.2480 | 0.8023 | 113.96 | 124.36 | 0.2444 | 0.6927 | 113.56 | 123.82 | 0.2411 |
| 140 | 0.9900 | 118.54 | 129.53 | 0.2559 | 0.8393 | 118.20 | 129.07 | 0.2524 | 0.7261 | 117.85 | 128.60 | 0.2492 |
| 160 | 1.0308 | 122.79 | 134.23 | 0.2636 | 0.8752 | 122.49 | 133.82 | 0.2601 | 0.7584 | 122.18 | 133.41 | 0.2570 |
| 180 | 1.0707 | 127.10 | 138.98 | 0.2712 | 0.9103 | 126.83 | 138.62 | 0.2678 | 0.7898 | 126.55 | 138.25 | 0.2647 |
| 200 | 1.1100 | 131.47 | 143.79 | 0.2786 | 0.9446 | 131.23 | 143.46 | 0.2752 | 0.8205 | 130.98 | 143.13 | 0.2722 |
| 220 | 1.1488 | 135.91 | 148.66 | 0.2859 | 0.9784 | 135.69 | 148.36 | 0.2825 | 0.8506 | 135.47 | 148.06 | 0.2796 |
| 240 | 1.1871 | 140.42 | 153.60 | 0.2930 | 1.0118 | 140.22 | 153.33 | 0.2897 | 0.8803 | 140.02 | 153.05 | 0.2868 |
| 260 | 1.2251 | 145.00 | 158.60 | 0.3001 | 1.0448 | 144.82 | 158.35 | 0.2968 | 0.9095 | 144.63 | 158.10 | 0.2940 |
| 280 | 1.2627 | 149.65 | 163.67 | 0.3070 | 1.0774 | 149.48 | 163.44 | 0.3038 | 0.9384 | 149.32 | 163.21 | 0.3010 |
| 300 | 1.3001 | 154.38 | 168.81 | 0.3139 | 1.1098 | 154.22 | 168.60 | 0.3107 | 0.9671 | 154.06 | 168.38 | 0.3079 |
| 320 | | | | | | | | | 0.9955 | 158.88 | 173.62 | 0.3147 |

**TABLE A-10E**

**Superheated refrigerant-134a (*Continued*)**

| $T$ °F | $v$ ft$^3$/lbm | $u$ Btu/lbm | $h$ Btu/lbm | $s$ Btu/(lbm·R) | $v$ ft$^3$/lbm | $u$ Btu/lbm | $h$ Btu/lbm | $s$ Btu/(lbm·R) | $v$ ft$^3$/lbm | $u$ Btu/lbm | $h$ Btu/lbm | $s$ Btu/(lbm·R) |
|---|---|---|---|---|---|---|---|---|---|---|---|---|
| | $P$ = **90 psia** ($T_{sat}$ = **72.83°F**) | | | | $P$ = **100 psia** ($T_{sat}$ = **79.17°F**) | | | | $P$ = **120 psia** ($T_{sat}$ = **90.54°F**) | | | |
| Sat. | 0.5278 | 102.89 | 111.68 | 0.2172 | 0.4747 | 103.68 | 112.46 | 0.2169 | 0.3941 | 105.06 | 113.82 | 0.2165 |
| 80 | 0.5408 | 104.46 | 113.47 | 0.2205 | 0.4761 | 103.87 | 112.68 | 0.2173 | | | | |
| 100 | 0.5751 | 108.82 | 118.39 | 0.2295 | 0.5086 | 108.32 | 117.73 | 0.2265 | 0.4080 | 107.26 | 116.32 | 0.2210 |
| 120 | 0.6073 | 113.15 | 123.27 | 0.2380 | 0.5388 | 112.73 | 122.70 | 0.2352 | 0.4355 | 111.84 | 121.52 | 0.2301 |
| 140 | 0.6380 | 117.50 | 128.12 | 0.2463 | 0.5674 | 117.13 | 127.63 | 0.2436 | 0.4610 | 116.37 | 126.61 | 0.2387 |
| 160 | 0.6675 | 121.87 | 132.98 | 0.2542 | 0.5947 | 121.55 | 132.55 | 0.2517 | 0.4852 | 120.89 | 131.66 | 0.2470 |
| 180 | 0.6961 | 126.28 | 137.87 | 0.2620 | 0.6210 | 125.99 | 137.49 | 0.2595 | 0.5082 | 125.42 | 136.70 | 0.2550 |
| 200 | 0.7239 | 130.73 | 142.79 | 0.2696 | 0.6466 | 130.48 | 142.45 | 0.2671 | 0.5305 | 129.97 | 141.75 | 0.2628 |
| 220 | 0.7512 | 135.25 | 147.76 | 0.2770 | 0.6716 | 135.02 | 147.45 | 0.2746 | 0.5520 | 134.56 | 146.82 | 0.2704 |
| 240 | 0.7779 | 139.82 | 152.77 | 0.2843 | 0.6960 | 139.61 | 152.49 | 0.2819 | 0.5731 | 139.20 | 151.92 | 0.2778 |
| 260 | 0.8043 | 144.45 | 157.84 | 0.2914 | 0.7201 | 144.26 | 157.59 | 0.2891 | 0.5937 | 143.89 | 157.07 | 0.2850 |
| 280 | 0.8303 | 149.15 | 162.97 | 0.2984 | 0.7438 | 148.98 | 162.74 | 0.2962 | 0.6140 | 148.63 | 162.26 | 0.2921 |
| 300 | 0.8561 | 153.91 | 168.16 | 0.3054 | 0.7672 | 153.75 | 167.95 | 0.3031 | 0.6339 | 153.43 | 167.51 | 0.2991 |
| 320 | 0.8816 | 158.73 | 173.42 | 0.3122 | 0.7904 | 158.59 | 173.21 | 0.3099 | 0.6537 | 158.29 | 172.81 | 0.3060 |

| $T$ °F | $v$ ft$^3$/lbm | $u$ Btu/lbm | $h$ Btu/lbm | $s$ Btu/(lbm·R) | $v$ ft$^3$/lbm | $u$ Btu/lbm | $h$ Btu/lbm | $s$ Btu/(lbm·R) | $v$ ft$^3$/lbm | $u$ Btu/lbm | $h$ Btu/lbm | $s$ Btu/(lbm·R) |
|---|---|---|---|---|---|---|---|---|---|---|---|---|
| | $P$ = **140 psia** ($T_{sat}$ = **100.56°F**) | | | | $P$ = **160 psia** ($T_{sat}$ = **109.55°F**) | | | | $P$ = **180 psia** ($T_{sat}$ = **117.74°F**) | | | |
| Sat. | 0.3358 | 106.25 | 114.95 | 0.2161 | 0.2916 | 107.28 | 115.91 | 0.2157 | 0.2569 | 108.18 | 116.74 | 0.2154 |
| 120 | 0.3610 | 110.90 | 120.25 | 0.2254 | 0.3044 | 109.88 | 118.89 | 0.2209 | 0.2595 | 108.77 | 117.41 | 0.2166 |
| 140 | 0.3846 | 115.58 | 125.24 | 0.2344 | 0.3269 | 114.73 | 124.41 | 0.2303 | 0.2814 | 113.83 | 123.21 | 0.2264 |
| 160 | 0.4066 | 120.21 | 130.74 | 0.2429 | 0.3474 | 119.49 | 129.78 | 0.2391 | 0.3011 | 118.74 | 128.77 | 0.2355 |
| 180 | 0.4274 | 124.82 | 135.89 | 0.2511 | 0.3666 | 124.20 | 135.06 | 0.2475 | 0.3191 | 123.56 | 134.19 | 0.2441 |
| 200 | 0.4474 | 129.44 | 141.03 | 0.2590 | 0.3849 | 128.90 | 140.29 | 0.2555 | 0.3361 | 128.34 | 139.53 | 0.2524 |
| 220 | 0.4666 | 134.09 | 146.18 | 0.2667 | 0.4023 | 133.61 | 145.52 | 0.2633 | 0.3523 | 133.11 | 144.84 | 0.2603 |
| 240 | 0.4852 | 138.77 | 151.34 | 0.2742 | 0.4192 | 138.34 | 150.75 | 0.2709 | 0.3678 | 137.90 | 150.15 | 0.2680 |
| 260 | 0.5034 | 143.50 | 156.54 | 0.2815 | 0.4356 | 143.11 | 156.00 | 0.2783 | 0.3828 | 142.71 | 155.46 | 0.2755 |
| 280 | 0.5212 | 148.28 | 161.78 | 0.2887 | 0.4516 | 147.92 | 161.29 | 0.2856 | 0.3974 | 147.55 | 160.79 | 0.2828 |
| 300 | 0.5387 | 153.11 | 167.06 | 0.2957 | 0.4672 | 152.78 | 166.61 | 0.2927 | 0.4116 | 152.44 | 166.15 | 0.2899 |
| 320 | 0.5559 | 157.99 | 172.39 | 0.3026 | 0.4826 | 157.69 | 171.98 | 0.2996 | 0.4256 | 157.38 | 171.55 | 0.2969 |
| 340 | 0.5730 | 162.93 | 177.78 | 0.3094 | 0.4978 | 162.65 | 177.39 | 0.3065 | 0.4393 | 162.36 | 177.00 | 0.3038 |
| 360 | 0.5898 | 167.93 | 183.21 | 0.3162 | 0.5128 | 167.67 | 182.85 | 0.3132 | 0.4529 | 167.40 | 182.49 | 0.3106 |

| $T$ °F | $v$ ft$^3$/lbm | $u$ Btu/lbm | $h$ Btu/lbm | $s$ Btu/(lbm·R) | $v$ ft$^3$/lbm | $u$ Btu/lbm | $h$ Btu/lbm | $s$ Btu/(lbm·R) | $v$ ft$^3$/lbm | $u$ Btu/lbm | $h$ Btu/lbm | $s$ Btu/(lbm·R) |
|---|---|---|---|---|---|---|---|---|---|---|---|---|
| | $P$ = **200 psia** ($T_{sat}$ = **125.28°F**) | | | | $P$ = **300 psia** ($T_{sat}$ = **156.17°F**) | | | | $P$ = **400 psia** ($T_{sat}$ = **179.95°F**) | | | |
| Sat. | 0.2288 | 108.98 | 117.44 | 0.2151 | 0.1424 | 111.72 | 119.62 | 0.2132 | 0.0965 | 112.77 | 119.91 | 0.2102 |
| 140 | 0.2446 | 112.87 | 121.92 | 0.2226 | | | | | | | | |
| 160 | 0.2636 | 117.94 | 127.70 | 0.2321 | 0.1462 | 112.95 | 121.07 | 0.2155 | | | | |
| 180 | 0.2809 | 122.88 | 133.28 | 0.2410 | 0.1633 | 118.93 | 128.00 | 0.2265 | 0.0965 | 112.79 | 119.93 | 0.2102 |
| 200 | 0.2970 | 127.76 | 138.75 | 0.2494 | 0.1777 | 124.47 | 134.34 | 0.2363 | 0.1143 | 120.14 | 128.60 | 0.2235 |
| 220 | 0.3121 | 132.60 | 144.15 | 0.2575 | 0.1905 | 129.79 | 140.36 | 0.2453 | 0.1275 | 126.35 | 135.79 | 0.2343 |
| 240 | 0.3266 | 137.44 | 149.53 | 0.2653 | 0.2021 | 134.99 | 146.21 | 0.2537 | 0.1386 | 132.12 | 142.38 | 0.2438 |
| 260 | 0.3405 | 142.30 | 154.90 | 0.2728 | 0.2130 | 140.12 | 151.95 | 0.2618 | 0.1484 | 137.65 | 148.64 | 0.2527 |
| 280 | 0.3540 | 147.18 | 160.28 | 0.2802 | 0.2234 | 145.23 | 157.63 | 0.2696 | 0.1575 | 143.06 | 154.72 | 0.2610 |
| 300 | 0.3671 | 152.10 | 165.69 | 0.2874 | 0.2333 | 150.33 | 163.28 | 0.2772 | 0.1660 | 148.39 | 160.67 | 0.2689 |
| 320 | 0.3799 | 157.07 | 171.13 | 0.2945 | 0.2428 | 155.44 | 168.92 | 0.2845 | 0.1740 | 153.69 | 166.57 | 0.2766 |
| 340 | 0.3926 | 162.07 | 176.60 | 0.3014 | 0.2521 | 160.57 | 174.56 | 0.2916 | 0.1816 | 158.97 | 172.42 | 0.2840 |
| 360 | 0.4050 | 167.13 | 182.12 | 0.3082 | 0.2611 | 165.74 | 180.23 | 0.2986 | 0.1890 | 164.26 | 178.26 | 0.2912 |
| 380 | | | | | 0.2699 | 170.94 | 185.92 | 0.3055 | 0.1962 | 169.57 | 184.09 | 0.2983 |
| 400 | | | | | 0.2786 | 176.18 | 191.64 | 0.3122 | 0.2032 | 174.90 | 189.94 | 0.3051 |

**FIGURE A-11**

*P-h* diagram for refrigerant-134a. (Reprinted by permission of the American Society of Heating, Refrigerating, and Air Conditioning Engineers, Inc., Atlanta, GA.)

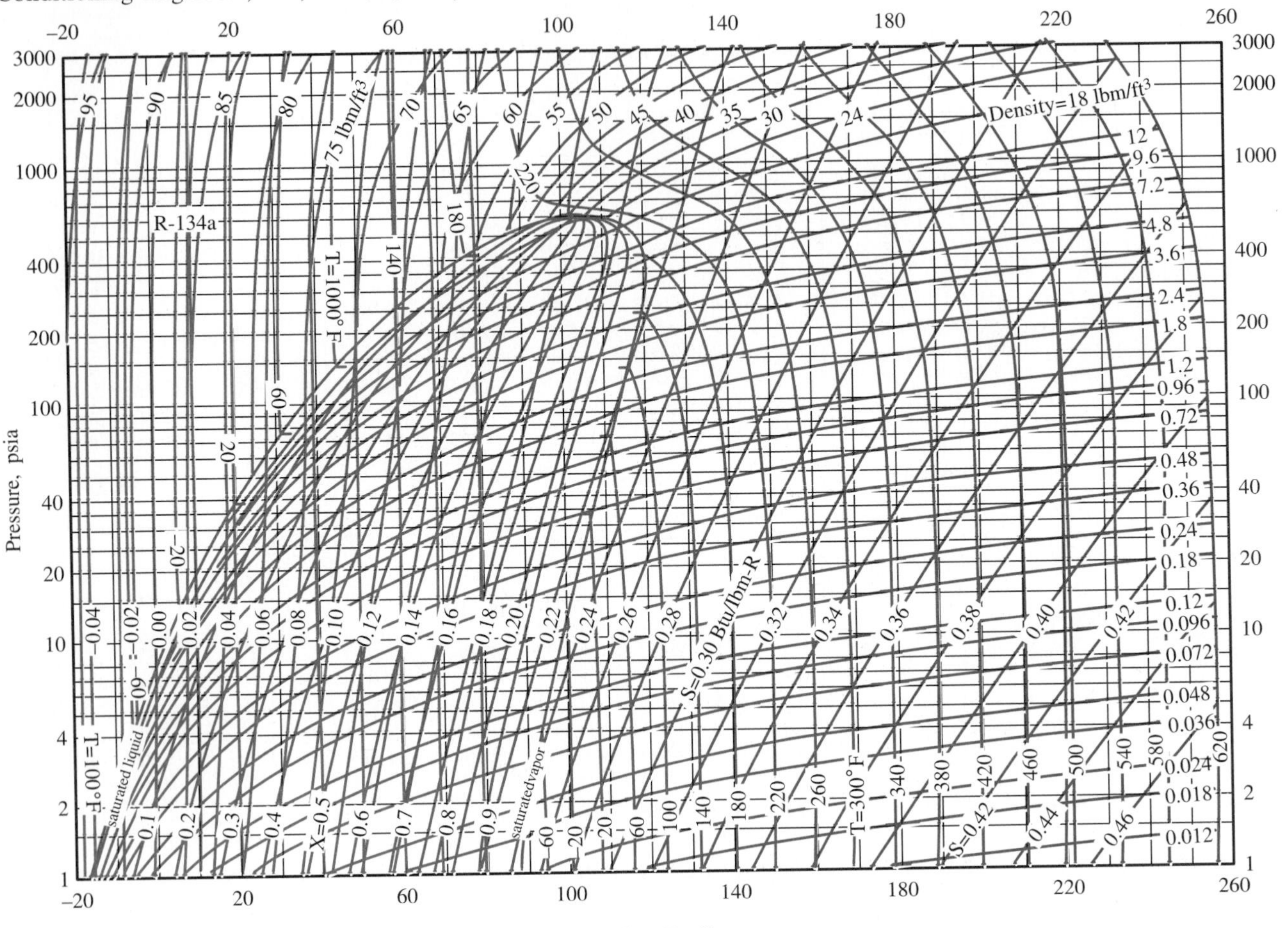

# TABLE A-12E

## Constants that appear in the Beattie–Bridgeman and the Benedict–Webb–Rubin equations of state

(*a*) The Beattie–Bridgeman equation of state is

$$P = \frac{R_u T}{\bar{v}^2}\left(1 - \frac{c}{\bar{v}T^3}\right)(\bar{v} + B) - \frac{A}{\bar{v}^2}, \quad \text{where } A = A_0\left(1 - \frac{a}{\bar{v}}\right) \quad \text{and} \quad B = B_0\left(1 - \frac{b}{\bar{v}}\right)$$

When $P$ is in psia, $\bar{v}$ is in ft³/lbmol, $T$ is in R, and $R_u = 10.73$ psia · ft³/(lbmol · R), the five constants in the Beattie–Bridgeman equation are as follows:

| Gas | $A_0$ | $a$ | $B_0$ | $b$ | $c$ |
|---|---|---|---|---|---|
| Air | 4,905.096 | 0.3093 | 0.7386 | −0.01764 | $4.054 \times 10^6$ |
| Argon, Ar | 4,865.515 | 0.3729 | 0.6297 | 0.0 | $5.596 \times 10^6$ |
| Carbon dioxide, $CO_2$ | 18,872.857 | 1.142 | 1.678 | 1.159 | $6.166 \times 10^7$ |
| Helium, He | 81.424 | 0.9587 | 0.2243 | 0.0 | $3.737 \times 10^3$ |
| Hydrogen, $H_2$ | 744.510 | −0.08105 | 0.3357 | −0.6982 | $4.708 \times 10^4$ |
| Nitrogen, $N_2$ | 5,068.324 | 0.4192 | 0.8083 | −0.1107 | $3.924 \times 10^6$ |
| Oxygen, $O_2$ | 5,620.956 | 0.4104 | 0.7407 | 0.06741 | $4.484 \times 10^6$ |

*Source:* Computed from Table A-12*a* by using the proper conversion factors.

(*b*) The Benedict–Webb–Rubin equation of state is

$$P = \frac{R_u T}{\bar{v}} + \left(B_0 R_u T - A_0 - \frac{C_0}{T^2}\right)\frac{1}{\bar{v}^2} + \frac{bR_u T - a}{\bar{v}^3} + \frac{a\alpha}{\bar{v}^6} + \frac{c}{\bar{v}^3 T^2}\left(1 + \frac{\gamma}{\bar{v}^2}\right)e^{-\gamma/\bar{v}^2}$$

When $P$ is in atm, $\bar{v}$ is in ft³/lbmol, $T$ is in R, and $R_u = 0.730$ atm · ft³/(lbmol · R), the eight constants in the Benedict–Webb–Rubin equation are as follows:

| Gas | $a$ | $A_0$ | $b$ | $B_0$ | $c$ | $C_0$ | $\alpha$ | $\gamma$ |
|---|---|---|---|---|---|---|---|---|
| *n*-Butane, $C_4H_{10}$ | 7747 | 2590 | 10.27 | 1.993 | $4.219 \times 10^9$ | $8.263 \times 10^8$ | 4.531 | 8.732 |
| Carbon dioxide, $CO_2$ | 563.1 | 703.0 | 1.852 | 0.7998 | $1.989 \times 10^8$ | $1.153 \times 10^8$ | 0.3486 | 1.384 |
| Carbon monoxide, CO | 150.7 | 344.5 | 0.676 | 0.8740 | $1.387 \times 10^7$ | $7.124 \times 10^6$ | 0.5556 | 1.541 |
| Methane, $CH_4$ | 203.1 | 476.4 | 0.868 | 0.6827 | $3.393 \times 10^7$ | $1.878 \times 10^7$ | 0.5120 | 1.541 |
| Nitrogen, $N_2$ | 103.2 | 270.6 | 0.598 | 0.6529 | $9.713 \times 10^6$ | $6.706 \times 10^6$ | 0.5235 | 1.361 |

*Source:* Kenneth Wark, *Thermodynamics,* 4th ed., McGraw-Hill, New York, 1983, p. 864, Table A-21. Originally published in H. W. Cooper and J. C. Goldfrank, *Hydrocarbon Processing,* Vol. 46, No. 12, p. 141, 1967.

**TABLE A-14E**
**Properties of Solid Metals**

| | Melting point | Properties at 540R | | | | Properties at various temperatures (R) $k$[Btu/(h · ft · R)]/$c_p$[Btu/(lbm · R)] | | | | | | |
|---|---|---|---|---|---|---|---|---|---|---|---|---|
| **Composition** | **R** | $\rho$ **lbm/ft$^3$** | $c_p$ **Btu/(lbm · R)** | $k$ **Btu/(h · ft · R)** | $\alpha \times 10^6$ **ft$^2$/s** | **180** | **360** | **720** | **1080** | **1440** | **1800** | **2160** |
| Aluminum<br>Pure | 1679 | 168 | 0.216 | 137 | 1045 | 174.5<br>0.115 | 137<br>0.191 | 138.6<br>0.226 | 133.4<br>0.246 | 126<br>0.273 | | |
| Alloy 2024-T6<br>(4.5% Cu, 1.5% Mg.<br>0.6% Mn) | 1395 | 173 | 0.209 | 102.3 | 785.8 | 37.6<br>0.113 | 94.2<br>0.188 | 107.5<br>0.22 | 107.5<br>0.249 | | | |
| Alloy 195, Cast<br>(4.5% Cu) | | 174.2 | 0.211 | 97 | 734 | | | 100.5<br>— | 106.9<br>— | | | |
| Beryllium | 2790 | 115.5 | 0.436 | 115.6 | 637.2 | 572<br>0.048 | 174<br>0.266 | 93<br>0.523 | 72.8<br>0.621 | 61.3<br>0.624 | 52.5<br>0.72 | 45.6<br>0.77 |
| Bismuth | 981 | 610.5 | 0.029 | 4.6 | 71 | 9.5<br>0.026 | 5.6<br>0.028 | 4.06<br>0.03 | | | | |
| Boron | 4631 | 156 | 0.264 | 15.6 | 105 | 109.7<br>0.03 | 32.06<br>0.143 | 9.7<br>0.349 | 6.1<br>0.451 | 5.5<br>0.515 | 5.7<br>0.558 | |
| Cadmium | 1069 | 540 | 0.055 | 55.6 | 521 | 117.3<br>0.047 | 57.4<br>0.053 | 54.7<br>0.057 | | | | |
| Chromium | 3812 | 447 | 0.107 | 54.1 | 313.2 | 91.9<br>0.045 | 64.1<br>0.091 | 52.5<br>0.115 | 46.6<br>0.129 | 41.2<br>0.138 | 37.8<br>0.147 | 35.8<br>0.162 |
| Cobalt | 3184 | 553.2 | 0.101 | 57.3 | 286.3 | 96.5<br>0.056 | 70.5<br>0.09 | 49.3<br>0.107 | 39<br>0.12 | 33.6<br>0.131 | 30.1<br>0.145 | 28.5<br>0.175 |
| Copper<br>Pure | 2445 | 559 | 0.092 | 231.7 | 1259.3 | 278.5<br>0.06 | 238.6<br>0.085 | 227.07<br>0.094 | 219<br>0.01 | 212<br>0.103 | 203.4<br>0.107 | 196<br>0.114 |
| Commercial bronze<br>(90% Cu, 10% Al) | 2328 | 550 | 0.1 | 30 | 150.7 | | 24.3<br>0.187 | 30<br>0.109 | 34<br>0.130 | | | |
| Phosphor gear bronze<br>(89% Cu, 11% Sn) | 1987 | 548.1 | 0.084 | 31.2 | 183 | | 23.7<br>— | 37.6<br>— | 42.8<br>— | | | |
| Cartridge brass<br>(70% Cu, 30% Zn) | 2139 | 532.5 | 0.09 | 63.6 | 364.9 | 43.3 | 54.9<br>0.09 | 79.2<br>0.09 | 86.0<br>0.101 | | | |
| Constantan<br>(55% Cu, 45% Ni) | 2687 | 557 | 0.092 | 13.3 | 72.3 | 9.8<br>0.06 | 1.1<br>0.09 | | | | | |
| Germanium | 2180 | 334.6 | 0.08 | 34.6 | 373.5 | 134<br>0.045 | 56<br>0.069 | 25<br>0.08 | 15.7<br>0.083 | 11.4<br>0.085 | 10.05<br>0.089 | 10.05<br>0.094 |
| Gold | 2180<br>2405 | 334.6<br>1205 | 0.08<br>0.03 | 34.6<br>183.2 | 373.5<br>1367 | 189<br>0.026 | 186.6<br>0.029 | 179.7<br>0.031 | 172.2<br>0.032 | 164.09<br>0.033 | 156<br>0.034 | 147.3<br>0.037 |
| Iridium | 4896 | 1404.6 | 0.031 | 85 | 541.4 | 99.4<br>0.021 | 88.4<br>0.029 | 83.2<br>0.031 | 79.7<br>0.032 | 76.3<br>0.034 | 72.8<br>0.036 | 69.3<br>0.038 |
| Iron<br>Pure | 3258 | 491.3 | 0.106 | 46.4 | 248.6 | 77.4<br>0.051 | 54.3<br>0.091 | 40.2<br>0.117 | 31.6<br>0.137 | 25.01<br>0.162 | 19<br>0.232 | 16.4<br>0.14 |
| Armco<br>(99.75% pure) | | 491.3 | 0.106 | 42 | 222.8 | 55.2<br>0.051 | 46.6<br>0.091 | 38<br>0.117 | 30.7<br>0.137 | 24.4<br>0.162 | 18.7<br>0.233 | 16.6<br>0.145 |
| Carbon steels<br>Plain carbon<br>(Mn ≤ 1%,<br>Si ≤ 0.1%) | | 490.3 | 0.103 | 35 | 190.6 | | | 32.8<br>0.116 | 27.7<br>0.113 | 22.7<br>0.163 | 17.4<br>0.279 | |
| AISI 1010 | | 489 | 0.103 | 37 | 202.4 | | | 33.9<br>0.116 | 28.2<br>0.133 | 22.7<br>0.163 | 18<br>0.278 | |

## TABLE A-14E
## Properties of Solid Metals (*Continued*)

| Composition | Melting point R | Properties at 540R: $\rho$ lbm/ft³ | $c_p$ Btu/(lbm · R) | $k$ Btu/(h · ft · R) | $\alpha \times 10^6$ ft²/s | Properties at various temperatures (R) $k$[Btu/(h · ft · R)]/$c_p$[Btu/(lbm · R)]: 180 | 360 | 720 | 1080 | 1440 | 1800 | 2160 |
|---|---|---|---|---|---|---|---|---|---|---|---|---|
| Carbon steels (*continued*) | | | | | | | | | | | | |
| Carbon-silicon (Mn ≤ 1%, 0.1% < Si ≤ 0.6%) | | 488 | 0.106 | 30 | 160.4 | | | 28.8<br>0.119 | 25.4<br>0.139 | 21.6<br>0.166 | 17<br>0.231 | |
| Carbon-manganese-silicon (1% < Mn ≤ 1.65%, 0.1% < Si ≤ 0.6%) | | 508 | 0.104 | 23.7 | 125 | | | 24.4<br>0.116 | 23<br>0.133 | 20.2<br>0.163 | 16<br>0.260 | |
| Chromium (low) steels $\frac{1}{2}$Cr-$\frac{1}{4}$Mo-Si (0.18% C, 0.65% Cr, 0.23% Mo, 0.6% Si) | | 488.3 | 0.106 | 21.8 | 117.4 | | | 22<br>0.117 | 21.2<br>0.137 | 19.3<br>0.164 | 15.6<br>0.231 | |
| 1Cr-$\frac{1}{2}$Mo (0.16% C, 1% Cr, 0.54% Mo, 0.39% Si) | | 490.6 | 0.106 | 24.5 | 131.3 | | | 24.3<br>0.117 | 22.6<br>0.137 | 20<br>0.164 | 15.8<br>0.231 | |
| 1 Cr-V (0.2% C, 1.02% Cr, 0.15% V) | | 489.2 | 0.106 | 28.3 | 151.8 | | | 27.0<br>0.117 | 24.3<br>0.137 | 21<br>0.164 | 16.3<br>0.231 | |
| Stainless steels AISI 302 | | 503 | 0.114 | 8.7 | 42 | | | 10<br>0.122 | 11.6<br>0.133 | 13.2<br>0.140 | 14.7<br>0.144 | |
| AISI | 3006 | 493.2 | 0.114 | 8.6 | 42.5 | 5.31<br>0.064 | 7.3<br>0.096 | 9.6<br>0.123 | 11.5<br>0.133 | 13<br>0.139 | 14.7<br>0.145 | 16.2<br>0.152 |
| AISI 316 | | 514.3 | 0.111 | 7.8 | 37.5 | | | 8.8<br>0.12 | 10.6<br>0.131 | 12.3<br>0.137 | 14<br>0.143 | |
| AISI 347 | | 498 | 0.114 | 8.2 | 40 | | | 9.1<br>0.122 | 1.1<br>0.133 | 12.7<br>0.14 | 14.3<br>0.144 | |
| Lead | 1082 | 708 | 0.03 | 20.4 | 259.4 | 23<br>0.028 | 21.2<br>0.029 | 19.7<br>0.031 | 18.1<br>0.034 | | | |
| Magnesium | 1661 | 109 | 0.245 | 90.2 | 943 | 87.9<br>0.155 | 91.9<br>0.223 | 88.4<br>0.256 | 86.0<br>0.279 | 84.4<br>0.302 | | |
| Molybdenum | 5209 | 639.3 | 0.06 | 79.7 | 578 | 1034<br>0.033 | 82.6<br>0.053 | 77.4<br>0.062 | 72.8<br>0.065 | 68.2<br>0.068 | 64.7<br>0.070 | 60.7<br>0.073 |
| Nickel Pure | 3110 | 555.6 | 0.106 | 52.4 | 247.6 | 94.8<br>0.055 | 61.8<br>0.091 | 46.3<br>0.115 | 37.9<br>0.141 | 39<br>0.126 | 41.4<br>0.134 | 44.0<br>0.141 |
| Nichrome (80% Ni, 20% Cr) | 3010 | 524.4 | 0.1 | 6.9 | 36.6 | | | 8.0<br>0.114 | 9.3<br>0.125 | 12.2<br>0.130 | | |
| Inconel X-750 (7.3% Ni, 15% Cr, 6.7% Fe) | 2997 | 531.3 | 0.104 | 6.8 | 33.4 | 5<br>— | 5.9<br>0.088 | 7.8<br>0.112 | 9.8<br>0.121 | 11.8<br>0.13 | 13.9<br>0.149 | 16.0 |
| Niobium | 4934 | 535 | 0.063 | 31 | 254 | 31.9<br>0.044 | 30.4<br>0.059 | 32<br>0.065 | 33.6<br>0.067 | 35.4<br>0.069 | 32.2<br>0.071 | 39.0<br>0.074 |
| Palladium | 3289 | 750.4 | 0.058 | 41.5 | 263.7 | 44.2<br>0.04 | 41.4<br>0.054 | 42.5<br>0.059 | 46<br>0.062 | 50<br>0.064 | 54.4<br>0.067 | 59.0<br>0.069 |
| Platinum Pure | 3681 | 1339 | 0.031 | 41.4 | 270 | 44.7<br>0.024 | 42<br>0.03 | 41.5<br>0.032 | 42.3<br>0.034 | 43.7<br>0.035 | 45.5<br>0.036 | 47.7<br>0.037 |
| Alloy 60Pt-40Rh (60% Pt, 40% Rh) | 3240 | 1038.2 | 0.038 | 27.2 | 187.3 | | | 30<br>— | 34<br>— | 37.5<br>— | 40<br>— | 42.2<br>— |

**TABLE A-14E**
**Properties of Solid Metals (*Continued*)**

| Composition | Melting point R | Properties at 540R: $\rho$ lbm/ft³ | $c_p$ Btu/(lbm · R) | $k$ Btu/(h · ft · R) | $\alpha \times 10^6$ ft²/s | Properties at various temperatures (R) $k$[Btu/(h · ft · R)]/$c_p$[Btu/(lbm · R)]: 180 | 360 | 720 | 1080 | 1440 | 1800 | 2160 |
|---|---|---|---|---|---|---|---|---|---|---|---|---|
| Rbenium | 6215 | 1317.2 | 0.032 | 27.7 | 180 | 34<br>0.023 | 30<br>0.03 | 26.6<br>0.033 | 25.5<br>0.034 | 25.4<br>0.036 | 25.8<br>0.037 | 26.6<br>0.038 |
| Rhodium | 4025 | 777.2 | 0.058 | 86.7 | 534 | 107.5<br>0.035 | 89<br>0.052 | 84.3<br>0.06 | 78.5<br>0.065 | 73.4<br>0.069 | 70<br>0.074 | 67.0<br>0.078 |
| Silicon | 3033 | 145.5 | 0.17 | 85.5 | 960.2 | 510.8<br>0.061 | 152.5<br>0.132 | 57.2<br>0.189 | 35.8<br>0.207 | 24.4<br>0.218 | 18.0<br>0.226 | 15.0<br>0.230 |
| Silver | 2223 | 656 | 0.056 | 248 | 1873 | 257<br>0.044 | 248.4<br>0.053 | 245.5<br>0.057 | 238<br>0.059 | 228.8<br>0.062 | 219<br>0.066 | 208.6<br>0.069 |
| Tantalum | 5884 | 1036.3 | 0.033 | 33.2 | 266 | 34.2<br>0.026 | 33.2<br>0.031 | 33.4<br>0.034 | 34<br>0.035 | 34.3<br>0.036 | 34.8<br>0.036 | 35.3<br>0.037 |
| Thorium | 3641 | 730.4 | 0.028 | 31.2 | 420.9 | 34.6<br>0.024 | 31.5<br>0.027 | 31.4<br>0.029 | 32.2<br>0.032 | 32.9<br>0.035 | 32.9<br>0.037 | 33.9<br>0.04 |
| Tin | 909 | 456.3 | 0.054 | 38.5 | 431.6 | 49.2<br>0.044 | 42.4<br>0.051 | 35.9<br>0.058 | | | | |
| Titanium | 3515 | 281 | 0.013 | 12.7 | 100.3 | 17.6<br>0.071 | 14.2<br>0.111 | 11.8<br>0.131 | 11.2<br>0.141 | 11.4<br>0.151 | 12<br>0.161 | 12.7<br>0.148 |
| Tungsten | 6588 | 1204.9 | 0.031 | 100.5 | 735.2 | 120.2<br>0.020 | 107.5<br>0.029 | 92<br>0.032 | 79.2<br>0.033 | 72.2<br>0.034 | 68.2<br>0.035 | 65.3<br>0.036 |
| Uranium | 2531 | 1190.5 | 0.027 | 16 | 134.5 | 12.5<br>0.022 | 14.5<br>0.026 | 17.1<br>0.029 | 19.6<br>0.035 | 22.4<br>0.042 | 25.4<br>0.043 | 28.3<br>0.038 |
| Vanadium | 3946 | 381 | 0.117 | 17.7 | 110.9 | 20.7<br>0.061 | 18<br>0.102 | 18<br>0.123 | 19.3<br>0.128 | 20.6<br>0.134 | 22.0<br>0.142 | 23.6<br>0.154 |
| Zinc | 1247 | 445.7 | 0.093 | 67 | 450 | 67.6<br>0.07 | 68.2<br>0.087 | 64.1<br>0.096 | 59.5<br>0.104 | | | |
| Zirconium | 3825 | 410.2 | 0.067 | 13.1 | 133.5 | 19.2<br>0.049 | 14.6<br>0.063 | 12.5<br>0.072 | 12<br>0.77 | 12.5<br>0.082 | 13.7<br>0.087 | 15.0<br>0.083 |

Tables A-14E through A-16E are obtained from the respective tables in SI units in Appendix 1 using proper conversion factors.

## TABLE A-15E
## Properties of solid nonmetals

| Composition | Melting point R | Properties at 540R: $\rho$ lbm/ft$^3$ | $c_p$ Btu/(lbm · R) | $k$ Btu/(h · ft · R) | $\alpha \times 10^6$ ft$^2$/s | Properties at various temperatures (R) $k$[Btu/(h · ft · R)]/$c_p$[Btu/(lbm · R)]: 180 | 360 | 720 | 1080 | 1440 | 1800 | 2160 |
|---|---|---|---|---|---|---|---|---|---|---|---|---|
| Aluminum oxide, sapphire | 4181 | 247.8 | 0.182 | 26.6 | 162.5 | 260 | 47.4 | 18.7 | 11 | 7.5 | 6 | |
| | | | | | | — | — | 0.224 | 0.265 | 0.281 | 0.293 | |
| Aluminum oxide, polycrystalline | 4181 | 247.8 | 0.182 | 20.8 | 128 | 76.8 | 31.7 | 15.3 | 9.3 | 6 | 4.5 | 3.8 |
| | | | | | | — | — | 0.244 | 0.265 | 0.281 | 0.293 | |
| Beryllium oxide | 4905 | 187.3 | 0.246 | 157.2 | 947.3 | | | 113.2 | 64.2 | 40.4 | 27.2 | 19 |
| | | | | | | | | 0.322 | 0.40 | 0.44 | 0.459 | 0.490 |
| Boron | 4631 | 156 | 0.264 | 16 | 107.5 | 109.8 | 30.3 | 10.8 | 6.5 | 4.6 | 3.6 | 3 |
| | | | | | | | | 0.355 | 0.445 | 0.509 | 0.561 | 0.610 |
| Boron fiber epoxy (30% vol) composite | 1062 | 130 | | | | | | | | | | |
| $k$, ∥ to fibers | | | | 1.3 | | 1.2 | 1.3 | 1.31 | | | | |
| $k$, ⊥ to fibers | | | | 0.34 | | 0.21 | 0.28 | 0.34 | | | | |
| $c_p$ | | | 0.268 | | | 0.086 | 0.18 | 0.34 | | | | |
| Carbon Amorphous | 2700 | 121.7 | — | 0.92 | — | 0.38 | 0.68 | 1.09 | 1.26 | 1.36 | 1.46 | 1.64 |
| Diamond, type IIa insulator | — | 219 | 0.121 | 1329 | — | 5778 | 2311.2 | 889.8 | | | | |
| | | | | | | | 0.005 | 0.046 | 0.203 | | | |
| Graphite, pyrolytic | 4091 | 138 | | | | | | | | | | |
| $k$, ∥ to layers | | | | 1126.7 | | 2871.6 | 1866.3 | 803.2 | 515.4 | 385.4 | 308.5 | 258.9 |
| $k$, ⊥ to layers | | | | 3.3 | | 9.7 | 5.3 | 2.4 | 1.5 | 1.16 | 0.92 | 0.77 |
| $c_p$ | | | 0.169 | | | 0.032 | 0.098 | 0.236 | 0.335 | 0.394 | 0.428 | 0.45 |
| Graphite fiber epoxy (25% vol) composite | 810 | 87.4 | | | | | | | | | | |
| $k$, heat flow ∥ to fibres | | | | 6.4 | | 3.3 | 5.0 | 7.5 | | | | |
| $k$, heat flow ⊥ to fibers | | | | 0.5 | 5 | 0.4 | 0.63 | | | | | |
| $c_p$ | | | 0.223 | | | 0.08 | 0.153 | 0.29 | | | | |
| Pyroceram, Corning 9606 | 2921 | 162.3 | 0.193 | 2.3 | 20.3 | 3.0 | 2.3 | 2.1 | 1.9 | 1.7 | 1.7 | 1.7 |
| Silicon carbide | 5580 | 197.3 | 0.161 | 283.1 | 2475.7 | | | — | — | — | 50.3 | 33.5 |
| | | | | | | | | 0.210 | 0.25 | 0.27 | 0.285 | 0.296 |
| Silicon dioxide, crystalline (quartz) | 3389 | 165.4 | | | | | | | | | | |
| $k$, ∥ to $c$ axis | | | | 6 | | 22.5 | 9.5 | 4.4 | 2.9 | 2.4 | | |
| $k$, ⊥ to $c$ axis | | | | 3.6 | | 12.0 | 5.9 | 2.7 | 2 | 1.8 | | |
| $c_p$ | | | 0.177 | | | — | — | 0.211 | 0.256 | 0.298 | | |
| Silicon dioxide, polycrystalline (fused silica) | 3389 | 138.6 | 0.177 | 0.79 | 9 | 0.4 | 0.65 | 0.87 | 1.01 | 1.25 | 1.65 | 2.31 |
| | | | | | | — | — | 0.216 | 0.248 | 0.264 | 0.276 | 0.286 |
| Silicon nitride | 3911 | 150 | 0.165 | 9.2 | 104 | — | — | 8.0 | 6.5 | 5.7 | 5.0 | 4.6 |
| | | | | | | — | 0.138 | 0.185 | 0.223 | 0.253 | 0.275 | 0.292 |
| Sulfur | 706 | 130 | 0.169 | 0.1 | 1.51 | 0.095 | 0.1 | | | | | |
| | | | | | | 0.962 | 0.144 | | | | | |
| Thorium dioxide | 6431 | 568.7 | 0.561 | 7.5 | 65.7 | | | 5.9 | 3.8 | 2.7 | 2.12 | 1.8 |
| | | | | | | | | 0.609 | 0.654 | 0.680 | 0.704 | 0.723 |
| Titanium dioxide, polycrystalline | 3840 | 259.5 | 0.170 | 4.9 | 30.1 | | | 4.0 | 2.9 | 2.3 | 2 | 1.9 |
| | | | | | | | | 0.192 | 0.210 | 0.217 | 0.222 | 0.225 |

**TABLE A-16E**
**Properties of common materials**
**(*a*) Building materials**

| Description/Composition | Typical properties at 540 R | | |
|---|---|---|---|
| | Density $\rho$ lbm/ft³ | Thermal conductivity $k$ Btu/(h · ft · R) | Specific Heat $c_p$ Btu/(lbm · R) |
| Building boards | | | |
| Asbestos-cement board | 119.8 | 0.33 | — |
| Gypsum or plaster board | 50 | 0.098 | — |
| Plywood | 34.0 | 0.07 | 0.290 |
| Sheathing, regular-density | 18.1 | 0.031 | 0.310 |
| Acoustic tile | 18.1 | 0.034 | 0.32 |
| Hardboard, siding | 39.9 | 0.054 | 0.279 |
| Hardboard, high-density | 63.0 | 0.086 | 0.329 |
| Particle board, low-density | 36.8 | 0.045 | 0.310 |
| Particle board, high-density | 62.4 | 0.098 | 0.310 |
| Woods | | | |
| Hardwoods (oak, maple) | 44.9 | 0.092 | 0.299 |
| Softwoods (fir, pine) | 31.8 | 0.069 | 0.329 |
| Masonry materials | | | |
| Cement mortar | 116.1 | 0.41 | 0.186 |
| Brick, common | 119.8 | 0.41 | 0.199 |
| Brick, face | 130 | 0.75 | — |
| Clay tile, hollow | | | |
| 1 cell deep, 10-cm-thick | — | 0.30 | — |
| 3 cells deep, 30-cm-thick | — | 0.39 | — |
| Concrete block, 3 oval cores | | | |
| sand/gravel, 20-cm-thick | — | 0.57 | — |
| cinder aggregate, 20-cm-thick | — | 0.38 | — |
| Concrete block, rectangular core | | | |
| 2 cores, 20-cm-thick, 16-kg | — | 0.63 | — |
| same with filled cores | — | 0.34 | — |
| Plastering materials | | | |
| Cement plaster, sand aggregate | 116.1 | 0.41 | — |
| Gypsum plaster, sand aggregate | 104.8 | 0.12 | 0.259 |
| Gypsum plaster, vermiculite aggregate | 44.9 | 0.14 | — |

**TABLE A-16E**
**Properties of common materials**
**(*b*) Insulating materials (*Continued*)**

| Description/Composition | Density $\rho$ lbm/ft³ | Thermal conductivity $k$ Btu/(h · ft · R) | Specific Heat $c_p$ Btu/(lbm · R) |
|---|---|---|---|
| | **Typical properties at 540 R** | | |
| Blanket and batt | | | |
| Glass fiber, paper faced | 1.0 | 0.026 | — |
| | 1.7 | 0.022 | — |
| | 2.5 | 0.02 | — |
| Glass fiber, coated; duct liner | 1.9 | 0.022 | 0.199 |
| Board and slab | | | |
| Cellular glass | 9.05 | 0.033 | 0.238 |
| Glass fiber, organic bonded | 6.55 | 0.02 | 0.189 |
| Polystyrene, expanded | | | |
| extruded (R-12) | 3.4 | 0.015 | 0.289 |
| molded beads | 1.0 | 0.023 | 0.289 |
| Mineral fiberboard; roofing material | 16.5 | 0.028 | — |
| Wood, shredded/cemented | 21.8 | 0.05 | 0.379 |
| Cork | 7.49 | 0.022 | 0.429 |
| Loose fill | | | |
| Cork, granulated | 9.9 | 0.026 | — |
| Diatomaceous silica, coarse | 21.8 | 0.039 | — |
| powder | 24.9 | 0.052 | — |
| Diatomaceous silica, fine powder | 12.5 | 0.03 | — |
| | 17.1 | 0.035 | — |
| Glass fiber, poured or blown | 1.0 | 0.024 | 0.199 |
| Vermiculite, flakes | 5 | 0.039 | 0.199 |
| | 9.9 | 0.036 | 0.238 |
| Formed/foamed-in-place | | | |
| Mineral wool granules with asbestos/inorganic binders, sprayed | 11.8 | 0.026 | — |
| Polyvinyl acetate cork mastic; sprayed or troweled | — | 0.057 | — |
| Urethane, two-part mixture; rigid foam | 4.3 | 0.015 | 0.249 |
| Reflective | | | |
| Aluminum foil separating fluffy glass mats; 10–12 layers; evaluated; for cryogenic applications (150 K) | 2.5 | 0.000 092 | — |
| Aluminum foil and glass paper laminate; 75–150 layers; evacuated; for cryogenic application (150 K) | 7.5 | 0.000 009 8 | — |
| Typical silica powder, evacuated | 9.9 | 0.000 98 | — |

**TABLE A-16E**
**Properties of common materials**
**(*c*) Other materials (*Continued*)**

| Description/composition | Temperature R | Density $\rho$ lbm/ft³ | Thermal conductivity $k$ Btu/(h · ft · R) | Specific heat $c_p$ Btu/lbm · R |
|---|---|---|---|---|
| Asphalt | 540 | 132 | 0.035 | 0.219 |
| Bakelite | 540 | 81.2 | 0.808 | 0.349 |
| Brick, refractory | | | | |
| Carborundum | 1569 | — | 10.7 | — |
| | 3009 | — | 6.4 | — |
| Chrome brick | 851 | 187.9 | 1.3 | 0.199 |
| | 1481 | | 1.4 | |
| | 2111 | | 1.2 | |
| Diatomaceous silica, fired | 860 | — | 0.14 | — |
| | 2061 | — | 0.17 | |
| Fire clay, burnt 2880 R | 1391 | 128 | 0.57 | 0.229 |
| | 1931 | — | 0.63 | |
| | 2471 | — | 0.63 | |
| Fire clay, burnt 3105 R | 1391 | 145.2 | 0.75 | 0.229 |
| | 1931 | | 0.8 | |
| | 2471 | | 0.8 | |
| Fire clay brick | 860 | 165.1 | 0.57 | 0.229 |
| | 1660 | | 0.86 | |
| | 2660 | | 1.04 | |
| Magnesite | 860 | — | 2.2 | 0.269 |
| | 1660 | — | 1.6 | |
| | 2660 | | 1.09 | |
| Clay | 540 | 91.1 | 0.75 | 0.210 |
| Coal, anthracite | 540 | 84.3 | 0.15 | 0.3 |
| Concrete (stone mix) | 540 | 143.6 | 0.8 | 0.210 |
| Cotton | 540 | 5 | 0.034 | 0.310 |
| Foodstuffs | | | | |
| Banana (75.7% water content) | 540 | 61.2 | 0.27 | 0.8 |
| Apple, red (75% water content) | 540 | 52.4 | 0.29 | 0.859 |
| Cake, batter | 540 | 45 | 0.128 | — |
| Cake, fully baked | 540 | 17.5 | 0.069 | — |
| Chicken meat, white | 356 | — | 0.924 | — |
| (74.4% water content) | 419 | — | 0.86 | |
| | 455 | | 0.78 | |
| | 473 | | 0.69 | |
| | 491 | | 0.275 | |
| | 329 | | 0.277 | |
| | 527 | | 0.282 | |

**TABLE A-16E**
**Properties of common materials**
**(*c*) Other materials (*Continued*)**

| Description/ composition | Temperature R | Density $\rho$ lbm/ft$^3$ | Thermal conductivity $k$ Btu/(h · ft · R) | Specific heat $c_p$ Btu/lbm · R |
|---|---|---|---|---|
| Glass | | | | |
| Plate (soda lime) | 540 | 156 | 0.8 | 0.179 |
| Pyrex | 540 | 139 | 0.8 | 0.199 |
| Ice | 491 | 57.4 | 1.08 | 0.487 |
| | 455 | — | 1.17 | 0.464 |
| Leather (sole) | 540 | 62.3 | 0.091 | — |
| Paper | 540 | 58 | 0.104 | 0.320 |
| Paraffin | 540 | 56.2 | 0.138 | 0.690 |
| Rock | | | | |
| Granite, Barre | 540 | 164.2 | 1.61 | 0.185 |
| Limestone, Salem | 540 | 144.8 | 1.24 | 0.193 |
| Marble, Halston | 540 | 167.3 | 1.61 | 0.198 |
| Quartzite, Sioux | 540 | 164.8 | 3.10 | 0.263 |
| Sandstone, Berea | 540 | 134.2 | 1.67 | 0.178 |
| Rubber, vulcanized | | | | |
| Soft | 540 | 68.6 | 0.075 | 0.48 |
| Hard | 540 | 74.3 | 0.092 | — |
| Sand | 540 | 94.6 | 0.156 | 0.191 |
| Soil | 540 | 128 | 0.3 | 0.439 |
| Snow | 540 | 6.9 | 0.028 | — |
| | 720 | 31.2 | 0.109 | — |
| Teflon | | 137.3 | 0.202 | — |
| | | | | — |
| Tissue, human | | | | |
| Skin | 540 | — | 0.213 | — |
| Fat layer (adipose) | 540 | — | 0.115 | — |
| Muscle | 540 | — | 0.236 | — |
| Wood, cross-grain | | | | |
| Balsa | 540 | 8.7 | 0.031 | — |
| Cypress | 540 | 29 | 0.056 | — |
| Fir | 540 | 25.9 | 0.063 | 0.649 |
| Oak | 540 | 34 | 0.098 | 0.569 |
| Yellow pine | 540 | 40 | 0.086 | 0.669 |
| White pine | 540 | 27.2 | 0.063 | — |
| Wood, radial | | | | |
| Oak | 540 | 34 | 0.109 | 0.569 |
| Fir | 540 | 26.2 | 0.08 | 0.649 |

**TABLE A-17E**
**Properties of some liquid metals**

| Metal | Temerature (°F) | Thermal conductivity Btu/(h · ft°F) | Density lbm/ft$^3$ | Heat capacity Btu/(lbm °F) | Absolute viscosity lbm/(ft · s) | Kinematic viscosity ft$^2$/s | Thermal diffusivity (ft$^2$/h) | Pr |
|---|---|---|---|---|---|---|---|---|
| Bismuth | 600 | 9.5 | 625 | 0.0345 | $1.09 \times 10^{-3}$ | $1.74 \times 10^{-6}$ | 0.44 | 0.014 |
| | 800 | 9.0 | 616 | 0.0357 | $0.90 \times 10^{-3}$ | $1.5 \times 10^{-6}$ | 0.41 | 0.013 |
| | 1000 | 9.0 | 608 | 0.0369 | $0.74 \times 10^{-3}$ | $1.2 \times 10^{-6}$ | 0.40 | 0.011 |
| | 1200 | 9.0 | 600 | 0.0381 | $0.62 \times 10^{-3}$ | $1.0 \times 10^{-6}$ | 0.39 | 0.0094 |
| | 1400 | 9.0 | 591 | 0.0393 | $0.53 \times 10^{-3}$ | $0.9 \times 10^{-6}$ | 0.39 | 0.0084 |
| Galium | 85 (m.p.) | 19.5 | 381 | 0.082 | $1.39 \times 10^{-3}$ | $3.6 \times 10^{-6}$ | 0.61 | 0.022 |
| | 200 | | 378 | 0.082 | $1.05 \times 10^{-3}$ | $2.8 \times 10^{-6}$ | | |
| | 500 | | 370 | 0.082 | $0.73 \times 10^{-3}$ | $2.0 \times 10^{-6}$ | | |
| | 800 | | 363 | | $0.58 \times 10^{-3}$ | $1.6 \times 10^{-6}$ | | |
| | 1200 | | 355 | | $0.47 \times 10^{-3}$ | $1.3 \times 10^{-6}$ | | |
| | 1600 | | 348 | | $0.44 \times 10^{-3}$ | $1.2 \times 10^{-6}$ | | |
| Lead | 700 | 9.3 | 658 | 0.038 | $1.61 \times 10^{-3}$ | $2.45 \times 10^{-6}$ | 0.37 | 0.024 |
| | 850 | 9.0 | 652 | 0.037 | $1.38 \times 10^{-3}$ | $2.12 \times 10^{-6}$ | 0.37 | 0.020 |
| | 1000 | 8.9 | 646 | 0.037 | $1.17 \times 10^{-3}$ | $1.81 \times 10^{-6}$ | 0.37 | 0.017 |
| | 1150 | 8.7 | 639 | 0.037 | $1.02 \times 10^{-3}$ | $1.60 \times 10^{-6}$ | 0.37 | 0.016 |
| | 1300 | 8.6 | 633 | | $0.92 \times 10^{-3}$ | $1.45 \times 10^{-6}$ | | |
| Lithium | 400 | 22 | 31.6 | 1.0 | $0.40 \times 10^{-3}$ | $13 \times 10^{-6}$ | 0.70 | 0.065 |
| | 600 | | 31.0 | 1.0 | $0.34 \times 10^{-3}$ | $11 \times 10^{-6}$ | | |
| | 800 | | 30.5 | 1.0 | $0.37 \times 10^{-3}$ | $12 \times 10^{-6}$ | | |
| | 1200 | | 29.4 | 1.0 | $0.29 \times 10^{-3}$ | $9.9 \times 10^{-6}$ | | |
| | 1800 | | 27.6 | 1.0 | $0.28 \times 10^{-3}$ | $10 \times 10^{-6}$ | | |
| Mercury | 50 | 4.7 | 847 | 0.033 | $1.07 \times 10^{-3}$ | $1.2 \times 10^{-6}$ | 0.17 | 0.027 |
| | 200 | 6.0 | 834 | 0.033 | $0.84 \times 10^{-3}$ | $1.0 \times 10^{-6}$ | 0.22 | 0.016 |
| | 300 | 6.7 | 826 | 0.033 | $0.74 \times 10^{-3}$ | $0.90 \times 10^{-6}$ | 0.25 | 0.012 |
| | 400 | 7.2 | 817 | 0.032 | $0.67 \times 10^{-3}$ | $0.82 \times 10^{-6}$ | 0.27 | 0.011 |
| | 600 | 8.1 | 802 | 0.032 | $0.58 \times 10^{-3}$ | $0.72 \times 10^{-6}$ | 0.31 | 0.0084 |
| Potassium | 300 | 26.0 | 50.4 | 0.19 | $0.25 \times 10^{-3}$ | $5.0 \times 10^{-6}$ | 2.7 | 0.1066 |
| | 500 | 24.7 | 48.7 | 0.19 | $0.16 \times 10^{-3}$ | $3.3 \times 10^{-6}$ | 2.7 | 0.0043 |
| | 800 | 22.8 | 46.3 | 0.18 | $0.12 \times 10^{-3}$ | $2.6 \times 10^{-6}$ | 2.7 | 0.0035 |
| | 1100 | 20.6 | 43.8 | 0.18 | $0.10 \times 10^{-3}$ | $2.3 \times 10^{-6}$ | 2.6 | 0.0032 |
| | 1300 | 19.1 | 42.1 | 0.18 | $0.090 \times 10^{-3}$ | $2.1 \times 10^{-6}$ | 2.5 | 0.0031 |
| Sodium | 200 | 49.8 | 58.0 | 0.33 | $0.47 \times 10^{-3}$ | $8.1 \times 10^{-6}$ | 2.6 | 0.011 |
| | 400 | 46.4 | 56.3 | 0.32 | $0.29 \times 10^{-3}$ | $5.1 \times 10^{-6}$ | 2.6 | 0.0072 |
| | 700 | 41.8 | 53.7 | 0.31 | $0.19 \times 10^{-3}$ | $3.5 \times 10^{-6}$ | 2.5 | 0.0050 |
| | 1000 | 37.8 | 51.2 | 0.30 | $0.14 \times 10^{-3}$ | $2.7 \times 10^{-6}$ | 2.4 | 0.0040 |
| | 1300 | 34.5 | 48.6 | 0.30 | $0.12 \times 10^{-3}$ | $2.5 \times 10^{-6}$ | 2.4 | 0.0038 |
| Tin | 500 | 19 | 433 | 0.0580 | $1.22 \times 10^{-3}$ | $2.82 \times 10^{-6}$ | 0.76 | 0.013 |
| | 700 | 19.4 | 428 | 0.0603 | $0.98 \times 10^{-3}$ | $2.3 \times 10^{-6}$ | 0.75 | 0.011 |
| | 850 | 19 | 425 | 0.0621 | $0.85 \times 10^{-3}$ | $2.0 \times 10^{-6}$ | 0.72 | 0.010 |
| | 1000 | 19 | 421 | 0.0639 | $0.76 \times 10^{-3}$ | $1.8 \times 10^{-6}$ | 0.71 | 0.093 |
| | 1200 | 19 | 417 | 0.0662 | $0.67 \times 10^{-3}$ | $1.6 \times 10^{-6}$ | 0.69 | 0.0084 |

*Source: Liquid Metals Handbook,* Atomic Energy Commission and Department of the Navy, NAVEXOS P-733 (rev.) 2d ed., Revised June 1954.

## Properties of some liquids

| Temperature T °F | Density, $\rho$ lbm/ft$^3$ | Specific heat $C_p$ Btu/(lbm °F) | Thermal conductivity, $k$ Btu/(h·ft·°F) | Dynamic viscosity, $\mu$ lbm/(ft·s) | Kinematic viscosity, $\nu$ ft$^2$/s | Thermal diffusivity, $\alpha$ ft$^2$/h | Volume expansivity, $\beta$ 1/R | Prandtl number Pr |
|---|---|---|---|---|---|---|---|---|
| | | | | *Water* | | | | |
| 32 | 62.4 | 1.01 | 0.319 | $120 \times 10^{-5}$ | $1.93 \times 10^{-5}$ | $5.07 \times 10^{-3}$ | $-0.37 \times 10^{-4}$ | 13.7 |
| 40 | 62.4 | 1.00 | 0.325 | $104 \times 10^{-5}$ | $1.67 \times 10^{-5}$ | $5.21 \times 10^{-3}$ | $0.20 \times 10^{-4}$ | 11.6 |
| 50 | 62.4 | 1.00 | 0.332 | $88 \times 10^{-5}$ | $1.40 \times 10^{-5}$ | $5.33 \times 10^{-3}$ | $0.49 \times 10^{-4}$ | 9.55 |
| 60 | 62.3 | 0.999 | 0.340 | $76 \times 10^{-5}$ | $1.22 \times 10^{-5}$ | $5.47 \times 10^{-3}$ | $0.85 \times 10^{-4}$ | 8.03 |
| 70 | 62.3 | 0.998 | 0.347 | $65.8 \times 10^{-5}$ | $1.06 \times 10^{-5}$ | $5.57 \times 10^{-3}$ | $1.2 \times 10^{-4}$ | 6.82 |
| 80 | 62.2 | 0.998 | 0.353 | $57.8 \times 10^{-5}$ | $0.93 \times 10^{-5}$ | $5.68 \times 10^{-3}$ | $1.5 \times 10^{-4}$ | 5.89 |
| 90 | 62.1 | 0.997 | 0.359 | $51.4 \times 10^{-5}$ | $0.825 \times 10^{-5}$ | $5.79 \times 10^{-3}$ | $1.8 \times 10^{-4}$ | 5.13 |
| 100 | 62.0 | 0.998 | 0.364 | $45.8 \times 10^{-5}$ | $0.740 \times 10^{-5}$ | $5.88 \times 10^{-3}$ | $2.0 \times 10^{-4}$ | 4.52 |
| 150 | 61.2 | 1.00 | 0.384 | $29.2 \times 10^{-5}$ | $0.477 \times 10^{-5}$ | $6.27 \times 10^{-3}$ | $3.1 \times 10^{-4}$ | 2.74 |
| 200 | 60.1 | 1.00 | 0.394 | $20.5 \times 10^{-5}$ | $0.341 \times 10^{-5}$ | $6.55 \times 10^{-3}$ | $4.0 \times 10^{-4}$ | 1.88 |
| 250 | 58.8 | 1.01 | 0.396 | $15.8 \times 10^{-5}$ | $0.269 \times 10^{-5}$ | $6.69 \times 10^{-3}$ | $4.8 \times 10^{-4}$ | 1.45 |
| 300 | 57.3 | 1.03 | 0.395 | $12.6 \times 10^{-5}$ | $0.220 \times 10^{-5}$ | $6.70 \times 10^{-3}$ | $6.0 \times 10^{-4}$ | 1.18 |
| 350 | 55.6 | 1.05 | 0.391 | $10.5 \times 10^{-5}$ | $0.189 \times 10^{-5}$ | $6.69 \times 10^{-3}$ | $6.9 \times 10^{-4}$ | 1.02 |
| 400 | 53.6 | 1.08 | 0.381 | $9.1 \times 10^{-5}$ | $0.170 \times 10^{-5}$ | $6.57 \times 10^{-3}$ | $8.0 \times 10^{-4}$ | 0.927 |
| 450 | 51.6 | 1.12 | 0.367 | $8.0 \times 10^{-5}$ | $0.155 \times 10^{-5}$ | $6.34 \times 10^{-3}$ | $9.0 \times 10^{-4}$ | 0.876 |
| 500 | 49.0 | 1.19 | 0.349 | $7.1 \times 10^{-5}$ | $0.145 \times 10^{-5}$ | $5.99 \times 10^{-3}$ | $10.0 \times 10^{-4}$ | 0.87 |
| 550 | 45.9 | 1.31 | 0.325 | $6.4 \times 10^{-5}$ | $0.139 \times 10^{-5}$ | $5.05 \times 10^{-3}$ | $11.0 \times 10^{-4}$ | 0.93 |
| 600 | 42.4 | 1.51 | 0.292 | $5.8 \times 10^{-5}$ | $0.137 \times 10^{-5}$ | $4.57 \times 10^{-3}$ | $12.0 \times 10^{-4}$ | 1.09 |
| | | | | *Ammonia* | | | | |
| −20 | 42.4 | 1.07 | 0.317 | $17.6 \times 10^{-5}$ | $0.417 \times 10^{-5}$ | $6.94 \times 10^{-3}$ | | 2.15 |
| 0 | 41.6 | 1.08 | 0.316 | $17.1 \times 10^{-5}$ | $0.410 \times 10^{-5}$ | $7.04 \times 10^{-3}$ | | 2.09 |
| 10 | 40.8 | 1.09 | 0.314 | $16.6 \times 10^{-5}$ | $0.407 \times 10^{-5}$ | $7.08 \times 10^{-3}$ | | 2.07 |
| 32 | 40.0 | 1.11 | 0.312 | $16.1 \times 10^{-5}$ | $0.402 \times 10^{-5}$ | $7.03 \times 10^{-3}$ | $1.2 \times 10^{-3}$ | 2.05 |
| 50 | 39.1 | 1.13 | 0.307 | $15.5 \times 10^{-5}$ | $0.396 \times 10^{-5}$ | $6.95 \times 10^{-3}$ | $1.3 \times 10^{-3}$ | 2.04 |
| 80 | 37.2 | 1.17 | 0.293 | $14.5 \times 10^{-5}$ | $0.386 \times 10^{-5}$ | $6.73 \times 10^{-3}$ | | 2.01 |
| 120 | 35.2 | 1.22 | 0.275 | $13.0 \times 10^{-5}$ | $0.355 \times 10^{-5}$ | $6.40 \times 10^{-3}$ | | 1.99 |
| | | | | *Light oil* | | | | |
| 60 | 57.0 | 0.43 | 0.077 | $5820 \times 10^{-5}$ | $102 \times 10^{-5}$ | $3.14 \times 10^{-3}$ | $0.38 \times 10^{-3}$ | 1170 |
| 80 | 56.8 | 0.44 | 0.077 | $2780 \times 10^{-5}$ | $49 \times 10^{-5}$ | $3.09 \times 10^{-3}$ | $0.38 \times 10^{-3}$ | 570 |
| 100 | 56.0 | 0.46 | 0.076 | $1530 \times 10^{-5}$ | $27.4 \times 10^{-5}$ | $2.95 \times 10^{-3}$ | $0.39 \times 10^{-3}$ | 340 |
| 150 | 54.3 | 0.48 | 0.075 | $530 \times 10^{-5}$ | $9.8 \times 10^{-5}$ | $2.88 \times 10^{-3}$ | $0.40 \times 10^{-3}$ | 122 |
| 200 | 54.0 | 0.51 | 0.074 | $250 \times 10^{-5}$ | $4.6 \times 10^{-5}$ | $2.69 \times 10^{-3}$ | $0.42 \times 10^{-3}$ | 62 |
| 250 | 53.0 | 0.52 | 0.074 | $139 \times 10^{-5}$ | $2.6 \times 10^{-5}$ | $2.67 \times 10^{-3}$ | $0.44 \times 10^{-3}$ | 35 |
| 300 | 51.8 | 0.54 | 0.073 | $830 \times 10^{-5}$ | $1.6 \times 10^{-5}$ | $2.62 \times 10^{-3}$ | $0.45 \times 10^{-3}$ | 22 |
| | | | | *Glycerin* | | | | |
| 50 | 79.3 | 0.554 | 0.165 | 2.56 | 0.0323 | $3.76 \times 10^{-3}$ | | 31,000 |
| 70 | 78.9 | 0.570 | 0.165 | 1.0 | 0.0127 | $3.67 \times 10^{-3}$ | $0.28 \times 10^{-3}$ | 12,500 |
| 85 | 78.5 | 0.584 | 0.164 | 0.424 | 0.0054 | $3.58 \times 10^{-3}$ | $0.30 \times 10^{-3}$ | 5,400 |
| 100 | 78.2 | 0.600 | 0.163 | 0.188 | 0.0024 | $3.45 \times 10^{-3}$ | | 2,500 |
| 120 | 77.7 | 0.617 | | 0.124 | 0.0016 | | | 1,600 |
| | | | | *Benzene* | | | | |
| 60 | 55.1 | 0.40 | 0.093 | $46.0 \times 10^{-5}$ | $0.835 \times 10^{-5}$ | $4.22 \times 10^{-3}$ | $0.60 \times 10^{-3}$ | 7.2 |
| 80 | 54.6 | 0.42 | 0.092 | $39.6 \times 10^{-5}$ | $0.725 \times 10^{-5}$ | $4.01 \times 10^{-3}$ | | 6.5 |
| 100 | 54.0 | 0.44 | 0.087 | $35.1 \times 10^{-5}$ | $0.650 \times 10^{-5}$ | $3.53 \times 10^{-3}$ | | 5.1 |
| 150 | 53.5 | 0.46 | | $26.0 \times 10^{-5}$ | $0.480 \times 10^{-5}$ | | | 4.5 |
| 200 | | | | $20.3 \times 10^{-5}$ | | | | 4.0 |
| | | | | *n-Butyl alcohol* | | | | |
| 60 | 50.5 | 0.55 | 0.097 | $226 \times 10^{-5}$ | $4.48 \times 10^{-5}$ | $3.49 \times 10^{-3}$ | | 46.6 |
| 100 | 49.7 | 0.61 | 0.096 | $129 \times 10^{-5}$ | $2.60 \times 10^{-5}$ | $3.16 \times 10^{-3}$ | $0.45 \times 10^{-3}$ | 29.5 |
| 150 | 48.5 | 0.68 | 0.095 | $67.5 \times 10^{-5}$ | $1.39 \times 10^{-5}$ | $2.88 \times 10^{-3}$ | $0.48 \times 10^{-3}$ | 17.4 |
| 200 | 47.2 | 0.77 | 0.094 | $38.6 \times 10^{-5}$ | $0.815 \times 10^{-5}$ | $2.58 \times 10^{-3}$ | | 11.3 |

*Source:* F Kreith, *Principles of Heat Transfer,* 3d ed., Addison-Wesley Educational Publishers, Inc., New York, 1973. Originally compiled from various sources. Reprinted by permission of Addison-Wesley Educational Publishers, Inc.

**TABLE A-19E**
**Properties of some gases at 1-atm pressure***

| Temperature $T$ °F | Density $\rho$ lbm/ft$^3$ | Specific heat $C_p$ Btu/(lbm·°F) | Thermal conductivity $k$ Btu/(h·ft·°F) | Dynamic viscosity $\mu$ lbm/(ft·s) | Kinematic viscosity $\nu$ ft$^2$/s | Thermal diffusivity $\alpha$ ft$^2$/h | Volume expansivity $\beta$ 1/R | Prandtl number Pr |
|---|---|---|---|---|---|---|---|---|
| | | | | *Air* | | | | |
| 0 | 0.086 | 0.239 | 0.0133 | $1.110 \times 10^{-5}$ | $0.130 \times 10^{-3}$ | 0.646 | $2.18 \times 10^{-3}$ | 0.73 |
| 32 | 0.081 | 0.240 | 0.0140 | $1.165 \times 10^{-5}$ | $0.145 \times 10^{-3}$ | 0.720 | $2.03 \times 10^{-3}$ | 0.72 |
| 60 | 0.077 | 0.240 | 0.0146 | $1.214 \times 10^{-5}$ | $0.159 \times 10^{-3}$ | 0.796 | $1.93 \times 10^{-3}$ | 0.72 |
| 80 | 0.074 | 0.240 | 0.0150 | $1.250 \times 10^{-5}$ | $0.170 \times 10^{-3}$ | 0.851 | $1.86 \times 10^{-3}$ | 0.72 |
| 100 | 0.071 | 0.240 | 0.0154 | $1.285 \times 10^{-5}$ | $0.180 \times 10^{-3}$ | 0.905 | $1.79 \times 10^{-3}$ | 0.72 |
| 120 | 0.069 | 0.240 | 0.0158 | $1.316 \times 10^{-5}$ | $0.192 \times 10^{-3}$ | 0.964 | $1.74 \times 10^{-3}$ | 0.72 |
| 140 | 0.067 | 0.241 | 0.0162 | $1.347 \times 10^{-5}$ | $0.204 \times 10^{-3}$ | 1.023 | $1.68 \times 10^{-3}$ | 0.72 |
| 160 | 0.064 | 0.241 | 0.0166 | $1.378 \times 10^{-5}$ | $0.215 \times 10^{-3}$ | 1.082 | $1.63 \times 10^{-3}$ | 0.72 |
| 180 | 0.062 | 0.241 | 0.0170 | $1.409 \times 10^{-5}$ | $0.227 \times 10^{-3}$ | 1.141 | $1.57 \times 10^{-3}$ | 0.72 |
| 200 | 0.060 | 0.241 | 0.0174 | $1.440 \times 10^{-5}$ | $0.239 \times 10^{-3}$ | 1.20 | $1.52 \times 10^{-3}$ | 0.72 |
| 300 | 0.052 | 0.243 | 0.0193 | $1.610 \times 10^{-5}$ | $0.306 \times 10^{-3}$ | 1.53 | $1.32 \times 10^{-3}$ | 0.71 |
| 400 | 0.046 | 0.245 | 0.0212 | $1.750 \times 10^{-5}$ | $0.378 \times 10^{-3}$ | 1.88 | $1.16 \times 10^{-3}$ | 0.689 |
| 500 | 0.0412 | 0.247 | 0.0231 | $1.890 \times 10^{-5}$ | $0.455 \times 10^{-3}$ | 2.27 | $1.04 \times 10^{-3}$ | 0.683 |
| 600 | 0.0373 | 0.250 | 0.0250 | $2.000 \times 10^{-5}$ | $0.540 \times 10^{-3}$ | 2.68 | $0.943 \times 10^{-3}$ | 0.685 |
| 700 | 0.0341 | 0.253 | 0.0268 | $2.14 \times 10^{-5}$ | $0.625 \times 10^{-3}$ | 3.10 | $0.862 \times 10^{-3}$ | 0.690 |
| 800 | 0.0314 | 0.256 | 0.0286 | $2.25 \times 10^{-5}$ | $0.717 \times 10^{-3}$ | 3.56 | $0.794 \times 10^{-3}$ | 0.697 |
| 900 | 0.0291 | 0.259 | 0.0303 | $2.36 \times 10^{-5}$ | $0.815 \times 10^{-3}$ | 4.02 | $0.735 \times 10^{-3}$ | 0.705 |
| 1000 | 0.0271 | 0.262 | 0.0319 | $2.47 \times 10^{-5}$ | $0.917 \times 10^{-3}$ | 4.50 | $0.685 \times 10^{-3}$ | 0.713 |
| 1500 | 0.0202 | 0.276 | 0.0400 | $3.00 \times 10^{-5}$ | $1.47 \times 10^{-3}$ | 7.19 | $0.510 \times 10^{-3}$ | 0.739 |
| 2000 | 0.0161 | 0.286 | 0.0471 | $3.45 \times 10^{-5}$ | $2.14 \times 10^{-3}$ | 10.2 | $0.406 \times 10^{-3}$ | 0.753 |
| | | | | *Water vapor* | | | | |
| 212 | 0.0372 | 0.451 | 0.0145 | $0.870 \times 10^{-5}$ | $0.234 \times 10^{-3}$ | 0.864 | $1.49 \times 10^{-3}$ | 0.96 |
| 300 | 0.0328 | 0.456 | 0.0171 | $1.000 \times 10^{-5}$ | $0.303 \times 10^{-3}$ | 1.14 | $1.32 \times 10^{-3}$ | 0.95 |
| 400 | 0.0288 | 0.462 | 0.0200 | $1.130 \times 10^{-5}$ | $0.395 \times 10^{-3}$ | 1.50 | $1.16 \times 10^{-3}$ | 0.94 |
| 500 | 0.0258 | 0.470 | 0.0228 | $1.265 \times 10^{-5}$ | $0.490 \times 10^{-3}$ | 1.88 | $1.04 \times 10^{-3}$ | 0.94 |
| 600 | 0.0233 | 0.477 | 0.0257 | $1.420 \times 10^{-5}$ | $0.610 \times 10^{-3}$ | 2.31 | $0.943 \times 10^{-3}$ | 0.94 |
| 700 | 0.0213 | 0.485 | 0.0288 | $1.555 \times 10^{-5}$ | $0.725 \times 10^{-3}$ | 2.79 | $0.862 \times 10^{-3}$ | 0.93 |
| 800 | 0.0196 | 0.494 | 0.0321 | $1.700 \times 10^{-5}$ | $0.855 \times 10^{-3}$ | 3.32 | $0.794 \times 10^{-3}$ | 0.92 |
| 900 | 0.0181 | 0.50 | 0.0355 | $1.810 \times 10^{-5}$ | $0.987 \times 10^{-3}$ | 3.93 | $0.735 \times 10^{-3}$ | 0.91 |
| 1000 | 0.0169 | 0.51 | 0.0388 | $1.920 \times 10^{-5}$ | $1.13 \times 10^{-3}$ | 4.50 | $0.685 \times 10^{-3}$ | 0.91 |
| 1200 | 0.0149 | 0.53 | 0.0457 | $2.14 \times 10^{-5}$ | $1.44 \times 10^{-3}$ | 5.80 | $0.603 \times 10^{-3}$ | 0.88 |
| 1400 | 0.0133 | 0.55 | 0.053 | $2.36 \times 10^{-5}$ | $1.78 \times 10^{-3}$ | 7.25 | $0.537 \times 10^{-3}$ | 0.87 |
| 1600 | 0.0120 | 0.56 | 0.061 | $2.58 \times 10^{-5}$ | $2.14 \times 10^{-3}$ | 9.07 | $0.485 \times 10^{-3}$ | 0.87 |
| 1800 | 0.0109 | 0.58 | 0.068 | $2.81 \times 10^{-5}$ | $2.58 \times 10^{-3}$ | 10.8 | $0.442 \times 10^{-3}$ | 0.87 |
| 2000 | 0.0100 | 0.60 | 0.076 | $3.03 \times 10^{-5}$ | $3.03 \times 10^{-3}$ | 12.7 | $0.406 \times 10^{-3}$ | 0.86 |
| | | | | *Oxygen* | | | | |
| 0 | 0.0955 | 0.2185 | 0.0131 | $1.215 \times 10^{-5}$ | $0.127 \times 10^{-3}$ | 0.627 | $2.18 \times 10^{-3}$ | 0.73 |
| 100 | 0.0785 | 0.2200 | 0.0159 | $1.420 \times 10^{-5}$ | $0.181 \times 10^{-3}$ | 0.880 | $1.79 \times 10^{-3}$ | 0.71 |
| 200 | 0.0666 | 0.2228 | 0.0179 | $1.610 \times 10^{-5}$ | $0.242 \times 10^{-3}$ | 1.20 | $1.52 \times 10^{-3}$ | 0.722 |
| 400 | 0.0511 | 0.2305 | 0.0228 | $1.955 \times 10^{-5}$ | $0.382 \times 10^{-3}$ | 1.94 | $1.16 \times 10^{-3}$ | 0.710 |
| 600 | 0.0415 | 0.2390 | 0.0277 | $2.26 \times 10^{-5}$ | $0.545 \times 10^{-3}$ | 2.79 | $0.943 \times 10^{-3}$ | 0.704 |
| 800 | 0.0349 | 0.2465 | 0.0324 | $2.53 \times 10^{-5}$ | $0.725 \times 10^{-3}$ | 3.76 | $0.794 \times 10^{-3}$ | 0.695 |
| 1000 | 0.0301 | 0.2528 | 0.0366 | $2.78 \times 10^{-5}$ | $0.924 \times 10^{-3}$ | 4.80 | $0.685 \times 10^{-3}$ | 0.690 |
| 1500 | 0.0224 | 0.2635 | 0.0465 | $3.32 \times 10^{-5}$ | $1.480 \times 10^{-3}$ | 7.88 | $0.510 \times 10^{-3}$ | 0.677 |
| | | | | *Nitrogen* | | | | |
| 0 | 0.0840 | 0.2478 | 0.0132 | $1.055 \times 10^{-5}$ | $0.125 \times 10^{-3}$ | 0.635 | $2.18 \times 10^{-3}$ | 0.713 |
| 100 | 0.0690 | 0.2484 | 0.0154 | $1.222 \times 10^{-5}$ | $0.177 \times 10^{-3}$ | 0.898 | $1.79 \times 10^{-3}$ | 0.71 |
| 200 | 0.0585 | 0.2490 | 0.0174 | $1.380 \times 10^{-5}$ | $0.236 \times 10^{-3}$ | 1.20 | $1.52 \times 10^{-3}$ | 0.71 |
| 400 | 0.0449 | 0.2515 | 0.0212 | $1.660 \times 10^{-5}$ | $0.370 \times 10^{-3}$ | 1.88 | $1.16 \times 10^{-3}$ | 0.71 |
| 600 | 0.0364 | 0.2564 | 0.0252 | $1.915 \times 10^{-5}$ | $0.526 \times 10^{-3}$ | 2.70 | $0.943 \times 10^{-3}$ | 0.70 |
| 800 | 0.0306 | 0.2623 | 0.0291 | $2.145 \times 10^{-5}$ | $0.702 \times 10^{-3}$ | 3.62 | $0.794 \times 10^{-3}$ | 0.70 |
| 1000 | 0.0264 | 0.2689 | 0.0330 | $2.355 \times 10^{-5}$ | $0.891 \times 10^{-3}$ | 4.65 | $0.685 \times 10^{-3}$ | 0.69 |
| 1500 | 0.0197 | 0.2835 | 0.0423 | $2.800 \times 10^{-5}$ | $1.420 \times 10^{-3}$ | 7.58 | $0.500 \times 10^{-3}$ | 0.676 |

**TABLE A-19E**
**Properties of some gases at 1-atm pressure* (*Continued*)**

| Temperature $T$ °F | Density $\rho$ lbm/ft$^3$ | Specific heat $C_p$ Btu/(lbm·°F) | Thermal conductivity $k$ Btu/(h·ft·°F) | Dynamic viscosity $\mu$ lbm/(ft·s) | Kinematic viscosity $\nu$ ft$^2$/s | Thermal diffusivity $\alpha$ ft$^2$/h | Volume expansivity $\beta$ 1/R | Prandtl number Pr |
|---|---|---|---|---|---|---|---|---|
| | | | | *Carbon dioxide* | | | | |
| 0 | 0.132 | 0.184 | 0.0076 | $0.88 \times 10^{-5}$ | $0.067 \times 10^{-3}$ | 0.313 | $2.18 \times 10^{-3}$ | 0.77 |
| 100 | 0.108 | 0.203 | 0.0100 | $1.05 \times 10^{-5}$ | $0.098 \times 10^{-3}$ | 0.455 | $1.79 \times 10^{-3}$ | 0.77 |
| 200 | 0.092 | 0.216 | 0.0125 | $1.22 \times 10^{-5}$ | $0.133 \times 10^{-3}$ | 0.63 | $1.52 \times 10^{-3}$ | 0.76 |
| 500 | 0.063 | 0.247 | 0.0198 | $1.67 \times 10^{-5}$ | $0.266 \times 10^{-3}$ | 1.27 | $1.04 \times 10^{-3}$ | 0.75 |
| 1000 | 0.0414 | 0.280 | 0.0318 | $2.30 \times 10^{-5}$ | $0.558 \times 10^{-3}$ | 2.75 | $0.685 \times 10^{-3}$ | 0.73 |
| 1500 | 0.0308 | 0.298 | 0.0420 | $2.86 \times 10^{-5}$ | $0.925 \times 10^{-3}$ | 4.58 | $0.510 \times 10^{-3}$ | 0.73 |
| 2000 | 0.0247 | 0.309 | 0.050 | $3.30 \times 10^{-5}$ | $1.34 \times 10^{-3}$ | 6.55 | $0.406 \times 10^{-3}$ | 0.735 |
| 3000 | 0.0175 | 0.322 | 0.061 | $3.92 \times 10^{-5}$ | $2.25 \times 10^{-3}$ | 10.8 | $0.289 \times 10^{-3}$ | 0.745 |
| | | | | *Carbon monoxide* | | | | |
| 0 | 0.0835 | 0.2482 | 0.0129 | $1.065 \times 10^{-5}$ | $0.128 \times 10^{-3}$ | 0.621 | | |
| 200 | 0.0582 | 0.2496 | 0.0169 | $1.390 \times 10^{-5}$ | $0.239 \times 10^{-3}$ | 1.16 | $2.18 \times 10^{-3}$ | 0.75 |
| 400 | 0.0446 | 0.2532 | 0.0208 | $1.670 \times 10^{-5}$ | $0.374 \times 10^{-3}$ | 1.84 | $1.52 \times 10^{-3}$ | 0.74 |
| 600 | 0.0362 | 0.2592 | 0.0246 | $1.910 \times 10^{-5}$ | $0.527 \times 10^{-3}$ | 2.62 | $1.16 \times 10^{-3}$ | 0.73 |
| 800 | 0.0305 | 0.2662 | 0.0285 | $2.134 \times 10^{-5}$ | $0.700 \times 10^{-3}$ | 3.50 | $0.943 \times 10^{-3}$ | 0.725 |
| 1000 | 0.0263 | 0.2730 | 0.0322 | $2.336 \times 10^{-5}$ | $0.887 \times 10^{-3}$ | 4.50 | $0.794 \times 10^{-3}$ | 0.72 |
| 1500 | 0.0196 | 0.2878 | 0.0414 | $2.783 \times 10^{-5}$ | $1.420 \times 10^{-3}$ | 7.33 | $0.685 \times 10^{-3}$ | 0.71 |
| | | | | | | | $0.510 \times 10^{-3}$ | 0.70 |
| | | | | *Hydrogen* | | | | |
| 0 | 0.0060 | 3.39 | 0.094 | $0.540 \times 10^{-5}$ | $0.89 \times 10^{-3}$ | 4.62 | | |
| 100 | 0.0049 | 3.42 | 0.110 | $0.620 \times 10^{-5}$ | $1.26 \times 10^{-3}$ | 6.56 | $2.18 \times 10^{-3}$ | 0.70 |
| 200 | 0.0042 | 3.44 | 0.122 | $0.692 \times 10^{-5}$ | $1.65 \times 10^{-3}$ | 8.45 | $1.79 \times 10^{-3}$ | 0.695 |
| 500 | 0.0028 | 3.47 | 0.160 | $0.884 \times 10^{-5}$ | $3.12 \times 10^{-3}$ | 16.5 | $1.52 \times 10^{-3}$ | 0.69 |
| 1000 | 0.0019 | 3.51 | 0.208 | $1.160 \times 10^{-5}$ | $6.2 \times 10^{-3}$ | 31.2 | $1.04 \times 10^{-3}$ | 0.69 |
| 1500 | 0.0014 | 3.62 | 0.260 | $1.415 \times 10^{-5}$ | $10.2 \times 10^{-3}$ | 51.4 | $0.685 \times 10^{-3}$ | 0.705 |
| 2000 | 0.0011 | 3.76 | 0.307 | $1.64 \times 10^{-5}$ | $14.4 \times 10^{-3}$ | 74.2 | $0.510 \times 10^{-3}$ | 0.71 |
| 3000 | 0.0008 | 4.02 | 0.380 | $1.72 \times 10^{-5}$ | $24.2 \times 10^{-3}$ | 118.0 | $0.406 \times 10^{-3}$ | 0.72 |
| | | | | | | | $0.289 \times 10^{-3}$ | 0.66 |
| | | | | *Helium* | | | | |
| 0 | 0.012 | 1.24 | 0.078 | $1.140 \times 10^{-5}$ | $0.950 \times 10^{-3}$ | 5.25 | | |
| 200 | 0.00835 | 1.24 | 0.097 | $1.480 \times 10^{-5}$ | $1.77 \times 10^{-3}$ | 9.36 | $2.18 \times 10^{-3}$ | 0.67 |
| 400 | 0.0064 | 1.24 | 0.115 | $1.780 \times 10^{-5}$ | $2.78 \times 10^{-3}$ | 14.5 | $1.52 \times 10^{-3}$ | 0.686 |
| 600 | 0.0052 | 1.24 | 0.129 | $2.02 \times 10^{-5}$ | $3.89 \times 10^{-3}$ | 20.0 | $1.16 \times 10^{-3}$ | 0.70 |
| 800 | 0.00436 | 1.24 | 0.138 | $2.285 \times 10^{-5}$ | $5.24 \times 10^{-3}$ | 25.5 | $0.943 \times 10^{-3}$ | 0.715 |
| 1000 | 0.00377 | 1.24 | — | $2.520 \times 10^{-5}$ | $6.69 \times 10^{-3}$ | — | $0.794 \times 10^{-3}$ | 0.73 |
| 1500 | 0.0028 | 1.24 | — | $3.160 \times 10^{-5}$ | $11.10 \times 10^{-3}$ | — | $0.685 \times 10^{-3}$ | — |
| | | | | | | | $0.510 \times 10^{-3}$ | — |

*Source:* F. Kreith, *Principles of Heat Transfer*, 3d ed., Addison-Wesley Educational Publishers, Inc., New York, 1973. Originally compiled from various sources. Reprinted by permission of Addison-Wesley Educational Publishers, Inc.

*For ideal gases, the properties $C_p$, $k$, $\mu$ and Pr are independent of pressure. The properties $\rho$, $\nu$, and $\alpha$ at a pressure $P$ other than 1 atm are determined by multiplying the values of $\rho$ at the given temperature by $P$, and by dividing the values of $\nu$ and $\alpha$ at the given temperature by $P$, where $P$ is in atm (1 atm = 101.325 kPa = 14.696 psi).

**TABLE A-20E**

**Properties of the atmosphere at high altitude**

| Altitude ft | Temperature °F | Pressure psia | Gravity $g$ ft/s$^2$ | Speed of sound ft/s | Density lbm/ft$^3$ | Viscosity $\mu$ lbm/(ft · s) | Thermal conductivity Btu/(h · ft · °F) |
|---|---|---|---|---|---|---|---|
| 0 | 59.00 | 14.7 | 32.174 | 1116 | 0.07647 | $1.202 \times 10^{-5}$ | 0.0146 |
| 500 | 57.22 | 14.4 | 32.173 | 1115 | 0.07536 | $1.199 \times 10^{-5}$ | 0.0146 |
| 1,000 | 55.43 | 14.2 | 32.171 | 1113 | 0.07426 | $1.196 \times 10^{-5}$ | 0.0146 |
| 1,500 | 53.65 | 13.9 | 32.169 | 1111 | 0.07317 | $1.193 \times 10^{-5}$ | 0.0145 |
| 2,000 | 51.87 | 13.7 | 32.168 | 1109 | 0.07210 | $1.190 \times 10^{-5}$ | 0.0145 |
| 2,500 | 50.09 | 13.4 | 32.166 | 1107 | 0.07104 | $1.186 \times 10^{-5}$ | 0.0144 |
| 3,000 | 48.30 | 13.2 | 32.165 | 1105 | 0.06998 | $1.183 \times 10^{-5}$ | 0.0144 |
| 3,500 | 46.52 | 12.9 | 32.163 | 1103 | 0.06985 | $1.180 \times 10^{-5}$ | 0.0143 |
| 4,000 | 44.74 | 12.7 | 32.162 | 1101 | 0.06792 | $1.177 \times 10^{-5}$ | 0.0143 |
| 4,500 | 42.96 | 12.5 | 32.160 | 1099 | 0.06690 | $1.173 \times 10^{-5}$ | 0.0142 |
| 5,000 | 41.17 | 12.2 | 32.159 | 1097 | 0.06590 | $1.170 \times 10^{-5}$ | 0.0142 |
| 5,500 | 39.39 | 12.0 | 32.157 | 1095 | 0.06491 | $1.167 \times 10^{-5}$ | 0.0141 |
| 6,000 | 37.61 | 11.8 | 32.156 | 1093 | 0.06393 | $1.164 \times 10^{-5}$ | 0.0141 |
| 6,500 | 35.83 | 11.6 | 32.154 | 1091 | 0.06296 | $1.160 \times 10^{-5}$ | 0.0141 |
| 7,000 | 34.05 | 11.3 | 32.152 | 1089 | 0.06200 | $1.157 \times 10^{-5}$ | 0.0140 |
| 7,500 | 32.26 | 11.1 | 32.151 | 1087 | 0.06105 | $1.154 \times 10^{-5}$ | 0.0140 |
| 8,000 | 30.48 | 10.9 | 32.149 | 1085 | 0.06012 | $1.150 \times 10^{-5}$ | 0.0139 |
| 8,500 | 28.70 | 10.7 | 32.148 | 1083 | 0.05919 | $1.147 \times 10^{-5}$ | 0.0139 |
| 9,000 | 26.92 | 10.5 | 32.146 | 1081 | 0.05828 | $1.144 \times 10^{-5}$ | 0.0138 |
| 9,500 | 25.14 | 10.3 | 32.145 | 1079 | 0.05738 | $1.140 \times 10^{-5}$ | 0.0138 |
| 10,000 | 23.36 | 10.1 | 32.145 | 1077 | 0.05648 | $1.137 \times 10^{-5}$ | 0.0137 |
| 11,000 | 19.79 | 9.72 | 32.140 | 1073 | 0.05473 | $1.130 \times 10^{-5}$ | 0.0136 |
| 12,000 | 16.23 | 9.34 | 32.137 | 1069 | 0.05302 | $1.124 \times 10^{-5}$ | 0.0136 |
| 13,000 | 12.67 | 8.99 | 32.134 | 1065 | 0.05135 | $1.117 \times 10^{-5}$ | 0.0135 |
| 14,000 | 9.12 | 8.63 | 32.131 | 1061 | 0.04973 | $1.110 \times 10^{-5}$ | 0.0134 |
| 15,000 | 5.55 | 8.29 | 32.128 | 1057 | 0.04814 | $1.104 \times 10^{-5}$ | 0.0133 |
| 16,000 | +1.99 | 7.97 | 32.125 | 1053 | 0.04659 | $1.097 \times 10^{-5}$ | 0.0132 |
| 17,000 | −1.58 | 7.65 | 32.122 | 1049 | 0.04508 | $1.090 \times 10^{-5}$ | 0.0132 |
| 18,000 | −5.14 | 7.34 | 32.119 | 1045 | 0.04361 | $1.083 \times 10^{-5}$ | 0.0130 |
| 19,000 | −8.70 | 7.05 | 32.115 | 1041 | 0.04217 | $1.076 \times 10^{-5}$ | 0.0129 |
| 20,000 | −12.2 | 6.76 | 32.112 | 1037 | 0.04077 | $1.070 \times 10^{-5}$ | 0.0128 |
| 22,000 | −19.4 | 6.21 | 32.106 | 1029 | 0.03808 | $1.056 \times 10^{-5}$ | 0.0126 |
| 24,000 | −26.5 | 5.70 | 32.100 | 1020 | 0.03553 | $1.042 \times 10^{-5}$ | 0.0124 |
| 26,000 | −33.6 | 5.22 | 32.094 | 1012 | 0.03311 | $1.028 \times 10^{-5}$ | 0.0122 |
| 28,000 | −40.7 | 4.78 | 32.088 | 1003 | 0.03082 | $1.014 \times 10^{-5}$ | 0.0121 |
| 30,000 | −47.8 | 4.37 | 32.082 | 995 | 0.02866 | $1.000 \times 10^{-5}$ | 0.0119 |
| 32,000 | −54.9 | 3.99 | 32.08 | 987 | 0.02661 | $0.986 \times 10^{-5}$ | 0.0117 |
| 34,000 | −62.0 | 3.63 | 32.07 | 978 | 0.02468 | $0.971 \times 10^{-5}$ | 0.0115 |
| 36,000 | −69.2 | 3.30 | 32.06 | 969 | 0.02285 | $0.956 \times 10^{-5}$ | 0.0113 |
| 38,000 | −69.7 | 3.05 | 32.06 | 968 | 0.02079 | $0.955 \times 10^{-5}$ | 0.0113 |
| 40,000 | −69.7 | 2.73 | 32.05 | 968 | 0.01890 | $0.955 \times 10^{-5}$ | 0.0113 |
| 45,000 | −69.7 | 2.148 | 32.04 | 968 | 0.01487 | $0.955 \times 10^{-5}$ | 0.0113 |
| 50,000 | −69.7 | 1.691 | 32.02 | 968 | 0.01171 | $0.955 \times 10^{-5}$ | 0.0113 |
| 55,000 | −69.7 | 1.332 | 32.00 | 968 | 0.00922 | $0.955 \times 10^{-5}$ | 0.0113 |
| 60,000 | −69.7 | 1.048 | 31.99 | 968 | 0.00726 | $0.955 \times 10^{-5}$ | 0.0113 |

*Source:* U.S. Standard Atmosphere Supplements, U.S. Government Printing Office, 1966. Based on year-around mean conditions at 45° latitute, and will vary with time of the year and the weather patterns. The conditions at sea level ($z = 0$) are taken to be $P = 14.696$ psia, $T = 59°F$, $\rho = 0.076474$ lbm/ft$^3$, $g = 32.1741$ ft$^2$/s.

# Index